THE GRASS GENERA
OF THE WORLD

THE GRASS GENERA OF THE WORLD

Revised Edition

Leslie Watson

Taxonomy Laboratory
Research School of Biological Sciences
Australian National University
GPO Box 475
Canberra ACT 2601
Australia

Michael J. Dallwitz

CSIRO Division of Entomology
GPO Box 1700
Canberra ACT 2601
Australia

CAB INTERNATIONAL

CAB INTERNATIONAL
Wallingford
Oxon OX10 8DE
UK

Tel: Wallingford (0491) 832111
Telex: 847964 (COMAGG G)
E-mail: cabi@cabi.org
Fax: (0491) 833508

A catalogue record for this book is available from the British Library.

ISBN   0 85198 802 4

First published 1992
Revised Edition published 1994

Printed and bound in the UK at the University Press, Cambridge

CONTENTS

1. INTRODUCTION

These detailed descriptions of grass genera have been generated by computer from a taxonomic data bank, and automatically typeset without further editing. It therefore seems desirable to provide some background information on the data and on the programs used.

One of us has been engaged for over twenty years in compiling data and making observations on grass genera, in order to investigate classificatory questions and to explore taxonomic applications of computer methods. The other has worked for a comparable time on computer key-making, and has developed the DELTA system for comprehensive representation and manipulation of taxonomic descriptions. We have both been especially concerned with the need to solve problems associated with computer handling of observations on morphology, anatomy, cytology, biochemistry, etc. (as opposed to more curatorial aspects of taxonomic work). A natural outcome of our association from 1973 was the establishment of an automated system for descriptions of grass genera, which was introduced in an article in *Taxon*, accompanied by microfiches carrying 324 prototype descriptions (Watson and Dallwitz 1981). We commenced with the grasses of Australia (cf. Watson and Dallwitz 1980, 1985), and became involved with those of Egypt (Watson 1977), Canada (Watson *et al.* 1986), Greece (Watson, Damanakis and Dallwitz 1988), and Southern Africa (Watson *et al.* 1989; Gibbs Russell *et al.* 1990), *en route* to the ultimate objective of world coverage.

Since 1988, we have been producing a continuously updated package for the grass genera of the world (Watson and Dallwitz 1988, 1989, 1991, 1994), comprising floppy disks with the program INTKEY and the data in a form accessible for interactive identification and information retrieval, a booklet with illustrations of character states, and the complete descriptions. The last were first circulated on five microfiches, but from mid-1990 we substituted screen-readable descriptions on floppy disks. The booklet will soon be superseded by copious, superior illustrations of characters and taxa for screen display.

The 1992 edition of the present book was the first hard-copy version of the world set of descriptions, which had been extensively improved since 1988 (in particular, more than 100 leaf anatomical descriptions had been added or significantly augmented). The present edition reflects numerous minor corrections and improvements made in the data during the past three years. It also includes eight additional (mostly recently described) genera, and the anatomical data for Chloridoideae have been greatly extended by new observations from the Taxonomy Lab (see Acknowledgements). We continue to exclude printed keys, because the INTKEY package (Watson and Dallwitz 1994) provides far better facilities for identification. It is an essential accompaniment to this book for persons wishing to conduct identifications, to obtain statistical details of character-state distributions, or to pursue other aspects of information retrieval on a large scale. Conversely, users of the automated package should find the printed descriptions useful, since they include highlighting of diagnostic features, and the book can be perused without exiting from the program.

1.1. The DELTA System

The DELTA system (Dallwitz 1980; Dallwitz, Paine and Zurcher 1993) provides facilities for accommodating the different kinds of descriptive data used by taxonomists, without information loss, in an easy-to-use format designed to minimize coding errors. An associated program, CONFOR, checks the data files for format and other errors, and translates the coded descriptions into natural language, offering a choice of formats. This book exemplifies the capacity of the system to store very complex descriptive data, and of CONFOR to reproduce them in tolerably readable form. CONFOR also produces summarized data for specified sets of taxa, giving for multistate characters the numbers of taxa exhibiting each character state, and for numeric characters providing the means, ranges, and names of taxa representing the extremes of ranges. Provision is made for automatically translating DELTA-coded data into the formats required by various other taxonomic programs, including KEY (Dallwitz 1974; Dallwitz, Paine and Zurcher 1993) for making printed keys; INTKEY (see below) for interactive identification and information retrieval; DIST (Dallwitz, Paine and Zurcher 1993) for generating distance matrices; and Hennig86 (Farris 1988), MacClade (Maddison and Maddison 1992), PAUP (Swofford 1984, 1991), and PHYLIP (Felsenstein 1993) for phylogenetic analysis. Another associated program, DELFOR, permits changing the sequences of characters and character states in the DELTA-coded data while automatically keeping all the files consistent with one another. DELTA format is also accepted by programs of the PANKEY package for key generation, interactive key construction, polyclaves, interactive identification, identification by matching, description printing, and preparing diagnostic descriptions (Pankhurst 1986).

DELTA was designated as a standard format for taxonomic descriptions at the 1988 meeting of TDWG (Taxonomic Databases Working Group for Plant Sciences) in St. Louis, Missouri. The DELTA programs are available accompanied by a Primer (Partridge, Dallwitz and Watson 1993), operational examples of real data and directives files, and a comprehensive User's Guide (Dallwitz, Paine, and Zurcher 1993).

The interactive identification/information retrieval program INTKEY has greatly extended the range of applications accessible through the DELTA format (see Watson *et al.* 1989), and offers professional research features which are lacking in simpler programs. These features include: entry and deletion of attributes in any order during an identification; calculation of the 'best' characters for use in identification; the ability to allow for errors (whether made by the user or in the data); the ability to express variability or uncertainty in attributes; optional display of notes on characters and character state definitions; direct handling of numeric values, including ranges of values and non-contiguous sets of values; the ability to alter the treatments of unknowns, inapplicables and overlapping values, as required for different applications (flexibility in this respect being particularly significant in relation to identification versus information retrieval); retrieving free-text information (that is, information not encoded in terms of the character list); freedom to carry out operations in any order (for example, displaying taxon descriptions or differences during the course of an identification); automatic handling of characters that become inapplicable when other characters take certain values; restricting operations to subsets of characters or taxa; defining keywords to represent subsets of characters and taxa; locating characters by included words, and taxa directly by name; no limits on numbers of taxa, characters, and character states; no limits on lengths of taxon names and character definitions; specifying 'character reliabilities' appropriate for particular purposes; obtaining lists of taxa possessing or lacking particular attributes or combinations of attributes; preparing lists of taxa uncoded for particular characters or sets of characters; listing similarities or differences between taxa, with the ability to vary the interpretations of 'similarity' and 'difference'; describing taxa in terms of nominated sets of characters; generating diagnosic descriptions for specimens or taxa, to specified degrees of redundancy; coalescing descriptions (e.g. to generate accurate generic descriptions from species descriptions); input of complex or lengthy sequences of commands from files; selective output of results to files; generating files suitable for input to other DELTA programs (for example, to highlight diagnostic features in printed descriptions, as in Chapter 4); screen display of illustrations of characters and taxa; complete on-line help (currently available in English, French, German, Malay, and Portuguese); and acceptable response times with large sets of data.

Linking of the DELTA system with the computer typesetting program TYPSET (Dallwitz and Zurcher 1988) permits the direct output of camera ready copy, as exemplified by this book.

The latest versions of the DELTA programs, and several INTKEY packages (including *The Grass Genera of the World*), are available by gopher or anonymous ftp from the following Internet hosts.

muse.bio.cornell.edu (directory: /pub/delta)
life.anu.edu.au (directory: /pub/biodiversity/delta)
spider.ento.csiro.au (directory: /delta)

The file Index.txt (note the upper-case I) contains a list of the available programs and data. To view this file during an ftp session, give the ftp commands 'get Index.txt' and '! more < Index.txt'. Most of the subdirectories of delta contain a file Download.1st which contains information about downloading and installing the files in that subdirectory. Files other than .1st, .txt, and .ps must be transmitted in binary mode (set by the ftp command 'binary').

Updates are announced on the Internet via the Taxacom mailing list. To subscribe to Taxacom, send the message

subscribe taxacom *your-first-name your-last-name*

by email to listserv@ucjeps.berkeley.edu. Listserv will automatically determine your email address.

1.2. The Data

1.2.1. Generic delimitation

Modern taxonomic treatments of the grass family (Poaceae, Gramineae) recognize about 10000 species and between 650 and 793 genera, the higher generic estimate being represented in the descriptions presented here. When development of the database commenced, the initial aim was to arrive at intellectually satisfying generic interpretations; but as time passed, the objectives of data banking became clearer, and the criteria for sampling changed. In particular, the inappropriateness in the present connection of *sensu lato* generic interpretations became apparent, because: (a) we hoped the database might achieve general acceptance as a useful facility, despite conflicting views among taxonomists over generic limits; (b) it is desirable to both maximize and specify the extent of sampling from large genera, especially in connection with esoteric characters; and (c) it is a simple matter to coalesce existing descriptions automatically, using the INTKEY package, but it is much harder to disentangle them. For these reasons, we employ the narrowest generic circumscriptions which seem likely to be demanded by at least some users, and which are at least to some extent taxonomically defensible (i.e., for which ostensibly adequate diagnostic descriptions can be prepared). This policy conflicts with some taxonomists' (especially teachers', and including Watson's) natural preference for relatively broad generic limits. On the other hand, it is consistent with our belief that, ultimately, the most useful databases will be those prepared at species level. Consequent on our approach, we include many entities which Clayton and Renvoize (1986) not unreasonably refer to the larger genera of which they are obviously satellites, or which might better be recognized at subgeneric rank. In all such cases, *sensu lato* generic affiliations are indicated in the descriptions. In a few cases (exemplified by *Koeleria* and *Trisetum*), we are obliged to provide *sensu lato* descriptions because absence of comparative data precludes compiling even minimally adequate descriptions of segregates in common usage.

Our compilation reveals that some large genera (notably *Panicum, Agrostis, Festuca, Poa, Eragrostis, Stipa, Bambusa, Arundinaria*) seem unduly heterogeneous. These are few in number, but it should be noted that they involve about 30% of the total species in the family. None of them have been monographed (and very few have even been critically reviewed in terms of tangible comparative data compiled at world level) for over a hundred years, making it impossible to reliably assess the worth of small segregates. Nevertheless, new genera continue to be proposed, sometimes with scant regard for the scientific prerequisites of adequate sampling of taxa and characters and the presentation of properly comparative observations. Many of the small genera 'accepted' here are based on geographically restricted sampling, and many supposed distinctions no doubt reflect ignorance of variation and character state correlations in the large genera. This problem, which bedevils the whole of flowering-plant taxonomy, will not begin to be alleviated until taxonomic research transcends political boundaries and is planned and funded at the world level rather than in terms of regional floras.

1.2.2. The descriptions

The primary initial aim was to have the grass generic character list cover all aspects of variation necessary, or potentially useful, for identification and/or classification (Clifford and Watson 1977); but it has continuously been extended, with a view to increasing the scope for general-purpose, research-orientated information retrieval. The earliest version in DELTA format (Watson and Dallwitz 1980) comprised 214 characters. As employed for this book (Chapter 2), it comprises 516 characters, dealing with nomenclature, general morphology, leaf anatomy and physiology, biochemistry, haploid and 2c DNA values, fruit and embryo structure, seedling form, cytology, intergeneric hybrids, phytogeography and distribution, ecology, pathogens, classification, and economic aspects. There is abundant, deliberate 'redundancy', to maximize the flexibility of the data for information retrieval, and for identifying complete material, fragmentary specimens, and fossils. It also carries information relevant to interpreting the descriptions, with clarification of terminology and references. The figure numbers refer to Watson and Dallwitz (1988). Characters 86, 144, 181, 249, 386, 387, 457, 463–469, 488–492, and 509–516 have been excluded from the descriptions in Chapter 4, because they are redundant in that connection, or because work on the corresponding data is still in progress.

The morphological characters and character states of the list are intended to encompass all the features customarily included in grass generic descriptions, plus others which deserve more attention than they usually receive. The morphological data in the descriptions represent information laboriously extracted from the literature (see References), checked and expanded via direct observations, and improved and augmented by colleagues and collaborators (see Acknowledgements).

The list of leaf anatomical characters derived originally mainly from Metcalfe (1960), but has been variously modified and extended to include more recently discovered aspects of variation. It naturally reflects the Taxonomy Lab's long-standing interest in features associated with variations in photosynthetic pathways. The leaf anatomical descriptions incorporate Metcalfe's observations, data obtained from slides of about 1500 species prepared and examined in the Taxonomy Lab, and assorted information from numerous published papers (see References and Acknowledgements).

DELTA descriptions (Table 1) are not limited to information conveyable directly via the character list, as provision is made for the inclusion of 'comments' (i.e., preparing a description need involve no information loss at all); but these data 'comments' are not at present directly accessible for automated key-making and interactive identification.

The references listed have mostly been used in preparing the generic descriptions, either as sources of descriptive, floristic, nomenclatural, economic, or ethnic data, in connection with generic circumscriptions and/or nomenclature, or as sources of background information. A few have not yet been used, but are informative on aspects which merit inclusion in the future. Individual references are annotated to indicate the kinds of data they have provided, but many fall under more than one category, and some assignments are fairly arbitrary. Note that 'indirect' descriptive data of the kind involving extrapolation from tribal or subfamily descriptions or deductions from printed keys have been largely eschewed; that inclusion in generic descriptions of particular taxonomic references does not necessarily mean that these have provided much of the *descriptive* data, or that taxonomic conclusions they contain have been accepted; and that nomenclatural inconsistencies between genera as described here and in the references pertaining to them result from taxonomic realignments. Many of the works listed are largely, or in part, compilations, which themselves list primary data sources: the latter are given here only when they have been consulted directly.

Table 1. A DELTA-format description. The program CONFOR uses the character list (Chapter 2) to translate this into natural language (Chapter 4).

Acamptoclados <Nash>/
2<~ *Eragrostis* (*E. sessilispica* Buckl.)> 4,2 5,2 6,3 9,(30–)40–50(–60)<?> 10,2 13,2
15,2 16,1<when young> 17,1<above the basal tuft of leaves> 18,2 28,2 29,2 30,1 31,2
32,1<the culm internodes very short> 33,1 34,2 35<villous in the auricle positions>
36,– 37,2 38<glabrous, pilose at the throat> 39,2 40,1<narrowly so, stiff and
subulate-tipped, almost acicular> 41,3 42,2 43,1–3 44,3 45,2 46,1/3<flat to loosely
involute> 47,2 48,1 50,2 51,3 52,2 53,2 55,1 56,3<the adjoining blade also hairy> 57,–
59,1<partial, of small hairs> 60,2 61,1 62,2 63,1 65,– 68,2 69,2 71,2 73,1 74,2 75,3
76,5<the long-internoded primary branches distant, stiffly spreading, mostly bearing
basal secondary branches, the latter mostly short and sometimes reduced to a single
spikelet, the spikelets appressed> 77,1 78,1<the axis curving or loosely spiral, pilose in
the axils> 80,2 81,2 82,3 84,1 85,1&2 86,2 87,2 88,2 92,– 93,– 94,2 97,– 99,– 101,–
102,1 106<distant,>,1 107,2 109,2 112,2 126,1 127,(7–)8–10(–12) 130,– 131,1 133,1
134,1 135,1 140,1 141,2<glabrous> 142,1 143,2 144,– 147,1 148,2 149,2 150,2 152,1
153,2 154,2 155,2 156,2 157,1 158,– 159,1 161,2 162,1 163,2 164,2<firmly
membranous to cartilaginous> 167,2 168,– 169,– 171,– 172,1 173,2 174,1 175,2 176,2
177,2 179,1 181,2 193,5–8 194,2 196,2 197,2 198,2<firmly membranous to
cartilaginous> 200<becoming somewhat indurated> 202,1 203,1 205,– 206,2 207,1
224,2 227,1 228,1<the lateral nerves also slightly raised> 229,2 231,2 232,3 234,1
235,1<prominently bowed out below> 236,2 238,2 239,1 240,2 241,2 242,2 243,2
245<slightly>,1<above> 247,1 248,2 249,2 250,2 251,1 252,2 253,– 255,3 256,1
257<medium sized> 258,2 259,2 260,1 261,2 263,2 266,2 267<apparently not white>
269,2 270,1<about 1.5 mm long> 273,2 274,3 276,– 277,2 278,1 279,1 280,1/2<?>
281,1 282,2 283,1 298,1 299,2 304,1<rather thick walled> 305,1 306,1 307,1 308,1
311,1 312,1 315,(45–)46–54(–61.5) 316,24–30 317,7.5–10.5 318,6.1–7.2
319,(21–)25.5–27(–30) 320,0.47–0.55 321,1 322,30–33 324,2&3 325,– 326,2
327,1<bordering the veins> 328,2<solitary> 329,2 330,1 332<prickles abundant
costally> 333,2<but costal prickles with round or rectangular pitted bases> 334,1
335,2<but many solitary> 337,1 338,3–4<a few>/6<predominating> 340,2 341,2 342,1
343,1 346,1 347,1 348,2 349,1 359,2 360,2 361,2 362,2 363,2 365,1 366,2<large, tall,
square- or round-topped, and infrequent small ones> 367,2 368,1 369,2 371,1 372,2
373,2 374,2 375,1<the adaxial fibre groups being contiguous with bundle sheath
extensions> 376,1 377,1 380,1 386,2 387,2 388,2 389,2 391,2 392,2 393,2 394,2 395,2
397,2 398,2 400,2 401,2 403,2 404,2 407,2 409,2 410,2 411,2 413,2 414,2 415,2 416,2
417,2 418,2 421,2 422,2 425,2 426,2 427,2 428,2 430,2 431,2 432,2 433,2 434,2 436,2
437,2 438,2 447,4 452,4 455,1 456<North America to Mexico> 457,6 458,2 459,3–4
460,2 462<plains and sandy prairies> 464,11 465,9 468,1 469,1 470,1 471,1 473,4
482,2 487,6 494,3 505<this project> 508,1 509<acampt50.gif acampt75.gif> 510,4
511,3 512,2 513,3 514,1

1.2.3. Shortcomings

Application of DELTA in its present form affords final products which fairly reflect
the scientific perspicacity, taxonomic skills, and literacy of users, as well as their will-
ingness to strive for expertise in handling the system itself. Some manifest deficiencies
in the present descriptions, however, are attributable to the fact that large-scale organization of the data preceded the inception of a prototype DELTA system, and the latter
underwent extensive modifications in the light of practical experience afforded by these
data. Inevitably, the character list was developed, and data were extensively gathered,
largely in ignorance of the recommendations we now give to people initiating databases.
Taking full advantage of DELTA requires users to define character states, encode data,
and prepare comments with subtlety and skill, to capitalize fully on facilities such as 'text

characters', 'implicit values,' 'dependent characters,' and so forth, which became available only relatively recently. Furthermore, the earlier data banking operations were conducted under severe technological constraints. Problems with computer storage necessitated minimizing numbers of characters and states (which, combined with the need to produce keys from an early stage, resulted in some highly undesirable, inclusive character states), and strictly limited the use and length of qualifying comments. Omission from the descriptions of literature references and detailed information on questions of sampling, and frequent failure to name species exhibiting exceptional character states, are attributable to this problem. The complex and clumsy operating systems of earlier mainframe computers, combined with a tendency for administrators to banish 'big' jobs to overnight batch operations, made it unduly onerous to produce the printed natural-language descriptions essential for routine checking of newly entered and revised descriptions. The clouds began to lift in 1983, when operations were transferred to a VAX 11/750, and disappeared in 1988, when suitable microcomputers were obtained. Our long-held optimism about the practical possibilities inherent in automated data banking was then finally justified; but the descriptions still bear residual scars. Despite strenuous and continuing efforts to keep them abreast of developments and to make improvements, they would obviously have been better, in content and expression, had the full DELTA system been available 20 years ago.

There will always be scope for extending the range of characters included, and for refining states. For example, a deficiency noticeable in the morphological coverage, especially noticeable in Andropogoneae and Olyreae, concerns the details of male and sterile spikelets (characters 115–123). To cope with these satisfactorily would involve grafting in separate suites of characters paralleling those employed for the 'female-fertile' spikelets; and in view of the shortage of information adequate for the purpose, it has not yet seemed worthwhile to inflate the character list to that extent. Other inadequacies arise from the need for improved understanding of comparative morphology, or from the need to back potentially useful concepts with comparative observations. For example, the fairly conventional characters used here to describe inflorescence form, and hence many of the inflorescence descriptions, are very unsatisfactory. Perhaps adoption of an inflorescence descriptive system such as that proposed by Allred (1982) would bring about a much-needed improvement, but it is easier to erect new terminologies and introduce such revisions into character lists than to implement them on a useful scale in the descriptions. Meanwhile, indequacies in the inflorescence terminology used here are especially apparent in Andropogoneae and Bambuseae; and taxonomic problems among the latter are further compounded by shortage of information on fruits.

The descriptions naturally contain out-and-out errors, which are corrected as they come to light. A few represent miscodings: coding errors were harder to avoid and detect before CONFOR accepted 'implicit' and 'dependent' values, and when computing operations were less smooth than they are now. Other kinds of errors and irritants range from misobservations (Watson's or other people's) to infelicitous wording of data comments. Too much descriptive information (especially anatomical) derives from dried material, and this must lead to some misobservations and to the overlooking of potentially interesting aspects of variation. Certain questionable aspects of descriptions arise from uncertainties over morphological interpretations and homologies, the significance of which will depend upon one's point of view and the uses to which the data are put. Some of these indicate interesting problems worthy of comparative ultrastructural, developmental, or experimental investigation. For example, are '*Enneapogon*-type' microhairs homologous with the other forms? Should the lateral sheath extensions of centothecoids be regarded as fusoids?

Information on more esoteric and inaccessible features naturally represents extrapolation from small samples, the extent of which has been discussed elsewhere (Watson 1987), and is also ascertainable from the automated database.

Finally, there are gaps in the data through simple lack of comparative observations on characters appearing in the formal list. In some cases, such as characters relating to habitat and economic aspects, the inputting of available data is nowhere near complete;

and the information on rusts and smuts needs bringing up to date. Gaps reflecting ignorance of published work no doubt exist, and will be filled when detected.

The generic circumscriptions and details of the descriptions could be improved everywhere — most obviously in Andropogoneae, Bambuseae, Stipeae, and Triticeae — by people with special knowledge. We would welcome input from persons willing and able to make useful contributions on any aspect, these to be acknowledged in the character list or under individual descriptions, as appropriate.

1.3. Acknowledgements

The following persons have provided data incorporated in the descriptions: S.G. Aiken, A. Van den Borre, H.T. Clifford, R.P. Ellis, G.E. Gibbs Russell, P.W. Hattersley, M.R. Henwood, E.A. Kellogg, T.D. Macfarlane, C.M. Weiller, H.D.V. Prendergast, B.K. Simon, and R.D. Webster. The nature and extent of individual contributions is detailed in Watson (1987), at appropriate places in the character list, and in individual generic descriptions. Generous loans of herbarium material, with permission to take anatomical specimens, have been made by the Australian National Herbarium; the Royal Botanic Gardens, Kew; the Botanical Research Institute (National Botanical Institute) of Pretoria; the National Herbarium of New South Wales; the Western Australian Herbarium; Botanischer Garten und Botanisches Museum, Berlin; and Museum national d'Histoire naturelle, Laboratoire de Phanérogamie, Paris (H. Scholz and Ph. Morat were very helpful in providing material of rare taxa). We are especially indebted to Chris Frylink, Gill Hines, Janette Lenz, Jill Hartley, and Eileen Dallwitz for dedicated and highly skilled technical and secretarial work associated with organizing and maintaining the database.

2. CHARACTER LIST

#1. <Synonyms: 'genera' included in the current description. SET MATCH OVERLAP to retrieve data with INTKEY>/
#2. <~ *Sensu lato* genus: i.e. genus in which this taxon is sometimes (not unreasonably) included. SET MATCH OVERLAP to retrieve data with INTKEY>/
#3. Excluding <segregate genera for which separate descriptions are provided. SET MATCH OVERLAP to retrieve data with INTKEY>/

Habit, vegetative morphology

#4. <Longevity of plants>/
    1. annual <or biennial, without remains of old sheaths or culms>/
    2. perennial <with remains of old sheaths and/or culms> <Figs 1, 2, 18>/
#5. <Reeds>/
    1. reeds <helophytic, tall, to (2–)3 m or more in height; culms woody and persistent, always leafy>/
    2. not reeds <implicit>/
#6. <Habit>/
    1. <long> rhizomatous/
    2. <long> stoloniferous/
    3. caespitose <Figs 1, 7>/
    4. decumbent <including 'rooting at the nodes'> <Fig. 2>/
#7. The flowering culms <whether having foliage leaves> <dimorphic culms — intended mainly for bambusoids>/
    1. leafless/
    2. leafy/
#8. Plants to <diameter: as yet uncoded>/
    cm in diameter/
#9. <Mature> culms <maximum height: data unreliable for large genera>/
    cm high/
#10. Culms <whether woody or herbaceous>/
    1. woody and persistent/
    2. herbaceous <not woody, not persistent>/
#11. Culms to <maximum diameter: note cm units, intended for bamboos>/
    cm in diameter/
#12. Culms <shape: intended for bamboos>/
    1. cylindrical/
    2. flattened on one side/
#13. Culms <whether scandent>/
    1. scandent/
    2. not scandent <self-supporting, scrambling or floating> <implicit>/
#14. Culms <habit: as yet uncoded>/
    1. self-supporting <implicit>/
    2. decumbent/
    3. scrambling/
    4. scandent/
    5. pendent/
    6. floating/
#15. Culms <whether branched above>/
    1. branched <vegetatively> above <Fig. 2>/
    2. unbranched <vegetatively> above <Figs 1, 7>/

#16. Culms <whether tuberous at base>/
    1. tuberous <at base> <Fig. 3>/
    2. not tuberous <at base — implicit>/
#17. Culms <number of aerial nodes: as yet uncoded>/
    noded/
#18. Culm nodes <exposure: few data>/
    1. exposed/
    2. hidden by leaf sheaths/
#19. Culm nodes <whether hairy or glabrous>/
    1. hairy <Figs 4, 33>/
    2. glabrous <Fig. 4>/
#20. Culm nodes <number of ridges: bamboos>/
    ridged/
#21. <Number of> primary branches/mid-culm node <intended mainly for bamboos>/
#22. Culm sheaths <persistence (intended mainly for bamboos)>/
    1. <or at least their bases> persistent/
    2. deciduous in their entirety/
#23. Culm leaves <presence: few data>/
    1. present/
    2. absent/
#24. Upper culm leaf blades <development: few data>/
    1. fully developed/
    2. reduced/
    3. vestigial <i.e. leaves reduced to sheaths — not to be confused with blade *abscission*>/
#25. <Mid> culm internodes <whether solid or hollow: avoid the 'peduncle'>/
    1. solid <or spongy>/
    2. <conspicuously> hollow/
#26. <Bambusoid habit, unicaespitose or pluricaespitose (intended for bamboos)>/
    1. unicaespitose/
    2. pluricaespitose/
#27. Rhizomes <form (intended mainly for bamboos)>/
    1. pachymorph <sympodial>/
    2. leptomorph <monopodial>/
#28. Plants <whether conspicuously armed>/
    1. conspicuously armed <specify how>/
    2. unarmed <implicit, except in bamboos>/
#29. Plants <presence of multicellular glands, with comments on location, form>/
    1. with multicellular glands/
    2. without multicellular glands <implicit>/
#30. Young <vegetative> shoots <whether extra- or intravaginal: poorly recorded>/
    1. extravaginal <bursting through the bases of subtending sheaths> <Fig. 6>/
    2. intravaginal <emerging from between subtending sheath and stem> <Fig. 5>/
#31. The <fresh> shoots <whether aromatic>/
    1. aromatic <when crushed>/
    2. not aromatic <when crushed — implicit>/
#32. Leaves <whether mainly basal, or mainly on the culms>/
    1. mostly basal <Figs 7, 14>/
    2. not <distinctly> basally aggregated <i.e., the culms leafy> <Figs 1, 2, 9, 16, 33>/
#33. Leaves <whether differentiated into sheath and blade>/
    1. clearly differentiated into sheath and blade <implicit>/
    2. not clearly differentiated into sheath and blade/
#34. Leaves <phyllotaxy>/
    1. spirally disposed <from intitiation> <Figs 8, 9>/
    2. <at least initially> distichous <though subject to subsequent displacement: the near-universal condition — implicit>/

#35. Leaves <whether auricles present or absent: see Clifford and Watson 1977, for
  definition>/
  1. auriculate <Fig. 10>/
  2. non-auriculate <Figs 12, 19 etc.>/
#36. Leaves <presence of auricular setae>/
  1. with auricular <'oral'> setae <Fig. 11>/
  2. without auricular <'oral'> setae <implicit, save in Bambusoideae> <Fig. 12
  etc.>/
#37. <Leaf> sheath margins <whether joined>/
  1. joined <to at least one-quarter of their length: 'sheaths tubular'>/
  2. free <implicit>/
#38. <Comments on sheaths>/
#39. Leaf blades <extreme reduction>/
  1. <all> greatly reduced <with main functions transferred elsewhere>/
  2. not all greatly reduced <implicit>/
#40. Leaf blades <shape: data incomplete>/
  1. linear/
  2. linear-lanceolate/
  3. lanceolate/
  4. ovate-lanceolate/
  5. ovate/
  6. elliptic <oblong>/
  7. obovate/
#41. Leaf blades <texture>/
  1. leathery/
  2. flimsy/
  3. neither leathery nor flimsy <to become implicit>/
#42. Leaf blades <whether broad or narrow (specify the true range)>/
  1. broad <maximum (flattened) width greater than 1 cm>/
  2. narrow <maximum (flattened) width less than 1 cm>/
#43. Leaf blades <mid-width: data very incomplete>/
  mm wide <in the middle>/
#44. Leaf blades <whether cordate or sagittate>/
  1. <at least some of them> cordate <Fig. 13>/
  2. <at least some of them> sagittate/
  3. not cordate, not sagittate <implicit>/
#45. Leaf blades <whether setaceous>/
  1. setaceous <i.e., fine and bristle-like: not to be confused with pungent, subulate
  etc.> <Fig. 14>/
  2. not setaceous <implicit>/
#46. Leaf blades <folded/rolled>/
  1. flat/
  2. folded <Fig. 42>/
  3. rolled/
  4. acicular/
#47. Leaf blades <whether needle-like>/
  1. hard, woody, needle-like <and plants prickly, e.g. *Triodia*> <Fig. 15>/
  2. not needle-like <plants not prickly> <implicit>/
#48. Leaf blades <presence abaxially of multicellular glands> <recorded only for
  Chloridoideae: Van den Borre 1994>/
  1. without abaxial multicellular glands/
  2. exhibiting multicellular glands abaxially/
#49. The abaxial glands <distribution of abaxial leaf blade glands> <recorded only for
  Chloridoideae: Van den Borre 1994>/
  1. on the blade margins/
  2. costal <usually basal to macrohairs>/
  3. intercostal/

#50. Leaf blades <whether pseudopetiolate>/
    1. pseudopetiolate <Figs 11, 42>/
    2. not pseudopetiolate <implicit>/
#51. Leaf blades <venation>/
    1. pinnately veined <Fig. 16>/
    2. palmately veined/
    3. <conventionally> parallel veined <venation neither pinnate nor palmate> <implicit>/
#52. Leaf blades <whether cross veined>/
    1. cross veined <with more than the occasional transverse vein detectable> <Fig. 17>/
    2. without cross venation/
#53. Leaf blades <whether disarticulating>/
    1. <or at least many of them, ultimately> disarticulating from the sheaths <Fig. 9>/
    2. persistent <not disarticulating>/
#54. Leaf blades <whether vernation rolled or folded>/
    1. rolled in bud/
    2. once-folded in bud/
    3. folded like a fan in bud/
#55. <Adaxial> ligule <presence>/
    1. <consistently> present <implicit>/
    2. absent <at least from upper leaves>/
#56. <Adaxial> ligule <form — avoid seedlings>/
    1. an unfringed membrane <may be variously hairy or ciliolate> <Fig. 19>/
    2. a fringed membrane <Figs 20, 21, 23>/
    3. a fringe of hairs <Fig. 22>/
    4. a rim of minute papillae/
#57. <Adaxial> ligule <shape of apex>/
    1. truncate/
    2. not truncate <acute, obtuse or rounded> <Fig. 19>/
#58. <Adaxial> ligule <length at middle: generally recorded only for membranous, unfringed forms>/
    mm long/
#59. <Outer> contra-ligule <presence: data very incomplete>/
    1. present <Figs 11, 24>/
    2. absent <Fig. 12 etc.>/

Reproductive organization

#60. Plants <whether monoecious, with bisexual spikelets, or dioecious>/
    1. <bisexual, but> monoecious with all the fertile spikelets unisexual/
    2. bisexual, with <at least some> bisexual spikelets <Plates 1:4, 1:5, 1:8, 2:10, 2:11, 2:13–2:17 etc.>/
    3. dioecious <with separate male and female-fertile individuals> <Figs 25, 26>/
#61. Plants <whether having hermaphrodite florets: not to be confused with presence or absence of hermaphrodite spikelets>/
    1. with <at least some> hermaphrodite florets <Plate 2:14 etc.>/
    2. without hermaphrodite florets/
#62. The spikelets <whether heterospiculate: exclusive of 'hidden' spikelets>/
    1. of <at least two> sexually distinct forms on the same plant <e.g., female or hermaphrodite and sterile or male-only; including 'heterogamous'. Vestigial spikelets represented only by their pedicels have here been regarded as spikelets> <Plates 1:3, 1:6; Figs 27–29, 33, 75, 76>/
    2. all alike in sexuality <on the same plant ('homogamous'): ignore hidden axillary spikelets, etc. Implicit>/

#63. The spikelets <sexuality>/
    1. hermaphrodite/
    2. female-only/
    3. male-only/
    4. sterile/
#64. The male and female-fertile spikelets <disposition on the plant>/
    1. in different inflorescences/
    2. on different <main> branches of the same inflorescence/
    3. segregated, in different parts of the same inflorescence branch <Plate 1:6>/
    4. mixed in the inflorescence/
#65. The spikelets <whether heteromorphic (intended mainly for heterospiculate andropogonoids)>/
    1. overtly heteromorphic <Plate 1:6; Figs 27, 28, 75, 76>/
    2. <externally> homomorphic/
#66. The spikelets <whether the spikelet combinations are all heterogamous: generally applied only to andropogonoids, inapplicable when exclusively homogamous>/
    1. in both homogamous and heterogamous combinations/
    2. all in heterogamous combinations/
#67. Plants <whether outbreeding or inbreeding — data extensively from Connor 1979>/
    1. outbreeding <allogamous>/
    2. inbreeding <autogamous>/
#68. <Cleistogamy — data mainly from Connor 1979: exposed spikelets>/
    1. exposed-cleistogamous <associated with varying degrees of spikelet and/or floret modification>/
    2. chasmogamous <unreliably implicit>/
#69. Plants <possession of hidden, cleistogamous spikelets in leaf axils or on specialised rhizomes>/
    1. with hidden cleistogenes <more or less hidden, usually conspicuously modified cleistogamous spikelets>/
    2. without hidden cleistogenes <implicit>/
#70. The hidden cleistogenes <location>/
    1. in the leaf sheaths/
    2. subterranean <rhizanthogenes>/
#71. <Whether apomixis observed, with data comment on pseudogamy/ non-pseudogamy: data mainly from Connor 1979>/
    1. apomictic/
    2. reproducing sexually <unreliably implicit>/
#72. <Occurrence of vivipary (poorly recorded)>/
    1. viviparous/
    2. not viviparous <unreliably implicit>/

Inflorescence

#73. Inflorescence <whether determinate (semelauctant) or indeterminate (iterauctant or with a seemingly indeterminate synflorescence): see McClure 1973, Calderon and Soderstrom 1973 etc. for definitions>/
    1. determinate <semelauctant — implicit>/
    2. indeterminate <iterauctant> <Fig. 41>/
#74. Inflorescence <whether possessing pseudospikelets: see McClure 1973 for definition>/
    1. with pseudospikelets <having basal bracts with axillary spikelets, in addition to or instead of the usual barren glumes> <Fig. 41>/
    2. without pseudospikelets <implicit>/

#75. Inflorescence <reduction>/
    1. reduced to a single spikelet/
    2. few spikeleted/
    3. <normally> many spikeleted <unreliably implicit>/
#76. Inflorescence <chasmogamous: overall form>/
    1. a single spike <Plate 1:9; Fig. 30>/
    2. of spicate <spike-like> main branches <of spikes, narrow racemes or narrow
       panicles> <Plate 1:4; Figs 31, 38, 53, 54>/
    3. a false spike, with spikelets <or spikelet clusters> on contracted axes <Plate
       1:5; Figs 42, 49, 50, 51, 77>/
    4. a single raceme <at least some of the spikelets clearly pedicellate> <Fig. 33,
       79>/
    5. paniculate <and not readily referable to any of the other states> <Plates 1:1,
       1:3, 1:8, 2:12; Figs 34, 35, 36, 37, 55>/
#77. Inflorescence <tumbleweeds>/
    1. deciduous in its entirety <including 'tumbleweeds'> <Fig. 26>/
    2. not deciduous <implicit>/
#78. Inflorescence <whether open or contracted: mainly applied to panicles>/
    1. open <Plates 1:8, 2:12; Figs 34, 35>/
    2. contracted <very compact, or narrow and spike-like> <Plates 1:1, 1:3; Figs 36,
       37>/
#79. Inflorescence <compact, solitary: form — mainly applied to panicles and solitary
      racemes>/
    1. capitate <more or less spherical> <Plate 1:3>/
    2. more or less ovoid/
    3. spicate <elongated-symmetrical, spike-like> <Fig. 37>/
    4. more or less irregular <neither capitate nor ovoid, not elongated-symmetrical>/
#80. Inflorescence <whether branches divaricate>/
    1. with conspicuously divaricate branchlets <Fig. 40>/
    2. without conspicuously divaricate branchlets <implicit>/
#81. Inflorescence <whether branchlets capillary: avoid INTKEY use with
      non-paniculate inflorescences, which have usually been scored as
      'inapplicable'>/
    1. with capillary branchlets <Fig. 35>/
    2. without capillary branchlets <implicit>/
#82. Inflorescence <whether digitate or subdigitate>/
    1. digitate <includes paired branches> <Figs 38, 54>/
    2. subdigitate/
    3. non-digitate <neither digitate nor 'subdigitate' — implicit>/
#83. Primary inflorescence branches <number: applied mainly to forms with spike-like
      main branches — data very incomplete>/
#84. Inflorescence <whether branches end in spikelets: caution required, notably in
      Paniceae where 'bristles' are interpreted as the naked tips of reduced axes>/
    1. with axes ending in spikelets <implicit> <Figs 30, 32, 54>/
    2. axes <often> not ending in spikelets <Figs 26, 31, 49>/
#85. Rachides <whether clearly (macroscopically) flattened, hollowed or winged (data
      fairly unreliable)>/
    1. hollowed <Plates 1:7, 1:9; Figs 30, 56>/
    2. flattened <Fig. 39>/
    3. winged/
    4. neither flattened nor hollowed, not winged <implicit>/
#86. Spikelets <whether embedded in the rachis>/
    1. all <more or less> partially embedded in the rachis <Plates 1:7, 1:9; Figs 30,
       42, 56>/
    2. not all embedded <implicit>/
#87. Inflorescence <whether spatheate (note: 'spatheate' not currently distinguished
      from 'spatheolate')>/
    1. spatheate <specify> <ignore mere early enclosure by an unmodified flag leaf>

<Figs 25, 26, 27, 43>/
 2. espatheate/

#88. Inflorescence <whether comprising a complex of 'partial inflorescences' and intervening foliar organs (= leaves, spathes, spatheoles>/
 1. a complex of 'partial inflorescences' and intervening foliar organs <i.e., a 'pseudo-inflorescence'> <Fig. 43>/
 2. not comprising 'partial inflorescences' and foliar organs/

#89. <Ultimate> spikelet-bearing axes <form> <intended mainly for andropogonoids and bamboos>/
 1. very much reduced <specify> <Plate 1:5; Figs 26, 42, 49, 50, 51>/
 2. spikes/
 3. 'racemes' <Plate 1:6; Fig. 43>/
 4. spikelike <evidently not spikes, on close examination. e.g. *Hemarthria*, *Phacelurus*> <Plate 1:7; Figs 27, 46, 47, 59, 60, 75, 76>/
 5. paniculate/
 6. capitate <= 1&5>/

#90. The spikelet-bearing axes <andropogonoid, number of spikelet-bearing 'articles' (joints)>/
 1. with only one spikelet-bearing 'article'/
 2. with 2–3 spikelet-bearing 'articles'/
 3. with 4–5 spikelet-bearing 'articles'/
 4. with 6–10 spikelet-bearing 'articles'/
 5. with more than 10 spikelet-bearing 'articles' <specify the approximate number>/

#91. The racemes <whether spikelet bearing to the base>/
 1. spikelet bearing to the base/
 2. without spikelets towards the base/

#92. <Ultimate> spikelet-bearing axes <grouping> <intended mainly for andropogonoids>/
 1. solitary/
 2. paired/
 3. clustered <in groups of three or more>/

#93. <Ultimate> spikelet-bearing axes <thickness of rachides> <intended mainly for andropogonoids>/
 1. with very slender rachides <Plate 1:6>/
 2. with substantial rachides <Plate 1:7; Fig. 27>/

#94. Spikelet-bearing axes <whether disarticulating. Note that spikelet-bearing axes may be greatly reduced>/
 1. disarticulating <often manifested in clearly articulated rachides. Excluding inflorescences falling whole (tumbleweeds)> <Figs 26, 27, 44, 45, 46, 47, 59, 76>/
 2. persistent <not disarticulating: implicit> <Figs 30, 38, 39, 53>/

#95. Spikelet-bearing axes <manner of disarticulation>/
 1. falling entire <Figs. 50, 51>/
 2. disarticulating at the joints <Figs 27, 44, 45, 46, 47, 59, 75, 76>/

#96. The pedicels and rachis internodes <*Bothriochloa*, *Dichanthium* and relatives>/
 1. with a longitudinal, translucent furrow/
 2. without a longitudinal, translucent furrow <implicit>/

#97. 'Articles' <('joints') of the spikelet-bearing rachis, shape (intended mainly for andropogonoids)>/
 1. linear/
 2. non-linear <Figs 44, 46, 47, 59, 75>/

#98. 'Articles' <of the spikelet-bearing rachis: whether bearing an elaiosome>/
 1. with a basal callus-knob <elaiosome> <Fig. 47>/
 2. without a basal callus-knob/

#99. 'Articles' <of the spikelet-bearing rachis, whether appendaged (intended mainly for andropogonoids)>/
    1. appendaged <Figs 45, 48>/
    2. not appendaged <Plate 1:7; Fig. 46>/

#100. 'Articles' <of the spikelet-bearing rachis, orientation of disarticulation (intended mainly for andropogonoids)>/
    1. disarticulating transversely <Plate 1:7; Figs 44, 46, 47, 59, 75>/
    2. disarticulating obliquely <Figs 27, 28>/

#101. 'Articles' <of the spikelet-bearing rachis, whether hairy (intended mainly for andropogonoids>/
    1. densely long-hairy/
    2. somewhat hairy/
    3. glabrous <Plate 1:7; Figs 46, 47, 59, 75>/

#102. Spikelets <and/or clusters, whether subtended by or associated with 'involucres' or bristles representing vestigial branches (note that 'bristles' must not be confused with hairs)>/
    1. <all> unaccompanied by bractiform involucres, not associated with setiform vestigial branches <implicit>/
    2. <at least some of them> subtended by solitary 'bristles' <vestigial branches>/
    3. <or clusters> with 'involucres' of 'bristles' <vestigial branches> <ignore true hairs> <Figs 49, 50, 51>/
    4. associated with bractiform involucres <Fig. 72>/

#103. The <reduced branch> 'bristles' <form, coalescence>/
    1. spiny, markedly coalescent basally <Fig. 50>/
    2. relatively slender, not spiny <Figs 49, 51>/

#104. The <reduced branch> 'bristles' <whether deciduous>/
    1. persisting on the axis <Fig. 49>/
    2. deciduous with the spikelets <Figs 50, 51>/

#105. The involucres <whether deciduous or persistent>/
    1. persistent on the rachis/
    2. shed with the fertile spikelets/

#106. Spikelets <grouping: recorded mainly in spikes and racemes>/
    1. <mainly> solitary <Plate 1:4; Figs 30, 38, 56>/
    2. <consistently> paired <Plate 1:6; Fig. 44>/
    3. <consistently> in triplets <Figs 46, 52, 59>/

#107. Spikelets <whether secund: currently a catch-all character, covering one-sidedness of inflorescence (e.g., *Dactylis*, dorsiventral rachides, etc.>/
    1. secund <Plates 1:2, 1:4, 2:15; Figs 31, 32, 38, 39, 42, 53, 54, 55>/
    2. not secund/

#108. Spikelets <insertion>/
    1. biseriate <on one side of the rachis> <Plate 1:4; Figs 31, 38, 39, 53, 54>/
    2. distichous <Fig. 30>/
    3. not two-ranked <not biseriate, not distichously arranged> <to become implicit>/

#109. Spikelets <insertion — revised version>/
    1. sessile <Plate 1:9; Figs 30, 56>/
    2. subsessile <Fig. 54>/
    3. pedicellate <Plates 1:1, 1:8, 1:5, 2:10–12 etc.; Figs 34, 35, 53, 63, etc.>/

#110. Pedicel apices <shape — recorded as yet only in Paniceae. Data mainly from R.D. Webster 1985>/
    1. oblique <Fig. 58>/
    2. truncate <Fig. 58>/
    3. discoid <Plate 2:12; Fig. 58>/
    4. cupuliform <Plates 1:8, 2:10; Fig. 57>/

#111. Spikelets <disposition, e.g. *Diplachne/Leptochloa*: not widely recorded>/
    1. imbricate/
    2. distant <not overlapping>/

#112. Spikelets <whether in regular 'long-and-short' combinations, as exemplified in typical andropogonoids>/
    1. consistently in 'long-and-short' combinations <i.e., pedicellate/sessile or long-pedicel/short-pedicel pairs or triplets: currently includes andropogonoid forms with the pedicellate 'spikelets' reduced to their pedicels> <Plates 1:6, 1:7; Figs 28, 44, 47, 59, 72, 76>/
    2. not <consistently> in distinct 'long-and-short' combinations <implicit>/
#113. Spikelets <detail of 'long-and-short' combinations (intended mainly for andropogonoids)>/
    1. in pedicellate/sessile combinations <Figs 28, 44, 47, 59>/
    2. unequally pedicellate in each combination/
#114. Pedicels of the 'pedicellate' spikelets <whether fused with the rachis: intended for andropogonoids>/
    1. discernible, but <extensively> fused with the rachis <Plate 1:7; Figs 47, 60>/
    2. free of the rachis <Fig. 44, 75>/
#115. The 'shorter' <andropogonoid> spikelets <sessile or shorter-pedicelled, sexuality>/
    1. hermaphrodite <Fig. 59>/
    2. female-only/
    3. male-only/
    4. sterile/
#116. The 'longer' <andropogonoid> spikelets <pedicelled or longer-pedicelled, sexuality>/
    1. hermaphrodite/
    2. female-only/
    3. male-only/
    4. sterile <comment if reduced to pedicels> <Figs 59, 60>/

Female-sterile spikelets

#117. <Comments on female-sterile spikelets>/
#118. Rachilla of male spikelets <whether prolonged beyond the uppermost male floret>/
    1. prolonged beyond the uppermost male floret/
    2. terminated by a male floret/
#119. The male spikelets <presence of glumes>/
    1. with glumes/
    2. without glumes/
#120. The male spikelets <whether with proximal sterile florets>/
    1. with proximal incomplete florets/
    2. without proximal incomplete florets/
#121. The male spikelets <number of florets>/
    floreted/
#122. The lemmas <of male spikelets, whether awned>/
    1. awnless/
    2. mucronate/
    3. awned/
#123. Male florets <number per male spikelet>/
#124. Male florets <of male-only spikelets, number of stamens>/
    staminate/
#125. The staminal filaments <in male-only spikelets, whether joined>/
    1. free <implicit>/
    2. joined/

Female-fertile spikelets

#126. <Female-fertile> spikelets <whether morphologically conventional>/
    1. morphologically 'conventional' <with readily identifiable glumes, lemmas and paleas> <implicit>/
    2. <very> unconventional <and hard to interpret>/
#127. <Female-fertile> spikelets <approximate length, excluding any awns: data unreliable for large genera>/
    mm long/
#128. <Female-fertile> spikelets <shape>/
    1. fusiform/
    2. subcylindrical/
    3. subspherical/
    4. suborbicular/
    5. cuneate/
    6. turbinate/
    7. oblong/
    8. elliptic/
    9. lanceolate/
    10. ovate/
    11. obovate/
    12. oblanceolate/
#129. <Female-fertile> spikelets <colour: few data>/
#130. <Female-fertile> spikelets <orientation of sessile to subsessile forms>/
    1. abaxial <G1 when present on the side away from the rachis; in panicoid forms having a proximal incomplete floret, the upper (female-fertile) lemma backs onto the rachis> <Plate 1:7; Figs 59, 70, 79>/
    2. adaxial <G1 when present against the rachis; in panicoid forms having a proximal incomplete floret, the upper (female-fertile) lemma is on the side away from the rachis> <Plate 1:4>/
#131. <Female-fertile> spikelets <plane of compression>/
    1. compressed laterally <lying on the side when placed on a flat surface> <Plates 1:2, 1:5, 1:9, 2:15, 2:17; Figs 30, 54, 56, 61, 67, 68, 73, 119>/
    2. not noticeably compressed <terete>/
    3. compressed <dorsally, ventrally or> dorsiventrally <lying on front or back when placed on a flat surface> <Plates 1:4, 1:7, 2:10, 2:11; Figs 70, 74, 75, 82, 97, 98>/
#132. <Female-fertile> spikelets <shape of 'dorsiventrally flattened' forms>/
    1. planoconvex/
    2. biconvex/
#133. <Female-fertile> spikelets <location of disarticulation positions>/
    1. <readily> disarticulating above the glumes <when mature>/
    2. disarticulating between the glumes/
    3. falling with the glumes <when mature> <pending data changes, including forms where the spikelets are shed by inflorescence disarticulation>/
    4. not disarticulating <common in cultivated cereals>/
#134. <Female-fertile> spikelets <whether rachilla disarticulates between the florets of spikelets with two or more fertile florets>/
    1. not disarticulating between the florets/
    2. disarticulating between the florets/
#135. <Female-fertile> spikelets <rachilla internode spacings: unsatisfactorily defined, and inadequately scored for treating state 1 as implicit>/
    1. with conventional internode spacings/
    2. with a distinctly elongated rachilla internode between the glumes <Fig. 62>/
    3. with a distinctly elongated rachilla internode above the glumes <Figs 63, 64, 81>/
    4. with distinctly elongated rachilla internodes between the florets/

#136. <Presence or absence of *Ichnanthus*-type stipe: Paniceae>/
    1. the upper floret conspicuously stipitate <Fig. 64>/
    2. the upper floret not stipitate/
#137. The stipe beneath the upper floret <thickness: *Ichnanthus* relatives>/
    1. filiform/
    2. not filiform <Fig. 64>/
#138. The stipe beneath the upper floret <shape: *Ichnanthus* relatives>/
    1. straight and swollen <Fig. 64>/
    2. curved, not swollen/
#139. The stipe beneath the upper floret <whether heterogeneous Zuloaga 1987>/
    1. heterogeneous <membranous towards the base of the palea, indurated on the lemma side>/
    2. homogeneous/
#140. Rachilla <of female-fertile spikelets, whether terminated by a female-fertile floret, or 'prolonged'>/
    1. prolonged beyond the uppermost female-fertile floret <i.e. not terminated by a female-fertile floret: note that 'racemose' spikelets with three or more female-fertile florets have all been awarded this state> <Figs 41, 56, 61, 65>/
    2. terminated by a female-fertile floret <not 'prolonged'>/
#141. Rachilla <of female-fertile spikelets, whether hairy>/
    1. hairy <between the female-fertile florets, or above the single one>/
    2. hairless/
#142. The rachilla extension <beyond the uppermost female-fertile floret of female-fertile spikelets, rudiments>/
    1. with incomplete florets/
    2. naked/
#143. Hairy callus <presence: an unsatisfactory catch-all character, but widely recorded and useful in keys>/
    1. present <Figs 28, 63, 72, 100>/
    2. absent/
#144. Callus hairs <presence, size: *Calamagrostis/Agrostis*>/
    1. present, more than 0.5 mm long/
    2. absent, or if present less than 0.5 mm long/
#145. Callus <presence/length: data very incomplete>/
    1. absent/
    2. short/
    3. long <Fig. 100>/
#146. Callus <whether blunt or pointed>/
    1. pointed <Figs 28, 100>/
    2. blunt/
#147. Glumes <of female-fertile spikelets, present or absent>/
    1. present <implicit>/
    2. absent/
#148. Glumes <of female-fertile spikelets, number: 'glumes' are barren, with neither axillary spikelets nor florets>/
    1. one per spikelet/
    2. two/
    3. several/
#149. Glumes <whether glumes of the female-fertile spikelets are all minute>/
    1. minute <relative to the rest of the spikelet> <Plate 1:2; Fig. 63>/
    2. relatively large <implicit>/
#150. Glumes <of female-fertile spikelets, whether markedly unequal in the intact spikelet; regardless of any differences in form>/
    1. very unequal <in length in the intact spikelet> <Plates 1:8, 2:10, 2:12; Figs 61, 68, 71>/
    2. more or less equal <in length in the intact spikelet: generally includes 'subequal'> <Plates 1:1, 1:7, 2:11; Figs 62, 66, 73, 79, 85, 89, 114, 119>/

#151. Glumes <length relative to the spikelet: poorly recorded>/
    1. <markedly> shorter than the spikelets <Figs 61, 68>/
    2. about equalling the spikelets <Plate 1:8; Figs 62, 93>/
    3. exceeding the spikelets <Figs 66, 67, 73, 85, 114>/
#152. Glumes <of female-fertile spikelets, lengths relative to proximal (adjacent)
    lemmas. Refers to the longer glume when glumes unequal>/
    1. <decidedly> shorter than the adjacent lemmas <in intact spikelets> <Figs 61,
    68, 89>/
    2. long relative to the adjacent lemmas <more or less equalling or exceeding
    them> <Plates 1:1, 1:7, 1:8, 2:15, 2:16; Figs 62, 66, 67, 71, 73, 79, 85, 114,
    119>/
#153. Glumes <of female-fertile spikelets, whether free or joined>/
    1. joined <at least basally>/
    2. free <implicit>/
#154. Glumes <of female-fertile spikelets, whether ventricose>/
    1. conspicuously ventricose <basally> <Fig. 69>/
    2. not ventricose <implicit>/
#155. Glumes <of sessile to subsessile female-fertile spikelets, position relative to
    rachis>/
    1. dorsiventral to the rachis <the entire spikelet orientated dorsiventrally to
    flatwise> <Plates 1:7, 1:9; Figs 56, 70, 79>/
    2. lateral to the rachis <the spikelets usually but not always borne flatwise>/
    3. displaced <e.g., lateral to each other on side away from rachis>/
#156. Glumes <of female-fertile spikelets, whether hairy>/
    1. hairy <Plates 1:1, 1:3, 1:8, 2:12, 2:13, 2:15, 2:16; Figs 52, 61>/
    2. hairless <Plate 1:5; Fig. 73>/
#157. Glumes <hairless, whether glabrous or scabrous>/
    1. glabrous <Plate 1:8; Fig. 73>/
    2. scabrous <Plate 1:5>/
#158. Glumes <of female-fertile spikelets, hair disposition>/
    1. with distinct hair tufts/
    2. with distinct rows of hairs/
    3. without conspicuous tufts or rows of hairs <implicit>/
#159. Glumes <of female-fertile spikelets, shape of apex>/
    1. pointed <Plates 1:3, 1:5, 2:13, 2:17; Figs 61, 67, 79, 85>/
    2. not pointed <blunt or incised>/
#160. Glumes <of female-fertile spikelets, shape>/
    1. subulate/
    2. not subulate <to become implicit>/
#161. Glumes <of female-fertile spikelets, whether awned>/
    1. awned <Plates 1:1, 2:15; Fig. 32>/
    2. awnless <Fig. 73>/
#162. Glumes <of female-fertile spikelets, whether carinate (i.e., one-keeled to middle
    or below)>/
    1. carinate <one-keeled> <Plates 1:5, 2:17; Figs 54, 67, 73, 89, 114>/
    2. non-carinate <includes forms with more than one keel, as well as those with
    non-keeled glumes> <Plates 1:5, 1:7>/
#163. Glumes <of female-fertile spikelets, whether conspicuously winged on the
    median keel>/
    1. with the keel conspicuously winged <Fig. 73>/
    2. without a median keel-wing <implicit>/
#164. Glumes <of female-fertile spikelets, whether markedly dissimilar in form or
    texture; ignore mere size difference>/
    1. very dissimilar <specify> <Plates 1:7, 2:10, 2:12; Figs 59, 61>/
    2. <more or less> similar <Plates 1:1, 2:11; Figs 66, 67, 73, 81, 85, 89>/
#165. Lower glume <*in situ* length relative to upper glume of female-fertile spikelet:
    not recorded if glumes more or less equal>/
    times the length of the upper glume/

#166. Lower glume <length relative to lowest lemma: not widely recorded>/
    1. shorter than the lowest lemma/
    2. about equalling the lowest lemma/
    3. much exceeding the lowest lemma/
#167. Lower glume <length relative to the lowest lemma (originally introduced to deal
    with *Colpodium/Catabrosa*)>/
    1. much shorter than half length of lowest lemma/
    2. longer than half length of lowest lemma/
#168. Lower glume <of female-fertile spikelets, whether distinctly two-keeled to the
    middle or below (intended mainly for andropogonoids)>/
    1. two-keeled <distinctly two-keeled to the middle or below> <Plate 1:7; Figs 74,
    75>/
    2. not two-keeled <not distinctly two-keeled, at least below the upper quarter>
    <implicit, save in Andropogonodae> <Fig. 72>/
#169. Lower glume <of female-fertile spikelets, shape of back (recorded mainly for
    andropogonoids)>/
    1. convex on the back <Plate 1:7; Fig. 72>/
    2. flattened on the back <Fig. 74>/
    3. concave <between the keels> on the back/
    4. sulcate on the back/
#170. Lower glume <of female-fertile spikelet, whether pitted with 1–3 pits, cf.
    Bothriochloa; not synonymous with lacunose, qv. (intended for
    andropogonoids)>/
    1. with a conspicuous pit <Fig. 74>/
    2. not pitted <implicit>/
#171. Lower glume <of female-fertile spikelet, texture (intended mainly for
    andropogonoids)>/
    1. relatively smooth <i.e. glabrous, scabrous or hairy, but not lacunose, rugose,
    tuberculate, muricate or spiny> <Plate 1:7; Fig. 60>/
    2. lacunose with <several-to-many> deep depressions <Figs 75, 76>/
    3. rugose/
    4. tuberculate <Fig. 59>/
    5. muricate/
    6. prickly <Fig. 77>/
#172. Lower glume <of female-fertile spikelet, mid-zone nerve number>/
    nerved/
#173. Upper glume <whether (asymmetrically) saccate: e.g. *Sacciolepis*>/
    1. distinctly saccate <Fig. 81>/
    2. not saccate <implicit>/
#174. Upper glume <(or the single glume) of female-fertile spikelets, mid-zone nerve
    number>/
    nerved/
#175. Upper glume <whether prickly>/
    1. prickly/
    2. not prickly <implicit>/
#176. <Female-fertile> spikelets <whether containing sterile or male-only florets in
    addition to female-fertile florets>/
    1. <normally> with female-fertile florets only/
    2. <or at least some of them, normally> with incomplete <sterile or male-only>
    florets <note that the situation at the apex of spikelets with more than three
    florets is often unknown or unclear> <Plates 1:8, 2:10, 2:12, 2:13, 2:16; Figs
    61, 64, 71, 78, 79>/
#177. The incomplete <male or sterile> florets <position in spikelet>/
    1. proximal to the female-fertile florets <Plates 1:8, 2:12, 2:13, 2:16; Figs 64, 71,
    79>/
    2. distal to the female-fertile florets <Figs 61, 78>/
    3. both distal and proximal to the female-fertile florets/
#178. The distal incomplete florets <number: few data>/

#179. The distal incomplete florets <specialisation>/
    1. merely underdeveloped <neither clearly specialised nor peculiarly modified in form> <Fig. 61>/
    2. clearly specialised and modified in form <Fig. 78>/
#180. The distal incomplete florets <whether awned: data very incomplete>/
    1. awned <Fig. 78>/
    2. awnless/
#181. <Female-fertile> spikelets <presence or absence of proximal incomplete florets. Strictly speaking, a redundant character, but universally recorded and very useful for key-making>/
    1. with proximal incomplete florets <includes empty lemmas> <Plates 1:8, 2:10, 2:12, 2:13, 2:16; Figs 64, 71, 79, 80, 84>/
    2. without proximal incomplete florets <and no proximal empty lemmas>/
#182. The proximal incomplete florets <of the female-fertile spikelets, when present, number (intended mainly for panicoids)>/
#183. The proximal incomplete florets <of the female-fertile spikelets, whether paleate>/
    1. paleate <Plate 2:13; Figs 71, 79, 80>/
    2. epaleate <Fig. 64>/
#184. Palea of the proximal incomplete florets <development>/
    1. fully developed <Fig. 84>/
    2. reduced <or vestigial> <Plate 2:13; Fig. 80>/
#185. Palea of the proximal incomplete florets <whether becoming hardened and enlarged laterally: Paniceae>/
    1. becoming conspicuously hardened and enlarged laterally/
    2. not becoming conspicuously hardened and enlarged laterally <implicit>/
#186. The proximal incomplete florets <of the female-fertile spikelets: sexuality>/
    1. male <Plate 2:13; Figs 71, 79, 80>/
    2. sterile <Plate 2:16>/
#187. The proximal <imperfect> lemmas <of the female-fertile spikelets: shape comments>/
#188. The proximal <imperfect> lemmas <of the female-fertile spikelets: whether awned>/
    1. awned/
    2. awnless/
#189. The proximal <imperfect> lemmas <of the female-fertile spikelets, mid-zone nerve number (intended mainly for panicoids)>/
nerved/
#190. The proximal <imperfect> lemmas <of the female-fertile spikelets, length relative to the female-fertile ones in the intact spikelet (intended mainly for panicoids)>/
    1. exceeded by the female-fertile lemmas <Fig. 80>/
    2. more or less equalling the female-fertile lemmas <Fig. 82>/
    3. decidedly exceeding the female-fertile lemmas <Plates 2:12, 2:16; Figs 64, 71, 84>/
#191. The proximal <imperfect> lemmas <of the female-fertile spikelets, firmness relative to the female-fertile ones (intended mainly for panicoids)>/
    1. less firm than the female-fertile lemmas <Plate 2:16; Figs 64, 71, 80>/
    2. similar in texture to the female-fertile lemmas/
    3. decidedly firmer than the female-fertile lemmas/
#192. The proximal <imperfect> lemmas <of the female-fertile spikelets, whether becoming indurated (intended mainly for panicoids)>/
    1. becoming indurated/
    2. not becoming indurated <Plate 2:16; Fig. 84>/
#193. <Number of> female-fertile florets <per female-fertile spikelet>/
#194. <Female-fertile> lemmas <insertion>/
    1. conspicuously non-distichous/
    2. not conspicuously non-distichous <implicit>/

#195. <Female-fertile> lemmas <shape comments>/
#196. <Female-fertile> lemmas <whether convolute>/
    1. convolute <and hiding the palea> <Fig. 85>/
    2. not convolute <implicit: but data not yet reliable>/
#197. <Female-fertile> lemmas <whether saccate>/
    1. saccate <Figs 81, 89>/
    2. not saccate <unreliably implicit>/
#198. <Female-fertile> lemmas <firmness, relative to the glumes>/
    1. less firm than the <firmer of the> glumes/
    2. similar in texture to the <firmer of the> glumes/
    3. decidedly firmer than the <firmer of the> glumes <Plate 2:16; Figs 64, 71, 81, 89, 114>/
#199. <Female-fertile> lemmas <texture: data provided for Australian Paniceae by R.D. Webster>/
    1. smooth <Figs 64, 81>/
    2. <longitudinally, minutely> striate <rugulose> <Fig. 82>/
    3. <transversely> rugose <Figs 83, 114>/
#200. <Female-fertile> lemmas <whether becoming indurated>/
    1. becoming indurated <cf. fingernails, when mature and dry> <Plates 2:11, 2:16; Figs 81, 83, 85, 114>/
    2. not becoming indurated <hyaline, membranous, leathery, cartilaginous etc.> <Fig. 71>/
#201. <Female-fertile> lemmas <of Paniceae, colour of mature L2: data from Webster 1986>/
    1. white in fruit/
    2. yellow in fruit/
    3. brown in fruit/
    4. black in fruit/
#202. <Female-fertile> lemmas <shape of apex>/
    1. entire <Figs 91, 97>/
    2. incised <Plate 2:18; Figs 66, 86, 87, 91>/
#203. <Female-fertile> lemmas <entire, whether pointed or blunt>/
    1. pointed <Fig. 97>/
    2. blunt <Figs 88, 101>/
#204. <Female-fertile> lemmas <number of lobes>/
    lobed/
#205. <Female-fertile> lemmas <whether deeply cleft>/
    1. deeply cleft <to a third or more> <Plate 2:18; Figs 66, 86, 87>/
    2. not deeply cleft <Fig. 91>/
#206. <Female-fertile> lemmas <whether crested, cf. *Cyrtococcum*>/
    1. crested at the tip <Fig. 89>/
    2. not crested <implicit>/
#207. <Female-fertile> lemmas <whether mucronate or awned>/
    1. awnless <neither mucronate nor awned> <Figs 68, 71, 73, 89 etc.>/
    2. mucronate <with a short, hard point or vestigial or incipient awn> <Plate 2:16; Fig. 104>/
    3. awned <Plates 1:5, 1:6, 2:18; Figs 56, 61, 66, 67, 85, 87, 90, 91, 114>/
#208. Awns <of female-fertile lemmas, form>/
    1. triple or trifid, commonly with a basal column <*Aristida* type> <Figs 92, 96>/
    2. not of the triple/trifid, basal column type <implicit>/
#209. Awns <of female-fertile lemmas, if present, number>/
#210. Awns <of female-fertile lemmas, position>/
    1. median <Figs 56, 61, 67, 91>/
    2. median and lateral <Plate 2:18; Figs 66, 86, 87>/
    3. lateral only/
#211. <The median> awns <whether different from the laterals in form>/
    1. different in form from the laterals <Figs 66, 86>/
    2. similar in form to the laterals <Plate 2:18; Fig. 87>/

#212. Awns <of female-fertile lemmas, position of (main, median)>/
    1. from a sinus <Figs 86, 91>/
    2. dorsal <Figs 56, 67, 90>/
    3. apical <Plate 1:5; Figs 61, 85, 87, 91, 92, 96, 114>/
#213. Awns <of dorsally awned female-fertile lemmas, position>/
    1. from near the top <from the upper quarter, or near the apex, or just behind an
       apical notch> <Fig. 56>/
    2. from well down the back <from near the middle, or below> <Figs 67, 90>/
#214. Awns <of female-fertile lemmas, whether hooked ('uncinate')>/
    1. hooked/
    2. not hooked <implicit>/
#215. Awns <of female-fertile lemmas, whether straight or geniculate when dry>/
    1. non-geniculate <straight or curved> <Plates 1:5, 2:18; Figs 56, 61, 87>/
    2. geniculate <usually twisted at the base> <Figs 66, 67, 86, 90>/
#216. Awns <of female-fertile lemmas, when non-geniculate, shape>/
    1. straight/
    2. recurving/
    3. flexuous/
#217. Awns <main, median of the female-fertile lemmas, hairiness>/
    1. hairless <glabrous or scabrous> <Figs 61, 67, 92>/
    2. hairy <but not long-plumose> <Fig. 91>/
    3. long-plumose <Plate 2:18; Figs 87, 96>/
#218. Awns <main, median of the female-fertile lemmas, relative length>/
    1. much shorter than the body of the lemma <Plate 1:5>/
    2. about as long as the body of the lemma/
    3. much longer than the body of the lemma <Plates 1:6, 2:18; Figs 61, 66, 67, 86,
       87, 91>/
#219. Awns <of female-fertile lemmas, number of veins entering base>/
    1. entered by one vein <Fig. 94>/
    2. entered by several <three or more> veins <Fig. 95>/
#220. Awns <of female-fertile lemmas, whether deciduous — e.g. *Stipa*/*Oryzopsis*>/
    1. deciduous/
    2. persistent <to become implicit>/
#221. The lateral awns <relative length>/
    1. shorter than the median <in the intact spikelet> <Figs 66, 86>/
    2. about equalling the median/
    3. exceeding the median/
#222. Awn bases <of female-fertile lemmas, whether twisted: few data>/
    1. twisted/
    2. not twisted/
#223. Awn bases <of female-fertile lemmas, whether flattened: few data>/
    1. flattened/
    2. not flattened/
#224. <Female-fertile> lemmas <whether hairy: excludes callus and awns>/
    1. <conspicuously> hairy <Plate 1:2; Figs 66, 86, 90>/
    2. hairless <glabrous, scabrous, sparsely puberulent, etc.> <Plate 2:16; Figs 61,
       64>/
#225. The hairs <of the female-fertile lemmas>/
    1. in tufts <Figs 66, 86>/
    2. not in tufts <implicit>/
#226. The hairs <of the female-fertile lemmas>/
    1. in transverse rows <Figs 66, 86>/
    2. not in transverse rows <implicit>/
#227. <Female-fertile> lemmas <hairless, whether glabrous or scabrous>/
    1. glabrous/
    2. scabrous/

#228. <Female-fertile> lemmas <whether carinate (i.e., one-keeled at least to the middle on the back>/
  1. carinate <with a single median keel> <Plate 1:2; Figs 54, 68>/
  2. non-carinate <rounded, flat, with two or more keels> <Figs 82, 83, 86, 90, 91>/
#229. The <lemma> keel <whether winged: few records>/
  1. winged/
  2. wingless/
#230. <Female-fertile> lemmas <whether margins *Digitaria* or *Paspalum* type (intended for Paniceae)>/
  1. having the margins <at least over the upper two-thirds> lying flat and exposed on the palea <*Digitaria*-type> <Plate 2:13; Fig. 97>/
  2. having the margins <at least over the lower two-thirds> tucked in onto the palea <*Paspalum*-type> <Plate 2:16; Fig. 98>/
#231. <Female-fertile> lemmas <presence of germination flap>/
  1. with a clear germination flap <when mature> <Figs 99, 100>/
  2. without a germination flap/
#232. <Female-fertile> lemmas <number of nerves traversing mid-region>/
  nerved/
#233. <Female-fertile> lemmas <confluence of nerves: data very incomplete>/
  1. with the nerves confluent <i.e. merging, not merely convergent> towards the tip/
  2. with the nerves non-confluent <Fig. 101>/
#234. Palea <female-fertile, presence in female-fertile florets>/
  1. present/
  2. absent/
#235. Palea <female-fertile, relative size>/
  1. relatively long <three-quarters or more of female-fertile lemma length> <Plates 2:13, 2:16; Figs 102, 104>/
  2. conspicuous but relatively short <less than three-quarters of female-fertile lemma length> <Fig. 88>/
  3. very reduced <or vestigial>/
#236. Palea <female-fertile, whether convolute>/
  1. convolute/
  2. not convolute <implicit: but data not yet reliable>/
#237. Palea <female-fertile, whether gaping: especially Aveneae>/
  1. gaping/
  2. tightly clasped by the lemma <not gaping>/
#238. Palea <female-fertile, whether incised>/
  1. entire <Fig. 106>/
  2. apically notched <Fig. 103>/
  3. deeply bifid/
#239. Palea <female-fertile, whether with awns or setae>/
  1. awnless, without apical setae <Figs 102, 103>/
  2. with apical setae <Fig. 104>/
  3. awned/
#240. Palea <texture: data very incomplete>/
  1. thinner than the lemma/
  2. textured like the lemma <Plates 2:11, 2:16; Figs 81, 97, 98>/
#241. Palea <female-fertile, whether indurated>/
  1. indurated <Plates 2:11, 2:16; Figs 81, 87>/
  2. not indurated <Figs 103, 105 etc.>/
#242. Palea <female-fertile, nerve number>/
  1. 1-nerved <truly 1-veined, or with two contiguous veins>/
  2. 2-nerved <with two well-separated nerves> <Figs 102, 103, 106>/
  3. several nerved <specify>/
  4. nerveless/

#243. Palea <female-fertile, whether dorsally 2-keeled, one-keeled (carinate), or
      keel-less>/
   1. one-keeled/
   2. 2-keeled <Figs 65, 88, 102, 103, 105, 106>/
   3. keel-less/
#244. Palea back <indumentum: as yet uncoded>/
   1. glabrous/
   2. scabrous/
   3. hairy/
#245. Palea keels <whether winged: data very incomplete>/
   1. winged <Figs 105, 106>/
   2. wingless <Fig. 102>/
#246. Palea keels <female-fertile, hairiness>/
   1. glabrous/
   2. scabrous <Fig. 105>/
   3. hairy <Figs 102, 103>/
#247. Lodicules <presence in female-fertile florets>/
   1. present/
   2. absent/
#248. Lodicules <number>/
#249. <Presence of third lodicule>/
   1. third lodicule present <Fig. 108>/
   2. no third lodicule/
#250. Lodicules <of female-fertile florets, whether anterior pair joined or free>/
   1. joined <at least basally> <Fig. 106>/
   2. free <Figs 103, 107–111>/
#251. Lodicules <of female-fertile florets, texture>/
   1. <distally> fleshy <'cuneate'; panicoid type> <Figs 103, 106, 107, 109>/
   2. <distally> membranous <i.e. pooid type> <Plate 2:14; Figs 108, 110, 111>/
#252. Lodicules <of female-fertile florets, whether hairy>/
   1. ciliate <or hairy> <Figs 103, 108, 110>/
   2. glabrous <Figs 106, 107, 109, 111>/
#253. Lodicules <of female-fertile florets, whether toothed: inapplicable when fleshy>/
   1. toothed/
   2. not toothed/
#254. Lodicules <of female-fertile florets, vascularization. Note: this fairly
      unsatisfactory character is not equivalent to 'presence or absence' of xylem>/
   1. heavily vascularized <cf. bamboos> <Figs 108, 112>/
   2. not or scarcely vascularized <i.e. the norm> <Figs 103, 106, 107, 109–111>/
#255. Stamens <number per female-fertile floret (not applicable to male spikelets or
      male florets)>/
#256. Stamens <whether filaments joined>/
   1. with free filaments <implicit>/
   2. monadelphous/
   3. diadelphous/
   4. triadelphous/
#257. Anthers <of female-fertile florets, length: data very incomplete, unreliable for
      large genera>/
   mm long/
#258. Anthers <whether penicillate>/
   1. penicillate <Fig. 114>/
   2. not penicillate <Fig. 113>/
#259. Anthers <whether connective apically prolonged>/
   1. with the connective apically prolonged/
   2. without an apically prolonged connective/
#260. Ovary <of female-fertile florets, whether apex glabrous or hairy>/
   1. glabrous <Plate 2:14; Figs 103, 106, 116, 118>/
   2. hairy <Figs 110, 112, 115, 117>/

#261. Ovary <whether with a conspicuous apical appendage>/
    1. with a conspicuous apical appendage <Fig. 115>/
    2. without a conspicuous apical appendage <implicit> <Figs 103, 116, 117>/
#262. The <ovary> appendage <form: intended for bamboos>/
    1. long, stiff and tapering <*Schizostachyum* type>/
    2. broadly conical, fleshy <*Bambusa* type>/
#263. Styles <whether fused>/
    1. fused <at least basally; each stigma assumed to represent one style> <Figs 107, 112, 117, 118>/
    2. free to their bases <Figs 103, 107, 110, 116>/
#264. Styles <fusion: as yet uncoded>/
    1. completely fused/
    2. joined below/
    3. free/
#265. Style bases <degree of separation: as yet uncoded>/
    1. adjacent/
    2. widely separated/
#266. Stigmas <number>/
#267. Stigmas <colour, in chasmogamous spikelets>/
    1. white <Plates 1:2, 1:5, 1:7, 2:14; Figs 110, 111, 115>/
    2. red <anthocyanin> pigmented <i.e. red, pink, purple or black> <Plates 1:3, 1:4, 1:6, 1:8, 2:10 etc.; Figs 79, 107>/
    3. <golden> brown <Plate 2:12>/

Fruit, embryo and seedling

#268. Disseminule <constitution: data not yet entered>/
    1. a naked seed/
    2. a free caryopsis/
    3. a caryopsis enclosed in but free of the lemma and palea/
    4. a caryopsis enclosed within and partially fused with the lemma and palea/
    5. consisting of the abscised spikelet/
    6. consisting of the abscised spikelet and its pedicel/
    7. comprising the rachis segment and associated structures/
    8. consisting of the disarticulated spikelet-bearing inflorescence unit/
    9. constituted by the complete, deciduous inflorescence/
#269. Fruit <adherence>/
    1. adhering to lemma and/or palea <Fig. 123>/
    2. free from both lemma and palea <but may be enclosed>/
#270. Fruit <length when mature>/
    1. small <less than 4 mm>/
    2. medium sized <4–10 mm>/
    3. large <more than 10 mm long>/
#271. Fruit <colour: few data>/
#272. Fruit <shape>/
    1. linear/
    2. fusiform/
    3. banana-shaped/
    4. ellipsoid <Fig. 121>/
    5. subglobose/
    6. pyriform/
#273. Fruit <whether grooved in transverse section>/
    1. longitudinally grooved <sulcate> <Fig. 121>/
    2. not grooved <includes terete, triangular in section, etc.: implicit> <Figs 122–124>/
#274. Fruit <plane of compression>/
    1. compressed laterally/
    2. compressed <dorsally, ventrally or> dorsiventrally <Figs 121–123>/

    3. not noticeably compressed <Figs 119–120>/
    4. trigonous/
#275. Fruit <indumentum: as yet uncoded>/
    1. glabrous/
    2. scabrous/
    3. hairy/
#276. Fruit <hair distribution>/
    1. with hairs confined to a terminal tuft <Fig. 121>/
    2. hairy on the body/
#277. Fruit <or grain surface pattern>/
    1. sculptured <Fig. 127>/
    2. <relatively> smooth <the near-universal condition: implicit> <Figs 121–126>/
#278. Hilum <form>/
    1. short <punctiform or shortly elliptical, less than half length of fruit> <Figs 122, 124>/
    2. long-linear <more than half as long as fruit> <Figs 121, 123>/
#279. Pericarp <texture>/
    1. thin <the usual condition: implicit> <Figs 119–126>/
    2. thick and hard/
    3. fleshy <fruit a berry>/
#280. Pericarp <whether fused or loose (or free)>/
    1. free <Figs 119, 120>/
    2. loosely adherent <fairly easily removable when soaked>/
    3. fused <implicit, except in Arundinoideae and Chloridoideae> <Figs 121–126>/
#281. Embryo <relative size>/
    1. large <at least one-third as long as fruit> <Fig. 125>/
    2. small <less than one-third as long as fruit> <Fig. 126>/
#282. Embryo <whether waisted in surface view>/
    1. waisted <Fig. 125>/
    2. not waisted <Fig. 126>/
#283. Seed <whether endospermic>/
    1. endospermic <implicit>/
    2. 'non-endospermic' <i.e. endosperm greatly reduced in the mature fruit: confined to some bamboos, where the condition is sometimes described as "liquid">/
#284. Endosperm <hard or liquid: data extensively from Terrell 1971, Rosengurtt *et al.* 1972>/
    1. liquid <soft or milky> in the mature fruit <confined to Pooideae>/
    2. hard/
#285. Endosperm <presence of lipid: data mainly from Rosengurtt *et al.* 1972>/
    1. with lipid/
    2. without lipid/
#286. Endosperm <form of starch grains: data mainly from Tateoka 1954, 1955, 1962>/
    1. containing only simple starch grains <each with only one hilum> <Fig. 129>/
    2. containing <at least some> compound starch grains <with at least some grains having two or more hila> <Fig. 128>/
#287. Embryo <presence of epiblast. Embro section data extensively from Reeder 1967, 1962 and Decker 1964>/
    1. with an epiblast <Fig. 132>/
    2. without an epiblast <Fig. 133>/
#288. Embryo <presence of scutellar tail>/
    1. with a scutellar tail <i.e. with a cleft between scutellum and coleorhiza> <Figs 132, 133>/
    2. without a scutellar tail/
#289. Embryo <relative length of mesocotyl internode>/
    1. with an elongated mesocotyl internode <Figs 132, 133>/
    2. with a negligible <short> mesocotyl internode/

#290. Embryo <number of scutellum bundles>/
    1. with one scutellum bundle <Fig. 131>/
    2. with more than one scutellum bundle/
#291. Embryonic leaf margins <whether overlapping or meeting>/
    1. meeting <Fig. 131>/
    2. overlapping <Fig. 130>/
#292. Seedling <relative length of mesocotyl: compiled data probably unreliable,
    because germination conditions should be standardized>/
    1. with a short mesocotyl <Figs 135, 136>/
    2. with a long mesocotyl <Figs 134, 137, 138>/
#293. Seedling <tightness of coleoptile: data extensively from Muller 1978>/
    1. with a loose coleoptile <at least near tip> <Fig. 135>/
    2. with a tight coleoptile/
#294. First seedling leaf <possession of lamina>/
    1. with a well-developed lamina/
    2. without a lamina/
#295. The <first seedling leaf> lamina <relative width: data on seedling leaf characters
    mainly from Kuwabara 1960, 1961 and H.T. Clifford (*pers. comm.*>/
    1. broad <length/breadth, ratio less than 20> <Figs 134, 137, 138>/
    2. narrow <length/breadth ratio 20 or more> <Fig. 136>/
#296. The <first seedling leaf> lamina <carriage>/
    1. erect <Fig. 136>/
    2. curved <Figs 134, 137>/
    3. supine <Fig. 138>/
#297. The <first seedling leaf> lamina <vein number, in middle>/
    veined/

Abaxial leaf blade epidermis

#298. <Whether abaxial leaf blade epidermis shows> costal/intercostal zonation/
    1. conspicuous <Plates 3:19, 3:21, 3:22; Figs 139–141, 143, 146, 147, 151–153,
    158, 169–174 etc.>/
    2. lacking <Figs 144, 185>/
#299. Papillae <presence in the abaxial leaf blade epidermis>/
    1. present <Plates 3:19, 3:22; Figs 139–141, 151–153, 160, 174, 179, 181>/
    2. absent/
#300. <Leaf blade abaxial epidermal> papillae <general location: data very
    incomplete>/
    1. costal <Figs 153, 160>/
    2. intercostal <Plate 3:22; Figs 151–153, 160>/
#301. Intercostal papillae <of the abaxial leaf blade epidermis, whether over-arching the
    stomata (at least at one end)>/
    1. <frequently> over-arching the stomata <Plate 3:22; Figs 140, 151, 152, 181>/
    2. not over-arching the stomata <Figs 174, 179>/
#302. Intercostal papillae <of the abaxial leaf blade epidermis, form, arrangement>/
    1. consisting of one oblique swelling per cell <Plate 3:22; Figs 140, 152, 181>/
    2. consisting of one symmetrical <conical or finger-like> projection per cell
    <Plate 3:19; Figs 151, 174>/
    3. several per cell <specify appearance> <Figs 139, 153, 179>/
#303. Long-cells <of abaxial leaf blade epidermis, whether similar in shape costally and
    intercostally>/
    1. similar in shape costally and intercostally <Figs 178, 180>/
    2. markedly different in shape costally and intercostally <Plate 3:21; Figs
    172–174, 176, 186>/
#304. Long-cells <of abaxial leaf blade epidermis, whether similar in thickness costally
    and intercostally>/
    1. of similar wall thickness costally and intercostally <Figs 178, 180, 186>/
    2. differing markedly in wall thickness costally and intercostally <Fig. 147>/

#305. Intercostal zones <of abaxial leaf blade epidermis, whether of typical long-cells>/
    1. with typical long-cells <more or less exclusively. Figs 139, 143, 144, 148, 152, 160, 169, 170, 172, 184, 186 etc.> <implicit when epidermis adequately described>/
    2. exhibiting many atypical long-cells <Figs 153, 174>/
    3. without typical long-cells <Fig. 146>/
#306. Mid-intercostal long-cells <of abaxial leaf blade epidermis, shape>/
    1. <more or less> rectangular <Plates 3:19, 3:21; Figs 139, 143, 145, 147, 149–152, 159, 160, 163, 164, 169, 170, 172, 173, 176, 177, 179, 180, 184–186>/
    2. <more or less> fusiform <narrowed at their ends> <Plate 3:20; Figs 144, 145, 148, 163, 176, 184>/
#307. Mid-intercostal long-cells <of abaxial leaf blade epidermis, whether walls straight or sinuous in (outer) optical section>/
    1. having markedly sinuous <tessellated> walls <Plates 3:19, 3:21; Figs 139, 143, 145, 147, 149, 150, 153, 159, 160, 164, 165, 168–175, 177–180, 182–186>/
    2. having straight or only gently undulating walls <Plate 3:20; Figs 144, 145, 148, 156, 163, 166, 181>/
#308. Microhairs <presence in abaxial leaf blade epidermis>/
    1. present <Plates 3:19, 3:22; Figs 139–142, 145, 149, 150, 152, 156, 160, 161, 164, 172, 174, 180, 182, 186 etc.>/
    2. absent/
#309. Microhairs <of abaxial leaf blade epidermis, shape>/
    1. more or less spherical/
    2. elongated <to become implicit>/
#310. Microhairs <of abaxial leaf blade epidermis, number of cells visible>/
    1. ostensibly one-celled <usually indicative of a sunken basal cell>/
    2. clearly two-celled <to become implicit>/
    3. uniseriate/
#311. Microhairs <of abaxial leaf blade epidermis, form>/
    1. panicoid-type <distal cell more or less parallel-sided or tapered to the apex; usually relatively elongated, thin-walled, often collapsed or missing> <Plate 3:22; Figs 139, 140, 149, 152, 156, 160–162, 164, 174, 177, 186>/
    2. chloridoid-type <distal cell inflated or more or less hemispherical, relatively short, usually thick-walled relative to the panicoid type, persistent> <Plate 3:19; Figs 141, 170, 180>/
    3. *Enneapogon*-type <long, with very long basal cell and relatively short, inflated apical cell> <Fig. 142>/
#312. <Abaxial leaf blade> microhair apical cell wall <thickness/firmness relative to that of the basal cell> <recorded only for Chloridoideae>/
    1. thinner than that of the basal cell and often collapsed/
    2. thinner than that of the basal cell but not tending to collapse/
    3. of similar thickness/rigidity to that of the basal cell/
#313. Microhairs <whether with 'partitioning membranes>/
    1. with 'partitioning membranes'/
    2. without 'partitioning membranes'/
#314. The 'partitioning membranes' <location>/
    1. in the basal cell/
    2. in the apical cell/
#315. Microhairs <of abaxial leaf blade, total external length: for species sample, see attached list plus Metcalfe 1960>/
    microns long/
#316. <Abaxial leaf blade> microhair basal cells <length> <recorded only for Chloridoideae: Van den Borre 1994>/
    microns long/
#317. Microhairs <of abaxial leaf blade, width at the septum: for species sample, see attached list>/
    microns wide at the septum/

#318. Microhair total length/width at septum <for species sample, see attached list.
   Useful ranges: 0.5–1.5 (more or less spherical); 1.5–3 (decidedly plump); 3–8
   (narrow); 8–40 (very narrow)>/
#319. Microhair apical cells <of abaxial leaf blade, length: for species sample, see the
   attached list plus Metcalfe 1960>/
   microns long/
#320. Microhair apical cell/total length ratio <for species sample, see attached list plus
   Metcalfe 1960. Useful approximations: 0–0.3 (a.c. markedly shorter than b.c.);
   0.3–0.7 (a.c. and b.c. about equal); 0.7–1.0 (a.c. markedly longer than b.c.)>/
#321. Stomata <abaxial, presence in the abaxial leaf blade epidermis>/
   1. common <abaxially>/
   2. absent or very rare <Fig. 183>/
#322. Stomata <of the abaxial leaf blade, end to end guard cell length: for species
   sample, see attached list>/
   microns long/
#323. <Stomatal> subsidiaries <of the abaxial leaf blade, whether papillate: data not
   comprehensively checked>/
   1. papillate <Fig. 139>/
   2. non-papillate <Figs 140, 152, 153>/
#324. <Stomatal> subsidiaries <of the abaxial leaf blade, shape>/
   1. parallel-sided <Plate 3:20; Figs 144, 148, 156, 163, 166, 167 (above)>/
   2. dome-shaped <Figs 143, 145, 164, 170, 182>/
   3. triangular <includes the triangular-truncate type> <Plates 3:21, 3:22; Figs 139,
      143, 147, 150, 152, 153, 160, 165, 167 (below), 168, 169, 171, 173, 174, 177,
      178, 179>/
#325. <Stomatal> subsidiaries <of the abaxial leaf blade, whether parallel-sided and
   triangular on the same leaf>/
   1. including both triangular and parallel-sided forms on the same leaf/
   2. not including both parallel-sided and triangular forms on the same leaf
      <implicit when epidermis adequately described>/
#326. <Stomatal> guard-cells <of the abaxial leaf blade, whether overlapped or
   overlapping (Watson & Johnston 1978: *Aust. J. Bot.* 26)>/
   1. overlapped by the interstomatals <Plate 3:20; Figs 166, 167>/
   2. overlapping to flush with the interstomatals <Plate 3:21; Figs 165, 167>/
#327. Intercostal short-cells <abaxial leaf blade, presence/abundance — prickles and
   hair bases not regarded as short-cells>/
   1. common <Figs 147, 149, 150, 152, 164, 169, 171–173, 175, 176, 179, 180,
      183, 184, 186>/
   2. absent or very rare <Figs 144, 148>/
#328. Intercostal short-cells <abaxial leaf blade epidermal, arrangement>/
   1. in cork/silica-cell pairs <Figs 152, 171–173, 175, 176, 180, 183, 184>/
   2. not paired <note that some short-cells recorded as 'solitary' probably represent
      superposed cork/silica-cell pairs> <Figs 152, 169, 179, 184>/
#329. Intercostal short-cells <abaxial leaf blade epidermal, whether silicified>/
   1. silicified <Figs 152, 171–173, 175, 176, 179, 180, 183>/
   2. not silicified <Figs 152, 169, 184>/
#330. Intercostal silica bodies <in the abaxial leaf blade epidermis> <recorded only for
   Chloridoideae: Van den Borre 1994>/
   1. absent/
   2. imperfectly developed/
   3. present and perfectly developed/
#331. Intercostal silica bodies <shape: data in process of organization, and not yet
   reliable>/
   1. rounded <currently including 'oblong' and 'elliptic'>/
   2. crescentic/
   3. tall-and-narrow/
   4. acutely-angled/
   5. cross-shaped/

6. vertically elongated-nodular <= vertically elongated-crenate, 'olyroid-type'>/
7. saddle shaped/
8. oryzoid-type/
9. <more or less> cubical/

#332. <Comments on macrohairs and prickles of the abaxial epidermis of the leaf blade>/

#333. Crown cells <presence in the abaxial leaf blade epidermis>/
1. present <Fig. 154>/
2. absent <implicit when epidermis adequately described>/

#334. Costal zones <of the abaxial leaf blade epidermis, possession of short-cells>/
1. with short-cells <implicit when epidermis adequately described>/
2. without short-cells/

#335. Costal short-cells <abaxial leaf blade epidermal, arrangement of short-cells; prickles, hair bases not counted as short-cells>/
1. conspicuously in long rows <of five or more cells> <Plates 3:19, 3:21, 3:22; Figs 140, 141, 143, 145, 146, 150–153, 156, 158, 160, 161, 164, 169, 171–175, 177, 180, 182, 184–186>/
2. predominantly paired <Figs 159, 170, 178>/
3. neither distinctly grouped into long rows nor predominantly paired <solitary; in short rows, mixtures of solitaries, pairs, short rows, etc.> <Figs 147, 148, 157, 162, 163, 176, 181>/

#336. Costal silica bodies <of the abaxial leaf blade epidermis, presence>/
1. present and well developed/
2. poorly developed/
3. absent/

#337. Costal silica bodies <distribution in the abaxial leaf blade epidermis, when present> <recorded only for Chloridoideae: Van den Borre 1994>/
1. present throughout the costal zones/
2. confined to the outer files of the costal zones/
3. confined to the central file(s) of the costal zones/
4. present in alternate cell files of the costal zones/

#338. Costal silica bodies <usual forms: preponderance often stated in data comments>/
1. horizontally-elongated crenate/sinuous <'pooid type'> <Plate 3:20; Figs 148, 155>/
2. horizontally-elongated smooth <Fig. 163>/
3. rounded <round to oval, potato shaped> <Figs 162, 178, 183>/
4. saddle shaped <Plate 3:19; Figs 150, 151, 158, 170>/
5. tall-and-narrow <Fig. 157>/
6. crescentic <Figs 159, 170>/
7. oryzoid <vertical dumb-bell> <Figs 139, 160, 164>/
8. 'panicoid-type' <cross shaped, dumb-bell shaped or nodular> <Plates 3:21,22; Figs 139–141, 143, 145, 149, 152, 156, 169, 171, 173, 175–177, 180–182, 184–186>/
9. acutely-angled <Isachne-type> <Figs 146, 161>/
10. rounded to upright-ovoid, with a raised prickly belt around the equator <'Scrotochloa-type'>/

#339. Costal silica bodies <panicoid, shape: variation likely to be underestimated — for identification, recommend MATCH I O U and 1–3/4 or 1–2/2–3/4>/
1. cross shaped <Figs 143, 145, 164, 174, 180>/
2. butterfly shaped <short dumb-bells with indented ends, intermediate between cross- and dumb-bell shaped, etc.> <Figs 143, 145, 177>/
3. dumb-bell shaped <Plate 3:21, 22; Figs 141, 143, 145, 152, 156, 177, 185>/
4. nodular <Figs 140, 149, 152, 156,176>/

#340. Costal silica bodies <of the abaxial leaf blade epidermis, presence of sharp points; including but not exclusively 'acutely-angled' and 'Isachne-type' sensu Metcalfe>/
1. <conspicuously> sharp-pointed <Figs 146, 161, 164, 173, 181>/
2. not sharp-pointed <implicit when epidermis adequately described>/

Transverse section of leaf blade, physiology, culm anatomy

#341. Leaf blades <whether the mid-laminar region consists largely of midrib>/
    1. <mainly> consisting of <reasonably interpretable as> midrib/
    2. 'laminar' <rather than consisting (largely) of midrib> <implicit>/
#342. <Maximum cells-distant count. Reliably indicative of photosynthetic pathway —
    see Hattersley and Watson 1975>/
    1. <showing a maximum cells-distant count of one, reliably predicting> C_4
       <Plates 3:23, 3:24; Figs 187, 188, 192–196, 198, 207–213, 222, 224–226>/
    2. <showing a maximum cells-distant count of two or more, reliably predicting>
       C_3 <Plates 3:25, 3:26; Figs 189–191, 197, 216, 218, 219, 223, 227>/
#343. The <C4> anatomical organization <of the leaf blade, whether conventional>/
    1. conventional <Plates 3:23, 3:24; Figs 187–188, 192–196, 210, 212, 222>
       <implicit when ts adequately described>/
    2. unconventional <Figs 198, 208–209, 211, 213>/
#344. Organization of <leaf blade> PCR tissue <when unconventional>/
    1. *Triodia* type <with the PCR cells forming a layer draping (at least in places)
       from one bundle to the next, rather than constituting discrete bundle sheaths>
       <Figs 198, 213>/
    2. *Alloteropsis* type <with two bundle sheaths, the inner being PCR> <Figs 208,
       209>/
    3. *Aristida* type <the PCR cells constituting a double bundle sheath>/
    4. *Arundinella* type <with single PCR files or groups in the mesophyll, in
       addition to the conventional PCR sheath> <Fig. 211>/
#345. <C_4> biochemical type <as determined by enzyme assay: see Hatch and Kagawa
    1974, Gutierrez et al. 1974(a) and 1974(b), Hatch, Kagawa and Craig 1975,
    and Prendergast, Hattersley and Stone 1987. Species samples in parentheses>/
    1. PCK/
    2. NAD–ME/
    3. NADP–ME/
#346. <Leaf blade XyMS: reliably indicative of C_4 type (Hattersley and Watson 1976.
    Note that this feature is ascertainable from primary vascular bundles only —
    see Character Note)>/
    1. XyMS+ <C_3, or C_4 'aspartate formers' type PCK or NAD–ME (exceptions:
       Eriachneae)> <Plates 3:24–26; Figs 187, 189–193, 195–198, 203–205, 212,
       214, 220, 223, 224>/
    2. XyMS– <C_4 'malate formers', type NADP–ME> <Plate 3:23; Figs 188, 199,
       208, 209, 222>/
#347. <Leaf blade> PCR sheath outlines <in C_4 forms> <data extensively from Ellis
    1977, and Prendergast and Hattersley 1987>/
    1. uneven <PCK or 'PCK-like', Figs 187, 193–195, 207, 210>/
    2. even <NAD-ME or 'NAD-ME-like'> <Figs 188, 192, 196, 212>/
#348. PCR sheaths of the primary vascular bundles <of the leaf blades, whether
    interrupted> <recorded only for Chloridoideae: Van den Borre 1994>/
    1. complete/
    2. interrupted/
#349. PCR sheaths of the primary vascular bundles <of the leaf blades, how
    interrupted> <recorded only for Chloridoideae: Van den Borre 1994>/
    1. interrupted abaxially only/
    2. interrupted adaxially only/
    3. interrupted both abaxially and adaxially/
    4. interrupted laterally <e.g. *Desmostachya*>/
#350. <Leaf blade> PCR sheath extensions <presence> <data mainly from Prendergast
    1987>/
    1. present <in at least some veins> <Figs 187, 193, 207>/
    2. absent <Figs 194–196>/
#351. Maximum number of <leaf blade PCR sheath> extension cells <data mainly from
    Prendergast 1987>/

#352. <Leaf blade> PCR cells <of C_4 forms, presence of a suberised lamella> <cf. Hattersley and Browning 1981>/
    1. with a suberised lamella <Figs 199–203>/
    2. without a suberised lamella <Figs 204–206>/
#353. <Leaf blade> PCR cell chloroplasts <of C_4 forms, shape> <data from Prendergast 1987, Prendergast, Hattersley and Stone 1987>/
    1. ovoid <Figs 193, 195>/
    2. elongated <Figs 194, 196, 205, 206>/
#354. <Leaf blade> PCR cell chloroplasts <of C_4 forms, whether granal. See Gutierrez et al. 1974, Carolin et al. 1973, Hattersley and Browning 1981>/
    1. with well developed grana <Figs 204–206>/
    2. with reduced grana <Figs 199–203>/
#355. <Leaf blade> PCR cell chloroplasts <position. Data extensively from Ellis 1977, Brown 1960, Prendergast and Hattersley 1987>/
    1. centrifugal/peripheral <sometimes NAD-ME, more often indicative of NADP–ME or PCK> <Plate 3:23; Figs 193, 195, 199, 201–203>/
    2. centripetal <NAD–ME: predominant in arid and semiarid species> <Figs 192, 194, 196, 204–206>/
#356. <Leaf blade> PBS cells <of C_3 forms, presence of suberised lamella>/
    1. with a suberised lamella/
    2. without a suberised lamella/
#357. <Leaf blade> mesophyll <whether chlorenchyma radiate: an ill-defined feature, not reliably indicative of photosynthetic pathway>/
    1. with radiate chlorenchyma <Plates 3:23, 3:24, 3:26; Figs 187, 192, 195–197, 207–210 etc.>/
    2. with non-radiate chlorenchyma <Plate 3:25; Figs 189–191, 213 etc.>/
#358. <Leaf blade> mesophyll <presence of palisade>/
    1. with <a clear> adaxial palisade <Figs 214, 216>/
    2. without <any obvious> adaxial palisade/
#359. <Leaf blade> mesophyll <presence of *Isachne*-type mesophyll>/
    1. *Isachne*-type <Plate 3:26; Fig. 197>/
    2. not *Isachne*-type <implicit when ts adequately described>/
#360. <Leaf blade> mesophyll <presence of 'circular cells' (i.e. isolated PCR cells or cell groups; 'distinctive cells')>/
    1. exhibiting 'circular cells' <Fig. 211>/
    2. without 'circular cells' <implicit when ts adequately described> <implicit when ts adequately described>/
#361. <Leaf blade> mesophyll <whether traversed by (at least some) columns of colourless mesophyll cells> <implicit when ts adequately described>/
    1. traversed by columns of colourless mesophyll cells <Plate 3:24; Figs 192, 210> <implicit when ts adequately described>/
    2. not traversed by colourless <mesophyll cell> columns <Figs 187, 188, 193, 212, 224, 225 etc.> <implicit when ts adequately described>/
#362. <Leaf blade> mesophyll <presence of arm cells (= 'ratchet' cells)>/
    1. with arm cells <Figs 213, 215, 216>/
    2. without arm cells <implicit when ts adequately described>/
#363. <Leaf blade> mesophyll <presence of fusoid cells>/
    1. with fusoids <Figs 214, 216–220>/
    2. without fusoids <implicit when ts adequately described>/
#364. The fusoids <whether part of the PBS>/
    1. an integral part of the PBS <Fig. 219>/
    2. external to <though contiguous with> the PBS <Figs 214–216, 218>/
#365. Leaf blade <ribbing>/
    1. with distinct, prominent adaxial ribs <only> <Figs 187, 189, 190, 227, 228>/
    2. 'nodular' in section <Plate 3:26; Fig. 191>/
    3. adaxially <more or less> flat <ignore mid-rib. Includes forms with abaxial ribs only> <Plates 3:23, 3:24; Figs 212, 224, 226>/

#366. Leaf blade <adaxial ribs, relative sizes>/
    1. with the ribs more or less constant in size <Fig. 187>/
    2. with the ribs very irregular in sizes <i.e. of two or more size orders; ignore the mid-rib> <Figs 189, 190, 227>/
#367. Midrib <of the leaf blade, prominence>/
    1. conspicuous <prominent in the outline, with distinctive sclerenchyma, etc.> <Plate 3:25; Figs 221, 222>/
    2. not readily distinguishable <other than by position>/
#368. Midrib <of the mid leaf blade, vascularization>/
    1. with one bundle only/
    2. having a conventional arc of bundles <i.e. at least three bundles> <Fig. 222>/
    3. having complex vascularization <i.e. with more than one bundle, not arranged in a conventional arc> <Fig. 221>/
#369. Midrib <and/or middle part of leaf blade, whether extensively of colourless mesophyll cells adaxially>/
    1. with <conspicuous> colourless mesophyll adaxially <Fig. 222>/
    2. without <conspicuous> colourless mesophyll adaxially <implicit when ts adequately described>/
#370. The lamina <in transverse section, symmetry around the midrib>/
    1. distinctly asymmetrical on either side of the midrib <usually involving marked asymmetry in the ribbing and/or the form of the margin; e.g. as in many bamboos>/
    2. symmetrical on either side of the midrib/
#371. Bulliforms <presence in the adaxial leaf blade of discrete adaxial groups: exclude 'hinge' groups flanking midribs>/
    1. present in discrete, regular adaxial groups <Plates 3:23, 3:24; Figs 187, 188, 191–197, 207, 208, 210, 212, 216, 221, 223–225. Presence probably under-recorded>/
    2. not in discrete, regular adaxial groups <bulliform cells absent or in ill defined or irregular groups, or constituting most of the epidermis> <Plate 3:25; Figs 189, 219, 226>/
#372. Bulliforms <form of groups: states 2–4 are better defined and more reliable for identification than state 1>/
    1. in simple <more or less> fans <i.e. without contiguous colourless mesophyll cells, deeply penetrating or not> <Figs 187, 190, 191, 196, 197, 207, 209, 220, 223; Metcalfe 1960, Fig. XV 2–4 and 6>/
    2. associated <internally> with colourless mesophyll cells to form deeply-penetrating fans <Plate 3:24; Figs 187, 192, 210, 212>/
    3. combining with colourless mesophyll cells to form narrow groups penetrating into the mesophyll <Fig. 224>/
#373. Bulliforms <whether associating with colourless mesophyll cells to form arches over minor leaf blade bundles>/
    1. associating with colourless mesophyll cells <sometimes> to form arches over small vascular bundles <Figs 188, 225>/
    2. nowhere involved in bulliform-plus-colourless mesophyll arches <implicit when ts adequately described>/
#374. <Presence in the leaf blade of small vascular bundles unaccompanied by sclerenchyma>/
    1. many of the smallest vascular bundles unaccompanied by sclerenchyma <Plate 3:23; Figs 222, 226>/
    2. all <or nearly all> the vascular bundles accompanied by sclerenchyma/
#375. Combined sclerenchyma girders <presence in the leaf blade of vascular bundles combining both adaxial and abaxial girders>/
    1. present <at least some, if only associated with midribs> <Plate 3:24, 3:25; Figs 187–195, 197, 212, 216, 220, 223–225, 227>/
    2. absent <cf. Figs 194, 208, 209>/

#376. Combined sclerenchyma girders <adaxial and abaxial sclerenchyma girders, whether forming 'anchors', I's or T's in one or more bundles of the leaf blade (include the midrib)>/
    1. forming 'figures' <'anchors', I's or T's> <in at least some bundles> <Plate 3:24; Figs 187–193, 195, 212, 216, 223, 227>/
    2. nowhere forming 'figures' <i.e. no 'anchors', I's or T's>/
#377. Sclerenchyma <whether all leaf blade sclerenchyma is bundle-associated>/
    1. all associated with vascular bundles <apart from any marginal fibres>/
    2. not all <obviously> bundle-associated <Plate 3:25; Fig. 227>/
#378. The 'extra' sclerenchyma <location of leaf blade sclerenchyma not associated with vascular bundles — exclusive of any in the midrib>/
    1. in abaxial groups/
    2. in a continuous abaxial layer/
    3. within the mesophyll/
    4. in adaxial groups/
#379. The 'extra' sclerenchyma <position of groups within the lamina — exclusive of midrib>/
    1. abaxial-hypodermal, the groups isolated <opposite bulliforms and/or furrows>/
    2. abaxial-hypodermal, the groups continuous with colourless columns/
    3. adaxial-hypodermal, contiguous with the bulliforms/
#380. The lamina margins <presence of marginal fibres, i.e. groups not associated with vascular bundles> <recorded only for Chloridoideae: Van den Borre 1994>/
    1. with fibres/
    2. without fibres/

Culm internodes

#381. Culm internode bundles <arrangement; poorly recorded, data mainly from Metcalfe 1960>/
    1. in one or two rings <ignore 'outer rings' of very few bundles>/
    2. in three or more rings/
    3. scattered/

Phytochemistry

#382. Tissues of the culm bases <whether accumulating abundant starch: data from Smith 1968, Smouter and Simpson 1989 and original observations>/
    1. with abundant starch/
    2. with little or no starch <implying fructans and/or sucrose>/
#383. Fructosans predominantly <short- or long-chain>/
    1. short-chain/
    2. long-chain/
#384. Leaves <presence of flavonoid sulphates: data from Harborne and Williams 1976>/
    1. containing flavonoid sulphates/
    2. without flavonoid sulphates/
#385. Leaf blade chlorophyll a:b ratio <data from Prendergast 1987>/
#386. Data on serological cross-reactions of pollen antigens <availability. To view comparative data from immunodiffusion and immunoelectrophoretic tests (photos of gels, histograms and tables), display the illustrations associated with this character — enter Alt+I, followed by PageDown. For further information, see Character Note (enter Alt+N)>/
    1. available/
    2. not available/
#387. RuBisCO $Km(CO_2)$ and Km(RUBP) data <availability. To view tables with comparative data, display the illustrations associated with this character: enter Alt+I, followed by PageDown>/
    1. available/
    2. not available <implicit>/

Special diagnostic features

#388. Spikelets <whether in much-reduced andropogonoid 'racemes', each of these reduced to a single triplet and enclosed at its base by a trumpet-like development of the peduncle tip — '*Anadelphia scyphofera*' etc.>/
    1. in much-reduced andropogonoid 'racemes', each of the latter reduced to a single triplet and enclosed at its base by a trumpet-like development of the peduncle tip/
    2. not borne as in '*Anadelphia scyphofera*' (q.v.) <implicit>/

#389. <Whether inflorescence of 2–3 glumeless, bracteate spikelets, the lodicules represented by a fringed annulus — *Anomochloa*>/
    1. inflorescence of 2–3 glumeless, bracteate spikelets, the lodicules represented by a fringed annulus/
    2. plant not as in *Anomochloa* (q.v.) <implicit>/

#390. <*Arundo/Phragmites*>/
    1. female-fertile lemmas conspicuously hairy; ligule hairs to 0.3 mm long, shorter than the membrane/
    2. female-fertile lemmas hairless; ligule hairs longer than 0.5 mm, longer than the membrane/

#391. <Whether the inflorescences of very peculiar pseudospikelets, characterized by development of rachides with long terminal segments, each of which serves as the pedicel of an abscissile spikelet — *Atractantha*>/
    1. the inflorescences of very peculiar pseudospikelets, characterized by development of rachides with long terminal segments, each of which serves as the pedicel of an abscissile spikelet/
    2. the inflorescences not as in *Atractantha* (q.v.) <implicit>/

#392. <Whether lemmas as broad as long, gibbous and umbonate, cordate at base — *Briza*>/
    1. lemmas as broad as long, gibbous and umbonate, cordate at base <*Briza*> <Fig. 88>/
    2. lemmas not as in *Briza* (q.v.) <implicit>/

#393. <Whether lemma awn winged, the wing extending down upper back of the lemma — *Brylkinia*>/
    1. lemma awn winged, the wing extending down the upper back of the lemma/
    2. lemma not wing-awned <implicit>/

#394. <Whether having male inflorescences elevated, with one to four spicate, unilateral branches, and the female spikelets in burr-like spatheate clusters — *Buchloë*>/
    1. the male inflorescences elevated, with one to four spicate, unilateral branches; female spikelets in burr-like clusters, usually two burrs per inflorescence, each burr on a short, stout rachis, partially enclosed in a broad, bractlike leaf sheath, falling entire with the indurate rachis united with the upper glumes/
    2. inflorescence not as in *Buchloë* (q.v.) <implicit>/

#395. <Whether upper glume extended downwards into a conspicuous spur — *Centrochloa*>/
    1. upper glume extended downwards into a conspicuous spur/
    2. upper glume not as in *Centrochloa* (q.v.) <implicit>/

#396. <Whether pedicels articulated and bearded with long hairs at and above the joint — *Chaetobromus*>/
    1. pedicels articulated and bearded with long hairs at and above the joint/
    2. pedicels not as in *Chaetobromus* (q.v.)/

#397. <Whether inflorescences in hard, globular 6–12 mm utricles — *Coix*>/
    1. inflorescences in hard, globular 6–12 mm utricles/
    2. inflorescences not as in *Coix* (q.v.) <implicit>/

#398. <Whether spikelets in numerous small, compact, short-branched panicles, each panicle at the tip of a stout, recurved peduncle and enclosed by a leathery, toothed involucre, the peduncles themselves subtended by the inflated sheaths of the (modified) upper leaves — *Cornucopiae*>/

1. spikelets in numerous small, compact, short-branched panicles, each panicle at the tip of a stout, recurved peduncle and enclosed by a leathery, toothed involucre, the peduncles themselves subtended by the inflated sheaths of the (modified) upper leaves/
2. spikelets not borne as in *Cornucopiae* (q.v.)/
#399. <*Cortaderia*/*Lamprothyrsus*>/
   1. the lemma awns lateral and median, the median strongly flattened/
   2. the median lemma awn not strongly flattened, laterals present or absent/
#400. <Whether lemmas awned, the awn bearing a ring of minute hairs at the middle, and apically clavate — *Corynephorus*>/
   1. lemmas awned, the awn bearing a ring of minute hairs at the middle, and apically clavate <Fig. 93>/
   2. lemmas without the characteristic *Corynephorus* awn <implicit>/
#401. <Whether the inflorescence of a few digitately-borne, bracteate spikelets, subtended by a spatheate leaf atop a single elongated culm internode, the plant very sedge-like in appearance — *Cyperochloa*>/
   1. the inflorescence of a few digitately-borne, bracteate spikelets, subtended by a spatheate leaf atop a single elongated culm internode, the plant very sedge-like in appearance/
   2. plants not as in *Cyperochloa* (q.v.) <implicit>/
#402. <Whether the lower glume exceeding the female-fertile lemma — *Diandrostachya*>/
   1. the lower glume exceeding the female-fertile lemma/
   2. the lower glume shorter than the female-fertile lemma/
#403. <Whether grain with a conspicuous whitish or yellowish, glossy beak — *Diarrhena*>/
   1. grain with a conspicuous whitish or yellowish, glossy beak/
   2. fruit not as in *Diarrhena* (q.v.) <implicit>/
#404. <Whether plants from a short rosette of winter leaves, the primary panicle producing secondary inflorescences with cleistogamous spikelets — *Dichanthelium*>/
   1. plants from a short rosette of winter leaves, the primary panicle producing secondary inflorescences with cleistogamous spikelets/
   2. plants not as in *Dichanthelium* (q.v.)/
#405. <Whether the pedicelled member of the sessile/pedicellate spikelet pairs much the larger, very striking, with a broad, flat, papery, reddish, long-awned lower glume — *Diectomis*>/
   1. the pedicelled member of the sessile/pedicellate spikelet pairs much the larger, very striking, with a broad, flat, papery, reddish, long-awned lower glume/
   2. without the sessile/pedicellate spikelet pairs characteristic of *Diectomis* <implicit>/
#406. <—*Enneapogon*/*Schmidtia*/*Cottea*/*Kaokochloa*>/
   1. female-fertile lemmas 9-lobed, each lobe terminating in an awn/
   2. female-fertile lemmas 6-lobed and 5-awned, with an awn arising between each pair of lobes/
   3. female-fertile lemmas irregularly lobed, the lobes produced into 7–11 awns/
   4. female-fertile lemmas with an incurved-emarginate apex, and a narrow awned lobe at each margin (sometimes with 1–2 shorter, additional lobes)/
#407. <Whether spikelets supported on a peculiar, hardened, cupuliform 'callus' — *Eriochloa*>/
   1. spikelets supported on a peculiar, hardened, cupuliform 'callus' <Plate 2:16>/
   2. no *Eriochloa*-type 'callus' <implicit>/
#408. <*Hackelochloa*/*Hemarthria*/*Rottboellia*>/
   1. lower glume of female-fertile spikelet globose, pitted/
   2. lower glume of female-fertile spikelet flattish, not pitted; 'pedicellate' spikelets similar to the female-fertile spikelets/
   3. lower glume of female-fertile spikelet flattish, not pitted; 'pedicellate' spikelets reduced, herbaceous/

#409. <Whether plants of wet places, the leaves remarkably thin and delicate —
Hubbardia>/
    1. plants of wet places, the leaves remarkably thin and delicate/
    2. plants not as in *Hubbardia* (q.v.) <implicit>/
#410. <Whether the adaxial surface of the leaf blade raised into sinuous lamellae —
Hydrothauma>/
    1. the adaxial surface of the leaf blade raised into sinuous lamellae/
    2. the adaxial surface of the leaf blade not as in *Hydrothauma* (q.v.)/
#411. <Whether plants aquatic, with inflated leaf sheaths serving as floats —
Hygroryza>/
    1. plants aquatic, with inflated leaf sheaths serving as floats/
    2. plants not as in *Hygroryza* (q.v.) <implicit>/
#412. <*Koeleria*/*Trisetum*>/
    1. panicle dense, cylindrical, ovoid, not interrupted: awns if present straight,
subterminal, inconspicuous in the inflorescence/
    2. panicle loose, or if dense then interrupted, neither cylindrical nor ovoid: awns
usually present, usually twisted, usually distinctly dorsal, conspicuous if
inflorescence compact/
#413. <Whether having female spikelets, with shell- or urn-shaped lemmas which are
closed save for an apical pore — *Leptaspis* and *Scrotochloa*>/
    1. having female spikelets, with shell- or urn-shaped lemmas which are closed
save for an apical pore/
    2. not having female spikelets as in *Leptaspis* and *Scrotochloa* (q.v.) <implicit>/
#414. <Whether female-fertile lemma very broad, with a conspicuous, succulent,
translucent region near the base of each wing — *Lombardochloa*>/
    1. female-fertile lemma very broad, with a conspicuous, succulent, translucent
region near the base of each wing/
    2. female-fertile lemma not as in *Lombardochloa* (q.v.) <implicit>/
#415. <Whether spikelets minute, shaped like cartoon birds' heads — *Lopholepis*>/
    1. spikelets minute, shaped like cartoon birds' heads/
    2. spikelets not as in *Lopholepis* (q.v.) <implicit>/
#416. <Whether plant coarsely tufted, with wiry leaf blades, the inflorescence of one
very peculiar spikelet enclosed in a sheath — *Lygeum*>/
    1. plant coarsely tufted, with wiry leaf blades, the inflorescence of one very
peculiar spikelet enclosed in a sheath/
    2. plant and inflorescence not as in *Lygeum* (q.v.) <implicit>/
#417. <Whether spikelets in 'false pairs', the pedicellate member of the andropogonoid
pair abscinding from its pedicel but remaining attached to the base of the
'article' above, alongside the sessile member of that 'article' — *Manisuris*>/
    1. spikelets in 'false pairs', the pedicellate member of the andropogonoid pair
abscinding from its pedicel but remaining attached to the base of the 'article'
above, alongside the sessile member of that 'article'/
    2. spikelets not arranged as in *Manisuris* (q.v.) <implicit>/
#418. <Whether spikelets with the distal incomplete florets and/or the rachilla apex
forming a terminal clavate appendage — *Melica* et al.>/
    1. spikelets with the distal incomplete florets and/or the rachilla apex forming a
terminal clavate appendage/
    2. spikelets without a terminal clavate appendage <implicit>/
#419. <*Merxmuellera*/*Karroochloa*/*Chaetobromus*/*Schismus*>/
    1. female-fertile lemmas with a bent awn, the awn twisted below/
    2. female-fertile lemmas awnless, mucronate or with a short straight awn/
#420. <*Merxmuellera*/*Karroochloa*>/
    1. spikelets 8–25 mm long, inflorescence longer than 60 mm long/
    2. spikelets 4–6(–7) mm long, inflorescence 10–60 mm long/
#421. <Whether spikelet with a single gibbous floret, the lemma awn placed off-centre
— *Nassella*>/
    1. spikelet with a single gibbous floret, the lemma awn placed off-centre/
    2. spikelet not as in *Nassella* (q.v.) <implicit>/

40

#422. <Whether the inflorescence a coarse, cylindrical 'raceme', apparently
representing a raceme of reduced 'glomerules', each glomerule shortly
pedunculate, comprising a single spikelet subtended crosswise by a lobed scale
forming an involucre-plus-bristle — *Odontelytrum*>/
 1. the inflorescence a coarse, cylindrical 'raceme', apparently representing a
raceme of reduced 'glomerules', each glomerule shortly pedunculate,
comprising a single spikelet subtended crosswise by a lobed scale forming an
involucre-plus-bristle/
 2. the inflorescence not as in *Odontelytrum* (q.v.) <implicit>/
#423. The lower lemma <whether combining a median translucent zone (cf. a
Bothriochloa pedicel) with a terminal pair of hygroscopically active setae —
Ophiochloa>/
 1. narrow, of peculiar form, having a central hyaline portion bordered by well
developed, conspicuously ciliate and terminally setose nerves (the whole being
reminiscent of a *Bothriochloa* pedicel), and exhibiting apically a pair of
cushion-based, 5–7 mm long, hygroscopically active and awnlike setae/
 2. not resembling a *Bothriochloa* pedicel in appearance, and without a pair of
hygroscopically active setae <implicit>/
#424. <*Paspalum/Echinochloa/Paspalidium*>/
 1. glumes and/or sterile lemmas awned or acuminate-mucronate/
 2. spikelets awnless, muticous/
 3. spikelets awnless, the female-fertile lemmas pointed or apiculate but not
mucronate/
#425. <Whether seed dark brown, with ruminate endosperm — *Phaenosperma*>/
 1. seed dark brown, with ruminate endosperm/
 2. seed not as in *Phaenosperma* (q.v.) <implicit>/
#426. <Whether spikelets borne on one side of a broad, leaflike rachis —
Phyllorhachis>/
 1. spikelets borne on one side of a broad, leaflike rachis/
 2. spikelets not borne on a broad, leaflike rachis <implicit>/
#427. <Whether scandent via leaf blades with retrorsely scabrid margins —
Prosphytochloa>/
 1. scandent via leaf blades with retrorsely scabrid margins/
 2. not scandent as in *Prosphytochloa* (q.v.) <implicit>/
#428. <Whether female-fertile lemma with its tip extended beyond the palea as a
conical, herbaceous beak (flotation device) composed of aerenchyma with
transverse septa, tapering into an awn — *Rhynchoryza*>/
 1. female-fertile lemma with its tip extended beyond the palea as a conical,
herbaceous beak (flotation device) composed of aerenchyma with transverse
septa, tapering into an awn/
 2. female-fertile lemma not as in *Rhynchoryza* (q.v.) <implicit>/
#429. <*Sorghastrum/Sorghum*>/
 1. spikelets ostensibly solitary, each accompanied by a barren pedicel/
 2. spikelets paired, all the pedicels spikelet-bearing/
#430. <Whether rush-like, with reduced leaf blades — *Spartochloa, Xerochloa*>/
 1. rush-like, with reduced leaf blades/
 2. not rush-like <implicit>/
#431. <Whether female inflorescence a large, deciduous globular head of sessile,
bristle-tipped racemes — *Spinifex*>/
 1. female inflorescence a large, deciduous globular head of sessile, bristle-tipped
racemes <Fig. 26>/
 2. inflorescence not as in *Spinifex* (q.v.) <implicit>/
#432. <Whether culms dimorphic, the fertile culms leafless, the vegetative culms each
with a single developed leaf, this being eligulate and with a terete, culm-like
'sheath' — *Steyermarkochloa*>/
 1. culms dimorphic, the fertile culms leafless, the vegetative culms each with a
single developed leaf, this being eligulate and with a terete, culm-like 'sheath'/
 2. plants not as in *Steyermarkochloa* (q.v.) <implicit>/

#433. <Whether flowering culms ultimately bending over, so as to enclose the ripening fruit — *Thuarea*>/
    1. flowering culms ultimately bending over, so as to enclose the ripening fruit/
    2. flowering culms not as in *Thuarea* (q.v.) <implicit>/
#434. <Whether lower glume <of female-fertile spikelet> with a rectangular window, surmounted by bristles — *Thyridolepis*>/
    1. lower glume <of female-fertile spikelet> with a rectangular window, surmounted by bristles <Fig. 70>/
    2. lower glume without a *Thyridolepis*-type window (q.v.) <implicit>/
#435. <Whether the lower glume of the pedicellate spikelet with a 5–10 mm (or longer) awn — *Urelytrum*>/
    1. the lower glume of the pedicellate spikelet with a 5–10 mm (or longer) awn/
    2. the lower glume of the pedicellate spikelet awnless/
#436. <Whether the inflorescence a spicate 'raceme', with each spikelet subtended at its base by a tiny hyaline bract (from Madagascar) — *Viguierella*>/
    1. the inflorescence a spicate 'raceme', with each spikelet subtended at its base by a tiny hyaline bract: Madagascar/
    2. the inflorescence not as in *Viguierella* (q.v.) <implicit>/
#437. <Whether fruiting inflorescence a massive, spatheate cob, the fruits in many rows — *Zea mays*>/
    1. fruiting inflorescence a massive, axillary, spatheate cob with a spongy axis, the fruits in many rows/
    2. fruiting inflorescence not as in maize (q.v.) <implicit>/
#438. <Whether stems cane-like, spikelets in bracteate, globular 1–3.5 cm heads — *Zygochloa*>/
    1. stems cane-like, spikelets in bracteate, globular 1–3.5 cm heads/
    2. plants not as in *Zygochloa* (q.v.) <implicit>/

Cytology

#439. Chromosome base number, $x =$/
#440. <Diploid chromosome numbers> $2n =$/
#441. <Recorded ploidy levels: data very incomplete>/
    ploid/
#442. Haplomic genome content <of Triticeae, as shown by genomic analysis: data incomplete>/
    1. A/
    2. B/
    3. C/
    4. D/
    5. E/
    6. F/
    7. G/
    8. H/
    9. I/
    10. J/
    11. K/
    12. L/
    13. M/
    14. N/
    15. O/
    16. P/
    17. Q/
    18. R/
    19. S/
    20. T/
    21. U/
    22. V/

      23. W/
      24. X/
      25. Y/
      26. Z/

#443. Chromosomes <size, cf. Avdulov 1931 — data not yet entered>/

#444. Haploid nuclear DNA content <2c value divided by ploidy: ranges and means>/
      pg/

#445. Mean diploid 2c DNA value <range and number of species studied in parenthesis: data mainly from Bennett and Smith 1976 and Bennett, Smith and Heslop-Harrison 1982>/
      pg/

#446. Nucleoli <in root tip meristems, whether or not persisting to metaphase>/
      1. persistent/
      2. disappearing before metaphase/

Taxonomy

#447. <Subfamily>/
      1. Pooideae/
      2. Bambusoideae/
      3. Arundinoideae/
      4. Chloridoideae/
      5. Panicoideae/

#448. <Supertribes of Watson *et al.* 1985, with name endings changed>/
      1. Triticodae/
      2. Poodae/
      3. Oryzodae/
      4. Bambusodae/
      5. Panicodae/
      6. Andropogonodae/

#449. <Tribe of Pooideae>/
      1. Triticeae/
      2. Brachypodieae/
      3. Bromeae/
      4. Aveneae <including Agrostideae, Phalarideae>/
      5. Meliceae/
      6. Seslerieae/
      7. Poeae <including Hainardieae, Monermeae>/

#450. <Tribe of Bambusoideae>/
      1. Oryzeae/
      2. Olyreae/
      3. Centotheceae/
      4. Anomochloeae/
      5. Brachyelytreae/
      6. Diarrheneae/
      7. Ehrharteae/
      8. Phaenospermateae/
      9. Phyllorhachideae/
      10. Phareae/
      11. Streptochaeteae/
      12. Streptogyneae/
      13. Guaduelleae/
      14. Puelieae/
      15. Bambuseae/

#451. <Tribe of Arundinoideae>/
      1. Steyermarkochloeae/
      2. Stipeae/
      3. Nardeae/

4. Lygeae/
5. Arundineae/
6. Danthonieae <and satellites>/
7. Cyperochloeae/
8. Micraireae/
9. Spartochloeae/
10. Aristideae/
11. Eriachneae/
12. Amphipogoneae/
#452. <Tribe/assemblage of Chloridoideae>/
   1. Pappophoreae/
   2. Orcuttieae/
   3. Triodieae/
   4. main chloridoid assemblage <including Chlorideae, Cynodonteae, Eragrosteae, Sporoboleae, Aeluropodeae, Jouveae, Unioleae, Leptureae, Lappagineae, Spartineae, Trageae, Perotideae, Pommereulleae>/
#453. <Tribe of Panicoideae>/
   1. Isachneae/
   2. Paniceae/
   3. Neurachneae/
   4. Arundinelleae/
   5. Andropogoneae/
   6. Maydeae/
#454. <Subtribe of Andropogoneae>/
   1. Andropogoninae <'awned Andropogoneae'>/
   2. Rottboelliinae <'awnless Andropogoneae'>/

Ecology, geography, regional floristic distribution

#455. <Number of species>/
   species/
#456. <Geographic distribution>/
#457. <World distribution: this 'character' is intended only for convenience in key-making — for more precise distributions, see 'geographical distribution'>/
   1. Western Eurasia, U.S.S.R. <includes Iran, Iraq, Turkey>/
   2. Mediterranean/
   3. Eastern Asia <Japan, China to India>/
   4. Africa <and Saudi Arabia>/
   5. Pacific <Malaysia, Indonesia, Australasia, Pacific Islands>/
   6. North America <Canada, Alaska, U.S.A., Mexico>/
   7. South and Central America, West Indies/
   8. Arctic/
#458. <Whether commonly adventive on an intercontinental scale>/
   1. commonly adventive/
   2. not commonly adventive <implicit>/
#459. <Habitat water requirement>/
   1. hydrophytic/
   2. helophytic <i.e., in marshy places>/
   3. mesophytic/
   4. xerophytic/
#460. <Habitat light requirement>/
   1. shade species/
   2. species of open habitats/
#461. <Salt tolerance, etc.>/
   1. halophytic/
   2. glycophytic <= not halophytic>/
#462. <Habitat notes: soil types, etc.>/

#463. <Number of species in Australia: mainly from Simon 1990>/
    species in Australia/
#464. <Geographical occurrence in Australasia — mainly after Simon 1978 (to be
      updated re. Simon 1990)>/
    1. Tasmania/
    2. New South Wales/
    3. Australian Capital Territory/
    4. Victoria/
    5. Western Australia/
    6. Queensland/
    7. Northern Territory/
    8. South Australia/
    9. New Guinea/
    10. New Zealand/
    11. not known in Australasia <implicit>/
#465. <Geographical distribution in southern Africa>/
    1. Namibia/
    2. Botswana/
    3. Transvaal/
    4. Orange Free State/
    5. Swaziland/
    6. Natal/
    7. Lesotho/
    8. Cape Province/
    9. not in southern Africa <implicit>/
#466. <Status and species numbers in southern Africa>/
    1. <number of> indigenous species/
    2. <number of> naturalized species <in southern Africa>/
    3. <number of> cultivated <species>/
#467. <Geographical distribution in North America: data not yet entered>/
#468. <Number of species in the Flora North America region: data from Kartesz and
      Kartesz 1980>/
    species in North America/
#469. <Status in North America: *Flora North America* region>/
    1. indigenous species/
    2. naturalized species <in North America>/
    3. cultivated/
#470. <Floristic Kingdoms: after Takhtajan 1969. Data deduced from information for
      Takhtajan's floristic regions (see below), provided by B. K. Simon 1987>/
    1. Holarctic/
    2. Paleotropical/
    3. Neotropical/
    4. Cape/
    5. Australian/
    6. Antarctic/
#471. <Holarctic Subkingdoms>/
    1. Boreal/
    2. Tethyan <ancient Mediterranean>/
    3. Madrean <Sonoran>/
#472. <Paleotropical Subkingdoms>/
    1. African/
    2. Madagascan/
    3. Indomalesian/
    4. Polynesian/
    5. Neocaledonian/
#473. <Boreal Subkingdom regions>/
    1. Arctic and Subarctic/
    2. Euro-Siberian/

    3. Eastern Asian/
    4. Atlantic North American/
    5. Rocky Mountains/
#474. <Tethyan Subkingdom regions>/
    1. Macaronesian/
    2. Mediterranean/
    3. Irano-Turanian/
#475. <African Subkingdom regions>/
    1. Saharo-Sindian/
    2. Sudano-Angolan/
    3. West African Rainforest/
    4. Namib-Karoo/
    5. Ascension and St. Helena/
#476. <Indomalesian Subkingdom regions>/
    1. Indian/
    2. Indo-Chinese/
    3. Malesian <Malayan>/
    4. Papuan/
#477. <Polynesian Subkingdom regions>/
    1. Hawaiian/
    2. Polynesian/
    3. Fijian/
#478. <Neotropical regions>/
    1. Caribbean/
    2. Venezuela and Surinam/
    3. Amazon/
    4. Central Brazilian/
    5. Pampas/
    6. Andean/
    7. Fernandezian/
#479. <Australian regions>/
    1. North and East Australian/
    2. South-West Australian/
    3. Central Australian/
#480. <Antarctic regions>/
    1. New Zealand/
    2. Patagonian/
    3. Antarctic and Subantarctic/
#481. <Euro-Siberian Subregions>/
    1. European/
    2. Siberian/
#482. <Atlantic North American Subregions>/
    1. Canadian-Appalachian/
    2. Southern Atlantic North American/
    3. Central Grasslands/
#483. <Sudano-Angolan Subregions>/
    1. Sahelo-Sudanian/
    2. Somalo-Ethiopian/
    3. South Tropical African/
    4. Kalaharian/
#484. <North and East Australian Subregions>/
    1. Tropical North and East Australian/
    2. Temperate and South-Eastern Australian/
#485. <Antarctic and Subantarctic Subregions>/
    1. South Temperate Oceanic Islands/
    2. Antarctic/

46

Hybrids

#486. <Intergeneric hybrids>/

Rusts and smuts

#487. Rusts — <genera>/
    1. *Dasturella*/
    2. *Phakopsora*/
    3. *Physopella*/
    4. *Stereostratum*/
    5. *Puccinia* <including *Uromyces*>/
    6. no rusts recorded <by Cummins 1971>/
#488. The *Puccinia* species from <morphological Group — after Cummins 1971,
    amended by D.B.O. Savile (pers. comm.)>/
    1. Group 1/
    2. Group 2/
    3. Group 5/
    4. Group 6/
    5. Group 7/
    6. Group 8/
#489. The *Puccinia* species from <the Group 1 species, Savile's subgroups>/
    1. subgroup 1(a)/
    2. subgroup 1(b)/
    3. subgroup 1(c)/
    4. subgroup 1(d)/
#490. The *Puccinia* species from <the Group 2 species, Savile's subgroups>/
    1. subgroup 2(a)/
    2. subgroup 2(b)/
#491. The *Puccinia* species from <the Group 5 species, Savile's subgroups>/
    1. subgroup 5(a)/
    2. subgroup 5(b)/
    3. subgroup 5(c)/
    4. subgroup 5(d)/
    5. subgroup 5(e)/
    6. subgroup 5(f)/
    7. subgroup 5(g)/
    8. subgroup 5(h)/
    9. subgroup 5(i)/
    10. subgroup 5(j)/
    11. subgroup 5(k)/
#492. The *Puccinia* species from <the Group 6 species, Savile's subgroups>/
    1. subgroup 6(a)/
    2. subgroup 6(b)/
    3. subgroup 6(c)/
    4. subgroup 6(d)/
    5. subgroup 6(e)/
#493. Taxonomically wide-ranging <rust> species: <wide-ranging here = recorded on 3
    or more host genera by Cummins 1971>/
    1. *Dasturella divina*/
    2. *Phakopsora incompleta*/
    3. *Physopella clemensiae*/
    4. *Stereostratum corticoides*/
    5. *Puccinia chaetochloae*/
    6. *Puccinia stenotaphri*/
    7. *Puccinia microspora*/
    8. *Puccinia polysora*/
    9. *Puccinia miscanthae*/
    10. *Puccinia nakanishikii*/

11. *Puccinia longicornis*/
12. *Puccinia kusanoi*/
13. *Puccinia eritraeensis*/
14. *Puccinia graminella*/
15. *Puccinia dolosa*/
16. *Puccinia orientalis*/
17. *Puccinia graminis*/
18. *Puccinia levis*/
19. *Puccinia substriata*/
20. '*Uromyces*' *setariae-italicae*/
21. '*Uromyces*' *schoenanthi*/
22. *Puccinia emaculata*/
23. *Puccinia cacabata*/
24. *Puccinia coronata*/
25. *Puccinia striiformis*/
26. *Puccinia montanensis*/
27. *Puccinia pygmaea*/
28. *Puccinia brachypodii-phoenicoidis*/
29. *Puccinia brachypodii*/
30. *Puccinia praegracilis*/
31. *Puccinia poarum*/
32. *Puccinia hordei*/
33. *Puccinia recondita*/
34. '*Uromyces*' *turcomanicum*/
35. '*Uromyces*' *fragilipes*/
36. '*Uromyces*' *dactylidis*/
37. '*Uromyces*' *hordeinus*/
38. *Puccinia monoica*/
39. *Puccinia versicolor*/
40. *Puccinia boutelouae*/
41. *Puccinia chloridis*/
42. *Puccinia schedonnardi*/
43. '*Uromyces*' *clignyi*/
44. '*Uromyces*' *eragrostidis*/
45. *Puccinia miyoshiana*/
46. *Puccinia cesatii*/
47. *Puccinia esclavensis*/
48. *Puccinia aristidae*/
49. no wide-ranging rust species <i.e. the positive records limited to rusts with restricted host ranges, as given by Cummins 1971: implicit>/
#494. Smuts <families: data not yet updated from Watson 1972, and *Panicum, Danthonia, Agropyron, Elymus* etc. omitted pending nomenclatural checking of records>/
    1. from Tilletiaceae/
    2. from Ustilaginaceae/
    3. not recorded <implicit: but see qualification>/
#495. <Smut genera> Tilletiaceae —/
    1. *Entyloma*/
    2. *Melanotaenium*/
    3. *Neovossia*/
    4. *Tilletia*/
    5. *Urocystis*/
#496. <Smut genera> Ustilaginaceae —/
    1. *Sorosporium*/
    2. *Sphacelotheca*/
    3. *Tolyposporella*/
    4. *Tolyposporium*/
    5. *Ustilago*/

Economic importance

#497. Significant weed species: <list extended from Häfliger and Scholtz 1980>/
#498. Cultivated fodder:/
#499. Important native pasture species:/
#500. Grain crop species:/
#501. Lawns and/or playing fields:/
#502. Commercial essential oils:/
#503. <Miscellaneous economic/ethnic data> <little yet entered>/

References, etc.

#504. Morphological/taxonomic: <references>/
#505. Leaf anatomical: <references>/

Special comments

#506. <Special comments>/
#507. <Adequacy of> fruit data <for reliability of 'implicit states'>/
    1. more or less satisfactory <implicit>/
    2. wanting <totally or seriously lacking>/
#508. <Adequacy of> anatomical data <for reliability of 'implicit states'>/
    1. more or less satisfactory <implicit>/
    2. epidermal only/
    3. for ts only/
    4. wanting <totally or seriously lacking>/

Illustrations

#509. <Illustrations for screen display>/

Additional characters, data incompletely prepared

#510. Colourless <mesophyll> cells <including any contiguous bulliforms, extent of penetration into the lamina> <recorded only for Chloridoideae: data under development — not recommended for current use>/
    1. penetrating the lamina to less than a quarter of its depth/
    2. penetrating the lamina to at least a quarter but less than a half of its depth/
    3. penetrating the lamina to about half its depth/
    4. penetrating the lamina to about 2/3 of its depth/
    5. traversing the lamina so as to link with the abaxial epidermis/
#511. Abaxial epidermis <of the leaf blade, presence of inflated cells> <recorded only for Chloridoideae: data under development — not recommended for current use>/
    1. without inflated cells <or groups>/
    2. with all the cells slightly inflated/
    3. with inflated cells confined to the furrows/
    4. with all the cells clearly bulliform <often irregular in size>/
#512. The bulliform groups <in the adaxial epidermis, excluding 'hinges', whether associated with colourless mesophyll cells> <recorded only for Chloridoideae: data under development — not recommended for current use >/
    1. not associated with colourless mesophyll cells/
    2. associated with colourless mesophyll cells/
#513. Intercostal short cells <in the abaxial leaf blade epidermis, presence/distribution> <recorded only for Chloridoideae>/
    1. absent/
    2. irregularly dispersed/
    3. regularly alternating with the long cells/

#514. Lamina mid-zone in transverse section <shape> <recorded only for
Chloridoideae: Van den Borre 1994>/
  1. open <exhibiting infolding or inrolling only under conditions of water stress>/
  2. infolded permanently <involving modified internal structure; involving some
'acicular', 'setaceous', 'filiform' and 'junciform' types>/
  3. terete <involving some 'acicular' and 'junciform' types>/
#515. Lamina mid-zone in transverse section <adaxial outline of permanently infolded
leaves, excluding terete blades> <recorded only for Chloridoideae: Van den
Borre 1994>/
  1. V-shaped/
  2. U-shaped/
  3. circular <disregarding adaxial channel>/
  4. triangular <base broad on either side of median bundle>/
#516. The adaxial channel <of the lamina, in transverse sections of permanently
infolded leaves, 'circular' forms> <recorded only for Chloridoideae:
Van den Borre 1994>/
  1. parallel-sided, with a digitate base/
  2. irregularly furrowed/
  3. rounded/
  4. triangular/
  5. reduced to a small groove/

3. CLASSIFICATION

3.1. Introduction

Of the 793 genera, about 390 are monotypic or ditypic. 23 have about 100 or more species, and contain approximately half of all grass species. Almost a third of the species are concentrated in the ten largest genera (*Agrostis, Aristida, Calamagrostis, Digitaria, Eragrostis, Festuca, Panicum, Paspalum, Poa* and *Stipa*), each with 200 or more species. In other words, the taxonomic structure of this family exhibits the all-pervading 'hollow curve' (data summarised by Willis, 1949), whose cause and significance in taxonomic theory has been the subject of much interesting discussion (e.g. Walters 1961, 1986). With a few highly significant exceptions (see Chapter 1), the genera are very homogeneous, the main bone of contention at generic level being the extent to which the delimitations as expressed here are too narrow. On the other hand, there is much room for dispute concerning the higher hierachical levels, where this huge and significant plant family presents further major challenges to taxonomists.

Taxonomists have long argued about classificatory methodology, notably regarding attitudes to 'phylogenetic' considerations. The current dispute concerns the relative merits of phenetics and cladistics, and the different methods within them. We subscribe to what is at present the minority view (cf. Sneath 1988), that valid cladistic inferences from taxonomic data of the traditional kind depend upon prior recognition of phenetic groups (without which there is no adequate basis for sampling). We also think that currently fashionable computer cladistic techniques are rather inappropriate for analysing taxon descriptions framed above the species level, because the results are confused by intra-taxon variability; furthermore, they are at present notoriously unpractical in relation to anything other than relatively small data sets, and are therefore especially prone in practice to errors consequent on superficial sampling. In the context of any large taxonomic group, sampling for cladistic studies should probably be conducted in terms of species descriptions, or even descriptions of individual organisms, *critically selected with reference to existing (non-cladistic) classifications*. Fortunately, however, functional taxonomy does not depend upon the revelation of incontrovertible phylogenies, which in any case may never be generally attainable. Even in circumstances where further phylogenetic inquiry seemed futile (for example, in some viruses, where the essential clues may be incomprehensibly scrambled or lost), few would doubt the necessity of taxonomic classifications or the need to continue refining them.

In the Poaceae, as in other groups of organisms, neither the phenetic nor the phylogenetic relationships of the hard core of problem taxa will ever be satisfactorily resolved by comparing genotypes indirectly in terms of their phenotypic (morphological, anatomical, and physiological) manifestations. Comparative molecular studies should provide more definitive answers for many of them, as well as providing the most reliable evidence about the details of group divergences. Indeed, comparative molecular studies of angiosperm taxonomy above the generic level are already giving challenging results. Regarding the main lines of grass classification and evolution, however, molecular contributions to date (summarised and discussed by Kellogg and Watson 1993) are exciting but exasperating.

Evidence from direct nucleic acid sequencing is starting to accumulate (Appels *et al.* 1987; Hamby and Zimmer 1988; Gibbs *et al.* 1990), and as sampling extends will presumably be the most convincingly interpretable. From a taxonomic standpoint, a phylogeny for Poaceae computed by Hamby and Zimmer from ribosomal RNA sequences seems remarkably sensible, considering that the data represent only nine genera. In fact, it proved almost exactly reproducible (Watson and Dallwitz, unpublished) via phenetic and cladistic analyses of the same minute sample of genera using the full range of descriptive information in the present database, right down to the (taxonomically unacceptable) clustering of *Hordeum* with *Avena* rather than with *Triticum*. Likewise, trees derived from the few available grass Rubisco LS sequences (Gibbs *et al.* 1990; Chase *et al.* 1993) are fairly consistent, as far as they go, with the subfamily classification given here (where they differ from taxonomic expectation, they also differ from one another).

In a near-molecular study, Esen and Hilu (1989) returned to serological comparisons of grass seed proteins (cf. Fairbrothers and Johnson 1961), applying modern techniques of enzyme-linked immunosorbent assays and computer analyses of data to prolamin fractions. Cross-referenced with the classification given here, their results conform in showing no or few discernible differences between species of *Triticum* or within the tribes Triticeae, Poeae, Chlorideae, Paniceae, and Andropogoneae, and they acknowledged distinctions between subfamilies. However, the results were further represented as showing, *inter alia*, lack of support for our dividing the subfamily Pooideae into the supertribes Triticodae and Poodae, and for associating *Bromus* with the former (cf. Macfarlane and Watson 1982); and as suggesting that the Oryzeae merit a subfamily separate from the Bambusoideae. Given the relatively small sample of grasses encompassed (19 species, with only 11 used for antiserum production and in reciprocal assays for taxonomic analyses), however, and taking note of quite serious discrepancies in the alternative hierarchies computed, as well as an alarming failure to persuade *Panicum* to cluster with *Pennisetum* rather than with *Tripsacum*, the results in relation to these more subtle questions of classification and evolution, though interesting, could scarcely be taken as definitive.

An approach aimed at comparing whole genomes via their thermal denaturation characteristics (King and Ingrouille 1987) had yielded quite different results. These made taxonomic sense within the sample from the subfamily Pooideae, in particular making a clear distinction between the Poodae and Triticodae and associating *Bromus* with the latter. However, the startling failure of the same procedure to distinguish *Phragmites*, *Spartina*, *Bambusa*, *Oryza*, *Saccharum*, *Sorghum* and *Zea* (representing the subfamilies Arundinoideae, Chloridoideae, Bambusoideae and Panicoideae) from the Pooideae necessitated ignoring them altogether for the purposes of taxonomic analysis and discussion. This probably exemplifies the situation where taxonomy is more informative about nucleic acid organization and the methodology of comparison than *vice versa*.

More recently, a 'phylogenetic analysis' of chloroplast DNA restriction-site variation in 28 species of the subfamily Pooideae by Soreng *et al.* (1990) has provided further support for the cladistic acceptability and taxonomic soundness of the Triticodae (again associating *Bromus* and *Brachypodium* with the Triticeae), and was also interpreted by them as supporting the monophyly of (versions of) the tribes Meliceae, Poeae and Aveneae.

A curious aspect of all this is that the Triticodae/Poodae distinction represents a rather sophisticated piece of non-molecular, non-cladistic taxonomic detective work. It is detectable by application of appropriate phenetic classificatory methods to the generic descriptions given in this book, yet has not been apparent in a thorough cladistic exploration *of the same data* (Kellogg and Watson 1993). The suggestion (e.g., Young and Watson 1970, Watson *et al.* 1985) that phenetic methods may sometimes be more informative *about phylogeny* than supposedly 'phylogenetic' approaches is generally ignored, and anyway the case is not yet proven with reference to grasses. Dependence of the molecular work on tiny samples (which are inconsistent from one study to the next), and doubts about the reliability of chloroplast genomes as phylogenetic indicators (summarized by Harris and Ingram 1991), serve to further complicate the issues. It has long been evident that comparative DNA studies are the key to major advances in taxonomic understanding. It is equally obvious that acceptable taxonomic and phylogenetic conclusions regarding genuinely contentious aspects of plant taxonomy will require extensive, very laborious and time-consuming sampling, and congruent results from ranges of genes (both nuclear and chloroplast) deemed adequate to represent whole genomes. Meanwhile, we note that by contrast with 'phylogenetic' classifications of the Dicotyledons by (e.g.) Hutchinson (1973), Cronquist (1968, 1981) and Takhtajan (1980), the pre-evolutionary and phenetic versions of Bentham and Hooker (1862, 1873) and Young and Watson (1970), respectively, compare rather well with the 'phylogeny' now being deduced from analyses of Rubisco LS sequences (e.g. Chase *et al.* 1993).

Fortunately, the method whereby a classification was produced, and the philosophy underlying it, will be largely irrelevant to users. Its worth will depend entirely on its usefulness, in particular its general reliability as a guide to sampling, and in facilitating accurate predictions.

The summarized classification given here is an extension to tribal level (to subtribal level in the case of Andropogoneae) of the classification detailed by Watson *et al.* (1985). Both the formal groupings and the sequences in which they appear represent attempts to portray overall similarities and differences between genera in a manner useful for practical purposes. This classification represents extensive and continuing taxonomic analyses of a large set of descriptive data, employing a variety of mainly phenetic numerical methods. The named groupings are for the most part presented with some confidence. They are based on application of a variety of classificatory strategies to alternative interpretations of the descriptive data, and on attempts to steer a sensible course between conflicting results. By contrast with this kind of product, the alternative classification of Clayton and Renvoize (1986) was avowedly 'phylogenetic'. The difference in approach seems rather fundamental, but allowing for the possibilities which exist for translating similar views on relationships into superficially different formal schemes, the resulting classifications are remarkably similar. Thus, the five subfamilies Pooideae, Bambusoideae, Arundinoideae, Chloridoideae, and Panicoideae are the same in name and, for the most part, in content. There is close correspondence regarding tribal delimitations, and our supertribes are detectable in sequences of genera and tribes as presented by Clayton and Renvoize (Bromeae adjoining, and *Brachypodium* within, the Triticeae; herbaceous bambusoids together; the eupanicoids and andropogonoids constituting well separated groups within the Panicoideae, with the Arundinelleae between them). The overtly cladistic analyses of Kellogg and Campbell (1987) produced an extensively compatible phylogenetic interpretation, and further confirmation of overall agreement among grass taxonomists will be seen in comparisons between our classification and recent treatments of the subfamilies by specialists (Conert 1987; Macfarlane 1987; Jacobs 1987). In short, there is at present something approaching consensus regarding the main outlines of grass classification, of which what follows is reasonably representative for practical purposes.

All the genera are here assigned to a subfamily, a tribe, and (where appropriate) to a supertribe, and the large genera are indicated. We still think the supertribes of Pooideae represent the most useful subdivision in that subfamily where, with the exceptions of Triticeae, Brachypodieae, Bromeae, and Meliceae, satisfactory tribes have yet to be defined. We believe them to be unattainable until the sampling problems posed especially by *Agrostis* and *Poa* (see Chapter 1) have been solved. The supertribes of Panicoideae acknowledge a rather clear distinction between the eu-panicoid and andropogonoid tribes, which is to some extent blurred by the Arundinelleae (these showing an interesting mixture of eu-panicoid, andropogonoid, and danthonioid features). Presentation of two subtribes within the Andropogoneae acknowledges the existence of a rather strong distinction among the members of that large and important assemblage. The tribe Paniceae as presented here offers abundant scope for further useful subdivision, but the taxonomic problem posed by the genus *Panicum* (see description, Chapter 4) obstructs progress in that direction. Perhaps the least satisfactory aspect of the tribes as presented here, however, is the *sensu lato* version of the Chlorideae ('Main chloridoid assemblage'). It is evidently desirable that this unwieldy assemblage be effectively broken down, but the time-honoured division into two weakly distinguished large tribes (Cynodonteae or Chlorideae *sensu stricto* and Eragrosteae) and a number of small tribes has little merit. Here too, a satisfactory classification is unattainable without better understanding of patterns of variation within the large genera, especially *Eragrostis*. Meanwhile, persons wanting a more detailed breakdown of this group are referred to Clayton and Renvoize (1986), who present many sequences of convincingly related genera as subtribes.

There remain genera whose position is uncertain, through lack of information (e.g., *Danthoniastrum*, *Euthryptochloa*, *Chandrasekharania*), or because they exhibit genuinely intermediate or highly peculiar features or combinations of features (e.g., *Lygeum*, *Nardus*, *Steyermarkochloa*, *Sartidia*, *Brachyelytrum*, *Ehrharta*, *Eriachne*). At subfamily level, the Arundinoideae constitute an unsatisfactory assemblage of convenience, which is not amenable to anything approaching a diagnostic description, and is probably polyphyletic (see Watson *et al.* (1985) and Kellogg and Campbell (1987) for detailed analyses couched in phenetic and phylogenetic terms, respectively). The individual arundinoid tribes are are coherent and useful, but their relationships to one another and to the other subfamilies remain controversial. Kellogg and Campbell interpreted their results as showing that 'the Pooideae, Bambusoideae, Panicoideae and perhaps the Chloridoideae are monophyletic and are derived from a highly polyphyletic Arundinoideae'. Elsewhere in the system, Soderstrom and Ellis (1987) have excluded from the Bambusoideae a number of isolated genera and small groups presented here among the herbaceous bambusoids (supertribe Oryzodae), without however suggesting alternative locations: i.e., *Brachyelytrum*, *Diarrhena*, *Phaenosperma*, the Ehrharteae, and the Centotheceae. The last are the least convincingly bambusoid of all these: they are sometimes presented as a separate subfamily (Soderstrom 1981; Clayton and Renvoize 1986), but might as readily be placed in the temporary rag-bag of the Arundinoideae. In any event, only about 12% of grass species fall into these areas of uncertainty at the highest hierarchical level, and most of them are contributed by the constellations around *Aristida*, *Stipa*, and *Danthonia*.

The brief descriptions of the larger groups given here are intended to indicate how they are distinguished, but also to illustrate the kinds of features upon which taxonomy imposes patterns. In fact, in Poaceae taxonomic bias at subtribal, tribal, supertribal or superfamily level is shown by most of the 360 morphological and anatomical characters represented in the descriptions (cf. Watson *et al.* 1985). It is also evident in the compiled data on physiology, cytology, haploid and 2c DNA values, ecology, geography, and in host ranges of rusts, smuts and viruses, but it should be remembered that information on esoteric and inaccessible features mostly represents extrapolation from small samples.

3.2. Classification

Subfamilies, supertribes, tribes and (for Andropogoneae only) subtribes. Genera with 40 or more species indicated by *; approximate species numbers for tribes, and for genera with 100 or more species, in parentheses. What follows is a summary of the classificatory information presented with the individual generic descriptions in Chapter 4. This is also accessible interactively (with the statistical details of character state distributions) via the INTKEY package.

POOIDEAE

Annual or perennial, herbaceous; culms unbranched above, usually with hollow internodes. Leaf blades not pseudopetiolate, without transverse veins. Ligule an unfringed membrane. Inflorescences various, but not comprising spikelike main branches; espatheate, espatheolate. Spikelets nearly always laterally compressed or terete, with 1–many hermaphrodite florets, rarely with incomplete proximal florets. Lemma without a germination flap. Palea usually 2-keeled and apically notched. Lodicules 2, usually membranous. Stigmas white. Hilum long-linear or short. Embryo small, usually with an epiblast, with neither mesocotyl internode nor scutellar tail, the embryonic leaf margins meeting. *Abaxial leaf blade epidermis.* Microhairs and papillae absent; costal silica bodies various, but hardly ever 'panicoid-type' or saddle-shaped; costal short-cells not in long rows. Stomatal guard-cells overlapped by the interstomatals. *Physiology, transverse section of leaf blade.* C_3. Adaxial surface usually ribbed. Mesophyll with neither fusoids nor arm cells, without colourless columns. Midrib usually with a single vascular bundle. All vascular bundles accompanied by sclerenchyma. *Cytology.* Basic chromosome number usually $x = 7$. Group mean diploid 2c DNA value 8.9 pg. Rusts: *Puccinia*. Smuts: *Entyloma, Tilletia, Urocystis, Ustilago.*

North and south temperate, tropical mountains (with a Laurasian bias).

Triticodae

Inflorescence usually spicate, commonly disarticulating at the joints. Glumes often lateral or 'displaced'. Lemma awns non-geniculate, usually entered by several veins. Ovary apex hairy, lodicules often ciliate. Endosperm containing only simple starch grains.

Abaxial leaf blade epidermis. Crown cells sometimes present; stomata often very large.
Cytology. Group mean diploid 2c DNA value 10.6 pg.

Triticeae (360): *Aegilops, Agropyron, Amblyopyrum, Australopyrum, Cockaynea, Crithopsis, Dasypyrum, *Elymus* (150), *Elytrigia, Eremopyrum, Festucopsis, Henrardia, Heteranthelium, Hordelymus, *Hordeum, Hystrix, Kengyilia, Leymus, Lophopyrum, Malacurus, Pascopyrum, Peridictyon, Psathyrostachys, Pseudoroegneria, Secale, Sitanion, Taeniatherum, Thinopyrum, Triticum.*

Brachypodieae (16): *Brachypodium.*

Bromeae (150): *Boissiera, *Bromus.*

Poodae

Inflorescence usually paniculate with persistent axes, or a persistent spike. Spikelets usually disarticulating above the glumes. Lemma awn geniculate or straight, entered by only one vein. Ovary apex usually glabrous; lodicules usually glabrous. Endosperm usually containing compound starch grains. Group mean 2c DNA value 7.9 pg.

Aveneae (including Agrostideae, Phalarideae; dubiously separable from the Poeae) (1050): **Agrostis* (220), *Aira, Airopsis, Alopecurus, Ammochloa, Ammophila, Amphibromus, Ancistragrostis, Aniselytron, Anthoxanthum, Antinoria, Apera, Arrhenatherum, Avellinia, Avena, Beckmannia, *Calamagrostis* (230), *Chaetopogon, Cinna, Cornucopiae, Corynephorus, Cyathopus, Danthoniastrum* (or Stipeae?), *Dasypoa, *Deschampsia, Deyeuxia, Dichelachne, Dielsiochloa, Duthiea* (or Danthonieae?), *Echinopogon, Euthryptochloa, Gastridium, Gaudinia, Gaudiniopsis, *Graphephorum, *Helictotrichon, Hierochloë, Holcus, Hyalopoa, Hypseochloa, *Koeleria, Lagurus, Leptagrostis, Libyella, Limnas, Limnodea, Linkagrostis, Maillea, Metcalfia* (or Danthonieae?), *Mibora, Milium, Nephelochloa, Pentapogon, Periballia, Peyritschia, Phalaris, Phleum, Pilgerochloa, Polypogon, Pseudarrhenatherum, Pseudodanthonia* (or Danthonieae?), *Pseudophleum, Relchela, Rhizocephalus, Scribneria, Sinochasea, Sphenopholis, Stephanachne, Stilpnophleum, Tovarochloa, Triplachne, *Trisetum, Vahlodea, Ventenata, Zingeria.*
Upper glume nearly always long relative to the adjacent lemma; female-fertile florets 1–2(–7).

Poeae (including Hainardieae, Monermeae; dubiously separable from the Aveneae) (1124): *Agropyropsis, Anthochloa, Aphanelytrum, Arctagrostis, Arctophila, Austrofestuca, Bellardiochloa, Briza, Calosteca, Castellia, Catabrosella, Catapodium, Coleanthus, Colpodium, Ctenopsis, Cutandia, Cynosurus, Dactylis, Desmazeria, Dissanthelium, Dryopoa, Dupontia, Eremopoa, Erianthecium, *Festuca* (360), Festucella, Gymnachne, Hainardia, Helleria, Hookerochloa, Lamarckia, Leucopoa, Lindbergella, Littledalea, Loliolum, Lolium, Lombardochloa, Megalachne, Microbriza, Micropyropsis, Micropyrum, Narduroides, Parafestuca, Parapholis, Phippsia, Pholiurus, *Poa* (500), Podophorus, Poidium, Pseudobromus, Psilurus, *Puccinellia, Rhomboelytrum, Sclerochloa, Scolochloa, Simplicia, Sphenopus, Torreyochloa, Tsvelevia, Vulpia, Vulpiella, Wangenheimia.*
Glumes usually short relative to adjacent lemma; female-fertile florets (1–)2–many.

Seslerieae (33): *Echinaria, Oreochloa, Psilathera, Sesleria, Sesleriella.*

Meliceae (136): *Brylkinia, Catabrosa, *Glyceria, Lycochloa, *Melica, Pleuropogon, Schizachne, Streblochaete, Triniochloa.*
Leaf sheath margins joined. Basic chromosome number $x = 9$ or 10.

BAMBUSOIDEAE

Mostly perennial, culms woody or herbaceous. Leaf blades often pseudopetiolate, often with transverse veins, often disarticulating. Inflorescence usually paniculate, often spatheate. Lemmas with non-geniculate awns or awnless. Palea keel-less, 1- or 2-keeled, notched or entire. Lodicules 1–10 (often 3). Hilum usually long-linear. Embryo usually small, with an epiblast, usually with a scutellar tail and overlapping leaf margins. *Abaxial leaf blade epidermis.* Microhairs present, panicoid type. Papillae often present. Costal silica bodies often 'panicoid-type', 'oryzoid-type' or saddle-shaped. Stomatal subsidiaries usually triangular or dome-shaped. *Physiology, transverse section of leaf blade.* C_3. Adaxial surface often flat. Mesophyll commonly with arm-cells and/or fusoids. Midrib usually with more than one vascular bundle, often with complex vasularization. All vascular bundles accompanied by sclerenchyma. *Cytology.* Chromosome base number usually $x = 10$, 11 or 12. Rusts: *Dasturella, Physopella, Stereostratum, Puccinia.* Smuts: *Entyloma, Tilletia, Sorosporium, Tolyposporium, Ustilago.*

Tropical/warm temperate, mostly forest/woodland and wet places.

Oryzodae

'Grasses', or to varying degrees more or less bambusoid in appearance; mostly herbaceous, culms commonly unbranched above. Lodicules commonly 2, rarely ciliate. Stigmas usually 2. Mesophyll with or without arm-cells and/or fusoids. Mostly diploids. Group mean diploid $2c$ DNA value 3.0 pg.

Oryzeae (73): *Chikusichloa, Hydrochloa, Hygroryza, Leersia, Luziola, Maltebrunia, Oryza, Porteresia, Potamophila, Prosphytochloa, Rhynchoryza, Zizania, Zizaniopsis.*

Spikelets laterally compressed, with one hermaphrodite floret and sometimes with a proximal sterile lemma. Glumes absent or reduced to a minute cupule. Hilum long-linear. Mostly hydrophytic or helophytic.

Olyreae (120): *Agnesia, Arberella, Buergersiochloa, Cryptochloa, Diandrolyra, Ekmanochloa, Froesiochloa, Lithachne, Maclurolyra, Mniochloa, Olyra, Pariana, Parodiolyra, Piresia, Piresiella, Raddia, Raddiella, Rehia, Reitzia, Sucrea.*

Monoecious, the spikelets mixed or in separate inflorescences. Female spikelets usually dorsally compressed, with well developed glumes and one floret; lemma hard, usually with a germination flap.

Centotheceae (33): *Bromuniola, Calderonella, Centotheca, Chasmanthium, Chevalierella, Gouldochloa, Lophatherum, Megastachya, Orthoclada, Pohlidium, Zeugites.*

Anomochloeae (1): *Anomochloa.*

Brachyelytreae (1): *Brachyelytrum.*

Diarrheneae (4–5): *Diarrhena.*

Ehrharteae (44): *Ehrharta, Microlaena, Petriella, Tetrarrhena.*

Spikelets compressed laterally or terete; with two proximal sterile florets and one hermaphrodite floret.

Phaenospermateae (1): *Phaenosperma.*

Phyllorhachideae (3): *Humbertochloa, Phyllorhachis.*

Phareae (14): *Leptaspis, Pharus, Scrotochloa, Suddia.*

Streptochaeteae (2): *Streptochaeta.*

Streptogyneae (2): *Streptogyna.*

Bambusodae

Woody bamboos with branching culms. Lodicules usually 3 or more, ciliate. Stigmas usually 3 or more. Usually tetraploid or hexaploid.

Guaduelleae (8): *Guaduella.*

Puelieae (6): *Puelia.*

Bambuseae (825): *Acidosasa, Actinocladum, Alvimia, Apoclada, Arthrostylidium, *Arundinaria, Athroostachys, Atractantha, Aulonemia, *Bambusa* (120), *Cephalostachyum, Chimonobambusa, *Chusquea* (100), *Colanthelia, Davidsea, Decaryochloa, Dendrocalamus, Dendrochloa, Dinochloa, Elytrostachys, Fargesia,*

*Gigantochloa, Glaziophyton, Greslania, Hickelia, Hitchcockella, Indocalamus, Indosasa, Melocalamus, Melocanna, *Merostachys, Metasasa, Monocladus, Myriocladus, Nastus, Neohouzeaua, Neurolepis, Ochlandra, Olmeca, Oreobambos, Otatea, Oxytenanthera, Perrierbambus, *Phyllostachys, Pseudocoix, Pseudosasa, Pseudostachyum, Pseudoxytenanthera, Racemobambos, Rhipidocladum, *Sasa, Schizostachyum, Semiarundinaria, Shibataea, *Sinarundinaria, Sinobambusa, Sphaerobambos, Swallenochloa, Teinostachyum, Thamnocalamus, Thyrsostachys, Vietnamosasa, Yushania.*

ARUNDINOIDEAE

Perennial herbs, often caespitose (sometimes 'bambusoid'). Leaf blades sometimes disarticulating. Ligule usually a fringed membrane or a fringe of hairs. Inflorescence usually paniculate, espatheate, axes persistent. Spikelets usually laterally compressed or terete and disarticulating above the glumes; with 1–many hermaphrodite florets, occasionally with proximal incomplete florets. Lemmas usually hairy, incised, usually awned, awns straight or geniculate. Palea usually 2-nerved and 2-keeled. Lodicules fleshy or membranous, ciliate or glabrous. Stigmas 2. Endosperm usually with compound starch grains. *Abaxial leaf blade epidermis.* Microhairs present or absent (then usually present somewhere on the plant). Non-papillate. Costal silica bodies of various forms, but hardly ever 'pooid-type'. Stomatal guard-cells usually not overlapped by the interstomatals; subsidiaries usually triangular to dome-shaped. *Physiology, transverse section of leaf blade.* C_3 or C_4 type NADP–ME. Adaxial surface usually ribbed. Mesophyll without fusoids, very rarely with arm-cells. Midrib usually with a single bundle. Smallest vascular bundles accompanied by sclerenchyma. *Cytology.* Chromosome base numbers variable. Rusts: *Dasturella, Puccinia.* Smuts: *Neovossia, Tilletia, Urocystis, Sorosporium, Sphacelotheca, Tolyposporium, Ustilago.*

Cosmopolitan, but with a marked Gondwanan bias.

Steyermarkochloeae (2): *Arundoclaytonia, Steyermarkochloa.*

Stipeae (419): *Aciachne, Ampelodesmos, Anemanthele, Anisopogon, Danthoniastrum* (or Aveneae?), *Lorenzochloa, Nassella, Orthachne, Oryzopsis, Piptatherum, Piptochaetium, Psammochloa, *Stipa (300), Trikeraia.*

Spikelets with one hermaphrodite floret only; lemma indurated, without a germination flap; with a geniculate apical awn entered by several veins. Palea keel-less. Lodicules often 3, glabrous. Hilum long-linear. *Abaxial leaf blade epidermis.* Microhairs absent (but a peculiar form sometimes found elsewhere on the plant). Costal silica bodies often 'panicoid-type', crescentic or rounded. *Physiology.* C_3.

Nardeae (1): *Nardus.*

Lygeae (1): *Lygeum.*

Arundineae (7): *Arundo, Phragmites, Thysanolaena.*

Mostly tall reeds. Spikelets with 2–many florets. Hilum short.

Danthonieae (315): *Alloeochaete, Centropodia, Chaetobromus, Chionochloa, Cortaderia, Crinipes, Danthonia, Danthonidium, Dichaetaria, Dregeochloa, Duthiea* (or Aveneae?), *Elytrophorus, Erythranthera, Gynerium, Habrochloa, Hakonechloa, Karroochloa, Lamprothyrsus, Merxmuellera, Metcalfia* (or Aveneae?), *Molinia, Monachather, Monostachya, Nematopoa, Notochloë, Pentameris, *Pentaschistis, Phaenanthoecium, Plinthanthesis, Poagrostis, Prionanthium, Pseudodanthonia* (or Aveneae?), *Pseudopentameris, Pyrrhanthera, *Rytidosperma, Schismus, Sieglingia, Styppeiochloa, Tribolium, Urochlaena, Zenkeria.*

Mostly less than 250 cm high. Spikelets with 2–many hermaphrodite florets, without proximal incomplete florets. Nearly all C_3 (exception *Centropodia*).

Cyperochloeae (1): *Cyperochloa*.

Micrairieae (13): *Micraira*.

Spartochloeae (1): *Spartochloa*.

Aristideae (344): **Aristida* (290), *Sartidia*, **Stipagrostis*.
Spikelets with one hermaphrodite floret, no incomplete florets. Lemma hardened, with or without a germination flap; with a characteristic trifid awn (or derivative of this). Palea reduced. C$_3$ (*Sartidia*) or C$_4$. Chromosome base number usually 11.

Eriachneae (42): **Eriachne, Pheidochloa*.
Ligule a fringe of hairs; inflorescence a panicle or raceme; lower glume (7–)9–15 nerved; spikelets without proximal incomplete florets; female-fertile florets (1–)2; hilum long-linear. XyMS+; C$_4$, type NADP-ME.

Amphipogoneae (9): *Amphipogon, Diplopogon*.
Inflorescence spicate or contracted; spikelets usually compressed dorsiventrally to terete; lemma deeply cleft into three apically awned lobes; palea with two apical awns or setae; pericarp free (opaque). Microhairs *Enneapogon*-type.

CHLORIDOIDEAE

Culms herbaceous, branched or unbranched above. Leaf blades not pseudopetiolate. Ligule nearly always a fringed membrane or of hairs. Inflorescences various, commonly of dorsiventral, spikelike main branches or paniculate, espatheate; axes usually persistent. Sometimes dioecious or monoecious with unisexual spikelets, usually with hermaphrodite spikelets and hermaphrodite florets. *Female-fertile spikelets* usually disarticulating above the glumes, compression lateral to dorsiventral; lower glume usually 1-nerved; very rarely with proximal incomplete florets; hermaphrodite (or female) florets 1–many. Lemmas without a germination flap, awn if present non-geniculate. Palea 2-nerved and 2-keeled. Lodicules 2, fleshy, glabrous. Pericarp often free or loose. Hilum short. Embryo large; usually with an epiblast; scutellar tail and mesocotyl internode present, embryonic leaf margins meeting. *Abaxial leaf blade epidermis*. Microhairs present, chloridoid type or (less often) panicoid type. Costal silica-bodies mostly panicoid-type or saddle shaped, the costal short-cells in long rows. Stomatal guard-cells not overlapped by the inter-stomatals; subsidiaries triangular to dome-shaped. *Physiology, transverse section of leaf blade*. C$_4$ (one known exception in *Eragrostis*), types PCK and NAD–ME (all XyMS+). Adaxial surface often flat. Mesophyll often traversed by colourless columns; fusoids absent, arm-cells (nearly always) absent. Bulliforms commonly combined with colourless cells to form deep-penetrating fans. Smallest vascular bundles accompanied by sclerenchyma. *Cytology*. Chromosome base number usually $x = 10$. Group mean 2c DNA value 1.1 pg. Rusts: *Physopella, Puccinia*. Smuts: *Entyloma, Melanotaenium, Tilletia, Sorosporium, Sphacelotheca, Tolyposporella, Ustilago*.

Mostly tropical and subtropical (especially Old World and Gondwanan), especially dry climates.

Triodieae (54): *Monodia, Plectrachne, Symplectrodia, Triodia*.

Pappophoreae (41): *Cottea, Enneapogon, Kaokochloa, Pappophorum Schmidtia*.

Orcuttieae (9): *Neostapfia, Orcuttia, Tuctoria*.

Main chloridoid assemblage (including Chlorideae, Cynodonteae, Eragrosteae, Sporoboleae, Aeluropodeae, Jouveae, Unioleae, Leptureae, Lappagineae, Spartineae, Trageae, Perotideae, Pommereulleae) (1305): *Acamptoclados, Acrachne, Aegopogon, Aeluropus, Afrotrichloris, Allolepis, Apochiton, Astrebla, Austrochloris, Bealia, Bewsia, Blepharidachne, Blepharoneuron, *Bouteloua, Brachyachne, Brachychloa, Buchloë, Buchlomimus, Calamovilfa, Catalepis, Cathestechum, Chaetostichium, Chaboissaea, *Chloris, Chrysochloa, Cladoraphis, Coelachyropsis, Coelachyrum, Craspedorhachis, Crypsis, Ctenium, Cyclostachya, Cynodon, Cypholepis, Dactyloctenium, Daknopholis, Dasyochloa, Decaryella, Desmostachya, Diandrochloa, Dignathia, Dinebra, Diplachne, Distichlis, Drake-Brockmania, Ectrosia, Ectrosiopsis, Eleusine,*

Enteropogon, Entoplocamia, Eragrostiella, *Eragrostis* (350), *Erioneuron, Eustachys,
Farrago, Fingerhuthia, Gouinia, Griffithsochloa, Gymnopogon, Halopyrum,
Harpachne, Harpochloa, Heterachne, Heterocarpha, Hilaria, Hubbardochloa, Indopoa,
Ischnurus, Jouvea, Kampochloa, Kengia, Leptocarydion, Leptochloa, Leptochloöpsis,
Leptothrium, Lepturella, Lepturidium, Lepturopetium, Lepturus, Lintonia, Lophacme,
Lopholepis, Lycurus, Melanocenchris, Microchloa, Monanthochloë, Monelytrum,
Mosdenia, *Muhlenbergia* (160), *Munroa, Myriostachya, Neeragrostis, Neesiochloa,
Neobouteloua, Neostapfiella, Neyraudia, Ochthochloa, Odyssea, Opizia, Orinus,
Oropetium, Oxychloris, Pentarrhaphis, Pereilema, Perotis, Piptophyllum, Planichloa,
Pogonarthria, Pogoneura, Pogonochloa, Polevansia, Pommereulla, Pringleochloa,
Psammagrostis, Pseudozoysia, Psilolemma, Pterochloris, Redfieldia, Reederochloa,
Rendlia, Richardsiella, Saugetia, Schaffnerella, Schedonnardus, Schenckochloa,
Schoenefeldia, Sclerodactylon, Scleropogon, Silentvalleya, Soderstromia, Sohnsia,
Spartina, *Sporobolus* (160), *Steirachne, Stiburus, Swallenia, Tetrachaete, Tetrachne,
Tetrapogon, Thellungia, Tragus, Trichoneura, Tridens, Triplasis, Tripogon, Triraphis,
Uniola, Urochondra, Vaseyochloa, Vietnamochloa, Viguierella, Willkommia, Zoysia.

PANICOIDEAE

Culms mostly herbaceous, usually branching above, internodes more often solid than hollow. Ligule a fringed or unfringed membrane, or a fringe of hairs. Inflorescence commonly paniculate, equally commonly of spikelike main branches (these often dorsiventral), spatheate or not; the axes disarticulating or persistent. Sometimes dioecious or monoecious, or with hermaphrodite spikelets. Female-fertile spikelets usually compressed dorsiventrally and falling with the glumes; rachilla not prolonged apically; nearly always with one proximal incomplete (male or sterile) floret and one hermaphrodite (rarely female-only) floret. Lodicules fleshy, usually glabrous. Stigmas usually red-pigmented. Hilum usually short. Endosperm starch grains usually simple. Embryo usually large; without an epiblast; scutellar tail and mesocotyl internode present; embryonic leaf margins overlapping. *Abaxial leaf blade epidermis.* Microhairs present, panicoid-type. Costal silica-bodies nearly always 'panicoid-type' (horizontally-elongated, cross-shaped, dumb-bell shaped or nodular); costal short-cells in long rows. Sometimes papillate. Stomatal guard-cells not overlapped by the subsidiaries; subsidiaries triangular or dome-shaped. *Physiology, transverse section of leaf blade.* C_3 or C_4. Adaxial surface often flat. Mesophyll without arm-cells, very rarely with fusoids. Midrib usually with an arc of bundles, often with adaxial colourless tissue. Smallest bundles commonly unaccompanied by sclerenchyma. *Cytology.* Chromosome base numbers mostly $x = 5$, 9 or 10. Group mean diploid $2c$ DNA value 3.0 pg. Rusts: *Dasturella, Phakopsora, Physopella, Puccinia.* Smuts: *Entyloma, Melanotaenium, Tilletia, Sorosporium, Sphacelotheca, Tolyposporella, Tolyposporium, Ustilago.*

Mainly tropical (especially Old World and Gondwanan), extending to temperate.

Panicodae

Inflorescence usually espatheate, the axes persistent or condensed into deciduous spikelet clusters. Spikelets sometimes in long-and-short combinations, but usually all alike in form and sexuality. Female-fertile lemma firm or indurated, with a germination flap, usually awnless and entire; the palea well developed. C_3 or C_4 (occasionally intermediate), types PCK, NAD–ME and NADP–ME (the latter all XyMS–). Basic chromosome number usually $x = 9$.

Pantropical to temperate; diverse habitat and rainfall requirements.

Isachneae (126): *Coelachne, Cyrtococcum, Heteranthoecia, Hubbardia, *Isachne* (100), *Limnopoa, Sphaerocaryum.*

Usually with both florets female-fertile; C_3.

Paniceae (about 2000): *Achlaena, Acostia, Acritochaete, Acroceras, Alexfloydia, Alloteropsis, Amphicarpum, Ancistrachne, Anthaenantiopsis, Anthenantia, Anthephora, Arthragrostis, Arthropogon, *Axonopus* (110), *Baptorhachis, Beckeropsis, Boivinella, *Brachiaria* (100), *Calyptochloa, Camusiella, Cenchrus, Centrochloa, Chaetium,*

Chaetopoa, Chamaeraphis, Chasechloa, Chloachne, Chlorocalymma, Cleistochloa, Cliffordiochloa, Commelinidium, Cymbosetaria, Cyphochlaena, Dallwatsonia, *Dichanthelium* (120), *Digitaria* (220), *Digitariopsis, Dimorphochloa, Dissochondrus, Eccoptocarpha,* *Echinochloa, Echinolaena, Entolasia, Eriochloa, Gerritea, Holcolemma, Homolepis, Homopholis, Hydrothauma, Hygrochloa, Hylebates, Hymenachne, Ichnanthus, Ixophorus, Lasiacis, Lecomtella, Leptocoryphium, Leptoloma, Leucophrys, Louisiella, Megaloprotachne, Melinis, Mesosetum, Microcalamus, Mildbraediochloa, Odontelytrum, Ophiochloa, Oplismenopsis, Oplismenus, Oryzidium, Otachyrium, Ottochloa,* *Panicum* (370), *Paratheria, Parectenium,* *Paspalidium,* *Paspalum* (320), *Pennisetum, Perulifera, Plagiantha, Plagiosetum, Poecilostachys, Pseudechinolaena, Pseudochaetochloa, Pseudoraphis, Reimarochloa, Reynaudia, Rhynchelytrum, Sacciolepis, Scutachne,* *Setaria* (110), *Setariopsis, Snowdenia, Spheneria, Spinifex, Steinchisma, Stenotaphrum, Stereochlaena, Streptolophus, Streptostachys, Taeniorhachis, Tarigidia, Tatianyx, Thrasya, Thrasyopsis, Thuarea, Thyridachne, Trachys, Tricholaena, Triscenia, Uranthoecium, Urochloa, Whiteochloa, Xerochloa, Yakirra, Yvesia, Zygochloa.*
C$_3$, or C$_4$ type PCK, NAD–ME and NADP–ME.

Neurachneae (10): *Neurachne, Paraneurachne, Thyridolepis.*
Inflorescence a single spicate raceme. C$_3$, C$_4$ type NADP–ME (XyMS–) or intermediate.

Arundinelleae (184): *Arundinella, Chandrasekharania, Danthoniopsis, Diandrostachya, Dilophotriche, Garnotia, Gilgiochloa, Isalus, Jansenella, Loudetia, Loudetiopsis, Trichopteryx, Tristachya, Zonotriche.*
Spikelets compressed laterally to terete, disarticulating above the glumes. Lemma usually bilobed or cleft, usually with a geniculate awn. Hilum usually long-linear. C$_4$ type NADP–ME (XyMS–, often with isolated PCR cells), or rarely C$_3$ (?). Chromosome base number $x = 10$ or 12.

Andropogonodae

Inflorescence often spatheate and/or spatheolate, the axes usually disarticulating. Spikelets usually in long-and-short combinations, often heterogamous, the members of the combinations different in sexuality (the longer-pedicelled members usually male or sterile). Glumes usually very dissimilar. The female-fertile lemma usually reduced to a stipe, insubstantial or hyaline; often bifid, with a geniculate awn; without a germination flap. Palea commonly reduced, vestigeal or absent. Exclusively C$_4$, type NADP–ME (XyMS–). Chromosome base number, x = mostly 5 or 10.
Tropical/warm temperate; mostly requiring seasonal high rainfall.

Andropogoneae (967).
Plants usually bisexual with hermaphrodite (upper) florets. Lodicules usually present.

Andropogoninae: *Agenium, Anadelphia,* *Andropogon* (100), *Andropterum, Apluda, Apocopis, Arthraxon, Asthenochloa, Bhidea, Bothriochloa, Capillipedium, Chrysopogon, Chumsriella, Clausospicula, Cleistachne,* *Cymbopogon, Dichanthium, Diectomis, Digastrium, Diheteropogon,* *Dimeria, Dybowskia, Eccoilopus, Elymandra, Eremopogon, Erianthus, Eriochrysis, Euclasta, Eulalia, Eulaliopsis, Exotheca, Germainia, Hemisorghum, Heteropogon, Homozeugos,* *Hyparrhenia, Hyperthelia, Hypogynium, Imperata,* *Ischaemum, Ischnochloa, Iseilema, Kerriochloa, Lasiorhachis, Leptosaccharum, Lophopogon, Microstegium, Miscanthidium, Miscanthus, Monium, Monocymbium, Narenga, Parahyparrhenia, Pleiadelphia, Pobeguinea, Pogonachne, Pogonatherum, Polliniopsis, Polytrias, Pseudanthistiria, Pseudodichanthium, Pseudopogonatherum, Pseudosorghum, Saccharum,* *Schizachyrium, Sclerostachya, Sehima, Sorghastrum, Sorghum, Spathia, Spodiopogon, Thelepogon, Themeda, Trachypogon, Triplopogon, Vetiveria, Ystia.*
Inflorescences mostly 'racemes' with slender rachides, articles without a basal callus knob and commonly hairy; dispersal unit usually with a basal, hairy callus; proximal incomplete floret usually sterile; female-fertile lemma very often incised or cleft, usually awned; female-fertile palea often absent, otherwise generally relatively short or much

reduced, commonly nerveless. Abaxial leaf blade epidermis commonly papillate; stomata 18–30–60 microns long; intercostal short-cells often absent or very rare, when present often solitary.

Rottboelliinae: *Chasmopodium, Coelorachis, Elionurus, Eremochloa, Glyphochloa, Hackelochloa, Hemarthria, Heteropholis, Jardinea, Lasiurus, Lepargochloa, Loxodera, Manisuris, Mnesithea, Ophiuros, Oxyrhachis, Phacelurus, Pseudovossia, Ratzeburgia, Rhytachne, Robynsiochloa, Rottboellia, Thaumastochloa, Thyrsia, Urelytrum, Vossia.*

Inflorescences with substantial rachides, mostly spikelike, commonly with spikelets more or less embedded and/or the pedicels more or less fused; articles mostly glabrous, nearly always non-linear and with a basal callus knob (elaiosome), usually without a hairy callus; proximal incomplete floret often male; female-fertile lemma entire, awnless; female-fertile palea present, often relatively long, often binerved. Abaxial leaf blade epidermis very rarely papillate; stomata 22–40–63 microns long; paired intercostal short-cells usually common.

Maydeae (32): *Chionachne, Coix, Euchlaena, Polytoca, Sclerachne, Trilobachne, Tripsacum, Zea.*

Plants monoecious, with all the fertile spikelets unisexual; without hermaphrodite florets. The male and female-fertile spikelets in different inflorescences, on different branches of the same inflorescence, or on different parts of the same branch. Lodicules absent.

4. DESCRIPTIONS

4.1. Implicit Character States

Unless indicated otherwise, the following character states are implicit throughout these descriptions, except where the characters concerned are inapplicable.

Habit, vegetative morphology. Not reeds. Culms not scandent. Culms not tuberous. Plants unarmed (except in Bambuseae, where unstated means unknown); without multicellular glands; the shoots not aromatic; leaves clearly differentiated into sheath and blade. Leaves distichous; without auricular setae (except in Bambusoideae, where unstated means unknown). Sheath margins free. Leaf blades not all greatly reduced. Leaf blades neither leathery nor flimsy. Leaf blades not cordate, not sagittate; not setaceous. Leaf blades not needle-like.

Leaf blades not pseudopetiolate; parallel veined. Ligule present (at least on lower leaves).

Reproductive organization. The spikelets all alike in sexuality (on the same plant: 'homogamous'); hermaphrodite; chasmogamous; plants without hidden cleistogenes; hidden cleistogenes (when present) in the leaf sheaths. Reproducing sexually; not viviparous.

Inflorescence. Inflorescence determinate; without pseudospikelets. *Inflorescence* many spikeleted. Inflorescence not deciduous. Inflorescence without conspicuously divaricate branchlets. Inflorescence without capillary branchlets; non-digitate. Inflorescence with axes ending in spikelets. Rachides neither flattened nor hollowed, not winged. Spikelets not all embedded. Spikelet-bearing axes persistent (not disarticulating); the pedicels and rachis internodes without a longitudinal, translucent furrow. Spikelets unaccompanied by bractiform involucres, not associated with setiform vestigial branches. Spikelets not in distinct 'long-and-short' combinations.

Female-sterile spikelets. The staminal filaments free.

Female-fertile spikelets. Spikelets morphologically 'conventional'.

Glumes present; relatively large. Glumes free; not ventricose. Glumes without conspicuous tufts or rows of hairs. *Glumes* without a median keel-wing; lower glume not two-keeled (except in Andropogonodae, where unstated means unknown); lower glume not pitted; upper glume not saccate; upper glume not prickly. Palea of the proximal incomplete florets not becoming conspicuously hardened and enlarged laterally.

Lemmas not conspicuously non-distichous; not convolute. Lemmas not crested; awns not of the triple/trifid, basal column type. Awns not hooked. The hairs not in tufts; not in transverse rows; palea not convolute. Stamens with free filaments. Ovary without a conspicuous apical appendage.

Fruit (when fruit data adequate). Fruit not grooved. Fruit smooth. Pericarp thin; fused (except in Arundinoideae and Chloridoideae, where unstated means unknown). Seed endospermic.

Abaxial leaf blade epidermis (when epidermal data adequate). Intercostal zones with typical long-cells. Subsidiaries not including both parallel-sided and triangular forms on the same leaf. Crown cells absent; costal zones with short-cells. Costal silica bodies not sharp-pointed.

Transverse section of leaf blade, physiology (when data adequate). Leaf blades 'laminar' (rather than consisting entirely of midrib); the anatomical organization when known to be C_4, conventional. Mesophyll not *Isachne*-type. Mesophyll without 'circular cells'. Mesophyll not traversed by colourless columns. Mesophyll without arm cells. Mesophyll without fusoids. *Midrib* without colourless mesophyll adaxially. *Bulliforms* nowhere involved in bulliform-plus-colourless mesophyll arches.

64

Special diagnostic features. (See Chapter 2 for descriptions of the corresponding 'positive' character states.) *Spikelets* not borne as in *'Anadelphia scyphofera'*. Plant not as in *Anomochloa*; the inflorescences not as in *Atractantha*. Lemmas not as in *Briza*. Lemma not wing-awned; inflorescence not as in *Buchloë*. Upper glume not as in *Centrochloa*. Inflorescences not as in *Coix*. Spikelets not borne as in *Cornucopiae*. Lemmas without the characteristic *Corynephorus* awn. Plants not as in *Cyperochloa*. Fruit not as in *Diarrhena*. Plants not as in *Dichanthelium*. Without the sessile/pedicellate spikelet pairs characteristic of *Diectomis*. No *Eriochloa*-type 'callus'. Plants not as in *Hubbardia*. The adaxial surface of the leaf blade not as in *Hydrothauma*. Plants not as in *Hygroryza*. Not having female spikelets as in *Leptaspis* and *Scrotochloa*. Female-fertile lemma not as in *Lombardochloa*. Spikelets not as in *Lopholepis*. Plant and inflorescence not as in *Lygeum*. Spikelets not arranged as in *Manisuris*. Spikelets without a terminal clavate appendage. Spikelet not as in *Nassella*. The inflorescence not as in *Odontelytrum*. The lower lemma not resembling a *Bothriochloa* pedicel in appearance, and without a pair of hygroscopically active setae. Seed not as in *Phaenosperma*. Spikelets not borne on a broad, leaflike rachis. Not scandent as in *Prosphytochloa*. Female-fertile lemma not as in *Rhynchoryza*. Not rush-like. Inflorescence not as in *Spinifex*. Plants not as in *Steyermarkochloa*. Flowering culms not as in *Thuarea*. Lower glume without a *Thyridolepis*-type window. The inflorescence not as in *Viguierella*. Fruiting inflorescence not as in maize. Plants not as in *Zygochloa*.

Rusts and smuts. Rusts — no rusts recorded (by Cummins, 1971). Smuts not recorded ((but see qualification in Chapter 2, character 477).

Special comments. Fruit data more or less satisfactory. Anatomical data more or less satisfactory.

4.2. The Genera

Acamptoclados Nash

~ *Eragrostis* (*E. sessilispica* Buckl.)

Habit, vegetative morphology. Perennial; caespitose. *Culms* (30–)40–50(–60) cm high (?); herbaceous; unbranched above; tuberous (when young); 1 noded (above the basal tuft of leaves). Culm nodes hidden by leaf sheaths. Plants unarmed. Young shoots extravaginal. *Leaves* mostly basal (the culm internodes very short); villous in the auricle positions. Glabrous, pilose at the throat. Leaf blades linear (narrowly so, stiff and subulate-tipped, almost acicular); narrow; 1–3 mm wide; flat, or rolled (flat to loosely involute); without abaxial multicellular glands; without cross venation; persistent. Ligule a fringe of hairs (the adjoining blade also hairy). *Contra-ligule* present (partial, of small hairs).

Reproductive organization. Plants bisexual, with bisexual spikelets; with hermaphrodite florets.

Inflorescence. *Inflorescence paniculate (the long-internoded primary branches distant, stiffly spreading, mostly bearing basal secondary branches, the latter mostly short and sometimes reduced to a single spikelet, the spikelets appressed); deciduous in its entirety*; open (the axis curving or loosely spiral, pilose in the axils). Rachides hollowed and flattened. Inflorescence espatheate; not comprising 'partial inflorescences' and foliar organs. Spikelets distant, solitary; not secund; subsessile.

Female-fertile spikelets. Spikelets (7–)8–10(–12) mm long; compressed laterally; disarticulating above the glumes; not disarticulating between the florets; with conventional internode spacings. Rachilla prolonged beyond the uppermost female-fertile floret; rachilla hairless (glabrous). The rachilla extension with incomplete florets. Hairy callus absent.

Glumes two; more or less equal; shorter than the adjacent lemmas; lateral to the rachis; hairless; glabrous; pointed; awnless; carinate; similar (firmly membranous to cartilaginous). Lower glume longer than half length of lowest lemma; 1 nerved. Upper glume 1 nerved. *Spikelets* with incomplete florets. The incomplete florets distal to the female-fertile florets. *The distal incomplete florets* merely underdeveloped.

Female-fertile florets 5–8. Lemmas similar in texture to the glumes (firmly membranous to cartilaginous); becoming somewhat indurated; entire; pointed; awnless; hairless; glabrous; carinate (the lateral nerves also slightly raised); without a germination flap; 3 nerved. **Palea** present; relatively long (prominently bowed out below); apically notched; awnless, without apical setae; *textured like the lemma*; not indurated; 2-nerved; 2-keeled. Palea keels slightly winged (above). *Lodicules* present; 2; free; fleshy; glabrous. *Stamens 3*. Anthers medium sized; not penicillate; without an apically prolonged connective. *Ovary* glabrous. Styles free to their bases. Stigmas 2; apparently not white.

Fruit, embryo and seedling. *Fruit* free from both lemma and palea; small (about 1.5 mm long); not noticeably compressed. Hilum short. Pericarp free, or loosely adherent (?). Embryo large; not waisted.

Abaxial leaf blade epidermis. *Costal/intercostal zonation* conspicuous. *Papillae* absent. *Long-cells* of similar wall thickness costally and intercostally (rather thick walled). Mid-intercostal long-cells rectangular; having markedly sinuous walls. *Microhairs* present; panicoid-type. Microhair apical cell wall thinner than that of the basal cell and often collapsed. Microhairs (45–)46–54(–61.5) microns long. Microhair basal cells 24–30 microns long. Microhairs 7.5–10.5 microns wide at the septum. Microhair total length/width at septum 6.1–7.2. Microhair apical cells (21–)25.5–27(–30) microns long. Microhair apical cell/total length ratio 0.47–0.55. *Stomata* common; 30–33 microns long. Subsidiaries dome-shaped and triangular. Guard-cells overlapping to flush with the interstomatals. *Intercostal short-cells* common (bordering the veins); not paired (solitary); not silicified. Intercostal silica bodies absent. Prickles abundant costally. *Crown cells* absent (but costal prickles with round or rectangular pitted bases). *Costal short-cells* predominantly paired (but many solitary). Costal silica bodies present throughout the costal zones; rounded to saddle shaped (a few), or crescentic (predominating).

Transverse section of leaf blade, physiology. C_4; XyMS+. PCR sheath outlines uneven. PCR sheaths of the primary vascular bundles interrupted; interrupted abaxially only. *Leaf blade* with distinct, prominent adaxial ribs; with the ribs very irregular in sizes (large, tall, square- or round-topped, and infrequent small ones). *Midrib* not readily distinguishable; with one bundle only. Bulliforms present in discrete, regular adaxial groups; associated with colourless mesophyll cells to form deeply-penetrating fans. All the vascular bundles accompanied by sclerenchyma. Combined sclerenchyma girders present (the adaxial fibre groups being contiguous with bundle sheath extensions); forming 'figures'. Sclerenchyma all associated with vascular bundles. The lamina margins with fibres.

Taxonomy. Chloridoideae; main chloridoid assemblage.

Ecology, geography, regional floristic distribution. 1 species. North America to Mexico. Mesophytic to xerophytic; species of open habitats. Plains and sandy prairies.

Holarctic. Boreal. Atlantic North American. Southern Atlantic North American.

References, etc. Leaf anatomical: this project.

Achlaena Griseb.

~ Arthropogon

Habit, vegetative morphology. Perennial; caespitose. **Culms 50–100 cm high**; herbaceous. Culm nodes glabrous. *Leaves* mostly basal; non-auriculate. Leaf blades linear-lanceolate; narrow; 2–6 mm wide; flat, or rolled (involute); without cross venation; persistent. *Ligule a fringed membrane*; 0.5–1 mm long.

Reproductive organization. Plants bisexual, with bisexual spikelets; with hermaphrodite florets.

Inflorescence. *Inflorescence paniculate*; open (long exserted, 7–15 cm long, the branches somewhat glutinous); espatheate; not comprising 'partial inflorescences' and foliar organs. Spikelet-bearing axes persistent. Spikelets solitary; pedicellate.

Female-fertile spikelets. Spikelets 6–8 mm long; compressed laterally; falling with the glumes. Rachilla terminated by a female-fertile floret. Hairy callus present (stipitate, minutely barbed). Callus pointed.

Glumes narrow, two; relatively large; very unequal (G_1 shorter); (the longer) long relative to the adjacent lemmas; without conspicuous tufts or rows of hairs; pointed;

awned (G₁ extending into, G₂ narrowed into slender, antrorse-scabrid awns); very dissimilar (G_1 subulate, G_2 broader, somewhat hardened). Lower glume 1 nerved. Upper glume 5 nerved, or 7 nerved. *Spikelets* with incomplete florets. The incomplete florets proximal to the female-fertile florets. *The proximal incomplete florets* 1; epaleate; sterile. The proximal lemmas awnless; 5 nerved, or 7 nerved; more or less equalling the female-fertile lemmas; decidedly firmer than the female-fertile lemmas (rather hard).

Female-fertile florets 1. **Lemmas** less firm than the glumes; not becoming indurated (thin, hyaline); entire; pointed; *awnless (but acuminate)*; without a germination flap; 1 nerved, or 3 nerved. *Palea* present; conspicuous but relatively short to very reduced; awnless, without apical setae; thinner than the lemma (hyaline); not indurated; 2-nerved. *Lodicules* present; 2; free. **Stamens 2.**

Fruit, embryo and seedling. Hilum short (oval). Embryo small (about 1/4 of the grain length).

Abaxial leaf blade epidermis. *Costal/intercostal zonation* conspicuous. *Papillae* absent. *Long-cells* similar in shape costally and intercostally to markedly different in shape costally and intercostally (the costals narrower); of similar wall thickness costally and intercostally. Mid-intercostal long-cells rectangular; having markedly sinuous walls. *Microhairs* present; elongated (plump); clearly two-celled; panicoid-type; 72–81 microns long; 18–21 microns wide at the septum. Microhair total length/width at septum 3.5–4.3. Microhair apical cells 43.5–49.5 microns long. Microhair apical cell/total length ratio 0.56–0.63. *Stomata* common; 48–52.5 microns long. Subsidiaries triangular (predominating), or dome-shaped. Guard-cells overlapping to flush with the interstomatals. *Intercostal short-cells* common; in cork/silica-cell pairs (in places), or not paired. *Costal short-cells* predominantly paired. Costal silica bodies oryzoid to 'panicoid-type'; when panicoid type, cross shaped (i.e. silica bodies vertical dumb-bell shaped to cross shaped).

Transverse section of leaf blade, physiology. C_4. The anatomical organization unconventional. Organization of PCR tissue *Arundinella* type (with extra PCR cells as lateral extensions from the sheaths of small bundles, as isolated clusters, and solitary). XyMS+. PCR sheath extensions absent. Mesophyll without adaxial palisade; exhibiting 'circular cells'. *Leaf blade* slightly 'nodular' in section; with the ribs more or less constant in size (low, flat to roundish-topped). *Midrib* conspicuous; having a conventional arc of bundles (with numerous major and minor bundles); with colourless mesophyll adaxially. *The lamina* symmetrical on either side of the midrib. Many of the smallest vascular bundles unaccompanied by sclerenchyma. Combined sclerenchyma girders present; forming 'figures' (most of the main bundles with large I's). Sclerenchyma all associated with vascular bundles.

Taxonomy. Panicoideae; Panicodae; Paniceae (Arthropogoneae).

Ecology, geography, regional floristic distribution. 1 species. Cuba and Jamaica. Species of open habitats. Open woods, dry slopes, savanna.

Neotropical. Caribbean.

References, etc. Morphological/taxonomic: e.g. Filgueiras 1982. Leaf anatomical: this project.

Special comments. Evidently close to *Arthropogon*, but seems fairly distinct morphologically and anatomically, as well as geographically. Fruit data wanting.

Aciachne Benth.

Habit, vegetative morphology. Perennial; cushion-forming and caespitose. **Culms *1–2 cm high*;** herbaceous; unbranched above. Young shoots intravaginal. *Leaves* not basally aggregated; non-auriculate; without auricular setae. Sheath margins free. The sheaths overlapping, whitish, shining. Leaf blades linear; narrow; 0.3–0.8 mm wide; setaceous (recurved); acicular (and involute, pungent); without cross venation; persistent. Ligule an unfringed membrane; 1–1.5 mm long.

Reproductive organization. Plants bisexual, with bisexual spikelets; with hermaphrodite florets. The spikelets of sexually distinct forms on the same plant, or all alike in sexuality; hermaphrodite, or hermaphrodite and sterile (the reduced inflorescence often exhibiting vestigial spikelets). Exposed-cleistogamous, or chasmogamous.

Inflorescence. *Inflorescence* reduced to a single spikelet, or few spikeleted; *greatly reduced, barely extruded from the sheaths at anthesis*; espatheate; not comprising 'partial inflorescences' and foliar organs. Spikelet-bearing axes persistent. Spikelets not secund; pedicellate.

Female-fertile spikelets. Spikelets about 3 mm long; compressed laterally; disarticulating above the glumes. Rachilla terminated by a female-fertile floret. Hairy callus absent.

Glumes present; two; more or less equal; shorter than the spikelets; shorter than the adjacent lemmas; not pointed (obtuse); awnless; non-carinate (rounded on the back); similar. Lower glume 3 nerved, or 5 nerved. Upper glume 5 nerved. *Spikelets with female-fertile florets only*.

Female-fertile florets 1. **Lemmas** fusiform; *convolute*; similar in texture to the glumes (hard); becoming indurated (shiny); entire; pointed; awned. *Awns* 1; median; apical (the lemma extending into the sharp, firm point); non-geniculate; hairless; much shorter than the body of the lemma to about as long as the body of the lemma; *persistent (with no line of demarcation from the body of the lemma)*. Lemmas hairless; non-carinate (abaxially rounded); 3 nerved. *Palea* present; relatively long; tightly clasped by the lemma; awnless, without apical setae; textured like the lemma; 2-nerved; keel-less. *Lodicules* present; 3; free; membranous; glabrous; not toothed; heavily vascularized (the anterior members with one strong vein). **Stamens 3**. Anthers not penicillate; without an apically prolonged connective. *Ovary* glabrous. Styles free to their bases. Stigmas 2; white.

Fruit, embryo and seedling. Fruit small (1.3–1.6 mm long); ellipsoid; not noticeably compressed (terete). Hilum long-linear. Embryo small; not waisted. Endosperm hard; containing compound starch grains. Embryo with an epiblast; without a scutellar tail; with a negligible mesocotyl internode; with one scutellum bundle. Embryonic leaf margins meeting.

Abaxial leaf blade epidermis. *Costal/intercostal zonation* lacking. *Papillae* absent. *Long-cells* similar in shape costally and intercostally; of similar wall thickness costally and intercostally. Mid-intercostal long-cells rectangular; having markedly sinuous walls. *Microhairs* absent (and also absent adaxially). *Stomata* absent or very rare. *Intercostal short-cells* common; in cork/silica-cell pairs. *Costal short-cells* predominantly paired, or neither distinctly grouped into long rows nor predominantly paired. Costal silica bodies rounded, or 'panicoid-type'.

Transverse section of leaf blade, physiology. C_3; XyMS+. *Mesophyll* with non-radiate chlorenchyma. *Leaf blade* adaxially flat. *Midrib* conspicuous; with one bundle only. Bulliforms present in discrete, regular adaxial groups, or not in discrete, regular adaxial groups; in simple fans (or these ill defined). All the vascular bundles accompanied by sclerenchyma. Combined sclerenchyma girders absent. Sclerenchyma not all bundle-associated. The 'extra' sclerenchyma in a continuous abaxial layer.

Cytology. Chromosome base number, $x = 11$. $2n = 22$. 2 ploid.

Taxonomy. Arundinoideae; Stipeae.

Ecology, geography, regional floristic distribution. 3 species. High Andes, northern Argentina to Costa Rica. Xerophytic; species of open habitats; glycophytic. High altitude grassland.

Neotropical. Andean.

References, etc. Morphological/taxonomic: Macfarlane and Watson 1980; Laegaard 1987. Leaf anatomical: this project.

Acidosasa Chu and Chao

Habit, vegetative morphology. Shrubby perennial. The flowering culms leafy. **Culms** tall; *woody and persistent*; branched above. Primary branches/mid-culm node 3. Rhizomes leptomorph. Plants unarmed. *Leaves* not basally aggregated. Leaf blades broad (large); pseudopetiolate; disarticulating from the sheaths; rolled in bud.

Reproductive organization. Plants bisexual, with bisexual spikelets; with hermaphrodite florets.

Inflorescence. *Inflorescence* determinate; without pseudospikelets; a single raceme, or paniculate; not comprising 'partial inflorescences' and foliar organs. *Spikelet-bearing axes* 'racemes', or paniculate (scanty); persistent. Spikelets pedicellate.

Female-fertile spikelets. *Spikelets* unconventional (having 4 glumes); compressed laterally; disarticulating above the glumes; disarticulating between the florets. Rachilla prolonged beyond the uppermost female-fertile floret.

Glumes present; *several (four)*.

Female-fertile florets *3–6 (? — 'several')*. **Stamens 6. Ovary without a conspicuous apical appendage**. Stigmas 3.

Taxonomy. Bambusoideae; Bambusodae; Bambuseae.

Ecology, geography, regional floristic distribution. 6 species. Mostly southern China, one species in Indo-China.

Paleotropical. Indomalesian. Indo-Chinese.

References, etc. Morphological/taxonomic: Chu and Chao 1979, Chao and Renvoize 1989.

Special comments. Fruit data wanting. Anatomical data wanting.

Acostia Swallen

~ *Digitaria* or *Panicum*?

Habit, vegetative morphology. Slender perennial; densely caespitose. *Culms* 25–40 cm high; herbaceous. Culm nodes hairy to glabrous. Sheaths keeled above. Leaf blades narrow; 3–4 mm wide (7–13 cm long, narrowed to both ends); flat; without cross venation; persistent. *Ligule a fringed membrane*; very short.

Reproductive organization. Plants bisexual, with bisexual spikelets; with hermaphrodite florets.

Inflorescence. *Inflorescence of spicate main branches*; *non-digitate*. Primary inflorescence branches 3–4 (these appressed, rather distant, up to 4.5 cm long). Inflorescence espatheate; not comprising 'partial inflorescences' and foliar organs. Spikelet-bearing axes persistent. Spikelets solitary and paired; secund; pedicellate (the pedicels shorter than the spikelets).

Female-fertile spikelets. *Spikelets 2–2.5 mm long*; abaxial; compressed dorsiventrally; *biconvex*; falling with the glumes; with conventional internode spacings. The upper floret not stipitate. Rachilla terminated by a female-fertile floret.

Glumes present; *one per spikelet (or the lower vestigial)*; (the upper) long relative to the adjacent lemmas; densely hairy; pointed (beyond the fruit); awnless; non-carinate. Upper glume 5 nerved. *Spikelets* with incomplete florets. The incomplete florets proximal to the female-fertile florets. *The proximal incomplete florets 1*. The proximal lemmas awnless; more or less equalling the female-fertile lemmas to decidedly exceeding the female-fertile lemmas (as long as the spikelet); densely hairy.

Female-fertile florets 1. Lemmas decidedly firmer than the glumes; smooth; entire; awnless; hairless; glabrous; non-carinate; having the margins tucked in onto the palea; with a clear germination flap (?). *Palea* present; awnless, without apical setae. *Ovary* glabrous (?). Stigmas 2 (?).

Fruit, embryo and seedling. Fruit small (1.7 mm long).

Taxonomy. Panicoideae; Panicodae; Paniceae.

Ecology, geography, regional floristic distribution. 1 species. Ecuador. Glycophytic (riverbanks).

Neotropical. Andean.

References, etc. Morphological/taxonomic: Swallen 1968.

Special comments. Fruit data wanting. Anatomical data wanting.

Acrachne Wright & Arn. ex Chiov.

Arthrochloa Lorch, *Camusia* Lorch, *Normanboria* Butzin

Habit, vegetative morphology. *Annual*; caespitose. *Culms* 12–80 cm high; herbaceous. Culm nodes glabrous. Culm internodes solid. *Leaves* not basally aggregated; non-auriculate. Leaf blades broadly linear (tapered to a hairlike tip); broad to narrow; flat; without abaxial multicellular glands; without cross venation; persistent. Ligule a fringed membrane to a fringe of hairs.

Reproductive organization. Plants bisexual, with bisexual spikelets; with hermaphrodite florets. Exposed-cleistogamous, or chasmogamous.

Inflorescence. *Inflorescence of spicate main branches*; subdigitate (usually with the lower spikes scattered, but becoming subdigitate above), or non-digitate (*A. racemosa*). Primary inflorescence branches 4–20. Rachides flattened. Inflorescence espatheate; not comprising 'partial inflorescences' and foliar organs. The racemes spikelet bearing to the base. Spikelet-bearing axes with substantial rachides; persistent. Spikelets solitary; secund (on the dorsiventral, slender, flattened rachis); biseriate; subsessile.

Female-fertile spikelets. *Spikelets* 5.5–13 mm long; adaxial; compressed laterally; *disarticulating above the glumes, or falling with the glumes, or not disarticulating (the lemmas falling acropetally from the rachilla, but the spikelet often falling wholly or in part before all the lemmas have been shed); not disarticulating between the florets to disarticulating between the florets (the rachilla tough or breaking irregularly, the paleas persistent)*. Rachilla prolonged beyond the uppermost female-fertile floret; rachilla hairless. The rachilla extension with incomplete florets. Hairy callus absent. *Callus* absent.

Glumes two; relatively large; *more or less equal*; shorter than the spikelets; shorter than the adjacent lemmas; dorsiventral to the rachis; awnless (but subulate via an excurrent mid-nerve); carinate; similar (thinly cartilaginous). Lower glume shorter than the lowest lemma; 1 nerved. Upper glume 1 nerved. *Spikelets* with incomplete florets. The incomplete florets distal to the female-fertile florets.

Female-fertile florets 8–20. Lemmas similar in texture to the glumes to decidedly firmer than the glumes (cartilaginous); not becoming indurated; incised; not deeply cleft (slightly notched via the slightly excurrent lateral nerves); mucronate (from the midnerve); hairless; glabrous; carinate; 3 nerved (the laterals closer to the margins than to the mid-nerve, and excurrent as small teeth). *Palea* present (lanceolate); 2-nerved. *Lodicules* present; 2; free; fleshy; glabrous. *Stamens* 3. Anthers not penicillate. *Ovary* glabrous. Styles free to their bases. Stigmas 2.

Fruit, embryo and seedling. Fruit free from both lemma and palea; small (0.8–1.1 mm long); ellipsoid; deeply *longitudinally grooved (on the hilar side)*; compressed dorsiventrally; sculptured. Hilum short. Pericarp free. Embryo large; not waisted. Endosperm containing compound starch grains. Embryo with an epiblast; with a scutellar tail; with an elongated mesocotyl internode.

Abaxial leaf blade epidermis. *Costal/intercostal zonation* conspicuous. *Papillae* absent. *Long-cells* similar in shape costally and intercostally; of similar wall thickness costally and intercostally. Mid-intercostal long-cells rectangular; having markedly sinuous walls. *Microhairs* present; chloridoid-type. Microhair apical cell wall of similar thickness/rigidity to that of the basal cell. Microhair basal cells 6–9 microns long. *Stomata* common. Subsidiaries dome-shaped (usually), or triangular (rarely). Guard-cells overlapping to flush with the interstomatals. *Intercostal short-cells* absent or very rare. Intercostal silica bodies absent. *Costal short-cells* conspicuously in long rows. Costal silica bodies present in alternate cell files of the costal zones; saddle shaped, or 'panicoid-type' (rarely).

Transverse section of leaf blade, physiology. C_4; XyMS+. PCR sheath outlines even. PCR sheaths of the primary vascular bundles complete. PCR cell chloroplasts centripetal. *Mesophyll* with radiate chlorenchyma. *Leaf blade* 'nodular' in section. *Midrib* conspicuous; having a conventional arc of bundles; with colourless mesophyll adaxially. All the vascular bundles accompanied by sclerenchyma. Combined sclerenchyma girders present (with the major bundles); forming 'figures', or nowhere forming 'figures'. Sclerenchyma all associated with vascular bundles.

Taxonomy. Chloridoideae; main chloridoid assemblage.

Ecology, geography, regional floristic distribution. 3 species. Abyssinia, southern Africa, Indochina, Indomalayan region, Australia. Mesophytic; shade species and species of open habitats; glycophytic. Sandy savanna.

Holarctic, Paleotropical, and Australian. Tethyan. African and Indomalesian. Irano-Turanian. Saharo-Sindian, Sudano-Angolan, and Namib-Karoo. Indian, Indo-Chinese, and Malesian. North and East Australian. Sahelo-Sudanian, Somalo-Ethiopian, South Tropical African, and Kalaharian. Tropical North and East Australian.

References, etc. Leaf anatomical: this project.

Acritochaete Pilger

Habit, vegetative morphology. Annual; decumbent. Culms herbaceous; branched above. Culm nodes glabrous. *Leaves* not basally aggregated; non-auriculate. Leaf blades linear to lanceolate (broadly linear to narrowly lanceolate); broad to narrow; 4–15 mm wide (to 14 cm long); without cross venation; persistent. Ligule an unfringed membrane (firm); truncate; 1 mm long.

Reproductive organization. Plants bisexual, with bisexual spikelets; with hermaphrodite florets.

Inflorescence. *Inflorescence of spicate main branches*. Primary inflorescence branches 2–4. Inflorescence espatheate; not comprising 'partial inflorescences' and foliar organs. Spikelet-bearing axes persistent. Spikelets solitary; secund; pedicellate.

Female-fertile spikelets. *Spikelets 5 mm long; compressed dorsiventrally*; falling with the glumes. Rachilla terminated by a female-fertile floret. Hairy callus absent.

Glumes two; very unequal; *shorter than the adjacent lemmas*; hairy; (the upper) *awned (attenuate into very long, filiform, flexuous awns)*; non-carinate; very dissimilar. Lower glume 0–1 nerved. Upper glume 2–4 nerved. *Spikelets* with incomplete florets. The incomplete florets proximal to the female-fertile florets. *Spikelets with proximal incomplete florets. The proximal incomplete florets* 1; epaleate; sterile. The proximal lemmas awned (identically with G_2); 6 nerved; more or less equalling the female-fertile lemmas; similar in texture to the female-fertile lemmas; not becoming indurated.

Female-fertile florets 1. *Lemmas* less firm than the glumes to similar in texture to the glumes; not becoming indurated; entire; pointed; *mucronate (with a terminal subule)*; hairless; non-carinate; having the margins lying flat and exposed on the palea; with a clear germination flap; 5 nerved. *Palea* present; relatively long; entire; 2-nerved; 2-keeled. *Lodicules* present; 2; free; fleshy; glabrous; not or scarcely vascularized. *Stamens* 3; with free filaments. Anthers not penicillate; without an apically prolonged connective. *Ovary* glabrous. Styles fused. Stigmas 2.

Fruit, embryo and seedling. *Fruit* free from both lemma and palea; small; compressed dorsiventrally. Hilum short. Embryo large; not waisted.

Abaxial leaf blade epidermis. *Costal/intercostal zonation* conspicuous. *Papillae* present. Intercostal papillae not over-arching the stomata; consisting of one oblique swelling per cell. Intercostal zones exhibiting many atypical long-cells. Mid-intercostal long-cells rectangular; having markedly sinuous walls. *Microhairs* present; panicoid-type. *Stomata* common (but confined to single files bordering veins). Subsidiaries triangular. Guard-cells overlapping to flush with the interstomatals. *Intercostal short-cells* absent or very rare. *Costal short-cells* conspicuously in long rows. Costal silica bodies 'panicoid-type'.

Transverse section of leaf blade, physiology. C_3; XyMS+. *Leaf blade* 'nodular' in section; with the ribs more or less constant in size. *Midrib* conspicuous; having a conventional arc of bundles. Bulliforms not in discrete, regular adaxial groups (these present only in irregular groups); in irregular groups. All the vascular bundles accompanied by sclerenchyma. Combined sclerenchyma girders present; forming 'figures'. Sclerenchyma all associated with vascular bundles.

Taxonomy. Panicoideae; Panicodae; Paniceae.

Ecology, geography, regional floristic distribution. 1 species. Tropical Africa. Mesophytic; shade species; glycophytic. Montane forest.

Paleotropical. African, Madagascan, and Indomalesian. Sudano-Angolan and West African Rainforest. Indian, Indo-Chinese, and Malesian. Sahelo-Sudanian, Somalo-Ethiopian, and South Tropical African.

References, etc. Leaf anatomical: this project.

Acroceras Stapf

Neohusnotia A. Camus

Excluding *Commelinidium*

Habit, vegetative morphology. Annual, or perennial; rhizomatous, or stoloniferous, or decumbent. *Culms* 10–125 cm high; herbaceous (often much-branched). Leaves non-auriculate. *Leaf blades* linear-lanceolate to ovate-lanceolate; *cordate (somewhat amplexicaul)*; pseudopetiolate, or not pseudopetiolate; cross veined, or without cross venation. *Ligule* present (mostly), or absent (rarely); a fringed membrane (very narrow), or a fringe of hairs.

Reproductive organization. Plants bisexual, with bisexual spikelets; with hermaphrodite florets. The spikelets all alike in sexuality.

Inflorescence. Inflorescence of spicate main branches (racemes or panicles), or paniculate; open; espatheate; not comprising 'partial inflorescences' and foliar organs. Spikelet-bearing axes persistent. *Spikelets* paired; secund; *pedicellate (the pairs of pedicels more or less connate below)*; consistently in 'long-and-short' combinations (at least in lower parts of panicle), or not in distinct 'long-and-short' combinations.

Female-fertile spikelets. Spikelets abaxial; compressed laterally to compressed dorsiventrally (terete below); falling with the glumes. Rachilla terminated by a female-fertile floret. Hairy callus absent.

Glumes two; very unequal; (the longer, upper) long relative to the adjacent lemmas (subequalling the L_1); dorsiventral to the rachis; awnless; similar (membranous). Lower glume 3 nerved. Upper glume 5–7 nerved, or 8–9 nerved (rarely). *Spikelets* with incomplete florets. The incomplete florets proximal to the female-fertile florets. *The proximal incomplete florets* 1; paleate. *Palea of the proximal incomplete florets fully developed.* The proximal incomplete florets male, or sterile. The proximal lemmas awnless; 5 nerved; more or less equalling the female-fertile lemmas to decidedly exceeding the female-fertile lemmas; less firm than the female-fertile lemmas to similar in texture to the female-fertile lemmas; becoming indurated to not becoming indurated (membranous to crustaceous).

Female-fertile florets 1. *Lemmas* decidedly firmer than the glumes; smooth to striate; becoming indurated to not becoming indurated; entire; *crested at the tip (the apex blunt, hard, laterally compressed)*; awnless; hairless (shiny); carinate to non-carinate; having the margins tucked in onto the palea; with a clear germination flap; 5 nerved. *Palea* present; relatively long; gaping (its tip reflexed); awnless, without apical setae; textured like the lemma (firm); indurated to not indurated; 2-nerved. *Lodicules* present; 2; free; fleshy; glabrous. *Stamens* 3. Anthers not penicillate. *Ovary* glabrous. Styles free to their bases. Stigmas 2.

Fruit, embryo and seedling. Fruit ellipsoid; flattened on one side. Hilum short, or long-linear (punctiform, oblong, or linear and half to two thirds the fruit length).

Abaxial leaf blade epidermis. *Costal/intercostal zonation* conspicuous. *Papillae* absent. Intercostal zones with typical long-cells, or without typical long-cells (long-cells cubical). Mid-intercostal long-cells having markedly sinuous walls, or having straight or only gently undulating walls. *Microhairs* present; panicoid-type and chloridoid-type; (36–)40–84 microns long; 3.9–6 microns wide at the septum. Microhair total length/width at septum 8–10.8. Microhair apical cells (24–)27–36(–38) microns long. Microhair apical cell/total length ratio 0.42–0.65. *Stomata* common; 25.5–30 microns long. Subsidiaries triangular. *Intercostal short-cells* common, or absent or very rare; when seen, in cork/silica-cell pairs; silicified. Intercostal silica bodies cross-shaped and oryzoid-type. *Costal short-cells* conspicuously in long rows, or neither distinctly grouped into long rows nor predominantly paired (rarely). Costal silica bodies 'panicoid-type'; cross shaped to dumb-bell shaped.

Transverse section of leaf blade, physiology. C_3; XyMS+. *Mesophyll* with radiate chlorenchyma; *Isachne*-type. *Leaf blade* with distinct, prominent adaxial ribs, or adaxially flat; with the ribs more or less constant in size. *Midrib* conspicuous; with one bundle only, or having a conventional arc of bundles. Bulliforms present in discrete, regular adaxial groups, or not in discrete, regular adaxial groups; in simple fans (or the fans ill defined). All the vascular bundles accompanied by sclerenchyma. Combined sclerenchyma girders present; forming 'figures', or nowhere forming 'figures'. Sclerenchyma all associated with vascular bundles.

Cytology. Chromosome base number, $x = 9$. $2n = 36$.

Taxonomy. Panicoideae; Panicodae; Paniceae.

Ecology, geography, regional floristic distribution. 15 species. Africa, Madagascar, Indomalayan region. Commonly adventive. Hydrophytic to mesophytic; shade species and species of open habitats; glycophytic. Shallow water, damp places and forests.

Paleotropical and Neotropical. African, Madagascan, and Indomalesian. Sudano-Angolan, West African Rainforest, and Namib-Karoo. Indian, Indo-Chinese, and Malesian. Caribbean, Venezuela and Surinam, Amazon, Central Brazilian, and Pampas. Sahelo-Sudanian, Somalo-Ethiopian, South Tropical African, and Kalaharian.

Economic importance. Significant weed species: *A. zizanioides*. Cultivated fodder: *A. macrum*. Important native pasture species: *A. macrum*.

References, etc. Morphological/taxonomic: Launert 1970; Zuloaga 1987. Leaf anatomical: Metcalfe 1960; this project.

Special comments. Fruit data wanting.

Actinocladum McClure ex Soderstrom

Habit, vegetative morphology. Perennial; caespitose. The flowering culms leafy. *Culms* 300–460 cm high; woody and persistent; to about 1.4 cm in diameter; branched above (and with rather different branches basally). Primary branches/mid-culm node 1 (apsidate). Culm sheaths persistent, or deciduous in their entirety. Culm internodes solid. Rhizomes pachymorph. Plants unarmed. *Leaves* not basally aggregated. Leaf blades broad; 10–40 mm wide; pseudopetiolate; disarticulating from the sheaths; rolled in bud. *Ligule* present; a minute ciliolate rim.

Reproductive organization. Plants bisexual, with bisexual spikelets; with hermaphrodite florets. The spikelets all alike in sexuality.

Inflorescence. *Inflorescence consisting of verticils of racemose or paniculate, few-flowered 'inflorescences'*; spatheate; a complex of 'partial inflorescences' and intervening foliar organs (with a spatheole at the first node of each ultimate inflorescence unit). *Spikelet-bearing axes* 'racemes', or paniculate; clustered; persistent. Spikelets not secund; long pedicellate.

Female-fertile spikelets. Spikelets 60–75 mm long; slightly compressed laterally; disarticulating above the glumes; disarticulating between the florets. Rachilla prolonged beyond the uppermost female-fertile floret; rachilla hairy (short-pilose). The rachilla extension with incomplete florets. Hairy callus present (of short hairs).

Glumes two; very unequal; shorter than the adjacent lemmas; free; hairless; glabrous; pointed; awnless (apiculate); non-carinate (rounded on the back); similar (leathery, triangular). Lower glume relatively smooth; 5 nerved. Upper glume 11 nerved. *Spikelets with incomplete florets. The incomplete florets distal to the female-fertile florets. The distal incomplete florets merely underdeveloped. Spikelets without proximal incomplete florets*.

Female-fertile florets 7–10. **Lemmas** ovate lanceolate; **less firm than the glumes**; not becoming indurated; entire; pointed; awnless; hairless; glabrous; non-carinate (rounded on the back); 13 nerved. *Palea* present; relatively long; minutely apically notched; awnless; without apical setae; not indurated; several nerved (6 between the keels, 5 on each wing); 2-keeled. *Lodicules* present; 3; free; membranous; ciliate, or glabrous; toothed; heavily vascularized. *Stamens* 3. Anthers not penicillate; without an apically prolonged connective. **Ovary** hairy; **with a conspicuous apical appendage**. The appendage broadly conical, fleshy. Styles fused (the top of the ovary being attenuate into a single style). Stigmas 2.

Fruit, embryo and seedling. Fruit large; not grooved (but the seed within it slightly grooved); slightly curved, compressed dorsiventrally; with hairs confined to a terminal tuft. Hilum long-linear. Pericarp dry, thick and hard; free. Embryo small.

Abaxial leaf blade epidermis. *Costal/intercostal zonation* conspicuous. *Papillae* present; costal and intercostal. Intercostal papillae over-arching the stomata (and almost completely covering them); several per cell (basically one median row per long-cell, large, thick walled, irregular, warty and sometimes bifurcated or branching). Intercostal zones probably with typical long-cells (but their shapes largely obscured but papillae, and hard to determine). Mid-intercostal long-cells rectangular; having markedly sinuous walls. *Microhairs* present; elongated; mostly clearly two-celled (but occasionally with an extra, short basal cell); panicoid-type; 40–48 microns long; 5.4–6 microns wide at the septum. Microhair total length/width at septum 7.5–8. Microhair apical cells 18–25.3 microns long. Microhair apical cell/total length ratio 0.44–0.53. *Stomata* common; 22.5–27 microns long. Guard-cells sunken. *Intercostal short-cells* absent or very rare. Numerous large costal prickles present, with abundant tiny pits and conspicuous basal rosettes. *Crown cells* absent. *Costal short-cells* predominantly paired. Costal silica bodies absent (in the material seen).

Transverse section of leaf blade, physiology. C_3. Mesophyll with adaxial palisade; with arm cells; with fusoids. The fusoids external to the PBS. *Leaf blade* with distinct, prominent adaxial ribs (with one large rib and a short series of smaller ones, near one margin), or adaxially flat (elsewhere). *Midrib* not readily distinguishable (unless the submarginal rib represents a highly asymmetrical midrib); with one bundle only. *The lamina* symmetrical on either side of the midrib (unless the large submarginal rib is interpreted as a highly asymmetrically placed midrib). Bulliforms present in discrete, regular adaxial groups (between all the adjoining veins); in simple fans. All the vascular bundles accompanied by sclerenchyma. Combined sclerenchyma girders present; forming 'figures' (all the bundles with narrow I's or 'anchors'). Sclerenchyma not all bundle-associated. The 'extra' sclerenchyma in abaxial groups and in adaxial groups; abaxial-hypodermal, the groups isolated and adaxial-hypodermal, contiguous with the bulliforms (there being small abaxial groups opposite the bulliforms, and adaxial hypodermal fibres adjoining and lateral to them).

Taxonomy. Bambusoideae; Bambusodae; Bambuseae.

Ecology, geography, regional floristic distribution. 1 species. Central Brazil. Neotropical. Central Brazilian.

References, etc. Morphological/taxonomic: Soderstrom 1981d. Leaf anatomical: this project.

Aegilops L.

Aegicon Adans., *Aegilemma* Löve, *Aegilonearum* Löve, *Aegilopodes* Löve, *Chennapyrum* Löve, *Comopyrum* Löve, *Cylindropyrum* (Jaub. & Spach) Löve, *Gastropyrum* (Jaub. & Spach) Löve, *Kiharapyrum* Löve, *Orrhopygium* Löve, *Patropyrum* Löve, *Perlaria* Fabric., *Sitopsis* (Jaub. & Spach) Löve

Excluding *Amblyopyrum*

Habit, vegetative morphology. *Annual*; rhizomatous, or caespitose. *Culms* 15–80 cm high; herbaceous. Culm internodes hollow. *Leaves* not basally aggregated; auriculate, or non-auriculate. Leaf blades linear to linear-lanceolate; 1.5–10 mm wide; usually flat; without cross venation. Ligule an unfringed membrane; truncate; 0.2–0.8 mm long.

Reproductive organization. Plants bisexual, with bisexual spikelets; with hermaphrodite florets. The spikelets of sexually distinct forms on the same plant, or all alike in sexuality; hermaphrodite, or hermaphrodite and sterile (there often being incomplete spikelets at base and tip of the spike).

Inflorescence. *Inflorescence a single spike.* Rachides hollowed. Inflorescence espatheate; not comprising 'partial inflorescences' and foliar organs. Spikelet-bearing axes with substantial rachides; disarticulating (usually); falling entire (occasionally — e.g. *Cylindropyrum*), or disarticulating at the joints. Spikelets solitary; not secund; distichous; sessile, or subsessile.

Female-fertile spikelets. Spikelets 5–12 mm long; not noticeably compressed (usually), or compressed laterally; disarticulating above the glumes, or falling with the glumes, or not disarticulating. Rachilla prolonged beyond the uppermost female-fertile floret; rachilla hairy. The rachilla extension with incomplete florets. Hairy callus absent. *Callus* very short; blunt.

Glumes two; more or less equal; shorter than the adjacent lemmas, or long relative to the adjacent lemmas; lateral to the rachis; without conspicuous tufts or rows of hairs; not pointed (usually with one or more teeth or awns); not subulate; awned (sometimes with more than one awn), or awnless; *non-carinate (generally rounded on the back)*; similar (leathery). *Lower glume 7–13 nerved.* Upper glume 7–13 nerved. *Spikelets* with incomplete florets. The incomplete florets distal to the female-fertile florets.

Female-fertile florets 2–8. *Lemmas* one to three toothed, or awned; *similar in texture to the glumes to decidedly firmer than the glumes*; becoming indurated to not becoming indurated; entire, or incised; awnless, or mucronate, or awned. Awns when present, 1, or 3; median, or median and lateral; the median similar in form to the laterals (when laterals present); from a sinus, or apical; non-geniculate; much shorter than the body of the lemma to much longer than the body of the lemma; entered by one vein. The lateral awns when present, shorter than the median, or about equalling the median. Lemmas hairy, or hairless; non-carinate (dorsally rounded); 5–7 nerved (usually), or 9–13 nerved (*Kiharapyrum*); with the nerves non-confluent. *Palea* present; relatively long; apically notched; not indurated (membranous); 2-nerved; 2-keeled. *Lodicules* present; 2; free; membranous; ciliate; not toothed. *Stamens* 3. Anthers 1.5–4.5 mm long; not penicillate. **Ovary hairy.** Stigmas 2.

Fruit, embryo and seedling. *Fruit* adhering to lemma and/or palea, or free from both lemma and palea; medium sized; longitudinally grooved; compressed dorsiventrally; with hairs confined to a terminal tuft. *Hilum long-linear.* Embryo large to small (to about 1/3 the caryopsis length); not waisted. Endosperm hard; without lipid; containing only simple starch grains.

Abaxial leaf blade epidermis. *Costal/intercostal zonation* conspicuous. *Papillae* absent. *Long-cells* similar in shape costally and intercostally (though the costals rather 270,4 smaller); of similar wall thickness costally and intercostally (thick walled). Mid-intercostal long-cells rectangular; having markedly sinuous walls (e.g. *A. cylindrica*), or having straight or only gently undulating walls. *Microhairs* absent. *Stomata* common; 42–45 microns long. Subsidiaries parallel-sided. Guard-cells overlapped by the interstomatals. *Intercostal short-cells* common, or absent or very rare; in cork/silica-cell pairs (in *A. cylindrica*); silicified (in *A. cylindrica*). Intercostal silica bodies rounded. *Crown cells* present (abundant costally, in *A. cylindrica*). *Costal short-cells* predominantly paired (e.g. in *A. cylindrica*), or neither distinctly grouped into long rows nor predominantly paired. Costal silica bodies horizontally-elongated crenate/sinuous, or rounded (exclusively, in *A. cylindrica*), or tall-and-narrow.

Transverse section of leaf blade, physiology. C_3; XyMS+. *Mesophyll* with non-radiate chlorenchyma; without adaxial palisade. *Leaf blade* with distinct, prominent adaxial ribs, or 'nodular' in section (the adaxial ribs wide and low); with the ribs more or less constant in size. *Midrib* not readily distinguishable; with one bundle only. Bulliforms present in discrete, regular adaxial groups; in simple fans (or the groups of fairly uniform cells). All the vascular bundles accompanied by sclerenchyma (except the extreme laterals, in *A. cylindrica*). Combined sclerenchyma girders present, or absent; nowhere forming 'figures'. Sclerenchyma all associated with vascular bundles.

Culm anatomy. *Culm internode bundles* in one or two rings.

Cytology. Chromosome base number, $x = 7$. $2n = 14$ and 28, or 42 (rarely). 2, 4, and 6 ploid. Haplomic genome content B, or C, or D, or L, or M, or U, or B and U, or C and D, or C and U, or D and M, or M and U, or D, M, and U. Haploid nuclear DNA content 2.3–7.5 pg (21 species, mean 5.6). Mean diploid 2c DNA value 11.2 pg (11 species, 7.2–14.3).

Taxonomy. Pooideae; Triticodae; Triticeae.

Ecology, geography, regional floristic distribution. 22 species. Western Mediterranean to central Asia. Commonly adventive. Xerophytic; species of open habitats.

Holarctic and Paleotropical. Boreal and Tethyan. African. Euro-Siberian. Macaronesian, Mediterranean, and Irano-Turanian. Saharo-Sindian. European.

Hybrids. Intergeneric hybrids with *Triticum* (×*Aegilotriticum* Wagner ex Tschermak), *Secale* (×*Aegilosecale* Ciferri & Giacom.), *Dasypyrum*, *Elytrigia*.

Rusts and smuts. Rusts — *Puccinia*. Taxonomically wide-ranging species: *Puccinia graminis*, *Puccinia striiformis*, *Puccinia hordei*, and *Puccinia recondita*. Smuts from Tilletiaceae and from Ustilaginaceae. Tilletiaceae — *Tilletia* and *Urocystis*. Ustilaginaceae — *Ustilago*.

Economic importance. Significant weed species: *A. cylindrica*, *A. geniculata*, *A. triuncialis*. Important native pasture species: several (e.g. *A. cylindrica*, *A. kotschyi*, *A. triuncialis* considered useful).

References, etc. Morphological/taxonomic: Löve 1984. Leaf anatomical: Metcalfe 1960; this project.

Aegopogon Humb. & Bonpl. ex Willd.

Atherophora Steud., *Hymenothecium* Lag., *Schellingia* Steud.

Habit, vegetative morphology. Delicate annual. *Culms* 6–30 cm high; herbaceous. *Leaves* not basally aggregated; non-auriculate. Sheaths loose. Leaf blades linear; narrow; 1–2 mm wide; flat; without abaxial multicellular glands; without cross venation; persistent. *Ligule a fringed membrane*.

Reproductive organization. Plants bisexual, with bisexual spikelets; with hermaphrodite florets. The spikelets of sexually distinct forms on the same plant; hermaphrodite, male-only, and sterile. The male and female-fertile spikelets mixed in the inflorescence (in triads, the central member hermaphrodite, the laterals male or sterile). The spikelets overtly heteromorphic.

Inflorescence. *Inflorescence a false spike, with spikelets on contracted axes (the main axis bearing triplets of spikelets with short flat pedicels, the triplets short-pedunculate, spreading, the rachis filiform)*; espatheate; not comprising 'partial inflorescences' and foliar organs. Spikelet-bearing axes disarticulating; falling entire (i.e. the triplets falling). *Spikelets* in triplets; *secund (the triplets on one side of the axis)*; sessile and pedicellate, or subsessile and pedicellate; *consistently in 'long-and-short' combinations*; in pedicellate/sessile combinations, or unequally pedicellate in each combination (the central hermaphrodite spikelet sessile or subsessile, the reduced laterals pedicellate). The 'shorter' spikelets hermaphrodite. The 'longer' spikelets male-only, or sterile.

Female-sterile spikelets. The longer-pedicelled male or neuter spikelets reduced.

Female-fertile spikelets. Spikelets 2 mm long; compressed laterally, or not noticeably compressed, or compressed dorsiventrally (?); falling with the glumes (the triplets disarticulating, with the pointed basal stipe). Rachilla terminated by a female-fertile floret.

Glumes present; two; more or less equal; shorter than the spikelets; shorter than the adjacent lemmas; free; not pointed (apically notched); awned (via the extended midnerve); similar (membranous, truncate or notched, awned). Lower glume 1 nerved (?). Upper glume 1 nerved (?). **Spikelets** with female-fertile florets only; *without proximal incomplete florets*.

Female-fertile florets 1. *Lemmas* similar in texture to the glumes (membranous); not becoming indurated; *incised*; with a narrow, nerveless lobe outside each lateral awn; not deeply cleft; awned. Awns 3; median and lateral; the median similar in form to the laterals; continued from the nerve; non-geniculate (delicate); hairless (scabrous); much longer than the body of the lemma; entered by one vein. The lateral awns shorter than the median. Lemmas hairless; without a germination flap; 3 nerved. *Palea* present; relatively long (but somewhat shorter than the lemma); apically notched; awned (two-awned); textured like the lemma; not indurated; 2-nerved. *Lodicules* present; 2; fleshy; glabrous. *Stamens* 3. Anthers small. *Ovary* glabrous. Stigmas 2.

Fruit, embryo and seedling. Fruit small (about 1.7 mm long in *A. bryophylus*); fusiform, or ellipsoid; compressed laterally. Hilum short. Pericarp fused. Embryo large; with an epiblast; with a scutellar tail; with an elongated mesocotyl internode. Embryonic leaf margins meeting.

Abaxial leaf blade epidermis. *Costal/intercostal zonation* conspicuous. *Papillae* present; intercostal. Intercostal papillae not over-arching the stomata (for the most part); consisting of one oblique swelling per cell (large). Mid-intercostal long-cells rectangular;

having straight or only gently undulating walls. *Microhairs* present; chloridoid-type. Microhair apical cell wall of similar thickness/rigidity to that of the basal cell. Microhair basal cells 30 microns long. *Stomata* common. Subsidiaries non-papillate; dome-shaped (sometimes approaching parallel). Guard-cells overlapping to flush with the interstomatals. *Intercostal short-cells* common; not paired (solitary); not silicified. Intercostal silica bodies absent to imperfectly developed. *Costal short-cells* conspicuously in long rows. Costal silica bodies present and well developed; confined to the central file(s) of the costal zones to present in alternate cell files of the costal zones; 'panicoid-type'; cross shaped to dumb-bell shaped.

Transverse section of leaf blade, physiology. C_4; XyMS+. PCR sheaths of the primary vascular bundles complete. PCR sheath extensions absent. *Leaf blade* with distinct, prominent adaxial ribs, or 'nodular' in section. *Midrib* not readily distinguishable; with one bundle only. Bulliforms present in discrete, regular adaxial groups; in simple fans (the large median cell of each group deeply inserted in the mesophyll). All the vascular bundles accompanied by sclerenchyma. Combined sclerenchyma girders present (with the main bundles); forming 'figures' (the main bundles with anchors). Sclerenchyma all associated with vascular bundles.

Taxonomy. Chloridoideae; main chloridoid assemblage.

Ecology, geography, regional floristic distribution. 3 species. Southern U.S.A. to Argentina.

Holarctic, Paleotropical, and Neotropical. Madrean. Indomalesian. Papuan. Caribbean, Venezuela and Surinam, and Andean.

Rusts and smuts. Rusts — *Puccinia*. Smuts from Ustilaginaceae. Ustilaginaceae — *Ustilago*.

References, etc. Leaf anatomical: this project.

Aeluropus Trin.

Aelbroeckia De Moor, *Chamaedactylis* T. Nees

Habit, vegetative morphology. Perennial; rhizomatous, or stoloniferous, or caespitose, or decumbent. **Culms** 5–40 cm high; *herbaceous*; branched above, or unbranched above. *Leaves* not basally aggregated; non-auriculate. Leaf blades linear to linear-lanceolate (with a cartilaginous, often pungent apex); narrow; 0.6–3 mm wide; flat, or folded; without abaxial multicellular glands; without cross venation. Ligule a fringed membrane to a fringe of hairs.

Reproductive organization. Plants bisexual, with bisexual spikelets; with hermaphrodite florets.

Inflorescence. *Inflorescence a single spike (ovoid to capitate), or of spicate main branches (then a 2-sided raceme of short, densely spiculate sessile spikes appressed to the main axis)*; espatheate; not comprising 'partial inflorescences' and foliar organs. Spikelet-bearing axes persistent. *Spikelets* solitary; *secund*.

Female-fertile spikelets. Spikelets 2.2–5 mm long; compressed laterally; disarticulating above the glumes. Rachilla prolonged beyond the uppermost female-fertile floret; rachilla hairless. Hairy callus absent. *Callus* short; blunt.

Glumes two; very unequal; shorter than the spikelets; shorter than the adjacent lemmas; lateral to the rachis; awnless; carinate; similar (membranous to leathery). Lower glume 1–3 nerved. Upper glume 5–7 nerved.

Female-fertile florets *4–18*. *Lemmas* similar in texture to the glumes; not becoming indurated; entire, or incised; when incised, not deeply cleft (emarginate); *mucronate*; hairy; carinate; *9–11 nerved*. *Palea* present; relatively long; entire to apically notched; awnless, without apical setae; not indurated (membranous); 2-nerved. *Lodicules* present; 2; free; fleshy; ciliate, or glabrous. *Stamens* 3. Anthers 0.8–1.6 mm long; not penicillate. *Ovary* glabrous. Styles free to their bases. Stigmas 2.

Fruit, embryo and seedling. *Fruit* free from both lemma and palea; ellipsoid; compressed dorsiventrally. Hilum short. *Pericarp fused*. Embryo large; not waisted. Endosperm hard; without lipid; containing compound starch grains. Embryo with an epiblast;

with a scutellar tail; with an elongated mesocotyl internode. Embryonic leaf margins meeting.

Abaxial leaf blade epidermis. *Costal/intercostal zonation* conspicuous. *Papillae* present (very abundant, large, thick walled); costal and intercostal. Intercostal papillae over-arching the stomata (so as to thoroughly obscure them); consisting of one oblique swelling per cell, or consisting of one symmetrical projection per cell. *Long-cells* markedly different in shape costally and intercostally (the costals conventionally shaped); of similar wall thickness costally and intercostally (quite thick walled). Intercostal zones exhibiting many atypical long-cells (all being relatively short). Mid-intercostal long-cells rectangular (to irregular); having markedly sinuous walls. *Microhairs* present; more or less spherical, or elongated; clearly two-celled; chloridoid-type. Microhair apical cell wall of similar thickness/rigidity to that of the basal cell. Microhairs 21–27 microns long. Microhair basal cells 9 microns long. Microhairs 13.5–18 microns wide at the septum. Microhair total length/width at septum 1.5–2. Microhair apical cells 14–18 microns long. Microhair apical cell/total length ratio 0.55–0.69. *Stomata* common (but almost invisible). Subsidiaries non-papillate; triangular. *Intercostal short-cells* common; hard to observe, but seemingly solitary, paired and even in short rows; not silicified. Intercostal silica bodies absent. *Costal short-cells* conspicuously in long rows. Costal silica bodies present in alternate cell files of the costal zones; 'panicoid-type'; mostly short butterfly shaped to dumb-bell shaped (a few almost square).

Transverse section of leaf blade, physiology. C_4; XyMS+. PCR sheath outlines even. PCR sheaths of the primary vascular bundles complete to interrupted; interrupted abaxially only. PCR sheath extensions absent. PCR cell chloroplasts centripetal. *Mesophyll* with radiate chlorenchyma; traversed by columns of colourless mesophyll cells (at least in places). *Leaf blade* with distinct, prominent adaxial ribs; with the ribs more or less constant in size. *Midrib* not readily distinguishable; with one bundle only. Bulliforms present in discrete, regular adaxial groups; in simple fans, or associated with colourless mesophyll cells to form deeply-penetrating fans (sometimes associated with colourless girders). All the vascular bundles accompanied by sclerenchyma. Combined sclerenchyma girders present; forming 'figures'. Sclerenchyma all associated with vascular bundles. The lamina margins with fibres.

Cytology. Chromosome base number, $x = 10$. $2n = 20$. 2 ploid.

Taxonomy. Chloridoideae; main chloridoid assemblage.

Ecology, geography, regional floristic distribution. 5 species. Mediterranean to India. Species of open habitats; halophytic. In sand of seashores and deserts.

Holarctic and Paleotropical. Boreal and Tethyan. African. Euro-Siberian. Mediterranean and Irano-Turanian. Saharo-Sindian and Sudano-Angolan. European and Siberian. Sahelo-Sudanian and Somalo-Ethiopian.

Rusts and smuts. Rusts — *Puccinia*. Smuts from Ustilaginaceae. Ustilaginaceae — *Sorosporium* and *Sphacelotheca*.

References, etc. Leaf anatomical: Metcalfe 1960; this project.

Afrotrichloris Chiov.

Habit, vegetative morphology. *Perennial*; caespitose. **Culms** 30–60 cm high; *herbaceous*; unbranched above. Plants unarmed. Leaf blades narrow; about 1.5 mm wide (to 25 cm long); rolled; without abaxial multicellular glands; without cross venation. Ligule scarious, long pilose.

Reproductive organization. Plants bisexual, with bisexual spikelets; with hermaphrodite florets.

Inflorescence. *Inflorescence a single spike (14–22 cm long, curved or flexuous)*. Inflorescence with axes ending in spikelets (?). Spikelet-bearing axes persistent.

Female-fertile spikelets. *Spikelets 15 mm long*; disarticulating above the glumes. Rachilla prolonged beyond the uppermost female-fertile floret. The rachilla extension with incomplete florets. Hairy callus present (?).

Glumes two; very unequal to more or less equal; (the upper) long relative to the adjacent lemmas; free; hairless; pointed; awnless (but acuminate); similar (divergent, narrowly lanceolate, hyaline). Lower glume 3–5 nerved. Upper glume 3–5 nerved. *Spike-*

lets with incomplete florets. The incomplete florets distal to the female-fertile florets. The distal incomplete florets several, clustered; incomplete florets awned (male or sterile, and lacking the palea).

Female-fertile florets *1*. *Lemmas* broadly rounded, thinly leathery below, with hyaline lobes; not becoming indurated; incised; 2 lobed; deeply cleft (to below the middle); *awned*. Awns 1, or 3; median, or median and lateral; the median similar in form to the laterals; from a sinus; non-geniculate; hairy; much longer than the body of the lemma. Lemmas hairy. The hairs in tufts (on the lateral nerves, the lobes glabrous), or not in tufts; not in transverse rows. Lemmas non-carinate (rounded on the back); 3 nerved. *Palea* present; relatively long; awnless, without apical setae; 2-nerved; 2-keeled. *Lodicules* present; 2; free; glabrous. *Stamens* 3. Anthers not penicillate; without an apically prolonged connective. *Ovary* glabrous. Styles free to their bases. Stigmas 2.

Fruit, embryo and seedling. Fruit small (2 mm long); ellipsoid (oblong-elliptical); compressed dorsiventrally. *Pericarp free.*

Abaxial leaf blade epidermis. *Costal/intercostal zonation* conspicuous. *Papillae* present; intercostal. Intercostal papillae over-arching the stomata; consisting of one oblique swelling per cell to consisting of one symmetrical projection per cell (the papillae large relative to the cells carrying them). Long-cells differing markedly in wall thickness costally and intercostally (intercostals thicker-walled). Intercostal zones with typical long-cells (but these rather short). Mid-intercostal long-cells rectangular; not readily observable in detail, because of the abundant papillae. *Microhairs* present; more or less spherical to elongated; clearly two-celled; chloridoid-type. Microhair apical cell wall of similar thickness/rigidity to that of the basal cell. Microhair basal cells 15 microns long. Microhair total length/width at septum 2. Microhair apical cell/total length ratio 0.37. *Stomata* common. Subsidiaries non-papillate; seemingly all dome-shaped (but extensively obscured by papillae). Guard-cells overlapping to flush with the interstomatals. *Intercostal short-cells* common; not paired (solitary); not silicified. Intercostal silica bodies absent. *Costal short-cells* conspicuously in long rows (but the short-cells rather long in some files). Costal silica bodies present in alternate cell files of the costal zones; exclusively saddle shaped (a rather round form of this type).

Transverse section of leaf blade, physiology. C_4; XyMS+. PCR sheaths of the primary vascular bundles interrupted; interrupted abaxially only and interrupted both abaxially and adaxially. PCR sheath extensions absent. *Mesophyll* with radiate chlorenchyma; traversed by columns of colourless mesophyll cells. *Leaf blade* with distinct, prominent adaxial ribs to 'nodular' in section. *Midrib* conspicuous (via a rather larger bundle, and its adaxial group of colourless cells); with one bundle only; with colourless mesophyll adaxially. Bulliforms present in discrete, regular adaxial groups; in simple fans, or associated with colourless mesophyll cells to form deeply-penetrating fans (these associated with colourless girders). All the vascular bundles accompanied by sclerenchyma. Combined sclerenchyma girders present (with most bundles); forming 'figures' (in the large bundles). Sclerenchyma all associated with vascular bundles. The lamina margins with fibres.

Taxonomy. Chloridoideae; main chloridoid assemblage.

Ecology, geography, regional floristic distribution. 2 species. Somalia. Paleotropical. African. Sudano-Angolan. Somalo-Ethiopian.

References, etc. Leaf anatomical: this project.

Special comments. Fruit data wanting.

Agenium Nees

Habit, vegetative morphology. Perennial; caespitose. Culms 50–80 cm high; herbaceous; branched above. *Leaves* not basally aggregated. Leaf blades linear to linear-lanceolate (acuminate); narrow; to 3 mm wide; without cross venation; disarticulating from the sheaths (at least on the lower culms). Ligule an unfringed membrane (minutely ciliolate only); truncate; 0.5 mm long. *Contra-ligule* absent.

Reproductive organization. Plants bisexual, with bisexual spikelets; with hermaphrodite florets. The spikelets of sexually distinct forms on the same plant; hermaphrodite and male-only, or hermaphrodite, male-only, and sterile, or female-only, male-only, and sterile. The male and female-fertile spikelets mixed in the inflorescence. The spikelets overtly heteromorphic (the upper pedicellate spikelets being awnless); in both homogamous and heterogamous combinations (with several conspicuous homogamous pairs at the raceme bases, these being awnless and male or neuter).

Inflorescence. *Inflorescence* of spicate main branches (of andropogonoid racemes), or a single raceme; usually *digitate, or subdigitate*. Primary inflorescence branches 1–3. Inflorescence espatheate; not comprising 'partial inflorescences' and foliar organs. *Spikelet-bearing axes* 'racemes'; the spikelet-bearing axes several-jointed; solitary, or paired, or clustered (the raceme bases, when digitate, filiform and flexuous); with very slender rachides; disarticulating; disarticulating at the joints. *'Articles'* non-linear; without a basal callus-knob; not appendaged; disarticulating obliquely; densely long-hairy (along the margins). Spikelets paired; not secund; sessile and pedicellate; consistently in 'long-and-short' combinations; in pedicellate/sessile combinations. Pedicels of the 'pedicellate' spikelets free of the rachis. The 'shorter' spikelets male-only to sterile (at the base of the racemes), or hermaphrodite to female-only (in the upper pairs). The 'longer' spikelets male-only, or sterile.

Female-sterile spikelets. The (upper) pedicelled spikelets male, larger. The male spikelets with glumes (the G_1 villous). The lemmas awnless.

Female-fertile spikelets. *Spikelets* 5 mm long; *compressed dorsiventrally (subterete)*; falling with the glumes (and with the adjacent joint and pedicel). Rachilla terminated by a female-fertile floret. Hairy callus present (the hairs shorter than in the pedicelled spikelet). Callus pointed.

Glumes two; more or less equal; long relative to the adjacent lemmas; dorsiventral to the rachis; hairy (G_1 more so); without conspicuous tufts or rows of hairs; not pointed (G_1 truncate, G_2 obtuse); awnless; very dissimilar (G_1 hairier, sulcate). *Lower glume not two-keeled*; *sulcate on the back (with a median, translucent groove)*; not pitted (but the sulcus looks like an extended pit); *6 nerved (without a median)*. Upper glume 3 nerved. *Spikelets* with incomplete florets. The incomplete florets proximal to the female-fertile florets. *The proximal incomplete florets* 1; epaleate; sterile. The proximal lemmas awnless (ciliate); 0 nerved; decidedly exceeding the female-fertile lemmas (the latter being reduced to its awn); not becoming indurated (hyaline, much shorter than the glumes).

Female-fertile florets 1. Lemmas consisting almost entirely of awn; entire; awned. Awns 1 (up to 40 mm long); geniculate; hairless to hairy (scabrid to shortly hairy); much longer than the body of the lemma (which is virtually non-existent); entered by one vein; deciduous. Lemmas 1 nerved (in the awn). **Palea absent**. Lodicules present; 2; free; fleshy; glabrous. *Ovary* glabrous. Stigmas 2; red pigmented.

Abaxial leaf blade epidermis. *Costal/intercostal zonation* conspicuous. *Papillae* present; intercostal. Intercostal papillae over-arching the stomata; consisting of one oblique swelling per cell, or consisting of one symmetrical projection per cell (large, rough). Intercostal zones with typical long-cells to exhibiting many atypical long-cells (and the interstomatals very short). Mid-intercostal long-cells having markedly sinuous walls. *Microhairs* present; panicoid-type; 25.5–27–27 microns long; 4–4.5(–5) microns wide at the septum. Microhair total length/width at septum 5–7.5. Microhair apical cells 15 microns long. Microhair apical cell/total length ratio 0.56–0.59. *Stomata* common; 18–21–21 microns long. Subsidiaries non-papillate; dome-shaped (mostly), or triangular (a few). Guard-cells overlapped by the interstomatals (sometimes, slightly), or overlapping to flush with the interstomatals. *Intercostal short-cells* absent or very rare (very scarce, ignoring the microhairs and a few small prickle bases). *Costal short-cells* conspicuously in long rows. Costal silica bodies 'panicoid-type'; mostly short dumb-bell shaped.

Transverse section of leaf blade, physiology. C_4; XyMS–. *Leaf blade* adaxially flat. *Midrib* conspicuous (via a large bundle with an I-shaped combination girder); limits not clearly defined. Bulliforms not in discrete, regular adaxial groups (the epidermis extensively bulliform). Many of the smallest vascular bundles unaccompanied by sclerenchyma. Combined sclerenchyma girders present (with all the primaries); forming 'figures' (most notably the mid-vein I). Sclerenchyma all associated with vascular bundles.

Taxonomy. Panicoideae; Andropogonodae; Andropogoneae; Andropogoninae.

Ecology, geography, regional floristic distribution. 4 species. Brazil to Argentina. Species of open habitats. Dry savanna.

Neotropical. Central Brazilian and Pampas.
References, etc. Leaf anatomical: this project.
Special comments. Description based mainly on *A. villosum*. Fruit data wanting.

Agnesia Zuloaga & Judziewicz

Habit, vegetative morphology. Delicate perennial; caespitose. The flowering culms leafy. **Culms *15–50 cm high***; herbaceous; *to* 0.1 cm in diameter; **unbranched above**. Culm nodes glabrous. Plants unarmed. *Leaves* not basally aggregated (the lower and middle culm leaves with reduced blades, the upper leaves fully developed in dense, pinnately presented complements of 3–15); auriculate (via a small, distal prolongation on one side of the sheath); without auricular setae. Sheaths slightly keeled. Leaf blades linear to lanceolate (acuminate); broad to narrow; (3–)10–30 mm wide (and (60–)80–160 mm long); flat; pseudopetiolate (the cuneate, slightly asymmetrical blade base attenuate to the 1–2 mm pseudopetiole); persistent (?). Ligule an unfringed membrane ('ciliolate'); truncate; 0.3–0.6 mm long.

Reproductive organization. *Plants monoecious with all the fertile spikelets unisexual*; without hermaphrodite florets. The spikelets of sexually distinct forms on the same plant; female-only and male-only. *The male and female-fertile spikelets on different branches of the same inflorescence, or segregated, in different parts of the same inflorescence branch (having a clavately pedicelled female spikelet terminating a racemose panicle which bears several to many male or male and female spikelets on erect filiform pedicels below, or occasionally with a short, erect branch from the lower part of the inflorescence bearing two male spikelets).*

Inflorescence. *Inflorescence* determinate; without pseudospikelets; few spikeleted to many spikeleted; a single raceme, or paniculate; spatheate (the male branch, when present, often bearing a minute bract), or espatheate (?); a complex of 'partial inflorescences' and intervening foliar organs, or not comprising 'partial inflorescences' and foliar organs (?). *Spikelet-bearing axes* 'racemes', or paniculate. Spikelets not secund (?); pedicellate.

Female-sterile spikelets. The male spikelets early deciduous, 7–8 mm long, linear, hyaline. The male spikelets without glumes; without proximal incomplete florets; 1 floreted. The lemmas awned (the awn flexuous, 1–2 mm long). Male florets 1.

Female-fertile spikelets. *Spikelets 11–15 mm long*; narrowly lanceolate; falling with the glumes (or the glumes tardily disarticulating); with conventional internode spacings. Rachilla terminated by a female-fertile floret. Hairy callus absent.

Glumes two; more or less equal (the lower slightly longer, even *in situ*); exceeding the spikelets; long relative to the adjacent lemmas (exceeding them); hairy to hairless (glabrous to puberulent); pointed; long *awned*; non-carinate; similar (firmly membranous, lanceolate, attenuate into the stiff, scabrous, 2.5–8 mm awns, with cross nerves). Lower glume 5 nerved. Upper glume 3 nerved. *Spikelets* with female-fertile florets only.

Female-fertile florets 1. Lemmas ellipsoidal or lanceolate; decidedly firmer than the glumes (leathery); not becoming indurated; pale, becoming mottled with dark spots; entire; pointed; mucronate to awned. Awns 1; median; apical; non-geniculate; straight; hairy (short-pilose); much shorter than the body of the lemma (1–1.5 mm long). Lemmas hairy (with appressed, silky hairs); non-carinate; having the margins tucked in onto the palea; perhaps with a clear germination flap (judging from the illustration in Zuloaga and Judziewicz 1993). *Palea* present; relatively long; tightly clasped by the lemma (and hidden by it, except at the base). Palea back hairy.

Taxonomy. Bambusoideae; Oryzodae; Olyreae.

Ecology, geography, regional floristic distribution. 1 species. Amazonian South America. Mesophytic; shade species; glycophytic. Wet lowland forests.

Neotropical. Amazon.

References, etc. Morphological/taxonomic: Zuloaga and Judziewicz 1993.

Special comments. Segregated from *Olyra* (*O. lancifolia* Mez). Available description poor. Fruit data wanting. Anatomical data wanting.

Agropyron Gaertn.

Costia Willkom, *Kratzmannia* Opiz
Excluding *Elymus, Leymus, Elytrigia, Australopyrum, Pascopyrum* etc.

Habit, vegetative morphology. Perennial; rhizomatous, or caespitose. **Culms** 15–150 cm high; *herbaceous*; unbranched above; tuberous, or not tuberous. Culm nodes hairy, or glabrous. Culm internodes solid, or hollow. *Leaves* not basally aggregated; auriculate, or non-auriculate. Leaf blades linear; narrow; 1.2–10 mm wide; not setaceous; flat, or rolled (convolute); without cross venation; persistent; rolled in bud. **Ligule** present; *an unfringed membrane*; truncate (or very short); 0.1–1 mm long (leathery to membranous). *Contra-ligule* absent.

Reproductive organization. Plants bisexual, with bisexual spikelets; with hermaphrodite florets; outbreeding.

Inflorescence. *Inflorescence a single spike (usually pectinate).* Rachides hollowed, or flattened, or winged, or neither flattened nor hollowed, not winged. Inflorescence espatheate; not comprising 'partial inflorescences' and foliar organs. Spikelet-bearing axes persistent. *Spikelets* solitary (diverging from the rachis); not secund; distichous (in rows); *pedicellate*; imbricate.

Female-fertile spikelets. Spikelets 6–12(–15) mm long; compressed laterally; disarticulating above the glumes; disarticulating between the florets. Rachilla prolonged beyond the uppermost female-fertile floret; rachilla hairy, or hairless (shortly pilose to scabrous). Hairy callus present, or absent. *Callus* very short; blunt.

Glumes two; more or less equal; shorter than the adjacent lemmas; lateral to the rachis; hairy (pilose), or hairless; when hairless, glabrous; pointed; not subulate; awned (the awn to 3 mm long), or awnless; *carinate (usually strongly keeled)*; similar (lanceolate ovate, somewhat asymmetric). Lower glume 2–5 nerved. Upper glume 2–5 nerved. *Spikelets* with female-fertile florets only (rarely), or with incomplete florets. The incomplete florets distal to the female-fertile florets.

Female-fertile florets 3–8 (rarely two, or up to ten). Lemmas similar in texture to the glumes (leathery); not becoming indurated; entire; pointed; awnless, or mucronate, or awned. Awns when present, 1; apical; non-geniculate; much shorter than the body of the lemma to about as long as the body of the lemma (rarely more than 5 mm long); entered by several veins. Lemmas hairy (pilose), or hairless; when hairless glabrous; carinate (at least slightly so); without a germination flap; 5 nerved; with the nerves confluent towards the tip. *Palea* present; relatively long; entire (truncate), or apically notched (emarginate); awnless, without apical setae; not indurated (membranous); nerveless (rarely), or 1-nerved (rarely), or 2-nerved; 2-keeled, or keel-less. *Lodicules* present; free; membranous; ciliate; not toothed (usually entire); not or scarcely vascularized. *Stamens* 3. Anthers (2.5–)3.5–6 mm long; not penicillate; without an apically prolonged connective. *Ovary* hairy. Styles free to their bases. Stigmas 2; white.

Fruit, embryo and seedling. *Fruit* adhering to lemma and/or palea; small to medium sized (3–5.5 mm long); longitudinally grooved; compressed dorsiventrally; with hairs confined to a terminal tuft. *Hilum long-linear.* Embryo small; not waisted. Endosperm hard; without lipid; containing only simple starch grains. Embryo with an epiblast, or without an epiblast; without a scutellar tail; with a negligible mesocotyl internode. Embryonic leaf margins meeting.

Seedling with a short mesocotyl. First seedling leaf with a well-developed lamina. The lamina narrow; erect; 3–5 veined.

Abaxial leaf blade epidermis. *Costal/intercostal zonation* conspicuous. *Papillae* absent. *Long-cells* markedly different in shape costally and intercostally (the costals narrower, more rectangular); differing markedly in wall thickness costally and intercostally (the costals with thicker, heavily pitted walls). Mid-intercostal long-cells mostly fusiform; having straight or only gently undulating walls. *Microhairs* absent. *Stomata* common; 46.5–51 microns long. Subsidiaries parallel-sided. Guard-cells overlapped by the interstomatals. *Intercostal short-cells* absent or very rare. Small prickles abundant. *Crown cells* absent. *Costal short-cells* neither distinctly grouped into long rows nor predominantly paired (mostly solitary, a few pairs). Costal silica bodies predominantly horizontally-elongated crenate/sinuous (but varying in length, regularity and sinuosity so as to approach panicoid forms at the one extreme, the horizontal-smooth form at the other). **Transverse section of leaf blade, physiology.** C_3; XyMS+. Mesophyll without adaxial palisade. *Leaf blade* with distinct, prominent adaxial ribs; with the ribs more or less constant in size (round topped). *Midrib* conspicuous (via a sharp keel and the large, heavily girdered bundle); with one bundle only. Bulliforms present in discrete, regular adaxial groups; exclusively in simple fans (in the furrows). All the vascular bundles accompanied by sclerenchyma. Combined sclerenchyma girders present (with all the primaries, the rest with adaxial girders or strands only); forming 'figures' (but at the most, only slight I's). Sclerenchyma all associated with vascular bundles.

Phytochemistry. Tissues of the culm bases with little or no starch. Fructosans predominantly short-chain.

Cytology. Chromosome base number, $x = 7$. $2n = 14$, 28, and 42. 2, 4, and 6 ploid. Haplomic genome content P.

Taxonomy. Pooideae; Triticodae; Triticeae.

Ecology, geography, regional floristic distribution. 15 species. Mediterranean to China and USSR. Commonly adventive. Mesophytic, or xerophytic; species of open habitats. Steppe, etc., on dry stony soils.

Holarctic. Boreal and Tethyan. Euro-Siberian. Irano-Turanian. European and Siberian.

Hybrids. Intergeneric hybrids with *Hordeum* (×*Agrohordeum* A. Camus), *Leymus* (×*Leymopyron* Tsvelev), *Elytrigia* (×*Agrotrigia* Tsvelev), *Secale*, *Triticum* (×*Agrotrisecale* Ciferri & Giacom., ×*Agrotriticum* Ciferri & Giacom.), *Sitanion* (×*Agrositanion* Bowden). See also ×*Agroelymus* A. Camus.

Economic importance. Significant weed species: *A. cristatum*. Cultivated fodder: *A. cristatum*.

References, etc. Morphological/taxonomic: Löve 1984. Leaf anatomical: this project.
Special comments. *Agropyron* sensu stricto.

Agropyropsis A. Camus

Habit, vegetative morphology. *Perennial*; stoloniferous, or caespitose. **Culms** 20–60 cm high; *herbaceous*; unbranched above; 1 noded. Culm nodes hidden by leaf sheaths. Young shoots intravaginal. *Leaves* mostly basal; non-auriculate; without auricular setae. Leaf blades linear to lanceolate; narrow; about 4 mm wide; rolled; without cross venation; persistent. Ligule an unfringed membrane (but ciliolate); truncate; about 0.5–1 mm long. *Contra-ligule* absent.

Reproductive organization. Plants bisexual, with bisexual spikelets; with hermaphrodite florets.

Inflorescence. *Inflorescence a single spike (tough)*. Rachides hollowed. Inflorescence espatheate; not comprising 'partial inflorescences' and foliar organs. Spikelets solitary; not secund; distichous; sessile; distant.

Female-fertile spikelets. Spikelets 5–9 mm long; compressed laterally to not noticeably compressed; disarticulating above the glumes; disarticulating between the florets. Rachilla prolonged beyond the uppermost female-fertile floret. The rachilla extension with incomplete florets. Hairy callus absent. *Callus* short; blunt (glabrous).

Glumes two; more or less equal; shorter than the spikelets; long relative to the adjacent lemmas; free; lateral to the rachis; hairless; glabrous; fairly not pointed; awnless; non-carinate; similar. Lower glume 3–5 nerved. Upper glume 3–5 nerved. *Spikelets* with incomplete florets. The incomplete florets distal to the female-fertile florets. The distal incomplete florets 1; *incomplete florets* merely underdeveloped; incomplete florets awnless.

Female-fertile florets 2–4(–6). Lemmas ovate; less firm than the glumes (somewhat cartilaginous); not becoming indurated; entire; awnless; hairless; glabrous; non-carinate (rounded on the back); without a germination flap; 3 nerved (lower florets), or 5 nerved (upper florets, in the material seen); with the nerves non-confluent. Palea present; relatively long (lanceolate); tightly clasped by the lemma; entire, or apically notched; awnless, without apical setae; textured like the lemma (membranous); 2-nerved; 2-keeled.

Palea keels wingless; hairy. *Lodicules* present; 2; free; membranous; not toothed; not or scarcely vascularized. *Stamens* 3. **Ovary** glabrous; *with a conspicuous apical appendage (this fleshy, glabrous)*. Styles free to their bases. Stigmas 2.

Fruit, embryo and seedling. *Fruit* free from both lemma and palea; small (2–3 mm long); fusiform; compressed dorsiventrally. Hilum short (shortly linear). Embryo small. Endosperm hard.

Abaxial leaf blade epidermis. *Costal/intercostal zonation* conspicuous. *Papillae* absent. *Long-cells* markedly different in shape costally and intercostally (the costals much narrower); of similar wall thickness costally and intercostally (the walls of medium thickness). Mid-intercostal long-cells rectangular; having markedly sinuous walls. *Microhairs* absent. *Stomata* common. Subsidiaries non-papillate; high dome-shaped and triangular. Guard-cells slightly overlapped by the interstomatals, or overlapping to flush with the interstomatals. *Intercostal short-cells* common; consistently in cork/silica-cell pairs; silicified. Intercostal silica bodies rounded. Costal silica bodies horizontally-elongated crenate/sinuous and horizontally-elongated smooth (a mixture of short-crenate forms and potato shapes).

Transverse section of leaf blade, physiology. C_3; XyMS+. Mesophyll without adaxial palisade. *Leaf blade* with distinct, prominent adaxial ribs; with the ribs more or less constant in size (these broad, flat-topped). *Midrib* conspicuous (by its large bundle, a fairly prominent, rounded keel and large bulliform 'hinges'); with one bundle only; without colourless mesophyll adaxially. *The lamina* symmetrical on either side of the midrib. Bulliforms present in discrete, regular adaxial groups; in simple fans (one at the base of each furrow). All the vascular bundles accompanied by sclerenchyma. Combined sclerenchyma girders present (with all the large bundles); forming 'figures' (mostly I's, the midrib with a large 'anchor'). Sclerenchyma all associated with vascular bundles (apart from large groups in the blade margins).

Taxonomy. Pooideae; Poodae; Poeae.

Ecology, geography, regional floristic distribution. 2 species. North Africa, Cape Verde Islands. Halophytic (in damp, saline soils).

Holarctic. Tethyan. Macaronesian, Mediterranean, and Irano-Turanian.

References, etc. Leaf anatomical: this project.

Agrostis L.

Agraulus P. Beauv., *Agrestis* Bub., *Anomalotis* Steud., *Bromidium* Nees, *Candollea* Steud., *Chaetotropis* Kunth, *Decandolea* Batard, *Didymochaeta* Steud., *Lachnagrostis* Trin., *Neoschischkinia* Tsvelev, *Notonema* Raf., *Pentatherum* Nabelek, *Podagrostis* (Griseb.) Scribn., *Senisetum* Koidz., *Trichodium* Michaux, *Vilfa* Adans.

Excluding *Linkagrostis* (*A. juressi*)

Habit, vegetative morphology. Annual, or perennial; rhizomatous, or stoloniferous, or caespitose, or decumbent. **Culms** mostly (3–)5–100 cm high; *herbaceous*; unbranched above. Culm nodes glabrous. Culm internodes hollow. *Leaves* mostly basal, or not basally aggregated; non-auriculate. *Sheath margins free*. Leaf blades linear; narrow; 0.2–10 mm wide; usually flat, or rolled (convolute, or canaliculate); without cross venation; persistent; rolled in bud. *Ligule an unfringed membrane*; truncate, or not truncate; 1–6 mm long.

Reproductive organization. *Plants* bisexual, with bisexual spikelets; *with hermaphrodite florets*; outbreeding. Exposed-cleistogamous, or chasmogamous.

Inflorescence. *Inflorescence paniculate*; deciduous in its entirety (*Lachnagrostis*), or not deciduous; open, or contracted (e.g., *Bromidium*); when contracted, spicate, or more or less irregular (not usually 'interrrupted'); with capillary branchlets, or without capillary branchlets; espatheate; not comprising 'partial inflorescences' and foliar organs. Spikelet-bearing axes persistent. Spikelets not secund; pedicellate.

Female-fertile spikelets. *Spikelets* 0.8–4 mm long; *compressed laterally*; disarticulating above the glumes. Rachilla prolonged beyond the uppermost female-fertile floret, or terminated by a female-fertile floret; rachilla hairy, or hairless. The rachilla extension when present, naked. Hairy callus present (the hairs less than 0.5 mm long), or absent. *Callus hairs absent, or if present less than 0.5 mm long*. *Callus* short (minute); blunt.

Glumes two; more or less equal; about equalling the spikelets to exceeding the spikelets (nearly always), or shorter than the spikelets (very rarely); long relative to the adjacent lemmas; pointed; awnless; carinate, or non-carinate; similar (usually narrow, membranous). Lower glume 1 nerved. *Upper glume 1 nerved*. *Spikelets* with female-fertile florets only.

Female-fertile florets *1*. *Lemmas less firm than the glumes (thinly membranous to hyaline)*; not becoming indurated; entire to incised (usually truncate or emarginate, sometimes toothed via excurrent veins); awnless, or mucronate, or awned. *Awns* when present, 1, or 3 (*Bromidium*), or 5 (rarely); median, or median and lateral (by extension of the lateral veins); the median different in form from the laterals (when laterals present); *dorsal*; *from well down the back*; geniculate; much shorter than the body of the lemma to about as long as the body of the lemma, or much longer than the body of the lemma (rarely); entered by one vein. The lateral awns (when present) shorter than the median to exceeding the median. Lemmas hairy, or hairless; non-carinate; without a germination flap; 3–5 nerved. *Palea* nearly always present; relatively long, or conspicuous but relatively short, or very reduced; entire, or apically notched; not indurated (hyaline or membranous); nerveless, or 2-nerved; 2-keeled, or keel-less. **Lodicules** *present*; 2; free; membranous; glabrous; not toothed; not or scarcely vascularized. *Stamens* 3. Anthers 0.3–2 mm long; not penicillate. *Ovary* glabrous. Styles free to their bases. Stigmas 2; white.

Fruit, embryo and seedling. *Fruit* free from both lemma and palea; small; longitudinally grooved, or not grooved (rarely); compressed dorsiventrally. Hilum short. Embryo small; not waisted. Endosperm liquid in the mature fruit, or hard; with lipid; containing compound starch grains. Embryo with an epiblast; without a scutellar tail; with a negligible mesocotyl internode. Embryonic leaf margins meeting.

Seedling with a short mesocotyl, or with a long mesocotyl; with a loose coleoptile, or with a tight coleoptile. First seedling leaf with a well-developed lamina. The lamina broad, or narrow; erect; 3–5 veined.

Abaxial leaf blade epidermis. *Costal/intercostal zonation* conspicuous. *Papillae* absent. *Long-cells* similar in shape costally and intercostally; of similar wall thickness costally and intercostally. Mid-intercostal long-cells fusiform; having straight or only gently undulating walls. *Microhairs* absent. *Stomata* common; 37–39 microns long (in *A. avenacea*). Subsidiaries low dome-shaped, or parallel-sided. Guard-cells overlapped by the interstomatals. *Intercostal short-cells* absent or very rare. *Costal short-cells* neither distinctly grouped into long rows nor predominantly paired. Costal silica bodies horizontally-elongated crenate/sinuous, or tall-and-narrow.

Transverse section of leaf blade, physiology. C_3; XyMS+. *Mesophyll* with non-radiate chlorenchyma. *Leaf blade* with distinct, prominent adaxial ribs, or 'nodular' in section; with the ribs more or less constant in size. *Midrib* conspicuous, or not readily distinguishable; with one bundle only. Bulliforms present in discrete, regular adaxial groups, or not in discrete, regular adaxial groups; when well defined, in simple fans (or the groups of fairly uniform cells). Many of the smallest vascular bundles unaccompanied by sclerenchyma, or all the vascular bundles accompanied by sclerenchyma. Combined sclerenchyma girders present, or absent; nowhere forming 'figures'. Sclerenchyma all associated with vascular bundles.

Culm anatomy. *Culm internode bundles* in one or two rings.

Phytochemistry. Tissues of the culm bases with little or no starch. Leaves without flavonoid sulphates (1 species).

Cytology. Chromosome base number, $x = 7$. $2n = 14, 16, 21, 28, 30, 32, 35, 42, 44, 46$, and 56 (and aneuploids). 2, 3, 4, 5, 6, 7, and 8 ploid (and aneuploids). Haploid nuclear DNA content 1.5 and 1.8 pg (2 species). Mean diploid 2c DNA value 6.9 pg (1 species).

Taxonomy. Pooideae; Poodae; Aveneae.

Ecology, geography, regional floristic distribution. About 220 species. Temperate. Commonly adventive. Helophytic, or mesophytic, or xerophytic (rarely); shade species and species of open habitats; nearly always glycophytic. Grassland, light woodland, rarely sand dunes.

Holarctic, Paleotropical, Neotropical, Cape, Australian, and Antarctic. Boreal, Tethyan, and Madrean. African, Madagascan, Indomalesian, and Polynesian. Arctic and Subarctic, Euro-Siberian, Eastern Asian, Atlantic North American, and Rocky Mountains. Macaronesian, Mediterranean, and Irano-Turanian. Saharo-Sindian, Sudano-Angolan, West African Rainforest, Namib-Karoo, and Ascension and St. Helena. Indian, Indo-Chinese, Malesian, and Papuan. Hawaiian. Caribbean, Central Brazilian, Pampas, Andean, and Fernandezian. North and East Australian and South-West Australian. New Zealand and Patagonian. European and Siberian. Canadian-Appalachian, Southern Atlantic North American, and Central Grasslands. Sahelo-Sudanian, Somalo-Ethiopian, and South Tropical African. Temperate and South-Eastern Australian.

Hybrids. Intergeneric hybrids with *Polypogon* (×*Agropogon* P. Fourn.), *Calamagrostis*.

Rusts and smuts. Rusts — *Puccinia*. Taxonomically wide-ranging species: *Puccinia graminis, Puccinia coronata, Puccinia striiformis, Puccinia pygmaea, Puccinia brachypodii, Puccinia praegracilis, Puccinia poarum, Puccinia recondita, 'Uromyces' fragilipes*, and *'Uromyces' dactylidis*. Smuts from Tilletiaceae and from Ustilaginaceae. Tilletiaceae — *Entyloma, Tilletia*, and *Urocystis*. Ustilaginaceae — *Sphacelotheca*.

Economic importance. Significant weed species: *A. canina, A. castellana, A. gigantea, A. stolonifera, A. tenuis*. Cultivated fodder: *A. palustris, A. tenuis* (only for poor, acid soils), etc. Important native pasture species: *A. exarata, A. oregonensis, A. tenuis, A. variabilis* etc. Lawns and/or playing fields: *A. canina, A. stolonifera, A. tenuis* etc.

References, etc. Morphological/taxonomic: Vickery 1941; Bjorkman 1960; Nicora 1962. Leaf anatomical: Metcalfe 1960; this project.

Aira L.

Airella (Dumort.) Dumort., *Aspris* Adans., *Caryophyllea* Opiz, *Fiorinia* Parl., *Fussia* Schur, *Salmasia* Bub.

Habit, vegetative morphology. Small, slender *annual*; caespitose. *Culms* 2–40 cm high; herbaceous; unbranched above. Culm nodes glabrous. Culm internodes hollow. *Leaves* mostly basal, or not basally aggregated; non-auriculate. Leaf blades linear; narrow; 0.3–2 mm wide; setaceous; flat, or folded, or rolled; without abaxial multicellular glands; without cross venation; persistent; rolled in bud, or once-folded in bud. *Ligule an unfringed membrane*; not truncate (acute); 2.4–5 mm long.

Reproductive organization. Plants bisexual, with bisexual spikelets; with hermaphrodite florets; inbreeding.

Inflorescence. Inflorescence paniculate; open, or contracted; when contracted, spicate to more or less irregular; with conspicuously divaricate branchlets, or without conspicuously divaricate branchlets; with capillary branchlets; espatheate; not comprising 'partial inflorescences' and foliar organs. Spikelet-bearing axes persistent. Spikelets not secund; pedicellate.

Female-fertile spikelets. *Spikelets 1.6–3.5 mm long*; compressed laterally; disarticulating above the glumes; disarticulating between the florets; with conventional internode spacings. The upper floret not stipitate. Rachilla prolonged beyond the uppermost female-fertile floret, or terminated by a female-fertile floret; rachilla hairless. The rachilla extension when present, naked. Hairy callus present, or absent. *Callus* short; blunt.

Glumes two; more or less equal; about equalling the spikelets; long relative to the adjacent lemmas; pointed; awnless; carinate, or non-carinate; similar (membranous, delicate). Lower glume 1–3 nerved. Upper glume 1–3 nerved. *Spikelets* with female-fertile florets only.

Female-fertile florets 2. *Lemmas decidedly firmer than the glumes (becoming papery)*; not becoming indurated; entire, or incised; awnless, or awned. *Awns* when present, 1; dorsal; from well down the back; *geniculate*; much shorter than the body of the lemma to much longer than the body of the lemma; entered by one vein. Lemmas hairless; carinate to non-carinate; 5 nerved. *Palea* present; relatively long; tightly clasped by the lemma; apically notched; awnless, without apical setae; 2-nerved. *Lodicules* present; 2; free; membranous; glabrous; not toothed. *Stamens* 3. Anthers 0.2–1.7 mm long; not penicillate. *Ovary* glabrous. Styles free to their bases. Stigmas 2; white.

Fruit, embryo and seedling. Fruit *adhering to lemma and/or palea*; small; fusiform; *longitudinally grooved*; compressed dorsiventrally. Hilum short. Embryo small; not waisted. Endosperm hard; with lipid; containing compound starch grains. Embryo with an epiblast; without a scutellar tail; with a negligible mesocotyl internode. Embryonic leaf margins meeting.

Seedling with a long mesocotyl; with a loose coleoptile, or with a tight coleoptile. First seedling leaf with a well-developed lamina. The lamina narrow; erect; 3–5 veined.

Abaxial leaf blade epidermis. *Costal/intercostal zonation* conspicuous. *Papillae* absent. *Long-cells* similar in shape costally and intercostally; of similar wall thickness costally and intercostally. Mid-intercostal long-cells fusiform; having markedly sinuous walls, or having straight or only gently undulating walls. *Microhairs* absent. *Stomata* absent or very rare, or common; 45–54 microns long (in *A. cupaniana*). Subsidiaries parallel-sided, or dome-shaped. Guard-cells overlapped by the interstomatals. *Intercostal short-cells* absent or very rare. Intercostal silica bodies absent. *Costal short-cells* conspicuously in long rows, or neither distinctly grouped into long rows nor predominantly paired. Costal silica bodies confined to the central file(s) of the costal zones; horizontally-elongated crenate/sinuous.

Transverse section of leaf blade, physiology. C_3; XyMS+. *Mesophyll* with non-radiate chlorenchyma; without adaxial palisade. *Leaf blade* with distinct, prominent adaxial ribs, or 'nodular' in section; with the ribs more or less constant in size. *Midrib* conspicuous; with one bundle only. Bulliforms not in discrete, regular adaxial groups. Many of the smallest vascular bundles unaccompanied by sclerenchyma, or all the vascular bundles accompanied by sclerenchyma. Combined sclerenchyma girders absent. Sclerenchyma all associated with vascular bundles.

Phytochemistry. Tissues of the culm bases with abundant starch, or with little or no starch. Leaves without flavonoid sulphates (1 species).

Cytology. Chromosome base number, $x = 7$. $2n = 14$ and 28. 2 and 4 ploid. Haploid nuclear DNA content 2.7–3.2 pg (4 species, mean 2.9). Mean diploid 2c DNA value 6 pg, or 5.9 pg (*A. elegantissima, A. praecox*).

Taxonomy. Pooideae; Poodae; Aveneae.

Ecology, geography, regional floristic distribution. 8 species. North and South temperate. Commonly adventive. Mesophytic to xerophytic; species of open habitats. Sandy soils.

Holarctic, Paleotropical, Cape, and Antarctic. Boreal and Tethyan. African. Euro-Siberian and Atlantic North American. Macaronesian, Mediterranean, and Irano-Turanian. Sudano-Angolan and West African Rainforest. Patagonian. European. Southern Atlantic North American and Central Grasslands. Sahelo-Sudanian, Somalo-Ethiopian, and South Tropical African.

Rusts and smuts. Rusts — *Puccinia*. Taxonomically wide-ranging species: *Puccinia graminis* and *Puccinia striiformis*. Smuts from Tilletiaceae. Tilletiaceae — *Tilletia*.

References, etc. Leaf anatomical: this project.

Airopsis Desv.

Aeropsis Aschers. & Graebn., *Sphaerella* Bub.

Habit, vegetative morphology. Very slender *annual*. *Culms* 3–25 cm high; herbaceous. Leaves non-auriculate. The uppermost sheath somewhat inflated. Leaf blades linear; narrow; to 1.5 mm wide; setaceous; rolled (convolute); without cross venation. Ligule an unfringed membrane; not truncate; 1–2 mm long.

Reproductive organization. Plants bisexual, with bisexual spikelets; with hermaphrodite florets.

Inflorescence. Inflorescence paniculate; open, or contracted; with capillary branchlets, or without capillary branchlets (?); espatheate; not comprising 'partial inflorescences' and foliar organs. Spikelet-bearing axes persistent. Spikelets not secund; pedicellate (the pedicels clavate).

Female-fertile spikelets. *Spikelets* 1.2–1.5 mm long; subspherical; *not noticeably compressed*; disarticulating above the glumes; with conventional internode spacings. Rachilla terminated by a female-fertile floret. Hairy callus absent.

Glumes two; relatively large; more or less equal; exceeding the spikelets; long relative to the adjacent lemmas; *conspicuously ventricose*; hairless (smooth, shining); pointed; awnless; non-carinate; similar. Lower glume 3 nerved (the laterals faint). Upper glume 3 nerved (the laterals faint). *Spikelets* with female-fertile florets only.

Female-fertile florets 2. Lemmas 3-dentate; similar in texture to the glumes; not becoming indurated; incised; not deeply cleft; awnless; hairy (puberulent); non-carinate; without a germination flap; 3 nerved. *Palea* present; relatively long; tightly clasped by the lemma; entire to apically notched (truncate to slightly denticulate); awnless, without apical setae; not indurated; 2-nerved; 2-keeled. *Lodicules* present; 2; free; membranous; glabrous; not toothed; not or scarcely vascularized. *Stamens* 3. Anthers 0.3 mm long. *Ovary* glabrous. Styles free to their bases. Stigmas 2.

Fruit, embryo and seedling. *Fruit* free from both lemma and palea; small (0.5–0.6 mm long); subglobose ('hemispherical'); compressed dorsiventrally. Hilum short (punctate). Embryo small. Endosperm liquid in the mature fruit.

Abaxial leaf blade epidermis. *Costal/intercostal zonation* conspicuous. *Papillae* absent. Mid-intercostal long-cells fusiform; having straight or only gently undulating walls. *Microhairs* absent. *Stomata* common. Subsidiaries parallel-sided. Guard-cells overlapped by the interstomatals. *Intercostal short-cells* absent or very rare. Prickles common. *Crown cells* absent. *Costal short-cells* neither distinctly grouped into long rows nor predominantly paired. Costal silica bodies absent to poorly developed; imperfectly horizontally-elongated crenate/sinuous.

Transverse section of leaf blade, physiology. C_3; XyMS+. *Mesophyll* with non-radiate chlorenchyma; without adaxial palisade. *Leaf blade* with distinct, prominent adaxial ribs; with the ribs very irregular in sizes. *Midrib* not readily distinguishable; with one bundle only. Bulliforms not in discrete, regular adaxial groups (seemingly, in the poor material seen). All the vascular bundles accompanied by sclerenchyma. Combined sclerenchyma girders absent (all the main bundles with adaxial strands and abaxial girders).

Cytology. Chromosome base number, $x = 4$. $2n = 8$. 2 ploid.

Taxonomy. Pooideae; Poodae; Aveneae.

Ecology, geography, regional floristic distribution. 1 species. Northwest Africa and southwest Europe to Sicily. Mesophytic to xerophytic; species of open habitats. Sandy places.

Holarctic. Boreal and Tethyan. Euro-Siberian. Mediterranean. European.

References, etc. Leaf anatomical: this project.

Alexfloydia B.K. Simon

Habit, vegetative morphology. Perennial; *stoloniferous*. **Culms** 18–24 cm high; *herbaceous*; not scandent; branched above; 3–4 noded. Culm nodes hidden by leaf sheaths. Plants unarmed; without multicellular glands. The shoots not aromatic. *Leaves* not basally aggregated; non-auriculate; without auricular setae. Sheath margins free. The sheaths keeled. Leaf blades linear; broad; 1–2 mm wide; flat to folded; not needle-like; without cross venation; persistent. Ligule a fringe of hairs.

Reproductive organization. Plants bisexual, with bisexual spikelets; with hermaphrodite florets. The spikelets hermaphrodite. Plants without hidden cleistogenes.

Inflorescence. *Inflorescence* few spikeleted; depauperate, *a single raceme, or paniculate*. Inflorescence with axes ending in spikelets. Inflorescence espatheate; not comprising 'partial inflorescences' and foliar organs. *Spikelets* unaccompanied by bractiform involucres, not associated with setiform vestigial branches; solitary; long pedicellate (the pedicels 4–7 mm long). Pedicel apices cupuliform. Spikelets distant.

Female-fertile spikelets. *Spikelets 3–3.5 mm long*; lanceolate to ovate; *compressed laterally*; falling with the glumes; with conventional internode spacings. The upper floret not stipitate. Rachilla terminated by a female-fertile floret. Hairy callus absent.

Glumes two; very unequal (the upper longer); (the upper) shorter than the spikelets to about equalling the spikelets; (the upper) shorter than the adjacent lemmas to long relative to the adjacent lemmas; scantily hairy (the upper, with a few cushion based hairs), or hairless (the lower); pointed; awnless; non-carinate; similar (papery, ovate to elliptic). *Lower glume* about 0.75 times the length of the upper glume; much shorter than the lowest lemma; 5 nerved, or 7 nerved. **Upper glume** not saccate; *9 nerved*. *Spikelets* with incomplete florets. The incomplete florets proximal to the female-fertile florets. *Spikelets with proximal incomplete florets*. *The proximal incomplete florets* 1; paleate. Palea of the proximal incomplete florets fully developed; not becoming conspicuously hardened and enlarged laterally. The proximal incomplete florets male. *The proximal lemmas* similar to the upper glume; awnless; *9 nerved*; decidedly exceeding the female-fertile lemmas; less firm than the female-fertile lemmas to similar in texture to the female-fertile lemmas (papery); not becoming indurated.

Female-fertile florets 1. Lemmas not saccate; firmly membranous or thinly cartilaginous; striate; not becoming indurated (no more than very thinly cartilaginous); yellow in fruit; entire; pointed (acute); not crested; awnless; hairless; non-carinate; having the margins lying flat and exposed on the palea; probably with a clear germination flap (the material seen disintegrating, but hinting at one); obscurely 5 nerved. *Palea* present; relatively long; gaping; entire; awnless, without apical setae; textured like the lemma; not indurated; 2-nerved; keel-less. *Lodicules* present; 2; free; membranous (quite large); ciliate; not or scarcely vascularized. *Stamens* 3. Anthers 0.8 mm long.

Abaxial leaf blade epidermis. *Costal/intercostal zonation* conspicuous. *Papillae* absent. *Long-cells* markedly different in shape costally and intercostally (the costals much narrower); of similar wall thickness costally and intercostally (thin walled). Mid-intercostal long-cells rectangular; having markedly sinuous walls. *Microhairs* present; elongated; clearly two-celled; panicoid-type; about 70 microns long; about 15 microns wide at the septum. Microhair total length/width at septum 3–6. Microhair apical cells about 50 microns long. Microhair apical cell/total length ratio 0.7–0.8. *Stomata* common; about 40 microns long. Subsidiaries non-papillate; parallel-sided, dome-shaped, and triangular (low to medium, predominantly more or less triangular with the apices truncated to various extents); including both triangular and parallel-sided forms on the same leaf. Guard-cells overlapping to flush with the interstomatals. *Intercostal short-cells* absent or very rare. Macrohairs infrequent, intergrading with long prickles, 1–2(–3) celled. *Costal short-cells* conspicuously in long rows. Costal silica bodies present and well developed; 'panicoid-type'; consistently elongated nodular.

Transverse section of leaf blade, physiology. C_4; XyMS–. PCR sheath outlines uneven. PCR sheath extensions absent. Mesophyll without 'circular cells'. *Leaf blade* adaxially flat. *Midrib* conspicuous (by virtue of a conspicuous abaxial keel, an arc of enlarged adaxial epidermal cells and some colourless mesophyll); with one bundle only, or having complex vascularization (depending on interpretation of the minor bundles flanking the median); with colourless mesophyll adaxially (in the form of a few large cells contiguous with the bulliform epidermis). *The lamina* symmetrical on either side of the midrib. Bulliforms not in discrete, regular adaxial groups (the adaxial epidermis mainly bulliform). Many of the smallest vascular bundles unaccompanied by sclerenchyma. Combined sclerenchyma girders absent (the sclerenchyma restricted to a large abaxial strand in the keel, and small adaxial and abaxial strands with the major laterals). Sclerenchyma all associated with vascular bundles.

Taxonomy. Panicoideae; Panicodae; Paniceae.

Ecology, geography, regional floristic distribution. 1 species. New South Wales. Mesophytic; halophytic to glycophytic. Australian. North and East Australian. Temperate and South-Eastern Australian.

References, etc. Morphological/taxonomic: B.K. Simon 1992. Leaf anatomical: this project.

Special comments. Fruit data wanting.

Alloeochaete (Rendle) C.E. Hubb.

Habit, vegetative morphology. Perennial; caespitose. **Culms** 40–200 cm high; herbaceous; **unbranched above**. Culm sheaths persistent. Young shoots intravaginal. *Leaves* mostly basal; non-auriculate. The sheath bases woolly-tomentose. Leaf blades broad (up to 2 cm in *A. oreogena*), or narrow; flat, or rolled; without cross venation; disarticulating from the sheaths. **Ligule a fringe of hairs**.

Reproductive organization. Plants bisexual, with bisexual spikelets; with hermaphrodite florets.

Inflorescence. Inflorescence paniculate; narrow; espatheate; not comprising 'partial inflorescences' and foliar organs. Spikelet-bearing axes persistent. Spikelets not secund; pedicellate.

Female-fertile spikelets. *Spikelets* 6–26 mm long; compressed laterally; disarticulating above the glumes; *disarticulating between the florets (above the persistent first floret)*. Rachilla prolonged beyond the uppermost female-fertile floret. The rachilla extension with incomplete florets. Hairy callus present. *Callus* short; blunt.

Glumes two; more or less equal; shorter than the spikelets; shorter than the adjacent lemmas; hairy (pilose), or hairless; sometimes glabrous; pointed (acute, acuminate or minutely bidentate); more or less *awned (mucronate to aristate, apically or from between short teeth)*; carinate; very dissimilar (with entire G_1 and bidentate G_2), or similar. Lower glume 3 nerved, or 5 nerved. Upper glume 3 nerved, or 5 nerved. *Spikelets* with incomplete florets. The incomplete florets both distal and proximal to the female-fertile florets (usually). *The distal incomplete florets* merely underdeveloped. **Spikelets with proximal incomplete florets**. *The proximal incomplete florets* 1; male (usually — though in three species the L_1 is occasionally hermaphrodite). *The proximal lemmas awned (from a slight or deep sinus, the awn short and straight to long and geniculate)*; 5 nerved; exceeded by the female-fertile lemmas; less firm than the female-fertile lemmas to similar in texture to the female-fertile lemmas (thinly membranous to papery); not becoming indurated.

Female-fertile florets *5–10*. Lemmas similar in texture to the glumes to decidedly firmer than the glumes; not becoming indurated; incised; 2 lobed; deeply cleft, or not deeply cleft (but always strongly bidentate); awned. Awns 1, or 3; median, or median and lateral (the lobes being aristate); the median different in form from the laterals (when laterals present); from a sinus; geniculate; hairless; much shorter than the body of the lemma to much longer than the body of the lemma. Lemmas usually hairy. The hairs in tufts (one on each side); not in transverse rows. Lemmas non-carinate (dorsally rounded); 5 nerved. *Palea* present; relatively long; entire to apically notched; awnless, without apical setae; not indurated (membranous); 2-nerved; 2-keeled. *Stamens* 3. Anthers 1–6.4 mm long. Stigmas 2.

Abaxial leaf blade epidermis. *Costal/intercostal zonation* conspicuous. *Papillae* absent. *Long-cells* markedly different in shape costally and intercostally (intercostals much broader); of similar wall thickness costally and intercostally (rather thin walled). Mid-intercostal long-cells rectangular; having markedly sinuous walls. *Microhairs* present (but scarce in material seen); panicoid-type (probably, but apical cells not seen); 36–45 microns long; 6–7 microns wide at the septum. Microhair total length/width at septum 5–6.5. Microhair apical cells 6–9 microns long. Microhair apical cell/total length ratio 0.15–0.2. *Stomata* common (but not abundant); 33–39 microns long. Subsidiaries low dome-shaped. Guard-cells overlapped by the interstomatals (sometimes; slightly), or overlapping to flush with the interstomatals. *Intercostal short-cells* common; in cork/silica-cell pairs; silicified. *Costal short-cells* conspicuously in long rows (but the short-cells often quite long). Costal silica bodies 'panicoid-type'; predominantly elongated dumb-bell shaped.

Transverse section of leaf blade, physiology. C_3 (indisputably); XyMS+. *Mesophyll* with non-radiate chlorenchyma; without adaxial palisade. *Leaf blade* with distinct, prominent adaxial ribs; with the ribs very irregular in sizes (small round-topped, alternating with large flat-topped). *Midrib* not readily distinguishable; with one bundle only. Bulliforms present in discrete, regular adaxial groups; in simple fans (in the furrows). All the vascular bundles accompanied by sclerenchyma. Combined sclerenchyma girders present (with the primaries); forming 'figures' (the primaries with massive I's). Sclerenchyma not all bundle-associated (with a small abaxial group opposite each furrow). The 'extra' sclerenchyma in abaxial groups; abaxial-hypodermal, the groups isolated.

Taxonomy. Arundinoideae; Danthonieae.

Ecology, geography, regional floristic distribution. 6 species. Angola, Tanzania, Malawi. Xerophytic; species of open habitats.

Paleotropical. African. Sudano-Angolan. South Tropical African.

References, etc. Morphological/taxonomic: Kabuye and Renvoize 1975. Leaf anatomical: this project.

Special comments. Fruit data wanting.

Allolepis Soderstrom & Decker

~ *Distichlis* (*D. texana*)

Habit, vegetative morphology. Perennial; rhizomatous and stoloniferous (the stolons long and stout). *Culms* 70–100 cm high; herbaceous. Culm nodes glabrous. Plants unarmed. *Leaves* not basally aggregated; non-auriculate. Leaf blades linear; narrow; 3–5 mm wide (to 30 cm long); flat; without abaxial multicellular glands; without cross venation; persistent. Ligule a fringe of hairs (very dense). *Contra-ligule* absent.

Reproductive organization. *Plants dioecious*; without hermaphrodite florets. The spikelets all alike in sexuality (on the same plant); female-only, or male-only. Plants outbreeding.

Inflorescence. Inflorescence paniculate (in both female and male plants); open; nondigitate; espatheate; not comprising 'partial inflorescences' and foliar organs. Spikelet-bearing axes persistent. Spikelets solitary; not secund; pedicellate.

Female-sterile spikelets. Male spikelets 10–15 mm long, with up to 20 florets; florets with three stamens. Rachilla of male spikelets prolonged beyond the uppermost male floret. *The male spikelets* with glumes; *10–20 floreted*. Male florets 3 staminate.

Female-fertile spikelets. Spikelets 9–15 mm long; compressed laterally; disarticulating above the glumes; disarticulating between the florets; with conventional internode spacings (though these longish). Rachilla prolonged beyond the uppermost female-fertile floret; rachilla hairless. The rachilla extension with incomplete florets. Hairy callus absent. *Callus* glabrous, spongy, somewhat distended.

Glumes two; relatively large; more or less equal; shorter than the spikelets; shorter than the adjacent lemmas; hairless (minutely scabrid on the keel); pointed; awnless; carinate; similar (broadly lanceolate, membranous). Lower glume 3 nerved. Upper glume 5 nerved. *Spikelets* with incomplete florets. The incomplete florets distal to the female-fertile florets. *The distal incomplete florets* merely underdeveloped.

Female-fertile florets 4–6. **Lemmas similar in texture to the glumes (membranous, the very wide hyaline margins enveloping the florets above)**; smooth; not becoming indurated; entire; pointed; awnless; hairless; glabrous; slightly carinate; without a germination flap; 3–4 nerved. **Palea** present; relatively long; *convolute (saccate below, closely convolute around the pistil)*; entire; awnless, without apical setae (but often delicately ciliate); textured like the lemma; not indurated; 2-nerved; 2-keeled. Palea keels winged (with conspicuous, hyaline wings); hairy (ciliate). *Lodicules* present; 2; free; fleshy to membranous (not 'cuneate', but apically thicker than the usual 'membranous' type); ciliate; lobed; heavily vascularized. *Stamens* 0 (represented by three minute, flattened, lodicule-like staminodes). *Ovary* glabrous. Styles fused (ovary attenuate into a long style). Stigmas 2 (exserted).

Fruit, embryo and seedling. Fruit ellipsoid.

Abaxial leaf blade epidermis. *Costal/intercostal zonation* conspicuous. *Papillae* absent. Mid-intercostal long-cells rectangular; having markedly sinuous walls. *Microhairs* present; more or less spherical to elongated; clearly two-celled; chloridoid-type (but apical cells thin-walled). Microhair apical cell wall thinner than that of the basal cell but not tending to collapse. Microhairs 27–31.5 microns long. Microhair basal cells 18–21 microns long. Microhairs 12–13.5 microns wide at the septum. Microhair total length/width at septum 2.2–2.6. Microhair apical cells 12–13.5 microns long. Microhair apical cell/total length ratio 0.4–0.44. *Stomata* common; 27–31.5 microns long. Subsidiaries mostly low to medium dome-shaped. Guard-cells overlapping to flush with the interstomatals. *Intercostal short-cells* common; in cork/silica-cell pairs, or not paired; silicified. Intercostal silica bodies present and perfectly developed; crescentic, or tall-and-narrow.

Costal short-cells conspicuously in long rows. Costal silica bodies present in alternate cell files of the costal zones; 'panicoid-type'; predominantly shortly cross shaped, dumbbell shaped, and nodular.

Transverse section of leaf blade, physiology. C_4; XyMS+. PCR sheaths of the primary vascular bundles complete. PCR sheath extensions absent. Mesophyll traversed by columns of colourless mesophyll cells. *Leaf blade* 'nodular' in section; with the ribs more or less constant in size. *Midrib* not readily distinguishable; with one bundle only. Bulliforms present in discrete, regular adaxial groups; associated with colourless mesophyll cells to form deeply-penetrating fans. All the vascular bundles accompanied by sclerenchyma. Combined sclerenchyma girders present; forming 'figures' (with most bundles). Sclerenchyma all associated with vascular bundles. The lamina margins with fibres.

Cytology. Chromosome base number, $x = 10$. $2n = 40$. 4 ploid.

Taxonomy. Chloridoideae; main chloridoid assemblage.

Ecology, geography, regional floristic distribution. 1 species. Southern United States. Species of open habitats; seemingly glycophytic (by contrast with the related *Distichlis*). Sandy places.

Holarctic. Boreal and Madrean. Atlantic North American. Southern Atlantic North American.

References, etc. Morphological/taxonomic: Soderstrom and Decker 1965. Leaf anatomical: this project.

Special comments. Fruit data wanting.

Alloteropsis Presl

Bluffia Nees, *Coridochloa* Nees, *Holosetum* Steud., *Mezochloa* Butzin, *Pterochlaena* Chiov.

Habit, vegetative morphology. Annual (rarely), or perennial; caespitose, or decumbent. **Culms** 20–150 cm high; *herbaceous*; unbranched above; tuberous, or not tuberous. Culm nodes hairy, or glabrous. Culm internodes hollow. The shoots aromatic (sometimes, coumarin-scented), or not aromatic. *Leaves* mostly basal; non-auriculate. Leaf blades linear to lanceolate; broad, or narrow; cordate, or not cordate, not sagittate; without cross venation; persistent; rolled in bud. Ligule a fringed membrane to a fringe of hairs. *Contraligule* present (in *A. semialata*), or absent (?).

Reproductive organization. Plants bisexual, with bisexual spikelets; with hermaphrodite florets.

Inflorescence. *Inflorescence* of spicate main branches; *digitate, or subdigitate (the branches in whorls on a short central axis)*; espatheate; not comprising 'partial inflorescences' and foliar organs. Spikelet-bearing axes persistent. Spikelets in triplets, or paired; secund. Pedicel apices cupuliform. Spikelets consistently in 'long-and-short' combinations.

Female-fertile spikelets. Spikelets 2.5–7 mm long; lanceolate, or ovate; abaxial; compressed dorsiventrally; falling with the glumes. Rachilla terminated by a female-fertile floret, or prolonged beyond the uppermost female-fertile floret (sometimes, minutely so). The rachilla extension when present, naked. Hairy callus absent.

Glumes two; very unequal; (the upper) long relative to the adjacent lemmas; dorsiventral to the rachis; hairy (G_2 ciliate marginally); pointed; awned, or awnless; very dissimilar (G_1 smaller, thinner, often mucronulate, the G_2 densely ciliate). Lower glume 3–5 nerved. *Upper glume 5 nerved. Spikelets* with incomplete florets. The incomplete florets proximal to the female-fertile florets. *Spikelets with proximal incomplete florets. The proximal incomplete florets* 1; paleate. *Palea of the proximal incomplete florets reduced (deeply bifid, 1-nerved). The proximal incomplete florets male. The proximal lemmas* awned (in '*Mezochloa*' = *A. paniculata*), or awnless; (3–)5 nerved (3 nerved in '*Mezochloa*'); *similar in texture to the female-fertile lemmas (but with a hyaline zone at the base)*; not becoming indurated.

Female-fertile florets 1. *Lemmas* long-attenuate into an awn or mucro; *similar in texture to the glumes*; smooth to striate; not becoming indurated; yellow in fruit; entire;

pointed; mucronate to awned. Awns when present, 1; apical; non-geniculate; hairless (scabrid); much shorter than the body of the lemma to about as long as the body of the lemma. Lemmas hairless; non-carinate; having the margins lying flat and exposed on the palea; with a clear germination flap; 5 nerved (usually), or 1–3 nerved ('*Mezochloa*'). *Palea* present (auriculate at base); relatively long; entire; awnless, without apical setae; 2-nerved. *Lodicules* present; 2; free; fleshy; glabrous. *Stamens* 3. Anthers not penicillate. *Ovary* glabrous. Styles free to their bases. Stigmas 2; red pigmented.

Fruit, embryo and seedling. Fruit small; compressed dorsiventrally. Hilum short. Embryo large; waisted. Endosperm hard; containing only simple starch grains. Embryo without an epiblast.

Seedling with a short mesocotyl. First seedling leaf with a well-developed lamina. The lamina broad; curved; 21–23 veined.

Abaxial leaf blade epidermis. *Costal/intercostal zonation* conspicuous. *Papillae* absent. Mid-intercostal long-cells rectangular; having markedly sinuous walls, or having straight or only gently undulating walls. *Microhairs* present; panicoid-type; (48–)50–66(–69) microns long; 6–7.5 microns wide at the septum. Microhair total length/width at septum 8–11. Microhair apical cells (21–)24–34(–36) microns long. Microhair apical cell/total length ratio 0.35–0.61. *Stomata* common; 34.5–36 microns long. Subsidiaries triangular. Guard-cells overlapping to flush with the interstomatals. *Intercostal short-cells* common (in *A. semialata*), or absent or very rare; not paired (solitary); not silicified. *Costal short-cells* conspicuously in long rows. Costal silica bodies 'panicoid-type'; butterfly shaped, or dumb-bell shaped, or nodular.

Transverse section of leaf blade, physiology. C_4 (in all the material examined except *A. semialata* ssp. *eckloniana*, including '*Coridochloa*'), or C_3 (*A. semialata* ssp. *eckloniana*). The anatomical organization when C4 unconventional. Organization of PCR tissue when C_4 *Alloteropsis* type (with an inner PCR sheath, and an outer sparsely chlorenchymatous sheath of unknown function). Biochemical type PCK (in Australian C_4 *A. semialata* ssp. *semialata*), or NADP–ME (in southern African C_4 *A. semialata*: evidently more biochemical typing is needed, given the intergrading C_4 anatomical forms, the NAD-ME anatomy of some species yet to be biochemically typed, and the problematical taxonomy); XyMS+ (*A. cimicina*, *A. quintasii*, *A. latifolia*, *A. paniculata*, *A. papillosa*), or XyMS– (in C_4 forms of *A. semialata*, plus *A. angusta*, *A. homblei*, *A. gwebiensis*). PCR sheath outlines when applicable uneven, or even. PCR sheath extensions when applicable present. Maximum number of extension cells 3. PCR cells of *A. semialata* with a suberised lamella. *PCR cell chloroplasts* of *A. semialata* ovoid; with well developed grana; centrifugal/peripheral (or evenly distributed, in *A. semialata* ssp. *semialata*, *A. angusta*, *A. homblei* and *A. gwebiensis*), or centripetal (in *A. cimicina*, *A. quintasii*, *A. latifolia*, *A. paniculata* and *A. papillosa*, which are anatomically indicated as NAD-ME). *Mesophyll* with radiate chlorenchyma, or with non-radiate chlorenchyma. *Leaf blade* 'nodular' in section to adaxially flat; with the ribs more or less constant in size. *Midrib* conspicuous, or not readily distinguishable; with one bundle only, or having a conventional arc of bundles; with colourless mesophyll adaxially, or without colourless mesophyll adaxially. Bulliforms present in discrete, regular adaxial groups; in simple fans and associated with colourless mesophyll cells to form deeply-penetrating fans. Many of the smallest vascular bundles unaccompanied by sclerenchyma. Combined sclerenchyma girders present, or absent; forming 'figures', or nowhere forming 'figures'. Sclerenchyma all associated with vascular bundles.

Phytochemistry. Leaf blade chlorophyll *a:b* ratio of Australian C_4 *A. semialata* 3.36–3.63.

Cytology. Chromosome base number, $x = 9$. $2n = 18$ (C_3 *A. semialata*), or 54 (C_4 *A. semialata*).

Taxonomy. Panicoideae; Panicodae; Paniceae.

Ecology, geography, regional floristic distribution. 5–8 species (with complexes around *A. semialata* and *A. paniculata* reflected in specific and generic synonyms). Tropical Africa, Asia & Australia. Helophytic, or mesophytic, or xerophytic; species of open habitats; glycophytic. Marshy and weedy places.

Paleotropical and Australian. African, Madagascan, Indomalesian, and Neocaledonian. Sudano-Angolan and West African Rainforest. Indian, Indo-Chinese, Malesian, and

Papuan. North and East Australian. Sahelo-Sudanian, Somalo-Ethiopian, and South Tropical African. Tropical North and East Australian.

Economic importance. Important native pasture species: *A. semialata*.

References, etc. Morphological/taxonomic: Gibbs Russell 1983. Leaf anatomical: Metcalfe 1960; Ellis 1974; Hattersley *et al*. 1977; Hattersley and Long 1990; this project.

Alopecurus L.

Alopecuropsis Opiz, *Colobachne* P. Beauv., *Tozzettia* Savi

Habit, vegetative morphology. Annual, or perennial; rhizomatous, or stoloniferous, or caespitose, or decumbent. *Culms* 10–110 cm high; herbaceous; unbranched above; tuberous (e.g. *A. bulbosus*), or not tuberous. Culm nodes glabrous. Culm internodes hollow. *Leaves* not basally aggregated; non-auriculate. *Leaf blades* linear; narrow; 0.7–10(–12) mm wide; flat, or rolled; *not pseudopetiolate*; without cross venation; persistent; rolled in bud. *Ligule an unfringed membrane*; truncate, or not truncate; 1–6 mm long.

Reproductive organization. Plants bisexual, with bisexual spikelets; with hermaphrodite florets. The spikelets all alike in sexuality. Plants outbreeding.

Inflorescence. *Inflorescence* paniculate; *contracted*; capitate, or more or less ovoid, or spicate; espatheate; not comprising 'partial inflorescences' and foliar organs. Spikelet-bearing axes persistent. Spikelets not secund; pedicellate.

Female-fertile spikelets. Spikelets 2–7 mm long; compressed laterally; falling with the glumes. Rachilla terminated by a female-fertile floret. Hairy callus absent.

Glumes two; more or less equal; long relative to the adjacent lemmas; joined (below), or free; pointed; awnless; carinate (the keels ciliate); similar (membranous). Lower glume 3 nerved. Upper glume 3 nerved. **Spikelets** with female-fertile florets only; *without proximal incomplete florets*.

Female-fertile florets 1. *Lemmas often with the margins connate below*; less firm than the glumes to similar in texture to the glumes (hyaline); not becoming indurated; entire; pointed, or blunt; awned. Awns 1; median; dorsal; from well down the back; geniculate; much shorter than the body of the lemma to much longer than the body of the lemma; entered by one vein. Lemmas hairy, or hairless; non-carinate; 5 nerved. **Palea** present (rarely), or absent (usually); when present, *very reduced*; awnless, without apical setae; not indurated. **Lodicules absent. Stamens** 3. Anthers 0.3–3.5 mm long; not penicillate. *Ovary* glabrous. Styles fused. Stigmas 2.

Fruit, embryo and seedling. *Fruit* free from both lemma and palea; small. Hilum short. Embryo small; not waisted. Endosperm liquid in the mature fruit, or hard; with lipid; containing compound starch grains. Embryo with an epiblast; without a scutellar tail; with a negligible mesocotyl internode. Embryonic leaf margins meeting.

Seedling with a long mesocotyl; with a loose coleoptile, or with a tight coleoptile. First seedling leaf with a well-developed lamina. The lamina broad, or narrow; erect; 3 veined.

Abaxial leaf blade epidermis. *Papillae* present, or absent. Mid-intercostal long-cells having markedly sinuous walls (rarely), or having straight or only gently undulating walls. *Microhairs* absent. *Stomata* common; 33–37.5 microns long (in *A. pratensis*). Subsidiaries parallel-sided. *Intercostal short-cells* absent or very rare. *Costal short-cells* neither distinctly grouped into long rows nor predominantly paired. Costal silica bodies horizontally-elongated crenate/sinuous to horizontally-elongated smooth.

Transverse section of leaf blade, physiology. C_3; XyMS+. *Mesophyll* with non-radiate chlorenchyma. Leaf blade with the ribs more or less constant in size. *Midrib* not readily distinguishable; with one bundle only. Bulliforms present in discrete, regular adaxial groups; in simple fans (or the cells fairly uniform in size within the groups). All the vascular bundles accompanied by sclerenchyma. Combined sclerenchyma girders present, or absent; nowhere forming 'figures'. Sclerenchyma all associated with vascular bundles.

Culm anatomy. *Culm internode bundles* in one or two rings, or in three or more rings.

Phytochemistry. Tissues of the culm bases with little or no starch. Fructosans predominantly long-chain. Leaves containing flavonoid sulphates (1 species).

Special diagnostic feature. *Spikelets not borne as in* **Cornucopiae** *(q.v.)*.

Cytology. Chromosome base number, $x = 7$. $2n = 14, 28, 42, 56$, and 100. 2, 4, 6, 8, 14, and 17 ploid. Haploid nuclear DNA content 3 pg (1 species).

Taxonomy. Pooideae; Poodae; Aveneae.

Ecology, geography, regional floristic distribution. 36 species. Eurasia & temperate South America. Commonly adventive. Helophytic, or mesophytic; species of open habitats; halophytic (*A. bulbosus*), or glycophytic. Damp meadows to stony slopes.

Holarctic, Paleotropical, Neotropical, and Antarctic. Boreal, Tethyan, and Madrean. African. Arctic and Subarctic, Euro-Siberian, Atlantic North American, and Rocky Mountains. Mediterranean and Irano-Turanian. Sudano-Angolan. Pampas and Andean. New Zealand and Patagonian. European. Canadian-Appalachian, Southern Atlantic North American, and Central Grasslands. Somalo-Ethiopian.

Rusts and smuts. Rusts — *Puccinia*. Taxonomically wide-ranging species: *Puccinia graminis*, *Puccinia coronata*, *Puccinia striiformis*, *Puccinia brachypodii*, *Puccinia recondita*, and *'Uromyces' dactylidis*. Smuts from Tilletiaceae and from Ustilaginaceae. Tilletiaceae — *Entyloma*, *Tilletia*, and *Urocystis*. Ustilaginaceae — *Sphacelotheca* and *Ustilago*.

Economic importance. Significant weed species: *A. aequalis*, *A. geniculatus*, *A. myosuroides* (Black Twitch, in temperate cereal crops). Cultivated fodder: *A. pratensis*. Important native pasture species: *A. pratensis*.

References, etc. Leaf anatomical: Metcalfe 1960; this project.

Alvimia Soderstrom & Londoño

Habit, vegetative morphology. Perennial. The flowering culms leafy. *Culms* ascending 8–25 metres into the vegetation; woody and persistent; to 1 cm in diameter; cylindrical; scandent; branched above. Primary branches/mid-culm node 1. Culm sheaths tardily deciduous. Culm internodes solid, or hollow. Rhizomes pachymorph. Plants unarmed. *Leaves* not basally aggregated; auriculate; with auricular setae (inconspicuous in *A. gracilis*). Leaf blades linear-lanceolate to ovate-lanceolate; broad to narrow; 3–30 mm wide; rolled; pseudopetiolate; seemingly without cross venation; disarticulating from the sheaths; rolled in bud. *Ligule* present; a ciliate or ciliolate rim; about 0.2–0.5 mm long. *Contra-ligule* present.

Reproductive organization. Plants bisexual, with bisexual spikelets; with hermaphrodite florets.

Inflorescence. Inflorescence indeterminate; *with pseudospikelets*; *of spicate main branches and paniculate (polytelic synflorescences, terminating leafy or leafless branches)*; spatheate (the prophyllate pseudospikelets subtended by bracts). Spikelet-bearing axes persistent. Spikelets solitary; not secund; pedicellate (via the elongated terminal segment of the rachis above the empty bract).

Female-fertile spikelets. *Spikelets* unconventional (being indeterminate); compressed laterally (?); disarticulating above the glumes; disarticulating between the florets. Rachilla prolonged beyond the uppermost female-fertile floret. The rachilla extension with incomplete florets (ending in a rudimentary floret, or in a bristle-like prolongation). Hairy callus absent.

Glumes several (0–3 gemmiparous bracts, 1 or 2 empty bracts, 0 or 1 gemmiparous glume); shorter than the adjacent lemmas; awnless. *Spikelets* with incomplete florets. The incomplete florets distal to the female-fertile florets. *The distal incomplete florets* merely underdeveloped.

Female-fertile florets 3–30. Lemmas not becoming indurated; entire; awnless (or apiculate), or mucronate; non-carinate; without a germination flap. *Palea* present; relatively long (as long as the lemma or longer); apically notched; awnless, without apical setae; 2-keeled. *Lodicules* present; 3; free; membranous; ciliate; heavily vascularized. **Stamens** 2 *(rarely 3)*. Anthers 3.3–5 mm long; not penicillate; without an apically prolonged connective. *Ovary* glabrous; without a conspicuous apical appendage. Styles fused (into one). Stigmas 2 (one of them sometimes branched).

Fruit, embryo and seedling. *Fruit* free from both lemma and palea (falling with the rachilla segment, the lemma and palea attached to its pointed base); large (10–20 mm long); reniform or olive-like; not noticeably compressed. Pericarp fleshy. Seed endospermic. Endosperm hard; containing only simple starch grains. Embryo with an epiblast; with a scutellar tail; with a negligible mesocotyl internode. Embryonic leaf margins overlapping.

Seedling with a short mesocotyl. First seedling leaf without a lamina.

Abaxial leaf blade epidermis. *Costal/intercostal zonation* conspicuous. *Papillae* present; costal and intercostal (but more abundant, and much more conspicuous, intercostally). Intercostal papillae over-arching the stomata; several per cell (variable in size, irregularly lobed, some more or less coronate). *Long-cells* similar in shape costally and intercostally to markedly different in shape costally and intercostally (the costals generally somewhat narrower); of similar wall thickness costally and intercostally (quite thick walled). Mid-intercostal long-cells rectangular; having markedly sinuous walls (the sinuosity coarse, with heavy pitting). *Microhairs* present; elongated; clearly two-celled; panicoid-type. *Stomata* common. Subsidiaries largely covered by overarching papillae; triangular (with truncated apices). Guard-cells seemingly but not certainly overlapping to flush with the interstomatals (the stomatal complexes sunken). *Intercostal short-cells* very common; not paired (solitary); not silicified (or the silica poorly developed, in the material seen). With a few small costal prickles. Costal zones with short-cells. *Costal short-cells* conspicuously in long rows (mostly solitary, a few pairs). Costal silica bodies poorly developed, or absent; not developed, the silica cells tall-and-narrow to saddle shaped.

Transverse section of leaf blade, physiology. C_3; XyMS+. Mesophyll without adaxial palisade; with arm cells; with fusoids. The fusoids external to the PBS. *Leaf blade* adaxially flat. Bulliforms present in discrete, regular adaxial groups; in simple fans (one in each of the slight furrows, where the fusoids from adjacent bundles approach oneanother). All the vascular bundles accompanied by sclerenchyma. Combined sclerenchyma girders present (with all the bundles); nowhere forming 'figures' (the girders slender). Sclerenchyma not all bundle-associated. The 'extra' sclerenchyma in abaxial groups and in adaxial groups; abaxial-hypodermal, the groups isolated and adaxial-hypodermal, contiguous with the bulliforms (opposite the bulliforms, and laterally adjacent to them).

Taxonomy. Bambusoideae; Bambusodae; Bambuseae.

Ecology, geography, regional floristic distribution. 3 species. Eastern coastal Bahia, Brazil.

Neotropical. Amazon.

References, etc. Morphological/taxonomic: Soderstrom and Londoño 1988. Leaf anatomical: this project.

Special comments. Fruit data wanting.

Amblyopyrum Eig

~ *Aegilops*

Habit, vegetative morphology. *Annual*. Culms herbaceous. Leaves auriculate, or non-auriculate. Leaf blades linear; narrow; 1.5–6 mm wide; usually flat; without cross venation. Ligule an unfringed membrane; truncate; 0.5 mm long.

Reproductive organization. Plants bisexual, with bisexual spikelets; with hermaphrodite florets. The spikelets of sexually distinct forms on the same plant, or all alike in sexuality; hermaphrodite, or hermaphrodite and sterile (rarely, sterile at the extremities of the inflorescence).

Inflorescence. *Inflorescence a single spike (glabrous, linear, usually very long)*. Rachides hollowed. Inflorescence espatheate; not comprising 'partial inflorescences' and foliar organs. *Spikelet-bearing axes* with substantial rachides; *disarticulating*; disarticulating at the joints. Spikelets solitary; not secund; distichous.

Female-fertile spikelets. Spikelets 8–15 mm long; compressed laterally; falling with the glumes. Rachilla prolonged beyond the uppermost female-fertile floret. The rachilla extension with incomplete florets. *Callus* very short; blunt.

Glumes two; more or less equal; long relative to the adjacent lemmas; lateral to the rachis; not pointed; not subulate; awnless; non-carinate; *similar (trapezoid or rectangular)*. Lower glume 3–9 nerved. Upper glume 4–9 nerved. *Spikelets* with incomplete florets. The incomplete florets distal to the female-fertile florets.

Female-fertile florets *4–8*. *Lemmas less firm than the glumes*; not becoming indurated; entire; blunt; awnless; hairy, or hairless; non-carinate; without a germination flap; 5 nerved; with the nerves non-confluent. *Palea* present; relatively long; 2-nerved; 2-keeled. *Lodicules* present; 2; free; membranous; ciliate; toothed, or not toothed; not or scarcely vascularized. *Stamens* 3. Anthers 3.5–4.8 mm long. *Ovary* hairy. Styles free to their bases. Stigmas 2.

Fruit, embryo and seedling. *Fruit* adhering to lemma and/or palea; small to medium sized (3.5–4 mm long); ellipsoid; shallowly longitudinally grooved; compressed dorsiventrally; with hairs confined to a terminal tuft. Hilum long-linear. Embryo large to small (to about 1/3 the caryposis length); with an epiblast.

Abaxial leaf blade epidermis. *Costal/intercostal zonation* conspicuous. *Papillae* absent. *Long-cells* similar in shape costally and intercostally, or markedly different in shape costally and intercostally; of similar wall thickness costally and intercostally, or differing markedly in wall thickness costally and intercostally. Mid-intercostal long-cells rectangular (mostly); having straight or only gently undulating walls. *Microhairs* absent. *Stomata* common; 39–41 microns long. Subsidiaries parallel-sided and dome-shaped. Guard-cells overlapped by the interstomatals. *Intercostal short-cells* absent or very rare (a few, solitary, not silicified, near veins). *Crown cells* present. *Costal short-cells* neither distinctly grouped into long rows nor predominantly paired. Costal silica bodies horizontally-elongated crenate/sinuous (predominantly), or horizontally-elongated smooth (few).

Transverse section of leaf blade, physiology. C_3; XyMS+. *Mesophyll* with non-radiate chlorenchyma. Bulliforms present in discrete, regular adaxial groups; in simple fans. Combined sclerenchyma girders present. Sclerenchyma all associated with vascular bundles.

Cytology. Chromosome base number, $x = 7$. $2n = 14$. 2 ploid. Haplomic genome content Z.

Taxonomy. Pooideae; Triticodae; Triticeae.

Ecology, geography, regional floristic distribution. 1 species. W. Asia. Holarctic. Boreal and Tethyan. Euro-Siberian. Mediterranean and Irano-Turanian. European.

References, etc. Morphological/taxonomic: Löve 1984. Leaf anatomical: this project.

Ammochloa Boiss.

Cephalochloa Coss. & Dur., *Dictyochloa* (Murbeck) E.G. Camus

Habit, vegetative morphology. Annual; caespitose. *Culms* 1–25 cm high; herbaceous; unbranched above. Culm internodes hollow. *Leaves* mostly basal; non-auriculate. Leaf blades linear to linear-lanceolate; narrow; 1–3 mm wide; flat, or rolled (convolute); without cross venation; persistent. Ligule an unfringed membrane; not truncate; 0.5–5 mm long.

Reproductive organization. Plants bisexual, with bisexual spikelets; with hermaphrodite florets. The spikelets of sexually distinct forms on the same plant; hermaphrodite and sterile (the latter reduced to small, sterile bracts at the base of the inflorescence); overtly heteromorphic.

Inflorescence. Inflorescence paniculate; contracted; capitate to more or less ovoid (reduced to a sub-globose head of close-packed spikelets); espatheate. Spikelet-bearing axes persistent. **Spikelets** *associated with bractiform involucres (these representing basal, sterile spikelets)*; not secund; subsessile.

Female-fertile spikelets. Spikelets 4–9 mm long; compressed laterally; disarticulating above the glumes. Rachilla prolonged beyond the uppermost female-fertile floret. The rachilla extension naked. Hairy callus absent. *Callus* short.

Glumes two (ovate-oblique); more or less equal; shorter than the spikelets; long relative to the adjacent lemmas; pointed; awnless; carinate (sometimes narrowly winged); similar (papery to membranous). Lower glume 1–2 nerved. Upper glume 1–2 nerved. *Spikelets* with female-fertile florets only.

Female-fertile florets 4–12. Lemmas similar in texture to the glumes to decidedly firmer than the glumes (papery to membranous, becoming leathery); not becoming indurated; entire; pointed; mucronate (the mucro recurved); hairless; carinate; 5–7 nerved. *Palea* present; relatively long; entire to apically notched; awnless, without apical setae; not indurated (membranous); 2-nerved; 2-keeled. *Lodicules* absent. *Stamens* 2–3. Anthers 0.6–0.8 mm long; not penicillate. **Ovary** glabrous; *with a conspicuous apical appendage (this membranous, associated with the style)*. Styles fused. Stigmas 2.

Fruit, embryo and seedling. **Fruit** free from both lemma and palea; *small to large (winged via the ovary appendage)*; compressed laterally, or not noticeably compressed. Hilum short. Embryo large, or small. Endosperm hard; with lipid. Embryo with an epiblast; without a scutellar tail; with a negligible mesocotyl internode. Embryonic leaf margins meeting.

Abaxial leaf blade epidermis. *Costal/intercostal zonation* conspicuous. *Papillae* absent. *Long-cells* markedly different in shape costally and intercostally; of similar wall thickness costally and intercostally. Mid-intercostal long-cells fusiform; having straight or only gently undulating walls. *Microhairs* absent. *Stomata* common. Subsidiaries parallel-sided. Guard-cells overlapped by the interstomatals. *Intercostal short-cells* absent or very rare. *Costal short-cells* neither distinctly grouped into long rows nor predominantly paired. Costal silica bodies horizontally-elongated crenate/sinuous.

Transverse section of leaf blade, physiology. C_3; XyMS+. *Mesophyll* with nonradiate chlorenchyma; without adaxial palisade. *Leaf blade* adaxially flat. *Midrib* conspicuous (rounded keel with larger bundle); with one bundle only. Bulliforms not in discrete, regular adaxial groups; nowhere involved in bulliform-plus-colourless mesophyll arches. All the vascular bundles accompanied by sclerenchyma. Combined sclerenchyma girders absent (strands only). Sclerenchyma all associated with vascular bundles.

Taxonomy. Pooideae; Poodae; Aveneae.

Ecology, geography, regional floristic distribution. 3 species. Mediterranean. Xerophytic; species of open habitats. Dry sandy places.

Holarctic and Paleotropical. Boreal and Tethyan. African. Euro-Siberian. Macaronesian, Mediterranean, and Irano-Turanian. Saharo-Sindian. European.

Rusts and smuts. Smuts from Tilletiaceae. Tilletiaceae — *Tilletia*.

References, etc. Leaf anatomical: this project.

Ammophila Host

Psamma P. Beauv.

Habit, vegetative morphology. Perennial; rhizomatous. *Culms* 20–130 cm high; herbaceous; unbranched above. Culm nodes glabrous. Culm internodes hollow. *Leaves* mostly basal; non-auriculate. *Leaf blades* linear; narrow; *2–5 mm wide (sharp-pointed, blue-green)*; rolled (convolute); *not pseudopetiolate*; without cross venation; persistent; rolled in bud. *Ligule an unfringed membrane*; not truncate; 1–30 mm long.

Reproductive organization. Plants bisexual, with bisexual spikelets; with hermaphrodite florets.

Inflorescence. *Inflorescence paniculate*; contracted; spicate; espatheate; not comprising 'partial inflorescences' and foliar organs. Spikelet-bearing axes persistent. Spikelets not secund; pedicellate.

Female-fertile spikelets. *Spikelets 9–15 mm long*; compressed laterally; disarticulating above the glumes. Rachilla prolonged beyond the uppermost female-fertile floret; rachilla hairy. The rachilla extension naked. Hairy callus present. *Callus* short; pointed.

Glumes two; more or less equal; about equalling the spikelets to exceeding the spikelets; long relative to the adjacent lemmas (exceeding them); pointed; awnless; carinate; similar. *Lower glume 1 nerved.* Upper glume 1–3 nerved. **Spikelets** with female-fertile florets only; *without proximal incomplete florets.*

Female-fertile florets 1. Lemmas similar in texture to the glumes; not becoming indurated; entire, or incised; when incised, 2 lobed; not deeply cleft; minutely awned, or mucronate. Awns 1; median; dorsal; from near the top (subterminal); non-geniculate; much shorter than the body of the lemma; entered by one vein. Lemmas hairless; carinate; without a germination flap; 5 nerved; with the nerves non-confluent. **Palea** present; relatively long; apically notched (minutely); awnless, without apical setae; not indurated (firm); *several nerved (often 4-nerved)*; keel-less. *Lodicules* present; 2; free; membranous; ciliate, or glabrous; not toothed; not or scarcely vascularized. *Stamens* 3; with free filaments. Anthers 4–5 mm long; not penicillate. *Ovary* glabrous. Styles free to their bases. Stigmas 2.

Fruit, embryo and seedling. Fruit medium sized; ellipsoid; longitudinally grooved; compressed dorsiventrally, or not noticeably compressed. Hilum long-linear (two thirds of the fruit length). Embryo small. Endosperm hard; with lipid; containing compound starch grains. Embryo with an epiblast; without a scutellar tail; with a negligible mesocotyl internode. Embryonic leaf margins meeting.

Seedling with a tight coleoptile.

Abaxial leaf blade epidermis. *Costal/intercostal zonation* lacking. *Papillae* absent. *Long-cells* similar in shape costally and intercostally; of similar wall thickness costally and intercostally (walls thick, pitted). Mid-intercostal long-cells rectangular; having markedly sinuous walls, or having straight or only gently undulating walls (different from one specimen to another). *Microhairs* absent. *Stomata* absent or very rare. *Intercostal short-cells* common; not paired; not silicified. *Costal short-cells* neither distinctly grouped into long rows nor predominantly paired (mainly solitary, a few paired). Costal silica bodies absent to poorly developed; in so far as recognisable horizontally-elongated crenate/sinuous (poorly developed, but the silica-cells mainly square, elongated-sinuous or elongated-crenate).

Transverse section of leaf blade, physiology. C_3; XyMS+. *Mesophyll* with non-radiate chlorenchyma. *Leaf blade* with distinct, prominent adaxial ribs; with the ribs very irregular in sizes. *Midrib* not readily distinguishable; with one bundle only. Bulliforms not in discrete, regular adaxial groups; in the furrows, in ill defined groups of small, irregularly sized cells. All the vascular bundles accompanied by sclerenchyma. Combined sclerenchyma girders present; forming 'figures' (each bundle with a large 'anchor' — the mesophyll being confined to lateral blocks in the ribs, and immediately beneath the furrows). Sclerenchyma not all bundle-associated (a continuous abaxial layer, linking with the 'anchors'). The 'extra' sclerenchyma in a continuous abaxial layer.

Cytology. Chromosome base number, $x = 7$. $2n = 14$, 28, and 56. 2, 4, and 8 ploid.

Taxonomy. Pooideae; Poodae; Aveneae.

Ecology, geography, regional floristic distribution. 2 species. North temperate. Commonly adventive. Xerophytic; species of open habitats; halophytic. Sand-binding and dune stabilizing.

Holarctic and Paleotropical. Boreal and Tethyan. African. Euro-Siberian and Atlantic North American. Mediterranean. Saharo-Sindian. European. Canadian-Appalachian.

Hybrids. *A. arenaria* hybridizes with *Calamagrostis epigejos* (×*Ammocalamagrostis* P. Fourn.; ×*Calamophila* O. Schwartz = ×*Ammocalamagrostis*, ×*Calammophila* Brand = ×*Ammocalamagrostis*).

Rusts and smuts. Rusts — *Puccinia*. Taxonomically wide-ranging species: *Puccinia graminis*, *Puccinia coronata*, and *Puccinia pygmaea*. Smuts from Ustilaginaceae. Ustilaginaceae — *Ustilago*.

Economic importance. *A. arenaria* and ×*Ammocalamagrostis* widely used as sand stabilizers.

References, etc. Leaf anatomical: Metcalfe 1960; this project.

Ampelodesmos Link

Habit, vegetative morphology. Robust perennial; rhizomatous. **Culms** 60–350 cm high; *herbaceous*. Culm internodes solid. Leaves non-auriculate. Leaf blades harsh; narrow; flat, or rolled; without cross venation. *Ligule a fringed membrane*; not truncate; 6–12 mm long.

Reproductive organization. Plants bisexual, with bisexual spikelets; with hermaphrodite florets.

Inflorescence. *Inflorescence paniculate*; open; espatheate; not comprising 'partial inflorescences' and foliar organs. Spikelet-bearing axes persistent. Spikelets not secund; pedicellate.

Female-fertile spikelets. Spikelets 10–15 mm long; compressed laterally; disarticulating above the glumes; disarticulating between the florets. Rachilla prolonged beyond the uppermost female-fertile floret; rachilla hairy (villous). Hairy callus present.

Glumes two; *more or less equal*; shorter than the spikelets; shorter than the adjacent lemmas; pointed (acuminate); awnless; carinate; similar (firmly membranous). Lower glume 3–5 nerved. Upper glume 3–5 nerved. *Spikelets* with female-fertile florets only, or with incomplete florets. The incomplete florets distal to the female-fertile florets.

Female-fertile florets 2–6 (*?*). Lemmas decidedly firmer than the glumes (leathery); not becoming indurated; incised; 2 lobed; not deeply cleft (shortly bidentate); mucronate, or awned. Awns when present, 1; from a sinus; non-geniculate; much shorter than the body of the lemma. Lemmas hairy (on the lower half); carinate to non-carinate; 5–7 nerved. *Palea* present; relatively long; apically notched (bidentate); awnless, without apical setae; 2-nerved; 2-keeled. *Lodicules* present; 3; free; membranous; ciliate (on the margins); not toothed; not or scarcely vascularized. *Stamens* 3. Anthers not penicillate. **Ovary** *hairy*. Styles free to their bases. Stigmas 2; white.

Fruit, embryo and seedling. Fruit medium sized (about 7 mm long); with hairs confined to a terminal tuft. Hilum long-linear. Embryo small. Endosperm hard. Embryo with an epiblast; without a scutellar tail; with a negligible mesocotyl internode. Embryonic leaf margins meeting.

Seedling with a tight coleoptile. First seedling leaf with a well-developed lamina. The lamina narrow (l/b ratio 70); curved; 7 veined.

Abaxial leaf blade epidermis. *Costal/intercostal zonation* lacking. *Papillae* absent. *Long-cells* similar in shape costally and intercostally; of similar wall thickness costally and intercostally. Mid-intercostal long-cells rectangular; having markedly sinuous walls. *Microhairs* absent. *Stomata* absent or very rare; 28.5–30 microns long. *Intercostal short-cells* common; in cork/silica-cell pairs. Intercostal silica bodies elliptic. *Costal short-cells* predominantly paired. Costal silica bodies rounded (mostly, elliptical), or tall-and-narrow.

Transverse section of leaf blade, physiology. C_3; XyMS+. *Mesophyll* with non-radiate chlorenchyma. *Leaf blade* with distinct, prominent adaxial ribs (flat-topped ribs); with the ribs very irregular in sizes. *Midrib* not readily distinguishable; with one bundle only. Bulliforms not in discrete, regular adaxial groups; in the furrows, in ill-defined groups of irregularly sized cells cf. *Ammophila*. All the vascular bundles accompanied by sclerenchyma. Sclerenchyma not all bundle-associated. The 'extra' sclerenchyma in a continuous abaxial layer.

Cytology. Chromosome base number, $x = 12$ (chromosomes small). $2n = 48$ and 96. 4 and 8 ploid.

Taxonomy. Arundinoideae; Stipeae.

Ecology, geography, regional floristic distribution. 1 species. Mediterranean. Xerophytic (mainly coastal).

Holarctic. Boreal and Tethyan. Euro-Siberian. Mediterranean. European.

Economic importance. A component of Esparto grass, used for papermaking.

References, etc. Morphological/taxonomic: Decker 1964b; Macfarlane and Watson 1980. Leaf anatomical: Metcalfe 1960; this project.

Amphibromus Nees

~ *Helictotrichon*

Habit, vegetative morphology. Perennial; rhizomatous, or caespitose, or decumbent. **Culms** 40–180 cm high; *herbaceous*. Culm nodes glabrous. Culm internodes hollow. *Leaves* not basally aggregated; non-auriculate. Sheath margins free. Leaf blades narrow; flat, or rolled (involute); without cross venation; persistent. *Ligule an unfringed membrane*; not truncate (elongated, becoming lacerated); 2–15 mm long.

Reproductive organization. Plants bisexual, with bisexual spikelets; with hermaphrodite florets. Exposed-cleistogamous, or chasmogamous. Plants with hidden cleistogenes, or without hidden cleistogenes. The hidden cleistogenes when present, in the leaf sheaths.

Inflorescence. *Inflorescence paniculate*; open (narrow, elongated); with capillary branchlets, or without capillary branchlets (often flexuose); espatheate; not comprising 'partial inflorescences' and foliar organs. Spikelet-bearing axes persistent. Spikelets not secund; pedicellate.

Female-fertile spikelets. Spikelets 7–15 mm long; compressed laterally; disarticulating above the glumes; disarticulating between the florets; with conventional internode spacings, or with distinctly elongated rachilla internodes between the florets. Rachilla prolonged beyond the uppermost female-fertile floret; rachilla hairy. *Hairy callus present (usually, silky)*.

Glumes two; more or less equal; shorter than the spikelets; *shorter than the adjacent lemmas*; pointed, or not pointed; awnless; carinate (slightly), or non-carinate; similar (G_2 broader). Lower glume 1–5 nerved. Upper glume 3–7 nerved. *Spikelets* with female-fertile florets only, or with incomplete florets. The incomplete florets distal to the female-fertile florets (the uppermost male). *The distal incomplete florets merely underdeveloped*.

Female-fertile florets *2–10*. *Lemmas* 4-toothed or distinctly bifid; decidedly firmer than the glumes; not becoming indurated (firm); incised; 2 lobed, or 4 lobed (toothed); not deeply cleft; *awned. Awns* 1, or 5; median, or median and lateral; the median different in form from the laterals (when laterals present); dorsal; *from well down the back (from about the middle)*; geniculate; entered by one vein. The lateral awns shorter than the median (straight, terminal). Lemmas hairless; non-carinate; without a germination flap; 3–15 nerved. *Palea* present; relatively long to conspicuous but relatively short; apically notched (the points acute); awnless; without apical setae; 1-nerved, or 2-nerved, or nerveless; 2-keeled (the keels ciliate), or keel-less. *Lodicules* present; 2; free; membranous; glabrous; not toothed. *Stamens* 3. Anthers not penicillate. **Ovary** *glabrous*. Styles free to their bases. Stigmas 2; white.

Fruit, embryo and seedling. Fruit small; not noticeably compressed. *Hilum long-linear*. Embryo small; not waisted. Endosperm hard; with lipid; containing compound starch grains.

First seedling leaf with a well-developed lamina.

Abaxial leaf blade epidermis. *Costal/intercostal zonation* conspicuous. *Papillae* absent. *Long-cells* similar in shape costally and intercostally; of similar wall thickness costally and intercostally. Mid-intercostal long-cells rectangular; having markedly sinuous walls, or having straight or only gently undulating walls. *Microhairs* absent. *Stomata* common. Subsidiaries parallel-sided. Guard-cells overlapped by the interstomatals, or overlapping to flush with the interstomatals. *Intercostal short-cells* absent or very rare. *Costal short-cells* neither distinctly grouped into long rows nor predominantly paired. Costal silica bodies horizontally-elongated crenate/sinuous, or horizontally-elongated smooth, or rounded, or tall-and-narrow.

Transverse section of leaf blade, physiology. C_3; XyMS+. *Mesophyll* with non-radiate chlorenchyma; without adaxial palisade; traversed by columns of colourless mesophyll cells. *Leaf blade* with distinct, prominent adaxial ribs; with the ribs more or less constant in size. Midrib with one bundle only. Bulliforms present in discrete, regular adaxial groups; associated with colourless mesophyll cells to form deeply-penetrating fans (these linked with traversing colourless columns or with abaxial sclerenchyma groups). All the vascular bundles accompanied by sclerenchyma. Combined sclerenchyma girders absent. Sclerenchyma not all bundle-associated. The 'extra' sclerenchyma in abaxial groups (opposite the furrows); abaxial-hypodermal, the groups continuous with colourless columns.

Phytochemistry. Tissues of the culm bases with little or no starch.

Taxonomy. Pooideae; Poodae; Aveneae.

Ecology, geography, regional floristic distribution. 6 species. Australia, New Zealand, South America. Helophytic, or mesophytic.

Neotropical, Australian, and Antarctic. Central Brazilian, Pampas, and Andean. North and East Australian and South-West Australian. Patagonian. Temperate and South-Eastern Australian.

Rusts and smuts. Rusts — *Puccinia*. Taxonomically wide-ranging species: *Puccinia graminis*. Smuts from Tilletiaceae. Tilletiaceae — *Urocystis*.

References, etc. Morphological/taxonomic: Jacobs and Lapinuro 1986. Leaf anatomical: this project.

Amphicarpum Kunth

Habit, vegetative morphology. Annual (erect), or perennial (the culms decumbent at the base). *Culms* 30–100 cm high; herbaceous. Culm nodes glabrous. Culm internodes hollow. Plants unarmed. *Leaves* mostly basal (*A. purshii*), or not basally aggregated; non-auriculate. Leaf blades broad, or narrow; 5–15 mm wide (10–15 cm long); flat; without cross venation; persistent. Ligule a fringe of hairs. *Contra-ligule* absent (but scattered tubercle-based hairs in that position in *A. muhlenbergianum*).

Reproductive organization. *Plants bisexual, with bisexual spikelets (but the 'chasmogamous' spikelets of the conspicuous terminal panicle not fruitful)*; with hermaphrodite florets. Exposed-cleistogamous and chasmogamous. *Plants with hidden cleistogenes. The hidden cleistogenes subterranean (on slender branches from the base of the culm, and sometimes also from lower nodes).*

Inflorescence. *Inflorescence* (i.e., the obvious, 'chasmogamous' but sterile inflorescence) *paniculate*; espatheate; not comprising 'partial inflorescences' and foliar organs. Spikelet-bearing axes persistent. Spikelets solitary; not secund; pedicellate.

Female-sterile spikelets. The exposed, sterile spikelets of the terminal panicle 4–7 mm long, G_1 sometimes obsolete, G_2 equalling L_1; L_2 and palea indurated, the lemma margins thin and flat.

Female-fertile spikelets. Spikelets 7–9 mm long (exclusively cleistogamous, plump, acuminate). Rachilla terminated by a female-fertile floret. Hairy callus absent.

Glumes present; one per spikelet to two; (when two) *very unequal (the lower obsolete or absent)*; long relative to the adjacent lemmas (G_2 equalling L_1); without conspicuous tufts or rows of hairs; awnless; non-carinate; (when two) very dissimilar (the lower vestigial). Upper glume strongly nerved. *Spikelets* with incomplete florets. The incomplete florets proximal to the female-fertile florets. *Spikelets with proximal incomplete florets. The proximal incomplete florets* 1; paleate. Palea of the proximal incomplete florets not becoming conspicuously hardened and enlarged laterally. The proximal lemmas awnless; exceeded by the female-fertile lemmas (at least in fruit); less firm than the female-fertile lemmas ('sub-rigid').

Female-fertile florets 1. *Lemmas acuminate*; decidedly firmer than the glumes; becoming indurated; entire; pointed; awnless; non-carinate; having the margins lying flat and exposed on the palea; with a clear germination flap. **Palea** present; awnless, without apical setae; *indurated. Stamens* 3.

Abaxial leaf blade epidermis. *Costal/intercostal zonation* conspicuous. *Papillae* absent. *Long-cells* markedly different in shape costally and intercostally (the costals narrower, 'normal'); differing markedly in wall thickness costally and intercostally (the costals thinner-walled). Intercostal zones exhibiting many atypical long-cells (large and short to isodiametric in the middle of the intercostal zone). Mid-intercostal long-cells rectangular to irregular in shape; having markedly sinuous walls. *Microhairs* present; panicoid-type; 54–60 microns long; 5.4 microns wide at the septum, or 8.4–9.6 microns wide at the septum. Microhair total length/width at septum 6.3–10. Microhair apical cells 31.5–34.5 microns long. Microhair apical cell/total length ratio 0.53–0.58. *Stomata* common (ranked alongside the veins); 18–21 microns long, or 37.5–43.5 microns long (in different specimens). Subsidiaries dome-shaped (mostly), or triangular (a few). Guard-cells overlapping to flush with the interstomatals. *Intercostal short-cells* common (especially alongside the veins); in cork/silica-cell pairs; silicified. *Costal short-cells* conspicuously in long rows. Costal silica bodies 'panicoid-type'; mostly shortish dumb-bell shaped.

Transverse section of leaf blade, physiology. C_3; XyMS+. *Mesophyll* with radiate chlorenchyma; *Isachne*-type (seemingly, in our poor material). *Leaf blade* 'nodular' in section to adaxially flat. *Midrib* not readily distinguishable; with one bundle only. Bulliforms present in discrete, regular adaxial groups; in simple fans. All the vascular bundles accompanied by sclerenchyma. Combined sclerenchyma girders present; forming 'figures' (with many I's). Sclerenchyma all associated with vascular bundles.

Special diagnostic feature. *Plants not as in* **Dichanthelium** *(q.v.)*.

Cytology. Chromosome base number, $x = 9$. $2n = 18$. 2 ploid.

Taxonomy. Panicoideae; Panicodae; Paniceae.

Ecology, geography, regional floristic distribution. 2 species. South-eastern U.S.A. Species of open habitats; glycophytic. Sandy pinewoods.

Holarctic. Boreal. Atlantic North American. Southern Atlantic North American.

References, etc. Leaf anatomical: Metcalfe 1960; this project.

Special comments. Fruit data wanting.

Amphipogon R.Br.

Gamelythrum Nees, *Pentacraspedon* Steud.

Habit, vegetative morphology. Perennial; caespitose (to short-rhizomatous). **Culms** 15–75 cm high; *herbaceous*; unbranched above. Culm nodes glabrous (and dark). Culm internodes hollow. Young shoots intravaginal. *Leaves* mostly basal, or not basally aggregated; non-auriculate. *Leaf blades* narrow; 1–3 mm wide; setaceous, or not setaceous (often acicular); flat, or rolled; without abaxial multicellular glands, or exhibiting multicellular glands abaxially (at the bases of macrohairs); without cross venation; *disarticulating from the sheaths*. Ligule a fringe of hairs. *Contra-ligule* absent.

Reproductive organization. Plants bisexual, with bisexual spikelets; with hermaphrodite florets. The spikelets of sexually distinct forms on the same plant, or all alike in sexuality; hermaphrodite, or hermaphrodite and sterile (sometimes reduced at the base of the inflorescence).

Inflorescence. Inflorescence a single spike (or approaching one), or paniculate; contracted; more or less ovoid, or spicate; espatheate; not comprising 'partial inflorescences' and foliar organs. Spikelet-bearing axes persistent. Spikelets not secund; sessile to pedicellate; consistently in 'long-and-short' combinations, or not in distinct 'long-and-short' combinations. The 'shorter' spikelets hermaphrodite. The 'longer' spikelets hermaphrodite.

Female-fertile spikelets. Spikelets 3–6 mm long; abaxial; not noticeably compressed to compressed dorsiventrally; disarticulating above the glumes. Rachilla terminated by a female-fertile floret. Hairy callus present. *Callus* short, or long (short and obtuse, or long and stipitate).

Glumes two; more or less equal; shorter than the adjacent lemmas to long relative to the adjacent lemmas; dorsiventral to the rachis; hairy, or hairless; usually entire, rarely 3-lobed; awned to awnless; non-carinate; similar (acute, obtuse, or tapering into an awn-like point, membranous or papery). *Lower glume 3 nerved.* Upper glume 3 nerved. **Spikelets** with female-fertile florets only; *without proximal incomplete florets*.

Female-fertile florets 1. **Lemmas** becoming decidedly firmer than the glumes; not becoming indurated; incised; 3 lobed; deeply cleft; *awned*. Awns 3; median and lateral (by bristles from the tips of the lobes); the median similar in form to the laterals; non-geniculate; hairless to hairy; about as long as the body of the lemma to much longer than the body of the lemma. The lateral awns about equalling the median. Lemmas hairy, or hairless; non-carinate; without a germination flap; 3 nerved. **Palea** present; relatively long; apically notched to deeply bifid; with apical setae, or awned; textured like the lemma; not indurated (hyaline); 2-nerved; *keel-less*. *Lodicules* present; 2; fleshy; glabrous; not or scarcely vascularized. *Stamens* 3. Anthers 1–5 mm long; not penicillate. *Ovary* glabrous. Styles fused. Stigmas 2.

Fruit, embryo and seedling. Fruit compressed dorsiventrally. Hilum short. *Pericarp free (and opaque)*. Embryo small. Endosperm containing compound starch grains. Embryo without an epiblast; with a scutellar tail; with an elongated mesocotyl internode. Embryonic leaf margins meeting.

Abaxial leaf blade epidermis. *Costal/intercostal zonation* conspicuous, or lacking (*A. laguroides*). *Papillae* present (usually very conspicuous and constituting volcano-like pits, but inconspicuous and in the form of less prominent pits in (e.g.) *A. amphipogonoides*); intercostal, or costal and intercostal. Intercostal papillae not over-arching the stomata; consisting of one symmetrical projection per cell (sometimes, costally), or several per cell (intercostal long-cells with rows). *Long-cells* markedly different in shape costally and intercostally; of similar wall thickness costally and intercostally (thick walled). Intercostal zones with typical long-cells (but some rather short). Mid-intercostal long-cells rectangular; having markedly sinuous walls. *Microhairs* present (sometimes?), or absent (from all material seen, but present adaxially and on lemmas, etc.); chloridoid-type and *Enneapogon*-type (having long basal cells and short, thin-walled, easily collapsing ovoid apices which are at least sometimes 2-celled, cf. *Enneapogon* — and ordinary chloridoid type microhairs seen on the lemmas of *A. caricinus*); with 'partitioning membranes' (*A. strictus*). The 'partitioning membranes' in the apical cell. *Stomata* absent or very rare, or common (*A. amphipogonoides*); (31.5–)33–42 microns long. Subsidiaries non-papillate; dome-shaped, or dome-shaped and triangular. Guard-cells overlapping to flush with the interstomatals. *Intercostal short-cells* common; in cork/silica-cell pairs, or not paired (solitary); silicified. Intercostal silica bodies absent, or present and perfectly developed. *Costal short-cells* neither distinctly grouped into long rows nor predominantly paired (all solitary). Costal silica bodies present and well developed, or poorly developed to absent (mostly); present throughout the costal zones; (these or at least the silica-cells) tall-and-narrow (almost exclusively, in all the species examined), or rounded (a few, notably in *A. strictus*), or saddle shaped (few at most).

Transverse section of leaf blade, physiology. C_3; XyMS+. *Mesophyll* with radiate chlorenchyma; *Isachne*-type (almost!), or not *Isachne*-type. *Leaf blade* with distinct, prominent adaxial ribs; with the ribs very irregular in sizes. *Midrib* not readily distinguishable; with one bundle only. Bulliforms present in discrete, regular adaxial groups (but the outer walls unusually thick), or not in discrete, regular adaxial groups; when present, in simple fans. All the vascular bundles accompanied by sclerenchyma. Combined sclerenchyma girders present (e.g *A. avenaceus*), or absent; when present, forming 'figures' (i.e. in *A. avenaceus*). Sclerenchyma all associated with vascular bundles, or not all bundle-associated (abaxial, continuous or interrupted). The 'extra' sclerenchyma when present, in abaxial groups to in a continuous abaxial layer. The lamina margins with fibres.

Taxonomy. Arundinoideae (?); Amphipogoneae.

Ecology, geography, regional floristic distribution. 8 species. Australia. Xerophytic; species of open habitats. Dry sandy grassland.

Australian. North and East Australian, South-West Australian, and Central Australian. Tropical North and East Australian and Temperate and South-Eastern Australian.

Rusts and smuts. Smuts from Ustilaginaceae. Ustilaginaceae — *Ustilago*.

References, etc. Morphological/taxonomic: Vickery 1950. Leaf anatomical: this project.

Special comments. Subfamilial relationships very problematical — the spikelet form (especially the lemma) and the microhairs are reminiscent of *Enneapogon*.

Anadelphia Hackel

Excluding *Monium, Pobeguinea, 'Anadelphia scyphofera'*

Habit, vegetative morphology. Annual, or perennial; caespitose (when perennial). *Culms* 20–200 cm high; herbaceous; branched above, or unbranched above. Culm nodes exposed; glabrous. Culm leaves present. Upper culm leaf blades fully developed. Young shoots intravaginal. *Leaves* mostly basal, or not basally aggregated; auriculate (or seemingly so, by virtue of the ligule), or non-auriculate. Leaf blades linear; narrow; 0.5–5 mm wide; setaceous (e.g. *A. pumila*), or not setaceous; flat, or acicular; without cross venation; persistent. **Ligule an unfringed membrane (sometimes laterally hardened, the margins auricle-like, when the blade narrow)**; truncate; about 1 mm long. *Contra-ligule* absent.

Reproductive organization. Plants bisexual, with bisexual spikelets; with hermaphrodite florets. The spikelets of sexually distinct forms on the same plant; hermaphrodite (occasionally seemingly all so, by suppression of the pedicelled members), or hermaphrodite and male-only, or hermaphrodite and sterile. The male and female-fertile spikelets mixed in the inflorescence. *The spikelets* overtly heteromorphic (pedicellate awnless, sessile awned); **all in heterogamous combinations**. Plants seemingly outbreeding, or inbreeding. Seemingly chasmogamous (the racemes usually long-peduncled and exserted from the spatheoles), or exposed-cleistogamous (seemingly, in *A. trepidaria*, in which the reduced 'raceme' is enclosed in the spatheole, cf. *Monium*).

Inflorescence. Inflorescence paniculate; open, or contracted; with capillary branchlets; spatheate (and spatheolate); *a complex of 'partial inflorescences' and intervening foliar organs*. **Spikelet-bearing axes** 'racemes' (usually), or very much reduced (to one sessile spikelet with one more or less vestigial partner, in *A. trepidaria* — cf. *Monium*); *the spikelet-bearing axes with 2–3 spikelet-bearing 'articles', or with 4–5 spikelet-bearing 'articles', or with only one spikelet-bearing 'article' (A. trepidaria)*; solitary; with very slender rachides; disarticulating; disarticulating at the joints. *'Articles'* linear; without a basal callus-knob; not appendaged; disarticulating transversely to disarticulating obliquely (less obviously oblique than in *Monium*); densely long-hairy, or somewhat hairy, or glabrous. Spikelets solitary (sometimes, at least seemingly, in *A. trepidaria*), or solitary and paired (the pedicelled members — including their pedicels — often missing, vestigial or much reduced and concealed in the callus hairs, at least in parts of the raceme); not secund; sessile and pedicellate (but the pedicellate members not always conspicuous); consistently in 'long-and-short' combinations (probably usually, if evidence of the pedicelled members is diligently sought), or not in distinct 'long-and-short' combinations (often ostensibly solitary, sometimes genuinely so); when paired, in pedicellate/sessile combinations. Pedicels of the 'pedicellate' spikelets free of the rachis. The 'shorter' spikelets hermaphrodite. The 'longer' spikelets male-only, or sterile.

Female-sterile spikelets. The pedicellate spikelets very variable in development and size, those of (e.g.) *A. pumila* and *A. leptocoma* larger than their sessile partner; awnless or the upper glume shortly awned. The lemmas awnless.

Female-fertile spikelets. Spikelets 3.5–5 mm long; compressed dorsiventrally; falling with the glumes. Rachilla terminated by a female-fertile floret. Hairy callus present. **Callus** usually *short*; blunt (usually), or pointed.

Glumes two; more or less equal; long relative to the adjacent lemmas; hairy, or hairless; without conspicuous tufts or rows of hairs; *awned (G_1 bidentate to bi-setaceous at tip, G_2 more or less long-awned)*; more or less carinate (G_2), or non-carinate (G_1); very dissimilar (both hardened, the G_1 dorsally rounded or flattened, acuminate and slightly bicarinate towards the tip, the G_2 naviculate and keeled towards the tip, its slender awn sometimes long and twisted around the more substantial one of the lemma). Lower glume not two-keeled (except towards the tip); convex on the back to concave on the back; not pitted; relatively smooth; 5–7 nerved (sometimes obscurely so, the median inconspicuous or absent). Upper glume 3 nerved. *Spikelets* with incomplete florets. The incomplete florets proximal to the female-fertile florets. *Spikelets with proximal incomplete florets*. *The proximal incomplete florets* 1; epaleate; sterile. The proximal lemmas awnless (truncate to pointed); 0 nerved, or 2 nerved (palea-like); more or less equalling the female-fertile lemmas to decidedly exceeding the female-fertile lemmas; similar in texture to the female-fertile lemmas (hyaline); not becoming indurated.

Female-fertile florets 1. Lemmas less firm than the glumes (i.e. the wings and lobes of the stipe hyaline); not becoming indurated; incised; 2 lobed; deeply cleft; awned. Awns 1; median; from a sinus; geniculate; hairless; much longer than the body of the lemma (8–50 mm long); entered by one vein. Awn bases twisted; not flattened. Lemmas hairless (sometimes ciliate); non-carinate; without a germination flap; emphatically only 1 nerved.

Palea *absent*. *Lodicules* present (small); 2; free; fleshy; glabrous; not or scarcely vascularized. *Stamens* 3. Anthers about 2 mm long; not penicillate; without an apically prolonged connective. *Ovary* glabrous. Styles fused (basally); joined below. Stigmas 2.

Fruit, embryo and seedling. Fruit compressed dorsiventrally (somewhat, at least in *A. bigeniculata*), or not noticeably compressed; glabrous. Hilum short. Embryo large.

Abaxial leaf blade epidermis. *Costal/intercostal zonation* conspicuous. *Papillae* present, or absent (*A. pumila*); intercostal. Intercostal papillae over-arching the stomata to not over-arching the stomata; consisting of one oblique swelling per cell, or consisting of one symmetrical projection per cell (large, fairly thick walled, surface-ornamented in *A. leptocoma*). *Long-cells* markedly different in shape costally and intercostally (the costals narrower and more regularly rectangular); of similar wall thickness costally and intercostally (fairly thin walled). Mid-intercostal long-cells rectangular; having markedly sinuous walls (*A. pumila*), or having straight or only gently undulating walls, or having markedly sinuous walls and having straight or only gently undulating walls (then sinuous only adjoining the costae). *Microhairs* present; elongated; clearly two-celled; panicoid-type; 33–45 microns long; 6–6.3 microns wide at the septum. Microhair total length/width at septum 5.5–7.1. Microhair apical cells 20–30 microns long. Microhair apical cell/total length ratio 0.59–0.67. *Stomata* common; 21–29 microns long. Subsidiaries non-papillate; variously dome-shaped and triangular. Guard-cells overlapping to flush with the interstomatals. *Intercostal short-cells* common (*A. pumila*), or absent or very rare; of *A. pumila* in cork/silica-cell pairs and not paired. *Costal short-cells* conspicuously in long rows. Costal silica bodies present and well developed; 'panicoid-type'; dumb-bell shaped.

Transverse section of leaf blade, physiology. C_4; XyMS– (but the presumed PCR sheath very thick walled and mestome-sheath like in *A. pumila* and *A. bigeniculata*). Mesophyll traversed by columns of colourless mesophyll cells (*A. leptocoma*), or not traversed by colourless columns. *Leaf blade* 'nodular' in section. *Midrib* conspicuous; with one bundle only to having a conventional arc of bundles (*A. bigeniculata*), or having a conventional arc of bundles; with colourless mesophyll adaxially (this represented by only a few cells in *A. bigeniculata*). *The lamina* symmetrical on either side of the midrib. Bulliforms present in discrete, regular adaxial groups (*A. leptocoma*), or not in discrete, regular adaxial groups; associated with colourless mesophyll cells to form deeply-penetrating fans (these linked with the traversing colourless columns, in *A. leptocoma*); associating with colourless mesophyll cells to form arches over small vascular bundles (seemingly, e.g. in *A. bigeniculata*), or nowhere involved in bulliform-plus-colourless mesophyll arches. Many of the smallest vascular bundles unaccompanied by sclerenchyma, or all the vascular bundles accompanied by sclerenchyma (*A. leptocoma*). Combined sclerenchyma girders present; forming 'figures', or nowhere forming 'figures'. Sclerenchyma all associated with vascular bundles.

Cytology. $2n = 20$.

Taxonomy. Panicoideae; Andropogonodae; Andropogoneae; Andropogoninae.

Ecology, geography, regional floristic distribution. 13 species. Tropical Africa. Savanna, often on shallow soils.

Paleotropical. African. Sudano-Angolan and West African Rainforest. Sahelo-Sudanian and South Tropical African.

References, etc. Leaf anatomical: this project.

Special comments. Contrast the treatments of Jacques-Félix 1962 and Clayton 1966: an effective resolution of generic limits (involving *Anadelphia, Monium* and *Pobeguinea*) awaits acqisition and proper analysis of dependable comparative data — meanwhile, the present descriptions are based on the incomplete sample examined (see the accompanying file).

Anadelphia scyphofera W.D. Clayton

Still referred to *Anadelphia* (but very distinct)

Habit, vegetative morphology. Annual. *Culms* 10–50 cm high; herbaceous; branched above to unbranched above. Culm nodes exposed; glabrous. Culm leaves present. *Leaves* not basally aggregated; ostensibly auriculate from the sheath, when the ligule is split. *Leaf blades* linear; narrow; about *0.5 mm wide*; acicular (triquetrous); without cross venation; persistent. *Ligule* present (of somewhat peculiar constitution, associated with the narrowness of the blade relative to the sheath); an unfringed membrane (leathery below and on the sides, mimicking auricles); not truncate; 2–3 mm long. *Contra-ligule* absent.

Reproductive organization. Plants bisexual, with bisexual spikelets; with hermaphrodite florets. The spikelets of sexually distinct forms on the same plant; hermaphrodite and sterile; overtly heteromorphic; all in heterogamous combinations. Plants presumably inbreeding. Exposed-cleistogamous (seemingly).

Inflorescence. Inflorescence paniculate; open; spatheate (and spatheolate); a complex of 'partial inflorescences' and intervening foliar organs. **Spikelet-bearing axes *very much reduced (each 'raceme' reduced to one triplet, clasped basally by the peculiar, trumpet-shaped tip of the peduncle)*;** the spikelet-bearing axes with only one spikelet-bearing 'article'; solitary; disarticulating; disarticulating at the joints (i.e. at the single articulation). 'Articles' without a basal callus-knob; disarticulating obliquely. Spikelets in triplets; sessile and pedicellate; consistently in 'long-and-short' combinations; in pedicellate/sessile combinations. Pedicels of the 'pedicellate' spikelets free of the rachis. The 'shorter' spikelets hermaphrodite. The 'longer' spikelets sterile.

Female-sterile spikelets. Sterile spikelets 2–4 mm long, reduced to the glumes.

Female-fertile spikelets. Spikelets 6–7 mm long; somewhat compressed dorsiventrally; falling with the glumes (and with the associated pedicellate spikelets). Rachilla terminated by a female-fertile floret. Hairy callus present. *Callus* long; pointed.

Glumes two; more or less equal; exceeding the spikelets; long relative to the adjacent lemmas; hairless; glabrous; awned; non-carinate; very dissimilar (the lower bidentate, awnless, canaliculate above, the upper round-backed to keeled, with a long, slender terminal awn). *Lower glume* two-keeled above, the keels closely apposed to form a groove; convex on the back and sulcate on the back; not pitted; relatively smooth; *6 nerved (with no median)*. Upper glume 3 nerved. *Spikelets* with incomplete florets. The incomplete florets proximal to the female-fertile florets. *The proximal incomplete florets* 1; epaleate; sterile. The proximal lemmas awnless; 2 nerved (palea-like); more or less equalling the female-fertile lemmas (about 3 mm long); less firm than the female-fertile lemmas to similar in texture to the female-fertile lemmas (hyaline); not becoming indurated.

Female-fertile florets 1. Lemmas stipitate below, with hyaline margins and lobes; less firm than the glumes; not becoming indurated; incised; 2 lobed; deeply cleft; awned. Awns 1; median; from a sinus; geniculate; hairless (scabridulous); much longer than the body of the lemma (to 50 mm long). Awn bases twisted; not flattened. Lemmas hairless; glabrous; non-carinate; without a germination flap; 3 nerved; with the nerves non-confluent. *Palea* present; conspicuous but relatively short (about 1.5 mm long); awnless, without apical setae; thinner than the lemma to textured like the lemma (hyaline); not indurated; 2-nerved, or nerveless. Palea back glabrous. *Lodicules* present; 2; fleshy; glabrous. *Ovary* glabrous.

Fruit, embryo and seedling. Fruit small to medium sized (3–4 mm long); ellipsoid; compressed dorsiventrally (slightly, above), or not noticeably compressed; glabrous. Hilum short. Embryo large. Endosperm hard.

Abaxial leaf blade epidermis. *Costal/intercostal zonation* conspicuous. *Papillae* absent. *Long-cells* markedly different in shape costally and intercostally (the costals much narrower); of similar wall thickness costally and intercostally (fairly thick walled). Mid-intercostal long-cells rectangular; having markedly sinuous walls (with conspicuous pits). *Microhairs* present; elongated; clearly two-celled; panicoid-type; 33–42 microns long; 4.5–5.4 microns wide at the septum. Microhair total length/width at septum 6.1–9.3. Microhair apical cells 15–21 microns long. Microhair apical cell/total length ratio 0.45–0.5. *Stomata* common; 24–27 microns long. Subsidiaries dome-shaped and triangular. Guard-cells overlapping to flush with the interstomatals (more or less flush). *Intercostal short-cells* absent or very rare (absent, except for a few pairs at costal/intercostal interfaces). *Costal short-cells* conspicuously in long rows (mostly), or predominantly paired (in some files, and exclusively over the midrib). Costal silica bodies present and well developed; 'panicoid-type'; consistently dumb-bell shaped.

Transverse section of leaf blade, physiology. Leaf blades entirely consisting of midrib. C_4; XyMS–. PCR sheath outlines uneven. PCR sheath extensions absent. Midrib having a conventional arc of bundles (the reduced blade with a large median bundle, two mid-lateral primaries and numerous minor bundles); with colourless mesophyll adaxially (the colourless tissue occupying the whole middle of the acicular blade). Bulliforms not in discrete, regular adaxial groups. All the vascular bundles accompanied by sclerenchyma (in the absence of a conventional lamina). Combined sclerenchyma girders absent (the large bundles with abaxial girders, the small ones with only strands). Sclerenchyma all associated with vascular bundles.

Special diagnostic feature. *Spikelets in much-reduced andropogonoid 'racemes', each of the latter reduced to a single triplet and enclosed at its base by a trumpet-like development of the peduncle tip.*

Taxonomy. Panicoideae; Andropogonodae; Andropogoneae; Andropogoninae.

Ecology, geography, regional floristic distribution. 1 species. Zambia. Mesophytic; species of open habitats; glycophytic.

Paleotropical. African. Sudano-Angolan. South Tropical African.

References, etc. Leaf anatomical: this project.

Special comments. Anomalous in *Anadelphia*; cf. *Clausospicula*.

Ancistrachne S. T. Blake

Habit, vegetative morphology. Perennial (with wiry culms); caespitose, or decumbent (sometimes shrubby, or scrambling). *Culms* 100–200 cm high; woody and persistent to herbaceous; branched above. Culm nodes glabrous. Culm internodes solid. Young shoots extravaginal. *Leaves* not basally aggregated; non-auriculate. *Leaf blades* narrow; without cross venation; *disarticulating from the sheaths*. Ligule present; *a fringe of hairs*.

Reproductive organization. Plants bisexual, with bisexual spikelets; with hermaphrodite florets.

Inflorescence. *Inflorescence paniculate*; *open (narrow, depauperate, the primary branches sometimes simple)*; espatheate; not comprising 'partial inflorescences' and foliar organs. Spikelet-bearing axes persistent. Spikelets not secund. Pedicel apices cupuliform.

Female-fertile spikelets. *Spikelets 2.2–3.5 mm long*; lanceolate; *abaxial*; *compressed dorsiventrally*; falling with the glumes. *Rachilla terminated by a female-fertile floret*. Hairy callus absent.

Glumes present; two; very unequal; (the upper) long relative to the adjacent lemmas; dorsiventral to the rachis (the lower abaxial); *hairy (hispid with hooked or flexuous hairs, the upper more so)*; pointed; awnless; non-carinate. Lower glume 0–5 nerved (from vestigial to half the spikelet length). Upper glume 5–11 nerved. *Spikelets with incomplete florets*. The incomplete florets proximal to the female-fertile florets. *Spikelets with proximal incomplete florets*. *The proximal incomplete florets* 1; paleate. Palea of the proximal incomplete florets reduced; not becoming conspicuously hardened and enlarged laterally. The proximal incomplete florets sterile. *The proximal lemmas hispid, the hairs hooked or flexuous*; awnless; *7–9 nerved*; more or less equalling the female-fertile lemmas to decidedly exceeding the female-fertile lemmas; less firm than the female-fertile lemmas (leathery); not becoming indurated.

Female-fertile florets 1. Lemmas decidedly firmer than the glumes; smooth; becoming indurated to not becoming indurated (cartilaginous to crustaceous); yellow in fruit, or brown in fruit; entire; pointed, or blunt; awnless to mucronate (the apex incurved); hairless; non-carinate; having the margins lying flat and exposed on the palea; with a clear germination flap; 5–7 nerved. *Palea* present; relatively long; entire; awnless, without apical setae; textured like the lemma; indurated, or not indurated; 2-nerved. *Lodicules* present; 2; free; fleshy; glabrous. *Stamens* 3. Anthers about 2 mm long; not penicillate. *Ovary* glabrous. Styles free to their bases. Stigmas 2.

Fruit, embryo and seedling. Fruit small; compressed dorsiventrally. Hilum short. Pericarp fused. Embryo large. Endosperm containing only simple starch grains. Embryo without an epiblast; with a scutellar tail; with an elongated mesocotyl internode. Embryonic leaf margins overlapping.

Abaxial leaf blade epidermis. *Costal/intercostal zonation* conspicuous. *Papillae* absent. *Long-cells* similar in shape costally and intercostally (but the intercostals shorter); of similar wall thickness costally and intercostally. Mid-intercostal long-cells rectangular; having markedly sinuous walls. *Microhairs* present; panicoid-type; 48–63 microns long; 6–7.5 microns wide at the septum. Microhair total length/width at septum 6.4–10.5. Microhair apical cells 21–24 microns long. Microhair apical cell/total length ratio 0.35–0.44. *Stomata* common; 28.5–31.5 microns long. Subsidiaries triangular. Guard-cells overlapping to flush with the interstomatals. *Intercostal short-cells* absent or very rare. *Costal short-cells* conspicuously in long rows. Costal silica bodies 'panicoid-type'.

Transverse section of leaf blade, physiology. C_3; XyMS+. *Mesophyll* with radiate chlorenchyma; without adaxial palisade; *Isachne*-type. *Leaf blade* adaxially flat. *Midrib* not readily distinguishable; with one bundle only. Bulliforms present in discrete, regular adaxial groups; in simple fans. All the vascular bundles accompanied by sclerenchyma. Combined sclerenchyma girders present; forming 'figures'. Sclerenchyma all associated with vascular bundles.

Taxonomy. Panicoideae; Panicodae; Paniceae.

Ecology, geography, regional floristic distribution. 4 species. Philippines, eastern Australia, New Caledonia. Shade species. Forests.

Paleotropical and Australian. Indomalesian, Polynesian, and Neocaledonian. Malesian. Fijian. North and East Australian. Tropical North and East Australian.

References, etc. Morphological/taxonomic: Blake 1941. Leaf anatomical: this project.

Special comments. Fruit data wanting.

Ancistragrostis S. T. Blake

~ *Deyeuxia* (*D. uncinioides* (S.T Blake) Royen & Veldk.)

Habit, vegetative morphology. Perennial; caespitose. **Culms 6–10 cm high; herbaceous**; unbranched above; 2 noded. Young shoots intravaginal. *Leaves* mostly basal; non-auriculate. Leaf blades linear; narrow; 0.4–0.6 mm wide; setaceous; rolled (convolute); without cross venation; persistent. Ligule an unfringed membrane; not truncate (ovate, lacerate); 1.1–1.5 mm long.

Reproductive organization. Plants bisexual, with bisexual spikelets; with hermaphrodite florets.

Inflorescence. Inflorescence paniculate; contracted; spicate; espatheate; not comprising 'partial inflorescences' and foliar organs. Spikelet-bearing axes persistent. Spikelets not secund; pedicellate (the pedicels 0.5–0.8 mm long); imbricate.

Female-fertile spikelets. Spikelets 3.5–4 mm long; compressed laterally; disarticulating above the glumes. Rachilla prolonged beyond the uppermost female-fertile floret; rachilla prolongation hairy. The rachilla extension naked. Hairy callus present. *Callus* short; blunt.

Glumes two; more or less equal (subequal); shorter than the spikelets; shorter than the adjacent lemmas; hairless; scabrous; subacute; awnless; carinate; similar (thinly cartilaginous). Lower glume shorter than the lowest lemma; 1 nerved. Upper glume 1 nerved, or 3 nerved. *Spikelets* with female-fertile florets only.

Female-fertile florets 1. *Lemmas convolute*; similar in texture to the glumes (thinly cartilaginous); incised; 2 lobed (minutely bilobed); not deeply cleft; awned. *Awns* 1; median; dorsal; from near the top; *hooked*; non-geniculate; hairless; much shorter than the body of the lemma (0.7 – 1.2 mm long); entered by one vein; persistent. *Lemmas* hairless; glabrous; *carinate*; 5 nerved; with the nerves non-confluent. *Palea* present; conspicuous but relatively short (about 2/3 of the lemma length); entire (obtuse); awnless, without apical setae; textured like the lemma; not indurated; 2-nerved; 2-keeled. Palea keels wingless; glabrous. *Lodicules* present; 2; free; membranous; glabrous; not toothed;

not or scarcely vascularized (hyaline). *Stamens* 3. Anthers 0.6 mm long; not penicillate; without an apically prolonged connective. *Ovary* glabrous. Styles fused. Stigmas 2.

Abaxial leaf blade epidermis. *Costal/intercostal zonation* conspicuous. *Papillae* absent. Mid-intercostal long-cells rectangular and fusiform; having markedly sinuous walls. *Microhairs* absent. *Stomata* common. Subsidiaries predominantly dome-shaped. Guard-cells overlapping to flush with the interstomatals (mostly more or less flush). *Intercostal short-cells* common; in cork/silica-cell pairs; silicified. Intercostal silica bodies mostly crescentic. *Costal short-cells* predominantly paired, or neither distinctly grouped into long rows nor predominantly paired (a few short rows). Costal silica bodies horizontally-elongated crenate/sinuous (predominating over most veins: a short variant of the crenate type), or tall-and-narrow to crescentic (a few).

Transverse section of leaf blade, physiology. C_3; XyMS+. *Mesophyll* with non-radiate chlorenchyma; without adaxial palisade. *Leaf blade* with distinct, prominent adaxial ribs (the tops rounded); with the ribs more or less constant in size. *Midrib* not readily distinguishable (except by its location); with one bundle only. Bulliforms present in discrete, regular adaxial groups; in simple fans (in each furrow). All the vascular bundles accompanied by sclerenchyma. Combined sclerenchyma girders absent (adaxially with strands only, abaxially with girders and strands). Sclerenchyma all associated with vascular bundles.

Taxonomy. Pooideae; Poodae; Aveneae.

Ecology, geography, regional floristic distribution. 2 species. Montane New Guinea. Mesophytic. Damp places and montane grassland.

Paleotropical. Indomalesian. Papuan.

References, etc. Morphological/taxonomic: Blake 1946. Leaf anatomical: this project.

Special comments. Fruit data wanting.

Andropogon L.

Anatherum P. Beauv., *Arthrostachys* Desv., *Arthrolophis* (Trin.) Chiov., *Dimeiostemon* Raf., *Eriopodium* Hochst., *Heterochloa* Desv., *Homoeatherum* Nees, *Leptopogon* Roberty

Excluding *Hypogynium*

Habit, vegetative morphology. Annual, or perennial; rhizomatous, or caespitose, or decumbent. *Culms* 8–250(–430) cm high; herbaceous; branched above, or unbranched above. Culm internodes solid. *The shoots not aromatic*. Leaves non-auriculate. Leaf blades linear; broad, or narrow; setaceous, or not setaceous; pseudopetiolate, or not pseudopetiolate; without cross venation; rolled in bud, or once-folded in bud. Ligule an unfringed membrane to a fringed membrane. *Contra-ligule* present (occasionally, of hairs, e.g. *A. gayanus*), or absent.

Reproductive organization. Plants bisexual, with bisexual spikelets; with hermaphrodite florets. The spikelets of sexually distinct forms on the same plant; hermaphrodite and male-only, or hermaphrodite and sterile (the pedicelled spiklet occasionally suppressed); overtly heteromorphic; in both homogamous and heterogamous combinations (the lowermost pair imperfect and homogamous), or all in heterogamous combinations. Plants outbreeding; with hidden cleistogenes, or without hidden cleistogenes. The hidden cleistogenes when present, in the leaf sheaths.

Inflorescence. *Inflorescence of spicate main branches, or paniculate (usually with paired or digitate 'racemes', these often spatheate and aggregated into false panicles)*; usually *spatheate*; a complex of 'partial inflorescences' and intervening foliar organs (often), or not comprising 'partial inflorescences' and foliar organs. **Spikelet-bearing axes** 'racemes'; *paired (nearly always), or clustered (very rarely only one, the raceme bases terete, often long, hardly ever deflexed)*; with very slender rachides, or with substantial rachides; disarticulating; disarticulating at the joints. 'Articles' without a basal callus-knob; appendaged, or not appendaged; densely long-hairy (plumose), or somewhat hairy (ciliate), or glabrous (rarely). *Spikelets paired*; sessile and pedicellate; consistently in 'long-and-short' combinations; in pedicellate/sessile combinations. Pedicels of the 'pedicellate' spikelets free of the rachis. The 'shorter' spikelets hermaphrodite. The 'longer' spikelets male-only, or sterile (sometimes reduced to their pedicels).

Female-sterile spikelets. The pedicelled spikelets male or barren, usually awnless, occasionally suppressed. The lemmas usually awnless.

Female-fertile spikelets. Spikelets compressed laterally, or not noticeably compressed, or compressed dorsiventrally; falling with the glumes. Rachilla terminated by a female-fertile floret. Hairy callus present. **Callus** *short*; blunt.

Glumes two; more or less equal; long relative to the adjacent lemmas; hairy, or hairless; without conspicuous tufts or rows of hairs; awned, or awnless (G_2 sometimes aristate); very dissimilar (subleathery to membranous, the lower flat, concave or canaliculate on the back, its margins folded and 2-keeled, the upper naviculate, carinate above). *Lower glume two-keeled*; flattened on the back to sulcate on the back; not pitted; relatively smooth; 1–11 nerved (sometimes lacking the midnerve). Upper glume 1–3 nerved. *Spikelets* with incomplete florets. The incomplete florets proximal to the female-fertile florets. *The proximal incomplete florets* 1; epaleate; sterile. The proximal lemmas awnless; 2 nerved; more or less equalling the female-fertile lemmas to decidedly exceeding the female-fertile lemmas; less firm than the female-fertile lemmas to similar in texture to the female-fertile lemmas (hyaline); not becoming indurated.

Female-fertile florets 1. **Lemmas** less firm than the glumes (hyaline to firm, sometimes substipitate beneath the awn); not becoming indurated; *incised (usually bifid)*; awned. Awns 1; median; from a sinus; geniculate; hairless (glabrous), or hairy (puberulous); about as long as the body of the lemma to much longer than the body of the lemma. Lemmas hairy, or hairless (often ciliate or ciliolate); non-carinate; 1–3 nerved. **Palea** present; *very reduced (hyaline)*; not indurated; nerveless. *Lodicules* present (tiny); 2; free; fleshy; ciliate, or glabrous. *Stamens* 1–3. Anthers penicillate, or not penicillate. *Ovary* glabrous. Stigmas 2.

Fruit, embryo and seedling. *Fruit* free from both lemma and palea; small; compressed laterally, or compressed dorsiventrally, or not noticeably compressed (subterete to planoconvex). Hilum short. Embryo large; waisted. Endosperm hard; without lipid. Embryo with an epiblast (e.g., *A. gayanus*), or without an epiblast; with a scutellar tail; with an elongated mesocotyl internode. Embryonic leaf margins overlapping.

Seedling with a long mesocotyl. First seedling leaf with a well-developed lamina.

Abaxial leaf blade epidermis. *Costal/intercostal zonation* conspicuous. *Papillae* absent. Mid-intercostal long-cells rectangular; having markedly sinuous walls, or having straight or only gently undulating walls. *Microhairs* present; panicoid-type; (24–)27–55 microns long; 6–7.5 microns wide at the septum. Microhair total length/width at septum 3.2–5. Microhair apical cells 15–30 microns long. Microhair apical cell/total length ratio 0.52–0.63. *Stomata* common; 27–30 microns long. Subsidiaries triangular, or dome-shaped and triangular. Guard-cells overlapping to flush with the interstomatals. *Intercostal short-cells* common, or absent or very rare; not paired (mainly solitary); not silicified. *Costal short-cells* conspicuously in long rows. Costal silica bodies 'panicoid-type'; cross shaped, or butterfly shaped, or dumb-bell shaped.

Transverse section of leaf blade, physiology. C_4; biochemical type NADP–ME (3 species); XyMS–. PCR cells with a suberised lamella. PCR cell chloroplasts with reduced grana; centrifugal/peripheral. *Mesophyll* with radiate chlorenchyma, or with non-radiate chlorenchyma; traversed by columns of colourless mesophyll cells, or not traversed by colourless columns. *Leaf blade* adaxially flat. *Midrib* conspicuous; having a conventional arc of bundles; with colourless mesophyll adaxially. Bulliforms present in discrete, regular adaxial groups, or not in discrete, regular adaxial groups; in irregular groups, or the epidermis extensively bulliform, or associated with colourless mesophyll cells to form deeply-penetrating fans (then sometimes associated with traversing colourless columns). Many of the smallest vascular bundles unaccompanied by sclerenchyma, or all the vascular bundles accompanied by sclerenchyma. Combined sclerenchyma girders present; nowhere forming 'figures'. Sclerenchyma all associated with vascular bundles.

Culm anatomy. *Culm internode bundles* scattered.

Phytochemistry. Tissues of the culm bases with abundant starch. Leaves without flavonoid sulphates (5 species).

Cytology. Chromosome base number, $x = 5$ and 10. $2n = 20$, 40, 60, 100, 120, and 180. 2–18 ploid. Nucleoli persistent.

Taxonomy. Panicoideae; Andropogonodae; Andropogoneae; Andropogoninae.

Ecology, geography, regional floristic distribution. About 100 species. Tropical. Commonly adventive. Mesophytic, or xerophytic. Mostly savanna, some in tropical highlands.

Holarctic, Paleotropical, Neotropical, and Cape. Boreal, Tethyan, and Madrean. African, Madagascan, and Indomalesian. Euro-Siberian, Eastern Asian, Atlantic North American, and Rocky Mountains. Macaronesian, Mediterranean, and Irano-Turanian. Saharo-Sindian, Sudano-Angolan, West African Rainforest, and Namib-Karoo. Indian, Indo-Chinese, and Papuan. Caribbean, Venezuela and Surinam, Amazon, Central Brazilian, Pampas, and Andean. European. Canadian-Appalachian, Southern Atlantic North American, and Central Grasslands. Sahelo-Sudanian, Somalo-Ethiopian, South Tropical African, and Kalaharian.

Rusts and smuts. Rusts — *Phakopsora* and *Puccinia*. Taxonomically wide-ranging species: *Phakopsora incompleta, Puccinia microspora, Puccinia nakanishikii, Puccinia eritraeensis, Puccinia versicolor*, and '*Uromyces*' *clignyi*. Smuts from Ustilaginaceae. Ustilaginaceae — *Sorosporium, Sphacelotheca, Tolyposporella*, and *Ustilago*.

Economic importance. Significant weed species: *A. bicornis, A. condensatus, A. gerardii, A. glomeratus, A. lateralis, A. scoparius, A. selloanus, A. ternarius, A. virginicus*. Cultivated fodder: *A. gayanus*. Important native pasture species: *A. gerardi, A. scoparius* (North America); *A. abyssinicus, A. chinensis, A. distachyos, A. lima* (Kenya); etc.

References, etc. Leaf anatomical: Metcalfe 1960; this project.

Andropterum Stapf

Habit, vegetative morphology. Perennial; decumbent. *Culms* about 20–100 cm high; herbaceous. *Leaves* not basally aggregated; non-auriculate. Leaf blades broadly linear (and acuminate); broad to narrow; to 15 mm wide; flat; without cross venation; persistent. *Ligule a fringe of hairs*.

Reproductive organization. Plants bisexual, with bisexual spikelets; with hermaphrodite florets. *The spikelets* of sexually distinct forms on the same plant; hermaphrodite and female-only, or male-only (sessile hermaphrodite, pedicelled male or sterile?); *overtly heteromorphic (the pedicelled spikelet larger, glabrous, asymmetrically winged via the G_1, awnless)*.

Inflorescence. *Inflorescence a single raceme*; espatheate; not comprising 'partial inflorescences' and foliar organs. **Spikelet-bearing axes** *'racemes' (with many spikelets)*; solitary; with substantial rachides; *disarticulating*; disarticulating at the joints. *'Articles'* non-linear (more or less clavate); ciliate. Spikelets paired; sessile and pedicellate; consistently in 'long-and-short' combinations; in pedicellate/sessile combinations. Pedicels of the 'pedicellate' spikelets free of the rachis. The 'shorter' spikelets hermaphrodite. The 'longer' spikelets hermaphrodite.

Female-fertile spikelets. *Spikelets* (the sessile members, to which this description is confined) 5–8 mm long; *compressed laterally (compressed between the internode and the pedicel)*; falling with the glumes. Rachilla terminated by a female-fertile floret.

Glumes two; long relative to the adjacent lemmas; awnless (G_1 bidentate, G_2 muticous or mucronate); very dissimilar (G_1 2-keeled and grooved, G_2 naviculate with a winged crest above). *Lower glume two-keeled*; sulcate on the back (the keels closely apposed). *Spikelets* with incomplete florets. The incomplete florets proximal to the female-fertile florets. *The proximal incomplete florets* 1; paleate; male. The proximal lemmas awnless.

Female-fertile florets 1. Lemmas incised; deeply cleft (bifid to 3/4 of their length); awned. Awns 1; median; from a sinus; geniculate; hairless (glabrous); much longer than the body of the lemma (8–15 mm long). Lemmas hairless; non-carinate; without a germination flap.

Abaxial leaf blade epidermis. *Costal/intercostal zonation* conspicuous. *Papillae* absent. *Long-cells* markedly different in shape costally and intercostally (the costals narrower); of similar wall thickness costally and intercostally. Mid-intercostal long-cells rectangular; having markedly sinuous walls. *Microhairs* present; elongated; clearly two-celled; panicoid-type; 42–45.6–48 microns long; 4.8–6–7.5 microns wide at the septum. Microhair total length/width at septum 5.6–7.7–9.4. Microhair apical cells 12–16.8–21 microns long. Microhair apical cell/total length ratio 0.29–0.37–0.44. *Stomata* common; 24–30 microns long. Subsidiaries dome-shaped and triangular. Guard-cells overlapping to flush with the interstomatals. *Intercostal short-cells* fairly common; not paired (solitary). Exhibiting abundant, short-tipped prickles with almost spherical bases, and large macrohairs with conspicuous basal rosettes. *Crown cells* absent. *Costal short-cells* conspicuously in long rows. Costal silica bodies absent to poorly developed; (these, or at least the silica cells) 'panicoid-type'; nodular.

Transverse section of leaf blade, physiology. C_4; XyMS–. PCR sheath outlines even. PCR sheath extensions absent. *Mesophyll* with radiate chlorenchyma. Midrib having a conventional arc of bundles (one large median, and several small ones on either side, nearer the abaxial surface); with colourless mesophyll adaxially (the large adaxial 'colourless' mass more or less lignified, and merging into fibre groups adaxial to the median bundle and beneath the adaxial epidermis). *The lamina* symmetrical on either side of the midrib. Bulliforms not in discrete, regular adaxial groups (for the most part — the epidermis extensively bulliform). Many of the smallest vascular bundles unaccompanied by sclerenchyma. Combined sclerenchyma girders present (but only in a few of the main bundles — the rest with strands only). Sclerenchyma all associated with vascular bundles.

Taxonomy. Panicoideae; Andropogonodae; Andropogoneae; Andropogoninae.

Ecology, geography, regional floristic distribution. 1 species. Tropical Africa. Mesophytic. Forest margins.

Paleotropical. African. Sudano-Angolan. South Tropical African.

References, etc. Leaf anatomical: this project.

Special comments. Note that spikelet details are given here only for the sessile spikelet. Fruit data wanting.

Anemanthele Veldk.

~ *Stipa*

Habit, vegetative morphology. Perennial; densely caespitose (from a short, cataphyllous rhizome). *Culms* 70–150 cm high; herbaceous; unbranched above. Culm nodes glabrous. Young shoots intravaginal. *Leaves* mostly basal; non-auriculate. Leaf blades narrow (up to 8 mm wide, to 40 cm long, stiff); flat, or rolled (flat to involute); without cross venation; persistent. *Ligule an unfringed membrane*; truncate; 1–1.5 mm long.

Reproductive organization. Plants bisexual, with bisexual spikelets; with hermaphrodite florets.

Inflorescence. *Inflorescence paniculate*; open (large, lax, the branches in verticils of 5 to 8); with capillary branchlets; espatheate; not comprising 'partial inflorescences' and foliar organs. Spikelet-bearing axes persistent. Spikelets not secund; pedicellate.

Female-fertile spikelets. *Spikelets 2.75–3.5 mm long*; greenish purple to leaden; compressed laterally to not noticeably compressed; disarticulating above the glumes. Rachilla terminated by a female-fertile floret. *Hairy callus present (small)*. Callus short (0.15–0.2 mm); blunt (obtuse, conical, straight).

Glumes two; very unequal to more or less equal (G_1 2.5–3 mm, G_2 2.75–3.5 mm); exceeding the spikelets; (the upper) long relative to the adjacent lemmas; hairless; glabrous (apart from the scabrous midrib); pointed (or the G_2 apically erose); awnless (G_2 with the midrib slightly excurrent); carinate; similar (scarious). *Lower glume 1 nerved.* Upper glume 1 nerved, or 3 nerved. **Spikelets** with female-fertile florets only; *without proximal incomplete florets*.

Female-fertile florets 1. Lemmas decidedly firmer than the glumes ('persistently membranous'); not becoming indurated; incised; 2 lobed; not deeply cleft (minutely bilobed at the tip); awned. Awns 1; median; from a sinus (i.e., from between the minute lobes); non-geniculate; entered by several veins (by all three); deciduous. Lemmas hairless; glabrous (to scaberulous); non-carinate (?); 3 nerved; with the nerves confluent towards the tip. *Palea* present (only its upper part covered, at anthesis, by the overlapping upper margins of the lemma); relatively long to conspicuous but relatively short (1.5–1.6 mm, compared with the 2–2.25 mm lemma); apically erose to slightly fimbriate; awnless, without apical setae; 2-nerved (weakly so); keel-less (neither keeled nor furrowed). *Lodicules* present; 2; free (but closely set and imbricate); membranous; glabrous; not toothed (lanceolate lingular); not or scarcely vascularized. **Stamens** *1*. Anthers about 1 mm long (yellow); not penicillate. *Ovary* glabrous. Styles free to their bases. Stigmas 2.

Fruit, embryo and seedling. *Fruit* free from both lemma and palea (exposed between the divergent lemma and palea); small (about 1.5 mm long, brown); fusiform; smooth (finely reticulate). Hilum short (in the lower 1/6, elliptic). Embryo large to small (about 0.3 times the grain length); with an epiblast; without a scutellar tail (the primary root bent away from the main axis); with a negligible mesocotyl internode. Embryonic leaf margins meeting.

Abaxial leaf blade epidermis. *Costal/intercostal zonation* lacking. *Papillae* absent. *Microhairs* absent. *Stomata* absent or very rare. *Intercostal short-cells* common; in cork/silica-cell pairs. *Costal short-cells* predominantly paired (with a few threes). Costal silica bodies horizontally-elongated smooth to rounded (more or less circular, irregularly isodiametric, oval to slightly elongated, and irregular).

Transverse section of leaf blade, physiology. C_3; XyMS+. *Mesophyll* with non-radiate chlorenchyma. *Leaf blade* with distinct, prominent adaxial ribs; with the ribs very irregular in sizes (of two size orders). *Midrib* conspicuous; with one bundle only. Bulliforms present in discrete, regular adaxial groups (lining the adaxial grooves). All the vascular bundles accompanied by sclerenchyma. Combined sclerenchyma girders present; forming 'figures' (the major bundles with T's).

Taxonomy. Arundinoideae; Stipeae.

Ecology, geography, regional floristic distribution. 1 species. New Zealand. Mesophytic (at the edges of streamlets, up to 460 m).

Antarctic. New Zealand.

References, etc. Morphological/taxonomic: Veldkamp 1985.

Aniselytron Merr.

Aulacolepis Hack., *Neoaulacolepis* Rauschert

~ *Calamagrostis*

Habit, vegetative morphology. Perennial; caespitose, or decumbent. *Culms* 40–100 cm high; herbaceous; unbranched above. Young shoots extravaginal. *Leaves* not basally aggregated; non-auriculate. Leaf blades linear; broad to narrow; 2.3–21 mm wide; flat; persistent. Ligule an unfringed membrane (sometimes ciliolate); truncate to not truncate (collar shaped to triangular); 0.3–7.5 mm long.

Reproductive organization. Plants bisexual, with bisexual spikelets; with hermaphrodite florets. Exposed-cleistogamous, or chasmogamous (?).

Inflorescence. *Inflorescence paniculate*; open; espatheate; not comprising 'partial inflorescences' and foliar organs. Spikelet-bearing axes persistent. Spikelets not secund; pedicellate; not in distinct 'long-and-short' combinations.

Female-fertile spikelets. *Spikelets* 2.4–4.6 mm long; at least somewhat compressed laterally; *disarticulating above the glumes*. Rachilla prolonged beyond the uppermost female-fertile floret; rachilla hairless (glabrous). The rachilla extension naked. *Hairy callus present*. *Callus* short; blunt.

Glumes two; very unequal (the G_1 being very small or rudimentary); shorter than the spikelets; shorter than the adjacent lemmas; hairless; glabrous to scabrous; pointed; awnless; carinate; *very dissimilar (the G_1 much shorter, sometimes a mere scale)*. Lower glume 0 nerved, or 1 nerved. Upper glume 1 nerved, or 3 nerved. **Spikelets** with female-fertile florets only; *without proximal incomplete florets*.

Female-fertile florets *1*. *Lemmas* acute; decidedly firmer than the glumes (leathery); not becoming indurated; entire; pointed; *awnless to mucronate (attenuate, sometimes into a hooked or filiform mucro)*; hairless; scabrous; *carinate*; without a germination flap; 5 nerved. *Palea* present; relatively long; entire (narrow); awnless, without apical setae; 2-nerved; 2-keeled (the rachilla prolongation residing between the keels). Palea keels wingless. *Lodicules* present; 2; free; membranous; glabrous; toothed, or not toothed; not or scarcely vascularized. **Stamens 3**. Anthers 0.6–2 mm long (yellow); not penicillate; without an apically prolonged connective. *Ovary* glabrous. Styles free to their bases. Stigmas 2.

Fruit, embryo and seedling. *Fruit* free from both lemma and palea; small (1.25–2 mm long); compressed laterally, or not noticeably compressed. Hilum short. Embryo small. Endosperm containing compound starch grains.

Abaxial leaf blade epidermis. *Costal/intercostal zonation* conspicuous. *Papillae* absent. *Long-cells* markedly different in shape costally and intercostally; differing markedly in wall thickness costally and intercostally (the costals thicker walled). Mid-intercostal long-cells fusiform; having straight or only gently undulating walls (to very minutely so). *Microhairs* absent. *Stomata* absent or very rare. *Intercostal short-cells* fairly common (adjoining the bases of adjacent prickles); not paired; not silicified. With numerous bulbous-based prickles. *Costal short-cells* neither distinctly grouped into long rows nor predominantly paired. Costal silica bodies horizontally-elongated smooth, or horizontally-elongated crenate/sinuous (a few being slightly sinuous).

Transverse section of leaf blade, physiology. C_3; XyMS+. *Mesophyll* with non-radiate chlorenchyma; without adaxial palisade. *Leaf blade* with distinct, prominent adaxial ribs to 'nodular' in section (the ribs round, low-topped); with the ribs more or less constant in size. *Midrib* conspicuous (via a large bundle and a slight abaxial keel); with one bundle only, or having a conventional arc of bundles (depending on interpretation — there being a small bundle on each side and fairly close to the large median). Bulliforms present in discrete, regular adaxial groups (in the broad furrows); in simple fans (the fans broad). Many of the smallest vascular bundles unaccompanied by sclerenchyma (and others with only negligible 'strands'). Combined sclerenchyma girders present (with the midrib and the other primaries); forming 'figures' (the median midrib bundle). Sclerenchyma all associated with vascular bundles.

Cytology. Chromosome base number, $x = 7$; 6 ploid.

Taxonomy. Pooideae; Poodae; Aveneae.

Ecology, geography, regional floristic distribution. 2 species. Sikkim, northern Burma, China, Japan, Taiwan, Luzon, Sabah, northern Sumatra. Mesophytic; shade species. Montane.

Holarctic and Paleotropical. Boreal. Indomalesian. Eastern Asian. Indo-Chinese and Malesian.

References, etc. Morphological/taxonomic: Korthoff and Veldkamp 1984. Leaf anatomical: this project.

Anisopogon R.Br.

Habit, vegetative morphology. Oat-like perennial; caespitose. **Culms** 60–110 cm high; *herbaceous*. Culm nodes glabrous. Culm internodes hollow. *Leaves* mostly basal; non-auriculate. Leaf blades linear-lanceolate; narrow; 2–3 mm wide; flat to rolled; without cross venation; persistent. Ligule a fringed membrane (the membrane short to relatively long, hairy on the back). *Contra-ligule* absent.

Reproductive organization. *Plants bisexual, with bisexual spikelets*; with hermaphrodite florets.

Inflorescence. Inflorescence few spikeleted (the spikelets large); paniculate; open; espatheate; not comprising 'partial inflorescences' and foliar organs. Spikelet-bearing axes persistent. Spikelets not secund; pedicellate.

Female-fertile spikelets. *Spikelets 40–60 mm long*; compressed laterally; disarticulating above the glumes. *Rachilla prolonged beyond the uppermost female-fertile floret (the slender prolongation 6–8 mm long, in the palea groove)*. Hairy callus present (about 6 mm long, with dense, white hairs).

Glumes two; relatively large; more or less equal (the upper slightly longer); exceeding the spikelets; long relative to the adjacent lemmas; pointed (lanceolate, attenuate); awnless; non-carinate; similar (3–5 cm long, rather flat and leafy). Lower glume 7–9 nerved. Upper glume 7–9 nerved. *Spikelets* with female-fertile florets only, or with incomplete florets (usually with a single, abortive floret). The incomplete florets distal to the female-fertile florets. *The distal incomplete florets* merely underdeveloped (minute).

Female-fertile florets *1. Lemmas decidedly firmer than the glumes*; not becoming indurated (but very firmly leathery); incised; 3 lobed; awned. Awns 3; median and lateral; the median different in form from the laterals; from a sinus; geniculate; hairless to hairy; much longer than the body of the lemma (up to 8 cm long); entered by several veins (5); persistent. The lateral awns shorter than the median (15–25 mm long, straight to twisted but non-geniculate, relatively slender). Lemmas hairy (with dense, white hairs); non-carinate (rounded on the back); without a germination flap; *3 nerved*. **Palea** present; relatively long (narrow, lanceolate); apically notched; awnless, without apical setae; textured like the lemma; firm, except for the membranous, bifid tip; 2-nerved; *2-keeled (the keels closely apposed below, diverging towards the tip)*. Palea keels wingless. *Lodicules* present; 3; free; membranous (two 'stipoid' and larger, the third smaller and flimsier); glabrous; not toothed. *Stamens* 3. Anthers 10–14 mm long; not penicillate; without an apically prolonged connective. **Ovary** *hairy (sparingly so, towards the apex)*. Styles free to their bases. Stigmas 2, or 3; white.

Fruit, embryo and seedling. Fruit medium sized; not noticeably compressed. Hilum long-linear. Pericarp free, or loosely adherent. Embryo small; waisted. Endosperm containing only simple starch grains.

Abaxial leaf blade epidermis. *Costal/intercostal zonation* conspicuous. *Papillae* absent. *Long-cells* similar in shape costally and intercostally (or the costals narrower); of similar wall thickness costally and intercostally (or the costals somewhat thicker-walled). Mid-intercostal long-cells rectangular; having markedly sinuous walls and having straight or only gently undulating walls (conspicuously pitted). *Microhairs* absent. *Stomata* absent or very rare; 18–27 microns long. Guard-cells overlapping to flush with the interstomatals (i.e., in the few seen). *Intercostal short-cells* common; in cork/silica-cell pairs. Heavily pitted prickle bases abundant. *Crown cells* absent. *Costal short-cells* varying in arrangement from file to file, conspicuously in long rows, or predominantly paired, or neither distinctly grouped into long rows nor predominantly paired. Costal silica bodies 'panicoid-type' (mostly), or tall-and-narrow (a few); predominantly short dumbbell shaped.

Transverse section of leaf blade, physiology. C_3; XyMS+. *Mesophyll* with non-radiate chlorenchyma; without adaxial palisade. *Leaf blade* with distinct, prominent adaxial ribs; with the ribs very irregular in sizes (small, round-topped ribs alternating with large, flat-topped ones). *Midrib* not readily distinguishable; with one bundle only. Bulliforms present in discrete, regular adaxial groups (one group in each furrow); in simple fans. All the vascular bundles accompanied by sclerenchyma. Combined sclerenchyma girders present; forming 'figures' (most bundles with a conspicuous 'I'). Sclerenchyma not all bundle-associated. The 'extra' sclerenchyma in abaxial groups, or in a continuous abaxial layer.

Taxonomy. Arundinoideae; Stipeae.

Ecology, geography, regional floristic distribution. 1 species. Australia. Xerophytic; species of open habitats; glycophytic. Light *Eucalyptus* forest and heathland.

Australian. North and East Australian. Temperate and South-Eastern Australian.

References, etc. Leaf anatomical: this project.

Anomochloa Brongn.

Habit, vegetative morphology. Perennial; rhizomatous and caespitose. *Culms* 30–50 cm high; herbaceous; unbranched above. Culm internodes solid. Rhizomes pachymorph. *Leaves* mostly basal; auriculate (on the sheaths); without auricular setae. *Leaf blades* narrowly lanceolate (to oblong-lanceolate); broad; (40–)60–100 mm wide (18–40 cm long); cordate; flat; *pseudopetiolate (the 'petioles' up to 25 cm)*; palmately veined; cross veined; seemingly persistent (but more or less 'articulated'); rolled in bud. Ligule a fringe of hairs. *Contra-ligule* absent.

Reproductive organization. Plants bisexual, with bisexual spikelets; with hermaphrodite florets.

Inflorescence. *Inflorescence spicate, cymoid, with 2–5 spikelets in the axil of each bract*; spatheate (and spatheolate); a complex of 'partial inflorescences' and intervening foliar organs. Spikelet-bearing axes persistent. Spikelets solitary; secund (in the partial inflorescences); pedicellate.

Female-fertile spikelets. Spikelets *unconventional (apparently lacking glumes, exhibiting an upper and a lower 'bract' and a 'perigonate annulus', all presenting interpretive problems)*. Rachilla terminated by a female-fertile floret.
Glumes absent. *Spikelets* with female-fertile florets only.
Female-fertile florets 1. Lemmas (if the lower 'bract' is seen as such) leathery, ovate, tessellate-nerved; not becoming indurated; non-carinate; without a germination flap; many nerved. **Palea** (if the 'upper bract' be so interpreted) present; *forming a closed tube below, enclosing the flower*; several nerved; keel-less. *Lodicules* present; joined (confluent, in the form of an annulus — if the annulus be so interpreted); ciliate (in that the annulus is fringed). *Stamens* 4. Anthers 3–5 mm long. *Ovary* glabrous. Stigmas 1.

Fruit, embryo and seedling. Fruit medium sized (up to 10 mm long); oblong to rectangular; shallowly longitudinally grooved; compressed laterally. Hilum long-linear. Embryo large (nearly as long as the fruit). Endosperm containing compound starch grains. Embryo with an epiblast; with a scutellar tail (inconspicuous); with a negligible mesocotyl internode. Embryonic leaf margins meeting.

Abaxial leaf blade epidermis. *Costal/intercostal zonation* conspicuous. *Papillae* absent. Intercostal zones exhibiting many atypical long-cells (rather short). Mid-intercostal long-cells having markedly sinuous walls. *Microhairs* present; more or less spherical to elongated; clearly two-celled; panicoid-type; (74–)80–90(–115) microns long. Microhair apical cells (45–)48–60(–63) microns long. Microhair apical cell/total length ratio about 0.64. *Stomata* common. Subsidiaries low dome-shaped and triangular. *Intercostal short-cells* common; in cork/silica-cell pairs; silicified. Intercostal silica bodies tall-and-narrow. *Costal short-cells* neither distinctly grouped into long rows nor predominantly paired (in series of 2–4). Costal silica bodies horizontally-elongated smooth and rounded.

Transverse section of leaf blade, physiology. C_3; XyMS+. *Mesophyll* with non-radiate chlorenchyma; with arm cells; with fusoids. The fusoids external to the PBS. *Leaf blade* blade with slight ribs and furrows adaxially. *Midrib* conspicuous (projecting adaxial rib); having complex vascularization (a large abaxial bundle, and 2 smaller ones in a line above it). Bulliforms present in discrete, regular adaxial groups to not in discrete, regular adaxial groups; in groups of the 'irregular' type. All the vascular bundles accompanied by sclerenchyma. Combined sclerenchyma girders present; forming 'figures' (incomplete anchors). Sclerenchyma all associated with vascular bundles.

Special diagnostic feature. *Inflorescence of 2–3 glumeless, bracteate spikelets, the lodicules represented by a fringed annulus*.

Cytology. Chromosome base number, $x = 9$. $2n = 36$.

Taxonomy. Bambusoideae; Oryzodae; Anomochloeae.

Ecology, geography, regional floristic distribution. 1 species. Tropical America. Shade species. In forests.
Neotropical. Central Brazilian.

References, etc. Morphological/taxonomic: Judziewcz and Soderstrom 1989. Leaf anatomical: Metcalfe 1960, Judziewcz and Soderstrom 1989.

Anthaenantiopsis Pilger

Habit, vegetative morphology. Perennial; caespitose. **Culms** 40–110 cm high; *herbaceous*; unbranched above. Culm nodes hairy, or glabrous. Culm internodes hollow. Young shoots intravaginal. *Leaves* not basally aggregated; non-auriculate. Sheaths hairy or glabrous, striate. *Leaf blades* linear to linear-lanceolate; narrow; (1–)1.5–9(–10) mm wide; setaceous (at least, tapering to filiform in *A. tristachya*), or not setaceous; *pseudopetiolate (narrow at the base, tapering to it in* **A. tristachya**); without cross venation. *Ligule an unfringed membrane (shortly laciniate), or a fringed membrane*; 0.3–1.2 mm long.

Reproductive organization. Plants bisexual, with bisexual spikelets; with hermaphrodite florets. *The spikelets all alike in sexuality*.

Inflorescence. *Inflorescence of spicate main branches (spiciform, of appressed or slightly divergent short racemes)*; non-digitate. *Inflorescence with axes ending in spikelets*. Inflorescence espatheate; not comprising 'partial inflorescences' and foliar organs. *Spikelet-bearing axes* 'racemes'. The racemes spikelet bearing to the base. Spikelet-bearing axes persistent. Spikelets secund (congested on the axes); biseriate; shortly pedicellate, or subsessile (the pedicels pilose).

Female-fertile spikelets. *Spikelets* 2.6–3.6 mm long; narrowly elliptic, or elliptic; adaxial; *compressed dorsiventrally*; biconvex; falling with the glumes. The upper floret not stipitate. Rachilla terminated by a female-fertile floret.

Glumes present; one per spikelet, or two (usually); very unequal (the lower being very small, said to be sometimes absent); about equalling the spikelets (i.e. the upper glumes); (the upper) long relative to the adjacent lemmas (sometimes enclosing the tip of the upper lemma); hairy (with tuberculate whitish hairs); awnless; non-carinate; very dissimilar (the lower reduced). Lower glume when present, much shorter than the lowest lemma; 0 nerved, or 1–3 nerved. Upper glume 5–7(–9) nerved. *Spikelets* with incomplete florets. The incomplete florets proximal to the female-fertile florets. *The proximal incomplete florets* 1; paleate. Palea of the proximal incomplete florets fully developed (hyaline); narrowly ovoid to ellipsoid. The proximal incomplete florets male. The proximal lemmas glumiform; awnless; 5–7(–9) nerved; more or less equalling the female-fertile lemmas to decidedly exceeding the female-fertile lemmas (as long as the spikelet); less firm than the female-fertile lemmas (and stiffly pilose).

Female-fertile florets 1. Lemmas decidedly firmer than the glumes (thinly crustaceous); smooth; becoming indurated; white in fruit (or whitish); entire; awnless; sometimes pubescent at the apex; non-carinate; having the margins tucked in onto the palea; with a clear germination flap. *Palea* present; relatively long; gaping; entire; awnless, without apical setae; thinner than the lemma to textured like the lemma (cartilaginous); indurated; 2-nerved. *Lodicules* present; 2; free; fleshy; glabrous. *Stamens* 3. Anthers not penicillate; without an apically prolonged connective. *Ovary* glabrous. Styles free to their bases. Stigmas 2.

Fruit, embryo and seedling. Fruit not grooved. Hilum short (punctifom or oblong).

Abaxial leaf blade epidermis. *Costal/intercostal zonation* conspicuous. *Papillae* absent. *Long-cells* markedly different in shape costally and intercostally (the costals narrower). Mid-intercostal long-cells rectangular and fusiform (and often rather irregular, sometimes with oblique end-walls); having markedly sinuous walls. *Microhairs* present; elongated; clearly two-celled; panicoid-type; 32–60 microns long. *Stomata* common; 19–32 microns long. Subsidiaries dome-shaped and triangular (variously 'truncated'). Guard-cells overlapping to flush with the interstomatals. *Intercostal short-cells* common; in cork/silica-cell pairs; silicified. *Costal short-cells* conspicuously in long rows, or conspicuously in long rows and neither distinctly grouped into long rows nor predominantly paired (in *A. tristachya*). Costal silica bodies 'panicoid-type'; dumb-bell shaped, or nodular (mostly, in *A tristachya*).

Transverse section of leaf blade, physiology. Leaf blades seemingly consisting of midrib (in the material of *A. trachystachya* seen, the lateral laminae being reduced to tiny flanges), or 'laminar'. C_4; XyMS–. PCR sheath outlines uneven. PCR sheath extensions absent. PCR cell chloroplasts centrifugal/peripheral. Mesophyll with arm cells, or without arm cells. *Leaf blade* with distinct, prominent adaxial ribs, or adaxially flat. *Midrib* conspicuous; having a conventional arc of bundles (or the bundles forming almost a circle in the reduced blade of *A. tristachya*); with colourless mesophyll adaxially (and in *A. tristachya* filling the centre of the vascular ring). Bulliforms present in discrete, regular adaxial groups (in the species other than *A tristachya*); in simple fans. Many of the smallest vascular bundles unaccompanied by sclerenchyma. Combined sclerenchyma girders present, or absent (*A. tristachya* exhibiting abaxial girders only, with the main bundles); when present, forming 'figures'. Sclerenchyma all associated with vascular bundles.

Cytology. Chromosome base number, $x = 10$. $2n = 20$, 40, and 41. 2 and 4 ploid.

Taxonomy. Panicoideae; Panicodae; Paniceae.

Ecology, geography, regional floristic distribution. 4 species. Brazil to Argentina. Species of open habitats; glycophytic. Savanna.

Neotropical. Central Brazilian.

References, etc. Morphological/taxonomic: Parodi 1938; Morrone *et al.* 1993. Leaf anatomical: this project; Morrone *et al.* 1993.

Special comments. Cf. *Brachiaria.* Fruit data wanting.

Anthenantia P. Beauv.

Anthaenantia P. Beauv. (alternative spelling), *Aulaxanthus* Elliott, *Aulaxia* Nutt

Habit, vegetative morphology. Perennial; caespitose. **Culms** 60–120 cm high; *herbaceous*; unbranched above. Culm nodes glabrous. Culm internodes hollow. *Leaves* mostly basal; non-auriculate. *Leaf blades* narrow (3–5 mm); without cross venation; *disarticulating from the sheaths (from the upper culms).* Ligule present; *a fringed membrane (short)*; 0.5 mm long.

Reproductive organization. Plants bisexual, with bisexual spikelets; with hermaphrodite florets.

Inflorescence. Inflorescence paniculate; open; with capillary branchlets; espatheate; not comprising 'partial inflorescences' and foliar organs. Spikelet-bearing axes persistent. Spikelets not secund; pedicellate.

Female-fertile spikelets. *Spikelets* 3 mm long; elliptic; *compressed dorsiventrally (but with the side of the lower lemma somewhat flattened)*; disarticulating above the glumes, or falling with the glumes, or not disarticulating. Rachilla terminated by a female-fertile floret. Hairy callus absent.

Glumes one per spikelet (the upper); relatively large; long relative to the adjacent lemmas; *hairy (with longitudinal rows of dense hairs)*; pointed; awnless; non-carinate. Upper glume 5 nerved. *Spikelets* with incomplete florets. The incomplete florets proximal to the female-fertile florets. *Spikelets with proximal incomplete florets. The proximal incomplete florets* 1; paleate; male. *The proximal lemmas* awnless; 5 nerved; more or less equalling the female-fertile lemmas; *less firm than the female-fertile lemmas (also with longitudinal rows of hairs)*; not becoming indurated.

Female-fertile florets 1. *Lemmas more or less boat-shaped and firm to the tip, by contrast with* **Leptocoryphium**; decidedly firmer than the glumes; not becoming indurated (cartilaginous, dark brown); entire; pointed; awnless; hairless; carinate to non-carinate; having the margins lying flat and exposed on the palea; with a clear germination flap; 3–5 nerved. *Palea* present; relatively long; awnless, without apical setae; keel-less. *Lodicules* present; 2; free; fleshy; glabrous; not or scarcely vascularized. *Stamens* 3. Anthers not penicillate; without an apically prolonged connective. *Ovary* glabrous. Styles free to their bases. Stigmas 2; red pigmented.

Fruit, embryo and seedling. Fruit small; compressed dorsiventrally. Hilum short. Embryo large. Endosperm hard.

Abaxial leaf blade epidermis. *Costal/intercostal zonation* conspicuous. *Papillae* absent. Mid-intercostal long-cells rectangular; having markedly sinuous walls. *Microhairs* present; panicoid-type. *Stomata* common. Subsidiaries triangular. Guard-cells overlapping to flush with the interstomatals. *Intercostal short-cells* common; not paired; silicified. *Costal short-cells* conspicuously in long rows. Costal silica bodies 'panicoid-type'.

Transverse section of leaf blade, physiology. C_4; XyMS–. *Mesophyll* with non-radiate chlorenchyma; traversed by columns of colourless mesophyll cells. *Leaf blade* with distinct, prominent adaxial ribs, or 'nodular' in section; with the ribs more or less constant in size. *Midrib* not readily distinguishable; having a conventional arc of bundles (in the blade, the smallest bundles are in an abaxial row, with the medium-sized ones above them in the ribs (phloem always abaxial): c.f. *Leptocoryphium*); with colourless mesophyll adaxially. Bulliforms present in discrete, regular adaxial groups; associated with colourless mesophyll cells to form deeply-penetrating fans. Many of the smallest vascular bundles unaccompanied by sclerenchyma. Combined sclerenchyma girders present; forming 'figures'. Sclerenchyma all associated with vascular bundles.

Cytology. Chromosome base number, $x = 10$ (?). $2n = 20$.

Taxonomy. Panicoideae; Panicodae; Paniceae.

Ecology, geography, regional floristic distribution. 2 species. Warm America. Pine barrens.

Neotropical. Venezuela and Surinam.
Economic importance. Important native pasture species: *A. rufa*.
References, etc. Leaf anatomical: this project.

Anthephora Schreber

Hypudaerus A. Br.

Habit, vegetative morphology. Annual, or perennial; rhizomatous, or caespitose to decumbent. *Culms* 15–150 cm high; herbaceous; branched above, or unbranched above. Culm nodes glabrous. Culm internodes hollow. Young shoots extravaginal, or intravaginal. *Leaves* not basally aggregated; non-auriculate. Leaf blades linear to lanceolate; narrow; 1–7 mm wide; not setaceous, or setaceous (when tightly rolled); flat, or rolled; without cross venation; persistent; rolled in bud. Ligule an unfringed membrane, or a fringed membrane; not truncate (often irregularly split almost to the base); 2–8 mm long. *Contra-ligule* absent.

Reproductive organization. Plants bisexual, with bisexual spikelets; with hermaphrodite florets. *The spikelets* of sexually distinct forms on the same plant (the involucres of the clusters being here interpreted as modified spikelets, rather than as modified lower glumes of the fertile spikelets); *hermaphrodite and sterile, or hermaphrodite and male-only (the glomerules comprising 1–11 central perfect spikelets with two or more outer sterile or male involucral spikelets, the latter mainly manifested as leathery, several nerved 'bracts')*; overtly heteromorphic (the involucral spikelets much modified, usually reduced to leathery 'bracts'). Apomictic, or reproducing sexually.

Inflorescence. Inflorescence a false spike, with spikelets on contracted axes (3–11 spikelets per glomerule, the glomerules conical, usually sessile); espatheate; not comprising 'partial inflorescences' and foliar organs. *Spikelet-bearing axes* very much reduced (to glomerules); disarticulating; falling entire (i.e. each glomerule falling from the persistent main axis). **Spikelets** *associated with bractiform involucres (these consisting of the leathery, bractlike vestiges of involucral spikelets, which are sometimes connate at the base)*. The involucres shed with the fertile spikelets. Spikelets not secund.

Female-sterile spikelets. *The outer, involucral spikelets mainly detected as broad, leathery 'bracts', these 2–15 nerved, sometimes basally connate; the bracts seemingly interpretable as modified lower glumes, being sometimes accompanied by a setaceous G_2 and a male floret.*

Female-fertile spikelets. Spikelets abaxial (in the sense that the backs of the glumes are to the inside of the glomerule); compressed dorsiventrally; falling with the glumes. Rachilla terminated by a female-fertile floret. *Callus* absent in the normal sense, but hairs at the base of the glomerule and on the involucre direct the resting position of the glomerule.

Glumes one per spikelet (the upper); long relative to the adjacent lemmas; pointed; awned; non-carinate. Upper glume 1 nerved. *Spikelets* with incomplete florets. The incomplete florets proximal to the female-fertile florets. **Spikelets with proximal incomplete florets**. *The proximal incomplete florets* 1; epaleate; sterile. The proximal lemmas awnless; 2–7 nerved; more or less equalling the female-fertile lemmas; less firm than the female-fertile lemmas; not becoming indurated.

Female-fertile florets 1. Lemmas lanceolate; not becoming indurated (membranous); entire; pointed; awnless; hairless; non-carinate; having the margins lying flat and exposed on the palea; with a clear germination flap, or without a germination flap; 3–5 nerved. *Palea* present; relatively long; entire; awnless, without apical setae; textured like the lemma; not indurated; 2-nerved. *Lodicules* absent. *Stamens* 3. *Ovary* glabrous. Styles fused, or free to their bases. Stigmas 2.

Fruit, embryo and seedling. Fruit compressed dorsiventrally. Hilum short. Endosperm hard; without lipid. Embryo without an epiblast; with a scutellar tail; with an elongated mesocotyl internode. Embryonic leaf margins overlapping.

Abaxial leaf blade epidermis. *Costal/intercostal zonation* conspicuous. *Papillae* absent. *Long-cells* markedly different in shape costally and intercostally (costals long, narrow, rectangular, intercostals broader, rather variable); of similar wall thickness costally and intercostally (thin walled). Intercostal zones with typical long-cells, or exhibiting many atypical long-cells (e.g. in *A. schinzii*, which has many of them nearly isodiametric). Mid-intercostal long-cells rectangular, or fusiform; having straight or only gently undulating walls (commonly), or having markedly sinuous walls (e.g. in *A. argentea*). *Microhairs* present; panicoid-type; (54–)57–84(–90) microns long; (24–)25–36(–45) microns wide at the septum. Microhair total length/width at septum 9.5–15. Microhair apical cells (3.6–)4.2–6(–9.3) microns long. Microhair apical cell/total length ratio 0.4–0.56. *Stomata* common; 18–27 microns long. Subsidiaries triangular. Guard-cells overlapped by the interstomatals, or overlapping to flush with the interstomatals. *Intercostal short-cells* absent or very rare. Costal prickles abundant. *Crown cells* absent. *Costal short-cells* conspicuously in long rows, or neither distinctly grouped into long rows nor predominantly paired. Costal silica bodies 'panicoid-type'; cross shaped, or dumb-bell shaped, or nodular.

Transverse section of leaf blade, physiology. C_4. The anatomical organization usually conventional, or unconventional (rarely, doubtfully). Organization of PCR tissue when unconventional, supposedly *Arundinella* type (see W.V. Brown 1977, quoting Johnson 1965). XyMS–. PCR sheath outlines even. PCR cell chloroplasts with reduced grana; centrifugal/peripheral. *Mesophyll* with radiate chlorenchyma; exhibiting 'circular cells', or without 'circular cells'. *Leaf blade* with distinct, prominent adaxial ribs, or adaxially flat (but usually with small, conspicuous abaxial ribs). *Midrib* conspicuous; having a conventional arc of bundles; with colourless mesophyll adaxially. Bulliforms present in discrete, regular adaxial groups, or not in discrete, regular adaxial groups (e.g. in *A. pubescens*, where the epidermis is mainly bulliform); in simple fans (when grouped); associating with colourless mesophyll cells to form arches over small vascular bundles, or nowhere involved in bulliform-plus-colourless mesophyll arches. Many of the smallest vascular bundles unaccompanied by sclerenchyma, or all the vascular bundles accompanied by sclerenchyma. Combined sclerenchyma girders present, or absent. Sclerenchyma all associated with vascular bundles.

Culm anatomy. *Culm internode bundles* in three or more rings.

Phytochemistry. Leaves without flavonoid sulphates (1 species).

Special diagnostic feature. *The inflorescence not as in* **Odontelytrum** *(q.v.).*

Cytology. Chromosome base number, $x = 9$. $2n = 18$, 36, and 40. 2 and 4 ploid.

Taxonomy. Panicoideae; Panicodae; Paniceae.

Ecology, geography, regional floristic distribution. 12 species. Tropical and southern Africa, Arabia, tropical America. Mesophytic to xerophytic; species of open habitats; glycophytic. In dry, sandy savanna.

Holarctic, Paleotropical, and Neotropical. Tethyan. African. Irano-Turanian. Saharo-Sindian, Sudano-Angolan, West African Rainforest, and Namib-Karoo. Caribbean, Venezuela and Surinam, Amazon, Central Brazilian, and Andean. Sahelo-Sudanian, Somalo-Ethiopian, South Tropical African, and Kalaharian.

Rusts and smuts. Rusts — *Puccinia*. Smuts from Ustilaginaceae. Ustilaginaceae — *Sorosporium* and *Sphacelotheca*.

References, etc. Leaf anatomical: Metcalfe 1960; this project; photographs of *A. argentea, A. pubescens* and *A. schinzii* provided by R. P. Ellis.

Special comments. Fruit data wanting.

Anthochloa Nees & Meyen ex Nees

Habit, vegetative morphology. Diminutive perennial; caespitose. *Culms* 5–10 cm high; herbaceous; unbranched above. Young shoots intravaginal. *Leaves* not basally aggregated; non-auriculate. Leaf blades linear; narrow; 1–3.5 mm wide; flat, or rolled (conduplicate); without cross venation; persistent. *Ligule an unfringed membrane*; truncate; 0.5 mm long.

Reproductive organization. Plants bisexual, with bisexual spikelets; with hermaphrodite florets (but with female and sterile florets distally in the spikelet).

Inflorescence. Inflorescence few spikeleted; paniculate; contracted. Inflorescence with axes ending in spikelets. Inflorescence espatheate; not comprising 'partial inflorescences' and foliar organs. Spikelet-bearing axes persistent. Spikelets not secund; shortly pedicellate.

Female-fertile spikelets. Spikelets 4.5–6.5 mm long; not noticeably compressed (terete); disarticulating above the glumes; disarticulating between the florets; with conventional internode spacings. Rachilla prolonged beyond the uppermost female-fertile floret. The rachilla extension with incomplete florets. Hairy callus absent.

Glumes present; two; more or less equal; shorter than the spikelets; shorter than the adjacent lemmas; hairless; not pointed (broadly rounded); awnless; non-carinate; similar (broad, membranous, irregularly dentate). Lower glume 3 nerved. Upper glume 5 nerved. **Spikelets *with incomplete florets (the lower florets hemaphrodite, female florets above, the uppermost sterile)*.** The incomplete florets distal to the female-fertile florets. **The distal incomplete florets *clearly specialised and modified in form (the rachilla terminating in a clavate clump of rudiments)*.**

Female-fertile florets 2–4. **Lemmas *flabelliform, irregularly toothed, often pinkish***; similar in texture to the glumes; not becoming indurated; incised (emarginate or split); awnless; hairless; glabrous; non-carinate; without a germination flap; 5 nerved, or 7 nerved. *Palea* present; relatively long; three lobed, the median lobe itself bilobed; awnless, without apical setae; 2-nerved; 2-keeled. *Lodicules* present; 2; free; membranous; glabrous; toothed; not or scarcely vascularized. *Stamens* 3. *Ovary* glabrous. Styles free to their bases. Stigmas 2.

Fruit, embryo and seedling. *Fruit* free from both lemma and palea; small (1.5 mm long). **Hilum short**. Embryo small. Endosperm hard; with lipid; containing compound starch grains. Embryo with an epiblast; without a scutellar tail; with a negligible mesocotyl internode. Embryonic leaf margins meeting.

Abaxial leaf blade epidermis. *Costal/intercostal zonation* conspicuous. *Papillae* absent. *Long-cells* similar in shape costally and intercostally, or markedly different in shape costally and intercostally; of similar wall thickness costally and intercostally. Midintercostal long-cells rectangular; having straight or only gently undulating walls. *Microhairs* absent. *Stomata* absent or very rare. *Intercostal short-cells* absent or very rare. *Costal short-cells* neither distinctly grouped into long rows nor predominantly paired. Costal silica bodies predominantly horizontally-elongated smooth (being at the most only slightly sinuate).

Transverse section of leaf blade, physiology. C_3; XyMS+. *Mesophyll* with nonradiate chlorenchyma; without adaxial palisade. *Leaf blade* adaxially flat. *Midrib* not readily distinguishable; with one bundle only. Bulliforms not in discrete, regular adaxial groups; nowhere involved in bulliform-plus-colourless mesophyll arches. All the vascular bundles accompanied by sclerenchyma. Combined sclerenchyma girders present (with all the bundles); the girders small. Sclerenchyma all associated with vascular bundles.

Special diagnostic feature. **Spikelets *with the distal incomplete florets and/or the rachilla apex forming a terminal clavate appendage*.**

Cytology. $2n = 42$.

Taxonomy. Pooideae; Poodae; Poeae.

Ecology, geography, regional floristic distribution. 2 species. Andes. Species of open habitats. In high mountains.

Holarctic and Neotropical. Madrean. Andean.

References, etc. Leaf anatomical: Metcalfe 1960; this project.

Anthoxanthum L.

Flavia Fabric., *Foenodorum* Krause, *Xanthonanthus* St-Lager
Excluding *Hierochloë*

Habit, vegetative morphology. Annual, or perennial; caespitose to decumbent. *Culms* 5–90 cm high; herbaceous; unbranched above. Culm nodes hairy, or glabrous. Culm internodes hollow. **The shoots aromatic (coumarin-scented)**. *Leaves* not basally aggregated; non-auriculate. Leaf blades linear to lanceolate; broad, or narrow (2–15); flat; without cross venation; persistent; rolled in bud. Ligule an unfringed membrane; not truncate; 2–5 mm long.

Reproductive organization. Plants bisexual, with bisexual spikelets; with hermaphrodite florets; outbreeding.

Inflorescence. Inflorescence a single raceme (rarely), or paniculate; contracted; more or less ovoid, or spicate; espatheate; not comprising 'partial inflorescences' and foliar organs. Spikelet-bearing axes persistent. Spikelets not secund; sessile to pedicellate.

Female-fertile spikelets. Spikelets 5–10 mm long; compressed laterally; disarticulating above the glumes. Rachilla terminated by a female-fertile floret.

Glumes two; very unequal; (the upper) long relative to the adjacent lemmas; pointed; awnless; carinate; similar (membranous). Lower glume 1 nerved. Upper glume 3 nerved. *Spikelets* with incomplete florets. The incomplete florets proximal to the female-fertile florets. *Spikelets with proximal incomplete florets. The proximal incomplete florets* 2; paleate (sometimes, the palea two-keeled), or epaleate; sterile. The proximal lemmas awned; 3 nerved; decidedly exceeding the female-fertile lemmas; less firm than the female-fertile lemmas (hairy); not becoming indurated.

Female-fertile florets 1. Lemmas decidedly firmer than the glumes; becoming indurated; entire, or incised; awnless, or awned. Awns when present, 1; geniculate; much shorter than the body of the lemma to much longer than the body of the lemma. Lemmas hairless; non-carinate; 1–7 nerved. *Palea* present; relatively long; entire; awnless, without apical setae; not indurated; 1-nerved, or 2-nerved, or nerveless; keel-less. **Lodicules absent**. Stamens 2, or 3 (rarely). Anthers 3.5–5 mm long; not penicillate. *Ovary* glabrous. Styles free to their bases. Stigmas 2.

Fruit, embryo and seedling. *Fruit* free from both lemma and palea; small. Hilum short. Embryo small; not waisted. Endosperm hard; with lipid; containing compound starch grains. Embryo with an epiblast; without a scutellar tail; with a negligible mesocotyl internode. Embryonic leaf margins meeting.

Seedling with a long mesocotyl; with a tight coleoptile. First seedling leaf with a well-developed lamina. The lamina narrow; 3 veined.

Abaxial leaf blade epidermis. *Costal/intercostal zonation* conspicuous. *Papillae* absent. *Long-cells* similar in shape costally and intercostally; of similar wall thickness costally and intercostally (thin walled). Mid-intercostal long-cells fusiform; having markedly sinuous walls. *Microhairs* absent. *Stomata* common; about 36 microns long (in *A. odoratum*). Subsidiaries parallel-sided. Guard-cells overlapped by the interstomatals. *Intercostal short-cells* absent or very rare. *Costal short-cells* neither distinctly grouped into long rows nor predominantly paired. Costal silica bodies horizontally-elongated crenate/sinuous, or horizontally-elongated smooth, or rounded.

Transverse section of leaf blade, physiology. C_3; XyMS+. *Mesophyll* with non-radiate chlorenchyma; without adaxial palisade. *Leaf blade* with distinct, prominent adaxial ribs, or 'nodular' in section; with the ribs more or less constant in size. *Midrib* conspicuous; with one bundle only. Bulliforms present in discrete, regular adaxial groups; in simple fans (or the cells fairly regular in size). All the vascular bundles accompanied by sclerenchyma. Combined sclerenchyma girders present; nowhere forming 'figures'. Sclerenchyma all associated with vascular bundles.

Culm anatomy. *Culm internode bundles* in one or two rings.

Phytochemistry. Tissues of the culm bases with little or no starch. Leaves without flavonoid sulphates (1 species).

Cytology. Chromosome base number, $x = 5$. $2n = 10$ and 20. 2 and 4 ploid. Mean diploid 2c DNA value 11.8 pg (*A. odoratum*).

Taxonomy. Pooideae; Poodae; Aveneae.

Ecology, geography, regional floristic distribution. 20 species. North temperate & mountains of tropical Africa & Asia. Commonly adventive. Mesophytic; shade species and species of open habitats. Meadows, grasslands and in light shade.

Holarctic, Paleotropical, Neotropical, Cape, and Antarctic. Boreal and Tethyan. African, Madagascan, and Indomalesian. Euro-Siberian, Eastern Asian, and Atlantic North American. Macaronesian, Mediterranean, and Irano-Turanian. Sudano-Angolan. Indian, Indo-Chinese, and Papuan. Caribbean and Andean. Patagonian. European and Siberian. Canadian-Appalachian. Sahelo-Sudanian, Somalo-Ethiopian, and South Tropical African.

Rusts and smuts. Rusts — *Puccinia*. Taxonomically wide-ranging species: *Puccinia graminis, Puccinia coronata, Puccinia brachypodii*, and *Puccinia recondita*. Smuts from Tilletiaceae and from Ustilaginaceae. Tilletiaceae — *Tilletia* and *Urocystis*. Ustilaginaceae — *Ustilago*.

Economic importance. Significant weed species: *A. aristatum, A. odoratum*.

References, etc. Morphological/taxonomic: Schouten and Veldkamp 1985. Leaf anatomical: Metcalfe 1960 and this project.

Special comments. Should perhaps include *Hierochloë*.

Antinoria Parl.

Habit, vegetative morphology. Annual, or perennial; caespitose, or decumbent. *Culms* 5–30 cm high; herbaceous; unbranched above. *Leaves* not basally aggregated; nonauriculate. Leaf blades linear; narrow; about 2.5 mm wide (to 12 cm long); flat; without cross venation; persistent. Ligule an unfringed membrane; usually lacerate; 1–3 mm long.

Reproductive organization. Plants bisexual, with bisexual spikelets; with hermaphrodite florets.

Inflorescence. *Inflorescence* paniculate; open (up to 10 by 7 cm); **with conspicuously divaricate branchlets**; with capillary branchlets; espatheate; not comprising 'partial inflorescences' and foliar organs. Spikelets not secund; pedicellate (the pedicels longer than the spikelets, clavate).

Female-fertile spikelets. Spikelets 1–2 mm long; compressed laterally; disarticulating above the glumes; disarticulating between the florets; with distinctly elongated rachilla internodes between the florets. The upper floret conspicuously stipitate. Rachilla terminated by a female-fertile floret; rachilla hairless.

Glumes two; more or less equal; **exceeding the spikelets**; long relative to the adjacent lemmas; hairless (scabridulous on the keel); 'subobtuse'; awnless; carinate; similar. Lower glume 3 nerved. Upper glume 3 nerved. *Spikelets* with female-fertile florets only.

Female-fertile florets 2. **Lemmas elliptic, widest near the tip by contrast with Airopsis, membranous**; similar in texture to the glumes; not becoming indurated; slightly incised; blunt; more or less truncate, but shallowly 3 lobed; not deeply cleft; awnless; hairless; glabrous; non-carinate; without a germination flap; 5 nerved. *Palea* present; relatively long; tightly clasped by the lemma; awnless, without apical setae; not indurated; 2-nerved; 2-keeled. *Stamens* 3. Anthers 0.5–1 mm long. *Ovary* glabrous. Styles free to their bases. Stigmas 2.

Fruit, embryo and seedling. Fruit pyriform; compressed dorsiventrally (rounded abaxially, flat ventrally). Hilum short. Embryo small.

Abaxial leaf blade epidermis. *Costal/intercostal zonation* conspicuous. *Papillae* present (very abundant); costal and intercostal (but mainly intercostal, being excluded from the costae by prickles). Intercostal papillae over-arching the stomata; several per cell (quite large, thick walled, circular, in one row of 2–5 per cell). Mid-intercostal long-cells rectangular to fusiform; having markedly sinuous walls and having straight or only gently undulating walls. *Microhairs* absent. *Stomata* common. Subsidiaries non-papillate; mostly more or less parallel-sided. Guard-cells overlapped by the interstomatals. *Intercostal short-cells* common; mostly apparently not paired (solitary). The costal zones dominated by chains of blunt prickles. *Costal short-cells* neither distinctly grouped into long rows nor predominantly paired (infrequent, amid the prickles). Costal silica bodies present and well developed (but examples few and far between); horizontally-elongated crenate/sinuous (short, often more or less imperfect).

Transverse section of leaf blade, physiology. C_3; XyMS+. Mesophyll without adaxial palisade. *Leaf blade* with distinct, prominent adaxial ribs; with the ribs more or less constant in size (tall). *Midrib* not readily distinguishable; with one bundle only. *The lamina* symmetrical on either side of the midrib. Bulliforms present in discrete, regular adaxial groups (perhaps, in the furrows), or not in discrete, regular adaxial groups (?); if present, in simple fans. All the vascular bundles accompanied by sclerenchyma. Combined sclerenchyma girders absent (the sclerenchyna confined to small adaxial strands, and small abaxial strands and girders). Sclerenchyma all associated with vascular bundles.

Cytology. Chromosome base number, $x = 7$. $2n = 14$. 2 ploid.

Taxonomy. Pooideae; Poodae; Aveneae.

Ecology, geography, regional floristic distribution. 2 species. Mediterranean. Mesophytic (in damp places); glycophytic.

Holarctic. Boreal and Tethyan. Euro-Siberian. Mediterranean. European.

References, etc. Leaf anatomical: this project.

Special comments. Morphological description compiled from *Flora Europaea* and Clayton & Renvoize (1987).

Apera Adans.

Anemagrostis Trin.

Habit, vegetative morphology. *Annual*; caespitose. *Culms* 10–120 cm high; herbaceous; unbranched above. Culm nodes glabrous. Culm internodes hollow. *Leaves* not basally aggregated; non-auriculate. Leaf blades narrow; 1–10 mm wide; flat, or rolled (convolute); without cross venation; persistent. *Ligule an unfringed membrane*; not truncate; 2–10 mm long.

Reproductive organization. Plants bisexual, with bisexual spikelets; with hermaphrodite florets; outbreeding.

Inflorescence. *Inflorescence paniculate*; open; with capillary branchlets; espatheate; not comprising 'partial inflorescences' and foliar organs. Spikelet-bearing axes persistent. Spikelets not secund; pedicellate.

Female-fertile spikelets. *Spikelets* 2–3.6 mm long; *compressed laterally*; disarticulating above the glumes. Rachilla prolonged beyond the uppermost female-fertile floret, or terminated by a female-fertile floret; rachilla hairless. The rachilla extension when present, with incomplete florets. Hairy callus absent. Callus blunt.

Glumes two; very unequal (usually), or more or less equal (rarely); (the upper) long relative to the adjacent lemmas; pointed; awnless; *non-carinate*; similar (membranous). Lower glume 1 nerved. Upper glume 3 nerved. *Spikelets* with female-fertile florets only, or with incomplete florets. The incomplete florets distal to the female-fertile florets.

Female-fertile florets 1 (usually), or 2–3. *Lemmas* bidentate; *decidedly firmer than the glumes (firmly membranous)*; not becoming indurated ('subindurate when mature'); incised; awned. *Awns* 1; median; dorsal; *from near the top*; non-geniculate, or geniculate (rarely); much longer than the body of the lemma; entered by one vein. Lemmas hairless; non-carinate; 5 nerved. *Palea* present; relatively long; apically notched; awnless, without apical setae; not indurated (hyaline); 2-nerved; 2-keeled, or keel-less. *Lodicules* present; 2; free; membranous; glabrous; toothed. *Stamens* 3. Anthers 0.3–2 mm long; not penicillate. *Ovary* glabrous. Styles free to their bases. Stigmas 2.

Fruit, embryo and seedling. **Fruit** free from both lemma and palea; small; *longitudinally grooved*; compressed dorsiventrally. *Hilum short*. Embryo small; not waisted. Endosperm liquid in the mature fruit, or hard; with lipid. Embryo with an epiblast; without a scutellar tail; with a negligible mesocotyl internode. Embryonic leaf margins meeting.

Seedling with a tight coleoptile. First seedling leaf with a well-developed lamina. The lamina narrow; erect; 3 veined.

Abaxial leaf blade epidermis. *Costal/intercostal zonation* conspicuous. *Papillae* absent. *Long-cells* similar in shape costally and intercostally (the costals narrower); of similar wall thickness costally and intercostally (thin walled). Mid-intercostal long-cells fusiform; having markedly sinuous walls. *Microhairs* absent. *Stomata* common. Subsidiaries parallel-sided. Guard-cells overlapped by the interstomatals. *Intercostal short-cells* absent or very rare (very rare); in cork/silica-cell pairs (also solitaries); silicified. *Costal short-cells* neither distinctly grouped into long rows nor predominantly paired (mainly solitary). Costal silica bodies horizontally-elongated crenate/sinuous, or horizontally-elongated smooth.

Transverse section of leaf blade, physiology. C_3; XyMS+. *Mesophyll* with non-radiate chlorenchyma. *Leaf blade* 'nodular' in section; with the ribs more or less constant in size. Bulliforms present in discrete, regular adaxial groups; in simple fans. All the vascular bundles accompanied by sclerenchyma. Combined sclerenchyma girders present; nowhere forming 'figures'. Sclerenchyma all associated with vascular bundles.

Cytology. Chromosome base number, $x = 7$. $2n = 14$. 2 ploid.

Taxonomy. Pooideae; Poodae; Aveneae.

Ecology, geography, regional floristic distribution. 3–4 species. Europe, western Asia. Commonly adventive. Mesophytic; species of open habitats. Dry sandy soils and arable land.

Holarctic and Antarctic. Boreal and Tethyan. Euro-Siberian. Mediterranean and Irano-Turanian. Patagonian. European and Siberian.

Rusts and smuts. Rusts — *Puccinia*. Taxonomically wide-ranging species: *Puccinia graminis* and *Puccinia coronata*. Smuts from Tilletiaceae. Tilletiaceae — *Tilletia* and *Urocystis*.

Economic importance. Significant weed species: *A. interrupta*, *A. spica-venti*.

References, etc. Leaf anatomical: this project.

Aphanelytrum Hackel

Habit, vegetative morphology. Perennial; stoloniferous and decumbent. Culms herbaceous; scrambling (over rocks and other vegetation). Leaves non-auriculate. *Sheath margins joined*. Leaf blades narrow; flat; without cross venation; persistent. Ligule an unfringed membrane; not truncate; 1 mm long.

Reproductive organization. Plants bisexual, with bisexual spikelets; with hermaphrodite florets.

Inflorescence. Inflorescence few spikeleted; paniculate; open; with capillary branchlets. Inflorescence with axes ending in spikelets. Inflorescence espatheate; not comprising 'partial inflorescences' and foliar organs. Spikelet-bearing axes persistent. Spikelets not secund; pedicellate.

Female-fertile spikelets. Spikelets *unconventional (peculiar, with very long rachilla internodes and tiny glumes)*; 4.5–7.5 mm long; not noticeably compressed; disarticulating above the glumes; with a distinctly elongated rachilla internode above the glumes and with distinctly elongated rachilla internodes between the florets. Rachilla prolonged beyond the uppermost female-fertile floret; rachilla hairless (scabrid). The rachilla extension naked. Hairy callus absent.

Glumes two; minute; very unequal; shorter than the spikelets; shorter than the adjacent lemmas; hairless; awnless; similar. Lower glume 0 nerved. *Upper glume 0 nerved*. *Spikelets* with female-fertile florets only.

Female-fertile florets 2–3. Lemmas lanceolate, broadly acuminate; not becoming indurated (membranous); incised; shortly 2 lobed; not deeply cleft; awned. Awns 1; median; from a sinus; non-geniculate; hairless; much shorter than the body of the lemma. Lemmas hairless; carinate; without a germination flap; 5 nerved. *Palea* present; relatively long; gaping; apically notched; awnless, without apical setae; not indurated; 2-nerved; 2-keeled. Palea keels wingless. *Lodicules* present; 2; free; membranous; glabrous; not toothed; not or scarcely vascularized. *Stamens* 3. Anthers 3.5–3.8 mm long; not penicillate; without an apically prolonged connective. *Ovary* glabrous. Styles free to their bases. Stigmas 2.

Fruit, embryo and seedling. *Fruit* free from both lemma and palea; small, or medium sized (3.5–4); compressed laterally. Hilum short. Embryo small. Endosperm hard. Embryo with an epiblast; without a scutellar tail; with a negligible mesocotyl internode. Embryonic leaf margins meeting.

Abaxial leaf blade epidermis. *Costal/intercostal zonation* conspicuous. *Papillae* absent. *Long-cells* markedly different in shape costally and intercostally (costals rectangular); differing markedly in wall thickness costally and intercostally (costals thicker- walled). Mid-intercostal long-cells exaggeratedly fusiform; having straight or only gently undulating walls. *Microhairs* absent. *Stomata* absent or very rare. *Intercostal short-cells* absent or very rare. *Costal short-cells* neither distinctly grouped into long rows nor predominantly paired. Costal silica bodies horizontally-elongated smooth (sometimes slightly sinuous).

Transverse section of leaf blade, physiology. C$_3$; XyMS+. *Mesophyll* with non-radiate chlorenchyma; without adaxial palisade. Midrib with one bundle only. Sclerenchyma all associated with vascular bundles.

Taxonomy. Pooideae; Poodae; Poeae.

Ecology, geography, regional floristic distribution. 1 species. Western tropical South America. Mesophytic; shade species; glycophytic. In humid montane forest. Neotropical. Andean.

References, etc. Leaf anatomical: this project.

Apluda L.

Calamina P. Beauv.

Habit, vegetative morphology. Perennial; decumbent (often scrambling). *Culms* 50–120 cm high; branched above. Culm nodes glabrous. Culm internodes solid. *Leaves* not basally aggregated; non-auriculate. *Leaf blades* broad (to 15 mm), or narrow; *pseudopetiolate*; without cross venation; persistent. Ligule an unfringed membrane.

Reproductive organization. *Plants bisexual, with bisexual spikelets*; with hermaphrodite florets. The spikelets of sexually distinct forms on the same plant; hermaphrodite, male-only, and sterile (in combination). Apomictic.

Inflorescence. Inflorescence paniculate; open; *spatheate (each naviculate spathe enclosing a raceme reduced to a single joint, with 3 spikelets)*; *a complex of 'partial inflorescences' and intervening foliar organs.* **Spikelet-bearing axes** very much reduced (to 1 sessile and 2 pedicelled spikelets); clustered (at the tips of the usually fascicled branchlets); with substantial rachides; *disarticulating (the whole triplet deciduous from the minute peduncle).* Spikelets in triplets; not secund; sessile and pedicellate; consistently in 'long-and-short' combinations; in pedicellate/sessile combinations (the pedicels flat and broad). Pedicels of the 'pedicellate' spikelets free of the rachis. The 'shorter' spikelets hermaphrodite. The 'longer' spikelets male-only and sterile (one male, the other reduced to a small glume).

Female-fertile spikelets. Spikelets compressed dorsiventrally; falling with the glumes. Rachilla terminated by a female-fertile floret. *Hairy callus absent.*

Glumes two; more or less equal; long relative to the adjacent lemmas; awnless; very dissimilar (lower leathery, upper thinner and naviculate). Lower glume 11 nerved. Upper glume 3–5 nerved. *Spikelets* with incomplete florets. The incomplete florets proximal to the female-fertile florets. *The proximal incomplete florets* 1; paleate; male. The proximal lemmas awnless; similar in texture to the female-fertile lemmas (thinly membranous to hyaline).

Female-fertile florets 1. Lemmas less firm than the glumes; not becoming indurated; entire, or incised; when incised 2 lobed; mucronate, or awned (entire and mucronate, or deeply bilobed/awned). Awns when present, 1; from a sinus; geniculate; much shorter than the body of the lemma to much longer than the body of the lemma. Lemmas hairless; non-carinate; 1 nerved. **Palea** present; *very reduced*; not indurated (hyaline); nerveless. *Lodicules* present; 2; free; fleshy; glabrous. *Stamens* 2–3. Anthers not penicillate. *Ovary* glabrous. Styles free to their bases. Stigmas 2; red pigmented.

Fruit, embryo and seedling. Fruit compressed dorsiventrally. Hilum short. Embryo large. Endosperm containing compound starch grains. Embryo without an epiblast; with a scutellar tail.

Abaxial leaf blade epidermis. *Costal/intercostal zonation* conspicuous. *Papillae* present. Intercostal papillae over-arching the stomata; consisting of one oblique swelling per cell (the interstomatal cells papilliform). *Long-cells* similar in shape costally and intercostally (the costals narrower); of similar wall thickness costally and intercostally (fairly thin walled). Intercostal zones with typical long-cells (but these rather short). Mid-intercostal long-cells rectangular; having markedly sinuous walls. *Microhairs* present; panicoid-type; 39–42 microns long; 6–7.5 microns wide at the septum. Microhair total length/width at septum 6.5–9. Microhair apical cells (12–)13.5–18 microns long. Microhair apical cell/total length ratio 0.31–0.43. *Stomata* common; 21–22.5(–24) microns long. Subsidiaries dome-shaped. *Intercostal short-cells* absent or very rare. *Costal short-cells* neither distinctly grouped into long rows nor predominantly paired. Costal silica bodies 'panicoid-type'; butterfly shaped, or dumb-bell shaped, or nodular.

Transverse section of leaf blade, physiology. C_4; XyMS–. *Mesophyll* with radiate chlorenchyma. *Leaf blade* 'nodular' in section, or adaxially flat; with the ribs more or less constant in size. *Midrib* conspicuous; having a conventional arc of bundles; with colourless mesophyll adaxially. Bulliforms present in discrete, regular adaxial groups; in simple fans (or in irregular groups). All the vascular bundles accompanied by sclerenchyma. Combined sclerenchyma girders present; forming 'figures'. Sclerenchyma all associated with vascular bundles.

Cytology. Chromosome base number, $x = 10$. $2n = 20$ and 40. 2 and 4 ploid.

Taxonomy. Panicoideae; Andropogonodae; Andropogoneae; Andropogoninae.

Ecology, geography, regional floristic distribution. 1 species. Mauritius & Socotra to Formosa & New Caledonia. Mesophytic. In thickets and forest margins.

Holarctic and Paleotropical. Tethyan. African, Madagascan, Indomalesian, and Neocaledonian. Irano-Turanian. Saharo-Sindian and Sudano-Angolan. Indian, Indo-Chinese, Malesian, and Papuan. Somalo-Ethiopian.

Rusts and smuts. Rusts — *Dasturella* and *Puccinia*. Taxonomically wide-ranging species: *'Uromyces' schoenanthi*. Smuts from Tilletiaceae and from Ustilaginaceae. Tilletiaceae — *Tilletia*. Ustilaginaceae — *Sorosporium* and *Sphacelotheca*.

References, etc. Leaf anatomical: Metcalfe 1960; this project.

Apochiton C.E. Hubb.

Habit, vegetative morphology. *Annual*; caespitose (loosely), or decumbent. *Culms* 30–70 cm high; herbaceous; unbranched above. Culm nodes glabrous. Plants unarmed. *Leaves* not basally aggregated; non-auriculate; without auricular setae. Leaf blades linear; narrow; to 4 mm wide; flat; without abaxial multicellular glands; without cross venation; persistent. Ligule a fringed membrane (but the fringe of short cilia); truncate; 0.5 mm long. *Contra-ligule* absent.

Reproductive organization. Plants bisexual, with bisexual spikelets; with hermaphrodite florets.

Inflorescence. *Inflorescence paniculate*; open (to 20 cm long); without capillary branchlets; espatheate; not comprising 'partial inflorescences' and foliar organs. Spikelet-bearing axes persistent. Spikelets not secund; pedicellate.

Female-fertile spikelets. Spikelets 6–8 mm long; compressed laterally; disarticulating above the glumes; disarticulating between the florets. Rachilla prolonged beyond the uppermost female-fertile floret; rachilla rachilla sparsely hairy. The rachilla extension with incomplete florets. Hairy callus present (the hairs white). Callus blunt.

Glumes present; two; very unequal to more or less equal; shorter than the adjacent lemmas; hairless; pointed (tapered into short points); awned (shortly), or awnless (then mucronate); carinate; similar (greyish, membranous, glabrous). Lower glume 3 nerved. Upper glume 3 nerved. *Spikelets* with incomplete florets. The incomplete florets distal to the female-fertile florets. The distal incomplete florets 1; *incomplete florets* merely underdeveloped.

Female-fertile florets 3–5. Lemmas similar in texture to the glumes (membranous); smooth; not becoming indurated; entire; awned. Awns 1; median; apical (attenuate from the lemma); non-geniculate; hairless (scabrid); much shorter than the body of the lemma to about as long as the body of the lemma (2–3 mm); entered by one vein; persistent. Lemmas hairy (with white hairs over the lower two thirds); somewhat carinate; without a germination flap; 3 nerved. Palea present; relatively long (but shorter than the lemma); *deeply bifid*; *awned (the two lobes attenuate into terminal awns about 2 mm long)*; not indurated (membranous); 2-nerved; 2-keeled. Palea keels hairy (with white hairs). *Lodicules* present; free; fleshy; glabrous; not or scarcely vascularized. *Stamens* 3. Anthers 1 mm long. *Ovary* glabrous. Styles free to their bases. Stigmas 2; red pigmented.

Fruit, embryo and seedling. *Fruit* free from both lemma and palea; small (about 2 mm long); ellipsoid; trigonous. Hilum short. *Pericarp free*. Embryo small (obscure).

Abaxial leaf blade epidermis. *Costal/intercostal zonation* conspicuous. *Papillae* present; costal and intercostal. Intercostal papillae not over-arching the stomata; consisting of one oblique swelling per cell (large but rather inconspicuous, at one end of many of the long-cells). *Long-cells* similar in shape costally and intercostally (the costals narrower); of similar wall thickness costally and intercostally (thin walled). Mid-intercostal long-cells rectangular; having markedly sinuous walls. *Microhairs* present; more or less spherical; clearly two-celled; chloridoid-type. Microhair apical cell wall of similar thickness/rigidity to that of the basal cell. Microhairs (22.5–)24–27(–30) microns long. Microhair basal cells 10–11 microns long. Microhairs 9–12 microns wide at the septum. Microhair total length/width at septum 1.9–3.3. Microhair apical cells 9–12 microns long. Microhair apical cell/total length ratio 0.35–0.5. *Stomata* common; 30–33 microns long. Subsidiaries mostly low dome-shaped. Guard-cells overlapping to flush with the interstomatals. *Intercostal short-cells* absent or very rare. Intercostal silica bodies absent. *Costal short-cells* conspicuously in long rows (but the 'short-cells' tending to be rather long). Costal silica bodies confined to the central file(s) of the costal zones; 'panicoid-type' (mostly), or saddle shaped (a few, mostly over the midrib); mostly cross shaped and butterfly shaped.

Transverse section of leaf blade, physiology. C_4; XyMS+. PCR sheaths of the primary vascular bundles complete. PCR sheath extensions absent. Mesophyll traversed by columns of colourless mesophyll cells. *Midrib* conspicuous; with one bundle only, or having a conventional arc of bundles (one large solitary bundle, or one large bundle with two smaller laterals, depending on interpretation); with colourless mesophyll adaxially. Bulliforms present in discrete, regular adaxial groups; in simple fans, or associated with colourless mesophyll cells to form deeply-penetrating fans. All the vascular bundles accompanied by sclerenchyma. Combined sclerenchyma girders present (but the girders short); nowhere forming 'figures'. Sclerenchyma all associated with vascular bundles.

Taxonomy. Chloridoideae; main chloridoid assemblage.

Ecology, geography, regional floristic distribution. 1 species. Tropical East Africa. Mesophytic; species of open habitats. Savanna, in seasonally wet soils.

Paleotropical. African. Sudano-Angolan. Somalo-Ethiopian and South Tropical African.

References, etc. Morphological/taxonomic: Hubbard 1936b. Leaf anatomical: this project.

Apoclada McClure

Habit, vegetative morphology. Perennial; caespitose. The flowering culms leafy. *Culms* 110 cm high (or more?); woody and persistent; branched above. *Primary branches/mid-culm node 2–5.* Culm internodes hollow. Unicaespitose. Rhizomes pachymorph. Plants unarmed. *Leaves* not basally aggregated; auricles minute, rudimentary or obsolete; with auricular setae. Leaf blades narrow; to about 1 mm wide; acicular (aculeiform, to 9 cm long); pseudopetiolate; without cross venation; disarticulating from the sheaths. *Ligule* present (very short, dorsally canescent); a fringed membrane to a fringe of hairs. *Contra-ligule* present.

Reproductive organization. Plants bisexual, with bisexual spikelets; with hermaphrodite florets. The spikelets all alike in sexuality.

Inflorescence. Inflorescence *determinate*; open; spatheate (inflorescence branches subtended by bracts, but no prophylls except when the 'racemes' are reduced to single spikelets); a complex of 'partial inflorescences' and intervening foliar organs. Spikelet-bearing axes persistent. Spikelets not secund; pedicellate.

Female-fertile spikelets. Spikelets morphologically 'conventional', or unconventional (the number of glumes variable); *70–400 mm long*; compressed laterally; disarticulating above the glumes; tardily disarticulating between the florets. Rachilla prolonged beyond the uppermost female-fertile floret; rachilla hairy, or hairless. The rachilla extension with incomplete florets. Hairy callus absent.

Glumes present, or absent; when present, one per spikelet to several (consisting of leaf sheaths with reduced blades, but varying from species to species in form, number, and spatial relation to the first lemma); relatively large; very unequal to more or less equal; shorter than the adjacent lemmas, or long relative to the adjacent lemmas; free; awnless. *Spikelets* with incomplete florets. The incomplete florets both distal and proximal to the female-fertile florets. *The distal incomplete florets* merely underdeveloped. **Spikelets with proximal incomplete florets.** *The proximal incomplete florets* 1–2 (?). The proximal lemmas awned, or awnless; exceeded by the female-fertile lemmas; similar in texture to the female-fertile lemmas.

Female-fertile florets (1–)3–15. Lemmas acuminate; not becoming indurated (papery); entire; pointed; awnless, or mucronate, or awned. Awns when present, 1; apical; non-geniculate; much shorter than the body of the lemma. Lemmas hairy, or hairless; carinate, or non-carinate (but then keeled towards the tip); 'several nerved'. *Palea* present; relatively long; apically notched (and 2 or 4 dentate); awnless, without apical setae; not indurated (papery); 2-nerved; 2-keeled. *Lodicules* present; free; membranous; ciliate; not toothed; heavily vascularized. *Stamens* 3 (occasionally 3–6 in *A. diversa*). Anthers not penicillate; without an apically prolonged connective. **Ovary** glabrous, or hairy; **with a conspicuous apical appendage.** The appendage broadly conical, fleshy. Styles fused. Stigmas 2 (rarely showing a vestigial third).

Fruit, embryo and seedling. Fruit longitudinally grooved; compressed dorsiventrally. Hilum long-linear. Embryo small.

Abaxial leaf blade epidermis. *Costal/intercostal zonation* conspicuous. *Papillae* present; costal and intercostal. Intercostal papillae from the ends of the interstomatals over-arching the stomata (and almost enclosing them); several per cell (a single, median row of large, circular or bifurcated papillae per long-cell). *Long-cells* markedly different in shape costally and intercostally (the costals much more regularly rectangular); of similar wall thickness costally and intercostally. Mid-intercostal long-cells rectangular; having markedly sinuous walls (with conspicuous pits). *Microhairs* present; elongated; clearly two-celled; panicoid-type; 42–51 microns long; 7.5–10.5 microns wide at the septum. Microhair total length/width at septum 5–6.8. Microhair apical cells 21.6–24 microns long. Microhair apical cell/total length ratio 0.42–0.57. *Stomata* common; 28.5–31.5 microns long. Subsidiaries obscured. *Intercostal short-cells* common; in cork/silica-cell pairs and not paired; not silicified (in the silica-deficient material seen). *Costal short-cells* predominantly paired. Costal silica bodies poorly developed; rounded (but most of the very few silica bodies in the material seen poorly developed).

Transverse section of leaf blade, physiology. C_3. Mesophyll without adaxial palisade; with arm cells (these very conspicuous); without fusoids (*sic*). *Leaf blade* more or less 'nodular' in section (the adaxial ribs low, round- to flat-topped). *Midrib* conspicuous (as indicated by a distinct longitudinal fold nearer to one margin of the lamina); with one bundle only. *The lamina* distinctly asymmetrical on either side of the midrib (assuming the fold is indicative of a laterally displaced midrib). Bulliforms present in discrete, regular adaxial groups (a conspicuous group in each adaxial groove); in simple fans (a few only), or associated with colourless mesophyll cells to form deeply-penetrating fans (most of the groups being associated internally with a large, deeply-penetrating median colourless cell). All the vascular bundles accompanied by sclerenchyma. Combined sclerenchyma girders present; forming 'figures' (nearly all the bundles with a conspicuous 'I' or 'anchor'). Sclerenchyma not all bundle-associated. The 'extra' sclerenchyma in abaxial groups and in adaxial groups (there being occasional abaxial groups opposite the bulliforms, and single layers of fibres lining the bulliforms —cf. *Atractantha*); abaxial-hypodermal, the groups isolated and adaxial-hypodermal, contiguous with the bulliforms.

Taxonomy. Bambusoideae; Bambusodae; Bambuseae.

Ecology, geography, regional floristic distribution. 4 species. Brazil. Neotropical. Central Brazilian and Pampas.

References, etc. Leaf anatomical: this project.

Apocopis Nees

Amblyachyrum Steud.

Habit, vegetative morphology. Annual, or perennial (low, delicate). Culms herbaceous; branched above, or unbranched above. Leaf blades narrow; without cross venation. Ligule an unfringed membrane (ciliolate).

Reproductive organization. Plants bisexual, with bisexual spikelets; with hermaphrodite florets (rarely), or without hermaphrodite florets. *The spikelets of sexually distinct forms on the same plant*; hermaphrodite, or hermaphrodite and sterile (when the pedicelled spikelet detectable); overtly heteromorphic (the pedicelled members much reduced or absent, the lowermost spikelets awnless); *all in heterogamous combinations*.

Inflorescence. Inflorescence of solitary 'racemes', or these 2–3(–4)nate but closely appressed, terminal, exserted; digitate, or non-digitate (when the 'racemes' solitary). Primary inflorescence branches 1–3(–4). Inflorescence spatheate, or espatheate; a complex of 'partial inflorescences' and intervening foliar organs, or not comprising 'partial inflorescences' and foliar organs. *Spikelet-bearing axes* spikes, or 'racemes'; solitary, or paired, or clustered; with substantial rachides; disarticulating; disarticulating at the joints. 'Articles' without a basal callus-knob; not appendaged; disarticulating obliquely; somewhat hairy (bearded at the apex, ciliate on the margins). Spikelets paired (or the pedicellate members suppressed); sessile and pedicellate; consistently in 'long-and-short' combinations; in pedicellate/sessile combinations. Pedicels of the 'pedicellate' spikelets free of the rachis. The 'shorter' spikelets hermaphrodite. *The 'longer' spikelets sterile (usually reduced to the pedicel, which is basally adnate to the lower glume of the sessile spikelet)*.

Female-fertile spikelets. *Spikelets* slightly *compressed dorsiventrally*; falling with the glumes. Rachilla terminated by a female-fertile floret. Hairy callus present. Callus blunt.

Glumes two (often dark coloured); *very unequal (G_2 much smaller)*; long relative to the adjacent lemmas (the lower only, the upper being much smaller); not pointed (truncate); awnless (G_2 sometimes with nerves excurrent as mucros); non-carinate; very dissimilar (G_2 2–3 keeled, G_1 flattened on the back and broadened above). Lower glume not two-keeled; convex on the back; not pitted; relatively smooth; obscurely 7–9 nerved. Upper glume 3 nerved. *Spikelets* with incomplete florets. The incomplete florets proximal to the female-fertile florets. *The proximal incomplete florets* 1; male. The proximal lemmas awnless; 0 nerved, or 1 nerved; more or less equalling the female-fertile lemmas; less firm than the female-fertile lemmas to similar in texture to the female-fertile lemmas (hyaline); not becoming indurated.

Female-fertile florets 1. Lemmas linear; less firm than the glumes; incised; not deeply cleft (notched); mucronate (from the sinus), or awned. Awns 1; median; from a sinus (or mucronate); geniculate; hairy (puberulous); much shorter than the body of the lemma to much longer than the body of the lemma. Lemmas hairless; non-carinate; without a germination flap; 1 nerved. **Palea** present; conspicuous but relatively short; entire (truncate or obtuse); awnless, without apical setae; not indurated (hyaline, glabrous); *nerveless*. *Lodicules* absent. **Stamens 2**. Anthers 0.75–3.5 mm long. *Ovary* glabrous. Stigmas 2.

Fruit, embryo and seedling. Hilum short. Embryo large.

Abaxial leaf blade epidermis. *Costal/intercostal zonation* conspicuous. *Papillae* present (large, scabrid). Intercostal papillae over-arching the stomata (laterally); consisting of one oblique swelling per cell to consisting of one symmetrical projection per cell. Long-cells of similar wall thickness costally and intercostally. Intercostal zones exhibiting many atypical long-cells (many being short and broad). Mid-intercostal long-cells intercostal long cells mostly hexagonal, many nearly isodiametric; having straight or only gently undulating walls. *Microhairs* present; panicoid-type; 30–45 microns long; 5.1–8.1 microns wide at the septum. Microhair total length/width at septum 3.2–8.3. Microhair apical cells 15–31.5 microns long. Microhair apical cell/total length ratio 0.48–0.75. *Stomata* common; 18–36 microns long. Subsidiaries variously dome-shaped (but the domes often more or less flat-topped), or triangular (in *A. mangalorensis*). Guard-cells overlapped by the interstomatals (slightly so, where not obscured by papillae). *Intercostal short-cells* absent or very rare. *Costal short-cells* conspicuously in long rows (but many pairs and short rows over midrib). Costal silica bodies horizontally-elongated crenate/sinuous (a few, over the midrib), or 'panicoid-type' (imperfect forms, especially over the midrib), or acutely-angled (in the form of numerous, sharp-pointed crosses over all veins but the midrib); often sharp-pointed.

Transverse section of leaf blade, physiology. C_4; XyMS–. *Mesophyll* with radiate chlorenchyma. *Leaf blade* adaxially flat. *Midrib* conspicuous (by virtue of its slight abaxial expansion, and adaxial colourless tissue); with one bundle only to having a conventional arc of bundles (depending on interpretation); with colourless mesophyll adaxially (small group, over midrib). Bulliforms not in discrete, regular adaxial groups (the adaxial epidermis mainly bulliform). Many of the smallest vascular bundles unaccompanied by sclerenchyma. Combined sclerenchyma girders present (with the primaries); nowhere forming 'figures'. Sclerenchyma all associated with vascular bundles.

Cytology. $2n = 20$ and 40.

Taxonomy. Panicoideae; Andropogonodae; Andropogoneae; Andropogoninae.

Ecology, geography, regional floristic distribution. 15 species. Burma, India, China and Southeast Asia. Xerophytic; species of open habitats. In dry shallow soils.

Paleotropical. Indomalesian. Indian and Indo-Chinese.

References, etc. Leaf anatomical: this project.

Arberella Soderstrom & Calderón

Habit, vegetative morphology. Perennial; caespitose. The flowering culms leafy (i.e. the culms not dimorphic). *Culms* 25–35 cm high; herbaceous; unbranched above. Culm internodes solid. Plants unarmed. **Leaves** not basally aggregated; *with auricular setae (but these tiny)*. Leaf blades linear-lanceolate to ovate-lanceolate; broad, or narrow; 5–30 mm wide (from 5–14 cm long, asymmetric at base); flat; pseudopetiolate; without cross venation; rolled in bud. Ligule an unfringed membrane (ciliolate), or a fringed membrane (ciliate); 0.4–0.8 mm long.

Reproductive organization. *Plants monoecious with all the fertile spikelets unisexual*; without hermaphrodite florets. The spikelets of sexually distinct forms on the same plant; female-only and male-only. The male and female-fertile spikelets segregated, in different parts of the same inflorescence branch (terminal female spikelets, smaller male ones below). The spikelets overtly heteromorphic.

Inflorescence. *Inflorescence* determinate, or indeterminate (according to interpretation); paniculate (several panicles (synflorescences) from each of several to many nodes); with capillary branchlets. Inflorescence (female) with axes ending in spikelets. *Inflorescence spatheate (prophylla projecting beyond the axillant sheaths)*; a complex of 'partial inflorescences' and intervening foliar organs. *Spikelet-bearing axes* paniculate; persistent. Spikelets solitary; not secund; pedicellate (the pedicels of female spikelets thickened at the tip).

Female-sterile spikelets. Male spikelets mainly paired, without glumes; 3.2–6.4 mm long, with one floret, lemma and palea membranous; lodicules 3, fleshy; stamens 3. The male spikelets without glumes; 1 floreted. The lemmas awnless. Male florets 3 staminate.

Female-fertile spikelets. *Spikelets* 8.5–22 mm long; *compressed dorsiventrally*; disarticulating above the glumes; with conventional internode spacings, or with a distinctly elongated rachilla internode above the glumes. Rachilla terminated by a female-fertile floret. Hairy callus absent. *Callus* absent, or short (columnar).

Glumes two; relatively large; very unequal, or more or less equal; about equalling the spikelets to exceeding the spikelets; long relative to the adjacent lemmas; hairless; glabrous; pointed; awnless; non-carinate; similar (membranous, ovate-lanceolate, attenuate). Lower glume 5–6 nerved, or 9–11 nerved (with transverse veinlets). Upper glume 5–9 nerved (with transverse veinlets). *Spikelets* with female-fertile florets only.

Female-fertile florets 1. Lemmas completely embracing the palea, ovate lanceolate; decidedly firmer than the glumes; becoming indurated (fleshy early, hardening later); entire; pointed; awnless (but apiculate); hairy (pilose basally and along margins); non-carinate; having the margins tucked in onto the palea; with a clear germination flap (apparently); 5 nerved. *Palea* present; relatively long; entire; awnless, without apical setae; indurated (textured like the lemma); several nerved (4–10); keel-less. *Lodicules* present; 3; free; fleshy; glabrous; heavily vascularized. *Stamens* 0. *Ovary* glabrous. Styles fused (into one long ribbon-like style). Stigmas 2 (plumose).

Fruit, embryo and seedling. Fruit subglobose (ovoid-spherical); not noticeably compressed. Hilum long-linear. Pericarp fused.

Abaxial leaf blade epidermis. *Costal/intercostal zonation* conspicuous (the costal zones narrow and distant from one another, the wide intercostal zones themselves subdivided into a broad, non-papillate, astomatal median zone, and narrow, stomatal, densely papillate zones bordering each costal zone). *Papillae* present; costal and intercostal (completely lacking from the wide median regions of the intercostal zones and near the blade margins, but abundant elsewhere). Intercostal papillae over-arching the stomata (except near the blade margins); several per cell (mostly of the coronate type or branched, large, in one or more irregular longitudinal rows per cell). Long-cells of similar wall thickness costally and intercostally. Mid-intercostal long-cells rectangular; having markedly sinuous walls. *Microhairs* present (abundant near the blade margins, seemingly absent elsewhere); elongated; clearly two-celled; panicoid-type; 45–60 microns long; 5.4–6.6 microns wide at the septum. Microhair total length/width at septum 5.4–6.8. Microhair apical cells (21–)30(–37.5) microns long. Microhair apical cell/total length ratio 0.5–0.53. *Stomata* common (bordering the costae, but sunken, overarched and hard to find except near the blade margins, where they are completely exposed); 24–27 microns long. *Intercostal short-cells* common; in cork/silica-cell pairs; silicified. Intercostal silica bodies vertically elongated-nodular. Large prickles without basal rosettes common costally and intercostally. *Crown cells* absent. *Costal short-cells* conspicuously in long rows. Costal silica bodies oryzoid (some, being vertically elongated crosses), or 'panicoid-type'; abundant, warty, mostly cross shaped.

Transverse section of leaf blade, physiology. C$_3$. Mesophyll seemingly without arm cells; with fusoids. The fusoids external to the PBS. *Leaf blade* with low, very wide adaxial ribs. *Midrib* conspicuous; having a conventional arc of bundles (one large, median primary bundle and a minor bundle on either side); with colourless mesophyll adaxially. *The lamina* symmetrical on either side of the midrib. Bulliforms not in discrete, regular adaxial groups (the adaxial epidermis extensively bulliform, with the cells gradually increasing in size towards the middle of each intercostal region). All the vascular bundles accompanied by sclerenchyma. Combined sclerenchyma girders present; forming 'figures' (all the bundles with slender girders, some of these constituting slight I's or 'anchors'). Sclerenchyma all associated with vascular bundles.

Cytology. Chromosome base number, $x = 11$. $2n = 22$. 2 ploid.

Taxonomy. Bambusoideae; Oryzodae; Olyreae.

Ecology, geography, regional floristic distribution. 3 species. Tropical America. Neotropical. Caribbean, Amazon, and Central Brazilian.

References, etc. Morphological/taxonomic: Soderstrom and Calderón 1979b; Soderstrom and Zuloaga 1988. Leaf anatomical: this project.

Arctagrostis Griseb.

Habit, vegetative morphology. Robust *perennial*; rhizomatous. **Culms** 10–120 cm high; *herbaceous*. Culm internodes hollow. Leaves non-auriculate. Leaf blades linear; narrow; 2–7 mm wide; flat; without cross venation. Ligule an unfringed membrane; truncate; 2–5 mm long.

Reproductive organization. *Plants bisexual, with bisexual spikelets*; with hermaphrodite florets. Exposed-cleistogamous, or chasmogamous (?).

Inflorescence. *Inflorescence paniculate*; open, or contracted; when contracted spicate, or more or less irregular; espatheate; not comprising 'partial inflorescences' and foliar organs. Spikelet-bearing axes persistent. Spikelets not secund; pedicellate.

Female-fertile spikelets. *Spikelets* 2.2–7 mm long; purplish; *not noticeably compressed*; disarticulating above the glumes. Rachilla terminated by a female-fertile floret. Hairy callus absent. *Callus* short.

Glumes two; more or less equal; *shorter than the adjacent lemmas*; hairless (usually glabrous); pointed; awnless; *carinate*; similar. Lower glume 1 nerved. *Upper glume 3 nerved*. Spikelets with female-fertile florets only.

Female-fertile florets 1. **Lemmas** similar in texture to the glumes to decidedly firmer than the glumes; not becoming indurated; entire, or incised; when entire pointed, or blunt; awnless, or mucronate; hairless; *carinate*; without a germination flap; 3–5 nerved. *Palea* present (similar to lemma); relatively long; textured like the lemma; not indurated; 2-nerved; obscurely 2-keeled, or keel-less. *Lodicules* present; 2; free; membranous; glabrous; not toothed; not or scarcely vascularized. *Stamens* 2, or 3. Anthers 1–3 mm long; not penicillate. *Ovary* glabrous. Styles free to their bases. Stigmas 2; white.

Fruit, embryo and seedling. *Fruit* free from both lemma and palea; small. Hilum short. Embryo small. Endosperm liquid in the mature fruit.

Abaxial leaf blade epidermis. *Costal/intercostal zonation* conspicuous. *Papillae* absent. *Long-cells* similar in shape costally and intercostally; differing markedly in wall thickness costally and intercostally. *Mid-intercostal long-cells* rectangular; having straight or only gently undulating walls. *Microhairs* absent. *Stomata* common. Subsidiaries parallel-sided. Guard-cells overlapped by the interstomatals. *Intercostal short-cells* absent or very rare. Costal silica bodies absent (in the material seen).

Transverse section of leaf blade, physiology. C_3; XyMS+. *Mesophyll* with non-radiate chlorenchyma.

Cytology. Chromosome base number, $x = 7$. $2n = 28, 42, 56,$ and 63. 4, 8, and 9 ploid.

Taxonomy. Pooideae; Poodae; Aveneae.

Ecology, geography, regional floristic distribution. 6 species. Arctic America, Eurasia. Helophytic, or mesophytic; species of open habitats. Arctic, marshy tundra.

Holarctic. Boreal. Arctic and Subarctic, Euro-Siberian, and Atlantic North American. European and Siberian. Canadian-Appalachian.

Rusts and smuts. Rusts — *Puccinia*. Taxonomically wide-ranging species: *Puccinia brachypodii*. Smuts from Tilletiaceae and from Ustilaginaceae. Tilletiaceae — *Urocystis*. Ustilaginaceae — *Ustilago*.

References, etc. Leaf anatomical: this project.

Arctophila Rupr. ex Andersson

Habit, vegetative morphology. Perennial; rhizomatous, or stoloniferous (with thick, brittle rhizomes). **Culms** robust, 30–80 cm high; *herbaceous*; unbranched above. Culm nodes glabrous. Culm internodes hollow. Young shoots extravaginal. **Leaves** *not basally aggregated*; non-auriculate. Leaf blades linear; narrow (purplish); 2–7(–13) mm wide; flat; without cross venation; persistent; rolled in bud. *Ligule an unfringed membrane*; not truncate (usually lacerate); 3–5 mm long.

Reproductive organization. Plants bisexual, with bisexual spikelets; with hermaphrodite florets.

Inflorescence. *Inflorescence paniculate*; open (with long branches, the lower often deflexed); espatheate; not comprising 'partial inflorescences' and foliar organs. Spikelet-bearing axes persistent. Spikelets not secund; pedicellate.

Female-fertile spikelets. Spikelets 2.5–8 mm long; compressed laterally to not noticeably compressed; disarticulating above the glumes. Rachilla prolonged beyond the uppermost female-fertile floret; rachilla hairless. The rachilla extension with incomplete florets. *Hairy callus present. Callus* short; blunt.

Glumes two; more or less equal; shorter than the spikelets; *long relative to the adjacent lemmas*; hairless; pointed; awnless; *non-carinate*; similar (membranous, ovate, apices scarious). Lower glume longer than half length of lowest lemma; 1 nerved. Upper glume 3 nerved. **Spikelets** *with incomplete florets*. The incomplete florets distal to the female-fertile florets. The distal incomplete florets 1; *incomplete florets* merely underdeveloped (vestigial, at the rachilla tip).

Female-fertile florets 2–7. *Lemmas* similar in texture to the glumes to decidedly firmer than the glumes; not becoming indurated; entire; blunt; *mucronate*; hairy (at the base), or hairless; non-carinate; without a germination flap; 3–5 nerved. *Palea* present; relatively long; apically notched; not indurated; 2-nerved; 2-keeled. *Lodicules* present; 2; free; membranous; glabrous; toothed, or not toothed; not or scarcely vascularized. *Stamens* 3. Anthers 1.2–2 mm long; not penicillate; without an apically prolonged connective. *Ovary* glabrous. Stigmas 2; white.

Fruit, embryo and seedling. *Fruit* free from both lemma and palea. Embryo small; without an epiblast; without a scutellar tail; with a negligible mesocotyl internode. Embryonic leaf margins meeting.

Abaxial leaf blade epidermis. *Costal/intercostal zonation* conspicuous. *Papillae* absent. *Long-cells* markedly different in shape costally and intercostally (the costals narrower, more regularly fusiform); differing markedly in wall thickness costally and intercostally (the costals with thick, pitted walls, the intercostal walls thin and unpitted). Mid-intercostal long-cells rectangular (occasionally), or fusiform (mostly); having straight or only gently undulating walls. *Microhairs* absent. *Stomata* common. Subsidiaries parallel-sided. Guard-cells overlapped by the interstomatals (but only slightly so). *Intercostal short-cells* absent or very rare. *Costal short-cells* neither distinctly grouped into long rows nor predominantly paired (solitary, infrequent). Costal silica bodies absent to poorly developed (i.e. in the infrequent costal short-cells in the material of *A. fulva* seen).

Transverse section of leaf blade, physiology. C_3; XyMS+. Mesophyll without adaxial palisade. *Leaf blade* 'nodular' in section (with broad, rounded adaxial and abaxial ribs); with the ribs more or less constant in size. *Midrib* not readily distinguishable; with one bundle only. Bulliforms present in discrete, regular adaxial groups (fairly ill defined), or not in discrete, regular adaxial groups; in simple fans (the 'groups' irregular). All the vascular bundles accompanied by sclerenchyma. Combined sclerenchyma girders present (heavy, with all the bundles); forming 'figures' (mostly I-shaped). Sclerenchyma all associated with vascular bundles.

Cytology. Chromosome base number, $x = 7$. $2n = 42$ and 63. 6 and 9 ploid.

Taxonomy. Pooideae; Poodae; Poeae.

Ecology, geography, regional floristic distribution. 1 species. Arctic and subarctic. Helophytic; species of open habitats; glycophytic. Marshy places and pool margins.

Holarctic. Boreal. Arctic and Subarctic, Euro-Siberian, Atlantic North American, and Rocky Mountains. European and Siberian. Canadian-Appalachian.

Hybrids. Intergeneric hybrids with *Dupontia* (×*Arctodupontia* Tsvelev).

References, etc. Leaf anatomical: this project.

Aristida L.

Aristopsis Catasus, *Arthratherum* P. Beauv., *Chaetaria* P. Beauv., *Curtopogon* P. Beauv., *Kielboul* Adans., *Moulinsia* Raf., *Streptachne* R. Br., *Trixostis* Raf.

Habit, vegetative morphology. Annual, or perennial; caespitose. *Culms* 10–100(–180) cm high; herbaceous; branched above, or unbranched above. Culm nodes glabrous. Culm internodes solid, or hollow. *Leaves* mostly basal, or not basally aggregated (in annual species); non-auriculate. *Leaf blades* linear, or linear-lanceolate; narrow; flat, or rolled; *not pseudopetiolate*; without cross venation; persistent; rolled in bud, or once-folded in bud. *Ligule* present; a fringed membrane to a fringe of hairs. *Contra-ligule* present (of hairs), or absent.

Reproductive organization. Plants bisexual, with bisexual spikelets; with hermaphrodite florets. Exposed-cleistogamous, or chasmogamous. Plants with hidden cleistogenes, or without hidden cleistogenes. The hidden cleistogenes in the leaf sheaths (when present).

Inflorescence. Inflorescence few spikeleted, or many spikeleted; paniculate; deciduous in its entirety, or not deciduous; open, or contracted. Rachides hollowed, or flattened, or winged, or neither flattened nor hollowed, not winged. Inflorescence espatheate; not comprising 'partial inflorescences' and foliar organs. Spikelet-bearing axes persistent. Spikelets not secund; pedicellate.

Female-fertile spikelets. Spikelets 4–30 mm long; compressed laterally to not noticeably compressed; disarticulating above the glumes. Rachilla terminated by a female-fertile floret. Hairy callus present. Callus *long*; pointed.

Glumes two; relatively large; very unequal to more or less equal; (at least the G_2) about equalling the spikelets, or exceeding the spikelets; usually pointed; awned, or awnless; carinate; very dissimilar, or similar (membranous to papery). Lower glume 1 nerved. Upper glume 1 nerved (usually), or 3 nerved (rarely). **Spikelets** with female-fertile florets only; *without proximal incomplete florets*.

Female-fertile florets 1. **Lemmas narrow, cylindrical**; convolute, or not convolute; *decidedly firmer than the glumes*; becoming indurated to not becoming indurated (leathery to indurated); entire; awned. *Awns triple or trifid, commonly with a basal* *column (usually), or not of the triple/trifid, basal column type (the column sometimes absent, the lateral branches sometimes reduced or absent)*; 1, or 3; apical; non-geniculate (at least, not geniculate in the normal sense); hairless (usually glabrous); much shorter than the body of the lemma to much longer than the body of the lemma; entered by several veins (usually with three veins in the column); deciduous, or persistent. Lemmas hairy (rarely), or hairless; glabrous, or scabrous; non-carinate; with a clear germination flap; 1–3 nerved. **Palea** present; *conspicuous but relatively short, or very reduced (enclosed by the lemma)*; entire; awnless, without apical setae; thinner than the lemma (hyaline); not indurated; 1-nerved, or 2-nerved, or nerveless. *Lodicules* present, or absent; when present, 2; free; membranous; glabrous; not toothed; heavily vascularized. *Stamens* 1–3. Anthers 0.7–2 mm long; not penicillate. *Ovary* glabrous. Styles free to their bases. Stigmas 2; red pigmented, or brown.

Fruit, embryo and seedling. *Fruit* free from both lemma and palea; small to large (3–11 mm long); fusiform; compressed dorsiventrally, or not noticeably compressed. Hilum short, or long-linear. Embryo large; waisted, or not waisted. Endosperm hard; without lipid; containing compound starch grains. Embryo without an epiblast; with a scutellar tail; with an elongated mesocotyl internode; with one scutellum bundle. Embryonic leaf margins meeting.

Seedling with a short mesocotyl. First seedling leaf with a well-developed lamina. The lamina narrow; curved; 3–5 veined.

Abaxial leaf blade epidermis. *Costal/intercostal zonation* conspicuous (or somewhat obscure in some species). *Papillae* absent. *Long-cells* similar in shape costally and intercostally; of similar wall thickness costally and intercostally (usually fairly thick walled). Mid-intercostal long-cells rectangular (and narrow); having markedly sinuous walls. *Microhairs* present; elongated; clearly two-celled; panicoid-type; 48–81(–87) microns long; 5–9.6 microns wide at the septum. Microhair total length/width at septum 8.4–12. Microhair apical cells 24–44 microns long. Microhair apical cell/total length ratio 0.4–0.6. *Stomata* common; 22.5–36 microns long. Subsidiaries dome-shaped, or triangular, or dome-shaped and triangular. Guard-cells overlapping to flush with the interstomatals. *Intercostal short-cells* common, or absent or very rare; in cork/silica-cell pairs (and solitaries); silicified (when paired), or not silicified (when solitary). Intercostal silica bodies variable in shape. *Costal short-cells* conspicuously in long rows, or neither distinctly grouped into long rows nor predominantly paired. Costal silica bodies horizontally-elongated crenate/sinuous, or horizontally-elongated smooth, or rounded, or saddle shaped, or tall-and-narrow, or crescentic, or oryzoid, or 'panicoid-type'; when panicoid type, variously cross shaped to dumb-bell shaped, or nodular (often exhibiting elongated dumb-bells with enlarged, bulbous ends, and/or very large, nodular forms).

Transverse section of leaf blade, physiology. C_4. The anatomical organization unconventional. Organization of PCR tissue *Aristida* type. **Biochemical type** NADP–ME (3 species); *XyMS– (with double PCR sheaths)*. PCR cells without a suberised lamella. *PCR cell chloroplasts* ovoid; with well developed grana (in the outer sheath), or with reduced grana (in the inner sheath); centrifugal/peripheral (in the inner sheath), or centripetal (in the outer sheath). *Mesophyll* with radiate chlorenchyma; traversed by columns of colourless mesophyll cells (usually), or not traversed by colourless columns (e.g. in *A. caput-medusae*). *Leaf blade* with distinct, prominent adaxial ribs; with the ribs very irregular in sizes. *Midrib* conspicuous, or not readily distinguishable; with one bundle only. Bulliforms associated with colourless mesophyll cells to form deeply-penetrating fans (these often linked with colourless girders, the latter fully traversing or not). All the vascular bundles accompanied by sclerenchyma. Combined sclerenchyma girders present; nowhere forming 'figures'. Sclerenchyma all associated with vascular bundles, or not all bundle-associated. The 'extra' sclerenchyma when present, in abaxial groups; abaxial-hypodermal, the groups continuous with colourless columns.

Culm anatomy. *Culm internode bundles* in three or more rings, or scattered.

Phytochemistry. Leaves without flavonoid sulphates (6 species). Leaf blade chlorophyll *a:b* ratio 4.46–4.65.

Cytology. Chromosome base number, $x = 11$ and 12. $2n = 22, 24, 36, 44, 48$, and 66. 2, 4, and 6 ploid. Nucleoli persistent.

Taxonomy. Arundinoideae; Aristideae.

Ecology, geography, regional floristic distribution. 290 species. Temperate and subtropical. Xerophytic.

Holarctic, Paleotropical, Neotropical, Cape, Australian, and Antarctic. Boreal, Tethyan, and Madrean. African, Madagascan, Indomalesian, Polynesian, and Neocaledonian. Euro-Siberian, Atlantic North American, and Rocky Mountains. Macaronesian, Mediterranean, and Irano-Turanian. Saharo-Sindian, Sudano-Angolan, West African Rainforest, Namib-Karoo, and Ascension and St. Helena. Indian, Indo-Chinese, Malesian, and Papuan. Fijian. Caribbean, Venezuela and Surinam, Amazon, Central Brazilian, Pampas, and Andean. North and East Australian and Central Australian. Patagonian. European and Siberian. Canadian-Appalachian, Southern Atlantic North American, and Central Grasslands. Sahelo-Sudanian, Somalo-Ethiopian, South Tropical African, and Kalaharian. Tropical North and East Australian and Temperate and South-Eastern Australian.

Rusts and smuts. Rusts — *Puccinia*. Taxonomically wide-ranging species: *Puccinia graminis* and *Puccinia aristidae*. Smuts from Tilletiaceae and from Ustilaginaceae. Tilletiaceae — *Tilletia*. Ustilaginaceae — *Sorosporium, Sphacelotheca, Tolyposporium*, and *Ustilago*.

Economic importance. Significant weed species: (e.g.) *A. adscensionis, A. dichotoma, A. longiseta, A. oligantha* — the awns of many species can injure livestock. Important native pasture species: none — generally poor value, but various species of minor grazing value in dry regions.

References, etc. Morphological/taxonomic: de Winter 1965; Lazarides 1980. Leaf anatomical: mainly from Metcalfe 1960 and this project.

Arrhenatherum P. Beauv.

Excluding *Pseudarrhenatherum*

Habit, vegetative morphology. Perennial; caespitose. *Culms* 30–200 cm high; herbaceous; unbranched above; tuberous, or not tuberous. Culm nodes glabrous. Culm internodes hollow. *Leaves* not basally aggregated; non-auriculate. Leaf blades linear; narrow; 2–7 mm wide; not setaceous; flat, or rolled (convolute); without cross venation; persistent; rolled in bud. *Ligule an unfringed membrane (sometimes puberulent)*; not truncate; 0.5–3 mm long.

Reproductive organization. Plants bisexual, with bisexual spikelets; with hermaphrodite florets, or without hermaphrodite florets (the upper floret being either hermaphrodite or female-only); outbreeding.

Inflorescence. Inflorescence paniculate; open; espatheate; not comprising 'partial inflorescences' and foliar organs. Spikelet-bearing axes persistent. Spikelets not secund; pedicellate.

Female-fertile spikelets. *Spikelets* 7–11 mm long; *compressed laterally*; *disarticulating above the glumes (the florets falling together)*; not disarticulating between the florets (persistent). Rachilla prolonged beyond the uppermost female-fertile floret; rachilla hairy. Hairy callus present. *Callus* short.

Glumes two; very unequal; about equalling the spikelets to exceeding the spikelets; long relative to the adjacent lemmas; pointed; awnless; carinate, or non-carinate; similar (membranous). Lower glume 1 nerved, or 3 nerved. *Upper glume 3 nerved. Spikelets* with incomplete florets. The incomplete florets proximal to the female-fertile florets, or both distal and proximal to the female-fertile florets. *The distal incomplete florets* merely underdeveloped (when present). *The proximal incomplete florets* when present, 1 (occasionally, both or all florets perfect); paleate; male. The proximal lemmas awned (the awn geniculate, from the lower back); 5–9 nerved; more or less equalling the female-fertile lemmas; similar in texture to the female-fertile lemmas; not becoming indurated.

Female-fertile florets *1 (or rarely 2–4)*. Lemmas decidedly firmer than the glumes; not becoming indurated; entire, or incised; when incised not deeply cleft (notched); awnless, or awned. Awns when present, 1; dorsal; non-geniculate (usually short and slender); much shorter than the body of the lemma to about as long as the body of the lemma; entered by one vein. Lemmas hairy, or hairless; non-carinate; without a germination flap;

5–9 nerved. *Palea* present; relatively long; apically notched (shortly bidentate); 2-nerved; 2-keeled. *Lodicules* present; 2; free; membranous; glabrous; not toothed. **Stamens 3**. Anthers 3.4–6.5 mm long; not penicillate. **Ovary hairy**. Styles free to their bases. Stigmas 2; white.

Fruit, embryo and seedling. *Fruit* small to large; *not grooved*; compressed dorsiventrally to not noticeably compressed; hairy on the body. Hilum long-linear. Embryo small; not waisted. Endosperm hard; with lipid; containing compound starch grains. Embryo with an epiblast; without a scutellar tail; with a negligible mesocotyl internode. Embryonic leaf margins meeting.

Seedling with a long mesocotyl; with a loose coleoptile, or with a tight coleoptile. First seedling leaf with a well-developed lamina. The lamina narrow; 3 veined.

Abaxial leaf blade epidermis. *Costal/intercostal zonation* conspicuous. *Papillae* absent. *Long-cells* similar in shape costally and intercostally; of similar wall thickness costally and intercostally (fairly thin walled). Mid-intercostal long-cells rectangular (long and narrow); having straight or only gently undulating walls. *Microhairs* absent. *Stomata* common. Subsidiaries parallel-sided. Guard-cells overlapped by the interstomatals. *Intercostal short-cells* absent or very rare. Prickles abundant costally and intercostally. *Costal short-cells* neither distinctly grouped into long rows nor predominantly paired. Costal silica bodies horizontally-elongated crenate/sinuous, or horizontally-elongated smooth (mostly), or rounded.

Transverse section of leaf blade, physiology. C_3; XyMS+. *Mesophyll* with non-radiate chlorenchyma. *Leaf blade* 'nodular' in section, or adaxially flat; with the ribs very irregular in sizes. *Midrib* not readily distinguishable; with one bundle only. Bulliforms in simple fans (the cells fairly uniformly sized, or occasionally in irregular groups). All the vascular bundles accompanied by sclerenchyma. Combined sclerenchyma girders present; nowhere forming 'figures'. Sclerenchyma all associated with vascular bundles.

Culm anatomy. *Culm internode bundles* in one or two rings.

Phytochemistry. Tissues of the culm bases with little or no starch. Fructosans predominantly long-chain. Leaves without flavonoid sulphates (1 species).

Cytology. Chromosome base number, $x = 7$. $2n = 14$, 28, and 42. 2, 4, and 6 ploid. Nucleoli disappearing before metaphase.

Taxonomy. Pooideae; Poodae; Aveneae.

Ecology, geography, regional floristic distribution. 4 species. Europe, Mediterranean. Commonly adventive. Mesophytic to xerophytic; species of open habitats. Dry grassland, edges of woods, disturbed ground.

Holarctic. Boreal and Tethyan. Euro-Siberian and Atlantic North American. Macaronesian, Mediterranean, and Irano-Turanian. European. Canadian-Appalachian.

Hybrids. Intergeneric hybrids with *Avena*.

Rusts and smuts. Rusts — *Puccinia*. Taxonomically wide-ranging species: *Puccinia graminis*, *Puccinia coronata*, *Puccinia striiformis*, *Puccinia hordei*, and *Puccinia recondita*. Smuts from Tilletiaceae and from Ustilaginaceae. Tilletiaceae — *Entyloma*, *Tilletia*, and *Urocystis*. Ustilaginaceae — *Ustilago*.

Economic importance. Significant weed species: *A. elatius*. Cultivated fodder: *A. elatius*.

References, etc. Leaf anatomical: Metcalfe 1960; this project.

Arthragrostis Lazarides

Habit, vegetative morphology. Annual. *Culms* 20–60 cm high; herbaceous. Culm nodes glabrous. Culm internodes hollow. *Leaves* not basally aggregated; non-auriculate. Leaf blades linear; narrow; 2–5 mm wide (to 10 cm long); flat; without cross venation. **Ligule** present; *a fringed membrane*.

Reproductive organization. Plants bisexual, with bisexual spikelets; with hermaphrodite florets.

Inflorescence. *Inflorescence paniculate*; *open (7–10 cm long)*; espatheate; not comprising 'partial inflorescences' and foliar organs. *Spikelet-bearing axes disarticulating*; falling entire (the primary inflorescence branches deciduous). Spikelets solitary; not secund; pedicellate. Pedicel apices truncate. Spikelets not in distinct 'long-and-short' combinations.

Female-fertile spikelets. Spikelets 4–4.9 mm long; lanceolate; compressed dorsiventrally; falling with the glumes; with a distinctly elongated rachilla internode between the glumes and with distinctly elongated rachilla internodes between the florets. The upper floret conspicuously stipitate. *The stipe beneath the upper floret filiform (not appendaged)*; homogeneous. Rachilla terminated by a female-fertile floret. Hairy callus absent.

Glumes two; very unequal; (the upper) long relative to the adjacent lemmas; hairless (G$_1$ glabrous, G$_2$ scabrid); pointed (G$_2$ acuminate), or not pointed (G$_1$ acute to rounded); awnless; non-carinate; very dissimilar. Lower glume 7 nerved. *Upper glume 11 nerved.* *Spikelets* with incomplete florets. The incomplete florets proximal to the female-fertile florets. *The proximal incomplete florets* 1; epaleate; sterile. The proximal lemmas awnless; 9–11 nerved; considerably longer than the upper lemmas; less firm than the female-fertile lemmas (membranous); not becoming indurated.

Female-fertile florets 1. *Lemmas* acute; *decidedly firmer than the glumes (cartilaginous)*; smooth; white in fruit; entire; pointed to blunt; awnless; hairless; glabrous; non-carinate; having the margins tucked in onto the palea; with a clear germination flap; obscurely nerved. *Palea* present; relatively long; entire; awnless, without apical setae; cartilaginous, like the lemma. *Lodicules* present; 2; free; fleshy; glabrous. *Stamens* 3. *Ovary* glabrous. Styles free to their bases. Stigmas 2.

Abaxial leaf blade epidermis. *Costal/intercostal zonation* conspicuous. *Papillae* absent. *Long-cells* markedly different in shape costally and intercostally (the costals very much narrower). Mid-intercostal long-cells rectangular; having markedly sinuous walls. *Microhairs* present; panicoid-type. *Stomata* common. Subsidiaries triangular. Guard-cells overlapping to flush with the interstomatals. *Intercostal short-cells* absent or very rare. With numerous large, cushion-based macrohairs. *Crown cells* absent. *Costal short-cells* conspicuously in long rows. Costal silica bodies 'panicoid-type'; mostly small butterfly shaped and dumb-bell shaped.

Transverse section of leaf blade, physiology. C$_4$; XyMS+. PCR sheath outlines uneven. PCR sheath extensions absent. *Leaf blade* adaxially flat. *Midrib* conspicuous (via a prominent, flat abaxial keel, and a large abaxial sclerenchyma girder associated with the median bundle); having a conventional arc of bundles (there being a smaller lateral on either side of the median). Bulliforms present in discrete, regular adaxial groups; in simple fans. Many of the smallest vascular bundles unaccompanied by sclerenchyma. Combined sclerenchyma girders absent (the larger bundles with small abaxial strands or girders only). Sclerenchyma all associated with vascular bundles.

Taxonomy. Panicoideae; Panicodae; Paniceae.

Ecology, geography, regional floristic distribution. 2 species. Australia. Mesophytic; species of open habitats. In scrub.

Australian. North and East Australian. Tropical North and East Australian.

References, etc. Morphological/taxonomic: Lazarides 1985. Leaf anatomical: this project.

Special comments. Fruit data wanting.

Arthraxon P. Beauv.

Alectoridia A. Rich., *Batratherum* Nees, *Lasiolytrum* Steud., *Lucaea* Kunth, *Pleuroplitis* Trin.

Habit, vegetative morphology. Annual, or perennial (often trailing); decumbent. *Culms* 10–100 cm high; herbaceous; branched above, or unbranched above; tuberous, or not tuberous. Culm nodes glabrous. Culm internodes solid, or hollow. Young shoots intravaginal. *Leaves* not basally aggregated; non-auriculate. *Leaf blades* linear-lanceolate to ovate-lanceolate; broad, or narrow; *cordate*; without cross venation; persistent. *Ligule a fringed membrane (short)*.

Reproductive organization. Plants bisexual, with bisexual spikelets; with hermaphrodite florets. The spikelets of sexually distinct forms on the same plant (usually); hermaphrodite, or hermaphrodite and male-only, or hermaphrodite and sterile; overtly heteromorphic, or homomorphic.

Inflorescence. *Inflorescence of spicate main branches*; usually subdigitate, or nondigitate (when the racemes solitary). Primary inflorescence branches (1–)2–30. Inflorescence espatheate; not comprising 'partial inflorescences' and foliar organs (but the inflorescences terminal and/or axillary). **Spikelet-bearing axes** spikes (rarely), or 'racemes' (usually); clustered; with very slender rachides; *disarticulating*; disarticulating at the joints. *'Articles'* linear; without a basal callus-knob; not appendaged; disarticulating transversely. Spikelets solitary (rarely), or paired; secund; sessile (the pedicelled spikelets sometimes completely suppressed), or sessile and pedicellate (but the latter often much reduced); consistently in 'long-and-short' combinations, or not in distinct 'long-and-short' combinations; when paired in pedicellate/sessile combinations. Pedicels of the 'pedicellate' spikelets (when present) free of the rachis. The 'shorter' spikelets hermaphrodite. The 'longer' spikelets male-only, or sterile.

Female-sterile spikelets. The pedicelled spikelets variable in form, male, sterile or vestigial, reduced to a sometimes microscopic pedicel, or totally suppressed.

Female-fertile spikelets. Spikelets 2–6.5 mm long; compressed laterally, or compressed dorsiventrally; falling with the glumes. Rachilla terminated by a female-fertile floret. Hairy callus absent.

Glumes two; more or less equal; long relative to the adjacent lemmas; without conspicuous tufts or rows of hairs; awnless; very dissimilar (the lower often leathery, rounded on the back, the upper less firm, laterally compressed). Lower glume more or less two-keeled, or not two-keeled; convex on the back; not pitted; rugose to prickly (sometimes with lateral rows of tubercles or spines); 5–20 nerved. Upper glume 1–5 nerved. *Spikelets with incomplete florets.* The incomplete florets proximal to the female-fertile florets. *The proximal incomplete florets* 1; epaleate; sterile. The proximal lemmas awnless; 0–1 nerved; more or less equalling the female-fertile lemmas; similar in texture to the female-fertile lemmas (membranous); not becoming indurated.

Female-fertile florets 1. Lemmas ovate-lanceolate; less firm than the glumes (hyaline); not becoming indurated; entire, or incised; pointed to blunt (acute to erose); when incised, not deeply cleft; awned (nearly always), or awnless (*A. submuticus*). *Awns* 1; median; dorsal; *from well down the back (near the base)*; geniculate; hairless (glabrous); much shorter than the body of the lemma to much longer than the body of the lemma. Lemmas hairless; non-carinate (rounded on the back); without a germination flap; 1 nerved, or 3 nerved. **Palea** present, or absent; when present, very reduced; entire (triangular, subacute); awnless, without apical setae; membranous; not indurated (hyaline); *nerveless*; keel-less. *Lodicules* present; 2; free; fleshy; glabrous. *Stamens* 2–3. Anthers not penicillate. *Ovary* glabrous. Styles free to their bases. Stigmas 2.

Fruit, embryo and seedling. *Fruit* free from both lemma and palea; fusiform, or ellipsoid; slightly compressed laterally. Hilum short. Embryo large. Endosperm hard; without lipid; containing compound starch grains. Embryo with an elongated mesocotyl internode.

Seedling with a long mesocotyl. First seedling leaf with a well-developed lamina. The lamina broad; supine.

Abaxial leaf blade epidermis. *Costal/intercostal zonation* conspicuous. *Papillae* present (though infrequent and often poorly defined in *A. hispidulus*). Intercostal papillae not over-arching the stomata; consisting of one oblique swelling per cell (at least in *A. hispidulus*). *Long-cells* markedly different in shape costally and intercostally (the costals much narrower, the intercostals broad); of similar wall thickness costally and intercostally (quite thin walled). Intercostal zones with typical long-cells (though these rather short, in places). Mid-intercostal long-cells rectangular; having markedly sinuous walls. *Microhairs* present; panicoid-type; (18–)22–37(–39) microns long; 4.5–6.3 microns wide at the septum. Microhair total length/width at septum 6.2–6.7. Microhair apical cells 10–21 microns long. Microhair apical cell/total length ratio 0.5–0.61. *Stomata* common; 30–33 microns long. Subsidiaries dome-shaped and triangular. Guard-cells overlapping to flush with the interstomatals. *Intercostal short-cells* absent or very rare. *Costal short-cells* conspicuously in long rows. Costal silica bodies 'panicoid-type'; butterfly shaped, or dumb-bell shaped.

Transverse section of leaf blade, physiology. C_4; XyMS–. *Mesophyll* with radiate chlorenchyma, or with non-radiate chlorenchyma (rarely). *Leaf blade* adaxially flat. *Midrib* conspicuous; with one bundle only. Bulliforms in irregular groups. Many of the smallest vascular bundles unaccompanied by sclerenchyma. Combined sclerenchyma girders absent (adaxial strands combined with abaxial girders or strands, or the bundles with abaxial sclerenchyma only). Sclerenchyma all associated with vascular bundles.

Culm anatomy. *Culm internode bundles* in one or two rings, or in three or more rings.

Phytochemistry. Leaves without flavonoid sulphates (1 species).

Cytology. Chromosome base number, $x = 9$ and 10. $2n = 18$, 20, 36, and 40. 2 and 4 ploid.

Taxonomy. Panicoideae; Andropogonodae; Andropogoneae; Andropogoninae.

Ecology, geography, regional floristic distribution. 7 species. Tropical Africa, Madagascar, Mauritius, Indomalayan region to Japan. Helophytic to mesophytic; shade species, or species of open habitats; glycophytic.

Holarctic, Paleotropical, and Australian. Boreal and Tethyan. African, Madagascan, and Indomalesian. Euro-Siberian, Eastern Asian, Atlantic North American, and Rocky Mountains. Macaronesian and Irano-Turanian. Saharo-Sindian, Sudano-Angolan, and West African Rainforest. Indian, Indo-Chinese, Malesian, and Papuan. North and East Australian. European. Canadian-Appalachian and Southern Atlantic North American. Sahelo-Sudanian, Somalo-Ethiopian, and South Tropical African. Tropical North and East Australian.

Rusts and smuts. Rusts — *Puccinia*. Smuts from Ustilaginaceae. Ustilaginaceae — *Sorosporium* and *Sphacelotheca*.

References, etc. Morphological/taxonomic: Van Welzen 1981. Leaf anatomical: Metcalfe 1960; this project.

Arthropogon Nees

Excluding *Achlaena*

Habit, vegetative morphology. Perennial; caespitose. **Culms** 60–100 cm high; herbaceous; **unbranched above**. Culm nodes glabrous. *Leaves* mostly basal, or not basally aggregated. Leaf blades linear, or linear-lanceolate; narrow; 1–6(–10) mm wide; setaceous (*A. filifolius*), or not setaceous; without cross venation. Ligule a fringed membrane, or a fringe of hairs.

Reproductive organization. Plants bisexual, with bisexual spikelets; with hermaphrodite florets. The spikelets hermaphrodite, or female-only.

Inflorescence. *Inflorescence paniculate*; open; with capillary branchlets (*A. filifolius*), or without capillary branchlets; espatheate; not comprising 'partial inflorescences' and foliar organs. Spikelet-bearing axes persistent. Spikelets not secund; pedicellate.

Female-fertile spikelets. Spikelets 3–11 mm long; compressed laterally; falling with the glumes. Rachilla terminated by a female-fertile floret. *Hairy callus present (not stipitate, by contrast with* **Achlaena***).*

Glumes two; very unequal (G_2 longer); *awned (the G_1 subulate)*; *very dissimilar (the lower linear to awnlike, the upper equalling the spikelet, entire, bidentate or bilobed, awned)*. Lower glume 1 nerved, or 3 nerved. Upper glume 3 nerved, or 5 nerved. *Spikelets* with incomplete florets. The incomplete florets proximal to the female-fertile florets. *Spikelets with proximal incomplete florets. The proximal incomplete florets* 1; paleate, or epaleate (e.g. *A. scaber*); male (e.g. *A. filifolius*), or sterile. *The proximal lemmas* awned (*A. xerachne*), or awnless; 3 nerved, or 5 nerved; *decidedly exceeding the female-fertile lemmas*; *decidedly firmer than the female-fertile lemmas (leathery)*; not becoming indurated.

Female-fertile florets 1. **Lemmas** less firm than the glumes (membranous); not becoming indurated; *awnless*; without a germination flap; 0 nerved, or 1 nerved, or 3 nerved. *Palea* present, or absent; conspicuous but relatively short to very reduced; awnless, without apical setae; thinner than the lemma, or textured like the lemma (membranous or hyaline); not indurated (scarious); 2-nerved, or nerveless. *Lodicules* present; 2. *Stamens* 3. *Ovary* glabrous. Stigmas 2.

Fruit, embryo and seedling. *Fruit* deciduous with lemma and palea; small to medium sized; chestnut brown in *A. villosus*. Embryo large.

Abaxial leaf blade epidermis. *Costal/intercostal zonation* conspicuous. *Papillae* absent. Mid-intercostal long-cells rectangular; having markedly sinuous walls (conspicuously pitted). *Microhairs* present; panicoid-type. *Stomata* common. Subsidiaries mostly triangular. Guard-cells overlapping to flush with the interstomatals. *Intercostal short-cells* common; in cork/silica-cell pairs; silicified. Intercostal silica bodies crescentic. With large cushion-based macrohairs. *Costal short-cells* conspicuously in long rows (but the short-cells often rather long, or irregular in length). Costal silica bodies rounded and 'panicoid-type'; shortly dumb-bell shaped, or nodular (and irregular forms, integrading with potato-shapes, etc.).

Transverse section of leaf blade, physiology. C_4. The anatomical organization conventional, or unconventional. Organization of PCR tissue when unconventional, *Arundinella* type. XyMS+ (? — see photo of *A. lanceolatus* in Filgueiras 1982), or XyMS– (*A. villosus*). *Mesophyll* with radiate chlorenchyma; exhibiting 'circular cells', or without 'circular cells' (but with many small bundles much reduced in this direction). *Midrib* not readily distinguishable; with one bundle only. Bulliforms present in discrete, regular adaxial groups, or not in discrete, regular adaxial groups (then epidermis seemingly extensively bulliform). All the vascular bundles accompanied by sclerenchyma (the small bundles and even some of the colourless groups with small abaxial strands). Combined sclerenchyma girders present; forming 'figures' (the primaries with I's). Sclerenchyma all associated with vascular bundles.

Taxonomy. Panicoideae; Panicodae; Paniceae (Arthropogoneae).

Ecology, geography, regional floristic distribution. 5 species. Brazil. Species of open habitats. Savanna.

Neotropical. Caribbean and Central Brazilian.

Rusts and smuts. Rusts — *Puccinia*.

References, etc. Morphological/taxonomic: Filgueiras 1982. Leaf anatomical: this project for *A. villosus*; supplemented from photos in Filgueiras 1982.

Special comments. Most of the anatomical description refers to *A. villosus*, which has been examined directly. A photo of *A. lanceolatus* in Filgueiras (1982) suggests there is variation in XyMS.

Arthrostylidium Rupr.

Habit, vegetative morphology. Perennial; caespitose. The flowering culms leafy. *Culms woody and persistent (the first culm internode often disproportionately elongated)*; scandent, or not scandent; branched above. Primary branches/mid-culm node 1. Unicaespitose. Rhizomes pachymorph. Plants unarmed. *Leaves* not basally aggregated; with auricular setae. Leaf blades pseudopetiolate; cross veined, or without cross venation; disarticulating from the sheaths.

Reproductive organization. Plants bisexual, with bisexual spikelets; with hermaphrodite florets.

Inflorescence. Inflorescence *determinate*; *of spicate main branches (of spicate racemes)*; digitate, or subdigitate, or non-digitate (?); *espatheate (with neither bracts nor prophylls)*; not comprising 'partial inflorescences' and foliar organs. Spikelet-bearing axes persistent. Spikelets solitary.

Female-fertile spikelets. Spikelets compressed laterally; disarticulating above the glumes. Rachilla prolonged beyond the uppermost female-fertile floret. The rachilla extension with incomplete florets. Hairy callus absent.

Glumes two; very unequal; shorter than the adjacent lemmas; pointed; carinate (apparently); similar. *Spikelets* with incomplete florets. The incomplete florets both distal and proximal to the female-fertile florets. *The distal incomplete florets* merely underdeveloped. *The proximal incomplete florets* 1. The proximal lemmas exceeded by the female-fertile lemmas.

Female-fertile florets 2–7 ('few to several'). Lemmas entire; pointed; awnless, or mucronate, or awned. Awns when present, 1; non-geniculate; much shorter than the body of the lemma. Lemmas carinate. *Palea* present; 2-nerved; 2-keeled (broadly sulcate). *Lodicules* nearly always present; 3; free; ciliate; not toothed; heavily vascularized. *Stamens* 3. Anthers not penicillate; without an apically prolonged connective. **Ovary** glabrous; *with a conspicuous apical appendage*. The appendage broadly conical, fleshy. Styles fused. Stigmas 2.

Fruit, embryo and seedling. Fruit longitudinally grooved; compressed dorsiventrally to not noticeably compressed. Hilum long-linear. Embryo small.

Abaxial leaf blade epidermis. *Costal/intercostal zonation* conspicuous. *Papillae* present, or absent. Long-cells of similar wall thickness costally and intercostally, or differing markedly in wall thickness costally and intercostally. Mid-intercostal long-cells having markedly sinuous walls. *Microhairs* present; panicoid-type. *Stomata* common. Subsidiaries low dome-shaped. *Intercostal short-cells* common (abundant); in cork/silica-cell pairs. *Costal short-cells* predominantly paired (and some short rows). Costal silica bodies saddle shaped.

Transverse section of leaf blade, physiology. C_3; XyMS+. Mesophyll with arm cells (these relatively inconspicuous); with fusoids (but these scarce). The fusoids external to the PBS. *Leaf blade* with distinct, prominent adaxial ribs, or 'nodular' in section; with the ribs more or less constant in size, or with the ribs very irregular in sizes. *Midrib* conspicuous; having complex vascularization. Bulliforms present in discrete, regular adaxial groups; in simple fans (mostly), or associated with colourless mesophyll cells to form deeply-penetrating fans. All the vascular bundles accompanied by sclerenchyma. Combined sclerenchyma girders present (nearly all the bundles); forming 'figures'.

Taxonomy. Bambusoideae; Bambusodae; Bambuseae.

Ecology, geography, regional floristic distribution. 20 species. New World. Neotropical. Caribbean, Venezuela and Surinam, Amazon, and Andean.

References, etc. Leaf anatomical: Metcalfe 1960.

Arundinaria Mich.

Bashania Keng f. & Yi, *Butania* Keng f., *Ludolphia* Willd., *Clavinodum* Wen, *Macronax* Raf, *Miegia* Pers., *Nipponocalamus* Nakai, *Oligostachyum* Wang & Ye, *Omeiocalamus* Keng f., *Pleioblastus* Nakai, *Triglossum* Roem. & Schult., *Tschompskia* Aschers. & Graebn.

Excluding *Pseudosasa*

Habit, vegetative morphology. Perennial. The flowering culms leafless, or leafy. *Culms* 200–800 cm high; woody and persistent; cylindrical; not scandent (usually). Primary branches/mid-culm node 1 (branching into 3–7 secondaries). Culm internodes hollow. Unicaespitose, or pluricaespitose. *Rhizomes leptomorph*. Plants unarmed. **Leaves** not basally aggregated; *with auricular setae*. Leaf blades broad, or narrow (relatively small); pseudopetiolate; cross veined; disarticulating from the sheaths. *Ligule a fringed membrane to a fringe of hairs*.

Reproductive organization. Plants bisexual, with bisexual spikelets; with hermaphrodite florets. The spikelets all alike in sexuality.

Inflorescence. *Inflorescence* determinate; without pseudospikelets; reduced to a single spikelet, or few spikeleted, or many spikeleted; of spicate main branches, or paniculate (i.e. variable, regardless of problems with terminology — generally open racemose or paniculate, sometimes both forms combined in the one plant, sometimes the 'branches' reduced to single spikelets); open; spatheate; a complex of 'partial inflorescences' and intervening foliar organs, or not comprising 'partial inflorescences' and foliar organs (?). *Spikelet-bearing axes* 'racemes', or paniculate; persistent. Spikelets pedicellate.

Female-fertile spikelets. Spikelets 10–80 mm long; compressed laterally; disarticulating above the glumes; disarticulating between the florets. Rachilla prolonged beyond the uppermost female-fertile floret; rachilla hairy. The rachilla extension with incomplete florets. Hairy callus absent.

Glumes one per spikelet, or two; very unequal, or more or less equal; shorter than the adjacent lemmas; awnless; similar. Lower glume 4–7 nerved. Upper glume 8–13 nerved. *Spikelets* with incomplete florets. The incomplete florets both distal and proximal to the female-fertile florets. *The distal incomplete florets* merely underdeveloped. **Spikelets with proximal incomplete florets**. The proximal incomplete florets 1 (or more?); male, or sterile (?). The proximal lemmas awnless; 9–11 nerved; exceeded by the female-fertile lemmas; similar in texture to the female-fertile lemmas; not becoming indurated.

Female-fertile florets 4–20 (?). Lemmas often acuminate or setaceous-tipped; similar in texture to the glumes; not becoming indurated (papery); entire; usually pointed; awnless, or mucronate, or awned. Awns when present, 1; apical; non-geniculate; much shorter than the body of the lemma to about as long as the body of the lemma (?). Lemmas 9–15 nerved. *Palea* present; relatively long, or conspicuous but relatively short; not indurated; several nerved (4 to 13 observed); 2-keeled (dorsally sulcate). *Lodicules* present (relatively large); 3; free; membranous; ciliate; not toothed; heavily vascularized. *Stamens* 3 (rarely 6?). *Ovary* glabrous, or hairy. Stigmas 2 (in *Bashania*), or 3.

Fruit, embryo and seedling. Fruit medium sized to large (1–1.2 cm long); longitudinally grooved. Hilum long-linear. Embryo small. Endosperm hard; without lipid; containing compound starch grains.

Abaxial leaf blade epidermis. *Costal/intercostal zonation* conspicuous. *Papillae* present. Intercostal papillae over-arching the stomata, or not over-arching the stomata; several per cell (in one or more rows per cell). Mid-intercostal long-cells having markedly sinuous walls (these thin). *Microhairs* present; panicoid-type (distal cells often variable in shape). *Stomata* common. Subsidiaries low to high dome-shaped (usually), or triangular (rarely). *Intercostal short-cells* common, or absent or very rare; in cork/silica-cell pairs; silicified, or not silicified. Intercostal silica bodies when present, often tall-and-narrow, or vertically elongated-nodular. *Costal short-cells* predominantly paired, or neither distinctly grouped into long rows nor predominantly paired, or conspicuously in long rows (sometimes, over main veins). Costal silica bodies saddle shaped (commonly), or oryzoid to 'panicoid-type' (sometimes tending to cross-shaped, then sometimes tending to be vertically elongated).

Transverse section of leaf blade, physiology. C_3; XyMS+. Mesophyll without adaxial palisade; with arm cells; with fusoids. The fusoids external to the PBS. *Leaf blade* with distinct, prominent adaxial ribs; with the ribs more or less constant in size, or with the ribs very irregular in sizes. *Midrib* conspicuous; having complex vascularization. *The lamina* distinctly asymmetrical on either side of the midrib. Bulliforms present in discrete, regular adaxial groups; in simple fans (mostly), or in simple fans and associated with colourless mesophyll cells to form deeply-penetrating fans. All the vascular bundles accompanied by sclerenchyma. Combined sclerenchyma girders present; forming 'figures'. Sclerenchyma all associated with vascular bundles.

Culm anatomy. *Culm internode bundles* in three or more rings to scattered.

Cytology. Chromosome base number, $x = 12$. $2n = 48$ (usually). 4 ploid (usually), or 2 ploid (rarely). Nucleoli persistent.

Taxonomy. Bambusoideae; Bambusodae; Bambuseae.

Ecology, geography, regional floristic distribution. About 50 species. In warm regions. Commonly adventive.

Holarctic, Paleotropical, Neotropical, and Cape. Boreal. African and Indomalesian. Eastern Asian and Atlantic North American. Sudano-Angolan and West African Rainforest. Indo-Chinese. Andean. Canadian-Appalachian and Southern Atlantic North American. Sahelo-Sudanian, Somalo-Ethiopian, and South Tropical African.

Rusts and smuts. Rusts — *Puccinia*. Smuts from Ustilaginaceae. Ustilaginaceae — *Ustilago*.

Economic importance. Culms of (e.g.) *A. amabilis* used for fishing rods, ski poles, umbrella shafts, furniture construction etc.; many others are cultivated as ornamentals or used in bonsai.

References, etc. Morphological/taxonomic: Chao and Renvoize 1989. Leaf anatomical: Metcalfe 1960; this project.

Arundinella Raddi

Acratherum Link, *Brandtia* Kunth, *Calamochloe* Reichenb., *Goldbachia* Trin., *Riedelia* Kunth, *Thysanachne* Presl

Habit, vegetative morphology. Annual, or perennial; mostly with tough, erect culms. *Culms* 30–150 cm high; herbaceous; branched above, or unbranched above. Culm nodes glabrous. Culm internodes solid, or hollow. *Leaves* not basally aggregated; non-auriculate. Leaf blades linear; narrow; not setaceous (rigid); flat, or rolled; without cross venation; disarticulating from the sheaths, or persistent; rolled in bud. *Ligule a fringed membrane (narrow)*; truncate.

Reproductive organization. Plants bisexual, with bisexual spikelets; with hermaphrodite florets. The spikelets of sexually distinct forms on the same plant, or all alike in sexuality; hermaphrodite, or hermaphrodite and sterile (sterile spikelets, when present, very reduced).

Inflorescence. *Inflorescence paniculate*; open, or contracted; with capillary branchlets, or without capillary branchlets; espatheate; not comprising 'partial inflorescences' and foliar organs. Spikelet-bearing axes persistent. Spikelets solitary, or paired; not secund; pedicellate; consistently in 'long-and-short' combinations, or not in distinct 'long-and-short' combinations. The 'shorter' spikelets hermaphrodite, or sterile (sometimes reduced to a glume). The 'longer' spikelets hermaphrodite.

Female-fertile spikelets. *Spikelets* 1.5–8 mm long; *compressed laterally*; disarticulating above the glumes, or disarticulating above the glumes and falling with the glumes (at least sometimes both); alays disarticulating between the florets (but not between the upper glume and the lower floret). *Rachilla terminated by a female-fertile floret. Hairy callus present. Callus* short; blunt.

Glumes two; very unequal; (the upper) long relative to the adjacent lemmas; pointed; awned, or awnless; very dissimilar to similar (membranous to papery, G_1 acute to mucronate, G_2 often caudate). *Lower glume 3 nerved.* Upper glume 5 nerved. *Spikelets with incomplete florets* (or rarely both florets perfect). The incomplete florets proximal to the female-fertile florets. *The proximal incomplete florets* 1; paleate. Palea of the proximal incomplete florets fully developed (narrow, two keeled). *The proximal incomplete florets male.* The proximal lemmas awnless; 3–7 nerved; less firm than the female-fertile lemmas to similar in texture to the female-fertile lemmas; not becoming indurated.

Female-fertile florets 1(–2). *Lemmas similar in texture to the glumes to decidedly firmer than the glumes (membranous to thinly leathery)*; not becoming indurated; entire, or incised; when entire pointed, or blunt; when incised, 2 lobed; not deeply cleft (entire, emarginate or bilobed); awnless, or awned. Awns 1 (usually), or 3; median, or median and lateral (via capillary bristles from the lobes); the median different in form from the laterals (when laterals present); from a sinus; geniculate; hairless; much shorter than the body of the lemma to much longer than the body of the lemma; persistent. Lemmas hairless (scabrid or scabridulous); non-carinate; having the margins lying flat and exposed on the palea, or having the margins tucked in onto the palea; with a clear germination flap; 1–7 nerved. *Palea* present; entire (narrow); awnless, without apical setae; not indurated; 2-nerved; 2-keeled. Palea keels wingless (the margins sometimes auriculate below). *Lodicules* present; 2; free; fleshy; glabrous. *Stamens* 3. Anthers 1.5–2 mm long; not penicillate. *Ovary* glabrous. Styles free to their bases. Stigmas 2; white, or red pigmented, or brown.

Fruit, embryo and seedling. *Fruit* free from both lemma and palea; small; compressed dorsiventrally, or not noticeably compressed. *Hilum short.* Embryo large; waisted. Endosperm hard; without lipid; containing compound starch grains. Embryo without an epiblast; with a scutellar tail; with an elongated mesocotyl internode. Embryonic leaf margins overlapping.

Seedling with a long mesocotyl.

Abaxial leaf blade epidermis. *Costal/intercostal zonation* conspicuous. *Papillae* absent. Mid-intercostal long-cells rectangular; having markedly sinuous walls. *Microhairs* present; panicoid-type; (36–)38–60(–62) microns long. Microhair apical cells (18–)21–42(–44) microns long. Microhair apical cell/total length ratio 0.56–0.67. *Stomata* common. Subsidiaries triangular. Guard-cells overlapping to flush with the interstomatals. *Intercostal short-cells* common; in cork/silica-cell pairs; silicified. Intercostal silica bodies tall-and-narrow, or crescentic. *Costal short-cells* predominantly paired. Costal silica bodies rounded (e.g. *A. nepalensis*), or crescentic (sometimes), or 'panicoid-type' (commonly); when panicoid type, cross shaped to dumb-bell shaped.

Transverse section of leaf blade, physiology. C_4. The anatomical organization conventional, or unconventional. Organization of PCR tissue when unconventional *Arundinella* type. Biochemical type NADP–ME (*A. nepalensis*); XyMS–. PCR sheath outlines uneven. PCR sheath extensions present. Maximum number of extension cells 1. PCR cells with a suberised lamella. *PCR cell chloroplasts* ovoid; with reduced grana (rudimentary); centrifugal/peripheral. *Mesophyll* with radiate chlorenchyma; exhibiting 'circular cells', or without 'circular cells'. *Midrib* conspicuous; with one bundle only, or having a conventional arc of bundles. Bulliforms present in discrete, regular adaxial groups; in simple fans. Many of the smallest vascular bundles unaccompanied by sclerenchyma. Combined sclerenchyma girders present; nowhere forming 'figures'. Sclerenchyma all associated with vascular bundles.

Culm anatomy. *Culm internode bundles* in one or two rings.

Phytochemistry. Leaf blade chlorophyll *a:b* ratio 4.45–4.51.

Cytology. Chromosome base number, $x = 7$, 10, 12, and 14. $2n = 14$, 20, 28, 36, and 56.

Taxonomy. Panicoideae; Panicodae; Arundinelleae.

Ecology, geography, regional floristic distribution. 55 species. In warm regions. Commonly adventive. Helophytic to mesophytic; species of open habitats; glycophytic. Marshy places, riverbanks and rocky slopes.

Holarctic, Paleotropical, Neotropical, Cape, and Australian. Boreal. African, Madagascan, and Indomalesian. Euro-Siberian and Eastern Asian. Saharo-Sindian, Sudano-Angolan, and West African Rainforest. Indian, Indo-Chinese, Malesian, and Papuan. Caribbean, Venezuela and Surinam, Central Brazilian, Pampas, and Andean. North and East Australian. Siberian. Sahelo-Sudanian, Somalo-Ethiopian, and South Tropical African. Tropical North and East Australian.

Rusts and smuts. Rusts — *Puccinia*. Taxonomically wide-ranging species: *Puccinia coronata*. Smuts from Tilletiaceae and from Ustilaginaceae. Tilletiaceae — *Tilletia*. Ustilaginaceae — *Sorosporium*, *Sphacelotheca*, and *Ustilago*.

Economic importance. Significant weed species: *A. bengalensis*, *A. leptochloa* (in North America). Important native pasture species: *A. setosa*.

References, etc. Leaf anatomical: Metcalfe 1960; this project.

Arundo L.

Amphidonax Nees, *Donacium* Fries, *Donax* P. Beauv., *Eudonax* Fries, *Scolochloa* Mert. & Koch

Habit, vegetative morphology. Perennial; mostly **reeds (*with long canes*)**; rhizomatous. The flowering culms leafy. *Culms* 200–600 cm high (or occasionally pendant from cliffs); woody and persistent; branched above (the main stems dominant). Culm nodes glabrous. Culm internodes hollow. Young shoots extravaginal. *Leaves* not basally aggregated; auriculate, or non-auriculate. **Leaf blades** linear-lanceolate to lanceolate; broad; **25–80 mm wide**; flat; not pseudopetiolate; without cross venation; disarticulating from the sheaths; rolled in bud. Ligule a fringed membrane (short). *Contra-ligule* absent.

Reproductive organization. Plants bisexual, with bisexual spikelets; with hermaphrodite florets.

Inflorescence. *Inflorescence paniculate (plumose)*; open; espatheate; not comprising 'partial inflorescences' and foliar organs. Spikelet-bearing axes persistent. Spikelets not secund; pedicellate.

Female-fertile spikelets. Spikelets 12–18 mm long; compressed laterally; disarticulating above the glumes; disarticulating between the florets. Rachilla prolonged beyond the uppermost female-fertile floret; rachilla hairless (save above the floret abscission zones — i.e., on the callus). Hairy callus present. *Callus* short; blunt.

Glumes two; more or less equal; about equalling the spikelets; long relative to the adjacent lemmas; pointed; awnless; carinate to non-carinate; similar (membranous). Lower glume 3–5 nerved. Upper glume 3–5 nerved. *Spikelets* with female-fertile florets only, or with incomplete florets. The incomplete florets distal to the female-fertile florets. *The distal incomplete florets* merely underdeveloped. **Spikelets without proximal incomplete florets**.

Female-fertile florets 2–7. Lemmas less firm than the glumes to similar in texture to the glumes (membranous or hyaline); not becoming indurated; entire, or incised; when incised, 2 lobed; not deeply cleft; awnless to awned. Awns when present, 1, or 3; median, or median and lateral; the median similar in form to the laterals (when laterals present); from a sinus; non-geniculate; much shorter than the body of the lemma; entered by one vein. The lateral awns (when present) shorter than the median. Lemmas villous hairy (on the back); non-carinate; 3–9 nerved. *Palea* present; relatively long to conspicuous but relatively short (from half to two thirds the lemma length); awnless, without apical setae; thinner than the lemma (delicately membranous), or textured like the lemma; 2-nerved; 2-keeled. *Lodicules* present; 2; free; fleshy; glabrous. *Stamens* 3. Anthers 2.5–3.5 mm long; not penicillate. *Ovary* glabrous. Styles free to their bases. Stigmas 2; green to greyish.

Fruit, embryo and seedling. *Fruit* free from both lemma and palea. Hilum short. Embryo large. Endosperm containing compound starch grains. Embryo without an epiblast; with a scutellar tail; with an elongated mesocotyl internode. Embryonic leaf margins meeting.

Abaxial leaf blade epidermis. *Costal/intercostal zonation* conspicuous. *Papillae* absent. *Long-cells* similar in shape costally and intercostally; of similar wall thickness costally and intercostally (thin walled). Mid-intercostal long-cells rectangular; having markedly sinuous walls. *Microhairs* present; panicoid-type; (54–)60–78(–81) microns long; 6.3–8.4 microns wide at the septum. Microhair total length/width at septum 9.6–11.9. Microhair apical cells 18–30 microns long. Microhair apical cell/total length ratio 0.33–0.35. *Stomata* common; 39–42–48 microns long. Subsidiaries dome-shaped, or dome-shaped and triangular. Guard-cells overlapping to flush with the interstomatals. *Intercostal short-cells* common, or absent or very rare; not paired (mainly solitary, sometimes in twos or threes); silicified (a few). Intercostal silica bodies tall-and-narrow, or cross-shaped, or saddle shaped, or oryzoid-type. *Costal short-cells* conspicuously in long rows (rarely), or neither distinctly grouped into long rows nor predominantly paired. Costal silica bodies 'panicoid-type'; cross shaped to dumb-bell shaped.

Transverse section of leaf blade, physiology. C_3; XyMS+. *Mesophyll* with radiate chlorenchyma (rarely), or with non-radiate chlorenchyma. *Leaf blade* adaxially flat. *Midrib* conspicuous; with one bundle only, or having a conventional arc of bundles. Bulliforms present in discrete, regular adaxial groups; combining with colourless mesophyll cells to form narrow groups penetrating into the mesophyll. All the vascular bundles accompanied by sclerenchyma. Combined sclerenchyma girders present; nowhere forming 'figures'. Sclerenchyma all associated with vascular bundles.

Culm anatomy. *Culm internode bundles* in three or more rings.

Phytochemistry. Tissues of the culm bases with little or no starch. Leaves without flavonoid sulphates (1 species).

Special diagnostic feature. Female-fertile lemmas conspicuously hairy; ligule hairs to 0.3 mm long, shorter than the membrane.

Cytology. Chromosome base number, $x = 12$. $2n = 60$, 72, 110, and 112. 6 and 9 ploid (and aneuploid). Nucleoli disappearing before metaphase.

Taxonomy. Arundinoideae; Arundineae.

Ecology, geography, regional floristic distribution. 3 species. Tropical and temperate. Commonly adventive. Helophytic to mesophytic.

Holarctic, Paleotropical, Neotropical, and Antarctic. Boreal and Tethyan. African and Indomalesian. Eastern Asian and Atlantic North American. Mediterranean and Irano-Turanian. Saharo-Sindian. Indo-Chinese. Central Brazilian, Pampas, and Andean. New Zealand. Central Grasslands.

Rusts and smuts. Rusts — *Dasturella* and *Puccinia*.

Economic importance. Significant weed species: *A. donax*, when blocking drainage ditches etc. Important native pasture species: *A. donax*. *A. donax* stems used for light construction work, and for making wood-wind reeds.

References, etc. Leaf anatomical: Metcalfe 1960; this project.

Arundoclaytonia Davidse & Ellis

Habit, vegetative morphology. Perennial (of peculiar habit — see discussion by Davidse and Ellis 1987); caespitose (but the 'basal' tufts become raised up to 70 cm by the elongation of a perennial 'trunk'). The flowering culms leafy. **Culms 200–300 cm high (the vegetative culms many-noded, covered for the lower 2–70 cm to a thickness of 1.5–6cm by appressed aerial roots and the remnants of sheath bases)**; woody and persistent (towards the base); to 1.5 cm in diameter. Culm internodes becoming hollow (above). Young shoots intravaginal. **Leaves** mostly basal (the cauline leaves more or less reduced); *spirally disposed (with 2/5 phyllotaxy)*. Leaf blades linear to linear-lanceolate; narrow to broad; those of the basal leaves 8–16 mm wide (and 45–80 cm long — those of the cauline leaves similar but smaller); flat (with involute margins and tips), or rolled (entirely involute); without cross venation; persistent. Ligule a fringed membrane; 0.3–0.9 mm long (the cilia 0.5–1.2 mm long). *Contra-ligule* absent.

Reproductive organization. *Plants monoecious with all the fertile spikelets unisexual*; without hermaphrodite florets. The spikelets of sexually distinct forms on the same plant (males and females in separate inflorescences, and reduced spikelets at the bases of the individual inflorescences); female-only, male-only, and sterile. The male and female-fertile spikelets in different inflorescences (the male inflorescences produced earlier than the females). The spikelets overtly heteromorphic.

Inflorescence. Inflorescence falsely paniculate (consisting of a false panicle of numerous pedunculate unisexual inflorescences); contracted (i.e. the individual inflorescences); capitate (each comprising a cluster of 7–20 spikelets: the male inflorescences usually longer-pedicellate, 9–13 mm wide and 15–23 mm high, the females 20–36 mm wide, 15–23 mm high); spatheate (each peduncle subtended by a sharp-pointed sheath). Spikelet-bearing axes persistent. *Spikelets* associated with bractiform involucres (the spikelet clusters each surrounded by one or two series of bracts and/ or rudimentary spikelets). The involucres persistent on the rachis. Spikelets not secund; sessile to pedicellate (the pedicels to 0.5 mm long).

Female-sterile spikelets. Male spikelets 3.5–7.5 mm long, rounded on the back, disarticulating below the glumes, 3–9 flowered, the uppermost florets reduced; glumes 2, unequal, usually with cross-veinlets; lemmas shorter than paleas, 3–9 nerved; paleas two-keeled, lodicules absent, stamens 2, anthers basifixed and 2.2–2.9 mm long. The male spikelets with glumes; 3–9 floreted. The lemmas mucronate. Male florets 2 staminate.

Female-fertile spikelets. Spikelets 7–19 mm long; compressed laterally; falling with the glumes; not disarticulating between the florets; with conventional internode spacings. Rachilla prolonged beyond the uppermost female-fertile floret (the uppermost floret rudimentary); rachilla hairy. The rachilla extension with incomplete florets. Hairy callus present (under the glumes). *Callus* short.

Glumes two; very unequal; shorter than the spikelets; shorter than the adjacent lemmas; hairy (at the base, scaberulous above); pointed (broadly acute); awnless; noncarinate; similar (herbaceous, ovate). *Lower glume* about 0.6–0.75 times the length of the upper glume; 1–3 nerved. Upper glume 3–5 nerved. *Spikelets* with incomplete florets. The incomplete florets both distal and proximal to the female-fertile florets. *The distal incomplete florets* merely underdeveloped; awnless. *The proximal incomplete florets* 1; paleate, or epaleate. Palea of the proximal incomplete florets when present, reduced. The proximal incomplete florets sterile. The proximal lemmas awnless; 7–9 nerved; exceeded by the female-fertile lemmas; similar in texture to the female-fertile lemmas; not becoming indurated.

Female-fertile florets 1. Lemmas not becoming indurated; entire; pointed (ovate-acute); awnless; hairy (pilose at the base and between the nerves above, scaberulous elsewhere); apparently carinate; without a germination flap; 9–11 nerved (with conspicuous cross-veinlets); with the nerves non-confluent. *Palea* present; relatively long (much longer than the lemma, curved in the upper half); convolute; entire (pointed); awnless, without apical setae; spongy-thickened and smooth-shiny below, herbaceous above; not indurated; several nerved (9–13 nerved); shallowly grooved on the back. *Lodicules* absent. *Stamens* 0 (absent, or represented by an anterior pair of rudimentary staminodes). *Ovary* glabrous. Styles fused (into one). Stigmas 2 (inconspicuously plumose, exserted through the apical orifice of the convolute palea).

Fruit, embryo and seedling. *Fruit* free from both lemma and palea; medium sized (6–7 mm long); fusiform (narrowing above); not noticeably compressed (terete). Hilum short (elliptic-punctiform). Embryo large to small (0.3–0.4 times the length of the grain).

Abaxial leaf blade epidermis. *Costal/intercostal zonation* lacking. *Papillae* absent. *Long-cells* similar in shape costally and intercostally; of similar wall thickness costally and intercostally. Mid-intercostal long-cells rectangular; having markedly sinuous walls (these conspicuously pitted). *Microhairs* absent. *Stomata* common. *Intercostal short-cells* common; consistently in cork/silica-cell pairs; silicified. *Costal short-cells* predominantly paired. Costal silica bodies tall-and-narrow.

Transverse section of leaf blade, physiology. C_3; XyMS+. *Mesophyll* with non-radiate chlorenchyma; without adaxial palisade; not *Isachne*-type (the cells isodiametric in ts, tightly packed). *Leaf blade* with distinct, prominent adaxial ribs; with the ribs more or less constant in size (flat-topped). *Midrib* not readily distinguishable; with one bundle only. *The lamina* symmetrical on either side of the midrib. Bulliforms present in discrete, regular adaxial groups; in simple fans (a group in each furrow, each with an inflated median cell). All the vascular bundles accompanied by sclerenchyma. Combined sclerenchyma girders present; forming 'figures' (all the bundles with conspicuous I's). Sclerenchyma not all bundle-associated (the abaxial girders linked to one another by a fibrous hypodermal layer). The 'extra' sclerenchyma in a continuous abaxial layer.

Special diagnostic feature. Plants not as in *Steyermarkochloa* (q.v.).

Taxonomy. Arundinoideae (?); Steyermarkochloeae.

Ecology, geography, regional floristic distribution. 1 species (*A. dissimilis*). South central Amazonian Brazil. Xerophytic; species of open habitats; glycophytic.

Neotropical. Amazon.

References, etc. Morphological/taxonomic: Davidse and Ellis 1987. Leaf anatomical: Davidse and Ellis 1987.

Asthenochloa Buese

Garnotiella Stapf

Habit, vegetative morphology. Decumbent annual, or perennial. Culms herbaceous (slender); branched above. *Leaves* not basally aggregated. Leaf blades linear-lanceolate to lanceolate; narrow; without cross venation.

Reproductive organization. Plants bisexual, with bisexual spikelets; with hermaphrodite florets. *The spikelets of sexually distinct forms on the same plant (but obscurely so, the 'pedicellate spikelet' being reduced to a minute pedicel hidden in the callus hairs)*; hermaphrodite, or hermaphrodite and sterile (but the latter rudimentary).

Inflorescence. Inflorescence paniculate (decompound); open; spatheate; a complex of 'partial inflorescences' and intervening foliar organs. **Spikelet-bearing axes** *very much reduced (pedicel-like, with cupuliform, long-hairy apices, bearing a plexus of a minute stipe (= rudiment of pedicelled spikelet), and one sessile spikelet)*; solitary; disarticulating; disarticulating at the joints (i.e., the spikelet and its vestigial companion (plus the bearded callus) articulate with the apex of the pedicel-like branchlet). 'Articles' with a basal callus-knob; densely long-hairy. Spikelets paired (but the pedicellate member suppressed); sessile and pedicellate (the latter reduced to their pedicels); consistently in 'long-and-short' combinations. Pedicels of the 'pedicellate' spikelets free of the rachis (but small and inconspicuous). The 'shorter' spikelets hermaphrodite. *The 'longer' spikelets sterile (reduced to tiny pedicels)*.

Female-fertile spikelets. Spikelets 2–2.5 mm long; compressed dorsiventrally (?); falling with the glumes. Rachilla terminated by a female-fertile floret. Hairy callus present (the hairs up to two thirds as long as spikelet).

Glumes two; more or less equal (G_2 somewhat longer); long relative to the adjacent lemmas; glumes with ciliate margins; awned (sometimes, G_2), or awnless; carinate (G_2), or non-carinate (G_1); very dissimilar (thinly herbaceous; G_1 2-toothed, dorsally flattened; G_2 acuminate or short-awned, naviculate). Lower glume flattened on the back; not pitted; relatively smooth; 2 nerved. Upper glume 3 nerved. **Spikelets** with female-fertile florets only; *without proximal incomplete florets*.

Female-fertile florets 1. Lemmas less firm than the glumes (thinly membranous); not becoming indurated; incised; awned. Awns 1; median; from a sinus; geniculate; hairless (scabrid); much longer than the body of the lemma. Lemmas hairless (the lobes shortly ciliolate); non-carinate; 1 nerved (?). **Palea** *absent. Lodicules* absent. *Stamens* 2. Anthers not penicillate; without an apically prolonged connective. *Ovary* glabrous. Stigmas 2.

Fruit, embryo and seedling. Hilum short. Embryo large.

Abaxial leaf blade epidermis. *Costal/intercostal zonation* conspicuous. *Papillae* present; costal and intercostal. Intercostal papillae not over-arching the stomata; several per cell (one median row of papillae per long-cell). *Long-cells* markedly different in shape costally and intercostally (the costals narrower). Mid-intercostal long-cells rectangular; having markedly sinuous walls. *Microhairs* present; elongated; clearly two-celled; panicoid-type; 39–51 microns long; 6–7.5 microns wide at the septum. Microhair total length/width at septum 6–7.5. Microhair apical cells 16–21 microns long. Microhair apical cell/total length ratio 0.33–0.43. *Stomata* common. Subsidiaries dome-shaped. Guard-cells overlapping to flush with the interstomatals. *Intercostal short-cells* absent or very rare. *Costal short-cells* conspicuously in long rows. Costal silica bodies 'panicoid-type'.

Transverse section of leaf blade, physiology. Probably C_4 (the material seen very poor). *Midrib* conspicuous; with one bundle only; with colourless mesophyll adaxially.

Taxonomy. Panicoideae; Andropogonodae; Andropogoneae; Andropogoninae.

Ecology, geography, regional floristic distribution. 1 species. Indonesian Archipelago. Mesophytic. In damp places.

Paleotropical. Indomalesian. Malesian.

References, etc. Leaf anatomical: this project.

Astrebla F. Muell.

Habit, vegetative morphology. Perennial; caespitose. *Culms* 30–90(–120) cm high; herbaceous; branched above. Culm nodes glabrous. Culm internodes solid. Young shoots intravaginal. *Leaves* not basally aggregated; non-auriculate. Leaf blades narrow; (3–)4–6(–8) mm wide; without abaxial multicellular glands; without cross venation; persistent; rolled in bud. *Ligule a fringe of hairs*.

Reproductive organization. *Plants bisexual, with bisexual spikelets*; with hermaphrodite florets. Exposed-cleistogamous, or chasmogamous.

Inflorescence. Inflorescence a single spike, or of spicate main branches, or a single raceme (with short pedicels); digitate, or non-digitate (when the racemes solitary). Primary inflorescence branches 1, or 2. Inflorescence espatheate; not comprising 'partial inflorescences' and foliar organs. Spikelet-bearing axes persistent. Spikelets solitary; secund; biseriate; sessile to subsessile, or pedicellate (with short pedicels); imbricate (cuneate).

Female-fertile spikelets. *Spikelets* 7–18 mm long (including the lobes); compressed laterally; *disarticulating above the glumes*; *not disarticulating between the florets*. Rachilla prolonged beyond the uppermost female-fertile floret. The rachilla extension with incomplete florets. Hairy callus absent.

Glumes two; very unequal to more or less equal; shorter than the adjacent lemmas, or long relative to the adjacent lemmas; pointed (acute); awnless; carinate; similar (membranous to papery). Lower glume 2–9 nerved. *Upper glume 7–13 nerved. Spikelets* with incomplete florets. The incomplete florets distal to the female-fertile florets.

Female-fertile florets *(2–)3–7*. Lemmas decidedly firmer than the glumes (leathery); not becoming indurated; incised; 3 lobed; deeply cleft; awned. Awns 1 (by terminal extension of the median lobe), or 3 (by extensions of all three lobes); median, or median and lateral; the median similar in form to the laterals (when laterals present); from a sinus, or apical; non-geniculate; often curved; hairless; much shorter than the body of the lemma to much longer than the body of the lemma. The lateral awns when present, shorter than the median to about equalling the median. Lemmas basally hairy; non-carinate (dorsally rounded); 3–11 nerved. *Palea* present; entire (acuminate); awnless, without apical setae; not indurated (firmly membranous to papery); 2-nerved; 2-keeled. Palea keels wingless (ciliate). *Lodicules* present; 2; free; fleshy; glabrous. *Stamens* 3. Anthers not penicillate. *Ovary* glabrous. Styles free to their bases. Stigmas 2.

Fruit, embryo and seedling. *Fruit* free from both lemma and palea; small; ellipsoid; longitudinally grooved; compressed dorsiventrally, or not noticeably compressed. Hilum short. **Pericarp free**. Embryo large. Endosperm containing compound starch grains. Embryo with an epiblast; with a scutellar tail; with an elongated mesocotyl internode. Embryonic leaf margins meeting.

Abaxial leaf blade epidermis. *Costal/intercostal zonation* conspicuous. *Papillae* present; intercostal. Intercostal papillae not over-arching the stomata; consisting of one symmetrical projection per cell (at least in *A. pectinata*). *Long-cells* similar in shape costally and intercostally; of similar wall thickness costally and intercostally (quite thin walled). Mid-intercostal long-cells rectangular; having markedly sinuous walls. *Microhairs* present; more or less spherical, or elongated; chloridoid-type (basal cells short). Microhair apical cell wall of similar thickness/rigidity to that of the basal cell. Microhairs 22–26 microns long. Microhair basal cells 12 microns long. Microhairs 12–13.5 microns wide at the septum. Microhair total length/width at septum 1.7–2. Microhair apical cells 9.6–15 microns long. Microhair apical cell/total length ratio 0.43–0.54. *Stomata* common; 21–24 microns long. Subsidiaries non-papillate; dome-shaped (mainly, in *A. pectinata*), or triangular, or dome-shaped and triangular. Guard-cells overlapping to flush with the interstomatals. *Intercostal short-cells* common; in cork/silica-cell pairs and not paired; silicified and not silicified. Intercostal silica bodies imperfectly developed; narrowly saddle shaped. *Costal short-cells* conspicuously in long rows. Costal silica bodies present in alternate cell files of the costal zones; large, saddle shaped.

Transverse section of leaf blade, physiology. C_4; biochemical type NAD–ME (4 species); XyMS+. PCR sheath outlines even. PCR sheaths of the primary vascular bundles complete to interrupted (only midrib); interrupted both abaxially and adaxially. PCR sheath extensions present, or absent. Maximum number of extension cells when present, 1–7. PCR cells without a suberised lamella. *PCR cell chloroplasts* elongated; with well developed grana; centripetal. *Mesophyll* with radiate chlorenchyma; not *Isachne*-type (but with very narrow-elongate PCA cells in *A. pectinata*); traversed by columns of colourless mesophyll cells. *Leaf blade* 'nodular' in section, or adaxially flat; with the ribs more or less constant in size. *Midrib* conspicuous; having a conventional arc of bundles; with colourless mesophyll adaxially, or without colourless mesophyll adaxially. Bulliforms present in discrete, regular adaxial groups (the groups large); associated with colourless mesophyll cells to form deeply-penetrating fans (these linking with the colourless columns). All the vascular bundles accompanied by sclerenchyma. Combined sclerenchyma girders present; forming 'figures', or nowhere forming 'figures'. Sclerenchyma all associated with vascular bundles. The lamina margins with fibres.

Phytochemistry. Leaf blade chlorophyll *a*:*b* ratio 4.17–4.53.

Cytology. Chromosome base number, $x = 10$. $2n = 40$.

Taxonomy. Chloridoideae; main chloridoid assemblage.

Ecology, geography, regional floristic distribution. 4 species. Australia. Xerophytic; species of open habitats. Dry sandy grassland.

Australian. North and East Australian and Central Australian. Tropical North and East Australian.

Rusts and smuts. Rusts — *Puccinia*.

Economic importance. Cultivated fodder: *A. triticoides*. Important native pasture species: all species palatable and valuable.

References, etc. Leaf anatomical: Metcalfe 1960 and this project.

Athroostachys Bentham

Habit, vegetative morphology. Perennial; rhizomatous. The flowering culms leafy. Culms woody and persistent; scandent; branched above. Primary branches/mid-culm node

3. Culm internodes hollow. Unicaespitose. Rhizomes pachymorph. Plants unarmed. *Leaves* not basally aggregated; with auricular setae. Leaf blades lanceolate; broad; pseudopetiolate; without cross venation; disarticulating from the sheaths. **Reproductive organization**. Plants bisexual, with bisexual spikelets; with hermaphrodite florets.

Inflorescence. *Inflorescence paniculate*; contracted; capitate; *spatheate (bracteate, the lower bracts liguliform caudate and foliar up to 3 cm long, the upper being short awned scales 0.5 mm long)*; not comprising 'partial inflorescences' and foliar organs. Spikelet-bearing axes persistent. Spikelets not secund; pedicellate.

Female-fertile spikelets. Spikelets readily interpreted, provided the subtending bracts are not regarded as glumes; 10–15 mm long; *compressed laterally*; disarticulating above the glumes; not disarticulating between the florets. Rachilla prolonged beyond the uppermost female-fertile floret. The rachilla extension with incomplete florets. Hairy callus absent.

Glumes two; very unequal; shorter than the adjacent lemmas; pointed; *awned*; similar. *Spikelets* with incomplete florets. The incomplete florets distal to the female-fertile florets. *The distal incomplete florets* merely underdeveloped.

Female-fertile florets 1. Lemmas entire; pointed; awnless to awned (?— subulate-acuminate); hairy (above). *Palea* present; relatively long; awnless, without apical setae; 2-nerved; 2-keeled. *Lodicules* present (their tips subulate-acuminate); 3; free; ciliate (towards their tips); not toothed; heavily vascularized. **Stamens 3**. Anthers not penicillate; with the connective apically prolonged (the thecae apiculate). **Ovary** glabrous; *with a conspicuous apical appendage*. The appendage broadly conical, fleshy. Styles fused. Stigmas 2.

Abaxial leaf blade epidermis. *Costal/intercostal zonation* conspicuous. *Papillae* present; costal and intercostal. Intercostal papillae over-arching the stomata (and almost completely covering them); several per cell (most long-cells with a median row of conspicuous, quite large, mainly bifurcated papillae). *Long-cells* similar in shape costally and intercostally; of similar wall thickness costally and intercostally. Mid-intercostal long-cells rectangular; having markedly sinuous walls. *Microhairs* present; elongated; clearly two-celled; panicoid-type; 60–69 microns long; 8.4–10.5 microns wide at the septum. Microhair total length/width at septum 5.9–7.7. Microhair apical cells 27–33 microns long. Microhair apical cell/total length ratio 0.4–0.48. *Stomata* common; 24–28.5 microns long. Guard-cells sunken. *Intercostal short-cells* fairly common (in 291–295,2 places), or absent or very rare (elsewhere); not paired (solitary); not silicified (in the material seen). Large costal prickles present, mostly without basal rosettes. *Crown cells* absent. *Costal short-cells* predominantly paired (superposed, often ostensibly solitary). Costal silica bodies poorly developed (and few).

Transverse section of leaf blade, physiology. C_3. Mesophyll without adaxial palisade; with arm cells; with fusoids. The fusoids external to the PBS. *Leaf blade* with distinct, prominent adaxial ribs, or adaxially flat (the ribs very low, wide and flat-topped to scarcely manifest). *Midrib* not readily distinguishable; with one bundle only. *The lamina* symmetrical on either side of the midrib. Bulliforms present in discrete, regular adaxial groups (between each bundle pair); in simple fans (the groups very large, with deeply penetrating median cells). All the vascular bundles accompanied by sclerenchyma. Combined sclerenchyma girders present (with all the bundles); forming 'figures' (some forming slender I's or 'anchors'). Sclerenchyma not all bundle-associated (there being small abaxial groups opposite the bulliforms, and adaxial hypodermal fibres lining their sides). The 'extra' sclerenchyma in abaxial groups and in adaxial groups; abaxial-hypodermal, the groups isolated and adaxial-hypodermal, contiguous with the bulliforms.

Taxonomy. Bambusoideae; Bambusodae; Bambuseae.

Ecology, geography, regional floristic distribution. 1 species. Brazil. Neotropical. Amazon and Central Brazilian.

References, etc. Leaf anatomical: this project.

Special comments. Fruit data wanting.

Atractantha McClure

Habit, vegetative morphology. Perennial. The flowering culms leafy. Culms woody and persistent; branched above. Culm nodes glabrous. Culm internodes solid, or hollow. Plants unarmed. *Leaves* not basally aggregated; auriculate (the auricles usually very weakly developed); with auricular setae. Leaf blades broad, or narrow; pseudopetiolate; without cross venation; disarticulating from the sheaths; rolled in bud. Ligule an unfringed membrane; truncate; 1 mm long (or less).

Reproductive organization. Plants bisexual, with bisexual spikelets; with hermaphrodite florets. Not viviparous.

Inflorescence. Inflorescence indeterminate; with pseudospikelets; *either capitate or diffuse, the branching pattern either distichous or sympodial, each axis representing the bracteate or prophyllate rachis of a pseudospikelet, the terminal segment of each rachis serving as the pedicel of an abscissile pseudospikelet*; spatheate (at least, 'bracteate' or prophyllate); not comprising 'partial inflorescences' and foliar organs. *Spikelet-bearing axes* paniculate to capitate; persistent. 'Articles' glabrous. Spikelets solitary; not secund.

Female-fertile spikelets. *Spikelets* unconventional (fairly, though the palea and lemma are recognizable in the pseudospikelet); 15–17 mm long; not noticeably compressed; disarticulating above the glumes (immediately below the lemma). Rachilla prolonged beyond the uppermost female-fertile floret; rachilla hairless. The rachilla extension with incomplete florets. Hairy callus absent.

Glumes absent (the spikelet subtended by a small prophyll, but no sterile bracts). *Spikelets* with incomplete florets. The incomplete florets distal to the female-fertile florets. *The distal incomplete florets* merely underdeveloped.

Female-fertile florets 1. Lemmas attenuately acuminate; not becoming indurated; entire; pointed; hairless; non-carinate; without a germination flap. *Palea* present; relatively long; apically notched to deeply bifid; awnless, without apical setae; not indurated; 2-keeled (with a narrow sulcus, or tubular). *Lodicules* present; 3; free; fleshy; ciliate; not toothed; heavily vascularized. *Stamens* 3. Anthers not penicillate; without an apically prolonged connective. *Ovary* hairy (antrorsely hispidulous); with a conspicuous apical appendage. The appendage broadly conical, fleshy. Styles fused. Stigmas 2(–3).

Abaxial leaf blade epidermis. *Costal/intercostal zonation* conspicuous. *Papillae* present; costal and intercostal. Intercostal papillae over-arching the stomata (and obscuring them); several per cell (mostly with a single, median row of large, circular or bifurcated papillae per long-cell). *Long-cells* similar in shape costally and intercostally to markedly different in shape costally and intercostally (the intercostals often shorter and wider); of similar wall thickness costally and intercostally. Mid-intercostal long-cells rectangular; having markedly sinuous walls. *Microhairs* present; elongated; clearly two-celled; panicoid-type; 51–79.5 microns long; 9.6–10.5 microns wide at the septum. Microhair total length/width at septum 4.9–7.6. Microhair apical cells 21–42 microns long. Microhair apical cell/total length ratio 0.4–0.55. *Stomata* common; 30–33 microns long. Guard-cells overlapped by the interstomatals. *Intercostal short-cells* common; in cork/silica-cell pairs; silicified. Intercostal silica bodies saddle shaped to oryzoid-type (cf. the costals). *Costal short-cells* predominantly paired. Costal silica bodies saddle shaped (a tall, narrowish form of these predominating), or oryzoid (sometimes with some of the saddles approaching this form).

Transverse section of leaf blade, physiology. C_3. Mesophyll with adaxial palisade; with arm cells; with fusoids. The fusoids external to the PBS. *Leaf blade* with distinct, prominent adaxial ribs; with the ribs more or less constant in size (low and flat-topped, save for a tall, narrow rib near one leaf margin). *Midrib* not readily distinguishable; with one bundle only. *The lamina* distinctly asymmetrical on either side of the midrib, or symmetrical on either side of the midrib (depending on whether the submarginal rib represents a highly asymmetrically placed midrib). Bulliforms present in discrete, regular adaxial groups; in simple fans (the groups large). All the vascular bundles accompanied by sclerenchyma. Combined sclerenchyma girders present (nearly all the bundles). Sclerenchyma not all bundle-associated (there being conspicuous abaxial groups opposite the bulliforms, and commonly small adaxial hypodermal arcs lining their sides). The 'extra' sclerenchyma in abaxial groups and in adaxial groups; abaxial-hypodermal, the groups isolated and adaxial-hypodermal, contiguous with the bulliforms.

Special diagnostic feature. *The inflorescences of very peculiar pseudospikelets, characterized by development of rachides with long terminal segments, each of which serves as the pedicel of an abscissile spikelet.*

Taxonomy. Bambusoideae; Bambusodae; Bambuseae.

Ecology, geography, regional floristic distribution. 2 species. Brazil. Neotropical. Amazon and Central Brazilian.

References, etc. Leaf anatomical: this project.

Special comments. Fruit data wanting.

Aulonemia Goudot

Matudacalamus Mackawa

Habit, vegetative morphology. Perennial; caespitose. The flowering culms leafy. *Culms woody and persistent*; branched above. *Primary branches/mid-culm node 1.* Culm sheaths persistent. Culm internodes hollow. Unicaespitose. Rhizomes pachymorph. Plants unarmed. *Leaves* not basally aggregated; with auricular setae, or without auricular setae. *Leaf blades broad (always?)*; pseudopetiolate; cross veined, or without cross venation; disarticulating from the sheaths; rolled in bud.

Reproductive organization. Plants bisexual, with bisexual spikelets; with hermaphrodite florets.

Inflorescence. Inflorescence paniculate; open; without capillary branchlets (but these slender); spatheate. *Spikelet-bearing axes* paniculate; persistent. Spikelets not secund; pedicellate.

Female-fertile spikelets. Spikelets 35–40 mm long (probably underestimating the true range); compressed laterally to not noticeably compressed; disarticulating above the glumes; disarticulating between the florets. Rachilla prolonged beyond the uppermost female-fertile floret. Hairy callus absent.

Glumes two; very unequal; shorter than the adjacent lemmas; apically *awned (both* G_1 *and* G_2*)*; similar. *Spikelets* with incomplete florets. The incomplete florets distal to the female-fertile florets, or both distal and proximal to the female-fertile florets. *The distal incomplete florets* merely underdeveloped. *The proximal incomplete florets* 1; sterile. The proximal lemmas awned, or awnless (then mucronate); exceeded by the female-fertile lemmas; similar in texture to the female-fertile lemmas; not becoming indurated.

Female-fertile florets *3–12 (?— 'few to many').* Lemmas similar in texture to the glumes; not becoming indurated; entire; blunt; mucronate, or awned. Awns when present, 1; apical; non-geniculate; hairless; much shorter than the body of the lemma to about as long as the body of the lemma. Lemmas non-carinate. *Palea* present; relatively long; entire (pointed); awnless, without apical setae; not indurated; 2-keeled. *Lodicules* present; 3; free; membranous; ciliate; not toothed; heavily vascularized. *Stamens* 3. Anthers not penicillate; without an apically prolonged connective. **Ovary** *hairy.* Styles fused. Stigmas 2.

Fruit, embryo and seedling. *Fruit* free from both lemma and palea; sub- fusiform, or ellipsoid; longitudinally grooved. Hilum long-linear. Embryo small.

Abaxial leaf blade epidermis. *Costal/intercostal zonation* conspicuous. *Papillae* absent (but present as inconspicuous rings around the stomata adaxially). *Long-cells* similar in shape costally and intercostally (but the costals somewhat smaller); of similar wall thickness costally and intercostally. Mid-intercostal long-cells rectangular (very regularly so); having markedly sinuous walls. *Microhairs* absent. *Stomata* common (abundant in broad bands adjoining the costal zones); 25.5–30 microns long. Subsidiaries consistently low to medium dome-shaped. Guard-cells slightly overlapped by the interstomatals. *Intercostal short-cells* common; in cork/silica-cell pairs; silicified. Intercostal silica bodies narrowly saddle shaped, or tall-and-narrow. Macrohairs and prickles absent. *Crown cells* absent. *Costal short-cells* predominantly paired. Costal silica bodies saddle shaped (predominating, a tall-and-narrow version), or tall-and-narrow (a few).

Transverse section of leaf blade, physiology. C_3; XyMS+. Mesophyll without adaxial palisade; with arm cells (but these conspicuous only around the fusoids in the poor material seen); with fusoids. The fusoids external to the PBS. *Leaf blade* with distinct, prominent adaxial ribs (especially towards the middle). *Midrib* conspicuous; having complex vascularization. *The lamina* distinctly asymmetrical on either side of the midrib (one side being more conspicuously ribbed in the region near the midrib). Bulliforms present in discrete, regular adaxial groups; in simple fans (with a large, conspicuous group in each intercostal zone). All the vascular bundles accompanied by sclerenchyma. Combined sclerenchyma girders present; forming 'figures' (every bundle with an I or an 'anchor'). Sclerenchyma all associated with vascular bundles (in *A. fulgor*).

Taxonomy. Bambusoideae; Bambusodae; Bambuseae.

Ecology, geography, regional floristic distribution. About 30 species. South America to Mexico and Costa Rica. Glycophytic.

Neotropical. Caribbean, Venezuela and Surinam, Central Brazilian, Pampas, and Andean.

References, etc. Leaf anatomical: this project.

Australopyrum (Tsvelev) A. Löve

~ *Agropyron*; cf.*Brachypodium*

Habit, vegetative morphology. Perennial; caespitose. Culms herbaceous; unbranched above. Young shoots intravaginal. *Leaves* not basally aggregated; auriculate. Sheath margins free. Leaf blades narrow; usually rolled; without cross venation; persistent. Ligule an unfringed membrane; truncate; short.

Reproductive organization. Plants bisexual, with bisexual spikelets; with hermaphrodite florets; inbreeding.

Inflorescence. *Inflorescence a single raceme (the spikelets somewhat spreading).* Rachides hollowed (notched against the spikelets). Inflorescence espatheate; not comprising 'partial inflorescences' and foliar organs. Spikelet-bearing axes persistent. Spikelets solitary; not secund; distichous; subsessile to pedicellate (i.e., the spikelets more or less distinctly stalked); imbricate.

Female-fertile spikelets. Spikelets 6–10 mm long; compressed laterally; disarticulating above the glumes; disarticulating between the florets. Rachilla prolonged beyond the uppermost female-fertile floret. The rachilla extension with incomplete florets.

Glumes present; two; more or less equal; *joined (slightly, at the base)*; lateral to the rachis; hairless; glabrous; pointed; not subulate; awnless; non-carinate; similar (rigid, broadly lanceolate, rather asymmetric). Lower glume 4–6 nerved. Upper glume 5–7 nerved (sic: glumes *not* '3 nerved'). *Spikelets* with incomplete florets. The incomplete florets distal to the female-fertile florets. *The distal incomplete florets* merely underdeveloped.

Female-fertile florets 5–7. Lemmas slightly involute; not becoming indurated; entire; pointed; awned. Awns 1; median; apical; hooked (in *A. uncinatum*), or not hooked; non-geniculate; much shorter than the body of the lemma to about as long as the body of the lemma. Lemmas without a germination flap; 5 nerved, or 7 nerved; with the nerves confluent towards the tip. *Palea* present; relatively long; 2-nerved; 2-keeled. *Lodicules* present; 2; ciliate, or glabrous. *Stamens* 3. Anthers 2–4 mm long (sic). **Ovary hairy.** Styles free to their bases. Stigmas 2; white.

Fruit, embryo and seedling. Fruit small (3–3.5 mm long). Hilum long-linear. Embryo small.

Abaxial leaf blade epidermis. *Costal/intercostal zonation* conspicuous. *Papillae* absent. *Long-cells* similar in shape costally and intercostally (the costals smaller); of similar wall thickness costally and intercostally. Mid-intercostal long-cells rectangular (a few almost fusiform in *A. pectinatum*); having markedly sinuous walls (exaggeratedly so in *A. velutinum*). *Microhairs* absent. *Stomata* common; 33–39 microns long (in *A. retrofractum* and *A. velutinum*), or 40.5–42 microns long (in *A. pectinatum*). Subsidiaries parallel-sided, or parallel-sided and dome-shaped, or parallel-sided, dome-shaped, and triangular (in *A. velutinum*); including both triangular and parallel-sided forms on the same leaf (and domes, in *A. velutinum*), or not including both parallel-sided and triangular forms on the same leaf. Guard-cells overlapped by the interstomatals. *Intercostal short-cells* common; in cork/silica-cell pairs; silicified. Prickles and macrohairs often common. *Crown cells* present, or absent. *Costal short-cells* predominantly paired. Costal silica bodies rounded (plus many irregularly isodiametric forms).

Transverse section of leaf blade, physiology. C_3; XyMS+. *Mesophyll* with non-radiate chlorenchyma; without adaxial palisade. *Leaf blade* with distinct, prominent adaxial ribs; with the ribs more or less constant in size (round topped). *Midrib* not readily distinguishable; with one bundle only. Bulliforms present in discrete, regular adaxial groups to not in discrete, regular adaxial groups (scarcely developed bulliform groups only, in some of the furrows); in groups of fairly evenly sized cells. All the vascular bundles accompanied by sclerenchyma. Combined sclerenchyma girders present (with all the bundles); forming 'figures' (most bundles with I's). Sclerenchyma all associated with vascular bundles.

Cytology. Chromosome base number, $x = 7$. $2n = 14$. 2 ploid. Haplomic genome content W.

Taxonomy. Pooideae; Triticodae; Triticeae.

Ecology, geography, regional floristic distribution. 2 species. Australia and New Guinea.

Paleotropical and Australian. Indomalesian. Papuan. North and East Australian. Temperate and South-Eastern Australian.

References, etc. Morphological/taxonomic: Löve 1984. Leaf anatomical: this project.

Austrochloris Lazarides

Habit, vegetative morphology. Perennial; caespitose. *Culms* 40–100 cm high; herbaceous. Culm nodes glabrous. Culm internodes solid. *Leaves* not basally aggregated; non-auriculate. Leaf blades narrow; 2–3 mm wide; without abaxial multicellular glands; without cross venation; persistent. Ligule a fringed membrane.

Reproductive organization. Plants bisexual, with bisexual spikelets; with hermaphrodite florets.

Inflorescence. *Inflorescence* of spicate main branches; **digitate**. Primary inflorescence branches 2, or 3. Rachides hollowed, or flattened (triquetrous). Inflorescence espatheate; not comprising 'partial inflorescences' and foliar organs. Spikelet-bearing axes persistent. Spikelets solitary; secund; biseriate.

Female-fertile spikelets. *Spikelets compressed dorsiventrally*; disarticulating above the glumes. Rachilla prolonged beyond the uppermost female-fertile floret; rachilla hairy. The rachilla extension with incomplete florets. Hairy callus absent.

Glumes two; more or less equal; long relative to the adjacent lemmas; pointed (acuminate); awnless; similar (narrow, membranous, divergent, the upper deciduous). Lower glume 1 nerved. *Upper glume 3 nerved*. Spikelets with incomplete florets. The incomplete florets distal to the female-fertile florets. The distal incomplete florets 1 (neuter, with a cuneate lemma). *Spikelets without proximal incomplete florets*.

Female-fertile florets 1. Lemmas broadly rounded; decidedly firmer than the glumes (cartilaginous); entire, or incised; when entire pointed, or blunt; awned. Awns 1; median; dorsal; from near the top; non-geniculate; about as long as the body of the lemma. Lemmas hairy; non-carinate (with 3 hairy keels); 3 nerved. *Palea* present; entire; awnless, without apical setae; thinner than the lemma (scarious); not indurated; 2-nerved. *Lodicules* present; 2; free; fleshy; glabrous. *Stamens* 3. Anthers 0.5 mm long; not penicillate. *Ovary* glabrous. Styles free to their bases. Stigmas 2; red pigmented.

Fruit, embryo and seedling. Fruit small; ellipsoid; compressed dorsiventrally. Hilum short. *Pericarp free*. Embryo large.

Abaxial leaf blade epidermis. *Papillae* present; intercostal. Intercostal papillae not over-arching the stomata; consisting of one oblique swelling per cell. Mid-intercostal long-cells having markedly sinuous walls. *Microhairs* present; more or less spherical; clearly two-celled; chloridoid-type. Microhair apical cell wall of similar thickness/rigidity to that of the basal cell. Microhairs 24–27 microns long. Microhair basal cells 12–16 microns long. Microhairs 10.5–12 microns wide at the septum. Microhair total length/

width at septum 2–2.4. Microhair apical cells (7.5–)9–11(–12) microns long. Microhair apical cell/total length ratio 0.31–0.47. *Stomata* common; 18–22.5 microns long. Subsidiaries non-papillate; triangular. Guard-cells overlapping to flush with the interstomatals. *Intercostal short-cells* absent or very rare. Intercostal silica bodies absent. *Costal short-cells* conspicuously in long rows, or neither distinctly grouped into long rows nor predominantly paired. Costal silica bodies present in alternate cell files of the costal zones; saddle shaped.

Transverse section of leaf blade, physiology. C_4; XyMS+. PCR sheath outlines even. PCR sheaths of the primary vascular bundles interrupted; interrupted abaxially only. PCR sheath extensions absent. PCR cell chloroplasts centripetal. *Mesophyll* with radiate chlorenchyma. *Leaf blade* 'nodular' in section; with the ribs more or less constant in size. *Midrib* conspicuous to not readily distinguishable; with one bundle only, or with one bundle only to having a conventional arc of bundles (?). Bulliforms present in discrete, regular adaxial groups; in simple fans. All the vascular bundles accompanied by sclerenchyma. Combined sclerenchyma girders present; forming 'figures'. Sclerenchyma all associated with vascular bundles.

Taxonomy. Chloridoideae; main chloridoid assemblage.

Ecology, geography, regional floristic distribution. 1 species. Australia. Species of open habitats. Savanna.

Australian. North and East Australian and Central Australian. Tropical North and East Australian.

References, etc. Morphological/taxonomic: Lazarides 1985. Leaf anatomical: this project.

Austrofestuca (Tsvel.) E.B. Alekseev

Schedonorus Beauv.
Excluding *Festucella, Hookerochloa*

Habit, vegetative morphology. Perennial; rhizomatous and caespitose. **Culms** 30–80 cm high; *herbaceous*; unbranched above. Culm nodes exposed, or hidden by leaf sheaths; glabrous. Culm internodes hollow. Plants unarmed. Young shoots intravaginal. *Leaves* mostly basal; non-auriculate. Sheaths terete. Leaf blades linear; narrow (cylindrical); 0.6–1.7 mm wide (8–50 cm long); involute, subulate, rigid; without cross venation; persistent; rolled in bud. *Ligule an unfringed membrane*; truncate, or not truncate; 0.8–2.1 mm long (basal leaves), or 3.8–5.5 mm long (culm leaves). *Contra-ligule* absent.

Reproductive organization. Plants bisexual, with bisexual spikelets; with hermaphrodite florets.

Inflorescence. *Inflorescence paniculate*; contracted; spicate; espatheate; not comprising 'partial inflorescences' and foliar organs. Spikelet-bearing axes persistent. Spikelets not secund; pedicellate.

Female-fertile spikelets. Spikelets 9–17 mm long; compressed laterally; disarticulating above the glumes; disarticulating between the florets. Rachilla prolonged beyond the uppermost female-fertile floret; rachilla hairy. The rachilla extension with incomplete florets. Hairy callus present, or absent. *Callus* short; blunt (truncate, glabrous, scabrous or villous).

Glumes two; more or less equal; shorter than the spikelets to about equalling the spikelets; *long relative to the adjacent lemmas*; hairless; pointed; awnless; carinate; similar (membranous to chartaceous). Lower glume 3 nerved, or 5 nerved (only at the base). Upper glume 3 nerved, or 5 nerved (only at the base). *Spikelets* with incomplete florets. The incomplete florets distal to the female-fertile florets. *The distal incomplete florets* merely underdeveloped.

Female-fertile florets *3–5*. *Lemmas* similar in texture to the glumes to decidedly firmer than the glumes (membranous to leathery); not becoming indurated; entire, or incised; not deeply cleft (no more than emarginate); awnless (muticous); hairless; *carinate*; without a germination flap; 5 nerved, or 7 nerved. *Palea* present; relatively long; apically notched; awnless, without apical setae; textured like the lemma; not indurated;

2-nerved; 2-keeled. Palea back scabrous, or hairy. Palea keels wingless; hairy. **Lodicules** present; 2; free; membranous; usually *ciliate*; toothed. *Stamens* 3. Anthers (3.8–)4.2–7.2 mm long; not penicillate. *Ovary* glabrous. Styles free to their bases. Stigmas 2; white.

Fruit, embryo and seedling. Fruit free from both lemma and palea; small (2.6–3.8 mm long); ventrally *longitudinally grooved*; compressed dorsiventrally (ventrally). Hilum short (usually oval, rarely punctiform). Embryo small; not waisted. Endosperm hard.

Abaxial leaf blade epidermis. *Costal/intercostal zonation* lacking. *Papillae* absent. *Long-cells* similar in shape costally and intercostally; of similar wall thickness costally and intercostally (walls thick and sinuous). Mid-intercostal long-cells fusiform (slightly); having markedly sinuous walls. *Microhairs* absent. *Stomata* absent or very rare. *Intercostal short-cells* common; not paired; silicified, or not silicified. *Costal short-cells* neither distinctly grouped into long rows nor predominantly paired. Costal silica bodies absent, or poorly developed; tall-and-narrow.

Transverse section of leaf blade, physiology. C_3; XyMS+. *Mesophyll* with non-radiate chlorenchyma. *Leaf blade* adaxially flat. *Midrib* conspicuous (in a small, central rib); with one bundle only. Bulliforms not in discrete, regular adaxial groups (with indistinct 'hinges' only). All the vascular bundles accompanied by sclerenchyma. Combined sclerenchyma girders present; forming 'figures'. Sclerenchyma not all bundle-associated. The 'extra' sclerenchyma in a continuous abaxial layer.

Taxonomy. Pooideae; Poodae; Poeae.

Ecology, geography, regional floristic distribution. 2 species. Extra-tropical Australasia. Xerophytic; species of open habitats; halophytic. *F. littoralis* and *F. pubinervis* are conspicous littoral sand and fore-dunes species.

Australian. North and East Australian and South-West Australian. Temperate and South-Eastern Australian.

References, etc. Morphological/taxonomic: see Jacobs (1990), *Telopea* 3, 601–603 for comments on *Austrofestuca/Festucella/Hookerochloa*. Leaf anatomical: this project.

Special comments. Clearly distinct from *Festuca*, but see *Festucella, Hookerochloa*.

Avellinia Parl.

~ Trisetum, Trisetaria

Habit, vegetative morphology. *Annual*; caespitose. *Culms* 2–30 cm high; herbaceous. Culm nodes glabrous. Culm internodes hollow. *Leaves* not basally aggregated; non-auriculate. Leaf blades linear; narrow; flat, or rolled (involute); without cross venation; persistent. Ligule an unfringed membrane to a fringed membrane (short, often hairy outside); truncate; 0.4–1 mm long.

Reproductive organization. Plants bisexual, with bisexual spikelets; with hermaphrodite florets.

Inflorescence. Inflorescence paniculate; open, or contracted; when contracted spicate, or more or less irregular; espatheate; not comprising 'partial inflorescences' and foliar organs. Spikelet-bearing axes persistent. Spikelets not secund; pedicellate.

Female-fertile spikelets. Spikelets 3–5 mm long; compressed laterally; disarticulating above the glumes. *Rachilla prolonged beyond the uppermost female-fertile floret*; rachilla hairy.

Glumes two; *very unequal*; (the longer) *long relative to the adjacent lemmas*; pointed; awnless; carinate; similar. Lower glume 1 nerved. Upper glume 3 nerved. *Spikelets* with female-fertile florets only, or with incomplete florets. The incomplete florets distal to the female-fertile florets.

Female-fertile florets 2–4. Lemmas apically bifid; similar in texture to the glumes; not becoming indurated; incised; awned. Awns 1; median; from a sinus; non-geniculate; much shorter than the body of the lemma; entered by one vein. Lemmas hairless; non-carinate; 3 nerved. **Palea** present; *conspicuous but relatively short; deeply bifid*; 2-nerved; 2-keeled. *Lodicules* present; 2; free; membranous; glabrous; not toothed. *Stamens* 3. Anthers 0.5 mm long; not penicillate. *Ovary* glabrous. Styles free to their bases. Stigmas 2.

Fruit, embryo and seedling. Fruit small. Hilum short. Embryo small. Endosperm liquid in the mature fruit; with lipid.

Abaxial leaf blade epidermis. *Papillae* absent. Mid-intercostal long-cells having straight or only gently undulating walls. *Microhairs* absent. *Stomata* common. Subsidiaries parallel-sided. Guard-cells overlapped by the interstomatals. *Intercostal short-cells* absent or very rare. *Costal short-cells* neither distinctly grouped into long rows nor predominantly paired. Costal silica bodies horizontally-elongated crenate/sinuous to horizontally-elongated smooth, or saddle shaped, or 'panicoid-type' (sometimes); sometimes indisputably nodular.

Transverse section of leaf blade, physiology. C_3; XyMS+. *Mesophyll* with radiate chlorenchyma. *Leaf blade* with distinct, prominent adaxial ribs; with the ribs more or less constant in size. *Midrib* not readily distinguishable; with one bundle only. Bulliforms inconspicuous in the poor material seen. All the vascular bundles accompanied by sclerenchyma. Combined sclerenchyma girders absent. Sclerenchyma all associated with vascular bundles.

Cytology. Chromosome base number, $x = 7$. $2n = 12$.

Taxonomy. Pooideae; Poodae; Aveneae.

Ecology, geography, regional floristic distribution. 2 species. Mediterranean. Not commonly adventive. Mesophytic, or xerophytic.

Holarctic. Boreal and Tethyan. Euro-Siberian. Mediterranean. European.

Rusts and smuts. Rusts — *Puccinia*. Taxonomically wide-ranging species: *Puccinia hordei*. Smuts from Tilletiaceae. Tilletiaceae — *Tilletia*.

References, etc. Leaf anatomical: this project.

Avena L.

Anelytrum Hack., *Preissia* Opiz

Habit, vegetative morphology. *Annual*; caespitose to decumbent. *Culms* 20–200 cm high; herbaceous; unbranched above. Culm nodes glabrous. Culm internodes hollow. *Leaves* not basally aggregated; non-auriculate. Leaf blades linear; narrow, or broad; 1.5–20 mm wide; flat (usually), or rolled (rarely convolute); without cross venation; persistent; rolled in bud, or once-folded in bud. *Ligule* present; an unfringed membrane; not truncate; 1–7 mm long.

Reproductive organization. *Plants bisexual, with bisexual spikelets*; with hermaphrodite florets; inbreeding. Exposed-cleistogamous, or chasmogamous.

Inflorescence. *Inflorescence paniculate*; open; with capillary branchlets, or without capillary branchlets; espatheate; not comprising 'partial inflorescences' and foliar organs. Spikelet-bearing axes persistent. Spikelets not secund; pedicellate; not in distinct 'long-and-short' combinations.

Female-fertile spikelets. Spikelets 10–45 mm long; compressed laterally; disarticulating above the glumes, or not disarticulating (cultivated forms); not disarticulating between the florets, or disarticulating between the florets. Rachilla prolonged beyond the uppermost female-fertile floret; rachilla hairy, or hairless. The rachilla extension with incomplete florets. Hairy callus present, or absent. Callus pointed.

Glumes two (lanceolate); more or less equal; about equalling the spikelets, or exceeding the spikelets (rarely shorter); long relative to the adjacent lemmas; pointed; awnless; non-carinate; similar (usually chaffy). Lower glume 3–11 nerved. Upper glume 3–11 nerved. *Spikelets* with incomplete florets. The incomplete florets distal to the female-fertile florets. *Spikelets without proximal incomplete florets*.

Female-fertile florets (1–)2–6. Lemmas similar in texture to the glumes (rarely), or decidedly firmer than the glumes (usually leathery to crustaceous); becoming indurated to not becoming indurated; incised; 2 lobed (bidentate or 2-aristulate); not deeply cleft; awnless, or awned. *Awns* when present, 1, or 3; median, or median and lateral; the median different in form from the laterals (when laterals present); dorsal; from well down the back; *geniculate*; much longer than the body of the lemma; entered by one vein. The lateral awns shorter than the median. Lemmas hairy, or hairless; non-carinate; without a germination flap; 5–9 nerved. *Palea* present; relatively long, or conspicuous but relatively short, or very reduced (but large); entire to apically notched; tough; 2-nerved; 2-keeled. *Lodicules* present; 2; free; membranous; glabrous; toothed, or not toothed. *Stamens* 3. Anthers 0.7–4 mm long; not penicillate. **Ovary *hairy***. Styles free to their bases. Stigmas 2; white.

Fruit, embryo and seedling. *Fruit* adhering to lemma and/or palea, or free from both lemma and palea; medium sized; longitudinally grooved; compressed dorsiventrally; hairy on the body. Hilum long-linear. Embryo small; not waisted. Endosperm hard; with lipid; containing compound starch grains. Embryo with an epiblast; without a scutellar tail; with a negligible mesocotyl internode. Embryonic leaf margins meeting.

Seedling with a long mesocotyl; with a loose coleoptile, or with a tight coleoptile. First seedling leaf with a well-developed lamina. The lamina broad, or narrow; erect; 7–15 veined.

Abaxial leaf blade epidermis. *Costal/intercostal zonation* conspicuous. *Papillae* absent. *Long-cells* similar in shape costally and intercostally (long and narrow); of similar wall thickness costally and intercostally (fairly thin walled). Mid-intercostal long-cells fusiform (slightly); having straight or only gently undulating walls. *Microhairs* absent. *Stomata* common. Subsidiaries parallel-sided. Guard-cells overlapped by the interstomatals. *Intercostal short-cells* absent or very rare. *Costal short-cells* predominantly paired, or neither distinctly grouped into long rows nor predominantly paired. Costal silica bodies horizontally-elongated crenate/sinuous, or horizontally-elongated smooth (or occasionally cubical).

Transverse section of leaf blade, physiology. C_3; XyMS+. *Mesophyll* with non-radiate chlorenchyma; without adaxial palisade. *Leaf blade* with distinct, prominent adaxial ribs, or 'nodular' in section, or adaxially flat; with the ribs more or less constant in size. *Midrib* conspicuous, or not readily distinguishable; with one bundle only. Bulliforms present in discrete, regular adaxial groups, or not in discrete, regular adaxial groups (the groups sometimes ill defined, or only hinge groups present); in simple fans (or in groups of fairly evenly sized cells). All the vascular bundles accompanied by sclerenchyma. Combined sclerenchyma girders present; nowhere forming 'figures'. Sclerenchyma all associated with vascular bundles.

Culm anatomy. *Culm internode bundles* in one or two rings.

Phytochemistry. Tissues of the culm bases with little or no starch.

Cytology. Chromosome base number, $x = 7$. $2n = 14$, 28, 42, 48, and 63. 2, 4, 6, 7, and 9 ploid. Haploid nuclear DNA content 4.3–5.5 pg (20 species, mean 4.7). Mean diploid 2c DNA value 9.9 pg (13 species, 8.8–11.0). Nucleoli disappearing before metaphase.

Taxonomy. Pooideae; Poodae; Aveneae.

Ecology, geography, regional floristic distribution. 27 species. Europe, Mediterranean, North Africa, western Asia. Commonly adventive. Mesophytic, or xerophytic; species of open habitats; glycophytic. Mostly in weedy places.

Holarctic, Paleotropical, Neotropical, and Cape. Boreal and Tethyan. African and Indomalesian. Euro-Siberian, Eastern Asian, and Atlantic North American. Macaronesian, Mediterranean, and Irano-Turanian. Saharo-Sindian and Sudano-Angolan. Indian and Indo-Chinese. Pampas. European and Siberian. Canadian-Appalachian, Southern Atlantic North American, and Central Grasslands. Somalo-Ethiopian.

Hybrids. Intergeneric hybrids with *Arrhenatherum*.

Rusts and smuts. Rusts — *Puccinia*. Taxonomically wide-ranging species: *Puccinia graminis*, *Puccinia coronata*, *Puccinia striiformis*, *Puccinia hordei*, and *Puccinia recondita*. Smuts from Tilletiaceae and from Ustilaginaceae. Tilletiaceae — *Entyloma*. Ustilaginaceae — *Ustilago*.

Economic importance. Significant weed species: *A. barbata*, *A. byzantina*, *A. fatua*, *A. ludoviciana*, *A. sterilis*, *A. strigosa*, *A. wiestii*. Grain crop species: mainly *A. sativa* (Oats).

References, etc. Morphological/taxonomic: Baum 1977. Leaf anatomical: Metcalfe 1960; this project.

Axonopus P. Beauv.

Anastrophus Schlecht., *Cabrera* Lag., *Lappogopsis* Steud.

Habit, vegetative morphology. Annual (rarely), or perennial; stoloniferous (sometimes mat-forming), or caespitose. **Culms** 15–100 cm high (or more?); *herbaceous*. Culm nodes hairy, or glabrous. Culm internodes solid, or hollow. *Leaves* mostly basal, or not basally aggregated; non-auriculate. Leaf blades linear-lanceolate to ovate-lanceolate; broad, or narrow; flat, or folded; without cross venation; persistent; rolled in bud. *Ligule an unfringed membrane*.

Reproductive organization. Plants bisexual, with bisexual spikelets; with hermaphrodite florets.

Inflorescence. *Inflorescence of spicate main branches*; digitate (rarely), or non-digitate. Rachides hollowed. Inflorescence espatheate; not comprising 'partial inflorescences' and foliar organs. Spikelet-bearing axes persistent. Spikelets solitary; secund; biseriate. Pedicel apices discoid.

Female-fertile spikelets. *Spikelets* 1.6–5.4 mm long; oblong, or elliptic, or lanceolate, or ovate, or obovate; *adaxial*; *compressed dorsiventrally*; biconvex; falling with the glumes. Rachilla terminated by a female-fertile floret. Hairy callus absent.

Glumes one per spikelet (membranous); long relative to the adjacent lemmas; dorsiventral to the rachis (the one glume (upper) outside); awnless. Upper glume 4–5 nerved. *Spikelets* with incomplete florets. The incomplete florets proximal to the female-fertile florets. *The proximal incomplete florets* 1; epaleate; sterile. The proximal lemmas awnless; 0 nerved, or 2 nerved (the median lacking); decidedly exceeding the female-fertile lemmas; less firm than the female-fertile lemmas (membranous); not becoming indurated.

Female-fertile florets 1. Lemmas decidedly firmer than the glumes; smooth to striate; becoming indurated to not becoming indurated (papery to crustaceous); yellow in fruit, or brown in fruit; entire; blunt; awnless; hairless; non-carinate; having the margins tucked in onto the palea; with a clear germination flap; 4 nerved. *Palea* present; relatively long; entire; awnless, without apical setae; textured like the lemma; indurated, or not indurated. *Lodicules* present; 2; free; fleshy; glabrous. *Stamens* 3. Anthers not penicillate. *Ovary* glabrous. Styles free to their bases; free. Stigmas 2; white (e.g. *A. rupestris*), or red pigmented (usually?).

Fruit, embryo and seedling. Fruit small; ellipsoid; compressed dorsiventrally. Hilum short. Embryo large; without an epiblast; with a scutellar tail; with an elongated mesocotyl internode. Embryonic leaf margins overlapping.

Abaxial leaf blade epidermis. *Costal/intercostal zonation* conspicuous. *Papillae* absent. *Long-cells* similar in shape costally and intercostally, or markedly different in shape costally and intercostally (the costals much narrower); of similar wall thickness costally and intercostally. Mid-intercostal long-cells rectangular, or fusiform (slightly); having markedly sinuous walls. *Microhairs* present; panicoid-type; 39–45 microns long; 7.5–9 microns wide at the septum. Microhair total length/width at septum 4.3–5.2. Microhair apical cells 21–30 microns long. Microhair apical cell/total length ratio 0.54–0.67. *Stomata* common; 36–45 microns long. Subsidiaries triangular. Guard-cells overlapping to flush with the interstomatals. *Intercostal short-cells* common; in cork/silica-cell pairs; silicified. *Costal short-cells* conspicuously in long rows. Costal silica bodies 'panicoid-type'; dumb-bell shaped.

Transverse section of leaf blade, physiology. C_4; biochemical type NADP–ME (1 species); XyMS–. PCR sheath outlines uneven. PCR sheath extensions present, or absent. Maximum number of extension cells 2. PCR cell chloroplasts with reduced grana; centrifugal/peripheral. *Mesophyll* with non-radiate chlorenchyma. *Leaf blade* 'nodular' in section; with the ribs more or less constant in size. *Midrib* conspicuous; having a conventional arc of bundles; with colourless mesophyll adaxially. Bulliforms present in discrete, regular adaxial groups; in simple fans. All the vascular bundles accompanied by sclerenchyma. Combined sclerenchyma girders present; nowhere forming 'figures'. Sclerenchyma all associated with vascular bundles.

Phytochemistry. Leaves without flavonoid sulphates (1 species).

Cytology. Chromosome base number, $x = 10$. $2n = 20, 40, 60,$ and 80. 2, 4, 6, and 8 ploid.

Taxonomy. Panicoideae; Panicodae; Paniceae.
Ecology, geography, regional floristic distribution. 114 species. Tropical South America. Commonly adventive. Helophytic to mesophytic; species of open habitats; glycophytic. Savanna, forest clearings, moist and weedy places.
Neotropical. Caribbean, Venezuela and Surinam, Amazon, Central Brazilian, Pampas, and Andean. Canadian-Appalachian, Southern Atlantic North American, and Central Grasslands. Sahelo-Sudanian, Somalo-Ethiopian, and South Tropical African.
Rusts and smuts. Rusts — *Physopella* and *Puccinia*. Taxonomically wide-ranging species: *Puccinia levis*. Smuts from Ustilaginaceae. Ustilaginaceae — *Sorosporium* and *Sphacelotheca*.
Economic importance. Significant weed species: *A. affinis*, *A. compressus*. Cultivated fodder: *A. affinis*, *A. compressus*. Important native pasture species: *A. affinis*, *A. flexuosus*. Lawns and/or playing fields: *A. compressus*.
References, etc. Leaf anatomical: this project.

Bambusa Schreber

Arundarbor Kuntze, *Bonia* Balansa, *Criciuma* Soderstrom & Londoño, *Dendrocalamopsis* (Chia & Fung) Keng f., *Eremocaulon* Soderstrom & Londoño, *Guadua* Kunth, *Holttumochloa* K.M. Wong, *Ischurochloa* Büse, *Kinabaluchloa* K.M. Wong, *Leleba* Nakai, *Lingnania* McClure, *Maclurochloa* K.M. Wong, *Soejatmia* K.M. Wong, and *Tetragonocalamus* Nakai

Habit, vegetative morphology. Perennial. The flowering culms leafless, or leafy. **Culms** (200–)500–3500 cm high; *woody and persistent*; to 15 cm in diameter; scandent, or not scandent; branched above. Culm nodes glabrous. Primary branches/mid-culm node 3–30 (several to many). Culm internodes hollow (sometimes nearly solid). Pluricaespitose. Rhizomes pachymorph. Plants conspicuously armed (with thorns from the nodes: *Guadua*), or unarmed. *Leaves* not basally aggregated; auriculate; with auricular setae, or without auricular setae. Leaf blades broad, or narrow (small to moderate-sized); pseudopetiolate; cross veined, or without cross venation; disarticulating from the sheaths; rolled in bud. Ligule an unfringed membrane to a fringed membrane; short or long.
Reproductive organization. Plants bisexual, with bisexual spikelets; with hermaphrodite florets. Not viviparous.
Inflorescence. **Inflorescence** indeterminate; *with pseudospikelets*; of pseudospikelets, these solitary or in tufts, fascicles or capitula on leafless branches; open; spatheate (with or without foliage leaves); a complex of 'partial inflorescences' and intervening foliar organs. *Spikelet-bearing axes* very much reduced, or paniculate to capitate; persistent. Spikelets not secund.
Female-fertile spikelets. Spikelets 10–80 mm long; compressed laterally to not noticeably compressed; disarticulating above the glumes; disarticulating between the florets. Rachilla prolonged beyond the uppermost female-fertile floret; rachilla hairless. Hairy callus absent.
Glumes (i.e. transitional, empty glumes, which are additional to bud-bearing basal bracts) present; one per spikelet to several (1–3); very unequal, or more or less equal; shorter than the spikelets; shorter than the adjacent lemmas; hairless; not pointed; awnless; carinate; similar. Lower glume longer than half length of lowest lemma; 7–18 nerved. Upper glume 7–18 nerved. *Spikelets* with incomplete florets. The incomplete florets distal to the female-fertile florets, or both distal and proximal to the female-fertile florets. *The distal incomplete florets* merely underdeveloped. *The proximal incomplete florets* 0–3 (? — fewer than 4); sterile. The proximal lemmas awnless; several-nerved; exceeded by the female-fertile lemmas; similar in texture to the female-fertile lemmas.
Female-fertile florets (1–)2–20 (–'many'). **Lemmas** similar in texture to the glumes; not becoming indurated; entire; *blunt*; awnless (usually), or mucronate (*Guadua*); hairless; carinate to non-carinate; 9–22 nerved. *Palea* present; relatively long; entire to apically notched; awnless, without apical setae; not indurated; several nerved (about 6–16); 2-keeled. Palea keels winged (*Guadua*), or wingless. **Lodicules present**; 3; free; membranous; ciliate; not toothed; heavily vascularized. *Stamens* 6. Anthers 3.5–8 mm long;

penicillate, or not penicillate; with the connective apically prolonged, or without an apically prolonged connective. **Ovary** *hairy*; *with a conspicuous apical appendage*. The appendage broadly conical, fleshy (except in *Guadua*? — see illustration in Londoño and Peterson 1992). Styles fused. Stigmas 3 (usually?).

Fruit, embryo and seedling. *Fruit* free from both lemma and palea; longitudinally grooved; not noticeably compressed; with hairs confined to a terminal tuft. Hilum long-linear. Embryo small; not waisted. Endosperm containing compound starch grains.

Abaxial leaf blade epidermis. *Costal/intercostal zonation* conspicuous. *Papillae* present. Intercostal papillae over-arching the stomata; several per cell. *Long-cells* similar in shape costally and intercostally; differing markedly in wall thickness costally and intercostally. Mid-intercostal long-cells rectangular; having markedly sinuous walls. *Microhairs* present; panicoid-type; 36–48 microns long (in *B. arnhemica*); 6–9 microns wide at the septum. Microhair total length/width at septum 4.5–8 (i.e. very variable, in *B. arnhemica*). Microhair apical cells 15–21 microns long. Microhair apical cell/total length ratio 0.37–0.46. *Stomata* common; 21–27 microns long. Subsidiaries parallel-sided, or dome-shaped, or triangular; including both triangular and parallel-sided forms on the same leaf. *Intercostal short-cells* common; in cork/silica-cell pairs. *Costal short-cells* conspicuously in long rows. Costal silica bodies saddle shaped, or oryzoid.

Transverse section of leaf blade, physiology. C_3; XyMS+. *Mesophyll* with non-radiate chlorenchyma; with arm cells; with fusoids. *Leaf blade* adaxially flat. *Midrib* conspicuous; having complex vascularization. *The lamina* distinctly asymmetrical on either side of the midrib. Bulliforms present in discrete, regular adaxial groups; in simple fans. All the vascular bundles accompanied by sclerenchyma. Combined sclerenchyma girders present; forming 'figures'.

Culm anatomy. *Culm internode bundles* in three or more rings.

Phytochemistry. Leaves without flavonoid sulphates (1 species).

Cytology. Chromosome base number, $x = 12$. $2n = 46$ (*Guadua*), or 24, 48, 70, and 72. 4 and 6 ploid (rarely diploid).

Taxonomy. Bambusoideae; Bambusodae; Bambuseae.

Ecology, geography, regional floristic distribution. About 120 species. Tropical and subtropical Asia, Africa, America. Commonly adventive.

Holarctic, Paleotropical, Neotropical, and Australian. Boreal. Indomalesian. Eastern Asian. Indian, Indo-Chinese, Malesian, and Papuan. Caribbean, Amazon, Central Brazilian, Pampas, and Andean. North and East Australian. Tropical North and East Australian.

Hybrids. Claimed intergeneric hybrids with *Saccharum* are erroneous, representing apomixis.

Rusts and smuts. Rusts — *Dasturella*, *Stereostratum*, and *Puccinia*. Taxonomically wide-ranging species: *Dasturella divina* and *Stereostratum corticoides*. Smuts from Tilletiaceae. Tilletiaceae — *Tilletia*.

Economic importance. *B. arundinacea*, *B. dissemulator*, *B. duriuscula*, *B. gibba*, *B. lapidea*, *B. malingensis*, *B. sinospinosa*, *B. tuldoides*, *B. vulgaris* etc. are widely used for general constructional work; others (e.g. *B. pervariabilis*) for fishing rods, ski poles, umbrella shafts, furniture etc.; others (e.g. *B. chungii*) for weaving; others (e.g. *B. flexuosa*, *B. gibba*, *B. sinospinosa*) are useful barrier plants; and the shoots of many (e.g. *B. beecheyana*, *B. gibboides*, *B. vulgaris*) are edible. Pulp for paper and rayon is obtained from (e.g.) *B. guadua*.

References, etc. Leaf anatomical: Metcalfe 1960; this project.

Special comments. This treatment follows Clayton and Renvoize (1986). Soderstrom and Ellis (1987) refer *Criciuma*, *Eremocaulon* and *Guadua* to their subtribe Guaduinae, along with *Olmeca*, and place *Tetragonocalamus* in the Arundinariinae, but revised generic descriptions adequate for the present purpose are not available. Further problematic 'genera' are *Kinabaluchloa* K.M. Wong (including *B. wrayi*), *Holttumochloa* K.M. Wong (including *B. magica*), *Maclurochloa* K.M. Wong (= *B. pauciflora* and *B. montana*), and *Soejatmia* K.M. Wong (= *B. ridleyi*). Sound generic circumscriptions depend on adequate comparative descriptions, and will be extensively unachievable for bamboos until the descriptive terminology of inflorescence and spikelet morphology has been standardized.

Baptorhachis Clayton & Renvoize

~ *Stereochlaena (S. foliosa)*

Habit, vegetative morphology. Annual. *Culms* 30–60 cm high; herbaceous; unbranched above. Culm nodes glabrous. Culm internodes hollow. *Leaves* not basally aggregated; auriculate. Leaf blades linear-lanceolate; narrow; 2–4 mm wide (and 2–5 cm long); flat; without cross venation; persistent. Ligule a fringed membrane; truncate; about 1 mm long. *Contra-ligule* absent.

Reproductive organization. *Plants bisexual, with bisexual spikelets*; with hermaphrodite florets.

Inflorescence. Inflorescence a single raceme. *Rachides* conspicuously *flattened and winged (the wing covering the spikelets, dark purple)*. Inflorescence espatheate; not comprising 'partial inflorescences' and foliar organs. *Spikelet-bearing axes* 'racemes'; persistent. Spikelets paired; secund; biseriate; sessile and pedicellate, or subsessile and pedicellate; densely imbricate; consistently in 'long-and-short' combinations (but these homogamous). The 'shorter' spikelets hermaphrodite. The 'longer' spikelets hermaphrodite.

Female-fertile spikelets. Spikelets 2.5–3 mm long; abaxial; compressed dorsiventrally; biconvex; falling with the glumes; with conventional internode spacings. Rachilla terminated by a female-fertile floret. Hairy callus absent (but the long hairs on the glume and first lemma direct the resting position of the spikelet). *Callus* absent.

Glumes one per spikelet (the G_1 suppressed); about equalling the spikelets (i.e. the G_2); long relative to the adjacent lemmas (i.e. the G_2 about equalling the L_1); dorsiventral to the rachis; hairy; with distinct hair tufts (the lateral keels with hair tufts below); pointed; awned (from a sinus or from the tip); non-carinate. Upper glume 3 nerved (the two laterals thickenced below). *Spikelets* with incomplete florets. The incomplete florets proximal to the female-fertile florets. *The proximal incomplete florets* 1; epaleate; sterile. The proximal lemmas bilobed; awned (from the sinus); 5 nerved, or 7 nerved; more or less equalling the female-fertile lemmas; less firm than the female-fertile lemmas (membranous between the leathery lateral keels, which are hair-tufted below).

Female-fertile florets 1. Lemmas decidedly firmer than the glumes (papery); smooth (and shining); not becoming indurated; entire; pointed (acute); awnless; hairless; glabrous; non-carinate; having the margins lying flat and exposed on the palea; without a germination flap; 3 nerved. *Palea* present; relatively long; tightly clasped by the lemma; entire; awnless, without apical setae; textured like the lemma (papery); not indurated; 2-nerved. *Lodicules* absent. *Stamens* 3. Anthers about 0.7 mm long; without an apically prolonged connective. *Ovary* glabrous. Styles free to their bases. Stigmas 2; white.

Abaxial leaf blade epidermis. *Microhairs* present; panicoid-type.

Transverse section of leaf blade, physiology. C_4; XyMS–. *Leaf blade* adaxially flat.

Special diagnostic feature. *Spikelets borne on one side of a broad, leaflike rachis (this colourful)*.

Taxonomy. Panicoideae; Panicodae; Paniceae.

Ecology, geography, regional floristic distribution. 1 species (*Stereochlaena foliacea*). Mozambique. Xerophytic; species of open habitats. On stony slopes.

Paleotropical. African. Sudano-Angolan. South Tropical African.

References, etc. Morphological/taxonomic: Clayton 1978b.

Special comments. Fruit data wanting. Anatomical data wanting.

Bealia Scribner

~ *Muhlenbergia (M. biloba* Hitchc.)

Habit, vegetative morphology. *Annual*; caespitose. *Culms* 9–35 cm high; herbaceous; unbranched above. Culm nodes glabrous. *Leaves* not basally aggregated; non-auriculate; without auricular setae. Sheaths keeled, often striate, longer than the internodes. Leaf blades linear; narrow; 0.6–1.4 mm wide (1–7 cm long); flat, or rolled (involute); without abaxial multicellular glands; without cross venation; persistent. *Ligule an unfringed membrane*; not truncate (apex acute or rounded, often irregularly toothed); 1.5–3.4 mm long. *Contra-ligule* absent.

Reproductive organization. Plants bisexual, with bisexual spikelets; with hermaphrodite florets. The spikelets hermaphrodite. Plants without hidden cleistogenes.

Inflorescence. *Inflorescence paniculate*; open; espatheate; not comprising 'partial inflorescences' and foliar organs. Spikelets solitary; not secund; pedicellate; imbricate.

Female-fertile spikelets. *Spikelets* about *4–4.5 mm long*; compressed laterally; disarticulating above the glumes. Rachilla terminated by a female-fertile floret (seemingly, from the description and illustrations seen). *Hairy callus present. Callus* short.

Glumes two; more or less equal; about equalling the spikelets to exceeding the spikelets; long relative to the adjacent lemmas; hairy (loosely pilose to villous, especially below); without conspicuous tufts or rows of hairs; pointed; awnless; similar. Lower glume faintly 1 nerved. *Upper glume* faintly *1 nerved*. *Spikelets* with female-fertile florets only.

Female-fertile florets 1. Lemmas not becoming indurated; incised; 2 lobed (the lobes rounded to obtuse); deeply cleft; awned. *Awns* 1; median; from a sinus; *non-geniculate*; flexuous (to crisp-curled); hairless (scabrous); much longer than the body of the lemma; entered by one vein. Lemmas hairy (tawny-villous below); without a germination flap; 3 nerved. *Palea* present; relatively long (about as long as the lemma, hairy below); entire (obtuse); awnless, without apical setae; not indurated; 2-nerved; 2-keeled. Palea keels wingless; hairy. *Lodicules* present; 2; free; fleshy; glabrous. *Stamens* 3. Anthers 1.2–2.3 mm long; not penicillate; without an apically prolonged connective. *Ovary* glabrous. Styles free to their bases. Stigmas 2.

Fruit, embryo and seedling. Fruit small (about 1.8 mm long); olive-brown; fusiform; not noticeably compressed. Hilum short. Pericarp fused. Embryo large.

Abaxial leaf blade epidermis. *Microhairs* present; elongated; clearly two-celled; chloridoid-type. Microhair apical cell wall of similar thickness/rigidity to that of the basal cell. Microhair basal cells 27–30 microns long. Microhair total length/width at septum 3.5. Microhair apical cell/total length ratio 0.25. *Stomata* common. Intercostal silica bodies absent. *Costal short-cells* conspicuously in long rows. Costal silica bodies present in alternate cell files of the costal zones; tall-and-narrow (to squarish).

Transverse section of leaf blade, physiology. C_4; XyMS+. PCR sheath outlines probably even. PCR sheaths of the primary vascular bundles interrupted; interrupted both abaxially and adaxially. PCR sheath extensions absent. PCR cell chloroplasts centripetal. *Mesophyll* with radiate chlorenchyma; traversed by columns of colourless mesophyll cells. *Midrib* conspicuous (via a bulbous abaxial keel); with one bundle only. *The lamina* symmetrical on either side of the midrib. Bulliforms present in discrete, regular adaxial groups to not in discrete, regular adaxial groups; in simple fans and associated with colourless mesophyll cells to form deeply-penetrating fans (the penetrating groups contributing to the colourless columns). Many of the smallest vascular bundles unaccompanied by sclerenchyma. Combined sclerenchyma girders present. Sclerenchyma all associated with vascular bundles. The lamina margins with fibres (and very large).

Cytology. Chromosome base number, $x = 8$. $2n = 16$.

Taxonomy. Chloridoideae; main chloridoid assemblage.

Ecology, geography, regional floristic distribution. 1 species (*B. mexicana = Muhlenbergia biloba* Hitchc.). Mexico. Mesophytic, or xerophytic; glycophytic. In shallow, sandy soils in coniferous woodland, at 2000–2300 m.

Neotropical. Caribbean.

References, etc. Morphological/taxonomic: Peterson 1989. Leaf anatomical: photo of mid-blade ts, Peterson *et al.* 1989; Van den Borre 1994.

Beckeropsis Figari & de Not.

~ Pennisetum

Habit, vegetative morphology. Annual, or perennial; rhizomatous, or caespitose. *Culms* 90–240 cm high; herbaceous; branched above. Culm nodes glabrous. Culm internodes hollow (in *B. uniseta*). Plants unarmed. Young shoots intravaginal. *Leaves* not basally aggregated; non-auriculate. Leaf blades linear-lanceolate; broad (to 12 mm wide in *B. uniseta*), or narrow; flat; pseudopetiolate, or not pseudopetiolate; without cross venation. Ligule a fringe of hairs. *Contra-ligule* absent.

Reproductive organization. Plants bisexual, with bisexual spikelets; with hermaphrodite florets.

Inflorescence. *Inflorescence* axillary, compound *paniculate (with racemelets each represented by a single spikelet-plus-bristle)*; espatheate (but sometimes closely subtended by a small leaf); not comprising 'partial inflorescences' and foliar organs. *Spikelet-bearing axes* very much reduced (reduced to a single spikelet with a long, reduced-branch 'bristle' at its base); (i.e. the much reduced racemelets) disarticulating; falling entire. **Spikelets *subtended by solitary 'bristles'***. The 'bristles' deciduous with the spikelets. Spikelets solitary; secund; pedicellate (but the 'pedicel' being presumably the base of the reduced branch); imbricate.

Female-fertile spikelets. Spikelets 2.5–3.5 mm long (in *B. uniseta*); abaxial; compressed dorsiventrally; planoconvex; falling with the glumes (spikelet and 'bristle' falling from the 'pedicel'); not disarticulating between the florets; with conventional internode spacings. Rachilla terminated by a female-fertile floret. Hairy callus absent. *Callus* absent.

Glumes two; very unequal; shorter than the spikelets; shorter than the adjacent lemmas; dorsiventral to the rachis; hairless; glabrous; not pointed; awnless; non-carinate; similar (hyaline, rounded or bilobed). Lower glume shorter than the lowest lemma; 0 nerved. *Upper glume 0 nerved. Spikelets* with incomplete florets. The incomplete florets proximal to the female-fertile florets. *The proximal incomplete florets* 1; epaleate; sterile. The proximal lemmas awnless; 5 nerved; similar in texture to the female-fertile lemmas (membranous); not becoming indurated.

Female-fertile florets 1. Lemmas decidedly firmer than the glumes; smooth; not becoming indurated; entire; pointed; awnless, or mucronate; hairless; glabrous; non-carinate; without a germination flap; 3–5 nerved. *Palea* present; relatively long; entire (truncate); awnless, without apical setae; thinner than the lemma (hyaline); not indurated; 2-nerved; 2-keeled. *Lodicules* present; 2; free; fleshy; glabrous. *Stamens* 3. Anthers about 1.5 mm long in *B. uniseta*; without an apically prolonged connective. *Ovary* glabrous. Styles fused (near the base). Stigmas 2; red pigmented (purple).

Fruit, embryo and seedling. *Fruit* free from both lemma and palea; small; compressed dorsiventrally. Hilum short. Embryo large.

Abaxial leaf blade epidermis. *Costal/intercostal zonation* conspicuous. *Papillae* absent. Mid-intercostal long-cells having markedly sinuous walls. *Microhairs* present; panicoid-type; 44–54 microns long. Microhair apical cells 24–36 microns long. Microhair apical cell/total length ratio 0.65. *Stomata* common. Subsidiaries dome-shaped and triangular. *Intercostal short-cells* absent or very rare. *Costal short-cells* conspicuously in long rows. Costal silica bodies 'panicoid-type'; cross shaped to dumb-bell shaped, or nodular.

Transverse section of leaf blade, physiology. C_4; XyMS–. PCR cell chloroplasts centrifugal/peripheral. *Mesophyll* with radiate chlorenchyma. *Leaf blade* with distinct, prominent adaxial ribs, or 'nodular' in section, or adaxially flat; with the ribs more or less constant in size. *Midrib* conspicuous; having a conventional arc of bundles; with colourless mesophyll adaxially. Bulliforms present in discrete, regular adaxial groups, or not in discrete, regular adaxial groups; in simple fans (or in groups of fairly evenly sized cells, or in irregular groups). Many of the smallest vascular bundles unaccompanied by sclerenchyma. Combined sclerenchyma girders absent. Sclerenchyma all associated with vascular bundles.

Culm anatomy. *Culm internode bundles* in three or more rings.

Phytochemistry. Leaves without flavonoid sulphates (*B. uniseta*).

Cytology. Chromosome base number, $x = 9$. $2n = 18$. 2 ploid.

Taxonomy. Panicoideae; Panicodae; Paniceae.

Ecology, geography, regional floristic distribution. 6 species. Tropical and southern Africa. Helophytic; shade species; glycophytic. Riverine woodland.

Paleotropical and Cape. African. Sudano-Angolan and West African Rainforest. Sahelo-Sudanian, Somalo-Ethiopian, and South Tropical African.

Rusts and smuts. Rusts — *Puccinia*. Taxonomically wide-ranging species: *Puccinia substriata*.

Economic importance. Important native pasture species: *B. uniseta*.
References, etc. Leaf anatomical: Metcalfe 1960; photos of *B. uniseta* provided by R.P. Ellis.

Beckmannia Host

Buchmannia Nutt, *Joachima* Ten.

Habit, vegetative morphology. Annual, or perennial; rhizomatous, or caespitose. *Culms* 30–150 cm high; herbaceous; tuberous (when perennial), or not tuberous. Culm internodes hollow. Leaves non-auriculate. Leaf blades narrow; 4–10 mm wide; flat; without cross venation; rolled in bud. *Ligule an unfringed membrane*; not truncate (acute); 5–10 mm long.

Reproductive organization. Plants bisexual, with bisexual spikelets; with hermaphrodite florets; outbreeding.

Inflorescence. *Inflorescence* of spicate main branches, or paniculate (the branches sometimes themselves branched); *open*; non-digitate (branches racemosely arranged); espatheate; not comprising 'partial inflorescences' and foliar organs. Spikelet-bearing axes persistent. *Spikelets secund (the branches unilateral)*; biseriate.

Female-fertile spikelets. *Spikelets* 1.5–4 mm long; suborbicular; *compressed laterally*; *falling with the glumes*. Rachilla prolonged beyond the uppermost female-fertile floret, or terminated by a female-fertile floret. The rachilla extension when present, with incomplete florets. Hairy callus absent.

Glumes two; more or less equal; long relative to the adjacent lemmas; conspicuously ventricose to not ventricose; pointed; awnless (but mucronate); *carinate*; *similar (herbaceous, navicular, more or less inflated, cross-veined)*. Lower glume 3 nerved. Upper glume 3 nerved. *Spikelets* with female-fertile florets only, or with incomplete florets. The incomplete florets when present, distal to the female-fertile florets. The distal incomplete florets 1 (male). *Spikelets without proximal incomplete florets*.

Female-fertile florets 1, or 2. Lemmas acuminate; cartilaginous; not becoming indurated; entire; pointed; awnless, or mucronate; hairy, or hairless; non-carinate (dorsally rounded); without a germination flap; 5 nerved. *Palea* present; relatively long; 2-nerved; 2-keeled, or keel-less. **Lodicules** present; 2; free; *membranous*; glabrous; toothed, or not toothed; not or scarcely vascularized. **Stamens 3**. Anthers 0.4–1.8 mm long. *Ovary* glabrous. Styles free to their bases. Stigmas 2; white.

Fruit, embryo and seedling. Fruit small (1.6–2 mm long); not noticeably compressed. *Hilum short*. Embryo small. Endosperm liquid in the mature fruit; with lipid. Embryo with an epiblast; without a scutellar tail; with a negligible mesocotyl internode. Embryonic leaf margins meeting.

Abaxial leaf blade epidermis. *Costal/intercostal zonation* conspicuous. *Papillae* absent. *Long-cells* markedly different in shape costally and intercostally; differing markedly in wall thickness costally and intercostally. Mid-intercostal long-cells fusiform; having straight or only gently undulating walls. *Microhairs* absent. *Stomata* common. Subsidiaries low dome-shaped, or parallel-sided. Guard-cells overlapped by the interstomatals (mostly, slightly). *Intercostal short-cells* absent or very rare. *Costal short-cells* neither distinctly grouped into long rows nor predominantly paired. Costal silica bodies horizontally-elongated crenate/sinuous, or rounded, or crescentic.

Transverse section of leaf blade, physiology. C_3; XyMS+. *Mesophyll* with non-radiate chlorenchyma.

Cytology. Chromosome base number, $x = 7$. $2n = 14$ (usually), or 16. 2 ploid.

Taxonomy. Pooideae; Poodae; Aveneae.

Ecology, geography, regional floristic distribution. 2 species. North Eurasia & North America. Helophytic, or mesophytic; species of open habitats. Meadows, etc.

Holarctic. Boreal, Tethyan, and Madrean. Arctic and Subarctic, Euro-Siberian, Eastern Asian, Atlantic North American, and Rocky Mountains. Mediterranean and Irano-Turanian. European and Siberian. Canadian-Appalachian, Southern Atlantic North American, and Central Grasslands.

Rusts and smuts. Rusts — *Puccinia*. Taxonomically wide-ranging species: *Puccinia graminis*, *Puccinia coronata*, and *Puccinia striiformis*. Smuts from Tilletiaceae and from Ustilaginaceae. Tilletiaceae — *Urocystis*. Ustilaginaceae — *Ustilago*.

Economic importance. Important native pasture species: *B. syzigachne*.

References, etc. Leaf anatomical: this project.

Bellardiochloa Chiov.

~ *Poa*

Habit, vegetative morphology. Perennial; caespitose (densely). **Culms** 8–50 cm high; *herbaceous*. Leaves non-auriculate. *Leaf blades* linear; narrow; 1 mm wide; *setaceous (glaucous)*; rolled; without cross venation; persistent. **Ligule** present; *an unfringed membrane*; not truncate; 2–7 mm long.

Reproductive organization. Plants bisexual, with bisexual spikelets; with hermaphrodite florets.

Inflorescence. *Inflorescence paniculate*; open, or contracted; espatheate; not comprising 'partial inflorescences' and foliar organs. Spikelet-bearing axes persistent. Spikelets not secund; pedicellate.

Female-fertile spikelets. Spikelets 3–7.5 mm long; compressed laterally; disarticulating above the glumes. *Rachilla* prolonged beyond the uppermost female-fertile floret; *rachilla hairy*. The rachilla extension with incomplete florets. *Hairy callus present*.

Glumes two; very unequal to more or less equal; shorter than the spikelets; shorter than the adjacent lemmas; pointed (acute); awnless; carinate; similar. Lower glume 3 nerved. Upper glume 3 nerved. *Spikelets* with incomplete florets. The incomplete florets distal to the female-fertile florets.

Female-fertile florets 2–8. Lemmas similar in texture to the glumes; not becoming indurated; entire, or incised; when entire pointed, or blunt; mucronate, or awned (very shortly so). Awns 1; median; from a sinus (or the mucro from a notch); non-geniculate; much shorter than the body of the lemma; entered by one vein. *Lemmas* hairless (shortly hairy at base, scabrid on keels, not lanate); *carinate*; without a germination flap; 5 nerved. *Palea* present; relatively long; 2-nerved; 2-keeled (hairy between the keels). *Lodicules* present; 2; free; membranous; glabrous; toothed, or not toothed; not or scarcely vascularized. *Stamens* 3. Anthers 1.4–2.5 mm long. *Ovary* glabrous. Styles free to their bases. Stigmas 2.

Fruit, embryo and seedling. *Fruit* adhering to lemma and/or palea; small (c. 1.8 mm long); compressed dorsiventrally. Hilum short. Embryo small. Endosperm hard; with lipid. Embryo with an epiblast; without a scutellar tail; with a negligible mesocotyl internode.

Abaxial leaf blade epidermis. *Costal/intercostal zonation* conspicuous. *Papillae* absent. *Long-cells* markedly different in shape costally and intercostally (the costals much narrower); of similar wall thickness costally and intercostally (thin walled). Mid-intercostal long-cells rectangular; having markedly sinuous walls (the sinuosity associated with conspicuous pitting). *Microhairs* absent. *Stomata* absent or very rare (none seen). *Intercostal short-cells* common; commonly in cork/silica-cell pairs; silicified. Intercostal silica bodies rounded to crescentic. With scattered costal and intercostal prickles. *Costal short-cells* predominantly paired. Costal silica bodies present and well developed; rounded and crescentic.

Transverse section of leaf blade, physiology. C_3; XyMS+. *Leaf blade* with distinct, prominent adaxial ribs; with the ribs more or less constant in size (round topped, one per bundle). *Midrib* conspicuous to not readily distinguishable; with one bundle only. *The lamina* symmetrical on either side of the midrib. Bulliforms present in discrete, regular adaxial groups, or not in discrete, regular adaxial groups (inconspicuous, in the material seen); if present, in simple fans (of small cells). All the vascular bundles accompanied by sclerenchyma. Combined sclerenchyma girders mostly absent (the material sectioned exhibiting only one combined girder, associated with the marginal bundle on one side of the blade, the others (including the midrib) with broad-footed abaxial girders and adaxial strands). Sclerenchyma all associated with vascular bundles, or not all bundle-associated (in one of the two specimens seen). The 'extra' sclerenchyma when present, in a continuous abaxial layer.

Cytology. Chromosome base number, $x = 7$. $2n = 14$. 2 ploid.

Taxonomy. Pooideae; Poodae; Poeae.

Ecology, geography, regional floristic distribution. 1 species. Southern Europe, western Asia.

Holarctic and Paleotropical. Boreal and Tethyan. African. Euro-Siberian. Mediterranean and Irano-Turanian. Sudano-Angolan and West African Rainforest. European. Sahelo-Sudanian.

References, etc. Leaf anatomical: this project.

Bewsia Goossens

Habit, vegetative morphology. Perennial; caespitose (with short, creeping rhizomes). *Culms* 26–93 cm high; herbaceous; to 0.2 cm in diameter; unbranched above. Culm nodes glabrous. Culm internodes solid. Plants unarmed. Young shoots intravaginal. *Leaves* mostly basal; non-auriculate; without auricular setae (but hairy at the mouth of the sheath and on the lower part of the blade). Leaf blades linear to linear-lanceolate; narrow; to 5 mm wide; flat, or rolled (the margins becoming involute under water stress); without abaxial multicellular glands; without cross venation. *Ligule an unfringed membrane (minutely ciliolate only)*; truncate; to 0.3 mm long.

Reproductive organization. Plants bisexual, with bisexual spikelets; with hermaphrodite florets. The spikelets all alike in sexuality.

Inflorescence. *Inflorescence of spicate main branches (these appressed to the central axis)*. Primary inflorescence branches about 10–15. Inflorescence espatheate; not comprising 'partial inflorescences' and foliar organs. *Spikelet-bearing axes* 'racemes'. The racemes spikelet bearing to the base. Spikelet-bearing axes persistent. Spikelets solitary; somewhat secund; biseriate; shortly pedicellate; imbricate.

Female-fertile spikelets. Spikelets 5.5–9 mm long; adaxial; strongly compressed laterally; disarticulating above the glumes; not disarticulating between the florets; with conventional internode spacings (rather long). Rachilla prolonged beyond the uppermost female-fertile floret; rachilla hairy (between L_1 and L_2). The rachilla extension with incomplete florets. Hairy callus present. *Callus* short; blunt.

Glumes two; more or less equal; about equalling the spikelets (a little shorter to a little longer); *long relative to the adjacent lemmas*; dorsiventral to the rachis; hairless (scabridulous); pointed (acuminate, often mucronate); awnless; strongly carinate; similar. Lower glume 1 nerved. Upper glume 1 nerved. *Spikelets* with incomplete florets. The incomplete florets distal to the female-fertile florets. *The distal incomplete florets* merely underdeveloped.

Female-fertile florets 2–6. *Lemmas* similar in texture to the glumes (membranous); not becoming indurated; entire, or incised; not deeply cleft (no more than minutely notched); *awned. Awns* 1; median; *dorsal*; from a quarter to a third of the way down; non-geniculate; hairless (scabrid); much shorter than the body of the lemma to about as long as the body of the lemma (1–4 mm long); entered by one vein. Lemmas hairy (the lower lemmas hairy below, beside the keel and on the margins); carinate; without a germination flap; 3 nerved; with the nerves non-confluent. *Palea* present; relatively long; minutely apically notched; awnless, without apical setae; textured like the lemma; not indurated; 2-nerved; 2-keeled. Palea keels hairy (minutely, densely ciliate). *Lodicules* present; 2; free; fleshy (long, narrow); glabrous. *Stamens* 3. Anthers 1.5–2.5 mm long; not penicillate; without an apically prolonged connective. *Ovary* glabrous. Styles free to their bases. Stigmas 2; red pigmented.

Fruit, embryo and seedling. Fruit small (about 2 mm long); linear (to oblong); not noticeably compressed (terete). Pericarp fused.

Abaxial leaf blade epidermis. *Costal/intercostal zonation* conspicuous. *Papillae* absent. Long-cells of similar wall thickness costally and intercostally (fairly thick walled). Mid-intercostal long-cells rectangular; having markedly sinuous walls. *Microhairs* absent. *Stomata* common (in rather distant single files). Subsidiaries parallel-sided to triangular (rather irregular — mostly high domes with truncated tops). Guard-cells overlapping to flush with the interstomatals. *Intercostal short-cells* common; not paired (solitary); not silicified (in material seen). Intercostal silica bodies absent. Prickles abundant. *Costal short-cells* conspicuously in long rows (the files interrupted by prickles). Costal silica bodies present and well developed (usually?), or absent (in some material seen, but the silica cells nodular); present in alternate cell files of the costal zones; 'panicoid-type'; dumb-bell shaped (with long, inconspicuous isthmuses), or nodular (a few).

Transverse section of leaf blade, physiology. C_4; XyMS+ (the MS thick-walled, sometimes double). PCR sheath outlines even. PCR sheaths of the primary vascular bundles interrupted; interrupted both abaxially and adaxially. PCR sheath extensions absent. PCR cell chloroplasts centripetal. *Leaf blade* adaxially flat (or with very low flat-topped ribs). *Midrib* conspicuous to not readily distinguishable; with one bundle only. Bulliforms present in discrete, regular adaxial groups; associated with colourless mesophyll cells to form deeply-penetrating fans. All the vascular bundles accompanied by sclerenchyma. Combined sclerenchyma girders present (with all the bundles); forming 'figures' (all bundles with anchors). Sclerenchyma all associated with vascular bundles to not all bundle-associated. The 'extra' sclerenchyma when present, in a continuous abaxial layer. The lamina margins with fibres.

Cytology. $2n = 30$.

Taxonomy. Chloridoideae; main chloridoid assemblage.

Ecology, geography, regional floristic distribution. 1 species. Southern tropical and South Africa. Mesophytic; species of open habitats. In grassveld, often on sandy soil.

Paleotropical. African. Sudano-Angolan. Sahelo-Sudanian and South Tropical African.

Rusts and smuts. Rusts — *Puccinia*.

References, etc. Leaf anatomical: this project; photos provided by R.P. Ellis.

Bhidea Stapf ex Bor

Habit, vegetative morphology. Annual. *Culms* 8–16 cm high; herbaceous. Culm nodes hairy, or glabrous. Leaves non-auriculate. Leaf blades linear; narrow; 3–4 mm wide (to 4 cm long); without cross venation. *Ligule an unfringed membrane*; 1 mm long.

Reproductive organization. Plants bisexual, with bisexual spikelets; with hermaphrodite florets. *The spikelets* of sexually distinct forms on the same plant (spikelet pairs homomorphic and sterile at the base of the raceme, heteromorphic heterogamous pairs above); hermaphrodite and male-only, or hermaphrodite and sterile; *overtly heteromorphic (the pedicelled and basal spikelets awnless)*; in both homogamous and heterogamous combinations (there being homomorphic, sterile pairs at the base of the raceme).

Inflorescence. Inflorescence of paired 'racemes', or these solitary (in *B. borii*); digitate, or non-digitate (*B. borii*). Primary inflorescence branches 1–2. *Inflorescence* spatheate (the raceme base(s) embraced by a spatheole); *a complex of 'partial inflorescences' and intervening foliar organs*. Spikelet-bearing axes 'racemes' (about 6-noded); solitary (*B. borii*), or paired; with very slender rachides (the raceme bases terete); disarticulating; disarticulating at the joints. '*Articles*' linear (slightly expanded above); without a basal callus-knob; not appendaged; disarticulating obliquely; densely long-hairy (down one side). Spikelets paired; sessile and pedicellate; consistently in 'long-and-short' combinations; in pedicellate/sessile combinations. Pedicels of the 'pedicellate' spikelets free of the rachis. The 'shorter' spikelets hermaphrodite, or sterile (reduced to glumes, at the base of the raceme). The 'longer' spikelets sterile (and awnless).

Female-sterile spikelets. The male spikelets with glumes. The lemmas awnless.

Female-fertile spikelets. Spikelets 7–20 mm long; compressed dorsiventrally; falling with the glumes. Rachilla terminated by a female-fertile floret. Hairy callus present. *Callus* short; blunt.

Glumes two; *very unequal (G_2 shorter, excluding the awn)*; (the longer) long relative to the adjacent lemmas; hairless; *awned (G_2, from a sinus)*; very dissimilar (G_1 narrow, pointed and 2-keeled above, awnless, G_2 trilobed and long-awned from a deep apical sinus). *Lower glume two-keeled (inrolled below, but asymmetrically keel-winged above)*; convex on the back to flattened on the back; not pitted; relatively smooth; 2 nerved (without intercarinal nerves). Upper glume 3 nerved. *Spikelets* with incomplete florets.

The incomplete florets proximal to the female-fertile florets. *The proximal incomplete florets* 1. The proximal lemmas awnless; 2 nerved; exceeded by the female-fertile lemmas; not becoming indurated.

Female-fertile florets 1. *Lemmas* less firm than the glumes; incised; not deeply cleft (i.e. to less than a quarter); *awned*. Awns 1; median; from a sinus; geniculate; hairless; much longer than the body of the lemma. Lemmas hairless; non-carinate; without a germination flap. *Palea* present; conspicuous but relatively short; awnless, without apical setae; not indurated. *Lodicules* present; 2; free; glabrous.

Abaxial leaf blade epidermis. *Costal/intercostal zonation* conspicuous. *Papillae* present (abundant); costal and intercostal. Intercostal papillae over-arching the stomata; several per cell (usually in a single row of 2–6 per cell, thick walled, circular, symmetrical to asymmetrical). *Long-cells* fairly markedly different in shape costally and intercostally (the costals narrower); of similar wall thickness costally and intercostally (fairly thin walled). Mid-intercostal long-cells rectangular; having markedly sinuous walls (the sinuosity coarse). *Microhairs* present; more or less spherical (sic); clearly two-celled; chloridoid-type (i.e., most unexpected). *Stomata* common. Subsidiaries papillate (each with a pair of inwardly directed papillae, one from each side); dome-shaped and triangular. Guard-cells overlapping to flush with the interstomatals. *Intercostal short-cells* absent or very rare. No macrohairs or prickles seen. *Costal short-cells* conspicuously in long rows. Costal silica bodies present and well developed; 'panicoid-type'; dumb-bell shaped and nodular.

Transverse section of leaf blade, physiology. C_4. The anatomical organization conventional. XyMS–. *Leaf blade* adaxially flat. *Midrib* conspicuous; having a conventional arc of bundles (a large median, with several smaller bundles on either side); with colourless mesophyll adaxially. *The lamina* symmetrical on either side of the midrib. Bulliforms not in discrete, regular adaxial groups (the epidermis mainly consisting of large, bulliform-like cells). All the vascular bundles accompanied by sclerenchyma. Combined sclerenchyma girders present (with all the bundles, even the tiniest); nowhere forming 'figures' (the fibre groups everywhere small, save the one abaxial to the median bundle). Sclerenchyma all associated with vascular bundles.

Taxonomy. Panicoideae; Andropogonodae; Andropogoneae; Andropogoninae.

Ecology, geography, regional floristic distribution. 3 species. India. Species of open habitats. Dry shallow soils.

Paleotropical. Indomalesian. Indian.

References, etc. Morphological/taxonomic: Bor 1948. Leaf anatomical: this project.

Special comments. Fruit data wanting.

Blepharidachne Hackel

Eremochloe S. Wats.

Habit, vegetative morphology. Usually perennial, or annual (rarely); branching above, with leafy inflorescence tufts on long internodes. *Culms* 8–20 cm high; herbaceous. *Leaves* mostly basal (in tufts); non-auriculate. Leaf blades narrow; acicular (involute, arcuate and rigid, less than 1 mm broad); without abaxial multicellular glands; without cross venation; persistent. *Ligule a fringe of hairs*.

Reproductive organization. Plants bisexual, with bisexual spikelets; with hermaphrodite florets.

Inflorescence. *Inflorescence* few spikeleted; *paniculate (of peduncled clusters, each invested by spathes)*; the individual clusters capitate; spatheate (the leaves subtending inflorescences with spathe-like sheaths); a complex of 'partial inflorescences' and intervening foliar organs. *Spikelet-bearing axes* very much reduced and paniculate (short, congested, usually surpassed by subtending leaves); persistent. **Spikelets *associated with bractiform involucres***; not secund (?); subsessile.

Female-fertile spikelets. Spikelets 6–8 mm long; compressed laterally; disarticulating above the glumes; not disarticulating between the florets; with conventional internode spacings. Rachilla prolonged beyond the uppermost female-fertile floret. The rachilla extension with incomplete florets. Hairy callus present (at the base of the unit as shed). *Callus* short; blunt.

Glumes two; more or less equal; about equalling the spikelets to exceeding the spikelets; long relative to the adjacent lemmas; pointed (acuminate); shortly awned, or awnless (then mucronate); similar (hyaline, shiny). Lower glume 1 nerved. *Upper glume 1 nerved. Spikelets* with incomplete florets. *The incomplete florets both distal and proximal to the female-fertile florets (L_1 and L_2 sterile or male, deeply lobed and awned, L_4 reduced to a 3-awned rudiment). The distal incomplete florets* clearly specialised and modified in form; awned (the terminal floret reduced to three awns). *Spikelets with proximal incomplete florets. The proximal incomplete florets* 2; paleate. Palea of the proximal incomplete florets usually reduced. The proximal incomplete florets male, or sterile. The proximal lemmas shortly awned (and hairy); 3 nerved; more or less equalling the female-fertile lemmas to decidedly exceeding the female-fertile lemmas; similar in texture to the female-fertile lemmas (and similar in form); not becoming indurated.

Female-fertile florets 1 (only L_3 fertile). Lemmas not becoming indurated; incised; 2 lobed; deeply cleft; awned. Awns 1, or 3; median, or median and lateral (via the awn-tipped lateral lobes); the median similar in form to the laterals (when laterals present); from a sinus; non-geniculate. The lateral awns when present, shorter than the median. Lemmas hairy (on the margins); non-carinate; without a germination flap; 3 nerved. *Palea* present; relatively long; apically notched; awnless, without apical setae; 2-nerved; 2-keeled. Palea keels hairy. *Lodicules* absent. *Stamens* (1–)2, or 3. *Ovary* glabrous. Stigmas 2.

Fruit, embryo and seedling. Fruit small (about 2 mm long); ellipsoid; compressed laterally. Hilum short. Pericarp fused. Embryo large; without an epiblast; with a scutellar tail; with an elongated mesocotyl internode. Embryonic leaf margins overlapping.

Abaxial leaf blade epidermis. *Costal/intercostal zonation* conspicuous. *Papillae* absent (though abundant adaxially). Mid-intercostal long-cells rectangular; having markedly sinuous walls. *Microhairs* absent (seemingly, from the material seen — though they would be easy to overlook in such material, with the intercostal zones in deep grooves and obscured by prickles or hairs. None seen adaxially either, in *B. kingii*). *Stomata* common; (21–)24–28(–30) microns long. Subsidiaries low dome-shaped and triangular. Guard-cells overlapping to flush with the interstomatals. *Intercostal short-cells* absent or very rare. Intercostal silica bodies absent. A few costal prickles forming a continuous series with the abundant macrohairs in *B. bigelovii*, and abundant prickles in *B. kingii*. *Crown cells* absent. *Costal short-cells* predominantly paired (in places, in both species), or neither distinctly grouped into long rows nor predominantly paired (elsewhere, in both species). Costal silica bodies present throughout the costal zones; rounded, or saddle shaped, or 'panicoid-type' (*B. bigelovii* with predominantly saddles merging into rounded forms and few 'panicoid' types, *B. kingii* with mainly 'panicoid types' and few saddles); when panicoid type, cross shaped (irregular), or dumb-bell shaped (short).

Transverse section of leaf blade, physiology. C_4; XyMS+. PCR sheath outlines even. PCR sheaths of the primary vascular bundles complete. PCR sheath extensions absent. PCR cell chloroplasts centripetal. *Leaf blade* 'nodular' in section; with the ribs more or less constant in size (round topped). *Midrib* fairly conspicuous, or not readily distinguishable; with one bundle only. *The lamina* symmetrical on either side of the midrib. Bulliforms present in discrete, regular adaxial groups; in simple fans. All the vascular bundles accompanied by sclerenchyma. Combined sclerenchyma girders absent (every bundle except the midrib with a massive abaxial strand and a smaller adaxial one, in *B. kingii*), or present (in *B. bigelovii*); forming 'figures' (the midrib in *B. kingii*, all the bundles in *B. bigelovii*). Sclerenchyma all associated with vascular bundles.

Cytology. Chromosome base number, $x = 7$. $2n = 14$. 2 ploid.

Taxonomy. Chloridoideae; main chloridoid assemblage.

Ecology, geography, regional floristic distribution. 3 species. U.S.A. and Argentina. Xerophytic; species of open habitats. Deserts, rocky slopes and xerophyllous scrub.

Holarctic and Neotropical. Boreal, Tethyan, and Madrean. Euro-Siberian. Irano-Turanian. Pampas. European and Siberian.

References, etc. Leaf anatomical: this project.

Blepharoneuron Nash

Habit, vegetative morphology. Annual (*B. shepherdi*), or perennial (*B. tricholepis*); caespitose, or decumbent. *Culms* (16–)20–60 cm high; herbaceous; branched above (*B. shepherdi*), or unbranched above. Culm nodes mostly hidden by leaf sheaths; glabrous. Culm internodes solid to hollow. Plants unarmed. Young shoots intravaginal. *Leaves* mostly basal; non-auriculate; without auricular setae. Leaf blades linear; narrow; to 2 mm wide; setaceous (*B. ticholepis*), or not setaceous; flat, or rolled (involute, conspicuously so in *B. tricholepis*); without abaxial multicellular glands; without cross venation; persistent. *Ligule an unfringed membrane*; not truncate (rounded, entire to lacerate); 0.7–3 mm long. *Contra-ligule* absent.

Reproductive organization. Plants bisexual, with bisexual spikelets; with hermaphrodite florets.

Inflorescence. Inflorescence paniculate; open (to somewhat contracted, greyish, elliptic, 5–20 cm long); with capillary branchlets (the pedicels wiry, flexuous); espatheate; not comprising 'partial inflorescences' and foliar organs. Spikelet-bearing axes persistent. Spikelets not secund; pedicellate (the long slender pedicels distended towards the tip).

Female-fertile spikelets. *Spikelets* (1.4–)1.8–3(–4) mm long; dark brown or purplish; slightly compressed laterally; *disarticulating above the glumes*. Rachilla terminated by a female-fertile floret. Hairy callus absent (callus glabrous). *Callus* short; blunt (truncate).

Glumes two; relatively large; very unequal to more or less equal (G_1 2/3 to 3/4 the length of G_2); about equalling the spikelets; (the upper) long relative to the adjacent lemmas; hairless (glabrous, glossy); glabrous; *not pointed (rounded to erose)*; awnless; non-carinate; similar (greyish green, delicately membranous, broadly ovate to lanceolate). Lower glume 1 nerved. Upper glume 1 nerved. **Spikelets** with female-fertile florets only; *without proximal incomplete florets*.

Female-fertile florets 1. Lemmas lanceolate, greyish green; similar in texture to the glumes to decidedly firmer than the glumes (hyaline to thinly membranous); smooth; not becoming indurated; entire (or erose); pointed to blunt; awnless, or mucronate (rarely); hairy (with long, silky hairs over the lower 2/3 of the midnerve and margins); slightly *carinate*; without a germination flap; 3 nerved. **Palea** present; relatively long; entire (pointed); awnless, without apical setae; textured like the lemma; not indurated; 2-nerved (though the nerves are much closer than is usual in paleae); *keel-less*. Palea back densely silky villous hairy (between the nerves below). *Lodicules* present; 2; free; fleshy (fairly); glabrous; not or scarcely vascularized. **Stamens 3**; with free filaments (these short). Anthers 0.8–2.1 mm long (relatively long); not penicillate; without an apically prolonged connective. *Ovary* glabrous. Styles free to their bases; free. Stigmas 2; apparently green.

Fruit, embryo and seedling. Fruit brownish; fusiform, or ellipsoid; glabrous. Hilum short. Pericarp fused. Embryo large. Endosperm hard. Embryo with an epiblast; with a scutellar tail; with an elongated mesocotyl internode. Embryonic leaf margins meeting.

Abaxial leaf blade epidermis. *Costal/intercostal zonation* conspicuous (despite the lack or scarcity of stomata in *B. tricholepis*). *Papillae* present (in *B. shepherdi*), or absent (in *B. tricholepis*); intercostal. Intercostal papillae not over-arching the stomata; consisting of one oblique swelling per cell (often adjacent to a stoma). *Long-cells* markedly different in shape costally and intercostally (costals narrower); of similar wall thickness costally and intercostally (thick walled). Intercostal zones with typical long-cells (all epidermal cells large). Mid-intercostal long-cells rectangular; having markedly sinuous walls. *Microhairs* present; elongated; clearly two-celled; chloridoid-type. Microhair apical cell wall thinner than that of the basal cell but not tending to collapse. Microhairs 30–31.5(–34.5) microns long; (10.5–)11.4–12.3(–14.4) microns wide at the septum. Microhair total length/width at septum 2–2.8. Microhair apical cells (10.5–)12–13.5(–15.6) microns long. Microhair apical cell/total length ratio 0.34–0.5. *Stomata* common (in *B. shepherdi*), or absent or very rare (*B. tricholepis*); 18–22.5 microns long (in *B. tricholepis*). Subsidiaries non-papillate; (*B. shepherdi* dome-shaped.

Intercostal short-cells common (in *B. tricholepis*), or absent or very rare (*B. shepherdi*); in *B. tricholepis* in cork/silica-cell pairs (also solitary and in short rows); silicified. Intercostal silica bodies imperfectly developed. Rows of costal prickles present. *Costal short-cells* conspicuously in long rows (interrupted by prickles). Costal silica bodies present in alternate cell files of the costal zones; relatively broad to narrow saddle shaped (sometimes approaching tall-and-narrow and crescentic).

Transverse section of leaf blade, physiology. C_4; XyMS+. PCR sheath outlines even. PCR sheaths of the primary vascular bundles interrupted; interrupted abaxially only. PCR sheath extensions present, or absent. Maximum number of extension cells when present, 1, or 2. PCR cell chloroplasts centripetal. *Mesophyll* with radiate chlorenchyma; traversed by columns of colourless mesophyll cells (between all bundles; very wide). *Leaf blade* with distinct, prominent adaxial ribs (only, in *B. tricholepis*), or 'nodular' in section (*B. shepherdi*); with the ribs more or less constant in size. *Midrib* conspicuous (with a somewhat larger bundle plus sclerenchyma and rib); with one bundle only. *The lamina* symmetrical on either side of the midrib. Bulliforms present in discrete, regular adaxial groups; associated with colourless mesophyll cells to form deeply-penetrating fans (these conrtibuting to the colourless intercostal columns). All the vascular bundles accompanied by sclerenchyma. Combined sclerenchyma girders absent (primaries with abaxial girders and adaxial strands (the PCA tissue encircling the top of each bundle); small bundles with strands only). Sclerenchyma all associated with vascular bundles. The lamina margins with fibres.

Cytology. Chromosome base number, $x = 8$. $2n = 16$. 2 ploid.

Taxonomy. Chloridoideae; main chloridoid assemblage.

Ecology, geography, regional floristic distribution. 2 species. Southwest U.S.A., Mexico. Xerophytic; species of open habitats; glycophytic. Dry upland coniferous forest. Holarctic. Madrean.

References, etc. Morphological/taxonomic: Peterson and Annable 1990. Leaf anatomical: this project, supplemented from Peterson and Annable 1990.

Boissiera Hochst. & Steud.

Euraphis (Trin.) Lindley, *Schnizleina* Steud., *Wiesta* Boiss.

~ *Bromus*

Habit, vegetative morphology. Annual; caespitose (or the culms solitary). *Culms* 5–15 cm high; herbaceous. Culm nodes hairy. Culm internodes hollow. Leaves auriculate. *Sheath margins joined*. Leaf blades linear; narrow; 1–6 mm wide; flat; without cross venation; persistent. Ligule an unfringed membrane.

Reproductive organization. Plants bisexual, with bisexual spikelets; with hermaphrodite florets.

Inflorescence. Inflorescence paniculate; contracted; capitate, or more or less ovoid; espatheate; not comprising 'partial inflorescences' and foliar organs. Spikelet-bearing axes persistent. Spikelets not secund; pedicellate.

Female-fertile spikelets. *Spikelets* 15–30 mm long; compressed laterally; disarticulating above the glumes; *not disarticulating between the florets*. Rachilla prolonged beyond the uppermost female-fertile floret. Hairy callus present, or absent.

Glumes two; very unequal to more or less equal; shorter than the spikelets; (the longer) long relative to the adjacent lemmas; awnless; similar (hyaline). Lower glume 3 nerved. Upper glume 5 nerved. *Spikelets* with female-fertile florets only, or with incomplete florets. The incomplete florets distal to the female-fertile florets. The distal incomplete florets usually several; incomplete florets awned (reduced to a bunch of awns).

Female-fertile florets 2–5. Lemmas decidedly firmer than the glumes (papery to leathery); not becoming indurated; incised; irregularly 5–9 lobed (toothed); not deeply cleft; awned. *Awns (5–)7(–9) (in a line across the lemma back)*; median and lateral; the median similar in form to the laterals; dorsal (from the upper part of the lemma back); from near the top; non-geniculate; straight to recurving; hairless (scabrid); about as long as the body of the lemma to much longer than the body of the lemma; entered by one vein. The lateral awns shorter than the median to about equalling the median (inserted at the same level on the lemma back, the outer members shorter). Awn bases twisted. Lemmas hairy; non-carinate; 9–13 nerved; with the nerves confluent towards the tip, or with the nerves non-confluent. *Palea* present; relatively long; palea apex minutely dentate; awnless, without apical setae; 2-nerved; 2-keeled. *Lodicules* present; 2; free; membranous. *Stamens* 3. Anthers 0.4–0.6 mm long. **Ovary** hairy; *with a conspicuous apical appendage (the styles lateral)*. Styles free to their bases. Stigmas 2.

Fruit, embryo and seedling. Fruit medium sized; longitudinally grooved; compressed dorsiventrally. Hilum long-linear. Embryo small. Endosperm hard; without lipid; containing only simple starch grains. Embryo with an epiblast; without a scutellar tail; with a negligible mesocotyl internode. Embryonic leaf margins meeting.

Abaxial leaf blade epidermis. *Costal/intercostal zonation* conspicuous. *Papillae* absent. *Long-cells* similar in shape costally and intercostally; of similar wall thickness costally and intercostally. Mid-intercostal long-cells fusiform; having straight or only gently undulating walls. *Microhairs* absent. *Stomata* common; 30–39 microns long. Subsidiaries parallel-sided. Guard-cells overlapped by the interstomatals. *Intercostal short-cells* common, or absent or very rare; not paired; not silicified. *Costal short-cells* neither distinctly grouped into long rows nor predominantly paired. Costal silica bodies horizontally-elongated crenate/sinuous to horizontally-elongated smooth, or rounded (rather irregular), or crescentic.

Transverse section of leaf blade, physiology. C_3; XyMS+. *Mesophyll* with non-radiate chlorenchyma; without adaxial palisade. *Leaf blade* 'nodular' in section; with the ribs more or less constant in size. Bulliforms present in discrete, regular adaxial groups (in the furrows); in simple fans. All the vascular bundles accompanied by sclerenchyma. Sclerenchyma all associated with vascular bundles.

Cytology. Chromosome base number, $x = 7$. $2n = 14$ and 28. 2 and 4 ploid. Mean diploid 2c DNA value 3.7 pg.

Taxonomy. Pooideae; Triticodae; Bromeae.

Ecology, geography, regional floristic distribution. 1 species. Western Asia. Mesophytic to xerophytic; species of open habitats. Dry stony soils.

Holarctic. Tethyan. Irano-Turanian.

Rusts and smuts. Rusts — *Puccinia*. Taxonomically wide-ranging species: *Puccinia striiformis*, *Puccinia hordei*, *Puccinia recondita*, and '*Uromyces*' *turcomanicum*.

Economic importance. Important native pasture species: *B. squarrosa*.

References, etc. Leaf anatomical: this project.

Boivinella A. Camus

~ *Cyphochlaena*

Habit, vegetative morphology. Culms 50–60 cm high; *herbaceous*. Plants unarmed. Sheath margins free. Sheaths basally inflated. Leaf blades ovate to elliptic (and attenuate); broad; 15–17 mm wide (and 8–9 cm long); cordate (or subcordate), or not cordate, not sagittate; without cross venation (or these not mentioned). Ligule 'membranous, long-pilose'.

Reproductive organization. Plants bisexual, with bisexual spikelets; with hermaphrodite florets. The spikelets of sexually distinct forms on the same plant; hermaphrodite and sterile; overtly heteromorphic.

Inflorescence. *Inflorescence of spicate main branches (of 'dorsiventral false spikes' at the apex of the culm)*; espatheate; not comprising 'partial inflorescences' and foliar organs. *Spikelet-bearing axes* 'racemes'; solitary; with substantial rachides; persistent. *Spikelets* paired (seemingly); secund (the rachis dorsiventral); biseriate (the pairs in two rows); sessile and pedicellate (the hermaphrodite spikelets short-pedicelled); *consistently in 'long-and-short' combinations*; in pedicellate/sessile combinations. Pedicels of the 'pedicellate' spikelets free of the rachis. The 'shorter' spikelets sterile (reduced to a glume). The 'longer' spikelets hermaphrodite.

Female-sterile spikelets. *The sterile spikelets sessile, reduced to one dorsally compressed glume*.

Female-fertile spikelets. Spikelets *unconventional (because of the 'extra' glume, representing the sessile spikelet)*; 2.2–2.5 mm long; abaxial; strongly *compressed laterally*; falling with the glumes. Rachilla terminated by a female-fertile floret.

Glumes two; more or less equal; about equalling the spikelets, or exceeding the spikelets; long relative to the adjacent lemmas; dorsiventral to the rachis; hairless; without conspicuous tufts or rows of hairs; awned to awnless; very dissimilar (G_1 subulate, and subaristate, G_2 obtuse and muticate). Lower glume relatively smooth; 1 nerved. Upper glume 3 nerved. *Spikelets* with incomplete florets. The incomplete florets proximal to the female-fertile florets. *The proximal incomplete florets* 1; paleate; male. The proximal lemmas saccate; awnless; decidedly exceeding the female-fertile lemmas; decidedly firmer than the female-fertile lemmas (cartilaginous, umbonate).

Female-fertile florets 1. Lemmas less firm than the glumes; not becoming indurated; white in fruit ('whitish'); awnless, or mucronate (?); hairless; apically subgibbous; having the margins lying flat and exposed on the palea. *Palea* present; not indurated (thin). **Stamens 6**. *Ovary* glabrous. Stigmas 2.

Fruit, embryo and seedling. Fruit small (1.5 mm long); compressed laterally (and subtriangular).

Abaxial leaf blade epidermis. *Costal/intercostal zonation* conspicuous. Mid-intercostal long-cells rectangular; having markedly sinuous walls. *Microhairs* present; panicoid-type. *Stomata* common. *Intercostal short-cells* common. *Costal short-cells* conspicuously in long rows. Costal silica bodies 'panicoid-type'.

Transverse section of leaf blade, physiology. C_3; XyMS+. *Mesophyll* with radiate chlorenchyma. Bulliforms present in discrete, regular adaxial groups; in simple fans. All the vascular bundles accompanied by sclerenchyma. Sclerenchyma all associated with vascular bundles.

Taxonomy. Panicoideae; Panicodae; Paniceae (Boivinelleae).

Ecology, geography, regional floristic distribution. 2 species. Madagascar. Paleotropical. Madagascan.

References, etc. Morphological/taxonomic: Camus 1925b; Bosser 1965.

Special comments. Fruit data wanting.

Bothriochloa Kuntze

Amphilophis Nash, *Gymnandropogon* (Nees) Duthie
~ *Dichanthium*

Habit, vegetative morphology. *Perennial*; rhizomatous, or stoloniferous, or caespitose, or decumbent. *Culms* 15–200 cm high; herbaceous; branched above, or unbranched above. Culm nodes hairy, or glabrous. Culm internodes solid. Young shoots intravaginal. The shoots aromatic, or not aromatic. *Leaves* not basally aggregated; non-auriculate. Leaf blades linear; narrow; flat; without cross venation; persistent; rolled in bud. *Ligule* present; an unfringed membrane to a fringed membrane; truncate, or not truncate. *Contraligule* absent.

Reproductive organization. Plants bisexual, with bisexual spikelets; with hermaphrodite florets. The spikelets of sexually distinct forms on the same plant; hermaphrodite and male-only, or hermaphrodite and sterile. The male and female-fertile spikelets mixed in the inflorescence. The spikelets overtly heteromorphic; all in heterogamous combinations. Exposed-cleistogamous, or chasmogamous. Apomictic, or reproducing sexually.

Inflorescence. Inflorescence of spicate main branches (many-jointed 'racemes'), or paniculate (rarely: the lower 'racemes' sometimes branched again at the base); digitate, or subdigitate (the racemes often almost palmate, towards the culm tips), or non-digitate; spatheate, or espatheate; a complex of 'partial inflorescences' and intervening foliar organs, or not comprising 'partial inflorescences' and foliar organs. **Spikelet-bearing axes** 'racemes'; *the spikelet-bearing axes* usually *with more than 10 spikelet-bearing 'articles'*. The racemes without spikelets towards the base. *Spikelet-bearing axes* solitary, or paired, or clustered; with very slender rachides; *disarticulating*; disarticulating at the joints. *The pedicels and rachis internodes with a longitudinal, translucent furrow*. 'Articles' linear; without a basal callus-knob; not appendaged; disarticulating transversely;

densely long-hairy (often villous), or somewhat hairy, or glabrous. Spikelets mainly paired (with a terminal triplet); secund; sessile and pedicellate; consistently in 'long-and-short' combinations; in pedicellate/sessile combinations. Pedicels of the 'pedicellate' spikelets free of the rachis. The 'shorter' spikelets hermaphrodite. The 'longer' spikelets male-only, or sterile.

Female-sterile spikelets. The pedicels of the male or sterile pedicellate spikelets often villous. The lemmas awnless.

Female-fertile spikelets. Spikelets compressed dorsiventrally; falling with the glumes (and with the joint). Rachilla terminated by a female-fertile floret. Hairy callus present. *Callus* short; blunt.

Glumes two; more or less equal; long relative to the adjacent lemmas; hairy (towards the base), or hairless; glabrous, or scabrous; pointed, or not pointed (blunt or minutely bifid); awnless; very dissimilar (the lower bicarinate and often with a pit on the back, the upper narrower, naviculate). *Lower glume two-keeled*; convex on the back to flattened on the back; with a conspicuous pit, or not pitted; relatively smooth; 5–9 nerved. Upper glume 1–4 nerved. *Spikelets* with incomplete florets. The incomplete florets proximal to the female-fertile florets. *The proximal incomplete florets* 1; epaleate; sterile. The proximal lemmas awnless; 0 nerved; more or less equalling the female-fertile lemmas to decidedly exceeding the female-fertile lemmas; similar in texture to the female-fertile lemmas; not becoming indurated.

Female-fertile florets 1. Lemmas linear, produced into the awn; less firm than the glumes (reduced to a hyaline stipe); not becoming indurated; entire; awned. Awns 1; median; apical; geniculate; hairless (glabrous); much longer than the body of the lemma. Lemmas hairless; glabrous; non-carinate; without a germination flap; 1–3 nerved. **Palea** present, or absent; when present, very reduced; not indurated; *nerveless*. *Lodicules* present; 2; free; fleshy; glabrous. *Stamens* 1–3. Anthers about 1.5 mm long; not penicillate; without an apically prolonged connective. *Ovary* glabrous. Styles free to their bases. Stigmas 2; red pigmented.

Fruit, embryo and seedling. *Fruit* free from both lemma and palea; small; compressed dorsiventrally. Hilum short. Embryo large. Endosperm hard; without lipid; containing only simple starch grains. Embryo without an epiblast; with a scutellar tail; with an elongated mesocotyl internode. Embryonic leaf margins overlapping.

Seedling with a long mesocotyl. First seedling leaf with a well-developed lamina. The lamina broad; curved, or supine; 21–30 veined.

Abaxial leaf blade epidermis. *Costal/intercostal zonation* conspicuous. *Papillae* present; intercostal. Intercostal papillae over-arching the stomata; consisting of one oblique swelling per cell. *Long-cells* similar in shape costally and intercostally; of similar wall thickness costally and intercostally. Mid-intercostal long-cells rectangular (long, narrow); having markedly sinuous walls. *Microhairs* present; panicoid-type; 38–66 microns long. Microhair apical cells 24–33 microns long. Microhair apical cell/total length ratio 0.5–0.64. *Stomata* common. Subsidiaries non-papillate; triangular. *Intercostal short-cells* absent or very rare. *Costal short-cells* conspicuously in long rows. Costal silica bodies 'panicoid-type'; cross shaped, or butterfly shaped, or dumb-bell shaped, or nodular.

Transverse section of leaf blade, physiology. C_4; XyMS–. PCR sheath outlines uneven. PCR cells with a suberised lamella. PCR cell chloroplasts with reduced grana; centrifugal/peripheral. *Mesophyll* with radiate chlorenchyma. *Leaf blade* adaxially flat (usually), or 'nodular' in section; with the ribs more or less constant in size (when present). *Midrib* conspicuous; with one bundle only, or having a conventional arc of bundles; with colourless mesophyll adaxially. Bulliforms present in discrete, regular adaxial groups, or not in discrete, regular adaxial groups; in simple fans (or in irregular groups), or associated with colourless mesophyll cells to form deeply-penetrating fans. Many of the smallest vascular bundles unaccompanied by sclerenchyma. Combined sclerenchyma girders present; forming 'figures', or nowhere forming 'figures'. Sclerenchyma all associated with vascular bundles.

Culm anatomy. *Culm internode bundles* in one or two rings, or in three or more rings.

Phytochemistry. Tissues of the culm bases with abundant starch. Leaves containing flavonoid sulphates (in at least some material of each of 6 species), or without flavonoid sulphates (at least sometimes, in 4 species).

Cytology. Chromosome base number, $x = 10$. $2n = 30$, 40, 50, and 120. 2, 4, 5, 6, and 12 ploid.

Taxonomy. Panicoideae; Andropogonodae; Andropogoneae; Andropogoninae.

Ecology, geography, regional floristic distribution. 35 species. Warm regions. Commonly adventive. Mesophytic; species of open habitats; glycophytic. Grassy places. Holarctic, Paleotropical, Neotropical, Australian, and Antarctic. Boreal, Tethyan, and Madrean. African, Madagascan, Indomalesian, Polynesian, and Neocaledonian. Eastern Asian and Atlantic North American. Irano-Turanian. Saharo-Sindian, Sudano-Angolan, West African Rainforest, and Namib-Karoo. Indian, Indo-Chinese, Malesian, and Papuan. Fijian. Caribbean, Central Brazilian, and Pampas. North and East Australian and Central Australian. Patagonian. Southern Atlantic North American and Central Grasslands. Sahelo-Sudanian, Somalo-Ethiopian, and South Tropical African. Tropical North and East Australian and Temperate and South-Eastern Australian.

Hybrids. Intergeneric hybrids with *Capillipedium, Dichanthium*.

Rusts and smuts. Rusts — *Puccinia*. Taxonomically wide-ranging species: *Puccinia nakanishikii, Puccinia graminis, Puccinia versicolor, 'Uromyces' clignyi*, and *Puccinia cesatii*. Smuts from Ustilaginaceae. Ustilaginaceae — *Sorosporium, Sphacelotheca*, and *Ustilago*.

Economic importance. Significant weed species: *B. ischaemum, B. pertusa, B. saccharoides*. Cultivated fodder: *B. insculpta*. Important native pasture species: *B. insculpta, B. ischaemum, B. radicans*.

References, etc. Leaf anatomical: Metcalfe 1960; this project.

Bouteloua Lag.

Actinochloa Roem. & Schult., *Antichloa* Steud., *Aristidium* (Endl.) Lindley, *Atheropogon* Willd., *Chondrosum* Desv., *Erucaria* Cerv., *Eutriana* Trin., *Heterosteca* Desv., *Nestlera* Steud., *Pleiodon* Reichenb., *Polyodon* Kunth, *Triaena* Kunth, *Triplathera* (Endl.) Lindley

Habit, vegetative morphology. Annual, or perennial; rhizomatous, or stoloniferous, or caespitose, or decumbent. *Culms* (10–)15–80 cm high; herbaceous; branched above, or unbranched above. Culm internodes solid, or hollow. *Leaves* mostly basal, or not basally aggregated; non-auriculate. Leaf blades narrow; flat, or folded, or rolled; exhibiting multicellular glands abaxially. The abaxial glands intercostal. Leaf blades without cross venation. *Ligule a fringe of hairs*.

Reproductive organization. Plants nearly always bisexual, with bisexual spikelets, or dioecious (and occasionally gynodioecious — *B. chondrosoides*); with hermaphrodite florets. The spikelets all alike in sexuality (usually), or of sexually distinct forms on the same plant; hermaphrodite, or female-only, or hermaphrodite and female-only. Plants outbreeding and inbreeding. Exposed-cleistogamous, or chasmogamous. Apomictic, or reproducing sexually.

Inflorescence. *Inflorescence of spicate main branches (the spikelets few to many), or a false spike, with spikelets on contracted axes (mainly in Section* **Atheropogon***). Inflorescence axes not ending in spikelets (the rachis of the 'spike' or cluster usually ending in a naked, straight or forked tip)*. Inflorescence espatheate; not comprising 'partial inflorescences' and foliar organs. Spikelet-bearing axes disarticulating (in Section *Atheropogon*), or persistent (Section *Chondrosum*); when falling, falling entire. **Spikelets** (when on much reduced 'branches') *subtended by solitary 'bristles'*; *solitary*; secund (on one side of rachis); biseriate.

Female-fertile spikelets. Spikelets compressed laterally; disarticulating above the glumes, or falling with the glumes (i.e., when the spikes shed). Rachilla prolonged beyond the uppermost female-fertile floret. The rachilla extension with incomplete florets.

Glumes two; very unequal (G_1 shorter and narrower); (the longer) shorter than the adjacent lemmas, or long relative to the adjacent lemmas; hairy, or hairless; pointed (acuminate or awn-tipped); awned (shortly so), or awnless. *Lower glume 1 nerved*. Upper glume 1 nerved. *Spikelets* usually with incomplete florets. The incomplete florets when present, which is usual, distal to the female-fertile florets. **The distal incomplete florets** *clearly specialised and modified in form (the rudiment(s) usually 3-awned, with the awns longer than those of the L_1).*

Female-fertile florets 1. Lemmas with nerves extending into short mucros or awns, with lobes or teeth between; not becoming indurated (membranous); incised; 1 lobed, or 2 lobed, or 3 lobed; not deeply cleft; awnless, or mucronate, or awned. Awns when present, 1 (sometimes with lateral mucros), or 3; median, or median and lateral; the median similar in form to the laterals (when laterals present); non-geniculate. The lateral awns when present, shorter than the median, or about equalling the median, or exceeding the median. Lemmas without a germination flap; 3 nerved. *Palea* present; awnless, without apical setae, or with apical setae, or awned (via excurrent nerves); 2-nerved. *Lodicules* present; 2; free; fleshy; glabrous; not or scarcely vascularized. *Stamens* 3. Anthers 0.4–4 mm long. *Ovary* glabrous. Stigmas 2.

Fruit, embryo and seedling. Fruit ellipsoid. Hilum short. Pericarp fused. Embryo large. Endosperm hard; without lipid. Embryo with an epiblast; with a scutellar tail; with an elongated mesocotyl internode. Embryonic leaf margins meeting.

Abaxial leaf blade epidermis. *Costal/intercostal zonation* conspicuous. *Papillae* absent. Long-cells differing markedly in wall thickness costally and intercostally (costals thicker walled). Mid-intercostal long-cells rectangular; having markedly sinuous walls. *Microhairs* present; more or less spherical to elongated; clearly two-celled; chloridoid-type. Microhair apical cell wall of similar thickness/rigidity to that of the basal cell. Microhairs without 'partitioning membranes' (in *B. curtipendula*); 15.6–29 microns long. Microhair basal cells 12–18 microns long. Microhairs 7.5–10.5 microns wide at the septum. Microhair total length/width at septum 1.9–2.8. Microhair apical cells 7.5–12 microns long. Microhair apical cell/total length ratio 0.33–0.48. *Stomata* common; 21–24 microns long. Subsidiaries dome-shaped, or triangular. *Intercostal short-cells* absent or very rare. Intercostal silica bodies absent. Prickles abundant in *B. curtipendula*. *Crown cells* absent. *Costal short-cells* predominantly paired, or neither distinctly grouped into long rows nor predominantly paired. Costal silica bodies present in alternate cell files of the costal zones; saddle shaped.

Transverse section of leaf blade, physiology. C_4; biochemical type PCK to NAD–ME, or NAD–ME (i.e. *B. curtipendula* is seemingly an NAD-ME/PCK intermediate while *B. gracilis* is NAD-ME); XyMS+. PCR sheath outlines even. PCR sheaths of the primary vascular bundles interrupted; interrupted both abaxially and adaxially. PCR sheath extensions absent. PCR cells without a suberised lamella. *PCR cell chloroplasts* ovoid, or elongated; with well developed grana; centrifugal/peripheral, or centripetal. *Mesophyll* with radiate chlorenchyma; traversed by columns of colourless mesophyll cells (linked with the bulliforms). *Leaf blade* 'nodular' in section (low ribs); with the ribs more or less constant in size. *Midrib* not readily distinguishable; with one bundle only. Bulliforms present in discrete, regular adaxial groups; associated with colourless mesophyll cells to form deeply-penetrating fans (these contributing to the colourless columns). All the vascular bundles accompanied by sclerenchyma. Combined sclerenchyma girders present (with the large bundles); forming 'figures' (anchors, in some large bundles). Sclerenchyma all associated with vascular bundles. The lamina margins with fibres.

Phytochemistry. Tissues of the culm bases with abundant starch. Leaf blade chlorophyll *a:b* ratio 3.13–4.3.

Cytology. Chromosome base number, $x = 10$. $2n = 20, 40, 41, 42, 56, 60, 70$, and 98. Nucleoli persistent.

Taxonomy. Chloridoideae; main chloridoid assemblage.

Ecology, geography, regional floristic distribution. 40 species. Canada to South America, especially southwest U.S.A. Species of open habitats. Dry plains and hillsides.

Holarctic, Neotropical, and Antarctic. Boreal and Madrean. Atlantic North American and Rocky Mountains. Caribbean, Venezuela and Surinam, Pampas, and Andean. Patagonian. Canadian-Appalachian, Southern Atlantic North American, and Central Grasslands.

Rusts and smuts. Rusts — *Puccinia*. Taxonomically wide-ranging species: *Puccinia cacabata, Puccinia boutelouae,* and *Puccinia chloridis*. Smuts from Ustilaginaceae. Ustilaginaceae — *Ustilago*.

Economic importance. Significant weed species: *B. aristidoides, B. barbata, B. gracilis*. Cultivated fodder: *B. curtipendula* etc. (Grama grasses). Important native pasture species: *B. curtipendula, B. gracilis, B. hirsuta*.

References, etc. Morphological/taxonomic: Griffiths 1912; Gould 1979a. Leaf anatomical: Metcalfe 1960; this project.

Brachiaria Griseb.

Pseudobrachiaria Launert
Excluding *Leucophrys*

Habit, vegetative morphology. Annual, or perennial; rhizomatous, or stoloniferous, or caespitose, or decumbent. **Culms** 7–200 cm high; *herbaceous*; branched above, or unbranched above. Culm nodes hairy, or glabrous. Culm internodes solid, or hollow. *Leaves* not basally aggregated; non-auriculate. Leaf blades linear to ovate-lanceolate; broad, or narrow; flat, or folded, or rolled; without cross venation; persistent; rolled in bud. *Ligule* present, or absent (occasionally); an unfringed membrane, or a fringed membrane, or a fringe of hairs. *Contra-ligule* absent.

Reproductive organization. Plants bisexual, with bisexual spikelets; with hermaphrodite florets; outbreeding. Apomictic, or reproducing sexually.

Inflorescence. *Inflorescence of spicate main branches (or occasionally with secondary racemelets)*. Primary inflorescence branches (1–)3–17(–30). Inflorescence with axes ending in spikelets, or axes not ending in spikelets. Rachides hollowed (or triquetrous), or flattened (ribbonlike), or winged (to filiform). Inflorescence espatheate; not comprising 'partial inflorescences' and foliar organs. Spikelet-bearing axes persistent. Spikelets solitary, or paired (or occasionally in fascicles); secund; biseriate; sessile to pedicellate. Pedicel apices discoid. Spikelets not in distinct 'long-and-short' combinations.

Female-fertile spikelets. *Spikelets* broadly *elliptic (plump, more or less obtuse, sometimes with the lowest internode accrescent to the sheathing base of G_1 to form a short, cylindrical stipe)*; *adaxial (or the orientation ambiguous)*; *not noticeably compressed (rarely), or compressed dorsiventrally*; biconvex; disarticulating above the glumes (in *. eruciformis*, where there is a secondary disarticulation beneath them), or falling with the glumes. Rachilla terminated by a female-fertile floret (nearly always), or prolonged beyond the uppermost female-fertile floret (occasionally, e.g. *B. glomerata*). The rachilla extension when present, naked. Hairy callus absent.

Glumes two; very unequal, or more or less equal (rarely); long relative to the adjacent lemmas; dorsiventral to the rachis; awnless; very dissimilar (the upper similar to the L_1). Lower glume 1–7 nerved. Upper glume 5–11 nerved. *Spikelets* with incomplete florets. The incomplete florets proximal to the female-fertile florets. *Spikelets with proximal incomplete florets. The proximal incomplete florets* 1; paleate, or epaleate; male, or sterile. *The proximal lemmas 5–7 nerved (the laterals distant from the median); more or less equalling the female-fertile lemmas*; less firm than the female-fertile lemmas; becoming indurated (rarely), or not becoming indurated.

Female-fertile florets 1. *Lemmas decidedly firmer than the glumes (crustaceous to subleathery)*; *striate, or rugose (and rarely smooth)*; becoming indurated to not becoming indurated; white in fruit; entire; pointed to blunt; not crested, or crested at the tip (in *B. paucispicata*); awnless to mucronate; *hairless (smooth or tuberculate)*; non-carinate; having the margins tucked in onto the palea; with a clear germination flap; 3–5 nerved. *Palea* present (the tip not reflexed); relatively long; gaping, or tightly clasped by the lemma; entire (apically rounded); awnless, without apical setae; textured like the lemma; indurated; 2-nerved. *Lodicules* present; 2; free; fleshy; glabrous. *Stamens* 3. Anthers not penicillate. *Ovary* glabrous. Styles free to their bases. Stigmas 2; red pigmented.

Fruit, embryo and seedling. Fruit small; compressed dorsiventrally. *Hilum short*. Embryo large. Endosperm hard; without lipid. Embryo without an epiblast; with a scutellar tail; with an elongated mesocotyl internode. Embryonic leaf margins overlapping.

Abaxial leaf blade epidermis. *Costal/intercostal zonation* conspicuous. *Papillae* present (rarely), or absent. Intercostal papillae not over-arching the stomata. Mid-intercostal long-cells rectangular; having markedly sinuous walls. *Microhairs* present; panicoid-type; (32–)40–67(–76) microns long. Microhair apical cells (18–)26–48 microns long. Microhair apical cell/total length ratio 0.52–0.69. *Stomata* common. Subsidiaries dome-shaped, or triangular. Guard-cells overlapping to flush with the interstomatals. *Intercostal short-cells* common (e.g. *B. decumbens*), or absent or very rare; in cork/silica-cell pairs; silicified. Intercostal silica bodies when present, cross-shaped. *Costal short-cells* conspicuously in long rows. Costal silica bodies 'panicoid-type'; cross shaped, or butterfly shaped, or dumb-bell shaped, or nodular.

Transverse section of leaf blade, physiology. C_4; *biochemical type* PCK (12 species, including *Pseudobrachiaria*); *XyMS+*. PCR sheath outlines uneven. PCR sheath extensions present, or absent. Maximum number of extension cells when present, 1. PCR cells with a suberised lamella. *PCR cell chloroplasts* ovoid; with well developed grana; centrifugal/peripheral. *Mesophyll* with radiate chlorenchyma. *Leaf blade* 'nodular' in section to adaxially flat. *Midrib* conspicuous, or not readily distinguishable; with one bundle only, or having a conventional arc of bundles; with colourless mesophyll adaxially (*B. decumbens*), or without colourless mesophyll adaxially. Bulliforms present in discrete, regular adaxial groups; associated with colourless mesophyll cells to form deeply-penetrating fans, or in simple fans and associated with colourless mesophyll cells to form deeply-penetrating fans. Many of the smallest vascular bundles unaccompanied by sclerenchyma, or all the vascular bundles accompanied by sclerenchyma (*B. decumbens*). Combined sclerenchyma girders present, or absent; nowhere forming 'figures'. Sclerenchyma all associated with vascular bundles.

Phytochemistry. Leaves without flavonoid sulphates (2 species). Leaf blade chlorophyll *a:b* ratio of *Pseudobrachiaria* 3.36–3.43.

Cytology. Chromosome base number, $x = 7$ and 9. $2n = 18, 28, 32, 36, 48, 54, 64$, and 72. 2–8 ploid. Nucleoli persistent.

Taxonomy. Panicoideae; Panicodae; Paniceae.

Ecology, geography, regional floristic distribution. About 100 species. Warm regions. Commonly adventive. Helophytic, or mesophytic, or xerophytic; mostly species of open habitats, or shade species. Diverse habitats, from semidesert to swamps.

Holarctic, Paleotropical, Neotropical, and Australian. Boreal, Tethyan, and Madrean. African, Madagascan, Indomalesian, Polynesian, and Neocaledonian. Euro-Siberian, Eastern Asian, and Atlantic North American. Macaronesian, Mediterranean, and Irano-Turanian. Saharo-Sindian, Sudano-Angolan, West African Rainforest, and Namib-Karoo. Indian, Indo-Chinese, Malesian, and Papuan. Fijian. Caribbean, Venezuela and Surinam, Amazon, Central Brazilian, Pampas, and Andean. North and East Australian and Central Australian. European. Southern Atlantic North American and Central Grasslands. Sahelo-Sudanian, Somalo-Ethiopian, South Tropical African, and Kalaharian. Tropical North and East Australian.

Rusts and smuts. Rusts — *Physopella* and *Puccinia*. Taxonomically wide-ranging species: *Puccinia orientalis*, *Puccinia levis*, and '*Uromyces*' *setariae-italicae*. Smuts from Tilletiaceae and from Ustilaginaceae. Tilletiaceae — *Melanotaenium* and *Tilletia*. Ustilaginaceae — *Sorosporium*, *Sphacelotheca*, *Tolyposporella*, and *Ustilago*.

Economic importance. Significant weed species: *B. ciliatissima*, *B. deflexa*, *B. distachya*, *B. eruciformis*, *B. fasciculata*, *B. lata*, *B. mutica*, *B. paspaloides*, *B. plantaginea*, *B. platyphylla*, *B. ramosa*, *B. reptans*, *B. texana*. Cultivated fodder: *B. arrecta* (Tanner), *B. brizantha* (Palisade), *B. decumbens* (Surinam), *B. mutica* (Para). Important native pasture species: *B. brizantha*, *B. bovonei*, *B. comata*, *B. decumbens*, *B. deflexa*, *B. distachya*, *B. jubata*, *B. nigropedata*, *B. serrata* etc. Grain crop species: *B. deflexa* — a minor cereal in West Africa.

References, etc. Leaf anatomical: Metcalfe 1960 and this project.

Special comments. All but the type species may be best referred to *Urochloa*: see Webster 1987. It is preferable, however, that generic circumscriptions around *Panicum* be reassessed on a basis of worldwide sampling before making formal realignments.

Brachyachne (Benth.) Stapf

Excluding *Lepturidium*

Habit, vegetative morphology. Annual, or perennial; stoloniferous, or caespitose. *Culms* 8–70 cm high; herbaceous. Culm nodes glabrous. Culm internodes solid. Leaves non-auriculate. Leaf blades linear; narrow; setaceous, or not setaceous; flat, or rolled (involute and filiform); without abaxial multicellular glands; without cross venation; persistent. Ligule a fringed membrane to a fringe of hairs.

Reproductive organization. Plants bisexual, with bisexual spikelets; with hermaphrodite florets. Exposed-cleistogamous, or chasmogamous.

Inflorescence. *Inflorescence* usually of spicate main branches (occasionally the racemes single); *digitate*. Rachides hollowed, or flattened (or triquetrous). Inflorescence espatheate; not comprising 'partial inflorescences' and foliar organs. Spikelet-bearing axes persistent. Spikelets solitary (appressed); secund (the racemes unilateral); biseriate; subsessile.

Female-fertile spikelets. *Spikelets* adaxial; strongly *compressed laterally*; disarticulating above the glumes. Rachilla usually prolonged beyond the uppermost female-fertile floret. The rachilla extension with incomplete florets, or naked. *Hairy callus present*. *Callus* short; pointed, or blunt.

Glumes two; more or less equal; *exceeding the spikelets (enclosing the floret)*; long relative to the adjacent lemmas; lateral to the rachis; pointed, or not pointed (sometimes minutely incised, with a tiny mucro); awnless; (at least the lower) carinate (the keel weak to very strong); with the keel conspicuously winged (especially the lower, in *B. convergens*), or without a median keel-wing; *very dissimilar (broad, usually thinly leathery, the lower curved in profile, the upper straighter)*. Lower glume 1 nerved. Upper glume 1 nerved. *Spikelets* with female-fertile florets only, or with incomplete florets. The incomplete florets when present, distal to the female-fertile florets. The distal incomplete florets 1; *incomplete florets* merely underdeveloped (tiny).

Female-fertile florets 1. *Lemmas* less firm than the glumes to similar in texture to the glumes (membranous to hyaline); not becoming indurated; entire, or incised; when entire, pointed, or blunt; when incised, 2 lobed; not deeply cleft; *awnless, or mucronate (rarely)*; hairy (long-haired on the nerves or all over); carinate (folded along the midnerve); 3 nerved. *Palea* present; relatively long to conspicuous but relatively short; awnless, without apical setae; textured like the lemma; not indurated; 2-nerved (with long hairs). *Lodicules* present; 2; joined, or free; fleshy; glabrous. *Stamens* 3. Anthers not penicillate. *Ovary* glabrous. Styles free to their bases. Stigmas 2.

Fruit, embryo and seedling. *Fruit* free from both lemma and palea; small; ellipsoid; compressed laterally. Hilum short. Pericarp fused. Embryo large; not waisted. Endosperm containing compound starch grains. Embryo without an epiblast; with a scutellar tail; with an elongated mesocotyl internode. Embryonic leaf margins overlapping.

Abaxial leaf blade epidermis. *Costal/intercostal zonation* conspicuous. *Papillae* present; intercostal. Intercostal papillae not over-arching the stomata; consisting of one symmetrical projection per cell. *Long-cells* the intercostals broader, often short; of similar wall thickness costally and intercostally. Intercostal zones with typical long-cells, or exhibiting many atypical long-cells. Mid-intercostal long-cells rectangular (to square); having markedly sinuous walls. *Microhairs* present; more or less spherical; clearly two-celled; chloridoid-type. Microhair apical cell wall of similar thickness/rigidity to that of the basal cell. Microhair basal cells 6 microns long. *Stomata* common. Subsidiaries non-papillate. Guard-cells overlapping to flush with the interstomatals. *Intercostal short-cells* common; not paired (mainly solitary); silicified (mostly), or not silicified (usually). Intercostal silica bodies imperfectly developed; tall-and-narrow (rare). *Costal short-cells* conspicuously in long rows. Costal silica bodies present in alternate cell files of the costal zones; saddle shaped.

Transverse section of leaf blade, physiology. C_4; XyMS+. PCR sheath outlines uneven, or even. PCR sheaths of the primary vascular bundles interrupted; interrupted abaxially only. PCR sheath extensions absent. PCR cell chloroplasts centrifugal/peripheral, or centripetal. *Mesophyll* with radiate chlorenchyma. *Leaf blade* 'nodular' in section; with the ribs more or less constant in size. *Midrib* not readily distinguishable; with one bundle only. Bulliforms present in discrete, regular adaxial groups. Many of the smallest vascular bundles unaccompanied by sclerenchyma. Combined sclerenchyma girders present; forming 'figures'. Sclerenchyma all associated with vascular bundles.

Taxonomy. Chloridoideae; main chloridoid assemblage.

Ecology, geography, regional floristic distribution. 10 species. Africa, Australia. Helophytic to mesophytic; species of open habitats; glycophytic. Seasonal swamps and moist rock crevices.

Paleotropical and Australian. African and Indomalesian. Sudano-Angolan and West African Rainforest. Malesian and Papuan. North and East Australian and Central Australian. Sahelo-Sudanian, Somalo-Ethiopian, and South Tropical African. Tropical North and East Australian.

Economic importance. Significant weed species: *B. convergens* (in North America).

References, etc. Leaf anatomical: this project.

Brachychloa Phillips

Habit, vegetative morphology. Decumbent annual, or perennial; stoloniferous. **Culms** 15–50 cm high; *herbaceous*; branched above (sometimes), or unbranched above (usually). Culm nodes glabrous. Culm internodes hollow. Young shoots intravaginal. *Leaves* not basally aggregated; non-auriculate. Leaf blades linear to lanceolate; narrow; 4–6 mm wide; flat; without abaxial multicellular glands; without cross venation; persistent. *Ligule a fringed membrane*; about 1 mm long.

Reproductive organization. *Plants bisexual, with bisexual spikelets*; with hermaphrodite florets.

Inflorescence. *Inflorescence of spicate main branches*; with the branches appressed and often crowded into a loose head in *B. schiemanniana*, spreading in *B. fragilis*; nondigitate. Primary inflorescence branches 6–16. Rachides flattened. Inflorescence espatheate; not comprising 'partial inflorescences' and foliar organs. The racemes spikelet bearing to the base. Spikelet-bearing axes disarticulating (*B. fragilis*), or persistent; when disarticulating, falling entire. Spikelets solitary; secund; biseriate; shortly pedicellate; imbricate.

Female-fertile spikelets. Spikelets 3.5–7 mm long; adaxial; compressed laterally; disarticulating above the glumes; disarticulating between the florets; with conventional internode spacings. Rachilla prolonged beyond the uppermost female-fertile floret; rachilla hairless. The rachilla extension with incomplete florets. *Callus* absent.

Glumes two; more or less equal; shorter than the spikelets; shorter than the adjacent lemmas; lateral to the rachis (by twisting of the pedicels); hairless; glabrous (to minutely scaberulous); pointed; awnless; somewhat carinate; similar (subcoriaceous). Lower glume 1–3 nerved. *Upper glume 3–7 nerved. Spikelets* with incomplete florets. The incomplete florets distal to the female-fertile florets. The distal incomplete florets awnless.

Female-fertile florets 3–7. *Lemmas not saccate*; similar in texture to the glumes (membranous); not becoming indurated; *incised*; blunt; 2 lobed; not deeply cleft (bidentate); very shortly mucronate (from between the lobes); hairy (on the margins, in *B. fragilis*), or hairless (in *B. schiemanniana*); carinate; without a germination flap; 3 nerved, or 5–7 nerved (in *B. schiemanniana*); with the nerves non-confluent. **Palea** present; relatively long (gibbous); *entire*; awnless, without apical setae; textured like the lemma; not indurated; 2-nerved; 2-keeled. Palea keels wingless. *Lodicules* present; 2; free; fleshy (?); glabrous; not or scarcely vascularized. *Stamens* 3. Anthers 0.4 mm long; not penicillate; without an apically prolonged connective. *Ovary* glabrous. Stigmas 2; red pigmented (dark purple).

Fruit, embryo and seedling. *Fruit* free from both lemma and palea; small (0.8 mm long); compressed laterally, or trigonous. Hilum short. *Pericarp free.*

Abaxial leaf blade epidermis. *Costal/intercostal zonation* conspicuous. *Papillae* present; intercostal. Intercostal papillae over-arching the stomata; consisting of one oblique swelling per cell (at the same end of each cell, thick walled and pitted). Long-cells differing markedly in wall thickness costally and intercostally (intercostals thicker walled). Intercostal zones with typical long-cells (on focusing through the papillae). Mid-intercostal long-cells rectangular; having markedly sinuous walls. *Microhairs* present;

more or less spherical; clearly two-celled; chloridoid-type (but well disguised in this papillate epidermis). Microhair apical cell wall of similar thickness/rigidity to that of the basal cell. Microhair basal cells 15 microns long. Microhair total length/width at septum 2. Microhair apical cell/total length ratio 0.3. *Stomata* common. Subsidiaries non-papillate; dome-shaped (mainly), or triangular. Guard-cells overlapping to flush with the interstomatals (on the side not obscured by a papilla). *Intercostal short-cells* common; not paired (solitary); not silicified. Intercostal silica bodies absent. *Costal short-cells* conspicuously in long rows. Costal silica bodies present in alternate cell files of the costal zones; exclusively saddle shaped (very regular).

Transverse section of leaf blade, physiology. C_4; XyMS+. PCR sheath outlines uneven and even. PCR sheaths of the primary vascular bundles interrupted; interrupted both abaxially and adaxially. PCR sheath extensions present (with some bundles only). Maximum number of extension cells 1. PCR cell chloroplasts centripetal. *Leaf blade* 'nodular' in section to adaxially flat. *Midrib* conspicuous; having a conventional arc of bundles (a large median, with a smaller lateral at either side); with colourless mesophyll adaxially. Bulliforms present in discrete, regular adaxial groups; seemingly exclusively in simple fans (but these deeply penetrating). All the vascular bundles accompanied by sclerenchyma. Combined sclerenchyma girders present; forming 'figures' (all the laterals with I's). Sclerenchyma all associated with vascular bundles (though the abaxial girders of the keel bundles form three heavy blocks). The lamina margins with fibres.

Taxonomy. Chloridoideae; main chloridoid assemblage.

Ecology, geography, regional floristic distribution. 2 species (*B. fragilis*, *B. schiemanniana*). Southern Mozambique, Natal. Mesophytic to xerophytic; shade species and species of open habitats; glycophytic. In coastal forests on sandy soil.

Paleotropical. African. Sudano-Angolan. South Tropical African.

References, etc. Leaf anatomical: this project; photos provided by R.P. Ellis.

Special comments. Generic circumscriptions around *Brachychloa*, *Drake-Brockmania* and *Heterocarpha* seem problematical, especially in the absence of anatomical data for *H. haareri* and *B. fragilis*.

Brachyelytrum P. Beauv.

Habit, vegetative morphology. Slender perennial; with short, knotty rhizomes. **Culms 50–100 cm high**; herbaceous. Culm nodes hairy, or glabrous. Culm internodes solid. Leaves non-auriculate. Leaf blades narrowly lanceolate; broad; 10–22 mm wide; pseudopetiolate (basally constricted), or not pseudopetiolate; cross veined, or without cross venation; persistent; rolled in bud. *Ligule* present; an unfringed membrane to a fringed membrane; 1.7–3.5 mm long.

Reproductive organization. Plants bisexual, with bisexual spikelets; with hermaphrodite florets. *The spikelets all alike in sexuality*.

Inflorescence. Inflorescence paniculate (scanty); espatheate; not comprising 'partial inflorescences' and foliar organs. Spikelet-bearing axes persistent. Spikelets not secund; pedicellate.

Female-fertile spikelets. Spikelets 7–12 mm long; not noticeably compressed to compressed dorsiventrally; disarticulating above the glumes. Rachilla prolonged beyond the uppermost female-fertile floret; rachilla hairy. The rachilla extension naked (bristle-like). Hairy callus present.

Glumes present; two, or one per spikelet (the lower sometimes vestigial or absent); (the upper, larger), minute to relatively large; very unequal; shorter than the adjacent lemmas (the upper no more than about 1/4 of the floret length); awnless. Lower glume 0–1 nerved. Upper glume 1 nerved. **Spikelets** with female-fertile florets only; *without proximal incomplete florets*.

Female-fertile florets 1. Lemmas tapering into the awn; herbaceous; not becoming indurated; entire; pointed; awned. Awns 1; median; apical; non-geniculate; hairless (scabrid); much longer than the body of the lemma; entered by several veins (3). Lemmas hairless; scabrous; non-carinate (rounded on the back); 5 nerved. **Palea** present; relatively long; *convolute*; 2-nerved; slightly 2-keeled. *Lodicules* present; 2; free; membranous; glabrous; not toothed; not or scarcely vascularized. **Stamens 3**. *Ovary* hairy; without a conspicuous apical appendage (but narrow and hairy above). Styles free to their bases. Stigmas 2; white.

Fruit, embryo and seedling. Fruit small to medium sized (5 mm long); linear; longitudinally grooved; compressed dorsiventrally. Hilum long-linear. Pericarp thick and hard; loosely adherent to fused (removable with difficulty). Embryo small; not waisted. Endosperm liquid in the mature fruit; containing only simple starch grains. Embryo with an epiblast; with a scutellar tail, or without a scutellar tail; with a negligible mesocotyl internode. Embryonic leaf margins overlapping.

Seedling with a short mesocotyl. First seedling leaf with a well-developed lamina. The lamina narrow; curved; 11–13 veined.

Abaxial leaf blade epidermis. *Costal/intercostal zonation* conspicuous. *Papillae* present; intercostal. Intercostal papillae not over-arching the stomata; often several per cell (large, irregular in shape). Mid-intercostal long-cells rectangular; having markedly sinuous walls. *Microhairs* absent. *Stomata* common. Subsidiaries non-papillate; parallelsided. Guard-cells overlapped by the interstomatals (slightly), or overlapping to flush with the interstomatals. *Intercostal short-cells* absent or very rare. *Costal short-cells* conspicuously in long rows. Costal silica bodies 'panicoid-type'.

Transverse section of leaf blade, physiology. C_3; XyMS+. *Mesophyll* with nonradiate chlorenchyma; without arm cells; without fusoids. *Leaf blade* adaxially flat. *Midrib* not readily distinguishable; with one bundle only. Bulliforms present in discrete, regular adaxial groups; in simple fans. Many of the smallest vascular bundles unaccompanied by sclerenchyma. Sclerenchyma all associated with vascular bundles.

Cytology. Chromosome base number, $x = 11$. $2n = 22$. 2 ploid.

Taxonomy. Bambusoideae; Oryzodae; Brachyelytreae.

Ecology, geography, regional floristic distribution. 1 species. North America, Japan & Korea. Shade species. Woodland.

Holarctic. Boreal. Eastern Asian and Atlantic North American. Canadian-Appalachian and Southern Atlantic North American.

References, etc. Morphological/taxonomic: Macfarlane and Watson 1980; Campbell, Garwood and Specht 1985. Leaf anatomical: this project.

Brachypodium P. Beauv.

Brevipodium A. & D. Löve, *Trachynia* Link, *Tragus* Panzer

Habit, vegetative morphology. Annual, or perennial; rhizomatous to caespitose. **Culms** 2–200 cm high; *herbaceous*; unbranched above. Culm nodes hairy. Culm internodes hollow. *Leaves* not basally aggregated; non-auriculate. Leaf blades linear; broad (rarely), or narrow; 3–12 mm wide; flat, or rolled (convolute); without cross venation; persistent. *Ligule an unfringed membrane*; truncate, or not truncate; 1–6 mm long.

Reproductive organization. Plants bisexual, with bisexual spikelets; with hermaphrodite florets; inbreeding.

Inflorescence. *Inflorescence* few spikeleted; *a single raceme, or paniculate (rarely)*; open. Rachides hollowed. Inflorescence espatheate; not comprising 'partial inflorescences' and foliar organs. Spikelet-bearing axes persistent. Spikelets solitary; secund (drooping to one side), or not secund; distichous; pedicellate.

Female-fertile spikelets. Spikelets 13–40 mm long; compressed laterally; disarticulating above the glumes. Rachilla prolonged beyond the uppermost female-fertile floret; rachilla hairless. The rachilla extension with incomplete florets. Hairy callus absent.

Glumes two; very unequal to more or less equal; shorter than the spikelets; shorter than the adjacent lemmas; dorsiventral to the rachis; pointed; awned, or awnless; noncarinate; similar (lanceolate). *Lower glume 5–7 nerved*. Upper glume 7–9 nerved. *Spikelets* with incomplete florets. The incomplete florets distal to the female-fertile florets. *The distal incomplete florets* merely underdeveloped; awned, or awnless.

Female-fertile florets 8–22. Lemmas ovate-lanceolate to acuminate; similar in texture to the glumes; not becoming indurated; entire; pointed; awned. *Awns* 1; median; *apical*; non-geniculate; much shorter than the body of the lemma to about as long as the body of the lemma; entered by several veins (3–5). Lemmas hairless; non-carinate; (5–)7–9 nerved. *Palea* present; relatively long; entire (truncate); awnless, without apical setae; 2-nerved; 2-keeled. *Lodicules* present; 2; free; membranous; ciliate; not toothed. *Stamens* 3. Anthers 0.4–4.5 mm long; not penicillate. **Ovary hairy**. Styles free to their bases. Stigmas 2; white.

Fruit, embryo and seedling. *Fruit* adhering to lemma and/or palea (*Trachynia*), or free from both lemma and palea; medium sized; longitudinally grooved; compressed dorsiventrally; with hairs confined to a terminal tuft. Hilum long-linear. Embryo small; not waisted. Endosperm hard; without lipid; containing only simple starch grains. Embryo with an epiblast, or without an epiblast; without a scutellar tail; with a negligible mesocotyl internode. Embryonic leaf margins meeting.

Seedling with a long mesocotyl; with a loose coleoptile, or with a tight coleoptile. First seedling leaf with a well-developed lamina. The lamina broad, or narrow; erect; 7–9 veined.

Abaxial leaf blade epidermis. *Costal/intercostal zonation* conspicuous. *Papillae* absent. *Long-cells* similar in shape costally and intercostally; of similar wall thickness costally and intercostally (thin walled). Mid-intercostal long-cells rectangular; having markedly sinuous walls (rarely), or having straight or only gently undulating walls. *Microhairs* absent. *Stomata* absent or very rare, or common; in *B. distachyon* 24–27 microns long. Subsidiaries parallel-sided. Guard-cells overlapped by the interstomatals. *Intercostal short-cells* common; not paired (mainly solitary); silicified (often, and with adjacent prickles), or not silicified. Intercostal silica bodies when present, rounded, or tall-and-narrow. *Crown cells* present, or absent. *Costal short-cells* conspicuously in long rows, or predominantly paired, or neither distinctly grouped into long rows nor predominantly paired. Costal silica bodies horizontally-elongated crenate/sinuous, rounded, and tall-and-narrow.

Transverse section of leaf blade, physiology. C_3; XyMS+. *Mesophyll* with non-radiate chlorenchyma; without adaxial palisade. *Leaf blade* with distinct, prominent adaxial ribs, or 'nodular' in section. *Midrib* conspicuous, or not readily distinguishable; with one bundle only. Bulliforms present in discrete, regular adaxial groups; in simple fans (or the groups of fairly uniform cells). All the vascular bundles accompanied by sclerenchyma. Combined sclerenchyma girders present; forming 'figures'. Sclerenchyma all associated with vascular bundles.

Culm anatomy. *Culm internode bundles* in three or more rings.

Phytochemistry. Leaves without flavonoid sulphates (1 species).

Cytology. Chromosome base number, $x = 5, 7, 9$, and 10. $2n = 10, 14, 16, 18, 28, 30, 42$, and 56. 2, 4, 6, and 8 ploid (and aneuploids).

Taxonomy. Pooideae; Triticodae; Brachypodieae.

Ecology, geography, regional floristic distribution. 16 species. Temperate, and tropical mountains. Commonly adventive. Mesophytic; shade species and species of open habitats.

Holarctic, Paleotropical, Neotropical, and Cape. Boreal, Tethyan, and Madrean. African, Madagascan, and Indomalesian. Arctic and Subarctic, Euro-Siberian, Eastern Asian, and Atlantic North American. Macaronesian, Mediterranean, and Irano-Turanian. Saharo-Sindian, Sudano-Angolan, and West African Rainforest. Indo-Chinese, Malesian, and Papuan. Caribbean, Pampas, and Andean. European and Siberian. Central Grasslands. Sahelo-Sudanian, Somalo-Ethiopian, and South Tropical African.

Rusts and smuts. Rusts — *Puccinia*. Taxonomically wide-ranging species: *Puccinia graminis*, *Puccinia coronata*, *Puccinia striiformis*, *Puccinia brachypodii-phoenicoidis*, and *Puccinia recondita*. Smuts from Tilletiaceae and from Ustilaginaceae. Tilletiaceae — *Tilletia*. Ustilaginaceae — *Ustilago*.

Economic importance. Significant weed species: *B. distachyon* (= *Trachynia*).

References, etc. Leaf anatomical: Metcalfe 1960; this project.

Briza L.

Brizochloa Jirásek & Chrtek (= *Briza humilis*), *Chascolytrum* Desv., *Chondrachyrum* Nees, *Macrobriza* (Tsvel.) Tsvel., *Tremularia* Fabric.

Excluding *Calosteca*, *Lombardochloa*

Habit, vegetative morphology. Annual, or perennial; rhizomatous, or caespitose. *Culms* 5–100 cm high; herbaceous; unbranched above. Culm nodes glabrous. Culm internodes hollow. Young shoots extravaginal, or intravaginal. *Leaves* not basally aggregated; non-auriculate. Leaf blades linear to linear-lanceolate; narrow; 1–10 mm wide; flat; without cross venation; persistent; rolled in bud. *Ligule an unfringed membrane*; not truncate; 0.8–2.5 mm long.

Reproductive organization. Plants bisexual, with bisexual spikelets; with hermaphrodite florets; outbreeding and inbreeding. Exposed-cleistogamous, or chasmogamous.

Inflorescence. *Inflorescence* paniculate; open; **with capillary branchlets (the spikelets drooping elegantly — 'quaking grass')**; espatheate; not comprising 'partial inflorescences' and foliar organs. Spikelet-bearing axes persistent. *Spikelets* unaccompanied by bractiform involucres, not associated with setiform vestigial branches; secund (drooping to one side), or not secund; pedicellate.

Female-fertile spikelets. Spikelets 2.5–25 mm long; compressed laterally; disarticulating above the glumes; with conventional internode spacings. Rachilla prolonged beyond the uppermost female-fertile floret; rachilla hairless. The rachilla extension with incomplete florets. Hairy callus absent. *Callus* short.

Glumes two; more or less equal; shorter than the spikelets; shorter than the adjacent lemmas; not pointed; awnless; carinate, or non-carinate; similar (broad and cordate, thin and papery). Lower glume 3–15 nerved. Upper glume 3–15 nerved. *Spikelets* with incomplete florets. The incomplete florets distal to the female-fertile florets. *The distal incomplete florets* merely underdeveloped; awnless.

Female-fertile florets 4–20. **Lemmas as broad as long, gibbous and umbonate, cordate**; similar in texture to the glumes; not becoming indurated; entire, or incised (obtuse, cuspidate, bidentate or mucronate); awnless, or mucronate (the mucro less than 1.5 mm long); hairy, or hairless; non-carinate; without a germination flap; 7–15 nerved. **Palea** present; **conspicuous but relatively short**; awnless, without apical setae (and with no apical appendage, by contrast with *Calosteca*); not indurated (hyaline); 2-nerved; 2-keeled. *Lodicules* present; 2; joined, or free; membranous; glabrous; toothed (rarely), or not toothed; not or scarcely vascularized. *Stamens* 3. Anthers 0.4–2.4 mm long; not penicillate. *Ovary* glabrous. Styles free to their bases. Stigmas 2; white.

Fruit, embryo and seedling. *Fruit* adhering to lemma and/or palea, or free from both lemma and palea; small; compressed dorsiventrally. Hilum short, or long-linear. Embryo small; not waisted. Endosperm liquid in the mature fruit (rarely), or hard; with lipid; containing compound starch grains. Embryo with an epiblast; without a scutellar tail; with a negligible mesocotyl internode. Embryonic leaf margins meeting.

Seedling with a long mesocotyl; with a tight coleoptile. First seedling leaf with a well-developed lamina. The lamina narrow; 3–5 veined.

Abaxial leaf blade epidermis. *Costal/intercostal zonation* conspicuous. *Papillae* absent. *Long-cells* markedly different in shape costally and intercostally; of similar wall thickness costally and intercostally. Mid-intercostal long-cells fusiform (contrasting with the smaller, narrower costals); having markedly sinuous walls (thin), or having straight or only gently undulating walls. *Microhairs* absent. *Stomata* common (usually), or absent or very rare. Subsidiaries parallel-sided. Guard-cells overlapped by the interstomatals. *Intercostal short-cells* common (e.g., *B. triloba*), or absent or very rare; when present, in cork/silica-cell pairs. *Costal short-cells* predominantly paired, or neither distinctly grouped into long rows nor predominantly paired. Costal silica bodies horizontally-elongated crenate/sinuous, or horizontally-elongated smooth, or rounded, or tall-and-narrow (or tending to rectangular).

Transverse section of leaf blade, physiology. C_3; XyMS+. *Mesophyll* with non-radiate chlorenchyma; without adaxial palisade. *Leaf blade* 'nodular' in section, or adaxially flat; with the ribs more or less constant in size. *Midrib* conspicuous, or not readily distinguishable; with one bundle only. Bulliforms present in discrete, regular adaxial groups (in the furrows); in simple fans. All the vascular bundles accompanied by sclerenchyma. Combined sclerenchyma girders present; forming 'figures', or nowhere forming 'figures'. Sclerenchyma all associated with vascular bundles.

Phytochemistry. Tissues of the culm bases with little or no starch. Leaves without flavonoid sulphates (several species).

Special diagnostic feature. *Lemmas as broad as long, gibbous and umbonate, cordate at base*.

Cytology. Chromosome base number, $x = 5$ and 7. $2n = 10$, 14, and 28. 2 and 4 ploid (and aneuploid). Haploid nuclear DNA content 5.2–10.8 pg (6 species, mean 7.3). Mean diploid 2c DNA value 17.7 pg (3 species, 14.6–21.6).

Taxonomy. Pooideae; Poodae; Poeae.

Ecology, geography, regional floristic distribution. 16 species. North temperate, South America. Commonly adventive. Mesophytic; mostly species of open habitats. On dry to moist soils.

Holarctic, Paleotropical, Neotropical, and Antarctic. Boreal, Tethyan, and Madrean. African and Indomalesian. Euro-Siberian and Atlantic North American. Macaronesian, Mediterranean, and Irano-Turanian. Saharo-Sindian. Indian and Papuan. Caribbean, Central Brazilian, Pampas, and Andean. Patagonian. European. Southern Atlantic North American and Central Grasslands.

Rusts and smuts. Rusts — *Puccinia*. Taxonomically wide-ranging species: *Puccinia graminis, Puccinia coronata, Puccinia striiformis, Puccinia recondita*, and '*Uromyces*' *dactylidis*. Smuts from Tilletiaceae and from Ustilaginaceae. Tilletiaceae — *Entyloma, Tilletia*, and *Urocystis*. Ustilaginaceae — *Ustilago*.

Economic importance. Significant weed species: *B. maxima, B. media, B. minor*. Cultivated ornamentals — especially *B. maxima*.

References, etc. Morphological/taxonomic: Mattei 1975, Rosengurtt and Arrillaga 1979, Nicora and Rúgolo de Agrasar 1981. Leaf anatomical: Metcalfe 1960; this project.

Bromuniola Stapf & C.E. Hubb.

Habit, vegetative morphology. Perennial; loosely caespitose. **Culms** 80–120 cm high; *herbaceous*. Culm internodes solid. Leaves without auricular setae. *Leaf blades* lanceolate; rolled; pseudopetiolate to not pseudopetiolate; *cross veined*. *Ligule a fringed membrane*.

Reproductive organization. Plants bisexual, with bisexual spikelets; with hermaphrodite florets.

Inflorescence. Inflorescence relatively few spikeleted; paniculate; open (broad, the branches thin and spreading); espatheate; not comprising 'partial inflorescences' and foliar organs. Spikelet-bearing axes persistent. Spikelets not secund.

Female-fertile spikelets. *Spikelets* 10–17 mm long; compressed laterally; *disarticulating above the glumes (the glumes and the sterile basal floret persisting on the pedicel)*; disarticulating between the florets. Rachilla prolonged beyond the uppermost female-fertile floret; rachilla hairy (and sinuous). Hairy callus absent.

Glumes present; two; very unequal to more or less equal; shorter than the adjacent lemmas; hairless; pointed (sometimes mucronate); awnless; carinate; similar (herbaceous, the margins hyaline). Lower glume 3–5 nerved. Upper glume 3–5 nerved. *Spikelets* with incomplete florets. The incomplete florets proximal to the female-fertile florets, or both distal and proximal to the female-fertile florets. *Spikelets with proximal incomplete florets*. *The proximal incomplete florets* 1; paleate, or epaleate (often); sterile. The proximal lemmas awned (subulate), or awnless; 5–7 nerved; exceeded by the female-fertile lemmas; similar in texture to the female-fertile lemmas; not becoming indurated.

Female-fertile florets *4–10*. Lemmas similar in texture to the glumes (herbaceous); not becoming indurated; entire, or incised (slightly, at the tip); not deeply cleft; awned. Awns 1; median; from a sinus, or apical; non-geniculate; hairless to hairy; much shorter than the body of the lemma to about as long as the body of the lemma. Lemmas hairless (keels scabrid); glabrous (the keels scabrid); carinate; 5–7 nerved. *Palea* present; conspicuous but relatively short; entire; awnless, without apical setae; 2-nerved; 2-keeled (and hunch-backed, the keels ciliate; projecting from the lemma into sinuations of the rachilla). *Lodicules* present; 2; free; fleshy. *Stamens* 3. *Ovary* glabrous. Styles free to their bases. Stigmas 2.

Fruit, embryo and seedling. *Fruit* free from both lemma and palea (but falling with them); compressed laterally (obliquely ovoid). Hilum short. Embryo large (about 1/3 of the fruit length); with an epiblast; with a scutellar tail; with an elongated mesocotyl internode. Embryonic leaf margins overlapping.

Abaxial leaf blade epidermis. *Costal/intercostal zonation* conspicuous. *Papillae* absent. Mid-intercostal long-cells having markedly sinuous walls. *Microhairs* present (partly sunken); panicoid-type; 33–46 microns long. Microhair apical cells 26–30(–41) microns long. Microhair apical cell/total length ratio 0.75. *Stomata* common. Subsidiaries triangular. *Intercostal short-cells* absent or very rare. *Costal short-cells* conspicuously in long rows. Costal silica bodies 'panicoid-type'; shortly dumb-bell shaped, or cross shaped.

Transverse section of leaf blade, physiology. C_3; XyMS+. *Mesophyll* with radiate chlorenchyma (indistinctly); without arm cells (?); without fusoids. *Leaf blade* adaxially flat. *Midrib* conspicuous (adaxially and abaxially projecting); having a conventional arc of bundles (the median flanked by 2 small laterals). Bulliforms present in discrete, regular adaxial groups. All the vascular bundles accompanied by sclerenchyma (exceptions near margins). Combined sclerenchyma girders present; forming 'figures' (in the midrib). Sclerenchyma all associated with vascular bundles.

Culm anatomy. *Culm internode bundles* scattered.

Taxonomy. Bambusoideae; Oryzodae; Centotheceae.

Ecology, geography, regional floristic distribution. 1 species. Angola. Shade species; glycophytic. Forests.

Paleotropical. African. Sudano-Angolan. South Tropical African.

References, etc. Leaf anatomical: Metcalfe 1960 and this project.

Bromus L.

Aechmorpha Steud., *Anisantha* Koch, *Avenaria* Fabrich., *Bromopsis* (Dumort.) Fourr., *Ceratochloa* P. Beauv., *Forasaccus* Bub., *Genea* (Dumort.) Dumort., *Libertia* Lejeune, *Michelaria* Dumort., *Nevskiella* Krecz & Vved., *Serrafalcus* Parl., *Stenofestuca* (Honda) Nakai, *Triniusa* Steud., *Trisetobromus* Nevski

Excluding *Boissiera*

Habit, vegetative morphology. Annual, or perennial; rhizomatous, or stoloniferous, or caespitose, or decumbent. *Culms* 3–190 cm high; herbaceous; unbranched above. Culm nodes hairy (rarely), or glabrous. Culm internodes solid (rarely), or hollow. *Leaves* not basally aggregated; auriculate, or non-auriculate. *Sheath margins joined*. Sheaths usually hairy. Leaf blades linear; broad, or narrow; 1–15 mm wide; usually flat, or rolled (somewhat involute, or convolute); without cross venation; persistent; rolled in bud. Ligule an unfringed membrane; not truncate; 1–9 mm long.

Reproductive organization. Plants bisexual, with bisexual spikelets; with hermaphrodite florets; outbreeding and inbreeding. Exposed-cleistogamous, or chasmogamous.

Inflorescence. Inflorescence reduced to a single spikelet (very rarely), or few spikeleted, or many spikeleted; a single raceme (rarely), or paniculate; open, or contracted; when contracted more or less ovoid, or spicate, or more or less irregular; with capillary branchlets, or without capillary branchlets; espatheate; not comprising 'partial inflorescences' and foliar organs. Spikelet-bearing axes persistent. Spikelets secund (falling to one side), or not secund; pedicellate.

Female-fertile spikelets. *Spikelets* (5–)10–70 mm long; compressed laterally; disarticulating above the glumes; *disarticulating between the florets*. Rachilla prolonged beyond the uppermost female-fertile floret; rachilla hairy, or hairless. The rachilla extension with incomplete florets. Hairy callus absent.

Glumes two; very unequal (usually), or more or less equal (rarely); shorter than the adjacent lemmas; free; pointed; awnless; carinate, or non-carinate; similar (herbaceous, persistent). Lower glume 1–5(–7) nerved. Upper glume 3–7(–9) nerved. *Spikelets* with incomplete florets. The incomplete florets distal to the female-fertile florets. *The distal incomplete florets* merely underdeveloped; usually awned, or awnless.

Female-fertile florets 3–30 (rarely 1–2). Lemmas similar in texture to the glumes to decidedly firmer than the glumes (herbaceous to subcoriaceous, the margins sometimes membranous); not becoming indurated; incised (usually), or entire (rarely); when incised,

2 lobed; when incised deeply cleft, or not deeply cleft; awnless, or mucronate, or awned. Awns when present, 1, or 3 (*B. danthoniae*, which may have two additional awnlets); median, or median and lateral (rarely); (the median) from a sinus, or dorsal; from near the top (generally 'subapical', but sometimes only very marginally so); non-geniculate; much shorter than the body of the lemma to much longer than the body of the lemma; entered by several veins. Lemmas hairy, or hairless; carinate (*Ceratochloa*), or non-carinate; without a germination flap; 5–15 nerved. *Palea* present; relatively long to conspicuous but relatively short; entire to apically notched; awnless, without apical setae; 2-nerved; 2-keeled. *Lodicules* present; 2; free; fleshy, or membranous; glabrous; not toothed; not or scarcely vascularized. *Stamens* 1–3. Anthers 0.3–7 mm long; not penicillate. **Ovary** hairy; **with a conspicuous apical appendage (the styles lateral to it)**. Styles free to their bases. Stigmas 2; white.

Fruit, embryo and seedling. *Fruit* adhering to lemma and/or palea; medium sized; longitudinally grooved; compressed laterally; with hairs confined to a terminal tuft. Hilum long-linear. Embryo small; waisted (rarely), or not waisted. Endosperm hard; without lipid; containing only simple starch grains. Embryo without an epiblast; without a scutellar tail; with a negligible mesocotyl internode. Embryonic leaf margins meeting.

Seedling with a short mesocotyl, or with a long mesocotyl; with a loose coleoptile, or with a tight coleoptile. First seedling leaf with a well-developed lamina. The lamina narrow; erect; 3–12 veined.

Abaxial leaf blade epidermis. *Costal/intercostal zonation* conspicuous. *Papillae* absent. Long-cells of similar wall thickness costally and intercostally (thin walled). Intercostal zones with typical long-cells (these often large). Mid-intercostal long-cells rectangular, or fusiform; having straight or only gently undulating walls. *Microhairs* absent. *Stomata* common; 43–48 microns long. Subsidiaries parallel-sided. Guard-cells overlapped by the interstomatals. *Intercostal short-cells* common, or absent or very rare; not paired (when present); silicified (often, when present). Intercostal silica bodies usually mainly tall-and-narrow. *Crown cells* present, or absent. *Costal short-cells* conspicuously in long rows, or predominantly paired, or neither distinctly grouped into long rows nor predominantly paired. Costal silica bodies horizontally-elongated crenate/sinuous, or horizontally-elongated smooth (commonly), or rounded, or tall-and-narrow, or crescentic.

Transverse section of leaf blade, physiology. C_3; XyMS+. *Mesophyll* with non-radiate chlorenchyma; without adaxial palisade. *Leaf blade* with distinct, prominent adaxial ribs, or 'nodular' in section, or adaxially flat; with the ribs more or less constant in size. *Midrib* conspicuous; with one bundle only. Bulliforms present in discrete, regular adaxial groups (in the furrows); in simple fans (or the groups of fairly uniform cells). Many of the smallest vascular bundles unaccompanied by sclerenchyma, or all the vascular bundles accompanied by sclerenchyma. Combined sclerenchyma girders present; forming 'figures', or nowhere forming 'figures'. Sclerenchyma all associated with vascular bundles.

Culm anatomy. *Culm internode bundles* in one or two rings.

Phytochemistry. Tissues of the culm bases with little or no starch. Fructosans predominantly short-chain. Leaves without flavonoid sulphates (4 species).

Cytology. Chromosome base number, $x = 7$. $2n = 14, 28, 42, 56$, and 70. 2, 4, 6, 8, and 10 ploid. Haploid nuclear DNA content 1.8–7 pg (39 species, mean 4.0). Mean diploid 2c DNA value 9.8 pg (11 species, (6.5–12.5). Nucleoli disappearing before metaphase.

Taxonomy. Pooideae; Triticodae; Bromeae.

Ecology, geography, regional floristic distribution. About 150 species. North temperate, tropical mountains, South America. Commonly adventive. Mesophytic, or xerophytic; shade species and species of open habitats.

Holarctic, Paleotropical, Neotropical, Cape, Australian, and Antarctic. Boreal, Tethyan, and Madrean. African and Indomalesian. Arctic and Subarctic, Euro-Siberian, Eastern Asian, Atlantic North American, and Rocky Mountains. Macaronesian, Mediterranean, and Irano-Turanian. Saharo-Sindian, Sudano-Angolan, and West African Rainforest. Indo-Chinese, Malesian, and Papuan. Caribbean, Venezuela and Surinam, Central Brazilian, Pampas, and Andean. North and East Australian and South-West Australian. New Zealand and Patagonian. European and Siberian. Canadian-Appalachian, Southern

Atlantic North American, and Central Grasslands. Sahelo-Sudanian, Somalo-Ethiopian, and South Tropical African. Temperate and South-Eastern Australian.

Hybrids. Supposed intergeneric hybrid with *Festuca*: ×*Bromofestuca* Prodan.

Rusts and smuts. Rusts — *Puccinia*. Taxonomically wide-ranging species: *Puccinia graminis, Puccinia coronata, Puccinia striiformis, Puccinia brachypodii-phoenicoidis, Puccinia hordei,* and *Puccinia recondita*. Smuts from Tilletiaceae and from Ustilaginaceae. Tilletiaceae — *Tilletia* and *Urocystis*. Ustilaginaceae — *Ustilago*.

Economic importance. Significant weed species: *B. arvensis, B. catharticus, B. commutatus, B. danthoniae, B. diandrus, B. erectus, B. hordeaceus, B. inermis, B. japonicus, B. lanceolatus, B. madritensis, B. pectinatus, B. rigidus, B. rubens, B. secalinus, B. squarrosus, B. sterilis, B. tectorum* (several species with injurious callus or awns. Cultivated fodder: *B. unioloides*; *B. inermis* cultivated for hay. Important native pasture species: *B. danthoniae, B. inermis, B. carinatus, B. catharticus, B. pectinatus, B. tectorum* etc. Grain crop species: *B. mango* — formerly grown as a cereal in Chile.

References, etc. Morphological/taxonomic: Wagnon 1952, Smith 1970. Leaf anatomical: Metcalfe 1960; this project.

Brylkinia Schmidt

Habit, vegetative morphology. Slender perennial; rhizomatous. *Culms* 25–70 cm high; herbaceous. Culm nodes glabrous. Leaves non-auriculate. Sheath margins joined. Leaf blades linear; narrow; 3–6 mm wide; flat; persistent. Ligule an unfringed membrane; 0.1 mm long.

Reproductive organization. Plants bisexual, with bisexual spikelets; with hermaphrodite florets.

Inflorescence. Inflorescence a single raceme; espatheate; not comprising 'partial inflorescences' and foliar organs. Spikelet-bearing axes persistent. Spikelets not secund (but pendulous, falling to one side); pedicellate.

Female-fertile spikelets. Spikelets 10–30 mm long; compressed laterally; falling with the glumes (from the base of the pedicel). Rachilla prolonged beyond the uppermost female-fertile floret. Hairy callus absent.

Glumes two (narrow); very unequal; shorter than the spikelets; shorter than the adjacent lemmas; pointed (acuminate); awnless; similar (herbaceous). Lower glume 3 nerved. Upper glume 5 nerved. *Spikelets* with incomplete florets. The incomplete florets proximal to the female-fertile florets. *The proximal incomplete florets* 2; epaleate; sterile. The proximal lemmas acuminate; awnless; similar in texture to the female-fertile lemmas; not becoming indurated.

Female-fertile florets 1. Lemmas similar in texture to the glumes to decidedly firmer than the glumes (firmly herbaceous to thinly leathery); not becoming indurated; awned. *Awns* 1; median; dorsal; from near the top; non-geniculate; **recurving (slightly bowed below, with a wing of green tissue extending from the upper back of the lemma to partway along awn)**. Lemmas carinate; 7 nerved. *Palea* present; relatively long; thinner than the lemma; 2-nerved (the keels closely adjacent); 2-keeled. *Lodicules* present (large, rectangular); 2; joined; membranous; glabrous; not toothed; not or scarcely vascularized. *Stamens* 3. Anthers 1.3–1.8 mm long. *Ovary* glabrous. Styles free to their bases. Stigmas 2; white.

Fruit, embryo and seedling. Fruit free from both lemma and palea; **with a thickened and glossy, umbonate cap.** Hilum long-linear. Embryo small. Endosperm containing compound starch grains. Embryo with an epiblast; without a scutellar tail; with a negligible mesocotyl internode. Embryonic leaf margins meeting.

Abaxial leaf blade epidermis. *Costal/intercostal zonation* conspicuous. *Papillae* absent. Mid-intercostal long-cells having markedly sinuous walls. *Microhairs* absent (and absent adaxially). *Stomata* absent or very rare. *Intercostal short-cells* common; in cork/silica-cell pairs (and solitary); silicified. Intercostal silica bodies tall-and-narrow. *Costal short-cells* neither distinctly grouped into long rows nor predominantly paired. Costal silica bodies horizontally-elongated crenate/sinuous to horizontally-elongated smooth.

Transverse section of leaf blade, physiology. C_3; XyMS+. *Leaf blade* with distinct, prominent adaxial ribs, or 'nodular' in section; with the ribs more or less constant in size.

Midrib conspicuous; with one bundle only. Bulliforms present in discrete, regular adaxial groups; in simple fans. All the vascular bundles accompanied by sclerenchyma.

Special diagnostic feature. *Lemma awn winged, the wing extending down the upper back of the lemma.*

Cytology. Chromosome base number, $x = 10$. $2n = 40$. 4 ploid.

Taxonomy. Pooideae; Poodae; Meliceae.

Ecology, geography, regional floristic distribution. 1 species. Japan, Manchuria, Sakhalin. Mesophytic; shade species; glycophytic. Woodland.

Holarctic. Boreal. Euro-Siberian and Eastern Asian. Siberian.

References, etc. Morphological/taxonomic: Macfarlane and Watson 1980. Leaf anatomical: Metcalfe 1960; this project.

Buchloë Engelm.

Bulbilis Raf., *Calanthera* Hook., *Lasiostega* Benth.

Habit, vegetative morphology. Grey-green, low, dense turf-forming perennial; stoloniferous. *Culms* 1–30 cm high; herbaceous; unbranched above. Culm internodes solid. *Leaves* not basally aggregated; non-auriculate. Leaf blades narrow; 1–2 mm wide (sparsely pilose); without abaxial multicellular glands; without cross venation; rolled in bud. Ligule a fringe of hairs; 0.5 mm long.

Reproductive organization. Plants monoecious with all the fertile spikelets unisexual (occasionally), or dioecious; without hermaphrodite florets. The spikelets of sexually distinct forms on the same plant, or all alike in sexuality; female-only, or male-only. The male and female-fertile spikelets in different inflorescences. Plants outbreeding (at least, when dioecious).

Inflorescence. *Inflorescence when male elevated, with 1–4 spicate, unilateral branches 6–14 mm long; when female with spikelets in burr-like clusters of 3–5(-7), usually 2 burrs per inflorescence, the burr on a short, stout rachis, partially enclosed in a broad, bracteate leaf sheath, falling entire with the indurate rachis united to the indurate G_2's*; spatheate (female inflorescences more or less hidden among culm leaves), or espatheate (male inflorescences); a complex of 'partial inflorescences' and intervening foliar organs (female inflorescence), or not comprising 'partial inflorescences' and foliar organs (male inflorescence). *Spikelet-bearing axes* of female inflorescences capitate; those of female plants disarticulating; falling entire. Spikelets secund (male spikelets in two rows on one side of rachis), or not secund (female spikelets in burrs).

Female-sterile spikelets. Male spikelets with 2 fertile florets, the lemmas 3-nerved, the lodicules 2, fleshy, heavily vascularised. The male spikelets with glumes (two, similar, of variable relative lengths, sometimes mucronate); 2 floreted.

Female-fertile spikelets. Spikelets adaxial; compressed dorsiventrally; falling with the glumes (the burrs falling entire), or not disarticulating. Rachilla terminated by a female-fertile floret. Hairy callus absent.

Glumes two; very unequal; (the longer) long relative to the adjacent lemmas; free (of one another); hairless; awnless; non-carinate; very dissimilar (G_1 narrow, thin, mucronate, well developed to obsolete, G_2 firm, thick, expanded in the middle with inflexed margins, enveloping the floret, abruptly contracted above, the apex with 3 rigid acuminate lobes). *Spikelets* with female-fertile florets only.

Female-fertile florets 1. Lemmas less firm than the glumes (firmly membranous); not becoming indurated; incised; 3 lobed (the middle lobe larger); the lobes mucronate; hairless; non-carinate (dorsally compressed); 3 nerved. *Palea* present; about equalling body of female-fertile lemma; 2-nerved. *Lodicules* present (reduced), or absent; when present, 2; membranous. *Ovary* glabrous. Stigmas 2; brown.

Fruit, embryo and seedling. *Fruit* free from both lemma and palea; ellipsoid. Hilum short. Pericarp fused. Embryo large; with an epiblast; with a scutellar tail; with an elongated mesocotyl internode. Embryonic leaf margins meeting.

Abaxial leaf blade epidermis. *Costal/intercostal zonation* conspicuous. *Papillae* absent. Mid-intercostal long-cells rectangular; having markedly sinuous walls (the sinuosity small-scale). *Microhairs* present; elongated; clearly two-celled; chloridoid-type (with elongated basal cells). Microhair apical cell wall of similar thickness/rigidity to that of the basal cell. Microhairs (34–)40–55(–75) microns long. Microhair basal cells 54 microns long. Microhairs 10.5–15 microns wide at the septum. Microhair total length/width at septum 4.25–5.1. Microhair apical cells (10.5–)12–21(–29) microns long. Microhair apical cell/total length ratio 0.21–0.39. *Stomata* common; 21–24 microns long. Subsidiaries slightly triangular, or dome-shaped. Guard-cells overlapping to flush with the interstomatals. *Intercostal short-cells* common; in cork/silica-cell pairs and not paired (some solitary); silicified. Intercostal silica bodies imperfectly developed; tall-and-narrow. *Costal short-cells* conspicuously in long rows (in some files), or predominantly paired (in others), or neither distinctly grouped into long rows nor predominantly paired (in a few places). Costal silica bodies present in alternate cell files of the costal zones; large, saddle shaped.

Transverse section of leaf blade, physiology. C_4; biochemical type NAD–ME (*dactyloides*); XyMS+. PCR sheath outlines even. PCR sheaths of the primary vascular bundles interrupted; interrupted abaxially only, or interrupted both abaxially and adaxially. PCR sheath extensions absent. PCR cells without a suberised lamella. PCR cell chloroplasts with well developed grana; centripetal. *Mesophyll* with radiate chlorenchyma; traversed by columns of colourless mesophyll cells. *Leaf blade* with distinct, prominent adaxial ribs (slight), or adaxially flat; with the ribs more or less constant in size. *Midrib* not readily distinguishable; with one bundle only. Bulliforms present in discrete, regular adaxial groups; in simple fans and associated with colourless mesophyll cells to form deeply-penetrating fans (the latter groups contributing to the colourless columns). All the vascular bundles accompanied by sclerenchyma. Combined sclerenchyma girders present. Sclerenchyma all associated with vascular bundles. The lamina margins with fibres (feww cells).

Phytochemistry. Tissues of the culm bases with abundant starch.

Special diagnostic feature. *The male inflorescences elevated, with one to four spicate, unilateral branches; female spikelets in burr-like clusters, usually two burrs per inflorescence, each burr on a short, stout rachis, partially enclosed in a broad, bractlike leaf sheath, falling entire with the indurate rachis united with the upper glumes.*

Cytology. Chromosome base number, $x = 10$. $2n = 20, 40, 56,$ and 60. $2, 4,$ and 6 ploid.

Taxonomy. Chloridoideae; main chloridoid assemblage.

Ecology, geography, regional floristic distribution. 1 species. North America. Xerophytic; species of open habitats. Dry plains.

Holarctic. Boreal and Madrean. Atlantic North American and Rocky Mountains. Southern Atlantic North American and Central Grasslands.

Rusts and smuts. Smuts from Tilletiaceae and from Ustilaginaceae. Tilletiaceae — *Tilletia*. Ustilaginaceae — *Ustilago*.

Economic importance. Significant weed species: *B. dactyloides*. Cultivated fodder: *B. dactyloides* (Buffalo). Important native pasture species: *B. dactyloides*.

References, etc. Leaf anatomical: Metcalfe 1960; this project.

Buchlomimus Reeder, Reeder & Rzedowski

Habit, vegetative morphology. Perennial; stoloniferous. *Culms* 5–15 cm high; herbaceous; unbranched above. Leaves non-auriculate. Leaf blades narrow; 1–3 mm wide (to 8 cm long); flat; exhibiting multicellular glands abaxially (at the bases of macrohairs). The abaxial glands intercostal. Leaf blades without cross venation; persistent. Ligule a fringe of hairs. *Contra-ligule* absent.

Reproductive organization. *Plants dioecious*; without hermaphrodite florets. The spikelets all alike in sexuality (on the same plant); female-only, or male-only. Plants outbreeding.

Inflorescence. *Inflorescence of spicate main branches (racemes straight to curved, pectinate, pedunculate, the female inflorescence prostrate, the male erect)*; digitate, or subdigitate, or non-digitate. Primary inflorescence branches (1–)2–3. Inflorescence axes not ending in spikelets (often ending in a scabrid bristle). Inflorescence espatheate; not comprising 'partial inflorescences' and foliar organs. Spikelet-bearing axes disarticulating (female inflorescence), or persistent (male infloresence); (of the female inflorescence) falling entire. Spikelets solitary; secund (pectinate); biseriate; subsessile; not in distinct 'long-and-short' combinations.

Female-sterile spikelets. Male racemes persistent, erect. Glumes equalling the floret, usually 1-nerved; one floreted, the lemma with three awn points. Rachilla of male spikelets prolonged beyond the uppermost male floret (the terminal rudiments, when present, consisting of scabrid awns which are sometimes very long). *The male spikelets 1 floreted (with or without rudiments)*. The lemmas awned. Male florets 1 (anthers bright orange).

Female-fertile spikelets. *Spikelets* 7 mm long; *compressed dorsiventrally*; falling with the glumes (and with the rachis: not disarticulating otherwise). Rachilla prolonged beyond the uppermost female-fertile floret; rachilla hairless. The rachilla extension with incomplete florets. Hairy callus with a tuft of hairs at the base of the glumes.

Glumes two; very unequal; (the upper) about equalling the spikelets; (the upper) long relative to the adjacent lemmas; lateral to the rachis; hairy (at the base); pointed; awnless; carinate (G_1), or non-carinate (G_2); narrow, the G_1 hyaline and keeled, the G_2 larger, thinly membranous, not keeled. Lower glume 1 nerved. Upper glume (3–)5 nerved. *Spikelets* with incomplete florets. The incomplete florets distal to the female-fertile florets. The distal incomplete florets 3; *incomplete florets* clearly specialised and modified in form; incomplete florets awned (the rudiments 3-awned, the awns long and scabrid, forming a terminal cluster).

Female-fertile florets 1. Lemmas decidedly firmer than the glumes (cartilaginous); smooth; not becoming indurated; incised; not deeply cleft (briefly triaristate); awned. Awns 3; median and lateral; the median similar in form to the laterals; from a sinus (from a short sinus between the two laterals); non-geniculate; recurving; hairless (scabrid); much shorter than the body of the lemma; entered by one vein. The lateral awns slightly shorter than the median. Lemmas hairless; glabrous; non-carinate (rounded on the back); without a germination flap; 3 nerved. *Palea* present; relatively long; apically notched (at the pointed tip); awnless, without apical setae (or the nerves slightly excurrent); not indurated (membranous); 2-nerved; 2-keeled. *Lodicules* present; 2; free; fleshy; glabrous; not or scarcely vascularized. *Stamens* 0 (but 3 staminodes). *Ovary* glabrous. Styles free to their bases. Stigmas 2.

Fruit, embryo and seedling. Fruit small to medium sized (about 4 mm long); ellipsoid; not noticeably compressed. Hilum short. Pericarp fused. Embryo large; waisted; with an epiblast; with a scutellar tail; with an elongated mesocotyl internode; with one scutellum bundle. Embryonic leaf margins meeting.

Abaxial leaf blade epidermis. *Costal/intercostal zonation* conspicuous. *Papillae* absent. *Long-cells* similar in shape costally and intercostally (narrow); of similar wall thickness costally and intercostally (walls of medium thickness). Mid-intercostal long-cells rectangular; having markedly sinuous walls (pitted). *Microhairs* present; elongated; clearly two-celled; panicoid-type (large, with long basal cells; the apical cells round-tipped, variable in length but all longer than in the chloridoid type). Microhair apical cell wall thinner than that of the basal cell and often collapsed. Microhair basal cells 30–42 microns long. Microhair total length/width at septum 5. Microhair apical cell/total length ratio 0.38. *Stomata* common. Subsidiaries dome-shaped (mostly), or triangular. Guard-cells overlapping to flush with the interstomatals. *Intercostal short-cells* common; in cork/ silica-cell pairs; silicified. Intercostal silica bodies present and perfectly developed; crescentic, tall-and-narrow, and saddle shaped. Cushion-based macrohairs present. *Costal short-cells* conspicuously in long rows (very consistently). Costal silica bodies confined to the central file(s) of the costal zones, or present in alternate cell files of the costal zones; homogeneously saddle shaped.

Transverse section of leaf blade, physiology. C_4; XyMS+. PCR sheaths of the primary vascular bundles interrupted; interrupted both abaxially and adaxially. PCR sheath extensions absent. Mesophyll traversed by columns of colourless mesophyll cells (between all bundles). *Leaf blade* adaxially flat. *Midrib* not readily distinguishable; with one bundle only. Bulliforms present in discrete, regular adaxial groups (between all bundles); associated with colourless mesophyll cells to form deeply-penetrating fans. Combined sclerenchyma girders present (with all the bundles); forming 'figures' (the primaries; minor bundles with scanty girders). Sclerenchyma all associated with vascular bundles. The lamina margins with fibres.

Cytology. Chromosome base number, $x = 10$ (n = 20). $2n = 40$.

Taxonomy. Chloridoideae; main chloridoid assemblage.

Ecology, geography, regional floristic distribution. 1 species. Mexico. Species of open habitats. Dry slopes.

Holarctic. Madrean.

References, etc. Morphological/taxonomic: Reeder, Reeder and Rzedowski 1965. Leaf anatomical: this project.

Buergersiochloa Pilger

Habit, vegetative morphology. *Perennial (with dimorphic culms)*. The flowering culms leafless. *Culms* 50–80 cm high. Plants unarmed. *Leaves* not basally aggregated; auriculate; with auricular setae. Leaf blades lanceolate, or ovate-lanceolate, or ovate; broad; 15–55 mm wide (6.5–27 cm long); pseudopetiolate; cross veined; rolled in bud. Ligule a fringed membrane (minute).

Reproductive organization. *Plants monoecious with all the fertile spikelets unisexual*; without hermaphrodite florets. The spikelets of sexually distinct forms on the same plant; female-only and male-only. The male and female-fertile spikelets segregated, in different parts of the same inflorescence branch (female spikelets above, male below). The spikelets overtly heteromorphic (males smaller).

Inflorescence. *Inflorescence* narrowly *paniculate*; espatheate; not comprising 'partial inflorescences' and foliar organs. Spikelet-bearing axes persistent. Spikelets not secund; pedicellate.

Female-sterile spikelets. Male spikelets 2.7–4.4 mm long. The male spikelets without glumes (or these vestigial); without proximal incomplete florets; 1 floreted. Male florets 1; 2–3 staminate. *The staminal filaments joined (monadelphous)*.

Female-fertile spikelets. Spikelets 4–9 mm long; not noticeably compressed to compressed dorsiventrally; disarticulating above the glumes. Rachilla terminated by a female-fertile floret. Hairy callus present. *Callus* short (minute, shortly bearded).

Glumes present; two (leathery); shorter than the adjacent lemmas (about half as long); not pointed (blunt); awnless. Lower glume 3–9 nerved. Upper glume 3–6 nerved. *Spikelets* with female-fertile florets only.

Female-fertile florets 1. Lemmas decidedly firmer than the glumes (leathery); not becoming indurated (or only slightly so); awned. Awns 1; median; apical; non-geniculate; about as long as the body of the lemma to much longer than the body of the lemma. Lemmas hairy (with appressed hairs); non-carinate; 5 nerved, or 7 nerved. *Palea* present (hairy); relatively long; awnless, without apical setae; 2-nerved. *Lodicules* present; 3; membranous (pointed); heavily vascularized. *Stamens* 0 (3 staminodes). *Ovary* glabrous. Styles fused. Stigmas 2.

Fruit, embryo and seedling. Hilum long-linear.

Abaxial leaf blade epidermis. *Costal/intercostal zonation* conspicuous. *Papillae* present; mostly intercostal. Intercostal papillae over-arching the stomata; several per cell (large, thick walled, clustered in rosettes around the stomata). Intercostal zones with typical long-cells (seemingly, the outlines obscured by papillae). Mid-intercostal long-cells having markedly sinuous walls. *Microhairs* present; elongated; panicoid-type. *Stomata* common (obscured by papillae). Subsidiaries non-papillate. *Intercostal short-cells* not observable. *Costal short-cells* conspicuously in long rows, or predominantly paired, or neither distinctly grouped into long rows nor predominantly paired (the disposition varying from place to place). Costal silica bodies absent to poorly developed (in material seen, but with panicoid type and saddle shaped silica cells).

Transverse section of leaf blade, physiology. C_3; XyMS+. *Mesophyll* with non-radiate chlorenchyma; without adaxial palisade; with arm cells; without fusoids. *Leaf blade* adaxially flat. *Midrib* seemingly not readily distinguishable (material seen poor). Bulliforms present in discrete, regular adaxial groups; in simple fans (these large). All the vascular bundles accompanied by sclerenchyma. Combined sclerenchyma girders present; forming 'figures' (in the main bundles). Sclerenchyma not all bundle-associated.

The 'extra' sclerenchyma in abaxial groups and in adaxial groups; abaxial-hypodermal, the groups isolated (opposite the bulliforms), or adaxial-hypodermal, contiguous with the bulliforms.

Taxonomy. Bambusoideae; Oryzodae; Olyreae ('Buergersiochloeae').

Ecology, geography, regional floristic distribution. 1 species. New Guinea. Paleotropical. Indomalesian. Papuan.

References, etc. Morphological/taxonomic: Blake 1946; Fitjen 1975. Leaf anatomical: this project.

Calamagrostis Adans.

Achaeta Fourn., *Amagris* Raf., *Ancistrochloa* Honda, *Anisachne* Keng, *Athernotus* Dulac, *Chamaecalamus* Meyen, *Cinnagrostis* Griseb., *Pteropodium* Steud., *Sclerodeuxia* Pilger

Excluding *Aniselytron, Deyeuxia, Dichelachne, Stilpnophleum*

Habit, vegetative morphology. Perennial (some species reedlike); rhizomatous, or stoloniferous, or caespitose, or decumbent. **Culms** 10–200 cm high; *herbaceous*; unbranched above. Culm nodes glabrous. Culm internodes hollow. *Leaves* not basally aggregated; non-auriculate. Leaf blades linear; narrow; 1–15 mm wide; setaceous, or not setaceous; flat (usually), or rolled (convolute); without cross venation; persistent; rolled in bud. *Ligule an unfringed membrane (sometimes erose-ciliate)*; truncate, or not truncate; 1–12 mm long. *Contra-ligule* present (a membrane), or absent.

Reproductive organization. Plants bisexual, with bisexual spikelets; with hermaphrodite florets; outbreeding. Exposed-cleistogamous, or chasmogamous. Apomictic (pseudogamous or non-pseudogamous), or reproducing sexually.

Inflorescence. *Inflorescence paniculate*; open (rarely), or contracted; when contracted capitate to spicate, or more or less irregular; espatheate; not comprising 'partial inflorescences' and foliar organs. Spikelet-bearing axes persistent. Spikelets not secund; pedicellate.

Female-fertile spikelets. *Spikelets* 3–7(–8) mm long; *compressed laterally*; disarticulating above the glumes. Rachilla prolonged beyond the uppermost female-fertile floret, or terminated by a female-fertile floret; rachilla hairy. The rachilla extension (when present) naked. *Hairy callus present (the hairs surrounding and often as long as or much exceeding the lemma). Callus hairs present, more than 0.5 mm long.*

Glumes two; more or less equal; exceeding the spikelets; long relative to the adjacent lemmas (the lemma only about 1/2–2/3 as long); pointed (acute to acuminate); awnless; carinate; similar. Lower glume 1 nerved. Upper glume 1–3 nerved. **Spikelets** with female-fertile florets only; *without proximal incomplete florets*.

Female-fertile florets 1. *Lemmas less firm than the glumes to similar in texture to the glumes (usually hyaline)*; not becoming indurated; *incised*; not deeply cleft (emarginate, bilobed or irregularly denticulate); awned. Awns 1; median; from a sinus, or dorsal; when dorsal, from well down the back; non-geniculate, or geniculate; much shorter than the body of the lemma to about as long as the body of the lemma, or much longer than the body of the lemma; entered by one vein. *Lemmas* hairless; *non-carinate*; 3–5 nerved. *Palea* present; relatively long; entire, or apically notched (via excurrent veins); awnless, without apical setae; textured like the lemma; not indurated (thin); 2-nerved; 2-keeled, or keel-less. **Lodicules** present; 2; free; *membranous*; glabrous; not toothed. *Stamens* 3. Anthers 1–4 mm long; not penicillate. *Ovary* glabrous. Styles free to their bases. Stigmas 2.

Fruit, embryo and seedling. *Fruit* free from both lemma and palea; small. *Hilum short*. Embryo small; not waisted. Endosperm liquid in the mature fruit, or hard; with lipid. Embryo with an epiblast; without a scutellar tail; with a negligible mesocotyl internode. Embryonic leaf margins meeting.

Seedling with a long mesocotyl; with a tight coleoptile. First seedling leaf with a well-developed lamina. The lamina narrow; 1 veined, or 3 veined.

Abaxial leaf blade epidermis. *Costal/intercostal zonation* conspicuous. *Papillae* absent. *Long-cells* similar in shape costally and intercostally (the costals narrower); of similar wall thickness costally and intercostally (thin walled). Mid-intercostal long-cells rectangular; having markedly sinuous walls. *Microhairs* absent. *Stomata* common. Subsidiaries low dome-shaped, or parallel-sided. Guard-cells overlapped by the interstomatals. *Intercostal short-cells* common; in cork/silica-cell pairs (sometimes), or not paired (but often paired with a prickle or hook); silicified (rarely), or not silicified (usually). Intercostal silica bodies when present, tall-and-narrow. *Costal short-cells* neither distinctly grouped into long rows nor predominantly paired. Costal silica bodies horizontally-elongated crenate/sinuous, or 'panicoid-type'; when 'panicoid type', nodular.

Transverse section of leaf blade, physiology. C_3; XyMS+. *Mesophyll* with non-radiate chlorenchyma. *Leaf blade* with distinct, prominent adaxial ribs; with the ribs more or less constant in size. *Midrib* conspicuous; having a conventional arc of bundles (*C. epigejos* exhibits a large median with a small lateral on either side). Bulliforms present in discrete, regular adaxial groups (in the furrows); in simple fans. All the vascular bundles accompanied by sclerenchyma. Combined sclerenchyma girders present; nowhere forming 'figures'. Sclerenchyma all associated with vascular bundles.

Culm anatomy. *Culm internode bundles* in three or more rings.

Phytochemistry. Tissues of the culm bases with little or no starch. Fructosans predominantly long-chain. Leaves without flavonoid sulphates (2 species).

Cytology. Chromosome base number, $x = 7$. $2n = 28$, 42, and 56, or 56–91. 4, 6, and 8 ploid (lowest $2n=28$?), or 8–13 ploid.

Taxonomy. Pooideae; Poodae; Aveneae.

Ecology, geography, regional floristic distribution. About 230 species. Temperate. Commonly adventive. Mostly helophytic to mesophytic; shade species to species of open habitats; glycophytic, or halophytic (rarely). Diverse habitats, including coastal sand — see ×*Ammocalamagrostis*.

Holarctic, Paleotropical, Neotropical, and Cape. Boreal, Tethyan, and Madrean. African and Polynesian. Arctic and Subarctic, Euro-Siberian, Eastern Asian, Atlantic North American, and Rocky Mountains. Mediterranean and Irano-Turanian. Sudano-Angolan and West African Rainforest. Hawaiian. Caribbean, Central Brazilian, Pampas, and Andean. European and Siberian. Canadian-Appalachian, Southern Atlantic North American, and Central Grasslands. Sahelo-Sudanian, Somalo-Ethiopian, and South Tropical African.

Hybrids. Intergeneric hybrids with *Agrostis*. *C. epigejos* hybridizes with *Ammophila arenaria* (×*Ammocalamagrostis* P. Fourn.; ×*Calamophila* O. Schwartz = ×*Ammocalamagrostis*, ×*Calammophila* Brand = ×*Ammocalamagrostis*).

Rusts and smuts. Rusts — *Puccinia*. Taxonomically wide-ranging species: *Puccinia graminis*, *Puccinia coronata*, *Puccinia striiformis*, *Puccinia pygmaea*, *Puccinia brachypodii*, *Puccinia poarum*, and *Puccinia recondita*. Smuts from Tilletiaceae and from Ustilaginaceae. Tilletiaceae — *Entyloma* and *Urocystis*. Ustilaginaceae — *Ustilago*.

Economic importance. Important native pasture species: *C. canadensis*, *C. inexpansa*, *C. rubescens*, *C. montanensis* etc. *Ammocalamagrostis* is a useful sandbinder.

References, etc. Leaf anatomical: Metcalfe 1960; this project.

Calamovilfa Hackel

Habit, vegetative morphology. Coarse perennial; rhizomatous, or caespitose (with short thick rhizomes). **Culms *50–200 cm high*; *herbaceous*;** unbranched above. Culm internodes solid. *Leaves* not basally aggregated; non-auriculate. Leaf blades narrow; rolled (firm); without abaxial multicellular glands; rolled in bud. Ligule a fringe of hairs.

Reproductive organization. Plants bisexual, with bisexual spikelets; with hermaphrodite florets.

Inflorescence. Inflorescence paniculate; open, or contracted; when contracted, spicate to more or less irregular; espatheate; not comprising 'partial inflorescences' and foliar organs. Spikelet-bearing axes persistent. Spikelets not secund; pedicellate.

Female-fertile spikelets. Spikelets 5–9 mm long; lanceolate; compressed laterally; disarticulating above the glumes. Rachilla terminated by a female-fertile floret. *Hairy callus present (the hairs sometimes longer than half lemma length).*

Glumes two; very unequal; shorter than the spikelets to about equalling the spikelets; shorter than the adjacent lemmas, or long relative to the adjacent lemmas; pointed (acute); awnless; carinate; similar (papery or membranous). Lower glume 1 nerved. Upper glume 1 nerved. **Spikelets** with female-fertile florets only; *without proximal incomplete florets.* *Female-fertile florets* 1 (rarely with a second). *Lemmas* similar in texture to the glumes; not becoming indurated; entire; pointed; *awnless*; hairy (pilose at base), or hairless; carinate (slightly), or non-carinate; *1 nerved*. *Palea* present; relatively long; 2-nerved; 2-keeled. *Lodicules* present; 2; free; fleshy; glabrous; heavily vascularized. *Stamens* 3. Anthers 3–5.5 mm long. *Ovary* glabrous. Stigmas 2.

Fruit, embryo and seedling. Fruit linear, or ellipsoid. Hilum short. *Pericarp free.* Endosperm hard; without lipid. Embryo with an epiblast; with a scutellar tail; with an elongated mesocotyl internode. Embryonic leaf margins meeting.

Abaxial leaf blade epidermis. *Costal/intercostal zonation* conspicuous. *Papillae* absent. *Long-cells* markedly different in shape costally and intercostally (the costals narrower, more regularly rectangular); of similar wall thickness costally and intercostally (the walls thick), or differing markedly in wall thickness costally and intercostally (the costals thicker walled). Mid-intercostal long-cells having markedly sinuous walls (heavily pitted). *Microhairs* present; elongated; clearly two-celled; chloridoid-type (with long basal cells). Microhair apical cell wall of similar thickness/rigidity to that of the basal cell. Microhairs 38–48(–60) microns long. Microhair basal cells 45 microns long. Microhairs 9.6–11.4 microns wide at the septum. Microhair total length/width at septum 3.9–5. Microhair apical cells 7–10(–12) microns long. Microhair apical cell/total length ratio 0.15–0.2. *Stomata* common; 22.5–30 microns long. Subsidiaries triangular, or dome-shaped (mostly, in *C. longifolia*). Guard-cells overlapping to flush with the interstomatals. *Intercostal short-cells* common; in cork/silica-cell pairs and not paired (some solitary); silicified. Intercostal silica bodies absent, or imperfectly developed; mostly saddle shaped. *Costal short-cells* neither distinctly grouped into long rows nor predominantly paired. Costal silica bodies present throughout the costal zones; mostly saddle shaped.

Transverse section of leaf blade, physiology. C_4; XyMS+. PCR sheath outlines even. PCR sheaths of the primary vascular bundles interrupted; interrupted both abaxially and adaxially. PCR sheath extensions absent. *Mesophyll* with radiate chlorenchyma; traversed by columns of colourless mesophyll cells. *Leaf blade* 'nodular' in section; with the ribs more or less constant in size. *Midrib* not readily distinguishable; with one bundle only. Bulliforms present in discrete, regular adaxial groups; associated with colourless mesophyll cells to form deeply-penetrating fans (incorporated in the traversing colourless columns). All the vascular bundles accompanied by sclerenchyma. Combined sclerenchyma girders present (with all the bundles); forming 'figures' (all the bundles with I's). Sclerenchyma all associated with vascular bundles. The lamina margins with fibres.

Cytology. Chromosome base number, $x = 10$. $2n = 40$.

Taxonomy. Chloridoideae; main chloridoid assemblage.

Ecology, geography, regional floristic distribution. 4 species. North America. Species of open habitats. Dry or marshy pine barrens, inland dunes, sandy prairie.

Holarctic. Boreal and Madrean. Atlantic North American and Rocky Mountains. Canadian-Appalachian, Southern Atlantic North American, and Central Grasslands.

Rusts and smuts. Rusts — *Puccinia.*

References, etc. Leaf anatomical: Metcalfe 1960; this project.

Calderonella Soderstrom & Decker

Habit, vegetative morphology. Perennial; stoloniferous and caespitose. The flowering culms leafy (the uppermost leaf with a reduced blade). *Culms* 70 cm high; herbaceous; unbranched above. Pluricaespitose. Plants unarmed. *Leaves* not basally aggregated. Leaf blades lanceolate; broad; 10–15 mm wide; flat; pseudopetiolate; cross veined (abaxially). *Ligule* present; an unfringed membrane (brown, papery, abaxially pilose); 0.1–0.2 mm long. *Contra-ligule* present (a hair fringe).

Reproductive organization. *Plants bisexual, with bisexual spikelets (3–5 dimorphous florets, the lowermost one (or two) pistillate and maturing first, the* *remainder staminate); without hermaphrodite florets (unless the second floret is some-times functionally hermaphrodite).*

Inflorescence. Inflorescence a single raceme (3–5 cm long, with 6 or 7 spikelets on short pedicels, the raceme terminating a long, filiform peduncle); espatheate (but uppermost leaf modified); not comprising 'partial inflorescences' and foliar organs. Spikelet-bearing axes solitary; persistent. Spikelets solitary; not secund; pedicellate (the pedicels 1–2 mm long).

Female-fertile spikelets. *Spikelets* 8 mm long; *compressed laterally (the lowermost floret becoming gibbous at maturity, causing the spikelet to bend at right angles to the rachis and become triangular in appearance);* falling with the glumes. Rachilla prolonged beyond the uppermost female-fertile floret. The rachilla extension with incomplete florets.

Glumes two; relatively large (G_1 about 6.5 mm long, G_2 about 4.5–5 mm); dorsiventral to the rachis; hairy (sparsely papillose-pilose intercostally); pointed; very dissimilar (G_1 lanceolate, G_2 parabolical in side view, both with transverse veinlets). Lower glume 13–16 nerved. Upper glume 10–11 nerved. *Spikelets* with incomplete florets. The incomplete florets distal to the female-fertile florets. *The distal incomplete florets clearly specialised and modified in form (i.e., they are male).*

Female-fertile florets 1 (the basal, which has 3 staminodes — but the second floret may have a (?)non-functional gynoecium). Lemmas with transverse veinlets; awnless; hairy (sparsely papillose-pilose between the veins); *15–19 nerved (but many nerves manifest only at base). Palea* present; awnless, without apical setae (truncate, the apex ciliolate); 2-nerved (with a connecting transverse nervelet at the summit); 2-keeled. Palea keels winged. *Lodicules* present; 2; joined (the inner edges partially fused below, in front of the ovary); fleshy; heavily vascularized. *Stamens* 3 (staminodal — 3 in male florets their anthers non-penicillate). *Ovary* glabrous. Styles fused. Stigmas 2.

Fruit, embryo and seedling. Fruit small (about 2.3 mm long); compressed laterally, ventrally and at the base. Hilum short. Embryo small.

Abaxial leaf blade epidermis. *Costal/intercostal zonation* conspicuous. *Papillae* absent. *Long-cells* markedly different in shape costally and intercostally (the costals narrower, more regularly rectangular, the walls less sinuous); of similar wall thickness costally and intercostally (quite thick walled). Mid-intercostal long-cells rectangular and fusiform; having markedly sinuous walls. *Microhairs* present; panicoid-type; 42–48 microns long; 4.5–6 microns wide at the septum. Microhair total length/width at septum 7–10.7. Microhair apical cells 27–30 microns long. Microhair apical cell/total length ratio 0.63–0.67. *Stomata* common; 27–33 microns long. Subsidiaries dome-shaped and triangular. Guard-cells overlapping to flush with the interstomatals. *Intercostal short-cells* common; silicified. Intercostal silica bodies cross-shaped. Prickles common. *Crown cells* absent. *Costal short-cells* conspicuously in long rows. Costal silica bodies 'panicoid-type'; cross shaped, butterfly shaped, and dumb-bell shaped.

Transverse section of leaf blade, physiology. C_3; XyMS+. Mesophyll with adaxial palisade; without arm cells; with fusoids. The fusoids an integral part of the PBS. *Midrib* conspicuous (via its large bundle with a heavy I-girder combination, and what appears to be a double PBS of irregularly shaped cells); with one bundle only. Bulliforms present in discrete, regular adaxial groups; in simple fans (large, wide groups, extending halfway into the mesophyll). All the vascular bundles accompanied by sclerenchyma. Combined sclerenchyma girders present (with the large bundles the rest having abaxial girders but adaxial strands); forming 'figures'. Sclerenchyma all associated with vascular bundles.

Taxonomy. Bambusoideae; Oryzodae; Centotheceae.

Ecology, geography, regional floristic distribution. 1 species. Panama. Shade species; glycophytic. Forests.

Neotropical. Caribbean and Andean.

References, etc. Morphological/taxonomic: Soderstrom and Decker 1973. Leaf anatomical: this project.

Calosteca Desv.

Calotheca P. Beauv.

~ *Briza*

Habit, vegetative morphology. Perennial (apparently); caespitose. *Culms* 40–80 cm high; herbaceous; unbranched above. Young shoots intravaginal. *Leaves* not basally aggregated; non-auriculate. Leaf blades linear; narrow; flat to folded; without cross venation; persistent; once-folded in bud. *Ligule an unfringed membrane*.

Reproductive organization. Plants bisexual, with bisexual spikelets; with hermaphrodite florets. Exposed-cleistogamous, or chasmogamous (?).

Inflorescence. *Inflorescence* paniculate; open; *with capillary branchlets (?)*; espatheate; not comprising 'partial inflorescences' and foliar organs. Spikelet-bearing axes persistent. Spikelets not secund; pedicellate.

Female-fertile spikelets. Spikelets 10–14 mm long; compressed laterally; disarticulating above the glumes; disarticulating between the florets; with conventional internode spacings. Rachilla prolonged beyond the uppermost female-fertile floret. The rachilla extension with incomplete florets. Hairy callus present.

Glumes two; more or less equal; shorter than the spikelets; shorter than the adjacent lemmas; hairless; awnless (to mucronate from a minutely bifid apex); *non-carinate (?)*; similar. *Spikelets* with incomplete florets. The incomplete florets distal to the female-fertile florets. *The distal incomplete florets* merely underdeveloped.

Female-fertile florets 5–10. *Lemmas cuneate at the base, broad in the middle, narrowed to the apex*; not becoming indurated; incised; 2 lobed; not deeply cleft (acuminate from the very broad base, but minutely incised at the tip); awned. Awns 1; median; from a sinus; non-geniculate; hairless; much shorter than the body of the lemma to about as long as the body of the lemma (2–5 mm long); entered by one vein. Lemmas non-carinate; without a germination flap; 7 nerved, or 9 nerved (?). **Palea** present; *conspicuous but relatively short (lanceolate)*; entire; awnless, without apical setae (but with a conspicuous, hyaline, apical appendage); thinner than the lemma (membranous); not indurated; 2-nerved; 2-keeled (hairy between the keels). Palea keels narrowly winged. *Lodicules* present; 2; free; membranous; glabrous; toothed. *Stamens* 3 (presumably?). *Ovary* glabrous. Stigmas 2.

Fruit, embryo and seedling. Fruit small (about 2 mm long); trigonous. *Hilum short*. Embryo small. Endosperm hard.

Abaxial leaf blade epidermis. *Costal/intercostal zonation* conspicuous (the costal zones narrow). *Papillae* absent. *Long-cells* markedly different in shape costally and intercostally (the costals much shorter and narrower, with markedly sinuous walls). Mid-intercostal long-cells rectangular to fusiform (very long, minimizing the tendency to be fusiform); having straight or only gently undulating walls. *Microhairs* absent. *Stomata* absent or very rare (confined to short, discontinuous single files adjoining the costal zones). Subsidiaries parallel-sided. Guard-cells overlapped by the interstomatals (but only very slightly so). *Intercostal short-cells* absent or very rare. *Costal short-cells* neither distinctly grouped into long rows nor predominantly paired. Costal silica bodies horizontally-elongated crenate/sinuous (crenate).

Transverse section of leaf blade, physiology. C_3; XyMS+. Mesophyll without adaxial palisade. *Leaf blade* with distinct, prominent adaxial ribs; with the ribs more or less constant in size (tall and narrow). *Midrib* somewhat conspicuous (with a larger rib); with one bundle only. *The lamina* symmetrical on either side of the midrib. Bulliforms present in discrete, regular adaxial groups; in simple fans. All the vascular bundles accompanied by sclerenchyma. Combined sclerenchyma girders absent (all bundles with a small adaxial strand, the main bundles with an abaxial strand as well). Sclerenchyma all associated with vascular bundles.

Phytochemistry. Leaves without flavonoid sulphates (1 species).

Special diagnostic feature. *Lemmas not as in* **Briza** *(q.v.)*.

Taxonomy. Pooideae; Poodae; Poeae.

Ecology, geography, regional floristic distribution. 1 species. South America. Neotropical. Andean.

References, etc. Morphological/taxonomic: Mattei 1975, Nicora and Rúgolo de Agrasar 1981. Leaf anatomical: this project.

Special comments. *Briza* sect. *Calotheca*, *Bromus brizoides*.

Calyptochloa C.E. Hubb.

Habit, vegetative morphology. Mat-forming perennial; decumbent. *Culms* 15–40 cm high; herbaceous. Culm nodes glabrous. Culm internodes hollow. Young shoots intravaginal. *Leaves* not basally aggregated; non-auriculate. Leaf blades narrow; 2–8 mm wide; folded; without cross venation; disarticulating from the sheaths; rolled in bud. *Ligule a fringe of hairs. Contra-ligule* absent.

Reproductive organization. Plants bisexual, with bisexual spikelets; with hermaphrodite florets. The spikelets all alike in sexuality. Exposed-cleistogamous, or chasmogamous. Plants with hidden cleistogenes, or without hidden cleistogenes. The hidden cleistogenes when present, in the leaf sheaths (borne singly, very modified).

Inflorescence. *Inflorescence* few spikeleted; *a single raceme (loose, spike-like)*; espatheate; not comprising 'partial inflorescences' and foliar organs. Spikelet-bearing axes persistent. Spikelets solitary; not secund; pedicellate. Pedicel apices truncate, or cupuliform.

Female-fertile spikelets. *Spikelets* 3.5 mm long; lanceolate, or ovate; adaxial; *compressed dorsiventrally*; falling with the glumes. Rachilla terminated by a female-fertile floret. Hairy callus absent.

Glumes two; very unequal; (the longer) long relative to the adjacent lemmas; (the upper) hairy (stiffly pilose dorsally and on the margins); awnless; very dissimilar (the lower vestigial, represented by a tiny scale). Lower glume 0 nerved. *Upper glume 7 nerved. Spikelets* with incomplete florets. The incomplete florets proximal to the female-fertile florets. *Spikelets with proximal incomplete florets. The proximal incomplete florets* 1; epaleate; sterile. The proximal lemmas awnless; 7 nerved; hairy, decidedly exceeding the female-fertile lemmas; less firm than the female-fertile lemmas to similar in texture to the female-fertile lemmas (chartaceous); not becoming indurated.

Female-fertile florets 1. *Lemmas* similar in texture to the glumes to decidedly firmer than the glumes; striate; becoming indurated (slightly); yellow in fruit, or brown in fruit; entire; pointed; *awned (attenuate into the awn)*. Awns 1; median; apical; non-geniculate; slightly curved; much shorter than the body of the lemma to about as long as the body of the lemma. Lemmas hairless; non-carinate; having the margins lying flat and exposed on the palea; 3 nerved. *Palea* present; relatively long; entire (pointed); awnless, without apical setae; textured like the lemma; 2-nerved. *Lodicules* present; 2; free; fleshy; glabrous. *Stamens* 3. Anthers 0.3 mm long; not penicillate. *Ovary* glabrous. Styles free to their bases. Stigmas 2.

Fruit, embryo and seedling. Fruit small; compressed dorsiventrally. Hilum short. Embryo large; waisted.

Abaxial leaf blade epidermis. *Costal/intercostal zonation* conspicuous. *Papillae* absent. *Long-cells* similar in shape costally and intercostally (narrow-rectangular); of similar wall thickness costally and intercostally (thin walled). Mid-intercostal long-cells rectangular; having markedly sinuous walls. *Microhairs* present; panicoid-type; 54–69 microns long; 6–9 microns wide at the septum. Microhair total length/width at septum 6.3–10. Microhair apical cells 21–33 microns long. Microhair apical cell/total length ratio 0.39–0.48. *Stomata* common; 33–36 microns long. Subsidiaries triangular. Guard-cells overlapping to flush with the interstomatals. *Intercostal short-cells* common; in cork/silica-cell pairs (mainly); silicified. *Costal short-cells* conspicuously in long rows. Costal silica bodies 'panicoid-type'.

Transverse section of leaf blade, physiology. C_3; XyMS+. *Mesophyll* with radiate chlorenchyma; *Isachne*-type. *Leaf blade* adaxially flat. *Midrib* not readily distinguishable; with one bundle only. Bulliforms present in discrete, regular adaxial groups; in simple fans. All the vascular bundles accompanied by sclerenchyma. Combined sclerenchyma girders present; forming 'figures', or nowhere forming 'figures'. Sclerenchyma all associated with vascular bundles.

Taxonomy. Panicoideae; Panicodae; Paniceae.

Ecology, geography, regional floristic distribution. 1 species. Australia. Shade species; glycophytic. Forests.

Australian. North and East Australian. Tropical North and East Australian.

References, etc. Morphological/taxonomic: Hubbard 1933c. Leaf anatomical: this project.

Camusiella Bosser

~ *Setaria*

Habit, vegetative morphology. Annual; caespitose. Culms unbranched above. Culm nodes glabrous. Culm internodes solid. Leaves non-auriculate. *Leaf blades pseudopetiolate (except those of upper culm leaves)*; without cross venation. Ligule a fringed membrane to a fringe of hairs.

Reproductive organization. Plants bisexual, with bisexual spikelets; with hermaphrodite florets. The spikelets of sexually distinct forms on the same plant, or all alike in sexuality; hermaphrodite, or hermaphrodite and sterile.

Inflorescence. *Inflorescence a false spike, with spikelets on contracted axes (bearing complex 'reduced branch-systems', each with a few spikelets and lots of bristles, cf.* Setaria*)*; non-digitate. Inflorescence axes not ending in spikelets. Inflorescence espatheate; not comprising 'partial inflorescences' and foliar organs. Spikelet-bearing axes persistent. Spikelets *with 'involucres' of 'bristles'*. The 'bristles' relatively slender, not spiny; persisting on the axis. Spikelets secund (dorsiventral racemes).

Female-fertile spikelets. *Spikelets* 3 mm long; *not noticeably compressed*; falling with the glumes. Rachilla terminated by a female-fertile floret. Hairy callus absent.

Glumes two; relatively large; very unequal; shorter than the adjacent lemmas; hairless; pointed; awnless; non-carinate; similar. Lower glume 3 nerved. Upper glume 5 nerved. *Spikelets* with incomplete florets. The incomplete florets proximal to the female-fertile florets. *The proximal incomplete florets* 1; sterile. The proximal lemmas awnless; 3 nerved; more or less equalling the female-fertile lemmas; less firm than the female-fertile lemmas; not becoming indurated.

Female-fertile florets 1. Lemmas decidedly firmer than the glumes; rugose; becoming indurated; entire; pointed (and beaked); awnless; hairless; non-carinate; having the margins lying flat and exposed on the palea; 5 nerved. *Palea* present; relatively long; awnless, without apical setae; 2-nerved; 2-keeled, or keel-less. *Lodicules* present; free; fleshy; glabrous; not or scarcely vascularized. *Stamens* 3. Anthers not penicillate. *Ovary* glabrous. Styles fused. Stigmas 2, or 3.

Fruit, embryo and seedling. Fruit small. Hilum short.

Abaxial leaf blade epidermis. *Costal/intercostal zonation* conspicuous. *Papillae* absent. Mid-intercostal long-cells having straight or only gently undulating walls. *Microhairs* present; panicoid-type (very narrow); (57–)60–72(–78) microns long; 30–40.5 microns wide at the septum. Microhair total length/width at septum 9.5–14.4. Microhair apical cells 5.4–6–6.3 microns long. Microhair apical cell/total length ratio 0.52–0.56. *Stomata* common; 30–33 microns long. Guard-cells overlapping to flush with the interstomatals. Prickles abundant, some with minute points which could be mistaken for papillae. *Costal short-cells* conspicuously in long rows. Costal silica bodies present and well developed (in addition to tiny birefringent crystals, which occur throughout the epidermis); 'panicoid-type', or acutely-angled (the panicoid forms often with acutely angled tips); sharp-pointed (the crosses and dumb-bells often with acutely-angled tips).

Transverse section of leaf blade, physiology. C_4; XyMS– (but some bundles XyMS+ — i.e. 'XyMS variable', cf. *Spinifex*). *Mesophyll* with radiate chlorenchyma. Leaf blade with the ribs more or less constant in size. *Midrib* conspicuous; having a conventional arc of bundles; with colourless mesophyll adaxially (the colourless tissue occurring across the blade). Bulliforms not in discrete, regular adaxial groups. Many of the smallest vascular bundles unaccompanied by sclerenchyma. Combined sclerenchyma girders present.

Taxonomy. Panicoideae; Panicodae; Paniceae.

Ecology, geography, regional floristic distribution. 2 species. Madagascar.

Paleotropical. Madagascan.
References, etc. Leaf anatomical: this project.

Capillipedium Stapf

Filipedium Raiz. & Jain

Habit, vegetative morphology. Annual, or perennial; caespitose to decumbent. Culms herbaceous; branched above, or unbranched above. Culm nodes hairy, or glabrous. Culm internodes solid. The shoots aromatic, or not aromatic. *Leaves* not basally aggregated; non-auriculate. Leaf blades narrow; flat; without cross venation; persistent; rolled in bud. Ligule an unfringed membrane to a fringed membrane. *Contra-ligule* absent.

Reproductive organization. Plants bisexual, with bisexual spikelets; with hermaphrodite florets. *The spikelets of sexually distinct forms on the same plant*; hermaphrodite and male-only, or hermaphrodite and sterile; overtly heteromorphic (the pedicelled spikelets awnless); all in heterogamous combinations. Apomictic, or reproducing sexually.

Inflorescence. *Inflorescence* paniculate; open; *with capillary branchlets*; espatheate; not comprising 'partial inflorescences' and foliar organs. **Spikelet-bearing axes** very much reduced, or 'racemes'; *the spikelet-bearing axes* with only one spikelet-bearing 'article' to with 6–10 spikelet-bearing 'articles' (racemes 1–2(-8) jointed); *with very slender rachides (the internodes and pedicels with a translucent median furrow)*; disarticulating (*C. parviflorum* may have only one triplet per raceme, the triplet being shed as a unit); falling entire (*C. parviflorum*), or disarticulating at the joints. *The pedicels and rachis internodes with a longitudinal, translucent furrow*. 'Articles' linear; without a basal callus-knob; densely long-hairy, or somewhat hairy, or glabrous. Spikelets in triplets (*C. parviflorum*), or paired; secund (pedicellate spikelets on one side of rachis, sessile ones on the other), or not secund (*C. parviflorum*); sessile and pedicellate; consistently in 'long-and-short' combinations; in pedicellate/sessile combinations. Pedicels of the 'pedicellate' spikelets free of the rachis. The 'shorter' spikelets hermaphrodite. The 'longer' spikelets male-only, or sterile.

Female-sterile spikelets. The male spikelets with glumes. The lemmas awnless.

Female-fertile spikelets. Spikelets compressed dorsiventrally; falling with the glumes. Rachilla terminated by a female-fertile floret. Hairy callus present.

Glumes two; more or less equal; long relative to the adjacent lemmas; awnless; very dissimilar. Lower glume two-keeled; convex on the back to concave on the back; not pitted; relatively smooth; 6–11 nerved. Upper glume 3 nerved (naviculate). *Spikelets* with incomplete florets. The incomplete florets proximal to the female-fertile florets. *The proximal incomplete florets* 1; sterile. The proximal lemmas awnless; 0 nerved; similar in texture to the female-fertile lemmas.

Female-fertile florets 1. Lemmas reduced to a hyaline stipe; less firm than the glumes; not becoming indurated; entire; awned. Awns 1; median; apical; geniculate; hairless (glabrous). Lemmas hairless; non-carinate; 1 nerved. **Palea absent**. *Lodicules* present; free; fleshy; glabrous. *Stamens* 3. Anthers not penicillate. *Ovary* glabrous. Styles free to their bases. Stigmas 2.

Fruit, embryo and seedling. *Fruit* free from both lemma and palea; small; compressed dorsiventrally. Hilum short. Embryo large. Endosperm containing only simple starch grains. Embryo without an epiblast; with a scutellar tail; with an elongated mesocotyl internode. Embryonic leaf margins overlapping.

Seedling with a long mesocotyl. First seedling leaf with a well-developed lamina. The lamina broad; supine; 13–20 veined.

Abaxial leaf blade epidermis. *Costal/intercostal zonation* conspicuous. *Papillae* present; costal and intercostal. Intercostal papillae over-arching the stomata; consisting of one oblique swelling per cell (intercostally), or several per cell (costally). *Long-cells* similar in shape costally and intercostally; of similar wall thickness costally and intercostally (fairly thin walled). Mid-intercostal long-cells rectangular; having markedly sinuous walls. *Microhairs* present; panicoid-type. *Stomata* common, or absent or very rare. Subsidiaries non-papillate; triangular. Guard-cells overlapping to flush with the interstomatals. *Intercostal short-cells* common; in cork/silica-cell pairs, or not paired (solitary); not silicified (usually). *Costal short-cells* conspicuously in long rows. Costal silica bodies 'panicoid-type'.

Transverse section of leaf blade, physiology. C_4; XyMS–. *Mesophyll* with radiate chlorenchyma; traversed by columns of colourless mesophyll cells. *Leaf blade* adaxially flat. *Midrib* conspicuous; having a conventional arc of bundles; with colourless mesophyll adaxially. Bulliforms present in discrete, regular adaxial groups (in the furrows); in simple fans and associated with colourless mesophyll cells to form deeply-penetrating fans; associating with colourless mesophyll cells to form arches over small vascular bundles. Many of the smallest vascular bundles unaccompanied by sclerenchyma. Combined sclerenchyma girders present; forming 'figures'. Sclerenchyma all associated with vascular bundles.

Phytochemistry. Leaves without flavonoid sulphates (1 species).

Cytology. Chromosome base number, $x = 10$. $2n = 20$, 40, and 60. 2, 4, and 6 ploid.

Taxonomy. Panicoideae; Andropogonodae; Andropogoneae; Andropogoninae.

Ecology, geography, regional floristic distribution. 14 species. Warm Old World. Species of open habitats.

Holarctic, Paleotropical, and Australian. Tethyan. African, Indomalesian, and Neo-caledonian. Irano-Turanian. Saharo-Sindian and Sudano-Angolan. Indian, Indo-Chinese, Malesian, and Papuan. North and East Australian. Somalo-Ethiopian and South Tropical African. Tropical North and East Australian.

Hybrids. Intergeneric hybrids with *Bothriochloa*.

Rusts and smuts. Rusts — *Puccinia*. Taxonomically wide-ranging species: *Puccinia nakanishikii*, *Puccinia eritraeensis*, *Puccinia versicolor*, *Puccinia miyoshiana*, and *Puccinia cesatii*. Smuts from Ustilaginaceae. Ustilaginaceae — *Sorosporium* and *Sphacelotheca*.

Economic importance. Significant weed species: *C. parviflorum*.

References, etc. Leaf anatomical: Metcalfe 1960; this project.

Castellia Tineo.

Habit, vegetative morphology. *Annual*. *Culms* 50–100 cm high; herbaceous. Leaves non-auriculate. Leaf blades linear; narrow; flat; without cross venation; persistent. *Ligule an unfringed membrane*; truncate; 1–2 mm long.

Reproductive organization. Plants bisexual, with bisexual spikelets; with hermaphrodite florets.

Inflorescence. *Inflorescence a single raceme (rarely), or paniculate*; open; espatheate; not comprising 'partial inflorescences' and foliar organs. Spikelet-bearing axes persistent. Spikelets not secund; pedicellate.

Female-fertile spikelets. Spikelets 9–20 mm long; compressed laterally; disarticulating above the glumes; disarticulating between the florets. Rachilla prolonged beyond the uppermost female-fertile floret. The rachilla extension with incomplete florets. Hairy callus absent.

Glumes two; relatively large; very unequal to more or less equal; shorter than the adjacent lemmas, or long relative to the adjacent lemmas; hairless; glabrous; pointed, or not pointed; awnless; non-carinate; similar. Lower glume 1–3 nerved. Upper glume 3–5 nerved. *Spikelets* with incomplete florets. The incomplete florets distal to the female-fertile florets.

Female-fertile florets *3–17*. *Lemmas less firm than the glumes*; not becoming indurated (membranous); entire; awnless; *densely tuberculate dorsally*; non-carinate; without a germination flap; 5 nerved. *Palea* present; relatively long; entire to apically notched; 2-nerved; 2-keeled. *Lodicules* present; 2; free; membranous; glabrous; toothed; not or scarcely vascularized. *Stamens* 3. Anthers 0.3–0.5 mm long; not penicillate; without an apically prolonged connective. *Ovary* glabrous. Styles free to their bases. Stigmas 2.

Fruit, embryo and seedling. *Fruit* adhering to lemma and/or palea (to the palea); small, or medium sized (3–4 mm long); compressed dorsiventrally. *Hilum long-linear*. Embryo small. Endosperm hard; without lipid; containing compound starch grains.

Embryo with an epiblast; without a scutellar tail; with a negligible mesocotyl internode. Embryonic leaf margins meeting.

Abaxial leaf blade epidermis. *Costal/intercostal zonation* conspicuous. *Papillae* absent. *Long-cells* markedly different in shape costally and intercostally (costals smaller); of similar wall thickness costally and intercostally. Mid-intercostal long-cells rectangular; having markedly sinuous walls. *Microhairs* absent. *Stomata* common; 33–36 microns long. Subsidiaries parallel-sided. Guard-cells overlapped by the interstomatals. *Intercostal short-cells* absent or very rare. *Costal short-cells* neither distinctly grouped into long rows nor predominantly paired. Costal silica bodies horizontally-elongated crenate/sinuous.

Transverse section of leaf blade, physiology. C_3; XyMS+. *Mesophyll* with non-radiate chlorenchyma; without adaxial palisade. *Leaf blade* with distinct, prominent adaxial ribs (small, rounded); with the ribs more or less constant in size. *Midrib* not readily distinguishable; with one bundle only; with colourless mesophyll adaxially. Bulliforms present in discrete, regular adaxial groups; in simple fans. Combined sclerenchyma girders present (in the primaries); forming 'figures' (narrow, in the primaries). Sclerenchyma all associated with vascular bundles.

Cytology. Chromosome base number, $x = 7$. $2n = 14$. 2 ploid.

Taxonomy. Pooideae; Poodae; Poeae.

Ecology, geography, regional floristic distribution. 1 species. Mediterranean to western Asia. Species of open habitats. Dry places.

Holarctic and Paleotropical. Tethyan. African. Macaronesian, Mediterranean, and Irano-Turanian. Saharo-Sindian.

References, etc. Leaf anatomical: this project.

Catabrosa P. Beauv.

Habit, vegetative morphology. Perennial; stoloniferous, or decumbent. *Culms* 5–70 cm high; herbaceous. Culm nodes glabrous. Culm internodes hollow. Leaves non-auriculate. *Sheath margins joined*. Leaf blades linear; narrow; to 3–10 mm wide; flat; without cross venation; persistent; once-folded in bud. Ligule an unfringed membrane; not truncate; 2–8 mm long.

Reproductive organization. Plants bisexual, with bisexual spikelets; with hermaphrodite florets.

Inflorescence. *Inflorescence* paniculate; *with capillary branchlets*; espatheate; not comprising 'partial inflorescences' and foliar organs. Spikelet-bearing axes persistent. Spikelets not secund; pedicellate.

Female-fertile spikelets. Spikelets 1.5–5 mm long; compressed laterally; disarticulating above the glumes. Rachilla prolonged beyond the uppermost female-fertile floret; rachilla hairless. The rachilla extension with incomplete florets, or naked. Hairy callus absent. *Callus* short (glabrous).

Glumes two; very unequal; shorter than the spikelets; shorter than the adjacent lemmas (the longer, upper glume less than 2/3 its length); free; not pointed (obtuse to truncate); awnless; *non-carinate*; similar (broad). Lower glume much shorter than half length of lowest lemma (less than 1/3 its length); 0–1 nerved. Upper glume 1–3 nerved. *Spikelets* with female-fertile florets only, or with incomplete florets. The incomplete florets if present, distal to the female-fertile florets. *The distal incomplete florets* merely underdeveloped.

Female-fertile florets (1–)2(–3). *Lemmas* decidedly firmer than the glumes (thinly membranous, with hyaline tips); not becoming indurated; entire (erose); blunt (obtuse to truncate); awnless; hairless; non-carinate; without a germination flap; prominently *3 nerved (these raised)*. *Palea* present; relatively long; awnless, without apical setae; not indurated (scarious); 2-nerved; 2-keeled. *Lodicules* present; 2; free; membranous; glabrous. *Stamens* 3. Anthers 0.7–1.8 mm long; not penicillate. *Ovary* glabrous. Styles free to their bases. Stigmas 2.

Fruit, embryo and seedling. *Fruit* free from both lemma and palea; small; compressed laterally. *Hilum short*. Embryo small; not waisted. Endosperm hard; without lipid. Embryo with an epiblast; without a scutellar tail; with a negligible mesocotyl internode. Embryonic leaf margins meeting.

Seedling with a tight coleoptile. First seedling leaf with a well-developed lamina. The lamina narrow; 3–4 veined.

Abaxial leaf blade epidermis. *Costal/intercostal zonation* conspicuous. *Papillae* present; costal and intercostal. Intercostal papillae over-arching the stomata (at one end); consisting of one oblique swelling per cell. Long-cells differing markedly in wall thickness costally and intercostally (the costals quite thick walled). Mid-intercostal long-cells rectangular and fusiform; having straight or only gently undulating walls. *Microhairs* absent. *Stomata* common; 22.5–24 microns long. Subsidiaries non-papillate; consistently parallel-sided. Guard-cells overlapped by the interstomatals. *Intercostal short-cells* common; not paired (solitary); not silicified. *Costal short-cells* neither distinctly grouped into long rows nor predominantly paired (mostly solitary). Costal silica bodies absent (in the material seen).

Transverse section of leaf blade, physiology. C_3; XyMS+. *Mesophyll* with non-radiate chlorenchyma; without adaxial palisade. *Leaf blade* adaxially flat. *Midrib* conspicuous (via an abaxial keel, and constrictions of the lamina on either side); with one bundle only. Bulliforms not in discrete, regular adaxial groups (apart from midrib hinges). All the vascular bundles accompanied by sclerenchyma (though all but the primaries are depauperate in sclerenchyma). Combined sclerenchyma girders absent (even the major bundles have only small adaxial and abaxial strands). Sclerenchyma all associated with vascular bundles (apart from conspicuous groups at the blade margins).

Cytology. Chromosome base number, $x = 5$. $2n = 10$ and 20. 2 and 4 ploid.

Taxonomy. Pooideae; Poodae; Meliceae.

Ecology, geography, regional floristic distribution. 2 species. North temperate. Helophytic to mesophytic; species of open habitats; glycophytic. In marshes and shallow water.

Holarctic and Antarctic. Boreal, Tethyan, and Madrean. Arctic and Subarctic, Euro-Siberian, Atlantic North American, and Rocky Mountains. Mediterranean and Irano-Turanian. Patagonian. European and Siberian. Canadian-Appalachian.

Rusts and smuts. Rusts — *Puccinia*. Taxonomically wide-ranging species: *Puccinia graminis*, *Puccinia coronata*, *Puccinia striiformis*, and *Puccinia brachypodii*. Smuts from Tilletiaceae and from Ustilaginaceae. Tilletiaceae — *Entyloma*. Ustilaginaceae — *Ustilago*.

References, etc. Leaf anatomical: Metcalfe 1960 and this project.

Catabrosella (Tzvelev) Tzvelev

~ *Colpodium*

Habit, vegetative morphology. Perennial; *caespitose*. Culms 5–15 cm high; herbaceous; tuberous, or not tuberous. Leaves non-auriculate. *Sheath margins joined (?)*. Leaf blades linear; narrow; 0.6–2 mm wide; usually flat; without cross venation; persistent. Ligule an unfringed membrane; 0.8–3.5 mm long.

Reproductive organization. Plants bisexual, with bisexual spikelets; with hermaphrodite florets.

Inflorescence. Inflorescence paniculate; open (lax, 2–5 branches per node); espatheate; not comprising 'partial inflorescences' and foliar organs. Spikelet-bearing axes persistent. Spikelets not secund; pedicellate.

Female-fertile spikelets. Spikelets 1.8–4 mm long; compressed laterally; disarticulating above the glumes. Rachilla prolonged beyond the uppermost female-fertile floret. Hairy callus absent. *Callus* short; blunt.

Glumes two; *very unequal*; shorter than the spikelets; shorter than the adjacent lemmas; awnless; non-carinate (?); similar (lanceolate to ovate). Lower glume 1 nerved. *Upper glume 3 nerved*.

Female-fertile florets (1–)2–5. *Lemmas* erose; not becoming indurated; *awnless*; hairy (in the lower half); *non-carinate*; without a germination flap; 5 nerved. *Palea* present; relatively long; 2-nerved; 2-keeled. *Lodicules* present; 2; free (?); membranous;

glabrous; not or scarcely vascularized. *Stamens* 3. Anthers 1–1.7 mm long. *Ovary* glabrous. Styles free to their bases. Stigmas 2.

Fruit, embryo and seedling. Fruit small. *Hilum short*. Embryo small.

Abaxial leaf blade epidermis. *Costal/intercostal zonation* conspicuous. *Papillae* absent. *Long-cells* similar in shape costally and intercostally; of similar wall thickness costally and intercostally (thin walled). Mid-intercostal long-cells rectangular; having straight or only gently undulating walls. *Microhairs* absent. *Stomata* common. Subsidiaries parallel-sided. Guard-cells overlapped by the interstomatals (very sunken). *Intercostal short-cells* common; not paired (solitary); not silicified. *Costal short-cells* neither distinctly grouped into long rows nor predominantly paired. Costal silica bodies horizontally-elongated smooth, or rounded, or crescentic.

Transverse section of leaf blade, physiology. C_3; XyMS+. Bulliforms present in discrete, regular adaxial groups; in simple fans. Combined sclerenchyma girders present. Sclerenchyma all associated with vascular bundles.

Cytology. Chromosome base number, $x = 5$, 6, and 9. $2n = 10$, 12, and 18. 2 ploid.

Taxonomy. Pooideae; Poodae; Poeae.

Ecology, geography, regional floristic distribution. 9 species. Southwestern to central Asia, Himalaya, western China.

Holarctic. Boreal and Tethyan. Euro-Siberian and Eastern Asian. Irano-Turanian. European and Siberian.

References, etc. Morphological/taxonomic: Bor 1970 (under *Colpodium*). Leaf anatomical: this project.

Catalepis Stapf & Stent

Habit, vegetative morphology. Perennial; caespitose. **Culms** 5–40 cm high; *herbaceous*; to about 0.1 cm in diameter; unbranched above. Culm nodes glabrous. Culm internodes solid. Plants unarmed. Young shoots intravaginal. *Leaves* mostly basal; non-auriculate. **Leaf blades** linear; narrow; to 1 mm wide; becoming *setaceous*; folded, or rolled (rarely flat); without abaxial multicellular glands; without cross venation; persistent. Ligule a fringe of hairs; 0.3 mm long. *Contra-ligule* absent.

Reproductive organization. Plants bisexual, with bisexual spikelets; with hermaphrodite florets.

Inflorescence. *Inflorescence of spicate main branches (spiciform, of numerous short racemes)*; contracted (very much so — the lateral branches short, sometimes reduced to 4 or 5 spikelets); non-digitate; espatheate; not comprising 'partial inflorescences' and foliar organs. Spikelet-bearing axes persistent. Spikelets solitary; not secund; shortly pedicellate.

Female-fertile spikelets. Spikelets 4–5 mm long; compressed laterally; falling with the glumes (seeming to disarticulate at the base of the pedicel). Rachilla prolonged beyond the uppermost female-fertile floret (the prolongation small). The rachilla extension naked. Hairy callus present (the i.e., the 'pedicel').

Glumes two; (the upper) relatively large; very unequal; the upper exceeding the spikelets; (the upper) long relative to the adjacent lemmas; lateral to the rachis; hairless (glabrous, ciliolate on the keel); pointed; awnless; (the upper) carinate; *very dissimilar (the lower reduced to a small subulate scale, the upper lanceolate)*. Lower glume much shorter than half length of lowest lemma; *0 nerved*. Upper glume 1 nerved. *Spikelets* with female-fertile florets only.

Female-fertile florets 1 (lanceolate). *Lemmas* similar in texture to the glumes (thin); not becoming indurated; entire; pointed; awnless; *hairy*. The hairs in tufts (one or two on either flank); not in transverse rows. Lemmas scabrous on the keels; carinate; without a germination flap; 3 nerved; with the nerves non-confluent. *Palea* present (broad); relatively long; apically notched (emarginate); awnless, without apical setae; textured like the lemma (hyaline, glabrous); not indurated; 2-nerved; 2-keeled. *Lodicules* present; 2; free; fleshy; glabrous; not or scarcely vascularized. *Stamens* 3. Anthers 1.5–2 mm long (i.e. relatively long); not penicillate; without an apically prolonged connective. *Ovary* glabrous. Styles fused. Stigmas 2; light brown.

Abaxial leaf blade epidermis. *Costal/intercostal zonation* conspicuous. *Papillae* absent. *Long-cells* markedly different in shape costally and intercostally (the costals much narrower); of similar wall thickness costally and intercostally (fairly thick walled). Mid-intercostal long-cells rectangular; having markedly sinuous walls (and pitted). *Microhairs* present; elongated; clearly two-celled; panicoid-type (very large, the apical cells narrow and often collapsed, but relatively short and round-tipped). Microhair apical cell wall thinner than that of the basal cell but not tending to collapse. Microhair basal cells 33 microns long. Microhair total length/width at septum 5. Microhair apical cell/total length ratio 0.3. *Stomata* common. Subsidiaries low to high dome-shaped and triangular. Guard-cells overlapping to flush with the interstomatals. *Intercostal short-cells* common; in cork/silica-cell pairs; silicified. Intercostal silica bodies present and perfectly developed; tall-and-narrow, or cross-shaped. *Costal short-cells* conspicuously in long rows. Costal silica bodies present in alternate cell files of the costal zones; assorted rounded, saddle shaped, and 'panicoid-type'; when panicoid type, cross shaped, or dumb-bell shaped, or nodular.

Transverse section of leaf blade, physiology. C_4; XyMS+. PCR sheath outlines uneven. PCR sheaths of the primary vascular bundles interrupted; interrupted both abaxially and adaxially. PCR sheath extensions present. Maximum number of extension cells 1. PCR cell chloroplasts centrifugal/peripheral. *Mesophyll* with radiate chlorenchyma. *Leaf blade* with distinct, prominent adaxial ribs to 'nodular' in section; with the ribs more or less constant in size (low, round-topped). *Midrib* conspicuous (somewhat: a larger bundle and keel, with heavier sclerenchyma); with one bundle only. Bulliforms present in discrete, regular adaxial groups (between all the bundles); in simple fans (each group with a large, deeply penetrating median cell). All the vascular bundles accompanied by sclerenchyma. Combined sclerenchyma girders present (in the primaries only, the small bundles with abaxial girders and adaxial strands); forming 'figures' (in the primary bundles). Sclerenchyma all associated with vascular bundles.

Taxonomy. Chloridoideae; main chloridoid assemblage.

Ecology, geography, regional floristic distribution. 1 species. South Africa. Mesophytic (locally abundant in mountain grassland); species of open habitats.

Paleotropical. African. Sudano-Angolan. South Tropical African.

References, etc. Leaf anatomical: this project; photos provided by R.P. Ellis.

Special comments. Fruit data wanting.

Catapodium Link

Scleropoa Griseb., *Synaphe* Dulac
~ *Desmazeria*

Habit, vegetative morphology. *Annual*; caespitose (or the culms solitary). *Culms* 10–50(–60) cm high; herbaceous; unbranched above; 2–7 noded. Culm nodes exposed, or hidden by leaf sheaths; glabrous. Culm internodes hollow. *Leaves* not basally aggregated; non-auriculate. Leaf blades linear; narrow; 1–4.8 mm wide; flat, or folded, or rolled (sometimes involute or convolute when dry); without cross venation; persistent. *Ligule an unfringed membrane*; truncate; 0.5–3 mm long.

Reproductive organization. Plants bisexual, with bisexual spikelets; with hermaphrodite florets; inbreeding. Exposed-cleistogamous, or chasmogamous.

Inflorescence. Inflorescence a single raceme, or paniculate (rigid, spikelike); open, or contracted (the branches with small adaxial pulvini); espatheate; not comprising 'partial inflorescences' and foliar organs. Spikelet-bearing axes persistent. *Spikelets secund (appressed to one side of the axis)*; biseriate; pedicellate (the pedicels short, thick).

Female-fertile spikelets. Spikelets 4–9(–10.2) mm long; oblong, or elliptic; compressed laterally; disarticulating above the glumes; disarticulating between the florets. Rachilla prolonged beyond the uppermost female-fertile floret; rachilla hairless. The rachilla extension with incomplete florets. Hairy callus absent.

Glumes present; two; more or less equal; shorter than the spikelets; shorter than the adjacent lemmas; *lateral to the rachis*; pointed; awnless; carinate; very dissimilar to similar (leathery, the lower lanceolate, the upper ovate). Lower glume 1–3 nerved. Upper glume 3 nerved, or 5 nerved. *Spikelets* with incomplete florets. The incomplete florets distal to the female-fertile florets. The distal incomplete florets 1; *incomplete florets* merely underdeveloped (much reduced); incomplete florets awnless.

Female-fertile florets *(3–)4–11(–12)*. *Lemmas dorsally rounded and glabrous basally, by contrast with* **Desmazeria**; less firm than the glumes to similar in texture to the glumes (membranous or leathery); not becoming indurated; entire; blunt; *awnless*; hairless; glabrous (at least towards the base); non-carinate; without a germination flap; 5 nerved. *Palea* present; relatively long; tightly clasped by the lemma; 2-nerved; 2-keeled. Palea keels wingless. *Lodicules* present; 2; free; membranous; glabrous; toothed, or not toothed. *Stamens* 3. Anthers 0.4–0.9 mm long; not penicillate. **Ovary** glabrous; *without a conspicuous apical appendage*. Styles free to their bases. Stigmas 2; white.

Fruit, embryo and seedling. *Fruit* free from both lemma and palea; small (1.5–2 mm long); ellipsoid; shallowly longitudinally grooved (ventrally); compressed dorsiventrally (ventrally). Hilum short. Embryo small; not waisted. Endosperm hard; with lipid. Embryo with an epiblast; without a scutellar tail; with a negligible mesocotyl internode. Embryonic leaf margins meeting.

Seedling with a tight coleoptile. First seedling leaf with a well-developed lamina. The lamina narrow; 3 veined.

Abaxial leaf blade epidermis. *Costal/intercostal zonation* conspicuous. *Papillae* absent. *Long-cells* similar in shape costally and intercostally; of similar wall thickness costally and intercostally (thin walled). Mid-intercostal long-cells rectangular; having straight or only gently undulating walls. *Microhairs* absent. *Stomata* absent or very rare *(C. loliaceum)*, or common *(C. rigidum)*; in *C. rigidum* 24–30 microns long. Subsidiaries parallel-sided. Guard-cells overlapped by the interstomatals, or overlapping to flush with the interstomatals. *Intercostal short-cells* absent or very rare; when present, in cork/silica-cell pairs; silicified. *Costal short-cells* neither distinctly grouped into long rows nor predominantly paired. Costal silica bodies horizontally-elongated crenate/sinuous, or horizontally-elongated smooth, or rounded.

Transverse section of leaf blade, physiology. C_3; XyMS+. *Mesophyll* with non-radiate chlorenchyma. *Leaf blade* with distinct, prominent adaxial ribs; with the ribs more or less constant in size. *Midrib* conspicuous; with one bundle only. Bulliforms present in discrete, regular adaxial groups (in the furrows); in simple fans. All the vascular bundles accompanied by sclerenchyma. Combined sclerenchyma girders absent. Sclerenchyma all associated with vascular bundles.

Special diagnostic feature. *Lemmas not as in* **Briza** *(q.v.)*.

Cytology. Chromosome base number, $x = 7$. $2n = 14$. 2 ploid. Mean diploid 2c DNA value 9.6 pg.

Taxonomy. Pooideae; Poodae; Poeae.

Ecology, geography, regional floristic distribution. 2 species. Europe, Mediterranean. Commonly adventive. Mesophytic to xerophytic; species of open habitats; halophytic, or glycophytic. In dry microhabitats, including maritime sand.

Holarctic. Boreal and Tethyan. Euro-Siberian. Mediterranean and Irano-Turanian. European.

Rusts and smuts. Rusts — *Puccinia*. Taxonomically wide-ranging species: *Puccinia graminis* and '*Uromyces*' *dactylidis*.

Economic importance. Significant weed species: *C. rigidum*.

References, etc. Leaf anatomical: Metcalfe 1960 and this project.

Cathestechum J. Presl

Excluding *Griffithsochloa*

Habit, vegetative morphology. Annual, or perennial; stoloniferous, or decumbent. *Culms* 10–40 cm high; herbaceous; branched above, or unbranched above. Culm internodes solid. Plants unarmed. Young shoots intravaginal. *Leaves* mostly basal; non-auriculate. Leaf blades linear; narrow; 0.5–2 mm wide; setaceous, or not setaceous; without abaxial multicellular glands; without cross venation; persistent. **Ligule** present; *a fringe of hairs (dense)*.

Reproductive organization. Plants bisexual, with bisexual spikelets; with hermaphrodite florets (the lower floret of the central spikelet commonly hermaphrodite), or without hermaphrodite florets. The spikelets of sexually distinct forms on the same plant; hermaphrodite, male-only, and sterile, or female-only, male-only, and sterile (in triplets, the two lower spikelets usually male or neuter, the central one female-fertile). The male and female-fertile spikelets mixed in the inflorescence. The spikelets usually overtly heteromorphic (the laterals more or less rudimentary).

Inflorescence. *Inflorescence a false spike, with spikelets on contracted axes (a raceme of cuneate 'spikes', each reduced to three spikelets). Inflorescence axes not ending in spikelets (in that the ostensibly terminal triplet is subtended by a more or less bifid bristle of a quarter to half its length).* Inflorescence espatheate; not comprising 'partial inflorescences' and foliar organs. Spikelet-bearing axes disarticulating; falling entire (the triplets fall entire, leaving the short pedicel on the rachis). **Spikelets *subtended by solitary 'bristles' (i.e. the rachis prolongation)*.** The 'bristles' deciduous with the spikelets. *Spikelets* in triplets; pedicellate; *consistently in 'long-and-short' combinations (the lateral spikelets shorter pedicelled, the central longer-pedicelled)*; unequally pedicellate in each combination. The 'shorter' spikelets male-only, or sterile. The 'longer' spikelets hermaphrodite, or female-only.

Female-sterile spikelets. The lower two spikelets of each triplet male or neuter. These with two florets, the lower often having three stamens with apiculate thecae, the upper usually much reduced. The male spikelets 2 floreted. Male florets 3 staminate.

Female-fertile spikelets. Spikelets about 4 mm long; compressed laterally; falling with the glumes (i.e. the triplets falling). Rachilla prolonged beyond the uppermost female-fertile floret; rachilla hairy. The rachilla extension with incomplete florets. Hairy callus present (the axis of the triplet being densely hairy, constituting a callus at the disarticulation).

Glumes two; very unequal; shorter than the spikelets; (the upper) long relative to the adjacent lemmas (approximately equalling them); hairy (G_2 only, the G_1 being glabrous); pointed (G_2 mucronate), or not pointed (G_1); awned (G_2 sometimes awn-tipped), or awnless; carinate (G_2), or non-carinate (G_1); very dissimilar (G_1 short, truncate, flabellate, glabrous; G_2 narrow, awn-tipped or lanceolate, hairy). *Lower glume* to about 0.25 times the length of the upper glume; 0 nerved. Upper glume 1 nerved. *Spikelets* with incomplete florets. The incomplete florets distal to the female-fertile florets (the lowermost of the three florets hermaphrodite or female-only, the upper two male or neuter). The distal incomplete florets 2; incomplete florets awned (but without the multiple awning of *Griffithsochloa*).

Female-fertile florets 1. Lemmas similar in texture to the glumes (thin); not becoming indurated; incised (the apical, sterile lemmas usually more deeply lobed); usually 4 lobed; deeply cleft (usually 4-partite); awned (the nerves usually extending into short awns). Awns 3; median and lateral; the median similar in form to the laterals; from a sinus; non-geniculate; hairless (scabrid); much shorter than the body of the lemma to about as long as the body of the lemma. The lateral awns shorter than the median to about equalling the median. Lemmas hairy (finely pilose on the back); non-carinate; without a germination flap; 3 nerved. *Palea* present; relatively long; apically notched; with apical setae to awned; not indurated (thin); 2-nerved; 2-keeled. *Lodicules* present; 2; free; fleshy; glabrous. *Stamens* 0, or 3. *Ovary* glabrous. Styles free to their bases (the stigmas densely plumose). Stigmas 2.

Fruit, embryo and seedling. Fruit ellipsoid.

Abaxial leaf blade epidermis. *Costal/intercostal zonation* conspicuous. *Papillae* absent. Long-cells of similar wall thickness costally and intercostally. Mid-intercostal long-cells rectangular; having markedly sinuous walls (pitted). *Microhairs* present; elongated; clearly two-celled; chloridoid-type (but thin walled, tending to collapse). Microhair apical cell wall thinner than that of the basal cell but not tending to collapse to of similar thickness/rigidity to that of the basal cell. Microhair basal cells 21–36 microns long. Microhair total length/width at septum 3–5. Microhair apical cell/total length ratio 0.3–0.4. *Stomata* common (in deep grooves). Subsidiaries dome-shaped. Guard-cells overlapping to flush with the interstomatals. *Intercostal short-cells* common; in cork/silica-cell pairs and not paired; silicified. Intercostal silica bodies present and perfectly developed; tall-and-narrow. *Costal short-cells* conspicuously in long rows. Costal silica bodies present in alternate cell files of the costal zones; very consistently saddle shaped.

Transverse section of leaf blade, physiology. C_4; XyMS+ (mestome sheath cells thick walled). PCR sheaths of the primary vascular bundles interrupted; interrupted abaxially only. PCR sheath extensions absent. Mesophyll traversed by columns of colourless mesophyll cells. *Leaf blade* 'nodular' in section to adaxially flat (large, rounded abaxial ribs); with the ribs more or less constant in size. *Midrib* not readily distinguishable; with one bundle only. Bulliforms present in discrete, regular adaxial groups; associated with colourless mesophyll cells to form deeply-penetrating fans. All the vascular bundles accompanied by sclerenchyma. Combined sclerenchyma girders absent (adaxially with strands only, but abaxially with big anchor-shaped girders). Sclerenchyma all associated with vascular bundles.

Cytology. Chromosome base number, $x = 10$. $2n = 60$.

Taxonomy. Chloridoideae; main chloridoid assemblage.

Ecology, geography, regional floristic distribution. 6 species. Southern U.S.A., Mexico. Species of open habitats. Scrub on dry hills.

Holarctic and Neotropical. Madrean. Caribbean.

Rusts and smuts. Rusts — *Puccinia*. Taxonomically wide-ranging species: *Puccinia cacabata* and *Puccinia boutelouae*. Smuts from Tilletiaceae. Tilletiaceae — *Tilletia*.

References, etc. Morphological/taxonomic: Griffiths 1912. Leaf anatomical: this project.

Special comments. Fruit data wanting.

Cenchrus L.

Echinaria Fabric., *Nastus* Lunell, *Raram* Adans.

Habit, vegetative morphology. Annual, or perennial ('sand-burrs'); rhizomatous, or stoloniferous, or caespitose, or decumbent. *Culms* 5–100(–150) cm high; herbaceous; branched above. Culm nodes glabrous. Culm internodes solid, or hollow. *Leaves* not basally aggregated; non-auriculate. Leaf blades linear, or linear-lanceolate; narrow; flat, or folded; without cross venation; persistent; rolled in bud. *Ligule a fringed membrane to a fringe of hairs*.

Reproductive organization. Plants bisexual, with bisexual spikelets; with hermaphrodite florets. The spikelets of sexually distinct forms on the same plant; hermaphrodite, or hermaphrodite and sterile (if the 'bristles' are taken to include vestigial spikelets). Apomictic, or reproducing sexually.

Inflorescence. *Inflorescence a false spike, with spikelets on contracted axes (the spikelets in prickly glomerules (burrs) composed of coalescing spines representing modified branchlets).* Rachides angular or compressed. Inflorescence espatheate; not comprising 'partial inflorescences' and foliar organs. *Spikelet-bearing axes* very much reduced (to burrs); disarticulating (but the main axis persistent); falling entire (i.e., the burrs falling). **Spikelets** *with 'involucres' of 'bristles' (the bristles coalescing, by contrast with* **Pennisetum***).* *The 'bristles'* nearly always *spiny, markedly coalescent basally (not spiny, merely ciliate, in* **C. ciliaris***)*; deciduous with the spikelets. Spikelets not secund.

Female-fertile spikelets. Spikelets lanceolate, or ovate; compressed dorsiventrally; falling with the glumes (i.e., in the burrs). Rachilla terminated by a female-fertile floret. Hairy callus absent.

Glumes present; two; very unequal; shorter than the adjacent lemmas, or long relative to the adjacent lemmas; awnless; very dissimilar, or similar (hyaline or membranous). Lower glume 1–5 nerved. Upper glume 1–7 nerved. *Spikelets* with incomplete florets. The incomplete florets proximal to the female-fertile florets. The proximal incomplete florets paleate, or epaleate (rarely). Palea of the proximal incomplete florets fully developed, or reduced (rarely). The proximal incomplete florets male, or sterile. The proximal lemmas awnless; 1–7 nerved; more or less equalling the female-fertile lemmas; less firm than the female-fertile lemmas; not becoming indurated.

Female-fertile florets 1. Lemmas similar in texture to the glumes to decidedly firmer than the glumes (firmly membranous, dull, papery or leathery); smooth; not becoming indurated; yellow in fruit; entire; awnless; hairless; non-carinate; having the margins lying flat and exposed on the palea; with a clear germination flap; 3–7 nerved. *Palea* present; relatively long; entire; awnless, without apical setae; textured like the lemma; not indurated; 2-nerved. **Lodicules *absent*.** *Stamens* 3. Anthers not penicillate. *Ovary* glabrous. Styles fused, or free to their bases. Stigmas 2.

Fruit, embryo and seedling. Fruit small; compressed dorsiventrally. Hilum short. Embryo large; waisted. Endosperm hard; without lipid; containing only simple starch grains. Embryo without an epiblast; with a scutellar tail; with an elongated mesocotyl internode. Embryonic leaf margins overlapping.

Seedling with a short mesocotyl. First seedling leaf with a well-developed lamina. The lamina broad; curved; 21–30 veined.

Abaxial leaf blade epidermis. *Costal/intercostal zonation* conspicuous. *Papillae* absent. *Long-cells* similar in shape costally and intercostally. Mid-intercostal long-cells rectangular; having markedly sinuous walls, or having straight or only gently undulating walls. *Microhairs* present; panicoid-type; (45–)54–84(–114) microns long; 6–11.5(–12.6) microns wide at the septum. Microhair total length/width at septum 6–10(–11.6). Microhair apical cells (27–)30–60(–72) microns long. Microhair apical cell/total length ratio 0.5–0.69. *Stomata* common; 30–45 microns long, or 36–52 microns long (*C. echinatus*). Subsidiaries markedly triangular. Guard-cells overlapping to flush with the interstomatals. *Intercostal short-cells* common, or absent or very rare. *Costal short-cells* conspicuously in long rows. Costal silica bodies 'panicoid-type'; cross shaped, or butterfly shaped, or dumb-bell shaped.

Transverse section of leaf blade, physiology. C_4; biochemical type NADP–ME (*C. pauciflorus*, *C. incertus*); XyMS–. PCR sheath outlines uneven. PCR sheath extensions absent. PCR cells with a suberised lamella. PCR cell chloroplasts with reduced grana; centrifugal/peripheral. *Mesophyll* with radiate chlorenchyma. *Midrib* conspicuous; having a conventional arc of bundles; with colourless mesophyll adaxially, or without colourless mesophyll adaxially (rarely). Bulliforms present in discrete, regular adaxial groups; in simple fans, or in simple fans and associated with colourless mesophyll cells to form deeply-penetrating fans. Many of the smallest vascular bundles unaccompanied by sclerenchyma. Combined sclerenchyma girders present, or absent; nowhere forming 'figures'. Sclerenchyma all associated with vascular bundles.

Culm anatomy. *Culm internode bundles* scattered.

Phytochemistry. Leaves without flavonoid sulphates (1 species).

Cytology. Chromosome base number, $x = 9$ and 12. $2n = 34, 35, 36, 40, 44, 45$, and 68. Mean diploid 2c DNA value 2.6 pg (?—ploidy of the one species studied unknown).

Taxonomy. Panicoideae; Panicodae; Paniceae.

Ecology, geography, regional floristic distribution. 22 species. Tropical and warm temperate. Commonly adventive. Mesophytic to xerophytic; shade species, or species of open habitats; halophytic, or glycophytic. Grassland, bush, sandy and weedy places.

Holarctic, Paleotropical, Neotropical, Australian, and Antarctic. Boreal, Tethyan, and Madrean. African, Madagascan, Indomalesian, Polynesian, and Neocaledonian. Atlantic North American and Rocky Mountains. Macaronesian, Mediterranean, and Irano-Turanian. Saharo-Sindian, Sudano-Angolan, West African Rainforest, and Namib-Karoo. Indian, Indo-Chinese, Malesian, and Papuan. Hawaiian and Fijian. Caribbean, Venezuela and Surinam, Amazon, Central Brazilian, Pampas, and Andean. North and East Australian. New Zealand. Canadian-Appalachian, Southern Atlantic North American, and Central Grasslands. Sahelo-Sudanian, Somalo-Ethiopian, South Tropical African, and Kalaharian. Tropical North and East Australian.

Rusts and smuts. Rusts — *Puccinia*. Smuts from Ustilaginaceae. Ustilaginaceae — *Sorosporium*, *Sphacelotheca*, *Tolyposporium*, and *Ustilago*.

Economic importance. Significant weed species: *C. biflorus*, *C. brownii*, *C. ciliaris*, *C. echinatus*, *C. incertus*, *C. longispinus*, *C. myosuroides*, *C. pauciflorus*, *C. tribuloides*. Cultivated fodder: *C. ciliaris* (drought resistant, tolerant of heavy grazing), *C. setiger*. Important native pasture species: *C. biflorus*, *C. ciliaris*, *C. pennisetiformis*, *C. setigerus*.

References, etc. Leaf anatomical: Metcalfe 1960 and this project.

Special comments. Some species overlapping with *Pennisetum*.

Centotheca Desv.

Centosteca Desv., *Ramosia* Merr.

Habit, vegetative morphology. Perennial; rhizomatous, or caespitose. The flowering culms leafy. **Culms** 25–125 cm high; *woody and persistent (at base)*. Culm nodes glabrous. Culm internodes solid. Plants unarmed. *Leaves* not basally aggregated; auriculate, or non-auriculate; without auricular setae. Leaf blades broadly linear to lanceolate; broad; pseudopetiolate to not pseudopetiolate; cross veined, or without cross venation (rarely); disarticulating from the sheaths, or persistent. *Ligule an unfringed membrane.*

Reproductive organization. Plants bisexual, with bisexual spikelets; with hermaphrodite florets. Viviparous, or not viviparous.

Inflorescence. Inflorescence paniculate, or of spicate main branches; open; espatheate; not comprising 'partial inflorescences' and foliar organs. Spikelet-bearing axes persistent. Spikelets not secund; pedicellate.

Female-fertile spikelets. Spikelets compressed laterally; disarticulating above the glumes (usually), or falling with the glumes; usually but tardily disarticulating between the florets; with a distinctly elongated rachilla internode between the glumes. Rachilla prolonged beyond the uppermost female-fertile floret; rachilla hairy. The rachilla extension with incomplete florets (usually), or naked. Hairy callus absent.

Glumes two; very unequal; shorter than the adjacent lemmas; free (widely separated); pointed; awnless; similar (membranous to herbaceous). Lower glume 3–5 nerved. Upper glume 3–5 nerved. *Spikelets* with female-fertile florets only, or with incomplete florets (usually with a rudimentary floret, and the upper 'fertile' florets sometimes female-only). The incomplete florets distal to the female-fertile florets. The distal incomplete florets usually 1. *Spikelets without proximal incomplete florets.*

Female-fertile florets (1–)2–3 (*C. uniflora* being 1-flowered). Lemmas not becoming indurated; entire to incised (obtuse to emarginate); mucronate; hairy (the upper lemmas with reflexed hairs); carinate; *9 nerved. Palea* present; awnless, without apical setae; 2-nerved. *Lodicules* present, or absent; when present, 2. *No third lodicule*. Lodicules free; fleshy; glabrous. *Stamens* 2–3. Anthers not penicillate. **Ovary** glabrous; *without a conspicuous apical appendage*. Styles free to their bases. Stigmas 2.

Fruit, embryo and seedling. *Fruit* free from both lemma and palea; small; compressed laterally. *Hilum short*. Embryo small. Endosperm containing only simple starch grains. Embryo with an epiblast; with a scutellar tail; with an elongated mesocotyl internode. Embryonic leaf margins meeting, or overlapping.

Seedling with a short mesocotyl. First seedling leaf with a well-developed lamina. The lamina broad; supine; 6–12 veined.

Abaxial leaf blade epidermis. *Papillae* absent. Mid-intercostal long-cells having markedly sinuous walls (thin walled). *Microhairs* present; panicoid-type; 31–36(–44) microns long. Microhair apical cells 18–22 microns long. Microhair apical cell/total length ratio 0.6. *Stomata* common; 30–36 microns long. Subsidiaries triangular. Guard-cells overlapping to flush with the interstomatals. *Intercostal short-cells* common to absent or very rare; when present, in cork/silica-cell pairs; silicified. Intercostal silica bodies when present, cross-shaped. *Costal short-cells* conspicuously in long rows. Costal silica bodies 'panicoid-type'; cross shaped, or butterfly shaped, or dumb-bell shaped.

Transverse section of leaf blade, physiology. C$_3$; XyMS+. Mesophyll with adaxial palisade; without arm cells; with fusoids, or without fusoids. The fusoids an integral part of the PBS. *Leaf blade* 'nodular' in section (rarely), or adaxially flat. *Midrib* conspicuous; with one bundle only. Bulliforms present in discrete, regular adaxial groups; in simple fans (the groups large and wide, and sometimes some irregular groups). All the vascular bundles accompanied by sclerenchyma. Combined sclerenchyma girders present; forming 'figures'. Sclerenchyma all associated with vascular bundles.

Cytology. Chromosome base number, $x = 12$. $2n = 24$. 2 ploid.

Taxonomy. Bambusoideae; Oryzodae; Centotheceae.

Ecology, geography, regional floristic distribution. 4 species. Tropical Africa, Asia, Polynesia. Shade species; glycophytic. Forests.

Paleotropical and Australian. African, Indomalesian, Polynesian, and Neocaledonian. Sudano-Angolan and West African Rainforest. Indian, Indo-Chinese, Malesian, and Papuan. Fijian. North and East Australian. Sahelo-Sudanian. Tropical North and East Australian.

Rusts and smuts. Rusts — *Puccinia*.

Economic importance. Significant weed species: *C. lappacea* (in North America).

References, etc. Leaf anatomical: Metcalfe 1960; this project.

Centrochloa Swallen

Habit, vegetative morphology. Annual; caespitose. *Culms* about 20–75 cm high; herbaceous; unbranched above. Culm nodes glabrous. *Leaves* mostly basal; non-auriculate. Leaf blades narrow; without cross venation; persistent. Ligule a fringed membrane.

Reproductive organization. Plants bisexual, with bisexual spikelets; with hermaphrodite florets.

Inflorescence. Inflorescence of spicate main branches (the rachides filiform); digitate to subdigitate. Rachides hollowed. Inflorescence espatheate; not comprising 'partial inflorescences' and foliar organs. Spikelet-bearing axes with very slender rachides; persistent. Spikelets solitary; secund.

Female-fertile spikelets. Spikelets 3 mm long; turbinate; adaxial; compressed dorsiventrally; falling with the glumes. Rachilla terminated by a female-fertile floret. *Hairy callus present (as a downward-pointing hairy projection, perhaps representing the lower glume, at the base of the spikelet — cf.* **Digitariopsis***)*.

Glumes one per spikelet (the upper); long relative to the adjacent lemmas; dorsiventral to the rachis; *hairy (densely straight-hairy, with dark brown hairs in 4 rows between and outside the veins, and with the rows terminating apically in tufts)*; pointed; awnless; non-carinate. Upper glume 3 nerved. *Spikelets* with incomplete florets. The incomplete florets proximal to the female-fertile florets. *The proximal incomplete florets* 1; sterile. The proximal lemmas awnless; 2–3 nerved; more or less equalling the female-fertile lemmas; less firm than the female-fertile lemmas; not becoming indurated.

Female-fertile florets 1. Lemmas decidedly firmer than the glumes (cartilaginous); not becoming indurated; entire; pointed; awnless; hairless; non-carinate; having the margins lying flat and exposed on the palea; with a clear germination flap; 5 nerved. *Palea* present; relatively long; 2-nerved; keel-less. *Lodicules* present; 2; free; fleshy; glabrous; not or scarcely vascularized. *Stamens* 3. *Ovary* glabrous. Stigmas 2.

Fruit, embryo and seedling. Fruit small; compressed dorsiventrally. Hilum short. Embryo large.

Abaxial leaf blade epidermis. *Costal/intercostal zonation* conspicuous. *Papillae* absent. *Long-cells* markedly different in shape costally and intercostally (the intercostals short and broad, the costals long and narrow); of similar wall thickness costally and intercostally (fairly thin walled). Mid-intercostal long-cells rectangular; having markedly sinuous walls. *Microhairs* present; panicoid-type. *Stomata* common. Subsidiaries triangular. Guard-cells overlapping to flush with the interstomatals. *Intercostal short-cells* absent or very rare. *Costal short-cells* conspicuously in long rows. Costal silica bodies 'panicoid-type'.

Transverse section of leaf blade, physiology. C_4; XyMS–. *Mesophyll* with radiate chlorenchyma. *Leaf blade* adaxially flat. *Midrib* conspicuous; having a conventional arc of bundles; with colourless mesophyll adaxially. Bulliforms not in discrete, regular adaxial groups (constituting most of the epidermis). Many of the smallest vascular bundles unaccompanied by sclerenchyma. Combined sclerenchyma girders present; forming 'figures'. Sclerenchyma all associated with vascular bundles.

Special diagnostic feature. *Upper glume extended downwards into a conspicuous spur*.

Taxonomy. Panicoideae; Panicodae; Paniceae.

Ecology, geography, regional floristic distribution. 1 species. Brazil. Species of open habitats. Savanna.

Neotropical. Central Brazilian.
References, etc. Leaf anatomical: this project.

Centropodia Reichenb.

Asthenatherum Nevski

Habit, vegetative morphology. Annual, or perennial (with glaucous stems and leaves); caespitose to decumbent. **Culms** 3–150 cm high; herbaceous; *unbranched above (but often branched near the base).* Culm nodes hairy, or glabrous. Culm internodes solid. Plants unarmed. Young shoots intravaginal. *Leaves* not basally aggregated; non-auriculate. *Leaf blades linear-lanceolate*; narrow; 1–5 mm wide; flat, or rolled (convolute, stiff, pungent); without cross venation; persistent. *Ligule a fringe of hairs*; about 1.5 mm long. *Contra-ligule* absent.

Reproductive organization. Plants bisexual, with bisexual spikelets; with hermaphrodite florets.

Inflorescence. *Inflorescence paniculate*; contracted; spicate; enclosed by spathe-like upper leaf sheaths; not comprising 'partial inflorescences' and foliar organs. Spikelet-bearing axes persistent. Spikelets not secund; pedicellate.

Female-fertile spikelets. *Spikelets* 7–10 mm long; *compressed laterally*; disarticulating above the glumes; disarticulating between the florets; with a distinctly elongated rachilla internode between the glumes. Rachilla prolonged beyond the uppermost female-fertile floret; rachilla hairless. Hairy callus present. *Callus* long; pungently pointed.

Glumes two; more or less equal; exceeding the spikelets; *long relative to the adjacent lemmas (enclosing the spikelet)*; hairless; pointed; awnless; carinate, or non-carinate; similar (papery to leathery). Lower glume much exceeding the lowest lemma; 5–11 nerved. Upper glume 5–11 nerved. *Spikelets* with female-fertile florets only, or with incomplete florets. The incomplete florets distal to the female-fertile florets. *The distal incomplete florets* merely underdeveloped; awned.

Female-fertile florets 2–5. Lemmas similar in texture to the glumes (papery); not becoming indurated; incised; 2 lobed; deeply cleft; awned. Awns 1, or 3; median, or median and lateral (by small straight extensions from the lobes); the median different in form from the laterals (when laterals present); from a sinus; geniculate; much shorter than the body of the lemma to about as long as the body of the lemma; entered by one vein. The lateral awns shorter than the median (and straight). Lemmas hairy. *The hairs* in tufts (6 to 8 bristle-tipped tufts); *in transverse rows (with a transverse row of tufts level with the base of the awn, as well as longitudinal rows of hairs)*. Lemmas non-carinate; without a germination flap; 7–11 nerved; with the nerves non-confluent. *Palea* present; relatively long (almost equalling the lemma); entire to apically notched; awnless, without apical setae; somewhat thinner than the lemma; not indurated; 2-nerved; 2-keeled. Palea keels winged (below); minutely scaberulous. *Lodicules* present; 2; free; fleshy; glabrous. *Stamens* 3. Anthers 0.7–1.2 mm long; not penicillate; without an apically prolonged connective. *Ovary* glabrous. Styles free to their bases. Stigmas 2; brown.

Fruit, embryo and seedling. *Fruit* free from both lemma and palea; longitudinally grooved; compressed dorsiventrally. *Hilum short.* Embryo large.

Abaxial leaf blade epidermis. *Costal/intercostal zonation* conspicuous. *Papillae* absent (but mimicked by prickle bases, especially in the costal zones). Mid-intercostal long-cells rectangular; having markedly sinuous walls, or having straight or only gently undulating walls. *Microhairs* present (but very hard to find — obscured by macrohairs and seemingly very rare), or absent; when seen panicoid-type to chloridoid-type (with short, rounded but thin-walled apical cells); in *C. forskalei* 36 microns long; 21 microns wide at the septum. Microhair total length/width at septum 1.7. Microhair apical cells 12 microns long. Microhair apical cell/total length ratio 0.33. *Stomata* common; about 27 microns long. Subsidiaries dome-shaped, or dome-shaped and triangular. *Intercostal short-cells* absent or very rare. *Costal short-cells* conspicuously in long rows, or neither distinctly grouped into long rows nor predominantly paired. Costal silica bodies tall-and-narrow, or 'panicoid-type'.

Transverse section of leaf blade, physiology. C_4; biochemical type biochemical type and ultrastructure need investigating, in view of the peculiarity of other C_4 arundinoids and especially in view of the variation in PCR sheath form; XyMS+. PCR sheath outlines uneven (e.g. *C. mossamedense*), or even (e.g. *C. glaucum*). PCR sheath extensions present (eg *C. mossamedense*), or absent. Maximum number of extension cells 2. PCR cell chloroplasts centripetal. *Mesophyll* with radiate chlorenchyma. *Leaf blade* 'nodular' in section; with the ribs more or less constant in size. *Midrib* not readily distinguishable; with one bundle only. Bulliforms present in discrete, regular adaxial groups; in simple fans (each group with a large, deeply penetrating median cell, cf. many chloridoids). All the vascular bundles accompanied by sclerenchyma. Combined sclerenchyma girders absent. Sclerenchyma all associated with vascular bundles.

Cytology. Chromosome base number, $x = 12$. $2n = 24$. 2 ploid.

Taxonomy. Arundinoideae; Danthonieae.

Ecology, geography, regional floristic distribution. 4 species. North Africa, South and South West Africa and Middle East. Xerophytic; species of open habitats.

Holarctic, Paleotropical, and Cape. Tethyan. African and Indomalesian. Irano-Turanian. Saharo-Sindian, Sudano-Angolan, and Namib-Karoo. Indian. Sahelo-Sudanian and South Tropical African.

Rusts and smuts. Rusts — *Puccinia*.

Economic importance. Important native pasture species: *C. forsskalii*.

References, etc. Morphological/taxonomic: Conert 1962. Leaf anatomical: this project; photographs of *Asthenatherum (=Centropodia) glaucum* and *A. mossamedense* provided by R. P. Ellis; Ellis 1984a.

Cephalostachyum Munro

~ Schizostachyum

Habit, vegetative morphology. Perennial. The flowering culms leafy. *Culms* 300–1300 cm high; woody and persistent; to 7.5 cm in diameter; scandent, or not scandent; branched above. Culm internodes hollow. Plants unarmed. Young shoots extravaginal. *Leaves* not basally aggregated; with auricular setae, or without auricular setae. Leaf blades linear-lanceolate to ovate-lanceolate (acuminate); broad; 15–100 mm wide (and 2.5–20 cm long); not cordate, not sagittate (base rounded or cuneate); flat; pseudopetiolate; without cross venation (but sometimes abaxially with pellucid dots simulating veinlets); disarticulating from the sheaths; rolled in bud. *Ligule* present (long).

Reproductive organization. Plants bisexual, with bisexual spikelets; with hermaphrodite florets. *The spikelets of sexually distinct forms on the same plant*; hermaphrodite, male-only, and sterile (perfect and variously imperfect or comprising empty glumes); overtly heteromorphic.

Inflorescence. Inflorescence indeterminate; *with pseudospikelets*; a terminal, globose head or an interrupted spike of heads; non-digitate; spatheate (the heads bracteate); a complex of 'partial inflorescences' and intervening foliar organs. *Spikelet-bearing axes* capitate; persistent. Spikelets not secund.

Female-fertile spikelets. Spikelets 12–25 mm long; compressed laterally to not noticeably compressed; disarticulating above the glumes; with conventional internode spacings, or with a distinctly elongated rachilla internode above the glumes. Rachilla conspicuously prolonged beyond the uppermost female-fertile floret. The rachilla extension with incomplete florets, or naked (terminated by a tiny aborted floret, in the material seen).

Glumes two to several; shorter than the adjacent lemmas; free; pointed; usually *awned (terminally)*; non-carinate; similar (broad, chaffy, many-nerved). Lower glume 11–13 nerved. Upper glume 11–15 nerved. *Spikelets* with female-fertile florets only, or with incomplete florets. The incomplete florets when detectable, distal to the female-fertile florets. The distal incomplete florets when detected, 1; *incomplete florets* merely underdeveloped (the rachilla prolongation terminated by a tiny aborted floret, in the material seen).

Female-fertile florets 1. Lemmas ovate-lanceolate; less firm than the glumes to similar in texture to the glumes; not becoming indurated; entire; pointed; awnless (e.g. *C. chapeleri*), or awned. Awns when present, 1; median; apical; non-geniculate; hairless; much shorter than the body of the lemma (and shorter than glume awns); entered by several veins. Lemmas hairy; non-carinate; without a germination flap; 14–18 nerved (in the species seen); with the nerves non-confluent. *Palea* present; relatively long; convolute; apically notched; bimucronate or awned; textured like the lemma; not indurated; several nerved (about 9–13); 2-keeled (sulcate between the close keels). Palea back hairy. Palea keels wingless; hairy. *Lodicules* present; 3; free; membranous; ciliate (or papillose); not toothed; heavily vascularized. *Stamens* 6. Anthers 4.5–5 mm long (in the material seen); penicillate, or not penicillate; with the connective apically prolonged, or without an apically prolonged connective. *Ovary* glabrous; with a conspicuous apical appendage. The appendage long, stiff and tapering. Styles fused; completely fused (the ovary apex attenuate into one long style). Stigmas 2, or 3 (short).

Fruit, embryo and seedling. *Fruit* free from both lemma and palea; longitudinally grooved, or not grooved. Pericarp thick and hard (crustaceous); free. Endosperm containing compound starch grains.

Abaxial leaf blade epidermis. *Costal/intercostal zonation* conspicuous. *Papillae* present. Intercostal papillae over-arching the stomata; several per cell. Intercostal zones exhibiting many atypical long-cells. Mid-intercostal long-cells having markedly sinuous walls. *Microhairs* present; panicoid-type (the cells of equal length); 64–69 microns long; 5–9 microns wide at the septum. Microhair total length/width at septum 6.1–11.9. Microhair apical cells 25.5–33 microns long. Microhair apical cell/total length ratio 0.39–0.5. *Stomata* common (obscured by the papillae); 15–21 microns long. *Intercostal short-cells* common. Costal silica bodies poorly developed (in the material seen); oryzoid (?).

Transverse section of leaf blade, physiology. C_3; XyMS+. Mesophyll with arm cells; with fusoids. *Leaf blade* adaxially flat (except on one side of midrib). *Midrib* conspicuous; having complex vascularization. *The lamina* distinctly asymmetrical on either side of the midrib. Bulliforms present in discrete, regular adaxial groups; in simple fans. All the vascular bundles accompanied by sclerenchyma. Combined sclerenchyma girders present; forming 'figures'.

Cytology. Chromosome base number, $x = 12$. $2n = 72$. 6 ploid.

Taxonomy. Bambusoideae; Bambusodae; Bambuseae.

Ecology, geography, regional floristic distribution. 12 species. Indomalaya, Madagascar.

Paleotropical. Madagascan and Indomalesian. Indian, Indo-Chinese, and Malesian.

References, etc. Leaf anatomical: Metcalfe 1960; this project.

Chaboissaea Fourn.

~ *Muhlenbergia* (*M. atacamensis* Parodi, *M. decumbens* Swallen, *M. ligulata* Fourn., *M. subbiflora* Hitchc.)

Habit, vegetative morphology. Annual, or perennial; caespitose, or decumbent. Culms (3–)5–70(–90) cm high; **herbaceous**; unbranched above (sometimes freely branched below). Culm nodes exposed; glabrous. Culm internodes hollow. *Leaves* not basally aggregated; non-auriculate; without auricular setae. Sheaths glabrous, usually shorter than the internodes, sometimes keeled. Leaf blades linear; narrow; 0.7–2.8(–3) mm wide; flat, or rolled (to loosely involute); without cross venation; persistent. *Ligule an unfringed membrane*; somewhat truncate, or not truncate; 1.5–3.2 mm long, or 6–10 mm long (*C. ligulata*). *Contra-ligule* absent.

Reproductive organization. Plants bisexual, with bisexual spikelets; with hermaphrodite florets. The spikelets hermaphrodite.

Inflorescence. *Inflorescence paniculate (terminal, narrow, the branches distant, alternate, subdivided, strongly appressed)*; open to contracted. Inflorescence with axes ending in spikelets. Inflorescence espatheate; not comprising 'partial inflorescences' and foliar organs. Spikelet-bearing axes persistent. Spikelets pedicellate (the pedicels 0.2–3 mm long); imbricate.

Female-fertile spikelets. Spikelets about 2.5–3.6 mm long; dark grey or lead coloured to greyish yellow; compressed laterally; disarticulating above the glumes; disarticulating between the florets (?). *Rachilla prolonged beyond the uppermost female-fertile floret*. The rachilla extension with incomplete florets. Hairy callus absent.

Glumes two; very unequal to more or less equal (to 'subequal'); shorter than the spikelets; shorter than the adjacent lemmas; hairless; glabrous (scabrous along the midnerve); pointed (acute or acuminate); awned, or awnless (often one or both awn-tipped); similar. Lower glume 1 nerved. Upper glume 1 nerved. *Spikelets* with incomplete florets. The incomplete florets distal to the female-fertile florets. The distal incomplete florets 1–2; *incomplete florets* merely underdeveloped.

Female-fertile florets *1*. Lemmas chartaceous; not becoming indurated; entire, or incised; pointed (acute to acuminate); sometimes minutely 2 lobed; not deeply cleft; awnless to awned. Awns 1; median; from a sinus, or dorsal, or apical (seemingly, from illustrations in Peterson and Annable 1992); if dorsally awned, from near the top; non-geniculate; straight, or flexuous; hairless; much shorter than the body of the lemma to much longer than the body of the lemma; entered by one vein. Lemmas hairy (with miute appressed hairs below, along the midnerve and margins); more or less *carinate (somewhat compressed-keeled, at least above)*; without a germination flap; 3 nerved. *Palea* present; relatively long; apically notched; awnless, without apical setae, or awned (the nerves often produced into two short awns); 2-nerved; 2-keeled. Palea back glabrous. Palea keels wingless. *Lodicules* present; 2; free; fleshy; glabrous. *Stamens* 3. Anthers 0.9–2 mm long; not penicillate; without an apically prolonged connective. *Ovary* glabrous. Styles free to their bases. Stigmas 2; 'dark gray'.

Fruit, embryo and seedling. Fruit small (1–2.5 mm long); brownish; fusiform. Hilum short (?). Embryo large. Seed endospermic. Endosperm hard. Embryo with an epiblast; with a scutellar tail; with an elongated mesocotyl internode. Embryonic leaf margins meeting.

Abaxial leaf blade epidermis. *Costal/intercostal zonation* conspicuous. *Papillae* present; intercostal. Intercostal papillae consisting of one symmetrical projection per cell. Mid-intercostal long-cells having markedly sinuous walls. *Microhairs* present; elongated; clearly two-celled; chloridoid-type. Microhair apical cell wall of similar thickness/rigidity to that of the basal cell. *Stomata* common. Subsidiaries dome-shaped. Intercostal short-cells not paired. Macrohairs absent, and prickles restricted to the blade margins. Costal silica bodies present and well developed; present in alternate cell files of the costal zones; saddle shaped.

Transverse section of leaf blade, physiology. C_4; XyMS+. PCR sheath outlines uneven. PCR sheaths of the primary vascular bundles interrupted; interrupted abaxially only, or interrupted both abaxially and adaxially. PCR sheath extensions absent. PCR cell chloroplasts centripetal. Mesophyll traversed by columns of colourless mesophyll cells. *Leaf blade* with distinct, prominent adaxial ribs, or 'nodular' in section. Midrib with one bundle only. Bulliforms present in discrete, regular adaxial groups; associated with colourless mesophyll cells to form deeply-penetrating fans (linked with the traversing columns of colourless cells). Combined sclerenchyma girders present (with primary bundles). Sclerenchyma all associated with vascular bundles.

Cytology. Chromosome base number, $x = 8$. $2n = 16$ and 32, or 14, 16, and 18 (*C. subbflora*). 2 and 4 ploid (*C. decumbens* being tetraploid).

Taxonomy. Chloridoideae; main chloridoid assemblage.

Ecology, geography, regional floristic distribution. 4 species. Mexico, Argentina. Helophytic to mesophytic; species of open habitats; glycophytic. Marshy meadows.

Neotropical. Caribbean.

References, etc. Morphological/taxonomic: Peterson, P.M. and Annable, C.R. (1992). A revision of *Chaboissaea* (Poaceae: Eragrostideae). *Madroño* **39**, 8–30 (from which the above is taken in its entirety). Leaf anatomical: exclusively from description and illustrations in Peterson and Annable 1992.

Chaetium Nees

Berchtoldia Presl

Habit, vegetative morphology. Perennial; caespitose. *Culms* herbaceous; *unbranched above*. Culm nodes glabrous. Culm internodes solid. *Leaves* mostly basal; non-auriculate. Leaf blades narrow; without cross venation; persistent.

Reproductive organization. Plants bisexual, with bisexual spikelets; with hermaphrodite florets. *The spikelets all alike in sexuality*.

Inflorescence. *Inflorescence a single raceme, or paniculate (of loose racemes from the main axis)*; espatheate; not comprising 'partial inflorescences' and foliar organs. Spikelet-bearing axes persistent. *Spikelets* solitary; not secund; pedicellate; *not in distinct 'long-and-short' combinations*.

Female-fertile spikelets. *Spikelets compressed dorsiventrally*; falling with the glumes. Rachilla terminated by a female-fertile floret. Hairy callus present (the spikelet tapering into a slender, bearded stipe).

Glumes two; more or less equal; long relative to the adjacent lemmas; hairless; scabrous; *awned (attenuate into very long, sinuous, scabrid awns)*; non-carinate; similar, or very dissimilar (the lower reduced to an awn in *C. bromoides*). Lower glume 3 nerved. *Upper glume 9–11 nerved. Spikelets* with incomplete florets. The incomplete florets proximal to the female-fertile florets. *Spikelets with proximal incomplete florets. The proximal incomplete florets* 1; sterile. The proximal lemmas acuminate to awned; awned to awnless; 7 nerved; more or less equalling the female-fertile lemmas; similar in texture to the female-fertile lemmas; not becoming indurated.

Female-fertile florets 1. Lemmas acuminate-apiculate to short awned; similar in texture to the glumes to decidedly firmer than the glumes; not becoming indurated; entire; awnless to mucronate, or awned (shortly); hairless (scabrid above); non-carinate; having the margins lying flat and exposed on the palea; with a clear germination flap; 5 nerved. *Palea* present; relatively long; apically notched (denticulate); 2-nerved; keel-less. *Lodicules* present; 2; free; fleshy; glabrous; not or scarcely vascularized. *Stamens* 3. Anthers not penicillate; without an apically prolonged connective. *Ovary* glabrous. Styles fused. Stigmas 2; brown.

Fruit, embryo and seedling. Embryo large.

Abaxial leaf blade epidermis. *Costal/intercostal zonation* conspicuous. *Papillae* absent. Mid-intercostal long-cells rectangular; having markedly sinuous walls. *Microhairs* present; panicoid-type; (42–)48–54(–60) microns long; (21–)28.5–31.2(–34.5) microns wide at the septum. Microhair total length/width at septum 3–7.1. Microhair apical cells 8.4–10.2(–14.1) microns long. Microhair apical cell/total length ratio 0.5–0.65. *Stomata* common; (36–)39–47(–51) microns long. Subsidiaries parallel-sided to triangular; including both triangular and parallel-sided forms on the same leaf. Guard-cells overlapping to flush with the interstomatals. *Intercostal short-cells* common; in cork/silica-cell pairs; silicified. *Costal short-cells* conspicuously in long rows (some veins), or predominantly paired (other veins). Costal silica bodies 'panicoid-type'.

Transverse section of leaf blade, physiology. C_4; XyMS+ (*C. bromoides*), or XyMS– (*C. festucoides, C. cubanum* — which Renvoize describes as 'anatomically intermediate C_3/C_4': see below). *Mesophyll* with radiate chlorenchyma, or with non-radiate chlorenchyma. *Leaf blade* 'nodular' in section; with the ribs more or less constant in size. *Midrib* conspicuous; having a conventional arc of bundles (one large, two small); with colourless mesophyll adaxially. Bulliforms present in discrete, regular adaxial groups; in simple fans. Many of the smallest vascular bundles unaccompanied by sclerenchyma. Combined sclerenchyma girders present; forming 'figures' (midribs only). Sclerenchyma all associated with vascular bundles.

Cytology. Chromosome base number, x = seemingly 13. $2n$ = 26. 2 ploid.

Taxonomy. Panicoideae; Panicodae; Paniceae.

Ecology, geography, regional floristic distribution. 3 species. Tropical America, Cuba. Species of open habitats. Grassland.

Neotropical. Caribbean and Central Brazilian.

Rusts and smuts. Rusts — *Puccinia*.

References, etc. Leaf anatomical: this project.

Special comments. Fruit data wanting.

Chaetobromus Nees

Habit, vegetative morphology. Perennial; rhizomatous (sometimes), or caespitose, or decumbent. **Culms** 15–75 cm high; *herbaceous*; to to 0.3 cm in diameter; branched above (but not profusely). Culm nodes black, glabrous. Culm internodes hollow. Plants unarmed. Young shoots intravaginal. *Leaves* not basally aggregated; non-auriculate (but hairy at the auricle positions). Leaf blades linear to linear-lanceolate; narrow; 1–5 mm wide; flat, or folded; without cross venation; persistent. *Ligule a fringe of hairs*.

Reproductive organization. *Plants bisexual, with bisexual spikelets*; with hermaphrodite florets.

Inflorescence. Inflorescence paniculate (rarely racemose, in depauperate plants); open, or contracted (sometimes with few spikelets); with capillary branchlets, or without capillary branchlets. Inflorescence with axes ending in spikelets. Inflorescence espatheate. Spikelet-bearing axes persistent. *Spikelets* solitary; not secund; *pedicellate (the pedicels articulated)*.

Female-fertile spikelets. *Spikelets* 10–17 mm long; compressed laterally; *falling with the glumes (the hairs on the persistent pedicel allowing the spikelet to move in only one direction)*; ultimately disarticulating between the florets (i.e., disarticulation occurring in the pedicel, above the glumes, and between the florets). Rachilla prolonged beyond the uppermost female-fertile floret; rachilla hairy. The rachilla extension with incomplete florets. *Hairy callus present (comprising the bearded upper part of pedicel, and the rachilla below the glumes)*. *Callus* long.

Glumes two; more or less equal; about equalling the spikelets, or exceeding the spikelets; long relative to the adjacent lemmas; free; hairy (puberulous), or hairless; scabrous (on the keels); pointed; awnless; carinate; similar (subherbaceous, with scarious margins, the upper narrower). Lower glume 5–10 nerved. Upper glume 3–5 nerved. *Spikelets* with incomplete florets. The incomplete florets distal to the female-fertile florets. *The distal incomplete florets* merely underdeveloped.

Female-fertile florets *(2–)3–4(–6)*. Lemmas less firm than the glumes to similar in texture to the glumes (membranous); not becoming indurated; incised; 2 lobed; deeply cleft; awned (but the L_1 sometimes with a reduced awn or awnless). Awns 1, or 3; median, or median and lateral (the lateral lemma lobes sometimes being bristle-tipped); the median different in form from the laterals (when laterals present); from a sinus (mostly), or apical (sometimes, in the L_1); geniculate (and twisted below); hairless; much longer than the body of the lemma; entered by one vein. The lateral awns when present, shorter than the median (and straight). Lemmas hairy (mostly), or hairless (L_1); non-carinate; 7–9 nerved. *Palea* present; relatively long; apically notched (obscurely 3-notched); awnless, without apical setae; textured like the lemma; not indurated; 2-nerved; 2-keeled. *Lodicules* present; free; fleshy; glabrous. *Stamens* 3. Anthers 5–6 mm long; not penicillate; without an apically prolonged connective. *Ovary* glabrous. Styles free to their bases (short). Stigmas 2; rusty red pigmented.

Fruit, embryo and seedling. *Fruit* tightly enclosed by lemma and palea; longitudinally grooved; compressed laterally (slightly). *Hilum long-linear*.

Abaxial leaf blade epidermis. *Costal/intercostal zonation* conspicuous. *Papillae* absent. *Long-cells* markedly different in shape costally and intercostally (the intercostals fusiform, the costals rectangular). Mid-intercostal long-cells fusiform; having markedly sinuous walls and having straight or only gently undulating walls (and the costals with sinuous walls). *Microhairs* absent (but panicoid type present adaxially). *Stomata* common (but apparently confined to two lateral files per intercostal zone); 36–42 microns long. Subsidiaries low dome-shaped, or parallel-sided. Guard-cells overlapping to flush with the interstomatals. *Intercostal short-cells* common; in cork/silica-cell pairs; silicified. Intercostal silica bodies often crescentic. *Costal short-cells* conspicuously in long rows. Costal silica bodies horizontally-elongated crenate/sinuous, or 'panicoid-type' (predominantly 'panicoid type', but the undulations sometimes amount to crenation).

Transverse section of leaf blade, physiology. C_4 (in *C. involucratus*, according to carbon isotope ratio — H. Ziegler, pers. comm.), or C_3 (seemingly, judging from the poor material seen of *C. dregeanus*, though the lateral cell count is low between all but a few bundles); XyMS+ (*C. dregeanus*). Mesophyll without adaxial palisade. *Leaf blade* adaxially flat (with marked abaxial ribs only). *Midrib* not readily distinguishable (the bundle and keel only very slightly larger); with one bundle only; with colourless mesophyll adaxially (in the form of a wide bundle-sheath extension). Bulliforms present in discrete, regular adaxial groups; in simple fans (the groups large). Many of the smallest vascular bundles unaccompanied by sclerenchyma (and these 'inserted' between the large veins, as in panicoid C_4 leaves). Combined sclerenchyma girders absent (the large bundles with adaxial strands and girders, linked to the bundles by colourless sheath extensions). Sclerenchyma all associated with vascular bundles.

Special diagnostic feature. Pedicels articulated and bearded with long hairs at and above the joint. Female-fertile lemmas with a bent awn, the awn twisted below.

Taxonomy. Arundinoideae; Danthonieae.

Ecology, geography, regional floristic distribution. 3 species. Southern Africa. Xerophytic; halophytic, or glycophytic. Commonly coastal, generally on sandy soil.

Paleotropical and Cape. African. Namib-Karoo.

References, etc. Leaf anatomical: this project.

Chaetopoa C.E. Hubb.

Habit, vegetative morphology. *Annual*; loosely caespitose. *Culms* 15–60 cm high (slender, geniculate-ascending); herbaceous; branched above. *Leaves* not basally aggregated; non-auriculate; without auricular setae. Sheath margins free. The sheaths with spreading to reflexed white hairs and tubercles. Leaf blades narrow; 2–6 mm wide (6.5–30 cm long); flat; without cross venation. Ligule an unfringed membrane; 4 mm long (lacerate above).

Reproductive organization. Plants bisexual, with bisexual spikelets; with hermaphrodite florets. The spikelets of sexually distinct forms on the same plant; hermaphrodite and male-only, or hermaphrodite, male-only, and sterile; homomorphic (the central fertile and outer sterile spikelets of the cluster superficially similar).

Inflorescence. *Inflorescence a false spike, with spikelets on contracted axes (dense, narrow, of numerous glomerules each comprising six outer sterile involucral spikelets and a central fertile one)*; espatheate; not comprising 'partial inflorescences' and foliar organs. *Spikelet-bearing axes* very much reduced (to glomerules); disarticulating; falling entire (i.e., the glomerules falling — the main axis persistent). *Spikelets* unaccompanied by bractiform involucres, not associated with setiform vestigial branches (but with 'false involucres' of sterile spikelets); not secund; pedicellate (all short-pedicelled); consistently in 'long-and-short' combinations; unequally pedicellate in each combination (one of the sterile spikelets is longer-pedicelled than the rest, and has a longer G_1). The 'shorter' spikelets hermaphrodite and sterile. The 'longer' spikelets sterile.

Female-sterile spikelets. *The 'involucral spikelets' occasionally male, usually sterile; glumes both bristle-like (by contrast with the involucral G_1 in* **Anthephora***); lower floret reduced to a membranous lemma, L_2 reduced or absent*.

Female-fertile spikelets. Spikelets 4–5 mm long; slightly compressed dorsiventrally; falling with the glumes (i.e., with the glomerules). Rachilla terminated by a female-fertile floret. Hairy callus absent.

Glumes two; relatively large; more or less equal; about equalling the spikelets to exceeding the spikelets; glumes as long as or longer than the florets; subulate; similar (both more or less bristle-like, the G_2 somewhat broader below). Lower glume 1 nerved. Upper glume 2 nerved. *Spikelets* with incomplete florets. The incomplete florets proximal to the female-fertile florets. *Spikelets with proximal incomplete florets*. *The proximal incomplete florets* 1; paleate. Palea of the proximal incomplete florets reduced. The proximal incomplete florets sterile. The proximal lemmas awnless (but apically acuminate and mucronate); 7–9 nerved; more or less equalling the female-fertile lemmas to decidedly exceeding the female-fertile lemmas; similar in texture to the female-fertile lemmas (chartaceous to membranous); not becoming indurated.

Female-fertile florets 1. **Lemmas** narrowly elliptical, acute; not becoming indurated; entire; pointed; *mucronate*; hairless; non-carinate (convex); having the margins lying flat and exposed on the palea; 3–5 nerved. *Palea* present; relatively long, or conspicuous but relatively short; entire (obtuse, ovate); awnless, without apical setae; not indurated (hyaline); 2-nerved. *Lodicules* present; 2; free; fleshy; glabrous. *Stamens* 3. Anthers not penicillate; without an apically prolonged connective. **Ovary hairy.** Styles fused. Stigmas 2; white.

Fruit, embryo and seedling. Fruit small (about 1.5 mm long); compressed dorsiventrally. Hilum short. Embryo large.

Abaxial leaf blade epidermis. *Costal/intercostal zonation* conspicuous. *Papillae* absent. *Long-cells* markedly different in shape costally and intercostally (the costals narrowly rectangular); of similar wall thickness costally and intercostally. Intercostal zones with typical long-cells (these rather irregular in shape). Mid-intercostal long-cells rectangular to fusiform; having markedly sinuous walls, or having straight or only gently undulating walls. *Microhairs* present (and fairly common alongside the veins, but hard to find); elongated; clearly two-celled; panicoid-type; 51–55.8–60 microns long; 5.4–6–7.5 microns wide at the septum. Microhair total length/width at septum 6.8–9.4–10.6. Microhair apical cells 30–33–36 microns long. Microhair apical cell/total length ratio 0.53–0.59–0.63. *Stomata* common; 27–30 microns long. Subsidiaries low to high dome-shaped, or triangular. Guard-cells overlapped by the interstomatals (slightly, in places), or overlapping to flush with the interstomatals. *Intercostal short-cells* absent or very rare (excluding the short cells associated with the bases of macrohairs). Macrohairs and prickles abundant. *Costal short-cells* conspicuously in long rows (the rows often interrupted by prickles), or neither distinctly grouped into long rows nor predominantly paired. Costal silica bodies absent to poorly developed; probably 'panicoid-type' (but the silica bodies themselves obscure in the material seen); cross shaped, or dumb-bell shaped (?).

Transverse section of leaf blade, physiology. C_4; XyMS–. PCR sheath outlines even. *Leaf blade* 'nodular' in section (the abaxial ribs rather more prominent); with the ribs more or less constant in size. *Midrib* conspicuous; having a conventional arc of bundles (the large, median keel bundle flanked on either side by 2—3 smaller bundles); with colourless mesophyll adaxially. *The lamina* symmetrical on either side of the midrib. Bulliforms not in discrete, regular adaxial groups (the epidermis extensively bulliform). Many of the smallest vascular bundles unaccompanied by sclerenchyma. Combined sclerenchyma girders absent. Sclerenchyma all associated with vascular bundles.

Taxonomy. Panicoideae; Panicodae; Paniceae (cf. *Anthephora*).

Ecology, geography, regional floristic distribution. 2 species. Tanzania. Species of open habitats. Soil pockets on rocks.

Paleotropical. African. Sudano-Angolan. South Tropical African.

References, etc. Morphological/taxonomic: Hubbard 1967; Clayton 1977. Leaf anatomical: this project.

Chaetopogon Janchen

Chaeturus Link

Habit, vegetative morphology. *Annual*. *Culms* 3–30 cm high; herbaceous. Leaves non-auriculate. Leaf blades linear; narrow; 0.5–1.5 mm wide; setaceous, or not setaceous; without cross venation. Ligule an unfringed membrane; not truncate (acute); 3–5 mm long.

Reproductive organization. Plants bisexual, with bisexual spikelets; with hermaphrodite florets.

Inflorescence. Inflorescence paniculate; contracted (narrow); espatheate; not comprising 'partial inflorescences' and foliar organs. Spikelet-bearing axes persistent. Spikelets not secund; shortly pedicellate.

Female-fertile spikelets. Spikelets 2.8–3 mm long; compressed laterally; falling with the glumes (and with the short pedicel). Rachilla terminated by a female-fertile floret. Hairy callus absent.

Glumes two; very unequal to more or less equal; exceeding the spikelets; long relative to the adjacent lemmas; hairless; scabrous; (the lower only) *awned*; carinate, or non-carinate; *very dissimilar (linear lanceolate, membranous, the lower produced into a long slender awn, the upper awnless)*. Lower glume 1–3 nerved. Upper glume 1–3 nerved. **Spikelets** with female-fertile florets only; *without proximal incomplete florets*.

Female-fertile florets 1. Lemmas less firm than the glumes to similar in texture to the glumes (hyaline); not becoming indurated; entire, or incised; pointed; not deeply cleft; awned. Awns 1; median; from a sinus, or apical; non-geniculate; much shorter than the body of the lemma to about as long as the body of the lemma (?); entered by one vein. Lemmas hairless; non-carinate; without a germination flap; 1 nerved, or 3 nerved. **Palea** present; conspicuous but relatively short (about half as long as the lemma); apically notched (denticulate); awnless, without apical setae; not indurated (hyaline); *1-nerved, or nerveless*; keel-less. *Lodicules* present; 2; free; membranous; glabrous; not toothed; not or scarcely vascularized. *Stamens* 3. *Ovary* glabrous. Styles free to their bases. Stigmas 2.

Fruit, embryo and seedling. Fruit small (about 2 mm long); compressed dorsiventrally. Hilum short. *Embryo small*. Endosperm hard; with lipid; containing compound starch grains.

Abaxial leaf blade epidermis. *Costal/intercostal zonation* conspicuous. *Papillae* absent. *Long-cells* similar in shape costally and intercostally; of similar wall thickness costally and intercostally. Mid-intercostal long-cells rectangular and fusiform; having straight or only gently undulating walls. *Microhairs* absent. *Stomata* fairly common; about 45 microns long. Subsidiaries parallel-sided. Guard-cells overlapped by the interstomatals. *Intercostal short-cells* absent or very rare. *Costal short-cells* neither distinctly grouped into long rows nor predominantly paired. Costal silica bodies horizontally-elongated crenate/sinuous (but so deeply 'crenate' as almost to qualify as 'nodular').

Transverse section of leaf blade, physiology. XyMS+.

Cytology. Chromosome base number, $x = 7$. $2n = 14$. 2 ploid.

Taxonomy. Pooideae; Poodae; Aveneae.

Ecology, geography, regional floristic distribution. 2 species. Mediterranean. Holarctic. Boreal and Tethyan. Euro-Siberian. Mediterranean. European.

References, etc. Leaf anatomical: this project.

Special comments. Anatomical data epidermal only.

Chaetostichium (Hochst.) C.E. Hubb.

~ *Oropetium*

Habit, vegetative morphology. Perennial; densely caespitose. *Culms* 5–12 cm high; herbaceous. Young shoots intravaginal. *Leaves* not basally aggregated. Leaf blades narrow; 0.3–1.2 mm wide (1–5 cm long); subsetaceous, rigid, flat or convolute; exhibiting multicellular glands abaxially. The abaxial glands intercostal. Leaf blades without cross venation; disarticulating from the sheaths (apparently). Ligule a fringe of hairs.

Reproductive organization. Plants bisexual, with bisexual spikelets; with hermaphrodite florets.

Inflorescence. *Inflorescence a single spike (slender, becoming recurved, to 8 cm long)*. Rachides hollowed (excavated to receive the spikelets). Inflorescence espatheate; not comprising 'partial inflorescences' and foliar organs. Spikelet-bearing axes persistent. *Spikelets* solitary; *secund*; *biseriate (the ranks adjacent)*; sessile.

Female-fertile spikelets. Spikelets 3–5 mm long; adaxial; compressed dorsiventrally; disarticulating above the glumes. Rachilla terminated by a female-fertile floret. Hairy callus present (minute, shortly bearded). *Callus* short.

Glumes one per spikelet, or two; (G_2) relatively large; when two, very unequal; (G_2) long relative to the adjacent lemmas; dorsiventral to the rachis; hairless; (the upper) awned (attenuate into a straight or curved awn to 7 mm long); non-carinate; *very dissimilar (G_1 hyaline, tiny or missing, G_2 linear-lanceolate, acuminate, awned, leathery)*. Lower glume 0 nerved. Upper glume 1 nerved. *Spikelets* with female-fertile florets only.

Female-fertile florets 1. Lemmas emarginate; less firm than the glumes (membranous); not becoming indurated; not deeply cleft; mucronate, or awned (minutely). Awns when present, 1; median; non-geniculate; much shorter than the body of the lemma; entered by one vein. Lemmas hairless; non-carinate; 3 nerved. **Palea** present; relatively long (slightly shorter than lemma); apically notched (emarginate); *with apical setae (via excurrent nerves)*; not indurated (membranous); 2-nerved; 2-keeled. *Lodicules* present; 2; free; glabrous. *Stamens* 3. Anthers 0.6–1 mm long; not penicillate; without an apically prolonged connective. *Ovary* glabrous. Styles free to their bases. Stigmas 2.

Fruit, embryo and seedling. *Fruit* free from both lemma and palea (included); small (2.5 mm long); fusiform; compressed laterally. Hilum short. *Pericarp loosely adherent (removable when soaked)*. Embryo small (about 1/4 the length of the fruit).

Abaxial leaf blade epidermis. *Costal/intercostal zonation* conspicuous. *Papillae* absent. *Long-cells* similar in shape costally and intercostally; of similar wall thickness costally and intercostally (medium-thin walled). Mid-intercostal long-cells rectangular; having markedly sinuous walls. *Microhairs* present; more or less spherical; clearly two-celled; chloridoid-type. Microhair apical cell wall of similar thickness/rigidity to that of the basal cell. Microhair basal cells 12 microns long. Microhair total length/width at septum 1.5. Microhair apical cell/total length ratio 0.3. *Stomata* common. Subsidiaries dome-shaped. Guard-cells overlapping to flush with the interstomatals (but the subsidiaries overlapped by the adjoining intercostals). *Intercostal short-cells* absent or very rare. Intercostal silica bodies absent. *Costal short-cells* conspicuously in long rows. Costal silica bodies present in alternate cell files of the costal zones; exclusively saddle shaped (a rather angular version of this type).

Transverse section of leaf blade, physiology. C_4; XyMS+. PCR sheaths of the primary vascular bundles interrupted; interrupted abaxially only. PCR sheath extensions absent. *Mesophyll* with radiate chlorenchyma; not traversed by colourless columns (but almost so). *Midrib* conspicuous (by virtue of a large, round abaxial keel, with a large anchor of sclerenchyma); with one bundle only. Bulliforms present in discrete, regular adaxial groups; associated with colourless mesophyll cells to form deeply-penetrating fans. All the vascular bundles accompanied by sclerenchyma. Combined sclerenchyma girders present; forming 'figures' (all the primaries with big anchors). Sclerenchyma all associated with vascular bundles (but the smaller bundles having their abaxial sclerenchyma in lateral groups, rather than directly opposite).

Taxonomy. Chloridoideae; main chloridoid assemblage.

Ecology, geography, regional floristic distribution. 1 species. Montane Northeast Africa.

Paleotropical. African. Saharo-Sindian.

References, etc. Morphological/taxonomic: Hubbard 1937b. Leaf anatomical: this project.

Chamaeraphis R.Br.

Setosa Ewart

Habit, vegetative morphology. Perennial; caespitose. *Culms* 20–50 cm high; herbaceous. Culm nodes glabrous. Culm internodes solid. *Leaves* mostly basal, or not basally aggregated; non-auriculate. Leaf blades narrow; not pseudopetiolate; without cross venation; persistent. Ligule an unfringed membrane.

Reproductive organization. Plants bisexual, with bisexual spikelets; with hermaphrodite florets (rarely), or without hermaphrodite florets (usually with L_1 male, L_2 female). *The spikelets all alike in sexuality*.

Inflorescence. *Inflorescence a false spike, with spikelets on contracted axes (densely spikelike, with distichously arranged, much reduced 'racemes')*; contracted. Inflorescence axes not ending in spikelets. Inflorescence espatheate; not comprising 'partial inflorescences' and foliar organs. **Spikelet-bearing axes very much reduced (to a single spikelet, a pungent callus, and a stout bristle)**; disarticulating; *falling entire (by contrast with* Pseudoraphis, *the reduced panicle branches disarticulate from the persistent main axis, complete with the pungent, bearded base, one spikelet and the awn-like tip)*. **Spikelets** *subtended by solitary 'bristles' (the terminal branch 'awns'* *being so interpreted)*. The 'bristles' deciduous with the spikelets. Spikelets solitary; not secund; pedicellate.

Female-fertile spikelets. Spikelets oblong, or elliptic, or lanceolate, or ovate, or obovate; abaxial; compressed dorsiventrally; falling with the glumes (and with the branchlet). Rachilla terminated by a female-fertile floret. Hairy callus present.

Glumes two; more or less equal; long relative to the adjacent lemmas; dorsiventral to the rachis (lower abax.); awnless. *Lower glume 9–11 nerved*. Upper glume 9–11 nerved. *Spikelets* with incomplete florets. The incomplete florets proximal to the female-fertile florets. *The proximal incomplete florets* 1; paleate. Palea of the proximal incomplete florets fully developed. The proximal incomplete florets male. The proximal lemmas awnless.

Female-fertile florets 1. Lemmas decidedly firmer than the glumes; smooth; white in fruit; awnless, or mucronate (?); without a germination flap; 7–9 nerved. *Palea* present; 2-nerved. *Lodicules* present; 2; free; fleshy; glabrous. *Stamens* 0, or 3. Anthers not penicillate. *Ovary* glabrous. Styles fused. Stigmas 2.

Fruit, embryo and seedling. Fruit small; compressed dorsiventrally. Hilum short. Embryo large; waisted.

Abaxial leaf blade epidermis. *Costal/intercostal zonation* conspicuous. *Papillae* present; intercostal. Intercostal papillae not over-arching the stomata; consisting of one symmetrical projection per cell. *Long-cells* markedly different in shape costally and intercostally; of similar wall thickness costally and intercostally (thick walled). Intercostal zones exhibiting many atypical long-cells. Mid-intercostal long-cells rectangular; having markedly sinuous walls. *Microhairs* present; panicoid-type. *Stomata* common. Subsidiaries non-papillate. Guard-cells overlapping to flush with the interstomatals. *Intercostal short-cells* common; in cork/silica-cell pairs; silicified. *Costal short-cells* conspicuously in long rows, or neither distinctly grouped into long rows nor predominantly paired (over the minor veins). Costal silica bodies 'panicoid-type'.

Transverse section of leaf blade, physiology. C_4; XyMS–. PCR sheath outlines uneven. PCR sheath extensions absent. *Mesophyll* with radiate chlorenchyma. *Leaf blade* adaxially flat. *Midrib* conspicuous; with one bundle only. Bulliforms not in discrete, regular adaxial groups (the epidermis extensively bulliform). Many of the smallest vascular bundles unaccompanied by sclerenchyma, or all the vascular bundles accompanied by sclerenchyma. Combined sclerenchyma girders present; forming 'figures'. Sclerenchyma all associated with vascular bundles.

Taxonomy. Panicoideae; Panicodae; Paniceae.

Ecology, geography, regional floristic distribution. 1 species. Australia. Mesophytic; species of open habitats. Subhumid open woodland and coastal grassland. Australian. North and East Australian. Tropical North and East Australian.

References, etc. Leaf anatomical: this project.

Chandrasekharania V.J. Nair, V.S. Ramachandran & P.V. Sreekumar

Habit, vegetative morphology. Annual; decumbent. *Culms* 20–40 cm high; herbaceous; unbranched above. Culm nodes glabrous. *Leaves* not basally aggregated; non-auriculate. *Leaf blades ovate*; narrow (but relatively broad); conspicuously *cordate*; *persistent*. Ligule an unfringed membrane (obscure).

Reproductive organization. Plants bisexual, with bisexual spikelets; with hermaphrodite florets.

Inflorescence. Inflorescence paniculate; contracted; capitate to more or less ovoid; not comprising 'partial inflorescences' and foliar organs. Spikelet-bearing axes persistent. Spikelets not secund; pedicellate (the scabrid pedicels up to 1.5 mm long).

Female-fertile spikelets. Spikelets 5–6 mm long; compressed laterally; disarticulating above the glumes; disarticulating between the florets (supposedly). Rachilla terminated by a female-fertile floret. *Callus* short.

Glumes present; two; very unequal; about equalling the spikelets; (G_2) long relative to the adjacent lemmas; hairy (G_1 densely hairy with tubercle-based hairs, on the upper half), or hairless (the G_2 merely scaberulous on the nerves); without conspicuous tufts or rows of hairs; pointed; shortly awned (each attenuate into a straight or curved, scabrous, 3 mm awn); very dissimilar (the smaller G_1 being hairy). Lower glume 7 nerved. Upper glume 7 nerved. *Spikelets* with female-fertile florets only.

Female-fertile florets 2. Lemmas leathery or firmly membranous; not becoming indurated; incised; 2 lobed; not deeply cleft (minutely bidentate); awned. Awns 1; median; from a sinus; non-geniculate; straight; hairless (scabrid); much shorter than the body of the lemma (1.5–2 mm long). Lemmas hairless; smooth, save for the scaberulous midrib; 5 nerved. *Palea* present; relatively long; apically notched; awnless, without apical setae (hairy below); thinner than the lemma (hyaline); not indurated; 2-nerved; 2-keeled. Palea keels wingless. *Lodicules* present; 2; free; fleshy; glabrous. *Stamens* 3. Anthers 1.25 mm long; not penicillate; without an apically prolonged connective. *Ovary* glabrous. Styles free to their bases. Stigmas 2.

Fruit, embryo and seedling. Fruit small (about 1 mm long). Hilum short (punctiform).

Transverse section of leaf blade, physiology. C_3 ('leaf anatomy almost identical to *Jansenella*', according to Clayton & Renvoize 1986).

Taxonomy. Panicoideae (?); Panicodae; Arundinelleae (? — not reliably classifiable without a better description).

Ecology, geography, regional floristic distribution. 1 species. India. Paleotropical. Indomalesian. Indian.

References, etc. Morphological/taxonomic: Nair, Ramachandran and Sreekumar 1982.

Special comments. Anatomical data wanting.

Chasechloa A. Camus

~ *Echinolaena*.

Habit, vegetative morphology. Culms herbaceous; branched above. Culm nodes glabrous. *Leaves* not basally aggregated; non-auriculate. *Leaf blades* broad; slightly *cordate*; without cross venation; persistent.

Reproductive organization. Plants bisexual, with bisexual spikelets; with hermaphrodite florets.

Inflorescence. *Inflorescence of spicate main branches (pectinate racemes)*; espatheate; not comprising 'partial inflorescences' and foliar organs. Spikelet-bearing axes persistent. Spikelets solitary; secund.

Female-fertile spikelets. Spikelets 5 mm long; compressed laterally; falling with the glumes. Rachilla terminated by a female-fertile floret. Hairy callus absent.

Glumes two; relatively large; more or less equal; shorter than the adjacent lemmas; lateral to the rachis; hairy (with stiff, tubercle-based hairs over the lower part of the keel); pointed; awnless (apiculate); carinate; *with the keel conspicuously winged (over its top third)*; similar. Lower glume 4–5 nerved. Upper glume 3–6 nerved. *Spikelets* with incomplete florets. The incomplete florets proximal to the female-fertile florets. *The proximal incomplete florets* 1; paleate; male. The proximal lemmas awnless; 11 nerved; decidedly exceeding the female-fertile lemmas; less firm than the female-fertile lemmas; not becoming indurated.

Female-fertile florets 1. *Lemmas* decidedly firmer than the glumes; not becoming indurated; entire; pointed; awnless; hairless; *carinate*; having the margins tucked in onto the palea; 5 nerved. *Palea* present; relatively long; 2-nerved; 2-keeled. *Lodicules* present; 2; free; fleshy; glabrous; not or scarcely vascularized. *Stamens* 3. Anthers not penicillate; without an apically prolonged connective. *Ovary* glabrous. Styles fused. Stigmas 2; white (seemingly, from dried material).

Abaxial leaf blade epidermis. *Costal/intercostal zonation* conspicuous. *Papillae* absent. Intercostal zones exhibiting many atypical long-cells. Mid-intercostal long-cells rectangular; having markedly sinuous walls. *Microhairs* present; panicoid-type; 51–54 microns long; 30–31.5 microns wide at the septum. Microhair total length/width at septum 6–6.1. Microhair apical cells 8.4–9 microns long. Microhair apical cell/total length ratio 0.58–0.59. *Stomata* common; 40–45 microns long. Subsidiaries triangular.

Guard-cells overlapping to flush with the interstomatals. *Intercostal short-cells* absent or very rare. *Costal short-cells* conspicuously in long rows. Costal silica bodies 'panicoid-type'; dumb-bell shaped (the cross-bars seemingly missing in the material seen, thus simulating saddles and crescents).

Transverse section of leaf blade, physiology. C_3; XyMS+. *Midrib* conspicuous; with one bundle only. All the vascular bundles accompanied by sclerenchyma. Combined sclerenchyma girders present. Sclerenchyma all associated with vascular bundles.

Taxonomy. Panicoideae; Panicodae; Paniceae.

Ecology, geography, regional floristic distribution. 3 species. Madagascar. Paleotropical. Madagascan.

References, etc. Leaf anatomical: this project.

Special comments. Fruit data wanting.

Chasmanthium Link

Excluding *Gouldochloa*

Habit, vegetative morphology. Erect perennial; rhizomatous, or caespitose. *Culms* 50–150 cm high; herbaceous; unbranched above. Culm nodes glabrous. Culm internodes hollow. Young shoots extravaginal. *Leaves* not basally aggregated; non-auriculate; without auricular setae. *Leaf blades* linear to lanceolate; *broad*; 10–20 mm wide (and 7 to 20 cm long); flat, or folded; not pseudopetiolate; *without cross venation*; *persistent*; rolled in bud. Ligule a fringed membrane; truncate. *Contra-ligule* absent.

Reproductive organization. Plants bisexual, with bisexual spikelets; with hermaphrodite florets. Exposed-cleistogamous, or chasmogamous.

Inflorescence. Inflorescence few spikeleted to many spikeleted; paniculate, or of spicate main branches (the primary branches reduced to racemes); open, or contracted; non-digitate; espatheate; not comprising 'partial inflorescences' and foliar organs. Spikelet-bearing axes persistent. Spikelets not secund; pedicellate.

Female-fertile spikelets. *Spikelets* morphologically 'conventional', or unconventional (in having supernumerary proximal sterile lemmas); 5–35 mm long; cuneate; markedly compressed laterally; disarticulating above the glumes (or above the sterile lemmas); disarticulating between the florets; with conventional internode spacings. Rachilla prolonged beyond the uppermost female-fertile floret. Hairy callus present, or absent. *Callus* absent, or short.

Glumes two; very unequal to more or less equal; shorter than the spikelets; shorter than the adjacent lemmas; hairless (apart from the keel); pointed (acute, acuminate, or rarely mucronate); awnless; carinate; similar (rigid, compressed). Lower glume 3–7 nerved. Upper glume 3–7 nerved. *Spikelets* with incomplete florets. The incomplete florets proximal to the female-fertile florets, or both distal and proximal to the female-fertile florets. *The distal incomplete florets* merely underdeveloped. *Spikelets with proximal incomplete florets. The proximal incomplete florets* 1–6; paleate, or epaleate; sterile. The proximal lemmas awnless; 5–11 nerved (?); exceeded by the female-fertile lemmas; similar in texture to the female-fertile lemmas; not becoming indurated.

Female-fertile florets *2–12 (or occasionally reduced to one, in few-spikeleted inflorescences)*. Lemmas acute to acuminate; less firm than the glumes to similar in texture to the glumes (papery); not becoming indurated; entire; pointed; awnless; hairless; glabrous; carinate (sometimes markedly flattened); without a germination flap; 7–11 nerved. *Palea* present (rigid, bowed at the base); relatively long; entire; awnless, without apical setae; textured like the lemma; 2-nerved; 2-keeled. Palea keels narrowly winged, or wingless. *Lodicules* present; 2; free; fleshy; glabrous; not or scarcely vascularized. *Stamens* 1(–3); with free filaments. *Ovary* glabrous. Styles free to their bases. Stigmas 2; red pigmented.

Fruit, embryo and seedling. *Fruit* free from both lemma and palea (loosely enclosed); dark brown to black; compressed laterally. Hilum short. Embryo large. Endosperm hard; without lipid. Embryo with an epiblast; with a scutellar tail; with an elongated mesocotyl internode. Embryonic leaf margins overlapping.

First seedling leaf with a well-developed lamina. The lamina 7 veined.

Abaxial leaf blade epidermis. *Costal/intercostal zonation* conspicuous. *Papillae* absent. Mid-intercostal long-cells rectangular, or rectangular and fusiform; having markedly sinuous walls, or having markedly sinuous walls and having straight or only gently undulating walls (in *C. laxum*). *Microhairs* present; panicoid-type; 42–54 microns long; 4.5–6 microns wide at the septum. Microhair total length/width at septum 7–11.3. Microhair apical cells 21–24(–33) microns long (*C. latifolium*), or 15–18 microns long (*C. laxum*). Microhair apical cell/total length ratio 0.31–0.33 (*C. laxum*), or 0.47–0.61 (*C. latifolium*). *Stomata* common; 15–21 microns long. Subsidiaries dome-shaped, or dome-shaped and triangular. Guard-cells overlapping to flush with the interstomatals. *Intercostal short-cells* common (fairly), or absent or very rare; not paired (solitary); not silicified. *Costal short-cells* conspicuously in long rows. Costal silica bodies 'panicoid-type'.

Transverse section of leaf blade, physiology. C_3; XyMS+. Mesophyll with adaxial palisade, or without adaxial palisade (this not apparent in the two species seen); without arm cells; without fusoids. *Leaf blade* with distinct, prominent adaxial ribs to adaxially flat. *Midrib* conspicuous (conspicuously keeled or not); with one bundle only (larger than the rest), or having a conventional arc of bundles (the large median accompanied by small laterals); with colourless mesophyll adaxially (e.g. *C. laxum*), or without colourless mesophyll adaxially (e.g. *C. latifolium*). Bulliforms present in discrete, regular adaxial groups; in simple fans. All the vascular bundles accompanied by sclerenchyma. Combined sclerenchyma girders present (with all the bundles); forming 'figures' (all the bundles with conspicuous 'anchors'). Sclerenchyma all associated with vascular bundles.

Culm anatomy. *Culm internode bundles* in one or two rings, or scattered.

Cytology. Chromosome base number, $x = 12$. $2n = 24$. 2 ploid. Nucleoli persistent.

Taxonomy. Bambusoideae; Oryzodae; Centotheceae.

Ecology, geography, regional floristic distribution. 6 species. Southeastern U.S.A. and northern Mexico. Shade species; glycophytic. Moist woodland to semiarid scrub.

Holarctic. Boreal and Madrean. Atlantic North American. Canadian-Appalachian, Southern Atlantic North American, and Central Grasslands.

References, etc. Morphological/taxonomic: Yates 1966b. Leaf anatomical: this project.

Chasmopodium Stapf

Habit, vegetative morphology. Robust annual, or perennial (rarely). *Culms* 90–550 cm high; herbaceous; branched above. *Leaves* not basally aggregated; non-auriculate. *Leaf blades* broad; 5–40 mm wide; *pseudopetiolate*; without cross venation. *Ligule* present; a fringe of hairs.

Reproductive organization. Plants bisexual, with bisexual spikelets; with hermaphrodite florets. The spikelets of sexually distinct forms on the same plant; hermaphrodite and male-only, or hermaphrodite and sterile; homomorphic (except that the terminal spikelet is sometimes very long and thin).

Inflorescence. Inflorescence of cylindrical dorsiventral 'racemes', terminating the sparsely branching culms; spatheate; *a complex of 'partial inflorescences' and intervening foliar organs (the 'racemes' distant from their spathes)*. Spikelet-bearing axes 'racemes'; solitary, or paired; with substantial rachides; disarticulating; disarticulating at the joints. *'Articles'* non-linear (narrowed at the base, concave at the apex, shorter than the spikelets); with a basal callus-knob; not appendaged; disarticulating transversely; somewhat hairy (ciliate on the back). Spikelets paired; secund (the axis dorsiventral, the sessile spikelets in two anterior rows and alternating, the pedicelled members posterior); sessile and pedicellate; consistently in 'long-and-short' combinations; in pedicellate/sessile combinations. Pedicels of the 'pedicellate' spikelets free of the rachis (stout). The 'shorter' spikelets hermaphrodite. The 'longer' spikelets male-only, or sterile.

Female-sterile spikelets. The pedicelled spikelet well developed. The lemmas awnless.

Female-fertile spikelets. Spikelets 6.3–7.5 mm long; compressed dorsiventrally; falling with the glumes. Rachilla terminated by a female-fertile floret. Hairy callus present (there being an annulus of hairs, above the disarticulation knob).

Glumes two; more or less equal; long relative to the adjacent lemmas; hairless; awnless; carinate (G₂), or non-carinate (G₁); G_2 **with the keel conspicuously winged (towards the top, the two keels of G_1 also somehwat winged)**; very dissimilar (leathery or crustaceous, the G₁ flat or concave dorsally, the G₂ naviculate). **Lower glume two-keeled**; flattened on the back to concave on the back; not pitted; relatively smooth; seemingly many nerved. Upper glume seemingly many nerved. *Spikelets* with incomplete florets. The incomplete florets proximal to the female-fertile florets. **Spikelets with proximal incomplete florets**. The proximal incomplete florets 1; paleate. Palea of the proximal incomplete florets fully developed. The proximal incomplete florets male. The proximal lemmas awnless; 3 nerved; more or less equalling the female-fertile lemmas; similar in texture to the female-fertile lemmas (hyaline); not becoming indurated.

Female-fertile florets 1. Lemmas less firm than the glumes (hyaline); not becoming indurated; entire; pointed, or blunt; awnless; hairless; non-carinate; without a germination flap; 3 nerved. *Palea* present; relatively long; entire; awnless, without apical setae; not indurated; 2-nerved; keel-less. *Lodicules* present; 2; free; fleshy; glabrous. *Stamens* 3. Anthers not penicillate; without an apically prolonged connective. *Ovary* glabrous. Stigmas 2.

Fruit, embryo and seedling. Embryo large (about 4/5 the length of the caryopsis).

Abaxial leaf blade epidermis. *Costal/intercostal zonation* conspicuous. Costal zones with short-cells. *Costal short-cells* conspicuously in long rows. Costal silica bodies present and well developed; 'panicoid-type'; dumb-bell shaped.

Transverse section of leaf blade, physiology. C₄ ('bundles crowded'). *Midrib* conspicuous; with colourless mesophyll adaxially. Bulliforms present in discrete, regular adaxial groups (between the vascular bundles).

Phytochemistry. Leaves without flavonoid sulphates (1 species).

Cytology. Chromosome base number, $x = 8$. $2n = 16$.

Taxonomy. Panicoideae; Andropogonodae; Andropogoneae; Rottboelliinae.

Ecology, geography, regional floristic distribution. 2 species. Tropical Africa. Paleotropical. African. Sudano-Angolan and West African Rainforest. Sahelo-Sudanian and South Tropical African.

References, etc. Leaf anatomical: Metcalfe 1960.

Special comments. Fruit data wanting.

Chevalierella A. Camus

Habit, vegetative morphology. Perennial; caespitose. The flowering culms leafy. **Culms 100–200 cm high**; herbaceous; unbranched above. Culm nodes hidden by leaf sheaths. Culm leaves present. Upper culm leaf blades fully developed. Culm internodes hollow. Plants unarmed. *Leaves* not basally aggregated; auriculate (on the sheath); without auricular setae. **Leaf blades** lanceolate, or elliptic; broad; 30–50 mm wide (and large); not cordate, not sagittate; flat; **pseudopetiolate (narrowed to the base)**; **cross veined**; persistent. *Contra-ligule* absent.

Reproductive organization. Plants bisexual, with bisexual spikelets; with hermaphrodite florets.

Inflorescence. Inflorescence of spicate main branches (in a terminal 'racemose panicle', about 40 cm long); espatheate; not comprising 'partial inflorescences' and foliar organs. *Spikelet-bearing axes* spikes; persistent. Spikelets solitary; secund; biseriate (on two sides of the slender, 3-sided axis); pedicellate.

Female-fertile spikelets. Spikelets about 5 mm long; adaxial; strongly compressed laterally; falling with the glumes; with a distinctly elongated rachilla internode between the glumes, with a distinctly elongated rachilla internode above the glumes, and with distinctly elongated rachilla internodes between the florets. Rachilla prolonged beyond the uppermost female-fertile floret; rachilla hairless. The rachilla extension with incomplete florets. Hairy callus present.

Glumes two; very unequal (by elongation of the internode between them, the G₁ and G₂ being actually about the same length); shorter than the spikelets; shorter than the adjacent lemmas; dorsiventral to the rachis; hairless; glabrous (scabrous on the midvein); *awned (the median nerve excurrent as a short subule)*; carinate; similar (membranous).

Lower glume shorter than the lowest lemma; 5 nerved. Upper glume 3–5 nerved. *Spikelets* with incomplete florets. The incomplete florets distal to the female-fertile florets. The distal incomplete florets 1, or 2; *incomplete florets* merely underdeveloped; incomplete florets awned.

Female-fertile florets 1. Lemmas ovate; similar in texture to the glumes to decidedly firmer than the glumes (papery); not becoming indurated; minutely incised, or entire; when detectably incised, minutely 2 lobed; not deeply cleft; awned. Awns 1; median; from a sinus, or dorsal to apical; from near the top (from just behind the apex or minute sinus — i.e. subapical); non-geniculate; flexuous; hairless (antrorsely scaberulous); much shorter than the body of the lemma to about as long as the body of the lemma; entered by one vein. Awn bases not twisted; not flattened. Lemmas hairless; glabrous; carinate; having the margins lying flat and exposed on the palea; weakly 3–5 nerved; with the nerves non-confluent. *Palea* present; relatively long; tightly clasped by the lemma; apically notched (minutely); awnless, without apical setae; thinner than the lemma to textured like the lemma; not indurated; 2-nerved; 2-keeled. Palea back glabrous. Palea keels winged; scabrous. **Lodicules** present; 2; *joined (represented by a single, anterior structure)*; glabrous. **Stamens** *2*. Anthers not penicillate; without an apically prolonged connective. *Ovary* glabrous. Styles free to their bases. Stigmas 2.

Fruit, embryo and seedling. Fruit small (2–4 mm long). Hilum short. Pericarp thin. Embryo small.

Abaxial leaf blade epidermis. *Costal/intercostal zonation* conspicuous. *Papillae* absent. *Long-cells* markedly different in shape costally and intercostally (the costals much narrower); of similar wall thickness costally and intercostally (of medium thickness). Mid-intercostal long-cells rectangular; having markedly sinuous walls (the sinuosity fairly coarse and irregular). *Microhairs* present; elongated; clearly two-celled; panicoid-type. *Stomata* common. Subsidiaries non-papillate; low dome-shaped to triangular. Guard-cells overlapping to flush with the interstomatals. *Intercostal short-cells* common; in cork/silica-cell pairs; silicified. Intercostal silica bodies irregularly cross-shaped. No macrohairs or prickles seen. Costal zones with short-cells. *Costal short-cells* conspicuously in long rows. Costal silica bodies present and well developed; 'panicoid-type'; cross shaped (a few), or dumb-bell shaped (mostly, short to elongated), or nodular (a few).

Transverse section of leaf blade, physiology. C_3; XyMS+. *Mesophyll* with non-radiate chlorenchyma; with adaxial palisade; without arm cells; with fusoids. The fusoids an integral part of the PBS (seemingly, in places), or external to the PBS (definitely, in places). *Leaf blade* adaxially flat. *Midrib* conspicuous (via its massive abaxial keel, heavy sclerenchyma and conspicuous lacunae); having a conventional arc of bundles (the bundles separated by lacunae); with colourless mesophyll adaxially (and with lacunae). *The lamina* symmetrical on either side of the midrib. Bulliforms present in discrete, regular adaxial groups; in simple fans (these large and wide). All the vascular bundles accompanied by sclerenchyma. Combined sclerenchyma girders present; forming 'figures' (most bundles with a small I or T). Sclerenchyma all associated with vascular bundles.

Taxonomy. Bambusoideae; Oryzodae; Centotheceae.

Ecology, geography, regional floristic distribution. 2 species. Congo. Shade species; glycophytic. Forests.

Paleotropical. African. West African Rainforest.

References, etc. Leaf anatomical: Jacques-Félix 1962; this project.

Chikusichloa Koidz.

Habit, vegetative morphology. *Culms* 70–175 cm high; herbaceous; unbranched above. *Leaves* not basally aggregated. Leaf blades linear to lanceolate (acuminate); broad, or narrow; 7–15 mm wide (30–50 cm long). Ligule truncate; 2–3 mm long.

Reproductive organization. Plants bisexual, with bisexual spikelets; with hermaphrodite florets. Exposed-cleistogamous, or chasmogamous (?).

Inflorescence. Inflorescence paniculate; open (about 40 cm long, 15 cm wide); espatheate; not comprising 'partial inflorescences' and foliar organs. Spikelet-bearing axes persistent. Spikelets not secund; pedicellate (the pedicels glabrous).

Female-fertile spikelets. Spikelets *unconventional (through reduction of the glumes, cf.* **Oryza***)*; 4–4.5 mm long; slightly *compressed dorsiventrally; borne on and falling with the long, slender stipe of the floret, by contrast with* **Leersia**. Rachilla terminated by a female-fertile floret.

Glumes *absent ('obsolete')*. *Spikelets* with female-fertile florets only.

Female-fertile florets 1. Lemmas lanceolate, acute to awned; not becoming indurated (membranous or papery); entire; pointed; awnless, or awned. Awns when present, 1; median; apical (by attenuation of the lemma); non-geniculate; about 4.5 mm long. Lemmas hairy, or hairless; strongly 5–7 nerved. *Palea* present; relatively long; entire (acute); awnless, without apical setae; not indurated (membranous); 2-nerved, or several nerved (sometimes 3); one-keeled. *Lodicules* present; 2. **Stamens** *1*. Anthers without an apically prolonged connective. *Ovary* glabrous. Stigmas 2.

Fruit, embryo and seedling. Fruit small (2 mm long). Hilum short. Pericarp very thin.

Abaxial leaf blade epidermis. *Costal/intercostal zonation* conspicuous. *Papillae* absent. *Long-cells* similar in shape costally and intercostally to markedly different in shape costally and intercostally (the costals tending to be narrower); of similar wall thickness costally and intercostally (the walls of medium thickness). Mid-intercostal long-cells rectangular; having markedly sinuous walls (the sinuosity fairly regular). *Microhairs* present; elongated; clearly two-celled; panicoid-type. *Stomata* common. Subsidiaries non-papillate; fairly high dome-shaped to triangular. Guard-cells overlapping to flush with the interstomatals. *Intercostal short-cells* common; in cork/silica-cell pairs. With a few intercostal prickles. *Costal short-cells* conspicuously in long rows. Costal silica bodies present and well developed; 'panicoid-type'; dumb-bell shaped (mainly), or butterfly shaped (a few).

Transverse section of leaf blade, physiology. C_3; XyMS+. Mesophyll with arm cells; with fusoids. The fusoids external to the PBS. *Leaf blade* 'nodular' in section, or adaxially flat. *Midrib* conspicuous; having complex vascularization (a large bundle abaxially in the keel, and a small adaxial one superposed, the pair embedded in either colourless tissue or large-celled, thin-walled sclerenchyma). Bulliforms present in discrete, regular adaxial groups; in simple fans (of large cells, between adjoining vascular bundles). All the vascular bundles accompanied by sclerenchyma. Combined sclerenchyma girders present (with all the bundles); forming 'figures' (I's). Sclerenchyma all associated with vascular bundles.

Cytology. Chromosome base number, $x = 12$. $2n = 24$. 2 ploid.

Taxonomy. Bambusoideae; Oryzodae; Oryzeae.

Ecology, geography, regional floristic distribution. 3 species. China, Riyukyu Is., Japan. Helophytic; shade species. Forests.

Holarctic and Paleotropical. Boreal. Indomalesian. Eastern Asian. Indo-Chinese and Malesian.

References, etc. Morphological/taxonomic: Koidzum 1925. Leaf anatomical: this project.

Special comments. Fruit data wanting.

Chimonobambusa Makino

Oreocalamus Keng, *Quiongzhuea* Hsueh & Yi

Habit, vegetative morphology. Perennial; rhizomatous, or stoloniferous. The flowering culms leafy. *Culms* 50–600 cm high; woody and persistent (monopodial); to 3 cm in diameter; branched above. *Primary branches/mid-culm node 3 (these subsequently giving rise to multiple twigs)*. Culm sheaths deciduous in their entirety (somewhat leathery, with very small blades). Culm internodes hollow. Rhizomes leptomorph. *Plants conspicuously armed (the two-ridged culm nodes usually thorny)*. *Leaves* not basally aggregated; with auricular setae. Leaf blades broad, or narrow; pseudopetiolate; cross veined; or without cross venation; disarticulating from the sheaths; rolled in bud.

Reproductive organization. Plants bisexual, with bisexual spikelets; with hermaphrodite florets.

Inflorescence. *Inflorescence* indeterminate (as indicated by bracts subtending the spikelets); seemingly without pseudospikelets; exhibiting 1–12 spikelets at each node of leafy or leafless branches; non-digitate; spatheate; a complex of 'partial inflorescences' and intervening foliar organs (inflorescence branches with or without foliage leaves). *Spikelet-bearing axes* 'racemes', or paniculate; persistent. Spikelets not secund; sessile.

Female-fertile spikelets. Spikelets disarticulating above the glumes; disarticulating between the florets (?); with distinctly elongated rachilla internodes between the florets ('the florets distant'). Rachilla prolonged beyond the uppermost female-fertile floret (?). The rachilla extension with incomplete florets. Hairy callus absent.

Glumes two to several; very unequal; shorter than the adjacent lemmas; awnless. Lower glume 3 nerved. Upper glume 7 nerved. *Spikelets* with incomplete florets. The incomplete florets distal to the female-fertile florets. *The distal incomplete florets* merely underdeveloped.

Female-fertile florets 2–8 (?). Lemmas ovate, membranous; entire; pointed; awnless; 8 nerved. *Palea* present; relatively long; entire; several nerved (7 observed); 2-keeled. *Lodicules* present; 3; free; membranous; ciliate; not toothed; heavily vascularized. *Stamens* 3. Anthers without an apically prolonged connective. *Ovary* glabrous. *Styles free to their bases (apparently)*. Stigmas 2 (feathery).

Fruit, embryo and seedling. Fruit not grooved; not noticeably compressed. Hilum short. Pericarp fleshy (at least, in Subg. *Oreocalamus* and *Quiongzhuea*); free. *Seed 'non-endospermic'*.

Abaxial leaf blade epidermis. *Costal/intercostal zonation* conspicuous. *Papillae* present; costal and intercostal. Intercostal papillae conspicuously over-arching the stomata; several per cell (mostly one row per cell, of the 'coronate' type). *Long-cells* similar in shape costally and intercostally; of similar wall thickness costally and intercostally. Mid-intercostal long-cells rectangular; having markedly sinuous walls. *Microhairs* present; panicoid-type (but large and fat); (54–)63–69 microns long (in *C. densifolia*); 6–9 microns wide at the septum. Microhair total length/width at septum 6–11.5 (i.e. very variable in *C. densifolia*). Microhair apical cells (21–)24–27 microns long. Microhair apical cell/total length ratio 0.35–0.43. *Stomata* common; about 27 microns long (in *C. densifolia*). Subsidiaries non-papillate; parallel-sided to dome-shaped. *Intercostal short-cells* common; not paired (mostly solitary); not silicified (mostly). *Costal short-cells* neither distinctly grouped into long rows nor predominantly paired (ones and twos). Costal silica bodies saddle shaped and oryzoid.

Transverse section of leaf blade, physiology. C_3; XyMS+. *Mesophyll* with non-radiate chlorenchyma; without adaxial palisade; with arm cells; with fusoids. The fusoids external to the PBS. Leaf blade with the ribs more or less constant in size. *Midrib* conspicuous (a larger bundle, with massive I-configuration of girders); with one bundle only. Bulliforms present in discrete, regular adaxial groups; in simple fans. All the vascular bundles accompanied by sclerenchyma. Combined sclerenchyma girders present (with all the bundles); forming 'figures' (all narrow I's, except midrib). Sclerenchyma all associated with vascular bundles.

Cytology. Chromosome base number, $x = 12$. $2n = 48$. 4 ploid.

Taxonomy. Bambusoideae; Bambusodae; Bambuseae.

Ecology, geography, regional floristic distribution. 10 species. Eastern Asia and Himalaya.

Holarctic and Paleotropical. Boreal. Indomalesian. Eastern Asian. Indian and Indo-Chinese.

Rusts and smuts. Rusts — *Stereostratum*. Taxonomically wide-ranging species: *Stereostratum corticoides*.

References, etc. Morphological/taxonomic: Chao and Renvoize 1989. Leaf anatomical: this project.

Special comments. Fruit data wanting.

Chionachne R.Br.

Habit, vegetative morphology. Annual (rarely), or perennial (reed-like); rhizomatous, or caespitose. *Culms* 60–200 cm high (?); woody and persistent, or herbaceous;

branched above. Culm nodes hairy, or glabrous. Culm internodes solid, or hollow. *Leaves* not basally aggregated; non-auriculate. Leaf blades broad; without cross venation; persistent. Ligule an unfringed membrane to a fringed membrane (ciliate).

Reproductive organization. *Plants monoecious with all the fertile spikelets unisexual*; without hermaphrodite florets. The spikelets of sexually distinct forms on the same plant; female-only and male-only. *The male and female-fertile spikelets segregated, in different parts of the same inflorescence branch (female spikelets few, near the bases of the spikes, the male spikelets distal).* The spikelets in both homogamous and heterogamous combinations (heterogamous below, male above).

Inflorescence. Inflorescence leafy, paniculate; open. Rachides hollowed. *Inflorescence* spatheate; *a complex of 'partial inflorescences' and intervening foliar organs*. Spikelet-bearing axes *'racemes' to spikelike*; solitary, or paired, or clustered; with substantial rachides; disarticulating; disarticulating at the joints. Spikelets solitary, or paired; secund; sessile and pedicellate; consistently in 'long-and-short' combinations; in pedicellate/sessile combinations. Pedicels of the 'pedicellate' spikelets discernible, but fused with the rachis, or free of the rachis. The 'shorter' spikelets female-only (towards bases of spikes), or male-only (in the upper parts of the spikes, where they are paired with pedicellate male spikelets). The 'longer' spikelets male-only (above), or sterile (below, where they are reduced to pedicels, more or less fused with the rachis and paired with female sessile spikelets).

Female-sterile spikelets. Male spikelets solitary and sessile, or in pairs with one sessile and one pedicelled.

Female-fertile spikelets. Spikelets compressed dorsiventrally; falling with the glumes. Rachilla terminated by a female-fertile floret.

Glumes two; very unequal; (the longer) long relative to the adjacent lemmas; awnless; very dissimilar (lower bigger, tough, rounded, with wings clasping the spikelet, the upper smaller, not tough, flat). Lower glume 7–20 nerved. Upper glume 7–9 nerved. *Spikelets with incomplete florets*. The incomplete florets proximal to the female-fertile florets. *Spikelets with proximal incomplete florets. The proximal incomplete florets* 1; sterile. The proximal lemmas awnless; 3 nerved; more or less equalling the female-fertile lemmas to decidedly exceeding the female-fertile lemmas; similar in texture to the female-fertile lemmas; not becoming indurated.

Female-fertile florets 1. Lemmas less firm than the glumes (membranous); not becoming indurated; awnless; hairless; non-carinate; 1–5 nerved. Palea present; relatively long; 2-nerved. *Lodicules* absent. *Ovary* glabrous. Styles fused. Stigmas 2; red pigmented.

Fruit, embryo and seedling. Fruit small; compressed dorsiventrally. Hilum short. Embryo large; waisted.

Abaxial leaf blade epidermis. *Costal/intercostal zonation* conspicuous. *Papillae* absent. *Long-cells* similar in shape costally and intercostally; of similar wall thickness costally and intercostally (thin walled). Mid-intercostal long-cells rectangular; having markedly sinuous walls. *Microhairs* present; panicoid-type; 54–57 microns long; 9–12 microns wide at the septum. Microhair total length/width at septum 4.75–6. Microhair apical cells 30–33 microns long. Microhair apical cell/total length ratio 0.53–0.61. *Stomata* common; 42–48 microns long. Subsidiaries triangular. Guard-cells overlapping to flush with the interstomatals. *Intercostal short-cells* common; in cork/silica-cell pairs; silicified. *Costal short-cells* conspicuously in long rows, or predominantly paired, or neither distinctly grouped into long rows nor predominantly paired. Costal silica bodies 'panicoid-type'.

Transverse section of leaf blade, physiology. C_4; XyMS–. *Mesophyll* with radiate chlorenchyma. *Leaf blade* adaxially flat. *Midrib* not readily distinguishable; having a conventional arc of bundles; with colourless mesophyll adaxially. Bulliforms present in discrete, regular adaxial groups; in simple fans. Many of the smallest vascular bundles unaccompanied by sclerenchyma. Combined sclerenchyma girders present; forming 'figures'. Sclerenchyma all associated with vascular bundles.

Cytology. Chromosome base number, $x = 10$. $2n = 20$. 2 ploid.

Taxonomy. Panicoideae; Andropogonodae; Maydeae.

Ecology, geography, regional floristic distribution. 7 species. Indomalayan region, Indochina, eastern Australia. Helophytic to mesophytic. Forest margins, streamsides.

Paleotropical and Australian. Indomalesian. Indian, Indo-Chinese, Malesian, and Papuan. North and East Australian. Tropical North and East Australian.
Rusts and smuts. Smuts from Ustilaginaceae. Ustilaginaceae — *Ustilago.*
Economic importance. Significant weed species: *C. hubbardiana* (in North America). Cultivated fodder: *C. semiteres.*
References, etc. Leaf anatomical: this project.

Chionochloa Zotov

~ *Danthonia* sensu lato, *Rytidosperma,* cf. *Cortaderia*

Habit, vegetative morphology. Perennial; caespitose. **Culms** 20–250 cm high (usually in coarse tussocks invested below with old sheaths); *herbaceous.* Culm nodes glabrous. Culm internodes hollow. Young shoots extravaginal, or intravaginal. *Leaves* mostly basal; non-auriculate. Sheaths with an apical hair tuft. *Leaf blades* linear; broad, or narrow; 0.8–10 mm wide (5–150cm long); setaceous, or not setaceous; flat, or rolled, or acicular; *not pseudopetiolate*; without cross venation; disarticulating from the sheaths (and sometimes fracturing into segments, cf. *Cortaderia*), or persistent. *Ligule a fringe of hairs*; usually about 1 mm long.

Reproductive organization. *Plants* nearly always *bisexual, with bisexual spikelets (with gynodioecism in* **C.** *bromoides)*; with hermaphrodite florets; inbreeding. Chasmogamous.

Inflorescence. *Inflorescence* few spikeleted to many spikeleted; *paniculate (usually hairy in the branch axils)*; open, or contracted; espatheate; not comprising 'partial inflorescences' and foliar organs. Spikelet-bearing axes persistent. Spikelets not secund; pedicellate.

Female-fertile spikelets. Spikelets 8–20 mm long; pale, golden or purpled; compressed laterally; disarticulating above the glumes; with conventional internode spacings. Rachilla prolonged beyond the uppermost female-fertile floret. Hairy callus present (the long hairs covering the lower part of the lemma). *Callus* short; blunt.

Glumes two; very unequal to more or less equal; shorter than the spikelets to exceeding the spikelets (but usually shorter); shorter than the adjacent lemmas to long relative to the adjacent lemmas; hairless (glabrous, occasionally prickle-toothed), or hairy (the upper, sometimes long-hairy on the lower margin); pointed; awned (rarely), or awnless; similar (membranous, the margins thin). Lower glume 1(–3) nerved. Upper glume 3 nerved, or 5 nerved, or 7 nerved. *Spikelets* with female-fertile florets only, or with incomplete florets. The incomplete florets distal to the female-fertile florets. *Spikelets without proximal incomplete florets.*

Female-fertile florets *2–8.* Lemmas not becoming indurated (membranous); incised; 2 lobed; deeply cleft to not deeply cleft (the lateral lobes usually shorter than th body); awned. Awns 1, or 3; median, or median and lateral (the lobes usually short-awned); the median different in form from the laterals (when laterals present); from a sinus; geniculate; entered by several veins (3 in *C. antarctica*). The lateral awns shorter than the median. Lemmas hairy (on the margins, or all over), or hairless. *The hairs not in tufts; in transverse rows, or not in transverse rows.* Lemmas non-carinate; 7–9 nerved. *Palea* present (hairy on the flanks below); relatively long (exceeding the lemma sinus); usually apically notched; awnless, without apical setae (usually), or awned (the keels being produced into 2mm awns in *C. beddiei*); 2-nerved; 2-keeled. Palea back glabrous, or scabrous, or hairy. Palea keels hairy. **Lodicules** present; 2; free; *membranous; ciliate;* toothed. *Stamens* 3. Anthers (2–)4–5(–6) mm long; not penicillate. *Ovary* glabrous. Styles free to their bases. Stigmas 2; white.

Fruit, embryo and seedling. *Fruit* small ((1.5-)2.5–3.5(-4) mm long); obovate, rugulose or smooth; *longitudinally grooved. Hilum long-linear (1/2 to 2/3 the length of the caryopsis).* Embryo large (1/3 to 1/2 the caryopsis length). Endosperm containing compound starch grains. Embryo without an epiblast; with a scutellar tail; with an elongated mesocotyl internode.

Abaxial leaf blade epidermis. *Costal/intercostal zonation* lacking. *Papillae* absent (but present adaxially). *Long-cells* similar in shape costally and intercostally (the intercostal zones being indistinguishable); of similar wall thickness costally and intercostally (thick to very thick walled). Mid-intercostal long-cells rectangular; having markedly sinuous walls. *Microhairs* absent (but present adaxially). *Stomata* absent or very rare; 30–43 microns long. Subsidiaries dome-shaped. *Intercostal short-cells* common; in cork/silica-cell pairs, or not paired; silicified, or not silicified. *Costal short-cells* predominantly paired, or neither distinctly grouped into long rows nor predominantly paired. Costal silica bodies rounded, or saddle shaped, or tall-and-narrow, or crescentic, or 'panicoid-type'.

Transverse section of leaf blade, physiology. C_3; XyMS+. *Mesophyll* with radiate chlorenchyma, or with non-radiate chlorenchyma; without adaxial palisade. *Leaf blade* with distinct, prominent adaxial ribs; with the ribs very irregular in sizes. *Midrib* conspicuous; with one bundle only; with colourless mesophyll adaxially, or without colourless mesophyll adaxially. Bulliforms present in discrete, regular adaxial groups (at the bases of the furrows, the groups sometimes small); in simple fans. All the vascular bundles accompanied by sclerenchyma. Combined sclerenchyma girders present; forming 'figures', or nowhere forming 'figures'. Sclerenchyma all associated with vascular bundles (*C. pallida*), or not all bundle-associated (*C. conspicua, C. frigida*). The 'extra' sclerenchyma in abaxial groups to in a continuous abaxial layer.

Phytochemistry. Leaves containing flavonoid sulphates (*C. conspicua, C. rigida*).

Cytology. Chromosome base number, $x = 6$ and 7. $2n = 42, 48, 72,$ and 96. 6, 12, and 16 ploid.

Taxonomy. Arundinoideae; Danthonieae.

Ecology, geography, regional floristic distribution. 19 species. New Zealand, southeastern Australia.

Australian and Antarctic. North and East Australian. New Zealand. Temperate and South-Eastern Australian.

Rusts and smuts. Rusts — *Puccinia*.

References, etc. Morphological/taxonomic: Zotov 1963; Vickery 1956; Connor 1991. Leaf anatomical: this project (from Australian material only).

Chloachne Stapf

~ *Poecilostachys*

Habit, vegetative morphology. Perennial; *rhizomatous, or stoloniferous*. *Culms* 30–100 cm high; herbaceous; branched above. Culm nodes glabrous. *Leaves* not basally aggregated; non-auriculate. Sheath margins free. *Leaf blades* lanceolate; broad; slightly cordate, or not cordate, not sagittate; pseudopetiolate, or not pseudopetiolate; *cross veined*. Ligule a fringed membrane.

Reproductive organization. *Plants* bisexual, with bisexual spikelets; *with hermaphrodite florets*. The spikelets of sexually distinct forms on the same plant, or all alike in sexuality; hermaphrodite, or hermaphrodite and sterile (sometimes sterile at the base of the inflorescence).

Inflorescence. Inflorescence of spicate main branches (mostly), or paniculate (the branches sometimes themselves branched at the base); espatheate; not comprising 'partial inflorescences' and foliar organs. Spikelet-bearing axes persistent. *Spikelets* in triplets (or small clusters), or paired; *secund (the racemes unilateral)*; pedicellate; not in distinct 'long-and-short' combinations.

Female-fertile spikelets. *Spikelets 6–8 mm long*; abaxial; *compressed laterally*; falling with the glumes. Rachilla prolonged beyond the uppermost female-fertile floret (the prolongation minute, vestigial), or terminated by a female-fertile floret. The rachilla extension when present, naked. Hairy callus absent.

Glumes two; very unequal; shorter than the adjacent lemmas, or long relative to the adjacent lemmas (i.e., the upper glumes); dorsiventral to the rachis; hairy (with very large, sparse hairs, or stiff, tubercle-based bristles); pointed (acuminate); awnless; non-carinate; similar (membranous-herbaceous). Lower glume 3–5 nerved. Upper glume 5 nerved. *Spikelets* with incomplete florets. The incomplete florets proximal to the female-fertile florets. *Spikelets with proximal incomplete florets*. *The proximal incomplete florets* 1; paleate. *Palea of the proximal incomplete florets reduced*. The proximal incomplete florets sterile. The proximal lemmas similar to the glumes, hairy or bristly; awnless; 7–8 nerved; decidedly exceeding the female-fertile lemmas; less firm than the female-fertile lemmas; not becoming indurated.

Female-fertile florets 1. Lemmas narrow; similar in texture to the glumes to decidedly firmer than the glumes (somewhat papyraceous); smooth (shining); not becoming indurated; entire; pointed; awnless to mucronate; hairless; glabrous (shiny); non-carinate; having the margins lying flat and exposed on the palea; with a clear germination flap; 5 nerved. *Palea* present; relatively long; entire (acute); awnless, without apical setae; textured like the lemma; not indurated; 2-nerved; folded on the margins. *Lodicules* present; 2; free; fleshy; glabrous; not or scarcely vascularized. *Stamens* 3. *Ovary* glabrous. Styles fused. Stigmas 2.

Fruit, embryo and seedling. Fruit small; compressed laterally (plano-convex). Hilum short. Embryo large; waisted.

Abaxial leaf blade epidermis. *Costal/intercostal zonation* conspicuous. *Papillae* absent. Intercostal zones exhibiting many atypical long-cells to without typical long-cells. Mid-intercostal long-cells rectangular; having markedly sinuous walls (coarsely, deeply). *Microhairs* present; panicoid-type; (69–)81–83 microns long; (30–)48–51(–53) microns wide at the septum. Microhair total length/width at septum 12.8–16.9. Microhair apical cells 4.8–5.7 microns long. Microhair apical cell/total length ratio 0.43–0.74. *Stomata* common; about 33 microns long. Guard-cells overlapping to flush with the interstomatals. *Intercostal short-cells* common; not paired (apparently solitary); silicified. Intercostal silica bodies tall-and-narrow. *Costal short-cells* conspicuously in long rows. Costal silica bodies 'panicoid-type'.

Transverse section of leaf blade, physiology. C_3; XyMS+. *Mesophyll* with non-radiate chlorenchyma. *Leaf blade* 'nodular' in section to adaxially flat; with the ribs more or less constant in size. *Midrib* conspicuous; with one bundle only. Bulliforms not in discrete, regular adaxial groups (bulliform-like cells constituting most of adaxial epidermis). All the vascular bundles accompanied by sclerenchyma. Combined sclerenchyma girders present; forming 'figures'. Sclerenchyma all associated with vascular bundles.

Taxonomy. Panicoideae; Panicodae; Paniceae.

Ecology, geography, regional floristic distribution. 2 species. Tropical Africa. Shade species.

Paleotropical. African. Sudano-Angolan and West African Rainforest. Somalo-Ethiopian and South Tropical African.

References, etc. Leaf anatomical: this project.

Chloris O. Swartz

Actinochloris Steud., *Agrostomia* Cerv., *Apogon* Steud., *Chloridopsis* Hack., *Chloropsis* Kuntze, *Chlorostis* Raf., *Geopogon* Steud., *Heterolepis* Boiss., *Leptochloris* Kuntze, *Phacellaria* Steud., *Trichloris* Benth.

Excluding *Daknopholis (C. boivinii)*, *Pterochloris*

Habit, vegetative morphology. Annual, or perennial; rhizomatous, or stoloniferous, or caespitose, or decumbent. *Culms* 10–300 cm high; herbaceous. Culm nodes glabrous. Culm internodes solid, or hollow. *Leaves* not basally aggregated; non-auriculate. Sheaths usually not compressed and keeled. Leaf blades linear; narrow; flat, or folded, or rolled; without abaxial multicellular glands; without cross venation; persistent; once-folded in bud. *Ligule a fringed membrane to a fringe of hairs*.

Reproductive organization. Plants bisexual, with bisexual spikelets; with hermaphrodite florets; outbreeding and inbreeding. Exposed-cleistogamous, or chasmogamous. Plants with hidden cleistogenes, or without hidden cleistogenes. The hidden cleistogenes when present, subterranean. Apomictic, or reproducing sexually.

Inflorescence. *Inflorescence of spicate main branches (rarely a single raceme)*; deciduous in its entirety, or not deciduous; *digitate, or subdigitate (except* **C. roxburghana***)*. Primary inflorescence branches 2–30 (pectinate or the spikelets appressed). Rachides hollowed, or flattened. Inflorescence espatheate; not comprising 'partial inflorescences' and foliar organs. Spikelet-bearing axes persistent. Spikelets solitary, or paired; secund (on the dorsiventral rachis); biseriate; subsessile, or pedicellate; not in distinct 'long-and-short' combinations.

Female-fertile spikelets. *Spikelets* 1.8–5.5 mm long; adaxial; *compressed laterally*; disarticulating above the glumes (the glumes usually persistent); with conventional internode spacings (or at least, without the modified internode of *Oxychloris*). Rachilla prolonged beyond the uppermost female-fertile floret; rachilla hairless. The rachilla extension with incomplete florets. *Hairy callus present (usually minute).* **Callus** short; *blunt.*

Glumes two; very unequal (divergent); (the longer) usually *shorter than the adjacent lemmas*; dorsiventral to the rachis to lateral to the rachis; pointed (acute or acuminate); awnless; similar to very dissimilar (narrow, membranous, or the lower sometimes subulate). Lower glume 1 nerved. Upper glume 1–4 nerved. *Spikelets* with incomplete florets. *The incomplete florets distal to the female-fertile florets (at least one of these conspicuous — the spikelet 'with two florets').* The distal incomplete florets 2–5; *incomplete florets* merely underdeveloped.

Female-fertile florets *1 (rarely 2).* Lemmas similar in texture to the glumes to decidedly firmer than the glumes (membranous or cartilaginous); not becoming indurated; entire (truncate), or incised; when incised 2 lobed; not deeply cleft; awned. Awns 1 (usually), or 3; median, or median and lateral (rarely, e.g. *Trichloris*); the median similar in form to the laterals (when laterals present); from a sinus, or apical; non-geniculate; entered by one vein. Lemmas hairy (at the margins, often decoratively), or hairless; carinate; without a germination flap; 1–7 nerved. *Palea* present; relatively long; entire to apically notched; awnless, without apical setae; not indurated (hyaline or membranous); 2-nerved; 2-keeled. *Lodicules* present; 2; free; fleshy; glabrous. *Stamens* 3. Anthers 0.2–1.3 mm long; not penicillate. *Ovary* glabrous. Styles free to their bases. Stigmas 2; white, or red pigmented.

Fruit, embryo and seedling. Fruit free from both lemma and palea; small; ellipsoid (to lanceolate); shallowly longitudinally grooved, or not grooved; *not noticeably compressed (subterete), or trigonous.* Hilum short. Pericarp free to fused. Embryo large (1/2 to 2/3 the grain length). Endosperm hard; without lipid; containing only simple starch grains, or containing compound starch grains. Embryo with an epiblast; with a scutellar tail; with an elongated mesocotyl internode. Embryonic leaf margins meeting, or overlapping.

Seedling with a long mesocotyl. First seedling leaf with a well-developed lamina. The lamina broad; curved; 6–12 veined.

Abaxial leaf blade epidermis. *Costal/intercostal zonation* conspicuous. *Papillae* present; intercostal. Intercostal papillae not over-arching the stomata; several per cell (as small finger-like projections). *Long-cells* similar in shape costally and intercostally. Midintercostal long-cells rectangular; having markedly sinuous walls. *Microhairs* present; usually more or less spherical; ostensibly one-celled, or clearly two-celled; chloridoidtype. Microhair apical cell wall of similar thickness/rigidity to that of the basal cell. Microhairs with 'partitioning membranes' (in *C. gayana*, also in ×*Cynochloris*). The 'partitioning membranes' in the basal cell. Microhairs 5–21 microns long. Microhair basal cells 9 microns long. Microhair total length/width at septum about 1.5. Microhair apical cells 5–18 microns long. Microhair apical cell/total length ratio about 0.5. *Stomata* common; about 18 microns long. Subsidiaries non-papillate; triangular (mostly), or domeshaped (low). Guard-cells overlapping to flush with the interstomatals. *Intercostal shortcells* common, or absent or very rare; in cork/silica-cell pairs, or not paired; silicified (usually). Intercostal silica bodies imperfectly developed; tall-and-narrow, or saddle shaped. *Costal short-cells* conspicuously in long rows, or neither distinctly grouped into long rows nor predominantly paired. Costal silica bodies present in alternate cell files of the costal zones; saddle shaped, or tall-and-narrow to crescentic.

Transverse section of leaf blade, physiology. C_4; biochemical type PCK (6 species); XyMS+. PCR sheath outlines uneven. PCR sheaths of the primary vascular bundles interrupted; interrupted both abaxially and adaxially. PCR sheath extensions usually absent, or present (in *C. virgata*). Maximum number of extension cells in *C. virgata* 2. PCR cells with a suberised lamella. *PCR cell chloroplasts* ovoid; with well developed grana; centrifugal/peripheral. *Mesophyll* with radiate chlorenchyma; traversed by columns of colourless mesophyll cells (and the colourless tissue often extended adaxially across the lamina), or not traversed by colourless columns. *Leaf blade* 'nodular' in section, or adaxially flat; with the ribs more or less constant in size. *Midrib* conspicuous; with one bundle only, or having a conventional arc of bundles; with colourless mesophyll adaxially, or without colourless mesophyll adaxially. Bulliforms present in discrete, regular adaxial groups (usually, but the groups sometimes restricted to midrib hinges); in simple fans, or associated with colourless mesophyll cells to form deeply-penetrating fans (these groups sometimes continuous with traversing colourless columns). Many of the smallest vascular bundles unaccompanied by sclerenchyma, or all the vascular bundles accompanied by sclerenchyma. Combined sclerenchyma girders present; nowhere forming 'figures'. Sclerenchyma all associated with vascular bundles. The lamina margins with fibres.

Phytochemistry. Tissues of the culm bases with abundant starch. Leaves containing flavonoid sulphates (1 species), or without flavonoid sulphates (3 species). Leaf blade chlorophyll *a:b* ratio 3.23–3.93.

Cytology. Chromosome base number, $x = 10$. $2n = 14, 20, 26, 30, 36, 40, 72, 80$, and 100. 2, 3, 4, 5, 8, and 10 ploid. Haploid nuclear DNA content 0.35 pg (1 species). Mean diploid 2c DNA value 0.7 pg (1 species). Nucleoli persistent.

Taxonomy. Chloridoideae; main chloridoid assemblage.

Ecology, geography, regional floristic distribution. About 55 species. Tropical and warm temperate. Commonly adventive. Mesophytic, or xerophytic; species of open habitats. Diverse habitats, mostly in short grassland on poor soil or disturbed ground.

Holarctic, Paleotropical, Neotropical, and Australian. Boreal, Tethyan, and Madrean. African, Madagascan, Indomalesian, and Neocaledonian. Euro-Siberian and Atlantic North American. Macaronesian and Irano-Turanian. Saharo-Sindian, Sudano-Angolan, West African Rainforest, and Namib-Karoo. Indian, Indo-Chinese, Malesian, and Papuan. Caribbean, Venezuela and Surinam, Central Brazilian, Pampas, and Andean. North and East Australian, South-West Australian, and Central Australian. Siberian. Southern Atlantic North American and Central Grasslands. Sahelo-Sudanian, Somalo-Ethiopian, South Tropical African, and Kalaharian. Tropical North and East Australian.

Hybrids. Intergeneric hybrids with *Cynodon* — ×*Cynochloris* Clifford & Everist: several species involved.

Rusts and smuts. Rusts — *Puccinia*. Taxonomically wide-ranging species: *Puccinia cacabata* and *Puccinia chloridis*. Smuts from Ustilaginaceae. Ustilaginaceae — *Sorosporium*, *Sphacelotheca*, and *Ustilago*.

Economic importance. Significant weed species: *C. barbata, C. gayana, C. halophila* (in North America), *C. pilosa, C. polydactyla, C. pycnothrix, C. virgata*. Cultivated fodder: *C. gayana* (Rhodes). Important native pasture species: *C. roxburghiana, C. virgata*.

References, etc. Morphological/taxonomic: Anderson 1974; Jacobs and Highet 1988. Leaf anatomical: Metcalfe 1960; this project.

Chlorocalymma W. Clayton

Habit, vegetative morphology. Annual. *Culms* 30 cm high; herbaceous; branched above. *Leaves* not basally aggregated; non-auriculate. Leaf blades linear (broadly); narrow; 3–7 mm wide (to 5 cm long); flat; without cross venation; persistent. *Ligule an unfringed membrane*; truncate (erose). *Contra-ligule* absent.

Reproductive organization. Plants bisexual, with bisexual spikelets; with hermaphrodite florets.

Inflorescence. Inflorescence a false spike, with spikelets on contracted axes (a spike bearing 2–3 little ovate pseudo-racemes, at the summit of the culm). Primary inflorescence branches 2–3 (i.e., the glomerules or 'pseudo-racemes'). *Rachides* of the pseudoracemes broadly *winged (the wings enfolding the several spikelets and their associated spiny involucres)*. Inflorescence espatheate; not comprising 'partial inflorescences' and foliar organs. *Spikelet-bearing axes* very much reduced (to pseudo-racemes); disarticulating; falling entire (the pseudo-racemes falling). **Spikelets** *associated with bractiform involucres ('each spikelet subtended and partly invested by a rigid,* *branched, spiny bract or involucre')*. The involucres shed with the fertile spikelets (i.e., as part of the pseudo-raceme complex). Spikelets secund (i.e., the spikelets on one side of the spikelet-bearing branch).

Female-fertile spikelets. *Spikelets* morphologically 'conventional' (in the last analysis); 5.5–6 mm long; falling with the glumes. Rachilla terminated by a female-fertile floret. Hairy callus absent.

Glumes two; relatively large; very unequal (G_1 about half length of G_2, and hyaline); *shorter than the adjacent lemmas*; hairless (G_2 scaberulous); pointed; awnless; non-carinate. *Lower glume 0 nerved*. Upper glume 3 nerved. *Spikelets with incomplete florets*. The incomplete florets proximal to the female-fertile florets. *The proximal incomplete florets* 1; epaleate (?); sterile. The proximal lemmas awnless; more or less equalling the female-fertile lemmas; less firm than the female-fertile lemmas (herbaceous); not becoming indurated.

Female-fertile florets 1. Lemmas similar in texture to the glumes to decidedly firmer than the glumes (leathery, whitish); not becoming indurated; entire; pointed; awnless; hairless; non-carinate; having the margins lying flat and exposed on the palea (seemingly). *Palea* present; relatively long; entire; awnless, without apical setae; not indurated. *Stamens* 3. *Ovary* glabrous. Styles free to their bases. Stigmas 2.

Fruit, embryo and seedling. Fruit small (2.5 mm long); compressed dorsiventrally. Hilum short. Embryo large.

Abaxial leaf blade epidermis. *Costal/intercostal zonation* conspicuous. *Papillae* absent. *Long-cells* markedly different in shape costally and intercostally (the costals much narrower); of similar wall thickness costally and intercostally (rather thin walled). Intercostal zones with typical long-cells. Mid-intercostal long-cells rectangular; having markedly sinuous walls. *Microhairs* present; panicoid-type; 36–48 microns long; 6 microns wide at the septum. Microhair total length/width at septum 7. Microhair apical cells 21–27 microns long. Microhair apical cell/total length ratio 0.53–0.67. *Stomata* common; 24–30 microns long. Subsidiaries dome-shaped (mostly), or triangular (in a few files). Guard-cells overlapping to flush with the interstomatals. *Intercostal short-cells* absent or very rare (very scarce); (the few seen) not paired (solitary); not silicified. Small prickles numerous. *Costal short-cells* conspicuously in long rows. Costal silica bodies 'panicoid-type'; cross shaped, butterfly shaped, and dumb-bell shaped (short).

Transverse section of leaf blade, physiology. C_4; XyMS–. *Leaf blade* adaxially flat. *Midrib* conspicuous (via its small keel, and adaxial colourless tissue); with one bundle only to having a conventional arc of bundles (depending on whether it is taken to include the small laterals flanking the median); with colourless mesophyll adaxially. Bulliforms not in discrete, regular adaxial groups (the adaxial epidermis extensively bulliform). Many of the smallest vascular bundles unaccompanied by sclerenchyma. Combined sclerenchyma girders absent (with relatively little sclerenchyma, save for adaxial and abaxial strands). Sclerenchyma all associated with vascular bundles.

Special diagnostic feature. *The inflorescence not as in* **Odontelytrum** *(q.v.)*.

Cytology. Chromosome base number, $x = 9$. $2n = 18$ (+ 3B).

Taxonomy. Panicoideae; Panicodae; Paniceae.

Ecology, geography, regional floristic distribution. 1 species. Tanzania. In dry bushland; species of open habitats. Bushland.

Paleotropical. African. Sudano-Angolan. South Tropical African.

References, etc. Morphological/taxonomic: Clayton 1970. Leaf anatomical: this project.

Chrysochloa Swallen

Bracteola Swallen

Habit, vegetative morphology. Annual, or perennial; stoloniferous, or caespitose. The flowering culms leafy. *Culms* 10–75 cm high; herbaceous. Culm nodes glabrous. Culm internodes solid. Plants unarmed. Young shoots intravaginal. *Leaves* mostly basal; non-auriculate. Leaf blades narrow; 3–6 mm wide (strap-shaped, the apices blunt); not setaceous (flat, folded or enrolled); without abaxial multicellular glands; without cross venation; persistent. Ligule a fringed membrane; truncate. *Contra-ligule* absent.

Reproductive organization. Plants bisexual, with bisexual spikelets; with hermaphrodite florets. The spikelets all alike in sexuality (save that terminal spikelets may be reduced).

Inflorescence. *Inflorescence* of spicate main branches; *digitate, or subdigitate (a terminal, spreading-ascending verticil of racemes)*. Primary inflorescence branches 2–6. Rachides trigonous. Inflorescence espatheate; not comprising 'partial inflorescences' and foliar organs. *Spikelet-bearing axes* spikelike; persistent. Spikelets solitary; secund; narrowly biseriate; subsessile (appressed); imbricate; not in distinct 'long-and-short' combinations.

Female-fertile spikelets. *Spikelets* 3–5 mm long; strongly compressed laterally; *disarticulating between the glumes (leaving G_1 on the rachis)*; disarticulating between the florets; with conventional internode spacings. Rachilla prolonged beyond the uppermost female-fertile floret; rachilla hairless. The rachilla extension with incomplete florets. Hairy callus absent.

Glumes two; more or less equal; *long relative to the adjacent lemmas*; hairless; pointed; shortly awned (G_2 sometimes), or awnless; carinate (and G_1 laterally compressed), or carinate and non-carinate (G_2 sometimes rounded on back); narrow, lanceolate, membranous. Lower glume 1 nerved. Upper glume 1 nerved, or 5 nerved (*C. orientalis*). *Spikelets* with incomplete florets. The incomplete florets distal to the female-fertile florets. The distal incomplete florets 1 (male or sterile); *incomplete florets* merely underdeveloped.

Female-fertile florets 1. Lemmas similar in texture to the glumes to decidedly firmer than the glumes (firmly membranous to leathery); not becoming indurated; entire; pointed; mucronate, or awned. Awns when present, 1; dorsal; from near the top (subapical); non-geniculate; straight, or flexuous; hairless (glabrous or scabrid); much shorter than the body of the lemma. Lemmas hairy (villous on the nerves); carinate; without a germination flap; 3 nerved. *Palea* present; relatively long; apically notched; awnless, without apical setae; thinner than the lemma; not indurated (very thin); 2-nerved; 2-keeled. *Lodicules* present; 2; free; fleshy; glabrous; not or scarcely vascularized. *Stamens* 3. Anthers not penicillate (relatively long); without an apically prolonged connective. *Ovary* glabrous. Styles free to their bases. Stigmas 2.

Fruit, embryo and seedling. *Fruit* free from both lemma and palea (but embraced); ellipsoid; trigonous. Pericarp fused.

Abaxial leaf blade epidermis. *Costal/intercostal zonation* conspicuous. *Papillae* present; intercostal. Intercostal papillae over-arching the stomata (slightly); consisting of one symmetrical projection per cell (one thick-walled papilla in the middle of each intercostal cell). *Long-cells* markedly different in shape costally and intercostally (costals normal); of similar wall thickness costally and intercostally (thick walled). Intercostal zones exhibiting many atypical long-cells (the 'long-cells' and interstomatals short). Mid-intercostal long-cells rectangular; having markedly sinuous walls. *Microhairs* present; more or less spherical (about 9µ in diameter); ostensibly one-celled; chloridoid-type; about 9 microns long. *Stomata* common; 18–21 microns long. Subsidiaries non-papillate; predominantly triangular. Guard-cells overlapping to flush with the interstomatals. *Intercostal short-cells* common; not paired (solitary); silicified. Intercostal silica bodies absent; small. *Costal short-cells* conspicuously in long rows. Costal silica bodies present in alternate cell files of the costal zones; exclusively saddle shaped.

Transverse section of leaf blade, physiology. C_4; XyMS+. PCR sheaths of the primary vascular bundles interrupted; interrupted both abaxially and adaxially. PCR sheath extensions absent. *Leaf blade* adaxially flat. *Midrib* conspicuous (via a large, narrow abaxial keel); having a conventional arc of bundles (a large median and a smaller lateral on either side); with colourless mesophyll adaxially (midrib and adjoining lamina adaxially colourless). Bulliforms present in discrete, regular adaxial groups (between all the lamina bundles); associated with colourless mesophyll cells to form deeply-penetrating fans (the fans large, rather irregularly shaped). All the vascular bundles accompanied by sclerenchyma. Combined sclerenchyma girders present (with all the bundles); forming 'figures' (most bundles with scanty girders; submarginal bundles with heavy I-shaped combinations). Sclerenchyma all associated with vascular bundles.

Cytology. $2n = 14$.

Taxonomy. Chloridoideae; main chloridoid assemblage.

Ecology, geography, regional floristic distribution. 5 species. Tropical Africa. Species of open habitats. Seasonally waterlogged ground.

Paleotropical. African. Sudano-Angolan and West African Rainforest. Sahelo-Sudanian, Somalo-Ethiopian, and South Tropical African.

References, etc. Leaf anatomical: this project.

Chrysopogon Trin.

Centrophorum Trin., *Chalcoelytrum* Lunell, *Pollinia* Spreng., *Raphis* Lour., *Trianthium* Desv.

Habit, vegetative morphology. Annual, or perennial; rhizomatous, or stoloniferous, or caespitose, or decumbent. *Culms* 15–150 cm high; herbaceous; usually unbranched above. Culm nodes hairy, or glabrous. Culm internodes solid. *Leaves* mostly basal; non-auriculate. Leaf blades narrow (often harsh and glaucous); without cross venation; persistent. Ligule a fringed membrane (short), or a fringe of hairs.

Reproductive organization. Plants bisexual, with bisexual spikelets; with hermaphrodite florets. *The spikelets of sexually distinct forms on the same plant*; hermaphrodite and male-only, or hermaphrodite and sterile; overtly heteromorphic (pedicellate spikelets flattened dorsally, awnless or not, often purple: often the sessile spikelet pallid or yellowish); all in heterogamous combinations.

Inflorescence. Inflorescence paniculate (the branches usually with terminal triads of spikelets, but sometimes with a long-pedicel/short-pedicel pair below the triad); open (with whorls of slender, persistent branches); with capillary branchlets; espatheate; not comprising 'partial inflorescences' and foliar organs. **Spikelet-bearing axes *very much reduced (usually to a single joint and the terminal triad, but sometimes with a long-pedicel/short-pedicel pair below)*; with very slender rachides; disarticulating; disarticulating at the joints (beneath the triad, and beneath the pairs when present). 'Articles' without a basal callus-knob; disarticulating obliquely. Spikelets in triplets, or in triplets and paired; not secund; sessile and pedicellate; consistently in 'long-and-short' combinations; in pedicellate/sessile combinations. Pedicels of the 'pedicellate' spikelets free of the rachis. The 'shorter' spikelets hermaphrodite. The 'longer' spikelets male-only, or sterile (with flat pedicels).

Female-sterile spikelets. Male spikelets often purplish, on slender pedicels, dorsally compressed, awned or awnless, L_1 empty, L_2 with a male floret. Rachilla of male spikelets terminated by a male floret. The male spikelets with glumes; with proximal incomplete florets; 2 floreted. The lemmas awnless, or awned. Male florets 1.

Female-fertile spikelets. *Spikelets* 5–8.5 mm long; *compressed laterally*; falling with the glumes. Rachilla terminated by a female-fertile floret. Hairy callus present (commonly fulvous).

Glumes two; more or less equal; long relative to the adjacent lemmas; awned and awnless (G_2 often awned), or awnless. Lower glume two-keeled, or not two-keeled; convex on the back (or keeled upwards, sometimes with spinulose margins); not pitted; relatively smooth; 5–7 nerved. Upper glume 3–5 nerved. *Spikelets* with incomplete florets. The incomplete florets proximal to the female-fertile florets. *Spikelets with proximal incomplete florets*. *The proximal incomplete florets* 1; epaleate; sterile. The proximal lemmas awnless; 2 nerved; more or less equalling the female-fertile lemmas; similar in texture to the female-fertile lemmas (hyaline); not becoming indurated.

Female-fertile florets 1. Lemmas less firm than the glumes (hyaline); not becoming indurated; entire, or incised; when two-lobed not deeply cleft (bidentate); awned. Awns 1; median; from a sinus, or apical; geniculate; hairless (glabrous), or hairy; much shorter than the body of the lemma to much longer than the body of the lemma (?). Lemmas hairless; non-carinate; 1–3 nerved. **Palea** present, or absent; when present, conspicuous but relatively short, or very reduced; not indurated; *nerveless*. Lodicules present; 2; free;

fleshy; glabrous. *Stamens* 3. Anthers not penicillate. *Ovary* glabrous. Styles free to their bases. Stigmas 2.

Fruit, embryo and seedling. *Fruit* free from both lemma and palea; compressed laterally. Hilum short. Embryo large. Endosperm hard; without lipid.

Abaxial leaf blade epidermis. *Costal/intercostal zonation* conspicuous. *Papillae* absent. *Long-cells* similar in shape costally and intercostally (but the intercostals with somewhat rounded ends); of similar wall thickness costally and intercostally. Mid-intercostal long-cells rectangular (with the 'corners' rounded); having markedly sinuous walls. *Microhairs* present; panicoid-type; (48–)50–60(–64) microns long. Microhair apical cells (25–)30–36 microns long. Microhair apical cell/total length ratio 0.6. *Stomata* common. Subsidiaries triangular. Guard-cells overlapping to flush with the interstomatals. *Intercostal short-cells* common; in cork/silica-cell pairs; silicified. Intercostal silica bodies crescentic and cross-shaped. *Costal short-cells* predominantly paired. Costal silica bodies 'panicoid-type'; cross shaped, or dumb-bell shaped.

Transverse section of leaf blade, physiology. C_4; XyMS–. PCR cells with a suberised lamella. PCR cell chloroplasts with reduced grana (rudimentary); centrifugal/peripheral. *Mesophyll* with radiate chlorenchyma. *Leaf blade* with distinct, prominent adaxial ribs, or 'nodular' in section; with the ribs more or less constant in size. *Midrib* not readily distinguishable; with one bundle only (rarely), or having a conventional arc of bundles. Bulliforms present in discrete, regular adaxial groups (or in more or less irregular groups), or not in discrete, regular adaxial groups (sometimes the groups confined to midrib hinges); in simple fans, or associated with colourless mesophyll cells to form deeply-penetrating fans. Many of the smallest vascular bundles unaccompanied by sclerenchyma. Combined sclerenchyma girders present; nowhere forming 'figures'. Sclerenchyma all associated with vascular bundles.

Phytochemistry. Leaves without flavonoid sulphates (1 species).

Cytology. Chromosome base number, $x = 5$ and 10. $2n = 20$ and 40.

Taxonomy. Panicoideae; Andropogonodae; Andropogoneae; Andropogoninae.

Ecology, geography, regional floristic distribution. 25 species. Tropical and subtropical. Mesophytic, or xerophytic (from subdesert to rainforest); species of open habitats; glycophytic. On poor soils, often in disturbed ground.

Holarctic, Paleotropical, Neotropical, and Australian. Boreal and Tethyan. African, Madagascan, Indomalesian, Polynesian, and Neocaledonian. Euro-Siberian. Mediterranean and Irano-Turanian. Saharo-Sindian, Sudano-Angolan, and West African Rainforest. Indian, Indo-Chinese, Malesian, and Papuan. Hawaiian. Caribbean. North and East Australian. European. Sahelo-Sudanian, Somalo-Ethiopian, South Tropical African, and Kalaharian. Tropical North and East Australian.

Rusts and smuts. Rusts — *Puccinia*. Taxonomically wide-ranging species: *Puccinia graminis*. Smuts from Ustilaginaceae. Ustilaginaceae — *Sorosporium*, *Sphacelotheca*, and *Ustilago*.

Economic importance. Significant weed species: *C. aciculatus* (sometimes, in pastures). Cultivated fodder: *C. fulvus*. Important native pasture species: *C. plumulosus*. Lawns and/or playing fields: *C. aciculatus* (in the humid tropics).

References, etc. Leaf anatomical: Metcalfe 1960; this project.

Chumsriella Bor

~ *Germainia*

Habit, vegetative morphology. Annual. *Culms* 10 cm high; herbaceous. *Leaves* not basally aggregated; non-auriculate. Leaf blades narrow; 2–3 mm wide (1–2 cm long); flat (hairy); without cross venation; persistent. **Ligule** present; *a fringe of hairs*.

Reproductive organization. *Plants monoecious with all the fertile spikelets unisexual*; without hermaphrodite florets. The spikelets of sexually distinct forms on the same plant; female-only and male-only; overtly heteromorphic (the pedicellate, female spikelets hairy and long-awned, the sessile males hairless and awnless); all in heterogamous combinations.

Inflorescence. *Inflorescence terminal, consisting of a pair of short 'racemes' on a long peduncle*. Primary inflorescence branches 2. Inflorescence spatheate (subtended by a spathiform sheath with reduced lamina). *Spikelet-bearing axes* 'racemes' (3–4 noded only); the spikelet-bearing axes with 2–3 spikelet-bearing 'articles', or with 4–5 spikelet-bearing 'articles'; paired (the pair fused by the lowest two joints); with substantial rachides (in proportion to their size); persistent. Spikelets paired (including the pair terminating the 'raceme'); sessile and pedicellate; consistently in 'long-and-short' combinations; in pedicellate/sessile combinations. Pedicels of the 'pedicellate' spikelets free of the rachis. The 'shorter' spikelets male-only. The 'longer' spikelets female-only.

Female-sterile spikelets. The sessile male spikelets 3–4 mm long, with 1–2 awnless florets, the lower sterile, the upper with three stamens. Rachilla of male spikelets terminated by a male floret. *The male spikelets* with glumes; with proximal incomplete florets; usually *2 floreted*. The lemmas awnless. Male florets 1; 3 staminate.

Female-fertile spikelets. Spikelets 2.5 mm long; falling with the glumes (deciduous from the cupular pedicel-tip). Rachilla terminated by a female-fertile floret.

Glumes two; more or less equal; long relative to the adjacent lemmas (exceeding them); hairy (densely brown-pilose); lanceolate, attenuate to more or less bifid apices; awnless; non-carinate; similar (but G_2 narrower, less densely hairy). Lower glume not two-keeled; not pitted. *Spikelets* with female-fertile florets only (i.e., lacking the sterile andropogonoid L_1).

Female-fertile florets 1. **Lemmas** lanceolate, produced into the awn; less firm than the glumes (hyaline below, papery beneath the awn); not becoming indurated; entire; *awned*. Awns 1; median; apical; geniculate (bigeniculate); hairy; much longer than the body of the lemma (30 mm or more long); entered by one vein. Lemmas non-carinate; without a germination flap; 1 nerved. *Palea* present; conspicuous but relatively short (about 1 mm long – i.e., half lemma length); entire (oblong-acute); awnless, without apical setae; not indurated; keel-less. *Stamens* 0. *Ovary* glabrous. Styles fused (briefly). Stigmas 2.

Fruit, embryo and seedling. Fruit small. Hilum short. Embryo large (3/4 as long as the grain).

Taxonomy. Panicoideae; Andropogonodae; Andropogoneae; Andropogoninae.

Ecology, geography, regional floristic distribution. 1 species. Thailand. Paleotropical. Indomalesian. Indo-Chinese.

References, etc. Morphological/taxonomic: Bor 1968b; Chaianan 1972.

Special comments. Anatomical data wanting.

Chusquea Kunth

Dendragrostis Jackson, *Rettbergia* Raddi
Excluding *Swallenochloa*

Habit, vegetative morphology. Perennial. The flowering culms leafy. *Culms woody and persistent*; scandent, or not scandent; branched above. Culm nodes glabrous. *Primary branches/mid-culm node 3–12 (or more — unrestricted pleioclade)*. Culm sheaths persistent, or deciduous in their entirety. Culm internodes solid (rarely with an irregular passage formed by shrinkage of pith). Unicaespitose, or pluricaespitose. Rhizomes pachymorph, or leptomorph. Plants conspicuously armed, or unarmed. Young shoots extravaginal. *Leaves* not basally aggregated; non-auriculate; without auricular setae. Leaf blades broad, or narrow; pseudopetiolate; cross veined, or without cross venation; disarticulating from the sheaths. Ligule an unfringed membrane.

Reproductive organization. Plants bisexual, with bisexual spikelets; with hermaphrodite florets.

Inflorescence. Inflorescence paniculate; open (rarely racemose or capitate); espatheate; not comprising 'partial inflorescences' and foliar organs. *Spikelet-bearing axes* paniculate. Spikelets solitary; not secund; pedicellate.

Female-fertile spikelets. *Spikelets* 8–10 mm long (*C. scandens*); *not noticeably compressed*; disarticulating above the glumes. Rachilla terminated by a female-fertile floret; rachilla hairless. Hairy callus absent.

Glumes two; more or less equal; shorter than the adjacent lemmas; hairless; not pointed; awnless; non-carinate; similar. *Spikelets* with incomplete florets. The incomplete florets proximal to the female-fertile florets. *Spikelets with proximal incomplete florets. The proximal incomplete florets* 1–3; sterile. The proximal lemmas awnless; exceeded by the female-fertile lemmas; similar in texture to the female-fertile lemmas; not becoming indurated.

Female-fertile florets 1. Lemmas similar in texture to the glumes; not becoming indurated; entire; pointed; awnless, or mucronate (?); hairless; non-carinate; without a germination flap. *Palea* present; relatively long; apically notched to deeply bifid; awnless, without apical setae; not indurated; keel-less. *Lodicules* present; 3; free; membranous; ciliate; not toothed. *Stamens* 3. Anthers not penicillate; without an apically prolonged connective. **Ovary** glabrous; *without a conspicuous apical appendage*. Styles fused. Stigmas 2.

Fruit, embryo and seedling. Fruit free from both lemma and palea; medium sized; slightly *longitudinally grooved*; not noticeably compressed. Hilum long-linear. Embryo small; not waisted.

Abaxial leaf blade epidermis. *Costal/intercostal zonation* conspicuous. *Papillae* present. Intercostal papillae several per cell. *Long-cells* similar in shape costally and intercostally, or markedly different in shape costally and intercostally; differing markedly in wall thickness costally and intercostally. Mid-intercostal long-cells rectangular; having markedly sinuous walls. *Microhairs* present; panicoid-type. *Stomata* common. Subsidiaries papillate; dome-shaped, or triangular, or parallel-sided (the triangles often with blunted/truncated apices); including both triangular and parallel-sided forms on the same leaf, or not including both parallel-sided and triangular forms on the same leaf. *Intercostal short-cells* common. *Costal short-cells* conspicuously in long rows. Costal silica bodies saddle shaped, or 'panicoid-type'; when panicoid type, cross shaped to dumb-bell shaped.

Transverse section of leaf blade, physiology. C_3; XyMS+. Mesophyll with arm cells; with fusoids. *Leaf blade* adaxially flat (adaxial ribs weakly developed). *Midrib* conspicuous; with one bundle only (rarely complex). Bulliforms present in discrete, regular adaxial groups; in simple fans (these wide or narrow). All the vascular bundles accompanied by sclerenchyma. Combined sclerenchyma girders present; forming 'figures'. Sclerenchyma all associated with vascular bundles.

Cytology. Chromosome base number, $x = 12$. $2n = 40, 44$, and 48. 4 ploid.

Taxonomy. Bambusoideae; Bambusodae; Bambuseae.

Ecology, geography, regional floristic distribution. More than 100 species. Mexico to Chile and Argentina. A very diverse genus with an altitudinal range from sea level to the lower limit of perpetual snow. Helophytic.

Neotropical and Antarctic. Caribbean, Venezuela and Surinam, Amazon, Central Brazilian, Pampas, Andean, and Fernandezian. Patagonian.

References, etc. Morphological/taxonomic: Soderstrom 1978; Soderstrom and Calderón 1978a, 1978b. Leaf anatomical: Metcalfe 1960; this project.

Cinna L.

Abola Adans., *Blyttia* Fries

Habit, vegetative morphology. Perennial (tall, of wet habitats); rhizomatous, or caespitose (laxly, or the culms solitary). *Culms* (20–)50–200(–220) cm high; herbaceous; tuberous (somewhat, in *C. arundinacea*), or not tuberous. Culm nodes glabrous (or somewhat scaberulous). Culm internodes hollow. Leaves non-auriculate. Sheath margins free. Sheaths glabrous. *Leaf blades linear*; broad to narrow; 1–20 mm wide (to 40 cm long); flat (with scabrous margins); without cross venation. *Ligule an unfringed membrane*; 2–10(–12) mm long.

Reproductive organization. Plants bisexual, with bisexual spikelets; with hermaphrodite florets.

Inflorescence. Inflorescence paniculate (with numerous spikelets); open; espatheate; not comprising 'partial inflorescences' and foliar organs. Spikelet-bearing axes persistent. Spikelets not secund; pedicellate.

Female-fertile spikelets. *Spikelets* (1.9–)2–6(–7.5) mm long; *compressed laterally*; *falling with the glumes.* Rachilla prolonged beyond the uppermost female-fertile floret (usually, as a stub or bristle), or terminated by a female-fertile floret (rarely). The rachilla extension with incomplete florets (usually), or naked. Hairy callus absent. *Callus* short.

Glumes two; more or less equal (the lower sometimes somewhat shorter); *long relative to the adjacent lemmas (a little shorter to a little longer)*; free; hairless; glabrous to scabrous (smooth to scaberulous); pointed (acute or acuminate); awnless (sometimes mucronate); carinate; similar (thinly membranous, the margins hyaline). Lower glume 1 nerved. Upper glume 1 nerved, or 3 nerved. *Spikelets* with female-fertile florets only (usually), or with incomplete florets (very rarely). The incomplete florets when present, distal to the female-fertile florets. The distal incomplete florets 1; *incomplete florets* merely underdeveloped. *Spikelets without proximal incomplete florets.*

Female-fertile florets 1 (nearly always, though the second floret when present is said to be occasionally 'fertile'). Lemmas similar to the glumes; similar in texture to the glumes to decidedly firmer than the glumes (firmly membranous); not becoming indurated; entire, or incised; when entire pointed, or blunt; awnless, or mucronate, or awned. Awns when present, 1; dorsal; from near the top (or subterminal); non-geniculate; straight; much shorter than the body of the lemma; entered by one vein. *Lemmas* hairless; *carinate*; without a germination flap; 3 nerved, or 5 nerved. *Palea* present; relatively long; thinner than the lemma (hyaline); 1-nerved, or 2-nerved (one veined in *C. arundinacea*, ostensibly so in the others by apposition of the two veins); one-keeled (literally so in *C. arundinacea*, ostensibly so by close apposition in the rest). *Lodicules* present; 2; free; membranous; ciliate; toothed, or not toothed (rarely); not or scarcely vascularized. **Stamens** *1, or 2.* Anthers (0.4–)0.6–1.2(–2.6) mm long. *Ovary* glabrous. Styles free to their bases. Stigmas 2; white.

Fruit, embryo and seedling. *Fruit* free from both lemma and palea; small (about 2.5 mm long); yellowish brown; compressed laterally. Hilum short. Embryo small. Endosperm liquid in the mature fruit, or hard; with lipid; containing compound starch grains. Embryo with an epiblast; without a scutellar tail; with a negligible mesocotyl internode. Embryonic leaf margins meeting.

Abaxial leaf blade epidermis. *Costal/intercostal zonation* conspicuous. *Papillae* absent. *Long-cells* markedly different in shape costally and intercostally; of similar wall thickness costally and intercostally. Mid-intercostal long-cells fusiform; having straight or only gently undulating walls. *Microhairs* absent. *Stomata* common; (30–)33–36(–38) microns long. Subsidiaries parallel-sided. Guard-cells overlapped by the interstomatals. *Intercostal short-cells* absent or very rare. Prickles abundant. *Costal short-cells* neither distinctly grouped into long rows nor predominantly paired. Costal silica bodies horizontally-elongated crenate/sinuous, or horizontally-elongated smooth.

Transverse section of leaf blade, physiology. C_3; XyMS+. *Mesophyll* with non-radiate chlorenchyma.

Cytology. Chromosome base number, $x = 7$. $2n = 28$. 4 ploid.

Taxonomy. Pooideae; Poodae; Aveneae.

Ecology, geography, regional floristic distribution. 3–4 species. Temperate Eurasia, North & South America. Helophytic, or mesophytic; shade species; glycophytic.

Holarctic and Neotropical. Boreal. Arctic and Subarctic, Euro-Siberian, Eastern Asian, Atlantic North American, and Rocky Mountains. Caribbean, Venezuela and Surinam, and Andean. European and Siberian. Canadian-Appalachian and Southern Atlantic North American.

Rusts and smuts. Rusts — *Puccinia*. Taxonomically wide-ranging species: *Puccinia graminis*, *Puccinia coronata*, and *Puccinia recondita*.

References, etc. Morphological/taxonomic: Brandenburg, Blackwell and Thieret 1991. Leaf anatomical: this project.

Cladoraphis Franch.

~ Eragrostis

Habit, vegetative morphology. Perennial; rhizomatous, or rhizomatous and stoloniferous (occasionally). **Culms** 20–80 cm high; *woody and persistent*; to 4 cm in diameter; branched above. Culm nodes glabrous. Culm internodes solid. *Plants conspicuously armed (with pungent tipped leaf blades and inflorescence axes)*. Young shoots intravaginal. *Leaves* mostly basal; non-auriculate. *Leaf blades* linear-lanceolate to lanceolate; narrow; to 6 mm wide; becoming rolled; *hard, woody, needle-like*; without abaxial multicellular glands; without cross venation; persistent. Ligule a fringe of hairs; to 2 mm long.

Reproductive organization. Plants bisexual, with bisexual spikelets; with hermaphrodite florets. The spikelets all alike in sexuality.

Inflorescence. Inflorescence paniculate (with distant branches, or reduced to a single branch or cluster); open (but the lateral branches compact). Primary inflorescence branches 4–25 (or more). *Inflorescence usually with axes not ending in spikelets (these usually pungent-tipped, but sometimes terminating in a cluster of spikelets)*. Inflorescence espatheate; not comprising 'partial inflorescences' and foliar organs. Spikelet-bearing axes persistent. Spikelets solitary; not secund; not two-ranked; pedicellate (the pedicels short); distant.

Female-fertile spikelets. Spikelets 7–16 mm long; adaxial; compressed laterally; disarticulating above the glumes; disarticulating between the florets (tardily); with conventional internode spacings. Rachilla prolonged beyond the uppermost female-fertile floret; rachilla hairy (puberulous), or hairless. The rachilla extension with incomplete florets. *Callus* absent.

Glumes two; more or less equal; shorter than the spikelets; shorter than the adjacent lemmas; dorsiventral to the rachis; hairless; glabrous; pointed (but sometimes split at the tip); awnless; carinate; similar. Lower glume 3 nerved. *Upper glume 3 nerved. Spikelets* with incomplete florets. The incomplete florets distal to the female-fertile florets. *The distal incomplete florets* merely underdeveloped; awnless.

Female-fertile florets 3–16. Lemmas not saccate; similar in texture to the glumes; not becoming indurated; entire; pointed to blunt; awnless (muticous); hairless (sometimes puberulent); finely scaberulous or very finely puberulous; carinate; without a germination flap; 3 nerved; with the nerves non-confluent. *Palea* present; relatively long (equalling the lemmas or slightly shorter); entire; awnless, without apical setae; textured like the lemma (membranous); not indurated; 2-nerved; 2-keeled. Palea keels wingless (but usually puberulous). *Lodicules* present; 2; free; fleshy. *Stamens* 3. Anthers 1.5–2 mm long; not penicillate; without an apically prolonged connective. *Ovary* glabrous. Styles free to their bases. Stigmas 2; white.

Fruit, embryo and seedling. *Fruit* free from both lemma and palea; small (1.3 to 2 mm long); compressed dorsiventrally. Hilum short. Pericarp free. Embryo large.

Abaxial leaf blade epidermis. *Costal/intercostal zonation* conspicuous. *Papillae* absent. Mid-intercostal long-cells rectangular. *Microhairs* present (but hidden in the deep intercostal grooves of the lamina); elongated; clearly two-celled; chloridoid-type. Microhair apical cell wall of similar thickness/rigidity to that of the basal cell. Microhair basal cells 30 microns long. Microhair total length/width at septum 3.4. Microhair apical cell/total length ratio 0.3. *Stomata* common. Intercostal short-cells not paired. Intercostal silica bodies absent. *Costal short-cells* neither distinctly grouped into long rows nor predominantly paired (the rows with many prickles). Costal silica bodies present in alternate cell files of the costal zones; saddle shaped (to squarish).

Transverse section of leaf blade, physiology. C_4. The anatomical organization fairly conventional. XyMS+. PCR sheath outlines uneven (owing to the adaxial extensions). PCR sheaths of the primary vascular bundles complete. PCR sheath extensions present (with most bundles). Maximum number of extension cells 4–5. PCR cell chloroplasts centripetal. *Mesophyll* with radiate chlorenchyma; traversed by columns of colourless mesophyll cells (between each pair of bundles). *Leaf blade* 'nodular' in section; with the ribs more or less constant in size (the abaxial ribs somewhat broader and flatter than their adaxial counterparts, the grooves between them narrow). *Midrib* not readily distinguishable; with one bundle only. Bulliforms present in discrete, regular adaxial groups (in each adaxial groove); associated with colourless mesophyll cells to form deeply-penetrating fans (these linking with the heavy colourless columns). All the vascular bundles accompanied by sclerenchyma. Combined sclerenchyma girders present; forming 'figures' (all the bundles with conspicuous I's or anchors). Sclerenchyma all associated with vascular bundles. The lamina margins with fibres.

Taxonomy. Chloridoideae; main chloridoid assemblage.

Ecology, geography, regional floristic distribution. 2 species. Southern Africa. Xerophytic; species of open habitats; halophytic, or glycophytic. *C. cyperoides* on beach dunes, *C. spinosa* on desert dunes and in sandy beds of dry watercourses.

Paleotropical. African. Namib-Karoo.

References, etc. Leaf anatomical: photos of *C. cyperoides* provided by R.P. Ellis; Van den Borre 1994 (*C. spinosa*).

Clausospicula M. Lazarides

~ *Anadelphia* sensu lato

Habit, vegetative morphology. Annual. Culms herbaceous; branched above. *Leaves* not basally aggregated. *Leaf blades* linear; *1.4–2.6 mm wide*; flat, or folded; without cross venation; persistent. *Ligule* present; an unfringed membrane; truncate; 0.3–0.5 mm long. *Contra-ligule* absent.

Reproductive organization. Plants bisexual, with bisexual spikelets; with hermaphrodite florets. The spikelets of sexually distinct forms on the same plant; hermaphrodite and sterile. The male and female-fertile spikelets mixed in the inflorescence. The spikelets overtly heteromorphic; all in heterogamous combinations. Plants inbreeding. Exposed-cleistogamous.

Inflorescence. Inflorescence paniculate; open; spatheate (and spatheolate); a complex of 'partial inflorescences' and intervening foliar organs. **Spikelet-bearing axes *very much reduced (each 'raceme' reduced to one 'triplet', clasped basally by the peculiar, trumpet-shaped tip of the peduncle)*;** the spikelet-bearing axes with only one spikelet-bearing 'article'; solitary; disarticulating; disarticulating at the joints (i.e. each at its only joint). The pedicels and rachis internodes without a longitudinal, translucent furrow. 'Articles' without a basal callus-knob; disarticulating obliquely. Spikelets in triplets (one of the pedicellate members sometimes missing); sessile and pedicellate; consistently in 'long-and-short' combinations; in pedicellate/sessile combinations. Pedicels of the 'pedicellate' spikelets free of the rachis. The 'shorter' spikelets hermaphrodite. The 'longer' spikelets sterile (sometimes reduced to the glabrous pedicels).

Female-sterile spikelets. The sterile spikelets, when present, 5–12 mm long, linear to lanceolate, awnless, disarticulating horizontally.

Female-fertile spikelets. Spikelets about 10 mm long; not noticeably compressed to compressed dorsiventrally; falling with the glumes. Rachilla terminated by a female-fertile floret. Hairy callus present. *Callus* long; pointed.

Glumes present; two; more or less equal; about equalling the spikelets, or exceeding the spikelets; long relative to the adjacent lemmas (more or less exceeding the L1); hairy and hairless (usually tubercled basally and towards the tip, otherwise glabrous); without conspicuous tufts or rows of hairs; awned; non-carinate; very dissimilar (both becoming indurated, the lower abruptly contracted above and with or without a short bristle from the narrow, truncate to notched apex, the upper attenuate into a long, slender scabrous awn 14–19 mm long which is often twisted around the lemma awn). Lower glume not two-keeled; convex on the back; not pitted; relatively smooth and tuberculate; 4 nerved (without a median). Upper glume 3 nerved. *Spikelets* with incomplete florets. The incomplete florets proximal to the female-fertile florets. *The proximal incomplete florets* 1; epaleate; sterile. The proximal lemmas awnless; 2 nerved, or 4 nerved; more or less equalling the female-fertile lemmas to decidedly exceeding the female-fertile lemmas; similar in texture to the female-fertile lemmas (hyaline); not becoming indurated.

Female-fertile florets 1. *Lemmas* less firm than the glumes; not becoming indurated; *entire (the hyaline margins of the stipe antero-laterally decurrent on the base of the awn, with no indication of lobes)*; awned. Awns 1; median; apical; geniculate; hairless to hairy; much longer than the body of the lemma (8–11 cm long); entered by several veins. Awn bases twisted; not flattened. Lemmas hairless; glabrous; non-carinate; without a germination flap; 3 nerved; with the nerves non-confluent. *Palea* present; conspicuous but relatively short; entire to apically notched; awnless, without apical setae; textured like the lemma (hyaline); not indurated; 2-nerved, or nerveless. *Lodicules* present; 2; free; fleshy; glabrous. *Stamens* 3 (short). Anthers 0.5–1.5 mm long; not penicillate; without an apically prolonged connective. *Ovary* glabrous. Styles free to their bases. Stigmas 2, or 3.

Fruit, embryo and seedling. *Fruit* free from both lemma and palea; medium sized (about 5.6 mm long); ellipsoid; compressed laterally (above), or compressed dorsiventrally to not noticeably compressed (below); glabrous. Hilum short. Embryo large.

Abaxial leaf blade epidermis. *Costal/intercostal zonation* conspicuous. *Papillae* absent. Long-cells of similar wall thickness costally and intercostally. Mid-intercostal long-cells rectangular to fusiform; having markedly sinuous walls. *Microhairs* present; panicoid-type; 39–47 microns long; 6–12 microns wide at the septum. Microhair total length/width at septum 3.25–7.5. Microhair apical cells 22–30 microns long. Microhair apical cell/total length ratio 0.57–0.67. *Stomata* common; 31–41 microns long. Subsidiaries triangular. Guard-cells overlapping to flush with the interstomatals. *Intercostal short-cells* common; in cork/silica-cell pairs; silicified. Intercostal silica bodies cross-shaped. Costal zones with short-cells. *Costal short-cells* conspicuously in long rows, or predominantly paired, or neither distinctly grouped into long rows nor predominantly paired. Costal silica bodies 'panicoid-type'; not sharp-pointed.

Transverse section of leaf blade, physiology. C_4; XyMS−. PCR sheath outlines uneven. PCR sheath extensions absent. *Leaf blade* with distinct, prominent adaxial ribs (the ribs low, variable in width). *Midrib* conspicuous; having a conventional arc of bundles (a large median, with four or five laterals on either side in a deep arc); with colourless mesophyll adaxially. Bulliforms present in discrete, regular adaxial groups (in places), or not in discrete, regular adaxial groups (mainly, the epidermis largely bulliform); in simple fans (in places), or associated with colourless mesophyll cells to form deeply-penetrating fans (mainly); associating with colourless mesophyll cells to form arches over small vascular bundles (noticeable only in Barritt 772), or nowhere involved in bulliform-plus-colourless mesophyll arches. Many of the smallest vascular bundles unaccompanied by sclerenchyma. Combined sclerenchyma girders present; nowhere forming 'figures'. Sclerenchyma all associated with vascular bundles.

Special diagnostic feature. *Spikelets in much-reduced andropogonoid 'racemes', each of the latter reduced to a single triplet and enclosed at its base by a trumpet-like development of the peduncle tip.*

Taxonomy. Panicoideae; Andropogonodae; Andropogoneae; Andropogoninae.

Ecology, geography, regional floristic distribution. 1 species. Northern Australia. Australian. North and East Australian.

References, etc. Morphological/taxonomic: Lazarides *et al*. 1990. Leaf anatomical: this project.

Cleistachne Benth.

Habit, vegetative morphology. *Annual. Culms* 60–250 cm high; herbaceous; usually unbranched above (sometimes with stilt roots). Plants unarmed. Leaves non-auriculate. Leaf blades linear; broad, or narrow; 4–15 mm wide; rolled; pseudopetiolate (often), or not pseudopetiolate; without cross venation. Ligule an unfringed membrane (scarious); not truncate (apically rounded).

Reproductive organization. Plants bisexual, with bisexual spikelets; with hermaphrodite florets.

Inflorescence. *Inflorescence paniculate (composed of racemes of greatly reduced 'racemes')*; large, terminal, linear to lanceolate; espatheate; not comprising 'partial inflorescences' and foliar organs. *Spikelet-bearing axes* ostensibly 'racemes' (these long, narrow, with many joints); with very slender rachides; persistent. 'Articles' densely longhairy, or somewhat hairy (the rachis and ostensible 'pedicels' with grey or brown hairs). *Spikelets solitary (each seemingly representing a 'raceme' reduced to a single spikelet,* *cf.* **Sorghum, Sorghastrum***); not secund; sessile (in that the ostensible 'pedicels' represent the peduncles of reduced lateral branches); *not in distinct 'long-and-short' combinations.*

Female-fertile spikelets. Spikelets (3–)4–5(–6) mm long; compressed dorsiventrally; falling with the glumes (disarticulating from the apex of the peduncle). Rachilla terminated by a female-fertile floret.

Glumes two; more or less equal; long relative to the adjacent lemmas; hairy; not pointed (truncate); awnless; non-carinate; similar (leathery, with inrolled margins). Lower glume not two-keeled; not pitted; relatively smooth; 7–9 nerved. *Upper glume 7–9 nerved. Spikelets* with incomplete florets. The incomplete florets proximal to the female-fertile florets. *The proximal incomplete florets* 1; epaleate; sterile. The proximal lemmas awnless; 2 nerved; decidedly exceeding the female-fertile lemmas; similar in texture to the female-fertile lemmas (hyaline); not becoming indurated.

Female-fertile florets 1. **Lemmas** less firm than the glumes (hyaline); not becoming indurated; *incised*; not deeply cleft (bidentate); awned. Awns 1; median; from a sinus; geniculate; hairless (glabrous); much longer than the body of the lemma. Lemmas hairy (ciliate apically); non-carinate; without a germination flap; 3 nerved. **Palea** present (but small); conspicuous but relatively short; entire; awnless, without apical setae (but ciliate); not indurated; seemingly *nerveless*; keel-less. *Lodicules* present; 2; free; fleshy; ciliate. *Stamens* 3. *Ovary* glabrous. Styles free to their bases. Stigmas 2.

Fruit, embryo and seedling. Hilum short. Embryo large.

Abaxial leaf blade epidermis. *Costal/intercostal zonation* conspicuous. *Papillae* present. Mid-intercostal long-cells having markedly sinuous walls (thin). *Microhairs* present; panicoid-type; (45–)48–54(–58) microns long. Microhair apical cells 30–36 microns long. Microhair apical cell/total length ratio 0.65. *Stomata* common. Subsidiaries triangular. *Intercostal short-cells* absent or very rare. *Costal short-cells* conspicuously in long rows. Costal silica bodies 'panicoid-type'; dumb-bell shaped and nodular.

Transverse section of leaf blade, physiology. C_4; XyMS–. *Mesophyll* with radiate chlorenchyma. *Leaf blade* adaxially flat. *Midrib* conspicuous; having a conventional arc of bundles (with numerous bundles); with colourless mesophyll adaxially. Bulliforms present in discrete, regular adaxial groups; in simple fans (these large-celled towards the middle of the blade, the groups irregular towards the margins). Many of the smallest vascular bundles unaccompanied by sclerenchyma. Combined sclerenchyma girders present (with the large bundles). Sclerenchyma all associated with vascular bundles.

Cytology. Chromosome base number, $x = 9$. $2n = 36$.

Taxonomy. Panicoideae; Andropogonodae; Andropogoneae; Andropogoninae.

Ecology, geography, regional floristic distribution. 1 species. Tropical Africa, India. Helophytic to mesophytic; species of open habitats; glycophytic. Riverbanks and old farmland.

Paleotropical. African and Indomalesian. Sudano-Angolan. Indian. South Tropical African.

References, etc. Leaf anatomical: Metcalfe 1960.

Cleistochloa C.E. Hubb.

Excluding *Dimorphochloa*

Habit, vegetative morphology. Perennial; caespitose. *Culms* 30–60 cm high; wiry; branched above. Culm nodes glabrous. Culm internodes hollow. Young shoots usually intravaginal. *Leaves* not basally aggregated; non-auriculate. *Leaf blades* narrow; flat, or rolled (rough, with tubercle-based hairs); without cross venation; *disarticulating from the sheaths*. Ligule a fringe of hairs.

Reproductive organization. Plants bisexual, with bisexual spikelets; with hermaphrodite florets. The exposed spikelets chasmogamous. *Plants with hidden cleistogenes*. The hidden cleistogenes in the leaf sheaths (borne singly, highly modified).

Inflorescence. *Inflorescence* few spikeleted; *a single raceme (spike-like, terminating culm branches)*; espatheate. Spikelet-bearing axes persistent. Spikelets solitary; not secund; pedicellate (the pedicels very short). Pedicel apices cupuliform.

Female-fertile spikelets. Spikelets 3.5–4.5 mm long; oblong, or elliptic, or obovate; adaxial; compressed dorsiventrally; biconvex; falling with the glumes. Rachilla terminated by a female-fertile floret. Hairy callus absent.

Glumes present; one per spikelet, or two (the lower minute or absent); (the upper) relatively large; very unequal; shorter than the adjacent lemmas; free; dorsiventral to the rachis; not pointed (rounded); awnless; very dissimilar (G₁ minute). Lower glume 0 nerved. Upper glume 5–7 nerved. *Spikelets* with incomplete florets. The incomplete florets proximal to the female-fertile florets. *The proximal incomplete florets* 1; epaleate; sterile. The proximal lemmas awnless (hairy on margins and apex); 7 nerved; more or less equalling the female-fertile lemmas; less firm than the female-fertile lemmas to similar in texture to the female-fertile lemmas; not becoming indurated.

Female-fertile florets 1. **Lemmas** similar to the L₁; striate; not becoming indurated; yellow in fruit, or brown in fruit; entire; ***blunt***; awnless; non-carinate; having the margins lying flat and exposed on the palea; with a clear germination flap; 5–7 nerved. *Palea* present; relatively long (margins hairy towards apex); entire (subacuminate); awnless, without apical setae; 2-nerved. *Lodicules* present; 2; free; fleshy. *Stamens* 3. Anthers 2.5 mm long; not penicillate. *Ovary* glabrous. Styles free to their bases. Stigmas 2.

Fruit, embryo and seedling. Fruit small; not noticeably compressed. Hilum short. Embryo large. Endosperm containing only simple starch grains. Embryo without an epiblast; with a scutellar tail; with an elongated mesocotyl internode. Embryonic leaf margins overlapping.

Abaxial leaf blade epidermis. *Costal/intercostal zonation* conspicuous. *Papillae* absent. *Long-cells* similar in shape costally and intercostally; of similar wall thickness costally and intercostally (fairly thin walled). Mid-intercostal long-cells rectangular; having markedly sinuous walls. *Microhairs* present; 48–54 microns long; 7.5–9 microns wide at the septum. Microhair total length/width at septum 5.3–7.2. Microhair apical cells 21–24 microns long. Microhair apical cell/total length ratio 0.44. *Stomata* common; 33–36 microns long. Subsidiaries dome-shaped (usually), or parallel-sided (rarely). Guard-cells overlapping to flush with the interstomatals. *Intercostal short-cells* common; in cork/silica-cell pairs; silicified. *Costal short-cells* conspicuously in long rows. Costal silica bodies 'panicoid-type'.

Transverse section of leaf blade, physiology. C₃; XyMS+. *Mesophyll* with radiate chlorenchyma; *Isachne*-type (or tending to this), or not *Isachne*-type. *Leaf blade* adaxially flat. *Midrib* not readily distinguishable; with one bundle only. Bulliforms present in discrete, regular adaxial groups; in simple fans. All the vascular bundles accompanied by sclerenchyma. Combined sclerenchyma girders present; forming 'figures'. Sclerenchyma all associated with vascular bundles.

Taxonomy. Panicoideae; Panicodae; Paniceae.

Ecology, geography, regional floristic distribution. 2 species. Australia. Xerophytic; species of open habitats. Dry sandstone.

Holarctic, Paleotropical, and Australian. Madrean. Indomalesian. Papuan. North and East Australian. Tropical North and East Australian.

References, etc. Morphological/taxonomic: Hubbard 1933b. Leaf anatomical: this project.

Cliffordiochloa B.K Simon

Habit, vegetative morphology. Weak perennial; caespitose. **Culms** 60–80 cm high; *herbaceous*; branched above; 4–6 noded. Culm nodes exposed. The shoots not aromatic. *Leaves* not basally aggregated. Sheaths compressed. Leaf blades linear; narrow; 2–3 mm wide; flat. Ligule an unfringed membrane, or a fringed membrane; 0.5 mm long.

Reproductive organization. Plants bisexual, with bisexual spikelets; with hermaphrodite florets. *The spikelets all alike in sexuality*; hermaphrodite.

Inflorescence. *Inflorescence of spicate main branches (the main axis 10–20 cm long, the branches spreading, not whorled)*; non-digitate. Inflorescence with axes ending in spikelets. Inflorescence espatheate. Spikelet-bearing axes persistent. *Spikelets* unaccompanied by bractiform involucres, not associated with setiform vestigial branches; paired (mostly), or in triplets (a few); secund (on one side of the branch); from two sides

250

of the three-sided rachis; pedicellate (the pedicels 0.5–1.5 mm long). Pedicel apices cupuliform. Spikelets consistently in 'long-and-short' combinations; unequally pedicellate in each combination. The 'shorter' spikelets hermaphrodite. The 'longer' spikelets hermaphrodite.

Female-fertile spikelets. *Spikelets* 1.5 mm long; elliptic; *compressed laterally*; falling with the glumes; with conventional internode spacings. The upper floret not stipitate. Rachilla terminated by a female-fertile floret. Hairy callus absent. *Callus* absent.

Glumes two; very unequal; (the upper) about equalling the spikelets (the lower about half as long); (the upper) long relative to the adjacent lemmas (about equalling the L_1); hairless; glabrous, or scabrous (on the keel); pointed (the lower), or not pointed (the upper notched); awnless; carinate (the lower strongly keeled, the upper only slightly so); *very dissimilar (membranous, the lower deltoid and almost keel-winged above, the upper less strongly keeled and slightly notched). Lower glume* about 0.5 times the length of the upper glume; shorter than the lowest lemma; longer than half length of lowest lemma; 1 nerved. *Upper glume* not saccate; 3 nerved, or 5 nerved. *Spikelets* with incomplete florets. The incomplete florets proximal to the female-fertile florets. *The proximal incomplete florets* 1; paleate. Palea of the proximal incomplete florets fully developed (2-keeled, hyaline); not becoming conspicuously hardened and enlarged laterally. The proximal incomplete florets sterile. The proximal lemmas resembling the upper glume in shape, size and texture; awnless; 3 nerved, or 5 nerved; more or less equalling the female-fertile lemmas to decidedly exceeding the female-fertile lemmas (slightly exceeding it); somewhat less firm than the female-fertile lemmas (membranous, like the glumes); not becoming indurated.

Female-fertile florets 1. **Lemmas** elliptic; not saccate; thin, but decidedly firmer than the glumes; striate; not becoming indurated (thinly cartilaginous); white in fruit; entire; blunt; not crested; awnless; hairless; glabrous; non-carinate; having the margins tucked in onto the palea; seemingly without a germination flap; *0 nerved. Palea* present; relatively long; tightly clasped by the lemma; entire; awnless, without apical setae; textured like the lemma; 2-nerved; keel-less. Palea back glabrous. *Lodicules* present; 2; fleshy; glabrous; not or scarcely vascularized. **Stamens 2**. Anthers not penicillate; without an apically prolonged connective. *Ovary* glabrous. Stigmas 2.

Abaxial leaf blade epidermis. *Costal/intercostal zonation* conspicuous. *Papillae* absent. *Long-cells* markedly different in shape costally and intercostally (the costals much narrower); of similar wall thickness costally and intercostally (walls of medium thickness). Intercostal zones with typical long-cells. Mid-intercostal long-cells rectangular; having markedly sinuous walls. *Microhairs* present; elongated; clearly two-celled; panicoid-type; 40–50 microns long; 7–10 microns wide at the septum. Microhair total length/width at septum 4–6. Microhair apical cells 25–30 microns long. Microhair apical cell/total length ratio 0.4–0.6. *Stomata* common; 20–30 microns long. Subsidiaries non-papillate; low dome-shaped to triangular, or parallel-sided (by extreme truncation of triangles); including both triangular and parallel-sided forms on the same leaf. Guard-cells overlapping to flush with the interstomatals. *Intercostal short-cells* common; in cork/silica-cell pairs and not paired (many solitary); silicified and not silicified. Intercostal silica bodies mostly cross-shaped. *Crown cells* absent. *Costal short-cells* conspicuously in long rows. Costal silica bodies present and well developed; 'panicoid-type'; nearly all dumb-bell shaped.

Transverse section of leaf blade, physiology. C_3; XyMS+. *Mesophyll* with radiate chlorenchyma; *Isachne*-type; without 'circular cells'; not traversed by colourless columns; without arm cells; without fusoids. *Midrib* conspicuous; with one bundle only, or having complex vascularization (depending on interpretation of the midrib). *The lamina* symmetrical on either side of the midrib. Bulliforms present in discrete, regular adaxial groups; in simple fans (these large). All the vascular bundles accompanied by sclerenchyma. Combined sclerenchyma girders present (with all the primary bundles, the minor bundles mostly with adaxial and abaxial strands). Sclerenchyma all associated with vascular bundles.

Taxonomy. Panicoideae; Panicodae; Paniceae.

Ecology, geography, regional floristic distribution. 1 species. Queensland. Helophytic.

Australian. North and East Australian. Tropical North and East Australian.
References, etc. Morphological/taxonomic: B.K. Simon 1992. Leaf anatomical: this project.
Special comments. Fruit data wanting.

Cockaynea Zotov

~ *Elymus* (Section *Stenostachys*)

Habit, vegetative morphology. Perennial; stoloniferous. Culms herbaceous. Culm nodes glabrous. Culm internodes hollow. Young shoots extravaginal. **Leaves** *mostly basal*; shortly auriculate, or non-auriculate (*C. laevis*). Leaf blades narrow; without cross venation; persistent. Ligule an unfringed membrane; truncate; 0.5 mm long.

Reproductive organization. Plants bisexual, with bisexual spikelets; with hermaphrodite florets. The spikelets all alike in sexuality. Plants inbreeding. Exposed-cleistogamous, or chasmogamous.

Inflorescence. *Inflorescence* many spikeleted; *a single spike (drooping or nodding)*. Rachides hollowed. *Inflorescence espatheate*; not comprising 'partial inflorescences' and foliar organs. Spikelet-bearing axes persistent. Spikelets solitary; not secund; sessile; imbricate.

Female-fertile spikelets. Spikelets 9–11 mm long; compressed laterally; disarticulating above the glumes. Rachilla prolonged beyond the uppermost female-fertile floret. The rachilla extension with incomplete florets. Hairy callus absent.

Glumes present, or absent; when present, *two (then reduced or awn-shaped)*; minute to relatively large (reduced to small stumps, or aristate and exceeding the rachis internodes); shorter than the adjacent lemmas; free; lateral to the rachis, or displaced (but the spikelets edgewise to the rachis); when present, subulate, or not subulate (then reduced); awned (awn-like). Lower glume 0 nerved, or 1 nerved. Upper glume 0 nerved, or 1 nerved. *Spikelets* with incomplete florets. The incomplete florets distal to the female-fertile florets.

Female-fertile florets *1–2*. Lemmas decidedly firmer than the glumes; awnless, or mucronate, or awned. Awns when present, 1; apical; non-geniculate; entered by one vein. Lemmas hairless; carinate to non-carinate; *9 nerved*. Palea present; relatively long; 2-nerved; 2-keeled. *Lodicules* present; 2; free; membranous; ciliate; toothed, or not toothed. *Stamens* 3. Anthers 5–10 mm long (? — 'long'); not penicillate. **Ovary** *hairy*. Styles free to their bases. Stigmas 2.

Fruit, embryo and seedling. *Fruit* adhering to lemma and/or palea (to the palea); longitudinally grooved. Hilum long-linear (nearly as long as the caryopsis). Embryo small.

Abaxial leaf blade epidermis. *Costal/intercostal zonation* conspicuous. *Papillae* absent. *Long-cells* similar in shape costally and intercostally (long rectangles); of similar wall thickness costally and intercostally (fairly thick walled). Mid-intercostal long-cells rectangular and fusiform; having straight or only gently undulating walls. *Microhairs* absent. *Stomata* absent or very rare; 42–46.5 microns long. Subsidiaries dome-shaped. Guard-cells overlapped by the interstomatals. *Intercostal short-cells* absent or very rare; silicified. *Costal short-cells* neither distinctly grouped into long rows nor predominantly paired. Costal silica bodies horizontally-elongated crenate/sinuous (and some rather square and crenate).

Transverse section of leaf blade, physiology. C_3; XyMS+. *Mesophyll* with non-radiate chlorenchyma. *Leaf blade* with distinct, prominent adaxial ribs; with the ribs more or less constant in size. *Midrib* conspicuous; with one bundle only. Combined sclerenchyma girders present; forming 'figures'.

Cytology. Chromosome base number, $x = 7$. $2n = 28$. 4 ploid. Haplomic genome content H and S.

Taxonomy. Pooideae; Triticodae; Triticeae.

Ecology, geography, regional floristic distribution. 2 species. New Zealand. Antarctic. New Zealand.

References, etc. Morphological/taxonomic: Löve and Connor 1982; Löve 1984. Leaf anatomical: this project.
Special comments. Fruit data wanting.

Coelachne R.Br.

Habit, vegetative morphology. Low annual, or perennial; caespitose, or decumbent. *Culms* 4–50 cm high; herbaceous. Culm nodes glabrous. Leaves non-auriculate. Leaf blades linear to ovate-lanceolate; narrow; cross veined, or without cross venation; persistent; rolled in bud. *Ligule a fringe of hairs*.

Reproductive organization. *Plants bisexual, with bisexual spikelets*; with hermaphrodite florets.

Inflorescence. Inflorescence of spicate main branches, or paniculate; open, or contracted; with capillary branchlets, or without capillary branchlets; espatheate; not comprising 'partial inflorescences' and foliar organs. Spikelet-bearing axes persistent. Spikelets solitary, or paired; not secund; pedicellate; not in distinct 'long-and-short' combinations.

Female-fertile spikelets. Spikelets ovate; not noticeably compressed to compressed dorsiventrally; disarticulating above the glumes; disarticulating between the florets; with distinctly elongated rachilla internodes between the florets. *Rachilla terminated by a female-fertile floret*; rachilla hairy. Hairy callus present, or absent.

Glumes two (persistent); very unequal to more or less equal; shorter than the spikelets (1/3 to 2/3 their length); shorter than the adjacent lemmas; hairless (usually glabrous); not pointed (obtuse); awnless; non-carinate; similar (broad, membranous). Lower glume 1–5 nerved. Upper glume 3–7 nerved. *Spikelets* with female-fertile florets only (but the upper floret female-only).

Female-fertile florets 2 *(dissimilar, separated by a slender rachilla segment)*. Lemmas decidedly firmer than the glumes (the lower becoming somewhat hardened, the upper remaining membranous); smooth; not becoming indurated; white in fruit; entire; blunt; awnless (obtuse); hairy (L_2 usually pubescent), or hairless (L_1 basally hairy or not); non-carinate; with a clear germination flap; *0 nerved*. *Palea* present; entire (obtuse); awnless, without apical setae; 2-nerved. *Lodicules* present; 2; free; fleshy; glabrous. *Stamens* 2–3. Anthers not penicillate. *Ovary* glabrous. Styles fused, or free to their bases. Stigmas 2.

Fruit, embryo and seedling. *Fruit* free from both lemma and palea; small; compressed dorsiventrally. Hilum short. Embryo large, or small; not waisted. Endosperm hard.

Abaxial leaf blade epidermis. *Costal/intercostal zonation* conspicuous. *Papillae* present, or absent. Intercostal papillae over-arching the stomata, or not over-arching the stomata; consisting of one symmetrical projection per cell (but not on every cell). *Long-cells* markedly different in shape costally and intercostally (the costals short, irregularly shaped); of similar wall thickness costally and intercostally (thin walled). Intercostal zones without typical long-cells (the cells mainly more or less isodiametric). Mid-intercostal long-cells rectangular (hexagonal to square); having straight or only gently undulating walls. *Microhairs* present; more or less spherical, or elongated; ostensibly one-celled, or clearly two-celled; chloridoid-type (apical cell somewhat pointed, but thick walled and broader than long); (7–)13–38(–42) microns long. Microhair apical cells 7–24(–26) microns long. Microhair apical cell/total length ratio 0.64–0.73. *Stomata* common; 24–27 microns long. Subsidiaries parallel-sided to triangular; including both triangular and parallel-sided forms on the same leaf. Guard-cells overlapping to flush with the interstomatals. *Intercostal short-cells* absent or very rare; not paired; not silicified. *Costal short-cells* conspicuously in long rows. Costal silica bodies acutely-angled (more or less cubical); sharp-pointed.

Transverse section of leaf blade, physiology. C_3; XyMS+. *Mesophyll* with radiate chlorenchyma; *Isachne*-type. *Leaf blade* with distinct, prominent adaxial ribs, or 'nodular' in section; with the ribs more or less constant in size. *Midrib* not readily distinguishable; with one bundle only. Bulliforms present in discrete, regular adaxial groups (in the furrows, the groups sometimes inconspicuous); in simple fans. All the vascular bundles accompanied by sclerenchyma. Combined sclerenchyma girders present; forming 'figures', or nowhere forming 'figures'. Sclerenchyma all associated with vascular bundles.

Cytology. Chromosome base number, $x = 10$. $2n = 40$.

Taxonomy. Panicoideae; Panicodae; Isachneae.

Ecology, geography, regional floristic distribution. 10 species. Palaeotropical. Helophytic; species of open habitats; glycophytic. Streamsides and marshes.

Holarctic, Paleotropical, and Australian. Boreal. African, Madagascan, and Indomalesian. Eastern Asian. Sudano-Angolan. Indian, Indo-Chinese, Malesian, and Papuan. North and East Australian. Somalo-Ethiopian and South Tropical African. Tropical North and East Australian.

References, etc. Leaf anatomical: Metcalfe 1960; this project.

Coelachyropsis Bor

~ Coelachyrum

Habit, vegetative morphology. Annual; decumbent. *Culms* 10–25 cm high; herbaceous; unbranched above. Culm nodes glabrous. *Leaves* not basally aggregated; non-auriculate. Leaf blades narrow; 2–6 mm wide; exhibiting multicellular glands abaxially. The abaxial glands on the blade margins, or costal (elongated, one cell file, with 4–8 cells). Leaf blades without cross venation; persistent. Ligule a fringed membrane.

Reproductive organization. Plants bisexual, with bisexual spikelets; with hermaphrodite florets.

Inflorescence. *Inflorescence* of spicate main branches (2–3 short spikes, in the material seen); *digitate*. Primary inflorescence branches 2–3. Inflorescence with axes ending in spikelets. Inflorescence espatheate; not comprising 'partial inflorescences' and foliar organs. Spikelet-bearing axes persistent. Spikelets solitary; secund; biseriate.

Female-fertile spikelets. Spikelets 5–7 mm long; compressed laterally; disarticulating above the glumes (?), or not disarticulating. Rachilla prolonged beyond the uppermost female-fertile floret; rachilla hairless. The rachilla extension with incomplete florets. Hairy callus absent.

Glumes two; more or less equal; shorter than the spikelets; shorter than the adjacent lemmas; lateral to the rachis; hairless; glabrous; pointed; awnless (but mucronate); carinate; similar (membranous-hyaline, broadly lanceolate acuminate). Lower glume 1 nerved. Upper glume 3–4 nerved. *Spikelets* with incomplete florets. The incomplete florets distal to the female-fertile florets. *The distal incomplete florets* merely underdeveloped.

Female-fertile florets 5–8. *Lemmas* broadly ovate, acuminate; *not saccate*; similar in texture to the glumes (membranous); not becoming indurated; entire; pointed; mucronate (attenuate into the mucro); *hairy (with long white hairs on the lower half along the nerves)*; carinate; having the margins lying flat and exposed on the palea; without a germination flap; 3 nerved. *Palea* present; relatively long; entire; awnless, without apical setae; not indurated (membranous); 2-nerved (with hairs along their lower two thirds); 2-keeled. *Lodicules* present; 2; free; fleshy; not or scarcely vascularized. *Stamens* 3 (with tiny anthers). Anthers not penicillate; without an apically prolonged connective. *Ovary* glabrous. Styles free to their bases. Stigmas 2.

Fruit, embryo and seedling. Fruit free from both lemma and palea; small (about 1 mm long); compressed dorsiventrally (angular when dry, and concave on the side away from the embryo); *sculptured (cf. Acrachne)*. Hilum short. Pericarp free. Embryo large. Endosperm hard.

Abaxial leaf blade epidermis. *Costal/intercostal zonation* conspicuous. *Papillae* present; intercostal. Intercostal papillae over-arching the stomata, or not over-arching the stomata; consisting of one oblique swelling per cell (large). *Long-cells* markedly different in shape costally and intercostally (the costals conventional, with sinuous walls). Intercostal zones with typical long-cells (long-cells thin walled, deforming, some may be short). Mid-intercostal long-cells having straight or only gently undulating walls. *Microhairs* present; more or less spherical; clearly two-celled; chloridoid-type. Microhair apical cell wall of similar thickness/rigidity to that of the basal cell. Microhair basal cells 9–12 microns long. Microhair total length/width at septum about 2. Microhair apical cell/total length ratio about 0.5. *Stomata* common. Subsidiaries non-papillate; including both triangular and parallel-sided forms on the same leaf. Guard-cells overlapping to flush with the interstomatals (except for overlapping by papillae). *Intercostal short-cells* absent or very rare. *Costal short-cells* conspicuously in long rows (although the short-cells are often rather long). Costal silica bodies present in alternate cell files of the costal zones; 'panicoid-type'; cross shaped and nodular (with points); sharp-pointed (the crosses and dumbbells with points).

Transverse section of leaf blade, physiology. C_4; XyMS+. PCR sheaths of the primary vascular bundles interrupted; interrupted both abaxially and adaxially. PCR sheath extensions absent. *Leaf blade* probably adaxially flat. *Midrib* not readily distinguishable; with one bundle only; with colourless mesophyll adaxially (there being colourless tissue adaxial to all the large vascular bundles). Bulliforms present in discrete, regular adaxial groups; in simple fans and associated with colourless mesophyll cells to form deeply-penetrating fans. All the vascular bundles accompanied by sclerenchyma. Combined sclerenchyma girders absent (colourless, large-celled bundle sheath extensions adaxially). Sclerenchyma all associated with vascular bundles. The lamina margins with fibres.

Taxonomy. Chloridoideae; main chloridoid assemblage.

Ecology, geography, regional floristic distribution. 1 species. Southern India, Ceylon.

Paleotropical. Indomalesian. Indian.

References, etc. Leaf anatomical: this project.

Coelachyrum Hochst. & Nees

Coeleochloa Steud.

Excluding *Coelachyropsis*, *Cypholepis*

Habit, vegetative morphology. Annual, or perennial; usually stoloniferous. *Culms* 15–100 cm high; herbaceous. Leaves non-auriculate. Leaf blades narrow; flat; without abaxial multicellular glands; without cross venation. *Ligule a fringed membrane*.

Reproductive organization. Plants bisexual, with bisexual spikelets; with hermaphrodite florets.

Inflorescence. Inflorescence of spicate main branches (slender racemes), or paniculate; digitate, or non-digitate; espatheate; not comprising 'partial inflorescences' and foliar organs. Spikelet-bearing axes persistent. Spikelets secund (the racemes unilateral), or not secund; biseriate; shortly pedicellate.

Female-fertile spikelets. Spikelets 5–7 mm long; compressed laterally; disarticulating above the glumes; disarticulating between the florets; with conventional internode spacings. Rachilla prolonged beyond the uppermost female-fertile floret. Hairy callus absent. *Callus* absent.

Glumes two; very unequal to more or less equal; shorter than the adjacent lemmas, or long relative to the adjacent lemmas; lateral to the rachis; awnless; similar (somewhat carinate, lanceolate). Lower glume 1 nerved, or 3 nerved. Upper glume 3 nerved. *Spikelets* with female-fertile florets only, or with incomplete florets. The incomplete florets distal to the female-fertile florets.

Female-fertile florets 2–10. *Lemmas saccate (below)*; not becoming indurated (membranous or scarious); awnless to mucronate; hairy, or hairless; lightly carinate, or non-carinate (as the fruit expands); 3 nerved. *Palea* present (broadly oval); entire; awnless, without apical setae; membranous; 2-nerved; 2-keeled (with narrow margins). *Lodicules* present; 2; free; fleshy; glabrous. *Stamens* 3. Anthers not penicillate. *Ovary* glabrous. Styles free to their bases. Stigmas 2.

Fruit, embryo and seedling. *Fruit* free from both lemma and palea; small (1.75–2 mm long); ellipsoid to subglobose (sub-orbicular); compressed dorsiventrally (concavo-convex); sculptured. Hilum short. Pericarp free. Embryo large; not waisted.

Abaxial leaf blade epidermis. *Costal/intercostal zonation* conspicuous. *Papillae* absent. Mid-intercostal long-cells rectangular; having markedly sinuous walls, or having straight or only gently undulating walls. *Microhairs* present; more or less spherical; ostensibly one-celled, or clearly two-celled (?); chloridoid-type. Microhair apical cells 6–11 microns long. *Stomata* common. Subsidiaries low dome-shaped and triangular. *Intercostal short-cells* absent or very rare; not paired. Intercostal silica bodies absent. *Costal short-cells* conspicuously in long rows. Costal silica bodies present in alternate cell files of the costal zones; saddle shaped, or oryzoid, or 'panicoid-type'; sometimes cross shaped, or butterfly shaped.

Transverse section of leaf blade, physiology. C_4; XyMS+. PCR sheaths of the primary vascular bundles complete. PCR sheath extensions absent. *Mesophyll* with radiate chlorenchyma; traversed by columns of colourless mesophyll cells. *Leaf blade* 'nodular' in section; with the ribs more or less constant in size. *Midrib* conspicuous; with one bundle only. Bulliforms present in discrete, regular adaxial groups; in simple fans and associated with colourless mesophyll cells to form deeply-penetrating fans (the simple groups each with a large, deeply-penetrating median cell). All the vascular bundles accompanied by sclerenchyma. Combined sclerenchyma girders present. Sclerenchyma all associated with vascular bundles. The lamina margins with fibres.

Taxonomy. Chloridoideae; main chloridoid assemblage.

Ecology, geography, regional floristic distribution. 6 species. North tropical Africa, tropical southwest Asia. Xerophytic; species of open habitats; halophytic, or glycophytic. Grassland, sand and semidesert.

Holarctic and Paleotropical. Tethyan. African and Indomalesian. Irano-Turanian. Saharo-Sindian and Sudano-Angolan. Indian. Sahelo-Sudanian and Somalo-Ethiopian.

References, etc. Morphological/taxonomic: Napper 1963. Leaf anatomical: Metcalfe 1960; this project; Van den Borre 1994.

Coelorachis Brongn.

Apogonia Nutt, *Coelorhachis* Brongn., *Cycloteria* Stapf
~ *Mnesithea* sensu lato

Habit, vegetative morphology. Perennial (mostly), or annual (*C. clarkei*); mostly robust, tall, forming clumps. *Culms* 70–400 cm high; herbaceous; branched above. Culm nodes hairy, or glabrous. Culm internodes solid. *Leaves* not basally aggregated; non-auriculate. Leaf blades linear; broad to narrow; flat (or rarely filiform); without cross venation; persistent. Ligule a fringed membrane to a fringe of hairs.

Reproductive organization. Plants bisexual, with bisexual spikelets; with hermaphrodite florets. The spikelets of sexually distinct forms on the same plant (usually), or all alike in sexuality (*C. parodiana*); hermaphrodite (rarely), or hermaphrodite and male-only, or hermaphrodite and sterile. The male and female-fertile spikelets mixed in the inflorescence. The spikelets overtly heteromorphic, or homomorphic; in both homogamous and heterogamous combinations, or all in heterogamous combinations. Exposed-cleistogamous, or chasmogamous (?).

Inflorescence. Inflorescence of spicate main branches, or paniculate (of long-peduncled, spikelike dorsiventral 'racemes', solitary at culm or branchlet apices, often in 'false panicles'). Rachides hollowed. *Inflorescence* spatheate; *a complex of 'partial inflorescences' and intervening foliar organs (the unit consisting of a 'raceme', its peduncle, the subtending leaf and the next internode (the peduncle of the unit))*. Spikelet-bearing axes spikelike; solitary, or clustered (fascicled); with substantial rachides; disarticulating; disarticulating at the joints. *'Articles'* non-linear (concave, clavate, shorter than the sessile spikelet); with a basal callus-knob; appendaged, or not appendaged; disarticulating transversely; glabrous. *Spikelets* paired, or in triplets (sometimes, then two sessile/one pedicellate); secund (the rachis dorsiventral, the sessile members alternating in two rows on one side); sessile and pedicellate; *consistently in 'long-and-short' combinations*; in pedicellate/sessile combinations. Pedicels of the 'pedicellate' spikelets free of the rachis (but closely contiguous). The 'shorter' spikelets hermaphrodite. The 'longer' spikelets hermaphrodite (*C. parodiana*), or male-only, or sterile.

Female-sterile spikelets. The pedicelled spikelets vestigial to well developed, bisexual in *C.parodiana*. The lemmas awnless.

Female-fertile spikelets. Spikelets 3–4.5 mm long; compressed dorsiventrally; falling with the glumes. Rachilla terminated by a female-fertile floret. Hairy callus absent. *Glumes* two; *more or less equal*; long relative to the adjacent lemmas; hairless; glabrous; awnless; very dissimilar (the lower two-keeled and two-winged above, the upper 1-keeled and wingless). *Lower glume two-keeled (the keels winged)*; convex on the back to flattened on the back; not pitted; relatively smooth, or lacunose with deep depressions, or rugose; 7 nerved. Upper glume 0–2 nerved. *Spikelets* with incomplete florets. The incomplete florets proximal to the female-fertile florets. *The proximal incomplete florets* 1; paleate, or epaleate (usually). Palea of the proximal incomplete florets when present, reduced. The proximal incomplete florets sterile. The proximal lemmas awnless; 0 nerved, or 2 nerved; more or less equalling the female-fertile lemmas; similar in texture to the female-fertile lemmas; not becoming indurated.

Female-fertile florets 1. **Lemmas** lanceolate; less firm than the glumes (hyaline); not becoming indurated; entire; pointed; awnless; hairless; non-carinate; *5 nerved*. *Palea* present; relatively long; entire; awnless, without apical setae; not indurated (hyaline); 2-nerved, or nerveless. *Lodicules* present; 2; free; fleshy; glabrous. *Stamens* 3. Anthers not penicillate. *Ovary* glabrous. Styles free to their bases. Stigmas 2; red pigmented.

Fruit, embryo and seedling. *Fruit* free from both lemma and palea; compressed dorsiventrally. Hilum short. Embryo large.

Abaxial leaf blade epidermis. *Costal/intercostal zonation* conspicuous. *Papillae* absent. *Long-cells* similar in shape costally and intercostally (long rectangles); of similar wall thickness costally and intercostally. Mid-intercostal long-cells rectangular; having markedly sinuous walls. *Microhairs* present; panicoid-type; 33–42 microns long; 6–7.5 microns wide at the septum. Microhair total length/width at septum 4.4–7. Microhair apical cells 18–22.5 microns long. Microhair apical cell/total length ratio 0.5–0.64. *Stomata* common; 27–30 microns long. Subsidiaries triangular. Guard-cells overlapping to flush with the interstomatals. *Intercostal short-cells* absent or very rare; in cork/silica-cell pairs; not silicified. *Costal short-cells* conspicuously in long rows. Costal silica bodies 'panicoid-type'.

Transverse section of leaf blade, physiology. C_4; XyMS–. *Mesophyll* with radiate chlorenchyma; traversed by columns of colourless mesophyll cells, or not traversed by colourless columns. *Leaf blade* 'nodular' in section; with the ribs more or less constant in size. *Midrib* conspicuous; having a conventional arc of bundles; with colourless mesophyll adaxially. Bulliforms present in discrete, regular adaxial groups; in simple fans and associated with colourless mesophyll cells to form deeply-penetrating fans; associating with colourless mesophyll cells to form arches over small vascular bundles. Many of the smallest vascular bundles unaccompanied by sclerenchyma. Combined sclerenchyma girders present; forming 'figures'.

Phytochemistry. Leaves without flavonoid sulphates (1 species).

Cytology. Chromosome base number, $x = 9$. $2n = 18$, 36, and 54. 2, 4, and 6 ploid.

Taxonomy. Panicoideae; Andropogonodae; Andropogoneae; Rottboelliinae.

Ecology, geography, regional floristic distribution. About 20 species. Mainly tropical. Helophytic to mesophytic; species of open habitats; glycophytic. Grassland and savanna, often on damp soils.

Holarctic, Paleotropical, Neotropical, Cape, and Australian. Boreal. African and Indomalesian. Atlantic North American. Sudano-Angolan and West African Rainforest. Indian, Indo-Chinese, Malesian, and Papuan. Caribbean. North and East Australian. Southern Atlantic North American and Central Grasslands. Sahelo-Sudanian, Somalo-Ethiopian, and South Tropical African. Tropical North and East Australian.

References, etc. Leaf anatomical: this project.

Coix L.

Lacryma Medik., *Lacryma-jobi* Ort., *Lacrymaria* Fabric., *Sphaerium* Kuntze

Habit, vegetative morphology. Annual to perennial; stems erect or straggling, with prop-roots from the lower nodes. *Culms* 70–400 cm high; herbaceous; branched above. Culm nodes glabrous. Culm internodes solid. *Leaves* not basally aggregated; non-auriculate. Leaf blades lanceolate; broad; 30–70 mm wide; cordate, or not cordate, not sagittate; flat; without cross venation; persistent; rolled in bud. *Ligule* present; an unfringed membrane to a fringed membrane.

Reproductive organization. Plants monoecious with all the fertile spikelets unisexual; without hermaphrodite florets. The spikelets of sexually distinct forms on the same plant; female-only, or male-only. The male and female-fertile spikelets on different branches of the same inflorescence (in different but apposed, spiciform racemes within the pistillate sheath). The spikelets overtly heteromorphic. Plants outbreeding. Apomictic, or reproducing sexually (?).

Inflorescence. Inflorescence paniculate (but peculiar — see below); spatheate; a complex of 'partial inflorescences' and intervening foliar organs (the partial inflorescences of peculiar form, on flattened peduncles, in leafy panicles). **Spikelet-bearing axes very much reduced (the female 'raceme' usually represented by three spikelets, enclosed in a globose, hardened involucre or 'pistillate sheath', separated from the male raceme by a prophyll at its base. The male raceme exserted on a peduncle through the apex of the involucre)**; disarticulating; falling entire (within its involucre). Spikelets not secund; sessile and pedicellate; consistently in 'long-and-short' combinations; in pedicellate/sessile combinations (in both male and female racemes). Pedicels of the 'pedicellate' spikelets free of the rachis. The 'shorter' spikelets of the female racemes female-only. The 'longer' spikelets of the female racemes female-only, or sterile.

Female-sterile spikelets. Male spikelets in pairs or triads, several per disarticulating raceme. Dorsally compressed, with two florets, both male or the lower sterile. Rachilla of male spikelets terminated by a male floret. The male spikelets with glumes; with proximal incomplete florets, or without proximal incomplete florets (the lower floret sterile or male); 2 floreted. The lemmas awnless. Male florets 1, or 2.

Female-fertile spikelets. Spikelets falling with the glumes. Rachilla terminated by a female-fertile floret.

Glumes two; more or less equal; long relative to the adjacent lemmas; pointed; awnless; very dissimilar (both beaked; the lower subglobose, hyaline below, subcartilaginous above; the upper narrower, strongly keeled, subhyaline). Lower glume not two-keeled; flattened on the back; not pitted; relatively smooth; many nerved. Upper glume 11 nerved. *Spikelets* with incomplete florets. The incomplete florets proximal to the female-fertile florets. *The proximal incomplete florets* 1; epaleate; sterile. The proximal lemmas awnless; 3–7 nerved; decidedly firmer than the female-fertile lemmas; not becoming indurated.

Female-fertile florets 1. Lemmas deltoid; less firm than the glumes to similar in texture to the glumes (similar to the upper glume, but less strongly keeled; very thin and hyaline beneath the beak); not becoming indurated; entire; pointed; mucronate (beaked); hairless; carinate (but less conspicuously so than the G_2); 3–5 nerved. *Palea* present; conspicuous but relatively short (broad, beaked); entire (subulate-beaked); not indurated (hyaline below); 2-nerved. *Lodicules* absent. *Stamens* 0 (or 3 staminodes). *Ovary* glabrous. Styles fused (into one very long style, exserted from the hard pistillate sheath). Stigmas 2.

Fruit, embryo and seedling. Fruit medium sized; compressed dorsiventrally. Hilum short (circular or elliptical, quite large). Embryo large. Endosperm hard; without lipid; containing only simple starch grains. Embryo without an epiblast; with a scutellar tail; with an elongated mesocotyl internode. Embryonic leaf margins overlapping.

Seedling with a long mesocotyl; with a loose coleoptile.

Abaxial leaf blade epidermis. *Costal/intercostal zonation* conspicuous. *Papillae* absent. *Long-cells* markedly different in shape costally and intercostally (the costals much narrower); of similar wall thickness costally and intercostally (fairly thick walled). Mid-intercostal long-cells rectangular; having markedly sinuous walls. *Microhairs* present; panicoid-type to chloridoid-type (the apical cells being flimsy, but relatively wide and round-tipped, and sometimes quite short); 31–42 microns long; 12–13.5 microns wide at the septum. Microhair total length/width at septum 2.9–3.25. Microhair apical cells 18–21(–28) microns long. Microhair apical cell/total length ratio 0.46–0.67. *Stomata* common; 31–36 microns long. Subsidiaries predominantly triangular. Guard-cells overlapping to flush with the interstomatals. *Intercostal short-cells* common; in cork/silica-cell pairs. *Costal short-cells* conspicuously in long rows. Costal silica bodies 'panicoid-type'; cross shaped to dumb-bell shaped.

Transverse section of leaf blade, physiology. C_4; XyMS– (but very ambiguously so: could have two PCR sheaths — need to see fresh material). PCR cell chloroplasts with reduced grana; centrifugal/peripheral. *Mesophyll* with radiate chlorenchyma, or with non-radiate chlorenchyma. *Leaf blade* adaxially flat. *Midrib* not readily distinguishable; with one bundle only. Bulliforms not in discrete, regular adaxial groups (apart from midrib hinges, the rest of epidermis mostly more or less bulliform). Many of the smallest vascular bundles unaccompanied by sclerenchyma. Combined sclerenchyma girders present. Sclerenchyma all associated with vascular bundles.

Culm anatomy. *Culm internode bundles* in three or more rings.

Phytochemistry. Leaves without flavonoid sulphates (1 species).

Special diagnostic feature. *Inflorescences in hard, globular 6–12 mm utricles*.

Cytology. Chromosome base number, $x = 5$. $2n = 10$, 20, and 40. 2, 4, and 8 ploid. Haploid nuclear DNA content 0.8 pg (1 species, $4x$).

Taxonomy. Panicoideae; Andropogonodae; Maydeae.

Ecology, geography, regional floristic distribution. 5 species. Tropical Asia. Commonly adventive. Helophytic to mesophytic; shade species, or species of open habitats; glycophytic. Forest margins and swamps.

Holarctic, Paleotropical, and Australian. Boreal and Tethyan. Madagascan, Indomalesian, and Neocaledonian. Eastern Asian. Irano-Turanian. Indo-Chinese, Malesian, and Papuan. North and East Australian. Tropical North and East Australian.

Rusts and smuts. Rusts — *Puccinia*. Smuts from Tilletiaceae and from Ustilaginaceae. Tilletiaceae — *Tilletia*. Ustilaginaceae — *Ustilago*.

Economic importance. Significant weed species: *C. lacryma-jobi*. Cultivated fodder: *C. lacryma-jobi* (Job's Tears). Grain crop species: utricles of *C. lacryma-jobi* sometimes made into flour.

References, etc. Leaf anatomical: Metcalfe 1960; this project.

Colanthelia McClure & Smith

Habit, vegetative morphology. Perennial (plants small to very tall, erect or decumbent). The flowering culms leafy. *Culms woody and persistent*; scandent, or not scandent; branched above. Primary branches/mid-culm node 1. *Culm sheaths deciduous in their entirety (midculm sheaths abscising from a conspicuous persistent girdle)*. Culm internodes hollow. Unicaespitose. *Rhizomes pachymorph*. Plants unarmed. **Leaves** not basally aggregated; *with auricular setae*. Leaf blades broad; pseudopetiolate; without cross venation; disarticulating from the sheaths; rolled in bud.

Reproductive organization. Plants bisexual, with bisexual spikelets; with hermaphrodite florets.

Inflorescence. Inflorescence spatheate; a complex of 'partial inflorescences' and intervening foliar organs (with groups of partial inflorescences at the nodes). *Spikelet-bearing axes* 'racemes', or paniculate; clustered; persistent. Spikelets not secund; pedicellate.

Female-fertile spikelets. Spikelets compressed laterally (?); disarticulating above the glumes (?); disarticulating between the florets. Rachilla prolonged beyond the uppermost female-fertile floret. The rachilla extension with incomplete florets.

Glumes two, or several (rarely); very unequal; shorter than the adjacent lemmas; awnless; similar. *Spikelets* with incomplete florets. The incomplete florets distal to the female-fertile florets. *The distal incomplete florets* merely underdeveloped.

Female-fertile florets 3–8 (or more). Lemmas not becoming indurated; awnless, or mucronate (?). *Palea* present; relatively long; entire to apically notched; not indurated; several nerved; 2-keeled (and sulcate). *Lodicules* present; 3; free; membranous; ciliate; heavily vascularized. *Stamens* 3. Anthers not penicillate; without an apically prolonged connective. **Ovary hairy**; *without a conspicuous apical appendage*. Stigmas 2.

Abaxial leaf blade epidermis. *Costal/intercostal zonation* conspicuous. *Papillae* present (though seemingly not towards the margins). Intercostal papillae over-arching the stomata; several per cell (irregular, sometimes branched, concentrated and largest around the stomata). *Long-cells* similar in shape costally and intercostally; of similar wall thickness costally and intercostally. Mid-intercostal long-cells rectangular; having markedly sinuous walls. *Microhairs* present; panicoid-type; in *cingulata* 60–78(–84) microns long; 8.4–9 microns wide at the septum. Microhair total length/width at septum 6.7–14. Microhair apical cells 24–31.5(–42) microns long. Microhair apical cell/total length ratio 0.38–0.5. *Stomata* common; in *C. cingulata* 25–30 microns long. Subsidiaries low dome-shaped. Guard-cells overlapped by the interstomatals (slightly, even when not covered by papillae). *Intercostal short-cells* common; not paired (solitary, when not paired with prickle bases); silicified. Intercostal silica bodies tall-and-narrow and saddle shaped. *Costal short-cells* neither distinctly grouped into long rows nor predominantly paired (mostly solitary). Costal silica bodies saddle shaped (predominantly), or tall-and-narrow.

Transverse section of leaf blade, physiology. C_3; XyMS+. Mesophyll with adaxial palisade (in places); without arm cells (apparently, in the material seen); with fusoids. The fusoids external to the PBS. *Leaf blade* adaxially flat. *Midrib* conspicuous (with a somewhat larger bundle, and a slight abaxial keel: on the other hand, one or two veins near the leaf margins are more strongly keeled); with one bundle only. Bulliforms present in discrete, regular adaxial groups; in simple fans (these large). All the vascular bundles accompanied by sclerenchyma. Combined sclerenchyma girders present; nowhere forming 'figures' (the girders mostly narrow). Sclerenchyma not all bundle-associated. The 'extra' sclerenchyma in abaxial groups; abaxial-hypodermal, the groups isolated (opposite the bulliform groups).

Taxonomy. Bambusoideae; Bambusodae; Bambuseae.
Ecology, geography, regional floristic distribution. 7 species. Brazil. Neotropical. Central Brazilian.
References, etc. Leaf anatomical: this project.
Special comments. Fruit data wanting.

Coleanthus Seidl.

Schmidtia Tratt., *Wilibalda* Roth

Habit, vegetative morphology. Dwarf annual; decumbent. *Culms* 2–10 cm high; herbaceous. Leaves non-auriculate. *Sheath margins joined*. Sheaths inflated. Leaf blades linear (usually recurved); narrow; about 1 mm wide; flat, or folded; without cross venation. *Ligule* present; an unfringed membrane; 1–1.5 mm long.

Reproductive organization. Plants bisexual, with bisexual spikelets; with hermaphrodite florets. The spikelets all alike in sexuality.

Inflorescence. Inflorescence paniculate (subtended by the expanded sheath of the upper leaf, the branches in compact, globular fascicles); contracted; espatheate; not comprising 'partial inflorescences' and foliar organs. Spikelet-bearing axes persistent (?). Spikelets not secund; pedicellate.

Female-fertile spikelets. *Spikelets* unconventional (in lacking glumes); 0.75–1.5 mm long; not noticeably compressed; disarticulating under the lemma. Rachilla terminated by a female-fertile floret. Hairy callus absent. *Callus* short.

Glumes *absent*. *Spikelets* with female-fertile florets only.

Female-fertile florets 1. Lemmas not becoming indurated (membranous or hyaline); entire; pointed; awned. Awns 1; median; apical (from the caudate- or aristate-acuminate lemma); non-geniculate; entered by one vein. Lemmas hairless (but ciliate on the keel); carinate; without a germination flap; 1 nerved. *Palea* present; conspicuous but relatively short; not indurated (membranous); 2-nerved; 2-keeled. *Lodicules* absent. *Stamens* 2. Anthers 0.3–0.4 mm long. *Ovary* glabrous. Styles free to their bases (very long). Stigmas 2; white.

Fruit, embryo and seedling. *Fruit* free from both lemma and palea; small (1.5 mm long). Hilum short (oval). Embryo small. Endosperm hard; without lipid; containing compound starch grains. Embryo with an epiblast; without a scutellar tail; with a negligible mesocotyl internode. Embryonic leaf margins meeting.

Abaxial leaf blade epidermis. *Costal/intercostal zonation* conspicuous. *Papillae* absent. Long-cells differing markedly in wall thickness costally and intercostally. Intercostal zones with typical long-cells (but seemingly lacking short-cells). Mid-intercostal long-cells having straight or only gently undulating walls. *Microhairs* absent. *Stomata* common. Subsidiaries parallel-sided. Guard-cells overlapped by the interstomatals. *Intercostal short-cells* absent or very rare. Costal zones seemingly without short-cells (in the material seen). Costal silica bodies absent.

Transverse section of leaf blade, physiology. C_3; XyMS+. Mesophyll without adaxial palisade. *Leaf blade* adaxially flat. *Midrib* conspicuous (mainly by its position); with one bundle only. *The lamina* symmetrical on either side of the midrib. Bulliforms not in discrete, regular adaxial groups (apart from midrib hinges). Many of the smallest vascular bundles unaccompanied by sclerenchyma (all bundles unaccompanied by sclerenchyma, in the material seen).

Cytology. Chromosome base number, $x = 7$. $2n = 14$. 2 ploid.

Taxonomy. Pooideae; Poodae; Poeae.

Ecology, geography, regional floristic distribution. 1 species. North temperate. Helophytic to mesophytic (in dried up lakes and reservoirs, at pond margins); glycophytic.

Holarctic. Boreal and Tethyan. Euro-Siberian. Mediterranean. European.

Rusts and smuts. Rusts — *Puccinia*. Taxonomically wide-ranging species: *Puccinia graminis*.

References, etc. Leaf anatomical: Metcalfe 1960; this project.

Colpodium Trin.

Paracolpodium Tsvelev., *Keniochloa* Melderis
Excluding *Catabrosella*, *Hyalopoa*

Habit, vegetative morphology. Perennial; rhizomatous, or stoloniferous, or caespitose, or decumbent. *Culms* 10–30 cm high; herbaceous; unbranched above. Leaves nonauriculate. Sheath margins joined to free. *Leaf blades* flat; *not pseudopetiolate*; without cross venation. *Ligule an unfringed membrane*; truncate, or not truncate; 3 mm long.

Reproductive organization. Plants bisexual, with bisexual spikelets; with hermaphrodite florets.

Inflorescence. Inflorescence paniculate; open; espatheate; not comprising 'partial inflorescences' and foliar organs. Spikelet-bearing axes persistent. Spikelets not secund; pedicellate.

Female-fertile spikelets. Spikelets 2–8 mm long; compressed laterally to not noticeably compressed; disarticulating above the glumes. Rachilla prolonged beyond the uppermost female-fertile floret, or terminated by a female-fertile floret. The rachilla extension when present, with incomplete florets, or naked. Hairy callus absent. *Callus* short; blunt.

Glumes two; relatively large; more or less equal; shorter than the adjacent lemmas, or long relative to the adjacent lemmas; pointed, or not pointed (broadly rounded or erose); awnless; non-carinate; similar. Lower glume longer than half length of lowest lemma; 1 nerved. Upper glume 1–3 nerved. *Spikelets* with female-fertile florets only, or with incomplete florets. The incomplete florets distal to the female-fertile florets. *The distal incomplete florets* merely underdeveloped; awnless. *Spikelets without proximal incomplete florets*.

Female-fertile florets 1. *Lemmas* with several teeth; similar in texture to the glumes to decidedly firmer than the glumes (thinly membranous, the tip hyaline); not becoming indurated; *incised*; 3–5 lobed (toothed); not deeply cleft; *awnless*; hairy, or hairless; carinate to non-carinate; without a germination flap; 3–5 nerved. *Palea* present; relatively long; 2-nerved; 2-keeled. *Lodicules* present; 2; free; membranous; glabrous; toothed; not or scarcely vascularized. *Stamens* 3. Anthers 1.5–3.5 mm long. *Ovary* glabrous. Styles free to their bases. Stigmas 2; white.

Fruit, embryo and seedling. Fruit free from both lemma and palea; *medium sized*. *Hilum short (oblong)*. Embryo small.

Abaxial leaf blade epidermis. *Costal/intercostal zonation* conspicuous. *Papillae* present (in *Keniochloa*), or absent; when present, costal and intercostal. Intercostal papillae over-arching the stomata; consisting of one oblique swelling per cell. *Long-cells* similar in shape costally and intercostally, or markedly different in shape costally and intercostally; of similar wall thickness costally and intercostally. Mid-intercostal long-cells rectangular, or fusiform; having straight or only gently undulating walls. *Microhairs* absent. *Stomata* common; 22.5–27 microns long (in *Keniochloa*), or 39–41 microns long (in *Paracolpodium*). Subsidiaries non-papillate; parallel-sided, or parallel-sided and dome-shaped (in *Keniochloa*). Guard-cells overlapped by the interstomatals. *Intercostal short-cells* common, or absent or very rare; not paired; not silicified. *Costal short-cells* neither distinctly grouped into long rows nor predominantly paired, or no costal short-cells in *Paracolpodium altaicum*. Costal silica bodies horizontally-elongated smooth, or rounded, or tall-and-narrow, or crescentic, or .

Transverse section of leaf blade, physiology. C$_3$; XyMS+. *Mesophyll* with non-radiate chlorenchyma. Midrib with one bundle only. Bulliforms present in discrete, regular adaxial groups; in simple fans. All the vascular bundles accompanied by sclerenchyma. Combined sclerenchyma girders present. Sclerenchyma all associated with vascular bundles.

Cytology. Chromosome base number, $x = 2$. $2n = 4$, or 8 (in *Keniochloa*), or 28 (*C. colchicum, Paracolpodium*). 2 ploid, or 4 ploid, or 14 ploid.

Taxonomy. Pooideae; Poodae; Poeae.

Ecology, geography, regional floristic distribution. Sensu stricto 3 species (?). High altitude North temperate. Montane.

Holarctic and Paleotropical. Boreal and Tethyan. African and Indomalesian. Arctic and Subarctic. Irano-Turanian. Sudano-Angolan. Indian. Somalo-Ethiopian and South Tropical African.

References, etc. Morphological/taxonomic: Bor 1970. Leaf anatomical: this project.

Commelinidium Stapf

~ *Acroceras*

Habit, vegetative morphology. Commelinaceous perennial; rhizomatous to stoloniferous, or decumbent. *Culms* 10–90 cm high; herbaceous. Culm nodes glabrous. Culm internodes solid. *Leaves* not basally aggregated; non-auriculate. Leaf blades ovate; broad; 15–40 mm wide; cordate, or not cordate, not sagittate; pseudopetiolate, or not pseudopetiolate; cross veined; disarticulating from the sheaths. Ligule a fringed membrane; short.

Reproductive organization. Plants bisexual, with bisexual spikelets; with hermaphrodite florets.

Inflorescence. Inflorescence of spicate main branches (spiciform racemes), or paniculate; open; espatheate; not comprising 'partial inflorescences' and foliar organs. Spikelet-bearing axes persistent. Spikelets solitary, or paired; somewhat secund; pedicellate.

Female-fertile spikelets. Spikelets 4–6 mm long; abaxial; compressed dorsiventrally; falling with the glumes; with a distinctly elongated rachilla internode between the glumes. Rachilla prolonged beyond the uppermost female-fertile floret (as a minute vestige), or terminated by a female-fertile floret. The rachilla extension when present, naked. Hairy callus absent.

Glumes two (separated by a marked internode); very unequal; (the longer) long relative to the adjacent lemmas; free; dorsiventral to the rachis; hairless; pointed (with a small apical callosity); awnless; carinate; similar (membranous-herbaceous). Lower glume 3–5 nerved. Upper glume 5 nerved. *Spikelets* with incomplete florets. The incomplete florets proximal to the female-fertile florets. *The proximal incomplete florets* 1; paleate. Palea of the proximal incomplete florets reduced. The proximal incomplete florets sterile. The proximal lemmas similar to the glumes; awnless; 5 nerved; more or less equalling the female-fertile lemmas to decidedly exceeding the female-fertile lemmas; less firm than the female-fertile lemmas; not becoming indurated.

Female-fertile florets 1. **Lemmas** decidedly firmer than the glumes (leathery); smooth; not becoming indurated; entire; *crested at the tip (cf.* **Acroceras**, **Cyrtococcum***);* awnless (but apiculate); hairless (shining); non-carinate; having the margins tucked in onto the palea; with a clear germination flap; *7–9 nerved*. *Palea* present; relatively long;

entire; awnless, without apical setae; textured like the lemma; not indurated; 2-nerved; keel-less. *Lodicules* present; 2; free; fleshy; glabrous; not or scarcely vascularized. *Stamens* 3. Anthers not penicillate; without an apically prolonged connective. *Ovary* glabrous. Styles free to their bases. Stigmas 2; red pigmented.

Fruit, embryo and seedling. Fruit small. Hilum short.

Abaxial leaf blade epidermis. *Costal/intercostal zonation* conspicuous. *Papillae* absent. Intercostal zones exhibiting many atypical long-cells. Mid-intercostal long-cells rectangular; having markedly sinuous walls. *Microhairs* present; panicoid-type; (60–)63–78(–79.5) microns long; 6.3–9 microns wide at the septum. Microhair total length/width at septum 7–12.4. Microhair apical cells 27–36 microns long. Microhair apical cell/total length ratio 0.42–0.46. *Stomata* common; 27–33 microns long. Subsidiaries high dome-shaped and triangular. Guard-cells overlapping to flush with the interstomatals. *Intercostal short-cells* absent or very rare. *Costal short-cells* conspicuously in long rows. Costal silica bodies 'panicoid-type'.

Transverse section of leaf blade, physiology. C_3; XyMS+. *Mesophyll* with radiate chlorenchyma. *Leaf blade* adaxially flat. *Midrib* conspicuous; with one bundle only. Bulliforms not in discrete, regular adaxial groups (constituting most of adaxial epidermis). All the vascular bundles accompanied by sclerenchyma. Combined sclerenchyma girders present; forming 'figures'. Sclerenchyma all associated with vascular bundles.

Cytology. $2n = 36$.

Taxonomy. Panicoideae; Panicodae; Paniceae.

Ecology, geography, regional floristic distribution. 3 species. West tropical Africa. Shade species.

Paleotropical. African. Sudano-Angolan and West African Rainforest. Sahelo-Sudanian, Somalo-Ethiopian, and South Tropical African.

References, etc. Leaf anatomical: this project.

Cornucopiae L.

Habit, vegetative morphology. Annual. *Culms* 10–40 cm high; herbaceous. *Leaves* not basally aggregated; non-auriculate. The upper sheaths inflated. Leaf blades linear (acuminate); narrow; flat. Ligule an unfringed membrane; not truncate; 1 mm long.

Reproductive organization. Plants bisexual, with bisexual spikelets; with hermaphrodite florets. The spikelets all alike in sexuality.

Inflorescence. *Inflorescence paniculate (but peculiar — the spikelets in numerous small, compact panicles, each of these enclosed by a leathery, toothed involucre at the tip of a stout, recurved peduncle; the peduncles themselves subtended by the inflated sheaths of the upper leaves)*; spatheate (the inflorescence units enclosed in tough, toothed involucres); a complex of 'partial inflorescences' and intervening foliar organs. *Spikelet-bearing axes* capitate; disarticulating; falling entire (the heads falling whole). *Spikelets* associated with bractiform involucres (each head surrounded by a cup-shaped involucral bract, which is is deciduous with it). The involucres shed with the fertile spikelets. Spikelets not secund; subsessile to pedicellate (the pedicels short).

Female-fertile spikelets. *Spikelets* more or less unconventional; 4–7 mm long; compressed laterally; falling with the glumes (in the heads). Rachilla terminated by a female-fertile floret. Hairy callus absent.

Glumes two; more or less equal; long relative to the adjacent lemmas; joined (in the lower third); hairless (but the keel ciliate below); not pointed (round or emarginate); awnless; carinate; similar. Lower glume 3 nerved. Upper glume 2–3 nerved. *Spikelets* with female-fertile florets only.

Female-fertile florets 1. Lemmas truncate; similar in texture to the glumes (the margins connate over the lower half); not becoming indurated; entire; blunt; awnless, or awned. Awns when present, 1; dorsal; from well down the back; non-geniculate; much shorter than the body of the lemma; entered by one vein. Lemmas hairless; non-carinate; without a germination flap; 5 nerved. *Palea* absent. *Lodicules* absent. *Stamens* 3. *Ovary* glabrous. Styles fused. Stigmas 2.

Fruit, embryo and seedling. *Fruit* free from both lemma and palea; small (1.5–2 mm long); compressed laterally. Hilum short. Embryo small. Endosperm hard; with lipid; containing compound starch grains.

Abaxial leaf blade epidermis. *Costal/intercostal zonation* conspicuous. *Papillae* absent. *Long-cells* similar in shape costally and intercostally, or markedly different in shape costally and intercostally (the costals bulbous between the silica cells); of similar wall thickness costally and intercostally, or differing markedly in wall thickness costally and intercostally (the costals thicker walled). Mid-intercostal long-cells rectangular and fusiform; having straight or only gently undulating walls. *Microhairs* absent. *Stomata* common, or absent or very rare; (27–)39(–60) microns long. Subsidiaries non-papillate; parallel-sided. Guard-cells overlapped by the interstomatals. *Intercostal short-cells* absent or very rare. With a few small costal prickles. *Costal short-cells* neither distinctly grouped into long rows nor predominantly paired (mostly solitary). Costal silica bodies present and well developed; horizontally-elongated crenate/sinuous (large and conspicuous).

Transverse section of leaf blade, physiology. C_3; XyMS+. *Mesophyll* with non-radiate chlorenchyma; without adaxial palisade. *Leaf blade* 'nodular' in section; with the ribs more or less constant in size. *Midrib* not readily distinguishable; with one bundle only. *The lamina* symmetrical on either side of the midrib. Bulliforms unclear, in the material seen. All the vascular bundles accompanied by sclerenchyma (but this scanty). Combined sclerenchyma girders absent (each bundle associated with small adaxial and abaxial strands). Sclerenchyma all associated with vascular bundles.

Special diagnostic feature. *Spikelets in numerous small, compact, short-branched panicles, each panicle at the tip of a stout, recurved peduncle and enclosed by a leathery, toothed involucre, the peduncles themselves subtended by the inflated sheaths of the (modified) upper leaves.*

Cytology. Chromosome base number, $x = 7$. $2n = 14$. 2 ploid.

Taxonomy. Pooideae; Poodae; Aveneae.

Ecology, geography, regional floristic distribution. 2 species. Eastern Mediterranean. Helophytic.

Holarctic. Tethyan. Mediterranean.

References, etc. Leaf anatomical: this project.

Cortaderia Stapf

Moorea Lemaire

Habit, vegetative morphology. Perennial; caespitose (mostly large, tussocky). *Culms* 100–400 cm high. Culm nodes glabrous. Culm internodes hollow. **Leaves *mostly basal***; non-auriculate. The sheaths disintegrating or rolling. Leaf blades linear (often harsh, with lacerating margins); broad, or narrow; without cross venation; disarticulating from the sheaths; once-folded in bud. **Ligule** present; *a fringe of hairs. Contra-ligule* present, or absent.

Reproductive organization. Plants bisexual, with bisexual spikelets, or dioecious; with hermaphrodite florets, or without hermaphrodite florets. The spikelets all alike in sexuality (i.e., on the same plant); hermaphrodite, or hermaphrodite and female-only, or female-only (mainly gynodioecious). Plants outbreeding and inbreeding. Apomictic (non-pseudogamous), or reproducing sexually.

Inflorescence. Inflorescence paniculate (large and plumose or small); open. Inflorescence with axes ending in spikelets. Inflorescence espatheate; not comprising 'partial inflorescences' and foliar organs. Spikelet-bearing axes persistent. Spikelets not secund; pedicellate.

Female-fertile spikelets. Spikelets 10–18 mm long; compressed laterally; disarticulating above the glumes; disarticulating between the florets. Rachilla prolonged beyond the uppermost female-fertile floret. The rachilla extension with incomplete florets. Hairy callus present. *Callus* long; pointed.

Glumes two; more or less equal; shorter than the spikelets to about equalling the spikelets (from 2/3 as long); *long relative to the adjacent lemmas*; hairless; glabrous; pointed; awnless; carinate; similar (narrow, hyaline). Lower glume 1(–3) nerved. Upper glume 1(–3) nerved. *Spikelets* with incomplete florets. The incomplete florets distal to the female-fertile florets. *The distal incomplete florets* merely underdeveloped. *Spikelets without proximal incomplete florets.*

Female-fertile florets *2–3(–5)*. Lemmas attenuate; similar in texture to the glumes to decidedly firmer than the glumes (membranous or hyaline); not becoming indurated; entire, or incised; when incised, 2 lobed; deeply cleft, or not deeply cleft; awned, or awnless. Awns when awned, 1, or 3; median, or median and lateral (via the lateral lobes); the median similar in form to the laterals (or somewhat more flattened, when laterals present); from a sinus, or apical; non-geniculate to geniculate; hairless; entered by several veins (3). The lateral awns when present, shorter than the median. Lemmas hairy. The hairs in tufts, or not in tufts; in transverse rows, or not in transverse rows. *Lemmas* noncarinate; *3 nerved*. Palea present (glabrous or hairy); relatively long; entire (truncate); awnless, without apical setae (but hairy); not indurated; 2-nerved; 2-keeled. Palea keels wingless. *Lodicules* present; 2; free; fleshy; ciliate. *Stamens* 3, or 0 (in female plants of dioecious species). Anthers not penicillate. *Ovary* glabrous. Styles free to their bases. Stigmas 2; brown.

Fruit, embryo and seedling. *Fruit* free from both lemma and palea. *Hilum long-linear*. Embryo large. Endosperm hard; without lipid; containing compound starch grains. Embryo without an epiblast; with a scutellar tail; with an elongated mesocotyl internode. Embryonic leaf margins meeting.

Abaxial leaf blade epidermis. *Costal/intercostal zonation* conspicuous. *Papillae* absent. *Long-cells* similar in shape costally and intercostally; of similar wall thickness costally and intercostally, or differing markedly in wall thickness costally and intercostally. Mid-intercostal long-cells rectangular; having markedly sinuous walls. *Microhairs* present, or absent; when present, panicoid-type. *Stomata* absent or very rare, or common. Subsidiaries dome-shaped, or parallel-sided and dome-shaped (*C. bifida* exhibiting a few parallels). Guard-cells when present, overlapped by the interstomatals (sic). *Intercostal short-cells* common; in cork/silica-cell pairs and not paired (solitary); when paired silicified. Intercostal silica bodies tall-and-narrow. *Costal short-cells* conspicuously in long rows and neither distinctly grouped into long rows nor predominantly paired (varying from vein to vein), or neither distinctly grouped into long rows nor predominantly paired (nearly all solitary in *C. bifida*, solitary and paired in *C. selloana*). Costal silica bodies variously horizontally-elongated crenate/sinuous, or tall-and-narrow, or 'panicoid-type'; when panicoid type, mostly short dumb-bell shaped.

Transverse section of leaf blade, physiology. C_3; XyMS+. *Mesophyll* with nonradiate chlorenchyma; without adaxial palisade. *Leaf blade* with distinct, prominent adaxial ribs to 'nodular' in section; with the ribs more or less constant in size (broad, flattopped). *Midrib* conspicuous (very much so in *C. selloana*), or not readily distinguishable (*C. archboldii*); with one bundle only (*C. archboldii*), or having a conventional arc of bundles. Bulliforms present in discrete, regular adaxial groups (at the bases of the furrows), or not in discrete, regular adaxial groups (the groups then irregular or ill defined, of small cells); in simple fans (e.g. *C. bifida*, or more often the groups ill-defined and of small cells cf. *Ammophila* —e.g. *C. selloana*). All the vascular bundles accompanied by sclerenchyma. Combined sclerenchyma girders present; forming 'figures' (in all the bundles). Sclerenchyma not all bundle-associated. The 'extra' sclerenchyma in abaxial groups to in a continuous abaxial layer.

Culm anatomy. *Culm internode bundles* in three or more rings.

Phytochemistry. Leaves containing flavonoid sulphates (4 species), or without flavonoid sulphates (*C. fulvida*).

Special diagnostic feature. The median lemma awn not strongly flattened, laterals present or absent.

Cytology. Chromosome base number, $x = 9$. $2n = 36, 72, 90,$ and 108. 4, 8, 10, and 12 ploid.

Taxonomy. Arundinoideae; Danthonieae.

Ecology, geography, regional floristic distribution. 24 species. New Zealand, South America. Commonly adventive. Mesophytic to xerophytic; species of open habitats. Hillsides, in scrub and in weedy places.

Neotropical and Antarctic. Caribbean, Venezuela and Surinam, Pampas, and Andean. New Zealand and Patagonian.

Rusts and smuts. Smuts from Ustilaginaceae. Ustilaginaceae — *Ustilago*.

Economic importance. Significant weed species: *C. selloana*, perhaps increasing in significance as such. Cultivated ornamentals, especially *C. selloana*.

References, etc. Morphological/taxonomic: Zotov 1963. Leaf anatomical: Metcalfe 1960; this project.

Corynephorus P. Beauv.

Anachortus Jirásek and Chrtek

Habit, vegetative morphology. Annual, or perennial; caespitose. *Culms* 10–60 cm high; herbaceous; unbranched above. Culm nodes glabrous. Culm internodes hollow. Leaves non-auriculate. Leaf blades linear; narrow; 0.3 to 0.8 mm in diameter; setaceous, or not setaceous; folded, or rolled and acicular; without cross venation; persistent. Ligule an unfringed membrane; not truncate; 2–4 mm long.

Reproductive organization. Plants bisexual, with bisexual spikelets; with hermaphrodite florets.

Inflorescence. Inflorescence paniculate; open, or contracted; with capillary branchlets, or without capillary branchlets; espatheate; not comprising 'partial inflorescences' and foliar organs. Spikelet-bearing axes persistent. Spikelets not secund; pedicellate.

Female-fertile spikelets. *Spikelets* 3–5 mm long; *compressed laterally*; disarticulating above the glumes; disarticulating between the florets. Rachilla prolonged beyond the uppermost female-fertile floret; rachilla hairy. The rachilla extension naked. Hairy callus present. *Callus* short.

Glumes two; more or less equal; about equalling the spikelets; long relative to the adjacent lemmas; pointed; awnless; carinate; similar (lanceolate). Lower glume 1 nerved. Upper glume 1 nerved, or 3 nerved (at base). *Spikelets* with female-fertile florets only.

Female-fertile florets 2. Lemmas similar in texture to the glumes (thinly membranous); not becoming indurated; incised; shortly 2 lobed; not deeply cleft (minutely bidenticulate); awned. *Awns* 1; median; dorsal; from well down the back; *geniculate (and peculiar, with a clavate apex enclosed by the glumes, and with a ring of minute hairs distal to the twisted lower half)*; much longer than the body of the lemma; entered by one vein. Lemmas hairless; non-carinate; 1 nerved. *Palea* present; relatively long; tightly clasped by the lemma; apically notched; awnless, without apical setae; 2-nerved. *Lodicules* present; 2; free; membranous; glabrous; toothed. *Stamens* 3. Anthers 0.5–1.5 mm long, or 0.4–0.5 mm long; not penicillate. *Ovary* glabrous. Styles free to their bases. Stigmas 2.

Fruit, embryo and seedling. *Fruit* usually, slightly adhering to lemma and/or palea; small; longitudinally grooved; compressed dorsiventrally. Hilum short. Embryo small; not waisted. Endosperm liquid in the mature fruit, or hard; with lipid.

Seedling with a short mesocotyl; with a loose coleoptile, or with a tight coleoptile. First seedling leaf with a well-developed lamina. The lamina narrow; erect; 3 veined.

Abaxial leaf blade epidermis. *Costal/intercostal zonation* lacking. *Papillae* absent. *Microhairs* absent. *Stomata* absent or very rare. *Intercostal short-cells* common; in cork/silica-cell pairs and not paired; silicified (when paired). Prickles abundant. *Costal short-cells* predominantly paired. Costal silica bodies rounded, tall-and-narrow, and crescentic.

Transverse section of leaf blade, physiology. C_3; XyMS+. *Mesophyll* with radiate chlorenchyma; without adaxial palisade. *Leaf blade* with distinct, prominent adaxial ribs (small, in the adaxial groove). *Midrib* not readily distinguishable (except by its position); with one bundle only. Bulliforms not in discrete, regular adaxial groups. Combined sclerenchyma girders absent. Sclerenchyma not all bundle-associated. The 'extra' sclerenchyma in a continuous abaxial layer.

Special diagnostic feature. *Lemmas awned, the awn bearing a ring of minute hairs at the middle, and apically clavate*.

Cytology. Chromosome base number, $x = 7$. $2n = 14$. 2 ploid. Haploid nuclear DNA content 1.2 pg (1 species). Mean diploid 2c DNA value 2.3 pg.

Taxonomy. Pooideae; Poodae; Aveneae.

Ecology, geography, regional floristic distribution. 5 species. Europe, Mediterranean. Commonly adventive. Xerophytic; species of open habitats. In sandy places, often coastal.

Holarctic and Paleotropical. Boreal and Tethyan. African. Euro-Siberian. Mediterranean and Irano-Turanian. Saharo-Sindian. European.

Rusts and smuts. Rusts — *Puccinia*. Taxonomically wide-ranging species: *Puccinia graminis*.

References, etc. Leaf anatomical: this project.

Cottea Kunth

Habit, vegetative morphology. Perennial; caespitose. **Culms** 30–60 cm high; herbaceous; branched above; *tuberous (the swellings associated with cleistogamous spikelets)*. Plants unarmed. *Leaves* not basally aggregated; non-auriculate. Sheaths pilose. Leaf blades linear; narrow; 3–7 mm wide; flat; exhibiting multicellular glands abaxially. The abaxial glands intercostal (at the base of macrohairs). Leaf blades without cross venation; persistent. Ligule a fringed membrane, or a fringe of hairs (?).

Reproductive organization. Plants bisexual, with bisexual spikelets; with hermaphrodite florets. Exposed-cleistogamous, or chasmogamous. *Plants with hidden cleistogenes*. The hidden cleistogenes in the leaf sheaths (one-flowered, in the basal sheaths).

Inflorescence. Inflorescence paniculate; open (rather loose, 8–15 cm long); espatheate; not comprising 'partial inflorescences' and foliar organs. Spikelet-bearing axes persistent. Spikelets not secund; pedicellate.

Female-fertile spikelets. Spikelets 5–7 mm long; disarticulating above the glumes; disarticulating between the florets; with conventional internode spacings. Rachilla prolonged beyond the uppermost female-fertile floret. The rachilla extension with incomplete florets. Hairy callus present. *Callus* short.

Glumes present; two; more or less equal; shorter than the spikelets; long relative to the adjacent lemmas (almost equalling them); hairy (pilose); acuminate or 3-toothed; awnless; similar (lanceolate). *Lower glume 7–13 nerved*. Upper glume 7–13 nerved. *Spikelets with incomplete florets*. The incomplete florets distal to the female-fertile florets. *The distal incomplete florets* merely underdeveloped; awned to awnless.

Female-fertile florets 4–8. Lemmas not becoming indurated; incised; irregularly 9–11 lobed; deeply cleft; awned. Awns 9, or 11; median and lateral; the median similar in form to the laterals; non-geniculate; recurving (spreading); hairless to hairy; much shorter than the body of the lemma to much longer than the body of the lemma. The lateral awns shorter than the median and about equalling the median. Lemmas hairy (villous below); non-carinate (dorsally rounded); 9–13 nerved. *Palea* present; relatively long (somewhat longer than lemma); awnless, without apical setae; 2-nerved; 2-keeled. Palea keels hairy. *Lodicules* present; 2; joined to the palea; glabrous. *Stamens* 3. *Ovary* glabrous.

Fruit, embryo and seedling. Fruit small (0.75 mm long); ellipsoid; compressed dorsiventrally to not noticeably compressed. Hilum short. Pericarp fused. Embryo large; waisted; with an epiblast; with a scutellar tail; with an elongated mesocotyl internode; with one scutellum bundle. Embryonic leaf margins overlapping.

Abaxial leaf blade epidermis. *Costal/intercostal zonation* conspicuous. *Papillae* absent. *Long-cells* similar in shape costally and intercostally (but the intercostals much larger); of similar wall thickness costally and intercostally (thin walled). Mid-intercostal long-cells rectangular; having markedly sinuous walls. *Microhairs* present; elongated; clearly two-celled; *Enneapogon*-type. Microhair apical cell wall of similar thickness/rigidity to that of the basal cell. Microhairs (138–)258–288(–300) microns long. Microhair basal cells 141–180 microns long. Microhairs (4.5–)10.5–12(–13.5) microns wide at the septum. Microhair total length/width at septum 22.2–30.7. Microhair apical cells (27–)42–48(–51) microns long. Microhair apical cell/total length ratio 0.14–0.19. *Stomata* common; (19.5–)21–24 microns long. Subsidiaries dome-shaped. Guard-cells overlapping to flush with the interstomatals. *Intercostal short-cells* absent or very rare. Intercostal silica bodies absent. Long macrohairs with basal rosettes present. *Costal short-*

cells conspicuously in long rows. Costal silica bodies present in alternate cell files of the costal zones; 'panicoid-type'; cross shaped (some), or dumb-bell shaped (mostly).

Transverse section of leaf blade, physiology. C_4; XyMS+. PCR sheaths of the primary vascular bundles interrupted; interrupted both abaxially and adaxially. PCR sheath extensions absent. *Leaf blade* 'nodular' in section to adaxially flat. *Midrib* not readily distinguishable; with one bundle only. Bulliforms present in discrete, regular adaxial groups; in simple fans (the large median cells deeply penetrating the mseophyll). All the vascular bundles accompanied by sclerenchyma. Combined sclerenchyma girders present (with all the bundles); forming 'figures' (I's with all bundles). Sclerenchyma all associated with vascular bundles.

Special diagnostic feature. Female-fertile lemmas irregularly lobed, the lobes produced into 7–11 awns.

Cytology. Chromosome base number, $x = 10$. $2n = 20$. 2 ploid.

Taxonomy. Chloridoideae; Pappophoreae.

Ecology, geography, regional floristic distribution. 1 species. Texas to Argentina. Xerophytic; species of open habitats.

Holarctic and Neotropical. Madrean. Andean.

References, etc. Leaf anatomical: this project.

Craspedorhachis Benth.

Habit, vegetative morphology. Perennial; often stoloniferous. *Culms* 20–100 cm high; herbaceous. Plants unarmed. Leaves non-auriculate (but with conspicuous hairs above the ligule). Leaf blades without abaxial multicellular glands; without cross venation; persistent. Ligule a fringed membrane to a fringe of hairs.

Reproductive organization. *Plants bisexual, with bisexual spikelets*; with hermaphrodite florets.

Inflorescence. *Inflorescence of spicate main branches (several slender spikes, usually on a long axis)*; digitate, or subdigitate, or non-digitate. Rachides flattened. Inflorescence espatheate; not comprising 'partial inflorescences' and foliar organs. *Spikelet-bearing axes* spikes; with substantial rachides (these zigzag, flattened); disarticulating (in *C. africana*, the peduncle bears a cupular disarticulation zone at the point of origin of the spikes, which seem to fall together), or persistent (?); if disarticulating, falling entire. Spikelets solitary; secund; biseriate.

Female-fertile spikelets. *Spikelets* about 3 mm long; adaxial; *compressed dorsiventrally*; disarticulating above the glumes. Rachilla terminated by a female-fertile floret. Hairy callus absent.

Glumes two; more or less equal; long relative to the adjacent lemmas (the upper considerably exceeding them); dorsiventral to the rachis; hairless; glabrous; pointed (acute); awnless; carinate (G_1, asymmetrically so), or non-carinate (G_2); (the lower) *with the keel conspicuously winged*; *very dissimilar (both long, membranous, the lower asymmetrically 1-keeled, the upper flat-backed, infolded with two keels)*. Lower glume G_2 two-keeled; 0–1 nerved. Upper glume 1–3 nerved. **Spikelets** with female-fertile florets only; *without proximal incomplete florets*.

Female-fertile florets 1. Lemmas less firm than the glumes (hyaline); not becoming indurated; awnless to mucronate; hairy (long-hairy on the veins); weakly *carinate*; without a germination flap; 3 nerved. *Palea* present; relatively long; apically notched; awnless, without apical setae; textured like the lemma; not indurated; 2-nerved; keel-less. *Lodicules* present; 2; free; fleshy; glabrous; not or scarcely vascularized. *Stamens* 3. Anthers not penicillate; without an apically prolonged connective. *Ovary* glabrous. Styles free to their bases. Stigmas 2.

Fruit, embryo and seedling. *Fruit* free from both lemma and palea; small (about 1 mm long); obovoid. Hilum short. Pericarp fused. Embryo large.

Abaxial leaf blade epidermis. *Costal/intercostal zonation* conspicuous. *Papillae* absent. Long-cells of similar wall thickness costally and intercostally. Mid-intercostal long-cells having markedly sinuous walls. *Microhairs* present; more or less spherical to elongated; clearly two-celled; chloridoid-type (but with an unusual apical, peg-like appendage). Microhair apical cell wall of similar thickness/rigidity to that of the basal cell. Microhairs 15–21 microns long. Microhair basal cells 9–12 microns long. Microhairs (4.5–)7.5 microns wide at the septum. Microhair total length/width at septum 2–3.3. Microhair apical cells 7.5–11 microns long. Microhair apical cell/total length ratio 0.45–0.6. *Stomata* common; 19.5–21 microns long. Subsidiaries mostly low dome-shaped. Guard-cells overlapping to flush with the interstomatals. *Intercostal short-cells* common; in cork/silica-cell pairs and not paired; not silicified. Intercostal silica bodies absent. *Costal short-cells* conspicuously in long rows. Costal silica bodies present in alternate cell files of the costal zones; 'panicoid-type'; consistently dumb-bell shaped.

Transverse section of leaf blade, physiology. C_4; XyMS+. PCR sheaths of the primary vascular bundles interrupted; interrupted abaxially only. PCR sheath extensions absent. PCR cell chloroplasts centripetal. Mesophyll traversed by columns of colourless mesophyll cells (in places, but most of the columns terminate one cell short of the abaxial epidermis). *Leaf blade* 'nodular' in section to adaxially flat. *Midrib* not readily distinguishable; with one bundle only. Bulliforms present in discrete, regular adaxial groups; associated with colourless mesophyll cells to form deeply-penetrating fans. All the vascular bundles accompanied by sclerenchyma. Combined sclerenchyma girders present; forming 'figures' (all bundles with anchor-shaped combined girders). Sclerenchyma all associated with vascular bundles. The lamina margins with fibres.

Cytology. $2n = 27$.

Taxonomy. Chloridoideae; main chloridoid assemblage.

Ecology, geography, regional floristic distribution. 5–6 species. Tropical Africa, North and South America. Mesophytic to xerophytic; species of open habitats; glycophytic. Sandy savanna.

Paleotropical. African and Madagascan. Sudano-Angolan and Namib-Karoo. South Tropical African and Kalaharian.

References, etc. Leaf anatomical: this project.

Crinipes Hochst.

Excluding *Crinipes (Triraphis) longipes* = *Nematopoa*

Habit, vegetative morphology. Perennial; caespitose. **Culms** 45–160 cm high; *herbaceous*; unbranched above. Young shoots extravaginal. **Leaves** *mostly basal*; non-auriculate. Leaf blades narrowly linear to linear-lanceolate; narrow; 5–14 mm wide (in *C. longifolius,* 'up to 6 mm' in *C. abyssinicus*); not setaceous; flat, or rolled (then convolute); without cross venation; disarticulating from the sheaths. *Ligule a fringe of hairs.*

Reproductive organization. Plants bisexual, with bisexual spikelets; with hermaphrodite florets.

Inflorescence. *Inflorescence paniculate (medium to large)*; open to contracted; more or less irregular; espatheate; not comprising 'partial inflorescences' and foliar organs. Spikelets not secund; pedicellate.

Female-fertile spikelets. Spikelets compressed laterally; disarticulating above the glumes; disarticulating between the florets. Rachilla prolonged beyond the uppermost female-fertile floret; rachilla hairy. The rachilla extension with incomplete florets. Hairy callus present. Callus blunt.

Glumes two; very unequal; shorter than the spikelets; shorter than the adjacent lemmas (1/2 to 3/4 as long); without conspicuous tufts or rows of hairs; pointed (acuminate); awned to awnless (usually mucronate or short-awned); similar (thinly membranous). Lower glume 1 nerved. Upper glume 1 nerved. *Spikelets* with incomplete florets. The incomplete florets distal to the female-fertile florets. *The distal incomplete florets* merely underdeveloped.

Female-fertile florets 2–3. Lemmas narrowly ovate to narrowly oblong-ovate; not becoming indurated (scarious-membranous); entire to incised; when incised, 2 lobed; not deeply cleft (bidenticulate); awned. Awns 1; median; from a sinus, or apical; non-geniculate; straight, or flexuous; hairless (scaberulous); much longer than the body of the lemma. *Lemmas hairy (but only between the lateral nerves and the margins)*; non-carinate (rounded on the back); *3 nerved*; with the nerves non-confluent. *Palea* present; relatively long; awnless, without apical setae; textured like the lemma (thinly membranous); not indurated; 2-nerved; 2-keeled. Palea back glabrous. Palea keels wingless. *Lodicules* present; 2; free; fleshy; ciliate, or glabrous. *Stamens* 3. Anthers 2–2.5 mm long. *Ovary* glabrous. Styles free to their bases. Stigmas 2; brown ('yellowish').

Fruit, embryo and seedling. *Fruit* free from both lemma and palea; small (1.8 to 2 mm long); compressed dorsiventrally (ventrally flat, dorsally convex). *Hilum short (linear, but only 1/4 to 1/3 the length of the grain).* Embryo small (1/4 to 1/3 the length of the grain).

Abaxial leaf blade epidermis. *Costal/intercostal zonation* conspicuous. *Papillae* absent. *Long-cells* markedly different in shape costally and intercostally (the costals narrower); of similar wall thickness costally and intercostally (fairly thick walled). Mid-intercostal long-cells regularly rectangular; having markedly sinuous walls (the sinuosity fine). *Microhairs* absent. *Stomata* absent or very rare (a few only seen, near the blade margin). Subsidiaries non-papillate; parallel-sided and dome-shaped. Guard-cells overlapping to flush with the interstomatals. *Intercostal short-cells* very common; not paired (solitary). Neither macrohairs nor prickles seen, except for prickles at the margins. Costal zones with short-cells. *Costal short-cells* conspicuously in long rows (but the short-cells fairly long). Costal silica bodies present and well developed; 'panicoid-type'; consistently, conspicuously nodular.

Transverse section of leaf blade, physiology. C_3; XyMS+ (the mestome sheath cells thick walled). *Mesophyll* with radiate chlorenchyma (especially around the smaller bundles); with adaxial palisade; *Isachne*-type (in places, especially around minor bundles). *Leaf blade* with distinct, prominent adaxial ribs; with the ribs very irregular in sizes (small over the smaller bundles, large and flat topped over the primaries). *Midrib* not readily distinguishable; with one bundle only. The lamina symmetrical on either side of the midrib. Bulliforms present in discrete, regular adaxial groups (in each furrow); in simple fans. All the vascular bundles accompanied by sclerenchyma. Combined sclerenchyma girders present (with every bundle); forming 'figures' ('anchors' with the smaller bundles, heavy I's with the primaries). Sclerenchyma not all bundle-associated. The 'extra' sclerenchyma in abaxial groups (or rather, in the form of single cells); abaxial-hypodermal, the groups isolated (opposite the bulliforms, but not of universal occurrence).

Taxonomy. Arundinoideae; Danthonieae.

Ecology, geography, regional floristic distribution. 2 species. Ethiopia, Sudan, Uganda. In rocky places.

Paleotropical. African. Saharo-Sindian and Sudano-Angolan. Sahelo-Sudanian and Somalo-Ethiopian.

References, etc. Morphological/taxonomic: Hubbard 1957. Leaf anatomical: this project.

Crithopsis Jaub & the Spach

Habit, vegetative morphology. *Annual*; caespitose. *Culms* 15–30 cm high; herbaceous; unbranched above. Culm nodes glabrous. *Leaves* not basally aggregated; auriculate. Leaf blades linear (acuminate); narrow; to 3 mm wide; flat; without cross venation; persistent. *Ligule an unfringed membrane.*

Reproductive organization. Plants bisexual, with bisexual spikelets; with hermaphrodite florets. *The spikelets all alike in sexuality.*

Inflorescence. *Inflorescence a false spike, with spikelets on contracted axes (the clusters each reduced to two spikelets).* Rachides flattened. Inflorescence espatheate; not comprising 'partial inflorescences' and foliar organs. Spikelet-bearing axes disarticulating; disarticulating at the joints. Spikelets paired; not secund; the pairs alternately distichous.

Female-fertile spikelets. Spikelets about 15 mm long; compressed laterally; falling with the glumes (and with the joint). Rachilla prolonged beyond the uppermost female-fertile floret; rachilla hairy. The rachilla extension with incomplete florets.

Glumes two; more or less equal; long relative to the adjacent lemmas; subulate; awned; non-carinate; similar (leathery, narrow, awn-like above). Lower glume 3 nerved. Upper glume 3 nerved. *Spikelets* with incomplete florets. The incomplete florets distal to the female-fertile florets. The distal incomplete florets 1; *incomplete florets* merely underdeveloped.

Female-fertile florets 1. Lemmas attenuate into the awn; similar in texture to the glumes (and similar to them in form); not becoming indurated; entire; pointed; awned. Awns 1; median; apical; non-geniculate; hairless (scabrid); about as long as the body of the lemma to much longer than the body of the lemma (4–7 mm long, erect); entered by several veins. Lemmas hairy, or hairless; non-carinate; 5 nerved. *Palea* present; relatively long; awnless, without apical setae; 2-nerved; 2-keeled. *Lodicules* present; 2; free; membranous; ciliate; toothed, or not toothed. *Stamens* 3. Anthers not penicillate. **Ovary *hairy***. Styles free to their bases. Stigmas 2; white.

Fruit, embryo and seedling. *Fruit* adhering to lemma and/or palea; medium sized; longitudinally grooved; compressed dorsiventrally; with hairs confined to a terminal tuft. Hilum long-linear. Embryo small.

Cytology. Chromosome base number, $x = 7$. $2n = 14$. 2 ploid. Haplomic genome content K.

Taxonomy. Pooideae; Triticodae; Triticeae.

Ecology, geography, regional floristic distribution. 1 species. North Africa & the Orient. Xerophytic; species of open habitats. In dry grassland.

Holarctic and Paleotropical. Tethyan. African. Mediterranean. Saharo-Sindian.

References, etc. Morphological/taxonomic: Löve 1984.

Special comments. Anatomical data wanting.

Crypsis Aiton

Antitragus Gaertn., *Ceytosis* Munro, *Heleochloa* Hort ex Roem., *Pallasia* Scop., *Pechea* Lapeyr., *Raddia* Mazziari, *Torgesia* Bornm.

Habit, vegetative morphology. **Annual**; caespitose to decumbent. *Culms* 5–30 cm high; herbaceous; branched above. Culm nodes glabrous. Culm internodes solid, or hollow. *Leaves* not basally aggregated; non-auriculate. Leaf blades linear (usually short); narrow; setaceous, or not setaceous; flat, or folded, or rolled; without abaxial multicellular glands; without cross venation; disarticulating from the sheaths; rolled in bud. Ligule a fringed membrane (narrow), or a fringe of hairs.

Reproductive organization. Plants bisexual, with bisexual spikelets; with hermaphrodite florets.

Inflorescence. *Inflorescence paniculate (a panicle of contracted, bracteate panicles or heads, or a single, elongated-contracted panicle)*; open (when many-headed), or contracted (when only one head); spatheate (via enlarged, upper leaf sheaths), or espatheate; a complex of 'partial inflorescences' and intervening foliar organs (e.g. *C. aculeata*), or not comprising 'partial inflorescences' and foliar organs (e.g. *C. alopecuroides*). Spikelet-bearing axes disarticulating (the heads falling), or persistent; falling entire. *Spikelets* not secund; *sessile*.

Female-fertile spikelets. *Spikelets* 2–6 mm long; strongly *compressed laterally*; disarticulating above the glumes, or falling with the glumes, or not disarticulating. Rachilla terminated by a female-fertile floret. Hairy callus absent.

Glumes two; more or less equal; long relative to the adjacent lemmas (almost equalling the floret); pointed; awnless; similar (narrow, complicate). Lower glume 0–1 nerved. Upper glume 0–1 nerved. *Spikelets* with female-fertile florets only.

Female-fertile florets 1. Lemmas not becoming indurated (membranous); entire; awnless to mucronate; hairless; carinate; 1 nerved. *Palea* present; relatively long; entire, or apically notched, or deeply bifid; awnless, without apical setae; not indurated (hyaline); 1-nerved (*Crypsis*), or 2-nerved (*Heleochloa*). **Lodicules *absent***. Stamens 2–3. Anthers 0.5–1.5 mm long, or 2–3 mm long; not penicillate. *Ovary* glabrous. Styles free to their bases. Stigmas 2; white.

Fruit, embryo and seedling. *Fruit* free from both lemma and palea; small (1.5 mm long); ellipsoid; compressed laterally, or not noticeably compressed; sculptured, or smooth. Hilum short. *Pericarp free (sometimes becoming swollen when wet, extruding* *the seed)*. Embryo large. Endosperm hard. Embryo with an epiblast; with a scutellar tail; with an elongated mesocotyl internode. Embryonic leaf margins meeting.

Abaxial leaf blade epidermis. *Costal/intercostal zonation* conspicuous. *Papillae* absent. *Long-cells* similar in shape costally and intercostally (but the intercostals larger); of similar wall thickness costally and intercostally. Mid-intercostal long-cells rectangular and fusiform; having markedly sinuous walls. *Microhairs* present; more or less spherical; clearly two-celled; chloridoid-type. Microhair apical cell wall of similar thickness/rigidity to that of the basal cell. Microhairs 19.5–24 microns long. Microhair basal cells 10–11 microns long. Microhairs 10.5–18 microns wide at the septum. Microhair total length/width at septum 1.3–2. Microhair apical cells 12–15 microns long. Microhair apical cell/total length ratio 0.57–0.63. *Stomata* common; 22–24 microns long. Subsidiaries dome-shaped. Guard-cells overlapping to flush with the interstomatals. *Intercostal short-cells* absent or very rare; in cork/silica-cell pairs and not paired (mainly paired, some solitary); silicified. Intercostal silica bodies present and perfectly developed; cross-shaped and oryzoid-type. *Costal short-cells* neither distinctly grouped into long rows nor predominantly paired. Costal silica bodies present throughout the costal zones to present in alternate cell files of the costal zones; 'panicoid-type'.

Transverse section of leaf blade, physiology. C_4; XyMS+. PCR sheaths of the primary vascular bundles complete. PCR sheath extensions absent. *Mesophyll* with radiate chlorenchyma. *Leaf blade* with distinct, prominent adaxial ribs, or 'nodular' in section; with the ribs more or less constant in size. *Midrib* not readily distinguishable; with one bundle only. Bulliforms present in discrete, regular adaxial groups. All the vascular bundles accompanied by sclerenchyma. Combined sclerenchyma girders absent. Sclerenchyma all associated with vascular bundles. The lamina margins with fibres.

Cytology. Chromosome base number, $x = 8$ and 9. $2n = 16$, 18, 32, 36, and 54. 2, 4, and 6 ploid.

Taxonomy. Chloridoideae; main chloridoid assemblage.

Ecology, geography, regional floristic distribution. 8 species. Mediterranean to North China. Commonly adventive. Species of open habitats; halophytic, or glycophytic. Wet soils.

Holarctic and Paleotropical. Boreal, Tethyan, and Madrean. African and Madagascan. Euro-Siberian. Mediterranean and Irano-Turanian. Saharo-Sindian and Sudano-Angolan. European and Siberian. Sahelo-Sudanian, Somalo-Ethiopian, and South Tropical African.

References, etc. Leaf anatomical: this project.

Cryptochloa Swallen

Habit, vegetative morphology. Perennial; caespitose (from short rhizomes). The flowering culms leafy. *Culms* 30–50 cm high; herbaceous; unbranched above. Plants unarmed. *Leaves* not basally aggregated; without auricular setae. Leaf blades broad; 1–2.5 mm wide; pseudopetiolate; cross veined; rolled in bud. *Contra-ligule* absent.

Reproductive organization. *Plants monoecious with all the fertile spikelets unisexual*; without hermaphrodite florets. The spikelets of sexually distinct forms on the same plant; hermaphrodite and male-only. *The male and female-fertile spikelets mixed in the inflorescence.* The spikelets overtly heteromorphic (male spikelets mostly shorter pedicelled, reduced). *Plants with inflorescences hardly exserted from the sheaths, inconspicuous*.

Inflorescence. Inflorescence determinate, or indeterminate (most species with synflorescences); *paniculate (often depauperate, having axillary partial inflorescences or synflorescences of variously reduced panicles)*. Spikelet-bearing axes paniculate; persistent. Spikelets not secund; pedicellate (the female pedicels thickened apically); consistently in 'long-and-short' combinations, or not in distinct 'long-and-short' combinations.

Female-sterile spikelets. Male spikelets reduced to lemma, palea and 2–3 free, non-penicillate stamens; sometimes awned. Rachilla of male spikelets terminated by a male floret. *The male spikelets without glumes*; without proximal incomplete florets; 1 floreted. The lemmas awnless, or awned. Male florets 1; 2 staminate, or 3 staminate. The staminal filaments free.

Female-fertile spikelets. *Spikelets 9–20 mm long*; *compressed dorsiventrally (not gibbous)*; disarticulating above the glumes; with a distinctly elongated rachilla internode above the glumes. Rachilla terminated by a female-fertile floret. *Callus* short (columnar).

Glumes two; more or less equal; long relative to the adjacent lemmas; hairless; awnless (sometimes acuminate); non-carinate; similar (thin, firm). Lower glume 5 nerved. Upper glume 5 nerved. *Spikelets* with female-fertile florets only.

Female-fertile florets 1. Lemmas decidedly firmer than the glumes (bony); becoming indurated; entire; pointed; awnless, or mucronate (?); hairy, or hairless; non-carinate; having the margins tucked in onto the palea; 5 nerved. *Palea* present (enclosed by lemma margins); relatively long; entire; awnless, without apical setae; indurated; 2-nerved. *Lodicules* present; 3; free; fleshy; glabrous; heavily vascularized. *Stamens* 0. *Ovary* glabrous. Styles fused (into one long style). Stigmas 2.

Fruit, embryo and seedling. *Fruit* free from both lemma and palea; compressed dorsiventrally. Hilum long-linear. Embryo small.

Abaxial leaf blade epidermis. *Costal/intercostal zonation* conspicuous. *Papillae* present (on the interstomatal cells, and also on the subsidiaries). Intercostal papillae overarching the stomata; several per cell (mostly around the edges of the interstomatals, and concentrated around the stomata). *Long-cells* markedly different in shape costally and intercostally (the costals much narrower); of similar wall thickness costally and intercostally (fairly thin walled). Mid-intercostal long-cells rectangular; having markedly sinuous walls. *Microhairs* present; panicoid-type. *Stomata* common. Subsidiaries papillate; triangular. *Intercostal short-cells* common; in cork/silica-cell pairs (and solitary). *Costal short-cells* conspicuously in long rows. Costal silica bodies saddle shaped, or oryzoid (at least, some approaching this), or 'panicoid-type'; when panicoid type, fat cross shaped.

Transverse section of leaf blade, physiology. C_3; XyMS+. *Mesophyll* with non-radiate chlorenchyma; with adaxial palisade (of large arm cells); with arm cells (these conspicuous); with fusoids. The fusoids external to the PBS. *Leaf blade* adaxially flat. *Midrib* conspicuous; with one bundle only; with colourless mesophyll adaxially. Bulliforms present in discrete, regular adaxial groups; in simple fans (the groups large). All the vascular bundles accompanied by sclerenchyma. Combined sclerenchyma girders present (with all the main bundles); forming 'figures' (most bundles with an anchor). Sclerenchyma all associated with vascular bundles.

Cytology. Chromosome base number, $x = 10$ and 11. $2n = 20$ and 22. 2 ploid.

Taxonomy. Bambusoideae; Oryzodae; Olyreae.

Ecology, geography, regional floristic distribution. 5 species. Central and South America.

Neotropical. Caribbean, Amazon, and Andean.

References, etc. Morphological/taxonomic: Soderstrom 1982a and 1982b; Soderstrom and Zuloaga 1988. Leaf anatomical: this project.

Ctenium Panzer

Aplocera Raf., *Campuloa* Desv., *Campulosus* Desv., *Monathera* Raf., *Monocera* Elliott, *Triatherus* Raf.

Habit, vegetative morphology. Perennial (usually), or annual; densely caespitose. **Culms** 40–100 cm high; *herbaceous*. Plants unarmed. The shoots aromatic (some species — e.g. *C. elegans* — smelling strongly of turpentine), or not aromatic. Leaf blades linear; narrow; setaceous, or not setaceous; flat, or rolled (convolute); without abaxial multicellular glands; without cross venation. *Ligule a fringed membrane (very short)*.

Reproductive organization. Plants bisexual, with bisexual spikelets; with hermaphrodite florets. The spikelets all alike in sexuality.

Inflorescence. Inflorescence a single spike, or of spicate main branches; non-digitate, or digitate (when of more than one spike). *Primary inflorescence branches 1–3 (the spikes pectinate, usually curved). Rachides hollowed and flattened (crescentic in section)*. Inflorescence espatheate; not comprising 'partial inflorescences' and foliar organs. *Spikelet-bearing axes* spikes; with substantial rachides; persistent. Spikelets solitary; secund; biseriate (along the midrib of the rachis).

Female-fertile spikelets. Spikelets 4–9 mm long; adaxial; compressed laterally; disarticulating above the glumes; not disarticulating between the florets. Rachilla prolonged beyond the uppermost female-fertile floret. The rachilla extension with incomplete florets.

Glumes present; two; very unequal; long relative to the adjacent lemmas; G_2 tubercled on the nerves; pointed (acute); G_2 *awned (shortly awn-tipped, and with a spreading awn from the middle of its back)*; carinate (G_1), or non-carinate (G_2 flat or rounded); very dissimilar (G_2 larger, firmer, awned). Lower glume 1 nerved. Upper glume 2 nerved, or 3 nerved. *Spikelets* with incomplete florets. **The incomplete florets both distal and proximal to the female-fertile florets.** The distal incomplete florets merely underdeveloped (male or barren). **Spikelets with proximal incomplete florets.** The proximal incomplete florets 2; paleate (the paleas hyaline, variously reduced), or epaleate; male, or sterile. The proximal lemmas awned (from just below tip); similar in texture to the female-fertile lemmas (thin, white).

Female-fertile florets 1. Lemmas less firm than the glumes (i.e., than G_2 — membranous); not becoming indurated; entire; pointed, or blunt; awned. Awns 1; median; dorsal; from near the top; non-geniculate; hairless (scabrid); about as long as the body of the lemma to much longer than the body of the lemma; entered by one vein. Lemmas hairy (ciliate on nerves); carinate; 3 nerved. *Palea* present; relatively long; not indurated; 2-nerved; 2-keeled. *Lodicules* present; 2; free; fleshy; glabrous. *Stamens* 3 (2 in male florets). *Ovary* glabrous. Styles free to their bases. Stigmas 2.

Fruit, embryo and seedling. *Fruit* free from both lemma and palea (embraced); ellipsoid. Hilum short. Pericarp fused. Embryo large; with an epiblast (rudimentary); with a scutellar tail; with an elongated mesocotyl internode. Embryonic leaf margins meeting.

Abaxial leaf blade epidermis. *Costal/intercostal zonation* conspicuous. *Papillae* absent. Mid-intercostal long-cells rectangular; having markedly sinuous walls. *Microhairs* present; more or less spherical to elongated; clearly two-celled; chloridoid-type (basal cell rather long). Microhair apical cell wall of similar thickness/rigidity to that of the basal cell. Microhairs 18–24 microns long. Microhair basal cells 13–18 microns long. Microhairs 9–10.5 microns wide at the septum. Microhair total length/width at septum 2–2.7. Microhair apical cells (7.5–)9 microns long. Microhair apical cell/total length ratio 0.38–0.43. *Stomata* common; 19.5–24 microns long. Subsidiaries triangular. *Intercostal short-cells* common (often paired with prickle bases). Intercostal silica bodies absent. Prickles very abundant in *C. polystachyum*. *Crown cells* absent. Costal silica bodies present in alternate cell files of the costal zones; 'panicoid-type'; cross shaped, or dumbbell shaped, or nodular.

Transverse section of leaf blade, physiology. C_4; XyMS+. PCR sheath outlines even. PCR sheaths of the primary vascular bundles interrupted; interrupted abaxially only. PCR sheath extensions present. Maximum number of extension cells 1, or 2. PCR cell chloroplasts centripetal. *Mesophyll* with radiate chlorenchyma; traversed by columns of colourless mesophyll cells; with arm cells (these rather conspicuous, in the material of *C. polystachyum* seen), or without arm cells (?). *Leaf blade* with distinct, prominent adaxial ribs to 'nodular' in section; with the ribs more or less constant in size. *Midrib* not readily distinguishable; with one bundle only. Bulliforms present in discrete, regular adaxial groups; associated with colourless mesophyll cells to form deeply-penetrating fans (these incorporated in the colourless columns). All the vascular bundles accompanied by sclerenchyma. Combined sclerenchyma girders present (with most bundles). Sclerenchyma all associated with vascular bundles.

Cytology. Chromosome base number, $x = 9$. $2n = 18, 36, 54$, and 160. 2 ploid, or 4 ploid, or 6 ploid, or 16 ploid (?).

Taxonomy. Chloridoideae; main chloridoid assemblage.

Ecology, geography, regional floristic distribution. 20 species. Tropical and subtropical America and Africa. Species of open habitats. Savanna.

Paleotropical and Neotropical. African and Madagascan. Sudano-Angolan and West African Rainforest. Caribbean. Sahelo-Sudanian, Somalo-Ethiopian, and South Tropical African.

References, etc. Leaf anatomical: Metcalfe 1960; this project.

Ctenopsis De Not

~ Vulpia

Habit, vegetative morphology. *Annual*; caespitose. *Culms* 10–40 cm high; herbaceous. Culm nodes glabrous. Culm internodes hollow. *Leaves* not basally aggregated; non-auriculate. Leaf blades linear; narrow; setaceous; rolled (convolute); without cross venation; persistent. Ligule an unfringed membrane; 0.7–1 mm long.

Reproductive organization. Plants bisexual, with bisexual spikelets; with hermaphrodite florets.

Inflorescence. Inflorescence a single raceme, or paniculate (sometimes sparingly branched); contracted; espatheate; not comprising 'partial inflorescences' and foliar organs. Spikelet-bearing axes persistent. *Spikelets* solitary; *secund (pectinate)*; pedicellate.

Female-fertile spikelets. Spikelets 3.5–12 mm long; compressed laterally; disarticulating above the glumes; disarticulating between the florets. Rachilla prolonged beyond the uppermost female-fertile floret; rachilla hairless. Hairy callus absent.

Glumes two; very unequal; (the longer) long relative to the adjacent lemmas; dorsiventral to the rachis; pointed; awnless; non-carinate (rounded on the back); very dissimilar (the lower minute). *Lower glume 0 nerved.* Upper glume 3 nerved. *Spikelets* with female-fertile florets only, or with incomplete florets. The incomplete florets distal to the female-fertile florets.

Female-fertile florets 3–14. Lemmas acuminate; less firm than the glumes (papery, the margins hyaline); not becoming indurated; entire; pointed; awnless, or mucronate (?); hairless; non-carinate; 5 nerved. *Palea* present; relatively long; apically notched; awnless, without apical setae; 2-nerved; 2-keeled. *Lodicules* present; 2; free; membranous; glabrous; toothed. *Stamens* 3. Anthers 0.5–2.5 mm long; not penicillate. *Ovary* glabrous. Styles free to their bases. Stigmas 2; white.

Fruit, embryo and seedling. *Fruit* free from both lemma and palea; small; longitudinally grooved; compressed dorsiventrally. *Hilum short.* Embryo small; not waisted.

Abaxial leaf blade epidermis. *Costal/intercostal zonation* conspicuous to lacking. *Papillae* absent. *Long-cells* similar in shape costally and intercostally; of similar wall thickness costally and intercostally. Mid-intercostal long-cells rectangular; having markedly sinuous walls. *Microhairs* absent. *Stomata* absent or very rare. *Intercostal short-cells* common; in cork/silica-cell pairs (mostly — some solitary); silicified (rarely), or not silicified. *Costal short-cells* predominantly paired. Costal silica bodies horizontally-elongated crenate/sinuous, or horizontally-elongated smooth, or rounded.

Transverse section of leaf blade, physiology. C_3; XyMS+. *Mesophyll* with non-radiate chlorenchyma; without adaxial palisade. *Midrib* conspicuous; with one bundle only. All the vascular bundles accompanied by sclerenchyma. Combined sclerenchyma girders present (*C. pectinella*), or absent (*C. patens*); in *C. pectinella* forming 'figures' (having tall, narrow I's with most bundles). Sclerenchyma all associated with vascular bundles.

Cytology. Chromosome base number, $x = 7$. $2n = 14$. 2 ploid.

Taxonomy. Pooideae; Poodae; Poeae.

Ecology, geography, regional floristic distribution. 4 species. Mediterranean & western Asia. Species of open habitats.

Holarctic, Paleotropical, and Cape. Tethyan. African. Mediterranean. Saharo-Sindian and Sudano-Angolan. South Tropical African.

References, etc. Leaf anatomical: this project.

Cutandia Willk.

Habit, vegetative morphology. Much-branched annual. *Culms* 10–40 cm high. Culm nodes glabrous. Culm internodes hollow. *Leaves* not basally aggregated; non-auriculate. Leaf blades linear; narrow; setaceous, or not setaceous; flat, or rolled (convolute); without cross venation; persistent. Ligule an unfringed membrane; truncate; 1.5–5 mm long.

Reproductive organization. Plants bisexual, with bisexual spikelets; with hermaphrodite florets.

275

Inflorescence. *Inflorescence* paniculate; *open (dichotomously-divaricately branched)*; *with conspicuously divaricate branchlets*; espatheate (but panicles partially enclosed by the upper leaf sheaths); not comprising 'partial inflorescences' and foliar organs. Spikelet-bearing axes disarticulating, or persistent; falling entire, or disarticulating at the joints (disarticulation occurring variously at the bases of the spikelets, between them, or at bases of branches). Spikelets not secund; pedicellate.

Female-fertile spikelets. *Spikelets 5–17 mm long*; compressed laterally; disarticulating above the glumes, or falling with the glumes; not disarticulating between the florets (proximally), or disarticulating between the florets (distally). Rachilla prolonged beyond the uppermost female-fertile floret. The rachilla extension with incomplete florets. Hairy callus absent. *Callus* short.

Glumes two; relatively large; very unequal; shorter than the spikelets; shorter than the adjacent lemmas; pointed, or not pointed; shortly awned, or awnless; carinate; similar (membranous). Lower glume 1–3 nerved. Upper glume 1–5 nerved. *Spikelets* with incomplete florets. The incomplete florets distal to the female-fertile florets.

Female-fertile florets 2–12. Lemmas similar in texture to the glumes (with hyaline margins); not becoming indurated; incised; usually 2 lobed (emarginate or bifid); awnless, or mucronate, or awned. Awns when present, 1; from a sinus, or dorsal; from near the top; much shorter than the body of the lemma; entered by one vein. Lemmas hairless; non-carinate (3-keeled); 3–5 nerved. *Palea* present; relatively long; entire (acute to truncate), or apically notched (bifid); awnless, without apical setae; 2-nerved; 2-keeled. **Lodicules** present; 2; free; membranous; *ciliate*; toothed. *Stamens* 3. Anthers 0.5–1.3 mm long; not penicillate. *Ovary* glabrous. Styles free to their bases. Stigmas 2.

Fruit, embryo and seedling. *Fruit* free from both lemma and palea; longitudinally grooved; trigonous. Hilum short (elongated, but short). Embryo small. Endosperm liquid in the mature fruit, or hard; without lipid. Embryo with an epiblast; without a scutellar tail; with a negligible mesocotyl internode. Embryonic leaf margins meeting.

Abaxial leaf blade epidermis. *Costal/intercostal zonation* conspicuous. *Papillae* absent. Long-cells of similar wall thickness costally and intercostally (very large). Mid-intercostal long-cells fusiform; having straight or only gently undulating walls. *Microhairs* absent. *Stomata* common. Subsidiaries parallel-sided. Guard-cells overlapped by the interstomatals. *Intercostal short-cells* absent or very rare. *Costal short-cells* neither distinctly grouped into long rows nor predominantly paired. Costal silica bodies horizontally-elongated crenate/sinuous to horizontally-elongated smooth, or 'panicoid-type'; often dumb-bell shaped.

Transverse section of leaf blade, physiology. C_3; XyMS+. *Mesophyll* with non-radiate chlorenchyma. *Leaf blade* with distinct, prominent adaxial ribs, or 'nodular' in section; with the ribs more or less constant in size. Midrib with one bundle only. Bulliforms not in discrete, regular adaxial groups (at least, bulliforms/groups inconspicuous in the poor material seen). All the vascular bundles accompanied by sclerenchyma. Combined sclerenchyma girders absent. Sclerenchyma all associated with vascular bundles.

Cytology. Chromosome base number, $x = 7$. $2n = 14$. 2 ploid.

Taxonomy. Pooideae; Poodae; Poeae.

Ecology, geography, regional floristic distribution. 6 species. Mediterranean, western Asia. Xerophytic; species of open habitats; halophytic, or glycophytic. Mostly in maritime sands or coastal rocky hills.

Holarctic and Paleotropical. Boreal and Tethyan. African. Euro-Siberian. Macaronesian, Mediterranean, and Irano-Turanian. Saharo-Sindian. European.

Rusts and smuts. Rusts — *Puccinia*. Taxonomically wide-ranging species: *Puccinia hordei*. Smuts from Ustilaginaceae. Ustilaginaceae — *Ustilago*.

Economic importance. Important native pasture species: *C. memphitica*.

References, etc. Morphological/taxonomic: Stace 1978b. Leaf anatomical: this project.

Cyathopus Stapf

Habit, vegetative morphology. Perennial. *Culms* 90–140 cm high; herbaceous. Leaves non-auriculate. Leaf blades narrow; flat; not pseudopetiolate; without cross venation. *Ligule an unfringed membrane*; not truncate.

Reproductive organization. Plants bisexual, with bisexual spikelets; with hermaphrodite florets. *The spikelets all alike in sexuality.*

Inflorescence. *Inflorescence paniculate (many spikeleted)*; open (large); *with capillary branchlets*; espatheate; not comprising 'partial inflorescences' and foliar organs. Spikelet-bearing axes persistent. Spikelets solitary; pedicellate. Pedicel apices minutely cupuliform.

Female-fertile spikelets. *Spikelets* about *3 mm long*; compressed laterally to not noticeably compressed; *falling with the glumes*. Rachilla terminated by a female-fertile floret.

Glumes present; two; more or less equal; long relative to the adjacent lemmas (slightly exceeding them); hairy; without conspicuous tufts or rows of hairs; pointed; awnless (but caudate); non-carinate (dorsally rounded); similar (chartaceous, each with a small apical beak). Lower glume strongly 3 nerved. Upper glume strongly 3 nerved. **Spikelets** with female-fertile florets only; *without proximal incomplete florets.*

Female-fertile florets 1. Lemmas ovate; less firm than the glumes (scarious); not becoming indurated; entire; pointed to blunt; awnless (muticous); hairless; carinate (weakly), or non-carinate (dorsally rounded); 5 nerved (the nerves obscure towards the apex). *Palea* present; relatively long; entire (pointed); awnless, without apical setae; not indurated (hyaline); 2-nerved; 2-keeled. Palea keels wingless. *Lodicules* present; 2; free; apparently not cuneate; glabrous; not toothed. *Stamens* 3. Anthers not penicillate; without an apically prolonged connective. *Ovary* glabrous. Stigmas 2.

Fruit, embryo and seedling. Fruit small (about 2 mm long). Hilum short. *Embryo large*.

Taxonomy. Pooideae (? — not reliably classifiable without better data); Poodae (?); Aveneae (?).

Ecology, geography, regional floristic distribution. 1 species. Eastern Himalayas. Mesophytic; shade species; glycophytic. In woods.

Holarctic. Boreal. Eastern Asian.

References, etc. Morphological/taxonomic: Stapf 1895.

Special comments. Fruit data wanting. Anatomical data wanting.

Cyclostachya J. & C. Reeder

Habit, vegetative morphology. Perennial; stoloniferous and caespitose. *Culms* 5–12 cm high; herbaceous; unbranched above. Culm nodes glabrous. Culm internodes solid. *Leaves* mostly basal; non-auriculate; hairy in the auricle positions. Leaf blades linear (setaceous at the tip); narrow; to 1 mm wide; without abaxial multicellular glands; without cross venation; persistent. *Ligule an unfringed membrane*; not truncate (jagged); 0.5–1 mm long. *Contra-ligule* absent.

Reproductive organization. *Plants dioecious*; without hermaphrodite florets. The spikelets all alike in sexuality (on the same plant); female-only, or male-only. Plants outbreeding.

Inflorescence. *Inflorescence* (male and female) *a single raceme (pectinate, curved, long pedunculate)*; espatheate; not comprising 'partial inflorescences' and foliar organs. *Spikelet-bearing axes disarticulating; falling entire (the raceme finally recurving into a circle).* Spikelets solitary; secund (pectinate, in two ranks); biseriate; pedicellate (on short pedicels).

Female-sterile spikelets. Male plants similar to females; male spikelets with one 3-stamened floret and 2–3 short-awned rudiments; glumes one-nerved; lemma 3-nerved, 3-toothed, short-awned; often with an abortive pistil. Rachilla of male spikelets prolonged beyond the uppermost male floret. The male spikelets with glumes; 1 floreted (plus rudiments). The lemmas short awned. Male florets 1; 3 staminate.

Female-fertile spikelets. Spikelets 8–9 mm long; compressed laterally to not noticeably compressed; disarticulating above the glumes (i.e., falling from the deciduous rachis). Rachilla prolonged beyond the uppermost female-fertile floret; rachilla hairy. The rachilla extension with incomplete florets. Hairy callus present.

Glumes two; very unequal (G_1 about two-thirds of G_2); (the upper) long relative to the adjacent lemmas; dorsiventral to the rachis; hairy (G_2); pointed (G_1 lanceolate-acuminate), or not pointed (G_2 apically notched); G_2 with 1–3 apical setae; carinate (G_1), or non-carinate (G_2); very dissimilar (hyaline, G_1 carinate-acuminate, glabrous, G_2 2-keeled, 2–3 lobed or notched at apex, with hairs on the back). Lower glume 1 nerved. Upper glume 2 nerved, or 3 nerved. *Spikelets* with incomplete florets. The incomplete florets distal to the female-fertile florets. The distal incomplete florets 2–3; *incomplete florets* clearly specialised and modified in form (clustered at the tip of the rachilla, their scabrid awns much longer than those of the fertile lemma); incomplete florets each three awned.

Female-fertile florets 1. Lemmas decidedly firmer than the glumes (membranous or cartilaginous); not becoming indurated; incised; 3 lobed; deeply cleft (into the awns); awned. Awns 3; median and lateral (via the acuminate lobes); the median more or less similar in form to the laterals (but not bending back); from a sinus, or apical (depending on interpretation, the three lobes acuminate into the awns); non-geniculate; hairless (scabrid); much shorter than the body of the lemma to about as long as the body of the lemma; entered by one vein. The lateral awns shorter than the median (bending back). Lemmas hairy; slightly keeled on the three nerves; without a germination flap; 3 nerved. *Palea* present (similar to the two-nerved examples of G_2); relatively long; apically notched; with apical setae (2, asymmetric, via the excurrent nerves); not indurated (hyaline to membranous, with green tissue); 2-nerved; 2-keeled. *Lodicules* present; 2; free; fleshy; glabrous; not or scarcely vascularized. *Stamens* 0 (3 staminodes only). *Ovary* glabrous. Styles free to their bases. Stigmas 2; stigmas pigmented.

Fruit, embryo and seedling. Fruit small (3 mm long); ellipsoid; not noticeably compressed. Hilum short. Pericarp fused. Embryo large; waisted; with an epiblast; with a scutellar tail; with an elongated mesocotyl internode. Embryonic leaf margins meeting.

Abaxial leaf blade epidermis. *Costal/intercostal zonation* conspicuous. *Papillae* absent. *Long-cells* markedly different in shape costally and intercostally (costals narrower); of similar wall thickness costally and intercostally (rather thick walled). Mid-intercostal long-cells rectangular; having markedly sinuous walls (and pitted). *Microhairs* present; elongated; clearly two-celled; chloridoid-type (very large). Microhair apical cell wall of similar thickness/rigidity to that of the basal cell. Microhairs (33–)34.5–36(–39) microns long. Microhair basal cells 21–24 microns long. Microhairs 12–15 microns wide at the septum. Microhair total length/width at septum 2.4–2.9. Microhair apical cells 10.5–16.5 microns long. Microhair apical cell/total length ratio 0.3–0.48. *Stomata* common; 24–25.5 microns long. Subsidiaries dome-shaped and triangular. Guard-cells overlapping to flush with the interstomatals. *Intercostal short-cells* common; in cork/silica-cell pairs; silicified. Intercostal silica bodies present and perfectly developed; tall-and-narrow and saddle shaped. *Costal short-cells* conspicuously in long rows. Costal silica bodies present in alternate cell files of the costal zones; predominantly saddle shaped.

Transverse section of leaf blade, physiology. C_4; XyMS+. PCR sheaths of the primary vascular bundles interrupted; interrupted abaxially only. PCR sheath extensions absent. Mesophyll traversed by columns of colourless mesophyll cells. *Leaf blade* 'nodular' in section (with prominent, flattish abaxial ribs); with the ribs more or less constant in size (slight). *Midrib* conspicuous (larger bundle and rib, and a large abaxial anchor-shaped girder); with one bundle only. Bulliforms present in discrete, regular adaxial groups; associated with colourless mesophyll cells to form deeply-penetrating fans (these associated with the colourless girders). All the vascular bundles accompanied by sclerenchyma. Combined sclerenchyma girders absent (adaxial strands only, the PCA tissue encircling the top of each bundle; abaxial strands or girders). Sclerenchyma all associated with vascular bundles. The lamina margins with fibres.

Taxonomy. Chloridoideae; main chloridoid assemblage.

Ecology, geography, regional floristic distribution. 1 species. Mexico. Species of open habitats. Dry places.

Holarctic. Madrean.

References, etc. Morphological/taxonomic: Reeder and Reeder 1963. Leaf anatomical: this project.

Cymbopogon Spreng.

Cymbanthelia Anderss., *Gymnanthelia* Schweinf.

Habit, vegetative morphology. Usually perennial (rarely annual); caespitose. **Culms** 15–300 cm high; herbaceous; usually **unbranched above**. Culm nodes glabrous. Culm internodes solid. **The shoots aromatic**. Leaves not basally aggregated; non-auriculate. Leaf blades linear (from broadly so to filiform); broad, or narrow; cordate, or not cordate, not sagittate; setaceous, or not setaceous; flat, or folded; without cross venation; persistent; rolled in bud. Ligule an unfringed membrane to a fringed membrane. *Contra-ligule* absent.

Reproductive organization. Plants bisexual, with bisexual spikelets; with hermaphrodite florets. *The spikelets* of sexually distinct forms on the same plant; hermaphrodite and male-only (usually), or hermaphrodite and sterile; overtly heteromorphic (the pedicelled spikelets not depressed abaxially, awnless); *in both homogamous and heterogamous combinations (the lowermost pair of the lowest raceme, or of each raceme, homogamous and imperfect)*. Plants outbreeding.

Inflorescence. Inflorescence paniculate (decompound, leafy). Rachides hollowed, or flattened, or winged, or neither flattened nor hollowed, not winged. *Inflorescence spatheate*; a complex of 'partial inflorescences' and intervening foliar organs. **Spikelet-bearing axes** 'racemes' (short, spikelike, each pair with a spatheole); *paired (the raceme bases short to more or less connate, flattened, often widely spreading or deflexed)*; with very slender rachides; disarticulating; disarticulating at the joints. *'Articles'* linear; appendaged, or not appendaged; densely long-hairy to somewhat hairy. *Spikelets* paired (or with a terminal triplet); *not secund*; sessile and pedicellate; consistently in 'long-and-short' combinations; in pedicellate/sessile combinations. Pedicels of the 'pedicellate' spikelets free of the rachis, or discernible, but fused with the rachis and free of the rachis (sometimes the pedicel of the homogamous pair being swollen and more or less fused with the internode). The 'shorter' spikelets hermaphrodite. The 'longer' spikelets male-only (usually), or sterile.

Female-sterile spikelets. Pedicellate spikelets never depressed or canaliculate on the back; only the L_1 present, hyaline, 2-nerved, its floret usually male but occasionally sterile or suppressed. The male spikelets 1 floreted. The lemmas awnless.

Female-fertile spikelets. Spikelets 3–7 mm long; compressed laterally, or not noticeably compressed, or compressed dorsiventrally; falling with the glumes. Rachilla terminated by a female-fertile floret. Hairy callus present. *Callus* short; blunt.

Glumes two; more or less equal; long relative to the adjacent lemmas; awnless; very dissimilar (the lower bicarinate, the upper naviculate). Lower glume two-keeled (the keels sometimes winged apically); flattened on the back to sulcate on the back; not pitted; relatively smooth; 1–5 nerved. Upper glume 1–5 nerved. *Spikelets* with incomplete florets. The incomplete florets proximal to the female-fertile florets. *The proximal incomplete florets* 1; epaleate; sterile. The proximal lemmas awnless; 2 nerved; less firm than the female-fertile lemmas to similar in texture to the female-fertile lemmas; not becoming indurated (hyaline).

Female-fertile florets 1. Lemmas hyaline to firm-stipitate beneath the awn; less firm than the glumes; not becoming indurated; apically incised; 2 lobed; awnless, or awned. Awns when present, 1; from a sinus; geniculate; hairless (glabrous); much shorter than the body of the lemma to much longer than the body of the lemma. Lemmas hairless; non-carinate; 1–3 nerved. *Palea* absent. *Lodicules* present; 2; free; fleshy; glabrous. *Stamens* 3. Anthers not penicillate. *Ovary* glabrous. Styles free to their bases. Stigmas 2; red pigmented.

Fruit, embryo and seedling. *Fruit* free from both lemma and palea; small; compressed dorsiventrally (subterete to planoconvex). Hilum short. Embryo large; waisted. Endosperm hard; without lipid; containing compound starch grains. Embryo without an epiblast; with a scutellar tail; with an elongated mesocotyl internode. Embryonic leaf margins overlapping.

Seedling with a long mesocotyl. First seedling leaf with a well-developed lamina. The lamina broad; curved; 21–30 veined.

Abaxial leaf blade epidermis. *Costal/intercostal zonation* conspicuous. *Papillae* present (rarely), or absent. Intercostal papillae when present, consisting of one oblique swelling per cell. *Long-cells* similar in shape costally and intercostally (narrow); of similar wall thickness costally and intercostally. Mid-intercostal long-cells rectangular (long); having markedly sinuous walls, or having straight or only gently undulating walls (rarely). *Microhairs* present; panicoid-type, or chloridoid-type (rarely); (26–)30–48(–60) microns long; 6–7.5 microns wide at the septum. Microhair total length/width at septum 4.8–7. Microhair apical cells (3–)13–22(–24) microns long. Microhair apical cell/total length ratio (0.17–)0.33–0.48–0.48. *Stomata* common; 21–22.5(–24) microns long. Subsidiaries variously low or high dome-shaped, or triangular, or dome-shaped and triangular. Guard-cells overlapping to flush with the interstomatals. *Intercostal short-cells* common, or absent or very rare; in cork/silica-cell pairs (and solitary); not silicified (usually), or silicified. Intercostal silica bodies when present, tall-and-narrow, or cross-shaped, or vertically elongated-nodular. *Costal short-cells* conspicuously in long rows, or neither distinctly grouped into long rows nor predominantly paired. Costal silica bodies 'panicoid-type'; cross shaped, or butterfly shaped, or dumb-bell shaped, or nodular (occasionally).

Transverse section of leaf blade, physiology. C_4; biochemical type NADP–ME (*C. citratus*); XyMS–. PCR sheath outlines uneven. PCR cell chloroplasts centrifugal/peripheral. *Mesophyll* with radiate chlorenchyma. *Leaf blade* adaxially flat. *Midrib* conspicuous; with one bundle only, or having a conventional arc of bundles; with colourless mesophyll adaxially (the colourless tissue often extending across the adaxial part of the blade). Bulliforms not in discrete, regular adaxial groups (in irregular groups); occasionally, irregularly associated with colourless mesophyll cells to form deeply-penetrating fans. Many of the smallest vascular bundles unaccompanied by sclerenchyma. Combined sclerenchyma girders present; forming 'figures'. Sclerenchyma all associated with vascular bundles.

Phytochemistry. Leaves containing flavonoid sulphates (1 species), or without flavonoid sulphates (4 species). Leaf blade chlorophyll *a:b* ratio 3.92–5.26.

Cytology. Chromosome base number, $x = 5$, or 10. $2n = 20, 22, 40$, and 60.

Taxonomy. Panicoideae; Andropogonodae; Andropogoneae; Andropogoninae.

Ecology, geography, regional floristic distribution. About 40 species. Tropical and subtropical Africa and Asia, Australia. Mesophytic to xerophytic; species of open habitats; glycophytic. Savanna.

Holarctic, Paleotropical, Cape, and Australian. Boreal and Tethyan. African, Madagascan, Indomalesian, Polynesian, and Neocaledonian. Eastern Asian. Irano-Turanian. Saharo-Sindian, Sudano-Angolan, West African Rainforest, and Namib-Karoo. Indian, Indo-Chinese, Malesian, and Papuan. Fijian. North and East Australian, South-West Australian, and Central Australian. Sahelo-Sudanian, Somalo-Ethiopian, South Tropical African, and Kalaharian. Tropical North and East Australian and Temperate and South-Eastern Australian.

Rusts and smuts. Rusts — *Puccinia*. Taxonomically wide-ranging species: *Puccinia nakanishikii, Puccinia eritraeensis, 'Uromyces' schoenanthi, Puccinia versicolor*, and *'Uromyces' clignyi*. Smuts from Ustilaginaceae. Ustilaginaceae — *C. refractus*.

Economic importance. Commercial essential oils: *C. nardus* and *C. winterianus* (citronella oil), *C. flexuosus* (East Indian Lemon-grass), *C. citratus* (West Indian Lemongrass), *C.martinii. C. citratus* used as a culinary herb.

References, etc. Morphological/taxonomic: Soenarko 1977. Leaf anatomical: Metcalfe 1960; this project.

Cymbosetaria Schweick.

~ *Setaria*

Habit, vegetative morphology. Annual. *Culms* 15–60 cm high; herbaceous; un-branched above (branching near the base). *Leaves* not basally aggregated; without auricular setae. **Leaf blades** lanceolate; broad; *sagittate*; flat; pseudopetiolate (except the uppermost); rolled in bud, or once-folded in bud (? — supposedly 'not pleated'). Ligule a fringed membrane (usually), or an unfringed membrane (rarely).

Reproductive organization. Plants bisexual, with bisexual spikelets; with hermaphrodite florets.

Inflorescence. *Inflorescence of spicate main branches*; open; non-digitate. Inflorescence with axes ending in spikelets, or axes not ending in spikelets (depending on interpretation of the 'bristle'). Inflorescence espatheate; not comprising 'partial inflorescences' and foliar organs. Spikelet-bearing axes persistent. **Spikelets *subtended by solitary 'bristles' (each subtended by a bristle)*. The 'bristles' persisting on the axis. Spikelets solitary; secund; pedicellate. Pedicel apices discoid.

Female-fertile spikelets. Spikelets abaxial; compressed dorsiventrally; falling with the glumes. Rachilla terminated by a female-fertile floret. Hairy callus absent.

Glumes two; relatively large; very unequal; shorter than the adjacent lemmas; dorsiventral to the rachis; hairless; awnless (G_2 apiculate); non-carinate; very dissimilar. Lower glume 3 nerved. Upper glume 5–7 nerved. *Spikelets* with incomplete florets. The incomplete florets proximal to the female-fertile florets. *The proximal incomplete florets* 1; paleate. Palea of the proximal incomplete florets fully developed. The proximal incomplete florets male. The proximal lemmas awnless; not becoming indurated (textured like the G_2).

Female-fertile florets 1. **Lemmas cymbiform**; decidedly firmer than the glumes; rugose; becoming indurated (crustaceous); entire; pointed, or blunt; awnless (but 'subapiculate'); hairless; carinate; having the margins tucked in onto the palea; with a clear germination flap; 3 nerved. *Palea* present; relatively long; entire; awnless, without apical setae (glabrous); textured like the lemma (crustaceous); indurated; 2-nerved; keel-less (flat backed). *Lodicules* present; free; fleshy; glabrous; not or scarcely vascularized. *Stamens* 3. Anthers not penicillate; without an apically prolonged connective. *Ovary* glabrous. Stigmas 2.

Fruit, embryo and seedling. Fruit small (1.25 mm long); compressed dorsiventrally (ventrally concave). Hilum short. Embryo large.

Abaxial leaf blade epidermis. *Costal/intercostal zonation* conspicuous. *Papillae* absent. *Long-cells* markedly different in shape costally and intercostally (the intercostals much broader). Mid-intercostal long-cells rectangular; having markedly sinuous walls. *Microhairs* present; panicoid-type. *Stomata* common. Subsidiaries low to high dome-shaped and triangular. Guard-cells overlapping to flush with the interstomatals. *Intercostal short-cells* absent or very rare. *Costal short-cells* conspicuously in long rows. Costal silica bodies 'panicoid-type'; dumb-bell shaped and nodular.

Transverse section of leaf blade, physiology. C_4; XyMS–. PCR cell chloroplasts centrifugal/peripheral. *Leaf blade* 'nodular' in section to adaxially flat. *Midrib* conspicuous (with a large abaxial keel); having a conventional arc of bundles (including 3 major bundles); with colourless mesophyll adaxially. Bulliforms not in discrete, regular adaxial groups (seemingly, in the poor material seen). Many of the smallest vascular bundles unaccompanied by sclerenchyma. Combined sclerenchyma girders present (with the major bundles). Sclerenchyma all associated with vascular bundles.

Taxonomy. Panicoideae; Panicodae; Paniceae.

Ecology, geography, regional floristic distribution. 1 species. Africa. Mesophytic; shade species; glycophytic.

Paleotropical. African and Madagascan. Sudano-Angolan and Namib-Karoo. Somalo-Ethiopian, South Tropical African, and Kalaharian.

References, etc. Morphological/taxonomic: Schweickerdt 1936. Leaf anatomical: this project.

Cynodon Rich.

Capriola Adans., *Dactilon* Vill., *Fibichia* Koel.

Habit, vegetative morphology. Perennial; rhizomatous and stoloniferous (often sward-forming). *Culms* 4–60(–100) cm high; herbaceous. Culm nodes glabrous. Culm internodes hollow. *Leaves* not basally aggregated; non-auriculate. Leaf blades linear; narrow; setaceous, or not setaceous; flat, or folded; without abaxial multicellular glands; without cross venation; persistent; rolled in bud, or once-folded in bud. *Ligule* present; a fringed membrane (very short), or a fringe of hairs.

Reproductive organization. Plants bisexual, with bisexual spikelets; with hermaphrodite florets; outbreeding.

Inflorescence. *Inflorescence of spicate main branches*; *digitate, or subdigitate (sometimes in two or more closely spaced whorls)*. Primary inflorescence branches 2–20. Rachides flattened. Inflorescence espatheate; not comprising 'partial inflorescences' and foliar organs. Spikelet-bearing axes persistent. Spikelets solitary; secund; biseriate.

Female-fertile spikelets. *Spikelets* 1.7–3 mm long; adaxial; *compressed laterally*; disarticulating above the glumes, or disarticulating between the glumes. Rachilla prolonged beyond the uppermost female-fertile floret, or terminated by a female-fertile floret (*C. incompletus*); rachilla hairless. The rachilla extension (when present) with incomplete florets (minute), or naked. Hairy callus absent.

Glumes two; very unequal to more or less equal; shorter than the adjacent lemmas; dorsiventral to the rachis to lateral to the rachis; pointed; awnless; carinate; *similar (narrow, lanceolate)*. Lower glume 1 nerved. Upper glume 1 nerved, or 1–3 nerved. *Spikelets* with female-fertile florets only (normally), or with incomplete florets. The incomplete florets (when present) distal to the female-fertile florets. *The distal incomplete florets* (if detectable) 1; **incomplete florets** *merely underdeveloped (usually minute and greatly reduced). Spikelets without proximal incomplete florets*.

Female-fertile florets *1*. *Lemmas* similar in texture to the glumes to decidedly firmer than the glumes (firmly cartilaginous); not becoming indurated; entire; pointed, or blunt; *awnless*; ciliate on the keel and lateral nerves, usually glabrous on the flanks; carinate; 1–4 nerved. *Palea* present; relatively long; entire to apically notched; awnless, without apical setae; textured like the lemma; not indurated; 2-nerved. *Lodicules* present; free; fleshy; glabrous. *Stamens* 3. Anthers 0.6–1.7 mm long; not penicillate. *Ovary* glabrous. Styles free to their bases. Stigmas 2; red pigmented.

Fruit, embryo and seedling. *Fruit* free from both lemma and palea; small; ellipsoid; compressed laterally, or trigonous. Hilum short. Pericarp fused. Embryo large. Endosperm containing compound starch grains. Embryo with an epiblast; with a scutellar tail; with an elongated mesocotyl internode. Embryonic leaf margins meeting.

Seedling with a long mesocotyl; with a loose coleoptile. First seedling leaf with a well-developed lamina. The lamina broad; curved; 7–9 veined.

Abaxial leaf blade epidermis. *Costal/intercostal zonation* conspicuous. *Papillae* present; costal and intercostal. Intercostal papillae over-arching the stomata, or not over-arching the stomata; consisting of one symmetrical projection per cell to several per cell (usually as finger-like projections). *Long-cells* similar in shape costally and intercostally; of similar wall thickness costally and intercostally. Mid-intercostal long-cells rectangular; having markedly sinuous walls. *Microhairs* present; more or less spherical, or elongated; clearly two-celled, or ostensibly one-celled (occasionally); chloridoid-type (sometimes sunken). Microhair apical cell wall of similar thickness/rigidity to that of the basal cell. Microhairs 15–24 microns long. Microhair basal cells 6 microns long. Microhairs 10.5–15 microns wide at the septum. Microhair total length/width at septum 1.4–1.9. Microhair apical cells 9–16 microns long. Microhair apical cell/total length ratio 0.52–0.8. *Stomata* common; 18–21 microns long. Subsidiaries non-papillate; low dome-shaped and triangular. Guard-cells overlapping to flush with the interstomatals. *Intercostal short-cells* common, or absent or very rare; not paired (usually solitary); silicified, or not silicified. Intercostal silica bodies absent, or imperfectly developed; when present, tall-and-narrow. *Costal short-cells* conspicuously in long rows. Costal silica bodies present in alternate cell files of the costal zones; saddle shaped, or tall-and-narrow.

Transverse section of leaf blade, physiology. C_4; biochemical type NAD–ME (2 species); XyMS+. PCR sheath outlines even. PCR sheaths of the primary vascular bundles interrupted; interrupted abaxially only. PCR sheath extensions absent. PCR cells without a suberised lamella. *PCR cell chloroplasts* elongated; with well developed grana; centripetal. *Mesophyll* with radiate chlorenchyma; traversed by columns of colourless mesophyll cells. **Leaf blade** *adaxially flat*. *Midrib* conspicuous; with one bundle only, or having a conventional arc of bundles. Bulliforms present in discrete, regular adaxial groups; associated with colourless mesophyll cells to form deeply-penetrating fans (these linked with the traversing colourless columns). All the vascular bundles accompanied by sclerenchyma. Combined sclerenchyma girders present; forming 'figures', or nowhere forming 'figures'. Sclerenchyma all associated with vascular bundles. The lamina margins with fibres.

Culm anatomy. *Culm internode bundles* in one or two rings.

Phytochemistry. Tissues of the culm bases with abundant starch. Leaves containing flavonoid sulphates (in some material of two species), or without flavonoid sulphates (3 species).

Cytology. Chromosome base number, $x = 9$ and 10. $2n = 16$, 18, 27, 36, 40, and 54. 2, 3, 4, and 6 ploid. Nucleoli persistent.

Taxonomy. Chloridoideae; main chloridoid assemblage.

Ecology, geography, regional floristic distribution. 10 species. Tropical and subtropical. Commonly adventive. Mesophytic, or xerophytic; species of open habitats; halophytic and glycophytic. Disturbed and arable land, weedy and sandy places, seashores.

Holarctic, Paleotropical, Neotropical, Cape, Australian, and Antarctic. Boreal, Tethyan, and Madrean. African, Madagascan, Indomalesian, and Neocaledonian. Euro-Siberian and Eastern Asian. Mediterranean and Irano-Turanian. Saharo-Sindian, Sudano-Angolan, West African Rainforest, and Namib-Karoo. Indian, Indo-Chinese, Malesian, and Papuan. Caribbean, Venezuela and Surinam, Amazon, Central Brazilian, Pampas, and Andean. North and East Australian. Patagonian. European and Siberian. Sahelo-Sudanian, Somalo-Ethiopian, South Tropical African, and Kalaharian. Tropical North and East Australian and Temperate and South-Eastern Australian.

Hybrids. Intergeneric hybrids with *Chloris* (×*Cynochloris* Clifford & Everist: several species involved).

Rusts and smuts. Rusts — *Puccinia*. Taxonomically wide-ranging species: *Puccinia graminis*. Smuts from Ustilaginaceae. Ustilaginaceae — *Sorosporium* and *Ustilago*.

Economic importance. Significant weed species: *C. dactylon* (sometimes, in arable land), *C. aethiopicus*, *C. incompletus*, *C. plectostachyus*. Important native pasture species: *C. dactylon*, *C. nlemfuensis*, *C. aethiopicus*, *C. plectostachyus*. Lawns and/or playing fields: *C. dactylon* (Bermuda, Couch), *C. transvaalensis* etc.

References, etc. Leaf anatomical: this project.

Cynosurus L.

Falonia Adans.

Habit, vegetative morphology. Annual, or perennial; caespitose. *Culms* 10–90 cm high; herbaceous; unbranched above; 3–10 noded. Culm nodes exposed; glabrous. Culm internodes hollow. *Leaves* not basally aggregated; non-auriculate. *Sheath margins free*. Sheaths not keeled, terete. Leaf blades linear; broad, or narrow; 0.5–9(–14) mm wide; flat; without cross venation; persistent; once-folded in bud. Ligule an unfringed membrane; truncate, or not truncate; 0.5–8.5 mm long.

Reproductive organization. Plants bisexual, with bisexual spikelets; with hermaphrodite florets. *The spikelets of sexually distinct forms on the same plant*; hermaphrodite and sterile; *overtly heteromorphic (fertile spikelets mixed with and more or less concealed by sterile ones consisting of rigid, lanceolate, awned glumes and lemmas)*. Plants outbreeding.

Inflorescence. *Inflorescence paniculate*; contracted; spicate, or more or less irregular; espatheate; not comprising 'partial inflorescences' and foliar organs. Spikelet-bearing axes persistent. *Spikelets secund*; shortly pedicellate, or subsessile. **Female-sterile spikelets**. The sterile spikelets consisting of rigid, lanceolate, awned glumes and lemmas occurring among the fertile ones. **Female-fertile spikelets**. Spikelets 2.8–10 mm long; compressed laterally; disarticulating above the glumes. *Rachilla prolonged beyond the uppermost female-fertile floret*; rachilla hairless. Hairy callus absent. *Callus* short; blunt.

Glumes two; more or less equal; shorter than the adjacent lemmas, or long relative to the adjacent lemmas; awnless; carinate; similar (narrow, thin). Lower glume 1 nerved. Upper glume 1 nerved. *Spikelets* with female-fertile florets only, or with incomplete florets. The incomplete florets distal to the female-fertile florets.

Female-fertile florets (1–)2–5. *Lemmas* similar in texture to the glumes to decidedly firmer than the glumes (papery or leathery); not becoming indurated; entire, or incised; *awned*. Awns 1; median; from a sinus, or apical; from near the top; non-geniculate; much shorter than the body of the lemma to much longer than the body of the lemma; entered by one vein. Lemmas hairy, or hairless; non-carinate; 5 nerved. *Palea* present; relatively long; tightly clasped by the lemma; apically notched (shortly bifid); thinner than the lemma (membranous); 2-nerved; 2-keeled. Palea keels wingless. *Lodicules* present; 2; free; membranous; glabrous; toothed. *Stamens* 3. Anthers 0.7–3.5 mm long; not penicillate. *Ovary* glabrous. Styles free to their bases. Stigmas 2; white.

Fruit, embryo and seedling. *Fruit* adhering to lemma and/or palea (to the palea); small to medium sized (2.2–4.3 mm long); ovoid or ellipsoid; longitudinally grooved, or not grooved; compressed dorsiventrally to not noticeably compressed. Hilum short, or long-linear (elliptic or linear). Embryo small; not waisted. Endosperm hard; with lipid, or without lipid; containing compound starch grains. Embryo with an epiblast; without a scutellar tail; with a negligible mesocotyl internode. Embryonic leaf margins meeting.

Seedling with a long mesocotyl; with a tight coleoptile. First seedling leaf with a well-developed lamina. The lamina narrow; 1 veined.

Abaxial leaf blade epidermis. *Costal/intercostal zonation* conspicuous. *Papillae* absent. *Long-cells* similar in shape costally and intercostally (the intercostals very large); differing markedly in wall thickness costally and intercostally (the costals thicker-walled, pitted). Mid-intercostal long-cells fusiform; having straight or only gently undulating walls. *Microhairs* absent. *Stomata* absent or very rare (*C. cristatus*), or common (*C. echinatus*); in *C. echinatus* (42–)45–50(–51) microns long. Subsidiaries parallel-sided. Guard-cells overlapped by the interstomatals. *Intercostal short-cells* absent or very rare; not paired (solitary); not silicified. *Costal short-cells* predominantly paired. Costal silica bodies horizontally-elongated crenate/sinuous, or horizontally-elongated smooth, or rounded.

Transverse section of leaf blade, physiology. C_3; XyMS+. *Mesophyll* with non-radiate chlorenchyma. *Leaf blade* with distinct, prominent adaxial ribs, or 'nodular' in section, or adaxially flat; with the ribs more or less constant in size, or with the ribs very irregular in sizes. *Midrib* conspicuous, or not readily distinguishable; with one bundle only. Bulliforms present in discrete, regular adaxial groups (in the furrows); in simple fans. All the vascular bundles accompanied by sclerenchyma. Combined sclerenchyma girders present; nowhere forming 'figures'. Sclerenchyma all associated with vascular bundles.

Culm anatomy. *Culm internode bundles* in one or two rings.

Cytology. Chromosome base number, $x = 7$. $2n = 14$. 2 ploid.

Taxonomy. Pooideae; Poodae; Poeae.

Ecology, geography, regional floristic distribution. 8 species. Europe, western Asia, North and South Africa. Commonly adventive. Mesophytic, or xerophytic; species of open habitats. Meadows, disturbed ground.

Holarctic, Paleotropical, and Neotropical. Boreal and Tethyan. African. Euro-Siberian and Atlantic North American. Macaronesian, Mediterranean, and Irano-Turanian. Saharo-Sindian. Caribbean. European. Canadian-Appalachian.

Rusts and smuts. Rusts — *Puccinia*. Taxonomically wide-ranging species: *Puccinia graminis, Puccinia coronata,* and *'Uromyces' dactylidis*. Smuts from Tilletiaceae and from Ustilaginaceae. Tilletiaceae — *Entyloma* and *Tilletia*. Ustilaginaceae — *Ustilago*.

Economic importance. Significant weed species: *C. cristatus, C. echinatus.* Cultivated fodder: *C. cristatus.* Lawns and/or playing fields: *C. cristatus.*

References, etc. Leaf anatomical: Metcalfe 1960; this project.

Cyperochloa Lazarides & L. Watson

Habit, vegetative morphology. Wiry *perennial (remarkably sedge-like in appearance, the plants forming colonies up to 2 m across)*; caespitose. *Culms* 7–50 cm high; herbaceous; unbranched above. Culm nodes erect, terete, glabrous or the topmost node pubescent. Culm internodes solid. Young shoots intravaginal. *Leaves* mostly basal; non-auriculate (but the ligule continuous with hairs at the lateral tips of the sheath). Leaf blades narrow; recurved, setaceous; tightly rolled; not pseudopetiolate; without cross venation; persistent. Ligule a fringe of hairs (short, dense). *Contra-ligule* absent.

Reproductive organization. Plants bisexual, with bisexual spikelets; with hermaphrodite florets.

Inflorescence. Inflorescence few spikeleted (generally 2–5); paniculate (borne on a very short peduncle, subtended by a short-sheathed terminal leaf with a reduced blade, atop the single elongated culm internode); contracted; capitate (consisting of 2–5 digitately-borne, bracteate spikelets); spatheate (the inflorescence subtended by the leaflike spathe, and each spikelet by a small, glume-like bract); not comprising 'partial inflorescences' and foliar organs. Spikelet-bearing axes persistent. Spikelets sessile to subsessile.

Female-fertile spikelets. Spikelets 7–9 mm long; broadly ovate; compressed laterally; disarticulating above the glumes; disarticulating between the florets; with conventional internode spacings. Rachilla prolonged beyond the uppermost female-fertile floret; rachilla hairless (each joint glabrous below the callus). The rachilla extension with incomplete florets. Hairy callus present. *Callus* short; blunt.

Glumes present; two; very unequal (the G$_2$ generally longer); shorter than the spikelets; shorter than the adjacent lemmas; hairy (hirsute or pubescent below and sometimes on the nerves, the margins ciliate); without conspicuous tufts or rows of hairs; pointed (acute), or not pointed (obtuse or emarginate); awnless (muticous); carinate to non-carinate (slightly keeled to rounded on the back); similar (ovate to lanceolate, membranous to cartilaginous). Lower glume 3 nerved, or 5 nerved. Upper glume 3 nerved, or 5 nerved. *Spikelets* with incomplete florets. The incomplete florets distal to the female-fertile florets. *The distal incomplete florets* merely underdeveloped (i.e., the upper 1–2).

Female-fertile florets 4–9. Lemmas similar in texture to the glumes; not becoming indurated; entire, or incised; blunt; not deeply cleft (at the most, minutely cleft); awnless (at the most, with a minute projection from the median nerve); hairy (conspicuously so below and on the margins, and also adaxially near the tip); carinate; without a germination flap; 5 nerved, or 7 nerved. *Palea* present; relatively long; entire to apically notched (no more than minutely so); awnless, without apical setae; thinner than the lemma (membranous with hyaline margins, hairy); not indurated; 2-nerved; 2-keeled. Palea keels wingless. *Lodicules* present; 2; free; fleshy; ciliate; not or scarcely vascularized. *Stamens* 3. Anthers 2.2–3 mm long (i.e., relatively long); not penicillate; without an apically prolonged connective. *Ovary* glabrous. Styles free to their bases. Stigmas 2; probably red pigmented.

Fruit, embryo and seedling. Fruit small (about 1.5 mm long); slightly indented on one side; not noticeably compressed; papillose, but not 'sculptured'. Hilum short. Embryo not visible through the opaque pericarp. Endosperm hard.

Abaxial leaf blade epidermis. *Costal/intercostal zonation* conspicuous. *Papillae* absent. *Long-cells* similar in shape costally and intercostally; of similar wall thickness costally and intercostally. Mid-intercostal long-cells rectangular; having markedly sinuous walls. *Microhairs* present; panicoid-type (but the apical cell quite broad); 33–36–39 microns long; (10.5–)12 microns wide at the septum. Microhair total length/ width at septum 2.75–3.43. Microhair apical cells 18–21–24 microns long. Microhair apical cell/total length ratio 0.55–0.62. *Stomata* common; 36(–39) microns long. Subsidiaries dome-shaped and triangular. Guard-cells overlapping to flush with the interstomatals. *Intercostal short-cells* common; in cork/silica-cell pairs. *Costal short-cells* predominantly paired. Costal silica bodies crescentic, or oryzoid, or 'panicoid-type'; mostly cross shaped (but rather irregular).

Transverse section of leaf blade, physiology. C_3; XyMS+. *Mesophyll* with non-radiate chlorenchyma; without adaxial palisade. *Leaf blade* with distinct, prominent adaxial ribs; with the ribs more or less constant in size (round topped). *Midrib* not readily distinguishable; with one bundle only. Bulliforms present in discrete, regular adaxial groups; in simple fans (the groups large). All the vascular bundles accompanied by sclerenchyma. Combined sclerenchyma girders present (with the primaries — some of the other bundles combining abaxial girders with adaxial strands); forming 'figures' (I's, in the primaries). Sclerenchyma all associated with vascular bundles.

Special diagnostic feature. *The inflorescence of a few digitately-borne, bracteate spikelets, subtended by a spatheate leaf atop a single elongated culm internode, the plant very sedge-like in appearance.*

Taxonomy. Arundinoideae; Cyperochloeae.

Ecology, geography, regional floristic distribution. 1 species. Australia. Xerophytic; species of open habitats. Dry sandy places. Australian. South-West Australian.

References, etc. Morphological/taxonomic: Lazarides and Watson 1986. Leaf anatomical: this project.

Cyphochlaena Hackel

Sclerochlaena A. Camus
Excluding *Boivinella*

Habit, vegetative morphology. Perennial. *Culms* 20–30 cm high; herbaceous; branched above. Culm nodes hidden by leaf sheaths. Culm leaves present. Upper culm leaf blades fully developed. *Leaves* not basally aggregated; non-auriculate. **Leaf blades** lanceolate to ovate-lanceolate; rather *flimsy*; relatively broad; 5–10 mm wide; flat; without cross venation; persistent. Ligule a fringed membrane. *Contra-ligule* absent.

Reproductive organization. Plants bisexual, with bisexual spikelets; with hermaphrodite florets (in the material seen). The spikelets of sexually distinct forms on the same plant; hermaphrodite and male-only, or hermaphrodite and sterile (this in the material seen). The male and female-fertile spikelets mixed in the inflorescence. The spikelets more or less overtly heteromorphic.

Inflorescence. Inflorescence of spicate main branches (a raceme of short, spicate branches). Inflorescence with axes ending in spikelets. Rachides narrowly winged. Inflorescence espatheate; not comprising 'partial inflorescences' and foliar organs. *Spikelet-bearing axes* short 'racemes'. The racemes spikelet bearing to the base. Spikelet-bearing axes solitary; with substantial rachides; persistent. *Spikelets* paired; secund; **biseriate (the pairs in two ranks)**; pedicellate, or subsessile and pedicellate (the pedicels winged, with long, stiff, cushion based hairs). Pedicel apices discoid to cupuliform. Spikelets imbricate; consistently in 'long-and-short' combinations; unequally pedicellate in each combination. Pedicels of the 'pedicellate' spikelets free of the rachis. The 'shorter' spikelets male-only, or sterile. The 'longer' spikelets hermaphrodite, or female-only.

Female-sterile spikelets. The short-pedicelled spikelets sterile or male, with a long-awned lower glume, membranous upper glume and membranous lower lemma. Rachilla of male spikelets terminated by a male floret. The male spikelets with glumes (these awned); without proximal incomplete florets; 2 floreted. The lemmas awned (the lower). Male florets 2; 3 staminate.

Female-fertile spikelets. Spikelets 1.5 mm long; abaxial; strongly compressed laterally; falling with the glumes; not disarticulating between the florets; with conventional internode spacings. Rachilla terminated by a female-fertile floret. *Callus* absent.

Glumes two; more or less equal (discounting awns); about equalling the spikelets to exceeding the spikelets; long relative to the adjacent lemmas; dorsiventral to the rachis;

hairless; glabrous; (the lower long) awned (the awn more or less deciduous); carinate; very dissimilar (the lower narrower, membranous, with a long, slender antrorsely scabrous awn from a slight sinus, the upper larger, saccate, hooded at the tip, awnless, leathery). Lower glume thinly 3 nerved. **Upper glume *distinctly saccate***; thinly 5 nerved. *Spikelets* with incomplete florets. The incomplete florets proximal to the female-fertile florets. *The proximal incomplete florets* 1; paleate (the palea saccate above). Palea of the proximal incomplete florets fully developed. The proximal incomplete florets male. The proximal lemmas saccate, hooded at the tip; awnless; 3 nerved; decidedly exceeding the female-fertile lemmas; decidedly firmer than the female-fertile lemmas (becoming cartilaginous).

Female-fertile florets 1. Lemmas less firm than the glumes (hyaline); not becoming indurated; entire; awnless; hairless; glabrous; non-carinate; having the margins lying flat and exposed on the palea; without a germination flap; 3 nerved. *Palea* present; relatively long; tightly clasped by the lemma; entire; awnless, without apical setae; textured like the lemma; not indurated; nerveless; keel-less. Palea back glabrous. *Lodicules* present; 2; free; fleshy; glabrous; not or scarcely vascularized. *Stamens* 3. *Ovary* glabrous. Styles free to their bases; free. Style bases adjacent. Stigmas 2.

Fruit, embryo and seedling. Fruit compressed laterally.

Abaxial leaf blade epidermis. *Costal/intercostal zonation* conspicuous. *Papillae* absent. *Long-cells* similar in shape costally and intercostally (the costals much narrower and more regularly rectangular). Intercostal zones with typical long-cells to exhibiting many atypical long-cells (these generally rather short, some very short). Mid-intercostal long-cells more or less isodiametric or irregular to rectangular; having markedly sinuous walls (coarsely sinuous). *Microhairs* present; elongated; clearly two-celled; panicoid-type. *Stomata* common (confined to rows alongside the costae). Subsidiaries non-papillate; dome-shaped and triangular. Guard-cells overlapping to flush with the interstomatals. *Intercostal short-cells* absent or very rare. With long, slender, cushion-based macrohairs intercostally. *Costal short-cells* conspicuously in long rows. Costal silica bodies present and well developed; 'panicoid-type'; cross shaped, dumb-bell shaped, and nodular.

Transverse section of leaf blade, physiology. C_3 (delta value: W.V. Brown 1977); XyMS+. Mesophyll without 'circular cells'; not traversed by colourless columns; without arm cells; without fusoids.

Taxonomy. Panicoideae; Panicodae; Paniceae (Boivinelleae).

Ecology, geography, regional floristic distribution. 2 species. Madagascar. In damp places.

Paleotropical. Madagascan.

References, etc. Morphological/taxonomic: Bosser 1965. Leaf anatomical: this project.

Special comments. Fruit data wanting.

Cypholepis Chiov.

~ *Coelachyrum*

Habit, vegetative morphology. *Perennial*; densely caespitose. **Culms** 30–100 cm high; ***herbaceous***. Culm nodes glabrous. Plants unarmed. *Leaves* not basally aggregated. Leaf blades linear; narrow; usually flat; without abaxial multicellular glands; without cross venation. Ligule a fringed membrane.

Reproductive organization. Plants bisexual, with bisexual spikelets; with hermaphrodite florets.

Inflorescence. *Inflorescence of spicate main branches*; non-digitate. Primary inflorescence branches 2–8 (these distant, erect). *Spikelet-bearing axes* spicate 'racemes'; persistent. Spikelets solitary; secund; biseriate; shortly pedicellate.

Female-fertile spikelets. *Spikelets* 5–10 mm long; compressed laterally; disarticulating above the glumes; ***disarticulating between the florets***. Rachilla prolonged beyond the uppermost female-fertile floret. The rachilla extension with incomplete florets.

Glumes two; more or less equal; *long relative to the adjacent lemmas*; hairless; not pointed (obtuse); awnless; carinate to non-carinate (rounded to slightly keeled on the back); similar (lanceolate, membranous). Lower glume 1 nerved. *Upper glume* usually *1 nerved*. *Spikelets* with incomplete florets. The incomplete florets distal to the female-fertile florets. *The distal incomplete florets* male.

Female-fertile florets *7–10*. *Lemmas* decidedly firmer than the glumes (membranous, becoming cartilaginous below); not becoming indurated; entire; pointed, or blunt (acute or obtuse); awnless; *hairy (pilose with club-shaped hairs on the lower back)*; 3 nerved. **Palea** present; *conspicuous but relatively short (about half the lemma length)*; entire (obtuse); awnless, without apical setae; not indurated (hyaline). *Lodicules* present (minute); 2; free; fleshy; glabrous. *Stamens* 3. *Ovary* glabrous (?). Styles free to their bases. Stigmas 2.

Fruit, embryo and seedling. Fruit small (1.2–1.4 mm long); compressed dorsiventrally (concavo-convex); smooth. Hilum short. Pericarp free. Embryo large.

Abaxial leaf blade epidermis. *Costal/intercostal zonation* conspicuous. *Papillae* present; intercostal (conspicuous). Intercostal papillae over-arching the stomata (at one end); consisting of one oblique swelling per cell. *Long-cells* markedly different in shape costally and intercostally (the costals narrower); of similar wall thickness costally and intercostally (medium thick walled). Mid-intercostal long-cells rectangular; having markedly sinuous walls to having straight or only gently undulating walls. *Microhairs* present; more or less spherical to elongated; clearly two-celled; chloridoid-type. Microhair apical cell wall of similar thickness/rigidity to that of the basal cell. Microhair basal cells 21 microns long. Microhair total length/width at septum 3. Microhair apical cell/total length ratio 0.4. *Stomata* common. Subsidiaries non-papillate; low to high dome-shaped. Guard-cells overlapped by the interstomatals (mostly, slightly). *Intercostal short-cells* absent or very rare. Intercostal silica bodies absent. Prickles abundant costally. *Crown cells* absent. *Costal short-cells* conspicuously in long rows. Costal silica bodies present in alternate cell files of the costal zones; large saddle shaped.

Transverse section of leaf blade, physiology. C_4; XyMS+. PCR sheath outlines uneven. PCR sheaths of the primary vascular bundles interrupted; interrupted both abaxially and adaxially. PCR sheath extensions absent. PCR cell chloroplasts centripetal. *Leaf blade* 'nodular' in section to adaxially flat; with the ribs very irregular in sizes. *Midrib* conspicuous; having a conventional arc of bundles (a large bundle, flanked on each side by two smaller ones); with colourless mesophyll adaxially. Bulliforms present in discrete, regular adaxial groups; in simple fans. All the vascular bundles accompanied by sclerenchyma. Combined sclerenchyma girders present (with the larger bundles); forming 'figures'. Sclerenchyma all associated with vascular bundles. The lamina margins with fibres.

Taxonomy. Chloridoideae; main chloridoid assemblage.

Ecology, geography, regional floristic distribution. 1 species. Northeast to southeast and southern Africa. Xerophytic; species of open habitats.

Paleotropical and Cape. African. Sudano-Angolan and Namib-Karoo. Somalo-Ethiopian, South Tropical African, and Kalaharian.

Rusts and smuts. Rusts — *Puccinia*. Taxonomically wide-ranging species: '*Uromyces*' *eragrostidis*.

References, etc. Leaf anatomical: this project.

Cyrtococcum Stapf

Habit, vegetative morphology. Perennial; stoloniferous, or decumbent. *Culms* 15–100 cm high; herbaceous. Culm nodes glabrous. Culm internodes hollow. *Leaves* not basally aggregated; non-auriculate. Leaf blades linear-lanceolate to ovate; broad, or narrow; cordate, or not cordate, not sagittate; pseudopetiolate, or not pseudopetiolate; cross veined, or without cross venation; persistent; rolled in bud. Ligule an unfringed membrane to a fringed membrane.

Reproductive organization. Plants bisexual, with bisexual spikelets; with hermaphrodite florets.

Inflorescence. Inflorescence paniculate; open, or contracted; espatheate; not comprising 'partial inflorescences' and foliar organs. Spikelet-bearing axes persistent. Spikelets not secund; pedicellate. Pedicel apices cupuliform.

Female-fertile spikelets. Spikelets 1–2.5 mm long; oblong, or elliptic, or obovate; gibbous, compressed laterally; falling with the glumes. Rachilla terminated by a female-fertile floret. Hairy callus absent.

Glumes two; more or less equal; shorter than the spikelets; shorter than the adjacent lemmas; awnless; carinate; similar (membranous, the upper blunt). Lower glume 1–3 nerved. *Upper glume 3 nerved. Spikelets* with incomplete florets. The incomplete florets proximal to the female-fertile florets. *The proximal incomplete florets* 1; paleate, or epaleate. Palea of the proximal incomplete florets when present, reduced. The proximal incomplete florets sterile. The proximal lemmas blunt; awnless; 3–5 nerved; more or less equalling the female-fertile lemmas; less firm than the female-fertile lemmas to decidedly firmer than the female-fertile lemmas; not becoming indurated.

Female-fertile florets 1. **Lemmas naviculate, gibbous on the back**; saccate; decidedly firmer than the glumes (papery to crustaceous); smooth, or striate; becoming indurated to not becoming indurated; white in fruit; entire; pointed; *crested at the tip*; awnless; hairless (smooth, shiny); carinate; having the margins tucked in onto the palea; with a clear germination flap; 5 nerved. *Palea* present; relatively long; entire; awnless, without apical setae; textured like the lemma; indurated, or not indurated; 2-nerved. *Lodicules* present; 2; free; fleshy; glabrous. *Stamens* 3. Anthers not penicillate. *Ovary* glabrous. Styles free to their bases. Stigmas 2.

Fruit, embryo and seedling. Fruit small; compressed laterally. Hilum short. Embryo large; not waisted.

Abaxial leaf blade epidermis. *Costal/intercostal zonation* conspicuous. *Papillae* present, or absent. Intercostal papillae not over-arching the stomata; consisting of one symmetrical projection per cell. *Long-cells* markedly different in shape costally and intercostally (the costals narrowly rectangular); of similar wall thickness costally and intercostally (very thin walled). Intercostal zones without typical long-cells (these being irregularly equidimensional). Mid-intercostal long-cells having markedly sinuous walls. *Microhairs* present; panicoid-type; 39–50 microns long; about 4.5 microns wide at the septum. Microhair total length/width at septum about 8.7. Microhair apical cells (17–)18–22(–27) microns long. Microhair apical cell/total length ratio 0.43–0.46. *Stomata* common. Subsidiaries low dome-shaped, or triangular; including both triangular and parallel-sided forms on the same leaf, or not including both parallel-sided and triangular forms on the same leaf. Guard-cells overlapping to flush with the interstomatals. *Intercostal short-cells* absent or very rare. *Costal short-cells* conspicuously in long rows. Costal silica bodies sometimes sharp-pointed, 'panicoid-type', or saddle shaped; cross shaped, or butterfly shaped, or dumb-bell shaped; sharp-pointed to not sharp-pointed.

Transverse section of leaf blade, physiology. C_3; XyMS+. *Mesophyll* with radiate chlorenchyma; without adaxial palisade; *Isachne*-type. *Leaf blade* with distinct, prominent adaxial ribs, or 'nodular' in section; with the ribs more or less constant in size. *Midrib* conspicuous, or not readily distinguishable (rarely); with one bundle only, or having a conventional arc of bundles. Bulliforms present in discrete, regular adaxial groups; in simple fans. All the vascular bundles accompanied by sclerenchyma. Combined sclerenchyma girders present; forming 'figures'. Sclerenchyma all associated with vascular bundles.

Culm anatomy. *Culm internode bundles* in one or two rings.

Taxonomy. Panicoideae; Panicodae; Isachneae.

Ecology, geography, regional floristic distribution. 12 species. Palaeotropical. Commonly adventive. Mesophytic; shade species. Forests.

Paleotropical and Australian. African, Madagascan, Indomalesian, Polynesian, and Neocaledonian. Sudano-Angolan and West African Rainforest. Indian, Indo-Chinese, Malesian, and Papuan. Fijian. North and East Australian. Sahelo-Sudanian, Somalo-Ethiopian, and South Tropical African. Tropical North and East Australian.

Rusts and smuts. Rusts — *Physopella* and *Puccinia*. Taxonomically wide-ranging species: *Physopella clemensiae, Puccinia orientalis*, and '*Uromyces*' *setariae-italicae*.

Economic importance. Significant weed species: *C. accrescens*, *C. oxyphyllum*, *C. patens*. Important native pasture species: e.g. *C. patens*.

References, etc. Leaf anatomical: Metcalfe 1960 and this project.

Dactylis L.

Amaxitis Adans., *Trachypoa* Bub.

Habit, vegetative morphology. Perennial; densely caespitose (with short, oblique rhizomes and/or stolons). *Culms* 15–200 cm high; herbaceous; unbranched above; tuberous, or not tuberous; 3–7 noded. Culm nodes exposed, or hidden by leaf sheaths; glabrous. Culm internodes hollow. *Leaves* not basally aggregated; non-auriculate. *Sheath margins joined (to halfway, at least in the upper leaves)*. Sheaths keeled, terete. Leaf blades linear to linear-lanceolate; broad, or narrow; 2–14 mm wide; flat, or folded, or rolled (involute); without cross venation; persistent; once-folded in bud. Ligule an unfringed membrane; not truncate; 2–17 mm long.

Reproductive organization. Plants bisexual, with bisexual spikelets; with hermaphrodite florets. The spikelets all alike in sexuality (usually), or of sexually distinct forms on the same plant (sometimes with rudimentary spikelets at the base of the inflorescence or at the bases of spikelet clusters). Plants outbreeding.

Inflorescence. *Inflorescence paniculate*; open, or contracted; espatheate; not comprising 'partial inflorescences' and foliar organs. Spikelet-bearing axes persistent. *Spikelets secund (in dense, one-sided clusters terminating the panicle branches)*; subsessile, or pedicellate (shortly so).

Female-fertile spikelets. Spikelets 4–8(–9.4) mm long; elliptic, or ovate; compressed laterally; disarticulating above the glumes. Rachilla prolonged beyond the uppermost female-fertile floret; rachilla hairless. The rachilla extension with incomplete florets. Hairy callus absent. *Callus* short; blunt.

Glumes two; relatively large; very unequal to more or less equal; shorter than the spikelets to about equalling the spikelets; shorter than the adjacent lemmas, or long relative to the adjacent lemmas (from about half to more than 3/4 as long); pointed; awned to awnless; *carinate*; similar (membranous, somewhat curved, often asymmetric about the midvein). Lower glume 1–3 nerved. Upper glume 3 nerved. *Spikelets* with female-fertile florets only, or with incomplete florets. The incomplete florets when present, distal to the female-fertile florets. The distal incomplete florets 1; *incomplete florets merely underdeveloped (rudimentary)*.

Female-fertile florets 2–5. Lemmas decidedly firmer than the glumes (papery); not becoming indurated; entire; pointed; awned. Awns 1; median; dorsal, or apical; from near the top; non-geniculate; much shorter than the body of the lemma; entered by one vein, or entered by several veins. Lemmas hairy, or hairless; scabrous (or scabrous and pilose); carinate; 5 nerved. *Palea* present; relatively long; tightly clasped by the lemma; apically notched; thinner than the lemma (membranous); 2-nerved; 2-keeled. Palea keels wingless. *Lodicules* present; 2; free; membranous; glabrous; toothed. *Stamens* 3. Anthers 1.5–4.5 mm long; not penicillate. *Ovary* glabrous. Styles free to their bases. Stigmas 2; white.

Fruit, embryo and seedling. *Fruit* adhering to lemma and/or palea (to the palea), or free from both lemma and palea; small; oblong to ellipsoid; not grooved; compressed dorsiventrally to not noticeably compressed. *Hilum short*. Embryo small; not waisted. Endosperm liquid in the mature fruit, or hard; with lipid; containing compound starch grains. Embryo with an epiblast; without a scutellar tail; with a negligible mesocotyl internode. Embryonic leaf margins meeting.

Seedling with a long mesocotyl; with a tight coleoptile. First seedling leaf with a well-developed lamina. The lamina narrow; erect; 3–5 veined.

Abaxial leaf blade epidermis. *Costal/intercostal zonation* conspicuous. *Papillae* absent. *Long-cells* similar in shape costally and intercostally; of similar wall thickness costally and intercostally. Mid-intercostal long-cells fusiform (slightly); having markedly sinuous walls, or having straight or only gently undulating walls (rarely). *Microhairs* absent. *Stomata* common. Subsidiaries parallel-sided. Guard-cells overlapped by the interstomatals. *Intercostal short-cells* absent or very rare; not paired; not silicified. *Costal* short-cells neither distinctly grouped into long rows nor predominantly paired. Costal silica bodies horizontally-elongated crenate/sinuous, or horizontally-elongated smooth (a few).

Transverse section of leaf blade, physiology. C_3; XyMS+. *Mesophyll* with non-radiate chlorenchyma. *Leaf blade* 'nodular' in section; with the ribs more or less constant in size. *Midrib* conspicuous (with a median bulliform group in its adaxial groove); with one bundle only. Bulliforms not in discrete, regular adaxial groups (apart from a large group above the median vascular bundle). All the vascular bundles accompanied by sclerenchyma. Combined sclerenchyma girders present; nowhere forming 'figures'. Sclerenchyma all associated with vascular bundles.

Phytochemistry. Tissues of the culm bases with little or no starch. Fructosans predominantly long-chain. Leaves without flavonoid sulphates (1 species).

Cytology. Chromosome base number, $x = 7$. $2n = 14$ and 28 (+ 0–4 B). 2 and 4 ploid (+0–4B). Haploid nuclear DNA content 4.9 pg (1 species, $2x$). Mean diploid 2c DNA value 9.8 pg (1 species).

Taxonomy. Pooideae; Poodae; Poeae.

Ecology, geography, regional floristic distribution. 1 species (or up to 5, by recognition of minor segregates). Temperate Eurasia. Commonly adventive. Mesophytic; shade species and species of open habitats. Meadows, woodlands and disturbed ground, in moist to dry places.

Holarctic, Paleotropical, and Neotropical. Boreal and Tethyan. African. Euro-Siberian, Atlantic North American, and Rocky Mountains. Macaronesian, Mediterranean, and Irano-Turanian. Saharo-Sindian. Pampas. European and Siberian. Canadian-Appalachian.

Rusts and smuts. Rusts — *Puccinia*. Taxonomically wide-ranging species: *Puccinia graminis, Puccinia coronata, Puccinia striiformis, Puccinia recondita*, and *'Uromyces' dactylidis*. Smuts from Tilletiaceae and from Ustilaginaceae. Tilletiaceae — *Entyloma, Tilletia*, and *Urocystis*. Ustilaginaceae — *Ustilago*.

Economic importance. Significant weed species: *D. glomerata*. Cultivated fodder: *D. glomerata* (Orchard Grass).

References, etc. Leaf anatomical: this project.

Dactyloctenium Willd.

Habit, vegetative morphology. Annual, or perennial; rhizomatous, or stoloniferous, or caespitose, or decumbent (mostly low, sometimes sward-forming). *Culms* 5–100(–160) cm high; herbaceous. Culm nodes glabrous. Culm internodes solid, or hollow. *Leaves* not basally aggregated; non-auriculate. Leaf blades linear to linear-lanceolate (sometimes pungent); narrow; flat, or rolled; without abaxial multicellular glands; without cross venation; persistent; once-folded in bud. Ligule a fringed membrane (narrow), or a fringe of hairs.

Reproductive organization. Plants bisexual, with bisexual spikelets; with hermaphrodite florets. The spikelets all alike in sexuality (or terminal spikelets sterile). Exposed-cleistogamous, or chasmogamous.

Inflorescence. *Inflorescence of spicate main branches (the mature spikelets sometimes almost at right-angles to the rachides); digitate.* Primary inflorescence branches 2–11. *Inflorescence axes not ending in spikelets (produced into a flattened point).* Rachides flattened. Inflorescence espatheate; not comprising 'partial inflorescences' and foliar organs. Spikelet-bearing axes with substantial rachides; disarticulating (sometimes very tardily); falling entire. Spikelets secund; biseriate; imbricate (sometimes becoming pectinate).

Female-fertile spikelets. Spikelets 2.3–8 mm long; adaxial; compressed laterally; disarticulating above the glumes; usually not disarticulating between the florets (the rachilla tough). Rachilla prolonged beyond the uppermost female-fertile floret; rachilla hairless. The rachilla extension with incomplete florets. Hairy callus absent.

Glumes two (persistent, membranous, laterally compressed); more or less equal; shorter than the spikelets; shorter than the adjacent lemmas; lateral to the rachis (the lower side-on); hairless; the upper awned (or mucronate), or awnless (the lower); carinate; *very dissimilar (membranous,the lower muticate, the upper obliquely awned or mucronate from just below the tip)*. Lower glume 1 nerved. Upper glume 1 nerved. *Spikelets* with incomplete florets. The incomplete florets distal to the female-fertile florets.

Female-fertile florets 3–6. Lemmas acute to shortly awned, often recurved at the tip; not becoming indurated (thinly membranous); entire, or incised; awnless, or mucronate, or awned. Awns when present, 1; median; apical; non-geniculate; straight, or recurving; much shorter than the body of the lemma. Lemmas hairless; glabrous; strongly carinate; having the margins lying flat and exposed on the palea; 1–3 nerved. *Palea* present; relatively long; apically notched; awnless, without apical setae; 2-nerved; 2-keeled. Palea keels winged (often), or wingless. *Lodicules* present; 2; fleshy. *Stamens* 3. Anthers 0.25–2.3 mm long; not penicillate. *Ovary* glabrous. Styles fused. Stigmas 2.

Fruit, embryo and seedling. Fruit free from both lemma and palea; small (0.7–1.1 mm long); ellipsoid to subglobose; obtusely triquetrous; *sculptured*. Hilum short. Pericarp thin (hyaline); free. Embryo large; not waisted. Endosperm containing compound starch grains. Embryo with an epiblast; with a scutellar tail; with an elongated mesocotyl internode. Embryonic leaf margins meeting.

Abaxial leaf blade epidermis. *Costal/intercostal zonation* conspicuous. *Papillae* present; costal and intercostal. Intercostal papillae over-arching the stomata; consisting of one oblique swelling per cell to consisting of one symmetrical projection per cell (large and thick walled). Mid-intercostal long-cells rectangular; having markedly sinuous walls and having straight or only gently undulating walls. *Microhairs* present; more or less spherical; ostensibly one-celled, or clearly two-celled; chloridoid-type. Microhair apical cell wall of similar thickness/rigidity to that of the basal cell. Microhairs with 'partitioning membranes' (in *D. aegyptium*). The 'partitioning membranes' in the basal cell. Microhairs 13–17 microns long. Microhair basal cells 15 microns long. Microhair total length/width at septum 1–1.5. Microhair apical cell/total length ratio 0.4. *Stomata* common. Subsidiaries low dome-shaped, or dome-shaped and triangular. Guard-cells overlapped by the interstomatals (regardless of the overlapping papillae). *Intercostal short-cells* common (rarely), or absent or very rare; not paired. Intercostal silica bodies absent. *Costal short-cells* conspicuously in long rows, or neither distinctly grouped into long rows nor predominantly paired. Costal silica bodies present in alternate cell files of the costal zones; consistently saddle shaped, or horizontally-elongated smooth and saddle shaped (sometimes, then the saddles predominating).

Transverse section of leaf blade, physiology. C_4; XyMS+. PCR sheath outlines uneven. PCR sheaths of the primary vascular bundles complete. PCR sheath extensions present. Maximum number of extension cells 1. PCR cell chloroplasts with well developed grana; centrifugal/peripheral. *Mesophyll* with radiate chlorenchyma; traversed by columns of colourless mesophyll cells, or not traversed by colourless columns. *Leaf blade* adaxially flat. *Midrib* conspicuous; having a conventional arc of bundles; with colourless mesophyll adaxially (the colourless tissue extending laterally across the adaxial part of the blade). Bulliforms present in discrete, regular adaxial groups; in simple fans, or associated with colourless mesophyll cells to form deeply-penetrating fans (the material of *D. aegyptium* seen having deep-penetrating simple fans only). All the vascular bundles accompanied by sclerenchyma. Combined sclerenchyma girders present; forming 'figures', or nowhere forming 'figures'. Sclerenchyma all associated with vascular bundles.

Culm anatomy. *Culm internode bundles* in one or two rings, or in three or more rings.

Phytochemistry. Leaves without flavonoid sulphates (1 species).

Cytology. Chromosome base number, $x = 10$ and 12. $2n = 20$, 36, and 48. 2, 3, and 4 ploid.

Taxonomy. Chloridoideae; main chloridoid assemblage.

Ecology, geography, regional floristic distribution. 13 species. In warm regions. Commonly adventive. Mesophytic to xerophytic; species of open habitats; halophytic and glycophytic. Sometimes in saline habitats or dunes, mostly in dry sandy soils.

Holarctic, Paleotropical, Neotropical, and Australian. Tethyan. African, Madagascan, Indomalesian, and Polynesian. Irano-Turanian. Saharo-Sindian, Sudano-Angolan, and West African Rainforest. Indian, Indo-Chinese, Malesian, and Papuan. Fijian. Caribbean, Venezuela and Surinam, Amazon, Central Brazilian, and Andean. North and East Australian and Central Australian. Sahelo-Sudanian, Somalo-Ethiopian, South Tropical African, and Kalaharian. Tropical North and East Australian and Temperate and South-Eastern Australian.

Rusts and smuts. Rusts — *Puccinia.* Smuts from Tilletiaceae and from Ustilaginaceae. Tilletiaceae — *Tilletia.* Ustilaginaceae — *Ustilago.*

Economic importance. Significant weed species: *D. aegyptium.* Cultivated fodder: *D. giganteum.* Important native pasture species: *D. aegyptium* (in arid and semiarid places). Lawns and/or playing fields: *D. aegyptium* (Crowfoot, Durban grass).

References, etc. Leaf anatomical: Metcalfe 1960; this project.

Daknopholis W. Clayton

Chloris boivinii

Habit, vegetative morphology. Prostrate annual; stoloniferous. *Culms* 10–30 cm high; herbaceous. Culm nodes glabrous. Plants unarmed. *Leaves* not basally aggregated; non-auriculate. Leaf blades narrow; 2–4 mm wide (and to 20 mm long); flat; without abaxial multicellular glands; without cross venation; persistent. Ligule a fringed membrane (short-fringed). *Contra-ligule* absent.

Reproductive organization. Plants bisexual, with bisexual spikelets; with hermaphrodite florets.

Inflorescence. *Inflorescence* of spicate main branches; *digitate.* Primary inflorescence branches 2–4. Rachides flattened. Inflorescence espatheate; not comprising 'partial inflorescences' and foliar organs. Spikelet-bearing axes persistent. Spikelets solitary; secund (on unilateral rachides); biseriate; subsessile; imbricate.

Female-fertile spikelets. *Spikelets 2 mm long*; adaxial; strongly compressed laterally (trigonous); disarticulating above the glumes. Rachilla briefly prolonged beyond the uppermost female-fertile floret. The rachilla extension naked (by contrast with *Chloris*). Hairy callus present.

Glumes two; relatively large; *more or less equal*; shorter than the spikelets; shorter than the adjacent lemmas; dorsiventral to the rachis to lateral to the rachis; hairless (glabrous, save for the scabrid keel and margins); awnless; carinate (G_1), or non-carinate (G_2); very dissimilar (membranous, thin). Lower glume 1 nerved. Upper glume 1 nerved. *Spikelets* with female-fertile florets only.

Female-fertile florets 1. Lemmas similar in texture to the glumes to decidedly firmer than the glumes (membranous to cartilaginous); not becoming indurated; incised; 2 lobed; not deeply cleft (bidentate); awned (usually), or awnless (rarely). Awns 1; median; from a sinus to dorsal; from near the top (just behind the slight notch); non-geniculate; hairless (scabrid); much longer than the body of the lemma; entered by one vein. *Lemmas* hairless (scabrid marginally and on the nerves); *carinate*; without a germination flap; 3 nerved. *Palea* present; relatively long; apically notched; awnless, without apical setae (ciliate on nerves); not indurated (thin); 2-nerved; 2-keeled. *Lodicules* present; 2; free; fleshy; glabrous; not or scarcely vascularized. *Stamens* 3. Anthers short; not penicillate; without an apically prolonged connective. *Ovary* glabrous. Styles free to their bases. Stigmas 2.

Fruit, embryo and seedling. Fruit ellipsoid; compressed laterally. *Pericarp free.*

Abaxial leaf blade epidermis. *Costal/intercostal zonation* conspicuous. *Papillae* present; intercostal. Intercostal papillae not over-arching the stomata (or scarcely so); several per cell (a row of large, cylindrical, thick-walled papillae along each long-cell and interstomatal). Long-cells of similar wall thickness costally and intercostally (walls of medium thickness). Mid-intercostal long-cells rectangular; having markedly sinuous walls. *Microhairs* present; more or less spherical; clearly two-celled; chloridoid-type (small). Microhair apical cell wall of similar thickness/rigidity to that of the basal cell. Microhair basal cells 9 microns long. Microhair total length/width at septum 2. Microhair apical cell/total cell length ratio 0.5. *Stomata* common. Subsidiaries irregularly dome-shaped and triangular. Guard-cells overlapping to flush with the interstomatals. *Intercostal short-cells* common; not paired (solitary). Intercostal silica bodies absent. *Costal short-cells* conspicuously in long rows. Costal silica bodies present in alternate cell files of the costal zones; saddle shaped (predominating), or 'panicoid-type' (a few).

Transverse section of leaf blade, physiology. C_4; XyMS+. PCR sheaths of the primary vascular bundles interrupted; interrupted abaxially only. PCR sheath extensions absent. *Leaf blade* adaxially flat (to slightly adaxially ribbed). *Midrib* conspicuous (by virtue of a large bulliform group above it); with one bundle only. Bulliforms present in discrete, regular adaxial groups (but inconspicuous save near the centre of the blade); in simple fans. All the vascular bundles accompanied by sclerenchyma. Combined sclerenchyma girders present (most bundles with minute girders). The lamina margins with fibres.

Taxonomy. Chloridoideae; main chloridoid assemblage.

Ecology, geography, regional floristic distribution. 1 species. Madagascar and Aldabra I. Species of open habitats; halophytic. Maritime sand.

Paleotropical. African and Madagascan. Sudano-Angolan. Somalo-Ethiopian and South Tropical African.

References, etc. Leaf anatomical: this project.

Dallwatsonia B.K. Simon

Habit, vegetative morphology. Perennial. *Culms* 40–130 cm high; herbaceous; branched above; 7 noded. Leaf blades linear, or linear-lanceolate. Ligule a fringed membrane.

Reproductive organization. Plants bisexual, with bisexual spikelets; with hermaphrodite florets. The spikelets all alike in sexuality.

Inflorescence. *Inflorescence of spicate main branches, or paniculate (the main axis to 22 cm long, the primary branches to 6 cm, the spikelet bearing branches reduced to 2–several spikelets)*; non-digitate; espatheate; not comprising 'partial inflorescences' and foliar organs. Spikelet-bearing axes persistent. *Spikelets* unaccompanied by bractiform involucres, not associated with setiform vestigial branches; more or less paired; secund to not secund (the spikelet bearing branches on two sides of the three-sided rachis); pedicellate (the pedicels 0.1–4 mm long). Pedicel apices cupuliform. *Spikelets* distant; somewhat *consistently in 'long-and-short' combinations*; unequally pedicellate in each combination.

Female-fertile spikelets. Spikelets 3.5–4 mm long; elliptic, or lanceolate; somewhat compressed laterally; falling with the glumes; not disarticulating between the florets; with conventional internode spacings. The upper floret not stipitate. Rachilla terminated by a female-fertile floret. Hairy callus absent. *Callus* absent.

Glumes two; *very unequal (the lower much shorter)*; (the upper) consistently somewhat *shorter than the spikelets*; *shorter than the adjacent lemmas*; hairless; glabrous; pointed; awnless; non-carinate (but with raised nerves); *very dissimilar (membranous, the lower ovate and much shorter, the upper lanceolate and almost as long as the L_1)*. Lower glume 0.3–0.5 times the length of the upper glume; shorter than the lowest lemma; much shorter than half length of lowest lemma; convex on the back; relatively smooth; *5 nerved. Upper glume* not saccate; 5–7 nerved. *Spikelets* with incomplete florets. The incomplete florets proximal to the female-fertile florets. *The proximal incomplete florets* 1; paleate. *Palea of the proximal incomplete florets reduced (no more than half the lemma length, linear or lanceolate, hyaline)*; not becoming conspicuously hardened and enlarged laterally. The proximal incomplete florets sterile. The proximal lemmas lanceolate, membranous, resembling the upper glume; awnless; 5 nerved; more or less equalling the female-fertile lemmas to decidedly exceeding the female-fertile lemmas; less firm than the female-fertile lemmas to similar in texture to the female-fertile lemmas; not becoming indurated.

Female-fertile florets 1. Lemmas lanceolate; not saccate; similar in texture to the glumes to decidedly firmer than the glumes (membranous to very thinly cartilaginous); smooth; not becoming indurated; white in fruit; entire; pointed; not deeply cleft; not crested; awnless; hairless; glabrous; non-carinate (rounded on the back); having the margins lying flat and exposed on the palea; with a clear germination flap; 5–7 nerved (the median lacking or basal only in the material seen); with the nerves non-confluent. *Palea* present; relatively long; gaping; entire; awnless, without apical setae; textured like the lemma; not indurated; 2-nerved; keel-less. Palea back glabrous. *Lodicules* present;

2; free; fleshy; glabrous; not or scarcely vascularized. **Stamens 3**. Anthers 1.4–2 mm long; not penicillate; without an apically prolonged connective. *Ovary* glabrous; without a conspicuous apical appendage. Styles free to their bases; free. Style bases widely separated. Stigmas 2.

Abaxial leaf blade epidermis. *Costal/intercostal zonation* conspicuous. *Papillae* absent. *Long-cells* markedly different in shape costally and intercostally (the costals much narrower); of similar wall thickness costally and intercostally (the walls of medium thickness). Mid-intercostal long-cells rectangular; having markedly sinuous walls. *Microhairs* present; elongated; clearly two-celled; panicoid-type (broad, almost balanoform); 50–55 microns long; about 10 microns wide at the septum. Microhair total length/width at septum 5–5.5. Microhair apical cells 35–40 microns long. Microhair apical cell/total length ratio 0.7–1. *Stomata* common; 15–20 microns long. Subsidiaries non-papillate; dome-shaped and triangular. Guard-cells overlapping to flush with the interstomatals. *Intercostal short-cells* common; in cork/silica-cell pairs (mostly), or not paired (some solitary); silicified and not silicified. Intercostal silica bodies mostly more or less cross-shaped. With a few small intercostal prickles. *Costal short-cells* conspicuously in long rows. Costal silica bodies present and well developed; 'panicoid-type'; short to medium dumb-bell shaped, or nodular (a few only).

Transverse section of leaf blade, physiology. C_3; XyMS+. Mesophyll seemingly *Isachne*-type (at least in places); not traversed by colourless columns; without fusoids (but most of the intercostal zones with a well defined aerenchymatous region in the middle). *Leaf blade* 'nodular' in section to adaxially flat. *Midrib* conspicuous (the keel large and abaxially prominent); having a conventional arc of bundles (the large median accompanied on either side by several small laterals); with colourless mesophyll adaxially. *The lamina* symmetrical on either side of the midrib. Bulliforms present in discrete, regular adaxial groups; in simple fans. All the vascular bundles accompanied by sclerenchyma. Combined sclerenchyma girders present (with all or most of the lateral bundles); forming 'figures' (some of the configurations somewhat I-shaped). Sclerenchyma all associated with vascular bundles.

Taxonomy. Panicoideae; Panicodae; Paniceae.

Ecology, geography, regional floristic distribution. 1 species. Queensland. Helophytic; glycophytic.

Australian. North and East Australian. Tropical North and East Australian.

References, etc. Morphological/taxonomic: B.K. Simon 1992. Leaf anatomical: this project.

Special comments. Fruit data wanting.

Danthonia DC.

Brachatera Desv., *Brachyathera* Kuntze, *Merathrepta* Raf., *Wilibald-Schmidtia* Conrad

Excluding *Chionochloa, Dregeochloa, Erythranthera, Karroochloa, Merxmuellera, Monachather, Monostachya, Plinthanthesis, Rytidosperma, Sieglingia*

Habit, vegetative morphology. Perennial; caespitose. **Culms** 10–100 cm high; *herbaceous*; *unbranched above*. Leaves non-auriculate. Leaf blades linear; narrow; without cross venation; persistent; once-folded in bud. *Ligule a fringe of hairs*.

Reproductive organization. *Plants bisexual, with bisexual spikelets*; with hermaphrodite florets; inbreeding. Exposed-cleistogamous, or chasmogamous. Plants with hidden cleistogenes, or without hidden cleistogenes. The hidden cleistogenes (when present) in the leaf sheaths (sometimes in the sheaths and very modified).

Inflorescence. *Inflorescence* few spikeleted; *paniculate (sometimes reduced almost to a raceme)*; *open*; espatheate; not comprising 'partial inflorescences' and foliar organs. Spikelet-bearing axes persistent. Spikelets not secund; pedicellate.

Female-fertile spikelets. Spikelets 4–25 mm long (?); compressed laterally; disarticulating above the glumes; disarticulating between the florets; with conventional internode spacings. Rachilla prolonged beyond the uppermost female-fertile floret. *Hairy callus present*. *Callus* short to long; pointed.

Glumes two; more or less equal; *about equalling the spikelets to exceeding the spikelets*; long relative to the adjacent lemmas; pointed; awnless; similar (papery). Lower glume 3–7 nerved. Upper glume 3–7 nerved. *Spikelets* with female-fertile florets only, or with incomplete florets. The incomplete florets when present, distal to the female-fertile florets.

Female-fertile florets 3–20 (?). *Lemmas* similar in texture to the glumes to decidedly firmer than the glumes; not becoming indurated; incised; 2 lobed (the lobes acute, acuminate or setaceous); deeply cleft, or not deeply cleft; *awned*. Awns 1, or 3; median, or median and lateral; the median different in form from the laterals (when laterals present); from a sinus; geniculate; much shorter than the body of the lemma to much longer than the body of the lemma; entered by one vein. The lateral awns when present, shorter than the median to exceeding the median (straight, terminating the lobes). Lemmas nearly always hairy. *The hairs not in tufts*; *not in transverse rows*. Lemmas non-carinate (rounded on the back); without a germination flap; 7–15 nerved (?). *Palea* present; relatively long to conspicuous but relatively short; entire to apically notched; awnless, without apical setae; not indurated; 2-nerved; 2-keeled. Palea keels wingless. **Lodicules** present (nearly always?); 2; free; fleshy, or membranous; *glabrous (usually — but with some exceptions)*. *Stamens* 3. Anthers not penicillate; without an apically prolonged connective. *Ovary* glabrous. Styles free to their bases. Stigmas 2.

Fruit, embryo and seedling. *Fruit* free from both lemma and palea; small; longitudinally grooved. *Hilum* usually *long-linear*. Embryo large. Endosperm hard; without lipid; containing compound starch grains. Embryo without an epiblast; with a scutellar tail; with an elongated mesocotyl internode. Embryonic leaf margins meeting.

First seedling leaf with a well-developed lamina. The lamina narrow; curved.

Abaxial leaf blade epidermis. *Costal/intercostal zonation* conspicuous. *Papillae* absent. Intercostal zones with typical long-cells, or exhibiting many atypical long-cells (these tending to be hexagonal). *Microhairs* present; panicoid-type. *Stomata* absent or very rare. *Intercostal short-cells* common; in cork/silica-cell pairs (and solitary); silicified. *Costal short-cells* conspicuously in long rows (sometimes in short series as well). Costal silica bodies tall-and-narrow, or 'panicoid-type'; when panicoid type, butterfly shaped, or dumb-bell shaped, or nodular.

Transverse section of leaf blade, physiology. C_3; XyMS+. *PBS cells* without a suberised lamella. *Mesophyll* with non-radiate chlorenchyma. *Leaf blade* with distinct, prominent adaxial ribs; with the ribs more or less constant in size. *Midrib* conspicuous (somewhat); with one bundle only. Bulliforms present in discrete, regular adaxial groups (at the bases of the furrows); in simple fans. All the vascular bundles accompanied by sclerenchyma (except at the leaf margins). Combined sclerenchyma girders present; forming 'figures' (e.g. in the midrib). Sclerenchyma all associated with vascular bundles.

Cytology. Chromosome base number, $x = 9$ (?). $2n = 18, 36$, and 48. 2 and 4 ploid. Nucleoli disappearing before metaphase.

Taxonomy. Arundinoideae; Danthonieae.

Ecology, geography, regional floristic distribution. About 20 species. The Northern hemisphere component of '*Danthonia*', very doubtfully distinguishable (i.e. as *Danthonia* sensu stricto) from the Southern component. Mesophytic to xerophytic; species of open habitats; glycophytic. Grasslands and open woodlands, often in hilly regions.

Holarctic. Boreal and Tethyan. Euro-Siberian, Atlantic North American, and Rocky Mountains. Irano-Turanian.

Hybrids. Intergeneric hybrid with *Sieglingia*: ×*Danthosieglingia* Domin.

Economic importance. Important native pasture species: *D. californica*, *D. compressa* etc.

References, etc. Morphological/taxonomic: Blake 1972a; Jacobs 1982. Leaf anatomical: Metcalfe 1960.

Special comments. *Sensu stricto*: the Northern hemisphere component of *Danthonia* sensu lato.

Danthoniastrum (J. Holub) J. Holub

~ *Metcalfia*

Habit, vegetative morphology. Perennial; caespitose (with slender stems). **Culms 10–50 cm high**; herbaceous. Culm nodes glabrous. Leaves non-auriculate. *Leaf blades* linear; narrow; 0.5–0.8 mm wide; *setaceous (rigid, pungent)*; rolled (convolute); without cross venation; disarticulating from the sheaths. *Ligule* present; an unfringed membrane, or a fringed membrane (?—ciliolate); truncate; 1–1.5 mm long.

Reproductive organization. Plants bisexual, with bisexual spikelets; with hermaphrodite florets.

Inflorescence. Inflorescence few spikeleted, or many spikeleted; a single raceme (with few spikelets), or paniculate (? sometimes); espatheate; not comprising 'partial inflorescences' and foliar organs. Spikelet-bearing axes persistent. Spikelets not secund; pedicellate.

Female-fertile spikelets. Spikelets 12–20(–25) mm long; compressed laterally; disarticulating above the glumes; disarticulating between the florets. Rachilla prolonged beyond the uppermost female-fertile floret; rachilla hairless (glabrous). The rachilla extension with incomplete florets. Hairy callus present. *Callus* fairly short.

Glumes two; very unequal; (the longer) long relative to the adjacent lemmas; hairless; pointed; awned (the upper, or both), or awnless; carinate to non-carinate (at the most, lightly keeled); similar (lanceolate, acuminate). Lower glume 3–5 nerved (rarely 1-nerved). Upper glume 5–7 nerved. *Spikelets* with incomplete florets. The incomplete florets distal to the female-fertile florets.

Female-fertile florets 2–5 (-7). *Lemmas* less firm than the glumes to similar in texture to the glumes (leathery); not becoming indurated; *incised*; 2 lobed (bidentate); deeply cleft, or not deeply cleft; awned. Awns 1 (the lateral lobes short, awnless), or 3 (the lateral lobes attenuate into awns); median, or median and lateral; the median different in form from the laterals (when laterals present); from a sinus; geniculate; about as long as the body of the lemma to much longer than the body of the lemma; entered by several veins (3). The lateral awns when present, shorter than the median (straight). Lemmas hairy (below); non-carinate; without a germination flap; 7 nerved. *Palea* present (linear-lanceolate); apically notched; 2-nerved; 2-keeled (the keels scabridulous). Palea keels winged. **Lodicules** present; *3*; free; membranous; ciliate; heavily vascularized (judging from Baum's illustration). *Stamens* 3. Anthers 3–4.5 mm long. **Ovary hairy**. Styles free to their bases (very short). Stigmas 2, or 3.

Fruit, embryo and seedling. *Fruit* free from both lemma and palea; small (2.7 mm long); compressed dorsiventrally; hairy on the body (over the upper 1/3). Hilum long-linear. Embryo small. Endosperm hard.

Abaxial leaf blade epidermis. *Costal/intercostal zonation* conspicuous to lacking (fairly obscure). *Papillae* absent. *Long-cells* fairly markedly different in shape costally and intercostally (narrower costally, at least in some zones); of similar wall thickness costally and intercostally (thick walled, with dense pitting). Intercostal zones exhibiting many atypical long-cells (many of them short to square or rounded). Mid-intercostal long-cells mainly rectangular; having markedly sinuous walls (the sinuosity fine, regular). *Microhairs* absent (and absent adaxially — militating against a danthonioid affinity). *Stomata* absent or very rare. *Intercostal short-cells* common; in cork/silica-cell pairs; silicified. Intercostal silica bodies tall-and-narrow. No macrohairs or prickles seen. *Costal short-cells* conspicuously in long rows. Costal silica bodies present and well developed; horizontally-elongated crenate/sinuous, rounded, and 'panicoid-type' (the silica bodies small, mostly shortly horizontally elongated, with a more or less pronounced isthmus which sometimes looks like a pair of crenations!); sometimes unambiguously dumb-bell shaped.

Transverse section of leaf blade, physiology. C_3; XyMS+. Mesophyll without adaxial palisade. *Leaf blade* with distinct, prominent adaxial ribs; with the ribs very irregular in sizes (with small, round-topped ribs separating the large, flat-topped ones). *Midrib* not readily distinguishable; with one bundle only. *The lamina* symmetrical on either side of the midrib. Bulliforms present in discrete, regular adaxial groups; in simple fans (of small cells, in the furrows). All the vascular bundles accompanied by sclerenchyma. Combined sclerenchyma girders present (with the primaries in the big ribs, the minor bundles with abaxial girders and adaxial strands); forming 'figures' (massive 'anchors'). Sclerenchyma not all bundle-associated. The 'extra' sclerenchyma in a continuous abaxial layer (linking the bases of the abaxial girders).

Taxonomy. Pooideae, or Arundinoideae; if pooid, Poodae; Aveneae; if arundinoid, Stipeae, or Danthonieae.

Ecology, geography, regional floristic distribution. 2 species. Balkan Peninsula. Holarctic. Boreal and Tethyan. Euro-Siberian. Mediterranean. European.

References, etc. Morphological/taxonomic: Baum 1973; Scholz 1982; Clayton 1985. Leaf anatomical: this project.

Special comments. The lack of microhairs (at least on the leaves) suggest Pooideae, but three (ciliate) lodicules, hairy ovary, 3-nerved lemma awn, etc., suggest otherwise. The available morphological descriptions are inadequate, there is no reliable information on presence or absence of genuine microhairs elsewhere on the plant (Baum enigmatically refers to 'one-celled microhairs' on the lodicules), and the embryo structure and cytological details are unknown.

Danthonidium C.E. Hubb.

Habit, vegetative morphology. *Annual*; caespitose. *Culms* 10–60 cm high; herbaceous; unbranched above. *Leaves* not basally aggregated; non-auriculate. Leaf blades narrow; 1–2.5 mm wide (to 7.5 cm long); without cross venation. Ligule a fringe of hairs.

Reproductive organization. Plants bisexual, with bisexual spikelets; with hermaphrodite florets.

Inflorescence. *Inflorescence a single raceme (short, many sided)*; contracted; espatheate; not comprising 'partial inflorescences' and foliar organs. Spikelet-bearing axes persistent. Spikelets solitary; not secund; very shortly pedicellate.

Female-fertile spikelets. Spikelets about 11 mm long; compressed laterally; disarticulating above the glumes. Rachilla prolonged beyond the uppermost female-fertile floret; rachilla hairy (above the floret, and spathulate). The rachilla extension with incomplete florets. Hairy callus present (slender, narrowly truncate).

Glumes two; very unequal (G_1 longer); exceeding the spikelets; long relative to the adjacent lemmas (the longer much exceeding it); hairless; glabrous; pointed; awned (both shortly awn-tipped); non-carinate; similar (papery, lanceolate). Lower glume 5 nerved. Upper glume 3 nerved. *Spikelets* with incomplete florets. The incomplete florets distal to the female-fertile florets. *The distal incomplete florets* merely underdeveloped.

Female-fertile florets 1. **Lemmas** cylindrical; **convolute (enfolding the palea)**; decidedly firmer than the glumes (cartilaginous or leathery); becoming indurated; shortly and broadly 2 lobed (the lobes glabrous); not deeply cleft; awned. Awns 3; median and lateral; the median different in form from the laterals; from a sinus (or just behind?); geniculate; hairless (scabrid); much longer than the body of the lemma. The lateral awns shorter than the median (straight, terminating the lobes). Lemmas hairy. *The hairs in transverse rows (one row, above the middle, and a dense basal row from the callus)*. Lemmas non-carinate; 9 nerved. *Palea* present; relatively long; apically notched; awnless, without apical setae to with apical setae; not indurated (thin); 2-nerved; 2-keeled. *Lodicules* present; 2; free; fleshy; glabrous. *Stamens* 3. Anthers not penicillate; without an apically prolonged connective. *Ovary* glabrous. Stigmas 2.

Abaxial leaf blade epidermis. *Costal/intercostal zonation* conspicuous. *Papillae* absent. Long-cells differing markedly in wall thickness costally and intercostally (the intercostals thicker walled). Mid-intercostal long-cells rectangular, or fusiform (mixed); having markedly sinuous walls. *Microhairs* present; panicoid-type (slender); 24–27–33 microns long; 3–4.5 microns wide at the septum. Microhair total length/width at septum 5.3–11. Microhair apical cells 18–24 microns long. Microhair apical cell/total length ratio 0.67–0.75. *Stomata* common; (34–)36 microns long. Guard-cells overlapping to flush with the interstomatals. *Intercostal short-cells* common; in cork/silica-cell pairs (and solitary); silicified. *Costal short-cells* conspicuously in long rows (mainly in very long, uninterrupted rows of very short cells, with a silica body in every other cell). Costal silica bodies exclusively oryzoid (small, very regular vertical dumb-bells).

Transverse section of leaf blade, physiology. C_3; XyMS+. *Midrib* conspicuous (as a larger bundle with a large flat-bottomed keel); with one bundle only; with colourless mesophyll adaxially (apparently with some colourless tissue between the adaxial fibres and the bundle). Bulliforms not in discrete, regular adaxial groups. All the vascular bundles accompanied by sclerenchyma. Combined sclerenchyma girders absent (each bundle with a wide adaxial strand and a small abaxial girder). Sclerenchyma all associated with vascular bundles.

Taxonomy. Arundinoideae; Danthonieae.

Ecology, geography, regional floristic distribution. 1 species. Bombay. Species of open habitats. Stony places.

Paleotropical. Indomalesian. Indian.

References, etc. Morphological/taxonomic: Hubbard 1937a. Leaf anatomical: this project.

Special comments. Fruit data wanting.

Danthoniopsis Stapf

Gazachloa Phipps, *Jacquesfelixia* Phipps, *Petrina* Phipps, *Pleioneura* (C. E. Hubb.) Phipps, *Rattraya* Phipps, *Xerodanthia* Phipps

Habit, vegetative morphology. Annual (rarely), or perennial; caespitose (sometimes densely so). **Culms** 25–200 cm high; *herbaceous*; branched above, or unbranched above. Culm nodes hairy, or glabrous. Culm internodes solid, or hollow. Plants unarmed. Young shoots intravaginal. *Leaves* not basally aggregated; non-auriculate. Leaf blades linear to lanceolate; broad, or narrow; 1.5–18 mm wide; flat, or rolled (but only slightly so); pseudopetiolate (*D. petiolata*), or not pseudopetiolate; without cross venation; persistent. Ligule a fringe of hairs; 0.5–2.5 mm long. *Contra-ligule* present (as an irregular line of hairs), or absent (rarely).

Reproductive organization. *Plants bisexual, with bisexual spikelets*; with hermaphrodite florets.

Inflorescence. Inflorescence paniculate (the spikelets in twos and threes); open, or contracted; espatheate; not comprising 'partial inflorescences' and foliar organs. Spikeletbearing axes persistent. Spikelets in triplets (rarely), or paired; not secund; pedicellate.

Female-fertile spikelets. *Spikelets* 5–20 mm long; purplish; compressed laterally; *disarticulating above the glumes*; disarticulating between the florets; with conventional internode spacings. Rachilla terminated by a female-fertile floret; rachilla hairy. Hairy callus present. *Callus* short; blunt (bidentate in *D. barbata*).

Glumes two; *very unequal*; (the longer) long relative to the adjacent lemmas; hairless; nearly always glabrous; awnless, or awned (G_2 sometimes aristate); non-carinate (abaxially rounded); *very dissimilar (G_1 acute to acuminate, G_2 with the tip extended)*. Lower glume shorter than the lowest lemma; 3–5 nerved. Upper glume 3–5 nerved. *Spikelets* with incomplete florets. The incomplete florets proximal to the female-fertile florets. *Spikelets with proximal incomplete florets*. *The proximal incomplete florets* 1; paleate. Palea of the proximal incomplete florets fully developed (membranous between its two narrowly winged keels). *The proximal incomplete florets male. The proximal lemmas awnless*; (3–)5–9 nerved.

Female-fertile florets 1. **Lemmas** usually but not always with 2–8 transverse rows of hairs or tufts; similar in texture to the glumes (to slightly firmer); *incised*; 2 lobed; deeply cleft; awned. Awns 1 (usually), or 3; median, or median and lateral (*D. chimanimaniensis*); the median different in form from the laterals (when laterals present); from a sinus; geniculate; hairless (scabrous); much longer than the body of the lemma; entered by several veins; deciduous. The lateral awns when present, shorter than the median (in the form of short, scaberulous bristles). Awn bases twisted; flattened. Lemmas hairy (usually), or hairless (sometimes glabrous). The hairs in tufts, or not in tufts; in transverse rows, or not in transverse rows (in Sect. Pleioneura). Lemmas non-carinate; having the margins tucked in onto the palea; with a clear germination flap (this just above the callus, often hidden by hairs); 7–9 nerved; with the nerves non-confluent. *Palea* present; relatively long (about equalling the lemma); apically notched; awnless, without apical setae; textured like the lemma; not indurated; 2-nerved; 2-keeled. *Palea keels winged (the wings clasped by the inrolled lemma margins, often terminating in a clavate swelling or auricle)*; glabrous, or scabrous. *Lodicules* present; 2; free; fleshy; glabrous. *Stamens* 3. Anthers 1.7–4 mm long; not penicillate; without an apically prolonged connective. *Ovary* glabrous. Styles free to their bases. Stigmas 2; white, or red pigmented (purple).

Fruit, embryo and seedling. *Hilum long-linear.* Embryo large.

Seedling with a loose coleoptile. First seedling leaf with a well-developed lamina. The lamina broad; curved.

Abaxial leaf blade epidermis. *Costal/intercostal zonation* conspicuous. *Papillae* absent. *Long-cells* similar in shape costally and intercostally, or markedly different in shape costally and intercostally (the costals narrower); of similar wall thickness costally and intercostally. Mid-intercostal long-cells rectangular; having markedly sinuous walls, or having straight or only gently undulating walls (the sinuosity sporadically apparent only at the outermost optical section, in *D. occidentalis*). *Microhairs* present; elongated; clearly two-celled; panicoid-type; 57–60 microns long; 6–7.5 microns wide at the septum. Microhair total length/width at septum 8–10. Microhair apical cells 24–30 microns long. Microhair apical cell/total length ratio 0.4–0.5. *Stomata* common; 30–34.5 microns long. Subsidiaries non-papillate; low to medium dome-shaped and triangular (in *D. occidentalis*, mostly with a small point). Guard-cells overlapping to flush with the interstomatals. *Intercostal short-cells* common; in cork/silica-cell pairs; silicified. Intercostal silica bodies in *D. occidentalis*, nearly all crescentic and oryzoid-type (small and irregularly shaped, but mostly more or less crescentic). *Costal short-cells* conspicuously in long rows. Costal silica bodies 'panicoid-type'; cross shaped (bordering the costal zones, in *D. occidentalis*), or dumb-bell shaped.

Transverse section of leaf blade, physiology. C_4. The anatomical organization conventional (usually), or unconventional. Organization of PCR tissue when unconventional, *Arundinella* type. XyMS+ (in *D. occidentalis*, the mestome sheath 2–3 cells wide at the XyMs position), or XyMS–. PCR sheath outlines even. PCR cell chloroplasts centrifugal/peripheral. *Mesophyll* with radiate chlorenchyma; without adaxial palisade; exhibiting 'circular cells' (?), or without 'circular cells' (*D. barbata, D. dinteri, D. occidentalis, D. pruinosa*); not traversed by colourless columns. *Leaf blade* 'nodular' in section (slightly), or adaxially flat (usually more conspicuously ribbed abaxially). *Midrib* conspicuous, or not readily distinguishable; with one bundle only, or having a conventional arc of bundles; with colourless mesophyll adaxially, or without colourless mesophyll adaxially (*D. occidentalis* with an adaxial mass of colourless cells between each pair of major bundles, over each of the intervening small bundles). Bulliforms present in discrete, regular adaxial groups (the groups large); in simple fans, or associated with colourless mesophyll cells to form deeply-penetrating fans; associating with colourless mesophyll cells to form arches over small vascular bundles (in *D. occidentalis*), or nowhere involved in bulliform-plus-colourless mesophyll arches. Many of the smallest vascular bundles unaccompanied by sclerenchyma (all of them, in *D. occidentalis*), or all the vascular bundles accompanied by sclerenchyma. Combined sclerenchyma girders present (with all the large bundles, in *D. occidentalis*), or absent; in *D. occidentalis* forming 'figures' (forming T's, in places). Sclerenchyma all associated with vascular bundles.

Cytology. Chromosome base number, $x = 9$, or 12 (?). $2n = 18$, 24, and 36. 2 and 4 ploid.

Taxonomy. Panicoideae; Panicodae; Arundinelleae.

Ecology, geography, regional floristic distribution. About 20 species. Africa, Arabia. Commonly adventive. Mesophytic to xerophytic; species of open habitats; glycophytic. Savanna woodland and desert fringes.

Holarctic and Paleotropical. Tethyan. African. Irano-Turanian. Saharo-Sindian, Sudano-Angolan, and Namib-Karoo. Sahelo-Sudanian, Somalo-Ethiopian, South Tropical African, and Kalaharian.

Rusts and smuts. Rusts — *Puccinia.*

References, etc. Morphological/taxonomic: Phipps 1967. Leaf anatomical: Metcalfe 1960; this project; photos of *D. dinteri* and *D. pruinosa* provided by R.P. Ellis.

Dasyochloa Willd. ex Rydberg

~ *Erioneuron* (*E. pulchellum*), *Tridens* (*T. pulchellus*)

Habit, vegetative morphology. Mop-like perennial; caespitose, or stoloniferous and caespitose. **Culms** 5–15 cm high; *herbaceous*; branched above. *Leaves* not basally aggregated. Leaf blades narrow; 0.3–0.5 mm wide; setaceous; rolled (involute); without abaxial multicellular glands; without cross venation. *Ligule* present; a fringe of hairs.

Reproductive organization. Plants bisexual, with bisexual spikelets; with hermaphrodite florets.

Inflorescence. *Inflorescence* few spikeleted; paniculate; *contracted (short peduncled, consisting of 3-spikeleted racemes)*; non-digitate; spatheate (each raceme subtended by a subulate-tipped sheath); *a complex of 'partial inflorescences' and intervening foliar organs*. Spikelet-bearing axes persistent.

Female-fertile spikelets. *Spikelets 4–6 mm long*; *compressed laterally (plump)*; disarticulating above the glumes; disarticulating between the florets; with conventional internode spacings. Rachilla prolonged beyond the uppermost female-fertile floret. The rachilla extension naked.

Glumes two; very unequal to more or less equal; about equalling the spikelets; (the longer) long relative to the adjacent lemmas; hairless; pointed; shortly awned (to mucronate); carinate; similar (membranous, with glands at the base). Lower glume 1 nerved. Upper glume 1 nerved. *Spikelets* with female-fertile florets only.

Female-fertile florets 2. Lemmas similar in texture to the glumes (thinly membranous); not becoming indurated; incised; 2 lobed; deeply cleft (to about halfway); awned. Awns 1; median; from a sinus; non-geniculate; hairless; much shorter than the body of the lemma to about as long as the body of the lemma; entered by one vein. Lemmas hairy; non-carinate; without a germination flap; 3 nerved (the laterals submarginal). *Palea* present; relatively long; entire; awnless, without apical setae (hairy below and on the keels); not indurated; 2-nerved; 2-keeled. *Stamens* 3 (?). *Ovary* glabrous. Stigmas 2; white.

Fruit, embryo and seedling. Fruit ellipsoid; compressed laterally. Hilum short. Pericarp fused. Embryo large; with an epiblast; with a scutellar tail; with an elongated mesocotyl internode. Embryonic leaf margins meeting.

Abaxial leaf blade epidermis. *Costal/intercostal zonation* conspicuous (the intercostal zones in furrows, and obscured by long prickles). *Papillae* absent. *Long-cells* markedly different in shape costally and intercostally (the costals much more regularly rectangular); of similar wall thickness costally and intercostally. Mid-intercostal long-cells rectangular; having markedly sinuous walls. *Microhairs* absent (none seen — but they would be hard to find in such material). *Stomata* common (hidden in the sides of the furrows, except towards the blade margins); 24–30 microns long. Subsidiaries low to high dome-shaped (mostly), or triangular (a few). Guard-cells overlapping to flush with the interstomatals. *Intercostal short-cells* absent or very rare; not paired. Intercostal silica bodies absent. Simple, thick walled prickles abundant costally and intercostally. *Crown cells* absent. *Costal short-cells* predominantly paired. Costal silica bodies present throughout the costal zones; 'panicoid-type', or crescentic (a few, plump), or rounded (a few, merging with crosses and crescents); basically cross shaped (often more or less malformed to simulate the other forms).

Transverse section of leaf blade, physiology. C_4; XyMS+. PCR sheath outlines even. PCR sheaths of the primary vascular bundles complete. PCR sheath extensions absent. PCR cell chloroplasts centripetal. *Mesophyll* with radiate chlorenchyma; traversed by columns of colourless mesophyll cells. *Leaf blade* 'nodular' in section to adaxially flat; with the ribs more or less constant in size (round topped). *Midrib* conspicuous to not readily distinguishable (having a somewhat larger abaxial strand); with one bundle only. *The lamina* symmetrical on either side of the midrib. Bulliforms present in discrete, regular adaxial groups; associated with colourless mesophyll cells to form deeply-penetrating fans (these incorporated in the traversing colourless columns). All the vascular bundles accompanied by sclerenchyma. Combined sclerenchyma girders absent (all the bundles with large adaxial and massive abaxial strands only). Sclerenchyma all associated with vascular bundles (except for large marginal groups). The lamina margins with fibres (large).

Cytology. Chromosome base number, $x = 8$. $2n = 16$. 2 ploid.

Taxonomy. Chloridoideae; main chloridoid assemblage.

Ecology, geography, regional floristic distribution. 1 species. Southern U.S.A. and Mexico. Xerophytic; species of open habitats. Rocky slopes.

Neotropical. Central Brazilian and Pampas.

References, etc. Morphological/taxonomic: Sanchez 1983. Leaf anatomical: this project.

Dasypoa Pilger

~ *Poa* (*P. scaberula*)

Habit, vegetative morphology. *Culms* 20–40 cm high; herbaceous; unbranched above. Culm nodes glabrous. Plants unarmed. *Leaves* not basally aggregated; non-auriculate. Leaf blades linear; narrow; to 3 mm wide; flat (except towards the tip); without cross venation; persistent. Ligule an unfringed membrane; not truncate; 2 mm long. *Contraligule* absent.

Reproductive organization. Plants bisexual, with bisexual spikelets; with hermaphrodite florets.

Inflorescence. Inflorescence paniculate; open; espatheate; not comprising 'partial inflorescences' and foliar organs. Spikelet-bearing axes persistent. Spikelets not secund; shortly pedicellate.

Female-fertile spikelets. *Spikelets* 4–5 mm long; *compressed laterally*; disarticulating above the glumes; disarticulating between the florets; with conventional internode spacings (or these slightly elongated). Rachilla prolonged beyond the uppermost female-fertile floret; rachilla hairless. *Hairy callus present (small, with long, fine, webby hairs more than 0.5 mm long)*. *Callus* short; blunt.

Glumes present; two; relatively large; more or less equal; shorter than the adjacent lemmas; hairy (scabrid-hairy on the keels); pointed; awnless; carinate. Lower glume longer than half length of lowest lemma; 1 nerved. **Upper glume** *distinctly saccate*; 3 nerved. *Spikelets* with incomplete florets (some of them). The incomplete florets distal to the female-fertile florets.

Female-fertile florets 3–5. Lemmas similar in texture to the glumes; not becoming indurated; entire; pointed; awnless; hairy; carinate (naviculate); without a germination flap; 5 nerved. *Palea* present; conspicuous but relatively short (about two-thirds lemma length); apically notched; awnless, without apical setae; not indurated (hyaline); 2-nerved; 2-keeled (scabrid-hairy on the keels). *Lodicules* present; 2; membranous; glabrous; toothed; not or scarcely vascularized. *Stamens* 3. Anthers short; not penicillate; without an apically prolonged connective. *Ovary* glabrous. Styles free to their bases. Stigmas 2.

Fruit, embryo and seedling. Fruit small. Hilum short. *Pericarp free*. Embryo small. Endosperm probably liquid in the mature fruit.

Abaxial leaf blade epidermis. *Costal/intercostal zonation* conspicuous. *Papillae* absent. Mid-intercostal long-cells fusiform; having straight or only gently undulating walls. *Microhairs* absent. *Stomata* common. Subsidiaries parallel-sided. Guard-cells overlapped by the interstomatals. *Intercostal short-cells* absent or very rare. Costal prickles abundant. *Crown cells* absent. *Costal short-cells* neither distinctly grouped into long rows nor predominantly paired (mostly solitary, a few short rows). Costal silica bodies horizontally-elongated crenate/sinuous (mostly crenate).

Transverse section of leaf blade, physiology. C_3; XyMS+. Mesophyll without adaxial palisade. Midrib with one bundle only. All the vascular bundles accompanied by sclerenchyma. Combined sclerenchyma girders present (but some small bundles with strands only). Sclerenchyma all associated with vascular bundles.

Taxonomy. Pooideae; Poodae; Aveneae.

Ecology, geography, regional floristic distribution. 1 species. Peru.

Neotropical. Andean.

References, etc. Leaf anatomical: this project.

Dasypyrum (Cosson & Durieu) Durand

Haynaldia Schur, *Pseudosecale* (Godron) Degen, *Secalidium* Schur

Habit, vegetative morphology. Annual, or perennial. *Culms* 20–100 cm high; herbaceous. *Leaves auriculate*. Sheath margins joined, or free. Leaf blades linear; narrow; 1–5 mm wide; flat; without cross venation. Ligule an unfringed membrane; truncate; 0.3–1 mm long.

Reproductive organization. Plants bisexual, with bisexual spikelets; with hermaphrodite florets.

Inflorescence. Inflorescence a single spike (compressed, dense). Rachides hollowed, or flattened, or winged, or neither flattened nor hollowed, not winged. Inflorescence espatheate; not comprising 'partial inflorescences' and foliar organs. Spikelet-bearing axes disarticulating; disarticulating at the joints. Spikelets solitary; not secund; distichous.

Female-fertile spikelets. Spikelets 7–22 mm long; compressed laterally; falling with the glumes. Rachilla prolonged beyond the uppermost female-fertile floret. The rachilla extension with incomplete florets. Hairy callus absent. *Callus* very short.

Glumes two; more or less equal; shorter than the adjacent lemmas; lateral to the rachis, or displaced (and adjacent to one another); *with distinct hair tufts*; not subulate; awned (both with long, scabrid awns); non-carinate; *similar (leathery with membranous margins, abruptly tapering into the awns, strongly 2-keeled, the keels bearing tufts of hair and convergent above)*. Lower glume two-keeled (with leathery, ciliate keels); 3–4 nerved. Upper glume 3–4 nerved (also bicarinate). *Spikelets* with incomplete florets. The incomplete florets distal to the female-fertile florets. The distal incomplete florets 1(–2) (sterile, long-stipitate).

Female-fertile florets *2(–3)*. Lemmas similar in texture to the glumes; not becoming indurated; entire, or incised; pointed; when incised, 2 lobed; not deeply cleft (bidentate); awned. Awns 1; median; apical (attenuate from the lemma), or from a sinus; non-geniculate; hairless (scabrid); much longer than the body of the lemma; entered by several veins. Lemmas hairy. The hairs in tufts (on the keel). Lemmas carinate; without a germination flap; 5 nerved. *Palea* present (narrowly lanceolate); relatively long; membranous; 2-nerved; 2-keeled. *Lodicules* present; 2; free; membranous; ciliate, or glabrous; not toothed; not or scarcely vascularized. *Stamens* 3. Anthers 5–7 mm long. *Ovary* hairy. Styles free to their bases. Stigmas 2.

Fruit, embryo and seedling. *Fruit* free from both lemma and palea (i.e. threshing free); small to medium sized (3.8–7 mm long); ellipsoid; shallowly longitudinally grooved; slightly to strongly compressed laterally; with hairs confined to a terminal tuft. Hilum long-linear. Embryo small. Endosperm hard; without lipid; containing only simple starch grains. Embryo with an epiblast.

Abaxial leaf blade epidermis. *Costal/intercostal zonation* conspicuous. *Papillae* absent. *Long-cells* markedly different in shape costally and intercostally (much narrower costally); fairly thin walled. Mid-intercostal long-cells rectangular to fusiform (some of them extraordinarily long); having straight or only gently undulating walls. *Microhairs* absent. *Stomata* common. Subsidiaries non-papillate; parallel-sided. Guard-cells overlapped by the interstomatals (slightly but consistently, but the apparatus not much sunken). *Intercostal short-cells* fairly common; not paired (solitary, large); not silicified. Small prickles intercostally, larger to very large ones costally. *Crown cells* absent (many of the small intercostal prickles with reduced or blunt 'points', but lacking the ring of pits). *Costal short-cells* neither distinctly grouped into long rows nor predominantly paired. Costal silica bodies present and well developed; horizontally-elongated crenate/sinuous to horizontally-elongated smooth.

Transverse section of leaf blade, physiology. C_3; XyMS+. Mesophyll without adaxial palisade. *Leaf blade* with distinct, prominent adaxial ribs, or 'nodular' in section; with the ribs more or less constant in size. *Midrib* not readily distinguishable; with one bundle only. *The lamina* symmetrical on either side of the midrib. Many of the smallest vascular bundles unaccompanied by sclerenchyma. Combined sclerenchyma girders present (with the primaries); forming 'figures' (a few I's). Sclerenchyma all associated with vascular bundles.

Cytology. Chromosome base number, $x = 7$. $2n = 14$ and 28. 2 and 4 ploid. Haplomic genome content V. Haploid nuclear DNA content 5.4 pg (1 species). Mean diploid 2c DNA value 10.7 pg (1 species).

Taxonomy. Pooideae; Triticodae; Triticeae.

Ecology, geography, regional floristic distribution. 3–5 species. Mediterranean. Xerophytic; species of open habitats. Stony slopes.

Holarctic. Boreal and Tethyan. Euro-Siberian. Mediterranean and Irano-Turanian. European.

Hybrids. Intergeneric hybrids with *Aegilops*.

Rusts and smuts. Rusts — *Puccinia*. Taxonomically wide-ranging species: *Puccinia striiformis* and *Puccinia recondita*. Smuts from Tilletiaceae and from Ustilaginaceae. Tilletiaceae — *Tilletia*. Ustilaginaceae — *Ustilago*.

References, etc. Morphological/taxonomic: Löve 1984. Leaf anatomical: this project.

Davidsea Soderstrom and Ellis

~ *Teinostachyum* (*T. attenuatum* (Thwaites) Munro)

Habit, vegetative morphology. Perennial. The flowering culms leafy. *Culms* 400–900 cm high; woody and persistent; to 2.5 cm in diameter; cylindrical; branched above. Primary branches/mid-culm node 3. Culm sheaths deciduous in their entirety (leaving a girdle). Culm internodes hollow. Unicaespitose. Rhizomes pachymorph. Plants unarmed. Young shoots intravaginal. *Leaves* not basally aggregated; with auricular setae. Leaf blades lanceolate (acuminate); broad; 20–30 mm wide (10–20 cm long); flat; pseudopetiolate; without cross venation; disarticulating from the sheaths; rolled in bud. Ligule truncate; 0.2–1 mm long. *Contra-ligule* in the form of a hard, glabrous rim.

Reproductive organization. Plants bisexual, with bisexual spikelets; with hermaphrodite florets.

Inflorescence. Inflorescence *indeterminate*; *with pseudospikelets (developing sympodially, by contrast with those of* **Pseudoxytenanthera** *etc.)*; terminating a leafy branch, consisting of a bracteate axis with capitate clusters of pseudospikelets along its length; spatheate. *Spikelet-bearing axes* capitate. Spikelets (i.e. the pseudospikelets) not secund; sessile.

Female-fertile spikelets. Spikelets about 20 mm long; falling with the glumes; not disarticulating between the florets. Rachilla prolonged beyond the uppermost female-fertile floret.

Glumes one per spikelet (this being empty, and preceded by two gemmiferous bracts and a long internode); shorter than the spikelets; shorter than the adjacent lemmas; hairless; glabrous (with ciliate margins below); awnless (apiculate). Upper glume (i.e. the one glume) 11–15 nerved. *Spikelets* with incomplete florets. The incomplete florets distal to the female-fertile florets. *The distal incomplete florets* merely underdeveloped; awnless.

Female-fertile florets 1. *Lemmas convolute*; similar in texture to the glumes (firm); not becoming indurated; entire; pointed; awnless (apiculate); hairless; glabrous; non-carinate; without a germination flap; 16 nerved (with transverse veinlets); with the nerves non-confluent. *Palea* present; relatively long (shorter than the lemma); apically notched; awnless, without apical setae; thinner than the lemma; not indurated; several nerved (5 — one between the keels and one down each side outside them); 2-keeled. Palea keels wingless; hairy. *Lodicules* present; 3; free; membranous; ciliate; not toothed; heavily vascularized. *Stamens* 6. Anthers about 2 mm long; not penicillate; with the connective apically prolonged (shortly apiculate). *Ovary* glabrous; with a conspicuous apical appendage. The appendage long, stiff and tapering. Styles fused (below). Stigmas 3 (shortly plumose).

Fruit, embryo and seedling. *Fruit* unknown.

Abaxial leaf blade epidermis. *Costal/intercostal zonation* conspicuous. *Papillae* present; costal and intercostal. Intercostal papillae over-arching the stomata (absent from the median, astomatal region of the intercostal zone); several per cell (often lobed or of the coronate type). *Long-cells* markedly different in shape costally and intercostally (the costals narrower). Mid-intercostal long-cells rectangular; having markedly sinuous walls. *Microhairs* present; elongated; clearly two-celled; panicoid-type. *Stomata* common. *Intercostal short-cells* common; in cork/silica-cell pairs; silicified. Intercostal silica bodies tall-and-narrow. *Costal short-cells* conspicuously in long rows. Costal silica bodies saddle shaped.

Transverse section of leaf blade, physiology. C_3; XyMS+. Mesophyll with adaxial palisade; with arm cells; with fusoids. The fusoids external to the PBS. *Leaf blade* with distinct, prominent adaxial ribs (these narrow and flat-topped except to one side of the midrib, where they are much more prominent). *Midrib* conspicuous; having complex vascularization. *The lamina* distinctly asymmetrical on either side of the midrib. Bulliforms present in discrete, regular adaxial groups; in simple fans. All the vascular bundles accompanied by sclerenchyma. Combined sclerenchyma girders present; forming 'figures'. Sclerenchyma all associated with vascular bundles.

Taxonomy. Bambusoideae; Bambusodae; Bambuseae.

Ecology, geography, regional floristic distribution. 1 species. Sri Lanka. Paleotropical. Indomalesian. Indian.

References, etc. Morphological/taxonomic: Soderstrom and Ellis 1988. Leaf anatomical: Soderstrom and Ellis 1988.

Special comments. Fruit data wanting.

Decaryella A. Camus

Habit, vegetative morphology. *Annual. Culms* 10–30 cm high; herbaceous. Culm nodes glabrous. *Leaves* not basally aggregated; non-auriculate. Leaf blades narrow; flat, or rolled; without cross venation. *Ligule a fringed membrane.*

Reproductive organization. Plants bisexual, with bisexual spikelets; with hermaphrodite florets. *The spikelets all alike in sexuality.*

Inflorescence. *Inflorescence a single raceme (loose, narrow, 4–12 cm long)*; espatheate; not comprising 'partial inflorescences' and foliar organs. Spikelet-bearing axes persistent. Spikelets solitary; not secund; long pedicellate; not in distinct 'long-and-short' combinations.

Female-fertile spikelets. *Spikelets* about 4.5 mm long; *compressed dorsiventrally; falling with the glumes (and with the pedicels).* Rachilla prolonged beyond the uppermost female-fertile floret, or terminated by a female-fertile floret (?). The rachilla extension when present, with incomplete florets, or naked. Hairy callus present (comprising the long, hairy pedicel). *Callus* long; pointed.

Glumes two; more or less equal; long relative to the adjacent lemmas; hairy; pointed; *awned (both having a terminal, subulate awn 3–7 mm long)*; non-carinate (dorsally rounded); similar (thick, leathery, smooth). Lower glume 5 nerved. Upper glume 5 nerved. *Spikelets* with female-fertile florets only, or with incomplete florets. The incomplete florets distal to the female-fertile florets. *The distal incomplete florets* merely underdeveloped. *Spikelets without proximal incomplete florets.*

Female-fertile florets 1–2. Lemmas less firm than the glumes (very thin); not becoming indurated; entire; pointed, or blunt; awnless; hairless; glabrous; non-carinate; 1 nerved, or 3 nerved. *Palea* present; entire; awnless, without apical setae; 2-nerved; 2-keeled.

Fruit, embryo and seedling. Embryo large.

Abaxial leaf blade epidermis. *Costal/intercostal zonation* conspicuous. *Papillae* absent. *Microhairs* present; chloridoid-type. *Stomata* common. Subsidiaries dome-shaped and triangular. *Intercostal short-cells* common; in cork/silica-cell pairs. Prickles and macrohairs present. *Costal short-cells* conspicuously in long rows. Costal silica bodies 'panicoid-type'; dumb-bell shaped.

Transverse section of leaf blade, physiology. C_4; XyMS+. *Mesophyll* with radiate chlorenchyma. Bulliforms present in discrete, regular adaxial groups; associated with colourless mesophyll cells to form deeply-penetrating fans. All the vascular bundles accompanied by sclerenchyma. Combined sclerenchyma girders present. Sclerenchyma all associated with vascular bundles.

Taxonomy. Chloridoideae; main chloridoid assemblage.

Ecology, geography, regional floristic distribution. 1 species. Madagascar. Species of open habitats. Dry bush.
Paleotropical. Madagascan.
Special comments. Spikelet description fairly inadequate. Fruit data wanting.

Decaryochloa A. Camus

Habit, vegetative morphology. Perennial. The flowering culms leafy. *Culms* woody and persistent; *scandent*; branched above. Primary branches/mid-culm node 3–7 ('several'). Rhizomes pachymorph. Leaf blades pseudopetiolate.

Reproductive organization. Plants bisexual, with bisexual spikelets; with hermaphrodite florets.

Inflorescence. *Inflorescence compound, with racemes of 1–3 spikelets grouped into spatheate fascicles*; spatheate; a complex of 'partial inflorescences' and intervening foliar organs. *Spikelet-bearing axes* 'racemes' (of one to three spikelets). Spikelets not secund.

Female-fertile spikelets. Rachilla prolonged beyond the uppermost female-fertile floret, or terminated by a female-fertile floret.
Glumes two to several (two to four).
Female-fertile florets 1. *Lemmas convolute*; leathery, not shiny. *Palea* present; convolute; textured like the lemma (leathery); keel-less. *Lodicules* present (small). Stamens 6; *diadelphous (in two groups of three)*. *Ovary* hairy; with a conspicuous apical appendage. The appendage broadly conical, fleshy. Stigmas 3.

Fruit, embryo and seedling. Pericarp fleshy.

Transverse section of leaf blade, physiology. C_3; XyMS+. *Mesophyll* with non-radiate chlorenchyma; with arm cells; with fusoids. The fusoids external to the PBS. *Leaf blade* with distinct, prominent adaxial ribs, or adaxially flat (the ribs very wide and low to negligible, save for one associated with one side of the midrib). *Midrib* conspicuous; with one bundle only; without colourless mesophyll adaxially. *The lamina* distinctly asymmetrical on either side of the midrib. Bulliforms present in discrete, regular adaxial groups; in simple fans. All the vascular bundles accompanied by sclerenchyma. Combined sclerenchyma girders present; forming 'figures' (I's or 'anchors' with most of the bundles). Sclerenchyma all associated with vascular bundles.

Taxonomy. Bambusoideae; Bambusodae; Bambuseae.

Ecology, geography, regional floristic distribution. 1 species. Madagascar. Shade species. In forest.
Paleotropical. Madagascan.
References, etc. Leaf anatomical: this project.
Special comments. Anatomical data for ts only.

Dendrocalamus Nees

Klemachloa Parker, *Neosinocalamus* Keng f., *Sinocalamus* McClure
Excluding *Gigantochloa, Oreobambos, Oxytenanthera*

Habit, vegetative morphology. Perennial; caespitose. The flowering culms leafy. *Culms* 600–4000 cm high (*D. giganteus* being probably the tallest grass); woody and persistent; to 30 cm in diameter; branched above. Culm nodes glabrous. Primary branches/mid-culm node numerous. Culm sheaths deciduous in their entirety. Culm internodes solid, or hollow. Pluricaespitose. Rhizomes pachymorph. Plants unarmed. Young shoots intravaginal. *Leaves* not basally aggregated; auriculate (but usually inconspicuously so). *Leaf blades* broad; pseudopetiolate; *without cross venation (but often with pellucid glands)*; disarticulating from the sheaths; rolled in bud. *Ligule* present; 3 mm long. *Contra-ligule* present.

Reproductive organization. Plants bisexual, with bisexual spikelets; with hermaphrodite florets. Not viviparous.

Inflorescence. Inflorescence indeterminate; *with pseudospikelets*; usually comprising spikes of spikelet tufts; *espatheate*. *Spikelet-bearing axes* very much reduced and paniculate, or very much reduced and capitate.

Female-fertile spikelets. *Spikelets* 8–20 mm long; *not noticeably compressed*. Rachilla prolonged beyond the uppermost female-fertile floret; rachilla hairless.

Glumes two, or several (?); relatively large; very unequal to more or less equal; shorter than the adjacent lemmas; pointed, or not pointed (ovate, acute or mucronate); awnless; non-carinate; similar. Lower glume longer than half length of lowest lemma; 17 nerved (in material seen). Upper glume 19 nerved (in material seen).

Female-fertile florets *2–8*. Lemmas similar in texture to the glumes; not becoming indurated; entire; pointed; awnless, or mucronate (?); hairless; non-carinate; without a germination flap; 29 nerved. *Palea* present; relatively long; entire, or apically notched, or deeply bifid; awnless, without apical setae; not indurated; several nerved (13 observed); 2-keeled (in lower florets), or keel-less (in upper florets). *Lodicules* absent ('or very scarce'). *Stamens* 6. Anthers 2.5–5 mm long, or 7–10 mm long (in *D. giganteus*); penicillate, or not penicillate; with the connective apically prolonged. *Ovary* hairy; with a conspicuous apical appendage. The appendage broadly conical, fleshy. Styles fused. Stigmas 1(–3); red pigmented.

Fruit, embryo and seedling. *Fruit* free from both lemma and palea; small to medium sized. Hilum long-linear. Pericarp thick and hard (crustaceous or hardened); free. Embryo not visible. Endosperm hard; containing compound starch grains. Embryo with an epiblast; with a scutellar tail; with a negligible mesocotyl internode; with more than one scutellum bundle. Embryonic leaf margins overlapping.

First seedling leaf without a lamina.

Abaxial leaf blade epidermis. *Costal/intercostal zonation* conspicuous. *Papillae* present; costal and intercostal. Intercostal papillae over-arching the stomata (sometimes conspicuously so), or not over-arching the stomata; several per cell. Mid-intercostal long-cells rectangular; having markedly sinuous walls. *Microhairs* present; panicoid-type; 48–51(–57) microns long (in *D. giganteus*); 4.5–7.5 microns wide at the septum. Microhair total length/width at septum 7.6–10.7. Microhair apical cells 24–25.5(–30) microns long. Microhair apical cell/total length ratio 0.47–0.53. *Stomata* common; 22.5–27 microns long. Subsidiaries non-papillate; high dome-shaped and triangular. Guard-cells overlapped by the interstomatals. *Intercostal short-cells* common, or absent or very rare. *Costal short-cells* conspicuously in long rows, or predominantly paired, or neither distinctly grouped into long rows nor predominantly paired. Costal silica bodies saddle shaped, or tall-and-narrow, or oryzoid.

Transverse section of leaf blade, physiology. C_3; XyMS+. *Mesophyll* with non-radiate chlorenchyma; with adaxial palisade; with arm cells; with fusoids. The fusoids external to the PBS. *Leaf blade* 'nodular' in section to adaxially flat; with the ribs more or less constant in size (low), or with the ribs very irregular in sizes. *Midrib* conspicuous; having complex vascularization. Bulliforms present in discrete, regular adaxial groups; in simple fans (sometimes large-celled, *Zea*-type). All the vascular bundles accompanied by sclerenchyma. Combined sclerenchyma girders present; forming 'figures'.

Cytology. Chromosome base number, $x = 12$. $2n = 48$, 64, and 72 (rarely 70). 4, 5, and 6 ploid.

Taxonomy. Bambusoideae; Bambusodae; Bambuseae.

Ecology, geography, regional floristic distribution. About 35 species. China, Indomalayan region. Commonly adventive.

Paleotropical. Indomalesian. Indian, Indo-Chinese, and Malesian.

Rusts and smuts. Rusts — *Dasturella*. Taxonomically wide-ranging species: *Dasturella divina*.

Economic importance. Significant weed species: *D. giganteus*. *D. strictus*, with solid culms, is a very important timber species; others (e.g. *D.asper, D. latiflorus*) have edible shoots; and splits of several species are used for weaving baskets etc.

References, etc. Leaf anatomical: Metcalfe 1960; this project.

Dendrochloa C.E. Parkinson

~ *Schizostachyum*

Habit, vegetative morphology. Perennial. The flowering culms leafy. *Culms* 1500–2000 cm high; woody and persistent; to 11 cm in diameter; not scandent; branched above. Pluricaespitose. Rhizomes leptomorph. Plants unarmed. *Leaves* not basally aggregated; without auricular setae. Leaf blades linear-lanceolate; broad (long-acuminate); 50–90 mm wide; rolled; pseudopetiolate (the pseudopetiole about 1 cm long); without cross venation (these inconspicuous and remote); disarticulating from the sheaths. Ligule erect and conspicuous; not truncate (obtuse); 3–7 mm long.

Reproductive organization. Plants bisexual, with bisexual spikelets; with hermaphrodite florets.

Inflorescence. *Inflorescence* determinate; without pseudospikelets; paniculate (about 75 cm long, with verticils of branches); spatheate; not comprising 'partial inflorescences' and foliar organs. *Spikelet-bearing axes* 'racemes', or paniculate (?); persistent. Spikelets solitary.

Female-fertile spikelets. Spikelets about 40–60 mm long; compressed laterally; disarticulating above the glumes; disarticulating between the florets; with distinctly elongated rachilla internodes between the florets (the segments about 2 cm long). Rachilla prolonged beyond the uppermost female-fertile floret. The rachilla extension with incomplete florets. Hairy callus absent.

Glumes present; two; shorter than the spikelets; shorter than the adjacent lemmas (3–10 mm long); hairless; awnless (mucronate); non-carinate; similar (rigid). *Spikelets* with incomplete florets. The incomplete florets distal to the female-fertile florets. ***Spikelets without proximal incomplete florets***.

Female-fertile florets 5–7. Lemmas about 1 cm wide; similar in texture to the glumes; not becoming indurated; entire; pointed; mucronate to awned (aristo-mucronate). Awns 1; median; apical; non-geniculate; much shorter than the body of the lemma. Lemmas hairless (but ciliate at the tip); glabrous; non-carinate; about 13 nerved. *Palea* present; relatively long (equalling the lemma, or longer); apically notched; awnless, without apical setae; several nerved (6 nerved between the keels, 5 nerved between them and the margins); 2-keeled. Palea keels wingless. *Lodicules* present; 3; membranous; ciliate; not toothed (narrowly obovate-oblong, 1 –1.2 cm long); heavily vascularized. **Stamens 6; triadelphous (one free, and a pair and a triplet with connate filaments)**. Anthers 14 mm long; without an apically prolonged connective. **Ovary** glabrous; *with a conspicuous apical appendage*. The appendage long, stiff and tapering. Styles fused (into one). Stigmas 3 (the 'style' apex obscurely trifid).

Fruit, embryo and seedling. Fruit large (about 2 cm long); not noticeably compressed (ellipso-cylindric). Seed endospermic.

Taxonomy. Bambusoideae; Bambusodae; Bambuseae.

Ecology, geography, regional floristic distribution. 1 species. Burma. Paleotropical. Indomalesian. Indo-Chinese.

References, etc. Morphological/taxonomic: Parkinson 1933.

Special comments. Anatomical data wanting.

Deschampsia P. Beauv.

Airidium Steud., *Aristavena* Albers & Butzin, *Avenella* Parl., *Campella* Link, *Czerniaevia* Ledeb., *Erioblastus* Honda, *Homoiachne* Pilger, *Lerchenfeldia* Schur, *Podinapus* Dulac

Excluding *Periballia*, *Vahlodea*

Habit, vegetative morphology. Annual, or perennial; rhizomatous, or stoloniferous, or caespitose, or decumbent (but usually caespitose). **Culms** 8–200 cm high; *herbaceous*; unbranched above. Culm nodes glabrous. Culm internodes hollow. Young shoots intravaginal. *Leaves* mostly basal; non-auriculate. *Sheath margins free*. Leaf blades linear; narrow; 0.3–6 mm wide; setaceous, or not setaceous; flat, or folded, or rolled (convolute); without cross venation; persistent; rolled in bud, or once-folded in bud. **Ligule** present; *an unfringed membrane*; not truncate; 3–15 mm long.

Reproductive organization. Plants bisexual, with bisexual spikelets; with hermaphrodite florets; outbreeding and inbreeding. Exposed-cleistogamous, or chasmogamous. Viviparous (sometimes), or not viviparous.

Inflorescence. *Inflorescence paniculate*; usually *open*; with capillary branchlets, or without capillary branchlets; espatheate; not comprising 'partial inflorescences' and foliar organs. Spikelet-bearing axes persistent. Spikelets not secund; pedicellate.

Female-fertile spikelets. *Spikelets* 3–9 mm long (small, delicate); *compressed laterally*; disarticulating above the glumes; *with distinctly elongated rachilla internodes between the florets (the lower floret sessile)*. Rachilla prolonged beyond the uppermost female-fertile floret (the terminal extension well developed); rachilla hairy (except *D. chapmanii*). The rachilla extension with incomplete florets, or naked (well developed beyond the uppermost floret). *Hairy callus present. Callus* short; blunt.

Glumes two; very unequal (rarely), or more or less equal; shorter than the spikelets to exceeding the spikelets; *long relative to the adjacent lemmas*; pointed; awnless; carinate, or non-carinate; similar (subscarious to membranous, with thin margins). *Lower glume 1 nerved. Upper glume 3 nerved.* Spikelets with female-fertile florets only (usually), or with incomplete florets. The incomplete florets when present, distal to the female-fertile florets. The distal incomplete florets when present, 1. *Spikelets without proximal incomplete florets*.

Female-fertile florets (1–)2(–3). Lemmas similar in texture to the glumes to decidedly firmer than the glumes (hyaline to cartilaginous); not becoming indurated; entire, or incised (2-lobed, 4-toothed or truncate); not deeply cleft; awned (usually), or awnless to mucronate (*D. tenella* and New Zealand relatives, distinguished from *Poa* by their denticulate lemma tip). Awns 1; median; dorsal; from well down the back (mostly from the base or lower half, a few from above the middle); slender, weakly geniculate, or nongeniculate; hairless (glabrous to scabridulous); much shorter than the body of the lemma to much longer than the body of the lemma; entered by one vein. Lemmas hairy (rarely), or hairless; non-carinate (rounded on the back); 4–7 nerved. *Palea* present; relatively long; tightly clasped by the lemma; awnless, without apical setae; not indurated (membranous); 2-nerved; 2-keeled. *Lodicules* present; 2; joined, or free; membranous; glabrous; toothed, or not toothed; not or scarcely vascularized. *Stamens* 3. Anthers 0.3–2.5 mm long; not penicillate. *Ovary* glabrous. Styles free to their bases. Stigmas 2.

Fruit, embryo and seedling. **Fruit** free from both lemma and palea; small; *longitudinally grooved*; compressed dorsiventrally, or not noticeably compressed. *Hilum short*. Pericarp thin (shining). Embryo large (rarely), or small; not waisted. Endosperm hard; with lipid; containing compound starch grains. Embryo with an epiblast; without a scutellar tail; with a negligible mesocotyl internode. Embryonic leaf margins meeting.

Seedling with a long mesocotyl; with a loose coleoptile, or with a tight coleoptile. First seedling leaf with a well-developed lamina. The lamina narrow; 3 veined.

Abaxial leaf blade epidermis. *Costal/intercostal zonation* conspicuous. *Papillae* absent. *Long-cells* similar in shape costally and intercostally; of similar wall thickness costally and intercostally (fairly thick walled). Mid-intercostal long-cells rectangular; having markedly sinuous walls. *Microhairs* absent. *Stomata* common, or absent or very rare; 36–39 microns long. Subsidiaries parallel-sided. Guard-cells overlapping to flush with the interstomatals. *Intercostal short-cells* common; in cork/silica-cell pairs, or not paired; when paired silicified. *Costal short-cells* predominantly paired. Costal silica bodies horizontally-elongated crenate/sinuous, or horizontally-elongated smooth, or rounded; sharp-pointed (rarely), or not sharp-pointed.

Transverse section of leaf blade, physiology. C_3; XyMS+. *Mesophyll* with non-radiate chlorenchyma; without adaxial palisade; traversed by columns of colourless mesophyll cells, or not traversed by colourless columns. *Leaf blade* with distinct, prominent adaxial ribs; with the ribs more or less constant in size. *Midrib* not readily distinguishable; with one bundle only. Bulliforms present in discrete, regular adaxial groups, or not in discrete, regular adaxial groups; when clearly grouped, in simple fans, or associated with colourless mesophyll cells to form deeply-penetrating fans. All the vascular bundles accompanied by sclerenchyma. Combined sclerenchyma girders absent. Sclerenchyma not all bundle-associated. The 'extra' sclerenchyma in abaxial groups (but sometimes forming an almost continuous layer); abaxial-hypodermal, the groups isolated (opposite the bulliforms), or abaxial-hypodermal, the groups continuous with colourless columns.

Phytochemistry. Leaves without flavonoid sulphates (3 species).

Special diagnostic feature. *Lemmas without the characteristic* **Corynephorus** *awn*.

Cytology. Chromosome base number, $x = 7$ and 13. $2n = 14$, 24, 26, 28, 32, 42, 52, and 56 (and 26+B). 2, 4, 6, and 8 ploid (and aneuploids: some complexes with $n = 13$, 14). Haploid nuclear DNA content 3.1–4.5 pg (3 species). Mean diploid 2c DNA value 10 pg (*D. antarctica*).

Taxonomy. Pooideae; Poodae; Aveneae.

Ecology, geography, regional floristic distribution. 40 species. North and South temperate, high altitude tropics. Commonly adventive. Helophytic, or mesophytic; shade species and species of open habitats. Meadows, upland grasslands and woods.

Holarctic, Paleotropical, Neotropical, Australian, and Antarctic. Boreal, Tethyan, and Madrean. African, Indomalesian, and Polynesian. Arctic and Subarctic, Euro-Siberian, Eastern Asian, Atlantic North American, and Rocky Mountains. Macaronesian, Mediterranean, and Irano-Turanian. Sudano-Angolan and West African Rainforest. Indo-Chinese and Papuan. Hawaiian. Caribbean, Pampas, and Andean. North and East Australian. New Zealand, Patagonian, and Antarctic and Subantarctic. European and Siberian. Canadian-Appalachian and Central Grasslands. Somalo-Ethiopian. Temperate and South-Eastern Australian. Antarctic.

Rusts and smuts. Rusts — *Puccinia*. Taxonomically wide-ranging species: *Puccinia graminis*, *Puccinia coronata*, *Puccinia pygmaea*, *Puccinia praegracilis*, *Puccinia hordei*, *Puccinia recondita*, and '*Uromyces*' *fragilipes*. Smuts from Tilletiaceae and from Ustilaginaceae. Tilletiaceae — *Entyloma* and *Tilletia*. Ustilaginaceae — *Ustilago*.

Economic importance. Significant weed species: *D. caespitosa*, *D. flexuosa*. Important native pasture species: *D. flexuosa*.

References, etc. Morphological/taxonomic: Koch 1979. Leaf anatomical: Metcalfe 1960; this project.

Desmazeria Dumort.

Brizopyrum Link
Excluding *Catapodium*

Habit, vegetative morphology. *Annual*. Culms herbaceous. Culm internodes hollow. Leaves non-auriculate. Leaf blades linear; narrow; flat, or rolled (convolute when dry), or acicular; without cross venation. *Ligule an unfringed membrane*.

Reproductive organization. Plants bisexual, with bisexual spikelets; with hermaphrodite florets; without hidden cleistogenes.

Inflorescence. *Inflorescence a single raceme, or paniculate (one-sided, with few, stiff branches)*; espatheate; not comprising 'partial inflorescences' and foliar organs. Spikelet-bearing axes persistent. Spikelets not secund; pedicellate (the pedicels stout).

Female-fertile spikelets. Spikelets 4–22 mm long; compressed laterally; disarticulating above the glumes; disarticulating between the florets. *Rachilla* prolonged beyond the uppermost female-fertile floret; *rachilla hairy (with capitate hairs)*. The rachilla extension with incomplete florets.

Glumes two; very unequal to more or less equal (the upper longer); (the longer) shorter than the adjacent lemmas, or long relative to the adjacent lemmas; dorsiventral to the rachis; hairless; awnless; carinate; similar (leathery). *Lower glume 3 nerved*. Upper glume 3–5 nerved. *Spikelets* with incomplete florets. The incomplete florets distal to the female-fertile florets.

Female-fertile florets 3–20. *Lemmas similar in texture to the glumes*; not becoming indurated; entire, or incised; when entire pointed, or blunt; *awnless*; *hairy (below, with capitate hairs)*; *carinate*; 5(–7) nerved (the intermediates sometimes faint). *Palea* present; relatively long; 2-nerved; 2-keeled. *Lodicules* present; 2; free; membranous; glabrous; toothed, or not toothed; not or scarcely vascularized. *Stamens* 3. Anthers 0.8–1.4 mm long; not penicillate. **Ovary** glabrous; *without a conspicuous apical appendage*. Styles free to their bases. Stigmas 2; white.

Fruit, embryo and seedling. Fruit small (1.8–2.0 mm long); compressed dorsiventrally. Hilum short. Embryo small. Endosperm hard; with lipid. Embryo with an epiblast.

Abaxial leaf blade epidermis. *Costal/intercostal zonation* conspicuous (the costal zones confined to a narrow median and a broader one at each margin of the blade). *Papillae* absent. *Long-cells* markedly different in shape costally and intercostally (the costals narrower and rectangular, with tessellated walls); differing markedly in wall thickness costally and intercostally (the costals thicker walled). Mid-intercostal long-cells fusiform; having straight or only gently undulating walls. *Microhairs* absent. *Stomata* absent or very rare (very few seen, confined to regions bordering the midrib); 37–51 microns long. Subsidiaries low dome-shaped, or parallel-sided. Guard-cells overlapped by the interstomatals. *Intercostal short-cells* absent or very rare. *Costal short-cells* predominantly paired (and solitary). Costal silica bodies poorly developed (in the material seen); tall-and-narrow (mostly, poorly developed and ill-defined).

Transverse section of leaf blade, physiology. C_3; XyMS+. Mesophyll without adaxial palisade. *Leaf blade* with distinct, prominent adaxial ribs; with the ribs more or less constant in size. *Midrib* not readily distinguishable; with one bundle only. *The lamina* symmetrical on either side of the midrib. Bulliforms probably present in discrete, regular adaxial groups (at the bases of the furrows — unclear in the material seen); in simple fans. All the vascular bundles accompanied by sclerenchyma (but the sclerenchyma very scanty). Combined sclerenchyma girders absent (small, inconspicuous adaxial strands only). Sclerenchyma all associated with vascular bundles.

Cytology. Chromosome base number, $x = 7$. $2n = 14$. 2 ploid.

Taxonomy. Pooideae; Poodae; Poeae.

Ecology, geography, regional floristic distribution. 5 species. Mediterranean, Europe. Commonly adventive. Xerophytic; species of open habitats. In sandy places.

Holarctic and Paleotropical. Boreal and Tethyan. African. Euro-Siberian. Macaronesian and Mediterranean. Saharo-Sindian. European.

References, etc. Morphological/taxonomic: Stace 1981, 1985; Brullo and Pavone 1985. Leaf anatomical: this project.

Special comments. No attempt here to account for *Flora Europaea* re-alignment (1980), involving *Catapodium* and *Scleropoa*.

Desmostachya (Hook. f.) Stapf

Stapfiola Kuntze

Habit, vegetative morphology. Perennial; rhizomatous and caespitose. Culms herbaceous. Culm nodes glabrous. Leaves non-auriculate. Leaf blades narrow; without abaxial multicellular glands; without cross venation; persistent. Ligule a fringed membrane to a fringe of hairs.

Reproductive organization. Plants bisexual, with bisexual spikelets; with hermaphrodite florets.

Inflorescence. *Inflorescence of spicate main branches (narrow, dense, of many racemes, on an elongate axis)*; contracted; spicate; espatheate; not comprising 'partial inflorescences' and foliar organs. Spikelet-bearing axes persistent. *Spikelets* secund; biseriate; *sessile*; densely imbricate.

Female-fertile spikelets. *Spikelets* 4–6 mm long; adaxial; *compressed laterally*; *falling with the glumes*; not disarticulating between the florets. Rachilla prolonged beyond the uppermost female-fertile floret; rachilla hairless. The rachilla extension with incomplete florets. Hairy callus absent.

Glumes two; very unequal; shorter than the spikelets; shorter than the adjacent lemmas; free; dorsiventral to the rachis; pointed; awnless; carinate; similar (membranous, oval). Lower glume 1 nerved. Upper glume 1 nerved. *Spikelets* with incomplete florets. The incomplete florets distal to the female-fertile florets.

Female-fertile florets 6–16. Lemmas deltoid; decidedly firmer than the glumes (papery to leathery); not becoming indurated; entire; pointed; awnless; hairless; glabrous; carinate; 3 nerved. *Palea* present; relatively long; apically notched; 2-nerved. *Lodicules* present; 2; free; fleshy; glabrous. *Stamens* 3. Anthers not penicillate. *Ovary* glabrous. Stigmas 2.

Fruit, embryo and seedling. *Fruit* free from both lemma and palea; ellipsoid; compressed dorsiventrally, or not noticeably compressed. Hilum short. Pericarp fused. Embryo large; not waisted.

Abaxial leaf blade epidermis. *Costal/intercostal zonation* conspicuous. *Papillae* absent. *Long-cells* markedly different in shape costally and intercostally (the costals much longer); of similar wall thickness costally and intercostally (fairly thick walled). Intercostal zones with typical long-cells (but the interstomatals very regularly cross shaped). Mid-intercostal long-cells rectangular; having markedly sinuous walls (pitted). *Microhairs* present; elongated; clearly two-celled; panicoid-type to chloridoid-type (basal cells fairly long, apical cells rather thin walled, variable, ranging from unexceptional chloridoid type to broad panicoid type — cf. *Eragrostis*?). Microhair apical cell wall thinner than that of the basal cell but not tending to collapse. Microhairs (27–)30–33(–42) microns long. Microhair basal cells 24 microns long. Microhairs 9–12 microns wide at the septum. Microhair total length/width at septum 2.5–3. Microhair apical cells 9–16 microns long. Microhair apical cell/total length ratio 0.35–0.4. *Stomata* common; 15–18 microns long. Subsidiaries triangular. *Intercostal short-cells* common; in cork/silica-cell pairs and not paired (i.e. in ones and twos); silicified. Intercostal silica bodies imperfectly developed; slightly saddle shaped, or tall-and-narrow, or crescentic. *Costal short-cells* conspicuously in long rows. Costal silica bodies present in alternate cell files of the costal zones; saddle shaped.

Transverse section of leaf blade, physiology. C_4; XyMS+. PCR sheath outlines even (despite conspicuous adaxial extension cells). PCR sheaths of the primary vascular bundles interrupted; interrupted both abaxially and adaxially and interrupted laterally. PCR sheath extensions present. Maximum number of extension cells not conventionally determinable, the PCR sheaths being interrupted by large colourless cells and sclerenchyma. PCR cell chloroplasts centrifugal/peripheral (seemingly, in *bipinnata*). *Mesophyll* with radiate chlorenchyma; traversed by columns of colourless mesophyll cells. *Leaf blade* with distinct, prominent adaxial ribs; with the ribs more or less constant in size. *Midrib* conspicuous; having a conventional arc of bundles; with colourless mesophyll adaxially. Bulliforms present in discrete, regular adaxial groups; associated with colourless mesophyll cells to form deeply-penetrating fans (these incorporated in the traversing colourless girders). All the vascular bundles accompanied by sclerenchyma. Combined sclerenchyma girders present (with all the bundles); forming 'figures' (all bundles with I's). Sclerenchyma not all bundle-associated. The 'extra' sclerenchyma in abaxial groups and in adaxial groups (comprising small groups at the bases of the colourless columns, and a more or less continuous adaxial layer). The lamina margins with fibres.

Cytology. Chromosome base number, $x = 10$. $2n = 20$.

Taxonomy. Chloridoideae; main chloridoid assemblage.

Ecology, geography, regional floristic distribution. 1 species. Northeast tropical Africa to India. Species of open habitats. Arid regions.

Holarctic and Paleotropical. Tethyan. African and Indomalesian. Irano-Turanian. Saharo-Sindian and Sudano-Angolan. Indian. Somalo-Ethiopian.

Rusts and smuts. Rusts — *Puccinia*. Taxonomically wide-ranging species: *Puccinia striiformis* and '*Uromyces*' *eragrostidis*.

Economic importance. Significant weed species: *D. bipinnata*.

References, etc. Leaf anatomical: Metcalfe 1960; this project.

Deyeuxia Clarion ex P. Beauv.

Sclerodeyeuxia (Stapf) Pilger
~ *Calamagrostis*. See also *Agrostis, Dichelachne*
Excluding *D. uncinioides* = *Ancistragrostis, Aniselytron, Stilpnophleum*

Habit, vegetative morphology. Perennial; caespitose. **Culms herbaceous**; unbranched above. Culm nodes glabrous. Culm internodes hollow. *Leaves* mostly basal, or not basally aggregated; non-auriculate. Leaf blades narrow; without cross venation; persistent. *Ligule an unfringed membrane*; truncate; 0.5–2 mm long.

Reproductive organization. Plants bisexual, with bisexual spikelets; with hermaphrodite florets; inbreeding.

Inflorescence. *Inflorescence paniculate*; open, or contracted; when contracted, spicate to more or less irregular; espatheate; not comprising 'partial inflorescences' and foliar organs. Spikelet-bearing axes persistent. Spikelets not secund; pedicellate.

Female-fertile spikelets. Spikelets 1–8 mm long; compressed laterally; disarticulating above the glumes. Rachilla prolonged beyond the uppermost female-fertile floret, or terminated by a female-fertile floret (rarely); rachilla hairy, or hairless. The rachilla extension when present, naked. Hairy callus present (the hairs sometimes 0.5 mm or more long, but shorter than the lemma), or absent.

Glumes two; more or less equal; about equalling the spikelets to exceeding the spikelets; *long relative to the adjacent lemmas (but the lemma at least 3/4 as long, by contrast with* Calamagrostis*)*; pointed, or not pointed; awnless; carinate; similar. Lower glume 1 nerved. *Upper glume 1 nerved. Spikelets* with female-fertile florets only.

Female-fertile florets *1. Lemmas decidedly firmer than the glumes (the main, equivocal distinction from* Agrostis*)*; not becoming indurated; *incised*; usually minutely 2–4 lobed; not deeply cleft (toothed); nearly always awned (rarely only mucronate). *Awns* when present, 1; *dorsal*; from near the top, or from well down the back; non-geniculate, or geniculate; *much shorter than the body of the lemma to about as long as the body of the lemma*; entered by one vein; deciduous, or persistent. Lemmas hairless; non-carinate; 4–5 nerved. *Palea* present; relatively long, or conspicuous but relatively short, or very reduced; 2-nerved; 2-keeled. *Lodicules* present; 2; free; membranous; glabrous; not toothed. *Stamens* 3. Anthers not penicillate. *Ovary* glabrous. Styles free to their bases. Stigmas 2; white.

Fruit, embryo and seedling. **Fruit** free from both lemma and palea; small; *longitudinally grooved*; with hairs confined to a terminal tuft. Hilum short. Embryo small.

Abaxial leaf blade epidermis. *Costal/intercostal zonation* conspicuous. *Papillae* absent. *Long-cells* similar in shape costally and intercostally (elongated); differing markedly in wall thickness costally and intercostally (the costals thicker walled). Mid-intercostal long-cells rectangular to fusiform; having straight or only gently undulating walls. *Microhairs* absent. *Stomata* common. Subsidiaries low dome-shaped, or parallel-sided. Guard-cells overlapped by the interstomatals. *Intercostal short-cells* absent or very rare; not paired; not silicified. *Costal short-cells* conspicuously in long rows, or neither distinctly grouped into long rows nor predominantly paired. Costal silica bodies horizontally-elongated crenate/sinuous, or horizontally-elongated smooth, or rounded.

Transverse section of leaf blade, physiology. C_3; XyMS+. *Mesophyll* with non-radiate chlorenchyma; without adaxial palisade. *Leaf blade* with distinct, prominent adaxial ribs; with the ribs more or less constant in size. *Midrib* not readily distinguishable; with one bundle only. Bulliforms not in discrete, regular adaxial groups (irregularly grouped in the furrows, of small cells — cf. *Ammophila*). All the vascular bundles accompanied by sclerenchyma. Combined sclerenchyma girders present; nowhere forming 'figures'. Sclerenchyma all associated with vascular bundles.

Cytology. Chromosome base number, $x = 7$. $2n = 28$.

Taxonomy. Pooideae; Poodae; Aveneae.

Ecology, geography, regional floristic distribution. 20 species. Temperate. Holarctic, Paleotropical, Neotropical, Australian, and Antarctic. Tethyan. Indomalesian. Irano-Turanian. Indian, Indo-Chinese, Malesian, and Papuan. Pampas. North and East Australian and South-West Australian. New Zealand and Patagonian. Temperate and South-Eastern Australian.

Rusts and smuts. Rusts — *Puccinia*. Taxonomically wide-ranging species: *Puccinia graminis*, *Puccinia hordei*, and *Puccinia recondita*. Smuts from Ustilaginaceae. Ustilaginaceae — *Ustilago*.

References, etc. Morphological/taxonomic: Vickery 1940. Leaf anatomical: Metcalfe 1960; this project.

Diandrochloa de Winter

~ Eragrostis

Habit, vegetative morphology. Annual, or perennial; caespitose. *Culms* 10–150 cm high; herbaceous (soft, geniculate or erect); branched above, or unbranched above. Culm internodes hollow. Plants unarmed. Leaves non-auriculate. Leaf blades linear to linear-lanceolate; narrow; sometimes tapered to a setaceous point; flat; without abaxial multicellular glands; without cross venation. *Ligule an unfringed membrane*.

Reproductive organization. Plants bisexual, with bisexual spikelets; with hermaphrodite florets. Exposed-cleistogamous, or chasmogamous (?).

Inflorescence. *Inflorescence paniculate*; open, or contracted; usually rigid, much longer than broad, contracted and dense or much branched; with conspicuously divaricate branchlets (when much-branched), or without conspicuously divaricate branchlets; nondigitate (the branches in pseudo-whorls on a central axis); espatheate; not comprising 'partial inflorescences' and foliar organs. Spikelet-bearing axes persistent. Spikelets solitary; not secund; pedicellate.

Female-fertile spikelets. *Spikelets 1–3.5 mm long*; compressed laterally; disarticulating above the glumes; disarticulating between the florets. Rachilla prolonged beyond the uppermost female-fertile floret; rachilla hairless (glabrous or scabrid). Hairy callus absent. *Callus* short; blunt.

Glumes two; very unequal to more or less equal; shorter than the spikelets; shorter than the adjacent lemmas to long relative to the adjacent lemmas; pointed (acute), or not pointed (apically rounded); awnless; carinate; similar (membranous or sub-hyaline, ovate to lanceolate, often green). Lower glume 1 nerved. Upper glume 1 nerved. *Spikelets* with female-fertile florets only, or with incomplete florets. The incomplete florets when present, distal to the female-fertile florets. *The distal incomplete florets* merely underdeveloped.

Female-fertile florets 2–14. Lemmas acute to obtuse; similar in texture to the glumes to decidedly firmer than the glumes (translucent or thinly leathery); not becoming indurated; entire, or incised; pointed, or blunt; not deeply cleft (sometimes emarginate or somewhat erose); awnless; hairless (the nerves sometimes scabrid); often with raised nerves; 3 nerved. *Palea* present; relatively long; entire (apically rounded or truncate), or apically notched (3-lobed); awnless, without apical setae; thinner than the lemma to textured like the lemma; not indurated (membranous); 2-nerved; 2-keeled. Palea keels glabrous, or scabrous. *Lodicules* present; 2; free; fleshy; glabrous. **Stamens 2**. Anthers 0.2–1 mm long; not penicillate; without an apically prolonged connective. *Ovary* glabrous. Styles free to their bases. Stigmas 2.

Fruit, embryo and seedling. *Fruit* free from both lemma and palea. Hilum short. Pericarp fused. Embryo large. Endosperm containing only simple starch grains.

Abaxial leaf blade epidermis. *Costal/intercostal zonation* conspicuous. *Papillae* absent. *Long-cells* markedly different in shape costally and intercostally (the intercostals much broader); of similar wall thickness costally and intercostally. Mid-intercostal long-cells rectangular; having markedly sinuous walls. *Microhairs* present; elongated; clearly two-celled; panicoid-type, or panicoid-type to chloridoid-type (i.e. paicoid type, or intermediate). Microhair apical cell wall of similar thickness/rigidity to that of the basal cell to thinner than that of the basal cell and often collapsed. Microhairs in *D. confertiflora*, 37.5–45 microns long. Microhair basal cells 24–27 microns long. Microhairs 8.4–9 microns wide at the septum. Microhair total length/width at septum 4.2–5. Microhair apical cells 36–37.5 microns long. Microhair apical cell/total length ratio 0.45–0.52. *Stomata* common; 21–24 microns long. Subsidiaries dome-shaped and triangular. Guard-cells overlapping to flush with the interstomatals. *Intercostal short-cells* common; not paired (solitary); not silicified. Intercostal silica bodies absent. *Costal short-cells* conspicuously in long rows. Costal silica bodies present in alternate cell files of the costal zones; horizontally-elongated smooth to rounded, or saddle shaped (a few), or 'panicoid-type'.

Transverse section of leaf blade, physiology. C_4; XyMS+. PCR sheath outlines uneven to even. PCR sheaths of the primary vascular bundles interrupted; interrupted abaxially only. PCR sheath extensions absent. PCR cell chloroplasts centrifugal/peripheral.

Mesophyll with radiate chlorenchyma. *Leaf blade* 'nodular' in section to adaxially flat. *Midrib* not readily distinguishable; with one bundle only. Bulliforms present in discrete, regular adaxial groups; in simple fans (each group with a large, deep-penetrating median cell). All the vascular bundles accompanied by sclerenchyma. Combined sclerenchyma girders present (with the primaries); forming 'figures' (in the primaries). Sclerenchyma all associated with vascular bundles. The lamina margins with fibres.

Cytology. Chromosome base number, $x = 10$.

Taxonomy. Chloridoideae; main chloridoid assemblage.

Ecology, geography, regional floristic distribution. 7 species. Americas, Australia, Asia, Africa. Helophytic; shade species, or species of open habitats; glycophytic.

Holarctic, Paleotropical, Neotropical, Cape, and Australian. Boreal and Tethyan. African and Indomalesian. Atlantic North American and Rocky Mountains. Irano-Turanian. Sudano-Angolan and Namib-Karoo. Indian and Indo-Chinese. Amazon, Central Brazilian, and Andean. Central Australian.

References, etc. Morphological/taxonomic: de Winter 1960; Koch (1978) argues for reduction to sectional level, on the grounds that only one character (the membranous, unfringed ligule) distinguishes *Diandrochloa* absolutely from *Eragrostis*. Leaf anatomical: this project.

Diandrolyra Stapf

Habit, vegetative morphology. Perennial; caespitose. The flowering culms leafy (but tending to carry the racemes on specialised, 1-leafed culms). *Culms* about 10–50 cm high; herbaceous; unbranched above. Culm nodes glabrous. Culm internodes solid. Plants unarmed. Young shoots extravaginal, or intravaginal. *Leaves* not basally aggregated; non-auriculate. Leaf blades broad; pseudopetiolate; cross veined; persistent; rolled in bud. *Ligule a fringed membrane*; 0.5–1.2 mm long.

Reproductive organization. *Plants monoecious with all the fertile spikelets unisexual*; without hermaphrodite florets. The spikelets of sexually distinct forms on the same plant; female-only and male-only. *The male and female-fertile spikelets mixed in the inflorescence (paired, male and female)*. The spikelets overtly heteromorphic. Not viviparous.

Inflorescence. *Inflorescence a short raceme, terminal on a normal leafy or a special 1-leafed culm, borne horizontally beneath the blade of the bent subtending leaf*. Spikelet-bearing axes persistent. 'Articles' not appendaged; glabrous. Spikelets paired (male and female); not secund; pedicellate; not in distinct 'long-and-short' combinations.

Female-sterile spikelets. Male spikelets with glumes reduced or lacking, having 3 stamens plus 3 staminodes and a sterile ovary with three stigmas. The male spikelets without glumes (or these vestigial); without proximal incomplete florets; 1 floreted. The lemmas awnless. Male florets 3 staminate, or 6 staminate (including the 3 staminodes). The staminal filaments free.

Female-fertile spikelets. *Spikelets* 9–11 mm long; *not noticeably compressed*; falling with the glumes. Rachilla terminated by a female-fertile floret.

Glumes two; more or less equal; long relative to the adjacent lemmas; hairless; awnless; non-carinate; similar (herbaceous, cuspidate). Lower glume 5–6 nerved. Upper glume 5–7 nerved. *Spikelets* with female-fertile florets only.

Female-fertile florets 1. **Lemmas similar in texture to the glumes (leathery)**; not becoming indurated; entire; blunt; awnless; hairless; glabrous; non-carinate; without a germination flap; 5–8 nerved. *Palea* present; relatively long; entire; awnless, without apical setae; not indurated; 2-nerved to several nerved; keel-less. *Lodicules* present; 3; free; fleshy; glabrous; toothed; not or scarcely vascularized. *Stamens* 0 (three staminodes). Styles fused (style base hairy). Stigmas 2.

Fruit, embryo and seedling. *Fruit* free from both lemma and palea; medium sized; compressed dorsiventrally. Hilum long-linear. Embryo small; not waisted. Endosperm hard; containing compound starch grains.

Seedling with a short mesocotyl; with a loose coleoptile. First seedling leaf without a lamina (two bladeless sheaths).

Abaxial leaf blade epidermis. *Costal/intercostal zonation* conspicuous. *Papillae* present. Intercostal papillae over-arching the stomata; several per cell. *Long-cells* markedly different in shape costally and intercostally; of similar wall thickness costally and intercostally. Mid-intercostal long-cells rectangular; having markedly sinuous walls. *Microhairs* present; panicoid-type. *Stomata* common. Subsidiaries triangular. Guard-cells overlapping to flush with the interstomatals. *Intercostal short-cells* common; in cork/silica-cell pairs; silicified. Intercostal silica bodies vertically elongated-nodular (?). *Costal short-cells* conspicuously in long rows. Costal silica bodies oryzoid.

Transverse section of leaf blade, physiology. C_3; XyMS+. *Mesophyll* with non-radiate chlorenchyma; with arm cells; with fusoids. *Leaf blade* adaxially flat. Midrib having a conventional arc of bundles. Bulliforms present in discrete, regular adaxial groups; in simple fans (the groups wide). All the vascular bundles accompanied by sclerenchyma. Combined sclerenchyma girders present; forming 'figures'. Sclerenchyma all associated with vascular bundles.

Cytology. Chromosome base number, $x = 11$.

Taxonomy. Bambusoideae; Oryzodae; Olyreae.

Ecology, geography, regional floristic distribution. 1 species. Panama. Shade species. Forest.

Neotropical. Central Brazilian.

References, etc. Leaf anatomical: this project.

Diandrostachya Jacq.-Fél.

~ Loudetiopsis

Habit, vegetative morphology. Annual, or perennial (with slender culms); caespitose. Culms herbaceous; unbranched above. Culm nodes glabrous. Culm internodes hollow. **Leaves** mostly basal; *distichous*; non-auriculate. Leaf blades linear; narrow; setaceous (at the tip), or not setaceous; often rolled (involute); without cross venation; persistent. *Ligule a fringe of hairs. Contra-ligule* absent.

Reproductive organization. Plants bisexual, with bisexual spikelets; with hermaphrodite florets.

Inflorescence. *Inflorescence* paniculate; *open*; espatheate; not comprising 'partial inflorescences' and foliar organs. Spikelet-bearing axes persistent. *Spikelets* in triplets (the triads terminating the panicle branches); not secund; *pedicellate (the pedicels free, short)*; not in distinct 'long-and-short' combinations (?).

Female-fertile spikelets. *Spikelets of the triad having their upper glumes backing onto one another*; compressed laterally (?); *disarticulating above the glumes*; with conventional internode spacings. Rachilla terminated by a female-fertile floret. Hairy callus present. *Callus* short; blunt (or emarginate, or bidentate).

Glumes two; very unequal (G_1 shorter, but usually more than half as long as the spikelet); (the upper) long relative to the adjacent lemmas; hairy (often apically penicillate); pointed (lanceolate to linear-lanceolate or acuminate); awnless; non-carinate (dorsally rounded); very dissimilar (G_2 much longer, more pointed and less hairy), or similar. Lower glume 3 nerved. Upper glume 3 nerved. *Spikelets* with incomplete florets. The incomplete florets proximal to the female-fertile florets. *Spikelets with proximal incomplete florets. The proximal incomplete florets* 1; paleate. Palea of the proximal incomplete florets fully developed (narrow, with two ciliate keels). The proximal incomplete florets male. The proximal lemmas awnless; 3 nerved; decidedly exceeding the female-fertile lemmas; less firm than the female-fertile lemmas (herbaceous to papery); not becoming indurated.

Female-fertile florets 1. Lemmas less firm than the glumes to decidedly firmer than the glumes; incised (at the summit, the lobes pointed); 2 lobed; deeply cleft (to more than 1/4 of their length), or not deeply cleft; awned. Awns 1; median; from a sinus; geniculate; hairy; much longer than the body of the lemma. *Lemmas* hairless (always glabrous?); non-carinate; having the margins tucked in onto the palea (the palea enclosed save at its summit); *with a clear germination flap*; 7 nerved. *Palea* present; apically notched (linear); awnless, without apical setae; 2-nerved; 2-keeled (channelled between the keels).

Palea keels wingless. *Lodicules* present; 2; free; fleshy; glabrous. **Stamens** 2. Anthers not penicillate; without an apically prolonged connective. *Ovary* glabrous. Styles free to their bases. Stigmas 2.

Fruit, embryo and seedling. *Fruit* free from both lemma and palea (but enclosed); longitudinally grooved. Hilum long-linear. Embryo large.

Abaxial leaf blade epidermis. *Costal/intercostal zonation* conspicuous. *Papillae* absent. *Long-cells* similar in shape costally and intercostally; of similar wall thickness costally and intercostally (rather thick walled). Mid-intercostal long-cells rectangular; having markedly sinuous walls (these heavily pitted). *Microhairs* present; panicoid-type. *Stomata* common. Subsidiaries consistently low dome-shaped. Guard-cells overlapping to flush with the interstomatals. *Intercostal short-cells* common (very regularly placed between all the long-cells); in cork/silica-cell pairs, or not paired (apparently solitary except when paired with prickle bases). With abundant small, short prickles. *Crown cells* absent. *Costal short-cells* conspicuously in long rows. Costal silica bodies 'panicoid-type'; predominantly elongated dumb-bell shaped.

Transverse section of leaf blade, physiology. C_4; XyMS−. PCR sheath outlines even. PCR cell chloroplasts centrifugal/peripheral. *Mesophyll* with radiate chlorenchyma; irregularly traversed by columns of colourless mesophyll cells. *Leaf blade* with distinct, prominent adaxial ribs; with the ribs very irregular in sizes. *Midrib* not readily distinguishable; with one bundle only. Bulliforms present in discrete, regular adaxial groups (in all the furrows); associated with colourless mesophyll cells to form deeply-penetrating fans; associating with colourless mesophyll cells to form arches over small vascular bundles (many of the bulliform groups associated with deeply penetrating or fully traversing lateral arms of colourless cells). All the vascular bundles accompanied by sclerenchyma. Combined sclerenchyma girders present; forming 'figures' (all the larger bundles with massive I's or T's). Sclerenchyma all associated with vascular bundles.

Special diagnostic feature. The lower glume exceeding the female-fertile lemma.

Taxonomy. Panicoideae; Panicodae; Arundinelleae.

Ecology, geography, regional floristic distribution. 5 species. Tropical Africa, South America.

Paleotropical and Neotropical. African. Sudano-Angolan and West African Rainforest. Amazon and Central Brazilian.

References, etc. Leaf anatomical: this project.

Diarrhena P. Beauv.

Diarina Raf., *Korycarpus* Lag., *Neomolinia* Honda, *Onoea* Franch. & Sav., *Roemeria* Roem. & Schult.

Habit, vegetative morphology. Large perennial; rhizomatous. *Culms* 50–150 cm high; herbaceous. Culm nodes nodes concealed. Leaves non-auriculate. Leaf blades broad; 1–2 mm wide; pseudopetiolate to not pseudopetiolate; without cross venation; persistent; rolled in bud. Ligule an unfringed membrane (but sometimes minutely ciliolate); 0.5–1 mm long.

Reproductive organization. Plants bisexual, with bisexual spikelets; with hermaphrodite florets. The spikelets all alike in sexuality.

Inflorescence. Inflorescence paniculate; open (diffuse); espatheate; not comprising 'partial inflorescences' and foliar organs. Spikelet-bearing axes persistent. Spikelets not secund; pedicellate.

Female-fertile spikelets. Spikelets 2.8–18 mm long (2.8–7 mm in *Neomolinia*, 10–18 mm in *Diarrhena* s.str.); compressed laterally; disarticulating above the glumes; disarticulating between the florets. Rachilla prolonged beyond the uppermost female-fertile floret; rachilla spinulose or smooth. Hairy callus absent.

Glumes two; very unequal; shorter than the spikelets; shorter than the adjacent lemmas; pointed, or not pointed (obtuse to acuminate); awnless; carinate. Lower glume 1–3 nerved. Upper glume 3–5 nerved. *Spikelets* with female-fertile florets only, or with incomplete florets. The incomplete florets distal to the female-fertile florets.

Female-fertile florets 2–3 (rarely 1 or 5). Lemmas decidedly firmer than the glumes; becoming indurated (horny); entire; blunt; awnless to mucronate; hairless (scabrous or glabrous); non-carinate (rounded on the back); 3–5 nerved. *Palea* present; relatively long; textured like the lemma (horny); indurated; 2-nerved; 2-keeled. *Lodicules* present; 2; free; membranous; apically ciliate; toothed, or not toothed; not or scarcely vascularized. *Stamens* (1–)2–3. Anthers 1.2–2.2 mm long; not penicillate. **Ovary** glabrous (though the rostellum is scabrous in *Neomolinia*); **with a conspicuous apical appendage (in the form of an apical beak, the styles attached terminally in its groove)**. Styles free to their bases. Stigmas 2.

Fruit, embryo and seedling. *Fruit* free from both lemma and palea; medium sized (about 5 mm long); compressed laterally, or not noticeably compressed (with a whitish or yellowish-glossy, sometimes enlarged and hardened beak); surface rough but not sculptured. Hilum short, or long-linear (short- or long-linear, from one third to as long as grain). Pericarp free. Embryo small (1/4 to 1/3 of grain length). Endosperm hard; containing compound starch grains. Embryo with an epiblast; with a scutellar tail; with a negligible mesocotyl internode. Embryonic leaf margins overlapping.

Seedling with a short mesocotyl. First seedling leaf with a well-developed lamina. The lamina narrow (broader towards the tip); erect; 7 veined.

Abaxial leaf blade epidermis. *Costal/intercostal zonation* conspicuous. *Papillae* absent. *Long-cells* markedly different in shape costally and intercostally (the costals relatively longer and narrower). Mid-intercostal long-cells rectangular to fusiform; having markedly sinuous walls. *Microhairs* present (*D. mandshurica* — see Renvoize 1985), or absent; as figured, more or less spherical (cf. *Hygroryza, Luziola* etc.?); ostensibly one-celled; obscure and hard to interpret. *Stomata* common. Subsidiaries low dome-shaped, or parallel-sided. Guard-cells slightly overlapped by the interstomatals. *Intercostal short-cells* common; in cork/silica-cell pairs; silicified. Intercostal silica bodies rounded, crescentic, tall-and-narrow, and cubical, or vertically elongated-nodular. *Costal short-cells* conspicuously in long rows, or neither distinctly grouped into long rows nor predominantly paired. Costal silica bodies horizontally-elongated crenate/sinuous, or 'panicoid-type'; when panicoid type, cross shaped, or butterfly shaped, or dumb-bell shaped, or nodular.

Transverse section of leaf blade, physiology. C_3; XyMS+. *Mesophyll* with non-radiate chlorenchyma; with arm cells; without fusoids. *Leaf blade* adaxially flat. *Midrib* conspicuous (with a big keel and massive blocks of sclerenchyma, at least in *D. americana*); having a conventional arc of bundles (the median being very large). Bulliforms present in discrete, regular adaxial groups; in simple fans. All the vascular bundles accompanied by sclerenchyma. Combined sclerenchyma girders present; forming 'figures', or nowhere forming 'figures'. Sclerenchyma all associated with vascular bundles.

Special diagnostic feature. *Grain with a conspicuous whitish or yellowish, glossy beak*.

Cytology. Chromosome base number, $x = 10$ and 19. $2n = 38$ and 60. 2 and 6 ploid.

Taxonomy. Bambusoideae; Oryzodae; Diarrheneae.

Ecology, geography, regional floristic distribution. 4–5 species. Eastern Asia, North America. Shade species.

Holarctic. Boreal. Euro-Siberian, Eastern Asian, and Atlantic North American. Siberian. Canadian-Appalachian and Southern Atlantic North American.

Rusts and smuts. Rusts — *Puccinia*. Taxonomically wide-ranging species: *Puccinia graminis*.

References, etc. Morphological/taxonomic: Tateoka 1957b; Schwab 1972; Macfarlane and Watson 1980. Leaf anatomical: Metcalfe 1960; this project.

Dichaetaria Nees

Habit, vegetative morphology. Perennial; caespitose (on a creeping, woody stock). **Culms** 75–100 cm high; *herbaceous*; unbranched above. Culm nodes glabrous. Culm internodes solid. Plants unarmed. Young shoots intravaginal. *Leaves* mostly basal; non-auriculate. Leaf blades narrow; 4–7 mm wide; flat; without cross venation. **Ligule *a fringed***

membrane. *Contra-ligule* present (the adaxial fringe continuing around the back of the leaf).

Reproductive organization. *Plants* bisexual, with bisexual spikelets; *with hermaphrodite florets*. The spikelets all alike in sexuality.

Inflorescence. Inflorescence paniculate; open; without capillary branchlets (but the axes very slender); espatheate; not comprising 'partial inflorescences' and foliar organs. Spikelet-bearing axes persistent. Spikelets not secund; shortly pedicellate.

Female-fertile spikelets. Spikelets morphologically 'conventional' (until the G_1 abscises); about 12 mm long; compressed laterally to not noticeably compressed; *disarticulating above the glumes (G_1 caducous, G_2 persisting after the spikelet has fallen)*. Rachilla prolonged beyond the uppermost female-fertile floret. The rachilla extension with incomplete florets. Hairy callus present. *Callus* long; pointed.

Glumes two; very unequal; shorter than the spikelets; shorter than the adjacent lemmas (G_2 about 2/3 of the lemma length); free; hairless; pointed; awnless (but sometimes sub-aristate); non-carinate (rounded on the back); similar (narrowly lanceolate). Lower glume 3 nerved. Upper glume 3 nerved. *Spikelets* with incomplete florets. The incomplete florets distal to the female-fertile florets. *The distal incomplete florets 1; **incomplete florets** clearly specialised and modified in form (the short length of rachilla distal to L_1 terminating in an empty, long, awn-like, recurved organ, which is minutely 2-aristate above the middle, very similar in form to the fertile lemma, with no sign of associated glumes)*.

Female-fertile florets 1. Lemmas decidedly firmer than the glumes; not becoming indurated (but firm, leathery); incised; awned. Awns 1; median; from a sinus (the long awn arising between/behind 2 minute setae); non-geniculate; hairless; about as long as the body of the lemma (up to 10 mm long). Lemmas hairless; glabrous; non-carinate (rounded on the back); without a germination flap; 3 nerved. *Palea* present (long and narrow); relatively long (equalling the lemma); entire (pointed); with apical setae; thinner than the lemma (membranous); not indurated; 2-nerved; 2-keeled. *Lodicules* present; 2; free; membranous; glabrous; not toothed; heavily vascularized. *Stamens* 3. **Ovary** glabrous; *with a conspicuous apical appendage (the styles terminal on this — see fruit description)*. Styles free to their bases. Stigmas 2.

Fruit, embryo and seedling. Fruit free from both lemma and palea; *small to medium sized (about 9 mm long, with a peculiar, small, glabrous, fleshy beak, pointed above, terminating in the styles)*; linear; longitudinally grooved. Hilum long-linear. Pericarp thin; fused. Embryo small; slightly waisted. Endosperm hard; containing compound starch grains. Embryo without an epiblast; with a scutellar tail.

Abaxial leaf blade epidermis. *Costal/intercostal zonation* conspicuous. *Papillae* absent. *Long-cells* similar in shape costally and intercostally (but the intercostals larger); of similar wall thickness costally and intercostally. Mid-intercostal long-cells rectangular; having markedly sinuous walls (these pitted). *Microhairs* present; panicoid-type (fairly broad, but very thin walled). *Stomata* common (in single files). Subsidiaries mostly low dome-shaped (a few approaching parallel-sided). Guard-cells overlapping to flush with the interstomatals. *Intercostal short-cells* common; in cork/silica-cell pairs; silicified. *Costal short-cells* conspicuously in long rows, or predominantly paired, or neither distinctly grouped into long rows nor predominantly paired (varying from vein to vein). Costal silica bodies saddle shaped, oryzoid, and 'panicoid-type' (a most unusual mixture, with intermediate forms); when identifiably panicoid type, cross shaped, butterfly shaped, and dumb-bell shaped.

Transverse section of leaf blade, physiology. C_3; XyMS+. *Mesophyll* with non-radiate chlorenchyma. *Leaf blade* with distinct, prominent adaxial ribs; with the ribs more or less constant in size. *Midrib* not readily distinguishable; with one bundle only. Bulliforms present in discrete, regular adaxial groups; in simple fans. All the vascular bundles accompanied by sclerenchyma. Combined sclerenchyma girders present (with all the bundles); forming 'figures'. Sclerenchyma all associated with vascular bundles.

Taxonomy. Arundinoideae; Danthonieae.

Ecology, geography, regional floristic distribution. 1 species. Southern India and Ceylon. Mesophytic; shade species. Woodland.

Paleotropical. Indomalesian. Indian.

References, etc. Leaf anatomical: this project.

Dichanthelium (A. Hitchc. & Chase) Gould

~ *Panicum*

Habit, vegetative morphology. *Perennial (with a rosette of short winter leaves)*; from a crown or with short matted rhizomes. *Culms* 20–70(–150) cm high; herbaceous. Culm internodes solid, or hollow. Leaves non-auriculate. Leaf blades broad, or narrow; usually flat; rolled in bud. Ligule an unfringed membrane to a fringe of hairs (?).

Reproductive organization. Plants bisexual, with bisexual spikelets; with hermaphrodite florets. *The spikelets of sexually distinct forms on the same plant (the first-formed spikelets rarely perfecting seed save in the Lanuginosa group and D.* clandestinum; *secondary branches from the primary panicle bearing cleistogamous, fertile spikelets)*; hermaphrodite, or hermaphrodite and sterile. Exposed-cleistogamous, or chasmogamous. Plants with hidden cleistogenes (often), or without hidden cleistogenes. The hidden cleistogenes in the leaf sheaths. Apomictic, or reproducing sexually.

Inflorescence. Inflorescence paniculate (primarily a terminal open panicle, with chasmogamous or abortive spikelets, subsequently complicated by branching from the culm nodes and production of secondary inflorescences); open; with capillary branchlets (often flexuous), or without capillary branchlets; ultimately a complex of 'partial inflorescences' and intervening foliar organs (the secondary branches repeatedly branching, the short branchlets fascicled and with very reduced leaves). *Spikelet-bearing axes* very much reduced (the grain bearing, cleistogamous ones), or paniculate; the terminal branches of the primary culm often falling. Spikelets not secund; pedicellate.

Female-fertile spikelets. *Spikelets* cleistogamous; 1.4–4.3 mm long; compressed dorsiventrally; falling with the glumes; with conventional internode spacings, or with distinctly elongated rachilla internodes between the florets. The upper floret conspicuously stipitate, or the upper floret not stipitate. The stipe beneath the upper floret when present, not filiform; straight and swollen; homogeneous. Rachilla terminated by a female-fertile floret. Hairy callus absent.

Glumes two; very unequal (the lower often minute); (the upper) long relative to the adjacent lemmas; hairless; awnless; non-carinate. Lower glume 0–7 nerved. Upper glume 3 nerved, or 7–9 nerved. *Spikelets* with incomplete florets. The incomplete florets proximal to the female-fertile florets. *The proximal incomplete florets* 1; male, or sterile. The proximal lemmas awnless; 3 nerved, or 5–9 nerved; less firm than the female-fertile lemmas; not becoming indurated.

Female-fertile florets 1. Lemmas similar in texture to the glumes to decidedly firmer than the glumes; smooth (shiny); becoming indurated (usually?); entire; blunt; awnless; hairless; non-carinate; having the margins tucked in onto the palea; with a clear germination flap; 3–11 nerved. *Palea* present; entire; awnless, without apical setae; textured like the lemma; indurated. *Lodicules* present; 2; free; fleshy. *Stamens* 3. *Ovary* glabrous. Styles free to their bases. Stigmas 2; red pigmented, or brown.

Fruit, embryo and seedling. *Fruit* free from both lemma and palea. Hilum short. Embryo large.

Abaxial leaf blade epidermis. *Costal/intercostal zonation* conspicuous. *Papillae* absent. Intercostal zones with typical long-cells, or exhibiting many atypical long-cells, or without typical long-cells. Mid-intercostal long-cells having markedly sinuous walls, or having straight or only gently undulating walls. *Microhairs* present; panicoid-type; (45–)51–81(–93) microns long; 4.5–10.5(–12) microns wide at the septum. Microhair total length/width at septum 4.25–17.3. Microhair apical cells 21–42(–48) microns long. Microhair apical cell/total length ratio 0.37–0.57. *Stomata* common; 30–48 microns long. Subsidiaries low dome-shaped to triangular. Guard-cells overlapping to flush with the interstomatals. *Intercostal short-cells* common, or absent or very rare; in cork/silica-cell pairs, or not paired. *Costal short-cells* conspicuously in long rows. Costal silica bodies 'panicoid-type' (sometimes sharp pointed); cross shaped to dumb-bell shaped; sharp-pointed, or not sharp-pointed.

Transverse section of leaf blade, physiology. C_3; XyMS+. *Mesophyll* with radiate chlorenchyma; *Isachne*-type, or not *Isachne*-type. *Leaf blade* with distinct, prominent adaxial ribs, or 'nodular' in section, or adaxially flat; with the ribs more or less constant in size. *Midrib* not readily distinguishable; with one bundle only. Bulliforms present in discrete, regular adaxial groups (at bases of furrows); in simple fans. All the vascular bundles accompanied by sclerenchyma. Combined sclerenchyma girders absent (there being colourless cells linking the bundles with adaxial strands). Sclerenchyma all associated with vascular bundles.

Phytochemistry. Leaves containing flavonoid sulphates (2 species).

Special diagnostic feature. *Plants from a short rosette of winter leaves, the primary panicle producing secondary inflorescences with cleistogamous spikelets*.

Cytology. Chromosome base number, $x = 9$. $2n = 18$ (mostly), or 36 (occasionally). Mostly 2 ploid.

Taxonomy. Panicoideae; Panicodae; Paniceae.

Ecology, geography, regional floristic distribution. 120 species. America. Holarctic and Paleotropical. Boreal and Madrean. Indomalesian and Polynesian. Atlantic North American. Malesian. Hawaiian. Southern Atlantic North American and Central Grasslands.

Economic importance. Significant weed species: *D. clandestinum*.

References, etc. Morphological/taxonomic: Gould and Clark 1978. Leaf anatomical: this project.

Dichanthium Willem.

Diplasanthum Desv., *Lepeocercis* Trin.

Excluding *Bothriochloa, Eremopogon*

Habit, vegetative morphology. Annual (rarely), or perennial; rhizomatous, or stoloniferous, or caespitose, or decumbent. *Culms* 15–200 cm high; herbaceous; branched above, or unbranched above. Culm nodes hairy, or glabrous. Culm internodes solid. The shoots aromatic, or not aromatic. *Leaves* not basally aggregated; non-auriculate. *Leaf blades* narrow; flat; without cross venation; *persistent*; rolled in bud. *Ligule* present; an unfringed membrane to a fringed membrane.

Reproductive organization. Plants bisexual, with bisexual spikelets; with hermaphrodite florets. *The spikelets* of sexually distinct forms on the same plant; hermaphrodite and male-only, or hermaphrodite and sterile; overtly heteromorphic (the pedicellate spikelets smaller, awnless), or homomorphic; *in both homogamous and heterogamous combinations (the lowest pair being imperfect and homogamous)*. Exposed-cleistogamous, or chasmogamous.

Inflorescence. *Inflorescence* of spicate main branches (many-jointed 'racemes'), or paniculate (the lower 'racemes' being sometimes branched again at the base); *digitate, or subdigitate (the racemes often almost palmate, towards the culm tips)*; spatheate, or espatheate; a complex of 'partial inflorescences' and intervening foliar organs, or not comprising 'partial inflorescences' and foliar organs. *Spikelet-bearing axes* 'racemes'; the spikelet-bearing axes usually with more than 10 spikelet-bearing 'articles'; solitary, or paired, or clustered; with very slender rachides; disarticulating; disarticulating at the joints. *The pedicels and rachis internodes without a longitudinal, translucent furrow*. '*Articles*' linear; without a basal callus-knob; not appendaged; disarticulating transversely; densely long-hairy, or somewhat hairy, or glabrous. Spikelets paired (with a terminal triplet); secund; sessile and pedicellate; consistently in 'long-and-short' combinations; in pedicellate/sessile combinations. Pedicels of the 'pedicellate' spikelets free of the rachis. The 'shorter' spikelets hermaphrodite (save at the raceme base, where the spikelet pairs are homogamous). The 'longer' spikelets male-only, or sterile.

Female-sterile spikelets. The pedicellate spikelets male or sterile, awnless. The lemmas awnless.

Female-fertile spikelets. Spikelets compressed dorsiventrally; falling with the glumes. Rachilla terminated by a female-fertile floret. *Hairy callus present*. Callus blunt.

Glumes two; relatively large; more or less equal; long relative to the adjacent lemmas; awnless; very dissimilar (the lower bicarinate, the upper narrower and naviculate). *Lower glume two-keeled*; convex on the back to flattened on the back; not pitted; relatively smooth; 5–9 nerved. Upper glume 1–4 nerved. *Spikelets* with incomplete florets. The incomplete florets proximal to the female-fertile florets. *The proximal incomplete florets* 1; epaleate; sterile. The proximal lemmas awnless; 0 nerved; similar in texture to the female-fertile lemmas; not becoming indurated.

Female-fertile florets 1. Lemmas reduced to a linear, hyaline stipe, produced into the awn; less firm than the glumes; not becoming indurated; entire; awned. *Awns* 1; median; *apical*; geniculate; hairless (glabrous); much longer than the body of the lemma. Lemmas hairless; non-carinate; 1–3 nerved. **Palea** present, or absent; when present, very reduced; not indurated; *nerveless. Lodicules* present; 2; free; fleshy; glabrous. *Stamens* 1–3. Anthers not penicillate. *Ovary* glabrous. Styles free to their bases. Stigmas 2; red pigmented.

Fruit, embryo and seedling. *Fruit* free from both lemma and palea; small; compressed dorsiventrally. Hilum short. Embryo large. Endosperm hard.

Abaxial leaf blade epidermis. *Costal/intercostal zonation* conspicuous. *Papillae* present. Intercostal papillae over-arching the stomata, or not over-arching the stomata; consisting of one oblique swelling per cell. Intercostal zones with typical long-cells (sometimes?), or exhibiting many atypical long-cells to without typical long-cells. Mid-intercostal long-cells rectangular; having markedly sinuous walls, or having markedly sinuous walls to having straight or only gently undulating walls. *Microhairs* present; panicoid-type; (42–)48–78(–84) microns long; (4.2–)5.1–6(–7.5) microns wide at the septum. Microhair total length/width at septum 9–14. Microhair apical cells (12–)21–28.5(–32) microns long. Microhair apical cell/total length ratio (0.29–)0.39–0.49(–0.51). *Stomata* common; 27–30 microns long. Subsidiaries low dome-shaped, or triangular (mostly). *Intercostal short-cells* common, or absent or very rare; when present, in cork/silica-cell pairs; silicified. Intercostal silica bodies vertically elongated-nodular, or cross-shaped, or oryzoid-type. *Costal short-cells* conspicuously in long rows. Costal silica bodies 'panicoid-type'; cross shaped, or dumb-bell shaped.

Transverse section of leaf blade, physiology. C_4; XyMS–. PCR cell chloroplasts with reduced grana; centrifugal/peripheral. *Mesophyll* with radiate chlorenchyma. *Leaf blade* 'nodular' in section (rarely), or adaxially flat; with the ribs more or less constant in size (rarely). *Midrib* conspicuous; with one bundle only, or having a conventional arc of bundles; with colourless mesophyll adaxially. Bulliforms present in discrete, regular adaxial groups (sometimes, in places), or not in discrete, regular adaxial groups (irregularly grouped, or the epidermis extensively bulliform); in simple fans (sometimes), or associated with colourless mesophyll cells to form deeply-penetrating fans (these mostly irregular). Many of the smallest vascular bundles unaccompanied by sclerenchyma. Combined sclerenchyma girders present; forming 'figures', or nowhere forming 'figures'. Sclerenchyma all associated with vascular bundles.

Culm anatomy. *Culm internode bundles* in one or two rings, or in three or more rings.

Phytochemistry. Leaves without flavonoid sulphates (1 species).

Cytology. Chromosome base number, $x = 10$. $2n = 20, 40, 50$, and 60. 2, 4, 5, and 6 ploid.

Taxonomy. Panicoideae; Andropogonodae; Andropogoneae; Andropogoninae.

Ecology, geography, regional floristic distribution. About 16 species. Old World Tropics. Commonly adventive. Helophytic to xerophytic; species of open habitats; glycophytic. Habitats from marshes to subdesert and disturbed ground.

Holarctic, Paleotropical, Cape, and Australian. Boreal and Tethyan. African, Madagascan, Indomalesian, and Neocaledonian. Euro-Siberian. Macaronesian, Mediterranean, and Irano-Turanian. Saharo-Sindian, Sudano-Angolan, and Namib-Karoo. Indian, Indo-Chinese, Malesian, and Papuan.

Hybrids. Intergeneric hybrids with *Bothriochloa*.

Rusts and smuts. Rusts — *Puccinia*. Taxonomically wide-ranging species: '*Uromyces*' *clignyi* and *Puccinia cesatii*. Smuts from Ustilaginaceae. Ustilaginaceae — *Sphacelotheca* and *Tolyposporella*.

Economic importance. Significant weed species: *D. annulatum, D. aristatum, D. caricosum, D. papillosum*. Important native pasture species: *D. caricosum, D. sericeum*.
References, etc. Leaf anatomical: Metcalfe 1960; this project.

Dichelachne Endl.

Habit, vegetative morphology. Perennial; caespitose. **Culms** 50–130 cm high; **herbaceous**; unbranched above. Culm nodes glabrous. Culm internodes hollow. *Leaves* not basally aggregated; non-auriculate. Leaf blades narrow; without cross venation; persistent. *Ligule an unfringed membrane*; truncate; 2–4 mm long.

Reproductive organization. Plants bisexual, with bisexual spikelets; with hermaphrodite florets; inbreeding. Exposed-cleistogamous, or chasmogamous.

Inflorescence. *Inflorescence paniculate*; contracted; *spicate, or more or less irregular*; espatheate; not comprising 'partial inflorescences' and foliar organs. Spikelet-bearing axes persistent. Spikelets not secund; pedicellate; not in distinct 'long-and-short' combinations.

Female-fertile spikelets. *Spikelets 5.5–9.5 mm long; compressed laterally*; disarticulating above the glumes. Rachilla minutely prolonged beyond the uppermost female-fertile floret, or terminated by a female-fertile floret. The rachilla extension when present, naked. *Hairy callus present (the hairs short)*. *Callus* short; blunt.

Glumes two; very unequal to more or less equal; (the upper) long relative to the adjacent lemmas (longer or slightly shorter than the floret); pointed (acuminate); awnless; carinate, or non-carinate; similar (hyaline to membranous). Lower glume 1 nerved. *Upper glume 1 nerved*. *Spikelets* with female-fertile florets only.

Female-fertile florets 1(–3). **Lemmas** linear to narrowly lanceolate; involute; *decidedly firmer than the glumes (the main, eqivocal distinction from Agrostis)*; not becoming indurated (thinly leathery); entire, or incised (tending to split); when incised, 2 lobed; not deeply cleft (no more than minutely notched); awned. *Awns* 1; median; from a sinus, or dorsal, or apical; from near the top; non-geniculate, or geniculate; usually recurving; *much longer than the body of the lemma (usually 1.5–3 cm long)*; entered by one vein. Lemmas hairless; faintly 5 nerved. *Palea* present; relatively long; entire (acute); awnless, without apical setae; not indurated; 2-nerved; 2-keeled (furrowed). *Lodicules* present; free; membranous; glabrous; not toothed. *Stamens* 1–3. Anthers penicillate, or not penicillate. *Ovary* glabrous. Styles free to their bases. Stigmas 2; white.

Fruit, embryo and seedling. **Fruit** free from both lemma and palea; small; *longitudinally grooved*; compressed dorsiventrally. Hilum short, or long-linear. Embryo small; not waisted. Endosperm liquid in the mature fruit; with lipid; containing only simple starch grains. Embryo with an epiblast; without a scutellar tail; with a negligible mesocotyl internode. Embryonic leaf margins meeting.

First seedling leaf with a well-developed lamina.

Abaxial leaf blade epidermis. *Costal/intercostal zonation* conspicuous. *Papillae* absent. *Long-cells* similar in shape costally and intercostally (narrow); differing markedly in wall thickness costally and intercostally (the costals thicker walled). Mid-intercostal long-cells fusiform; having straight or only gently undulating walls. *Microhairs* absent. *Stomata* absent or very rare, or common. Subsidiaries parallel-sided. *Intercostal short-cells* absent or very rare; usually in cork/silica-cell pairs; silicified, or not silicified. Intercostal silica bodies when present, rounded. *Costal short-cells* neither distinctly grouped into long rows nor predominantly paired. Costal silica bodies horizontally-elongated crenate/sinuous, or horizontally-elongated smooth, or rounded.

Transverse section of leaf blade, physiology. C_3; XyMS+. *PBS cells* without a suberised lamella. *Mesophyll* with non-radiate chlorenchyma. *Leaf blade* with distinct, prominent adaxial ribs, or 'nodular' in section; with the ribs more or less constant in size. *Midrib* conspicuous, or not readily distinguishable; with one bundle only. Bulliforms present in discrete, regular adaxial groups (in the furrows); in simple fans. All the vascular bundles accompanied by sclerenchyma. Combined sclerenchyma girders present; nowhere forming 'figures'. Sclerenchyma all associated with vascular bundles.

Culm anatomy. *Culm internode bundles* in one or two rings.

Cytology. Chromosome base number, $x = 7$. $2n = 70$. 10 ploid.

Taxonomy. Pooideae; Poodae; Aveneae.

Ecology, geography, regional floristic distribution. 5 species. Australia, New Zealand, New Guinea, Timor, Pacific. Mesophytic, or xerophytic; species of open habitats. Forest margins and upland grasslands.

Paleotropical, Australian, and Antarctic. Indomalesian. Papuan. North and East Australian and South-West Australian. New Zealand. Tropical North and East Australian and Temperate and South-Eastern Australian.

Rusts and smuts. Rusts — *Puccinia*. Taxonomically wide-ranging species: *Puccinia graminis*.

References, etc. Morphological/taxonomic: Veldkamp 1974. Leaf anatomical: Metcalfe 1960; this project.

Special comments. Separable only with difficulty from *Calamagrostis*, *Agrostis* and *Deyeuxia*.

Diectomis Kunth

~ *Andropogon* (*A. fastigiatus*)

Habit, vegetative morphology. Annual (becoming red on drying); caespitose. *Culms* 30–150 cm high; herbaceous; branched above. Culm nodes exposed; glabrous (dark). Culm internodes solid. The shoots not aromatic. *Leaves* not basally aggregated; non-auriculate. Leaf blades linear; narrow; 1–4 mm wide; flat (rather flaccid); without cross venation. Ligule an unfringed membrane (tough); not truncate; up to 10–20 mm long.

Reproductive organization. Plants bisexual, with bisexual spikelets; with hermaphrodite florets. The spikelets of sexually distinct forms on the same plant; hermaphrodite and sterile. The male and female-fertile spikelets mixed in the inflorescence. The spikelets overtly heteromorphic (the pedicelled member much larger); all in heterogamous combinations.

Inflorescence. Inflorescence compound paniculate; open; spatheate; a complex of 'partial inflorescences' and intervening foliar organs. **Spikelet-bearing axes** 'racemes'; *the spikelet-bearing axes* with more than 10 spikelet-bearing 'articles' ('many jointed'); solitary (one per spathe); *disarticulating*; disarticulating at the joints. *'Articles'* non-linear (trumpet-shaped); without a basal callus-knob; appendaged (the appendage terminal, small, bidentate); densely long-hairy (long-villous, with white hairs sometimes to 7 mm long). *Spikelets* paired; sessile and pedicellate; *consistently in 'long-and-short' combinations*; in pedicellate/sessile combinations. Pedicels of the 'pedicellate' spikelets free of the rachis (trumpet-shaped). The 'shorter' spikelets hermaphrodite. The 'longer' spikelets sterile.

Female-sterile spikelets. *The pedicel translucent down the middle and expanded at the tip, the pedicellate spikelet much the larger, up to 8 mm long, its lower glume large, flat, papery, reddish, with a 5–7 mm awn.*

Female-fertile spikelets. Spikelets 4–5 mm long; compressed laterally (between the rachis and the pedicel); falling with the glumes (and with the 'article' and the accompanying pedicelled member). Rachilla terminated by a female-fertile floret. Hairy callus present. *Callus* short.

Glumes two; more or less equal; exceeding the spikelets; long relative to the adjacent lemmas; the upper awned (with a long, fine bristle from the sinus); *very dissimilar (the lower two-keeled, membranous in the sulcus, with firm keels and awnless or mucronate, the upper deeply naviculate, with a long slender awn from its apical sinus)*. Lower glume two-keeled; sulcate on the back (deeply and narrowly grooved between the keels, cf. *Andropogon* Sect. Piestium etc.); not pitted; relatively smooth; 2 nerved (without a median nerve, the sulcus membranous). Upper glume 3 nerved. *Spikelets* with incomplete florets. The incomplete florets proximal to the female-fertile florets. *The proximal incomplete florets* 1; epaleate; sterile. The proximal lemmas ciliolate; awnless; 2 nerved; more or less equalling the female-fertile lemmas; similar in texture to the female-fertile lemmas (hyaline); not becoming indurated.

Female-fertile florets 1. Lemmas broad, deeply naviculate-complanate; less firm than the glumes (hyaline); not becoming indurated; incised; 2 lobed; not deeply cleft (bidentate); awned. Awns 1; median; from a sinus; geniculate; hairless (glabrous); much longer than the body of the lemma. *Lemmas* ciliolate; non-carinate; without a germination flap; *2 nerved*. *Palea* present; conspicuous but relatively short; entire; awnless, without apical setae (ciliate); textured like the lemma (hyaline); nerveless; keel-less. Palea back glabrous. *Lodicules* present; 2; fleshy; glabrous. *Stamens* 3. Anthers relatively long; not penicillate; without an apically prolonged connective. *Ovary* glabrous. Styles free to their bases. Stigmas 2.

Fruit, embryo and seedling. Fruit ellipsoid; not grooved; compressed laterally; glabrous. Hilum short. Embryo large (about half the grain length).

Abaxial leaf blade epidermis. *Costal/intercostal zonation* conspicuous. *Papillae* present; intercostal (at least, absent over the midrib). Intercostal papillae not over-arching the stomata (but associated with them); consisting of one oblique swelling per cell (each approaching one end of a stoma). *Long-cells* fairly similar in shape costally and intercostally; differing markedly in wall thickness costally and intercostally (the costals thicker walled, at least over the midrib). Intercostal zones with typical long-cells. Mid-intercostal long-cells rectangular; having markedly sinuous walls and having straight or only gently undulating walls (the long-cell walls thin and hard to observe). *Microhairs* present; elongated; clearly two-celled. *Stomata* common. Subsidiaries non-papillate; dome-shaped (medium domes). Guard-cells overlapping to flush with the interstomatals. *Intercostal short-cells* common; not paired (solitary, associated with prickle bases); not silicified. Prickles abundant costally and intercostally, variable in size. *Crown cells* absent. *Costal short-cells* conspicuously in long rows. Costal silica bodies present and well developed; 'panicoid-type'; dumb-bell shaped.

Transverse section of leaf blade, physiology. C_4; XyMS–. PCR sheath outlines even. PCR sheath extensions absent. *Mesophyll* with non-radiate chlorenchyma; not *Isachne*-type; without 'circular cells'; not traversed by colourless columns. *Leaf blade* with distinct, prominent adaxial ribs; with the ribs very irregular in sizes. *Midrib* conspicuous; having a conventional arc of bundles (with one median primary and one to several minor bundles on either side); with colourless mesophyll adaxially. *The lamina* symmetrical on either side of the midrib. Bulliforms present in discrete, regular adaxial groups (but these infrequent), or not in discrete, regular adaxial groups (the bulliform cells mostly irregular, occupying much of the intercostal zones); in simple fans (in places). Many of the smallest vascular bundles unaccompanied by sclerenchyma. Combined sclerenchyma girders present (with the primary bundles); forming 'figures' (anchors, I's and T's). Sclerenchyma all associated with vascular bundles.

Special diagnostic feature. *The pedicelled member of the sessile/pedicellate spikelet pairs much the larger, very striking, with a broad, flat, papery, reddish, long-awned lower glume*.

Cytology. Chromosome base number, $x = 5$, or 10. $2n = 20$. 2 ploid.

Taxonomy. Panicoideae; Andropogonodae; Andropogoneae; Andropogoninae.

Ecology, geography, regional floristic distribution. 1 species (*D. fastigiata* (Sw.) Kunth). Pantropical. Mesophytic; species of open habitats; glycophytic. Savanna in sandy soils, on rocky slopes or near streams in grassland, open forest etc.

Holarctic, Paleotropical, and Neotropical. African, Madagascan, Indomalesian, Polynesian, and Neocaledonian (?). Indian, Indo-Chinese, Malesian, and Papuan (?). Caribbean.

Economic importance. Important native pasture species: *D. fastigiata* provides good fodder until the awns form.

Special comments. Fruit data wanting.

Dielsiochloa Pilger

Habit, vegetative morphology. Perennial; caespitose. Culms herbaceous. Leaves non-auriculate (?). *Leaf blades* narrow; *setaceous*; without cross venation. *Ligule an unfringed membrane*.

Reproductive organization. Plants bisexual, with bisexual spikelets; with hermaphrodite florets. The spikelets all alike in sexuality.

Inflorescence. Inflorescence paniculate; oblong, contracted; espatheate; not comprising 'partial inflorescences' and foliar organs. Spikelet-bearing axes persistent. Spikelets not secund; pedicellate.

Female-fertile spikelets. *Spikelets* compressed laterally; disarticulating above the glumes; *not disarticulating between the florets*. Rachilla prolonged beyond the uppermost female-fertile floret. The rachilla extension with incomplete florets. Hairy callus absent.

Glumes two; very unequal; shorter than the adjacent lemmas; pointed; awnless; *carinate*. Lower glume 1 nerved. *Upper glume 3 nerved*. *Spikelets* with incomplete florets. The incomplete florets distal to the female-fertile florets. *The distal incomplete florets 4–7*; **incomplete florets** *clearly specialised and modified in form*; *incomplete florets awned (the prominent awns of the sterile lemmas apparently functioning to aid disperal)*.

Female-fertile florets *2–3(–4)*. Lemmas firmly membranous; incised; 2 lobed; awned. Awns 1; median; dorsal; from near the top, or from well down the back; non-geniculate. Lemmas hairless; carinate to non-carinate; without a germination flap; 5 nerved. *Palea* present; gaping; 2-nerved. *Lodicules* present; 2; free; membranous; glabrous; toothed; not or scarcely vascularized. *Stamens* 2–3. *Ovary* glabrous. Styles free to their bases. Stigmas 2.

Fruit, embryo and seedling. Fruit small, or medium sized (4 mm long); compressed dorsiventrally. *Hilum long-linear*. Embryo small. Endosperm hard; containing compound starch grains.

Abaxial leaf blade epidermis. *Papillae* absent. *Long-cells* markedly different in shape costally and intercostally (the costals much smaller and narrower); of similar wall thickness costally and intercostally (walls thick, pitted). Mid-intercostal long-cells rectangular; having markedly sinuous walls. *Microhairs* absent. *Stomata* absent or very rare (present in the furrows adaxially — with parallel-sided subsidiaries). *Intercostal short-cells* absent or very rare. Thick walled prickles common costally. *Crown cells* absent. Costal silica bodies poorly developed; perhaps tall-and-narrow, or crescentic, or poorly developed in material seen.

Transverse section of leaf blade, physiology. C_3; XyMS+. Mesophyll without adaxial palisade. *Leaf blade* with distinct, prominent adaxial ribs; with the ribs very irregular in sizes. *Midrib* not readily distinguishable; with one bundle only. *The lamina* symmetrical on either side of the midrib. Bulliforms unclear in the poor material seen. All the vascular bundles accompanied by sclerenchyma. Combined sclerenchyma girders absent (some bundles with an abaxial strand only, others with a small adaxial strand as well). Sclerenchyma all associated with vascular bundles (except at the blade margins).

Taxonomy. Pooideae; Poodae; Aveneae.

Ecology, geography, regional floristic distribution. 1 species. Peru. Species of open habitats. In high altitude grassland.

Neotropical. Caribbean and Andean.

References, etc. Leaf anatomical: this project.

Digastrium (Hackel) A. Camus

~ *Ischaemum*

Habit, vegetative morphology. Perennial. Culms herbaceous. Culm nodes hairy, or glabrous. *Leaves* not basally aggregated; non-auriculate. Leaf blades broad, or narrow; without cross venation; persistent. *Ligule* present; an unfringed membrane.

Reproductive organization. Plants bisexual, with bisexual spikelets; with hermaphrodite florets. *The spikelets of sexually distinct forms on the same plant*; hermaphrodite and sterile; overtly heteromorphic; all in heterogamous combinations.

Inflorescence. *Inflorescence* of spicate main branches, or a single raceme (the spikelike racemes with thick, inflated joints and pedicels); often *digitate*; espatheate; not comprising 'partial inflorescences' and foliar organs. **Spikelet-bearing axes** 'racemes'; solitary to clustered; *with substantial rachides (the joints thick, inflated)*; disarticulating; disarticulating at the joints. *'Articles'* non-linear. *Spikelets* paired; *not secund*; sessile and pedicellate; consistently in 'long-and-short' combinations; in pedicellate/sessile combinations. Pedicels of the 'pedicellate' spikelets free of the rachis. The 'shorter' spikelets hermaphrodite. The 'longer' spikelets sterile.

Female-sterile spikelets. The pedicellate spikelets sterile, reduced to glumes.

Female-fertile spikelets. Spikelets compressed dorsiventrally; falling with the glumes. Rachilla terminated by a female-fertile floret. Hairy callus absent.

Glumes two; more or less equal; long relative to the adjacent lemmas; awnless. Lower glume not pitted; relatively smooth; 5–7 nerved. Upper glume 5–7 nerved. *Spikelets* with incomplete florets. The incomplete florets proximal to the female-fertile florets. *The proximal incomplete florets* 1; paleate; male. The proximal lemmas awnless; not becoming indurated.

Female-fertile florets 1. **Lemmas** less firm than the glumes; *incised*; 2 lobed (bifid); awned. Awns 1; median; from a sinus; geniculate. Lemmas hairless; non-carinate; 3 nerved. *Palea* present; 2-nerved (?). **Lodicules present**; 2; free; fleshy; ciliate, or glabrous. *Stamens* 3. Anthers not penicillate. *Ovary* glabrous. Styles fused, or free to their bases. Stigmas 2.

Fruit, embryo and seedling. Fruit small; compressed dorsiventrally. Hilum short.

Abaxial leaf blade epidermis. *Costal/intercostal zonation* conspicuous. *Papillae* present. Intercostal papillae over-arching the stomata; many or several per cell. Long-cells of similar wall thickness costally and intercostally (rather thin walled). Intercostal zones mainly with typical long-cells. Mid-intercostal long-cells rectangular; having markedly sinuous walls. *Microhairs* present; panicoid-type and chloridoid-type (the basal cells fairly long, the apical cells sometimes partially collapsing but quite thick walled, hemispherical to somewhat elongated); (24–)25.5–27(–30) microns long; (7.5–)8.4–9 microns wide at the septum. Microhair total length/width at septum 2.7–3.6. Microhair apical cells 7.5–9–9.6 microns long. Microhair apical cell/total length ratio 0.29–0.37. *Stomata* common; (27–)28.5–30(–36) microns long. Subsidiaries papillate; triangular. *Intercostal short-cells* common to absent or very rare; not paired (solitary). *Costal short-cells* conspicuously in long rows. Costal silica bodies 'panicoid-type'; mostly dumb-bell shaped; sharp-pointed.

Transverse section of leaf blade, physiology. C_4. *Mesophyll* with radiate chlorenchyma. *Leaf blade* adaxially flat. *Midrib* conspicuous; having a conventional arc of bundles; with colourless mesophyll adaxially. Bulliforms present in discrete, regular adaxial groups; in simple fans. Many of the smallest vascular bundles unaccompanied by sclerenchyma. Combined sclerenchyma girders present (with the major bundles); forming 'figures'. Sclerenchyma all associated with vascular bundles.

Taxonomy. Panicoideae; Andropogonodae; Andropogoneae; Andropogoninae.

Ecology, geography, regional floristic distribution. 2 species. Australasia.

Paleotropical and Australian. Indomalesian. Papuan. North and East Australian. Tropical North and East Australian.

References, etc. Leaf anatomical: this project.

Special comments. Fruit data wanting.

Digitaria Haller

Acicarpa Raddi, *Digitariella* De Winter, *Elytroblepharum* Steud., *Elytroblepharum* (Steud.) Schlecht., *Eriachne* Phil., *Gramerium* Desv., *Sanguinaria* Bub., *Sanguinella* Gleichen, *Syntherisma* Walt., *Trichachne* Nees, *Valota* Adans.

Excluding *Acostia*, *Digitariopsis*

Habit, vegetative morphology. Annual, or perennial; rhizomatous, or stoloniferous, or caespitose, or decumbent (sometimes sward forming). **Culms** (6–)15–300 cm high (or more?); *herbaceous*; branched above, or unbranched above. Culm nodes glabrous. Culm internodes solid, or hollow. *Leaves* not basally aggregated; non-auriculate. Leaf blades linear to lanceolate; broad, or narrow; flat, or folded, or rolled; without cross venation; persistent; rolled in bud. *Ligule usually an unfringed membrane*. *Contra-ligule* absent.

Reproductive organization. Plants bisexual, with bisexual spikelets; with hermaphrodite florets. *The spikelets all alike in sexuality*; overtly heteromorphic (sometimes, notably regarding indumentum), or homomorphic. Plants inbreeding. Exposed-cleistogamous, or chasmogamous. Plants with hidden cleistogenes (very rarely), or without hidden cleistogenes. The hidden cleistogenes when present, in the leaf sheaths.

Inflorescence. *Inflorescence* nearly always *of spicate main branches (rarely with some secondary branchlets or a single raceme)*; open, or contracted; digitate, or subdigitate, or non-digitate. Primary inflorescence branches 2–50. Rachides hollowed, or flattened, or winged. Inflorescence espatheate; not comprising 'partial inflorescences' and foliar organs. *Spikelet-bearing axes persistent*. *Spikelets* solitary, or paired to in triplets (usually in groups of 2–3); secund; *pedicellate (except for two Malaysian species)*. Pedicel apices truncate, or discoid, or cupuliform. Spikelets imbricate, or distant; consistently in 'long-and-short' combinations (and the longer-pedicelled members sometimes larger and/or hairier), or not in distinct 'long-and-short' combinations. The 'shorter' spikelets hermaphrodite. The 'longer' spikelets hermaphrodite.

Female-fertile spikelets. *Spikelets* elliptic, or lanceolate, or ovate, or obovate; abaxial; *compressed dorsiventrally*; falling with the glumes. Rachilla terminated by a female-fertile floret. *Hairy callus absent*.

Glumes one per spikelet, or two; *very unequal (the lower tiny or suppressed)*; shorter than the adjacent lemmas, or long relative to the adjacent lemmas (i.e., the upper, sometimes); free; dorsiventral to the rachis; pointed, or not pointed; awnless; non-carinate; very dissimilar. *Lower glume* 0–0.4 times the length of the upper glume; when present, much shorter than half length of lowest lemma; 0–1 nerved, or 3 nerved (at the base). *Upper glume* gibbous or not; 3–7 nerved (rarely nerveless). *Spikelets* with incomplete florets. The incomplete florets proximal to the female-fertile florets. *Spikelets with proximal incomplete florets*. *The proximal incomplete florets* 1; paleate, or epaleate. Palea of the proximal incomplete florets when present, reduced. The proximal incomplete florets sterile. *The proximal lemmas usually as long as the spikelet, usually hairy, the hairs between the first and second lateral nerves and along the margins (contrast with* **Stereochlaena***)*; awnless; 3–7(–11) nerved; exceeded by the female-fertile lemmas to decidedly exceeding the female-fertile lemmas; less firm than the female-fertile lemmas; not becoming indurated.

Female-fertile florets 1. Lemmas similar in texture to the glumes to decidedly firmer than the glumes (cartilaginous); smooth to striate; yellow in fruit, or brown in fruit, or black in fruit; entire; pointed (mostly subacute to acuminate); awnless (but often apiculate); hairless (no more than minutely striate-papillate); non-carinate, or carinate (Section *Monodactylae*, where some species have a pronounced median keel); *having the margins lying flat and exposed on the palea*; *with a clear germination flap*; obscurely *1–3 nerved*. *Palea* present; relatively long (about equalling the lemma); tightly clasped by the lemma; entire; awnless, without apical setae; textured like the lemma; 2-nerved. *Lodicules* present; 2; joined, or free; fleshy; glabrous. *Stamens* 3. Anthers not penicillate. *Ovary* glabrous. Styles fused, or free to their bases. Stigmas 2; red pigmented.

Fruit, embryo and seedling. Fruit small; compressed dorsiventrally. Hilum short. Embryo large; waisted. Endosperm hard; without lipid; containing compound starch grains. Embryo without an epiblast; with a scutellar tail; with an elongated mesocotyl internode. Embryonic leaf margins overlapping.

Seedling with a long mesocotyl; with a loose coleoptile. First seedling leaf with a well-developed lamina. The lamina broad; erect, or curved, or supine; 13–20 veined.

Abaxial leaf blade epidermis. *Costal/intercostal zonation* conspicuous. *Papillae* absent. *Long-cells* similar in shape costally and intercostally; of similar wall thickness costally and intercostally. Mid-intercostal long-cells rectangular; having markedly sinuous walls. *Microhairs* present; panicoid-type; (34–)42–75(–78) microns long; 3.6–7.5 microns wide at the septum. Microhair total length/width at septum 8–18.3. Microhair apical cells (17–)22–48(–50) microns long. Microhair apical cell/total length ratio 0.44–0.67. *Stomata* common; 24–36 microns long. Subsidiaries high to low dome-shaped and triangular. Guard-cells overlapping to flush with the interstomatals. *Intercostal short-cells* absent or very rare (scarce); not paired (solitary when seen); when seen, not silicified. Prickles sometimes common. *Costal short-cells* conspicuously in long rows. Costal silica bodies 'panicoid-type'; cross shaped, or butterfly shaped, or dumb-bell shaped.

Transverse section of leaf blade, physiology. C_4; biochemical type NADP–ME (*D. sanguinalis*); XyMS–. PCR sheath outlines uneven. PCR sheath extensions absent. PCR cells with a suberised lamella. PCR cell chloroplasts with reduced grana; centrifugal/peripheral. *Mesophyll* with radiate chlorenchyma. *Leaf blade* with distinct, prominent adaxial ribs, or 'nodular' in section, or adaxially flat; with the ribs more or less constant in size. *Midrib* conspicuous, or not readily distinguishable; with one bundle only, or having a conventional arc of bundles; with colourless mesophyll adaxially, or without colourless mesophyll adaxially. Bulliforms present in discrete, regular adaxial groups, or not in discrete, regular adaxial groups (commonly in irregular groups); when regularly grouped, in simple fans. Many of the smallest vascular bundles unaccompanied by sclerenchyma. Combined sclerenchyma girders present (rarely), or absent; nowhere forming 'figures'. Sclerenchyma all associated with vascular bundles.

Culm anatomy. *Culm internode bundles* in one or two rings.

Phytochemistry. Tissues of the culm bases with little or no starch (*D. sanguinalis*). Leaves without flavonoid sulphates (4 species).

Cytology. Chromosome base number, $x = 9$, 15, and 17. $2n = 18$, 30, 36, 45, 54, 60, 70, 72, and 76, or 108. Nucleoli persistent.

Taxonomy. Panicoideae; Panicodae; Paniceae.

Ecology, geography, regional floristic distribution. 220 species. Mainly in warm regions. Commonly adventive. Mesophytic, or xerophytic; mostly species of open habitats. Diverse habitats, including weedy ground and sandy beaches.

Holarctic, Paleotropical, Neotropical, Australian, and Antarctic. Boreal, Tethyan, and Madrean. African, Madagascan, Indomalesian, Polynesian, and Neocaledonian. Euro-Siberian, Eastern Asian, and Atlantic North American. Macaronesian, Mediterranean, and Irano-Turanian. Saharo-Sindian, Sudano-Angolan, West African Rainforest, and Namib-Karoo. Indian, Indo-Chinese, Malesian, and Papuan. Polynesian. Caribbean, Venezuela and Surinam, Amazon, Central Brazilian, and Pampas. North and East Australian and Central Australian. Patagonian. European and Siberian. Canadian-Appalachian, Southern Atlantic North American, and Central Grasslands. Sahelo-Sudanian, Somalo-Ethiopian, South Tropical African, and Kalaharian. Tropical North and East Australian and Temperate and South-Eastern Australian.

Rusts and smuts. Rusts — *Physopella* and *Puccinia*. Taxonomically wide-ranging species:. *Puccinia levis*, *Puccinia substriata*, and *Puccinia esclavensis*. Smuts from Tilletiaceae and from Ustilaginaceae. Tilletiaceae — *Tilletia*. Ustilaginaceae — *Sorosporium*, *Sphacelotheca*, and *Ustilago*.

Economic importance. Significant weed species: 'crabgrasses' —*D. adscendens*, *D. decumbens*, *D. ciliaris*, *D. fuscescens*, *D. horizontalis*, *D. ischaemum*, *D. longiflora*, *D. radicosa*, *D. sanguinalis*, *D. scalarum*, *D. setigera*, *D. ternata*, *D. violascens*. Cultivated fodder: *D. decumbens*, *D. eriantha* (Pangola). Important native pasture species: *D. aridicola*, *D. didactyla*, *D. gazensis*, *D. longiflora*, *D. milanjiana*, *D. nodosa*, *D. pearsonii*, *D. velutina* and others. Grain crop species: minor west African cereals on infertile soils: *D. exilis* (Achna, Fonio), *D. iburua*. Lawns and/or playing fields: *D. didactyla*, *D. longiflora*, *D. swazilandensis*, *D. timorensis*.

References, etc. Leaf anatomical: Metcalfe 1960; this project.

Digitariopsis C.E. Hubb.

~ *Digitaria*

Excluding *Leptoloma*

Habit, vegetative morphology. Annual, or perennial; caespitose. Culms herbaceous; unbranched above. Culm nodes glabrous. *Leaves* not basally aggregated; non-auriculate. Leaf blades narrow; without cross venation; persistent. Ligule a fringed membrane (minutely fringed).

Reproductive organization. Plants bisexual, with bisexual spikelets; with hermaphrodite florets. The spikelets all alike in sexuality.

Inflorescence. *Inflorescence of spicate main branches*; with capillary branchlets (i.e. capillary rachides). Primary inflorescence branches 12–20. Inflorescence espatheate;

not comprising 'partial inflorescences' and foliar organs. Spikelet-bearing axes persistent. Spikelets solitary, or paired; secund; biseriate; pedicellate.

Female-fertile spikelets. *Spikelets* about 3 mm long; abaxial; *compressed dorsiventrally*; falling with the glumes. Rachilla terminated by a female-fertile floret. Hairy callus present (associated with a characteristic basal prolongation of the solitary glume). *Callus* long.

Glumes one per spikelet (the upper); shorter than the adjacent lemmas; dorsiventral to the rachis; hairy; *with distinct hair tufts and with distinct rows of hairs (with dark brown hairs in 4 rows, between and outside the veins, prolonged into tufts apically)*; pointed; awnless; non-carinate. **Upper glume** *prolonged downwards into a short spur, this seemingly the extreme of a tendency seen in* **Digitaria** *species such as* **D. flaccida**; 3 nerved. *Spikelets* with incomplete florets. The incomplete florets proximal to the female-fertile florets. *The proximal incomplete florets* 1; paleate. Palea of the proximal incomplete florets reduced. The proximal incomplete florets sterile. The proximal lemmas awnless; 5 nerved; more or less equalling the female-fertile lemmas; less firm than the female-fertile lemmas to similar in texture to the female-fertile lemmas (hyaline, densely hairy below and on the margins); not becoming indurated.

Female-fertile florets 1. **Lemmas** similar in texture to the glumes to decidedly firmer than the glumes; not becoming indurated; entire; pointed; awnless; hairless; non-carinate; having the margins lying flat and exposed on the palea; *with a clear germination flap*; 3 nerved. *Palea* present; relatively long; entire; awnless, without apical setae; textured like the lemma; not indurated; 2-nerved; 2-keeled. *Lodicules* present; 2; free; fleshy; glabrous; not or scarcely vascularized. *Stamens* 3. Anthers not penicillate; without an apically prolonged connective. *Ovary* glabrous. *Styles* basally *fused*. Stigmas 2; red pigmented.

Fruit, embryo and seedling. Fruit compressed dorsiventrally. Hilum short. Pericarp fused. Embryo large.

Abaxial leaf blade epidermis. *Costal/intercostal zonation* conspicuous. *Papillae* absent. Mid-intercostal long-cells fusiform; having straight or only gently undulating walls (to slightly sinuous in high focus). *Microhairs* present; panicoid-type; (24–)27–32(–45) microns long; (15–)16–23(–33) microns wide at the septum. Microhair total length/width at septum 5–10.7. Microhair apical cells (4.2–)4.5–4.8(–5.4) microns long. Microhair apical cell/total length ratio 0.55–0.73. *Stomata* common; 30–36 microns long. Subsidiaries dome-shaped and triangular. Guard-cells overlapping to flush with the interstomatals. *Intercostal short-cells* absent or very rare. *Costal short-cells* conspicuously in long rows. Costal silica bodies 'panicoid-type' and acutely-angled, or tall-and-narrow (representing incomplete crosses); mostly cross shaped; sometimes sharp-pointed.

Transverse section of leaf blade, physiology. C_4; XyMS–. *Mesophyll* with radiate chlorenchyma. *Leaf blade* adaxially flat. *Midrib* not readily distinguishable; with one bundle only. Bulliforms not in discrete, regular adaxial groups (constituting most of the adaxial epidermis). Many of the smallest vascular bundles unaccompanied by sclerenchyma. Combined sclerenchyma girders present; forming 'figures'. Sclerenchyma all associated with vascular bundles.

Taxonomy. Panicoideae; Panicodae; Paniceae.

Ecology, geography, regional floristic distribution. 2 species. Tropical Africa. Paleotropical. African. Sudano-Angolan. South Tropical African.

References, etc. Leaf anatomical: this project.

Dignathia Stapf

Habit, vegetative morphology. Annual, or perennial; caespitose. **Culms** (1–)5–50 cm high; *herbaceous*. Plants unarmed. *Leaves* not basally aggregated; non-auriculate. Leaf blades narrow; 1.5–3 mm wide (to 8 cm long); not setaceous (rigid); without cross venation; rolled in bud. *Ligule a fringed membrane*.

Reproductive organization. Plants bisexual, with bisexual spikelets; with hermaphrodite florets. The spikelets of sexually distinct forms on the same plant; hermaphrodite and sterile; overtly heteromorphic.

Inflorescence. *Inflorescence a false spike, with spikelets on contracted axes (the pedunculate branchlets each with 2–3 spikelets, the terminal one sterile and reduced to glumes, the rest fertile)*; contracted; espatheate; not comprising 'partial inflorescences' and foliar organs. *Spikelet-bearing axes* very much reduced (to 1–2 spikelets plus a rudiment on the curved rachis); disarticulating; falling entire (the branchlets falling complete with fertile and sterile spikelets). Spikelets not in distinct 'long-and-short' combinations.

Female-fertile spikelets. Spikelets 3–7 mm long; the G_1 laterally contiguous with rachis; obliquely compressed laterally; falling with the glumes. Rachilla terminated by a female-fertile floret.

Glumes two; very unequal; (the upper) long relative to the adjacent lemmas (exceeding the lemma); lateral to the rachis; hairy; pointed (acute, acuminate or subulate-beaked); *awned (one or both acuminate into a curved or flexuous subule)*; non-carinate; *very dissimilar (thickened-cartilaginous dorsally with hyaline margins, G_1 shorter, G_2 broader and marked with green lines)*. Lower glume relatively smooth; several nerved. Upper glume several nerved. *Spikelets* with female-fertile florets only.

Female-fertile florets 1. Lemmas less firm than the glumes (hyaline-membranous); not becoming indurated; entire; pointed; awned. Awns 1; median; apical; non-geniculate; much shorter than the body of the lemma (a short slender aristule, about 1 mm long). Lemmas hairy, or hairless; *carinate*; 3 nerved. *Palea* present; relatively long; awnless, without apical setae; 2-nerved; 2-keeled. *Lodicules* present; 2; free; fleshy; glabrous. *Stamens* 3. Anthers 1–2 mm long; not penicillate; without an apically prolonged connective. *Ovary* glabrous. Styles free to their bases. Stigmas 2.

Fruit, embryo and seedling. Fruit small (1.5–2 mm long). Hilum short. Pericarp fused. Embryo large.

Abaxial leaf blade epidermis. *Costal/intercostal zonation* conspicuous. *Papillae* present; intercostal (but on only some cells). Intercostal papillae not over-arching the stomata; consisting of one oblique swelling per cell, or consisting of one symmetrical projection per cell. Long-cells differing markedly in wall thickness costally and intercostally (the costals thinner). Mid-intercostal long-cells rectangular; having markedly sinuous walls. *Microhairs* present; elongated; clearly two-celled; chloridoid-type; 27–31.5 microns long; 7.2–9 microns wide at the septum. Microhair total length/width at septum 3–4.2. Microhair apical cells 7.5–10.5 microns long. Microhair apical cell/total length ratio 0.24–0.36. *Stomata* common; (18–)21–24 microns long. Subsidiaries non-papillate; mostly low to high dome-shaped, or triangular (some, fairly). Guard-cells overlapping to flush with the interstomatals. *Intercostal short-cells* common; not paired (mostly solitary); silicified. With a few costal prickles. *Crown cells* absent. *Costal short-cells* conspicuously in long rows. Costal silica bodies 'panicoid-type'; dumb-bell shaped.

Transverse section of leaf blade, physiology. C_4; XyMS+. PCR sheath extensions absent. *Mesophyll* with radiate chlorenchyma. *Leaf blade* 'nodular' in section to adaxially flat; with the ribs more or less constant in size (slight). *Midrib* not readily distinguishable; with one bundle only. Bulliforms present in discrete, regular adaxial groups; in simple fans (these deeply penetrating). All the vascular bundles accompanied by sclerenchyma. Combined sclerenchyma girders present (with all the bundles); forming 'figures' (all bundles). Sclerenchyma all associated with vascular bundles.

Taxonomy. Chloridoideae; main chloridoid assemblage.

Ecology, geography, regional floristic distribution. 5 species. East Africa and India. Species of open habitats. Dry bushland.

Paleotropical. African and Indomalesian. Saharo-Sindian and Sudano-Angolan. Indian. Somalo-Ethiopian and South Tropical African.

References, etc. Morphological/taxonomic: Stapf 1911. Leaf anatomical: this project.

Diheteropogon (Hack.) Stapf

Habit, vegetative morphology. Annual, or perennial (slender); caespitose. **Culms** 15–230 cm high; herbaceous (to woody at the base); mainly *unbranched above*. Culm internodes solid. *Leaves* not basally aggregated. Sheath margins free. *Leaf blades* greatly reduced (sometimes, to their midribs in *D. filifolius*), or not all greatly reduced; linear, or linear-lanceolate; broad to narrow; cordate, or not cordate, not sagittate; flat, or acicular; without cross venation. Ligule an unfringed membrane; not truncate.

Reproductive organization. Plants bisexual, with bisexual spikelets; with hermaphrodite florets. *The spikelets* of sexually distinct forms on the same plant (all male or sterile at the bases of the 'racemes', heterogamous above); hermaphrodite and male-only, or hermaphrodite and sterile; overtly heteromorphic; *in both homogamous and heterogamous combinations (the 3–9 proximal pairs homogamous, imperfect, hardly heteromorphic)*.

Inflorescence. Inflorescence of spicate main branches, or paniculate (of paired 'racemes', terminal or in a scanty false panicle); spatheate; *a complex of 'partial inflorescences' and intervening foliar organs*. *Spikelet-bearing axes* 'racemes'; paired (not deflexed the bases terete); with substantial rachides; disarticulating; disarticulating at the joints. **'Articles'** *non-linear (thickened and hollowed at the summit)*; not appendaged; disarticulating obliquely; densely long-hairy to somewhat hairy. Spikelets paired; sessile and pedicellate; consistently in 'long-and-short' combinations; in pedicellate/sessile combinations. Pedicels of the 'pedicellate' spikelets free of the rachis. The 'shorter' spikelets hermaphrodite (save at the base of the raceme). The 'longer' spikelets male-only, or sterile (?).

Female-sterile spikelets. The pedicelled spikelets awnless or aristulate, larger, callus glabrous.

Female-fertile spikelets. Spikelets 5–9 mm long; compressed dorsiventrally (subterete); falling with the glumes (or with a slight tendency to disarticulate above them). Rachilla terminated by a female-fertile floret. Hairy callus present. *Callus* long; pointed.

Glumes two; more or less equal; G_2 with ciliate margins; awnless; carinate (G_2), or non-carinate (G_1); very dissimilar (somewhat leathery, G_1 bicarinate and grooved between the keels, G_2 not bicarinate). *Lower glume two-keeled*; sulcate on the back; not pitted; relatively smooth; many nerved. Upper glume 3 nerved. *Spikelets* with incomplete florets. The incomplete florets proximal to the female-fertile florets. *The proximal incomplete florets* 1; epaleate; sterile. The proximal lemmas awnless; 2 nerved; more or less equalling the female-fertile lemmas; similar in texture to the female-fertile lemmas (thin); not becoming indurated.

Female-fertile florets 1. **Lemmas** less firm than the glumes (hyaline); not becoming indurated; *incised*; awned. Awns 1; median; from a sinus; twice geniculate; hairless (scabrid), or hairy; much longer than the body of the lemma (and sturdy). Lemmas with ciliate lobes; non-carinate; without a germination flap; 1 nerved (?). **Palea** present; conspicuous but relatively short (small); entire; awnless, without apical setae (ciliolate); not indurated (hyaline); *2-nerved*; keel-less. *Lodicules* present; 2; free. *Stamens* 3. *Ovary* glabrous. Stigmas 2.

Fruit, embryo and seedling. *Fruit* free from both lemma and palea.

Abaxial leaf blade epidermis. *Costal/intercostal zonation* conspicuous. *Papillae* absent. Long-cells of similar wall thickness costally and intercostally. Mid-intercostal long-cells mostly rectangular; having markedly sinuous walls, or having straight or only gently undulating walls. *Microhairs* present; panicoid-type. *Stomata* common. Subsidiaries dome-shaped and triangular. *Intercostal short-cells* common; in cork/silica-cell pairs, or not paired (some solitary); silicified, or not silicified. Intercostal silica bodies infrequent and variable in shape. *Crown cells* absent. *Costal short-cells* conspicuously in long rows, or conspicuously in long rows and predominantly paired. Costal silica bodies 'panicoid-type'; cross shaped.

Transverse section of leaf blade, physiology. Leaf blades consisting of midrib, or 'laminar'. C_4; XyMS–. PCR sheath extensions absent. PCR cell chloroplasts centrifugal/peripheral. Mesophyll traversed by columns of colourless mesophyll cells. *Leaf blade* (when the blade not acicular and reduced almost to its midrib) with distinct, prominent adaxial ribs, or adaxially flat; when present, with the ribs more or less constant in size. *Midrib* conspicuous (or the blade reduced to the midrib); with one bundle only, or having a conventional arc of bundles; with colourless mesophyll adaxially. Bulliforms present in discrete, regular adaxial groups, or not in discrete, regular adaxial groups (the adaxial part of the leaf extensively colourless in *D. filifolius*, the adaxial epidermis of irregularly grouped bulliform cells in *D. amplectens*). Many of the smallest vascular bundles unaccompanied by sclerenchyma. Sclerenchyma all associated with vascular bundles.

Phytochemistry. Leaves containing flavonoid sulphates (1 species).

Special diagnostic feature. *Spikelets not borne as in* 'Anadelphia scyphofera' *(q.v.).*

Cytology. Chromosome base number, $x = 10$. $2n = 20$ and 40. 2 and 4 ploid.

Taxonomy. Panicoideae; Andropogonodae; Andropogoneae; Andropogoninae.

Ecology, geography, regional floristic distribution. 5 species. Tropical Africa. Helophytic; species of open habitats; glycophytic. Savanna.

Paleotropical. African and Madagascan. Sudano-Angolan, West African Rainforest, and Namib-Karoo. Sahelo-Sudanian, Somalo-Ethiopian, South Tropical African, and Kalaharian.

References, etc. Morphological/taxonomic: Stapf 1922. Leaf anatomical: Metcalfe 1960 (*Andropogon amplectens*); photos of *D. filifolius* and *D. amplectens* provided by R.P. Ellis.

Special comments. Fruit data wanting.

Dilophotriche Jacq.-Fél.

~ *Loudetiopsis p.p.*

Habit, vegetative morphology. Annual, or perennial (culms erect, rather slender); caespitose. **Culms** 100–120 cm high; *herbaceous*; cylindrical; unbranched above; 4–6 noded. Culm nodes exposed; glabrous. Culm leaves present. Upper culm leaf blades fully developed. Culm internodes hollow. *Leaves* not basally aggregated; non-auriculate. Leaf blades linear; narrow; often becoming involute setaceous, or not setaceous; flat, or rolled (becoming involute); without cross venation. Ligule a fringe of hairs. *Contra-ligule* absent.

Reproductive organization. Plants bisexual, with bisexual spikelets; with hermaphrodite florets.

Inflorescence. *Inflorescence* few spikeleted to many spikeleted; paniculate (diffuse, the branches often setulose); *open*; with capillary branchlets (often), or without capillary branchlets (but then the branchlets slender); espatheate; not comprising 'partial inflorescences' and foliar organs. Spikelet-bearing axes persistent. Spikelets in triplets, or paired (rarely); not secund; pedicellate; not in distinct 'long-and-short' combinations (but the pedicels varying in length).

Female-fertile spikelets. *Spikelets* 7–12 mm long; compressed laterally; *disarticulating above the glumes*; disarticulating between the florets (above the L_1 only in *D. pobeguinii*, above the L_1 and above G_2 in *D. occidentalis*). The upper floret conspicuously stipitate. Rachilla terminated by a female-fertile floret. Hairy callus present. **Callus** *long (with long hairs)*; blunt (at the tip).

Glumes two; *very unequal*; (the upper) long relative to the adjacent lemmas; hairy (with long tubercle-based hairs), or hairless (glabrous); pointed; awnless; non-carinate; very dissimilar to similar (more or less lanceolate, the lower shorter and sometimes hairier with more prominent nerves). Lower glume shorter than the lowest lemma; convex on the back to flattened on the back; in *D. pobeguinii* tuberculate; 3 nerved. Upper glume 3 nerved. *Spikelets* with incomplete florets. The incomplete florets proximal to the female-fertile florets. *Spikelets with proximal incomplete florets. The proximal incomplete florets* 1; paleate. Palea of the proximal incomplete florets fully developed (two keeled). The proximal incomplete florets male. The proximal lemmas lanceolate; awnless; 3 nerved; decidedly exceeding the female-fertile lemmas; less firm than the female-fertile lemmas (herbaceous-membranous); not becoming indurated.

Female-fertile florets 1. *Lemmas* decidedly firmer than the glumes; not becoming indurated (becoming leathery); *incised*; 2 lobed (the lobes finally setaceous); deeply cleft; awned. Awns 1, or 3; median, or median and lateral (if the lobes are considered awn-tipped); the median different in form from the laterals (if laterals are considered present); from a sinus; geniculate; hairless, or hairy; much longer than the body of the lemma; entered by several veins; deciduous (seemingly, in *D. pobeguinii*). The lateral awns (if considered as such) shorter than the median. Awn bases twisted; flattened. Lemmas hairy. *The hairs* in tufts, or not in tufts; *in transverse rows (the lemma variously pilose, tufted on either side in the middle or with an fringe of long hairs on either side beneath the sinus)*. Lemmas non-carinate; 7 nerved; with the nerves confluent towards the tip. *Palea* present; relatively long; entire; awnless, without apical setae; textured like the lemma; 2-nerved; 2-keeled (the keels fairly thick and hard). Palea back glabrous. Palea keels wingless; glabrous. *Lodicules* present; 2; free; fleshy; glabrous; not toothed; not or scarcely vascularized. *Stamens* 3. *Ovary* glabrous. Styles free to their bases; free. Style bases widely separated. Stigmas 2; white, or brown.

Fruit, embryo and seedling. *Fruit* free from both lemma and palea; small; longitudinally grooved. *Hilum long-linear*. Embryo large.

Abaxial leaf blade epidermis. *Costal/intercostal zonation* conspicuous. *Papillae* absent. *Long-cells* markedly different in shape costally and intercostally (the costals much narrower); of similar wall thickness costally and intercostally (thin walled). Mid-intercostal long-cells somewhat fusiform, or rectangular; having markedly sinuous walls (the sinuosity fairly coarse). *Microhairs* present; elongated; clearly two-celled; panicoid-type. *Stomata* common. Subsidiaries non-papillate; dome-shaped (*D. occidentalis*), or triangular (with sharp points, in *D. pobeguinii*). Guard-cells overlapping to flush with the interstomatals. *Intercostal short-cells* common (abundant); mostly in cork/silica-cell pairs; silicified. Intercostal silica bodies irregularly cross-shaped (mostly more or less distorted, in *D. pobeguinii*). No microhairs or prickles seen in *D. pobeguinii*. *Costal short-cells* conspicuously in long rows. Costal silica bodies present and well developed; 'panicoid-type'; long and short dumb-bell shaped (mostly with rather rough outlines, in *D. pobeguinii*).

Transverse section of leaf blade, physiology. C_4; XyMS+ (? — see Jacques-Félix 1962), or XyMS– (indisputably, in *D. pobeguinii*). PCR sheath outlines uneven. PCR sheath extensions absent. Mesophyll without 'circular cells'; traversed by columns of colourless mesophyll cells, or not traversed by colourless columns. *Leaf blade* adaxially flat. *Midrib* not readily distinguishable; with one bundle only. Bulliforms present in discrete, regular adaxial groups, or not in discrete, regular adaxial groups (the epidermis extensively bulliform, the groups mostly irregular); associating with colourless mesophyll cells to form arches over small vascular bundles. Many of the smallest vascular bundles unaccompanied by sclerenchyma. Combined sclerenchyma girders present (*D. occidentalis*), or absent (the abaxial fibres in *D. pobeguinii* as strands only); in *D. occidentalis* forming 'figures'. Sclerenchyma all associated with vascular bundles.

Taxonomy. Panicoideae; Panicodae; Arundinelleae.

Ecology, geography, regional floristic distribution. 3 species. West Africa. Mesophytic. In damp pockets on rock outcrops.

Paleotropical. African. Sudano-Angolan and West African Rainforest. Sahelo-Sudanian.

References, etc. Morphological/taxonomic: Jacques-Félix 1962. Leaf anatomical: drawing of *D. occidentalis*, *D. purpurea* and *D. tristachya* in Jacques-Félix 1962; *D. pobeguinii* and *D. occidentalis*, this project.

Dimeria R.Br.

Didactylon Zoll. & Mor., *Haplachne* Presl, *Psilostachys* Steud., *Pterostachyum* Steud., *Woodrowia* Stapf

Habit, vegetative morphology. Annual, or perennial (often pigmented when mature); caespitose. Culms herbaceous. Culm nodes hairy (usually?). Culm internodes hollow. Leaves auriculate, or non-auriculate. Leaf blades linear (acuminate); narrow; without cross venation; persistent. Ligule an unfringed membrane.

Reproductive organization. Plants bisexual, with bisexual spikelets; with hermaphrodite florets.

Inflorescence. *Inflorescence* of spicate main branches, or a single raceme (the raceme or racemes spiciform); deciduous in its entirety (in *D. woodrowii*, the two mature racemes coil into a ball and are shed together), or not deciduous; when of more than one raceme, *digitate*. Primary inflorescence branches 1–9. Rachides hollowed (triquetrous), or flattened. Inflorescence espatheate; not comprising 'partial inflorescences' and foliar organs. *Spikelet-bearing axes* 'racemes'; with very slender rachides, or with substantial rachides (filiform or flattened); persistent (the rachis tough). *Spikelets solitary*; secund; biseriate; shortly pedicellate (the pedicels usually but not always shortly clavate); imbricate, or distant; emphatically *not in distinct 'long-and-short' combinations (despite being obviously 'andropogonoid')*.

Female-fertile spikelets. Spikelets strongly compressed laterally; falling with the glumes, or not disarticulating (*D. woodrowii*). Hairy callus present (shortly bearded).

Glumes two; more or less equal; awnless; carinate; with the keel conspicuously winged, or without a median keel-wing; similar. Lower glume not two-keeled; not pitted; relatively smooth; 1–3 nerved. Upper glume 1–3 nerved. *Spikelets* with incomplete florets. The incomplete florets proximal to the female-fertile florets. *The proximal incomplete florets* 1; epaleate; sterile. The proximal lemmas awnless; 0 nerved; similar in texture to the female-fertile lemmas (thinly membranous to hyaline).

Female-fertile florets 1. Lemmas less firm than the glumes; not becoming indurated; incised (or rarely entire); awnless, or awned. Awns when present, 1; usually from a sinus; geniculate; hairless (glabrous); much longer than the body of the lemma. Lemmas hairless; without a germination flap; 1–3 nerved. **Palea absent**. *Lodicules* absent. *Stamens* usually 2 (the filaments short). Anthers not penicillate. *Ovary* glabrous. Styles fused. Stigmas 2.

Fruit, embryo and seedling. Fruit small; compressed laterally (linear). Hilum short (elliptical). Embryo large. Endosperm containing compound starch grains. Embryo without an epiblast; with a scutellar tail; with an elongated mesocotyl internode. Embryonic leaf margins overlapping.

Seedling with a long mesocotyl.

Abaxial leaf blade epidermis. *Costal/intercostal zonation* conspicuous. *Papillae* present; costal and intercostal. Intercostal papillae not over-arching the stomata; several per cell (one longitudinal row per cell). *Long-cells* similar in shape costally and intercostally (narrowly so); of similar wall thickness costally and intercostally (thin walled). Mid-intercostal long-cells rectangular; having markedly sinuous walls. *Microhairs* present; more or less spherical, or elongated; ostensibly one-celled, or clearly two-celled; panicoid-type and chloridoid-type; 9–18 microns long; (7.2–)7.5–8.4(–9) microns wide at the septum. Microhair total length/width at septum 1–3. Microhair apical cells 7.5–9(–10.5) microns long. Microhair apical cell/total length ratio 0.5–0.6. *Stomata* common; 18–19.5 microns long. Subsidiaries papillate; triangular. Guard-cells overlapping to flush with the interstomatals. *Intercostal short-cells* absent or very rare; not paired (very rare); when seen not silicified. *Costal short-cells* conspicuously in long rows. Costal silica bodies horizontally-elongated crenate/sinuous (almost), or 'panicoid-type' (mostly); cross shaped, or dumb-bell shaped, or nodular.

Transverse section of leaf blade, physiology. C_4; XyMS–. *Mesophyll* with radiate chlorenchyma. *Leaf blade* adaxially flat. *Midrib* conspicuous; having a conventional arc of bundles; with colourless mesophyll adaxially. Bulliforms not in discrete, regular adaxial groups (irregularly grouped in the largely bulliform epidermis). Many of the smallest vascular bundles unaccompanied by sclerenchyma. Combined sclerenchyma girders present; forming 'figures'. Sclerenchyma all associated with vascular bundles.

Culm anatomy. *Culm internode bundles* in one or two rings, or in three or more rings.

Phytochemistry. Leaves without flavonoid sulphates (1 species).

Cytology. $2n = 14$.

Taxonomy. Panicoideae; Andropogonodae; Andropogoneae; Andropogoninae.

Ecology, geography, regional floristic distribution. 40 species. Mascarene Is., Southeast Asia, Indomalayan region, Australia, Polynesia. Mesophytic; species of open habitats. Forest glades and margins.

Holarctic, Paleotropical, and Australian. Boreal. African, Madagascan, Indomalesian, and Polynesian. Eastern Asian. Saharo-Sindian. Indian, Indo-Chinese, Malesian, and Papuan. Polynesian. North and East Australian. Tropical North and East Australian.

Rusts and smuts. Rusts — *Phakopsora*. Taxonomically wide-ranging species: *Phakopsora incompleta*. Smuts from Ustilaginaceae. Ustilaginaceae — *Sorosporium*.

References, etc. Leaf anatomical: Metcalfe 1960; this project.

Dimorphochloa S. T. Blake

~ *Cleistochloa*

Habit, vegetative morphology. Wiry, bushy perennial; caespitose. *Culms* 40–100 cm high; branched above. Culm nodes hairy. Culm internodes hollow. Young shoots extravaginal. *Leaves* not basally aggregated; non-auriculate. The sheaths short. Leaf blades narrow; not setaceous; rolled (short); without cross venation; disarticulating from the sheaths; rolled in bud. *Ligule a fringe of hairs.*

Reproductive organization. Plants bisexual, with bisexual spikelets; with hermaphrodite florets. *The spikelets of sexually distinct forms on the same plant (dimorphic, chasmogamous/cleistogamous).* Exposed-cleistogamous and chasmogamous (the cleistogamous spikelets borne singly on short leafy shoots at the previous season's nodes). Plants without hidden cleistogenes.

Inflorescence. *Inflorescence conspicuously of two kinds, terminal and axillary, with chasmogamous and cleistogamous spikelets respectively, the terminal inflorescence spikelike, a raceme or sparse panicle*; spatheate, or espatheate (according to interpretation of the limits of the inflorescence); a complex of 'partial inflorescences' and intervening foliar organs, or not comprising 'partial inflorescences' and foliar organs (according to interpretation). *Spikelet-bearing axes* 'racemes'; with very slender rachides (these triquetous or compressed); persistent. Spikelets solitary; not secund; shortly pedicellate.

Female-fertile spikelets. Spikelets 4.7–5.2 mm long; adaxial; compressed dorsiventrally; falling with the glumes. Rachilla terminated by a female-fertile floret. Hairy callus absent.

Glumes one per spikelet, or two; very unequal; (the upper) long relative to the adjacent lemmas; free; dorsiventral to the rachis; hairy (G_2, with short hairs on the back); awnless; very dissimilar (the lower minute to flimsy or absent, upper substantial and nearly equalling spikelet). Lower glume 0–1 nerved. Upper glume 5–7 nerved. *Spikelets* with incomplete florets. The incomplete florets proximal to the female-fertile florets. *Spikelets with proximal incomplete florets. The proximal incomplete florets* 1; sterile. The proximal lemmas awnless; 7 nerved; more or less equalling the female-fertile lemmas; becoming indurated, or not becoming indurated (tending to crustaceous).

Female-fertile florets 1. **Lemmas** decidedly firmer than the glumes; smooth; *becoming indurated*; entire; pointed; *mucronate (the mucro incurved)*; hairless; non-carinate; with a clear germination flap; 5–7 nerved. *Palea* present; entire; 2-nerved. *Lodicules* present; 2; free; fleshy; glabrous. *Stamens* 3. Anthers about 2 mm long; not penicillate. *Ovary* glabrous. Styles free to their bases. Stigmas 2; red pigmented.

Fruit, embryo and seedling. Fruit small (3–3.5 mm long); compressed dorsiventrally. Hilum short. Embryo large. Endosperm containing only simple starch grains. Embryo without an epiblast; with a scutellar tail; with an elongated mesocotyl internode.

Abaxial leaf blade epidermis. *Costal/intercostal zonation* conspicuous. *Papillae* absent. *Long-cells* similar in shape costally and intercostally; of similar wall thickness costally and intercostally (fairly thin walled). Mid-intercostal long-cells rectangular; having markedly sinuous walls. *Microhairs* present; panicoid-type; 57–72 microns long; 8.4–10.5 microns wide at the septum. Microhair total length/width at septum 6.3–8.6. Microhair apical cells 27–34.5 microns long. Microhair apical cell/total length ratio 0.42–0.5. *Stomata* common; 36–39 microns long. Subsidiaries dome-shaped. Guard-cells overlapping to flush with the interstomatals. *Intercostal short-cells* common; not silicified. *Costal short-cells* neither distinctly grouped into long rows nor predominantly paired. Costal silica bodies 'panicoid-type'.

Transverse section of leaf blade, physiology. C_3; XyMS+. *Mesophyll* with radiate chlorenchyma, or with non-radiate chlorenchyma (rarely). *Leaf blade* 'nodular' in section; with the ribs more or less constant in size. *Midrib* not readily distinguishable; with one bundle only. Bulliforms present in discrete, regular adaxial groups; in simple fans. All the vascular bundles accompanied by sclerenchyma. Combined sclerenchyma girders present; forming 'figures'. Sclerenchyma all associated with vascular bundles.

Taxonomy. Panicoideae; Panicodae; Paniceae.

Ecology, geography, regional floristic distribution. 1 species. Australia.

Australian. North and East Australian. Tropical North and East Australian.
References, etc. Morphological/taxonomic: Blake 1941b. Leaf anatomical: this project.

Dinebra Jacq.

Habit, vegetative morphology. *Annual*; caespitose to decumbent. *Culms* 15–120 cm high; herbaceous; unbranched above. Culm nodes glabrous. Culm internodes hollow. Young shoots intravaginal. *Leaves* not basally aggregated; non-auriculate. Leaf blades broad (rarely), or narrow; usually flat; without abaxial multicellular glands; without cross venation; persistent; rolled in bud. *Ligule a fringed membrane (very narrow)*. *Contra-ligule* absent.

Reproductive organization. Plants bisexual, with bisexual spikelets; with hermaphrodite florets.

Inflorescence. *Inflorescence of spicate main branches, or paniculate (a raceme of numerous small spikes which become deflexed at maturity, the lower spikelets of each spike often replaced by small deciduous branchlets)*; *non-digitate*; espatheate; not comprising 'partial inflorescences' and foliar organs. Spikelet-bearing axes with substantial rachides; disarticulating, or persistent; when disarticulating, falling entire (the smaller laterals deciduous). *Spikelets* solitary; *secund*; biseriate; sessile.

Female-fertile spikelets. Spikelets 3.5–10 mm long; cuneate; adaxial; compressed laterally; disarticulating above the glumes; when with two or more florets, disarticulating between the florets. *Rachilla prolonged beyond the uppermost female-fertile floret*; rachilla hairless. The rachilla extension with incomplete florets. Hairy callus absent.

Glumes present; two; more or less equal; exceeding the spikelets; long relative to the adjacent lemmas (much exceeding them); dorsiventral to the rachis; pointed; *awned (acuminate-aristate)*; carinate (subulate); very dissimilar, or similar (leathery or membranous, the lower often very asymmetrical). Lower glume much exceeding the lowest lemma; 1–2 nerved. Upper glume 1–2 nerved. *Spikelets* with incomplete florets. The incomplete florets distal to the female-fertile florets. *Spikelets without proximal incomplete florets*.

Female-fertile florets 1–2. Lemmas pointed to incised; less firm than the glumes to similar in texture to the glumes (thinly membranous); not becoming indurated; not deeply cleft (acute to emarginate); awnless to mucronate; hairy (usually, pilose on the nerves), or hairless (glabrous in *D. retroflexa*); carinate (slightly), or non-carinate; 3 nerved. *Palea* present; awnless, without apical setae; 2-nerved. *Lodicules* present; 2; free; fleshy; glabrous; toothed. *Stamens* 3. Anthers not penicillate. *Ovary* glabrous. Styles free to their bases. Stigmas 2; red pigmented.

Fruit, embryo and seedling. **Fruit** free from both lemma and palea; small; ellipsoid; shallowly concave on the hilar side; *trigonous*. Hilum short. Pericarp fused. Embryo large. Endosperm hard; without lipid; containing compound starch grains. Embryo with an epiblast; with a scutellar tail; with an elongated mesocotyl internode; with one scutellum bundle. Embryonic leaf margins meeting.

Seedling with a long mesocotyl; with a loose coleoptile. First seedling leaf with a well-developed lamina. The lamina broad; curved; 9 veined.

Abaxial leaf blade epidermis. *Costal/intercostal zonation* conspicuous. *Papillae* present; intercostal. Intercostal papillae not over-arching the stomata; consisting of one oblique swelling per cell to consisting of one symmetrical projection per cell (most intercostal long-cells with a papilla at one end). *Long-cells* similar in shape costally and intercostally; of similar wall thickness costally and intercostally (thin walled). Mid-intercostal long-cells rectangular; having markedly sinuous walls. *Microhairs* present; more or less spherical to elongated; clearly two-celled; chloridoid-type. Microhair apical cell wall of similar thickness/rigidity to that of the basal cell. Microhairs 30–34.5 microns long. Microhair basal cells 24 microns long. Microhairs 12(–13.5) microns wide at the septum. Microhair total length/width at septum 2.5–2.9. Microhair apical cells 7.5–10.5 microns long. Microhair apical cell/total length ratio 0.2–0.27. *Stomata* common; 45–46.5 microns long. Subsidiaries non-papillate; triangular. Guard-cells overlapping to flush with the interstomatals. *Intercostal short-cells* absent or very rare; not paired. Intercostal silica bodies absent. Prickles common. *Costal short-cells* conspicuously in long rows. Costal silica bodies present in alternate cell files of the costal zones; 'panicoid-type'; cross shaped, or dumb-bell shaped, or nodular.

Transverse section of leaf blade, physiology. C_4; XyMS+. PCR sheath outlines uneven. PCR sheaths of the primary vascular bundles interrupted; interrupted abaxially only. PCR sheath extensions absent. PCR cell chloroplasts centrifugal/peripheral. *Mesophyll* with radiate chlorenchyma. *Leaf blade* adaxially flat. *Midrib* not readily distinguishable; with one bundle only. Bulliforms in simple fans. All the vascular bundles accompanied by sclerenchyma. Combined sclerenchyma girders present; nowhere forming 'figures'. Sclerenchyma all associated with vascular bundles.

Cytology. Chromosome base number, $x = 10$. $2n = 20$. 2 ploid.

Taxonomy. Chloridoideae; main chloridoid assemblage.

Ecology, geography, regional floristic distribution. 3 species. Tropical Africa, Asia. Commonly adventive. Helophytic to mesophytic (in seasonally wet places); shade species, or species of open habitats; glycophytic. Savanna.

Paleotropical. African, Madagascan, and Indomalesian. Saharo-Sindian and Sudano-Angolan. Indian. Sahelo-Sudanian, Somalo-Ethiopian, South Tropical African, and Kalaharian.

Economic importance. Significant weed species: *D. retroflexa.*

References, etc. Leaf anatomical: this project.

Dinochloa Buese

Habit, vegetative morphology. Perennial. The flowering culms leafy. *Culms* 1000–3000 cm high (sympodial, zigzag); woody and persistent; to 5 cm in diameter; scandent (high-climbing); branched above. Primary branches/mid-culm node several to many. Culm sheaths persistent (with a rugose girdle at the base). Culm internodes solid, or hollow (flexuous). Plants conspicuously armed (thorny), or unarmed (but the culms rough). *Leaves* not basally aggregated; auriculate (with bristles adjoining). Leaf blades broad; 25–100 mm wide (by 8 to 45 cm long); not cordate, not sagittate (bases cuneate); pseudopetiolate; cross veined; disarticulating from the sheaths; rolled in bud. Ligule a fringed membrane ('shortly ciliate').

Reproductive organization. Plants bisexual, with bisexual spikelets; with hermaphrodite florets.

Inflorescence. Inflorescence indeterminate; with pseudospikelets; *a large fan of slender, leafless inflorescence branches up to 3 m long, with spikelets in pairs and groups at their nodes*; spatheate; a complex of 'partial inflorescences' and intervening foliar organs (but without foliage leaves). *Spikelet-bearing axes* paniculate; persistent. Spikelets not secund (the spikelet pairs and clusters alternate); pedicellate; consistently in 'long-and-short' combinations, or not in distinct 'long-and-short' combinations. The 'shorter' spikelets hermaphrodite. The 'longer' spikelets hermaphrodite.

Female-fertile spikelets. Spikelets 2–4(–5) mm long; slightly compressed laterally; falling with the glumes. Rachilla terminated by a female-fertile floret.

Glumes two, or several (2–3); very unequal (G_1 smaller); shorter than the adjacent lemmas; free; *conspicuously ventricose*; not pointed (broad, obtuse, convolute); awnless; similar. Lower glume 9 nerved (in material seen). Upper glume 9 nerved (in material seen). *Spikelets* with female-fertile florets only.

Female-fertile florets 1. Lemmas similar in texture to the glumes; mucronate (or mucronulate); 9 nerved (in the species seen). *Palea* present (glabrous); relatively long (longer than lemma); apically notched (slightly 2-pointed, the points touching); awnless, without apical setae (mucronate); thinner than the lemma; not indurated; several nerved (11 in the species seen); keel-less. *Lodicules* absent. *Stamens* 6 (short). Anthers 4 mm long; penicillate, or not penicillate; with the connective apically prolonged (acute). *Ovary* glabrous; with a conspicuous apical appendage. The appendage broadly conical, fleshy. Styles fused (into a single style with a wide solid base). Stigmas 3 (slightly plumose).

Fruit, embryo and seedling. Fruit subglobose (spheroidal); not noticeably compressed. Pericarp thin, or fleshy (i.e. pericarp sometimes thick, wrinkled when dry). Embryo large (1/2 to 1/3 as long as fruit). *Seed 'non-endospermic'.*

Abaxial leaf blade epidermis. *Costal/intercostal zonation* conspicuous. *Papillae* present; intercostal. Intercostal papillae over-arching the stomata; several per cell (variously shaped, sometimes exclusively forming rings over and around the stomata). *Long-cells* similar in shape costally and intercostally; of similar wall thickness costally and intercostally. Mid-intercostal long-cells rectangular; having markedly sinuous walls (thin). *Microhairs* present; clearly two-celled, or uniseriate (occasionally 3-celled); panicoid-type (rather variable in form); (60–)66–75(–78) microns long (in *D. pubivanea*); 5.4–19.5 microns wide at the septum. Microhair total length/width at septum 6.7–14.4 (i.e. very variable, in *D. pubivanea*). Microhair apical cells (31.5–)40.5–42(–51) microns long. Microhair apical cell/total length ratio 0.53–0.65. *Stomata* common (obscured by papillae); 28.5–30 microns long (in *D. pubivanea*). Subsidiaries non-papillate. Guard-cells overlapped by the interstomatals. *Intercostal short-cells* common; in cork/silica-cell pairs; silicified. Intercostal silica bodies tall-and-narrow. *Costal short-cells* conspicuously in long rows, or predominantly paired, or neither distinctly grouped into long rows nor predominantly paired. Costal silica bodies saddle shaped, or tall-and-narrow, or crescentic, or oryzoid.

Transverse section of leaf blade, physiology. C_3; XyMS+. *Mesophyll* with non-radiate chlorenchyma; with adaxial palisade; with arm cells; with fusoids (very large). The fusoids external to the PBS. *Leaf blade* adaxially flat. *Midrib* conspicuous; having complex vascularization. Bulliforms present in discrete, regular adaxial groups; in simple fans. All the vascular bundles accompanied by sclerenchyma. Combined sclerenchyma girders present; forming 'figures' (most bundles).

Cytology. Chromosome base number, $x = 12$. $2n = 72$.

Taxonomy. Bambusoideae; Bambusodae; Bambuseae.

Ecology, geography, regional floristic distribution. About 25 species. Southeast Asia, Indo-Malaya.

Paleotropical. Indomalesian. Indo-Chinese and Malesian.

References, etc. Leaf anatomical: Metcalfe 1960; this project.

Diplachne P. Beauv.

~ *Leptochloa*

Excluding *Kengia*

Habit, vegetative morphology. Perennial; stoloniferous, or caespitose (some tall). **Culms** 30–270 cm high; ***herbaceous***; unbranched above. Culm nodes glabrous. Culm internodes hollow. *Leaves* mostly basal, or not basally aggregated; non-auriculate. Leaf blades linear; broad, or narrow; flat, or rolled (often involute); without abaxial multicellular glands; without cross venation; persistent; rolled in bud. Ligule an unfringed membrane, or a fringed membrane, or a fringe of hairs (*D. chloridiformis, D. monticola*). *Contra-ligule* present (of hairs), or absent.

Reproductive organization. *Plants bisexual, with bisexual spikelets*; with hermaphrodite florets. Exposed-cleistogamous, or chasmogamous. Plants with hidden cleistogenes, or without hidden cleistogenes. The hidden cleistogenes when present, in the leaf sheaths.

Inflorescence. *Inflorescence* of spicate main branches (a contracted panicle of spikelike racemes), or a single raceme (rarely), or paniculate; ***open***; with capillary branchlets, or without capillary branchlets; digitate, or non-digitate; espatheate; not comprising 'partial inflorescences' and foliar organs. The racemes spikelet bearing to the base. Spikelet-bearing axes persistent. *Spikelets not secund (or scarcely so)*; biseriate; subsessile to pedicellate; usually more or less distant.

Female-fertile spikelets. *Spikelets* 6–15 mm long (narrow); *not noticeably compressed to compressed dorsiventrally (more or less terete)*; disarticulating above the glumes; disarticulating between the florets. Rachilla prolonged beyond the uppermost female-fertile floret; rachilla hairless (glabrous or scaberulous, the internode apices widened and oblique).

Glumes two; relatively large; very unequal to more or less equal; shorter than the spikelets; shorter than the adjacent lemmas; dorsiventral to the rachis, or lateral to the rachis, or displaced; hairless; glabrous; awnless; carinate; similar (membranous). Lower glume shorter than the lowest lemma; 1 nerved. Upper glume 1 nerved. *Spikelets* with female-fertile florets only, or with incomplete florets. The incomplete florets distal to the female-fertile florets. The distal incomplete florets awnless.

Female-fertile florets 5–20. Lemmas not becoming indurated; entire, or incised; blunt; when incised, 2 lobed (bidentate); not deeply cleft; awnless, or mucronate (from the sinus), or awned. Awns when present, 1; from a sinus, or dorsal; from near the top; non-geniculate; much shorter than the body of the lemma. *Lemmas hairy (the margins ciliate below, and usually with a tuft of hairs below the middle of the mid-nerve), or hairless.* The hairs in tufts (usually), or not in tufts. *Lemmas* non-carinate (rarely keeled); *3 nerved. Palea* present; entire (e.g. *D. eleusine*), or apically notched; not indurated (membranous); 2-nerved; 2-keeled. Palea keels hairy (ciliate, below). **Lodicules** present; 2; free; *fleshy*; glabrous. *Stamens* 3. Anthers not penicillate. *Ovary* glabrous. Styles free to their bases. Stigmas 2; white.

Fruit, embryo and seedling. *Fruit* free from both lemma and palea; small; compressed laterally, or compressed dorsiventrally. Hilum short. Pericarp free, or loosely adherent, or fused. Embryo large; not waisted. Endosperm hard; without lipid; containing compound starch grains. Embryo with an epiblast; with a scutellar tail; with an elongated mesocotyl internode. Embryonic leaf margins meeting.

Seedling with a short mesocotyl. First seedling leaf with a well-developed lamina. The lamina broad; curved; 7 veined.

Abaxial leaf blade epidermis. *Costal/intercostal zonation* conspicuous. *Papillae* present; costal, or intercostal, or costal and intercostal. Intercostal papillae not overarching the stomata; consisting of one symmetrical projection per cell (finger-like). *Long-cells* similar in shape costally and intercostally (narrowly so); of similar wall thickness costally and intercostally (thin walled). Mid-intercostal long-cells rectangular; having markedly sinuous walls. *Microhairs* present; more or less spherical; clearly two-celled; chloridoid-type. Microhair apical cell wall of similar thickness/rigidity to that of the basal cell. Microhairs 12–18–19.5 microns long. Microhair basal cells 9 microns long. Microhairs 9 microns wide at the septum. Microhair total length/width at septum 1.3–2.2. Microhair apical cells (6.6–)9–9.6(–10.5) microns long. Microhair apical cell/total length ratio 0.46–0.58. *Stomata* common; 21–24 microns long. Subsidiaries triangular. Guard-cells overlapping to flush with the interstomatals. *Intercostal short-cells* common; not paired (mostly solitary); silicified and not silicified. Intercostal silica bodies absent, or imperfectly developed. *Costal short-cells* conspicuously in long rows. Costal silica bodies present in alternate cell files of the costal zones; saddle shaped.

Transverse section of leaf blade, physiology. C_4; XyMS+. PCR sheath outlines even. PCR sheaths of the primary vascular bundles interrupted; interrupted both abaxially and adaxially. PCR sheath extensions absent. PCR cell chloroplasts centripetal. *Mesophyll* with non-radiate chlorenchyma. *Leaf blade* 'nodular' in section; with the ribs very irregular in sizes. *Midrib* conspicuous; having a conventional arc of bundles; with colourless mesophyll adaxially. Bulliforms present in discrete, regular adaxial groups; in simple fans. All the vascular bundles accompanied by sclerenchyma. Combined sclerenchyma girders present; forming 'figures'. Sclerenchyma all associated with vascular bundles. The lamina margins with fibres.

Culm anatomy. *Culm internode bundles* in one or two rings.

Phytochemistry. Leaves without flavonoid sulphates (1 species).

Cytology. Chromosome base number, $x = 10$. $2n = 40$ and 60. 4 and 6 ploid.

Taxonomy. Chloridoideae; main chloridoid assemblage.

Ecology, geography, regional floristic distribution. 18 species. Tropical and subtropical. Commonly adventive. Helophytic, or mesophytic, or xerophytic; shade species and species of open habitats; halophytic and glycophytic. Woodland, savanna, dry and swampy soils.

Holarctic, Paleotropical, Neotropical, Cape, Australian, and Antarctic. Boreal, Tethyan, and Madrean. African, Madagascan, and Indomalesian. Euro-Siberian, Eastern Asian, and Atlantic North American. Irano-Turanian. Saharo-Sindian, Sudano-Angolan, West African Rainforest, and Namib-Karoo. Indian, Indo-Chinese, and Malesian. Central Brazilian and Pampas. North and East Australian and Central Australian. Patagonian.

European and Siberian. Canadian-Appalachian and Southern Atlantic North American. Sahelo-Sudanian, Somalo-Ethiopian, South Tropical African, and Kalaharian. Tropical North and East Australian and Temperate and South-Eastern Australian.

Rusts and smuts. Smuts from Ustilaginaceae. Ustilaginaceae — *Ustilago*.

Economic importance. Significant weed species: *D. fascicularis, D. fusca*.

References, etc. Leaf anatomical: this project.

Diplopogon R.Br.

Dipogonia P. Beauv.

~ *Amphipogon*

Habit, vegetative morphology. Perennial; caespitose. *Culms* 30–60 cm high; herbaceous; unbranched above. Culm nodes glabrous. Leaves non-auriculate. Leaf blades narrow; setaceous (and convolute); without cross venation; persistent. Ligule a fringe of hairs.

Reproductive organization. *Plants* bisexual, with bisexual spikelets; *with hermaphrodite florets*. The spikelets of sexually distinct forms on the same plant; hermaphrodite and sterile; overtly heteromorphic (the sterile spikelets reduced to setaceous bracts).

Inflorescence. Inflorescence paniculate; contracted; capitate to more or less ovoid; espatheate; not comprising 'partial inflorescences' and foliar organs. Spikelet-bearing axes persistent. **Spikelets *associated with bractiform involucres (constituted by the glumes of the lowermost spikelets)***; not secund; shortly pedicellate, or subsessile.

Female-sterile spikelets. *The sterile spikelets forming a fairly inconspicuous involucre of linear, convolute, subulate or setaceous bracts at the base of the panicle*.

Female-fertile spikelets. Spikelets compressed laterally; disarticulating above the glumes. Rachilla terminated by a female-fertile floret.

Glumes two; more or less equal (G_1 being slightly longer than G_2); long relative to the adjacent lemmas (exceeding them); pointed (acuminate); awned (the G_1s usually being acuminate into a long straight point); non-carinate (rounded on the back); similar (lanceolate and hyaline). Lower glume 1 nerved. Upper glume 1 nerved. *Spikelets* with female-fertile florets only.

Female-fertile florets 1. *Lemmas* narrow bodied; *convolute (below)*; decidedly firmer than the glumes (firmly chartaceous); not becoming indurated; incised; 3 lobed; deeply cleft; awned. *Awns* 3; median and lateral (the lateral lobes acuminate into short, erect, scabrid awns, the median much longer); the median different in form from the laterals; from a sinus (i.e., from between the lateral lobes); *non-geniculate (but much flattened below, twisted and laterally displacing)*; hairless; about as long as the body of the lemma to much longer than the body of the lemma; entered by one vein. The lateral awns shorter than the median (erect, scabrid). Lemmas hairy (with silky hairs below); non-carinate; 3 nerved. *Palea* present; relatively long; awned (apically 2-awned); not indurated (membranous); 2-nerved. *Lodicules* present; 2; free; fleshy. *Stamens* 3. Anthers not penicillate. *Ovary* glabrous. Styles free to their bases. Stigmas 2.

Fruit, embryo and seedling. Fruit small. Hilum short. Embryo small.

Abaxial leaf blade epidermis. *Costal/intercostal zonation* lacking. *Papillae* present. Intercostal papillae not over-arching the stomata. Mid-intercostal long-cells rectangular; having markedly sinuous walls. *Microhairs* absent (but present adaxially); panicoid type. *Stomata* absent or very rare. Subsidiaries non-papillate. *Intercostal short-cells* common. *Costal short-cells* neither distinctly grouped into long rows nor predominantly paired. Costal silica bodies tall-and-narrow and crescentic.

Transverse section of leaf blade, physiology. C_3; XyMS+. *Mesophyll* with non-radiate chlorenchyma. *Leaf blade* with distinct, prominent adaxial ribs, or 'nodular' in section; with the ribs more or less constant in size. *Midrib* not readily distinguishable; with one bundle only. Bulliforms present in discrete, regular adaxial groups to not in discrete, regular adaxial groups; in simple fans (poorly differentiated fans, to *Ammophila*-type aggregations of small cells, in the furrows). All the vascular bundles accompanied by sclerenchyma. Combined sclerenchyma girders absent. Sclerenchyma not all bundle-associated. The 'extra' sclerenchyma in abaxial groups; abaxial-hypodermal, the groups isolated (opposite the adaxial furrows).

Taxonomy. Arundinoideae (?); Amphipogoneae.

Ecology, geography, regional floristic distribution. 1 species. Western Australia. Australian. South-West Australian.

References, etc. Morphological/taxonomic: Macfarlane and Watson 1980. Leaf anatomical: this project.

Dissanthelium Trin.

Graminiastrum Krause, *Phalaridium* Nees & Meyen, *Stenochloa* Nutt

Habit, vegetative morphology. Annual, or perennial (mostly dwarf); rhizomatous, or stoloniferous, or caespitose, or decumbent. *Culms herbaceous*. Leaves non-auriculate. Leaf blades narrow; setaceous, or not setaceous; without cross venation. **Ligule** present; *an unfringed membrane*; not truncate; 2–6 mm long.

Reproductive organization. Plants bisexual, with bisexual spikelets; with hermaphrodite florets.

Inflorescence. Inflorescence paniculate; open, or contracted; often spicate; espatheate; not comprising 'partial inflorescences' and foliar organs. Spikelet-bearing axes persistent. Spikelets not secund; pedicellate.

Female-fertile spikelets. Spikelets 2.5–5 mm long; compressed laterally; disarticulating above the glumes; disarticulating between the florets. Rachilla prolonged beyond the uppermost female-fertile floret. The rachilla extension naked. *Hairy callus absent*.

Glumes two; more or less equal; about equalling the spikelets; *long relative to the adjacent lemmas (often exceeding all the florets)*; pointed (acute); awnless; carinate; similar (firm, membranous to herbaceous). Lower glume 1–3 nerved. Upper glume 3 nerved. *Spikelets with female-fertile florets only*.

Female-fertile florets *2(–3)*. *Lemmas less firm than the glumes (membranous)*; not becoming indurated; entire, or incised; when entire pointed, or blunt; awnless; hairy, or hairless; carinate; without a germination flap; 3–5 nerved. **Palea** present; *conspicuous but relatively short*; tightly clasped by the lemma; 2-nerved; 2-keeled. *Lodicules* present; 2; free; membranous; glabrous; toothed; not or scarcely vascularized. *Stamens* 3. *Ovary* glabrous. Styles free to their bases. Stigmas 2; white.

Fruit, embryo and seedling. *Fruit* free from both lemma and palea; small (1.3 mm long); compressed dorsiventrally. Hilum short. Embryo small. Endosperm hard; with lipid; containing compound starch grains. Embryo with an epiblast; without a scutellar tail; with a negligible mesocotyl internode. Embryonic leaf margins meeting.

Abaxial leaf blade epidermis. *Costal/intercostal zonation* conspicuous. *Papillae* absent. *Long-cells* markedly different in shape costally and intercostally (the costals narrower); of similar wall thickness costally and intercostally. Mid-intercostal long-cells rectangular; having markedly sinuous walls. *Microhairs* absent. *Stomata* common. Subsidiaries parallel-sided, dome-shaped, and triangular (low to high); including both triangular and parallel-sided forms on the same leaf (and high and low domes). Guard-cells overlapped by the interstomatals (slightly). *Intercostal short-cells* common; in cork/silica-cell pairs; silicified. Intercostal silica bodies mostly crescentic, or tall-and-narrow. *Costal short-cells* predominantly paired. Costal silica bodies rounded, or crescentic (crescents predominating).

Transverse section of leaf blade, physiology. C_3; XyMS+. *Mesophyll* with non-radiate chlorenchyma; without adaxial palisade. *Leaf blade* adaxially flat. *Midrib* conspicuous (by its 'hinges', abaxial keel, and the large vascular bundle possessing only an abaxial strand); with one bundle only. *The lamina* symmetrical on either side of the midrib (save for the conspicuous 'midrib hinges'). Bulliforms not in discrete, regular adaxial groups. All the vascular bundles accompanied by sclerenchyma. Combined sclerenchyma girders present; forming 'figures' (with the largest bundles, except that of the midrib). Sclerenchyma all associated with vascular bundles.

Taxonomy. Pooideae; Poodae; Poeae.

Ecology, geography, regional floristic distribution. 17 species. Mostly Andean, also California Islands, Mexico, Argentina. Species of open habitats. Mostly in High Andean puna.

Holarctic and Neotropical. Madrean. Andean.

References, etc. Leaf anatomical: this project.

Dissochondrus (Hillebr.) Kuntze

Habit, vegetative morphology. Perennial; caespitose. Culm nodes hairy. *Leaves* not basally aggregated; auriculate. *Leaf blades broad*; pseudopetiolate; without cross venation; disarticulating from the sheaths.

Reproductive organization. Plants bisexual, with bisexual spikelets; with hermaphrodite florets.

Inflorescence. Inflorescence paniculate; contracted; spicate. Inflorescence axes not ending in spikelets. Inflorescence espatheate; not comprising 'partial inflorescences' and foliar organs. Spikelet-bearing axes persistent. **Spikelets** *subtended by solitary 'bristles', or with 'involucres' of 'bristles' (these long, sinuous, antrorsely scabrid).* The 'bristles' persisting on the axis. Spikelets not secund; pedicellate. Pedicel apices cupuliform.

Female-fertile spikelets. Spikelets 3 mm long; compressed dorsiventrally; falling with the glumes. Rachilla terminated by a female-fertile floret. Hairy callus absent.

Glumes two; very unequal; (the longer) long relative to the adjacent lemmas; hairless; scabrous; pointed; awnless; non-carinate; similar. Lower glume 5 nerved. Upper glume 7–9 nerved. *Spikelets with female-fertile florets only.*

Female-fertile florets 2 *(i.e. both the florets hermaphrodite).* Lemmas decidedly firmer than the glumes; becoming indurated, or not becoming indurated (leathery or crustaceous); entire; pointed; awnless; hairless; non-carinate; having the margins tucked in onto the palea; with a clear germination flap; 3–5 nerved. *Palea* present; relatively long; textured like the lemma; 2-nerved; keel-less. *Lodicules* present; 2; free; fleshy; glabrous. *Stamens* 3. Anthers not penicillate; without an apically prolonged connective (but the anthers divaricate). *Ovary* glabrous. Styles free to their bases. Stigmas 2.

Fruit, embryo and seedling. Fruit small.

Abaxial leaf blade epidermis. *Costal/intercostal zonation* conspicuous. *Papillae* absent. Mid-intercostal long-cells fusiform; having markedly sinuous walls. *Microhairs* present (equalling the long-cells in length); panicoid-type; (60–)66–81(–87) microns long; (27–)30–44(–57) microns wide at the septum. Microhair total length/width at septum 6.1–19.3. Microhair apical cells (4.2–)5.4–6 microns long. Microhair apical cell/total length ratio 0.45–0.66. *Stomata* common; 21–24 microns long. Subsidiaries dome-shaped. Guard-cells overlapping to flush with the interstomatals. *Intercostal short-cells* common (fairly); in cork/silica-cell pairs. *Costal short-cells* conspicuously in long rows. Costal silica bodies 'panicoid-type'; cross shaped, dumb-bell shaped, and nodular (mostly nodular).

Transverse section of leaf blade, physiology. C_4. The anatomical organization unconventional. Organization of PCR tissue *Arundinella* type. XyMS–. *Mesophyll* with radiate chlorenchyma (slightly); exhibiting 'circular cells' (solitaries, small groups and some obviously 'reduced bundles'). *Leaf blade* adaxially flat. Midrib with one bundle only to having a conventional arc of bundles (a small bundle under each midrib 'hinge'). Bulliforms present in discrete, regular adaxial groups; in simple fans. All the vascular bundles accompanied by sclerenchyma. Combined sclerenchyma girders present; forming 'figures'. Sclerenchyma all associated with vascular bundles.

Taxonomy. Panicoideae; Panicodae; Paniceae.

Ecology, geography, regional floristic distribution. 1 species. Hawaii. Shade species. On slopes.

Paleotropical. Polynesian. Hawaiian.

References, etc. Leaf anatomical: this project.

Special comments. Fruit data wanting.

Distichlis Raf.

Trisiola Raf. (1825)

Excluding *Allolepis (D. texana)*

Habit, vegetative morphology. Perennial; *saltgrass, with thick, conspicuously distichous leaves*. Culms 10–70 cm high; herbaceous. Culm nodes glabrous. Culm internodes solid. **Leaves *not basally aggregated***; non-auriculate. *Leaf blades* narrow; without abaxial multicellular glands; without cross venation; *disarticulating from the sheaths*; rolled in bud. Ligule a fringed membrane to a fringe of hairs (?).

Reproductive organization. *Plants monoecious with all the fertile spikelets unisexual (rarely), or dioecious*; without hermaphrodite florets. The spikelets of sexually distinct forms on the same plant (when monoecious), or all alike in sexuality (on the same plant, i.e., when dioecious); female-only, or male-only, or female-only and male-only. Plants outbreeding (at least when dioecious).

Inflorescence. Inflorescence reduced to a single spikelet (*D. australis*), or few spikeleted, or many spikeleted; of spicate main branches (of short racemes on a central axis), or a single raceme, or paniculate (or reduced); contracted; more or less ovoid, or spicate, or more or less irregular; non-digitate; espatheate; not comprising 'partial inflorescences' and foliar organs. Spikelet-bearing axes persistent. Spikelets solitary; not secund; pedicellate.

Female-sterile spikelets. Male spikelets similar to the females, but of thinner texture. Male florets 3 staminate.

Female-fertile spikelets. *Spikelets* 6–28 mm long; compressed laterally; tardily disarticulating above the glumes; *disarticulating between the florets*. Rachilla prolonged beyond the uppermost female-fertile floret; rachilla hairless. Hairy callus absent.

Glumes two; very unequal; shorter than the spikelets; shorter than the adjacent lemmas; pointed; awnless; similar (firm). Lower glume 1–5 nerved. Upper glume 4–9 nerved.

Female-fertile florets *3–20*. Lemmas not becoming indurated; entire; pointed; awnless; hairless; glabrous; carinate to non-carinate; (3–)7–9(–11) nerved. *Palea* present; relatively long; leathery; 2-nerved; 2-keeled. Palea keels more or less winged. *Lodicules* present; 2; free; fleshy, or membranous; glabrous; toothed. *Stamens* 0. *Ovary* glabrous. Styles free to their bases. Stigmas 2.

Fruit, embryo and seedling. Fruit ellipsoid. *Hilum short*. Endosperm hard; containing only simple starch grains. Embryo with an epiblast; with a scutellar tail; with an elongated mesocotyl internode. Embryonic leaf margins meeting.

Abaxial leaf blade epidermis. *Costal/intercostal zonation* conspicuous. *Papillae* present (a few, usually near stomata), or absent (but abundant adaxially); intercostal. Intercostal papillae over-arching the stomata to not over-arching the stomata; consisting of one oblique swelling per cell. *Long-cells* similar in shape costally and intercostally; of similar wall thickness costally and intercostally (fairly thick walled). Mid-intercostal long-cells rectangular; having markedly sinuous walls. *Microhairs* present, or absent; more or less spherical, or elongated; ostensibly one-celled, or clearly two-celled; chloridoid-type. Microhair apical cell wall of similar thickness/rigidity to that of the basal cell. Microhairs with 'partitioning membranes' (in *D. stricta*). The 'partitioning membranes' in the basal cell. Microhairs when seen 30–32 microns long. Microhair basal cells 24 microns long. Microhairs about 13 microns wide at the septum. Microhair total length/width at septum 2.3. Microhair apical cells 7.5 microns long. Microhair apical cell/total length ratio 0.24. *Stomata* common; 21–24 microns long. Subsidiaries dome-shaped, or dome-shaped and triangular. Guard-cells overlapped by the interstomatals, or overlapping to flush with the interstomatals. *Intercostal short-cells* common; in cork/silica-cell pairs (occasionally), or not paired (solitary); silicified (rarely), or not silicified. Intercostal silica bodies absent, or imperfectly developed; when present, tall-and-narrow. *Costal short-cells* predominantly paired, or neither distinctly grouped into long rows nor predominantly paired. Costal silica bodies present throughout the costal zones; rounded (common), or saddle shaped, or tall-and-narrow (or more or less cuboid).

Transverse section of leaf blade, physiology. C_4; XyMS+. PCR sheath outlines even. PCR sheaths of the primary vascular bundles interrupted; interrupted abaxially only.

PCR sheath extensions present. Maximum number of extension cells 1. PCR cell chloroplasts centripetal. *Mesophyll* with radiate chlorenchyma; traversed by columns of colourless mesophyll cells. *Leaf blade* with distinct, prominent adaxial ribs; with the ribs more or less constant in size. *Midrib* not readily distinguishable; with one bundle only. Bulliforms present in discrete, regular adaxial groups; in simple fans (irregular, of small cells), or associated with colourless mesophyll cells to form deeply-penetrating fans (sometimes connected with traversing colourless columns). All the vascular bundles accompanied by sclerenchyma. Combined sclerenchyma girders present; forming 'figures'. Sclerenchyma all associated with vascular bundles. The lamina margins with fibres.

Phytochemistry. Tissues of the culm bases with abundant starch. Leaves without flavonoid sulphates (1 species).

Cytology. Chromosome base number, $x = 10$. $2n = 40$ (in *D. stricta*). 4 ploid.

Taxonomy. Chloridoideae; main chloridoid assemblage.

Ecology, geography, regional floristic distribution. About 6 species. 12 in America, 1 in Australia. Species of open habitats; halophytic. Seashores and deserts.

Holarctic, Neotropical, Australian, and Antarctic. Boreal and Madrean. Atlantic North American and Rocky Mountains. Caribbean, Pampas, and Andean. North and East Australian. Patagonian. Canadian-Appalachian, Southern Atlantic North American, and Central Grasslands. Temperate and South-Eastern Australian.

Rusts and smuts. Rusts — *Puccinia*. Smuts from Ustilaginaceae. Ustilaginaceae — *Ustilago*.

Economic importance. Significant weed species: *D. spicata*, *D. stricta*.

References, etc. Leaf anatomical: Metcalfe 1960; this project.

Special comments. Fruit data wanting.

Drake-Brockmania Stapf

Excluding *Heterocarpha*

Habit, vegetative morphology. Annual, or perennial; caespitose, or decumbent. *Culms* 5–65 cm high; herbaceous; branched above, or unbranched above. Plants unarmed. Young shoots intravaginal. *Leaves* not basally aggregated; non-auriculate. Leaf blades linear; narrow; flat; exhibiting multicellular glands abaxially (at the base of hairs). The abaxial glands intercostal. Leaf blades without cross venation. Ligule an unfringed membrane; truncate; 0.7–1.5 mm long.

Reproductive organization. Plants bisexual, with bisexual spikelets; with hermaphrodite florets.

Inflorescence. Inflorescence of spicate main branches (of short, broad spikes on a central axis); contracted; non-digitate (the axis short), or subdigitate. Primary inflorescence branches 3–6. Inflorescence with axes ending in spikelets. Inflorescence espatheate; not comprising 'partial inflorescences' and foliar organs. *Spikelet-bearing axes disarticulating*; falling entire (the spikes deciduous). Spikelets secund; biseriate; subsessile; imbricate.

Female-fertile spikelets. Spikelets 6–14 mm long; strongly compressed laterally; disarticulating above the glumes (but the spikes falling first); eventually disarticulating between the florets. Rachilla prolonged beyond the uppermost female-fertile floret. The rachilla extension with incomplete florets. Hairy callus absent.

Glumes two; very unequal (G_2 conspicuous, spreading); shorter than the spikelets; (the upper) long relative to the adjacent lemmas (almost as long); awnless; carinate; glumes lanceolate. Lower glume much exceeding the lowest lemma; 1–5 nerved. Upper glume 7–17 nerved. *Spikelets* with incomplete florets. The incomplete florets distal to the female-fertile florets. *The distal incomplete florets merely underdeveloped*.

Female-fertile florets *5–18*. Lemmas lemma papery; not becoming indurated; entire; pointed; mucronate (or cuspidate); hairy (villous below on keel and margins); carinate; without a germination flap; (3–)5–7 nerved. *Palea* present; relatively long; entire to apically notched; awnless, without apical setae; 2-nerved; 2-keeled. Palea keels winged, or wingless. *Lodicules* present; 2; glabrous. *Stamens* 3. Anthers not penicillate; without an apically prolonged connective. *Ovary* glabrous. Styles free to their bases. Stigmas 2.

Fruit, embryo and seedling. Fruit ellipsoid; trigonous. Hilum short. *Pericarp loosely adherent (removable when wet)*. Embryo large.

Abaxial leaf blade epidermis. *Costal/intercostal zonation* conspicuous. *Papillae* present; costal and intercostal. Intercostal papillae over-arching the stomata (slightly), or not over-arching the stomata; consisting of one oblique swelling per cell to consisting of one symmetrical projection per cell (thick-walled). *Long-cells* markedly different in shape costally and intercostally (the costals longer and narrower). Intercostal zones with typical long-cells to exhibiting many atypical long-cells (many of them quite short). Mid-intercostal long-cells rectangular; having markedly sinuous walls to having straight or only gently undulating walls. *Microhairs* present; more or less spherical; ostensibly one-celled; chloridoid-type (obviously glandular). Microhair apical cell wall of similar thickness/rigidity to that of the basal cell. Microhairs 21–22.5(–26) microns long. Microhair basal cells 9 microns long. Microhair total length/width at septum 1. Microhair apical cell/total length ratio 0.5. *Stomata* common; 15–18 microns long. Subsidiaries dome-shaped (mostly), or triangular (a few). Guard-cells overlapping to flush with the interstomatals. *Intercostal short-cells* absent or very rare (but lots of short 'long-cells'); not paired. Intercostal silica bodies absent. With a few costal prickles. *Crown cells* absent. *Costal short-cells* conspicuously in long rows. Costal silica bodies present in alternate cell files of the costal zones; 'panicoid-type'; mostly dumb-bell shaped.

Transverse section of leaf blade, physiology. C_4; XyMS+. PCR sheaths of the primary vascular bundles interrupted; interrupted both abaxially and adaxially. PCR sheath extensions absent. *Leaf blade* 'nodular' in section (with large and small 'nodules'). *Midrib* not readily distinguishable; with one bundle only. Bulliforms present in discrete, regular adaxial groups (these infrequent); in simple fans. All the vascular bundles accompanied by sclerenchyma. Combined sclerenchyma girders present (with most bundles); forming 'figures' (with most bundles). Sclerenchyma all associated with vascular bundles.

Taxonomy. Chloridoideae; main chloridoid assemblage.

Ecology, geography, regional floristic distribution. 1 species (*D. somalensis*). Eastern Africa. Mesophytic; species of open habitats. Damp places in savanna.

Paleotropical. African. Sudano-Angolan. Sahelo-Sudanian and Somalo-Ethiopian.

References, etc. Morphological/taxonomic: Phillips 1974. Leaf anatomical: this project.

Special comments. Generic circumscriptions around *Brachychloa*, *Drake-Brockmania* and *Heterocarpha* seem problematical, especially in the absence of anatomical data for *H. haareri* and *B. fragilis*.

Dregeochloa Conert

~ *Rytidosperma, Danthonia* sensu lato

Habit, vegetative morphology. Perennial; stoloniferous (sometimes), or caespitose (with short, often branched creeping rhizomes). **Culms** 40–250 cm high; *herbaceous*; unbranched above (but usually considerably branched just below the soil surface). Culm nodes glabrous. Culm internodes solid. Plants unarmed. Young shoots intravaginal. *Leaves* mostly basal, or not basally aggregated; non-auriculate. Hair-tufted at the mouth of the sheath. Leaf blades linear, or ovate-lanceolate to ovate; narrow (very rigid); to 3 mm wide; setaceous, or not setaceous; usually folded; without cross venation; persistent. *Ligule a fringe of hairs*; minute, to 1 mm long.

Reproductive organization. *Plants bisexual, with bisexual spikelets*; with hermaphrodite florets.

Inflorescence. Inflorescence few spikeleted (4–12); a single raceme, or paniculate (rarely); when paniculate, contracted (and reduced); espatheate; not comprising 'partial inflorescences' and foliar organs. Spikelet-bearing axes persistent. Spikelets solitary; not secund; not two-ranked; pedicellate (the rachis and pedicels hairy); distant.

Female-fertile spikelets. Spikelets 10–13 mm long; somewhat compressed laterally; disarticulating above the glumes; disarticulating between the florets; with conventional internode spacings. Rachilla prolonged beyond the uppermost female-fertile floret; rachilla hairy, or hairless. The rachilla extension with incomplete florets. Hairy callus present (with a beard on each side). *Callus* long; pointed.

Glumes two; more or less equal; about equalling the spikelets to exceeding the spikelets; *long relative to the adjacent lemmas*; minutely hairy, or hairless; when hairless glabrous; pointed (acute); awnless; non-carinate; similar (lanceolate, scarious or herbaceous below, G_1 narrower). Lower glume 5 nerved. Upper glume 5–7 nerved. *Spikelets* with incomplete florets. The incomplete florets distal to the female-fertile florets. *The distal incomplete florets* merely underdeveloped; awned. *Spikelets without proximal incomplete florets*.

Female-fertile florets *3–8*. Lemmas similar in texture to the glumes (membranous); not becoming indurated; incised; 2 lobed (the lobes acute or bristle-tipped); deeply cleft (to about 1/3); awned. Awns 1; median; from a sinus; geniculate; hairless; about as long as the body of the lemma; entered by one vein; persistent. Lemmas hairy. *The hairs in tufts; in transverse rows (the lobes minutely hairy, a row of tufts at their base, and larger marginal tufts beneath)*. Lemmas non-carinate; 7–9 nerved; with the nerves non-confluent. *Palea* present; relatively long; entire to apically notched; awnless, without apical setae; thinner than the lemma; not indurated (hyaline); 2-nerved; 2-keeled. Palea keels wingless; hairy. *Lodicules* present; 2; free; fleshy; glabrous. *Stamens* 3. Anthers 3 mm long; not penicillate; without an apically prolonged connective. **Ovary** sparsely *hairy*. Styles free to their bases. Stigmas 2; red pigmented.

Fruit, embryo and seedling. *Fruit* free from both lemma and palea; small; longitudinally grooved; compressed dorsiventrally. Hilum short (punctiform). *Pericarp thick and hard*; free.

Abaxial leaf blade epidermis. *Costal/intercostal zonation* conspicuous. *Papillae* absent. Long-cells differing markedly in wall thickness costally and intercostally (the costals very thick-walled). *Microhairs* present (perhaps?), or absent (apparently, though the hairy grooves are inaccessible: present adaxially); panicoid-type. *Stomata* hidden. *Costal short-cells* predominantly paired. Costal silica bodies tall-and-narrow.

Transverse section of leaf blade, physiology. C_3 (obviously so in *D. pumila*, but but the anatomy of *D. calvinensis* is equivocal, to say the least: most mesophyll cells are no more than one cell distant, and the only seeming exceptions are at the tops of the adaxial ribs. A candidate for intermediacy); XyMS+. *Leaf blade* with round-topped adaxial ribs, and broad, flat-topped abaxial ribs with deep intercostal furrows; with the ribs more or less constant in size (adaxially). *Midrib* not readily distinguishable; with one bundle only. Bulliforms present in discrete, regular adaxial groups (in the adaxial furrows); in simple fans (but these deeply penetrating in *D. pumila*). All the vascular bundles accompanied by sclerenchyma. Combined sclerenchyma girders absent (the sclerenchyma 'strands' in hypodermal bands in the adaxial and abaxial ribs). Sclerenchyma all associated with vascular bundles.

Taxonomy. Arundinoideae; Danthonieae.

Ecology, geography, regional floristic distribution. 2 species. Drier parts of southern Africa. Xerophytic; species of open habitats; halophytic (sometimes), or glycophytic (usually). *D. pumila* in blown sand over rocks.

Paleotropical and Cape. African. Namib-Karoo.

References, etc. Leaf anatomical: this project; photos of *D. pumila* provided by R.P. Ellis.

Dryopoa Vick.

Habit, vegetative morphology. Perennial; caespitose. *Culms* 115–500 cm high; herbaceous; unbranched above; 4–8 noded. Culm nodes exposed; glabrous. Culm internodes hollow. *Leaves* not basally aggregated; non-auriculate. Sheaths not keeled, terete, striate. Leaf blades linear-lanceolate (and long-acuminate); broad to narrow; 7–18(–24) mm wide; flat; without cross venation; persistent. *Ligule* an unfringed membrane; not truncate; *6–18(–20) mm long (striate-veined, cartilaginous)*.

Reproductive organization. Plants bisexual, with bisexual spikelets; with hermaphrodite florets. The spikelets all alike in sexuality.

Inflorescence. *Inflorescence* paniculate; open (broadly pyramidal, 20–50 cm long, up to 50 cm wide); *with capillary branchlets (towards the extremities)*. Inflorescence with axes ending in spikelets. Inflorescence espatheate; not comprising 'partial inflorescences' and foliar organs. Spikelet-bearing axes persistent. Spikelets not secund; long pedicellate.

Female-fertile spikelets. Spikelets 6–11.9 mm long; broadly elliptic; compressed laterally; disarticulating above the glumes; disarticulating between the florets; with distinctly elongated rachilla internodes between the florets. Rachilla prolonged beyond the uppermost female-fertile floret; rachilla hairy. The rachilla extension with incomplete florets. Hairy callus present. *Callus* short; blunt.

Glumes two; more or less equal; shorter than the spikelets; shorter than the adjacent lemmas (half as long to somewhat shorter); hairless; pointed; awnless; carinate (the keels scabrid); similar (membranous). Lower glume 3 nerved. Upper glume 3 nerved. *Spikelets* with incomplete florets. The incomplete florets distal to the female-fertile florets. The distal incomplete florets 1; *incomplete florets* merely underdeveloped (rudimentary).

Female-fertile florets *3–5(–6)*. Lemmas narrowly elliptic; decidedly firmer than the glumes (thinly cartilaginous); very firm; entire, or incised; pointed; entire or shortly 2 lobed; not deeply cleft (at the most minutely incised); awnless to awned. Awns when present, 1; median; from a sinus (subapically), or dorsal; from near the top; non-geniculate; straight; hairless (scabrous); much shorter than the body of the lemma (to 2 mm long); entered by one vein. Awn bases not twisted; not flattened. Lemmas hairless; scabrous; slightly carinate; without a germination flap; 5 nerved (the nerves raised and scaberulous); with the nerves non-confluent. *Palea* present; relatively long (slightly exceeding the lemma); tightly clasped by the lemma; entire (and pointed, but easily splitting); awnless, without apical setae; textured like the lemma; not indurated (thinly cartilaginous); 1-nerved (ciliolate), or 2-nerved; 2-keeled. Palea keels wingless; minutely, densely hairy. *Lodicules* present; 2; free; membranous; glabrous; toothed, or not toothed; not or scarcely vascularized. *Stamens* 3. Anthers 2.5–3.4 mm long; not penicillate; without an apically prolonged connective. **Ovary** *hairy (the hairs apical, minute, stiff)*. Styles free to their bases. Stigmas 2; white.

Fruit, embryo and seedling. *Fruit* free from both lemma and palea; small (about 2.5 mm long); brown, green beneath the pericarp; obovoid; longitudinally grooved to not grooved (sometimes with a shallow depression); slightly compressed dorsiventrally (ventrally); hairy (hispid); with hairs confined to a terminal tuft. Hilum short to long-linear (0.5–0.8 mm long, narrowly elliptical). *Pericarp loosely adherent (readily removeable)*. Embryo small; not waisted. Endosperm hard. Embryo with an epiblast.

Abaxial leaf blade epidermis. *Costal/intercostal zonation* conspicuous. *Papillae* absent. *Long-cells* markedly different in shape costally and intercostally (the costals elongated-rectangular, narrower); of similar wall thickness costally and intercostally (thin walled). Mid-intercostal long-cells fusiform; having straight or only gently undulating walls. *Microhairs* absent. *Stomata* fairly common (but confined to the zones adjoining the costae). Subsidiaries parallel-sided. Guard-cells overlapped by the interstomatals. *Intercostal short-cells* common; in cork/silica-cell pairs and not paired (often solitary); not silicified. *Costal short-cells* neither distinctly grouped into long rows nor predominantly paired. Costal silica bodies horizontally-elongated crenate/sinuous (a few), or horizontally-elongated smooth (a few), or rounded (e.g. some potato shaped, many malformed), or crescentic (a few, malformed).

Transverse section of leaf blade, physiology. C_3; XyMS+. *Mesophyll* with non-radiate chlorenchyma; without adaxial palisade; with a suspicion of arm-cells in some of the poor material seen. *Leaf blade* with distinct, prominent adaxial ribs to 'nodular' in section; with the ribs more or less constant in size (these wide and low). *Midrib* conspicuous; having a conventional arc of bundles (three, corresponding with ridges on the midrib), or with one bundle only (towards the upper third of the blade, or throughout the narrower blades). Bulliforms present in discrete, regular adaxial groups (in each furrow); in simple fans (these wide). All the vascular bundles accompanied by sclerenchyma. Combined sclerenchyma girders present (with all the bundles); nowhere forming 'figures'. Sclerenchyma all associated with vascular bundles.

Taxonomy. Pooideae; Poodae; Poeae.

Ecology, geography, regional floristic distribution. 1 species. Southeast Australia, Tasmania.

Australian. North and East Australian. Temperate and South-Eastern Australian.
References, etc. Morphological/taxonomic: Vickery 1963. Leaf anatomical: this project.

Dupontia R.Br.

Habit, vegetative morphology. Perennial; rhizomatous, or stoloniferous. *Culms* 10–50 cm high; herbaceous. Leaves non-auriculate. *Sheath margins joined*. Sheaths often purplish. Leaf blades linear; narrow; 1–5 mm wide; flat, or folded; without cross venation. Ligule an unfringed membrane; truncate, or not truncate; 1–3 mm long.

Reproductive organization. Plants bisexual, with bisexual spikelets; with hermaphrodite florets.

Inflorescence. *Inflorescence paniculate*; open, or contracted; espatheate; not comprising 'partial inflorescences' and foliar organs. Spikelet-bearing axes persistent. Spikelets not secund; pedicellate.

Female-fertile spikelets. Spikelets 4–9 mm long; compressed laterally to not noticeably compressed; disarticulating above the glumes. Rachilla prolonged beyond the uppermost female-fertile floret. The rachilla extension naked. Hairy callus present, or absent. *Callus* short; blunt.

Glumes two; relatively large; very unequal to more or less equal; (the upper) about equalling the spikelets; (the upper) *long relative to the adjacent lemmas (exceeding them)*; pointed, or not pointed; awnless; non-carinate; similar (membranous, with wide hyaline margins). Lower glume 1–3 nerved. Upper glume 1–3 nerved. *Spikelets* with female-fertile florets only.

Female-fertile florets (1–)2(–4). **Lemmas** acute; less firm than the glumes to similar in texture to the glumes (scarious); not becoming indurated; *entire*; pointed; *awnless, or mucronate (via the excurrent mid-vein)*; hairy (at least at the base), or hairless; non-carinate; without a germination flap; 3 nerved, or 5 nerved. *Palea* present; relatively long (but shorter than the lemma); 2-nerved; 2-keeled. **Lodicules** present; 2; *free*; membranous; glabrous; toothed, or not toothed; not or scarcely vascularized. *Stamens* 3. Anthers 1.5–3 mm long. *Ovary* glabrous. Styles free to their bases (occasionally 3). Stigmas 2; white.

Fruit, embryo and seedling. *Fruit* free from both lemma and palea. Pericarp fused (?). Embryo without an epiblast (Decker 1964); without a scutellar tail; with a negligible mesocotyl internode. Embryonic leaf margins meeting.

Abaxial leaf blade epidermis. *Costal/intercostal zonation* lacking (save that the midrib zone is barely distinguishable, by rather narrower long-cells). *Papillae* absent. *Long-cells* similar in shape costally and intercostally; of similar wall thickness costally and intercostally (fairly thick walled). Mid-intercostal long-cells rectangular; having markedly sinuous walls. *Microhairs* absent. *Stomata* absent or very rare. *Intercostal short-cells* common; not paired (solitary). *Costal short-cells* neither distinctly grouped into long rows nor predominantly paired (mostly solitary). Costal silica bodies horizontally-elongated crenate/sinuous (very short with few crenations, or imperfect), or rounded, or tall-and-narrow to crescentic (a few).

Transverse section of leaf blade, physiology. C_3; XyMS+. *Mesophyll* with non-radiate chlorenchyma; without adaxial palisade. *Leaf blade* with a constriction on either side of the midrib, otherwise scarcely ribbed. *Midrib* conspicuous (with a rib and a slight keel); with one bundle only. Bulliforms not in discrete, regular adaxial groups (other than the conspicuous 'hinges' flanking the midrib). All the vascular bundles accompanied by sclerenchyma. Combined sclerenchyma girders present (with all the bundles); nowhere forming 'figures' (the girders slender). Sclerenchyma all associated with vascular bundles.

Cytology. Chromosome base number, $x = 7$. $2n = 42, 44, 88,$ and 132. 6, 12, and 19 ploid (?).

Taxonomy. Pooideae; Poodae; Poeae.

Ecology, geography, regional floristic distribution. 2 species. Arctic. Helophytic; species of open habitats; glycophytic.

Holarctic. Boreal. Arctic and Subarctic, Euro-Siberian, and Atlantic North American. European and Siberian. Canadian-Appalachian.

Hybrids. Intergeneric hybrid with *Arctophila*: ×*Arctodupontia* Tsvelev. With *Arctopoa* (= *Poa*): ×*Dupontopoa* N.S. Probatova (exemplified by *Poa labradorica* Steudel, with ×*Dupontopoa dezhnevii* representing taxonomic errors: see Darbyshire *et al.* 1992, Darbyshire and Cayouette 1992).

References, etc. Leaf anatomical: Metcalfe 1960; this project.

Special comments. Fruit data wanting.

Duthiea Hackel

Thrixgyne Keng, *Triavenopsis* Candargy

Habit, vegetative morphology. Perennial; rhizomatous, or caespitose. *Culms* 20–70 cm high; herbaceous; unbranched above. Young shoots intravaginal. *Leaves* not basally aggregated (often overtopping the inflorescence); non-auriculate. *Leaf blades* linear; narrow; 2–6 mm wide; flat, or rolled (usually convolute); *not pseudopetiolate*; without cross venation. Ligule an unfringed membrane; not truncate; 5–8 mm long (lacerate).

Reproductive organization. *Plants bisexual, with bisexual spikelets*; with hermaphrodite florets.

Inflorescence. Inflorescence few spikeleted, or many spikeleted; a single raceme, or paniculate (the lower nodes with well developed collars); contracted (the spikelets often gathered into a short unilateral cluster); *spatheate (the lowest spikelet and sometimes some others subtended by a deciduous, membranous scale of unknown homology)*; not comprising 'partial inflorescences' and foliar organs. Spikelets secund (the raceme unilateral), or not secund; pedicellate.

Female-fertile spikelets. Spikelets 15–25 mm long; compressed laterally to not noticeably compressed; disarticulating above the glumes; disarticulating between the florets. Rachilla prolonged beyond the uppermost female-fertile floret; rachilla hairy, or hairless. The rachilla extension with incomplete florets, or naked. Hairy callus present.

Glumes two; more or less equal; about equalling the spikelets; long relative to the adjacent lemmas; pointed (acute to acuminate); slightly awned to awnless; non-carinate; similar (membranous with thinner margins). Lower glume 5–9 nerved. Upper glume 7–13 nerved. *Spikelets* with female-fertile florets only, or with incomplete florets. The incomplete florets when present, distal to the female-fertile florets. *The distal incomplete florets* merely underdeveloped.

Female-fertile florets (1–)2–3(–9). Lemmas ovate; similar in texture to the glumes to decidedly firmer than the glumes (herbaceous to leathery, with thin margins and tip); incised; 2 lobed; deeply cleft to not deeply cleft (from bidentate to cleft to a third or more); awned. Awns 1; median; from a sinus; geniculate; much longer than the body of the lemma; entered by one vein. Lemmas hairy. *The hairs* in tufts; *in transverse rows (the tufts in a median horizontal row)*. Lemmas non-carinate; without a germination flap; 9–13 nerved. *Palea* present; relatively long; apically notched; awnless, without apical setae, or with apical setae to awned (sometimes 2-aristulate); 2-nerved; 2-keeled. Palea keels wingless. **Lodicules absent. Stamens** 3. Anthers penicillate. **Ovary hairy**. Styles fused. Stigmas 2.

Fruit, embryo and seedling. Fruit beaked via the persistent style; hairy on the body. Hilum long-linear. Embryo small.

Abaxial leaf blade epidermis. *Costal/intercostal zonation* fairly conspicuous. *Papillae* absent. *Long-cells* similar in shape costally and intercostally (the costals smaller); of similar wall thickness costally and intercostally. Mid-intercostal long-cells rectangular; having markedly sinuous walls (and pitted). *Microhairs* present (possibly — one doubtful example seen), or absent (?). *Stomata* absent or very rare. *Intercostal short-cells* absent or very rare. Costal prickles present. *Crown cells* absent. *Costal short-cells* conspicuously in long rows. Costal silica bodies few (though abundant adaxially); 'panicoid-type'; more or less nodular.

Transverse section of leaf blade, physiology. C_3; XyMS+. *Mesophyll* with non-radiate chlorenchyma. *Leaf blade* with distinct, prominent adaxial ribs; with the ribs very irregular in sizes. *Midrib* seemingly not readily distinguishable; with one bundle only. Bulliforms present in discrete, regular adaxial groups (at bases of furrows); in simple fans (these small). All the vascular bundles accompanied by sclerenchyma. Combined sclerenchyma girders present; forming 'figures' (the large bundles with T's and anchors). Sclerenchyma not all bundle-associated. The 'extra' sclerenchyma in abaxial groups (of 1–3 cells); abaxial-hypodermal, the groups isolated (opposite the furrows).

Cytology. Chromosome base number, $x = 7$. Haploid nuclear DNA content unknown, but the chromosomes 'large.

Taxonomy. Pooideae, or Arundinoideae; if pooid, Poodae; Aveneae (?); if arundinoid, Danthonieae (?).

Ecology, geography, regional floristic distribution. 3 species. Himalayas. Species of open habitats. Mountain slopes.

Holarctic and Paleotropical. Tethyan. African. Irano-Turanian. Saharo-Sindian.

References, etc. Leaf anatomical: this project.

Dybowskia Stapf

~ Hyparrhenia

Habit, vegetative morphology. *Annual (with prop roots)*. The flowering culms leafy. *Culms* about 20–100 cm high; herbaceous; to 1 cm in diameter; unbranched above; about 7 noded. Culm nodes exposed; glabrous. Culm leaves present. Upper culm leaf blades fully developed. Culm internodes solid. *Leaves* not basally aggregated; non-auriculate. Leaf blades linear; broad; 6–14 mm wide; without cross venation; persistent. Ligule an unfringed membrane; not truncate; 1.5–4 mm long. *Contra-ligule* absent.

Reproductive organization. Plants bisexual, with bisexual spikelets; with hermaphrodite florets. *The spikelets* of sexually distinct forms on the same plant; hermaphrodite and male-only; overtly heteromorphic (the pedicellate spikelets awnless); *in both homogamous and heterogamous combinations (the raceme with two basal, large homogamous pairs, forming an involucre for the triad above).*

Inflorescence. *Inflorescence* spatheate; *a complex of 'partial inflorescences' and intervening foliar organs (the short 'racemes' incompletely exserted from the spathes).* Spikelet-bearing axes *very much reduced (2–2.5 cm long, of few spikelets, the adjacent joints of the basal homogamous pairs entering into the formation of the 'raceme' base; pedicels of the trio basally adnate laterally to the G_1 of the female-fertile (sessile) member)*; paired (permanently continuous); with substantial rachides; disarticulating; disarticulating at the joints (the terminal, female-fertile spikelet falling with the pedicels, the homogamous pairs tardily separating later). 'Articles' non-linear; without a basal callus-knob; appendaged (the terminal disarticulation auriculiform); disarticulating obliquely. Spikelets paired and in triplets (with a terminal triad); not secund; sessile and pedicellate; consistently in 'long-and-short' combinations (the basal two homogenous pairs sessile, the triad comprising 1 sessile and 2 pedicellate spikelets); in pedicellate/sessile combinations (in the triad). Pedicels of the 'pedicellate' spikelets free of the rachis (but fused to the G_1 of the sessile spikelet). The 'shorter' spikelets hermaphrodite (terminal), or male-only (basal). The 'longer' spikelets male-only.

Female-sterile spikelets. Rachilla of male spikelets terminated by a male floret. The male spikelets with glumes; without proximal incomplete florets; 1 floreted. The lemmas awnless. Male florets 1; 3 staminate.

Female-fertile spikelets. *Spikelets* 15–19 mm long; *not noticeably compressed to compressed dorsiventrally*; falling with the glumes (and with the pedicels). Rachilla terminated by a female-fertile floret. Hairy callus present (densely bearded). *Callus* long (with long hairs); pointed.

Glumes two; more or less equal; exceeding the spikelets; long relative to the adjacent lemmas; hairy (G_1 densely villous at the base); not pointed; awnless; non-carinate; very dissimilar. *Lower glume* much exceeding the lowest lemma; not two-keeled (below, but shortly two-keeled towards the short broad beak); convex on the back; not pitted; relatively smooth; *11 nerved, or 13 nerved*. Upper glume 3 nerved. *Spikelets* with incomplete florets. The incomplete florets proximal to the female-fertile florets. *The proximal incomplete florets* 1; epaleate; sterile. *The proximal lemmas* awnless; *2 nerved*; hyaline; not becoming indurated.

Female-fertile florets 1. Lemmas linear, rolled so that the margins meet, but not enclosing other organs, extending into the awn above; less firm than the glumes (but reduced almost to the awn); not becoming indurated; incised; 2 lobed; not deeply cleft (minutely toothed, the teeth hyaline); awned. Awns 1; median; from a sinus (from between the tiny teeth); twice geniculate; hairy (silky-pubescent); much longer than the body of the lemma; deciduous (or at least falling with the deciduous L₂). Awn bases flattened. Lemmas hairy; non-carinate; without a germination flap; 1 nerved. *Palea* present; conspicuous but relatively short; awnless, without apical setae; thinner than the lemma; not indurated (hyaline); nerveless; 2-keeled. Palea back glabrous. Palea keels wingless; glabrous. *Lodicules* present; 2; free; fleshy; ciliate, or glabrous; not or scarcely vascularized. *Stamens* 3. Anthers about 5 mm long; not penicillate; without an apically prolonged connective. *Ovary* glabrous. Stigmas 2.

Abaxial leaf blade epidermis. *Costal/intercostal zonation* conspicuous. *Papillae* present; intercostal (large, on the interstomatals, conspicuous only in places). Intercostal papillae when present, consisting of one oblique swelling per cell. *Long-cells* markedly different in shape costally and intercostally (the costals much narrower, more regularly rectangular); of similar wall thickness costally and intercostally (fairly thin walled). Mid-intercostal long-cells fairly regularly rectangular; having markedly sinuous walls. *Microhairs* present; elongated; clearly two-celled; panicoid-type. *Stomata* common. Subsidiaries non-papillate; low dome-shaped. Guard-cells overlapping to flush with the interstomatals. *Intercostal short-cells* common (often paired with prickle bases); sometimes in cork/silica-cell pairs; sometimes silicified. Intercostal silica bodies crescentic, cross-shaped, and oryzoid-type. Numerous small, bulbous prickles and occasional cushion-based macrohairs, in the intercostal zones. *Costal short-cells* conspicuously in long rows. Costal silica bodies present and well developed; 'panicoid-type'; butterfly shaped and dumb-bell shaped.

Transverse section of leaf blade, physiology. C₄; XyMS+ to XyMS− (XyMS 'variable'). PCR sheath outlines uneven. PCR sheath extensions absent. *Leaf blade* adaxially flat. *Midrib* conspicuous; having a conventional arc of bundles (three primaries, with numerous small bundles between); with colourless mesophyll adaxially (and a narrow median adaxial-hypodermal layer of sclerenchyma). *The lamina* symmetrical on either side of the midrib. Bulliforms not in discrete, regular adaxial groups (the adaxial epidermis largely bulliform). Many of the smallest vascular bundles unaccompanied by sclerenchyma. Combined sclerenchyma girders present (with the primaries and larger minor bundles). Sclerenchyma all associated with vascular bundles.

Taxonomy. Panicoideae; Andropogonodae; Andropogoneae; Andropogoninae.

Ecology, geography, regional floristic distribution. 1 species. Tropical Africa. Paleotropical. African. West African Rainforest.

References, etc. Leaf anatomical: this project.

Special comments. Fruit data wanting.

Eccoilopus Steud.

~ *Spodiopogon*

Habit, vegetative morphology. Rather tall perennial; caespitose. Leaf blades cordate (often), or not cordate, not sagittate; pseudopetiolate (often), or not pseudopetiolate. *Ligule an unfringed membrane*.

Reproductive organization. Plants bisexual, with bisexual spikelets; with hermaphrodite florets. *The spikelets all alike in sexuality*; homomorphic.

Inflorescence. *Inflorescence of spicate main branches, or paniculate (the pedunculate 'racemes' often whorled, sometimes basally branched)*; not comprising 'partial inflorescences' and foliar organs. **Spikelet-bearing axes** 'racemes', or paniculate; with very slender rachides; *persistent (despite being jointed)*. 'Articles' not appendaged.

Spikelets paired; pedicellate; *consistently in 'long-and-short' combinations*; unequally pedicellate in each combination. Pedicels of the 'pedicellate' spikelets free of the rachis. The 'shorter' spikelets hermaphrodite. The 'longer' spikelets hermaphrodite.

Female-fertile spikelets. *Spikelets* 4–6 mm long; *not noticeably compressed (terete)*; falling with the glumes (from the pedicels). Rachilla terminated by a female-fertile floret. Hairy callus present (as a tuft of short hairs). *Callus* short.

Glumes two; more or less equal; long relative to the adjacent lemmas; without conspicuous tufts or rows of hairs; awnless; *very dissimilar (the lower chartaceous, pallid, convex with raised nerves,the upper cymbiform)*. Lower glume not two-keeled; convex on the back; not pitted; relatively smooth; prominently 7 nerved, or 9 nerved (ridged). Upper glume 7 nerved. *Spikelets* with incomplete florets. The incomplete florets proximal to the female-fertile florets. *The proximal incomplete florets* 1; paleate, or epaleate; sterile. The proximal lemmas awnless; 1 nerved; more or less equalling the female-fertile lemmas; similar in texture to the female-fertile lemmas (hyaline).

Female-fertile florets 1. Lemmas less firm than the glumes (hyaline); not becoming indurated; incised; 2 lobed; deeply cleft; awned, or mucronate (in that the awn may be 'imperfect'). Awns when present, 1; from a sinus; geniculate. Lemmas without a germination flap. *Palea* present; conspicuous but relatively short; nerveless (hyaline). *Ovary* glabrous. Stigmas 2.

Fruit, embryo and seedling. Fruit small (2 mm long); not noticeably compressed (cylindrical). Hilum short. Embryo large.

Transverse section of leaf blade, physiology. C_4.

Cytology. $2n = 40$.

Taxonomy. Panicoideae; Andropogonodae; Andropogoneae; Andropogoninae.

Ecology, geography, regional floristic distribution. 4 species. Asia.

Holarctic and Paleotropical. Boreal. Indomalesian. Eastern Asian. Indian and Indo-Chinese.

Rusts and smuts. Rusts — *Puccinia*. Taxonomically wide-ranging species: *Puccinia miyoshiana*.

Special comments. Anatomical data wanting.

Eccoptocarpha Launert

Habit, vegetative morphology. *Annual*; loosely caespitose (or culms single). *Culms* 40–75 cm high; herbaceous; unbranched above. *Leaves* not basally aggregated; non-auriculate. Leaf blades narrow; 2–9 mm wide (2.5–10 cm long); flat; without cross venation; persistent. **Ligule** present; *a fringe of hairs*.

Reproductive organization. Plants bisexual, with bisexual spikelets; with hermaphrodite florets. *The spikelets all alike in sexuality.*

Inflorescence. Inflorescence determinate; without pseudospikelets; *of spicate main branches*; non-digitate. Primary inflorescence branches 2–7. Inflorescence espatheate; not comprising 'partial inflorescences' and foliar organs. Spikelet-bearing axes persistent. Spikelets solitary; secund (the racemes dorsiventral); shortly pedicellate. Pedicel apices discoid.

Female-fertile spikelets. *Spikelets* 3.2–3.6 mm long; obovate; adaxial; compressed dorsiventrally; falling with the glumes; *with a distinctly elongated rachilla internode between the glumes and with distinctly elongated rachilla internodes between the florets (the L_2 being borne on a long, s-shaped stalk which straightens on maturity, and the glumes being slightly separated)*. The upper floret conspicuously stipitate. Rachilla terminated by a female-fertile floret; rachilla hairless. Hairy callus absent.

Glumes two; relatively large; more or less equal (G_1 only slightly shorter, about 3/4 the length of the spikelet); long relative to the adjacent lemmas; dorsiventral to the rachis; hairy (G_1 with glandular hairs, G_2 with apical, rigid hairs); not pointed; awnless; non-carinate; *very dissimilar (G_1 membranous, glandular-hairy, G_2 firmer and prominently net-veined)*. Lower glume 5 nerved. *Upper glume* not saccate; 5 nerved. *Spikelets* with incomplete florets. The incomplete florets proximal to the female-fertile florets. *Spikelets with proximal incomplete florets*. *The proximal incomplete florets* 1; paleate. Palea of the proximal incomplete florets reduced. The proximal incomplete florets male. *The prox-*

imal lemmas reticulately veined, similar in form and texture to G$_2$; awnless; 5 nerved; decidedly exceeding the female-fertile lemmas; less firm than the female-fertile lemmas; not becoming indurated.

Female-fertile florets 1 (very small relative to the rest of the spikelet). *Lemmas* decidedly firmer than the glumes; smooth (shiny); becoming indurated (crustaceous); entire; blunt; awnless; *hairy (at the tip)*; non-carinate; having the margins tucked in onto the palea; obscurely 5 nerved. *Palea* present; relatively long; entire (rounded at the tip); awnless, without apical setae; textured like the lemma (smooth); not indurated; 2-nerved. *Lodicules* present. *Stamens* 3. *Ovary* glabrous. Styles free to their bases. Stigmas 2.

Fruit, embryo and seedling. Fruit small (1.75 mm long); compressed dorsiventrally. Hilum short. Embryo large.

Abaxial leaf blade epidermis. *Costal/intercostal zonation* conspicuous. *Papillae* absent. Mid-intercostal long-cells rectangular; having markedly sinuous walls. *Microhairs* present; panicoid-type; 33–39 microns long; 3.9–6.6 microns wide at the septum. Microhair total length/width at septum 5.5–10. Microhair apical cells 27–30 microns long. Microhair apical cell/total length ratio 0.75–0.82. *Stomata* common; 33–36 microns long. Subsidiaries low dome-shaped (mostly), or triangular. Guard-cells overlapping to flush with the interstomatals. *Intercostal short-cells* common; in cork/silica-cell pairs; silicified. Intercostal silica bodies cross-shaped. *Costal short-cells* predominantly paired (and a few short rows). Costal silica bodies 'panicoid-type'; nearly all short dumb-bell shaped, or cross shaped.

Transverse section of leaf blade, physiology. C$_4$; XyMS+. PCR sheath outlines uneven. *Midrib* not readily distinguishable (the main bundle slightly larger, with a somewhat heavier girder); with one bundle only, or having a conventional arc of bundles (if adjoining tiny bundles are counted as part of the midrib). Bulliforms the epidermis extensively 'bulliform'. Many of the smallest vascular bundles unaccompanied by sclerenchyma. Combined sclerenchyma girders present (with the primaries); nowhere forming 'figures'. Sclerenchyma all associated with vascular bundles.

Taxonomy. Panicoideae; Panicodae; Paniceae.

Ecology, geography, regional floristic distribution. 1 species. Tropical Africa. Species of open habitats. Savanna.

Paleotropical. African. Sudano-Angolan. South Tropical African.

References, etc. Morphological/taxonomic: Launert 1965. Leaf anatomical: this project.

Echinaria Desf.

Panicastrella Moench

Habit, vegetative morphology. Annual; erect or ascending. *Culms* 1.5–25 cm high; herbaceous. Culm nodes glabrous. Leaves non-auriculate. *Sheath margins joined*. Leaf blades linear; narrow; 1.5–5 mm wide; flat; without cross venation. Ligule an unfringed membrane (but ciliolate); truncate; 0.5 mm long.

Reproductive organization. Plants bisexual, with bisexual spikelets; with hermaphrodite florets. The spikelets of sexually distinct forms on the same plant (with sterile, reduced members at the base of the inflorescence), or all alike in sexuality; often hermaphrodite and sterile; overtly heteromorphic (sterile-reduced *versus* fertile), or homomorphic.

Inflorescence. *Inflorescence paniculate*; contracted (prickly); *capitate, or more or less ovoid (5–15 mm long)*; espatheate; not comprising 'partial inflorescences' and foliar organs. Spikelet-bearing axes persistent. *Spikelets* associated with bractiform involucres (constituted by the basal, sterile spikelets), or unaccompanied by bractiform involucres, not associated with setiform vestigial branches; not secund; shortly pedicellate, or subsessile.

Female-sterile spikelets. The sterile spikelets, when present, basal and bractlike.

Female-fertile spikelets. Spikelets 4–12 mm long; compressed laterally to not noticeably compressed; disarticulating above the glumes. Rachilla prolonged beyond the uppermost female-fertile floret. *Hairy callus present. Callus* short.

Glumes two; relatively large (membranous); very unequal to more or less equal; shorter than the spikelets; long relative to the adjacent lemmas; mucronate to awned (the lower with 2–5 veins excurrent as awns, the upper with an excurrent midrib); carinate. Lower glume 2–5 nerved. Upper glume 1 nerved. *Spikelets* with female-fertile florets only, or with incomplete florets. The incomplete florets distal to the female-fertile florets. *Female-fertile florets* (1–)3–4. Lemmas decidedly firmer than the glumes (leathery); not becoming indurated; awned. *Awns* 5, or 7; *median and lateral (the 5–7 strong lemma veins being produced as flattened awns which become deflexed at maturity, the middle awn the longest)*; the median similar in form to the laterals; apical; non-geniculate; re-curving (deflexing); much longer than the body of the lemma (lemma 2 mm long, the middle awn 4–6 mm long); entered by one vein. The lateral awns shorter than the median. Lemmas hairy; non-carinate; without a germination flap; 5–7 nerved. **Palea** present; relatively long; *awned (two or rarely five veins produced as flattened awns)*; 2-nerved, or several nerved (up to 5); 2-keeled. *Lodicules* present, or absent; when present, 2; free; membranous; glabrous; not toothed; not or scarcely vascularized. *Stamens* 3. Anthers 0.8–1.3 mm long. *Ovary* hairy. *Styles fused.* Stigmas 2.

Fruit, embryo and seedling. *Fruit* free from both lemma and palea; small (2 mm long); not noticeably compressed; with hairs confined to a terminal tuft. Hilum short. Embryo small. Endosperm hard; without lipid; containing compound starch grains. Embryo with an epiblast.

Abaxial leaf blade epidermis. *Costal/intercostal zonation* conspicuous. *Papillae* absent. *Long-cells* similar in shape costally and intercostally; of similar wall thickness costally and intercostally. Mid-intercostal long-cells rectangular and fusiform; having markedly sinuous walls. *Microhairs* absent. *Stomata* common (fairly, but mostly alongside the veins); 36–39 microns long. Subsidiaries parallel-sided. Guard-cells over-lapped by the interstomatals. *Intercostal short-cells* absent or very rare. Prickle bases abundant. *Costal short-cells* neither distinctly grouped into long rows nor predominantly paired (mostly solitary). Costal silica bodies horizontally-elongated crenate/sinuous (mostly), or horizontally-elongated smooth.

Transverse section of leaf blade, physiology. C_3; XyMS+. Mesophyll without ad-axial palisade. *Midrib* not readily distinguishable; with one bundle only. *The lamina* sym-metrical on either side of the midrib. Bulliforms present in discrete, regular adaxial groups; in simple fans. All the vascular bundles accompanied by sclerenchyma. Combined sclerenchyma girders present (but the largest bundles with abaxial strands only); forming 'figures' (I's). Sclerenchyma all associated with vascular bundles.

Cytology. Chromosome base number, $x = 7$ and 9. $2n = 14$ and 18. 2 ploid.

Taxonomy. Pooideae; Poodae; Seslerieae.

Ecology, geography, regional floristic distribution. 1 species. Mediterranean, eastern Asia. Xerophytic; species of open habitats.

Holarctic. Boreal and Tethyan. Euro-Siberian. Mediterranean and Irano-Turanian. European.

Rusts and smuts. Rusts — *Puccinia.* Taxonomically wide-ranging species: *Puccinia hordei.*

References, etc. Leaf anatomical: this project.

Echinochloa P. Beauv.

Ornithospermum Dumoulin, *Tema* Adans.

Habit, vegetative morphology. Annual, or perennial; caespitose to decumbent (or floating). **Culms** 40–360 cm high; *herbaceous*; sometimes floating; branched above, or unbranched above; tuberous, or not tuberous. Culm nodes glabrous. Culm internodes solid, or hollow. *Leaves* not basally aggregated; non-auriculate. **Leaf blades broad**; flat; without cross venation; persistent; rolled in bud. **Ligule** present, or absent; when present, *a fringe of hairs.* *Contra-ligule* present (of hairs), or absent.

Reproductive organization. Plants bisexual, with bisexual spikelets; with hermaph-rodite florets. *The spikelets all alike in sexuality.* Plants outbreeding and inbreeding. Exposed-cleistogamous, or chasmogamous.

Inflorescence. *Inflorescence of spicate main branches (the spikelets often hispid). Inflorescence with axes ending in spikelets.* Inflorescence espatheate; not comprising 'partial inflorescences' and foliar organs. Spikelet-bearing axes persistent. *Spikelets* paired, or clustered in little secondary racemelets; secund; typically and characteristically in four dense rows, but occasionally in two; shortly pedicellate, or subsessile. Pedicel apices oblique, or truncate, or discoid, or cupuliform. Spikelets imbricate.

Female-fertile spikelets. *Spikelets* 2.3–7 mm long; elliptic, or lanceolate, or ovate; *adaxial (probably best interpreted as adaxial relative to the reduced, spikelet-bearing branch); compressed dorsiventrally; planoconvex;* falling with the glumes. Rachilla terminated by a female-fertile floret. Hairy callus absent.

Glumes two; very unequal; (the upper) shorter than the adjacent lemmas to long relative to the adjacent lemmas; dorsiventral to the rachis; pointed; awned, or awnless; with the keel conspicuously winged, or without a median keel-wing; *very dissimilar (the G_1 usually much shorter, ovate, often mucronate, the G_2 strongly concave, acute, cuspidate or awned).* Lower glume 0–3 nerved. Upper glume 5 nerved, or 7 nerved. *Spikelets* with incomplete florets. The incomplete florets proximal to the female-fertile florets. *Spikelets with proximal incomplete florets.* **The proximal incomplete florets** 1; *paleate.* Palea of the proximal incomplete florets fully developed, or reduced (e.g. *E. kimberleyensis*). The proximal incomplete florets male (rarely), or sterile. **The proximal lemmas** similar to the G_2 but flattened on the back, often with a large cusp or awned; awned, or awnless; *5 nerved*; less firm than the female-fertile lemmas to similar in texture to the female-fertile lemmas; not becoming indurated.

Female-fertile florets 1. **Lemmas usually blunt-apiculate, with a laterally compressed, incurved beak**; decidedly firmer than the glumes; smooth; becoming indurated to not becoming indurated (subleathery to crustaceous); yellow in fruit; entire; awnless (obtuse to apiculate); hairless (shiny); non-carinate; having the margins tucked in onto the palea; 5 nerved. *Palea* present (the tip reflexed); relatively long; entire; awnless, without apical setae; textured like the lemma; indurated, or not indurated; 2-nerved. *Lodicules* present; joined, or free; fleshy; glabrous. *Stamens* 3. Anthers 0.4–1.2 mm long; not penicillate. *Ovary* glabrous. Styles free to their bases. Stigmas 2; red pigmented.

Fruit, embryo and seedling. Fruit small; compressed dorsiventrally. Hilum short. Embryo large; waisted, or not waisted. Endosperm hard; without lipid; containing only simple starch grains. Embryo without an epiblast; with a scutellar tail; with an elongated mesocotyl internode. Embryonic leaf margins overlapping.

Seedling with a long mesocotyl; with a loose coleoptile. First seedling leaf with a well-developed lamina. The lamina broad; curved; 13–20 veined (?).

Abaxial leaf blade epidermis. *Costal/intercostal zonation* conspicuous. *Papillae* present. Intercostal papillae over-arching the stomata, or not over-arching the stomata; consisting of one oblique swelling per cell. *Long-cells* similar in shape costally and intercostally to markedly different in shape costally and intercostally (the costals narrower); of similar wall thickness costally and intercostally. Mid-intercostal long-cells rectangular; having markedly sinuous walls, or having straight or only gently undulating walls. *Microhairs* present; panicoid-type; 36–60 microns long; 6–6.6 microns wide at the septum. Microhair total length/width at septum 7.3–9. Microhair apical cells (16–)22–30 microns long. Microhair apical cell/total length ratio 0.48–0.57. *Stomata* common; 39–46.5 microns long. Subsidiaries low dome-shaped, or triangular. Guard-cells overlapping to flush with the interstomatals. *Intercostal short-cells* absent or very rare; in cork/silica-cell pairs, or not paired; silicified (when paired), or not silicified. Intercostal silica bodies when present, cross-shaped. *Costal short-cells* conspicuously in long rows. Costal silica bodies 'panicoid-type'; cross shaped to nodular.

Transverse section of leaf blade, physiology. C_4; biochemical type NADP–ME (3 species); XyMS–. PCR sheath outlines uneven. PCR sheath extensions absent. PCR cell chloroplasts with reduced grana; centrifugal/peripheral. *Mesophyll* with radiate chlorenchyma. *Leaf blade* with distinct, prominent adaxial ribs, or adaxially flat; with the ribs more or less constant in size. *Midrib* conspicuous; having a conventional arc of bundles; with colourless mesophyll adaxially. Bulliforms not in discrete, regular adaxial groups (mostly in irregular groups, or the epidermis extensively bulliform); sometimes, irregularly in simple fans. Many of the smallest vascular bundles unaccompanied by sclerenchyma. Combined sclerenchyma girders absent. Sclerenchyma all associated with vascular bundles.

Culm anatomy. *Culm internode bundles* in one or two rings.

Phytochemistry. Tissues of the culm bases with abundant starch. Leaves without flavonoid sulphates (2 species).

Special diagnostic feature. Glumes and/or sterile lemmas awned or acuminate-mucronate.

Cytology. Chromosome base number, x = 9. $2n$ = 27, 36, 42, 48, 54, 72, and 108. Mean diploid 2c DNA value 2.7 pg (?—*E. frumentacea*, ploidy unknown).

Taxonomy. Panicoideae; Panicodae; Paniceae.

Ecology, geography, regional floristic distribution. 30–40 species. In warm regions. Commonly adventive. Hydrophytic, helophytic, and mesophytic; mostly species of open habitats; glycophytic. In water and moist or marshy places, also in disturbed ground and weedy places.

Holarctic, Paleotropical, Neotropical, and Australian. Boreal, Tethyan, and Madrean. African, Madagascan, Indomalesian, Polynesian, and Neocaledonian. Euro-Siberian, Eastern Asian, and Atlantic North American. Macaronesian, Mediterranean, and Irano-Turanian. Saharo-Sindian, Sudano-Angolan, West African Rainforest, and Namib-Karoo. Indian, Indo-Chinese, Malesian, and Papuan. Fijian. Caribbean, Venezuela and Surinam, Amazon, Central Brazilian, Pampas, and Andean. North and East Australian and Central Australian. European and Siberian. Canadian-Appalachian, Southern Atlantic North American, and Central Grasslands. Sahelo-Sudanian, Somalo-Ethiopian, South Tropical African, and Kalaharian. Tropical North and East Australian.

Rusts and smuts. Rusts — *Puccinia*. Smuts from Tilletiaceae and from Ustilaginaceae. Tilletiaceae — *Entyloma* and *Tilletia*. Ustilaginaceae — *Sorosporium*, *Sphacelotheca*, *Tolyposporium*, and *Ustilago*.

Economic importance. Significant weed species: *E. oryzoides* and *E. stagnina* (in rice), *E. crusgalli* (especially in arable land), *E. colonum*, *E. crus-pavonis*, *E. pyramidalis*. Cultivated fodder: *E. frumentacea*, *E. stagnina*. Important native pasture species: all more or less palatable to stock. Grain crop species: minor cereals: *E. frumentacea* (Sawa — India), *E. utilis* (China and Japan), *E. colona* (Shama millet), *E. pyramidalis*.

References, etc. Leaf anatomical: Metcalfe 1960; this project.

Echinolaena Desv.

Excluding *Chasechloa*

Habit, vegetative morphology. Annual, or perennial; rhizomatous to stoloniferous. Culms herbaceous; branched above. Culm nodes hairy, or glabrous. Culm internodes solid. *Leaves* not basally aggregated; non-auriculate. *Leaf blades* linear to lanceolate; broad, or narrow; *cordate*; tending to pseudopetiolate, or not pseudopetiolate; without cross venation.

Reproductive organization. *Plants bisexual, with bisexual spikelets*; with hermaphrodite florets. The spikelets of sexually distinct forms on the same plant, or all alike in sexuality; hermaphrodite, or hermaphrodite and sterile (the sessile members often aborted).

Inflorescence. Inflorescence of spicate main branches, or a single raceme. Primary inflorescence branches 1–5 (of pectinate 'racemes'). *Inflorescence with axes ending in spikelets (but the lower glume of the terminal spikelet simulating a rachis extension)*. Rachides hollowed. Inflorescence espatheate; not comprising 'partial inflorescences' and foliar organs. Spikelet-bearing axes persistent. *Spikelets* solitary (ostensibly, by abortion), or paired; secund; sessile and pedicellate; *consistently in 'long-and-short' combinations*; in pedicellate/sessile combinations. The 'shorter' spikelets hermaphrodite, or female-only, or sterile. The 'longer' spikelets hermaphrodite.

Female-fertile spikelets. Spikelets 6–10 mm long; compressed laterally; falling with the glumes. The upper floret conspicuously stipitate (its callus with narrow wings or scars at the base of the lemma). Rachilla terminated by a female-fertile floret. Hairy callus absent.

Glumes two; very unequal to more or less equal; (the longer) long relative to the adjacent lemmas; *lateral to the rachis*; pointed; awned to awnless (acute to shortly awned); carinate to non-carinate (the nerves forming ribs); *similar (tuberculate-bristly)*. Lower glume tuberculate (with tubercle based bristles); 3–9 nerved. Upper glume 7–9 nerved. *Spikelets* with incomplete florets. The incomplete florets proximal to the female-fertile florets. *The proximal incomplete florets* 1; paleate; male. The proximal lemmas awnless; 5 nerved; decidedly exceeding the female-fertile lemmas; similar in texture to the female-fertile lemmas; not becoming indurated.

Female-fertile florets 1. Lemmas similar in texture to the glumes to decidedly firmer than the glumes; smooth; not becoming indurated; entire; blunt; awnless; hairless; glabrous; non-carinate; having the margins tucked in onto the palea; with a clear germination flap; 3–5 nerved. *Palea* present; relatively long; 2-nerved; keel-less. *Lodicules* present; 2; free; fleshy; glabrous; not or scarcely vascularized. *Stamens* 3. Anthers not penicillate; without an apically prolonged connective. *Ovary* glabrous. Styles free to their bases. Stigmas 2.

Fruit, embryo and seedling. Fruit small. Hilum long-linear.

Abaxial leaf blade epidermis. *Costal/intercostal zonation* conspicuous. *Papillae* absent. *Long-cells* similar in shape costally and intercostally; of similar wall thickness costally and intercostally (fairly thick walled). Mid-intercostal long-cells rectangular; having markedly sinuous walls. *Microhairs* present (large); panicoid-type; 75–90 microns long, or (48–)51–60(–63) microns long (in *E. gracilis*); 9–12 microns wide at the septum, or (7.5–)8.4–9(–9.3) microns wide at the septum (*E. gracilis*). Microhair total length/ width at septum 5.3–8.7. Microhair apical cells 39–48 microns long, or 27–36 microns long (*E. gracilis*). Microhair apical cell/total length ratio 0.5–0.7. *Stomata* common; 57–63 microns long, or (39–)45–48(–51) microns long (in *E. gracilis*). Subsidiaries triangular. Guard-cells overlapping to flush with the interstomatals. *Intercostal short-cells* common; not paired; not silicified. *Costal short-cells* conspicuously in long rows. Costal silica bodies 'panicoid-type'; cross shaped and dumb-bell shaped.

Transverse section of leaf blade, physiology. C_3; XyMS+. *Mesophyll* with radiate chlorenchyma; *Isachne*-type. *Leaf blade* adaxially flat. Midrib with one bundle only. Bulliforms present in discrete, regular adaxial groups; in simple fans. All the vascular bundles accompanied by sclerenchyma. Combined sclerenchyma girders absent. Sclerenchyma all associated with vascular bundles.

Phytochemistry. Leaves without flavonoid sulphates (1 species).

Cytology. Chromosome base number, $x = 9$. $2n = 60$.

Taxonomy. Panicoideae; Panicodae; Paniceae.

Ecology, geography, regional floristic distribution. 6 species. Central & South America, 1 Madagascar. Helophytic, or mesophytic; species of open habitats. Savanna.

Paleotropical and Neotropical. Madagascan. Caribbean, Central Brazilian, and Andean.

References, etc. Leaf anatomical: this project.

Echinopogon P. Beauv.

Hystericina Steud.

Habit, vegetative morphology. Perennial; rhizomatous, or caespitose. Culms herbaceous. Culm nodes glabrous. Culm internodes solid, or hollow. *Leaves* not basally aggregated; non-auriculate. *Leaf blades* narrow; *not pseudopetiolate*; without cross venation; persistent. Ligule an unfringed membrane; truncate; 1.5–3 mm long.

Reproductive organization. Plants bisexual, with bisexual spikelets; with hermaphrodite florets.

Inflorescence. *Inflorescence paniculate*; contracted; *capitate, or more or less ovoid*; espatheate; not comprising 'partial inflorescences' and foliar organs. Spikelet-bearing axes persistent. Spikelets not secund; pedicellate.

Female-fertile spikelets. Spikelets 2.5–10 mm long; compressed laterally; disarticulating above the glumes. Rachilla prolonged beyond the uppermost female-fertile floret. The rachilla extension naked. Hairy callus present.

Glumes two; relatively large; more or less equal; long relative to the adjacent lemmas; pointed (acute to acuminate); awnless; carinate (the keels stiffly ciliate); similar (membranous). Lower glume 1 nerved. *Upper glume 1 nerved*. *Spikelets* with female-fertile florets only.

Female-fertile florets *1*. *Lemmas* decidedly firmer than the glumes (thinly leathery); not becoming indurated; *incised*; *2 lobed (the apex narrow, with two slender, erect, acuminate or setiform lateral lobes)*; mucronate (*E. phleoides*), or awned. Awns 1; median; from a sinus (or slightly behind it); non-geniculate; hairless (scabrid); entered by one vein. Lemmas hairless; non-carinate; distinctly 5–7(–11) nerved. *Palea* present; relatively long; minutely apically notched (3-toothed); 2-nerved; 2-keeled, or keel-less. **Lodicules** present; 2; free; membranous; *ciliate*; not toothed. *Stamens* 3. Anthers not penicillate. *Ovary* glabrous. Styles free to their bases. Stigmas 2; white.

Fruit, embryo and seedling. *Fruit* free from both lemma and palea; small; longitudinally grooved; not noticeably compressed; with hairs confined to a terminal tuft. *Hilum long-linear*. Embryo small; not waisted. Endosperm liquid in the mature fruit, or hard; containing compound starch grains. Embryo with an epiblast; without a scutellar tail; with a negligible mesocotyl internode. Embryonic leaf margins meeting.

Seedling with a short mesocotyl. First seedling leaf with a well-developed lamina. The lamina narrow; erect; 3–5 veined.

Abaxial leaf blade epidermis. *Costal/intercostal zonation* conspicuous. *Papillae* absent. *Long-cells* similar in shape costally and intercostally (elongated); of similar wall thickness costally and intercostally. Mid-intercostal long-cells rectangular to fusiform; having straight or only gently undulating walls. *Microhairs* absent. *Stomata* common. Subsidiaries parallel-sided, or dome-shaped. Guard-cells overlapped by the interstomatals. *Intercostal short-cells* common (mainly adjacent to the costal regions), or absent or very rare; in cork/silica-cell pairs (usually), or not paired (sometimes solitary); silicified (when paired), or not silicified. *Costal short-cells* conspicuously in long rows, or neither distinctly grouped into long rows nor predominantly paired. Costal silica bodies horizontally-elongated crenate/sinuous, or horizontally-elongated smooth (a few, rectangular), or tall-and-narrow.

Transverse section of leaf blade, physiology. C_3; XyMS+. *Mesophyll* with non-radiate chlorenchyma. *Midrib* not readily distinguishable; with one bundle only. Bulliforms present in discrete, regular adaxial groups; in simple fans. All the vascular bundles accompanied by sclerenchyma. Combined sclerenchyma girders present; nowhere forming 'figures'. Sclerenchyma all associated with vascular bundles.

Cytology. Chromosome base number, $x = 7$. $2n = 42$.

Taxonomy. Pooideae; Poodae; Aveneae.

Ecology, geography, regional floristic distribution. 7 species. Australia, New Zealand, New Guinea. Mesophytic; glycophytic. In open woodland.

Paleotropical, Australian, and Antarctic. Indomalesian. Papuan. North and East Australian. New Zealand. Tropical North and East Australian and Temperate and South-Eastern Australian.

Rusts and smuts. Rusts — *Puccinia*. Taxonomically wide-ranging species: *Puccinia graminis*.

References, etc. Morphological/taxonomic: Hubbard 1935. Leaf anatomical: this project.

Ectrosia R.Br.

Excluding *Ectrosiopsis, Planichloa*

Habit, vegetative morphology. Annual, or perennial; caespitose. **Culms** 10–60 cm high; *herbaceous*; branched above, or unbranched above. Culm nodes glabrous. Culm internodes hollow. *Leaves* mostly basal, or not basally aggregated; non-auriculate. Leaf blades narrow; setaceous, or not setaceous; from setaceous in *E. agrostoides* to flat and up to 3 mm wide in *E. leporina*; without abaxial multicellular glands; without cross venation; persistent. **Ligule a fringed membrane (very narrow), or a fringe of hairs**. *Contraligule* absent.

Reproductive organization. Plants bisexual, with bisexual spikelets; with hermaphrodite florets. Exposed-cleistogamous, or chasmogamous.

Inflorescence. *Inflorescence paniculate (or almost reduced to a raceme of racemes)*; open to contracted; with capillary branchlets (*E. eragrostoides*), or without capillary branchlets; non-digitate; espatheate; not comprising 'partial inflorescences' and foliar organs. Spikelet-bearing axes persistent. Spikelets solitary; secund, or not secund; pedicellate.

Female-fertile spikelets. *Spikelets* 1.5–13 mm long; compressed laterally; disarticulating above the glumes; *not disarticulating between the florets*; with distinctly elongated rachilla internodes between the florets (the one above the glumes shorter than the rest). Rachilla prolonged beyond the uppermost female-fertile floret; rachilla hairless. The rachilla extension with incomplete florets. Hairy callus absent.

Glumes two; *very unequal*; shorter than the spikelets; shorter than the adjacent lemmas; hairless; glabrous; pointed, or not pointed; awnless; carinate; similar. Lower glume 1 nerved. Upper glume 1 nerved. *Spikelets* with incomplete florets. The incomplete florets distal to the female-fertile florets. **The distal incomplete florets *clearly specialised and modified in form (long-awned, or reduced to awns). Spikelets without proximal incomplete florets*.

Female-fertile florets 1–20. *Lemmas* similar in texture to the glumes (thinly membranous); not becoming indurated; entire, or incised; when entire pointed, or blunt; when incised, 2 lobed; not deeply cleft; *awned*. Awns 1; median; from a sinus, or apical; non-geniculate; hairless (scabrous); much shorter than the body of the lemma to much longer than the body of the lemma. *Lemmas* hairless; glabrous, or scabrous; *carinate*; without a germination flap; *1–3 nerved*. *Palea* present; relatively long; entire; awnless; without apical setae; not indurated; 2-nerved; strongly 2-keeled. Palea keels wingless; hairy (ciliate). *Lodicules* present; 2; fleshy; glabrous; not or scarcely vascularized. *Stamens* 3. Anthers very short; not penicillate. *Ovary* glabrous. Styles free to their bases. Stigmas 2; white.

Fruit, embryo and seedling. Fruit small; ellipsoid; not noticeably compressed. Hilum short. Pericarp fused. Embryo large. Endosperm hard; containing compound starch grains. Embryo with an epiblast; with a scutellar tail; with an elongated mesocotyl internode. Embryonic leaf margins meeting.

Abaxial leaf blade epidermis. *Costal/intercostal zonation* conspicuous. *Papillae* absent. Mid-intercostal long-cells rectangular; having markedly sinuous walls. *Microhairs* present; elongated; clearly two-celled; panicoid-type. Microhair apical cell wall thinner than that of the basal cell and often collapsed. Microhairs (45–)48–66(–69) microns long. Microhair basal cells 21 microns long. Microhairs (4.5–)5.4–6(–6.6) microns wide at the septum. Microhair total length/width at septum 7.5–11.1. Microhair apical cells (19.5–)25.5–36(–39) microns long. Microhair apical cell/total length ratio 0.43–0.64. *Stomata* common; 24–27 microns long. Subsidiaries low dome-shaped, or triangular. Guard-cells overlapping to flush with the interstomatals. *Intercostal short-cells* common; in cork/silica-cell pairs (*E. leporina*), or not paired (solitary, in the other species seen); silicified (*E. leporina*, with cork/silica-cell pairs), or not silicified. Intercostal silica bodies present and perfectly developed; in *E. leporina* tall-and-narrow and oryzoid-type. *Costal short-cells* predominantly paired. Costal silica bodies present throughout the costal zones; crescentic (mostly), or saddle shaped (a few), or tall-and-narrow (a few).

Transverse section of leaf blade, physiology. C_4; XyMS+. PCR sheath outlines uneven. PCR sheaths of the primary vascular bundles interrupted; interrupted both abaxially and adaxially. PCR sheath extensions present. Maximum number of extension cells 1–6. *Mesophyll* with radiate chlorenchyma. *Leaf blade* with distinct, prominent adaxial ribs to adaxially flat; with the ribs more or less constant in size. *Midrib* not readily distinguishable; with one bundle only. Bulliforms present in discrete, regular adaxial groups; in simple fans (the median cells large and deeply penetrating). All the vascular bundles accompanied by sclerenchyma. Combined sclerenchyma girders present (with all the primaries); forming 'figures' (anchors and I's in the primaries). Sclerenchyma all associated with vascular bundles. The lamina margins with fibres.

Taxonomy. Chloridoideae; main chloridoid assemblage.

Ecology, geography, regional floristic distribution. 12 species. Tropical Australia. Species of open habitats. Poor sandy soils.

Paleotropical and Australian. Indomalesian. Malesian and Papuan. North and East Australian. Tropical North and East Australian.

References, etc. Leaf anatomical: this project.

Ectrosiopsis (Ohwi) Jansen

~ *Ectrosia* (*E. lasioclada* (Merr.) S.T. Blake, = *Eragrostis* sect. Ectrosiopsis)

Habit, vegetative morphology. Perennial; caespitose. **Culms** 30–40 cm high; *herbaceous*; unbranched above. Plants unarmed. **Leaves** *mostly basal*; non-auriculate. *Leaf blades* narrow; *2–2.5 mm wide (to 2 cm long, sometimes with tuberculate hairs)*; setaceous, or not setaceous; without abaxial multicellular glands; without cross venation. **Ligule** present; *a fringe of hairs*.

Reproductive organization. *Plants bisexual, with bisexual spikelets*; with hermaphrodite florets.

Inflorescence. *Inflorescence paniculate*; *contracted (elongate, 10–15 cm long)*; espatheate; not comprising 'partial inflorescences' and foliar organs. Spikelet-bearing axes persistent. Spikelets not secund; shortly pedicellate.

Female-fertile spikelets. *Spikelets* 3–7 mm long; compressed laterally; disarticulating above the glumes (but the glumes deciduous); *disarticulating between the florets*; *with distinctly elongated rachilla internodes between the florets (?)*. Rachilla prolonged beyond the uppermost female-fertile floret; rachilla hairless. The rachilla extension with incomplete florets.

Glumes two; very unequal; shorter than the spikelets; shorter than the adjacent lemmas (the G_2 somewhat shorter than the L_1); pointed (acute); awnless; similar (membranous, naviculate). Lower glume 1 nerved. Upper glume 1 nerved. *Spikelets* with incomplete florets. The incomplete florets distal to the female-fertile florets. *The distal incomplete florets* merely underdeveloped.

Female-fertile florets *3–7*. Lemmas attenuate-acuminate; similar in texture to the glumes (membranous); not becoming indurated; entire; pointed; awnless to awned (setaceously acuminate to shortly awned). Awns when present, 1; median; apical; non-geniculate; much shorter than the body of the lemma. *Lemmas hairless*; carinate ('lightly'); *3 nerved. Palea* present; 2-keeled. Palea keels wingless. *Lodicules* present; 2; free; fleshy. *Stamens* 3. Anthers 0.3–0.4 mm long.

Fruit, embryo and seedling. Fruit ellipsoid. Embryo with an epiblast; with a scutellar tail; with an elongated mesocotyl internode.

Abaxial leaf blade epidermis. *Costal/intercostal zonation* conspicuous (but clearly different only in the presence of stomata intercostally). *Papillae* absent. *Long-cells* similar in shape costally and intercostally; of similar wall thickness costally and intercostally. Mid-intercostal long-cells rectangular; having markedly sinuous walls (fairly thick). *Microhairs* present; elongated; clearly two-celled; panicoid-type (large, broad in the middle). Microhair apical cell wall thinner than that of the basal cell and often collapsed. Microhairs (45–)51 microns long. Microhair basal cells 27–30 microns long. Microhairs 6–9 microns wide at the septum. Microhair total length/width at septum 5–8.5. Microhair apical cells 21–30 microns long. Microhair apical cell/total length ratio 0.41–0.59. *Stomata* common; 24–27 microns long. Subsidiaries dome-shaped (mostly), or triangular (a few). Guard-cells overlapping to flush with the interstomatals. *Intercostal short-cells* common; in cork/silica-cell pairs; silicified. Intercostal silica bodies present and perfectly developed; crescentic, tall-and-narrow, and saddle shaped (crescents predominating). *Costal short-cells* predominantly paired. Costal silica bodies present throughout the costal zones; crescentic (predominating), or tall-and-narrow, or saddle shaped (a few referable to narrow versions of this type).

Transverse section of leaf blade, physiology. C_4; XyMS+. PCR sheath outlines uneven. PCR sheaths of the primary vascular bundles interrupted; interrupted abaxially only. PCR sheath extensions present. Maximum number of extension cells 1. PCR cell chloroplasts probably centrifugal/peripheral (judging from poorly preserved material). *Leaf blade* adaxially flat. *Midrib* not readily distinguishable; with one bundle only. Bulliforms present in discrete, regular adaxial groups; in simple fans (the groups large, each with a large and deeply penetrating median cell). All the vascular bundles accompanied by sclerenchyma. Combined sclerenchyma girders present (with all the bundles); forming 'figures' (anchors or I's with all of them). Sclerenchyma all associated with vascular bundles.

Taxonomy. Chloridoideae; main chloridoid assemblage.

Ecology, geography, regional floristic distribution. 1 species (*E. lasioclada*). Australia, New Guinea, Caroline Is. Species of open habitats. Damp sandy depressions.

Paleotropical and Australian. Indomalesian. Papuan. North and East Australian. Tropical North and East Australian.

References, etc. Morphological/taxonomic: Jansen 1952. Leaf anatomical: this project.

Special comments. Fruit data wanting.

Ehrharta Thunb.

Diplax Bennett, *Trochera* L. Rich.

Excluding *Microlaena, Petriella, Tetrarrhena*

Habit, vegetative morphology. Annual, or perennial; rhizomatous, or stoloniferous, or caespitose, or decumbent. *Culms* 6–150 cm high; woody and persistent, or herbaceous; branched above, or unbranched above; tuberous, or not tuberous. Culm nodes hairy, or glabrous. Culm internodes solid, or hollow. Young shoots intravaginal. *Leaves* mostly basal, or not basally aggregated; auriculate, or non-auriculate. *Leaf blades* linear to linear-lanceolate; broad, or narrow; 0.5–12 mm wide; setaceous, or not setaceous; flat, or folded, or rolled; *not pseudopetiolate*; without cross venation; disarticulating from the sheaths, or persistent. Ligule an unfringed membrane, or a fringed membrane, or a fringe of hairs; truncate; 0.5–3 mm long. *Contra-ligule* absent.

Reproductive organization. *Plants bisexual, with bisexual spikelets*; with hermaphrodite florets; outbreeding.

Inflorescence. Inflorescence few spikeleted to many spikeleted; a single raceme, or paniculate (then narrow, with slender branches); open, or contracted; with capillary branchlets, or without capillary branchlets; espatheate (though in two species the mature inflorescence base is enclosed in the uppermost leaf sheath); not comprising 'partial inflorescences' and foliar organs. Spikelet-bearing axes persistent. Spikelets solitary; secund, or not secund; pedicellate; imbricate, or distant; not in distinct 'long-and-short' combinations.

Female-fertile spikelets. Spikelets 2–17 mm long; compressed laterally, or not noticeably compressed; disarticulating above the glumes; not disarticulating between the florets; with conventional internode spacings. Rachilla inconspicuously prolonged beyond the uppermost female-fertile floret, or terminated by a female-fertile floret; rachilla hairy, or hairless. The rachilla extension when present, naked. Hairy callus absent. *Callus* absent.

Glumes two; very unequal, or more or less equal; shorter than the spikelets to exceeding the spikelets; shorter than the adjacent lemmas, or long relative to the adjacent lemmas; hairless; glabrous, or scabrous; pointed, or not pointed (blunt to obtuse); awnless; carinate, or non-carinate; similar (membranous). **Lower glume** 0.5–1 times the length of the upper glume; *5 nerved*. Upper glume 5 nerved. *Spikelets* with incomplete florets. The incomplete florets proximal to the female-fertile florets. *Spikelets with proximal incomplete florets*. The proximal incomplete florets 2; epaleate; sterile. *The proximal lemmas curiously varied in form, often hardened and transversely ridged, the lower sometimes bearing a knob-like appendage (elaieosome?) at its base*; awned (abruptly from the back, or the lemma tapering into the awn), or awnless; exceeded by the female-fertile lemmas to decidedly exceeding the female-fertile lemmas; less firm than the female-fertile lemmas to similar in texture to the female-fertile lemmas; becoming indurated, or becoming indurated and not becoming indurated (the first sterile lemma being sometimes non-indurated).

Female-fertile florets 1. Lemmas decidedly firmer than the glumes; usually smooth (and often polished), or striate; becoming indurated; entire; pointed, or blunt (truncate); awnless (usually), or mucronate (occasionally); rarely sparsely hairy, or hairless; usually glabrous, or scabrous; carinate; without a germination flap; 5–7 nerved; with the nerves non-confluent. **Palea** present; relatively long (narrow); entire; awnless, without apical setae; thinner than the lemma; not indurated; 1-nerved to several nerved, or nerveless (rarely); *2-keeled, or one-keeled (by apposition of the two)*. Palea keels wingless. *Lodicules* present; 2; free; membranous; ciliate, or glabrous; toothed. *Stamens* 3, or 4, or 6. Anthers not penicillate; without an apically prolonged connective. *Ovary* glabrous. Styles fused to free to their bases. Stigmas 2; white, or brown.

Fruit, embryo and seedling. *Fruit* free from both lemma and palea; small; compressed laterally. **Hilum long-linear.** Embryo small; waisted. Endosperm hard; without lipid; containing compound starch grains. Embryo without an epiblast; with a scutellar tail; with a negligible mesocotyl internode. Embryonic leaf margins overlapping.

Seedling with a short mesocotyl, or with a long mesocotyl. First seedling leaf with a well-developed lamina. The lamina broad, or narrow; erect; 6–12 veined.

Abaxial leaf blade epidermis. *Costal/intercostal zonation* conspicuous. *Papillae* absent. *Long-cells* similar in shape costally and intercostally (but the intercostals much larger); of similar wall thickness costally and intercostally. Mid-intercostal long-cells rectangular, or fusiform; having markedly sinuous walls, or having straight or only gently undulating walls. *Microhairs* present, or absent; panicoid-type; 32–44 microns long (e.g. *E. pusilla*), or 39–60 microns long; 4.9–7.3 microns wide at the septum (in *E. rehmannii*, *E. pusilla*), or (7–)8–10(–12) microns wide at the septum. Microhair total length/width at septum 2.7–10. Microhair apical cells 17–49 microns long. Microhair apical cell/total length ratio 0.42–0.71. *Stomata* common; 24–44(–49) microns long. Subsidiaries low to high dome-shaped. Guard-cells overlapped by the interstomatals. *Intercostal short-cells* common, or absent or very rare; when seen not paired; when seen silicified, or not silicified. Intercostal silica bodies when present, rounded, or crescentic. *Costal short-cells* conspicuously in long rows, or predominantly paired, or neither distinctly grouped into long rows nor predominantly paired. Costal silica bodies rounded, or saddle shaped (to almost cubical), or crescentic, or 'panicoid-type'; when panicoid type, cross shaped, or butterfly shaped, or dumb-bell shaped.

Transverse section of leaf blade, physiology. C_3; XyMS+. *Mesophyll* with non-radiate chlorenchyma; without adaxial palisade; traversed by columns of colourless mesophyll cells, or not traversed by colourless columns (rarely); with arm cells (sometimes, in southern African species), or without arm cells; without fusoids. *Leaf blade* with distinct, prominent adaxial ribs, or adaxially flat; with the ribs more or less constant in size, or with the ribs very irregular in sizes. *Midrib* conspicuous, or not readily distinguishable; with one bundle only; with colourless mesophyll adaxially (rarely), or without colourless mesophyll adaxially. Bulliforms present in discrete, regular adaxial groups; in simple fans. All the vascular bundles accompanied by sclerenchyma. Combined sclerenchyma girders present. Sclerenchyma all associated with vascular bundles.

Phytochemistry. Tissues of the culm bases with abundant starch, or with little or no starch. Leaves without flavonoid sulphates (1 species).

Cytology. Chromosome base number, $x = 12$. $2n = 24$ and 48. 2 and 4 ploid.

Taxonomy. Bambusoideae; Oryzodae; Ehrharteae.

Ecology, geography, regional floristic distribution. 27 species. Southern and tropical Africa, Mascarene Is., New Zealand. Commonly adventive. Helophytic (most of the annuals), or mesophytic; shade species (*E. erecta*), or species of open habitats; halophytic (*E. villosa*), or glycophytic. Diverse habitats, with *E. villosa* in coastal sand dunes.

Paleotropical, Cape, and Antarctic. African and Indomalesian. Saharo-Sindian, Sudano-Angolan, and Namib-Karoo. Indian. New Zealand. Sahelo-Sudanian, Somalo-Ethiopian, and South Tropical African.

Rusts and smuts. Smuts from Tilletiaceae and from Ustilaginaceae. Tilletiaceae — *Tilletia*. Ustilaginaceae — *Ustilago*.

Economic importance. Significant weed species: *E. brevifolia, E. calycina, E. dura, E. erecta, E. longiflora, E. villosa*. Cultivated fodder: *E. erecta, E. calycina*. Important native pasture species: *E. erecta*.

References, etc. Morphological/taxonomic: Willemse 1982; Gibbs Russell and Ellis 1987, 1988; Gibbs Russell 1987. Leaf anatomical: Ellis 1987 and this project.

Ekmanochloa Hitchcock

Habit, vegetative morphology. Erect perennial (with dimorphic culms); caespitose. The flowering culms leafless (at least, with 2–3 bladeless leaves). **Culms 30–100 cm high**; herbaceous; unbranched above; *tuberous (with small 'corms')*. Plants unarmed. *Leaves* not basally aggregated; non-auriculate; without auricular setae. Sheaths shorter than the internodes, striate, glabrous or pilose distally. *Leaf blades* of the vegetative culms greatly reduced (sometimes, in *E. subaphylla?*), or not all greatly reduced; linear-lanceolate, or lanceolate; narrow; 1–4 mm wide (8–55 cm long); not cordate, not sagittate (slightly asymmetric at the base); flat; shortly pseudopetiolate; cross veined (the transverse veins seen to be abundant only at high magnification); disarticulating from the sheaths; rolled in bud. *Ligule* present (minute); an unfringed membrane (laciniate); truncate; 0.1–0.2 mm long.

Reproductive organization. *Plants monoecious with all the fertile spikelets unisexual*; without hermaphrodite florets. The spikelets of sexually distinct forms on the same plant; female-only and male-only. *The male and female-fertile spikelets on different branches of the same inflorescence (there being a pair of slender racemes, one male the other female)*. The spikelets overtly heteromorphic.

Inflorescence. Inflorescence of spicate main branches (a pair of conjugate, appressed, slender, unisexual racemes, the male raceme slightly shorter than the female); digitate. Primary inflorescence branches 2. Inflorescence espatheate; not comprising 'partial inflorescences' and foliar organs. *Spikelet-bearing axes* 'racemes'; with very slender rachides (these filiform, slightly flexuose); persistent. Spikelets solitary; secund; shortly pedicellate; not in distinct 'long-and-short' combinations.

Female-sterile spikelets. The male spikelets 2.7–3.3 mm long, smaller than the females, ellipsoid, hyaline, reduced to the subequal lemma and palea and flower, the lemma awnless. Rachilla of male spikelets terminated by a male floret. The male spikelets without glumes; with proximal incomplete florets; 1 floreted. The lemmas awnless. Male florets 1; 2 staminate (*E. subaphylla*), or 3 staminate (the anthers 0.8 mm long).

Female-fertile spikelets. Spikelets 6–9 mm long (0.9 mm wide); lanceolate; compressed dorsiventrally; disarticulating above the glumes (the floret 'early deciduous'); with conventional internode spacings, or with a distinctly elongated rachilla internode above the glumes (0.5 mm long in *A. subaphylla*). Rachilla terminated by a female-fertile floret. Hairy callus absent.

Glumes two; very unequal, or more or less equal; shorter than the spikelets, or about equalling the spikelets; shorter than the adjacent lemmas, or long relative to the adjacent lemmas (*E. subaphylla* having short glumes); hairy, or hairless; awnless; non-carinate; similar (membranous). Lower glume 5 nerved. Upper glume 3 nerved. *Spikelets* with female-fertile florets only.

Female-fertile florets 1. Lemmas terete, narrowing gradually into the slender awn; decidedly firmer than the glumes (leathery); becoming indurated; entire; pointed (attenuate into the awn); long awned. Awns 1; median; apical; non-geniculate; straight, or flexuous; hairless; much longer than the body of the lemma. Lemmas hairy, or hairless; non-carinate; having the margins tucked in onto the palea; with a clear germination flap (?); 5 nerved, or 7 nerved. *Palea* present; relatively long; tightly clasped by the lemma; awnless, without apical setae; textured like the lemma; 2-nerved. *Lodicules* present; 3; membranous (flabellate); not or scarcely vascularized. *Stamens* 0 (no staminodes mentioned). *Ovary* glabrous. Styles free to their bases. Stigmas 2.

Fruit, embryo and seedling. Disseminule a free caryopsis. *Fruit* free from both lemma and palea; small to medium sized (4.2 mm long in *E. aristata*); brown; fusiform;

longitudinally grooved; glabrous. Hilum long-linear (as long as the caryopsis). Embryo small.

Abaxial leaf blade epidermis. *Costal/intercostal zonation* conspicuous. *Papillae* present (abundant); costal and intercostal. Intercostal papillae over-arching the stomata; several per cell (fairly large, more or less circular or branched, thick walled, in one or two rows per long-cell and forming rings overlooking the stomatal crypts). Long-cells of similar wall thickness costally and intercostally (fairly thin walled). Mid-intercostal long-cells rectangular; having markedly sinuous walls (the sinuosity coarse). *Microhairs* present; elongated; clearly two-celled; panicoid-type (the basal cells much elongated); 45–50 microns long. Microhair apical cells 5–14 microns long. *Stomata* common (but lacking in the mid-intercostal regions); 17–28 microns long. Subsidiaries obscured by the papillae. *Intercostal short-cells* common; in cork/silica-cell pairs; silicified. Intercostal silica bodies tall-and-narrow and vertically elongated-nodular (slim). With scattered prickles costally and intercostally. *Costal short-cells* conspicuously in long rows. Costal silica bodies present and well developed; 'panicoid-type' (mostly), or oryzoid (a few); cross shaped to dumb-bell shaped (large, short); sharp-pointed, or not sharp-pointed.

Transverse section of leaf blade, physiology. C_3; XyMS+. Mesophyll with adaxial palisade (*E. aristata*), or without adaxial palisade (*E. ekmanochloa*); with arm cells (but fairly inconspicuous); most definitely without fusoids (in the material seen, cf. Zuloaga *et al*. 1993). *Leaf blade* more or less adaxially flat. *Midrib* conspicuous, or not readily distinguishable; with one bundle only; without colourless mesophyll adaxially. *The lamina* symmetrical on either side of the midrib. Bulliforms present in discrete, regular adaxial groups; in simple fans (these wide, large celled). All the vascular bundles accompanied by sclerenchyma. Combined sclerenchyma girders present; forming 'figures' (I's with the main bundles). Sclerenchyma all associated with vascular bundles.

Taxonomy. Bambusoideae; Oryzodae; Olyreae.

Ecology, geography, regional floristic distribution. 2 species. Cuba. Probably shade species; glycophytic. On limestone rocks and in pinewoods.

Neotropical. Caribbean.

References, etc. Morphological/taxonomic: Zuloaga *et al*. 1993. Leaf anatomical: this project; Zuloaga *et al*. 1993.

Eleusine Gaertn.

Habit, vegetative morphology. Annual, or perennial (the culms flattened); caespitose (or mat-forming). *Culms* 10–150 cm high; herbaceous. Culm nodes glabrous. Culm internodes solid, or hollow. Young shoots intravaginal. *Leaves* not basally aggregated; non-auriculate. Sheath margins free. The sheaths keeled. Leaf blades linear; narrow; flat, or folded; without abaxial multicellular glands; without cross venation; persistent; rolled in bud, or once-folded in bud. Ligule a fringed membrane. *Contra-ligule* absent.

Reproductive organization. Plants bisexual, with bisexual spikelets; with hermaphrodite florets; inbreeding. Exposed-cleistogamous, or chasmogamous. Plants without hidden cleistogenes. Viviparous (sometimes), or not viviparous.

Inflorescence. *Inflorescence* of spicate main branches; open, or contracted (sometimes forming a capitulum); *digitate, or subdigitate (then shortly racemose, but clustered at the top of the culm)*. Primary inflorescence branches 1–16. Rachides flattened. Inflorescence espatheate; not comprising 'partial inflorescences' and foliar organs. The racemes spikelet bearing to the base. Spikelet-bearing axes with substantial rachides; persistent. *Spikelets* secund; biseriate; *sessile to subsessile*; imbricate.

Female-fertile spikelets. Spikelets 3.5–11 mm long; adaxial; compressed laterally; disarticulating above the glumes, or not disarticulating (*E. coracana*); disarticulating between the florets (except in *E. coracana*). Rachilla prolonged beyond the uppermost female-fertile floret; rachilla hairy, or hairless. Hairy callus absent. *Callus* absent.

Glumes two; *very unequal*; shorter than the spikelets; shorter than the adjacent lemmas; dorsiventral to the rachis to lateral to the rachis; awnless; carinate; with the keel conspicuously winged, or without a median keel-wing. Lower glume 1 nerved. Upper glume 3–5(–7) nerved. *Spikelets* with female-fertile florets only, or with incomplete florets. The incomplete florets distal to the female-fertile florets.

Female-fertile florets *3–15*. Lemmas not becoming indurated; entire; pointed, or blunt; awnless to mucronate; hairless; glabrous; carinate; 3 nerved. *Palea* present; apically notched; awnless, without apical setae; 2-nerved; 2-keeled. Palea keels winged, or wingless. *Lodicules* present; 2; joined, or free; fleshy, or membranous; glabrous; toothed, or not toothed. *Stamens* 3. Anthers 0.5–0.8 mm long; not penicillate. *Ovary* glabrous. Styles free to their bases. Stigmas 2; white, or brown.

Fruit, embryo and seedling. Fruit free from both lemma and palea; small (0.9–2 mm long); ellipsoid to subglobose; longitudinally grooved (*E. multiflora*), or not grooved; not noticeably compressed (globose), or trigonous; *sculptured*. Hilum short. Pericarp thin (hyaline); free. Embryo large; not waisted. Endosperm hard; without lipid; containing compound starch grains. Embryo with an epiblast; with a scutellar tail; with an elongated mesocotyl internode. Embryonic leaf margins meeting.

Seedling with a long mesocotyl. First seedling leaf with a well-developed lamina. The lamina broad; curved; 6–12 veined.

Abaxial leaf blade epidermis. *Costal/intercostal zonation* conspicuous. *Papillae* absent. *Long-cells* similar in shape costally and intercostally; of similar wall thickness costally and intercostally. Mid-intercostal long-cells rectangular; having markedly sinuous walls. *Microhairs* present; more or less spherical, or elongated; ostensibly one-celled, or clearly two-celled; chloridoid-type (often with hidden bases). Microhair apical cell wall of similar thickness/rigidity to that of the basal cell. Microhairs with 'partitioning membranes' (in *E. indica*). The 'partitioning membranes' in the basal cell. Microhairs 19–21 microns long. Microhair basal cells 12–14 microns long. Microhairs 7.5–9 microns wide at the septum. Microhair total length/width at septum 2.17–2.8. Microhair apical cells 13.5–16.5 microns long. Microhair apical cell/total length ratio 0.64–0.8. *Stomata* common; 30–33 microns long. Subsidiaries triangular. Guard-cells overlapping to flush with the interstomatals. *Intercostal short-cells* common; not paired. Intercostal silica bodies absent. *Costal short-cells* conspicuously in long rows. Costal silica bodies present in alternate cell files of the costal zones; rounded to saddle shaped.

Transverse section of leaf blade, physiology. C_4; biochemical type NAD–ME (2 species); XyMS+. PCR sheath outlines even. PCR sheaths of the primary vascular bundles complete. PCR sheath extensions absent. PCR cells without a suberised lamella. PCR cell chloroplasts with well developed grana; centripetal. *Mesophyll* with radiate chlorenchyma; traversed by columns of colourless mesophyll cells. *Leaf blade* adaxially flat; with the ribs more or less constant in size. *Midrib* conspicuous; with one bundle only. Bulliforms present in discrete, regular adaxial groups; associated with colourless mesophyll cells to form deeply-penetrating fans, or in simple fans and associated with colourless mesophyll cells to form deeply-penetrating fans (often linked with traversing colourless girders). All the vascular bundles accompanied by sclerenchyma. Combined sclerenchyma girders present; forming 'figures', or nowhere forming 'figures'. Sclerenchyma all associated with vascular bundles. The lamina margins with fibres.

Culm anatomy. *Culm internode bundles* in one or two rings, or in three or more rings.

Phytochemistry. Leaf blade chlorophyll *a:b* ratio 3.43–4.3.

Cytology. Chromosome base number, $x = 9$. $2n = 18, 20, 36$, and 40. 2 and 4 ploid. Haploid nuclear DNA content 0.7–0.8 pg (2 species, $2x$ and $4x$). Mean diploid 2c DNA value 1.4 pg (1 species). Nucleoli persistent.

Taxonomy. Chloridoideae; main chloridoid assemblage.

Ecology, geography, regional floristic distribution. 9 species. Tropical and subtropical. Commonly adventive. Mesophytic, or xerophytic; species of open habitats. Savanna, grassland, weedy places.

Holarctic, Paleotropical, Neotropical, Cape, and Antarctic. Boreal, Tethyan, and Madrean. African, Madagascan, Indomalesian, and Polynesian. Euro-Siberian, Eastern Asian, and Atlantic North American. Irano-Turanian. Saharo-Sindian, Sudano-Angolan, West African Rainforest, and Namib-Karoo. Indian, Malesian, and Papuan. Fijian. Caribbean, Venezuela and Surinam, Amazon, Central Brazilian, and Pampas. New Zealand and Patagonian. Siberian. Canadian-Appalachian. Sahelo-Sudanian, Somalo-Ethiopian, South Tropical African, and Kalaharian.

Rusts and smuts. Smuts from Ustilaginaceae. Ustilaginaceae — *Ustilago*.

Economic importance. Significant weed species: *E. africana, E. coracana, E. indica, E. tristachya.* Grain crop species: *E. coracana* (Finger Millet, Ragi) widely grown in uplands of India, China and Africa.

References, etc. Leaf anatomical: Metcalfe 1960; this project.

Elionurus Humb. & Bonpl.

Callichloea Steud., *Habrurus* Hochst.

Habit, vegetative morphology. Annual, or perennial; caespitose. *Culms* 10–150 cm high; herbaceous; unbranched above. Culm nodes glabrous. Culm internodes solid. The shoots aromatic (with a bitter taste), or not aromatic. *Leaves* not basally aggregated; non-auriculate. Leaf blades narrow; setaceous, or not setaceous; sometimes flat, or folded (tightly); without cross venation; persistent. *Ligule a fringed membrane (very short), or a fringe of hairs.*

Reproductive organization. Plants bisexual, with bisexual spikelets; with hermaphrodite florets. *The spikelets* of sexually distinct forms on the same plant; hermaphrodite and male-only; *homomorphic*; all in heterogamous combinations.

Inflorescence. Inflorescence a single raceme, or paniculate (of single 'racemes', terminal or sometimes axillary and gathered into false panicles). Rachides flattened. Inflorescence spatheate; a complex of 'partial inflorescences' and intervening foliar organs, or not comprising 'partial inflorescences' and foliar organs. *Spikelet-bearing axes* spikelike (flexuous); solitary; with substantial rachides (these flattened); disarticulating; disarticulating at the joints. *'Articles'* linear, or non-linear; appendaged (with a scarious lobe, in *E. elegans* and relatives), or not appendaged; disarticulating obliquely; densely long-hairy. Spikelets paired; secund (all the sessile spikelets on one side); sessile and pedicellate; consistently in 'long-and-short' combinations; in pedicellate/sessile combinations. Pedicels of the 'pedicellate' spikelets free of the rachis (resembling the internode). The 'shorter' spikelets hermaphrodite. The 'longer' spikelets male-only.

Female-sterile spikelets. The pedicelled, male spikelets well developed, muticous or aristulate. The male spikelets with glumes; 2 floreted. The lemmas awnless.

Female-fertile spikelets. Spikelets compressed dorsiventrally; falling with the glumes. Rachilla terminated by a female-fertile floret. Hairy callus present.

Glumes two; relatively large; very unequal to more or less equal; (the longer) long relative to the adjacent lemmas; hairy, or hairless; *with distinct hair tufts, or with distinct rows of hairs; awned (the G₁ often cuspidate to a bifid tip, with tails several mm long), or awnless*; very dissimilar (the lower tougher, carinate on the edges, its keels generally glandular or with tufts of hairs, the upper membranous, lanceolate, not 2-keeled), or similar (rarely, then both subulate). Lower glume two-keeled (keels not winged, but often ciliate and bordered by a dense oil streak); convex on the back to flattened on the back; not pitted; relatively smooth, or tuberculate (then with ciliate or hair-tufted tubercles on the keels); 9 nerved. Upper glume 3 nerved. *Spikelets* with incomplete florets. The incomplete florets proximal to the female-fertile florets. *Spikelets with proximal incomplete florets. The proximal incomplete florets* 1; epaleate; sterile. The proximal lemmas awnless; decidedly exceeding the female-fertile lemmas; similar in texture to the female-fertile lemmas (hyaline); not becoming indurated.

Female-fertile florets 1. *Lemmas* less firm than the glumes (hyaline); not becoming indurated; entire; awnless; *0 nerved. Palea* present, or absent; when present, very reduced; not indurated; nerveless. *Lodicules* present; 2; free; fleshy; glabrous. *Stamens* 3. Anthers not penicillate. *Ovary* glabrous. Styles free to their bases. Stigmas 2; red pigmented.

Fruit, embryo and seedling. *Fruit* free from both lemma and palea; small; compressed dorsiventrally. Hilum short. Embryo large; waisted. Endosperm hard; without lipid.

Abaxial leaf blade epidermis. *Papillae* absent. Mid-intercostal long-cells having markedly sinuous walls. *Microhairs* present; panicoid-type; 45–80 microns long. Microhair apical cells 22–51 microns long. Microhair apical cell/total length ratio 0.53–0.6. *Stomata* common. Subsidiaries tall dome-shaped, or triangular. Guard-cells overlapping to flush with the interstomatals. *Intercostal short-cells* common, or absent or very rare; when present, in cork/silica-cell pairs, or in cork/silica-cell pairs and not paired; silicified. Intercostal silica bodies when present, rounded, or crescentic, or tall-and-narrow. *Costal short-cells* predominantly paired. Costal silica bodies tall-and-narrow, or 'panicoid-type'; cross shaped, or butterfly shaped, or dumb-bell shaped, or nodular; sharp-pointed.

Transverse section of leaf blade, physiology. C_4; XyMS–. PCR cell chloroplasts centrifugal/peripheral. *Mesophyll* with radiate chlorenchyma. *Leaf blade* 'nodular' in section; with the ribs more or less constant in size. *Midrib* conspicuous (sometimes with an adaxial bulliform group); having a conventional arc of bundles; with colourless mesophyll adaxially (the colourless tissue sometimes extending across the lamina adaxially). Bulliforms not in discrete, regular adaxial groups (in irregular groups, the epidermis extensively bulliform). Many of the smallest vascular bundles unaccompanied by sclerenchyma. Combined sclerenchyma girders present; forming 'figures' (rarely), or nowhere forming 'figures'. Sclerenchyma all associated with vascular bundles.

Phytochemistry. Leaves without flavonoid sulphates (2 species).

Cytology. Chromosome base number, $x = 5$ and 10. $2n = 10$ and 20.

Taxonomy. Panicoideae; Andropogonodae; Andropogoneae; Rottboelliinae.

Ecology, geography, regional floristic distribution. 15 species. Tropical and subtropical. Mesophytic to xerophytic; species of open habitats; glycophytic. Savanna, often on dry soils.

Holarctic, Paleotropical, Neotropical, and Australian. Boreal, Tethyan, and Madrean. African, Madagascan, and Indomalesian. Atlantic North American. Macaronesian, Mediterranean, and Irano-Turanian. Saharo-Sindian, Sudano-Angolan, West African Rainforest, and Namib-Karoo. Indian and Papuan. Caribbean, Venezuela and Surinam, Central Brazilian, Pampas, and Andean. North and East Australian. Southern Atlantic North American and Central Grasslands. Sahelo-Sudanian, Somalo-Ethiopian, South Tropical African, and Kalaharian. Tropical North and East Australian.

Rusts and smuts. Smuts from Ustilaginaceae. Ustilaginaceae — *Sphacelotheca* and *Ustilago*.

Economic importance. Important native pasture species: several, especially in tropical America and Africa (e.g. *E. royleanus*, *E. tripsacoides*).

References, etc. Leaf anatomical: Metcalfe 1960; this project.

Elymandra Stapf

Excluding *Pleiadelphia*

Habit, vegetative morphology. Coarse annual, or perennial; caespitose. **Culms** 50–250 cm high; *herbaceous*; branched above. *Leaves* not basally aggregated; non-auriculate; without auricular setae. Leaf blades linear; narrow; 2–9 mm wide ((3-)10–60 cm long); not setaceous (rolled or with revolute margins); without cross venation. Ligule an unfringed membrane; truncate; (0.5–)1–3 mm long. *Contra-ligule* absent.

Reproductive organization. Plants bisexual, with bisexual spikelets; with hermaphrodite florets. *The spikelets* of sexually distinct forms on the same plant; hermaphrodite and male-only, or hermaphrodite and sterile; overtly heteromorphic (only the female-fertile spikelets awned); *in both homogamous and heterogamous combinations (each raceme with 1–6 or more male-only or sterile pairs at the base, then one or more heterogamous pairs above and a heterogamous terminal triad).* Plants seemingly inbreeding (the long-pedunculate racemes enclosed in the spatheoles), or outbreeding (?).

Inflorescence. Inflorescence paniculate (of long-exserted 'racemes' gathered into a false panicle); spatheate (and spatheolate); *a complex of 'partial inflorescences' and intervening foliar organs (the spathes and spatheoles narrow, with subulate or setaceous tips).* Spikelet-bearing axes elongated 'racemes'; *the spikelet-bearing axes* with 4–5 spikelet-bearing 'articles' to with more than 10 spikelet-bearing 'articles' ((4-)-18 pairs); *paired (in each spatheole)*; *with substantial rachides*; disarticulating; disarticulating at the joints. *'Articles'* linear; without a basal callus-knob; not appendaged; disarticulating obliquely; somewhat hairy to glabrous. Spikelets paired and in triplets (pairs below, with a terminal heterogamous triad); sessile and pedicellate; consistently in 'long-and-short' combinations (in the upper part of the raceme, those of the lower homogamous pairs all being sessile); in pedicellate/sessile combinations. Pedicels of the 'pedicellate' spikelets free of the rachis. The 'shorter' spikelets hermaphrodite (i.e., in the heterogamous combinations). The 'longer' spikelets male-only, or sterile (rarely).

Female-sterile spikelets. The pedicellate spikelets male or rarely sterile, awnless, with an elongated linear callus. The male spikelets with glumes; 2 floreted. The lemmas awnless.

Female-fertile spikelets. Spikelets 7–11(–12) mm long; not noticeably compressed to compressed dorsiventrally; falling with the glumes. Rachilla terminated by a female-fertile floret. Hairy callus present (shortly white bearded). Callus pointed to blunt.

Glumes two; more or less equal; long relative to the adjacent lemmas; hairy (with white or yellowish hairs); without conspicuous tufts or rows of hairs; *awned (the G$_2$, except sometimes in* **E. androphila***)*; non-carinate; very dissimilar (leathery, the G$_1$ obtuse or truncate, the thinner G$_2$ pointed with an awn or subule, dorsally rounded and grooved). Lower glume not two-keeled; convex on the back; not pitted; relatively smooth; 7 nerved. Upper glume 3 nerved. *Spikelets* with incomplete florets. The incomplete florets proximal to the female-fertile florets. *The proximal incomplete florets* 1; epaleate; sterile. The proximal lemmas awnless; 2 nerved; more or less equalling the female-fertile lemmas; similar in texture to the female-fertile lemmas (hyaline); not becoming indurated.

Female-fertile florets 1. Lemmas stipitate, with hyaline wings and teeth; less firm than the glumes; not becoming indurated; incised; 2 lobed; deeply cleft; awned. Awns 1; median; from a sinus; geniculate; hairy; much longer than the body of the lemma. Lemmas with the hyaline part ciliate; non-carinate; without a germination flap; 1 nerved. **Palea** *absent. Lodicules* present; 2; free (small); fleshy; glabrous. *Stamens* 3. Anthers not penicillate; without an apically prolonged connective. *Ovary* glabrous. Styles basally fused. Stigmas 2.

Abaxial leaf blade epidermis. *Costal/intercostal zonation* conspicuous. *Papillae* absent. *Long-cells* mostly markedly different in shape costally and intercostally (the costals generally narrower and more regularly rectangular); of similar wall thickness costally and intercostally. Intercostal zones with typical long-cells to exhibiting many atypical long-cells. Mid-intercostal long-cells rectangular, or rectangular to fusiform; having markedly sinuous walls to having straight or only gently undulating walls. *Microhairs* present; elongated; clearly two-celled; panicoid-type; of *E. grallata*, 48–54(–60) microns long; 6 microns wide at the septum. Microhair total length/width at septum 8–10. Microhair apical cells 15–18 microns long. Microhair apical cell/total length ratio 0.3–0.4. *Stomata* common; of *E. grallata* 34.5–43.5 microns long. Subsidiaries tall triangular, or dome-shaped. Guard-cells overlapping to flush with the interstomatals. *Intercostal short-cells* common (in *E. archaelymandra*), or absent or very rare (but *E. grallata* with numerous small prickles); of *E. archaelymandra* in cork/silica-cell pairs; silicified. Intercostal silica bodies crescentic, or tall-and-narrow. *Costal short-cells* conspicuously in long rows. Costal silica bodies present and well developed; 'panicoid-type'; mostly dumb-bell shaped.

Transverse section of leaf blade, physiology. C$_4$; XyMS–. PCR sheath extensions absent. *Mesophyll* with radiate chlorenchyma; traversed by columns of colourless mesophyll cells (associated with the 'arches'), or not traversed by colourless columns. *Leaf blade* with distinct, prominent adaxial ribs, or adaxially flat. *Midrib* very conspicuous (and abaxially prominent); having a conventional arc of bundles (with several small laterals each side of a usually very large median); with colourless mesophyll adaxially. Bulliforms present in discrete, regular adaxial groups to not in discrete, regular adaxial groups (the epidermis often largely, more or less irregularly bulliform between the costae); sometimes, irregularly in simple fans, or associated with colourless mesophyll cells to form deeply-penetrating fans; associating with colourless mesophyll cells to form arches over small vascular bundles (these especially spectacular in *E. androphila*). Many of the smallest vascular bundles unaccompanied by sclerenchyma, or all the vascular bundles accompanied by sclerenchyma. Combined sclerenchyma girders present (primaries); forming 'figures' (these usually slender, massive in *E. androphila*). Sclerenchyma all associated with vascular bundles.

Cytology. $2n = 20$.

Taxonomy. Panicoideae; Andropogonodae; Andropogoneae; Andropogoninae.
Ecology, geography, regional floristic distribution. 4 species. Tropical Africa. Mesophytic; shade species, or species of open habitats; glycophytic. Savanna woodland. Paleotropical and Neotropical. African. Sudano-Angolan, West African Rainforest, and Namib-Karoo. Central Brazilian. Sahelo-Sudanian, Somalo-Ethiopian, and South Tropical African.
Rusts and smuts. Smuts from Tilletiaceae and from Ustilaginaceae. Tilletiaceae — *Tilletia*. Ustilaginaceae — *Sphacelotheca*.
References, etc. Leaf anatomical: this project — most of the material seen poor.

Elymus L.

Anthosachne Steud., *Asperella* Humb., *Campeiostachys* Drob., *Clinelymus* (Griseb.) Nevski, *Goulardia* Husn., *Gymnostichum* Schreb., *Roegneria* C. Koch, *Semeiostachys* Drob., *Terellia* Lunell, *Zeia* Lunell

Excluding *Cockaynea* (*Elymus* sect. *Stenostachys*), *Elytrigia, Festucopsis, Hystrix, Lophopyrum, Leymus, Pascopyrum, Pseudoroegneria, Sitanion, Taeniatherum, Thinopyrum*

Habit, vegetative morphology. *Perennial*; *caespitose (or turf-forming)*. Culms 20–150(–200) cm high; *herbaceous*; unbranched above. Culm nodes glabrous. Culm internodes hollow. Young shoots intravaginal. *Leaves* usually not basally aggregated; auriculate (the basal leaves), or non-auriculate (the stem leaves, sometimes). Sheath margins joined, or free. Leaf blades linear; broad, or narrow; 1–18 mm wide; flat, or rolled (involute or convolute); without cross venation; persistent; rolled in bud. Ligule an unfringed membrane; truncate; 0.1–2 mm long. *Contra-ligule* absent.

Reproductive organization. Plants bisexual, with bisexual spikelets; with hermaphrodite florets. The spikelets all alike in sexuality. Plants inbreeding. Exposed-cleistogamous, or chasmogamous. Apomictic (in *E. scaber* ('*E. scabrus*') = '*Agropyron scabrum*'), or reproducing sexually.

Inflorescence. *Inflorescence a single spike, or a false spike, with spikelets on contracted axes (erect or drooping)*; espatheate; not comprising 'partial inflorescences' and foliar organs. Spikelet-bearing axes persistent (i.e. with *Hystrix* and *Sitanion* excluded). Spikelets solitary, or paired, or in triplets (rarely up to four); not secund; distichous, or not two-ranked (sometimes not in regular rows, e.g. *Clinelymus*); sessile to subsessile (the pedicels 0.3 — 1.5 mm long); usually imbricate.

Female-fertile spikelets. *Spikelets* 8–25(–35) mm long; *compressed laterally*; disarticulating above the glumes; not disarticulating between the florets, or disarticulating between the florets (usually, the joints well developed). Rachilla prolonged beyond the uppermost female-fertile floret; rachilla hairy (shortly pilose), or hairless (scabrous). The rachilla extension with incomplete florets. Hairy callus present (usually), or absent. **Callus** fairly *long*; blunt (obtuse-triangular).

Glumes two (well developed); relatively large; very unequal to more or less equal; shorter than the adjacent lemmas, or long relative to the adjacent lemmas; joined (basally), or free; when scoreable, lateral to the rachis; hairless; glabrous (or scabrid on the veins); without conspicuous tufts or rows of hairs; pointed (often), or not pointed (sometimes blunt); not subulate; awned (often, up to 8 mm long), or awnless; non-carinate (but the veins usually raised); similar (from lanceolate-ovate to narrow lanceolate). *Lower glume 3–7 nerved.* Upper glume 3–7 nerved. *Spikelets* with incomplete florets. The incomplete florets distal to the female-fertile florets. *The distal incomplete florets* merely underdeveloped.

Female-fertile florets *(2–)3–7(–9)*. *Lemmas* lanceolate-oblongate; less firm than the glumes to similar in texture to the glumes; not becoming indurated; entire; *pointed*; awnless, or mucronate, or awned. Awns when present, 1; median; apical; non-geniculate; straight, or recurving; much shorter than the body of the lemma to much longer than the body of the lemma (usually longer, up to 50 mm long, erect or more often divergent); entered by several veins. Lemmas hairy (shortly pilose), or hairless; when hairless glabrous, or scabrous; non-carinate; without a germination flap; 5 nerved; with the nerves confluent towards the tip. *Palea* present; relatively long; awnless, without apical setae; 2-nerved; 2-keeled. Palea keels usually scabrous, or hairy. *Lodicules* present; 2; free; membranous; usually ciliate; toothed, or not toothed (usually entire). *Stamens* 3. *Anthers 1–4 mm long (rarely to 5 mm, nearly always 'short')*; not penicillate. **Ovary** *hairy*. Styles free to their bases. Stigmas 2; white.

Fruit, embryo and seedling. *Fruit* usually adhering to lemma and/or palea; medium sized (5–9 mm long); longitudinally grooved; compressed dorsiventrally; with hairs confined to a terminal tuft. Hilum long-linear. Embryo small. Endosperm hard; without lipid; containing only simple starch grains. Embryo without an epiblast; without a scutellar tail; with a negligible mesocotyl internode; with one scutellum bundle. Embryonic leaf margins meeting.

Seedling with a long mesocotyl; with a loose coleoptile. First seedling leaf with a well-developed lamina. The lamina narrow; 3–5 veined.

Abaxial leaf blade epidermis. *Costal/intercostal zonation* conspicuous, or lacking. *Papillae* absent. Mid-intercostal long-cells having markedly sinuous walls, or having straight or only gently undulating walls (thin). *Microhairs* absent. *Stomata* absent or very rare (e.g. *E. caninus*), or common. Subsidiaries parallel-sided, or dome-shaped, or parallel-sided and dome-shaped. *Intercostal short-cells* common, or absent or very rare (e.g. *E. interruptus*); when present, in cork/silica-cell pairs, or not paired; silicified, or not silicified. Intercostal silica bodies when present, rounded (or oval), or crescentic, or tall-and-narrow. *Crown cells* present, or absent. *Costal short-cells* predominantly paired, or neither distinctly grouped into long rows nor predominantly paired (mostly solitary). Costal silica bodies horizontally-elongated crenate/sinuous, or horizontally-elongated smooth, or rounded, or tall-and-narrow (usually), or crescentic, or saddle shaped (a tendency, in *E. caninus*).

Transverse section of leaf blade, physiology. C_3; XyMS+. *Leaf blade* with distinct, prominent adaxial ribs (low, rounded), or adaxially flat; when present, with the ribs more or less constant in size. *Midrib* conspicuous, or not readily distinguishable; with one bundle only, or having a conventional arc of bundles (e.g. *E. interruptus*). Bulliforms present in discrete, regular adaxial groups; in simple fans. Many of the smallest vascular bundles unaccompanied by sclerenchyma (e.g. *E. caninus*), or all the vascular bundles accompanied by sclerenchyma. Combined sclerenchyma girders present. Sclerenchyma all associated with vascular bundles.

Phytochemistry. Tissues of the culm bases with little or no starch. Fructosans predominantly short-chain.

Cytology. Chromosome base number, $x = 7$. $2n = 28$ and 42, or 56 (rarely). 4, 6, and 8 ploid (self pollinated). Haplomic genome content H and S, or S and Y (*Roegneria*). Nucleoli disappearing before metaphase.

Taxonomy. Pooideae; Triticodae; Triticeae.

Ecology, geography, regional floristic distribution. About 150 species. Widespread extratropically in both hemispheres. Commonly adventive. Mesophytic, or xerophytic.

Holarctic, Paleotropical, Australian, and Antarctic. Boreal and Tethyan. Indomalesian. Euro-Siberian, Eastern Asian, Atlantic North American, and Rocky Mountains. Mediterranean and Irano-Turanian. Papuan. North and East Australian. New Zealand. European and Siberian. Canadian-Appalachian and Central Grasslands. Tropical North and East Australian and Temperate and South-Eastern Australian.

Hybrids. Intergeneric hybrids with *Sitanion*, *Triticum*.

Rusts and smuts. Rusts — *Puccinia*. Taxonomically wide-ranging species: *Puccinia striiformis*.

References, etc. Morphological/taxonomic: Löve 1984. Leaf anatomical: Metcalfe 1960; this project.

Elytrigia Desv.

Trichopyrum (Nevski) Löve
~ *Agropyron, Elymus*
Excluding *Lophopyrum, Pascopyrum, Pseudoroegneria, Thinopyrum*

Habit, vegetative morphology. Perennial; *rhizomatous (or densely turf-forming)*. **Culms** 20–150 cm high; *herbaceous*; unbranched above. Culm internodes solid, or hollow. *Leaves* not basally aggregated; auriculate, or non-auriculate. Sheath margins joined (often, on vegetative shoots), or free. Leaf blades linear; narrow; 1.2–10 mm wide; flat, or rolled (convolute); without cross venation; persistent; rolled in bud, or once-folded in bud. Ligule an unfringed membrane; truncate; 0.2–1 mm long (tough membranous). *Contra-ligule* absent.

Reproductive organization. Plants bisexual, with bisexual spikelets; with hermaphrodite florets. The spikelets of sexually distinct forms on the same plant, or all alike in sexuality; hermaphrodite, or hermaphrodite and sterile (sterile spikelets, when present, localised at the tip of the rachis). Plants outbreeding.

Inflorescence. *Inflorescence a single spike (erect or drooping, linear)*; espatheate; not comprising 'partial inflorescences' and foliar organs. *Spikelet-bearing axes persistent. Spikelets solitary*; not secund; distichous; *sessile to subsessile (the pedicels less than 0.3 mm long)*.

Female-fertile spikelets. Spikelets 7–23 mm long; compressed laterally to not noticeably compressed; disarticulating above the glumes, or falling with the glumes; not disarticulating between the florets, or disarticulating between the florets (the joints poorly developed). Rachilla prolonged beyond the uppermost female-fertile floret; rachilla hairy (rarely), or hairless (usually scabrous). The rachilla extension with incomplete florets. Hairy callus present, or absent. *Callus* short.

Glumes present; *two*; very unequal to more or less equal; *shorter than the adjacent lemmas*; free; lateral to the rachis; hairy (rarely), or hairless (glabrous); pointed, or not pointed; not subulate; awned, or awnless; non-carinate (or slightly so only towards the tip); *similar (ovate, oblongate or lanceolate, not awnlike)*. Lower glume 3–11 nerved. Upper glume 3–11 nerved. *Spikelets* with incomplete florets. The incomplete florets distal to the female-fertile florets. *The distal incomplete florets* merely underdeveloped.

Female-fertile florets *3–7(–10)*. Lemmas similar in texture to the glumes (leathery, lanceolate); entire, or incised; pointed, or blunt; awnless, or mucronate, or awned. Awns when present, 1; from a sinus, or apical; non-geniculate; much shorter than the body of the lemma to much longer than the body of the lemma (to 20 mm long); entered by several veins. Lemmas hairy (somewhat pilose), or hairless; when hairless glabrous, or scabrous; non-carinate; without a germination flap; 5 nerved; with the nerves confluent towards the tip. *Palea* present; relatively long; 2-nerved; 2-keeled (the keels scabrous or ciliate). *Lodicules* present; 2; free; membranous; ciliate; usually not toothed; not or scarcely vascularized. *Stamens* 3. Anthers 5–8 mm long (relatively long). *Ovary* hairy. Styles free to their bases. Stigmas 2; white.

Fruit, embryo and seedling. *Fruit* usually adhering to lemma and/or palea; medium sized (4–6 mm long); longitudinally grooved; compressed dorsiventrally; with hairs confined to a terminal tuft. *Hilum long-linear*. Embryo small. Endosperm hard; without lipid; containing only simple starch grains. Embryo with an epiblast (cf. Reeder's illustration of *E. repens*); without a scutellar tail; with a negligible mesocotyl internode. Embryonic leaf margins meeting.

Seedling with a tight coleoptile. First seedling leaf with a well-developed lamina. The lamina narrow; 2 veined, or 3–5 veined.

Abaxial leaf blade epidermis. *Costal/intercostal zonation* conspicuous. *Papillae* absent. *Long-cells* similar in shape costally and intercostally; of similar wall thickness costally and intercostally. Mid-intercostal long-cells rectangular and fusiform; having markedly sinuous walls and having straight or only gently undulating walls. *Microhairs* absent. *Stomata* common. Subsidiaries low dome-shaped, or parallel-sided (a few). Guard-cells overlapped by the interstomatals. *Intercostal short-cells* common; not paired (mostly solitary, a few pairs); silicified. Intercostal silica bodies tall-and-narrow, or saddle shaped, or cubical, or rounded. *Crown cells* present, or absent. *Costal short-cells* predominantly paired, or neither distinctly grouped into long rows nor predominantly paired (e.g. mainly solitary in *E. repens*). Costal silica bodies horizontally-elongated crenate/sinuous, or rounded (commonly), or saddle shaped (sometimes more or less cubical), or tall-and-narrow, or crescentic, or 'panicoid-type' (some in *E. repens*).

Transverse section of leaf blade, physiology. C_3; XyMS+. *Mesophyll* with non-radiate chlorenchyma; without adaxial palisade. *Leaf blade* with distinct, prominent adaxial ribs; with the ribs very irregular in sizes. *Midrib* not readily distinguishable; with one bundle only. Bulliforms present in discrete, regular adaxial groups (in the furrows); in simple fans. Many of the smallest vascular bundles unaccompanied by sclerenchyma (occasionally), or all the vascular bundles accompanied by sclerenchyma. Combined sclerenchyma girders present; forming 'figures'. Sclerenchyma all associated with vascular bundles.

Phytochemistry. Tissues of the culm bases with little or no starch. Fructosans predominantly short-chain.

Cytology. Chromosome base number, $x = 7$. $2n = 42$, or 56, or 84 (rarely). 6, 8, and 12 ploid. Haplomic genome content E, J, and S, or S and X, or E and S (*Trichopyrum*). Haploid nuclear DNA content 4.3–5.9 pg (2 species).

Taxonomy. Pooideae; Triticodae; Triticeae.

Ecology, geography, regional floristic distribution. 8 species. North and south temperate. Commonly adventive. Mesophytic, or xerophytic; halophytic, or glycophytic. Diverse habitats, including sand dunes.

Holarctic. Boreal and Tethyan. Euro-Siberian. Mediterranean and Irano-Turanian.

Hybrids. Intergeneric hybrids with *Agropyron* (×*Agrotrigia* Tsvelev), *Hordeum* (×*Elytrordeum* Hylander, ×*Elyhordeum* Zizan & Petrowa), *Aegilops*, *Leymus* (×*Leymotrigia* Tsvelev), *Lophopyrum*, *Secale*, *Triticum* (×*Trititrigia* Tsvelev), *Thinopyrum*.

Economic importance. Significant weed species: *E. repens* (Quick Grass, Scutch, Couch).

References, etc. Morphological/taxonomic: Löve 1984. Leaf anatomical: Metcalfe 1960; this project.

Elytrophorus P. Beauv.

Echinalysium Trin.

Habit, vegetative morphology. Annual; caespitose. *Culms* 10–50 cm high; herbaceous. Culm nodes glabrous. Culm internodes solid. *Leaves* not basally aggregated; non-auriculate. Leaf blades linear; narrow; flat; without abaxial multicellular glands; without cross venation; persistent; rolled in bud. Ligule an unfringed membrane to a fringed membrane.

Reproductive organization. Plants bisexual, with bisexual spikelets; with hermaphrodite florets. The spikelets of sexually distinct forms on the same plant; hermaphrodite, or hermaphrodite and sterile (reduced, sterile spikelets often present at the bases of the spikelet clusters).

Inflorescence. Inflorescence a false spike, with spikelets on contracted axes (the glomerules sometimes confluent to form a cylinder); *espatheate (but the glomerules and the clusters within them subtended by the enlarged, spreading glumes of the lower spikelets)*; not comprising 'partial inflorescences' and foliar organs. *Spikelet-bearing axes* very much reduced (to glomerules); persistent. **Spikelets** *associated with bractiform involucres (these constituted by the enlarged glumes of the lower spikelets)*; not secund.

Female-fertile spikelets. Spikelets strongly compressed laterally; disarticulating above the glumes; disarticulating between the florets; with conventional internode spacings. Rachilla prolonged beyond the uppermost female-fertile floret. The rachilla extension with incomplete florets. Hairy callus absent.

Glumes two; more or less equal; shorter than the spikelets, or about equalling the spikelets; long relative to the adjacent lemmas; hairy (sparsely hispid on the margins); pointed (acuminate); awned (shortly aristulate), or awnless (muticous); carinate; similar (narrowly lanceolate, persistent, membranous). Lower glume 1 nerved. Upper glume 1 nerved. *Spikelets* with incomplete florets. The incomplete florets distal to the female-fertile florets.

Female-fertile florets 2–6. Lemmas drawn out into the awn; similar in texture to the glumes (membranous, granular, ovate); not becoming indurated; entire; pointed; awned.

Awns 1; median; apical (lemma becoming setaceous at the summit); non-geniculate; much shorter than the body of the lemma; entered by one vein. *Lemmas* hairless (or scabrid ciliate on the keel and margins); *carinate (naviculate)*; 3 nerved. **Palea** present; *conspicuous but relatively short*; deeply bifid; awnless, without apical setae; not indurated (hyaline); 2-nerved (or more?); 2-keeled. Palea keels winged (the wings narrow or broad, dorsal). *Lodicules* present; 1, or 2; free; fleshy; glabrous; not toothed. *Stamens* 1–3. Anthers 0.5 mm long; not penicillate. *Ovary* glabrous. Styles fused, or free to their bases. Stigmas 2.

Fruit, embryo and seedling. *Fruit* free from both lemma and palea; small; not noticeably compressed. Hilum short. Pericarp free. Embryo large (half the fruit length); not waisted. Endosperm containing compound starch grains. Embryo without an epiblast; with a scutellar tail; with an elongated mesocotyl internode. Embryonic leaf margins meeting.

First seedling leaf with a well-developed lamina. The lamina narrow; erect.

Abaxial leaf blade epidermis. *Costal/intercostal zonation* conspicuous. *Papillae* absent. *Long-cells* markedly different in shape costally and intercostally (the costals long-rectangular, the intercostals much longer and fusiform); of similar wall thickness costally and intercostally. Intercostal zones with typical long-cells (though these very long). Mid-intercostal long-cells fusiform; having straight or only gently undulating walls. *Microhairs* present; elongated; clearly two-celled; panicoid-type. Microhair apical cell wall thinner than that of the basal cell and often collapsed. Microhair basal cells 12–15 microns long. *Stomata* common. Subsidiaries parallel-sided. Guard-cells overlapping to flush with the interstomatals. *Intercostal short-cells* absent or very rare. Intercostal silica bodies absent. *Costal short-cells* conspicuously in long rows. Costal silica bodies present in alternate cell files of the costal zones; 'panicoid-type'; nodular.

Transverse section of leaf blade, physiology. C_3; XyMS+. *Mesophyll* with radiate chlorenchyma; without fusoids (but with intercostal lacunae). *Leaf blade* with distinct, prominent adaxial ribs; with the ribs more or less constant in size. *Midrib* not readily distinguishable; with one bundle only. Bulliforms present in discrete, regular adaxial groups (in some species, between the vascular bundles), or not in discrete, regular adaxial groups (the groups inconpicuous, or restricted to midrib hinges); in simple fans. Many of the smallest vascular bundles unaccompanied by sclerenchyma. Combined sclerenchyma girders absent. Sclerenchyma all associated with vascular bundles, or not all bundle-associated. The 'extra' sclerenchyma when present, in abaxial groups; abaxial-hypodermal, the groups isolated (opposite the furrows). The lamina margins without fibres.

Culm anatomy. *Culm internode bundles* in one or two rings.

Cytology. Chromosome base number, $x = 12$, or 13 (?). $2n = 24$, or 26 (?). 2 ploid.

Taxonomy. Arundinoideae; Danthonieae (?).

Ecology, geography, regional floristic distribution. 2–4 species. Tropical Africa, tropical Asia, Australia. Commonly adventive. Helophytic.

Paleotropical and Australian. African and Indomalesian. Saharo-Sindian, Sudano-Angolan, West African Rainforest, and Namib-Karoo. Indian and Indo-Chinese. North and East Australian and Central Australian. Sahelo-Sudanian, Somalo-Ethiopian, and South Tropical African. Tropical North and East Australian and Temperate and South-Eastern Australian.

Rusts and smuts. Smuts from Tilletiaceae. Tilletiaceae — *Tilletia*.

Economic importance. Significant weed species: *E. articulatus* (in North America).

References, etc. Leaf anatomical: Metcalfe 1960; this project.

Elytrostachys McClure

Habit, vegetative morphology. Scrambling, sympodial perennial. The flowering culms leafy. Culms woody and persistent; scandent to not scandent ('culms self-supporting below, typically weak and pendulous or clambering above'); branched above. Primary branches/mid-culm node numerous, with one dominant. Culm internodes hollow. Rhizomes pachymorph. Plants unarmed. *Leaves* not basally aggregated; with auricular setae. Leaf blades broad; pseudopetiolate; cross veined, or without cross venation; disarticulating from the sheaths; rolled in bud.

Reproductive organization. Plants bisexual, with bisexual spikelets; with hermaphrodite florets. Viviparous.

Inflorescence. **Inflorescence** indeterminate; *with pseudospikelets*; of pseudospikelet clusters terminating leafy or leafless axes; spatheate; a complex of 'partial inflorescences' and intervening foliar organs. *Spikelet-bearing axes* paniculate to capitate; persistent. Spikelets not secund.

Female-fertile spikelets. *Spikelets* unconventional; compressed laterally (?); disarticulating above the glumes; with a distinctly elongated rachilla internode above the glumes. Rachilla prolonged beyond the uppermost female-fertile floret. The rachilla extension with incomplete florets. Hairy callus absent.

Glumes two (in accordance with the definition of glumes applied here); very unequal; shorter than the adjacent lemmas; awnless; similar. *Spikelets* with incomplete florets. The incomplete florets distal to the female-fertile florets. *The distal incomplete florets* merely underdeveloped.

Female-fertile florets 1, or 2. Lemmas not becoming indurated; entire; pointed, or blunt; awnless, or mucronate (?). *Palea* present; relatively long; awnless, without apical setae; not indurated; 2-keeled (sulcate). *Lodicules* present; 3; free; membranous; ciliate; not toothed; heavily vascularized. *Stamens* 6. **Ovary** *glabrous*; with a conspicuous apical appendage. *The appendage broadly conical, fleshy*. Styles fused (the ovary attenuate to the stigmas). *Stigmas 2*.

Fruit, embryo and seedling. Fruit longitudinally grooved. Hilum long-linear. *Pericarp thick and hard (leathery, thickened at the apex)*.

Abaxial leaf blade epidermis. *Costal/intercostal zonation* conspicuous (and the intercostal zones with broad stomatal and astomatal bands). *Papillae* present; costal and intercostal. Intercostal papillae over-arching the stomata (almost covering them); several per cell (mostly with a single median row of 6–10 circular or somewhat bifurcated papillae per long-cell). *Long-cells* markedly different in shape costally and intercostally (the costals narrower); of similar wall thickness costally and intercostally. Mid-intercostal long-cells rectangular; having markedly sinuous walls. *Microhairs* present; elongated; clearly two-celled; panicoid-type; 57–60 microns long; 6–6.3 microns wide at the septum. Microhair total length/width at septum 9.5–10. Microhair apical cells 36–40.5 microns long. Microhair apical cell/total length ratio 0.6–0.7. *Stomata* common; 21–24 microns long. Subsidiaries triangular. *Intercostal short-cells* common (in the astomatal regions); not paired (solitary); not silicified. Macrohairs present costally, a few prickles present costally and intercostally. *Crown cells* absent. *Costal short-cells* neither distinctly grouped into long rows nor predominantly paired. Costal silica bodies scarce, tall-and-narrow to crescentic.

Transverse section of leaf blade, physiology. C_3; XyMS+. Mesophyll with arm cells; with fusoids. The fusoids external to the PBS. *Leaf blade* adaxially flat. *Midrib* conspicuous; having complex vascularization (one large bundle, and assorted small ones). *The lamina* symmetrical on either side of the midrib. Bulliforms present in discrete, regular adaxial groups (with a large group in the middle of each intercostal zone); in simple fans. All the vascular bundles accompanied by sclerenchyma. Combined sclerenchyma girders present; forming 'figures' (all the bundles with a conspicuous 'anchor'). Sclerenchyma not all bundle-associated (there being adaxial hypodermal fibres lining the sides of the bulliforms, but no abaxial groups opposite them). The 'extra' sclerenchyma in adaxial groups; adaxial-hypodermal, contiguous with the bulliforms.

Taxonomy. Bambusoideae; Bambusodae; Bambuseae.

Ecology, geography, regional floristic distribution. 2 species. South America: Honduras to Venezuela. Mesophytic; glycophytic. Rain forest.

Neotropical. Caribbean, Venezuela and Surinam, and Andean.

References, etc. Leaf anatomical: this project.

Enneapogon Desv. ex P. Beauv.

Calotheria Wight and Arn.

Habit, vegetative morphology. Annual (rarely), or perennial; caespitose. **Culms** (3–)5–100(–110) cm high; *herbaceous*. Culm nodes hairy. Culm internodes hollow. *Leaves* not basally aggregated; auriculate, or non-auriculate. Leaf blades linear; narrow; setaceous, or not setaceous; flat, or rolled (often convolute); without abaxial multicellular glands; without cross venation; disarticulating from the sheaths, or persistent; rolled in bud. *Ligule a fringe of hairs*.

Reproductive organization. *Plants bisexual, with bisexual spikelets*; with hermaphrodite florets. Exposed-cleistogamous, or chasmogamous. Plants with hidden cleistogenes, or without hidden cleistogenes. The hidden cleistogenes (when present) in the leaf sheaths (often basal and very modified).

Inflorescence. Inflorescence paniculate; feathery, contracted; capitate, or more or less ovoid, or spicate, or more or less irregular; espatheate; not comprising 'partial inflorescences' and foliar organs. Spikelet-bearing axes persistent. Spikelets not secund; shortly pedicellate.

Female-fertile spikelets. Spikelets 3.5–11 mm long; compressed laterally, or not noticeably compressed, or compressed dorsiventrally; disarticulating above the glumes; not disarticulating between the florets. Rachilla prolonged beyond the uppermost female-fertile floret; rachilla hairy. The rachilla extension with incomplete florets. *Callus* short.

Glumes two; very unequal to more or less equal; about equalling the spikelets; long relative to the adjacent lemmas; hairy, or hairless; awnless; similar (membranous). Lower glume 5–21 nerved. Upper glume 5–21 nerved. *Spikelets* with incomplete florets. The incomplete florets distal to the female-fertile florets. *The distal incomplete florets* merely underdeveloped (sterile or rudimentary), or clearly specialised and modified in form (sometimes reduced to awns and forming a brushlike appendage).

Female-fertile florets 1–3. *Lemmas* decidedly firmer than the glumes (firm or leathery); not becoming indurated; incised; 9 lobed; *deeply cleft (into awns)*; *awned*. *Awns 9*; median and lateral; the median similar in form to the laterals; apical; non-geniculate; hairless, or hairy, or long-plumose; about as long as the body of the lemma to much longer than the body of the lemma. The lateral awns about equalling the median. Lemmas villous hairy; non-carinate; 9 nerved (smooth or ribbed). *Palea* present; relatively long (longer than the body of the lemma); entire (pointed); not indurated (thinly membranous); 2-nerved; 2-keeled (the keels ciliate). Palea keels hairy. *Lodicules* present; 2; free; fleshy, or membranous; glabrous; not toothed. *Stamens* 3. Anthers 0.2–1.5 mm long; not penicillate. *Ovary* glabrous. Styles free to their bases. Stigmas 2; white.

Fruit, embryo and seedling. *Fruit* free from both lemma and palea; small; compressed dorsiventrally. Hilum short. *Pericarp thick and hard*; fused. Embryo large; waisted. Endosperm hard; without lipid; containing compound starch grains. Embryo with an epiblast; with a scutellar tail; with an elongated mesocotyl internode. Embryonic leaf margins overlapping.

Seedling with a short mesocotyl. First seedling leaf with a well-developed lamina. The lamina narrow.

Abaxial leaf blade epidermis. *Costal/intercostal zonation* conspicuous. *Papillae* absent. *Long-cells* similar in shape costally and intercostally; of similar wall thickness costally and intercostally (the costals only slightly thicker walled). Mid-intercostal long-cells rectangular; having markedly sinuous walls. *Microhairs* present (but sometimes rare — e.g., in *E. clelandii*, where they are, however, abundant on the glumes and lemmas); elongated; clearly two-celled, or clearly two-celled to uniseriate; *Enneapogon*-type (sometimes up to 8 times the length of an average panicoid-type microhair, and occasionally exhibiting a thin transverse septum in the elongated basal 'cell'). Microhair apical cell wall of similar thickness/rigidity to that of the basal cell. Microhairs with 'partitioning membranes' (in *E. nigricans*). The 'partitioning membranes' in the apical cell. Microhair basal cells 60 microns long. Microhair total length/width at septum 8. Microhair apical cell/total length ratio 0.6. *Stomata* common. Subsidiaries low dome-shaped, or triangular. Guard-cells overlapping to flush with the interstomatals. *Intercostal short-cells* common, or absent or very rare; not paired (when present); not silicified. Intercostal silica bodies absent. *Costal short-cells* conspicuously in long rows. Costal silica bodies present in alternate cell files of the costal zones; 'panicoid-type'; cross shaped to dumbbell shaped, or nodular (occasionally).

Transverse section of leaf blade, physiology. C_4; biochemical type NAD–ME (7 species); XyMS+. PCR sheath outlines uneven. PCR sheaths of the primary vascular bundles interrupted; interrupted abaxially only. PCR sheath extensions present. Maximum number of extension cells 1 (usually). PCR cells with a suberised lamella. *PCR cell chloroplasts* elongated; with well developed grana; centrifugal/peripheral to centripetal. *Mesophyll* with radiate chlorenchyma; traversed by columns of colourless mesophyll cells. *Leaf blade* 'nodular' in section; with the ribs more or less constant in size. *Midrib* not readily distinguishable; with one bundle only. Bulliforms present in discrete, regular adaxial groups; associated with colourless mesophyll cells to form deeply-penetrating fans, or in simple fans and associated with colourless mesophyll cells to form deeply-penetrating fans (often linked with traversing colourless columns). All the vascular bundles accompanied by sclerenchyma. Combined sclerenchyma girders present; forming 'figures'. Sclerenchyma all associated with vascular bundles. The lamina margins with fibres.

Culm anatomy. *Culm internode bundles* in one or two rings.

Phytochemistry. Leaf blade chlorophyll *a*:*b* ratio 3.66–4.44.

Special diagnostic feature. Female-fertile lemmas 9-lobed, each lobe terminating in an awn.

Cytology. Chromosome base number, $x = 9$ and 10. $2n = 18$, 36, and 40. 2 and 4 ploid.

Taxonomy. Chloridoideae; Pappophoreae.

Ecology, geography, regional floristic distribution. 30 species. In warm regions. Xerophytic; species of open habitats; glycophytic. Bushland and semidesert.

Holarctic, Paleotropical, Neotropical, and Australian. Boreal, Tethyan, and Madrean. African, Madagascan, and Indomalesian. Atlantic North American. Macaronesian, Mediterranean, and Irano-Turanian. Saharo-Sindian, Sudano-Angolan, and Namib-Karoo. Indian, Indo-Chinese, and Papuan. Caribbean. North and East Australian and Central Australian. Central Grasslands. Sahelo-Sudanian, Somalo-Ethiopian, South Tropical African, and Kalaharian. Tropical North and East Australian and Temperate and South-Eastern Australian.

Rusts and smuts. Smuts from Ustilaginaceae. Ustilaginaceae — *Sphacelotheca* and *Ustilago*.

Economic importance. Significant weed species: *E. cenchroides*, *E. scoparius*. Important native pasture species: several important in dry regions, e.g. *E. gracilis*, *E. nigricans*, *E. desvauxii*.

References, etc. Morphological/taxonomic: Burbidge 1941. Leaf anatomical: Metcalfe 1960; this project.

Enteropogon Nees

Macrostachya A. Rich.
Excluding *Saugetia*

Habit, vegetative morphology. *Perennial*; caespitose. **Culms** 20–120 cm high; *herbaceous*. Culm nodes glabrous. Culm internodes solid, or hollow. *Leaves* mostly basal, or not basally aggregated; non-auriculate. Leaf blades linear; narrow; setaceous, or not setaceous; flat, or rolled (then involute-filiform); without abaxial multicellular glands; not pseudopetiolate; without cross venation; persistent; rolled in bud. Ligule a fringed membrane (short).

Reproductive organization. *Plants bisexual, with bisexual spikelets*; with hermaphrodite florets; with hidden cleistogenes (in *E. chlorideus*), or without hidden cleistogenes. The hidden cleistogenes of *E. chlorideus* subterranean.

Inflorescence. *Inflorescence a single spike, or of spicate main branches, or a single raceme (with short pedicels)*; digitate, or subdigitate, or non-digitate (when consisting of a single raceme). Primary inflorescence branches 1–15. Inflorescence espatheate; not comprising 'partial inflorescences' and foliar organs. Spikelet-bearing axes persistent. Spikelets solitary; secund; biseriate; sessile to subsessile.

Female-fertile spikelets. *Spikelets* 5–8 mm long; adaxial; *compressed dorsiventrally*; disarticulating above the glumes; disarticulating between the florets. *Rachilla prolonged beyond the uppermost female-fertile floret.* The rachilla extension with incomplete florets. Hairy callus present. *Callus* short; blunt to pointed.

Glumes two (hyaline); very unequal; (the longer) long relative to the adjacent lemmas; dorsiventral to the rachis; hairless; glabrous; awned (G_2, often), or awnless; carinate; subulate to lanceolate, minutely bidentate, membranous, divergent. Lower glume 1 nerved. *Upper glume 1 nerved. Spikelets* with incomplete florets. The incomplete florets distal to the female-fertile florets. *The distal incomplete florets* 1, or 2–4; **incomplete florets** merely underdeveloped, or clearly specialised and modified in form (sometimes represented by a cluster of awned rudiments); *incomplete florets awned.*

Female-fertile florets 1, or 2 (the L_2 male or with a hermaphrodite floret). Lemmas decidedly firmer than the glumes (leathery, rigid); entire, or incised; when incised, 2 lobed; not deeply cleft (bidentate); awned. Awns 1; median; from a sinus, or apical; nongeniculate; about as long as the body of the lemma to much longer than the body of the lemma; entered by one vein, or entered by several veins (rarely, a pair of spikes). Lemmas hairless (glabrous or scaberulous); glabrous to scabrous; non-carinate (rounded to flat on the back); having the margins tucked in onto the palea; 3 nerved (the median raised). *Palea* present; relatively long; entire to apically notched; awnless, without apical setae; not indurated (membranous or scarious); 2-nerved; 2-keeled. *Lodicules* present; 2; free; fleshy; glabrous. *Stamens* 3. Anthers not penicillate. *Ovary* glabrous. Styles free to their bases. Stigmas 2.

Fruit, embryo and seedling. Fruit small; ellipsoid; shallowly longitudinally grooved; compressed dorsiventrally (concavo-convex). Hilum short. Pericarp free (according to Clayton and Renvoize 1986), or fused. Embryo small to large (up to 1/3 of the grain length). Endosperm containing only simple starch grains.

Abaxial leaf blade epidermis. *Costal/intercostal zonation* conspicuous. *Papillae* present; intercostal. Intercostal papillae over-arching the stomata; consisting of one symmetrical projection per cell, or consisting of one oblique swelling per cell. *Long-cells* similar in shape costally and intercostally; of similar wall thickness costally and intercostally. Mid-intercostal long-cells rectangular; having markedly sinuous walls. *Microhairs* present; elongated; clearly two-celled; chloridoid-type. Microhair apical cell wall of similar thickness/rigidity to that of the basal cell. Microhairs 21.6–24 microns long. Microhair basal cells 12–15 microns long. Microhairs (6–)9–11.4(–12) microns wide at the septum. Microhair total length/width at septum 1.8–4. Microhair apical cells 9.6–15 microns long. Microhair apical cell/total length ratio 0.43–0.63. *Stomata* common; about 15 microns long. Subsidiaries triangular. Guard-cells overlapping to flush with the interstomatals. *Intercostal short-cells* common, or absent or very rare; not paired (mainly solitary); not silicified. Intercostal silica bodies absent. *Costal short-cells* conspicuously in long rows, or neither distinctly grouped into long rows nor predominantly paired. Costal silica bodies present in alternate cell files of the costal zones; rounded (some), or saddle shaped (mainly), or crescentic.

Transverse section of leaf blade, physiology. C_4; XyMS+. PCR sheath outlines even. PCR sheaths of the primary vascular bundles interrupted; interrupted both abaxially and adaxially. PCR sheath extensions present, or absent. Maximum number of extension cells when present, 1–2. PCR cell chloroplasts centripetal. *Mesophyll* with radiate chlorenchyma; traversed by columns of colourless mesophyll cells, or not traversed by colourless columns. *Leaf blade* 'nodular' in section, or adaxially flat; with the ribs more or less constant in size. *Midrib* conspicuous (rarely), or not readily distinguishable; with one bundle only, or having a conventional arc of bundles; with colourless mesophyll adaxially, or without colourless mesophyll adaxially. Bulliforms present in discrete, regular adaxial groups; in simple fans, or associated with colourless mesophyll cells to form deeply-penetrating fans (linked or not with traversing colourless columns). Many of the smallest vascular bundles unaccompanied by sclerenchyma, or all the vascular bundles accompanied by sclerenchyma. Combined sclerenchyma girders present; forming 'figures', or nowhere forming 'figures'. Sclerenchyma all associated with vascular bundles. The lamina margins with fibres.

Phytochemistry. Leaves without flavonoid sulphates (2 species).

Cytology. Chromosome base number, $x = 10$. $2n = 20$. 2 ploid.

Taxonomy. Chloridoideae; main chloridoid assemblage.

Ecology, geography, regional floristic distribution. 6 species. Africa, Seychelles, India, Formosa, Australia, Pacific. Mesophytic to xerophytic; shade species, or species of open habitats; glycophytic. Savanna on sand or clay.

Holarctic, Paleotropical, and Australian. Tethyan. African, Madagascan, and Indomalesian. Macaronesian. Saharo-Sindian, Sudano-Angolan, West African Rainforest, and Namib-Karoo. Indian, Indo-Chinese, and Malesian. North and East Australian and Central Australian. Sahelo-Sudanian, Somalo-Ethiopian, South Tropical African, and Kalaharian. Tropical North and East Australian and Temperate and South-Eastern Australian.

Rusts and smuts. Rusts — *Puccinia*. Taxonomically wide-ranging species: *Puccinia eritraeensis*.

Economic importance. Important native pasture species: *E. acicularis*, in arid/semiarid places; *E. macrostachyus*.

References, etc. Morphological/taxonomic: Jacobs and Highet 1988. Leaf anatomical: Metcalfe 1960; this project.

Entolasia Stapf

Habit, vegetative morphology. *Perennial*; rhizomatous, or caespitose. *Culms* 20–120 cm high; woody and persistent (wiry, bushy), or herbaceous; scandent (sometimes straggling/climbing), or not scandent; branched above, or unbranched above; tuberous, or not tuberous. Culm nodes hairy, or glabrous. Culm internodes hollow. *Leaves* not basally aggregated; non-auriculate. Leaf blades broad, or narrow; cordate, or not cordate, not sagittate; pseudopetiolate, or not pseudopetiolate; without cross venation; disarticulating from the sheaths (in Australia), or persistent (in southern Africa — evidence of shaky generic circumscription); rolled in bud. *Ligule a fringe of hairs*.

Reproductive organization. Plants bisexual, with bisexual spikelets; with hermaphrodite florets.

Inflorescence. *Inflorescence of spicate main branches, or paniculate (but usually with sessile spiciform racemes, appressed to the common axis)*; open, or contracted. Rachides hollowed to flattened. Inflorescence espatheate; not comprising 'partial inflorescences' and foliar organs. Spikelet-bearing axes persistent. Spikelets solitary, or paired; second. Pedicel apices discoid. Spikelets consistently in 'long-and-short' combinations (rarely), or not in distinct 'long-and-short' combinations.

Female-fertile spikelets. *Spikelets* 2.5–6 mm long; elliptic, or lanceolate, or oblanceolate; adaxial; usually *compressed dorsiventrally*; falling with the glumes. Rachilla terminated by a female-fertile floret. Hairy callus absent.

Glumes present; two; very unequal; (the longer) *long relative to the adjacent lemmas*; dorsiventral to the rachis; awnless; non-carinate; very dissimilar (the lower tiny, hyaline, the upper membranous and equalling the spikelet). Lower glume 0–3 nerved. Upper glume 3–7 nerved. *Spikelets* with incomplete florets. The incomplete florets proximal to the female-fertile florets. *Spikelets with proximal incomplete florets*. **The proximal incomplete florets** 1; *epaleate*; sterile. The proximal lemmas awnless; 5 nerved; more or less equalling the female-fertile lemmas to decidedly exceeding the female-fertile lemmas; less firm than the female-fertile lemmas to similar in texture to the female-fertile lemmas (but shorter than the G_2, which it resembles much more than does the L_2); not becoming indurated.

Female-fertile florets 1. *Lemmas* similar in texture to the glumes to decidedly firmer than the glumes (membranous to leathery); smooth to striate; not becoming indurated; white in fruit, or yellow in fruit; entire; blunt; awnless; densely silky *hairy*; non-carinate; having the margins tucked in onto the palea; with a clear germination flap; 3–5 nerved. *Palea* present; relatively long; entire; awnless, without apical setae; 2-nerved; 2-keeled (the margins inrolled). Palea back hairy (between the keels). *Lodicules* present; 2; fleshy;

glabrous. *Stamens* 3. Anthers not penicillate. **Ovary** glabrous; *without a conspicuous apical appendage*. Styles free to their bases. Stigmas 2; red pigmented.

Fruit, embryo and seedling. Fruit small; compressed dorsiventrally. Hilum short. Embryo large. Endosperm containing only simple starch grains. Embryo without an epiblast; with a scutellar tail; with an elongated mesocotyl internode. Embryonic leaf margins meeting.

First seedling leaf with a well-developed lamina. The lamina broad; curved; 6–12 veined.

Abaxial leaf blade epidermis. *Costal/intercostal zonation* conspicuous. *Papillae* absent. *Long-cells* markedly different in shape costally and intercostally; of similar wall thickness costally and intercostally. Intercostal zones sometimes exhibiting many atypical long-cells. Mid-intercostal long-cells rectangular to fusiform (and sometimes very short); having markedly sinuous walls. *Microhairs* present; panicoid-type; 45–75(–78) microns long; 6–9 microns wide at the septum. Microhair total length/width at septum 5.4–10. Microhair apical cells (19.5–)21–30(–33) microns long. Microhair apical cell/total length ratio 0.33–0.65. *Stomata* common; 33–51 microns long. Subsidiaries triangular. Guardcells overlapping to flush with the interstomatals. *Intercostal short-cells* absent or very rare; not paired (usually solitary); not silicified. *Costal short-cells* conspicuously in long rows. Costal silica bodies 'panicoid-type'; dumb-bell shaped.

Transverse section of leaf blade, physiology. C_3; XyMS+. *Mesophyll* with radiate chlorenchyma; without adaxial palisade; *Isachne*-type. *Leaf blade* adaxially flat. *Midrib* not readily distinguishable; with one bundle only. Bulliforms present in discrete, regular adaxial groups; in simple fans. All the vascular bundles accompanied by sclerenchyma. Combined sclerenchyma girders present. Sclerenchyma all associated with vascular bundles.

Cytology. Chromosome base number, $x = 9$. $2n = 18$. 2 ploid.

Taxonomy. Panicoideae; Panicodae; Paniceae.

Ecology, geography, regional floristic distribution. 5 species. Tropical Africa, eastern Australia. Helophytic, mesophytic, and xerophytic; shade species and species of open habitats; glycophytic. Marshy places, damp grassland and dry forest.

Paleotropical and Australian. African, Indomalesian, and Neocaledonian. Sudano-Angolan, West African Rainforest, and Namib-Karoo. Papuan. North and East Australian. Sahelo-Sudanian, Somalo-Ethiopian, South Tropical African, and Kalaharian. Tropical North and East Australian and Temperate and South-Eastern Australian.

Rusts and smuts. Rusts — *Puccinia*. Taxonomically wide-ranging species: *Puccinia levis*.

References, etc. Leaf anatomical: this project.

Entoplocamia Stapf

Habit, vegetative morphology. Robust *annual*. Culms (20–)40–110 cm high; herbaceous; to to 0.3 cm in diameter; unbranched above. Culm nodes glabrous. Culm internodes solid. Plants unarmed. Young shoots intravaginal. *Leaves* mostly basal; non-auriculate. Sheath margins free. Leaf blades linear-lanceolate; narrow; 2–10 mm wide (to about 30 cm long); flat, or rolled; without abaxial multicellular glands; without cross venation; persistent. Ligule a fringe of hairs; about 1 mm long.

Reproductive organization. Plants bisexual, with bisexual spikelets; with hermaphrodite florets.

Inflorescence. *Inflorescence a false spike, with spikelets on contracted axes, or a single spike, or a single raceme (the spikelets solitary or in clusters or short secondary spikes, on the rachis of a simple or compound spike)*; espatheate; not comprising 'partial inflorescences' and foliar organs. Spikelet-bearing axes persistent. Spikelets secund.

Female-fertile spikelets. Spikelets 9–20 mm long; compressed laterally (becoming twisted when mature); falling with the glumes; not disarticulating between the florets (falling whole); with conventional internode spacings. Rachilla prolonged beyond the uppermost female-fertile floret; rachilla hairless. The rachilla extension with incomplete florets. *Callus* absent.

Glumes two; very unequal to more or less equal; shorter than the spikelets; shorter than the adjacent lemmas; hairy (at the margins), or hairless; pointed (acute); awnless; carinate (the keel scabrid); similar (thin, membranous, ovate). Lower glume 3 nerved. Upper glume 5 nerved. *Spikelets* with incomplete florets. The incomplete florets both distal and proximal to the female-fertile florets. *The distal incomplete florets* merely underdeveloped. **Spikelets with proximal incomplete florets.** The proximal incomplete florets 2; epaleate; sterile. The proximal lemmas mucronate to awned, or awnless; 6–8 nerved; less firm than the female-fertile lemmas to similar in texture to the female-fertile lemmas (intermediate in texture between glumes and fertile lemmas).

Female-fertile florets 4–20. Lemmas decidedly firmer than the glumes (cartilaginous at the base, chartaceous above, hyaline at the margins); entire; pointed; mucronate to awned (via the excurrent mid-nerve). Awns 1; median; stout, apical; hairless; much shorter than the body of the lemma. Lemmas hairy (villous along the lower margins); carinate; without a germination flap; 9–11 nerved; with the nerves non-confluent. *Palea* present; relatively long; apically notched; thinner than the lemma; not indurated; 2-nerved; 2-keeled. Palea keels winged (the margins and wings villous). **Lodicules absent.** *Stamens* 3. Anthers 4–5 mm long; not penicillate; without an apically prolonged connective. *Ovary* glabrous. Styles fused. Stigmas 2; white.

Fruit, embryo and seedling. *Fruit* free from both lemma and palea (loosely enclosed); small (2 mm long); ellipsoid; compressed laterally. Hilum short. Pericarp free. Embryo large; with an epiblast; with a scutellar tail; with an elongated mesocotyl internode. Embryonic leaf margins overlapping.

Abaxial leaf blade epidermis. *Costal/intercostal zonation* conspicuous. *Papillae* absent. *Long-cells* markedly different in shape costally and intercostally (intercostals much broader); of similar wall thickness costally and intercostally (fairly thick walled). Mid-intercostal long-cells rectangular; having markedly sinuous walls (and pitted). *Microhairs* present; elongated; clearly two-celled; chloridoid-type. Microhair apical cell wall thinner than that of the basal cell and often collapsed to thinner than that of the basal cell but not tending to collapse. Microhairs (34.5–)36–39 microns long. Microhair basal cells 18–21 microns long. Microhairs 9.6–12 microns wide at the septum. Microhair total length/width at septum 3–4.1. Microhair apical cells 16.5–21 microns long. Microhair apical cell/total length ratio 0.42–0.56. *Stomata* common; 22.5–27 microns long. Subsidiaries dome-shaped to triangular. Guard-cells overlapping to flush with the interstomatals. *Intercostal short-cells* common; in cork/silica-cell pairs; silicified. Intercostal silica bodies present and perfectly developed; tall-and-narrow, or saddle shaped. *Costal short-cells* conspicuously in long rows, or predominantly paired. Costal silica bodies present in alternate cell files of the costal zones; predominantly large, broad saddle shaped.

Transverse section of leaf blade, physiology. C_4; XyMS+. PCR sheath outlines even. PCR sheaths of the primary vascular bundles interrupted; interrupted abaxially only to interrupted both abaxially and adaxially. PCR sheath extensions absent. PCR cell chloroplasts centripetal. *Mesophyll* with radiate chlorenchyma; traversed by columns of colourless mesophyll cells (in places). *Leaf blade* adaxially flat (with no more than slight abaxial ribs over primary bundles). *Midrib* rather conspicuous (with a slight, rounded, abaxial keel); having a conventional arc of bundles (the median with a smaller lateral on either side); with colourless mesophyll adaxially, or without colourless mesophyll adaxially. Bulliforms present in discrete, regular adaxial groups; associated with colourless mesophyll cells to form deeply-penetrating fans (these sometimes linked with traversing colourless columns). All the vascular bundles accompanied by sclerenchyma. Combined sclerenchyma girders present (with the primaries only); forming 'figures' (in the large bundles). Sclerenchyma all associated with vascular bundles. The lamina margins with fibres.

Cytology. Chromosome base number, $x = 10$.

Taxonomy. Chloridoideae; main chloridoid assemblage.

Ecology, geography, regional floristic distribution. 1 species. Angola, southwest and southern Africa. Xerophytic (but often grows in depressions where moisture collects); species of open habitats; halophytic (sometimes), or glycophytic (usually).

Paleotropical and Cape. African. Namib-Karoo.

References, etc. Leaf anatomical: this project; photos of *E. africana* provided by R.P. Ellis.

Eragrostiella Bor

Habit, vegetative morphology. Perennial; compactly caespitose. *Culms* 20–50 cm high; herbaceous. Culm internodes hollow. **Leaves** usually *mostly basal*; non-auriculate. Leaf blades narrow; setaceous, or not setaceous (often filiform, rarely flat); without abaxial multicellular glands; without cross venation; persistent. *Ligule a fringed membrane*; short.

Reproductive organization. Plants bisexual, with bisexual spikelets; with hermaphrodite florets. The spikelets all alike in sexuality.

Inflorescence. Inflorescence a single spike, or a single raceme (long-pedunculate); espatheate; not comprising 'partial inflorescences' and foliar organs. Spikelet-bearing axes persistent. Spikelets secund; biseriate; subsessile, or pedicellate.

Female-fertile spikelets. *Spikelets* 6–20 mm long; compressed laterally; disarticulating above the glumes; tardily *disarticulating between the florets, or not disarticulating between the florets (the paleas persistent)*. Rachilla prolonged beyond the uppermost female-fertile floret; rachilla hairy. The rachilla extension with incomplete florets. Hairy callus absent.

Glumes present; two; more or less equal; shorter than the adjacent lemmas, or long relative to the adjacent lemmas; awnless; carinate (the lower), or non-carinate (the upper being keeled only at the tip); *very dissimilar*. Lower glume 1 nerved. Upper glume 3 nerved. *Spikelets* with incomplete florets. The incomplete florets distal to the female-fertile florets. *The distal incomplete florets* merely underdeveloped.

Female-fertile florets 6–50. Lemmas less firm than the glumes to similar in texture to the glumes (membranous to cartilaginous); not becoming indurated; entire to incised (acute to emarginate); not deeply cleft; awnless; hairless; glabrous; carinate. The keel wingless. Lemmas without a germination flap; 3 nerved. *Palea* present (persistent); relatively long; entire; awnless, without apical setae; 2-nerved; 2-keeled. Palea keels narrowly to broadly winged. *Lodicules* present; 2; fleshy. Anthers not penicillate. *Ovary* glabrous. Stigmas 2.

Fruit, embryo and seedling. *Fruit* ellipsoid; not noticeably compressed (terete), or trigonous; *sculptured (reticulate-striate)*. Pericarp fused (?). Embryo large; with an epiblast; with a scutellar tail; with an elongated mesocotyl internode. Embryonic leaf margins meeting.

Abaxial leaf blade epidermis. *Costal/intercostal zonation* conspicuous. *Papillae* absent. *Long-cells* similar in shape costally and intercostally; of similar wall thickness costally and intercostally. Mid-intercostal long-cells rectangular; having markedly sinuous walls. *Microhairs* present; more or less spherical; clearly two-celled; chloridoid-type. Microhair apical cell wall of similar thickness/rigidity to that of the basal cell. Microhairs 14–18 microns long. Microhair basal cells 9 microns long. Microhairs 12(–13.5) microns wide at the septum. Microhair total length/width at septum 1.4–1.8. Microhair apical cells 7–12 microns long. Microhair apical cell/total length ratio 0.4–0.73. *Stomata* common; 22.5–25.5 microns long. Subsidiaries low dome-shaped, or triangular. Guard-cells overlapping to flush with the interstomatals. *Intercostal short-cells* common; in cork/silica-cell pairs; silicified. Intercostal silica bodies imperfectly developed; saddle shaped (small), or tall-and-narrow. *Costal short-cells* conspicuously in long rows, or neither distinctly grouped into long rows nor predominantly paired. Costal silica bodies present in alternate cell files of the costal zones; saddle shaped.

Transverse section of leaf blade, physiology. C_4; XyMS+. PCR sheath outlines even. PCR sheaths of the primary vascular bundles interrupted; interrupted abaxially only. PCR sheath extensions present, or absent. Maximum number of extension cells 1–2. *Mesophyll* with radiate chlorenchyma; traversed by columns of colourless mesophyll cells. *Leaf blade* with distinct, prominent adaxial ribs, or adaxially flat; with the ribs more or less constant in size. *Midrib* not readily distinguishable; with one bundle only. Bulliforms present in discrete, regular adaxial groups; associated with colourless mesophyll cells to form deeply-penetrating fans (these linked with traversing columns of colourless cells). All the vascular bundles accompanied by sclerenchyma. Combined sclerenchyma girders present; forming 'figures'. Sclerenchyma all associated with vascular bundles. The lamina margins with fibres.

Taxonomy. Chloridoideae; main chloridoid assemblage.

Ecology, geography, regional floristic distribution. 5 species. India, Burma, Ceylon, Australia. Xerophytic; species of open habitats. Dry grassland and bush, on thin soils.

Paleotropical and Australian. African and Indomalesian. Saharo-Sindian and Sudano-Angolan. Indian and Indo-Chinese. North and East Australian. Somalo-Ethiopian. Tropical North and East Australian.

Rusts and smuts. Smuts from Tilletiaceae. Tilletiaceae — *Tilletia*.

References, etc. Leaf anatomical: Metcalfe 1960; this project.

Eragrostis N. M. Wolf

Boriskerella Terekhov, *Erochloe* Raf., *Erosion* Lunell, *Exagrostis* Steud., *Macroblepharus* Philippi, *Psilantha* (K. Koch) Tzvelev, *Roshevitzia* Tsvelev, *Triphlebia* Stapf, *Vilfagrostis* Doell

Excluding *Acamptoclados*, *Cladoraphis*, *Diandrochloa*, *Neeragrostis*, *Stiburus*, *Thellungia*

Habit, vegetative morphology. Annual, or perennial; caespitose (sometimes shrubby), or decumbent, or stoloniferous (rarely — e.g. *E. hypnoides*, *E. barbinodis*). *Culms* 10–300 cm high; herbaceous (usually), or woody and persistent (occasionally); branched above (sometimes), or unbranched above (mostly); tuberous (rarely), or not tuberous; 1–5(–20) noded. Culm nodes nearly always glabrous, or hairy (very rarely, e.g. *E. annulata*, *E. kennedyae*). Culm internodes solid, or hollow. Plants without multicellular glands, or with multicellular glands (commonly, on various organs: buttonlike, elevated pitted glands mostly on leaf blades (e.g. *E. cilianensis*), crateriform glands (e.g. *E. annulata*), spot- or band-shaped glands often beneath nodes (e.g. *E. trichophora*), gland-based macrohairs common on leaf sheaths, blades or collars (e.g. *E. trichocolea*)). *Leaves* mostly basal, or not basally aggregated; auriculate (rarely), or non-auriculate (mostly). Sheath margins free. The sheaths keeled or not, glabrous or hairy and glandular. Leaf blades linear; broad to narrow (mostly narrow); flat, or folded, or rolled; without abaxial multicellular glands, or exhibiting multicellular glands abaxially (sometimes). The abaxial glands on the blade margins, or costal. *Leaf blades not pseudopetiolate*; without cross venation; disarticulating from the sheaths (rarely, e.g. *E. australasica*), or persistent (usually); rolled in bud. **Ligule** present; *a fringe of hairs (perhaps always, with* **Diandrochloa** *etc. excluded)*.

Reproductive organization. Plants bisexual, with bisexual spikelets (with the exception of *E. contrerasii* Pohl — see *Neeragrostis*); with hermaphrodite florets. The spikelets of sexually distinct forms on the same plant (occasionally), or all alike in sexuality (usually); sometimes hermaphrodite and male-only (e.g. *E. superba*, with reduced spikelets towards the extremities of the panicle). Plants outbreeding and inbreeding. Exposed-cleistogamous, or chasmogamous. Plants with hidden cleistogenes (rarely, e.g. *E. scaligera*, *E. riobrancensis*), or without hidden cleistogenes. The hidden cleistogenes when present, in the leaf sheaths. Apomictic, or reproducing sexually. Viviparous (sometimes), or not viviparous.

Inflorescence. *Inflorescence few spikeleted, or many spikeleted; a false spike, with spikelets on contracted axes (occasionally, e.g.* **E. chapelieri***)), or paniculate (mostly a decompound panicle, often with multicellular glands, characteristically scented)*; deciduous in its entirety (rarely, e.g. *E. curtipedicellata*), or not deciduous; open, or contracted; with conspicuously divaricate branchlets, or without conspicuously divaricate branchlets; with capillary branchlets, or without capillary branchlets (mostly). Inflorescence with axes ending in spikelets (nearly always), or axes not ending in spikelets (ending in bristles in no more than two or three species, e.g. *E. canescens*, *E. hispida*). Inflorescence espatheate; not comprising 'partial inflorescences' and foliar organs. Spikelet-bearing axes persistent. Spikelets not secund; subsessile, or pedicellate.

Female-fertile spikelets. *Spikelets* 1–6.5–25(–40) mm long (and 0.5–3–8 mm wide, often lead coloured or purple tinged); *compressed laterally (usually strongly so, but sometimes subterete in 'Section Cylindrostachya')*; disarticulating above the glumes (usually), or falling with the glumes (e.g *E. superba*); *not disarticulating between the florets (commonly, the paleas persistent), or disarticulating between the florets*; *with conventional internode spacings*. Rachilla prolonged beyond the uppermost female-fertile floret; rachilla hairy (occasionally), or hairless (usually). The rachilla extension with incomplete florets (mostly), or naked (sometimes). *Hairy callus absent. Callus* absent (usually), or short; when present, blunt.

Glumes two (persistent most frequently in subgenus Caesia, deciduous most frequently in subgenus Eragrostis); very unequal to more or less equal; shorter than the spikelets; shorter than the adjacent lemmas; free; hairless; glabrous, or scabrous (on the veins); pointed (mostly), or not pointed (occasionally blunt); awnless; carinate (mostly), or non-carinate (a few); similar (mostly, membranous), or very dissimilar (*E. stenostachya*). Lower glume shorter than the lowest lemma to about equalling the lowest lemma; 1 nerved. *Upper glume 1 nerved. Spikelets* with female-fertile florets only, or with incomplete florets. The incomplete florets (when present) nearly always distal to the female-fertile florets (a very few species with 1–3 incomplete lower florets). The distal incomplete florets 1–2(–3); *incomplete florets* merely underdeveloped (mostly with reduced palea and 1–2–3 veined lemma); incomplete florets awnless. *Spikelets* nearly always *without proximal incomplete florets*.

Female-fertile florets *(2–)3–10–45(–50). Lemmas* similar in texture to the glumes to decidedly firmer than the glumes (narrow, membranous to papery); not becoming indurated; entire, or incised; when incised, not deeply cleft (emarginate); awnless, or mucronate (very rarely almost awned, e.g. *E. walteri*); *hairless (usually glabrous)*; glabrous, or scabrous; carinate (mostly), or non-carinate. *The keel* when present, *wingless*. Lemmas without a germination flap; (1–)3(–5) nerved (1–3 in *E. kennedyae*, 3–5 in *E. walteri*); with the nerves non-confluent (the lateral veins more often conspicuous than inconspicuous, usually with only the midvein reaching the margin). **Palea** present (often persistent on the rachilla after the lemma and fruit have dropped); relatively long (usually, but shorter than the lemma), or conspicuous but relatively short; entire (mostly), or apically notched (then usually bifid, rarely trifid); awnless, without apical setae; *thinner than the lemma (membranous/hyaline)*; not indurated; 2-nerved; 2-keeled. Palea keels winged (rarely), or wingless; hairy (usually), or glabrous to scabrous. *Lodicules* present (usually), or absent; when present, 2; free; fleshy; glabrous. *Stamens* (2–)3 (mostly 3, but sometimes 2–3 or 2 (*E. ciliaris*). Anthers (0.1–)0.2–0.6–1.8(–2.5) mm long; not penicillate; without an apically prolonged connective. *Ovary* glabrous. Styles free to their bases. Stigmas 2; white, or red pigmented.

Fruit, embryo and seedling. *Fruit* free from both lemma and palea; small (0.4–0.8–2.4 mm long); ellipsoid, or subglobose (or obovate); not grooved (mostly), or longitudinally grooved; compressed laterally (commonly, e.g. *E. pilgerana*), or compressed dorsiventrally (rarely), or not noticeably compressed (mostly), or trigonous (rarely); scabrous; smooth, or sculptured (sometimes finely striate or reticulate). *Hilum short.* Pericarp fused (usually), or loosely adherent (rather readily detachable in some species), or free (e.g. in *E. megalosperma*, *E. stapfiana*). Embryo large (0.3–0.5–0.7 times the length of the caryopsis); waisted (usually), or not waisted. Endosperm hard; without lipid; containing compound starch grains. Embryo with an epiblast; with a scutellar tail; with an elongated mesocotyl internode. Embryonic leaf margins meeting.

Seedling with a long mesocotyl; with a loose coleoptile. First seedling leaf with a well-developed lamina. The lamina broad; curved; 3–5 veined.

Abaxial leaf blade epidermis. *Costal/intercostal zonation* conspicuous. *Papillae* absent (nearly always), or present (*E. obtusiflora*). *Long-cells* similar in shape costally and intercostally; of similar wall thickness costally and intercostally. Mid-intercostal long-cells rectangular; having markedly sinuous walls (nearly always), or having straight or only gently undulating walls (rarely). *Microhairs* present; elongated; clearly two-celled; panicoid-type (subgenus Caesia), or chloridoid-type (subgenus Eragrostis, or some species with more or less intermediate forms), or *Enneapogon*-type (known only in *E. annulata*). Microhair apical cell wall thinner than that of the basal cell and often collapsed (subgenus Caesia), or thinner than that of the basal cell but not tending to collapse (few), or of similar thickness/rigidity to that of the basal cell (subgenus Eragrostis). Microhairs without 'partitioning membranes' (*E. cilianensis, E. parviflora, E. elongata*, despite the first two being 'chloridoid-type'); (30–)33–75(–156) microns long; (6–)7.5–11.5–15(–22) microns wide at the septum. Microhair total length/width at septum 2.75–12.7. Microhair apical cells (10–)12–41(–48) microns long. Microhair apical cell/total length ratio 0.3–0.61. *Stomata* common; (16.8–)19–36(–45) microns long. Subsidiaries non-papillate; low dome-shaped (usually), or triangular. Guard-cells overlapping to flush with the interstomatals. *Intercostal short-cells* common; in cork/silica-cell pairs, or not paired (solitary); silicified (when paired), or not silicified. Intercostal silica bodies absent, or imperfectly developed, or present and perfectly developed; when present, rounded, or crescentic, or tall-and-narrow (most frequently), or saddle shaped. Intercostal macrohairs sometimes present, often associated with rosettes of specialized cells. Marginal and/or costal prickles often present, intercostal prickles sometimes present, these sometimes paired with short-cells. *Costal short-cells* conspicuously in long rows, or predominantly paired, or neither distinctly grouped into long rows nor predominantly paired (not infrequently). Costal silica bodies present and well developed; present throughout the costal zones (subgenus Caesia), or confined to the central file(s) of the costal zones (rare), or present in alternate cell files of the costal zones (subgenus Eragrostis); saddle shaped (commonly), or rounded, or tall-and-narrow (rarely), or crescentic, or 'panicoid-type'; sometimes cross shaped, or nodular.

Transverse section of leaf blade, physiology. C_4 (with the startling exception of *E. walteri*: see Ellis 1984); biochemical type NAD–ME (14 species); XyMS+. PCR sheath outlines uneven, or even, or uneven to even. PCR sheath extensions present (subgenus Caesia), or absent (subgenus Eragrostis). Maximum number of extension cells 1–3–7. PCR cells with a suberised lamella, or without a suberised lamella. *PCR cell chloroplasts* ovoid, or elongated; with well developed grana; centrifugal/peripheral, or centripetal, or centrifugal/peripheral to centripetal. *Mesophyll* with radiate chlorenchyma, or with non-radiate chlorenchyma; without adaxial palisade; traversed by columns of colourless mesophyll cells (rarely), or not traversed by colourless columns. *Leaf blade* with distinct, prominent adaxial ribs (commonly), or 'nodular' in section (usually), or adaxially flat (*E. chapelieri*); when ribbed, with the ribs more or less constant in size (but larger above the primary bundles), or with the ribs very irregular in sizes (rarely). *Midrib* conspicuous, or not readily distinguishable (mostly); with one bundle only, or having a conventional arc of bundles (rarely). Bulliforms present in discrete, regular adaxial groups (nearly always), or not in discrete, regular adaxial groups (*E. pergracilis*, no bulliforms in *E. plana*); associated with colourless mesophyll cells to form deeply-penetrating fans (mostly), or in simple fans and associated with colourless mesophyll cells to form deeply-penetrating fans (these sometimes linked to traversing columns of colourless cells). Usually all the vascular bundles accompanied by sclerenchyma, or many of the smallest vascular bundles unaccompanied by sclerenchyma (not uncommonly, especially towards the blade margins). Combined sclerenchyma girders present; forming 'figures'. Sclerenchyma all associated with vascular bundles. The lamina margins without fibres, or with fibres.

Culm anatomy. *Culm internode bundles* in one or two rings.

Phytochemistry. Tissues of the culm bases with abundant starch. Leaves containing flavonoid sulphates (*E. horizontalis*), or without flavonoid sulphates (4 species). Leaf blade chlorophyll *a:b* ratio 3.39–4.48.

Cytology. Chromosome base number, $x = 10$. $2n = 20, 40, 50, 60, 80, 100,$ and 108. 2, 4, 5, 6, 8, and 10 ploid. Haploid nuclear DNA content 0.32 pg (1 species, $4x$). Nucleoli persistent.

Taxonomy. Chloridoideae; main chloridoid assemblage.

Ecology, geography, regional floristic distribution. 350 species. Cosmopolitan, mostly subtropical. Commonly adventive. Helophytic (rarely), or mesophytic, or xerophytic; mostly species of open habitats; halophytic, or glycophytic. Often on poor or sandy soils or disturbed ground.

Holarctic, Paleotropical, Neotropical, Cape, Australian, and Antarctic. Boreal, Tethyan, and Madrean. African, Madagascan, Indomalesian, Polynesian, and Neocaledonian. Euro-Siberian, Eastern Asian, Atlantic North American, and Rocky Mountains. Macaronesian, Mediterranean, and Irano-Turanian. Saharo-Sindian, Sudano-Angolan, West African Rainforest, Namib-Karoo, and Ascension and St. Helena. Indian, Indo-Chinese, Malesian, and Papuan. Hawaiian and Fijian. Caribbean, Venezuela and Surinam, Amazon, Central Brazilian, Pampas, and Andean. North and East Australian, South-West Australian, and Central Australian. New Zealand and Patagonian. European and Siberian. Canadian-Appalachian, Southern Atlantic North American, and Central Grasslands. Sahelo-Sudanian, Somalo-Ethiopian, South Tropical African, and Kalaharian. Tropical North and East Australian and Temperate and South-Eastern Australian.

Rusts and smuts. Rusts — *Physopella* and *Puccinia*. Taxonomically wide-ranging species: '*Uromyces*' *eragrostidis*. Smuts from Tilletiaceae and from Ustilaginaceae. Tilletiaceae — *Tilletia*. Ustilaginaceae — *Sorosporium, Sphacelotheca, Tolyposporella*, and *Ustilago*.

Economic importance. Significant weed species: *E. amabilis, E. aspera, E. aethiopica, E. atrovirens, E. barrelieri, E. cilianensis, E. curvula, E. diffusa, E. diarrhena, E. ferruginea, E. japonica, E. malayana, E. megastachya, E. mexicana, E. minor, E. multicaulis, E. neomexicana, E. pectinacea, E. pilosa, E. superba, E. tenella, E. tenuifolia, E. tremula, E. unioloides, E. virescens, E. viscosa, E. zeylanica*. Cultivated fodder: *E. tef*; *E. curvula* and *E. superba* drought resistant, used for reseeding denuded land. Important native pasture species: mostly more or less unpalatable, but a few useful — e.g. *E. capensis, E. cilianensis, E. ciliaris, E. curvula, E. patens, E. pilosa, E. rigidior, E. superba*. Grain crop species: *E. tef* (Teff) a staple cereal in Ethiopia, potentially of wide interest.

References, etc. Morphological/taxonomic: Koch 1978; Van den Borre and Watson 1994. Leaf anatomical: Metcalfe 1960; Prendergast *et al.* 1986; Van den Borre and Watson 1994; and this project.

Eremochloa Buese

Pectinaria (Benth.) Hack.

Habit, vegetative morphology. Perennial; caespitose. Culm nodes hairy. Culm internodes solid. Young shoots extravaginal. *Leaves* mostly basal; non-auriculate. Leaf blades narrow; without cross venation; persistent; once-folded in bud. Ligule an unfringed membrane.

Reproductive organization. Plants bisexual, with bisexual spikelets; with hermaphrodite florets. The spikelets of sexually distinct forms on the same plant; hermaphrodite and sterile; overtly heteromorphic (the 'pedicelled spikelets' very reduced); all in heterogamous combinations. Plants outbreeding.

Inflorescence. Inflorescence comprising axillary or terminal spike-like, slender flattened 'racemes'; espatheate; not comprising 'partial inflorescences' and foliar organs (or not usually interpreted as such). **Spikelet-bearing axes** *spikelike (exserted on long slender peduncles, curved, with imbricate spikelets)*; solitary, or paired (rarely); with substantial rachides (these swollen, 3-angled); tardily *disarticulating*; disarticulating at the joints. '*Articles*' non-linear (clavate); with a basal callus-knob; densely long-hairy to somewhat hairy. Spikelets paired; secund (the sessile members in two alternating rows, on one side of the rachis); sessile and pedicellate; imbricate; consistently in 'long-and-short' combinations; in pedicellate/sessile combinations. *Pedicels of the 'pedicellate' spikelets free of the rachis (flattened)*. The 'shorter' spikelets hermaphrodite (or those near the tip o sterile). The 'longer' spikelets sterile (reduced to a pedicel only, or with rudiments of a spikelet which may take the form of a bristle or mucro).

Female-sterile spikelets. Vestigial.

Female-fertile spikelets. Spikelets compressed dorsiventrally; falling with the glumes (and with the adjacent internode and pedicel). Rachilla terminated by a female-fertile floret. Hairy callus present, or absent.

Glumes two; very unequal; (the longer) long relative to the adjacent lemmas; awnless; very dissimilar (the lower 2-keeled and spiny on the margins, the upper naviculate, smooth, glabrous). *Lower glume two-keeled*; not pitted; *prickly (the lower keels with filiform or flattened scabrid curved or horizontal spines)*; 5–9 nerved. Upper glume 3 nerved. *Spikelets* with incomplete florets. The incomplete florets proximal to the female-fertile florets. **The proximal incomplete florets** 1; paleate; *male*. The proximal lemmas awnless; more or less equalling the female-fertile lemmas to decidedly exceeding the female-fertile lemmas; similar in texture to the female-fertile lemmas (hyaline); not becoming indurated.

Female-fertile florets 1. **Lemmas** less firm than the glumes (hyaline); not becoming indurated; entire; pointed; awnless; hairless; non-carinate; *0 nerved*. *Palea* present; textured like the lemma (hyaline); not indurated; nerveless. *Lodicules* present; 2; free; fleshy; glabrous. *Stamens* 3. Anthers not penicillate. *Ovary* glabrous. Styles free to their bases. Stigmas 2; red pigmented.

Fruit, embryo and seedling. Fruit small; compressed dorsiventrally. Hilum short. Embryo large. Endosperm containing only simple starch grains. Embryo without an epiblast; with a scutellar tail; with an elongated mesocotyl internode.

Abaxial leaf blade epidermis. *Costal/intercostal zonation* conspicuous. *Papillae* absent. *Long-cells* similar in shape costally and intercostally; of similar wall thickness costally and intercostally. Mid-intercostal long-cells rectangular (but sometimes with rounded ends); having markedly sinuous walls. *Microhairs* present; panicoid-type; 53.7–58.6 microns long; 10.7–12.2 microns wide at the septum. Microhair total length/width at septum 4.4–5.5. Microhair apical cells 29–39 microns long. Microhair apical cell/total length ratio 0.55–0.67. *Stomata* common; 36.6–41.5 microns long. Subsidiaries triangular. Guard-cells overlapping to flush with the interstomatals. *Intercostal short-cells* common; in cork/silica-cell pairs; silicified. *Costal short-cells* conspicuously in long rows. Costal silica bodies 'panicoid-type'.

Transverse section of leaf blade, physiology. C_4; XyMS–. PCR sheath outlines uneven. PCR cell chloroplasts with reduced grana; centrifugal/peripheral. *Mesophyll* with radiate chlorenchyma. *Leaf blade* adaxially flat. *Midrib* conspicuous (with a large adaxial bulliform group); having a conventional arc of bundles; with colourless mesophyll adaxially. Bulliforms not in discrete, regular adaxial groups (the epidermis extensively bulliform). Many of the smallest vascular bundles unaccompanied by sclerenchyma. Combined sclerenchyma girders present; forming 'figures'. Sclerenchyma all associated with vascular bundles.

Cytology. Chromosome base number, $x = 9$. $2n = 18$. 2 ploid. Nucleoli persistent.

Taxonomy. Panicoideae; Andropogonodae; Andropogoneae; Rottboelliinae.

Ecology, geography, regional floristic distribution. 9 species. India, Ceylon, southern China, Southeast Asia, western Malaysia, Australia. Mesophytic; species of open habitats. Short grassland.

Paleotropical and Australian. Indomalesian. Indian, Indo-Chinese, Malesian, and Papuan. North and East Australian. Tropical North and East Australian.

Rusts and smuts. Smuts from Ustilaginaceae. Ustilaginaceae — *Sorosporium* and *Sphacelotheca*.

Economic importance. Lawns and/or playing fields: *E. ophiuroides*.

References, etc. Leaf anatomical: this project.

Eremopoa Roshev.

Habit, vegetative morphology. *Annual*; caespitose (or the culms solitary). *Culms* 5–60 cm high; herbaceous; unbranched above. Culm nodes glabrous. Culm internodes hollow. Leaves non-auriculate. Leaf blades linear; narrow; 1–4 mm wide; flat, or rolled; without cross venation. Ligule an unfringed membrane; not truncate; 1–2.5 mm long.

Reproductive organization. Plants bisexual, with bisexual spikelets; with hermaphrodite florets. The spikelets all alike in sexuality.

Inflorescence. *Inflorescence paniculate (the branches whorled, the lower sometimes sterile)*; open; espatheate; not comprising 'partial inflorescences' and foliar organs. Spikelet-bearing axes persistent. Spikelets not secund; pedicellate.

Female-fertile spikelets. Spikelets 3–8 mm long; compressed laterally to not noticeably compressed; disarticulating above the glumes. Rachilla prolonged beyond the uppermost female-fertile floret. The rachilla extension with incomplete florets. Hairy callus absent. *Callus* short; blunt.

Glumes present; two; very unequal; shorter than the spikelets; *shorter than the adjacent lemmas*; pointed; awnless; similar (membranous). Lower glume 1 nerved. Upper glume 3 nerved. *Spikelets* with incomplete florets. The incomplete florets distal to the female-fertile florets.

Female-fertile florets (1–)2–6. *Lemmas lanceolate to narrowly oblong in lateral view, by contrast with* **Poa**; not becoming indurated; entire; pointed (acute to acuminate), or blunt; *awnless to mucronate*; *hairy*; non-carinate (keeled only towards the tip); without a germination flap; 5 nerved. *Palea* present; relatively long; 2-nerved; 2-keeled. *Lodicules* present; 2; free; membranous; glabrous; toothed, or not toothed; not or scarcely vascularized. *Stamens* 3. Anthers 0.4–2.5 mm long. *Ovary* glabrous. Styles free to their bases. Stigmas 2; white.

Fruit, embryo and seedling. **Fruit** adhering to lemma and/or palea; *medium sized (5 mm long)*; compressed dorsiventrally. *Hilum short*. Embryo small. Endosperm containing compound starch grains.

Abaxial leaf blade epidermis. *Costal/intercostal zonation* conspicuous. *Papillae* absent. *Long-cells* markedly different in shape costally and intercostally (the costals smaller, more nearly rectangular); differing markedly in wall thickness costally and intercostally (the intercostals thin walled, the costals quite thick walled). Mid-intercostal long-cells fusiform; having straight or only gently undulating walls. *Microhairs* absent. *Stomata* common. Subsidiaries parallel-sided. Guard-cells overlapped by the interstomatals. *Intercostal short-cells* absent or very rare (there being a few solitary cells apparently representing uncompleted stomatal apparati). *Costal short-cells* neither distinctly grouped into long rows nor predominantly paired (mostly solitary). Costal silica bodies exclusively horizontally-elongated crenate/sinuous.

Transverse section of leaf blade, physiology. C_3; XyMS+. *Mesophyll* with non-radiate chlorenchyma; without adaxial palisade. *Leaf blade* adaxially flat (apart from a constriction on either side of the midrib). *Midrib* conspicuous; with one bundle only. Bulliforms not in discrete, regular adaxial groups (save for the 'midrib hinges'). Many of the smallest vascular bundles unaccompanied by sclerenchyma (the blade being very depauperate of sclerenchyma). Combined sclerenchyma girders absent (only small abaxial strands seen). Sclerenchyma all associated with vascular bundles.

Cytology. Chromosome base number, $x = 7$. $2n = 14$, 28, and 42. 2, 4, and 6 ploid.

Taxonomy. Pooideae; Poodae; Poeae.

Ecology, geography, regional floristic distribution. 4 species. Western and central Asia to the Himalayas. Mesophytic.

Holarctic and Paleotropical. Boreal and Tethyan. African. Euro-Siberian. Mediterranean and Irano-Turanian. Saharo-Sindian. European.

References, etc. Leaf anatomical: this project.

Eremopogon Stapf

~ *Dichanthium*

Habit, vegetative morphology. Annual (rarely), or perennial; caespitose. Culms herbaceous; branched above. Leaves non-auriculate. Leaf blades without cross venation. *Ligule* present.

Reproductive organization. Plants bisexual, with bisexual spikelets; with hermaphrodite florets. *The spikelets of sexually distinct forms on the same plant*; hermaphrodite and male-only, or hermaphrodite and sterile; overtly heteromorphic (the pedicellate members awnless); all in heterogamous combinations.

Inflorescence. *Inflorescence of 'racemes', each branch with a basal bladeless sheath and a terminal 'raceme'*; spatheate; a complex of 'partial inflorescences' and intervening foliar organs. **Spikelet-bearing axes** *'racemes'*; solitary; with very slender rachides; disarticulating; disarticulating at the joints. *The pedicels and rachis internodes without a longitudinal, translucent furrow.* *'Articles'* linear; not appendaged; disarticulating transversely. Spikelets paired; sessile and pedicellate; consistently in 'long-and-short' combinations; in pedicellate/sessile combinations. Pedicels of the 'pedicellate' spikelets free of the rachis. The 'shorter' spikelets hermaphrodite. The 'longer' spikelets male-only, or sterile.

Female-sterile spikelets. The lemmas awnless.

Female-fertile spikelets. Spikelets compressed dorsiventrally; falling with the glumes. Rachilla terminated by a female-fertile floret. Hairy callus present.

Glumes two; more or less equal; long relative to the adjacent lemmas; awnless; very dissimilar (the lower bicarinate, the upper naviculate). *Lower glume* two-keeled (the keels not winged); *with a conspicuous pit*; relatively smooth. Upper glume 2 nerved. *Spikelets* with incomplete florets. The incomplete florets proximal to the female-fertile florets. *The proximal incomplete florets* 1; epaleate; sterile. The proximal lemmas awnless; 0 nerved; not becoming indurated (hyaline).

Female-fertile florets 1. Lemmas reduced to a linear stipe beneath the awn; less firm than the glumes; entire, or incised (sometimes); sometimes 2 lobed (bidentate); awned. Awns 1; median; from a sinus, or apical; geniculate; hairless (glabrous); much longer than the body of the lemma. Lemmas hairless; 0–1 nerved. **Palea *absent***. *Lodicules* present; 2. *Stamens* 3. *Ovary* glabrous. Styles free to their bases. Stigmas 2.

Fruit, embryo and seedling. Hilum short. Embryo large.

Abaxial leaf blade epidermis. *Costal/intercostal zonation* conspicuous (on focusing through the overlying papillae). *Papillae* present. Intercostal papillae over-arching the stomata (and obscuring all else in high focus); consisting of one symmetrical projection per cell (unusually large, bulbous, overlapping one another as well as the stomata). *Long-cells* similar in shape costally and intercostally; of similar wall thickness costally and intercostally (thin walled). Mid-intercostal long-cells rectangular; having markedly sinuous walls (but the sinuosity very small scale). *Microhairs* present; panicoid-type; 51–66 microns long; 6.9–9 microns wide at the septum. Microhair total length/width at septum 5.7–8.6. Microhair apical cells 27–39 microns long. Microhair apical cell/total length ratio 0.53–0.6. *Stomata* common; 27–30 microns long. Subsidiaries dome-shaped and triangular. Guard-cells with the guard-cell insertion obscured. *Intercostal short-cells* absent or very rare. A few small prickles present. *Costal short-cells* conspicuously in long rows. Costal silica bodies 'panicoid-type'.

Transverse section of leaf blade, physiology. C_4; XyMS–. *Mesophyll* with radiate chlorenchyma. *Leaf blade* adaxially flat. *Midrib* conspicuous (via an abaxial keel and the conspicuous sclerenchyma group beneath its main bundle); with one bundle only to having a conventional arc of bundles (the large bundle flanked on either side by one or two small ones, these arguably within the limits of the midrib); without colourless mesophyll adaxially (though there appear to be one or two colourless cells adjoining the adaxial sclerenchyma of the main bundle). Bulliforms not in discrete, regular adaxial groups (the entire epidermis being 'bulliform', all its cells tanniniferous except over the midrib). Many of the smallest vascular bundles unaccompanied by sclerenchyma. Combined sclerenchyma girders present (with all the primaries); forming 'figures' (the midrib with an anchor). Sclerenchyma all associated with vascular bundles.

Cytology. Chromosome base number, $x = 10$. $2n = 40$. 4 ploid.

Taxonomy. Panicoideae; Andropogonodae; Andropogoneae; Andropogoninae.

Ecology, geography, regional floristic distribution. 4 species. Warm Old World. Paleotropical. African and Indomalesian. Sudano-Angolan and West African Rainforest. Indian, Indo-Chinese, and Papuan.

Rusts and smuts. Rusts — *Puccinia*. Taxonomically wide-ranging species: '*Uromyces*' *clignyi*. Smuts from Ustilaginaceae. Ustilaginaceae — *Sphacelotheca* and *Ustilago*.

References, etc. Leaf anatomical: this project.

Eremopyrum (Ledeb.) Jaub. & Spach

Habit, vegetative morphology. Annual; caespitose. *Culms* 20–40 cm high; herbaceous. Culm nodes glabrous. Culm internodes hollow. *Leaves* not basally aggregated; auriculate. Sheath margins joined, or free. Leaf blades linear to linear-lanceolate; 1–7 mm wide; flat; without cross venation; persistent; rolled in bud. Ligule an unfringed membrane; truncate; 0.4–2 mm long.

Reproductive organization. Plants bisexual, with bisexual spikelets; with hermaphrodite florets.

Inflorescence. *Inflorescence a single spike (the spikelets often divergent from the rachis)*; espatheate; not comprising 'partial inflorescences' and foliar organs. *Spikelet-bearing axes* usually *disarticulating*; usually either *falling entire, or disarticulating at the joints*. Spikelets solitary; not secund; alternately distichous; sessile.

Female-fertile spikelets. Spikelets 9–25 mm long; compressed laterally; disarticulating above the glumes; disarticulating between the florets. Rachilla prolonged beyond the uppermost female-fertile floret. The rachilla extension with incomplete florets. Callus blunt.

Glumes two; relatively large; more or less equal; shorter than the adjacent lemmas, or long relative to the adjacent lemmas; joined, or free; lateral to the rachis; pointed; not subulate; shortly awned, or awnless; carinate; similar (becoming very hard at maturity). Lower glume 1–4 nerved. Upper glume 1–4 nerved. *Spikelets* with incomplete florets. The incomplete florets distal to the female-fertile florets.

Female-fertile florets *3–6*. Lemmas similar in texture to the glumes (leathery); not becoming indurated; awnless, or mucronate, or awned. Awns when present, 1; apical; non-geniculate; entered by several veins. Lemmas hairy, or hairless; carinate; 5–7 nerved; with the nerves confluent towards the tip. **Palea** present; *conspicuous but relatively short*; entire to apically notched; awnless, without apical setae, or with apical setae (via prolongation of the keels); thinner than the lemma (membranous); not indurated; 2-nerved; 2-keeled. *Lodicules* present; 2; free; membranous; ciliate; toothed, or not toothed. *Stamens* 3. Anthers 0.5–1.3 mm long; not penicillate. *Ovary* hairy. Styles free to their bases. Stigmas 2; white.

Fruit, embryo and seedling. *Fruit* somewhat adhering to lemma and/or palea; small to large; longitudinally grooved; slightly compressed dorsiventrally; with hairs confined to a terminal tuft. Hilum long-linear. Embryo small.

Abaxial leaf blade epidermis. *Costal/intercostal zonation* conspicuous. *Papillae* absent. Mid-intercostal long-cells rectangular and fusiform; having straight or only gently undulating walls (these thin). *Microhairs* absent. *Stomata* common; 41.5–56 microns long. Subsidiaries parallel-sided. Guard-cells overlapped by the interstomatals. *Intercostal short-cells* absent or very rare (excluding macrohair bases). Macrohair bases abundant. *Costal short-cells* neither distinctly grouped into long rows nor predominantly paired (solitary). Costal silica bodies horizontally-elongated crenate/sinuous, or horizontally-elongated smooth.

Transverse section of leaf blade, physiology. C_3; XyMS+. *Mesophyll* with non-radiate chlorenchyma; without adaxial palisade. *Leaf blade* 'nodular' in section; with the ribs more or less constant in size. *Midrib* not readily distinguishable; with one bundle only. Bulliforms not in discrete, regular adaxial groups (the epidermis large-celled). All the vascular bundles accompanied by sclerenchyma. Combined sclerenchyma girders absent (strands only). Sclerenchyma all associated with vascular bundles.

Cytology. Chromosome base number, $x = 7$. $2n = 14$ and 28. 2 and 4 ploid. Haplomic genome content F. Haploid nuclear DNA content 5.5 pg (1 species, $2x$). Mean diploid 2c DNA value 11 pg (1 species).

Taxonomy. Pooideae; Triticodae; Triticeae.

Ecology, geography, regional floristic distribution. 5 species. Mediterranean to central Asia. Xerophytic; species of open habitats. Stony slopes, steppe, semi-desert.

Holarctic. Boreal and Tethyan. Euro-Siberian. Mediterranean and Irano-Turanian. European.

Rusts and smuts. Smuts from Tilletiaceae and from Ustilaginaceae. Tilletiaceae — *Tilletia*. Ustilaginaceae — *Ustilago*.

Economic importance. Important native pasture species: *E. bonaepartis*.

References, etc. Morphological/taxonomic: Löve 1984. Leaf anatomical: this project.

Eriachne R.Br.

Achneria P. Beauv., *Massia* Bal.

Habit, vegetative morphology. Annual, or perennial; rhizomatous, or caespitose. **Culms** 5–100 cm high; *herbaceous*; tuberous (rarely), or not tuberous. Culm nodes hairy, or glabrous. Culm internodes solid, or hollow. *Leaves* not basally aggregated; non-auriculate. *Leaf blades* narrow; often more or less *setaceous*; flat, or rolled (sometimes pungent); without cross venation; persistent; once-folded in bud. *Ligule a fringe of hairs*. *Contra-ligule* present (of hairs), or absent.

Reproductive organization. Plants bisexual, with bisexual spikelets; with hermaphrodite florets. Exposed-cleistogamous, or chasmogamous.

Inflorescence. *Inflorescence paniculate, or a single raceme (rarely)*; open, or contracted; sometimes spicate; espatheate; not comprising 'partial inflorescences' and foliar organs. Spikelet-bearing axes persistent. Spikelets solitary; not secund; pedicellate.

Female-fertile spikelets. Spikelets 2–15 mm long; compressed laterally to compressed dorsiventrally; disarticulating above the glumes; not disarticulating between the florets, or disarticulating between the florets; with conventional internode spacings. Rachilla prolonged beyond the uppermost female-fertile floret (rarely), or terminated by a female-fertile floret. The rachilla extension (when present) with incomplete florets. Hairy callus present. **Callus *short***; blunt.

Glumes two; more or less equal; shorter than the spikelets to exceeding the spikelets; shorter than the adjacent lemmas, or long relative to the adjacent lemmas; awned, or awnless; carinate (rarely), or non-carinate (usually, rounded on the back); similar (persistent, broad, scarious to subhyaline or cartilaginous). *Lower glume (7–)9–15 nerved*. Upper glume (7–)9–15 nerved. *Spikelets* with female-fertile florets only, or with incomplete florets. The incomplete florets distal to the female-fertile florets (i.e. the upper of the two florets sometimes reduced). *Spikelets without proximal incomplete florets*.

Female-fertile florets (1–)2. Lemmas similar in texture to the glumes to decidedly firmer than the glumes (scarious to cartilaginous); becoming indurated to not becoming indurated; entire, or incised; awnless, or mucronate, or awned. Awns when present, 1; from a sinus, or apical; non-geniculate; much shorter than the body of the lemma to much longer than the body of the lemma. Lemmas hairy (with long white hairs); carinate to non-carinate; having the margins tucked in onto the palea; 5–7 nerved. *Palea* present; entire (pointed), or apically notched; awnless, without apical setae, or with apical setae, or awned (sometimes with 1 or 2 apical setae or awns); textured like the lemma; indurated, or not indurated; 2-nerved; 2-keeled to keel-less (the lateral 'keels' blunt). *Lodicules* present; 2; free; fleshy; glabrous. *Stamens* (2–)3. Anthers not penicillate. *Ovary* glabrous. Styles free to their bases. Stigmas 2; white, or red pigmented.

Fruit, embryo and seedling. *Fruit* free from both lemma and palea; small, or medium sized, or large; ellipsoid; compressed dorsiventrally (plano-convex). *Hilum long-linear*. Embryo large, or small; waisted, or not waisted. Endosperm hard; containing only simple starch grains, or containing compound starch grains. Embryo without an epiblast; with a scutellar tail; with an elongated mesocotyl internode. Embryonic leaf margins slightly overlapping.

Seedling with a long mesocotyl; with a loose coleoptile. First seedling leaf with a well-developed lamina. The lamina broad; erect; 6–12 veined.

Abaxial leaf blade epidermis. *Costal/intercostal zonation* conspicuous. *Papillae* absent. *Long-cells* similar in shape costally and intercostally; of similar wall thickness costally and intercostally. Mid-intercostal long-cells rectangular; having markedly sinuous walls. *Microhairs* present; panicoid-type; 45–68 microns long; (5.1–)5.4–9(–9.8) microns wide at the septum. Microhair total length/width at septum 5.7–11.7. Microhair apical cells 25.5–36(–44) microns long. Microhair apical cell/total length ratio 0.42–0.62. *Stomata* common; 19–48 microns long. Subsidiaries dome-shaped. Guard-cells overlapping to flush with the interstomatals. *Intercostal short-cells* common, or absent or very rare; in cork/silica-cell pairs, or not paired (solitary); silicified. Intercostal silica bodies oryzoid-type. *Costal short-cells* conspicuously in long rows. Costal silica bodies oryzoid, or 'panicoid-type'; sharp-pointed (the oryzoid forms, sometimes), or not sharp-pointed.

Transverse section of leaf blade, physiology. C_4; biochemical type NADP–ME (5 species); XyMS+. PCR sheath outlines uneven (usually), or even, or uneven to even. PCR sheath extensions present, or absent. Maximum number of extension cells when present, 1–3. PCR cells without a suberised lamella. *PCR cell chloroplasts* ovoid; with well developed grana; usually centrifugal/peripheral, or centripetal (e.g. *E. pulchella*). *Mesophyll* with radiate chlorenchyma, or with non-radiate chlorenchyma. *Leaf blade* with distinct, prominent adaxial ribs, or 'nodular' in section; with the ribs more or less constant in size, or with the ribs very irregular in sizes. *Midrib* not readily distinguishable; with one bundle only, or having a conventional arc of bundles (rarely). Bulliforms present in discrete, regular adaxial groups (in the furrows); in simple fans. All the vascular bundles accompanied by sclerenchyma. Combined sclerenchyma girders present; forming 'figures'. Sclerenchyma all associated with vascular bundles.

Phytochemistry. Leaf blade chlorophyll *a:b* ratio 4.12–5.

Taxonomy. Arundinoideae (or Panicoideae); Eriachneae.

Ecology, geography, regional floristic distribution. 40 species. China, Indomalayan region, Australia. Species of open habitats. Savanna.

Holarctic, Paleotropical, and Australian. Boreal. Indomalesian. Eastern Asian. Indian, Indo-Chinese, Malesian, and Papuan. North and East Australian and Central Australian. Tropical North and East Australian.

Rusts and smuts. Smuts from Ustilaginaceae. Ustilaginaceae — *Sorosporium*.

References, etc. Morphological/taxonomic: Van Eck-Borsboom 1980. Leaf anatomical: Metcalfe 1960; this project.

Erianthecium L. Parodi

Habit, vegetative morphology. Perennial; caespitose. **Culms** 30–45 cm high; herbaceous; *tuberous*. Culm nodes glabrous. Leaves non-auriculate. Leaf blades linear; narrow; 3–5 mm wide; flat; without cross venation. *Ligule an unfringed membrane*; truncate to not truncate; 3–4 mm long.

Reproductive organization. Plants bisexual, with bisexual spikelets; with hermaphrodite florets. Exposed-cleistogamous.

Inflorescence. Inflorescence few spikeleted (8–12); paniculate; contracted; espatheate; not comprising 'partial inflorescences' and foliar organs. Spikelet-bearing axes persistent. Spikelets not secund; pedicellate.

Female-fertile spikelets. Spikelets 7–8 mm long; compressed laterally; disarticulating above the glumes; disarticulating between the florets; with conventional internode spacings. Rachilla prolonged beyond the uppermost female-fertile floret; rachilla hairless. Hairy callus absent.

Glumes two; relatively large; very unequal to more or less equal (the upper somewhat longer); shorter than the spikelets; long relative to the adjacent lemmas; hairless; glabrous; pointed (acute); awnless; carinate; similar (lanceolate-acute, papery). Lower glume somewhat shorter than the lowest lemma; 3 nerved, or 5 nerved. *Upper glume 7 nerved*. *Spikelets* with female-fertile florets only, or with incomplete florets. The incomplete florets if present, distal to the female-fertile florets. *The distal incomplete florets* if present, merely underdeveloped.

Female-fertile florets 3–5. Lemmas similar in texture to the glumes (papery or leathery); not becoming indurated; entire (according to the original description), or incised (according to Clayton and Renvoize 1986); if incised, 2 lobed; not deeply cleft (bidentate, according to Clayton and Renvoize); awned. Awns 1; median; dorsal (according to the original description), or from a sinus (or from behind it?); from near the top; non-geniculate; hairless; much shorter than the body of the lemma to about as long as the body of the lemma; entered by one vein. Lemmas densely hairy; non-carinate; without a germination flap; 7 nerved, or 9 nerved. *Palea* present; relatively long; tightly clasped by the lemma; apically notched; awnless, without apical setae; textured like the lemma; not indurated; 2-nerved; 2-keeled. Palea back hairy. Palea keels wingless; hairy (long-ciliate). *Lodicules* present; 2; free; membranous; glabrous; not toothed; not or scarcely vascularized. *Stamens* 3. Anthers about 1.5 mm long; not penicillate; without an apically prolonged connective. *Ovary* glabrous. Styles free to their bases. Stigmas 2 ('pubescent, terminally exserted').

Fruit, embryo and seedling. Fruit small (1 mm long); ellipsoid to subglobose; compressed dorsiventrally (dorsally convex, ventrally flat). *Hilum short*. Embryo small. Endosperm liquid in the mature fruit, or hard; with lipid; containing compound starch grains. Embryo with an epiblast.

Abaxial leaf blade epidermis. *Costal/intercostal zonation* conspicuous. *Papillae* absent. *Long-cells* markedly different in shape costally and intercostally; of similar wall thickness costally and intercostally. *Mid-intercostal long-cells* fusiform; having straight or only gently undulating walls. *Microhairs* absent. *Stomata* absent or very rare. *Intercostal short-cells* absent or very rare. Macrohair bases common. *Costal short-cells* neither distinctly grouped into long rows nor predominantly paired. Costal silica bodies horizontally-elongated crenate/sinuous, or rounded (a few), or 'panicoid-type' (some); when panicoid type, mainly cross shaped and nodular (the latter elongated).

Transverse section of leaf blade, physiology. C_3; XyMS+. *Mesophyll* with non-radiate chlorenchyma. *Leaf blade* 'nodular' in section to adaxially flat; with the ribs more or less constant in size. Bulliforms present in discrete, regular adaxial groups; in simple fans. All the vascular bundles accompanied by sclerenchyma. Combined sclerenchyma girders present. Sclerenchyma all associated with vascular bundles.

Taxonomy. Pooideae; Poodae; Poeae.

Ecology, geography, regional floristic distribution. 1 species. Uruguay. Species of open habitats. Stony slopes.

Neotropical. Pampas.

References, etc. Morphological/taxonomic: Parodi 1938. Leaf anatomical: this project.

Erianthus Michx.

Ripidium Trin.
~ *Saccharum*

Habit, vegetative morphology. Perennial; sometimes *reeds (or reedlike)*; caespitose. *Culms* 200–400 cm high; unbranched above. Culm nodes glabrous. Culm internodes solid. *Leaves* not basally aggregated; non-auriculate. Leaf blades broad (always?); without cross venation; persistent. Ligule an unfringed membrane to a fringed membrane.

Reproductive organization. Plants bisexual, with bisexual spikelets; with hermaphrodite florets. The spikelets homomorphic (save in glume nervation). Exposed-cleistogamous, or chasmogamous.

Inflorescence. Inflorescence paniculate (silky-hairy); open; espatheate; not comprising 'partial inflorescences' and foliar organs. **Spikelet-bearing axes** 'racemes'; with very slender rachides; *disarticulating*; disarticulating at the joints. *'Articles'* linear; not appendaged; disarticulating transversely; densely long-hairy (plumose), or somewhat hairy. Spikelets paired; not secund; sessile and pedicellate; consistently in 'long-and-short' combinations; in pedicellate/sessile combinations. Pedicels of the 'pedicellate' spikelets free of the rachis. The 'shorter' spikelets hermaphrodite. The 'longer' spikelets hermaphrodite.

Female-fertile spikelets. Spikelets compressed dorsiventrally; falling with the glumes (the pedicelled member falling from its pedicel, the sessile falling with the joint and pedicel). Rachilla terminated by a female-fertile floret. Hairy callus present.

Glumes two; more or less equal; long relative to the adjacent lemmas; awnless; very dissimilar (the lower bicarinate, the upper naviculate). Lower glume two-keeled; not pitted; relatively smooth; 2 nerved. Upper glume 3 nerved. *Spikelets* with incomplete florets. The incomplete florets proximal to the female-fertile florets. *The proximal incomplete florets* 1; epaleate; sterile. The proximal lemmas awnless; 1 nerved; more or less equalling the female-fertile lemmas; similar in texture to the female-fertile lemmas (thinly membranous); not becoming indurated.

Female-fertile florets 1. *Lemmas* less firm than the glumes (hyaline); not becoming indurated; entire, or incised; when entire pointed (to aristate), or blunt; not deeply cleft; *mucronate, or awned*. Awns 1; median; apical; non-geniculate; hairless; much shorter than the body of the lemma to much longer than the body of the lemma. Lemmas ciliate; non-carinate; 3 nerved. **Palea** present; (when present) conspicuous but relatively short, or very reduced; entire; awnless, without apical setae; not indurated (hyaline); *nerveless*. *Lodicules* present; 2; free; fleshy; glabrous. *Stamens* 3. Anthers not penicillate. *Ovary* glabrous. Styles free to their bases. Stigmas 2.

Fruit, embryo and seedling. *Fruit* free from both lemma and palea; small. Hilum short. Endosperm hard; without lipid; containing compound starch grains. Embryo without an epiblast; with a scutellar tail; with an elongated mesocotyl internode. Embryonic leaf margins overlapping.

Abaxial leaf blade epidermis. *Costal/intercostal zonation* conspicuous. *Papillae* absent. *Long-cells* similar in shape costally and intercostally; of similar wall thickness costally and intercostally. Mid-intercostal long-cells rectangular; having markedly sinuous walls. *Microhairs* present; panicoid-type; 42–54 microns long; 3.3–5.4 microns wide at the septum. Microhair total length/width at septum 8.5–13.8. Microhair apical cells (10–)19.5–25 microns long. Microhair apical cell/total length ratio 0.39–0.47. *Stomata* common; 24–25.5(–27) microns long. Subsidiaries triangular. Guard-cells overlapping to flush with the interstomatals. *Intercostal short-cells* absent or very rare; not paired (solitary); not silicified. *Costal short-cells* conspicuously in long rows. Costal silica bodies 'panicoid-type'; cross shaped to dumb-bell shaped.

Transverse section of leaf blade, physiology. C_4; XyMS–. PCR cell chloroplasts with reduced grana. *Mesophyll* with radiate chlorenchyma. *Leaf blade* with distinct, prominent adaxial ribs, or 'nodular' in section; with the ribs very irregular in sizes. *Midrib* conspicuous; having a conventional arc of bundles; with colourless mesophyll adaxially. Bulliforms present in discrete, regular adaxial groups; in simple fans, or associated with colourless mesophyll cells to form deeply-penetrating fans; associating with colourless mesophyll cells to form arches over small vascular bundles, or nowhere involved in bulliform-plus-colourless mesophyll arches. All the vascular bundles accompanied by sclerenchyma. Combined sclerenchyma girders present; nowhere forming 'figures'. Sclerenchyma all associated with vascular bundles.

Phytochemistry. Leaves containing flavonoid sulphates (*E. maximus*), or without flavonoid sulphates (8 species).

Cytology. Chromosome base number, $x = 5$ and 10. $2n = 10, 15, 20, 30,$ and 60.

Taxonomy. Panicoideae; Andropogonodae; Andropogoneae; Andropogoninae.

Ecology, geography, regional floristic distribution. 28 species. Tropical America, southeast Europe to eastern Asia, Indomalayan region, Polynesia, Sahara, Madagascar. Helophytic, or mesophytic.

Holarctic, Paleotropical, and Neotropical. Boreal and Tethyan. African, Madagascan, Indomalesian, and Polynesian. Euro-Siberian. Irano-Turanian. Saharo-Sindian. Sudano-Angolan, and West African Rainforest. Indian, Indo-Chinese, Malesian, and Papuan. Polynesian and Fijian. Venezuela and Surinam, Amazon, and Andean.

Hybrids. Intergeneric hybrids with *Saccharum*.

Rusts and smuts. Rusts — *Puccinia*. Taxonomically wide-ranging species: *Puccinia microspora*. Smuts from Ustilaginaceae. Ustilaginaceae — *Sphacelotheca* and *Ustilago*.

References, etc. Leaf anatomical: Metcalfe 1960; this project.

Eriochloa Kunth

Aglycia Steud., *Glandiloba* (Raf.) Steud., *Helopus* Trin., *Oedipachne* Link

Habit, vegetative morphology. Annual, or perennial; stoloniferous, or caespitose to decumbent. *Culms* 20–170 cm high; herbaceous; branched above, or unbranched above. Culm nodes hairy. Culm internodes hollow. *Leaves* mostly basal, or not basally aggregated; non-auriculate. Leaf blades narrow; setaceous (sometimes?), or not setaceous; usually flat; without cross venation; persistent; rolled in bud. Ligule a fringed membrane (very reduced), or a fringe of hairs.

Reproductive organization. Plants bisexual, with bisexual spikelets; with hermaphrodite florets. *The spikelets all alike in sexuality*. Apomictic, or reproducing sexually.

Inflorescence. Inflorescence of spicate main branches, or paniculate (usually of simple or compound racemes, very rarely a single raceme (*E. distachya*) or a contracted panicle (*E. meyeriana*)); open; espatheate; not comprising 'partial inflorescences' and foliar organs. Spikelet-bearing axes persistent. Spikelets solitary, or paired (or in clusters); secund. Pedicel apices discoid. Spikelets consistently in 'long-and-short' combinations, or not in distinct 'long-and-short' combinations.

Female-fertile spikelets. *Spikelets* elliptic, or lanceolate, or ovate; *adaxial (nearly always with a small basal swelling representing fusion of the usually vestigial G_1 to the swollen internode); compressed dorsiventrally*; biconvex; falling with the glumes. Rachilla terminated by a female-fertile floret. Hairy callus absent.

Glumes two (but G_1 nearly always very reduced); very unequal (nearly always), or more or less equal (the lower usually vestigial, but as long as the spikelet in *E. rovumensis*); (the upper) long relative to the adjacent lemmas; dorsiventral to the rachis; pointed (G_2 pointed to aristulate); awned (G_2, when aristulate), or awnless; non-carinate; *very dissimilar (the lower nearly always reduced to a small cupuliform strip adherent to the thickened rachilla internode)*. Upper glume 5 nerved. *Spikelets* with incomplete florets. The incomplete florets proximal to the female-fertile florets. *The proximal incomplete florets* 1; paleate, or epaleate. Palea of the proximal incomplete florets (when present) fully developed to reduced. The proximal incomplete florets male, or sterile. The proximal lemmas awnless; 5 nerved (similar to the G_2, or shorter); decidedly exceeding the female-fertile lemmas; less firm than the female-fertile lemmas (membranous); not becoming indurated.

Female-fertile florets 1. *Lemmas* decidedly firmer than the glumes; rugose; becoming indurated to not becoming indurated (papery to crustaceous); yellow in fruit; entire; *mucronate to awned (the mucro or awn barbellate)*. Awns apical; non-geniculate; much shorter than the body of the lemma. Lemmas hairless (glabrous or apically puberulous); non-carinate; having the margins tucked in onto the palea; with a clear germination flap; 5 nerved. *Palea* present; relatively long; entire (rounded apically); awnless, without apical setae; textured like the lemma; 2-nerved; 2-keeled (the keels thickened). *Lodicules* present; 2; free; fleshy; glabrous. *Stamens* 3. Anthers not penicillate. *Ovary* glabrous. Styles fused, or free to their bases. Stigmas 2; red pigmented.

Fruit, embryo and seedling. Fruit small; compressed dorsiventrally. Hilum short. Embryo large. Endosperm hard; containing only simple starch grains. Embryo without an epiblast; with a scutellar tail; with an elongated mesocotyl internode. Embryonic leaf margins overlapping.

Seedling with a long mesocotyl.

Abaxial leaf blade epidermis. *Costal/intercostal zonation* conspicuous. *Papillae* absent. *Long-cells* similar in shape costally and intercostally; of similar wall thickness costally and intercostally. Mid-intercostal long-cells rectangular; having markedly sinuous walls. *Microhairs* present; panicoid-type; 54–64.5 microns long; 6–6.6 microns wide at the septum. Microhair total length/width at septum 9–10. Microhair apical cells 22.5–27 microns long. Microhair apical cell/total length ratio 0.4–0.42. *Stomata* common; 33–36 microns long. Subsidiaries dome-shaped, or triangular. Guard-cells overlapping to flush with the interstomatals. *Intercostal short-cells* common; in cork/silica-cell pairs, or not paired (solitary); silicified (when paired), or not silicified. *Costal short-cells* conspicuously in long rows. Costal silica bodies 'panicoid-type'; cross shaped, or dumb-bell shaped, or nodular.

Transverse section of leaf blade, physiology. C_4; biochemical type PCK (5 species); XyMS+. PCR sheath outlines uneven. PCR sheath extensions absent. PCR cells with a suberised lamella. PCR cell chloroplasts with well developed grana; centrifugal/peripheral. *Mesophyll* with radiate chlorenchyma, or with non-radiate chlorenchyma. *Leaf blade* adaxially flat; with the ribs more or less constant in size. *Midrib* conspicuous; having a conventional arc of bundles; with colourless mesophyll adaxially. Bulliforms present in discrete, regular adaxial groups; in simple fans. Many of the smallest vascular bundles unaccompanied by sclerenchyma. Combined sclerenchyma girders present; nowhere forming 'figures'. Sclerenchyma all associated with vascular bundles.

Culm anatomy. *Culm internode bundles* in one or two rings.

Special diagnostic feature. *Spikelets supported on a peculiar, hardened, cupuliform 'callus' (representing the G_1)*.

Cytology. Chromosome base number, $x = 9$. $2n = 36, 54$, and 72. 4, 6, and 8 ploid.

Taxonomy. Panicoideae; Panicodae; Paniceae.

Ecology, geography, regional floristic distribution. 30 species. Subtropical. Commonly adventive. Helophytic to mesophytic; species of open habitats; glycophytic. Damp ground and weedy places.

Holarctic, Paleotropical, Neotropical, and Australian. Boreal, Tethyan, and Madrean. African, Madagascan, Indomalesian, Polynesian, and Neocaledonian. Euro-Siberian, Eastern Asian, and Atlantic North American. Irano-Turanian. Saharo-Sindian, Sudano-Angolan, West African Rainforest, and Namib-Karoo. Indian, Indo-Chinese, Malesian, and Papuan. Fijian. Caribbean, Venezuela and Surinam, Central Brazilian, Pampas, and Andean. North and East Australian and Central Australian. European and Siberian. Southern Atlantic North American and Central Grasslands. Sahelo-Sudanian, Somalo-Ethiopian, and South Tropical African. Tropical North and East Australian and Temperate and South-Eastern Australian.

Rusts and smuts. Rusts — *Puccinia*. Taxonomically wide-ranging species: *Puccinia levis* and '*Uromyces*' *setariae-italicae*. Smuts from Tilletiaceae and from Ustilaginaceae. Tilletiaceae — *Tilletia*. Ustilaginaceae — *Sorosporium* and *Ustilago*.

Economic importance. Significant weed species: *E. gracilis, E. punctata*. Cultivated fodder: *E. polystachya*. Important native pasture species: *E. gracilis, E. procera*.

References, etc. Leaf anatomical: this project.

Eriochrysis P. Beauv.

Plazerium Kunth

Excluding *Leptosaccharum*

Habit, vegetative morphology. Perennial; caespitose. **Culms *40–120 cm high***; herbaceous; unbranched above. Culm nodes hairy, or glabrous. Culm internodes hollow. *Leaves* mostly basal; non-auriculate. Leaf blades linear; narrow; flat (usually), or folded (rarely); without cross venation. Ligule a fringed membrane, or a fringe of hairs.

Reproductive organization. Plants bisexual, with bisexual spikelets; with hermaphrodite florets. *The spikelets of sexually distinct forms on the same plant*; hermaphrodite and female-only; more or less homomorphic (but the pedicelled members smaller); all in heterogamous combinations.

Inflorescence. *Inflorescence of spicate main branches, or paniculate (of spiciform 'racemes', in a raceme or panicle, with tawny-red hairs)*; contracted (narrow); non-digitate. Primary inflorescence branches 6–12. Inflorescence espatheate; not comprising 'partial inflorescences' and foliar organs. *Spikelet-bearing axes* 'racemes'; solitary; with substantial rachides; disarticulating; disarticulating at the joints. '*Articles*' linear to non-linear (clavate); not appendaged; disarticulating transversely; densely long-hairy (at the disarticulation line). Spikelets paired; sessile and pedicellate; consistently in 'long-and-short' combinations; in pedicellate/sessile combinations. Pedicels of the 'pedicellate' spikelets free of the rachis. The 'shorter' spikelets hermaphrodite. *The 'longer' spikelets female-only*.

Female-fertile spikelets. Spikelets compressed dorsiventrally; falling with the glumes. Rachilla terminated by a female-fertile floret. Hairy callus present.

Glumes two; more or less equal; long relative to the adjacent lemmas; free; hairy; awnless; non-carinate; very dissimilar (the G_2 thinner, not 2-keeled). Lower glume two-keeled; convex on the back; not pitted; relatively smooth; indistinctly many nerved. Upper glume 1 nerved, or 3 nerved. *Spikelets* with incomplete florets. The incomplete florets proximal to the female-fertile florets. *The proximal incomplete florets* 1; epaleate; sterile. The proximal lemmas awnless; 0 nerved; decidedly exceeding the female-fertile lemmas; similar in texture to the female-fertile lemmas (hyaline); not becoming indurated.

Female-fertile florets 1. Lemmas less firm than the glumes (hyaline); not becoming indurated; entire to incised (truncate to serrate); not deeply cleft; awnless; non-carinate; without a germination flap; 0 nerved. **Palea *absent***. *Lodicules* present; 2; free; fleshy;

glabrous. *Stamens* 3 (in the sessile spikelets, rudimentary in the pedicellate spikelets). Anthers not penicillate; without an apically prolonged connective. *Ovary* glabrous. Stigmas 2; red pigmented, or brown.

Fruit, embryo and seedling. Hilum short. Embryo large.

Abaxial leaf blade epidermis. *Costal/intercostal zonation* conspicuous. *Papillae* absent. *Long-cells* similar in shape costally and intercostally to markedly different in shape costally and intercostally (the costals narrower); of similar wall thickness costally and intercostally. Mid-intercostal long-cells rectangular and fusiform; having markedly sinuous walls. *Microhairs* absent. *Stomata* common; 24–28.5 microns long. Subsidiaries triangular. Guard-cells overlapped by the interstomatals (sic). *Intercostal short-cells* absent or very rare (unless the rows over minor veins called 'intercostal'). With numerous complex-based macrohairs over the veins. *Costal short-cells* predominantly paired. Costal silica bodies 'panicoid-type', or acutely-angled (a few); short dumb-bell shaped, or cross shaped; sharp-pointed (a few 'crosses' of *Isachne* type).

Transverse section of leaf blade, physiology. C_4; XyMS– (but the PCR sheath cell walls very thick and pitted, resembling a mestome sheath). PCR cell chloroplasts centrifugal/peripheral. *Leaf blade* adaxially flat. *Midrib* conspicuous; having a conventional arc of bundles (one large keel bundle, 3 small ones on each side); with colourless mesophyll adaxially (this 'colourless tissue' lignified). Bulliforms present in discrete, regular adaxial groups (the groups of large cells); in simple fans. All the vascular bundles accompanied by sclerenchyma. Combined sclerenchyma girders present (nearly all bundles); forming 'figures' (mid-rib constituting a large anchor. Other main bundles with I's). Sclerenchyma all associated with vascular bundles.

Cytology. $2n = 20$.

Taxonomy. Panicoideae; Andropogonodae; Andropogoneae; Andropogoninae.

Ecology, geography, regional floristic distribution. 7 species. Tropical America and tropical Africa. Helophytic; species of open habitats; glycophytic. In swamps and moist places.

Paleotropical and Neotropical. African. Sudano-Angolan and West African Rainforest. Caribbean, Venezuela and Surinam, Central Brazilian, Pampas, and Andean. Sahelo-Sudanian, Somalo-Ethiopian, and South Tropical African.

References, etc. Leaf anatomical: this project.

Erioneuron Nash

Habit, vegetative morphology. Perennial; densely caespitose. Culms 15–40 cm high; *herbaceous*. Leaves mostly basal; non-auriculate. Leaf blades narrow; 1–1.5 mm wide; flat, or folded, or rolled; exhibiting multicellular glands abaxially (at the base of macrohairs). The abaxial glands intercostal. Leaf blades without cross venation. *Ligule a fringe of hairs*.

Reproductive organization. Plants bisexual, with bisexual spikelets; with hermaphrodite florets.

Inflorescence. *Inflorescence* few spikeleted; *paniculate*; contracted; *capitate, or more or less ovoid*. Inflorescence with axes ending in spikelets. Inflorescence espatheate; not comprising 'partial inflorescences' and foliar organs. Spikelet-bearing axes persistent. Spikelets not secund; subsessile to pedicellate.

Female-fertile spikelets. Spikelets 5.5–20 mm long; compressed laterally (plump); disarticulating above the glumes; disarticulating between the florets; with conventional internode spacings. Rachilla prolonged beyond the uppermost female-fertile floret. The rachilla extension with incomplete florets.

Glumes two; relatively large; very unequal to more or less equal; shorter than the spikelets to about equalling the spikelets; shorter than the adjacent lemmas to long relative to the adjacent lemmas; free; hairless; glabrous; pointed; awned to awnless; similar (membranous to leathery). Lower glume 1 nerved. *Upper glume 1 nerved. Spikelets* with incomplete florets. The incomplete florets distal to the female-fertile florets. **The distal incomplete florets** merely underdeveloped; *awned*.

Female-fertile florets *4–18*. Lemmas not becoming indurated; entire, or incised; when incised, 2 lobed; not deeply cleft; awned. Awns 1 (often accompanied by two lateral mucros); median; from a sinus (or from between the lobes); non-geniculate; hairless; much shorter than the body of the lemma to about as long as the body of the lemma; entered by one vein. Lemmas hairy (on the nerves, at least below); *non-carinate (dorsally rounded)*; without a germination flap; 3 nerved (the laterals marginal). **Palea** present (pilose below); *conspicuous but relatively short*; entire; awnless, without apical setae; 2-nerved; 2-keeled. Palea back hairy (villous below). Palea keels hairy (ciliolate to ciliate). *Lodicules* present; 2; sometimes joined to the palea; fleshy; glabrous. *Stamens* 1–3. Anthers 0.5–1.2 mm long. *Ovary* glabrous. Stigmas 2; white.

Fruit, embryo and seedling. Fruit small (about 1.6 mm long); ellipsoid; compressed laterally; smooth (translucent). Hilum short. Pericarp fused. Embryo large (projecting downwards); waisted. Endosperm hard. Embryo with an epiblast; with a scutellar tail; with an elongated mesocotyl internode. Embryonic leaf margins meeting.

Abaxial leaf blade epidermis. *Costal/intercostal zonation* conspicuous. *Papillae* absent. Long-cells of similar wall thickness costally and intercostally (fairly thick walled). Mid-intercostal long-cells rectangular; having markedly sinuous walls. *Microhairs* present; more or less spherical to elongated; clearly two-celled; chloridoid-type (large, the apical cells massive, bulbous, thick walled). Microhair apical cell wall of similar thickness/rigidity to that of the basal cell. Microhairs (25.5–)28.5–34.5(–39) microns long. Microhair basal cells 15 microns long. Microhairs 12–15 microns wide at the septum. Microhair total length/width at septum 2.1–2.6. Microhair apical cells 15–22.5 microns long. Microhair apical cell/total length ratio 0.5–0.75. *Stomata* common; 24–28.5 microns long. Subsidiaries predominantly low to high dome-shaped. Guard-cells overlapping to flush with the interstomatals. *Intercostal short-cells* absent or very rare. Intercostal silica bodies absent. *Costal short-cells* conspicuously in long rows, or predominantly paired. Costal silica bodies present throughout the costal zones, or confined to the central file(s) of the costal zones; rounded to saddle shaped (predominating over the midrib), or tall-and-narrow (a few, over minor veins), or 'panicoid-type' (predominating over the minor veins); when panicoid type, short dumb-bell shaped, or cross shaped.

Transverse section of leaf blade, physiology. C_4; XyMS+. PCR sheath outlines even. PCR sheaths of the primary vascular bundles complete. PCR sheath extensions absent. PCR cell chloroplasts centripetal. *Mesophyll* with radiate chlorenchyma; traversed by columns of colourless mesophyll cells. *Leaf blade* 'nodular' in section to adaxially flat. *Midrib* conspicuous (via the rounded keel and a large abaxial sclerenchyma group, and lack of an adaxial strand); with one bundle only. Bulliforms present in discrete, regular adaxial groups; associated with colourless mesophyll cells to form deeply-penetrating fans (these incorporated in the traversing colourless columns). Many of the smallest vascular bundles unaccompanied by sclerenchyma. Combined sclerenchyma girders absent (adaxial strands only, and abaxial strands or girders). Sclerenchyma all associated with vascular bundles. The lamina margins with fibres (large).

Cytology. Chromosome base number, $x = 8$. $2n = 16$. 2 ploid.

Taxonomy. Chloridoideae; main chloridoid assemblage.

Ecology, geography, regional floristic distribution. 5 species. Southwest U.S.A. and Mexico. Species of open habitats. Rocky slopes.

Holarctic and Antarctic. Boreal and Madrean. Atlantic North American. Patagonian. Southern Atlantic North American and Central Grasslands.

References, etc. Leaf anatomical: this project.

Special comments. A segregate of *Tridens*.

Erythranthera Zotov

~ *Rytidosperma, Danthonia* sensu lato

Habit, vegetative morphology. Perennial; rhizomatous, or caespitose, or decumbent (often sward-forming). **Culms *3–12 cm high*; *herbaceous*.** Culm nodes glabrous. Culm internodes hollow. *Leaves* mostly basal; non-auriculate. **Leaf blades** narrow; setaceous; flat, or folded (to 3 cm long); without cross venation; *disarticulating from the sheaths*. *Ligule a fringed membrane (very short), or a fringe of hairs.*

Reproductive organization. Plants bisexual, with bisexual spikelets; with hermaphrodite florets.

Inflorescence. Inflorescence few spikeleted; a single raceme, or paniculate; open, or contracted; espatheate; not comprising 'partial inflorescences' and foliar organs. Spikelet-bearing axes persistent. Spikelets not secund; pedicellate.

Female-fertile spikelets. *Spikelets* 2.5–5 mm long; compressed laterally; disarticulating above the glumes; *with distinctly elongated rachilla internodes between the florets (internodes up to 1/3 as long as the lemmas)*. Rachilla prolonged beyond the uppermost female-fertile floret. Hairy callus absent. *Callus* short; blunt (flattened).

Glumes two; more or less equal; *long relative to the adjacent lemmas*; free; awnless; similar. Lower glume 3–5 nerved. Upper glume 3 nerved. *Spikelets* with female-fertile florets only, or with incomplete florets. The incomplete florets distal to the female-fertile florets.

Female-fertile florets *3–5*. *Lemmas* not becoming indurated; incised; minutely 3 lobed; not deeply cleft (with 3 minute apical teeth); *awnless*; hairy, or hairless (glabrous). The hairs when present, not in tufts; not in transverse rows (scattered). Lemmas carinate to non-carinate; 7–9 nerved. *Palea* present; 2-nerved; 2-keeled. Palea keels hairy (ciliate). *Lodicules* present; 2; joined, or free; fleshy, or membranous; glabrous; toothed. *Stamens* 3. Anthers not penicillate. *Ovary* glabrous. Styles free to their bases. Stigmas 2; brown.

Fruit, embryo and seedling. Fruit small. Hilum short. Embryo large. Endosperm containing compound starch grains. Embryo without an epiblast; with a scutellar tail; with an elongated mesocotyl internode.

Abaxial leaf blade epidermis. *Costal/intercostal zonation* conspicuous. *Papillae* absent. *Long-cells* similar in shape costally and intercostally (long and narrow); of similar wall thickness costally and intercostally (thick walled). Mid-intercostal long-cells rectangular; having markedly sinuous walls. *Microhairs* present; panicoid-type; 45–55.5(–60) microns long; 9.6–14.4 microns wide at the septum. Microhair total length/width at septum 3.1–5.8. Microhair apical cells 24–36 microns long. Microhair apical cell/total length ratio 0.43–0.59. *Stomata* absent or very rare. Subsidiaries dome-shaped, or triangular. *Intercostal short-cells* common; in cork/silica-cell pairs; silicified. Intercostal silica bodies mainly crescentic. *Costal short-cells* conspicuously in long rows, or neither distinctly grouped into long rows nor predominantly paired. Costal silica bodies rounded (including potato-shaped), or crescentic, or 'panicoid-type'.

Transverse section of leaf blade, physiology. C_3; XyMS+. *Mesophyll* with radiate chlorenchyma, or with non-radiate chlorenchyma; without adaxial palisade. *Leaf blade* with distinct, prominent adaxial ribs, or adaxially flat; with the ribs more or less constant in size. *Midrib* conspicuous, or not readily distinguishable; with one bundle only. Bulliforms present in discrete, regular adaxial groups (in the furrows); in simple fans. All the vascular bundles accompanied by sclerenchyma. Combined sclerenchyma girders present; forming 'figures'. Sclerenchyma all associated with vascular bundles.

Taxonomy. Arundinoideae; Danthonieae.

Ecology, geography, regional floristic distribution. 3 species. Australia, New Zealand. Species of open habitats. Alpine.

Australian and Antarctic. North and East Australian. New Zealand. Temperate and South-Eastern Australian.

References, etc. Morphological/taxonomic: Zotov 1963. Leaf anatomical: this project.

Euchlaena Schrad.

~ *Zea*

Habit, vegetative morphology. Robust annual, or perennial. **Culms 200–500 cm high**; herbaceous; unbranched above. Culm internodes solid. *Leaves* not basally aggregated; non-auriculate; without auricular setae. Leaf blades broadly linear-lanceolate; broad; 20–80 mm wide; flat; without cross venation; persistent; rolled in bud. Ligule an unfringed membrane. *Contra-ligule* absent.

Reproductive organization. *Plants monoecious with all the fertile spikelets unisexual*; without hermaphrodite florets. The spikelets of sexually distinct forms on the same plant; female-only, or male-only. The male and female-fertile spikelets in different inflorescences. The spikelets overtly heteromorphic.

Inflorescence. *Inflorescence comprising terminal panicles of male spikelets, and axillary spikes of pistillate spikelets*. Rachides of female inflorescence hollowed. Inflorescence spatheate. *Spikelet-bearing axes* spikes (female, these borne two to several, enclosed in foliaceous spathes or husks, in the leaf sheaths), or 'racemes' (male); (the females) with substantial rachides; disarticulating (i.e., the female rachides); disarticulating at the joints. *'Articles'* non-linear; without a basal callus-knob; not appendaged; disarticulating obliquely; somewhat hairy, or glabrous. Spikelets solitary (females), or paired (males); secund (male pairs on one side of the rachis), or not secund (females); biseriate (the male pairs), or distichous (female); sessile (female), or sessile and pedicellate (the males in sessile-pedicellate pairs); consistently in 'long-and-short' combinations (male), or not in distinct 'long-and-short' combinations (female).

Female-sterile spikelets. *Male inflorescence and spikelets like those of Zea*. Rachilla of male spikelets terminated by a male floret. The male spikelets with glumes (two, many nerved); without proximal incomplete florets; 2 floreted. The lemmas awnless. Male florets 2; 3 staminate.

Female-fertile spikelets. *Spikelets* abaxial (the G_1 covering the rachis cavity); *compressed laterally*; falling with the glumes (and with the joint). Rachilla terminated by a female-fertile floret. Hairy callus absent.

Glumes two; more or less equal; about equalling the spikelets to exceeding the spikelets; long relative to the adjacent lemmas; dorsiventral to the rachis; hairless; glabrous; apiculate; awnless; non-carinate; very dissimilar (G_1 indurate, G_2 membranous). Lower glume not two-keeled; convex on the back; not pitted; relatively smooth; about 11–13 nerved. Upper glume 9 nerved. *Spikelets* with incomplete florets. The incomplete florets proximal to the female-fertile florets. *Spikelets with proximal incomplete florets. The proximal incomplete florets* 1; epaleate; sterile. The proximal lemmas awnless; 2 nerved, or 3 nerved; similar in texture to the female-fertile lemmas (hyaline); not becoming indurated.

Female-fertile florets 1. **Lemmas** less firm than the glumes (hyaline); not becoming indurated; entire; awnless; hairless; non-carinate; without a germination flap; *1 nerved*. *Palea* present; relatively long; apically notched; textured like the lemma (hyaline); not indurated; 2-nerved; keel-less. *Lodicules* absent. *Stamens* 0. *Ovary* glabrous. Styles fused (nearly to the tip). Stigmas 2.

Fruit, embryo and seedling. Hilum short. Embryo large.

Abaxial leaf blade epidermis. *Costal/intercostal zonation* conspicuous. *Papillae* absent. Mid-intercostal long-cells having markedly sinuous walls (thin). *Microhairs* present; panicoid-type (but the distal cells rather inflated); 60–72 microns long. Microhair apical cells 36–48 microns long. Microhair apical cell/total length ratio 0.64. *Stomata* common. Subsidiaries triangular. *Intercostal short-cells* common; in cork/silica-cell pairs; silicified. Intercostal silica bodies cross-shaped, tall-and-narrow, and vertically elongated-nodular. *Costal short-cells* conspicuously in long rows. Costal silica bodies 'panicoid-type'; cross shaped.

Transverse section of leaf blade, physiology. C_4; biochemical type NADP–ME (*E. mexicana*); XyMS–. PCR cell chloroplasts with reduced grana; centrifugal/peripheral. *Mesophyll* with non-radiate chlorenchyma. *Leaf blade* adaxially flat. *Midrib* conspicuous; having a conventional arc of bundles (a large median, flanked on either side by a large lateral and smaller bundles of various sizes); with colourless mesophyll adaxially. Bulliforms present in discrete, regular adaxial groups; in simple fans (the cells large). Many of the smallest vascular bundles unaccompanied by sclerenchyma. Combined sclerenchyma girders present (with the larger bundles). Sclerenchyma all associated with vascular bundles.

Special diagnostic feature. *Fruiting inflorescence not as in maize (q.v.) (the female spikelets free of one another, falling, not forming a cob)*.

Cytology. Chromosome base number, $x = 10$. $2n = 20$ and 40. 2 and 4 ploid.

Taxonomy. Panicoideae; Andropogonodae; Maydeae.

Ecology, geography, regional floristic distribution. 4 species. Mexico. Mesophytic. Field margins.

Holarctic and Neotropical. Madrean. Caribbean.

Hybrids. Intergeneric hybrids with *Zea* — ×*Euchlaezea* Janaki ex Bor.

Rusts and smuts. Rusts — *Physopella* and *Puccinia*. Taxonomically wide-ranging species: *Puccinia polysora*. Smuts from Ustilaginaceae. Ustilaginaceae — *Ustilago*.

Economic importance. Cultivated fodder: *E. mexicana* (Teosinte).

References, etc. Leaf anatomical: Metcalfe 1960.

Special comments. The descriptions of maize relatives require updating, to conform with modern taxonomic views (e.g. Doebley and Iltis 1980, Iltis 1987). The older and convenient but artificial treatment, separating the readily distinguishable *Zea mays* from the rest, is retained here pending acqisition of detailed comparative morphological data.

Euclasta Franch.

Indochloa Bor

Habit, vegetative morphology. Rambling annual. **Culms** 30–200 cm high; herbaceous; *branched above*. Culm nodes hairy. *The shoots aromatic*. *Leaves* not basally aggregated. Leaf blades narrow; 2–8 mm wide; flat; without cross venation.

Reproductive organization. Plants bisexual, with bisexual spikelets; with hermaphrodite florets. *The spikelets* of sexually distinct forms on the same plant; hermaphrodite and male-only, or hermaphrodite and sterile; overtly heteromorphic; *in both homogamous and heterogamous combinations (the proximal 1–3 pairs imperfect and homogamous)*.

Inflorescence. *Inflorescence of spicate main branches (terminal and axillary, the 'racemes' delicate, solitary or clustered, peduncled, 3–6 cm long)*; non-digitate (corymbiform or sub-paniculate), or subdigitate. Primary inflorescence branches 1–15. Rachides flattened. Inflorescence espatheate; not comprising 'partial inflorescences' and foliar organs. *Spikelet-bearing axes* 'racemes'; solitary to clustered; with very slender rachides (these flattened, sinuous); disarticulating; disarticulating at the joints (above, but the basal homogamous spikelets persistent). The pedicels and rachis internodes with a longitudinal, translucent furrow. *'Articles' linear (hyaline and glandular down the middle)*; not appendaged; disarticulating transversely; densely long-hairy. Spikelets paired; sessile and pedicellate; consistently in 'long-and-short' combinations; in pedicellate/sessile combinations. Pedicels of the 'pedicellate' spikelets free of the rachis. The 'shorter' spikelets hermaphrodite (except at the base of the raceme, where both sessile and pedicellate spikelets are male or neuter and persistent). The 'longer' spikelets male-only, or sterile.

Female-sterile spikelets. The pedicelled spikelets large, herbaceous.

Female-fertile spikelets. Spikelets 3.6–4 mm long; compressed dorsiventrally; falling with the glumes. Rachilla terminated by a female-fertile floret. Hairy callus present (minutely bearded). *Callus* short.

Glumes two; relatively large; more or less equal; long relative to the adjacent lemmas (much exceeding them); hairy (G_1); awnless; very dissimilar (G_1 truncate or obtuse and 2-keeled, the G_2 naviculate and 1-keeled). Lower glume two-keeled (truncate or obtuse); not pitted; 9 nerved. Upper glume 3 nerved. *Spikelets* with incomplete florets. The incomplete florets proximal to the female-fertile florets. *The proximal incomplete florets* 1; epaleate; sterile. The proximal lemmas awnless; 0 nerved; not becoming indurated (very small, hyaline).

Female-fertile florets 1. Lemmas reduced to a linear, hyaline stipe beneath the awn; less firm than the glumes; not becoming indurated; entire; not deeply cleft; awned. *Awns* 1; median; apical (continuous from the stipe); geniculate; *hairy*; much longer than the body of the lemma. Lemmas hairless; non-carinate; without a germination flap; 1 nerved. *Palea* absent. *Lodicules* present; 2; free; glabrous; not or scarcely vascularized. *Stamens* 3; with free filaments. Anthers as broad as long; not penicillate; without an apically prolonged connective. *Ovary* glabrous. Styles free to their bases. Stigmas 2.

Fruit, embryo and seedling. Hilum short. Embryo large.

Abaxial leaf blade epidermis. *Costal/intercostal zonation* conspicuous. *Papillae* present. Intercostal papillae consisting of one oblique swelling per cell. Mid-intercostal long-cells having straight or only gently undulating walls (mostly, the walls thin). *Microhairs* present; panicoid-type; 40–54 microns long. Microhair apical cells 12–18 microns long. Microhair apical cell/total length ratio 0.32. *Stomata* common. Subsidiaries dome-shaped to triangular. *Intercostal short-cells* absent or very rare. Cushion-based macrohairs present. *Costal short-cells* conspicuously in long rows. Costal silica bodies 'panicoid-type'; dumb-bell shaped (mostly), or nodular (some, and some tending to saddle-shaped).

Transverse section of leaf blade, physiology. C_4; XyMS–. *Mesophyll* with radiate chlorenchyma. *Leaf blade* adaxially flat. *Midrib* conspicuous; having a conventional arc of bundles (one large median, 2–3 small laterals on either side). Bulliforms not in discrete, regular adaxial groups (irregularly grouped). Many of the smallest vascular bundles unaccompanied by sclerenchyma. Combined sclerenchyma girders present. Sclerenchyma all associated with vascular bundles.

Cytology. Chromosome base number, $x = 10$. $2n = 40$.

Taxonomy. Panicoideae; Andropogonodae; Andropogoneae; Andropogoninae.

Ecology, geography, regional floristic distribution. 2 species. Tropical America, Africa, Madagascar, India. Mesophytic; shade species, or species of open habitats. In partial shade, or among taller grasses.

Paleotropical. African and Madagascan. Sudano-Angolan and West African Rain-forest. Somalo-Ethiopian and South Tropical African.

References, etc. Leaf anatomical: Metcalfe 1960.

Eulalia Kunth

Puliculum Haines
Excluding *Pseudopogonatherum*

Habit, vegetative morphology. Perennial (usually), or annual; caespitose, or decumbent. *Culms* 10–150 cm high; herbaceous; unbranched above; tuberous, or not tuberous. Culm nodes hairy, or glabrous. Culm internodes solid. *Leaves* not basally aggregated; non-auriculate. Leaf blades linear; narrow; flat; without cross venation; persistent. *Ligule* present; an unfringed membrane, or a fringed membrane.

Reproductive organization. Plants bisexual, with bisexual spikelets; with hermaphrodite florets. *The spikelets all alike in sexuality*; homomorphic.

Inflorescence. *Inflorescence of spicate main branches (very hairy or silky, often brown or purple); digitate, or subdigitate (usually, with a short axis).* Primary inflorescence branches 1–20. Inflorescence espatheate; not comprising 'partial inflorescences' and foliar organs. **Spikelet-bearing axes** 'racemes' (spiciform); *with very slender rachides*; disarticulating; disarticulating at the joints. '**Articles**' linear; not appendaged; *disarticulating obliquely*; densely long-hairy. Spikelets paired; secund, or not secund; sessile and pedicellate; consistently in 'long-and-short' combinations; in pedicellate/sessile combinations. Pedicels of the 'pedicellate' spikelets free of the rachis. The 'shorter' spikelets hermaphrodite. The 'longer' spikelets hermaphrodite.

Female-fertile spikelets. Spikelets compressed dorsiventrally; falling with the glumes (the pedicelled falling from the pedicel, the sessile falling with the joint and pedicel). Rachilla terminated by a female-fertile floret. Hairy callus present. *Callus* short; blunt.

Glumes two; more or less equal; long relative to the adjacent lemmas; hairy; without conspicuous tufts or rows of hairs; awned (the G_1 being rarely bilobed or 2-awned), or awnless; very dissimilar (both villous and rigid to leathery, the lower flattened to depressed on the back and more or less bicarinate, the upper naviculate). *Lower glume two-keeled (the keels not winged)*; flattened on the back to concave on the back; not pitted; relatively smooth; 1–9 nerved. Upper glume 1–5 nerved. *Spikelets* with female-fertile florets only, or with incomplete florets. The incomplete florets proximal to the female-fertile florets. *The proximal incomplete florets* when present, 1; when present, epaleate; when present, sterile. The proximal lemmas awnless; exceeded by the female-fertile lemmas; similar in texture to the female-fertile lemmas (hyaline); not becoming indurated.

Female-fertile florets 1. Lemmas concurrent with the awn, or bilobed; less firm than the glumes (hyaline); not becoming indurated; entire, or incised; when incised, 2 lobed; when incised, deeply cleft, or not deeply cleft; awned (nearly always), or awnless (in *E. manipurensis*). Awns 1; usually from a sinus; geniculate; usually hairless; much longer than the body of the lemma. *Lemmas* hairless; non-carinate; without a germination flap; *1–3 nerved.* **Palea** present, or absent; when present, very reduced; entire; awnless, without apical setae; not indurated (hyaline); *nerveless. Lodicules* present; 2; free; fleshy; ciliate, or glabrous. *Stamens* (2–)3. Anthers not penicillate. *Ovary* glabrous. Styles fused, or free to their bases. Stigmas 2; red pigmented.

Fruit, embryo and seedling. *Fruit* free from both lemma and palea; small; compressed dorsiventrally, or not noticeably compressed. Hilum short. Embryo large. Endosperm containing only simple starch grains. Embryo without an epiblast; with a scutellar tail; with an elongated mesocotyl internode.

Abaxial leaf blade epidermis. *Costal/intercostal zonation* conspicuous. *Papillae* present. Intercostal papillae not over-arching the stomata; several per cell (finger-like). *Long-cells* similar in shape costally and intercostally; of similar wall thickness costally and intercostally. Mid-intercostal long-cells rectangular; having markedly sinuous walls. *Microhairs* present; panicoid-type; 24–48 microns long; (4.5–)5.4 microns wide at the septum. Microhair total length/width at septum 5.6–7.8. Microhair apical cells 12–29 microns long. Microhair apical cell/total length ratio 0.56–0.8. *Stomata* common; 19.5–24 microns long. Subsidiaries dome-shaped (low to high), or triangular. Guard-cells overlapping to flush with the interstomatals. *Intercostal short-cells* absent or very rare; not paired (solitary); not silicified. *Costal short-cells* conspicuously in long rows, or neither distinctly grouped into long rows nor predominantly paired. Costal silica bodies 'panicoid-type'; cross shaped, or butterfly shaped, or dumb-bell shaped, or nodular.

Transverse section of leaf blade, physiology. C_4; XyMS–. PCR cell chloroplasts with reduced grana; centrifugal/peripheral. *Mesophyll* with radiate chlorenchyma; traversed by columns of colourless mesophyll cells, or not traversed by colourless columns. *Leaf blade* with distinct, prominent adaxial ribs, or 'nodular' in section, or adaxially flat; with the ribs more or less constant in size. *Midrib* conspicuous; with one bundle only, or having a conventional arc of bundles. Bulliforms present in discrete, regular adaxial groups; associated with colourless mesophyll cells to form deeply-penetrating fans, or in simple fans and associated with colourless mesophyll cells to form deeply-penetrating fans (sometimes linked with traversing colourless columns, sometimes with large-celled groups of the *Zea* type). Many of the smallest vascular bundles unaccompanied by sclerenchyma. Combined sclerenchyma girders present, or absent; forming 'figures', or nowhere forming 'figures'. Sclerenchyma all associated with vascular bundles.

Culm anatomy. *Culm internode bundles* in three or more rings.

Phytochemistry. Leaves without flavonoid sulphates (2 species).

Cytology. Chromosome base number, $x = 5$ and 10. $2n = 20$ and 40.

Taxonomy. Panicoideae; Andropogonodae; Andropogoneae; Andropogoninae.

Ecology, geography, regional floristic distribution. 30 species. Tropical and subtropical Africa, Asia, Australia. Helophytic to mesophytic; species of open habitats; halophytic, or glycophytic. Grassland, sometimes in moist places, and *E. ridleyi* in maritime sand.

Holarctic, Paleotropical, Neotropical, Cape, and Australian. Boreal. African, Madagascan, and Indomalesian. Eastern Asian. Sudano-Angolan and Namib-Karoo. Indian, Indo-Chinese, Malesian, and Papuan. Andean. North and East Australian and Central Australian. Sahelo-Sudanian, Somalo-Ethiopian, South Tropical African, and Kalaharian. Tropical North and East Australian and Temperate and South-Eastern Australian.

Rusts and smuts. Rusts — *Puccinia*. Smuts from Ustilaginaceae. Ustilaginaceae — *Sorosporium*.

Economic importance. Significant weed species: *E. villosa.* Important native pasture species: *E. aurea.*

References, etc. Leaf anatomical: Metcalfe 1960; this project.

Eulaliopsis Honda

Pollinidium Haines

Habit, vegetative morphology. Perennial; caespitose. *Culms* 40–100 cm high; herbaceous (with thickened, woolly rootstock). Leaf blades linear; narrow (long, rigid); 2–3 mm wide; folded, or rolled; without cross venation. Ligule a fringed membrane, or a fringe of hairs (the fringe dense).

Reproductive organization. Plants bisexual, with bisexual spikelets; with hermaphrodite florets. The spikelets all alike in sexuality; hermaphrodite; homomorphic.

Inflorescence. *Inflorescence* axillary, of spicate main branches ('racemes'); *subdigitate*. Primary inflorescence branches (1–)2–4. Inflorescence espatheate; not comprising 'partial inflorescences' and foliar organs. *Spikelet-bearing axes* 'racemes' (spicate, longer than the common peduncle). The racemes without spikelets towards the base (pedunculate). Spikelet-bearing axes paired to clustered (very rarely solitary); with very slender rachides; disarticulating; disarticulating at the joints. 'Articles' densely long-hairy (with basal tufts of brown hairs). Spikelets paired; sessile and pedicellate; consistently in 'long-and-short' combinations; in pedicellate/sessile combinations. Pedicels of the 'pedicellate' spikelets free of the rachis. The 'shorter' spikelets hermaphrodite. The 'longer' spikelets hermaphrodite (i.e., the spikelets homogamous).

Female-fertile spikelets. Spikelets 3.5–7 mm long (silky-villous); compressed dorsiventrally; falling with the glumes (the pedicellate spikelet falling from the pedicel, the sessile one falling with the joint). Rachilla terminated by a female-fertile floret. Hairy callus present (with tufts of brown hairs on the sessile spikelet/joint unit and at the base of the spikelet). Callus blunt.

Glumes two; more or less equal (the G_2 somewhat longer); long relative to the adjacent lemmas; rufously hairy; *with distinct hair tufts*; awned (the upper shortly so, sometimes), or awnless; carinate (G_2), or non-carinate (G_1); very dissimilar (the G_1 rounded dorsally, 2–3 toothed, muticous, the G_2 keeled, notched or cuspidate or aristulate). Lower glume not two-keeled; convex on the back (with raised nerves); not pitted; relatively smooth. *Spikelets* with incomplete florets. The incomplete florets proximal to the female-fertile florets. **The proximal incomplete florets 1 (better developed than in Eulalia)**; paleate. Palea of the proximal incomplete florets fully developed. The proximal incomplete florets male, or sterile. The proximal lemmas awnless; 0 nerved; not becoming indurated (hyaline).

Female-fertile florets 1. Lemmas linear; less firm than the glumes (hyaline); not becoming indurated; incised; minutely 2 lobed; not deeply cleft (2-toothed); awned. Awns 1; median; from a sinus; geniculate; much longer than the body of the lemma. Lemmas non-carinate; without a germination flap; 1 nerved. *Palea* present; conspicuous but relatively short (short, broadly ovate); entire (obtuse); awnless, without apical setae (ciliate or glabrous); nerveless; keel-less. **Lodicules** present; 2; free; *ciliate (each with a tuft of long hairs)*. *Stamens* 3. *Ovary* glabrous. Stigmas 2.

Fruit, embryo and seedling. Fruit ellipsoid.

Abaxial leaf blade epidermis. *Costal/intercostal zonation* conspicuous. *Papillae* absent. *Long-cells* markedly different in shape costally and intercostally (the costals narrower, more regularly rectangular); of similar wall thickness costally and intercostally (the walls of medium thickness). Mid-intercostal long-cells rectangular to fusiform; having markedly sinuous walls to having straight or only gently undulating walls (the sinuosity fairly fine to negligible). *Microhairs* present; elongated; clearly two-celled; panicoid-type. *Stomata* common. Subsidiaries non-papillate; dome-shaped (mostly low domes). Guard-cells overlapping to flush with the interstomatals. *Intercostal short-cells* common; in cork/silica-cell pairs, or not paired. No macrohairs or prickles seen. *Costal short-cells* conspicuously in long rows. Costal silica bodies present and well developed; 'panicoid-type'; butterfly shaped to dumb-bell shaped.

Transverse section of leaf blade, physiology. C_4. The anatomical organization somewhat unconventional (in that the minor bundles are so crowded as to be almost superposed). XyMS–. *Leaf blade* adaxially flat (with slight abaxial ribs beneath the primary bundles). *Midrib* conspicuous (with a pointed keel, an adaxial group of bulliforms, and midrib 'hinges'); having a conventional arc of bundles; with colourless mesophyll adaxially (in that there may be a few colourless cells beneath the adaxial bulliform group), or without colourless mesophyll adaxially (?). *The lamina* symmetrical on either side of the midrib. Bulliforms present in discrete, regular adaxial groups; associating with colourless mesophyll cells to form arches over small vascular bundles. Many of the smallest vascular bundles unaccompanied by sclerenchyma. Combined sclerenchyma girders present (with the primaries); forming 'figures' (large 'anchors' and I's). Sclerenchyma all associated with vascular bundles.

Taxonomy. Panicoideae; Andropogonodae; Andropogoneae; Andropogoninae.

Ecology, geography, regional floristic distribution. 2 species. India and Southeast Asia. Species of open habitats. Dry hillsides.

Holarctic and Paleotropical. Tethyan. Indomalesian. Irano-Turanian. Indian, Indo-Chinese, and Malesian.

Rusts and smuts. Smuts from Ustilaginaceae. Ustilaginaceae — *Ustilago*.

Economic importance. *E. binata* used for making paper, mats, rope and string.

References, etc. Morphological/taxonomic: Bor 1957a. Leaf anatomical: this project.

Special comments. Fruit data wanting.

Eustachys Desf.

Chloroides Regel, *Langsdorffia* Regel, *Schultesia* Spreng.

Habit, vegetative morphology. Annual, or perennial; caespitose. *Culms* 20–100 cm high; herbaceous. Culm nodes glabrous. Culm internodes solid. *Leaves* not basally aggregated; non-auriculate. The sheaths laterally folded and keeled. Leaf blades linear; broad, or narrow; flat, or folded; without abaxial multicellular glands; pseudopetiolate, or not pseudopetiolate; without cross venation; persistent. *Ligule a fringe of hairs*.

Reproductive organization. Plants bisexual, with bisexual spikelets; with hermaphrodite florets.

Inflorescence. *Inflorescence of spicate main branches (pectinate spikes or spicate racemes)*; *digitate*. Primary inflorescence branches 2–45. Rachides hollowed, or flattened (or angular). Inflorescence espatheate; not comprising 'partial inflorescences' and foliar organs. Spikelet-bearing axes persistent. Spikelets solitary; secund; biseriate.

Female-fertile spikelets. Spikelets 1.7–5 mm long; adaxial; compressed laterally to not noticeably compressed; disarticulating above the glumes. Rachilla prolonged beyond the uppermost female-fertile floret (but terminating in the incomplete floret); rachilla hairless. The rachilla extension with incomplete florets. Hairy callus absent.

Glumes two; relatively large (thinly membranous); very unequal to more or less equal; shorter than the adjacent lemmas; dorsiventral to the rachis; hairless; scabrous; not pointed (notched or blunt); the upper *awned (from below its apex)*; carinate; very dissimilar (both membranous, but the G_2 broader and awned). Lower glume 1 nerved. Upper glume 1 nerved. *Spikelets* with incomplete florets. The incomplete florets distal to the female-fertile florets. The distal incomplete florets usually 2; *incomplete florets* merely underdeveloped (greatly reduced, or the lower male). *Spikelets without proximal incomplete florets*.

Female-fertile florets *1*. Lemmas decidedly firmer than the glumes (firmly membranous to leathery); not becoming indurated; dark brown in fruit; entire, or incised; when incised, 2 lobed; not deeply cleft (notched); awnless, or mucronate; hairy (on margins and keel); carinate; 3 nerved. *Palea* present; entire (oblanceolate); awnless, without apical setae; thinner than the lemma (scarious); not indurated; 2-nerved; 2-keeled. Palea keels winged (and ciliolate). *Lodicules* present; 2; free; fleshy; glabrous. *Stamens* 3. Anthers not penicillate. *Ovary* glabrous. Styles free to their bases. Stigmas 2.

Fruit, embryo and seedling. *Fruit* free from both lemma and palea; small; ellipsoid; trigonous. Hilum short. Pericarp loosely adherent, or fused. Embryo large; not waisted.

Abaxial leaf blade epidermis. *Costal/intercostal zonation* conspicuous. *Papillae* present; intercostal. Intercostal papillae over-arching the stomata; consisting of one oblique swelling per cell, or several per cell (usually with two diverging, oblique swellings). *Long-cells* similar in shape costally and intercostally; of similar wall thickness costally and intercostally. Intercostal zones with typical long-cells (though these rather short). Mid-intercostal long-cells rectangular; having markedly sinuous walls. *Microhairs* present; more or less spherical; ostensibly one-celled; chloridoid-type; 18–22 microns long. *Stomata* common; (18–)28–30 microns long. Subsidiaries triangular. Guard-cells overlapping to flush with the interstomatals. *Intercostal short-cells* common; not paired; not silicified. Intercostal silica bodies absent. *Costal short-cells* conspicuously in long rows. Costal silica bodies present in alternate cell files of the costal zones; saddle shaped.

Transverse section of leaf blade, physiology. C_4; biochemical type NAD–ME (*E. distichophylla*); XyMS+. PCR sheath outlines uneven, or even. PCR sheaths of the primary vascular bundles interrupted; interrupted both abaxially and adaxially. PCR sheath extensions absent. PCR cell chloroplasts centrifugal/peripheral, or centripetal. *Mesophyll* with radiate chlorenchyma. *Leaf blade* adaxially flat. *Midrib* conspicuous (with a large adaxial bulliform group); with one bundle only; with colourless mesophyll adaxially (this extending across the lamina and adaxial to most of the bundles). Bulliforms present in discrete, regular adaxial groups (more or less, but the groups large, occupying most of the intercostal regions, and separated by few smaller epidermal cells from the colourless and bulliform tissue adaxial to the vascular bundles); associated with colourless mesophyll cells to form deeply-penetrating fans. All the vascular bundles accompanied by sclerenchyma. Combined sclerenchyma girders absent. Sclerenchyma all associated with vascular bundles.

Cytology. Chromosome base number, $x = 10$. $2n = 40$. 4 ploid.

Taxonomy. Chloridoideae; main chloridoid assemblage.

Ecology, geography, regional floristic distribution. 10 species. Tropical America, West Indies, tropical and South Africa. Commonly adventive. Mesophytic; species of open habitats; glycophytic. Savanna, on a variety of soils.

Holarctic, Paleotropical, and Neotropical. Boreal. African and Indomalesian. Atlantic North American. Sudano-Angolan and Namib-Karoo. Indian, Indo-Chinese, and Malesian. Caribbean. Southern Atlantic North American and Central Grasslands. Sahelo-Sudanian, Somalo-Ethiopian, South Tropical African, and Kalaharian.

Economic importance. Important native pasture species: *E. paspaloides*.

References, etc. Leaf anatomical: Metcalfe 1960; this project.

Euthryptochloa Cope

Habit, vegetative morphology. Perennial; loosely caespitose. *Culms* 20–100 cm high; herbaceous; unbranched above. *Leaves* not basally aggregated; non-auriculate. *Leaf blades* lanceolate to elliptic; broad; up to 25 mm wide; *pseudopetiolate*. Ligule an unfringed membrane; not truncate (herbaceous, many nerved, blunt); 20–30 mm long.

Reproductive organization. Plants bisexual, with bisexual spikelets; with hermaphrodite florets.

Inflorescence. Inflorescence paniculate; open to contracted. Inflorescence with axes ending in spikelets. Inflorescence espatheate; not comprising 'partial inflorescences' and foliar organs. Spikelets not secund; shortly pedicellate.

Female-fertile spikelets. *Spikelets 3.6–4.3 mm long*; *compressed dorsiventrally (according to the available description)*; falling with the glumes. Rachilla terminated by a female-fertile floret. Hairy callus absent. *Callus* short; blunt.

Glumes two; very unequal; (the upper) about equalling the spikelets; (the upper) long relative to the adjacent lemmas; hairless; pointed; awnless; similar (thinly membranous). *Lower glume 1 nerved*. Upper glume 3 nerved. **Spikelets** with female-fertile florets only; *without proximal incomplete florets*.

Female-fertile florets 1. Lemmas oblong-ovate; similar in texture to the glumes to decidedly firmer than the glumes (membranous); not becoming indurated; entire; pointed (sub-acute); awnless; hairless; 3 nerved. *Palea* present; relatively long; entire; awnless, without apical setae; 2-nerved (?); 2-keeled (with a dorsal sulcus). Palea keels wingless. *Lodicules* present; 2 (?); membranous. **Stamens 3**.

Abaxial leaf blade epidermis. *Microhairs* probably absent (the leaf anatomy being described merely as 'pooid').

Transverse section of leaf blade, physiology. Said to have pooid leaf anatomy, so presumably C_3.

Taxonomy. Pooideae (according to Cope), or Bambusoideae (? — at least equally likely, from the available description); Poodae (if really pooid), or Oryzodae (if bambusoid); if pooid Aveneae (?); if bambusoid Phaenospermateae (? — the only available description comes quite close to that of *Phaenosperma*).

Ecology, geography, regional floristic distribution. 1 species. China. Mesophytic; shade species. In montane forest.

Holarctic. Boreal. Eastern Asian.

References, etc. Morphological/taxonomic: T.A. Cope 1987. *Kew Bull*. **42**, 707–709.

Special comments. Data limited to the original, inadequate description and illustration. Fruit data wanting. Anatomical data wanting.

Exotheca Anderss.

Habit, vegetative morphology. Perennial; densely caespitose. *Culms* 30–200 cm high; herbaceous; branched above, or unbranched above. Culm nodes glabrous. **Leaves** not basally aggregated; *auriculate (on the sheaths)*. Leaf blades linear; narrow; 2–4 mm wide; without cross venation. Ligule an unfringed membrane (adnate to the sheath auricles).

Reproductive organization. Plants bisexual, with bisexual spikelets; with hermaphrodite florets. *The spikelets of sexually distinct forms on the same plant*; hermaphrodite, male-only, and sterile; overtly heteromorphic (the imperfect spikelets awnless); *in both homogamous and heterogamous combinations (with two proximal male or sterile, homogamous and homomorphic pairs and a terminal heterogamous triad)*.

Inflorescence. Inflorescence with long-pedunculate racemes of few spikelets, terminating culms on their distantly spaced branches; spatheate; a complex of 'partial inflorescences' and intervening foliar organs. **Spikelet-bearing axes** 'racemes' (short, of few spikelets); *paired (the upper raceme base terete, 15–25 mm long so that the pair align end to end, by contrast with* **Hyparrhenia***); disarticulating*; disarticulating at the joints (disarticulating readily beneath the triad, persistent or tardily disarticulating between the basal homogamous pairs). 'Articles' somewhat auriculate beneath the terminal triad; disarticulating obliquely; glabrous. Spikelets in triplets (at the raceme tip), or paired (proximally); sessile and pedicellate; consistently in 'long-and-short' combinations; in pedicellate/sessile combinations. Pedicels of the 'pedicellate' spikelets free of the rachis. The 'shorter' spikelets hermaphrodite (the solitary sessile spikelet of the triad), or male-only to sterile (the rest). The 'longer' spikelets male-only, or sterile.

Female-sterile spikelets. The lemmas awnless.

Female-fertile spikelets. *Spikelets* 12–16 mm long; *not noticeably compressed (terete)*; falling with the glumes. Rachilla terminated by a female-fertile floret (the triad falling). Hairy callus present.

Glumes two; more or less equal; long relative to the adjacent lemmas; free; hairy; awnless; non-carinate; very dissimilar (both somewhat leathery, the G_1 ending in a 2-keeled, bidentate, herbaceous beak, the G_2 thinner, rounded on the back, 1-keeled at its membranous tip). *Lower glume* not two-keeled; convex on the back; not pitted; relatively smooth; *9 nerved*. Upper glume 3 nerved. *Spikelets* with incomplete florets. The incomplete florets proximal to the female-fertile florets. *The proximal incomplete florets* 1; epaleate; sterile. The proximal lemmas awnless; faintly 2–3 nerved; more or less equalling the female-fertile lemmas to decidedly exceeding the female-fertile lemmas; similar in texture to the female-fertile lemmas (hyaline); not becoming indurated.

Female-fertile florets 1. Lemmas stipitate, the teeth hyaline; less firm than the glumes; not becoming indurated; incised; 2 lobed; not deeply cleft (2-toothed); awned. Awns 1; median; from a sinus; geniculate; hairy (pilose); much longer than the body of the lemma. Lemmas lemmas ciliate; non-carinate; without a germination flap; 1 nerved. **Palea** present; *conspicuous but relatively short*; entire; awnless, without apical setae (but long-ciliate); not indurated (hyaline); nerveless; keel-less. *Lodicules* present; 2; free; glabrous. *Stamens* 3. *Ovary* glabrous. Stigmas 2.

Fruit, embryo and seedling. Fruit claviform-oblong.

Abaxial leaf blade epidermis. *Costal/intercostal zonation* conspicuous. *Papillae* absent. *Long-cells* markedly different in shape costally and intercostally (the costals much narrower); of similar wall thickness costally and intercostally (thin walled). Mid-intercostal long-cells rectangular; having markedly sinuous walls. *Microhairs* present; elongated; clearly two-celled; panicoid-type; 54–60–66 microns long; 5.4–6–6.6 microns wide at the septum. Microhair total length/width at septum 8.6–9.9–11.7. Microhair apical cells 25.5–29–33 microns long. Microhair apical cell/total length ratio 0.45–0.48–0.53. *Stomata* common; 25.5–26.4–27 microns long. Subsidiaries predominantly high dome-shaped and triangular. Guard-cells overlapping to flush with the interstomatals. *Intercostal short-cells* common; not paired (mostly ostensibly solitary). *Costal short-cells* conspicuously in long rows. Costal silica bodies 'panicoid-type'; predominantly short dumbbell shaped.

Transverse section of leaf blade, physiology. C_4; XyMS–. PCR sheath outlines even. PCR sheath extensions absent. *Leaf blade* adaxially flat. *Midrib* conspicuous; having a conventional arc of bundles (the large median flanked on either side by several smaller bundles, all displaced towards the abaxial surface); with colourless mesophyll adaxially. *The lamina* symmetrical on either side of the midrib. Bulliforms not in discrete, regular adaxial groups (in that the epidermis is extensively bulliform). Many of the smallest vascular bundles unaccompanied by sclerenchyma. Combined sclerenchyma girders present; forming 'figures' (the primaries with I's and T's). Sclerenchyma all associated with vascular bundles.

Phytochemistry. Leaves without flavonoid sulphates (1 species).

Taxonomy. Panicoideae; Andropogonodae; Andropogoneae; Andropogoninae.

Ecology, geography, regional floristic distribution. 1 species. Tropical Africa. Species of open habitats. Upland grassland.

Paleotropical. African and Indomalesian. Sudano-Angolan. Indo-Chinese. Sahelo-Sudanian, Somalo-Ethiopian, and South Tropical African.

Rusts and smuts. Rusts — *Phakopsora* and *Puccinia*. Taxonomically wide-ranging species: *Phakopsora incompleta* and *'Uromyces' clignyi*.

References, etc. Morphological/taxonomic: Clayton 1966. Leaf anatomical: this project.

Special comments. Fruit data wanting.

Fargesia Franch. (*sensu stricto*)

~ *Thamnocalamus*

Excluding *Ampelocalamus, Chimonocalamus, Drepanostachyum, Sinarundinaria* (see comments)

Habit, vegetative morphology. Perennial. *Culms woody and persistent*; *cylindrical (by contrast with* **Phyllostachys***)*. Primary branches/mid-culm node 5–20 (? — 'numerous'). Plants unarmed. *Leaves* not basally aggregated. Leaf blades pseudopetiolate; cross veined; disarticulating from the sheaths; rolled in bud.

Reproductive organization. Plants bisexual, with bisexual spikelets; with hermaphrodite florets.

Inflorescence. Inflorescence dense paniculate; spatheate. Spikelet-bearing axes persistent. *Spikelets secund (inflorescence 'almost one-sided')*; pedicellate.

Female-fertile spikelets. Spikelets compressed laterally; disarticulating above the glumes. Rachilla prolonged beyond the uppermost female-fertile floret. The rachilla extension with incomplete florets.

Glumes two; hairy ('slightly pilose'); pointed (acuminate); awnless; similar. Lower glume 5 nerved. Upper glume 7–9 nerved. *Spikelets* with incomplete florets. The incomplete florets distal to the female-fertile florets. The distal incomplete florets usually 2; *incomplete florets* merely underdeveloped.

Female-fertile florets usually 2. Lemmas acuminate; similar in texture to the glumes; not becoming indurated; entire; pointed; without tessellate venation. *Palea* present (without tessellate venation); 2-keeled. *Lodicules* present; 3; membranous; ciliate. **Stamens 3.** Anthers with the connective apically prolonged. Ovary without a conspicuous apical appendage. Styles fused (into one). *Stigmas 3.*

Fruit, embryo and seedling. Fruit longitudinally grooved. Hilum long-linear. Endosperm containing compound starch grains.

Taxonomy. Bambusoideae; Bambusodae; Bambuseae.

Ecology, geography, regional floristic distribution. 2 species. China. Holarctic. Boreal. Eastern Asian.

Special comments. Clayton and Renvoize (1986) and Soderstrom and Ellis (1987) propose very different generic interpretations of the species in this circle of affinity, but there are no available descriptions adequate for the present purpose. Soderstrom and Ellis (1987) propose including *Ampelocalamus* Chen, Wen and Shang, *Chimonocalamus* Hsueh and Yi, *Drepanostachyum* Keng f. and *Sinarundinaria* Nakai, and speak elsewhere (1988) of 'fargesioid' *Arundinaria* species. Clayton and Renvoize include the first three in *Sinarundinaria*, and include *Fargesia* in *Thamnocalamus*. To further complicate matters, *Sinarundinaria* is sometimes presented as a synonym of *Phyllostachys* Sieb. and Zucc. Fruit data wanting. Anatomical data wanting.

Farrago W. Clayton

Habit, vegetative morphology. Wiry annual; caespitose. *Culms* 25–50 cm high; herbaceous; to 0.07 cm in diameter; unbranched above. Culm nodes glabrous. Plants unarmed. *Leaves* not basally aggregated; non-auriculate. Leaf blades linear (acuminate); narrow; about 1.5 mm wide (and 3–8 cm long, glabrous); setaceous, or not setaceous; without abaxial multicellular glands; without cross venation; persistent. *Ligule a fringe of hairs*. *Contra-ligule* absent.

Reproductive organization. Plants bisexual, with bisexual spikelets; with hermaphrodite florets. *The spikelets* of sexually distinct forms on the same plant; *hermaphrodite and sterile (the basal spikelet of the triplet reduced to a long-aristate glume, the middle one fertile, the upper reduced to a linear, long-aristate glume)*.

Inflorescence. Inflorescence a false spike, with spikelets on contracted axes (a 'pseudo-raceme'); espatheate; not comprising 'partial inflorescences' and foliar organs. *Spikelet-bearing axes* very much reduced (to glomerules); disarticulating; falling entire (the glomerules falling). **Spikelets** *with 'involucres' of 'bristles' (these constituted by the basal sterile spikelets)*. The 'bristles' deciduous with the spikelets (?). *Spikelets* in triplets; secund (the glomerules in two ranks on one side); sessile and pedicellate; *consistently in 'long-and-short' combinations*; in pedicellate/sessile combinations (the sessile, fertile spikelet accompanied by two bristle-like sterile spikelets, the one sessile, the other pedicellate).

Female-fertile spikelets. *Spikelets* unconventional (the present description and interpretation after Clayton (1967)); 2.5–3.5 mm long; compressed dorsiventrally; falling with the glumes. Rachilla terminated by a female-fertile floret. Hairy callus present (at the bases of the glomerule and of each spikelet).

Glumes two; more or less equal; exceeding the spikelets; long relative to the adjacent lemmas (much exceeding them); (the lower) awned (with an 8 mm scabrid awn); carinate; very dissimilar (membranous — G_1 asymmetric, ciliate-keeled, long-aristate, G_2 symmetrical, scabridulous, shortly aristulate). Lower glume 1 nerved. Upper glume 1 nerved. *Spikelets* with female-fertile florets only.

Female-fertile florets 1. Lemmas only about 2 mm long; less firm than the glumes (hyaline); not becoming indurated; entire; blunt; awnless; hairy (dorsally pilose above); 0 nerved. *Palea* present; conspicuous but relatively short (about half the lemma length); entire; awnless, without apical setae; not indurated; nerveless; keel-less. *Lodicules* present; 2; free; glabrous; not or scarcely vascularized. *Stamens* 3. Anthers not penicillate; without an apically prolonged connective. *Ovary* glabrous. Styles basally fused. Stigmas 2.

Fruit, embryo and seedling. Fruit small (1.6 mm long); ellipsoid; compressed dorsiventrally. Hilum short (?). Pericarp fused. Embryo large (?); with an epiblast; with a scutellar tail; with an elongated mesocotyl internode; with one scutellum bundle. Embryonic leaf margins meeting.

Abaxial leaf blade epidermis. *Costal/intercostal zonation* conspicuous. *Papillae* absent. *Microhairs* present; elongated; clearly two-celled; chloridoid-type. Microhair apical cell wall of similar thickness/rigidity to that of the basal cell. Microhair basal cells 12 microns long. Microhair total length/width at septum 3. Microhair apical cell/total length ratio 0.5. *Stomata* common. Subsidiaries triangular. Intercostal silica bodies absent. *Costal short-cells* conspicuously in long rows. Costal silica bodies present in alternate cell files of the costal zones; 'panicoid-type'; mostly dumb-bell shaped.

Transverse section of leaf blade, physiology. C_4; XyMS+. PCR sheaths of the primary vascular bundles interrupted; interrupted abaxially only. PCR sheath extensions absent. *Mesophyll* with radiate chlorenchyma. *Leaf blade* with distinct, prominent adaxial ribs. *Midrib* conspicuous; with one bundle only. Bulliforms present in discrete, regular adaxial groups. All the vascular bundles accompanied by sclerenchyma. Sclerenchyma all associated with vascular bundles. The lamina margins without fibres.

Taxonomy. Chloridoideae; main chloridoid assemblage.

Ecology, geography, regional floristic distribution. 1 species. Tanzania. Xerophytic; species of open habitats. Rock crevices.

Paleotropical. African. Sudano-Angolan. South Tropical African.

References, etc. Morphological/taxonomic: Clayton 1968. Leaf anatomical: Van den Borre 1994.

Festuca L.

Amphigenes Janka, *Anatherum* Nabelek, *Argillochloa* Weber, *Bucetum* Parnell, *Drymochloa* Holub, *Drymonaetes* Fourr., *Festucaria* Fabric., *Gnomonia* Lunell, *Gramen* Krause, *Hellerochloa* Rauschert, *Leiopoa* Ohwi, *Lojaconoa* Gand., *Nabelekia* Roshev., *Wasatchia* M. E. Jones

Excluding *Austrofestuca, Helleria, Leucopoa, Parafestuca, Pseudobromus, Tsvelevia*

Habit, vegetative morphology. Perennial; rhizomatous, or stoloniferous, or caespitose, or decumbent. **Culms** 2–200 cm high; *herbaceous*; unbranched above; tuberous, or not tuberous. Culm nodes glabrous. Culm internodes solid, or hollow. Young shoots extravaginal, or intravaginal. *Leaves* mostly basal, or not basally aggregated; auriculate, or non-auriculate. Sheath margins joined, or free. Leaf blades linear to linear-lanceolate; narrow; 0.2–15 mm wide; setaceous, or not setaceous; flat, or folded, or rolled (convolute or involute); without cross venation; persistent; rolled in bud, or once-folded in bud. *Ligule an unfringed membrane (sometimes ciliolate)*; truncate; 0.1–1.5(–5.5) mm long (usually less than 1 mm).

Reproductive organization. Plants bisexual, with bisexual spikelets (*Leucopoa* being excluded); with hermaphrodite florets. The spikelets all alike in sexuality. Plants outbreeding and inbreeding (by cleistogamy). Exposed-cleistogamous, or chasmogamous. Viviparous (sometimes), or not viviparous.

Inflorescence. *Inflorescence paniculate*; open (usually), or contracted (rarely); when contracted *spicate, or more or less irregular*; with capillary branchlets, or without capillary branchlets; espatheate; not comprising 'partial inflorescences' and foliar organs. Spikelet-bearing axes persistent. *Spikelets not secund*; pedicellate.

Female-fertile spikelets. *Spikelets* 3–20 mm long; *compressed laterally*; disarticulating above the glumes; disarticulating between the florets. Rachilla prolonged beyond the uppermost female-fertile floret; rachilla hairy, or hairless. The rachilla extension with incomplete florets. *Hairy callus absent*.

Glumes two; very unequal; shorter than the spikelets; *shorter than the adjacent lemmas*; pointed; awnless; carinate to non-carinate; similar (usually narrow to ovate-lanceolate). *Lower glume 1–3 nerved*. Upper glume (1–)3–5 nerved. *Spikelets* with incomplete florets. The incomplete florets distal to the female-fertile florets. *The distal incomplete florets* merely underdeveloped.

Female-fertile florets *2–14 (rarely 1)*. Lemmas similar in texture to the glumes to decidedly firmer than the glumes; not becoming indurated; entire, or incised; when entire pointed, or blunt; when incised, not deeply cleft; awnless, or mucronate, or awned. Awns when present, 1; from a sinus, or apical; non-geniculate; much shorter than the body of the lemma (usually), or about as long as the body of the lemma (sometimes, rarely somewhat longer); entered by one vein. Lemmas hairy (rarely), or hairless; non-carinate;

3–7 nerved. **Palea** present; *relatively long*; tightly clasped by the lemma; apically notched; awnless, without apical setae; *textured like the lemma*; not indurated (submembranous); 2-nerved; 2-keeled. *Lodicules* present; 2; free; membranous; ciliate, or glabrous; toothed. *Stamens* 3. Anthers 0.4–6 mm long; not penicillate. **Ovary** glabrous, or hairy; *without a conspicuous apical appendage*. Styles free to their bases. Stigmas 2; white.

Fruit, embryo and seedling. Fruit adhering to lemma and/or palea, or free from both lemma and palea; small, or medium sized, or large; fusiform, or ellipsoid; *longitudinally grooved*; compressed dorsiventrally; with hairs confined to a terminal tuft. *Hilum long-linear (usually about as long as the grain, but sometimes elliptical and only half as long)*. Embryo small; not waisted. Endosperm hard; without lipid; containing compound starch grains. Embryo with an epiblast; without a scutellar tail; with a negligible meso-cotyl internode. Embryonic leaf margins meeting.

Seedling with a long mesocotyl; with a loose coleoptile, or with a tight coleoptile. First seedling leaf with a well-developed lamina. The lamina narrow; erect; 3–5 veined.

Abaxial leaf blade epidermis. *Costal/intercostal zonation* conspicuous. *Papillae* absent. *Long-cells* similar in shape costally and intercostally, or markedly different in shape costally and intercostally; of similar wall thickness costally and intercostally. Mid-intercostal long-cells rectangular, or fusiform; having markedly sinuous walls (rarely almost straight). *Microhairs* absent. *Stomata* absent or very rare, or common; when stomata present, 39–51 microns long. Subsidiaries low dome-shaped, or parallel-sided, or parallel-sided and dome-shaped. Guard-cells overlapped by the interstomatals, or over-lapping to flush with the interstomatals. *Intercostal short-cells* common; in cork/silica-cell pairs; silicified. Intercostal silica bodies rounded, or tall-and-narrow, or crescentic. *Costal short-cells* predominantly paired, or neither distinctly grouped into long rows nor predominantly paired. Costal silica bodies rounded (commonly), or tall-and-narrow, or crescentic, or horizontally-elongated crenate/sinuous (occasionally), or horizontally-elon-gated smooth (occasionally).

Transverse section of leaf blade, physiology. C_3; XyMS+. *Mesophyll* with non-radiate chlorenchyma. *Leaf blade* with distinct, prominent adaxial ribs, or adaxially flat; with the ribs more or less constant in size to with the ribs very irregular in sizes. *Midrib* conspicuous, or not readily distinguishable; with one bundle only. Bulliforms present in discrete, regular adaxial groups (usually, in the furrows), or not in discrete, regular adaxial groups; usually in simple fans. Many of the smallest vascular bundles unaccompanied by sclerenchyma, or all the vascular bundles accompanied by sclerenchyma. Combined sclerenchyma girders present, or absent; forming 'figures', or nowhere forming 'figures'. Sclerenchyma all associated with vascular bundles, or not all bundle-associated. The 'extra' sclerenchyma when present, in abaxial groups, or in a continuous abaxial layer (often).

Culm anatomy. *Culm internode bundles* in one or two rings.

Phytochemistry. Tissues of the culm bases with little or no starch. Fructosans pre-dominantly short-chain. Leaves without flavonoid sulphates (4 species).

Cytology. Chromosome base number, $x = 7$. $2n = 14, 28, 35, 42, 56,$ and 70. 2, 4, 5, 6, 8, and 10 ploid. Haploid nuclear DNA content (1.5–)1.8–3.6 pg (13 species, mean 2.6 — the lowest values representing high polyploids). Mean diploid 2c DNA value 6.5 pg (9 species, 3.4–9.5).

Taxonomy. Pooideae; Poodae; Poeae.

Ecology, geography, regional floristic distribution. 360 species (or more). Worldwide temperate & mountains. Commonly adventive. Helophytic (rarely), or meso-phytic (mostly), or xerophytic (rarely); halophytic, or glycophytic. Hillsides, mountains, plains, meadows.

Holarctic, Paleotropical, Neotropical, Cape, Australian, and Antarctic. Boreal, Tethyan, and Madrean. African, Madagascan, Indomalesian, and Polynesian. Arctic and Subarctic, Euro-Siberian, Eastern Asian, Atlantic North American, and Rocky Mountains. Macaronesian, Mediterranean, and Irano-Turanian. Saharo-Sindian, Sudano-Angolan, and West African Rainforest. Indian, Indo-Chinese, Malesian, and Papuan. Hawaiian. Caribbean, Central Brazilian, Pampas, and Andean. North and East Australian. New Zealand and Patagonian. European and Siberian. Canadian-Appalachian, Southern

Atlantic North American, and Central Grasslands. Sahelo-Sudanian, Somalo-Ethiopian, and South Tropical African. Temperate and South-Eastern Australian.

Hybrids. Intergeneric hybrids with *Vulpia* (×*Festulpia* Melderis ex Stace & R. Cotton), with *Lolium* (×*Festulolium* Aschers. & Graebn.) and supposedly with *Bromus* (×*Bromofestuca* Prodan. — *Bull. Grad. Bot. Univ. Cluj* **16**, 93 (1936)).

Rusts and smuts. Rusts — *Puccinia*. Taxonomically wide-ranging species: *Puccinia graminis*, *Puccinia coronata*, *Puccinia striiformis*, *Puccinia pygmaea*, *Puccinia brachypodii*, *Puccinia poarum*, *Puccinia recondita*, 'Uromyces' turcomanicum, 'Uromyces' dactylidis, and 'Uromyces' hordeinus. Smuts from Tilletiaceae and from Ustilaginaceae. Tilletiaceae — *Entyloma*, *Tilletia*, and *Urocystis*. Ustilaginaceae — *Ustilago*.

Economic importance. Significant weed species: *F. arundinacea, F. ovina, F. rubra, F. tenuifolia.* Cultivated fodder: *F. arundinacea, F. pratensis, F. rubra* etc. (Fescues). Lawns and/or playing fields: several fine-leaved species — *F. rubra, F. tenuifolia, F. ovina* etc.

References, etc. Morphological/taxonomic: Vickery 1939; Alekseev 1984. Leaf anatomical: Metcalfe 1960; this project; Aiken *et al.* 1985.

Festucella E. Alekseev

~ *Austrofestuca (Festuca) eriopoda*

Habit, vegetative morphology. Perennial; densely caespitose. **Culms** 50–160 cm high; *herbaceous*; unbranched above; 2–3 noded. Culm nodes exposed; glabrous. Culm internodes hollow. Young shoots intravaginal. *Leaves* mostly basal; non-auriculate. Sheath margins joined (in the basal leaves), or free (upper leaves). Sheaths terete. *Leaf blades* linear; narrow; *0.5–0.7 mm wide*; tightly folded, or rolled to acicular; without cross venation. *Ligule an unfringed membrane (ciliolate)*; 2–3 mm long.

Reproductive organization. Plants bisexual, with bisexual spikelets; with hermaphrodite florets.

Inflorescence. *Inflorescence paniculate*; open; espatheate; not comprising 'partial inflorescences' and foliar organs. Spikelet-bearing axes persistent. Spikelets not secund; pedicellate.

Female-fertile spikelets. Spikelets 6.8–12 mm long; compressed laterally; disarticulating above the glumes; disarticulating between the florets. Rachilla prolonged beyond the uppermost female-fertile floret. The rachilla extension with incomplete florets. Hairy callus present. *Callus* short; blunt (villous).

Glumes two; more or less equal; shorter than the spikelets; shorter than the adjacent lemmas; pointed; awnless; carinate to non-carinate; similar (lanceolate, membranous with thin margins). *Lower glume 3 nerved.* Upper glume 3 nerved, or 5 nerved. *Spikelets* with female-fertile florets only, or with incomplete florets. The incomplete florets distal to the female-fertile florets. *The distal incomplete florets* merely underdeveloped.

Female-fertile florets 3–5. *Lemmas* similar in texture to the glumes; not becoming indurated; more or less incised; 2 lobed; not deeply cleft (minutely bidenticulate or erose); *awned. Awns* 1; median; *from a sinus*; non-geniculate; hairless (scabrous); much shorter than the body of the lemma (0.5–2 mm long). *Lemmas* hairless; scabrous (on the keel and below); *carinate*; 5 nerved; with the nerves non-confluent. *Palea* present; relatively long; entire to apically notched; awnless, without apical setae; textured like the lemma; not indurated; 2-nerved; 2-keeled. Palea back glabrous to hairy. Palea keels wingless; scabrous to hairy. *Lodicules* present; 2; free; membranous; sparsely ciliate, or glabrous; toothed; not or scarcely vascularized. *Stamens* 3. Anthers 3.6–5.5 mm long. *Ovary* glabrous. Styles free to their bases. Stigmas 2; white.

Fruit, embryo and seedling. *Fruit* free from both lemma and palea; medium sized (3.8–4.5 mm long); fusiform, or ellipsoid; ventrally longitudinally grooved; slightly compressed dorsiventrally. *Hilum short (oval)*. Embryo small; not waisted.

Abaxial leaf blade epidermis. *Costal/intercostal zonation* conspicuous. *Papillae* absent. *Long-cells* markedly different in shape costally and intercostally (the costals much narrower, longer, more rectangular). Mid-intercostal long-cells fusiform; having markedly sinuous walls. *Microhairs* absent. *Stomata* absent or very rare. *Intercostal short-cells* absent or very rare. Abundant costal prickles present, with pitted bases. *Crown cells* absent. *Costal short-cells* predominantly paired (and a few short rows). Costal silica bodies horizontally-elongated smooth (almost exclusively, some quite short), or horizontally-elongated crenate/sinuous (very few).

Transverse section of leaf blade, physiology. C$_3$; XyMS+. *Mesophyll* with non-radiate chlorenchyma; without adaxial palisade. *Midrib* conspicuous; with one bundle only. Bulliforms not in discrete, regular adaxial groups (confined to the conspicuous 'midrib hinges' in the near acicular blade). All the vascular bundles accompanied by sclerenchyma. Combined sclerenchyma girders absent (abaxial strands only, with all the bundles). Sclerenchyma all associated with vascular bundles.

Taxonomy. Pooideae; Poodae; Poeae.

Ecology, geography, regional floristic distribution. 1 species. Eastern Australia. Helophytic, or mesophytic (?); glycophytic.

Australian. North and East Australian.

References, etc. Morphological/taxonomic: Alekseev 1985, Jacobs 1990. Leaf anatomical: this project.

Festucopsis (C.E. Hubb.) Melderis

~ *Elymus*

Excluding *Peridictyon*

Habit, vegetative morphology. Perennial; caespitose. *Culms* 25–60 cm high; herbaceous; 2 noded, or 2–3 noded. Culm nodes glabrous. Young shoots intravaginal. Leaves non-auriculate. Sheath margins joined (innovation leaves), or free (culm leaves). *The connate basal sheaths disintegrating into more or less discrete fibres. Leaf blades narrow*; 0.4–1 mm wide; filiform or *setaceous*; folded, or rolled (involute); without cross venation. Ligule an unfringed membrane (but ciliolate); truncate; very short, about 0.1–0.5 mm long.

Reproductive organization. Plants bisexual, with bisexual spikelets; with hermaphrodite florets.

Inflorescence. *Inflorescence* rather few spikeleted (about 8–12); *a single spike (5–15 cm long).* Rachides hollowed (concave against the spikelets). Inflorescence espatheate; not comprising 'partial inflorescences' and foliar organs. Spikelet-bearing axes persistent. Spikelets solitary; not secund; sessile.

Female-fertile spikelets. Spikelets 10–25(–26) mm long; compressed laterally to not noticeably compressed; disarticulating above the glumes; disarticulating between the florets. Rachilla prolonged beyond the uppermost female-fertile floret. Hairy callus absent. *Callus* very short; blunt (to truncate).

Glumes two; more or less equal; shorter than the spikelets; shorter than the adjacent lemmas; free; dorsiventral to the rachis; pointed (acute); not subulate; awnless; non-carinate; *similar (somewhat asymmetrical, leathery, with narrow membranous margins).* Lower glume 1–3 nerved, or 3–4 nerved. Upper glume 3–5(–6) nerved. *Spikelets* with female-fertile florets only, or with incomplete florets. The incomplete florets distal to the female-fertile florets.

Female-fertile florets (2–)3–8(–12). Lemmas similar in texture to the glumes (leathery); not becoming indurated; entire; pointed; awnless (*F. festucoides*), or awned (at least the lower ones, in *F. serpentini*). Awns when present, 1; median; apical; non-geniculate; much shorter than the body of the lemma to about as long as the body of the lemma. Lemmas hairy (puberulent basally), or hairless; non-carinate; without a germination flap; 3–5 nerved. *Palea* present; conspicuous but relatively short (up to two thirds the lemma length), or very reduced; apically notched; 2-nerved; 2-keeled. Palea keels hairy (stiffly ciliate above). *Lodicules* present; 2; free; membranous; ciliate; not toothed; not or scarcely vascularized. *Stamens* 3. Anthers (1.6–)3–5.5 mm long. **Ovary *hairy.*** Styles free to their bases. Stigmas 2.

Fruit, embryo and seedling. Fruit medium sized (4–6 mm long); with hairs confined to a terminal tuft. Hilum long-linear. Embryo small. Endosperm containing only simple starch grains.

Transverse section of leaf blade, physiology. Probably C_3. Mesophyll not traversed by colourless columns; without fusoids. *Leaf blade* with distinct, prominent adaxial ribs, or 'nodular' in section. Midrib with one bundle only; without colourless mesophyll adaxially. All the vascular bundles accompanied by sclerenchyma. Combined sclerenchyma girders present; forming 'figures' (I's and anchors). Sclerenchyma all associated with vascular bundles.

Cytology. Chromosome base number, $x = 7$. $2n = 14$. 2 ploid. Haplomic genome content G. Mean diploid 2c DNA value unknown, but chromosomes 'large'.

Taxonomy. Pooideae; Triticodae; Triticeae.

Ecology, geography, regional floristic distribution. 2 species (*F. serpentini* (C.E. Hubb.) Melderis, and the poorly known *F. festucoides* (Maire) Löve). Albania, High Atlas of Morocco. Species of open habitats; glycophytic. Rocky or stony slopes or gullies.

Holarctic. Boreal and Tethyan. Euro-Siberian. Mediterranean. European.

References, etc. Morphological/taxonomic: Löve 1984, Seberg *et al.* 1991. Leaf anatomical: scanty – Seberg *et al.* 1991.

Special comments. Description scarcely adequate, e.g. lacking information on spikelet orientation. Anatomical data wanting.

Fingerhuthia Nees

Lasiotrichos Lehm.

Habit, vegetative morphology. Perennial (usually), or annual (rarely, in desert areas); caespitose. **Culms** 5–117 cm high; herbaceous; ***unbranched above***. Culm nodes glabrous. Culm internodes hollow. Plants unarmed. Young shoots intravaginal. *Leaves* mostly basal; non-auriculate. ***Sheath margins free***. Leaf blades long linear; narrow; 2–5 mm wide; flat, or folded; without abaxial multicellular glands; without cross venation; persistent. *Ligule a fringe of hairs*; 0.1–0.7 mm long (and the base of the blade sometimes hairy above the ligule).

Reproductive organization. Plants bisexual, with bisexual spikelets; with hermaphrodite florets. The spikelets of sexually distinct forms on the same plant, or all alike in sexuality; hermaphrodite, or hermaphrodite and sterile (the lowest sometimes barren). Apomictic (peculiarly so), or reproducing sexually (?).

Inflorescence. Inflorescence a single raceme, or paniculate (to 12 cm long); densely contracted; more or less ovoid to spicate; espatheate; not comprising 'partial inflorescences' and foliar organs. Spikelet-bearing axes persistent. *Spikelets* solitary; *not secund*; pedicellate; imbricate.

Female-fertile spikelets. *Spikelets* 4–7 mm long; strongly ***compressed laterally***; *falling with the glumes (disarticulating from persistent pedicels)*; not disarticulating between the florets; with conventional internode spacings. ***Rachilla prolonged beyond the uppermost female-fertile floret***; rachilla tough. The rachilla extension with incomplete florets. Callus of spikelet blunt.

Glumes two; more or less equal; exceeding the spikelets; long relative to the adjacent lemmas; hairy (with straight hairs on the keels and margins); pointed; awned, or awnless (shortly awned or mucronate); carinate (ciliate on keel); similar (narrow, folded, thin). Lower glume 1 nerved. Upper glume 1 nerved. *Spikelets* with incomplete florets. The incomplete florets distal to the female-fertile florets. The distal incomplete florets 1–3; *incomplete florets* merely underdeveloped (male or rudimentary).

Female-fertile florets 1. *Lemmas* similar in texture to the glumes to decidedly firmer than the glumes (rather firmly membranous); not becoming indurated; entire; pointed, or blunt; *mucronate (via the excurrent median nerve)*; hairy (long-ciliate on the keels and margins), or hairless (*F. sesleriformis*); carinate; without a germination flap; (3–)5 nerved, or 7 nerved; with the nerves confluent towards the tip. *Palea* present; relatively long; apically notched; awnless, without apical setae; textured like the lemma; basally indurated to not indurated; 2-nerved; 2-keeled. Palea keels wingless. *Lodicules* present;

2; free; fleshy; glabrous. *Stamens* 3. Anthers 2–2.5 mm long; without an apically pro-longed connective. *Ovary* glabrous. Styles free to their bases. Stigmas 2; white.

Fruit, embryo and seedling. *Fruit* free from both lemma and palea; ellipsoid; somewhat compressed laterally. Embryo with an epiblast; with a scutellar tail; with an elongated mesocotyl internode. Embryonic leaf margins overlapping.

Abaxial leaf blade epidermis. *Costal/intercostal zonation* conspicuous. *Papillae* absent. *Long-cells* similar in shape costally and intercostally (but the costals smaller); of similar wall thickness costally and intercostally (quite thick walled). Mid-intercostal long-cells rectangular; having markedly sinuous walls. *Microhairs* present; elongated; clearly two-celled; chloridoid-type to *Enneapogon*-type (suggestive of a small version of the latter). Microhair apical cell wall thinner than that of the basal cell but not tending to collapse to of similar thickness/rigidity to that of the basal cell. Microhairs 52.5–67 microns long. Microhair basal cells 33–36 microns long. Microhairs 9–10.5 microns wide at the septum. Microhair total length/width at septum 5.1–6.3. Microhair apical cells 18–24 microns long. Microhair apical cell/total length ratio 0.34–0.44. *Stomata* common; 24–27 microns long. Subsidiaries triangular. Guard-cells overlapped by the interstomatals (very slightly). *Intercostal short-cells* common; in cork/silica-cell pairs (and some solitary); silicified. Intercostal silica bodies imperfectly developed; rounded, or saddle shaped, or crescentic, or cubical. *Costal short-cells* predominantly paired. Costal silica bodies present throughout the costal zones; saddle shaped (a round version, predominating everywhere), or rounded (a few, merging with the saddles), or crescentic (a few, plus a few more or less cubical).

Transverse section of leaf blade, physiology. C_4; XyMS+. PCR sheath outlines even. PCR sheaths of the primary vascular bundles interrupted; interrupted both abaxially and adaxially. PCR sheath extensions absent. PCR cell chloroplasts centripetal. *Mesophyll* with radiate chlorenchyma; traversed by columns of colourless mesophyll cells. *Leaf blade* 'nodular' in section to adaxially flat; with the ribs more or less constant in size (somewhat larger over large bundles). *Midrib* not readily distinguishable; with one bundle only. Bulliforms present in discrete, regular lateral groups; associated with colourless mesophyll cells to form deeply-penetrating fans (these incorporated in the traversing columns of colourless cells). All the vascular bundles accompanied by sclerenchyma. Combined sclerenchyma girders present (with the primaries); forming 'figures' (anchors, but only in the primaries: the smaller bundles with strands only). Sclerenchyma all associated with vascular bundles. The lamina margins with fibres (large).

Cytology. Chromosome base number, $x = 10$. $2n = 20$ and 40. 2 and 4 ploid.

Taxonomy. Chloridoideae; main chloridoid assemblage.

Ecology, geography, regional floristic distribution. 2 species. Southern Africa, Afghanistan and Arabia. Helophytic, or mesophytic, or xerophytic; species of open habitats; glycophytic.

Holarctic, Paleotropical, and Cape. Tethyan. African. Irano-Turanian. Saharo-Sindian and Sudano-Angolan. South Tropical African and Kalaharian.

Rusts and smuts. Smuts from Ustilaginaceae. Ustilaginaceae — *Ustilago*.

References, etc. Leaf anatomical: Metcalfe 1960; this project.

Special comments. Fruit data wanting.

Froesiochloa G.A. Black

Habit, vegetative morphology. Perennial. The flowering culms leafy. Culms herbaceous. **Leaves** not basally aggregated; *without auricular setae*. Leaf blades lanceolate to ovate (?); pseudopetiolate; rolled in bud. Ligule an unfringed membrane.

Reproductive organization. *Plants monoecious with all the fertile spikelets unisexual*; without hermaphrodite florets. The spikelets of sexually distinct forms on the same plant; female-only and male-only. *The male and female-fertile spikelets mixed in the inflorescence (a central female surrounded by 6–10 males, in each cluster)*. The spikelets overtly heteromorphic (the female spikelets lanceolate).

Inflorescence. Inflorescence without pseudospikelets; a false spike, with spikelets on contracted axes (of 2–4 deciduous clusters on a central axis). Spikelet-bearing axes disarticulating; falling entire (the clusters falling).

Female-sterile spikelets. Male spikelets 1 flowered, lemma membranous and 3 nerved, lodicules 3, 6 stamens with joined filaments. Rachilla of male spikelets terminated by a male floret. The male spikelets without glumes; without proximal incomplete florets; 1 floreted. The lemmas awnless. Male florets 1; 6 staminate. *The staminal filaments joined*.

Female-fertile spikelets. Spikelets lanceolate; compressed dorsiventrally; falling with the glumes (i.e. with the clusters). Rachilla terminated by a female-fertile floret.

Glumes two; about equalling the spikelets to exceeding the spikelets; long relative to the adjacent lemmas; pointed (acuminate, herbaceous); awnless; non-carinate; similar. Lower glume 3–7 nerved. Upper glume 3–7 nerved. *Spikelets* with female-fertile florets only.

Female-fertile florets 1. Lemmas decidedly firmer than the glumes (leathery); entire; pointed ('subacute'); mucronate; non-carinate; 5–7 nerved. *Palea* present; awnless, without apical setae; 2-nerved, or several nerved (?). *Lodicules* present; 3. *Stamens* 0. *Ovary* glabrous.

Fruit, embryo and seedling. Disseminule consisting of the disarticulated spikelet-bearing inflorescence unit.

Taxonomy. Bambusoideae; Oryzodae; Olyreae.

Ecology, geography, regional floristic distribution. 2 species. Brazil, Guiana. Mesophytic; shade species. In forests.

Neotropical. Caribbean, Venezuela and Surinam, Amazon, and Central Brazilian.

References, etc. Morphological/taxonomic: G.A. Black (1950: not yet seen — data here from Clayton and Renvoize 1986).

Special comments. Fruit data wanting. Anatomical data wanting.

Garnotia Brongn.

Berghausia Endl., *Miquelia* Arn. & Nees

Habit, vegetative morphology. Annual (rarely), or perennial; rhizomatous, or caespitose (mostly), or decumbent. Culms unbranched above. Culm nodes hairy (usually), or glabrous. Culm internodes solid, or hollow. *Leaves* mostly basal, or not basally aggregated; non-auriculate. Leaf blades narrow; without cross venation; disarticulating from the sheaths, or persistent. **Ligule** present; *a fringed membrane*; short.

Reproductive organization. Plants bisexual, with bisexual spikelets; with hermaphrodite florets. Exposed-cleistogamous, or chasmogamous. Plants with hidden cleistogenes, or without hidden cleistogenes. The hidden cleistogenes (when present) in the leaf sheaths (in concealed, contracted panicles).

Inflorescence. *Inflorescence of spicate main branches, or paniculate*; open, or contracted; espatheate; not comprising 'partial inflorescences' and foliar organs. Spikelet-bearing axes persistent. Spikelets solitary (rarely), or in triplets (rarely), or paired; not secund; pedicellate; consistently in 'long-and-short' combinations, or not in distinct 'long-and-short' combinations; unequally pedicellate in each combination. The 'shorter' spikelets hermaphrodite. The 'longer' spikelets hermaphrodite.

Female-fertile spikelets. *Spikelets* 2–5 mm long (?); *not noticeably compressed to compressed dorsiventrally; falling with the glumes*. Rachilla terminated by a female-fertile floret. Hairy callus present, or absent.

Glumes two; relatively large (membranous); more or less equal (the G_1 usually slightly longer than the G_2); about equalling the spikelets; long relative to the adjacent lemmas; awned, or awnless; *very dissimilar (the G_1 more or less convex, the G_2 flattened and often furrowed on the back)*. Lower glume convex on the back; 3–5 nerved. Upper glume 3–5 nerved. **Spikelets** with female-fertile florets only; *without proximal incomplete florets*.

Female-fertile florets 1. **Lemmas decidedly firmer than the glumes (cartilaginous)**; not becoming indurated; entire, or incised; when incised, 2 lobed; when incised, not deeply cleft (shortly bidentate or bisetaceous); awnless (rarely), or mucronate (rarely), or awned (usually). Awns when present, 1; from a sinus, or apical; non-geniculate, or geniculate; much shorter than the body of the lemma to much longer than the body of the lemma. Lemmas hairless; glabrous; non-carinate (convex); without a germination flap; 3–5 nerved. **Palea** present; *relatively long (with ciliate margins, and auricles near the base)*; 2-nerved. *Lodicules* present; 2; free; fleshy; ciliate, or glabrous. **Stamens 3**. Anthers not penicillate. *Ovary* glabrous. Styles free to their bases. Stigmas 2.

Fruit, embryo and seedling. *Hilum short (round)*. Embryo large, or small.

Abaxial leaf blade epidermis. *Costal/intercostal zonation* conspicuous. *Papillae* absent. *Long-cells* similar in shape costally and intercostally, or markedly different in shape costally and intercostally (with the costals longer and narrower); of similar wall thickness costally and intercostally. Mid-intercostal long-cells rectangular (narrow); having markedly sinuous walls. *Microhairs* present; panicoid-type; 48–98 microns long; 9–11.4 microns wide at the septum. Microhair total length/width at septum 7.4–10.2. Microhair apical cells 24–48 microns long. Microhair apical cell/total length ratio 0.4–0.52. *Stomata* common; 40–42 microns long. Subsidiaries triangular. Guard-cells overlapping to flush with the interstomatals. *Intercostal short-cells* common; in cork/silica-cell pairs; silicified. *Costal short-cells* conspicuously in long rows, or predominantly paired, or neither distinctly grouped into long rows nor predominantly paired. Costal silica bodies 'panicoid-type' (mostly), or crescentic, or saddle shaped (rarely); when panicoid, cross shaped to dumb-bell shaped.

Transverse section of leaf blade, physiology. C_4. The anatomical organization conventional, or unconventional. Organization of PCR tissue when unconventional, *Arundinella* type. XyMS–. PCR sheath outlines uneven. PCR sheath extensions absent. *Mesophyll* with non-radiate chlorenchyma; exhibiting 'circular cells', or without 'circular cells'. *Leaf blade* with distinct, prominent adaxial ribs, or adaxially flat; with the ribs more or less constant in size. *Midrib* conspicuous; having a conventional arc of bundles; with colourless mesophyll adaxially, or without colourless mesophyll adaxially. Bulliforms present in discrete, regular adaxial groups; in simple fans. Many of the smallest vascular bundles unaccompanied by sclerenchyma. Combined sclerenchyma girders present; forming 'figures', or nowhere forming 'figures'. Sclerenchyma all associated with vascular bundles.

Phytochemistry. Leaves without flavonoid sulphates (1 species).

Cytology. $2n = 20$.

Taxonomy. Panicoideae; Panicodae; Arundinelleae.

Ecology, geography, regional floristic distribution. 30 species. Eastern Asia, Northeast Australia, Pacific. Mesophytic. Light woodland and moist rocky slopes.

Paleotropical and Australian. Indomalesian and Polynesian. Indian, Indo-Chinese, Malesian, and Papuan. Hawaiian and Fijian. North and East Australian. Tropical North and East Australian.

References, etc. Morphological/taxonomic: Gould 1973. Leaf anatomical: Metcalfe 1960; this project.

Gastridium P. Beauv.

Habit, vegetative morphology. Annual; caespitose (or the culms solitary). *Culms* 10–60 cm high; herbaceous. Culm nodes glabrous. Culm internodes hollow. *Leaves* not basally aggregated; non-auriculate. Leaf blades linear; narrow; 1.5–4 mm wide; flat; without cross venation; persistent. Ligule an unfringed membrane; not truncate (veined); 4–6 mm long.

Reproductive organization. Plants bisexual, with bisexual spikelets; with hermaphrodite florets.

Inflorescence. Inflorescence paniculate; contracted; spicate; espatheate; not comprising 'partial inflorescences' and foliar organs. Spikelet-bearing axes persistent. Spikelets not secund; pedicellate.

Female-fertile spikelets. Spikelets 3–6.5 mm long; compressed laterally; disarticulating above the glumes. *Rachilla* shortly *prolonged beyond the uppermost female-fertile floret (or rarely not prolonged?). The rachilla extension* when present, *naked*. Hairy callus absent. *Callus* short; blunt.

Glumes two; relatively large; very unequal to more or less equal (the lower somewhat longer); exceeding the spikelets; long relative to the adjacent lemmas (the floret enclosed by their swollen bases); *conspicuously ventricose*; pointed (acuminate); awnless; carinate; *similar (swollen, globular and more or less cartilaginous at the base, membranous above)*. Lower glume 1 nerved. Upper glume 1 nerved. *Spikelets* with female-fertile florets only.

Female-fertile florets 1. Lemmas less firm than the glumes; not becoming indurated; incised (more or less dentate); mucronate to awned (the midvein usually excurrent). Awns when present, 1; dorsal; from near the top; geniculate; much shorter than the body of the lemma to much longer than the body of the lemma; entered by one vein. Lemmas hairy, or hairless; non-carinate; 5 nerved. *Palea* present; relatively long; apically notched; awnless, without apical setae; not indurated (hyaline); 2-nerved; 2-keeled. *Lodicules* present; 2; free; membranous; glabrous; not toothed. *Stamens* 3. Anthers 0.5–0.7 mm long; not penicillate. *Ovary* glabrous. Styles free to their bases. Stigmas 2.

Fruit, embryo and seedling. *Fruit* slightly adhering to lemma and/or palea; small; compressed dorsiventrally (slightly). Hilum short. Embryo small; not waisted. Endosperm hard; with lipid. Embryo with an epiblast; without a scutellar tail; with a negligible mesocotyl internode. Embryonic leaf margins meeting.

Seedling with a short mesocotyl. First seedling leaf with a well-developed lamina. The lamina narrow; erect; 3 veined.

Abaxial leaf blade epidermis. *Costal/intercostal zonation* conspicuous. *Papillae* absent. *Long-cells* similar in shape costally and intercostally (the costals shorter); of similar wall thickness costally and intercostally. Mid-intercostal long-cells rectangular; having straight or only gently undulating walls. *Microhairs* absent. *Stomata* common; 36–54 microns long. Subsidiaries parallel-sided. Guard-cells overlapped by the interstomatals. *Intercostal short-cells* absent or very rare; when present, not paired (solitary); silicified. *Costal short-cells* neither distinctly grouped into long rows nor predominantly paired. Costal silica bodies horizontally-elongated crenate/sinuous.

Transverse section of leaf blade, physiology. C_3; XyMS+. *Mesophyll* with non-radiate chlorenchyma; without adaxial palisade. *Leaf blade* with distinct, prominent adaxial ribs, or 'nodular' in section; with the ribs more or less constant in size. *Midrib* not readily distinguishable; with one bundle only. Bulliforms present in discrete, regular adaxial groups (in the furrows); in simple fans. All the vascular bundles accompanied by sclerenchyma. Combined sclerenchyma girders present; forming 'figures'. Sclerenchyma all associated with vascular bundles.

Cytology. Chromosome base number, $x = 7$. $2n = 14$ and 28. 2 ploid.

Taxonomy. Pooideae; Poodae; Aveneae.

Ecology, geography, regional floristic distribution. 2 species. Canaries, western Europe, Mediterranean. Commonly adventive. Mesophytic to xerophytic; species of open habitats. Grassy places and arable land.

Holarctic and Paleotropical. Boreal, Tethyan, and Madrean. African. Euro-Siberian. Macaronesian and Irano-Turanian. Saharo-Sindian and Sudano-Angolan. European. Sahelo-Sudanian and Somalo-Ethiopian.

References, etc. Leaf anatomical: this project.

Gaudinia P. Beauv.

Arthrostachya Link, *Falmiria* Reichenb., *Cyclichnium* Dulac, *Meringurus* Murbeck

Habit, vegetative morphology. Biennial or *annual*; caespitose. *Culms* 15–120 cm high; herbaceous. Culm nodes glabrous. Culm internodes hollow. *Leaves* not basally aggregated; non-auriculate. Leaf blades linear; narrow; 1.5–4 mm wide; flat; without cross venation; persistent. Ligule an unfringed membrane; truncate (minutely serrated); 0.5–0.9 mm long.

Reproductive organization. Plants bisexual, with bisexual spikelets; with hermaphrodite florets.

Inflorescence. *Inflorescence a single spike*. Rachides hollowed. Inflorescence espatheate; not comprising 'partial inflorescences' and foliar organs. Spikelet-bearing axes disarticulating, or persistent (variable); when disarticulating, disarticulating at the joints (above the insertions of the spikelets). *Spikelets solitary (appressed to the concave rachis)*; not secund; distichous; sessile.

Female-fertile spikelets. Spikelets 5.5–20 mm long; compressed laterally; disarticulating above the glumes; disarticulating between the florets. Rachilla prolonged beyond the uppermost female-fertile floret; rachilla hairless. The rachilla extension with incomplete florets. Hairy callus absent. *Callus* short; blunt.

Glumes two; very unequal to more or less equal; shorter than the spikelets to about equalling the spikelets; (the longer) long relative to the adjacent lemmas; lateral to the rachis; pointed; awnless; non-carinate; similar. Lower glume 3–5 nerved. Upper glume 5–11 nerved. *Spikelets* with incomplete florets. The incomplete florets distal to the female-fertile florets. *The distal incomplete florets* merely underdeveloped.

Female-fertile florets *3–11*. Lemmas similar in texture to the glumes (leathery); not becoming indurated; entire to incised (or bicuspid); awned, or awnless (*G. hispanica*). *Awns* 1; median; dorsal; from near the top; *geniculate*; entered by one vein. Lemmas hairy, or hairless; weakly carinate; 5 nerved, or 7 nerved, or 9 nerved. *Palea* present; relatively long; entire to apically notched; awnless, without apical setae; 2-nerved. *Lodicules* present; 2; free; membranous; glabrous; toothed. *Stamens* 3. Anthers 2.5–6 mm long; not penicillate. Ovary *hairy*. Styles free to their bases. Stigmas 2; white.

Fruit, embryo and seedling. Fruit with hairs confined to a terminal tuft. *Hilum short (round)*. Embryo small. Endosperm liquid in the mature fruit; with lipid; containing compound starch grains.

Abaxial leaf blade epidermis. *Costal/intercostal zonation* conspicuous. *Papillae* absent. *Long-cells* similar in shape costally and intercostally; of similar wall thickness costally and intercostally. Mid-intercostal long-cells fusiform; having straight or only gently undulating walls. *Microhairs* absent. *Stomata* absent or very rare (*G. fragilis*), or common; in *G. maroccana*, 28–33 microns long. Subsidiaries parallel-sided. Guard-cells overlapped by the interstomatals. *Intercostal short-cells* absent or very rare; in cork/silica-cell pairs, or not paired; silicified, or not silicified. *Costal short-cells* neither distinctly grouped into long rows nor predominantly paired. Costal silica bodies horizontally-elongated crenate/sinuous, or horizontally-elongated smooth.

Transverse section of leaf blade, physiology. C_3; XyMS+. *Mesophyll* with non-radiate chlorenchyma; without adaxial palisade. *Leaf blade* with distinct, prominent adaxial ribs, or 'nodular' in section; with the ribs more or less constant in size. *Midrib* conspicuous (rarely), or not readily distinguishable; with one bundle only. Bulliforms present in discrete, regular adaxial groups (between the bundles); in simple fans. All the vascular bundles accompanied by sclerenchyma. Combined sclerenchyma girders absent. Sclerenchyma all associated with vascular bundles.

Cytology. Chromosome base number, $x = 7$. $2n = 14$ (and 14+1). 2 ploid.

Taxonomy. Pooideae; Poodae; Aveneae.

Ecology, geography, regional floristic distribution. 4 species. Mediterranean, Azores. Commonly adventive. Mesophytic to xerophytic; species of open habitats; glycophytic. Weedy places.

Holarctic and Paleotropical. Boreal and Tethyan. African. Euro-Siberian. Mediterranean. Saharo-Sindian. European.

Rusts and smuts. Rusts — *Puccinia*. Taxonomically wide-ranging species: *Puccinia striiformis, Puccinia hordei*, and *Puccinia recondita*. Smuts from Tilletiaceae. Tilletiaceae — *Entyloma*.

References, etc. Leaf anatomical: this project.

Gaudiniopsis (Boiss.) Eig

~ *Ventenata*

Habit, vegetative morphology. *Annual*; erect or ascending. *Culms* 10–50 cm high; herbaceous. Leaves non-auriculate. Leaf blades linear; 1–3.5 mm wide; often rolled (convolute); without cross venation. Ligule an unfringed membrane; not truncate (lacerate); 3–6 mm long.

Reproductive organization. Plants bisexual, with bisexual spikelets; with hermaphrodite florets.

Inflorescence. Inflorescence paniculate; contracted (the ends of the branches minutely hispidulous); spicate; espatheate; not comprising 'partial inflorescences' and foliar organs. Spikelet-bearing axes persistent. Spikelets not secund; pedicellate.

Female-fertile spikelets. Spikelets 9–13 mm long; compressed laterally; disarticulating above the glumes; disarticulating between the florets. Rachilla prolonged beyond the uppermost female-fertile floret. Hairy callus present, or absent.

Glumes two; very unequal; shorter than the spikelets; shorter than the adjacent lemmas; free; pointed (acute); awnless; carinate, or non-carinate; similar (lanceolate). *Lower glume 3 nerved*. Upper glume 3–5 nerved. *Spikelets* with female-fertile florets only, or with incomplete florets. The incomplete florets distal to the female-fertile florets. *The distal incomplete florets* merely underdeveloped.

Female-fertile florets 4–10. *Lemmas* similar in texture to the glumes (papery); not becoming indurated; entire, or incised; when entire, pointed; *awned (apart from the lowest, which is acute and awnless)*. Awns 1; median; dorsal; from well down the back; geniculate; much longer than the body of the lemma (about 10 mm long); entered by one vein. Lemmas hairy, or hairless; non-carinate; without a germination flap; 5 nerved. **Palea** present; *conspicuous but relatively short (about two thirds the lemma length)*; 2-nerved; 2-keeled. *Lodicules* present; 2; free; membranous; glabrous; toothed, or not toothed; not or scarcely vascularized. *Stamens* 3. *Ovary* glabrous. Styles free to their bases. Stigmas 2.

Fruit, embryo and seedling. *Fruit* free from both lemma and palea; small (2 mm long); compressed dorsiventrally. **Hilum short**. Embryo small. Endosperm liquid in the mature fruit.

Abaxial leaf blade epidermis. *Costal/intercostal zonation* conspicuous. *Papillae* absent. *Long-cells* similar in shape costally and intercostally; of similar wall thickness costally and intercostally. Mid-intercostal long-cells rectangular; having markedly sinuous walls (pitted and rather thick). *Microhairs* absent. *Stomata* common; 37.5–52 microns long. Subsidiaries parallel-sided. Guard-cells mostly only slightly overlapped by the interstomatals. *Intercostal short-cells* absent or very rare. *Crown cells* present. *Costal short-cells* predominantly paired. Costal silica bodies horizontally-elongated crenate/sinuous to horizontally-elongated smooth (few, shortish), or rounded (mainly).

Transverse section of leaf blade, physiology. C_3; XyMS+.

Taxonomy. Pooideae; Poodae; Aveneae.

Ecology, geography, regional floristic distribution. 1 species. Asia Minor. Holarctic. Boreal and Tethyan. Euro-Siberian. Mediterranean and Irano-Turanian. European.

References, etc. Leaf anatomical: this project.

Germainia Bal. & Poitr.

Balansochloa Kuntze, *Sclerandrium* Stapf & Hubbard
Excluding *Chumsriella*

Habit, vegetative morphology. Annual, or perennial; stoloniferous (rarely), or caespitose. *Culms* 6–100 cm high; herbaceous. Culm internodes hollow (or variable?). *Leaves* mostly basal, or not basally aggregated; non-auriculate. Leaf blades narrow; without cross venation. *Ligule an unfringed membrane, or a fringed membrane (?)*; truncate; 0.5–2 mm long.

Reproductive organization. Plants monoecious with all the fertile spikelets unisexual; without hermaphrodite florets. The spikelets of sexually distinct forms on the same plant; female-only and male-only, or female-only, male-only, and sterile. The male and female-fertile spikelets mixed in the inflorescence. *The spikelets* overtly heteromorphic (the sessile and involucral spikelets awnless, with dissimilar glumes); *in both homogamous and heterogamous combinations (each raceme with one or two basal, homogamous pairs)*.

Inflorescence. *Inflorescence peduncled, often long-exserted, solitary, capitate to elongate and consisting of few- to many-jointed racemes*; *digitate (the racemes appressed or divergent)*; espatheate (but the peduncle sometimes enclosed in the uppermost sheath); not comprising 'partial inflorescences' and foliar organs. *Spikelet-bearing axes* very much reduced (to a peduncle, with two sessile spikelets and a pedicellate one on a short axis), or 'racemes'; solitary, or paired to clustered (1–2(–6) racemes, fused at their bases); tardily disarticulating; finally disarticulating at the joints. **Spikelets** *associated with bractiform involucres (the basal, sessile, homogamous spikelets forming an involucre round the heterogamous pairs)*; paired; sessile and pedicellate; consistently in 'long-and-short' combinations; in pedicellate/sessile combinations. Pedicels of the 'pedicellate' spikelets free of the rachis. The 'shorter' spikelets male-only, or sterile (when 'involucral'). The 'longer' spikelets female-only.

Female-sterile spikelets. The sessile spikelet larger, male or barren, dorsally compressed. Male spikelets with 2 stamens in the upper floret. Rachilla of male spikelets terminated by a male floret. The male spikelets with glumes; with proximal incomplete florets; 2 floreted. The lemmas awnless. Male florets 2; (the upper floret) 2 staminate.

Female-fertile spikelets. Spikelets somewhat compressed dorsiventrally (subterete); falling with the glumes. Rachilla terminated by a female-fertile floret. Hairy callus present (brown to yellow). Callus pointed, or blunt.

Glumes two; more or less equal; long relative to the adjacent lemmas; hairy, or hairless; pointed, or not pointed; awnless; non-carinate; similar (papery, narrow, blunt). Lower glume not two-keeled; not pitted; relatively smooth; 0–5 nerved. Upper glume 0–3 nerved. *Spikelets* with female-fertile florets only, or with incomplete florets (*G. truncatiglumis*). The incomplete florets when present, proximal to the female-fertile florets (sometimes suppressed). *The proximal incomplete florets* when present, 1; sterile. The proximal lemmas awnless.

Female-fertile florets 1. Lemmas without apical teeth, reduced to the membranous or stipitate awn-base; less firm than the glumes; entire; awned. Awns 1; median; apical; geniculate; hairy. Lemmas hairless; 1 nerved. *Palea* present; relatively long; entire, or apically notched, or deeply bifid; awnless, without apical setae; not indurated (hyaline); nerveless. *Lodicules* absent. *Stamens* 0. *Ovary* glabrous. Styles free to their bases. Stigmas 2.

Fruit, embryo and seedling. *Fruit* free from both lemma and palea; small (0.6–3.2 mm long). Hilum short. Embryo large.

Abaxial leaf blade epidermis. *Costal/intercostal zonation* conspicuous. *Papillae* present. Intercostal papillae over-arching the stomata. Mid-intercostal long-cells having markedly sinuous walls. *Microhairs* present; panicoid-type. *Stomata* common. Subsidiaries papillate; triangular. Guard-cells overlapping to flush with the interstomatals. *Intercostal short-cells* common. *Costal short-cells* neither distinctly grouped into long rows nor predominantly paired. Costal silica bodies 'panicoid-type'.

Transverse section of leaf blade, physiology. C_4; XyMS–. *Mesophyll* with radiate chlorenchyma. *Leaf blade* adaxially flat. *Midrib* conspicuous; having a conventional arc of bundles; with colourless mesophyll adaxially. Bulliforms present in discrete, regular adaxial groups; in simple fans, or associated with colourless mesophyll cells to form deeply-penetrating fans. Many of the smallest vascular bundles unaccompanied by sclerenchyma. Combined sclerenchyma girders present; forming 'figures'. Sclerenchyma all associated with vascular bundles.

Cytology. $2n = 14$.

Taxonomy. Panicoideae; Andropogonodae; Andropogoneae; Andropogoninae.

Ecology, geography, regional floristic distribution. 8 species. Asia, Malesia & North Australia.

Paleotropical and Australian. Indomalesian. Indo-Chinese, Malesian, and Papuan. North and East Australian. Tropical North and East Australian.

References, etc. Morphological/taxonomic: Chaianan 1972. Leaf anatomical: this project.

Gerritea Zuloaga & Morrone

Habit, vegetative morphology. Perennial; caespitose. *Culms* 20–100 cm high (procumbent, to 120 cm long); herbaceous; profusely branched above; many. Culm nodes hidden by leaf sheaths; hairy (pilose, with long whitish hairs). Culm internodes hollow. Young shoots intravaginal. *Leaves* not basally aggregated (mostly cauline); minutely auriculate; without auricular setae. Strongly keeled. *Leaf blades* linear-lanceolate; broad to narrow; 7–15 mm wide; flat; *pseudopetiolate (attenuate to a 2 cm long pseudopetiole)*; without cross venation; disarticulating from the sheaths. Ligule a fringed membrane (abaxially pilose); truncate; 0.3–0.6 mm long.

Reproductive organization. Plants bisexual, with bisexual spikelets; with hermaphrodite florets. The spikelets all alike in sexuality; hermaphrodite.

Inflorescence. *Inflorescence* many spikeleted; paniculate (terminal, exserted, lax, pyramindal); open (14–30 cm long, 4–12 cm wide with divergent first-order branches, the peduncle up to 10 cm long); *with capillary branchlets (?)*; espatheate; not comprising 'partial inflorescences' and foliar organs. Spikelets solitary; not secund; long pedicellate (the pedicels flexuous, to 8 mm long).

Female-fertile spikelets. *Spikelets 1.8–2.2 mm long (0.5 mm wide)*; elliptic; greenish or tinged with purple; slightly compressed laterally; falling with the glumes (?); disarticulating between the florets; with conventional internode spacings. The upper floret not stipitate. Rachilla terminated by a female-fertile floret. Hairy callus absent (?).

Glumes two; very unequal to more or less equal (subequal, or the lower somewhat shorter); long relative to the adjacent lemmas; hairy (with long, papillose-based hairs towards the margins); pointed (acuminate); awnless; non-carinate; very dissimilar to similar (the lower acuminate, the upper acute). *Lower glume* 0.75–1 times the length of the upper glume; longer than half length of lowest lemma; 3 nerved. *Spikelets* with incomplete florets. The incomplete florets proximal to the female-fertile florets. *Spikelets with proximal incomplete florets*. *The proximal incomplete florets* 1; paleate. Palea of the proximal incomplete florets more or less fully developed (0.7–1 mm long, hyaline, 2 nerved). The proximal incomplete florets sterile. The proximal lemmas glumiform, acute; awnless; more or less equalling the female-fertile lemmas, or decidedly exceeding the female-fertile lemmas (?); similar in texture to the female-fertile lemmas (?); not becoming indurated.

Female-fertile florets 1. *Lemmas* narrowly ellipsoid; *similar in texture to the glumes (? — 'membranous')*; smooth; not becoming indurated; 'pale'; entire; pointed; awnless; hairless; glabrous; non-carinate; *having the margins lying flat and exposed on the palea*; 3 nerved. *Palea* present; relatively long; gaping (free of the lemma at the tip); awnless, without apical setae; not indurated; 2-nerved. Palea back glabrous. *Lodicules* present; 2; free; fleshy ('truncate'); glabrous; not or scarcely vascularized. *Stamens* 3. Anthers 0.6–0.7 mm long; not penicillate; without an apically prolonged connective. *Ovary* glabrous. Styles free to their bases. Stigmas 2; white ('whitish').

Fruit, embryo and seedling. Disseminule unknown.

Abaxial leaf blade epidermis. *Costal/intercostal zonation* conspicuous. *Papillae* absent. Intercostal zones with typical long-cells. Mid-intercostal long-cells rectangular; having markedly sinuous walls. *Microhairs* present; elongated; clearly two-celled; panicoid-type (the distal cells sharply pointed); 70–98 microns long. *Stomata* common; 26–32.5 microns long. Subsidiaries non-papillate; triangular. *Intercostal short-cells* common (irregularly dispersed); in cork/silica-cell pairs and not paired (solitary and paired); silicified. Intercostal silica bodies cross-shaped (or irregular). Small hooks and macrohairs present. *Costal short-cells* conspicuously in long rows. Costal silica bodies present and well developed; 'panicoid-type'; dumb-bell shaped.

Transverse section of leaf blade, physiology. C_3; XyMS+. *Mesophyll* with radiate chlorenchyma; more or less *Isachne*-type. *Leaf blade* slightly 'nodular' in section; with the ribs more or less constant in size (low round topped). *Midrib* conspicuous (by the large median bundle and keel); with one bundle only, or having a conventional arc of bundles (sometimes with two small laterals). *The lamina* symmetrical on either side of the midrib. Bulliforms present in discrete, regular adaxial groups (in each furrow); in simple fans. All the vascular bundles accompanied by sclerenchyma. Combined sclerenchyma girders present (with all the bundles); forming 'figures' (anchors). Sclerenchyma all associated with vascular bundles.

Taxonomy. Panicoideae; Panicodae; Paniceae.

Ecology, geography, regional floristic distribution. 1 species. La Paz, Bolivia. Mesophytic; shade species; glycophytic.

Neotropical. Andean.

References, etc. Morphological/taxonomic: Zuloaga and Morrone 1993. Leaf anatomical: Zuloaga and Morrone 1993.

Special comments. Fruit data wanting.

Gigantochloa Kurtz ex Munro

~ *Dendrocalamus*

Habit, vegetative morphology. Arborescent perennial; sympodial, caespitose. The flowering culms leafless, or leafy. **Culms** 300–3000 cm high; *woody and persistent*; to 14 cm in diameter; scandent, or not scandent; branched above. Primary branches/mid-culm node several, with one dominant. Culm internodes hollow. Rhizomes pachymorph. Plants unarmed. *Leaves* not basally aggregated; auriculate (low sheath auricles, bristly or not); with auricular setae, or without auricular setae. Leaf blades broad (and the leaves rather large); pseudopetiolate; cross veined; disarticulating from the sheaths; rolled in bud. Ligule an unfringed membrane, or a fringed membrane.

Reproductive organization. Plants bisexual, with bisexual spikelets; with hermaphrodite florets. The spikelets of sexually distinct forms on the same plant (?), or all alike in sexuality.

Inflorescence. Inflorescence indeterminate; *with pseudospikelets (the spikelets with 'several basal bracts and glumes')*; of paniculate pseudo-inflorescences, in alternate clusters along leafless branches; spatheate; a complex of 'partial inflorescences' and intervening foliar organs. *Spikelet-bearing axes* paniculate; persistent. Spikelets not secund.

Female-fertile spikelets. Spikelets 9–50 mm long; compressed laterally; falling with the glumes ('spikelets falling whole'). Rachilla prolonged beyond the uppermost female-fertile floret.

Glumes present; two to several; shorter than the adjacent lemmas; awnless (mucronate); similar. *Spikelets* with incomplete florets. The incomplete florets distal to the female-fertile florets, or both distal and proximal to the female-fertile florets (if upper 'glumes' (which may subtend young spikelets) are regarded as proximal lemmas). *The distal incomplete florets* merely underdeveloped. *Spikelets with proximal incomplete florets*. The proximal lemmas awnless (mucronate); exceeded by the female-fertile lemmas.

Female-fertile florets 2–5. Lemmas increasing in length acropetally; mucronate (or at least apiculate); with more than 5 nerves. *Palea* present (much narrower than lemmas); relatively long, or conspicuous but relatively short; entire (in upper florets), or apically notched (in lower florets); awnless, without apical setae; thinner than the lemma (thin, translucent); not indurated; several nerved (2–5 between the keels, 0–3 on either side); 2-keeled. *Lodicules* usually absent. **Stamens** 6; *monadelphous (the tube longer than the palea)*. Anthers 4–9 mm long; penicillate; with the connective apically prolonged. *Ovary* hairy; with a conspicuous apical appendage. The appendage broadly conical, fleshy. Styles fused (into one long style, not rigid, not hollow). Stigmas 1(–3).

Fruit, embryo and seedling. Fruit narrowly fusiform; longitudinally grooved; not noticeably compressed; with hairs confined to a terminal tuft. Hilum long-linear. Embryo small.

Abaxial leaf blade epidermis. *Costal/intercostal zonation* conspicuous. *Papillae* present. Intercostal papillae over-arching the stomata; several per cell (of various shapes). Mid-intercostal long-cells having markedly sinuous walls. *Microhairs* present; panicoid-type. *Stomata* common (obscured by papillae). *Intercostal short-cells* common; in cork/silica-cell pairs; silicified. Intercostal silica bodies tall-and-narrow. *Costal short-cells* conspicuously in long rows (by amalgamation of pairs), or predominantly paired. Costal silica bodies saddle shaped.

Transverse section of leaf blade, physiology. C_3; XyMS+. Mesophyll traversed by columns of colourless mesophyll cells; with arm cells; with fusoids. *Leaf blade* with distinct, prominent adaxial ribs; with the ribs more or less constant in size. *Midrib* conspicuous; having complex vascularization. Bulliforms present in discrete, regular adaxial groups; associated with colourless mesophyll cells to form deeply-penetrating fans (these linking with traversing colourless cells). All the vascular bundles accompanied by sclerenchyma. Combined sclerenchyma girders present. Sclerenchyma all associated with vascular bundles.

Phytochemistry. Leaves without flavonoid sulphates.

Cytology. Chromosome base number, $x = 12$. $2n = 70$ and 72. 6 ploid.

Taxonomy. Bambusoideae; Bambusodae; Bambuseae.

Ecology, geography, regional floristic distribution. 20 species. Indomalayan. Forests.

Paleotropical. Indomalesian. Malesian.

Economic importance. Culms of *G. apus*, *G. maxima*, *G. nigrociliata* etc. used for house construction, as well as for furniture, household appliances etc.

References, etc. Leaf anatomical: Metcalfe 1960; this project.

Gilgiochloa Pilger

Habit, vegetative morphology. *Annual. Culms* 30–90 cm high; herbaceous. Culm nodes hairy. Plants unarmed. *Leaf blades lanceolate*; narrow; flat; without cross venation. Ligule a fringe of hairs.

Reproductive organization. Plants bisexual, with bisexual spikelets; with hermaphrodite florets.

Inflorescence. Inflorescence paniculate; contracted; spicate; espatheate; not comprising 'partial inflorescences' and foliar organs. Spikelet-bearing axes persistent. Spikelets pedicellate (the pedicels accrescent to the axes).

Female-fertile spikelets. Spikelets morphologically 'conventional'; 7–10 mm long; purplish; compressed laterally; *disarticulating above the glumes*; disarticulating between the florets. Rachilla terminated by a female-fertile floret. Hairy callus present. *Callus* short; blunt.

Glumes two; very unequal; (the upper) long relative to the adjacent lemmas (as long as the spikelet); hairy (the G_1 usually sparsely hairy), or hairless (glabrous); pointed; *awned (the G_1 with a 4–5 mm awn, the G_2 pointed or continued into a bristle)*; thin-membranous. Lower glume 3 nerved. Upper glume 3 nerved. *Spikelets* with incomplete florets. The incomplete florets proximal to the female-fertile florets. *Spikelets with proximal incomplete florets. The proximal incomplete florets* 1; paleate. *Palea of the proximal incomplete florets fully developed (becoming conspicuously indurated, thickened between the keels). The proximal incomplete florets sterile.* The proximal lemmas awnless (pointed); 5–7 nerved.

Female-fertile florets 1. Lemmas similar in texture to the glumes to decidedly firmer than the glumes (thinly leathery); not becoming indurated; incised; deeply cleft; awned. Awns 3; median and lateral; the median different in form from the laterals; from a sinus; geniculate; much longer than the body of the lemma. The lateral awns shorter than the median (straight). Lemmas hairy. The hairs in tufts (on the lemma margins and at the awn base); in transverse rows. Lemmas non-carinate; 5 nerved. *Palea* present; relatively long; apically notched; 2-nerved; 2-keeled. *Palea keels narrowly winged (the wings terminating in clavate appendages). Lodicules* present; 2; free; fleshy; glabrous. *Stamens* 3. *Ovary* glabrous. Stigmas 2.

Fruit, embryo and seedling. *Hilum long-linear.*

Abaxial leaf blade epidermis. *Costal/intercostal zonation* conspicuous. *Papillae* absent. Mid-intercostal long-cells rectangular; having markedly sinuous walls. *Microhairs* present; panicoid-type; 58–72 microns long; 3.6–5.4 microns wide at the septum. Microhair total length/width at septum 11.7–17.1. Microhair apical cells 28–36 microns long. Microhair apical cell/total length ratio 0.48–0.52. *Stomata* common; 25–29 microns long. Subsidiaries dome-shaped and triangular. Guard-cells overlapping to flush with the interstomatals. *Intercostal short-cells* common (in places); not paired; not silicified.

Costal short-cells conspicuously in long rows. Costal silica bodies 'panicoid-type'; mostly shortish dumb-bell shaped.

Transverse section of leaf blade, physiology. C_4; XyMS–. *Mesophyll* with radiate chlorenchyma; seemingly without 'circular cells' (in the material seen). *Midrib* conspicuous, or not readily distinguishable; with one bundle only. Many of the smallest vascular bundles unaccompanied by sclerenchyma. Sclerenchyma all associated with vascular bundles (rather depauperate).

Taxonomy. Panicoideae; Panicodae; Arundinelleae.

Ecology, geography, regional floristic distribution. 1 species. East tropical Africa. Mesophytic; species of open habitats. Clearings in thickets and woodland.

Paleotropical. African. Sudano-Angolan. Somalo-Ethiopian and South Tropical African.

References, etc. Morphological/taxonomic: Phipps 1967. Leaf anatomical: this project.

Glaziophyton Franch.

Habit, vegetative morphology. Sympodial, juncoid perennial; *caespitose (the first internode of the primary culm disproportionately elongate, the upper nodes condensed).* The flowering culms leafless (but after burning, small secondary culms arise by tillering from the stumps). Culms woody and persistent. Primary branches/mid-culm node 1. Culm internodes hollow (but partitioned by numerous thin septa of pith). Pluricaespitose. Rhizomes pachymorph. Plants unarmed. Leaves auriculate (the auricles weakly developed), or non-auriculate; without auricular setae. *Leaf blades* lanceolate (acuminate); *broad*; about 10 mm wide; pseudopetiolate; cross veined; disarticulating from the sheaths; rolled in bud.

Reproductive organization. Plants bisexual, with bisexual spikelets; with hermaphrodite florets.

Inflorescence. Inflorescence falsely paniculate (to 1 m long, never terminal to a culm); open (effusely branched, the main axis often trigonous); spatheate (with bracts and prophylls); a complex of 'partial inflorescences' and intervening foliar organs. Spikelet-bearing axes persistent. Spikelets not secund; not in distinct 'long-and-short' combinations.

Female-fertile spikelets. Spikelets to 24 mm long (maximum); compressed laterally to not noticeably compressed; disarticulating above the glumes; disarticulating between the florets; with a distinctly elongated rachilla internode above the glumes. Rachilla prolonged beyond the uppermost female-fertile floret; rachilla hairless. Hairy callus absent.

Glumes two; very unequal; shorter than the adjacent lemmas; hairless; glabrous; awnless (somewhat apiculate); non-carinate; similar. *Spikelets* with incomplete florets. The incomplete florets distal to the female-fertile florets, or both distal and proximal to the female-fertile florets. *The distal incomplete florets* merely underdeveloped. *The proximal incomplete florets* 0, or 1; male, or sterile. The proximal lemmas awnless; exceeded by the female-fertile lemmas; similar in texture to the female-fertile lemmas.

Female-fertile florets (1–)2, or 3. Lemmas not becoming indurated; entire; pointed, or blunt; awnless, or mucronate (?); hairless; glabrous; non-carinate. *Palea* present; relatively long; apically notched; awnless, without apical setae; 2-keeled (dorsally sulcate, the keels ciliate). *Lodicules* present; 3; free; membranous; glabrous; not toothed; heavily vascularized. *Stamens* 3. Anthers not penicillate; without an apically prolonged connective. *Ovary* glabrous. Styles fused. Stigmas 2.

Special diagnostic feature. *Rush-like, with reduced leaf blades (but the juncoid primary culms sprouting typical bamboo twigs from their bases after burning).*

Taxonomy. Bambusoideae; Bambusodae; Bambuseae.

Ecology, geography, regional floristic distribution. 1 species. Brazil. Neotropical. Amazon and Central Brazilian.

Special comments. Fruit data wanting. Anatomical data wanting.

Glyceria R.Br.

Devauxia Kunth, *Exydra* Endl., *Heleochloa* Fries, *Hemibromus* Steud., *Hydrochloa* Hartm., *Hydropoa* (Dumort.) Dumort., *Nevroloma* Raf., *Plotia* Steud., *Porroteranthe* Steud.

Habit, vegetative morphology. Glabrous, perennial; rhizomatous, or stoloniferous, or caespitose, or decumbent. *Culms* 25–200 cm high; herbaceous. Culm nodes glabrous. Culm internodes hollow. *Leaves* mostly basal, or not basally aggregated; non-auriculate. *Sheath margins joined*. Leaf blades linear; broad, or narrow; 1.5–18 mm wide; flat, or folded; cross veined, or without cross venation (when narrower); persistent; once-folded in bud. Ligule an unfringed membrane; not truncate.

Reproductive organization. Plants bisexual, with bisexual spikelets; with hermaphrodite florets; outbreeding and inbreeding.

Inflorescence. Inflorescence paniculate; open; espatheate; not comprising 'partial inflorescences' and foliar organs. Spikelet-bearing axes persistent. Spikelets secund, or not secund.

Female-fertile spikelets. *Spikelets* 2–35 mm long; *compressed laterally*; disarticulating above the glumes; disarticulating between the florets. Rachilla prolonged beyond the uppermost female-fertile floret; rachilla hairless. The rachilla extension with incomplete florets. Hairy callus absent. *Callus* short; blunt.

Glumes two; very unequal to more or less equal; nearly always shorter than the spikelets; shorter than the adjacent lemmas; pointed, or not pointed; awnless; non-carinate; similar (herbaceous). Lower glume 1(–3) nerved. *Upper glume 1(–3) nerved*. Spikelets with incomplete florets. The incomplete florets distal to the female-fertile florets. *The distal incomplete florets* merely underdeveloped.

Female-fertile florets 3–16. **Lemmas** similar in texture to the glumes to decidedly firmer than the glumes (membranous to thinly leathery); not becoming indurated; entire, or incised; when entire pointed, or blunt; when incised, 3 lobed, or 5 lobed; not deeply cleft (toothed); awnless; hairless; *non-carinate (rounded on the back)*; (5–)7(–11) nerved; with the nerves non-confluent. *Palea* present; relatively long; entire to apically notched; awnless, without apical setae, or with apical setae; thinner than the lemma; not indurated (membranous); 2-nerved; 2-keeled. Palea keels sometimes, slightly winged, or wingless. **Lodicules** present; joined; *fleshy*; glabrous; toothed, or not toothed. *Stamens* 2, or 3. Anthers 0.3–3 mm long; not penicillate. **Ovary** glabrous; *without a conspicuous apical appendage*. Styles free to their bases. Stigmas 2; white.

Fruit, embryo and seedling. *Fruit* free from both lemma and palea; small; longitudinally grooved; compressed dorsiventrally. Hilum long-linear. Embryo small; not waisted. Endosperm hard; without lipid; containing compound starch grains. Embryo with an epiblast; without a scutellar tail; with a negligible mesocotyl internode. Embryonic leaf margins meeting.

Seedling with a long mesocotyl; with a tight coleoptile. First seedling leaf with a well-developed lamina. The lamina narrow; 3 veined.

Abaxial leaf blade epidermis. *Costal/intercostal zonation* conspicuous. *Papillae* present, or absent. Intercostal papillae over-arching the stomata, or not over-arching the stomata; consisting of one oblique swelling per cell. *Long-cells* similar in shape costally and intercostally, or markedly different in shape costally and intercostally (the costals often fusiform); of similar wall thickness costally and intercostally. Intercostal zones with typical long-cells, or exhibiting many atypical long-cells (*G. australis* having many short long-cells). Mid-intercostal long-cells rectangular; having straight or only gently undulating walls. *Microhairs* absent. *Stomata* common; 24–36 microns long. Subsidiaries low dome-shaped, or parallel-sided, or parallel-sided and dome-shaped. Guard-cells overlapped by the interstomatals (at least in *G. australis*). *Intercostal short-cells* common, or absent or very rare. *Costal short-cells* neither distinctly grouped into long rows nor predominantly paired. Costal silica bodies horizontally-elongated crenate/sinuous.

Transverse section of leaf blade, physiology. C_3; XyMS+. *Mesophyll* with non-radiate chlorenchyma; without adaxial palisade; without fusoids (but at least sometimes with large cavities). *Leaf blade* with distinct, prominent adaxial ribs, or adaxially flat; with the ribs more or less constant in size. *Midrib* conspicuous to not readily distinguishable; with one bundle only. Bulliforms present in discrete, regular adaxial groups, or not in discrete, regular adaxial groups (sometimes restricted to midrib 'hinges'); when grouped, in simple fans. All the vascular bundles accompanied by sclerenchyma. Combined sclerenchyma girders present, or absent; forming 'figures' (e.g. *G. australis*), or nowhere forming 'figures'. Sclerenchyma all associated with vascular bundles.

Culm anatomy. *Culm internode bundles* in one or two rings.

Cytology. Chromosome base number, $x = 10$. $2n = 20, 28, 40, 56$, and 60. 2, 4, and 6 ploid (and aneuploids).

Taxonomy. Pooideae; Poodae; Meliceae.

Ecology, geography, regional floristic distribution. 40 species. Cosmopolitan. Commonly adventive. *Hydrophytic to helophytic*; glycophytic.

Holarctic, Paleotropical, Neotropical, Australian, and Antarctic. Boreal, Tethyan, and Madrean. Indomalesian. Arctic and Subarctic, Euro-Siberian, Eastern Asian, Atlantic North American, and Rocky Mountains. Macaronesian, Mediterranean, and Irano-Turanian. Indian and Indo-Chinese. Pampas and Andean. North and East Australian and South-West Australian. Patagonian. European and Siberian. Canadian-Appalachian, Southern Atlantic North American, and Central Grasslands. Temperate and South-Eastern Australian.

Rusts and smuts. Rusts — *Puccinia*. Taxonomically wide-ranging species: *Puccinia graminis*, *Puccinia coronata*, *Puccinia striiformis*, *Puccinia brachypodii*, and *Puccinia recondita*. Smuts from Tilletiaceae and from Ustilaginaceae. Tilletiaceae — *Entyloma* and *Urocystis*. Ustilaginaceae — *Ustilago*.

Economic importance. Important native pasture species: in swampy places, *G. elata*, *G. grandis*, *G. maxima*, *G. pauciflora*, *G. striata*.

References, etc. Leaf anatomical: Metcalfe 1960; this project.

Glyphochloa W. D. Clayton

Habit, vegetative morphology. *Annual*. Culms *10–60 cm high*; herbaceous; *branched above*. Leaves not basally aggregated. Leaf blades linear; broad to narrow; 5–15 mm wide. *Ligule an unfringed membrane*.

Reproductive organization. *Plants bisexual, with bisexual spikelets*; with hermaphrodite florets. The spikelets of sexually distinct forms on the same plant; hermaphrodite and male-only, or hermaphrodite and sterile. The male and female-fertile spikelets mixed in the inflorescence. The spikelets overtly heteromorphic; all in heterogamous combinations.

Inflorescence. Inflorescence of single, dorsiventral 'racemes' terminating the culm branches. Rachides hollowed, or flattened. Inflorescence spatheate. **Spikelet-bearing axes** *spikelike*; solitary; with substantial rachides; *disarticulating*; disarticulating at the joints. *'Articles'* non-linear (inflated, clavate or turbinate); with a basal callus-knob; disarticulating transversely, or disarticulating obliquely. Spikelets paired; secund (the raceme dorsiventral, with the sessile members in two alternating rows on one side of the rachis); consistently in 'long-and-short' combinations; in pedicellate/sessile combinations. Pedicels of the 'pedicellate' spikelets discernible, but fused with the rachis. The 'shorter' spikelets hermaphrodite. The 'longer' spikelets male-only, or sterile (?).

Female-sterile spikelets. *The pedicelled spikelets as large as the sessile, male or neuter, G_1 smooth and asymmetrically or unilaterally winged, G_2 variously keel-winged*.

Female-fertile spikelets. Spikelets morphologically 'conventional' (but the G_1 curiously winged, ornamented and awned); abaxial; compressed dorsiventrally; *falling with the glumes (and with the joint plus the pedicelled spikelet from the same joint, by contrast with* **Manisuris** *sensu stricto)*. Rachilla terminated by a female-fertile floret. Hairy callus absent.

Glumes present; two; long relative to the adjacent lemmas; hairy, or hairless; without conspicuous tufts or rows of hairs; awned (the G_1 1–2 aristate or tailed), or awnless (*G. clarkei*); non-carinate; *very dissimilar (the G_1 hardened, curiously ornamented and winged above, and aristate)*. *Lower glume two-keeled*; convex on the back; lacunose with deep depressions, or rugose, or tuberculate, or prickly (and laterally winged above the middle). *Spikelets* with incomplete florets. The incomplete florets proximal to the female-fertile florets. *The proximal incomplete florets* 1; sterile. The proximal lemmas awnless; 0 nerved, or 2 nerved; similar in texture to the female-fertile lemmas (hyaline).

Female-fertile florets 1. **Lemmas** less firm than the glumes (hyaline); not becoming indurated; entire; awnless; *0 nerved, or 2 nerved*. *Palea* present. *Lodicules* present; 2; fleshy. *Stamens* 3.

Fruit, embryo and seedling. Hilum short. Embryo large.

Transverse section of leaf blade, physiology. C_4; XyMS–.

Taxonomy. Panicoideae; Andropogonodae; Andropogoneae; Rottboelliinae.

Ecology, geography, regional floristic distribution. 8 species. Central and peninsular India. On rocks.

Paleotropical. Indomalesian. Indian.

References, etc. Morphological/taxonomic: Jain 1970.

Special comments. Anatomical data wanting.

Gouinia Fourn.

Pogochloa S. Moore

Habit, vegetative morphology. Perennial. *Culms* 30–350 cm high; woody and persistent (sometimes, somewhat), or herbaceous; scandent (in one species), or not scandent; branched above, or unbranched above. Culm nodes glabrous. Culm internodes solid. *Leaves* not basally aggregated; non-auriculate. **Leaf blades** linear-lanceolate; broad, or narrow; flat, or rolled (involute); without abaxial multicellular glands; *not pseudopetiolate*; without cross venation. Ligule a fringed membrane, or a fringe of hairs. *Contraligule* absent.

Reproductive organization. Plants bisexual, with bisexual spikelets; with hermaphrodite florets.

Inflorescence. *Inflorescence of spicate main branches*; open; non-digitate (the branches racemosely arranged); espatheate; not comprising 'partial inflorescences' and foliar organs. The racemes without spikelets towards the base. Spikelet-bearing axes with very slender rachides; persistent. Spikelets solitary; secund; biseriate; subsessile to pedicellate.

Female-fertile spikelets. *Spikelets 6–15 mm long*; adaxial; *compressed laterally*; disarticulating above the glumes; *disarticulating between the florets*. Rachilla prolonged beyond the uppermost female-fertile floret; rachilla hairy, or hairless. The rachilla extension with incomplete florets. Hairy callus present. *Callus* short; blunt.

Glumes present; two; very unequal to more or less equal; shorter than the spikelets; shorter than the adjacent lemmas; dorsiventral to the rachis; pointed (to attenuate); awnless; carinate; similar (lanceolate). Lower glume shorter than the lowest lemma; 1–5 nerved. Upper glume (1–)3–7 nerved. *Spikelets* with incomplete florets. The incomplete florets distal to the female-fertile florets. *The distal incomplete florets* 1; **incomplete florets** merely underdeveloped; *incomplete florets awned (awned or reduced to the awn, by contrast with* **Leptochloa***)*.

Female-fertile florets 2–5. Lemmas similar in texture to the glumes (thin); not becoming indurated; entire, or incised; pointed, or blunt; when incised, 2 lobed; not deeply cleft (bidentate); awned. Awns 1; median; from a sinus, or apical; non-geniculate; hairless (scabrid); much shorter than the body of the lemma (rarely), or about as long as the body of the lemma to much longer than the body of the lemma; persistent. Lemmas hairy (densely ciliate below adjoining the midnerve, and often ciliate or pilose on the margins); carinate; without a germination flap; 3 nerved, or 5(–7) nerved; with the nerves confluent towards the tip. *Palea* present; relatively long, or conspicuous but relatively short; apically notched (two-toothed); awnless, without apical setae, or with apical setae to awned (via the teeth); textured like the lemma; not indurated; 2-nerved; 2-keeled. Palea keels somewhat winged. *Lodicules* present; 2 (tiny); free; fleshy; glabrous; not or scarcely vascularized. *Stamens* 3. Anthers about 0.5 mm long; not penicillate; without an apically prolonged connective. *Ovary* glabrous. Stigmas 2.

Fruit, embryo and seedling. *Fruit* free from both lemma and palea; small (3–4 mm long); brown or black; ellipsoid; strongly longitudinally grooved (ventrally); compressed laterally (slightly); smooth. *Hilum long-linear*. Pericarp fused. Embryo small (i.e. generally less than a third of the fruit length); not waisted. Endosperm hard.

Abaxial leaf blade epidermis. *Costal/intercostal zonation* conspicuous. *Papillae* present (in *G. latifolia*, Diaz 1993), or absent (mostly); in *G. latifolia*, intercostal. Intercostal papillae not over-arching the stomata. *Long-cells* similar in shape costally and intercostally; of similar wall thickness costally and intercostally (fairly thick walled). Midintercostal long-cells rectangular; having markedly sinuous walls (these pitted). *Microhairs* present; elongated; clearly two-celled; panicoid-type to chloridoid-type (but fairly narrow, with longish basal cells). Microhair apical cell wall thinner than that of the basal cell but not tending to collapse. Microhairs (28.5–)31–36(–37.5) microns long. Microhair basal cells 21 microns long. Microhairs (7.5–)8.4–11.4(–11.7) microns wide at the septum. Microhair total length/width at septum 2.9–5. Microhair apical cells 10–13.5 microns long. Microhair apical cell/total length ratio 0.32–0.43. *Stomata* common; 18–22.5 microns long. Subsidiaries dome-shaped, or triangular (but their apices truncated). Guard-cells overlapping to flush with the interstomatals. *Intercostal short-cells* common; in cork/silica-cell pairs; silicified. Intercostal silica bodies present and perfectly developed; tall-and-narrow and oryzoid-type. *Costal short-cells* conspicuously in long rows. Costal silica bodies present in alternate cell files of the costal zones; 'panicoid-type'; mostly dumb-bell shaped (short to elongated).

Transverse section of leaf blade, physiology. C_4; XyMS+. PCR sheath outlines uneven (the abaxial 2 or 3 PCR cells very small, the lower laterals much wider than the upper ones). PCR sheaths of the primary vascular bundles interrupted; interrupted abaxially only. PCR sheath extensions absent. *Leaf blade* adaxially flat (or the ribs broad and flat topped, scarcely prominent). *Midrib* conspicuous; having a conventional arc of bundles (one large median bundle, with two laterals on either side); with colourless mesophyll adaxially. Bulliforms present in discrete, regular adaxial groups; in simple fans and associated with colourless mesophyll cells to form deeply-penetrating fans (only one example of the latter seen, but the simple fans all with deeply penetrating median cells). All the vascular bundles accompanied by sclerenchyma. Combined sclerenchyma girders present (with all but those in the midrib); forming 'figures' (all the lateral bundles with conspicuous I's). Sclerenchyma all associated with vascular bundles. The lamina margins with fibres.

Cytology. Chromosome base number, $x = 10$. $2n = 40$ (where available, but with n counts of 20, 32, 38 and 40).

Taxonomy. Chloridoideae; main chloridoid assemblage.

Ecology, geography, regional floristic distribution. 9 species (according to Diaz, 1993). Mexico to Argentina, West Indies. Shade species and species of open habitats. Thickets and open places on hill slopes.

Neotropical. Caribbean, Amazon, Central Brazilian, and Andean.

Rusts and smuts. Rusts — *Puccinia*.

References, etc. Morphological/taxonomic: Swallen 1935. Leaf anatomical: this project.

Gouldochloa Valdés, Morden & Hatch

~ *Chasmanthium*

Habit, vegetative morphology. Perennial; caespitose. *The flowering culms leafy (but differing from the non-flowering culms in having elongated internodes)*. Culms 30–60 cm high (usually erect, sometimes decumbent to ascending); herbaceous; to 0.26 cm in diameter; unbranched above (seemingly). **Leaves** not basally aggregated; auriculate; *with auricular setae (or at least, the auricles pilose)*. Leaf blades linear; narrow; 2–3.1 mm wide (and 3–7 cm long); flat, or rolled (inrolled or involute); without cross venation; persistent (the senescent tips becoming spirally coiled). *Ligule* present; a fringed membrane.

Reproductive organization. Plants bisexual, with bisexual spikelets; with hermaphrodite florets.

Inflorescence. *Inflorescence* paniculate; *contracted (spikelike, with short, rigid, closely appressed primary branches)*; spicate; spatheate; *a complex of 'partial inflorescences' and intervening foliar organs (in that the lowermost spikelets of primary branches are often subtended by a 0.5–3.5 mm scale)*. Spikelet-bearing axes persistent. Spikelets solitary; not secund.

Female-fertile spikelets. Spikelets 8–16 mm long; compressed laterally; disarticulating above the glumes; disarticulating between the florets; with conventional internode spacings. Rachilla prolonged beyond the uppermost female-fertile floret; rachilla hairless.

Glumes two; more or less equal (subequal, the G_1 being slightly longer); shorter than the adjacent lemmas; hairless; awnless; seemingly non-carinate (but the nerves 'protruding abaxially'); similar (pale green, chartaceous, with hyaline margins). Lower glume 7 nerved, or 9 nerved. Upper glume 7 nerved, or 9 nerved. *Spikelets* with incomplete florets. The incomplete florets distal to the female-fertile florets, or both distal and proximal to the female-fertile florets (the lowermost floret staminate or perfect, the upper florets male or neuter). The distal incomplete florets usually several; *incomplete florets* merely underdeveloped. *The proximal incomplete florets* when present, 1; when present, paleate; when present, male. The proximal lemmas awnless; 7 nerved, or 9 nerved; exceeded by the female-fertile lemmas; similar in texture to the female-fertile lemmas (rigid, chartaceous); not becoming indurated.

Female-fertile florets 1, or 2 (the second, and sometimes the third). Lemmas similar in texture to the glumes (rigid, chartaceous, with hyaline margins); not becoming indurated; entire; awnless; glabrous except for a tuft near the callus; non-carinate (seemingly, though the nerves prominent); without a germination flap (or at least, no flap mentioned); 7 nerved, or 9 nerved. *Palea* present; relatively long to conspicuous but relatively short (subequal with to 3mm shorter than the lemma); apically notched (the teeth 0.3–0.4 mm long); awnless, without apical setae; texture unstated; not indurated; 2-nerved; sharply 2-keeled. *Lodicules* present; 2; free; glabrous; regularly 3-lobed; not or scarcely vascularized (obscurely 3 nerved). **Stamens** *1 (anterior to the pistil)*. Anthers 1.5–2.1 mm long; not penicillate; without an apically prolonged connective. *Ovary* glabrous. Stigmas 2; red pigmented.

Fruit, embryo and seedling. Fruit small (about 3 mm long); compressed laterally (ovate to triangular in section). Hilum short (probably, but unstated). Pericarp fused (yellow brown above, golden brown below). Embryo small (1/5 of the caryopsis length); with an epiblast; with a scutellar tail; with an elongated mesocotyl internode. Embryonic leaf margins overlapping.

Abaxial leaf blade epidermis. *Costal/intercostal zonation* conspicuous. *Papillae* absent. *Long-cells* markedly different in shape costally and intercostally (the costals much narrower). Mid-intercostal long-cells rectangular; having markedly sinuous walls. *Microhairs* present; panicoid-type. *Stomata* common. Subsidiaries triangular. *Intercostal short-cells* absent or very rare (seemingly). *Costal short-cells* conspicuously in long rows. Costal silica bodies 'panicoid-type'; cross shaped and dumb-bell shaped.

Transverse section of leaf blade, physiology. C_3; XyMS+. *Mesophyll* with non-radiate chlorenchyma; seemingly without arm cells; without fusoids. *Leaf blade* with distinct, prominent adaxial ribs to 'nodular' in section. Bulliforms present in discrete, regular adaxial groups; in simple fans. All the vascular bundles accompanied by sclerenchyma. Combined sclerenchyma girders present; forming 'figures'. Sclerenchyma all associated with vascular bundles.

Cytology. Chromosome base number, $x = 12$. $2n = 24$. 2 ploid.

Taxonomy. Bambusoideae; Oryzodae; Centotheceae.

Ecology, geography, regional floristic distribution. 1 species. Mexico. Xerophytic; species of open habitats. In desert scrub.

Holarctic. Madrean.

References, etc. Morphological/taxonomic: Valdés, Morden and Hatch 1986.

Graphephorum Desv.

~ *Trisetum*

Habit, vegetative morphology. *Perennial*; caespitose. **Culms** 50–100 cm high; *herbaceous*; unbranched above. Young shoots intravaginal. Leaves non-auriculate; without auricular setae. Leaf blades linear; narrow; 2–8 mm wide; flat; without cross venation. *Ligule an unfringed membrane (?).*

Reproductive organization. Plants bisexual, with bisexual spikelets; with hermaphrodite florets. The spikelets all alike in sexuality; hermaphrodite.

Inflorescence. *Inflorescence paniculate*; open to contracted (8–20 cm long, nodding or erect); espatheate; not comprising 'partial inflorescences' and foliar organs. Spikelet-bearing axes persistent. Spikelets not secund; pedicellate; imbricate.

Female-fertile spikelets. Spikelets 5–7 mm long; compressed laterally; disarticulating above the glumes; disarticulating between the florets. Rachilla prolonged beyond the uppermost female-fertile floret; rachilla copiously hairy (pilose). *Hairy callus present. Callus* short.

Glumes two; very unequal to more or less equal; (the upper) *about equalling the spikelets*; *long relative to the adjacent lemmas (about as long)*; hairless; scabrous; pointed; awnless; carinate; similar. *Spikelets* with female-fertile florets only, or with incomplete florets (?). The incomplete florets (if they occur) distal to the female-fertile florets. *The distal incomplete florets* merely underdeveloped.

Female-fertile florets 2–3(–4) *('two to several').* **Lemmas** not becoming indurated; *entire*; pointed to blunt; awnless (usually), or mucronate (rarely with a dorsal, subapical awn-point); hairless (above the callus); scabrous; *carinate*; without a germination flap; 5 nerved (or more?); with the nerves non-confluent. **Palea** present; relatively long; *tightly clasped by the lemma*; awnless, without apical setae; 2-nerved; 2-keeled. **Ovary** glabrous; *without a conspicuous apical appendage.*

Fruit, embryo and seedling. *Fruit not grooved*; smooth. *Hilum short (?).* Pericarp thin. Embryo small (?).

Taxonomy. Pooideae; Triticodae; Aveneae.

Ecology, geography, regional floristic distribution. 3 species (Clayton and Renvoize 1986). North and central America. Helophytic, or mesophytic; species of open habitats; glycophytic. Boggy meadows and moist ground, or montane.

Holarctic and Neotropical. Boreal and Madrean. Atlantic North American and Rocky Mountains. Caribbean.

References, etc. Morphological/taxonomic: Hitchcock and Chase (1950: *Trisetum melicoides*, *T.wolfii*); Clayton anf Renvoize (1986).

Special comments. Description very inadequate. Fruit data wanting. Anatomical data wanting.

Greslania Bal.

Habit, vegetative morphology. Shrubby perennial. **Culms** 100–300 cm high; *woody and persistent*; *unbranched above (while vegetative).* Rhizomes pachymorph. Leaf blades pseudopetiolate.

Reproductive organization. Plants bisexual, with bisexual spikelets; with hermaphrodite florets.

Inflorescence. Inflorescence *determinate*; large, falsely paniculate (apical, of aggregated 1-spikelet racemes and their bracts and prophylls); non-digitate; spatheate; a complex of 'partial inflorescences' and intervening foliar organs. *Spikelet-bearing axes* very much reduced; persistent.

Female-fertile spikelets. Rachilla prolonged beyond the uppermost female-fertile floret. The rachilla extension with incomplete florets.

Glumes two; awnless; similar. *Spikelets* with incomplete florets. The incomplete florets distal to the female-fertile florets. *The distal incomplete florets* merely underdeveloped (rudimentary, pistillate).

Female-fertile florets 1. Lemmas not becoming indurated; entire; blunt; awnless; 'lemma nerves obscure'. *Palea* present; awnless, without apical setae; 2-keeled. *Lodicules* present; 3. *Stamens* 6. **Ovary** glabrous; **with a conspicuous apical appendage**. The appendage broadly conical, fleshy. Styles fused (short). Stigmas 3.

Fruit, embryo and seedling. Pericarp fleshy; free.

Abaxial leaf blade epidermis. *Costal/intercostal zonation* conspicuous. *Papillae* present; intercostal (in the stomatal bands only). Intercostal papillae over-arching the stomata; several per cell (various shapes, cuticularised). Mid-intercostal long-cells having markedly sinuous walls (these pitted, thick). *Microhairs* present; panicoid-type. *Stomata* common (obscured by papillae). *Intercostal short-cells* common; in cork/silica-cell pairs; silicified. Intercostal silica bodies tall-and-narrow. *Costal short-cells* conspicuously in long rows, or predominantly paired. Costal silica bodies saddle shaped.

Transverse section of leaf blade, physiology. C_3; XyMS+. *Mesophyll* with non-radiate chlorenchyma; with arm cells; with fusoids (large). *Leaf blade* with distinct, prominent adaxial ribs; with the ribs more or less constant in size. *Midrib* conspicuous; having complex vascularization. Bulliforms present in discrete, regular adaxial groups; in simple fans (in the furrows). All the vascular bundles accompanied by sclerenchyma. Combined sclerenchyma girders present (with most bundles).

Taxonomy. Bambusoideae; Bambusodae; Bambuseae.

Ecology, geography, regional floristic distribution. 4 species. New Caledonia. Paleotropical. Neocaledonian.

References, etc. Leaf anatomical: Metcalfe 1960.

Special comments. Fruit data wanting.

Griffithsochloa G.J. Pierce

~ Cathestechum

Habit, vegetative morphology. Short, erect, delicate perennial; caespitose. *Culms* about 40–50 cm high; herbaceous; unbranched above. Culm nodes minutely hairy. *Leaves* mostly basal; non-auriculate. Leaf blades linear; narrow; about 1–2 mm wide (and 5–6 cm long); flat; without abaxial multicellular glands; without cross venation. Ligule a fringe of hairs.

Reproductive organization. Plants bisexual, with bisexual spikelets; with hermaphrodite florets (at least the first floret of the central spikelet of the triplet being usually hermaphrodite), or without hermaphrodite florets. The spikelets of sexually distinct forms on the same plant to all alike in sexuality; hermaphrodite, or hermaphrodite and male-only, or hermaphrodite, male-only, and sterile, or female-only, male-only, and sterile (the lateral spikelets of the triplets tending to be staminate or sterile, the central one hermaphrodite or occasionally female); homomorphic.

Inflorescence. *Inflorescence a false spike, with spikelets on contracted axes (a raceme of 'spikes', each reduced to three spikelets). Inflorescence axes not ending in spikelets (the main branches bare-tipped, the clusters with the rachis prolonged as a short, sometimes winged awn)*. Inflorescence espatheate; not comprising 'partial inflorescences' and foliar organs. Spikelet-bearing axes disarticulating; falling entire (the triplets falling). **Spikelets *subtended by solitary 'bristles' (i.e. by the rachis prolongation); in triplets*; secund (the main branches on one side of the culm).

Female-fertile spikelets. Spikelets 2–2.5 mm long; compressed laterally; falling with the glumes (the triplets falling whole); disarticulating between the florets. Rachilla prolonged beyond the uppermost female-fertile floret; rachilla hairy. The rachilla extension with incomplete florets. Hairy callus present (constituted by the hairy axis of the triplet).

Glumes two; *very unequal (but the G_1 longer than in* **Cathesticum** *sensu stricto)*; shorter than the spikelets; (the upper) long relative to the adjacent lemmas; hairy (the G_1 hirsute); pointed (G_1 acuminate, G_2 trullate); awnless; very dissimilar (G_1 acuminate, G_2 trullate). **Lower glume** about 0.7 times the length of the upper glume; *1 nerved (?)*. Upper glume 1 nerved (?). *Spikelets* with incomplete florets. The incomplete florets distal to the female-fertile florets (the first floret being female or hermaphrodite, the second hermaphrodite or male or sterile, the remaining 2–3 florets sterile). The distal incomplete florets 2–4; *incomplete florets clearly specialised and modified in form (for dispersal)*; incomplete florets awned (the terminal florets being reduced to multiple-awned lemmas).

Female-fertile florets 1–2. Lemmas similar in texture to the glumes (thin); not becoming indurated; incised; 3 lobed; deeply cleft to not deeply cleft (less deeply than in the upper, sterile florets); awned, or mucronate. Awns 3, or 5 (the lowermost lemma 3 awned, the others 3 to 9 awned, the awn number increasing with the degree of sterility); median and lateral; the median similar in form to the laterals; apical; non-geniculate; much shorter than the body of the lemma. The lateral awns shorter than the median. Lemmas hairy (with a few scattered hairs); non-carinate; 3 nerved. *Palea* present; relatively long; apically notched; with apical setae, or awned; not indurated; 2-nerved; 2-keeled. Palea keels wingless. *Lodicules* present; 2; free; fleshy; glabrous. *Stamens* 0, or 3. *Ovary* glabrous. Stigmas 2.

Fruit, embryo and seedling. Fruit ellipsoid; compressed dorsiventrally. Hilum short. Pericarp fused. Embryo large; with an epiblast; with a scutellar tail; with an elongated mesocotyl internode. Embryonic leaf margins meeting.

Abaxial leaf blade epidermis. *Costal/intercostal zonation* conspicuous. *Papillae* absent (?). Mid-intercostal long-cells rectangular; having markedly sinuous walls. *Microhairs* present; elongated; clearly two-celled; chloridoid-type. Microhair apical cell wall of similar thickness/rigidity to that of the basal cell. Microhair basal cells 15–18 microns long. Microhair total length/width at septum 5–6. Microhair apical cell/total length ratio about 0.4. *Stomata* common. Subsidiaries mostly dome-shaped (?). Intercostal short-cells not paired. Intercostal silica bodies absent. *Costal short-cells* conspicuously in long rows. Costal silica bodies present in alternate cell files of the costal zones; saddle shaped.

Transverse section of leaf blade, physiology. C_4. PCR sheath outlines even. PCR sheaths of the primary vascular bundles interrupted; interrupted both abaxially and adaxially. PCR sheath extensions absent. Mesophyll traversed by columns of colourless mesophyll cells. Bulliforms present in discrete, regular adaxial groups; associated with colourless mesophyll cells to form deeply-penetrating fans (these linked with traversing columns of colourless cells). The lamina margins with fibres.

Cytology. Chromosome base number, $x = 10$; 2 ploid.

Taxonomy. Chloridoideae; main chloridoid assemblage.

Ecology, geography, regional floristic distribution. 1 species (. μϑλτιφιδϑμ). Mexico. Xerophytic; species of open habitats. Dry scrub.

Neotropical. Caribbean.

References, etc. Morphological/taxonomic: Pierce 1978. Leaf anatomical: Pierce 1978.

Guaduella Franch.

Microbambus K. Schum.

Habit, vegetative morphology. Perennial (of zingiberaceous aspect); rhizomatous. The flowering culms leafless, or leafy (with some segregation into fertile/sterile-vegetative). *Culms herbaceous*; branched above to unbranched above. Culm nodes exposed, or hidden by leaf sheaths; glabrous. Primary branches/mid-culm node 1 (scarce). Culm sheaths persistent. Culm leaves present. Upper culm leaf blades fully developed. Culm internodes hollow. Rhizomes pachymorph. Plants unarmed. *Leaves* not basally aggregated; non-auriculate; without auricular setae. Leaf blades ovate-lanceolate to elliptic; broad; 25–75 mm wide; flat; pseudopetiolate; without cross venation; disarticulating from the sheaths. *Ligule a fringed membrane. Contra-ligule* absent.

Reproductive organization. Plants bisexual, with bisexual spikelets; with hermaphrodite florets.

Inflorescence. Inflorescence a single raceme, or paniculate (branching at the base); open; espatheate; not comprising 'partial inflorescences' and foliar organs. Spikelets solitary; not secund; shortly pedicellate, or sessile to subsessile.

Female-fertile spikelets. *Spikelets* morphologically 'conventional'; 20–90 mm long; compressed laterally to not noticeably compressed; disarticulating above the glumes; disarticulating between the florets; with distinctly elongated rachilla internodes between the florets. Rachilla prolonged beyond the uppermost female-fertile floret. *Callus* absent.

Glumes two; very unequal; shorter than the adjacent lemmas; free; hairy, or hairless; glabrous, or scabrous; pointed, or not pointed; awnless; carinate to non-carinate; similar. *Lower glume* about 0.5–0.75 times the length of the upper glume; 3 nerved, or 5 nerved, or 7 nerved. Upper glume 5 nerved, or 7 nerved, or 9 nerved, or 11 nerved. *Spikelets* with incomplete florets. The incomplete florets proximal to the female-fertile florets, or both distal and proximal to the female-fertile florets. *The distal incomplete florets* merely underdeveloped. **Spikelets with proximal incomplete florets.** *The proximal incomplete florets* 1–3; paleate; male. The proximal lemmas awnless; more or less equalling the female-fertile lemmas; similar in texture to the female-fertile lemmas.

Female-fertile florets 5–12. Lemmas ovate-lanceolate; decidedly firmer than the glumes; smooth; not becoming indurated; entire; pointed; awnless; hairy, or hairless; carinate to non-carinate; without a germination flap; 9 nerved, or 11–12 nerved; with the nerves non-confluent. *Palea* present; relatively long; entire (truncate), or apically notched; awnless, without apical setae (apically fringed); thinner than the lemma to textured like the lemma; not indurated; several nerved (5–11 nerved); 2-keeled. Palea back hairy. Palea keels more or less winged, or wingless; hairy. *Lodicules* present; 3; free; membranous; ciliate; not toothed. **Stamens 6.** Anthers 4–6 mm long; penicillate; with the connective apically prolonged. *Ovary* hairy. Styles free to their bases. Stigmas 2.

Fruit, embryo and seedling. Hilum long-linear. Embryo small.

Abaxial leaf blade epidermis. *Costal/intercostal zonation* conspicuous. *Papillae* present (in *G. oblonga* (Metcalfe 1960)), or absent (in the three species seen). Intercostal papillae over-arching the stomata. *Long-cells* markedly different in shape costally and intercostally (the costals much narrower); of similar wall thickness costally and intercostally (of medium thickness). Mid-intercostal long-cells rectangular; having markedly sinuous walls (coarsely, fairly evenly sinuous). *Microhairs* present; elongated; consistently uniseriate (all three celled, very large in *G. foliosa*); ostensibly panicoid type, but with two collapsing apical cells. *Stomata* common. Subsidiaries non-papillate; dome-shaped. Guard-cells overlapping to flush with the interstomatals (when recordable). *Intercostal short-cells* common; in cork/silica-cell pairs; silicified. Intercostal silica bodies saddle shaped (smaller and narrower than the costals). Small macrohairs and bulbous prickles abundant intercostally in *G. zenkeri*, scarce in *G. marantifolia*, absent in *G. foliosa*. *Costal short-cells* conspicuously in long rows. Costal silica bodies present and well developed; saddle shaped.

Transverse section of leaf blade, physiology. C_3; XyMS+. Mesophyll with adaxial palisade (fairly clear in *G. foliosa* and *G. marantifolia*), or without adaxial palisade; with arm cells (very conspicuous in *G. marantifolia*), or without arm cells (seemingly, in *G. foliosa*); with fusoids. The fusoids external to the PBS. *Leaf blade* adaxially flat; with the ribs more or less constant in size. *Midrib* conspicuous; with one bundle only (this large, in *G. zenkeri*), or having a conventional arc of bundles (three bundles, in *G. foliosa* and *G. marantifolia*), or having complex vascularization (*G. oblonga*); with colourless mesophyll adaxially. *The lamina* symmetrical on either side of the midrib. Bulliforms present in discrete, regular adaxial groups; in simple fans. All the vascular bundles accompanied by sclerenchyma. Combined sclerenchyma girders present; forming 'figures'.

Taxonomy. Bambusoideae; Bambusodae; Guaduelleae.

Ecology, geography, regional floristic distribution. 8 species. Tropical Africa. Shade species. Rainforest.

Paleotropical. African. West African Rainforest.

References, etc. Leaf anatomical: Metcalfe 1960; this project.

Special comments. Fruit data wanting.

Gymnachne L. Parodi

~ *Rhomboelytrum*

Habit, vegetative morphology. Perennial; caespitose. *Culms* 25–40 cm high; herbaceous; unbranched above. Young shoots intravaginal. *Leaves* not basally aggregated; non-auriculate; without auricular setae. Leaf blades linear; narrow; flat, or rolled (subconvolute); without cross venation; persistent. *Ligule an unfringed membrane*; truncate, or not truncate; 1.5–2 mm long.

Reproductive organization. Plants bisexual, with bisexual spikelets; with hermaphrodite florets. Exposed-cleistogamous.

Inflorescence. *Inflorescence paniculate*; contracted; spicate to more or less irregular; espatheate; not comprising 'partial inflorescences' and foliar organs. Spikelet-bearing axes persistent. Spikelets not secund.

Female-fertile spikelets. Spikelets 4–6.7 mm long; compressed laterally; disarticulating above the glumes; disarticulating between the florets; with conventional internode spacings. Rachilla prolonged beyond the uppermost female-fertile floret; rachilla hairless (glabrous). Hairy callus present. *Callus* short; blunt.

Glumes two; more or less equal; shorter than the spikelets; shorter than the adjacent lemmas, or long relative to the adjacent lemmas; hairless; scabrous; pointed to not pointed; awnless; carinate; similar. Lower glume longer than half length of lowest lemma; 3 nerved. Upper glume 3 nerved. *Spikelets* with female-fertile florets only, or with incomplete florets. The incomplete florets distal to the female-fertile florets. *The distal incomplete florets* (if present) merely underdeveloped; awnless.

Female-fertile florets 3–6. *Lemmas* lanceolate; similar in texture to the glumes; not becoming indurated; *incised*; briefly 2 lobed; not deeply cleft; awnless, or mucronate (in the notch); hairless; glabrous; carinate to non-carinate; without a germination flap; *(1–)3(–4) nerved*. Palea present; relatively long; tightly clasped by the lemma; minutely apically notched; awnless, without apical setae; not indurated; 2-nerved; 2-keeled. Palea keels wingless; hairy. *Lodicules* present; 2; free; membranous; glabrous; not toothed; not or scarcely vascularized. **Stamens 1**. Anthers about 0.6 mm long; not penicillate; without an apically prolonged connective. *Ovary* glabrous. Styles free to their bases. Stigmas 2.

Fruit, embryo and seedling. Fruit small (1.5 mm long); fusiform; longitudinally grooved; somewhat trigonous. Hilum short. Embryo small; with an epiblast; without a scutellar tail; with a negligible mesocotyl internode. Embryonic leaf margins meeting.

Abaxial leaf blade epidermis. *Costal/intercostal zonation* conspicuous. *Papillae* absent. *Long-cells* markedly different in shape costally and intercostally (the costals smaller, narrower, more regularly rectangular); of similar wall thickness costally and intercostally (fairly thick walled). Mid-intercostal long-cells rectangular and fusiform; having markedly sinuous walls (these conspicuously pitted). *Microhairs* absent. *Stomata* absent or very rare. *Intercostal short-cells* common; mostly in cork/silica-cell pairs (a few solitary, and a few short rows); silicified. Intercostal silica bodies crescentic. *Costal short-cells* conspicuously in long rows and neither distinctly grouped into long rows nor predominantly paired (some costae with long rows and high concentrations of silica bodies and only occasional interrrupting prickles, other costae with predominently short rows). Costal silica bodies horizontally-elongated crenate/sinuous (these predominating, but variable in lenght and shape to approach panicoid forms), or 'panicoid-type' (some unambiguously referable to these); when panicoid type, cross shaped, or dumb-bell shaped, or nodular.

Transverse section of leaf blade, physiology. C_3; XyMS+. *Mesophyll* with radiate chlorenchyma (in places); without adaxial palisade. *Leaf blade* with distinct, prominent adaxial ribs; with the ribs more or less constant in size (round topped). *Midrib* not readily distinguishable; with one bundle only. *The lamina* symmetrical on either side of the midrib. Bulliforms present in discrete, regular adaxial groups (a large group in every furrow); in simple fans. All the vascular bundles accompanied by sclerenchyma. Combined sclerenchyma girders present (with all the primaries); forming 'figures' (the larger bundles with slight I's). Sclerenchyma all associated with vascular bundles.

Taxonomy. Pooideae; Poodae; Poeae.

Ecology, geography, regional floristic distribution. 1 species. Chile. Neotropical. Andean.

References, etc. Leaf anatomical: this project.

Special comments. Nomenclatural and taxonomic confusion involves this genus and *Rhomboelytrum*. This morphological description includes *R. berteroanum* = *G. jaffuelii* or *G. filiformis*.

Gymnopogon P. Beauv.

Anthopogon Nutt., *Alloiantheros* Elliott ex Raf., *Biatherium* Desv., *Sciadonardus* Steud., *Monochaete* Doell., *Doellochloa* Kuntze

Habit, vegetative morphology. Annual (rarely), or perennial; caespitose (with short, scaly rhizomes). **Culms 20–60 cm high**; *herbaceous*; unbranched above. *Leaves* not basally aggregated; conspicuously distichous; non-auriculate. Sheaths keel-less. Leaf blades linear to ovate-lanceolate; broad, or narrow; cordate, or not cordate, not sagittate; not setaceous (short, stiff, often folded when dry); without abaxial multicellular glands; pseudopetiolate, or not pseudopetiolate. *Ligule an unfringed membrane, or a fringed membrane (?)*.

Reproductive organization. *Plants bisexual, with bisexual spikelets*; with hermaphrodite florets. The spikelets all alike in sexuality (except that the lowermost spikelets of the branches may be sterile). Exposed-cleistogamous, or chasmogamous.

Inflorescence. *Inflorescence of spicate main branches*; open; subdigitate to non-digitate; espatheate; not comprising 'partial inflorescences' and foliar organs. The racemes spikelet bearing to the base, or without spikelets towards the base. Spikelet-bearing axes persistent. Spikelets solitary (appressed); secund (in two rows along one side of rachis); biseriate; distant.

Female-fertile spikelets. *Spikelets 3–6 mm long; compressed laterally to not noticeably compressed*; disarticulating above the glumes; *not disarticulating between the florets*. Rachilla prolonged beyond the uppermost female-fertile floret. The rachilla extension with incomplete florets. Hairy callus present (short or linear), or absent (?).

Glumes two; more or less equal; about equalling the spikelets to exceeding the spikelets; *long relative to the adjacent lemmas*; free; dorsiventral to the rachis; pointed; awned, or awnless; *similar (narrow, acuminate)*. Lower glume 1 nerved. Upper glume 1 nerved. *Spikelets* with incomplete florets. The incomplete florets distal to the female-fertile florets. The distal incomplete florets 1, or 2; incomplete florets awned (the terminal rudiment sometimes 2–3 awned). *Spikelets without proximal incomplete florets*.

Female-fertile florets 1(–3). *Lemmas* not becoming indurated (membranous); *incised*; 2 lobed; not deeply cleft (minutely notched); mucronate (rarely), or awned (usually). *Awns* when present, 1; *from a sinus*; non-geniculate (?); much shorter than the body of the lemma to about as long as the body of the lemma. Lemmas hairy, or hairless; 3 nerved. *Palea* present; relatively long; 2-nerved. *Ovary* glabrous. Stigmas probably 2.

Fruit, embryo and seedling. Fruit ellipsoid; compressed dorsiventrally to not noticeably compressed. *Hilum short*. Endosperm hard; without lipid; containing only simple starch grains. Embryo with an epiblast; with a scutellar tail; with an elongated mesocotyl internode. Embryonic leaf margins meeting.

Abaxial leaf blade epidermis. *Costal/intercostal zonation* conspicuous (the intercostal zones very narrow). *Papillae* present; costal and intercostal. Intercostal papillae not over-arching the stomata; several per cell (on some costal and intercostal long-cells and some interstomatals, small, sometimes obscure). Long-cells of similar wall thickness costally and intercostally (thick walled). Mid-intercostal long-cells rectangular; having straight or only gently undulating walls (distant sinuations). *Microhairs* present; more or less spherical, or elongated; clearly two-celled; chloridoid-type. Microhair apical cell wall of similar thickness/rigidity to that of the basal cell. Microhairs about 19 microns long. Microhair basal cells 6–9 microns long. Microhairs about 7.5 microns wide at the septum. Microhair total length/width at septum 0.25. Microhair apical cells about 12.3 microns long. Microhair apical cell/total length ratio 0.66. *Stomata* common; 18–21 microns long. Subsidiaries predominantly triangular, or dome-shaped and triangular, or parallel-sided (in *G. ambiguus*, according to Metcalfe). Guard-cells overlapping to flush with the interstomatals. *Intercostal short-cells* absent or very rare (rare, solitary or paired). Intercostal silica bodies absent. *Costal short-cells* conspicuously in long rows. Costal silica bodies present in alternate cell files of the costal zones; 'panicoid-type'; dumb-bell shaped.

Transverse section of leaf blade, physiology. C_4; XyMS+. PCR sheath outlines even. PCR sheaths of the primary vascular bundles interrupted; interrupted abaxially only. PCR sheath extensions present, or absent. Maximum number of extension cells 1, or 2.

PCR cell chloroplasts centripetal. *Mesophyll* with radiate chlorenchyma; traversed by columns of colourless mesophyll cells (*G. ambiguus*), or not traversed by colourless columns (in *G. foliosus*, none of the columns seems to fully traverse). *Leaf blade* with distinct, prominent adaxial ribs (the ribs low, round topped); with the ribs more or less constant in size (slight ribs only). *Midrib* not readily distinguishable; with one bundle only. Bulliforms present in discrete, regular adaxial groups; in simple fans and associated with colourless mesophyll cells to form deeply-penetrating fans (sometimes linked with traversing colourless columns). All the vascular bundles accompanied by sclerenchyma. Combined sclerenchyma girders present (with all the bundles); forming 'figures' (all the bundles with conspicuous anchors). Sclerenchyma all associated with vascular bundles (but in *G. foliosus* the abaxial anchor groups are so wide as to form an almost continuous abaxial layer).

Phytochemistry. Leaves without flavonoid sulphates (2 species).

Cytology. Chromosome base number, $x = 10$.

Taxonomy. Chloridoideae; main chloridoid assemblage.

Ecology, geography, regional floristic distribution. 15 species. America, West Indies. Xerophytic; shade species and species of open habitats. Open places or light woodland on dry sandy soil.

Holarctic, Paleotropical, and Neotropical. Boreal. Indomalesian. Atlantic North American. Indian and Indo-Chinese. Caribbean, Venezuela and Surinam, Amazon, Central Brazilian, Pampas, and Andean. Southern Atlantic North American and Central Grasslands.

Rusts and smuts. Rusts — *Puccinia*. Taxonomically wide-ranging species: *Puccinia boutelouae*.

References, etc. Leaf anatomical: Metcalfe 1960 and this project.

Special comments. Lacking reliable data on ligule form, spikelet orientation, lemma form, number of stigmas, etc. Fruit data somewhat wanting.

Gynerium P. Beauv.

Gynerium Humb. & Bonpl.

Habit, vegetative morphology. Very large, soil-binding, perennial; *reeds*; rhizomatous. *Culms* 200–1000 cm high. Culm internodes solid. *Leaves* not basally aggregated (clustered towards the culm tips); non-auriculate; without auricular setae (but with tufts of longer hairs at the auricle positions). Leaf blades linear-lanceolate to lanceolate; broad; 4–8 mm wide (and to 2 m long, sharply serrulate); flat; without cross venation; disarticulating from the sheaths (the still-laminate leaves forming large flabelliform crowns on the culms). Ligule a fringed membrane to a fringe of hairs; to 4mm long. *Contra-ligule* present (a membrane), or absent.

Reproductive organization. *Plants dioecious*; without hermaphrodite florets. The spikelets all alike in sexuality (on the same plant); female-only, or male-only. Plants outbreeding.

Inflorescence. Inflorescence paniculate (very large); open; with capillary branchlets; espatheate; not comprising 'partial inflorescences' and foliar organs. Spikelet-bearing axes persistent. Spikelets not secund.

Female-sterile spikelets. Male spikelets with glabrous lemmas, short paleas, fleshy lodicules and two stamens. Male florets 2 staminate.

Female-fertile spikelets. Spikelets 6–9 mm long (approximately 8 mm); compressed laterally; disarticulating above the glumes; disarticulating between the florets. Rachilla prolonged beyond the uppermost female-fertile floret; rachilla hairless. The rachilla extension naked. Hairy callus present. *Callus* short.

Glumes two; very unequal; (the upper) long relative to the adjacent lemmas (exceeding them); hairless; pointed; awnless (but long attenuate), or awned; carinate (G_1), or non-carinate (G_2 rounded on the back); very dissimilar (the lower hyaline, the upper acuminate-subulate and recurved). *Lower glume* about 0.3 times the length of the upper glume; 1 nerved. Upper glume 3 nerved. *Spikelets* with female-fertile florets only.

Female-fertile florets 2. Lemmas similar in texture to the glumes; not becoming indurated; entire; pointed; awnless (but long-attenuate, the tip with long villous hairs); silky hairy; non-carinate (but keeled above); 5 nerved. *Palea* present; relatively long; apically notched; awnless, without apical setae; not indurated; 2-nerved; 2-keeled (and ciliate). *Lodicules* present; 2; free; membranous (poorly developed); ciliate; not toothed. *Stamens* 0 (2 staminodes). *Ovary* glabrous. Styles free to their bases. Stigmas 2; red pigmented.

Abaxial leaf blade epidermis. *Costal/intercostal zonation* conspicuous (but impossible to distinguish between costal and astomatal intercostal regions). *Papillae* absent. Mid-intercostal long-cells rectangular; having markedly sinuous walls. *Microhairs* present; elongated; clearly two-celled; panicoid-type (at least adaxially — the abaxial ones mostly damaged in material seen). *Stomata* common; 33–39 microns long. Subsidiaries dome-shaped (but becoming somewhat triangular at lower levels of focus). Guard-cells overlapping to flush with the interstomatals. *Costal short-cells* predominantly paired. Costal silica bodies rounded, saddle shaped, and crescentic.

Transverse section of leaf blade, physiology. C$_3$; XyMS+ (the mestome sheath often several cells thick, heavily lignified). *Mesophyll* with non-radiate chlorenchyma; without adaxial palisade; traversed by columns of colourless mesophyll cells (these very broad, extending from the adaxial bulliforms to join with conspicuous abaxial sclerenchyma columns, rather than extending from epidermis to epidermis); with arm cells (very conspicuous). *Leaf blade* with distinct, prominent adaxial ribs (the ribs low and broad). *Midrib* conspicuous (large, with a wide abaxial keel and big air spaces between the major vascular bundles); having complex vascularization (comprising an arc of several large primary bundles situated towards the adaxial surface, and an abaxial arc of numerous minor bundles some of which occur in the abaxial sclerenchyma girders of the primaries); with colourless mesophyll adaxially. *The lamina* symmetrical on either side of the midrib. Bulliforms present in discrete, regular adaxial groups (in each furrow); associated with colourless mesophyll cells to form deeply-penetrating fans (these linked with the traversing colourless columns). All the vascular bundles accompanied by sclerenchyma. Combined sclerenchyma girders present (with all the bundles, save some in the midrib); forming 'figures' (most bundles with very tall, narrow I's). Sclerenchyma not all bundle-associated. The 'extra' sclerenchyma in abaxial groups and in adaxial groups; abaxial-hypodermal, the groups continuous with colourless columns and adaxial-hypodermal, contiguous with the bulliforms (there being conspicuous intercostal abaxial 'girders' linked with the colourless columns, which constitute false costae in the abaxial epidermis, and some adaxial groups lining the sides of the bulliforms).

Phytochemistry. Leaves containing flavonoid sulphates (*G. sagittatum*).

Cytology. 2n = 72 and 76.

Taxonomy. Arundinoideae; Danthonieae.

Ecology, geography, regional floristic distribution. 1 species. Mexico to subtropical South America. Helophytic (streamsides and wet places).

Neotropical. Caribbean, Amazon, and Andean.

Economic importance. Used for construction work, basketry etc.

References, etc. Leaf anatomical: Metcalfe 1960; this project.

Special comments. Fruit data wanting.

Habrochloa C.E. Hubb.

Habit, vegetative morphology. Slender *annual*. *Culms* 5–50 cm high; herbaceous; unbranched above. *Leaves* not basally aggregated; non-auriculate. Leaf blades narrow; 2–2.5 mm wide (and 1–8 cm long); flat; without cross venation. *Ligule a fringe of hairs*.

Reproductive organization. Plants bisexual, with bisexual spikelets; with hermaphrodite florets. Exposed-cleistogamous.

Inflorescence. Inflorescence paniculate; delicate, open; with capillary branchlets; non-digitate; espatheate; not comprising 'partial inflorescences' and foliar organs. Spikelet-bearing axes persistent. Spikelets not secund.

Female-fertile spikelets. *Spikelets 2–2.5 mm long*; compressed laterally; disarticulating above the glumes; disarticulating between the florets; with conventional internode spacings. Rachilla prolonged beyond the uppermost female-fertile floret; rachilla hairless. The rachilla extension with incomplete florets. Hairy callus present (small). *Callus* short.

Glumes two; relatively large; more or less equal (the G_2 slightly shorter); shorter than the spikelets to about equalling the spikelets; *long relative to the adjacent lemmas (exceeding them)*; hairless; pointed (acute to acuminate); awnless; carinate; similar (linear-lanceolate, hyaline). Lower glume 1 nerved. Upper glume 1 nerved. *Spikelets* with incomplete florets. The incomplete florets distal to the female-fertile florets. *The distal incomplete florets* merely underdeveloped.

Female-fertile florets 2–5. Lemmas similar in texture to the glumes to decidedly firmer than the glumes (thinly membranous, translucent); not becoming indurated; incised; very shortly 2 lobed (the lobes themselves obtuse or minutely bidentate); not deeply cleft; awned. *Awns* 1; median; from a sinus; non-geniculate; gently recurving, or flexuous; hairless (scabrid); *much longer than the body of the lemma (to about 7 mm long, very slender)*. Lemmas hairy; carinate; *3 nerved. Palea* present; relatively long; entire to apically notched; awnless, without apical setae; not indurated (thinly membranous); 2-nerved; 2-keeled. *Lodicules* present; 2; free; fleshy; glabrous. *Stamens* 3. Anthers 0.1–0.3 mm long; not penicillate; without an apically prolonged connective. *Ovary* glabrous. Styles free to their bases. Stigmas 2.

Fruit, embryo and seedling. Fruit small (0.8–1 mm long); linear, or ellipsoid; trigonous. Hilum short. Embryo small (about 1/4 of the fruit length).

Abaxial leaf blade epidermis. *Costal/intercostal zonation* conspicuous. *Papillae* absent. Mid-intercostal long-cells rectangular and fusiform; having markedly sinuous walls (but the sinuations gentle). *Microhairs* present; panicoid-type (the apical cells all missing in material seen); 30(–37.5) microns long; 5.4 microns wide at the septum. Microhair total length/width at septum 5.56–6.94. Microhair apical cells 16.5–21 microns long. Microhair apical cell/total length ratio 0.55–0.6. *Stomata* absent or very rare (though surprisingly so, in this relatively thin-walled epidermis); 21–22.5 microns long. *Intercostal short-cells* absent or very rare (none seen). *Costal short-cells* conspicuously in long rows (though some files have relatively long short-cells). Costal silica bodies 'panicoid-type'; more or less exclusively dumb-bell shaped (large).

Transverse section of leaf blade, physiology. Almost certainly C_3 (though the material seen poor); XyMS+. *Leaf blade* adaxially flat. *Midrib* not readily distinguishable; with one bundle only. Bulliforms not in discrete, regular adaxial groups. Combined sclerenchyma girders present; forming 'figures' (small anchors in the primaries only, the other bundles depauperate in sclerenchyma). Sclerenchyma all associated with vascular bundles.

Taxonomy. Arundinoideae; Danthonieae.

Ecology, geography, regional floristic distribution. 1 species. East tropical Africa. Shade species; glycophytic. On slopes.

Paleotropical. African. Sudano-Angolan. South Tropical African.

References, etc. Morphological/taxonomic: Hubbard 1967a. Leaf anatomical: this project.

Hackelochloa Kuntze

Rytilix Rafin.
~ *Mnesithea*

Habit, vegetative morphology. *Annual*; caespitose. *Culms* (5–)30–100 cm high; herbaceous; branched above. Culm nodes hairy, or glabrous. Culm internodes solid. *Leaves* not basally aggregated; non-auriculate. Leaf blades linear-lanceolate; broad, or narrow; up to 10–15 mm wide; flat; without cross venation; persistent. Ligule a fringed membrane.

Reproductive organization. *Plants bisexual, with bisexual spikelets*; with hermaphrodite florets. The spikelets of sexually distinct forms on the same plant, or all alike in sexuality (the pairs sometimes homogamous); hermaphrodite, or hermaphrodite and male-only, or hermaphrodite, male-only, and sterile. The male and female-fertile spikelets mixed in the inflorescence. The spikelets overtly heteromorphic (the pedicellate spikelets narrowly ovate and winged, herbaceous); in both homogamous and heterogamous combinations, or all in heterogamous combinations.

Inflorescence. Inflorescence paniculate (the numerous 'racemes' solitary in their spathes, usually in fascicles). Rachides hollowed. *Inflorescence spatheate*; a complex of 'partial inflorescences' and intervening foliar organs. *Spikelet-bearing axes* spikelike; solitary (in their spathes); with substantial rachides; disarticulating; disarticulating at the joints (the sessile spikelets falling with the joint and the pedicelled spikelet). **'Articles'** non-linear (grooved, concave); with a basal callus-knob; *disarticulating obliquely*; glabrous. Spikelets paired; secund (the sessile members in two alternating rows, on one side of the rachis); consistently in 'long-and-short' combinations; in pedicellate/sessile combinations. Pedicels of the 'pedicellate' spikelets discernible, but fused with the rachis. The 'shorter' spikelets hermaphrodite. The 'longer' spikelets hermaphrodite (the glumes herbaceous and similar, the lower lemma present or absent), or male-only, or sterile.

Female-sterile spikelets. The pedicellate spikelets sometimes male-only or sterile.

Female-fertile spikelets. Spikelets 1–3 mm long; abaxial; compressed dorsiventrally (to globose); falling with the glumes. Rachilla terminated by a female-fertile floret.

Glumes two; relatively large; not greatly different in length, but G$_2$ hooded by G$_1$; long relative to the adjacent lemmas; awnless; *very dissimilar (the lower cartilaginous, globose, reticulate-pitted, the upper thinner and embedded in the axis)*. **Lower glume not two-keeled**; convex on the back; lacunose with deep depressions; 7–10 nerved. Upper glume 3 nerved. *Spikelets* with incomplete florets. The incomplete florets proximal to the female-fertile florets. *The proximal incomplete florets* 1; epaleate; sterile. The proximal lemmas awnless; 0 nerved; decidedly exceeding the female-fertile lemmas; similar in texture to the female-fertile lemmas (hyaline); not becoming indurated.

Female-fertile florets 1. **Lemmas** ovate-obtuse; less firm than the glumes (hyaline, flimsy); not becoming indurated; entire; blunt; awnless; hairless; non-carinate; without a germination flap; *0 nerved*. *Palea* present, or absent; when present, relatively long, or conspicuous but relatively short, or very reduced; entire; awnless, without apical setae; not indurated (hyaline); nerveless (hyaline, flimsy). *Lodicules* present; fleshy. *Stamens* 3. Anthers not penicillate. *Ovary* glabrous. Styles free to their bases. Stigmas 2.

Fruit, embryo and seedling. Fruit compressed dorsiventrally. Hilum short. Embryo large; without an epiblast; with a scutellar tail; with an elongated mesocotyl internode. Embryonic leaf margins overlapping.

Abaxial leaf blade epidermis. *Costal/intercostal zonation* conspicuous. *Papillae* absent. *Long-cells* similar in shape costally and intercostally; of similar wall thickness costally and intercostally (fairly thin walled). Mid-intercostal long-cells rectangular; having markedly sinuous walls. *Microhairs* present; panicoid-type; 43.5–54 microns long; 4.5–7.5 microns wide at the septum. Microhair total length/width at septum 6.4–10.3. Microhair apical cells 21–34.5 microns long. Microhair apical cell/total length ratio 0.44–0.72. *Stomata* common; 24–39 microns long. Subsidiaries triangular. Guard-cells overlapping to flush with the interstomatals. *Intercostal short-cells* common; in cork/silica-cell pairs; silicified. Intercostal silica bodies cross-shaped. *Costal short-cells* conspicuously in long rows. Costal silica bodies 'panicoid-type'.

Transverse section of leaf blade, physiology. C$_4$; XyMS–. *Mesophyll* with radiate chlorenchyma. *Leaf blade* adaxially flat. *Midrib* conspicuous; with one bundle only, or having a conventional arc of bundles (rarely). Bulliforms present in discrete, regular adaxial groups (and also in irregular groups); in simple fans (some of these of the large celled, *Zea*-type). Many of the smallest vascular bundles unaccompanied by sclerenchyma. Combined sclerenchyma girders present; forming 'figures'. Sclerenchyma all associated with vascular bundles.

Special diagnostic feature. Lower glume of female-fertile spikelet globose, pitted.

Cytology. Chromosome base number, $x = 7$ (?). $2n = 14$ (?). 2 ploid.

Taxonomy. Panicoideae; Andropogonodae; Andropogoneae; Rottboelliinae.

Ecology, geography, regional floristic distribution. 2 species. Tropics, southern China and southern U.S.A. Helophytic to mesophytic; species of open habitats; glycophytic. Grassland and disturbed ground.

Paleotropical, Neotropical, and Australian. African, Madagascan, and Indomalesian. Saharo-Sindian, Sudano-Angolan, and West African Rainforest. Indian, Indo-Chinese,

Malesian, and Papuan. Caribbean, Central Brazilian, and Andean. North and East Australian. Sahelo-Sudanian, Somalo-Ethiopian, and South Tropical African. Tropical North and East Australian.

Rusts and smuts. Rusts — *Puccinia*. Taxonomically wide-ranging species: *Puccinia levis*. Smuts from Ustilaginaceae. Ustilaginaceae — *Sphacelotheca*.

References, etc. Leaf anatomical: this project.

Hainardia Greuter

Monerma (Willd.) Coss. & Dur.

Habit, vegetative morphology. Annual; caespitose. *Culms* 5–40(–47) cm high; herbaceous; branched above; 2–5 noded. Culm nodes exposed; glabrous. Culm internodes solid. Young shoots intravaginal. *Leaves* not basally aggregated; non-auriculate. Sheaths keeled or not, terete to inflated. Leaf blades linear; narrow; 1.5–2.5 mm wide; flat, or rolled (convolute); without cross venation; persistent. Ligule an unfringed membrane; truncate; (0.2–)0.6–1 mm long.

Reproductive organization. Plants bisexual, with bisexual spikelets; with hermaphrodite florets. The spikelets of sexually distinct forms on the same plant (with rudiments at the base of the spikelet), or all alike in sexuality; hermaphrodite, or hermaphrodite and sterile.

Inflorescence. *Inflorescence a single spike (with a hard, cylindrical, articulated rachis, the spikelets embedded in alternate notches)*. Rachides hollowed (notched). Inflorescence espatheate; not comprising 'partial inflorescences' and foliar organs. *Spikelet-bearing axes disarticulating*; disarticulating at the joints. Spikelets solitary; not secund; distichous.

Female-fertile spikelets. *Spikelets* 4–8 mm long; adaxial (with reference to the missing G_1); *compressed dorsiventrally*; falling with the glumes. Rachilla prolonged beyond the uppermost female-fertile floret. The rachilla extension with incomplete florets, or naked. Hairy callus absent. *Callus* absent.

Glumes one per spikelet (the leathery G_2, which closes the hollow of the rachis); relatively large (firm); long relative to the adjacent lemmas; dorsiventral to the rachis; hairless; pointed (acuminate); awnless; non-carinate. Upper glume 3–7(–9) nerved (the nerves raised). *Spikelets* with female-fertile florets only, or with incomplete florets. The incomplete florets when present, distal to the female-fertile florets. The distal incomplete florets 1; *incomplete florets merely underdeveloped (rudimentary)*.

Female-fertile florets 1. Lemmas lanceolate; less firm than the glumes (membranous); not becoming indurated; entire; pointed; awnless; hairless; non-carinate; 3(–5) nerved; with the nerves non-confluent (the laterals short). *Palea* present; relatively long; entire to apically notched; awnless, without apical setae; thinner than the lemma (hyaline); not indurated; 2-nerved; weakly 2-keeled. Palea keels wingless. *Lodicules* present; 2; free; membranous; glabrous; toothed, or not toothed. *Stamens* 1–3. Anthers 1.8–3.5 mm long; not penicillate. *Ovary* glabrous. Styles free to their bases. Stigmas 2; white.

Fruit, embryo and seedling. *Fruit* free from both lemma and palea; small to medium sized (3–4.3 mm long); oblong to ellipsoid; not grooved; compressed dorsiventrally to not noticeably compressed. Hilum short (ovate to elliptic, about 0.6 mm long). Embryo small; not waisted. Endosperm hard; with lipid; containing compound starch grains. Embryo with an epiblast.

First seedling leaf with a well-developed lamina. The lamina narrow; erect.

Abaxial leaf blade epidermis. *Costal/intercostal zonation* conspicuous. *Papillae* absent. *Long-cells* similar in shape costally and intercostally; of similar wall thickness costally and intercostally (fairly thick walled). Mid-intercostal long-cells rectangular; having markedly sinuous walls. *Microhairs* absent. *Stomata* common. Subsidiaries parallel-sided. Guard-cells overlapped by the interstomatals. *Intercostal short-cells* common; in cork/silica-cell pairs; silicified. *Costal short-cells* predominantly paired. Costal silica bodies rounded and tall-and-narrow.

Transverse section of leaf blade, physiology. C_3; XyMS+. *Mesophyll* with non-radiate chlorenchyma; without adaxial palisade. *Leaf blade* with distinct, prominent adaxial ribs, or 'nodular' in section; with the ribs more or less constant in size. *Midrib* not readily distinguishable; with one bundle only. Bulliforms present in discrete, regular adaxial groups (at the bases of the furrows); in simple fans. All the vascular bundles accompanied by sclerenchyma. Combined sclerenchyma girders present; forming 'figures'. Sclerenchyma all associated with vascular bundles.

Cytology. Chromosome base number, $x = 13$. $2n = 26$ and 52. 2 and 4 ploid.

Taxonomy. Pooideae; Poodae; Poeae.

Ecology, geography, regional floristic distribution. 1 species. Mediterranean to Iraq. Commonly adventive. Mesophytic; species of open habitats. Meadows, etc., often coastal.

Holarctic. Boreal and Tethyan. Euro-Siberian. Macaronesian and Mediterranean. European.

Rusts and smuts. Smuts from Ustilaginaceae. Ustilaginaceae — *Ustilago*.

References, etc. Leaf anatomical: this project.

Hakonechloa Makino

Habit, vegetative morphology. Perennial; rhizomatous. **Culms** 30–60 cm high; *herbaceous*; to 2 cm in diameter; unbranched above. Culm nodes glabrous. Rhizomes leptomorph. Plants unarmed. *Leaves* not basally aggregated; non-auriculate. Leaf blades narrow; 2–8 mm wide; without cross venation. *Ligule a fringed membrane.* Contraligule present (the adaxial ligule continued as a fringe of hairs around the back).

Reproductive organization. Plants bisexual, with bisexual spikelets; with hermaphrodite florets.

Inflorescence. *Inflorescence paniculate (small)*; open; with capillary branchlets; espatheate; not comprising 'partial inflorescences' and foliar organs. Spikelet-bearing axes persistent. Spikelets not secund.

Female-fertile spikelets. *Spikelets 9–11 mm long*; compressed laterally; disarticulating above the glumes; disarticulating between the florets. Rachilla prolonged beyond the uppermost female-fertile floret; rachilla hairy (with long white hairs, above the disarticulation zones). The rachilla extension with incomplete florets. Hairy callus present (slender).

Glumes two; very unequal; shorter than the adjacent lemmas; hairless; glabrous; pointed; awnless; non-carinate; *very dissimilar*. Lower glume 3 nerved. Upper glume 3 nerved. *Spikelets* with incomplete florets. The incomplete florets distal to the female-fertile florets. *The distal incomplete florets* merely underdeveloped.

Female-fertile florets 3–6. Lemmas acuminate; similar in texture to the glumes (herbaceous/membranous); not becoming indurated; entire (or sometimes with a slight notch); pointed; awned. Awns 1; median; apical; non-geniculate; hairless (scabrid); much shorter than the body of the lemma to about as long as the body of the lemma. Lemmas hairy (along the margins); non-carinate (rounded); having the margins lying flat and exposed on the palea; without a germination flap; 3 nerved. *Palea* present; relatively long; slightly apically notched; awnless, without apical setae; not indurated (thinly membranous); 2-nerved; 2-keeled. *Lodicules* present; free; fleshy; glabrous; heavily vascularized. *Stamens* 3. Anthers not penicillate; without an apically prolonged connective. *Ovary* glabrous. Styles free to their bases. Stigmas 2; brown (apparently).

Fruit, embryo and seedling. *Fruit* free from both lemma and palea. *Pericarp free.* Embryo without an epiblast; with a scutellar tail; with an elongated mesocotyl internode. Embryonic leaf margins meeting.

Abaxial leaf blade epidermis. *Costal/intercostal zonation* conspicuous. *Papillae* absent. Long-cells of similar wall thickness costally and intercostally. Mid-intercostal long-cells rectangular; having markedly sinuous walls. *Microhairs* present; panicoid-type (both cells long, the apical very flimsy); (44–)48–60(–62) microns long. Microhair apical cells 29–39 microns long. Microhair apical cell/total length ratio 0.62. *Stomata* absent or very rare. *Intercostal short-cells* common; not paired (solitary); not silicified. Costal prickles present. *Crown cells* absent. *Costal short-cells* conspicuously in long rows (some veins), or neither distinctly grouped into long rows nor predominantly paired (the long-cells often short). Costal silica bodies 'panicoid-type'; cross shaped, butterfly shaped, and dumb-bell shaped, or nodular.

Transverse section of leaf blade, physiology. C_3; XyMS+. *Mesophyll* with non-radiate chlorenchyma; with fusoids (seemingly), or without fusoids (need to see fresh material). *Leaf blade* adaxially flat (low abaxial ribs). *Midrib* conspicuous (a larger bundle, with a slight double keel); with one bundle only. Bulliforms present in discrete, regular adaxial groups; in simple fans. All the vascular bundles accompanied by sclerenchyma. Combined sclerenchyma girders present (with all the bundles); forming 'figures' (in the midrib). Sclerenchyma all associated with vascular bundles.

Phytochemistry. Leaves without flavonoid sulphates (*H. macra*).

Cytology. $2n = 50$.

Taxonomy. Arundinoideae; Danthonieae.

Ecology, geography, regional floristic distribution. 1 species. Japan. Mesophytic; glycophytic. On wet cliffs.

Holarctic. Boreal. Eastern Asian.

References, etc. Leaf anatomical: Metcalfe 1960; this project.

Special comments. Fruit data wanting.

Halopyrum Stapf

Habit, vegetative morphology. Perennial; stoloniferous and caespitose. *Culms herbaceous (tough)*. Leaves non-auriculate. *Leaf blades* narrow (junciform); rolled (convolute, stiff, filiform-tipped); without abaxial multicellular glands; without cross venation; *disarticulating from the sheaths*. *Ligule a fringe of hairs*.

Reproductive organization. Plants bisexual, with bisexual spikelets; with hermaphrodite florets.

Inflorescence. Inflorescence of spicate main branches (the short, erect branches being themselves unbranched), or paniculate (elongated, contracted); non-digitate; espatheate; not comprising 'partial inflorescences' and foliar organs. Spikelet-bearing axes persistent. Spikelets solitary; not secund.

Female-fertile spikelets. Spikelets 10–30 mm long; compressed laterally; disarticulating above the glumes; disarticulating between the florets. Rachilla prolonged beyond the uppermost female-fertile floret; rachilla hairy (the internodes bearded at the tip, with long hairs up to half the lemma length). The rachilla extension with incomplete florets. *Hairy callus present*.

Glumes two; *more or less equal (subequal)*; shorter than the spikelets; shorter than the adjacent lemmas to long relative to the adjacent lemmas; hairless; pointed (acuminate); awnless; carinate; similar (leathery, ovate-lanceolate). Lower glume shorter than the lowest lemma to about equalling the lowest lemma; 3–5 nerved. Upper glume 5–7 nerved. *Spikelets* with incomplete florets. The incomplete florets distal to the female-fertile florets. *The distal incomplete florets* merely underdeveloped.

Female-fertile florets 6–20. *Lemmas* ovate-oblong, acute; similar in texture to the glumes (leathery); entire, or incised; pointed; if incised 2 lobed; not deeply cleft (bidenticulate); *mucronate*; hairy (asperulous or minutely hairy); carinate; 3 nerved. *Palea* present; relatively long; entire; awnless, without apical setae; 2-nerved; 2-keeled (the keels scaberulous). *Lodicules* present; 2; free; fleshy; glabrous. *Stamens* 3. *Ovary* glabrous. Styles free to their bases. Stigmas 2.

Fruit, embryo and seedling. *Fruit* ellipsoid; *longitudinally grooved*; compressed dorsiventrally (concavo-convex). *Hilum short*. Pericarp fused. Embryo large; with an epiblast; with a scutellar tail; with an elongated mesocotyl internode. Embryonic leaf margins meeting.

Abaxial leaf blade epidermis. *Costal/intercostal zonation* conspicuous. *Papillae* present (in the deep grooves, and abundant adaxially); intercostal. Intercostal papillae over-arching the stomata; consisting of one oblique swelling per cell, or consisting of one symmetrical projection per cell (finger-like). Long-cells differing markedly in wall thickness costally and intercostally (the costals thick-walled). *Microhairs* present (in deep intercostal grooves); more or less spherical; clearly two-celled; chloridoid-type. Microhair apical cell wall of similar thickness/rigidity to that of the basal cell. Microhairs

27–33–34.5 microns long; 21–22.5 microns wide at the septum. Microhair total length/ width at septum 1.2–1.6. Microhair apical cells 17.4–27 microns long. Microhair apical cell/total length ratio 0.54–0.78. *Stomata* absent or very rare. Intercostal silica bodies absent. *Costal short-cells* neither distinctly grouped into long rows nor predominantly paired. Costal silica bodies present throughout the costal zones; saddle shaped (mainly), or tall-and-narrow, or 'panicoid-type' (a few).

Transverse section of leaf blade, physiology. C_4; XyMS+. PCR sheaths of the primary vascular bundles interrupted; interrupted abaxially only. PCR sheath extensions present. Maximum number of extension cells 1–5. *Mesophyll* with radiate chlorenchyma; traversed by columns of colourless mesophyll cells. *Leaf blade* 'nodular' in section; with the ribs very irregular in sizes. *Midrib* not readily distinguishable; with one bundle only. Bulliforms present in discrete, regular adaxial groups; in simple fans and associated with colourless mesophyll cells to form deeply-penetrating fans (sometimes linked with traversing columns of colourless cells). All the vascular bundles accompanied by sclerenchyma. Combined sclerenchyma girders present; forming 'figures'. Sclerenchyma all associated with vascular bundles. The lamina margins with fibres.

Taxonomy. Chloridoideae; main chloridoid assemblage.

Ecology, geography, regional floristic distribution. 1 species. Coastal Indian Ocean. Xerophytic; species of open habitats; halophytic. A coastal sand stabilizer.

Holarctic and Paleotropical. Tethyan. African, Madagascan, and Indomalesian. Irano-Turanian. Saharo-Sindian and Sudano-Angolan. Indian. Sahelo-Sudanian, Somalo-Ethiopian, and South Tropical African.

References, etc. Morphological/taxonomic: Stapf 1896. Leaf anatomical: this project.

Harpachne Hochst.

Habit, vegetative morphology. Perennial; caespitose. *Culms* 20–75 cm high; herbaceous; unbranched above. Culm nodes glabrous. Plants unarmed. Young shoots intravaginal. *Leaves* not basally aggregated; non-auriculate. Leaf blades linear; narrow; 2–5 mm wide; exhibiting multicellular glands abaxially (maybe at base of hairs?). The abaxial glands intercostal. Leaf blades without cross venation; persistent. Ligule a fringe of hairs. *Contra-ligule* present (of inconspicuous hairs, more pronounced laterally).

Reproductive organization. *Plants bisexual, with bisexual spikelets*; with hermaphrodite florets.

Inflorescence. *Inflorescence a single raceme, or paniculate, or a false spike, with spikelets on contracted axes (bottlebrush-like, spiciform, the lowermost pedicels sometimes branching, the solitary or paired spikelets interpretable as reduced 'clusters')*; espatheate; not comprising 'partial inflorescences' and foliar organs. Spikelet-bearing axes persistent (the main axis persistent), or disarticulating (if the deciduous spikelet-plus-pedicel is interpreted as a reduced branch); (if interpreted as reduced branches) falling entire. *Spikelets* almost in verticils on the rachis; *secund (symmetrically disposed, but turned to one side and deflexing)*.

Female-fertile spikelets. *Spikelets* 5–11 mm long; purplish; strongly compressed laterally; *falling with the glumes (abscising near the bases of the 'pedicels', to leave hairy stumps (true pedicels?) on the rachis)*; disarticulating between the florets; with distinctly elongated rachilla internodes between the florets (curved). Rachilla prolonged beyond the uppermost female-fertile floret; rachilla hairless (glabrous, flexuous). Hairy callus present (consisting of the main part of the 'pedicel'). *Callus* long; pointed.

Glumes present; two; very unequal (G_1 about two-thirds the length of G_2); shorter than the spikelets; shorter than the adjacent lemmas; hairless; glabrous; not pointed (rounded); awnless; carinate; similar (membranous, narrow, oblong). Lower glume not two-keeled; 1 nerved. Upper glume 1 nerved. *Spikelets* with incomplete florets. The incomplete florets distal to the female-fertile florets, or both distal and proximal to the female-fertile florets (the L_1 being smaller than those above, and sometimes lacking stamens in the material seen). *The distal incomplete florets* merely underdeveloped. *The proximal incomplete florets* when present, 1.

Female-fertile florets *4–16 (increasing in size acropetally)*. Lemmas enlarged at the base, then acuminate to the tip; similar in texture to the glumes to decidedly firmer than the glumes (membranous, or cartilaginous below); smooth; not becoming indurated; entire; pointed; mucronate, or awned. Awns when present, 1; apical; geniculate; much shorter than the body of the lemma. Lemmas hairless; glabrous; carinate; without a germination flap; 3 nerved. **Palea** present; *conspicuous but relatively short (to about two-thirds of the lemma length)*; entire; awnless, without apical setae; not indurated (membranous); 2-nerved (and deeply hollowed between); 2-keeled. Palea keels winged (the wings ciliolate). *Lodicules* present; 2; free; fleshy; glabrous. *Stamens* 3; with free filaments (short). Anthers relatively long; not penicillate; without an apically prolonged connective. *Ovary* glabrous. Styles free to their bases. Stigmas 2; white.

Fruit, embryo and seedling. *Fruit* small (0.75–1.3 mm long); *compressed laterally*. Hilum short. Pericarp fused. Embryo large.

Abaxial leaf blade epidermis. *Costal/intercostal zonation* conspicuous. *Papillae* absent. *Long-cells* similar in shape costally and intercostally; of similar wall thickness costally and intercostally (medium thick walled). Mid-intercostal long-cells rectangular; having markedly sinuous walls (pitted). *Microhairs* present; elongated; clearly two-celled; panicoid-type to chloridoid-type (the apical cells round-tipped, not collapsing, short (chlorioid) to quite elongated (panicoid)). Microhair apical cell wall thinner than that of the basal cell but not tending to collapse. Microhair basal cells 18–21 microns long. Microhair total length/width at septum 4. Microhair apical cell/total length ratio 0.5. *Stomata* common. Subsidiaries high dome-shaped and triangular. Guard-cells overlapping to flush with the interstomatals. *Intercostal short-cells* common; not paired (mostly solitary); not silicified (apparently). Intercostal silica bodies absent. *Costal short-cells* predominantly paired. Costal silica bodies present throughout the costal zones (mostly), or present in alternate cell files of the costal zones; saddle shaped (plus various imperfect forms), or rounded (merging into broad saddles), or crescentic (a few).

Transverse section of leaf blade, physiology. C_4; XyMS+. PCR sheaths of the primary vascular bundles interrupted; interrupted abaxially only. PCR sheath extensions present. Maximum number of extension cells 1–2. *Leaf blade* adaxially flat. *Midrib* conspicuous (somewhat: a wider abaxial rib and a large anchor of abaxial sclerenchyma); with one bundle only. Bulliforms present in discrete, regular adaxial groups (between all the bundles); in simple fans (but the median cels of the simple fans deeply penetrating). All the vascular bundles accompanied by sclerenchyma. Combined sclerenchyma girders present (with all the main bundles); forming 'figures' (all the main bundles). Sclerenchyma all associated with vascular bundles.

Taxonomy. Chloridoideae; main chloridoid assemblage.

Ecology, geography, regional floristic distribution. 2 species. Tropical Africa. Species of open habitats. Savanna.

Paleotropical. African. Saharo-Sindian and Sudano-Angolan. Sahelo-Sudanian, Somalo-Ethiopian, and South Tropical African.

References, etc. Leaf anatomical: this project.

Harpochloa Kunth

Habit, vegetative morphology. Perennial; densely caespitose. *Culms* 30–90 cm high; herbaceous; unbranched above. Culm nodes glabrous. Culm internodes solid. Plants unarmed. Young shoots intravaginal. *Leaves* mostly basal; non-auriculate (but with auricular hairs). The basal sheaths persistent, keeled. Leaf blades stiffly linear; narrow; 1–4 mm wide (by 7–25 cm long); flat, or rolled; without abaxial multicellular glands; without cross venation; persistent. *Ligule a fringed membrane*; truncate; 0.2–0.3 mm long. *Contra-ligule* absent.

Reproductive organization. *Plants bisexual, with bisexual spikelets*; with hermaphrodite florets.

Inflorescence. *Inflorescence a single spike (very rarely two)*; usually non-digitate. *Inflorescence axes not ending in spikelets (conspicuously bare-tipped). Rachides hollowed and flattened (crescentic in section, the edges ciliate).* Inflorescence espatheate; not comprising 'partial inflorescences' and foliar organs. Spikelet-bearing axes persistent. Spikelets solitary; secund; alternately biseriate (along the midrib of the rachis); very densely imbricate.

Female-fertile spikelets. *Spikelets* 6–7 mm long (darkly pigmented); adaxial; compressed laterally; *disarticulating above the glumes*; not disarticulating between the florets; with conventional internode spacings. Rachilla prolonged beyond the uppermost female-fertile floret; rachilla hairless. The rachilla extension with incomplete florets. Hairy callus absent. *Callus* absent.

Glumes two (dark grey-green); very unequal (G_2 much larger); about equalling the spikelets (i.e. the upper glumes); (the upper) long relative to the adjacent lemmas; lateral to the rachis; hairless (G_2 keels scabrid); glabrous; without conspicuous tufts or rows of hairs; pointed; awnless; carinate (G_1), or non-carinate (the G_2 being 2-keeled); very dissimilar (the G_1 smaller, 1-keeled, thinner, the G_2 2-keeled, firm). Lower glume 1 nerved. Upper glume 3 nerved. *Spikelets* with incomplete florets. *The incomplete florets distal to the female-fertile florets (the second and third male, the fourth if present sterile, these all enclosed in the lemma of the lower male floret and not exceeding the L_1).* The distal incomplete florets 2–3; *incomplete florets* merely underdeveloped; incomplete florets awnless.

Female-fertile florets 1. Lemmas similar in texture to the glumes (firmly membranous); not becoming indurated (membranous to papery, grey); entire; blunt; awnless; hairy (the L_1 being long-ciliate on its keel and margins); carinate (folded); without a germination flap; 3 nerved; with the nerves non-confluent. *Palea* present (hairy near its tip); relatively long; slightly apically notched; awnless, without apical setae; textured like the lemma; not indurated (membranous); 2-nerved; 2-keeled. Palea keels slightly winged; glabrous. *Lodicules* present; free; fleshy (winged); glabrous; not or scarcely vascularized. *Stamens* 3 (in hermaphrodite and male florets). Anthers to 2.5 mm long (larger than in the male florets); not penicillate; without an apically prolonged connective. *Ovary* glabrous. Styles free to their bases. Stigmas 2; brown.

Fruit, embryo and seedling. *Fruit* free from both lemma and palea (but embraced); small (3 mm long); ellipsoid; obtusely triquetrous. Hilum short. Pericarp fused. Embryo large.

Abaxial leaf blade epidermis. *Costal/intercostal zonation* conspicuous. *Papillae* present; costal and intercostal. Intercostal papillae over-arching the stomata; consisting of one symmetrical projection per cell (long-cells and interstomatals mostly each with one large, tall, thick-walled cylindrical papilla). *Long-cells* similar in shape costally and intercostally; of similar wall thickness costally and intercostally (walls thick, pitted). Intercostal zones with typical long-cells (though these rather short). Mid-intercostal long-cells rectangular; having markedly sinuous walls. *Microhairs* present; more or less spherical; ostensibly one-celled; chloridoid-type. *Stomata* common. Subsidiaries triangular (detectable despite the papillae). Guard-cells overlapping to flush with the interstomatals. *Intercostal short-cells* common; not paired (solitary). Intercostal silica bodies absent. *Costal short-cells* conspicuously in long rows. Costal silica bodies confined to the central file(s) of the costal zones; saddle shaped.

Transverse section of leaf blade, physiology. C_4. The anatomical organization somewhat unconventional. Organization of PCR tissue almost *Triodia* type (though no linking across adjacent PCR sheaths seen). XyMS+. PCR sheaths of the primary vascular bundles interrupted; interrupted both abaxially and adaxially. PCR sheath extensions absent. PCR cell chloroplasts centripetal. Mesophyll perhaps with arm cells (?). *Leaf blade* adaxially flat (except over the midrib, but with deep abaxial clefts between the bundles). *Midrib* conspicuous (by its position, and its small adaxial ribs); having a conventional arc of bundles; with colourless mesophyll adaxially (the entire adaxial two-thirds of the blade comprising 'colourless tissue' of large (apparently lignified) cells, which broadly truncates the PCR sheaths). Bulliforms not in discrete, regular adaxial groups (no bulliforms seen). All the vascular bundles accompanied by sclerenchyma. Combined sclerenchyma girders absent (all bundles with abaxial girders only). Sclerenchyma not all bundle-associated. The 'extra' sclerenchyma in adaxial groups (there being an irregularly discontinuous adaxial hypodermis of thick-walled fibres).

Cytology. Chromosome base number, $x = 10$. $2n = 40$.

Taxonomy. Chloridoideae; main chloridoid assemblage.

Ecology, geography, regional floristic distribution. 1 species. Southern Africa. Mesophytic; species of open habitats; glycophytic. Grassland.

Paleotropical. African. Sudano-Angolan. South Tropical African.
References, etc. Leaf anatomical: this project; photos provided by R.P. Ellis.

Helictotrichon Besser ex Roem. & Schult.

Avenastrum Opiz, *Avenochloa* Holub, *Avenula* (Dumort.) Dumort., *Danthorhiza* Ten., *Heuffelia* Schur, *Stipavena* Vierh.
Excluding *Amphibromus*

Habit, vegetative morphology. Perennial; caespitose. **Culms** 15–150 cm high; *herbaceous*; unbranched above. Culm internodes hollow. *Leaves* not basally aggregated; non-auriculate. Leaf blades linear; narrow; 1–10 mm wide; flat, or folded, or rolled (convolute); without cross venation; persistent; rolled in bud, or once-folded in bud. *Ligule an unfringed membrane (sometimes puberulent)*; not truncate; 2–8 mm long.

Reproductive organization. Plants bisexual, with bisexual spikelets; with hermaphrodite florets. Exposed-cleistogamous, or chasmogamous.

Inflorescence. *Inflorescence paniculate*; open (usually narrow); espatheate; not comprising 'partial inflorescences' and foliar organs. Spikelet-bearing axes persistent. Spikelets not secund.

Female-fertile spikelets. Spikelets 8–25 mm long; compressed laterally; disarticulating above the glumes; disarticulating between the florets. Rachilla prolonged beyond the uppermost female-fertile floret; rachilla hairy. The rachilla extension with incomplete florets. *Hairy callus present. Callus* short (0.5–1.2 mm long); pointed.

Glumes two; very unequal (usually), or more or less equal; usually shorter than the spikelets; (the upper usually) *shorter than the adjacent lemmas*; pointed; awnless; carinate; similar (persistent, hyaline to scarious or firm and herbaceous). Lower glume 1–3 nerved. Upper glume 3(–5) nerved. *Spikelets* with incomplete florets. The incomplete florets distal to the female-fertile florets. *The distal incomplete florets* merely underdeveloped; awned.

Female-fertile florets *2–7*. Lemmas decidedly firmer than the glumes (firmly membranous to leathery); not becoming indurated; usually incised; 2 lobed (usually), or 4 lobed; not deeply cleft (bidentate or bisetulose, rarely 4-dentate); awned. Awns 1; median; dorsal; from near the top, or from well down the back (but from above the middle); geniculate (usually), or non-geniculate; occasionally straight, or recurving; much longer than the body of the lemma (always?); entered by one vein. Lemmas hairy, or hairless; carinate to non-carinate (weakly keeled to dorsally rounded); 5–7 nerved. *Palea* present; relatively long; tightly clasped by the lemma; apically notched; awnless, without apical setae; not indurated (membranous); 2-nerved; 2-keeled. *Lodicules* present; 2; free; membranous; glabrous; toothed; not or scarcely vascularized. *Stamens* 3. Anthers not penicillate. **Ovary hairy**. Styles free to their bases. Stigmas 2.

Fruit, embryo and seedling. Fruit medium sized; longitudinally grooved; compressed laterally, or not noticeably compressed; with hairs confined to a terminal tuft. Hilum long-linear. Embryo small; not waisted. *Endosperm liquid in the mature fruit*; with lipid; containing compound starch grains. Embryo with an epiblast; with a negligible mesocotyl internode.

Seedling with a long mesocotyl; with a tight coleoptile. First seedling leaf with a well-developed lamina. The lamina narrow; 3–5 veined.

Abaxial leaf blade epidermis. *Costal/intercostal zonation* conspicuous. *Papillae* absent. *Long-cells* similar in shape costally and intercostally; of similar wall thickness costally and intercostally. Mid-intercostal long-cells fusiform; having markedly sinuous walls, or having straight or only gently undulating walls. *Microhairs* absent. *Stomata* common. Subsidiaries low dome-shaped, or parallel-sided. Guard-cells overlapped by the interstomatals. *Intercostal short-cells* absent or very rare (none seen). *Costal short-cells* predominantly paired, or neither distinctly grouped into long rows nor predominantly paired. Costal silica bodies horizontally-elongated crenate/sinuous, or horizontally-elongated smooth, or rounded.

Transverse section of leaf blade, physiology. C_3; XyMS+. *Mesophyll* with non-radiate chlorenchyma; without adaxial palisade. *Leaf blade* adaxially flat. *Midrib* conspicuous; with one bundle only. Bulliforms not in discrete, regular adaxial groups (limited to midrib 'hinges'). All the vascular bundles accompanied by sclerenchyma. Combined sclerenchyma girders present. Sclerenchyma all associated with vascular bundles.

Phytochemistry. Leaves without flavonoid sulphates (1 species).

Cytology. Chromosome base number, $x = 7$. $2n = 14$, 28, 42, 70, 81, 98, 112, 126, 133, and 147. 2, 4, 6, 10, 12, 14, 16, 18, 19, and 21 ploid. Haploid nuclear DNA content 3–5 pg (2 species). Mean diploid 2c DNA value 10.1 pg (*pubescens*).

Taxonomy. Pooideae; Poodae; Aveneae.

Ecology, geography, regional floristic distribution. About 90 species. Europe, Africa, Southeast Asia, North & South America. Mesophytic to xerophytic, or helophytic (rarely); mostly species of open habitats. Dry hillsides, meadows, margins of woods.

Holarctic, Paleotropical, Neotropical, and Cape. Boreal, Tethyan, and Madrean. African, Madagascan, and Indomalesian. Euro-Siberian, Eastern Asian, Atlantic North American, and Rocky Mountains. Mediterranean and Irano-Turanian. Sudano-Angolan and West African Rainforest. Indo-Chinese and Malesian. Pampas and Andean. European. Canadian-Appalachian. Sahelo-Sudanian, Somalo-Ethiopian, and South Tropical African.

Rusts and smuts. Rusts — *Puccinia*. Taxonomically wide-ranging species: *Puccinia graminis* and *Puccinia coronata*. Smuts from Tilletiaceae and from Ustilaginaceae. Tilletiaceae — *Urocystis*. Ustilaginaceae — *Ustilago*.

Economic importance. Important native pasture species: *H. milanjianum* in Kenya.

References, etc. Leaf anatomical: Metcalfe 1960; this project.

Helleria Fourn.

~ Festuca

Habit, vegetative morphology. Perennial; caespitose. Culms herbaceous. *Leaves non-auriculate*. *Sheath margins free*. Leaf blades narrow; setaceous; without cross venation. **Ligule** present; *an unfringed membrane*; truncate; 0.5 mm long.

Reproductive organization. Plants bisexual, with bisexual spikelets; with hermaphrodite florets.

Inflorescence. *Inflorescence paniculate*; open; espatheate; not comprising 'partial inflorescences' and foliar organs. Spikelet-bearing axes persistent. Spikelets not secund.

Female-fertile spikelets. *Spikelets* 10–15 mm long; *not noticeably compressed*; disarticulating above the glumes. Rachilla prolonged beyond the uppermost female-fertile floret. The rachilla extension with incomplete florets. Hairy callus absent.

Glumes present; two; more or less equal; long relative to the adjacent lemmas (exceeding the L_1); pointed (acute); awnless; non-carinate; similar. Lower glume 3 nerved. *Upper glume 3 nerved*. *Spikelets* with incomplete florets. The incomplete florets distal to the female-fertile florets.

Female-fertile florets 2–5. *Lemmas* similar in texture to the glumes; not becoming indurated; *incised*; awned. Awns 1; median; from a sinus; non-geniculate; entered by one vein. Lemmas hairless; non-carinate; without a germination flap; 5 nerved. *Palea* present; relatively long; 2-nerved; 2-keeled. *Lodicules* present; 2; free; membranous; glabrous; toothed; not or scarcely vascularized. *Stamens* 3. *Ovary* glabrous. Styles free to their bases. Stigmas 2.

Fruit, embryo and seedling. Fruit small (5 mm long); deep purple; compressed dorsiventrally. *Hilum long-linear*. Embryo small. Endosperm hard; with lipid; containing compound starch grains. Embryo with an epiblast.

Abaxial leaf blade epidermis. *Costal/intercostal zonation* lacking. *Papillae* absent. *Long-cells* similar in shape costally and intercostally; of similar wall thickness costally and intercostally. Mid-intercostal long-cells rectangular; having markedly sinuous walls (walls thick, conspicuously pitted). *Microhairs* absent. Stomata absent or very rare. *Intercostal short-cells* common; in cork/silica-cell pairs (in *H. fragilis*), or in cork/silica-cell pairs and not paired (commonly solitary, in *H. livida*); silicified. Intercostal silica bodies tall-and-narrow, or crescentic, or oryzoid-type, or rounded. *Costal short-cells* predominantly paired (in *H. fragilis*), or neither distinctly grouped into long rows nor predominantly paired (mostly solitary but some paired, in *H. livida*). Costal silica bodies rounded

(H. fragilis), or tall-and-narrow (especially *H. livida*), or crescentic (conspicuous in *IH. fragilis*), or oryzoid (tall-and-narrows expanded top and bottom, in *H. livida*).

Transverse section of leaf blade, physiology. C_3; XyMS+. Mesophyll without adaxial palisade. *Leaf blade* with distinct, prominent adaxial ribs; with the ribs more or less constant in size. *Midrib* not readily distinguishable; with one bundle only. *The lamina* symmetrical on either side of the midrib. Bulliforms present in discrete, regular adaxial groups; in simple fans. All the vascular bundles accompanied by sclerenchyma. Combined sclerenchyma girders absent (most with a small adaxial strand in the top of the overlying rib, and none in contact with the abaxial fibres). Sclerenchyma not all bundle-associated. The 'extra' sclerenchyma in abaxial groups, or in a continuous abaxial layer (this 2–3 cells thick); (when non-continuous) abaxial-hypodermal, the groups isolated (opposite the furrows).

Taxonomy. Pooideae; Poodae; Poeae.

Ecology, geography, regional floristic distribution. 2 species. Mexico, Venezuela. Neotropical. Caribbean and Venezuela and Surinam.

References, etc. Morphological/taxonomic: Alekseev 1984. Leaf anatomical: this project.

Hemarthria R.Br.

Lodicularia P. Beauv.

Habit, vegetative morphology. Perennial; stoloniferous, or caespitose, or decumbent. *Culms* 30–150 cm high; herbaceous; branched above. Culm nodes glabrous. Culm internodes hollow. *Leaves* not basally aggregated; non-auriculate. Leaf blades linear-lanceolate (usually), or linear; narrow; flat; without cross venation; persistent; once-folded in bud. Ligule a fringed membrane.

Reproductive organization. Plants bisexual, with bisexual spikelets; with hermaphrodite florets. The spikelets all alike in sexuality; overtly heteromorphic (the sunken, 'sessile' spikelets with dissimilar glumes, the non-sunken 'pedicelled' spikelets with similar glumes). Exposed-cleistogamous, or chasmogamous.

Inflorescence. Inflorescence of spicate main branches, or paniculate (of flattened, dorsiventral spicate 'racemes', arising one or more from the sheaths of each of the upper leaves). Rachides hollowed (the sessile spikelet embedded). *Inflorescence spatheate (usually)*; a complex of 'partial inflorescences' and intervening foliar organs. **Spikelet-bearing axes** spikelike (often curved); solitary, or clustered (fascicled); with substantial rachides; tardily *disarticulating (the rachis initially tough)*; ultimately disarticulating at the joints. *'Articles'* non-linear (clavate); with a basal callus-knob (rarely), or without a basal callus-knob (usually); not appendaged; disarticulating transversely to disarticulating obliquely; glabrous. *Spikelets paired (each pair comprising a sessile spikelet and the 'pedicelled' member of the 'pair' below)*; secund (the sessile members in two alternating rows, more or less on one side of the rachis); consistently in 'long-and-short' combinations (the fused pedicels discernible). Pedicels of the 'pedicellate' spikelets discernible, but fused with the rachis. The 'shorter' spikelets hermaphrodite. *The 'longer' spikelets hermaphrodite*.

Female-fertile spikelets. Spikelets 3–7 mm long; abaxial; compressed dorsiventrally; falling with the glumes. Rachilla terminated by a female-fertile floret. Hairy callus absent. *Callus* short (obtuse or truncate, in the sessile spikelet), or absent (in the pedicelled spikelet).

Glumes present; two; more or less equal; long relative to the adjacent lemmas; dorsiventral to the rachis; hairless; glabrous; awned (the lower biaristate in *H. debilis*, the G_2 sometimes long-acuminate/aristate), or awnless; non-carinate; very dissimilar (in the embedded spikelet, the outer tough, the inner membranous), or similar (both tough in the pedicelled spikelets). *Lower glume two-keeled (the keels not winged)*; convex on the back to flattened on the back; not pitted; relatively smooth; 5–7 nerved. Upper glume 5–7 nerved. *Spikelets* with incomplete florets. The incomplete florets proximal to the female-fertile florets. *The proximal incomplete florets* 1; epaleate; sterile. The proximal lemmas awnless; 2 nerved; decidedly exceeding the female-fertile lemmas; similar in texture to the female-fertile lemmas; not becoming indurated.

Female-fertile florets 1. **Lemmas** not indented; less firm than the glumes (hyaline); not becoming indurated; entire; awnless; hairless; non-carinate; *0 nerved*. *Palea* present; conspicuous but relatively short; entire; awnless, without apical setae; not indurated (hyaline); nerveless. *Lodicules* present; 2; free; fleshy; glabrous. *Stamens* 3. Anthers not penicillate. *Ovary* glabrous. Styles free to their bases. Stigmas 2; white.

Fruit, embryo and seedling. *Fruit* free from both lemma and palea; small; compressed dorsiventrally. Hilum short. Embryo large. Endosperm containing compound starch grains. Embryo with an elongated mesocotyl internode.

Seedling with a long mesocotyl.

Abaxial leaf blade epidermis. *Costal/intercostal zonation* conspicuous. *Papillae* absent. *Long-cells* markedly different in shape costally and intercostally (the costals smaller and narrower than the unusually large intercostals, at least in *H. uncinata*); differing markedly in wall thickness costally and intercostally (the intercostals thicker walled, in *H. uncinata*). Mid-intercostal long-cells rectangular; having markedly sinuous walls. *Microhairs* present; panicoid-type; 48–84 microns long; 11.4–14.4 microns wide at the septum. Microhair total length/width at septum 4.8–6.5. Microhair apical cells 20–51 microns long. Microhair apical cell/total length ratio 0.46–0.65. *Stomata* common; 33–39 microns long. Subsidiaries mostly triangular (in *H. uncinata*). Guard-cells overlapping to flush with the interstomatals. *Intercostal short-cells* common; in cork/silica-cell pairs; silicified. Intercostal silica bodies mostly cross-shaped, or vertically elongated-nodular. Heavily pitted, thick walled prickle bases present in *H. uncinata*. *Crown cells* absent. *Costal short-cells* neither distinctly grouped into long rows nor predominantly paired (sic). Costal silica bodies 'panicoid-type'; cross shaped and butterfly shaped, or dumbbell shaped.

Transverse section of leaf blade, physiology. C_4; XyMS–. PCR cell chloroplasts with reduced grana; centrifugal/peripheral. *Mesophyll* with radiate chlorenchyma. *Leaf blade* adaxially flat. *Midrib* conspicuous; having a conventional arc of bundles; with colourless mesophyll adaxially (the colourless tissue extending across the adaxial part of the blade, contiguous with the epidermis). Bulliforms not in discrete, regular adaxial groups (the epidermis extensively bulliform). Many of the smallest vascular bundles unaccompanied by sclerenchyma. Combined sclerenchyma girders present (with the main bundles); forming 'figures', or nowhere forming 'figures'. Sclerenchyma all associated with vascular bundles.

Culm anatomy. *Culm internode bundles* in one or two rings.

Phytochemistry. Leaves without flavonoid sulphates (1 species).

Special diagnostic feature. Lower glume of female-fertile spikelet flattish, not pitted; 'pedicellate' spikelets similar to the female-fertile spikelets.

Cytology. Chromosome base number, $x = 9$, or 10. $2n = 18$, or 20 (18+2B), or 36, or 54. 2, 4, and 6 ploid (and aneuploids).

Taxonomy. Panicoideae; Andropogonodae; Andropogoneae; Rottboelliinae.

Ecology, geography, regional floristic distribution. 12 species. Tropical Africa, Madagascar, eastern Asia, Indomalayan region, Australia. Commonly adventive. Hydrophytic to helophytic; species of open habitats; glycophytic. In water or in wet places.

Holarctic, Paleotropical, Neotropical, Cape, and Australian. Boreal, Tethyan, and Madrean. African, Madagascan, and Indomalesian. Euro-Siberian, Eastern Asian, and Atlantic North American. Macaronesian, Mediterranean, and Irano-Turanian. Saharo-Sindian, Sudano-Angolan, and Namib-Karoo. Indian, Indo-Chinese, Malesian, and Papuan. Pampas and Andean. North and East Australian and South-West Australian. Siberian. Southern Atlantic North American and Central Grasslands. Sahelo-Sudanian, Somalo-Ethiopian, South Tropical African, and Kalaharian. Tropical North and East Australian and Temperate and South-Eastern Australian.

Rusts and smuts. Rusts — *Puccinia*. Taxonomically wide-ranging species: *Puccinia microspora*, *Puccinia levis*, and '*Uromyces*' *clignyi*. Smuts from Ustilaginaceae. Ustilaginaceae — *Sphacelotheca* and *Ustilago*.

Economic importance. Significant weed species: *H. altissima*, *H. compressa*.

References, etc. Leaf anatomical: Metcalfe 1960; this project.

Hemisorghum C.E. Hubb.

Habit, vegetative morphology. Stout perennial; caespitose. Culms herbaceous. Culm internodes solid. *Leaf blades linear-lanceolate*; *broad*; flat. *Ligule a fringed membrane (?)*.

Reproductive organization. *Plants bisexual, with bisexual spikelets*; with hermaphrodite florets. The spikelets of sexually distinct forms on the same plant; hermaphrodite and sterile; overtly heteromorphic (the pedicelled smaller, lacking the L_1); all in heterogamous combinations.

Inflorescence. Inflorescence paniculate (a large decompound panicle with whorled, triquetrous or angular, scabrid branches); open; espatheate; not comprising 'partial inflorescences' and foliar organs (?). *Spikelet-bearing axes* 'racemes' (few to several-noded); solitary; with very slender rachides; disarticulating; disarticulating at the joints. *'Articles'* linear (slender); not appendaged; disarticulating transversely. Spikelets paired; consistently in 'long-and-short' combinations; in pedicellate/sessile combinations. Pedicels of the 'pedicellate' spikelets free of the rachis. The 'shorter' spikelets hermaphrodite. The 'longer' spikelets sterile.

Female-sterile spikelets. The sterile pedicelled spikelets without an L_1.

Female-fertile spikelets. Spikelets compressed dorsiventrally (?); falling with the glumes (falling with the joint and the pedicelled spikelet). Rachilla terminated by a female-fertile floret. Hairy callus present (with a ring of minute hairs). *Callus* short; blunt.

Glumes two; more or less equal; long relative to the adjacent lemmas; G_2 with ciliate margins; awnless; *very dissimilar (leathery, the G_1 obtuse, dorsally flat or slightly convex, 2-keeled for most of its length with the margins inflexed towards the base, the G_2 lanceolate, acute, dorsally rounded, becoming 1-keeled upwards)*. *Lower glume two-keeled*; convex on the back to flattened on the back; not pitted; relatively smooth; 8–11 nerved. *Upper glume 7 nerved*. *Spikelets* with incomplete florets. The incomplete florets proximal to the female-fertile florets. *The proximal incomplete florets* 1; sterile. The proximal lemmas awnless; 2 nerved; more or less equalling the female-fertile lemmas to decidedly exceeding the female-fertile lemmas; similar in texture to the female-fertile lemmas (thinly membranous to hyaline); not becoming indurated.

Female-fertile florets 1. Lemmas less firm than the glumes (thin); not becoming indurated; incised; minutely 2 lobed; not deeply cleft (minutely bidentate); awnless, or mucronate (between the teeth); ciliolate; non-carinate; without a germination flap; 1 nerved. **Palea** present; relatively long (about as long as the lemma); linear oblong; awnless, without apical setae; textured like the lemma; not indurated; *nerveless*; keel-less. *Lodicules* present; 2; free; fleshy; glabrous. *Stamens* 3. *Ovary* glabrous. Styles fused (at base). Stigmas 2; red pigmented.

Fruit, embryo and seedling. Fruit compressed dorsiventrally (plano-convex). Hilum short. Embryo large.

Taxonomy. Panicoideae; Andropogonodae; Andropogoneae; Andropogoninae.

Ecology, geography, regional floristic distribution. 2 species. India, Burma, Southeast Asia. On hillsides and riverbanks.

Paleotropical. Indomalesian. Indo-Chinese.

Special comments. Anatomical data wanting.

Henrardia C.E. Hubb.

Habit, vegetative morphology. Annual. Culms herbaceous. Leaves auriculate, or non-auriculate. Leaf blades linear; narrow; 1–3 mm wide; flat, or rolled (convolute); without cross venation. Ligule an unfringed membrane; unevenly truncate; 0.2–0.5 mm long.

Reproductive organization. Plants bisexual, with bisexual spikelets; with hermaphrodite florets. *The spikelets all alike in sexuality*.

Inflorescence. *Inflorescence a single spike (with embedded spikelets)*. Rachides hollowed. Inflorescence espatheate; not comprising 'partial inflorescences' and foliar organs. *Spikelet-bearing axes disarticulating*; disarticulating at the joints. Spikelets solitary; not secund; distichous.

Female-fertile spikelets. *Spikelets* 7–10 mm long; *not noticeably compressed*; falling with the glumes. Rachilla prolonged beyond the uppermost female-fertile floret. Hairy callus absent.

Glumes two; more or less equal; exceeding the spikelets; long relative to the adjacent lemmas; *displaced (side by side, both abaxial)*; without conspicuous tufts or rows of hairs; pointed, or not pointed (acute to obtuse); not subulate; awnless; non-carinate; similar (asymmetrical). Lower glume 3–7 nerved. Upper glume 5–9 nerved. *Spikelets* with female-fertile florets only, or with incomplete florets. The incomplete florets distal to the female-fertile florets.

Female-fertile florets 1–2. Lemmas acute; less firm than the glumes (very soft); not becoming indurated; entire; pointed; awnless; hairy; non-carinate; without a germination flap; 3–5 nerved. *Palea* present; relatively long; 2-nerved; 2-keeled. *Lodicules* present; 2; free; membranous; ciliate; toothed, or not toothed; not or scarcely vascularized. *Stamens* 3. Anthers 0.5–2.2 mm long. *Ovary* hairy. Styles free to their bases. Stigmas 2.

Fruit, embryo and seedling. *Fruit* adhering to lemma and/or palea (slightly); medium sized (4–6.2 mm long); shallowly longitudinally grooved; compressed dorsiventrally; with hairs confined to a terminal tuft. Hilum long-linear. Embryo large to small (up to 1/3 the length of the caryopsis). Endosperm containing only simple starch grains. Embryo with an epiblast.

Abaxial leaf blade epidermis. *Costal/intercostal zonation* conspicuous. *Papillae* absent. *Long-cells* similar in shape costally and intercostally; differing markedly in wall thickness costally and intercostally (costals with thicker walls). Mid-intercostal long-cells rectangular; having markedly sinuous walls. *Microhairs* absent. *Stomata* common; 60–63 microns long. Subsidiaries low dome-shaped, or parallel-sided. Guard-cells overlapped by the interstomatals (mostly, slightly). *Intercostal short-cells* absent or very rare. Prickle/macrohair bases abundant. *Costal short-cells* neither distinctly grouped into long rows nor predominantly paired (mostly solitary). Costal silica bodies horizontally-elongated crenate/sinuous (some almost qualifying as 'nodular'), or rounded (a few).

Transverse section of leaf blade, physiology. C_3; XyMS+. Mesophyll without adaxial palisade. *Leaf blade* with distinct, prominent adaxial ribs; with the ribs more or less constant in size. *Midrib* not readily distinguishable; with one bundle only. *The lamina* symmetrical on either side of the midrib. Bulliforms present in discrete, regular adaxial groups (in the furrows); in simple fans. All the vascular bundles accompanied by sclerenchyma. Combined sclerenchyma girders present (large bundles only); forming 'figures'. Sclerenchyma all associated with vascular bundles (except for massive groups in the blade margins).

Cytology. Chromosome base number, $x = 7$. $2n = 14$. 2 ploid. Haplomic genome content O.

Taxonomy. Pooideae; Triticodae; Triticeae.

Ecology, geography, regional floristic distribution. 2 species. Southwest & Central Asia. Species of open habitats. Dry slopes.

Holarctic. Boreal and Tethyan. Euro-Siberian. Mediterranean and Irano-Turanian. European.

References, etc. Morphological/taxonomic: Löve 1984. Leaf anatomical: this project.

Heterachne Benth.

Habit, vegetative morphology. *Annual*; caespitose. *Culms* 10–50 cm high; herbaceous; unbranched above. Culm nodes glabrous. Culm internodes hollow. Young shoots intravaginal. *Leaves* not basally aggregated; non-auriculate (but sometimes hairy at the auricle position). Leaf blades narrow; setaceous, or not setaceous; rolled (when dry); without abaxial multicellular glands; without cross venation; persistent. *Ligule a fringed membrane (very reduced), or a fringe of hairs*. *Contra-ligule* absent.

Reproductive organization. Plants bisexual, with bisexual spikelets; with hermaphrodite florets. The spikelets of sexually distinct forms on the same plant, or all alike in sexuality; hermaphrodite, or hermaphrodite and sterile (there being abortive spikelets in all the clusters, and a lowermost cluster of imperfect spikelets, in the material examined of *H. abortiva*). Exposed-cleistogamous, or chasmogamous.

Inflorescence. *Inflorescence a false spike, with spikelets on contracted axes (several globular heads in an interrupted spike), or paniculate (with sessile heads of spikelets in the upper leaf sheaths)*; contracted. *Inflorescence with axes ending in spikelets*. Inflorescence espatheate; not comprising 'partial inflorescences' and foliar organs. *Spikelet-bearing axes* very much reduced (to globular heads); persistent. Spikelets solitary; not secund.

Female-fertile spikelets. Spikelets about 3–4 mm long; suborbicular, or oblong; very much compressed laterally; disarticulating above the glumes (these deciduous); not disarticulating between the florets. *Rachilla prolonged beyond the uppermost female-fertile floret (and the prolongation sharply bent)*; rachilla hairless. The rachilla extension with incomplete florets. Hairy callus absent. *Callus* absent.

Glumes two; relatively large; very unequal to more or less equal; shorter than the spikelets; shorter than the adjacent lemmas; hairless; glabrous; pointed; awnless; carinate; similar (very compressed-complicate). Lower glume 1 nerved. Upper glume 1–3 nerved. *Spikelets* with incomplete florets. The incomplete florets distal to the female-fertile florets. The distal incomplete florets 1–4 (in a fan-shaped cluster); *incomplete florets merely underdeveloped*.

Female-fertile florets 1(–2). *Lemmas* similar in texture to the glumes (membranous); not becoming indurated; entire; blunt; awnless; hairless; glabrous; *carinate. The keel winged*. Lemmas without a germination flap; 1–3 nerved. *Palea* present; relatively long; entire; awnless, without apical setae; 2-nerved; 2-keeled. Palea keels winged (the wings broad, ciliate). *Lodicules* present; 2; fleshy; glabrous; not or scarcely vascularized. *Stamens* 3. Anthers minute; not penicillate; without an apically prolonged connective. *Ovary* glabrous. Styles free to their bases (short). Stigmas 2.

Fruit, embryo and seedling. *Fruit* free from both lemma and palea (but enclosed); small; ellipsoid; compressed laterally. Hilum short. Pericarp fused. Embryo large; not waisted. Endosperm containing compound starch grains. Embryo with an epiblast; with a scutellar tail; with an elongated mesocotyl internode. Embryonic leaf margins meeting.

Abaxial leaf blade epidermis. *Costal/intercostal zonation* conspicuous (or the zonation fairly obscure, in *H. abortiva*). *Papillae* absent. *Long-cells* similar in shape costally and intercostally; of similar wall thickness costally and intercostally (fairly thick walled). Mid-intercostal long-cells rectangular; having markedly sinuous walls. *Microhairs* present; elongated; clearly two-celled; panicoid-type. Microhair apical cell wall thinner than that of the basal cell and often collapsed. Microhairs 54–63 microns long. Microhair basal cells 24 microns long. Microhairs 9–12.5 microns wide at the septum. Microhair total length/width at septum 4.5–6.3. Microhair apical cells (25.5–)29.4–30 microns long. Microhair apical cell/total length ratio 0.47–0.54. *Stomata* common; 30–33 microns long. Subsidiaries predominantly triangular (in *H. abortiva*). Guard-cells overlapping to flush with the interstomatals. *Intercostal short-cells* common (e.g. in *H. abortiva*), or absent or very rare; in cork/silica-cell pairs to not paired; silicified. Intercostal silica bodies absent, or imperfectly developed; tall-and-narrow to crescentic (in *H.abortiva*). *Costal short-cells* predominantly paired. Costal silica bodies present throughout the costal zones; rounded (a few), or saddle shaped, or tall-and-narrow, or crescentic.

Transverse section of leaf blade, physiology. C_4; XyMS+. PCR sheath outlines uneven. PCR sheaths of the primary vascular bundles interrupted; interrupted abaxially only. PCR sheath extensions present. Maximum number of extension cells 3–5. *Leaf blade* with distinct, prominent adaxial ribs; with the ribs more or less constant in size. *Midrib* not readily distinguishable; with one bundle only. Bulliforms present in discrete, regular adaxial groups; in simple fans (their large median cells deeply penetrating), or associated with colourless mesophyll cells to form deeply-penetrating fans. All the vascular bundles accompanied by sclerenchyma. Combined sclerenchyma girders present; forming 'figures'. Sclerenchyma all associated with vascular bundles. The lamina margins with fibres.

Taxonomy. Chloridoideae; main chloridoid assemblage.

Ecology, geography, regional floristic distribution. 3 species. Northern Australia. Mesophytic. In damp depressions.

Australian. North and East Australian. Tropical North and East Australian.

References, etc. Morphological/taxonomic: Hubbard 1935. Leaf anatomical: this project.

Heteranthelium Hochst.

Habit, vegetative morphology. *Annual. Culms* 5–30 cm high; herbaceous. Culm nodes with short bristles below the nodes. Leaves auriculate. Sheath margins joined, or free. Leaf blades narrow; flat; without cross venation. Ligule an unfringed membrane; truncate; 0.2–0.8 mm long.

Reproductive organization. Plants bisexual, with bisexual spikelets; with hermaphrodite florets. *The spikelets* of sexually distinct forms on the same plant; *hermaphrodite and sterile (each spikelet cluster with the lowest one or two members fertile, the other sterile).*

Inflorescence. *Inflorescence a false spike, with spikelets on contracted axes (in clusters)*; espatheate; not comprising 'partial inflorescences' and foliar organs. Spikelet-bearing axes disarticulating; falling entire (i.e., the clusters falling). Spikelets in clusters of 1–2 fertile and 1–3 sterile; not secund; distichous.

Female-fertile spikelets. *Spikelets* 8–15 mm long; *compressed laterally*; falling with the glumes (i.e., in the clusters); not disarticulating between the florets. Rachilla prolonged beyond the uppermost female-fertile floret; rachilla somewhat spinulose. The rachilla extension with incomplete florets. *Callus* very short.

Glumes two; relatively large; displaced (side by side away from the rachis); subulate; awned (both awn-like); *similar (thinly leathery, linear-subulate, with basal capitate hairs)*. Lower glume 1 nerved. Upper glume 1 nerved. *Spikelets* with incomplete florets. The incomplete florets distal to the female-fertile florets. The distal incomplete florets 1–3; *incomplete florets* clearly specialised and modified in form (forming a tuft of awns).

Female-fertile florets 1–2 (usually the first two). Lemmas modified into the awn; decidedly firmer than the glumes (cartilaginous to leathery); entire; pointed; awned. Awns 1; median; apical; non-geniculate; much shorter than the body of the lemma; entered by several veins. Lemmas hairy (tuberculate on the back, with capitate hairs on the margins); non-carinate; without a germination flap; 5–7 nerved. **Palea** present; relatively long; apically notched; *with apical setae (apical cusps, about 1 mm long)*; 2-nerved; 2-keeled. *Lodicules* present; 2; free; membranous; ciliate; not toothed; not or scarcely vascularized. *Stamens* 3. Anthers 0.8–1.2 mm long. **Ovary hairy.** Styles free to their bases. Stigmas 2.

Fruit, embryo and seedling. *Fruit* somewhat adhering to lemma and/or palea; medium sized (4.5–7.5 mm long); ellipsoid; longitudinally grooved; compressed laterally; with hairs confined to a terminal tuft. Hilum long-linear. Embryo large to small (up to 1/3 the caryopsis length). Endosperm hard; without lipid; containing only simple starch grains. Embryo with an epiblast.

Abaxial leaf blade epidermis. *Costal/intercostal zonation* conspicuous. *Papillae* absent. *Long-cells* markedly different in shape costally and intercostally (costal 'long-cells' often short). Mid-intercostal long-cells mostly rectangular; having straight or only gently undulating walls. *Microhairs* absent. *Stomata* common; 39–45 microns long. Subsidiaries parallel-sided, or parallel-sided and dome-shaped. Guard-cells overlapped by the interstomatals. *Intercostal short-cells* absent or very rare. Abundant macrohairs present. *Costal short-cells* neither distinctly grouped into long rows nor predominantly paired. Costal silica bodies horizontally-elongated crenate/sinuous, or horizontally-elongated smooth, or rounded (few).

Transverse section of leaf blade, physiology. C_3; XyMS+. Mesophyll without adaxial palisade. *Midrib* not readily distinguishable; with one bundle only. *The lamina* symmetrical on either side of the midrib. Bulliforms present in discrete, regular adaxial groups (in the furrows); in simple fans. All the vascular bundles accompanied by sclerenchyma (but the smallest with minute strands only). Combined sclerenchyma girders present; forming 'figures' (I's and anchors). Sclerenchyma all associated with vascular bundles.

Cytology. Chromosome base number, $x = 7$. $2n = 14$. 2 ploid. Haplomic genome content Q.

Taxonomy. Pooideae; Triticodae; Triticeae.

Ecology, geography, regional floristic distribution. 1 species. Southwest Asia. Xerophytic; species of open habitats. On dry slopes.

Holarctic. Tethyan. Irano-Turanian.

Rusts and smuts. Rusts — *Puccinia*. Taxonomically wide-ranging species: *Puccinia graminis* and *Puccinia striiformis*.

References, etc. Morphological/taxonomic: Löve 1984. Leaf anatomical: Metcalfe 1960; this project.

Heteranthoecia Stapf

Habit, vegetative morphology. Annual; decumbent (mat-forming). *Culms* 20–30 cm high; herbaceous; branched above. *Leaves* not basally aggregated; non-auriculate. *Leaf blades* lanceolate; narrow; 2–5 mm wide; *cross veined*. Ligule a fringed membrane (very narrow), or a fringe of hairs.

Reproductive organization. Plants bisexual, with bisexual spikelets; with hermaphrodite florets (the lower floret hermaphrodite, the upper female-only).

Inflorescence. Inflorescence of spicate main branches (a panicle of short, unilateral spikes). *Inflorescence axes not ending in spikelets (the rachides ending in blunt naked tips)*. Inflorescence espatheate; not comprising 'partial inflorescences' and foliar organs. Spikelet-bearing axes persistent. Spikelets secund; biseriate; contiguous.

Female-fertile spikelets. Spikelets 1.7–2.3 mm long; abaxial; compressed dorsiventrally; disarticulating above the glumes; disarticulating between the florets. Rachilla terminated by a female-fertile floret. Hairy callus absent.

Glumes two; more or less equal; shorter than the adjacent lemmas; hairless; pointed; awnless; similar (firmly membranous, with hyaline margins). Lower glume 4–7 nerved. Upper glume 4–7 nerved. *Spikelets* with female-fertile florets only.

Female-fertile florets *2 (but the upper usually female-only, much shorter and blunter)*. Lemmas similar in texture to the glumes to decidedly firmer than the glumes (papery); becoming indurated (L_1), or not becoming indurated (L_2); entire; pointed (L1 acuminate), or blunt (L2 obtuse to subacute); awnless; hairy; non-carinate; having the margins tucked in onto the palea; with a clear germination flap; 5 nerved. *Palea* present; relatively long; entire, or apically notched (slightly); awnless, without apical setae; not indurated; 2-nerved. *Lodicules* present; 2; free; fleshy; glabrous. *Stamens* 3. *Ovary* glabrous. Styles free to their bases. Stigmas 2.

Fruit, embryo and seedling. Fruit small; slightly compressed dorsiventrally. Hilum short. Embryo large. Endosperm containing only simple starch grains. Embryo without an epiblast; with a scutellar tail; with an elongated mesocotyl internode.

Abaxial leaf blade epidermis. *Costal/intercostal zonation* conspicuous. *Papillae* present. Intercostal zones without typical long-cells (these being cubical). Mid-intercostal long-cells having straight or only gently undulating walls. *Microhairs* present; panicoid-type; 15–22 microns long. Microhair apical cells 9–16 microns long. Microhair apical cell/total length ratio 0.68. *Stomata* common. Subsidiaries parallel-sided to triangular; including both triangular and parallel-sided forms on the same leaf. *Intercostal short-cells* absent or very rare. *Costal short-cells* conspicuously in long rows. Costal silica bodies acutely-angled; sharp-pointed.

Transverse section of leaf blade, physiology. C_3; XyMS+. *Mesophyll* with radiate chlorenchyma; *Isachne*-type. *Leaf blade* with distinct, prominent adaxial ribs; with the ribs more or less constant in size. *Midrib* not readily distinguishable; with one bundle only. Bulliforms present in discrete, regular adaxial groups (in the furrows); in simple fans. All the vascular bundles accompanied by sclerenchyma. Combined sclerenchyma girders present. Sclerenchyma all associated with vascular bundles.

Taxonomy. Panicoideae; Panicodae; Isachneae.

Ecology, geography, regional floristic distribution. 1 species. Tropical Africa. Helophytic (swamps and shallow water).

Paleotropical. African. Sudano-Angolan and West African Rainforest. Sahelo-Sudanian and Somalo-Ethiopian.

References, etc. Leaf anatomical: Metcalfe 1960.

Heterocarpha Stapf & C.E. Hubb.

~ *Drake-Brockmania*

Habit, vegetative morphology. Perennial; stoloniferous and caespitose. *Culms* 15–63 cm high; herbaceous; unbranched above. Culm nodes glabrous. Plants unarmed. *Leaves* not basally aggregated; non-auriculate; without auricular setae. Sheath margins free. *Leaf blades lanceolate*; narrow; (2–)3–7(–9.5) mm wide (and 3–18 cm long); flat; without abaxial multicellular glands; without cross venation; *persistent*. Ligule present; a fringed membrane (a short rim, with a conspicuous fringe); 1–1.5 mm long. *Contra-ligule* absent.

Reproductive organization. Plants bisexual, with bisexual spikelets; with hermaphrodite florets.

Inflorescence. *Inflorescence of spicate main branches (3.5–11 cm long, rather narrow, with short, dense racemes which are reflexed terminally)*. Primary inflorescence branches (5–)8–12. Inflorescence espatheate; not comprising 'partial inflorescences' and foliar organs. The racemes without spikelets towards the base. Spikelet-bearing axes persistent. Spikelets solitary; secund; biseriate (in two ranks on one side of rachis); subsessile; imbricate.

Female-fertile spikelets. Spikelets 6–10(–14) mm long; adaxial; markedly compressed laterally; disarticulating above the glumes; disarticulating between the florets; with conventional internode spacings. Rachilla prolonged beyond the uppermost female-fertile floret; rachilla hairless (glabrous). The rachilla extension with incomplete florets. Hairy callus absent.

Glumes two; more or less equal; shorter than the spikelets; shorter than the adjacent lemmas; dorsiventral to the rachis to lateral to the rachis; hairless (scabrid on the keels); pointed (acute); awnless; carinate; similar (ovate, membranous). Lower glume 1 nerved, or 2 nerved. *Upper glume 5–8 nerved. Spikelets* with incomplete florets. The incomplete florets distal to the female-fertile florets. *The distal incomplete florets* merely underdeveloped.

Female-fertile florets *(4–)5–10*. Lemmas similar in texture to the glumes (membranous); smooth; not becoming indurated; entire to incised; not deeply cleft (blunt to slightly notched); mucronate (minutely, from the slight apical sinus); hairy (villous on margins and keels), or hairless (?); glabrous; carinate; without a germination flap; 5–7 nerved (with 2–4 very thin nerves between the three heavy ones). **Palea** present (bent); relatively long; *apically notched*; awnless, without apical setae; not indurated (membranous); 2-nerved; 2-keeled (hairy between the keels, deeply furrowed). *Lodicules* present; 2; free; fleshy; glabrous; not or scarcely vascularized. *Stamens* 3. Anthers short; not penicillate; without an apically prolonged connective. *Ovary* glabrous. Styles free to their bases. Stigmas 2 (but sometimes with a vestigial third); red pigmented.

Fruit, embryo and seedling. *Fruit* free from both lemma and palea; small (1 mm long); somewhat compressed laterally. Hilum short. **Pericarp free (thin)**. Embryo large.

Taxonomy. Chloridoideae; main chloridoid assemblage.

Ecology, geography, regional floristic distribution. 1 species (*H. haareri*). Tropical East Africa. Species of open habitats; halophytic. In maritime sands, in dry rather saline soils fringing *Suaeda* bush and in dry *Acacia* bush.

Paleotropical. African. Sudano-Angolan. Somalo-Ethiopian and South Tropical African.

References, etc. Morphological/taxonomic: Stapf and Hubbard 1929.

Special comments. Referred to *Drake-Brockmania* by Phillips (1974), but in view of the morphological differences, anatomical data are badly needed for *H. haareri* and *B. fragilis*. Fruit data wanting. Anatomical data wanting.

Heteropholis C.E. Hubb.

~ *Mnesithea*

Habit, vegetative morphology. Annual, or perennial; mostly decumbent. *Culms* slender, 60–170 cm high; herbaceous; branched above. Culm internodes solid. *Leaves* not basally aggregated; auriculate (with sheath auricles), or non-auriculate. Leaf blades linear to lanceolate; broad, or narrow; cordate, or not cordate, not sagittate; flat, or folded (conduplicate when young, later flat); without cross venation. *Ligule an unfringed membrane*.

Reproductive organization. *Plants bisexual, with bisexual spikelets*; with hermaphrodite florets. The spikelets of sexually distinct forms on the same plant; hermaphrodite and male-only, or hermaphrodite and sterile. The male and female-fertile spikelets mixed in the inflorescence. The spikelets overtly heteromorphic; all in heterogamous combinations.

Inflorescence. Inflorescence of single, cylindrical to flattened, dorsiventral 'racemes' terminating culms and branches. Rachides hollowed (with small cavities). *Inflorescence* spatheate; *a complex of 'partial inflorescences' and intervening foliar organs (the 'racemes' spatheate)*. **Spikelet-bearing axes** *spikelike (the spikelets apparently opposite, by fusion of joints and pedicels)*; solitary; with substantial rachides; *disarticulating*; disarticulating at the joints. 'Articles' non-linear (clavate, with 2 small cavities); with a basal callus-knob; not appendaged; disarticulating transversely; glabrous. Spikelets paired; secund (dorsiventral 'racemes'); consistently in 'long-and-short' combinations; in pedicellate/sessile combinations. Pedicels of the 'pedicellate' spikelets discernible, but fused with the rachis. The 'shorter' spikelets hermaphrodite. The 'longer' spikelets male-only, or sterile.

Female-sterile spikelets. The pedicelled spikelet thinly leathery, the G_1 not lacunose. The male spikelets with glumes. The lemmas awnless.

Female-fertile spikelets. Spikelets abaxial; compressed dorsiventrally; falling with the glumes (and with the joint). Rachilla terminated by a female-fertile floret. Hairy callus absent.

Glumes two; more or less equal; long relative to the adjacent lemmas; dorsiventral to the rachis; hairless; glabrous; awnless; G_2 one-keeled, often obscurely so; *very dissimilar (the G_1 leathery, transversely rugose or pitted and with two keel-wings above, the G_2 smooth, membranous, more or less naviculate)*. *Lower glume not two-keeled (winged only at the tip)*; convex on the back; not pitted; *lacunose with deep depressions, or rugose (transversely)*; 7–9 nerved. Upper glume 3–5 nerved. *Spikelets* with incomplete florets. The incomplete florets proximal to the female-fertile florets. *The proximal incomplete florets* 1; paleate, or epaleate. Palea of the proximal incomplete florets when present, reduced. The proximal incomplete florets male, or sterile. The proximal lemmas awnless; 2 nerved; more or less equalling the female-fertile lemmas; similar in texture to the female-fertile lemmas (hyaline); not becoming indurated.

Female-fertile florets 1. Lemmas less firm than the glumes (hyaline); not becoming indurated; entire; pointed, or blunt; awnless; hairless; non-carinate; without a germination flap; 0 nerved, or 3 nerved. *Palea* present; relatively long, or conspicuous but relatively short, or very reduced; entire; awnless, without apical setae; not indurated (hyaline); 2-nerved, or nerveless; keel-less. *Lodicules* present; 2; free; fleshy; glabrous. *Stamens* 3. *Ovary* glabrous. Styles free to their bases. Stigmas 2; red pigmented.

Fruit, embryo and seedling. Fruit compressed dorsiventrally. Hilum short. Embryo large.

Abaxial leaf blade epidermis. *Costal/intercostal zonation* conspicuous. *Papillae* absent. Long-cells of similar wall thickness costally and intercostally. Mid-intercostal long-cells rectangular; having markedly sinuous walls. *Microhairs* present; more or less panicoid-type (but relatively broad); 52.5–72 microns long; 6–9 microns wide at the septum. Microhair total length/width at septum 7–10. Microhair apical cells 30–45 microns long. Microhair apical cell/total length ratio 0.57–0.63. *Stomata* common; 43.5–46.5 microns long. Subsidiaries triangular (with tall points). Guard-cells overlapping to flush with the interstomatals. *Intercostal short-cells* common; in cork/silica-cell pairs; silicified (sometimes). *Costal short-cells* conspicuously in long rows. Costal silica bodies 'panicoid-type'; cross shaped, dumb-bell shaped, and nodular.

Transverse section of leaf blade, physiology. C_4; XyMS–. *Leaf blade* adaxially flat. *Midrib* conspicuous; having a conventional arc of bundles (a large keel bundle, with several smaller laterals); with colourless mesophyll adaxially. Bulliforms not in discrete, regular adaxial groups (the adaxial epidermis mainly bulliform). Many of the smallest vascular bundles unaccompanied by sclerenchyma. Combined sclerenchyma girders present (each primary bundle). Sclerenchyma all associated with vascular bundles.

Taxonomy. Panicoideae; Andropogonodae; Andropogoneae; Rottboelliinae.

Ecology, geography, regional floristic distribution. 5 species. Tropical Africa, Asia, Australia. In wooded grassland.

Paleotropical and Australian. African and Indomalesian. Sudano-Angolan. Indian. North and East Australian. South Tropical African. Tropical North and East Australian.

References, etc. Morphological/taxonomic: Hubbard 1956. Leaf anatomical: this project.

Heteropogon Pers.

Spirotheros Raf.

Habit, vegetative morphology. Annual, or perennial; caespitose. *Culms* 20–100 cm high; herbaceous; branched above, or unbranched above. Culm nodes glabrous. Culm internodes solid. *Leaves* not basally aggregated; non-auriculate. Sheath margins free. The sheaths keeled. Leaf blades linear; narrow; flat; without cross venation; persistent. *Ligule a fringed membrane*.

Reproductive organization. Plants bisexual, with bisexual spikelets; with hermaphrodite florets, or without hermaphrodite florets. *The spikelets of sexually distinct forms on the same plant (heterogamous, but only in the upper parts of the raceme)*; hermaphrodite, male-only, and sterile, or female-only, male-only, and sterile; overtly heteromorphic; in both homogamous and heterogamous combinations (lower pairs homogamous and homomorphic, male or sterile). Apomictic, or reproducing sexually.

Inflorescence. Inflorescence a single raceme, or paniculate (the single 'racemes' sometimes in false panicles); spatheate, or espatheate; a complex of 'partial inflorescences' and intervening foliar organs (when having axillary fascicles), or not comprising 'partial inflorescences' and foliar organs. **Spikelet-bearing axes** 'racemes' (of several to many joints); *solitary*; disarticulating; *disarticulating at the joints (between the heterogamous upper spikelet pairs)*. 'Articles' linear; *disarticulating obliquely. Spikelets* paired; secund; consistently in 'long-and-short' combinations; *in pedicellate/sessile combinations (but the pedicel reduced to a short stump, the spikelet being supported on a long, slender callus)*. Pedicels of the 'pedicellate' spikelets free of the rachis. The 'shorter' spikelets hermaphrodite (in upper regions of spike-like panicles only), or female-only. The 'longer' spikelets male-only, or sterile.

Female-sterile spikelets. *The pedicellate male or sterile spikelets larger, with a pedicel-like callus (the true pedicel represented by a stump); awnless, dorsally flattened, rather asymmetric. G_1 herbaceous, many nerved, winged above*. The male spikelets with glumes. The lemmas awnless.

Female-fertile spikelets. Spikelets not noticeably compressed to compressed dorsiventrally (subterete); falling with the glumes. Rachilla terminated by a female-fertile floret. Hairy callus present. *Callus* long; pointed.

Glumes two; more or less equal; long relative to the adjacent lemmas; hairy; without conspicuous tufts or rows of hairs; awnless; very dissimilar (the upper with deep longitudinal grooves). Lower glume not two-keeled; convex on the back; not pitted; relatively smooth, or tuberculate; 5–9 nerved. Upper glume 3 nerved. *Spikelets* with incomplete florets. The incomplete florets proximal to the female-fertile florets. *The proximal incomplete florets* 1; epaleate; sterile. The proximal lemmas awnless; 0 nerved; similar in texture to the female-fertile lemmas (hyaline); not becoming indurated.

Female-fertile florets 1. **Lemmas** stipitate-cartilaginous below, produced into the awn; less firm than the glumes (hyaline); not becoming indurated; entire; *awned*. Awns 1; median; apical; geniculate; hairy; much longer than the body of the lemma. Lemmas non-carinate; 3 nerved. **Palea** present, or absent; when present, very reduced; *nerveless*. *Lodicules* present, or absent; when present, 2; free; fleshy; glabrous. *Stamens* 0–3. Anthers not penicillate. *Ovary* glabrous. Styles free to their bases. Stigmas 2; red pigmented.

Fruit, embryo and seedling. Fruit free from both lemma and palea; *longitudinally grooved (channelled on one side)*; compressed dorsiventrally. Hilum short. Embryo large. Endosperm containing only simple starch grains. Embryo without an epiblast; with a scutellar tail; with an elongated mesocotyl internode. Embryonic leaf margins overlapping.

First seedling leaf with a well-developed lamina. The lamina broad; curved; 21–30 veined.

Abaxial leaf blade epidermis. *Costal/intercostal zonation* conspicuous. *Papillae* present (e.g. *H. contortus*), or absent (e.g. *H. triticeus*). Intercostal papillae consisting of one oblique swelling per cell (in *H. contortus*). Long-cells of similar wall thickness costally and intercostally. Mid-intercostal long-cells rectangular; having markedly sinuous walls. *Microhairs* present; panicoid-type; 61.5–82 microns long; 7.2–9.6 microns wide at the septum. Microhair total length/width at septum 6.4–8.8. Microhair apical cells 24–38 microns long. Microhair apical cell/total length ratio 0.41–0.51. *Stomata* common; 42–45 microns long. Subsidiaries triangular. Guard-cells overlapping to flush with the interstomatals. *Intercostal short-cells* common; in cork/silica-cell pairs, or not paired (many solitary, some twos, some short rows — being an unusual feature in Poaceae); not silicified. *Costal short-cells* conspicuously in long rows. Costal silica bodies 'panicoid-type'; cross shaped to dumb-bell shaped.

Transverse section of leaf blade, physiology. C_4; XyMS–. PCR sheath outlines even (the walls thick). PCR sheath extensions absent. PCR cells with a suberised lamella. PCR cell chloroplasts with reduced grana; centrifugal/peripheral. *Mesophyll* with radiate chlorenchyma; traversed by columns of colourless mesophyll cells (in places, apparently, in *H. triticeus*), or not traversed by colourless columns. *Leaf blade* adaxially flat (the furrows scarcely detectable). *Midrib* conspicuous; having a conventional arc of bundles; with colourless mesophyll adaxially (e.g. in *H. triticeus*), or without colourless mesophyll adaxially (e.g. in *H. contortus*). Bulliforms present in discrete, regular adaxial groups; associated with colourless mesophyll cells to form deeply-penetrating fans (sometimes linked with traversing columns of colourless cells, or in places irregularly grouped); associating with colourless mesophyll cells to form arches over small vascular bundles (in places,in *H. triticeus*), or nowhere involved in bulliform-plus-colourless mesophyll arches. Many of the smallest vascular bundles unaccompanied by sclerenchyma. Combined sclerenchyma girders present; nowhere forming 'figures'. Sclerenchyma all associated with vascular bundles.

Culm anatomy. *Culm internode bundles* scattered.

Phytochemistry. Leaves without flavonoid sulphates (2 species).

Cytology. Chromosome base number, $x = 10$ and 11. $2n = 20, 22, 40, 44, 50, 60$, and 80. 2, 4, 6, and 8 ploid.

Taxonomy. Panicoideae; Andropogonodae; Andropogoneae; Andropogoninae.

Ecology, geography, regional floristic distribution. 7 species. Tropical. Mesophytic to xerophytic; species of open habitats; glycophytic. Dry places, often on poor soils.

Holarctic, Paleotropical, Neotropical, Cape, and Australian. Boreal, Tethyan, and Madrean. African, Madagascan, Indomalesian, and Neocaledonian. Euro-Siberian and Atlantic North American. Mediterranean and Irano-Turanian. Saharo-Sindian, Sudano-Angolan, West African Rainforest, and Namib-Karoo. Indian, Indo-Chinese, Malesian, and Papuan. Caribbean, Central Brazilian, and Andean. North and East Australian. European. Southern Atlantic North American and Central Grasslands. Sahelo-Sudanian, Somalo-Ethiopian, South Tropical African, and Kalaharian. Tropical North and East Australian.

Rusts and smuts. Rusts — *Puccinia*. Taxonomically wide-ranging species: *Puccinia versicolor* and '*Uromyces*' *clignyi*. Smuts from Ustilaginaceae. Ustilaginaceae — *Sorosporium*, *Sphacelotheca*, and *Tolyposporium*.

Economic importance. Significant weed species: *H. contortus* (with needle-sharp, penetrative callus). Important native pasture species: *H. contortus*.

References, etc. Leaf anatomical: Metcalfe 1960; this project.

Hickelia A. Camus

Excluding *Pseudocoix*

Habit, vegetative morphology. Perennial. The flowering culms leafy. Culms woody and persistent (slender); scandent; branched above. *Leaves* not basally aggregated; without auricular setae. Leaf blades lanceolate (acuminate); broad; to 10–15 mm wide (to 10 cm long); rounded at the base; pseudopetiolate; disarticulating from the sheaths (presumably). *Ligule* present; small.

Reproductive organization. Plants bisexual, with bisexual spikelets; with hermaphrodite florets.

Inflorescence. Inflorescence *indeterminate*; with stellate spikelet clusters around the nodes. *Spikelet-bearing axes* capitate; clustered (the dense capitula aggregated); persistent.

Female-fertile spikelets. Spikelets 10–18 mm long; becoming ventricose; disarticulating above the glumes (the glumes persistent). Rachilla prolonged beyond the uppermost female-fertile floret.

Glumes several (3–5); relatively large; hairy (long-pilose); pointed; awnless; **very dissimilar (the lower short and membranous, the two upper crustaceous and like the lemma)**. *Spikelets* with incomplete florets. The incomplete florets proximal to the female-fertile florets. *The proximal incomplete florets* 2. **The proximal lemmas** awnless; more or less equalling the female-fertile lemmas to decidedly exceeding the female-fertile lemmas; similar in texture to the female-fertile lemmas; **becoming indurated (shining, ventricose)**.

Female-fertile florets 1. **Lemmas** broad, convolute, ovate; **decidedly firmer than the glumes**; becoming indurated (and inflated); entire; awnless; non-carinate; many-nerved. **Palea** present; relatively long; palea apically pilose; indurated (shining); several nerved (16–20 nerves); **keel-less. Lodicules** present ('acute, small'). **Stamens** 6. **Ovary** hairy; **with a conspicuous apical appendage**. The appendage broadly conical, fleshy. Stigmas 3.

Fruit, embryo and seedling. Fruit medium sized (9 mm long); longitudinally grooved. Pericarp thin (becoming dilated at the base of the style).

Taxonomy. Bambusoideae; Bambusodae; Bambuseae.

Ecology, geography, regional floristic distribution. 1 species. Madagascar. Glycophytic. Forest.

Paleotropical. Madagascan.

Special comments. Anatomical data wanting.

Hierochloë (Gmel.) R.Br.

Ataxia R. Br., *Dimeria* Raf., *Disarrenum* Labill., *Savastana* Schrank, *Torresia* Ruiz & Pavon

~ *Anthoxanthum*

Habit, vegetative morphology. Perennial; rhizomatous, or caespitose. *Culms* 7–120 cm high; herbaceous. Culm nodes glabrous. Culm internodes hollow. Young shoots extravaginal. *The shoots aromatic (coumarin-scented)*. *Leaves* not basally aggregated; non-auriculate. Leaf blades linear to lanceolate; narrow; flat, or folded, or rolled; without cross venation; persistent; rolled in bud. Ligule an unfringed membrane; not truncate; 2–5.5 mm long.

Reproductive organization. Plants bisexual, with bisexual spikelets; with hermaphrodite florets. Apomictic, or reproducing sexually.

Inflorescence. *Inflorescence* paniculate; **open**; with capillary branchlets, or without capillary branchlets; espatheate; not comprising 'partial inflorescences' and foliar organs. Spikelet-bearing axes persistent. Spikelets not secund.

Female-fertile spikelets. Spikelets 3.5–7 mm long; compressed laterally; disarticulating above the glumes; not disarticulating between the florets. Rachilla terminated by a female-fertile floret. Hairy callus absent.

Glumes two; very unequal to more or less equal; about equalling the spikelets; shorter than the adjacent lemmas, or long relative to the adjacent lemmas; pointed; awnless; carinate; similar (membranous, ovate). Lower glume 1–5 nerved. Upper glume 3–5 nerved. *Spikelets* with incomplete florets. The incomplete florets proximal to the female-fertile florets. **Spikelets with proximal incomplete florets.** **The proximal incomplete florets** 2; paleate; *male (with 3 stamens, or rarely neuter)*. The proximal lemmas awned, or awnless; 5 nerved; more or less equalling the female-fertile lemmas to decidedly exceeding the female-fertile lemmas; less firm than the female-fertile lemmas (membranous); not becoming indurated.

Female-fertile florets 1. Lemmas becoming indurated (usually), or not becoming indurated (in only 1 species); entire; blunt; awnless, or mucronate, or awned. Awns when present, 1; dorsal; from near the top; non-geniculate; much shorter than the body of the lemma; entered by one vein. Lemmas hairy, or hairless; non-carinate; 3–5 nerved. *Palea* present; relatively long; 1-nerved; keel-less (2-keeled in male florets). *Lodicules* present; 2; free; membranous; glabrous; toothed, or not toothed. *Stamens* 2 (3 in male florets). Anthers 1–3.5 mm long; not penicillate. *Ovary* glabrous. Styles fused, or free to their bases. Stigmas 2; white, or brown.

Fruit, embryo and seedling. *Fruit* free from both lemma and palea; small; compressed laterally. Hilum short. *Embryo small*; not waisted. Endosperm hard; with lipid; containing compound starch grains. Embryo with an epiblast; without a scutellar tail; with a negligible mesocotyl internode. Embryonic leaf margins meeting.

First seedling leaf with a well-developed lamina. The lamina narrow; 3–5 veined.

Abaxial leaf blade epidermis. *Costal/intercostal zonation* conspicuous. *Papillae* absent. *Long-cells* similar in shape costally and intercostally; differing markedly in wall thickness costally and intercostally (the costals thicker). Mid-intercostal long-cells rectangular; having markedly sinuous walls. *Microhairs* absent. *Stomata* common, or absent or very rare. Subsidiaries low dome-shaped, or parallel-sided. *Intercostal short-cells* common, or absent or very rare. *Costal short-cells* conspicuously in long rows (rarely), or neither distinctly grouped into long rows nor predominantly paired. Costal silica bodies horizontally-elongated crenate/sinuous, or horizontally-elongated smooth.

Transverse section of leaf blade, physiology. C_3; XyMS+. *Mesophyll* with non-radiate chlorenchyma. *Leaf blade* 'nodular' in section; with the ribs more or less constant in size. *Midrib* conspicuous, or not readily distinguishable; with one bundle only, or having a conventional arc of bundles. Bulliforms present in discrete, regular adaxial groups (in the furrows); in simple fans (and sometimes in ill-defined groups of small cells, cf. *Ammophila*). All the vascular bundles accompanied by sclerenchyma. Combined sclerenchyma girders present; forming 'figures', or nowhere forming 'figures'. Sclerenchyma all associated with vascular bundles.

Phytochemistry. Leaves without flavonoid sulphates (1 species).

Cytology. Chromosome base number, $x = 7$. $2n = 14, 28, 42, 56, 64, 66, 68, 71$, and 72, or 74–78. 2, 4, 6, 8, 9, 10, and 11 ploid (and aneuploids).

Taxonomy. Pooideae; Poodae; Aveneae.

Ecology, geography, regional floristic distribution. 30 species. Temperate & cold regions. Helophytic to mesophytic; shade species and species of open habitats. Woods, marshes, grasslands, tundra.

Holarctic, Paleotropical, Neotropical, Australian, and Antarctic. Boreal, Tethyan, and Madrean. Indomalesian. Arctic and Subarctic, Euro-Siberian, Eastern Asian, Atlantic North American, and Rocky Mountains. Mediterranean and Irano-Turanian. Indo-Chinese, Malesian, and Papuan. Caribbean and Andean. North and East Australian. New Zealand and Patagonian. European and Siberian. Canadian-Appalachian. Temperate and South-Eastern Australian.

Rusts and smuts. Rusts — *Puccinia*. Taxonomically wide-ranging species: *Puccinia graminis, Puccinia coronata, Puccinia praegracilis*, and *Puccinia recondita*. Smuts from Tilletiaceae and from Ustilaginaceae. Tilletiaceae — *Urocystis*. Ustilaginaceae — *Ustilago*.

References, etc. Morphological/taxonomic: Schouten and Veldkamp 1985. Leaf anatomical: Metcalfe 1960; this project.

Hilaria Kunth

Hexarrhena Presl, *Pleuraphis* Torrey, *Scleropelta* Buckley, *Symbasiandra* Steud.

Habit, vegetative morphology. Perennial; rhizomatous, or stoloniferous, or caespitose. **Culms 65–100 cm high**; herbaceous; branched above, or unbranched above. Culm internodes solid. *Leaves* not basally aggregated; auriculate (e.g. *H. jamesii*), or non-auriculate. Leaf blades narrow; 1.5–5 mm wide; flat, or folded, or rolled; without abaxial multicellular glands; tending to pseudopetiolate, or not pseudopetiolate; without cross venation; persistent. Ligule a fringed membrane to a fringe of hairs. *Contra-ligule* absent.

Reproductive organization. *Plants bisexual, with bisexual spikelets*; with hermaphrodite florets. *The spikelets* of sexually distinct forms on the same plant; *hermaphrodite and male-only (in triplets, two staminate and one hermaphrodite)*; overtly heteromorphic. Apomictic, or reproducing sexually.

Inflorescence. Inflorescence a false spike, with spikelets on contracted axes; contracted; espatheate; not comprising 'partial inflorescences' and foliar organs. *Spikelet-bearing axes* very much reduced (to triplets of spikelets); disarticulating (though the main axis persists); falling entire (i.e., the triplets falling). **Spikelets** *associated with bractiform involucres (constituted by the glumes of the three spikelets, which surround the triplet and are shed with it)*. The involucres shed with the fertile spikelets. Spikelets in triplets; not secund; sessile and subsessile (the central member subsessile, the others sessile).

Female-sterile spikelets. The lateral spikelets male, dorsally compressed, 2–3 flowered, their asymmetric glumes 4–5 nerved and awned at the side. Lodicules present. The male spikelets with glumes; 2–3 floreted.

Female-fertile spikelets. *Spikelets* 6–10 mm long; *asymmetrically compressed, the floret being flattened laterally, and the glumes being displaced and flattened against the rachis*; falling with the glumes (the clusters disarticulate). Rachilla terminated by a female-fertile floret.

Glumes two; more or less equal; shorter than the spikelets, or about equalling the spikelets; shorter than the adjacent lemmas, or long relative to the adjacent lemmas; joined, or free; displaced (against the rachis); hairy (ciliate); not pointed (cleft and ciliate); awned (dorsally from the mid-nerve, and in some species terminally from the laterals as well); carinate; similar. Lower glume 5–7 nerved. Upper glume 5–7 nerved. *Spikelets* with female-fertile florets only.

Female-fertile florets 1. Lemmas less firm than the glumes; not becoming indurated (membranous or hyaline); entire, or incised; when incised, 2 lobed; not deeply cleft; awnless, or mucronate, or awned. Awns when present, 1; median; dorsal; from near the top; non-geniculate; hairless; much shorter than the body of the lemma; entered by one vein. *Lemmas carinate*; without a germination flap; 3 nerved; with the nerves non-confluent. *Palea* present; relatively long; entire to apically notched; awnless, without apical setae; textured like the lemma (thin, hyaline); not indurated; 2-nerved; 2-keeled. Palea keels wingless. *Lodicules* absent (at least, in the material of *H. jamesii* seen). *Stamens* 3. Anthers 2.3–4 mm long (i.e. relatively long); not penicillate; without an apically prolonged connective. *Ovary* glabrous. Styles fused. Stigmas 2; red pigmented.

Fruit, embryo and seedling. Fruit ellipsoid. Hilum short. Embryo large; with an epiblast; with a scutellar tail; with an elongated mesocotyl internode. Embryonic leaf margins overlapping.

Abaxial leaf blade epidermis. *Costal/intercostal zonation* conspicuous. *Papillae* present; intercostal. Intercostal papillae over-arching the stomata; consisting of one symmetrical projection per cell, or consisting of one oblique swelling per cell (small, very thick walled, mostly towards one end of each long-cell and interstomatal). Long-cells of similar wall thickness costally and intercostally (quite thick walled). Mid-intercostal long-cells rectangular; having markedly sinuous walls (these pitted). *Microhairs* present; elongated; clearly two-celled; chloridoid-type (small, the basal cells relatively long, the distal cells very short). Microhair apical cell wall of similar thickness/rigidity to that of the basal cell. Microhairs 27–33 microns long. Microhair basal cells 24 microns long. Microhairs 7.5–10.5 microns wide at the septum. Microhair total length/width at septum 3–4.2. Microhair apical cells 7.5–10.5 microns long. Microhair apical cell/total length ratio 0.23–0.35. *Stomata* common; about 24 microns long. Subsidiaries dome-shaped and triangular. Guard-cells overlapped by the interstomatals (very slightly). *Intercostal short-cells* common; not paired (solitary, except that they are frequently paired with the bases of the abundant prickles). Intercostal silica bodies absent. *Costal short-cells* conspicuously in long rows. Costal silica bodies present in alternate cell files of the costal zones; 'panicoid-type'; consistently dumb-bell shaped (large).

Transverse section of leaf blade, physiology. C_4; biochemical type PCK (1 species); XyMS+. PCR sheath outlines even. PCR sheaths of the primary vascular bundles interrupted; interrupted abaxially only, or interrupted both abaxially and adaxially. PCR sheath extensions present. Maximum number of extension cells 1. PCR cell chloroplasts seemingly centripetal (in the poorish material seen). *Mesophyll* with radiate chlorenchyma; traversed by columns of colourless mesophyll cells (broad and conspicuous, between all the veins). *Leaf blade* 'nodular' in section; with the ribs more or less constant in size (broadly round topped). *Midrib* not readily distinguishable; with one bundle only. Bulliforms present in discrete, regular adaxial groups (in each groove); associated with colourless mesophyll cells to form deeply-penetrating fans (these linked with traversing columns of colourless cells). All the vascular bundles accompanied by sclerenchyma. Combined sclerenchyma girders present (with all the bundles); forming 'figures' (anchors or I's with all the bundles). Sclerenchyma all associated with vascular bundles. The lamina margins with fibres.

Cytology. Chromosome base number, $x = 9$. $2n = 36, 72, 86, 90,$ and 120. $4, 8,$ and 10 ploid (and higher). Nucleoli persistent.

Taxonomy. Chloridoideae; main chloridoid assemblage.

Ecology, geography, regional floristic distribution. 9 species. Southwest U.S.A. and Mexico to Venezuela. Xerophytic; species of open habitats. Arid and semi-arid plains. Holarctic, Paleotropical, and Neotropical. Boreal and Madrean. Madagascan. Atlantic North American. Caribbean. Southern Atlantic North American and Central Grasslands.

Rusts and smuts. Rusts — *Puccinia*. Taxonomically wide-ranging species: *Puccinia aristidae*. Smuts from Ustilaginaceae. Ustilaginaceae — *Ustilago*.

Economic importance. Important native pasture species: *H. belangeri*.

References, etc. Leaf anatomical: this project.

Special comments. Fruit data wanting.

Hitchcockella A. Camus

Habit, vegetative morphology. Perennial (shrub). The flowering culms leafy. *Culms woody and persistent (slender)*; not scandent (?); branched above (the branches in fascicles, more or less curved). Primary branches/mid-culm node several. Rhizomes leptomorph. *Leaves* not basally aggregated. Leaf blades lanceolate to elliptic (acuminate and mucronate); narrow; rounded at the base; pseudopetiolate (one supposes); without cross venation (apparently); disarticulating from the sheaths. *Ligule* present ('nearly none').

Reproductive organization. *Plants bisexual, with bisexual spikelets*; with hermaphrodite florets.

Inflorescence. Inflorescence of 1–3 spikelets at the apices of branchlets and rather hidden by leaves; spatheate; *a complex of 'partial inflorescences' and intervening foliar organs (presumably)*. *Spikelet-bearing axes* presumably very much reduced. Spikelets solitary; not secund; not in distinct 'long-and-short' combinations.

Female-fertile spikelets. Spikelets 12 mm long; strongly compressed laterally; disarticulating above the glumes (the assumed proximal lemma(s) falling with the floret); not disarticulating between the florets. Rachilla terminated by a female-fertile floret (?).

Glumes two, or several (?—with abundant scope for confusing glumes and proximal sterile lemmas, at least in the available description); shorter than the adjacent lemmas; (G_2) hairy; pointed; carinate; similar (lanceolate, acuminate, subulate). Lower glume 7 nerved. Upper glume 7–9 nerved. *Spikelets* with incomplete florets (probably?). The incomplete florets (if to be interpreted as such) proximal to the female-fertile florets. *Spikelets with proximal incomplete florets (?)*. *The proximal incomplete florets* 1; sterile. The proximal lemmas lemma-like — subulate, compressed, keeled, hairy; 13 nerved; exceeded by the female-fertile lemmas; similar in texture to the female-fertile lemmas.

Female-fertile florets 1. Lemmas acuminate; incised (obscurely bidenticulate); awnless; hairy (ciliate); carinate. *Palea* present. **Stamens 6**. **Ovary** glabrous (rostrate); *without a conspicuous apical appendage*. Styles fused (into one long style). Stigmas 2.

Taxonomy. Bambusoideae; Bambusodae; Bambuseae.

Ecology, geography, regional floristic distribution. 1 species. Madagascar. Paleotropical. Madagascan.

References, etc. Morphological/taxonomic: Camus 1925a.

Special comments. From the original, very inadequate description. Fruit data wanting. Anatomical data wanting.

Holcolemma Stapf & C.E. Hubb.

Habit, vegetative morphology. Annual, or perennial; loosely caespitose (with a knotty rootstock and geniculately ascending culms). *Culms* 30–60 cm high; herbaceous. *Leaves* not basally aggregated; non-auriculate. Leaf blades narrow; without cross venation; persistent. *Ligule a fringed membrane*; very short.

Reproductive organization. Plants bisexual, with bisexual spikelets; with hermaphrodite florets.

Inflorescence. Inflorescence paniculate; contracted; spicate. *Inflorescence axes not ending in spikelets (at least the larger branchlets terminated by a usually fairly inconspicuous bristle, which subtends the 'terminal' spikelet)*. Inflorescence espatheate; not comprising 'partial inflorescences' and foliar organs. Spikelet-bearing axes persistent. **Spikelets** (at least at the branch tips) *subtended by solitary 'bristles'*. The 'bristles' persisting on the axis. Spikelets not secund.

Female-fertile spikelets. Spikelets 2.5–3.5 mm long; abaxial; compressed dorsiventrally; falling with the glumes. Rachilla terminated by a female-fertile floret. Hairy callus absent.

Glumes two; very unequal (G_1 a quarter, G_2 a half the spikelet length); shorter than the spikelets; shorter than the adjacent lemmas; hairless; pointed, or not pointed; awnless; non-carinate; similar (membranous or chartaceous). *Lower glume 3 nerved*. Upper glume 3–5 nerved, or 13 nerved. *Spikelets* with incomplete florets. The incomplete florets proximal to the female-fertile florets. *The proximal incomplete florets* 1; paleate. *Palea of the proximal incomplete florets* fully developed; *becoming conspicuously hardened and enlarged laterally (after anthesis, when the large leathery keels come to clasp the side of the upper floret)*. The proximal incomplete florets male, or sterile. *The proximal lemmas doubly saccate, with a deep, grooved fold down the back*; awnless; 5–13 nerved; more or less equalling the female-fertile lemmas to decidedly exceeding the female-fertile lemmas (the L_2 sometimes much the shorter); less firm than the female-fertile lemmas to similar in texture to the female-fertile lemmas (membranous or herbaceous); not becoming indurated.

Female-fertile florets 1. Lemmas decidedly firmer than the glumes; rugose; becoming indurated (thinly crustaceous), or not becoming indurated; entire; pointed; awnless, or mucronate; hairless; non-carinate; having the margins lying flat and exposed on the palea; with a clear germination flap; 5 nerved. *Palea* present; relatively long; textured like the lemma; indurated, or not indurated; 2-nerved; 2-keeled. *Lodicules* present; free; fleshy; glabrous; not or scarcely vascularized. *Stamens* 3. Anthers not penicillate; without an apically prolonged connective. *Ovary* glabrous. Styles free to their bases. Stigmas 2; red pigmented.

Fruit, embryo and seedling. Fruit small; compressed dorsiventrally. Hilum short. Embryo large.

Abaxial leaf blade epidermis. *Costal/intercostal zonation* conspicuous. *Papillae* absent. Intercostal zones exhibiting many atypical long-cells (typical long-cells only alongside the veins: the rest rounded, irregular or isodiametric). Mid-intercostal long-cells rectangular, or fusiform; long-cells bordering veins sinuous, most intercostal cells not sinuous. *Microhairs* present; panicoid-type; (52.5–)54–57(–60) microns long; 7.5–9 microns wide at the septum. Microhair total length/width at septum 6.2–7.2. Microhair apical cells 30–34.5 microns long. Microhair apical cell/total length ratio 0.53–0.61. *Stomata* common (but only alongside veins); 40–46.5 microns long. Subsidiaries triangu-

464

lar. Guard-cells overlapping to flush with the interstomatals. *Intercostal short-cells* common; in cork/silica-cell pairs. *Costal short-cells* conspicuously in long rows. Costal silica bodies 'panicoid-type'; mostly dumb-bell shaped.

Transverse section of leaf blade, physiology. C_3; XyMS+. *Mesophyll* with radiate chlorenchyma; *Isachne*-type. *Leaf blade* 'nodular' in section to adaxially flat. Midrib with one bundle only. Bulliforms present in discrete, regular adaxial groups; in simple fans. All the vascular bundles accompanied by sclerenchyma. Combined sclerenchyma girders absent. Sclerenchyma all associated with vascular bundles.

Taxonomy. Panicoideae; Panicodae; Paniceae.

Ecology, geography, regional floristic distribution. 4 species. East Africa, southern India, Ceylon, Queensland. Shade species, or species of open habitats. Weedy places.

Paleotropical and Australian. African and Indomalesian. Sudano-Angolan. Indian and Indo-Chinese. North and East Australian. Somalo-Ethiopian and South Tropical African. Tropical North and East Australian.

References, etc. Morphological/taxonomic: Stapf and Hubbard 1929; Clayton 1978a. Leaf anatomical: this project.

Holcus L.

Arthrochloa R. Br., *Ginannia* Bub., *Homalachna* Kuntze, *Nothoholcus* Nash, *Notholcus* Hitchc., *Sorghum* Adans.

Habit, vegetative morphology. Annual (rarely), or perennial; rhizomatous to stoloniferous, or caespitose. **Culms** 8–150 cm high; *herbaceous*; unbranched above. Culm nodes hairy, or glabrous. Culm internodes hollow. *Leaves* not basally aggregated; non-auriculate. Leaf blades linear to linear-lanceolate; broad, or narrow; 2–16 mm wide; flat; without cross venation; persistent; rolled in bud. Ligule an unfringed membrane to a fringed membrane; not truncate (rounded); 1–5 mm long.

Reproductive organization. *Plants* bisexual, with bisexual spikelets; *with hermaphrodite florets*; outbreeding.

Inflorescence. *Inflorescence paniculate*; fairly open to contracted; when contracted spicate, or more or less irregular; espatheate; not comprising 'partial inflorescences' and foliar organs. Spikelet-bearing axes persistent. Spikelets not secund.

Female-fertile spikelets. *Spikelets* 3–8 mm long; compressed laterally; *falling with the glumes*; usually with a distinctly elongated rachilla internode above the glumes (with the lower floret elevated on a curved internode). *Rachilla prolonged beyond the uppermost female-fertile floret (and sometimes prolonged beyond the upper, male-only floret)*; rachilla hairy, or hairless. *The rachilla extension usually with incomplete florets (in the sense that the upper floret is usually male-only, regardless of whether there is an extension beyond it).* Hairy callus present, or absent. *Callus* very short.

Glumes two; more or less equal; about equalling the spikelets to exceeding the spikelets (enclosing them); long relative to the adjacent lemmas; pointed; *carinate*; similar (membranous). Lower glume 1 nerved. Upper glume 3 nerved. *Spikelets* with incomplete florets. The incomplete florets distal to the female-fertile florets (spikelets 2-flowered, the lower hermaphrodite, the upper usually male-only). *The distal incomplete florets* 1; incomplete florets *clearly specialised and modified in form (male, falling entire); incomplete florets awned (with a conspicuous short, hooked or geniculate dorsal awn from above the middle).*

Female-fertile florets 1 (very rarely, both florets hermaphrodite). *Lemmas decidedly firmer than the glumes (leathery or cartilaginous, shiny)*; entire, or incised; when entire pointed, or blunt; when incised, 2 lobed; not deeply cleft (bidentate); awnless, or awned. Awns when present, 1; dorsal; from near the top, or from well down the back; geniculate. Lemmas hairy, or hairless; *carinate*; indistinctly 3–5 nerved. *Palea* present; relatively long; tightly clasped by the lemma; textured like the lemma; not indurated (thin); 2-nerved; 2-keeled. *Lodicules* present; 2; free; membranous; glabrous; toothed, or not toothed. *Stamens* 3. Anthers not penicillate. *Ovary* glabrous. Styles free to their bases. Stigmas 2; white.

Fruit, embryo and seedling. *Fruit* adhering to lemma and/or palea, or free from both lemma and palea; small; longitudinally grooved (rarely), or not grooved; compressed laterally. Hilum short, or long-linear (rarely). Embryo small; not waisted. Endosperm liquid in the mature fruit, or hard; with lipid. Embryo with an epiblast; without a scutellar tail; with a negligible mesocotyl internode. Embryonic leaf margins meeting.

Seedling with a short mesocotyl, or with a long mesocotyl; with a loose coleoptile. First seedling leaf with a well-developed lamina. The lamina narrow; erect; 3 veined.

Abaxial leaf blade epidermis. *Costal/intercostal zonation* conspicuous. *Papillae* absent. *Long-cells* similar in shape costally and intercostally (but the intercostals much larger); of similar wall thickness costally and intercostally. Mid-intercostal long-cells rectangular; having straight or only gently undulating walls. *Microhairs* absent. *Stomata* common. Subsidiaries parallel-sided. Guard-cells overlapped by the interstomatals. *Intercostal short-cells* absent or very rare. Prickles abundant. *Costal short-cells* neither distinctly grouped into long rows nor predominantly paired. Costal silica bodies horizontally-elongated crenate/sinuous, or horizontally-elongated smooth, or rounded.

Transverse section of leaf blade, physiology. C_3; XyMS+. *Mesophyll* with radiate chlorenchyma (rarely), or with non-radiate chlorenchyma. *Leaf blade* with distinct, prominent adaxial ribs, or 'nodular' in section; with the ribs more or less constant in size. *Midrib* not readily distinguishable; with one bundle only; with colourless mesophyll adaxially (sometimes, in *H. lanatus*), or without colourless mesophyll adaxially. Bulliforms present in discrete, regular adaxial groups; in simple fans. All the vascular bundles accompanied by sclerenchyma. Combined sclerenchyma girders present; nowhere forming 'figures'. Sclerenchyma all associated with vascular bundles.

Culm anatomy. *Culm internode bundles* in one or two rings.

Phytochemistry. Tissues of the culm bases with little or no starch. Leaves without flavonoid sulphates (1 species).

Cytology. Chromosome base number, $x = 4$ and 7. $2n = 8$, 14, 28, 35, 42, and 49. 2, 4, 5, 6, and 7 ploid.

Taxonomy. Pooideae; Poodae; Aveneae.

Ecology, geography, regional floristic distribution. 6 species. 8 Canary Is., North Africa, Europe to Asia Minor & Caucasus; 1 South Africa. Commonly adventive. Mesophytic; shade species and species of open habitats; glycophytic. Grassland, open woodland, disturbed ground.

Holarctic, Paleotropical, Neotropical, and Cape. Boreal and Tethyan. African. Euro-Siberian and Atlantic North American. Macaronesian and Mediterranean. Sudano-Angolan. Andean. European. Canadian-Appalachian, Southern Atlantic North American, and Central Grasslands. South Tropical African.

Rusts and smuts. Rusts — *Puccinia*. Taxonomically wide-ranging species: *Puccinia coronata*, *Puccinia striiformis*, and *Puccinia hordei*. Smuts from Tilletiaceae and from Ustilaginaceae. Tilletiaceae — *Entyloma* and *Tilletia*. Ustilaginaceae — *Ustilago*.

Economic importance. Significant weed species: *H. lanatus*, *H. mollis*. Cultivated fodder: *H. lanatus*.

References, etc. Leaf anatomical: Metcalfe 1960; this project.

Homolepis Chase

Habit, vegetative morphology. Annual, or perennial; rooting at the nodes and then erect, rarely caespitose (*H. longispicula*). **Culms** 15–120 cm high; *herbaceous*; scandent, or not scandent; branched above. Culm nodes glabrous. Culm internodes hollow. Plants unarmed. **Leaves** mostly basal (*H. longispicula*), or not basally aggregated; *auriculate (the auricles small, on the sheath)*. The sheath margins with stiff hairs. Leaf blades linear (in *H. longispicula*), or linear-lanceolate to ovate-lanceolate (usually, and acuminate); broad (usually), or narrow (in *H. longispicula*); cordate (usually), or not cordate, not sagittate (*H. longispicula*); pseudopetiolate (usually), or not pseudopetiolate (*H. longispicula*); without cross venation; seemingly persistent. *Ligule* present; an unfringed membrane (*H. glutinosa*), or a fringed membrane (*H. isocalycina*), or a fringe of hairs (*H. longispicula*). *Contra-ligule* present (*H. aturensis*), or absent.

Reproductive organization. *Plants bisexual, with bisexual spikelets*; with hermaphrodite florets.

Inflorescence. *Inflorescence paniculate*; open; *with capillary branchlets*; espatheate; not comprising 'partial inflorescences' and foliar organs. Spikelet-bearing axes persistent. Spikelets not secund; not in distinct 'long-and-short' combinations.

Female-fertile spikelets. *Spikelets* 2.7–11 mm long; lanceolate; *compressed dorsiventrally*; falling with the glumes; with conventional internode spacings to with a distinctly elongated rachilla internode between the glumes. Rachilla terminated by a female-fertile floret. Hairy callus absent.

Glumes two; *more or less equal*; exceeding the spikelets; long relative to the adjacent lemmas; sparsely hairy, or hairless; pointed; awnless; non-carinate; *similar*. Lower glume 5–9 nerved. Upper glume 7–9 nerved. *Spikelets* with incomplete florets. The incomplete florets proximal to the female-fertile florets. *The proximal incomplete florets* 1; paleate. Palea of the proximal incomplete florets fully developed (membranous). The proximal incomplete florets male, or sterile. *The proximal lemmas* awnless; *5–7 nerved (with dense long stiff white hairs on the back, the folded margins enclosing the L_2 in palea-like fashion)*; decidedly exceeding the female-fertile lemmas; ultimately less firm than the female-fertile lemmas; not becoming indurated.

Female-fertile florets 1. Lemmas decidedly firmer than the glumes (ultimately); smooth; becoming indurated (cartilaginous); awnless; hairless; non-carinate; having the margins lying flat and exposed on the palea; with a clear germination flap; 5 nerved. *Palea* present; relatively long; entire (pointed); awnless, without apical setae; thinly cartilaginous, with hyaline infolded margins above; 2-nerved; keel-less. *Lodicules* present; 2; free; fleshy; glabrous; not or scarcely vascularized. *Stamens* 3. Anthers not penicillate; without an apically prolonged connective. *Ovary* glabrous. Styles basally fused, or free to their bases. Stigmas 2; red pigmented.

Fruit, embryo and seedling. Fruit small; compressed dorsiventrally. *Hilum long-linear*. Embryo large; without an epiblast; with a scutellar tail; with an elongated mesocotyl internode; with more than one scutellum bundle (3). Embryonic leaf margins overlapping.

Abaxial leaf blade epidermis. *Costal/intercostal zonation* conspicuous. *Papillae* absent. *Long-cells* markedly different in shape costally and intercostally (the costals narrower). Mid-intercostal long-cells rectangular; having markedly sinuous walls. *Microhairs* present; panicoid-type (but the distal cells large and fat); (52–)54–69(–70.5) microns long, or 36–42 microns long (in *H. longispicula*); (6.6–)7.5–12(–13.5) microns wide at the septum. Microhair total length/width at septum 5.5–9.3, or 3.1–3.6 (*H. longispicula*). Microhair apical cells (30–)33–39(–41) microns long, or 18–24 microns long (*H. longispicula*). Microhair apical cell/total length ratio 0.5–0.7. *Stomata* common; 27–43.5 microns long. Subsidiaries triangular. Guard-cells overlapping to flush with the interstomatals. *Intercostal short-cells* common; in cork/silica-cell pairs; silicified. *Costal short-cells* conspicuously in long rows (e.g., *H. aturensis*), or predominantly paired (*H. glutinosum*, *H. longispicula*), or conspicuously in long rows and neither distinctly grouped into long rows nor predominantly paired (*H. isocalycia*). Costal silica bodies basically 'panicoid-type', or tall-and-narrow to crescentic (intergrading with narrow crosses); mainly cross shaped, dumb-bell shaped, and nodular.

Transverse section of leaf blade, physiology. C_3; XyMS+. *Mesophyll* with non-radiate chlorenchyma; without adaxial palisade; with fusoids (these very large and regular in *H. aturensis*, *H. isocalycia*, *H. glutinosum*), or without fusoids (*H. longispicula*). The fusoids external to the PBS (contiguous with laterally extended PBS cells). *Leaf blade* 'nodular' in section to adaxially flat; when present, with the ribs more or less constant in size (broad, low). *Midrib* conspicuous (by virtue of its large bundle, with heavier abaxial sclerenchyma); with one bundle only. Bulliforms present in discrete, regular adaxial groups; in simple fans (the groups large). All the vascular bundles accompanied by sclerenchyma. Combined sclerenchyma girders present (with all the bundles); forming 'figures' (in the primaries, but the sclerenchyma masses small except in *H. longispicula*). Sclerenchyma all associated with vascular bundles.

Cytology. $2n = 22$, 24, and 40.

Taxonomy. Panicoideae; Panicodae; Paniceae.

Ecology, geography, regional floristic distribution. 4 species. Tropical South America. Helophytic to mesophytic; shade species and species of open habitats; glycophytic. In moist woods and wet places.

Neotropical. Caribbean, Venezuela and Surinam, Amazon, Central Brazilian, and Andean.

References, etc. Leaf anatomical: this project.

Homopholis C.E. Hubb.

Habit, vegetative morphology. Perennial; stoloniferous (often), or caespitose. *Culms* 20–50 cm high; herbaceous. Culm nodes glabrous. Culm internodes hollow. *Leaves* not basally aggregated; non-auriculate. Leaf blades narrow; 2–2.5 mm wide; without cross venation; persistent; rolled in bud. *Ligule an unfringed membrane*; truncate; 1.5 mm long.

Reproductive organization. Plants bisexual, with bisexual spikelets; with hermaphrodite florets.

Inflorescence. *Inflorescence paniculate*; *open (up to 25 cm long, the branches to 15 cm with one or few spikelets towards their ends)*; with capillary branchlets; espatheate; not comprising 'partial inflorescences' and foliar organs. Spikelet-bearing axes persistent. Spikelets not secund. Pedicel apices cupuliform.

Female-fertile spikelets. *Spikelets* 4.5–6 mm long; lanceolate; *compressed dorsiventrally*; falling with the glumes. Rachilla terminated by a female-fertile floret. Hairy callus absent.

Glumes two; *more or less equal (the G₁ slightly longer)*; long relative to the adjacent lemmas; pointed; awnless; very dissimilar to similar (the upper shortly hairy on the back). Lower glume 5–7 nerved. *Upper glume 7 nerved. Spikelets* with incomplete florets. The incomplete florets proximal to the female-fertile florets. *Spikelets with proximal incomplete florets*. *The proximal incomplete florets* 1; paleate, or epaleate. Palea of the proximal incomplete florets (when present) reduced (nerveless). The proximal incomplete florets sterile. *The proximal lemmas* similar to G₂; awnless; 7 nerved; decidedly exceeding the female-fertile lemmas; *less firm than the female-fertile lemmas*; not becoming indurated.

Female-fertile florets 1. *Lemmas* about half the spikelet length, shortly beaked; *decidedly firmer than the glumes*; smooth; becoming indurated (slightly so, and shiny); yellow in fruit; entire; pointed; awnless (but rostrate); hairless; non-carinate (rounded on the back); having the margins lying flat and exposed on the palea; with a clear germination flap; 5–7 nerved. *Palea* present (shortly auriculate at the base); relatively long; entire; awnless, without apical setae; thinner than the lemma (membranous); not indurated; 2-nerved. *Lodicules* present; 2; free; fleshy; glabrous. *Stamens* 3. Anthers 1.5 mm long; not penicillate. *Ovary* glabrous. Styles free to their bases. Stigmas 2; red pigmented.

Fruit, embryo and seedling. Fruit small; compressed dorsiventrally. *Hilum short*. Embryo large.

Abaxial leaf blade epidermis. *Papillae* absent (except on the subsidiaries). Mid-intercostal long-cells having markedly sinuous walls. *Microhairs* present; panicoid-type; 33–39 microns long; 13–17.4 microns wide at the septum. Microhair total length/width at septum 5–6.1. Microhair apical cells 5.4–7.2 microns long. Microhair apical cell/total length ratio 0.39–0.45. *Stomata* common; 29–33 microns long. Subsidiaries papillate; triangular. Costal silica bodies 'panicoid-type'.

Transverse section of leaf blade, physiology. C₃; XyMS+. *Mesophyll* with radiate chlorenchyma. *Leaf blade* 'nodular' in section; with the ribs more or less constant in size. *Midrib* not readily distinguishable; with one bundle only. Bulliforms present in discrete, regular adaxial groups; in simple fans. All the vascular bundles accompanied by sclerenchyma. Combined sclerenchyma girders present; forming 'figures'. Sclerenchyma all associated with vascular bundles.

Taxonomy. Panicoideae; Panicodae; Paniceae.

Ecology, geography, regional floristic distribution. 2 species. Australia. Shade species; glycophytic. In forests.

Australian. North and East Australian. Tropical North and East Australian.

Economic importance. Important native pasture species: *H. belsonii*.

References, etc. Morphological/taxonomic: Hubbard 1934; Webster 1987. Leaf anatomical: this project.

Homozeugos Stapf

Habit, vegetative morphology. Perennial; caespitose. *Culms* 60–200 cm high; herbaceous. Culm nodes hairy. Young shoots intravaginal. Leaves auriculate (with sheath auricles), or non-auriculate. Leaf blades narrowly linear (or junciform); narrow; without cross venation. *Ligule an unfringed membrane*; truncate, or not truncate.

Reproductive organization. Plants bisexual, with bisexual spikelets; with hermaphrodite florets. *The spikelets all alike in sexuality (in homogamous pairs).*

Inflorescence. *Inflorescence of spicate main branches (of few fascicled 'racemes', or a solitary 'raceme')*; digitate (usually), or non-digitate (when the 'raceme' solitary). Primary inflorescence branches 1–6. Inflorescence espatheate; not comprising 'partial inflorescences' and foliar organs. **Spikelet-bearing axes** slender 'racemes'; clustered; with very slender rachides; *disarticulating*; disarticulating at the joints. *'Articles'* linear; not appendaged; disarticulating obliquely; densely long-hairy. Spikelets paired; consistently in 'long-and-short' combinations; in pedicellate/sessile combinations. Pedicels of the 'pedicellate' spikelets free of the rachis. The 'shorter' spikelets hermaphrodite. The 'longer' spikelets hermaphrodite.

Female-fertile spikelets. *Spikelets* 6–14 mm long; *not noticeably compressed (almost cylindrical)*; falling with the glumes. Rachilla terminated by a female-fertile floret. Hairy callus present (spiny). *Callus* long; pointed.

Glumes two; more or less equal; long relative to the adjacent lemmas; free; hairy; without conspicuous tufts or rows of hairs; awnless; very dissimilar. Lower glume not two-keeled (except towards the tip); convex on the back (leathery); relatively smooth; 7–8 nerved (these evenly spaced). Upper glume 3 nerved. *Spikelets* with incomplete florets. The incomplete florets proximal to the female-fertile florets. *The proximal incomplete florets* 1; sterile. The proximal lemmas awnless; less firm than the female-fertile lemmas (hyaline); not becoming indurated.

Female-fertile florets 1. Lemmas linear; less firm than the glumes (scarious below, leathery above); not becoming indurated; incised; 2 lobed; not deeply cleft (bidentate); awned. Awns 1; median; from a sinus, or apical; geniculate; hairy to long-plumose; much longer than the body of the lemma. Lemmas hairy (ciliate).

Abaxial leaf blade epidermis. *Costal/intercostal zonation* conspicuous. *Papillae* present; intercostal. Intercostal papillae over-arching the stomata (slightly, at one end); consisting of one oblique swelling per cell (mostly on the interstomatals, in the form of one large, thick-walled papilla to each cell). *Long-cells* markedly different in shape costally and intercostally (the costals much narrower); of similar wall thickness costally and intercostally (walls medium thickness). Mid-intercostal long-cells rectangular; having markedly sinuous walls. *Microhairs* present; elongated; clearly two-celled; panicoid-type; 36–41–45 microns long; 4.5–4.8–5.4 microns wide at the septum. Microhair total length/width at septum 8–8.8. Microhair apical cells 16.5–20–22.5 microns long. Microhair apical cell/total length ratio 0.46–0.5. *Stomata* common; 27(–30) microns long. Subsidiaries non-papillate; low dome-shaped and triangular. Guard-cells overlapping to flush with the interstomatals. *Intercostal short-cells* absent or very rare (ignoring the prickles). Small prickles abundant. *Crown cells* absent (but with abundant costal prickles). *Costal short-cells* conspicuously in long rows. Costal silica bodies 'panicoid-type'; mostly nodular.

Transverse section of leaf blade, physiology. C_4; XyMS–. PCR sheath outlines even. PCR sheath extensions absent. *Leaf blade* with distinct, prominent adaxial ribs (over the primary bundles); with the ribs more or less constant in size (round-topped). *Midrib* very conspicuous; having a conventional arc of bundles (several small bundles on either side, between the large median and two large laterals, all the bundles towards the abaxial surface); with colourless mesophyll adaxially. *The lamina* symmetrical on either side of the midrib. Bulliforms present in discrete, regular adaxial groups; in simple fans and associated with colourless mesophyll cells to form deeply-penetrating fans; associating with colourless mesophyll cells to form arches over small vascular bundles. All the vascular bundles accompanied by sclerenchyma (the numerous small bundles each with an abaxial strand). Combined sclerenchyma girders present (with all the primary laterals, and some of the lower order bundles); forming 'figures' (especially T's). Sclerenchyma all associated with vascular bundles.

Taxonomy. Panicoideae; Andropogonodae; Andropogoneae; Andropogoninae.

Ecology, geography, regional floristic distribution. 5 species. Tropical Africa. Mesophytic. Savanna.

Paleotropical. African. Sudano-Angolan. South Tropical African.

References, etc. Morphological/taxonomic: Stapf 1915. Leaf anatomical: this project.

Special comments. Fruit data wanting.

Hookerochloa E. Alekseev

~ Austrofestuca (Festuca) hookeriana

Habit, vegetative morphology. Perennial; caespitose. **Culms** 55–180 cm high; *herbaceous*; unbranched above; 2–3 noded. Culm nodes exposed, or hidden by leaf sheaths; glabrous. Culm internodes hollow. Young shoots intravaginal (in our material). *Leaves* mostly basal; non-auriculate. Sheath margins joined (to 2/3 of their length, in the lower leaves), or free (upper leaves). Sheaths terete, not splitting longitudinally into fibres. Leaf blades linear; narrow; 3–6.8 mm wide (11–70 cm long); flat to folded; without cross venation; persistent; once-folded in bud. *Ligule an unfringed membrane*; truncate, or not truncate; 0.2–2.5 mm long (basal leaves), or 2.7–6.8 mm long (culm leaves). *Contraligule* absent.

Reproductive organization. Plants bisexual, with bisexual spikelets; with hermaphrodite florets.

Inflorescence. *Inflorescence* paniculate; *open (diffuse)*; espatheate; not comprising 'partial inflorescences' and foliar organs. Spikelet-bearing axes persistent. Spikelets not secund.

Female-fertile spikelets. Spikelets 7.7–14 mm long; compressed laterally; disarticulating above the glumes; disarticulating between the florets; with conventional internode spacings. Rachilla prolonged beyond the uppermost female-fertile floret; rachilla usually hairy. The rachilla extension naked. Hairy callus present. *Callus* short; blunt (densely villous).

Glumes two; very unequal to more or less equal; shorter than the spikelets; *shorter than the adjacent lemmas*; hairless; scabrous; pointed; awnless; carinate; similar (membranous, with thin margins). Lower glume 3 nerved. Upper glume 3 nerved, or 5 nerved. *Spikelets* with incomplete florets. The incomplete florets distal to the female-fertile florets. *The distal incomplete florets* merely underdeveloped.

Female-fertile florets 3–5. Lemmas lanceolate; similar in texture to the glumes (membranous to chartaceous); not becoming indurated; entire (usually), or incised; when incised, 2 lobed; not deeply cleft (sometimes minutely bidenticulate or erose); mucronate to awned (the lower lemmas usually conspicuously awned, the upper lemmas sometimes awnless). Awns 1; median; from a sinus, or apical (subapical); non-geniculate; hairless (scabrous); much shorter than the body of the lemma (0.2–3.4 mm long); entered by one vein. *Lemmas* hairless; scabrous (below); *carinate*; without a germination flap; 5–7 nerved; *with the nerves confluent towards the tip*. *Palea* present; relatively long; tightly clasped by the lemma; entire to apically notched; awnless, without apical setae; textured like the lemma; not indurated; 2-nerved; 2-keeled. Palea back glabrous, or scabrous. Palea keels wingless; hairy (densely puberulous). *Lodicules* present; 2; free; membranous; sparsely ciliate, or glabrous; toothed, or not toothed; not or scarcely vascularized. *Stamens* 3. Anthers 3.4–4.7 mm long; not penicillate; without an apically prolonged connective. *Ovary* glabrous. Styles free to their bases. Stigmas 2; white.

Fruit, embryo and seedling. **Fruit** free from both lemma and palea; small (2.6–3.4 mm long); ellipsoid; ventrally *longitudinally grooved*; compressed dorsiventrally; glabrous. *Hilum short (oval)*. Embryo small; not waisted. Endosperm hard. Embryo with an epiblast.

Abaxial leaf blade epidermis. *Costal/intercostal zonation* conspicuous. *Papillae* absent. *Long-cells* similar in shape costally and intercostally; of similar wall thickness costally and intercostally. Mid-intercostal long-cells rectangular; having markedly sinuous walls. *Microhairs* absent. *Stomata* common; 45–54 microns long. Subsidiaries mostly dome-shaped (but some truncated and approaching parallel-sided). Guard-cells overlapped by the interstomatals. Intercostal short-cells in cork/silica-cell pairs; silicified. Intercostal silica bodies irregularly rounded (potato shaped), or crescentic. With a few costal prickles. *Crown cells* absent. *Costal short-cells* neither distinctly grouped into long rows nor predominantly paired (mostly in pairs and short rows). Costal silica bodies rounded (predominating), or crescentic (a few, intergrading with the rounded forms).

Transverse section of leaf blade, physiology. C_3; XyMS+. *Mesophyll* with non-radiate chlorenchyma; without adaxial palisade. *Leaf blade* adaxially flat. *Midrib* conspicuous (by its abaxial keel and flanking hinges); with one bundle only. Bulliforms not in discrete, regular adaxial groups (save for the midrib 'hinges'). All the vascular bundles accompanied by sclerenchyma. Combined sclerenchyma girders present; forming 'figures' (with all the bundles). Sclerenchyma all associated with vascular bundles.

Taxonomy. Pooideae; Poodae; Poeae.

Ecology, geography, regional floristic distribution. 1 species. Eastern Australia. Helophytic; glycophytic. Alpine and subalpine swamps.

Australian. North and East Australian.

References, etc. Morphological/taxonomic: Alekseev 1985; Jacobs 1990. Leaf anatomical: this project.

Hordelymus (Jessen) C.O. Harz

Cuviera Koel.

Habit, vegetative morphology. Perennial; caespitose (from short, creeping rhizomes). *Culms* 40–110 cm high; herbaceous. Culm nodes hairy. Leaves auriculate. Leaf blades linear; broad, or narrow; 4–14 mm wide; flat; without cross venation. Ligule an unfringed membrane; truncate; 0.7–1 mm long.

Reproductive organization. Plants bisexual, with bisexual spikelets; with hermaphrodite florets. The spikelets of sexually distinct forms on the same plant; hermaphrodite and male-only, or hermaphrodite and sterile (the central spikelet of each triplet imperfect).

Inflorescence. *Inflorescence a false spike, with spikelets on contracted axes (the spikelets in triplets on reduced axes, like* **Hordeum** *but all pedicellate)*; contracted; espatheate; not comprising 'partial inflorescences' and foliar organs. Spikelet-bearing axes persistent (the rachis tough). *Spikelets* in triplets; not secund; distichous; *consistently in 'long-and-short' combinations*; unequally pedicellate in each combination. The 'shorter' spikelets (i.e. the central member of each triplet) male-only, or sterile. The 'longer' spikelets (i.e. the lateral members) hermaphrodite.

Female-fertile spikelets. Spikelets about 12 mm long; compressed dorsiventrally; disarticulating above the glumes. Rachilla prolonged beyond the uppermost female-fertile floret. The rachilla extension usually with incomplete florets (the upper of the two florets usually being imperfect, and with bristle beyond it).

Glumes two; more or less equal; long relative to the adjacent lemmas; *joined*; displaced (side by side); hairless; scabrous; subulate to not subulate (linear subulate to linear); *awned (but not awnlike)*; carinate, or non-carinate; similar. Lower glume 3 nerved. Upper glume 3 nerved. *Spikelets* usually with incomplete florets. The incomplete florets distal to the female-fertile florets. The distal incomplete florets 1; *incomplete florets* merely underdeveloped.

Female-fertile florets 1(–2) (the upper floret occasionally being also perfect). Lemmas lanceolate acuminate to the awn; similar in texture to the glumes; entire; pointed; awned. Awns 1; median; apical; non-geniculate; hairless; about as long as the body of the lemma to much longer than the body of the lemma; entered by several veins. Lemmas hairless; scabrous; non-carinate; without a germination flap; 5–7 nerved. *Palea* present (narrow); relatively long; 2-nerved; 2-keeled. *Lodicules* present; 2; free; membranous; ciliate; not toothed; not or scarcely vascularized. *Stamens* 3. Anthers 3–4 mm long. *Ovary* hairy. Styles free to their bases. Stigmas 2.

Fruit, embryo and seedling. *Fruit* slightly adhering to lemma and/or palea; medium sized (7–8 mm long); longitudinally grooved; compressed dorsiventrally; with hairs confined to a terminal tuft. Hilum long-linear. Embryo small.

Abaxial leaf blade epidermis. *Costal/intercostal zonation* conspicuous. *Papillae* absent. Long-cells differing markedly in wall thickness costally and intercostally (costals thicker-walled). Mid-intercostal long-cells rectangular; having markedly sinuous walls, or having straight or only gently undulating walls. *Microhairs* absent. *Stomata* absent or very rare. Subsidiaries low dome-shaped, or parallel-sided. *Intercostal short-cells* absent or very rare. Small prickles numerous. *Costal short-cells* neither distinctly grouped into long rows nor predominantly paired (the few seen mostly solitary). Costal silica bodies (the feww seen) horizontally-elongated crenate/sinuous, or horizontally-elongated smooth.

Transverse section of leaf blade, physiology. C_3; XyMS+. Leaf blade with the ribs very irregular in sizes. *Midrib* conspicuous (the girders more massive). Bulliforms present in discrete, regular adaxial groups; in simple fans. Combined sclerenchyma girders present. Sclerenchyma all associated with vascular bundles.

Cytology. Chromosome base number, $x = 7$. $2n = 28$ and 70. 4 and 10 ploid. Haplomic genome content H and T.

Taxonomy. Pooideae; Triticodae; Triticeae.

Ecology, geography, regional floristic distribution. 1 species. Europe, Western Asia. Shade species. Woodland.

Holarctic. Boreal and Tethyan. Euro-Siberian. Irano-Turanian. European.

Rusts and smuts. Rusts — *Puccinia*. Taxonomically wide-ranging species: *Puccinia graminis* and *Puccinia striiformis*.

References, etc. Morphological/taxonomic: Löve 1984. Leaf anatomical: Metcalfe 1960; this project.

Hordeum L.

Critesion Raf., *Critho* Meyer, *Zeocrithon* P. Beauv., *Zeocriton* Wolf

Habit, vegetative morphology. Annual, or perennial; caespitose (or the culms solitary). *Culms* 5–130 cm high; herbaceous; unbranched above; tuberous, or not tuberous. Culm nodes glabrous. Culm internodes hollow. *Leaves* not basally aggregated; auriculate, or non-auriculate. Leaf blades linear; broad (rarely), or narrow; 1.5–15 mm wide; not setaceous; usually flat, or rolled (convolute); without cross venation; persistent; rolled in bud. Ligule an unfringed membrane; truncate; 0.5–1 mm long.

Reproductive organization. *Plants bisexual, with bisexual spikelets*; with hermaphrodite florets. *The spikelets of sexually distinct forms on the same plant*; hermaphrodite and male-only, or hermaphrodite and sterile (the lateral spikelets sterile in *Critesion*, male in *Hordeum* sensu stricto). Plants outbreeding and inbreeding. Exposed-cleistogamous, or chasmogamous.

Inflorescence. *Inflorescence a false spike, with spikelets on contracted axes (the spikelets in triads)*; contracted. Inflorescence with axes ending in spikelets, or axes not ending in spikelets (the main axis lacking a terminal spikelet). Rachides hollowed. Inflorescence espatheate; not comprising 'partial inflorescences' and foliar organs. Spikelet-bearing axes disarticulating, or persistent (in cultivated forms); falling entire (the triplets shed, in *Hordeum* sensu stricto), or disarticulating at the joints (the main axis fragile, e.g. in *Critesion*). *Spikelets in triplets (the triplets shed together)*; not secund; distichous, or not two-ranked (in 2–6 rows); consistently in 'long-and-short' combinations, or not in distinct 'long-and-short' combinations (rarely, the laterals also sessile); usually in pedicellate/sessile combinations (the central spikelet of each triplet sessile, the laterals pedicellate). The 'shorter' spikelets hermaphrodite. The 'longer' spikelets hermaphrodite, or male-only, or sterile (in *Critesion*).

Female-sterile spikelets. The lateral spikelets usually smaller, when sterile sometimes reduced to clusters of awns.

Female-fertile spikelets. Spikelets compressed laterally to compressed dorsiventrally; falling with the glumes (in the deciduous triplets), or not disarticulating (in cultivated forms). Rachilla usually prolonged beyond the uppermost female-fertile floret (at least in the central spikelets of the triplets); rachilla hairy, or hairless. *The rachilla extension naked.*

Glumes two (unless the side by side pair in such forms represent a split G_1, with G_2 lacking (Butzin 1965)); more or less equal; free; displaced (side by side); subulate to not subulate; awned; similar (persistent, awn- or bristle-like above). Lower glume 1(–3) nerved. Upper glume 1(–3) nerved. **Spikelets** with female-fertile florets only; *without proximal incomplete florets.*

Female-fertile florets 1. Lemmas acuminate to an awn or awn point; similar in texture to the glumes (leathery); not becoming indurated; entire, or incised; when incised, 2 lobed, or 3 lobed; awned (nearly always), or mucronate (in some South American species). Awns 1; median; apical; non-geniculate; much shorter than the body of the lemma to much longer than the body of the lemma; entered by several veins. Lemmas hairless; non-carinate; 5 nerved. *Palea* present; relatively long; entire, or apically notched, or deeply bifid; awnless, without apical setae; 2-nerved; 2-keeled. *Lodicules* present; free; membranous; ciliate; toothed, or not toothed. *Stamens* 3. Anthers 0.2–5 mm long; not penicillate. **Ovary hairy.** Styles free to their bases. Stigmas 2; white.

Fruit, embryo and seedling. *Fruit* adhering to lemma and/or palea, or free from both lemma and palea; small to large; longitudinally grooved; compressed dorsiventrally; with hairs confined to a terminal tuft. Hilum long-linear. Embryo small; not waisted. Endosperm hard; without lipid; containing only simple starch grains. Embryo with an epiblast; without a scutellar tail; with a negligible mesocotyl internode. Embryonic leaf margins meeting.

Seedling with a short mesocotyl; with a loose coleoptile, or with a tight coleoptile. First seedling leaf with a well-developed lamina. The lamina narrow; erect; 3–7 veined.

Abaxial leaf blade epidermis. *Costal/intercostal zonation* conspicuous. *Papillae* absent. *Long-cells* the intercostals much larger; of similar wall thickness costally and intercostally. Mid-intercostal long-cells rectangular; having markedly sinuous walls, or having straight or only gently undulating walls. *Microhairs* absent. *Stomata* common. Subsidiaries parallel-sided, or parallel-sided and dome-shaped. Guard-cells overlapped by the interstomatals. *Intercostal short-cells* common, or absent or very rare; when present, in cork/silica-cell pairs, or not paired (solitary); silicified. Intercostal silica bodies when present, rounded, or crescentic. *Costal short-cells* conspicuously in long rows, or predominantly paired, or neither distinctly grouped into long rows nor predominantly paired. Costal silica bodies horizontally-elongated crenate/sinuous, or horizontally-elongated smooth, or rounded, or crescentic.

Transverse section of leaf blade, physiology. C_3; XyMS+. *PBS cells* without a suberised lamella. *Mesophyll* with non-radiate chlorenchyma. *Leaf blade* with distinct, prominent adaxial ribs, or 'nodular' in section, or adaxially flat; with the ribs more or less constant in size. *Midrib* conspicuous, or not readily distinguishable; with one bundle only. Bulliforms present in discrete, regular adaxial groups (in the furrows); in simple fans. All the vascular bundles accompanied by sclerenchyma. Combined sclerenchyma girders present; forming 'figures', or nowhere forming 'figures'. Sclerenchyma all associated with vascular bundles.

Culm anatomy. *Culm internode bundles* in one or two rings.

Phytochemistry. Tissues of the culm bases with little or no starch. Fructosans predominantly short-chain. Leaves without flavonoid sulphates (2 species).

Cytology. Chromosome base number, $x = 7$. $2n = 14$, 28, and 42 (the polyploids supposedly only in '*Critesion*'). 2, 4, and 6 ploid. Haplomic genome content H (*Critesion*), or I (*Hordeum* sensu stricto). Haploid nuclear DNA content 5.4–5.6 pg (15 species, mean 5.5). Mean diploid 2c DNA value 11 pg (11 species, 10.8–11.1). Nucleoli disappearing before metaphase.

Taxonomy. Pooideae; Triticodae; Triticeae.

Ecology, geography, regional floristic distribution. About 40 species. North temperate & South America. Commonly adventive. Mesophytic, or xerophytic; species of open habitats; halophytic and glycophytic. Open weedy or sandy places, mostly in dry soils.

Holarctic and Paleotropical. Boreal and Tethyan. African. Euro-Siberian. Mediterranean and Irano-Turanian. Sudano-Angolan. European. Somalo-Ethiopian.

Hybrids. Intergeneric hybrids with *Elytrigia* (×*Elytrordeum* Hylander, ×*Elyhordeum* Zizan & Petrowa), *Agropyron* (×*Agrohordeum* A. Camus), *Secale* (×*Hordale* Ciferri & Giacom.), *Sitanion* (×*Sitordeum* Bowden), *Triticum* (×*Tritordeum* Aschers. & Graebn.).

Rusts and smuts. Rusts — *Puccinia*. Taxonomically wide-ranging species: *Puccinia graminis*, *Puccinia coronata*, *Puccinia striiformis*, *Puccinia montanensis*, *Puccinia hordei*, *Puccinia recondita*, '*Uromyces*' *turcomanicum*, '*Uromyces*' *fragilipes*, and '*Uromyces*' *hordeinus*. Smuts from Tilletiaceae and from Ustilaginaceae. Tilletiaceae — *Entyloma*, *Tilletia*, and *Urocystis*. Ustilaginaceae — *Ustilago*.

Economic importance. Significant weed species: *H. jubatum*, *H. leporinum*, *H. marinum*, *H. murinum* (the fruiting inflorescence parts of these causing eye and other damage to livestock, and problems with wool). Grain crop species: *H. vulgare* (Barley); including (e.g.) *H. distichon*, *H. aegiceras*.

References, etc. Morphological/taxonomic: Löve 1984. Leaf anatomical: Metcalfe 1960; this project.

Hubbardia Bor

Habit, vegetative morphology. Delicate annual; stoloniferous. Culms herbaceous. *Leaves* not basally aggregated; non-auriculate. Leaf blades narrowly elliptic; flimsy; relatively broad; 3–7 mm wide (and about 1.5–4 cm long); flat; without cross venation. *Ligule* absent.

Reproductive organization. Plants bisexual, with bisexual spikelets; with hermaphrodite florets.

Inflorescence. Inflorescence paniculate (small, scanty, terminating the leafy branches); open; with capillary branchlets; espatheate; not comprising 'partial inflorescences' and foliar organs. Spikelet-bearing axes persistent. Spikelets solitary; not secund; not in distinct 'long-and-short' combinations.

Female-fertile spikelets. Spikelets *unconventional (said to have two lemmas, both lacking paleas)*; 2.7 mm long; not noticeably compressed to compressed dorsiventrally; disarticulating above the glumes (the glumes persistent); not disarticulating between the florets. Rachilla terminated by a female-fertile floret (?). Hairy callus absent.

Glumes two; more or less equal; about equalling the spikelets; long relative to the adjacent lemmas; hairless; awnless; non-carinate (lanceolate, round-backed); similar (thinly membranous). Lower glume 5–7 nerved. Upper glume 5–7 nerved. *Spikelets* with incomplete florets. The incomplete florets proximal to the female-fertile florets. *The proximal incomplete florets* 1; epaleate; sterile. The proximal lemmas lanceolate; awnless; 7–9 nerved; more or less equalling the female-fertile lemmas; similar in texture to the female-fertile lemmas; not becoming indurated.

Female-fertile florets 1. Lemmas not becoming indurated (membranous); entire; pointed, or blunt; awnless; hairless; non-carinate; 7–9 nerved. **Palea** *absent (unless the 'L_2' is a many-nerved palea)*. *Lodicules* present; 2; free; fleshy; glabrous. *Stamens* 3. *Ovary* glabrous. Styles free to their bases. Stigmas 2.

Fruit, embryo and seedling. Fruit small (1.5 mm long); fusiform; compressed dorsiventrally. Hilum short (very shortly linear). Embryo large (about a third of the fruit length). Endosperm containing compound starch grains.

Abaxial leaf blade epidermis. *Costal/intercostal zonation* conspicuous. *Papillae* absent. Intercostal zones exhibiting many atypical long-cells. Mid-intercostal long-cells having straight or only gently undulating walls. *Microhairs* absent (but present adaxially); panicoid type. *Stomata* absent or very rare. *Intercostal short-cells* absent or very rare. Costal silica bodies acutely-angled (cuboid); sharp-pointed (square to rectangular).

Transverse section of leaf blade, physiology. C_3 (seemingly with only two layers of mesophyll, the cells of upper layer irregularly lobed, those of the lower forming a palisade).

Special diagnostic feature. *Plants of wet places, the leaves remarkably thin and delicate*.

Taxonomy. Panicoideae; Panicodae; Isachneae.

Ecology, geography, regional floristic distribution. 1 species. Southern India. Helophytic (discovered in wet soil and high humidity, in the spray from a waterfall). Paleotropical. Indomalesian. Indian.

References, etc. Leaf anatomical: Clifford 1967.

Special comments. Extinct?.

Hubbardochloa Auquier

~ *Muhlenbergia*?

Habit, vegetative morphology. Slender annual. *Culms* 4–12(–17) cm high; herbaceous; branched above. *Leaves* not basally aggregated; non-auriculate. *Leaf blades* linear-lanceolate; narrow; 2–5 mm wide; flat, or rolled (convolute when dry); without abaxial multicellular glands; *pseudopetiolate (the base very abruptly contracted)*; without cross venation. Ligule a fringe of hairs.

Reproductive organization. Plants bisexual, with bisexual spikelets; with hermaphrodite florets. The spikelets all alike in sexuality. Plants seemingly inbreeding. Exposed-cleistogamous, or chasmogamous (?).

Inflorescence. *Inflorescence* paniculate; open; *with capillary branchlets*; espatheate; not comprising 'partial inflorescences' and foliar organs. Spikelet-bearing axes persistent. Spikelets solitary; secund, or not secund.

Female-fertile spikelets. Spikelets 1.2–1.4 mm long; fusiform; somewhat compressed laterally; disarticulating above the glumes. Rachilla terminated by a female-fertile floret. Hairy callus present (minute). *Callus* short; blunt.

Glumes two; more or less equal; exceeding the spikelets; long relative to the adjacent lemmas; hairless; glabrous to scabrous; pointed (the upper being acute), or not pointed (the lower being obtuse-erose); awnless; carinate (the upper), or non-carinate (the lower); slightly dissimilar. Lower glume 1 nerved. Upper glume 1 nerved. *Spikelets* with female-fertile florets only.

Female-fertile florets 1. Lemmas less firm than the glumes (hyaline-membranous); not becoming indurated; entire; blunt (truncate); awned. Awns 1; median; apical; non-geniculate; flexuous; hairless (scaberulous); much longer than the body of the lemma; entered by one vein. Lemmas hairy (on the margins); non-carinate (dorsally rounded); 1 nerved. **Palea** *absent*. *Lodicules* absent. *Stamens* 3. Anthers 0.3–0.4 mm long. *Ovary* glabrous. Styles free to their bases. Stigmas 2.

Fruit, embryo and seedling. Fruit small (0.9–1 mm long); fusiform; not noticeably compressed (almost cylindrical). Hilum short. Pericarp fused. Embryo small.

Abaxial leaf blade epidermis. *Costal/intercostal zonation* fairly conspicuous. *Papillae* present; very abundant, costal and intercostal. Intercostal papillae several per cell (nearly every cell with one or two rows of smallish, round or branch-paired papillae). *Long-cells* similar in shape costally and intercostally; of similar wall thickness costally and intercostally (rather thin walled). Mid-intercostal long-cells rectangular; having markedly sinuous walls (the sinuosity coarse). *Microhairs* present; elongated; clearly two-celled; chloridoid-type. Microhair apical cell wall of similar thickness/rigidity to that of the basal cell. Microhair basal cells 12 microns long. Microhair total length/width at septum 3. Microhair apical cell/total length ratio 0.33. *Stomata* absent or very rare. *Intercostal short-cells* absent or very rare; not paired. Intercostal silica bodies absent. No macrohairs or prickles seen. *Costal short-cells* conspicuously in long rows. Costal silica bodies present and well developed; present in alternate cell files of the costal zones; 'panicoid-type'; cross shaped, butterfly shaped, and dumb-bell shaped (mostly dumb-bells).

Transverse section of leaf blade, physiology. C_4; XyMS+. PCR sheath outlines even. PCR sheaths of the primary vascular bundles interrupted; interrupted abaxially only. PCR sheath extensions absent. *Mesophyll* with radiate chlorenchyma; not traversed by colourless columns. *Leaf blade* with distinct, prominent adaxial ribs; with the ribs more or less constant in size (round topped, one per bundle). *Midrib* not readily distinguishable; with one bundle only. Bulliforms present in discrete, regular adaxial groups (in every furrow); in simple fans (deeply penetrating). All the vascular bundles accompanied by sclerenchyma. Sclerenchyma all associated with vascular bundles (scanty). The lamina margins with fibres.

Taxonomy. Chloridoideae; main chloridoid assemblage.

Ecology, geography, regional floristic distribution. 1 species. Rwanda, Burundi, Zambia. Species of open habitats. Stony slopes, in savanna.

Paleotropical. African. West African Rainforest. South Tropical African.

References, etc. Morphological/taxonomic: Auquier 1980. Leaf anatomical: this project.

Humbertochloa A. Camus & Stapf

Habit, vegetative morphology. Perennial. The flowering culms leafy. **Culms** about 20–100 cm high; *woody and persistent (slender, bamboo-like)*; branched above. Culm internodes hollow. Plants unarmed. *Leaves* not basally aggregated; non-auriculate; without auricular setae. Leaf blades broad; 10–30 mm wide (and 2–9 cm long); conspicuously cordate; pseudopetiolate; cross veined; disarticulating from the sheaths; rolled in bud. *Ligule* present; truncate (short).

Reproductive organization. Plants monoecious with all the fertile spikelets unisexual; without hermaphrodite florets. The spikelets of sexually distinct forms on the same plant; female-only and male-only (separate male and female spikelets in the conspicuous terminal inflorescences, and sometimes with small, few-spikeleted female inflorescences in the upper leaf sheaths). The male and female-fertile spikelets in different inflorescences. The spikelets overtly heteromorphic (the males much smaller). Plants with hidden cleistogenes, or without hidden cleistogenes. The hidden cleistogenes (when present) in the leaf sheaths (in the form of few-spikeleted female inflorescences, in the upper leaf sheaths).

Inflorescence. *Inflorescence with the spikelets in secondary, short spiciform subsessile 'racemes', the latter borne alternately along the thickened midrib on one side of the thickened, leaf-like primary axis*. Rachides flattened and winged (spathiform). Inflorescence spatheate. **Spikelet-bearing axes** *very much reduced (the female racemes with one spikelet, the males with 1–4)*; disarticulating. *Spikelets* with 'involucres' of 'bristles' (the glumes of vestigial spikelets?). The 'bristles' persisting on the axis. Spikelets secund (i.e., on one side of the main axis).

Female-sterile spikelets. Male spikelets borne 1–4 per raceme, smaller than the females, the upper (fertile) floret with 3–6 stamens and a vestigial ovary. Male florets 3–6 staminate.

Female-fertile spikelets. Spikelets 10–11 mm long; abaxial; compressed laterally (asymmetric); falling with the glumes. Rachilla prolonged beyond the uppermost female-fertile floret. Hairy callus absent.

Glumes two; very unequal; shorter than the adjacent lemmas; very dissimilar (G_1 shorter, rigid and subulate, G_2 membranous, keeled, folded, lanceolate or ovate-oblong). Lower glume 0 nerved. Upper glume 5–7 nerved. *Spikelets* with incomplete florets. The incomplete florets proximal to the female-fertile florets. *The proximal incomplete florets* 1; sterile. The proximal lemmas awnless; rigid, thick.

Female-fertile florets 1. Lemmas lanceolate-acuminate; thinly leathery; entire; pointed; awnless; hairless; non-carinate (dorsally rounded); 7–11 nerved (the nerves anastomosing above). *Palea* present; conspicuous but relatively short (about two-thirds the lemma length); entire (lanceolate); awnless, without apical setae; not indurated (thinly leathery); 2-nerved. *Lodicules* present; 2; membranous ('hyaline'); heavily vascularized. *Stamens* 0, or 3–6 (staminodal). *Ovary* glabrous. Styles fused (into one long style). Stigmas 2.

Fruit, embryo and seedling. *Fruit* free from both lemma and palea. Hilum long-linear. Embryo small. Endosperm containing compound starch grains.

Abaxial leaf blade epidermis. *Costal/intercostal zonation* conspicuous. *Papillae* absent. *Long-cells* markedly different in shape costally and intercostally (the costals narrower); of similar wall thickness costally and intercostally (fairly thin walled). Mid-intercostal long-cells rectangular; having markedly sinuous walls (the sinuosity coarse). *Microhairs* present; elongated; clearly two-celled; panicoid-type; 57–69 microns long;

7.5–9 microns wide at the septum. Microhair total length/width at septum 6.3–9.2. Microhair apical cells 30–37.5 microns long. Microhair apical cell/total length ratio 0.5–0.57. *Stomata* common; 24–27 microns long. Subsidiaries mostly triangular. Guard-cells overlapping to flush with the interstomatals. *Intercostal short-cells* common; in cork/silica-cell pairs (these abundant, small); silicified. Intercostal silica bodies crescentic, or tall-and-narrow, or oryzoid-type (small). A few bulbous-based prickles present. *Crown cells* absent. *Costal short-cells* neither distinctly grouped into long rows nor predominantly paired (seemingly nearly all solitary). Costal silica bodies saddle shaped, or oryzoid (predominating, intergrading with the saddles).

Transverse section of leaf blade, physiology. C$_3$ (the bundles very widely separated); XyMS+ (the mestome sheath itself double in the midrib bundle). *Mesophyll* with non-radiate chlorenchyma; probably with adaxial palisade (judging from the distorted material seen); with arm cells; without fusoids. *Leaf blade* 'nodular' in section. *Midrib* conspicuous (by virtue of its large bundle and small, blunt, abaxial keel); with one bundle only. *The lamina* symmetrical on either side of the midrib. Bulliforms present in discrete, regular adaxial groups (these large and wide); in simple fans. All the vascular bundles accompanied by sclerenchyma. Combined sclerenchyma girders present; forming 'figures' (the median with a large, pronounced I, the other bundles with slight 'anchors' or I's). Sclerenchyma all associated with vascular bundles.

Special diagnostic feature. *Spikelets borne on one side of a broad, leaflike rachis.*

Taxonomy. Bambusoideae; Oryzodae; Phyllorhachideae.

Ecology, geography, regional floristic distribution. 2 species. Tropical East Africa, Madagascar. Shade species. In forests.

Paleotropical. African. Sudano-Angolan. South Tropical African.

References, etc. Morphological/taxonomic: Hubbard 1939. Leaf anatomical: this project.

Hyalopoa (Tzvelev) Tzvelev

~ *Colpodium*

Habit, vegetative morphology. Perennial; caespitose. **Culms** 15–70 cm high; *herbaceous*; unbranched above. Culm nodes glabrous. *Leaves* not basally aggregated. Leaf blades narrow; 1–6 mm wide (by 4–15 cm long); flat; without cross venation; persistent.

Reproductive organization. *Plants bisexual, with bisexual spikelets*; with hermaphrodite florets.

Inflorescence. *Inflorescence* paniculate; *open (effuse, nodding, the branches flexuose)*; espatheate; not comprising 'partial inflorescences' and foliar organs. Spikelet-bearing axes persistent. Spikelets not in distinct 'long-and-short' combinations.

Female-fertile spikelets. *Spikelets 5–10 mm long*; compressed laterally; disarticulating above the glumes. Rachilla prolonged beyond the uppermost female-fertile floret. Hairy callus absent. *Callus* short; blunt.

Glumes two; very unequal; shorter than the spikelets; shorter than the adjacent lemmas; *not pointed (rounded-denticulate)*; awnless; non-carinate; similar (cuneate to lanceolate, often purple-tipped). Lower glume longer than half length of lowest lemma; 1 nerved. Upper glume 1 nerved, or 3 nerved. *Spikelets* with female-fertile florets only, or with incomplete florets. The incomplete florets distal to the female-fertile florets. *The distal incomplete florets* merely underdeveloped.

Female-fertile florets 2–4. Lemmas decidedly firmer than the glumes (leathery below); not becoming indurated; entire, or incised (very slightly); pointed, or blunt; not deeply cleft; awnless; hairy (pilose below, between the ciliate nerves); *carinate. The keel* slightly *winged*. Lemmas without a germination flap; 3 nerved, or 5 nerved. *Palea* present; relatively long; awnless, without apical setae; 2-nerved; 2-keeled. Palea keels wingless. *Lodicules* present; 2; free; membranous; glabrous; not toothed; not or scarcely vascularized. *Stamens* 3. Anthers 1.5–3.5 mm long. *Ovary* glabrous. Styles free to their bases. Stigmas 2; white.

Fruit, embryo and seedling. *Fruit* free from both lemma and palea; small (1.8–2.5 mm long). Hilum short. Embryo small.

Cytology. Chromosome base number, $x = 7$. $2n = 42$. 6 ploid.

Taxonomy. Pooideae; Poodae; Aveneae.

Ecology, geography, regional floristic distribution. 4 species. Caucasus, Turkey, northwest Himalaya, northeast Siberia. Helophytic (moist upland habitats); species of open habitats; glycophytic.

Holarctic. Boreal and Tethyan. Euro-Siberian and Eastern Asian. Irano-Turanian. Siberian.

References, etc. Morphological/taxonomic: Bor 1970 (under *Colpodium*).

Special comments. Anatomical data wanting.

Hydrochloa P. Beauv.

~ Luziola (L. fluitans, = H. caroliensis

Habit, vegetative morphology. Perennial; a slender aquatic, floating or with trailing culms 30–100 cm long. Culms herbaceous; branched above. **Leaves** not basally aggregated; *with auricular setae (usually 2 or 3 on either side)*. Leaf blades narrow; 2–5 mm wide (by 2–6 cm long); flat. Ligule an unfringed membrane; 0.5–1 mm long.

Reproductive organization. *Plants monoecious with all the fertile spikelets unisexual*; without hermaphrodite florets. The spikelets of sexually distinct forms on the same plant; female-only and male-only. *The male and female-fertile spikelets in different inflorescences (male spikelets in small few-flowered terminal panicles or racemes, female spikelets axillary, solitary or in few-flowered racemes)*. The spikelets overtly heteromorphic.

Inflorescence. Inflorescence reduced to a single spikelet, or few spikeleted; a single raceme, or paniculate; espatheate; not comprising 'partial inflorescences' and foliar organs. Spikelet-bearing axes persistent.

Female-sterile spikelets. Male spikelets about 4 mm long, 1-flowered, the lemma thin and 7-nerved the palea thin and 2-nerved, the floret with 6 free stamens. The male spikelets without glumes; 1 floreted. Male florets 6 staminate. The staminal filaments free.

Female-fertile spikelets. Spikelets unconventional (through lacking organs, presumably glumes); *2 mm long*; not noticeably compressed to compressed dorsiventrally; *abscising below the spikelet*. Rachilla terminated by a female-fertile floret.

Glumes *absent*. *Spikelets* with female-fertile florets only.

Female-fertile florets 1. Lemmas not becoming indurated (thin); entire; pointed; hairless; without a germination flap; 5–7 nerved. *Palea* present; relatively long (thin); not indurated; several nerved (4–7); one-keeled. *Stamens* 0. Styles fused (?). Stigmas 2 (long).

Abaxial leaf blade epidermis. *Costal/intercostal zonation* conspicuous. *Papillae* present; costal. Intercostal papillae complex — one large papilla on each long-cell and interstomatal, itself covered by the numerous, minute papillae which extend over most of the cell surface. *Long-cells* markedly different in shape costally and intercostally (the costals much narrower and more regular); of similar wall thickness costally and intercostally (thin walled). Mid-intercostal long-cells irregular, but rectangular; having markedly sinuous walls. *Microhairs* present (detectable by the bases, but hard to locate and none seen in good condition); ostensibly one-celled (cf. *Luziola*?); minute. *Stomata* fairly common (in single files near the costal zones); 18–21 microns long. Subsidiaries often papillate (two papillae each, near their ends). Guard-cells deeply sunken amongst the surrounding papillate cells. *Intercostal short-cells* common; silicified. Intercostal silica bodies oryzoid-type. *Costal short-cells* conspicuously in long rows. Costal silica bodies 'panicoid-type'; exclusively dumb-bell shaped (short).

Transverse section of leaf blade, physiology. C_3; XyMS+. Mesophyll without adaxial palisade; probably with arm cells (but the material seen very poor); without fusoids. *Midrib* fairly conspicuous (by its larger bundle); with one bundle only. *The lamina* symmetrical on either side of the midrib. Bulliforms present in discrete, regular adaxial groups (a large group in each intercostal zone); in simple fans. All the vascular bundles accompanied by sclerenchyma. Combined sclerenchyma girders present (with all the bundles). Sclerenchyma all associated with vascular bundles.

Cytology. Chromosome base number, $x = 12$. $2n = 24$.

Taxonomy. Bambusoideae; Oryzodae; Oryzeae.

Ecology, geography, regional floristic distribution. 1 species. Southeast U.S.A. Hydrophytic.

Holarctic. Boreal. Atlantic North American. Southern Atlantic North American.

References, etc. Leaf anatomical: this project.

Special comments. Fruit data wanting.

Hydrothauma C.E. Hubb.

Habit, vegetative morphology. Aquatic annual. Culms herbaceous; branched above. *Leaves* not basally aggregated. *Leaf blades linear (the emergent blades with sinuous longitudinal lamellae adaxially)*; narrow; *pseudopetiolate (the blades of the lower leaves floating on their long 'petioles')*. Ligule an unfringed membrane.

Reproductive organization. Plants bisexual, with bisexual spikelets; with hermaphrodite florets.

Inflorescence. Inflorescence short, slender, a single raceme (with paired spikelets), or paniculate; spicate; non-digitate; espatheate; not comprising 'partial inflorescences' and foliar organs. Spikelet-bearing axes persistent. Spikelets paired; secund; consistently in 'long-and-short' combinations. The 'shorter' spikelets hermaphrodite. The 'longer' spikelets hermaphrodite.

Female-fertile spikelets. Spikelets abaxial; compressed dorsiventrally (asymmetrical in profile); falling with the glumes. Rachilla terminated by a female-fertile floret. Hairy callus absent.

Glumes two; very unequal; (the upper) about equalling the spikelets; (the upper) long relative to the adjacent lemmas; dorsiventral to the rachis; hairless; awnless; non-carinate; very dissimilar (the lower short, orbicular-truncate, membranous-hyaline, the upper elongated, gibbous, equalling the spikelet, membranous-herbaceous). Lower glume 0 nerved. **Upper glume** *distinctly saccate (cf.* **Sacciolepis***)*; 7–9 nerved. *Spikelets* with incomplete florets. The incomplete florets proximal to the female-fertile florets. *The proximal incomplete florets* 1; paleate. Palea of the proximal incomplete florets reduced. The proximal incomplete florets male, or sterile. The proximal lemmas awnless; 7 nerved; decidedly exceeding the female-fertile lemmas (resembling G_2); less firm than the female-fertile lemmas; not becoming indurated.

Female-fertile florets 1. Lemmas decidedly firmer than the glumes (leathery to crustaceous); becoming indurated; entire; awnless; hairless; non-carinate; obscurely nerved. *Palea* present; relatively long (somewhat leathery); tightly clasped by the lemma; entire (pointed); awnless, without apical setae; textured like the lemma; indurated. *Lodicules* present; 2; free; fleshy. *Stamens* 3. Anthers not penicillate; without an apically prolonged connective (anthers divaricate). *Ovary* glabrous. Styles free to their bases. Stigmas 2.

Abaxial leaf blade epidermis. *Costal/intercostal zonation* conspicuous. *Papillae* absent (though very abundant adaxially, both intercostally and over the lamellae). Long-cells of similar wall thickness costally and intercostally (fairly thin walled). Mid-intercostal long-cells rectangular; having markedly sinuous walls (the sinuosity coarse and irregular). *Microhairs* present; elongated; clearly two-celled; panicoid-type; without 'partitioning membranes' (the costals very much narrower). *Stomata* absent or very rare. *Intercostal short-cells* common (in places — exclusive of abundant microhair bases); not paired (seemingly solitary); not silicified. *Costal short-cells* conspicuously in long rows. Costal silica bodies present and well developed; 'panicoid-type'; nearly all nodular.

Transverse section of leaf blade, physiology. Leaf blades 'laminar' (very unusual, with high densely papillate lamella-ribs running longitudinally in association with each vascular bundle). C_3; XyMS+. Mesophyll without adaxial palisade; not *Isachne*-type; without fusoids (but with rounded or angular lacunae in the abaxial part of the mesophyll). *Leaf blade* with distinct, prominent adaxial ribs (i.e., the lamellae). Combined sclerenchyma girders absent (the sclerenchyma very scanty). Sclerenchyma not all bundle-associated. The 'extra' sclerenchyma in adaxial groups (there being small groups of hypodermal fibres, and single large lignified hypodermal cells, in the lamellae).

Special diagnostic feature. *The adaxial surface of the leaf blade raised into sinuous lamellae.*

Taxonomy. Panicoideae; Panicodae; Paniceae.

Ecology, geography, regional floristic distribution. 1 species. Southern tropical Africa. Hydrophytic (in shallow pools).

Paleotropical. African. Sudano-Angolan. South Tropical African.

References, etc. Morphological/taxonomic: Hubbard 1947. Leaf anatomical: Metcalfe 1960; this project.

Special comments. Worthy of detailed anatomical study based on better material than that seen. The presence of abundant microhairs on the submerged, astomatal abaxial blade surface is odd, and many of the vascular bundles are associated with an abaxial sclerenchyma group incorporating an extraordinarily large, lignified element. Fruit data wanting.

Hygrochloa Lazarides

Habit, vegetative morphology. Annual, or perennial (aquatic to hydrophytic); caespitose (*H. cravenii*, erect, not submerged, to 1 m high), or decumbent (*H. aquatica*, floating from submerged tufts). *Culms* 50–100 cm high; herbaceous. Culm internodes hollow (and aerenchymatous). *Leaves* not basally aggregated; non-auriculate. Leaf blades narrow; without cross venation; persistent. Ligule a fringe of hairs.

Reproductive organization. *Plants monoecious with all the fertile spikelets unisexual*; without hermaphrodite florets. The spikelets of sexually distinct forms on the same plant; female-only and male-only. The male and female-fertile spikelets on different branches of the same inflorescence (female spikelets in appressed spiciform 'racemes' below, male spikelets on the main axis above). The spikelets overtly heteromorphic.

Inflorescence. Inflorescence of spicate main branches. *Inflorescence axes not ending in spikelets (the branch rachides and the panicle axis briefly bare-tipped)*. Rachides hollowed. Inflorescence espatheate; not comprising 'partial inflorescences' and foliar organs. *Spikelet-bearing axes persistent*. Spikelets solitary; secund; biseriate. Pedicel apices discoid.

Female-sterile spikelets. Male spikelets 3.5–5 mm long, with 2 florets, both staminate with three stamens or one neuter. The male spikelets with glumes; 2 floreted. The lemmas awnless. Male florets 1, or 2; 3 staminate.

Female-fertile spikelets. Spikelets 2–2.75 mm long; oblong, or elliptic, or obovate; abaxial; compressed dorsiventrally; falling with the glumes. Rachilla terminated by a female-fertile floret. Hairy callus absent.

Glumes two; very unequal; shorter than the adjacent lemmas, or long relative to the adjacent lemmas; dorsiventral to the rachis; hairless; not pointed; awnless; non-carinate; very dissimilar (both membranous, the upper much larger and more substantial). Lower glume 0–3 nerved. Upper glume 5–7 nerved. *Spikelets* with incomplete florets. The incomplete florets proximal to the female-fertile florets. *The proximal incomplete florets* 1; epaleate; sterile. The proximal lemmas awnless; 5–7 nerved; more or less equalling the female-fertile lemmas; less firm than the female-fertile lemmas (membranous); not becoming indurated.

Female-fertile florets 1. Lemmas decidedly firmer than the glumes (leathery to crustaceous); striate (or striolate); becoming indurated; yellow in fruit, or brown in fruit; entire; pointed; awnless (apiculate); hairless; non-carinate; having the margins lying flat and exposed on the palea; with a clear germination flap; 5 nerved. *Palea* present; relatively long; entire; awnless, without apical setae; textured like the lemma; indurated, or not indurated; 1-nerved; 2-keeled. *Lodicules* present; 2; free; fleshy; glabrous. *Stamens* 0. *Ovary* glabrous. Styles fused. Stigmas 2; red pigmented.

Fruit, embryo and seedling. Fruit small (1.5–1.75 mm long); compressed dorsiventrally. Hilum short. Embryo large.

Abaxial leaf blade epidermis. *Costal/intercostal zonation* conspicuous. *Papillae* absent. *Long-cells* similar in shape costally and intercostally (long-rectangular); of similar wall thickness costally and intercostally (thin walled). Mid-intercostal long-cells having markedly sinuous walls. *Microhairs* present; panicoid-type (distal cells very flimsy and often missing); 34.5–36 microns long; 5.4–7.5 microns wide at the septum. Microhair total length/width at septum 4.8–6.7. Microhair apical cells 18–21 microns long. Microhair apical cell/total length ratio 0.52–0.58. *Stomata* common; 33–39 microns long. Subsidiaries triangular. Guard-cells overlapping to flush with the interstomatals (flush). *Intercostal short-cells* common; in cork/silica-cell pairs (mostly superposed); silicified, or not silicified. *Costal short-cells* conspicuously in long rows, or neither distinctly grouped into long rows nor predominantly paired. Costal silica bodies 'panicoid-type'; nodular.

Transverse section of leaf blade, physiology. C_4; XyMS–. PCR sheath outlines uneven. PCR sheath extensions absent. *Leaf blade* adaxially with numerous large, inflated hairs, clustered on the ribs and less frequent intercostally. *Midrib* not readily distinguishable; with one bundle only. Bulliforms not in discrete, regular adaxial groups (the epidermis extensively 'bulliform'). Many of the smallest vascular bundles unaccompanied by sclerenchyma. Combined sclerenchyma girders absent (with the adaxial sclerenchyma confined to tiny strands, associated with the larger bundles). Sclerenchyma all associated with vascular bundles.

Taxonomy. Panicoideae; Panicodae; Paniceae.

Ecology, geography, regional floristic distribution. 2 species. North Australia. Hydrophytic to helophytic.

Australian. North and East Australian. Tropical North and East Australian.

References, etc. Morphological/taxonomic: Lazarides 1979. Leaf anatomical: this project.

Hygroryza Nees

Potamochloa Griff.

Habit, vegetative morphology. Aquatic perennial; stoloniferous (with adventitious roots). The flowering culms leafy. Culms herbaceous; unbranched above. Culm nodes glabrous. *Leaves* not basally aggregated; non-auriculate; without auricular setae. Sheath margins free. Sheaths inflated, tessellate-veined. Leaf blades ovate-lanceolate to elliptic; broad; to 20 mm wide; somewhat cordate; flat; pseudopetiolate (floating); cross veined; persistent. Ligule an unfringed membrane; truncate; 0.8 mm long. *Contra-ligule* absent.

Reproductive organization. Plants bisexual, with bisexual spikelets; with hermaphrodite florets.

Inflorescence. Inflorescence relatively few spikeleted (partly enclosed in the upper leaf sheath); paniculate; open; espatheate (but the uppermost leaf quite reduced); not comprising 'partial inflorescences' and foliar organs. Spikelet-bearing axes persistent. Spikelets solitary; not secund; imbricate.

Female-fertile spikelets. *Spikelets* unconventional (by virtue of the vestigial glumes); compressed laterally; falling with the glumes (at least, disarticulating well below the annular rim which presumably represents them: the floret falling on its long, slender stipe). Rachilla terminated by a female-fertile floret. Hairy callus absent.

Glumes more or less *absent (being represented at the most by an annular rim).* *Spikelets* with female-fertile florets only.

Female-fertile florets 1. Lemmas attenuate; not becoming indurated (papery or firmly membranous); entire; pointed; awned. Awns 1; median; apical; non-geniculate; hairless; about as long as the body of the lemma to much longer than the body of the lemma. Lemmas hairy (hispid, especially on the veins); carinate; without a germination flap; 5 nerved (the laterals close to the margins); with the nerves confluent towards the tip. *Palea* present; relatively long (about equalling the lemma); entire (acute); awnless, without apical setae; textured like the lemma; not indurated (papery); several nerved (3); one-keeled. *Lodicules* present; 2; free; membranous; glabrous; not toothed; heavily vascularized. *Stamens* 6. Anthers about 1 mm long; not penicillate; without an apically prolonged connective. *Ovary* glabrous. Styles free to their bases. Stigmas 2.

Fruit, embryo and seedling. *Fruit* free from both lemma and palea; small (about 3.5mm long); not noticeably compressed (terete). Hilum long-linear. Embryo small. Endosperm hard.

Abaxial leaf blade epidermis. *Costal/intercostal zonation* conspicuous. *Papillae* present; costal and intercostal. Intercostal papillae over-arching the stomata (but only very slightly, at the ends); several per cell (each long- and interstomatal cell with one large central papilla, this and the entire cell surface ornamented by numerous small cuticular papillae — only the guard cells and subsidiaries lack ornamentation). Long-cells of similar wall thickness costally and intercostally (walls of medium thickness). Intercostal zones exhibiting many atypical long-cells (most being short to more or less isodiametric). Mid-intercostal long-cells having markedly sinuous walls. *Microhairs* present; elongated; ostensibly one-celled; these in the form of minute, seemingly 1-celled trichomes — their bases presenting the appearance of those of conventional microhairs at first sight. *Stomata* common. Subsidiaries non-papillate; dome-shaped and triangular. Guard-cells overlapped by the interstomatals (very slightly), or overlapping to flush with the interstomatals. *Intercostal short-cells* common; not paired (solitary); silicified. Intercostal silica bodies minutely oryzoid-type. *Costal short-cells* conspicuously in long rows. Costal silica bodies oryzoid (exclusively, a large version).

Transverse section of leaf blade, physiology. C_3. *Mesophyll* with non-radiate chlorenchyma; with arm cells, or without arm cells (? — these not apparent in the poor material seen); with fusoids. The fusoids external to the PBS. *Leaf blade* with distinct, prominent adaxial ribs. Midrib with one bundle only. Combined sclerenchyma girders present. Sclerenchyma all associated with vascular bundles.

Special diagnostic feature. *Plants aquatic, with inflated leaf sheaths serving as floats.*

Cytology. Chromosome base number, $x = 12$. $2n = 24$. 2 ploid.

Taxonomy. Bambusoideae; Oryzodae; Oryzeae.

Ecology, geography, regional floristic distribution. 1 species. Eastern Asia. Hydrophytic.

Paleotropical. Indomalesian. Indian and Indo-Chinese.

Economic importance. Significant weed species: *H. aristata* (in rice).

References, etc. Leaf anatomical: this project.

Hylebates Chippindall

Habit, vegetative morphology. Annual, or perennial; decumbent. **Culms** 40–300 cm high; *herbaceous*. *Leaves* not basally aggregated. *Leaf blades* linear to lanceolate; *broad*; 10–40 mm wide; cordate (amplexicaul, in *H. cordatus*), or not cordate, not sagittate; *not pseudopetiolate*. Ligule present; *a fringed membrane*; very short.

Reproductive organization. Plants bisexual, with bisexual spikelets; with hermaphrodite florets.

Inflorescence. *Inflorescence paniculate*; open; with capillary branchlets; espatheate; not comprising 'partial inflorescences' and foliar organs. Spikelet-bearing axes persistent. Spikelets not secund.

Female-fertile spikelets. *Spikelets* 2–3.5 mm long; *compressed dorsiventrally*; falling with the glumes; with a distinctly elongated rachilla internode between the glumes. Rachilla terminated by a female-fertile floret. Hairy callus absent.

Glumes two; very unequal (G_1 about a third the length of G_2); (the upper) about equalling the spikelets; (the upper) long relative to the adjacent lemmas; hairless; awnless; non-carinate; very dissimilar (G_1 much smaller, ovate, basally clasping the spikelet). *Spikelets* with incomplete florets. The incomplete florets proximal to the female-fertile florets. *Spikelets with proximal incomplete florets.* *The proximal incomplete florets* 1; sterile. *The proximal lemmas* shortly *awned (terminally)*; less firm than the female-fertile lemmas to similar in texture to the female-fertile lemmas (membranous); not becoming indurated.

Female-fertile florets 1. Lemmas similar in texture to the glumes to decidedly firmer than the glumes (thinly papery); smooth (? — dull); not becoming indurated; entire; pointed, or blunt; awnless; hairless; non-carinate; having the margins lying flat and exposed on the palea (enfolding and concealing it). *Palea* present; tightly clasped by the lemma; awnless, without apical setae; 2-nerved. *Lodicules* present; 2; fleshy. *Stamens* 3. Stigmas 2.

Abaxial leaf blade epidermis. *Costal/intercostal zonation* conspicuous. *Papillae* absent. *Long-cells* markedly different in shape costally and intercostally (the costals narrow, rectangular); of similar wall thickness costally and intercostally (very thin walled). Intercostal zones without typical long-cells (the cells more or less irregularly isodiametric). Mid-intercostal long-cells having markedly sinuous walls and having straight or only gently undulating walls (often very coarsely and irregularly sinuate). *Microhairs* present; elongated; clearly two-celled; panicoid-type (but unusual, with a vase-shaped basal cell which narrows abruptly towards the apical cell). *Stomata* common. Subsidiaries non-papillate; low dome-shaped to triangular. Guard-cells overlapping to flush with the interstomatals. *Intercostal short-cells* absent or very rare. Large, cushion-based macrohairscommon intercostally. *Costal short-cells* conspicuously in long rows. Costal silica bodies present and well developed; 'panicoid-type'; dumb-bell shaped and nodular.

Transverse section of leaf blade, physiology. C_3. *Midrib* conspicuous; with one bundle only, or having a conventional arc of bundles (depending on the interpretation of its limits, which might take in a small lateral on eiter side); with colourless mesophyll adaxially. *The lamina* symmetrical on either side of the midrib. Bulliforms present in discrete, regular adaxial groups; in simple fans. Sclerenchyma all associated with vascular bundles.

Phytochemistry. Leaves without flavonoid sulphates (*H. chlorochloë*).

Taxonomy. Panicoideae; Panicodae; Paniceae.

Ecology, geography, regional floristic distribution. 2 species. Central Africa. Mesophytic; shade species. In woodland and riverine forest.

Paleotropical. African. Sudano-Angolan. Somalo-Ethiopian and South Tropical African.

Economic importance. Important native pasture species: *H. chlorochloë*.

References, etc. Leaf anatomical: this project.

Special comments. Fruit data wanting.

Hymenachne P. Beauv.

Habit, vegetative morphology. *Perennial*; decumbent aquatics. **Culms** 50–200 cm high; *herbaceous*. Culm nodes glabrous. Culm internodes solid (but aerenchymatous). *Leaves* not basally aggregated; non-auriculate. Leaf blades linear to lanceolate; broad, or narrow; cordate, or not cordate, not sagittate; cross veined, or without cross venation; persistent. **Ligule** present; *an unfringed membrane*.

Reproductive organization. Plants bisexual, with bisexual spikelets; with hermaphrodite florets. The spikelets all alike in sexuality.

Inflorescence. Inflorescence of spicate main branches, or paniculate (spiciform or contracted, often with appressed branches); espatheate; not comprising 'partial inflorescences' and foliar organs. *Spikelet-bearing axes persistent. Spikelets not secund.* Pedicel apices truncate, or discoid.

Female-fertile spikelets. *Spikelets* narrowly lanceolate; abaxial; *not noticeably compressed to compressed dorsiventrally*; falling with the glumes; with a distinctly elongated rachilla internode between the glumes. Rachilla terminated by a female-fertile floret. Hairy callus absent.

Glumes present; two; *very unequal*; (the upper) shorter than the spikelets to about equalling the spikelets (G_1 about half as long); shorter than the adjacent lemmas to long relative to the adjacent lemmas; free (widely separated); awnless; *similar (hyaline, not gibbous)*. *Lower glume* about 0.5 times the length of the upper glume; 1–3 nerved. Upper glume 3–7 nerved (usually acuminate). *Spikelets* with incomplete florets. The incomplete florets proximal to the female-fertile florets. *Spikelets with proximal incomplete florets*. *The proximal incomplete florets* 1; epaleate; sterile. *The proximal lemmas* acute, acuminate or awnlike; more or less *awned, or awnless*; 3–5 nerved; more or less equalling the female-fertile lemmas to decidedly exceeding the female-fertile lemmas; less firm than the female-fertile lemmas to similar in texture to the female-fertile lemmas (hyaline, lanceolate, acuminate); not becoming indurated.

Female-fertile florets 1. **Lemmas** similar in texture to the glumes to decidedly firmer than the glumes (membranous to leathery); smooth; not becoming indurated; white in fruit; entire; pointed; awnless; hairless; non-carinate; having the margins lying flat and exposed on the palea (except at the base); **without a germination flap**; 2 nerved, or 3 nerved. *Palea* present; relatively long; gaping; entire; awnless, without apical setae; textured like the lemma; not indurated. *Lodicules* present; 2; free; fleshy; glabrous. *Stamens* 3. Anthers not penicillate. *Ovary* glabrous. Styles fused. Stigmas 2.

Fruit, embryo and seedling. *Fruit* free from both lemma and palea; small; compressed dorsiventrally. **Hilum short**. Embryo large; waisted.

Abaxial leaf blade epidermis. *Costal/intercostal zonation* conspicuous. *Papillae* absent. *Long-cells* similar in shape costally and intercostally; of similar wall thickness costally and intercostally. Mid-intercostal long-cells rectangular to fusiform; having markedly sinuous walls. *Microhairs* present; more or less chloridoid-type (though the apical cells fairly thin walled); 27–36 microns long; 7.5–9 microns wide at the septum. Microhair total length/width at septum 2–4.3. Microhair apical cells (12–)13.5–15(–18) microns long. Microhair apical cell/total length ratio 0.44–0.54. *Stomata* common; 30–34.5 microns long. Subsidiaries triangular. Guard-cells overlapping to flush with the interstomatals. *Intercostal short-cells* common; in cork/silica-cell pairs, or not paired (solitary); silicified (when paired), or not silicified. *Costal short-cells* conspicuously in long rows. Costal silica bodies 'panicoid-type'.

Transverse section of leaf blade, physiology. C_3; XyMS+. *Mesophyll* with non-radiate chlorenchyma. *Leaf blade* 'nodular' in section, or adaxially flat; with the ribs more or less constant in size. *Midrib* conspicuous; having a conventional arc of bundles (a small lateral on each side). Bulliforms present in discrete, regular adaxial groups; in simple fans. All the vascular bundles accompanied by sclerenchyma. Combined sclerenchyma girders present; nowhere forming 'figures'. Sclerenchyma all associated with vascular bundles.

Cytology. $2n = 24$.

Taxonomy. Panicoideae; Panicodae; Paniceae.

Ecology, geography, regional floristic distribution. 5 species. Tropical. Hydrophytic to helophytic (in swamps); glycophytic.

Paleotropical, Neotropical, and Australian. Indomalesian. Indian, Indo-Chinese, Malesian, and Papuan. Caribbean, Venezuela and Surinam, Amazon, Central Brazilian, Pampas, and Andean. North and East Australian. Tropical North and East Australian.

Economic importance. Important native pasture species: *H. acutigluma*, in swamps.

References, etc. Leaf anatomical: this project.

Hyparrhenia Anderss.

Excluding *Dybowskia, Hyperthelia*

Habit, vegetative morphology. Annual (rarely), or perennial (usually large); caespitose. *Culms* 30–300(–400) cm high; herbaceous. Culm nodes glabrous. Culm internodes solid. **The shoots not aromatic**. *Leaves* not basally aggregated; non-auriculate. Leaf blades linear; narrow; setaceous, or not setaceous; usually flat, or folded (sometimes); without cross venation; persistent; rolled in bud. Ligule an unfringed membrane.

Reproductive organization. *Plants bisexual, with bisexual spikelets*; with hermaphrodite florets. *The spikelets of sexually distinct forms on the same plant (the lower spikelet pairs homogamous, the upper pairs heterogamous)*; hermaphrodite and male-only, or hermaphrodite and sterile; overtly heteromorphic (imperfect spikelets sometimes with awned glumes, the L_2 awnless); *in both homogamous and heterogamous combinations (the proximal 1–2 pairs homogamous, male or neuter)*. Plants inbreeding. Apomictic, or reproducing sexually.

Inflorescence. *Inflorescence* falsely paniculate (leafy, often with coloured spatheoles); *with capillary branchlets (i.e., the articles of the racemes, and the peduncles slender)*; spatheate; *a complex of 'partial inflorescences' and intervening foliar organs*. Spikelet-bearing axes 'racemes'; *paired (the racemes with a common peduncle, often deflexed, their bases terete or flattened, that of the upper usually much shorter than 9 mm by contrast with **Exotheca**)*; with very slender rachides; disarticulating; disarticulating at the joints. *'Articles'* linear; appendaged, or not appendaged; disarticulating obliquely; densely long-hairy, or somewhat hairy, or glabrous. Spikelets paired (with terminal triplets); somewhat secund, or not secund; consistently in 'long-and-short' combinations; in pedicellate/sessile combinations. Pedicels of the 'pedicellate' spikelets free of the rachis. The 'shorter' spikelets hermaphrodite (in the upper pairs only). The 'longer' spikelets male-only, or sterile.

Female-sterile spikelets. The pedicelled spikelets male or sterile, without a callus, usually longer than the sessile, G_1 often mucronate or aristate. L_2 awnless, sometimes suppressed. The male spikelets with glumes; 1 floreted, or 2 floreted. The lemmas awnless.

Female-fertile spikelets. Spikelets 3.5–10 mm long; not noticeably compressed to compressed dorsiventrally; falling with the glumes. Rachilla terminated by a female-fertile floret. Hairy callus present. Callus pointed to blunt.

Glumes two; more or less equal; long relative to the adjacent lemmas; free; not pointed (truncate); *awnless*; very dissimilar (the lower dorsally rounded or flattened, the upper narrower, shallowly naviculate). Lower glume not two-keeled (striate or grooved); convex on the back (rarely slightly concave or with two or more shallow depressions); not pitted; relatively smooth; 7–9 nerved. Upper glume 3 nerved. *Spikelets* with incomplete florets. The incomplete florets proximal to the female-fertile florets. **Spikelets with proximal incomplete florets**. *The proximal incomplete florets* 1; epaleate; sterile. The proximal lemmas awnless; 0 nerved, or 2 nerved; similar in texture to the female-fertile lemmas (hyaline); not becoming indurated.

Female-fertile florets 1. **Lemmas** hyaline below, stipitate to the awn; less firm than the glumes (hyaline, but hardening into the awn); not becoming indurated; *incised*; 2 lobed (usually minutely bidentate); not deeply cleft; awned. Awns 1; median; from a sinus (flanked by tiny teeth); geniculate; hairy, or long-plumose; much longer than the body of the lemma. Lemmas hairy, or hairless; non-carinate; without a germination flap; 1 nerved. *Palea* present, or absent; when present, relatively long, or conspicuous but relatively short, or very reduced; not indurated; 2-nerved, or nerveless. *Lodicules* present; 2; free; fleshy, or membranous; glabrous. *Stamens* 3. Anthers not penicillate. *Ovary* glabrous. Styles free to their bases. Stigmas 2; red pigmented.

Fruit, embryo and seedling. *Fruit* free from both lemma and palea; compressed dorsiventrally, or not noticeably compressed. Hilum short. Embryo large. Endosperm containing only simple starch grains. Embryo without an epiblast; with a scutellar tail; with an elongated mesocotyl internode. Embryonic leaf margins overlapping.

Abaxial leaf blade epidermis. *Costal/intercostal zonation* conspicuous. *Papillae* present. Intercostal papillae over-arching the stomata; consisting of one oblique swelling per cell. *Long-cells* similar in shape costally and intercostally; of similar wall thickness costally and intercostally. Mid-intercostal long-cells rectangular; having markedly sinuous walls. *Microhairs* present; panicoid-type; (46.5–)52–66(–72) microns long; 6–7 microns wide at the septum. Microhair total length/width at septum 6–8.5. Microhair apical cells 15–30 microns long. Microhair apical cell/total length ratio 0.33–0.55. *Stomata* common; 21–25.5 microns long. Subsidiaries triangular, or dome-shaped and triangular. Guard-cells overlapping to flush with the interstomatals. *Intercostal short-cells* common; in cork/silica-cell pairs (a few), or not paired (mainly solitary); silicified (when paired), or not silicified. *Costal short-cells* conspicuously in long rows. Costal silica bodies 'panicoid-type'; cross shaped, or butterfly shaped, or dumb-bell shaped, or nodular.

Transverse section of leaf blade, physiology. C_4; biochemical type NADP–ME (*H. hirta*); XyMS–. PCR cells with a suberised lamella. PCR cell chloroplasts with reduced grana; centrifugal/peripheral. *Mesophyll* with radiate chlorenchyma. *Leaf blade* adaxially flat. *Midrib* conspicuous; having a conventional arc of bundles; with colourless mesophyll adaxially. Bulliforms present in discrete, regular adaxial groups; in simple fans (the groups of large cells (cf. *Zea*) or of small cells, sometimes the bulliforms irregularly grouped). Many of the smallest vascular bundles unaccompanied by sclerenchyma. Combined sclerenchyma girders present; forming 'figures'. Sclerenchyma all associated with vascular bundles.

Phytochemistry. Leaves without flavonoid sulphates (2 species).

Cytology. Chromosome base number, $x = 10$ and 15. $2n = 20, 30, 40, 44, 45$, and 60. 2, 4, and 6 ploid. Nucleoli persistent.

Taxonomy. Panicoideae; Andropogonodae; Andropogoneae; Andropogoninae.

Ecology, geography, regional floristic distribution. About 55 species. Mediterranean, Africa, Arabia, America. Commonly adventive. Mesophytic, or xerophytic; species of open habitats; glycophytic. Savanna.

Holarctic, Paleotropical, Cape, and Australian. Boreal and Tethyan. African, Madagascan, and Indomalesian. Euro-Siberian. Macaronesian, Mediterranean, and Irano-Turanian. Saharo-Sindian, Sudano-Angolan, West African Rainforest, and Namib-Karoo. Indian, Indo-Chinese, Malesian, and Papuan. North and East Australian. European. Sahelo-Sudanian, Somalo-Ethiopian, South Tropical African, and Kalaharian. Tropical North and East Australian.

Rusts and smuts. Rusts — *Puccinia*. Taxonomically wide-ranging species: *Puccinia levis*, *Puccinia versicolor*, and '*Uromyces*' *clignyi*. Smuts from Tilletiaceae and from Ustilaginaceae. Tilletiaceae — *Tilletia*. Ustilaginaceae — *Sorosporium*, *Sphacelotheca*, and *Ustilago*.

Economic importance. Significant weed species: *H. hirta*, *H. rufa*. Cultivated fodder: *H. rufa*. Important native pasture species: most more or less unpalatable, but sometimes useful, especially when young: *H. anamesa*, *H. cymbaria*, *H. dregeana*, *H. filipendula*, *H. poecilotricha*, etc.

References, etc. Morphological/taxonomic: Clayton 1969. Leaf anatomical: Metcalfe 1960; this project.

Hyperthelia W. Clayton

~ *Hyparrhenia* (including *H. dissoluta*)

Habit, vegetative morphology. Annual, or perennial; caespitose. *Culms* 100–750 cm high; herbaceous; to 1 cm in diameter; branched above (to form compound inflorescences). Culm nodes glabrous. Culm sheaths persistent. Culm internodes hollow. Young shoots intravaginal. The shoots not aromatic. *Leaves* not basally aggregated; nonauriculate. Leaf blades linear; broad, or narrow; 3–40 mm wide; flat, or rolled (on drying); pseudopetiolate (pseudopetioles to 16 cm long), or not pseudopetiolate; without cross venation; persistent. Ligule an unfringed membrane (usually), or a fringed membrane (rarely); truncate, or not truncate (rounded to acute, the upper edges of the sheath adnate to its sides); 3–4 mm long. *Contra-ligule* absent.

Reproductive organization. *Plants* bisexual, with bisexual spikelets; **with hermaphrodite florets**. The spikelets of sexually distinct forms on the same plant; hermaphrodite and male-only. The male and female-fertile spikelets mixed in the inflorescence. *The spikelets* overtly heteromorphic; **in both homogamous and heterogamous combinations (with one homogamous pair, at the base of the lower raceme)**.

Inflorescence. Inflorescence falsely paniculate (large, leafy); spatheate; *a complex of 'partial inflorescences' and intervening foliar organs*. Spikelet-bearing axes very much reduced (usually), or 'racemes' (rarely); the spikelet-bearing axes usually with only one spikelet-bearing 'article' (rarely with 4 or 5); paired (the pair subtended by a linear to lanceolate spatheole, the bases terete, sometimes deflexed); with very slender rachides; disarticulating; disarticulating at the joints. '**Articles**' linear; *appendaged (the raceme-base with a long scarious appendage at the tip, 3–20 mm long, flat or rolled into a funnel around the raceme base)*; disarticulating obliquely. *Spikelets* paired, or in triplets (sometimes having one female-fertile spikelet with a pair of pedicelled male spikelets, the triplet disarticulating in its entirety); *consistently in 'long-and-short' combinations*; in pedicellate/sessile combinations. Pedicels of the 'pedicellate' spikelets free of the rachis. The 'shorter' spikelets hermaphrodite (in the heterogamous combinations), or male-only (in the homogamous combinations). The 'longer' spikelets male-only.

Female-sterile spikelets. The homogamous and pedicellate spikelets male, linear-lanceolate, with two hyaline lemmas; pedicellate spikelets with a short basal callus. The male spikelets 2 floreted.

Female-fertile spikelets. Spikelets 8–35 mm long; abaxial; not noticeably compressed to compressed dorsiventrally; biconvex; falling with the glumes; not disarticulating between the florets; with conventional internode spacings. Rachilla terminated by a female-fertile floret. Hairy callus present. *Callus* long; pointed.

Glumes two; more or less equal; exceeding the spikelets; long relative to the adjacent lemmas; dorsiventral to the rachis; hairy (G_1, towards the tip), or hairless (G_2); without conspicuous tufts or rows of hairs; not pointed (G_1 bifurcate, G_2 blunt); awned (G_2, sometimes), or awnless; non-carinate; very dissimilar. *Lower glume* much exceeding the lowest lemma; *not two-keeled*; sulcate on the back; not pitted; relatively smooth. *Upper glume 1 nerved*. *Spikelets* with incomplete florets. The incomplete florets proximal to the female-fertile florets. *The proximal incomplete florets* 1; epaleate; sterile. The proximal lemmas awnless; 0 nerved; similar in texture to the female-fertile lemmas (hyaline); not becoming indurated.

Female-fertile florets 1. *Lemmas* less firm than the glumes (hyaline at margins and tips); becoming indurated along a central sulcate column below the awn; *incised*; shortly 2 lobed; not deeply cleft; awned. Awns 1; median; from a sinus; geniculate; hairy; much longer than the body of the lemma. Lemmas hairless (or with a few hairs only, at the edges); non-carinate (flat, sulcate); without a germination flap; 1 nerved. *Palea* present, or absent; when present, conspicuous but relatively short, or very reduced. *Lodicules* present; 2; free; fleshy; glabrous. *Stamens* 3. Anthers 3–4 mm long; not penicillate; without an apically prolonged connective. *Ovary* glabrous. Styles free to their bases. Stigmas 2; dark red pigmented.

Fruit, embryo and seedling. *Fruit* free from both lemma and palea; narrowly ellipsoid. Hilum short. Embryo large.

Abaxial leaf blade epidermis. *Stomata* common. Subsidiaries dome-shaped and triangular. *Intercostal short-cells* absent or very rare. *Costal short-cells* conspicuously in long rows. Costal silica bodies 'panicoid-type'.

Transverse section of leaf blade, physiology. C_4; XyMS–. PCR cell chloroplasts centrifugal/peripheral. Mesophyll without adaxial palisade. *Leaf blade* adaxially flat (practically smooth). *Midrib* conspicuous; having complex vascularization. Bulliforms somtimes present in discrete, regular adaxial groups (but mostly in irregular groups); sometimes in simple fans. Many of the smallest vascular bundles unaccompanied by sclerenchyma. Combined sclerenchyma girders present (with most large bundles).

Special diagnostic feature. Spikelets in much-reduced andropogonoid 'racemes', each of the latter reduced to a single triplet and enclosed at its base by a trumpet-like development of the peduncle tip (sometimes), or not borne as in 'Anadelphia scyphofera' (q.v.).

Cytology. Chromosome base number, $x = 10$.

Taxonomy. Panicoideae; Andropogonodae; Andropogoneae; Andropogoninae.

Ecology, geography, regional floristic distribution. 6 species. Tropical and southern Africa. Mesophytic; species of open habitats; glycophytic. Grasslands and savanna.

Paleotropical. African and Madagascan. Sudano-Angolan, West African Rainforest, and Namib-Karoo. Sahelo-Sudanian, Somalo-Ethiopian, South Tropical African, and Kalaharian.

Economic importance. Grain crop species: *Hyperthelia edulis* harvested wild.

References, etc. Morphological/taxonomic: Clayton 1966. Leaf anatomical: this project.

Hypogynium Nees

~ *Andropogon*

Habit, vegetative morphology. Perennial; densely caespitose. *Culms* about 20–100 cm high; herbaceous. Culm nodes glabrous. Plants unarmed. Young shoots intravaginal. *Leaves* not basally aggregated; non-auriculate. Sheaths glabrous. Leaf blades linear; narrow; 2–4 mm wide; flat (at the base), or rolled (convolute above); without cross venation; persistent. Ligule an unfringed membrane; not truncate (rounded to acute).

Reproductive organization. *Plants monoecious with all the fertile spikelets unisexual*; without hermaphrodite florets. The spikelets of sexually distinct forms on the same plant; female-only and male-only. The male and female-fertile spikelets mixed in the inflorescence. *The spikelets homomorphic*; all in heterogamous combinations (?).

Inflorescence. Inflorescence paniculate; spatheate; a complex of 'partial inflorescences' and intervening foliar organs (often decompound). *Spikelet-bearing axes* short 'racemes'; solitary; with very slender rachides (filiform); disarticulating; disarticulating at the joints. 'Articles' not appendaged; disarticulating obliquely (or at least, sub-obliquely); glabrous. Spikelets paired; consistently in 'long-and-short' combinations; in pedicellate/sessile combinations. Pedicels of the 'pedicellate' spikelets free of the rachis. The 'shorter' spikelets female-only. The 'longer' spikelets male-only.

Female-sterile spikelets. The pedicelled spikelets very similar in form to the sessile, but glumes stronger nerved, lemmas slightly shorter, and with 3 functional stamens. The male spikelets with glumes; 2 floreted. The lemmas awnless. Male florets 3 staminate.

Female-fertile spikelets. Spikelets 3–4 mm long; compressed dorsiventrally; falling with the glumes. Rachilla terminated by a female-fertile floret. *Hairy callus present (small, glabrous on the back, bearded on sides and front)*.

Glumes two; more or less equal; long relative to the adjacent lemmas (exceeding them); without conspicuous tufts or rows of hairs; awnless; very dissimilar (both papery, G_1 bicarinate, G_2 navicular). Lower glume two-keeled; not pitted; relatively smooth. Upper glume 1–3 nerved. *Spikelets* with incomplete florets. The incomplete florets proximal to the female-fertile florets. *Spikelets with proximal incomplete florets. The proximal incomplete florets* 1; epaleate; sterile. The proximal lemmas awnless; 0 nerved; more or less equalling the female-fertile lemmas to decidedly exceeding the female-fertile lemmas; hyaline; not becoming indurated.

Female-fertile florets 1. Lemmas less firm than the glumes (hyaline); not becoming indurated; entire; awnless; hairless; non-carinate; without a germination flap; 1 nerved. *Palea* present, or absent (in *H. virgatum*); when present, conspicuous but relatively short, or very reduced; entire; awnless; without apical setae; not indurated; nerveless. **Lodicules present**; 2; fleshy; glabrous. *Stamens* 0 (but with 3 staminodes). *Ovary* glabrous. Stigmas 2.

Fruit, embryo and seedling. Fruit small (about 2 mm long in *H. virgatum*); brown; fusiform; smooth. Hilum short. Embryo large; not waisted.

Abaxial leaf blade epidermis. *Costal/intercostal zonation* conspicuous. *Papillae* absent. *Long-cells* markedly different in shape costally and intercostally (the costals much narrower); of similar wall thickness costally and intercostally. Mid-intercostal long-cells rectangular, or rectangular and fusiform; having markedly sinuous walls. *Microhairs* present; panicoid-type. *Stomata* common. Subsidiaries triangular. *Intercostal short-cells* common. *Costal short-cells* conspicuously in long rows. Costal silica bodies 'panicoid-type'; mostly dumb-bell shaped.

Transverse section of leaf blade, physiology. C_4; XyMS–. PCR sheath outlines uneven. PCR sheath extensions absent. PCR cell chloroplasts centrifugal/peripheral. *Leaf blade* adaxially flat. *Midrib* conspicuous (via a conspicuous abaxial keel, and a large adaxial bulliform group); with one bundle only, or having a conventional arc of bundles (if two small flanking bundles are regarded as part of the midrib); with colourless mesophyll adaxially, or without colourless mesophyll adaxially. *The lamina* symmetrical on either side of the midrib. Bulliforms not in discrete, regular adaxial groups (restricted to a single group over the midrib). Many of the smallest vascular bundles unaccompanied by sclerenchyma. Combined sclerenchyma girders present (with the larger bundles); forming 'figures'. Sclerenchyma all associated with vascular bundles.

Cytology. Chromosome base number, $x = 5$. $2n = 30$.

Taxonomy. Panicoideae; Andropogonodae; Andropogoneae; Andropogoninae.

Ecology, geography, regional floristic distribution. 2 species. Tropical America and Africa. Mesophytic; species of open habitats; glycophytic. Savanna.

Paleotropical and Neotropical. African. Sudano-Angolan. Venezuela and Surinam, Amazon, and Central Brazilian. South Tropical African.

References, etc. Leaf anatomical: based mainly on photos of *H. festuciformis* supplied by R.P. Ellis, but see also Metcalfe 1960.

Hypseochloa C.E. Hubb.

Habit, vegetative morphology. Slender *annual*. *Culms* 4–16 cm high; herbaceous. Leaves non-auriculate. Leaf blades narrow; flat, or rolled; without cross venation; persistent. Ligule an unfringed membrane; not truncate; 1–3 mm long.

Reproductive organization. Plants bisexual, with bisexual spikelets; with hermaphrodite florets. Exposed-cleistogamous, or chasmogamous.

Inflorescence. *Inflorescence paniculate*; open; *with capillary branchlets*; espatheate; not comprising 'partial inflorescences' and foliar organs. Spikelet-bearing axes persistent. Spikelets not secund.

Female-fertile spikelets. Spikelets 2–3 mm long; compressed laterally to not noticeably compressed; disarticulating above the glumes. Rachilla prolonged beyond the uppermost female-fertile floret. The rachilla extension naked (minute). Hairy callus present. *Callus* short; blunt.

Glumes present; two; more or less equal; exceeding the spikelets; long relative to the adjacent lemmas (exceeding them); pointed (acuminate); awnless; *carinate*; similar (membranous). Lower glume 3 nerved, or 5 nerved. Upper glume 3 nerved, or 5 nerved. **Spikelets** with female-fertile florets only; *without proximal incomplete florets*.

Female-fertile florets 1. **Lemmas** decidedly firmer than the glumes (thinly leathery); *becoming indurated (at least by contrast with* **Agrostis***)*; incised; 2 lobed (shortly bidentate); not deeply cleft; awned. Awns 1; median; dorsal; from near the top, or from well down the back; geniculate; much longer than the body of the lemma; entered by one vein. Lemmas hairless; glabrous, or scabrous; non-carinate; having the margins tucked in onto the palea; without a germination flap; 5–7 nerved. *Palea* present; relatively long; tightly clasped by the lemma; minutely apically notched; awnless, without apical setae; not indurated (hyaline/membranous); 2-nerved; 2-keeled. *Lodicules* present; 2; free; membranous; glabrous; toothed, or not toothed; not or scarcely vascularized. *Stamens* 3. *Ovary* glabrous. Styles free to their bases. Stigmas 2.

Fruit, embryo and seedling. Fruit small (1.2 mm long); compressed dorsiventrally, or not noticeably compressed. Hilum short. Embryo small. Endosperm hard; with lipid; containing compound starch grains.

Abaxial leaf blade epidermis. *Costal/intercostal zonation* conspicuous. *Papillae* absent. *Long-cells* markedly different in shape costally and intercostally (costals shorter, bulbous); of similar wall thickness costally and intercostally (thin-walled). Mid-intercostal long-cells fusiform; having straight or only gently undulating walls. *Microhairs* absent. *Stomata* common; 42–57 microns long. Subsidiaries parallel-sided. Guard-cells overlapped by the interstomatals. *Intercostal short-cells* absent or very rare. *Costal short-cells* neither distinctly grouped into long rows nor predominantly paired (mostly solitary or paired). Costal silica bodies horizontally-elongated crenate/sinuous (many deeply crenate, some almost 'nodular', sometimes with pointed corners); sharp-pointed (many of the crenate/nodular silica-bodies have pointed 'corners').

Transverse section of leaf blade, physiology. C_3; XyMS+. *Mesophyll* with radiate chlorenchyma (in places); without adaxial palisade. *Leaf blade* with distinct, prominent adaxial ribs to 'nodular' in section; with the ribs more or less constant in size (low, round-topped). *Midrib* not readily distinguishable; with one bundle only. Bulliforms present in discrete, regular adaxial groups (in the furrows); in simple fans (these broad). All the vascular bundles accompanied by sclerenchyma. Combined sclerenchyma girders absent (mostly strands only). Sclerenchyma all associated with vascular bundles.

Taxonomy. Pooideae; Poodae; Aveneae.

Ecology, geography, regional floristic distribution. 2 species. On mountains of Tropical West Africa. Mesophytic; species of open habitats; glycophytic. Mountain grassland.

Paleotropical. African. West African Rainforest.

References, etc. Leaf anatomical: this project.

Hystrix Moench

~ *Elymus*

Habit, vegetative morphology. *Perennial*; caespitose. *Culms* 60–200 cm high; herbaceous. Culm nodes hairy (puberulent), or glabrous. Culm internodes hollow. *Leaves* not basally aggregated; auriculate. *Sheath margins joined*. Leaf blades linear to linear-lanceolate; broad, or narrow; 3–30 mm wide; flat; without cross venation. Ligule an unfringed membrane (tough); truncate; 0.3–1 mm long.

Reproductive organization. Plants bisexual, with bisexual spikelets; with hermaphrodite florets. The spikelets of sexually distinct forms on the same plant, or all alike in sexuality; hermaphrodite, or hermaphrodite and sterile (sometimes with sterile spikelets at the tip of the rachis).

Inflorescence. *Inflorescence a single spike, or a false spike, with spikelets on contracted axes*. Rachides flattened (continuous). Inflorescence espatheate; not comprising 'partial inflorescences' and foliar organs. Spikelet-bearing axes persistent. Spikelets solitary to paired; not secund; distichous (in regular rows).

Female-fertile spikelets. Spikelets 8–18 mm long; adaxial; compressed laterally; disarticulating above the glumes; disarticulating between the florets. Rachilla prolonged beyond the uppermost female-fertile floret. The rachilla extension with incomplete florets. Hairy callus present. Callus blunt.

Glumes present (in the lower spikelets, but often missing from the upper spikelets); two (but small), or one per spikelet (the G_1 minute or absent); minute, or relatively large; *very unequal*; shorter than the spikelets; shorter than the adjacent lemmas; free; *displaced (to one side)*; subulate; awned (the outer one awn-like), or awnless; non-carinate. Lower glume 0–1 nerved. Upper glume 1–2 nerved. *Spikelets* with incomplete florets. The incomplete florets distal to the female-fertile florets.

Female-fertile florets 2–4. Lemmas convex, tapering into long awns; entire; pointed; awned. Awns 1; median; apical; non-geniculate; much longer than the body of the lemma; entered by several veins. Lemmas hairy, or hairless; non-carinate; without a germination flap; 5–7 nerved; with the nerves confluent towards the tip. *Palea* present; relatively long; 2-nerved; 2-keeled. *Lodicules* present; free; membranous; ciliate; not toothed; not or scarcely vascularized. *Stamens* 3. Anthers 2–7 mm long. *Ovary* hairy. Styles free to their bases. Stigmas 2.

Fruit, embryo and seedling. *Fruit* adhering to lemma and/or palea; small to medium sized (4–6 mm long); longitudinally grooved; compressed dorsiventrally; with hairs confined to a terminal tuft. Hilum long-linear. Embryo small. Endosperm hard; without lipid; containing only simple starch grains. Embryo without an epiblast; without a scutellar tail; with a negligible mesocotyl internode. Embryonic leaf margins meeting.

Abaxial leaf blade epidermis. *Costal/intercostal zonation* conspicuous. *Papillae* absent. *Long-cells* markedly different in shape costally and intercostally (the costals smaller, more regularly rectangular); of similar wall thickness costally and intercostally (thin walled). Mid-intercostal long-cells rectangular to fusiform; having straight or only gently undulating walls. *Microhairs* absent. *Stomata* absent or very rare. *Intercostal short-cells* absent or very rare. Prickles common. *Crown cells* absent. *Costal short-cells* neither distinctly grouped into long rows nor predominantly paired (mostly solitary, or adjoining prickle-bases, but occasionally in short rows). Costal silica bodies horizontally-elongated crenate/sinuous.

Transverse section of leaf blade, physiology. C_3; XyMS+. Mesophyll without adaxial palisade. Combined sclerenchyma girders present; forming 'figures'. Sclerenchyma all associated with vascular bundles.

Cytology. Chromosome base number, $x = 7$. $2n = 28$ and 56. 4 and 8 ploid. Haplomic genome content H and S.

Taxonomy. Pooideae; Triticodae; Triticeae.

Ecology, geography, regional floristic distribution. 9 species. Asia, North America, New Zealand. Mesophytic; shade species and species of open habitats. Woodland and meadows.

Holarctic. Boreal and Tethyan. Atlantic North American and Rocky Mountains. Irano-Turanian. Canadian-Appalachian.

Rusts and smuts. Rusts — *Puccinia*. Taxonomically wide-ranging species: *Puccinia coronata, Puccinia striiformis, Puccinia montanensis,* and *Puccinia recondita.* Smuts from Ustilaginaceae. Ustilaginaceae — *Ustilago.*
References, etc. Morphological/taxonomic: Löve 1984. Leaf anatomical: this project.

Ichnanthus P. Beauv.

Ischnanthus Roem. & Schult., *Navicularia* Radd.

Habit, vegetative morphology. Annual, or perennial; caespitose, or decumbent. *Culms herbaceous.* Culm nodes hairy, or glabrous. Culm internodes solid, or hollow. *Leaves* not basally aggregated; non-auriculate. Leaf blades linear to ovate; broad, or narrow; pseudopetiolate, or not pseudopetiolate; cross veined, or without cross venation; persistent. *Ligule* short *a fringed membrane, or a fringe of hairs.*

Reproductive organization. Plants bisexual, with bisexual spikelets; with hermaphrodite florets. *The spikelets all alike in sexuality.*

Inflorescence. *Inflorescence* paniculate, or of spicate main branches (the primary branches sometimes unbranched); open; *with capillary branchlets*; espatheate; not comprising 'partial inflorescences' and foliar organs. Spikelet-bearing axes persistent. *Spikelets not secund.* Pedicel apices truncate, or cupuliform.

Female-fertile spikelets. *Spikelets 4–5 mm long*; oblong, or elliptic, or lanceolate, or ovate, or obovate, or oblanceolate; adaxial (or not orientated); compressed laterally to compressed dorsiventrally (the L_2 dorsally compressed); disarticulating above the glumes, or falling with the glumes; with distinctly elongated rachilla internodes between the florets. The upper floret conspicuously stipitate. *The stipe beneath the upper floret* not filiform; *curved, not swollen (about 1 mm long, bearing a small, white, annular, membranous, minutely two-lobed appendage below the palea)*; homogeneous. Rachilla terminated by a female-fertile floret. Hairy callus absent.

Glumes two; very unequal (the lower usually about 1/2 to 3/4 of the spikelet length); (the longer) long relative to the adjacent lemmas; pointed; awnless; similar (membranous-herbaceous). *Lower glume* 0.5–0.75(–1) times the length of the upper glume; 3–7 nerved. Upper glume 5–9 nerved. *Spikelets* with incomplete florets. The incomplete florets proximal to the female-fertile florets. *Spikelets with proximal incomplete florets.* The proximal incomplete florets 1; paleate. Palea of the proximal incomplete florets fully developed. *The proximal incomplete florets male. The proximal lemmas* awnless; *7 nerved*; decidedly exceeding the female-fertile lemmas; less firm than the female-fertile lemmas; not becoming indurated.

Female-fertile florets 1. Lemmas decidedly firmer than the glumes (papery to leathery); smooth; becoming indurated; white in fruit, or yellow in fruit; awnless; hairless (shiny); non-carinate; having the margins tucked in onto the palea; with a clear germination flap; 5–7 nerved. *Palea* present; relatively long; tightly clasped by the lemma; entire; awnless, without apical setae; textured like the lemma; indurated; 2-nerved. *Lodicules* present; 2; free; fleshy; glabrous. *Stamens* 3. Anthers not penicillate. *Ovary* glabrous. Styles free to their bases. Stigmas 2; red pigmented.

Fruit, embryo and seedling. *Disseminule often deciduous before the rest of the spikelet, comprising the fruit and the P_2 on the stipe-like rachilla joint, enclosed by the L_1.* Fruit small; compressed dorsiventrally. *Hilum short.* Embryo large.

Abaxial leaf blade epidermis. *Costal/intercostal zonation* conspicuous. *Papillae* absent. Mid-intercostal long-cells rectangular; having markedly sinuous walls (these very thin). *Microhairs* present; panicoid-type; 88–92.7 microns long; 5.8–8.8 microns wide at the septum. Microhair total length/width at septum 10–15.4. Microhair apical cells 46.4–53.7 microns long. Microhair apical cell/total length ratio 0.53–0.58. *Stomata* common (in rows adjoining the veins); 51–70 microns long. Subsidiaries low dome-shaped and triangular. Guard-cells overlapping to flush with the interstomatals. *Intercostal short-cells* absent or very rare (very few); not paired. *Costal short-cells* conspicuously in long rows. Costal silica bodies 'panicoid-type'; mostly dumb-bell shaped.

Transverse section of leaf blade, physiology. C_3; XyMS+. *Leaf blade* 'nodular' in section. *Midrib* conspicuous (as a larger bundle and adaxial rib); with one bundle only.

Bulliforms present in discrete, regular adaxial groups; in simple fans. All the vascular bundles accompanied by sclerenchyma. Combined sclerenchyma girders present (but most sclerenchyma in strands); forming 'figures' (few). Sclerenchyma all associated with vascular bundles.

Phytochemistry. Leaves containing flavonoid sulphates (1 species), or without flavonoid sulphates (1 species).

Cytology. $2n = 18$, 20, 40, and 54.

Taxonomy. Panicoideae; Panicodae; Paniceae.

Ecology, geography, regional floristic distribution. 33 species. Tropical America, Indomalayan region, Australia. Mesophytic; shade species and species of open habitats; glycophytic. Forest, grassland and disturbed ground.

Paleotropical, Neotropical, and Australian. African, Indomalesian, and Polynesian. Sudano-Angolan and West African Rainforest. Indian, Indo-Chinese, Malesian, and Papuan. Polynesian. Caribbean, Venezuela and Surinam, Amazon, Central Brazilian, Pampas, and Andean. North and East Australian. Sahelo-Sudanian. Tropical North and East Australian.

Rusts and smuts. Rusts — *Puccinia*. Taxonomically wide-ranging species: *Puccinia levis*.

References, etc. Leaf anatomical: this project.

Imperata Cyr.

Syllepis Fourn.

Habit, vegetative morphology. Perennial; rhizomatous. *Culms* 10–150 cm high; herbaceous; unbranched above. Culm nodes hairy, or glabrous. Culm internodes solid. *Leaves* mostly basal; non-auriculate. Leaf blades broad, or narrow; pseudopetiolate (attenuate), or not pseudopetiolate; without cross venation; persistent; rolled in bud. *Ligule a fringed membrane*.

Reproductive organization. Plants bisexual, with bisexual spikelets; with hermaphrodite florets. The spikelets all alike in sexuality; homomorphic.

Inflorescence. Inflorescence paniculate (silky, spiciform or loosely contracted, the branches with numerous short 'racemes'); contracted; spicate; espatheate; not comprising 'partial inflorescences' and foliar organs. *Spikelet-bearing axes* short 'racemes'; with very slender rachides; persistent. *'Articles'* linear; densely long-hairy (the hairs silvery). *Spikelets* paired; not secund; *consistently in 'long-and-short' combinations*; unequally pedicellate in each combination. Pedicels of the 'pedicellate' spikelets free of the rachis. The 'shorter' spikelets hermaphrodite. The 'longer' spikelets hermaphrodite.

Female-fertile spikelets. Spikelets 2.5–5 mm long; not noticeably compressed to compressed dorsiventrally; falling with the glumes (falling entire from their pedicels). Rachilla terminated by a female-fertile floret. Hairy callus present (silky, silvery).

Glumes two; more or less equal; long relative to the adjacent lemmas; without conspicuous tufts or rows of hairs; pointed, or not pointed; awnless; *similar (membranous, with long silvery hairs especially towards the base)*. *Lower glume* not two-keeled; convex on the back; not pitted; relatively smooth; *5–7 nerved*. Upper glume 3–5 nerved. *Spikelets* with female-fertile florets only (rarely), or with incomplete florets. The incomplete florets proximal to the female-fertile florets. *The proximal incomplete florets* 1; paleate (rarely), or epaleate; male (rarely), or sterile. The proximal lemmas awnless; 0 nerved, or 1 nerved; decidedly exceeding the female-fertile lemmas (rarely, the latter may even be absent); similar in texture to the female-fertile lemmas; not becoming indurated.

Female-fertile florets 1. *Lemmas lanceolate to oblanceolate, reduced, sometimes absent*; less firm than the glumes (hyaline); not becoming indurated; entire, or incised (denticulate); when entire pointed, or blunt; awnless; hairless; non-carinate; 0–1 nerved. *Palea* present; relatively long, or conspicuous but relatively short, or very reduced (broad); apically notched; awnless, without apical setae; not indurated (hyaline); nerveless. *Lodicules* absent. **Stamens 1 (Section Imperatella), or 2 (Section Eriopogon).** Anthers not penicillate. *Ovary* glabrous. Styles fused. Stigmas 2; red pigmented.

Fruit, embryo and seedling. *Fruit* free from both lemma and palea; small; not notice-ably compressed. Hilum short. Embryo large. Endosperm containing only simple starch grains. Embryo without an epiblast; with a scutellar tail; with an elongated mesocotyl internode. Embryonic leaf margins overlapping.

Seedling with a long mesocotyl.

Abaxial leaf blade epidermis. *Costal/intercostal zonation* conspicuous. *Papillae* absent. *Long-cells* similar in shape costally and intercostally; of similar wall thickness costally and intercostally. Mid-intercostal long-cells rectangular; having markedly sinuous walls. *Microhairs* present; panicoid-type; (40–)42–54(–60) microns long; 5.4–6 microns wide at the septum. Microhair total length/width at septum 7.8–8.9. Microhair apical cells (16.5–)18.6–22.5(–24) microns long. Microhair apical cell/total length ratio 0.34–0.47. *Stomata* common; 24–27 microns long. Subsidiaries triangular. Guard-cells overlapping to flush with the interstomatals. *Intercostal short-cells* common; in cork/silica-cell pairs, or not paired (solitaries and triplets); silicified, or not silicified. Intercostal silica bodies when present, cross-shaped, or saddle shaped. *Costal short-cells* conspicuously in long rows. Costal silica bodies 'panicoid-type'; dumb-bell shaped, or nodular.

Transverse section of leaf blade, physiology. C_4; XyMS–. PCR cell chloroplasts with reduced grana; centrifugal/peripheral. *Mesophyll* with radiate chlorenchyma; traversed by columns of colourless mesophyll cells (rarely), or not traversed by colourless columns. *Leaf blade* adaxially flat. *Midrib* conspicuous; having a conventional arc of bundles. Bulliforms present in discrete, regular adaxial groups; associated with colourless mesophyll cells to form deeply-penetrating fans (sometimes linked to the abaxial epidermis by colourles cells); associating with colourless mesophyll cells to form arches over small vascular bundles. All the vascular bundles accompanied by sclerenchyma. Combined sclerenchyma girders present; nowhere forming 'figures'. Sclerenchyma all associated with vascular bundles, or not all bundle-associated.

Phytochemistry. Leaves without flavonoid sulphates (2 species).

Cytology. Chromosome base number, $x = 5$ and 10. $2n = 20, 40, 50$, and 60.

Taxonomy. Panicoideae; Andropogonodae; Andropogoneae; Andropogoninae.

Ecology, geography, regional floristic distribution. 8 species. Tropical and subtropical. Commonly adventive (*I. cylindrica* being one of the world's worst weeds). Helophytic, or mesophytic, or xerophytic; species of open habitats; halophytic, or glycophytic. Often in damp or weedy places, some forms of *I. cylindrica* in coastal sand.

Holarctic, Paleotropical, Neotropical, Cape, Australian, and Antarctic. Boreal, Tethyan, and Madrean. African, Madagascan, Indomalesian, Polynesian, and Neocaledonian. Euro-Siberian, Eastern Asian, and Atlantic North American. Mediterranean and Irano-Turanian. Saharo-Sindian, Sudano-Angolan, West African Rainforest, and Namib-Karoo. Indian, Indo-Chinese, Malesian, and Papuan. Fijian. Caribbean, Venezuela and Surinam, Amazon, Central Brazilian, Pampas, and Andean. North and East Australian. New Zealand and Patagonian. European. Central Grasslands. Sahelo-Sudanian, Somalo-Ethiopian, and South Tropical African. Tropical North and East Australian and Temperate and South-Eastern Australian.

Hybrids. Intergeneric hybrids procured with *Saccharum*.

Rusts and smuts. Rusts — *Puccinia*. Taxonomically wide-ranging species: *Puccinia microspora* and *Puccinia miscanthae*. Smuts from Ustilaginaceae. Ustilaginaceae — *Sphacelotheca* and *Ustilago*.

Economic importance. Significant weed species: *I. brasiliensis*, *I. cylindrica*. *I. cylindrica* var. *major* cultivated for papermaking.

References, etc. Leaf anatomical: Metcalfe 1960; this project.

Indocalamus Nakai

Gelidocalamus Wen, *Ferrocalamus* Hsueh & Keng f.

Excluding *Monocladus*

Habit, vegetative morphology. Perennial (shrubs). The flowering culms leafy. *Culms* slender, 60–300 cm high; woody and persistent; branched above. **Primary branches/mid-culm node 1 (but producing many branches later)**. Culm sheaths persistent. ***Rhizomes leptomorph***. Plants unarmed. **Leaves** not basally aggregated; ***with auricular setae***. Leaf blades broad, or narrow; 5–25 mm wide; cordate, or not cordate, not sagittate; pseudopetiolate; cross veined, or without cross venation; disarticulating from the sheaths; rolled in bud. **Ligule** present; ***a fringed membrane***; short. *Contra-ligule* present (in the species seen).

Reproductive organization. Plants bisexual, with bisexual spikelets; with hermaphrodite florets.

Inflorescence. Inflorescence paniculate; spatheate, or espatheate (the panicles spatheate or not, terminating leafy or leafless shoots); a complex of 'partial inflorescences' and intervening foliar organs, or not comprising 'partial inflorescences' and foliar organs. **Spikelet-bearing axes** *paniculate*; persistent. Spikelets not secund.

Female-fertile spikelets. Spikelets 8–18 mm long; with distinctly elongated rachilla internodes between the florets. Rachilla prolonged beyond the uppermost female-fertile floret. The rachilla extension with incomplete florets.

Glumes two; very unequal; shorter than the adjacent lemmas; awnless; similar (membranous). Lower glume 9 nerved (in material seen). Upper glume 9 nerved (in material seen). *Spikelets* with incomplete florets. The incomplete florets distal to the female-fertile florets. ***Spikelets without proximal incomplete florets***.

Female-fertile florets *3–8*. Lemmas sometimes tessellate; 7 nerved, or 9 nerved (in material seen). *Palea* present; relatively long; several nerved (3–5 nerved in material seen); 2-keeled. *Stamens* 3. Stigmas 2.

Fruit, embryo and seedling. Fruit longitudinally grooved, or not grooved. Pericarp fleshy (at least, in *Ferrocalamus*, Wen and He 1989), or thin (?); fused (*Gelidocalamus* and others), or free (*Ferrocalamus*). Seed endospermic (*Gelidocalamus* and others), or 'non-endospermic' (*Ferrocalamus*). Endosperm when present, containing compound starch grains.

Abaxial leaf blade epidermis. *Costal/intercostal zonation* conspicuous. *Papillae* present (large, thick-walled); costal and intercostal. Intercostal papillae over-arching the stomata; several per cell (one row per cell). Mid-intercostal long-cells rectangular; having markedly sinuous walls (these thin). *Microhairs* present; panicoid-type; 45–52.5 microns long (in *I. debilis*); 5.1–6 microns wide at the septum. Microhair total length/width at septum 7.5–10.2. Microhair apical cells 21–25.5 microns long. Microhair apical cell/total length ratio 0.47–0.5. *Stomata* common (alongside the veins); 24–28.5 microns long (in *I. debilis*). Subsidiaries parallel-sided and dome-shaped. Guard-cells overlapped by the interstomatals. *Intercostal short-cells* common; not paired; silicified. *Costal short-cells* neither distinctly grouped into long rows nor predominantly paired. Costal silica bodies 'panicoid-type'.

Transverse section of leaf blade, physiology. C_3; XyMS+. *Mesophyll* with non-radiate chlorenchyma; with adaxial palisade; with arm cells; with fusoids. The fusoids external to the PBS. *Midrib* conspicuous (larger bundle); with one bundle only. Bulliforms present in discrete, regular adaxial groups; in simple fans. All the vascular bundles accompanied by sclerenchyma. Combined sclerenchyma girders present; forming 'figures' (with most bundles). Sclerenchyma all associated with vascular bundles.

Cytology. Chromosome base number, $x = 12$. $2n = 48$. 4 ploid.

Taxonomy. Bambusoideae; Bambusodae; Bambuseae.

Ecology, geography, regional floristic distribution. 6 species. Tropical Asia. Woodland, and forming thickets in open country.

Paleotropical. Indomalesian. Indian and Indo-Chinese.

Economic importance. Leaves of *I. longiauritus* and *I. sinicus* used for wrapping Chinese tamales, for lining baskets, for roofing boats etc.

References, etc. Leaf anatomical: this project.

Special comments. Fruit data wanting.

Indopoa Bor

Habit, vegetative morphology. *Annual (sometimes epiphytic)*. *Culms* 10–16 cm high; herbaceous; unbranched above. Leaves non-auriculate. Leaf blades linear; narrow; 1–2 mm wide (to 12 cm long); not cordate, not sagittate; setaceous (involute); without abaxial multicellular glands; without cross venation; persistent. *Ligule* present; an unfringed membrane; not truncate.

Reproductive organization. Plants bisexual, with bisexual spikelets; with hermaphrodite florets; presumed inbreeding. Seemingly exposed-cleistogamous.

Inflorescence. **Inflorescence** determinate; without pseudospikelets; *a single spike (to 8 cm long)*; without capillary branchlets; non-digitate; espatheate (but the upper culm leaves sometimes reduced); not comprising 'partial inflorescences' and foliar organs. Spikelet-bearing axes persistent. Spikelets solitary; not secund (alternate).

Female-fertile spikelets. *Spikelets* morphologically 'conventional' (but seemingly cleistogamous); 7–9 mm long; adaxial; compressed laterally; disarticulating above the glumes; disarticulating between the florets. Rachilla prolonged beyond the uppermost female-fertile floret. The rachilla extension with incomplete florets. Hairy callus present. Callus pointed.

Glumes two; relatively large; very unequal; (the upper) long relative to the adjacent lemmas; free; dorsiventral to the rachis; hairless (glabrous, the nerves and mucro scaberulous); pointed; shortly awned, or awnless (but then mucronate); carinate; similar (thin, the G_1 shorter and asymmetric at the tip). Lower glume 1 nerved. Upper glume 1 nerved. *Spikelets* with incomplete florets. The incomplete florets distal to the female-fertile florets. *The distal incomplete florets* merely underdeveloped.

Female-fertile florets 4–6. Lemmas similar in texture to the glumes to decidedly firmer than the glumes; not becoming indurated; incised; 2 lobed; deeply cleft; awned. Awns 3; median and lateral; the median different in form from the laterals; from a sinus; geniculate (the base twisting); hairless (scabrid); much longer than the body of the lemma; entered by one vein. The lateral awns shorter than the median. *Lemmas* hairless; without a germination flap; *3 nerved (the median very broad, later curving round so as almost to longitudinally enclose the slender caryopsis)*; with the nerves non-confluent (extending into the 3 awns). *Palea* present; conspicuous but relatively short (about 4 mm long, compared with the 6 mm lemma); entire (oblanceolate); awnless, without apical setae; not indurated; 2-nerved; 2-keeled (the keels very shortly ciliate). *Lodicules* present; 2; free; glabrous. *Stamens* 3. Anthers minute, 0.25–0.3 mm long; not penicillate; without an apically prolonged connective. **Ovary** glabrous; *with a conspicuous apical appendage (this glabrous, emarginate, fleshy)*. Stigmas 2; white.

Fruit, embryo and seedling. *Fruit small to medium sized (3–4.5 mm long, very slender and needlelike, with an apical, glabrous, emarginate, fleshy appendage)*; linear (needle-like). Hilum short. Embryo small (about one-fifth of grain length). Seed endospermic. *Endosperm hard (of a peculiar, flinty texture)*.

Abaxial leaf blade epidermis. *Costal/intercostal zonation* conspicuous. *Papillae* absent. *Long-cells* markedly different in shape costally and intercostally (the costals more regularly rectangular, and much narrower). Mid-intercostal long-cells rectangular and fusiform; having markedly sinuous walls and having straight or only gently undulating walls. *Microhairs* present; more or less spherical; clearly two-celled; chloridoid-type (of the sunken type). Microhair apical cell wall of similar thickness/rigidity to that of the basal cell. Microhairs 15–18 microns long. Microhair basal cells 6 microns long. Microhairs 9–12 microns wide at the septum. Microhair total length/width at septum 1.3–1.9. Microhair apical cell/total length ratio 0.76. *Stomata* common; 36–43.5 microns long. Subsidiaries low dome-shaped, or triangular. Guard-cells overlapping to flush with the interstomatals. *Intercostal short-cells* absent or very rare. Intercostal silica bodies absent. *Costal short-cells* conspicuously in long rows. Costal silica bodies present in alternate cell files of the costal zones; saddle shaped (almost exclusively), or oryzoid (some tending to this).

Transverse section of leaf blade, physiology. C_4; XyMS+. PCR sheaths of the primary vascular bundles interrupted; interrupted both abaxially and adaxially. PCR sheath extensions absent. *Midrib* not readily distinguishable; with one bundle only. Bulliforms present in discrete, regular adaxial groups (between the vascular bundles, more conspicuous nearer the middle of the blade); in simple fans. All the vascular bundles accompanied by sclerenchyma. Combined sclerenchyma girders present; forming 'figures' (with the primaries). Sclerenchyma all associated with vascular bundles.

Taxonomy. Chloridoideae; main chloridoid assemblage.

Ecology, geography, regional floristic distribution. 1 species. India. On rocks.

Paleotropical. Indomalesian. Indian.

References, etc. Morphological/taxonomic: Bor 1957b. Leaf anatomical: Metcalfe 1960; this project.

Special comments. *Tripogon pauperculus.*

Indosasa McLure

Habit, vegetative morphology. Perennial (shrub). The flowering culms leafy. Culms woody and persistent; cylindrical; branched above. Culm nodes 2 ridged. Primary branches/mid-culm node usually 3. *Rhizomes leptomorph. Leaves* not basally aggregated; with auricular setae. *Leaf blades broad (large)*; pseudopetiolate; disarticulating from the sheaths.

Reproductive organization. Plants bisexual, with bisexual spikelets; with hermaphrodite florets.

Inflorescence. **Inflorescence** weakly indeterminate; *with pseudospikelets*; 'comprising short espatheate branches loosely grouped about a node', the spikelets in tight clusters; spatheate, or espatheate; a complex of 'partial inflorescences' and intervening foliar organs, or not comprising 'partial inflorescences' and foliar organs ('*I. hispidula* approaches a compound inflorescence, with spathiform bracts'). Spikelet-bearing axes persistent. Spikelets solitary; sessile.

Female-fertile spikelets. Spikelets compressed laterally (?); disarticulating above the glumes; disarticulating between the florets. Rachilla prolonged beyond the uppermost female-fertile floret. The rachilla extension with incomplete florets.

Glumes two; similar. *Spikelets* with incomplete florets. The incomplete florets distal to the female-fertile florets.

Female-fertile florets 3–20 ('several to many'). Lemmas leathery; awnless (?). *Lodicules* present; 3; membranous; ciliate to glabrous; heavily vascularized. **Stamens 6**. Anthers without an apically prolonged connective. *Ovary without a conspicuous apical appendage*. Styles fused. Stigmas 3.

Fruit, embryo and seedling. Fruit large (8 mm long in *I. sinica*); ellipsoid.

Taxonomy. Bambusoideae; Bambusodae; Bambuseae.

Ecology, geography, regional floristic distribution. 12 species. Asia, especially China and Vietnam. Shade species; glycophytic. Forest and roadsides.

Paleotropical. Indomalesian. Indo-Chinese.

References, etc. Morphological/taxonomic: Chao and Chu 1983, Chao and Renvoize 1989.

Special comments. Morphological description poor. Fruit data wanting. Anatomical data wanting.

Isachne R.Br.

Habit, vegetative morphology. Annual (rarely), or perennial (often aquatic); rhizomatous, or stoloniferous, or caespitose, or decumbent. *Culms* 5–100 cm high; herbaceous (rarely almost shrubby); scandent (to 6 m in *I. arundinacea*), or not scandent; branched above, or unbranched above. Culm nodes glabrous. Culm internodes hollow. *Leaves* not basally aggregated; non-auriculate. *Leaf blades* linear-lanceolate to ovate, or linear (rarely); broad, or narrow; 0.5–12(–20) mm wide; cordate (occasionally — *I. polygonoides*), or not cordate, not sagittate; without cross venation; *disarticulating from the sheaths*. Ligule a fringed membrane (very narrow), or a fringe of hairs. *Contra-ligule* present (of hairs), or absent.

Reproductive organization. Plants bisexual, with bisexual spikelets; with hermaphrodite florets, or without hermaphrodite florets.

Inflorescence. Inflorescence paniculate; open; with capillary branchlets, or without capillary branchlets; espatheate; not comprising 'partial inflorescences' and foliar organs. Spikelet-bearing axes persistent. Spikelets not secund. Pedicel apices cupuliform. *Spikelets not in distinct 'long-and-short' combinations*.

Female-fertile spikelets. *Spikelets 0.6–2 mm long*; elliptic, or ovate, or obovate; *compressed dorsiventrally*; *disarticulating above the glumes*; not disarticulating between the florets, or disarticulating between the florets. Rachilla terminated by a female-fertile floret; rachilla hairless. Hairy callus absent.

Glumes two; more or less equal; long relative to the adjacent lemmas; awnless; similar (membranous). Lower glume 3–9 nerved. Upper glume 5–9 nerved. *Spikelets* with female-fertile florets only, or with incomplete florets. The incomplete florets (when present) proximal to the female-fertile florets, or distal to the female-fertile florets (the distal floret may be female-only, and rarely the proximal floret is male-only). *The proximal incomplete florets* when present, 1; when present, paleate. Palea of the proximal incomplete florets fully developed. The proximal incomplete florets when present, male. The proximal lemmas variable — similar to L_2 or much longer and membranous; awnless.

Female-fertile florets *1 (occasionally), or 2 (usually — typically with both florets perfect)*. Lemmas orbicular to broadly elliptic; similar in texture to the glumes to decidedly firmer than the glumes (chartaceous to leathery); smooth to striate; not becoming indurated; entire; pointed, or blunt; awnless; hairy (pubescent), or hairless (glabrous); non-carinate; having the margins tucked in onto the palea; with a clear germination flap; 5–7 nerved. *Palea* present; relatively long; entire; awnless, without apical setae; textured like the lemma; not indurated; 2-nerved. *Lodicules* present; 2; free; fleshy; glabrous. **Stamens** *3*. Anthers not penicillate. *Ovary* glabrous. Styles free to their bases. Stigmas 2; red pigmented.

Fruit, embryo and seedling. Fruit small; compressed dorsiventrally. *Hilum long-linear*. Embryo large. Endosperm containing only simple starch grains. Embryo without an epiblast; with a scutellar tail; with an elongated mesocotyl internode. Embryonic leaf margins meeting.

Seedling with a short mesocotyl, or with a long mesocotyl. First seedling leaf with a well-developed lamina. The lamina broad; curved; 6–12 veined.

Abaxial leaf blade epidermis. *Costal/intercostal zonation* conspicuous. *Papillae* present, or absent. Intercostal papillae not over-arching the stomata; consisting of one symmetrical projection per cell (finger-like). *Long-cells* markedly different in shape costally and intercostally (the intercostals more or less hexagonal); of similar wall thickness costally and intercostally. Intercostal zones without typical long-cells (the long-cells more or less cubical or even vertically elongated). Mid-intercostal long-cells having markedly sinuous walls, or having straight or only gently undulating walls. *Microhairs* present; panicoid-type (though the distal cell is rather thick-walled). *Stomata* common. Subsidiaries parallel-sided to triangular; including both triangular and parallel-sided forms on the same leaf. Guard-cells overlapped by the interstomatals, or overlapping to flush with the interstomatals. *Intercostal short-cells* absent or very rare. *Costal short-cells* conspicuously in long rows. Costal silica bodies acutely-angled (more or less cubical); sharp-pointed.

Transverse section of leaf blade, physiology. C_3; XyMS+. *PBS cells* without a suberised lamella. *Mesophyll* with radiate chlorenchyma; *Isachne*-type. *Leaf blade* 'nodular' in section; with the ribs more or less constant in size. *Midrib* conspicuous; with one bundle only. Bulliforms present in discrete, regular adaxial groups (in the shallow furrows); in simple fans. All the vascular bundles accompanied by sclerenchyma. Combined sclerenchyma girders present; forming 'figures', or nowhere forming 'figures'. Sclerenchyma all associated with vascular bundles.

Culm anatomy. *Culm internode bundles* in one or two rings.

Phytochemistry. Leaves without flavonoid sulphates (2 species).

Cytology. Chromosome base number, $x = 10$. $2n = 20$, 50, and 60. 2, 5, and 6 ploid.

Taxonomy. Panicoideae; Panicodae; Isachneae.

Ecology, geography, regional floristic distribution. About 100 species. Tropical and subtropical. Helophytic, or mesophytic; shade species, or species of open habitats; glycophytic. Mostly in marshy ground.

Holarctic, Paleotropical, Neotropical, Australian, and Antarctic. Boreal and Tethyan. African, Madagascan, Indomalesian, Polynesian, and Neocaledonian. Eastern Asian. Irano-Turanian. Saharo-Sindian, Sudano-Angolan, and West African Rainforest. Indian, Indo-Chinese, Malesian, and Papuan. Hawaiian and Fijian. Caribbean, Amazon, Central Brazilian, and Andean. North and East Australian. New Zealand. Sahelo-Sudanian, Somalo-Ethiopian, and South Tropical African. Tropical North and East Australian and Temperate and South-Eastern Australian.

Rusts and smuts. Rusts — *Puccinia*. Smuts from Ustilaginaceae. Ustilaginaceae — *Sphacelotheca* and *Ustilago*.

Economic importance. Significant weed species: *I. globosa* (in rice). Important native pasture species: in wet places, e.g. *I. albens, I. globosa*.

References, etc. Leaf anatomical: Metcalfe 1960; this project.

Isalus J. Phipps

~ *Tristachya*

Habit, vegetative morphology. Perennial; caespitose. *Culms* 35–70 cm high; herbaceous; unbranched above; about 4 noded. Culm nodes exposed; glabrous. Culm leaves present. Upper culm leaf blades fully developed. Young shoots intravaginal. *Leaves* not basally aggregated; non-auriculate. Leaf blades linear to linear-lanceolate; narrow; 1.5–2 mm wide; flat, or rolled; without cross venation; persistent. *Ligule a fringe of hairs*. *Contra-ligule* absent.

Reproductive organization. Plants bisexual, with bisexual spikelets; with hermaphrodite florets.

Inflorescence. Inflorescence paniculate; open (with the triads of spikelets terminating the branches); espatheate; not comprising 'partial inflorescences' and foliar organs. Spikelet-bearing axes persistent. Spikelets in triplets; not secund; sessile; not in distinct 'long-and-short' combinations.

Female-fertile spikelets. Spikelets 9–14 mm long; compressed laterally; disarticulating above the glumes; not disarticulating between the florets; with conventional internode spacings. The upper floret not stipitate. Rachilla terminated by a female-fertile floret. Hairy callus present. Callus pointed (with long hairs).

Glumes two; more or less equal; about equalling the spikelets; long relative to the adjacent lemmas; hairy (at least the lower, with tubercle-based hairs), or hairless; *with distinct rows of hairs*; pointed (to subulate); awnless (at least the upper) carinate; similar (lanceolate-acuminate, firmly membranous to leathery). Lower glume about equalling the lowest lemma; tuberculate; 3 nerved. Upper glume 3 nerved. *Spikelets* with incomplete florets. The incomplete florets proximal to the female-fertile florets. *Spikelets with proximal incomplete florets*. *The proximal incomplete florets* 1; paleate. Palea of the proximal incomplete florets fully developed. The proximal incomplete florets male (with three stamens), or sterile. The proximal lemmas similar to the upper glume in shape and texture, sometimes with tubercle-based hairs at the summit; awnless; 5 nerved, or 7 nerved; decidedly exceeding the female-fertile lemmas; less firm than the female-fertile lemmas to similar in texture to the female-fertile lemmas (firmly membranous to leathery); not becoming indurated.

Female-fertile florets 1. Lemmas similar in texture to the glumes to decidedly firmer than the glumes; smooth; not becoming indurated; incised; 2 lobed; deeply cleft; awned. Awns 1; median; from a sinus (between the setaceous-pointed lateral lobes); geniculate; hairless; much longer than the body of the lemma; entered by several veins. Awn bases not twisted; not flattened. Lemmas hairy. *The hairs* in tufts; *in transverse rows (the lemma hairy at the base and with 6–8 tufts aligned transversely in the upper third)*. Lemmas non-carinate; having the margins tucked in onto the palea; without a germination flap; 7 nerved, or 9 nerved; with the nerves confluent towards the tip. *Palea* present; relatively long; tightly clasped by the lemma; apically notched; awnless, without apical setae; thinner than the lemma; not indurated; 2-nerved; 2-keeled. Palea back hairy. Palea keels wingless (but thickened); hairy. *Stamens* 3. Anthers about 3 mm long (in *I. isalensis*); without an apically prolonged connective. **Ovary** sparsely *hairy*. Styles free to their bases; free. Style bases adjacent. Stigmas 2; brown.

Abaxial leaf blade epidermis. *Costal/intercostal zonation* conspicuous. *Papillae* absent. *Long-cells* markedly different in shape costally and intercostally (the costals much narrower); of similar wall thickness costally and intercostally (fairly thin walled). Intercostal zones with typical long-cells to exhibiting many atypical long-cells (*I. humbertii* having very short long-cells in some files). Mid-intercostal long-cells rectangular to fusiform (mostly more or less rectangular, but the ends often fairly rounded); having markedly sinuous walls (the sinuosity fine or coarse, heavily pitted). *Microhairs* present; elongated; clearly two-celled; panicoid-type (narrow). *Stomata* common. Subsidiaries non-papillate; low dome-shaped and triangular (often with a small point). Guard-cells overlapping to flush with the interstomatals. *Intercostal short-cells* common; in cork/silica-cell pairs and not paired (many solitary); mostly not silicified (in the material of two species seen). *I. isalensis* with large, cushion based macrohairs in the intercostal zones, and prickles bordering the costae, *I. humbertii* with neither. *Costal short-cells* conspicuously in long rows. Costal silica bodies present and well developed, or absent (*I. humbertii*); in *I. isalensis* 'panicoid-type'; consistently short to medium dumb-bell shaped (in *I. isalensis*, the silica cells dumb-bell shaped in *I. humbertii*).

Transverse section of leaf blade, physiology. C_4. The anatomical organization conventional. XyMS–. PCR sheath extensions absent. Mesophyll without 'circular cells'; traversed by columns of colourless mesophyll cells (in places). *Leaf blade* with distinct, prominent adaxial ribs to 'nodular' in section, or adaxially flat; with the ribs more or less constant in size (low). *Midrib* not readily distinguishable; with one bundle only, or having a conventional arc of bundles (depending on interpretation). *The lamina* symmetrical on either side of the midrib. Bulliforms present in discrete, regular adaxial groups (a large group in every furrow, overlying a small bundle); in simple fans (in places), or associated with colourless mesophyll cells to form deeply-penetrating fans; associating with colourless mesophyll cells to form arches over small vascular bundles (these colourless cells sometimes seeming to make traversing columns). Many of the smallest vascular bundles unaccompanied by sclerenchyma. Combined sclerenchyma girders present (with all the main bundles); forming 'figures' (the large bundles with massive I's and 'anchors'). Sclerenchyma all associated with vascular bundles.

Taxonomy. Panicoideae; Panicodae; Arundinelleae.

Ecology, geography, regional floristic distribution. 3 species. Madagascar. Paleotropical. Madagascan.

References, etc. Leaf anatomical: this project.

Special comments. Fruit data wanting.

Ischaemum L.

Argopogon Mimeur, *Collardoa* Cav., *Ischaemopogon* Griseb., *Meoschium* P. Beauv., *Schoenanthus* Adans.

Excluding *Digastrium*

Habit, vegetative morphology. Annual, or perennial; rhizomatous, or stoloniferous, or caespitose, or decumbent. *Culms* 10–350 cm high; herbaceous; branched above, or unbranched above. Culm nodes hairy, or glabrous. Culm internodes solid, or hollow. *Leaves* not basally aggregated; non-auriculate, or auriculate (with sheath auricles). Leaf blades linear (usually), or linear-lanceolate to lanceolate; broad (rarely), or narrow; cordate to sagittate, or not cordate, not sagittate; flat; pseudopetiolate, or not pseudopetiolate; without cross venation; persistent; rolled in bud. Ligule an unfringed membrane.

Reproductive organization. Plants bisexual, with bisexual spikelets; with hermaphrodite florets. The spikelets of sexually distinct forms on the same plant, or all alike in sexuality; hermaphrodite, or hermaphrodite and male-only (rarely), or hermaphrodite and sterile (rarely); overtly heteromorphic (the pedicelled spikelet sometimes much smaller, often asymmetric), or homomorphic.

Inflorescence. *Inflorescence* terminal or axillary, of spicate main branches; usually *digitate. Primary inflorescence branches (1–)2(–14) (usually paired, one-sided and locked back to back to simulate a single spike). Inflorescence* spatheate (the uppermost leaf reduced to a spatheate sheath), or espatheate; *not comprising 'partial inflorescences' and foliar organs. Spikelet-bearing axes* 'racemes'; paired, or clustered (rarely solitary); with substantial rachides (these stout, triangular); disarticulating; disarticulating at the joints. *'Articles'* non-linear (clavate or inflated); glabrous, or more often hairy. *Spikelets* paired; *secund (the raceme dorsiventral, its segments usually appearing U or V-shaped from the back owing to the thick internodes and pedicels)*; consistently in 'long-and-short' combinations; in pedicellate/sessile combinations, or unequally pedicellate in each combination. Pedicels of the 'pedicellate' spikelets free of the rachis (usually stoutly linear to obovoid, sometimes very short). The 'shorter' spikelets hermaphrodite. The 'longer' spikelets hermaphrodite, or male-only (rarely), or sterile (rarely).

Female-sterile spikelets. The pedicelled spikelet as large as the sessile or much reduced, variously compressed, often asymmetrical.

Female-fertile spikelets. Spikelets compressed dorsiventrally; falling with the glumes. Rachilla terminated by a female-fertile floret. Hairy callus present, or absent.

Glumes two; more or less equal; long relative to the adjacent lemmas; awned (the upper, sometimes), or awnless; very dissimilar (the lower leathery and usually 2-keeled, the upper 1-keeled above and sometimes awned). *Lower glume* usually *two-keeled (winged or not)*; convex on the back to concave on the back; not pitted; relatively smooth (rarely), or rugose (transversely), or tuberculate (on the margins); 7–11 nerved. Upper glume 5–11 nerved. *Spikelets* with incomplete florets. The incomplete florets proximal to the female-fertile florets. *The proximal incomplete florets* 1; paleate. Palea of the proximal incomplete florets fully developed. *The proximal incomplete florets male*. The proximal lemmas awnless; similar in texture to the female-fertile lemmas; not becoming indurated.

Female-fertile florets 1. **Lemmas** less firm than the glumes (firmly membranous); not becoming indurated; *incised*; 2 lobed; awned (usually), or mucronate, or awnless (rarely). Awns when present, 1; or mucros from a sinus; geniculate; hairless (glabrous); much shorter than the body of the lemma to much longer than the body of the lemma. Lemmas hairless; non-carinate; without a germination flap; 1–5 nerved. **Palea** present; relatively long; apically notched; awnless, without apical setae; thinner than the lemma (hyaline); not indurated; *nerveless*. Lodicules present; 2; free; fleshy; glabrous. *Stamens* 3. Anthers not penicillate. *Ovary* glabrous. Styles free to their bases. Stigmas 2; red pigmented.

Fruit, embryo and seedling. *Fruit* free from both lemma and palea; compressed dorsiventrally, or not noticeably compressed. Hilum short. Embryo large. Endosperm hard; without lipid; containing compound starch grains. Embryo without an epiblast; with a scutellar tail; with an elongated mesocotyl internode. Embryonic leaf margins overlapping.

Seedling with a long mesocotyl.

Abaxial leaf blade epidermis. *Costal/intercostal zonation* conspicuous. *Papillae* present; costal and intercostal. Intercostal papillae over-arching the stomata (but only from the subsidiary cells); several per cell (costal and intercostal long-cells with 1–2 irregular rows of small, thick walled papillae, the subsidiaries each with a small papilla at each end). *Long-cells* markedly different in shape costally and intercostally (the costals much narrower); of similar wall thickness costally and intercostally (quite thin walled). Mid-intercostal long-cells rectangular and fusiform; having markedly sinuous walls. *Microhairs* present; more or less spherical, or elongated; ostensibly one-celled, or clearly two-celled; panicoid-type (some in *I. commutatum* being 'balanoform'), or chloridoid-type; 18–48 microns long; 7.2–9.6 microns wide at the septum. Microhair total length/width at septum 1–4.4. Microhair apical cells 4–19.5 microns long. Microhair apical cell/total length ratio 0.29–0.61. *Stomata* common; 34.5–42 microns long. Subsidiaries papillate; predominantly triangular. Guard-cells overlapping to flush with the interstomatals. *Intercostal short-cells* absent or very rare (ignoring the bulbous bases of short prickles). *Costal short-cells* conspicuously in long rows, or neither distinctly grouped into long rows nor predominantly paired. Costal silica bodies 'panicoid-type'; cross shaped, or butterfly shaped, or dumb-bell shaped, or nodular; sharp-pointed.

Transverse section of leaf blade, physiology. C_4; XyMS–. PCR sheath outlines even. PCR cell chloroplasts centrifugal/peripheral. *Mesophyll* with radiate chlorenchyma. *Leaf blade* adaxially flat. *Midrib* conspicuous; with one bundle only, or having a conventional arc of bundles; with colourless mesophyll adaxially. Bulliforms not in discrete, regular adaxial groups (the epidermis extensively bulliform, the cells irregularly grouped). Many of the smallest vascular bundles unaccompanied by sclerenchyma. Combined sclerenchyma girders present; nowhere forming 'figures'. Sclerenchyma all associated with vascular bundles.

Culm anatomy. *Culm internode bundles* in one or two rings.

Phytochemistry. Leaves without flavonoid sulphates (2 species).

Cytology. Chromosome base number, $x = 9$, or 10. $2n = 18$, 20, 40, 54, 56, and 68. 2, 4, 6, and 8 ploid (?).

Taxonomy. Panicoideae; Andropogonodae; Andropogoneae; Andropogoninae.

Ecology, geography, regional floristic distribution. 60 species. Tropical and subtropical. Commonly adventive. Helophytic (mostly), or mesophytic, or xerophytic; shade species, or species of open habitats; halophytic, or glycophytic. Some in damp or shady places, some (e.g. *I. muticum, I. triticeum*) in coastal sand.

Holarctic, Paleotropical, Neotropical, and Australian. Boreal. African, Madagascan, Indomalesian, Polynesian, and Neocaledonian. Eastern Asian. Saharo-Sindian, Sudano-Angolan, West African Rainforest, and Namib-Karoo. Indian, Indo-Chinese, Malesian, and Papuan. Hawaiian and Fijian. Caribbean, Amazon, Central Brazilian, Pampas, and Andean. North and East Australian. Sahelo-Sudanian, Somalo-Ethiopian, South Tropical African, and Kalaharian. Tropical North and East Australian.

Rusts and smuts. Rusts — *Phakopsora* and *Puccinia*. Taxonomically wide-ranging species: *Phakopsora incompleta* and *Puccinia versicolor*. Smuts from Ustilaginaceae. Ustilaginaceae — *Sorosporium, Sphacelotheca*, and *Ustilago*.

Economic importance. Significant weed species: *I. indicum, I. muticum, I. rugosum* (in ricefields), *I. timorense*. Important native pasture species: *I. muticum, I. rugosum*.

References, etc. Leaf anatomical: Metcalfe 1960 and this project.

Ischnochloa J.D. Hook.

~ Microstegium

Habit, vegetative morphology. Small, delicate annual. *Culms* 10–20 cm high; herbaceous. *Leaves* not basally aggregated. Leaf blades oblong elliptic; narrow (but relatively broad); not setaceous. Ligule a fringed membrane.

Reproductive organization. Plants bisexual, with bisexual spikelets; with hermaphrodite florets. The spikelets all alike in sexuality (homogamous).

Inflorescence. *Inflorescence a single terminal 'raceme'*. Rachides flattened. Inflorescence espatheate; not comprising 'partial inflorescences' and foliar organs. **Spikelet-bearing axes** 'racemes'; solitary; with very slender rachides (these flattened); *persistent*. *Spikelets* paired; not secund; sessile and pedicellate; *consistently in 'long-and-short' combinations*; in pedicellate/sessile combinations. Pedicels of the 'pedicellate' spikelets free of the rachis. The 'shorter' spikelets hermaphrodite. The 'longer' spikelets hermaphrodite.

Female-fertile spikelets. Spikelets small; compressed dorsiventrally. Rachilla terminated by a female-fertile floret. Hairy callus present. *Callus* short.

Glumes two; more or less equal; long relative to the adjacent lemmas; free; without conspicuous tufts or rows of hairs; not pointed; awnless. *Lower glume two-keeled (the keels ciliolate)*; deeply or shallowly sulcate on the back; not pitted; relatively smooth; 5–6 nerved. Upper glume 3 nerved. **Spikelets** with female-fertile florets only; *without proximal incomplete florets*.

Female-fertile florets 1. Lemmas less firm than the glumes (glumes almost leathery, lemma hyaline); not becoming indurated; incised; 2 lobed; deeply cleft; awned. Awns 1; median; from a sinus; geniculate; much longer than the body of the lemma. **Lemmas** hairless; non-carinate; without a germination flap; *1 nerved*. *Palea* present; not indurated (scarious). *Lodicules* present; 2; free; fleshy; glabrous. *Stamens* 3. Anthers not penicillate; without an apically prolonged connective. *Ovary* glabrous. Styles fused. Stigmas 2.

Taxonomy. Panicoideae; Andropogonodae; Andropogoneae; Andropogoninae.

Ecology, geography, regional floristic distribution. 1 species. Himalayas. Growing in moss at 1800–2100 m altitude.

Holarctic and Paleotropical. Boreal. Indomalesian. Eastern Asian. Indian.

References, etc. Morphological/taxonomic: Hooker 1896.

Special comments. Fruit data wanting. Anatomical data wanting.

Ischnurus Balf.

~ *Lepturus*

Habit, vegetative morphology. Small annual; caespitose. *Culms* 15 cm high; herbaceous. Young shoots extravaginal. Leaf blades narrow; without abaxial multicellular glands; without cross venation. Ligule a fringed membrane.

Reproductive organization. Plants bisexual, with bisexual spikelets; with hermaphrodite florets. The spikelets all alike in sexuality.

Inflorescence. Inflorescence a single spike (cylindrical); non-digitate. Rachides hollowed and winged. Inflorescence espatheate; not comprising 'partial inflorescences' and foliar organs. *Spikelet-bearing axes disarticulating*; disarticulating at the joints. '**Articles**' non-linear (hollowed to receive the spikelets); *appendaged (with small, lateral auriculate wings)*. Spikelets solitary; not secund; alternately distichous; sessile.

Female-fertile spikelets. Spikelets small; adaxial; compressed dorsiventrally; falling with the glumes. Rachilla terminated by a female-fertile floret.

Glumes one per spikelet (the G$_2$); (the G$_2$) exceeding the spikelets (covering the sunken floret); long relative to the adjacent lemmas (shorter than the joint); *not pointed (obtuse)*; awnless; non-carinate. Upper glume (i.e. the only one) 5 nerved, or 7 nerved (?).

Female-fertile florets 1. Lemmas less firm than the glumes (scarious); not becoming indurated; entire; pointed, or blunt; awnless; the nerves ciliate; without a germination flap; 3 nerved. *Palea* present; not indurated (scarious); 2-nerved. Styles free to their bases. Stigmas 2.

Fruit, embryo and seedling. Fruit small (about 2 mm long). Hilum short. *Pericarp free*. Embryo large.

Abaxial leaf blade epidermis. *Costal/intercostal zonation* conspicuous. *Papillae* present; nearly all intercostal. Intercostal papillae over-arching the stomata; consisting of one symmetrical projection per cell, or several per cell (mostly large, thick walled, more or less symmetrical). *Long-cells* markedly different in shape costally and intercostally (the costals narrower); of similar wall thickness costally and intercostally (fairly thick walled). Mid-intercostal long-cells rectangular; having markedly sinuous walls (the sinuosity irregular, coarse). *Microhairs* present; more or less spherical; clearly two-celled; chloridoid-type. Microhair apical cell wall of similar thickness/rigidity to that of the basal cell. Microhair basal cells 12 microns long. Microhair total length/width at septum 1.6. Microhair apical cell/total length ratio 0.5. *Stomata* common. Subsidiaries non-papillate; dome-shaped and triangular. Guard-cells overlapping to flush with the interstomatals. *Intercostal short-cells* common; in cork/silica-cell pairs to not paired; silicified. Intercostal silica bodies imperfectly developed; tall-and-narrow to acutely-angled. Prickles present, confined to the midrib. *Costal short-cells* conspicuously in long rows. Costal silica bodies present and well developed; present in alternate cell files of the costal zones; saddle shaped.

Transverse section of leaf blade, physiology. Leaf blades consisting of midrib to 'laminar' (largely dominated by the midrib, with only 2–3 bundles in each small lateral flange, in the material seen). C$_4$. The anatomical organization conventional. XyMS+. PCR sheath outlines even. PCR sheaths of the primary vascular bundles interrupted; interrupted both abaxially and adaxially. PCR sheath extensions absent. *Midrib* conspicuous; having a conventional arc of bundles (a large median primary, a smaller primary on either side, and 2–3 minor bundles in between); with colourless mesophyll adaxially. *The lamina* symmetrical on either side of the midrib. Bulliforms represented only by a pair of large midrib 'hinges'. All the vascular bundles accompanied by sclerenchyma. Combined sclerenchyma girders present (in the lamina bundles, but the fibre groups scanty). Sclerenchyma all associated with vascular bundles. The lamina margins without fibres.

Taxonomy. Chloridoideae; main chloridoid assemblage.

Ecology, geography, regional floristic distribution. 1 species. Socotra.
Paleotropical. African. Sudano-Angolan. Somalo-Ethiopian.

References, etc. Leaf anatomical: Metcalfe 1960; this project.

Iseilema Anderss.

Habit, vegetative morphology. Annual, or perennial (usually delicate, often with coloured leaves, culms and spathes); caespitose. Culms branched above. Culm nodes glabrous. Culm internodes hollow. The shoots aromatic (from conspicuous glands on the leaves and spathes), or not aromatic (rarely). *Leaves* not basally aggregated; non-auriculate. Leaf blades narrow; flat, or folded; without cross venation; persistent. *Ligule a fringed membrane*.

Reproductive organization. Plants bisexual, with bisexual spikelets; with hermaphrodite florets (usually), or without hermaphrodite florets (rarely). *The spikelets* of sexually distinct forms on the same plant; hermaphrodite and male-only, or hermaphrodite and sterile, or female-only and male-only, or female-only and sterile; overtly heteromorphic; *in both homogamous and heterogamous combinations (with two basal pairs of homogamous, male or sterile spikelets, connate to form a false tetramerous whorl beneath the single heterogamous triplet)*.

Inflorescence. Inflorescence falsely paniculate (the single racemes or tight clusters gathered with their spatheoles into a narrow false panicle); open; spatheate; a complex of 'partial inflorescences' and intervening foliar organs. **Spikelet-bearing axes** very much reduced (with small, short spike-like 'racemes', similar to *Themeda* but disarticulating below the involucral spikelets); disarticulating; *falling entire (complete with the involucral spikelets, and sometimes with the investing spatheole, which may then be indeurated)*. **Spikelets** *associated with bractiform involucres (comprising the involucral spikelets)*. The involucres (i.e. the involucral spikelets) shed with the fertile spikelets. Spikelets distal to the involucral whorl in triplets; not secund; sessile and pedicellate (two pedicellate members per heterogamous triplet, the pedicels filiform); consistently in 'long-and-short' combinations; in pedicellate/sessile combinations. The 'shorter' spikelets hermaphrodite, or female-only. The 'longer' spikelets male-only, or sterile.

Female-sterile spikelets. *The two basal pairs of spikelets homogamous, male or sterile, all more or less pedicelled, connate into a 'tetramerous' false whorl beneath the single heterogamous triplet, often indurated*.

Female-fertile spikelets. Spikelets compressed dorsiventrally; falling with the glumes (and with the whole racemelet. the upper pedicellate pair usually smaller than the sessile, sometimes much reduced, deciduous or not). Rachilla terminated by a female-fertile floret. Hairy callus absent. *Callus* absent.

Glumes two; more or less equal; long relative to the adjacent lemmas; awnless. Lower glume not two-keeled; flattened on the back; not pitted; relatively smooth; 8–10 nerved. Upper glume 3 nerved. *Spikelets* with incomplete florets. The incomplete florets proximal to the female-fertile florets. *The proximal incomplete florets* 1; epaleate; sterile. The proximal lemmas awnless; 0 nerved, or 3 nerved; exceeded by the female-fertile lemmas; not becoming indurated (hyaline).

Female-fertile florets 1. Lemmas stipitate beneath the awn; less firm than the glumes; not becoming indurated; entire, or incised; when incised, 2 lobed; awned. Awns 1; median; from a sinus, or apical; geniculate. Lemmas hairless; non-carinate; without a germination flap; 1 nerved. **Palea** *absent*. *Lodicules* present; 2; fleshy. *Stamens* 1–3. Anthers not penicillate. *Ovary* glabrous. Styles fused. Stigmas 2; red pigmented.

Fruit, embryo and seedling. *Fruit* free from both lemma and palea; small; compressed dorsiventrally. Hilum short. Embryo large.

Abaxial leaf blade epidermis. *Costal/intercostal zonation* conspicuous. *Papillae* present. Intercostal papillae over-arching the stomata; consisting of one oblique swelling per cell (not all the cells being papillate). *Long-cells* similar in shape costally and intercostally; of similar wall thickness costally and intercostally. Mid-intercostal long-cells rectangular; having markedly sinuous walls. *Microhairs* present; panicoid-type; 57–63 microns long; 5.4–6 microns wide at the septum. Microhair total length/width at septum 10.5–11.1. Microhair apical cells 31.5–37.5 microns long. Microhair apical cell/total length ratio 0.51–0.63. *Stomata* common; 24–25.5 microns long. Subsidiaries triangular. Guard-cells overlapping to flush with the interstomatals. *Intercostal short-cells* absent or very rare; in cork/silica-cell pairs, or not paired (solitary); silicified (when paired), or not silicified. *Costal short-cells* conspicuously in long rows. Costal silica bodies 'panicoid-type'; cross shaped and dumb-bell shaped.

Transverse section of leaf blade, physiology. C_4; XyMS–. *Mesophyll* with radiate chlorenchyma. *Leaf blade* adaxially flat. *Midrib* conspicuous; having a conventional arc of bundles; with colourless mesophyll adaxially, or without colourless mesophyll adaxially. Bulliforms present in discrete, regular adaxial groups; in simple fans. Many of the smallest vascular bundles unaccompanied by sclerenchyma. Combined sclerenchyma girders present; forming 'figures'. Sclerenchyma all associated with vascular bundles.

Cytology. Chromosome base number, $x = 4$, or 5 (?). $2n = 6, 8, 18, 28$, and 36. 2, 7, and 9 ploid (?).

Taxonomy. Panicoideae; Andropogonodae; Andropogoneae; Andropogoninae.

Ecology, geography, regional floristic distribution. 20 species. Indomalayan region, Australia. Species of open habitats. Grassland.

Paleotropical and Australian. Indomalesian. Indian, Indo-Chinese, and Malesian. North and East Australian and Central Australian. Tropical North and East Australian.

Rusts and smuts. Smuts from Ustilaginaceae. Ustilaginaceae — *Sorosporium*, *Sphacelotheca*, and *Ustilago*.

Economic importance. Important native pasture species: *I. vaginiflorum*.

References, etc. Leaf anatomical: Metcalfe 1960; this project.

Ixophorus Schlechtd.

Habit, vegetative morphology. Annual, or perennial; caespitose. Culms herbaceous; unbranched above. Culm nodes glabrous. Culm internodes solid. *Leaves* not basally aggregated; non-auriculate. Leaf blades broad; without cross venation; persistent. Ligule a fringed membrane; truncate; 1.5 mm long.

Reproductive organization. Plants bisexual, with bisexual spikelets; without hermaphrodite florets (the second floret being female-only).

Inflorescence. Inflorescence of spicate main branches; open. Inflorescence axes not ending in spikelets. Rachides hollowed. Inflorescence espatheate; not comprising 'partial inflorescences' and foliar organs. Spikelet-bearing axes persistent. **Spikelets *subtended by solitary 'bristles' (each subtended and exceeded by a single, sticky bristle)***. The 'bristles' persisting on the axis. Spikelets solitary; secund (the branches dorsiventral).

Female-fertile spikelets. Spikelets 3.5 mm long; abaxial; compressed dorsiventrally; falling with the glumes. Rachilla terminated by a female-fertile floret. Hairy callus absent.

Glumes two; very unequal; (the longer) long relative to the adjacent lemmas (almost equalling the spikelet, the lower much shorter); dorsiventral to the rachis; hairless; glabrous; pointed; awnless; non-carinate; similar (herbaceous). Lower glume 3 nerved. *Upper glume 11 nerved*. *Spikelets* with incomplete florets. The incomplete florets proximal to the female-fertile florets (the latter being female-only). *The proximal incomplete florets* 1; paleate. *Palea of the proximal incomplete florets* fully developed; *becoming conspicuously hardened and enlarged laterally (hyaline at first, then developing expanded leathery flanks and winged keels clasping the upper floret)*. The proximal incomplete florets male. The proximal lemmas awnless; 5 nerved; decidedly exceeding the female-fertile lemmas; less firm than the female-fertile lemmas (cartilaginous); not becoming indurated.

Female-fertile florets 1. Lemmas decidedly firmer than the glumes; minutely striate; not becoming indurated (tough); mucronate; hairless; non-carinate; having the margins tucked in onto the palea; with a clear germination flap; 5 nerved. *Palea* present (firm, rugulose between the veins); relatively long; tightly clasped by the lemma (enclosed by it at its tip); 2-nerved; 2-keeled. *Lodicules* present; 2; free; fleshy; glabrous; not or scarcely vascularized. *Stamens* 0. *Ovary* glabrous. Styles basally fused. Stigmas 2; red pigmented.

Fruit, embryo and seedling. *Fruit* free from both lemma and palea; small (about 3 mm long); compressed dorsiventrally. Hilum short. Embryo large; waisted. Endosperm hard.

Abaxial leaf blade epidermis. *Costal/intercostal zonation* conspicuous. *Papillae* absent. *Long-cells* markedly different in shape costally and intercostally; of similar wall thickness costally and intercostally. Intercostal zones exhibiting many atypical long-cells (many being quite short). Mid-intercostal long-cells intercostal long-cells variable — some rectangular, some fusiform, many short; having markedly sinuous walls. *Microhairs* present; panicoid-type; 49–51 microns long; 8.4–9 microns wide at the septum. Microhair total length/width at septum 5.7–6.1. Microhair apical cells 31.5–36 microns long. Microhair apical cell/total length ratio 0.62–0.71. *Stomata* common; 37.5–48 microns long. Subsidiaries mostly dome-shaped. Guard-cells overlapped by the interstomatals, or overlapping to flush with the interstomatals. *Intercostal short-cells* absent or very rare. *Costal short-cells* conspicuously in long rows. Costal silica bodies 'panicoid-type'; short butterfly shaped and dumb-bell shaped, or cross shaped.

Transverse section of leaf blade, physiology. C_4; XyMS– (median midrib bundle XyMS+, the rest XyMS-). *Leaf blade* 'nodular' in section; with the ribs more or less constant in size. *Midrib* conspicuous; having a conventional arc of bundles; with colourless mesophyll adaxially. Bulliforms present in discrete, regular adaxial groups (in places, but the epidermis mostly irregularly bulliform); occasionally in simple fans. Many of the smallest vascular bundles unaccompanied by sclerenchyma. Combined sclerenchyma girders present.

Taxonomy. Panicoideae; Panicodae; Paniceae.

Ecology, geography, regional floristic distribution. 2–3 species. Mexico. Species of open habitats. Weedy places.

Neotropical. Caribbean.

Rusts and smuts. Rusts — *Puccinia*. Taxonomically wide-ranging species: *Puccinia chaetochloae*.

References, etc. Leaf anatomical: this project.

Jansenella Bor

Habit, vegetative morphology. *Annual*. Culms herbaceous; unbranched above. *Leaves* not basally aggregated. *Leaf blades* lanceolate; narrow; 3–7 mm wide (to about 3.5 cm long); slightly *cordate*; flat; without cross venation. Ligule an unfringed membrane.

Reproductive organization. Plants bisexual, with bisexual spikelets; with hermaphrodite florets.

Inflorescence. Inflorescence paniculate; contracted; spicate; espatheate; not comprising 'partial inflorescences' and foliar organs. Spikelet-bearing axes persistent. Spikelets paired (usually), or in triplets (rarely); not secund.

Female-fertile spikelets. Spikelets 6–9 mm long; purplish; compressed laterally; disarticulating above the glumes. Rachilla terminated by a female-fertile floret. Hairy callus present. *Callus* short; blunt.

Glumes two; very unequal; (the upper) long relative to the adjacent lemmas; hairless; pointed; awned (aristulate, the points 1–2 mm long), or awnless. Lower glume 3 nerved. Upper glume 5 nerved. *Spikelets* with incomplete florets. The incomplete florets proximal to the female-fertile florets. *The proximal incomplete florets* when present, 1; paleate. Palea of the proximal incomplete florets fully developed (two keeled, narrowly two winged). The proximal incomplete florets male, or sterile (or even female). The proximal lemmas awned (with a short, dorsal capillary awn); 5–9 nerved; decidedly exceeding the female-fertile lemmas (similar to G_2); less firm than the female-fertile lemmas (membranous, leathery); not becoming indurated.

Female-fertile florets 1(–2). *Lemmas* decidedly firmer than the glumes; becoming leathery; incised; 2 lobed; deeply cleft; *awned*. Awns 1, or 3; median, or median and lateral (the two lobes aristulate); the median different in form from the laterals (if laterals considered present); from a sinus; geniculate. The lateral awns shorter than the median. Lemmas hairy. *The hairs* in tufts (these thick, near the margins at the bases of the lobes); *in transverse rows (transversely bearded with brown hairs in the lower half)*. *Lemmas* non-carinate; *7–9 nerved*. *Palea* present; relatively long; apically notched; 2-nerved; 2-keeled. *Palea keels winged (and with peculiar unicellular, turgid hairs between the keels)*. *Lodicules* present; 2; free; glabrous. *Stamens* 3. Anthers not penicillate; without an apically prolonged connective. *Ovary* glabrous. Stigmas 2.

Fruit, embryo and seedling. Fruit small (1.3 mm long). Hilum short. Embryo large.

Abaxial leaf blade epidermis. *Costal/intercostal zonation* conspicuous. *Microhairs* present; panicoid-type. Costal silica bodies 'panicoid-type'; dumb-bell shaped.

Transverse section of leaf blade, physiology. Supposedly C_3 (but needing further study: certainly the anatomical drawings of Türpe (1970) show neither 'circular cells' nor the expected small bundles); XyMS− (from Türpe's detailed drawings, which if accurate reveal this as the only known C3 XyMS− combination). *Mesophyll* with radiate chlorenchyma. *Leaf blade* adaxially flat. *Midrib* not readily distinguishable.

Cytology. $2n = 20$.

Taxonomy. Panicoideae; Panicodae; Arundinelleae.

Ecology, geography, regional floristic distribution. 1 species. India, Burma. Mesophytic. Moist places in hills.

Paleotropical. Indomalesian. Indian.

References, etc. Leaf anatomical: Metcalfe 1960, Türpe 1970.

Jardinea Steud.

~ Phacelurus

Habit, vegetative morphology. Coarse, robust perennial; caespitose. *Culms* 200–400 cm high; herbaceous; to almost 1 cm in diameter. Culm leaves present. Upper culm leaf blades fully developed. Culm internodes solid. *Leaves* not basally aggregated; non-auriculate. Leaf blades linear, or linear-lanceolate; broad to narrow; 7–15 mm wide (or more); flat (long); without cross venation; persistent. Ligule a fringed membrane; 0.2–0.5 mm long. *Contra-ligule* absent.

Reproductive organization. Plants bisexual, with bisexual spikelets; with hermaphrodite florets. The spikelets of sexually distinct forms on the same plant, or all alike in sexuality (the pedicellate member hermaphrodite, male, female, sterile or vestigial, increasingly reduced towards the raceme tips); hermaphrodite, or hermaphrodite and female-only, or hermaphrodite, female-only, and male-only, or hermaphrodite, female-only, and sterile, or hermaphrodite, female-only, male-only, and sterile. The male and female-fertile spikelets (when both represented) mixed in the inflorescence. The spikelets overtly heteromorphic to homomorphic; in both homogamous and heterogamous combinations, or all in heterogamous combinations.

Inflorescence. Inflorescence of spicate main branches, or paniculate (long, slender, spiciform 'racemes', borne racemosely on a common axis or some branched); open; subdigitate (the axis sometimes quite short), or non-digitate. Rachides hollowed (to receive the sessile spikelets and the more or less flattened pedicels). Inflorescence espatheate; not comprising 'partial inflorescences' and foliar organs. *Spikelet-bearing axes* spikelike (by appression/embedding of the spikelets and pedicels); the spikelet-bearing axes with more than 10 spikelet-bearing 'articles'. The racemes spikelet bearing to the base. Spikelet-bearing axes clustered (mostly numerous, the occasional one solitary); with substantial rachides; disarticulating; disarticulating at the joints. *'Articles'* non-linear (hollowed below, expanded at the apex); with a basal callus-knob; not appendaged; disarticulating transversely; glabrous. *Spikelets* paired; secund (the racemes somewhat dorsiventral); sessile and pedicellate; imbricate; *consistently in 'long-and-short' combinations*; in pedicellate/sessile combinations. Pedicels of the 'pedicellate' spikelets free of the rachis (more or less flattened, appressed). The 'shorter' spikelets hermaphrodite. The 'longer' spikelets hermaphrodite, or female-only, or male-only, or sterile.

Female-sterile spikelets. The pedicellate spikelets usually smaller, hermaphrodite or variously reduced (often to female-only rather than male-only in the material seen) or vestigial. Rachilla of male spikelets represented only in *J. angolensis*, in the specimens seen, prolonged beyond the uppermost male floret. The male spikelets with glumes; with proximal incomplete florets; 1 floreted. The lemmas awnless. Male florets 1; 4 staminate (sic).

Female-fertile spikelets. *Spikelets* 5–10 mm long; *compressed dorsiventrally*; planoconvex; falling with the glumes (the sessile spikelets falling with the adjacent joint and pedicel). Rachilla terminated by a female-fertile floret. Hairy callus absent. *Callus* short (beneath the sessile spikelet); pointed.

Glumes two; very unequal (G_1 longer), or more or less equal; exceeding the spikelets; (the longer) long relative to the adjacent lemmas (exceeding them); hairless; scabrous; pointed; awnless; carinate (the upper), or non-carinate (the lower); with the keel conspicuously winged; very dissimilar (G_1 dorsally flattened with incurved margins, leathery and much firmer than G_2, G_2 smaller, naviculate and cartilaginous). Lower glume much exceeding the lowest lemma; two-keeled to not two-keeled (keeled above, incurved below); convex on the back to flattened on the back; not pitted; muricate to prickly (mainly near the margins); 4–6 nerved. *Upper glume* 3 nerved; *prickly (mainly along the keel)*. *Spikelets* with incomplete florets. The incomplete florets proximal to the female-fertile florets. *The proximal incomplete florets* 1; paleate, or epaleate (to a scabrous, awnlike structure). Palea of the proximal incomplete florets fully developed, or reduced. The proximal incomplete florets sterile. The proximal lemmas lanceolate, one or two keeled with ciliate margins and scabrous keels; awnless; 2 nerved, or 3 nerved; more or less equalling the female-fertile lemmas to decidedly exceeding the female-fertile lemmas; similar in texture to the female-fertile lemmas (hyaline); not becoming indurated.

Female-fertile florets 1. Lemmas less firm than the glumes (hyaline); smooth; not becoming indurated; entire; pointed; awnless; hairy, or hairless (with ciliate margins); carinate. The keel wingless. Lemmas without a germination flap; 3 nerved. *Palea* present, or absent; when present, relatively long, or conspicuous but relatively short, or very reduced; entire; awnless, without apical setae; textured like the lemma; not indurated; 2-nerved, or nerveless; keel-less. Palea back glabrous, or scabrous. *Lodicules* present; 2; free; fleshy; glabrous; not or scarcely vascularized. *Stamens* 3. Anthers 2.5–3.5 mm long; not penicillate; without an apically prolonged connective. *Ovary* glabrous. Styles free to their bases; free. Style bases adjacent. Stigmas 2; red pigmented.

Fruit, embryo and seedling. Fruit compressed dorsiventrally.

Abaxial leaf blade epidermis. *Costal/intercostal zonation* conspicuous. *Papillae* absent. *Long-cells* narrower costally; differing markedly in wall thickness costally and intercostally (the costals thicker walled). Mid-intercostal long-cells rectangular; having markedly sinuous walls. *Microhairs* present; elongated; clearly two-celled; panicoid-type. *Stomata* common. Subsidiaries non-papillate; high to medium dome-shaped (mostly), or triangular (a few, slightly). Guard-cells overlapping to flush with the interstomatals. *Intercostal short-cells* common; mostly in cork/silica-cell pairs. No macrohairs or prickles seen. Costal zones with short-cells. *Costal short-cells* predominantly paired, or neither distinctly grouped into long rows nor predominantly paired (the pairs in places supplemented by short rows). Costal silica bodies 'panicoid-type'; shortly dumb-bell shaped.

Transverse section of leaf blade, physiology. C_4; XyMS–. PCR sheath outlines uneven. PCR sheath extensions absent. *Mesophyll* with radiate chlorenchyma. *Leaf blade* with distinct, prominent adaxial ribs; with the ribs very irregular in sizes (mostly low, encompassing more than one bundle). *Midrib* very conspicuous; having a conventional arc of bundles (these abaxial, with a large median primary, a large primary on each flank, and smaller bundles in between); with colourless mesophyll adaxially (and a narrow, adaxial lignified hypodermal layer). Bulliforms present in discrete, regular adaxial groups (in places, in the furrows), or not in discrete, regular adaxial groups (in places); in simple fans and associated with colourless mesophyll cells to form deeply-penetrating fans; associating with colourless mesophyll cells to form arches over small vascular bundles (in places). All the vascular bundles accompanied by sclerenchyma (even the tiniest bundles nearly all associated with at least one abaxial fibre). Combined sclerenchyma girders present (with all but the smallest bundles); forming 'figures' (I's and anchors). Sclerenchyma all associated with vascular bundles.

Taxonomy. Panicoideae; Andropogonodae; Andropogoneae; Rottboelliinae.

Ecology, geography, regional floristic distribution. 3 species. Tropical Africa. Helophytic, or mesophytic.

Paleotropical. African. Sudano-Angolan and West African Rainforest. Somalo-Ethiopian and South Tropical African.

References, etc. Leaf anatomical: this project.

Special comments. Fruit data wanting.

Jouvea Fourn.

Rhachidospermum Vasey

Habit, vegetative morphology. Perennial; stoloniferous and caespitose. *Culms* 50 cm high; herbaceous. *Leaves* not basally aggregated; non-auriculate. Leaf blades narrow; 2–4 mm wide (glabrous); flat, or rolled (involute); without abaxial multicellular glands; without cross venation; disarticulating from the sheaths. Ligule a fringed membrane (the membrane very short, with a conspicuous fringe).

Reproductive organization. Plants monoecious with all the fertile spikelets unisexual, or dioecious; without hermaphrodite florets. The spikelets of sexually distinct forms on the same plant (when monoecious), or all alike in sexuality (when dioecious); female-only, or male-only. The male and female-fertile spikelets in different inflorescences.

Inflorescence. *Inflorescence with male spikelets in long-peduncled, dorsiventral spikes, with a spikelet terminating the rachis and the spikes in axillary fascicles (J. pilosa) or groups of 1–3 (J.* straminea*); the female spikelets thornlike, each with an associated prophyll and spathe, these units in axillary umbels (J. pilosa, cf. female* Spinifex*) or in small spiny clusters (J.* straminea*); spatheate (the female spikelet clusters with umbels of spathes), or espatheate (the male inflorescences). **Spikelets *associated with bractiform involucres (i.e., the spathes)*; secund (males), or not secund (females).

Female-sterile spikelets. Male spikelets sessile, disarticulating above the glumes, their 3–5 fertile florets with 2 large fleshy lodicules and 3 stamens. The male spikelets with glumes; without proximal incomplete florets; 3–7 floreted. Male florets 3–5; 3 staminate.

Female-fertile spikelets. Spikelets so *unconventional (as to be only dubiously interpretable as spikelets rather than spikes)*; 20–30 mm long, or 40–50 mm long; *not noticeably compressed (cylindrical, thorn-like)*; falling whole; not disarticulating between the florets. Rachilla prolonged beyond the uppermost female-fertile floret (as a subulate point, which becomes a thorn at maturity); rachilla hairless. The rachilla extension with incomplete florets. Hairy callus absent.

**Glumes *absent*. *Spikelets* with incomplete florets. The incomplete florets distal to the female-fertile florets. The distal incomplete florets 1–3; *incomplete florets* merely underdeveloped.

**Female-fertile florets 2–4(–8) *(completely embedded in the thick, spongy, cylindrical rachilla). Lemmas forming a tube around the floret, adnate to and embedded in the rachilla for most of their length, with an apical pore through which the stigmas protrude*; becoming indurated; awnless (?); hairless; glabrous; non-carinate. **Palea *present (adaxially, but detected only on sectioning the 'spikelet')*; (in sections) conspicuous but relatively short (not much longer than the ovary); awnless, without apical setae; thinner than the lemma (very flimsy); not indurated; nerveless. *Lodicules* absent. *Stamens* 0 (reduced to three vestiges). *Ovary* glabrous. Styles fused (ovary attenuate into the long style). Stigmas 2 (long, exserted); red pigmented, or brown (?).

Fruit, embryo and seedling. Fruit ellipsoid. Embryo with an epiblast; with a scutellar tail; with an elongated mesocotyl internode. Embryonic leaf margins meeting.

Abaxial leaf blade epidermis. *Costal/intercostal zonation* conspicuous. *Papillae* absent. Mid-intercostal long-cells rectangular; having markedly sinuous walls (and pitted, with pits also in the outer walls). *Microhairs* present; more or less spherical; clearly two-celled; chloridoid-type (basal cell embedded in *J. straminea*, normal in *J. pilosa*). Microhair apical cell wall of similar thickness/rigidity to that of the basal cell. Microhairs 30–39 microns long (*J. pilosa*), or 18–21 microns long (*J. straminea*). Microhair basal cells 6 microns long. Microhairs 18.6–21.6 microns wide at the septum (*J. pilosa*), or 13.5–16.5 microns wide at the septum (*J. straminea*). Microhair total length/width at septum 1.4–2.1. Microhair apical cells (10.5–)12–16.5(–17.4) microns long. Microhair apical cell/total length ratio 0.32–0.52, or 0.64–0.67. *Stomata* common; 43.5–48 microns long

(J. pilosa), or 28.5–33 microns long *(J. straminea)*. Subsidiaries high dome-shaped *(J. straminea)*, or parallel-sided and dome-shaped *(J. pilosa*, where the domes are mainly low). Guard-cells overlapping to flush with the interstomatals. *Intercostal short-cells* common; in cork/silica-cell pairs; silicified. Intercostal silica bodies present and perfectly developed; saddle shaped (tiny), or crescentic. *Costal short-cells* predominantly paired. Costal silica bodies present throughout the costal zones; crescentic (predominating), or saddle shaped (a few, more or less imperfect).

Transverse section of leaf blade, physiology. C_4; XyMS+. PCR sheaths of the primary vascular bundles complete. PCR sheath extensions present. Maximum number of extension cells 2. Mesophyll traversed by columns of colourless mesophyll cells. *Leaf blade* adaxially flat. *Midrib* conspicuous (with a large abaxial anchor-shaped strand in *J. straminea*), or not readily distinguishable; with one bundle only. Bulliforms present in discrete, regular adaxial groups; associated with colourless mesophyll cells to form deeply-penetrating fans (these linked with traversing columns of colourless cells). All the vascular bundles accompanied by sclerenchyma. Combined sclerenchyma girders absent (most bundles with scant abaxial strands, but many altogether lacking adaxial sclerenchyma). Sclerenchyma all associated with vascular bundles. The lamina margins with fibres.

Taxonomy. Chloridoideae; main chloridoid assemblage.

Ecology, geography, regional floristic distribution. 2 species. California to Central America. Xerophytic; species of open habitats; halophytic, or glycophytic. Sandy places, coastal dunes and mud flats.

Neotropical. Caribbean.

References, etc. Morphological/taxonomic: Weatherwax 1939. Leaf anatomical: Metcalfe 1960; this project.

Special comments. Fruit data wanting.

Kampochloa W. Clayton

Habit, vegetative morphology. Perennial; densely caespitose. *Culms* 15–30 cm high; herbaceous; unbranched above. Plants unarmed. *Leaves* mostly basal; non-auriculate. Leaf blades narrow; 1–3 mm wide (and 3–5 cm long); without abaxial multicellular glands; without cross venation; persistent. Ligule hyaline; 1 mm long.

Reproductive organization. Plants bisexual, with bisexual spikelets; with hermaphrodite florets.

Inflorescence. *Inflorescence a single raceme (with very short pedicels)*; the raceme curved, 1–2.4 cm long. *Inflorescence axes not ending in spikelets (in that the rachis tip appears to be a naked point)*. Inflorescence not comprising 'partial inflorescences' and foliar organs. Spikelet-bearing axes persistent. Spikelets secund; biseriate; shortly pedicellate.

Female-fertile spikelets. *Spikelets* 3.5 mm long; compressed laterally; *disarticulating above the glumes (?)*. Rachilla prolonged beyond the uppermost female-fertile floret; rachilla hairless. The rachilla extension with incomplete florets. Hairy callus present.

Glumes two; very unequal; shorter than the adjacent lemmas; free; *awned (the G_1 with 1 mm subule from a sinus, the G_2 also incised, but with a 2.7 mm dorsal awn)*; non-carinate; very dissimilar (the G_2 green, puberulent, its longer awn from low on the back). Lower glume 1 nerved. *Spikelets* with incomplete florets. The incomplete florets distal to the female-fertile florets. *The distal incomplete florets* clearly specialised and modified in form (the second floret male or sterile, succeeded distally by several florets represented by a bunch of vestigial lemmas in the form of hyaline scales and long awns).

Female-fertile florets 1. Lemmas less firm than the glumes to similar in texture to the glumes (hyaline); not becoming indurated; incised; not deeply cleft; awned. Awns 1; median; from a sinus; non-geniculate; hairless; much shorter than the body of the lemma (very short, scarcely more than a mucro). Lemmas hairy; *carinate (long-ciliate along the nerves)*; 3 nerved. *Palea* present; relatively long; apically notched; awnless, without apical setae (glabrous); not indurated (hyaline); 2-nerved; 2-keeled. *Lodicules* present; 2;

free; fleshy; glabrous. *Stamens* 3. Anthers not penicillate; without an apically prolonged connective. *Ovary* glabrous. Styles free to their bases. Stigmas 2.

Abaxial leaf blade epidermis. *Costal/intercostal zonation* conspicuous. *Papillae* absent. Mid-intercostal long-cells rectangular; having markedly sinuous walls. *Microhairs* present; more or less spherical; ostensibly one-celled, or clearly two-celled; chloridoid-type (the basal cell often sunken). Microhair apical cell wall of similar thickness/rigidity to that of the basal cell. Microhairs 12–15 microns long. Microhair basal cells 4–5 microns long. Microhairs 6–8.4 microns wide at the septum. Microhair total length/width at septum 1.8–2.3. Microhair apical cell/total length ratio 0.6–1. *Stomata* common; 25.5–31.5 microns long. Subsidiaries obscured by overlaps from the adjoining cells. Guard-cells overlapped by the interstomatals (in some files), or overlapping to flush with the interstomatals. *Intercostal short-cells* common (in some files); not paired (solitary); not silicified. Intercostal silica bodies absent. *Costal short-cells* conspicuously in long rows. Costal silica bodies present in alternate cell files of the costal zones; 'panicoid-type'; mostly butterfly shaped and dumb-bell shaped.

Transverse section of leaf blade, physiology. C_4; XyMS+. PCR sheaths of the primary vascular bundles interrupted; interrupted abaxially only. PCR sheath extensions absent. Mesophyll traversed by columns of colourless mesophyll cells. *Leaf blade* 'nodular' in section to adaxially flat. *Midrib* not readily distinguishable; with one bundle only. Bulliforms present in discrete, regular adaxial groups; associated with colourless mesophyll cells to form deeply-penetrating fans (these linking with the traversing colourless cell columns). All the vascular bundles accompanied by sclerenchyma. Combined sclerenchyma girders present (with all the bundles); forming 'figures' (all the bundles with I's or anchors). Sclerenchyma all associated with vascular bundles. The lamina margins without fibres.

Taxonomy. Chloridoideae; main chloridoid assemblage.

Ecology, geography, regional floristic distribution. 1 species. Angola and Zambia. Paleotropical. African. Sudano-Angolan. South Tropical African.

References, etc. Leaf anatomical: this project.

Special comments. Fruit data wanting.

Kaokochloa de Winter

Habit, vegetative morphology. Pilose *annual*; culms geniculate or prostrate at base, rooting at the nodes. *Culms* 15–80 cm high; herbaceous; branched above. Culm nodes hairy. Culm internodes hollow. *Leaves* not basally aggregated; non-auriculate; without auricular setae (but with auricular hairs). Leaf blades linear-lanceolate to lanceolate; broad, or narrow; 5–15 mm wide (5–12 cm long); flat, or rolled; exhibiting multicellular glands abaxially (at the bases of the microhairs). The abaxial glands intercostal. Leaf blades without cross venation; persistent. Ligule a fringe of hairs; truncate; 0.5–1.5 mm long. *Contra-ligule* absent.

Reproductive organization. Plants bisexual, with bisexual spikelets; with hermaphrodite florets.

Inflorescence. *Inflorescence paniculate*; open to contracted; espatheate; not comprising 'partial inflorescences' and foliar organs. Spikelet-bearing axes persistent. Spikelets not secund; shortly pedicellate, or subsessile; imbricate.

Female-fertile spikelets. *Spikelets* 5–8 mm long; not noticeably compressed ('subglobose'); *disarticulating between the glumes (the upper glume falling with the spikelet)*; not disarticulating between the florets; with conventional internode spacings. Rachilla prolonged beyond the uppermost female-fertile floret; rachilla hairy. The rachilla extension with incomplete florets. Hairy callus absent. *Callus* absent.

Glumes two; more or less equal; about equalling the spikelets; long relative to the adjacent lemmas; hairy (villous and glandular); without conspicuous tufts or rows of hairs; pointed (subacute); awnless; non-carinate (broadly rounded abaxially); similar. Lower glume about equalling the lowest lemma; 9 nerved, or 11 nerved. Upper glume 9 nerved, or 11 nerved. *Spikelets* with incomplete florets. The incomplete florets distal to the female-fertile florets. *The distal incomplete florets* merely underdeveloped; awnless (or with only minute vestiges of awns).

Female-fertile florets *3–6. Lemmas with an incurved, emarginate apex and a narrow awned lobe at each margin, and sometimes with 1–2 shorter, additional lobes;* becoming indurated; incised (between excurrent nerves); 3–8 lobed; not deeply cleft; awned. Awns 2, or 3, or 5; median and lateral (sometimes), or lateral only (the two marginal nerves excurrent into large awns, the median and other nerves occasionally contributing smaller awns); the median (when present) similar in form to the laterals; apical; non-geniculate; hairy (and glandular); when present, much shorter than the body of the lemma to about as long as the body of the lemma; entered by one vein. The lateral awns exceeding the median. Lemmas hairy (the lower two profusely so between the nerves, the upper ones hairy only at the margins); non-carinate (abaxially rounded to flattened); without a germination flap; 9 nerved; with the nerves non-confluent. *Palea* present; relatively long (but narrower than the lemma); minutely apically notched; thinner than the lemma; not indurated (thinly leathery); 2-nerved; 2-keeled. Palea keels hairy. *Lodicules* present; 2; free; fleshy; glabrous. *Stamens* 3. Anthers 3–4 mm long (yellow). *Ovary* glabrous. Stigmas 2.

Fruit, embryo and seedling. Fruit compressed dorsiventrally (ventrally flattened). Hilum short. Pericarp fused. Embryo large; with an epiblast; with a scutellar tail; with an elongated mesocotyl internode. Embryonic leaf margins overlapping.

Abaxial leaf blade epidermis. *Costal/intercostal zonation* conspicuous. *Papillae* seemingly absent. *Long-cells* markedly different in shape costally and intercostally (the costals much narrower). Mid-intercostal long-cells rectangular; having markedly sinuous walls. *Microhairs* present; elongated; clearly two-celled; *Enneapogon*-type. Microhair apical cell wall of similar thickness/rigidity to that of the basal cell. Microhair basal cells 210–480 microns long. Microhair total length/width at septum 30. Microhair apical cell/total length ratio 0.01. *Stomata* common. Subsidiaries dome-shaped and triangular. *Intercostal short-cells* seemingly absent or very rare; not paired. Intercostal silica bodies absent. *Costal short-cells* conspicuously in long rows. Costal silica bodies present in alternate cell files of the costal zones; predominantly 'panicoid-type'; mostly dumb-bell shaped.

Transverse section of leaf blade, physiology. C_4; XyMS+. PCR sheath outlines uneven. PCR sheaths of the primary vascular bundles complete. PCR sheath extensions absent. PCR cell chloroplasts centrifugal/peripheral. *Mesophyll* with radiate chlorenchyma; traversed by columns of colourless mesophyll cells (perhaps, in places), or not traversed by colourless columns. *Leaf blade* adaxially flat. *Midrib* not readily distinguishable; with one bundle only. Bulliforms present in discrete, regular adaxial groups (between adjacent bundles, and each group corresponding with an abaxial bulliform group); in simple fans and associated with colourless mesophyll cells to form deeply-penetrating fans (some of the simple fans of the deeply-penetrating type, some of the others seeming to link with traversing colourless columns). Many of the smallest vascular bundles unaccompanied by sclerenchyma, or all the vascular bundles accompanied by sclerenchyma. Combined sclerenchyma girders absent. Sclerenchyma all associated with vascular bundles. The lamina margins with fibres.

Special diagnostic feature. Female-fertile lemmas with an incurved-emarginate apex, and a narrow awned lobe at each margin (sometimes with 1–2 shorter, additional lobes).

Taxonomy. Chloridoideae; Pappophoreae.

Ecology, geography, regional floristic distribution. 1 species. Southern Africa. Xerophytic; species of open habitats; glycophytic. In semi-desert.

Paleotropical. African. Namib-Karoo.

References, etc. Morphological/taxonomic: de Winter 1961. Leaf anatomical: photos provided by R.P. Ellis.

Karroochloa Conert & Túrpe

~ *Rytidosperma, Danthonia* sensu lato

Habit, vegetative morphology. Annual, or perennial; stoloniferous, or caespitose. **Culms** 4–40 cm high; *herbaceous*; unbranched above. Culm nodes glabrous. Culm internodes solid. Plants unarmed. Young shoots extravaginal (rarely), or intravaginal. *Leaves* mostly basal to not basally aggregated; non-auriculate. Hair-tufted at the mouth of the sheath. Leaf blades linear; narrow; to 2 mm wide; flat, or folded, or rolled; without cross venation; persistent. **Ligule** present; *a fringe of hairs. Contra-ligule* present (as a line of hairs).

Reproductive organization. *Plants bisexual, with bisexual spikelets*; with hermaphrodite florets.

Inflorescence. Inflorescence paniculate; contracted (1–6 cm long); more or less ovoid; espatheate; not comprising 'partial inflorescences' and foliar organs. Spikelet-bearing axes persistent. Spikelets solitary; not secund; pedicellate.

Female-fertile spikelets. Spikelets 4–6(–7) mm long; compressed laterally; disarticulating above the glumes; disarticulating between the florets; with conventional internode spacings. Rachilla prolonged beyond the uppermost female-fertile floret; rachilla hairy. The rachilla extension with incomplete florets. *Hairy callus present. Callus* short, or long; blunt.

Glumes two; more or less equal (subequal); about equalling the spikelets; long relative to the adjacent lemmas; hairless; glabrous; pointed; awnless; carinate; similar (membranous, the margins and apices hyaline). Lower glume much exceeding the lowest lemma; 3–5 nerved. Upper glume 3–5(–7) nerved. *Spikelets* with incomplete florets. The incomplete florets distal to the female-fertile florets. *The distal incomplete florets* merely underdeveloped; awned.

Female-fertile florets 3–7. *Lemmas with fringes or tufts of white hairs, except in* **K. curva**; similar in texture to the glumes (membranous); not becoming indurated; incised; 2 lobed; deeply cleft; awned. Awns 1, or 3; median, or median and lateral (by small extensions from the lobes); the median different in form from the laterals (when laterals present); from a sinus; geniculate; hairless; about as long as the body of the lemma to much longer than the body of the lemma. The lateral awns shorter than the median (and straight). Lemmas hairy. The hairs in tufts, or not in tufts; in transverse rows, or not in transverse rows. *Lemmas* non-carinate; without a germination flap; *9 nerved*; with the nerves non-confluent. *Palea* present; relatively long (almost equalling the lemma); entire to apically notched; awnless, without apical setae (glabrous or pilose); textured like the lemma; not indurated (membranous, the margins hyaline); 2-nerved; 2-keeled. Palea keels wingless; glabrous, or scabrous. **Lodicules** present; 2; free; fleshy; *ciliate. Stamens* 3. Anthers 1.4–2.2 mm long; not penicillate; without an apically prolonged connective. *Ovary* glabrous. Styles free to their bases. Stigmas 2.

Fruit, embryo and seedling. *Fruit* free from both lemma and palea (but enclosed); small (0.8–1 mm long). *Hilum short. Embryo small*.

Abaxial leaf blade epidermis. *Costal/intercostal zonation* conspicuous. *Papillae* absent. Mid-intercostal long-cells rectangular, or rectangular and fusiform; having markedly sinuous walls. *Microhairs* present; panicoid-type. *Stomata* common, or absent or very rare. Subsidiaries dome-shaped. *Intercostal short-cells* common; in cork/silica-cell pairs, or not paired; silicified. *Costal short-cells* usually conspicuously in long rows, or neither distinctly grouped into long rows nor predominantly paired (in *K. purpurea*). Costal silica bodies tall-and-narrow to crescentic (*K. purpurea*), or 'panicoid-type' (usually); usually cross shaped, or dumb-bell shaped.

Transverse section of leaf blade, physiology. C_3; XyMS+. *Mesophyll* with non-radiate chlorenchyma; *Isachne*-type (or at least tending to this), or not *Isachne*-type. *Leaf blade* with distinct, prominent adaxial ribs to 'nodular' in section, or adaxially flat; when ribbed with the ribs more or less constant in size. *Midrib* not readily distinguishable (apart from position); with one bundle only. Bulliforms present in discrete, regular adaxial groups, or not in discrete, regular adaxial groups; sometimes in simple fans. All the vascular bundles accompanied by sclerenchyma. Combined sclerenchyma girders absent. Sclerenchyma all associated with vascular bundles.

Special diagnostic feature. Female-fertile lemmas with a bent awn, the awn twisted below. *Spikelets 4–6(–7) mm long, inflorescence 10–60 mm long*.

Cytology. Chromosome base number, $x = 6$. $2n = 12$ and 24. 2 and 4 ploid.

Taxonomy. Arundinoideae; Danthonieae.

Ecology, geography, regional floristic distribution. 4 species. Southern Africa. Mesophytic; species of open habitats; glycophytic. Grassland and among rocks.

Paleotropical and Cape. African. Namib-Karoo.

References, etc. Morphological/taxonomic: Conert 1969. Leaf anatomical: Conert and Türpe 1969; photos of *K. purpurea* provided by R.P. Ellis.

Kengia Packer

Cleistogenes Keng, *Moliniopsis* Gand.

~ *Diplachne*

Habit, vegetative morphology. Perennial. *Culms* 10–100 cm high; herbaceous. Plants unarmed. *Leaves* not basally aggregated. **Leaf blades linear-lanceolate**; narrow; flat, or rolled (convolute); without abaxial multicellular glands; without cross venation. Ligule a fringed membrane to a fringe of hairs.

Reproductive organization. Plants bisexual, with bisexual spikelets; with hermaphrodite florets. Exposed-cleistogamous, or chasmogamous. *Plants with hidden cleistogenes*. The hidden cleistogenes in the leaf sheaths (especially associated with the upper leaves).

Inflorescence. Inflorescence of spicate main branches (of loose racemes on a central axis), or paniculate; open to contracted; espatheate; not comprising 'partial inflorescences' and foliar organs. Spikelets not secund; pedicellate; distant.

Female-fertile spikelets. Spikelets 4–13 mm long; compressed laterally; disarticulating above the glumes; disarticulating between the florets; with distinctly elongated rachilla internodes between the florets. Rachilla prolonged beyond the uppermost female-fertile floret. The rachilla extension with incomplete florets. Hairy callus present. Callus blunt.

Glumes two; very unequal; shorter than the spikelets; shorter than the adjacent lemmas; pointed to not pointed; awnless; similar (membranous). Lower glume 1 nerved, or 3–5 nerved (at base). Upper glume 1 nerved, or 3–5(–7) nerved (at base). *Spikelets with incomplete florets*. The incomplete florets distal to the female-fertile florets. *The distal incomplete florets* merely underdeveloped; awnless.

Female-fertile florets *2–6 (to many?)*. Lemmas not becoming indurated; usually incised; minutely 2 lobed; not deeply cleft (two-toothed); mucronate to awned. Awns 1; median; from a sinus; non-geniculate. Lemmas hairy (pilose along margins), or hairless; carinate; 3 nerved, or 5(–7) nerved. *Palea* present; relatively long; apically notched (two-toothed); awnless, without apical setae; textured like the lemma; not indurated; 2-nerved; 2-keeled. **Lodicules absent (usually, most florets being cleistogamous)**. Stamens 3. Anthers 0.7–3.5 mm long.

Fruit, embryo and seedling. *Fruit* free from both lemma and palea. Hilum short. Embryo large. Endosperm hard; without lipid. Embryo with an epiblast; with a scutellar tail; with an elongated mesocotyl internode. Embryonic leaf margins meeting.

Abaxial leaf blade epidermis. *Papillae* present; costal and intercostal. Intercostal papillae consisting of one oblique swelling per cell. Mid-intercostal long-cells rectangular. *Microhairs* present; elongated; clearly two-celled; chloridoid-type. Microhair apical cell wall of similar thickness/rigidity to that of the basal cell. Microhair basal cells 21–27 microns long. Microhair total length/width at septum 2.2. Microhair apical cell/total length ratio 0.3. *Stomata* common. Intercostal short-cells not paired; silicified. Intercostal silica bodies present and perfectly developed (but blurred by papillae); rounded, tall-and-narrow, and oryzoid-type. *Costal short-cells* conspicuously in long rows. Costal silica bodies present in alternate cell files of the costal zones; 'panicoid-type'.

Transverse section of leaf blade, physiology. C_4; XyMS+. PCR sheaths of the primary vascular bundles interrupted; interrupted both abaxially and adaxially. PCR sheath extensions present. Maximum number of extension cells 2. Mesophyll not traversed by colourless columns. *Leaf blade* with distinct, prominent adaxial ribs to adaxially flat. *Midrib* not readily distinguishable; with one bundle only; without colourless mesophyll adaxially, or with colourless mesophyll adaxially (at least, sometimes with a few colourless cells between the bundles and the sclerenchyma). Bulliforms present in discrete, regular adaxial groups. All the vascular bundles accompanied by sclerenchyma. Combined sclerenchyma girders absent. Sclerenchyma all associated with vascular bundles. The lamina margins with fibres.

Cytology. Chromosome base number, $x = 10$. $2n = 40$. 4 ploid.

Taxonomy. Chloridoideae; main chloridoid assemblage.

Ecology, geography, regional floristic distribution. 10 species. Temperate Eurasia. Mesophytic, or xerophytic (mainly); species of open habitats; glycophytic.

Holarctic, Paleotropical, and Antarctic. Boreal and Tethyan. Indomalesian. Euro-Siberian. Mediterranean and Irano-Turanian. Indian. New Zealand and Antarctic and Subantarctic. European. South Temperate Oceanic Islands.

Rusts and smuts. Rusts — *Puccinia*.

References, etc. Leaf anatomical: Van den Borre 1994.

Kengyilia Yen & J.L Yang

Habit, vegetative morphology. Perennial; caespitose. *Culms* 60 cm high (erect); herbaceous; unbranched above; 2–3 noded. Culm nodes glabrous (the culm white pubescent just beneath the spike). Young shoots intravaginal. *Leaves* not basally aggregated. Leaf blades linear; narrow; about 3 mm wide ((6–)7–8 cm long); acicular ('has short, rodlike leaves that form tubelike structures', 'complanate or involute', adaxially white pubescent or villous, abaxially glabrous); without cross venation; persistent. Ligule an unfringed membrane; truncate; 0.5 mm long.

Reproductive organization. Plants bisexual, with bisexual spikelets; with hermaphrodite florets.

Inflorescence. *Inflorescence a single spike ((7–8–12 cm long, 4–6 mm wide, erect, the densely pilose internodes 4–10 mm long)*; espatheate; not comprising 'partial inflorescences' and foliar organs. Spikelet-bearing axes persistent. *Spikelets* probably solitary; probably not secund; probably distichous; presumably *sessile*.

Female-fertile spikelets. Spikelets 15–20 mm long; ovate; green to purpling; compressed laterally; disarticulating above the glumes; disarticulating between the florets (? — the rachilla articulated). Rachilla prolonged beyond the uppermost female-fertile floret; rachilla hairy (puberulent). The rachilla extension with incomplete florets. *Hairy callus present*.

Glumes two; very unequal to more or less equal (? — 'subequal', but the illustration depicting them very unequal); shorter than the spikelets; long relative to the adjacent lemmas (the longer glume about equalling the lowest lemma, in the illustration seen); lateral to the rachis (?); hairy (sometimes, on the nerves), or hairless; mostly glabrous; pointed (sometimes asymmetrically dentate); awned to awnless (mucronate); non-carinate (?); similar ('herbaceous'). Lower glume 3–5 nerved (? — 'glumes 3–5 nerved'). Upper glume 3–5 nerved (?). *Spikelets* probably with incomplete florets. The incomplete florets distal to the female-fertile florets. *The distal incomplete florets* merely underdeveloped; awned, or awnless.

Female-fertile florets *(5–)7–8*. *Lemmas* entire; **blunt**; shortly **awned**. Awns 1; median; apical; non-geniculate; straight; much shorter than the body of the lemma (1–4 mm long). Lemmas hairy (white villous, the hairs 1 mm long); non-carinate; without a germination flap. *Palea* present; relatively long (equalling or slightly shorter than the lemma); tightly clasped by the lemma; entire ('truncate'); awnless, without apical setae; 2-nerved; 2-keeled. Palea back hairy (white-pilose). Palea keels wingless; hairy. **Anthers 2–3 mm long**. *Ovary* hairy (judging from a poor drawing of the fruit).

Fruit, embryo and seedling. Fruit medium sized (about 7 mm long); dark brown; presumably longitudinally grooved (?); with hairs confined to a terminal tuft. Embryo small.

Cytology. Chromosome base number, $x = 7$. $2n = 42$. 6 ploid. Haplomic genome content P, S, and Y (? — L.W. being insufficiently knowledgeable about genomic analysis to interpret the garbled published description).

Taxonomy. Pooideae; Triticodae; Triticeae.

Ecology, geography, regional floristic distribution. 1 species (*K. gobicola*). Gobi Desert, West china. Xerophytic; species of open habitats. In dry, stony, high elevation desert (stony gobi).

Holarctic. Boreal. Eastern Asian.

References, etc. Morphological/taxonomic: Yen and Yang (1990).

Special comments. Based on the very poor original description and grotty illustrations in *Can. J. Bot.* **68** (1990): no information on spikelet orientation, lodicules, fruit form, etc. Fruit data wanting. Anatomical data wanting.

Kerriochloa C.E. Hubb.

Habit, vegetative morphology. Slender perennial; decumbent. *Culms* 20–25 cm high; herbaceous; branched above. *Leaves* not basally aggregated; non-auriculate. Leaf blades flat (short); without cross venation. Ligule an unfringed membrane.

Reproductive organization. Plants bisexual, with bisexual spikelets; with hermaphrodite florets. The spikelets of sexually distinct forms on the same plant; hermaphrodite and sterile; overtly heteromorphic (the pedicelled member very reduced); all in heterogamous combinations. Viviparous.

Inflorescence. Inflorescence of solitary, shortly peduncled terminal 'racemes'; non-digitate; *spatheate (the fragile, spicate racemes partially enclosed by narrow spathes)*. *Spikelet-bearing axes* 'racemes'; solitary; with substantial rachides; disarticulating; disarticulating at the joints. *'Articles'* non-linear (cuneate); not appendaged; disarticulating transversely; densely long-hairy. *Spikelets* paired; somewhat secund; sessile and pedicellate; *consistently in 'long-and-short' combinations*; in pedicellate/sessile combinations. Pedicels of the 'pedicellate' spikelets free of the rachis. The 'shorter' spikelets hermaphrodite. The 'longer' spikelets sterile (reduced to the G_1).

Female-sterile spikelets. The pedicelled spikelet much smaller, reduced to its membranous, nerveless G_1.

Female-fertile spikelets. *Spikelets* 5–6 mm long; slightly *compressed laterally*; falling with the glumes (deciduous with the adjacent internode and pedicel). Rachilla terminated by a female-fertile floret. Hairy callus present. *Callus* short; blunt (truncate).

Glumes two; more or less equal; long relative to the adjacent lemmas; free; hairy; without conspicuous tufts or rows of hairs; not pointed (the (G_1) obtuse, the G_2 notched); the G_2 *awned (from its sinus)*; carinate (G_2), or non-carinate (G_1); very dissimilar (G_1 dorsally convex and awnless, G_2 cymbiform, awned). *Lower glume not two-keeled (wingless, the margins incurved)*; convex on the back; with a conspicuous pit; relatively smooth; 5 nerved. Upper glume 3 nerved. *Spikelets* with incomplete florets. The incomplete florets proximal to the female-fertile florets. *Spikelets with proximal incomplete florets. The proximal incomplete florets* 1; sterile (much shorter than the glumes). The proximal lemmas awnless; 3 nerved; decidedly exceeding the female-fertile lemmas; similar in texture to the female-fertile lemmas to decidedly firmer than the female-fertile lemmas (thinly membranous, with broad hyaline margins, entire); not becoming indurated.

Female-fertile florets 1. Lemmas less firm than the glumes (hyaline); not becoming indurated; incised; 2 lobed; not deeply cleft; awned. Awns 1; median; from a sinus; geniculate; hairless; much longer than the body of the lemma. Lemmas hairless; non-carinate; without a germination flap; 3 nerved. *Palea* present; shorter than the L_2; entire (acuminate); awnless, without apical setae; not indurated (hyaline); 2-nerved. *Lodicules* present; 2; free; glabrous; toothed (conspicuously bidentate). *Stamens* 3. Anthers not penicillate; without an apically prolonged connective. *Ovary* glabrous. Styles fused. Stigmas 2; red pigmented.

Fruit, embryo and seedling. Fruit not noticeably compressed. Hilum short. Embryo large.

Taxonomy. Panicoideae; Andropogonodae; Andropogoneae; Andropogoninae.

Ecology, geography, regional floristic distribution. 1 species. Thailand and Cambodia. Species of open habitats.

Paleotropical. Indomalesian. Indo-Chinese.

References, etc. Morphological/taxonomic: Hubbard 1951.

Special comments. Anatomical data wanting.

Koeleria Pers.

Aegialina Schult., *Aegialitis* Trin., *Airochloa* Link, *Brachystylus* Dulac, *Ktenosachne* Steud., *Leptophyllochloa* Cald., *Lophochloa* Reichenb., *Poarion* Reichenb., *Rostraria* Trin., *Wilhelmsia* Koch

Excluding *Avellinia*

Habit, vegetative morphology. Annual (*Lophochloa, Rostraria*), or perennial; caespitose (usually), or rhizomatous (rarely). **Culms** 5–120 cm high; *herbaceous*; tuberous, or not tuberous. Culm nodes glabrous. Culm internodes hollow. Leaves non-auriculate. Leaf blades linear; narrow; 0.5–6 mm wide; flat, or folded, or rolled (convolute); without cross venation; persistent. **Ligule** present; *an unfringed membrane (sometimes puberulent and ciliolate)*; truncate, or not truncate.

Reproductive organization. Plants bisexual, with bisexual spikelets; with hermaphrodite florets; outbreeding and inbreeding.

Inflorescence. *Inflorescence paniculate*; *contracted (not interrupted)*; more or less ovoid, or spicate; espatheate; not comprising 'partial inflorescences' and foliar organs. Spikelet-bearing axes persistent. Spikelets not secund; pedicellate.

Female-fertile spikelets. Spikelets (2.7–)4–7 mm long; compressed laterally; disarticulating above the glumes; disarticulating between the florets. Rachilla prolonged beyond the uppermost female-fertile floret; rachilla hairy, or hairless. Hairy callus present, or absent. *Callus* short.

Glumes two; very unequal, or more or less equal; shorter than the spikelets, or about equalling the spikelets; shorter than the adjacent lemmas (rarely), or long relative to the adjacent lemmas (the G_2 usually equalling the first lemma); pointed; awnless; *carinate*; very dissimilar (sometimes, e.g. *K. pumila*), or similar. Lower glume 1 nerved. *Upper glume 3 nerved*. *Spikelets* with female-fertile florets only, or with incomplete florets. The incomplete florets distal to the female-fertile florets.

Female-fertile florets *2–4*. *Lemmas* similar in texture to the glumes; not becoming indurated; entire (usually), or incised (e.g. in *Lophochloa*); *blunt*; *awnless, or mucronate, or awned (but then the awns relatively inconspicuous, by contrast with* **Trisetum***)*. Awns 1; median; from a sinus, or dorsal; from near the top; non-geniculate; much shorter than the body of the lemma to about as long as the body of the lemma; entered by one vein. Lemmas hairy, or hairless; *carinate*; 3–5 nerved. **Palea** present; relatively long (usually), or conspicuous but relatively short (e.g. *K. cristata, K. pumila*); *gaping*; apically notched; thinner than the lemma (hyaline or membranous); not indurated; 2-nerved; 2-keeled. *Lodicules* present; 2; free; membranous; glabrous; toothed. *Stamens* 3. Anthers 0.3–0.7 mm long (*Rostraria*), or 1.3–3 mm long; not penicillate. *Ovary* glabrous. Styles free to their bases. Stigmas 2.

Fruit, embryo and seedling. Fruit small; slightly compressed laterally. Hilum short, or long-linear. Embryo small; not waisted. *Endosperm liquid in the mature fruit*; with lipid; containing compound starch grains. Embryo with an epiblast; without a scutellar tail; with a negligible mesocotyl internode. Embryonic leaf margins meeting.

Seedling with a short mesocotyl, or with a long mesocotyl; with a tight coleoptile. First seedling leaf with a well-developed lamina. The lamina narrow; erect; 3 veined.

Abaxial leaf blade epidermis. *Costal/intercostal zonation* conspicuous. *Papillae* absent. *Long-cells* similar in shape costally and intercostally, or markedly different in shape costally and intercostally; of similar wall thickness costally and intercostally (thin walled). Intercostal zones with typical long-cells, or exhibiting many atypical long-cells. Mid-intercostal long-cells fusiform; having straight or only gently undulating walls. *Microhairs* absent. Stomata absent or very rare, or common. Subsidiaries low domeshaped, or parallel-sided. Guard-cells overlapped by the interstomatals. *Intercostal short-cells* absent or very rare; in cork/silica-cell pairs, or not paired (some solitaries, and abundant prickles); silicified (when paired), or not silicified. *Costal short-cells* neither distinctly grouped into long rows nor predominantly paired. Costal silica bodies horizontally-elongated crenate/sinuous to horizontally-elongated smooth, or rounded, or saddle shaped (occasionally).

Transverse section of leaf blade, physiology. C_3; XyMS+. *Mesophyll* with non-radiate chlorenchyma. *Leaf blade* with distinct, prominent adaxial ribs, or 'nodular' in section; with the ribs more or less constant in size. *Midrib* not readily distinguishable; with one bundle only. Bulliforms present in discrete, regular adaxial groups (between the vascular bundles), or not in discrete, regular adaxial groups (e.g. *K. cristata, K. pumila*); in simple fans. All the vascular bundles accompanied by sclerenchyma. Combined sclerenchyma girders present; forming 'figures'. Sclerenchyma all associated with vascular bundles.

Phytochemistry. Leaves without flavonoid sulphates (1 species).

Special diagnostic feature. *Panicle dense, cylindrical, ovoid, not interrupted: awns if present straight, subterminal, inconspicuous in the inflorescence.*

Cytology. Chromosome base number, $x = 7$. $2n = 14, 26, 28, 40, 42, 43, 56, 70, 84, 112$, and 126. $2, 4, 6, 8, 10, 12, 16$, and 18 ploid (and aneuploids).

Taxonomy. Pooideae; Poodae; Aveneae.

Ecology, geography, regional floristic distribution. About 60 species. North and south temperate. Commonly adventive. Mesophytic, or xerophytic; mostly species of open habitats. In dry grassland and rocky places.

Holarctic, Paleotropical, Neotropical, Cape, and Antarctic. Boreal, Tethyan, and Madrean. African. Arctic and Subarctic, Euro-Siberian, Eastern Asian, and Atlantic North American. Mediterranean and Irano-Turanian. Sudano-Angolan and West African Rainforest. Pampas, Andean, and Fernandezian. New Zealand and Patagonian. European and Siberian. Canadian-Appalachian and Southern Atlantic North American. Sahelo-Sudanian, Somalo-Ethiopian, and South Tropical African.

Hybrids. Intergeneric hybrids with *Trisetum*: ×*Trisetokoeleria* Tsvelev.

Rusts and smuts. Rusts — *Puccinia*. Taxonomically wide-ranging species: *Puccinia graminis, Puccinia coronata, Puccinia striiformis, Puccinia poarum, Puccinia hordei, Puccinia recondita*, and *Puccinia monoica*. Smuts from Tilletiaceae and from Ustilaginaceae. Tilletiaceae — *Entyloma, Tilletia*, and *Urocystis*. Ustilaginaceae — *Ustilago*.

Economic importance. Significant weed species: *K. phleoides, K. pyramidata*. Important native pasture species: e.g. *K. cristata*.

References, etc. Leaf anatomical: Metcalfe 1960 and this project.

Special comments. The taxonomic situation around *Koeleria* and *Trisetum* (involving *Trisetaria, Graphephorum, Rostraria, Lophochloa* and *Peyritschia*) is hopelessly unsatisfactory, and demands either a world monograph of the approximately 150 species involved, or critical assessment via extensive worldwide sampling and an adequate character list. This *sensu lato* version of *Koeleria* reflects the impossibility of preparing adequate descriptions for segregate genera, and is poorly separable from that of *Trisetum*.

Lagurus L.

Avena Scop.

Habit, vegetative morphology. Slender *annual*; caespitose. *Culms* 8–50(–60) cm high; herbaceous; unbranched above. Culm nodes hairy, or glabrous. Culm internodes hollow. *Leaves* not basally aggregated; non-auriculate. *Leaf blades linear-lanceolate*; narrow; 1.5–10 mm wide; not setaceous (pubescent); flat; without cross venation; persistent; rolled in bud. Ligule an unfringed membrane, or a fringed membrane (rarely); irregularly truncate; 2.5–3 mm long.

Reproductive organization. Plants bisexual, with bisexual spikelets; with hermaphrodite florets; inbreeding.

Inflorescence. *Inflorescence paniculate*; contracted; *more or less ovoid (silky-white hairy and bristly)*; espatheate; not comprising 'partial inflorescences' and foliar organs. Spikelet-bearing axes persistent. *Spikelets not secund*; pedicellate.

Female-fertile spikelets. Spikelets 5–10 mm long; compressed laterally; disarticulating above the glumes. *Rachilla prolonged beyond the uppermost female-fertile floret*; rachilla hairy. The rachilla extension naked. Hairy callus absent. Callus blunt.

Glumes two; more or less equal; long relative to the adjacent lemmas; hairy (villous); pointed; *awned*; non-carinate; similar (narrowly lanceolate, membranous, hairy, tapering into fine bristles, thinly membranous). Lower glume 1 nerved. Upper glume 1 nerved. *Spikelets* with female-fertile florets only.

Female-fertile florets *1*. Lemmas similar in texture to the glumes to decidedly firmer than the glumes (membranous); not becoming indurated; incised; 2 lobed (narrowed into 2 awn-tipped teeth); not deeply cleft; awned. Awns 3; median and lateral (with two short terminal laterals in addition to the longer median); the median different in form from the laterals; dorsal; from near the top, or from well down the back; geniculate; hairless; much longer than the body of the lemma; entered by one vein. The lateral awns shorter than the median (terminal, straight). Lemmas hairy, or hairless; non-carinate; 5 nerved. *Palea* present; relatively long (but shorter than the lemma); apically notched; awnless, without apical setae; not indurated (thin); 2-nerved; 2-keeled, or keel-less. *Lodicules* present; 2; free; membranous; glabrous; toothed. *Stamens* 3. Anthers 1.5–2 mm long; not penicillate. *Ovary* glabrous. Styles fused, or free to their bases. Stigmas 2; white.

Fruit, embryo and seedling. *Fruit* free from both lemma and palea; small. Hilum short. Embryo small; not waisted. Endosperm liquid in the mature fruit; with lipid.

Abaxial leaf blade epidermis. *Costal/intercostal zonation* conspicuous. *Papillae* absent. *Long-cells* similar in shape costally and intercostally; of similar wall thickness costally and intercostally. Mid-intercostal long-cells rectangular to fusiform; having straight or only gently undulating walls. *Microhairs* absent. *Stomata* common; 37–45 microns long. Subsidiaries parallel-sided. Guard-cells overlapped by the interstomatals. *Intercostal short-cells* absent or very rare; in cork/silica-cell pairs, or not paired (solitaries, and pairs adjacent to the prickles); not silicified. *Costal short-cells* conspicuously in long rows. Costal silica bodies horizontally-elongated smooth (predominating), or rounded.

Transverse section of leaf blade, physiology. C_3; XyMS+. *Mesophyll* with non-radiate chlorenchyma; without adaxial palisade. *Leaf blade* with distinct, prominent adaxial ribs, or adaxially flat; with the ribs more or less constant in size. *Midrib* conspicuous; with one bundle only. Bulliforms not in discrete, regular adaxial groups (ill defined, in material seen). All the vascular bundles accompanied by sclerenchyma. Combined sclerenchyma girders present; nowhere forming 'figures'. Sclerenchyma all associated with vascular bundles.

Culm anatomy. *Culm internode bundles* in one or two rings.

Phytochemistry. Tissues of the culm bases with little or no starch.

Cytology. Chromosome base number, $x = 7$. $2n = 14$. 2 ploid.

Taxonomy. Pooideae; Poodae; Aveneae.

Ecology, geography, regional floristic distribution. 1 species. Mediterranean. Commonly adventive. Xerophytic; species of open habitats. Especially maritime sands.

Holarctic, Paleotropical, and Cape. Boreal and Tethyan. African. Euro-Siberian and Atlantic North American. Macaronesian, Mediterranean, and Irano-Turanian. Saharo-Sindian. European. Canadian-Appalachian.

Rusts and smuts. Rusts — *Puccinia*. Taxonomically wide-ranging species: *Puccinia coronata* and *Puccinia hordei*.

Economic importance. Significant weed species: *L. ovatus*. Cultivated as an ornamental.

References, etc. Leaf anatomical: Metcalfe 1960; this project.

Lamarckia Moench mut. Koeler

Achyrodes Boehmer, *Chrysurus* Pers., *Pterium* Desv., *Tinaea* Garzia

Habit, vegetative morphology. Annual; caespitose. *Culms* 5–200(–300) cm high; herbaceous; branched above to unbranched above; 1–5 noded. Culm nodes exposed, or hidden by leaf sheaths; glabrous. Culm internodes hollow. *Leaves* not basally aggregated;

non-auriculate. *Sheath margins joined (for up to two thirds their length)*. Sheaths keeled. Leaf blades narrow; 2.6–7.5 mm wide; not setaceous; flat; without cross venation; persistent. Ligule an unfringed membrane; not truncate; (4–)5–10.2 mm long.

Reproductive organization. Plants bisexual, with bisexual spikelets; with hermaphrodite florets. *The spikelets of sexually distinct forms on the same plant; hermaphrodite, male-only, and sterile (only the terminal spikelet in each fascicle being hermaphrodite, the other 3–4 male-only or with 3–6 empty, awnless, truncate lemmas)*.

Inflorescence. *Inflorescence a false spike, with spikelets on contracted axes (the spikelets in peduncled fascicles)*; contracted; espatheate; not comprising 'partial inflorescences' and foliar organs. Spikelet-bearing axes disarticulating; falling entire (the clusters of 3–5 spikelets falling whole). Spikelets secund; pedicellate (the pedicels villous).

Female-sterile spikelets. The lateral, sterile spikelets reduced to the glumes and lemmas, narrow-elongated, concealing the hermaphrodite spikelets save for the awns.

Female-fertile spikelets. Spikelets 2.5–4.3 mm long; cuneate; compressed laterally; falling with the glumes (in the clusters); with a distinctly elongated rachilla internode above the glumes and with distinctly elongated rachilla internodes between the florets. Rachilla prolonged beyond the uppermost female-fertile floret. The rachilla extension with incomplete florets. Hairy callus absent. *Callus* short; blunt (glabrous).

Glumes two; more or less equal; about equalling the spikelets to exceeding the spikelets; long relative to the adjacent lemmas; hairless; pointed; awned to awnless (acuminate to shortly aristate); carinate; similar (membranous, linear-lanceolate, hyaline). Lower glume 1 nerved. Upper glume 1 nerved. *Spikelets* with incomplete florets. The incomplete florets distal to the female-fertile florets. The distal incomplete florets 1; incomplete florets awned (the sterile rudiment with a long awn).

Female-fertile florets 1. Lemmas papery; not becoming indurated; entire to incised; when incised, 2 lobed; not deeply cleft (bidentate); awned. Awns 1; median; dorsal; from near the top; non-geniculate; straight, or recurving; much longer than the body of the lemma (4.7–7.2 mm long); entered by one vein. Lemmas hairless; non-carinate; 4–5 nerved. *Palea* present; relatively long; tightly clasped by the lemma; textured like the lemma (membranous); 2-nerved; 2-keeled. Palea keels wingless. *Lodicules* present; 2; membranous; glabrous; toothed. *Stamens* 3. Anthers 0.5–0.8 mm long; not penicillate. *Ovary* glabrous. Styles free to their bases. Stigmas 2; white.

Fruit, embryo and seedling. *Fruit* adhering to lemma and/or palea (slightly, to the palea); small (1.6–2 mm long); ovoid or ellipsoid; ventrally longitudinally grooved; glabrous. *Hilum short (elliptic)*. Embryo small. Endosperm liquid in the mature fruit, or hard; with lipid. Embryo with an epiblast; without a scutellar tail; with a negligible mesocotyl internode. Embryonic leaf margins meeting.

Abaxial leaf blade epidermis. *Costal/intercostal zonation* conspicuous. *Papillae* absent (from material seen, but see Metcalfe's description). *Long-cells* markedly different in shape costally and intercostally (many of the costals quite rectangular); of similar wall thickness costally and intercostally (quite thin walled). Mid-intercostal long-cells strongly fusiform; having straight or only gently undulating walls. *Microhairs* absent. *Stomata* common; 24–28.5 microns long, or 33–36 microns long (in different specimens of *L. aurea*). Subsidiaries low dome-shaped, or parallel-sided. Guard-cells overlapped by the interstomatals (mostly, but only very slightly), or overlapping to flush with the interstomatals. *Intercostal short-cells* absent or very rare. *Costal short-cells* conspicuously in long rows. Costal silica bodies horizontally-elongated crenate/sinuous.

Transverse section of leaf blade, physiology. C_3; XyMS+. *Mesophyll* with non-radiate chlorenchyma; without adaxial palisade. *Leaf blade* adaxially flat. *Midrib* conspicuous (via a prominent abaxial keel); with one bundle only; with colourless mesophyll adaxially. Bulliforms not in discrete, regular adaxial groups (restricted to midrib 'hinges'). Many of the smallest vascular bundles unaccompanied by sclerenchyma (and many others with only 1-celled strands). Combined sclerenchyma girders present (some of the primaries only); nowhere forming 'figures'. Sclerenchyma all associated with vascular bundles.

Culm anatomy. *Culm internode bundles* in one or two rings.

Phytochemistry. Leaves without flavonoid sulphates (1 species).

Cytology. Chromosome base number, $x = 7$. $2n = 14$. 2 ploid.

Taxonomy. Pooideae; Poodae; Poeae.

Ecology, geography, regional floristic distribution. 1 species. Mediterranean to Pakistan. Commonly adventive. Mesophytic, or xerophytic; species of open habitats. Dry places.

Holarctic and Paleotropical. Boreal and Tethyan. African. Euro-Siberian. Mediterranean. Saharo-Sindian. European.

Rusts and smuts. Rusts — *Puccinia*. Taxonomically wide-ranging species: *Puccinia graminis*, *Puccinia coronata*, and *Puccinia striiformis*. Smuts from Tilletiaceae. Tilletiaceae — *Entyloma*.

Economic importance. Significant weed species: *L. aurea*.

References, etc. Leaf anatomical: Metcalfe 1960; this project.

Lamprothyrsus Pilger

Habit, vegetative morphology. *Cortaderia*-like perennial; caespitose. *Culms* 100–230 cm high; herbaceous. Young shoots intravaginal. *Leaves* not basally aggregated. Leaf blades linear; narrow; rolled (long, involute); without cross venation; disarticulating from the sheaths. Ligule a fringe of hairs.

Reproductive organization. *Plants* (gyno?) *dioecious (males rarer than females)*; without hermaphrodite florets. The spikelets hermaphrodite and female-only (?), or female-only, or male-only. Plants outbreeding. Apomictic (non-pseudogamous).

Inflorescence. *Inflorescence* paniculate; *open to contracted (large, to 50 cm long, with many spikelets)*; non-digitate; espatheate; not comprising 'partial inflorescences' and foliar organs. Spikelet-bearing axes persistent. Spikelets not secund; pedicellate.

Female-fertile spikelets. *Spikelets compressed dorsiventrally*; disarticulating above the glumes; disarticulating between the florets (at the bases of the rachilla joints). Rachilla prolonged beyond the uppermost female-fertile floret; rachilla hairy (long-villous). The rachilla extension with incomplete florets. Hairy callus present. *Callus* linear.

Glumes two; very unequal; shorter than the spikelets (1/2–2/3 as long); long relative to the adjacent lemmas (slightly longer); hairless; glabrous; pointed; awned to awnless; similar (hyaline, acuminate to an awn point). Lower glume 0 nerved, or 1 nerved. Upper glume 0 nerved, or 1 nerved. *Spikelets* with incomplete florets. The incomplete florets distal to the female-fertile florets. *The distal incomplete florets* merely underdeveloped.

Female-fertile florets 4–10. Lemmas hyaline; not becoming indurated; incised; 2 lobed; deeply cleft; awned. *Awns* 3; median and lateral (the lemma teeth acuminate into long delicate awns); the median different in form from the laterals; from a sinus; non-geniculate to geniculate; *recurving (flat, spreading from near the scarcely twisted base, flexuous or loosely twisted above)*. Lemmas hairy (with silvery hairs); non-carinate; 5 nerved. *Palea* present; relatively long (longer than the body of the lemma); apically notched; awnless, without apical setae; 2-keeled. Palea back hairy (pilose). Palea keels hairy. *Lodicules* present; 2; free; ciliate, or glabrous. *Stamens* 0 (3 staminodes). *Ovary* glabrous. Styles free to their bases. Stigmas 2.

Fruit, embryo and seedling. Hilum long-linear. Embryo without an epiblast; with a scutellar tail; with an elongated mesocotyl internode. Embryonic leaf margins meeting.

Abaxial leaf blade epidermis. *Costal/intercostal zonation* conspicuous, or lacking (*L. peruvianus*). *Papillae* absent. Mid-intercostal long-cells rectangular; having markedly sinuous walls (rather thick). *Microhairs* present, or absent (*L. peruvianus*); panicoid-type. *Stomata* absent or very rare (*L. peruvianus*), or common (*L. hieronymi*). Subsidiaries low dome-shaped, or triangular (slightly). *Intercostal short-cells* common. *Costal short-cells* conspicuously in long rows (*L. hieronymi*), or neither distinctly grouped into long rows nor predominantly paired (short-cells regularly spaced, solitary in *L. peruvianus*). Costal silica bodies absent (*L. peruvianus*), or present and well developed (*L. hieronymi*); of *L. hieronymi* 'panicoid-type'; somewhat dumb-bell shaped; not sharp-pointed.

Transverse section of leaf blade, physiology. C_3; XyMS+. *Mesophyll* with non-radiate chlorenchyma; without adaxial palisade. *Leaf blade* with distinct, prominent adaxial ribs (round- and flat-topped); with the ribs very irregular in sizes. *Midrib* conspicuous (the large bundle having a slight abaxial keel); with one bundle only. Bulliforms present in discrete, regular adaxial groups; in simple fans. All the vascular bundles accompanied by sclerenchyma. Combined sclerenchyma girders present (with adaxial 'sheath extensions'); forming 'figures' (I's or anchors with all the bundles). Sclerenchyma not all bundle-associated (there being an abaxial, more or less interrupted lignified hypodermis). The 'extra' sclerenchyma in abaxial groups.

Special diagnostic feature. The lemma awns lateral and median, the median strongly flattened.

Taxonomy. Arundinoideae; Danthonieae.

Ecology, geography, regional floristic distribution. 3 species. Tropical South America.

Neotropical. Pampas and Andean.

References, etc. Leaf anatomical: Metcalfe 1960; this project.

Special comments. Fruit data wanting.

Lasiacis A. Hitchc.

Pseudolasiacis A. Camus

Habit, vegetative morphology. Perennial; rhizomatous to stoloniferous, or decumbent. Culms woody and persistent (usually), or herbaceous (10%); scandent (to 10 m in some species), or not scandent; branched above. Culm nodes glabrous. Culm internodes solid, or hollow. *Leaves* not basally aggregated; auriculate (with sheath auricles), or non-auriculate. Sheath margins free. *Leaf blades* linear to elliptic; *broad*; cordate, or not cordate, not sagittate; pseudopetiolate, or not pseudopetiolate; without cross venation; disarticulating from the sheaths; rolled in bud. *Ligule an unfringed membrane to a fringed membrane*; 0.1–9 mm long. *Contra-ligule* absent.

Reproductive organization. Plants bisexual, with bisexual spikelets; with hermaphrodite florets.

Inflorescence. *Inflorescence paniculate (with the spikelets borne obliquely on their pedicels)*; open, or contracted; espatheate; not comprising 'partial inflorescences' and foliar organs. Spikelet-bearing axes persistent. Spikelets not secund; pedicellate; not in distinct 'long-and-short' combinations.

Female-fertile spikelets. Spikelets morphologically 'conventional' (usually), or unconventional (*L. anomala*, with an extra sterile lemma); 2.6–5 mm long; slightly *compressed dorsiventrally (subglobose)*; falling with the glumes; with conventional internode spacings, or with distinctly elongated rachilla internodes between the florets (in the form of a conical boss, in *L. grisebachii* and *L. ruscifolia*). Rachilla prolonged beyond the uppermost female-fertile floret (occasionally), or terminated by a female-fertile floret. The rachilla extension when present, naked. Hairy callus absent.

Glumes two; relatively large; very unequal; (the upper) about equalling the spikelets; (the upper) long relative to the adjacent lemmas; hairy (with webby white hairs at the apex); pointed, or not pointed; awnless; non-carinate; similar (membranous, abruptly apiculate, blackening and accumulating oil when mature). Lower glume 5–13 nerved. Upper glume 7–15 nerved. *Spikelets* with incomplete florets. The incomplete florets proximal to the female-fertile florets. *Spikelets with proximal incomplete florets. The proximal incomplete florets* 1, or 2 (in *L. anomala*); paleate. *Palea of the proximal incomplete florets not becoming conspicuously hardened and enlarged laterally*. The proximal incomplete florets male, or sterile. *The proximal lemmas apically pubescent, blackening and accumulating oil like the glumes*; awnless; 7–15 nerved; more or less equalling the female-fertile lemmas (and the upper glume); less firm than the female-fertile lemmas; not becoming indurated.

Female-fertile florets 1. *Lemmas* with a shallow excavation and a tuft of hair at the tip; decidedly firmer than the glumes; smooth; *becoming indurated*; usually turning dark brown in fruit; entire; blunt; awnless; *hairy (with webby white hairs at the apex)*; non-carinate; having the margins tucked in onto the palea; with a clear germination flap; *7 nerved*. *Palea* present; relatively long; indurated; 2-nerved; 2-keeled (apically excavated, like the lemma). *Lodicules* present; 2; free; fleshy; glabrous; not or scarcely vascularized. *Stamens* 3. Anthers not penicillate; without an apically prolonged connective. **Ovary** glabrous;

without a conspicuous apical appendage. Styles fused (basally), or free to their bases. Stigmas 2; white, or red pigmented.

Fruit, embryo and seedling. Fruit small; ellipsoid; broadly longitudinally grooved, or not grooved; compressed dorsiventrally. Hilum short. Embryo large; not waisted. Endosperm hard. Embryo without an epiblast; with a scutellar tail; with an elongated mesocotyl internode. Embryonic leaf margins overlapping.

First seedling leaf with a well-developed lamina. The lamina broad; curved to supine.

Abaxial leaf blade epidermis. *Costal/intercostal zonation* conspicuous. *Papillae* absent. Intercostal zones exhibiting many atypical long-cells to without typical long-cells (the long-cells relatively short). Mid-intercostal long-cells rectangular; having markedly sinuous walls. *Microhairs* present; panicoid-type; 48–55.5 microns long; 6–6.6 microns wide at the septum. Microhair total length/width at septum 7.6–9.3. Microhair apical cells 16.5–22.5 microns long. Microhair apical cell/total length ratio 0.34–0.47. *Stomata* common; 30–33 microns long. Subsidiaries triangular. Guard-cells overlapped by the interstomatals (e.g. *L. sorghoides*), or overlapping to flush with the interstomatals (e.g. *L. procerrima*). *Intercostal short-cells* absent or very rare. *Costal short-cells* conspicuously in long rows. Costal silica bodies 'panicoid-type'; cross shaped, or butterfly shaped, or dumb-bell shaped, or nodular; not sharp-pointed.

Transverse section of leaf blade, physiology. C_3; XyMS+. *Mesophyll* with radiate chlorenchyma; with adaxial palisade; *Isachne*-type. *Leaf blade* with distinct, prominent adaxial ribs, or 'nodular' in section, or adaxially flat; with the ribs more or less constant in size. *Midrib* not readily distinguishable; with one bundle only, or having a conventional arc of bundles. Bulliforms present in discrete, regular adaxial groups (these broad, between main bundles); in simple fans. All the vascular bundles accompanied by sclerenchyma. Combined sclerenchyma girders present; forming 'figures'. Sclerenchyma all associated with vascular bundles.

Phytochemistry. Leaves without flavonoid sulphates (1 species).

Cytology. Chromosome base number, $x = 9$. $2n = 18$ and 36.

Taxonomy. Panicoideae; Panicodae; Paniceae.

Ecology, geography, regional floristic distribution. 20 species. Tropical and subtropical America. Mesophytic. Forest margins.

Neotropical. Caribbean, Venezuela and Surinam, Amazon, Central Brazilian, Pampas, and Andean.

Rusts and smuts. Rusts — *Physopella* and *Puccinia*.

References, etc. Morphological/taxonomic: Parodi 1938; Davidse 1978. Leaf anatomical: this project.

Lasiorhachis (Hack.) Stapf

Lasiorrachis Stapf
~ *Saccharum*

Habit, vegetative morphology. Perennial; caespitose. *Culms* 50–100 cm high; herbaceous; unbranched above. Culm nodes exposed; hairy. Culm leaves present. Upper culm leaf blades fully developed. Culm internodes solid. Young shoots extravaginal. *Leaves* mostly basal; non-auriculate (hairy at the auricle positions). Leaf blades linear to linear-lanceolate; harsh; narrow; 5–10 mm wide; flat, or rolled (involute); pseudopetiolate (attenuate to the base, cf. *Imperata cylindrica*); without cross venation; persistent. Ligule an unfringed membrane; not truncate; 0.7–1 mm long. *Contra-ligule* absent.

Reproductive organization. Plants bisexual, with bisexual spikelets; with hermaphrodite florets. The spikelets of sexually distinct forms on the same plant (at least towards the tips of the inflorescence branches); hermaphrodite, male-only, and sterile (in the material seen), or hermaphrodite and male-only (in other collections, according to Stapf 1926); overtly heteromorphic, or overtly heteromorphic to homomorphic; in both homogamous and heterogamous combinations (homogamous below, heterogamous above).

Inflorescence. Inflorescence paniculate (pilose, with white hairs); open; espatheate; not comprising 'partial inflorescences' and foliar organs. **Spikelet-bearing axes** paniculate; *the spikelet-bearing axes* with more than 10 spikelet-bearing 'articles'; with very slender rachides (these not channelled); *disarticulating*; disarticulating at the joints. '*Articles*' linear; without a basal callus-knob; not appendaged; disarticulating transversely; densely long-hairy. Spikelets paired; not secund; sessile and pedicellate; consistently in 'long-and-short' combinations; in pedicellate/sessile combinations. Pedicels of the 'pedicellate' spikelets free of the rachis. The 'shorter' spikelets hermaphrodite. The 'longer' spikelets hermaphrodite, male-only, and sterile (in the material seen, the much reduced members at the tips of the branches), or hermaphrodite and male-only (all or nearly all male, in some specimens).

Female-sterile spikelets. The pedicelled spikelets falling from their pedicels, apparently very variable in sexuality and completeness, hermaphrodite to male or rudimentary, sometimes with two functional male florets in the material seen.

Female-fertile spikelets. Spikelets 4–6 mm long; elliptic to lanceolate; compressed dorsiventrally; planoconvex; falling with the glumes (and in the case of sessile spikelets, with the associated rachis joint and pedicel). Rachilla terminated by a female-fertile floret. Hairy callus present. *Callus* short (hairy in both sessile and pedicellate spikelets); blunt (truncate).

Glumes two; more or less equal; exceeding the spikelets; long relative to the adjacent lemmas; hairy (both pilose); (the upper) pointed; awnless; (the upper) carinate; very dissimilar (both firmly membranous or papery, G_1 bicarinate, G_2 naviculate). Lower glume two-keeled; convex on the back to flattened on the back; relatively smooth; 6–9 nerved. Upper glume 5–7 nerved. *Spikelets* with incomplete florets. The incomplete florets proximal to the female-fertile florets. *The proximal incomplete florets* 1; epaleate; sterile. The proximal lemmas awnless; 2 nerved, or 3 nerved, or 5 nerved; more or less equalling the female-fertile lemmas, or decidedly exceeding the female-fertile lemmas; decidedly firmer than the female-fertile lemmas (membranous with hyaline wings); not becoming indurated.

Female-fertile florets 1. Lemmas broad; less firm than the glumes (membranous or hyaline); not becoming indurated; entire to incised; when incised, 2 lobed; not deeply cleft (no more than minutely incised); mucronate to awned. Awns when present, 1; median; from a sinus, or apical; non-geniculate; straight; hairless; much shorter than the body of the lemma; entered by one vein. Awn bases not twisted; not flattened. Lemmas hairy (along the margins only); non-carinate; without a germination flap; 1 nerved, or 3 nerved. *Palea* present; conspicuous but relatively short; apically notched; awnless; without apical setae (but ciliate); textured like the lemma; not indurated; inconspicuously 2-nerved, or nerveless; keel-less. *Lodicules* present; 2; free; fleshy; ciliate (on their points, with deciduous needlike hairs, in the material seen). *Stamens* 3. Anthers about 2 mm long; minutely penicillate; without an apically prolonged connective. **Ovary** hairy (on the appendage (q.v.), the hairs resembling those on the lodicules); *with a conspicuous apical appendage (all those examined with a rather to very conspicuous, linear, hairy appendage, perhaps representing a vestigial style but originating distinctly lower on the apex than the normal pair)*. Styles free to their bases. Stigmas 2 (the putative vestigial style without stigmatic papillae); red pigmented.

Abaxial leaf blade epidermis. *Costal/intercostal zonation* conspicuous. *Papillae* absent. *Long-cells* markedly different in shape costally and intercostally (the costals narrower and more uniform); of similar wall thickness costally and intercostally (thin walled). Intercostal zones with typical long-cells, or exhibiting many atypical long-cells (isodiametric to square, in places). Mid-intercostal long-cells rectangular (to rhomboid); mostly having markedly sinuous walls. *Microhairs* present (large); elongated; clearly two-celled; panicoid-type. Microhair apical cell/total length ratio about 0.5. *Stomata* common. Subsidiaries non-papillate; high dome-shaped to triangular. Guard-cells overlapping to flush with the interstomatals. *Intercostal short-cells* common; not paired (solitary); not silicified. No macrohairs or prickles seen. Costal zones with short-cells. *Costal short-cells* neither distinctly grouped into long rows nor predominantly paired (pairs or short rows). Costal silica bodies present and well developed (in places, but lacking in others); 'panicoid-type'; short dumb-bell shaped (mostly), or cross shaped.

Transverse section of leaf blade, physiology. C_4. The anatomical organization conventional. XyMS– (the PCR cell walls often thick, mestome sheath-like). PCR sheath outlines uneven. PCR sheath extensions absent. *Mesophyll* with radiate chlorenchyma; traversed by columns of colourless mesophyll cells (in places, associated with the bulliform-plus-colourless cell arches). *Leaf blade* adaxially flat. *Midrib* very conspicuous; having a conventional arc of bundles (these numerous, abaxial); with colourless mesophyll adaxially. *The lamina* symmetrical on either side of the midrib. Bulliforms present in discrete, regular adaxial groups; associated with colourless mesophyll cells to form deeply-penetrating fans; associating with colourless mesophyll cells to form arches over small vascular bundles. Many of the smallest vascular bundles unaccompanied by sclerenchyma. Combined sclerenchyma girders present (with the major bundles); forming 'figures' (large I's). Sclerenchyma all associated with vascular bundles.

Taxonomy. Panicoideae; Andropogonodae; Andropogoneae; Andropogoninae.

Ecology, geography, regional floristic distribution. 1 species. Madagascar. Paleotropical. Madagascan.

References, etc. Leaf anatomical: this project.

Special comments. Description restricted to *L. hildebrandtii* (Hack.) Stapf. Stapf's (1926) dismissal of the hairy ovary apex is untenable, and Bosser's association of two *Miscanthidium* species with *L. hildebrandtii* is unconvincing. The material seen (Hildebrandt 3755) has most of the pedicellate spikelets male, except near the baranch tips. 'Glabrous' lodicules (seen in some of the florets examined, cf. Bosser 1968) may reflect easily dislodged hairs (observed in other florets). Fruit data wanting.

Lasiurus Boiss.

Habit, vegetative morphology. Perennial; caespitose. Culms woody and persistent to herbaceous (woody and persistent at the base). Young shoots intravaginal. Leaves non-auriculate. Leaf blades without cross venation. *Ligule* present; a fringed membrane to a fringe of hairs.

Reproductive organization. Plants bisexual, with bisexual spikelets; with hermaphrodite florets. The spikelets of sexually distinct forms on the same plant; hermaphrodite and male-only, or hermaphrodite and sterile; overtly heteromorphic to homomorphic (the 1 of the pedicelled members not tailed); all in heterogamous combinations.

Inflorescence. *Inflorescence of spicate main branches (of terminal, silky-hairy spiciform racemes)*; open; digitate, or subdigitate, or non-digitate; spatheate; *a complex of 'partial inflorescences' and intervening foliar organs (with spathaceous sheaths)*. Spikelet-bearing axes 'racemes'; solitary; with substantial rachides; *disarticulating*; disarticulating at the joints. *'Articles'* non-linear (stoutly clavate); with a basal callus-knob; appendaged (the appendages suborbicular); disarticulating transversely; densely long-hairy (bearded at the nodes). *Spikelets* paired and in triplets (mostly triplets); not secund (the sessile spikelets alternating); sessile and pedicellate (the pedicels resembling the internodes); *consistently in 'long-and-short' combinations (two sessile and one pedicelled spikelet per triplet)*; in pedicellate/sessile combinations. Pedicels of the 'pedicellate' spikelets free of the rachis. The 'shorter' spikelets hermaphrodite. The 'longer' spikelets male-only, or sterile.

Female-sterile spikelets. The pedicellate spikelets well developed, with tailless G_1.

Female-fertile spikelets. Spikelets compressed dorsiventrally; falling with the glumes. Rachilla terminated by a female-fertile floret. Hairy callus present.

Glumes two; very unequal; (the longer) long relative to the adjacent lemmas; hairy (G_1 densely ciliate); awnless; *very dissimilar (the lower flat-backed, bicarinate above, subleathery and hairy with a short bidentate or flattened apical extension, the upper naviculate and membranous)*. Lower glume flattened on the back; not pitted; relatively smooth. *Spikelets* with incomplete florets. The incomplete florets proximal to the female-fertile florets. *The proximal incomplete florets* 1; paleate. Palea of the proximal incomplete florets fully developed. *The proximal incomplete florets male*. The proximal lemmas awnless; 3 nerved; similar in texture to the female-fertile lemmas; not becoming indurated.

Female-fertile florets 1. **Lemmas** less firm than the glumes (hyaline); not becoming indurated; entire; pointed; awnless; hairless; *carinate*; 3 nerved. *Palea* present; relatively long; entire; awnless, without apical setae; not indurated (hyaline); 2-nerved. *Lodicules* present; 2; free; fleshy; glabrous. *Stamens* 3. *Ovary* glabrous. Styles free to their bases. Stigmas 2.

Fruit, embryo and seedling. *Fruit* free from both lemma and palea; compressed dorsiventrally. Hilum short. Embryo large.

Abaxial leaf blade epidermis. *Costal/intercostal zonation* conspicuous. *Papillae* absent. *Long-cells* similar in shape costally and intercostally to markedly different in shape costally and intercostally (the costals bulbous between the silica cells, elsewhere more regularly rectangular); of similar wall thickness costally and intercostally to differing markedly in wall thickness costally and intercostally (thick walled, the costals tending to be thicker). Intercostal zones with typical long-cells, or exhibiting many atypical long-cells (in places). Mid-intercostal long-cells rectangular; having markedly sinuous walls (with conspicuous pitting, including in the outer wall). *Microhairs* present, or absent (seemingly, in the material seen); when recorded, panicoid-type. *Stomata* common. Subsidiaries non-papillate; triangular. Guard-cells overlapping to flush with the interstomatals. *Intercostal short-cells* common; not paired (mostly, seemingly solitary), or in cork/silica-cell pairs (sometimes, adjoining the costal zones); silicified (in places), or not silicified (mainly). The costal zones with small to large prickles, grading into macrohairs. *Costal short-cells* predominantly paired (mostly), or neither distinctly grouped into long rows nor predominantly paired (in places, solitary or in short rows). Costal silica bodies present and well developed; more or less 'panicoid-type', or rounded; mostly cross shaped (but with only slight furrowing).

Transverse section of leaf blade, physiology. C_4. The anatomical organization conventional. XyMS−. PCR sheath extensions absent. Mesophyll traversed by columns of colourless mesophyll cells (perhaps, in places), or not traversed by colourless columns. *Leaf blade* with distinct, prominent adaxial ribs; with the ribs more or less constant in size (low, wide). *Midrib* very conspicuous; having a conventional arc of bundles (three primaries, separated by minor bundles); with colourless mesophyll adaxially (this extending outwards beneath the bulliform 'hinges'). *The lamina* symmetrical on either side of the midrib. Bulliforms present in discrete, regular adaxial groups (with a large group in each furrow, overlying small vascular bundles); associated with internal colourless tissue, the latter thinner walled and extending inwards asymmetrically; associating with colourless mesophyll cells to form arches over small vascular bundles. Many of the smallest vascular bundles unaccompanied by sclerenchyma (most but not all of the smallest bundles being associated with a small sclerenchyma group). Combined sclerenchyma girders present to absent (perhaps present with a few primaries, but in most cases there is only an adaxial strand, with colourless tissue intervening). Sclerenchyma all associated with vascular bundles.

Phytochemistry. Leaves without flavonoid sulphates (1 species).

Cytology. Chromosome base number, $x = 7$, or 9. $2n = 18$ and 56. 2 and 6 ploid.

Taxonomy. Panicoideae; Andropogonodae; Andropogoneae; Rottboelliinae.

Ecology, geography, regional floristic distribution. 3 species. Tropical East Africa, India. Xerophytic; species of open habitats. Subdesert.

Holarctic and Paleotropical. Tethyan. African and Indomalesian. Irano-Turanian. Saharo-Sindian and Sudano-Angolan. Indian. Sahelo-Sudanian and Somalo-Ethiopian.

Rusts and smuts. Smuts from Ustilaginaceae. Ustilaginaceae — *Sorosporium*.

Economic importance. Important native pasture species: *L. hirsutus*, in arid conditions.

References, etc. Leaf anatomical: Metcalfe 1960; this project.

Lecomtella A. Camus

Habit, vegetative morphology. Bamboo-like perennial. *Culms* about 50–200 cm high; herbaceous; branched above. Culm leaves present. Upper culm leaf blades fully developed. Culm internodes hollow. Plants unarmed. *Leaves* not basally aggregated; non-auriculate; without auricular setae. Leaf blades lanceolate (-acuminate); broad; 10–20 mm wide (12–22 cm long); without cross venation. ***Ligule a fringe of hairs***. *Contra-ligule* present (as a fringe of hairs).

Reproductive organization. *Plants bisexual, with bisexual spikelets; without hermaphrodite florets (the upper floret of the hermaphrodite spikelet female-only)*. The spikelets of sexually distinct forms on the same plant; hermaphrodite and male-only. The male and female-fertile spikelets segregated, in different parts of the same inflorescence branch (the short branches with the spikelets male-only below, hermaphrodite above). The spikelets homomorphic.

Inflorescence. Inflorescence paniculate; contracted; more or less irregular; espatheate; not comprising 'partial inflorescences' and foliar organs. Spikelet-bearing axes with very slender rachides; persistent. Spikelets not secund; pedicellate. Pedicel apices cupuliform. Spikelets imbricate.

Female-sterile spikelets. Male spikelets similar to the hermaphrodite, but both florets male with membranous lemmas. Rachilla of male spikelets terminated by a male floret. The male spikelets with glumes; without proximal incomplete florets; 2 floreted (both male). The lemmas awnless. Male florets 2; 3 staminate.

Female-fertile spikelets. *Spikelets* 10 mm long (or more); weakly *compressed laterally*; falling with the glumes; disarticulating between the florets; *with distinctly elongated rachilla internodes between the florets (the upper floret borne on a short stipe with two apical wings)*. The upper floret conspicuously stipitate. The stipe beneath the upper floret not filiform; straight and swollen; homogeneous. Rachilla terminated by a female-fertile floret; rachilla hairless. Hairy callus absent. *Callus* absent.

Glumes two; very unequal to more or less equal; shorter than the spikelets; (the longer) shorter than the adjacent lemmas; hairless; glabrous; pointed (acuminate); awnless; carinate; similar (herbaceous, ovate-lanceolate). *Lower glume* about 0.7 times the length of the upper glume; 3 nerved. Upper glume 7 nerved. *Spikelets* with incomplete florets. The incomplete florets proximal to the female-fertile florets. *Spikelets with proximal incomplete florets*. *The proximal incomplete florets* 1; paleate. Palea of the proximal incomplete florets fully developed (hyaline, two keeled). The proximal incomplete florets male (with 3 stamens). The proximal lemmas lanceolate; awnless; 7 nerved, or 9 nerved; decidedly exceeding the female-fertile lemmas (equalling the spikelet); less firm than the female-fertile lemmas (herbaceous); not becoming indurated.

Female-fertile florets 1 (female-only). *Lemmas* ovate; *decidedly firmer than the glumes (crustaceous, tuberculate at the tip, with 3 thickened cushions apically on the nerves)*; becoming indurated; entire; blunt; awnless; hairy; non-carinate; having the margins lying flat and exposed on the palea; with a clear germination flap; 5 nerved; with the nerves non-confluent. *Palea* present; relatively long; tightly clasped by the lemma; entire (tuberculate at the tip, with thickened cushions apically on the nerves); awnless, without apical setae; textured like the lemma; thickened apically; 2-nerved; keel-less (with incurved margins). *Lodicules* present; 2; free; fleshy (quadrangular); glabrous; not or scarcely vascularized. *Ovary* glabrous. Styles free to their bases. Stigmas 2 (very long, slender).

Fruit, embryo and seedling. Pericarp thin.

Abaxial leaf blade epidermis. *Costal/intercostal zonation* conspicuous. *Papillae* absent. Long-cells of similar wall thickness costally and intercostally (of medium thickness). Intercostal zones exhibiting many atypical long-cells (these and the interstomatals frequently very short). Mid-intercostal long-cells rectangular; having markedly sinuous walls (conspicuously pitted). *Microhairs* seemingly absent (but these would be very hard to detect amid the nemerous prickles). *Stomata* common (unusually abundant). Subsidiaries non-papillate; high dome-shaped and triangular. Guard-cells overlapping to flush with the interstomatals. *Intercostal short-cells* common; not paired (solitary, small); not silicified. With long, almost continuous chains of large, bulbous prickles costally, and with vey numerous smaller to tiny prickles intercostally; no macrohairs seen. Costal zones with short-cells. *Costal short-cells* predominantly paired (in places), or neither distinctly grouped into long rows nor predominantly paired (the pairs and short rows separated by prickles). Costal silica bodies present and well developed (but relatively infrequent); 'panicoid-type'; very shortly dumb-bell shaped, or cross shaped.

Transverse section of leaf blade, physiology. C_3; XyMS+. *Mesophyll* with non-radiate chlorenchyma; without adaxial palisade; without 'circular cells'. *Leaf blade* 'nodular' in section; with the ribs more or less constant in size. *Midrib* conspicuous to not readily distinguishable (perhaps represented by a slightly enlarged primary bundle); with one bundle only. Bulliforms present in discrete, regular adaxial groups; in simple fans (these large, in each furrow). All the vascular bundles accompanied by sclerenchyma. Combined sclerenchyma girders present; forming 'figures' (all the bundles with I's). Sclerenchyma all associated with vascular bundles.

Taxonomy. Panicoideae; Panicodae; Paniceae (Boivinelleae).

Ecology, geography, regional floristic distribution. 1 species. Madagascar. Mesophytic. Forest margins.

Paleotropical. Madagascan.

References, etc. Morphological/taxonomic: Camus 1925b. Leaf anatomical: this project.

Special comments. Anatomical data wanting.

Leersia Soland.

Aplexia Faf., *Asprella* Schreb., *Blepharochloa* Endl., *Ehrhartia* Weber, *Endodia* Raf., *Homalocenchrus* Mieg, *Laertia* Gromov, *Pseudoryza* Griff., *Turraya* Wall.

Habit, vegetative morphology. Annual (rarely), or perennial; rhizomatous, or stoloniferous, or caespitose. *Culms* 30–150 cm high; herbaceous. Culm nodes hairy, or glabrous. Culm internodes solid, or hollow. Young shoots intravaginal. *Leaves* not basally aggregated; non-auriculate; without auricular setae. *Leaf blades* linear; *narrow*; flat, or folded, or rolled; without cross venation; persistent; rolled in bud. Ligule an unfringed membrane; truncate. *Contra-ligule* absent.

Reproductive organization. *Plants bisexual, with bisexual spikelets*; with hermaphrodite florets. The spikelets all alike in sexuality. Exposed-cleistogamous, or chasmogamous. Plants with hidden cleistogenes, or without hidden cleistogenes. The hidden cleistogenes when present, in the leaf sheaths.

Inflorescence. Inflorescence paniculate, or of spicate main branches (the primary branches sometimes simple); open; with capillary branchlets, or without capillary branchlets; non-digitate; espatheate; not comprising 'partial inflorescences' and foliar organs. Spikelet-bearing axes persistent. Spikelets solitary; secund, or not secund; pedicellate.

Female-fertile spikelets. Spikelets unconventional (the dilated pedicel apices may or may not represent 'glumes', the palea may 'really' be a lemma, etc. — c.f. *Oryza*); 3–6 mm long; strongly *compressed laterally*; disarticulating above the glumes (or at least, above the rim assumed to represent them); *with no more than a short stipe beneath the floret, by contrast with* **Chikusichloa**. Rachilla terminated by a female-fertile floret. Hairy callus absent.

Glumes *absent (apparently represented by a narrow rim at the tip of the pedicel)*. Spikelets with female-fertile florets only; *without proximal incomplete florets*.

Female-fertile florets *1*. Lemmas not becoming indurated (membranous to papyraceous); awnless to awned (often caudate). Awns when present, 1; median; apical; non-geniculate; much shorter than the body of the lemma to much longer than the body of the lemma. *Lemmas* usually *hairy (hispid on the nerves)*; carinate; without a germination flap; 3–5 nerved. *Palea* present; relatively long (but much narrower than the lemma); entire; awnless, without apical setae; tough; several nerved (3); one-keeled. *Lodicules* present; 2; free; fleshy, or membranous; glabrous; not toothed; heavily vascularized. *Stamens* 1–3, or 6. Anthers 0.5–2.4 mm long; not penicillate; without an apically prolonged connective. *Ovary* glabrous. Styles free to their bases. Stigmas 2; white.

Fruit, embryo and seedling. *Fruit* free from both lemma and palea; small; compressed laterally. Hilum long-linear. Embryo small; not waisted. Endosperm hard; without lipid; containing compound starch grains. Embryo with an epiblast; without a scutellar tail; with a negligible mesocotyl internode. Embryonic leaf margins overlapping.

Seedling with a short mesocotyl, or with a long mesocotyl. First seedling leaf without a lamina.

Abaxial leaf blade epidermis. *Costal/intercostal zonation* conspicuous. *Papillae* present; costal and intercostal. Intercostal papillae over-arching the stomata; several per cell. Mid-intercostal long-cells rectangular; having markedly sinuous walls. *Microhairs* present; panicoid-type; 22–30 microns long (*L. hexandra*), or 32–42 microns long (*L. oryzoides*); (3.9–)4.2–5.7 microns wide at the septum. Microhair total length/width at septum 4.7–7.1. Microhair apical cells 10–16 microns long (*L. hexandra*), or 16–24 microns long (*L. oryzoides*). Microhair apical cell/total length ratio 0.52–0.54. *Stomata* common; in *L. hexandra*, 22–30 microns long. Subsidiaries papillate; triangular. Guard-cells overlapping to flush with the interstomatals. *Intercostal short-cells* common, or absent or very rare; silicified, or not silicified. Intercostal silica bodies where present, crescentic, or tall-and-narrow, or oryzoid-type, or cross-shaped, or oryzoid-type. *Costal short-cells* conspicuously in long rows. Costal silica bodies oryzoid; not sharp-pointed.

Transverse section of leaf blade, physiology. C_3; XyMS+. *Mesophyll* with non-radiate chlorenchyma; traversed by columns of colourless mesophyll cells, or not traversed by colourless columns; with arm cells, or without arm cells; without fusoids. *Leaf blade* with distinct, prominent adaxial ribs; with the ribs more or less constant in size. *Midrib* conspicuous, or not readily distinguishable; with one bundle only, or having complex vascularization (rarely); with colourless mesophyll adaxially (rarely), or without colourless mesophyll adaxially. Bulliforms present in discrete, regular adaxial groups (in the furrows); in simple fans (*L. hexandra*), or associated with colourless mesophyll cells to form deeply-penetrating fans (*L. oryzoides*). All the vascular bundles accompanied by sclerenchyma. Combined sclerenchyma girders present; forming 'figures'. Sclerenchyma all associated with vascular bundles.

Culm anatomy. *Culm internode bundles* in one or two rings.

Phytochemistry. Leaves without flavonoid sulphates (1 species).

Cytology. Chromosome base number, $x = 12$. $2n = 24, 48,$ and 60. 2, 4, and 5 ploid.

Taxonomy. Bambusoideae; Oryzodae; Oryzeae.

Ecology, geography, regional floristic distribution. 18 species. Tropical and warm temperate. Commonly adventive. Helophytic; shade species and species of open habitats.

Holarctic, Paleotropical, Neotropical, and Australian. Boreal, Tethyan, and Madrean. African, Madagascan, and Indomalesian. Arctic and Subarctic, Euro-Siberian, Eastern Asian, Atlantic North American, and Rocky Mountains. Macaronesian, Mediterranean, and Irano-Turanian. Saharo-Sindian, Sudano-Angolan, West African Rainforest, and Namib-Karoo. Indian, Indo-Chinese, Malesian, and Papuan. Caribbean, Venezuela and Surinam, Central Brazilian, Pampas, and Andean. North and East Australian. European. Canadian-Appalachian, Southern Atlantic North American, and Central Grasslands. Sahelo-Sudanian, Somalo-Ethiopian, South Tropical African, and Kalaharian. Tropical North and East Australian and Temperate and South-Eastern Australian.

Rusts and smuts. Rusts — *Puccinia*. Taxonomically wide-ranging species: *Puccinia graminis*, *Puccinia striiformis*, and *Puccinia recondita*. Smuts from Tilletiaceae and from Ustilaginaceae. Tilletiaceae — *Tilletia*. Ustilaginaceae — *Sorosporium*, *Tolyposporium*, and *Ustilago*.

Economic importance. Significant weed species: *L. hexandra* (in rice), *L. oryzoides*. Important native pasture species: *L. hexandra*.

References, etc. Leaf anatomical: Metcalfe 1960; this project.

Lepargochloa Launert

~ *Loxodera* (*L. rhytachnoides*

Habit, vegetative morphology. Perennial; decumbent. *Culms* about 20–90 cm high; herbaceous; unbranched above. Culm nodes hairy. *Leaves* not basally aggregated; non-auriculate. Leaf blades narrowly linear; narrow; to 6 mm wide; flat, or rolled (involute); without cross venation. *Ligule an unfringed membrane*; 2 mm long.

Reproductive organization. Plants bisexual, with bisexual spikelets; with hermaphrodite florets. The spikelets homomorphic.

Inflorescence. *Inflorescence* a terminal, spiciform 'raceme'; *non-digitate*; espatheate (seemingly); not comprising 'partial inflorescences' and foliar organs. **Spikelet-bearing axes** 'racemes' and spikelike; solitary; with substantial rachides; *persistent (seemingly)*. '*Articles*' non-linear (subclavate); not appendaged (but the tips bristly-hairy); transversely jointed; densely long-hairy. Spikelets paired; sessile and pedicellate; consistently in 'long-and-short' combinations; in pedicellate/sessile combinations. Pedicels of the 'pedicellate' spikelets free of the rachis. The 'shorter' spikelets hermaphrodite. The 'longer' spikelets hermaphrodite.

Female-fertile spikelets. Spikelets 7–9 mm long; compressed dorsiventrally; falling with the glumes (presumably?). Rachilla terminated by a female-fertile floret. Hairy callus present. *Callus* short.

Glumes two; more or less equal; long relative to the adjacent lemmas; hairy (G_1 with setaceous hairs, G_2 ciliate); **with distinct hair tufts (the setaceous hairs of the G_1 in tufts)**; awnless; very dissimilar (G_1 leathery, rugose, convex to flattened on the back, G_2 thinner and boat-shaped). Lower glume not two-keeled; convex on the back to flattened on the back; not pitted; rugose to tuberculate; 7 nerved. Upper glume 3 nerved. *Spikelets* with incomplete florets. The incomplete florets proximal to the female-fertile florets. *Spikelets with proximal incomplete florets*. The proximal incomplete florets 1; paleate; male. The proximal lemmas awnless; 2(–3) nerved; decidedly exceeding the female-fertile lemmas; similar in texture to the female-fertile lemmas (thinly membranous); not becoming indurated.

Female-fertile florets 1. Lemmas ovate-elliptic; less firm than the glumes (thinly membranous); not becoming indurated; entire; pointed, or blunt; awnless; hairless (with ciliate margins); non-carinate; without a germination flap; 3 nerved. *Palea* present; relatively long; entire (acute or sub-acute); awnless, without apical setae; not indurated; 2-nerved; 2-keeled. *Lodicules* present; 2; free; fleshy; glabrous. *Stamens* 3. Anthers not penicillate; without an apically prolonged connective. *Ovary* glabrous. Stigmas 2.

Taxonomy. Panicoideae; Andropogonodae; Andropogoneae; Rottboelliinae.

Ecology, geography, regional floristic distribution. 1 species. Tropical southern Africa.

Paleotropical. African. Sudano-Angolan. South Tropical African.

References, etc. Morphological/taxonomic: Launert 1961.

Special comments. Fruit data wanting. Anatomical data wanting.

Leptagrostis C.E. Hubb.

Habit, vegetative morphology. Perennial; caespitose. *Culms* 22 cm high (slender); herbaceous; unbranched above. Plants unarmed. Leaves non-auriculate. Leaf blades narrow; 2–6 mm wide (10–25 cm long); setaceous at the tip; without cross venation. *Ligule a fringe of hairs*.

Reproductive organization. Plants bisexual, with bisexual spikelets; with hermaphrodite florets.

Inflorescence. *Inflorescence paniculate*; contracted (narrow, 5–12.5 cm long); **with capillary branchlets**; espatheate; not comprising 'partial inflorescences' and foliar organs. Spikelet-bearing axes persistent. Spikelets not secund; shortly pedicellate.

Female-fertile spikelets. Spikelets 3.5–4 mm long; purple-tinged; compressed laterally; disarticulating above the glumes. Rachilla prolonged beyond the uppermost female-fertile floret; rachilla hairy, or hairless. The rachilla extension naked. **Hairy callus present (the white hairs to 3.5 mm long, half as long as the lemma)**. Callus blunt.

Glumes two; very unequal (G_1 shorter); (the upper) long relative to the adjacent lemmas (the lower half as long); hairless; pointed; awnless; keeled above; similar (lanceolate, acuminate, thinly membranous). Lower glume 1 nerved. *Upper glume 3 nerved*. Spikelets with female-fertile florets only; **without proximal incomplete florets**.

Female-fertile florets 1. *Lemmas* broadly lanceolate, flattened; similar in texture to the glumes (thinly membranous); not becoming indurated; entire, or incised; when entire, pointed (acuminate into the awn); when incised 2 lobed; when incised, not deeply cleft (bidenticulate); *awned*. Awns 1; median; from a sinus (i.e., from between minute teeth), or apical; non-geniculate; straight; much shorter than the body of the lemma (little more than 1 mm long); entered by one vein. Lemmas hairless; glabrous; non-carinate (dorsally rounded); 3–5 nerved. *Palea* present; relatively long; awnless, without apical setae (apically ciliolate); not indurated (membranous); 2-nerved. **Lodicules** present; 2; free;

membranous; glabrous. *Stamens* 3. Anthers 1.5 mm long. *Ovary* glabrous. Styles free to their bases. Stigmas 2.

Taxonomy. Pooideae (?); Poodae (?); Aveneae (?).

Ecology, geography, regional floristic distribution. 1 species. Abyssinia. Paleotropical. African. Sudano-Angolan. Somalo-Ethiopian.

References, etc. Morphological/taxonomic: Hubbard 1939b.

Special comments. Description hopelessly inadequate for reliable classificatory assignment. Fruit data wanting. Anatomical data wanting.

Leptaspis R.Br.

Excluding *Scrotochloa*

Habit, vegetative morphology. Perennial; rhizomatous. The flowering culms leafy. *Culms* 30–100 cm high; herbaceous. Culm nodes glabrous. Culm internodes solid. *Leaves* not basally aggregated; non-auriculate; without auricular setae. Leaf blades broad; pseudopetiolate; pinnately veined (the laterals slanting obliquely from the midrib); cross veined; persistent; rolled in bud. **Ligule** present (short); *a fringed membrane. Contraligule* absent.

Reproductive organization. Plants monoecious with all the fertile spikelets unisexual; without hermaphrodite florets. The spikelets of sexually distinct forms on the same plant; female-only and male-only. The male and female-fertile spikelets mixed in the inflorescence. The spikelets overtly heteromorphic (the male spikelets smaller).

Inflorescence. Inflorescence paniculate; open (with 1–3 primary branches at each of (1–)3–7 nodes). Rachides hollowed, or flattened, or winged, or neither flattened nor hollowed, not winged. Inflorescence spatheate (the spikelet branchlets often subtended by bracts), or espatheate; not comprising 'partial inflorescences' and foliar organs. Spikelet-bearing axes persistent. Spikelets not secund; pedicellate (the female pedicels filiform, not clavate); consistently in 'long-and-short' combinations. The 'shorter' spikelets female-only. The 'longer' spikelets male-only.

Female-sterile spikelets. Male spikelets smaller than the females, narrower, the floret with 6 stamens, the anthers non-penicillate, the floret caducous. The male spikelets with glumes; 1 floreted. The lemmas awnless, or mucronate (?). Male florets 6 staminate.

Female-fertile spikelets. Spikelets compressed dorsiventrally; disarticulating above the glumes. Rachilla terminated by a female-fertile floret; rachilla hairless. Hairy callus absent.

Glumes two, or several (often 3); minute, or relatively large; more or less equal; shorter than the adjacent lemmas; pointed (cuspidate); awnless; similar (herbaceous, persistent). Lower glume 1–3 nerved. Upper glume 1–3 nerved.

Female-fertile florets 1. *Lemmas inflated, sacciform, cochleate, their margins joined to form utricles which are closed save for a lateral pore from which the stigmas emerge*; decidedly firmer than the glumes; becoming indurated; awnless, or mucronate (?); hairy; non-carinate; 9 nerved. *Palea* present (enclosed within the lemma, linear); apically notched (bidentate); indurated (when mature); 2-nerved; 2-keeled. *Lodicules* present (but small), or absent; when present, 2, or 3; free; membranous; glabrous; not toothed. *Stamens* 0 (and no staminodes). *Ovary* glabrous. Styles fused (into one). Stigmas 3.

Fruit, embryo and seedling. *Fruit* free from both lemma and palea. Pericarp fused. Endosperm hard.

Abaxial leaf blade epidermis. *Costal/intercostal zonation* conspicuous. *Papillae* absent. *Long-cells* markedly different in shape costally and intercostally (the costals much narrower); of similar wall thickness costally and intercostally. Mid-intercostal long-cells rectangular; having markedly sinuous walls. *Microhairs* absent. *Stomata* absent or very rare (in places), or common. Subsidiaries tall dome-shaped to triangular. Guard-cells overlapped by the interstomatals to overlapping to flush with the interstomatals. *Intercostal short-cells* absent or very rare. *Costal short-cells* conspicuously in long rows. Costal silica bodies small, horizontally-elongated smooth (a few, almost), or 'panicoid-type' (predominantly, but much reduced); seemingly 'smoothed out' dumb-bell shaped and nodular; not sharp-pointed.

Transverse section of leaf blade, physiology. C_3; XyMS+. Mesophyll with adaxial palisade (of cubical cells and an abaxial one of tall cells); traversed by columns of colourless mesophyll cells; without arm cells; with fusoids. The fusoids external to the PBS. *Leaf blade* 'nodular' in section; with the ribs more or less constant in size. *Midrib* conspicuous; having complex vascularization; with colourless mesophyll adaxially. Bulliforms not in discrete, regular adaxial groups (ill defined or absent). All the vascular bundles accompanied by sclerenchyma. Combined sclerenchyma girders present; forming 'figures'. Sclerenchyma all associated with vascular bundles.

Special diagnostic feature. *Having female spikelets, with shell- or urn-shaped lemmas which are closed save for an apical pore*.

Cytology. Chromosome base number, $x = 12$. $2n = 24$. 2 ploid.

Taxonomy. Bambusoideae; Oryzodae; Phareae.

Ecology, geography, regional floristic distribution. 5 species. Tropical west Africa, Mascarene Is., Ceylon, Fiji, New Guinea. Shade species. In forest.

Paleotropical and Australian. African, Madagascan, Indomalesian, Polynesian, and Neocaledonian. Saharo-Sindian, Sudano-Angolan, and West African Rainforest. Indian, Indo-Chinese, Malesian, and Papuan. Fijian. North and East Australian. Sahelo-Sudanian, Somalo-Ethiopian, and South Tropical African. Tropical North and East Australian.

References, etc. Morphological/taxonomic: Judziewicz 1984. Leaf anatomical: Metcalfe 1960; this project.

Special comments. Fruit data wanting.

Leptocarydion Stapf

Habit, vegetative morphology. *Annual*; loosely caespitose, or decumbent (rarely, rooting at the nodes). *Culms* 13–130 cm high; herbaceous; branched above, or unbranched above. Culm nodes glabrous. Culm internodes solid. Plants unarmed. Young shoots intravaginal. *Leaves* not basally aggregated; non-auriculate, or auriculate (at the base of the blade). *Leaf blades lanceolate to ovate*; *broad*; 4–15 mm wide; cordate, or not cordate, not sagittate; flat, or rolled; without abaxial multicellular glands; without cross venation; persistent. Ligule a fringed membrane; truncate; 0.3–0.4 mm long. *Contra-ligule* absent.

Reproductive organization. Plants bisexual, with bisexual spikelets; with hermaphrodite florets.

Inflorescence. *Inflorescence of spicate main branches*; contracted (to about 20 cm long, the thin spicate laterals appressed). Primary inflorescence branches 10–20 (or more —i.e. many). Rachides neither flattened nor hollowed, not winged (very slender). Inflorescence espatheate; not comprising 'partial inflorescences' and foliar organs. The racemes spikelet bearing to the base (or nearly so). Spikelet-bearing axes persistent. Spikelets solitary; secund; biseriate; subsessile.

Female-fertile spikelets. Spikelets 5–11 mm long; lanceolate; compressed laterally; disarticulating above the glumes; disarticulating between the florets; with conventional internode spacings. Rachilla prolonged beyond the uppermost female-fertile floret; rachilla hairy. The rachilla extension with incomplete florets. Hairy callus present (slender). *Callus* short; pointed.

Glumes two; relatively large; very unequal; shorter than the spikelets; (the upper) long relative to the adjacent lemmas; hairless; scabrous (on the keels); pointed (G_1 acuminate-mucronate); awnless; carinate; similar (reddish, subhyaline, very narrow). *Lower glume* 0.5–0.75 times the length of the upper glume; shorter than the lowest lemma; 1 nerved. Upper glume 1 nerved. *Spikelets* with incomplete florets. The incomplete florets distal to the female-fertile florets. *The distal incomplete florets* merely underdeveloped; awned.

Female-fertile florets *6–12*. Lemmas similar in texture to the glumes to decidedly firmer than the glumes (thin); smooth; not becoming indurated; incised; 4 lobed (minutely 4 toothed); not deeply cleft; awned. Awns 1; median (the midnerve excurrent); from a sinus; non-geniculate (very slender); hairless (scabrid); about as long as the body of the lemma to much longer than the body of the lemma; entered by one vein. Lemmas hairy (pubescent below the middle, the edges ciliate); carinate; without a germination flap; 3 nerved; with the nerves non-confluent. *Palea* present (linear-oblong); relatively long, or conspicuous but relatively short; entire; awnless, without apical setae; textured like the lemma; not indurated (thin); 2-nerved; 2-keeled. Palea keels wingless; hairy. *Lodicules* present; 2; free; fleshy; glabrous; not toothed; not or scarcely vascularized. *Stamens* 2, or 3. Anthers 0.4 mm long; not penicillate; without an apically prolonged connective. *Ovary* glabrous. Styles free to their bases. Stigmas 2; brown.

Fruit, embryo and seedling. *Fruit* free from both lemma and palea (tightly embraced); small (to 1 mm long); linear; obtusely triquetrous. Hilum short. Pericarp fused. Embryo large.

Abaxial leaf blade epidermis. *Costal/intercostal zonation* conspicuous. *Papillae* present; costal and intercostal (more conspicuous intercostally). Intercostal papillae overarching the stomata; several per cell. Mid-intercostal long-cells rectangular; having markedly sinuous walls (coarsely so). *Microhairs* present; elongated; clearly two-celled; chloridoid-type. Microhair apical cell wall of similar thickness/rigidity to that of the basal cell. Microhairs 18–20 microns long. Microhair basal cells 12–14 microns long. Microhair apical cells 4–8 microns long. Microhair apical cell/total length ratio 0.3. *Stomata* common. Subsidiaries high dome-shaped to triangular. *Intercostal short-cells* common; mostly not paired (but with a few pairs and threes); not silicified. Intercostal silica bodies absent. *Costal short-cells* conspicuously in long rows (the files often interrupted by prickles). Costal silica bodies present in alternate cell files of the costal zones; 'panicoid-type'; mostly dumb-bell shaped (sometimes short); not sharp-pointed.

Transverse section of leaf blade, physiology. C_4; XyMS+. PCR sheath outlines even. PCR sheaths of the primary vascular bundles interrupted; interrupted abaxially only. PCR sheath extensions absent. PCR cell chloroplasts centripetal. *Mesophyll* with radiate chlorenchyma. *Leaf blade* adaxially flat. *Midrib* not readily distinguishable; with one bundle only. Bulliforms present in discrete, regular adaxial groups; in simple fans (mostly), or associated with colourless mesophyll cells to form deeply-penetrating fans (a few only, in the material seen). All the vascular bundles accompanied by sclerenchyma. Combined sclerenchyma girders present (with most bundles). Sclerenchyma all associated with vascular bundles. The lamina margins without fibres.

Taxonomy. Chloridoideae; main chloridoid assemblage.

Ecology, geography, regional floristic distribution. 1 species. East and southern Africa. Mesophytic.

Paleotropical. African and Madagascan. Sudano-Angolan and Namib-Karoo. Somalo-Ethiopian, South Tropical African, and Kalaharian.

Economic importance. Important native pasture species: *L. vulpiastrum*.

References, etc. Leaf anatomical: Metcalfe 1960; this project; photos provided by R.P. Ellis.

Leptochloa P. Beauv.

Anoplia Steud., *Baldomiria* Herter, *Diachroa* Nutt, *Diacisperma* Kuntze, *Disakisperma* Steud., *Ipnum* Phil., *Leptostachys* Meyer, *Oxydenia* Nutt, *Rabdochloa* P. Beauv.

Excluding *Diplachne*, *Kengia*

Habit, vegetative morphology. Annual, or perennial; rhizomatous, or stoloniferous, or caespitose, or decumbent. Culms woody and persistent, or herbaceous; branched above, or unbranched above; tuberous, or not tuberous. Culm nodes glabrous. Culm internodes solid, or hollow. *Leaves* mostly basal, or not basally aggregated; non-auriculate. *Leaf blades* linear, or linear-lanceolate; narrow; flat, or rolled; without abaxial multicellular glands; *not pseudopetiolate*; without cross venation; persistent; rolled in bud. **Ligule** present; *an unfringed membrane to a fringed membrane*.

Reproductive organization. Plants bisexual, with bisexual spikelets; with hermaphrodite florets. The spikelets all alike in sexuality. Exposed-cleistogamous, or chasmogamous. *Plants without hidden cleistogenes*.

Inflorescence. *Inflorescence of spicate main branches (spiciform racemes)*; *open*; non-digitate (the racemes often whorled), or subdigitate. Rachides hollowed, or flattened, or neither flattened nor hollowed, not winged. Inflorescence espatheate; not comprising 'partial inflorescences' and foliar organs. The racemes spikelet bearing to the base. **Spikelet-bearing axes** with very slender rachides; *persistent*. Spikelets solitary; clearly secund; shortly pedicellate; usually imbricate; not in distinct 'long-and-short' combinations.

Female-fertile spikelets. Spikelets 1–5 mm long (rarely up to 7 mm); adaxial; compressed laterally to not noticeably compressed; disarticulating above the glumes; disarticulating between the florets. Rachilla prolonged beyond the uppermost female-fertile floret; rachilla hairless. Hairy callus present, or absent. *Callus* short (minute); blunt.

Glumes two; very unequal to more or less equal; shorter than the spikelets; usually *shorter than the adjacent lemmas*; free (narrow, membranous); dorsiventral to the rachis (the lower often asymmetric); hairless; pointed; awnless; carinate. Lower glume shorter than the lowest lemma; 1 nerved. **Upper glume 1 nerved**. *Spikelets* with female-fertile florets only, or with incomplete florets. The incomplete florets distal to the female-fertile florets. The distal incomplete florets awnless. **Spikelets without proximal incomplete florets**.

Female-fertile florets *(1–)3–6*. Lemmas less firm than the glumes to similar in texture to the glumes (membranous to hyaline); not becoming indurated; entire (rarely), or incised; when entire, blunt; when incised, 2 lobed (bidentate); when incised, not deeply cleft; awnless, or mucronate, or awned. *Awns* when present, 1; from a sinus; non-geniculate; *much shorter than the body of the lemma*. Lemmas hairy (with appressed hairs on the lateral nerves), or hairless; usually more or less *carinate*; 3 nerved. *Palea* present; entire to apically notched; awnless, without apical setae; not indurated (membranous); 2-nerved; 2-keeled (often glabrous between the keels, but hairy outside them). *Lodicules* present; 2; free; fleshy. *Stamens* 2–3. Anthers not penicillate. *Ovary* glabrous. Styles free to their bases. Stigmas 2.

Fruit, embryo and seedling. Fruit adhering to lemma and/or palea, or free from both lemma and palea; small (0.5–2 mm long); *compressed laterally, or compressed dorsiventrally*; smooth (or striolate). **Hilum short**. Pericarp free, or loosely adherent, or fused. Embryo large; not waisted. Endosperm hard; without lipid; containing compound starch grains. Embryo with an epiblast; with a scutellar tail; with an elongated mesocotyl internode. Embryonic leaf margins meeting.

Seedling with a long mesocotyl. First seedling leaf with a well-developed lamina. The lamina broad; curved; 3–5 veined.

Abaxial leaf blade epidermis. *Costal/intercostal zonation* conspicuous. *Papillae* nearly always present; intercostal. Intercostal papillae over-arching the stomata, or not over-arching the stomata; consisting of one oblique swelling per cell, or consisting of one symmetrical projection per cell. *Long-cells* similar in shape costally and intercostally; differing markedly in wall thickness costally and intercostally (thicker *intercostally*). Intercostal zones with typical long-cells, or exhibiting many atypical long-cells (many long-cells very short). Mid-intercostal long-cells rectangular; having markedly sinuous walls. *Microhairs* present; more or less spherical to elongated; nearly always clearly two-celled, or ostensibly one-celled (rarely); chloridoid-type (of various forms, but very rarely of the ostensibly one-celled type). Microhair apical cell wall of similar thickness/rigidity to that of the basal cell. Microhairs with 'partitioning membranes' (in *L. digitata*). The 'partitioning membranes' in the basal cell. Microhairs (13–)19.5–36(–42) microns long. Microhair basal cells 27–30 microns long. Microhairs (5.4–)9–10(–18) microns wide at the septum. Microhair total length/width at septum 1.3–3.3. Microhair apical cells (6–)7–15(–16.5) microns long. Microhair apical cell/total length ratio 0.21–0.7(–1). *Stomata* common; 15–25.5 microns long. Subsidiaries low dome-shaped to triangular. Guard-cells overlapping to flush with the interstomatals. *Intercostal short-cells* common; in cork/silica-cell pairs, or not paired (solitary); silicified, or not silicified. Intercostal silica bodies absent, or imperfectly developed. *Costal short-cells* conspicuously in long rows, or neither distinctly grouped into long rows nor predominantly paired. Costal silica bodies present in alternate cell files of the costal zones; saddle shaped, or tall-and-narrow to crescentic, or 'panicoid-type'; not sharp-pointed.

Transverse section of leaf blade, physiology. C_4; biochemical type PCK (*L. ciliolata*), or NAD–ME (*L. digitata*); XyMS+. PCR sheath outlines uneven, or even. PCR sheaths of the primary vascular bundles interrupted; interrupted both abaxially and adaxially. PCR sheath extensions present, or absent. Maximum number of extension cells when present, 1. PCR cells with a suberised lamella, or without a suberised lamella. *PCR cell chloroplasts* ovoid, or elongated; with well developed grana; centrifugal/peripheral (*L. ciliolata*), or centripetal. *Mesophyll* with radiate chlorenchyma; traversed by columns of colourless mesophyll cells, or not traversed by colourless columns. **Leaf blade *'nodular' in section***; with the ribs more or less constant in size. *Midrib* conspicuous, or not readily distinguishable; with one bundle only, or having a conventional arc of bundles; with colourless mesophyll adaxially, or without colourless mesophyll adaxially. Bulliforms present in discrete, regular adaxial groups, or not in discrete, regular adaxial groups (?); at least sometimes associated with colourless mesophyll cells to form deeply-penetrating fans. All the vascular bundles accompanied by sclerenchyma. Combined sclerenchyma girders present; forming 'figures'. Sclerenchyma all associated with vascular bundles. The lamina margins with fibres.

Phytochemistry. Leaves without flavonoid sulphates (1 species). Leaf blade chlorophyll *a*:*b* ratio 3.41–4.26.

Cytology. Chromosome base number, $x = 10$. $2n = 20$, 40, and 60. 2, 4, and 6 ploid. Nucleoli persistent.

Taxonomy. Chloridoideae; main chloridoid assemblage.

Ecology, geography, regional floristic distribution. 27 species. Tropical and subtropical. Commonly adventive. Helophytic, mesophytic, and xerophytic; shade species and species of open habitats; halophytic and glycophytic. Woodland, savanna, dry and swampy soils.

Holarctic, Paleotropical, Neotropical, and Australian. Boreal, Tethyan, and Madrean. African, Madagascan, Indomalesian, and Neocaledonian. Eastern Asian and Atlantic North American. Mediterranean. Saharo-Sindian, Sudano-Angolan, and West African Rainforest. Indian, Indo-Chinese, Malesian, and Papuan. Caribbean, Amazon, Central Brazilian, Pampas, and Andean. North and East Australian and Central Australian. Southern Atlantic North American and Central Grasslands. Sahelo-Sudanian, Somalo-Ethiopian, South Tropical African, and Kalaharian. Tropical North and East Australian.

Rusts and smuts. Rusts — *Puccinia*. Smuts from Ustilaginaceae. Ustilaginaceae — *Ustilago*.

Economic importance. Significant weed species: *L. chinensis*, *L. coerulescens*, *L. fascicularis*, *L. filiformis*, *L. panicea*, *L scabra*, *L. uninervia*, *L. virgata*. Important native pasture species: e.g. *L. chinensis*, *L. dubia*, *L. obtusiflora*, *L. panicea*.

References, etc. Morphological/taxonomic: Lazarides 1980a. Leaf anatomical: Metcalfe 1960; this project.

Special comments. Note that there are serious taxonomic problems involving *Leptochloa*, *Trichoneura*, *Diplachne*, etc.

Leptochloöpsis Yates

~ *Uniola* (*U. condensata* and *U. virgata*)

Habit, vegetative morphology. Perennial (robust); caespitose. **Culms** 100–200 cm high; *herbaceous*. Plants unarmed. Leaf blades attenuated from about 8 mm wide; narrow; flat, or rolled (involute); without abaxial multicellular glands; without cross venation.

Reproductive organization. Plants bisexual, with bisexual spikelets; with hermaphrodite florets.

Inflorescence. *Inflorescence of spicate main branches*; contracted (the stiff branches narrowly ascending). Primary inflorescence branches numerous. Inflorescence espatheate; not comprising 'partial inflorescences' and foliar organs. Spikelet-bearing axes slender; persistent. Spikelets secund (conspicuously on one side of the rachis, usually closely imbricate); biseriate; shortly pedicellate; imbricate.

Female-fertile spikelets. Spikelets 1.5–7 mm long; oblong, or elliptic; compressed laterally; falling with the glumes; not disarticulating between the florets. Rachilla prolonged beyond the uppermost female-fertile floret.

Glumes present; two; very unequal (the upper longer); shorter than the adjacent lemmas; hairless; pointed (acuminate); awnless; carinate; similar. *Spikelets* with incomplete florets. The incomplete florets proximal to the female-fertile florets, or both distal and proximal to the female-fertile florets (?). *Spikelets with proximal incomplete florets*. *The proximal incomplete florets* 1–3 (?); sterile. The proximal lemmas awnless; 5 nerved; exceeded by the female-fertile lemmas; similar in texture to the female-fertile lemmas.

Female-fertile florets *3–6*. Lemmas acuminate; not becoming indurated (firm, smooth, shining); entire; pointed; awnless, or mucronate (?); carinate (compressed-keeled, especially above); 5 nerved. *Palea* present; awnless, without apical setae. *Lodicules* present; free; fleshy; glabrous. *Stamens* 3. Stigmas 2.

Fruit, embryo and seedling. *Fruit* longitudinally grooved, or not grooved; *not noticeably compressed, or trigonous*. Hilum short. Pericarp fused (?). Embryo large; with an epiblast; with a scutellar tail; with an elongated mesocotyl internode. Embryonic leaf margins meeting.

First seedling leaf with a well-developed lamina. The lamina narrow; erect to curved.

Abaxial leaf blade epidermis. *Costal/intercostal zonation* conspicuous (but obviously distinguished only by the distribution of stomatal bands). *Papillae* absent. *Long-cells* similar in shape costally and intercostally; of similar wall thickness costally and intercostally. Mid-intercostal long-cells rectangular; having markedly sinuous walls. *Microhairs* present (but scarce, and no good examples seen); chloridoid-type (?). *Stomata* common; 24–27 microns long. Subsidiaries low dome-shaped. *Intercostal short-cells* common; in cork/silica-cell pairs (some solitary); silicified. Intercostal silica bodies absent; rounded and crescentic. *Costal short-cells* predominantly paired (some solitary). Costal silica bodies present throughout the costal zones; rounded, or saddle shaped (or rectangular), or crescentic; not sharp-pointed.

Transverse section of leaf blade, physiology. C_4; XyMS+. PCR sheaths of the primary vascular bundles interrupted; interrupted both abaxially and adaxially. PCR sheath extensions present. Maximum number of extension cells 6–10. PCR cells with a suberised lamella, or without a suberised lamella. Mesophyll traversed by columns of colourless mesophyll cells (connected to abaxial sclerenchyma groups). *Leaf blade* with distinct, prominent adaxial ribs. Bulliforms present in discrete, regular adaxial groups; associated with colourless mesophyll cells to form deeply-penetrating fans (these linked with traversing columns of colourless cells). All the vascular bundles accompanied by sclerenchyma. Combined sclerenchyma girders present; forming 'figures' (in the large bundles). Sclerenchyma not all bundle-associated. The 'extra' sclerenchyma in abaxial groups; abaxial-hypodermal, the groups isolated, or abaxial-hypodermal, the groups continuous with colourless columns.

Cytology. Chromosome base number, $x = 10$ (n = 20). $2n = 40$.

Taxonomy. Chloridoideae; main chloridoid assemblage.

Ecology, geography, regional floristic distribution. 2 species. Puerto Rico and South to Central America. Xerophytic; species of open habitats; halophytic, or glycophytic.

Neotropical. Caribbean.

References, etc. Leaf anatomical: this project.

Leptocoryphium Nees

Habit, vegetative morphology. Perennial; caespitose. **Culms** 30–100 cm high; herbaceous; unbranched above; somewhat *tuberous*. Culm nodes glabrous. Young shoots intravaginal. *Leaves* mostly basal; non-auriculate. Leaf blades narrow; about 5 mm wide; not setaceous (tending to subulate); flat, or folded; without cross venation; persistent. Ligule a fringed membrane.

Reproductive organization. Plants bisexual, with bisexual spikelets; with hermaphrodite florets.

Inflorescence. Inflorescence paniculate; open; with capillary branchlets; espatheate; not comprising 'partial inflorescences' and foliar organs. Spikelet-bearing axes persistent. Spikelets not secund; pedicellate.

Female-fertile spikelets. Spikelets *unconventional (as a eu-panicoid, because either a glume or the L_1 is missing)*; about 4.8 mm long; compressed dorsiventrally; falling with the glumes. Rachilla terminated by a female-fertile floret. Hairy callus absent.

Glumes one per spikelet (if the second organ is assumed to be the L₁); shorter than the adjacent lemmas to long relative to the adjacent lemmas; hairy; not pointed; awnless; non-carinate. Upper glume (as interpreted here) 5 nerved. *Spikelets* with incomplete florets. The incomplete florets proximal to the female-fertile florets. *The proximal incomplete florets* 1 (hairy); epaleate; sterile. The proximal lemmas awnless; 5 nerved; more or less equalling the female-fertile lemmas; similar in texture to the female-fertile lemmas; not becoming indurated.

Female-fertile florets 1. Lemmas flat, becoming hyaline towards the tip; similar in texture to the glumes; smooth; becoming indurated; brown in fruit; entire; pointed; awnless; hairless (but ciliate on the upper edges); non-carinate; having the margins lying flat and exposed on the palea (unusual in not gripping it); with a clear germination flap; 5 nerved. *Palea* present; relatively long; gaping; entire; awnless, without apical setae; textured like the lemma; 2-nerved; keel-less. *Lodicules* present; 2; free; fleshy; glabrous; not or scarcely vascularized. *Stamens* 3. Anthers not penicillate. *Ovary* glabrous. Stigmas 2; red pigmented.

Fruit, embryo and seedling. *Fruit* free from both lemma and palea; small (2.5 mm long); ellipsoid. Hilum short. Embryo large; not waisted.

Abaxial leaf blade epidermis. *Costal/intercostal zonation* conspicuous. *Papillae* absent. Mid-intercostal long-cells rectangular; having markedly sinuous walls. *Microhairs* present (but very scarce); panicoid-type; 39–46.5 microns long; 9–9.6 microns wide at the septum. Microhair total length/width at septum 4.3–4.8. Microhair apical cells 24–27 microns long. Microhair apical cell/total length ratio 0.58–0.62. *Stomata* common; 34.5–36 microns long. Subsidiaries triangular. Guard-cells overlapping to flush with the interstomatals. *Intercostal short-cells* common; in cork/silica-cell pairs and not paired; not silicified. *Costal short-cells* conspicuously in long rows. Costal silica bodies 'panicoid-type'; cross shaped and dumb-bell shaped; not sharp-pointed.

Transverse section of leaf blade, physiology. C₄. The anatomical organization unconventional (somewhat, in that the minor vascular bundles are scattered across the thickish blade portions between the major bundles). Organization of PCR tissue approaching the *Arundinella* type. XyMS+. Mesophyll without 'circular cells' (but the PCR cells of numerous tiny bundles could be mistaken for these). *Leaf blade* 'nodular' in section; with the ribs more or less constant in size. *Midrib* conspicuous; having a conventional arc of bundles (large median, small laterals); with colourless mesophyll adaxially. Bulliforms not in discrete, regular adaxial groups (*Ammophila*-type groups of small cells in the furrows, save for the 'midrib hinges'). All the vascular bundles accompanied by sclerenchyma. Combined sclerenchyma girders present; forming 'figures'. Sclerenchyma all associated with vascular bundles.

Phytochemistry. Leaves without flavonoid sulphates (1 species).

Cytology. Chromosome base number, $x = 10$. $2n = 40$. 4 ploid.

Taxonomy. Panicoideae; Panicodae; Paniceae.

Ecology, geography, regional floristic distribution. 1 species. Warm America, West Indies. Species of open habitats. Savanna.

Neotropical and Antarctic. Caribbean, Venezuela and Surinam, Amazon, Central Brazilian, Pampas, and Andean. Patagonian.

References, etc. Morphological/taxonomic: Parodi 1969. Leaf anatomical: this project.

Leptoloma Chase

~ *Digitaria*

Habit, vegetative morphology. Perennial; rhizomatous, or caespitose (with a knotty rhizomatous base). *Culms* 25–80 cm high; herbaceous. Culm nodes glabrous. *Leaves* not basally aggregated; non-auriculate. Leaf blades narrow; 2–6 mm wide (rather rigid); without cross venation; persistent. Ligule an unfringed membrane; truncate; 1 mm long.

Reproductive organization. Plants bisexual, with bisexual spikelets; with hermaphrodite florets.

Inflorescence. *Inflorescence* paniculate; *deciduous in its entirety*; open; with capillary branchlets; espatheate; not comprising 'partial inflorescences' and foliar organs. Spikelet-bearing axes persistent. Spikelets not secund; pedicellate.

Female-fertile spikelets. Spikelets 2–4 mm long; compressed dorsiventrally; falling with the glumes. Rachilla terminated by a female-fertile floret. Hairy callus absent (although there may be hair tufts from bases of G_2 and L_1).

Glumes two (the lower minute and vestigial), or one per spikelet; the G_2 relatively large; (when both present), very unequal; (the upper) long relative to the adjacent lemmas; (the upper) often hairy (between the veins); awnless; non-carinate. *Lower glume 0 nerved.* Upper glume 3–5 nerved. *Spikelets* with incomplete florets. The incomplete florets proximal to the female-fertile florets. *The proximal incomplete florets* 1; paleate, or epaleate. Palea of the proximal incomplete florets when present, reduced. The proximal incomplete florets sterile. The proximal lemmas awnless; 5–7 nerved; more or less equalling the female-fertile lemmas; less firm than the female-fertile lemmas (membranous, often hairy between the veins); not becoming indurated.

Female-fertile florets 1. Lemmas decidedly firmer than the glumes (thinly cartilaginous); striate (minutely rugulose); not becoming indurated; brown in fruit; entire; pointed; awnless; hairless; non-carinate; having the margins lying flat and exposed on the palea (hyaline); with a clear germination flap; 3 nerved. *Palea* present; relatively long; awnless, without apical setae; textured like the lemma; not indurated; 2-nerved. *Lodicules* present; 2; joined, or free; fleshy; glabrous; not or scarcely vascularized. *Stamens* 3. *Ovary* glabrous. Styles free to their bases. Stigmas 2; red pigmented.

Abaxial leaf blade epidermis. *Costal/intercostal zonation* conspicuous. *Papillae* absent. *Long-cells* markedly different in shape costally and intercostally; of similar wall thickness costally and intercostally. Mid-intercostal long-cells rectangular and fusiform; having markedly sinuous walls. *Microhairs* present; panicoid-type; (51–)60–66(–75) microns long; 3.6–4.5(–5.4) microns wide at the septum. Microhair total length/width at septum 11.3–18.3. Microhair apical cells (16.5–)22.5–31.5(–39) microns long. Microhair apical cell/total length ratio 0.32–0.52. *Stomata* common; 25.5–28.5 microns long. Subsidiaries parallel-sided, dome-shaped, and triangular; including both triangular and parallel-sided forms on the same leaf. Guard-cells overlapped by the interstomatals (mostly, but a few more or less flush). *Intercostal short-cells* absent or very rare. *Costal short-cells* conspicuously in long rows. Costal silica bodies 'panicoid-type'; mostly dumb-bell shaped (some with points); sharp-pointed, or not sharp-pointed (some dumb-bells have points).

Transverse section of leaf blade, physiology. C_4; XyMS–. *Leaf blade* 'nodular' in section; with the ribs more or less constant in size. *Midrib* not readily distinguishable; with one bundle only. Bulliforms not in discrete, regular adaxial groups (constituting most of epidermis). Many of the smallest vascular bundles unaccompanied by sclerenchyma. Combined sclerenchyma girders present. Sclerenchyma all associated with vascular bundles.

Phytochemistry. Tissues of the culm bases with abundant starch.

Cytology. $2n = 72$, or 36, or 70 (mostly 72).

Taxonomy. Panicoideae; Panicodae; Paniceae.

Ecology, geography, regional floristic distribution. 1 species. North America. Holarctic. Boreal and Madrean. Atlantic North American. Canadian-Appalachian, Southern Atlantic North American, and Central Grasslands.

Rusts and smuts. Smuts from Tilletiaceae and from Ustilaginaceae. Tilletiaceae — *Entyloma.* Ustilaginaceae — *Sorosporium.*

References, etc. Leaf anatomical: this project.

Special comments. Fruit data wanting.

Leptosaccharum (Hackel) A. Camus

~ *Eriochrysis*

Habit, vegetative morphology. Perennial; caespitose. *Culms* 40–70 cm high; herbaceous; unbranched above. Culm nodes hairy. *Leaves* mostly basal; non-auriculate. Sheath margins free. Leaf blades narrow; 1.5–2 mm wide; setaceous; without cross venation. Ligule a fringe of hairs; 1–1.5 mm long.

Reproductive organization. Plants bisexual, with bisexual spikelets; with hermaphrodite florets.

Inflorescence. Inflorescence paniculate; contracted (narrow). Inflorescence with axes ending in spikelets. Inflorescence espatheate; not comprising 'partial inflorescences' and foliar organs. *Spikelet-bearing axes* 'racemes'; solitary; persistent (distinctly articulated, but tenacious?). *'Articles' densely long-hairy (with conspicuous tufts of golden hairs at each joint and beneath each spikelet). Spikelets* not secund; *consistently in 'long-and-short' combinations (but irregularly so)*. The 'shorter' spikelets hermaphrodite. The 'longer' spikelets hermaphrodite.

Female-fertile spikelets. Spikelets *unconventional (because they have only one glume, with reduced L_2 and P2)*; 4–7 mm long; compressed dorsiventrally; falling with the glumes. Rachilla terminated by a female-fertile floret. Hairy callus present.

Glumes one per spikelet; long relative to the adjacent lemmas; hairy; not pointed; awnless; non-carinate. Upper glume 3 nerved. *Spikelets* with incomplete florets. The incomplete florets proximal to the female-fertile florets. *The proximal incomplete florets* 1; epaleate; sterile. The proximal lemmas awnless; 3–5 nerved; decidedly exceeding the female-fertile lemmas; decidedly firmer than the female-fertile lemmas (hairy, similar to the glume); not becoming indurated.

Female-fertile florets 1. **Lemmas greatly reduced**; less firm than the glumes (flimsy); not becoming indurated; entire; pointed; awnless; hairy; non-carinate; having the margins lying flat and exposed on the palea; without a germination flap; 0–1 nerved. **Palea** present (but reduced, similar to the L_2); conspicuous but relatively short; awnless, without apical setae; textured like the lemma (flimsy); not indurated; *nerveless*; keel-less. *Lodicules* present; 2; free; fleshy (large); glabrous; not or scarcely vascularized. *Stamens* 3 (very small, shorter than the lodicules). Anthers not penicillate; without an apically prolonged connective. *Ovary* glabrous. Styles fused to free to their bases. Stigmas 2; red pigmented.

Fruit, embryo and seedling. Fruit small. Pericarp fused. Embryo large.

Abaxial leaf blade epidermis. *Costal/intercostal zonation* lacking. *Papillae* absent. Mid-intercostal long-cells rectangular; having markedly sinuous walls. *Microhairs* absent. *Stomata* absent or very rare; 22.5–27 microns long. Subsidiaries triangular. Guard-cells overlapping to flush with the interstomatals. *Intercostal short-cells* common; in cork/silica-cell pairs (no zonation, but pairs everywhere). *Costal short-cells* predominantly paired. Costal silica bodies tall-and-narrow, crescentic, and oryzoid (intergrading); not sharp-pointed.

Transverse section of leaf blade, physiology. C_4; XyMS– (but bundles variable). *Mesophyll* with radiate chlorenchyma. *Leaf blade* with distinct, prominent adaxial ribs; with the ribs very irregular in sizes. *Midrib* conspicuous; having a conventional arc of bundles; with colourless mesophyll adaxially. Bulliforms not in discrete, regular adaxial groups. All the vascular bundles accompanied by sclerenchyma. Sclerenchyma not all bundle-associated. The 'extra' sclerenchyma in a continuous abaxial layer.

Taxonomy. Panicoideae; Andropogonodae; Andropogoneae; Andropogoninae.

Ecology, geography, regional floristic distribution. 1 species. Brazil & Paraguay. Neotropical. Central Brazilian and Pampas.

References, etc. Morphological/taxonomic: Camus 1923. Leaf anatomical: this project.

Leptothrium Kunth

Latipes Kunth

Habit, vegetative morphology. Perennial; caespitose. Culms 10–75 cm high; herbaceous; branched above. Culm nodes glabrous. Culm internodes solid. Plants unarmed. *Leaves* not basally aggregated; non-auriculate. Sheath margins free. Leaf blades narrow; 1–2 mm wide (to 8 cm long); flat, or rolled; without abaxial multicellular glands; without cross venation; persistent. Ligule a fringe of hairs. *Contra-ligule* absent.

Reproductive organization. Plants bisexual, with bisexual spikelets; with hermaphrodite florets. *The spikelets all alike in sexuality*.

Inflorescence. *Inflorescence a false spike, with spikelets on contracted axes (an elongated false raceme), or a single raceme (the 'clusters' each reduced to one peduncled spikelet in L.* rigidum); *contracted (the 'pedicels'(peduncles) flattened, cuneate, fringed, the rachis long and slender)*; espatheate; not comprising 'partial inflorescences' and foliar organs. **Spikelet-bearing axes** very much reduced (i.e., the panicle branches reduced to the pedicel-like peduncle and one or two spikelets); *disarticulating*; falling entire (the peduncles falling). Spikelets solitary (*L. rigidum*), or paired (*L. senegalensis*); not secund; sessile.

Female-fertile spikelets. Spikelets 3–8 mm long; compressed laterally (except for the larger asymmetric spikelet in *L. senegalensis*); falling with the glumes (and the pedicels); with conventional internode spacings. Rachilla terminated by a female-fertile floret. Hairy callus absent.

Glumes two; relatively large (asymmetric in *L. senegalensis*); very unequal; (the upper) long relative to the adjacent lemmas (exceeding the lemma); pointed (subulate-acuminate); awnless (subulate, muticate, leathery, shining); with the keel conspicuously winged (in upper part of G_2 in *L. rigidum*), or without a median keel-wing; *very dissimilar (G_1 of L. rigidum and of one spikelet of the pair in L. senegalensis dorsally flattened, ciliate towards the tip, recurving, more or less free of the rest of the spikelet; G_2 shorter, laterally flattened towards the tip, tuberculate-spiny or not)*. Lower glume tuberculate, or prickly; 1 nerved, or 3–5 nerved. *Upper glume 5 nerved*; prickly, or not prickly. *Spikelets* with female-fertile florets only.

Female-fertile florets 1. **Lemmas** less firm than the glumes (hyaline); not becoming indurated; entire; pointed; awnless; hairless; glabrous; *carinate*; without a germination flap; 1 nerved, or 3 nerved. *Palea* present; very reduced (and delicate); entire; awnless, without apical setae; nerveless; keel-less. *Lodicules* present; 2; free; fleshy, or membranous; glabrous; not or scarcely vascularized. *Stamens* 3. Anthers relatively long; not penicillate; without an apically prolonged connective. *Ovary* glabrous. Styles free to their bases. Stigmas 2.

Fruit, embryo and seedling. *Fruit* free from both lemma and palea; small (2 mm long); compressed laterally. Hilum short. *Pericarp loosely adherent*. Embryo large; waisted.

Abaxial leaf blade epidermis. *Costal/intercostal zonation* conspicuous. *Papillae* present, or absent; when seen, intercostal. Intercostal papillae consisting of one oblique swelling per cell. *Long-cells* the costal cells longer; of similar wall thickness costally and intercostally (medium thin walled). Mid-intercostal long-cells rectangular; having markedly sinuous walls. *Microhairs* present; more or less spherical to elongated; clearly two-celled; chloridoid-type. Microhair apical cell wall of similar thickness/rigidity to that of the basal cell. Microhairs 21–33 microns long. Microhair basal cells 18 microns long. Microhairs 11.4–13.5 microns wide at the septum. Microhair total length/width at septum 1.7–2.8. Microhair apical cells 7.5–8.4 microns long (*L. rigidum*), or 9.6–12.6 microns long (*L. senegalensis*). Microhair apical cell/total length ratio 0.23–0.48. *Stomata* common; 16.5–18 microns long (*L. rigidum*), or 21–34.5 microns long (*L. senegalensis*). Subsidiaries low dome-shaped and triangular. Guard-cells overlapping to flush with the interstomatals. *Intercostal short-cells* common, or absent or very rare (very rare in *L. senegalensis*); in cork/silica-cell pairs (and solitary); silicified. Intercostal silica bodies absent. Costal prickles abundant. *Costal short-cells* conspicuously in long rows. Costal silica bodies present in alternate cell files of the costal zones; 'panicoid-type'; mostly dumb-bell shaped; not sharp-pointed.

Transverse section of leaf blade, physiology. C_4; XyMS+. PCR sheaths of the primary vascular bundles interrupted; interrupted abaxially only. PCR sheath extensions absent. *Mesophyll* with radiate chlorenchyma. *Leaf blade* with distinct, prominent adaxial ribs; with the ribs more or less constant in size (round-topped). *Midrib* not readily distinguishable; with one bundle only. Bulliforms present in discrete, regular adaxial groups (the groups large, between each bundle pair); in simple fans (the large median cells deeply penetrating). All the vascular bundles accompanied by sclerenchyma. Combined sclerenchyma girders present (with most bundles); forming 'figures'. Sclerenchyma all associated with vascular bundles.

Cytology. Chromosome base number, $x = 10$.

Taxonomy. Chloridoideae; main chloridoid assemblage.

Ecology, geography, regional floristic distribution. 2 species. Caribbean & Senegal to Pakistan. Xerophytic; halophytic, or glycophytic. Dry bushland, open seashore, sandy thickets.

Paleotropical and Neotropical. African. Saharo-Sindian and Sudano-Angolan. Caribbean. Sahelo-Sudanian and Somalo-Ethiopian.

Economic importance. Cultivated fodder: *L. senegalense* used for reseeding denuded ground. Important native pasture species: *L. senegalense*.

References, etc. Leaf anatomical: this project.

Lepturella Stapf

~ *Oropetium* (*O. aristatum* (Stapf) Pilger, = sect. *Lepturella*)

Habit, vegetative morphology. Dwarf, cushion forming annual; caespitose. **Culms 2–9 cm high**; herbaceous; branched above. Culm nodes glabrous. Culm internodes hollow. Plants unarmed. Young shoots intravaginal. *Leaves* not basally aggregated; non-auriculate. Leaf blades linear; narrow; about 2 mm wide (to 4 cm long); usually rolled; exhibiting multicellular glands abaxially (base of macrohairs). The abaxial glands on the blade margins. Leaf blades without cross venation; disarticulating from the sheaths, or persistent. Ligule an unfringed membrane (laciniate, with only a few hairs), or a fringed membrane; truncate; about 0.2 mm long. *Contra-ligule* absent.

Reproductive organization. Plants bisexual, with bisexual spikelets; with hermaphrodite florets.

Inflorescence. *Inflorescence a single spike (straight or curved, cylindrical, slender, short)*. Primary inflorescence branches 1 (i.e., solitary spikes). Rachides hollowed. Inflorescence espatheate; not comprising 'partial inflorescences' and foliar organs. **Spikelet-bearing axes** spikes; *disarticulating*; disarticulating at the joints (into segments with one spikelet). *'Articles'* non-linear; disarticulating transversely; glabrous. Spikelets solitary; not secund; distichous (the ranks opposite); sessile.

Female-fertile spikelets. *Spikelets* 2.5–3.5 mm long; adaxial; *compressed laterally*; falling with the glumes (and with the joint); with conventional internode spacings. Rachilla terminated by a female-fertile floret. Hairy callus present (short). *Callus* short.

Glumes two, or one per spikelet (the G_1 reduced or missing, save in the terminal spikelet); very unequal (save in the terminal spikelets, which have equal glumes); exceeding the spikelets; (the upper) long relative to the adjacent lemmas; dorsiventral to the rachis; hairless; *awned (G_2 tapered into a long subule)*; carinate; very dissimilar (G_1 a scarious scale or missing, G_2 large, covering the floret, awned, rigid). Lower glume 0 nerved. Upper glume 1 nerved. *Spikelets* with female-fertile florets only.

Female-fertile florets 1. Lemmas less firm than the glumes (membranous); not becoming indurated; incised; 2 lobed; not deeply cleft; awned. Awns 1; median; from a sinus; non-geniculate; hairless (scabrous); about as long as the body of the lemma to much longer than the body of the lemma; entered by one vein. Lemmas hairless; without a germination flap; 3 nerved. *Palea* present (oblong); relatively long; entire; with apical setae (the keels shortly excurrent); membranous; not indurated; 2-nerved; 2-keeled. Palea keels wingless; glabrous. *Lodicules* present; 2; free; glabrous. *Stamens* 3. Anthers not penicillate; without an apically prolonged connective. *Ovary* glabrous. Stigmas 2; brown.

Fruit, embryo and seedling. Fruit compressed laterally. Hilum short. Pericarp fused. Embryo large.

Abaxial leaf blade epidermis. *Costal/intercostal zonation* conspicuous. *Papillae* absent. Mid-intercostal long-cells rectangular; having markedly sinuous walls. *Microhairs* present; more or less spherical; clearly two-celled; chloridoid-type. Microhair apical cell wall of similar thickness/rigidity to that of the basal cell. Microhair basal cells 12 microns long. Microhair total length/width at septum 2. Microhair apical cell/total length ratio 0.5. *Stomata* common. Intercostal short-cells in cork/silica-cell pairs; silicified. Intercostal silica bodies imperfectly developed; tall-and-narrow. *Costal short-cells* conspicuously in long rows. Costal silica bodies present in alternate cell files of the costal zones; saddle shaped; not sharp-pointed (to squarish).

Transverse section of leaf blade, physiology. C_4; XyMS+. PCR sheaths of the primary vascular bundles interrupted; interrupted abaxially only. PCR sheath extensions absent. *Mesophyll* with radiate chlorenchyma; traversed by columns of colourless mesophyll cells. *Leaf blade* with distinct, prominent adaxial ribs. *Midrib* not readily distinguishable; with one bundle only. Bulliforms present in discrete, regular adaxial groups; associated with colourless mesophyll cells to form deeply-penetrating fans (these linked with the traversing columns of colourless cells). All the vascular bundles accompanied by sclerenchyma. Combined sclerenchyma girders present; forming 'figures'. Sclerenchyma all associated with vascular bundles. The lamina margins without fibres.

Taxonomy. Chloridoideae; main chloridoid assemblage.

Ecology, geography, regional floristic distribution. 1 species (*L. aristata*). West tropical Africa. Mesophytic to xerophytic; species of open habitats; glycophytic.

Paleotropical. African. Sudano-Angolan and West African Rainforest. Somalo-Ethiopian and South Tropical African.

References, etc. Leaf anatomical: Metcalfe 1960; Van den Borre 1994.

Lepturidium Hitchc. and Ekman

~ Brachyachne

Habit, vegetative morphology. Wiry perennial; *caespitose, or decumbent. Culms* 10–30 cm high; herbaceous; unbranched above. *Leaves* not basally aggregated; conspicuously distichous; non-auriculate. Sheaths overlapping, pilose or villous above and around the base of the blade. *Leaf blades* linear-lanceolate; narrow; 1–2 mm wide (and 1–2 cm long); flat, or rolled (becoming involute, the tips hard, obtuse, boat-shaped); without abaxial multicellular glands; without cross venation; *persistent*. Ligule a fringed membrane, or a fringe of hairs (?).

Reproductive organization. *Plants dioecious (according to the original description — no pistils found in the specimens examined)*; without hermaphrodite florets. The spikelets female-only, or male-only (?). Plants presumably outbreeding.

Inflorescence. *Inflorescence a single raceme (with appressed spikelets)*. Inflorescence with axes ending in spikelets. Rachides triquetrous. Inflorescence espatheate; not comprising 'partial inflorescences' and foliar organs. Spikelet-bearing axes persistent. Spikelets solitary; secund; biseriate; subsessile; imbricate (overlapping to about half their length).

Female-sterile spikelets. The exclusively male spikelets with two equal, awnless, 1-nerved glumes. One proximal male floret with three stamens, its lemma 3 nerved, mucronate between two very short teeth. Second floret more or less reduced, sometimes paleate but sterile. Rachilla of male spikelets prolonged beyond the uppermost male floret. *The male spikelets* with glumes (two); without proximal incomplete florets; *2 floreted (the lower fertile)*. The lemmas mucronate. Male florets 1; 3 staminate.

Female-fertile spikelets. *Spikelets* (presumably?) *without proximal incomplete florets*.

Fruit, embryo and seedling. Pericarp and female-fertile plants unknown.

Abaxial leaf blade epidermis. *Costal/intercostal zonation* conspicuous. *Papillae* absent. *Long-cells* markedly different in shape costally and intercostally (the costals much narrower); of similar wall thickness costally and intercostally (fairly thick walled). Mid-intercostal long-cells rectangular; having markedly sinuous walls (the sinuosity fine to coarse, regular to irregular). *Microhairs* present; more or less spherical to elongated; clearly two-celled; chloridoid-type. Microhair apical cell wall of similar thickness/rigidity to that of the basal cell. Microhair basal cells 15 microns long. Microhair total length/width at septum 2. Microhair apical cell/total length ratio 0.5. *Stomata* common. Subsidiaries non-papillate; triangular (mostly of the truncated type). Guard-cells overlapping to flush with the interstomatals. *Intercostal short-cells* fairly common; in cork/silica-cell pairs; silicified. Intercostal silica bodies imperfectly developed; short dumb-bell shaped, like some of the costals. No macrohairs or prickles seen. *Costal short-cells* conspicuously in long rows. Costal silica bodies present and well developed; present in alternate cell files of the costal zones; 'panicoid-type' (large); dumb-bell shaped (often with a quite long isthmus).

Transverse section of leaf blade, physiology. C_4; XyMS+. PCR sheath outlines uneven to even (the lowermost, abaxial cell on each side often much enlarged). PCR sheaths of the primary vascular bundles interrupted; interrupted abaxially only. PCR sheath extensions absent. Mesophyll traversed by columns of colourless mesophyll cells (perhaps, in places?), or not traversed by colourless columns (?). *Leaf blade* with distinct, prominent adaxial ribs; with the ribs more or less constant in size. *Midrib* not readily distinguishable; with one bundle only. *The lamina* symmetrical on either side of the midrib. Bulliforms present in discrete, regular adaxial groups; associated with colourless mesophyll cells to form deeply-penetrating fans. All the vascular bundles accompanied by sclerenchyma. Combined sclerenchyma girders present; forming 'figures' ('anchors' with all the bundles, the abaxial girders interrupting the PCR sheaths, the adaxial girders not doing so). Sclerenchyma all associated with vascular bundles. The lamina margins with fibres.

Taxonomy. Chloridoideae; main chloridoid assemblage.

Ecology, geography, regional floristic distribution. 1 species (rare). Cuba. Species of open habitats; halophytic (on salt flats).

Neotropical. Caribbean.

References, etc. Morphological/taxonomic: Hitchcock and Ekman (1936), in 'Manual of the Grasses of the West Indies', USDA Miscellaneous Publ. 243, Washington. Leaf anatomical: this project.

Special comments. Fruit data wanting.

Lepturopetium Morat

Habit, vegetative morphology. Wiry perennial; decumbent. **Culms** 30–50 cm high (slender); *herbaceous*; *branched above*. Culm nodes glabrous. Plants unarmed. *Leaves* not basally aggregated; non-auriculate. Leaf blades linear; narrow; 2.8–4 mm wide (5–7 cm long); flat, or folded, or rolled; without abaxial multicellular glands; without cross venation; persistent. *Ligule a fringed membrane*.

Reproductive organization. Plants bisexual, with bisexual spikelets; with hermaphrodite florets.

Inflorescence. *Inflorescence a single spike* (**L. kuniense**), *or of spicate main branches (mostly paired, in* **L. marshallense**); digitate, or non-digitate (when the racemes solitary). Primary inflorescence branches 1–3 (each with 10–16 spikelets). Rachides hollowed. Inflorescence espatheate; not comprising 'partial inflorescences' and foliar organs. Spikelet-bearing axes persistent. *Spikelets* solitary; secund to not secund (somewhat so, according to Morat); distichous; *sessile*; imbricate.

Female-fertile spikelets. Spikelets 4.5–7 mm long; adaxial; compressed laterally; disarticulating above the glumes, or falling with the glumes (?); not disarticulating between the florets (the rachilla not articulated). Rachilla prolonged beyond the uppermost female-fertile floret; rachilla hairy. The rachilla extension with incomplete florets. Hairy callus present (beneath the L_1).

Glumes two; very unequal; (the longer) long relative to the adjacent lemmas; dorsiventral to the rachis; pointed (both or only the G_2 acuminate); awned (the G_2 sometimes shortly awn tipped), or awnless; non-carinate; very dissimilar (the G_2 covering the spikelet, rigid, leathery, the G_1 shorter and thinner — except in the terminal spikelet, where both glumes resemble a normal G_2). Lower glume longer than half length of lowest lemma; 1 nerved. *Upper glume 5–7 nerved*. *Spikelets* with incomplete florets. The incomplete florets distal to the female-fertile florets. The distal incomplete florets 1 (sterile); *incomplete florets* merely underdeveloped; incomplete florets awnless. *Spikelets without proximal incomplete florets*.

Female-fertile florets 1 (*L. kuniense*), or 2 (*L. marshallense*). *Lemmas* less firm than the glumes (papery); not becoming indurated; incised (minutely so); 2 lobed; not deeply cleft; *awned*. Awns 1; median; from a sinus; non-geniculate; hairless (scabrid); much shorter than the body of the lemma to about as long as the body of the lemma (1.5–4.5

mm long); entered by one vein; persistent. Lemmas hairy (shortly pilose, at least apically); non-carinate; 3 nerved. *Palea* present; relatively long; entire to apically notched; awnless, without apical setae (truncate-ciliate); thinner than the lemma; not indurated (hyaline); 2-nerved; 2-keeled. Palea keels wingless. *Lodicules* present ('reduced'); 2. *Stamens* 3. Anthers not penicillate; without an apically prolonged connective. *Ovary* glabrous. Stigmas 2; red pigmented.

Abaxial leaf blade epidermis. *Costal/intercostal zonation* conspicuous. *Papillae* present (and very abundant); costal and intercostal. Intercostal papillae over-arching the stomata; consisting of one symmetrical projection per cell, or several per cell (the interstomatals mostly with one large papilla, the long-cells often with a median row). Mid-intercostal long-cells rectangular; having markedly sinuous walls. *Microhairs* present (hard to find amongst the abundant papillae); more or less spherical; ostensibly one-celled to clearly two-celled; chloridoid-type. Microhair apical cell wall of similar thickness/rigidity to that of the basal cell. Microhair basal cells 6 microns long. Microhair total length/width at septum 1.5. Microhair apical cell/total length ratio 0.6. *Stomata* common. Subsidiaries non-papillate; dome-shaped and triangular. Guard-cells overlapping to flush with the interstomatals. *Intercostal short-cells* common; not paired. Intercostal silica bodies absent. *Costal short-cells* conspicuously in long rows. Costal silica bodies confined to the central file(s) of the costal zones; predominantly saddle shaped (but many incompletely formed); not sharp-pointed.

Transverse section of leaf blade, physiology. C_4; XyMS+. PCR sheath outlines conspicuously uneven (the minor bundles), or even (the primaries). PCR sheaths of the primary vascular bundles interrupted; interrupted abaxially only. PCR sheath extensions absent. *Leaf blade* adaxially flat. *Midrib* conspicuous; having a conventional arc of bundles; with colourless mesophyll adaxially. *The lamina* symmetrical on either side of the midrib. Bulliforms present in discrete, regular adaxial groups (between all bundles); in simple fans (the median cells deeply penetrating), or associated with colourless mesophyll cells to form deeply-penetrating fans. All the vascular bundles accompanied by sclerenchyma. Combined sclerenchyma girders present (with all the bundles); forming 'figures' (the primaries with conspicuous I's). Sclerenchyma all associated with vascular bundles. The lamina margins with fibres.

Taxonomy. Chloridoideae; main chloridoid assemblage.

Ecology, geography, regional floristic distribution. 2 species. New Caledonia, Marshall Islands, Cocos Islands. Species of open habitats; presumably halophytic. Calcareous soil, in costal hinterland.

Paleotropical and Neotropical. Polynesian and Neocaledonian. Polynesian. Caribbean.

References, etc. Morphological/taxonomic: Morat 1981; Fosberg and Sachet 1982. Leaf anatomical: this project.

Special comments. Possible intergeneric hybrids, involving *Lepturus* as one parent. Fruit data wanting.

Lepturus R.Br.

Lepiurus Dum., *Leptocercus* Raf., *Monerma* P. Beauv.

Habit, vegetative morphology. Perennial; stoloniferous and caespitose. *Culms* 10–60 cm high; herbaceous; branched above, or unbranched above. Culm nodes glabrous. Culm internodes solid. *Leaves* not basally aggregated; non-auriculate. Leaf blades linear to linear-lanceolate; narrow; not setaceous; flat, or rolled (involute); without abaxial multicellular glands; without cross venation; persistent. Ligule an unfringed membrane.

Reproductive organization. Plants bisexual, with bisexual spikelets; with hermaphrodite florets.

Inflorescence. *Inflorescence a single spike (almost cylindrical, the joints striate)*. Rachides hollowed. Inflorescence espatheate; not comprising 'partial inflorescences' and foliar organs. *Spikelet-bearing axes* with substantial rachides; *disarticulating*; disarticulating at the joints. *'Articles'* non-linear (hollowed to receive the spikelets); not appendaged; disarticulating transversely, or disarticulating obliquely; glabrous. Spikelets solitary; not secund; distichous; sessile.

Female-fertile spikelets. Spikelets 3–15(–20) mm long; adaxial; compressed dorsiventrally; falling with the glumes. *Rachilla prolonged beyond the uppermost female-fertile floret.* The rachilla extension with incomplete florets. Hairy callus absent.

Glumes one per spikelet (the lower usually missing), or two; very unequal; (the upper) long relative to the adjacent lemmas; free; dorsiventral to the rachis; pointed; awnless, or awned (the G_2 sometimes tapered into a short awn); *very dissimilar (the lower, if present, reduced to a minute triangular scale, the upper exceeding the spikelet, thickened, with rows of minute bristles on the back).* Lower glume 1 nerved. Upper glume 5–12 nerved. *Spikelets* with incomplete florets. The incomplete florets distal to the female-fertile florets.

Female-fertile florets 1–2. Lemmas less firm than the glumes (membranous); not becoming indurated; entire; pointed; awnless; hairy (minutely so, at the base), or hairless (glabrous); non-carinate; 3 nerved. *Palea* present (lanceolate); relatively long; not indurated (hyaline); 2-nerved; 2-keeled. *Lodicules* present; 2; free; fleshy; glabrous. *Stamens* 3. Anthers not penicillate. *Ovary* glabrous. Styles free to their bases. Stigmas 2; red pigmented.

Fruit, embryo and seedling. *Fruit* free from both lemma and palea; ellipsoid; compressed dorsiventrally. Hilum short. *Pericarp free.* Embryo large; not waisted. Endosperm hard; containing only simple starch grains.

Abaxial leaf blade epidermis. *Costal/intercostal zonation* conspicuous. *Papillae* present; costal and intercostal. Intercostal papillae not over-arching the stomata; several per cell (finger-like). *Long-cells* similar in shape costally and intercostally; of similar wall thickness costally and intercostally (fairly thin walled). Mid-intercostal long-cells rectangular; having markedly sinuous walls. *Microhairs* present; more or less spherical; clearly two-celled; chloridoid-type. Microhair apical cell wall of similar thickness/rigidity to that of the basal cell. Microhairs (22.5–)24–25.5(–28.5) microns long; (9.6–)10.5–11.4(–13.8) microns wide at the septum. Microhair total length/width at septum 1.6–2.7. Microhair apical cells (7.5–)9–11(–12) microns long. Microhair apical cell/total length ratio 0.29–0.47. *Stomata* common; 19.5–22.5 microns long. Subsidiaries triangular. Guard-cells overlapping to flush with the interstomatals. *Intercostal short-cells* common; in cork/silica-cell pairs, or not paired; silicified (a few), or not silicified. Intercostal silica bodies imperfectly developed; tall-and-narrow. *Costal short-cells* neither distinctly grouped into long rows nor predominantly paired. Costal silica bodies present in alternate cell files of the costal zones; saddle shaped; not sharp-pointed.

Transverse section of leaf blade, physiology. C_4; XyMS+. PCR sheath outlines uneven, or even. PCR sheaths of the primary vascular bundles interrupted; interrupted abaxially only. PCR sheath extensions absent. *Mesophyll* with radiate chlorenchyma; traversed by columns of colourless mesophyll cells, or not traversed by colourless columns. *Leaf blade* with distinct, prominent adaxial ribs; with the ribs more or less constant in size. *Midrib* conspicuous; having a conventional arc of bundles; with colourless mesophyll adaxially (adaxial colourless tissue sometimes also present in other parts of the blade). Bulliforms present in discrete, regular adaxial groups; in simple fans, or in simple fans and associated with colourless mesophyll cells to form deeply-penetrating fans (sometimes linked with girders of colourless cells). Many of the smallest vascular bundles unaccompanied by sclerenchyma. Combined sclerenchyma girders present; forming 'figures'. Sclerenchyma all associated with vascular bundles. The lamina margins with fibres.

Cytology. $2n = 14, 18, 26, 36, 52,$ and 54.

Taxonomy. Chloridoideae; main chloridoid assemblage.

Ecology, geography, regional floristic distribution. About 8 species. Coastal east Africa, Madagascar, to Australia, & Polynesia. Xerophytic; species of open habitats; halophytic. Sandy beaches and coastal hinterland, usually good sandbinders.

Holarctic, Paleotropical, Cape, and Australian. Boreal. African, Madagascan, Indomalesian, Polynesian, and Neocaledonian. Eastern Asian. Saharo-Sindian and Sudano-Angolan. Indian, Indo-Chinese, Malesian, and Papuan. Hawaiian, Polynesian, and Fijian.

North and East Australian. Somalo-Ethiopian and South Tropical African. Tropical North and East Australian.

Hybrids. *Lepturopetium* (q.v.) may be an intergeneric hybrid involving *Lepturus* as one parent.

Rusts and smuts. Rusts — *Puccinia*.

References, etc. Leaf anatomical: Metcalfe 1960; this project.

Leucophrys Rendle

~ *Brachiaria*

Habit, vegetative morphology. Perennial. **Culms** 7–100 cm high; *woody and persistent (stiffly geniculate)*; branched above (the plants bushy). Culm nodes glabrous. Culm internodes solid. Young shoots intravaginal. *Leaves* not basally aggregated; non-auriculate. Leaf blades linear-lanceolate; narrow; 2–4 mm wide; flat, or rolled; hard, woody, needle-like (sometimes spiny, hard and brittle), or not needle-like; without cross venation; disarticulating from the sheaths (leaving the culms as the main organs of photosynthesis). Ligule a fringe of hairs; 0.6–0.8 mm long. *Contra-ligule* absent.

Reproductive organization. Plants bisexual, with bisexual spikelets; with hermaphrodite florets.

Inflorescence. Inflorescence paniculate; contracted; espatheate; not comprising 'partial inflorescences' and foliar organs. Spikelet-bearing axes persistent. *Spikelets* solitary, or paired; not secund; *pedicellate (and substipitate, with a short stalk fitting into the pedicel apex)*. Pedicel apices cupuliform. Spikelets consistently in 'long-and-short' combinations, or not in distinct 'long-and-short' combinations.

Female-fertile spikelets. Spikelets 4–6 mm long; abaxial to adaxial (the orientation variable); compressed dorsiventrally; biconvex; falling with the glumes; not disarticulating between the florets; with conventional internode spacings. Rachilla terminated by a female-fertile floret. Hairy callus absent. *Callus* absent.

Glumes two; relatively large; very unequal to more or less equal; about equalling the spikelets to exceeding the spikelets; (the longer) long relative to the adjacent lemmas; hairy; *with distinct rows of hairs (the upper with a transverse row above the middle, the lower glabrous save at the base)*; pointed (the tips minutely truncate); awnless (but the tips caudate, inrolled, membranous); non-carinate; *very dissimilar (the lower obtuse or notched at the apex, pilose at the base, the upper tapering, caudate, dorsally long-villous with a transverse fringe just above the middle)*. *Lower glume* 0.75–1 times the length of the upper glume; about equalling the lowest lemma; 3 nerved. Upper glume 5–7 nerved. *Spikelets* with incomplete florets. The incomplete florets proximal to the female-fertile florets. *The proximal incomplete florets* 1; paleate. Palea of the proximal incomplete florets fully developed; not becoming conspicuously hardened and enlarged laterally. The proximal incomplete florets male. *The proximal lemmas awnless (the tip inrolled, membranous, sometimes slightly caudate, the back with a transverse fringe of hairs)*; 3 nerved; decidedly exceeding the female-fertile lemmas (about twice as long); less firm than the female-fertile lemmas; not becoming indurated.

Female-fertile florets 1. Lemmas decidedly firmer than the glumes; smooth; becoming indurated (glossy); entire; pointed, or blunt; awnless; hairless; glabrous; non-carinate; having the margins tucked in onto the palea; with a clear germination flap (basal); 5 nerved. **Palea** present; relatively long; tightly clasped by the lemma; entire; awnless, without apical setae; textured like the lemma; *indurated*; 2-nerved; 2-keeled. Palea keels glabrous. *Lodicules* present; 2; free; fleshy; glabrous; not or scarcely vascularized. *Stamens* 3. Anthers 2–2.5 mm long; not penicillate; without an apically prolonged connective. *Ovary* glabrous. Styles free to their bases. Stigmas 2.

Abaxial leaf blade epidermis. *Costal/intercostal zonation* conspicuous. *Papillae* absent. *Long-cells* markedly different in shape costally and intercostally (the costals relatively longer and narrower). Intercostal zones with typical long-cells. Mid-intercostal long-cells rectangular; having markedly sinuous walls. *Microhairs* present; panicoid-type. *Stomata* common. Subsidiaries mostly dome-shaped. *Intercostal short-cells* common; seemingly not paired. With rosettes of isodiametric cells around each of the abundant macrohairs. *Costal short-cells* conspicuously in long rows. Costal silica bodies 'panicoid-type'; not sharp-pointed.

Transverse section of leaf blade, physiology. C_4; XyMS+. PCR sheath extensions absent. PCR cell chloroplasts centrifugal/peripheral. *Mesophyll* with radiate chlorenchyma. *Leaf blade* adaxially flat. *Midrib* not readily distinguishable; with one bundle only. Bulliforms present in discrete, regular adaxial groups (and also some rather irregular groups); in simple fans. Many of the smallest vascular bundles unaccompanied by sclerenchyma. Combined sclerenchyma girders present (with the major ones only); nowhere forming 'figures'. Sclerenchyma all associated with vascular bundles.

Taxonomy. Panicoideae; Panicodae; Paniceae.

Ecology, geography, regional floristic distribution. 1 species. Tropical and southern Africa. Helophytic, or xerophytic; species of open habitats; glycophytic. Sandy riverbeds in semidesert.

Paleotropical. African. Namib-Karoo.

References, etc. Morphological/taxonomic: Rendle 1922; Launert 1970. Leaf anatomical: photos of *L. mesocoma* provided by R.P. Ellis.

Special comments. Fruit data wanting.

Leucopoa Griseb.

Hesperochloa Rydb.

~ *Festuca*

Habit, vegetative morphology. Perennial; rhizomatous, or caespitose. **Culms** 30–120 cm high; *herbaceous*. Culm internodes hollow. Leaves non-auriculate. Leaf blades narrow; flat; without cross venation. *Ligule an unfringed membrane*; truncate; 0.1–5 mm long.

Reproductive organization. Plants bisexual, with bisexual spikelets, or dioecious (sporadically); with hermaphrodite florets. The spikelets all alike in sexuality; hermaphrodite, or female-only, or male-only.

Inflorescence. *Inflorescence paniculate*; open, or contracted; espatheate; not comprising 'partial inflorescences' and foliar organs. Spikelet-bearing axes persistent. Spikelets not secund; pedicellate.

Female-fertile spikelets. *Spikelets 6–10 mm long*; compressed laterally; disarticulating above the glumes; disarticulating between the florets. Rachilla prolonged beyond the uppermost female-fertile floret. The rachilla extension with incomplete florets. Hairy callus absent.

Glumes two; very unequal; shorter than the spikelets; shorter than the adjacent lemmas; pointed, or not pointed; awnless; carinate; similar. Lower glume 1 nerved. Upper glume 3 nerved. *Spikelets* with incomplete florets. The incomplete florets distal to the female-fertile florets. *The distal incomplete florets* merely underdeveloped.

Female-fertile florets 3–9. *Lemmas less firm than the glumes*; not becoming indurated; entire to incised (irregularly serrate); not deeply cleft; awnless; hairless; carinate to non-carinate; without a germination flap; 5 nerved. *Palea* present; relatively long; 2-nerved; 2-keeled. *Lodicules* present; 2; free; membranous; glabrous; not toothed; not or scarcely vascularized. *Stamens* 3. Anthers 2.5–4.5 mm long. **Ovary hairy.** Styles free to their bases. Stigmas 2.

Fruit, embryo and seedling. Hilum long-linear. Embryo small. Endosperm hard; containing compound starch grains. Embryo with an epiblast.

First seedling leaf with a well-developed lamina. The lamina narrow; erect; 3 veined.

Abaxial leaf blade epidermis. *Costal/intercostal zonation* conspicuous. *Papillae* absent. *Long-cells* similar in shape costally and intercostally; of similar wall thickness costally and intercostally. Mid-intercostal long-cells rectangular (mostly), or fusiform (a few); having markedly sinuous walls (thick and pitted). *Microhairs* absent. *Stomata* absent or very rare. *Intercostal short-cells* common; in cork/silica-cell pairs (but the 'silica cells' without silica); not silicified. *Costal short-cells* neither distinctly grouped into long rows nor predominantly paired. Costal silica bodies rounded (mostly), or crescentic (some); not sharp-pointed.

Transverse section of leaf blade, physiology. C$_3$; XyMS+. *Mesophyll* with non-radiate chlorenchyma. *Leaf blade* with distinct, prominent adaxial ribs; with the ribs very irregular in sizes. *Midrib* not readily distinguishable; with one bundle only, or having a conventional arc of bundles (minor bundles flanking midrib). Bulliforms present in discrete, regular adaxial groups; in simple fans. Many of the smallest vascular bundles unaccompanied by sclerenchyma. Combined sclerenchyma girders present; forming 'figures'. Sclerenchyma all associated with vascular bundles.

Phytochemistry. Leaves without flavonoid sulphates (1 species).

Cytology. Chromosome base number, $x = 7$.

Taxonomy. Pooideae; Poodae; Poeae.

Ecology, geography, regional floristic distribution. 6 species. Western & Central Asia to Himalayas, North America.

Holarctic and Paleotropical. Boreal, Tethyan, and Madrean. Indomalesian. Euro-Siberian. Irano-Turanian. Indian. Siberian.

References, etc. Leaf anatomical: this project.

Leymus Hochst.

Aneurolepidium Nevski
Excluding *Malacurus*

Habit, vegetative morphology. Perennial; *rhizomatous (or turf-forming)*. **Culms** 20–150 cm high; *herbaceous*; unbranched above. Culm internodes hollow. Young shoots extravaginal. *Leaves* mostly basal (often), or not basally aggregated; auriculate (the auricles lanceolate-crescentic), or non-auriculate. Sheath margins free. Leaf blades linear (harsh, glaucus, pungent); broad, or narrow; 2–15 mm wide; flat, or rolled (convolute); without cross venation; persistent; rolled in bud. Ligule an unfringed membrane; truncate, or not truncate; 0.2–1(–3) mm long. *Contra-ligule* absent.

Reproductive organization. Plants bisexual, with bisexual spikelets; with hermaphrodite florets; outbreeding.

Inflorescence. *Inflorescence a single spike, or a false spike, with spikelets on contracted axes (erect, linear, rarely oblongate)*; espatheate; not comprising 'partial inflorescences' and foliar organs. *Spikelet-bearing axes persistent. Spikelets paired to in triplets (rarely solitary or in clusters of up to six)*; not secund; distichous; *subsessile*; usually imbricate.

Female-fertile spikelets. Spikelets (6–)8–20(–25) mm long; compressed laterally to not noticeably compressed; disarticulating above the glumes; disarticulating between the florets (the joints well developed). Rachilla prolonged beyond the uppermost female-fertile floret; rachilla hairy (shortly pilose), or hairless (scabrous). The rachilla extension with incomplete florets. Hairy callus present, or absent. Callus blunt (triangular or rounded).

Glumes two; very unequal to more or less equal; shorter than the adjacent lemmas to long relative to the adjacent lemmas; joined (basally), or free; when scoreable, lateral to the rachis, or displaced (side by side); hairy (short pilose), or hairless (scabrous or rarely glabrous); when hairless glabrous (rarely), or scabrous; pointed (linear lanceolate to almost setiform); subulate, or not subulate; awned to awnless; carinate (often), or non-carinate; similar. Lower glume 1–3 nerved. Upper glume 1–3 nerved. *Spikelets* with incomplete florets. The incomplete florets distal to the female-fertile florets. *The distal incomplete florets* merely underdeveloped.

Female-fertile florets *(2–)3–7(–12)*. Lemmas acute, with a cusp or awn; similar in texture to the glumes (leathery); not becoming indurated; entire; pointed; mucronate, or awned. Awns when present, 1; median; apical; non-geniculate; much shorter than the body of the lemma (up to 4 mm long); entered by several veins. Lemmas hairy (pilose), or hairless; when hairless glabrous, or scabrous; non-carinate; without a germination flap; 5 nerved, or 7 nerved; with the nerves confluent towards the tip. *Palea* present; relatively long (almost equalling the lemma); awnless, without apical setae; not indurated; 2-nerved; 2-keeled (pilose or scabrous along the keels, and often between them). *Lodicules* present; 2; free; membranous; ciliate; not toothed (usually entire). *Stamens* 3. Anthers 5–7 mm long (relatively long); not penicillate; without an apically prolonged connective. **Ovary hairy**. Styles free to their bases. Stigmas 2; white.

Fruit, embryo and seedling. *Fruit* adhering to lemma and/or palea; small to medium sized (2.5–9 mm long); longitudinally grooved; compressed dorsiventrally; usually with hairs confined to a terminal tuft. Hilum long-linear. Embryo small. Endosperm hard; without lipid; containing only simple starch grains. Embryo without an epiblast; without a scutellar tail; with a negligible mesocotyl internode; with one scutellum bundle. Embryonic leaf margins meeting.

First seedling leaf with a well-developed lamina. The lamina narrow.

Abaxial leaf blade epidermis. *Costal/intercostal zonation* conspicuous, or lacking. *Papillae* absent. Long-cells of similar wall thickness costally and intercostally, or differing markedly in wall thickness costally and intercostally. Mid-intercostal long-cells having markedly sinuous walls. *Microhairs* absent. *Stomata* common. Subsidiaries dome-shaped. *Intercostal short-cells* common; in cork/silica-cell pairs (and solitary); silicified. *Costal short-cells* predominantly paired (and solitary). Costal silica bodies rounded, or tall-and-narrow (or more or less rectangular); not sharp-pointed.

Transverse section of leaf blade, physiology. C_3; XyMS+. *Leaf blade* with distinct, prominent adaxial ribs; with the ribs more or less constant in size (e.g., *L. condensatus*), or with the ribs very irregular in sizes (e.g. *L. arenarius*). Midrib not readily distinguishable; with one bundle only. Bulliforms present in discrete, regular adaxial groups, or not in discrete, regular adaxial groups (in the furrows, often small and variable in size, cf. *Ammophila*); sometimes in simple fans. All the vascular bundles accompanied by sclerenchyma. Combined sclerenchyma girders present (with the primaries); forming 'figures', or nowhere forming 'figures'. Sclerenchyma all associated with vascular bundles.

Cytology. Chromosome base number, $x = 7$. $2n = 28, 42, 56$, and 84. 4, 6, 8, and 12 ploid. Haplomic genome content J and N.

Taxonomy. Pooideae; Triticodae; Triticeae.

Ecology, geography, regional floristic distribution. About 30 species. In the non-tropical southern hemisphere and especially mountains of central Asia and North America. Commonly adventive. Mesophytic, or xerophytic (often littoral or halophytic); species of open habitats; halophytic, or glycophytic. With a wide habitat range, including coastal sand dunes.

Holarctic. Boreal. Euro-Siberian, Atlantic North American, and Rocky Mountains. European and Siberian. Canadian-Appalachian.

Hybrids. Intergeneric hybrids with *Agropyron* (×*Leymopyron* Tsvelev), *Elytrigia* (×*Leymotrigia* Tsvelev), *Psathyrostachys* (×*Leymostachys* Tsvelev), *Thinopyrum* (several species involved). See also ×*Elyleymus* Baum.

Rusts and smuts. Rusts — *Puccinia*.

Economic importance. Including useful sandbinders.

References, etc. Morphological/taxonomic: Löve 1984; Atkins, Barkworth and Dewey 1984. Leaf anatomical: Metcalfe 1960 (after nomenclatural adjustments) and this project.

Libyella Pamp.

Habit, vegetative morphology. Diminutive annual; caespitose. *Culms* 2–5 cm high, or 0 cm high (the internodes reduced); herbaceous. *Leaves* mostly basal; non-auriculate. *Sheath margins joined*. Leaf blades linear to linear-lanceolate; narrow; 0.4–1 mm wide; folded; without cross venation; persistent. *Ligule* present (its margins connate); an unfringed membrane; not truncate; 0.5–1 mm long. *Contra-ligule* absent.

Reproductive organization. Plants bisexual, with bisexual spikelets; with hermaphrodite florets. The spikelets of sexually distinct forms on the same plant; hermaphrodite and female-only, or hermaphrodite, female-only, and male-only (the lowest spikelet of the terminal inflorescence female and hidden in the uppermost sheath, the rest hermaphrodite or rarely male). The male and female-fertile spikelets segregated, in different parts of the same inflorescence branch (the lowest spikelet of the inflorescence, hidden in the uppermost leaf sheath, being female). *Plants with hidden cleistogenes. The hidden* cleistogenes *in the leaf sheaths and subterranean (in the radical leaf sheaths: solitary, lacking glumes, with stigmas up to 25 mm long projecting from the sheaths)*.

Inflorescence. *Inflorescence* few spikeleted; *a single spike*. Inflorescence with axes ending in spikelets. Inflorescence espatheate; not comprising 'partial inflorescences' and foliar organs. Spikelet-bearing axes persistent. Spikelets solitary; not secund; sessile to subsessile (occasionally), or pedicellate; distant.

Female-fertile spikelets. Spikelets 3–4 mm long; compressed laterally; disarticulating above the glumes. Rachilla terminated by a female-fertile floret. Hairy callus absent.

Glumes one per spikelet (in the hermaphrodite/male spikelets), or two (in the single female spikelet); very small, scarious; very unequal to more or less equal; shorter than the spikelets; shorter than the adjacent lemmas; lateral to the rachis; hairless; glabrous; not pointed; awnless; non-carinate; similar. Lower glume 0 nerved, or 1 nerved. Upper glume 0 nerved, or 1 nerved. *Spikelets* with female-fertile florets only.

Female-fertile florets 1 (but cleistogamous spikelets of *L. cyrenaica* may have two). Lemmas lanceolate; similar in texture to the glumes (scarious); not becoming indurated; more or less incised; not deeply cleft (irregularly serrate or erose); awnless; hairy (shortly villous on the lower half); carinate; without a germination flap; 5 nerved (though the laterals may be very short). *Palea* present; relatively long; irregularly serrate or erose; awnless, without apical setae; textured like the lemma; not indurated; 2-nerved; 2-keeled. Palea back hairy. Palea keels wingless; hairy. *Lodicules* present, or absent; when present, 2; free; membranous; glabrous; toothed; not or scarcely vascularized. *Stamens* 3. *Ovary* glabrous. Styles free to their bases. Stigmas 2.

Fruit, embryo and seedling. *Fruit* free from both lemma and palea; small (1.4–2.5 mm long); compressed laterally. Hilum short. Embryo small. Endosperm hard; containing compound starch grains. Embryo with an epiblast; without a scutellar tail; with a negligible mesocotyl internode. Embryonic leaf margins meeting.

Abaxial leaf blade epidermis. *Costal/intercostal zonation* lacking. *Papillae* absent. *Long-cells* similar in shape costally and intercostally; of similar wall thickness costally and intercostally (rather thick walled). Intercostal zones with typical long-cells. Mid-intercostal long-cells fusiform (these, and all the long-cells); having straight or only gently undulating walls. *Microhairs* absent. *Stomata* fairly common. Subsidiaries non-papillate; parallel-sided. Guard-cells conspicuously overlapped by the interstomatals. *Intercostal short-cells* absent or very rare. No prickles or macrohairs seen. Costal zones without short-cells. Costal silica bodies absent.

Transverse section of leaf blade, physiology. C_3; XyMS+. *Mesophyll* with non-radiate chlorenchyma; without adaxial palisade. *Midrib* fairly conspicuous (via the large bundle and a fairly prominent adaxial rib); with one bundle only (this being the only primary bundle in the blade). *The lamina* symmetrical on either side of the midrib. Bulliforms seemingly not in discrete, regular adaxial groups. Many of the smallest vascular bundles unaccompanied by sclerenchyma (the blade apparently lacking bundle-associated sclerenchyma). Combined sclerenchyma girders absent. Sclerenchyma confined to a tiny group in each blade margin.

Taxonomy. Pooideae; Poodae; Aveneae.

Ecology, geography, regional floristic distribution. 1 species. Libya, Morocco. In coastal sandy places.

Holarctic. Tethyan. Mediterranean.

References, etc. Leaf anatomical: this project.

Limnas Trin.

Habit, vegetative morphology. Perennial. *Culms* 15–60 cm high; herbaceous. Leaves non-auriculate. *Leaf blades* linear; narrow; 0.6–4 mm wide; setaceous, or not setaceous; flat, or folded; *not pseudopetiolate*; without cross venation. **Ligule** present; *an unfringed membrane*; not truncate; 1–3 mm long.

Reproductive organization. Plants bisexual, with bisexual spikelets; with hermaphrodite florets.

Inflorescence. Inflorescence paniculate; fairly open (rather narrow, with short branchlets); espatheate; not comprising 'partial inflorescences' and foliar organs. Spikelet-bearing axes persistent. Spikelets not secund; pedicellate.

Female-fertile spikelets. *Spikelets* 3–5 mm long; *not noticeably compressed*; falling with the glumes. Rachilla terminated by a female-fertile floret. Hairy callus present (beneath the glumes).

Glumes two; more or less equal; long relative to the adjacent lemmas; pointed (acute); awnless; carinate; similar (leathery). Lower glume 3 nerved. Upper glume 3 nerved. Spikelets with female-fertile florets only; *without proximal incomplete florets*.

Female-fertile florets 1. Lemmas similar in texture to the glumes (leathery); not becoming indurated; entire; pointed (acute); awned. Awns 1; median; dorsal; from well down the back; geniculate; entered by one vein. Lemmas hairless; non-carinate; without a germination flap; prominently 5 nerved. **Palea** present; conspicuous but relatively short (about 1/4 of the lemma length); *1-nerved*; one-keeled. *Lodicules* present; 2; free; membranous; glabrous; not toothed; not or scarcely vascularized. *Stamens* 2, or 3. Anthers 2–2.5 mm long. *Ovary* glabrous. Styles fused. Stigmas 2.

Fruit, embryo and seedling. *Fruit* free from both lemma and palea.

Abaxial leaf blade epidermis. *Costal/intercostal zonation* lacking. *Papillae* absent. *Long-cells* similar in shape costally and intercostally; of similar wall thickness costally and intercostally. Mid-intercostal long-cells rectangular; having markedly sinuous walls. *Microhairs* absent. *Stomata* absent or very rare. *Intercostal short-cells* common; in cork/silica-cell pairs (many also in short rows, interspersed with prickles); silicified. *Crown cells* present (common, in addition to the numerous prickles). *Costal short-cells* neither distinctly grouped into long rows nor predominantly paired (in short rows and paired). Costal silica bodies rounded, tall-and-narrow, and crescentic; sharp-pointed (many having sharp points, including 'tall, narrow' and 'round' forms).

Transverse section of leaf blade, physiology. *C₃*; XyMS+. *Mesophyll* with non-radiate chlorenchyma; without adaxial palisade. *Leaf blade* with distinct, prominent adaxial ribs (and the leaf margins inrolling); with the ribs very irregular in sizes. *Midrib* conspicuous (by a slight abaxial keel); with one bundle only. Bulliforms not in discrete, regular adaxial groups (bulliforms not apparent in the poor material seen). All the vascular bundles accompanied by sclerenchyma. Sclerenchyma not all bundle-associated. The 'extra' sclerenchyma in a continuous abaxial layer (constituting a lignified hypodermis).

Cytology. Chromosome base number, $x = 7$. $2n = 28$.

Taxonomy. Pooideae; Poodae; Aveneae.

Ecology, geography, regional floristic distribution. 2 species. Central Asia to Northeast Siberia. Mesophytic to xerophytic; glycophytic. Open woods and stony slopes.

Holarctic. Boreal. Euro-Siberian. Siberian.

References, etc. Leaf anatomical: this project.

Special comments. Fruit data wanting.

Limnodea Dewey ex Coult.

Greenia Nutt, *Sclerachne* Trin., *Thurberia* Benth.

Habit, vegetative morphology. *Annual*; caespitose. *Culms* 20–50 cm high; herbaceous; branching basally. *Leaves* not basally aggregated; non-auriculate. Leaf blades narrow; flat (pubescent); without cross venation. Ligule an unfringed membrane; not truncate (lacerate); 1–2 mm long.

Reproductive organization. Plants bisexual, with bisexual spikelets; with hermaphrodite florets.

Inflorescence. Inflorescence paniculate; open (narrow); espatheate; not comprising 'partial inflorescences' and foliar organs. Spikelet-bearing axes persistent. Spikelets not secund; pedicellate.

Female-fertile spikelets. *Spikelets* 3–4 mm long; *not noticeably compressed; falling with the glumes. Rachilla prolonged beyond the uppermost female-fertile floret (as a short bristle)*. The rachilla extension naked. Hairy callus absent. *Callus* conspicuous, beneath the glumes.

Glumes two; more or less equal; exceeding the spikelets; long relative to the adjacent lemmas; often densely hairy (hispidulous or pilose); pointed (acute); awnless; non-carinate; similar (firm, leathery). Lower glume 3 nerved. Upper glume 3 nerved. *Spikelets* with female-fertile florets only.

Female-fertile florets 1. Lemmas less firm than the glumes (membranous); not becoming indurated; incised; 2 lobed; not deeply cleft (bidentate); awned. Awns 1; median; dorsal; from near the top; geniculate; hairless; much longer than the body of the lemma (slender, 8–10 mm long); entered by one vein. Lemmas hairless; weakly carinate, or non-carinate; without a germination flap; inconspicuously 3 nerved. **Palea** present; **conspicuous but relatively short**; thinner than the lemma (hyaline); 2-nerved; keel-less. *Lodicules* present; 2; free; membranous; glabrous; toothed, or not toothed; not or scarcely vascularized. *Stamens* 3. *Ovary* glabrous. Styles free to their bases. Stigmas 2.

Fruit, embryo and seedling. Fruit small (2.5 mm long). Embryo small. Endosperm hard; containing compound starch grains. Embryo with an epiblast; without a scutellar tail; with a negligible mesocotyl internode. Embryonic leaf margins meeting.

Abaxial leaf blade epidermis. *Costal/intercostal zonation* conspicuous. *Papillae* absent. *Long-cells* markedly different in shape costally and intercostally; of similar wall thickness costally and intercostally, or differing markedly in wall thickness costally and intercostally. Mid-intercostal long-cells rectangular and fusiform (mostly); having straight or only gently undulating walls. *Microhairs* absent. *Stomata* common (adjoining the veins). Subsidiaries low dome-shaped, or parallel-sided. Guard-cells overlapped by the interstomatals. *Intercostal short-cells* absent or very rare. *Costal short-cells* conspicuously in long rows, or neither distinctly grouped into long rows nor predominantly paired. Costal silica bodies horizontally-elongated crenate/sinuous (with a few deeply crenate examples approaching 'nodular'); not sharp-pointed.

Transverse section of leaf blade, physiology. C_3; XyMS+. *Mesophyll* with non-radiate chlorenchyma. Leaf blade with the ribs more or less constant in size. Midrib with one bundle only. All the vascular bundles accompanied by sclerenchyma. Combined sclerenchyma girders present. Sclerenchyma all associated with vascular bundles.

Cytology. Chromosome base number, $x = 7$. $2n = 14$. 2 ploid. Nucleoli disappearing before metaphase.

Taxonomy. Pooideae; Poodae; Aveneae.

Ecology, geography, regional floristic distribution. 1 species. Southern U.S.A. Xerophytic; species of open habitats. Prairie.

Holarctic. Boreal and Madrean. Atlantic North American. Southern Atlantic North American and Central Grasslands.

Rusts and smuts. Rusts — *Puccinia*. Taxonomically wide-ranging species: *Puccinia graminis*.

References, etc. Leaf anatomical: this project.

Limnopoa C.E. Hubb.

Habit, vegetative morphology. Annual, or perennial; forming floating mats. *Culms* 5–25 cm high; herbaceous; branched above. Culm nodes glabrous. *Leaves* not basally aggregated; non-auriculate. Sheath margins free. The sheaths rather inflated. Leaf blades narrow (short); setaceous (towards the tips); without cross venation; persistent. Ligule a fringe of hairs.

Reproductive organization. *Plants bisexual, with bisexual spikelets*; *without hermaphrodite florets (the lower of the two in each spikelet male, the upper female)*.

Inflorescence. *Inflorescence a single raceme (unilateral)*; contracted (reduced, spicate, terminal, the sessile spikelets embedded, the pedicellate ones closely applied to the hollowed rachis). Rachides hollowed. Inflorescence espatheate; not comprising 'partial inflorescences' and foliar organs. Spikelet-bearing axes persistent. Spikelets paired; secund; pedicellate; consistently in 'long-and-short' combinations. The 'shorter' spikelets hermaphrodite. The 'longer' spikelets hermaphrodite.

Female-fertile spikelets. Spikelets 3–4 mm long; abaxial; compressed dorsiventrally; disarticulating above the glumes; disarticulating between the florets. Rachilla terminated by a female-fertile floret. Hairy callus absent.

Glumes two; more or less equal; shorter than the adjacent lemmas (but exceeding the small L_2); dorsiventral to the rachis; hairless; awnless; non-carinate; similar (ovate, membranous). Lower glume 5–9 nerved (with three main veins). Upper glume 5–9 nerved (3–4 main veins). *Spikelets* with incomplete florets (with 2 florets, the lower male, the upper female). The incomplete florets proximal to the female-fertile florets. *The proximal incomplete florets* 1; paleate. Palea of the proximal incomplete florets fully developed. The proximal incomplete florets male. The proximal lemmas narrowly ovate, glabrous; awnless; 5–7 nerved; decidedly exceeding the female-fertile lemmas; less firm than the female-fertile lemmas (membranous); not becoming indurated.

Female-fertile florets 1. Lemmas elliptic, puberulous, about 2/3 the length of L_1; decidedly firmer than the glumes; not becoming indurated (cartilaginous); entire; blunt; awnless; hairy (woolly); non-carinate; having the margins tucked in onto the palea; with a clear germination flap; 5–7 nerved. *Palea* present; relatively long; entire; awnless, without apical setae; cartilaginous, hairy; about 5; 2-keeled. *Lodicules* present; 2; free; fleshy; glabrous; not or scarcely vascularized. *Stamens* 0 (three with long non-penicillate anthers, in the lower floret only). *Ovary* glabrous. Styles fused to free to their bases. Stigmas 2; red pigmented.

Fruit, embryo and seedling. Fruit small; compressed dorsiventrally. ***Hilum long-linear***. Embryo small; not waisted.

Abaxial leaf blade epidermis. *Costal/intercostal zonation* conspicuous. *Papillae* present. Intercostal papillae over-arching the stomata; consisting of one symmetrical projection per cell. Intercostal zones exhibiting many atypical long-cells. Mid-intercostal long-cells rectangular, or fusiform; having straight or only gently undulating walls. *Microhairs* present (but very scarce); panicoid-type; (51–)57–60(–66) microns long; 9–11.4 microns wide at the septum. Microhair total length/width at septum 5–6.9. Microhair apical cells (34–)36–41(–45) microns long. Microhair apical cell/total length ratio 0.6–0.68. *Stomata* common; 21–27 microns long. Subsidiaries mostly dome-shaped. Guard-cells overlapping to flush with the interstomatals. *Intercostal short-cells* absent or very rare. *Costal short-cells* conspicuously in long rows. Costal silica bodies 'panicoid-type'; cross shaped and dumb-bell shaped (with sharp points — prickly); sharp-pointed (prickly crosses & dumb-bells).

Transverse section of leaf blade, physiology. C_3; XyMS+. Mesophyll *Isachne*-type. *Leaf blade* with distinct, prominent adaxial ribs; with the ribs more or less constant in size. *Midrib* not readily distinguishable (but mid-zone of leaf distinguished by presence of air-spaces); with one bundle only. Bulliforms present in discrete, regular adaxial groups; in simple fans. All the vascular bundles accompanied by sclerenchyma. Combined sclerenchyma girders absent. Sclerenchyma all associated with vascular bundles.

Taxonomy. Panicoideae; Panicodae; Isachneae.

Ecology, geography, regional floristic distribution. 1 species. India. Hydrophytic (in tanks, forming a thick mass of tangled stems on the surface).

Paleotropical. Indomalesian. Indian.

References, etc. Morphological/taxonomic: Hubbard 1943. Leaf anatomical: this project.

Lindbergella Bor

Habit, vegetative morphology. *Annual*. Culms herbaceous. Leaves non-auriculate. Leaf blades narrow; without cross venation. *Ligule an unfringed membrane*; not truncate; 2–3 mm long.

Reproductive organization. Plants bisexual, with bisexual spikelets; with hermaphrodite florets. The spikelets all alike in sexuality.

Inflorescence. *Inflorescence* paniculate; open; *without conspicuously divaricate branchlets*; espatheate; not comprising 'partial inflorescences' and foliar organs. Spikelet-bearing axes persistent. Spikelets not secund; pedicellate (the pedicels terete).

Female-fertile spikelets. Spikelets 3–4.5 mm long; compressed laterally; disarticulating above the glumes. Rachilla prolonged beyond the uppermost female-fertile floret. The rachilla extension with incomplete florets. Hairy callus absent.

Glumes two; very unequal to more or less equal; shorter than the adjacent lemmas; pointed (acute to acuminate); awnless; carinate; similar. *Lower glume 3 nerved*. Upper glume 3 nerved, or 5 nerved. *Spikelets* with incomplete florets. The incomplete florets distal to the female-fertile florets.

Female-fertile florets *2–5*. *Lemmas* similar in texture to the glumes (leathery); not becoming indurated; *incised*; 2 lobed; not deeply cleft (notched); *mucronate (with a tiny 'awn' in the apical notch)*; *hairy (but asperulous, not 'webby', with appressed hairs on the nerves)*; carinate; without a germination flap; *3 nerved*. *Palea* present; relatively long; 2-nerved; 2-keeled. *Lodicules* present; 2; joined, or free; membranous; glabrous; toothed; not or scarcely vascularized. *Stamens* 3. *Ovary* glabrous. Styles free to their bases. Stigmas 2.

Fruit, embryo and seedling. Fruit small (2.5–2.8 mm long); compressed laterally. Hilum short. Embryo small.

Abaxial leaf blade epidermis. *Costal/intercostal zonation* conspicuous. *Papillae* absent. *Long-cells* markedly different in shape costally and intercostally (the costals narrower, rectangular); differing markedly in wall thickness costally and intercostally (the intercostals thin walled, the costals quite thick walled). Mid-intercostal long-cells fusiform; having straight or only gently undulating walls. *Microhairs* absent. *Stomata* fairly common. Subsidiaries non-papillate; parallel-sided. Guard-cells overlapped by the interstomatals. *Intercostal short-cells* absent or very rare. Strongly pointed prickles common costally. *Costal short-cells* neither distinctly grouped into long rows nor predominantly paired (mostly solitary, fairly infrequent). Costal silica bodies present and well developed; horizontally-elongated crenate/sinuous and horizontally-elongated smooth.

Transverse section of leaf blade, physiology. C_3; XyMS+. *Leaf blade* adaxially flat, or with distinct, prominent adaxial ribs; with the ribs more or less constant in size (only slight ribs apparent in the collapsed material seen). *Midrib* conspicuous to not readily distinguishable (perhaps slightly larger and more prominent abaxially); with one bundle only. *The lamina* symmetrical on either side of the midrib. Bulliforms not determinable in the material seen. All the vascular bundles accompanied by sclerenchyma. Combined sclerenchyma girders absent (all the bundles seeming to have adaxial and abaxial strands). Sclerenchyma all associated with vascular bundles.

Taxonomy. Pooideae; Poodae; Poeae.

Ecology, geography, regional floristic distribution. 1 species. Cyprus. Species of open habitats. Rocky slopes.

Holarctic. Tethyan. Mediterranean.

References, etc. Leaf anatomical: this project.

Linkagrostis Garcia, Blanca & Torres

~ *Agrostis* (*A. juressi* Link)

Habit, vegetative morphology. Perennial; stoloniferous. **Culms** 40–80 cm high; *herbaceous*; unbranched above; 2–4 noded. Culm nodes exposed. Culm leaves present. Young shoots intravaginal. *Leaves* not basally aggregated; non-auriculate. **Leaf blades** linear to lanceolate; narrow; *(4–)5–8(–10) mm wide (4–15 cm long)*; flat; without cross venation; persistent. Ligule an unfringed membrane; truncate; 0.2–0.5 mm long.

Reproductive organization. Plants bisexual, with bisexual spikelets; with hermaphrodite florets. The spikelets hermaphrodite.

Inflorescence. Inflorescence paniculate; contracted (with short, erect branches); spicate (interrupted); espatheate; not comprising 'partial inflorescences' and foliar organs. Spikelet-bearing axes persistent. *Spikelets* not secund; *pedicellate (the pedicels much shorter than the spikelets, not clavate)*; imbricate.

Female-fertile spikelets. Spikelets 2.5–3.2 mm long; compressed laterally; disarticulating above the glumes. Hairy callus absent. *Callus* glabrous.

Glumes two; more or less equal; *about equalling the spikelets to exceeding the spikelets*; long relative to the adjacent lemmas (somewhat exceeding it); puberulent; not pointed; awnless; carinate; similar (herbaceous, lanceolate). Lower glume 1 nerved. *Upper* glume *1 nerved*. **Spikelets** with female-fertile florets only; *without proximal incomplete florets*.

Female-fertile florets *1. Lemmas similar in texture to the glumes (leathery, green)*; not becoming indurated; entire to incised; more or less linear, narrowed above; not deeply cleft (obtuse, truncate or denticulate); awnless to mucronate; hairless; non-carinate; without a germination flap; *5 nerved*; with the nerves non-confluent. **Palea** present; *conspicuous but relatively short to very reduced (no more than 1/5 the lemma length)*; tightly clasped by the lemma; apically notched; awnless, without apical setae. *Lodicules* present; 2; free; membranous (falcate); glabrous. *Stamens* 3. Anthers about 1.2 mm long. *Ovary* glabrous. Stigmas 2; white.

Fruit, embryo and seedling. Fruit small (about 1.5 mm long); ellipsoid; longitudinally grooved; compressed dorsiventrally. Hilum short. Embryo small.

Abaxial leaf blade epidermis. *Costal/intercostal zonation* conspicuous. *Papillae* absent. Intercostal zones with typical long-cells. Mid-intercostal long-cells fusiform; having straight or only gently undulating walls. *Microhairs* absent. *Stomata* common. Costal silica bodies horizontally-elongated crenate/sinuous (?).

Transverse section of leaf blade, physiology. C$_3$; XyMS+. Mesophyll not *Isachne*-type; not traversed by colourless columns. *Leaf blade* adaxially flat. *Midrib* not readily distinguishable; with one bundle only; without colourless mesophyll adaxially. Bulliforms present in discrete, regular adaxial groups; in simple fans. Combined sclerenchyma girders present. Sclerenchyma all associated with vascular bundles.

Cytology. Chromosome base number, $x = 7$. $2n = 14$.

Taxonomy. Pooideae; Poodae; Aveneae.

Ecology, geography, regional floristic distribution. 1 species. Iberian Peninsula and northeast Africa. Mesophytic; shade species; glycophytic. In damp, shady places. Holarctic. Tethyan. Mediterranean.

References, etc. Morphological/taxonomic: Garcia, Blanca and Torres 1987. Leaf anatomical: Garcia, Blanca and Torres 1987.

Lintonia Stapf

Joannegria Chiov., *Negria* Chiov.

Habit, vegetative morphology. Perennial; caespitose, or rhizomatous and caespitose. **Culms** 20–90 cm high; *herbaceous*; branched above. Culm nodes glabrous. Culm internodes hollow. Plants unarmed. Young shoots intravaginal. *Leaves* not basally aggregated; non-auriculate. Leaf blades linear (tapered to a fine, acuminate tip); narrow; 2–6 mm wide; flat; without abaxial multicellular glands; without cross venation; persistent. **Ligule** *an unfringed membrane (minutely ciliolate, with long hairs at the auricle positions)*; truncate; 1 mm long. *Contra-ligule* absent.

Reproductive organization. Plants bisexual, with bisexual spikelets; with hermaphrodite florets.

Inflorescence. *Inflorescence of spicate main branches*; open; digitate, or non-digitate (*L. brizoides*). Primary inflorescence branches 2–8. Inflorescence espatheate (but often enveloped below by the sheath of the uppermost culm leaf); not comprising 'partial inflorescences' and foliar organs. *Spikelet-bearing axes* 'racemes'. The racemes without spikelets towards the base. Spikelet-bearing axes sinuous, with very slender rachides; persistent. Spikelets solitary; somewhat secund, or not secund; biseriate, or not two-ranked; subsessile to pedicellate.

Female-fertile spikelets. Spikelets 4–11 mm long; plump, cuneate, or elliptic; adaxial; compressed laterally; disarticulating above the glumes; not disarticulating between the florets (the rachilla tough); with conventional internode spacings. Rachilla prolonged beyond the uppermost female-fertile floret; rachilla hairy (below the lowest floret), or hairless (between the upper florets). The rachilla extension with incomplete florets. Hairy callus present (i.e., below the lowermost floret). *Callus* on the lower floret short; blunt.

Glumes two; very unequal (G$_1$ shorter); shorter than the spikelets; shorter than the adjacent lemmas; free; dorsiventral to the rachis; hairless; glabrous (the keel sometimes scabrous); pointed (the nerve slightly excurrent), or not pointed (slightly notched); awnless (sub-mucronate); carinate (slightly), or non-carinate (rounded on the back); similar (persistent, hyaline-membranous). Lower glume 1 nerved. Upper glume 1 nerved. *Spikelets* with incomplete florets. The incomplete florets distal to the female-fertile florets. The distal incomplete florets 1–4; *incomplete florets* merely underdeveloped; incomplete florets shortly awned.

Female-fertile florets 2–4. Lemmas decidedly firmer than the glumes (tough and cartilaginous, at least in part); not becoming indurated; shortly incised; 2 lobed; not deeply cleft (emarginate or 2-toothed); awned. *Awns* 1; median; *dorsal*; from near the top; non-geniculate (curved); hairless (scabrid); much shorter than the body of the lemma to about as long as the body of the lemma; entered by one vein. *Lemmas hairy (the hairs clavate and apiculate, in 5–6 longitudinal rows on the nerves)*; non-carinate (abaxially rounded); without a germination flap; 5–9 nerved; with the nerves non-confluent. *Palea* present; relatively long (about 3/4 the length of the lemma); entire, or apically notched (between the prolonged nerve tips); awnless, without apical setae, or with apical setae (comprising the short scabrid nerve-tips); thinner than the lemma; not indurated (hyaline); 2-nerved; 2-keeled (sharply complanate). Palea keels wingless; hairy (conspicuously ciliate). *Lodicules* present; 2; free; fleshy (cylindrical rather than cuneate); glabrous. **Stamens 3; *with free filaments (these attached at the base of the ovary, the androecium and gynoecium being borne on a short stipe between the lodicules)*.** Anthers 1 mm long; not penicillate; without an apically prolonged connective. *Ovary* glabrous. Styles free to their bases. Stigmas 2; red pigmented (dark purple).

Fruit, embryo and seedling. *Fruit* free from both lemma and palea; small (1.3–2.2 mm long); ellipsoid; strongly compressed dorsiventrally. Hilum short (elliptical). *Pericarp free*. Embryo large (about half the length of the fruit); not waisted.

Abaxial leaf blade epidermis. *Costal/intercostal zonation* conspicuous. *Papillae* present; intercostal. Intercostal papillae over-arching the stomata (at one end); consisting of one symmetrical projection per cell (to almost spherical). *Long-cells* similar in shape costally and intercostally; of similar wall thickness costally and intercostally (thin walled). Mid-intercostal long-cells rectangular; having markedly sinuous walls. *Microhairs* present; more or less spherical; clearly two-celled; chloridoid-type (large, with a relatively short basal cell and a hemispherical distal cell). Microhair apical cell wall of similar thickness/rigidity to that of the basal cell. Microhairs 18–21 microns long. Microhair basal cells 6 microns long. Microhairs (9–)11.4–12(–13.5) microns wide at the septum. Microhair total length/width at septum 1.5–2.2. Microhair apical cells 12–13.5 microns long. Microhair apical cell/total length ratio 0.6–0.75. *Stomata* common; 22.5–25.5 microns long. Subsidiaries mostly triangular. Guard-cells overlapping to flush with the interstomatals. *Intercostal short-cells* absent or very rare; where detected, not paired. Intercostal silica bodies absent. *Crown cells* absent (and prickles at leaf margins only). *Costal short-cells* conspicuously in long rows (but the short-cells sometimes rather long). Costal silica bodies present in alternate cell files of the costal zones; saddle shaped; not sharp-pointed.

Transverse section of leaf blade, physiology. C_4; XyMS+. PCR sheath outlines uneven. PCR sheaths of the primary vascular bundles interrupted; interrupted both abaxially and adaxially. PCR sheath extensions absent. PCR cell chloroplasts centrifugal/peripheral. *Mesophyll* with radiate chlorenchyma. *Midrib* conspicuous; having a conventional arc of bundles (large median, and 4 small bundles on each side); with colourless mesophyll adaxially. Bulliforms present in discrete, regular adaxial groups; in simple fans and associated with colourless mesophyll cells to form deeply-penetrating fans (a few of the latter only, with small internal colourless cells). All the vascular bundles accompanied by sclerenchyma. Combined sclerenchyma girders present (with the primaries); forming 'figures'. Sclerenchyma all associated with vascular bundles. The lamina margins with fibres.

Cytology. $2n = 30$.

Taxonomy. Chloridoideae; main chloridoid assemblage.

Ecology, geography, regional floristic distribution. 2 species. Tropical east Africa. Helophytic, or mesophytic; shade species, or species of open habitats; glycophytic. Savanna, heavy soils in seasonally wet places.

Paleotropical. African. Sudano-Angolan. Sahelo-Sudanian, Somalo-Ethiopian, South Tropical African, and Kalaharian.

References, etc. Leaf anatomical: this project; photos of *L. nutans* provided by R.P. Ellis.

Lithachne P. Beauv.

Habit, vegetative morphology. Perennial; caespitose. The flowering culms leafy. *Culms* 15–60 cm high; woody and persistent, or herbaceous (wiry); unbranched above. Plants unarmed. *Leaves* not basally aggregated; auriculate (in the form of tiny, membranous, erect appendages in *L. pauciflora*), or non-auriculate; without auricular setae. Leaf blades ovate-lanceolate; broad, or narrow; 8–30 mm wide; not cordate, not sagittate (the base asymmetric); flat, or rolled (reflexing at night); pseudopetiolate; without cross venation; disarticulating from the sheaths to persistent; rolled in bud. *Ligule* present (or with a tuft of hairs in its place); an unfringed membrane (minutely ciliolate, the pseudopetiole hairy above it at least in *L. pauciflora*); truncate. *Contra-ligule* absent.

Reproductive organization. *Plants monoecious with all the fertile spikelets unisexual*; without hermaphrodite florets. The spikelets of sexually distinct forms on the same plant; female-only and male-only. The male and female-fertile spikelets in different inflorescences, or in different inflorescences and segregated, in different parts of the same inflorescence branch (with a terminal male panicle (sometimes absent), and axillary infloresences reduced to a single female spikelet or the latter with males below). The spikelets overtly heteromorphic.

Inflorescence. Inflorescence determinate, or indeterminate (2–many inflorescences per node in *L. pauciflora*); *a single raceme, or paniculate (few-flowered axillary racemes or panicles, terminated by female spikelets, or reduced to two spikelets, female above and male below. Terminal panicle if present male)*; spatheate; a complex of 'partial inflorescences' and intervening foliar organs. *Spikelet-bearing axes* 'racemes', or paniculate; persistent. Spikelets not secund; pedicellate (male pedicels shorter than female, the latter expanded above).

Female-sterile spikelets. The male spikelets smaller than the female, reduced to lemma, palea and 2–3 free stamens. Rachilla of male spikelets terminated by a male floret. The male spikelets without glumes; without proximal incomplete florets; 1 floreted. Male florets 1; 2–3 staminate.

Female-fertile spikelets. *Spikelets* 6–8 mm long (triangular); *compressed laterally*; disarticulating above the glumes; with a distinctly elongated rachilla internode above the glumes (the segment under the floret being distinctly swollen). Rachilla terminated by a female-fertile floret. Hairy callus absent. *Callus* short (columnar).

Glumes two; more or less equal; about equalling the spikelets; long relative to the adjacent lemmas; pointed (ovate-acuminateto caudate); awnless; non-carinate (rounded on the back); similar (membranous). Lower glume 9–11 nerved. Upper glume 9–11 nerved. *Spikelets* with female-fertile florets only.

Female-fertile florets 1. *Lemmas bony, dorsally gibbous and asymmetrical; saccate*; decidedly firmer than the glumes; smooth (shiny); becoming indurated; entire; blunt (hooded); awnless; hairless; glabrous; non-carinate (dorsally rounded); with a clear germination flap; 7 nerved; with the nerves confluent towards the tip. *Palea* present; relatively long; entire; awnless, without apical setae; textured like the lemma; indurated; several nerved (3); keel-less. *Lodicules* present (tiny, in *L. pauciflora*); 3; free; glabrous; heavily vascularized. *Stamens* 0. *Ovary* glabrous. Styles fused (into one).

Fruit, embryo and seedling. *Fruit* free from both lemma and palea; small; somewhat compressed laterally. Hilum long-linear. Embryo small. Endosperm hard.

Seedling with a short mesocotyl. First seedling leaf without a lamina (the seedling with two sheaths).

Abaxial leaf blade epidermis. *Costal/intercostal zonation* conspicuous. *Papillae* present (and abundant intercostally). Intercostal papillae over-arching the stomata (but only from the subsidiaries); several per cell (small, circular, thick walled — several rows per cell, and a pair on each subsidiary). Mid-intercostal long-cells rectangular; having markedly sinuous walls. *Microhairs* present (very abundant); panicoid-type; 52–57 microns long (in *pauciflora*); 4.5–5.7 microns wide at the septum. Microhair apical cells 22.5–25.5 microns long. Microhair apical cell/total length ratio 0.39–0.47. *Stomata* common; (24–)27–30 microns long (in *L. pauciflora*). Subsidiaries papillate; triangular (sharp pointed). Guard-cells noticeably sunken, but the guard cells not noticeably overlapped. *Intercostal short-cells* common; not paired (mostly solitary). *Costal short-cells* conspicuously in long rows. Costal silica bodies intergrading oryzoid and 'panicoid-type'; cross shaped (many fat, almost square), or dumb-bell shaped; not sharp-pointed.

Transverse section of leaf blade, physiology. C$_3$; XyMS+. *Mesophyll* with non-radiate chlorenchyma; without adaxial palisade; with arm cells (almost certainly, but the material seen very poor); with fusoids. The fusoids external to the PBS. *Leaf blade* ad-axially flat. *Midrib* conspicuous (via the large bundle with its heavy sclerenchyma girders); with one bundle only. Bulliforms present in discrete, regular adaxial groups (these wide); in simple fans. All the vascular bundles accompanied by sclerenchyma. Combined sclerenchyma girders present; forming 'figures' (all the bundles with an anchor or an I). Sclerenchyma all associated with vascular bundles.

Cytology. Chromosome base number, $x = 11$. $2n = 22$ and 44. 2 and 4 ploid.

Taxonomy. Bambusoideae; Oryzodae; Olyreae.

Ecology, geography, regional floristic distribution. 4 species. West Indies, Central and South America. Shade species.

Neotropical. Caribbean, Central Brazilian, Pampas, and Andean.

References, etc. Morphological/taxonomic: Soderstrom 1980. Leaf anatomical: this project.

Littledalea Hemsley

Habit, vegetative morphology. Robust perennial; rhizomatous. *Culms* 12–35 cm high; herbaceous. Leaves non-auriculate. *Sheath margins free (for at least three quarters of the lengths of the sheaths)*. Leaf blades linear; narrow; 1–4 mm wide; flat, or rolled (convolute); without cross venation. *Ligule an unfringed membrane*; not truncate; 0.5–3.5 mm long.

Reproductive organization. Plants bisexual, with bisexual spikelets; with hermaph-rodite florets.

Inflorescence. Inflorescence paniculate; open; espatheate; not comprising 'partial in-florescences' and foliar organs. Spikelet-bearing axes persistent. *Spikelets* not secund; *pedicellate*.

Female-fertile spikelets. Spikelets 18–35 mm long (i.e., large); purplish, papery; compressed laterally; disarticulating above the glumes; disarticulating between the florets; with conventional internode spacings. Rachilla prolonged beyond the uppermost female-fertile floret; rachilla hairless. The rachilla extension with incomplete florets. Hairy callus absent. *Callus* short; blunt.

Glumes two; very unequal; shorter than the adjacent lemmas; hairless; glabrous; pointed to not pointed; awnless; non-carinate; similar. Lower glume 1–3 nerved. Upper glume 5 nerved, or 7 nerved. *Spikelets* with incomplete florets. The incomplete florets distal to the female-fertile florets. *The distal incomplete florets* merely underdeveloped; awnless.

Female-fertile florets 7–9. Lemmas apically rounded or erose; similar in texture to the glumes; not becoming indurated; incised; pointed to blunt; when incised, 2 lobed; not deeply cleft; awnless, or mucronate to awned. Awns when present, 1; median; from a sinus; non-geniculate; hairless; much shorter than the body of the lemma; entered by one vein. Lemmas hairless; scabrous; non-carinate; without a germination flap; 5–7 nerved; with the nerves non-confluent. **Palea** present; *conspicuous but relatively short (about half the lemma length)*; tightly clasped by the lemma; apically notched; awnless, without apical setae; textured like the lemma; not indurated; 2-nerved; 2-keeled. Palea keels wing-less; scabrous. *Lodicules* present; 2; free; membranous; ciliate, or glabrous; toothed; not or scarcely vascularized. *Stamens* 3. Anthers 6–7 mm long; not penicillate; without an apically prolonged connective. **Ovary** apically *hairy*; *without a conspicuous apical ap-pendage (not at all Bromus-like, in material seen)*. Styles free to their bases. Stigmas 2.

Fruit, embryo and seedling. Endosperm containing only simple starch grains.

Abaxial leaf blade epidermis. *Costal/intercostal zonation* lacking. *Papillae* absent. *Long-cells* similar in shape costally and intercostally; of similar wall thickness costally and intercostally. *Mid-intercostal long-cells* rectangular; having markedly sinuous walls (rather thick, pitted). *Microhairs* absent. *Stomata* absent or very rare (none seen). *Intercostal short-cells* common; in cork/silica-cell pairs (frequent pairs, but no silica bodies in the 'silica cells'); not silicified. *Costal short-cells* predominantly paired (some solitary). Costal silica bodies absent.

Transverse section of leaf blade, physiology. C_3; XyMS+. *Mesophyll* with non-radiate chlorenchyma; without adaxial palisade. *Leaf blade* with distinct, prominent adaxial ribs; with the ribs more or less constant in size to with the ribs very irregular in sizes. *Midrib* not readily distinguishable; with one bundle only. Bulliforms not in discrete, regular adaxial groups. All the vascular bundles accompanied by sclerenchyma. Combined sclerenchyma girders present (with the primaries only, the rest with strands); forming 'figures' (in the primaries). Sclerenchyma all associated with vascular bundles.

Taxonomy. Pooideae; Poodae (or Triticodae?); Poeae (?).

Ecology, geography, regional floristic distribution. 3 species. Central Asia to western China. Mesophytic to xerophytic; species of open habitats. Stony slopes.

Holarctic. Tethyan. Irano-Turanian.

References, etc. Leaf anatomical: this project.

Special comments. Morphological description revised by E.A. Kellogg, November 1988. Fruit data wanting.

Loliolum Krecz. & Bobr.

Habit, vegetative morphology. *Annual*. *Culms* 5–20 cm high; herbaceous. Leaves non-auriculate. Leaf blades linear; narrow; 0.6–1.2 mm wide; not setaceous; rolled (convolute); without cross venation. Ligule an unfringed membrane; truncate; 1.5 mm long.

Reproductive organization. Plants bisexual, with bisexual spikelets; with hermaphrodite florets.

Inflorescence. *Inflorescence* unilateral, *a single spike, or a single raceme*; espatheate; not comprising 'partial inflorescences' and foliar organs. Spikelet-bearing axes persistent. *Spikelets* solitary; secund; *subsessile*.

Female-fertile spikelets. Spikelets 3–6 mm long; compressed laterally; disarticulating above the glumes. Rachilla prolonged beyond the uppermost female-fertile floret. The rachilla extension with incomplete florets. Hairy callus absent.

Glumes two; very unequal to more or less equal; (the upper) about equalling the spikelets to exceeding the spikelets; (the upper) long relative to the adjacent lemmas; *displaced (the lower displaced to the outside)*; pointed (subulate-acuminate); awnless; non-carinate. Lower glume 3 nerved. *Upper glume 3 nerved*. *Spikelets* with incomplete florets. The incomplete florets distal to the female-fertile florets.

Female-fertile florets 3–6(–9). *Lemmas less firm than the glumes (leathery)*; not becoming indurated; entire; pointed (acute); awned (save for the lowest one or two, which are awnless). Awns 1; median; apical; non-geniculate; entered by one vein. Lemmas hairy, or hairless (sometimes, the lower lemmas only are hairy); non-carinate; without a germination flap; 5 nerved. *Palea* present; relatively long (subequalling the lemma); apically notched (bidentate); 2-nerved; 2-keeled. *Lodicules* present; 2; free; membranous; glabrous; toothed; not or scarcely vascularized. *Stamens* 3. Anthers 0.2–0.3 mm long. *Ovary* glabrous. Styles free to their bases. Stigmas 2.

Fruit, embryo and seedling. *Fruit* slightly adhering to lemma and/or palea; small (2–2.5 mm long); compressed dorsiventrally. Hilum short. Embryo small. Endosperm hard.

Abaxial leaf blade epidermis. *Costal/intercostal zonation* conspicuous. *Papillae* absent. *Long-cells* markedly different in shape costally and intercostally (the costals narrow, rectangular); of similar wall thickness costally and intercostally (the walls of medium thickness). *Mid-intercostal long-cells* more or less rectangular, or fusiform (especially mid-intercostally); having straight or only gently undulating walls. *Microhairs* absent. *Stomata* common (mainly adjacent to the costae). Subsidiaries non-papillate; parallel-sided to dome-shaped. Guard-cells overlapped by the interstomatals (and the stomata quite deeply sunken). *Intercostal short-cells* absent or very rare. Neither prickles nor macrohairs seen, except for the latter at the margin. *Costal short-cells* neither distinctly grouped into long rows nor predominantly paired (solitary, in pairs and short rows — the intervening long-cells sometimes fairly short). Costal silica bodies present and well developed; horizontally-elongated crenate/sinuous.

Transverse section of leaf blade, physiology. C$_3$; XyMS+. Mesophyll without adaxial palisade. *Midrib* conspicuous (mainly by its position in the infolded, almost acicular blade, which has only three bundles in the material seen); with one bundle only. *The lamina* symmetrical on either side of the midrib. Bulliforms not in discrete, regular adaxial groups (absent). All the vascular bundles accompanied by sclerenchyma. Combined sclerenchyma girders absent (each bundle — the midrib and the two laterals — with small adaxial strands only). Sclerenchyma all associated with vascular bundles.

Taxonomy. Pooideae; Poodae; Poeae.

Ecology, geography, regional floristic distribution. 1 species. Turkey & Caucasus, East to Baluchistan (Pakistan). Species of open habitats. Dry places.

Holarctic. Tethyan. Irano-Turanian.

References, etc. Leaf anatomical: this project.

Lolium L.

Arthrochortus Lowe, *Craepalia* Schrank, *Crypturus* Link

Habit, vegetative morphology. Annual, or perennial; rhizomatous, or stoloniferous, or caespitose, or decumbent. **Culms** 10–130 cm high; ***herbaceous***; unbranched above. Culm nodes glabrous. Culm internodes hollow. **Leaves** not basally aggregated; ***auriculate***. ***Sheath margins free***. Leaf blades linear; usually narrow; 2–12 mm wide; flat, or folded, or rolled; without cross venation; persistent; rolled in bud, or once-folded in bud. Ligule an unfringed membrane; truncate; 0.8–2 mm long.

Reproductive organization. Plants bisexual, with bisexual spikelets; with hermaphrodite florets; outbreeding and inbreeding.

Inflorescence. Inflorescence a single spike (with partially embedded spikelets). Rachides hollowed. Inflorescence espatheate; not comprising 'partial inflorescences' and foliar organs. Spikelet-bearing axes persistent. Spikelets solitary; not secund; conspicuously distichous; sessile.

Female-fertile spikelets. Spikelets 7–26 mm long; compressed laterally; disarticulating above the glumes; disarticulating between the florets. Rachilla prolonged beyond the uppermost female-fertile floret; rachilla hairless. The rachilla extension with incomplete florets. Hairy callus absent. *Callus* short.

Glumes one per spikelet (except that the terminal spikelet has two); shorter than the adjacent lemmas, or long relative to the adjacent lemmas; dorsiventral to the rachis; pointed, or not pointed; awnless; non-carinate. Upper glume (i.e. the only glume) 3–7 nerved (membranous). *Spikelets* with incomplete florets. The incomplete florets distal to the female-fertile florets. *The distal incomplete florets* merely underdeveloped.

Female-fertile florets 2–22. Lemmas less firm than the glumes to decidedly firmer than the glumes (membranous to papery, sometimes turgid or hardening in fruit); becoming indurated to not becoming indurated; entire, or incised; when entire pointed, or blunt; awnless, or awned. Awns when present, 1; from a sinus, or dorsal; when dorsal, from near the top; non-geniculate; hairless; much shorter than the body of the lemma; entered by one vein. Lemmas hairless; non-carinate; 5–7 nerved. *Palea* present; relatively long (usually ciliate); apically notched; awnless, without apical setae; not indurated; 2-nerved; 2-keeled. *Lodicules* present; 2; free; membranous; glabrous; toothed, or not toothed; not or scarcely vascularized. *Stamens* 3. Anthers 1.3–4.5 mm long; not penicillate. *Ovary* glabrous. Styles free to their bases. Stigmas 2; white.

Fruit, embryo and seedling. *Fruit* somewhat adhering to lemma and/or palea; small, or medium sized, or large; longitudinally grooved; compressed dorsiventrally. Hilum long-linear. Embryo small; not waisted. Endosperm hard; without lipid; containing compound starch grains. Embryo with an epiblast; without a scutellar tail; with a negligible mesocotyl internode. Embryonic leaf margins meeting.

Seedling with a short mesocotyl, or with a long mesocotyl; with a tight coleoptile. First seedling leaf with a well-developed lamina. The lamina narrow; erect; 3–7 veined.

Abaxial leaf blade epidermis. *Costal/intercostal zonation* conspicuous. *Papillae* absent. *Long-cells* markedly different in shape costally and intercostally (costals rectangular, intercostals longer, fusiform); of similar wall thickness costally and intercostally (but the walls of the costals sinuous). Mid-intercostal long-cells fusiform; having straight or only gently undulating walls (those bordering the veins sinuous, by contrast). *Microhairs* absent. *Stomata* common. Subsidiaries low dome-shaped, or parallel-sided. Guard-cells overlapped by the interstomatals. *Intercostal short-cells* absent or very rare. *Costal short-cells* predominantly paired, or neither distinctly grouped into long rows nor predominantly paired. Costal silica bodies horizontally-elongated crenate/sinuous, or horizontally-elongated smooth, or rounded (some almost cubical); not sharp-pointed.

Transverse section of leaf blade, physiology. C_3; XyMS+. *PBS cells* without a suberised lamella. *Mesophyll* with non-radiate chlorenchyma. *Leaf blade* with distinct, prominent adaxial ribs; with the ribs more or less constant in size. *Midrib* conspicuous; with one bundle only. Bulliforms present in discrete, regular adaxial groups; in simple fans. All the vascular bundles accompanied by sclerenchyma. Combined sclerenchyma girders present, or absent; nowhere forming 'figures'. Sclerenchyma all associated with vascular bundles.

Phytochemistry. Tissues of the culm bases with little or no starch. Fructosans predominantly short-chain. Leaves without flavonoid sulphates (1 species).

Cytology. Chromosome base number, $x = 7$. $2n = 14$ and 28. 2 and 4 ploid. Haploid nuclear DNA content (2.2–)3.2–6.9 pg (8 species, mean 5.0). Mean diploid 2c DNA value 9.9 pg (5 species, (4.3–)6.4–13.6).

Taxonomy. Pooideae; Poodae; Poeae.

Ecology, geography, regional floristic distribution. 8 species. Temperate Eurasia, north Africa. Commonly adventive. Mesophytic; species of open habitats.

Holarctic, Paleotropical, Neotropical, and Cape. Boreal, Tethyan, and Madrean. Indomalesian. Euro-Siberian, Eastern Asian, and Atlantic North American. Macaronesian and Irano-Turanian. Indian. Pampas and Andean. European and Siberian. Canadian-Appalachian, Southern Atlantic North American, and Central Grasslands.

Hybrids. Intergeneric hybrids with *Festuca* — ×*Festulolium* Aschers. & Graebn. (several species of each genus involved).

Rusts and smuts. Rusts — *Puccinia*. Taxonomically wide-ranging species: *Puccinia graminis*, *Puccinia coronata*, *Puccinia striiformis*, *Puccinia brachypodii*, *Puccinia hordei*, and *Puccinia recondita*. Smuts from Tilletiaceae and from Ustilaginaceae. Tilletiaceae — *Tilletia* and *Urocystis*. Ustilaginaceae — *Ustilago*.

Economic importance. Significant weed species: *L. multiflorum*, *L. perenne*, *L. persicum*, *L. remotum*, *L. rigidum*, *L. temulentum* (darnel — with toxic grain). Cultivated fodder: *L. multiflorum*, *L. perenne*. Lawns and/or playing fields: *L. perenne*.

References, etc. Leaf anatomical: Metcalfe 1960; this project.

Lombardochloa Rosengurtt & Arillaga

Poidium rufum (Presl) Mattei
~ *Briza*

Habit, vegetative morphology. Perennial; caespitose. *Culms* 50–150 cm high (?); herbaceous. Young shoots intravaginal. Leaves non-auriculate. The fibrous remains of the sheaths persisting. Leaf blades narrow; flat, or folded; without cross venation; persistent. Ligule an unfringed membrane.

Reproductive organization. Plants bisexual, with bisexual spikelets; with hermaphrodite florets. Exposed-cleistogamous, or chasmogamous (?).

Inflorescence. Inflorescence paniculate; open; espatheate; not comprising 'partial inflorescences' and foliar organs. Spikelet-bearing axes persistent. Spikelets not secund; pedicellate.

Female-fertile spikelets. Spikelets compressed laterally; disarticulating above the glumes; disarticulating between the florets; with a distinctly elongated rachilla internode above the glumes. Rachilla prolonged beyond the uppermost female-fertile floret. The rachilla extension with incomplete florets. Hairy callus present. *Callus* short.

Glumes two; more or less equal; about equalling the spikelets; long relative to the adjacent lemmas; hairless; glabrous; awnless; carinate; similar (naviculate). Lower glume 3 nerved. Upper glume 3 nerved. *Spikelets* with incomplete florets. The incomplete florets distal to the female-fertile florets. *The distal incomplete florets* merely underdeveloped.

Female-fertile florets 2–4. **Lemmas broad, with a succulent, translucent region near the base of each wing**; not becoming indurated; entire; pointed to blunt; awnless; hairless; carinate to non-carinate (slightly keeled when mature); 5 nerved. *Palea* present; relatively long; entire (acute or obtuse); awnless, without apical setae; not indurated; 2-nerved; 2-keeled. Palea keels wingless. *Lodicules* present; 2; free; membranous; glabrous; not toothed. *Stamens* 3, or 1 (in cleistogamous florets). *Ovary* glabrous. Stigmas 2.

Fruit, embryo and seedling. *Fruit* adhering to lemma and/or palea to free from both lemma and palea; longitudinally grooved. Hilum short. Endosperm with lipid.

Phytochemistry. Leaves without flavonoid sulphates (1 species).

Special diagnostic feature. Lemmas not as in *Briza* (q.v.). *Female-fertile lemma very broad, with a conspicuous, succulent, translucent region near the base of each wing*.

Taxonomy. Pooideae; Poodae; Poeae.

Ecology, geography, regional floristic distribution. 1 species. South America. Neotropical. Andean.

References, etc. Morphological/taxonomic: Rosengurtt and Arrillaga de Maffei 1979.
Special comments. Anatomical data wanting.

Lophacme Stapf

Habit, vegetative morphology. Perennial; caespitose. *Culms* 13–57 cm high; herbaceous; unbranched above (though branching below). Culm nodes glabrous. Culm leaves present. Upper culm leaf blades reduced. Culm internodes solid. Plants unarmed. Young shoots intravaginal. *Leaves* mostly basal; non-auriculate. **Leaf blades** linear-lanceolate; *narrow*; 2–4 mm wide (to 6 cm long); rolled (involute); without abaxial multicellular glands; without cross venation; persistent. Ligule an unfringed membrane (in *L. parva*), or a fringed membrane (in *L. digitata*); when unfringed truncate; 0.1–0.4 mm long. *Contra-ligule* absent.

Reproductive organization. Plants bisexual, with bisexual spikelets; with hermaphrodite florets. The spikelets of sexually distinct forms on the same plant (sometimes? — see below), or all alike in sexuality.

Inflorescence. *Inflorescence* of spicate main branches (consisting of slender spikelike racemes); *digitate*. Primary inflorescence branches 2–8. Inflorescence espatheate, or spatheate (in that the spikelets are sometimes subtended by very minute, sparsely hairy, hyaline scales — vestigial bracts or spikelets?); not comprising 'partial inflorescences' and foliar organs. The racemes spikelet bearing to the base. Spikelet-bearing axes with very slender rachides; persistent. Spikelets solitary; secund; biseriate; shortly pedicellate, or subsessile; somewhat distant.

Female-fertile spikelets. *Spikelets* 3.5–6 mm long; adaxial; *compressed laterally*; disarticulating above the glumes; disarticulating between the florets; with distinctly elongated rachilla internodes between the florets (between L_1 and L_2 and above L2). Rachilla prolonged beyond the uppermost female-fertile floret; rachilla hairy, or hairless. The rachilla extension with incomplete florets. Hairy callus present (minutely bearded). *Callus* short; pointed.

Glumes two; very unequal (G_1 shorter), or more or less equal; shorter than the spikelets to about equalling the spikelets; shorter than the adjacent lemmas (G_1, sometimes), or long relative to the adjacent lemmas; hairless; glabrous (the keel scabrous); pointed (acute or acuminate); awnless; carinate; *similar (linear-lanceolate, membranous)*. Lower glume shorter than the lowest lemma to about equalling the lowest lemma; 1 nerved. Upper glume 1 nerved. *Spikelets* with incomplete florets. The incomplete florets distal to the female-fertile florets. The distal incomplete florets about 4; *incomplete florets* clearly specialised and modified in form (forming a tuft which remains attached to the upper fertile floret); incomplete florets awned (reduced to awns).

Female-fertile florets 1, or 2 (the second floret sometimes male-only). Lemmas less firm than the glumes (thinly membranous); not becoming indurated; incised; 2 lobed (two-toothed); not deeply cleft; awned. Awns 1; median; from a sinus; non-geniculate; straight, or recurving (fine); hairless (scabrous); about as long as the body of the lemma to much longer than the body of the lemma; entered by one vein. Lemmas hairy; carinate; without a germination flap; 3 nerved. *Palea* present; relatively long (reaching the bases of the lemma lobes); entire; awnless, without apical setae; not indurated (thinly membranous or hyaline); 2-nerved; 2-keeled. Palea keels wingless; scabrous (or sparsely ciliate). *Lodicules* present; free; somewhat fleshy; glabrous; conspicuously toothed; not or scarcely vascularized. *Stamens* 3. Anthers 1.5 mm long (i.e. relatively large); not penicillate; without an apically prolonged connective. *Ovary* glabrous. Styles free to their bases. Stigmas 2; brown.

Fruit, embryo and seedling. *Fruit* free from both lemma and palea; small (1.8 mm long); fusiform; not noticeably compressed. Hilum short. **Pericarp loosely adherent** *(easily removable after soaking).* Embryo large (a little more than 1/3 the length of the fruit); not waisted.

Abaxial leaf blade epidermis. *Costal/intercostal zonation* conspicuous. *Papillae* absent. *Long-cells* similar in shape costally and intercostally; of similar wall thickness costally and intercostally. Mid-intercostal long-cells rectangular; having markedly sinuous walls. *Microhairs* present; elongated; clearly two-celled; panicoid-type to chloridoid-type (intermediate: apical cells broad and rounded at tip, not tending to collapse, rather variable in length but mostly longer than the normal chloridoid type). Microhair apical cell wall thinner than that of the basal cell but not tending to collapse. Microhairs (15–)18–21 microns long; 5.4–6.9 microns wide at the septum. Microhair total length/ width at septum 2.2–3.3. Microhair apical cells 9–12 microns long. Microhair apical cell/ total length ratio 0.57–0.67. *Stomata* common; 19.5–21.6 microns long. Subsidiaries mostly low triangular (with small points), or dome-shaped (a few). Guard-cells overlapping to flush with the interstomatals. *Intercostal short-cells* common (in some regions, rare or missing in others); in cork/silica-cell pairs; silicified. Intercostal silica bodies imperfectly developed; cross-shaped (perfect to imperfect). *Costal short-cells* conspicuously in long rows. Costal silica bodies present in alternate cell files of the costal zones; 'panicoid-type'; large dumb-bell shaped; not sharp-pointed.

Transverse section of leaf blade, physiology. C_4; XyMS+. PCR sheath outlines even. PCR sheaths of the primary vascular bundles interrupted; interrupted abaxially only. PCR sheath extensions absent. PCR cell chloroplasts centripetal. Mesophyll traversed by columns of colourless mesophyll cells (in places, but most of the columns fall short of the abaxial surface). *Leaf blade* adaxially flat. *Midrib* not readily distinguishable; with one bundle only. Bulliforms present in discrete, regular adaxial groups (between each pair of bundles); associated with colourless mesophyll cells to form deeply-penetrating fans (mostly, the groups large, linked with colourless columns which sometimes reach the abaxial epidermis), or in simple fans (a few of these, with large, deeply penetrating median cells). All the vascular bundles accompanied by sclerenchyma. Combined sclerenchyma girders present (with all the bundles); forming 'figures' (all the bundles). Sclerenchyma all associated with vascular bundles. The lamina margins with fibres.

Taxonomy. Chloridoideae; main chloridoid assemblage.

Ecology, geography, regional floristic distribution. 2 species. Southern tropical and southern Africa. Helophytic, or mesophytic (open grassland or streamsides); shade species (*L. parva*), or species of open habitats (*L. digitata*); glycophytic.

Paleotropical. African. Sudano-Angolan. South Tropical African.

References, etc. Leaf anatomical: this project; photos of *L. digitata* provided by R.P. Ellis.

Lophatherum Brongn.

Acroelytrum Steud., *Allelotheca* Steud.

Habit, vegetative morphology. Perennial; caespitose. *Culms* 50–120 cm high; herbaceous; not tuberous (but *L. gracile* with root tubers). Culm nodes glabrous. Culm internodes solid, or hollow. *Leaves* not basally aggregated; non-auriculate; without auricular setae. *Leaf blades* linear-lanceolate to ovate-lanceolate; broad; not cordate, not sagittate (rather oblique at the base); pseudopetiolate (the pseudopetiole often winged); *cross veined*; persistent. **Ligule** present; *an unfringed membrane (ciliolate)*; truncate.

Reproductive organization. Plants bisexual, with bisexual spikelets; with hermaphrodite florets.

Inflorescence. Inflorescence of spicate main branches, or paniculate (the unilateral racemes solitary to paired on the common axis); open; espatheate; not comprising 'partial inflorescences' and foliar organs. Spikelet-bearing axes persistent. Spikelets not secund; pedicellate.

Female-fertile spikelets. Spikelets 8–13 mm long; compressed laterally (to subterete); falling with the glumes; not disarticulating between the florets. *Rachilla prolonged beyond the uppermost female-fertile floret*; rachilla hairless. The rachilla extension with incomplete florets. Hairy callus absent.

Glumes two; more or less equal; shorter than the adjacent lemmas; free; lower often with tubercle-based hairs submarginally near the apex; not pointed (obtuse); awnless; non-carinate (rounded on the back); membranous with hyaline margins. Lower glume 5 nerved. *Upper glume 7 nerved. Spikelets* with incomplete florets. The incomplete florets distal to the female-fertile florets (reduced to their recurved, retrorsely-scabrid lemma awns). The distal incomplete florets 3–4(–9); *incomplete florets* clearly specialised and modified in form (the fan of awns forming an adhering apparatus).

Female-fertile florets 1. Lemmas not becoming indurated; entire; blunt; awned (the awns retrorsely scabrid). Awns 1; median; apical; non-geniculate; hairless; much shorter than the body of the lemma. Lemmas hairless; glabrous; carinate; strongly 7–9 nerved. *Palea* present; 2-nerved; strongly 2-keeled. *Lodicules* present; 2; free; fleshy; glabrous; heavily vascularized. **Stamens 2**. Anthers not penicillate. Styles fused.

Fruit, embryo and seedling. Embryo with an epiblast; with a scutellar tail; with an elongated mesocotyl internode. Embryonic leaf margins overlapping.

Abaxial leaf blade epidermis. *Costal/intercostal zonation* conspicuous. *Papillae* absent. *Long-cells* similar in shape costally and intercostally; of similar wall thickness costally and intercostally (thin walled). Mid-intercostal long-cells rectangular; having markedly sinuous walls. *Microhairs* present; panicoid-type; 31–54 microns long; (5.4–)6 microns wide at the septum. Microhair total length/width at septum 8.2–9. Microhair apical cells (18–)22–30 microns long. Microhair apical cell/total length ratio 0.5–0.6. *Stomata* common; (33–)36–42(–45) microns long. Subsidiaries triangular. Guard-cells overlapping to flush with the interstomatals. *Intercostal short-cells* common; in cork/silica-cell pairs (usually), or not paired (some solitary); silicified. Intercostal silica bodies cross-shaped, or vertically elongated-nodular. *Costal short-cells* conspicuously in long rows. Costal silica bodies 'panicoid-type'; dumb-bell shaped; not sharp-pointed.

Transverse section of leaf blade, physiology. C_3; XyMS+. Mesophyll with adaxial palisade (at least over the bundles); without arm cells; with fusoids (in the form of short, lateral PBS extensions, c.f. *Orthoclada, Centotheca* etc.). The fusoids an integral part of the PBS. *Leaf blade* adaxially flat. *Midrib* conspicuous; with one bundle only. Bulliforms not in discrete, regular adaxial groups (much of the intercostal epidermis irregularly bulliform). All the vascular bundles accompanied by sclerenchyma. Combined sclerenchyma girders present; forming 'figures'. Sclerenchyma all associated with vascular bundles.

Cytology. Chromosome base number, $x = 12$. $2n = 48$. 4 ploid.

Taxonomy. Bambusoideae; Oryzodae; Centotheceae.

Ecology, geography, regional floristic distribution. 2 species. Eastern Asia, Indomalayan region, tropical Australia. Shade species to species of open habitats.

Holarctic, Paleotropical, and Australian. Boreal. Indomalesian. Eastern Asian. Indian, Indo-Chinese, Malesian, and Papuan. North and East Australian. Tropical North and East Australian.

Rusts and smuts. Rusts — *Puccinia*.
References, etc. Leaf anatomical: Metcalfe 1960 and this project.
Special comments. Fruit data wanting.

Lopholepis Decne.

Holboellia Hook.

Habit, vegetative morphology. Annual; caespitose. *Culms* 15–45 cm high; herbaceous; to 0.15 cm in diameter; unbranched above. Culm nodes glabrous. Plants unarmed. *Leaves* not basally aggregated; non-auriculate. Leaf blades narrow; 3–6 mm wide; somewhat cordate (or amplexicaul); without abaxial multicellular glands; without cross venation; persistent. Ligule an unfringed membrane; 0.3–0.5 mm long. *Contra-ligule* absent.

Reproductive organization. Plants bisexual, with bisexual spikelets; with hermaphrodite florets.

Inflorescence. *Inflorescence* 5–10 cm long, cylindric, dense, ostensibly *a false spike, with spikelets on contracted axes (best interpreted as a raceme of one-spikeleted racemelets)*; espatheate; not comprising 'partial inflorescences' and foliar organs. *Spikelet-bearing axes* very much reduced (to a short, cuneate rachis with one spikelet); solitary; disarticulating (interpreting the 'pedicel' or 'callus' as a reduced branch, which is shed); falling entire. Spikelets solitary; not secund (the racemelets spiralled); sessile.

Female-fertile spikelets. Spikelets *unconventional (at least at first sight — of bizarre appearance, resembling a caricatured bird's head in outline)*; about 2 mm long (excluding the 2 mm 'callus'); compressed laterally; *falling with the glumes (the disarticulation occurring at the base of a 1.5 mm hispidulous 'pedicel' or 'callus', close to the main axis)*. Rachilla terminated by a female-fertile floret. Hairy callus present (i.e. the 'pedicel').

Glumes two; more or less equal (although the G_1 is larger, enclosing the G_2); long relative to the adjacent lemmas; hairy (with spicule-hairs on the keels); awnless; both strongly carinate; with the keel conspicuously winged; very dissimilar (both beak- shaped, the G_1 leathery with globose base, narrow, and boat-shaped above, with a narrow, 2-lobed cartilaginous keel-wing having pectinate spinules, the G_2 shorter and pectinate-keeled, only the upper half with a pectinate keel-wing). Lower glume 1 nerved. Upper glume 1 nerved. *Spikelets* with female-fertile florets only.

Female-fertile florets 1 (small). Lemmas rounded; saccate; less firm than the glumes (hyaline); not becoming indurated; entire; blunt; awnless; hairless; glabrous; non-carinate; having the margins lying flat and exposed on the palea; without a germination flap; 1 nerved. *Palea* present; relatively long; entire; awnless, without apical setae; not indurated (hyaline); 2-nerved (except at the base); 2-keeled. *Lodicules* absent. *Ovary* glabrous. Styles free to their bases. Stigmas 2.

Fruit, embryo and seedling. Fruit free from both lemma and palea; *small (about 1 mm long, shaped like a bird's head, hollowed above both hilum and embryo, curved and beaked above)*. Hilum short. Pericarp thin; fused. Embryo large. Endosperm hard.

Abaxial leaf blade epidermis. *Costal/intercostal zonation* conspicuous. *Papillae* absent. *Long-cells* markedly different in shape costally and intercostally (the costals almost straight-walled); of similar wall thickness costally and intercostally. Mid-intercostal long-cells rectangular; having markedly sinuous walls. *Microhairs* present; more or less spherical to elongated; clearly two-celled; chloridoid-type (the apical cells usually 'apiculate'). Microhair apical cell wall of similar thickness/rigidity to that of the basal cell. Microhairs 18–21 microns long. Microhair basal cells 9 microns long. Microhairs 6–6.6 microns wide at the septum. Microhair total length/width at septum 2.7–3.5. Microhair apical cells (7.5–)9 microns long. Microhair apical cell/total length ratio 0.43–0.5. *Stomata* common; 21–25.5 microns long. Subsidiaries triangular, or parallel-sided (in that the apices of the triangles are to varying extents truncated); including both triangular and parallel-sided forms on the same leaf. Guard-cells overlapped by the interstomatals (slightly). *Intercostal short-cells* absent or very rare. Intercostal silica bodies absent. Prickles absent. *Crown cells* absent. *Costal short-cells* conspicuously in long rows. Costal silica bodies present in alternate cell files of the costal zones; 'panicoid-type'; cross shaped (a few), or dumb-bell shaped (mostly); not sharp-pointed.

Transverse section of leaf blade, physiology. C_4; XyMS+. PCR sheaths of the primary vascular bundles interrupted; interrupted abaxially only. PCR sheath extensions absent. *Leaf blade* adaxially flat. *Midrib* not readily distinguishable; with one bundle only. Bulliforms present in discrete, regular adaxial groups; associated with colourless mesophyll cells to form deeply-penetrating fans. All the vascular bundles accompanied by sclerenchyma. Combined sclerenchyma girders absent (seemingly abaxial girders only). Sclerenchyma all associated with vascular bundles.

Special diagnostic feature. *Spikelets minute, shaped like cartoon birds' heads.*

Taxonomy. Chloridoideae; main chloridoid assemblage.

Ecology, geography, regional floristic distribution. 1 species. India, Ceylon. Xerophytic; species of open habitats. Disturbed places in sandy ground: a sand stabilizer.

Paleotropical. Indomalesian. Indian.

References, etc. Leaf anatomical: this project.

Lophopogon Hackel

Habit, vegetative morphology. Low, delicate annual, or perennial; caespitose. Culms herbaceous; branched above, or unbranched above.

Reproductive organization. Plants bisexual, with bisexual spikelets; with hermaphrodite florets. *The spikelets* of sexually distinct forms on the same plant; hermaphrodite and male-only; overtly heteromorphic (the pedicellate spikelets awned, the sessile members of the lower pairs awnless); *in both homogamous and heterogamous combinations (the upper pairs homogamous, the lower pairs with the sessile spikelets male).*

Inflorescence. Inflorescence of spicate main branches; contracted; digitate. *Primary inflorescence branches 2 (the two short 'racemes' closely appressed to form an ovate head).* Inflorescence espatheate; not comprising 'partial inflorescences' and foliar organs. *Spikelet-bearing axes* 'racemes'; paired; with very slender rachides; disarticulating; disarticulating at the joints. 'Articles' disarticulating obliquely; densely long-hairy (the hairs reddish). Spikelets paired; sessile and pedicellate (the pedicel tips oblique); consistently in 'long-and-short' combinations; in pedicellate/sessile combinations. Pedicels of the 'pedicellate' spikelets free of the rachis. The 'shorter' spikelets male-only (in the lower part of the inflorescence), or hermaphrodite (above). The 'longer' spikelets hermaphrodite.

Female-sterile spikelets. The lower sessile spikelets male-only, dorsally compressed, awnless. The male spikelets with glumes; 2 floreted. The lemmas awnless.

Female-fertile spikelets. *Spikelets compressed laterally*; falling with the glumes. Rachilla terminated by a female-fertile floret. Callus pointed.

Glumes two; very unequal; *with distinct hair tufts (G_1 truncate, with transversely placed tufts, at least in upper spikelets)*; awned (G_2 only, the awn short); very dissimilar (G_1 hair-tufted, truncate, 3-toothed and awnless, the G_2 short-awned). Lower glume not two-keeled; convex on the back; not pitted; distinctly nerved. *Spikelets* with incomplete florets. The incomplete florets proximal to the female-fertile florets. *Spikelets with proximal incomplete florets. The proximal incomplete florets 1.*

Female-fertile florets 1. Lemmas linear; less firm than the glumes; incised; 2 lobed; deeply cleft; awned. Awns 1; median; from a sinus; geniculate. Lemmas non-carinate; without a germination flap. *Lodicules* absent. **Stamens 2.**

Taxonomy. Panicoideae; Andropogonodae; Andropogoneae; Andropogoninae.

Ecology, geography, regional floristic distribution. 2 species. India. Species of open habitats.

Paleotropical. Indomalesian. Indian.

Special comments. Fruit data wanting. Anatomical data wanting.

Lophopyrum A. Löve

~ Elymus, Elytrigia

Habit, vegetative morphology. Perennial (with stout, firm glabrous culms); *caespitose*. **Culms** 30–100 cm high; *herbaceous*; unbranched above. Culm nodes glabrous. Culm internodes hollow. Young shoots intravaginal. *Leaves* not basally aggregated. Leaf blades narrow (glaucous, stiff, with thick veins); rolled; without cross venation; persistent. *Ligule an unfringed membrane*; truncate. *Contra-ligule* absent.

Reproductive organization. Plants bisexual, with bisexual spikelets; with hermaphrodite florets.

Inflorescence. *Inflorescence a single spike (the spikelets appressed early, but becoming divaricate)*. Rachides flattened (against the spikelets). Inflorescence espatheate; not comprising 'partial inflorescences' and foliar organs. Spikelet-bearing axes persistent (usually), or disarticulating; when disarticulating, disarticulating at the joints (the spikelet falling with the internode above). Spikelets solitary; not secund; distichous; sessile.

Female-fertile spikelets. Spikelets compressed laterally; disarticulating above the glumes, or falling with the glumes; not disarticulating between the florets to disarticulating between the florets (?). Rachilla prolonged beyond the uppermost female-fertile floret. The rachilla extension with incomplete florets. Hairy callus absent. **Callus** short; *pointed (glabrous)*.

Glumes two; more or less equal (or slightly unequal); shorter than the spikelets; shorter than the adjacent lemmas; lateral to the rachis; not pointed (obtuse to truncate); not subulate; awnless; non-carinate; similar (oblong, leathery to hardened). Lower glume 3–9 nerved. Upper glume 3–9 nerved. *Spikelets* with incomplete florets. The incomplete florets distal to the female-fertile florets. *The distal incomplete florets merely underdeveloped*.

Female-fertile florets 3–12 ('few to many'). *Lemmas* entire; blunt; *awnless*; sparsely hairy, or hairless; non-carinate; without a germination flap; 5 nerved; with the nerves confluent towards the tip. *Palea* present; relatively long (slightly shorter than the lemma); awnless, without apical setae; 2-nerved; 2-keeled. *Lodicules* present; 2; free; membranous; ciliate; not toothed. *Stamens* 3. Anthers 4–10 mm long. *Ovary* hairy. Styles free to their bases. Stigmas 2; white.

Fruit, embryo and seedling. *Fruit* adhering to lemma and/or palea; medium sized; longitudinally grooved; compressed dorsiventrally; with hairs confined to a terminal tuft. *Hilum long-linear*. Embryo small. Endosperm containing only simple starch grains.

Abaxial leaf blade epidermis. *Costal/intercostal zonation* lacking (for the most part). *Papillae* absent. *Long-cells* similar in shape costally and intercostally; of similar wall thickness costally and intercostally (indistinguishable in form, fairly thick walled). Mid-intercostal long-cells rectangular; having markedly sinuous walls. *Microhairs* absent. *Stomata* absent or very rare (in the material of *L. elongatum* seen), or common (*L. elongatum* according to Metcalfe). Subsidiaries low dome-shaped (recorded by Metcalfe). *Intercostal short-cells* common; in cork/silica-cell pairs; silicified. *Crown cells* present (common, fairly small). *Costal short-cells* predominantly paired (some solitary). Costal silica bodies saddle shaped, or tall-and-narrow, or crescentic, or rounded (integrading with the other forms); not sharp-pointed.

Transverse section of leaf blade, physiology. C_3; XyMS+. *Mesophyll* with non-radiate chlorenchyma; without adaxial palisade. *Leaf blade* with distinct, prominent adaxial ribs; with the ribs very irregular in sizes (the large ribs varying in shape, and the blade asymmetric in section). *Midrib* not readily distinguishable; with one bundle only. Bulliforms present in discrete, regular adaxial groups (the groups very small, in the furrows); in simple fans. All the vascular bundles accompanied by sclerenchyma. Combined sclerenchyma girders present (with the main bundles — the minor bundles with abaxial girders only); forming 'figures' (the main bundles with massive I's). Sclerenchyma all associated with vascular bundles.

Cytology. Chromosome base number, $x = 7$. $2n = 14, 28, 56$, and 70. 2, 4, 8, and 10 ploid. Haplomic genome content E. Haploid nuclear DNA content 5.6 pg (1 species). Mean diploid 2c DNA value 11.2 pg (1 species).

Taxonomy. Pooideae; Triticodae; Triticeae.
Ecology, geography, regional floristic distribution. 11 species. Mediterranean to Iran. Commonly adventive. Xerophytic; species of open habitats; halophytic. Coastal sands.
Holarctic. Boreal and Tethyan. Euro-Siberian. Mediterranean and Irano-Turanian. European.
Hybrids. Intergeneric hybrids with *Elytigia, Leymus, Sitanion, Triticum*.
References, etc. Morphological/taxonomic: Löve 1984. Leaf anatomical: this project.

Lorenzochloa J. & C. Reeder

Parodiella J. & C. Reeder
~ *Orthachne*

Habit, vegetative morphology. Perennial; densely caespitose. **Culms** 15–38 cm high; *herbaceous*; unbranched above. *Leaves* mostly basal; non-auriculate. Leaf blades narrow; acicular (erect, rigid, terete, pungent); without cross venation; persistent. *Ligule an unfringed membrane*; not truncate (acute); 2–4 mm long.
Reproductive organization. Plants bisexual, with bisexual spikelets; with hermaphrodite florets.
Inflorescence. *Inflorescence paniculate*; open (but narrow and rather few flowered, the branches conspicuously pulvinate); espatheate; not comprising 'partial inflorescences' and foliar organs. Spikelet-bearing axes persistent. Spikelets not secund; pedicellate.
Female-fertile spikelets. *Spikelets 2.3–2.6 mm long*; compressed laterally, or not noticeably compressed (?); disarticulating above the glumes. Rachilla terminated by a female-fertile floret. Hairy callus present. *Callus* short.
Glumes two; more or less equal; shorter than the spikelets; *shorter than the adjacent lemmas*; hairless; not pointed (truncate or emarginate); awnless; non-carinate; similar (firm). *Lower glume 3 nerved*. Upper glume 3 nerved. *Spikelets* with female-fertile florets only.
Female-fertile florets 1. *Lemmas* tapered into the awn; similar in texture to the glumes; not becoming indurated; entire; pointed; *awned*. Awns 1; median; apical; non-geniculate; straight, or flexuous (erect or slightly flexuose, stout with a distinct line of demarcation from the rest of the lemma); hairless (scabrous); about as long as the body of the lemma to much longer than the body of the lemma (to 5 mm long). Lemmas hairy (pubescent on the margins and along the lower half of the mid-nerve); non-carinate; without a germination flap; 5 nerved. *Palea* present; relatively long; awnless, without apical setae; not indurated (firm); 2-nerved; 2-keeled (often pubescent between the keels). Palea keels wingless. *Lodicules* present; 3. *Third lodicule present*. Lodicules free; membranous; glabrous; not toothed; not or scarcely vascularized. *Stamens* 3. Anthers not penicillate; without an apically prolonged connective. *Ovary* glabrous. Styles free to their bases. Stigmas 2.
Fruit, embryo and seedling. Fruit small (1.2 to 1.4 mm long); broadly fusiform; compressed dorsiventrally to not noticeably compressed. Hilum long-linear. Embryo small; not waisted; with an epiblast; without a scutellar tail; with a negligible mesocotyl internode. Embryonic leaf margins meeting.
Abaxial leaf blade epidermis. *Costal/intercostal zonation* lacking. *Papillae* absent. *Long-cells* similar in shape costally and intercostally; of similar wall thickness costally and intercostally. Mid-intercostal long-cells rectangular; having markedly sinuous walls. *Microhairs* absent. *Stomata* absent or very rare. Prickles abundant in the silica-cell files. *Crown cells* absent. *Costal short-cells* conspicuously in long rows (the rows interrupted only by prickles). Costal silica bodies 'panicoid-type'; dumb-bell shaped (mostly, short), or cross shaped (a few), or nodular (a few, short); not sharp-pointed.
Transverse section of leaf blade, physiology. C_3; XyMS+. *Mesophyll* with non-radiate chlorenchyma; without adaxial palisade. *Leaf blade* with distinct, prominent adaxial ribs; with the ribs very irregular in sizes. *Midrib* not readily distinguishable; with one bundle only. *The lamina* symmetrical on either side of the midrib. Bulliforms not in discrete, regular adaxial groups (*Ammophila*-type groups of small cells in the furrows).

All the vascular bundles accompanied by sclerenchyma. Combined sclerenchyma girders present; forming 'figures' (each main bundle with a large 'anchor'). Sclerenchyma not all bundle-associated (the feet of the 'anchors' joining). The 'extra' sclerenchyma in a continuous abaxial layer.

Cytology. Chromosome base number, $x = 11$. $2n = 22$. 2 ploid.

Taxonomy. Arundinoideae; Stipeae.

Ecology, geography, regional floristic distribution. 1 species. High Andes of Venezuela, Colombia, Peru.

Neotropical. Caribbean.

References, etc. Leaf anatomical: this project.

Loudetia Hochst.

Excluding *Loudetiopsis* s.l.

Habit, vegetative morphology. Annual (rarely), or perennial; caespitose. **Culms** (25–)40–500 cm high; *herbaceous (usually erect, slender or robust)*; branched above, or unbranched above. Culm nodes hairy, or glabrous. Culm internodes solid, or hollow. Young shoots intravaginal. *Leaves* mostly basal, or not basally aggregated; non-auriculate; without auricular setae (but often with auricular hair tufts). *Leaf blades* linear (often rigid); *neither leathery nor flimsy*; narrow; not cordate, not sagittate; flat, or rolled (convolute); pseudopetiolate, or not pseudopetiolate; without cross venation; persistent. Ligule a fringed membrane (narrow), or a fringe of hairs; 0.5–3 mm long. *Contra-ligule* present (of hairs), or absent.

Reproductive organization. *Plants bisexual, with bisexual spikelets*; with hermaphrodite florets.

Inflorescence. Inflorescence paniculate; open, or contracted (rarely more or less spiciform); more or less ovoid; espatheate; not comprising 'partial inflorescences' and foliar organs. Spikelet-bearing axes persistent. *Spikelets solitary, or paired (long-pedicelled triplets in* **L. togoensis**); not secund; pedicellate; consistently in 'long-and-short' combinations, or not in distinct 'long-and-short' combinations.

Female-fertile spikelets. *Spikelets* 6–25 mm long; brown; compressed laterally to not noticeably compressed; *disarticulating above the glumes*; *disarticulating between the florets (disarticulating between the two florets, less readily or not at all between* G_2 *and* L_1); with conventional internode spacings. Rachilla prolonged beyond the uppermost female-fertile floret (very rarely), or terminated by a female-fertile floret; rachilla hairless. The rachilla extension when present, naked. Hairy callus present. *Callus* short, or long; pointed, or blunt (oblong to linear, bidentate or obliquely pungent).

Glumes two; relatively large; very unequal; (the upper) long relative to the adjacent lemmas (nearly as long as the spikelet); coarsely hairy (with brown or black hair cushions), or hairless; without conspicuous tufts or rows of hairs; pointed, or not pointed; awned (the G_2 may be setaceous-acuminate), or awnless; non-carinate; similar. Lower glume shorter than the lowest lemma; 3 nerved. Upper glume 3(–5) nerved. *Spikelets* with incomplete florets. The incomplete florets proximal to the female-fertile florets. *Spikelets with proximal incomplete florets. The proximal incomplete florets* 1; paleate, or epaleate (*L. togoensis*). Palea of the proximal incomplete florets when present, fully developed (membranous, two keeled). The proximal incomplete florets male, or sterile. *The proximal lemmas awnless (muticous)*; 3 nerved, or 5 nerved (*L. togoensis*); more or less equalling the female-fertile lemmas to decidedly exceeding the female-fertile lemmas; less firm than the female-fertile lemmas to similar in texture to the female-fertile lemmas; not becoming indurated (papery to leathery, similar to G_2).

Female-fertile florets 1. **Lemmas** similar in texture to the glumes to decidedly firmer than the glumes; not becoming indurated (more or less leathery); usually *incised (shortly so, rarely entire)*; usually 2 lobed; *not deeply cleft*; awned. Awns 1; median; from a sinus; geniculate; hairy; much longer than the body of the lemma; entered by one vein; deciduous (usually), or persistent (*L. phragmitoides*). Lemmas hairy to hairless (pilose to glabrescent). The hairs not in tufts; not in transverse rows. Lemmas non-carinate (rounded on the back); having the margins tucked in onto the palea; without a germination flap;

5–9 nerved; with the nerves non-confluent. *Palea* present (linear); relatively long; apically notched; awnless, without apical setae; thinner than the lemma; not indurated (membranous); 2-nerved; 2-keeled (the keels thick). Palea keels wingless. *Lodicules* present; 2; free; fleshy; glabrous; not or scarcely vascularized. *Stamens* 2(–3). Anthers penicillate (*L. togoensis*), or not penicillate; without an apically prolonged connective. *Ovary* glabrous. Styles free to their bases. Stigmas 2; brown.

Fruit, embryo and seedling. Fruit slightly longitudinally grooved. **Hilum long-linear**. Embryo large.

Abaxial leaf blade epidermis. *Costal/intercostal zonation* conspicuous. *Papillae* absent. Intercostal zones with typical long-cells, or exhibiting many atypical long-cells, or without typical long-cells (long-cells normal to cubical). Mid-intercostal long-cells rectangular; having markedly sinuous walls. *Microhairs* present; panicoid-type; 60–90(–106) microns long; 5.4–6.6 microns wide at the septum. Microhair total length/width at septum 12–14.5. Microhair apical cells 24–50 microns long. Microhair apical cell/total length ratio 0.37–0.5. *Stomata* common; 24–30 microns long. Subsidiaries low to high dome-shaped (mostly), or triangular. Guard-cells overlapping to flush with the interstomatals. *Intercostal short-cells* common, or absent or very rare (ignoring cells associated with macrohair bases). *Costal short-cells* conspicuously in long rows. Costal silica bodies 'panicoid-type'; mostly dumb-bell shaped; not sharp-pointed.

Transverse section of leaf blade, physiology. C_4. The anatomical organization conventional (usually), or unconventional. Organization of PCR tissue when unconventional, *Arundinella* type. XyMS–. PCR sheath outlines uneven. PCR sheath extensions present. Maximum number of extension cells 1–5. PCR cell chloroplasts centrifugal/peripheral. *Mesophyll* with radiate chlorenchyma; exhibiting 'circular cells', or without 'circular cells' (absent from *L. flavida* and *L. simplex*); traversed by columns of colourless mesophyll cells. *Leaf blade* with distinct, prominent adaxial ribs, or 'nodular' in section, or adaxially flat; with the ribs more or less constant in size. *Midrib* conspicuous, or not readily distinguishable; with one bundle only, or having a conventional arc of bundles; with colourless mesophyll adaxially, or without colourless mesophyll adaxially. Bulliforms present in discrete, regular adaxial groups; in simple fans, or associated with colourless mesophyll cells to form deeply-penetrating fans (commonly linking with traversing columns of colourless cells); associating with colourless mesophyll cells to form arches over small vascular bundles. Many of the smallest vascular bundles unaccompanied by sclerenchyma (in *L. simplex*), or all the vascular bundles accompanied by sclerenchyma. Combined sclerenchyma girders present. Sclerenchyma all associated with vascular bundles.

Phytochemistry. Leaves without flavonoid sulphates (2 species).

Special diagnostic feature. The lower glume shorter than the female-fertile lemma.

Cytology. Chromosome base number, $x = 6$ and 12. $2n = 20, 24, 40$, and 60.

Taxonomy. Panicoideae; Panicodae; Arundinelleae.

Ecology, geography, regional floristic distribution. About 26 species. Tropical and southern Africa, Madagascar, with 1 in South America. Helophytic, or mesophytic, or xerophytic; species of open habitats; glycophytic. In savanna woodland, often on poor shallow soils.

Paleotropical. African and Madagascan. Saharo-Sindian, Sudano-Angolan, West African Rainforest, and Namib-Karoo. Sahelo-Sudanian, Somalo-Ethiopian, South Tropical African, and Kalaharian.

Rusts and smuts. Rusts — *Phakopsora* and *Puccinia*. Smuts from Ustilaginaceae. Ustilaginaceae — *Sorosporium*.

Economic importance. Significant weed species: *L. simplex*. Grain crop species: wild *L. esculenta* harvested in Sudan.

References, etc. Leaf anatomical: Metcalfe 1960; this project; photos of *L. flavida* and *L. simplex* provided by R.P. Ellis.

Loudetiopsis Conert

~ *Loudetia* Section *Pseudotristachya* C.E. Hubb., cf. Jacques-Felix 1962
Excluding *Diandrostachya, Dilophotriche*

Habit, vegetative morphology. Usually perennial. *Culms* 30–120 cm high; herbaceous (usually erect); unbranched above. Culm nodes hairy (*L. kerstingii*), or glabrous. Culm internodes hollow. Plants unarmed. *Leaves* not basally aggregated; non-auriculate. Leaf blades linear, or lanceolate (*L. tristachyoides*); narrow (rigid); 2–7 mm wide; flat, or rolled (convolute); without cross venation; persistent. *Ligule a fringe of hairs (and the blade pilose immediately above it)*; 0.3–0.5 mm long. *Contra-ligule* absent.

Reproductive organization. *Plants bisexual, with bisexual spikelets*; with hermaphrodite florets.

Inflorescence. Inflorescence paniculate; usually open, or contracted (rarely, but not spiciform); with capillary branchlets, or without capillary branchlets; espatheate; not comprising 'partial inflorescences' and foliar organs. Spikelet-bearing axes persistent. *Spikelets* in triplets; not secund; *pedicellate (the pedicels of all three spikelets short, arising together at the tip of the peduncle, which is often hooked)*; not in distinct 'long-and-short' combinations (but the pedicels somewhat unequal).

Female-fertile spikelets. *Spikelets* 8–20 mm long; brown; compressed laterally, or not noticeably compressed; *disarticulating above the glumes*; *disarticulating between the florets (more readily than between G_2 and L_1, cf.* Loudetia*)*; with conventional internode spacings. Rachilla terminated by a female-fertile floret. Hairy callus present. *Callus* short; blunt (truncate or slightly bifid).

Glumes two; very unequal; (the longer) long relative to the adjacent lemmas; hairy (usually with tubercle-based hairs); without conspicuous tufts or rows of hairs; pointed, or not pointed; *awnless*; *very dissimilar (the G_1 shorter, wider and hairier)*. Lower glume shorter than the lowest lemma; 3 nerved, or 5 nerved. Upper glume 3 nerved. *Spikelets* with incomplete florets. The incomplete florets proximal to the female-fertile florets. *Spikelets with proximal incomplete florets*. *The proximal incomplete florets* 1; paleate. Palea of the proximal incomplete florets fully developed (two keeled, membranous). The proximal incomplete florets male. The proximal lemmas awnless; 3 nerved; decidedly exceeding the female-fertile lemmas; similar in texture to the female-fertile lemmas; not becoming indurated.

Female-fertile florets 1. Lemmas similar in texture to the glumes to decidedly firmer than the glumes (leathery); not becoming indurated; incised (apically); 2 lobed; not deeply cleft; awned. Awns 1; median; from a sinus; geniculate; hairless (scabrous), or hairy (with short, appressed hairs); much longer than the body of the lemma; entered by one vein. Lemmas hairy, or hairless; non-carinate (dorsally rounded); without a germination flap; 5–9 nerved. *Palea* present; relatively long (linear); apically notched; awnless, without apical setae; thinner than the lemma; not indurated (membranous to leathery); 2-nerved; 2-keeled (the keels leathery). Palea keels wingless; glabrous. *Lodicules* present; 2; free; fleshy; not or scarcely vascularized. **Stamens** 2. Anthers 2.5 mm long; not penicillate; without an apically prolonged connective. *Ovary* glabrous. Styles free to their bases. Stigmas 2; red pigmented (purple).

Fruit, embryo and seedling. Fruit longitudinally grooved, or not grooved. Hilum long-linear. Embryo large.

Abaxial leaf blade epidermis. *Costal/intercostal zonation* conspicuous. *Papillae* absent. *Long-cells* markedly different in shape costally and intercostally (in *L. chrysothrix*, the costals narrower); of similar wall thickness costally and intercostally. Mid-intercostal long-cells rectangular; having markedly sinuous walls. *Microhairs* present; more or less spherical; panicoid-type; 66–75 microns long; 6.6–8.4 microns wide at the septum. Microhair total length/width at septum 7.9–11.3. Microhair apical cells 28–33 microns long. Microhair apical cell/total length ratio 0.42–0.46. *Stomata* common; 30–40.5 microns long. Subsidiaries low dome-shaped (consistently, in *L. chrysothrix*). Guard-cells overlapping to flush with the interstomatals. *Intercostal short-cells* common; not paired (but often adjoining small prickles, in *L. chrysothrix*). *Costal short-cells* conspicuously in long rows. Costal silica bodies 'panicoid-type'; dumb-bell shaped (but in the material of *L. chrysothrix* seen, these all seem to lack the isthmus, so that the actual silica takes the form of more or less round, saddle-shaped or crescentic pieces, occurring two per cell); not sharp-pointed.

Transverse section of leaf blade, physiology. C_4; XyMS+ (e.g. *L. chrysothrix*), or XyMS− (e.g. *L. ambiens*). PCR sheath outlines uneven. PCR sheath extensions present.

Maximum number of extension cells 3. PCR cell chloroplasts centrifugal/peripheral. Mesophyll in places traversed by columns of colourless mesophyll cells. *Leaf blade* with distinct, prominent adaxial ribs; with the ribs very irregular in sizes. *Midrib* not readily distinguishable; with one bundle only, or having a conventional arc of bundles (the large bundle accompanied on either side by a minor one). Bulliforms present in discrete, regular adaxial groups (in each furrow); associated with colourless mesophyll cells to form deeply-penetrating fans (commonly linked with traversing columns of colourless cells); associating with colourless mesophyll cells to form arches over small vascular bundles. All the vascular bundles accompanied by sclerenchyma. Combined sclerenchyma girders present (with all the major bundles — the minor ones with only abaxial girders, or with strands only); forming 'figures'. Sclerenchyma all associated with vascular bundles.

Special diagnostic feature. *The lower glume shorter than the female-fertile lemma.*

Cytology. Chromosome base number, $x = 10$. $2n = 20$.

Taxonomy. Panicoideae; Panicodae; Arundinelleae.

Ecology, geography, regional floristic distribution. 11 species. West Africa. Helophytic (sometimes), or mesophytic; species of open habitats; glycophytic. Savanna, usually in woodland or shallow pockets on bare rocks, rarely in swamps.

Paleotropical. African. Sudano-Angolan and West African Rainforest. Sahelo-Sudanian.

References, etc. Leaf anatomical: Metcalfe 1960; this project.

Louisiella C.E. Hubb. & Léonard

Habit, vegetative morphology. Perennial; rhizomatous, or decumbent. *Culms herbaceous*; branched above. Culm nodes hairy. Culm internodes spongy. *Leaves* not basally aggregated; non-auriculate. Leaf blades narrow; without cross venation; persistent. *Ligule a fringed membrane (very short)*.

Reproductive organization. Plants bisexual, with bisexual spikelets; with hermaphrodite florets.

Inflorescence. Inflorescence paniculate; open; espatheate; not comprising 'partial inflorescences' and foliar organs. *Spikelet-bearing axes persistent.* Spikelets paired or in short appressed racemelets; secund (?); pedicellate.

Female-fertile spikelets. *Spikelets 8–9 mm long*; lanceolate; *compressed dorsiventrally*; falling with the glumes. Rachilla terminated by a female-fertile floret. Hairy callus absent.

Glumes two; very unequal; (the longer) long relative to the adjacent lemmas; hairless; awnless; non-carinate; very dissimilar (the lower very small, nerveless and truncate, the upper equalling the spikelet, long acuminate, strongly nerved). *Lower glume 0 nerved*. Upper glume 5–7 nerved. *Spikelets* with incomplete florets. The incomplete florets proximal to the female-fertile florets. *Spikelets with proximal incomplete florets. The proximal incomplete florets* 1; epaleate; sterile. The proximal lemmas long acuminate; awnless; 7–9 nerved; decidedly exceeding the female-fertile lemmas; similar in texture to the female-fertile lemmas; not becoming indurated.

Female-fertile florets 1. Lemmas lanceolate; similar in texture to the glumes to decidedly firmer than the glumes; not becoming indurated (membranous to leathery); entire; pointed; awnless; hairy to hairless (glabrous, with a scabrid or hairy tip); non-carinate; having the margins lying flat and exposed on the palea; with a clear germination flap; 5–7 nerved. *Palea* present; relatively long; entire (pointed); awnless; without apical setae; textured like the lemma; not indurated; 2-nerved (near the margins); keel-less. *Lodicules* present; 2; free; fleshy; glabrous; not or scarcely vascularized. *Stamens* 3. Anthers not penicillate; without an apically prolonged connective. *Ovary* glabrous. Styles fused to free to their bases. Stigmas 2; red pigmented.

Fruit, embryo and seedling. Fruit small (2.5 mm long). *Hilum long-linear*. Embryo large to small (up to 1/3 as long as the grain).

Abaxial leaf blade epidermis. *Costal/intercostal zonation* conspicuous. *Papillae* present. Intercostal papillae over-arching the stomata; several per cell (numerous per cell, in 1-several (average 2) rows, and large and ornamented at the blade margin). Mid-intercostal long-cells rectangular; having markedly sinuous walls. *Microhairs* present; panicoid-type; 27–36 microns long; 15–21 microns wide at the septum. Microhair total length/width at septum 5.3–7.7. Microhair apical cells (3.9–)4.5–5.7(–6) microns long. Microhair apical cell/total length ratio 0.5–0.7. *Stomata* common; 19–24 microns long. Subsidiaries high dome-shaped and triangular; including both triangular and parallel-sided forms on the same leaf. Guard-cells overlapping to flush with the interstomatals. *Intercostal short-cells* common; in cork/silica-cell pairs. *Costal short-cells* neither distinctly grouped into long rows nor predominantly paired ('short-cells' much longer than usual, and the files interrupted by prickles). Costal silica bodies 'panicoid-type'; small, cross shaped and dumb-bell shaped; not sharp-pointed.

Transverse section of leaf blade, physiology. C_4; XyMS+. PCR sheath outlines uneven. *Mesophyll* with radiate chlorenchyma. *Leaf blade* 'nodular' in section; with the ribs more or less constant in size. *Midrib* not readily distinguishable (one slightly larger bundle); with one bundle only. Bulliforms present in discrete, regular adaxial groups; in simple fans. All the vascular bundles accompanied by sclerenchyma. Combined sclerenchyma girders present; nowhere forming 'figures'. Sclerenchyma all associated with vascular bundles.

Taxonomy. Panicoideae; Panicodae; Paniceae.

Ecology, geography, regional floristic distribution. 1 species. Tropical Africa. Hydrophytic, or helophytic (floating, or decumbent on mud).

Paleotropical. African. West African Rainforest.

References, etc. Leaf anatomical: this project.

Special comments. Fruit data wanting.

Loxodera Launert

Plagiarthron Duv.

Excluding *Lepargochloa*

Habit, vegetative morphology. Perennial; caespitose. *Culms* 100 cm high; herbaceous; unbranched above. **Culm nodes glabrous.** *Leaves* not basally aggregated; non-auriculate. Leaf blades linear; narrow; without cross venation. **Ligule an unfringed membrane**; 0.8–1 mm long.

Reproductive organization. *Plants bisexual, with bisexual spikelets*; with hermaphrodite florets. The spikelets of sexually distinct forms on the same plant; hermaphrodite and male-only, or hermaphrodite and sterile; overtly heteromorphic (the pedicellate spikelet sometimes reduced to its glumes), or homomorphic; all in heterogamous combinations.

Inflorescence. Inflorescence a single apical 'raceme'. Rachides hollowed, or flattened. Inflorescence espatheate (seemingly); not comprising 'partial inflorescences' and foliar organs. **Spikelet-bearing axes** *spikelike (erect, subcylindrical)*; solitary; with substantial rachides; disarticulating; disarticulating at the joints. 'Articles' non-linear (hollowed to receive the sessile spikelet); appendaged; disarticulating obliquely; densely long-hairy. *Spikelets* paired; not secund; sessile and pedicellate (the pedicels resembling the internodes); **consistently in 'long-and-short' combinations**; in pedicellate/sessile combinations. Pedicels of the 'pedicellate' spikelets free of the rachis. The 'shorter' spikelets hermaphrodite. The 'longer' spikelets hermaphrodite, or male-only, or sterile.

Female-sterile spikelets. The pedicelled spikelet well developed to much reduced, its lower glume sometimes attenuate into a long, curved awn.

Female-fertile spikelets. Spikelets 7.5 mm long; compressed dorsiventrally; falling with the glumes (attached to the joint and pedicel). Rachilla terminated by a female-fertile floret. *Hairy callus present (the hairs encircling each node).* *Callus* short (and broad); blunt.

Glumes two; more or less equal; long relative to the adjacent lemmas; hairy; without conspicuous tufts or rows of hairs; awnless; very dissimilar (G_1 leathery and keel-less, G_2 thinner and boat-shaped). **Lower glume** not two-keeled (seemingly); convex on the back to flattened on the back; not pitted; **7–9 nerved.** Upper glume 3 nerved. *Spikelets* with incomplete florets. The incomplete florets proximal to the female-fertile florets. *The proximal incomplete florets* 1; paleate; male. The proximal lemmas awnless; 3(–5) nerved; decidedly exceeding the female-fertile lemmas; similar in texture to the female-fertile lemmas (thinly membranous); not becoming indurated.

Female-fertile florets 1. Lemmas less firm than the glumes (thinly membranous); not becoming indurated; entire; pointed, or blunt; awnless; hairless; non-carinate; without a germination flap; 3 nerved. *Palea* present; relatively long; entire; awnless, without apical setae; not indurated; 2-nerved. *Lodicules* present; 2; fleshy; glabrous. *Stamens* 3. *Ovary* glabrous. Styles fused. Stigmas 2.

Fruit, embryo and seedling. Hilum short. Pericarp unknown in mature state.

Abaxial leaf blade epidermis. *Costal/intercostal zonation* conspicuous. *Papillae* present (in *L. ledermannii*), or absent; intercostal. Intercostal papillae over-arching the stomata (conspicuously, at one end); consisting of one oblique swelling per cell (at one end of each interstomatal). Intercostal zones with typical long-cells (in *L. caespitosa*), or exhibiting many atypical long-cells (in *L. ledermannii*, where perturbation is associated with the basal rosettes of abundant macrohairs). Mid-intercostal long-cells rectangular (irregularly so in *L. ledermannii*); having markedly sinuous walls to having straight or only gently undulating walls. *Microhairs* present; elongated; clearly two-celled; panicoid-type; (33–)40–62(–69) microns long; 8.4–11.4 microns wide at the septum. Microhair total length/width at septum 2.9–5.4 (*L. ledermanni*), or 5.1–7.2 (*L. caespitosa*). Microhair apical cells (19.5–)22–30(–33) microns long. Microhair apical cell/total length ratio 0.47–0.59. *Stomata* common; 30–36 microns long (*L. ledermanni*), or 36.6–40.5 microns long (*L. caespitosa*). Subsidiaries non-papillate; predominantly triangular (often with truncated apices, in *L. caespitosa*). Guard-cells overlapped by the interstomatals (slightly), or overlapping to flush with the interstomatals. *Intercostal short-cells* common (solitary, associated with prickle bases, in *L. caespitosa*), or absent or very rare. *Costal short-cells* conspicuously in long rows, or predominantly paired, or neither distinctly grouped into long rows nor predominantly paired (the arrangement varying from place to place in both species seen). Costal silica bodies 'panicoid-type'; short, dumb-bell shaped, or cross shaped; not sharp-pointed.

Transverse section of leaf blade, physiology. C_4; XyMS–. PCR sheath outlines even. PCR sheath extensions absent. Mesophyll traversed by columns of colourless mesophyll cells (occasionally, from colourless arches), or not traversed by colourless columns. *Leaf blade* adaxially flat. *Midrib* conspicuous; having a conventional arc of bundles (the large median flanked on either side by several smaller bundles, and with large laterals at the outer margins of the midrib zone); with colourless mesophyll adaxially (comprising two large bulliform groups, and a mass of underlying colourless tissue contiguous with them which extends across the blade). *The lamina* symmetrical on either side of the midrib. Bulliforms present in discrete, regular adaxial groups (between pairs of main bundles); in simple fans and associated with colourless mesophyll cells to form deeply-penetrating fans (with colourless masses formed from bulliforms and contiguous underlying colourless tissue occupying the adaxial part of the blade between each pair of main bundles); associating with colourless mesophyll cells to form arches over small vascular bundles. Many of the smallest vascular bundles unaccompanied by sclerenchyma. Combined sclerenchyma girders present (with all the main laterals); forming 'figures' (mostly slender T's or I's). Sclerenchyma all associated with vascular bundles.

Taxonomy. Panicoideae; Andropogonodae; Andropogoneae; Rottboelliinae.

Ecology, geography, regional floristic distribution. 3 species. Tropical southern Africa. Helophytic; glycophytic. Savanna, in damp drainage hollows.

Paleotropical. African. Sudano-Angolan and West African Rainforest. Sahelo-Sudanian and South Tropical African.

References, etc. Morphological/taxonomic: Launert 1961, 1963, 1965a; Clayton 1977. Leaf anatomical: this project.

Special comments. Fruit data wanting.

Luziola A.L. Juss.

Arrozia Kunth, *Caryochloa* Trin.
Excluding *Hydrochloa*

Habit, vegetative morphology. Annual, or perennial; stoloniferous, or decumbent. The flowering culms leafy. *Culms* 10–30 cm high; herbaceous; branched above. Plants unarmed. *Leaves* not basally aggregated; without auricular setae. Leaf blades broad, or narrow; 1–10(–20) mm wide; flat. *Ligule an unfringed membrane*; not truncate (obtuse or acuminate, entire or laciniate); long.

Reproductive organization. *Plants monoecious with all the fertile spikelets unisexual*; without hermaphrodite florets. The spikelets of sexually distinct forms on the same plant; female-only and male-only. The male and female-fertile spikelets in different inflorescences (the staminate and pistillate spikelets in separate panicles), or on different branches of the same inflorescence (then with the male spikelets uppermost).

Inflorescence. Inflorescence paniculate (the panicles terminal and axillary); with capillary branchlets; espatheate; not comprising 'partial inflorescences' and foliar organs. Spikelet-bearing axes persistent. Spikelets solitary (at least, not paired); not secund; pedicellate.

Female-sterile spikelets. Male spikelets with one floret, a pointed lemma and 6–18 stamens. The male spikelets without glumes; 1 floreted. The lemmas awnless. Male florets 6–18 staminate.

Female-fertile spikelets. *Spikelets* unconventional (assumed to lack glumes); 2–5 mm long; compressed laterally to not noticeably compressed; disarticulating from the pedicel. Rachilla terminated by a female-fertile floret.

Glumes absent. *Spikelets* with female-fertile florets only.

Female-fertile florets 1. Lemmas not becoming indurated (thin); entire; pointed; awnless, or mucronate (?); lemmas several to many nerved. *Palea* present; relatively long (about equalling the lemma); thinner than the lemma (thin); not indurated; several nerved; one-keeled. *Stamens* 0. *Ovary* glabrous. Stigmas 2.

Fruit, embryo and seedling. *Fruit* free from both lemma and palea; small (1–2 mm long); ellipsoid to subglobose; not noticeably compressed. Hilum long-linear. *Pericarp* thick and hard; *free*. Embryo small. Endosperm hard; without lipid.

Abaxial leaf blade epidermis. *Costal/intercostal zonation* conspicuous. *Papillae* present; costal and intercostal. Intercostal papillae not over-arching the stomata; several per cell (or rather, many, on all cells save the guard-cells and subsidiaries). *Long-cells* markedly different in shape costally and intercostally (the costals narrower, more regularly rectangular); of similar wall thickness costally and intercostally (thin walled). Intercostal zones with typical long-cells to exhibiting many atypical long-cells. Mid-intercostal long-cells rectangular and fusiform; having markedly sinuous walls (the sinuosity fairly coarse). *Microhairs* present; elongated; ostensibly one-celled (cf. those of *Hygroryza*); minute; 19.5–21 microns long. *Stomata* common; 27–30 microns long. Subsidiaries non-papillate; triangular (a few), or dome-shaped (mostly). Guard-cells overlapped by the interstomatals (very slightly). *Intercostal short-cells* common; not paired (mostly seemingly solitary); silicified. Intercostal silica bodies oryzoid-type. Large prickle bases common. *Crown cells* absent. *Costal short-cells* conspicuously in long rows. Costal silica bodies 'panicoid-type' (predominating), or oryzoid (a few, especially alongside the veins); when panicoid type, short dumb-bell shaped, or cross shaped (the latter sometimes intergrading with oryzoid type); not sharp-pointed.

Transverse section of leaf blade, physiology. C_3; XyMS+. *Mesophyll* with non-radiate chlorenchyma; without adaxial palisade; with arm cells (seemingly, but not very conspicuous in the material seen); without fusoids. *Midrib* conspicuous (by a keel, and a large lacuna or region of colourless tissue above); having a conventional arc of bundles (a large median, with two smaller bundles on either side); with colourless mesophyll adaxially (probably). Bulliforms not in discrete, regular adaxial groups; in simple fans. All the vascular bundles accompanied by sclerenchyma. Combined sclerenchyma girders present; forming 'figures'. Sclerenchyma all associated with vascular bundles.

Taxonomy. Bambusoideae; Oryzodae; Oryzeae.

Ecology, geography, regional floristic distribution. 11 species. Southern U.S.A. to tropical South America. Hydrophytic, or helophytic.

Neotropical. Caribbean, Venezuela and Surinam, Amazon, Central Brazilian, Pampas, and Andean.

Economic importance. Significant weed species: *L. spruceana* (in North America).

References, etc. Leaf anatomical: this project.

Lycochloa Samuelsson

Habit, vegetative morphology. Perennial; rhizomatous. Culms herbaceous. Leaves non-auriculate. Leaf blades without cross venation. *Ligule an unfringed membrane*; truncate; 4–5 mm long.

Reproductive organization. Plants bisexual, with bisexual spikelets; with hermaphrodite florets. The spikelets all alike in sexuality.

Inflorescence. *Inflorescence a single raceme (lax)*; espatheate; not comprising 'partial inflorescences' and foliar organs. Spikelet-bearing axes persistent. Spikelets solitary; secund (nodding to one side); shortly pedicellate; distant.

Female-fertile spikelets. Spikelets 15–20 mm long; compressed laterally; disarticulating above the glumes; not disarticulating between the florets. Rachilla prolonged beyond the uppermost female-fertile floret. The rachilla extension with incomplete florets. Hairy callus present. Callus blunt.

Glumes two; very unequal; shorter than the adjacent lemmas; pointed (acute); awnless; similar (papery). Lower glume 5 nerved. Upper glume 5 nerved. *Spikelets* with incomplete florets. The incomplete florets distal to the female-fertile florets. The distal incomplete florets 3–4; *incomplete florets* clearly specialised and modified in form (forming a narrowly lanceolate cluster).

Female-fertile florets 2. *Lemmas* decidedly firmer than the glumes (leathery); not becoming indurated; incised; 2 lobed; not deeply cleft (bidentate); *awned*. *Awns* 1; median; *dorsal*; from near the top (or at about 1/3 the way down); geniculate. Lemmas hairy; non-carinate; without a germination flap; 7–11 nerved (and ribbed). *Palea* present; 2-nerved. **Lodicules** present; 2; *joined*; fleshy; not toothed; not or scarcely vascularized. *Ovary* glabrous (?); without a conspicuous apical appendage (?). Stigmas 2 (?).

Fruit, embryo and seedling. *Fruit* free from both lemma and palea; medium sized (5–8 mm long). Hilum long-linear. Embryo small. Endosperm hard; without lipid.

Abaxial leaf blade epidermis. *Costal/intercostal zonation* conspicuous. *Papillae* absent. Intercostal zones exhibiting many atypical long-cells (some tending to be hexagonal). Mid-intercostal long-cells having markedly sinuous walls. *Microhairs* absent. *Stomata* absent or very rare. *Intercostal short-cells* common; not paired; silicified. Intercostal silica bodies tall-and-narrow (elliptical). *Costal short-cells* conspicuously in long rows, or neither distinctly grouped into long rows nor predominantly paired (mostly in short or long rows, some solitary). Costal silica bodies 'panicoid-type'; mostly nodular; not sharp-pointed.

Transverse section of leaf blade, physiology. C_3; XyMS+. *Mesophyll* with non-radiate chlorenchyma. *Leaf blade* with distinct, prominent adaxial ribs to adaxially flat (the abaxial ribs more conspicuous); with the ribs more or less constant in size. *Midrib* conspicuous; with one bundle only. Bulliforms present in discrete, regular adaxial groups; in simple fans and associated with colourless mesophyll cells to form deeply-penetrating fans. All the vascular bundles accompanied by sclerenchyma. Combined sclerenchyma girders present; forming 'figures'. Sclerenchyma all associated with vascular bundles.

Taxonomy. Pooideae; Poodae; Meliceae.

Ecology, geography, regional floristic distribution. 1 species. Lebanon. Xerophytic (among rocks).

Holarctic. Tethyan. Irano-Turanian.

References, etc. Leaf anatomical: Metcalfe 1960.

Lycurus Kunth

Pleopogon Nutt

Habit, vegetative morphology. Perennial; caespitose. *Culms* 20–100 cm high; herbaceous. Culm internodes solid. *Leaves* mostly basal, or not basally aggregated; non-auriculate. Sheath margins free. Sheaths laterally compressed, keeled. Leaf blades linear;

narrow; 0.5–3 mm wide; flat, or folded; without abaxial multicellular glands; without cross venation. *Ligule* present; an unfringed membrane; not truncate (pointed).

Reproductive organization. Plants bisexual, with bisexual spikelets; with hermaphrodite florets. The spikelets of sexually distinct forms on the same plant; hermaphrodite and male-only, or hermaphrodite and sterile (the lower of each pair sterile or staminate).

Inflorescence. Inflorescence a false spike, with spikelets on contracted axes, or paniculate; contracted; spicate (and bristly, the branches short); espatheate; not comprising 'partial inflorescences' and foliar organs. *Spikelet-bearing axes disarticulating*; falling entire (the short branches shed). *Spikelets paired (the lower sterile or staminate, the other hermaphrodite)*; not secund; shortly pedicellate; consistently in 'long-and-short' combinations, or not in distinct 'long-and-short' combinations.

Female-sterile spikelets. The lower spikelet similar to the upper, sterile or male.

Female-fertile spikelets. Spikelets subcylindrical; compressed laterally to not noticeably compressed; falling with the glumes (deciduous in pairs). *Rachilla terminated by a female-fertile floret*.

Glumes present; two; more or less equal; shorter than the adjacent lemmas, or long relative to the adjacent lemmas; *awned*; very dissimilar (the lower usually with 2–3 slender awns, the upper 1-awned). Lower glume 2 nerved, or 3 nerved. Upper glume 1 nerved. **Spikelets** with female-fertile florets only; *without proximal incomplete florets*.

Female-fertile florets 1 (the sterile spikelets also 1-flowered). *Lemmas* tapering into the awn; *decidedly firmer than the glumes*; not becoming indurated (firm); entire; pointed; awned. Awns 1; median; apical; non-geniculate (slender); much shorter than the body of the lemma to much longer than the body of the lemma. Lemmas without a germination flap; 3 nerved. *Palea* present; relatively long; awnless, without apical setae; thinner than the lemma. **Lodicules** *present*; 2; free; not or scarcely vascularized. *Stamens* 3. Anthers 1 mm long. *Ovary* glabrous. Styles free to their bases. Stigmas 2 (short).

Fruit, embryo and seedling. Fruit fusiform. Pericarp fused. Embryo large; with an epiblast; with a scutellar tail; with an elongated mesocotyl internode. Embryonic leaf margins meeting.

Abaxial leaf blade epidermis. *Costal/intercostal zonation* conspicuous. *Papillae* present; intercostal. Intercostal papillae over-arching the stomata (in places, at one end or from the side); consisting of one oblique swelling per cell (mostly at one end of each interstomatal, and sometimes in the middle of a long-cell). *Long-cells* markedly different in shape costally and intercostally (the costals being much narrower and more regularly rectangular). Intercostal zones exhibiting many atypical long-cells (many being quite short). Mid-intercostal long-cells rectangular and fusiform; having straight or only gently undulating walls. *Microhairs* present; elongated; clearly two-celled; chloridoid-type (the basal cells somewhat the longer). Microhair apical cell wall of similar thickness/rigidity to that of the basal cell. Microhairs (37.5–)45–46.5(–49.5) microns long. Microhair basal cells 30 microns long. Microhairs 15.6–16.5–17.4 microns wide at the septum. Microhair total length/width at septum 2.3–2.9. Microhair apical cells (10.5–)11.4–12(–18) microns long. Microhair apical cell/total length ratio 0.28–0.36. *Stomata* common; 21–24 microns long. Subsidiaries dome-shaped (mostly), or triangular (a few). Guard-cells overlapping to flush with the interstomatals. *Intercostal short-cells* absent or very rare (ignoring the hair bases); not paired. Intercostal silica bodies absent. Short macrohairs common. *Crown cells* absent. *Costal short-cells* conspicuously in long rows. Costal silica bodies present in alternate cell files of the costal zones; saddle shaped (squarish to tall-and-narrow versions of this form predominating), or tall-and-narrow; not sharp-pointed.

Transverse section of leaf blade, physiology. C_4; XyMS+. PCR sheath outlines even. PCR sheaths of the primary vascular bundles interrupted; interrupted both abaxially and adaxially. PCR sheath extensions absent. PCR cell chloroplasts centripetal. *Mesophyll* with radiate chlorenchyma; traversed by columns of colourless mesophyll cells (conspicuously, from every furrow). *Leaf blade* 'nodular' in section, or adaxially flat; with the ribs more or less constant in size (low). *Midrib* conspicuous; with one bundle only. Bulliforms present in discrete, regular adaxial groups; associated with colourless mesophyll cells to form deeply-penetrating fans (these linked with traversing columns of colourless cells). All the vascular bundles accompanied by sclerenchyma. Combined sclerenchyma girders present (with the primaries only — the rest with strands only);

forming 'figures' (the primaries with I's). Sclerenchyma all associated with vascular bundles. The lamina margins with fibres.

Cytology. Chromosome base number, $x = 10$. $2n = 28$ and 40.

Taxonomy. Chloridoideae; main chloridoid assemblage.

Ecology, geography, regional floristic distribution. 3 species. Southern U.S.A., Hawaii to north tropical South America. Species of open habitats. Plains and rocky hills.

Holarctic and Neotropical. Boreal and Madrean. Atlantic North American. Caribbean and Andean. Central Grasslands.

Rusts and smuts. Rusts — *Puccinia*. Taxonomically wide-ranging species: *Puccinia schedonnardi*. Smuts from Ustilaginaceae. Ustilaginaceae — *Ustilago*.

Economic importance. Important native pasture species: *L. phleoides*.

References, etc. Leaf anatomical: this project.

Lygeum Loefl ex L.

Linospartum Adans., *Spartum* P. Beauv.

Habit, vegetative morphology. Wiry perennial; rhizomatous (the rhizomes scaly). *Culms* 30–70 cm high; herbaceous; branched above. Culm nodes glabrous. Culm internodes solid. *Leaves* not basally aggregated; non-auriculate. Leaf blades narrow (junciform); without cross venation; persistent; once-folded in bud. *Ligule* present; an unfringed membrane; not truncate (acute); 7 mm long.

Reproductive organization. Plants bisexual, with bisexual spikelets; with hermaphrodite florets.

Inflorescence. *Inflorescence* reduced to a single spikelet; *of one terminal spikelet, enclosed in an indurated spathe until it protrudes from the side at maturity*; spatheate (the hard spathes ovate, acute or with a laminate tip). *Spikelet-bearing axes* very much reduced (to a single spikelet).

Female-fertile spikelets. Spikelets *unconventional (through apparent absence of glumes, and congenital fusion of lemmas and paleas)*; 35–45 mm long; compressed laterally; the spikelet falling whole. Rachilla terminated by a female-fertile floret.

Glumes absent. *Spikelets* with female-fertile florets only.

Female-fertile florets 2. *Lemmas leathery, 2 cm long, the two connate by their margins over the lower half into a rigid, hairy tube, the upper halves free*; becoming indurated; awnless; hairy (below, glabrous above); non-carinate; 9 nerved. *Palea* present (the two connate by their backs below); relatively long (3–4 cm: much exceeding the lemms); awnless, without apical setae; 2-nerved. *Lodicules* absent. *Stamens* 3. Anthers about 15 mm long; not penicillate. *Ovary* glabrous. Styles fused (into a single style). Stigmas 1.

Fruit, embryo and seedling. Fruit medium sized (8–9 mm long); fusiform. Hilum long-linear. Embryo small. Endosperm containing compound starch grains.

Seedling with a short mesocotyl. First seedling leaf with a well-developed lamina. The lamina narrow; erect; 7 veined.

Abaxial leaf blade epidermis. *Costal/intercostal zonation* conspicuous. *Papillae* absent. *Long-cells* similar in shape costally and intercostally; of similar wall thickness costally and intercostally (rather thick walled). Mid-intercostal long-cells rectangular; having markedly sinuous walls. *Microhairs* present (but infrequent); chloridoid-type (sic: large and spectacular, in the material seen); 72–93 microns long; 21.6–24 microns wide at the septum. Microhair total length/width at septum 3–4.3. Microhair apical cells 24–25.5 microns long. Microhair apical cell/total length ratio 0.27–0.33. *Stomata* common; (45–)54–57 microns long. Subsidiaries triangular. Guard-cells overlapped by the interstomatals. *Intercostal short-cells* common; silicified. Intercostal silica bodies rounded. *Costal short-cells* predominantly paired, or neither distinctly grouped into long rows nor predominantly paired (rarely). Costal silica bodies exclusively rounded; not sharp-pointed.

Transverse section of leaf blade, physiology. C_3; XyMS+. *Mesophyll* with non-radiate chlorenchyma. *Leaf blade* with distinct, prominent adaxial ribs; with the ribs more or less constant in size. *Midrib* not readily distinguishable; with one bundle only. Bulliforms not in discrete, regular adaxial groups (in the furrows, small and variable in size cf. *Ammophila*). All the vascular bundles accompanied by sclerenchyma. Combined sclerenchyma girders present, or absent; forming 'figures'. Sclerenchyma not all bundle-associated. The 'extra' sclerenchyma in abaxial groups (these columnar); opposite some of the furrows.

Phytochemistry. Leaves without flavonoid sulphates.

Special diagnostic feature. *Plant coarsely tufted, with wiry leaf blades, the inflorescence of one very peculiar spikelet enclosed in a sheath*.

Cytology. Chromosome base number, $x = 10$. $2n = 40$ (the chromosomes 'large'). 4 ploid (the 40 chromosomes large).

Taxonomy. Arundinoideae; Lygeae.

Ecology, geography, regional floristic distribution. 1 species. Mediterranean. Xerophytic; species of open habitats.

Holarctic and Paleotropical. Boreal and Tethyan. African. Euro-Siberian. Mediterranean. Saharo-Sindian. European.

Rusts and smuts. Rusts — *Puccinia*. Taxonomically wide-ranging species: '*Uromyces' dactylidis*. Smuts from Ustilaginaceae. Ustilaginaceae — *Ustilago*.

Economic importance. Esparto grass, yielding fibre for paper, rope etc.

References, etc. Leaf anatomical: Metcalfe 1960; this project.

Maclurolyra Calderón and Soderstrom

Habit, vegetative morphology. Perennial; caespitose. The flowering culms leafy (but with only 1–3 developed blades). *Culms* 20–50 cm high; unbranched above. Culm nodes hairy. Culm internodes solid (the middle internodes much elongated). Rhizomes pachymorph. Plants unarmed. Young shoots extravaginal and intravaginal. **Leaves** mostly basal; *with auricular setae*. Leaf blades ovate-lanceolate, or elliptic (acuminate); broad; 30–50 mm wide (and 10–21 cm long); not cordate, not sagittate (asymmetrically rounded at base); pseudopetiolate (pulvinate, twisted through 180°); cross veined; rolled in bud. *Ligule* present (thick); 0.5–1.2 mm long. *Contra-ligule* absent.

Reproductive organization. *Plants monoecious with all the fertile spikelets unisexual*; without hermaphrodite florets. The spikelets of sexually distinct forms on the same plant (male and female); female-only and male-only. *The male and female-fertile spikelets mixed in the inflorescence (each infloresence unit with 2–several female spikelets and 1–2 males)*. The spikelets overtly heteromorphic; all in heterogamous combinations.

Inflorescence. Inflorescence *indeterminate (a synflorescence)*; paniculate (terminal, of reduced racemelike panicles); contracted; rigid-fusiform; spatheate (spatheoles, etc. hidden within within the uppermost sheaths); a complex of 'partial inflorescences' and intervening foliar organs (but not obviously so, without dissection). *Spikelet-bearing axes* reduced, mixed-sex 'racemes'; persistent. Spikelets paired; not secund; subsessile, or pedicellate; consistently in 'long-and-short' combinations; unequally pedicellate in each combination. The 'shorter' spikelets female-only. The 'longer' spikelets male-only.

Female-sterile spikelets. *Male spikelets smaller than the females, without glumes, lemmas 7 nerved, florets with 3 fertile stamens and non-penicillate anthers and 3 staminodes*. Rachilla of male spikelets terminated by a male floret. The male spikelets without glumes; without proximal incomplete florets; 1 floreted. Male florets 1; 3 staminate, or 6 staminate (if the 3 staminodes are called stamens).

Female-fertile spikelets. *Spikelets* 9.5–11.5 mm long; *compressed laterally to not noticeably compressed (? 'depressed')*; disarticulating above the glumes; with conventional internode spacings. Rachilla terminated by a female-fertile floret (and no elongated internode above the glumes). Hairy callus absent. *Callus* absent.

Glumes two; more or less equal; long relative to the adjacent lemmas (about equalling them); free; hairless (glabrous, short-hispid at tip); pointed; awnless; non-carinate; similar (lanceolate-acute, leathery). Lower glume 5–6 nerved. Upper glume 5–7 nerved. *Spikelets* with female-fertile florets only.

Female-fertile florets 1. Lemmas narrowly lanceolate-acuminate; decidedly firmer than the glumes; becoming indurated (leathery, becoming crustaceous); entire; pointed;

awless; hairy (covered with long appressed hairs); non-carinate; having the margins tucked in onto the palea (and enclosing it); (5–)7(–8) nerved. *Palea* present (but enclosed, elliptic-acuminate); relatively long; entire; awnless, without apical setae; indurated; several nerved (2–4); keel-less (convolute, villous). *Lodicules* present; 3, or 4; free; glabrous; not toothed; heavily vascularized. *Stamens* 0 (3 minute staminodes, alternating with the lodicules). *Ovary* glabrous (but the style hairy below middle). Styles fused (the ovary attenuate into one long slender style). Stigmas 2.

Fruit, embryo and seedling. *Fruit* free from both lemma and palea (but enclosed); medium sized (5.5 mm long); compressed dorsiventrally (ventrally). Hilum long-linear. Embryo small. Endosperm containing compound starch grains.

Seedling with a short mesocotyl. First seedling leaf without a lamina (the seedling with two sheaths).

Abaxial leaf blade epidermis. *Costal/intercostal zonation* conspicuous. *Papillae* present. Intercostal papillae over-arching the stomata; several per cell (more than 1 row per cell). Mid-intercostal long-cells rectangular; having markedly sinuous walls. *Microhairs* present; clearly two-celled, or uniseriate (i.e. some 3-celled); panicoid-type. *Stomata* common (in bands alongside veins). Subsidiaries papillate; triangular. *Intercostal short-cells* common; in cork/silica-cell pairs; silicified. Intercostal silica bodies tall-and-narrow and oryzoid-type. *Costal short-cells* conspicuously in long rows (a few pairs and short rows). Costal silica bodies saddle shaped, or oryzoid (some tending to this form), or 'panicoid-type' (some); when panicoid type, tending to cross shaped; not sharp-pointed.

Transverse section of leaf blade, physiology. C_3; XyMS+. Mesophyll with arm cells; with fusoids (large). *Leaf blade* with distinct, prominent adaxial ribs (wide, rounded); with the ribs more or less constant in size. *Midrib* conspicuous (larger rib); having a conventional arc of bundles (1 or 2 small bundles each side of median). Bulliforms present in discrete, regular adaxial groups. All the vascular bundles accompanied by sclerenchyma. Combined sclerenchyma girders present (with all the bundles); forming 'figures'.

Cytology. Chromosome base number, $x = 11$. $2n = 22$. 2 ploid.

Taxonomy. Bambusoideae; Oryzodae; Olyreae.

Ecology, geography, regional floristic distribution. 1 species. Panama. Neotropical. Caribbean.

References, etc. Morphological/taxonomic: Calderón and Soderstrom 1973.

Maillea Parl.

~ *Phleum* (IP. crypsoides)

Habit, vegetative morphology. *Annual.* *Culms* 2–10 cm high; herbaceous. Leaves non-auriculate. Sheath margins free. The upper sheaths inflated. Leaf blades linear; narrow; flat; without cross venation. Ligule an unfringed membrane; truncate; 1.5–2.5 mm long.

Reproductive organization. Plants bisexual, with bisexual spikelets; with hermaphrodite florets.

Inflorescence. *Inflorescence paniculate (sessile among the leaves)*; contracted; more or less ovoid (to 5 cm long); espatheate; not comprising 'partial inflorescences' and foliar organs. Spikelet-bearing axes persistent. Spikelets not secund.

Female-fertile spikelets. Spikelets 2–4.5 mm long; compressed laterally; disarticulating above the glumes. Rachilla prolonged beyond the uppermost female-fertile floret. The rachilla extension naked. Hairy callus absent.

Glumes two; more or less equal; exceeding the spikelets; long relative to the adjacent lemmas (about three times as long); free; pointed (apiculate); awnless; carinate; **with the keel conspicuously winged**; similar. Lower glume 1 nerved. Upper glume 1 nerved. *Spikelets* with female-fertile florets only.

Female-fertile florets 1. Lemmas truncate; less firm than the glumes to similar in texture to the glumes (and only about 1/3 of glume length); not becoming indurated; entire; blunt; awnless; hairless; non-carinate; without a germination flap; 1(–3) nerved. **Palea** present; relatively long (exceeding the lemma); *1-nerved*; keel-less. *Lodicules* absent. **Stamens** *2*. *Ovary* glabrous. Styles free to their bases. Stigmas 2.

Fruit, embryo and seedling. *Fruit* free from both lemma and palea; not noticeably compressed. Hilum short. Embryo small.

Abaxial leaf blade epidermis. *Costal/intercostal zonation* conspicuous. *Papillae* present; costal and intercostal. Intercostal papillae over-arching the stomata (over-arching them laterally); consisting of one oblique swelling per cell, or consisting of one symmetrical projection per cell. *Long-cells* similar in shape costally and intercostally; of similar wall thickness costally and intercostally. Intercostal zones exhibiting many atypical long-cells (many of them short to more or less round). Mid-intercostal long-cells fusiform (to round); having straight or only gently undulating walls. *Microhairs* absent. *Stomata* common; (42–)45–52(–66) microns long. Subsidiaries non-papillate; parallel-sided. Guard-cells overlapped by the interstomatals (the stomata distinctly sunken). *Intercostal short-cells* common to absent or very rare; not paired (solitary, large, rounded, merging into the range of long-cell forms); not silicified. *Costal short-cells* neither distinctly grouped into long rows nor predominantly paired. Costal silica bodies present and well developed; horizontally-elongated crenate/sinuous (mostly), or horizontally-elongated smooth (very few); not sharp-pointed.

Transverse section of leaf blade, physiology. C_3; XyMS+. *Mesophyll* with non-radiate chlorenchyma; without adaxial palisade. *Leaf blade* with distinct, prominent adaxial ribs; with the ribs more or less constant in size. *Midrib* not readily distinguishable; with one bundle only. *The lamina* symmetrical on either side of the midrib. All the vascular bundles accompanied by sclerenchyma (except for the outermost bundle on either side, in the material seen). Combined sclerenchyma girders absent (the sclerenchyma confined to small adaxial and abaxial strands). Sclerenchyma all associated with vascular bundles.

Taxonomy. Pooideae; Poodae; Aveneae.

Ecology, geography, regional floristic distribution. 1 species. Eastern Mediterranean islands and Greece.

Holarctic. Tethyan. Mediterranean.

References, etc. Leaf anatomical: this project.

Malacurus Nevski

~ Leymus

Habit, vegetative morphology. Perennial; caespitose. *Culms herbaceous*. Leaves auriculate. Leaf blades without cross venation. Ligule an unfringed membrane; truncate; 0.7–8 mm long.

Reproductive organization. Plants bisexual, with bisexual spikelets; with hermaphrodite florets. The spikelets of sexually distinct forms on the same plant, or all alike in sexuality; hermaphrodite, or hermaphrodite and sterile (with sterile spikelets at the tip of the rachis).

Inflorescence. *Inflorescence a false spike, with spikelets on contracted axes (the clusters reduced to pairs)*; espatheate; not comprising 'partial inflorescences' and foliar organs. Spikelet-bearing axes persistent. Spikelets paired; not secund.

Female-fertile spikelets. *Spikelets* 8–16 mm long; *not noticeably compressed*; disarticulating above the glumes; *disarticulating between the florets (the rachilla very fragile)*. Rachilla prolonged beyond the uppermost female-fertile floret. Hairy callus absent.

Glumes two (much reduced); very unequal to more or less equal; joined (those of adjacent spikelets sometimes joined), or free; lateral to the rachis, or displaced (much reduced); hairless; glabrous; pointed; awned (awn-like); similar (subulate-setiform). Lower glume 1 nerved. Upper glume 1 nerved. *Spikelets* with female-fertile florets only, or with incomplete florets. The incomplete florets distal to the female-fertile florets.

Female-fertile florets *2–4*. Lemmas not becoming indurated; entire, or incised; awnless, or awned. Awns when present, 1; median; from a sinus, or apical; non-geniculate; much shorter than the body of the lemma (no more than 2 mm long); when present, deciduous. Lemmas hairy (densely lanate); non-carinate; without a germination flap; 7–9 nerved. *Palea* present; relatively long; 2-nerved; 2-keeled. *Lodicules* present; 2; free; membranous; ciliate; not toothed; not or scarcely vascularized. *Stamens* 3. *Ovary* hairy. Styles free to their bases. Stigmas 2.

Fruit, embryo and seedling. **Fruit** adhering to lemma and/or palea; *not grooved*. Hilum long-linear. Embryo small.

Abaxial leaf blade epidermis. *Costal/intercostal zonation* conspicuous. *Papillae* absent. Mid-intercostal long-cells rectangular; having markedly sinuous walls (and pitted). *Microhairs* absent. *Stomata* common; 69–75 microns long. Subsidiaries low dome-shaped, or parallel-sided. Guard-cells overlapped by the interstomatals (clearly sunken). *Intercostal short-cells* common; intercostal short-cells solitary and paired. Costal silica bodies absent to poorly developed (in the material seen); horizontally-elongated smooth, or rounded, or tall-and-narrow.

Transverse section of leaf blade, physiology. C_3; XyMS+. Mesophyll without adaxial palisade. *Leaf blade* with distinct, prominent adaxial ribs. *Midrib* not readily distinguishable; with one bundle only. *The lamina* symmetrical on either side of the midrib. Bulliforms present in discrete, regular adaxial groups; in simple fans. All the vascular bundles accompanied by sclerenchyma. Combined sclerenchyma girders present; forming 'figures'. Sclerenchyma all associated with vascular bundles.

Cytology. Chromosome base number, $x = 7$. $2n = 42$. 6 ploid.

Taxonomy. Pooideae; Triticodae; Triticeae.

Ecology, geography, regional floristic distribution. 1 species. Central Asia. Holarctic. Boreal. Euro-Siberian. Siberian.

References, etc. Morphological/taxonomic: Löve 1984. Leaf anatomical: this project.

Maltebrunia Kunth

Habit, vegetative morphology. Erect perennial; stoloniferous. *Culms* 30–150 cm high (?); herbaceous; not scandent; unbranched above. *Leaves* not basally aggregated; auriculate (with sheath auricles adnate to the ligule); without auricular setae. *Leaf blades lanceolate to ovate*; broad; 15–50 mm wide; flat; usually pseudopetiolate; cross veined (?), or without cross venation (in *M. leersioides*). Ligule an unfringed membrane; not truncate (toothed). *Contra-ligule* absent.

Reproductive organization. Plants bisexual, with bisexual spikelets; with hermaphrodite florets.

Inflorescence. Inflorescence paniculate; open; with capillary branchlets; espatheate; not comprising 'partial inflorescences' and foliar organs. Spikelet-bearing axes persistent. Spikelets not secund.

Female-fertile spikelets. Spikelets unconventional (owing to the reduced glumes, cf. *Oryza*); 7–10 mm long; lightly compressed laterally; disarticulating above the glumes (i.e. above their vestiges); *with only a short stipe above the sterile lemmas*. Rachilla terminated by a female-fertile floret. Hairy callus absent. *Callus* short; blunt.

Glumes present to absent (vestigial); two; *minute (reduced to a bilobed annulus)*; more or less equal; shorter than the spikelets; shorter than the adjacent lemmas; joined; awnless. Lower glume 0 nerved. Upper glume 0 nerved. *Spikelets* with incomplete florets. The incomplete florets proximal to the female-fertile florets. **The proximal incomplete florets 2**; epaleate; sterile. The proximal lemmas subulate; exceeded by the female-fertile lemmas (about 1/5 to 1/10 the length of the spikelet); less firm than the female-fertile lemmas (thinly membranous).

Female-fertile florets 1. Lemmas ovate-oblong or elliptic, acute, the margins clasping the lateral nerves of the palea; not becoming indurated (firmly chartaceous); entire; pointed; awnless; hairy (with thin, appressed white hairs), or hairless; carinate; without a germination flap; 5 nerved; *with the nerves non-confluent*. *Palea* present; relatively long (equalling the spikelet); tightly clasped by the lemma; entire (narrow, acute); awnless, without apical setae; textured like the lemma; not indurated; several nerved (three nerved); one-keeled. Palea keels wingless. *Lodicules* present; 2; free; membranous (but relatively substantial, in the material seen); glabrous. **Stamens 6**. Anthers about 3–4 mm long (relatively long, linear); not penicillate; without an apically prolonged connective. *Ovary* glabrous. Styles free to their bases. Stigmas 2 (plumose).

Abaxial leaf blade epidermis. *Costal/intercostal zonation* conspicuous. *Papillae* present; intercostal (mostly in the stomatal zones, adjoining the costae — not to be confused with the abundant, bulbous prickle bases). Intercostal papillae not over-arching the stomata (for the most part); several per cell (small, round, scattered on the interstomatals). *Long-cells* similar in shape costally and intercostally (the costals rather smaller); of similar wall thickness costally and intercostally (of medium thickness). Mid-intercostal long-cells rectangular; having markedly sinuous walls (the sinuosity coarse). *Microhairs* present; elongated; clearly two-celled; panicoid-type. *Stomata* common (but absent from the mid-intercostal regions). Subsidiaries non-papillate; dome-shaped to tri-angular. Guard-cells overlapping to flush with the interstomatals. *Intercostal short-cells* common (hard to observe in the material seen); in cork/silica-cell pairs and not paired; seemingly not silicified. Large bulbous-based prickles with reduced points, abundant both costally and intercostally. *Costal short-cells* conspicuously in long rows (in some files), or predominantly paired (in other files). Costal silica bodies oryzoid.

Transverse section of leaf blade, physiology. C_3; XyMS+. Mesophyll without ad-axial palisade; with arm cells; with fusoids. The fusoids external to the PBS. *Leaf blade* adaxially flat. *Midrib* conspicuous; having complex vascularization (there being an abaxial arc of three with a median one adaxially, in the material seen); with colourless mesophyll adaxially. *The lamina* symmetrical on either side of the midrib. Bulliforms constituting most of the adaxial epidermis, in very wide fans. All the vascular bundles accompanied by sclerenchyma. Combined sclerenchyma girders present (with most of the bundles); forming 'figures' (I's with the smaller bundles, 'anchors' with the large ones). Sclerenchyma all associated with vascular bundles.

Taxonomy. Bambusoideae; Oryzodae; Oryzeae.

Ecology, geography, regional floristic distribution. 3–5 species. Gabon, Tanzania, Madagascar. Shade species; glycophytic.

Paleotropical. African. Sudano-Angolan. South Tropical African.

References, etc. Morphological/taxonomic: Hubbard 1962. Leaf anatomical: this project.

Special comments. Fruit data wanting.

Manisuris L.

Peltophorus Desv.

Excluding all but *Manisuris myuros* L.

Habit, vegetative morphology. Perennial. *Culms* 15–70 cm high; herbaceous; branched above. Leaf blades linear; narrow.

Reproductive organization. Plants bisexual, with bisexual spikelets; with hermaph-rodite florets. The spikelets of sexually distinct forms on the same plant; hermaphrodite and male-only, or hermaphrodite and sterile (?). The male and female-fertile spikelets mixed in the inflorescence. The spikelets overtly heteromorphic; all in heterogamous combinations.

Inflorescence. Inflorescence of spiciform 'racemes', solitary or fascicled. Rachides hollowed. Inflorescence spatheate. **Spikelet-bearing axes** spikelike; solitary (axillary and terminal); with substantial rachides; disarticulating; *disarticulating at the joints (in a peculiar and characteristic fashion)*. 'Articles' non-linear (stoutly clavate); with a basal callus-knob. Spikelets paired; secund (the sessile members in two rows, on one side of the rachis); sessile (by fusion of the pedicels); consistently in 'long-and-short' combina-tions; in pedicellate/sessile combinations. *Pedicels of the 'pedicellate' spikelets discern-ible, but fused with the rachis (the 'pedicelled' spikelet abscinds from its own pedicel, but remains fused to the base of the internode above, so that there are 'false pairs' of spikelets comprising the pedicelled member of one segment plus the sessile member of the segment above — by contrast with the 'normal' arrangement in* **Glyphochloa***)*. The 'shorter' spikelets hermaphrodite. The 'longer' spikelets male-only, or sterile (?).

Female-sterile spikelets. 'Pedicelled' spikelets male or neuter (?), about 4 mm long, striate, awnless, the G_1 winged on one margin, the G_2 naviculate and keel-winged at the tip.

Female-fertile spikelets. Spikelets 2.5–3 mm long; abaxial; compressed dorsiventrally; falling with the glumes (and with the joint plus the pedicelled spikelet from the joint below — by contrast with *Glyphochloa*). Rachilla terminated by a female-fertile floret. Hairy callus absent.

Glumes present; two; long relative to the adjacent lemmas; hairless; without conspicuous tufts or rows of hairs; awnless; non-carinate; very dissimilar (G_1 hardened, laterally winged, with a deep transverse groove interrupting the wings). Lower glume two-keeled; not pitted; relatively smooth. *Spikelets* with incomplete florets. The incomplete florets proximal to the female-fertile florets. *The proximal incomplete florets* 1; epaleate; sterile. The proximal lemmas awnless; 0 nerved, or 2 nerved; similar in texture to the female-fertile lemmas (hyaline).

Female-fertile florets 1. Lemmas entire; less firm than the glumes (i.e., than the G_1 — hyaline); not becoming indurated; entire; awnless; 0 nerved, or 2 nerved. *Palea* present. *Lodicules* present; 2; fleshy. *Stamens* 3.

Fruit, embryo and seedling. Hilum short. Embryo large.

Abaxial leaf blade epidermis. *Costal/intercostal zonation* conspicuous. *Papillae* absent. *Long-cells* markedly different in shape costally and intercostally (the costals much narrower); of similar wall thickness costally and intercostally (fairly thin walled). Mid-intercostal long-cells rectangular to fusiform; having markedly sinuous walls (the sinuosity fine to coarse, fairly irregular). *Microhairs* present; elongated; clearly two-celled; panicoid-type. *Stomata* common. Subsidiaries non-papillate; dome-shaped to triangular. Guard-cells overlapping to flush with the interstomatals. *Intercostal short-cells* common; in cork/silica-cell pairs. No prickles or macrohairs seen. *Costal short-cells* conspicuously in long rows. Costal silica bodies present and well developed; 'panicoid-type'; butterfly shaped and dumb-bell shaped, or nodular (mostly short dumb-bells).

Transverse section of leaf blade, physiology. C_4. The anatomical organization conventional. XyMS–. *Midrib* conspicuous; having a conventional arc of bundles; with colourless mesophyll adaxially. Many of the smallest vascular bundles unaccompanied by sclerenchyma. Combined sclerenchyma girders present (with the primaries). Sclerenchyma all associated with vascular bundles.

Special diagnostic feature. *Spikelets in 'false pairs', the pedicellate member of the andropogonoid pair abscinding from its pedicel but remaining attached to the base of the 'article' above, alongside the sessile member of that 'article'.*

Cytology. Chromosome base number, $x = 9$. $2n = 18$ and 36. 2 and 4 ploid.

Taxonomy. Panicoideae; Andropogonodae; Andropogoneae; Rottboelliinae.

Ecology, geography, regional floristic distribution. 1 species. India. Dry places. Paleotropical. Indomalesian. Indian.

References, etc. Morphological/taxonomic: Jain 1970. Leaf anatomical: this project.

Special comments. Sensu stricto, = *M. myuros*.

Megalachne Steud.

Pantathera Phil.

Habit, vegetative morphology. Perennial; caespitose. **Culms** 40–100 cm high; *herbaceous*. Young shoots intravaginal. *Leaves* not basally aggregated; distichous; non-auriculate. Sheaths glabrous. Leaf blades linear; narrow; flat, or rolled (convolute when dry); without cross venation; persistent. Ligule an unfringed membrane; not truncate (elongated, laciniate); 5–10 mm long.

Reproductive organization. Plants bisexual, with bisexual spikelets; with hermaphrodite florets. The spikelets all alike in sexuality.

Inflorescence. *Inflorescence* rather few spikeleted; *paniculate*; fairly open; espatheate; not comprising 'partial inflorescences' and foliar organs. Spikelet-bearing axes persistent. Spikelets not secund; pedicellate.

Female-fertile spikelets. Spikelets 10–26 mm long; compressed laterally; disarticulating above the glumes; disarticulating between the florets; with conventional internode spacings. Rachilla prolonged beyond the uppermost female-fertile floret. The rachilla extension with incomplete florets. Hairy callus present (well developed, with stiff hairs).

Glumes present; two; more or less equal; long relative to the adjacent lemmas; hairless; pointed; *awned (both, or only the lower)*; carinate to non-carinate; very dissimilar to similar (herbaceous, linear-lanceolate, aristulate to awned). Lower glume 3–5 nerved. Upper glume 3–5 nerved. *Spikelets* with incomplete florets. The incomplete florets distal to the female-fertile florets. *The distal incomplete florets* merely underdeveloped; awned. *Spikelets without proximal incomplete florets.*

Female-fertile florets 2–6. Lemmas lanceolate-acuminate; similar in texture to the glumes to decidedly firmer than the glumes (subleathery); not becoming indurated; entire; awned. Awns 1; median; apical; non-geniculate; recurving and flexuous (divaricating at maturity); hairless; about as long as the body of the lemma to much longer than the body of the lemma; entered by several veins; persistent. *Lemmas* hairless; scabrous; somewhat *carinate*; without a germination flap; 5 nerved, or 7 nerved. *Palea* present; relatively long; tightly clasped by the lemma; apically notched; awnless, without apical setae to with apical setae; thinner than the lemma (membranous); not indurated; 2-nerved; 2-keeled. Palea keels wingless; scabrous. *Lodicules* present; 2; free; membranous; glabrous; toothed; not or scarcely vascularized. **Stamens 3**. Anthers about 0.6 mm long; not penicillate; without an apically prolonged connective. **Ovary hairy**. Styles fused to free to their bases. Stigmas 2–3.

Fruit, embryo and seedling. Fruit medium sized (6–7 mm long); compressed dorsiventrally; with hairs confined to a terminal tuft. Hilum long-linear. Embryo small. Endosperm containing compound starch grains. Embryo with an epiblast; without a scutellar tail; with a negligible mesocotyl internode.

Abaxial leaf blade epidermis. *Costal/intercostal zonation* lacking. *Papillae* absent. *Long-cells* similar in shape costally and intercostally; of similar wall thickness costally and intercostally. Mid-intercostal long-cells rectangular; having markedly sinuous walls (thick, pitted). *Microhairs* absent. *Stomata* absent or very rare. *Intercostal short-cells* common; in cork/silica-cell pairs; silicified. *Costal short-cells* predominantly paired. Costal silica bodies rounded (mostly), or crescentic (a few); not sharp-pointed.

Transverse section of leaf blade, physiology. C_3; XyMS+. *Mesophyll* with non-radiate chlorenchyma; without adaxial palisade. *Leaf blade* with distinct, prominent adaxial ribs; with the ribs more or less constant in size. *Midrib* not readily distinguishable; with one bundle only. Bulliforms present in discrete, regular adaxial groups (the groups large). All the vascular bundles accompanied by sclerenchyma. Combined sclerenchyma girders present (with all the bundles); forming 'figures' (all bundles with strong T's to I's). Sclerenchyma not all bundle-associated. The 'extra' sclerenchyma in abaxial groups (in addition to strong marginal groups); abaxial-hypodermal, the groups isolated (of 1–few cells, opposite the bulliforms).

Taxonomy. Pooideae; Poodae; Poeae.

Ecology, geography, regional floristic distribution. 2 species. Juan Fernandez Is. (Chile). Species of open habitats. Rocky slopes.

Neotropical. Fernandezian.

References, etc. Leaf anatomical: this project.

Megaloprotachne C.E. Hubb.

Habit, vegetative morphology. Annual; caespitose, or decumbent (sometimes rooting at the lower nodes). *Culms* 15–90 cm high; herbaceous; branched above, or unbranched above. Culm nodes hairy. Culm internodes hollow. Young shoots intravaginal. *Leaves* not basally aggregated; non-auriculate. Leaf blades linear; narrow; 3–7 mm wide; flat; without cross venation; persistent. *Ligule a fringed membrane*; not truncate (acute); about 1 mm long (excluding the hairs, which are 1 to 2 mm long). *Contra-ligule* absent.

Reproductive organization. Plants bisexual, with bisexual spikelets; with hermaphrodite florets.

Inflorescence. Inflorescence of spicate main branches (spike-like racemes or narrow panicles); subdigitate (usually with several racemes from below the apex), or non-digitate. Primary inflorescence branches (1–)5–8(–15). Inflorescence espatheate; not comprising 'partial inflorescences' and foliar organs. The racemes spikelet bearing to the base. Spikelet-bearing axes persistent. Spikelets paired; secund; pedicellate. Pedicel apices discoid.

Spikelets loosely imbricate; *consistently in 'long-and-short' combinations (but the spikelets homogamous)*; unequally pedicellate in each combination. The 'shorter' spikelets hermaphrodite. The 'longer' spikelets hermaphrodite.

Female-fertile spikelets. *Spikelets 4–5 mm long*; abaxial; compressed dorsiventrally; falling with the glumes; with conventional internode spacings. The upper floret not stipitate. Rachilla terminated by a female-fertile floret. Hairy callus absent. *Callus* absent.

Glumes two; more or less equal; about equalling the spikelets; long relative to the adjacent lemmas; dorsiventral to the rachis; hairy and hairless; not pointed (obtuse or emarginate); awnless; non-carinate; *very dissimilar (the lower hairless, the upper with four dense rows of long, green to dark purple hairs between the veins). Lower glume 5–7 nerved. Upper glume 3 nerved.* *Spikelets* with incomplete florets. The incomplete florets proximal to the female-fertile florets. *Spikelets with proximal incomplete florets. The proximal incomplete florets* 1; paleate. Palea of the proximal incomplete florets fully developed (two keeled). The proximal incomplete florets male. *The proximal lemmas* awnless; *3–5 nerved (glabrous on the back, with dense long red-brown hairs at the margins)*; decidedly exceeding the female-fertile lemmas; less firm than the female-fertile lemmas; becoming indurated (on the margins, remaining membranous on the back).

Female-fertile florets 1. *Lemmas decidedly firmer than the glumes*; striate; becoming indurated to not becoming indurated (cartilaginous-crustaceous); entire; pointed (acuminate); awnless; hairless; glabrous (smooth and shining); non-carinate; having the margins lying flat and exposed on the palea; with a clear germination flap; 3 nerved (the nerves obscure). *Palea* present; relatively long; entire; awnless, without apical setae; textured like the lemma; indurated; 2-nerved; 2-keeled. Palea keels wingless. *Lodicules* present; 2; free; fleshy; glabrous; not or scarcely vascularized. *Stamens* 3. Anthers 3 mm long; not penicillate; without an apically prolonged connective. *Ovary* glabrous. Stigmas 2; red pigmented.

Fruit, embryo and seedling. *Fruit* free from both lemma and palea; small (almost 2 mm long); compressed dorsiventrally. Hilum short. Embryo large; waisted.

Abaxial leaf blade epidermis. *Costal/intercostal zonation* conspicuous. *Long-cells* markedly different in shape costally and intercostally (the costals much narrower). Mid-intercostal long-cells rectangular and fusiform; having markedly sinuous walls and having straight or only gently undulating walls. *Microhairs* present; panicoid-type (some quite ordinary, but also with numerous long, narrow, warty 'macrohairs', many of which appear to have a basal cell); i.e. the normal ones 45–54 microns long; 5.5–7.5 microns wide at the septum. Microhair total length/width at septum 6–10. Microhair apical cells 24–27 microns long. Microhair apical cell/total length ratio 0.5–0.58. *Stomata* common; 39–45 microns long. Subsidiaries parallel-sided and dome-shaped. Guard-cells overlapping to flush with the interstomatals. *Intercostal short-cells* hair bases impede interpretation; not recordable because macrohair bases impede interpretation. Costal prickles abundant. *Crown cells* absent. *Costal short-cells* conspicuously in long rows. Costal silica bodies 'panicoid-type'; short dumb-bell shaped (with sharp points), or cross shaped; sharp-pointed (sharp corners on dumb-bells).

Transverse section of leaf blade, physiology. C_4; XyMS–. PCR sheath outlines uneven. PCR cell chloroplasts centrifugal/peripheral. *Mesophyll* with radiate chlorenchyma. *Leaf blade* 'nodular' in section to adaxially flat; with the ribs more or less constant in size (and slight). *Midrib* conspicuous; having a conventional arc of bundles; with colourless mesophyll adaxially. Bulliforms present in discrete, regular adaxial groups (in places, but mainly irregularly grouped and constituting most of the epidermis); in simple fans (the more or less irregular fans large), or associated with colourless mesophyll cells to form deeply-penetrating fans; associating with colourless mesophyll cells to form arches over small vascular bundles (especially towards the midrib). Many of the smallest vascular bundles unaccompanied by sclerenchyma. Combined sclerenchyma girders absent. Sclerenchyma all associated with vascular bundles.

Taxonomy. Panicoideae; Panicodae; Paniceae.

Ecology, geography, regional floristic distribution. 1 species (supposedly 2). Southern tropical and South Africa. Mesophytic to xerophytic; shade species, or species of open habitats; glycophytic. In open *Acacia* and mopane savanna.

Paleotropical and Cape. African. Sudano-Angolan and Namib-Karoo. Kalaharian. **References, etc.** Morphological/taxonomic: Hubbard 1929. Leaf anatomical: this project; photos of *M. albescens* provided by R.P. Ellis.

Megastachya P. Beauv.

Habit, vegetative morphology. Annual (tall, erect), or perennial (weakly); sometimes stoloniferous, or decumbent (forming secondary shoots from the rooting nodes). *Culms* 30–100 cm high; herbaceous; branched above. Culm nodes glabrous. Culm internodes solid. Young shoots intravaginal. *Leaves* not basally aggregated; non-auriculate; without auricular setae. *Leaf blades* linear-lanceolate to lanceolate; broad; to 20 mm wide; *cordate (amplexicaul)*; flat; cross veined (abaxially); persistent. Ligule an unfringed membrane; not truncate; about 0.5 mm long. *Contra-ligule* absent.

Reproductive organization. Plants bisexual, with bisexual spikelets; with hermaphrodite florets. The spikelets all alike in sexuality. Viviparous, or not viviparous.

Inflorescence. Inflorescence paniculate; open (large, racemose); with capillary branchlets; espatheate; not comprising 'partial inflorescences' and foliar organs. Spikelet-bearing axes persistent. Spikelets not secund; pedicellate (the pedicels long and slender).

Female-fertile spikelets. Spikelets 7–15 mm long (becoming smaller acropetally in the spikelet); compressed laterally; disarticulating above the glumes; disarticulating between the florets; with conventional internode spacings. Rachilla prolonged beyond the uppermost female-fertile floret; rachilla hairless. The rachilla extension with incomplete florets. Hairy callus absent. *Callus* absent.

Glumes two; very unequal; shorter than the spikelets; shorter than the adjacent lemmas; hairless; glabrous (on the sides), or scabrous (on the mid-nerve); shortly awned (or mucronate, from the excurrent mid-nerve); carinate; similar (membranous-herbaceous, broadly ovate). *Lower glume* 0.75 times the length of the upper glume; shorter than the lowest lemma; 3–4 nerved. Upper glume 3–4 nerved. *Spikelets* with incomplete florets. The incomplete florets distal to the female-fertile florets. The distal incomplete florets 1; *incomplete florets* merely underdeveloped; incomplete florets awnless.

Female-fertile florets *12–17*. Lemmas similar in texture to the glumes; not becoming indurated; incised; 2 lobed; not deeply cleft (bidentate); awnless, or mucronate (the mucro from between the lobes, via the excurrent mid-nerve); hairless; glabrous (on the sides), or scabrous (on the mid-nerve); carinate; without a germination flap; obscurely 5–7 nerved. *Palea* present (narrower than lemma); relatively long; tightly clasped by the lemma; entire; awnless, without apical setae; textured like the lemma; not indurated; 2-nerved; 2-keeled. Palea keels hairy. **Lodicules *absent***. *Stamens* 2–3. Anthers not penicillate; without an apically prolonged connective. *Ovary* glabrous. Styles free to their bases. Stigmas 2 (long).

Fruit, embryo and seedling. *Fruit* free from both lemma and palea (but falling with them); small (about 1 mm long); subglobose; triquetrous. Hilum short. Embryo small. Endosperm containing only simple starch grains. Embryo with an epiblast; with a scutellar tail; with an elongated mesocotyl internode.

Abaxial leaf blade epidermis. *Costal/intercostal zonation* conspicuous. *Papillae* absent. Mid-intercostal long-cells having markedly sinuous walls (coarsely so, the walls thin). *Microhairs* present; panicoid-type (balanoform); (48–)54–68 microns long. Microhair apical cells 37–58 microns long. Microhair apical cell/total length ratio 0.78. *Stomata* common. Subsidiaries triangular. *Intercostal short-cells* absent or very rare. *Costal short-cells* conspicuously in long rows. Costal silica bodies 'panicoid-type'; cross shaped and dumb-bell shaped, or butterfly shaped; not sharp-pointed.

Transverse section of leaf blade, physiology. C_3; XyMS+. Mesophyll without adaxial palisade; without arm cells (according to Metcalfe 1960), or with arm cells (?); with fusoids (as represented by laterally extended PBS cells). The fusoids an integral part of the PBS. *Leaf blade* adaxially flat. *Midrib* conspicuous; having complex vascularization (1 large median with 2 tiny laterals, all enclosed in a common sheath). Bulliforms present in discrete, regular adaxial groups; in simple fans (the fans very wide, occupying most of the epidermis of each intercostal zone). Combined sclerenchyma girders present. Sclerenchyma all associated with vascular bundles.

Cytology. Chromosome base number, $x = 12$. $2n = 48$. 4 ploid.

Taxonomy. Bambusoideae; Oryzodae; Centotheceae.

Ecology, geography, regional floristic distribution. 1 species. Tropical and southern Africa. Mesophytic; shade species; glycophytic. In forests.

Paleotropical. African and Madagascan. Sudano-Angolan and West African Rainforest. Somalo-Ethiopian and South Tropical African.

References, etc. Leaf anatomical: Metcalfe 1960; photos of *M. mucronata* provided by R.P. Ellis.

Melanocenchris Nees

Gracilea Hook. f., *Ptiloneilema* Steud., *Roylea* Steud.

Habit, vegetative morphology. Annual, or perennial; caespitose. *Culms* 5–20 cm high; herbaceous. Culm nodes glabrous. Culm internodes hollow. *Leaves* not basally aggregated; non-auriculate. Leaf blades narrow; 5–6 mm wide; setaceous, or not setaceous; exhibiting multicellular glands abaxially (base of macrohairs). The abaxial glands intercostal. *Leaf blades cross veined*; persistent. Ligule a fringe of hairs.

Reproductive organization. Plants bisexual, with bisexual spikelets; with hermaphrodite florets. The spikelets of sexually distinct forms on the same plant; hermaphrodite and male-only, or hermaphrodite and sterile. The male and female-fertile spikelets segregated, in different parts of the same inflorescence branch, or mixed in the inflorescence (each head commonly with 1–2 hermaphrodite spikelets and 2–3 progressively smaller male or sterile members).

Inflorescence. *Inflorescence a false spike, with spikelets on contracted axes (the spikelets in turbinate heads)*. *Inflorescence axes not ending in spikelets (the reduced raceme terminated by a forked bristle)*. Inflorescence espatheate; not comprising 'partial inflorescences' and foliar organs. *Spikelet-bearing axes* very much reduced (to heads of spikelets); disarticulating; falling entire (the heads falling — the main axis persistent). Spikelets secund (the heads secund on the spikes); pedicellate.

Female-sterile spikelets. Sterile and (when present) male spikelets similar to the hermaphrodites, but more or less reduced.

Female-fertile spikelets. Spikelets 4 mm long; compressed laterally, or compressed dorsiventrally (?); falling with the glumes; not disarticulating between the florets. Rachilla prolonged beyond the uppermost female-fertile floret. The rachilla extension with incomplete florets (the second floret male or sterile, with an extension above). Hairy callus present.

Glumes two; very unequal to more or less equal; shorter than the adjacent lemmas to long relative to the adjacent lemmas; hairy (G_1 hairy on the back and at the base of the awn, G_2 hairy on the nerves and at the base of the awn); awned (both G_1 and G_2 with their mid-nerve produced into a purplish, scabrid awn); very dissimilar (the lower linear, the upper with broad hyaline wings). Lower glume 1 nerved. Upper glume 3 nerved. *Spikelets* with incomplete florets. The incomplete florets distal to the female-fertile florets. The distal incomplete florets 1.

Female-fertile florets 1. Lemmas lobed/awned, except for the L_1 of *M. monoica*; not becoming indurated; incised; 3 lobed (the lobes acuminate-awned); deeply cleft (above); awned. Awns 3; median and lateral; the median similar in form to the laterals; from a sinus; non-geniculate; much shorter than the body of the lemma. The lateral awns somewhat shorter than the median. *Lemmas hairy (covered with short clavate hairs)*; carinate; 3 nerved. *Palea* present; relatively long; apically notched to deeply bifid; with apical setae to awned; 2-nerved; 2-keeled (abaxially covered with clavate hairs). *Lodicules* present; 2; free; fleshy; glabrous. *Stamens* 3. Anthers not penicillate. *Ovary* glabrous. Styles free to their bases. Stigmas 2.

Fruit, embryo and seedling. *Fruit* free from both lemma and palea; small; ellipsoid; compressed dorsiventrally. Hilum short. Pericarp fused. Embryo large; not waisted.

Abaxial leaf blade epidermis. *Costal/intercostal zonation* conspicuous. *Papillae* present; intercostal. Intercostal papillae over-arching the stomata; consisting of one symmetrical projection per cell (finger-like). *Long-cells* similar in shape costally and intercostally; of similar wall thickness costally and intercostally. Mid-intercostal long-cells rectangular; having markedly sinuous walls. *Microhairs* present; elongated; clearly two-celled; chloridoid-type. Microhair apical cell wall of similar thickness/rigidity to that of the basal cell. Microhairs 61.5–66 microns long. Microhair basal cells 54 microns long. Microhairs 15–17.5 microns wide at the septum. Microhair total length/width at septum 3.2–4.2. Microhair apical cells 10.5–12.9 microns long. Microhair apical cell/total length ratio 0.17–0.21. *Stomata* common; (19.5–)21 microns long. Subsidiaries triangular. Guard-cells overlapping to flush with the interstomatals. *Intercostal short-cells* absent or very rare (none seen). Intercostal silica bodies absent. *Costal short-cells* conspicuously in long rows. Costal silica bodies present in alternate cell files of the costal zones; saddle shaped, tall-and-narrow, and crescentic; not sharp-pointed.

Transverse section of leaf blade, physiology. C_4; XyMS+. PCR sheaths of the primary vascular bundles interrupted; interrupted abaxially only to interrupted both abaxially and adaxially. PCR sheath extensions absent. *Mesophyll* with radiate chlorenchyma. *Leaf blade* 'nodular' in section; with the ribs more or less constant in size. *Midrib* not readily distinguishable; with one bundle only. Bulliforms present in discrete, regular adaxial groups; in simple fans. All the vascular bundles accompanied by sclerenchyma. Combined sclerenchyma girders present; forming 'figures'. Sclerenchyma all associated with vascular bundles. The lamina margins without fibres.

Taxonomy. Chloridoideae; main chloridoid assemblage.

Ecology, geography, regional floristic distribution. 3 species. Northeastern tropical Africa, India, Ceylon. Species of open habitats. Dry plains and hillsides.

Holarctic and Paleotropical. Tethyan. African and Indomalesian. Irano-Turanian. Saharo-Sindian and Sudano-Angolan. Indian. Somalo-Ethiopian.

References, etc. Leaf anatomical: this project.

Melica L.

Beckeria Bernh., *Bromelica* (Thurber) Farw., *Claudia* Opiz, *Dalucum* Adans., *Verinea* Merino

Habit, vegetative morphology. Perennial; rhizomatous. *Culms* 10–150(–200) cm high; herbaceous; scandent (*M. sarmentosa*, via filiform, retrorsely scabrid leaf blade tips), or not scandent; unbranched above; tuberous (often), or not tuberous. Culm internodes hollow. Leaves auriculate, or non-auriculate. **Sheath margins joined.** Leaf blades linear; broad to narrow; 1.3–15 mm wide; flat, or rolled (convolute); cross veined (rarely), or without cross venation; rolled in bud. Ligule an unfringed membrane to a fringed membrane, or a fringe of hairs (rarely); truncate, or not truncate; 0.1–4 mm long.

Reproductive organization. Plants bisexual, with bisexual spikelets; with hermaphrodite florets. Exposed-cleistogamous, or chasmogamous.

Inflorescence. Inflorescence few spikeleted to many spikeleted; a single raceme, or paniculate; open, or contracted; with capillary branchlets, or without capillary branchlets; espatheate; not comprising 'partial inflorescences' and foliar organs. Spikelet-bearing axes persistent. Spikelets secund, or not secund; pedicellate.

Female-fertile spikelets. Spikelets 4–20 mm long; compressed laterally to not noticeably compressed; disarticulating above the glumes, or falling with the glumes (and sometimes disarticulating both above and below them); not disarticulating between the florets, or disarticulating between the florets (tardily or reluctantly); with conventional internode spacings, or with a distinctly elongated rachilla internode between the glumes. Rachilla prolonged beyond the uppermost female-fertile floret; rachilla hairless. The rachilla extension with incomplete florets. Hairy callus absent. *Callus* short (glabrous).

Glumes two; relatively large; very unequal to more or less equal; shorter than the spikelets, or about equalling the spikelets (usually?); shorter than the adjacent lemmas, or long relative to the adjacent lemmas (usually?); pointed, or not pointed; awnless; non-carinate; very dissimilar (the lower sometimes greatly enlarged), or similar (often coloured, papery, thin-tipped). Lower glume 1–7 nerved. Upper glume (3–)5–7 nerved. *Spikelets* with incomplete florets. The incomplete florets distal to the female-fertile florets. **The distal incomplete florets *merely underdeveloped, or clearly specialised and*** *modified in form (often, modified as a ball of successively enveloped lemmas or as a swollen rachilla extension).*

Female-fertile florets 1–7. Lemmas similar in texture to the glumes to decidedly firmer than the glumes (usually leathery, sometimes membranous); entire, or incised; when entire pointed, or blunt; awnless, or mucronate, or awned. Awns when present, 1; from a sinus, or apical; non-geniculate; straight; hairless; much shorter than the body of the lemma to about as long as the body of the lemma. Lemmas hairy, or hairless; non-carinate; without a germination flap; 5–9(–15) nerved. *Palea* present; relatively long, or conspicuous but relatively short, or very reduced; entire, or apically notched (bidentate); thinner than the lemma; 2-nerved; 2-keeled. **Lodicules** present; 2; *joined*; *fleshy*; glabrous; not or scarcely vascularized. *Stamens* 3. Anthers 0.6–3.5 mm long; not penicillate. **Ovary** glabrous; *without a conspicuous apical appendage*. Styles free to their bases. Stigmas 2.

Fruit, embryo and seedling. Fruit free from both lemma and palea; small; sometimes red; *longitudinally grooved*; compressed dorsiventrally, or not noticeably compressed. Hilum long-linear. Pericarp thin. Embryo small. Endosperm hard; without lipid; containing compound starch grains. Embryo with an epiblast; without a scutellar tail; with a negligible mesocotyl internode. Embryonic leaf margins meeting.

Seedling with a long mesocotyl; with a loose coleoptile, or with a tight coleoptile. First seedling leaf with a well-developed lamina. The lamina broad, or narrow; erect; 3–5 veined.

Abaxial leaf blade epidermis. *Costal/intercostal zonation* conspicuous. *Papillae* absent. *Long-cells* similar in shape costally and intercostally, or markedly different in shape costally and intercostally; of similar wall thickness costally and intercostally, or differing markedly in wall thickness costally and intercostally. Mid-intercostal long-cells rectangular, or fusiform; having markedly sinuous walls, or having straight or only gently undulating walls. *Microhairs* absent. *Stomata* absent or very rare, or common. Subsidiaries parallel-sided. *Intercostal short-cells* common, or absent or very rare; not paired; silicified, or not silicified. Intercostal silica bodies when seen tall-and-narrow. *Costal short-cells* conspicuously in long rows, or predominantly paired, or neither distinctly grouped into long rows nor predominantly paired. Costal silica bodies horizontally-elongated crenate/sinuous, or horizontally-elongated smooth, or rounded; not sharp-pointed.

Transverse section of leaf blade, physiology. C_3; XyMS+. *Mesophyll* with non-radiate chlorenchyma. *Leaf blade* with distinct, prominent adaxial ribs, or adaxially flat; with the ribs more or less constant in size. *Midrib* conspicuous, or not readily distinguishable; with one bundle only. Bulliforms present in discrete, regular adaxial groups (at the bases of the furrows); in simple fans (and sometimes also in groups of small, irregularly sized cells, cf. *Ammophila*). All the vascular bundles accompanied by sclerenchyma. Combined sclerenchyma girders present; forming 'figures', or nowhere forming 'figures'. Sclerenchyma all associated with vascular bundles.

Phytochemistry. Leaves without flavonoid sulphates (2 species).

Special diagnostic feature. Spikelets with the distal incomplete florets and/or the rachilla apex forming a terminal clavate appendage, or spikelets without a terminal clavate appendage.

Cytology. Chromosome base number, $x = 9$. $2n = 14$ (rarely), or 18, or 36. 2 and 4 ploid. Nucleoli disappearing before metaphase.

Taxonomy. Pooideae; Poodae; Meliceae.

Ecology, geography, regional floristic distribution. About 80 species. North temperate, southern Africa and South America. *Mesophytic to xerophytic*; shade species and species of open habitats.

Holarctic, Paleotropical, Neotropical, Cape, and Antarctic. Boreal, Tethyan, and Madrean. African and Indomalesian. Euro-Siberian, Eastern Asian, Atlantic North American, and Rocky Mountains. Macaronesian, Mediterranean, and Irano-Turanian. Saharo-Sindian and Sudano-Angolan. Indian and Indo-Chinese. Caribbean, Pampas, and Andean. Patagonian. European and Siberian. Canadian-Appalachian, Southern Atlantic North American, and Central Grasslands. South Tropical African.

Rusts and smuts. Rusts — *Puccinia*. Taxonomically wide-ranging species: *Puccinia graminis, Puccinia coronata, Puccinia brachypodii, Puccinia poarum*, and *Puccinia*

schedonnardi. Smuts from Tilletiaceae and from Ustilaginaceae. Tilletiaceae — *Urocystis*. Ustilaginaceae — *Sphacelotheca* and *Ustilago*.
References, etc. Leaf anatomical: Metcalfe 1960; this project.

Melinis P. Beauv.

Suaria Schrank, *Tristegis* Nees

Habit, vegetative morphology. Annual, or perennial; decumbent. *Culms* 30–120 cm high; herbaceous. Culm nodes hairy. Culm internodes solid to hollow. *The shoots aromatic. Leaves* not basally aggregated; non-auriculate. Leaf blades narrow; without cross venation; persistent. Ligule a fringed membrane (very narrow), or a fringe of hairs.

Reproductive organization. Plants bisexual, with bisexual spikelets; with hermaphrodite florets.

Inflorescence. *Inflorescence paniculate*; open; with capillary branchlets (flexuose); espatheate; not comprising 'partial inflorescences' and foliar organs. Spikelet-bearing axes persistent. Spikelets not secund; pedicellate; not in distinct 'long-and-short' combinations.

Female-fertile spikelets. Spikelets 1–4 mm long; oblong; not noticeably compressed to compressed dorsiventrally; falling with the glumes; with conventional internode spacings. Rachilla terminated by a female-fertile floret. Hairy callus absent.

Glumes one per spikelet, or two; very unequal; (the upper) long relative to the adjacent lemmas; awned, or awnless; *very dissimilar (the lower tiny, membranous or reduced to a rim, the upper equalling spikelet, truncate, emarginate or bifid at summit, often awned from the sinus, straight on the back). Lower glume 0 nerved.* Upper glume 7 nerved. *Spikelets* with incomplete florets. The incomplete florets proximal to the female-fertile florets. *The proximal incomplete florets* 1; paleate, or epaleate. Palea of the proximal incomplete florets when present, fully developed. The proximal incomplete florets male, or sterile. The proximal lemmas awned (usually, with a subulate awn from the bifid apex), or awnless (rarely); 3–5 nerved; similar in texture to the female-fertile lemmas, or decidedly firmer than the female-fertile lemmas; not becoming indurated.

Female-fertile florets 1. Lemmas less firm than the glumes (hyaline or thinly membranous); smooth; not becoming indurated; white in fruit; entire, or incised; awnless; hairless; carinate; having the margins lying flat and exposed on the palea; without a germination flap; 1–3 nerved. *Palea* present; relatively long; textured like the lemma; not indurated; 2-nerved; keel-less (or flattened). *Lodicules* present; 2; fleshy, or membranous. *Stamens* 3. Anthers not penicillate. *Ovary* glabrous. Styles free to their bases. Stigmas 2.

Fruit, embryo and seedling. Hilum short. Embryo large. Endosperm containing only simple starch grains. Embryo without an epiblast; with a scutellar tail; with an elongated mesocotyl internode. Embryonic leaf margins overlapping.

Seedling with a long mesocotyl. First seedling leaf with a well-developed lamina. The lamina broad; curved; 13–20 veined.

Abaxial leaf blade epidermis. *Costal/intercostal zonation* conspicuous. *Papillae* absent. Mid-intercostal long-cells rectangular; having markedly sinuous walls. *Microhairs* present; panicoid-type; 63–70.5 microns long; 5.4–6 microns wide at the septum. Microhair total length/width at septum 11–13. Microhair apical cells 30–34.5 microns long. Microhair apical cell/total length ratio 0.47–0.5. *Stomata* common; 30–33 microns long. Subsidiaries low dome-shaped and triangular. Guard-cells overlapping to flush with the interstomatals. *Intercostal short-cells* absent or very rare. Prickle bases abundant. *Costal short-cells* conspicuously in long rows. Costal silica bodies 'panicoid-type'; cross shaped, or butterfly shaped, or butterfly shaped, or dumb-bell shaped; not sharp-pointed.

Transverse section of leaf blade, physiology. C_4; biochemical type PCK (*M. minutiflora*); XyMS+. PCR sheath outlines uneven. *PCR cell chloroplasts* ovoid; centrifugal/peripheral. *Mesophyll* with radiate chlorenchyma. *Leaf blade* adaxially flat. *Midrib* conspicuous; having a conventional arc of bundles. Bulliforms present in discrete, regular adaxial groups; associated with colourless mesophyll cells to form deeply-penetrating fans. Many of the smallest vascular bundles unaccompanied by sclerenchyma. Combined sclerenchyma girders present. Sclerenchyma all associated with vascular bundles.

Culm anatomy. *Culm internode bundles* in three or more rings.

Phytochemistry. Leaves without flavonoid sulphates (2 species). Leaf blade chlorophyll *a:b* ratio 3.31–3.46.

Cytology. Chromosome base number, $x = 9$. $2n = 36$.

Taxonomy. Panicoideae; Panicodae; Paniceae (Melinideae).

Ecology, geography, regional floristic distribution. About 12 species. 1 in tropical South America, West Indies,; 17 in tropical and South Africa, Madagascar. Commonly adventive. Helophytic, or mesophytic; shade species and species of open habitats; glycophytic. Savanna woodland, open grassland and disturbed ground.

Paleotropical. African, Madagascan, and Polynesian. Sudano-Angolan and West African Rainforest. Fijian. Sahelo-Sudanian, Somalo-Ethiopian, and South Tropical African.

Rusts and smuts. Rusts — *Physopella* and *Puccinia*. Taxonomically wide-ranging species: '*Uromyces' setariae-italicae.*

Economic importance. Cultivated fodder: *M. minutiflora*. Important native pasture species: *M. minutiflora.*

References, etc. Leaf anatomical: Metcalfe 1960 and this project.

Melocalamus Benth.

Habit, vegetative morphology. Perennial. The flowering culms leafy. **Culms** 500–3000 cm high; woody and persistent; *to* 2.5 cm in diameter; *scandent (tufted, spreading, arching)*; branched above. Primary branches/mid-culm node several to many. Culm sheaths persistent (brittle). Culm internodes hollow. Rhizomes pachymorph. Plants unarmed. *Leaves* not basally aggregated; auriculate (the auricles hirsute, reflexed, caducous); with auricular setae. Leaf blades lanceolate to elliptic; broad; 22–50 mm wide (and 14–26 cm long); not cordate, not sagittate (rounded at base); pseudopetiolate (the petiolule hairy); without cross venation; disarticulating from the sheaths; rolled in bud. *Ligule* present (narrow, entire).

Reproductive organization. Plants bisexual, with bisexual spikelets; with hermaphrodite florets. *The spikelets of sexually distinct forms on the same plant*; hermaphrodite and sterile (panicle 'with several fertile and many sterile spikelets'). Sometimes viviparous.

Inflorescence. *Inflorescence* indeterminate; with pseudospikelets; paniculate (a large, compound, interrupted panicle of small sub-globose heads); spatheate; a complex of 'partial inflorescences' and intervening foliar organs (the panicle often leafy). *Spikelet-bearing axes* capitate; persistent. Spikelets not secund; not in distinct 'long-and-short' combinations.

Female-fertile spikelets. Spikelets 1.7–2.5 mm long; disarticulating above the glumes (?); disarticulating between the florets (?); with conventional internode spacings. Rachilla prolonged beyond the uppermost female-fertile floret. The rachilla extension with incomplete florets.

Glumes two; more or less equal; shorter than the adjacent lemmas; hairless; awnless (shortly mucronate); similar (broadly oval, ventricose). *Spikelets* with incomplete florets. The incomplete florets distal to the female-fertile florets. The distal incomplete florets 1; *incomplete florets* merely underdeveloped.

Female-fertile florets 2. Lemmas similar in texture to the glumes; not becoming indurated; entire; awnless, or mucronate (?); fertile lemma often ciliate on margins. *Palea* present; relatively long; apically notched; awnless, without apical setae; not indurated; several nerved (4–5 nerves between the keels); 2-keeled. Palea keels hairy (ciliate). *Lodicules* present (large); 3; free; membranous; ciliate; not toothed (blunt); heavily vascularized. *Stamens* 6; with free filaments (these short). *Ovary* glabrous; with a conspicuous apical appendage. The appendage broadly conical, fleshy. Styles fused. Stigmas 2, or 3 (short, plumose).

Fruit, embryo and seedling. Fruit large (2.25–3 cm in diameter); subglobose. Pericarp fleshy; free. *Seed 'non-endospermic'.*

Abaxial leaf blade epidermis. *Costal/intercostal zonation* conspicuous. *Papillae* present. Intercostal papillae over-arching the stomata (four or more overarching each stoma). *Long-cells* markedly different in shape costally and intercostally (the costals rather longer). Intercostal zones exhibiting many atypical long-cells (a short form). Mid-intercostal long-cells having markedly sinuous walls. *Microhairs* present; panicoid-type. *Stomata* common (obscured by papillae). *Intercostal short-cells* common; in cork/silica-cell pairs; silicified. Intercostal silica bodies vertically elongated-nodular. *Costal short-cells* predominantly paired. Costal silica bodies saddle shaped (mostly, tall), or tall-and-narrow; not sharp-pointed.

Transverse section of leaf blade, physiology. C_3; XyMS+. Mesophyll with arm cells; with fusoids. *Leaf blade* adaxially flat (save near midrib). *Midrib* conspicuous; having complex vascularization. Bulliforms present in discrete, regular adaxial groups; in simple fans. All the vascular bundles accompanied by sclerenchyma. Combined sclerenchyma girders present (with all the bundles); forming 'figures' (larger bundles with this tendency).

Taxonomy. Bambusoideae; Bambusodae; Bambuseae.

Ecology, geography, regional floristic distribution. 1 species. India, Burma. Forests.

Paleotropical. Indomalesian. Indo-Chinese.

References, etc. Leaf anatomical: Metcalfe 1960; this project.

Special comments. Fruit data wanting.

Melocanna Trin.

Beesha Kunth

Habit, vegetative morphology. Perennial (shrubs or trees); rhizomatous. The flowering culms leafy. **Culms *1000–2000 cm high***; woody and persistent; to 9 cm in diameter; branched above. Primary branches/mid-culm node many. Culm sheaths often persistent (leathery). Culm internodes hollow. Rhizomes pachymorph (metamorph type II). Plants unarmed. *Leaves* not basally aggregated; auricles inconspicuous; without auricular setae. Leaf blades broad; 20–90 mm wide; pseudopetiolate; without cross venation; disarticulating from the sheaths; rolled in bud.

Reproductive organization. Plants bisexual, with bisexual spikelets; with hermaphrodite florets. Sometimes viviparous.

Inflorescence. *Inflorescence* indeterminate; with pseudospikelets; paniculate (large, compound, with spatheate, oblong, tight groups of few to several spikelets imbricate and secund on the branches); spatheate (the spikelet fascicles bracteate); a complex of 'partial inflorescences' and intervening foliar organs. *Spikelet-bearing axes* paniculate; persistent. Spikelets secund (the fascicles secund on the spicate inflorescence branches).

Female-fertile spikelets. Spikelets 10–15 mm long; disarticulating above the glumes (?); disarticulating between the florets (?). Rachilla prolonged beyond the uppermost female-fertile floret, or terminated by a female-fertile floret.

Glumes two to several (to four); shorter than the adjacent lemmas; pointed; awnless; similar (ovate-lanceolate). *Spikelets* presumably? with incomplete florets. The incomplete florets proximal to the female-fertile florets (assuming the 'several glumes' may represent sterile lemmas). *Spikelets with proximal incomplete florets (?)*. *The proximal incomplete florets* 1–2 (?); sterile.

Female-fertile florets 1. Lemmas ovate-lanceolate, pungent; less firm than the glumes to similar in texture to the glumes; entire; pointed; awnless. *Palea* present; relatively long; several nerved (?); keel-less. *Lodicules* present; 2; free; membranous (narrow); ciliate; heavily vascularized (3–5 nerved). *Stamens* 5–7; with free filaments, or monadelphous to triadelphous (irregularly united). Anthers not penicillate; without an apically prolonged connective. *Ovary* glabrous; with a conspicuous apical appendage. The appendage long, stiff and tapering. Styles fused (ovary attenuate into a long style). Stigmas 2–4 (feathery, recurved).

Fruit, embryo and seedling. *Fruit large (reaching 7–12 cm long in* **M. baccifera***)*; pyriform. *Pericarp fleshy*. Embryo large (but not visible). *Seed 'non-endospermic' (or* *hardly so, when mature: the abnormally large scutellum absorbs nearly all the endosperm, and the seed germinates while fruit still attached to parent plant)*. Embryo with an epiblast; with a scutellar tail; with a negligible mesocotyl internode. Embryonic leaf margins overlapping.

Abaxial leaf blade epidermis. *Costal/intercostal zonation* conspicuous. *Papillae* present (confined to stomatal bands). Intercostal papillae over-arching the stomata. Intercostal zones with typical long-cells to exhibiting many atypical long-cells (the long-cells short). Mid-intercostal long-cells having markedly sinuous walls (thin). *Microhairs* present; panicoid-type; 63–66 microns long (in *M. bambusoides*); 6–7.5 microns wide at the septum. Microhair total length/width at septum 8.4–11. Microhair apical cells 33.6–37.5 microns long. Microhair apical cell/total length ratio 0.52–0.57. *Stomata* common (obscured by papillae); 24–27 microns long (in *M. bambusoides*). *Intercostal short-cells* common; in cork/silica-cell pairs; silicified. Intercostal silica bodies tall-and-narrow. *Costal short-cells* predominantly paired (occasionally in short rows). Costal silica bodies saddle shaped; not sharp-pointed.

Transverse section of leaf blade, physiology. C_3; XyMS+. Mesophyll with arm cells; with fusoids (conspicuous). The fusoids external to the PBS. *Leaf blade* with distinct, prominent adaxial ribs (ribs slight, widely spaced), or adaxially flat; with the ribs more or less constant in size. *Midrib* conspicuous; having complex vascularization. Bulliforms present in discrete, regular adaxial groups; in simple fans (mostly), or associated with colourless mesophyll cells to form deeply-penetrating fans. All the vascular bundles accompanied by sclerenchyma. Combined sclerenchyma girders present (with most bundles); forming 'figures' (many bundles, somewhat so).

Taxonomy. Bambusoideae; Bambusodae; Bambuseae.

Ecology, geography, regional floristic distribution. 3 species. Eastern Asia. Paleotropical. Indomalesian. Indo-Chinese.

Economic importance. Culms of *M. baccifera* yield pulp for high quality papers; splits are used in handicrafts; and the large fruits are edible.

References, etc. Leaf anatomical: Metcalfe 1960; this project.

Merostachys Spreng.

Brasilocalamus Nakai

Habit, vegetative morphology. Arborescent, shrubby or scrambling perennial. The flowering culms leafy. *Culms woody and persistent*; scandent, or not scandent; branched above. Primary branches/mid-culm node 1 (forming a fan). Culm sheaths with constricted, more or less petiolate, reflexed blades. Culm internodes solid, or hollow. Unicaespitose. Rhizomes pachymorph. Plants unarmed. Leaves with auricular setae. Leaf blades broad; pseudopetiolate; without cross venation; disarticulating from the sheaths. Ligule an unfringed membrane.

Reproductive organization. Plants bisexual, with bisexual spikelets; with hermaphrodite florets. The spikelets of sexually distinct forms on the same plant, or all alike in sexuality; hermaphrodite, or hermaphrodite and male-only, or hermaphrodite and sterile (with imperfect spikelets towards the tips of the racemes).

Inflorescence. Inflorescence a single spike, or a single raceme (spicate, pectinate, terminal on leafy twigs). *Inflorescence axes not ending in spikelets (the tips excurrent, with rudimentary spikelets)*. Rachides flattened. Inflorescence spatheate (with 'bracts' subtending at least the lowest spikelets); not comprising 'partial inflorescences' and foliar organs. *Spikelet-bearing axes* spikes, or 'racemes'; solitary; persistent. Spikelets solitary (usually), or paired, or in triplets; secund; sessile to subsessile.

Female-fertile spikelets. Spikelets compressed laterally to not noticeably compressed; disarticulating above the glumes; when more than one-floreted, disarticulating between the florets. Rachilla prolonged beyond the uppermost female-fertile floret. The rachilla extension with incomplete florets.

Glumes one per spikelet (the upper); shorter than the adjacent lemmas; hairy; pointed; non-carinate. Upper glume 1 nerved (in material seen). *Spikelets* with incomplete florets. The incomplete florets both distal and proximal to the female-fertile florets. *The distal incomplete florets* merely underdeveloped. *The proximal incomplete florets* 1; sterile (but sometimes with a rudimentary floret – hence its interpretation as an L_1 rather than a G_2). The proximal lemmas awnless; 3 nerved; similar in texture to the female-fertile lemmas (fragile, brittle); not becoming indurated.

Female-fertile florets 1–2(–10). Lemmas similar in texture to the glumes; not becoming indurated; awnless, or mucronate (?); hairy; non-carinate; without a germination flap; 9–11 nerved (in material seen). *Palea* present; relatively long; entire; awnless, without apical setae; several nerved (6 in material seen); 2-keeled (and sulcate between). *Lodicules* present; 3; free; membranous; ciliate; heavily vascularized. *Stamens* 3. **Ovary** glabrous; **with a conspicuous apical appendage**. The appendage broadly conical, fleshy. Stigmas 2.

Fruit, embryo and seedling. *Fruit* free from both lemma and palea; not noticeably compressed. Hilum with the hilum not visible externally. Pericarp thin, or thick and hard (leathery or crustaceous); fused, or loosely adherent. Embryo embryo not visible externally.

Abaxial leaf blade epidermis. *Costal/intercostal zonation* conspicuous. *Papillae* present. Intercostal papillae over-arching the stomata; several per cell (mainly one row per long-cell, variable in size). *Long-cells* similar in shape costally and intercostally; of similar wall thickness costally and intercostally. Mid-intercostal long-cells rectangular; having markedly sinuous walls (thin). *Microhairs* present; panicoid-type (large, the basal cells variable in length); 45–54 microns long (in *M. anceps*); 9–12 microns wide at the septum. Microhair total length/width at septum 4.4–5.7. Microhair apical cells 24–28.5 microns long. Microhair apical cell/total length ratio 0.46–0.57. *Stomata* common (obscured); 27–30 microns long (in *M. anceps*). Guard-cells sunken, and obscured by papillae. *Intercostal short-cells* common; silicified. Intercostal silica bodies tall-and-narrow. *Costal short-cells* neither distinctly grouped into long rows nor predominantly paired (mostly solitary, sometimes associated with large, bulbous prickles). Costal silica bodies present and well developed to poorly developed; tall-and-narrow.

Transverse section of leaf blade, physiology. C_3; XyMS+. Mesophyll with adaxial palisade; with arm cells (Metcalfe), or without arm cells (very inconspicuous in *M. anceps*); with fusoids. The fusoids external to the PBS. *Leaf blade* adaxially flat. *Midrib* not readily distinguishable; with one bundle only. Bulliforms present in discrete, regular adaxial groups; in simple fans, or associated with colourless mesophyll cells to form deeply-penetrating fans (if associated lignified cells are interpreted as 'colourless cells'). All the vascular bundles accompanied by sclerenchyma. Combined sclerenchyma girders present (with most bundles); forming 'figures' (anchors with most bundles). Sclerenchyma not all bundle-associated. The 'extra' sclerenchyma in abaxial groups (opposite the bulliforms), or in adaxial groups (the 'groups' mostly reduced to one or few cells, in *M. anceps*); abaxial-hypodermal, the groups isolated and adaxial-hypodermal, contiguous with the bulliforms.

Taxonomy. Bambusoideae; Bambusodae; Bambuseae.

Ecology, geography, regional floristic distribution. About 40 species. South America. Forests.

Neotropical. Caribbean, Central Brazilian, Pampas, and Andean.

References, etc. Leaf anatomical: Metcalfe 1960; this project.

Merxmuellera Conert

~ Rytidosperma, Danthonia sensu lato

Habit, vegetative morphology. Perennial; caespitose. Culms 15–200 cm high; **herbaceous**; unbranched above; tuberous (in *M. rufa*, in the sense that the sheaths form a bulbous structure below the soil surface), or not tuberous. Culm nodes glabrous. Culm internodes solid, or hollow. Plants conspicuously armed (*M. drakensbergensis* having pungent leaf blades), or unarmed. Young shoots intravaginal. *Leaves* mostly basal (usually), or not basally aggregated (in *M. arundinacea*); non-auriculate. The mouths and bases of the sheaths often wooly. Leaf blades linear; broad (rarely), or narrow; 4–15 mm wide; nearly always rolled; disarticulating from the sheaths, or persistent (often disarticulating only when old). *Ligule a fringe of hairs*.

Reproductive organization. *Plants bisexual, with bisexual spikelets*; with hermaphrodite florets.

Inflorescence. Inflorescence a single raceme (rarely — *M. disticha*), or paniculate; contracted (narrow, occasionally spike-like, usually longer than 6 cm by contrast with *Karroochloa*); espatheate; not comprising 'partial inflorescences' and foliar organs. Spikelet-bearing axes persistent. Spikelets secund, or not secund; pedicellate.

Female-fertile spikelets. Spikelets 8–25 mm long; compressed laterally; disarticulating above the glumes; disarticulating between the florets; with conventional internode spacings. Rachilla prolonged beyond the uppermost female-fertile floret; rachilla hairy (above). The rachilla extension with incomplete florets. Hairy callus present (0.6–2 mm long, with bearded margins). *Callus short, or long*.

Glumes two; more or less equal (G_2 somewhat shorter); about equalling the spikelets to exceeding the spikelets; *long relative to the adjacent lemmas*; hairless; glabrous (shining); pointed; awnless; carinate; similar (papery, the margins and apex hyaline, the midnerve percurrent). Lower glume much exceeding the lowest lemma; 1 nerved, or 3(–5) nerved. Upper glume 1 nerved, or 3(–5) nerved. *Spikelets* with incomplete florets. The incomplete florets distal to the female-fertile florets. *The distal incomplete florets merely underdeveloped*.

Female-fertile florets *3–10*. Lemmas similar in texture to the glumes; not becoming indurated; incised; 2 lobed; deeply cleft to not deeply cleft; awned. Awns 1, or 3; median, or median and lateral (the lobes sometimes finely awn-tipped); the median different in form from the laterals (when laterals present); from a sinus (the lobes sometimes basally adherent to it); geniculate; hairless; much longer than the body of the lemma. The lateral awns (when present) shorter than the median. Lemmas hairy, or hairless (rarely glabrous). *The hairs* when present, *in tufts*; *not in transverse rows*. Lemmas non-carinate (dorsally rounded); without a germination flap; (7–)9 nerved; with the nerves non-confluent. *Palea* present; relatively long (lanceolate); tightly clasped by the lemma; apically notched (bidentate); awnless, without apical setae; textured like the lemma; not indurated (hyaline); 2-nerved; 2-keeled. Palea keels wingless; hairy. **Lodicules** present; 2; free; *membranous*; ciliate (often), or glabrous. *Stamens* 3. Anthers without an apically prolonged connective. *Ovary* glabrous. Stigmas 2 (long, plumose); white.

Fruit, embryo and seedling. *Fruit* free from both lemma and palea; small (2–3 mm long); not noticeably compressed.

Abaxial leaf blade epidermis. *Costal/intercostal zonation* lacking. *Papillae* absent. *Long-cells* similar in shape costally and intercostally; of similar wall thickness costally and intercostally. Mid-intercostal long-cells rectangular; having markedly sinuous walls. *Microhairs* absent. *Stomata* absent or very rare. *Intercostal short-cells* common; in cork/silica-cell pairs; silicified. *Costal short-cells* predominantly paired. Costal silica bodies rounded and tall-and-narrow; not sharp-pointed.

Transverse section of leaf blade, physiology. C_3; XyMS+. *Mesophyll* with non-radiate chlorenchyma. *Leaf blade* 'nodular' in section; with the ribs very irregular in sizes. *Midrib* not readily distinguishable (apart from its position); with one bundle only. Bulliforms not in discrete, regular adaxial groups. All the vascular bundles accompanied by sclerenchyma. Combined sclerenchyma girders present; forming 'figures' (anchors). Sclerenchyma not all bundle-associated (the anchors abaxially linked forming a continuous subepidermal band). The 'extra' sclerenchyma in a continuous abaxial layer.

Special diagnostic feature. Female-fertile lemmas with a bent awn, the awn twisted below. Spikelets 8–25 mm long, inflorescence longer than 60 mm long.

Cytology. Chromosome base number, $x = 6$. $2n = 12$. 2 ploid.

Taxonomy. Arundinoideae; Danthonieae.

Ecology, geography, regional floristic distribution. 16 species. South and South West Africa. Mesophytic to xerophytic (often in mountains); species of open habitats; glycophytic.

Paleotropical and Cape. African. Sudano-Angolan and Namib-Karoo. South Tropical African.

References, etc. Morphological/taxonomic: Conert 1970, 1971. Leaf anatomical: Ellis 1981.
Special comments. Fruit data wanting.

Mesosetum Steud.

Bifaria (Hack.) Kuntze, *Peniculus* Swallen

Habit, vegetative morphology. Annual, or perennial; stoloniferous, or caespitose. Culms herbaceous. Culm nodes hairy. Culm internodes hollow. *Leaves* mostly basal, or not basally aggregated; non-auriculate. Leaf blades narrow; flat, or rolled (convolute); without cross venation; persistent. Ligule a fringed membrane, or a fringe of hairs.

Reproductive organization. Plants bisexual, with bisexual spikelets; with hermaphrodite florets.

Inflorescence. *Inflorescence a single spike, or a single raceme (with subsessile spikelets).* Rachides hollowed to flattened, or winged (membranously, in *M. ansatum*). Inflorescence espatheate; not comprising 'partial inflorescences' and foliar organs. Spikelet-bearing axes persistent. *Spikelets* solitary; *not secund*; biseriate; *subsessile.*

Female-fertile spikelets. *Spikelets* adaxial; *compressed dorsiventrally*; falling with the glumes. Rachilla terminated by a female-fertile floret. Hairy callus present, or absent.

Glumes two; very unequal; (the upper) long relative to the adjacent lemmas; dorsiventral to the rachis; hairy (the upper only, or both), or hairless; (the lower) awned (or mucronate, sometimes), or awnless; carinate (the upper strongly keeled, slightly winged). Lower glume 3 nerved. Upper glume (3–)5(–7) nerved. *Spikelets* with incomplete florets. The incomplete florets proximal to the female-fertile florets. *Spikelets with proximal incomplete florets. The proximal incomplete florets* 1; paleate, or epaleate. Palea of the proximal incomplete florets when present, reduced (hyaline). The proximal incomplete florets sterile. The proximal lemmas usually hyaline down the middle; awnless; 5 nerved; decidedly exceeding the female-fertile lemmas; less firm than the female-fertile lemmas; not becoming indurated.

Female-fertile florets 1. Lemmas similar in texture to the glumes to decidedly firmer than the glumes (leathery); smooth; not becoming indurated; entire; acute to mucronate; awnless; hairless; glabrous; non-carinate; having the margins lying flat and exposed on the palea; with a clear germination flap; 5 nerved. **Palea** present; relatively long; entire; awnless, without apical setae; textured like the lemma (apart from the hyaline margins); *2-nerved*; keel-less. *Lodicules* present; 2; free; fleshy; glabrous; not or scarcely vascularized. *Stamens* 3. Anthers not penicillate. *Ovary* glabrous. Styles free to their bases. Stigmas 2; red pigmented.

Fruit, embryo and seedling. *Fruit* free from both lemma and palea; small; longitudinally grooved; compressed laterally (slightly), or not noticeably compressed. **Hilum long-linear (at least in M. pittieri).** Embryo large to small; not waisted. Endosperm hard.

Abaxial leaf blade epidermis. *Costal/intercostal zonation* conspicuous. *Papillae* absent. Mid-intercostal long-cells having markedly sinuous walls. *Microhairs* present; ostensibly one-celled (the apical cells bulbous-based with pointed tips, seemingly lying horizontally over short, embedded basal cells); small, peculiar and hard to interpret in both species seen — cf. *Coelachne, Coelachyrum*?. *Stomata* common; 43.5–54 microns long. Subsidiaries triangular. Guard-cells overlapping to flush with the interstomatals. *Intercostal short-cells* common; in cork/silica-cell pairs; silicified. *Costal short-cells* conspicuously in long rows (*M. loliiforme*), or neither distinctly grouped into long rows nor predominantly paired (*M. pittieri*). Costal silica bodies 'panicoid-type'; mostly cross shaped; not sharp-pointed.

Transverse section of leaf blade, physiology. C_4; XyMS– (but variable in bundles of *M. loliiforme*). *Mesophyll* with radiate chlorenchyma. *Leaf blade* 'nodular' in section; with the ribs more or less constant in size. *Midrib* not readily distinguishable (slightly larger); having a conventional arc of bundles (1 large, 2 small). Bulliforms present in discrete, regular adaxial groups; in simple fans. Many of the smallest vascular bundles unaccompanied by sclerenchyma. Combined sclerenchyma girders present (*M. loliiforme*), or absent (*M. pittieri*); nowhere forming 'figures'. Sclerenchyma all associated with vascular bundles (excluding the blade margins).

Cytology. $2n = 16$.

Taxonomy. Panicoideae; Panicodae; Paniceae.

Ecology, geography, regional floristic distribution. 35 species. C. & tropical South America, West Indies. Species of open habitats. Savanna.

Neotropical. Caribbean, Venezuela and Surinam, Central Brazilian, and Andean.

Rusts and smuts. Smuts from Ustilaginaceae. Ustilaginaceae — *Sphacelotheca*.

References, etc. Leaf anatomical: this project.

Metasasa W.T. Lin

Habit, vegetative morphology. Perennial; rhizomatous. The flowering culms leafy. **Culms *300–700 cm high***; woody and persistent; to 3 cm in diameter; cylindrical; branched above. ***Primary branches/mid-culm node 2***. Culm sheaths deciduous in their entirety. Rhizomes leptomorph. Plants unarmed. *Leaves* not basally aggregated; non-auriculate; without auricular setae. *Leaf blades* lanceolate to ovate-lanceolate; broad; 25–50 mm wide (and 13–26 cm long); rolled; pseudopetiolate; *cross veined*; disarticulating from the sheaths; rolled in bud. *Ligule* present; about 3 mm long.

Reproductive organization. Plants bisexual, with bisexual spikelets; with hermaphrodite florets.

Inflorescence. *Inflorescence* determinate; without pseudospikelets; few spikeleted (6–8); paniculate; open. Spikelet-bearing axes persistent. Spikelets solitary; not secund; pedicellate.

Female-fertile spikelets. Spikelets 30–70 mm long; compressed laterally; disarticulating above the glumes; disarticulating between the florets; with conventional internode spacings. Rachilla prolonged beyond the uppermost female-fertile floret; rachilla hairy. The rachilla extension with incomplete florets. Hairy callus absent.

Glumes two; very unequal (judging from the original illustration); shorter than the spikelets; shorter than the adjacent lemmas; awnless; carinate; similar. *Spikelets* with incomplete florets. The incomplete florets distal to the female-fertile florets.

Female-fertile florets *7–9*. Lemmas not becoming indurated; entire; pointed (acuminate); awnless; hairy; without a germination flap; 13–18 nerved. *Palea* present; relatively long; entire to apically notched; awnless, without apical setae; several nerved (4–5 nerved between the keels); 2-keeled. *Lodicules* present; 3; free; membranous; glabrous; not toothed (lanceolate); heavily vascularized. **Stamens *6***. Anthers 5–5.5 mm long. Ovary without a conspicuous apical appendage. Styles fused (into one). Stigmas 2, or 3 (one of the two stigmas illustrated as being deeply two-lobed).

Fruit, embryo and seedling. Fruit medium sized (about 8mm long). Hilum seemingly long-linear.

Taxonomy. Bambusoideae; Bambusodae; Bambuseae.

Ecology, geography, regional floristic distribution. 1 species. Guangdong, China. Paleotropical. Indomalesian. Indo-Chinese.

References, etc. Morphological/taxonomic: Lin 1988.

Special comments. Anatomical data wanting.

Metcalfia Conert

Excluding *Danthoniastrum*

Habit, vegetative morphology. Perennial; caespitose. *Culms* 70–90 cm high; herbaceous; unbranched above. Culm nodes glabrous. Young shoots intravaginal. *Leaves auriculate (with long sheath auricles)*. Leaf blades narrow; setaceous; rolled (pungent); without cross venation. *Ligule an unfringed membrane*; truncate, or not truncate; 1–4 mm long.

Reproductive organization. Plants bisexual, with bisexual spikelets; with hermaphrodite florets.

Inflorescence. Inflorescence few spikeleted; paniculate (spiciform); contracted; espatheate; not comprising 'partial inflorescences' and foliar organs. Spikelet-bearing axes persistent. Spikelets paired; not secund; pedicellate.

Female-fertile spikelets. *Spikelets* 12–19 mm long; *not noticeably compressed*; disarticulating above the glumes. Rachilla prolonged beyond the uppermost female-fertile floret. The rachilla extension with incomplete florets. Hairy callus present (the hairs long, white). *Callus* short (1–1.2 mm long); blunt.

Glumes two; very unequal, or more or less equal; (the longer) long relative to the adjacent lemmas; hairless; pointed (acuminate); awned, or awnless; non-carinate; similar (membranous). Lower glume 7–9 nerved. Upper glume 7–9 nerved. *Spikelets* with incomplete florets. The incomplete florets distal to the female-fertile florets. *The distal incomplete florets* merely underdeveloped.

Female-fertile florets 2–3. Lemmas similar in texture to the glumes; not becoming indurated; incised; 2 lobed; deeply cleft (notched); awned. Awns 1 (but with short setae terminating the lateral lobes); median; from a sinus (the notch deep); geniculate; hairless (scabrid); about as long as the body of the lemma to much longer than the body of the lemma. Lemmas hairy (beneath the insertion of the awn); non-carinate; without a germination flap; 7 nerved. *Palea* present; relatively long; apically notched; awnless, without apical setae (but apically ciliate); 2-nerved; 2-keeled. *Lodicules* present; 3; free; membranous; glabrous; not toothed; heavily vascularized (the anterior pair). **Stamens 2**. Anthers 1.5–2 mm long. *Ovary* hairy. Styles free to their bases. Stigmas 2.

Fruit, embryo and seedling. Hilum long-linear. Embryo small; with an epiblast; without a scutellar tail; with a negligible mesocotyl internode. Embryonic leaf margins overlapping.

Abaxial leaf blade epidermis. *Costal/intercostal zonation* lacking. *Papillae* absent. *Long-cells* similar in shape costally and intercostally; of similar wall thickness costally and intercostally. Mid-intercostal long-cells rectangular; having markedly sinuous walls (thick, pitted). *Microhairs* absent. *Stomata* absent or very rare. *Intercostal short-cells* common; mostly in cork/silica-cell pairs; silicified. *Costal short-cells* predominantly paired. Costal silica bodies rounded, or tall-and-narrow, or crescentic (mostly rounded, tall-and-narrow or crescentic, but the *silica cells* often cross- to dumb-bell shaped); not sharp-pointed.

Transverse section of leaf blade, physiology. C_3; XyMS+. *Mesophyll* with radiate chlorenchyma; without adaxial palisade. *Leaf blade* with distinct, prominent adaxial ribs; with the ribs very irregular in sizes. *Midrib* not readily distinguishable; with one bundle only. All the vascular bundles accompanied by sclerenchyma. Combined sclerenchyma girders present (with all the bundles); forming 'figures' (all bundles). Sclerenchyma not all bundle-associated. The 'extra' sclerenchyma in a continuous abaxial layer.

Cytology. Chromosome base number, $x =$ and chromosome 'size' seemingly unknown.

Taxonomy. Pooideae, or Arundinoideae; if pooid, Poodae; Aveneae; if arundinoid, Stipeae, or Danthonieae.

Ecology, geography, regional floristic distribution. 1 species. Mexico. Species of open habitats. Stony hillsides.

Holarctic. Madrean.

References, etc. Morphological/taxonomic: Tateoka 1964b; Conert 1960; Clayton 1985. Leaf anatomical: this project.

Special comments. A taxonomic puzzle — cf. *Danthoniastrum*? The embryo structure as described by Tateoka (1964) is *not* 'pooid', and the leaf anatomical evidence (see above) is ambiguous. Reliable taxonomic assignment requires that adaxial leaf epidermis, glumes, lemmas and lodicules be examined for microhairs, and that cytological information be obtained.

Mibora Adans.

Chamagrostis Borkh., *Knappia* Sm., *Micagrostis* Juss., *Rothia* Borkh., *Sturmia* Hoppe

Habit, vegetative morphology. Diminutive *annual*; caespitose. *Culms* 2–15 cm high (filiform); herbaceous. *Leaves* mostly basal; non-auriculate. Leaf blades linear; narrow; 0.3–1 mm wide; flat, or folded; without cross venation; persistent. Ligule an unfringed membrane; truncate; 0.5–1.5 mm long.

Reproductive organization. Plants bisexual, with bisexual spikelets; with hermaphrodite florets.

Inflorescence. *Inflorescence a single raceme (unilateral, spikelike)*; espatheate; not comprising 'partial inflorescences' and foliar organs. Spikelet-bearing axes persistent. Spikelets solitary; secund; distichous; pedicellate.

Female-fertile spikelets. *Spikelets 1.8–3 mm long*; compressed laterally; disarticulating above the glumes. Rachilla terminated by a female-fertile floret. Hairy callus absent. *Callus* very short.

Glumes two; more or less equal; exceeding the spikelets; long relative to the adjacent lemmas; not pointed (obtuse to emarginate); awnless; non-carinate; similar (thinly membranous). Lower glume 1 nerved. *Upper glume 1 nerved*. *Spikelets* with female-fertile florets only.

Female-fertile florets 1. Lemmas less firm than the glumes (hyaline); not becoming indurated; entire to incised (truncate-denticulate); awnless; hairy; non-carinate; *5 nerved*. *Palea* present; relatively long; awnless, without apical setae (hairy); 2-nerved; keel-less. *Lodicules* present; 2. *Stamens* 3. Anthers 1.4–1.7 mm long; not penicillate. *Ovary* glabrous. Styles fused. Stigmas 2.

Fruit, embryo and seedling. *Fruit* free from both lemma and palea; small. Hilum short. Embryo small. Endosperm liquid in the mature fruit, or hard; with lipid.

Seedling with a loose coleoptile. First seedling leaf with a well-developed lamina. The lamina narrow; 1–3 veined.

Abaxial leaf blade epidermis. *Costal/intercostal zonation* lacking. *Papillae* absent. Mid-intercostal long-cells fusiform; having straight or only gently undulating walls. *Microhairs* absent. *Stomata* common. Subsidiaries parallel-sided. Guard-cells overlapped by the interstomatals. *Intercostal short-cells* absent or very rare (none seen). Costal silica bodies absent.

Transverse section of leaf blade, physiology. C_3; XyMS+. *Mesophyll* with non-radiate chlorenchyma. *Leaf blade* with distinct, prominent adaxial ribs; with the ribs more or less constant in size. *Midrib* conspicuous; with one bundle only. Bulliforms not in discrete, regular adaxial groups (no bulliform cells). Many of the smallest vascular bundles unaccompanied by sclerenchyma. Combined sclerenchyma girders absent. Sclerenchyma all associated with vascular bundles.

Cytology. Chromosome base number, $x = 7$. $2n = 14$. 2 ploid. Haploid nuclear DNA content 2.8 pg (1 species, $2x$). Mean diploid 2c DNA value 5.5 pg.

Taxonomy. Pooideae; Poodae; Aveneae.

Ecology, geography, regional floristic distribution. 2 species. Mediterranean. Commonly adventive. Mesophytic; species of open habitats. Damp, sandy soils.

Holarctic. Boreal and Tethyan. Euro-Siberian. Mediterranean. European.

References, etc. Leaf anatomical: this project.

Micraira F. Muell.

Habit, vegetative morphology. Mat-forming *perennial (polytrichoid in appearance)*. Culms 1–9 cm high; *herbaceous*. **Leaves** not basally aggregated; *spirally disposed*; non-auriculate. Leaf blades narrow; cordate, or not cordate, not sagittate; setaceous to not setaceous (pungent or muticous); pseudopetiolate, or not pseudopetiolate; without cross venation; disarticulating from the sheaths. Ligule a fringed membrane to a fringe of hairs.

Reproductive organization. Plants bisexual, with bisexual spikelets; with hermaphrodite florets.

Inflorescence. Inflorescence a single spike, or a single raceme, or paniculate; open, or contracted; with capillary branchlets; espatheate; not comprising 'partial inflorescences' and foliar organs. Spikelet-bearing axes persistent. Spikelets not secund; pedicellate; not in distinct 'long-and-short' combinations.

Female-fertile spikelets. *Spikelets 0.75–2 mm long*; compressed laterally; disarticulating above the glumes. Rachilla prolonged beyond the uppermost female-fertile floret. Hairy callus absent.

Glumes two; more or less equal; shorter than the spikelets to exceeding the spikelets; shorter than the adjacent lemmas to long relative to the adjacent lemmas; pointed; awnless (muticous, mucronate or mucronulate); similar (membranous). Lower glume 1 nerved. Upper glume 1 nerved. *Spikelets* with female-fertile florets only, or with incomplete florets. The incomplete florets proximal to the female-fertile florets. *The proximal incomplete florets* 1; male, or sterile. The proximal lemmas awnless; 5–9 nerved; more or less equalling the female-fertile lemmas; similar in texture to the female-fertile lemmas; not becoming indurated.

Female-fertile florets 1–2. **Lemmas almost rectangular**; less firm than the glumes (thinly membranous to hyaline); not becoming indurated; entire; blunt; awnless; hairless; carinate to non-carinate; 5–9 nerved. *Palea* present; relatively long, or conspicuous but relatively short, or very reduced (almost rectangular); entire to deeply bifid; awnless, without apical setae; thinner than the lemma to textured like the lemma; not indurated; 2-nerved, or several nerved (2, or 5–7); 2-keeled. *Lodicules* absent. *Stamens* 2. Anthers 0.5–1.2 mm long; not penicillate. *Ovary* glabrous. Stigmas 2; red pigmented.

Fruit, embryo and seedling. Fruit small (0.3–0.75 mm long); ellipsoid; compressed dorsiventrally. Hilum long-linear. Embryo small. Endosperm containing only simple starch grains. Embryo without an epiblast; with a scutellar tail.

First seedling leaf with a well-developed lamina. The lamina narrow; curved.

Abaxial leaf blade epidermis. *Costal/intercostal zonation* conspicuous. *Papillae* present, or absent (e.g., *M. subulifolia*). Intercostal papillae not over-arching the stomata; several per cell (small, up to 15 per cell). *Long-cells* similar in shape costally and intercostally (but the costals rather smaller); of similar wall thickness costally and intercostally (thin walled). Mid-intercostal long-cells rectangular; having markedly sinuous walls, or having straight or only gently undulating walls. *Microhairs* present; panicoid-type; (36–)45–51 microns long; 3.6–6.5(–8.4) microns wide at the septum. Microhair total length/width at septum 6.1–13.3. Microhair apical cells (16.5–)22.5–24(–25.5) microns long. Microhair apical cell/total length ratio 0.46–0.55. *Stomata* absent or very rare, or common; 18–24 microns long. Subsidiaries dome-shaped. Guard-cells overlapping to flush with the interstomatals. *Intercostal short-cells* absent or very rare; not paired (rare); not silicified. *Costal short-cells* conspicuously in long rows. Costal silica bodies 'panicoid-type'; short, angular, butterfly shaped, or dumb-bell shaped, or cross shaped; sharp-pointed (all the silica-bodies angular, many more or less rectangular — cf. *Pheidochloa*, but not vertically elongated).

Transverse section of leaf blade, physiology. C_3; XyMS+. *Mesophyll* with non-radiate chlorenchyma; *Isachne*-type, or not *Isachne*-type (e.g., *M. subulifolia*). Leaf blade with the ribs more or less constant in size. *Midrib* not readily distinguishable; with one bundle only. Bulliforms present in discrete, regular adaxial groups; in simple fans. All the vascular bundles accompanied by sclerenchyma. Combined sclerenchyma girders present; forming 'figures'. Sclerenchyma all associated with vascular bundles.

Taxonomy. Arundinoideae; Micraireae.

Ecology, geography, regional floristic distribution. 13 species. Australia. Rocky places, in shallow soils.

Australian. North and East Australian. Tropical North and East Australian.

References, etc. Morphological/taxonomic: Clifford 1964; Lazarides 1979, 1985. Leaf anatomical: Metcalfe 1960; this project.

Microbriza Parodi ex Nicora et Rúg.

Monostemon Henr.

Habit, vegetative morphology. Perennial; caespitose. *Culms* 40–80 cm high; herbaceous; unbranched above. Young shoots intravaginal. *Leaves* not basally aggregated; non-auriculate; without auricular setae. Sheaths glabrous or pilose, becoming fibrous when dead. Leaf blades linear; narrow; flat; without cross venation; persistent (the fibrous remains of the sheaths persisting). *Ligule an unfringed membrane*; truncate; 1–3 mm long.

Reproductive organization. Plants bisexual, with bisexual spikelets; with hermaphrodite florets. Exposed-cleistogamous, or chasmogamous.

Inflorescence. Inflorescence paniculate; open (the spikelets clustered at the extremities); with capillary branchlets; espatheate; not comprising 'partial inflorescences' and foliar organs. Spikelet-bearing axes persistent. Spikelets not secund; pedicellate; imbricate.

Female-fertile spikelets. *Spikelets 1.2–2.5 mm long*; compressed laterally (the individual florets dorsiventrally compressed); disarticulating above the glumes; *not disarticulating between the florets (at least, the florets tending to fall together)*; with conventional internode spacings. Rachilla prolonged beyond the uppermost female-fertile floret. Hairy callus absent.

Glumes two; more or less equal; about equalling the spikelets; long relative to the adjacent lemmas; hairless; scabrous (to echinulate); awnless; carinate to non-carinate; similar (broadly lanceolate-navicular). Lower glume 3 nerved. Upper glume 3 nerved. *Spikelets* with female-fertile florets only, or with incomplete florets (the presence of a distal rudiment variable). The incomplete florets (when present) distal to the female-fertile florets. *The distal incomplete florets* when present, merely underdeveloped; awnless.

Female-fertile florets 2–3. *Lemmas navicular, dorsally gibbous but without the* **Briza** *umbo, the margins infolded on the palea*; similar in texture to the glumes; not becoming indurated (leathery); entire; pointed to blunt; awnless to mucronate; hairless; verrucose to echinulate; non-carinate; without a germination flap; 5 nerved. *Palea* present; relatively long; tightly clasped by the lemma; entire (lanceolate, acute); awnless, without apical setae; textured like the lemma (leathery); not indurated; 2-nerved; 2-keeled (and flat between the keels). Palea keels wingless; glabrous to scabrous. *Lodicules* present; 2; free; membranous; glabrous; not toothed (lanceolate). **Stamens 1**. *Ovary* glabrous. Styles free to their bases. Stigmas 2.

Fruit, embryo and seedling. *Fruit* adhering to lemma and/or palea (adhering to the palea along the two lateral grooves); small; ellipsoid; with two lateral, longitudinal grooves, wherein lie the margins of the palea; compressed dorsiventrally (ventrally flattened, dorsally convex). Hilum short. Embryo small. Endosperm hard.

Abaxial leaf blade epidermis. *Costal/intercostal zonation* conspicuous. *Papillae* absent. *Long-cells* markedly different in shape costally and intercostally (the costals much narrower, rectangular). Mid-intercostal long-cells fusiform; having straight or only gently undulating walls. *Microhairs* absent. *Stomata* fairly common. Subsidiaries parallel-sided. Guard-cells overlapped by the interstomatals. *Intercostal short-cells* absent or very rare. *Costal short-cells* conspicuously in long rows (the long rows interrupted only occasionally, by prickles). Costal silica bodies horizontally-elongated crenate/sinuous, or 'panicoid-type' (the bodies ill defined, in the material seen); if panicoid type, nodular; not sharp-pointed.

Transverse section of leaf blade, physiology. C_3; XyMS+. *Leaf blade* with distinct, prominent adaxial ribs, or 'nodular' in section; with the ribs more or less constant in size. *Midrib* not readily distinguishable; with one bundle only. Bulliforms present in discrete, regular adaxial groups (in all the furrows); in simple fans. All the vascular bundles accompanied by sclerenchyma. Combined sclerenchyma girders present; forming 'figures' (in all the main bundles). Sclerenchyma all associated with vascular bundles.

Special diagnostic feature. Lemmas not as in *Briza* (q.v.).

Taxonomy. Pooideae; Poodae; Poeae.

Ecology, geography, regional floristic distribution. 2 species. South America. Helophytic; glycophytic.

Neotropical. Central Brazilian and Pampas.

References, etc. Leaf anatomical: this project.

Microcalamus Franch.

Habit, vegetative morphology. Perennial; rhizomatous and stoloniferous. *Culms* 15–60 cm high; herbaceous; unbranched above. Culm nodes hairy. Young shoots extravaginal. *Leaves* not basally aggregated; non-auriculate. Leaf blades lanceolate to ovate; broad; 15–50 mm wide; pseudopetiolate; cross veined; persistent. *Ligule* present; an unfringed membrane, or a fringed membrane (laciniate, minutely fringed).

Reproductive organization. Plants bisexual, with bisexual spikelets; with hermaphrodite florets.

Inflorescence. Inflorescence paniculate; open, or contracted (about the primary branches); espatheate; not comprising 'partial inflorescences' and foliar organs. Spikelet-bearing axes persistent. Spikelets not secund; pedicellate.

Female-fertile spikelets. Spikelets weakly compressed laterally; falling with the glumes. Rachilla terminated by a female-fertile floret. Hairy callus present.

Glumes two; very unequal; shorter than the adjacent lemmas; hairless; pointed; awnless; non-carinate; similar (ovate, papery). Lower glume 5–7 nerved. Upper glume 5–7 nerved. *Spikelets* with incomplete florets. The incomplete florets proximal to the female-fertile florets. *The proximal incomplete florets* 1; paleate. Palea of the proximal incomplete florets fully developed. The proximal incomplete florets sterile. The proximal lemmas awnless; 7 nerved; somewhat exceeded by the female-fertile lemmas; less firm than the female-fertile lemmas (papery, similar to the glumes); not becoming indurated.

Female-fertile florets 1. *Lemmas* rostrate or subulate; similar in texture to the glumes to decidedly firmer than the glumes (subleathery); not becoming indurated; entire; pointed; *crested at the tip*; awnless (but beaked); *hairy*; non-carinate; having the margins lying flat and exposed on the palea; without a germination flap; 5 nerved. *Palea* present; relatively long; entire; 2-nerved; 2-keeled. *Lodicules* present; 2; free; fleshy; glabrous; not or scarcely vascularized. *Stamens* 3. Anthers not penicillate; without an apically prolonged connective. *Ovary* glabrous. Styles fused. Stigmas 2; white (in dried material).

Fruit, embryo and seedling. Hilum short.

Abaxial leaf blade epidermis. *Costal/intercostal zonation* conspicuous. *Papillae* absent. Intercostal zones exhibiting many atypical long-cells. Mid-intercostal long-cells having straight or only gently undulating walls. *Microhairs* present; panicoid-type. *Stomata* common (but very localised, in one or two rows adjoining veins). Subsidiaries triangular. Guard-cells overlapping to flush with the interstomatals. *Intercostal short-cells* common; in cork/silica-cell pairs; silicified. *Costal short-cells* conspicuously in long rows. Costal silica bodies 'panicoid-type'; mostly cross shaped; not sharp-pointed.

Transverse section of leaf blade, physiology. C_3; XyMS+. *Mesophyll* with non-radiate chlorenchyma. *Leaf blade* adaxially flat. *Midrib* conspicuous (keel, large bundle); with one bundle only. Bulliforms not in discrete, regular adaxial groups (constituting most of epidermis). All the vascular bundles accompanied by sclerenchyma. Combined sclerenchyma girders present (midrib bundle only); forming 'figures' (in the midrib). Sclerenchyma all associated with vascular bundles.

Taxonomy. Panicoideae; Panicodae; Paniceae.

Ecology, geography, regional floristic distribution. 4 species. Tropical west Africa. Shade species; glycophytic. In forest.

Paleotropical. African. West African Rainforest.

References, etc. Leaf anatomical: this project.

Microchloa R.Br.

Micropogon Pfeiffer
Excluding *Rendlia*

Habit, vegetative morphology. Annual (rarely), or perennial; caespitose (low), or decumbent (mat-forming). *Culms* 5–60 cm high; herbaceous; unbranched above. Culm nodes glabrous. Culm internodes solid. *Leaves* mostly basal, or not basally aggregated; non-auriculate. Leaf blades narrow (stiff, often convolute); setaceous, or not setaceous; without abaxial multicellular glands; without cross venation; persistent. Ligule a fringed membrane (narrow), or a fringe of hairs.

Reproductive organization. Plants bisexual, with bisexual spikelets; with hermaphrodite florets. Exposed-cleistogamous, or chasmogamous (?).

Inflorescence. *Inflorescence a single spike (slender, often curved, the spikelets inclined to pectinate)*. Rachides hollowed (crescentic in section), or hollowed and winged. Inflorescence espatheate (but often embraced by the uppermost sheath); not comprising 'partial inflorescences' and foliar organs. Spikelet-bearing axes persistent (tough, narrow). Spikelets solitary; secund; biseriate; sessile.

Female-fertile spikelets. Spikelets 1.7–5.5 mm long; adaxial (but twisted); not noticeably compressed to compressed dorsiventrally; disarticulating above the glumes. Rachilla terminated by a female-fertile floret. Hairy callus present. *Callus* short; pointed.

Glumes two; more or less equal; exceeding the spikelets; long relative to the adjacent lemmas; lateral to the rachis; hairless; glabrous; pointed (lanceolate-acute); awnless; very dissimilar (the lower asymmetric, cymbiform, keeled, twisted at the base, the upper flat). Lower glume 1 nerved. Upper glume 1 nerved. **Spikelets** *with female-fertile florets only*.

Female-fertile florets 1. Lemmas not becoming indurated (membranous or hyaline); entire to incised; when entire pointed, or blunt; not deeply cleft (no more than emarginate); awnless, or mucronate; hairy (ciliate on the nerves); carinate to non-carinate; without a germination flap; *2 nerved*. Palea present; entire to apically notched; awnless, without apical setae; textured like the lemma; not indurated; 2-nerved; 2-keeled. Palea keels hairy. *Lodicules* present; 2; free; fleshy; glabrous. *Stamens* 3. Anthers relatively long; not penicillate. *Ovary* glabrous. Styles free to their bases. Stigmas 2.

Fruit, embryo and seedling. *Fruit* free from both lemma and palea; small (0.9–1.5 mm long); ellipsoid; compressed dorsiventrally, or not noticeably compressed. Hilum short. Pericarp fused. Embryo large; not waisted. Endosperm hard; without lipid. Embryo with an epiblast; with a scutellar tail; with an elongated mesocotyl internode. Embryonic leaf margins meeting.

Abaxial leaf blade epidermis. *Costal/intercostal zonation* conspicuous. *Papillae* absent. *Long-cells* similar in shape costally and intercostally; of similar wall thickness costally and intercostally. Mid-intercostal long-cells rectangular; having markedly sinuous walls. *Microhairs* present; more or less spherical, or elongated; ostensibly one-celled, or clearly two-celled (?); chloridoid-type. *Stomata* common; 21.6–27 microns long. Subsidiaries triangular. Guard-cells overlapping to flush with the interstomatals. *Intercostal short-cells* common; in cork/silica-cell pairs (rarely), or not paired (usually solitary); silicified, or not silicified. Intercostal silica bodies absent. *Costal short-cells* conspicuously in long rows. Costal silica bodies present in alternate cell files of the costal zones; saddle shaped; not sharp-pointed.

Transverse section of leaf blade, physiology. C_4; XyMS+. PCR sheath outlines uneven, or even. PCR sheaths of the primary vascular bundles complete to interrupted; interrupted abaxially only. PCR sheath extensions absent. PCR cell chloroplasts centrifugal/peripheral (usually), or centripetal (in some individuals of *M. caffra*?). *Mesophyll* with radiate chlorenchyma; traversed by columns of colourless mesophyll cells (rarely), or not traversed by colourless columns. *Leaf blade* adaxially flat. *Midrib* not readily distinguishable; with one bundle only. Bulliforms present in discrete, regular adaxial groups; associated with colourless mesophyll cells to form deeply-penetrating fans (these sometimes linked with traversing columns of colourless cells). All the vascular bundles accompanied by sclerenchyma. Combined sclerenchyma girders present; forming 'figures'. Sclerenchyma all associated with vascular bundles. The lamina margins without fibres.

Cytology. Chromosome base number, $x = 10$. $2n = 40$.

Taxonomy. Chloridoideae; main chloridoid assemblage.

Ecology, geography, regional floristic distribution. 4 species. 3 in Africa, 1 pantropical. Mesophytic to xerophytic; species of open habitats; glycophytic. Savanna, in shallow hard soils.

Holarctic, Paleotropical, Neotropical, and Australian. Madrean. African, Madagascan, and Indomalesian. Saharo-Sindian, Sudano-Angolan, West African Rainforest, and Namib-Karoo. Indian, Indo-Chinese, and Malesian. Caribbean, Central Brazilian, Pampas, and Andean. North and East Australian. Sahelo-Sudanian, Somalo-Ethiopian, South Tropical African, and Kalaharian. Tropical North and East Australian.

Rusts and smuts. Rusts — *Puccinia*. Smuts from Ustilaginaceae. Ustilaginaceae — *Ustilago*.

References, etc. Leaf anatomical: Metcalfe 1960; this project.

Microlaena R.Br.

~ *Ehrharta*

Habit, vegetative morphology. Perennial; stoloniferous and caespitose. *Culms* 30–200 cm high; woody and persistent, or herbaceous. Culm nodes glabrous. Culm internodes hollow. Young shoots extravaginal. *Leaves* not basally aggregated; auriculate, or non-auriculate; without auricular setae. Leaf blades linear to linear-lanceolate; narrow; flat (or concave); without cross venation; persistent; rolled in bud. Ligule an unfringed membrane to a fringed membrane (a hyaline rim, with caducous cilia). *Contra-ligule* absent.

Reproductive organization. Plants bisexual, with bisexual spikelets; with hermaphrodite florets. The spikelets all alike in sexuality (but often cleistogamous, leading to reduced paleas and lodicules, and indehiscent stamens). Plants inbreeding. Exposed-cleistogamous, or chasmogamous. Plants with hidden cleistogenes, or without hidden cleistogenes. The hidden cleistogenes when present, in the leaf sheaths.

Inflorescence. Inflorescence a single raceme, or paniculate; open; espatheate; not comprising 'partial inflorescences' and foliar organs. Spikelet-bearing axes persistent. Spikelets not secund; pedicellate.

Female-fertile spikelets. *Spikelets compressed laterally*; disarticulating above the glumes; **with a distinctly elongated rachilla internode above the glumes (i.e., beneath the empty lemmas)**. Rachilla terminated by a female-fertile floret; rachilla hairless. Hairy callus present.

Glumes two; very *minute*; very unequal (the G_2 longer); shorter than the adjacent lemmas; awnless; similar (membranous). Lower glume 0–1 nerved. Upper glume 0–1 nerved. *Spikelets* with incomplete florets. The incomplete florets proximal to the female-fertile florets. *Spikelets with proximal incomplete florets. The proximal incomplete florets* 2 (similar); epaleate; sterile. The proximal lemmas awned (long acuminate, tapered into long slender awns); 5–9 nerved; decidedly exceeding the female-fertile lemmas; decidedly firmer than the female-fertile lemmas (cartilaginous, usually ribbed and grooved).

Female-fertile florets 1. Lemmas decidedly firmer than the glumes (cartilaginous, with thin margins); becoming indurated to not becoming indurated; entire; pointed; awnless, or mucronate, or awned (tapered into the stout awn). Awns (when present) 1; median; apical; non-geniculate; hairless (scabrid); much shorter than the body of the lemma. Lemmas hairless; carinate (scabrid-ciliate on the keel); 5–7 nerved. **Palea** usually present; when present, relatively long, or conspicuous but relatively short, or very reduced; entire; awnless, without apical setae; thinner than the lemma; not indurated (thinly membranous); *1-nerved (or nerveless)*; one-keeled, or keel-less. Palea keels wingless. *Lodicules* present; 2; free; membranous; glabrous. *Stamens* 2–6. Anthers not penicillate. *Ovary* glabrous. Styles free to their bases. Stigmas 2; white.

Fruit, embryo and seedling. *Fruit* free from both lemma and palea; medium sized; oblong-linear; compressed laterally. Hilum long-linear. Embryo small; not waisted. Endosperm containing compound starch grains. Embryo with an epiblast; with a scutellar tail; with an elongated mesocotyl internode. Embryonic leaf margins overlapping.

Seedling with a short mesocotyl; with a tight coleoptile. First seedling leaf with a well-developed lamina. The lamina broad; curved; 7–9 veined.

Abaxial leaf blade epidermis. *Costal/intercostal zonation* conspicuous. *Papillae* absent. *Long-cells* similar in shape costally and intercostally; of similar wall thickness costally and intercostally. Mid-intercostal long-cells rectangular and fusiform; having markedly sinuous walls (rectangular), or having straight or only gently undulating walls (fusiform). *Microhairs* present; panicoid-type; 36–54 microns long; 6–9.3 microns wide at the septum. Microhair total length/width at septum 4.8–9. Microhair apical cells (16–)21–33(–36) microns long. Microhair apical cell/total length ratio 0.53–0.71. *Stomata* common; 21–24 microns long. Subsidiaries dome-shaped, or triangular. Guard-cells overlapping to flush with the interstomatals. *Intercostal short-cells* common, or absent or very rare; not silicified. *Costal short-cells* conspicuously in long rows. Costal silica bodies 'panicoid-type'; cross shaped, or butterfly shaped, or dumb-bell shaped; not sharp-pointed.

Transverse section of leaf blade, physiology. C_3; XyMS+. *Mesophyll* with non-radiate chlorenchyma; without adaxial palisade; without arm cells; without fusoids. *Leaf blade* 'nodular' in section; with the ribs more or less constant in size. *Midrib* conspicuous; with one bundle only. Bulliforms present in discrete, regular adaxial groups; in simple fans. All the vascular bundles accompanied by sclerenchyma. Combined sclerenchyma girders present; forming 'figures'. Sclerenchyma all associated with vascular bundles.

Culm anatomy. *Culm internode bundles* in one or two rings.

Phytochemistry. Tissues of the culm bases with abundant starch. Leaves containing flavonoid sulphates (*M. stipoides*).

Cytology. Chromosome base number, $x = 10$.

Taxonomy. Bambusoideae; Oryzodae; Ehrharteae.

Ecology, geography, regional floristic distribution. 10 species. Philippines, Java to Australasia. Helophytic to mesophytic; shade species and species of open habitats.

Paleotropical, Australian, and Antarctic. Indomalesian and Polynesian. Malesian and Papuan. Hawaiian and Fijian. North and East Australian and South-West Australian. New Zealand. Tropical North and East Australian and Temperate and South-Eastern Australian.

Rusts and smuts. Rusts — *Puccinia*.

References, etc. Morphological/taxonomic: Willemse 1982. Leaf anatomical: Metcalfe 1960; this project.

Micropyropsis Zarco & Cabezudo

Habit, vegetative morphology. Perennial. **Culms** 30–110 cm high; herbaceous; un-branched above; *tuberous*. Culm nodes glabrous. *Leaves* not basally aggregated. Leaf blades linear; narrow; 1–5 mm wide (10–25 cm long); without cross venation; persistent. *Ligule an unfringed membrane*.

Reproductive organization. Plants bisexual, with bisexual spikelets; with hermaphrodite florets.

Inflorescence. *Inflorescence a single raceme (spikelike)*. Rachides hollowed (to receive the spikelets). Inflorescence espatheate; not comprising 'partial inflorescences' and foliar organs. Spikelet-bearing axes persistent. Spikelets solitary; not secund; distichous (seemingly); subsessile to pedicellate (the pedicels about 0.5 mm long); distant.

Female-fertile spikelets. Spikelets 10–26 mm long; adaxial; compressed laterally; disarticulating above the glumes; disarticulating between the florets. Rachilla prolonged beyond the uppermost female-fertile floret; rachilla hairless. The rachilla extension with incomplete florets. Hairy callus absent. *Callus* short; blunt (glabrous).

Glumes two; very unequal; shorter than the spikelets; shorter than the adjacent lemmas; dorsiventral to the rachis; hairless; not subulate; awnless; similar. Lower glume 1 nerved. Upper glume 3 nerved. *Spikelets* with incomplete florets. The incomplete florets distal to the female-fertile florets. *The distal incomplete florets* merely underdeveloped.

Female-fertile florets 3–13. Lemmas not becoming indurated; entire; pointed; awned. Awns 1; median; dorsal; from near the top; non-geniculate; hairless; much shorter than the body of the lemma to about as long as the body of the lemma (2–6 mm long, slender). Lemmas hairless; glabrous; non-carinate (dorsally rounded); 5 nerved. *Palea* present; relatively long; entire to apically notched; awnless, without apical setae; not indurated; 2-nerved; 2-keeled. *Lodicules* present; 2; free; membranous; glabrous; toothed (bilobed). *Stamens* 3. Anthers 3 mm long. *Ovary* glabrous; without a conspicuous apical appendage. Styles free to their bases. Stigmas 2.

Fruit, embryo and seedling. Fruit *adhering to lemma and/or palea*; small to medium sized (about 4 mm long).

Transverse section of leaf blade, physiology. *Leaf blade* with distinct, prominent adaxial ribs; with the ribs more or less constant in size. *Midrib* not readily distinguishable; with one bundle only. Combined sclerenchyma girders present; forming 'figures'. Sclerenchyma all associated with vascular bundles.

Taxonomy. Pooideae; Poodae; Poeae.

Ecology, geography, regional floristic distribution. 1 species. Spain. In damp sands.

Holarctic. Tethyan. Mediterranean.

References, etc. Morphological/taxonomic: Zarco and Cabezudo 1983.

Special comments. Fruit data wanting. Anatomical data wanting.

Micropyrum Link

Habit, vegetative morphology. *Annual. Culms* 50–100 cm high; herbaceous. Leaves non-auriculate. Leaf blades linear; narrow; flat, or rolled (convolute when dry); without cross venation. *Ligule an unfringed membrane*; truncate; 0.2–1 mm long.

Reproductive organization. Plants bisexual, with bisexual spikelets; with hermaphrodite florets. The spikelets all alike in sexuality.

Inflorescence. *Inflorescence a single raceme (spiciform), or paniculate (sparingly branched)*; espatheate; not comprising 'partial inflorescences' and foliar organs. Spikelet-bearing axes persistent. *Spikelets not secund*; when in a raceme, distichous; pedicellate.

Female-fertile spikelets. Spikelets 4–16 mm long; compressed laterally; disarticulating above the glumes; disarticulating between the florets. Rachilla prolonged beyond the uppermost female-fertile floret. The rachilla extension with incomplete florets. Hairy callus absent.

Glumes two; very unequal to more or less equal; shorter than the adjacent lemmas; dorsiventral to the rachis; pointed, or not pointed (acute to rounded); awnless; carinate, or non-carinate; similar (leathery). Lower glume 1–3 nerved. Upper glume 3–5 nerved. *Spikelets* with incomplete florets. The incomplete florets distal to the female-fertile florets.

Female-fertile florets 3–9(–14). *Lemmas decidedly firmer than the glumes (papery)*; not becoming indurated; entire to incised (obtuse to emarginate); awnless (but apiculate), or mucronate, or awned. *Awns* when present, 1; *from a sinus, or apical*; nongeniculate; much shorter than the body of the lemma to about as long as the body of the lemma; entered by one vein. Lemmas hairless; non-carinate (dorsally rounded); without a germination flap; 5 nerved. *Palea* present; relatively long; apically notched (shortly bifid); awnless, without apical setae; 2-nerved; 2-keeled. *Lodicules* present; 2; free; membranous; glabrous; toothed; not or scarcely vascularized. *Stamens* 3. Anthers 0.5–3.2 mm long. *Ovary* glabrous; *without a conspicuous apical appendage*. Styles free to their bases. Stigmas 2.

Fruit, embryo and seedling. *Fruit* adhering to lemma and/or palea; medium sized (2.5–3.2 mm long); compressed dorsiventrally. *Hilum long-linear*. Embryo small. Endosperm hard; without lipid; containing compound starch grains. Embryo with an epiblast.

Abaxial leaf blade epidermis. *Costal/intercostal zonation* conspicuous. *Papillae* absent. *Long-cells* similar in shape costally and intercostally; of similar wall thickness costally and intercostally. Mid-intercostal long-cells rectangular; having markedly sinuous walls (and pitted). *Microhairs* absent. *Stomata* absent or very rare. *Intercostal short-cells* absent or very rare. *Costal short-cells* predominantly paired. Costal silica bodies rounded (numerous); not sharp-pointed.

Transverse section of leaf blade, physiology. C_3; XyMS+. *Leaf blade* with distinct, prominent adaxial ribs; with the ribs more or less constant in size. *Midrib* not readily distinguishable; with one bundle only. Bulliforms poorly defined, in the poor material seen. Combined sclerenchyma girders absent. Sclerenchyma all associated with vascular bundles.

Cytology. Chromosome base number, $x = 7$. $2n = 14$. 2 ploid.

Taxonomy. Pooideae; Poodae; Poeae.

Ecology, geography, regional floristic distribution. 3 species. Central Europe, Mediterranean. Species of open habitats. Dry places.

Holarctic. Boreal and Tethyan. Euro-Siberian. Mediterranean. European.

References, etc. Leaf anatomical: this project.

Microstegium Nees

Coelarthron Hook.f., *Ephebopogon* Steud., *Leptatherum* Nees, *Nemastachys* Steud., *Psilopogon* Hochst.

Excluding *Ischnochloa*

Habit, vegetative morphology. Creeping or rambling annual, or perennial; decumbent. **Culms** 30–60 cm high; *herbaceous*; unbranched above. *Leaves* not basally aggregated; non-auriculate. Leaf blades linear to lanceolate; broad, or narrow; flat; pseudopetiolate to not pseudopetiolate; persistent. *Ligule an unfringed membrane*.

Reproductive organization. Plants bisexual, with bisexual spikelets; with hermaphrodite florets. *The spikelets* usually *all alike in sexuality (rarely, the pedicellate spikelet male)*; overtly heteromorphic to homomorphic (the pedicelled member with a less concave lower glume, and sometimes slightly smaller). Exposed-cleistogamous, or chasmogamous.

Inflorescence. *Inflorescence of spicate main branches (flexuous, fragile racemes, these not villous)*; digitate, or subdigitate (to scattered on a short axis), or non-digitate (sometimes solitary). Primary inflorescence branches 1–25. Inflorescence espatheate; not comprising 'partial inflorescences' and foliar organs. Spikelet-bearing axes with very slender rachides; disarticulating; disarticulating at the joints. '*Articles*' linear, or non-linear (then clavate or pyriform); not appendaged; disarticulating transversely; densely long-hairy (rarely, villous), or somewhat hairy, or glabrous. Spikelets paired; sessile and pedicellate; consistently in 'long-and-short' combinations; in pedicellate/sessile combinations. Pedicels of the 'pedicellate' spikelets free of the rachis. The 'shorter' spikelets hermaphrodite. The 'longer' spikelets hermaphrodite.

Female-fertile spikelets. *Spikelets compressed dorsiventrally*; falling with the glumes (the pedicelled spikelet falling from its pedicel, the sessile falling with the adjacent internode and pedicel). Rachilla terminated by a female-fertile floret. Hairy callus present. *Callus* short.

Glumes two; more or less equal; long relative to the adjacent lemmas; awned (the upper, sometimes), or awnless; very dissimilar (the lower bicarinate and chanelled, the upper laterally compressed, naviculate). *Lower glume* two-keeled (via sharply inflexed margins, the keels not winged); *sulcate on the back (with a deep groove or broadly concave median channel)*; not pitted; relatively smooth; 4–6 nerved. Upper glume 3 nerved. *Spikelets* with incomplete florets (though these are often more or less suppressed in some spikelets). The incomplete florets (when present) proximal to the female-fertile florets. *The proximal incomplete florets* 1; paleate, or epaleate. Palea of the proximal incomplete florets reduced. The proximal incomplete florets sterile. The proximal lemmas exceeded by the female-fertile lemmas to decidedly exceeding the female-fertile lemmas; similar in texture to the female-fertile lemmas (hyaline); not becoming indurated.

Female-fertile florets 1. **Lemmas** less firm than the glumes (hyaline or membranous); not becoming indurated; *incised*; 2 lobed (or 2-toothed); deeply cleft to not deeply cleft (usually bidentate to bifid, often minute); awned (nearly always), or awnless (rarely). Awns 1; median; usually from a sinus; geniculate; hairless (glabrous); much longer than the body of the lemma. Lemmas hairless; non-carinate. **Palea** present, or absent; relatively long, or conspicuous but relatively short to very reduced (always small, but sometimes exceeding the body of the L_2); not indurated (hyaline); *nerveless*. *Lodicules* present; 2; free; fleshy; glabrous. *Stamens* (2–)3. Anthers not penicillate. *Ovary* glabrous. Styles free to their bases. Stigmas 2.

Fruit, embryo and seedling. *Fruit* free from both lemma and palea. Hilum short. Embryo large. Endosperm hard; containing only simple starch grains. Embryo with an elongated mesocotyl internode.

Seedling with a long mesocotyl.

Abaxial leaf blade epidermis. *Costal/intercostal zonation* conspicuous. *Papillae* present. Intercostal papillae not over-arching the stomata; several per cell (at least in *M. spectabile*, where the papillae are mostly confined to costal long-cells and those adjoining the costal zones). Mid-intercostal long-cells rectangular; having markedly sinuous walls. *Microhairs* present; panicoid-type; 42–54 microns long; 6.6–7.5 microns wide at the septum. Microhair total length/width at septum 6.4–8.1. Microhair apical cells (18–)20–26(–32) microns long. Microhair apical cell/total length ratio 0.42–0.53. *Stomata* common; 28.5–34.5 microns long. Subsidiaries triangular (mostly), or dome-shaped (some,low). Guard-cells overlapping to flush with the interstomatals. *Intercostal* short-cells absent or very rare. Prickle bases abundant. *Costal short-cells* conspicuously in long rows. Costal silica bodies 'panicoid-type'; cross shaped, or butterfly shaped, or dumb-bell shaped, or nodular; not sharp-pointed.

Transverse section of leaf blade, physiology. C_4; biochemical type NADP–ME (1 species); XyMS–. PCR cell chloroplasts centrifugal/peripheral. *Mesophyll* with radiate chlorenchyma. *Leaf blade* adaxially flat. *Midrib* conspicuous; having a conventional arc of bundles; with colourless mesophyll adaxially, or without colourless mesophyll adaxially. Bulliforms not in discrete, regular adaxial groups (most of the epidermis being irregularly bulliform, apart from the midrib 'hinge groups'). Many of the smallest vascular bundles unaccompanied by sclerenchyma. Combined sclerenchyma girders present.

Cytology. Chromosome base number, $x = 10$. $2n = 20$ and 40.

Taxonomy. Panicoideae; Andropogonodae; Andropogoneae; Andropogoninae.

Ecology, geography, regional floristic distribution. About 15 species. Tropical and subtropical Africa and Asia. Commonly adventive. Mesophytic; shade species; glycophytic.

Holarctic, Paleotropical, Cape, and Australian. Boreal and Tethyan. African, Indomalesian, Polynesian, and Neocaledonian. Eastern Asian. Irano-Turanian. Sudano-Angolan and West African Rainforest. Indian, Indo-Chinese, Malesian, and Papuan. Fijian. North and East Australian. Sahelo-Sudanian, Somalo-Ethiopian, and South Tropical African. Tropical North and East Australian.

Rusts and smuts. Rusts — *Phakopsora* and *Puccinia*. Taxonomically wide-ranging species: *Phakopsora incompleta*. Smuts from Ustilaginaceae. Ustilaginaceae — *Sphacelotheca*.

Economic importance. Significant weed species: *M. vimineum*.

References, etc. Leaf anatomical: Metcalfe 1960; this project.

Mildbraediochloa Butzin

Habit, vegetative morphology. *Perennial. Culms herbaceous.* Leaf blades without cross venation.

Reproductive organization. *Plants bisexual, with bisexual spikelets*; with hermaphrodite florets.

Inflorescence. *Inflorescence paniculate*; contracted ('panicle dense'); espatheate; not comprising 'partial inflorescences' and foliar organs. Spikelet-bearing axes persistent. Spikelets pedicellate.

Female-fertile spikelets. *Spikelets compressed laterally.* Rachilla terminated by a female-fertile floret.

Glumes two; *very unequal*; long relative to the adjacent lemmas; the upper *awned*; very dissimilar (G_1 short and awnless, G_2 bilobed with the median nerve excurrent into an awn). Upper glume obscurely 5 nerved. *Spikelets* with incomplete florets. The incomplete florets proximal to the female-fertile florets. *Spikelets with proximal incomplete florets. The proximal incomplete florets* 1; paleate; male. *The proximal lemmas awned (bilobed, the median nerve excurrent between).*

Female-fertile florets 1. *Lemmas* less firm than the glumes (thinly membranous, by contrast with the firmly membranous G_2); not becoming indurated; incised; 2 lobed; *awned*. Awns 1; median; from a sinus. *Palea* present. Stigmas 2.

Abaxial leaf blade epidermis. *Costal/intercostal zonation* conspicuous. *Papillae* absent. *Long-cells* more or less markedly different in shape costally and intercostally (the costals much narrower); of similar wall thickness costally and intercostally (the walls of medium thickness). Intercostal zones exhibiting many atypical long-cells (mainly in the mid-intercostal regions). Mid-intercostal long-cells rectangular; having markedly sinuous walls (the sinuosity fairly coarse to fine, irregular). *Microhairs* present; elongated; clearly two-celled; panicoid-type. *Stomata* common. Subsidiaries non-papillate; low dome-shaped to triangular. Guard-cells overlapping to flush with the interstomatals. *Intercostal short-cells* fairly common (but not to be confused with small prickles); not paired (seemingly solitary); not obviously silicified. With numerous very small intercostal prickles. *Costal short-cells* conspicuously in long rows (but the 'short-cells' often quite long). Costal silica bodies present and well developed; 'panicoid-type'; cross shaped and dumb-bell shaped; sharp-pointed and not sharp-pointed.

Transverse section of leaf blade, physiology. C_4. The anatomical organization conventional. XyMS+. PCR sheath outlines uneven. PCR sheath extensions absent. *Mesophyll* with radiate chlorenchyma; traversed by columns of colourless mesophyll cells (these columns infrequent, of small cells, one cell wide). *Leaf blade* adaxially flat. *Midrib* not readily distinguishable; with one bundle only. *The lamina* symmetrical on either side of the midrib. Bulliforms present in discrete, regular adaxial groups; in simple fans and associated with colourless mesophyll cells to form deeply-penetrating fans (the fans very large, usually simple but occasionally with one or two small clourless cells contiguous internally, rarely these forming a linking column). Many of the smallest vascular bundles unaccompanied by sclerenchyma. Combined sclerenchyma girders present (with a few of the primaries); nowhere forming 'figures' (the sclerenchyma scanty). Sclerenchyma all associated with vascular bundles.

Taxonomy. Panicoideae; Panicodae; Paniceae.

Ecology, geography, regional floristic distribution. 1 species. Tropical Africa: Annobon Is. On cliffs.

Paleotropical. African. West African Rainforest.

References, etc. Morphological/taxonomic: Butzin 1971. Leaf anatomical: this project.

Special comments. Fruit data wanting.

Milium L.

Milearium Moench

Habit, vegetative morphology. Annual, or perennial; stoloniferous, or caespitose. *Culms* 10–180 cm high; herbaceous; unbranched above. Culm nodes glabrous. Culm internodes hollow. Young shoots extravaginal. Leaves non-auriculate. *Leaf blades* linear to linear-lanceolate; broad, or narrow; 2–18 mm wide; flat; *not pseudopetiolate*; without cross venation; persistent; rolled in bud. *Ligule an unfringed membrane*; not truncate; 4–5 mm long.

Reproductive organization. Plants bisexual, with bisexual spikelets; with hermaphrodite florets.

Inflorescence. Inflorescence paniculate; open; with capillary branchlets; espatheate; not comprising 'partial inflorescences' and foliar organs. Spikelet-bearing axes persistent. Spikelets not secund; pedicellate.

Female-fertile spikelets. *Spikelets* 2–4 mm long; somewhat *compressed dorsiventrally*; disarticulating above the glumes. Rachilla terminated by a female-fertile floret. *Hairy callus absent. Callus* short; blunt.

Glumes two; more or less equal; about equalling the spikelets to exceeding the spikelets; long relative to the adjacent lemmas; free; pointed (acute); awnless; non-carinate; similar (membranous). Lower glume 3 nerved. Upper glume 3 nerved. **Spikelets** with female-fertile florets only; *without proximal incomplete florets*.

Female-fertile florets 1. Lemmas rounded; not convolute; decidedly firmer than the glumes (leathery); becoming indurated (shiny in fruit); entire; blunt; awnless; hairless; non-carinate; 5 nerved. **Palea** present; relatively long; entire; textured like the lemma (leathery); indurated; 2-nerved; *keel-less. Lodicules* present; 2; free; membranous; glabrous; toothed, or not toothed; not or scarcely vascularized. *Stamens* 3. Anthers 1.3–3 mm long; not penicillate. *Ovary* glabrous. Styles free to their bases. Stigmas 2; white.

Fruit, embryo and seedling. Fruit small (about 2 mm long); compressed dorsiventrally, or not noticeably compressed. *Hilum short (1/5 to almost 1/2 grain length).* Embryo small; not waisted. Endosperm hard; with lipid; containing compound starch grains. Embryo with an epiblast; without a scutellar tail; with a negligible mesocotyl internode. Embryonic leaf margins meeting.

Seedling with a long mesocotyl; with a loose coleoptile. First seedling leaf with a well-developed lamina. The lamina broad; erect; 3–5 veined.

Abaxial leaf blade epidermis. *Costal/intercostal zonation* conspicuous. *Papillae* absent. *Long-cells* markedly different in shape costally and intercostally (the costals narrower, rectangular); of similar wall thickness costally and intercostally (thin walled). Mid-intercostal long-cells fusiform; having straight or only gently undulating walls. *Microhairs* absent. *Stomata* common. Subsidiaries parallel-sided. Guard-cells overlapped by the interstomatals. *Intercostal short-cells* absent or very rare. *Costal short-cells* neither distinctly grouped into long rows nor predominantly paired. Costal silica bodies absent (in the material seen).

Transverse section of leaf blade, physiology. C_3; XyMS+. *Mesophyll* with non-radiate chlorenchyma; without adaxial palisade. *Leaf blade* adaxially flat. *Midrib* conspicuous, or not readily distinguishable; with one bundle only; with colourless mesophyll adaxially, or without colourless mesophyll adaxially. Bulliforms present in discrete, regular adaxial groups (between the vascular bundles). All the vascular bundles accompanied by sclerenchyma. Combined sclerenchyma girders present; forming 'figures'.

Cytology. Chromosome base number, $x = 4$, 5, 7, and 9. $2n = 8$, 10, 28, and 42. 2, 4, and 6 ploid.

Taxonomy. Pooideae; Poodae; Aveneae.

Ecology, geography, regional floristic distribution. 3–4 species. North temperate. Commonly adventive. Mesophytic and xerophytic; shade species and species of open habitats.

Holarctic. Boreal and Tethyan. Euro-Siberian, Eastern Asian, and Atlantic North American. Mediterranean and Irano-Turanian. European and Siberian. Canadian-Appalachian.

Rusts and smuts. Rusts — *Puccinia*. Taxonomically wide-ranging species: *Puccinia graminis*, *Puccinia coronata*, *Puccinia striiformis*, *Puccinia brachypodii*, *Puccinia recondita*, and '*Uromyces*' *dactylidis*. Smuts from Tilletiaceae and from Ustilaginaceae. Tilletiaceae — *Urocystis*. Ustilaginaceae — *Ustilago*.

References, etc. Morphological/taxonomic: Macfarlane and Watson 1980. Leaf anatomical: this project.

Miscanthidium Stapf

~ *Miscanthus*

Habit, vegetative morphology. Erect perennial; sometimes rhizomatous. *Culms* 100–400 cm high; herbaceous; unbranched above. Leaf blades linear; broad to narrow; flat, or acicular; pseudopetiolate (attenuate to the sheath), or not pseudopetiolate; without cross venation. *Ligule an unfringed membrane.*

Reproductive organization. Plants bisexual, with bisexual spikelets; with hermaphrodite florets. The spikelets homomorphic.

Inflorescence. *Inflorescence of spicate main branches, or paniculate (the panicle often large, branched, silky, red or brown, the central axis longer and the racemes shorter than in* Miscanthus *sensu stricto)*; open, or contracted; espatheate; not comprising 'partial inflorescences' and foliar organs. **Spikelet-bearing axes** 'racemes' (slender, flexuous); with very slender rachides; ultimately *disarticulating*; tardily disarticulating at the joints. '*Articles*' linear; not appendaged; disarticulating transversely. *Spikelets* paired; pedicellate; consistently in 'long-and-short' combinations; *unequally pedicellate in each combination*. Pedicels of the 'pedicellate' spikelets free of the rachis. The 'shorter' spikelets hermaphrodite. The 'longer' spikelets hermaphrodite.

Female-fertile spikelets. Spikelets compressed dorsiventrally; falling with the glumes (disarticulating from the pedicels before break-up of the rachis). Rachilla terminated by a female-fertile floret. Hairy callus present (spikelets with an involucre of very long hairs at the base). Callus blunt.

Glumes two; more or less equal; long relative to the adjacent lemmas; awnless; carinate (G_2), or non-carinate (G_1); very dissimilar (papery to leathery, the G_1 flat-backed, 2-keeled with inflexed margins and nerves between the keels, the G_2 naviculate). *Lower glume two-keeled*; flattened on the back; not pitted; relatively smooth; vein number variable. Upper glume 1 nerved, or 3 nerved. *Spikelets* with incomplete florets. The incomplete florets proximal to the female-fertile florets. *The proximal incomplete florets* 1; epaleate; sterile. The proximal lemmas awnless (muticous or mucronate); similar in texture to the female-fertile lemmas (hyaline); not becoming indurated.

Female-fertile florets 1. Lemmas less firm than the glumes (hyaline); not becoming indurated; entire (seemingly); not deeply cleft; awned. Awns 1; median; apical; geniculate (twisted, slightly bent); about as long as the body of the lemma to much longer than the body of the lemma. **Lemmas hairy (marginally)**; non-carinate; without a germination flap. **Palea** present; conspicuous but relatively short; entire; awnless, without apical setae (ciliate); not indurated (hyaline); **nerveless**; keel-less. *Lodicules* present; 2; free; fleshy. *Stamens* 3. Stigmas 2.

Fruit, embryo and seedling. Hilum short. Embryo large; without a scutellar tail.

Abaxial leaf blade epidermis. *Costal/intercostal zonation* conspicuous. *Papillae* present (in *M. sorghum*), or absent; intercostal. Intercostal papillae over-arching the stomata; consisting of one oblique swelling per cell (elongated, almost clavate, thick walled). Mid-intercostal long-cells having straight or only gently undulating walls (in *M. sorghum*), or having markedly sinuous walls. *Microhairs* present (though none seen in *M. sorghum*, where the intercostal zones are extensively obscured by prickles and papillae); panicoid-type; (34–)36–42(–44) microns long. Microhair apical cells 13–22 microns long. Microhair apical cell/total length ratio 0.49. *Stomata* common. Subsidiaries non-papillate; low dome-shaped, or triangular (mostly domes in *M. sorghum*). *Intercostal short-cells* absent or very rare. *Costal short-cells* predominantly paired and neither distinctly grouped into long rows nor predominantly paired (mostly solitary and paired, but the costal 'long-cells' sometimes rather short, giving the appearance of long rows). Costal silica bodies 'panicoid-type'; cross shaped to dumb-bell shaped, or nodular (a few); not sharp-pointed.

Transverse section of leaf blade, physiology. Leaf blades consisting of midrib (in *M. teretifolium*, where the centre consists wholly of colourless tissue, with the chlorenchyma and vb's confined to the peripheral zone), or 'laminar'. C_4; XyMS–. PCR sheath outlines uneven. PCR cell chloroplasts centrifugal/peripheral. *Leaf blade of M. sorghum* 'nodular' in section; with the ribs very irregular in sizes (in *M. sorghum*, especially near the midrib). *Midrib* conspicuous (or the blade reduced to the midrib); in *M. sorghum* having a conventional arc of bundles; with colourless mesophyll adaxially (at least in the lower part of the blade). Bulliforms present in discrete, regular adaxial groups (these large, in each furrow), or not in discrete, regular adaxial groups (*M. teretifolium*); in simple fans and associated with colourless mesophyll cells to form deeply-penetrating fans (the fans in *M. sorghum*, more or less deformed); associating with colourless mesophyll cells to form arches over small vascular bundles. All the vascular bundles accompanied by sclerenchyma. Combined sclerenchyma girders present, or absent (abaxial girders only in *M. teretifolium*); in *M. sorghum* forming 'figures' (the large bundles with massive T's or I's). Sclerenchyma all associated with vascular bundles.

Cytology. Chromosome base number, $x = 15$. $2n = 28$ and 30. 2 ploid.

Taxonomy. Panicoideae; Andropogonodae; Andropogoneae; Andropogoninae.

Ecology, geography, regional floristic distribution. 6–7 species. Tropical and southern Africa. Helophytic; shade species, or species of open habitats; glycophytic. Streamsides and forest margins.

Paleotropical. African. Saharo-Sindian and Sudano-Angolan.

Hybrids. Intergeneric hybrids procured with *Saccharum*.

References, etc. Leaf anatomical: Metcalfe 1960 and this project.

Miscanthus Anderss.

Triarrhena (Maxim.) Nakai, *Xiphagrostis* Cov.

Excluding *Miscanthidium, Sclerostachya*

Habit, vegetative morphology. Perennial; rhizomatous and caespitose, or caespitose (usually tall, cane-like or reed-like). **Culms 200–350 cm high**. Culm nodes hairy, or glabrous. Culm internodes solid. *Leaves* not basally aggregated; non-auriculate. Leaf blades broad, or narrow; flat; without cross venation; persistent; rolled in bud. *Ligule a fringed membrane.*

Reproductive organization. Plants bisexual, with bisexual spikelets; with hermaphrodite florets. The spikelets all alike in sexuality; homomorphic.

Inflorescence. Inflorescence of spicate main branches and paniculate; open (the panicles large, fan-shaped or corymbiform, plumose); espatheate; not comprising 'partial inflorescences' and foliar organs. **Spikelet-bearing axes** 'racemes'; *persistent*. *Spikelets* paired; not secund; pedicellate; *consistently in 'long-and-short' combinations*; unequally pedicellate in each combination. The 'shorter' spikelets hermaphrodite. The 'longer' spikelets hermaphrodite.

Female-fertile spikelets. *Spikelets compressed dorsiventrally*; falling with the glumes (falling from their pedicels). Rachilla terminated by a female-fertile floret. Hairy callus present (with long fine hairs). *Callus* short; blunt.

Glumes two (papery to membranous); more or less equal; long relative to the adjacent lemmas; awnless. *Lower glume 3–4 nerved.* Upper glume 1–5 nerved. *Spikelets* with incomplete florets. The incomplete florets proximal to the female-fertile florets. *The proximal incomplete florets* 1; sterile. The proximal lemmas awnless; decidedly exceeding the female-fertile lemmas; not becoming indurated (hyaline).

Female-fertile florets 1. Lemmas usually becoming stiptate beneath an awn, but rarely reduced, hyaline and awnless; less firm than the glumes; not becoming indurated; entire, or incised; when incised, 2 lobed; not deeply cleft (bidentate); awnless (rarely), or awned. Awns 1; median; usually from a sinus; geniculate; hairless (glabrous); much shorter than the body of the lemma to much longer than the body of the lemma. Lemmas hairless; non-carinate; without a germination flap; 0–3 nerved. **Palea** present; conspicuous but relatively short; awnless, without apical setae; not indurated; *nerveless. Lodicules* present; 2; free; fleshy; glabrous. *Stamens* 2–3. Anthers not penicillate. *Ovary* glabrous. Styles free to their bases. Stigmas 2; red pigmented.

Fruit, embryo and seedling. Hilum short. Embryo large. Endosperm containing compound starch grains. Embryo without an epiblast; with a scutellar tail; with an elongated mesocotyl internode. Embryonic leaf margins overlapping.

Seedling with a long mesocotyl.

Abaxial leaf blade epidermis. *Costal/intercostal zonation* conspicuous. *Papillae* present. Intercostal papillae not over-arching the stomata; consisting of one oblique swelling per cell, or several per cell (finger-like). *Long-cells* similar in shape costally and intercostally; of similar wall thickness costally and intercostally. Mid-intercostal long-cells rectangular; having markedly sinuous walls. *Microhairs* present; panicoid-type; (48–)52–72(–77) microns long; 6.3–7.5 microns wide at the septum. Microhair total length/width at septum 8.2–9.4. Microhair apical cells 18–39 microns long. Microhair apical cell/total length ratio 0.42–0.53. *Stomata* common; 22–29 microns long. Subsidiaries low dome-shaped to triangular. Guard-cells overlapping to flush with the interstomatals. *Intercostal short-cells* absent or very rare; in cork/silica-cell pairs, or not paired (solitary); silicified, or not silicified. Intercostal silica bodies when present, tall-and-narrow. *Costal short-cells* conspicuously in long rows. Costal silica bodies 'panicoid-type'; cross shaped to dumb-bell shaped, or nodular; not sharp-pointed.

Transverse section of leaf blade, physiology. C_4; XyMS–. PCR cell chloroplasts with reduced grana. *Mesophyll* with radiate chlorenchyma; traversed by columns of colourless mesophyll cells. *Leaf blade* with distinct, prominent adaxial ribs (rarely), or adaxially flat; with the ribs more or less constant in size. *Midrib* conspicuous; having a conventional arc of bundles; with colourless mesophyll adaxially. Bulliforms present in discrete, regular adaxial groups; in simple fans and associated with colourless mesophyll cells to form deeply-penetrating fans; associating with colourless mesophyll cells to form arches over small vascular bundles. All the vascular bundles accompanied by sclerenchyma. Combined sclerenchyma girders present; nowhere forming 'figures'. Sclerenchyma all associated with vascular bundles.

Phytochemistry. Leaves without flavonoid sulphates (3 species).

Cytology. Chromosome base number, $x = 19$. $2n = 35$–43, or 57, 76, 95, and 114.

Taxonomy. Panicoideae; Andropogonodae; Andropogoneae; Andropogoninae.

Ecology, geography, regional floristic distribution. 20 species. Japan & Philippines. Commonly adventive. Helophytic, or mesophytic; species of open habitats. Hillsides and marshes.

Holarctic and Paleotropical. Boreal. Indomalesian, Polynesian, and Neocaledonian. Euro-Siberian, Eastern Asian, and Atlantic North American. Indian, Indo-Chinese, Malesian, and Papuan. Polynesian and Fijian. Siberian. Canadian-Appalachian.

Hybrids. Intergeneric hybrids procured with *Saccharum*.

Rusts and smuts. Rusts — *Puccinia*. Taxonomically wide-ranging species: *Puccinia miscanthae*. Smuts from Ustilaginaceae. Ustilaginaceae — *Sphacelotheca* and *Ustilago*.

Economic importance. Significant weed species: *M. floridulus*. Some (e.g. *M. sinensis*) cultivated a ornamentals; *M. sinensis* is the subject of extensive field trials as fuel and raw material for making paper and chipboard.

References, etc. Leaf anatomical: Metcalfe 1960; this project.

Mnesithea Kunth

Dipterum Desv., *Thyridostachyum* Nees
Excluding *Hackelochloa, Heteropholis, Coelorachis*

Habit, vegetative morphology. Perennial; caespitose. *Culms* 50–170 cm high; herbaceous. Culm internodes solid. Leaves non-auriculate. Leaf blades broad, or narrow; 5–25 mm wide; flat, or rolled (finally involute); without cross venation. **Ligule** present; *an unfringed membrane*; truncate; short.

Reproductive organization. *Plants bisexual, with bisexual spikelets*; with hermaphrodite florets. The spikelets of sexually distinct forms on the same plant; hermaphrodite and sterile; overtly heteromorphic (the pedicelled members reduced to the two glumes, or vestigial); all in heterogamous combinations.

Inflorescence. Inflorescence of subterete, spike-like 'racemes'; non-digitate. Rachides hollowed. *Inflorescence* spatheate; *a complex of 'partial inflorescences' and intervening foliar organs*. **Spikelet-bearing axes** *spikelike*; solitary; with substantial rachides; *disarticulating*; disarticulating at the joints. '*Articles*' non-linear (shortly clavate, concave, with embedded sessile spikelets and cupuliform apices); with a basal callus-knob; not appendaged; disarticulating transversely (the cupuliform apex fitting the basal knob of the next higher joint); glabrous (sometimes striate). *Spikelets in triplets (2 sessile, 1 pedicelled in the lower part of the spikelike 'raceme'), or paired (towards the tips, or throughout depauperate racemes)*; secund (the sessile members in two alternating rows, on one side of the dorsiventral rachis); sessile and pedicellate (the pedicels slender). Pedicel apices cupuliform. Spikelets consistently in 'long-and-short' combinations; in pedicellate/sessile combinations. Pedicels of the 'pedicellate' spikelets discernible, but fused with the rachis (in association with spikelet pairs, towards the inflorescence tips), or free of the rachis (with the triplets). The 'shorter' spikelets hermaphrodite. The 'longer' spikelets sterile (reduced to two minute glumes, or rudimentary).

Female-sterile spikelets. The pedicelled spikelet very greatly reduced or vestigial.

Female-fertile spikelets. Spikelets 3.5–4.5 mm long; abaxial (the G_2 embedded in rachis); compressed dorsiventrally; falling with the glumes. Rachilla terminated by a female-fertile floret. Hairy callus absent.

Glumes present; two; more or less equal; long relative to the adjacent lemmas; free; dorsiventral to the rachis; hairy (e.g., *M. mollicoma*, with tubercle-based short white hairs), or hairless; without conspicuous tufts or rows of hairs; awnless; non-carinate; very dissimilar (G_1 obliquely ovate with narrowly incurved margins, indurated, the G_2 convex and much thinner). *Lower glume not two-keeled*; convex on the back, or flattened on the back; not pitted; *relatively smooth, or tuberculate*. Upper glume 3 nerved. *Spikelets* with incomplete florets. The incomplete florets proximal to the female-fertile florets. *The proximal incomplete florets* 1; epaleate; sterile. The proximal lemmas awnless; 2 nerved; more or less equalling the female-fertile lemmas to decidedly exceeding the female-fertile lemmas; similar in texture to the female-fertile lemmas (hyaline); not becoming indurated.

Female-fertile florets 1. Lemmas oblong; less firm than the glumes (hyaline); not becoming indurated; entire; pointed, or blunt; awnless; hairless; non-carinate; without a germination flap; 3 nerved. *Palea* present; awnless, without apical setae; not indurated (hyaline); 2-nerved, or nerveless; keel-less. *Lodicules* present; free; fleshy; glabrous. *Stamens* 3. *Ovary* glabrous. Styles free to their bases. Stigmas 2; red pigmented.

Fruit, embryo and seedling. Fruit compressed dorsiventrally.

Abaxial leaf blade epidermis. *Costal/intercostal zonation* conspicuous. *Papillae* absent. Long-cells of similar wall thickness costally and intercostally. Intercostal zones exhibiting many atypical long-cells (many very short ones). Mid-intercostal long-cells rectangular (in *M. laevis*, many long-cells are almost square); having markedly sinuous walls (exaggeratedly so in *M. laevis*). *Microhairs* present; panicoid-type (but relatively broad in *laevis*); 45–54 microns long (*M. laevis*), or 69–84 microns long (*M. mollicoma*); 8.4–11.4 microns wide at the septum (*M. laevis*), or 5.4–6.6 microns wide at the septum (*M. mollicoma*). Microhair total length/width at septum 4.2–5.4 (*M. laevis*), or 10.4–15.6 (*M. mollicoma*). Microhair apical cells 25.5–39 microns long (*M. laevis*), or 39–51 microns long (*M. mollicoma*). Microhair apical cell/total length ratio 0.57–0.62 (*M. laevis*), or 0.5–0.63 (*M. mollicoma*). *Stomata* common; 54–63 microns long (*M. laevis*), or 30–33 microns long (*M. mollicoma*). Subsidiaries triangular. Guard-cells overlapping to flush with the interstomatals. *Intercostal short-cells* common; in cork/silica-cell pairs; silicified (*M. laevis*), or not silicified. Intercostal silica bodies when present, oryzoid-type. Cushion-based macrohairs present in *M. mollicoma*. *Costal short-cells* predominantly paired. Costal silica bodies oryzoid (all, in *M.laevis*), or 'panicoid-type' (*M. mollicoma*); not sharp-pointed.

Transverse section of leaf blade, physiology. C_4; XyMS– (but occasional cells in ms position in *M. laevis*). *Leaf blade* 'nodular' in section. *Midrib* conspicuous (a smaller primary bundle in *M. laevis* with the closely opposed hinge-cells forming a single bulliform group above it, conventional in *M. mollicoma*); having a conventional arc of bundles (a median keel bundle and several small laterals); with colourless mesophyll adaxially (over the midrib in *M. mollicoma*, over the primary bundles except the midrib in *M. laevis*). Bulliforms not in discrete, regular adaxial groups (the epidermis of large cells); associating with colourless mesophyll cells to form arches over small vascular bundles. Many of the smallest vascular bundles unaccompanied by sclerenchyma. Combined sclerenchyma girders present (with the primaries); nowhere forming 'figures'. Sclerenchyma all associated with vascular bundles.

Cytology. Chromosome base number, $x = 9$. $2n = 18$. 2 ploid.

Taxonomy. Panicoideae; Andropogonodae; Andropogoneae; Rottboelliinae.

Ecology, geography, regional floristic distribution. 5 species. India, Ceylon to Southeast Asia. Mesophytic. Damp grassland.

Paleotropical. Indomalesian. Indian, Indo-Chinese, and Malesian.

Rusts and smuts. Smuts from Ustilaginaceae. Ustilaginaceae — *Sorosporium* and *Sphacelotheca*.

References, etc. Leaf anatomical: this project.

Special comments. Fruit data wanting.

Mniochloa Chase

Habit, vegetative morphology. Slender perennial (with dimorphic culms); caespitose. The flowering culms leafless (with 1–3 bladeless sheaths). **Culms** 10–25 cm high (the vegetative culms 3–12 cm tall, the flowering culms longer); herbaceous; unbranched above; *tuberous*. Rhizomes pachymorph. Plants unarmed. Young shoots intravaginal. *Leaves* not basally aggregated; without auricular setae. *Leaf blades elliptic*; narrow (but relatively broad); 3–5 mm wide (and 8–18 mm long); not cordate, not sagittate (base cuneate, no more than slightly asymmetric); flat; shortly pseudopetiolate; persistent; rolled in bud. Ligule an unfringed membrane, or a fringed membrane (?— 'a hyaline, lacerate, ciliate membrane'); truncate; about 0.1 mm long.

Reproductive organization. *Plants monoecious with all the fertile spikelets unisexual*; without hermaphrodite florets. The spikelets of sexually distinct forms on the same plant; male-only and sterile. The male and female-fertile spikelets on different branches of the same inflorescence. The spikelets overtly heteromorphic.

Inflorescence. *Inflorescence of spicate main branches (of two spike-like racemes, the one male, the somewhat longer one female)*; digitate. Primary inflorescence branches 2. Inflorescence espatheate; not comprising 'partial inflorescences' and foliar organs. *Spikelet-bearing axes* 'racemes'; persistent. Spikelets solitary; secund (on one side of the slender, triquetrous rachis); pedicellate (with very short, clavate pedicels); not in distinct 'long-and-short' combinations.

Female-sterile spikelets. The male racemes slightly shorter; male spikelets 1.3—1.7 mm long, much shorter than females, ellipsoid, acute; lemma 1–nerved, palea 2–nerved, lemma and palea membranous; 3 free, non-penicillate stamens. Rachilla of male spikelets terminated by a male floret. The male spikelets without glumes; without proximal incomplete florets; 1 floreted. The lemmas awnless. Male florets 1; 3 staminate. The staminal filaments free.

Female-fertile spikelets. Spikelets 2–4.5 mm long; narrowly elliptic; compressed dorsiventrally; with conventional internode spacings (i.e. no stipe above the glumes). Rachilla terminated by a female-fertile floret. Hairy callus absent.

Glumes two; relatively large; very unequal to more or less equal (the lower somewhat shorter); shorter than the spikelets; shorter than the adjacent lemmas; hairless; pointed to not pointed; awnless; non-carinate; similar (delicately membranous, ovate-elliptical). *Spikelets* with female-fertile florets only.

Female-fertile florets 1. Lemmas similar in texture to the glumes, or decidedly firmer than the glumes (membranous); becoming indurated (white-cartilaginous); entire; pointed, or blunt; awnless (and not apiculate); hairless; glabrous; non-carinate; having the margins tucked in onto the palea. *Palea* present; relatively long; entire; awnless, without apical setae (glabrous); textured like the lemma (which enfolds it); indurated. *Lodicules* present; 3. *Stamens* 0 (and no vestiges).

Fruit, embryo and seedling. Disseminule a free caryopsis. *Fruit* free from both lemma and palea; small (1.5–2 mm long); fusiform; longitudinally grooved; compressed dorsiventrally. Hilum long-linear. Embryo small.

Abaxial leaf blade epidermis. *Costal/intercostal zonation* conspicuous. *Papillae* present; costal and intercostal. Intercostal papillae over-arching the stomata, or not over-arching the stomata; several per cell (small, numerous, and often forming rings around the stomata). *Long-cells* markedly different in shape costally and intercostally (the costals much narrower); of similar wall thickness costally and intercostally (thin walled). Mid-intercostal long-cells rectangular; having markedly sinuous walls. *Microhairs* present; elongated; clearly two-celled; panicoid-type; 36–46 microns long; 6.3–7.5 microns wide at the septum. Microhair total length/width at septum 9–10.2. Microhair apical cells 8–15 microns long. Microhair apical cell/total length ratio 0.2–0.4. *Stomata* common; 13–17 microns long. Subsidiaries non-papillate; parallel-sided. Guard-cells overlapped by the interstomatals (slightly). *Intercostal short-cells* absent or very rare. No macrohairs, but hooks and prickles common. *Costal short-cells* conspicuously in long rows. Costal silica bodies 'panicoid-type'; predominantly short dumb-bell shaped.

Transverse section of leaf blade, physiology. C_3; XyMS+. Mesophyll with adaxial palisade; with arm cells; without fusoids. *Leaf blade* adaxially flat. *Midrib* conspicuous (by virtue of its large bundle and relatively massive 'anchor' of sclerenchyma); with one bundle only. *The lamina* symmetrical on either side of the midrib. Bulliforms present in discrete, regular adaxial groups; in simple fans. All the vascular bundles accompanied by sclerenchyma. Combined sclerenchyma girders present; forming 'figures' (most bundles with an 'anchor'). Sclerenchyma all associated with vascular bundles.

Taxonomy. Bambusoideae; Oryzodae; Olyreae.

Ecology, geography, regional floristic distribution. 1 species. Cuba. Mesophytic; shade species. On lowland limestone cliffs.

Neotropical. Caribbean.

References, etc. Morphological/taxonomic: Chase 1908; Zuloaga *et al.* 1993. Leaf anatomical: Zuloaga *et al.* 1993.

Special comments. Fruit data wanting.

Molinia Schrank

Amblytes Dulac., *Enodium* Gaud., *Moliniopsis* Hayata

Habit, vegetative morphology. Perennial; caespitose. *Culms* 15–250 cm high; herbaceous; tuberous to not tuberous (the lowest internodes short, swollen, persisting for several years). Culm nodes glabrous. Culm internodes hollow. *Leaves* mostly basal (the intermediate internodes condensed, the uppermost constituting most of the elevated culm); non-auriculate. *Leaf blades* narrow; 3–10 mm wide; flat, or rolled; without cross venation; *disarticulating from the sheaths*; rolled in bud. Ligule a fringe of hairs.

Reproductive organization. Plants bisexual, with bisexual spikelets; with hermaphrodite florets; outbreeding.

Inflorescence. *Inflorescence* paniculate; open, or contracted; **with capillary branchlets**; espatheate; not comprising 'partial inflorescences' and foliar organs. Spikelet-bearing axes persistent. Spikelets not secund; pedicellate.

Female-fertile spikelets. *Spikelets* 3.2–14 mm long; **compressed laterally**; disarticulating above the glumes; disarticulating between the florets; with distinctly elongated rachilla internodes between the florets. Rachilla prolonged beyond the uppermost female-fertile floret; rachilla hairless (rough). Hairy callus present, or absent (*M. coerulea*). *Callus* short; blunt.

Glumes two; very unequal to more or less equal; shorter than the spikelets; shorter than the adjacent lemmas; pointed; awnless; carinate; similar (membranous). Lower glume 1–3 nerved. Upper glume 1–5 nerved. *Spikelets* with female-fertile florets only, or with incomplete florets. The incomplete florets distal to the female-fertile florets. *The distal incomplete florets* merely underdeveloped.

Female-fertile florets (1–)2–5(–6). **Lemmas decidedly firmer than the glumes (firmly membranous)**; not becoming indurated; entire; pointed (acute); awnless, or mucronate; hairless (somewhat scabrous above); non-carinate (rounded on the back); 3(–5) nerved. *Palea* present; relatively long; entire to apically notched (truncate to emarginate); awnless, without apical setae; 2-nerved. *Lodicules* present; 2; joined (at the base), or free; fleshy; glabrous. *Stamens* 3. Anthers 1.5–3 mm long; not penicillate. *Ovary* glabrous. Styles free to their bases. Stigmas 2.

Fruit, embryo and seedling. *Fruit* free from both lemma and palea; small. *Hilum long-linear. Pericarp loosely adherent (reluctantly separable)*. Embryo large; not waisted. Endosperm hard; without lipid; containing compound starch grains. Embryo without an epiblast; with a scutellar tail; with an elongated mesocotyl internode. Embryonic leaf margins meeting.

Seedling with a loose coleoptile. First seedling leaf with a well-developed lamina. The lamina narrow; 5 veined.

Abaxial leaf blade epidermis. *Costal/intercostal zonation* conspicuous. *Papillae* absent. Mid-intercostal long-cells rectangular and fusiform; having markedly sinuous walls and having straight or only gently undulating walls. *Microhairs* present (in *Moliniopsis*), or absent; panicoid-type; in *Moliniopsis* (45–)48–54(–57) microns long; 6–8.4 microns wide at the septum. Microhair total length/width at septum 5.7–8.6. Microhair apical cells 24–27 microns long. Microhair apical cell/total length ratio 0.42–0.57. *Stomata* absent or very rare (*Moliniopsis*), or common; in *Molinia* s.str. (27–)30–34.5 microns long. Subsidiaries triangular (few), or dome-shaped (mostly, low). *Intercostal short-cells* common; in cork/silica-cell pairs; silicified. Intercostal silica bodies crescentic to tall-and-narrow, or vertically elongated-nodular. *Costal short-cells* conspicuously in long rows (but some of the 'short cells' rather long). Costal silica bodies 'panicoid-type'; cross shaped, or butterfly shaped, or dumb-bell shaped; not sharp-pointed.

Transverse section of leaf blade, physiology. C_3; XyMS+. *Mesophyll* with non-radiate chlorenchyma. *Leaf blade* with distinct, prominent adaxial ribs; with the ribs more or less constant in size. *Midrib* conspicuous; with one bundle only, or having a conventional arc of bundles. Bulliforms present in discrete, regular adaxial groups (at the bases of the shallow furrows); in simple fans. Combined sclerenchyma girders present; forming 'figures' (at least in the mid-rib). Sclerenchyma all associated with vascular bundles.

Phytochemistry. Leaves without flavonoid sulphates (*M. coerulea*).

Cytology. Chromosome base number, $x = 9$. $2n = 18$, 36, and 90. 2, 4, and 10 ploid.

Taxonomy. Arundinoideae; Danthonieae.

Ecology, geography, regional floristic distribution. 2–5 species. Temperate Eurasia. Helophytic (calcifuge); species of open habitats; glycophytic. Wet moorland, heaths.

Holarctic. Boreal and Tethyan. Euro-Siberian. Mediterranean and Irano-Turanian. European and Siberian.

Rusts and smuts. Rusts — *Puccinia*. Taxonomically wide-ranging species: *Puccinia graminis* and *Puccinia coronata*. Smuts from Tilletiaceae. Tilletiaceae — *Neovossia*.

References, etc. Morphological/taxonomic: Jirásek 1966. Leaf anatomical: Metcalfe 1960; this project.

Monachather Steud.

Habit, vegetative morphology. Perennial; caespitose. **Culms** 50–80 cm high; herbaceous; unbranched above; *tuberous (the bases somewhat swollen, covered with scale-like sheaths bearing dense woolly hairs)*. Culm nodes glabrous. Leaves non-auriculate. Leaf blades narrow; 3–4 mm wide; flat; without cross venation; persistent. *Ligule an unfringed membrane (jagged, often with dense marginal tufts of cilia)*; truncate; 1–3 mm long.

Reproductive organization. *Plants bisexual, with bisexual spikelets*; with hermaphrodite florets.

Inflorescence. Inflorescence few spikeleted; paniculate; contracted (linear, nearly racemose); espatheate; not comprising 'partial inflorescences' and foliar organs. Spikelet-bearing axes persistent. Spikelets not secund; pedicellate; mostly not in distinct 'long-and-short' combinations (though sometimes there are few such pairs).

Female-fertile spikelets. Spikelets (8–)11–14(–17) mm long; compressed laterally; disarticulating above the glumes; disarticulating between the florets; with distinctly elongated rachilla internodes between the florets. Rachilla prolonged beyond the uppermost female-fertile floret; rachilla hairless (scabrid). The rachilla extension with incomplete florets. Hairy callus present (the hairs long). *Callus* short; blunt.

Glumes two; more or less equal; exceeding the spikelets; long relative to the adjacent lemmas (much exceeding them); hairless; glabrous; pointed; awnless; non-carinate; similar (papery). *Lower glume 13–19 nerved*. Upper glume 11–13 nerved. *Spikelets* with incomplete florets. The incomplete florets distal to the female-fertile florets.

Female-fertile florets (3–)5–6(–8). Lemmas decidedly firmer than the glumes; not becoming indurated; incised; 2 lobed; deeply cleft; awned. Awns 1; median; from a sinus (from between the long, glabrous lemma lobes); geniculate; much shorter than the body of the lemma to about as long as the body of the lemma. Lemmas hairy. *The hairs in transverse rows (one row basal, one row beneath the lobes)*. Lemmas non-carinate; with a clear germination flap; 9–13 nerved. Palea *present (becoming hard and shining, bent across the fruit)*; relatively long (longer than body of lemma); entire (obtuse or truncate), or apically notched (emarginate); awnless, without apical setae; leathery; indurated; 2-nerved; 2-keeled. Palea keels winged; hairy (ciliate). *Lodicules* present; 2; free; fleshy; glabrous; not or scarcely vascularized. *Stamens* 3. Anthers about 0.3 mm long; not penicillate; without an apically prolonged connective. *Ovary* glabrous. Styles free to their bases. Stigmas 2.

Fruit, embryo and seedling. *Fruit* free from both lemma and palea; small (1–2 mm long); longitudinally grooved (depressed beneath the palea margins, concave along the hilum face); compressed dorsiventrally. Hilum short. Embryo large; waisted. Endosperm hard.

First seedling leaf with a well-developed lamina. The lamina broad; curved; 11 veined.

Abaxial leaf blade epidermis. *Costal/intercostal zonation* conspicuous. *Papillae* absent. *Long-cells* similar in shape costally and intercostally; differing markedly in wall thickness costally and intercostally. Mid-intercostal long-cells rectangular; having markedly sinuous walls. *Microhairs* present; panicoid-type; (40.5–)45–60(–66) microns long; (6.6–)8.1–9(–12) microns wide at the septum. Microhair total length/width at septum 5–9.1. Microhair apical cells 16.5–25.5(–28.5) microns long. Microhair apical cell/total length ratio 0.36–0.48. *Stomata* common; 31.5–36 microns long. Subsidiaries dome-shaped to triangular, or parallel-sided (by varying degrees of truncation of the apices of 'triangles'); including both triangular and parallel-sided forms on the same leaf (and a variety of domes). Guard-cells overlapping to flush with the interstomatals. *Intercostal short-cells* absent or very rare. *Costal short-cells* predominantly paired. Costal silica bodies 'panicoid-type'; mostly cross shaped, butterfly shaped, and dumb-bell shaped; not sharp-pointed.

Transverse section of leaf blade, physiology. C_3; XyMS+. *Mesophyll* with non-radiate chlorenchyma. *Leaf blade* 'nodular' in section; with the ribs more or less constant in size. *Midrib* not readily distinguishable; with one bundle only. Bulliforms present in discrete, regular adaxial groups to not in discrete, regular adaxial groups (mostly *Ammophila*-type arrangements of small cells, in the bases of the furrows); in places, inconspicuously in simple fans. All the vascular bundles accompanied by sclerenchyma. Combined sclerenchyma girders present; forming 'figures'. Sclerenchyma all associated with vascular bundles.

Cytology. Chromosome base number, $x = 6$. $2n = 72$. 12 ploid.

Taxonomy. Arundinoideae; Danthonieae (but with several eu-panicoid features).

Ecology, geography, regional floristic distribution. 1 species. Australia. Xerophytic; species of open habitats. Arid grassland.

Australian. Central Australian.

Economic importance. Important native pasture species: *M. paradoxa*, in dry places.

References, etc. Morphological/taxonomic: Blake 1972a; Vickery 1956. Leaf anatomical: this project.

Monanthochloë Engelm.

Halochloa Griseb., *Solenophyllum* Baillon

Habit, vegetative morphology. 'Ericoid' perennial; rhizomatous and stoloniferous (decumbent, mat-forming). **Culms 5–20 cm high**; herbaceous; branched above. *Leaves* not basally aggregated; conspicuously distichous; non-auriculate; without auricular setae. Leaf blades linear-lanceolate to lanceolate; narrow; 1–4 mm wide (to 1 cm long); thick, firm, folded or involute; acicular (subulate); without abaxial multicellular glands; without cross venation; persistent. Ligule a fringed membrane; minute. *Contra-ligule* absent.

Reproductive organization. *Plants dioecious*; without hermaphrodite florets. The spikelets all alike in sexuality (on the same plant); female-only, or male-only. Plants outbreeding.

Inflorescence. *Inflorescence reduced to a single spikelet*; inconspicuous, the spikelets borne singly, more or less concealed by leaf sheaths; spatheate, or espatheate (depending on interpretation); a complex of 'partial inflorescences' and intervening foliar organs, or not comprising 'partial inflorescences' and foliar organs (depending on interpretation).

Female-sterile spikelets. Male spikelets without glumes, with 2–5 three-stamened florets and incomplete distal florets. The male spikelets without glumes; with proximal incomplete florets; 2–5 floreted. Male florets 3 staminate.

Female-fertile spikelets. *Spikelets* unconventional (being without glumes); 8–10 mm long; compressed laterally to not noticeably compressed; disarticulating only tardily, at the lowest rachilla node; not disarticulating between the florets; with conventional internode spacings. Rachilla prolonged beyond the uppermost female-fertile floret. The rachilla extension with incomplete florets. Hairy callus absent.

Glumes absent. *Spikelets* with incomplete florets. The incomplete florets distal to the female-fertile florets. *The distal incomplete florets* merely underdeveloped; awnless.

Female-fertile florets 2–5. Lemmas convolute, oblong-lanceolate, leathery or leaflike in texture; not becoming indurated; entire; pointed; awnless; hairless; non-carinate (rounded on the back); perhaps with a germination flap — see illustration by Nicora and de Agrasar (1987); 9 nerved. *Palea* present; relatively long; tightly clasped by the lemma; entire; awnless, without apical setae; textured like the lemma (leathery); 2-nerved (narrow); 2-keeled. Palea keels winged (to embrace the floret above); ciliate. *Lodicules* absent. *Stamens* 0 (three staminodes). *Ovary* glabrous. Stigmas 2.

Fruit, embryo and seedling. *Fruit* free from both lemma and palea; ellipsoid. Hilum short. Pericarp fused. Embryo with an epiblast; with a scutellar tail; with an elongated mesocotyl internode. Embryonic leaf margins overlapping.

Abaxial leaf blade epidermis. *Costal/intercostal zonation* conspicuous. *Papillae* present; intercostal. Intercostal papillae seemingly consisting of one oblique swelling per cell, or consisting of one symmetrical projection per cell. Intercostal zones seemingly without typical long-cells (not observable in detail, being obscured by large papillae and prickles). *Microhairs* present; chloridoid-type (overarched by semicircles of prickles). *Stomata* common (but largely obscured). Prickles abundant. *Costal short-cells* predominantly paired (in places), or neither distinctly grouped into long rows nor predominantly paired (then solitary). Costal silica bodies present throughout the costal zones; rounded (mostly, but mostly more or less vertically elongated), or saddle shaped (few); not sharp-pointed.

Transverse section of leaf blade, physiology. C_4; XyMS+. PCR sheath outlines seemingly even (in the poor material seen). PCR sheaths of the primary vascular bundles complete. PCR sheath extensions seemingly absent. PCR cell chloroplasts centripetal. *Leaf blade* 'nodular' in section; with the ribs more or less constant in size (except for the lower one over the midrib). *Midrib* conspicuous (in lacking an adaxial rib, and having no adaxial strand); with one bundle only. *The lamina* symmetrical on either side of the midrib. Bulliforms present in discrete, regular adaxial groups (in the furrows); seemingly exclusively in simple fans (these quite small). All the vascular bundles accompanied by sclerenchyma. Combined sclerenchyma girders absent. Sclerenchyma all associated with vascular bundles (the bundles other than that of the midrib all with massive adaxial strands and abaxial strands or girders). The lamina margins with fibres.

Cytology. Chromosome base number, $x = 10$. $2n = 40$.

Taxonomy. Chloridoideae; main chloridoid assemblage.

Ecology, geography, regional floristic distribution. 3 species. America, West Indies. Helophytic; species of open habitats; halophytic. Inland salt pans.

Holarctic and Neotropical. Boreal and Madrean. Atlantic North American. Caribbean. Southern Atlantic North American.

Rusts and smuts. Rusts — *Puccinia*.

References, etc. Leaf anatomical: this project.

Monelytrum Hackel

Habit, vegetative morphology. Annual, or perennial; stoloniferous (each 'stolon' being a single, bare internode), or caespitose, or decumbent. *Culms* 8–80 cm high; herbaceous; branched above, or unbranched above. Culm nodes glabrous. Culm internodes solid. Plants unarmed. Young shoots intravaginal. *Leaves* not basally aggregated; non-auriculate. *Leaf blades* narrow; 2–7 mm wide (their margins thickened, with tubercle-based hairs); somewhat *cordate*; flat, or rolled (convolute); without abaxial multicellular glands; without cross venation; persistent. Ligule a fringed membrane; 1 mm long. *Contra-ligule* absent.

Reproductive organization. Plants bisexual, with bisexual spikelets; with hermaphrodite florets. The spikelets of sexually distinct forms on the same plant; hermaphrodite and sterile (there being 1–3 sterile spikelets at the tips of the reduced inflorescence branches); overtly heteromorphic (the sterile spikelets more or less awn-like); all in heterogamous combinations.

Inflorescence. Inflorescence bristly, a false spike, with spikelets on contracted axes. *Inflorescence axes not ending in spikelets (but rather, ending in bristles representing the uppermost 1–3 spikelets)*. Inflorescence espatheate; not comprising 'partial inflorescences' and foliar organs. *Spikelet-bearing axes* very much reduced (to shortly pedunculate spikelet clusters); disarticulating; falling entire (i.e., the clusters shed). Spikelets solitary; not secund; sessile to pedicellate; imbricate.

Female-sterile spikelets. The sterile spikelets awnlike, at the tips of the reduced branches.

Female-fertile spikelets. Spikelets 3–4 mm long; adaxial; compressed dorsiventrally; planoconvex; falling with the glumes (in the glomerules); with conventional internode spacings. Rachilla terminated by a female-fertile floret. *Hairy callus present (at the base of the cluster)*. *Callus* long (formed from the rachis base); blunt.

Glumes one per spikelet (G_1 sometimes absent), or two; (the upper) relatively large; very unequal; (the upper) long relative to the adjacent lemmas; hairy; (the upper) awned (with a slightly recurved awn at least as long as itself); non-carinate; very dissimilar (G_1 reduced to a minute scale, G_2 flat, elliptic-lanceolate, herbaceous). Lower glume much shorter than half length of lowest lemma (minute); 0 nerved. Upper glume 5–7 nerved. *Spikelets* with female-fertile florets only.

Female-fertile florets 1. Lemmas less firm than the glumes (membranous); not becoming indurated; entire to incised; pointed; when incised, minutely 3 lobed; not deeply cleft; mucronate to awned (from the mid-nerve). Awns when present, 1; median; apical; non-geniculate; hairless (scaberulous); much shorter than the body of the lemma; entered by one vein. Lemmas sparsely hairy; non-carinate; without a germination flap; 3 nerved. *Palea* present (broadly lanceolate); relatively long; entire (truncate); awnless, without apical setae (with scattered hairs); slightly thinner than the lemma; not indurated; 2-nerved; 2-keeled. *Lodicules* present; 2; free; fleshy; glabrous; not or scarcely vascularized. *Stamens* 3. Anthers 2 mm long; not penicillate; without an apically prolonged connective. *Ovary* glabrous. Styles free to their bases. Stigmas 2; red pigmented.

Fruit, embryo and seedling. *Fruit* free from both lemma and palea; small (2 mm long); ellipsoid; compressed dorsiventrally (dorsally). Hilum short (elliptical). Pericarp fused. Embryo large (about 1/3 the length of the fruit); not waisted.

Abaxial leaf blade epidermis. *Costal/intercostal zonation* conspicuous. *Papillae* present; intercostal. Intercostal papillae consisting of one oblique swelling per cell (at one end of each interstomatal cell). Long-cells differing markedly in wall thickness costally and intercostally (intercostal walls thicker). Mid-intercostal long-cells rectangular (very); having markedly sinuous walls. *Microhairs* present; elongated; clearly two-celled; chloridoid-type. Microhair apical cell wall of similar thickness/rigidity to that of the basal cell. Microhairs (39–)42–45(–51) microns long. Microhair basal cells 33 microns long. Microhairs 12–14.4 microns wide at the septum. Microhair total length/width at septum 2.9–4.3. Microhair apical cells (9–)10.5–12(–15) microns long. Microhair apical cell/total length ratio 0.23–0.33. *Stomata* common; 27–33 microns long. Subsidiaries dome-shaped and triangular. Guard-cells stomata sunken below the raised (concave) ends of the interstomatals, but not overlapped. *Intercostal short-cells* absent or very rare. Intercostal silica bodies absent. *Costal short-cells* conspicuously in long rows (but the 'short-cells' often rather long, giving an impression of some pairs and short rows). Costal silica bodies present in alternate cell files of the costal zones; almost exclusively saddle shaped; not sharp-pointed.

Transverse section of leaf blade, physiology. C_4; XyMS+. PCR sheath outlines even. PCR sheaths of the primary vascular bundles complete to interrupted; interrupted adaxially only. PCR sheath extensions absent. PCR cell chloroplasts centripetal. *Leaf blade* adaxially flat (with low abaxial ribs). *Midrib* not readily distinguishable; with one bundle only. Bulliforms present in discrete, regular adaxial groups; in simple fans (the large median cells deeply penetrating), or associated with colourless mesophyll cells to form deeply-penetrating fans. All the vascular bundles accompanied by sclerenchyma. Combined sclerenchyma girders present (with the primaries — some minor bundles with strands only); forming 'figures'. Sclerenchyma all associated with vascular bundles. The lamina margins with fibres.

Taxonomy. Chloridoideae; main chloridoid assemblage.

Ecology, geography, regional floristic distribution. 2 species. Southwest Africa to southern Angola. Xerophytic; species of open habitats; glycophytic. In seasonally moist locations?.

Paleotropical. African. Namib-Karoo.

References, etc. Leaf anatomical: Metcalfe 1960; this project; photos of *M. luderitzianum* provided by R.P. Ellis.

Monium Stapf

~ *Anadelphia*

Habit, vegetative morphology. Annual, or perennial; when perennial, caespitose. **Culms** 20–90 cm high; *herbaceous*; branched above, or unbranched above. Culm nodes exposed; glabrous. Culm leaves present. *Leaves* not basally aggregated. The sheaths with a strong median nerve. Leaf blades linear; narrow; 0.25–3 mm wide; setaceous, or not setaceous; flat; without cross venation; persistent. Ligule an unfringed membrane; truncate to not truncate; 1–2 mm long. *Contra-ligule* absent.

Reproductive organization. Plants bisexual, with bisexual spikelets; with hermaphrodite florets. *The spikelets* of sexually distinct forms on the same plant (but the pedicelled spikelet may be so reduced as to be hidden among the callus hairs of the fertile spikelet); hermaphrodite and sterile, or hermaphrodite and male-only; overtly heteromorphic (the pedicelled spikelet much reduced); *all in heterogamous combinations*. Exposed-cleistogamous (? — the greatly reduced 'raceme' inclosed within its spathe), or chasmogamous (?).

Inflorescence. Inflorescence paniculate; open; with capillary branchlets to without capillary branchlets; *spatheate (and spatheolate)*; a complex of 'partial inflorescences' and intervening foliar organs. **Spikelet-bearing axes** *very much reduced (the 'raceme' shorter than the spatheole, bearing only the one fertile spikelet and its pedicelled satellite)*; the spikelet-bearing axes with only one spikelet-bearing 'article'; solitary; disarticulating; disarticulating at the joints (i.e., each 'raceme' disarticulating at the only joint). 'Articles' without a basal callus-knob; disarticulating obliquely (the single 'joint' obliquely articulated on the somewhat swollen peduncle tip). Spikelets paired (no triplets); sessile and pedicellate; consistently in 'long-and-short' combinations; in pedicellate/sessile combinations. Pedicels of the 'pedicellate' spikelets free of the rachis. The 'shorter' spikelets hermaphrodite. The 'longer' spikelets male-only, or sterile (and often reduced to a vestigial pedicel).

Female-sterile spikelets. Pedicellate spikelets often reduced to a minute pedicel, which may be concealed among the callus hairs. When developed, acuminate with a glabrous callus.

Female-fertile spikelets. Spikelets 5–10 mm long; not noticeably compressed to compressed dorsiventrally; falling with the glumes. Rachilla terminated by a female-fertile floret. Hairy callus present. **Callus** *long*; pointed.

Glumes two; more or less equal; exceeding the spikelets; long relative to the adjacent lemmas (exceeding them); free; hairless; glabrous; without conspicuous tufts or rows of hairs; awned (the G2 only, or sometimes the G1 two-awned), or awnless (the G_2 then subulate-tipped or apiculate); non-carinate; very dissimilar (both leathery, the G_1 apically truncate to bidentate and round-backed, the G_2 subulate or apiculate at tip, naviculate). *Lower glume* not two-keeled (except towards the tip); convex on the back; *not pitted*; relatively smooth; 5 nerved, or 7 nerved (with a median). Upper glume 3 nerved. *Spikelets* with incomplete florets. The incomplete florets proximal to the female-fertile florets. *Spikelets with proximal incomplete florets*. The proximal incomplete florets 1; epaleate; sterile. The proximal lemmas awnless (pointed to bilobed); 0 nerved, or 2 nerved; more or less equalling the female-fertile lemmas; similar in texture to the female-fertile lemmas (hyaline); not becoming indurated.

Female-fertile florets 1. *Lemmas* somewhat displaced; less firm than the glumes (hyaline); not becoming indurated; incised; 2 lobed; deeply cleft to not deeply cleft; awned. Awns 1; median; from a sinus; geniculate; hairless, or hairy; much longer than the body of the lemma (5–16 cm long). Awn bases twisted; not flattened. Lemmas hairy, or hairless; glabrous; non-carinate; without a germination flap; 1 nerved. **Palea** *absent (according to Jacques-Félix 1962, and in the material seen — by contrast with* 'Anadelphia scyphofera'). Lodicules present; 2; fleshy; glabrous. *Stamens* 3. *Ovary* glabrous. Styles basally fused; joined below. Stigmas 2.

Fruit, embryo and seedling. Fruit ellipsoid; not noticeably compressed (more or less cylindrical); glabrous. Embryo large. Endosperm hard.

Abaxial leaf blade epidermis. *Costal/intercostal zonation* conspicuous. *Papillae* present; costal and intercostal (but not over the midrib). Intercostal papillae not over-arching the stomata; consisting of one oblique swelling per cell, or consisting of one symmetrical projection per cell (large, thick walled). Mid-intercostal long-cells basically rectangular; having straight or only gently undulating walls. *Microhairs* present; elongated; clearly two-celled; panicoid-type; 31–36 microns long; 5.1–5.4 microns wide at the septum. Microhair total length/width at septum 6.1–6.7. Microhair apical cells 19–24 microns long. Microhair apical cell/total length ratio 0.59–0.67. *Stomata* common; 24–26 microns long. Subsidiaries non-papillate; triangular. Guard-cells overlapping to flush with the interstomatals. *Intercostal short-cells* seemingly absent or very rare. *Costal short-cells* conspicuously in long rows. Costal silica bodies present and well developed (in places, but extensively obscured by papillae); 'panicoid-type'; not sharp-pointed.

Transverse section of leaf blade, physiology. C_4; XyMS–. PCR sheath outlines uneven. *Leaf blade* with distinct, prominent adaxial ribs (the ribs very low). *Midrib* conspicuous; having a conventional arc of bundles (one median, with smaller laterals); with colourless mesophyll adaxially (e.g. *M. macrochaeta*), or without colourless mesophyll adaxially (but with the overlying adaxial epidermis bulliform). Bulliforms present in discrete, regular adaxial groups (the groups wide), or not in discrete, regular adaxial groups (if the same configuration is interpreted as 'epidermis extensively bulliform'); in simple fans; associating with colourless mesophyll cells to form arches over small vascular bundles (seemingly, in places, in *M. funerea*), or nowhere involved in bulliform-plus-colourless mesophyll arches. All the vascular bundles accompanied by sclerenchyma (in *M. funerea*), or many of the smallest vascular bundles unaccompanied by sclerenchyma (in *M. trichaeta*). Combined sclerenchyma girders present (with all the primaries and some of the smaller bundles, other small bundles with small strands only); forming 'figures' (the main bundles). Sclerenchyma all associated with vascular bundles.

Taxonomy. Panicoideae; Andropogonodae; Andropogoneae; Andropogoninae.

Ecology, geography, regional floristic distribution. 7 species. Tropical West Africa.

Paleotropical. African. West African Rainforest.

References, etc. Leaf anatomical: this project.

Special comments. See comment under *Anadelphia* regarding unsatisfactory taxonomy involving *Monium, Anadelphia* and *Pobeguinea*.

Monocladus Chia, Fung & Yang

Habit, vegetative morphology. Perennial. The flowering culms leafy. *Culms* 100–500 cm high; woody and persistent; to 1.5 cm in diameter; branched above. Primary branches/mid-culm node 1. Culm internodes solid. Unicaespitose. *Rhizomes pachymorph*. *Leaves* not basally aggregated; auriculate; with auricular setae. Leaf blades linear-lanceolate to lanceolate; subcoriaceous; broad; 35–80 mm wide (and 20–40 cm long); pseudopetiolate; minutely cross veined; disarticulating from the sheaths.

Reproductive organization. Plants bisexual, with bisexual spikelets; with hermaphrodite florets.

Inflorescence. *Inflorescence* indeterminate; *with pseudospikelets*; spatheate.

Female-fertile spikelets. Spikelets disarticulating above the glumes; disarticulating between the florets. Rachilla prolonged beyond the uppermost female-fertile floret.

Glumes two; shorter than the spikelets; shorter than the adjacent lemmas; pointed; similar. *Spikelets* with incomplete florets. The incomplete florets distal to the female-fertile florets. *The distal incomplete florets* merely underdeveloped.

Lemmas entire; pointed; awnless. *Palea* present; relatively long (in the first floret, and leathery), or conspicuous but relatively short (in the other florets, and membranous). *Lodicules* present; 3; free; glabrous. *Stamens* 6. Anthers not penicillate; without an apically prolonged connective. **Ovary** glabrous; *without a conspicuous apical appendage*. Styles basally fused. Stigmas 3.

Taxonomy. Bambusoideae; Bambusodae; Bambuseae.

Ecology, geography, regional floristic distribution. 4 species. Southern China. Glycophytic.

Paleotropical. Indomalesian. Indo-Chinese.

References, etc. Morphological/taxonomic: Chia, Fung and Yang 1988.

Special comments. Cf. *Indosasa, Schizostachyum*. The ovary apex (not mentioned in the original, unsatisfactory Latin descriptions) is assumed to be unappendaged. The only other difference from *Schizostachyum* is the glabrous lodicules. Fruit data wanting. Anatomical data wanting.

Monocymbium Stapf

Habit, vegetative morphology. Perennial; caespitose. *Culms* 30–120 cm high; herbaceous; branched above, or unbranched above. *Leaves* not basally aggregated; non-auriculate. Leaf blades linear; narrow; flat (tapering to a sharp point); without cross venation. *Ligule an unfringed membrane*; truncate.

Reproductive organization. Plants bisexual, with bisexual spikelets; with hermaphrodite florets. The spikelets of sexually distinct forms on the same plant; hermaphrodite and male-only; overtly heteromorphic (the pedicellate spikelets awnless); all in heterogamous combinations.

Inflorescence. Inflorescence paniculate (of solitary 'racemes', loosely gathered into a false panicle); with capillary branchlets; conspicuously spatheate (each raceme embraced by a reddish spatheole); *a complex of 'partial inflorescences' and intervening foliar organs*. *Spikelet-bearing axes* dense 'racemes' (the spikelets more or less concealing the short internodes); the spikelet-bearing axes with 6–10 spikelet-bearing 'articles', or with more than 10 spikelet-bearing 'articles' (with at least 6 spikelet pairs); solitary; with very slender rachides (filiform); disarticulating; disarticulating at the joints. *'Articles'* linear; not appendaged; disarticulating obliquely; densely long-hairy. Spikelets paired; secund; sessile and pedicellate; consistently in 'long-and-short' combinations; in pedicellate/sessile combinations. Pedicels of the 'pedicellate' spikelets free of the rachis. The 'shorter' spikelets hermaphrodite. The 'longer' spikelets male-only.

Female-sterile spikelets. Pedicellate spikelets male, similar in form to the sessile members but are awnless, with a linear callus. The lemmas awnless.

Female-fertile spikelets. Spikelets compressed dorsiventrally (flattened dorsally, the sides rounded); falling with the glumes (deciduous with the adjacent joint and pedicel). Rachilla terminated by a female-fertile floret. **Callus** *short (but not clearly distinct from G_1, obscurely bearded)*; blunt.

Glumes two; more or less equal; long relative to the adjacent lemmas (exceeding them); hairy, or hairless; without conspicuous tufts or rows of hairs; (the upper) *awned (from a notch)*; very dissimilar (thinly cartilaginous, only the G_2 awned). Lower glume not two-keeled (naviculate, laterally compressed and keeled over its upper third); convex on the back; not pitted; relatively smooth; 7 nerved (?). Upper glume 3 nerved. *Spikelets* with incomplete florets. The incomplete florets proximal to the female-fertile florets. *Spikelets with proximal incomplete florets*. *The proximal incomplete florets* 1; epaleate; sterile. The proximal lemmas awnless; more or less equalling the female-fertile lemmas to decidedly exceeding the female-fertile lemmas; similar in texture to the female-fertile lemmas (hyaline); not becoming indurated.

Female-fertile florets 1. Lemmas less firm than the glumes (except for the cartilaginous median zone); not becoming indurated; incised; 2 lobed; deeply cleft; awned. Awns 1; median; from a sinus; geniculate; hairless; much longer than the body of the lemma. Lemmas hairless; non-carinate; without a germination flap; 1 nerved. **Palea** *absent*. Lodicules present; 2; free; fleshy; glabrous. *Stamens* 3. *Ovary* glabrous. *Styles free to their bases*. Stigmas 2.

Fruit, embryo and seedling. Fruit compressed dorsiventrally. Hilum short. Embryo large.

Abaxial leaf blade epidermis. *Costal/intercostal zonation* conspicuous. *Papillae* absent. Long-cells differing markedly in wall thickness costally and intercostally (costals thinner). Mid-intercostal long-cells rectangular; having markedly sinuous walls (mostly). *Microhairs* present; panicoid-type; 39–51 microns long; 5.4–6 microns wide at the septum. Microhair total length/width at septum 6.5–8.5. Microhair apical cells 21–26 microns long. Microhair apical cell/total length ratio 0.44–0.54. *Stomata* common; 27–30 microns long. Subsidiaries dome-shaped (mostly), or triangular. Guard-cells overlapping to flush with the interstomatals. *Intercostal short-cells* common; in cork/silica-cell pairs (and solitary); silicified (but rarely so). *Costal short-cells* conspicuously in long rows. Costal silica bodies 'panicoid-type'; cross shaped, butterfly shaped, and dumb-bell shaped; not sharp-pointed.

Transverse section of leaf blade, physiology. C_4; XyMS–. PCR cell chloroplasts centrifugal/peripheral. *Mesophyll* with radiate chlorenchyma; traversed by columns of colourless mesophyll cells (a few of the arches reaching the abaxial epidermis, near midrib). *Leaf blade* adaxially flat. *Midrib* conspicuous (via a larger bundle and abaxial rib); with one bundle only. *The lamina* symmetrical on either side of the midrib. Bulliforms not in discrete, regular adaxial groups (the epidermis mainly bulliform); associating with colourless mesophyll cells to form arches over small vascular bundles (especially near the midrib, in *M. ceresiiforme*). Many of the smallest vascular bundles unaccompanied by sclerenchyma. Combined sclerenchyma girders present. Sclerenchyma all associated with vascular bundles.

Cytology. Chromosome base number, $x = 5$, or 10. $2n = 20$. 2 ploid, or 4 ploid.

Taxonomy. Panicoideae; Andropogonodae; Andropogoneae; Andropogoninae.

Ecology, geography, regional floristic distribution. 4 species. Tropical and southern Africa. Mesophytic; species of open habitats; glycophytic. Savanna.

Paleotropical and Cape. African. Sudano-Angolan and West African Rainforest. Somalo-Ethiopian and South Tropical African.

Rusts and smuts. Rusts — *Puccinia*. Taxonomically wide-ranging species: *Puccinia versicolor* and '*Uromyces*' *clignyi*. Smuts from Ustilaginaceae. Ustilaginaceae — *Sphacelotheca*.

Economic importance. Significant weed species: *M. ceresiiforme*.

References, etc. Leaf anatomical: this project.

Monodia S.W.L. Jacobs

Habit, vegetative morphology. *Stipa*-like, hummock-forming perennial; caespitose. *Culms* 70–150 cm high; herbaceous. Leaves non-auriculate. *Leaf blades* narrow (about 60 cm long); flat (at first), or folded (later); *hard, woody, needle-like (pungent)*; without abaxial multicellular glands; without cross venation; persistent (the persistent, older blades becoming recurved). *Ligule a fringe of hairs*; up to 3 mm long.

Reproductive organization. Plants bisexual, with bisexual spikelets; with hermaphrodite florets.

Inflorescence. Inflorescence of spicate main branches (?), or paniculate; contracted (to about 50 cm long); espatheate; not comprising 'partial inflorescences' and foliar organs. The racemes spikelet bearing to the base. Spikelet-bearing axes persistent. Spikelets not secund; shortly pedicellate.

Female-fertile spikelets. Spikelets 16–24 mm long; compressed laterally; disarticulating above the glumes; with conventional internode spacings. The upper floret not stipitate. Rachilla prolonged beyond the uppermost female-fertile floret. The rachilla extension naked. Hairy callus present. *Callus* long; pointed (oblique).

Glumes two; more or less equal; exceeding the spikelets; long relative to the adjacent lemmas; hairless; glabrous to scabrous; pointed (acute); awnless; similar. Lower glume 9 nerved, or 11 nerved. Upper glume 3 nerved, or 5 nerved. *Spikelets* with female-fertile florets only.

Female-fertile florets 1. **Lemmas convolute**; not becoming indurated; incised; 2 lobed; not deeply cleft (the lobes short, membranous, unequal); awned. Awns 1; median; from a sinus; geniculate (the column lightly twisted, the bristle falcate); hairless; much longer than the body of the lemma (5–6 cm long); entered by one vein; persistent. *Lemmas hairy (with a line of appressed hairs on each margin, but otherwise glabrous)*; non-carinate; 3 nerved. *Palea* present; relatively long; awnless, without apical setae (and glabrous); not indurated; 2-nerved; 2-keeled (the rachilla borne within the groove between). Palea keels wingless. *Lodicules* present; 2; free; fleshy; glabrous. *Stamens* 3. Anthers 4–6 mm long; not penicillate; without an apically prolonged connective. *Ovary* glabrous. Stigmas 2; pale.

Fruit, embryo and seedling. Fruit ellipsoid.

Abaxial leaf blade epidermis. *Costal/intercostal zonation* conspicuous (in the central part, although the grooves are easily overlooked in surface view), or lacking (on the wide lateral flanges). *Papillae* present; intercostal (mainly confined to the grooves and their margins). Intercostal papillae consisting of one oblique swelling per cell. *Long-cells* similar in shape costally and intercostally; of similar wall thickness costally and intercostally (thick walled, and pitted). Mid-intercostal long-cells rectangular; having markedly sinuous walls. *Microhairs* present (but hidden in the very narrow grooves); chloridoid-type. *Stomata* not visible. Subsidiaries non-papillate. *Costal short-cells* conspicuously in long rows. Costal silica bodies 'panicoid-type'; mostly small, dumb-bell shaped; not sharp-pointed.

Transverse section of leaf blade, physiology. C$_4$. The anatomical organization unconventional. Organization of PCR tissue *Triodia* type (with 'draping' PCR tissue). XyMS+. Mesophyll traversed by columns of colourless mesophyll cells (linking the bases of the deep and narrow adaxial and abaxial grooves); with arm cells (the PCA tissue exhibiting conspicuous ingrowths, cf. *Triodia* etc.). Leaf blade with the ribs very irregular in sizes (and shapes). *Midrib* conspicuous (by its position); with one bundle only; with colourless mesophyll adaxially (this and most of the large bundles with a large, colourless more or less lignified group between the top of the bundle and the adaxial sclerenchyma group). Bulliforms present in discrete, regular adaxial groups (at the bases of the central adaxial grooves); associated with colourless mesophyll cells to form deeply-penetrating fans (these continuous with the traversing columns of colourless cells). All the vascular bundles accompanied by sclerenchyma. Combined sclerenchyma girders present (if the colourless groups are regarded as part of the sclerenchyma); forming 'figures' (massive). Sclerenchyma not all bundle-associated. The 'extra' sclerenchyma in a continuous abaxial layer (the massive lateral flanges of the leaf blade with a heavy, continuous layer of fibres, beneath a mass of large-celled sclerenchyma abaxial to the vascular bundles and mesophyll — cf. *Triodia*).

Taxonomy. Chloridoideae; Triodieae.

Ecology, geography, regional floristic distribution. 1 species. Western Australia. Xerophytic; species of open habitats. Stony ground.

Australian. North and East Australian. Tropical North and East Australian.

References, etc. Morphological/taxonomic: Jacobs 1984. Leaf anatomical: this project.

Special comments. Fruit data wanting.

Monostachya Merr.

~ *Rytidosperma, Danthonia* sensu lato

Habit, vegetative morphology. Perennial; low, mat-forming alpines. *Culms* 2–20 cm high (?); herbaceous. Culm nodes glabrous. Culm internodes hollow. *Leaves* mostly basal; non-auriculate. *Leaf blades* narrow; *setaceous*; without cross venation; *persistent*. *Ligule a fringe of hairs (?)*.

Reproductive organization. Plants bisexual, with bisexual spikelets; with hermaphrodite florets. The spikelets of sexually distinct forms on the same plant, or all alike in sexuality; hermaphrodite, or hermaphrodite and sterile (having vestigial spikelets, much reduced, beneath the terminal perfect one).

Inflorescence. *Inflorescence* reduced to a single spikelet, or few spikeleted; if of more than one spikelet *a single raceme*; espatheate; not comprising 'partial inflorescences' and foliar organs. Spikelet-bearing axes persistent. Spikelets not secund; pedicellate.

Female-fertile spikelets. Spikelets compressed laterally; disarticulating above the glumes; with conventional internode spacings. Rachilla prolonged beyond the uppermost female-fertile floret; rachilla hairless. Hairy callus present, or absent.

Glumes two; more or less equal; long relative to the adjacent lemmas; awnless; similar. Lower glume 3 nerved. Upper glume 3 nerved. *Spikelets* with female-fertile florets only, or with incomplete florets. The incomplete florets distal to the female-fertile florets.

Female-fertile florets 1–10. **Lemmas** not becoming indurated; *incised*; 2 lobed; mucronate, or awned. Awns when present, from a sinus; non-geniculate; when present, much shorter than the body of the lemma. Lemmas hairless; non-carinate; 3–5 nerved. *Palea* present; 2-nerved. *Lodicules* present; free; fleshy; ciliate. *Stamens* 3. Anthers not penicillate. *Ovary* glabrous. Styles free to their bases. Stigmas 2.

Fruit, embryo and seedling. *Fruit* small; *compressed laterally*. Hilum short. Embryo small; not waisted.

Abaxial leaf blade epidermis. *Costal/intercostal zonation* conspicuous (with paired short-cells costally). *Papillae* absent. *Long-cells* similar in shape costally and intercostally (long-rectangular); of similar wall thickness costally and intercostally (walls thick and sinuous). Mid-intercostal long-cells having markedly sinuous walls. *Microhairs* absent (but present adaxially); panicoid type. *Stomata* absent or very rare (abaxially, but present adaxially). Subsidiaries not available abaxially, but triangles and domes present adaxially. *Intercostal short-cells* common; in cork/silica-cell pairs; not silicified (usually). *Costal short-cells* predominantly paired. Costal silica bodies rounded; not sharp-pointed.

Transverse section of leaf blade, physiology. C$_3$; XyMS+. *Mesophyll* with non-radiate chlorenchyma. *Leaf blade* with distinct, prominent adaxial ribs; with the ribs more or less constant in size. *Midrib* not readily distinguishable; with one bundle only. All the vascular bundles accompanied by sclerenchyma. Combined sclerenchyma girders present; forming 'figures'. Sclerenchyma all associated with vascular bundles.

Cytology. Chromosome base number, $x = 5$. $2n = 20$. 4 ploid.

Taxonomy. Arundinoideae; Danthonieae.

Ecology, geography, regional floristic distribution. About 4 species. Philippines, New Guinea.

Paleotropical. Indomalesian. Papuan.

References, etc. Morphological/taxonomic: Jacobs 1982. Leaf anatomical: this project.

Mosdenia Stent

Habit, vegetative morphology. Perennial; *stoloniferous (the stolons with densely imbricate cataphylls)*. Culms 10–90 cm high; *herbaceous*; unbranched above. Culm internodes solid. Plants unarmed. Young shoots intravaginal. *Leaves* mostly basal (those on the culms with reduced blades), or not basally aggregated; non-auriculate. Leaf blades linear to linear-lanceolate; narrow; to 5 mm wide (and to 12 cm long); without abaxial multicellular glands; without cross venation; persistent. **Ligule** present; *an unfringed membrane (laciniate)*; 0.3–0.5 mm long.

Reproductive organization. Plants bisexual, with bisexual spikelets; with hermaphrodite florets. The spikelets of sexually distinct forms on the same plant, or all alike in sexuality; hermaphrodite, or hermaphrodite and sterile (those at the tip of the inflorescence sometimes reduced).

Inflorescence. *Inflorescence a single spike (a dense, continuous, elongated bottlebrush, the spikelets spreading at right angles to the axis)*; espatheate; not comprising 'partial inflorescences' and foliar organs. *Spikelet-bearing axes persistent.* Spikelets solitary; not secund; not two-ranked (in whorls or spirals); sessile, or subsessile ('on minutely tomentose notches); imbricate.

Female-fertile spikelets. Spikelets 2.5–3.75 mm long (sub-falcate); falling with the glumes; with conventional internode spacings. *Rachilla terminated by a female-fertile floret.* Hairy callus present. *Callus* short; blunt.

Glumes two; more or less equal (the G$_1$ slightly longer and broader); about equalling the spikelets (or somewhat longer); long relative to the adjacent lemmas (exceeding them); lateral to the rachis; hairless (glabrous save at callus); pointed; awnless; carinate (G$_1$), or non-carinate; very dissimilar (the G$_2$ narrower, flat-backed). Lower glume about equalling the lowest lemma (slightly exceeding it); 1 nerved. *Upper glume 1 nerved.* *Spikelets* with female-fertile florets only.

Female-fertile florets 1. Lemmas small, ovate-lanceolate; less firm than the glumes; delicate, hyaline; entire; pointed; awnless; hairless; carinate; without a germination flap; 1 nerved, or 3 nerved. *Palea* present; relatively long; apically notched (minutely); awnless, without apical setae; textured like the lemma (delicate, hyaline); not indurated; 2-nerved; 2-keeled. *Lodicules* present; 2; free; fleshy; glabrous. *Stamens* 3. Anthers 1.3–1.5 mm long; not penicillate; without an apically prolonged connective. *Ovary* glabrous. Styles free to their bases. Stigmas 2; red pigmented (purple), or brown.

Fruit, embryo and seedling. *Fruit* free from both lemma and palea; small (about 1.5 mm long); ellipsoid; longitudinally grooved; compressed dorsiventrally (dorsally). Hilum short (elliptical). Pericarp fused. Embryo large (about 1/3 the length of the fruit).

Abaxial leaf blade epidermis. *Costal/intercostal zonation* conspicuous. *Papillae* absent. *Long-cells* markedly different in shape costally and intercostally (the costals more regularly narrow). Mid-intercostal long-cells rectangular; having markedly sinuous walls. *Microhairs* present; chloridoid-type. *Stomata* common. Subsidiaries low dome-shaped. *Intercostal short-cells* absent or very rare (few). *Costal short-cells* conspicuously in long rows. Costal silica bodies present in alternate cell files of the costal zones; 'panicoid-type'; cross shaped (to rectangular, near the leaf margins), or dumb-bell shaped (elsewhere); not sharp-pointed.

Transverse section of leaf blade, physiology. C_4; XyMS+. PCR sheaths of the primary vascular bundles interrupted; interrupted both abaxially and adaxially. PCR sheath extensions absent. PCR cell chloroplasts centripetal. *Mesophyll* with radiate chlorenchyma; traversed by columns of colourless mesophyll cells (seemingly, in places), or not traversed by colourless columns (but the bulliform-plus-colourless cell groups very deeply penetrating). *Leaf blade* 'nodular' in section to adaxially flat. *Midrib* not readily distinguishable; with one bundle only. Bulliforms present in discrete, regular adaxial groups; in simple fans and associated with colourless mesophyll cells to form deeply-penetrating fans. All the vascular bundles accompanied by sclerenchyma. Combined sclerenchyma girders present (with all the bundles); forming 'figures' (most bundles with rectangular adaxial and winged-crescentic abaxial girders).

Cytology. $2n = 40$.

Taxonomy. Chloridoideae; main chloridoid assemblage.

Ecology, geography, regional floristic distribution. 1 species. South Africa. Mesophytic; species of open habitats; glycophytic. Dry savanna.

Paleotropical. African. Sudano-Angolan. South Tropical African.

References, etc. Morphological/taxonomic: Stent 1922. Leaf anatomical: photos of *M. leptostachys* provided by R.P. Ellis.

Muhlenbergia Schreber

Acroxis Steud., *Anthipsimus* Raf., *Calycodon* Nutt, *Clomena* P. Beauv., *Crypsinna* Fourn., *Dactylogramma* Link, *Dilepyrum* Michaux, *Epicampes* Presl, *Lepyroxis* Fourn., *Podosemum* Desv., *Sericrostis* Raf., *Tosagris* P. Beauv., *Trichochloa* DC., *Vaseya* Thurber

Excluding *Chaboissaea, Bealia, Hubbardochloa*

Habit, vegetative morphology. Annual (rarely), or perennial; rhizomatous, or stoloniferous, or caespitose, or decumbent. *Culms* 2–100 cm high, or 100–200 cm high (rarely); herbaceous; branched above, or unbranched above. Culm nodes glabrous. Culm internodes solid, or hollow. *Leaves* not basally aggregated; non-auriculate. Leaf blades narrow; setaceous, or not setaceous; flat, or rolled; without abaxial multicellular glands; without cross venation; persistent; rolled in bud. Ligule an unfringed membrane (e.g., in Section *Epicampes*), or a fringed membrane.

Reproductive organization. Plants bisexual, with bisexual spikelets; with hermaphrodite florets. *The spikelets all alike in sexuality*. Exposed-cleistogamous, or chasmogamous. Plants with hidden cleistogenes, or without hidden cleistogenes. The hidden cleistogenes when present, in the leaf sheaths (sometimes very modified).

Inflorescence. *Inflorescence paniculate*; open, or contracted (then narrow, sometimes spikelike); with capillary branchlets, or without capillary branchlets; espatheate; not comprising 'partial inflorescences' and foliar organs. Spikelet-bearing axes persistent. Spikelets not secund; pedicellate.

Female-fertile spikelets. *Spikelets 1.7–3.5 mm long; compressed laterally;* disarticulating above the glumes. *Rachilla terminated by a female-fertile floret (nearly always, the upper floret when present usually being fertile). Hairy callus present (very short).* Callus short.

Glumes two (though rarely the G_1 is obsolete); very unequal to more or less equal; shorter than the spikelets to about equalling the spikelets (rarely longer); shorter than the adjacent lemmas (usually), or long relative to the adjacent lemmas; obtuse to acuminate; awned, or awnless; carinate, or non-carinate; very dissimilar, or similar. *Lower glume 0–2 nerved.* Upper glume 0–3 nerved. **Spikelets** nearly always with female-fertile florets only; *without proximal incomplete florets.*

Female-fertile florets 1, or 2 (rarely). Lemmas similar in texture to the glumes to decidedly firmer than the glumes; not becoming indurated (firmly membranous); entire, or incised; when incised 2 lobed (minutely bifid); not deeply cleft (at the most shortly bidentate); mucronate, or awned. Awns 1; from a sinus, or dorsal, or apical; when dorsal, from near the top; non-geniculate; straight, or flexuous; much shorter than the body of the lemma to much longer than the body of the lemma. Lemmas hairy (rarely long-pilose, usually minutely pilose), or hairless; non-carinate; without a germination flap; *3 nerved.* *Palea* present; 2-nerved. *Lodicules* present; 2; free; fleshy; glabrous. *Stamens* 3. Anthers 0.3–2.7 mm long; not penicillate. *Ovary* glabrous. Styles free to their bases. Stigmas 2; red pigmented, or brown.

Fruit, embryo and seedling. *Fruit* free from both lemma and palea (but the lemma tightly enclosing the grain); small; fusiform, or ellipsoid; not noticeably compressed. Hilum short. Pericarp fused. Embryo large. Endosperm hard; without lipid. Embryo with an epiblast; with a scutellar tail; with an elongated mesocotyl internode. Embryonic leaf margins meeting.

Abaxial leaf blade epidermis. *Costal/intercostal zonation* conspicuous. *Papillae* present; intercostal. Intercostal papillae over-arching the stomata; consisting of one oblique swelling per cell. *Long-cells* similar in shape costally and intercostally; of similar wall thickness costally and intercostally. Mid-intercostal long-cells rectangular; having straight or only gently undulating walls. *Microhairs* present; elongated; clearly two-celled; chloridoid-type. Microhair apical cell wall of similar thickness/rigidity to that of the basal cell. Microhairs (16.5–)18–50(–52.5) microns long. Microhair basal cells 36 microns long. Microhairs (4.5–)5.4–12(–12.3) microns wide at the septum. Microhair total length/width at septum 1.5–11.2. Microhair apical cells (3–)4.5–10.5(–12) microns long. Microhair apical cell/total length ratio 0.14–0.33. *Stomata* common; 16.5–22.5 microns long. Subsidiaries low to high dome-shaped. Guard-cells overlapping to flush with the interstomatals. *Intercostal short-cells* absent or very rare. Intercostal silica bodies absent. *Costal short-cells* conspicuously in long rows, or neither distinctly grouped into long rows nor predominantly paired. Costal silica bodies present in alternate cell files of the costal zones; saddle shaped, or 'panicoid-type' (sometimes pointed); when panicoid type, cross shaped to dumb-bell shaped; sharp-pointed, or not sharp-pointed.

Transverse section of leaf blade, physiology. C_4; biochemical type PCK (1 species); XyMS+. PCR sheath outlines uneven (e.g. *M. diversiglumis, M. ciliata*), or even (e.g. *M. breviculmis, M. pusilla, M. ramulosa*). PCR sheaths of the primary vascular bundles interrupted; interrupted abaxially only. PCR sheath extensions absent. PCR cells without a suberised lamella. PCR cell chloroplasts with well developed grana; centrifugal/peripheral (e.g. *M. microsperma*, and seemingly *M. ciliata* and *M. diversiglumis*), or centripetal (e.g. *M. ramulosa*). *Mesophyll* with radiate chlorenchyma, or with non-radiate chlorenchyma (rarely); traversed by columns of colourless mesophyll cells (e.g. *M. confusa, M. breviculmis, M. pusilla*), or not traversed by colourless columns (e.g. *M. ciliata, M. diversiglumis, M. microsperma*). *Leaf blade* 'nodular' in section, or adaxially flat; when present, with the ribs more or less constant in size (low). *Midrib* conspicuous; with one bundle only. *The lamina* symmetrical on either side of the midrib. Bulliforms present in discrete, regular adaxial groups; in simple fans (predominant in e.g. *M. diversiglumis, M. microsperma*), or associated with colourless mesophyll cells to form deeply-penetrating fans (often linked with traversing colourless columns). All the vascular bundles accompanied by sclerenchyma. Combined sclerenchyma girders present, or absent (e.g. *M.*

confusa); forming 'figures', or nowhere forming 'figures'. Sclerenchyma all associated with vascular bundles.

Phytochemistry. Tissues of the culm bases with abundant starch.

Cytology. Chromosome base number, $x = 10$. $2n = 20, 40, 42$, and 60. 2 and 4 ploid.

Taxonomy. Chloridoideae; main chloridoid assemblage.

Ecology, geography, regional floristic distribution. 160 species. Himalayas to Japan, North America to Andes. Commonly adventive. Mesophytic to xerophytic; species of open habitats. Diverse habitats, commonly in open grassland, important in arid and semi-arid regions.

Holarctic, Paleotropical, and Neotropical. Boreal, Tethyan, and Madrean. Indomalesian. Eastern Asian, Atlantic North American, and Rocky Mountains. Irano-Turanian. Indian, Indo-Chinese, Malesian, and Papuan. Caribbean, Pampas, and Andean. Canadian-Appalachian, Southern Atlantic North American, and Central Grasslands.

Rusts and smuts. Rusts — *Puccinia*. Taxonomically wide-ranging species: *Puccinia graminis*, *Puccinia striiformis*, and *Puccinia schedonnardi*. Smuts from Tilletiaceae and from Ustilaginaceae. Tilletiaceae — *Entyloma* and *Tilletia*. Ustilaginaceae — *Sphacelotheca* and *Ustilago*.

Economic importance. Significant weed species: *M. schreberi*. Important native pasture species: *M. emersleyi, M. montana, M. pauciflora, M. wrightii*.

References, etc. Morphological/taxonomic: Soderstrom 1967. Leaf anatomical: Metcalfe 1960; this project; and for illustrations of leaf transverse sections showing photosynthetic pathway-related features, see Peterson *et al.* 1989.

Munroa J. Torr.

Hemimunroa Parodi

Habit, vegetative morphology. Prostrate *annual*; stoloniferous (low, spreading, much-branched, exhibiting long naked internodes, with fascicles of leaves and inflorescences). *Culms* 5–20 cm high; herbaceous. Culm internodes solid, or hollow. Leaves non-auriculate. The sheaths expanded. Leaf blades short, acuminate, pungent; narrow; flat, or folded, or rolled, or acicular; without abaxial multicellular glands; without cross venation; persistent. Ligule a fringe of hairs.

Reproductive organization. Plants bisexual, with bisexual spikelets; with hermaphrodite florets.

Inflorescence. Inflorescence reduced to a single spikelet, or few spikeleted; paniculate (of terminal, leafy clusters); contracted; capitate. Inflorescence with axes ending in spikelets. *Inflorescence* spatheate (i.e., the inflorescence leaves with broad sheaths); *a complex of 'partial inflorescences' and intervening foliar organs (in that the reduced panicles of small spikelet clusters are almost hidden among the fascicles of leaves)*. *Spikelet-bearing axes* very much reduced and paniculate (comprising small subsessile clusters of 2–4 spikelets); persistent, or disarticulating (the complete inflorescences deciduous in some species). Spikelets not secund; subsessile.

Female-fertile spikelets. *Spikelets* 6–8 mm long; *compressed dorsiventrally*; disarticulating above the glumes, or falling with the glumes (in species where the inflorescence falls entire); not disarticulating between the florets (when the inflorescence is shed), or disarticulating between the florets. Rachilla prolonged beyond the uppermost female-fertile floret. The rachilla extension with incomplete florets. Hairy callus absent.

Glumes two (lower spikelets), or one per spikelet (upper spikelets); very unequal, or more or less equal; shorter than the spikelets; shorter than the adjacent lemmas; hairless; pointed (acute, narrow); awnless. Lower glume 1 nerved. Upper glume 1 nerved. *Spikelets* with incomplete florets. The incomplete florets distal to the female-fertile florets (the proximal 1–2 florets female or hermaphrodite, those above hermaphrodite, the uppermost reduced or sterile). *The distal incomplete florets merely underdeveloped.*

Female-fertile florets 2–3 (upper spikelets), or 3–8 (lower spikelets). Lemmas less firm than the glumes to similar in texture to the glumes (lower leathery, upper membranous); not becoming indurated; incised; 4 lobed, or 2 lobed; mucronate, or awned (one or three mucronate or awned, from the nerves between the lobes). Awns when present, 1, or 3; median, or median and lateral; the median from a sinus, or apical; non-geniculate; hairless; much shorter than the body of the lemma (to 2 mm long). Lemmas hairy. The hairs in tufts (from the middle of each margin). Lemmas non-carinate; without a germination flap; 3 nerved. *Palea* present; relatively long; tightly clasped by the lemma; apically notched; awnless, without apical setae; not indurated (membranous); 2-nerved; 2-keeled. *Lodicules* present (small), or absent; 2; joined to the palea; membranous. *Stamens* 2, or 3. Anthers 1–1.5 mm long; not penicillate; without an apically prolonged connective. *Ovary* glabrous. Stigmas 2; white.

Fruit, embryo and seedling. Fruit small; ellipsoid; compressed laterally. Hilum short. Pericarp thin (translucent, oval); fused. Embryo large; with an epiblast; with a scutellar tail; with an elongated mesocotyl internode. Embryonic leaf margins meeting.

Abaxial leaf blade epidermis. *Costal/intercostal zonation* conspicuous. *Papillae* absent. Mid-intercostal long-cells rectangular (rather irregularly so, but not fusiform); having markedly sinuous walls (the walls quite thick). *Microhairs* present; more or less spherical; clearly two-celled; chloridoid-type (of the short basal cell type). Microhair apical cell wall of similar thickness/rigidity to that of the basal cell. Microhairs 24(–25.5) microns long. Microhair basal cells 9 microns long. Microhairs 11.4–14.4 microns wide at the septum. Microhair total length/width at septum 1.7–2.2. Microhair apical cells 10.5–15 microns long. Microhair apical cell/total length ratio 0.44–0.63. *Stomata* common; 24–27 microns long. Subsidiaries dome-shaped and triangular. Guard-cells overlapping to flush with the interstomatals. *Intercostal short-cells* common; in cork/silica-cell pairs, or not paired; silicified. Intercostal silica bodies imperfectly developed; tall-and-narrow. *Costal short-cells* conspicuously in long rows (over some veins), or predominantly paired (in places, especially near the leaf margins), or neither distinctly grouped into long rows nor predominantly paired (with some short rows, or the long rows irregularly broken by long 'short-cells'). Costal silica bodies present throughout the costal zones to present in alternate cell files of the costal zones; 'panicoid-type' (mostly), or saddle shaped (common only near the leaf margins); predominantly plump cross shaped and dumb-bell shaped; not sharp-pointed.

Transverse section of leaf blade, physiology. C_4; XyMS+. PCR sheaths of the primary vascular bundles complete. PCR sheath extensions absent. *Mesophyll* with radiate chlorenchyma. *Leaf blade* 'nodular' in section, or adaxially flat. *Midrib* conspicuous (via its fairly conspicuous rounded keel, and the lack of adaxial sclerenchyma); with one bundle only. Bulliforms present in discrete, regular adaxial groups; in simple fans (these sometimes deeply penetrating via a large median cell), or associated with colourless mesophyll cells to form deeply-penetrating fans. All the vascular bundles accompanied by sclerenchyma. Combined sclerenchyma girders present (with all bundles save the median); forming 'figures' (all but the midrib with massive anchors). Sclerenchyma not all bundle-associated. The 'extra' sclerenchyma in abaxial groups; abaxial-hypodermal, the groups isolated (apart from massive marginal sclerenchyma groups, there is a conspicuous group abaxially opposite each bulliform hinge group — i.e. flanking the midrib). The lamina margins with fibres.

Cytology. Chromosome base number, $x = 7$ and 8. $2n = 14$ and 16. 2 ploid.

Taxonomy. Chloridoideae; main chloridoid assemblage.

Ecology, geography, regional floristic distribution. 5 species. 1 in the western U.S.A., 4 in the central Andes. Xerophytic; species of open habitats. Dry plains.

Holarctic and Neotropical. Boreal and Madrean. Atlantic North American and Rocky Mountains. Andean. Central Grasslands.

References, etc. Leaf anatomical: Metcalfe 1960; this project.

Myriocladus Swallen

Habit, vegetative morphology. Perennial. The flowering culms leafy. *Culms woody and persistent (the lowermost internode usually disproportionately elongate, usually with the others condensed so that the leaves cluster atop the canes)*; not scandent; branched above. Primary branches/mid-culm node 1. Culm sheaths persistent (thickened, indurate). Culm internodes solid, or hollow. Unicaespitose. Rhizomes pachymorph. Plants unarmed. *Leaves* not basally aggregated; with auricular setae. *Leaf blades leathery*;

broad; pseudopetiolate, or not pseudopetiolate; cross veined, or without cross venation; disarticulating from the sheaths; rolled in bud.

Reproductive organization. Plants bisexual, with bisexual spikelets; with hermaphrodite florets.

Inflorescence. Inflorescence many spikeleted, or few spikeleted (rarely); of spicate main branches, or a single raceme, or paniculate (terminating leafy or leafless branches, sometimes terminating the culms, generally long and narrow, the branching racemose to paniculate); espatheate; not comprising 'partial inflorescences' and foliar organs. *Spikelet-bearing axes* 'racemes', or spikelike, or paniculate; persistent. Spikelets usually secund; sessile to pedicellate.

Female-fertile spikelets. Spikelets 3–4 mm long; disarticulating above the glumes; not disarticulating between the florets, or disarticulating between the florets. Rachilla prolonged beyond the uppermost female-fertile floret. The rachilla extension with incomplete florets. Hairy callus absent.

Glumes two; shorter than the adjacent lemmas; not pointed; awnless. *Spikelets* with incomplete florets. The incomplete florets both distal and proximal to the female-fertile florets. *The distal incomplete florets* merely underdeveloped. **Spikelets with proximal incomplete florets**. *The proximal incomplete florets* 1. The proximal lemmas awnless; similar in texture to the female-fertile lemmas.

Female-fertile florets *2–5*. Lemmas not becoming indurated; entire; pointed, or blunt; awnless. *Palea* present; relatively long; entire; awnless, without apical setae; 2-keeled (and sulcate). *Lodicules* present; 3; membranous; glabrous; not toothed; heavily vascularized. *Stamens* 3. Anthers not penicillate; without an apically prolonged connective. Styles fused (short). Stigmas 2.

Fruit, embryo and seedling. Hilum long-linear.

Abaxial leaf blade epidermis. *Costal/intercostal zonation* conspicuous. *Papillae* present; intercostal, or costal and intercostal (in places). Intercostal papillae over-arching the stomata (in rings); several per cell. *Long-cells* markedly different in shape costally and intercostally (the costals regular, elongated-rectangular, the intercostals less regular and generally relatively short); of similar wall thickness costally and intercostally. Intercostal zones with typical long-cells (though these largely interstomatal and relatively short, the stomata being abundant in most places). Mid-intercostal long-cells having markedly sinuous walls. *Microhairs* present; elongated; clearly two-celled; panicoid-type (large). *Stomata* common. Subsidiaries seemingly non-papillate; seemingly mostly triangular. Guard-cells with outlines obscured by the papillae. *Intercostal short-cells* absent or very rare. *Costal short-cells* predominantly paired. Costal silica bodies rounded (mostly), or saddle shaped (a few clear examples); not sharp-pointed.

Transverse section of leaf blade, physiology. C_3; XyMS+. Mesophyll with adaxial palisade; with arm cells; with fusoids. The fusoids external to the PBS. *Leaf blade* adaxially flat. *Midrib* not readily distinguishable. *The lamina* symmetrical on either side of the midrib. Bulliforms present in discrete, regular adaxial groups; in simple fans (these large, the median cell deep-penetrating). All the vascular bundles accompanied by sclerenchyma. Combined sclerenchyma girders present; forming 'figures' (a few bundles with I's, the adaxial arms slender). Sclerenchyma not all bundle-associated. The 'extra' sclerenchyma in abaxial groups and in adaxial groups; abaxial-hypodermal, the groups isolated and adaxial-hypodermal, contiguous with the bulliforms (there being fibres flanking the bulliform groups, and an abaxial fibre group opposite each bulliform group).

Taxonomy. Bambusoideae; Bambusodae; Bambuseae.

Ecology, geography, regional floristic distribution. 20 species. Venezuela. Helophytic, or mesophytic. Sandstone tablelands, over 1000 m.

Neotropical. Venezuela and Surinam.

References, etc. Leaf anatomical: this project.

Myriostachya J.D. Hook.

Habit, vegetative morphology. Perennial; with thick rhizomes. **Culms *150–300 cm high (often floating and thickened)*;** herbaceous. Culm internodes solid. Rhizomes pachymorph. Plants unarmed. **Leaves *mostly basal*;** non-auriculate. Leaf blades broad;

10–26 mm wide (?—up to 2.6 cm broad and 80 cm long); without abaxial multicellular glands; without cross venation; persistent. *Ligule a fringed membrane*. Contra-ligule absent.

Reproductive organization. Plants bisexual, with bisexual spikelets; with hermaphrodite florets.

Inflorescence. Inflorescence of spicate main branches and paniculate (a narrow panicle of numerous spike-like branches, tending to verticils); espatheate; not comprising 'partial inflorescences' and foliar organs. *Spikelet-bearing axes* spikelike; disarticulating, or persistent; if deciduous falling entire. Spikelets solitary; not secund; pedicellate.

Female-fertile spikelets. Spikelets 6–15 mm long; adaxial; compressed laterally; falling with the glumes and disarticulating above the glumes (falling with the pedicels, but the rachilla also disarticulating tardily); not disarticulating between the florets (or only tardily so). Rachilla prolonged beyond the uppermost female-fertile floret; rachilla hairless. The rachilla extension with incomplete florets. Hairy callus absent.

Glumes two; more or less equal; long relative to the adjacent lemmas; dorsiventral to the rachis; hairless (minutely scabrid); pointed (lanceolate-acuminate); **awned (acuminate into the straight awns)**; carinate; similar. Lower glume 1 nerved. Upper glume 1 nerved. *Spikelets* with incomplete florets. The incomplete florets distal to the female-fertile florets. *The distal incomplete florets* merely underdeveloped.

Female-fertile florets 2–20. Lemmas similar in texture to the glumes; not becoming indurated (leathery); entire; pointed; awnless (but cuspidate); hairless; glabrous; carinate; without a germination flap; 3 nerved. *Palea* present; relatively long; entire; awnless, without apical setae; not indurated (leathery or membranous); 2-nerved; 2-keeled. *Lodicules* present (minute); 2; free; fleshy; glabrous; not or scarcely vascularized. *Stamens* 3. Anthers very small; not penicillate; without an apically prolonged connective. *Ovary* glabrous. Styles fused. Stigmas 2.

Fruit, embryo and seedling. Fruit subterete. Hilum short (presumably). Pericarp fused (presumably). Embryo large.

Abaxial leaf blade epidermis. *Costal/intercostal zonation* conspicuous. *Papillae* absent. Long-cells of similar wall thickness costally and intercostally. Mid-intercostal long-cells rectangular; having markedly sinuous walls. *Microhairs* present; more or less spherical to elongated; clearly two-celled; chloridoid-type. Microhair apical cell wall of similar thickness/rigidity to that of the basal cell. Microhairs (27–)28.5–30(–33) microns long. Microhair basal cells 21 microns long. Microhairs 12.6–15.6 microns wide at the septum. Microhair total length/width at septum 1.8–2.3. Microhair apical cells 6–9 microns long. Microhair apical cell/total length ratio 0.22–0.29. *Stomata* common (very abundant, closely packed); 18–21 microns long. Subsidiaries dome-shaped (mostly), or triangular (a few). Guard-cells overlapping to flush with the interstomatals (but the ends of the subsidiaries tending to be overlapped by the 'corners' of the interstomatals). *Intercostal short-cells* common; in cork/silica-cell pairs and not paired; silicified. Intercostal silica bodies present and perfectly developed; saddle shaped. *Costal short-cells* predominantly paired (and in short rows), or neither distinctly grouped into long rows nor predominantly paired. Costal silica bodies present in alternate cell files of the costal zones; saddle shaped (mostly), or crescentic (a few, over the larger veins); not sharp-pointed.

Transverse section of leaf blade, physiology. C_4; XyMS+ (the mestome sheath cells very thick-walled). PCR sheaths of the primary vascular bundles interrupted; interrupted both abaxially and adaxially. PCR sheath extensions present. Maximum number of extension cells 6. Mesophyll traversed by columns of colourless mesophyll cells. *Leaf blade* adaxially flat. *Midrib* conspicuous (with adaxial lacunae); having a conventional arc of bundles (several); with colourless mesophyll adaxially (and lacunae). Bulliforms present in discrete, regular adaxial groups; associated with colourless mesophyll cells to form deeply-penetrating fans (these linking with traversing columns of colourless cells). All the vascular bundles accompanied by sclerenchyma. Combined sclerenchyma girders present; forming 'figures' (anchors). Sclerenchyma not all bundle-associated. The 'extra' sclerenchyma in abaxial groups.

Taxonomy. Chloridoideae; main chloridoid assemblage.

Ecology, geography, regional floristic distribution. 1 species. Indo-China, Thailand, Malay Peninsula, Indonesia, also India, Burma and Ceylon. Hydrophytic to helophytic; halophytic (in brackish water and tidal river margins).

Paleotropical. Indomalesian. Indian and Malesian.

References, etc. Leaf anatomical: this project.

Narduroides Rouy

Habit, vegetative morphology. Slender, erect, rigid *annual*. *Culms* 20–40 cm high; herbaceous. Leaves non-auriculate. Leaf blades linear; narrow; flat, or rolled (convolute when dry); without cross venation. Ligule an unfringed membrane; truncate; 0.7 mm long.

Reproductive organization. Plants bisexual, with bisexual spikelets; with hermaphrodite florets. *The spikelets all alike in sexuality*.

Inflorescence. Inflorescence unilateral, a single spike, or a single raceme (the pedicels to 0.5 mm long), or paniculate (with few branches below); rigid, the spikelets appressed, set in depressions in the rachis. Rachides hollowed. Inflorescence espatheate; not comprising 'partial inflorescences' and foliar organs. Spikelet-bearing axes persistent. *Spikelets* solitary; *secund*; *distichous*; subsessile (the pedicels no more than 0.5 mm long).

Female-fertile spikelets. Spikelets 3–7 mm long; compressed laterally; disarticulating above the glumes; disarticulating between the florets. Rachilla prolonged beyond the uppermost female-fertile floret. The rachilla extension naked. Hairy callus absent.

Glumes two; more or less equal; *long relative to the adjacent lemmas*; dorsiventral to the rachis; *not pointed (broadly rounded to emarginate)*; awnless; non-carinate; similar (with a prominent midrib). Lower glume 1–5 nerved. Upper glume 3–5 nerved. *Spikelets* with female-fertile florets only.

Female-fertile florets 4–6. Lemmas similar in texture to the glumes (papery, the margins hyaline); not becoming indurated; entire to incised (emarginate to shortly bifid, sometimes apiculate); when incised, 2 lobed; not deeply cleft; awnless; hairless; glabrous; non-carinate (to very slightly so); without a germination flap; 5 nerved. *Palea* present; relatively long; apically notched (shortly 2-fid); 2-nerved; 2-keeled. *Lodicules* present; 2; free; membranous; glabrous; toothed; not or scarcely vascularized. *Stamens* 3. *Ovary* glabrous. Styles free to their bases. Stigmas 2.

Fruit, embryo and seedling. *Fruit* adhering to lemma and/or palea (slightly, to the palea); small (1–1.8 mm long); not noticeably compressed. Hilum short. Embryo small. Endosperm hard; without lipid; containing compound starch grains. Embryo with an epiblast; without a scutellar tail; with a negligible mesocotyl internode. Embryonic leaf margins meeting.

Abaxial leaf blade epidermis. *Costal/intercostal zonation* conspicuous. *Papillae* present to absent (the cuticle seeming to undulate with reference to the lateral sinuations, in places forming low, oblique papillae). *Long-cells* markedly different in shape costally and intercostally (the costals much smaller and narrower); of similar wall thickness costally and intercostally (fairly thick walled). Mid-intercostal long-cells rectangular (quite regular); having markedly sinuous walls (the sinuosity coarse, associated with conspicuous pitting). *Microhairs* absent. *Stomata* absent or very rare. *Intercostal short-cells* absent or very rare (very few); in cork/silica-cell pairs (a few pairs seen); silicified. *Costal short-cells* neither distinctly grouped into long rows nor predominantly paired (mostly in pairs and short rows). Costal silica bodies present and well developed; horizontally-elongated crenate/sinuous and horizontally-elongated smooth, or rounded (many of these being more or less irregular).

Transverse section of leaf blade, physiology. C_3; XyMS+. *Leaf blade* with distinct, prominent adaxial ribs. *Midrib* fairly conspicuous (with a large, flat topped adaxial rib and an adaxial strand); with one bundle only. *The lamina* symmetrical on either side of the midrib. Bulliforms present in discrete, regular adaxial groups, or not in discrete, regular adaxial groups (? — not apparent in the poor material seen); if present, in simple fans. All the vascular bundles accompanied by sclerenchyma. Combined sclerenchyma girders absent (all the bundles with small abaxial strands, apart from the midrib with its adaxial strand as well). Sclerenchyma all associated with vascular bundles.

Cytology. Chromosome base number, $x = 7$. $2n = 14$. 2 ploid.

Taxonomy. Pooideae; Poodae; Poeae.

Ecology, geography, regional floristic distribution. 1 species. Mediterranean. Xerophytic.

Holarctic. Boreal and Tethyan. Euro-Siberian. Mediterranean. European.

References, etc. Leaf anatomical: this project.

Nardus L.

Natschia Bub.

Habit, vegetative morphology. Tough, wiry perennial; caespitose. *Culms* 10–50 cm high; herbaceous. Culm nodes glabrous. Culm internodes hollow. Young shoots extravaginal. *Leaves* mostly basal; non-auriculate. Smooth. Leaf blades linear (sharp pointed); narrow (about 0.5 mm in diameter, 4–30 cm long); setaceous; tightly rolled; without cross venation; persistent; rolled in bud. Ligule an unfringed membrane; not truncate (blunt); 2 mm long.

Reproductive organization. Plants bisexual, with bisexual spikelets; with hermaphrodite florets. Apomictic (non-pseudogamous), or reproducing sexually.

Inflorescence. Inflorescence a single spike (slender, one-sided). Inflorescence axes not ending in spikelets (terminating in a bristle up to 1 cm long). Rachides hollowed. Inflorescence espatheate; not comprising 'partial inflorescences' and foliar organs. Spikelet-bearing axes persistent. Spikelets solitary; secund; biseriate; sessile; loosely to closely imbricate.

Female-fertile spikelets. Spikelets 5–9 mm long (narrow); (the larger) more or less abaxial; compressed laterally; disarticulating above the glumes. Rachilla terminated by a female-fertile floret. Hairy callus absent. *Callus* short; blunt.

Glumes *present*; one per spikelet, or two; *minute*; very unequal; shorter than the adjacent lemmas (G₁ a small scale, G₂ minute or absent); *joined (their rudiments joined with one another and with the rachis)*; hairless; awnless. Lower glume 0 nerved. Upper glume 0 nerved. *Spikelets* with female-fertile florets only.

Female-fertile florets 1. Lemmas linear-lanceolate or lanceolate-oblong; decidedly firmer than the glumes; not becoming indurated (papery); entire; pointed; awned. Awns 1; median; apical; non-geniculate; hairless; much shorter than the body of the lemma (1–3 mm long). Lemmas hairless; non-carinate (2–3 keeled); 3 nerved. *Palea* present; relatively long; entire; awnless, without apical setae; thinner than the lemma; 2-nerved. *Lodicules* absent. *Stamens* 3. Anthers 2.5–4 mm long; not penicillate. *Ovary* glabrous. Styles fused. *Stigmas 1*.

Fruit, embryo and seedling. Fruit small. Hilum short (but linear). Embryo small; not waisted. Endosperm hard; without lipid; containing compound starch grains. Embryo with an epiblast; without a scutellar tail; with a negligible mesocotyl internode. Embryonic leaf margins meeting.

Seedling with a short mesocotyl; with a tight coleoptile. First seedling leaf with a well-developed lamina. The lamina narrow; erect; 3 veined.

Abaxial leaf blade epidermis. *Costal/intercostal zonation* conspicuous. *Papillae* absent. Long-cells of similar wall thickness costally and intercostally. Mid-intercostal long-cells rectangular; having markedly sinuous walls. *Microhairs* present; panicoid-type; (68–)72–96(–105) microns long; 6.6–10.5 microns wide at the septum. Microhair total length/width at septum 9–15.6. Microhair apical cells (34–)42–52(–54) microns long. Microhair apical cell/total length ratio 0.45–0.61. *Stomata* absent or very rare, or common; when present, 25.5–30–33 microns long. Subsidiaries triangular. Guard-cells overlapping to flush with the interstomatals. *Intercostal short-cells* common; in cork/silica-cell pairs (and solitary); silicified. Intercostal silica bodies tall-and-narrow and cubical. With abundant intercostal macrohairs. *Costal short-cells* neither distinctly grouped into long rows nor predominantly paired (mostly 1–5, but in a few places the intervening 'long-cells' are short). Costal silica bodies rounded, saddle shaped, and tall-and-narrow (intergrading, sometimes almost cubical); not sharp-pointed.

Transverse section of leaf blade, physiology. C_3; XyMS+. *Mesophyll* with non-radiate chlorenchyma. *Leaf blade* with distinct, prominent adaxial ribs; with the ribs more or less constant in size. *Midrib* conspicuous; with one bundle only. Bulliforms not in discrete, regular adaxial groups (small, variable in size, in the bases of the furrows). All the vascular bundles accompanied by sclerenchyma. Combined sclerenchyma girders present; forming 'figures'. Sclerenchyma all associated with vascular bundles.

Cytology. Chromosome base number, $x = 13$. $2n = 26$–30 (the chromosomes 'large'). 2 ploid.

Taxonomy. Arundinoideae; Nardeae.

Ecology, geography, regional floristic distribution. 1 species. Europe, western Asia. Commonly adventive. Xerophytic; species of open habitats; glycophytic. Moorland.

Holarctic. Boreal and Tethyan. Arctic and Subarctic, Euro-Siberian, and Atlantic North American. Macaronesian and Mediterranean. European. Canadian-Appalachian.

Economic importance. Very unpalatable to stock.

References, etc. Leaf anatomical: Metcalfe 1960; this project.

Narenga Bor

~ *Saccharum*

Habit, vegetative morphology. Robust perennial; stoloniferous. *Culms* 200–300 cm high; herbaceous. Culm internodes solid. *Leaf blades* broad to narrow; 4–30 mm wide (up to 1 m long); flat, or folded; *pseudopetiolate*; without cross venation. Ligule an unfringed membrane.

Reproductive organization. Plants bisexual, with bisexual spikelets; with hermaphrodite florets. *The spikelets all alike in sexuality*; homomorphic.

Inflorescence. *Inflorescence paniculate*; contracted (narrow, dense, woolly); espatheate; not comprising 'partial inflorescences' and foliar organs. **Spikelet-bearing axes** 'racemes'; with very slender rachides; *disarticulating*; disarticulating at the joints. **'Articles' non-linear (clavate)**; not appendaged; disarticulating transversely; somewhat hairy (at base, ciliolate on the lower margins, glabrous above). Spikelets paired; not secund; sessile and pedicellate; consistently in 'long-and-short' combinations; in pedicellate/sessile combinations. Pedicels of the 'pedicellate' spikelets free of the rachis. The 'shorter' spikelets hermaphrodite. The 'longer' spikelets hermaphrodite.

Female-fertile spikelets. Spikelets 2–3 mm long; compressed dorsiventrally; falling with the glumes (pedicellate spikelets falling from the pedicels, sessile ones falling with the adjacent joint and pedicel). Rachilla terminated by a female-fertile floret. Hairy callus present.

Glumes two; more or less equal; long relative to the adjacent lemmas; free; hairless (shiny, brown); awnless; carinate (G_2), or non-carinate (G_1); very dissimilar (leathery, the G_1 dorsally flattened, the G_2 keeled). Lower glume not two-keeled; flattened on the back; not pitted; relatively smooth. *Spikelets* with incomplete florets. The incomplete florets proximal to the female-fertile florets. *The proximal incomplete florets* 1; epaleate; sterile. The proximal lemmas awned, or awnless; similar in texture to the female-fertile lemmas (hyaline); not becoming indurated.

Female-fertile florets 1. **Lemmas** truncate; less firm than the glumes; entire; blunt; non-carinate; **0 nerved, or 1 nerved.** *Lodicules* present; ciliate (?). *Stamens* 3. *Ovary* glabrous.

Fruit, embryo and seedling. Fruit not noticeably compressed. Hilum short.

Phytochemistry. Leaves without flavonoid sulphates (1 species).

Cytology. Chromosome base number, $x = 5$ (?). $2n = 30$. 6 ploid.

Taxonomy. Panicoideae; Andropogonodae; Andropogoneae; Andropogoninae.

Ecology, geography, regional floristic distribution. 2 species. Indo-Malaysia. Paleotropical. Indomalesian. Indo-Chinese and Malesian.

Hybrids. Intergeneric hybrids with *Saccharum*, *Sclerostachya*.

Rusts and smuts. Smuts from Ustilaginaceae. Ustilaginaceae — *Sphacelotheca*.

Special comments. Anatomical data wanting.

Nassella Desv.

~ *Stipa* p.p.

Habit, vegetative morphology. Perennial; caespitose. *Culms* 25–65 cm high; herbaceous. Culm nodes glabrous. *Leaves* mostly basal. Leaf blades narrow; setaceous, or not setaceous; without cross venation; persistent; rolled in bud. Ligule an unfringed membrane; truncate. *Contra-ligule* absent.

Reproductive organization. Plants bisexual, with bisexual spikelets; with hermaphrodite florets. Exposed-cleistogamous, or chasmogamous.

Inflorescence. Inflorescence paniculate; deciduous in its entirety, or not deciduous; open; with capillary branchlets; espatheate; not comprising 'partial inflorescences' and foliar organs. Spikelet-bearing axes persistent. Spikelets not secund; pedicellate.

Female-fertile spikelets. *Spikelets* 1–3 mm long; *compressed laterally (but plump and gibbous)*; disarticulating above the glumes. Rachilla terminated by a female-fertile floret. Hairy callus present. *Callus* short; blunt.

Glumes two; more or less equal; exceeding the spikelets; long relative to the adjacent lemmas; hairless; glabrous; pointed; awned (acuminate into an awn), or awnless; noncarinate (rounded on the back); similar. Lower glume 3 nerved. Upper glume 3 nerved. *Spikelets* with female-fertile florets only.

Female-fertile florets 1. Lemmas convolute; saccate (above); decidedly firmer than the glumes; becoming indurated; entire; awned. *Awns* 1; *located asymmetrically*; dorsal; from near the top; geniculate; hairless (scabrid); much longer than the body of the lemma; entered by several veins; deciduous. Lemmas hairy, or hairless; when 'hairless', glabrous, or scabrous (or tuberculate); non-carinate; without a germination flap; 3 nerved (obscurely); with the nerves confluent towards the tip. *Palea* present; conspicuous but relatively short; entire; awnless, without apical setae; thinner than the lemma (scarious); not indurated; nerveless; keel-less. *Lodicules* present; 2 (in material seen); fleshy ('stipoid'); glabrous; not toothed; not or scarcely vascularized. *Stamens* 3. Anthers penicillate (in material seen), or not penicillate (?); without an apically prolonged connective. *Ovary* glabrous. Styles free to their bases. Stigmas 2.

Fruit, embryo and seedling. Fruit small; oblong to pyriform; compressed laterally. Hilum long-linear. Embryo large; waisted. Endosperm containing compound starch grains. Embryo with an epiblast; without a scutellar tail; with a negligible mesocotyl internode. Embryonic leaf margins meeting.

First seedling leaf with a well-developed lamina. The lamina narrow; curved.

Abaxial leaf blade epidermis. *Costal/intercostal zonation* conspicuous, or lacking. *Papillae* absent. *Long-cells* similar in shape costally and intercostally; of similar wall thickness costally and intercostally (thick-walled and pitted). Mid-intercostal long-cells rectangular; having markedly sinuous walls. *Microhairs* absent. *Stomata* absent or very rare; when seen 28.5–30 microns long. *Intercostal short-cells* common; in cork/silica-cell pairs; silicified. *Costal short-cells* conspicuously in long rows. Costal silica bodies 'panicoid-type', or tall-and-narrow; not sharp-pointed.

Transverse section of leaf blade, physiology. C_3; XyMS+. *Mesophyll* with non-radiate chlorenchyma; without adaxial palisade. *Leaf blade* with distinct, prominent adaxial ribs; with the ribs more or less constant in size. *Midrib* not readily distinguishable; with one bundle only. Bulliforms present in discrete, regular adaxial groups (in *Ammophila*-type groups of small cells, in the bases of the furrows). All the vascular bundles accompanied by sclerenchyma. Combined sclerenchyma girders present; forming 'figures'. Sclerenchyma not all bundle-associated. The 'extra' sclerenchyma in a continuous abaxial layer.

Phytochemistry. Tissues of the culm bases with abundant starch.

Special diagnostic feature. *Spikelet with a single gibbous floret, the lemma awn placed off-centre*.

Cytology. $2n = 38$.

Taxonomy. Arundinoideae; Stipeae.

Ecology, geography, regional floristic distribution. 15 species. Andes. Commonly adventive. Mesophytic to xerophytic; species of open habitats; glycophytic.

Neotropical and Antarctic. Caribbean and Andean. Patagonian.

Rusts and smuts. Rusts —*Puccinia*. Taxonomically wide-ranging species: *Puccinia graminella*.

Economic importance. Significant weed species: *N. trichotoma* (in pastures).

References, etc. Leaf anatomical: this project.

Nastus Juss.

Chloothamnus Büse, *Oreiostachys* Gamble, *Stemmatospermum* P. Beauv.

Habit, vegetative morphology. Arboresent or scrambling perennial. *Culms* woody and persistent; *to* 4.5 cm in diameter; *scandent*; branched above. Primary branches/mid-culm node many from middle and upper nodes. Rhizomes pachymorph. Plants unarmed. Leaf blades broad, or narrow; 3–30 mm wide; pseudopetiolate; disarticulating from the sheaths; rolled in bud. Ligule an unfringed membrane to a fringe of hairs.

Reproductive organization. Plants bisexual, with bisexual spikelets; with hermaphrodite florets.

Inflorescence. Inflorescence without pseudospikelets (the 'glumes' all barren); reduced to a single spikelet, or few spikeleted, or many spikeleted; a single raceme, or paniculate (terminating leafy branches); spatheate (the panicles bracteate), or espatheate; not comprising 'partial inflorescences' and foliar organs. *Spikelet-bearing axes* 'racemes', or paniculate; persistent. Spikelets not secund; subsessile, or pedicellate; consistently in 'long-and-short' combinations, or not in distinct 'long-and-short' combinations. The 'shorter' spikelets hermaphrodite. The 'longer' spikelets hermaphrodite.

Female-fertile spikelets. *Spikelets* 8.5–25 mm long; compressed laterally; *disarticulating between the glumes (2–3 of them deciduous with the floret)*. Rachilla prolonged beyond the uppermost female-fertile floret, or terminated by a female-fertile floret. The rachilla extension when present, with incomplete florets (with a rudiment).

Glumes present; several (4–6 'barren glumes'); increasing in size acropetally; shorter than the adjacent lemmas; hairy, or hairless; often apiculate. Lower glumes sometimes with setiform. Lower glume 9–13 nerved (in material seen). Upper glume 9–13 nerved (in material seen). *Spikelets* with female-fertile florets only, or with incomplete florets. The incomplete florets distal to the female-fertile florets. *The distal incomplete florets merely underdeveloped.*

Female-fertile florets 1. Lemmas similar in texture to the glumes (firmly membranous); entire; pointed; awnless, or mucronate (?); hairy, or hairless; 11–15 nerved (in material seen). *Palea* present; relatively long; apically notched; awnless, without apical setae; several nerved (13–17 in material seen); 2-keeled, or keel-less (when rachilla extension present). *Lodicules* present; 3 (very large); free; membranous; ciliate; heavily vascularized. **Stamens** 6; *with free filaments*. Anthers not penicillate; without an apically prolonged connective. **Ovary** hairy; *with a conspicuous apical appendage*. The appendage broadly conical, fleshy (expanded, fleshy, sometimes hollow, not attenuated into the style). Styles fused. Stigmas 3.

Fruit, embryo and seedling. Pericarp fleshy.

Abaxial leaf blade epidermis. *Costal/intercostal zonation* conspicuous. *Papillae* present. Intercostal papillae over-arching the stomata (which are sunken); several per cell (mostly large, around the stomata). *Long-cells* markedly different in shape costally and intercostally (the costals longer); of similar wall thickness costally and intercostally. Mid-intercostal long-cells rectangular; having markedly sinuous walls (thin). *Microhairs* present; panicoid-type; (54–)57–78(–81) microns long; 8.4–9.6 microns wide at the septum. Microhair total length/width at septum 6–9. Microhair apical cells (24–)25.5–39(–40.5) microns long. Microhair apical cell/total length ratio 0.43–0.56. *Stomata* common (obscured by papillae); 24–30(–34.5) microns long. *Intercostal short-cells* common, or absent or very rare; in cork/silica-cell pairs; silicified. Intercostal silica bodies tall-and-narrow, or saddle shaped. With large short-pointed, bulbous intercostal prickles, their bases ringed by pits in the adjoining cells. *Crown cells* absent. *Costal short-cells* conspicuously in long rows, or predominantly paired, or neither distinctly grouped into long rows nor predominantly paired (but often paired). Costal silica bodies saddle shaped, tall-and-narrow, crescentic, and oryzoid (predominantly saddles, sometimes merging into the others); not sharp-pointed.

Transverse section of leaf blade, physiology. C_3; XyMS+ (the MS sometimes double). Mesophyll with adaxial palisade; with arm cells; with fusoids. The fusoids external to the PBS. *Leaf blade* with distinct, prominent adaxial ribs, or 'nodular' in section; with the ribs more or less constant in size. *Midrib* conspicuous; with one bundle only (large, embedded in lignified tissue), or having complex vascularization. Bulliforms present in discrete, regular adaxial groups (these very large, in every intercostal zone); in simple fans, or in simple fans and associated with colourless mesophyll cells to form deeply-penetrating fans. All the vascular bundles accompanied by sclerenchyma. Combined sclerenchyma girders present (with most bundles); forming 'figures' (in many bundles). Sclerenchyma all associated with vascular bundles.

Taxonomy. Bambusoideae; Bambusodae; Bambuseae.

Ecology, geography, regional floristic distribution. 7 species. Malesia, New Guinea.

Paleotropical. Madagascan and Indomalesian. Malesian and Papuan.

References, etc. Leaf anatomical: Metcalfe 1960; this project.

Special comments. Fruit data wanting.

Neeragrostis Bush

~ *Eragrostis* ('*E. reptans*')

Habit, vegetative morphology. *Annual*; stoloniferous, or decumbent. *Culms* 5–15 cm high; herbaceous. *Leaves* not basally aggregated; non-auriculate. Leaf blades linear (acuminate); narrow; 1.5–3 mm wide (1–3 cm long); flat; without abaxial multicellular glands; without cross venation. Ligule a fringe of hairs; 0.5–1 mm long.

Reproductive organization. *Plants dioecious*; without hermaphrodite florets. The spikelets all alike in sexuality (on the one plant); female-only, or male-only. Plants outbreeding.

Inflorescence. Inflorescence of spicate main branches (in the male), or paniculate (in the female); very contracted (in the female); capitate (in the female, 'subcapitate' in the male); digitate (in the male), or non-digitate (in the densely contracted female panicles); espatheate; not comprising 'partial inflorescences' and foliar organs. Spikelets secund (male), or not secund (female); biseriate (male), or not two-ranked (female); subsessile to pedicellate.

Female-sterile spikelets. Male spikelets about 8 mm long, many-floreted, glumes unequal, lemmas 3 nerved; palea almost as long as lemma, 2-keeled; 2 fleshy lodicules, 3 stamens 1.5–2.5 mm long. Rachilla of male spikelets prolonged beyond the uppermost male floret. The male spikelets with glumes; without proximal incomplete florets; about 8–15 floreted. Male florets about 8–12; 3 staminate.

Female-fertile spikelets. *Spikelets* about 5–6 mm long; compressed laterally; disarticulating above the glumes; *not disarticulating between the florets (the rachilla and paleas persistent)*. Rachilla prolonged beyond the uppermost female-fertile floret. The rachilla extension with incomplete florets. Hairy callus absent. *Callus* absent.

Glumes two; very unequal; shorter than the spikelets; shorter than the adjacent lemmas; hairless; pointed; awnless; similar. Lower glume 1 nerved. Upper glume 1–2 nerved. *Spikelets* with incomplete florets. The incomplete florets distal to the female-fertile florets. *The distal incomplete florets* merely underdeveloped; awnless.

Female-fertile florets about 8–12. Lemmas similar in texture to the glumes; not becoming indurated; entire; pointed (acuminate); awnless; often hairy (sparsely villous); carinate; 3 nerved. *Palea* present; conspicuous but relatively short (about half as long as the lemma); awnless, without apical setae; not indurated; 2-nerved; 2-keeled. Palea keels wingless; hairy. *Lodicules* present; 2; free; fleshy; glabrous. **Stamens 0.** *Ovary* glabrous. Styles basally fused (forming a persistent beak on the fruit). Stigmas 2 (conspicuously emergent at anthesis).

Fruit, embryo and seedling. Fruit small (0.8–0.9 mm long); ellipsoid; compressed laterally. Hilum short. Pericarp fused. Embryo large (0.3–0.4 mm long); waisted. Endosperm hard. Embryo with an epiblast.

Abaxial leaf blade epidermis. *Costal/intercostal zonation* conspicuous. *Papillae* absent. Mid-intercostal long-cells rectangular; having markedly sinuous walls. *Microhairs* present; elongated; clearly two-celled; *Enneapogon*-type. Microhair apical cell wall of similar thickness/rigidity to that of the basal cell. Microhairs 90–250 microns long. Microhair basal cells 130–230 microns long. Microhair total length/width at septum 10–15. Microhair apical cells 15–25 microns long. Microhair apical cell/total length ratio about 0.1–0.2. *Stomata* common. *Intercostal short-cells* common; in cork/silica-cell pairs; silicified. Intercostal silica bodies present and perfectly developed; tall-and-narrow, or cubical. *Costal short-cells* neither distinctly grouped into long rows nor predominantly paired. Costal silica bodies present in alternate cell files of the costal zones; 'panicoid-type'; not sharp-pointed.

Transverse section of leaf blade, physiology. C_4; XyMS+. PCR sheaths of the primary vascular bundles interrupted; interrupted abaxially only. PCR sheath extensions absent. *Leaf blade* with distinct, prominent adaxial ribs to 'nodular' in section; with the ribs more or less constant in size. *Midrib* not readily distinguishable; with one bundle only. Bulliforms present in discrete, regular adaxial groups; seemingly exclusively in simple fans. All the vascular bundles accompanied by sclerenchyma. Sclerenchyma all associated with vascular bundles. The lamina margins with fibres.

Taxonomy. Chloridoideae; main chloridoid assemblage.

Ecology, geography, regional floristic distribution. 1 species. Southern U.S.A., Mexico to Surinam. Mesophytic; species of open habitats; glycophytic. River banks, sandy and open ground.

Holarctic and Neotropical. Boreal and Madrean. Caribbean.

References, etc. Morphological/taxonomic: Nicora 1962. Leaf anatomical: Nicora 1962.

Special comments. Very distinct from typical *Eragrostis*, judging from Nicora's thorough description; but the distinction is blurred by (e.g.) *E. hypnoides* (Lam.) B.S.P. and especially by *E. contrerasii* Pohl, which is also dioecious with creeping habit.

Neesiochloa Pilger

Habit, vegetative morphology. *Annual*; caespitose. *Culms* 26 cm high; herbaceous; unbranched above. Plants unarmed. *Leaves* not basally aggregated; non-auriculate. Leaf blades narrow; 0.5–2 mm wide (?—to 2 mm wide, the margins cartilaginous, with tubercular hairs marginally and abaxially); not setaceous; flat; without cross venation; persistent. Ligule a fringed membrane. *Contra-ligule* absent.

Reproductive organization. Plants bisexual, with bisexual spikelets; with hermaphrodite florets.

Inflorescence. Inflorescence few spikeleted; very sparse, a single raceme (with long, capillary pedicels), or paniculate (the lower branches sometimes branched again); open; espatheate; not comprising 'partial inflorescences' and foliar organs. Spikelet-bearing axes persistent. Spikelets solitary; not secund; pedicellate (the long pedicels glandular).

Female-fertile spikelets. *Spikelets 5–9 mm long (red tinged, reminiscent of* **Briza**, *but not pendulous)*; suborbicular; compressed laterally (but plump); disarticulating above the glumes; disarticulating between the florets; with conventional internode spacings. Rachilla prolonged beyond the uppermost female-fertile floret; rachilla very hairy. The rachilla extension with incomplete florets. *Hairy callus present*.

Glumes two; more or less equal; shorter than the spikelets; long relative to the adjacent lemmas; free; hairless; glabrous; pointed; awned (attenuate into short awns); carinate; similar (thinly membranous, ovate acuminate). Lower glume 1 nerved. Upper glume 1 nerved. *Spikelets* with incomplete florets. The incomplete florets distal to the female-fertile florets. *The distal incomplete florets* merely underdeveloped.

Female-fertile florets 4–8. *Lemmas fan-shaped, very broad above — much broader than long*; similar in texture to the glumes (membranous); not becoming indurated; incised; not deeply cleft (emarginate); awned. Awns 1; median; from a sinus; non-geniculate; hairless (scabrid); about as long as the body of the lemma to much longer than the body of the lemma; entered by one vein. Lemmas hairy (along the nerves); carinate; without a germination flap; 3 nerved. **Palea** present; *relatively long (bent sharply inwards over the pistil, with hair tufts at the elbows)*; entire; awnless, without apical setae; not indurated (membranous); 2-nerved; 2-keeled. Palea keels hairy. *Lodicules* present; 2; free; small, rather thin (not the usual cuneate type); glabrous. *Stamens* 3; with free filaments (fairly long). Anthers very short; not penicillate; without an apically prolonged connective. *Ovary* glabrous. Styles free to their bases. Stigmas 2; red pigmented.

Fruit, embryo and seedling. *Fruit* small (about 1.5 mm long); *longitudinally grooved*; compressed laterally. Hilum short. *Pericarp free*. Embryo large; slightly waisted.

Abaxial leaf blade epidermis. *Costal/intercostal zonation* conspicuous. *Papillae* absent. *Long-cells* markedly different in shape costally and intercostally (costals longer and narrower); of similar wall thickness costally and intercostally (walls of medium thickness). Mid-intercostal long-cells rectangular; having markedly sinuous walls. *Microhairs* present; more or less spherical to elongated; clearly two-celled; chloridoid-type; (30–)31.5–36(–39) microns long; (12.6–)14.4–15 microns wide at the septum. Microhair total length/width at septum 2–2.7. Microhair apical cells (9–)9.6–10.5(–12) microns long. Microhair apical cell/total length ratio 0.27–0.36. *Stomata* common; 24–30 microns long. Subsidiaries mostly dome-shaped. Guard-cells overlapping to flush with the interstomatals. *Intercostal short-cells* common (fairly); not paired (solitary); not silicified. Cushion-based macrohairs present. *Costal short-cells* conspicuously in long rows (the short-cells rather long in some files). Costal silica bodies saddle shaped (predominating, save over the midrib and blade margins), or rounded (common over some veins), or oryzoid (a few), or 'panicoid-type'; including a few short dumb-bell shaped, or cross shaped; not sharp-pointed.

Transverse section of leaf blade, physiology. C_4; XyMS+. PCR sheath extensions absent. *Mesophyll* with radiate chlorenchyma. *Leaf blade* adaxially flat. *Midrib* not readily distinguishable; with one bundle only. Bulliforms present in discrete, regular adaxial groups; associated with colourless mesophyll cells to form deeply-penetrating fans. All the vascular bundles accompanied by sclerenchyma. Combined sclerenchyma girders present (with all bundles); forming 'figures' (all the large bundles anchor-shaped). Sclerenchyma all associated with vascular bundles.

Taxonomy. Chloridoideae; main chloridoid assemblage.

Ecology, geography, regional floristic distribution. 1 species. Brazil. Mesophytic; species of open habitats. Disturbed ground.

Neotropical. Central Brazilian.

References, etc. Leaf anatomical: this project.

Nematopoa C.E. Hubb.

Triraphis longipes, Crinipes longipes

Habit, vegetative morphology. Perennial; loosely caespitose. Culms herbaceous; unbranched above. Young shoots extravaginal. Leaves non-auriculate. Leaf blades narrow; setaceous (filiform); without cross venation. **Ligule a fringe of hairs**.

Reproductive organization. Plants bisexual, with bisexual spikelets; with hermaphrodite florets. The spikelets all alike in sexuality.

Inflorescence. *Inflorescence paniculate*; open (ovate, oblong); **with capillary branchlets**; espatheate; not comprising 'partial inflorescences' and foliar organs. Spikelet-bearing axes persistent. Spikelets not secund; pedicellate.

Female-fertile spikelets. *Spikelets compressed laterally*; disarticulating above the glumes; disarticulating between the florets. Rachilla prolonged beyond the uppermost female-fertile floret; rachilla hairless (scabridulous). The rachilla extension with incomplete florets. Hairy callus present (densely bearded). *Callus* short.

Glumes two; very unequal; shorter than the spikelets; shorter than the adjacent lemmas (small); awnless; similar (thinly membranous, hyaline, G_1 narrowly lanceolate, G_2 broader). Lower glume 0 nerved, or 1 nerved. Upper glume 0 nerved, or 1 nerved.

Spikelets with incomplete florets. The incomplete florets distal to the female-fertile florets. *The distal incomplete florets* merely underdeveloped.

Female-fertile florets *4–7*. **Lemmas** similar in texture to the glumes to decidedly firmer than the glumes (membranous); not becoming indurated; *incised*; 2 lobed (shortly and narrowly bilobed, the lobes commonly mucronulate); not deeply cleft; awned. Awns 1; median; from a sinus; non-geniculate; flexuous (very thin). Lemmas hairy; non-carinate (convex); *3 nerved (the laterals near the margins)*. *Palea* present (narrowly oblong); relatively long; entire (obtuse); awnless, without apical setae; not indurated; 2-nerved; 2-keeled. *Lodicules* present (minute); 2; free; fleshy; glabrous. *Stamens* 3. *Ovary* glabrous. Styles free to their bases. Stigmas 2; red pigmented.

Fruit, embryo and seedling. *Fruit* broadly *fusiform*; not noticeably compressed. Hilum short. Embryo large; not waisted. Endosperm hard.

Abaxial leaf blade epidermis. *Costal/intercostal zonation* conspicuous. *Papillae* absent. Mid-intercostal long-cells rectangular; having markedly sinuous walls. *Microhairs* present; panicoid-type; 34–50 microns long; 5.5–9 microns wide at the septum. Microhair total length/width at septum 5.2–6.9. Microhair apical cells 16–30 microns long. Microhair apical cell/total length ratio 0.45–0.6. *Stomata* absent or very rare. *Intercostal short-cells* common; in cork/silica-cell pairs; silicified. *Costal short-cells* conspicuously in long rows (but the files frequently interrupted by large prickles). Costal silica bodies 'panicoid-type', or oryzoid (mostly panicoid type, but a few vertically elongated crosses); nearly all cross shaped; not sharp-pointed.

Taxonomy. Arundinoideae; Danthonieae (?).

Ecology, geography, regional floristic distribution. 1 species. South Africa. Mesophytic (in moist places).

Paleotropical. African. Sudano-Angolan. South Tropical African.

References, etc. Morphological/taxonomic: Hubbard 1957. Leaf anatomical: this project.

Special comments. Anatomical data epidermal only.

Neobouteloua Gould

Habit, vegetative morphology. Perennial; rhizomatous and decumbent. *Culms* 12–40 cm high; herbaceous; unbranched above. Culm nodes glabrous. Culm internodes solid. Plants unarmed. *Leaves* not basally aggregated; non-auriculate. Sheaths longer than the internodes. *Leaf blades* linear-lanceolate to lanceolate; narrow; to 2.5 mm wide; flat (rigid, pungent-tipped); without abaxial multicellular glands; *not pseudopetiolate*; without cross venation; persistent. *Ligule* present; a fringed membrane to a fringe of hairs. *Contra-ligule* absent.

Reproductive organization. *Plants bisexual, with bisexual spikelets*; with hermaphrodite florets. The spikelets all alike in sexuality (except perhaps for depauperate spikelets at the inflorescence tip).

Inflorescence. *Inflorescence of spicate main branches*; fairly open. Primary inflorescence branches 12–20 (or more). *Inflorescence with axes ending in spikelets*. Rachides triquetrous, ridged. Inflorescence espatheate; not comprising 'partial inflorescences' and foliar organs. Spikelet-bearing axes with very slender rachides; persistent. Spikelets solitary; secund; biseriate.

Female-fertile spikelets. *Spikelets 2 mm long*; adaxial; compressed laterally (but the individual florets dorsally compressed); disarticulating above the glumes; the rachilla somewhat elongated beneath the crown of sterile florets. Rachilla prolonged beyond the uppermost female-fertile floret; rachilla hairless. The rachilla extension with incomplete florets. Hairy callus present. *Callus* short.

Glumes two; very unequal; shorter than the adjacent lemmas (somewhat); dorsiventral to the rachis; hairless; glabrous; pointed; awnless (but G_2 mucronate); carinate; *very dissimilar (both hyaline, G_1 pointed, G_2 acuminate)*. Lower glume 1 nerved. Upper glume 1 nerved. *Spikelets* with incomplete florets. The incomplete florets distal to the female-fertile florets. *The distal incomplete florets 3–5*; **incomplete florets** *clearly specialised and modified in form (clustered at rachilla tip, reduced to their lemmas, these* *each deeply incised into three long, divergent, equal awns and the awns of all the lemmas forming a crown). Spikelets without proximal incomplete florets.*

Female-fertile florets 1. **Lemmas** decidedly firmer than the glumes (thinly membranous); not becoming indurated; incised; *3 lobed (the lobes attenuate into the awns)*; deeply cleft; awned. Awns 3; median and lateral; the median different in form from the laterals (curved back); apical; non-geniculate; recurving; hairless (scabrid); about as long as the body of the lemma; entered by one vein. The lateral awns shorter than the median (upright). Lemmas hairy (on the back and along lateral nerves); non-carinate; without a germination flap; 3 nerved. *Palea* present; relatively long; tightly clasped by the lemma; entire (spathulate); awnless, without apical setae (glabrous); thinner than the lemma; not indurated (hyaline); 2-nerved; 2-keeled. Palea keels wingless; scabrous. *Lodicules* present; 2; free; tiny and delicate; glabrous; not or scarcely vascularized. *Stamens* 3. Anthers very short; not penicillate; without an apically prolonged connective. *Ovary* glabrous. Styles free to their bases. Stigmas 2; red pigmented.

Fruit, embryo and seedling. *Fruit* free from both lemma and palea; small (to about 1.5 mm long); ellipsoid; trigonous. Hilum short. Pericarp fused. Embryo large.

Abaxial leaf blade epidermis. *Costal/intercostal zonation* conspicuous (the intercostal regions sunken). *Papillae* present; costal and intercostal. Intercostal papillae over-arching the stomata (slightly, at one end, sometimes), or not over-arching the stomata; consisting of one oblique swelling per cell (involving interstomatals), or consisting of one symmetrical projection per cell (elsewhere, these small, circular, usually displaced towards on end of the cell). Mid-intercostal long-cells having markedly sinuous walls. *Microhairs* present; more or less spherical to elongated; clearly two-celled; chloridoid-type. Microhair apical cell wall of similar thickness/rigidity to that of the basal cell. Microhairs 24–27(–30) microns long. Microhair basal cells 12 microns long. Microhairs 9–10.5–12 microns wide at the septum. Microhair total length/width at septum 2.3–2.9. Microhair apical cells 7.5–9(–12) microns long. Microhair apical cell/total length ratio 0.28–0.38. *Stomata* common; 19.5–21–22.5 microns long. Subsidiaries dome-shaped and triangular. Guard-cells overlapped by the interstomatals (slightly). *Intercostal short-cells* absent or very rare (apparently). Intercostal silica bodies absent. *Costal short-cells* conspicuously in long rows (but the 'short-cells' quite long in some files). Costal silica bodies present in alternate cell files of the costal zones; almost exclusively saddle shaped; not sharp-pointed.

Transverse section of leaf blade, physiology. C_4; XyMS+. PCR sheaths of the primary vascular bundles interrupted; interrupted abaxially only. PCR sheath extensions absent. *Leaf blade* 'nodular' in section; with the ribs more or less constant in size. *Midrib* conspicuous; having a conventional arc of bundles (with a small bundle on either side of the large median); with colourless mesophyll adaxially. Bulliforms present in discrete, regular adaxial groups (the groups not very conspicuous); seemingly exclusively in simple fans. All the vascular bundles accompanied by sclerenchyma. Combined sclerenchyma girders present (with all the bundles); forming 'figures' (all the bundles). Sclerenchyma all associated with vascular bundles. The lamina margins with fibres.

Taxonomy. Chloridoideae; main chloridoid assemblage.

Ecology, geography, regional floristic distribution. 1 species. Argentina. Species of open habitats. Dry plains and hillsides.

Neotropical. Pampas.

References, etc. Leaf anatomical: this project.

Neohouzeaua A. Camus

~ Shizostachyum

Habit, vegetative morphology. Perennial (large bushes). Culms woody and persistent; scandent, or not scandent; branched above. Rhizomes pachymorph. *Leaves* not basally aggregated. Leaf blades pseudopetiolate; disarticulating from the sheaths.

Reproductive organization. Plants bisexual, with bisexual spikelets; with hermaphrodite florets.

Inflorescence. **Inflorescence** indeterminate (presumably?); *with pseudospikelets (presumably?)*; paniculate (dense, oblong). Spikelet-bearing axes persistent. Spikelets not secund; pedicellate.

Female-fertile spikelets. Rachilla prolonged beyond the uppermost female-fertile floret (?).

Glumes present; '3–4'; awnless (mucronate); similar.

Female-fertile florets 1. Lemmas twisted; similar in texture to the glumes; not becoming indurated; mucronate to awned (or almost awned). **Palea *absent*. *Lodicules* absent ('or very small'). **Stamens** 6; *monadelphous*. *Ovary with a conspicuous apical appendage*. The appendage long, stiff and tapering. Styles fused (into one prolonged style). Stigmas 3.

Fruit, embryo and seedling. Pericarp thick and hard; free (the fruit a nut).

Taxonomy. Bambusoideae; Bambusodae; Bambuseae.

Ecology, geography, regional floristic distribution. 4 species. Southeast Asia. Paleotropical. Indomalesian. Indo-Chinese.

Special comments. Description very inadequate. Fruit data wanting. Anatomical data wanting.

Neostapfia Davy

Davyella Hack., *Stapfia* Davy

Habit, vegetative morphology. Viscid annual; culms ascendng from the decumbent base. *Culms* 7–30 cm high; herbaceous; unbranched above. Culm nodes glabrous. Culm internodes solid. Plants unarmed. The shoots aromatic. *Leaves* not basally aggregated; not clearly differentiated into sheath and blade; spirally disposed (seemingly, in material seen); non-auriculate. Leaf blades broad to narrow; 5–12 mm wide (not distinctly separated from the sheaths, 5–10 cm long, the leaves about 12 mm wide in middle); exhibiting multicellular glands abaxially. The abaxial glands on the blade margins. Leaf blades not pseudopetiolate; without cross venation; persistent. *Ligule* absent. *Contra-ligule* absent.

Reproductive organization. Plants bisexual, with bisexual spikelets; with hermaphrodite florets. The spikelets of sexually distinct forms on the same plant (interpreting the terminal inflorescence bracts as abortivente spikelets); hermaphrodite and sterile.

Inflorescence. *Inflorescence a single raceme (dense, spicate, the spikelets spiralled)*; contracted; soft, dense, 3–7 cm long. Inflorescence axes not ending in spikelets (the upper part of the inflorescence axis with linear-lanceolate, empty 'bracts' instead of spikelets). *Inflorescence espatheate*; not comprising 'partial inflorescences' and foliar organs. Spikelet-bearing axes persistent. Spikelets not secund; shortly pedicellate (the pedicels persistent).

Female-fertile spikelets. Spikelets *unconventional (not only lacking glumes, but having the lemmas spiralled on the rachilla)*; 6–7 mm long; compressed dorsiventrally; falling from their pedicels, above the glume vestiges; disarticulating between the florets. Rachilla prolonged beyond the uppermost female-fertile floret; rachilla hairless. The rachilla extension with incomplete florets. Hairy callus absent.

Glumes absent. *Spikelets* with incomplete florets. The incomplete florets distal to the female-fertile florets (the rachilla ending in an incomplete floret). *The distal incomplete florets* merely underdeveloped. *Spikelets without proximal incomplete florets*.

Female-fertile florets 2–5. **Lemmas** *conspicuously non-distichous*; *flabellate, broad, fringed*; not becoming indurated (thinly membranous); entire; blunt; awnless; hairless; non-carinate; without a germination flap; 9–15 nerved. *Palea* present (narrower than the lemma); relatively long (somewhat shorter than the lemma); apically notched; awnless, without apical setae; not indurated (hyaline); 2-nerved; 2-keeled. Palea keels glabrous. *Lodicules* present; 2; free; fleshy; ciliate, or glabrous; not or scarcely vascularized. *Stamens* 3 (the filaments short and thick). Anthers long; not penicillate; without an apically prolonged connective. *Ovary* glabrous. Styles free to their bases. Stigmas 2.

Fruit, embryo and seedling. Fruit compressed laterally. Hilum short (large). Pericarp with a thick, viscid pericarp, obscuring the embryo. Embryo large (but obscured);

with an epiblast; with a scutellar tail; with an elongated mesocotyl internode; with one scutellum bundle. Embryonic leaf margins meeting.

Abaxial leaf blade epidermis. *Costal/intercostal zonation* lacking (and no short-cell/long-cell organisation). *Papillae* absent. Intercostal zones without typical long-cells. Mid-intercostal long-cells epidermal walls straight. *Microhairs* present; more or less spherical; clearly two-celled; chloridoid-type (button-mushroom type, cf. *Orcuttia*). Microhair apical cell wall of similar thickness/rigidity to that of the basal cell. Microhair basal cells 6–9 microns long. Microhair total length/width at septum 3.1. Microhair apical cell/total length ratio 0.48. *Stomata* common; (21–)22.5–24 microns long. Subsidiaries tall dome-shaped (but the guard-cells are not typically poaceous). Guard-cells overlapped by the interstomatals. *Intercostal short-cells* absent or very rare. Intercostal silica bodies absent. With 'crozier' macrohairs, plus large capitate multicellular glands, especially at the blade margins. Costal zones without short-cells. *Costal short-cells* lacking. Costal silica bodies absent.

Transverse section of leaf blade, physiology. C_4; XyMS+. PCR sheaths of the primary vascular bundles complete. PCR sheath extensions absent. *Mesophyll* with radiate chlorenchyma. *Leaf blade* 'nodular' in section to adaxially flat. *Midrib* not readily distinguishable; with one bundle only. Bulliforms present in discrete, regular adaxial groups; in simple fans (and in irregular groups). Many of the smallest vascular bundles unaccompanied by sclerenchyma (in fact, all the bundles lack sclerenchyma accompaniment). Combined sclerenchyma girders absent.

Culm anatomy. *Culm internode bundles* scattered.

Cytology. Chromosome base number, $x = 10$. $2n = 40$. 4 ploid.

Taxonomy. Chloridoideae; Orcuttieae.

Ecology, geography, regional floristic distribution. 1 species. Southwestern U.S.A. Helophytic.

Holarctic. Boreal. Atlantic North American. Southern Atlantic North American.

Rusts and smuts. Rusts — *Puccinia*. Taxonomically wide-ranging species: *Puccinia graminis*.

References, etc. Leaf anatomical: Metcalfe 1960; this project.

Special comments. Fruit data wanting.

Neostapfiella A. Camus

Habit, vegetative morphology. Annual; stoloniferous. *Culms* 15–30 cm high; herbaceous; unbranched above. Culm nodes glabrous. **Leaves** *mostly basal*; non-auriculate. The sheaths flattened. *Leaf blades* narrow; 3–4 mm wide; without abaxial multicellular glands; *pseudopetiolate*; without cross venation. Ligule a fringed membrane.

Reproductive organization. Plants bisexual, with bisexual spikelets; with hermaphrodite florets.

Inflorescence. *Inflorescence* of spicate main branches; *digitate*. Primary inflorescence branches 2–4. Inflorescence espatheate; not comprising 'partial inflorescences' and foliar organs. Spikelet-bearing axes persistent. Spikelets solitary; subsessile.

Female-fertile spikelets. Spikelets 4.5–5.5 mm long; cuneate; adaxial; compressed laterally; disarticulating above the glumes. Rachilla prolonged beyond the uppermost female-fertile floret (when the L_2 has an incomplete floret), or terminated by a female-fertile floret. The rachilla extension when present, with incomplete florets. Hairy callus present. Callus pointed (oblong to linear).

Glumes two; shorter than the spikelets; shorter than the adjacent lemmas; dorsiventral to the rachis; hairless; pointed; awnless; carinate; similar (membranous). Lower glume 1 nerved. Upper glume 1 nerved. *Spikelets* with female-fertile florets only, or with incomplete florets. The incomplete florets distal to the female-fertile florets. *The distal incomplete florets* merely underdeveloped.

Female-fertile florets *2 (usually), or 1 (the L_2 being sometimes male or sterile)*. Lemmas decidedly firmer than the glumes (leathery); becoming indurated; incised; 2 lobed; not deeply cleft (bidentate); awned. Awns 1; median; dorsal; from near the top; non-geniculate; hairless (scabrid); much longer than the body of the lemma. Lemmas hairy (pubescent on the back, ciliate near the margins); carinate; prominently 3 nerved (the outer nerves not flanked by wings). *Palea* present; relatively long; awnless, without apical setae.

Fruit, embryo and seedling. Fruit lanceolate; trigonous. Pericarp fused.

Abaxial leaf blade epidermis. *Costal/intercostal zonation* conspicuous. *Papillae* present (in *N. humbertiana*), or absent (*N. perrieri*); in *N. humbertiana*, costal and intercostal (but much more conspicuous intercostally). Intercostal papillae over-arching the stomata (at one end, at least in places); several per cell (small, circular and thick walled, or larger and oblique at the ends of interstomatals). *Long-cells* similar in shape costally and intercostally to markedly different in shape costally and intercostally (the costals mostly narrower); of similar wall thickness costally and intercostally (of medium thickness). Intercostal zones with typical long-cells. Mid-intercostal long-cells rectangular; having markedly sinuous walls. *Microhairs* present; more or less spherical to elongated; clearly two-celled; chloridoid-type. Microhair apical cell wall of similar thickness/rigidity to that of the basal cell. Microhair basal cells 6–7 microns long. Microhair total length/width at septum 1.6. Microhair apical cell/total length ratio 0.6. *Stomata* common. Subsidiaries non-papillate; mostly more or less triangular (in *N. perrieri*), or dome-shaped and triangular (*N. humbertiana*). Guard-cells overlapping to flush with the interstomatals. *Intercostal short-cells* common (inconspicuous in *N. humbertiana*); in cork/silica-cell pairs; silicified (*N.perrieri*), or not silicified (? *N. humbertiana*). Intercostal silica bodies imperfectly developed, or present and perfectly developed; rounded and tall-and-narrow (or more or less misshapen, mostly very small). No prickles or macrohairs seen. *Costal short-cells* conspicuously in long rows. Costal silica bodies present and well developed; present in alternate cell files of the costal zones; very consistently saddle shaped.

Transverse section of leaf blade, physiology. C_4; XyMS+. PCR sheath outlines even (judging from poorly preserved material of *N. perrieri*), or uneven (judging from poorly preserved material of *N. humbertiana*). PCR sheaths of the primary vascular bundles interrupted; interrupted both abaxially and adaxially. PCR sheath extensions absent. *Leaf blade* with distinct, prominent adaxial ribs, or adaxially flat (more or less, in *N. humbertiana*); with the ribs more or less constant in size. *Midrib* conspicuous (by the large, rounded abaxial keel, an adaxial bulliform group and conspicuous 'hinges'); with one bundle only; with colourless mesophyll adaxially (*N. perrieri*), or without colourless mesophyll adaxially (*N. humbertiana*). *The lamina* symmetrical on either side of the midrib. Bulliforms present in discrete, regular adaxial groups, or not in discrete, regular adaxial groups (the groups, if present, small and indistinct in the material seen); seemingly in simple fans (only). Many of the smallest vascular bundles unaccompanied by sclerenchyma, or all the vascular bundles accompanied by sclerenchyma. Combined sclerenchyma girders present; forming 'figures' (heavy I's with the large bundles in *N. perrieri*, light ones in *N. humbertiana*). Sclerenchyma all associated with vascular bundles.

Taxonomy. Chloridoideae; main chloridoid assemblage.

Ecology, geography, regional floristic distribution. 3 species. Madagascar. Helophytic, or mesophytic; halophytic, or glycophytic. Savanna.

Paleotropical. Madagascan.

References, etc. Leaf anatomical: this project.

Special comments. Fruit data wanting.

Nephelochloa Boiss.

Habit, vegetative morphology. *Annual*. Culms herbaceous. Leaves non-auriculate. Leaf blades without cross venation. Ligule an unfringed membrane; not truncate; 3–4 mm long.

Reproductive organization. Plants bisexual, with bisexual spikelets; with hermaphrodite florets. The spikelets all alike in sexuality.

Inflorescence. *Inflorescence paniculate (with many slender branches at each node)*; *open (the basal branchlets sterile)*; espatheate; not comprising 'partial inflorescences' and foliar organs. Spikelet-bearing axes persistent. Spikelets not secund; pedicellate.

Female-fertile spikelets. *Spikelets 3–3.3 mm long*; compressed laterally; disarticulating above the glumes; disarticulating between the florets. Rachilla prolonged beyond the uppermost female-fertile floret; rachilla hairy (beneath the spikelets). *Hairy callus present*.

Glumes two; very unequal to more or less equal; shorter than the adjacent lemmas, or long relative to the adjacent lemmas; pointed (acute); awnless; *non-carinate*; similar. Lower glume 1 nerved. Upper glume 3 nerved. *Spikelets* with female-fertile florets only, or with incomplete florets. The incomplete florets distal to the female-fertile florets.

Female-fertile florets 2–5. Lemmas similar in texture to the glumes to decidedly firmer than the glumes (membranous); not becoming indurated; incised; 2 lobed (bidentate); awned. *Awns* 1; median; *from a sinus (from a deep notch)*; non-geniculate (fine). Lemmas hairless; glabrous; non-carinate; without a germination flap; 5 nerved. *Palea* present; relatively long; 2-nerved; 2-keeled. **Lodicules** present; 2; free; *membranous*; glabrous; toothed; not or scarcely vascularized. *Stamens* 3. *Ovary* glabrous. Styles free to their bases. Stigmas 2.

Fruit, embryo and seedling. Fruit small (0.8 mm long). Hilum short. Embryo small. Endosperm containing compound starch grains. Embryo with an epiblast; without a scutellar tail; with a negligible mesocotyl internode. Embryonic leaf margins meeting.

Abaxial leaf blade epidermis. *Costal/intercostal zonation* conspicuous. *Papillae* absent. Long-cells differing markedly in wall thickness costally and intercostally (costals somewhat thicker-walled). Mid-intercostal long-cells rectangular; having markedly sinuous walls. *Microhairs* absent. *Stomata* common; 24–30–32 microns long. Subsidiaries parallel-sided, or dome-shaped (low). Guard-cells slightly overlapped by the interstomatals. *Intercostal short-cells* absent or very rare. *Costal short-cells* neither distinctly grouped into long rows nor predominantly paired (but often adjoining prickles). Costal silica bodies horizontally-elongated crenate/sinuous (mostly), or rounded (a few), or tall-and-narrow (a few); not sharp-pointed.

Transverse section of leaf blade, physiology. C_3; XyMS+. *Mesophyll* with non-radiate chlorenchyma; without adaxial palisade. *Leaf blade* with distinct, prominent adaxial ribs; with the ribs more or less constant in size (large, round-topped). *Midrib* not readily distinguishable; with one bundle only. Bulliforms not in discrete, regular adaxial groups. All the vascular bundles accompanied by sclerenchyma. Combined sclerenchyma girders absent (adaxial and abaxial strands only, or adaxial strands and abaxial girders). Sclerenchyma all associated with vascular bundles.

Taxonomy. Pooideae; Poodae; Aveneae.

Ecology, geography, regional floristic distribution. 1 species. Western Asia. Species of open habitats. Dry places.

Holarctic. Boreal and Tethyan. Eastern Asian. Irano-Turanian.

References, etc. Leaf anatomical: this project.

Neurachne R.Br.

Habit, vegetative morphology. Perennial; caespitose (from short rhizomes). *Culms* 15–50 cm high; herbaceous. Culm nodes hairy, or glabrous. Culm internodes solid. Young shoots extravaginal and intravaginal (the primary shoots extravaginal, each terminating in a tuft of intravaginal culm shoots). *Leaves* mostly basal; non-auriculate. Leaf blades linear (to very narrowly so); narrow; 0.9–3.5 mm wide; flat, or rolled; exhibiting multicellular glands abaxially. The abaxial glands intercostal. *Leaf blades* without cross venation; *persistent*; rolled in bud. *Ligule a fringe of hairs*.

Reproductive organization. Plants bisexual, with bisexual spikelets; with hermaphrodite florets. *The spikelets of sexually distinct forms on the same plant*; hermaphrodite and sterile (the lowermost being reduced).

Inflorescence. *Inflorescence a single spike (almost, in N. munroi), or a single raceme (spike-like, or even capitate to ovoid in N. alopecuroidea and N. minor)*; contracted; capitate, or more or less ovoid, or spicate; espatheate; not comprising 'partial inflorescences' and foliar organs. Spikelet-bearing axes persistent. *Spikelets solitary*; not secund; not two-ranked (spiralled); sessile (or almost so), or subsessile, or pedicellate (the pedicels short). Pedicel apices oblique, or discoid, or cupuliform.

Female-sterile spikelets. Several of the lowermost spikelets generally much reduced, closer together and more persistent.

Female-fertile spikelets. *Spikelets* 5–13 mm long (erect); lanceolate; abaxial; ***compressed dorsiventrally***; falling with the glumes. Rachilla terminated by a female-fertile floret. ***Hairy callus present (with very conspicuous long white hairs)***.

Glumes two; more or less equal; long relative to the adjacent lemmas (exceeding them); dorsiventral to the rachis; hairy (especially marginally); pointed (acuminate); awned, or awnless; very dissimilar to similar (both rigidly membranous, ovate-acuminate to lanceolate-subulate and becoming hardened towards the base, the lower ciliate or not, the upper with a dense narrow sub-marginal beard of long white hairs on each side below). Lower glume 3–7 nerved. Upper glume 7–13 nerved. *Spikelets* with incomplete florets. The incomplete florets proximal to the female-fertile florets. *The proximal incomplete florets* 1; paleate. Palea of the proximal incomplete florets fully developed (except occasionally in *N. alopecuroidea*). The proximal incomplete florets male. The proximal lemmas awnless; 5–7 nerved; more or less equalling the female-fertile lemmas; similar in texture to the female-fertile lemmas to decidedly firmer than the female-fertile lemmas; not becoming indurated.

Female-fertile florets 1. **Lemmas less firm than the glumes (membranous to hyaline, in N. munroi** *smaller and flimsier than the palea)*; smooth; not becoming indurated (hyaline or membranous); white in fruit; awnless; hairless; non-carinate; having the margins lying flat and exposed on the palea; without a germination flap; 0–5 nerved. *Palea* present; relatively long; entire; awnless, without apical setae; textured like the lemma; not indurated; 2-nerved. *Lodicules* present; 2; free; fleshy; glabrous. *Stamens* 3. Anthers not penicillate. *Ovary* glabrous. Styles fused. ***Stigmas 2 (but three styles, one usually reduced but forming an appendage on the grain)***; white, or red pigmented.

Fruit, embryo and seedling. Fruit small; trigonous. Hilum short. Embryo large. Endosperm containing only simple starch grains. Embryo without an epiblast; with a scutellar tail; with a negligible mesocotyl internode.

First seedling leaf with a well-developed lamina. The lamina broad; supine.

Abaxial leaf blade epidermis. *Costal/intercostal zonation* conspicuous. *Papillae* absent. Long-cells of similar wall thickness costally and intercostally. Mid-intercostal long-cells rectangular (or at least, not fusiform); having markedly sinuous walls. *Microhairs* present; elongated; clearly two-celled; panicoid-type. Microhair apical cell wall thinner than that of the basal cell and often collapsed. Microhairs (39–)45–68(–69) microns long. Microhair basal cells 21 microns long. Microhairs (6–)7.2–12(–15) microns wide at the septum. Microhair total length/width at septum 4.6–9.1, or 11 (*N. tenuifolia*). Microhair apical cells (15–)22–33(–34.5) microns long. Microhair apical cell/ total length ratio 0.36–0.61. *Stomata* common; 27–42 microns long, or 21–27 microns long (*N. tenuifolia*). Subsidiaries dome-shaped and triangular. Guard-cells overlapping to flush with the interstomatals (a few slightly overlapped in *N. alopecuroidea*). *Intercostal short-cells* common; in cork/silica-cell pairs; silicified, or not silicified. Intercostal silica bodies absent. Often with cushion-based macrohairs and/or prickles. *Crown cells* present (or having structures strongly reminiscent of them), or absent. *Costal short-cells* conspicuously in long rows (*N. alopecuroidea, N. lanigera*), or predominantly paired (*N. munroi, N. tenuifolia*). Costal silica bodies present in alternate cell files of the costal zones; 'panicoid-type' (usually), or rounded (the common form in *N. tenuifolia*, scarce in the other species); not sharp-pointed.

Transverse section of leaf blade, physiology. C_4 (*N. munroi*, which has an 'extra' sheath, cf. *Alloteropsis*), or C_3 (*N. alopecuroidea, N. lanigera, N. queenslandica, N. tenuifolia*; *N. minor* is a C_3/C_4 intermediate). The anatomical organization when C_4 unconventional. Organization of PCR tissue when C_4 *Alloteropsis* type. Biochemical type when C_4, NADP–ME (*N. munroi*); XyMS+ (*N. alopecuroidea, N. tenuifolia, N. lanigera, N. queenslandica*), or XyMS– (*N. munroi*). PCR sheath outlines in *N. munroi* uneven to even. PCR sheath extensions present. Maximum number of extension cells 4. PCR cells with a suberised lamella. PCR cell chloroplasts with well developed grana; centrifugal/ peripheral. *PBS cells* of the C_3 species with a suberised lamella (*lanigera?*), or without a suberised lamella. *Mesophyll* with radiate chlorenchyma, or with non-radiate chlorenchyma (*N. tenuifolia*); *Isachne*-type (e.g. *N. lanigera*), or not *Isachne*-type (e.g. *N. alopecuroidea*). *Leaf blade* 'nodular' in section (the abaxial ribs more prominent); with the ribs more or less constant in size. *Midrib* not readily distinguishable; with one bundle only. Bulliforms present in discrete, regular adaxial groups (sometimes associated with/comprising hair cushions); in simple fans, or in simple fans and associated with colourless mesophyll cells to form deeply-penetrating fans. Many of the smallest vascular bundles unaccompanied by sclerenchyma (*N. tenuifolia*), or all the vascular bundles accompanied by sclerenchyma. Combined sclerenchyma girders present (*N. tenuifolia*), or absent; forming 'figures' (*N. tenuifolia*), or nowhere forming 'figures'. Sclerenchyma all associated with vascular bundles (apart from strong marginal fibre groups). The lamina margins with fibres.

Phytochemistry. Leaf blade chlorophyll *a:b* ratio of *N. munroi* 4.06–4.97.

Cytology. Chromosome base number, $x = 9$. $2n = 18$, 36, and 54 (rarely 37, 53). 2, 4, and 6 ploid (and some aneuploids).

Taxonomy. Panicoideae; Panicodae; Neurachneae.

Ecology, geography, regional floristic distribution. 6 species. Australia. Xerophytic; species of open habitats; glycophytic. Heath, sandstone, light woodland and scrub.

Australian. Central Australian.

Rusts and smuts. Smuts from Ustilaginaceae. Ustilaginaceae — *Ustilago*.

References, etc. Morphological/taxonomic: Blake 1972b. Leaf anatomical: for leaf anatomy and photosynthetic pathways, see Hattersley *et al.* 1982, Hattersley and Roksandic 1983.

Neurolepis Meissner

Planotia Munro, *Platonia* Kunth

Habit, vegetative morphology. Grasslike perennial; caespitose. The flowering culms leafy. *Culms* 100–550 cm high; woody and persistent to herbaceous; to 2.6 cm in diameter; unbranched above. Culm nodes glabrous. Culm internodes hollow. Unicaespitose (usually), or pluricaespitose (e.g. *N. aristata*). Rhizomes pachymorph. Plants unarmed. *Leaves* not basally aggregated. Leaf blades linear; broad, or narrow (typically coarse, grass-like); sometimes very large: up to 30 cm wide and 5 m long in *N. elata*; pseudopetiolate, or not pseudopetiolate; cross veined, or without cross venation; disarticulating from the sheaths, or persistent. *Ligule a fringed membrane (and sometimes double, with the usual short membrane backed by a larger one). Contra-ligule* present.

Reproductive organization. Plants bisexual, with bisexual spikelets; with hermaphrodite florets.

Inflorescence. Inflorescence paniculate (each terminating an unbranched culm); open, or contracted (elongated, broadly to narrowly paniculate, with a strong rachis). Inflorescence with axes ending in spikelets, or axes not ending in spikelets. Inflorescence espatheate; not comprising 'partial inflorescences' and foliar organs. *Spikelet-bearing axes* 'racemes' to spikelike; persistent. Spikelets solitary; secund to not secund ('commonly only one order of the branches...showing secund orientation'); pedicellate.

Female-fertile spikelets. *Spikelets not noticeably compressed*; disarticulating above the glumes; not disarticulating between the florets, or disarticulating between the florets. Rachilla terminated by a female-fertile floret; rachilla hairless. Hairy callus absent.

Glumes present; two; more or less equal; shorter than the adjacent lemmas; hairless; not pointed; awnless; non-carinate; similar. *Spikelets* with incomplete florets. The incomplete florets proximal to the female-fertile florets. *Spikelets with proximal incomplete florets. The proximal incomplete florets 2–3*; sterile. The proximal lemmas awnless; more or less equalling the female-fertile lemmas; similar in texture to the female-fertile lemmas.

Female-fertile florets 1. Lemmas entire; blunt; awnless; hairless; non-carinate; 3–7 nerved. **Palea** present; relatively long; entire to apically notched; awnless, without apical setae; *2-nerved*; keel-less. *Lodicules* present; 3; free; membranous; ciliate to glabrous; not toothed. *Stamens* 3–6. Anthers not penicillate; without an apically prolonged connective. **Ovary** glabrous; *without a conspicuous apical appendage*. Styles fused. Stigmas 2.

Fruit, embryo and seedling. *Fruit* free from both lemma and palea; small. Hilum long-linear. Embryo small ('basal, clearly to weakly manifest').

Abaxial leaf blade epidermis. *Papillae* present. Intercostal papillae over-arching the stomata, or not over-arching the stomata; several per cell. *Long-cells* similar in shape costally and intercostally. *Microhairs* present; panicoid-type; 42–66(–70) microns long. Microhair apical cells 16–21 microns long (*N. aristata*), or 10–40 microns long (*N. nobilis*). Microhair apical cell/total length ratio 0.3 (*N. aristata*), or 0.46 (*N. nobilis*). *Stomata* common. Subsidiaries low to high dome-shaped. *Intercostal short-cells* absent or very rare (rarely). *Costal short-cells* predominantly paired (and solitary). Costal silica bodies rounded, or tall-and-narrow (sometimes tall-narrow-crenate), or crescentic; not sharp-pointed.

Transverse section of leaf blade, physiology. C_3; XyMS+. *Mesophyll* with nonradiate chlorenchyma; with arm cells; with fusoids. The fusoids external to the PBS. *Midrib* not readily distinguishable; having complex vascularization (sometimes obscure). Bulliforms present in discrete, regular adaxial groups; associated with colourless mesophyll cells to form deeply-penetrating fans. All the vascular bundles accompanied by sclerenchyma. Combined sclerenchyma girders present; forming 'figures'.

Cytology. Chromosome base number, $x = 12$. $2n = 48$. 4 ploid.

Taxonomy. Bambusoideae; Bambusodae; Bambuseae.

Ecology, geography, regional floristic distribution. 9 species. South America (Venezuela to Peru), Trinidad.

Neotropical. Venezuela and Surinam and Andean.

References, etc. Morphological/taxonomic: Soderstrom 1969. Leaf anatomical: Metcalfe 1960.

Neyraudia Hook.f.

Habit, vegetative morphology. Perennial; *reeds (resembling* **Arundo** *and* **Phragmites** *in aspect)*; forming loose tufts. *Culms* 75–500 cm high; woody and persistent; branched above. Culm nodes glabrous. Culm internodes solid. Rhizomes pachymorph. Plants unarmed. *Leaves* not basally aggregated; non-auriculate. Leaf blades broad; 15–40 mm wide (often splitting longitudinally); without abaxial multicellular glands; without cross venation; disarticulating from the sheaths. *Ligule a fringe of hairs (long)*. *Contra-ligule* present.

Reproductive organization. Plants bisexual, with bisexual spikelets; with hermaphrodite florets.

Inflorescence. Inflorescence paniculate (large, plumose); open; with capillary branchlets; non-digitate (the branches more or less verticillate); espatheate; not comprising 'partial inflorescences' and foliar organs. Spikelet-bearing axes persistent. Spikelets not secund; pedicellate (with thin pedicels).

Female-fertile spikelets. Spikelets 5–9 mm long; compressed laterally; disarticulating above the glumes; disarticulating between the florets (somewhat below the lemmas). Rachilla prolonged beyond the uppermost female-fertile floret; rachilla hairy (but glabrous except above the abscission zone of each floret, the conspicuous hair tuft being on the 'callus'). The rachilla extension with incomplete florets. Hairy callus present (with long white hairs).

Glumes two; very unequal to more or less equal; shorter than the spikelets; shorter than the adjacent lemmas; hairless; pointed (acute to acuminate); awnless; carinate; similar. Lower glume 1–3 nerved. Upper glume 1–3 nerved. *Spikelets* with incomplete florets. The incomplete florets both distal and proximal to the female-fertile florets. *The distal incomplete florets* merely underdeveloped. *The proximal incomplete florets* 1; sterile. The proximal lemmas awnless (acuminate, larger than but similar to the glumes); 3 nerved; exceeded by the female-fertile lemmas; similar in texture to the female-fertile lemmas; not becoming indurated.

Female-fertile florets 3–8. *Lemmas* similar in texture to the glumes (membranous); not becoming indurated; *incised*; 2 lobed; not deeply cleft (setaceously bidentate); awned (the distal florets with longer awns). Awns 1, or 3; median, or median and lateral (in that the lateral nerves may be excurrent into very short awns); the median similar in form to the laterals (when laterals present); from a sinus; non-geniculate; recurving; hairless; much shorter than the body of the lemma to about as long as the body of the lemma. The lateral awns when present, shorter than the median. Lemmas hairy (bearded marginally and on the lateral nerves); non-carinate (dorsally rounded); having the margins lying flat and exposed on the palea (hyaline); without a germination flap; 3 nerved. *Palea* present; relatively long; apically notched; awnless, without apical setae; not indurated; 2-nerved; 2-keeled. *Lodicules* present; 2; free; fleshy; glabrous; not or scarcely vascularized. *Stamens* 3. Anthers not penicillate; without an apically prolonged connective. *Ovary* glabrous. Styles free to their bases. Stigmas 2.

Fruit, embryo and seedling. *Fruit* free from both lemma and palea; small (1.5–2 mm long); linear; subteretete to angular. Hilum short. Pericarp fused. Embryo large; with an epiblast; with a scutellar tail; with an elongated mesocotyl internode. Embryonic leaf margins meeting.

Abaxial leaf blade epidermis. *Costal/intercostal zonation* conspicuous. *Papillae* absent. *Long-cells* markedly different in shape costally and intercostally (the costals narrower and relatively longer); of similar wall thickness costally and intercostally. Mid-intercostal long-cells rectangular; and pitted. *Microhairs* absent. *Stomata* common; 18–21–22.5 microns long. Subsidiaries high dome-shaped (mostly), or triangular. Guard-cells overlapping to flush with the interstomatals. *Intercostal short-cells* common; in cork/silica-cell pairs (and a few solitary); silicified. Intercostal silica bodies present and perfectly developed; crescentic, saddle shaped, and oryzoid-type. *Costal short-cells* conspicuously in long rows (but many of the short-cells quite long). Costal silica bodies present in alternate cell files of the costal zones; rounded and saddle shaped; not sharp-pointed.

Transverse section of leaf blade, physiology. C_4; XyMS+. PCR sheaths of the primary vascular bundles interrupted; interrupted both abaxially and adaxially. PCR sheath extensions absent. *Leaf blade* with distinct, prominent adaxial ribs; with the ribs more or less constant in size. *Midrib* not readily distinguishable (its bundle somewhat larger); with one bundle only. Bulliforms present in discrete, regular adaxial groups; in simple fans and associated with colourless mesophyll cells to form deeply-penetrating fans. All the vascular bundles accompanied by sclerenchyma. Combined sclerenchyma girders present (with all the bundles); forming 'figures'. Sclerenchyma not all bundle-associated. The 'extra' sclerenchyma in abaxial groups; abaxial-hypodermal, the groups isolated (in small abaxial groups, opposite the bulliforms). The lamina margins with fibres.

Cytology. Chromosome base number, $x = 10$ (?). $2n = 40$.

Taxonomy. Chloridoideae; main chloridoid assemblage.

Ecology, geography, regional floristic distribution. 2 species. Tropical Africa, Madagascar, China, Indomalayan region. Helophytic to mesophytic.

Paleotropical. African, Madagascan, and Indomalesian. Sudano-Angolan. Indian, Indo-Chinese, and Malesian. Somalo-Ethiopian and South Tropical African.

Rusts and smuts. Rusts — *Puccinia*. Smuts from Ustilaginaceae. Ustilaginaceae — *Ustilago*.

References, etc. Leaf anatomical: this project.

Notochloë Domin.

Habit, vegetative morphology. Perennial; caespitose. **Culms** 30–50 cm high; *herbaceous*. Culm nodes glabrous. Leaves non-auriculate. Leaf blades narrow; 2–3 mm wide; flat, or rolled (inrolled); without cross venation. *Ligule a rim of minute papillae*; 0.1–1 mm long.

Reproductive organization. Plants bisexual, with bisexual spikelets; with hermaphrodite florets.

Inflorescence. *Inflorescence* few spikeleted (5–10); *paniculate (short)*; open; with capillary branchlets; espatheate; not comprising 'partial inflorescences' and foliar organs. Spikelet-bearing axes persistent. Spikelets not secund; pedicellate.

Female-fertile spikelets. Spikelets 12–22 mm long; compressed laterally; disarticulating above the glumes. Rachilla prolonged beyond the uppermost female-fertile floret. Hairy callus present.

Glumes present; two; very unequal; shorter than the spikelets; (the longer) shorter than the adjacent lemmas; hairless; glabrous; pointed (acute); awnless; carinate. Lower glume 3 nerved. Upper glume 5 nerved. *Spikelets* with female-fertile florets only, or with incomplete florets. The incomplete florets distal to the female-fertile florets.

Female-fertile florets *8–14*. Lemmas similar in texture to the glumes (thinly leathery); not becoming indurated; incised; 3 lobed; not deeply cleft (shortly 3-toothed); mucronate (the teeth with minute terminal mucros); hairless; glabrous; non-carinate; 5–7 nerved (and ridged). *Palea* present. Stigmas 2.

Abaxial leaf blade epidermis. *Costal/intercostal zonation* conspicuous to lacking. *Papillae* absent. *Long-cells* similar in shape costally and intercostally; of similar wall thickness costally and intercostally. Mid-intercostal long-cells fusiform; having markedly sinuous walls. *Microhairs* present; panicoid-type; (75–)78–99(–111) microns long; (15–)20.4–21.6(–24) microns wide at the septum. Microhair total length/width at septum 3.75–5.1. Microhair apical cells (34.5–)36–42(–60) microns long. Microhair apical cell/total length ratio 0.42–0.54. *Stomata* absent or very rare; 38.4–43 microns long. Subsidiaries parallel-sided. Guard-cells overlapping to flush with the interstomatals. *Intercostal short-cells* common; not paired (solitary); not silicified. *Costal short-cells* neither distinctly grouped into long rows nor predominantly paired. Costal silica bodies tall-and-narrow (a few), or crescentic; not sharp-pointed.

Transverse section of leaf blade, physiology. C_3; XyMS+. *Mesophyll* with non-radiate chlorenchyma; without adaxial palisade. *Leaf blade* 'nodular' in section; with the ribs very irregular in sizes. *Midrib* conspicuous (by virtue of large 'hinge groups'); with one bundle only. Bulliforms present in discrete, regular adaxial groups; in simple fans (in addition to the large 'hinges'). All the vascular bundles accompanied by sclerenchyma. Combined sclerenchyma girders present; forming 'figures'. Sclerenchyma all associated with vascular bundles.

Taxonomy. Arundinoideae; Danthonieae.

Ecology, geography, regional floristic distribution. 1 species. Australia. Helophytic; glycophytic (upland swamps).

Australian. North and East Australian. Temperate and South-Eastern Australian.

References, etc. Leaf anatomical: this project.

Special comments. Fruit data wanting.

Ochlandra Thwaites

Beesha Munro, *Irulia* Bedd.

Habit, vegetative morphology. Perennial; reeds, or not reeds (shrubs). The flowering culms leafy. *Culms* 200–600 cm high; woody and persistent; to 5 cm in diameter. Culm sheaths persistent. Culm internodes hollow. Plants unarmed. **Leaves** not basally aggregated; *auriculate (with pellucid dots)*; without auricular setae. Leaf blades broad; pseudopetiolate; disarticulating from the sheaths. Ligule an unfringed membrane; truncate; short.

Reproductive organization. Plants bisexual, with bisexual spikelets; with hermaphrodite florets.

Inflorescence. *Inflorescence* indeterminate; with pseudospikelets; a false spike, with spikelets on contracted axes (the clusters stellate); spatheate, or espatheate. **Spikelet-bearing axes** *capitate*. Spikelets not secund.

Female-fertile spikelets. Spikelets 18–50 mm long; not noticeably compressed; falling with the glumes. Rachilla terminated by a female-fertile floret; rachilla hairless.

Glumes present; two; more or less equal; shorter than the adjacent lemmas; hairless; pointed; awnless (but shortly mucronate); non-carinate; similar. Lower glume much shorter than half length of lowest lemma; 40 nerved (or more, in material seen). Upper glume 40 nerved (or more, in material seen). *Spikelets* with incomplete florets. The incomplete florets proximal to the female-fertile florets. *Spikelets with proximal incomplete florets. The proximal incomplete florets* 1–3; sterile. The proximal lemmas awnless (but shortly mucronate); 40 nerved (or more, in material seen); exceeded by the female-fertile lemmas; similar in texture to the female-fertile lemmas; not becoming indurated.

Female-fertile florets 1. Lemmas similar in texture to the glumes; not becoming indurated; entire; pointed; non-carinate; without a germination flap; 40 nerved (or more, in material seen). *Palea* present; relatively long; entire, or apically notched, or deeply bifid;

awless, without apical setae; not indurated; several nerved (9 or more in material seen); keel-less. *Lodicules* present; 1–5(–15); free; membranous; glabrous; toothed, or not toothed; heavily vascularized (sometimes up to 15 mm long). **Stamens 6–120**; with free filaments to monadelphous. Anthers not penicillate; with the connective apically prolonged. *Ovary* glabrous; with a conspicuous apical appendage. The appendage long, stiff and tapering. Styles fused. Stigmas 3–6.

Fruit, embryo and seedling. *Fruit* free from both lemma and palea; large (2–5 cm long); ellipsoid, or subglobose, or pyriform; not noticeably compressed. Hilum not visible. Pericarp fleshy. Embryo not visible. *Seed 'non-endospermic'*.

Seedling with a short mesocotyl; with a loose coleoptile. First seedling leaf without a lamina.

Abaxial leaf blade epidermis. *Costal/intercostal zonation* conspicuous. *Papillae* present. Intercostal papillae over-arching the stomata (and obscuring them); several per cell (often one row per long-cell). *Long-cells* similar in shape costally and intercostally; of similar wall thickness costally and intercostally. Mid-intercostal long-cells rectangular; having markedly sinuous walls. *Microhairs* present; panicoid-type; 54–60 microns long; 7.5–9–10.5 microns wide at the septum. Microhair total length/width at septum 5.1–7.4. Microhair apical cells 25.5–30 microns long. Microhair apical cell/total length ratio 0.47–0.54. *Stomata* common (obscured papillae and sunken); 22.5–27 microns long. *Intercostal short-cells* common; in cork/silica-cell pairs; silicified. *Costal short-cells* predominantly paired, or neither distinctly grouped into long rows nor predominantly paired. Costal silica bodies saddle shaped, or crescentic (e.g. predominant in *O. stridula*); not sharp-pointed.

Transverse section of leaf blade, physiology. C_3; XyMS+. *Mesophyll* with non-radiate chlorenchyma; without adaxial palisade; with arm cells; with fusoids. The fusoids external to the PBS. *Leaf blade* adaxially flat. *Midrib* conspicuous; having complex vascularization. *The lamina* distinctly asymmetrical on either side of the midrib. Bulliforms present in discrete, regular adaxial groups; in simple fans (the median cells deeply penetrating). All the vascular bundles accompanied by sclerenchyma. Combined sclerenchyma girders present (with most bundles); forming 'figures' (with most bundles). Sclerenchyma all associated with vascular bundles.

Cytology. Chromosome base number, $x = 12$. $2n = 72$. 6 ploid.

Taxonomy. Bambusoideae; Bambusodae; Bambuseae.

Ecology, geography, regional floristic distribution. 12 species. Madagascar, India, Ceylon. Forest thickets.

Paleotropical. Madagascan and Indomalesian. Indian.

References, etc. Leaf anatomical: Metcalfe 1960; this project.

Ochthochloa Edgwe.

Habit, vegetative morphology. Perennial; stoloniferous and caespitose. **Culms** 10–40 cm high; herbaceous; branched above; *tuberous*. Plants unarmed. *Leaves* not basally aggregated; non-auriculate. Leaf blades linear-lanceolate; narrow; without abaxial multicellular glands; without cross venation; persistent. Ligule a fringed membrane.

Reproductive organization. Plants bisexual, with bisexual spikelets; with hermaphrodite florets.

Inflorescence. *Inflorescence* of spicate main branches; *digitate*. Primary inflorescence branches (2–)3–5. Inflorescence espatheate; not comprising 'partial inflorescences' and foliar organs. *Spikelet-bearing axes* spikes; disarticulating; falling entire (i.e., the spikes falling entire). Spikelets solitary; secund (the spikes dorsiventral).

Female-fertile spikelets. Spikelets 4–6.5(–8) mm long; strongly compressed laterally; disarticulating above the glumes; not disarticulating between the florets. Rachilla prolonged beyond the uppermost female-fertile floret. Hairy callus absent.

Glumes two; very unequal (G_1 smaller); shorter than the spikelets; (the upper) shorter than the adjacent lemmas, or long relative to the adjacent lemmas; pointed; awned (G_2, shortly), or awnless; carinate (the keel scabrid); *very dissimilar (membranous, elliptic to oblanceolate, the upper tapering to a cuspidate or shortly-subulate tip and with a thickened, three-nerved keel)*. Lower glume much exceeding the lowest lemma; 1 nerved.

Upper glume 3 nerved (in the keel). *Spikelets* with female-fertile florets only, or with incomplete florets. The incomplete florets if present, distal to the female-fertile florets. *The distal incomplete florets merely underdeveloped.*

Female-fertile florets 3–6. Lemmas acute; not becoming indurated (membranous); entire; pointed; mucronate; hairy (villous below on the margins and keel); 3 nerved. *Palea* present; relatively long; entire (pointed); awnless, without apical setae (hairy); not indurated; 2-nerved; 2-keeled. Palea keels hairy (ciliate). *Ovary* glabrous. Styles free to their bases. Stigmas 2.

Fruit, embryo and seedling. Fruit small (1.5–2 mm long); ellipsoid. Hilum short. *Pericarp free.* Embryo large.

Abaxial leaf blade epidermis. *Costal/intercostal zonation* conspicuous. *Papillae* present; intercostal. Intercostal papillae over-arching the stomata; consisting of one symmetrical projection per cell to several per cell. *Long-cells* markedly different in shape costally and intercostally (the costals longer); of similar wall thickness costally and intercostally (medium walled). Intercostal zones with typical long-cells (or tending to be rather short). Mid-intercostal long-cells rectangular; having markedly sinuous walls and conspicuous pits). *Microhairs* present; more or less spherical; clearly two-celled; chloridoid-type (with sunken basal cell). Microhair apical cell wall of similar thickness/rigidity to that of the basal cell. Microhairs (18–)21 microns long. Microhair basal cells 6–9 microns long. Microhairs 9.6–11.4–12 microns wide at the septum. Microhair total length/width at septum 1.6–2.2. Microhair apical cells (8.4–)9 microns long. Microhair apical cell/total length ratio 0.43–0.47. *Stomata* common; (21–)22.5–24(–25.5) microns long. Subsidiaries dome-shaped and triangular. Guard-cells overlapping to flush with the interstomatals (apart from the papillae). *Intercostal short-cells* common; not paired (solitary); not silicified. Intercostal silica bodies absent. *Costal short-cells* conspicuously in long rows. Costal silica bodies present in alternate cell files of the costal zones; exclusively saddle shaped; not sharp-pointed.

Transverse section of leaf blade, physiology. C_4; XyMS+. PCR sheaths of the primary vascular bundles interrupted; interrupted abaxially only, or interrupted both abaxially and adaxially. PCR sheath extensions present. Maximum number of extension cells 1–3. Mesophyll traversed by columns of colourless mesophyll cells. *Leaf blade* adaxially flat. *Midrib* conspicuous; having a conventional arc of bundles (a large median, with two smaller laterals on each side); with colourless mesophyll adaxially. Bulliforms present in discrete, regular adaxial groups; in simple fans and associated with colourless mesophyll cells to form deeply-penetrating fans (linking with traversing columns of colourless cells). All the vascular bundles accompanied by sclerenchyma. Combined sclerenchyma girders present (at least the primary laterals); forming 'figures' (some of the primaries). Sclerenchyma all associated with vascular bundles. The lamina margins with fibres.

Taxonomy. Chloridoideae; main chloridoid assemblage.

Ecology, geography, regional floristic distribution. 1 species. North Africa, Iran to India. Xerophytic; species of open habitats. Semidesert.

Paleotropical. African. Saharo-Sindian.

References, etc. Morphological/taxonomic: Hilu 1981. Leaf anatomical: this project.

Odontelytrum Hackel

Habit, vegetative morphology. Perennial; stoloniferous. *Culms* 60–100 cm high (standing 30–40 cm above the water); herbaceous; branched above. Culm nodes glabrous. Culm internodes hollow. Plants unarmed. Young shoots intravaginal. *Leaves* not basally aggregated; non-auriculate. Leaf blades linear, or linear-lanceolate; narrow; to 9 mm wide; flat, or rolled; without cross venation; persistent. Ligule an unfringed membrane to a fringed membrane; truncate (entire or laciniate); 1–1.5 mm long. *Contra-ligule* absent.

Reproductive organization. Plants bisexual, with bisexual spikelets; with hermaphrodite florets. The spikelets all alike in sexuality.

Inflorescence. *Inflorescence a false spike, with spikelets on contracted axes, or a single raceme (a coarse, cylindrical 'raceme', apparently representing a raceme of* reduced 'glomerules', each glomerule shortly pedunculate, comprising a single spikelet subtended by a lobed scale forming an involucre-plus-bristle). Inflorescence morphologically difficult — comparable with that of **Anthepora, Cenchrus** or **Pseudoraphis?**. Inflorescence espatheate (but enveloped below by the uppermost leaf sheath, whose blade is at least as long as the inflorescence); not comprising 'partial inflorescences' and foliar organs. **Spikelet-bearing axes very much reduced (each reduced to one spikelet, with a herbaceous, lobed scale and bristle)**; disarticulating; falling entire (i.e., the reduced 'glomerules' deciduous — the main axis persistent). **Spikelets associated with bractiform involucres and subtended by solitary 'bristles' (each spikelet with a purplish, irregularly 4–6 lobed involucre, this being herbaceous except for one lobe, which is almost free, awnlike, scabrid and 12–25 mm long)**. The 'bristles' deciduous with the spikelets. The involucres shed with the fertile spikelets. Spikelets solitary; not secund; sessile.

Female-fertile spikelets. Spikelets 10–14 mm long; abaxial; compressed dorsiventrally; falling with the glumes; not disarticulating between the florets; with conventional internode spacings. Rachilla terminated by a female-fertile floret; rachilla hairless. Hairy callus absent. *Callus* absent.

Glumes one per spikelet (the G_1 missing), or two; very unequal (the G_1 when present very small); shorter than the spikelets; shorter than the adjacent lemmas (the G_2 about half length of spikelet); free; dorsiventral to the rachis; hairless; pointed (G_2 narrowly ovate); awnless. Lower glume much shorter than half length of lowest lemma; when present, 0 nerved. Upper glume 1–3 nerved (scarious). *Spikelets* with incomplete florets. The incomplete florets proximal to the female-fertile florets. *The proximal incomplete florets* 1; paleate. Palea of the proximal incomplete florets fully developed (as long as the lemma). The proximal incomplete florets male. The proximal lemmas awnless; 9 nerved; more or less equalling the female-fertile lemmas; less firm than the female-fertile lemmas (membranous).

Female-fertile florets 1. Lemmas decidedly firmer than the glumes (cartilaginous below, herbaceous above); smooth; not becoming indurated; entire; pointed; awnless (the tip caudate, membranous); hairless; glabrous; non-carinate; having the margins lying flat and exposed on the palea; without a germination flap; 7 nerved. *Palea* present; relatively long (equalling the lemma); tip caudate; textured like the lemma; 2-nerved; keel-less (dorsally rounded). Palea back glabrous. *Lodicules* absent. *Stamens* 3. Anthers 5.5 mm long; not penicillate; without an apically prolonged connective. Styles fused (to the bases of the stigmas). Stigmas 2; red pigmented.

Fruit, embryo and seedling. Fruit compressed dorsiventrally. Embryo large (about 1/3 the length of the fruit).

Abaxial leaf blade epidermis. *Costal/intercostal zonation* conspicuous. *Papillae* present; costal and intercostal. Intercostal papillae consisting of one oblique swelling per cell. *Long-cells* similar in shape costally and intercostally (but the costals rather smaller). Intercostal zones without typical long-cells (of more or less isodiametric cells). Mid-intercostal long-cells rectangular (mostly quite square); having markedly sinuous walls. *Microhairs* present; panicoid-type. *Stomata* seemingly absent or very rare. *Costal short-cells* conspicuously in long rows. Costal silica bodies 'panicoid-type'; not sharp-pointed.

Transverse section of leaf blade, physiology. C_4; XyMS+. PCR cell chloroplasts centrifugal/peripheral. *Leaf blade* with distinct, prominent adaxial ribs to 'nodular' in section; with the ribs more or less constant in size (low, round-topped). *Midrib* conspicuous (very); having a conventional arc of bundles (a large median with about 8 smaller laterals on either side, all abaxial); with colourless mesophyll adaxially (the colourless mass with a lacuna in the centre). Bulliforms present in discrete, regular adaxial groups; associating with colourless mesophyll cells to form arches over small vascular bundles (in places, apparently). Many of the smallest vascular bundles unaccompanied by sclerenchyma. Combined sclerenchyma girders present (with the primaries). Sclerenchyma all associated with vascular bundles.

Special diagnostic feature. *The inflorescence a coarse, cylindrical 'raceme', apparently representing a raceme of reduced 'glomerules', each glomerule shortly pedunculate, comprising a single spikelet subtended crosswise by a lobed scale forming an involucre-plus-bristle.*

Taxonomy. Panicoideae; Panicodae; Paniceae.

Ecology, geography, regional floristic distribution. 1 species (*O. abyssinicum*). Abyssinia and southern Africa. Helophytic (in flowing or standing water); species of open habitats; glycophytic.

Paleotropical. African. Sudano-Angolan. Somalo-Ethiopian.

References, etc. Morphological/taxonomic: Stapf 1916. Leaf anatomical: photos of *O. abyssinicum* provided by R.P. Ellis.

Odyssea Stapf

Habit, vegetative morphology. Glaucous, creeping perennial; rhizomatous. The flowering culms leafy. *Culms* 5–75 cm high; herbaceous; branched above, or unbranched above. Plants conspicuously armed (leaf blades short, rigid and very pungent-tipped). *Leaves* not basally aggregated; conspicuously distichous; non-auriculate. The sheaths imbricate. *Leaf blades* narrow; 2–4 mm wide; flat and rolled (inrolled from the flat base); *hard, woody, needle-like*; without abaxial multicellular glands; without cross venation; *disarticulating from the sheaths*. Ligule a fringe of hairs. *Contra-ligule* absent.

Reproductive organization. Plants bisexual, with bisexual spikelets; with hermaphrodite florets.

Inflorescence. *Inflorescence paniculate (of short, crowded branches)*; fairly to very contracted; capitate to more or less ovoid, or more or less irregular; espatheate; not comprising 'partial inflorescences' and foliar organs. Spikelet-bearing axes persistent. Spikelets solitary; somewhat secund, or not secund; pedicellate.

Female-fertile spikelets. Spikelets 5–9 mm long; adaxial; somewhat compressed laterally; disarticulating above the glumes; disarticulating between the florets. Rachilla prolonged beyond the uppermost female-fertile floret; rachilla hairy (to 'subglabrous'). The rachilla extension with incomplete florets. Hairy callus present (with lateral hair tufts, not conspicuous). *Callus* short.

Glumes two; very unequal; shorter than the spikelets; shorter than the adjacent lemmas; hairless; glabrous; glume apices pointed or irregularly toothed; awnless (but sometimes with the nerve tip constituting a tiny mucro); carinate; similar (thinly membranous to hyaline). Lower glume 1 nerved. Upper glume 1 nerved. *Spikelets* with incomplete florets. The incomplete florets distal to the female-fertile florets. *The distal incomplete florets* merely underdeveloped.

Female-fertile florets 2–8. Lemmas similar in texture to the glumes to decidedly firmer than the glumes (membranous with scarious margins, or scarious); not becoming indurated; incised; not deeply cleft (bidenticulate); mucronate; hairy (with silky white hairs on the lower half of the mid-nerve, and along the laterals); somewhat carinate, or non-carinate; without a germination flap; 3 nerved. *Palea* present; relatively long; slightly apically notched; awnless, without apical setae, or with apical setae (the nerves sometimes slightly prolonged); textured like the lemma; not indurated; 2-nerved; 2-keeled. Palea keels silky hairy. *Lodicules* present; 2; free; fleshy; glabrous. *Stamens* 3. Anthers 2 mm long; not penicillate; without an apically prolonged connective. *Ovary* glabrous. Styles free to their bases. Stigmas 2.

Fruit, embryo and seedling. Fruit small (1.1–1.5 mm long); ellipsoid; slightly concave on the hilar side; compressed laterally. Hilum short. *Pericarp free*. Embryo large (around 1/3 grain length).

Abaxial leaf blade epidermis. *Costal/intercostal zonation* conspicuous. *Papillae* absent. Mid-intercostal long-cells having markedly sinuous walls. *Microhairs* present (very large); elongated; clearly two-celled; chloridoid-type. Microhair apical cell wall thinner than that of the basal cell but not tending to collapse. Microhairs $(39-)42-48(-51)$ microns long. Microhair basal cells 39 microns long. Microhairs $(16.5-)20.4-21$ microns wide at the septum. Microhair total length/width at septum 2–3.1. Microhair apical cells $(15-)16.5-20(-21)$ microns long. Microhair apical cell/total length ratio 0.31–0.42. *Stomata* common; 21–22.5–24 microns long. Subsidiaries triangular. Guard-cells overlapping to flush with the interstomatals. Intercostal short-cells not paired. Intercostal silica bodies absent. *Costal short-cells* conspicuously in long rows. Costal silica bodies present in alternate cell files of the costal zones; 'panicoid-type'; cross shaped, butterfly shaped, and dumb-bell shaped; not sharp-pointed.

Transverse section of leaf blade, physiology. C_4; XyMS+. PCR sheaths of the primary vascular bundles complete. PCR sheath extensions present. Maximum number of extension cells 1–3. PCR cell chloroplasts centripetal. Mesophyll not traversed by colourless columns (but deeply penetrating bulliform groups almost traversing the leaf). *Leaf blade* 'nodular' in section (with deep adaxial and abaxial furrows, the abaxial ribs broader and flat-bottomed); with the ribs more or less constant in size. *Midrib* not readily distinguishable; with one bundle only. Bulliforms present in discrete, regular adaxial groups; associated with colourless mesophyll cells to form deeply-penetrating fans. All the vascular bundles accompanied by sclerenchyma. Combined sclerenchyma girders present (including bundle sheath extensions); forming 'figures' (I's with all bundles). Sclerenchyma all associated with vascular bundles.

Taxonomy. Chloridoideae; main chloridoid assemblage.

Ecology, geography, regional floristic distribution. 2 species. Coastal Red Sea, tropical and southwest Africa. Xerophytic; species of open habitats; halophytic. A sandbinder.

Paleotropical and Cape. African. Sudano-Angolan and Namib-Karoo. Somalo-Ethiopian, South Tropical African, and Kalaharian.

References, etc. Morphological/taxonomic: Stapf 1922. Leaf anatomical: this project.

Olmeca Soderstrom

Habit, vegetative morphology. Arborescent perennial; rhizomatous (the stems solitary, the rhizome necks up to 8 m long). The flowering culms leafy. *Culms* 150 cm high; woody and persistent; to 5 cm in diameter; branched above. Culm nodes glabrous. Primary branches/mid-culm node 1. Culm internodes solid, or hollow. Unicaespitose. *Rhizomes pachymorph.* Plants unarmed. *Leaves* not basally aggregated; non-auriculate; with auricular setae. Leaf blades broad; pseudopetiolate, or not pseudopetiolate; without cross venation; rolled in bud. Ligule truncate, or not truncate; 1–4 mm long. *Contra-ligule* present, or absent.

Reproductive organization. Plants bisexual, with bisexual spikelets; with hermaphrodite florets. Not viviparous.

Inflorescence. Inflorescence *determinate*; paniculate; open; espatheate; not comprising 'partial inflorescences' and foliar organs. *Spikelet-bearing axes* paniculate; with very slender rachides. 'Articles' glabrous. Spikelets pedicellate.

Female-fertile spikelets. Spikelets 30–40 mm long; not noticeably compressed; disarticulating above the glumes; disarticulating between the florets. Rachilla prolonged beyond the uppermost female-fertile floret; rachilla hairless. The rachilla extension with incomplete florets. Hairy callus absent.

Glumes two; very unequal; shorter than the adjacent lemmas; hairless; not pointed; awnless; non-carinate; similar. Lower glume longer than half length of lowest lemma. *Spikelets* with incomplete florets. The incomplete florets distal to the female-fertile florets. *The distal incomplete florets* merely underdeveloped.

Female-fertile florets 3–12 (variable in number). Lemmas not becoming indurated; entire; pointed; awnless, or mucronate (?); hairless; non-carinate; without a germination flap; 7–11 nerved. *Palea* present; relatively long; apically notched; awnless, without apical setae; not indurated; 2-nerved, or several nerved (2–6); 2-keeled. *Lodicules* present; 3; free; membranous; ciliate; toothed, or not toothed; heavily vascularized. *Stamens* 3. Anthers not penicillate; without an apically prolonged connective. *Ovary* glabrous. Styles fused. Stigmas 2–3.

Fruit, embryo and seedling. Fruit large (1.5–2.5 cm in diameter); subglobose. Pericarp fleshy. *Seed 'non-endospermic'.*

Seedling with a short mesocotyl. First seedling leaf without a lamina.

Abaxial leaf blade epidermis. *Costal/intercostal zonation* conspicuous. *Papillae* absent. *Long-cells* similar in shape costally and intercostally (the costals rather smaller); of similar wall thickness costally and intercostally. Mid-intercostal long-cells rectangular; having markedly sinuous walls. *Microhairs* present; elongated; clearly two-celled; panicoid-type; 48–60 microns long; 6.9–9.6 microns wide at the septum. Microhair total length/width at septum 5.6–7.7. Microhair apical cells 24–33 microns long. Microhair apical cell/total length ratio 0.5–0.55. *Stomata* common; 25–30 microns long. Subsidiaries high dome-shaped (mostly), or triangular (some). Guard-cells overlapping to flush with the interstomatals. *Intercostal short-cells* common; in cork/silica-cell pairs; silicified, or not silicified. Intercostal silica bodies when present, narrowly saddle shaped. *Costal short-cells* predominantly paired. Costal silica bodies saddle shaped (a narrowish version predominating), or oryzoid (some of the saddles being almost interpretable as such); not sharp-pointed.

Transverse section of leaf blade, physiology. C_3; XyMS+. Mesophyll with adaxial palisade; with arm cells; with fusoids. The fusoids external to the PBS. *Leaf blade* adaxially flat. *Midrib* conspicuous; having complex vascularization. *The lamina* symmetrical on either side of the midrib. Bulliforms present in discrete, regular adaxial groups (a large group in each intercostal zone); in simple fans. All the vascular bundles accompanied by sclerenchyma. Combined sclerenchyma girders present; forming 'figures' (the primaries and some of the smaller bundles with large 'anchors' or I's). Sclerenchyma all associated with vascular bundles.

Taxonomy. Bambusoideae; Bambusodae; Bambuseae.

Ecology, geography, regional floristic distribution. 2 species. Mexico.

Holarctic and Neotropical. Madrean. Caribbean.

References, etc. Morphological/taxonomic: Sodertrom 1981e. Leaf anatomical: this project.

Special comments. Fruit data wanting.

Olyra L.

Mapira Adans.

Excluding *Parodiolyra*

Habit, vegetative morphology. Perennial; caespitose. The flowering culms leafy. **Culms** 50–500 cm high; *woody and persistent*; scandent (twining), or not scandent; self-supporting, or scrambling, or scandent; branched above. Culm nodes hairy, or glabrous. Culm internodes hollow. *Leaves* not basally aggregated; auriculate, or non-auriculate; without auricular setae. **Leaf blades** lanceolate to ovate; broad; (12–)30–130 mm wide; *cordate (sometimes basally asymmetric), or sagittate*; flat; pseudopetiolate; cross veined (these rarely inconspicuous?); disarticulating from the sheaths; rolled in bud. Ligule an unfringed membrane, or a fringed membrane; truncate.

Reproductive organization. *Plants monoecious with all the fertile spikelets unisexual (the male spikelets immediately beneath the females, or the lower parts of the panicle exclusively male)*; without hermaphrodite florets. The spikelets of sexually distinct forms on the same plant; female-only and male-only. The male and female-fertile spikelets on different branches of the same inflorescence, or segregated, in different parts of the same inflorescence branch (female spikelets terminating the branches, the smaller male spikelets below, or the upper branches all female, the lower all male). The spikelets overtly heteromorphic.

Inflorescence. *Inflorescence* determinate, or indeterminate (?synflorescences); paniculate; open, or contracted; spatheate, or espatheate (?); a complex of 'partial inflorescences' and intervening foliar organs (spatheolate synflorescences), or not comprising 'partial inflorescences' and foliar organs (?). Spikelet-bearing axes persistent. Spikelets solitary; not secund; pedicellate (female spikelets on stout, clavate peduncles, males on slender pedicels).

Female-sterile spikelets. The male spikelets usually shorter than the females; without glumes; stamens 3, free. Rachilla of male spikelets terminated by a male floret. The male spikelets without glumes; without proximal incomplete florets; 1 floreted. Male florets 1; 3 staminate. The staminal filaments free.

Female-fertile spikelets. *Spikelets* (5.5–)7–50 mm long; elliptic, or ovate; compressed dorsiventrally; nearly always *disarticulating above the glumes (the glumes persistent, except in* O. micrantha*)*; with conventional internode spacings, or with a distinctly elongated rachilla internode above the glumes. The upper floret conspicuously stipitate, or the upper floret not stipitate. Rachilla terminated by a female-fertile floret. Hairy callus absent. *Callus* absent.

Glumes two; very unequal (the lower longer), or more or less equal; about equalling the spikelets to exceeding the spikelets; long relative to the adjacent lemmas; not ventricose (by contrast with *Parodiolyra*); hairy to hairless; pointed (acuminate to aristate); awnless, or awned (often caudate-acuminate); non-carinate; similar (papery or herbaceous, often with transverse veinlets). Lower glume 5–15 nerved. Upper glume (3–)5–11 nerved. *Spikelets* with female-fertile florets only.

Female-fertile florets 1 (the anthecium recalling that of Paniceae). Lemmas finally decidedly firmer than the glumes; becoming indurated; entire; blunt; awnless; hairy, or hairless; non-carinate; having the margins tucked in onto the palea; with a clear germination flap. *Palea* present; relatively long; tightly clasped by the lemma; entire; awnless, without apical setae; indurated; 2-nerved, or several nerved (2–4); keel-less. *Lodicules* present; 3; free; membranous; glabrous; heavily vascularized. *Stamens* 0. *Ovary* glabrous. Styles fused (into one); completely fused. Stigmas 2.

Fruit, embryo and seedling. *Fruit* free from both lemma and palea; small to medium sized (3.8–9 mm long); fusiform, or ellipsoid; compressed dorsiventrally, or not noticeably compressed. Hilum long-linear (as long as the caryopsis). Embryo small (a fifth or less of the caryopsis length). Endosperm hard; without lipid. Embryo with an epiblast; with a scutellar tail; with a negligible mesocotyl internode. Embryonic leaf margins overlapping.

Seedling with a short mesocotyl; with a loose coleoptile. First seedling leaf without a lamina.

Abaxial leaf blade epidermis. *Costal/intercostal zonation* conspicuous. *Papillae* present. Intercostal papillae not over-arching the stomata; several per cell (most long-cells and interstomatals with one or two rows of smallish, round papillae in *O. latifolia*). Intercostal zones with typical long-cells (though these rather short). Mid-intercostal long-cells rectangular; having markedly sinuous walls. *Microhairs* present; panicoid-type; (48–)51–72(–77) microns long; (3.6–)3.9–4.2(–5.4) microns wide at the septum. Microhair total length/width at septum 10.6–13.3. Microhair apical cells 19.5–21–36 microns long. Microhair apical cell/total length ratio 0.39–0.45. *Stomata* common; 21–25.5 microns long. Subsidiaries predominantly triangular. Guard-cells overlapping to flush with the interstomatals. *Intercostal short-cells* common; in cork/silica-cell pairs; silicified. Intercostal silica bodies vertically elongated-nodular. *Costal short-cells* conspicuously in long rows. Costal silica bodies saddle shaped (common over some veins, in *O. latifolia*), or oryzoid, or 'panicoid-type'; in *O. latifolia* commonly plump cross shaped and butterfly shaped; not sharp-pointed.

Transverse section of leaf blade, physiology. C_3; XyMS+. *Mesophyll* with non-radiate chlorenchyma; without adaxial palisade; with arm cells; with fusoids. The fusoids external to the PBS. *Leaf blade* adaxially flat. *Midrib* conspicuous (adaxial and abaxial projections); with one bundle only, or having a conventional arc of bundles, or having complex vascularization. Bulliforms present in discrete, regular adaxial groups; in simple fans (the fans wide). All the vascular bundles accompanied by sclerenchyma. Combined sclerenchyma girders present; forming 'figures'. Sclerenchyma all associated with vascular bundles.

Phytochemistry. Leaves without flavonoid sulphates (*O. latifolia*).

Cytology. $2n = 14, 20, 22, 30, 40$, and 44.

Taxonomy. Bambusoideae; Oryzodae; Olyreae.

Ecology, geography, regional floristic distribution. 23 species. Tropical America, Africa. Commonly adventive. Mesophytic; shade species; glycophytic. In forests.

Paleotropical and Neotropical. African and Madagascan. Sudano-Angolan and West African Rainforest. Caribbean, Venezuela and Surinam, Amazon, Central Brazilian, and Andean. Somalo-Ethiopian and South Tropical African.

Rusts and smuts. Rusts — *Physopella* and *Puccinia*.

References, etc. Morphological/taxonomic: Soderstrom and Zuloaga 1989. Leaf anatomical: Metcalfe 1960; this project.

Ophiochloa Filgueiras, Davidse & Zuloaga

Habit, vegetative morphology. Perennial; densely caespitose. *Culms* 50–80 cm high; herbaceous (erect, stiff); unbranched above; 6–10 noded. Culm nodes exposed; glabrous (dark). Culm internodes hollow (glabrous, stramineous). Young shoots intravaginal. *Leaves* not basally aggregated (basal and cauline); non-auriculate. Sheaths dorsally rounded, glabrous. Leaf blades linear; narrow; 0.5–1 mm wide (10–20 cm long); setaceous; rolled to acicular (U-shaped in section below, nearly cylindrical distally, the apex subpungent); without cross venation; disarticulating from the sheaths to persistent. *Ligule a fringe of hairs (minute)*; 0.5 mm long (with hairs to 6.5 mm long behind).

Reproductive organization. Plants bisexual, with bisexual spikelets; with hermaphrodite florets. The spikelets hermaphrodite.

Inflorescence. Inflorescence of spicate main branches, or a single raceme (the 6–10 cm long, delicate, erect or drooping, spikelike racemes one terminal and usually one axillary at the tops of the culms, well exserted on 6–22 cm long peduncles); digitate, or non-digitate. Primary inflorescence branches 1, or 2. *Inflorescence axes not ending in spikelets (the rachis apex extending beyond the spikelets as a sterile projection to 5 mm long)*. Rachides flattened and winged (the membranous wings 0.–1.1 mm wide). Inflorescence spatheate (the axillary raceme prophyllate), or espatheate; a complex of 'partial inflorescences' and intervening foliar organs, or not comprising 'partial inflorescences' and foliar organs. *Spikelet-bearing axes persistent. Spikelets* solitary; secund (the rachides unilateral); biseriate (borne alternately on each side of the midrib); *subsessile (the pedicels adnate to the rachis midrib for most of their length)*; imbricate.

Female-fertile spikelets. *Spikelets* somewhat unconventional (owing to the lack of a lower glume, combined with the peculiar structure of the lemma of the lower floret); 2.3–3.1 mm long (and 0.4–0.5 mm wide); narrowly elliptic to lanceolate; adaxial (i.e. the only, upper glume being abaxial); compressed dorsiventrally; falling with the glumes; not disarticulating between the florets; with conventional internode spacings. The upper floret not stipitate. Rachilla terminated by a female-fertile floret. Hairy callus absent. Callus blunt.

Glumes one per spikelet; relatively large; about equalling the spikelets; *long relative to the adjacent lemmas (the colourless glume embracing the upper floret, but leaving the lower lemma free)*; dorsiventral to the rachis; hairless; scabrous (with conspicuous prickle hairs); pointed (acute); awnless; non-carinate. Upper glume i.e. the single glume, indistinctly 2 nerved. *Spikelets* with incomplete florets. The incomplete florets proximal to the female-fertile florets. *Spikelets with proximal incomplete florets. The proximal incomplete florets* 1; epaleate (this seeming the best interpretation, despite resulting in a binerved lemma); sterile. *The proximal lemmas of peculiar, narrow form, having a central hyaline portion bordered by well developed, conspicuously ciliate and terminally setose nerves, the whole being reminiscent of a* **Bothriochloa** *pedicel; awnless (but with a striking pair of terminal, cushion-based, 5–7 mm long setae, which are hygroscopically active and awnlike)*; 2 nerved; somewhat exceeded by the female-fertile lemmas; similar in texture to the female-fertile lemmas to decidedly firmer than the female-fertile lemmas; not becoming indurated.

Female-fertile florets 1. Lemmas similar in texture to the glumes; not becoming indurated (membranous save for the hyaline apex, by contrast with the heavy margins of the lower lemma); copper-coloured; entire; pointed (acute); awnless; hairless (with a few prickle hairs only); non-carinate; having the margins tucked in onto the palea (below, but flat and free from the palea at the tip); without a germination flap; 'indistinctly nerved'. *Palea* present; relatively long (about equalling the lemma); gaping (at the tip); apically notched; awnless, without apical setae; thinner than the lemma (hyaline); not indurated; 2-nerved. *Lodicules* present; 2; free; fleshy; glabrous (?). *Stamens* 2 (usually), or 3 (rarely). Anthers 1.3–1.7 mm long (purple); not penicillate; without an apically prolonged connective. *Ovary* glabrous. Styles free to their bases. Stigmas 2; white (or rather, yellow), or red pigmented (varying in the same inflorescence).

Fruit, embryo and seedling. Fruit small (1–1.2 mm long); wine-red; narrowly obovate to clavate; not grooved. Hilum short. Embryo large (about half the caryopsis length); not waisted.

Transverse section of leaf blade, physiology. Leaf blades seemingly consisting of midrib. C_4; XyMS–. PCR sheath outlines even. PCR sheath extensions absent. Mesophyll without adaxial palisade. *Midrib* conspicuous, in that the blade appears largely reduced to midrib, with a single group of large bulliforms occupying the shallow adaxial groove; having a conventional arc of bundles; with colourless mesophyll adaxially. Many of the smallest vascular bundles unaccompanied by sclerenchyma. Combined sclerenchyma girders absent (the sclerenchyma seemingly limited to abaxial girders with the larger byndles). Sclerenchyma all associated with vascular bundles.

Special diagnostic feature. *The lower lemma narrow, of peculiar form, having a central hyaline portion bordered by well developed, conspicously ciliate and terminally setose nerves (the whole being reminiscent of a* **Bothriochloa** *pedicel), and exhibiting apically a pair of cushion-based, 5–7 mm long, hygroscopically active and awnlike setae*.

Taxonomy. Panicoideae; Panicodae; Paniceae.

Ecology, geography, regional floristic distribution. 1 species (*O. hydrolithica*). Goiás, Brazil. Glycophytic. On serpentine (ophiolite) rocks, in seasonally running water. Neotropical. Central Brazilian.

References, etc. Morphological/taxonomic: Filgueiras, Davidse and Zuloaga 1993. Leaf anatomical: Filgueiras, Davidse and Zuloaga 1993.

Special comments. Probably another odd *Digitaria* relative, as assigned by Filgueiras *et al.*, although a preliminary INTKEY exploration based on this description suggests there is as much in common with *Paspalum* and even with some Andropogoneae (Rottboelliinae, e.g. *Phacelurus*). Anatomical data for ts only.

Ophiuros Gaertn. f.

Habit, vegetative morphology. Annual, or perennial; caespitose (forming tall clumps). *Culms* 100–250 cm high; branched above; tuberous, or not tuberous. Culm nodes glabrous. Culm internodes solid. *Leaves* not basally aggregated; non-auriculate. Leaf blades broad, or narrow; flat, or folded, or rolled (convolute); without cross venation; persistent. Ligule a fringed membrane.

Reproductive organization. Plants bisexual, with bisexual spikelets; with hermaphrodite florets. The spikelets all alike in sexuality (ostensibly), or of sexually distinct forms on the same plant (if the vestigial pedicellate members are perceived as such); hermaphrodite, or hermaphrodite and sterile; overtly heteromorphic (if the pedicellate members are perceived as spikelets).

Inflorescence. Inflorescence compound paniculate (with slender, cylindrical, distichous spikes, terminal on culm branches). Rachides hollowed. *Inflorescence* spatheate; *a complex of 'partial inflorescences' and intervening foliar organs*. Spikelet-bearing axes spikelike *(the spikelets sunk in hollows of the terete rachis)*; solitary, or clustered (fascicled); with substantial rachides; *disarticulating*; disarticulating at the joints. *'Articles'* non-linear; with a basal callus-knob; disarticulating transversely; glabrous. *Spikelets* solitary (ostensibly solitary), or paired (acknowledging an 'adnate pedicel'); *not secund (the sessile spikelets in two opposite rows)*; distichous; sessile and pedicellate; consistently in 'long-and-short' combinations (disguisedly so); in pedicellate/sessile combinations. Pedicels of the 'pedicellate' spikelets discernible, but fused with the rachis. The 'shorter' spikelets hermaphrodite. *The 'longer' spikelets sterile (and reduced to the adnate pedicel)*.

Female-sterile spikelets. The pedicelled spikelet absent, being represented only by its pedicel, which is fused to the internode and sometimes barely recognisable.

Female-fertile spikelets. Spikelets abaxial; compressed dorsiventrally; falling with the glumes (and with the joint). Rachilla terminated by a female-fertile floret. Hairy callus absent.

Glumes two; more or less equal; long relative to the adjacent lemmas; awnless; very dissimilar (the lower thickly leathery or crustaceous, the upper thinly hyaline). *Lower glume* not two-keeled (with or without narrow wings at the tip); convex on the back to flattened on the back; not pitted; *lacunose with deep depressions, or rugose (with a basal transverse groove)*; 5–7 nerved. Upper glume 3 nerved. *Spikelets* with incomplete florets.

The incomplete florets proximal to the female-fertile florets. *Spikelets with proximal incomplete florets*. *The proximal incomplete florets* 1; paleate; male, or sterile. The proximal lemmas awnless; 0 nerved, or 2 nerved; not becoming indurated.

Female-fertile florets 1. Lemmas less firm than the glumes (thinly membranous); not becoming indurated; awnless; hairless; non-carinate; 0 nerved, or 2 nerved, or 3 nerved. *Palea* present; entire; not indurated (thin); 2-nerved. *Lodicules* present; 2; free; fleshy; glabrous. *Stamens* 3. Anthers not penicillate. *Ovary* glabrous. Styles free to their bases. Stigmas 2.

Fruit, embryo and seedling. Fruit slightly compressed dorsiventrally. Embryo small (about 1/4 of the fruit length).

Abaxial leaf blade epidermis. *Costal/intercostal zonation* conspicuous. *Papillae* absent. *Long-cells* similar in shape costally and intercostally; of similar wall thickness costally and intercostally. Mid-intercostal long-cells rectangular (with rounded corners), or fusiform (or tending to this); having markedly sinuous walls. *Microhairs* present; panicoid-type; (51–)52.5–54(–58.5) microns long; 10.5–12 microns wide at the septum. Microhair total length/width at septum 4.25–5.6. Microhair apical cells 24–40.5 microns long. Microhair apical cell/total length ratio 0.69–0.74. *Stomata* common; 24–29 microns long. Subsidiaries triangular. Guard-cells overlapping to flush with the interstomatals. *Intercostal short-cells* common; in cork/silica-cell pairs; silicified. *Costal short-cells* predominantly paired. Costal silica bodies 'panicoid-type'; mainly cross shaped; not sharp-pointed.

Transverse section of leaf blade, physiology. C_4; XyMS–. *Mesophyll* with radiate chlorenchyma. *Leaf blade* adaxially flat. *Midrib* conspicuous; having a conventional arc of bundles; with colourless mesophyll adaxially. Bulliforms present in discrete, regular adaxial groups (in the large-celled epidermis); in simple fans (the groups large celled, *Zea*-type). Many of the smallest vascular bundles unaccompanied by sclerenchyma. Combined sclerenchyma girders present; forming 'figures'. Sclerenchyma all associated with vascular bundles.

Phytochemistry. Leaves without flavonoid sulphates (1 species).

Taxonomy. Panicoideae; Andropogonodae; Andropogoneae; Rottboelliinae.

Ecology, geography, regional floristic distribution. 4 species. 1 in northeast tropical Africa, 6 in the Indomalayan region, Australia. Damp places in savanna.

Paleotropical and Australian. African and Indomalesian. Saharo-Sindian and Sudano-Angolan. Indian, Indo-Chinese, Malesian, and Papuan. North and East Australian. Somalo-Ethiopian. Tropical North and East Australian.

Rusts and smuts. Smuts from Ustilaginaceae. Ustilaginaceae — *Sphacelotheca* and *Ustilago*.

References, etc. Leaf anatomical: this project.

Special comments. Fruit data wanting.

Opizia J. & C. Presl

Casiostega Galeotti

Habit, vegetative morphology. Perennial; stoloniferous (sward forming). *Culms* 10 cm high; herbaceous. Culm internodes solid. *Leaves* not basally aggregated; non-auriculate; without auricular setae. Leaf blades linear to linear-lanceolate (acuminate); narrow; 1–2 mm wide; flat, or folded; without abaxial multicellular glands; without cross venation. *Ligule a fringed membrane (hairy on the back, with a short fringe)*; not truncate (rounded); 0.5–1 mm long. *Contra-ligule* absent.

Reproductive organization. *Plants dioecious*; without hermaphrodite florets. The spikelets all alike in sexuality (on the same plant); female-only, or male-only. Plants outbreeding.

Inflorescence. Inflorescence of spicate main branches (male plants with short panicles of 2–3 spike-like racemes, each raceme pulvinate at the base, reflexing and deciduous; with minute bristles associated with raceme bases), or a single raceme (female inflorescences solitary, spikelike). Inflorescence with axes ending in spikelets (female racemes), or axes not ending in spikelets (male racemes, which terminate in a short bristle). Inflorescence espatheate; not comprising 'partial inflorescences' and foliar organs. Spikelet-bearing axes disarticulating (males), or persistent (females); males falling entire. **Spikelets *subtended by solitary 'bristles' (each female raceme having short bristles near the tip, with a spikelet beyond).*** The 'bristles' persisting on the axis. Spikelets solitary; secund (in male racemes), or not secund (female racemes); very shortly pedicellate. Pedicel apices slightly cupuliform.

Female-sterile spikelets. Male spikelets 3–4 mm long, shortly pedicellate, secund, awnless; glumes, lemma and palea thinly membranous; glumes 1-nerved, lemma 3-nerved, palea 2-nerved; two fleshy lodicules; no prolonged rachilla, no rudimentary floret. Rachilla of male spikelets terminated by a male floret. The male spikelets with glumes; without proximal incomplete florets; 1 floreted. Male florets 1.

Female-fertile spikelets. Spikelets 3–4 mm long; adaxial; compressed laterally to not noticeably compressed; falling with the glumes; with conventional internode spacings. Rachilla prolonged beyond the uppermost female-fertile floret; rachilla hairy. The rachilla extension with incomplete florets. Hairy callus present. Callus blunt.

Glumes two; very unequal (G_1 about a third the length of G_2); (the upper) long relative to the adjacent lemmas; dorsiventral to the rachis; hairy (all over the back); awnless; non-carinate; dorsally hairy, hyaline, the upper with a minutely-toothed apex, the lower pointed. Lower glume 1 nerved. Upper glume 3 nerved. *Spikelets* with incomplete florets. The incomplete florets distal to the female-fertile florets. ***The distal incomplete florets*** 1; **incomplete florets *clearly specialised and modified in form (the terminal sterile floret reduced to the three long awns and the lobes tiny compared with the L_1, the lower part strongly adnate to the back of the palea)***.

Female-fertile florets 1. Lemmas decidedly firmer than the glumes (firmly membranous or leathery); becoming indurated; incised; 4 lobed (the lobes hyaline); deeply cleft; awned. Awns 3; median and lateral; the median non-geniculate; hairless (scabrid); much longer than the body of the lemma. ***The lateral awns shorter than the median (somewhat, arising from the inner sides of the outermost lobes)***. Lemmas hairy; somewhat carinate; without a germination flap; 3 nerved. **Palea *present (strongly winged above, bowed out below)***; relatively long; deeply bifid; awnless, without apical setae; not indurated (but very firm, cf. the lemma); 2-nerved; 2-keeled. Palea keels strongly winged (the wings projecting above the bases of the lemma awns). *Lodicules* present; 2; free; fleshy; glabrous; not or scarcely vascularized. *Stamens* 0. *Ovary* glabrous. *Styles fused (the ovary asymmetrically beaked, and saccate to fit the palea)*. Stigmas 2; white (?).

Fruit, embryo and seedling. Fruit small (about 1 mm long); ellipsoid; compressed laterally. Hilum short. Pericarp rather tough; free. Embryo large.

Abaxial leaf blade epidermis. *Costal/intercostal zonation* conspicuous. *Papillae* absent. *Long-cells* similar in shape costally and intercostally; of similar wall thickness costally and intercostally. Mid-intercostal long-cells rectangular; having markedly sinuous walls (thick, pitted). *Microhairs* present; elongated; clearly two-celled; panicoid-type (rather large, the apical cell very thin-walled, about equalling the basal cell in length). Microhair apical cell wall thinner than that of the basal cell and often collapsed. Microhairs (45–)54–60(–69) microns long. Microhair basal cells 30 microns long. Microhairs (6.6–)7.2–8.4(–9) microns wide at the septum. Microhair total length/width at septum 6.4–8.2. Microhair apical cells (22.5–)28.5–31.5(–35.4) microns long. Microhair apical cell/total length ratio 0.5–0.53. *Stomata* common; 24–25.5–27 microns long. Subsidiaries low to medium dome-shaped. Guard-cells overlapping to flush with the interstomatals. *Intercostal short-cells* common; in cork/silica-cell pairs; silicified. Intercostal silica bodies imperfectly developed; mainly crescentic (small). *Costal short-cells* predominantly paired. Costal silica bodies present throughout the costal zones to present in alternate cell files of the costal zones; crescentic (mainly), or saddle shaped (especially over the main veins), or tall-and-narrow (some); not sharp-pointed.

Transverse section of leaf blade, physiology. C_4; XyMS+. PCR sheaths of the primary vascular bundles interrupted; interrupted abaxially only. PCR sheath extensions absent. *Leaf blade* adaxially flat. *Midrib* conspicuous (with a large anchor-shaped girder abaxially, and bulliform epidermis adaxially); with one bundle only. Bulliforms present in discrete, regular adaxial groups; in simple fans and associated with colourless mesophyll cells to form deeply-penetrating fans. All the vascular bundles accompanied by sclerenchyma. Combined sclerenchyma girders present (one primary bundle towards the margin on each side, the rest (except the midrib) with scanty adaxial and abaxial strands); forming 'figures' (the two aforementioned). Sclerenchyma all associated with vascular bundles. The lamina margins with fibres.

Taxonomy. Chloridoideae; main chloridoid assemblage.

Ecology, geography, regional floristic distribution. 1 species. Southern Mexico, Cuba. Species of open habitats. Shallow soils on dry hillsides.

Neotropical. Caribbean.

References, etc. Leaf anatomical: this project.

Oplismenopsis L. Parodi

Habit, vegetative morphology. Aquatic perennial; rhizomatous. Culms herbaceous; to 1.2 cm in diameter; branched above. Culm nodes glabrous. Culm internodes hollow. *Leaves* not basally aggregated; non-auriculate. Sheaths glabrous, with cross-veins. *Leaf blades* lanceolate to ovate; broad; 10–27 mm wide; *cordate*; flat, or rolled (convolute); shortly pseudopetiolate; without cross venation; persistent. Ligule a fringed membrane; truncate; 1–1.5 mm long. *Contra-ligule* absent.

Reproductive organization. Plants bisexual, with bisexual spikelets; with hermaphrodite florets.

Inflorescence. Inflorescence of spicate main branches, or paniculate (of loose, non-digitate racemes); open; espatheate; not comprising 'partial inflorescences' and foliar organs. Spikelet-bearing axes persistent. Spikelets solitary; secund; biseriate; pedicellate.

Female-fertile spikelets. *Spikelets* about *9 mm long*; lanceolate; adaxial; compressed dorsiventrally; falling with the glumes. Rachilla terminated by a female-fertile floret. Hairy callus absent.

Glumes two; very unequal; (the upper) about equalling the spikelets (the lower about 1/2–1/3 its length); (the upper) long relative to the adjacent lemmas; dorsiventral to the rachis; hairless; pointed; *awned (the upper caudate-aristate)*; non-carinate; herbaceous. Lower glume 3 nerved. Upper glume 5–7 nerved. *Spikelets* with incomplete florets. The incomplete florets proximal to the female-fertile florets. *The proximal incomplete florets* 1; paleate. Palea of the proximal incomplete florets fully developed. The proximal incomplete florets male. The proximal lemmas acuminate to shortly awned; awned, or awnless (but then attenuate-mucronate); 5 nerved; herbaceous.

Female-fertile florets 1. Lemmas ovate-lanceolate; papyraceous; smooth to striate; entire; not crested; awnless; hairless; glabrous (shining); non-carinate; with a clear germination flap; 5 nerved. **Palea** present; relatively long; entire; awnless, without apical setae; textured like the lemma; 2-nerved; *2-keeled*. Palea keels wingless. *Lodicules* present; 2; free; fleshy; glabrous; not or scarcely vascularized. *Stamens* 3. Anthers not penicillate. *Ovary* glabrous. Styles basally fused. Stigmas 2; red pigmented.

Fruit, embryo and seedling. Fruit small to medium sized (about 4 mm long); ellipsoid; longitudinally grooved; compressed dorsiventrally (slightly). **Hilum long-linear** *(nearly as long as the grain)*. Embryo large; not waisted. Endosperm hard.

Abaxial leaf blade epidermis. *Costal/intercostal zonation* conspicuous. *Papillae* absent. Mid-intercostal long-cells having markedly sinuous walls. *Microhairs* present; panicoid-type. *Stomata* common. Subsidiaries triangular. Guard-cells overlapping to flush with the interstomatals. *Intercostal short-cells* common; in cork/silica-cell pairs. *Costal short-cells* conspicuously in long rows. Costal silica bodies 'panicoid-type'; cross shaped (mostly), or nodular (few); not sharp-pointed.

Transverse section of leaf blade, physiology. C$_3$; XyMS+. *Mesophyll* with radiate chlorenchyma; *Isachne*-type. *Leaf blade* with distinct, prominent adaxial ribs to 'nodular' in section; with the ribs more or less constant in size. *Midrib* not readily distinguishable; with one bundle only. Bulliforms present in discrete, regular adaxial groups; in simple fans. All the vascular bundles accompanied by sclerenchyma. Combined sclerenchyma girders present; forming 'figures'. Sclerenchyma all associated with vascular bundles.

Cytology. 2n = 20.

Taxonomy. Panicoideae; Panicodae; Paniceae.

Ecology, geography, regional floristic distribution. 1 species. South America. Hydrophytic, or helophytic.

Neotropical. Pampas.

Rusts and smuts. Smuts from Ustilaginaceae. Ustilaginaceae — *Sorosporium*.

References, etc. Morphological/taxonomic: Parodi 1937. Leaf anatomical: this project.

Oplismenus P. Beauv.

Hekaterosachne Steud., *Hippagrostis* Kuntze, *Orthopogon* R. Br.

Habit, vegetative morphology. Annual, or perennial; decumbent. *Culms* 10–100 cm high; herbaceous; freely branched above. Culm nodes hairy. Culm internodes solid, or hollow. *Leaves* not basally aggregated; non-auriculate. Leaf blades linear to ovate; broad, or narrow; not setaceous; flat (thin); pseudopetiolate, or not pseudopetiolate; cross veined, or without cross venation; persistent; rolled in bud. Ligule a fringed membrane (very short), or a fringe of hairs.

Reproductive organization. Plants bisexual, with bisexual spikelets; with hermaphrodite florets.

Inflorescence. *Inflorescence of spicate main branches (short unilateral racemes along a central axis)*; open; espatheate; not comprising 'partial inflorescences' and foliar organs. Spikelet-bearing axes persistent. *Spikelets* solitary, or paired (or the lower member reduced, or in clusters); *secund*; biseriate; shortly pedicellate; imbricate, or distant; not in distinct 'long-and-short' combinations.

Female-fertile spikelets. Spikelets elliptic, or lanceolate, or ovate; abaxial; weakly compressed laterally, or not noticeably compressed dorsiventrally; falling with the glumes. Rachilla terminated by a female-fertile floret. *Hairy callus present*.

Glumes two; more or less equal; shorter than the adjacent lemmas, or long relative to the adjacent lemmas; dorsiventral to the rachis; not pointed; *awned (both or at least the lower, the awn of the lower always longer)*; carinate; *similar (herbaceous, the awns often viscid)*. Lower glume 3–5 nerved. Upper glume 5 nerved. *Spikelets* with incomplete florets. The incomplete florets proximal to the female-fertile florets. *Spikelets with proximal incomplete florets*. *The proximal incomplete florets* 1; paleate. Palea of the proximal incomplete florets fully developed to reduced. The proximal incomplete florets male, or sterile. The proximal lemmas shortly awned, or awnless; 5–9 nerved; not becoming indurated.

Female-fertile florets 1. Lemmas dorsally compressed; similar in texture to the glumes to decidedly firmer than the glumes (papery to leathery); smooth (glossy); becoming indurated to not becoming indurated; white in fruit, or yellow in fruit; entire; indistinctly crested at the tip, or not crested; awnless; hairless; carinate to non-carinate; having the margins tucked in onto the palea; with a clear germination flap; 3–5 nerved. *Palea* present; relatively long; entire (acute); awnless, without apical setae; textured like the lemma (smooth, glossy); indurated, or not indurated; 2-nerved. *Lodicules* present; 2; fleshy. *Stamens* 3. Anthers 1–2 mm long; not penicillate. *Ovary* glabrous. Styles free to their bases. Stigmas 2.

Fruit, embryo and seedling. Fruit ellipsoid; longitudinally grooved (slightly), or not grooved; compressed dorsiventrally. Hilum short to long-linear (oblong, up to a half as long as the fruit). Embryo large. Endosperm hard; without lipid; containing only simple starch grains. Embryo without an epiblast; with a scutellar tail; with an elongated mesocotyl internode. Embryonic leaf margins overlapping.

Seedling with a long mesocotyl. First seedling leaf with a well-developed lamina. The lamina broad; supine; 13–20 veined.

Abaxial leaf blade epidermis. *Costal/intercostal zonation* conspicuous. *Papillae* absent. *Long-cells* markedly different in shape costally and intercostally (the costals narrow-rectangular); of similar wall thickness costally and intercostally (thin walled). Intercostal zones exhibiting many atypical long-cells, or without typical long-cells (incorporating or comprising squarish long-cells). Mid-intercostal long-cells rectangular (usually short, some irregular in shape); having markedly sinuous walls. *Microhairs* present; panicoid-type; (38–)46–59(–60) microns long; 4.5–6 microns wide at the septum. Microhair total length/width at septum 10–11.3. Microhair apical cells (20–)25–34.5(–36) microns long. Microhair apical cell/total length ratio 0.52–0.55. *Stomata* absent or very rare, or common; (34–)36–39(–42) microns long. Subsidiaries low dome-shaped, or triangular. Guard-cells overlapping to flush with the interstomatals. *Intercostal short-cells* common, or absent or very rare; not paired (mainly solitary); silicified, or not silicified. Intercostal silica bodies when present, cross-shaped. *Costal short-cells* conspicuously in long rows. Costal silica bodies 'panicoid-type'; cross shaped to dumb-bell shaped, or nodular; not sharp-pointed.

Transverse section of leaf blade, physiology. C_3; XyMS+. *Mesophyll* with radiate chlorenchyma; *Isachne*-type to not *Isachne*-type. *Leaf blade* with distinct, prominent adaxial ribs, or adaxially flat; with the ribs more or less constant in size. *Midrib* conspicuous; with one bundle only. Bulliforms present in discrete, regular adaxial groups (these wide); in simple fans. All the vascular bundles accompanied by sclerenchyma. Combined sclerenchyma girders present; forming 'figures'. Sclerenchyma all associated with vascular bundles.

Culm anatomy. *Culm internode bundles* in one or two rings.

Phytochemistry. Leaves without flavonoid sulphates (*O. hirtellus*).

Cytology. Chromosome base number, $x = 9$, 10, and 11. $2n = 18$, 36, 54, 72, and 90.

Taxonomy. Panicoideae; Panicodae; Paniceae.

Ecology, geography, regional floristic distribution. 5 species. Tropical and subtropical. Mesophytic; shade species; glycophytic. In forest.

Holarctic, Paleotropical, Neotropical, Cape, Australian, and Antarctic. Boreal and Tethyan. African, Madagascan, Indomalesian, Polynesian, and Neocaledonian. Euro-Siberian, Eastern Asian, and Atlantic North American. Mediterranean and Irano-Turanian. Saharo-Sindian, Sudano-Angolan, West African Rainforest, and Namib-Karoo. Indian, Indo-Chinese, Malesian, and Papuan. Fijian. Caribbean, Venezuela and Surinam, Amazon, Central Brazilian, Pampas, and Andean. North and East Australian. New Zealand. European. Southern Atlantic North American and Central Grasslands. Sahelo-Sudanian, Somalo-Ethiopian, South Tropical African, and Kalaharian. Tropical North and East Australian and Temperate and South-Eastern Australian.

Rusts and smuts. Rusts — *Phakopsora* and *Puccinia*. Smuts from Tilletiaceae and from Ustilaginaceae. Tilletiaceae — *Tilletia*. Ustilaginaceae — *Ustilago*.

Economic importance. Significant weed species: *O. burmanii*, *O. compositus*, *O. hirtellus*, *O. undulatifolius*. Important native pasture species: *O. compositus*.

References, etc. Leaf anatomical: Metcalfe 1960; this project.

Orcuttia Vasey

Habit, vegetative morphology. Aquatic *annual (producing long, juvenile, floating basal leaves)*. *Culms* 5–20 cm high; herbaceous; unbranched above. Culm internodes solid. The shoots aromatic. **Leaves** not basally aggregated; *not clearly differentiated into sheath and blade*; non-auriculate. Sheath margins free (but no clear distinction between sheath and blade). Leaf blades narrow (short); exhibiting multicellular glands abaxially. The abaxial glands intercostal. Leaf blades without cross venation; persistent. *Ligule* absent.

Reproductive organization. Plants bisexual, with bisexual spikelets; with hermaphrodite florets. The spikelets all alike in sexuality.

Inflorescence. Inflorescence a single spike, or a single raceme (then spike-like); espatheate; not comprising 'partial inflorescences' and foliar organs. Spikelet-bearing axes persistent. Spikelets solitary; not secund; distichous; subsessile to pedicellate.

Female-fertile spikelets. *Spikelets* 8–15 mm long; *compressed laterally*; disarticulating above the glumes; disarticulating between the florets (when wet). Rachilla prolonged beyond the uppermost female-fertile floret. The rachilla extension with incomplete florets.

Glumes two; more or less equal; shorter than the spikelets; shorter than the adjacent lemmas; not pointed; awnless; *similar (broad, 2–5 toothed)*. Lower glume 9–15 nerved.

Upper glume 9–15 nerved. *Spikelets* with incomplete florets. The incomplete florets distal to the female-fertile florets. *The distal incomplete florets* merely underdeveloped.

Female-fertile florets 5–11 (?). *Lemmas* firm; not becoming indurated; incised; 5 lobed; *deeply cleft*; mucronate (with mucronate teeth), or awned (the teeth awn-tipped). Awns when present, 5; when present, median and lateral; the median similar in form to the laterals; non-geniculate. Lemmas hairy, or hairless; 13–15 nerved. *Palea* present (broad); relatively long (as long as the lemma); 2-nerved; 2-keeled. Palea keels glabrous. *Lodicules* absent. *Stamens* 3. Stigmas 2.

Fruit, embryo and seedling. Fruit compressed laterally. Hilum short (large). Embryo large; without an epiblast; with a scutellar tail; with an elongated mesocotyl internode; with one scutellum bundle. Embryonic leaf margins meeting.

Abaxial leaf blade epidermis. *Costal/intercostal zonation* conspicuous. *Papillae* present; intercostal. Intercostal papillae not over-arching the stomata; consisting of one oblique swelling per cell. Mid-intercostal long-cells having straight or only gently undulating walls (or only slightly so). *Microhairs* present; elongated; clearly two-celled; chloridoid-type (*Neostapfia* type: 'button mushroom' shaped, and secretory). Microhair apical cell wall of similar thickness/rigidity to that of the basal cell. Microhairs 24–33(–36) microns long. Microhair basal cells 12 microns long. Microhairs 13.5–18.6 microns wide at the septum. Microhair total length/width at septum 1.3–2.7. Microhair apical cells 10.5–13.5 microns long. Microhair apical cell/total length ratio 0.32–0.44. *Stomata* common; 21–27 microns long (*O. inaequalis*), or 30–37.5 microns long (*O. californica*). Subsidiaries mostly high dome-shaped. *Intercostal short-cells* absent or very rare. Intercostal silica bodies absent. *Costal short-cells* conspicuously in long rows. Costal silica bodies present throughout the costal zones; 'panicoid-type'; cross shaped to dumb-bell shaped, or nodular; not sharp-pointed.

Transverse section of leaf blade, physiology. C_4; XyMS+. PCR sheaths of the primary vascular bundles complete. PCR sheath extensions absent. *Mesophyll* with radiate chlorenchyma; traversed by columns of colourless mesophyll cells. *Leaf blade* adaxially flat. *Midrib* not readily distinguishable; with one bundle only. Bulliforms present in discrete, regular adaxial groups; in simple fans and associated with colourless mesophyll cells to form deeply-penetrating fans (linking with traversing columns of colourless cells). Many of the smallest vascular bundles unaccompanied by sclerenchyma. Combined sclerenchyma girders present, or absent. Sclerenchyma all associated with vascular bundles.

Cytology. Chromosome base number, $x = 10$ (probably). $2n = 24$–32. 2 and 4 ploid (? — and aneuploids).

Taxonomy. Chloridoideae; Orcuttieae.

Ecology, geography, regional floristic distribution. 5 species. California and Mexico. Helophytic.

Holarctic. Madrean.

Rusts and smuts. Rusts — *Puccinia*. Taxonomically wide-ranging species: *Puccinia graminis*.

References, etc. Morphological/taxonomic: Reeder 1965. Leaf anatomical: Metcalfe 1960 and this project.

Oreobambos K. Schum.

~ *Dendrocalamus*

Habit, vegetative morphology. Arborescent perennial. The flowering culms leafy. *Culms* 450–1800 cm high; woody and persistent; to 10 cm in diameter. Culm nodes glabrous. Primary branches/mid-culm node several, on dominant. Culm internodes hollow. Pluricaespitose. Rhizomes pachymorph. Plants unarmed. *Leaves* not basally aggregated; without auricular setae. Leaf blades lanceolate to elliptic; broad; 25–60 mm wide; pseudopetiolate; without cross venation; disarticulating from the sheaths; rolled in bud.

Reproductive organization. Plants bisexual, with bisexual spikelets; with hermaphrodite florets. Not viviparous.

Inflorescence. *Inflorescence* indeterminate; *with pseudospikelets*; a false spike, with spikelets on contracted axes (with involucrate, cupuliform clusters of pseudospikelets on bare branches); spatheate (each cluster subtended by a deciduous spathe, and associated with broad, leathery bracts). *Spikelet-bearing axes* very much reduced (the spikelet clusters of 3 or more spikelets or pseudo-spikelets each subtended by a pair of leafy bracts); persistent. **Spikelets** *associated with bractiform involucres*.

Female-fertile spikelets. Spikelets 12–15 mm long; compressed laterally to not noticeably compressed; falling with the glumes. Rachilla prolonged beyond the uppermost female-fertile floret; rachilla hairless. The rachilla extension naked. Hairy callus absent.

Glumes one per spikelet; long relative to the adjacent lemmas; hairless (papery); glabrous; not pointed; awnless; non-carinate. Upper glume 11–18 nerved. **Spikelets** with female-fertile florets only; *without proximal incomplete florets*.

Female-fertile florets 2 (subequal). Lemmas similar in texture to the glumes to decidedly firmer than the glumes (papery to thinly leathery); not becoming indurated; entire; blunt; awnless; hairless; glabrous; without a germination flap; 11–23 nerved. *Palea* present; relatively long; apically notched; awnless, without apical setae; not indurated; several nerved (5–11); 2-keeled. *Lodicules* absent. *Stamens* 6. Anthers not penicillate; with the connective apically prolonged. *Ovary* hairy; with a conspicuous apical appendage. The appendage broadly conical, fleshy. Styles fused (into one). Stigmas 1.

Fruit, embryo and seedling. Pericarp thick and hard; free (and with a crustaceous, hairy apical appendage).

Abaxial leaf blade epidermis. *Costal/intercostal zonation* conspicuous. *Papillae* present. Intercostal papillae over-arching the stomata; several per cell. *Long-cells* similar in shape costally and intercostally; differing markedly in wall thickness costally and intercostally. Mid-intercostal long-cells having markedly sinuous walls. *Microhairs* present; panicoid-type. *Stomata* common (obscured by papillae). Subsidiaries dome-shaped, or triangular (?—outlines obscure). *Intercostal short-cells* common; in cork/silica-cell pairs. *Costal short-cells* predominantly paired. Costal silica bodies saddle shaped (rather variable); not sharp-pointed.

Transverse section of leaf blade, physiology. C_3; XyMS+. *Mesophyll* with non-radiate chlorenchyma; with arm cells; with fusoids. *Leaf blade* adaxially flat. *Midrib* conspicuous; having complex vascularization. Bulliforms present in discrete, regular adaxial groups; in simple fans. All the vascular bundles accompanied by sclerenchyma. Combined sclerenchyma girders present; forming 'figures'.

Taxonomy. Bambusoideae; Bambusodae; Bambuseae.

Ecology, geography, regional floristic distribution. 1 species. Tropical East Africa. Forest.

Paleotropical. African. Sudano-Angolan. Somalo-Ethiopian and South Tropical African.

References, etc. Leaf anatomical: Metcalfe 1960.

Special comments. Fruit data wanting.

Oreochloa Link

~ *Sesleria*

Habit, vegetative morphology. Montane perennial; rhizomatous, or caespitose. *Culms* 15–40 cm high; herbaceous. Leaves non-auriculate. **Sheath margins joined**. Leaf blades linear; narrow; 0.5–2 mm wide; setaceous, or not setaceous; flat, or rolled (involute); without cross venation. Ligule an unfringed membrane; not truncate; 1–6 mm long.

Reproductive organization. Plants bisexual, with bisexual spikelets; with hermaphrodite florets. The spikelets of sexually distinct forms on the same plant, or all alike in sexuality; hermaphrodite, or hermaphrodite and sterile (sometimes with small sterile bracts, representing reduced spikelets, at the base of the inflorescence); sometimes overtly heteromorphic (reduced and sterile/fertile).

Inflorescence. *Inflorescence paniculate*; contracted; more or less ovoid, or spicate; espatheate (ignoring any basal 'bracts'); not comprising 'partial inflorescences' and foliar organs. Spikelet-bearing axes persistent. Spikelets secund, or not secund; unilateral or distichous; pedicellate.

Female-fertile spikelets. Spikelets 3–6 mm long; compressed laterally; disarticulating above the glumes. Rachilla prolonged beyond the uppermost female-fertile floret. The rachilla extension with incomplete florets. Hairy callus absent.

Glumes two; very unequal, or more or less equal; shorter than the adjacent lemmas; pointed (acute); awnless; *non-carinate*; similar. Lower glume 1(–3) nerved. Upper glume 1(–3) nerved. *Spikelets* with incomplete florets. The incomplete florets distal to the female-fertile florets.

Female-fertile florets 3–7. Lemmas similar in texture to the glumes (membranous); not becoming indurated; entire, or incised; when incised, 3 lobed (3-toothed); not deeply cleft; awnless, or mucronate; hairy; carinate; without a germination flap; obscurely 5–7 nerved. *Palea* present; relatively long; 2-nerved; 2-keeled. Palea keels hairy (ciliate). *Lodicules* present; 2; free; membranous; glabrous; toothed, or not toothed; not or scarcely vascularized. *Stamens* 3. Anthers 2–3.5 mm long. *Ovary* glabrous. *Styles fused*. Stigmas 2.

Fruit, embryo and seedling. *Fruit* free from both lemma and palea; small (2 mm long). Hilum short. Embryo small; with an epiblast; without a scutellar tail; with a negligible mesocotyl internode. Embryonic leaf margins meeting.

Abaxial leaf blade epidermis. *Costal/intercostal zonation* lacking. *Papillae* absent. *Long-cells* similar in shape costally and intercostally; of similar wall thickness costally and intercostally. Mid-intercostal long-cells rectangular; having markedly sinuous walls (thick and pitted). *Microhairs* absent. *Stomata* absent or very rare. *Intercostal short-cells* common; not paired; not silicified. *Costal short-cells* neither distinctly grouped into long rows nor predominantly paired (mostly solitary). Costal silica bodies absent (in the material seen).

Transverse section of leaf blade, physiology. C_3; XyMS+. *Mesophyll* with non-radiate chlorenchyma. *Leaf blade* with distinct, prominent adaxial ribs, or 'nodular' in section; with the ribs more or less constant in size. *Midrib* conspicuous (larger bundle and rib); with one bundle only. Bulliforms present in discrete, regular adaxial groups (as well as the 'midrib hinges' and *Ammophila*-type arrangements of small cells in the furrows); in simple fans (sin some of the furrows). All the vascular bundles accompanied by sclerenchyma. Combined sclerenchyma girders absent (each bundle with adaxial and abaxial strands). Sclerenchyma all associated with vascular bundles, or not all bundle-associated.

Cytology. Chromosome base number, $x = 7$. $2n = 14$. 2 ploid.

Taxonomy. Pooideae; Poodae; Seslerieae.

Ecology, geography, regional floristic distribution. 4 species. Southern Europe. Species of open habitats. Mountain rocks, calcifuge.

Holarctic. Boreal and Tethyan. Euro-Siberian. Mediterranean. European.

Rusts and smuts. Smuts from Tilletiaceae. Tilletiaceae — *Entyloma*.

References, etc. Leaf anatomical: this project.

Orinus A. Hitchc.

Habit, vegetative morphology. Perennial; rhizomatous and caespitose. **Culms** 30–50 cm high; *herbaceous*. Rhizomes pachymorph. Plants unarmed. Leaves non-auriculate. Leaf blades narrow; 2–5 mm wide (by 3–10 cm long, sparsely pilose); becoming involute; with setiform, slightly pungent tips; exhibiting multicellular glands abaxially. The abaxial glands intercostal. Leaf blades without cross venation. Ligule thin, lacerate; 1 mm long.

Reproductive organization. Plants bisexual, with bisexual spikelets; with hermaphrodite florets. The spikelets all alike in sexuality. *Plants without hidden cleistogenes*.

Inflorescence. *Inflorescence of spicate main branches (erect or ascending racemes)*; non-digitate (the axis elongate). Primary inflorescence branches 5–8. Inflorescence espatheate; not comprising 'partial inflorescences' and foliar organs. Spikelet-bearing axes persistent. *Spikelets* solitary; secund (the rachides dorsiventral); *not two-ranked (in one row, on one side of the rachis, appressed)*; shortly pedicellate.

Female-fertile spikelets. *Spikelets* 6 mm long; pale or leaden-purplish tinged; disarticulating above the glumes; *disarticulating between the florets*; with distinctly elongated rachilla internodes between the florets (a 1 mm internode between the first and second florets). Rachilla prolonged beyond the uppermost female-fertile floret; rachilla hairless (glabrous).

Glumes two; relatively large (4–5 mm long); more or less equal (only slightly unequal); shorter than the adjacent lemmas; hairy (sparsely villous to nearly glabrous); without conspicuous tufts or rows of hairs; pointed (acute); awnless; similar (pale, membranous). Lower glume shorter than the lowest lemma; 1 nerved. *Upper glume 3 nerved. Spikelets* with female-fertile florets only, or with incomplete florets. The incomplete florets distal to the female-fertile florets. *The distal incomplete florets* merely underdeveloped.

Female-fertile florets 2–3 (–4). *Lemmas* similar in texture to the glumes (the tips hyaline); not becoming indurated; entire; blunt, or pointed; awnless, or mucronate (but only slightly so); **hairy (with villous hairs all over, not confined to the nerves as in Leptochloa and Trichoneura)**; indistinctly carinate; **3 nerved (somewhat concave between the nerves)**. *Palea* present; relatively long; awnless, without apical setae; 2-keeled. Palea keels hairy. *Stamens* 3; with free filaments. Anthers 2.5–2.75 mm long. Stigmas 2.

Fruit, embryo and seedling. Fruit small (2.5 mm long); not noticeably compressed (cylindrical). Hilum short. Pericarp fused. Embryo large (about 1/3 the grain length).

Abaxial leaf blade epidermis. *Costal/intercostal zonation* conspicuous. *Papillae* present (very abundant); costal and intercostal. Intercostal papillae over-arching the stomata; consisting of one symmetrical projection per cell. Intercostal zones with typical long-cells (or these rather short). Mid-intercostal long-cells rectangular; having markedly sinuous walls. *Microhairs* present; elongated; clearly two-celled; chloridoid-type. Microhair apical cell wall of similar thickness/rigidity to that of the basal cell. Microhairs 21–24–25.5 microns long. Microhair basal cells 24 microns long. Microhairs (9.3–)10.5–11.4(–15) microns wide at the septum. Microhair total length/width at septum 1.6–2.6. Microhair apical cells 10.5–12–16.5 microns long. Microhair apical cell/total length ratio 0.41–0.68. *Stomata* common (alongside the veins); 21–22.5(–25.5) microns long. Subsidiaries triangular (though not readily observable). *Intercostal short-cells* common; in cork/silica-cell pairs and not paired (some solitary); not silicified. Intercostal silica bodies absent. *Costal short-cells* conspicuously in long rows. Costal silica bodies present in alternate cell files of the costal zones; 'panicoid-type'; limited to cross shaped and butterfly shaped; not sharp-pointed.

Transverse section of leaf blade, physiology. C_4; XyMS+. PCR sheaths of the primary vascular bundles interrupted; interrupted abaxially only. PCR sheath extensions absent. *Leaf blade* 'nodular' in section to adaxially flat. *Midrib* not readily distinguishable; with one bundle only. Bulliforms present in discrete, regular adaxial groups; in simple fans (these predominating, with large, deeply penetrating median cells), or associated with colourless mesophyll cells to form deeply-penetrating fans (a few). All the vascular bundles accompanied by sclerenchyma. Combined sclerenchyma girders present (with all the bundles); forming 'figures' (all bundles with small I's). Sclerenchyma all associated with vascular bundles. The lamina margins with fibres.

Taxonomy. Chloridoideae; main chloridoid assemblage.

Ecology, geography, regional floristic distribution. 2 species. Western Himalayas. Xerophytic; species of open habitats. Desert sand dunes at high altitude.

Holarctic. Tethyan. Irano-Turanian.

References, etc. Morphological/taxonomic: Hitchcock 1933. Leaf anatomical: this project.

Oropetium Trin.

Kralikella Coss. and Dur.

~ *Tripogon*

Excluding *Chaetostichium, Lepturella*

Habit, vegetative morphology. Annual, or perennial; caespitose (dwarf, cushion-forming). **Culms** 2–15(–17) cm high; *herbaceous*; branched above, or unbranched above.

Culm nodes glabrous. Culm internodes hollow. Plants unarmed. Young shoots intravaginal. *Leaves* mostly basal, or not basally aggregated; non-auriculate. Leaf blades linear; narrow; about 2 mm wide (to 4 cm long); setaceous, or not setaceous; flat, or folded, or rolled; exhibiting multicellular glands abaxially. The abaxial glands intercostal and on the blade margins. Leaf blades without cross venation; disarticulating from the sheaths, or persistent. Ligule an unfringed membrane, or a fringed membrane; truncate; about 0.2 mm long. *Contra-ligule* absent.

Reproductive organization. Plants bisexual, with bisexual spikelets; with hermaphrodite florets.

Inflorescence. *Inflorescence a single spike (straight, curved, sinuous or coiled).* Rachides hollowed. Inflorescence espatheate; not comprising 'partial inflorescences' and foliar organs. *Spikelet-bearing axes* spikes; with substantial rachides (though these slender, herbaceous or spongy); persistent, or disarticulating; when fragile, disarticulating at the joints (or fracturing into segments of 1–4 spikelets). *'Articles'* non-linear; disarticulating transversely; glabrous. *Spikelets* solitary; not secund; *distichous*; sessile.

Female-fertile spikelets. *Spikelets 2.5–3.5 mm long*; adaxial; *compressed laterally*; disarticulating above the glumes, or falling with the glumes (and with the joint); with conventional internode spacings. Rachilla prolonged beyond the uppermost female-fertile floret, or terminated by a female-fertile floret. The rachilla extension when present, with incomplete florets. Hairy callus present, or absent. *Callus* absent, or short; when present, blunt.

Glumes two, or one per spikelet (the G_1 sometimes vestigial or missing); relatively large (G_2); very unequal (except in terminal spikelets); exceeding the spikelets; (the upper) long relative to the adjacent lemmas; dorsiventral to the rachis; hairless; pointed; awnless; carinate, or non-carinate; very dissimilar (G_1 reduced and scarious or missing, G_2 covering the florets, hardened). Lower glume if present, much shorter than half length of lowest lemma; 0 nerved. Upper glume 1 nerved, or 3 nerved. *Spikelets* with female-fertile florets only, or with incomplete florets. The incomplete florets when present, distal to the female-fertile florets. The distal incomplete florets 1 (this male or sterile); *incomplete florets* merely underdeveloped.

Female-fertile florets 1 (or the second floret rarely also hermaphrodite?). Lemmas less firm than the glumes (hyaline); not becoming indurated; incised; 2 lobed (bidentate); not deeply cleft; mucronate, or awned. Awns when present, 1; (or mucro) from a sinus; non-geniculate; hairless (scabrous); much shorter than the body of the lemma; entered by one vein. Lemmas hairy (at the base or on the nerves), or hairless; carinate to non-carinate; without a germination flap; 3 nerved. *Palea* present; relatively long (oblong); entire, or apically notched, or deeply bifid; awnless, without apical setae; not indurated (hyaline); 2-nerved; 2-keeled. Palea keels wingless; glabrous. *Lodicules* present; 2; free; fleshy; glabrous. *Stamens* 3; with free filaments. Anthers about 0.7 mm long; not penicillate; without an apically prolonged connective. *Ovary* glabrous. Styles free to their bases. Stigmas 2; brown.

Fruit, embryo and seedling. *Fruit* free from both lemma and palea (but included); small (about 1.5 mm long); fusiform; compressed laterally. Hilum short. *Pericarp loosely adherent (removable when soaked)*. Embryo small (about 1/4 the length of the fruit).

Abaxial leaf blade epidermis. *Costal/intercostal zonation* conspicuous. *Papillae* absent. *Long-cells* similar in shape costally and intercostally; of similar wall thickness costally and intercostally. Mid-intercostal long-cells rectangular; having markedly sinuous walls (moderately thick, pitted). *Microhairs* present; more or less spherical; clearly two-celled; chloridoid-type. Microhair apical cell wall thinner than that of the basal cell but not tending to collapse. Microhairs (16–)19.5–21(–23) microns long. Microhair basal cells 12 microns long. Microhairs (9–)11.4–12(–12.6) microns wide at the septum. Microhair total length/width at septum 1.5–2. Microhair apical cells (6–)8.4–9(–10.5) microns long. Microhair apical cell/total length ratio 0.36–0.47. *Stomata* common; 22.5–24 microns long. Subsidiaries triangular. *Intercostal short-cells* absent or very rare. Intercostal silica bodies absent. *Costal short-cells* conspicuously in long rows. Costal silica bodies present in alternate cell files of the costal zones; saddle shaped; not sharp-pointed.

Transverse section of leaf blade, physiology. C_4; XyMS+. PCR sheath outlines even. PCR sheaths of the primary vascular bundles interrupted; interrupted abaxially only. PCR sheath extensions absent. PCR cell chloroplasts centripetal. *Mesophyll* with radiate chlorenchyma; without adaxial palisade; traversed by columns of colourless mesophyll cells (in places). *Leaf blade* with distinct, prominent adaxial ribs to 'nodular' in section; with the ribs more or less constant in size (rib apices rounded). *Midrib* not readily distinguishable; with one bundle only. Bulliforms present in discrete, regular adaxial groups; associated with colourless mesophyll cells to form deeply-penetrating fans (these sometimes linking with traversing columns of clourless cells). All the vascular bundles accompanied by sclerenchyma. Combined sclerenchyma girders present; forming 'figures' (most bundles with anchors). Sclerenchyma all associated with vascular bundles.

Cytology. Chromosome base number, $x = 10$. $2n = 20$.

Taxonomy. Chloridoideae; main chloridoid assemblage.

Ecology, geography, regional floristic distribution. 3–4 species. Arid subtropical Africa and mountains. Mesophytic to xerophytic; species of open habitats; glycophytic. In shallow soil between or over rocks and in outwashes.

Paleotropical. African and Indomalesian. Saharo-Sindian, Sudano-Angolan, West African Rainforest, and Namib-Karoo. Indian and Indo-Chinese. Sahelo-Sudanian, Somalo-Ethiopian, South Tropical African, and Kalaharian.

References, etc. Morphological/taxonomic: Phillips 1974. Leaf anatomical: Metcalfe 1960; this project; photos of *O. capense* provided by R.P. Ellis.

Orthachne Nees ex Steud.

~ *Stipa*, cf. *Lorenzochloa*
Excluding *Lorenzochloa*

Habit, vegetative morphology. Perennial. *Culms* 5–35 cm high (?); herbaceous; unbranched above. Leaf blades filiform; narrow; rolled and acicular; without cross venation.

Reproductive organization. Plants bisexual, with bisexual spikelets; with hermaphrodite florets.

Inflorescence. *Inflorescence* many spikeleted (at least, usually of 10 or more, by contrast with *Aciachne*); *paniculate (borne well above the leaves)*; open; espatheate; not comprising 'partial inflorescences' and foliar organs. Spikelet-bearing axes persistent. Spikelets not secund; pedicellate.

Female-fertile spikelets. *Spikelets not noticeably compressed (the floret cylindrical)*; disarticulating above the glumes. Rachilla terminated by a female-fertile floret. Hairy callus present. *Callus* short; pointed.

Glumes two; shorter than the spikelets; shorter than the adjacent lemmas; pointed, or not pointed (truncate or acute); awnless; similar (membranous). Lower glume 1 nerved. *Upper glume 1 nerved.* Spikelets with female-fertile florets only; *without proximal incomplete florets*.

Female-fertile florets 1. *Lemmas covering only the margins of the palea*; not convolute; not saccate; similar in texture to the glumes (membranous); not becoming indurated; entire; awned. Awns 1; median; apical (from the acuminate lemma); non-geniculate; persistent. Awn bases slightly twisted. Lemmas non-carinate; 5 nerved. **Palea** present; relatively long; membranous; 2-nerved; *keel-less*. **Lodicules** present; *3*. **Stamens 3**.

Abaxial leaf blade epidermis. *Costal/intercostal zonation* lacking. *Papillae* absent (though abundant adaxially). *Long-cells* similar in shape costally and intercostally; of similar wall thickness costally and intercostally (thick walled). Mid-intercostal long-cells rectangular; having markedly sinuous walls (these very heavily pitted). *Microhairs* absent. *Stomata* absent or very rare (present adaxially only, with parallel-sided subsidiaries). *Intercostal short-cells* common. *Costal short-cells* neither distinctly grouped into long rows nor predominantly paired (frequently in short rows). Costal silica bodies seemingly 'panicoid-type' (but very hard to observe in the material available); not sharp-pointed.

Transverse section of leaf blade, physiology. C_3; XyMS+. *Leaf blade* with distinct, prominent adaxial ribs; with the ribs very irregular in sizes. *Midrib* not readily distinguishable; with one bundle only. Bulliforms not in discrete, regular adaxial groups (not seen). All the vascular bundles accompanied by sclerenchyma. Combined sclerenchyma girders present (with the primaries); forming 'figures'. Sclerenchyma not all bundle-associated. The 'extra' sclerenchyma in a continuous abaxial layer (this submerging the 'feet' of the 'anchors').

Cytology. Chromosome base number, $x = 11$. $2n = 22$. 2 ploid.

Taxonomy. Arundinoideae; Stipeae.

Ecology, geography, regional floristic distribution. 2 species. Chile. Xerophytic; species of open habitats. Montane.

Antarctic. Patagonian.

References, etc. Morphological/taxonomic: Hughes 1923; Reeder and Reeder 1968. Leaf anatomical: this project.

Special comments. Fruit data wanting.

Orthoclada P. Beauv.

Habit, vegetative morphology. Perennial; rhizomatous, or caespitose. The flowering culms leafy. *Culms* 80–180 cm high; herbaceous; to 0.4 cm in diameter; branched above. Culm nodes glabrous. Culm internodes hollow. Plants unarmed. **Leaves** not basally aggregated; auriculate; *with auricular setae*. Leaf blades lanceolate to elliptic; broad; pseudopetiolate; cross veined; rolled in bud. *Ligule* present; an unfringed membrane (minutely ciliolate); truncate; 0.5–1 mm long. *Contra-ligule* absent.

Reproductive organization. *Plants* bisexual, with bisexual spikelets; *with hermaphrodite florets*.

Inflorescence. *Inflorescence* paniculate, or of spicate main branches (then the primary branches reduced to open racemes); open; *with capillary branchlets*; espatheate; not comprising 'partial inflorescences' and foliar organs. Spikelet-bearing axes persistent. Spikelets solitary; pedicellate.

Female-fertile spikelets. *Spikelets* 4–12 mm long; compressed laterally; *falling with the glumes*. Rachilla prolonged beyond the uppermost female-fertile floret (adherent to the lower back of the palea); rachilla hairy (minutely). The rachilla extension with incomplete florets. Hairy callus absent.

Glumes two; very unequal (G_1 about 2/3 the length of G_2); shorter than the adjacent lemmas; free; hairless (sparsely scabrid); pointed; awnless; carinate (somewhat); similar (lanceolate, herbaceous-membranous). Lower glume 3–5 nerved. Upper glume 3–5 nerved. *Spikelets* with incomplete florets. The incomplete florets distal to the female-fertile florets. *The distal incomplete florets* merely underdeveloped.

Female-fertile florets 1–5 (decreasing in size acropetally). Lemmas acuminate; similar in texture to the glumes to decidedly firmer than the glumes; not becoming indurated; entire; pointed; awnless; hairless (sparsely scabrid above); slightly carinate; 5–6 nerved. *Palea* present; relatively long; apically notched (minutely); awnless, without apical setae; textured like the lemma; not indurated; 2-nerved (but folded, bringing the keels together); 2-keeled (the keels joined to the lower part of the internode above). Palea keels minutely winged (above, and the internode above situated in the groove between them). *Lodicules* present; 2; free; fleshy; glabrous; not toothed; heavily vascularized. *Stamens* 2 (*O. laxa*), or 3. Anthers not penicillate; without an apically prolonged connective. *Ovary* glabrous. Styles free to their bases. Stigmas 2.

Fruit, embryo and seedling. *Fruit* free from both lemma and palea; small; compressed laterally. Hilum short. Embryo large; with an epiblast; with a scutellar tail; with an elongated mesocotyl internode. Embryonic leaf margins overlapping.

Abaxial leaf blade epidermis. *Costal/intercostal zonation* conspicuous. *Papillae* absent. *Long-cells* markedly different in shape costally and intercostally (the costals relatively much longer); of similar wall thickness costally and intercostally (thin walled). Intercostal zones with typical long-cells (though with some rather short). Mid-intercostal long-cells rectangular; having markedly sinuous walls (very deeply so). *Microhairs* present; panicoid-type; (46.5–)57–60 microns long; 3–3.6 microns wide at the septum. Microhair total length/width at septum 12.9–20. Microhair apical cells (28.5–)34.5–37.5(–39) microns long. Microhair apical cell/total length ratio 0.58–0.66.

Stomata common; (30–)36–39 microns long. Subsidiaries triangular. Guard-cells overlapping to flush with the interstomatals. *Intercostal short-cells* absent or very rare. *Costal short-cells* conspicuously in long rows. Costal silica bodies 'panicoid-type'; mostly cross shaped, butterfly shaped, and dumb-bell shaped; not sharp-pointed.

Transverse section of leaf blade, physiology. C_3; XyMS+. Mesophyll with adaxial palisade; seemingly without arm cells (but material seen poor); with fusoids (shortish). The fusoids an integral part of the PBS. Leaf blade with the ribs more or less constant in size. *Midrib* conspicuous; having a conventional arc of bundles (one large bundle, two small laterals); with colourless mesophyll adaxially. Bulliforms present in discrete, regular adaxial groups (these large); in simple fans. All the vascular bundles accompanied by sclerenchyma. Combined sclerenchyma girders absent (adaxial strand-abaxial girder combinations only). Sclerenchyma all associated with vascular bundles.

Cytology. Chromosome base number, $x = 12$. $2n = 24$. 2 ploid.

Taxonomy. Bambusoideae; Oryzodae; Centotheceae.

Ecology, geography, regional floristic distribution. 2 species. 1 in southern Mexico to tropical South America and West Indies; 1 in southeastern tropical Africa. Shade species; glycophytic. In forests.

Paleotropical and Neotropical. African. Sudano-Angolan and West African Rainforest. Caribbean, Amazon, Central Brazilian, and Andean. South Tropical African.

References, etc. Leaf anatomical: this project.

Oryza L.

Padia Moritzi

Habit, vegetative morphology. Annual, or perennial; rhizomatous, or caespitose. *Culms* 30–300 cm high; herbaceous. Culm nodes glabrous. Culm internodes hollow. *Leaves* not basally aggregated; usually auriculate; with auricular setae (auricles with up to 4 setae), or without auricular setae. *Leaf blades linear to linear-lanceolate; neither leathery nor flimsy*; broad, or narrow; flat; pseudopetiolate, or not pseudopetiolate; without cross venation; persistent. Ligule an unfringed membrane; 3–45 mm long.

Reproductive organization. Plants bisexual, with bisexual spikelets; with hermaphrodite florets. *The spikelets all alike in sexuality*. Plants outbreeding and inbreeding. Exposed-cleistogamous, or chasmogamous.

Inflorescence. Inflorescence paniculate (axes usually wavy, the spikelets appressed), or of spicate main branches (the primary branches often reduced to racemes); open; with capillary branchlets, or without capillary branchlets; espatheate; not comprising 'partial inflorescences' and foliar organs. Spikelet-bearing axes persistent. Spikelets not secund; pedicellate.

Female-fertile spikelets. Spikelets unconventional (disarticulating above a cupuliform pedicel apex, which is taken to represent glumes, so that the organ serving as a palea may 'really' be a lemma); 4–12 mm long; strongly compressed laterally; disarticulating above the glumes (i.e. above the pedicel cup representing them); *with conventional internode spacings (i.e. the floret not stipitate)*. Rachilla terminated by a female-fertile floret. Hairy callus absent.

Glumes *present to absent (represented only by a small 2-lobed cupule)*; if present, two; *minute*; more or less equal; shorter than the adjacent lemmas; joined; awnless. Lower glume 0 nerved. Upper glume 0 nerved. *Spikelets* with incomplete florets. The incomplete florets proximal to the female-fertile florets. *Spikelets with proximal incomplete florets*. *The proximal incomplete florets* 2 (small, vestigial, no more than half the spikelet length, sometimes only bristles); epaleate; sterile. The proximal lemmas awnless; exceeded by the female-fertile lemmas (usually between 1/8 and 1/2 the length of the spikelet).

Female-fertile florets 1. Lemmas becoming indurated to not becoming indurated (leathery to indurated); entire; pointed; awnless, or mucronate, or awned. Awns when present, 1; apical; non-geniculate; much shorter than the body of the lemma to much longer than the body of the lemma. Lemmas hairy (hispid-ciliate on the nerves), or hairless (glabrous); strongly carinate; 3–9 nerved. *Palea* present; relatively long (but narrower than the lemma); tightly clasped by the lemma (along the lateral nerves); entire (mucronate to subulate); awnless, without apical setae, or with apical setae, or awned (sometimes with an apical awn or seta); textured like the lemma; not indurated (leathery); several nerved; one-keeled. *Lodicules* present; 2; membranous (but the membranous flange may be narrow); glabrous; toothed, or not toothed; heavily vascularized. **Stamens 6**. Anthers 2–3 mm long; not penicillate. *Ovary* glabrous. Styles fused (basally), or free to their bases. Stigmas 2.

Fruit, embryo and seedling. *Fruit* adhering to lemma and/or palea, or free from both lemma and palea; small, or medium sized, or large; compressed laterally. Hilum long-linear. Embryo small. Endosperm hard; without lipid; containing compound starch grains. Embryo with an epiblast; with a scutellar tail; with a negligible mesocotyl internode. Embryonic leaf margins overlapping.

Seedling with a long mesocotyl. First seedling leaf without a lamina.

Abaxial leaf blade epidermis. *Costal/intercostal zonation* conspicuous. *Papillae* present. Intercostal papillae over-arching the stomata (to some extent); several per cell (*O. australiensis* having long-cells variously with a single longitudinal row of large papillae, a double row of half-sized ones, or something in between; and the guard-cells over-arched at their ends by tiny projections, seemingly from the subsidiaries). Mid-intercostal long-cells rectangular; having markedly sinuous walls. *Microhairs* present; panicoid-type; without 'partitioning membranes' (in *O. sativa*); (24–)26–42(–45) microns long; 3.6–5.4 microns wide at the septum. Microhair total length/width at septum 7.1–12.5. Microhair apical cells (14–)19–22.5(–25.5) microns long. Microhair apical cell/total length ratio 0.46–0.57. *Stomata* common; (22.5–)24–28.5(–30) microns long. Subsidiaries papillate; dome-shaped, or triangular (mostly). Guard-cells overlapped by the interstomatals (very slightly). *Intercostal short-cells* absent or very rare. *Costal short-cells* conspicuously in long rows. Costal silica bodies oryzoid (mostly), or rounded, or crescentic (the rounded and crescentic forms associated with the short-cell pairs which occur over some veins in *O. sativa*); not sharp-pointed.

Transverse section of leaf blade, physiology. C$_3$; XyMS+. *Mesophyll* with non-radiate chlorenchyma; without adaxial palisade; with arm cells; without fusoids. Leaf blade with the ribs more or less constant in size (except for enlarged ribs associated with the leaf margins), or with the ribs very irregular in sizes. *Midrib* conspicuous; having complex vascularization. Bulliforms present in discrete, regular adaxial groups; in simple fans, or in simple fans and associated with colourless mesophyll cells to form deeply-penetrating fans (a few *Sporobolus*-type groups in at least some species, e.g. *O. australiensis*). All the vascular bundles accompanied by sclerenchyma. Combined sclerenchyma girders present; forming 'figures' (in most bundles). Sclerenchyma all associated with vascular bundles.

Culm anatomy. *Culm internode bundles* in one or two rings.

Phytochemistry. Tissues of the culm bases with abundant starch. Leaves without flavonoid sulphates (2 species).

Special diagnostic feature. *Not scandent as in* **Prosphytochloa** *(q.v.)*. *Female-fertile lemma not as in* **Rhynchoryza** *(q.v.)*.

Cytology. Chromosome base number, $x = 12$. $2n = 24$ and 48. 2 and 4 ploid. Haploid nuclear DNA content 0.6–1.1 pg (12 species, mean 0.8). Mean diploid 2c DNA value 1.7 pg (10 species, 1.1–2.5). Nucleoli persistent.

Taxonomy. Bambusoideae; Oryzodae; Oryzeae.

Ecology, geography, regional floristic distribution. 25 species. Tropical. Commonly adventive. Hydrophytic, or helophytic; shade species, or species of open habitats; glycophytic.

Holarctic, Paleotropical, and Neotropical. Boreal and Tethyan. African, Madagascan, and Indomalesian. Euro-Siberian and Atlantic North American. Irano-Turanian. Saharo-Sindian, Sudano-Angolan, West African Rainforest, and Namib-Karoo. Indian, Indo-Chinese, Malesian, and Papuan. Caribbean, Amazon, Pampas, and Andean. European. Southern Atlantic North American and Central Grasslands. Sahelo-Sudanian, Somalo-Ethiopian, South Tropical African, and Kalaharian.

Hybrids. Intergeneric hybrid claimed with *Triticum*: ×*Oryticum* Wang & Tang in *Acta Phytotax. Sin.* **20**, 179 (1982).

Rusts and smuts. Rusts — *Puccinia.* Taxonomically wide-ranging species: *Puccinia graminis.* Smuts from Tilletiaceae. Tilletiaceae — *Entyloma* and *Tilletia.*

Economic importance. Significant weed species: *O. barthii, O. perennis* (in rice), *O. punctata, O. rufipogon.* Important native pasture species: e.g. *O. longistaminata*(Kenya), *O. ridleyi* (Malaya). Grain crop species: *O. sativa* (Rice); also *O. glaberrima* (west Africa).

References, etc. Leaf anatomical: Metcalfe 1960; this project.

Oryzidium C.E. Hubb. & Schweick.

Habit, vegetative morphology. Floating perennial; stoloniferous. *Culms* 40–120 cm high (the lower internodes trailing in water or floating); herbaceous (spongy); branched above. Culm nodes glabrous. Culm sheaths persistent (even when permanently submerged). Culm internodes solid (spongy). Plants unarmed. Young shoots intravaginal. *Leaves* not basally aggregated; non-auriculate. Leaf blades linear; narrow; 6–10 mm wide; flat; without cross venation; persistent. Ligule a fringe of hairs; 3–4 mm long. *Contra-ligule* absent.

Reproductive organization. *Plants bisexual, with bisexual spikelets; without hermaphrodite florets (the lower floret male, the upper female).*

Inflorescence. *Inflorescence paniculate*; narrow, the branches nearly erect; espatheate; not comprising 'partial inflorescences' and foliar organs. Spikelet-bearing axes persistent. Spikelets solitary; not secund; pedicellate; somewhat imbricate.

Female-fertile spikelets. Spikelets 8–10 mm long; lanceolate; compressed dorsiventrally; biconvex; falling with the glumes; not disarticulating between the florets; with distinctly elongated rachilla internodes between the florets (i.e., the upper floret stipitate). Rachilla terminated by a female-fertile floret; rachilla hairless. Hairy callus absent.

Glumes two; very unequal; (the upper) long relative to the adjacent lemmas; hairless; glabrous (G_1), or scabrous (G_2, on the veins); (the upper) *awned (attenuate into a long straight awn)*; non-carinate; very dissimilar (the G_1 a small, membranous, truncate scale, the G_2 large, firm, awned). Lower glume shorter than the lowest lemma; 0 nerved, or 3 nerved (faint). Upper glume 5–7 nerved. *Spikelets* with incomplete florets. The incomplete florets proximal to the female-fertile florets. *The proximal incomplete florets* 1; paleate. Palea of the proximal incomplete florets fully developed; not becoming conspicuously hardened and enlarged laterally. The proximal incomplete florets male (with 3 stamens). The proximal lemmas acuminate; awned (attenuate); 5 nerved; decidedly exceeding the female-fertile lemmas; similar in texture to the female-fertile lemmas; not becoming indurated.

Female-fertile florets 1. Lemmas similar in texture to the glumes to decidedly firmer than the glumes (thinly leathery); smooth; not becoming indurated; entire; pointed; mucronate (or mucronulate); hairless; non-carinate; having the margins lying flat and exposed on the palea; with a clear germination flap; 7 nerved. *Palea* present; relatively long; tightly clasped by the lemma; entire; awnless, without apical setae; textured like the lemma; not indurated; 2-nerved; keel-less. *Lodicules* present; 2; free; fleshy; glabrous; not or scarcely vascularized. *Stamens* 0 (the male floret with non-penicillate anthers 2.5–3 mm long). *Ovary* glabrous. Styles free to their bases. Stigmas 2; white.

Fruit, embryo and seedling. *Fruit* free from both lemma and palea (but clasped below by the flaps of the palea); small (3–3.5 mm long); ellipsoid; compressed dorsiventrally. Hilum short. Embryo large; not waisted.

Abaxial leaf blade epidermis. *Costal/intercostal zonation* conspicuous. *Papillae* absent. Mid-intercostal long-cells rectangular; having markedly sinuous walls. *Microhairs* present; panicoid-type; (72–)75–84(–90) microns long; (4.8–)5.7–6 microns wide at the septum. Microhair total length/width at septum 12–18.8. Microhair apical cells (46–)51–56(–57) microns long. Microhair apical cell/total length ratio 0.63–0.68. *Stomata* common (easy to find, but thinly spread); 27–33 microns long. Guard-cells overlapping to flush with the interstomatals. *Intercostal short-cells* common; in cork/silica-cell pairs. *Costal short-cells* conspicuously in long rows (some veins), or predominantly paired (some veins), or neither distinctly grouped into long rows nor predominantly paired (some veins). Costal silica bodies 'panicoid-type'; all very small, dumb-bell shaped and nodular; not sharp-pointed. **Transverse section of leaf blade, physiology.** C_4; XyMS+. PCR sheath outlines even. PCR cell chloroplasts seemingly centripetal. *Leaf blade* with distinct, prominent adaxial ribs; with the ribs more or less constant in size. *Midrib* conspicuous (with a large air-space on either side of the large bundle); having a conventional arc of bundles (one large bundle, two small); with colourless mesophyll adaxially. Bulliforms present in discrete, regular adaxial groups; in simple fans. All the vascular bundles accompanied by sclerenchyma. Combined sclerenchyma girders absent. Sclerenchyma all associated with vascular bundles.

Taxonomy. Panicoideae; Panicodae; Paniceae.

Ecology, geography, regional floristic distribution. 1 species. Southern tropical Africa. Hydrophytic (in permanent water); species of open habitats; glycophytic.

Paleotropical. African. Sudano-Angolan and Namib-Karoo. Kalaharian.

References, etc. Morphological/taxonomic: Hubbard and Schweickerdt 1936. Leaf anatomical: this project; photos of *O. barnardii* provided by R.P. Ellis.

Oryzopsis Michx.

Dilepyrum Raf., *Eriocoma* Nutt., *Fendleria* Steud., *Piptatherum* P. Beauv., *Urachne* Trin.

Habit, vegetative morphology. Perennial; caespitose. *Culms* 10–150 cm high; unbranched above. Culm nodes glabrous. Culm internodes solid, or hollow. Young shoots extravaginal, or intravaginal. Leaves non-auriculate. *Leaf blades* broad, or narrow; 0.7–15 mm wide; flat, or folded, or rolled (involute); *not pseudopetiolate*; without cross venation; persistent; rolled in bud, or once-folded in bud. *Ligule an unfringed membrane, or a fringed membrane (?)*; truncate to not truncate; 0.2–15 mm long.

Reproductive organization. *Plants bisexual, with bisexual spikelets*; with hermaphrodite florets; outbreeding. Exposed-cleistogamous, or chasmogamous.

Inflorescence. Inflorescence paniculate; open; with capillary branchlets, or without capillary branchlets; espatheate; not comprising 'partial inflorescences' and foliar organs. Spikelet-bearing axes persistent. Spikelets not secund; pedicellate.

Female-fertile spikelets. Spikelets 2–10 mm long; compressed laterally to not noticeably compressed (*Oryzopsis* sensu stricto), or not noticeably compressed to compressed dorsiventrally (*Piptatherum*); disarticulating above the glumes. Rachilla terminated by a female-fertile floret. Hairy callus present, or absent (*Piptatherum*). *Callus* short; blunt (conical, straight or usually incurved in *Piptatherum*).

Glumes two; more or less equal; about equalling the spikelets, or exceeding the spikelets; long relative to the adjacent lemmas; pointed; awnless; non-carinate; similar (membranous). Lower glume 3–7 nerved. Upper glume 3–7 nerved. **Spikelets** with female-fertile florets only; *without proximal incomplete florets*.

Female-fertile florets 1. *Lemmas* convolute (rarely), or not convolute; not saccate; decidedly firmer than the glumes; *becoming indurated*; incised (apically); 2 lobed (obscurely); not deeply cleft; awned. *Awns* 1; median; *from a sinus*; non-geniculate (*Piptatherum*), or geniculate; hairless, or hairy; much shorter than the body of the lemma to much longer than the body of the lemma; deciduous (usually?). Lemmas hairy, or hairless; non-carinate; 3–5 nerved (rarely to 9 nerved). **Palea** present (covered by the lemma only marginally in *Piptatherum*, completely in *Oryzopsis* sensu stricto); relatively long; awnless, without apical setae; *textured like the lemma (leathery)*; indurated; 2-nerved; *keel-less*. *Lodicules* present; 2, or 3 (*Piptatherum*); free; membranous; ciliate, or glabrous (*Piptatherum*); not toothed; heavily vascularized (the stipoid pair), or not or scarcely vascularized (the third, when present). *Stamens* 3. Anthers 1.5–5 mm long; penicillate (usually), or not penicillate. *Ovary* glabrous. Styles fused (*Oryzopsis* sensu stricto), or free to their bases (*Piptatherum*). Stigmas 2.

Fruit, embryo and seedling. *Fruit* free from both lemma and palea; small to medium sized (1.5–4.0 mm long). *Hilum long-linear*. Embryo large to small (1/4 to 1/2 the grain length). Endosperm hard; without lipid. Embryo with an epiblast; without a scutellar tail; with a negligible mesocotyl internode. Embryonic leaf margins meeting.

Seedling with a long mesocotyl. First seedling leaf with a well-developed lamina. The lamina narrow; erect.

Abaxial leaf blade epidermis. *Costal/intercostal zonation* conspicuous. *Papillae* absent. *Long-cells* similar in shape costally and intercostally (narrow-rectangular); of similar wall thickness costally and intercostally (fairly thick-walled, the costals somewhat less so). Mid-intercostal long-cells rectangular; having markedly sinuous walls. *Microhairs* present (sometimes?), or absent (present adaxially in at least some species); elongated; ostensibly one-celled; cf. those of *Stipa*. *Stomata* common, or absent or very rare; 21–24 microns long. Subsidiaries low to high dome-shaped. Guard-cells overlapping to flush with the interstomatals. *Intercostal short-cells* common; in cork/silica-cell pairs, or not paired (solitary); silicified (when paired). Intercostal silica bodies when present, rounded (elliptical or oval), or crescentic, or tall-and-narrow. *Costal short-cells* conspicuously in long rows. Costal silica bodies rounded (including potato shapes), or tall-and-narrow (to more or less cubical), or crescentic, or 'panicoid-type'; sometimes butterfly shaped, or dumb-bell shaped (e.g. *O. racemosa*); not sharp-pointed.

Transverse section of leaf blade, physiology. C_3; XyMS+. *Mesophyll* with non-radiate chlorenchyma; without adaxial palisade. *Leaf blade* 'nodular' in section; with the ribs very irregular in sizes. *Midrib* conspicuous, or not readily distinguishable; with one bundle only. Bulliforms present in discrete, regular adaxial groups (in the furrows); in simple fans. All the vascular bundles accompanied by sclerenchyma. Combined sclerenchyma girders present; forming 'figures'. Sclerenchyma all associated with vascular bundles.

Culm anatomy. *Culm internode bundles* in one or two rings.

Phytochemistry. Tissues of the culm bases with abundant starch.

Cytology. Chromosome base number, $x = 11$, 12, and 14; 2, 4, 5, and 10 ploid, or 11 ploid.

Taxonomy. Arundinoideae; Stipeae.

Ecology, geography, regional floristic distribution. About 35 species. North temperate & subtropical. Commonly adventive. Mesophytic to xerophytic; shade species, or species of open habitats (mainly).

Holarctic, Paleotropical, and Neotropical. Boreal, Tethyan, and Madrean. African and Indomalesian. Euro-Siberian, Atlantic North American, and Rocky Mountains. Irano-Turanian. Saharo-Sindian and Sudano-Angolan. Indo-Chinese. Andean. European. Canadian-Appalachian and Central Grasslands. Somalo-Ethiopian.

Hybrids. Intergeneric hybrids with *Stipa* — ×*Stiporyzopsis* B.L. Johnson & Rogler.

Rusts and smuts. Rusts — *Puccinia*. Smuts from Tilletiaceae and from Ustilaginaceae. Tilletiaceae — *Urocystis*. Ustilaginaceae — *Ustilago*.

Economic importance. Significant weed species: *O. miliacea*. Cultivated fodder: *O. miliacea*.

References, etc. Morphological/taxonomic: Freitag 1975. Leaf anatomical: Metcalfe 1960; this project.

Otachyrium Nees

Habit, vegetative morphology. *Perennial*; rhizomatous and caespitose. *Culms* 20–150 cm high; herbaceous; branched above, or unbranched above. Culm nodes hairy (usually). The shoots not aromatic. *Leaves* not basally aggregated; non-auriculate. Leaf blades linear to linear-lanceolate; narrow; 1–9 mm wide; setaceous, or not setaceous; flat, or rolled, or acicular; without cross venation. *Ligule a fringed membrane*; truncate.

Reproductive organization. Plants bisexual, with bisexual spikelets; with hermaphrodite florets, or without hermaphrodite florets (the upper floret may be female-only in two species).

Inflorescence. Inflorescence of spicate main branches, or paniculate; open, or contracted; espatheate; not comprising 'partial inflorescences' and foliar organs. Spikelet-bearing axes persistent. Spikelets paired (usually); pedicellate. Pedicel apices cupuliform.

Spikelets consistently in 'long-and-short' combinations; unequally pedicellate in each combination.

Female-fertile spikelets. Spikelets 2–8.5 mm long; compressed laterally to compressed dorsiventrally; falling with the glumes; with conventional internode spacings. The upper floret not stipitate. Rachilla terminated by a female-fertile floret (though one species occasionally has a third, terminal, abortive floret). Hairy callus absent.

Glumes two; more or less equal (subequal); shorter than the spikelets (1/4 to 1/2 as long); *shorter than the adjacent lemmas*; hairy, or hairless; awnless; carinate; *similar (membranous)*. Lower glume 1–3 nerved. Upper glume 1–7 nerved. *Spikelets* with incomplete florets. The incomplete florets proximal to the female-fertile florets. *Spikelets with proximal incomplete florets. The proximal incomplete florets* 1; paleate. *Palea of the proximal incomplete florets* fully developed; *becoming conspicuously hardened and enlarged laterally (greatly enlarged and often winged, expanding widely beyond the spikelet when mature, and adhering to the upper floret)*. The proximal incomplete florets male. The proximal lemmas sometimes sulcate; awnless; 3 nerved; more or less equalling the female-fertile lemmas to decidedly exceeding the female-fertile lemmas; less firm than the female-fertile lemmas (membranous); not becoming indurated.

Female-fertile florets 1. Lemmas acute; decidedly firmer than the glumes; smooth (and shining), or striate; becoming indurated to not becoming indurated (from papery to crustaceous); pallid to black; entire; pointed; awnless; hairless; non-carinate (convex or gibbous); having the margins tucked in onto the palea; without a germination flap; 3–5 nerved. *Palea* present; relatively long; entire; awnless, without apical setae; textured like the lemma; indurated, or not indurated; 2-nerved; 2-keeled. Palea keels winged (in four of the species), or wingless. *Lodicules* present; 2; free; fleshy; glabrous. *Stamens* 3. Anthers not penicillate; without an apically prolonged connective. *Ovary* glabrous. Styles free to their bases. Stigmas 2.

Fruit, embryo and seedling. Fruit small (1.5 — 2.5 mm long); ellipsoid; compressed dorsiventrally (plano-convex). Hilum short. Embryo large; not waisted.

Abaxial leaf blade epidermis. *Costal/intercostal zonation* conspicuous. *Papillae* absent. *Long-cells* markedly different in shape costally and intercostally (the costals longer and narrower); of similar wall thickness costally and intercostally (quite thin walled). Mid-intercostal long-cells rectangular; having markedly sinuous walls. *Microhairs* present; panicoid-type; 34.5–36–42 microns long; 7.5–9 microns wide at the septum. Microhair total length/width at septum 4–5.6. Microhair apical cells 21–24–27 microns long. Microhair apical cell/total length ratio 0.57–0.75. *Stomata* common; (30–)36–42 microns long. Subsidiaries mostly triangular (low, with small points). Guard-cells overlapping to flush with the interstomatals. *Intercostal short-cells* common; not paired (solitary); not silicified. Intercostal prickles abundant. *Costal short-cells* conspicuously in long rows. Costal silica bodies 'panicoid-type'; cross shaped, dumb-bell shaped, and nodular; not sharp-pointed.

Transverse section of leaf blade, physiology. C_3; XyMS+. Mesophyll without adaxial palisade; with fusoids (?— a suspicion of these, not confirmable in available material). *Leaf blade* 'nodular' in section to adaxially flat. *Midrib* conspicuous; with one bundle only, or having a conventional arc of bundles (depending on the limits set on the midrib); with colourless mesophyll adaxially. Bulliforms present in discrete, regular adaxial groups; in simple fans. All the vascular bundles accompanied by sclerenchyma. Combined sclerenchyma girders present (with the main bundles); forming 'figures' (the main bundles with small I's). Sclerenchyma all associated with vascular bundles.

Cytology. Chromosome base number, $x = 9$; 2 ploid.

Taxonomy. Panicoideae; Panicodae; Paniceae.

Ecology, geography, regional floristic distribution. 7 species. Tropical South America, West Indies. Helophytic, or mesophytic; glycophytic.

Neotropical. Caribbean, Venezuela and Surinam, Amazon, Central Brazilian, and Pampas.

References, etc. Morphological/taxonomic: Sendulsky and Soderstrom 1984. Leaf anatomical: this project.

Otatea McClure & Smith

~ *Sinarundinaria*

Habit, vegetative morphology. Perennial. The flowering culms leafy. *Culms woody and persistent*; branched above. Primary branches/mid-culm node 3. Culm sheaths persistent ('lacking a well-marked basal girdle'). Culm internodes hollow. Unicaespitose, or pluricaespitose. Rhizomes pachymorph. Plants unarmed. *Leaves* not basally aggregated; with auricular setae (but small), or without auricular setae. Leaf blades lanceolate (acuminate), or linear-lanceolate; broad to narrow; pseudopetiolate; cross veined, or without cross venation; disarticulating from the sheaths. *Contra-ligule* present.

Reproductive organization. Plants bisexual, with bisexual spikelets; with hermaphrodite florets.

Inflorescence. Inflorescence paniculate; open (with numerous spikelets, by contrast with *Yushania*); non-digitate; spatheate; *a complex of 'partial inflorescences' and intervening foliar organs*. *Spikelet-bearing axes* paniculate; persistent. Spikelets not secund; pedicellate.

Female-fertile spikelets. Spikelets 35–40 mm long; compressed laterally; disarticulating above the glumes; disarticulating between the florets. Rachilla prolonged beyond the uppermost female-fertile floret; rachilla hairy (ciliate at the dilated segment apices). The rachilla extension with incomplete florets. Hairy callus absent.

Glumes two; very unequal to more or less equal; shorter than the adjacent lemmas; pointed; *awned (the terminal scaberulous subule to 2.5 mm long)*; carinate; similar (lanceolate, subule-tipped). *Spikelets* with incomplete florets. The incomplete florets distal to the female-fertile florets. *The distal incomplete florets* merely underdeveloped.

Female-fertile florets 3–7. Lemmas similar in texture to the glumes; not becoming indurated; entire; pointed; awned (acuminate into the subule). Awns 1; median; apical; non-geniculate; much shorter than the body of the lemma (about 2.5 mm long). *Lemmas carinate*. *Palea* present; relatively long; entire (truncate); awnless, without apical setae; not indurated; 2-keeled. *Lodicules* present; 3; free; membranous; ciliate; not toothed; heavily vascularized. **Stamens 3**. Anthers not penicillate; without an apically prolonged connective. *Ovary* glabrous; without a conspicuous apical appendage. Styles fused. Stigmas 2.

Fruit, embryo and seedling. *Fruit* free from both lemma and palea; longitudinally grooved. Hilum long-linear. Embryo small; not waisted.

Abaxial leaf blade epidermis. *Costal/intercostal zonation* conspicuous (and an even more conspicuous distinction between the epidermes on either side of the midrib — the one smooth and epapillate, the other with abundant prickles and papillate). *Papillae* present (on the one side of the blade); intercostal. Intercostal papillae several per cell (mostly one median row of small, circular papillae per long-cell). *Long-cells* similar in shape costally and intercostally to markedly different in shape costally and intercostally (the costals tending to be smaller and narrower); of similar wall thickness costally and intercostally (thin walled). Mid-intercostal long-cells rectangular; having markedly sinuous walls. *Microhairs* present (on one side of the blade only); elongated; clearly two-celled; panicoid-type (more or less, but the apical cells consistently blunt). *Stomata* absent or very rare (confined to the adaxial surface, where each is associated with a ring of overarching papillae). *Intercostal short-cells* common; in cork/silica-cell pairs; silicified. Intercostal silica bodies narrowly saddle shaped, or crescentic. *Crown cells* absent (but costal and intercostal prickles abundant over the papillate half). *Costal short-cells* conspicuously in long rows (but the costal 'short-cells' often relatively long, and the files frequently interrupted by prickles). Costal silica bodies saddle shaped (abundant and predominating), or 'panicoid-type' (confined to the vicinity of the midrib); when panicoid type, short dumb-bell shaped; not sharp-pointed.

Transverse section of leaf blade, physiology. C_3; XyMS+. Mesophyll with arm cells; with fusoids. The fusoids external to the PBS. *Leaf blade* adaxially flat. *Midrib* conspicuous (by virtue of its large bundle and abaxially prominent keel); with one bundle only. *The lamina* symmetrical on either side of the midrib (in transverse section — despite the epidermal distinction). Bulliforms present in discrete, regular adaxial groups (these large, in the middle of each intercostal zone); in simple fans. All the vascular bundles accompanied by sclerenchyma. Combined sclerenchyma girders present (with all the bundles); forming 'figures' (most bundles with I's or 'anchors'). Sclerenchyma all associated with vascular bundles.

Taxonomy. Bambusoideae; Bambusodae; Bambuseae.

Ecology, geography, regional floristic distribution. 2 species. Mexico and Central America.

Neotropical. Caribbean.

References, etc. Leaf anatomical: this project.

Special comments. See Clayton and Renvoize (1986) and Soderstrom and Ellis (1987) for very different generic interpretations of the species in this circle of affinity. There are no available generic descriptions adequate for the present purpose.

Ottochloa Dandy

Hemigymnia Stapf

Habit, vegetative morphology. Perennial; decumbent (slender). *Culms herbaceous.* Culm nodes glabrous. Culm internodes solid. *Leaves* not basally aggregated; non-auriculate. *Leaf blades lanceolate*; broad, or narrow; flat (thin); pseudopetiolate, or not pseudopetiolate; cross veined, or without cross venation; *disarticulating from the sheaths*; rolled in bud. Ligule an unfringed membrane to a fringed membrane.

Reproductive organization. Plants bisexual, with bisexual spikelets; with hermaphrodite florets.

Inflorescence. *Inflorescence paniculate (the branches weakly unilateral, with appressed secondary racemelets); open*; with capillary branchlets; non-digitate; espatheate; not comprising 'partial inflorescences' and foliar organs. Spikelet-bearing axes persistent. Spikelets solitary, or paired; not secund; pedicellate (the pedicels widened upwards). Pedicel apices cupuliform.

Female-fertile spikelets. *Spikelets 1.75–2 mm long*; elliptic; abaxial or not orientated; *compressed dorsiventrally*; falling with the glumes. Rachilla terminated by a female-fertile floret. Hairy callus absent.

Glumes two; more or less equal; shorter than the spikelets (about 1/2 to 2/3 as long); *shorter than the adjacent lemmas*; hairless (usually glabrous); pointed; awnless; similar (membranous). Lower glume 3 nerved. Upper glume 3–5 nerved. *Spikelets* with incomplete florets. The incomplete florets proximal to the female-fertile florets. *Spikelets with proximal incomplete florets. The proximal incomplete florets* 1; epaleate; male. The proximal lemmas awnless; 5–9 nerved; more or less equalling the female-fertile lemmas; less firm than the female-fertile lemmas; not becoming indurated.

Female-fertile florets 1. Lemmas decidedly firmer than the glumes (subleathery); smooth to striate; becoming indurated to not becoming indurated; yellow in fruit; entire; pointed; awnless to mucronate (or rather, apiculate to mucronulate); hairless; non-carinate; having the margins lying flat and exposed on the palea; with a clear germination flap; 3–5 nerved. *Palea* present; relatively long; entire; awnless, without apical setae; textured like the lemma; 2-nerved. *Lodicules* present; 2; free; fleshy; glabrous. *Stamens* 3. Anthers not penicillate. *Ovary* glabrous. Styles free to their bases. Stigmas 2; red pigmented.

Fruit, embryo and seedling. Fruit small; compressed dorsiventrally (strongly). Hilum short. Embryo large; not waisted.

Abaxial leaf blade epidermis. *Costal/intercostal zonation* conspicuous. *Papillae* absent. *Long-cells* similar in shape costally and intercostally; of similar wall thickness costally and intercostally (the costals slightly thicker walled). Mid-intercostal long-cells rectangular; having markedly sinuous walls. *Microhairs* present; elongated; clearly two-celled; panicoid-type; 39–69 microns long; 5–8.4 microns wide at the septum. Microhair total length/width at septum 6.5–10. Microhair apical cells 21–36 microns long. Microhair apical cell/total length ratio 0.48–0.62. *Stomata* common; 24–37.5 microns long. Subsidiaries dome-shaped. Guard-cells overlapping to flush with the interstomatals. *Intercostal short-cells* common; in cork/silica-cell pairs, or not paired (solitary); silicified, or not silicified. *Costal short-cells* conspicuously in long rows. Costal silica bodies 'panicoid-type'; not sharp-pointed.

Transverse section of leaf blade, physiology. C_3; XyMS+. *Mesophyll* with radiate chlorenchyma. *Leaf blade* 'nodular' in section; with the ribs more or less constant in size. *Midrib* conspicuous, or not readily distinguishable; with one bundle only. Bulliforms present in discrete, regular adaxial groups; in simple fans. All the vascular bundles accompanied by sclerenchyma. Combined sclerenchyma girders present; forming 'figures'. Sclerenchyma all associated with vascular bundles.

Culm anatomy. *Culm internode bundles* in one or two rings.

Cytology. Chromosome base number, $x = 9$. $2n = 18$. 2 ploid.

Taxonomy. Panicoideae; Panicodae; Paniceae.

Ecology, geography, regional floristic distribution. 4 species. Africa, Indomalayan region, Australia. Shade species and species of open habitats. Damp and shady places.

Paleotropical and Australian. African, Indomalesian, and Neocaledonian. West African Rainforest. Indian, Indo-Chinese, Malesian, and Papuan. North and East Australian. Tropical North and East Australian.

Rusts and smuts. Rusts — *Physopella* and *Puccinia*. Taxonomically wide-ranging species: *Physopella clemensiae*, *Puccinia orientalis*, and '*Uromyces*' *setariae-italicae*.

Economic importance. Significant weed species: *O. nodosa*.

References, etc. Leaf anatomical: this project.

Oxychloris Lazarides

Habit, vegetative morphology. Annual, or perennial (short-lived); caespitose. *Culms* 15–50 cm high; herbaceous; unbranched above (5–7 noded). Culm nodes glabrous. Culm internodes solid (spongy). Young shoots intravaginal. *Leaves* not basally aggregated; non-auriculate. Leaf blades narrow (less than 3.5 mm wide); flat, or rolled; without abaxial multicellular glands; without cross venation; persistent. Ligule a fringed membrane (short). *Contra-ligule* absent.

Reproductive organization. *Plants bisexual, with bisexual spikelets*; with hermaphrodite florets.

Inflorescence. *Inflorescence* of spicate main branches; *digitate*. Primary inflorescence branches 3–6. Rachides hollowed and flattened (triqetrous). Inflorescence espatheate; not comprising 'partial inflorescences' and foliar organs. Spikelet-bearing axes persistent. Spikelets solitary; secund (the rachis dorsiventral); biseriate; subsessile to pedicellate.

Female-fertile spikelets. *Spikelets* 1.8–4.5 mm long (-6); compressed laterally; *falling with the glumes (the spikelets falling whole)*; not disarticulating between the florets; with distinctly elongated rachilla internodes between the florets (the sterile florets in a terminal cluster, separated from the basal floret by a thickened, elongated internode). Rachilla prolonged beyond the uppermost female-fertile floret (and modified with spongy tissue to form a conspicuous thickened inernode, separating the clustered sterile florets from the basal fertile one); rachilla glabrous apart fom a few basal hairs. The rachilla extension with incomplete florets. Hairy callus present (2.5–3mm long, being elongated and pungent). *Callus* long; pointed.

Glumes two; very unequal; long relative to the adjacent lemmas; lateral to the rachis; hairless; glabrous; not pointed (G_1 entire, G_2 2-lobed or truncate); awnless; carinate; similar (thinly membranous to hyaline, often purple-tinged). Lower glume 1 nerved. Upper glume 1 nerved. *Spikelets* with incomplete florets. The incomplete florets distal to the female-fertile florets (without paleas). The distal incomplete florets 3–5; *incomplete florets* clearly specialised and modified in form (the clustered sterile lemmas large, winged and awned, the lower ones broad, flaring and 7 nerved). *Spikelets without proximal incomplete florets*.

Female-fertile florets 1 (basal). Lemmas decidedly firmer than the glumes (cartilaginous to indurated); not becoming indurated (but dark brown when mature); incised; 2 lobed; not deeply cleft (2-toothed); awned. Awns 1; median; dorsal; from near the top; non-geniculate; hairless; about as long as the body of the lemma to much longer than the body of the lemma; entered by one vein. Lemmas hairy. The hairs in tufts (from near the apices of the lateral nerves and lower on the mid-nerve). Lemmas carinate (the mid-nerve ribbed); without a germination flap; 3 nerved (the laterals submarginal); with the nerves confluent towards the tip. *Palea* present; relatively long; apically notched; awnless, without apical setae; not indurated (membranous); 2-nerved; 2-keeled. Palea keels winged (the wings ciliate). **Lodicules absent**. *Stamens* 3. Anthers very small; not penicillate. *Ovary* glabrous. Styles free to their bases. Stigmas 2; red pigmented.

Fruit, embryo and seedling. *Fruit* free from both lemma and palea; small (1.3–2 mm long); tigonous. Hilum short. Pericarp fused. Embryo large (almost as long as the fruit). Endosperm hard.

Abaxial leaf blade epidermis. *Costal/intercostal zonation* conspicuous. *Papillae* present; intercostal. Intercostal papillae over-arching the stomata; consisting of one symmetrical projection per cell (mostly). *Long-cells* markedly different in shape costally and intercostally (the non-papillate costals narrower and more regular). Mid-intercostal long-cells rectangular; having markedly sinuous walls. *Microhairs* present; more or less spherical; clearly two-celled; chloridoid-type. Microhair apical cell wall of similar thickness/rigidity to that of the basal cell. Microhairs 22.5–24–27 microns long. Microhair basal cells 12 microns long. Microhairs 9.6–12–15 microns wide at the septum. Microhair total length/width at septum 1.8–2.5. Microhair apical cells (9–)10.5–12(–13.5) microns long. Microhair apical cell/total length ratio 0.33–0.56. *Stomata* common; 16.5–19.5 microns long. Subsidiaries dome-shaped and triangular. *Intercostal short-cells* common; not paired (solitary); not silicified. Intercostal silica bodies absent. *Costal short-cells* conspicuously in long rows. Costal silica bodies present in alternate cell files of the costal zones; exclusively saddle shaped; not sharp-pointed.

Transverse section of leaf blade, physiology. C_4; biochemical type NAD–ME (1 species); XyMS+. PCR sheath outlines even. PCR sheaths of the primary vascular bundles interrupted; interrupted both abaxially and adaxially. PCR sheath extensions present. Maximum number of extension cells 1. PCR cells without a suberised lamella. *PCR cell chloroplasts* elongated; with well developed grana; centripetal. *Leaf blade* 'nodular' in section; with the ribs more or less constant in size. *Midrib* not readily distinguishable; with one bundle only. Bulliforms not in discrete, regular adaxial groups (seemingly, in the poor material seen). All the vascular bundles accompanied by sclerenchyma. Combined sclerenchyma girders present; forming 'figures'. Sclerenchyma all associated with vascular bundles. The lamina margins with fibres.

Phytochemistry. Leaf blade chlorophyll *a:b* ratio 3.88–4.5.

Taxonomy. Chloridoideae; main chloridoid assemblage.

Ecology, geography, regional floristic distribution. 1 species. Australia. Xerophytic; species of open habitats; halophytic to glycophytic. Dry savanna.

Australian. North and East Australian and Central Australian. Tropical North and East Australian.

References, etc. Morphological/taxonomic: Lazarides 1985. Leaf anatomical: this project.

Oxyrhachis Pilger

Habit, vegetative morphology. Perennial; caespitose. *Culms* 20–80 cm high; herbaceous; unbranched above. Culm nodes glabrous (seemingly fragile). Culm internodes solid. Young shoots intravaginal. *Leaves* mostly basal; non-auriculate. Leaf blades linear; narrow (filiform, conduplicate-involute, rigid); about 0.5–0.7 mm wide; setaceous; folded, or rolled; without cross venation; persistent; rolled in bud. Ligule a fringed membrane, or a fringe of hairs (short); truncate; 0.2–0.3 mm long. *Contra-ligule* absent.

Reproductive organization. Plants bisexual, with bisexual spikelets; with hermaphrodite florets. *The spikelets all alike in sexuality*.

Inflorescence. *Inflorescence a single spike (narrow, cylindrical, terminating the culm)*. Rachides hollowed. Inflorescence espatheate; not comprising 'partial inflorescences' and foliar organs. **Spikelet-bearing axes** cylindrical spikes; solitary; with substantial rachides; *disarticulating*; disarticulating at the joints. '*Articles*' linear; without a basal callus-knob; slightly appendaged (at the upper end); disarticulating obliquely (very much so); glabrous. *Spikelets solitary (i.e. only theoretically in pairs, the 'pedicel' fused* *with and indistinguishable from the rachis)*; not secund (the sessile spikelets in two opposite rows); distichous; sessile; distant.

Female-fertile spikelets. Spikelets 4–6 mm long; abaxial; compressed dorsiventrally; planoconvex; falling with the glumes (and with the adjacent joint); with conventional internode spacings. Rachilla terminated by a female-fertile floret. Hairy callus absent. Callus blunt (and thick).

Glumes two; more or less equal; exceeding the spikelets; long relative to the adjacent lemmas; dorsiventral to the rachis; hairless; glabrous; not pointed (the tips rounded); awnless; non-carinate (rounded on back); very dissimilar (G_1 obtuse, leathery, G_2 apically notched or entire, membranous-hyaline). Lower glume much exceeding the lowest lemma; not two-keeled (and wingless); convex on the back; not pitted; relatively smooth; 6–7 nerved. Upper glume 2 nerved. *Spikelets* with incomplete florets. The incomplete florets proximal to the female-fertile florets. *The proximal incomplete florets* 1; epaleate; sterile. The proximal lemmas awnless (obtuse); 0 nerved, or 2 nerved; more or less equalling the female-fertile lemmas to decidedly exceeding the female-fertile lemmas; similar in texture to the female-fertile lemmas (hyaline); not becoming indurated.

Female-fertile florets 1. **Lemmas** less firm than the glumes; entire; blunt (truncate-obtuse); awnless; hairless; glabrous; non-carinate; without a germination flap; *2 nerved*. *Palea* present, or absent; when present, very reduced (adherent to the lodicules); entire (truncate), or apically notched (emarginate or bilobed); awnless, without apical setae; textured like the lemma; not indurated (hyaline); nerveless; keel-less. *Lodicules* present; 2; free; fleshy; glabrous. *Stamens* 3. Anthers about 2.5 mm long; without an apically prolonged connective. *Ovary* glabrous. Styles free to their bases. Stigmas 2; brown.

Fruit, embryo and seedling. Fruit compressed dorsiventrally. Hilum short. Embryo large.

Abaxial leaf blade epidermis. *Costal/intercostal zonation* conspicuous to lacking (the epidermis very featureless, but the costal zones — especially over the midrib — with narrower long-cells). *Papillae* absent. *Long-cells* similar in shape costally and intercostally; of similar wall thickness costally and intercostally (fairly thick walled). Intercostal zones mainly with typical long-cells (with a few short to almost circular). Mid-intercostal long-cells with shortly rounded ends; having markedly sinuous walls (and conspicuous pits). *Microhairs* absent. *Stomata* absent or very rare. *Intercostal short-cells* common; mostly in cork/silica-cell pairs (mostly somewhat superposed); silicified. *Costal short-cells* also predominantly paired (the short-cell configurations and silica bodies not noticeably different from those of the intercostal zones). Costal silica bodies all small, intergrading rounded, tall-and-narrow, and crescentic; not sharp-pointed.

Transverse section of leaf blade, physiology. C_4; XyMS–. PCR sheath outlines even. PCR sheath extensions absent. *Mesophyll* with radiate chlorenchyma. *Leaf blade* with distinct, prominent adaxial ribs (the rather slight ribs emphasised by tufts of large, inflated macrohairs, cf. *Hygrochloa*); with the ribs more or less constant in size. *Midrib* conspicuous (by its position, the narrowing of the lamina, the smaller abaxial epidermal cells and smaller abaxial sclerenchyma group associated with it, and by the smaller macrohairs above it); with one bundle only. *The lamina* symmetrical on either side of the midrib. Bulliforms bulliforms indistinct. All the vascular bundles accompanied by sclerenchyma. Combined sclerenchyma girders present (with the main laterals); forming 'figures' (the main laterals with 'anchors'). Sclerenchyma not all bundle-associated. The 'extra' sclerenchyma in a continuous abaxial layer (there being a conspicuous, lignified, large-celled hypodermis abaxially between the main lateral bundles).

Taxonomy. Panicoideae; Andropogonodae; Andropogoneae; Rottboelliinae.

Ecology, geography, regional floristic distribution. 1 species. Tropical Africa, Madagascar. Helophytic; species of open habitats; glycophytic. Upland streamsides and marshy places.

Paleotropical. African and Madagascan. Sudano-Angolan and West African Rainforest. Sahelo-Sudanian and South Tropical African.

References, etc. Leaf anatomical: this project.

Oxytenanthera Munro

Houzeaubambus Mattei, *Scirpobambus* Kuntze
~ *Dendrocalamus*

Habit, vegetative morphology. Arborescent shrubby perennial; caespitose. The flowering culms leafy (but the culm leaves deciduous). *Culms* 300–1300 cm high (somewhat crooked, bending over to the ground); woody and persistent (forming dense clumps); to 10 cm in diameter; cylindrical; branched above (at the nodal line). Culm nodes glabrous. Primary branches/mid-culm node 1 (with subsidiary branches from these forming clusters). Culm sheaths persistent. Culm internodes solid, or hollow. Pluricaespitose. Rhizomes pachymorph. Plants unarmed. Young shoots intravaginal. *Leaves* not basally aggregated; non-auriculate; with auricular setae (these deciduous). Leaf blades linear-lanceolate to lanceolate; broad; 10–30 mm wide; flat; pseudopetiolate; without cross venation; disarticulating from the sheaths. Ligule an unfringed membrane; truncate; 0.3–0.5 mm long. *Contra-ligule* present.

Reproductive organization. Plants bisexual, with bisexual spikelets; with hermaphrodite florets. The spikelets of sexually distinct forms on the same plant (there being numerous sterile spikelets); hermaphrodite and sterile.

Inflorescence. Inflorescence indeterminate; *with pseudospikelets*; a false spike, with spikelets on contracted axes (the clusters sometimes confluent or reduced to a single terminal cluster like that of female *Spinifex*); spatheate (each spikelet cluster subtended by a papery sheath, and individual spikelets by several short, papery 'bracts'); not comprising 'partial inflorescences' and foliar organs. *Spikelet-bearing axes* capitate. **Spikelets associated with bractiform involucres**.

Female-fertile spikelets. Spikelets 15–45 mm long; compressed laterally to not noticeably compressed; falling with the glumes; not disarticulating between the florets. Rachilla terminated by a female-fertile floret; rachilla hairless. Hairy callus absent.

Glumes two (cross-veined); very unequal; shorter than the adjacent lemmas; hairy (shortly hispidulous); pointed, or not pointed (obtuse to acute); awnless; non-carinate; similar (papery to leathery). Lower glume much shorter than half length of lowest lemma; 17–30 nerved. Upper glume 17–30 nerved. *Spikelets* with incomplete florets. The incomplete florets proximal to the female-fertile florets. *Spikelets with proximal incomplete florets. The proximal incomplete florets* 1–3; male, or sterile (the paleas when present two-keeled). The proximal lemmas awned, or awnless (then mucronate); 26–32 nerved; exceeded by the female-fertile lemmas; similar in texture to the female-fertile lemmas; not becoming indurated.

Female-fertile florets 1. Lemmas similar in texture to the glumes (papery to thinly leathery); not becoming indurated; entire; pointed; mucronate to awned. Awns 1; median; apical; non-geniculate; to 7 mm long. Lemmas hairy (hispid); non-carinate; without a germination flap; 11–23 nerved (with cross-nerves). *Palea* present; relatively long (may exceed the lemma); entire (pointed); awnless, without apical setae; not indurated; several nerved (16–19); keel-less (convolute). *Lodicules* absent. *Stamens* 6; monadelphous. Anthers not penicillate; with the connective apically prolonged. *Ovary* glabrous (but the style mostly shortly hairy); with a conspicuous apical appendage. The appendage long, stiff and tapering. Styles fused (the ovary attenuate into the single, hollow style). Stigmas 3.

Fruit, embryo and seedling. *Fruit* free from both lemma and palea; large; not noticeably compressed. Hilum long-linear. Embryo small. Endosperm containing compound starch grains.

Seedling with a short mesocotyl; with a loose coleoptile. First seedling leaf without a lamina.

Abaxial leaf blade epidermis. *Costal/intercostal zonation* conspicuous. *Papillae* present. Intercostal papillae over-arching the stomata; several per cell. *Long-cells* markedly different in shape costally and intercostally; differing markedly in wall thickness costally and intercostally. Mid-intercostal long-cells having markedly sinuous walls. *Microhairs* present; panicoid-type; 42–46.5 microns long; 5.4–6 microns wide at the septum. Microhair total length/width at septum 7–8.8. Microhair apical cells 21–24 microns long. Microhair apical cell/total length ratio 0.45–0.52. *Stomata* common; 27–30 microns long.

Subsidiaries low to high dome-shaped, or triangular. *Intercostal short-cells* absent or very rare. *Costal short-cells* predominantly paired. Costal silica bodies saddle shaped and oryzoid; not sharp-pointed.

Transverse section of leaf blade, physiology. C_3; XyMS+. *Mesophyll* with non-radiate chlorenchyma; without adaxial palisade; without arm cells; with fusoids. The fusoids external to the PBS. *Leaf blade* adaxially flat; with the ribs more or less constant in size. *Midrib* conspicuous; having complex vascularization. Bulliforms present in discrete, regular adaxial groups; in simple fans. All the vascular bundles accompanied by sclerenchyma. Combined sclerenchyma girders present; forming 'figures'. Sclerenchyma all associated with vascular bundles.

Cytology. Chromosome base number, $x = 12$. $2n = 72$. 6 ploid.

Taxonomy. Bambusoideae; Bambusodae; Bambuseae.

Ecology, geography, regional floristic distribution. 1 species. Africa. Mesophytic; shade species; glycophytic. Growing in the protection of larger trees.

Paleotropical. African. Sudano-Angolan and West African Rainforest. Sahelo-Sudanian, Somalo-Ethiopian, and South Tropical African.

Rusts and smuts. Rusts — *Dasturella*. Taxonomically wide-ranging species: *Dasturella divina*.

Economic importance. *O. abyssinica* stems used for light construction and fencing.

References, etc. Leaf anatomical: Metcalfe 1960; this project.

Panicum L.

Chasea Nieuw., *Coleataenia* Griseb., *Dileucaden* (Raf.) Steud., *Eatonia* Raf., *Eriolytrum* Kunth, *Milium* Adans., *Monachne* P. Beauv., *Phanopyrum* (Raf.) Nash, *Polyneura* Peter, *Psilochloa* Launert, *Setiacis* S.L. Chen & Y.X. Jin (?— original description inadequate)

Excluding *Dichanthelium, Steinchisma*

Habit, vegetative morphology. Annual, or perennial (but no overwintering rosette — by contrast with *Dichanthelium*); rhizomatous, or stoloniferous, or caespitose, or decumbent. *Culms* 20–400 cm high; woody and persistent, or herbaceous; branched above, or unbranched above; tuberous (rarely), or not tuberous. Culm nodes hairy, or glabrous. Culm internodes solid, or hollow. Plants with multicellular glands (rarely, then stalked, e.g. in the *Clavelligera* group), or without multicellular glands (usually). *Leaves* mostly basal, or not basally aggregated; auriculate (rarely), or non-auriculate. Leaf blades broad, or narrow; cordate, or not cordate, not sagittate; setaceous, or not setaceous; flat (usually); pseudopetiolate (rarely), or not pseudopetiolate; without cross venation; disarticulating from the sheaths (occasionally), or persistent; rolled in bud. Ligule an unfringed membrane, or a fringed membrane to a fringe of hairs. *Contra-ligule* present (of hairs), or absent.

Reproductive organization. *Plants bisexual, with bisexual spikelets*; with hermaphrodite florets. *The spikelets all alike in sexuality*. Plants inbreeding. Exposed-cleistogamous, or chasmogamous. Apomictic, or reproducing sexually.

Inflorescence. *Inflorescence paniculate (except in the* **Stolonifera** *group and some species of section Laxa, where it consists of racemes and the distiction from* **Brachiaria** *breaks down)*; deciduous in its entirety, or not deciduous; open, or contracted; with capillary branchlets, or without capillary branchlets. Rachides hollowed, or flattened, or winged, or neither flattened nor hollowed, not winged. Inflorescence espatheate; not comprising 'partial inflorescences' and foliar organs. Spikelet-bearing axes persistent. *Spikelets not secund (except in the American Agrostoidea group, '*Psilochloa*', some species in section Laxa, etc.)*; pedicellate. *Pedicel apices cupuliform. Spikelets not in distinct 'long-and-short' combinations*.

Female-fertile spikelets. *Spikelets* 1.4–6 mm long; elliptic, or lanceolate, or ovate, or obovate; adaxial (in the few cases where the orientation is ascertainable); *compressed dorsiventrally (with very few exceptions: e.g.* **P. hemitomum**); falling with the glumes, or not disarticulating; with conventional internode spacings, or with a distinctly elongated rachilla internode between the glumes, or with distinctly elongated rachilla internodes between the florets, or with a distinctly elongated rachilla internode between the glumes and with distinctly elongated rachilla internodes between the florets. The upper floret conspicuously stipitate (e.g. sections Phanopyrum, Rudgeana), or the upper floret not stipitate. *The stipe beneath the upper floret* when present, not filiform; *straight and swollen*; heterogeneous (section Rudgeana), or homogeneous. Rachilla terminated by a female-fertile floret (very rarely prolonged, e.g. occasionally in *P. heliophilum*: Zuloaga and Morone 1991). *Hairy callus absent.*

Glumes two; nearly always *very unequal*; (the longer) long relative to the adjacent lemmas; *without conspicuous tufts or rows of hairs*; nearly always awnless (the G_2 truncate to pointed, very rarely shortly awn-tipped); very dissimilar, or similar (herbaceous-membranous, the lower sometimes very short and nerveless). *Lower glume* 0.15–0.8 times the length of the upper glume; 1–7 nerved. Upper glume 3–9 nerved. *Spikelets* nearly always with incomplete florets. The incomplete florets proximal to the female-fertile florets. *Spikelets with proximal incomplete florets. The proximal incomplete florets* nearly always 1 (rarely 2); paleate, or epaleate. *Palea of the proximal incomplete florets* when present, fully developed to reduced; *not becoming conspicuously hardened and enlarged laterally.* The proximal incomplete florets male, or sterile. *The proximal lemmas* awnless; 3 nerved (rarely), or 5–9 nerved, or 11 nerved (rarely); more or less equalling the female-fertile lemmas to decidedly exceeding the female-fertile lemmas; *less firm than the female-fertile lemmas*; not becoming indurated.

Female-fertile florets 1. *Lemmas* similar in texture to the glumes to decidedly firmer than the glumes; smooth (rarely rugose: subgenus Megathyrsus (*P. maximum*)); becoming indurated to not becoming indurated (leathery, bony or cartilaginous, usually becoming indurated, but membranous in a feww species of section Laxa, associated with other morphological peculiarities); yellow in fruit, or brown in fruit; entire; pointed, or blunt; usually not crested; awnless (rarely minutely apiculate); *hairless*; non-carinate; *having the margins tucked in onto the palea (nearly always, but no doubt there are exceptions among the forms with non-indurated lemmas)*; with a clear germination flap; 3–11 nerved. *Palea* present; relatively long; tightly clasped by the lemma (except in *P. discrepans*, where it is free at the tip); entire; awnless, without apical setae; textured like the lemma; indurated, or not indurated; 2-nerved. *Lodicules* present; 2; free; fleshy; glabrous. *Stamens* 3. Anthers 0.3–2 mm long; not penicillate. **Ovary** glabrous; *without a conspicuous apical appendage.* Styles free to their bases. Stigmas 2; red pigmented.

Fruit, embryo and seedling. *Fruit* free from both lemma and palea; small; compressed dorsiventrally. *Hilum* nearly always *short (but linear in (e.g.)* **P. glutinosum, P. macranthum, P. pilgerianum = Psilochloa***).* Embryo large. Endosperm hard; without lipid; containing only simple starch grains, or containing compound starch grains. Embryo without an epiblast; with a scutellar tail; with an elongated mesocotyl internode. Embryonic leaf margins overlapping.

Seedling with a long mesocotyl. First seedling leaf with a well-developed lamina. The lamina broad; erect, or curved; 6–12 veined.

Abaxial leaf blade epidermis. *Costal/intercostal zonation* conspicuous. *Papillae* absent. *Long-cells* similar in shape costally and intercostally; of similar wall thickness costally and intercostally. Intercostal zones with typical long-cells, or exhibiting many atypical long-cells, or without typical long-cells. Mid-intercostal long-cells rectangular; having markedly sinuous walls. *Microhairs* present; clearly two-celled (nearly always), or uniseriate (3(–4) celled in *P. validum*, Zuloaga *et al.* 1989); panicoid-type; without 'partitioning membranes' (in *P. virgatum*); (38–)42–78(–85) microns long; (5.4–)6–6.3(–6.6) microns wide at the septum. Microhair total length/width at septum 6.3–9. Microhair apical cells (25.5–)26–48(–55) microns long. Microhair apical cell/total length ratio 0.57–0.62. *Stomata* common; 30–33 microns long. Subsidiaries low dome-shaped, or triangular, or dome-shaped and triangular. Guard-cells overlapping to flush with the interstomatals. *Intercostal short-cells* common, or absent or very rare; in cork/silica-cell pairs, or not paired (solitary); silicified, or not silicified. Intercostal silica bodies when present, cross-shaped, or rounded, or tall-and-narrow. *Costal short-cells* conspicuously in long rows. Costal silica bodies 'panicoid-type', or tall-and-narrow to crescentic (rarely); mostly cross shaped to dumb-bell shaped, or nodular; not sharp-pointed.

Transverse section of leaf blade, physiology. C_4, or C_3. The anatomical organization when C_4 conventional, or unconventional. Organization of PCR tissue in a few C_4 species *Alloteropsis* type. Biochemical type PCK (5 species), or NAD–ME (14 species), or NADP–ME (4 species); when biochemically tested, XyMS+ (C_3, or C_4 NAD-ME or PCK), or XyMS– (NADP-ME). PCR sheath outlines uneven, or even. PCR sheath extensions present (rarely), or absent (usually). Maximum number of extension cells when present, 1. PCR cells with a suberised lamella, or without a suberised lamella. *PCR cell chloroplasts* ovoid, or elongated; with well developed grana, or with reduced grana; centrifugal/peripheral, or centripetal. *PBS cells* without a suberised lamella. *Mesophyll* with radiate chlorenchyma; *Isachne*-type, or not *Isachne*-type; traversed by columns of colourless mesophyll cells, or not traversed by colourless columns. *Leaf blade* 'nodular' in section, or adaxially flat; with the ribs more or less constant in size, or with the ribs very irregular in sizes. *Midrib* conspicuous, or not readily distinguishable; with one bundle only, or having a conventional arc of bundles; with colourless mesophyll adaxially, or without colourless mesophyll adaxially. Bulliforms present in discrete, regular adaxial groups; in simple fans and associated with colourless mesophyll cells to form deeply-penetrating fans (sometimes linking with traversing columns of colourless cells). Many of the smallest vascular bundles unaccompanied by sclerenchyma, or all the vascular bundles accompanied by sclerenchyma (rarely). Combined sclerenchyma girders present; forming 'figures', or nowhere forming 'figures'. Sclerenchyma all associated with vascular bundles.

Phytochemistry. Tissues of the culm bases with abundant starch. Leaves containing flavonoid sulphates (2 species), or without flavonoid sulphates (8 species). Leaf blade chlorophyll $a{:}b$ ratio 3.2–3.56 (PCK), or 3.5–4.82 (NAD-ME), or 3.86–4.64 (NADP-ME).

Cytology. Chromosome base number, $x = 7, 9,$ and 10. $2n = 18$ (seemingly rarely), or 36, or 37, or 54, or 72. Haploid nuclear DNA content 0.5 pg (1 species, $6x$). Nucleoli persistent.

Taxonomy. Panicoideae; Panicodae; Paniceae.

Ecology, geography, regional floristic distribution. About 370 species. Tropical, subtropical and warm temperate. Commonly adventive. Mesophytic, or xerophytic; shade species and species of open habitats; halophytic (rarely), or glycophytic. Diverse habitats: *P. pinifolium* sandbinding.

Holarctic, Paleotropical, Neotropical, Australian, and Antarctic. Boreal, Tethyan, and Madrean. African, Madagascan, Indomalesian, Polynesian, and Neocaledonian. Euro-Siberian, Eastern Asian, Atlantic North American, and Rocky Mountains. Macaronesian, Mediterranean, and Irano-Turanian. Saharo-Sindian, Sudano-Angolan, West African Rainforest, and Namib-Karoo. Indian, Indo-Chinese, Malesian, and Papuan. Hawaiian and Fijian. Caribbean, Venezuela and Surinam, Amazon, Central Brazilian, Pampas, and Andean. North and East Australian and Central Australian. New Zealand and Patagonian. European and Siberian. Canadian-Appalachian, Southern Atlantic North American, and Central Grasslands. Sahelo-Sudanian, Somalo-Ethiopian, South Tropical African, and Kalaharian. Tropical North and East Australian and Temperate and South-Eastern Australian.

Rusts and smuts. Rusts — *Puccinia*. Taxonomically wide-ranging species: *Puccinia dolosa*, '*Uromyces*' *setariae-italicae*, and *Puccinia esclavensis*.

Economic importance. Significant weed species: *P. antidotale*, *P. barbipulvinatum*, *P. bisulcatum*, *P. brevifolium*, *P. capillare*, *P. dichotomoflorum*, *P. gattingeri*, *P. laevifolium*, *P. maximum*, *P. miliaceum*, *P. natalense*, *P. obtusum*, *P. repens*, *P. sarmentosum*, *P. trichoides*, *P. turgidum*, *P. virgatum*, etc. Cultivated fodder: *P. coloratum* (Buffalo), *P. maximum* (Guinea grass), *P. miliaceum*, *P. purpurascens*, *P. schinzii*. Important native pasture species: many species, e.g. *P. bulbosum*, *P. coloratum*, *P. maximum*, *P. merkeri*, *P. obtusum*, *P. poaeoides*, *P. repens*, *P. stipitatum*, *P. trichocladum*, *P. trichoides*, *P. texanum*, *P. virgatum*. Grain crop species: *P. miliaceum* (Proso millet); also *P. sonorum* (Sauwi), *P. sumatrense* (Sama).

References, etc. Morphological/taxonomic: Hitchcock and Chase 1910, 1915; Hsu 1965; Brown 1977; Zuloaga and Soderstrom 1985; Zuloaga 1987; Ellis 1988. Leaf anatomical: mainly Metcalfe 1960 and this project.

Special comments. Generic limits among the relatives of *Panicum* (*Ancistrachne, Brachiaria, Dichanthelium, Digitaria, Eriochloa, Homolepis, Hylebates, Hymenachne, Ichnanthus, Paspalidium, Sacciolepis, Setaria, Tricholaena, Urochloa, Whiteochloa* etc., and many small segregates) need critical revision in terms of comparative data recorded at world level. Consequent re-alignments of species might reduce the variability here attributed to *Panicum*. See W. V. Brown (1977) for discussion, and Zuloaga (1987) for a detailed treatment of the *New World* species. Occasional species with a rachilla prolongation or a second sterile floret, and small suites of species with laterally compressed spikelets, secund one-sided inflorescence branches, stipitate upper florets, linear hila, etc., pose major hazards for printed generic keys; and for use in in that context, the description of *Panicum* as encoded here will generally require editing at regional level. The agriculturally important species *P. maximum* well illustrates the situation. It clearly belongs with species currently referred to *Urochloa* and/or *Brachiaria* (PCK, rugose upper lemma, etc.), but changes of this kind cannot be effectively implemented until the generic circumscriptions have been clarified (cf. Webster 1987, Zuloaga 1987).

Pappophorum Schreber

Polyraphis (Trin.) Lindley

Habit, vegetative morphology. Perennial; caespitose. *Culms* 30–150(–200) cm high; herbaceous. Young shoots intravaginal. *Leaves* not basally aggregated; non-auriculate. Leaf blades linear (glabrous); narrow; 2–6 mm wide; flat, or rolled (rigid); exhibiting multicellular glands abaxially (at the base of macrohairs). The abaxial glands intercostal. Leaf blades without cross venation. Ligule a fringe of hairs.

Reproductive organization. Plants bisexual, with bisexual spikelets; with hermaphrodite florets. Exposed-cleistogamous, or chasmogamous. Plants with hidden cleistogenes, or without hidden cleistogenes. The hidden cleistogenes when present, in the leaf sheaths (sometimes basal and highly modified).

Inflorescence. Inflorescence paniculate; open, or contracted; when contracted spicate, or more or less irregular; espatheate; not comprising 'partial inflorescences' and foliar organs. Spikelet-bearing axes persistent. Spikelets not secund; pedicellate.

Female-fertile spikelets. Spikelets compressed laterally, or not noticeably compressed, or compressed dorsiventrally; disarticulating above the glumes; not disarticulating between the florets (or only tardily so). Rachilla prolonged beyond the uppermost female-fertile floret. The rachilla extension with incomplete florets. *Hairy callus present.*

Glumes two; more or less equal; about equalling the spikelets; long relative to the adjacent lemmas; pointed; awnless; carinate; similar (thinly membranous, acute or mucronate). Lower glume 1 nerved. Upper glume 1 nerved. *Spikelets* with incomplete florets. The incomplete florets distal to the female-fertile florets. *The distal incomplete florets* reduced to a brush-like appendage.

Female-fertile florets (1–)3–5. Lemmas decidedly firmer than the glumes (leathery); not becoming indurated; incised; 13–23 lobed; deeply cleft (dissected above into awns); awned. *Awns 13–23*; median and lateral (spreading, unequal, together forming a pappus-like crown to the spikelet); the median similar in form to the laterals; non-geniculate; hairless to long-plumose; much longer than the body of the lemma (simple, or branched at the base). Lemmas hairy (at least below); non-carinate (dorsally rounded); without a germination flap; (5–)7–11 nerved (or more). *Palea* present; relatively long; apically notched; awnless, without apical setae; textured like the lemma (papyraceous); 2-nerved; 2-keeled. Palea keels wingless; scabrous to hairy. *Lodicules* present; 2; free; glabrous. *Stamens* 3. *Ovary* glabrous. Stigmas 2.

Fruit, embryo and seedling. *Fruit* free from both lemma and palea; small; ellipsoid; compressed dorsiventrally. Hilum short (1/5 as long as the grain). Pericarp fused. Embryo large. Endosperm hard; without lipid. Embryo with an epiblast; with a scutellar tail; with an elongated mesocotyl internode. Embryonic leaf margins meeting.

Abaxial leaf blade epidermis. *Costal/intercostal zonation* conspicuous. *Papillae* absent. *Long-cells* markedly different in shape costally and intercostally (the costals relatively long and narrow); of similar wall thickness costally and intercostally. Mid-intercostal long-cells rectangular; having markedly sinuous walls. *Microhairs* present; more or less spherical to elongated; clearly two-celled; chloridoid-type (and demonstrated to secrete salt). Microhair apical cell wall of similar thickness/rigidity to that of the basal cell. Microhairs $(27-)28.5-30$ microns long. Microhair basal cells 18 microns long. Microhairs $14.4-16.5$ microns wide at the septum. Microhair total length/width at septum $1.7-2$. Microhair apical cells $10.5-13.5$ microns long. Microhair apical cell/total length ratio $0.35-0.45$. *Stomata* common; $25.5-31.5$ microns long. Guard-cells overlapping to flush with the interstomatals. *Intercostal short-cells* common; not paired (solitary). Intercostal silica bodies absent. *Costal short-cells* conspicuously in long rows. Costal silica bodies present in alternate cell files of the costal zones; saddle shaped (the commonest form), or 'panicoid-type' (common in places); when panicoid type, cross shaped and butterfly shaped; not sharp-pointed.

Transverse section of leaf blade, physiology. C_4; XyMS+. PCR sheaths of the primary vascular bundles interrupted; interrupted abaxially only. PCR sheath extensions present. Maximum number of extension cells 1, or 2–3. PCR cell chloroplasts centripetal. *Mesophyll* with radiate chlorenchyma; traversed by columns of colourless mesophyll cells (seemingly, occasionally). *Leaf blade* with distinct, prominent adaxial ribs, or 'nodular' in section; with the ribs more or less constant in size. *Midrib* not readily distinguishable; with one bundle only. Bulliforms present in discrete, regular adaxial groups; associated with colourless mesophyll cells to form deeply-penetrating fans (these sometimes connecting with traversing colourless columns). All the vascular bundles accompanied by sclerenchyma. Combined sclerenchyma girders present; forming 'figures'. Sclerenchyma all associated with vascular bundles. The lamina margins with fibres.

Special diagnostic feature. *Spikelets with the distal incomplete florets and/or the rachilla apex forming a terminal clavate appendage.*

Cytology. Chromosome base number, $x = 10$. $2n = 40$, or 60, or 100. 4, 6, and 10 ploid. Nucleoli persistent.

Taxonomy. Chloridoideae; Pappophoreae, or main chloridoid assemblage (relationships dubious).

Ecology, geography, regional floristic distribution. 8 species. U.S.A., South America. Xerophytic; species of open habitats. Grassland and bushland.

Holarctic, Neotropical, and Antarctic. Boreal, Tethyan, and Madrean. Euro-Siberian and Atlantic North American. Irano-Turanian. Venezuela and Surinam, Amazon, Central Brazilian, Pampas, and Andean. Patagonian. Siberian. Southern Atlantic North American and Central Grasslands.

Rusts and smuts. Rusts — *Puccinia*. Smuts from Ustilaginaceae. Ustilaginaceae — *Sphacelotheca* and *Ustilago*.

References, etc. Morphological/taxonomic: Reeder 1965; Reeder and Toolin 1989. Leaf anatomical: Metcalfe 1960; this project.

Parafestuca E. Alekseev

~ *Festuca* (*F. albida*)

Habit, vegetative morphology. Perennial; densely caespitose. **Culms** 25–100 cm high; *herbaceous*; unbranched above. Young shoots intravaginal. Leaves non-auriculate. The sheaths splitting longitudinally into fibres. Leaf blades linear; broad to narrow; 3–10 mm wide (–12); flat, or rolled (convolute); without cross venation; persistent. *Ligule an unfringed membrane*; truncate; 0.8–2 mm long.

Reproductive organization. Plants bisexual, with bisexual spikelets; with hermaphrodite florets.

Inflorescence. *Inflorescence paniculate*; open to contracted; espatheate; not comprising 'partial inflorescences' and foliar organs. Spikelet-bearing axes persistent. Spikelets not secund; pedicellate.

Female-fertile spikelets. *Spikelets 6–8(–10) mm long*; compressed laterally; disarticulating above the glumes; disarticulating between the florets. Rachilla prolonged beyond the uppermost female-fertile floret. The rachilla extension with incomplete florets. Hairy callus present. *Callus* short.

Glumes two; very unequal; shorter than the spikelets; *long relative to the adjacent lemmas*; hairless; scabrous (marginally and below); pointed; awnless; carinate; similar (lanceolate). Lower glume 1 nerved. Upper glume 3 nerved. *Spikelets* with incomplete florets. The incomplete florets distal to the female-fertile florets. *The distal incomplete florets* merely underdeveloped.

Female-fertile florets *2–3(–4)*. *Lemmas* similar in texture to the glumes; not becoming indurated; entire; pointed; awnless; hairless; scabrous (on the keel); *carinate*; *3 nerved*. *Palea* present; relatively long to conspicuous but relatively short; apically notched; awnless, without apical setae; not indurated; 2-nerved; 2-keeled. Palea keels wingless; scabrous to hairy (below). *Lodicules* present; 2; free; membranous; glabrous; toothed (the teeth equal); not or scarcely vascularized. **Stamens** *3*. Anthers 2.5–3 mm long. *Ovary* glabrous. Stigmas 2.

Fruit, embryo and seedling. *Fruit* small to medium sized (3.5 – 4 mm long); fusiform; somewhat *compressed dorsiventrally (ventrally)*. Hilum short. Embryo small.

Transverse section of leaf blade, physiology. C$_3$. *Leaf blade* with distinct, prominent adaxial ribs; with the ribs very irregular in sizes. *Midrib* conspicuous; with one bundle only. Bulliforms present in discrete, regular adaxial groups; in simple fans. All the vascular bundles accompanied by sclerenchyma. Combined sclerenchyma girders present (with the main bundles only); forming 'figures' (in the main bundles).

Taxonomy. Pooideae; Poodae; Poeae.

Ecology, geography, regional floristic distribution. 1 species. Madeira. Holarctic. Tethyan. Macaronesian.

References, etc. Morphological/taxonomic: Alekseev 1985. Leaf anatomical: Alekseev 1985.

Special comments. Anatomical data for ts only.

Parahyparrhenia A. Camus

Habit, vegetative morphology. Annual, or perennial. Culms herbaceous. Leaf blades narrow (very); not setaceous (flat or involute); flat, or rolled (involute); without cross venation. Ligule an unfringed membrane.

Reproductive organization. Plants bisexual, with bisexual spikelets; with hermaphrodite florets. The spikelets of sexually distinct forms on the same plant; hermaphrodite and male-only, or hermaphrodite and sterile; overtly heteromorphic; in both homogamous and heterogamous combinations (one male or neuter pair, inconspicuous, at the base of the lower raceme).

Inflorescence. *Inflorescence* of spicate main branches (a single raceme or raceme-pair), or paniculate (then a scanty, spatheate false panicle of racemes); *with capillary branchlets (i.e., the articles of racemes and peduncles)*; *spatheate*; a complex of 'partial inflorescences' and intervening foliar organs, or not comprising 'partial inflorescences' and foliar organs (when comprising only two racemes). **Spikelet-bearing axes** 'racemes' (the rachis few to many-jointed); *paired (not deflexed, the bases terete, unequal, the upper filiform)*; with very slender rachides; disarticulating; disarticulating at the joints. *'Articles'* linear; not appendaged (by contrast with *Hyperthelia*); disarticulating obliquely; somewhat hairy (ciliate on both sides). Spikelets paired; sessile and pedicellate; consistently in 'long-and-short' combinations; in pedicellate/sessile combinations. Pedicels of the 'pedicellate' spikelets free of the rachis. The 'shorter' spikelets hermaphrodite (sometimes male or sterile in the lowermost pair). The 'longer' spikelets male-only, or sterile.

Female-sterile spikelets. The pedicelled spikelets about as long as the sessile, narrowly lanceolate, male with two hyaline lemmas or empty. Callus small and oblong or absent. Pedicel ciliate on both sides.

Female-fertile spikelets. Spikelets 4.5–12 mm long; not noticeably compressed to compressed dorsiventrally (subterete); falling with the glumes (falling with adjacent joint and pedicel). Rachilla terminated by a female-fertile floret. Hairy callus present (long, pointed, curved or oblique). *Callus* long; pointed.

Glumes two; more or less equal; long relative to the adjacent lemmas; hairless (glabrous, or the margins ciliate); (the upper) *awned (or at least aristulate)*; carinate (G$_2$), or non-carinate (G_1); very dissimilar (the lower with a dorsal median groove and narrow involute margins, sub-keeled upwards and sometimes winged, bidentate and bimucronate, the upper cymbiform, awned or aristulate). *Lower glume two-keeled (with a dorsal median groove)*; sulcate on the back (without a conspicuous herbaceous tip, by contrast with *Hyperthelia*); not pitted; relatively smooth. *Spikelets* with incomplete florets. The incomplete florets proximal to the female-fertile florets. *The proximal incomplete florets* 1; sterile. The proximal lemmas awnless (narrow-lanceolate); similar in texture to the female-fertile lemmas (hyaline); not becoming indurated.

Female-fertile florets 1. Lemmas stipitate below; less firm than the glumes (hyaline); not becoming indurated; incised; 2 lobed (bidentate); awned. Awns 1; median; from a sinus; geniculate; hairy; much longer than the body of the lemma. Lemmas non-carinate; without a germination flap; 1 nerved. **Palea** present, or absent; when present, *very reduced*. *Lodicules* present; 2; free; glabrous. *Stamens* 3 (the anthers rather long). *Ovary* glabrous. Stigmas 2.

Abaxial leaf blade epidermis. *Costal/intercostal zonation* conspicuous (the intercostal zones very narrow). *Papillae* present; intercostal (on each interstomatal cell). Intercostal papillae over-arching the stomata (slightly, in places), or not over-arching the stomata (for the most part); consisting of one oblique swelling per cell, or consisting of one symmetrical projection per cell. Long-cells of similar wall thickness costally and intercostally (thin walled). Mid-intercostal long-cells rectangular; having markedly sinuous walls and having straight or only gently undulating walls (the sinuosity very fine). *Microhairs* present; elongated; clearly two-celled; panicoid-type. *Stomata* common. Subsidiaries non-papillate; dome-shaped. Guard-cells overlapping to flush with the interstomatals. *Intercostal short-cells* common; in cork/silica-cell pairs; silicified. Intercostal silica bodies cross-shaped (and short dumb-bell shaped). No macrohairs or prickles seen. *Costal short-cells* conspicuously in long rows. Costal silica bodies present and well developed; 'panicoid-type'; dumb-bell shaped (mostly), or nodular (a few).

Transverse section of leaf blade, physiology. Leaf blades largely consisting of midrib, or 'laminar'. C_4. The anatomical organization conventional. XyMS−. PCR sheath outlines uneven. Mesophyll traversed by columns of colourless mesophyll cells (in places). *Midrib* conspicuous; having a conventional arc of bundles (an abaxial arc, with a median primary and small bundles on either side, or three primaries with small bundles between them); with colourless mesophyll adaxially. *The lamina* symmetrical on either side of the midrib. Bulliforms present in discrete, regular adaxial groups; associated with colourless mesophyll cells to form deeply-penetrating fans; associating with colourless mesophyll cells to form arches over small vascular bundles. Many of the smallest vascular bundles unaccompanied by sclerenchyma (though many of the tiniest have minute abaxial sclerenchyma groups). Combined sclerenchyma girders present (with the primaries). Sclerenchyma all associated with vascular bundles.

Taxonomy. Panicoideae; Andropogonodae; Andropogoneae; Andropogoninae.

Ecology, geography, regional floristic distribution. 5 species. Tropical Africa and Thailand. Species of open habitats. Savanna grassland, pools on rock outcrops.

Paleotropical. African and Indomalesian. Sudano-Angolan and West African Rainforest. Indo-Chinese. Sahelo-Sudanian.

References, etc. Morphological/taxonomic: Clayton 1966. Leaf anatomical: this project.

Special comments. Fruit data wanting.

Paraneurachne S. T. Blake

Habit, vegetative morphology. Perennial; stoloniferous, or caespitose, or decumbent. *Culms* 10–45 cm high; branched above. Culm nodes hairy. Culm internodes solid. Young shoots culm branches intravaginal. *Leaves* not basally aggregated; non-auriculate. *Leaf blades* linear to ovate; narrow; 2–4 mm wide; without cross venation; *disarticulating from the sheaths*; rolled in bud. *Ligule a fringe of hairs*.

Reproductive organization. Plants bisexual, with bisexual spikelets; with hermaphrodite florets. The spikelets of sexually distinct forms on the same plant, or all alike in sexuality; hermaphrodite, or hermaphrodite and sterile (lowest spikelet reduced). Not viviparous.

Inflorescence. *Inflorescence a single raceme (spike-like)*; the racemes contracted, the spikelets spiralled; *espatheate*; not comprising 'partial inflorescences' and foliar organs. Spikelet-bearing axes persistent. Spikelets solitary; not secund; pedicellate (the pedicels short, persistent). Pedicel apices oblique, or discoid.

Female-fertile spikelets. *Spikelets* 7–12 mm long; lanceolate; abaxial; *compressed dorsiventrally*; falling with the glumes. Rachilla terminated by a female-fertile floret; rachilla hairless. *Hairy callus present (very conspicuous, with long white hairs)*.

Glumes two; more or less equal; long relative to the adjacent lemmas; dorsiventral to the rachis; hairy (with dense white hairs); pointed (subulate-acuminate); awned (acuminate into the short subule), or awnless (acuminate, hardly truly awned); non-carinate; very dissimilar (lower membranous, flat on back, not bearded. Upper convex and hardened towards the base, with a dense narrow beard of long white submarginal hairs on each side below). *Lower glume two-keeled (above)*; 5–7 nerved. Upper glume 11–13 nerved. *Spikelets* with incomplete florets. The incomplete florets proximal to the female-fertile florets. *The proximal incomplete florets* 1; paleate. Palea of the proximal incomplete florets reduced. The proximal incomplete florets male. The proximal lemmas gibbous below; awnless; 7 nerved; exceeded by the female-fertile lemmas to decidedly exceeding the female-fertile lemmas.

Female-fertile florets 1. Lemmas similar in texture to the glumes; smooth to striate; not becoming indurated (L_2 thinly rigid. cf. *Thyridolepis*, contrast *Neurachne*); yellow in fruit; entire; pointed; awnless; hairy, or hairless; non-carinate; having the margins lying flat and exposed on the palea; with a clear germination flap; 3–7 nerved. *Palea* present (acuminate); relatively long; apically notched; awnless, without apical setae; textured like the lemma; indurated; 2-nerved; 2-keeled. *Lodicules* present; 2; free; fleshy; glabrous; not or scarcely vascularized. *Stamens* 3. Anthers about 2.5 mm long; not penicillate; without an apically prolonged connective. *Ovary* glabrous. Styles fused, or free to their bases. Stigmas 2 (and 2 styles); red pigmented.

Fruit, embryo and seedling. Fruit small; compressed dorsiventrally. Hilum short. Embryo large; waisted. Endosperm containing only simple starch grains. Embryo without an epiblast; with a scutellar tail; with a negligible mesocotyl internode.

First seedling leaf with a well-developed lamina. The lamina broad; curved.

Abaxial leaf blade epidermis. *Costal/intercostal zonation* conspicuous. *Papillae* absent. Long-cells of similar wall thickness costally and intercostally. Mid-intercostal long-cells rectangular; having markedly sinuous walls. *Microhairs* present; 51–63 microns long; 9–9.9 microns wide at the septum. Microhair total length/width at septum 5.2–7. Microhair apical cells 28.5–33 microns long. Microhair apical cell/total length ratio 0.52–0.59. *Stomata* common; 27–36 microns long. Guard-cells overlapping to flush with the interstomatals (slightly overlapping). *Intercostal short-cells* common; in cork/silica-cell pairs; silicified. Many prickles present, mostly reduced to their bases. *Crown cells* present (or at least, many structures (obviously 'reduced prickles') approaching these). *Costal short-cells* conspicuously in long rows and neither distinctly grouped into long rows nor predominantly paired (mostly pairs and short rows, but some long rows — quite a mixture). Costal silica bodies 'panicoid-type'; mostly cross shaped, butterfly shaped, and dumb-bell shaped (short); not sharp-pointed.

Transverse section of leaf blade, physiology. C_4. The anatomical organization unconventional. Organization of PCR tissue *Alloteropsis* type. Biochemical type NADP–ME; XyMS– (the 'inner sheath' of large, thin walled cells is PCR, but there is another sheath of smaller, sparsely chlorenchymatous cells outside it. c.f.*Alloteropsis*, *Neurachne*). PCR sheath outlines even. PCR sheath extensions present. Maximum number of extension cells 1. PCR cells with a suberised lamella. PCR cell chloroplasts with well developed grana; centrifugal/peripheral. *Mesophyll* with radiate chlorenchyma. *Leaf blade* with distinct, prominent adaxial ribs (with broad, flat-topped adaxial ribs, each including several vascular bundles); with the ribs more or less constant in size. *Midrib* not readily distinguishable; with one bundle only to having a conventional arc of bundles (with a small bundle beneath each bulliform group). Bulliforms present in discrete, regular adaxial groups; in simple fans, or in simple fans and associated with colourless mesophyll cells to form deeply-penetrating fans. All the vascular bundles accompanied by sclerenchyma. Combined sclerenchyma girders absent (combining broad abaxial girders with adaxial strands). Sclerenchyma all associated with vascular bundles.

Cytology. Chromosome base number, $x = 9$. $2n = 36$. 4 ploid.

Taxonomy. Panicoideae; Panicodae; Neurachneae.

Ecology, geography, regional floristic distribution. 1 species. Northern Australia. Xerophytic; species of open habitats. Dry grassland.

Australian. Central Australian.

References, etc. Morphological/taxonomic: Blake 1972b. Leaf anatomical: Hattersley *et al.* 1982.

Parapholis C.E. Hubb.

Lepidurus Janchen

Habit, vegetative morphology. Slender annual (erect or more or less prostrate); caespitose. *Culms* 2–50 cm high; herbaceous; branched above, or unbranched above; 3–9 noded. Culm nodes exposed, or hidden by leaf sheaths; glabrous. Culm internodes hollow. Young shoots intravaginal. *Leaves* not basally aggregated; auriculate, or non-auriculate. Sheaths not keeled. Leaf blades linear; narrow; 0.4–3 mm wide; setaceous, or not setaceous; flat, or rolled (convolute); without cross venation; persistent; once-folded in bud. Ligule an unfringed membrane; truncate; 0.3–0.9 mm long.

Reproductive organization. Plants bisexual, with bisexual spikelets; with hermaphrodite florets. The spikelets all alike in sexuality, or of sexually distinct forms on the same plant (sometimes with rudiments at the base of the inflorescence); hermaphrodite, or hermaphrodite and sterile.

Inflorescence. *Inflorescence a single spike (cylindrical, rigid, often curved)*. Rachides hollowed. Inflorescence espatheate; not comprising 'partial inflorescences' and foliar organs. *Spikelet-bearing axes disarticulating*; disarticulating at the joints. Spikelets solitary; not secund; distichous; sessile.

Female-fertile spikelets. *Spikelets 4–8.5 mm long*; more or less abaxial (closing off the hollowed internode); *compressed laterally*; falling with the glumes (shed with rachis joints). *Rachilla prolonged beyond the uppermost female-fertile floret*; rachilla hairless. The rachilla extension with incomplete florets, or naked. Hairy callus absent. *Callus* absent.

Glumes two; more or less equal; long relative to the adjacent lemmas; *lateral to the rachis to displaced (side by side)*; pointed; awnless; sometimes asymmetric and winged; similar (leathery). Lower glume 3–5 nerved (the nerves raised). Upper glume 3–5 nerved (the nerves raised). *Spikelets* with female-fertile florets only, or with incomplete florets. The incomplete florets if present, distal to the female-fertile florets. The distal incomplete florets 1; *incomplete florets* merely underdeveloped (a rudiment).

Female-fertile florets *1*. Lemmas less firm than the glumes (membranous, side-on to the rachis); not becoming indurated; awnless; hairless; glabrous; non-carinate; 1 nerved, or 3 nerved (the laterals very short). *Palea* present; relatively long; tightly clasped by the lemma; entire to apically notched; awnless, without apical setae; thinner than the lemma (hyaline); 2-nerved; weakly 2-keeled. Palea keels wingless. *Lodicules* present; 2; free; membranous; glabrous; not toothed. *Stamens* 3. Anthers 0.5–4 mm long; not penicillate. *Ovary* glabrous. Styles free to their bases. Stigmas 2; white.

Fruit, embryo and seedling. *Fruit* free from both lemma and palea; small (3–3.6 mm long); narrowly ovoid or ellipsoid; ventrally longitudinally grooved, or not grooved; compressed dorsiventrally to not noticeably compressed. *Hilum short*. Embryo small. Endosperm liquid in the mature fruit, or hard; with lipid; containing compound starch grains. Embryo with an epiblast.

Seedling with a loose coleoptile. First seedling leaf with a well-developed lamina. The lamina narrow; 3 veined.

Abaxial leaf blade epidermis. *Costal/intercostal zonation* conspicuous. *Papillae* absent. *Long-cells* similar in shape costally and intercostally; of similar wall thickness costally and intercostally. Mid-intercostal long-cells rectangular; having markedly sinuous walls, or having straight or only gently undulating walls. *Microhairs* absent. *Stomata* absent or very rare, or common; (39–)42–45(–48) microns long. Subsidiaries low dome-shaped, or parallel-sided. Guard-cells overlapped by the interstomatals. *Intercostal short-cells* common; in cork/silica-cell pairs (mainly), or not paired (a few solitaries); silicified (when paired). Intercostal silica bodies rounded, or crescentic. *Costal short-cells* predominantly paired, or neither distinctly grouped into long rows nor predominantly paired. Costal silica bodies rounded (mostly), or tall-and-narrow to crescentic; not sharp-pointed.

Transverse section of leaf blade, physiology. C_3; XyMS+. *Mesophyll* with non-radiate chlorenchyma; without adaxial palisade. *Leaf blade* with distinct, prominent adaxial ribs; with the ribs more or less constant in size. *Midrib* conspicuous (rarely), or not readily distinguishable; with one bundle only. Bulliforms present in discrete, regular adaxial groups, or not in discrete, regular adaxial groups (*P. strigosa* with small bulliforms of variable size, irregularly grouped in the furrows, cf. *Ammophila*); sometimes in simple fans (e.g. *P. incurva*). All the vascular bundles accompanied by sclerenchyma. Combined sclerenchyma girders absent. Sclerenchyma all associated with vascular bundles.

Cytology. Chromosome base number, $x = 7$, 9, and 19. $2n = 14$, 28, 32, 36, 38, and 42. 2, 4, and 6 ploid.

Taxonomy. Pooideae; Poodae; Poeae.

Ecology, geography, regional floristic distribution. 6 species. Western Europe, Mediterranean to India. Commonly adventive. Species of open habitats; halophytic, or glycophytic. Sandy maritime soils and saltmarshes.

Holarctic and Paleotropical. Boreal and Tethyan. African and Indomalesian. Euro-Siberian. Mediterranean and Irano-Turanian. Saharo-Sindian. Indian. European.

References, etc. Leaf anatomical: Metcalfe 1960; this project.

Paratheria Griseb.

Habit, vegetative morphology. Perennial; geniculate ascending, rooting at the nodes. *Culms* 15–80 cm high; herbaceous; branched above. Culm nodes hairy, or glabrous. Culm internodes hollow. Young shoots intravaginal. *Leaves* not basally aggregated; non-auriculate. Leaf blades linear; narrow; 3–5 mm wide; flat; without cross venation; persistent. *Ligule* present; a fringed membrane (very narrow), or a fringe of hairs; about 0.3 mm long. *Contra-ligule* present (a line of hairs).

Reproductive organization. Plants bisexual, with bisexual spikelets; with hermaphrodite florets. The spikelets all alike in sexuality (but sometimes with cleistogamous spikelets lacking bristles at the base of the inflorescence). *Plants with hidden cleistogenes.* The hidden cleistogenes in the leaf sheaths (the upper sheaths).

Inflorescence. *Inflorescence a false spike, with spikelets on contracted axes (loosely spicate, with appressed, deciduous, one-spikeleted racemelets).* Inflorescence axes not ending in spikelets (the racemelet ending in the bristle, which subtends the spikelet and extends beyond it). Inflorescence espatheate; not comprising 'partial inflorescences' and foliar organs. **Spikelet-bearing axes *very much reduced (to the long, pungent stipe, one spikelet and its subtending bristle)*; disarticulating; falling entire (i.e., the racemelets disarticulating, complete with spikelet and bristle, and contributing a pointed callus beneath the spikelet). **Spikelets *subtended by solitary 'bristles'*.** The 'bristles' deciduous with the spikelets. Spikelets solitary (one per racemelet); not secund; subsessile.

Female-fertile spikelets. Spikelets 8–13 mm long; abaxial; compressed dorsiventrally; falling with the glumes (and the racemelet); with conventional internode spacings. Rachilla terminated by a female-fertile floret. Hairy callus absent. *Callus* long (constituted by the proximal part of the branch beneath the spikelet); pointed.

Glumes two; minute; more or less equal; shorter than the spikelets; shorter than the adjacent lemmas; dorsiventral to the rachis; hairless; not pointed; awnless; non-carinate; similar (hyaline). Lower glume 0 nerved. Upper glume 0 nerved. *Spikelets* with incomplete florets. The incomplete florets proximal to the female-fertile florets. *The proximal incomplete florets* 1; epaleate; sterile. The proximal lemmas awnless (but acuminate-subulate); 7–11 nerved (these anastomosing above); more or less equalling the female-fertile lemmas; similar in texture to the female-fertile lemmas (firmly membranous); not becoming indurated.

Female-fertile florets 1. Lemmas decidedly firmer than the glumes; smooth; not becoming indurated; entire; pointed; awnless (but acuminate-subulate, like the L_1); hairless; glabrous; non-carinate; having the margins lying flat and exposed on the palea; without a germination flap; 7 nerved. *Palea* present; relatively long (about equalling the lemma); entire; awnless, without apical setae; textured like the lemma; not indurated; 2-nerved (linear-lanceolate, acuminate, membranous); keel-less (abaxially rounded). *Lodicules* present; 2; free; fleshy; glabrous; not or scarcely vascularized. *Stamens* 3. Anthers 2.5 mm long; not penicillate; without an apically prolonged connective. *Ovary* glabrous. Styles fused. Stigmas 2; white.

Fruit, embryo and seedling. Fruit compressed dorsiventrally. Hilum short. Embryo large.

Abaxial leaf blade epidermis. *Costal/intercostal zonation* conspicuous. *Papillae* absent. *Long-cells* similar in shape costally and intercostally. Mid-intercostal long-cells rectangular; having markedly sinuous walls. *Microhairs* present; panicoid-type; 57–63–66 microns long; 4.5–5.4–6 microns wide at the septum. Microhair total length/ width at septum 9.5–14.7. Microhair apical cells 30–34.5 microns long. Microhair apical cell/total length ratio 0.52–0.6. *Stomata* common; (24–)27–28.5(–33) microns long. Subsidiaries low dome-shaped (mostly), or triangular. Guard-cells overlapping to flush with the interstomatals. *Intercostal short-cells* common; in cork/silica-cell pairs. *Costal short-cells* conspicuously in long rows. Costal silica bodies 'panicoid-type'; mostly dumb-bell shaped; not sharp-pointed.

Transverse section of leaf blade, physiology. C_4; XyMS–. *Mesophyll* with non-radiate chlorenchyma. *Leaf blade* 'nodular' in section; with the ribs more or less constant in size. *Midrib* conspicuous; having a conventional arc of bundles; with colourless mesophyll adaxially. Bulliforms present in discrete, regular adaxial groups to not in discrete, regular adaxial groups; occasionally in simple fans (but the epidermis largely irregularly bulliform). All the vascular bundles accompanied by sclerenchyma. Combined sclerenchyma girders present; nowhere forming 'figures'. Sclerenchyma all associated with vascular bundles.

Taxonomy. Panicoideae; Panicodae; Paniceae.

Ecology, geography, regional floristic distribution. 2 species. Africa, Madagascar, Cuba, Brazil. Commonly adventive. Hydrophytic, or helophytic; species of open habitats; glycophytic. Swamps and lakes.

Paleotropical and Neotropical. African and Madagascan. Sudano-Angolan, West African Rainforest, and Namib-Karoo. Caribbean, Amazon, and Central Brazilian. Sahelo-Sudanian, Somalo-Ethiopian, and South Tropical African.

References, etc. Leaf anatomical: this project.

Parectenium P. Beauv. corr. Stapf

Paractaenium P. Beauv.
~ *Plagiosetum*

Habit, vegetative morphology. Annual; caespitose. *Culms* 30–90 cm high; herbaceous; branched above. Culm nodes glabrous. Culm internodes solid. Leaves non-auriculate. Leaf blades narrow; 2–5 mm wide; flat; without cross venation; persistent. *Ligule a fringed membrane*; short.

Reproductive organization. Plants bisexual, with bisexual spikelets; with hermaphrodite florets.

Inflorescence. *Inflorescence of spicate main branches (with up to 5 spikelets per raceme, but reducing to one or two)*; the racemes finally deflexing; non-digitate. Inflorescence axes not ending in spikelets (each short lateral raceme ending in a bristle). Rachides hollowed. Inflorescence espatheate; not comprising 'partial inflorescences' and foliar organs. *Spikelet-bearing axes* very much reduced, or spikelike; disarticulating (but the main axis persistent); falling entire (the short branches articulated at their bases, falling whole). **Spikelets** *subtended by solitary 'bristles' (individually, at least at the base of* *the racemelets, by contrast with* **Plagiosetum***). The 'bristles' deciduous with the spike-lets.* Spikelets solitary; secund; shortly pedicellate. Pedicel apices truncate.

Female-fertile spikelets. Spikelets 3.5–5 mm long; oblong, or elliptic, or lanceolate, or ovate, or obovate; abaxial; compressed dorsiventrally; falling with the glumes (and with the branch). Rachilla terminated by a female-fertile floret.

Glumes two; relatively large; very unequal; (the G_2, which equals the spikelet) long relative to the adjacent lemmas; dorsiventral to the rachis; not pointed; awnless; non-carinate. Lower glume 3–5 nerved. Upper glume 9–13 nerved. *Spikelets* with incomplete florets. The incomplete florets proximal to the female-fertile florets. *The proximal incomplete florets* 1; paleate. Palea of the proximal incomplete florets reduced. The proximal incomplete florets sterile. The proximal lemmas awnless; 9–11 nerved; decidedly exceeding the female-fertile lemmas; less firm than the female-fertile lemmas (herbaceous-membranous, similar to the G_2); not becoming indurated.

Female-fertile florets 1. Lemmas decidedly firmer than the glumes (leathery or carti-laginous); rugose; becoming indurated (slightly); yellow in fruit, or brown in fruit; entire; pointed, or blunt; awnless; hairless (shortly ciliolate at the apex); non-carinate; having the margins lying flat and exposed on the palea; 3–5 nerved. *Palea* present; relatively long (elliptical); entire; awnless, without apical setae; textured like the lemma; slightly indurated (like the lemma). *Lodicules* present; 2; fleshy. *Stamens* 3. Anthers about 2 mm long; not penicillate. *Ovary* glabrous. Styles free to their bases. Stigmas 2.

Fruit, embryo and seedling. Fruit small; compressed dorsiventrally. Hilum short. Embryo small. Endosperm containing only simple starch grains. Embryo without an epi-blast; with a scutellar tail; with an elongated mesocotyl internode. Embryonic leaf margins overlapping.

Abaxial leaf blade epidermis. *Costal/intercostal zonation* conspicuous. *Papillae* absent. Mid-intercostal long-cells having markedly sinuous walls. *Microhairs* present; panicoid-type; (36–)47–50(–63) microns long; 5.4–7.5 microns wide at the septum. Microhair total length/width at septum 7.5–10.5. Microhair apical cells (22.5–)28.5–42(–43.5) microns long. Microhair apical cell/total length ratio 0.59–0.7. *Stomata* common; 21–25.5 microns long. Subsidiaries dome-shaped. Guard-cells overlap-ping to flush with the interstomatals. *Intercostal short-cells* common, or absent or very rare. *Costal short-cells* conspicuously in long rows. Costal silica bodies 'panicoid-type'; not sharp-pointed.

Transverse section of leaf blade, physiology. C_4; XyMS–. PCR sheath outlines uneven. PCR sheath extensions present. Maximum number of extension cells 1. PCR cell chloroplasts centrifugal/peripheral. *Mesophyll* with radiate chlorenchyma. *Leaf blade* with distinct, prominent adaxial ribs; with the ribs more or less constant in size. *Midrib* not readily distinguishable; with one bundle only. Bulliforms not in discrete, regular ad-axial groups (the epidermis extensively bulliform). Many of the smallest vascular bundles unaccompanied by sclerenchyma. Combined sclerenchyma girders absent. Sclerenchyma all associated with vascular bundles.

Taxonomy. Panicoideae; Panicodae; Paniceae.

Ecology, geography, regional floristic distribution. 1 species. Australia. Species of open habitats; halophytic, or glycophytic. Sandy places, sometimes maritime.

Australian. Central Australian.

References, etc. Leaf anatomical: this project.

Pariana Aub.

Eremitis Doell

Habit, vegetative morphology. Perennial; from a creeping rootstock with crowded stems. The flowering culms leafless, or leafy. **Culms** about 50–200 cm high; *woody and persistent*; to 0.5 cm in diameter; unbranched above. Culm nodes glabrous. Plants unarmed. *Leaves* not basally aggregated; auriculate; with auricular setae. Sheath margins free. The sheaths with (*Pariana* s. str.) or without (*Eremitis*) scarlike marks. Leaf blades linear to elliptic; broad; shortly pseudopetiolate; cross veined. *Ligule* present (often very short); not truncate; up to 5 mm long.

Reproductive organization. *Plants monoecious with all the fertile spikelets unisexual*; without hermaphrodite florets. The spikelets of sexually distinct forms on the same plant; female-only and male-only. *The male and female-fertile spikelets mixed in the inflorescence (in verticils, each comprising a central female surrounded by (4-)5(-6) males)*. The spikelets overtly heteromorphic. Plants probably outbreeding (with evidence of entomophily, in *Pariana* s. str.), or inbreeding (*Eremitis*?); with hidden cleistogenes (*Eremitis*), or without hidden cleistogenes (*Pariana*). The hidden cleistogenes in *Eremitis* subterranean. Not viviparous.

Inflorescence. *Inflorescence a false spike, with spikelets on contracted axes (the spikelets in verticils, the whole reduced to a single terminal verticil with vestiges below it in* **Eremitis***)*; espatheate; not comprising 'partial inflorescences' and foliar organs. Spikelet-bearing axes disarticulating; disarticulating at the joints (of several segments which separate at maturity (*Pariana*), or only the terminal segment disarticulating (*Eremitis*)). Spikelets in verticils of (5-)6(-7); not secund; sessile and pedicellate; consistently in 'long-and-short' combinations; in pedicellate/sessile combinations (the female spikelet sessile, the surrounding males on short pedicels). The 'shorter' spikelets female-only. The 'longer' spikelets male-only.

Female-sterile spikelets. *The male spikelets with their short pedicels flattened and coalescent, the florets having 10–40 stamens with their filaments free or joined*. The male spikelets with glumes (but some sometimes suppressed); 1 floreted. Male florets 10–40 staminate. The staminal filaments free, or joined.

Female-fertile spikelets. Spikelets 5–13 mm long; not noticeably compressed; disarticulating above the glumes. Rachilla terminated by a female-fertile floret. Hairy callus absent.

Glumes two; more or less equal; long relative to the adjacent lemmas; hairless; pointed; awnless; similar. Lower glume 1–3 nerved. Upper glume 1–3 nerved. *Spikelets with female-fertile florets only*.

Female-fertile florets 1. Lemmas decidedly firmer than the glumes; becoming indurated; hairless; non-carinate; 3 nerved. *Palea* present; relatively long; entire; awnless, without apical setae; indurated; several nerved; keel-less. *Lodicules* present; 3; free; ciliate, or glabrous; not toothed. *Stamens* 0. *Ovary* glabrous. Styles fused, or free to their bases. Stigmas 2 (plumose in *Pariana*, not so in *Eremitis*).

Fruit, embryo and seedling. *Fruit* free from both lemma and palea; small to medium sized. Embryo with an epiblast; with a scutellar tail; with a negligible mesocotyl internode. Embryonic leaf margins overlapping.

Abaxial leaf blade epidermis. *Papillae* present. Intercostal papillae several per cell. *Long-cells* similar in shape costally and intercostally (fairly short). Mid-intercostal long-cells rectangular; having markedly sinuous walls. *Microhairs* present; panicoid-type; 66–90 microns long. Microhair apical cells 18–38 microns long. Microhair apical cell/ total length ratio 0.26–0.52. *Stomata* common. Subsidiaries tall dome-shaped, or triangular. *Intercostal short-cells* common; silicified. Intercostal silica bodies vertically elongated-nodular, or cross-shaped and vertically elongated-nodular. *Costal short-cells* neither distinctly grouped into long rows nor predominantly paired. Costal silica bodies oryzoid and 'panicoid-type' (occasionally tall-narrow-crenate); commonly cross shaped; not sharp-pointed.

Transverse section of leaf blade, physiology. C_3; XyMS+. *Mesophyll* with non-radiate chlorenchyma; with arm cells; with fusoids. *Leaf blade* adaxially flat. *Midrib* conspicuous; seemingly having a conventional arc of bundles (?—a large central bundle and two laterals, but whether there is a complex is not clear from published descriptions); with the vascular bundles embedded in large-celled, colourless ground tissue. Bulliforms present in discrete, regular adaxial groups; in simple fans. All the vascular bundles accompanied by sclerenchyma. Combined sclerenchyma girders present; forming 'figures'. Sclerenchyma all associated with vascular bundles.

Cytology. Chromosome base number, $x = 11$, or 12. $2n = 44$ and 48.

Taxonomy. Bambusoideae; Oryzodae; Olyreae (?).

Ecology, geography, regional floristic distribution. 34 species. Tropical South America. Shade species. Forest floor plants.

Neotropical. Caribbean, Venezuela and Surinam, Amazon, and Andean.

References, etc. Leaf anatomical: Metcalfe 1960.
Special comments. Fruit data wanting.

Parodiolyra Soderstrom and Zuloaga

~ *Olyra*

Habit, vegetative morphology. Perennial; caespitose, or decumbent (or vine-like, or clambering). The flowering culms leafy. *Culms* 30–160 cm high (or trailing and climbing and up to 10 m long); woody and persistent to herbaceous; scandent, or not scandent; decumbent, scrambling, and scandent; branched above. Culm nodes hairy (pilose, with retrorse hairs). Culm internodes hollow. *Leaves* not basally aggregated; non-auriculate; without auricular setae. Sheaths glabrous to pilose. Leaf blades lanceolate to ovate; broad; 25–130 mm wide; slightly cordate, or not cordate, not sagittate; flat; pseudo-petiolate; cross veined, or without cross venation (?); disarticulating from the sheaths; rolled in bud. *Ligule an unfringed membrane to a fringed membrane*; truncate.

Reproductive organization. *Plants monoecious with all the fertile spikelets unisexual*; without hermaphrodite florets. The spikelets of sexually distinct forms on the same plant; female-only and male-only. *The male and female-fertile spikelets on different branches of the same inflorescence, or segregated, in different parts of the same inflorescence branch (the lower branches with male spikelets only, the upper branches male basally and female terminally or female-only: cf.* **Olyra***).* The spikelets overtly heteromorphic.

Inflorescence. *Inflorescence* determinate, or indeterminate; paniculate; open (lax and diffuse); probably spatheate (cf. *Olyra*). *Spikelets* solitary; not secund; *pedicellate (the pedicels filiform, by contrast with* **Olyra***)*.

Female-sterile spikelets. The male spikelets without glumes, stamens 3. Rachilla of male spikelets terminated by a male floret. The male spikelets without glumes; without proximal incomplete florets; 1 floreted. Male florets 1; 3 staminate.

Female-fertile spikelets. *Spikelets* 2–5.8 mm long; compressed dorsiventrally; falling with the glumes; *with a distinctly elongated rachilla internode between the glumes (this prominent and thickened)*. Rachilla terminated by a female-fertile floret. Hairy callus absent. *Callus* absent.

Glumes *present*; two; more or less equal; about equalling the spikelets to exceeding the spikelets; long relative to the adjacent lemmas; *conspicuously ventricose ('inflated to indurate at maturity')*; hairy (shortly pilose); awnless; non-carinate; similar (membranous, inflated). Lower glume 5–9 nerved. Upper glume 3–6 nerved. *Spikelets* with female-fertile florets only.

Female-fertile florets *1 (the anthecium resembling that typical of Paniceae)*. Lemmas decidedly firmer than the glumes; smooth (shining); becoming indurated; entire; pointed, or blunt; awnless; hairy, or hairless; non-carinate; having the margins tucked in onto the palea; with a clear germination flap; 5 nerved. *Palea* present; relatively long; tightly clasped by the lemma; entire; awnless; without apical setae; textured like the lemma; indurated; 2-nerved; keel-less. *Lodicules* present; 3; free; glabrous; heavily vascularized. *Stamens* 0. *Ovary* glabrous. Styles fused (into one). Stigmas 2.

Fruit, embryo and seedling. Disseminule a caryopsis enclosed in but free of the lemma and palea. *Fruit* free from both lemma and palea; small (1–2.5 mm long); brownish; ellipsoid; compressed dorsiventrally. *Hilum long-linear (but markedly shorter than the caryopsis, by contrast with* **Olyra***)*. Pericarp fused. Embryo small. Endosperm hard.

Abaxial leaf blade epidermis. *Costal/intercostal* zonation conspicuous. *Papillae* present; costal and intercostal. Intercostal papillae not over-arching the stomata; several per cell (small, circular, with one or two irregular rows per long-cell in *P. ramosissima*, larger in *P. lateralis*, very variable in size and number and some very large and branched in *P. leutzelburgii*). Mid-intercostal long-cells rectangular; having markedly sinuous walls (with coarse sinuosity). *Microhairs* present; elongated; clearly two-celled; panicoid-type. *Stomata* common. Subsidiaries papillate; high dome-shaped (mostly), or triangular. Guard-cells overlapping to flush with the interstomatals. *Intercostal short-cells* common;

in cork/silica-cell pairs, or not paired (solitary in *P. leutzelburgii*); silicified. Intercostal silica bodies vertically elongated-nodular. Bulbous-based prickles common. *Crown cells* absent. *Costal short-cells* conspicuously in long rows. Costal silica bodies 'panicoid-type'; consistently angular cross shaped.

Transverse section of leaf blade, physiology. C_3; XyMS+. Mesophyll with adaxial palisade; with arm cells; with fusoids, or without fusoids (none seen in *P. lateralis*). The fusoids external to the PBS. *Leaf blade* adaxially flat. *Midrib* conspicuous to not readily distinguishable (sometimes with a somewhat raised, rounded adaxial rib); with one bundle only. *The lamina* symmetrical on either side of the midrib. Bulliforms present in discrete, regular adaxial groups (the groups large, in each mid-intercostal region); in simple fans. All the vascular bundles accompanied by sclerenchyma. Combined sclerenchyma girders present; forming 'figures' (all the bundles with 'anchors'). Sclerenchyma all associated with vascular bundles.

Cytology. $2n = 36$.

Taxonomy. Bambusoideae; Oryzodae; Olyreae.

Ecology, geography, regional floristic distribution. 3 species. From Costa Rica south to Bolivia and Bahia, Brazil. Mesophytic; shade species and species of open habitats; glycophytic. In low and mid-elevation forests and savannas.

Neotropical. Caribbean, Venezuela and Surinam, Amazon, and Central Brazilian.

References, etc. Morphological/taxonomic: Soderstrom and Zuloaga 1989. Leaf anatomical: this project.

Pascopyrum A. Löve

'Agropyron smithii', ~ *Elymus, Elytrigia*

Habit, vegetative morphology. Strongly glaucous *perennial*; *rhizomatous*. **Culms** 30–90 cm high; *herbaceous*; unbranched above. *Leaves* not basally aggregated; auriculate (slender, pointed, on some of the sheaths). Sheath margins free. Leaf blades narrow; 2–7 mm wide; flat, or rolled (stiff, involute when dry); without cross venation; persistent. *Ligule an unfringed membrane*; truncate; 1 mm long.

Reproductive organization. Plants bisexual, with bisexual spikelets; with hermaphrodite florets; outbreeding.

Inflorescence. *Inflorescence a single spike, or a false spike, with spikelets on contracted axes (erect, 7–15 cm long, the 'clusters' reduced to one or two spikelets per node)*; espatheate; not comprising 'partial inflorescences' and foliar organs. Spikelet-bearing axes persistent. *Spikelets* solitary (at all nodes), or solitary and paired (with a few pairs at the middle nodes); not secund; distichous; *sessile*; imbricate (not pectinate).

Female-fertile spikelets. Spikelets mostly 15–25 mm long; compressed laterally; disarticulating above the glumes; disarticulating between the florets. Rachilla prolonged beyond the uppermost female-fertile floret. The rachilla extension with incomplete florets.

Glumes two; more or less equal (to slightly unequal); shorter than the spikelets; *long relative to the adjacent lemmas (the G_2 generally equalling or exceeding the L_1)*; lateral to the rachis; hairless; glabrous; pointed (tapering); not subulate (lanceolate and tapering, by contrast with those of *Leymus* —but subulate when dry); awned, or awnless (tapering into an acute or short-awned apex); non-carinate; similar (rigid, with scarious margins). Lower glume 3–7 nerved. Upper glume 3–7 nerved. *Spikelets* with incomplete florets. The incomplete florets distal to the female-fertile florets. *The distal incomplete florets merely underdeveloped.*

Female-fertile florets 4–9. Lemmas similar in texture to the glumes (firm, pale); not becoming indurated; entire; pointed, or blunt; awnless, or mucronate, or awned. Awns when present, 1; median; apical; non-geniculate; much shorter than the body of the lemma; entered by several veins. Lemmas glabrous, or basally pubescent; non-carinate; without a germination flap; 'obscurely nerved'; with the nerves confluent towards the tip. *Palea* present; relatively long; 2-nerved; 2-keeled. Palea back hairy (pubescent). *Stamens* 3. Anthers 4–6 mm long. Styles free to their bases. Stigmas 2; white.

Abaxial leaf blade epidermis. *Costal/intercostal zonation* fairly conspicuous. *Papillae* absent. *Long-cells* similar in shape costally and intercostally; of similar wall thickness costally and intercostally (fairly thick walled, pitted). Mid-intercostal long-cells rectangular; having markedly sinuous walls. *Microhairs* absent. *Stomata* common; (72–)75–81(–84) microns long. Subsidiaries low dome-shaped, or parallel-sided. Guard-cells overlapped by the interstomatals (slightly). *Intercostal short-cells* common; in cork/silica-cell pairs (and solitary); silicified (occasionally only). Intercostal silica bodies rounded and tall-and-narrow. *Crown cells* absent (from the material seen, though it shows a few crown cell-like prickles near the margins). *Costal short-cells* neither distinctly grouped into long rows nor predominantly paired (solitary, paired and a few threes). Costal silica bodies crescentic; not sharp-pointed.

Transverse section of leaf blade, physiology. C_3; XyMS+. *Mesophyll* with non-radiate chlorenchyma; without adaxial palisade. *Leaf blade* with distinct, prominent adaxial ribs; with the ribs very irregular in sizes (there being a decidedly smaller rib intercalated between the large, round-topped ones towards the margin on either side). *Midrib* not readily distinguishable; with one bundle only. Bulliforms present in discrete, regular adaxial groups (in the furrows); exclusively in simple fans. All the vascular bundles accompanied by sclerenchyma. Combined sclerenchyma girders present (with all the main veins, the minor ones having small abaxial strands only); forming 'figures' (I's to T's in the main bundles). Sclerenchyma all associated with vascular bundles.

Phytochemistry. Tissues of the culm bases with little or no starch. Fructosans predominantly short-chain.

Cytology. Chromosome base number, $x = 7$. $2n = 56$. 8 ploid. Nucleoli disappearing before metaphase.

Taxonomy. Pooideae; Triticodae; Triticeae.

Ecology, geography, regional floristic distribution. 1 species. North America. Mesophytic; halophytic (in heavy, saline-alkaline soils).

Holarctic. Boreal. Atlantic North American and Rocky Mountains. Central Grasslands.

References, etc. Morphological/taxonomic: Löve 1984. Leaf anatomical: this project.

Special comments. Fruit data wanting.

Paspalidium Stapf

Habit, vegetative morphology. Annual, or perennial (often aquatic); rhizomatous, or caespitose to decumbent. Culms herbaceous. Culm nodes glabrous. Culm internodes hollow. Young shoots extravaginal (usually?), or intravaginal. *Leaves* not basally aggregated; non-auriculate. Leaf blades broad, or narrow; flat, or rolled; without cross venation; persistent; once-folded in bud. *Ligule* present; a fringed membrane (very narrow), or a fringe of hairs.

Reproductive organization. Plants bisexual, with bisexual spikelets; with hermaphrodite florets.

Inflorescence. *Inflorescence of spicate main branches to a false spike, with spikelets on contracted axes (the branches generally appressed to the rachis, and sometimes greatly reduced). Inflorescence axes not ending in spikelets (each terminating in a conspicuous or more or less inconspicuous bristle).* Inflorescence espatheate; not comprising 'partial inflorescences' and foliar organs. **Spikelet-bearing axes** 'racemes', or very much reduced; *persistent.* **Spikelets** *unaccompanied by bractiform involucres, not associated with setiform vestigial branches (this being the 'distiction' from* **Setaria**: *however, the terminal spikelet of each branch is associated with the branch-tip bristle, and since the 'branches' may be reduced to single spikelets, the separation is scarcely adequate)*; solitary, or paired; secund (the inflorescence branches dorsiventral); (when the racemes not greatly reduced) biseriate. Pedicel apices discoid.

Female-fertile spikelets. *Spikelets* elliptic, or lanceolate, or ovate; *abaxial*; compressed dorsiventrally; falling with the glumes. Rachilla terminated by a female-fertile floret. Hairy callus absent.

Glumes two; very unequal; (the upper) about equalling the spikelets; (the upper) shorter than the adjacent lemmas, or long relative to the adjacent lemmas; dorsiventral to the rachis; awnless; non-carinate. Lower glume 1–5 nerved. Upper glume 5–11 nerved. *Spikelets* with incomplete florets. The incomplete florets proximal to the female-fertile florets. *Spikelets with proximal incomplete florets. The proximal incomplete florets* 1; paleate, or epaleate. Palea of the proximal incomplete florets when present, fully developed to reduced. The proximal incomplete florets male, or sterile. The proximal lemmas resembling the upper glume; awnless; more or less equalling the female-fertile lemmas; less firm than the female-fertile lemmas; not becoming indurated (membranous).

Female-fertile florets 1. *Lemmas* decidedly firmer than the glumes; *rugose*; becoming indurated (crustaceous); yellow in fruit, or brown in fruit; entire; pointed; awnless (often apiculate); hairless; non-carinate; having the margins tucked in onto the palea; with a clear germination flap; 5 nerved. *Palea* present; relatively long; entire; awnless, without apical setae; firm, like the lemma; 2-nerved. *Lodicules* present; 2; free; fleshy; glabrous. *Stamens* 3. Anthers not penicillate. *Ovary* glabrous. Styles fused, or free to their bases. Stigmas 2.

Fruit, embryo and seedling. Fruit small; ellipsoid to subglobose; compressed dorsiventrally. Hilum short. Embryo large. Endosperm containing only simple starch grains. Embryo without an epiblast; with a scutellar tail; with an elongated mesocotyl internode. Embryonic leaf margins overlapping.

First seedling leaf with a well-developed lamina. The lamina broad; curved; 13–20 veined.

Abaxial leaf blade epidermis. *Costal/intercostal zonation* conspicuous. *Papillae* absent. *Long-cells* similar in shape costally and intercostally; of similar wall thickness costally and intercostally (fairly thin walled). Mid-intercostal long-cells rectangular; having markedly sinuous walls. *Microhairs* present; panicoid-type; (40.5–)42–72(–81) microns long; (5.4–)5.7–7.5 microns wide at the septum. Microhair total length/width at septum 7.1–12.3. Microhair apical cells (21–)22.5–42(–51) microns long. Microhair apical cell/total length ratio 0.47–0.63. *Stomata* common; (24–)25.5–30 microns long. Subsidiaries low dome-shaped (sometimes, a few), or triangular. Guard-cells overlapping to flush with the interstomatals. *Intercostal short-cells* common, or absent or very rare; in cork/silica-cell pairs (usually), or not paired (a few solitaries); silicified (when paired). Intercostal silica bodies cross-shaped, or crescentic. *Costal short-cells* conspicuously in long rows. Costal silica bodies 'panicoid-type'; cross shaped, or butterfly shaped, or dumb-bell shaped; not sharp-pointed.

Transverse section of leaf blade, physiology. C_4; XyMS–. PCR sheath outlines uneven. PCR sheath extensions absent. PCR cell chloroplasts with reduced grana; centrifugal/peripheral. *Mesophyll* with radiate chlorenchyma. *Leaf blade* with distinct, prominent adaxial ribs, or 'nodular' in section. *Midrib* conspicuous, or not readily distinguishable (rarely); with one bundle only to having a conventional arc of bundles (a small bundle under each midrib 'hinge'). Bulliforms present in discrete, regular adaxial groups; in simple fans. Many of the smallest vascular bundles unaccompanied by sclerenchyma. Combined sclerenchyma girders present, or absent; nowhere forming 'figures'. Sclerenchyma all associated with vascular bundles.

Culm anatomy. *Culm internode bundles* in one or two rings.

Phytochemistry. Leaves without flavonoid sulphates (1 species).

Special diagnostic feature. Spikelets awnless, the female-fertile lemmas pointed or apiculate but not mucronate.

Cytology. Chromosome base number, $x = 9$. $2n = 18$, 36, and 54. 2, 4, and 6 ploid.

Taxonomy. Panicoideae; Panicodae; Paniceae.

Ecology, geography, regional floristic distribution. About 40 species. In warm regions. Hydrophytic to mesophytic; shade species and species of open habitats; glycophytic. Swamps, forests, dry slopes.

Holarctic, Paleotropical, Neotropical, and Australian. Boreal and Tethyan. African, Madagascan, Indomalesian, and Neocaledonian. Atlantic North American. Mediterranean and Irano-Turanian. Saharo-Sindian, Sudano-Angolan, West African Rainforest, and Namib-Karoo. Indian, Indo-Chinese, Malesian, and Papuan. Caribbean, Amazon, Central Brazilian, Pampas, and Andean. North and East Australian and Central Australian. Southern Atlantic North American. Sahelo-Sudanian, Somalo-Ethiopian, South Tropical

African, and Kalaharian. Tropical North and East Australian and Temperate and South-Eastern Australian.

Rusts and smuts. Rusts — *Puccinia*. Taxonomically wide-ranging species: '*Uromyces*' *setariae-italicae*. Smuts from Ustilaginaceae. Ustilaginaceae — *Sorosporium* and *Ustilago*.

Economic importance. Significant weed species: *P. flavidum*, *P. geminatum*, *P. punctatum*. Important native pasture species: *P. desertorum*, *P. geminatum*, *P. punctatum*.

References, etc. Leaf anatomical: Metcalfe 1960 and this project.

Special comments. Somewhat marginally separable from *Setaria*.

Paspalum L.

Anachyris Nees, *Cerea* Schlecht., *Ceresia* Pers., *Cleachne* Roland. ex Rottb., *Cymotochloa* Schlecht., *Dichromus* Schlecht., *Digitaria* Fabric., *Dimorphostachys* Fourn., *Maizilla* Schlecht., *Moenchia* Steud., *Paspalanthium* Desv., *Reimaria* Fluegge, *Sabsab* Adans., *Wirtgenia* Doell

Habit, vegetative morphology. Perennial (usually), or annual; rhizomatous, or stoloniferous, or caespitose, or decumbent. **Culms** 10–300 cm high (rarely taller, sometimes with culms trailing to 2 m or more); *herbaceous*. Culm nodes hairy, or glabrous. Culm internodes solid, or hollow. *Leaves* not basally aggregated; non-auriculate. Leaf blades linear, or linear to linear-lanceolate; broad, or narrow; (1–)3–25(–30) mm wide; flat, or folded, or rolled; without abaxial multicellular glands; without cross venation; persistent; rolled in bud. Ligule an unfringed membrane to a fringe of hairs.

Reproductive organization. Plants bisexual, with bisexual spikelets; with hermaphrodite florets. *The spikelets all alike in sexuality*. Plants outbreeding, or inbreeding. Exposed-cleistogamous, or chasmogamous. Plants with hidden cleistogenes, or without hidden cleistogenes. The hidden cleistogenes when present, subterranean. Apomictic, or reproducing sexually.

Inflorescence. *Inflorescence of spicate main branches*; digitate, or subdigitate, or non-digitate. Primary inflorescence branches (1–)2–20(–60). Inflorescence with axes ending in spikelets (usually), or axes not ending in spikelets (rachides naked-tipped in (e.g.) *P. repens*). Rachides hollowed, or flattened, or winged. Inflorescence espatheate; not comprising 'partial inflorescences' and foliar organs. Spikelet-bearing axes disarticulating (e.g., *P. repens*), or persistent; when disarticulating falling entire. Spikelets solitary, or paired; secund; biseriate (or in 3 or four rows); subsessile, or pedicellate. Pedicel apices oblique, or truncate, or discoid, or cupuliform. Spikelets consistently in 'long-and-short' combinations, or not in distinct 'long-and-short' combinations; when in long/short pairs unequally pedicellate in each combination. Pedicels of the 'pedicellate' spikelets free of the rachis. The 'shorter' spikelets hermaphrodite. The 'longer' spikelets hermaphrodite.

Female-fertile spikelets. *Spikelets* (1.2–)1.5–4.2(–4.5) mm long; suborbicular, or elliptic, or lanceolate, or ovate, or obovate; *abaxial*; *compressed dorsiventrally*; *plano-convex*; falling with the glumes. Rachilla terminated by a female-fertile floret. *Hairy callus absent*.

Glumes present (usually), or absent (in Section *Anachyris*); when present, one per spikelet (in species with an 'andropogonoid' spikelet arrangement), or two; *very unequal*; (the upper) long relative to the adjacent lemmas; free; dorsiventral to the rachis; awnless; very dissimilar (the G_1 usually much reduced). *Lower glume 0–1 nerved*. Upper glume 3–6 nerved. *Spikelets* with incomplete florets. The incomplete florets proximal to the female-fertile florets. *Spikelets with proximal incomplete florets*. **The proximal incomplete florets** 1; *epaleate*; sterile. The proximal lemmas awnless; 3–5 nerved (usually), or 2 nerved (*P. ceresia*); more or less equalling the female-fertile lemmas; less firm than the female-fertile lemmas to similar in texture to the female-fertile lemmas; not becoming indurated.

Female-fertile florets 1. **Lemmas** apiculate; similar in texture to the glumes to decidedly firmer than the glumes (papery to crustaceous); smooth to striate; becoming indurated to not becoming indurated; yellow in fruit, or brown in fruit; entire; *blunt*; awnless; hairless; non-carinate; having the margins tucked in onto the palea; with a clear germination flap; 3–5 nerved. *Palea* present; relatively long; entire; awnless, without apical setae; textured like the lemma (usually glossy); indurated, or not indurated; 2-nerved. *Lodicules* present; 2; free; fleshy; glabrous. *Stamens* 3. Anthers not penicillate. *Ovary* glabrous. Styles free to their bases. Stigmas 2; red pigmented.

Fruit, embryo and seedling. Fruit small; compressed dorsiventrally. Hilum short. Embryo large, or small (rarely). Endosperm hard; without lipid; containing only simple starch grains. Embryo without an epiblast; with a scutellar tail; with an elongated mesocotyl internode. Embryonic leaf margins overlapping.

Seedling with a long mesocotyl. First seedling leaf with a well-developed lamina. The lamina broad; curved.

Abaxial leaf blade epidermis. *Costal/intercostal zonation* conspicuous. *Papillae* absent. *Long-cells* similar in shape costally and intercostally; of similar wall thickness costally and intercostally. Mid-intercostal long-cells rectangular; having markedly sinuous walls. *Microhairs* present; elongated; ostensibly one-celled (the basal cell sometimes completely embedded), or clearly two-celled; panicoid-type; (28–)34–45(–46.5) microns long; 5.7–6.3 microns wide at the septum. Microhair total length/width at septum 6.2–6.4. Microhair apical cells 12–30 microns long. Microhair apical cell/total length ratio when applicable 0.49–0.62. *Stomata* common; (30–)31–36(–42) microns long. Subsidiaries low dome-shaped and triangular, or triangular. Guard-cells overlapping to flush with the interstomatals. *Intercostal short-cells* common, or absent or very rare; in cork/silica-cell pairs (usually), or not paired (a few solitaries); silicified (when paired). Intercostal silica bodies imperfectly developed; when present, cross-shaped, or cross-shaped and vertically elongated-nodular. *Costal short-cells* conspicuously in long rows. Costal silica bodies present throughout the costal zones, or confined to the central file(s) of the costal zones to present in alternate cell files of the costal zones; 'panicoid-type'; cross shaped to dumb-bell shaped, or nodular; not sharp-pointed.

Transverse section of leaf blade, physiology. C_4; biochemical type NADP–ME (*P. notatum, P. dilatatum*); XyMS–. PCR sheath outlines uneven. PCR sheath extensions absent. PCR cell chloroplasts with reduced grana; centrifugal/peripheral. *Mesophyll* with radiate chlorenchyma, or with non-radiate chlorenchyma. *Leaf blade* adaxially flat. *Midrib* conspicuous; having a conventional arc of bundles; with colourless mesophyll adaxially, or without colourless mesophyll adaxially. Bulliforms present in discrete, regular adaxial groups, or not in discrete, regular adaxial groups (sometimes irregularly grouped); often in simple fans. Many of the smallest vascular bundles unaccompanied by sclerenchyma. Combined sclerenchyma girders present; nowhere forming 'figures'. Sclerenchyma all associated with vascular bundles. The lamina margins with fibres (large).

Culm anatomy. *Culm internode bundles* in three or more rings.

Phytochemistry. Tissues of the culm bases with abundant starch. Leaves containing flavonoid sulphates (2 species), or without flavonoid sulphates (2 species).

Special diagnostic feature. Spikelets awnless, muticous.

Cytology. Chromosome base number, $x = 10$ and 12. $2n = 20, 40, 48, 50, 60, 63$, and 80. 2, 4, and 6 ploid. Mean diploid 2c DNA value 1.6 pg (2 species of unknown ploidy, 1.2 and 2.1). Nucleoli persistent.

Taxonomy. Panicoideae; Panicodae; Paniceae.

Ecology, geography, regional floristic distribution. 320 species. In warm regions. Commonly adventive. Mostly helophytic, or mesophytic, or xerophytic; mostly species of open habitats; halophytic (a few, e.g. *P. distichum*), or glycophytic. In diverse habitats — savanna, damp places, forest margins, weedy ground, coastal sands (including useful sandbinders), coastal and inland saltmarshes.

Holarctic, Paleotropical, Neotropical, Cape, Australian, and Antarctic. Boreal, Tethyan, and Madrean. African, Madagascan, Indomalesian, Polynesian, and Neocaledonian. Euro-Siberian, Eastern Asian, and Atlantic North American. Macaronesian, Mediterranean, and Irano-Turanian. Saharo-Sindian, Sudano-Angolan, West African Rainforest, and Namib-Karoo. Indian, Indo-Chinese, Malesian, and Papuan. Polynesian and Fijian. Caribbean, Venezuela and Surinam, Amazon, Central Brazilian, Pampas, and Andean. North and East Australian. New Zealand and Patagonian. European. Canadian-Appalachian, Southern Atlantic North American, and Central Grasslands. Sahelo-Sudanian, Somalo-Ethiopian, South Tropical African, and Kalaharian. Tropical North and East Australian and Temperate and South-Eastern Australian.

Rusts and smuts. Rusts — *Physopella* and *Puccinia*. Taxonomically wide-ranging species: *Puccinia chaetochloae, Puccinia dolosa, Puccinia levis, Puccinia substriata, Puccinia emaculata, Puccinia coronata*, and *Puccinia esclavensis*. Smuts from Tilletiaceae and from Ustilaginaceae. Tilletiaceae — *Tilletia*. Ustilaginaceae — *Sorosporium, Sphacelotheca, Tolyposporium*, and *Ustilago*.

Economic importance. Significant weed species: *P. ciliatifolium, P. conjugatum, P. dilatatum, P. fimbriatum, P. fluitans, P. laeve, P. lividum, P. longifolium, P. notatum, P. paspaloides, P. plicatulum, P. scrobiculatum, P. thunbergii, P. urvillei, P. vaginatum, P. virgatum*. Cultivated fodder: *P. dilatatum* (Dallis), *P. notatum* (Bahia), *P. plicatulum*. Important native pasture species: several, e.g. *P. auriculatum, P. glumaceum, P. notatum, P. paniculatum, P. scrobiculatum*. Grain crop species: *P. scrobiculatum* (Kodo — India).

References, etc. Leaf anatomical: Metcalfe 1960; this project.

Pennisetum Rich.

Amphochaeta Anderss., *Catatherophora* Steud., *Eriochaeta* Fig. & De Not, *Gymnotrix* P. Beauv., *Loydia* Delile, *Macrochaeta* Steud., *Penicillaria* Willd., *Pentastachya* Steud., *Sericura* Hassk.

Excluding *Beckeropsis, Pseudochaetochloa*

Habit, vegetative morphology. Annual (rarely), or perennial; stoloniferous, or caespitose, or decumbent. *Culms* 15–800 cm high; herbaceous; branched above, or unbranched above. Culm internodes solid, or hollow. Young shoots extravaginal, or intravaginal, or extravaginal and intravaginal. *Leaves* mostly basal, or not basally aggregated; auriculate, or non-auriculate. Leaf blades broad, or narrow; 3–35(–50) mm wide (and up to 1 m or more long in some species); without cross venation; disarticulating from the sheaths, or persistent; rolled in bud, or once-folded in bud. *Ligule a fringed membrane to a fringe of hairs*. *Contra-ligule* present (of hairs), or absent.

Reproductive organization. *Plants bisexual, with bisexual spikelets*; with hermaphrodite florets. The spikelets of sexually distinct forms on the same plant, or all alike in sexuality; hermaphrodite, or hermaphrodite and male-only (peripheral spikelets of the glomerules may be male-only); overtly heteromorphic (section Heterecetachya, where the outer, male spikelets are laterally compressed and keeled, with larger upper glume and lower lemma), or homomorphic (mostly). Plants outbreeding; with hidden cleistogenes (e.g. *P. clandestinum*, which lacks 'normal' inflorescences), or without hidden cleistogenes. The hidden cleistogenes when present, in the leaf sheaths. Apomictic, or reproducing sexually.

Inflorescence. *Inflorescence a false spike, with spikelets on contracted axes, or paniculate (the spikelets fascicled in false spikes, in small groups or apparently solitary, but always surrounded at their bases by reduced-branch bristles)*; contracted (into false spikes); espatheate; not comprising 'partial inflorescences' and foliar organs. *Spikelet-bearing axes* very much reduced (to one or few spikelets plus bristles); disarticulating (but the main axis persistent); falling entire (the false spikes or spikelet-plus-bristle clusters falling). **Spikelets** *with 'involucres' of 'bristles' (these sessile to shortly stalked, the bristles relatively slender, basally free or scarcely united, by contrast with* **Cenchrus***). The 'bristles' relatively slender, not spiny*; deciduous with the spikelets. Spikelets secund (when the inflorescence is branched), or not secund; subsessile, or pedicellate.

Female-fertile spikelets. *Spikelets compressed dorsiventrally (to subterete)*; falling with the glumes, or disarticulating above the glumes and falling with the glumes, or not disarticulating (in cultivated forms); not disarticulating between the florets, or disarticulating between the florets (i.e the upper floret sometimes readily disarticulating from the rest of the spikelet). Rachilla terminated by a female-fertile floret. Hairy callus absent.

Glumes two; very unequal (G_1 often minute or vestigial); shorter than the adjacent lemmas, or long relative to the adjacent lemmas (G_2 very short to as long as the spikelet); free; awnless; very dissimilar, or similar (hyaline or membranous). Lower glume 0–5 nerved. Upper glume 0–11 nerved. *Spikelets* with incomplete florets. The incomplete florets proximal to the female-fertile florets. ***Spikelets with proximal incomplete florets.*** *The proximal incomplete florets* 1; paleate, or epaleate. Palea of the proximal incomplete florets when present, fully developed to reduced. The proximal incomplete florets male, or sterile. The proximal lemmas awnless; 3–9(–15) nerved; exceeded by the female-fertile lemmas to more or less equalling the female-fertile lemmas (mostly), or decidedly exceeding the female-fertile lemmas (section Brevivalvula); less firm than the female-fertile lemmas to similar in texture to the female-fertile lemmas (membranous); not becoming indurated.

Female-fertile florets 1. ***Lemmas*** similar in texture to the glumes to decidedly firmer than the glumes; ***smooth, or striate***; not becoming indurated (membranous to chartaceous or subleathery, the texture hardened or unchanged at maturity); white in fruit, or yellow in fruit; entire; pointed, or blunt; awnless, or mucronate; hairy (near the margins), or hairless (glabrous); non-carinate; having the margins lying flat and exposed on the palea; with a clear germination flap; 5–7 nerved. *Palea* present; relatively long; entire; awnless, without apical setae; textured like the lemma (apart from the thinner margins); not indurated; 2-nerved. *Lodicules* present (then often very minute), or absent; when present, 2; glabrous. *Stamens* 3. Anthers penicillate (conspicuously so in section Penicillaria), or not penicillate. *Ovary* glabrous. Styles fused, or free to their bases. Stigmas 2; white.

Fruit, embryo and seedling. Fruit small; compressed dorsiventrally. Hilum short. Embryo large; waisted. Endosperm hard; without lipid; containing only simple starch grains. Embryo without an epiblast; with a scutellar tail; with an elongated mesocotyl internode. Embryonic leaf margins overlapping.

Seedling with a long mesocotyl. First seedling leaf with a well-developed lamina. The lamina broad; erect, or curved; 21–30 veined.

Abaxial leaf blade epidermis. *Costal/intercostal zonation* conspicuous. *Papillae* absent. *Long-cells* similar in shape costally and intercostally; of similar wall thickness costally and intercostally. Mid-intercostal long-cells rectangular; having markedly sinuous walls. *Microhairs* present; panicoid-type; (30–)33–70(–102) microns long; (5.4–)6–6.9(–7.2) microns wide at the septum. Microhair total length/width at septum 4.5–6.1. Microhair apical cells (18.6–)21–46(–54) microns long. Microhair apical cell/total length ratio 0.53–0.73. *Stomata* common; 24–25.5–27 microns long. Subsidiaries triangular, or dome-shaped and triangular. Guard-cells overlapping to flush with the interstomatals. *Intercostal short-cells* common; in cork/silica-cell pairs; silicified. Intercostal silica bodies tall-and-narrow, or vertically elongated-nodular. *Costal short-cells* predominantly paired, or neither distinctly grouped into long rows nor predominantly paired. Costal silica bodies 'panicoid-type'; cross shaped, or butterfly shaped, or dumb-bell shaped, or nodular; not sharp-pointed.

Transverse section of leaf blade, physiology. C_4; biochemical type NADP–ME (2 species); XyMS–. PCR sheath outlines uneven. PCR sheath extensions absent. PCR cell chloroplasts with reduced grana; centrifugal/peripheral. *Mesophyll* with radiate chlorenchyma; without adaxial palisade. *Leaf blade* with distinct, prominent adaxial ribs, or 'nodular' in section, or adaxially flat; with the ribs very irregular in sizes. *Midrib* conspicuous; with one bundle only, or having a conventional arc of bundles; with colourless mesophyll adaxially (this sometimes mirrored by adaxial colourless tissue above the lateral vacular bundles), or without colourless mesophyll adaxially (rarely). Bulliforms present in discrete, regular adaxial groups; in simple fans, or in simple fans and associated with colourless mesophyll cells to form deeply-penetrating fans. Many of the smallest vascular bundles unaccompanied by sclerenchyma. Combined sclerenchyma girders present, or absent; nowhere forming 'figures'. Sclerenchyma all associated with vascular bundles.

Culm anatomy. *Culm internode bundles* scattered.

Phytochemistry. Leaves containing flavonoid sulphates (some material only of 1 species), or without flavonoid sulphates (2 species).

Cytology. Chromosome base number, $x = 9$. $2n = 14, 18, 22, 34, 35, 36, 45, 52$, and 54, or 32–54. Haploid nuclear DNA content 0.6–2.5 pg (2 species, $2x$ and $4x$).

Taxonomy. Panicoideae; Panicodae; Paniceae.

Ecology, geography, regional floristic distribution. About 80 species. In warm regions. Commonly adventive. Helophytic, mesophytic, and xerophytic; shade species and species of open habitats. Savanna, woodland, weedy ground.

Holarctic, Paleotropical, Neotropical, Cape, and Australian. Boreal and Tethyan. African, Madagascan, Indomalesian, Polynesian, and Neocaledonian. Euro-Siberian and Eastern Asian. Mediterranean and Irano-Turanian. Saharo-Sindian, Sudano-Angolan, West African Rainforest, and Namib-Karoo. Indian, Indo-Chinese, Malesian, and Papuan. Polynesian. Caribbean, Venezuela and Surinam, Amazon, Central Brazilian, Pampas, and Andean. North and East Australian. European. Sahelo-Sudanian, Somalo-Ethiopian, South Tropical African, and Kalaharian. Tropical North and East Australian and Temperate and South-Eastern Australian.

Rusts and smuts. Rusts — *Phakopsora* and *Puccinia*. Taxonomically wide-ranging species: *Puccinia chaetochloae*, *Puccinia stenotaphri*, *Puccinia levis*, *Puccinia substriata*, '*Uromyces*' *setariae-italicae*, and *Puccinia esclavensis*. Smuts from Tilletiaceae and from Ustilaginaceae. Tilletiaceae — *Tilletia*. Ustilaginaceae — *Sorosporium*, *Sphacelotheca*, *Tolyposporium*, and *Ustilago*.

Economic importance. Significant weed species: *P. alopecuroides*, *P. americanum*, *P. clandestinum*, *P. pedicellatum*, *P. polystachyon*, *P. purpureum*, *P. setaceum*, *P. sieberianum* (in *P. glaucum*), *P. villosum*. Cultivated fodder: *P. clandestinum* (Kikuyu), *P. purpureum* (Elephant). Important native pasture species: *P. donsonii*, *P. massaicum*, *P. trachyphyllum* etc. Grain crop species: *P. glaucum* (Pearl Millet — the most drought tolerant tropical cereal). Lawns and/or playing fields: *P. clandestinum*.

References, etc. Leaf anatomical: Metcalfe 1960; this project.

Special comments. Some of the species closely approach *Cenchrus*.

Pentameris P. Beauv.

Habit, vegetative morphology. Perennial; caespitose. *Culms* 25–200 cm high; woody and persistent, or herbaceous (from a woody or suffrutescent base); branched above, or unbranched above. Culm nodes glabrous. Culm internodes hollow. Plants unarmed. Young shoots intravaginal. *Leaves* mostly basal, or not basally aggregated; auriculate (*P. thuarii*), or non-auriculate (but hairy in the auricle positions). **Leaf blades** linear to linear-lanceolate; narrow (rigid or wiry, usually long and filiform, sometimes curled or squarrose); 1–4 mm wide; *not pseudopetiolate*; without cross venation; disarticulating from the sheaths, or persistent. Ligule a fringe of hairs; 1.5–2.5 mm long. *Contraligule* absent.

Reproductive organization. Plants bisexual, with bisexual spikelets; with hermaphrodite florets.

Inflorescence. Inflorescence few spikeleted to many spikeleted; paniculate; open, or contracted (sometimes scanty); with capillary branchlets, or without capillary branchlets; non-digitate (the branching sometimes trichotomous); espatheate; not comprising 'partial inflorescences' and foliar organs. Spikelet-bearing axes persistent. *Spikelets* solitary; not secund; *pedicellate (the pedicels usually long-capillary, expanded under the spikelets)*.

Female-fertile spikelets. Spikelets 13–25(–30) mm long; compressed laterally; disarticulating above the glumes; disarticulating between the florets. Rachilla prolonged beyond the uppermost female-fertile floret (as a small extension). Hairy callus present. *Callus* short; blunt.

Glumes two; more or less equal; exceeding the spikelets; long relative to the adjacent lemmas (much exceeding them); hairy (puberulous, in *P. dregeana*), or hairless; other than in *P. dregeana*, glabrous; pointed (acute or acuminate); awned (setaceously acuminate), or awnless; carinate; similar (ovate-lanceolate, thin). Lower glume much exceeding the lowest lemma; 1 nerved, or 3 nerved. Upper glume 1 nerved, or 3 nerved. *Spikelets* occasionally with incomplete florets, or with female-fertile florets only (usually). The incomplete florets when present, distal to the female-fertile florets. *The distal incomplete florets* merely underdeveloped.

Female-fertile florets 2. Lemmas similar in texture to the glumes to decidedly firmer than the glumes; not becoming indurated (membranous); incised; 2 lobed; deeply cleft; awned. *Awns* 3; *median and lateral (the lateral lemma lobes each with a 1–7 mm bristle* *from the inner side, more or less adnate below)*; the median different in form from the laterals; from a sinus; geniculate (near the middle); hairless (scabrid); much longer than the body of the lemma; entered by several veins (3). The lateral awns shorter than the median (being loosely twisted or straight bristles, from the inner margins of the lobes). Lemmas hairy. *The hairs* not in tufts; *not in transverse rows (in longitudinal rows between the veins)*. Lemmas non-carinate; having the margins lying flat and exposed on the palea; without a germination flap; 7 nerved, or 9 nerved (-11 nerved); with the nerves non-confluent. *Palea* present (hairy); relatively long; apically notched, or deeply bifid; awnless, without apical setae; textured like the lemma (membranous); 2-nerved; 2-keeled. *Lodicules* present; 2; joined (at the base), or free; fleshy; ciliate, or glabrous. *Stamens* 3. Anthers 4.5–5 mm long; not penicillate; without an apically prolonged connective. **Ovary hairy (with a deciduous apical tomentum of branched hairs)**. Styles free to their bases (short). Stigmas 2; white.

Fruit, embryo and seedling. *Fruit* free from both lemma and palea (but enclosed by slightly hardened lemma and palea); small (3 mm long); ellipsoid, or subglobose. Hilum long-linear. *Pericarp thick and hard (crustaceous)*; free. Embryo small.

Abaxial leaf blade epidermis. *Costal/intercostal zonation* lacking (or the costal long-cells narrower). *Papillae* absent. *Long-cells* similar in shape costally and intercostally; of similar wall thickness costally and intercostally (walls heavily thickened and pitted). Mid-intercostal long-cells rectangular, or rectangular to fusiform (in *P. longiglumis*); having markedly sinuous walls, or having straight or only gently undulating walls (in *P. longiglumis*). *Microhairs* absent. *Stomata* absent or very rare. *Intercostal short-cells* common; in cork/silica-cell pairs; silicified. Intercostal silica bodies rounded. *Costal short-cells* predominantly paired. Costal silica bodies rounded, or tall-and-narrow (a few, in *P. longiglumis*); not sharp-pointed.

Transverse section of leaf blade, physiology. C_3; XyMS+. *Mesophyll* with non-radiate chlorenchyma; without adaxial palisade. *Leaf blade* 'nodular' in section; with the ribs very irregular in sizes. *Midrib* not readily distinguishable (except via its position); with one bundle only. Bulliforms present in discrete, regular adaxial groups to not in discrete, regular adaxial groups (there being small, more or less distinct groups in the bases of the deep furrows); more or less in simple fans. All the vascular bundles accompanied by sclerenchyma. Combined sclerenchyma girders present; forming 'figures' (all the main bundles with I's or 'anchors'). Sclerenchyma not all bundle-associated. The 'extra' sclerenchyma in a continuous abaxial layer.

Taxonomy. Arundinoideae; Danthonieae.

Ecology, geography, regional floristic distribution. 5 species. South Africa. Mesophytic; species of open habitats; glycophytic.

Paleotropical and Cape. African. Namib-Karoo.

References, etc. Leaf anatomical: Ellis 1985.

Pentapogon R.Br.

Habit, vegetative morphology. Annual, or perennial; caespitose. *Culms* 20–60 cm high; herbaceous. Culm nodes hairy. Culm internodes hollow. Leaves auriculate, or non-auriculate. Leaf blades narrow; not setaceous (inrolled); without cross venation; persistent. *Ligule an unfringed membrane*; truncate, or not truncate; 0.5–2 mm long.

Reproductive organization. Plants bisexual, with bisexual spikelets; with hermaphrodite florets.

Inflorescence. Inflorescence paniculate; open (the branches scabrid); espatheate; not comprising 'partial inflorescences' and foliar organs. Spikelet-bearing axes persistent. Spikelets not secund; pedicellate.

Female-fertile spikelets. *Spikelets 5–15 mm long*; compressed laterally; disarticulating above the glumes. *Rachilla terminated by a female-fertile floret*. Hairy callus present.

Glumes two; very unequal to more or less equal; (the longer) exceeding the spikelets; (the longer) long relative to the adjacent lemmas (exceeding them); hairless (scabrid, especially on the keels); pointed; shortly awned, or awnless; carinate; similar (membranous, acuminate). Lower glume 1–3 nerved. Upper glume 3–5 nerved. *Spikelets* with female-fertile florets only.

Female-fertile florets 1. **Lemmas** decidedly firmer than the glumes (leathery); not becoming indurated; *incised*; 4 lobed; deeply cleft (deeply incised into 4 lanceolate, awn-tipped lobes, with a long median awn); *awned*. *Awns 5*; median and lateral; the median different in form from the laterals; dorsal; from near the top (i.e., arising just behind the sinus); geniculate; hairless (scabrid); much longer than the body of the lemma; entered by one vein. The lateral awns shorter than the median (straight, slender). Lemmas hairless; glabrous; non-carinate; 5 nerved. **Palea** present; relatively long, or conspicuous but relatively short, or very reduced; tightly clasped by the lemma; 2-nerved; *2-keeled*. *Lodicules* present; 2; free; membranous; glabrous; not toothed. *Stamens* 3. Anthers not penicillate. *Ovary* glabrous. Styles free to their bases. Stigmas 2.

Fruit, embryo and seedling. Fruit small; not noticeably compressed. Hilum short. Embryo small. Endosperm liquid in the mature fruit (Clayton and Renvoize 1986).

Abaxial leaf blade epidermis. *Costal/intercostal zonation* conspicuous. *Papillae* absent. *Long-cells* similar in shape costally and intercostally (but the costals much smaller); of similar wall thickness costally and intercostally. Mid-intercostal long-cells rectangular to fusiform; having straight or only gently undulating walls. *Microhairs* absent. *Stomata* absent or very rare. Guard-cells overlapped by the interstomatals. *Inter-costal short-cells* absent or very rare. *Costal short-cells* conspicuously in long rows, or neither distinctly grouped into long rows nor predominantly paired. Costal silica bodies horizontally-elongated crenate/sinuous and horizontally-elongated smooth, or rounded (a few); not sharp-pointed.

Transverse section of leaf blade, physiology. C_3; XyMS+. *Mesophyll* with non-radiate chlorenchyma; without adaxial palisade. *Leaf blade* 'nodular' in section; with the ribs more or less constant in size. *Midrib* not readily distinguishable; with one bundle only. Bulliforms present in discrete, regular adaxial groups; in simple fans. All the vascular bundles accompanied by sclerenchyma. Combined sclerenchyma girders present, or absent; nowhere forming 'figures'. Sclerenchyma all associated with vascular bundles.

Taxonomy. Pooideae; Poodae; Aveneae.

Ecology, geography, regional floristic distribution. 1 species. Australia, Tasmania. Mesophytic, or xerophytic; glycophytic. Open woodland.

Australian. North and East Australian. Temperate and South-Eastern Australian.

References, etc. Leaf anatomical: Metcalfe 1960 and this project.

Pentarrhaphis Kunth

Polyschistis Presl, *Strombodurus* Steud.

Habit, vegetative morphology. Perennial; caespitose. **Culms *30–60 cm high*;** herbaceous; unbranched above (slender). Culm nodes glabrous. Plants unarmed. Young shoots intravaginal. *Leaves* mostly basal; non-auriculate. Leaf blades linear; narrow; to 2 mm wide; setaceous (at the tips); without cross venation; persistent. **Ligule *a fringe of hairs*.** *Contra-ligule* absent.

Reproductive organization. Plants bisexual, with bisexual spikelets; with hermaphrodite florets. *The spikelets* of sexually distinct forms on the same plant, or all alike in sexuality (*P. polymorpha*); *hermaphrodite, or hermaphrodite and sterile (2-spikeleted racemelets, P. polymorpha having both well developed, the others with one well developed and the other reduced to its awnlike glumes)*; all in heterogamous combinations.

Inflorescence. *Inflorescence a false spike, with spikelets on contracted axes (the clusters cuneate, bristly). Inflorescence axes not ending in spikelets (with a forked rachis prolongation).* Inflorescence espatheate; not comprising 'partial inflorescences' and foliar organs. **Spikelet-bearing axes *very much reduced (to the clusters, which in two species are reduced to a single fertile spikelet and a vestigial one reduced to its awns, the rachis terminating in a forked bristle (5 bristles in total); in P. polymorpha, both spikelets are well developed)*;** disarticulating; falling entire (i.e., the reduced glomerules falling). *Spikelets* unaccompanied by bractiform involucres, not associated with setiform vestigial branches, or with 'involucres' of 'bristles'. The 'bristles' when present, deciduous with the spikelets. Spikelets not secund; shortly pedicellate.

Female-fertile spikelets. Spikelets more or less morphologically 'conventional' (there is scope to argue that the structure regarded here as the 'fertile spikelet' is really two spikelets, but this would necessitate arguing that the 'upper one' has no glume); about 4 mm long; *compressed dorsiventrally*; falling with the glumes (and with the associated rudiments); not disarticulating between the florets. Rachilla prolonged beyond the uppermost female-fertile floret; rachilla hairy (beneath the sterile floret, glabrous above it). The rachilla extension with incomplete florets.

Glumes present; two (the G_1 represented by one of the bristles, hard to recognise as such); shorter than the adjacent lemmas; (the upper) hairy (having long white hairs over the lower half); apically two-lobed; awned (the lower reduced to an awn, the upper with a 1.5 mm awn from a sinus); (the upper) non-carinate (rounded, hyaline); very dissimilar (the lower a hirsute bristle, the upper acicular to lanceolate). Upper glume 1 nerved. *Spikelets* with incomplete florets. The incomplete florets distal to the female-fertile florets. The distal incomplete florets 1 (male or sterile).

Female-fertile florets 1. **Lemmas** membranous or cartilaginous; not becoming indurated; incised; 4 lobed (two large lateral lobes, and the central awn flanked by two teeth); deeply cleft (three cleft, laterally to halfway and shallowly in association with the central awn); *awned*. Awns 3 (all recurved-spreading); median and lateral; the median similar in form to the laterals; from a sinus (from between the central teeth); non-geniculate; recurving; hairless (scabrid); about as long as the body of the lemma (all three awns of similar lengths). The lateral awns about equalling the median (from the inner edges of the outer lobes). Lemmas hairy; non-carinate (rounded); without a germination flap; 3 nerved. *Palea* present; relatively long; apically notched (rather deeply); awnless, without apical setae, or with apical setae (via excurrent nerves); not indurated (hyaline, glabrous); 2-nerved; 2-keeled. *Lodicules* present; 2; free; fleshy; glabrous; not or scarcely vascularized. *Stamens* 3; with free filaments (these short). Anthers long; not penicillate; without an apically prolonged connective. *Ovary* glabrous. Styles free to their bases. Stigmas 2; white.

Fruit, embryo and seedling. Fruit ellipsoid.

Abaxial leaf blade epidermis. *Costal/intercostal zonation* conspicuous. *Papillae* absent. *Long-cells* similar in shape costally and intercostally. Mid-intercostal long-cells rectangular; having markedly sinuous walls (and pitted). *Microhairs* present; elongated; clearly two-celled; panicoid-type to chloridoid-type (the apical cell thin-walled, often collapsed); (34.5–)36–39(–42) microns long; (6–)7.5–8.4 microns wide at the septum. Microhair total length/width at septum 4.3–5.8. Microhair apical cells (13.5–)14.4–18(–19.5) microns long. Microhair apical cell/total length ratio 0.39–0.43. *Stomata* common; (27–)28.5–30(–33) microns long. Subsidiaries dome-shaped and triangular (mostly dome-shaped). Guard-cells overlapping to flush with the interstomatals. *Intercostal short-cells* common; in cork/silica-cell pairs; silicified. Intercostal silica bodies crescentic and saddle shaped (small). *Costal short-cells* conspicuously in long rows. Costal silica bodies saddle shaped (larger towards the blade margins); not sharp-pointed.

Transverse section of leaf blade, physiology. C_4; XyMS+ (the sheath cells thick-walled). PCR sheath extensions absent. Mesophyll traversed by columns of colourless mesophyll cells (but not in the marginal flanges). *Leaf blade* adaxially flat (with expanded leaf margins, each bearing two large bundles and a small one on the outside). *Midrib* not readily distinguishable; with one bundle only. Bulliforms present in discrete, regular adaxial groups; associated with colourless mesophyll cells to form deeply-penetrating fans. All the vascular bundles accompanied by sclerenchyma. Combined sclerenchyma girders present (with the large marginal bundles, the rest mostly having only scanty adaxial and abaxial strands); forming 'figures' (the marginal bundles). Sclerenchyma all associated with vascular bundles.

Taxonomy. Chloridoideae; main chloridoid assemblage.

Ecology, geography, regional floristic distribution. 3 species. Mexico to Columbia. Species of open habitats. Dry scrub.

Neotropical. Caribbean and Andean.

References, etc. Leaf anatomical: this project.

Special comments. Fruit data wanting.

Pentaschistis (Nees) Spach

Achneria Benth., *Afrachneria* Sprague
Excluding *Poagrostis*

Habit, vegetative morphology. Perennial (usually), or annual (less commonly); usually caespitose. *Culms* 10–150 cm high; herbaceous; branched above, or unbranched above. Culm nodes glabrous. Culm internodes hollow. Plants unarmed; with multicellular glands, or without multicellular glands (present in various forms, sometimes sunken, in about half the species). Young shoots intravaginal. The shoots aromatic (or foetid), or not aromatic. *Leaves* mostly basal (usually), or not basally aggregated; non-auriculate. Leaf blades linear to lanceolate (or filiform, often with multicellular, stalked or saucer-shaped glands); narrow; 0.5–5 mm wide; setaceous, or not setaceous; rolled (usually), or flat; exhibiting multicellular glands abaxially (at the base of macrohairs). The abaxial glands intercostal. Leaf blades without cross venation; persistent. *Ligule a fringe of hairs*. *Contra-ligule* present (rarely, as a fringe of hairs), or absent.

Reproductive organization. Plants bisexual, with bisexual spikelets; with hermaphrodite florets; without hidden cleistogenes.

Inflorescence. Inflorescence few spikeleted, or many spikeleted; paniculate (the branches often with glands); open, or contracted (sometimes spicate); with capillary branchlets, or without capillary branchlets; espatheate; not comprising 'partial inflorescences' and foliar organs. Spikelet-bearing axes persistent. Spikelets not secund; pedicellate.

Female-fertile spikelets. Spikelets 1–19 mm long; compressed laterally; disarticulating above the glumes; disarticulating between the florets; with conventional internode spacings. Rachilla prolonged beyond the uppermost female-fertile floret (as a minute bristle, or with a rudimentary third floret), or terminated by a female-fertile floret. The rachilla extension (when present) with incomplete florets, or naked. Hairy callus present. *Callus* short; blunt.

Glumes two; more or less equal; about equalling the spikelets to exceeding the spikelets; *long relative to the adjacent lemmas (enclosing the spikelet)*; hairy, or hairless; glabrous; pointed; awnless; carinate; similar (narrow to lanceolate, green or scarious, rarely hyaline, shining, often with glands). Lower glume much exceeding the lowest lemma (usually), or about equalling the lowest lemma; 1 nerved (above the base). *Upper glume 1 nerved (above the base)*. *Spikelets* occasionally with incomplete florets, or with female-fertile florets only (usually). The incomplete florets when present, distal to the female-fertile florets. *The distal incomplete florets* merely underdeveloped.

Female-fertile florets 2. *Lemmas* similar in texture to the glumes (membranous); not becoming indurated; incised (bifid, rarely 3–4-fid); 2 lobed, or 4–5 lobed (rarely); not deeply cleft; *awnless, or mucronate, or awned (generally awned from the central sinus, and with a point or straight awn on each lobe)*. Awns when present, 1 (rarely), or 3 (usually), or 5 (rarely); median, or median and lateral (usually); the median different in form from the laterals (when laterals present); from a sinus; usually geniculate; hairless; much longer than the body of the lemma; entered by several veins (two or three). The lateral awns when present, shorter than the median (straight, inserted in the sinuses and partially fused to the lateral lemma lobes). Lemmas hairy, or hairless; non-carinate (dorsally rounded); without a germination flap; *5–7 nerved*; with the nerves non-confluent. *Palea* present; relatively long; tightly clasped by the lemma; apically notched; awnless, without apical setae; textured like the lemma; not indurated (thinly membranous); 2-nerved; 2-keeled. Palea keels wingless; glabrous. *Lodicules* present; 2; free; fleshy; ciliate, or glabrous. *Stamens* 3. Anthers not penicillate; without an apically prolonged connective. *Ovary* glabrous. Styles free to their bases. Stigmas 2; white.

Fruit, embryo and seedling. Fruit small; longitudinally grooved. *Hilum short (but linear-oblong)*. Pericarp free, or fused, or loosely adherent. Embryo large to small.

Abaxial leaf blade epidermis. *Costal/intercostal zonation* conspicuous. *Papillae* absent. *Long-cells* similar in shape costally and intercostally; of similar wall thickness costally and intercostally (but the costals slightly thicker walled). Mid-intercostal long-cells rectangular; having markedly sinuous walls. *Microhairs* present; elongated; clearly two-celled; panicoid-type. Microhair apical cell wall thinner than that of the basal cell and often collapsed. Microhairs 42–45(–54) microns long. Microhair basal cells 24–27 microns long. Microhairs 5.4–6–8.4 microns wide at the septum. Microhair total length/width at septum 5–10. Microhair apical cells (16.5–)19.5–21(–22.5) microns long. Microhair apical cell/total length ratio 0.39–0.49. *Stomata* common; 24–27–30 microns long. Subsidiaries dome-shaped. Guard-cells overlapping to flush with the interstomatals. *Intercostal short-cells* absent or very rare. Intercostal silica bodies absent. *Costal short-cells* conspicuously in long rows. Costal silica bodies present in alternate cell files of the costal zones; 'panicoid-type'; dumb-bell shaped; not sharp-pointed.

Transverse section of leaf blade, physiology. C_3; XyMS+. *Mesophyll* with non-radiate chlorenchyma; without adaxial palisade. *Leaf blade* with distinct, prominent ad-axial ribs; with the ribs more or less constant in size. *Midrib* conspicuous; with one bundle only. Bulliforms present in discrete, regular adaxial groups; in simple fans. All the vascular bundles accompanied by sclerenchyma. Combined sclerenchyma girders present; forming 'figures'. Sclerenchyma all associated with vascular bundles. The lamina margins with fibres (few cells).

Cytology. Chromosome base number, $x = 7$ and 13. $2n = 14, 26, 28, 42$, and 52.

Taxonomy. Arundinoideae; Danthonieae (?).

Ecology, geography, regional floristic distribution. About 65 species. Africa, Madagascar. Commonly adventive. Xerophytic, or mesophytic; species of open habitats; glycophytic.

Paleotropical and Cape. African and Madagascan. Sudano-Angolan, West African Rainforest, and Namib-Karoo. Sahelo-Sudanian, Somalo-Ethiopian, and South Tropical African.

Economic importance. Significant weed species: *P. thunbergii*. Important native pasture species: *P. borussica*.

References, etc. Leaf anatomical: mainly this project.

Pereilema J. & C. Presl

Habit, vegetative morphology. Annual; caespitose. *Culms* 30–60 cm high; herbaceous; unbranched above. Culm nodes glabrous. Plants unarmed. **Leaves** not basally aggregated; *auriculate*. Leaf blades narrow; 2–5 mm wide; flat; without abaxial multicellular glands; without cross venation; persistent. Ligule an unfringed membrane; 0.5 mm long. *Contra-ligule* absent.

Reproductive organization. Plants bisexual, with bisexual spikelets; with hermaphrodite florets. *The spikelets* of sexually distinct forms on the same plant; hermaphrodite and sterile, or hermaphrodite, male-only, and sterile; *overtly heteromorphic (in bunches, some of each group bisexual and the rest incomplete or 'reduced to glumes or bristles').*

Inflorescence. *Inflorescence of spicate main branches to a false spike, with spikelets on contracted axes (the spikelet clusters being subsessile on appressed spike-like laterals, which decrease in length towards the tip of the main axis — cf.* Sporobolus, *apart from the awns)*; contracted (plumose by the long lemma awns). Primary inflorescence branches 16–25 (or more). Inflorescence espatheate; not comprising 'partial inflorescences' and foliar organs. *Spikelet-bearing axes* very much reduced, or 'racemes'; disarticulating (*P. ciliatum*), or persistent; falling entire (the main branches falling, in *P. ciliatum*). Spikelets *with 'involucres' of 'bristles' (representing reduced spikelets — the 'involucre' to one side)*. The 'bristles' deciduous with the spikelets. Spikelets secund (the clusters one-sided).

Female-fertile spikelets. *Spikelets* morphologically 'conventional' (when the spikelet cluster is dissected out); 2 mm long (to 3 cm with the awns); compressed laterally; disarticulating above the glumes (when the inflorescence branches fall entire, the florets also disarticulate). Rachilla terminated by a female-fertile floret. Hairy callus present.

Glumes two; more or less equal; shorter than the spikelets; shorter than the adjacent lemmas; when not reduced to bristles, hairless (the keel scabrid); awned (the awns long, straight, scabrid); carinate; similar (hyaline, emarginate, or sometimes reduced to ciliate bristles in *P. ciliatum*). Lower glume 1 nerved. Upper glume 1 nerved. *Spikelets* with female-fertile florets only.

Female-fertile florets 1. Lemmas similar in texture to the glumes to decidedly firmer than the glumes (becoming hardened in the fruit); not deeply cleft; mucronate to awned. Awns 1; median; apical; non-geniculate; hairless (scabrid, slender); usually much longer than the body of the lemma (to 3 cm long); persistent. *Lemmas* hairless; scabrous; *carinate*; without a germination flap; 3 nerved. *Palea* present; relatively long; apically notched; awnless, without apical setae to with apical setae (the nerves slightly excurrent); not indurated (hyaline); 2-nerved; 2-keeled. *Lodicules* absent. *Stamens* 3. Anthers not penicillate; without an apically prolonged connective. *Ovary* glabrous. Styles free to their bases. Stigmas 2.

Fruit, embryo and seedling. Fruit ellipsoid. Pericarp fused. Embryo with an epiblast; with a scutellar tail; with an elongated mesocotyl internode. Embryonic leaf margins meeting.

Abaxial leaf blade epidermis. *Costal/intercostal zonation* conspicuous. *Papillae* present; intercostal. Intercostal papillae several per cell. Long-cells intercostal long-cells very thin-walled. Mid-intercostal long-cells having straight or only gently undulating walls. *Microhairs* present; elongated; clearly two-celled; chloridoid-type. Microhair apical cell wall of similar thickness/rigidity to that of the basal cell. Microhairs (24–)25.5–27(–30) microns long. Microhair basal cells 9–12 microns long. Microhairs 10.5–12 microns wide at the septum. Microhair total length/width at septum 2.1–2.6. Microhair apical cells 6–9 microns long. Microhair apical cell/total length ratio 0.22–0.35. *Stomata* common; 19–24 microns long. Subsidiaries parallel-sided and dome-shaped. Guard-cells overlapping to flush with the interstomatals. Intercostal silica bodies absent. Prickles abundant. Costal silica bodies present in alternate cell files of the costal zones; 'panicoid-type'; not sharp-pointed.

Transverse section of leaf blade, physiology. C_4; XyMS+. PCR sheaths of the primary vascular bundles interrupted; interrupted abaxially only. PCR sheath extensions absent. Mesophyll traversed by columns of colourless mesophyll cells. *Leaf blade* 'nodular' in section. *Midrib* conspicuous; having a conventional arc of bundles (the large median flanked by a small lateral on each side); with colourless mesophyll adaxially. Bulliforms present in discrete, regular adaxial groups (in addition to large 'midrib hinges'); associated with colourless mesophyll cells to form deeply-penetrating fans (these incorporated in the wide colourless girders). All the vascular bundles accompanied by sclerenchyma. Sclerenchyma all associated with vascular bundles.

Cytology. Chromosome base number, $x = 10$. $2n = 20$.

Taxonomy. Chloridoideae; main chloridoid assemblage.

Ecology, geography, regional floristic distribution. 3 species. Mexico to tropical South America. Species of open habitats. Hills and weedy places in savanna.

Neotropical. Caribbean and Andean.

Rusts and smuts. Rusts — *Puccinia*.

References, etc. Leaf anatomical: this project.

Special comments. Fruit data wanting.

Periballia Trin.

Molineria Parl., *Molineriella* Rouy
~ *Deschampsia*

Habit, vegetative morphology. Annual; caespitose (or the culms solitary). *Culms* 3–25 cm high; herbaceous. Culm nodes glabrous. Leaves non-auriculate. Leaf blades linear; narrow; 0.5–2 mm wide; setaceous, or not setaceous; flat, or rolled (convolute); without cross venation; persistent. *Ligule an unfringed membrane*.

Reproductive organization. Plants bisexual, with bisexual spikelets; with hermaphrodite florets.

Inflorescence. Inflorescence paniculate (the lower branches sterile in *P. involucrata*); open; with capillary branchlets; espatheate; not comprising 'partial inflorescences' and foliar organs. Spikelet-bearing axes persistent. Spikelets not secund; pedicellate.

Female-fertile spikelets. Spikelets 1.75–2 mm long; compressed laterally; disarticulating above the glumes; with distinctly elongated rachilla internodes between the florets. *Rachilla terminated by a female-fertile floret*; rachilla hairy, or hairless. Hairy callus present (the hairs 1/5–2/3 the lemma length).

Glumes two; more or less equal; shorter than the spikelets; *shorter than the adjacent lemmas*; free; awnless; similar (membranous). Lower glume 1–3 nerved. Upper glume 1–3 nerved. *Spikelets* with female-fertile florets only.

Female-fertile florets 2. *Lemmas lanceolate*; less firm than the glumes (hyaline), or similar in texture to the glumes; not becoming indurated; entire to incised; not deeply cleft (blunt, truncate or irregularly toothed); awnless, or awned. Awns if present, 1; median; dorsal; from near the top to from well down the back; non-geniculate; straight. Lemmas hairless; non-carinate; 3–7 nerved. *Palea* present; tightly clasped by the lemma; 2-nerved. *Lodicules* present; 2; free; membranous; glabrous; not toothed. *Stamens* 3. Anthers 0.8–1.2 mm long; not penicillate. *Ovary* glabrous. Styles free to their bases. Stigmas 2.

Fruit, embryo and seedling. **Fruit** slightly adhering to lemma and/or palea; small; narrowly ellipsoid; *longitudinally grooved (or flattened, on one face)*. Hilum short. Embryo small; not waisted. Endosperm liquid in the mature fruit; without lipid.

Abaxial leaf blade epidermis. *Costal/intercostal zonation* conspicuous. *Papillae* absent. *Long-cells* markedly different in shape costally and intercostally (the costals very narrow, parallel-sided); differing markedly in wall thickness costally and intercostally (the costals thicker-walled). Mid-intercostal long-cells fusiform; having straight or only gently undulating walls (the walls thin). *Microhairs* absent. *Stomata* common. Subsidiaries parallel-sided. Guard-cells overlapped by the interstomatals. *Intercostal short-cells* absent or very rare. *Costal short-cells* neither distinctly grouped into long rows nor predominantly paired. Costal silica bodies horizontally-elongated crenate/sinuous, or horizontally-elongated smooth; not sharp-pointed.

Transverse section of leaf blade, physiology. C_3; XyMS+.

Cytology. Chromosome base number, $x = 4$ and 7. $2n = 8$, 14, and 18. 2 ploid. Haploid nuclear DNA content 3.3 pg (1 species). Mean diploid 2c DNA value 6.6 pg (*P. involucrata*).

Taxonomy. Pooideae; Poodae; Aveneae.

Ecology, geography, regional floristic distribution. 3 species. Mediterranean. Commonly adventive. Xerophytic; species of open habitats. Dry sandy places.

Holarctic. Boreal and Tethyan. Euro-Siberian. Mediterranean. European.

References, etc. Leaf anatomical: this project.

Special comments. Anatomical data largely epidermal only.

Peridictyon O. Seberg, S. Frederiksen & C. Baden

Festucopsis sancta (Janka) Meld.

Habit, vegetative morphology. Perennial; loosely caespitose. *Culms* (37–)47–80(–110) cm high; unbranched above; 2–3 noded. Culm nodes glabrous. Young shoots intravaginal. Leaves auriculate (the auricles (0.1-)0.3–0.7(-0.8) mm long). Sheath margins joined (the innovation leaves), or free. *The basal sheaths disintegrating into a brownish mass of reticulate fibres*. *Leaf blades* linear; narrow; (1–)1.6–2.6(–3.1) mm wide; not setaceous (junciform); rolled; *disarticulating from the sheaths ('with an abscission layer at junction of sheath and blade')*. *Ligule an unfringed membrane*; truncate; 'extremely short'.

Reproductive organization. Plants bisexual, with bisexual spikelets; with hermaphrodite florets.

Inflorescence. *Inflorescence* rather few spikeleted ((6–)7–13(–15)); *a single spike ((5–)8–14(–17 cm long)*. Rachides flattened. Inflorescence espatheate. *Spikelet-bearing axes persistent (?)*. Spikelets solitary; not secund; distichous (?); sessile.

Female-fertile spikelets. Spikelets (11–)12–19(–21) mm long; green; compressed laterally, or not noticeably compressed (?); disarticulating above the glumes; disarticulating between the florets. Rachilla prolonged beyond the uppermost female-fertile floret. Hairy callus absent. *Callus* very short.

Glumes two; more or less equal; shorter than the spikelets; shorter than the adjacent lemmas; hairless; glabrous; pointed; *awned (the awn apical, 0.5–1.3 mm long)*; non-carinate; *similar (lanceolate, leathery)*. Lower glume 2–4 nerved. Upper glume 3–6 nerved. *Spikelets* presumably with incomplete florets. The incomplete florets distal to the female-fertile florets. *The distal incomplete florets* merely underdeveloped.

Female-fertile florets *5–10*. *Lemmas* entire (?); *awned*. Awns 1; median; apical (?); non-geniculate; much shorter than the body of the lemma (3–9 mm long). Lemmas hairless; glabrous; non-carinate. *Palea* present; relatively long (equalling to slightly shorter than the lemma); 2-nerved; 2-keeled. Palea keels hairy (ciliate above). *Lodicules* present; 2; presumably membranous; ciliate. *Stamens* 3. Anthers 4.2–6(–6.3) mm long. *Ovary* hairy.

Fruit, embryo and seedling. *Fruit* adhering to lemma and/or palea (to the lemma); medium sized (4.5–5.5 mm long); with hairs confined to a terminal tuft. Hilum long-linear. Endosperm containing only simple starch grains.

Abaxial leaf blade epidermis. *Costal/intercostal zonation* lacking. *Papillae* absent. *Long-cells* similar in shape costally and intercostally (indistinguishable); of similar wall thickness costally and intercostally (thick walled). Mid-intercostal long-cells rectangular; having markedly sinuous walls (the sinuosity quite coarse, regular, associated with conspicuous pitting, and pits also conspicuous in the outside walls). *Microhairs* absent. *Stomata* absent or very rare. *Intercostal short-cells* absent or very rare; in cork/silica-cell pairs; silicified. Intercostal silica bodies rounded and crescentic. No macrohairs or prickles seen. *Crown cells* absent. *Costal short-cells* predominantly paired. Costal silica bodies present and well developed; rounded and crescentic.

Transverse section of leaf blade, physiology. C_3; XyMS+. Mesophyll without adaxial palisade. *Leaf blade* with distinct, prominent adaxial ribs; with the ribs more or less constant in size (broadly round topped). *Midrib* conspicuous (by the rather pointed abaxial keel, and its larger abaxial girder); with one bundle only. *The lamina* symmetrical on either side of the midrib. Bulliforms present in discrete, regular adaxial groups; in simple fans (these small-celled, in the furrows). All the vascular bundles accompanied by sclerenchyma. Combined sclerenchyma girders present (with the primaries, the minor bundles with small adaxial strands and abaxial strands or girders); forming 'figures' (the primaries with massive I's). Sclerenchyma not all bundle-associated. The 'extra' sclerenchyma in a continuous abaxial layer (linking the abaxial sclerenchyma of the bundles).

Cytology. Chromosome base number, $x = 7$. $2n = 14$. 2 ploid. Mean diploid 2c DNA value unknown, but chromosomes 'large'.

Taxonomy. Pooideae; Triticodae; Triticeae.

Ecology, geography, regional floristic distribution. 1 species. Greece and Bulgaria. Species of open habitats; glycophytic. Calcareous, stony or rocky slopes.

Holarctic. Tethyan. Mediterranean. European.

References, etc. Morphological/taxonomic: Seberg *et al.* 1991. Leaf anatomical: this project.

Special comments. Morphological description scarcely adequate — e.g. spikelet orientation unknown. Fruit data wanting.

Perotis Aiton

Xystidium Trin.

Habit, vegetative morphology. Annual, or perennial (rarely); caespitose. **Culms** 12–100 cm high; *herbaceous*. Culm nodes glabrous. Culm internodes solid. *Leaves* not basally aggregated; non-auriculate. Leaf blades narrow (but relatively broad at the base); cordate, or not cordate, not sagittate; flat or folded; without abaxial multicellular glands; without cross venation; persistent. Ligule an unfringed membrane to a fringed membrane.

Reproductive organization. *Plants bisexual, with bisexual spikelets*; with hermaphrodite florets.

Inflorescence. *Inflorescence a single spike, or a single raceme (a narrow 'bottlebrush', bearded by the long glume awns)*; espatheate; not comprising 'partial inflorescences' and foliar organs. Spikelet-bearing axes persistent. Spikelets solitary (often reflexing when mature); not secund; subsessile to pedicellate.

Female-fertile spikelets. *Spikelets* 1.2–5.5 mm long; *compressed laterally*; falling with the glumes. Rachilla terminated by a female-fertile floret. Hairy callus present, or absent.

Glumes two; more or less equal; long relative to the adjacent lemmas (considerably exceeding them); hairy (often), or hairless; awned; carinate; *similar (narrow, membranous to cartilaginous, tipped by long capillary awns)*. Lower glume 1 nerved. Upper glume 1 nerved. **Spikelets** with female-fertile florets only; *without proximal incomplete florets*.

Female-fertile florets *1*. Lemmas lanceolate; less firm than the glumes (hyaline); not becoming indurated; entire; pointed, or blunt; awnless; hairless; glabrous; carinate; 1 nerved. **Palea** present; conspicuous but relatively short (but almost equalling the lemma); not indurated (hyaline); *nerveless*. *Lodicules* present; 2; fleshy; glabrous. *Stamens* 3. Anthers not penicillate. *Ovary* glabrous. Styles fused. Stigmas 2.

Fruit, embryo and seedling. Fruit small to medium sized (almost as long as the glumes); longitudinally grooved; compressed dorsiventrally, or not noticeably compressed. Hilum short. Pericarp fused. Embryo large. Endosperm hard; without lipid; containing compound starch grains. Embryo with an epiblast; with a scutellar tail; with an elongated mesocotyl internode. Embryonic leaf margins meeting.

Abaxial leaf blade epidermis. *Costal/intercostal zonation* conspicuous. *Papillae* absent. Mid-intercostal long-cells having markedly sinuous walls. *Microhairs* present; elongated; clearly two-celled; chloridoid-type. Microhair apical cell wall of similar thickness/rigidity to that of the basal cell. Microhairs 10–18 microns long. Microhair basal cells 7–8 microns long. Microhairs (5.4–)6–8.5(–9) microns wide at the septum. Microhair total length/width at septum 2–3.3. Microhair apical cells 4–9 microns long. Microhair apical cell/total length ratio 0.38–0.6. *Stomata* common; 21–27 microns long. Subsidiaries low to high dome-shaped and triangular. *Intercostal short-cells* absent or very rare. Intercostal silica bodies absent. *Costal short-cells* conspicuously in long rows. Costal silica bodies present in alternate cell files of the costal zones; 'panicoid-type'; cross shaped, or dumb-bell shaped; not sharp-pointed.

Transverse section of leaf blade, physiology. C_4; XyMS+. PCR sheath outlines even. PCR sheaths of the primary vascular bundles interrupted; interrupted abaxially only. PCR sheath extensions absent. PCR cell chloroplasts centripetal. *Mesophyll* with radiate chlorenchyma; traversed by columns of colourless mesophyll cells. *Leaf blade* adaxially flat. *Midrib* not readily distinguishable; with one bundle only. Bulliforms present in discrete, regular adaxial groups; in simple fans and associated with colourless mesophyll cells to form deeply-penetrating fans (often linked with traversing colourless girders). Combined sclerenchyma girders present; forming 'figures'. Sclerenchyma all associated with vascular bundles. The lamina margins with fibres.

Culm anatomy. *Culm internode bundles* in one or two rings, or in three or more rings.

Cytology. Chromosome base number, $x = 10$. $2n = 36$, or 40.

Taxonomy. Chloridoideae; main chloridoid assemblage.

Ecology, geography, regional floristic distribution. 10 species. Africa, India, Ceylon, eastern Asia, Australia. Mesophytic, or xerophytic; species of open habitats; glycophytic. Savanna and grassland, often ruderal.

Paleotropical and Australian. African, Madagascan, and Indomalesian. Saharo-Sindian, Sudano-Angolan, West African Rainforest, and Namib-Karoo. Indian, Indo-Chinese, Malesian, and Papuan. North and East Australian. Sahelo-Sudanian, Somalo-Ethiopian, South Tropical African, and Kalaharian. Tropical North and East Australian.

Rusts and smuts. Rusts — *Puccinia*. Smuts from Tilletiaceae. Tilletiaceae — *Tilletia*.

References, etc. Leaf anatomical: Metcalfe 1960; this project.

Perrierbambus A. Camus

Habit, vegetative morphology. Bushy perennial. *Culms woody and persistent*; branched above. Primary branches/mid-culm node several. Rhizomes leptomorph. **Leaves** not basally aggregated; *with auricular setae*. Leaf blades pseudopetiolate; disarticulating from the sheaths.

Reproductive organization. Plants bisexual, with bisexual spikelets; with hermaphrodite florets.

Inflorescence. *Inflorescence* reduced to a single spikelet to few spikeleted, or many spikeleted (depending how the 'inflorescence' is delimited); with spikelets solitary or 2–3, terminating simple branches which are borne in fairly distant, dense semi-whorls; the slender individual branches each bearing many small-bladed leaves, the latter clustered towards the tip and almost involucrate around the spikelet(s); *espatheate (the inflorescence surrounded by involucral leaf blades, instead of spathiform sheaths)*; not comprising 'partial inflorescences' and foliar organs. *Spikelet-bearing axes* not racemose; persistent. Spikelets not secund; shortly pedicellate.

Female-fertile spikelets. *Spikelets* unconventional (or at least, hard to interpret in the inadequate material seen); compressed laterally to not noticeably compressed. Rachilla prolonged beyond the uppermost female-fertile floret (in all the few spikelets dissected); rachilla hairless. The rachilla extension with incomplete florets. Hairy callus absent.

Glumes two; very unequal; hairless; glabrous; pointed; awnless (apiculate); noncarinate (dorsally rounded); similar (lanceolate-ovate). Lower glume 9–11 nerved. Upper glume 13 nerved. *Spikelets* with incomplete florets (in the material seen). The incomplete florets distal to the female-fertile florets. The distal incomplete florets 1; *incomplete florets* merely underdeveloped; incomplete florets awnless.

Female-fertile florets 1. Lemmas lanceolate-ovate; similar in texture to the glumes to decidedly firmer than the glumes (firmly membranous); not becoming indurated; entire; awnless; hairless; glabrous; without a germination flap; about 13–15 nerved. Palea present; relatively long; entire; awnless, without apical setae; textured like the lemma; not indurated; several nerved; 2-keeled (but the keels closely apposed). Palea keels wingless. *Lodicules* present; 3; free; membranous; glabrous; heavily vascularized. *Stamens 5*, or 6 (one seemingly separate from the whorl of five, in some of the spikelets seen). Anthers not penicillate; without an apically prolonged connective. **Ovary** inconspicuously hairy (at the tip); seemingly *without a conspicuous apical appendage (but the apex hardening to form a shiny beak on the fruit)*. Styles fused. Stigmas 3.

Fruit, embryo and seedling. *Pericarp fleshy*; free.

Taxonomy. Bambusoideae; Bambusodae; Bambuseae.

Ecology, geography, regional floristic distribution. 2 species. Madagascar. Paleotropical. Madagascan.

Special comments. The spikelet morphology is evidently 'difficult', and this description reflects very tentative interpretation of depauperate and inadequate (type) material. Fruit data wanting. Anatomical data wanting.

Perulifera A. Camus

~ *Pseudechinolaena*

Habit, vegetative morphology. Annual. *Culms* 10–15 cm high; herbaceous; branched above. Culm nodes exposed; hairy. Plants unarmed. *Leaves* not basally aggregated; non-auriculate; without auricular setae. Sheaths slightly inflated, not keeled. Leaf blades lanceolate; broad; 3–5 mm wide (1.5–4 cm long); rounded at base; flat; more or less pseudopetiolate; without cross venation. *Ligule an unfringed membrane (ciliolate)*; truncate. *Contra-ligule* absent.

Reproductive organization. Plants bisexual, with bisexual spikelets; with hermaphrodite florets. *The spikelets* of sexually distinct forms on the same plant; hermaphrodite and sterile; *overtly heteromorphic (but the sterile spikelets so 'reduced' as to be confused with the glumes)*.

Inflorescence. Inflorescence of spicate main branches (a 3.5–6 cm raceme of spike-like racemes). Primary inflorescence branches 4–7. Inflorescence espatheate; not comprising 'partial inflorescences' and foliar organs. Spikelet-bearing axes persistent. *Spikelets* solitary (apparently), or paired ('really'); secund; biseriate; shortly pedicellate, or sessile to subsessile (the sterile member); imbricate; *consistently in 'long-and-short' combinations (but not obviously so until the sterile spikelet is identified as such). The 'shorter' spikelets sterile (reduced to a lanceolate, glabrous glume contiguous with the fertile spikelet)*. The 'longer' spikelets hermaphrodite.

Female-sterile spikelets. The sterile spikelet of each pair reduced to an elongate glume.

Female-fertile spikelets. *Spikelets* unconventional (because of the reduced sterile spikelet continguous with the G_1); 2–2.2 mm long; purple; abaxial; strongly compressed laterally; falling with the glumes. Rachilla terminated by a female-fertile floret.

Glumes two; relatively large; (the upper) long relative to the adjacent lemmas; dorsiventral to the rachis; *awned (the G_1 attenuate into a 2.5–4 mm awn)*; very dissimilar (both membranous, the G_1 smaller, narrower, attenuate-aristate and asperulous, the G_2 basally gibbous, verrucose and mucronate). Lower glume 3 nerved. **Upper glume** *distinctly saccate (swelled out 'like a wallet')*; 7 nerved; probably becoming more or less prickly, or not prickly (?). *Spikelets* with incomplete florets. The incomplete florets proximal to the female-fertile florets. *The proximal incomplete florets* 1; paleate. Palea of the proximal incomplete florets fully developed. The proximal incomplete florets male (with three stamens). The proximal lemmas carinate, subgibbous above; awnless (but rostrate); 3 nerved; decidedly exceeding the female-fertile lemmas; less firm than the female-fertile lemmas (membranous); not becoming indurated.

Female-fertile florets 1. Lemmas laterally compressed, acute; very thinly cartilaginous; smooth; not becoming indurated; entire; pointed; awnless; hairless (shiny); non-carinate (convex); having the margins lying flat and exposed on the palea; 3 nerved, or 5 nerved. *Palea* present; relatively long; somewhat thinner than the lemma (hyaline, less rigid); not indurated; 2-nerved; slightly 2-keeled. *Lodicules* present; 2; fleshy (lobed); glabrous. *Stamens* 3 (the anthers deeply sagittate basally and divaricate above, the thecae easily separating to simulate six stamens). Anthers about 1 mm long (relatively long, yellow, the filaments short and thick); not penicillate; without an apically prolonged connective. *Ovary* glabrous. Stigmas 2; red pigmented.

Fruit, embryo and seedling. *Fruit* free from both lemma and palea; small (0.9 mm long); elongated pyriform; not grooved; compressed laterally, or not noticeably compressed; glabrous. Hilum short. Embryo large; waisted. Endosperm hard.

Abaxial leaf blade epidermis. *Costal/intercostal zonation* conspicuous. *Papillae* absent. *Long-cells* markedly different in shape costally and intercostally (the costals much smaller and narrower). Intercostal zones with typical long-cells to exhibiting many atypical long-cells (many of them rather short, some very short). Mid-intercostal long-cells more or less isodiametric or somewhat irregular to rectangular; having markedly sinuous walls (coarsely sinuous). *Microhairs* present; elongated; clearly two-celled; panicoid-type. *Stomata* common (but confined to rows alongside the costae). Subsidiaries non-papillate; low to high dome-shaped. Guard-cells overlapping to flush with the interstomatals. *Intercostal short-cells* absent or very rare. With numerous long, slender, cushion-based macrohairs and a few small prickles intercostally. *Costal short-cells* mostly conspicuously in long rows. Costal silica bodies present and well developed; 'panicoid-type'; dumb-bell shaped and nodular.

Transverse section of leaf blade, physiology. C_3; XyMS+. Mesophyll not *Isachne*-type; without arm cells; without fusoids. *Midrib* conspicuous; with one bundle only.

Taxonomy. Panicoideae; Panicodae; Paniceae.

Ecology, geography, regional floristic distribution. 1 species. Madagascar. Paleotropical. Madagascan.

References, etc. Morphological/taxonomic: Camus 1927b.

Special comments. Fruit data wanting.

Petriella Zotov

~ *Ehrharta*

Habit, vegetative morphology. Perennial; rhizomatous. *Culms* 1–40 cm high; woody and persistent, or herbaceous; unbranched above (but branching near the base). Culm nodes glabrous. Plants unarmed. Young shoots intravaginal. **Leaves** not basally aggregated; *auriculate*. Leaf blades linear-lanceolate to ovate-lanceolate; narrow; 2–3 mm wide; flat, or folded; without cross venation; disarticulating from the sheaths (from the lower sheaths), or persistent. Ligule a fringed membrane; not truncate; to 0.5 mm long. *Contra-ligule* absent.

Reproductive organization. Plants bisexual, with bisexual spikelets; with hermaphrodite florets.

Inflorescence. *Inflorescence* a single raceme, or paniculate; *contracted*; espatheate; not comprising 'partial inflorescences' and foliar organs. Spikelet-bearing axes persistent. Spikelets solitary; not secund; pedicellate; imbricate.

Female-fertile spikelets. Spikelets 3–5 mm long; compressed laterally; disarticulating above the glumes; with conventional internode spacings. Rachilla terminated by a female-fertile floret. *Callus* absent.

Glumes present; two; more or less equal; shorter than the spikelets; *shorter than the adjacent lemmas (2.0–3.5 mm long)*; hairless; glabrous; pointed, or not pointed; awnless; carinate; very dissimilar, or similar (G_1 or G_2 may be rounded, truncate or acute at the tip). Lower glume shorter than the lowest lemma; much shorter than half length of lowest lemma; 1–3 nerved. Upper glume 3–5 nerved. *Spikelets* with incomplete florets. The incomplete florets proximal to the female-fertile florets. *Spikelets with proximal incomplete florets*. **The proximal incomplete florets 2** *(the lemmas glabrous except at the base, without basal appendages, with prominent longitudinal veins but without transverse ribs)*; epaleate; sterile. The proximal lemmas awned (the apex acuminate, contracted into a short terminal awn); 0–7 nerved; more or less equalling the female-fertile lemmas to decidedly exceeding the female-fertile lemmas; decidedly firmer than the female-fertile lemmas; not becoming indurated.

Female-fertile florets 1. Lemmas decidedly firmer than the glumes (chartaceous); not becoming indurated; entire; blunt (truncate); awnless; hairless; glabrous; carinate; without a germination flap; 1–5 nerved. *Palea* present; relatively long (3/4 the length of the lemma); entire; awnless, without apical setae; thinner than the lemma (hyaline-membranous); not indurated; 1-nerved (i.e. the two veins contiguous); one-keeled. *Lodicules* present; 2; free; membranous; slightly toothed; not or scarcely vascularized. **Stamens 2**. Anthers 1–1.5 mm long; without an apically prolonged connective. *Ovary* glabrous. Styles free to their bases. Stigmas 2; white.

Fruit, embryo and seedling. *Fruit* free from both lemma and palea; small (2–3.5 mm long); ellipsoid; strongly compressed laterally. Hilum long-linear. Embryo small.

Taxonomy. Bambusoideae; Oryzodae; Ehrharteae.

Ecology, geography, regional floristic distribution. 2 species. New Zealand. Helophytic to mesophytic; species of open habitats; glycophytic. Mountains.

Antarctic. New Zealand.

Special comments. Seems scarcely distinguishable from *Ehrharta* and *Tetrarrhena*. Anatomical data wanting.

Peyritschia Fourn.

~ *Trisetum*: *P.koelerioides* (Peyr.) Fourn. (= *Trisetum altijugum* (Fourn.) Scribn.) and *P. pringlei* Scribn. (= *Trisetum kochianum* I. Hdz. T.). See Hernández and Koch 1987

Habit, vegetative morphology. Perennial. Culms herbaceous. Culm internodes hollow. Young shoots extravaginal. Leaves non-auriculate. Leaf blades narrow; without cross venation; persistent. *Ligule an unfringed membrane*; truncate.

Reproductive organization. Plants bisexual, with bisexual spikelets; with hermaphrodite florets.

Inflorescence. *Inflorescence paniculate*; open to contracted; espatheate; not comprising 'partial inflorescences' and foliar organs. Spikelet-bearing axes persistent. Spikelets not secund; pedicellate.

Female-fertile spikelets. Spikelets compressed laterally; disarticulating above the glumes (?). Rachilla prolonged beyond the uppermost female-fertile floret; rachilla hairy to hairless.

Glumes two; more or less equal; about equalling the spikelets; *long relative to the adjacent lemmas*; pointed; awnless; carinate; similar. *Spikelets* with female-fertile florets only, or with incomplete florets. The incomplete florets if present, distal to the female-fertile florets.

Female-fertile florets 2. Lemmas similar in texture to the glumes; not becoming indurated; incised; 2 lobed (the lobes rounded or erose); not deeply cleft; mucronate (the awn sometimes rudimentary), or awned. Awns when present, 1; median; dorsal; from near the top, or from well down the back; non-geniculate, or geniculate; much shorter than the body of the lemma to about as long as the body of the lemma; entered by one vein. **Palea** present; relatively long; tightly clasped by the lemma; *with apical setae*; 2-nerved; 2-keeled. *Lodicules* present; 2; free; membranous; toothed. **Stamens 2 (Clayton and Renvoize 1986)**. Styles free to their bases. Stigmas 2.

Fruit, embryo and seedling. Fruit compressed dorsiventrally. Hilum short. Pericarp thin (shining). Embryo small. Endosperm liquid in the mature fruit.

Abaxial leaf blade epidermis. *Costal/intercostal zonation* conspicuous. *Papillae* absent. *Long-cells* markedly different in shape costally and intercostally (the costals narrower); differing markedly in wall thickness costally and intercostally (the costals thicker walled). Mid-intercostal long-cells fusiform; having straight or only gently undulating walls. *Microhairs* absent. *Stomata* common to absent or very rare. Subsidiaries parallel-sided to dome-shaped. Guard-cells overlapped by the interstomatals. *Intercostal short-cells* absent or very rare (excluding the bases of prickles). *Costal short-cells* neither distinctly grouped into long rows nor predominantly paired (solitary, paired and in short rows), or conspicuously in long rows (in places, in *P. koelerioides*). Costal silica bodies present and well developed; horizontally-elongated crenate/sinuous (crenate); not sharp-pointed.

Transverse section of leaf blade, physiology. C_3; XyMS+. *Mesophyll* with non-radiate chlorenchyma; without adaxial palisade. *Leaf blade* with distinct, prominent adaxial ribs; with the ribs more or less constant in size (low, round topped). *Midrib* not readily distinguishable; with one bundle only. *The lamina* symmetrical on either side of the midrib. Bulliforms present in discrete, regular adaxial groups (one in each furrow); in simple fans. All the vascular bundles accompanied by sclerenchyma. Combined sclerenchyma girders present; forming 'figures' (all the major bundles with substantial I's or T's). Sclerenchyma all associated with vascular bundles.

Special diagnostic feature. Panicle loose, or if dense then interrupted, neither cylindrical nor ovoid: awns usually present, usually twisted, usually distinctly dorsal, conspicuous if inflorescence compact.

Cytology. Chromosome base number, $x = 7$. $2n = 28$. 4 ploid.

Taxonomy. Pooideae; Poodae; Aveneae.

Ecology, geography, regional floristic distribution. 2 species. Mexico. Holarctic and Neotropical. Madrean. Caribbean.

Rusts and smuts. Rusts — *Puccinia*. Taxonomically wide-ranging species: *Puccinia poarum*.

References, etc. Morphological/taxonomic: Koch 1979. Leaf anatomical: this project.

Phacelurus Griseb.

Pseudophacelurus (Steud.) A. Camus

Excluding *Jardinea, Pseudovossia, Thyrsia*

Habit, vegetative morphology. Robust perennial. *Culms* 20–150 cm high; herbaceous (tough); branched above, or unbranched above. Leaf blades linear; narrow; flat, or folded, or acicular (rarely, when represented by the midrib); without cross venation. *Ligule an unfringed membrane.*

Reproductive organization. *Plants bisexual, with bisexual spikelets*; with hermaphrodite florets. *The spikelets of sexually distinct forms on the same plant*; hermaphrodite (rarely), or hermaphrodite and male-only, or hermaphrodite and sterile; overtly heteromorphic (the pedicellate spikelets usually smaller or vestigial), or homomorphic; in both homogamous and heterogamous combinations, or all in heterogamous combinations.

Inflorescence. *Inflorescence of spicate main branches (usually terminal, of flattened spicate 'racemes')*; usually *digitate, or subdigitate (the racemes rarely solitary, or on an elongated axis)*; espatheate; not comprising 'partial inflorescences' and foliar organs. **Spikelet-bearing axes** spicate 'racemes' to spikelike; *clustered (on a common axis, rarely solitary)*; with substantial rachides; disarticulating; disarticulating at the joints. 'Articles' non-linear (clavate to inflated, hollowed at the apex); with a basal callus-knob; disarticulating transversely; glabrous. Spikelets paired; sessile and pedicellate (the pedicels resembling the internodes); consistently in 'long-and-short' combinations; in pedicellate/sessile combinations. *Pedicels of the 'pedicellate' spikelets free of the rachis (not articulated, by contrast with* **Pseudovossia***).* The 'shorter' spikelets hermaphrodite. The 'longer' spikelets hermaphrodite (rarely), or male-only, or sterile.

Female-sterile spikelets. Pedicelled spikelets usually more or less resembling the sessile but usually smaller, rarely bisexual and sometimes vestigial.

Female-fertile spikelets. Spikelets compressed dorsiventrally (dorsally flat, convex or rarely concave); falling with the glumes. Rachilla terminated by a female-fertile floret. Hairy callus absent. *Callus* short (minute); blunt (truncate).

Glumes two; more or less equal; long relative to the adjacent lemmas; hairless; awnless; carinate (G_2), or non-carinate (G_1); very dissimilar (leathery to membranous, the G_1 2-keeled and flat, the G_2 cymbiform). Lower glume two-keeled (sometimes winged); flattened on the back; not pitted; relatively smooth. *Spikelets* with incomplete florets. The incomplete florets proximal to the female-fertile florets. *The proximal incomplete florets* 1; male, or sterile. The proximal lemmas awnless; 2 nerved; similar in texture to the female-fertile lemmas (hyaline); not becoming indurated.

Female-fertile florets 1. *Lemmas* less firm than the glumes (hyaline); not becoming indurated; *entire*; pointed, or blunt; *awnless*; non-carinate; without a germination flap; 3 nerved. *Lodicules* present; 2; free; fleshy; glabrous. *Stamens* 3. *Ovary* glabrous. Stigmas 2.

Fruit, embryo and seedling. Fruit compressed dorsiventrally. Embryo large.

Abaxial leaf blade epidermis. *Costal/intercostal zonation* conspicuous. *Papillae* absent. *Long-cells* markedly different in shape costally and intercostally (the costals narrower, more regularly rectangular); of similar wall thickness costally and intercostally. Intercostal zones with typical long-cells, or exhibiting many atypical long-cells (some of them short, in places, in *P. huillensis*). Mid-intercostal long-cells rectangular, or rectangular and fusiform (in *P. huillensis*); having markedly sinuous walls. *Microhairs* present; elongated; clearly two-celled; panicoid-type; 57–60 microns long; 7.5–9 microns wide at the septum. Microhair total length/width at septum 6.3–7.1. Microhair apical cells 30–32 microns long. Microhair apical cell/total length ratio 0.5–0.55. *Stomata* common; 36–39 microns long. Subsidiaries triangular (predominantly an extreme form, in *P. huillensis*), or dome-shaped. Guard-cells overlapping to flush with the interstomatals (clearly overlapping them in places, in *P. huillensis*). *Intercostal short-cells* common; in cork/silica-cell pairs; silicified. *Costal short-cells* predominantly paired. Costal silica bodies 'panicoid-type'; not sharp-pointed.

Transverse section of leaf blade, physiology. Leaf blades consisting of midrib (e.g. *P. franksae*), or 'laminar'. C_4; XyMS–. PCR sheath outlines even. PCR sheath extensions absent. PCR cell chloroplasts centrifugal/peripheral. *Leaf blade* adaxially flat (or with only slight ribs, over the main bundles, in *P. huillensis*). *Midrib* very conspicuous (or the blade more or less reduced to its midrib); having a conventional arc of bundles (abaxially); with colourless mesophyll adaxially (and the colourless tissue extending laterally, about halfway to the margins, in the lamina of *P. huillensis*). Bulliforms not in discrete, regular adaxial groups (the epidermis mostly more or less bulliform). Many of the smallest vascular bundles unaccompanied by sclerenchyma. Combined sclerenchyma girders present (with all the primaries, in *P. huillensis*); forming 'figures' (forming 'anchors', in places).

Special diagnostic feature. *The lower glume of the pedicellate spikelet awnless.*

Cytology. Chromosome base number, $x = 10$. $2n = 20$ and 40. 2 and 4 ploid.

Taxonomy. Panicoideae; Andropogonodae; Andropogoneae; Rottboelliinae.

Ecology, geography, regional floristic distribution. 7 species. Africa to Indo-China and Japan. Helophytic to mesophytic; shade species, or species of open habitats; glycophytic. Woodland and grassland, in moist places.

Holarctic and Paleotropical. Boreal and Tethyan. African and Indomalesian. Euro-Siberian. Mediterranean and Irano-Turanian. Sudano-Angolan. Indian and Indo-Chinese. European. South Tropical African.

Rusts and smuts. Rusts — *Puccinia*.

References, etc. Morphological/taxonomic: Clayton 1978b. Leaf anatomical: this project; photos of *P. franksae* provided by R.P. Ellis.

Phaenanthoecium C.E. Hubb.

Danthonia köstlinni, Streblochaete köstlinii

Habit, vegetative morphology. Perennial; caespitose (culms decumbent or ascending, sometimes trailing to 2 m). *Culms herbaceous*; *branched above*. Leaves not basally aggregated; non-auriculate. Leaf blades narrowly linear; narrow; flat, or rolled (convolute); without cross venation. **Ligule** present; *a fringe of hairs*.

Reproductive organization. Plants bisexual, with bisexual spikelets; with hermaphrodite florets.

Inflorescence. Inflorescence determinate; without pseudospikelets; few spikeleted; *a single raceme*; espatheate; not comprising 'partial inflorescences' and foliar organs. Spikelet-bearing axes persistent. Spikelets shortly pedicellate.

Female-fertile spikelets. Spikelets slightly compressed laterally; disarticulating above the glumes; disarticulating between the florets. Rachilla prolonged beyond the uppermost female-fertile floret; rachilla hairless (glabrous). Hairy callus present. *Callus* very short; blunt.

Glumes two; relatively large; more or less equal (or slightly unequal); *shorter than the spikelets*; free; not pointed (emarginate, bilobed or obtuse); awnless; carinate (G_1, somewhat), or non-carinate; similar (narrowly oblong to oblong-elliptic, membranous). Lower glume 1 nerved, or 3 nerved. Upper glume 3 nerved, or 4 nerved. *Spikelets* with female-fertile florets only, or with incomplete florets. The incomplete florets distal to the female-fertile florets. *The distal incomplete florets* merely underdeveloped.

Female-fertile florets 5–9. Lemmas similar in texture to the glumes (membranous); not becoming indurated; incised; shortly 1 lobed; not deeply cleft; awned. Awns 3; median and lateral (the lateral lobes attenuate into long setae); the median different in form from the laterals; from a sinus; geniculate. Lemmas hairy (on the margins, otherwise glabrous). The hairs in tufts (along the margins). Lemmas non-carinate (dorsally rounded); 9 nerved. *Palea* present; linear-oblong; entire (obtuse or truncate); awnless; without apical setae; not indurated (membranous); 2-nerved; 2-keeled. *Lodicules* present (minute); 2; fleshy (cuneate); ciliate (or at least ciliolate). *Stamens* 3. *Ovary* glabrous; without a conspicuous apical appendage. Styles free to their bases. Stigmas 2.

Fruit, embryo and seedling. *Fruit* free from both lemma and palea (loosely included); concavo-convex. *Hilum long-linear (about 4/5 of the grain length)*. Embryo small (about a quarter of the grain length).

Abaxial leaf blade epidermis. *Costal/intercostal zonation* conspicuous. *Papillae* absent. *Long-cells* similar in shape costally and intercostally (the intercostals larger); of similar wall thickness costally and intercostally (walls of medium thickness). Mid-intercostal long-cells rectangular; having markedly sinuous walls. *Microhairs* present; panicoid-type (large); (52.5–)54–60(–64.5) microns long; (6–)6.3–6.6(–8) microns wide at the septum. Microhair total length/width at septum 6.7–9.8. Microhair apical cells 21–22.5–27 microns long. Microhair apical cell/total length ratio 0.38–0.42. *Stomata* common; 25.5–28.5 microns long. Subsidiaries mostly high dome-shaped. Guard-cells overlapping to flush with the interstomatals. *Intercostal short-cells* common; not paired (solitary); not silicified. *Costal short-cells* conspicuously in long rows. Costal silica bodies 'panicoid-type'; almost exclusively cross shaped, butterfly shaped, and dumb-bell shaped (short); not sharp-pointed.

Transverse section of leaf blade, physiology. C_3; XyMS+. Mesophyll without adaxial palisade. *Leaf blade* with distinct, prominent adaxial ribs; with the ribs more or less constant in size. *Midrib* not readily distinguishable; with one bundle only. Bulliforms present in discrete, regular adaxial groups; in simple fans. All the vascular bundles accompanied by sclerenchyma. Combined sclerenchyma girders present; forming 'figures' (all the bundles with I's). Sclerenchyma all associated with vascular bundles.

Taxonomy. Arundinoideae; Danthonieae.

Ecology, geography, regional floristic distribution. 1 species. Northeast Africa. Shade species. Shady cliffs.

Paleotropical. African. Sudano-Angolan. Somalo-Ethiopian.

References, etc. Morphological/taxonomic: Hubbard 1936c. Leaf anatomical: this project.

Phaenosperma Munro ex Benth.

Habit, vegetative morphology. Perennial; robust, rhizomatous. *Culms* 100–300 cm high; herbaceous; to 0.03 cm in diameter. Culm nodes glabrous. Culm internodes solid, or hollow. *Leaves* not basally aggregated; non-auriculate. Leaf blades broadly linear; broad; 10–30 mm wide; flat; pseudopetiolate (at least the basal ones, becoming inverted by twisting of the 'petiole'); pinnately veined (with nerves slanting obliquely from the midrib); cross veined; disarticulating from the sheaths (?). Ligule an unfringed membrane; not truncate (acute to rounded); 6–10 mm long. *Contra-ligule* absent.

Reproductive organization. Plants bisexual, with bisexual spikelets; with hermaphrodite florets. The spikelets all alike in sexuality.

Inflorescence. Inflorescence paniculate; open; espatheate; not comprising 'partial inflorescences' and foliar organs. Spikelet-bearing axes persistent. Spikelets not secund; pedicellate.

Female-fertile spikelets. Spikelets 3–3.5 mm long; compressed dorsiventrally; falling with the glumes; with conventional internode spacings. Rachilla terminated by a female-fertile floret. Hairy callus absent.

Glumes two; very unequal; shorter than the spikelets; shorter than the adjacent lemmas; basally joined; hairless; glabrous; awnless; non-carinate; very dissimilar (membranous, G_1 deltoid-lanceolate, G_2 oblong). Lower glume 1 nerved. Upper glume 3 nerved. *Spikelets* with female-fertile florets only.

Female-fertile florets 1. Lemmas similar in texture to the glumes to decidedly firmer than the glumes (membranous); not becoming indurated; entire (but splitting); blunt; awnless; hairless; glabrous; non-carinate (rounded on the back); without a germination flap; 3–7 nerved; with the nerves confluent towards the tip. Palea present; relatively long (slightly exceeding the lemma); entire; awnless, without apical setae; not indurated; 2-nerved; 2-keeled. Palea keels wingless. *Lodicules* present; 3 (fairly large); free; membranous; glabrous; not toothed. *Stamens* 3. Anthers not penicillate. *Ovary* glabrous. Styles free to their bases. Stigmas 2.

Fruit, embryo and seedling. *Fruit* free from both lemma and palea (and protruding conspicuously when mature); small (2.5–3 mm long); subglobose; slightly longitudinally grooved (along the hilum); sculptured (rugose). Hilum long-linear. Pericarp free (the seed swollen and dark brown). Embryo small. *Endosperm hard (ruminate)*; containing only simple starch grains. Embryo with an epiblast; with a scutellar tail; with a negligible mesocotyl internode; with one scutellum bundle. Embryonic leaf margins overlapping.

Abaxial leaf blade epidermis. *Costal/intercostal zonation* conspicuous. *Papillae* absent. Mid-intercostal long-cells having straight or only gently undulating walls. *Microhairs* absent. Stomata common; 19.5–21–22.5 microns long. Subsidiaries very low to high dome-shaped. *Intercostal short-cells* common; in cork/silica-cell pairs; silicified. Intercostal silica bodies rounded (elliptical), or crescentic. *Costal short-cells* predominantly paired (a few short rows). Costal silica bodies rounded and crescentic; not sharp-pointed.

Transverse section of leaf blade, physiology. C$_3$; XyMS+. *Mesophyll* with non-radiate chlorenchyma; without arm cells; without fusoids. *Leaf blade* adaxially flat. *Midrib* conspicuous; having a conventional arc of bundles (3 bundles); with colourless mesophyll adaxially (the material seen giving the appearance of a proliferation of PBS tissue between the large median bundle and its adaxial strand of sclerenchyma). Bulliforms present in discrete, regular adaxial groups (in addition to the large 'midrib hinges'); in simple fans (the fans wide, in the shallow furrows). All the vascular bundles accompanied by sclerenchyma. Combined sclerenchyma girders present; nowhere forming 'figures'. Sclerenchyma all associated with vascular bundles.

Culm anatomy. *Culm internode bundles* in three or more rings.

Special diagnostic feature. *Seed dark brown, with ruminate endosperm.*

Cytology. Chromosome base number, $x = 12$. $2n = 24$. 2 ploid.

Taxonomy. Bambusoideae; Oryzodae; Phaenospermateae.

Ecology, geography, regional floristic distribution. 1 species. Japan, China, Korea. Shade species; glycophytic. In warm temperate forest.

Holarctic. Boreal. Eastern Asian.

References, etc. Morphological/taxonomic: Conert 1959; Macfarlane and Watson 1980. Leaf anatomical: Metcalfe 1960; this project.

Phalaris L.

Baldingera Gaertn., Meyer & Scherb., *Digraphis* Trin., *Endallex* Raf., *Phalaridantha* St-Lager, *Phalaroides* Wolf, *Typhoides* Moench

Habit, vegetative morphology. Annual, or perennial; some species reedlike; rhizomatous, or caespitose, or decumbent. *Culms* 10–200 cm high; herbaceous; unbranched above; tuberous, or not tuberous. Culm internodes hollow. *Leaves* not basally aggregated; non-auriculate. Leaf blades linear to linear-lanceolate; broad, or narrow; 2–20 mm wide; flat; without cross venation; persistent; rolled in bud. **Ligule** present; *an unfringed membrane*; truncate, or not truncate; 2–12 mm long.

Reproductive organization. Plants bisexual, with bisexual spikelets; with hermaphrodite florets. The spikelets of sexually distinct forms on the same plant (rarely), or all alike in sexuality; hermaphrodite, or hermaphrodite and sterile (clusters of one fertile and several deformed-sterile in *P. paradoxa* and *P. caerulescens*). Plants outbreeding and inbreeding.

Inflorescence. Inflorescence a false spike, with spikelets on contracted axes, or paniculate; open (rarely), or contracted; when contracted capitate, or more or less ovoid, or spicate, or more or less irregular; espatheate; not comprising 'partial inflorescences' and foliar organs. Spikelet-bearing axes disarticulating, or persistent; when disarticulating, falling entire (the clusters falling in *P. paradoxa* and *P. caerulescens*). Spikelets not secund; pedicellate.

Female-fertile spikelets. Spikelets 3.5–9.5 mm long; strongly compressed laterally; disarticulating above the glumes, or falling with the glumes, or not disarticulating. Rachilla terminated by a female-fertile floret. Hairy callus absent.

Glumes two; more or less equal; about equalling the spikelets to exceeding the spikelets; *long relative to the adjacent lemmas*; pointed; awnless; carinate; with the keel conspicuously winged, or without a median keel-wing; *similar (papery)*. Lower glume 1–5 nerved. Upper glume 1–5 nerved. *Spikelets* usually with incomplete florets. The incomplete florets proximal to the female-fertile florets. *Spikelets with proximal incomplete florets. The proximal incomplete florets* 1, or 2; usually epaleate; usually sterile. *The proximal lemmas* awnless; exceeded by the female-fertile lemmas; *less firm than the female-fertile lemmas (reduced to scales)*; not becoming indurated.

Female-fertile florets 1. *Lemmas decidedly firmer than the glumes*; becoming indurated; entire; pointed; awnless; hairy, or hairless; *carinate*; 5 nerved. **Palea** present; relatively long; entire to apically notched; awnless, without apical setae; *textured like the lemma*; indurated, or not indurated; 1-nerved, or 2-nerved; keel-less. *Lodicules* present; 2; free; membranous; glabrous; not toothed. *Stamens* 3. Anthers 1.2–6 mm long; not penicillate. *Ovary* glabrous. Styles fused, or free to their bases. Stigmas 2; white.

Fruit, embryo and seedling. *Fruit* free from both lemma and palea; small; compressed laterally. *Hilum long-linear*. Embryo large (up to a third of the grain length), or small; not waisted. Endosperm hard; with lipid; containing compound starch grains. Embryo with an epiblast; without a scutellar tail; with a negligible mesocotyl internode. Embryonic leaf margins meeting.

Seedling with a long mesocotyl; with a loose coleoptile, or with a tight coleoptile. First seedling leaf with a well-developed lamina. The lamina narrow; erect; 3–5 veined.

Abaxial leaf blade epidermis. *Costal/intercostal zonation* conspicuous. *Papillae* absent. *Long-cells* similar in shape costally and intercostally; of similar wall thickness costally and intercostally (fairly thin walled). Mid-intercostal long-cells fusiform; having straight or only gently undulating walls. *Microhairs* absent. *Stomata* common; (30–)33–39(–42) microns long. Subsidiaries low dome-shaped, or parallel-sided. Guard-cells overlapped by the interstomatals. *Intercostal short-cells* common (rarely), or absent or very rare (absent); in cork/silica-cell pairs, or not paired; silicified, or not silicified. Intercostal silica bodies when present, rounded (oval). *Costal short-cells* neither distinctly grouped into long rows nor predominantly paired (usually), or conspicuously in long rows and neither distinctly grouped into long rows nor predominantly paired (*P. angusta*). Costal silica bodies horizontally-elongated crenate/sinuous, or horizontally-elongated smooth, or rounded (or cuboid — varying from species to species), or 'panicoid-type' (occasionally); occasionally cross shaped, or butterfly shaped; not sharp-pointed.

Transverse section of leaf blade, physiology. C_3; XyMS+. *PBS cells* without a suberised lamella. *Mesophyll* with radiate chlorenchyma (rarely), or with non-radiate chlorenchyma. *Leaf blade* adaxially flat; with the ribs more or less constant in size. *Midrib* conspicuous; with one bundle only. Bulliforms present in discrete, regular adaxial groups; in simple fans. All the vascular bundles accompanied by sclerenchyma. Combined sclerenchyma girders present, or absent; nowhere forming 'figures'. Sclerenchyma all associated with vascular bundles.

Phytochemistry. Tissues of the culm bases with little or no starch. Fructosans predominantly long-chain. Leaves without flavonoid sulphates (1 species).

Cytology. Chromosome base number, $x = 6$ and 7. $2n = 12, 14, 28, 35, 42$, and 56 (and aneuploids). 2, 4, 6, and 8 ploid (and aneuploids). Haploid nuclear DNA content 1.7–3.7 pg (13 species, mean 2.4). Mean diploid 2c DNA value 5.8 pg (9 species, (2.8–)3.5–9.0). Nucleoli disappearing before metaphase.

Taxonomy. Pooideae; Poodae; Aveneae.

Ecology, geography, regional floristic distribution. 16 species. North temperate, South America. Commonly adventive. Helophytic, or mesophytic; species of open habitats. In weedy places, damp soils and swamps.

Holarctic, Paleotropical, Neotropical, and Antarctic. Boreal, Tethyan, and Madrean. African and Indomalesian. Arctic and Subarctic, Euro-Siberian, Eastern Asian, Atlantic North American, and Rocky Mountains. Macaronesian, Mediterranean, and Irano-Turanian. Saharo-Sindian and Sudano-Angolan. Indian. Caribbean, Pampas, and Andean. Patagonian. European and Siberian. Canadian-Appalachian, Southern Atlantic North American, and Central Grasslands. Somalo-Ethiopian.

Rusts and smuts. Rusts — *Puccinia*. Taxonomically wide-ranging species: *Puccinia graminis*, *Puccinia coronata*, and *Puccinia striiformis*. Smuts from Tilletiaceae and from Ustilaginaceae. Tilletiaceae — *Entyloma* and *Tilletia*. Ustilaginaceae — *Ustilago*.

Economic importance. Significant weed species: *P. aquatica*, *P. arundincacea*, *P. brachystachys*, *P. canariensis*, *P. minor*, *P. paradoxa*. Cultivated fodder: *P. arundinacea*, *P. aquatica*. Important native pasture species: *P. arundinacea*. Grain crop species: *P. canariensis* (Canary grass), mainly for bird seed.

References, etc. Morphological/taxonomic: Anderson 1961. Leaf anatomical: Metcalfe 1960; this project.

Pharus P.Browne

Habit, vegetative morphology. Perennial; rhizomatous. The flowering culms leafy. Culms herbaceous; unbranched above. Culm internodes solid. Rhizomes leptomorph. Plants unarmed. *Leaves* not basally aggregated. *Leaf blades* elliptic, or obovate; broad;

pseudopetiolate (becoming inverted by twising of the 'petiole'); *pinnately veined (the nerves running obliquely from mid-rib to margin)*; cross veined.

Reproductive organization. Plants monoecious with all the fertile spikelets unisexual; without hermaphrodite florets. The spikelets of sexually distinct forms on the same plant; female-only and male-only. The male and female-fertile spikelets mixed in the inflorescence. The spikelets overtly heteromorphic (the females larger).

Inflorescence. Inflorescence paniculate ((1–)3–8 noded, fragile); open; non-digitate. Inflorescence axes not ending in spikelets (apically), or with axes ending in spikelets. Inflorescence espatheate (no subtending bracts with the spikelet branchlets); not comprising 'partial inflorescences' and foliar organs. *Spikelet-bearing axes disarticulating*; falling entire (the primary branches breaking away from the main rachis). Spikelets paired; not secund; sessile and pedicellate, or subsessile and pedicellate (the pistillate pedicels filiform); consistently in 'long-and-short' combinations. The 'shorter' spikelets female-only. The 'longer' spikelets male-only.

Female-sterile spikelets. Male spikelets smaller, laterally compressed, with 2 glumes; stamens 6 per floret. The floret persistent. The male spikelets with glumes; 1 floreted. Male florets 6 staminate.

Female-fertile spikelets. *Spikelets not noticeably compressed (terete)*; falling with the glumes. Rachilla terminated by a female-fertile floret.

Glumes two to several (occasionally 3); more or less equal; shorter than the adjacent lemmas to long relative to the adjacent lemmas; pointed (acute); awnless; similar (membranous, persisitent). Lower glume 5 nerved. Upper glume 7 nerved. *Spikelets* with female-fertile florets only.

Female-fertile florets 1. **Lemmas cylindrical, twisted, often with a minute, bent beak, in one species with fused margins**; decidedly firmer than the glumes; becoming indurated (rather thin at anthesis, but hardening in fruit); entire; awnless, or mucronate; **hairy (with hooked hairs, becoming adhesive in fruit)**; non-carinate (becoming involute in fruit); having the margins tucked in onto the palea; 5–7 nerved. *Palea* present (very narrow, folded around the long style); relatively long; tightly clasped by the lemma; awnless, without apical setae; not indurated (membranous); 2-nerved; 2-keeled. *Lodicules* present; 3; free. *Stamens* 0. *Ovary* glabrous. Styles fused. Stigmas 3.

Fruit, embryo and seedling. Hilum long-linear. Endosperm hard; without lipid. Embryo with an epiblast; with a scutellar tail, or without a scutellar tail. Embryonic leaf margins overlapping.

Abaxial leaf blade epidermis. *Costal/intercostal zonation* conspicuous. *Papillae* absent. *Long-cells* markedly different in shape costally and intercostally (the costals very long, narrow, straight-walled, fibre-like with tapered ends). Mid-intercostal long-cells rectangular; having markedly sinuous walls (thin). *Microhairs* absent. *Stomata* common (but relatively infrequent, alongside the veins); 24–25.5–27 microns long. Subsidiaries parallel-sided to dome-shaped (high domes to tall parallels). Guard-cells overlapping to flush with the interstomatals. *Intercostal short-cells* absent or very rare. *Costal short-cells* conspicuously in long rows. Costal silica bodies 'panicoid-type'; cross shaped to dumbbell shaped; not sharp-pointed.

Transverse section of leaf blade, physiology. C_3; XyMS+. Mesophyll with arm cells; with fusoids. The fusoids external to the PBS. *Leaf blade* adaxially flat. *Midrib* conspicuous; having complex vascularization; with colourless mesophyll adaxially. Bulliforms present in discrete, regular adaxial groups (also in rather irregular groups); in simple fans. All the vascular bundles accompanied by sclerenchyma. Combined sclerenchyma girders present (some; others with only strands). Sclerenchyma all associated with vascular bundles.

Culm anatomy. *Culm internode bundles* scattered.

Cytology. Chromosome base number, $x = 12$. $2n = 24$ and 48. 2 and 4 ploid.

Taxonomy. Bambusoideae; Oryzodae; Phareae.

Ecology, geography, regional floristic distribution. 6 species. Tropical America, West Indies. Shade species.

Neotropical. Caribbean, Venezuela and Surinam, Amazon, Pampas, and Andean.

References, etc. Leaf anatomical: Metcalfe 1960; this project.

Special comments. Fruit data wanting.

Pheidochloa S. T. Blake

Habit, vegetative morphology. Slender annual; caespitose. *Culms* 20–50 cm high (filiform); herbaceous; unbranched above. Culm nodes glabrous. Culm internodes solid. Plants unarmed. Young shoots intravaginal. Leaves non-auriculate. Leaf blades narrow; setaceous (short, convolute); without cross venation; persistent. *Ligule a fringe of hairs*.

Reproductive organization. *Plants bisexual, with bisexual spikelets*; with hermaphrodite florets. The spikelets all alike in sexuality. Exposed-cleistogamous.

Inflorescence. Inflorescence few spikeleted (4–6); a single raceme, or paniculate (a raceme or depauperate panicle of few spikelets); espatheate; not comprising 'partial inflorescences' and foliar organs. Spikelet-bearing axes persistent. Spikelets not secund; pedicellate.

Female-fertile spikelets. *Spikelets* 6–13 mm long; compressed laterally; disarticulating above the glumes; **with a distinctly elongated rachilla internode between the glumes (glabrous)**. Rachilla terminated by a female-fertile floret. Hairy callus present. *Callus* long; pointed.

Glumes present; two (separated by an unusually long internode); very unequal (G_2 twice the length of G_1); exceeding the spikelets; (the longer) long relative to the adjacent lemmas (much longer); hairless; pointed; awnless; carinate; similar (herbaceous-membranous). *Lower glume 7 nerved.* Upper glume 7 nerved. *Spikelets* with female-fertile florets only.

Female-fertile florets 2 (*similar*). Lemmas similar in texture to the glumes to decidedly firmer than the glumes; not becoming indurated (thinly cartilaginous, often purplish); entire; pointed; awned. Awns 1; median; apical; non-geniculate (purple); straight (bristle-like); about as long as the body of the lemma to much longer than the body of the lemma; entered by several veins (3 or 5). Lemmas hairy; non-carinate; without a germination flap; 7 nerved. *Palea* present (glabrous); relatively long; awnless, without apical setae; not indurated; 2-nerved; 2-keeled. *Lodicules* present; free; membranous (hyaline, minute); glabrous; not toothed. **Stamens 2.** Anthers minute; not penicillate; without an apically prolonged connective. *Ovary* glabrous. Styles free to their bases. Stigmas 2.

Fruit, embryo and seedling. *Fruit* free from both lemma and palea; small; linear; longitudinally grooved (concavo-convex); compressed dorsiventrally. Hilum long-linear. Embryo small; not waisted. Endosperm hard. Embryo without an epiblast; with a scutellar tail; with an elongated mesocotyl internode; with one scutellum bundle. Embryonic leaf margins meeting.

First seedling leaf with a well-developed lamina. The lamina fairly broad; curved.

Abaxial leaf blade epidermis. *Costal/intercostal zonation* conspicuous. *Papillae* absent. *Long-cells* similar in shape costally and intercostally; of similar wall thickness costally and intercostally (fairly thick walled). Mid-intercostal long-cells rectangular, or fusiform (in the files containing the stomata); having markedly sinuous walls. *Microhairs* present; panicoid-type (the basal cell tending to be plump and distally inflated, the apical cell rather broad, but tapered and thin-walled); (42–)46–57(–58.5) microns long; (9–)9.6–14.4(–15) microns wide at the septum. Microhair total length/width at septum 3.6–6.3. Microhair apical cells (20.4–)27–33(–34.5) microns long. Microhair apical cell/total length ratio 0.48–0.63. *Stomata* common; (30–)33–36(–39) microns long. Subsidiaries triangular. Guard-cells overlapping to flush with the interstomatals. *Intercostal short-cells* absent or very rare. *Costal short-cells* conspicuously in long rows. Costal silica bodies oryzoid and 'panicoid-type'; mostly cross shaped (but many of the crosses more or less 'oryzoid', and most having points); sharp-pointed (many of the crosses are 'oryzoid' in form, and most have many points).

Transverse section of leaf blade, physiology. C_4; biochemical type NADP–ME (*P. gracilis*); XyMS+. PCR sheath outlines uneven. PCR sheath extensions absent. PCR cells without a suberised lamella. *PCR cell chloroplasts* ovoid; with well developed grana; centrifugal/peripheral. *Mesophyll* with radiate chlorenchyma. *Leaf blade* adaxially flat. *Midrib* not readily distinguishable; with one bundle only; with colourless mesophyll adaxially. Bulliforms not in discrete, regular adaxial groups (epidermis of large cells, with no bulliform groups other than the small bulliforms-plus-colourless cells group over the median vascular bundle). All the vascular bundles accompanied by sclerenchyma.

Combined sclerenchyma girders absent. Sclerenchyma all associated with vascular bundles.

Phytochemistry. Leaf blade chlorophyll *a:b* ratio 4.85–4.93.

Taxonomy. Arundinoideae (or Panicoideae); Eriachneae.

Ecology, geography, regional floristic distribution. 2 species. Australia & New Guinea. In damp sandy heaths.

Paleotropical and Australian. Indomalesian. Papuan. North and East Australian. Tropical North and East Australian.

References, etc. Morphological/taxonomic: Blake 1944a. Leaf anatomical: this project.

Phippsia R.Br.

Vilfa Beauv.

Habit, vegetative morphology. Dwarf *perennial*; caespitose, or decumbent. *Culms* 2–25 cm high; herbaceous. Leaves non-auriculate. *Sheath margins joined*. The upper sheaths somewhat expanded. Leaf blades linear; narrow; 1–3 mm wide; flat; without cross venation; rolled in bud. Ligule an unfringed membrane; not truncate; 1–1.5 mm long.

Reproductive organization. Plants bisexual, with bisexual spikelets; with hermaphrodite florets. Exposed-cleistogamous, or chasmogamous (?). Plants with hidden cleistogenes, or without hidden cleistogenes. The hidden cleistogenes when present, in the leaf sheaths.

Inflorescence. Inflorescence paniculate; open, or contracted (to spike-like); espatheate; not comprising 'partial inflorescences' and foliar organs. Spikelet-bearing axes persistent. Spikelets not secund; pedicellate.

Female-fertile spikelets. Spikelets 1–2.5 mm long; compressed laterally to not noticeably compressed (terete); disarticulating above the glumes (but the glumes also often caducous). Rachilla terminated by a female-fertile floret. Hairy callus absent. *Callus* short.

Glumes one per spikelet, or two; minute; very unequal; *shorter than the adjacent lemmas (the longer less than a third as long)*; free; awnless; similar (more or less caducous). Lower glume much shorter than half length of lowest lemma; 0 nerved. *Upper glume 0 nerved. Spikelets* with female-fertile florets only.

Female-fertile florets *1*. Lemmas decidedly firmer than the glumes (membranous); not becoming indurated; entire; pointed, or blunt; awnless; hairy, or hairless; carinate to non-carinate; without a germination flap; 3–5 nerved. *Palea* present; relatively long; 2-nerved; 2-keeled. Palea keels hairy (often spinulose), or scabrous (?). *Lodicules* present; 2; free; membranous; glabrous; not toothed; not or scarcely vascularized. *Stamens* 1, or 2(–3). Anthers 0.3–0.6 mm long. *Ovary* glabrous. Styles free to their bases. Stigmas 2.

Fruit, embryo and seedling. *Fruit* free from both lemma and palea; ellipsoid (protruding from the floret when mature). Hilum short (oval). Embryo small. Endosperm hard. Embryo with an epiblast; without a scutellar tail; with a negligible mesocotyl internode. Embryonic leaf margins meeting.

Abaxial leaf blade epidermis. *Costal/intercostal zonation* lacking (and without short-cells). *Papillae* absent. Mid-intercostal long-cells fusiform; having straight or only gently undulating walls. *Microhairs* absent. *Stomata* common (fairly); (36–)39–42(–45) microns long. Subsidiaries parallel-sided. Guard-cells overlapped by the interstomatals (the subsidiaries also sunken). *Intercostal short-cells* absent or very rare. Costal zones without short-cells. Costal silica bodies absent.

Transverse section of leaf blade, physiology. C_3; XyMS+. Mesophyll without adaxial palisade. *Leaf blade* adaxially flat. *Midrib* conspicuous (prominent adaxially, and flanked by conspicuous bulliform 'hinges'); with one bundle only. *The lamina* symmetrical on either side of the midrib. Bulliforms not in discrete, regular adaxial groups (apart from the midrib 'hinges'). Many of the smallest vascular bundles unaccompanied by sclerenchyma (and no sclerenchyma with any of the other bundles either, in the material of *P. algida* seen).

Cytology. Chromosome base number, $x = 7$. $2n = 28$. 4 ploid.

Taxonomy. Pooideae; Poodae; Poeae.

Ecology, geography, regional floristic distribution. 3 species. Arctic circumpolar, South American mountains. Helophytic; species of open habitats.

Holarctic. Boreal. Arctic and Subarctic, Euro-Siberian, and Rocky Mountains. European.

Hybrids. Intergeneric hybrids with *Puccinellia* (×*Pucciphippsia* Tsvelev).

Rusts and smuts. Rusts — *Puccinia*. Taxonomically wide-ranging species: *Puccinia brachypodii*.

References, etc. Leaf anatomical: this project.

Phleum L.

Achnodon Link, *Achnodonton* P. Beauv., *Chilochloa* P. Beauv., *Heleochloa* P. Beauv., *Phalarella* Boiss., *Plantinia* Bub., *Stelephuros* Adans.

Excluding *Maillea, Pseudophleum*

Habit, vegetative morphology. Annual, or perennial; rhizomatous, or stoloniferous, or caespitose, or decumbent. *Culms* 4–150 cm high; herbaceous; tuberous, or not tuberous. Culm nodes glabrous. Culm internodes hollow. *Leaves* not basally aggregated; non-auriculate. *Leaf blades* linear; narrow; 1–10 mm wide; usually flat; *not pseudopetiolate*; without cross venation; persistent; rolled in bud. **Ligule** present; *an unfringed membrane*; truncate, or not truncate; 1.5–6 mm long.

Reproductive organization. Plants bisexual, with bisexual spikelets; with hermaphrodite florets; outbreeding and inbreeding.

Inflorescence. Inflorescence paniculate; contracted; capitate, or more or less ovoid, or spicate; espatheate; not comprising 'partial inflorescences' and foliar organs. Spikelet-bearing axes persistent. Spikelets not secund.

Female-fertile spikelets. *Spikelets* 1.6–5 mm long; strongly compressed laterally; *disarticulating above the glumes*. Rachilla prolonged beyond the uppermost female-fertile floret, or terminated by a female-fertile floret. The rachilla extension when present, naked. Hairy callus absent. *Callus* short.

Glumes two; more or less equal; long relative to the adjacent lemmas (exceeding them); pointed; shortly *awned*; carinate; similar (membranous, the margins overlapping for most of their length). Lower glume 3 nerved. Upper glume 3 nerved. *Spikelets* with female-fertile florets only.

Female-fertile florets *1*. *Lemmas* less firm than the glumes (membranous); not becoming indurated; entire; blunt (truncate or obtuse); *awnless*; hairy, or hairless (glabrous to densely ciliate); non-carinate; 5–7 nerved. **Palea** present; relatively long; entire to apically notched; awnless, without apical setae; not indurated; 1-nerved, or 2-nerved; *keelless*. *Lodicules* present; 2; free; membranous; glabrous; toothed; not or scarcely vascularized. *Stamens* 3. Anthers 0.3–2.3 mm long; not penicillate. *Ovary* glabrous. Styles fused, or free to their bases. Stigmas 2.

Fruit, embryo and seedling. Fruit small. Hilum short. Embryo small; not waisted. Endosperm liquid in the mature fruit; containing compound starch grains. Embryo with an epiblast; without a scutellar tail; with a negligible mesocotyl internode. Embryonic leaf margins meeting.

Seedling with a long mesocotyl; with a loose coleoptile, or with a tight coleoptile. First seedling leaf with a well-developed lamina. The lamina broad, or narrow; erect; 3 veined.

Abaxial leaf blade epidermis. *Costal/intercostal zonation* conspicuous. *Papillae* absent. *Long-cells* similar in shape costally and intercostally; of similar wall thickness costally and intercostally. Mid-intercostal long-cells fusiform; having markedly sinuous walls. *Microhairs* absent. *Stomata* common; (36–)42–45 microns long. Subsidiaries parallel-sided. Guard-cells overlapped by the interstomatals. *Intercostal short-cells* absent or very rare (rare); not paired (usually solitary); not silicified. *Crown cells* absent (but some reduced prickles approach these). *Costal short-cells* conspicuously in long rows. Costal silica bodies horizontally-elongated crenate/sinuous, or horizontally-elongated smooth, or rounded; not sharp-pointed.

Transverse section of leaf blade, physiology. C_3; XyMS+. *Mesophyll* with non-radiate chlorenchyma; without adaxial palisade. *Leaf blade* with distinct, prominent adaxial ribs; with the ribs more or less constant in size. *Midrib* conspicuous; with one bundle only. Bulliforms present in discrete, regular adaxial groups; in simple fans. Many of the smallest vascular bundles unaccompanied by sclerenchyma, or all the vascular bundles accompanied by sclerenchyma. Combined sclerenchyma girders present; nowhere forming 'figures'. Sclerenchyma all associated with vascular bundles.

Phytochemistry. Tissues of the culm bases with little or no starch. Fructosans predominantly long-chain. Leaves without flavonoid sulphates (2 species).

Cytology. Chromosome base number, $x = 7$. $2n = 10$ (rarely), or 14, or 28, or 42. 2, 4, and 6 ploid. Haploid nuclear DNA content 1.6–1.7 pg (2 species). Mean diploid 2c DNA value 3.4 pg (1 species).

Taxonomy. Pooideae; Poodae; Aveneae.

Ecology, geography, regional floristic distribution. 15 species. Temperate Eurasia, America. Commonly adventive. Mostly mesophytic; species of open habitats; mostly glycophytic, or halophytic. Meadows and dry places, *P. arenarium* in coastal sand.

Holarctic, Paleotropical, and Neotropical. Boreal, Tethyan, and Madrean. African and Indomalesian. Arctic and Subarctic, Euro-Siberian, Atlantic North American, and Rocky Mountains. Mediterranean and Irano-Turanian. Saharo-Sindian. Indian. Venezuela and Surinam and Andean. European and Siberian. Canadian-Appalachian.

Rusts and smuts. Rusts — *Puccinia*. Taxonomically wide-ranging species: *Puccinia graminis*, *Puccinia coronata*, *Puccinia striiformis*, *Puccinia brachypodii*, *Puccinia poarum*, and '*Uromyces*' *dactylidis*. Smuts from Tilletiaceae and from Ustilaginaceae. Tilletiaceae — *Entyloma*, *Tilletia*, and *Urocystis*. Ustilaginaceae — *Ustilago*.

Economic importance. Cultivated fodder: *P. bertolonii*, *P. pratense* (Timothy). Lawns and/or playing fields: *P. bertolonii*.

References, etc. Leaf anatomical: Metcalfe 1960 and this project.

Pholiurus Trin.

Habit, vegetative morphology. Annual; caespitose. *Culms* 5–40 cm high; herbaceous. Culm nodes glabrous. *Leaves* not basally aggregated; non-auriculate. Leaf blades linear; narrow; 1–3 mm wide; not setaceous; flat, or rolled (convolute). Ligule an unfringed membrane; not truncate; 3–4 mm long.

Reproductive organization. Plants bisexual, with bisexual spikelets; with hermaphrodite florets.

Inflorescence. *Inflorescence a single spike (tough, cylindrical)*. Rachides hollowed. Inflorescence espatheate; not comprising 'partial inflorescences' and foliar organs. Spikelet-bearing axes persistent. Spikelets solitary; not secund; distichous; sessile.

Female-fertile spikelets. Spikelets 4–7 mm long; somewhat compressed laterally; falling with the glumes; not disarticulating between the florets. *Rachilla terminated by a female-fertile floret*. Hairy callus absent.

Glumes two; more or less equal; long relative to the adjacent lemmas (exceeding them); lateral to the rachis; hairless; not pointed (obtuse, ovate); awnless; non-carinate; similar (leathery, strongly veined). Lower glume 5 nerved. Upper glume 5 nerved. *Spikelets* with female-fertile florets only.

Female-fertile florets 2. Lemmas less firm than the glumes (membranous); not becoming indurated; entire; pointed; awnless; hairless; non-carinate; 0–2 nerved. *Palea* present; relatively long; 2-nerved; 2-keeled. *Lodicules* present; 2; free; membranous; not toothed. *Stamens* 3. Anthers 1.8–2.5 mm long. *Ovary* glabrous. Stigmas 2.

Fruit, embryo and seedling. *Fruit* free from both lemma and palea. Endosperm liquid in the mature fruit to hard; containing compound starch grains.

Abaxial leaf blade epidermis. *Costal/intercostal zonation* conspicuous. *Papillae* absent. Mid-intercostal long-cells fusiform; having straight or only gently undulating walls (thin). *Microhairs* absent. *Stomata* common; (31.5–)36–39(–45) microns long. Subsidiaries parallel-sided. Guard-cells overlapped by the interstomatals. *Intercostal short-cells* absent or very rare. *Costal short-cells* neither distinctly grouped into long rows nor predominantly paired (solitary). Costal silica bodies horizontally-elongated crenate/sinuous; not sharp-pointed.

Transverse section of leaf blade, physiology. C_3; XyMS+. *Mesophyll* with non-radiate chlorenchyma; without adaxial palisade. Leaf blade with the ribs more or less constant in size. *Midrib* conspicuous; with one bundle only. Bulliforms not apparent in the poor material seen. All the vascular bundles accompanied by sclerenchyma. Combined sclerenchyma girders present (in the midrib only); forming 'figures' (midrib only). Sclerenchyma all associated with vascular bundles.

Special diagnostic feature. *Plant and inflorescence not as in* **Lygeum** *(q.v.).*

Cytology. Chromosome base number, $x = 7$ (?). $2n = 14$ and 36.

Taxonomy. Pooideae; Poodae; Poeae.

Ecology, geography, regional floristic distribution. 1 species. Southeast Europe. Commonly adventive. Species of open habitats; halophytic, or glycophytic. In coastal sands and saline soils.

Holarctic. Boreal and Tethyan. Euro-Siberian. Mediterranean and Irano-Turanian. European and Siberian.

Rusts and smuts. Smuts from Tilletiaceae. Tilletiaceae — *Tilletia*.

References, etc. Leaf anatomical: this project.

Special comments. Fruit data wanting.

Phragmites Adans.

Czernya Presl, *Miphragtes* Nieuwland, *Oxyanthe* Steud., *Trichoon* Roth, *Xenochloa* Roem. & Schult.

Habit, vegetative morphology. Perennial; *reeds (often forming dense stands)*; rhizomatous and stoloniferous. *Culms* 60–400 cm high (–1000 cm); woody and persistent to herbaceous (often somewhat persistent); branched above (especially when main culm damaged), or unbranched above. Culm nodes glabrous. Culm internodes hollow. Young shoots extravaginal. *Leaves* not basally aggregated; auriculate (scarcely), or non-auriculate. Leaf blades linear-lanceolate to lanceolate; broad; 6–50 mm wide; flat, or rolled (convolute); not pseudopetiolate; without cross venation; disarticulating from the sheaths; rolled in bud. *Ligule a fringe of hairs. Contra-ligule* absent.

Reproductive organization. Plants bisexual, with bisexual spikelets; with hermaphrodite florets.

Inflorescence. *Inflorescence* paniculate; *open (20–60 cm long, plumose, the fertile lemmas surrounded by long white silky hairs)*; with capillary branchlets (towards the spikelets); espatheate; not comprising 'partial inflorescences' and foliar organs. Spikelet-bearing axes persistent. Spikelets not secund; pedicellate.

Female-fertile spikelets. Spikelets 9–16 mm long; compressed laterally; disarticulating above the glumes (at least above the L_1); disarticulating between the florets. *Rachilla* prolonged beyond the uppermost female-fertile floret; *rachilla hairy (with very long white hairs, but only above the disarticulation zones of the florets)*. The rachilla extension with incomplete florets. Hairy callus present (with long fine silky hairs). *Callus* long (and slender); blunt.

Glumes two; very unequal; shorter than the spikelets; shorter than the adjacent lemmas; pointed; awnless; non-carinate (rounded on the back); similar (membranous). Lower glume 3–5 nerved. Upper glume 3–5 nerved. *Spikelets* with incomplete florets. The incomplete florets both distal and proximal to the female-fertile florets. *The distal incomplete florets* merely underdeveloped. *The proximal incomplete florets* 1; paleate; male (the stamens often 2), or sterile. The proximal lemmas awnless; 3(–7) nerved; more or less equalling the female-fertile lemmas; similar in texture to the female-fertile lemmas (membranous); not becoming indurated.

Female-fertile florets (2–)3–10. Lemmas narrow, lanceolate; similar in texture to the glumes (membranous); not becoming indurated; entire; pointed (acute to acuminate or aristulate); awnless, or awned (narrow-attenuate, muticous to aristulate). Awns (if lemmas aristulate) 1; median; apical; non-geniculate; much shorter than the body of the lemma. *Lemmas hairless*; glabrous; non-carinate; 1–3 nerved. *Palea* present; conspicuous but relatively short; entire; awnless, without apical setae (glabrous); 2-nerved. *Lodicules* present; 2; free; fleshy; ciliate, or glabrous. *Stamens* 3. Anthers 0.6–2.5 mm long; not penicillate. *Ovary* glabrous. Styles fused. Stigmas 2; brown.

Fruit, embryo and seedling. *Fruit* free from both lemma and palea; small; compressed dorsiventrally (to sub-terete). Hilum short. Embryo large; not waisted. Endosperm hard; without lipid; containing compound starch grains. Embryo without an epiblast; with a scutellar tail; with an elongated mesocotyl internode. Embryonic leaf margins meeting.

Seedling with a short mesocotyl, or with a long mesocotyl. First seedling leaf with a well-developed lamina. The lamina broad; curved; 13–20 veined.

Abaxial leaf blade epidermis. *Costal/intercostal zonation* conspicuous. *Papillae* absent. *Long-cells* similar in shape costally and intercostally; of similar wall thickness costally and intercostally (fairly thick walled). Intercostal zones with typical long-cells, or exhibiting many atypical long-cells (with many of them short). Mid-intercostal long-cells rectangular; having markedly sinuous walls. *Microhairs* present, or absent; when present, panicoid-type; 24–28.5 microns long; 4.5–6.3 microns wide at the septum. Microhair total length/width at septum 4.5–5.3. Microhair apical cells 12–13.5 microns long. Microhair apical cell/total length ratio 0.47–0.5. *Stomata* common; (30–)33–34.5(–36) microns long. Subsidiaries low dome-shaped. Guard-cells overlapped by the interstomatals, or overlapping to flush with the interstomatals. *Intercostal short-cells* common; in cork/silica-cell pairs; silicified. Intercostal silica bodies rounded, or crescentic. *Costal short-cells* neither distinctly grouped into long rows nor predominantly paired. Costal silica bodies rounded, or horizontally-elongated smooth to tall-and-narrow (or cuboid), or crescentic; not sharp-pointed.

Transverse section of leaf blade, physiology. C_3; XyMS+. *Mesophyll* with non-radiate chlorenchyma; without adaxial palisade; with arm cells. *Leaf blade* adaxially flat. *Midrib* conspicuous; having a conventional arc of bundles. Bulliforms present in discrete, regular adaxial groups; in simple fans (these with deeply penetrating median cells), or associated with colourless mesophyll cells to form deeply-penetrating fans (a few), or combining with colourless mesophyll cells to form narrow groups penetrating into the mesophyll. All the vascular bundles accompanied by sclerenchyma. Combined sclerenchyma girders present; nowhere forming 'figures'. Sclerenchyma all associated with vascular bundles.

Culm anatomy. *Culm internode bundles* in one or two rings, or in three or more rings.

Phytochemistry. Tissues of the culm bases with abundant starch. Leaves without flavonoid sulphates (2 species).

Special diagnostic feature. Female-fertile lemmas hairless; ligule hairs longer than 0.5 mm, longer than the membrane.

Cytology. Chromosome base number, $x = 12$. $2n = 36, 44, 46, 48, 49, 50, 51, 52, 54$, and 96. 3, 4, and 8 ploid (and aneuploids).

Taxonomy. Arundinoideae; Arundineae.

Ecology, geography, regional floristic distribution. 3 species. Cosmopolitan. Helophytic.

Holarctic, Paleotropical, Neotropical, Australian, and Antarctic. Boreal, Tethyan, and Madrean. African, Madagascan, Indomalesian, and Neocaledonian. Euro-Siberian, Eastern Asian, and Atlantic North American. Macaronesian, Mediterranean, and Irano-Turanian. Saharo-Sindian, Sudano-Angolan, West African Rainforest, and Namib-Karoo. Indian, Indo-Chinese, Malesian, and Papuan. Caribbean, Pampas, and Andean. North and East Australian and Central Australian. Patagonian. European and Siberian. Canadian-Appalachian and Southern Atlantic North American. Sahelo-Sudanian, Somalo-Ethiopian, South Tropical African, and Kalaharian. Tropical North and East Australian and Temperate and South-Eastern Australian.

Rusts and smuts. Rusts — *Puccinia*. Smuts from Tilletiaceae and from Ustilaginaceae. Tilletiaceae — *Neovossia*. Ustilaginaceae — *Ustilago*.

Economic importance. Significant weed species: *P. australis*, *P. karka* (causing water loss and stagnation, and impeding navigation). Used for mats and thatching.

References, etc. Leaf anatomical: Metcalfe 1960; this project.

Phyllorhachis Trimen

Habit, vegetative morphology. Perennial (with slender culms); rhizomatous. *Culms* scrambling, 50–130 cm high (or more); herbaceous; to 0.2 cm in diameter; branched above. Culm nodes hairy, or glabrous. Culm sheaths persistent (with the blades, above), or deciduous in their entirety (from the bases of the old culms). Culm internodes hollow. Plants unarmed. Young shoots extravaginal. *Leaves* not basally aggregated; non-auriculate; without auricular setae. *Leaf blades* lanceolate to ovate-lanceolate; broad; 10–25 mm wide (and to 12 cm long); *sagittate*; flat; pseudopetiolate (the pseudopetiole to 2.5 mm long); without cross venation; disarticulating from the sheaths (in association with culm branching). Ligule a fringe of hairs; 0.2 mm long. *Contra-ligule* absent.

Reproductive organization. Plants monoecious with all the fertile spikelets unisexual; without hermaphrodite florets. The spikelets of sexually distinct forms on the same plant; female-only and male-only. The male and female-fertile spikelets mixed in the inflorescence (single female spikelets basal to male spikelets in the very contracted branches — glomerules — of the terminal inflorescences, and with female spikelets in small, homogamous axillary inflorescences of 1 or 2 spikelets each). The spikelets overtly heteromorphic.

Inflorescence. *Inflorescence terminal, consisting of heterogamous racemules each of 3–4 spikelets, borne on flattened secondary rachides arranged in two series on one side of an expanded, herbaceous, primary rachis; and sometimes also with axillary, homogamous racemules of 2–3 female spikelets on slender rachides.* Inflorescence axes not ending in spikelets (with a scabrid bristle in each glomerule seeming to represent the branch tip). Rachides flattened and winged (spathe-like). Inflorescence espatheate (but the rachis flattened, spathe-like); not comprising 'partial inflorescences' and foliar organs. Spikelet-bearing axes disarticulating; falling entire (i.e., the racemules fall from the persistent main axis). Spikelets consistently in 'long-and-short' combinations (in the terminal inflorescence). The 'shorter' spikelets female-only and male-only. The 'longer' spikelets male-only and sterile.

Female-sterile spikelets. Male spikelets with glumes 2, short, similar, keel-less, oblong or oval, awnless; G_1 nerveless, G_2 1-nerved; one proximal, 3-nerved sterile lemma; one oblong, 5–7 nerved fertile lemma enclosing a 2-nerved palea and 6 free stamens with 2.5 mm anthers. Rachilla of male spikelets terminated by a male floret. The male spikelets with glumes; with proximal incomplete florets; 2 floreted (the proximal sterile). Male florets 1; 6 staminate. The staminal filaments free.

Female-fertile spikelets. *Spikelets* unconventional (the 'G_1' and palea highly peculiar); 10–16 mm long; abaxial (with respect to the main axis); compressed laterally, or not noticeably compressed, or compressed dorsiventrally (?); falling with the glumes (the glomerules disarticulating). Rachilla terminated by a female-fertile floret. Hairy callus absent. *Callus* absent.

Glumes two; more or less equal; shorter than the adjacent lemmas; awned (G_1 awn-like); carinate (G_2); *very dissimilar (the G_1 subulate, rigid and awnlike, the G_2 oblong and firm)*. Lower glume 0–1 nerved. Upper glume 5–9 nerved. *Spikelets* with incomplete florets. The incomplete florets proximal to the female-fertile florets. **The proximal incomplete florets** 1; *sterile (reduced, rigid, thick)*. The proximal lemmas awnless; many nerved, broadly lanceolate, sulcate in the back; more or less equalling the female-fertile lemmas to decidedly exceeding the female-fertile lemmas; decidedly firmer than the female-fertile lemmas.

Female-fertile florets 1. Lemmas caudate-acuminate; not becoming indurated; entire; pointed; awnless; hairless; non-carinate; without a germination flap; 11–17 nerved. *Palea* present (sulcate between the two principal nerves); relatively long; entire (pointed); awnless, without apical setae; several nerved (8–12). *Lodicules* present; 2; free; membranous; glabrous. *Stamens* 0 (6 rudiments only). *Ovary* glabrous. Styles fused (into one, long). Stigmas 2; creamy white.

Fruit, embryo and seedling. *Fruit* free from both lemma and palea; medium sized (6–7 mm long); fusiform; slightly longitudinally grooved; not noticeably compressed. Hilum long-linear. Embryo small (about 1/6 the length of the fruit).

Abaxial leaf blade epidermis. *Costal/intercostal zonation* conspicuous. *Papillae* absent. Intercostal zones exhibiting many atypical long-cells (many being rather short). Mid-intercostal long-cells having markedly sinuous walls (thin walled). *Microhairs* present; panicoid-type; 42–50 microns long. Microhair apical cells 25–30 microns long. Microhair apical cell/total length ratio 0.6. *Stomata* common. Subsidiaries triangular. *Intercostal short-cells* common; in cork/silica-cell pairs; silicified. Intercostal silica bodies tall-and-narrow, or tall-and-narrow and oryzoid-type. Macrohairs present. *Costal short-cells* conspicuously in long rows. Costal silica bodies oryzoid (mostly), or saddle shaped (tendency), or 'panicoid-type' (tendency); when panicoid type, tending to cross shaped; not sharp-pointed.

Transverse section of leaf blade, physiology. C_3; XyMS+. *Mesophyll* with non-radiate chlorenchyma; with arm cells; without fusoids. *Leaf blade* adaxially flat. *Midrib* conspicuous; having a conventional arc of bundles (a median bundle and 2 small laterals). Bulliforms present in discrete, regular adaxial groups; in simple fans (the groups wide). All the vascular bundles accompanied by sclerenchyma. Combined sclerenchyma girders present; forming 'figures'. Sclerenchyma all associated with vascular bundles.

Special diagnostic feature. *Spikelets borne on one side of a broad, leaflike rachis.*

Cytology. Chromosome base number, $x = 12$. $2n = 24$. 2 ploid.

Taxonomy. Bambusoideae; Oryzodae; Phyllorhachideae.

Ecology, geography, regional floristic distribution. 1 species. Southern tropical Africa. Mesophytic; shade species; glycophytic. In riverine forests.

Paleotropical. African. Sudano-Angolan. South Tropical African.

References, etc. Leaf anatomical: Metcalfe 1960.

Phyllostachys Sieb & Zucc.

Sinoarundinaria Ohwi

Habit, vegetative morphology. Arborescent or shrubby perennial; rhizomatous. The flowering culms leafy. **Culms *300–2000 cm high*;** woody and persistent; to 20 cm in diameter; flattened on one side; branched above. *Primary branches/mid-culm node 2 (but rebranching).* Culm sheaths deciduous in their entirety. Culm internodes hollow. Rhizomes leptomorph. Plants unarmed. *Leaves* not basally aggregated; auriculate, or non-auriculate; with auricular setae. *Leaf blades* broad; pseudopetiolate; *cross veined*; disarticulating from the sheaths; rolled in bud.

Reproductive organization. Plants bisexual, with bisexual spikelets; with hermaphrodite florets.

Inflorescence. Inflorescence compound paniculate (of spicate, 1-spikeleted branchlets, aggregated into spatheate clusters); spatheate; a complex of 'partial inflorescences' and intervening foliar organs (with or without foliage leaves). *Spikelet-bearing axes* 'racemes' and very much reduced; clustered; persistent. Spikelets not secund.

Female-fertile spikelets. Spikelets 18–80 mm long; compressed laterally to not noticeably compressed; disarticulating above the glumes; disarticulating between the florets. Rachilla prolonged beyond the uppermost female-fertile floret; rachilla hairless (glabrous). Hairy callus absent.

Glumes one per spikelet, or two (or 3, and the lateral spikelets with an outer bract at the base); shorter than the adjacent lemmas; pointed; awnless. Lower glume many-nerved. Upper glume many nerved. *Spikelets* with female-fertile florets only, or with incomplete florets. The incomplete florets if present, distal to the female-fertile florets.

Female-fertile florets 1–4. Lemmas entire; pointed; awnless, or mucronate (?); carinate to non-carinate; lemmas many veined. *Palea* present; relatively long; apically notched; awnless, without apical setae to with apical setae; several nerved; 2-keeled. *Lodicules* present; 2, or 3; free; membranous; ciliate, or glabrous; heavily vascularized. **Stamens 3.** Anthers penicillate, or not penicillate; with the connective apically prolonged, or without an apically prolonged connective. **Ovary** glabrous; *with a conspicuous apical appendage.* The appendage broadly conical, fleshy. Styles fused (into one, long). Stigmas 2–3.

Fruit, embryo and seedling. Fruit longitudinally grooved. Hilum long-linear. Embryo small. Endosperm containing compound starch grains. Embryo with an epiblast; with a scutellar tail; with a negligible mesocotyl internode. Embryonic leaf margins overlapping.

Abaxial leaf blade epidermis. *Costal/intercostal zonation* conspicuous. *Papillae* present. Intercostal papillae over-arching the stomata; several per cell. Mid-intercostal long-cells having markedly sinuous walls (thin walled). *Microhairs* present; panicoid-type (but variable in shape). *Stomata* common (outlines often more or less obscured by papillae). Subsidiaries low to high dome-shaped. *Intercostal short-cells* common, or absent or very rare; in cork/silica-cell pairs, or not paired; silicified. Intercostal silica bodies tall-and-narrow. *Costal short-cells* conspicuously in long rows, or neither distinctly grouped into long rows nor predominantly paired. Costal silica bodies saddle shaped; not sharp-pointed.

Transverse section of leaf blade, physiology. C_3; XyMS+. Mesophyll with arm cells; with fusoids (rarely), or without fusoids (usually, or fusoids if present rare and/or inconspicuous). *Leaf blade* with distinct, prominent adaxial ribs (these low), or adaxially flat; with the ribs more or less constant in size. *Midrib* conspicuous; having complex vascularization. Bulliforms present in discrete, regular adaxial groups; in simple fans and associated with colourless mesophyll cells to form deeply-penetrating fans. All the vascular bundles accompanied by sclerenchyma. Combined sclerenchyma girders present (with most bundles); forming 'figures' (with most bundles).

Cytology. Chromosome base number, $x = 12$. $2n = 24$ (rarely), or 48, or 72. 2 ploid, or 4 ploid, or 6 ploid.

Taxonomy. Bambusoideae; Bambusodae; Bambuseae.

Ecology, geography, regional floristic distribution. About 50 species. Eastern Asia. Holarctic, Paleotropical, and Neotropical. Boreal. Indomalesian. Eastern Asian. Indian and Indo-Chinese. Caribbean.

Rusts and smuts. Rusts — *Stereostratum* and *Puccinia*. Taxonomically wide-ranging species: *Stereostratum corticoides*, *Puccinia longicornis*, and *Puccinia kusanoi*. Smuts from Ustilaginaceae. Ustilaginaceae — *Ustilago*.

Economic importance. Culms of *P. aurea*, *P. bambusoides*, *P. glauca*, *P. nigra*, *P. vivax* used for walking sticks, fishing rods, furniture, handicrafts etc; young shoots of *P. aurea*, *P. bambusoides*, *P. glauca*, *P. nidularia*, *P. vivax* eaten as vegetables.

References, etc. Leaf anatomical: Metcalfe 1960; this project.

Pilgerochloa Eig

~ Ventenata

Habit, vegetative morphology. *Annual*. Culms herbaceous. Leaves non-auriculate. Leaf blades without cross venation. *Ligule an unfringed membrane*; not truncate; 2–4 mm long.

Reproductive organization. Plants bisexual, with bisexual spikelets; with hermaphrodite florets.

Inflorescence. *Inflorescence* paniculate; *open*; espatheate; not comprising 'partial inflorescences' and foliar organs. Spikelet-bearing axes persistent. Spikelets not secund; pedicellate.

Female-fertile spikelets. Spikelets 9–11 mm long; compressed laterally; disarticulating above the glumes. Rachilla prolonged beyond the uppermost female-fertile floret. The rachilla extension with incomplete florets. Hairy callus absent.

Glumes two; *very unequal*; shorter than the adjacent lemmas; pointed (acuminate); awnless; carinate; similar. Lower glume 1 nerved. Upper glume 3 nerved. *Spikelets* with incomplete florets. The incomplete florets distal to the female-fertile florets.

Female-fertile florets 2–4. *Lemmas* similar in texture to the glumes to decidedly firmer than the glumes; not becoming indurated; all *awned (including the lowest)*. *Awns* 1; median; dorsal; from well down the back (from the middle third); *geniculate*; entered by one vein. Lemmas hairy; non-carinate; without a germination flap; 5–6 nerved. **Palea** present; *conspicuous but relatively short (about one half the lemma length)*; 2-nerved;

2-keeled. *Lodicules* present; 2; free; membranous; glabrous; toothed; not or scarcely vascularized. *Stamens* 3. *Ovary* glabrous. Styles free to their bases. Stigmas 2.

Fruit, embryo and seedling. *Fruit* free from both lemma and palea. Embryo small. *Endosperm liquid in the mature fruit.*

Abaxial leaf blade epidermis. *Costal/intercostal zonation* conspicuous (seemingly involving only the midrib). *Papillae* absent. *Long-cells* markedly different in shape costally and intercostally (the costals much narrower); of similar wall thickness costally and intercostally to differing markedly in wall thickness costally and intercostally (the walls of medium thickness, those of the costals tending to be thicker). Mid-intercostal long-cells rectangular (very long, rarely tending slightly to fusiform); having markedly sinuous walls (the sinuosity fairly fine and even, associated with conspicuous pitting). *Microhairs* absent. *Stomata* fairly common (but mostly confined to a single file on either side of the midrib); 45–54 microns long. Subsidiaries non-papillate; parallel-sided to dome-shaped. Guard-cells slightly but consistently overlapped by the interstomatals (the apparatus not noticeably sunken). *Intercostal short-cells* fairly common; in cork/silica-cell pairs (often), or not paired (a few solitary); silicified, or not silicified. *Costal short-cells* predominantly paired (both cells horizontally quite elongated). Costal silica bodies present and well developed; horizontally-elongated crenate/sinuous to horizontally-elongated smooth, or rounded (sometimes numerous, mostly more or less irregular), or crescentic (few).

Transverse section of leaf blade, physiology. C_3; XyMS+. *Leaf blade* with distinct, prominent adaxial ribs; with the ribs more or less constant in size. *Midrib* conspicuous (by virtue of its large abaxial keel); with one bundle only. *The lamina* symmetrical on either side of the midrib. Bulliforms not apparent in the poor material seen. All the vascular bundles accompanied by sclerenchyma. Combined sclerenchyma girders absent (only small adaxial strands, save for the small abaxial one with the midrib). Sclerenchyma all associated with vascular bundles.

Taxonomy. Pooideae; Poodae; Aveneae.

Ecology, geography, regional floristic distribution. 1 species. Asia Minor. Xerophytic.

Holarctic. Tethyan. Irano-Turanian.

References, etc. Leaf anatomical: this project.

Piptatherum P. Beauv.

~ *Oryzopsis*

Habit, vegetative morphology. Perennial; densely or loosely caespitose. *Culms* 10–140 cm high; herbaceous; unbranched above. Culm nodes glabrous. Culm internodes solid, or hollow (?). Young shoots extravaginal, or intravaginal. Leaves non-auriculate. Leaf blades broad to narrow; 0.7–15 mm wide; not setaceous; flat, or folded, or rolled (involute); without cross venation; persistent; rolled in bud, or once-folded in bud (?). Ligule an unfringed membrane; truncate to not truncate; 0.2–15 mm long.

Reproductive organization. Plants bisexual, with bisexual spikelets; with hermaphrodite florets.

Inflorescence. Inflorescence paniculate; open (usually with 1–3(–5) branches at the lower nodes); with capillary branchlets, or without capillary branchlets; espatheate; not comprising 'partial inflorescences' and foliar organs. Spikelet-bearing axes persistent. Spikelets not secund; pedicellate.

Female-fertile spikelets. Spikelets 2–9 mm long; compressed dorsiventrally (dorsally compressed to subterete); disarticulating above the glumes. Rachilla terminated by a female-fertile floret. Hairy callus absent. *Callus* short; blunt (glabrous).

Glumes present; two; more or less equal (subequal); about equalling the spikelets, or exceeding the spikelets; long relative to the adjacent lemmas; generally with small asperities, at least above; pointed; awnless; non-carinate; similar. Lower glume 3–5 nerved (–7). Upper glume 3–7 nerved. *Spikelets* with female-fertile florets only.

Female-fertile florets 1. Lemmas not saccate; decidedly firmer than the glumes; becoming indurated; usually dark brown to black and shiny, brown in *P. miliaceum*; inconspicuously incised; obscurely 2 lobed; not deeply cleft; awned. Awns 1 (fine);

median; from a sinus; non-geniculate; usually straight (erect, neither twisted nor curved); hairless (scabrous); much shorter than the body of the lemma to much longer than the body of the lemma; deciduous (except in *P. virescens*). Lemmas hairy (the hairs white or golden to brown), or hairless; non-carinate; 3–5 nerved (rarely more?). *Palea* present (only marginally covered by the lemma); relatively long; awnless, without apical setae; textured like the lemma; indurated; 2-nerved; keel-less. *Lodicules* present; 3; free; membranous (two 'stipoid' and two-nerved, one thinner and nerveless); glabrous; not toothed; heavily vascularized (the stipoid pair), or not or scarcely vascularized (the third). *Stamens* 3. Anthers 1.5–4.5 mm long; penicillate (usually), or not penicillate. *Ovary* glabrous. Styles free to their bases. Stigmas 2.

Fruit, embryo and seedling. *Fruit* free from both lemma and palea; small to medium sized (1.5–4.0 mm long); compressed dorsiventrally. Hilum long-linear. Embryo large to small (usually about a third of the grain length). Endosperm hard; without lipid. Embryo with an epiblast; without a scutellar tail; with a negligible mesocotyl internode. Embryonic leaf margins meeting (unreliably — these embryo data, from Reeder, may refer to *Oryzopsis*).

Seedling with a long mesocotyl. First seedling leaf with a well-developed lamina. The lamina narrow; erect.

Abaxial leaf blade epidermis. *Costal/intercostal zonation* conspicuous. *Papillae* absent. *Long-cells* similar in shape costally and intercostally (narrowly rectangular); of similar wall thickness costally and intercostally (fairly thick walled, the costals somewhat less so). Mid-intercostal long-cells rectangular; having markedly sinuous walls. *Microhairs* absent (but present adaxially in at least some species); one-celled?; cf. those of *Stipa*. Stomata common; 21–24 microns long. Subsidiaries low to high dome-shaped. Guard-cells overlapping to flush with the interstomatals. *Intercostal short-cells* common; in cork/silica-cell pairs and not paired (some solitary); silicified (in the pairs). *Costal short-cells* conspicuously in long rows (mostly, with a few pairs). Costal silica bodies rounded (a few, including potato shapes), or tall-and-narrow to crescentic, or 'panicoid-type'; when panicoid type, cross shaped, or dumb-bell shaped; not sharp-pointed.

Transverse section of leaf blade, physiology. C_3; XyMS+. *Mesophyll* with non-radiate chlorenchyma; without adaxial palisade. *Leaf blade* 'nodular' in section; with the ribs very irregular in sizes. *Midrib* conspicuous to not readily distinguishable; with one bundle only. Bulliforms present in discrete, regular adaxial groups; in simple fans. Many of the smallest vascular bundles unaccompanied by sclerenchyma, or all the vascular bundles accompanied by sclerenchyma. Combined sclerenchyma girders present; forming 'figures'. Sclerenchyma all associated with vascular bundles.

Culm anatomy. *Culm internode bundles* in one or two rings.

Cytology. Chromosome base number, $x = 12$. $2n = 24$. 2 ploid.

Taxonomy. Arundinoideae; Stipeae.

Ecology, geography, regional floristic distribution. 25 species, or 26 species. Old World subtropics. Commonly adventive. Mesophytic to xerophytic.

Holarctic. Boreal and Tethyan. Euro-Siberian. Macaronesian and Mediterranean. European.

Economic importance. Significant weed species: *P. miliaceum*. Cultivated fodder: *P. miliaceum*.

References, etc. Morphological/taxonomic: Freitag 1975. Leaf anatomical: Metcalfe 1960 and this project.

Special comments. Not separable from the *sensu lato* description of *Oryzopsis* in this data set.

Piptochaetium Presl

Caryochloa Spreng., *Podopogon* Raf.

Habit, vegetative morphology. Perennial; caespitose. *Culms* 30–150 cm high; herbaceous; unbranched above. Young shoots intravaginal. *Leaves* mostly basal; non-auriculate (but with thickenings at the auricle positions). Leaf blades broad, or narrow (narrow, usually involute); without cross venation; persistent. Ligule an unfringed membrane; 0.3–3(–8) mm long. *Contra-ligule* absent.

Reproductive organization. Plants bisexual, with bisexual spikelets; with hermaphrodite florets. Exposed-cleistogamous, or chasmogamous.

Inflorescence. Inflorescence few spikeleted to many spikeleted; paniculate; open (usually), or contracted; with capillary branchlets, or without capillary branchlets; espatheate; not comprising 'partial inflorescences' and foliar organs. Spikelet-bearing axes persistent. Spikelets not secund; usually long pedicellate.

Female-fertile spikelets. Spikelets 2–24 mm long; compressed laterally to not noticeably compressed; disarticulating above the glumes. Rachilla terminated by a female-fertile floret. Hairy callus usually present. *Callus* short; acute.

Glumes two; more or less equal; about equalling the spikelets, or exceeding the spikelets; long relative to the adjacent lemmas (usually exceeding the floret); hairless; pointed (abruptly acuminate); awnless; non-carinate (convex on the back); similar (thin). Lower glume 3–7 nerved. Upper glume 3–7 nerved. *Spikelets* with female-fertile florets only.

Female-fertile florets 1. **Lemmas** often dark-pigmented, sometimes cylindrical, usually asymmetrically obovoid to subglobose or lenticular; not convolute; *saccate*; decidedly firmer than the glumes; becoming indurated (smooth, verrucose, striate or tuberculate); awned. **Awns** 1; median; *apical (the summit of the lemma often expanded into a corona)*; geniculate (often bi-geniculate), or geniculate; sometimes non-geniculate and curved; hairless (glabrous), or hairy; much longer than the body of the lemma, or much shorter than the body of the lemma (much reduced, in *P. cucullatum*); deciduous, or persistent. Lemmas hairy, or hairless; *carinate (usually somewhat compressed and keeled)*; without a germination flap; 5 nerved. **Palea** present; *relatively long (the apex of the keel projecting as a minute point above the summit of the lemma)*; entire; pointed, sometimes with an apical mucro; indurated (except marginally); 2-nerved (these fairly adjacent); *2-keeled (with a narrow sulcus between the two close indurated keels, which are not enclosed by the lemma)*. Lodicules present; 2, or 3; free; membranous; glabrous; not or scarcely vascularized. *Stamens* 3. Anthers not penicillate; without an apically prolonged connective. *Ovary* glabrous. Styles free to their bases. Stigmas 2.

Fruit, embryo and seedling. Fruit longitudinally grooved. *Hilum long-linear*. Endosperm hard; without lipid.

Abaxial leaf blade epidermis. *Costal/intercostal zonation* conspicuous. *Papillae* absent. Long-cells of similar wall thickness costally and intercostally. Mid-intercostal long-cells rectangular, or rectangular and fusiform; having markedly sinuous walls. *Microhairs* absent. Stomata absent or very rare. *Intercostal short-cells* common; in cork/silica-cell pairs; silicified. Intercostal silica bodies tall-and-narrow, or crescentic, or cross-shaped. Small prickles abundant. *Crown cells* absent. *Costal short-cells* conspicuously in long rows. Costal silica bodies 'panicoid-type'; not sharp-pointed.

Transverse section of leaf blade, physiology. C$_3$; XyMS+. Mesophyll traversed by columns of colourless mesophyll cells (at the 'hinge positions'). *Leaf blade* with distinct, prominent adaxial ribs to 'nodular' in section (having only 3 similar sized veins, the lamina very thin and lacking photosynthetic tissue between the 3 'ribs'). Midrib with one bundle only. Bulliforms present in discrete, regular adaxial groups (the groups broad, at the bases of the furrows). All the vascular bundles accompanied by sclerenchyma. Combined sclerenchyma girders absent (combining adaxial and abaxial strands, or adaxial strands and abaxial girders). Sclerenchyma not all bundle-associated. The 'extra' sclerenchyma in abaxial groups; abaxial-hypodermal, the groups continuous with colourless columns.

Cytology. Chromosome base number, $x = 11$, or 12 (?). $2n = 22$ and 44. 2 and 4 ploid.

Taxonomy. Arundinoideae; Stipeae.

Ecology, geography, regional floristic distribution. About 30 species. North & (mainly) South America. Mesophytic to xerophytic; species of open habitats. Steppe.

Holarctic, Neotropical, and Antarctic. Madrean. Caribbean, Pampas, Andean, and Fernandezian. Patagonian.

Rusts and smuts. Rusts — *Puccinia*. Taxonomically wide-ranging species: *Puccinia graminella*. Smuts from Tilletiaceae and from Ustilaginaceae. Tilletiaceae — *Tilletia*. Ustilaginaceae — *Ustilago*.

Economic importance. Important native pasture species: *P. fimbriatum*.
References, etc. Leaf anatomical: Metcalfe 1960; this project.

Piptophyllum C.E. Hubb.

'*Pentaschistis welwitschii*'
Cf. *Triraphis, Crinipes*

Habit, vegetative morphology. Slender, erect perennial; caespitose (the base tomentose and fibrous). Culms herbaceous; unbranched above. Leaves non-auriculate. *Leaf blades* narrow; setaceous (folded); without cross venation; *disarticulating from the sheaths*. Ligule present; *a fringe of hairs*.

Reproductive organization. Plants bisexual, with bisexual spikelets; with hermaphrodite florets.

Inflorescence. Inflorescence determinate; without pseudospikelets; paniculate; open (lanceolate-oblong or narrowly oblong); *with capillary branchlets*; non-digitate; espatheate; not comprising 'partial inflorescences' and foliar organs. Spikelet-bearing axes persistent. Spikelets not secund; pedicellate.

Female-fertile spikelets. Spikelets compressed laterally; disarticulating above the glumes; disarticulating between the florets; with conventional internode spacings (the internodes very short). Rachilla prolonged beyond the uppermost female-fertile floret. The rachilla extension naked.

Glumes two; relatively large; more or less equal (only slightly unequal); shorter than the spikelets; *shorter than the adjacent lemmas*; acute or the apex minutely toothed; awnless (but mucronate); similar (thinly membranous). Lower glume 1 nerved. Upper glume 1 nerved. **Spikelets** with female-fertile florets only; *without proximal incomplete florets*.

Female-fertile florets 2. *Lemmas* not becoming indurated (herbaceous membranous, becoming firm); incised; *2 lobed*; not deeply cleft (the lobes narrow, terminating in short setae); *awned*. Awns 1; median; from a sinus; non-geniculate to geniculate (very slender, curved or flexuose above but flattened and twisted at the base). Lemmas hairy (pilose on the margins); non-carinate (convex); 5–9 nerved. *Palea* present; relatively long; entire (obtuse), or apically notched; awnless, without apical setae; not indurated (thinly membranous); 2-nerved; 2-keeled. *Lodicules* present; 2; fleshy (minute); minutely ciliate, or glabrous. *Stamens* 3. *Ovary* glabrous. Styles free to their bases. Stigmas 2.

Fruit, embryo and seedling. Fruit not noticeably compressed (terete). *Hilum short (but linear)*. Embryo large.

Taxonomy. Arundinoideae, or Chloridoideae; main chloridoid assemblage (?).

Ecology, geography, regional floristic distribution. 1 species. Angola. Mesophytic; species of open habitats. Damp rocky places.

Paleotropical. African. Sudano-Angolan. South Tropical African.

References, etc. Morphological/taxonomic: Hubbard 1957.

Special comments. Anatomical data wanting.

Piresia Swallen

Habit, vegetative morphology. Perennial. *The flowering culms mostly leafless (and decumbent in the leaf litter, but erect foliage culms also sometimes having a terminal inflorescence)*. Culms 8–40 cm high; herbaceous; unbranched above. Plants unarmed. Leaves not basally aggregated; without auricular setae. Leaf blades pseudopetiolate; cross veined. *Contra-ligule* absent.

Reproductive organization. *Plants monoecious with all the fertile spikelets unisexual*; without hermaphrodite florets. The spikelets of sexually distinct forms on the same plant; female-only and male-only. *The male and female-fertile spikelets mixed in the inflorescence (one or two males beneath each female)*. The spikelets overtly heteromorphic.

Inflorescence. Inflorescence paniculate (depauperate, racemelike); spatheate; a complex of 'partial inflorescences' and intervening foliar organs (with spatheoles?).

Spikelet-bearing axes 'racemes', or paniculate; persistent. *Spikelets* in triplets, or paired; not secund; *not in distinct 'long-and-short' combinations*.

Female-sterile spikelets. Male spikelets usually lacking glumes, with 3 free, non-penicillate stamens. Rachilla of male spikelets terminated by a male floret. The male spikelets usually without glumes; without proximal incomplete florets. Male florets 1; 3 staminate. The staminal filaments free.

Female-fertile spikelets. Spikelets 5–7 mm long; narrowly elliptic; with convention-al internode spacings. Rachilla terminated by a female-fertile floret.

Glumes two; more or less equal; long relative to the adjacent lemmas; pointed; awn-less (apiculate); non-carinate; similar (herbaceous). Lower glume 3 nerved, or 5 nerved. Upper glume 3 nerved, or 5 nerved. *Spikelets* with female-fertile florets only.

Female-fertile florets 1. Lemmas decidedly firmer than the glumes (leathery); bec-oming indurated; awnless; hairy (appressed pilose); non-carinate; having the margins tucked in onto the palea. *Palea* present; relatively long; entire; awnless, without apical setae; textured like the lemma; indurated; 2-nerved. *Stamens* 0 (staminodes 0 or three). Styles fused. Stigmas 2.

Fruit, embryo and seedling. Fruit compressed dorsiventrally. *Hilum long-linear.* Embryo small.

Abaxial leaf blade epidermis. *Costal/intercostal zonation* conspicuous. *Papillae* present (and very abundant); costal and intercostal (but far more conspicuous intercostally). Intercostal papillae over-arching the stomata; several per cell (one or two irregular rows of irregularly shaped, rather angular papillae per long-cell). Mid-intercostal long-cells rectangular; having markedly sinuous walls. *Microhairs* present; elongated; clearly two-celled; panicoid-type; 24–39 microns long; 3.6–4.5 microns wide at the septum. Microhair total length/width at septum 7.7–10.8. Microhair apical cells 15–17.5 microns long. Microhair apical cell/total length ratio 0.4–0.45. *Stomata* common; 18–21 microns long. Subsidiaries papillate (commonly two on each); high dome-shaped (mostly), or triangular. *Intercostal short-cells* common; in cork/silica-cell pairs; silicified. Intercostal silica bodies vertically elongated-nodular. Large, cushion-based macrohairs common. *Crown cells* absent. *Costal short-cells* conspicuously in long rows. Costal silica bodies 'panicoid-type'; mostly Maltese- cross shaped (sometimes almost 'oryziod'), or butterfly shaped to dumb-bell shaped (a few, short); not sharp-pointed.

Transverse section of leaf blade, physiology. C_3; XyMS+. Mesophyll with adaxial palisade; with arm cells (the cells adjoining the fusoids with a few large and conspicuous ingrowths); with fusoids. The fusoids external to the PBS. *Leaf blade* adaxially flat. *Midrib* conspicuous (by virtue of its large bundle and adaxially prominent rib); with one bundle only. *The lamina* symmetrical on either side of the midrib. Bulliforms present in discrete, regular adaxial groups; in simple fans (the groups large and wide, one in each intercostal zone). All the vascular bundles accompanied by sclerenchyma. Combined scle-renchyma girders present; forming 'figures' (all the bundles with a small 'anchor', except the midrib with strands only). Sclerenchyma all associated with vascular bundles.

Special diagnostic feature. *Not having female spikelets as in* **Leptaspis** *and* **Scrotochloa** *(q.v.).*

Cytology. Chromosome base number, $x = 11$. $2n = 22$. 2 ploid.

Taxonomy. Bambusoideae; Oryzodae; Olyreae.

Ecology, geography, regional floristic distribution. 4 species. Tropical America. Mesophytic; shade species.

Neotropical. Caribbean, Venezuela and Surinam, Amazon, and Central Brazilian.

References, etc. Morphological/taxonomic: Swallen 1964; Soderstrom 1982b. Leaf anatomical: this project.

Piresiella Judziewicz, Zuloaga & Morone

Habit, vegetative morphology. Delicate perennial (with dimorphic culms); stolonif-erous and caespitose. The flowering culms leafless (with 1–2 bladeless sheaths). **Culms** 3–20 cm high (the flowering culms shorter than the vegetative culms); herbaceous; un-branched above; *tuberous*; 3–9 noded (the leaves concentrated towards the tips of the vegetative culms). Culm nodes exposed; hairy (retrorsely slightly bearded). Rhizomes pachymorph. Plants unarmed. *Leaves* not basally aggregated; non-auriculate; without auricular setae. Sheaths glabrous except on one margin, ciliate at the summit. *Leaf blades ovate (strongly asymmetrical at the base)*; broad to narrow; 6–10 mm wide (14–20 mm long); not cordate, not sagittate (the base truncate); flat; shortly pseudopetiolate; persistent; rolled in bud. Ligule an unfringed membrane (ciliolate); truncate; about 0.1 mm long.

Reproductive organization. *Plants monoecious with all the fertile spikelets unisexual*; without hermaphrodite florets. The spikelets of sexually distinct forms on the same plant; female-only and male-only. The male and female-fertile spikelets on different branches of the same inflorescence. The spikelets overtly heteromorphic.

Inflorescence. **Inflorescence** determinate; *of spicate main branches (terminal, exserted, each consisting of two conjugate, ascending, appressed, subequal, unisexual racemes each with 6–9 spikelets, the male raceme somewhat shorter)*; digitate; espatheate; not comprising 'partial inflorescences' and foliar organs. *Spikelet-bearing axes* 'racemes'; paired; with very slender rachides. Spikelets solitary; secund (the spikelets on one side of the slender raceme); shortly pedicellate, the male pedicels 0.3–1.5 mm long, the females 0.2–0.6 mm.

Female-sterile spikelets. The male spikelets on slightly shorter racemes, 1.3–1.8 mm long, ellipsoid, glabrous, persistent. Rachilla of male spikelets terminated by a male floret. The male spikelets without glumes; without proximal incomplete florets; 1 floreted. The lemmas awnless. Male florets 1; 2 staminate.

Female-fertile spikelets. Spikelets 4–4.8 mm long; ovoid; compressed dorsiventrally; falling with the glumes ('or the glumes tardily deciduous'); with a distinctly elongated rachilla internode above the glumes (this 0.3–0.5 mm long, elaiosome-like). Rachilla terminated by a female-fertile floret. Hairy callus absent.

Glumes two; more or less equal; exceeding the spikelets; long relative to the adjacent lemmas; hairless; glabrous; pointed (acute to acuminate); awnless; non-carinate; similar (narrowly ovate, chartaceous, sulcate, greenish or whitish, sometimes purple-tipped). Lower glume 3(–5) nerved (with a few cross-veins). Upper glume 3(–5) nerved (with a few cross-veins). *Spikelets* with female-fertile florets only.

Female-fertile florets 1. Lemmas similar in texture to the glumes, or decidedly firmer than the glumes (?); becoming indurated, or not becoming indurated (?); more or less mucronate (cucullate or apiculate); hairy (covered with delicate, appressed, white silky hairs); non-carinate; seemingly having the margins tucked in onto the palea; 'not evidently nerved'. *Palea* present; relatively long (about equalling the lemma); tightly clasped by the lemma; entire; awnless, without apical setae; 2-nerved; 2-keeled. Palea keels wingless. *Lodicules* present; 3; free. *Stamens* 0 (no staminodes mentioned). *Ovary* glabrous. Styles free to their bases. Stigmas and styles 3.

Fruit, embryo and seedling. Disseminule a free caryopsis. *Fruit* free from both lemma and palea; small (2 mm long); ellipsoid; longitudinally grooved; compressed dorsiventrally. Hilum long-linear (as long as the caryopsis). Embryo small (0.2–0.3 mm long).

Abaxial leaf blade epidermis. *Costal/intercostal zonation* conspicuous. *Papillae* present; intercostal (small, circular, confined to the ends of the interstomatals and the stomatal sides of the long-cells adjoining the subsidiaries, and arranged so as to form a beautiful ring around each stomatal apparatus). Intercostal papillae not over-arching the stomata. *Long-cells* markedly different in shape costally and intercostally (the costals much narrower); of similar wall thickness costally and intercostally (thin walled). Mid-intercostal long-cells rectangular; having markedly sinuous walls. *Microhairs* present; elongated; clearly two-celled; panicoid-type; (49–)57–69(–75) microns long; 6.3–7.5 microns wide at the septum. Microhair total length/width at septum 9–10.2. Microhair apical cells 24–28.5 microns long. Microhair apical cell/total length ratio 0.4–0.45. *Stomata* common; (19–)24–27 microns long. Subsidiaries non-papillate; parallel-sided. Guard-cells overlapped by the interstomatals (slightly). *Intercostal short-cells* common; in cork/silica-cell pairs; silicified. Intercostal silica bodies oryzoid-type. Macrohairs present near the blade margins. *Costal short-cells* conspicuously in long rows. Costal silica bodies present and well developed; 'panicoid-type'; predominantly short dumb-bell shaped.

Transverse section of leaf blade, physiology. C_3; XyMS+. Mesophyll with adaxial palisade (one layer); with arm cells (but very obscure in the poor material seen); with fusoids. The fusoids external to the PBS. *Leaf blade* adaxially flat. *Midrib* conspicuous (by virtue of its large bundle and relatively massive 'anchor' of sclerenchyma); with one bundle only. *The lamina* symmetrical on either side of the midrib. Bulliforms in simple fans (between all the bundles). All the vascular bundles accompanied by sclerenchyma. Combined sclerenchyma girders present; forming 'figures' (most bundles with an 'anchor'). Sclerenchyma all associated with vascular bundles.

Taxonomy. Bambusoideae; Oryzodae; Olyreae.

Ecology, geography, regional floristic distribution. 1 species (*P. strephioides*). Cuba. Mesophytic; shade species; glycophytic. Lowland ravines, palm savannas and streambanks.

Neotropical. Caribbean.

References, etc. Morphological/taxonomic: Zuloaga *et al.* (1993). Leaf anatomical: Zuloaga *et al.* (1993); this project.

Special comments. Segregated from *Mniochloa* (*M. strephioides*).

Plagiantha Renvoize

Habit, vegetative morphology. *Annual (or short-lived)*; loosely caespitose. *Culms* 30–80 cm high; herbaceous. *Leaves* not basally aggregated. Leaf blades lanceolate; narrow; 3–7 mm wide; without cross venation; persistent.

Reproductive organization. Plants bisexual, with bisexual spikelets; with hermaphrodite florets.

Inflorescence. *Inflorescence paniculate*; open; with capillary branchlets; espatheate; not comprising 'partial inflorescences' and foliar organs. *Spikelet-bearing axes persistent*. Spikelets not secund; pedicellate. Pedicel apices oblique.

Female-fertile spikelets. *Spikelets 2.2–2.8 mm long*; *compressed dorsiventrally*; falling with the glumes. The upper floret not stipitate. Rachilla terminated by a female-fertile floret.

Glumes two; very unequal; (the upper) about equalling the spikelets; (the upper) long relative to the adjacent lemmas; hairless; not pointed; awnless; non-carinate. *Lower glume* about 0.25 times the length of the upper glume; 3–5 nerved. Upper glume 5 nerved. *Spikelets* with incomplete florets. The incomplete florets proximal to the female-fertile florets. *Spikelets with proximal incomplete florets. The proximal incomplete florets* 1; paleate. *Palea of the proximal incomplete florets* fully developed; *becoming conspicuously hardened and enlarged laterally. The proximal lemmas two-keeled, sulcate and nerveless between them*; awnless; *4 nerved*; slightly exceeded by the female-fertile lemmas; less firm than the female-fertile lemmas (membranous); not becoming indurated.

Female-fertile florets 1. *Lemmas* papillose; not becoming indurated; pallid; entire (and apiculate); awnless; hairless; non-carinate; *having the margins tucked in onto the palea*. *Palea* present; relatively long; tightly clasped by the lemma; awnless; without apical setae; textured like the lemma.

Abaxial leaf blade epidermis. *Costal/intercostal zonation* conspicuous. *Papillae* absent. *Long-cells* markedly different in shape costally and intercostally (the costals much narrower); of similar wall thickness costally and intercostally (thin walled). Mid-intercostal long-cells rectangular; having markedly sinuous walls. *Microhairs* present; elongated; clearly two-celled; panicoid-type. *Stomata* common. Subsidiaries low dome-shaped. Guard-cells overlapping to flush with the interstomatals. *Intercostal short-cells* common; not paired (solitary, rectangular, often large); not silicified. *Costal short-cells* conspicuously in long rows. Costal silica bodies 'panicoid-type'; butterfly shaped to nodular (ranging from short to elongated); not sharp-pointed.

Transverse section of leaf blade, physiology. C_3; XyMS+. *Mesophyll* somewhat noticeably, in places, with radiate chlorenchyma; without adaxial palisade; *Isachne*-type to not *Isachne*-type. *Leaf blade* 'nodular' in section to adaxially flat. *Midrib* conspicuous; with one bundle only, or having a conventional arc of bundles (a large median, with a small bundle on either side which could be considered part of the midrib); with colourless mesophyll adaxially. *The lamina* symmetrical on either side of the midrib. Bulliforms present in discrete, regular adaxial groups (these wide, one in each of the slight furrows); in simple fans. All the vascular bundles accompanied by sclerenchyma. Combined sclerenchyma girders present (primaries only); forming 'figures' (the midrib median and some other primaries with anchors). Sclerenchyma all associated with vascular bundles. **Taxonomy**. Panicoideae; Panicodae; Paniceae.

Ecology, geography, regional floristic distribution. 1 species. Brazil. Mesophytic; shade species; glycophytic.

Neotropical. Central Brazilian.

References, etc. Morphological/taxonomic: Renvoize 1982. *Kew Bull*. **37**, 323. Leaf anatomical: this project.

Special comments. Fruit data wanting.

Plagiosetum Benth.

Excluding *Parectenium*

Habit, vegetative morphology. Annual; caespitose to decumbent. *Culms* 20–50 cm high; herbaceous; branched above. Culm nodes glabrous. Culm internodes hollow. Leaves non-auriculate. Leaf blades narrow; not setaceous; flat, or rolled (convolute); without cross venation; persistent; rolled in bud. Ligule a fringed membrane to a fringe of hairs; 1 mm long.

Reproductive organization. Plants bisexual, with bisexual spikelets; with hermaphrodite florets. The spikelets all alike in sexuality.

Inflorescence. *Inflorescence a false spike, with spikelets on contracted axes (racemose, the cuneate, flattened branches each reduced to 1–2 spikelets and several variously united bristles, reflexing at maturity)*. Inflorescence axes not ending in spikelets (the racemelet ending in a fan of bristles). Rachides hollowed to flattened. Inflorescence espatheate; not comprising 'partial inflorescences' and foliar organs. **Spikelet-bearing axes** very much reduced (with 1–2(-3) spikelets, plus bristles); *disarticulating*; falling entire (the short branches articulated at the base, falling whole). **Spikelets with 'involucres' of 'bristles' (or at least with the single bristle branched, cf. Parectenium)**. The 'bristles' relatively slender, not spiny; deciduous with the spikelets. Spikelets solitary; secund (the racemes to one side of the main axis).

Female-fertile spikelets. Spikelets 6–7.5 mm long; compressed dorsiventrally; falling with the glumes (the branches falling whole). Rachilla terminated by a female-fertile floret.

Glumes present; two; very unequal; (the longer) long relative to the adjacent lemmas (equalling the spikelet); awnless. Lower glume 3–5 nerved. *Upper glume 9–17 nerved*. *Spikelets* with incomplete florets. The incomplete florets proximal to the female-fertile florets. *The proximal incomplete florets* 1. *The proximal lemmas* awnless; *18–20 nerved*; less firm than the female-fertile lemmas (similar to G_2); not becoming indurated.

Female-fertile florets 1. Lemmas decidedly firmer than the glumes; rugose (shining); becoming indurated; entire; pointed, or blunt; awnless; hairless; non-carinate; having the margins tucked in onto the palea; with a clear germination flap; 3–7 nerved. *Palea* present; relatively long; entire (apiculate); awnless, without apical setae; textured like the lemma; indurated. *Lodicules* present; 2; fleshy. *Stamens* 3. Anthers not penicillate. *Ovary* glabrous. Styles free to their bases. Stigmas 2.

Fruit, embryo and seedling. Fruit small; compressed dorsiventrally. Hilum short. Embryo large. Endosperm containing compound starch grains. Embryo without an epiblast; with a scutellar tail; with an elongated mesocotyl internode. Embryonic leaf margins overlapping.

Abaxial leaf blade epidermis. *Costal/intercostal zonation* conspicuous. *Papillae* absent. *Long-cells* similar in shape costally and intercostally; of similar wall thickness costally and intercostally (fairly thin walled). Mid-intercostal long-cells rectangular; having markedly sinuous walls. *Microhairs* present; panicoid-type; 45–51–57 microns long; (6.3–)6.9–8.4(-9) microns wide at the septum. Microhair total length/width at septum 5.6–7.4. Microhair apical cells (27–)30–35(-40.5) microns long. Microhair apical cell/total length ratio 0.59–0.71. *Stomata* common; (24–)27–28.5(-30) microns long. Subsidiaries parallel-sided and dome-shaped. Guard-cells overlapping to flush with the interstomatals. *Intercostal short-cells* common; not paired (mainly solitary); not silicified.

Costal short-cells conspicuously in long rows. Costal silica bodies 'panicoid-type'; not sharp-pointed.

Transverse section of leaf blade, physiology. C_4; XyMS–. PCR sheath outlines uneven. PCR sheath extensions absent. PCR cell chloroplasts centrifugal/peripheral. *Mesophyll* with radiate chlorenchyma, or with non-radiate chlorenchyma (rarely). *Leaf blade* adaxially flat. *Midrib* conspicuous; with one bundle only; with colourless mesophyll adaxially. Bulliforms not in discrete, regular adaxial groups (the epidermis irregularly bulliform). Many of the smallest vascular bundles unaccompanied by sclerenchyma. Combined sclerenchyma girders present; nowhere forming 'figures'. Sclerenchyma all associated with vascular bundles.

Taxonomy. Panicoideae; Panicodae; Paniceae.

Ecology, geography, regional floristic distribution. 1 species. Australia. Xerophytic; species of open habitats. Dry sandhills.

Australian. Central Australian.

References, etc. Leaf anatomical: this project.

Planichloa B. Simon

~ Ectrosia

Habit, vegetative morphology. Erect *annual*. *Culms* 12–40 cm high; herbaceous; unbranched above; 1–3 noded (terete). *Leaves* not basally aggregated; non-auriculate. Sheaths hispid, with tubercle based hairs. Leaf blades linear (hispid); narrow; 1–4 mm wide (to 10 cm long); flat, or folded; without abaxial multicellular glands; without cross venation; persistent. *Ligule a fringed membrane*; 0.2–0.3 mm long.

Reproductive organization. Plants bisexual, with bisexual spikelets; with hermaphrodite florets.

Inflorescence. Inflorescence paniculate; contracted to open; when contracted, more or less irregular; espatheate; not comprising 'partial inflorescences' and foliar organs. Spikelet-bearing axes persistent. Spikelets not secund; pedicellate (the pedicels 0.5–1 mm long, hairy).

Female-fertile spikelets. *Spikelets* 5–11 mm long (3–6 mm wide); strongly compressed laterally; disarticulating above the glumes (but the glumes falling later); *not disarticulating between the florets (the rachilla rigid)*. Rachilla prolonged beyond the uppermost female-fertile floret. The rachilla extension with incomplete florets. Hairy callus absent. *Callus* short.

Glumes two; very unequal; shorter than the spikelets; shorter than the adjacent lemmas; hairless; scaberulous on the margins and keel; pointed; awnless; carinate; similar (lanceolate, acuminate, usually mauve). Lower glume 1 nerved. Upper glume 3 nerved. *Spikelets* with incomplete florets. The incomplete florets distal to the female-fertile florets. The distal incomplete florets awned.

Female-fertile florets 2–8. Lemmas lemmas rigid, flattened, yellowish green or the apices infused with mauve, dark-pigmented on the lateral nerves; similar in texture to the glumes (leathery); not becoming indurated; entire; pointed; awnless (acute to acuminate), or awned (the awns increasingly emphasized acropetally in the spikelet). Awns 1; median; apical; non-geniculate; hairless (scabrous); much shorter than the body of the lemma to about as long as the body of the lemma; persistent. Lemmas hairless; scaberulous on the keel and margins; carinate; without a germination flap; *5 nerved, or 7 nerved (the laterals more or less grouped together)*. **Palea** present; *conspicuous but relatively short (about half the lemma length, 'comma shaped')*; awnless, without apical setae; hyaline, with leathery and scaberulous margins; not indurated; 2-nerved; 2-keeled. Palea keels winged; scabrous. *Lodicules* present; 2; free; fleshy; glabrous. *Stamens* 3. Anthers about 0.5 mm long (mauvish red). *Ovary* glabrous. Stigmas 2; pale yellow, plumose.

Fruit, embryo and seedling. Fruit small (about 1.4 mm long); ellipsoid; slightly compressed laterally, or not noticeably compressed. Hilum short. Pericarp fused. Embryo large (about half the length of the caryopsis).

Abaxial leaf blade epidermis. *Costal/intercostal zonation* conspicuous. *Papillae* absent. *Long-cells* markedly different in shape costally and intercostally (the costals smaller and narrower); of similar wall thickness costally and intercostally. Mid-intercostal long-cells rectangular; having markedly sinuous walls. *Microhairs* present; elongated; clearly two-celled; panicoid-type. Microhair apical cell wall thinner than that of the basal cell and often collapsed. Microhair basal cells 21 microns long. Microhair total length/ width at septum 5. Microhair apical cell/total length ratio 0.5. *Stomata* common. Subsidiaries triangular (predominantly), or dome-shaped. Guard-cells overlapping to flush with the interstomatals. *Intercostal short-cells* common; in cork/silica-cell pairs (a few only, with crescentic silica bodies), or not paired (mostly solitary and unsilicified); mostly not silicified. Intercostal silica bodies absent, or imperfectly developed. *Costal short-cells* predominantly paired. Costal silica bodies present throughout the costal zones; saddle shaped (a very small version, predominating), or tall-and-narrow to crescentic (intergrading with the saddles); not sharp-pointed.

Transverse section of leaf blade, physiology. C_4; XyMS+. PCR sheath outlines uneven. PCR sheaths of the primary vascular bundles interrupted; interrupted abaxially only. PCR sheath extensions present. Maximum number of extension cells 2–3. *Mesophyll* with radiate chlorenchyma. *Leaf blade* with distinct, prominent adaxial ribs; with the ribs more or less constant in size (low). *Midrib* fairly conspicuous (via a widish, flat abaxial keel and the larger bundle); with one bundle only. *The lamina* symmetrical on either side of the midrib. Bulliforms present in discrete, regular adaxial groups (the groups very large, one in each furrow); in simple fans (these predominating, the median cells large and deeply penetrating), or associated with colourless mesophyll cells to form deeply-penetrating fans (these infrequent). All the vascular bundles accompanied by sclerenchyma. Combined sclerenchyma girders present (nearly all the bundles); forming 'figures' (nearly every bundle with an 'I' or an 'anchor'). Sclerenchyma all associated with vascular bundles. The lamina margins with fibres.

Taxonomy. Chloridoideae; main chloridoid assemblage.

Ecology, geography, regional floristic distribution. 1 species. Restricted to the Cook pastoral district of northern Queensland. Mesophytic to xerophytic; species of open habitats; glycophytic. In sandy soils.

Australian. North and East Australian. Tropical North and East Australian.

References, etc. Morphological/taxonomic: Simon 1986. Leaf anatomical: this project.

Plectrachne Henrard

Habit, vegetative morphology. Perennial; caespitose. *Culms* 20–175 cm high. Culm nodes glabrous. *The shoots aromatic*. *Leaves* mostly basal; non-auriculate. *Leaf blades* narrow; acicular; *hard, woody, needle-like*; without abaxial multicellular glands; without cross venation; persistent. Ligule a fringe of hairs (usually very short).

Reproductive organization. Plants bisexual, with bisexual spikelets; with hermaphrodite florets.

Inflorescence. Inflorescence paniculate; open, or contracted; when contracted spicate; non-digitate; espatheate; not comprising 'partial inflorescences' and foliar organs. Spikelet-bearing axes persistent. Spikelets not secund; pedicellate.

Female-fertile spikelets. *Spikelets* compressed laterally; disarticulating above the glumes; *with conventional internode spacings (by contrast with* Symplectrodia*)*. Rachilla prolonged beyond the uppermost female-fertile floret; rachilla hairy. The rachilla extension with incomplete florets. Hairy callus present.

Glumes two; more or less equal; *about equalling the spikelets to exceeding the spikelets*; long relative to the adjacent lemmas; awnless; carinate, or non-carinate. Lower glume 3–7 nerved. Upper glume 3–7 nerved. *Spikelets* with incomplete florets. The incomplete florets distal to the female-fertile florets.

Female-fertile florets 3–10. *Lemmas* becoming indurated to not becoming indurated (leathery to indurated); incised; 3 lobed; *deeply cleft (the lobes longer than the body of the lemma)*; awned. Awns 1, or 3; median, or median and lateral (the lobes awnlike); the median similar in form to the laterals (when laterals present); from a sinus; non-geniculate. The lateral awns shorter than the median. Lemmas hairy, or hairless; non-carinate; 9 nerved. *Palea* present; relatively long; 2-nerved. *Lodicules* present; 2; free; fleshy; glabrous; heavily vascularized. *Stamens* 3. Anthers not penicillate. *Ovary* glabrous. Styles free to their bases. Stigmas 2.

Fruit, embryo and seedling. Fruit small; ellipsoid. Hilum short. Pericarp fused.

Abaxial leaf blade epidermis. *Costal/intercostal zonation* conspicuous and lacking (centre of leaf/margins). *Papillae* absent (but abundant adaxially). *Microhairs* present (? — probably, but well hidden in the grooves); (adaxial) elongated; clearly two-celled; chloridoid-type (where seen, adaxially). Microhair apical cell wall of similar thickness/rigidity to that of the basal cell. Microhair basal cells 21 microns long. *Stomata* absent or very rare (or if present, hidden in the narrow grooves). *Costal short-cells* conspicuously in long rows. Costal silica bodies present and well developed to absent; 'panicoid-type'; not sharp-pointed.

Transverse section of leaf blade, physiology. C₄. The anatomical organization unconventional. Organization of PCR tissue *Triodia* type. XyMS+. PCR sheath outlines even. PCR cell chloroplasts centrifugal/peripheral. *Mesophyll* with radiate chlorenchyma; traversed by columns of colourless mesophyll cells (these linking the adaxial and abaxial grooves, at least in the middle part of the blade where the PCR cells do not 'drape'); with arm cells. *Leaf blade* 'nodular' in section, or with distinct, prominent adaxial ribs and 'nodular' in section (the abaxial grooves sometimes lacking towards the leaf margins); with the ribs very irregular in sizes. *Midrib* conspicuous (by its position); with one bundle only; without colourless mesophyll adaxially (but the blade often with blocks of abaxial colourless tissue towards the margins). Bulliforms present in discrete, regular adaxial groups (as part of the traversing columns), or not in discrete, regular adaxial groups (bulliforms often not recognisable as such); if recognisable, associated with colourless mesophyll cells to form deeply-penetrating fans (incorporated in the traversing columns of colourless cells). All the vascular bundles accompanied by sclerenchyma. Combined sclerenchyma girders present, or absent; forming 'figures'. Sclerenchyma all associated with vascular bundles, or not all bundle-associated (sometimes with a continuous abaxial layer). The 'extra' sclerenchyma when present, in abaxial groups, or in a continuous abaxial layer.

Taxonomy. Chloridoideae; Triodieae.

Ecology, geography, regional floristic distribution. 16 species. Australia. Xerophytic; species of open habitats. Sandy or stony, arid soils.

Australian. Central Australian.

References, etc. Morphological/taxonomic: Jacobs 1971. Leaf anatomical: Metcalfe 1960; this project.

Pleiadelphia Stapf

~ *Elymandra (E. gossweileri)*

Habit, vegetative morphology. Annual. *Culms* 80–190 cm high; herbaceous; branched above. Culm nodes glabrous. Culm leaves present. Plants unarmed. *Leaves* not basally aggregated; with tough, auricle-like extensions of the sheath along the sides of the ligule. Leaf blades linear (attenuated); narrow; 1–4 mm wide (and up to 30 cm long, scabrid above and on the margins); setaceous-tipped; without cross venation; persistent. Ligule an unfringed membrane; truncate; about 0.5 mm long. *Contra-ligule* absent.

Reproductive organization. Plants bisexual, with bisexual spikelets; with hermaphrodite florets. *The spikelets* of sexually distinct forms on the same plant; hermaphrodite and sterile; overtly heteromorphic (only the fertile spikelet awned); *in both homogamous and heterogamous combinations (the raceme with 1–4 basal, homogamous sterile pairs and a terminal heterogamous triad)*. Plants inbreeding (seemingly, the 'racemes' enclosed in the spatheoles). Exposed-cleistogamous (?).

Inflorescence. Inflorescence paniculate; open; spatheate; *a complex of 'partial inflorescences' and intervening foliar organs*. Spikelet-bearing axes 'racemes' and very much reduced (reduced to 1–4 basal homogamous pairs and the terminal triad); *solitary (one per spatheole)*; disarticulating; disarticulating at the joints (i.e. at the only joint).

'*Articles*' linear; without a basal callus-knob; not appendaged; disarticulating obliquely. Spikelets in triplets (with a terminal triad), or paired (below); sessile and pedicellate; consistently in 'long-and-short' combinations; in pedicellate/sessile combinations. Pedicels of the 'pedicellate' spikelets free of the rachis. The 'shorter' spikelets hermaphrodite (in the terminal triad). The 'longer' spikelets sterile.

Female-sterile spikelets. The sterile spikelets awnless, tending not to disarticulate. The lemmas awnless.

Female-fertile spikelets. Spikelets (12–)14–16 mm long; not noticeably compressed to compressed dorsiventrally; falling with the glumes. Rachilla terminated by a female-fertile floret. Hairy callus present (densely fulvous-bearded). *Callus* long; pointed.

Glumes two; more or less equal; long relative to the adjacent lemmas; at least the G_1 hairy; without conspicuous tufts or rows of hairs; *awned (G_2)*; non-carinate; very dissimilar (the G_1 leathery, obtuse or truncate and awnless, the G_2 thinner, truncate to emarginate with a long slender awn, weakly impressed by the margins of G_1). *Lower glume* 2-grooved to accommodate the pedicels of the sterile spikelets; convex on the back; not pitted; relatively smooth; *5–7 nerved (the median present but inconspicuous)*. Upper glume 3 nerved. *Spikelets* with incomplete florets. The incomplete florets proximal to the female-fertile florets. *Spikelets with proximal incomplete florets. The proximal incomplete florets* 1; epaleate; sterile. The proximal lemmas awnless; thinly 2 nerved; more or less equalling the female-fertile lemmas to decidedly exceeding the female-fertile lemmas; similar in texture to the female-fertile lemmas (hyaline); not becoming indurated.

Female-fertile florets 1. Lemmas less firm than the glumes (hyaline); not becoming indurated; incised; 2 lobed; deeply cleft; awned. Awns 1; median; from a sinus; geniculate; hairless to hairy (glabrescent); much longer than the body of the lemma (8–9 cm long); entered by one vein. Lemmas hairless; non-carinate; without a germination flap; 1 nerved. **Palea** *absent. Lodicules* present; 2; free; fleshy; glabrous. *Stamens* 3. Anthers about 3 mm long (the filaments short); not penicillate; without an apically prolonged connective. *Ovary* glabrous. Styles free to their bases; free. Style bases adjacent. Stigmas 2.

Fruit, embryo and seedling. Fruit not noticeably compressed. Hilum short. Embryo large.

Abaxial leaf blade epidermis. *Costal/intercostal zonation* conspicuous. *Papillae* absent. *Long-cells* markedly different in shape costally and intercostally (the costals much narrower); of similar wall thickness costally and intercostally (thin walled). Mid-intercostal long-cells rectangular and fusiform; having markedly sinuous walls and having straight or only gently undulating walls. *Microhairs* present; elongated; clearly two-celled; panicoid-type. *Stomata* common. Subsidiaries mostly high triangular. Guard-cells overlapping to flush with the interstomatals. *Intercostal short-cells* fairly common; in cork/silica-cell pairs and not paired (paired and solitary); silicified and not silicified. Intercostal silica bodies tall-and-narrow. *Costal short-cells* conspicuously in long rows. Costal silica bodies present and well developed; saddle shaped and 'panicoid-type'; dumb-bell shaped; not sharp-pointed.

Transverse section of leaf blade, physiology. C_4; XyMS–. PCR sheath outlines uneven. PCR sheath extensions absent. *Mesophyll* with radiate chlorenchyma. *Leaf blade* adaxially flat. *Midrib* conspicuous; having a conventional arc of bundles (a large median, with three or four smaller bundles on either side); with colourless mesophyll adaxially. *The lamina* symmetrical on either side of the midrib. Bulliforms present in discrete, regular adaxial groups (in places, but the epidermis mainly bulliform); in simple fans (in places), or associated with colourless mesophyll cells to form deeply-penetrating fans (in places); associating with colourless mesophyll cells to form arches over small vascular bundles (there often being a colourless cell adjacent to the epidermis on either side of a small bundle). Many of the smallest vascular bundles unaccompanied by sclerenchyma. Combined sclerenchyma girders present (with the larger bundles); forming 'figures' (the larger bundles with I's). Sclerenchyma all associated with vascular bundles.

Taxonomy. Panicoideae; Andropogonodae; Andropogoneae; Andropogoninae.

Ecology, geography, regional floristic distribution. 1–2 species. Tropical Africa. Glycophytic. On shallow sandy soils.

Paleotropical. African. West African Rainforest.

References, etc. Leaf anatomical: this project.

Pleuropogon R.Br.

Lepitoma Steud., *Lophochlaena* Nees

Habit, vegetative morphology. Annual (rarely), or perennial; rhizomatous, or stoloniferous, or caespitose, or decumbent. *Culms* 10–150 cm high; herbaceous. Culm internodes hollow. *Leaves* not basally aggregated; non-auriculate. *Sheath margins joined.* Leaf blades linear; narrow; 1–3 mm wide; usually flat; without cross venation; rolled in bud, or once-folded in bud. *Ligule* present; an unfringed membrane; not truncate; 2–6 mm long.

Reproductive organization. Plants bisexual, with bisexual spikelets; with hermaphrodite florets.

Inflorescence. *Inflorescence a single raceme (rarely paniculate)*; open (rarely); espatheate; not comprising 'partial inflorescences' and foliar organs. Spikelet-bearing axes persistent. Spikelets solitary; secund (the raceme unilateral); shortly pedicellate (often deflexed).

Female-fertile spikelets. Spikelets 8–15 mm long; compressed laterally; disarticulating above the glumes; disarticulating between the florets. Rachilla prolonged beyond the uppermost female-fertile floret. The rachilla extension with incomplete florets. Hairy callus absent.

Glumes two; very unequal; shorter than the spikelets; shorter than the adjacent lemmas; pointed, or not pointed; awnless; non-carinate; similar (membranous). *Lower glume 1 nerved.* Upper glume 3 nerved. *Spikelets* with incomplete florets. The incomplete florets distal to the female-fertile florets. *The distal incomplete florets* merely underdeveloped.

Female-fertile florets 6–20. Lemmas similar in texture to the glumes to decidedly firmer than the glumes (membranous to thinly leathery); not becoming indurated; entire to incised; when incised, not deeply cleft (bidentate, irregularly toothed or erose); mucronate to awned. Awns when present, 1; from a sinus, or apical; non-geniculate; much shorter than the body of the lemma; entered by one vein. Lemmas hairy, or hairless; 7 nerved. **Palea** present; *awnless, without apical setae, or awned (from the base of the keels, and sometimes from higher on the keels as well)*; 2-nerved; 2-keeled. Palea keels winged (below). **Lodicules** present; 2; *membranous. Stamens* 3. Anthers 1.2–2 mm long. **Ovary** glabrous; *without a conspicuous apical appendage.* Stigmas 2.

Fruit, embryo and seedling. Fruit free from both lemma and palea; *deep red*. Pericarp fused. Embryo with an epiblast; without a scutellar tail; with a negligible mesocotyl internode. Embryonic leaf margins meeting.

Abaxial leaf blade epidermis. *Costal/intercostal zonation* conspicuous. *Papillae* present. Intercostal papillae over-arching the stomata; several per cell. *Long-cells* similar in shape costally and intercostally; of similar wall thickness costally and intercostally to differing markedly in wall thickness costally and intercostally (the costals somewhat thicker). Mid-intercostal long-cells rectangular; having markedly sinuous walls (rarely), or having straight or only gently undulating walls. *Microhairs* absent. *Stomata* common; 36–39 microns long. Subsidiaries parallel-sided. Guard-cells overlapped by the interstomatals. *Intercostal short-cells* absent or very rare. *Costal short-cells* neither distinctly grouped into long rows nor predominantly paired. Costal silica bodies horizontally-elongated crenate/sinuous; not sharp-pointed.

Transverse section of leaf blade, physiology. C_3; XyMS+. *Mesophyll* with non-radiate chlorenchyma. Sclerenchyma not all bundle-associated. The 'extra' sclerenchyma in abaxial groups, or in abaxial groups and in adaxial groups.

Cytology. Chromosome base number, $x = 9$, or 10. $2n = 40$ and 42. 4 ploid.

Taxonomy. Pooideae; Poodae; Meliceae.

Ecology, geography, regional floristic distribution. 6 species. Western North America & circumpolar. Helophytic; glycophytic.

Holarctic. Boreal. Arctic and Subarctic and Euro-Siberian. European and Siberian.

References, etc. Leaf anatomical: Metcalfe 1960; this project.

Special comments. Fruit data wanting.

Plinthanthesis Steud.

Blakeochloa Veldkamp
~ *Rytidosperma, Danthonia* sensu lato

Habit, vegetative morphology. Perennial; caespitose. **Culms** 20–70 cm high; *herbaceous*; unbranched above. Culm nodes glabrous. Leaves non-auriculate. Leaf blades narrow; setaceous to not setaceous; flat, or rolled; without cross venation; persistent; once-folded in bud. *Ligule a fringe of hairs (short)*.

Reproductive organization. *Plants bisexual, with bisexual spikelets*; with hermaphrodite florets.

Inflorescence. *Inflorescence* few spikeleted to many spikeleted; *paniculate*; open, or contracted; with capillary branchlets to without capillary branchlets; espatheate; not comprising 'partial inflorescences' and foliar organs. Spikelet-bearing axes persistent. Spikelets not secund; pedicellate.

Female-fertile spikelets. Spikelets 4–9 mm long; compressed laterally; disarticulating above the glumes; disarticulating between the florets (the disarticulation horizontal, leading to a short, obtuse callus); with distinctly elongated rachilla internodes between the florets (the internode more than half the lemma length). Rachilla prolonged beyond the uppermost female-fertile floret; rachilla hairless. The rachilla extension with incomplete florets. *Hairy callus absent. Callus* short; blunt.

Glumes two; more or less equal; shorter than the spikelets to exceeding the spikelets; long relative to the adjacent lemmas (exceeding them); hairless; glabrous, or scabrous; pointed (acute); awnless; keeled above; similar (translucent or firm). Lower glume 1 nerved, or 3 nerved. Upper glume 1 nerved, or 3 nerved. *Spikelets* with incomplete florets. The incomplete florets distal to the female-fertile florets. *The distal incomplete florets merely underdeveloped.*

Female-fertile florets 3–5. Lemmas similar in texture to the glumes to decidedly firmer than the glumes (firmly membranous to leathery); not becoming indurated; incised; 2 lobed; deeply cleft (to about halfway), or not deeply cleft (the lobes minute, or up to 4 mm long); awnless, or mucronate, or awned (shortly). Awns when present, 1; median; when present, from a sinus; non-geniculate (flat, curved), or geniculate; when present, much shorter than the body of the lemma to about as long as the body of the lemma. Lemmas evenly hairy (at least below). *The hairs not in tufts*; *not in transverse rows*. Lemmas non-carinate (rounded on the back); without a germination flap; 7 nerved, or 9 nerved. *Palea* present; relatively long (equalling the lemma); entire to apically notched; awnless, without apical setae (densely white-hairy below); thinner than the lemma to textured like the lemma; not indurated; 2-nerved; 2-keeled. Palea keels wingless. **Lodicules** present; 2; *membranous*; glabrous; not toothed. *Stamens* 3. Anthers 1.5–3 mm long. *Ovary* glabrous. Styles free to their bases. Stigmas 2.

Fruit, embryo and seedling. *Fruit* free from both lemma and palea; small; longitudinally grooved. *Hilum long-linear*. Embryo large. Endosperm hard.

Abaxial leaf blade epidermis. *Costal/intercostal zonation* conspicuous. *Papillae* absent. *Long-cells* markedly different in shape costally and intercostally (in *P. paradoxa* the intercostals less regular). Mid-intercostal long-cells rectangular (*P. rodwayi*), or rectangular to fusiform (*P. paradoxa*); having markedly sinuous walls. *Microhairs* present; panicoid-type (conventional in *P. rodwayi*, in *P. paradoxa* the basal cell fat, the apical cell conical); (54–)60–72(–81) microns long (*P. paradoxa*), or 60–69 microns long (*P. rodwayi*); 18–21–25.5 microns wide at the septum (*P. paradoxa*), or 11.4–12 microns wide at the septum (*P. rodwayi*). Microhair total length/width at septum 2.5–4 (*P. paradoxa*), or 5.3–6.1 (*P. rodwayi*). Microhair apical cells (22.5–)30–37.5(–46.5) microns long (*P. paradoxa*), or (17.4–)22.5–24(–28.5) microns long (*P. rodwayi*). Microhair apical cell/total length ratio 0.42–0.57 (*P. paradoxa*), or 0.26–0.41 (*P. rodwayi*). *Stomata* absent or very rare (*P. rodwayi*), or common (*P. paradoxa*, in places); in *P. paradoxa* (36–)45–48(–51) microns long. Subsidiaries of *P. paradoxa* dome-shaped, or parallel-sided (more or less). Guard-cells overlapping to flush with the interstomatals (in *P. paradoxa*, in which many stomata are accompanied at one end by a third 'subsidiary', in the form of a very short intercostal cell — a most unusual feature). *Intercostal short-cells* common; in cork/silica-cell pairs (and solitary); silicified. *Costal short-cells* conspicuously in long rows (very regular). Costal silica bodies 'panicoid-type'; very regular cross shaped and butterfly shaped; not sharp-pointed.

Transverse section of leaf blade, physiology. C_3; XyMS+. *Mesophyll* with non-radiate chlorenchyma; without adaxial palisade. *Leaf blade* with distinct, prominent adaxial ribs (*P. rodwayi*), or 'nodular' in section (*P. paradoxa*); with the ribs very irregular in sizes. *Midrib* not readily distinguishable; with one bundle only. Bulliforms not in discrete, regular adaxial groups (or at least, only inconspicuously grouped). All the vascular bundles accompanied by sclerenchyma. Combined sclerenchyma girders present (with the primaries — the smaller bundles often with strands only); forming 'figures'. Sclerenchyma all associated with vascular bundles.

Taxonomy. Arundinoideae; Danthonieae.

Ecology, geography, regional floristic distribution. 3 species. Southeastern Australia. Mesophytic to xerophytic; species of open habitats; glycophytic. Upland heaths.

Australian. North and East Australian. Temperate and South-Eastern Australian.

References, etc. Morphological/taxonomic: Blake 1972a, Vickery 1956. Leaf anatomical: this project.

Poa L.

Arctopoa (Griseb.) Probat., *Neuropoa* Clayton, *Oreopoa* Grand., *Paneion* Lunell, *Parodiochloa* C.E. Hubb., *Poagrostis* Raf.

Excluding *Bellardiochloa, Dasypoa, Poidium*

Habit, vegetative morphology. Annual, or perennial; rhizomatous, or stoloniferous, or caespitose, or decumbent. **Culms** (1–)4–150 cm high; *herbaceous*; unbranched above; tuberous, or not tuberous. Culm internodes hollow. Young shoots extravaginal, or intravaginal. *Leaves* mostly basal; non-auriculate. Sheath margins joined, or free. Leaf blades linear, or linear-lanceolate (often ending in a boat-shaped tip); nearly always narrow; 0.2–12 mm wide (rarely wider); setaceous, or not setaceous; flat, or folded (or canaliculate), or rolled (involute or convolute); without cross venation; persistent (usually), or disarticulating from the sheaths (very rarely); once-folded in bud. *Ligule an unfringed membrane, or a fringed membrane (rarely)*; truncate, or not truncate; (0.1–)0.5–6(–15) mm long. *Contra-ligule* absent.

Reproductive organization. Plants bisexual, with bisexual spikelets, or dioecious (or gynodioecious, in subgenus *Andinae*); with hermaphrodite florets, or without hermaphrodite florets (rarely). The spikelets hermaphrodite, or female-only, or hermaphrodite and female-only, or male-only. Plants outbreeding and inbreeding. Exposed-cleistogamous, or chasmogamous. Apomictic (pseudogamous and non-pseudogamous), or reproducing sexually. Viviparous (sometimes), or not viviparous.

Inflorescence. *Inflorescence* reduced to a single spikelet to few spikeleted (very rarely), or many spikeleted (nearly always); *paniculate*; open, or contracted; with capillary branchlets, or without capillary branchlets; espatheate; not comprising 'partial inflorescences' and foliar organs. Spikelet-bearing axes persistent. Spikelets not secund; pedicellate.

Female-fertile spikelets. *Spikelets* 2–11 mm long; *compressed laterally*; disarticulating above the glumes; disarticulating between the florets. *Rachilla prolonged beyond the uppermost female-fertile floret; rachilla hairless (nearly always glabrous).* Hairy callus present, or absent. *Callus* short; blunt.

Glumes two; more or less equal, or very unequal to more or less equal (nearly always 'subequal', with the G_1 somewhat shorter); shorter than the spikelets; *shorter than the adjacent lemmas*; free; pointed, or not pointed (rarely rounded); awnless; *carinate*; similar (membranous). Lower glume 1 nerved, or 3 nerved. Upper glume (1–)3(–5) nerved. *Spikelets* with female-fertile florets only, or with incomplete florets. The incomplete florets distal to the female-fertile florets. *The distal incomplete florets merely underdeveloped. Spikelets without proximal incomplete florets.*

Female-fertile florets (1–)2–13(–15) (the one-floreted species very unusual, perhaps restricted to Malesia). *Lemmas similar in texture to the glumes*; not becoming indurated;

nearly always *entire (very rarely tridenticulate)*; pointed; awnless, or mucronate (2–3 mm terminal 'awns' in a very few species, e.g. in the southern South American *P. flabellata*); **hairy (often with web-like hairs), or hairless (rarely)**; *carinate*; without a germination flap; (3–)5 nerved, or 7–11 nerved (rarely, e.g. in the Australian *Neuropoa*); with the nerves non-confluent. *Palea* present; relatively long, or conspicuous but relatively short; usually tightly clasped by the lemma; entire to apically notched (emarginate); awnless, without apical setae; textured like the lemma; 2-nerved; 2-keeled. Palea keels wingless; glabrous to hairy (glabrous, aculeolate or ciliate). **Lodicules** present; 2; free; *membranous*; nearly always glabrous (occasionally ciliolate); toothed, or not toothed; not or scarcely vascularized. *Stamens* 3. Anthers 0.2–3 mm long; not penicillate; without an apically prolonged connective. *Ovary* glabrous. Styles free to their bases. Stigmas 2; white.

Fruit, embryo and seedling. Fruit free from both lemma and palea; small; generally brown?; *fusiform*; longitudinally grooved (rarely?), or not grooved (usually); compressed laterally (occasionally?), or compressed dorsiventrally to not noticeably compressed; glabrous. *Hilum short*. Embryo small; not waisted. *Endosperm hard*; with lipid; containing compound starch grains. Embryo with an epiblast; without a scutellar tail; with a negligible mesocotyl internode. Embryonic leaf margins meeting.

Seedling with a short mesocotyl, or with a long mesocotyl; with a loose coleoptile, or with a tight coleoptile. First seedling leaf with a well-developed lamina. The lamina narrow; erect; 3 veined, or 5 veined.

Abaxial leaf blade epidermis. *Costal/intercostal zonation* conspicuous (usually), or lacking. *Papillae* absent (usually), or present (e.g. *P. sieberana*, *P. helmsii*); when present, intercostal, or costal and intercostal. Intercostal papillae not over-arching the stomata; several per cell (of the coronate-pit type). *Long-cells* similar in shape costally and intercostally; of similar wall thickness costally and intercostally (thick or thin walled). Mid-intercostal long-cells rectangular, or fusiform; having markedly sinuous walls, or having straight or only gently undulating walls. *Microhairs* absent. *Stomata* absent or very rare, or common; when present, (27–)30–54(–57) microns long. Subsidiaries non-papillate; parallel-sided, or dome-shaped (low), or parallel-sided and dome-shaped. Guard-cells overlapped by the interstomatals. *Intercostal short-cells* common, or absent or very rare; in cork/silica-cell pairs, or not paired; silicified, or not silicified. Intercostal silica bodies when present, tall-and-narrow, or crescentic. *Costal short-cells* predominantly paired, or neither distinctly grouped into long rows nor predominantly paired. Costal silica bodies horizontally-elongated crenate/sinuous, or horizontally-elongated smooth, or rounded, or tall-and-narrow, or crescentic; not sharp-pointed.

Transverse section of leaf blade, physiology. C_3; XyMS+. *Mesophyll* with non-radiate chlorenchyma; without adaxial palisade. *Leaf blade* 'nodular' in section, or adaxially flat; with the ribs more or less constant in size. *Midrib* conspicuous; with one bundle only. Bulliforms present in discrete, regular adaxial groups, or not in discrete, regular adaxial groups (commonly restricted to the 'midrib hinge' pair); in simple fans. Many of the smallest vascular bundles unaccompanied by sclerenchyma, or all the vascular bundles accompanied by sclerenchyma. Combined sclerenchyma girders present, or absent; forming 'figures', or nowhere forming 'figures'. Sclerenchyma all associated with vascular bundles, or not all bundle-associated. The 'extra' sclerenchyma when present, in abaxial groups, or in a continuous abaxial layer (e.g. *P. labillardieri*).

Phytochemistry. Tissues of the culm bases with little or no starch. Fructosans predominantly long-chain.

Cytology. Chromosome base number, $x = 7$. $2n = 14$, or 28, or 35, or 38, or 38–117, or 42, or 43, or 44, or 56, or 50–56, or 63, or 65, or 70–72, or 76 (etc). 2, 4, 5, 6, 7, 8, 9, 10, 11, and 12 ploid (etc., and aneuploids). Haploid nuclear DNA content 0.9–2.8 pg (6 species, mean 1.5). Mean diploid 2c DNA value 3.6 pg (3 species, 2.4–5.6). Nucleoli disappearing before metaphase.

Taxonomy. Pooideae; Poodae; Poeae.

Ecology, geography, regional floristic distribution. About 500 species. Cosmopolitan. Commonly adventive. Helophytic (rarely), or mesophytic (mostly), or xerophytic (rarely); shade species and species of open habitats; mostly glycophytic, or halophytic (a few). Typically in grasslands and meadows, a few (e.g. *P. macrantha, P. confinis*) in coastal sand.

Holarctic, Paleotropical, Neotropical, Australian, and Antarctic. Boreal, Tethyan, and Madrean. African, Madagascan, Indomalesian, Polynesian, and Neocaledonian. Arctic and Subarctic, Euro-Siberian, Eastern Asian, Atlantic North American, and Rocky Mountains. Macaronesian, Mediterranean, and Irano-Turanian. Saharo-Sindian, Sudano-Angolan, West African Rainforest, and Namib-Karoo. Indian, Indo-Chinese, Malesian, and Papuan. Hawaiian. Caribbean, Central Brazilian, Pampas, and Andean. North and East Australian and South-West Australian. New Zealand and Patagonian. European and Siberian. Canadian-Appalachian, Southern Atlantic North American, and Central Grasslands. Sahelo-Sudanian, Somalo-Ethiopian, and South Tropical African. Tropical North and East Australian and Temperate and South-Eastern Australian.

Hybrids. Intergeneric hybrid of *Arctopoa* (= *Poa*) with *Dupontia*: ×*Dupontopoa* N.S. Probatova (exemplified by *Poa labradorica* Steudel, with ×*Dupontopoa dezhnevii* representing taxonomic errors: see Darbyshire *et al.* 1992, Darbyshire and Cayouette 1992).

Rusts and smuts. Rusts — *Puccinia*. Taxonomically wide-ranging species: *Puccinia graminis, Puccinia coronata, Puccinia striiformis, Puccinia brachypodii, Puccinia recondita, 'Uromyces' dactylidis*, and *Puccinia monoica*. Smuts from Tilletiaceae and from Ustilaginaceae. Tilletiaceae — *Entyloma, Tilletia*, and *Urocystis*. Ustilaginaceae — *Ustilago*.

Economic importance. Significant weed species: *P. annua, P. bulbosa, P. compressa, P. infirma, P. pratensis, P. sphondylodes, P. sylvicola, P. trivialis*. Cultivated fodder: notably *P. palustris, P. pratensis* (Meadow, Bluegrass), *P. compressa*. Important native pasture species: many valuable: e.g. *P. alpina, P. arctica, P. arida, P. compressa, P. epilis, P. interior, P. gracillima, P. juncifolia, P. palustris, P. pratensis, P. rupicola, P. schimperana, P. trivialis*. Lawns and/or playing fields: *P. nemoralis, P. pratensis, P. trivialis* etc. — the generally inoffensive *P. annua* being commonly present as a weed.

References, etc. Morphological/taxonomic: Vickery 1970. Leaf anatomical: mainly Metcalfe 1960; this project.

Poagrostis Stapf

~ *Pentaschistis*

Habit, vegetative morphology. *Annual (delicate)*; loosely caespitose. *Culms* 4–25 cm high; herbaceous; branched above. Culm nodes glabrous. Culm internodes hollow. Plants unarmed. Young shoots intravaginal. *Leaves* mostly basal, or not basally aggregated; non-auriculate (but a few hairs at the auricular positions). Leaf blades linear (acute); narrow; 1–1.75 mm wide (and 1.2–2.5 cm long); flat; without cross venation; persistent. *Ligule a fringe of hairs*; about 1 mm long. *Contra-ligule* absent.

Reproductive organization. Plants bisexual, with bisexual spikelets; with hermaphrodite florets.

Inflorescence. Inflorescence reduced to a single spikelet, or few spikeleted, or many spikeleted; paniculate; open; with capillary branchlets; espatheate; not comprising 'partial inflorescences' and foliar organs. Spikelet-bearing axes persistent. Spikelets not secund; pedicellate (the pedicels long-capillary).

Female-fertile spikelets. *Spikelets 2 mm long*; compressed laterally; disarticulating above the glumes. Rachilla very shortly prolonged beyond the uppermost female-fertile floret. The rachilla extension naked. Hairy callus absent. *Callus* absent, or short; when present, blunt.

Glumes two; more or less equal; about equalling the spikelets, or exceeding the spikelets; long relative to the adjacent lemmas (slightly exceeding them); hairless; glabrous; pointed (acute); awnless; carinate; similar (membranous, lanceolate). Lower glume 3 nerved. Upper glume 3 nerved. **Spikelets** with female-fertile florets only; *without proximal incomplete florets*.

Female-fertile florets 1. Lemmas acute, lanceolate; less firm than the glumes to similar in texture to the glumes (delicately membranous); not becoming indurated; entire;

pointed; awnless; hairy (finely and sparsely silky-villous); non-carinate; without a germination flap; obscurely 7 nerved. *Palea* present; relatively long (subequalling the lemma); entire; awnless, without apical setae; textured like the lemma; not indurated; 2-nerved (below the middle). **Lodicules** present; 2; free; ***membranous***; glabrous. *Stamens* 3. Anthers about 1.5 mm long; not penicillate; without an apically prolonged connective. *Ovary* glabrous. Styles free to their bases. Stigmas 2.

Fruit, embryo and seedling. Fruit small (1.75 mm, oblong); compressed dorsiventrally (slightly, dorsally). Hilum short (? hilum obscure). Pericarp thin (sub-crustaceous), or thick and hard (?). ***Embryo small (but 1/4–1/3 grain length).***

Abaxial leaf blade epidermis. *Costal/intercostal zonation* conspicuous. *Papillae* absent. Mid-intercostal long-cells rectangular and fusiform; having markedly sinuous walls (these quite thin). *Microhairs* present; panicoid-type; 54–57 microns long; 10.5–11.4 microns wide at the septum. Microhair total length/width at septum 5–5.1. Microhair apical cells 28.5–34.5 microns long. Microhair apical cell/total length ratio 0.53–0.6. *Stomata* absent or very rare; (25.5–)33–36(–39) microns long. *Intercostal short-cells* fairly common; in cork/silica-cell pairs (and solitary). *Costal short-cells* conspicuously in long rows. Costal silica bodies 'panicoid-type'; mostly dumb-bell shaped and nodular; not sharp-pointed.

Transverse section of leaf blade, physiology. C_3; XyMS+. *Midrib* not readily distinguishable; with one bundle only. Sclerenchyma all associated with vascular bundles.

Taxonomy. Arundinoideae; Danthonieae.

Ecology, geography, regional floristic distribution. 1–2 species. South Africa. Helophytic, or mesophytic (on mountain slopes); shade species (of rocks); glycophytic. Cape and Antarctic. Patagonian.

References, etc. Leaf anatomical: this project.

Pobeguinea Jacques-Félix

~ *Anadelphia*

Habit, vegetative morphology. Annual (rarely), or perennial; caespitose. *Culms* 50–100 cm high; branched above, or unbranched above. Culm nodes exposed; glabrous. Culm leaves present. Upper culm leaf blades fully developed. Young shoots intravaginal. *Leaves* not basally aggregated. Leaf blades linear; narrow; 2–5 mm wide; flat; without cross venation; persistent. Ligule an unfringed membrane; truncate; about 0.5 mm long. *Contra-ligule* absent.

Reproductive organization. Plants bisexual, with bisexual spikelets; with hermaphrodite florets. *The spikelets of sexually distinct forms on the same plant*; hermaphrodite and male-only, or hermaphrodite and sterile. The male and female-fertile spikelets mixed in the inflorescence. The spikelets overtly heteromorphic; all in heterogamous combinations. Plants seemingly inbreeding (the stamens enclosed within the spikelets, and the 'racemes' enclosed in the spatheoles). Seemingly exposed-cleistogamous.

Inflorescence. Inflorescence paniculate; open; with capillary branchlets; spatheate (and spatheolate); a complex of 'partial inflorescences' and intervening foliar organs. Spikelet-bearing axes *very much reduced (consisting of a single triplet of which the pedicellate members may be much reduced or rarely missing (or concealed among the callus hairs?), or the triplet preceded by a heterogamous pair)*; the spikelet-bearing axes with only one spikelet-bearing 'article', or with 2–3 spikelet-bearing 'articles' (one or two 'articles'); solitary; disarticulating; disarticulating at the joints (or at the single articulation). *'Articles'* linear; without a basal callus-knob; not appendaged; disarticulating obliquely; densely long-hairy to somewhat hairy. Spikelets paired and in triplets, or in triplets (i.e. seemingly always with a more or less readily recognizable terminal triplet); not secund; sessile and pedicellate; consistently in 'long-and-short' combinations; in pedicellate/sessile combinations. Pedicels of the 'pedicellate' spikelets free of the rachis. The 'shorter' spikelets hermaphrodite. The 'longer' spikelets male-only, or sterile (with hairy pedicels).

Female-sterile spikelets. The male/sterile spikelets linear-lanceolate, awnless, glabrous, disarticulating transversely. Rachilla of male spikelets terminated by a male floret. The male spikelets with glumes; with proximal incomplete florets. The lemmas awnless. Male florets 1; 3 staminate.

Female-fertile spikelets. Spikelets 4.5–7.5 mm long; compressed dorsiventrally; falling with the glumes. Rachilla terminated by a female-fertile floret. Hairy callus present. *Callus* fairly long; pointed.

Glumes two; more or less equal; exceeding the spikelets; long relative to the adjacent lemmas (exceeding them); hairy (towards the apex), or hairless; glabrous; without conspicuous tufts or rows of hairs; awned to awnless (the lower muticous or bisubulate, the upper apiculate to long-awned); very dissimilar (both hardened and shiny, the lower rather flat-backed, truncate to bidentate or bisubulate, the upper more or less carinate by accommodation of the pedicels, truncate to emarginate and apiculate to long-awned). *Lower glume convex on the back to flattened on the back; with a conspicuous pit (at the base, at least in* **P. afzeliana** *and* **P hamata**); relatively smooth; *4–7 nerved (sometimes conspicuously lacking the median).* Upper glume 3 nerved. *Spikelets* with incomplete florets. The incomplete florets proximal to the female-fertile florets. *The proximal incomplete florets* 1; epaleate; sterile. The proximal lemmas awnless; 0 nerved, or 2 nerved (palea-like); decidedly exceeding the female-fertile lemmas; similar in texture to the female-fertile lemmas (hyaline); not becoming indurated.

Female-fertile florets 1. Lemmas less firm than the glumes (stipitate-winged below, the wings and lobes hyaline); not becoming indurated; incised; 2 lobed; deeply cleft; awned. Awns 1; median; from a sinus; geniculate; much longer than the body of the lemma (12–35 mm long); entered by one vein. Awn bases twisted; not flattened. Lemmas hairless; non-carinate; without a germination flap; 1 nerved. **Palea absent.** *Lodicules* present; 2; free; fleshy; ciliate. *Stamens* 3 (short, included). Anthers about 2.5 mm long; not penicillate; without an apically prolonged connective. *Ovary* glabrous. Styles basally fused; joined below. Stigmas 2.

Fruit, embryo and seedling. Fruit ellipsoid; glabrous. Hilum short. Embryo large.

Abaxial leaf blade epidermis. *Costal/intercostal zonation* conspicuous. *Papillae* present; costal and intercostal (*P. hamata*), or intercostal (*P. afzeliana*, confined to the interstomatals). Intercostal papillae over-arching the stomata, or not over-arching the stomata; consisting of one oblique swelling per cell to consisting of one symmetrical projection per cell, or consisting of one symmetrical projection per cell (very large, fairly thick walled). Long-cells of similar wall thickness costally and intercostally (fairly thin walled). Intercostal zones exhibiting many atypical long-cells. Mid-intercostal long-cells rectangular to fusiform (or variously irregular); having straight or only gently undulating walls. *Microhairs* present; elongated; clearly two-celled; panicoid-type; 30–51 microns long; 4–8.4 microns wide at the septum. Microhair total length/width at septum 3.6–8.5. Microhair apical cells 12–22.5 microns long. Microhair apical cell/total length ratio 0.3–0.6. *Stomata* common; 22–27 microns long. Subsidiaries non-papillate; triangular. Guard-cells overlapping to flush with the interstomatals. *Intercostal short-cells* seemingly absent or very rare (but numerous short long-cells present). *Costal short-cells* conspicuously in long rows. Costal silica bodies present and well developed; 'panicoid-type'; dumb-bell shaped; not sharp-pointed.

Transverse section of leaf blade, physiology. C_4; XyMS–. PCR sheath outlines uneven. PCR sheath extensions absent. *Leaf blade* adaxially flat. *Midrib* conspicuous; having a conventional arc of bundles (a large median, and 2–3 smaller bundles on either side); with colourless mesophyll adaxially. *The lamina* symmetrical on either side of the midrib. Bulliforms present in discrete, regular adaxial groups (in places, most obviously in *P. afzeliana*), or not in discrete, regular adaxial groups (the epidermis mostly more or less bulliform except over the larger bundles, and the 'groups' irregular in form); in simple fans (in places, especially in *P. afzeliana*), or associated with colourless mesophyll cells to form deeply-penetrating fans (especially in *P. afzeliana*); associating with colourless mesophyll cells to form arches over small vascular bundles (often with single colourless cells adjoining the bulliform epidermis on either side of a small bundle). Many of the smallest vascular bundles unaccompanied by sclerenchyma. Combined sclerenchyma girders present (with the primaries only); forming 'figures' (the primaries, in *P. hamata*), or nowhere forming 'figures'. Sclerenchyma all associated with vascular bundles.

Taxonomy. Panicoideae; Andropogonodae; Andropogoneae; Andropogoninae.
Ecology, geography, regional floristic distribution. 4 species. Tropical West Africa. Mesophytic; species of open habitats; glycophytic.
Paleotropical. African. Sudano-Angolan and West African Rainforest. South Tropical African and Kalaharian.
References, etc. Leaf anatomical: this project.

Podophorus Phil.

Habit, vegetative morphology. Perennial (?). Culms herbaceous. *Leaves* not basally aggregated; non-auriculate. *Leaf blades* linear to linear-lanceolate; narrow; about 4–6 mm wide; flat (pilose); *not pseudopetiolate*; without cross venation. Ligule an unfringed membrane (but abaxially hairy); not truncate (laciniate); about 5 mm long.
Reproductive organization. Plants bisexual, with bisexual spikelets; with hermaphrodite florets.
Inflorescence. *Inflorescence* few spikeleted; *paniculate*; open; espatheate; not comprising 'partial inflorescences' and foliar organs. Spikelet-bearing axes persistent. Spikelets not secund; pedicellate.
Female-fertile spikelets. Spikelets 10–13 mm long; compressed laterally; disarticulating above the glumes; not disarticulating between the florets. Rachilla prolonged beyond the uppermost female-fertile floret. The rachilla extension with incomplete florets. Hairy callus present. *Callus* short.
Glumes two; very unequal; shorter than the spikelets; shorter than the adjacent lemmas; hairless; pointed (acute); awnless; non-carinate; fairly similar (lanceolate, herbaceous). Lower glume 3 nerved. Upper glume 5 nerved. *Spikelets* with incomplete florets. The incomplete florets distal to the female-fertile florets. The distal incomplete florets 1, or 2; *incomplete florets* merely underdeveloped; incomplete florets awned. *Spikelets without proximal incomplete florets*.
Female-fertile florets *1*. Lemmas subcylindrical; decidedly firmer than the glumes (leathery); rigid; entire; awned. Awns 1; median; apical; non-geniculate, or geniculate (? — see illustration by Nicora and Rúgolo de Agrasar 1987); flexuous; hairless; much longer than the body of the lemma; persistent. Lemmas hairless; scabrous; non-carinate; without a germination flap; 5 nerved, or 7 nerved. *Palea* present; relatively long; tightly clasped by the lemma; apically notched; awnless, without apical setae; 2-nerved; 2-keeled. *Lodicules* present; 2; free; membranous; glabrous; toothed; not or scarcely vascularized. **Stamens 2**. **Ovary** *hairy (on the top)*; without a conspicuous apical appendage (not at all appendaged, just hairy). Styles free to their bases. Stigmas 2; white.
Fruit, embryo and seedling. Fruit medium sized (5 mm long); with hairs confined to a terminal tuft. Hilum long-linear. Embryo small.
Abaxial leaf blade epidermis. *Costal/intercostal zonation* conspicuous. *Papillae* absent. *Long-cells* costal long-cells much smaller than intercostals. Mid-intercostal long-cells rectangular and fusiform; having markedly sinuous walls (and pitted, fairly thin-walled). *Microhairs* absent. *Stomata* absent or very rare. *Intercostal short-cells* common; in cork/silica-cell pairs (commonly); silicified. Numerous macrohairs present costally and intercostally. *Costal short-cells* predominantly paired. Costal silica bodies horizontally-elongated crenate/sinuous (a few), or rounded (numerous), or tall-and-narrow (a few); not sharp-pointed.
Transverse section of leaf blade, physiology. C_3; XyMS+. *Mesophyll* with non-radiate chlorenchyma; without adaxial palisade. *Leaf blade* adaxially flat; with the ribs more or less constant in size. *Midrib* conspicuous (a larger bundle and an abaxially rounded keel); with one bundle only. Bulliforms present in discrete, regular adaxial groups; in simple fans (these large, mostly distorted by macrohairs). All the vascular bundles accompanied by sclerenchyma. Combined sclerenchyma girders present; forming 'figures' (all the bundles). Sclerenchyma all associated with vascular bundles.
Taxonomy. Pooideae; Poodae; Poeae.
Ecology, geography, regional floristic distribution. 1 species. Juan Fernandez Is. (Chile).
Neotropical. Fernandezian.

References, etc. Morphological/taxonomic: Macfarlane and Watson 1980. Leaf anatomical: this project.

Poecilostachys Hackel

Excluding *Chloachne*

Habit, vegetative morphology. Perennial; decumbent (trailing or scrambling). *Culms herbaceous*; branched above. Culm nodes glabrous. Culm internodes solid. *Leaves* not basally aggregated; non-auriculate. Leaf blades linear to lanceolate; broad, or narrow; without cross venation; persistent.

Reproductive organization. Plants bisexual, with bisexual spikelets; with hermaphrodite florets.

Inflorescence. Inflorescence of spicate main branches, or paniculate (with several short, unilateral, spicate racemes); open; espatheate; not comprising 'partial inflorescences' and foliar organs. Spikelet-bearing axes persistent. *Spikelets* unaccompanied by bractiform involucres, not associated with setiform vestigial branches, or subtended by solitary 'bristles'. The 'bristles' when present, persisting on the axis. *Spikelets* paired; *secund*; pedicellate.

Female-fertile spikelets. *Spikelets compressed laterally*; falling with the glumes. Rachilla terminated by a female-fertile floret. Hairy callus absent.

Glumes two; more or less equal; shorter than the spikelets; shorter than the adjacent lemmas; free; pointed; awned, or awnless (or mucronate); carinate to non-carinate; *similar*. Lower glume 5 nerved. Upper glume 5 nerved. *Spikelets* with incomplete florets. The incomplete florets proximal to the female-fertile florets. *Spikelets with proximal incomplete florets*. *The proximal incomplete florets* 1; sterile. The proximal lemmas awnless; 7–9 nerved; decidedly exceeding the female-fertile lemmas; similar in texture to the female-fertile lemmas; not becoming indurated.

Female-fertile florets 1. Lemmas strongly laterally compressed; similar in texture to the glumes to decidedly firmer than the glumes (membranous to cartilaginous); not becoming indurated; entire; pointed; awnless; non-carinate; having the margins lying flat and exposed on the palea, or having the margins tucked in onto the palea; without a germination flap; 5 nerved. *Palea* present; relatively long; 2-nerved; keel-less. **Lodicules** present; 2; *joined*; fleshy, or membranous; glabrous; heavily vascularized, or not or scarcely vascularized. *Stamens* 3. Anthers not penicillate; without an apically prolonged connective. *Ovary* glabrous. Styles fused. Stigmas 2.

Fruit, embryo and seedling. *Fruit* small, or medium sized; *not noticeably compressed. Hilum long-linear*. Embryo large; not waisted. Endosperm hard.

Abaxial leaf blade epidermis. *Costal/intercostal zonation* conspicuous. *Papillae* absent. *Long-cells* markedly different in shape costally and intercostally (the costals much narrower). Intercostal zones exhibiting many atypical long-cells (many being almost square in *P. festucaceus*). Mid-intercostal long-cells rectangular; having markedly sinuous walls. *Microhairs* present; elongated; clearly two-celled; panicoid-type; about 54 microns long; 3.3 microns wide at the septum. Microhair total length/width at septum 16.4. Microhair apical cells about 33 microns long. Microhair apical cell/total length ratio 0.61. *Stomata* common (but only bordering on the veins); 36–39 microns long. Subsidiaries low triangular, or dome-shaped, or dome-shaped and triangular. Guard-cells overlapping to flush with the interstomatals. *Intercostal short-cells* common, or absent or very rare (very rare in *P. festucaceus*); not paired; silicified. *Costal short-cells* conspicuously in long rows. Costal silica bodies 'panicoid-type'; mostly nodular; not sharp-pointed.

Transverse section of leaf blade, physiology. C_3; XyMS+. *Mesophyll* with non-radiate chlorenchyma. *Leaf blade* with distinct, prominent adaxial ribs to 'nodular' in section; with the ribs more or less constant in size (distant, low, round topped). *Midrib* fairly conspicuous; with one bundle only. Bulliforms not in discrete, regular adaxial groups (most of the epidermis between the ribs 'bulliform', at least in *P. festucaceus*). All the vascular bundles accompanied by sclerenchyma. Combined sclerenchyma girders present; forming 'figures' (in the major bundles). Sclerenchyma all associated with vascular bundles.

Taxonomy. Panicoideae; Panicodae; Paniceae.
Ecology, geography, regional floristic distribution. 20 species. Madagascar. Shade species (forest).
Paleotropical. Madagascan.
References, etc. Leaf anatomical: this project, from poor material of *P. festucaceus* and *P. viguieri*.

Pogonachne Bor

Habit, vegetative morphology. Stilt-rooted annual. *Culms* 100 cm high; herbaceous. *Leaf blades* absent from the upper culm leaves; lower on the culms, broad; 8–15 mm wide (to 25 cm long); without cross venation. **Ligule** present; *an unfringed membrane (laciniate)*; 3 mm long.

Reproductive organization. Plants bisexual, with bisexual spikelets; with hermaphrodite florets. The spikelets of sexually distinct forms on the same plant; hermaphrodite and sterile (hermaphrodite and 'reduced' - but the latter are somewhat theoretical); overtly heteromorphic (though the vestigial, sessile spikelets are hardly recognisable as such).

Inflorescence. *Inflorescence* falsely *paniculate (scanty, of single racemes and their spatheoles)*; *spatheate*; not comprising 'partial inflorescences' and foliar organs. *Spikelet-bearing axes* 'racemes'; solitary; eventually disarticulating; tardily disarticulating at the joints (after the spikelets have fallen). *'Articles'* linear; not appendaged. Spikelets paired (accepting the vestiges as spikelets); pedicellate, or sessile and pedicellate (accepting the vestiges, which are sometimes totally suppressed, as 'sessile spikelets'); consistently in 'long-and-short' combinations (in vestigial form). Pedicels of the 'pedicellate' spikelets free of the rachis. The 'shorter' spikelets sterile (the 'sessile spikelets', when present, being reduced to small scales between internode and pedicel). The 'longer' spikelets hermaphrodite.

Female-sterile spikelets. *The 'sessile spikelets' vestigial and reduced to small scales between the internode and the pedicel, or completely suppressed.*

Female-fertile spikelets. *Spikelets* 7–10 mm long; strongly *compressed laterally*; falling with the glumes. Rachilla terminated by a female-fertile floret. Hairy callus present (the hairs down one side).

Glumes two; more or less equal; long relative to the adjacent lemmas; hairy; (the upper) *with distinct hair tufts (with a tuft of hairs in the middle third)*; awnless; non-carinate (rounded on the back); *very dissimilar (G₁ ovate-lanceolate, entire and sparsely hairy towards the tip, G₂ asymmetric, with a tuft of hairs above the middle)*. Lower glume not two-keeled (and wingless); convex on the back; not pitted; G₁ several-nerved. Upper glume 5 nerved. *Spikelets* with incomplete florets. The incomplete florets proximal to the female-fertile florets. *The proximal incomplete florets* 1; sterile. The proximal lemmas awnless; 3 nerved; more or less equalling the female-fertile lemmas; similar in texture to the female-fertile lemmas (hyaline); not becoming indurated.

Female-fertile florets 1. Lemmas less firm than the glumes; incised; not deeply cleft; awned. Awns 1; median; from a sinus; geniculate; hairless (scabrid); much longer than the body of the lemma. *Lemmas* hairless; glabrous; non-carinate; without a germination flap; *1 nerved. Palea* present; relatively long; entire; awnless, without apical setae; not indurated; 2-nerved; keel-less. *Lodicules* present; 2; free; fleshy; glabrous. *Stamens* 3. Anthers not penicillate; without an apically prolonged connective. *Ovary* glabrous. Styles free to their bases. Stigmas 2.

Taxonomy. Panicoideae; Andropogonodae; Andropogoneae; Andropogoninae (seemingly).

Ecology, geography, regional floristic distribution. 1 species. Bombay.
Paleotropical. Indomalesian. Indian.

References, etc. Morphological/taxonomic: Bor 1949.

Special comments. Fruit data wanting. Anatomical data wanting.

Pogonarthria Stapf

Habit, vegetative morphology. Annual, or perennial; caespitose. *Culms* 13–250 cm high; herbaceous; branched above, or unbranched above. Culm nodes glabrous. Culm internodes solid. Plants unarmed. Young shoots intravaginal. *Leaves* mostly basal, or not basally aggregated; non-auriculate; without auricular setae (but the mouth of the sheath sometimes hairy). Leaf blades linear; narrow; 2–5 mm wide; setaceous, or not setaceous; flat, or rolled (convolute); without abaxial multicellular glands; without cross venation; persistent. Ligule a fringed membrane, or a fringe of hairs; 0.25–0.5 mm long. *Contraligule* absent.

Reproductive organization. Plants bisexual, with bisexual spikelets; with hermaphrodite florets.

Inflorescence. *Inflorescence of spicate main branches (a raceme of numerous, upcurved, spike-like branches)*; with conspicuously divaricate branchlets, or without conspicuously divaricate branchlets (when immature); non-digitate (the branches tending to whorls). Rachides the slender spikelet bearing rachides flattened and sinuous. Inflorescence espatheate; not comprising 'partial inflorescences' and foliar organs. *Spikelet-bearing axes* spikelike. The racemes without spikelets towards the base. *Spikelet-bearing axes disarticulating*; falling entire (the racemes falling after the spikelets have broken up). Spikelets solitary; secund (the spikelet bearing rachides one-sided); biseriate; shortly pedicellate; imbricate.

Female-fertile spikelets. *Spikelets* 3.3–7.8 mm long; compressed laterally; disarticulating above the glumes; *not disarticulating between the florets, or disarticulating between the florets (disarticulating between the lemmas, or the glumes and lemmas falling irregularly to leave the paleas on the persistent rachilla)*; with conventional internode spacings. Rachilla prolonged beyond the uppermost female-fertile floret; rachilla hairy (ciliate at the disarticulating apices). The rachilla extension with incomplete florets. Hairy callus absent. *Callus* absent.

Glumes two; *very unequal (G₁ about 2/3 of G₂)*; shorter than the spikelets; shorter than the adjacent lemmas; lateral to the rachis; hairless; glabrous (scabrid on the keel); pointed (subacuminate); awnless; carinate; similar (rigidly membranous). Lower glume shorter than the lowest lemma; 1 nerved. Upper glume 1 nerved. *Spikelets* with incomplete florets. The incomplete florets distal to the female-fertile florets. *The distal incomplete florets* merely underdeveloped; awnless.

Female-fertile florets 2–8 *(decreasing in size upwards)*. *Lemmas* acute to acuminate; similar in texture to the glumes; not becoming indurated; entire; pointed; awnless (but sometimes subaristate); *hairless*; glabrous (scabrid on the median nerve); more or less carinate; without a germination flap; 3 nerved; with the nerves non-confluent. *Palea* present; relatively long; minutely apically notched; awnless, without apical setae; textured like the lemma (membranous); not indurated; 2-nerved; 2-keeled. Palea keels wingless; scabrous. *Lodicules* present; 2; free; fleshy (but narrow); glabrous; not or scarcely vascularized. *Stamens* 3. Anthers 0.6–1.3 mm long; not penicillate; without an apically prolonged connective. *Ovary* glabrous. Styles free to their bases. Stigmas 2; brown.

Fruit, embryo and seedling. *Fruit* free from both lemma and palea; small (0.5–1 mm long); ellipsoid; obtusely trigonous. Hilum short. Pericarp fused. Embryo large (about 1/2 grain length).

Abaxial leaf blade epidermis. *Costal/intercostal zonation* conspicuous. *Papillae* absent. *Long-cells* similar in shape costally and intercostally, or markedly different in shape costally and intercostally (the costals smaller); of similar wall thickness costally and intercostally. Mid-intercostal long-cells rectangular; having markedly sinuous walls (deeply, and pitted). *Microhairs* present, or absent (from some material of *P. squarrosa*); more or less spherical to elongated; clearly two-celled; when present, chloridoid-type (rather variable in shape, cf. *Eragrostis*). Microhair apical cell wall of similar thickness/rigidity to that of the basal cell. Microhairs without 'partitioning membranes' (in *P. squarrosa*); (27–)28.5–30 microns long (*P. fleckii*), or 36–54 microns long (*P. squarrosa*). Microhair basal cells 21–24 microns long. Microhairs 11.4–15 microns wide at the septum (*P. fleckii*). Microhair total length/width at septum 1.9–2.5 (*P. fleckii*).

Microhair apical cells (5.4–)6–7.5 microns long (*P. fleckii*), or 14–22 microns long (*P. squarrosa*). Microhair apical cell/total length ratio 0.2–0.28 (*P. fleckii*), or 0.43 (*P. squarrosa*). *Stomata* common; (22.5–)25.5–28.5(–30) microns long. Subsidiaries triangular (a few), or dome-shaped (mostly). Guard-cells overlapping to flush with the interstomatals. *Intercostal short-cells* common; in cork/silica-cell pairs; silicified. Intercostal silica bodies absent, or imperfectly developed; rounded, or crescentic, or saddle shaped. *Costal short-cells* conspicuously in long rows (*P. fleckii*), or predominantly paired and neither distinctly grouped into long rows nor predominantly paired (over different veins). Costal silica bodies present in alternate cell files of the costal zones; saddle shaped to crescentic (*P. squarrosa*), or 'panicoid-type' (*P. fleckii*); not sharp-pointed.

Transverse section of leaf blade, physiology. C_4; XyMS+. PCR sheath outlines uneven to even (more even in *P. fleckii* than in *P. squarrosa*). PCR sheaths of the primary vascular bundles interrupted; interrupted both abaxially and adaxially. PCR sheath extensions present, or absent. Maximum number of extension cells when present, 1. PCR cell chloroplasts centripetal. *Mesophyll* with radiate chlorenchyma; traversed by columns of colourless mesophyll cells, or not traversed by colourless columns. *Leaf blade* 'nodular' in section to adaxially flat (the ribs very low). *Midrib* conspicuous (bundle somewhat larger), or not readily distinguishable; with one bundle only. Bulliforms present in discrete, regular adaxial groups; in simple fans (a few, with large median cells), or associated with colourless mesophyll cells to form deeply-penetrating fans (often linked to the abaxial epidermis by girders of colourless cells). All the vascular bundles accompanied by sclerenchyma. Combined sclerenchyma girders present (nearly all the bundles); forming 'figures' (all I's or anchors). Sclerenchyma all associated with vascular bundles. The lamina margins with fibres.

Cytology. $2n = 40$, or 42.

Taxonomy. Chloridoideae; main chloridoid assemblage.

Ecology, geography, regional floristic distribution. 4 species. Tropical and southern Africa. Mesophytic to xerophytic; species of open habitats; glycophytic. Savanna grasslands, often in shallow or sandy soils or in disturbed places.

Paleotropical. African and Madagascan. Sudano-Angolan, West African Rainforest, and Namib-Karoo. Sahelo-Sudanian, Somalo-Ethiopian, South Tropical African, and Kalaharian.

Rusts and smuts. Rusts — *Puccinia*.

Economic importance. Significant weed species: *P. squarrosa*.

References, etc. Morphological/taxonomic: Rendle 1899. Leaf anatomical: Metcalfe 1960; this project; photos of *P. fleckii* and *P. squarrosa* provided by R.P. Ellis.

Pogonatherum P. Beauv.

Homoplitis Trin., *Pogonopsis* Presl

Habit, vegetative morphology. Delicate perennial; caespitose, or decumbent. *Culms* 12–60 cm high; branched above. Culm nodes hairy, or glabrous. Culm internodes solid. Young shoots intravaginal. *Leaves* not basally aggregated; non-auriculate. *Leaf blades* narrow; usually flat; *pseudopetiolate*; without cross venation; persistent. Ligule a fringed membrane.

Reproductive organization. Plants bisexual, with bisexual spikelets; with hermaphrodite florets (in the sessile spikelets). *The spikelets of sexually distinct forms on the same plant*; hermaphrodite and female-only (hermaphrodite-sessile and female-pedicelled); more or less *homomorphic (the pedicelled slightly smaller)*.

Inflorescence. *Inflorescence a solitary, spicate, pedunculate, axillary 'raceme'*; espatheate; not comprising 'partial inflorescences' and foliar organs. *Spikelet-bearing axes* 'racemes'; solitary; disarticulating; disarticulating at the joints. Spikelets paired; not secund; sessile and pedicellate; consistently in 'long-and-short' combinations; in pedicellate/sessile combinations. The 'shorter' spikelets hermaphrodite. The 'longer' spikelets female-only.

Female-fertile spikelets. Spikelets spikelets small; compressed laterally; falling with the glumes (the pedicelled falling from its pedicel, the sessile falling with the internode and pedicel). Rachilla terminated by a female-fertile floret. Hairy callus present, or absent. Callus blunt.

Glumes two; more or less equal; long relative to the adjacent lemmas; hairy, or hairless; without conspicuous tufts or rows of hairs; (the upper) **awned (with a capillary awn from the mid-nerve)**; very dissimilar (G_1 cartilaginous, rounded on the back and truncate, G_2 compressed-keeled and awned). **Lower glume** strongly convex on the back; not pitted; relatively smooth; **0–1 nerved**. Upper glume 1 nerved. *Spikelets* with female-fertile florets only, or with incomplete florets. The incomplete florets proximal to the female-fertile florets. *The proximal incomplete florets* 1; male, or sterile. The proximal lemmas 0 nerved, or 1 nerved; not becoming indurated.

Female-fertile florets 1. Lemmas less firm than the glumes (thinly membranous to hyaline); not becoming indurated; incised; 2 lobed; deeply cleft to not deeply cleft (to 1/3 to 1/2 its length); awned. Awns 1; median; from a sinus; geniculate; much longer than the body of the lemma (filiform). Lemmas hairless; non-carinate; 1 nerved. *Palea* present; nerveless. *Lodicules* absent. **Stamens 1–2**. Anthers not penicillate. *Ovary* glabrous. Styles fused. Stigmas 2.

Fruit, embryo and seedling. Fruit small; compressed dorsiventrally (flattened on the front). Hilum short. Embryo large.

Abaxial leaf blade epidermis. *Costal/intercostal zonation* conspicuous. *Papillae* present. Intercostal papillae over-arching the stomata; several per cell (often slightly irregular in shape). *Long-cells* similar in shape costally and intercostally; of similar wall thickness costally and intercostally. Mid-intercostal long-cells rectangular; having markedly sinuous walls. *Microhairs* present; panicoid-type; 24–27(–30) microns long; (3.6–)4.8–5.4 microns wide at the septum. Microhair total length/width at septum 4.4–7.5. Microhair apical cells (10.5–)15–16.5 microns long. Microhair apical cell/total length ratio 0.5–0.63. *Stomata* common; (18–)21–22.5(–24) microns long. Subsidiaries triangular (or obscured by the papillae). Guard-cells overlapping to flush with the interstomatals. *Intercostal short-cells* absent or very rare. *Costal short-cells* conspicuously in long rows. Costal silica bodies acutely-angled (very acutely rectangular); sharp-pointed.

Transverse section of leaf blade, physiology. C_4; XyMS–. *Mesophyll* with radiate chlorenchyma. *Leaf blade* with distinct, prominent adaxial ribs, or 'nodular' in section; with the ribs more or less constant in size. *Midrib* conspicuous, or not readily distinguishable; with one bundle only. Bulliforms present in discrete, regular adaxial groups (and in irregular groups); in simple fans. All the vascular bundles accompanied by sclerenchyma. Combined sclerenchyma girders present; forming 'figures'. Sclerenchyma all associated with vascular bundles.

Cytology. $2n = 20$.

Taxonomy. Panicoideae; Andropogonodae; Andropogoneae; Andropogoninae.

Ecology, geography, regional floristic distribution. 3 species. India to Japan. Holarctic, Paleotropical, and Australian. Boreal and Tethyan. Indomalesian. Eastern Asian. Irano-Turanian. Indian, Indo-Chinese, Malesian, and Papuan. North and East Australian. Tropical North and East Australian.

Rusts and smuts. Rusts — *Puccinia*.

References, etc. Leaf anatomical: Metcalfe 1960; this project.

Pogoneura Napper

Habit, vegetative morphology. *Annual*. *Culms* 30–60 cm high; herbaceous; unbranched above. Culm nodes glabrous. Plants unarmed. *Leaves* not basally aggregated. Leaf blades narrow; 2–4 mm wide (about 4 cm long); exhibiting multicellular glands abaxially. The abaxial glands on the blade margins. Leaf blades without cross venation; persistent. *Ligule* present; very short, lacerate' or 'a line of hairs', in different descriptions.

Reproductive organization. Plants bisexual, with bisexual spikelets; with hermaphrodite florets.

Inflorescence. *Inflorescence of spicate main branches (of slender racemes from the main rachis)*; non-digitate. Primary inflorescence branches 12–25 (about 6 cm long, with few spikelets). Inflorescence espatheate; not comprising 'partial inflorescences' and foliar organs. Spikelet-bearing axes persistent. Spikelets solitary.

Female-fertile spikelets. *Spikelets* 5–6 mm long; *not noticeably compressed (terete)*; disarticulating above the glumes; *disarticulating between the florets*. Rachilla prolonged beyond the uppermost female-fertile floret. Hairy callus present.

Glumes two; very unequal to more or less equal; *exceeding the spikelets (enfolding the florets)*; long relative to the adjacent lemmas; hairless (glabrous, with scabridulous keels); pointed (acute or acuminate); awnless; similar (narrowly lanceolate, membranous). Lower glume much exceeding the lowest lemma; 1 nerved. Upper glume 1 nerved. *Spikelets* with female-fertile florets only, or with incomplete florets. The incomplete florets distal to the female-fertile florets. The distal incomplete florets 1; *incomplete florets* merely underdeveloped (vestigial).

Female-fertile florets 2. Lemmas not becoming indurated; incised; 2 lobed; not deeply cleft (bidentate); shortly awned. Awns 1; median; from a sinus; non-geniculate; hairless; much shorter than the body of the lemma (about 1 mm long). Lemmas hairy (villous on the lateral nerves, the nerves grey-green); non-carinate; 3 nerved. *Palea* present; relatively long; entire to apically notched (emarginate); awnless, without apical setae; not indurated; 2-nerved; 2-keeled. Palea keels hairy (conspicuously fringed, the hairs grey-green). *Lodicules* 'minute or missing'. *Stamens* 3. Anthers 0.4–0.5 mm long. *Ovary* glabrous. Stigmas 2.

Fruit, embryo and seedling. Fruit ellipsoid; slightly compressed dorsiventrally.

Abaxial leaf blade epidermis. *Costal/intercostal zonation* conspicuous. *Papillae* absent. *Long-cells* markedly different in shape costally and intercostally (the costals narrower and longer); of similar wall thickness costally and intercostally (walls of medium thickness). Mid-intercostal long-cells rectangular; having markedly sinuous walls. *Microhairs* present; elongated; clearly two-celled; chloridoid-type (the basal cell quite long and expanding distally). Microhair apical cell wall of similar thickness/rigidity to that of the basal cell. Microhairs (27–)28.5–33(–36) microns long. Microhair basal cells 21–24 microns long. Microhairs 6.9–8.4 microns wide at the septum. Microhair total length/width at septum 3.2–4.8. Microhair apical cells 7.5–9 microns long. Microhair apical cell/total length ratio 0.23–0.33. *Stomata* common; 19.5–24–27 microns long. Subsidiaries all dome-shaped. Guard-cells overlapping to flush with the interstomatals. *Intercostal short-cells* common; in cork/silica-cell pairs and not paired (some solitary); silicified. Intercostal silica bodies present and perfectly developed; mostly saddle shaped (small). *Costal short-cells* conspicuously in long rows. Costal silica bodies present in alternate cell files of the costal zones; exclusively saddle shaped; not sharp-pointed.

Transverse section of leaf blade, physiology. C_4; XyMS+. PCR sheaths of the primary vascular bundles interrupted; interrupted abaxially only. PCR sheath extensions absent. *Leaf blade* adaxially flat. *Midrib* conspicuous (via a larger bundle with more abaxial sclerenchyma, and a small sharp keel); with one bundle only. Bulliforms present in discrete, regular adaxial groups; in simple fans (the large median cell deeply penetrating). All the vascular bundles accompanied by sclerenchyma. Combined sclerenchyma girders present (but few: most bundles with a broad abaxial girder and a small adaxial strand); forming 'figures' (a few primaries with I's). Sclerenchyma all associated with vascular bundles. The lamina margins with fibres.

Taxonomy. Chloridoideae; main chloridoid assemblage (cf. *Leptochloa*).

Ecology, geography, regional floristic distribution. 1 species. East Africa. Paleotropical. African. Sudano-Angolan. Sahelo-Sudanian and Somalo-Ethiopian.

References, etc. Morphological/taxonomic: Napper 1963. Leaf anatomical: this project.

Special comments. Fruit data wanting.

Pogonochloa C.E. Hubb.

Habit, vegetative morphology. Perennial; caespitose. **Culms** 30–90 cm high; *herbaceous*; unbranched above. Culm nodes glabrous. Culm internodes solid. Plants unarmed. Young shoots intravaginal. *Leaves* mostly basal to not basally aggregated; non-auriculate. Leaf blades linear (tough, glaucous); narrow; usually folded; without abaxial multicellular glands; without cross venation; rolled in bud. *Ligule* present; a fringed membrane (narrow); truncate; 0.3 mm long. *Contra-ligule* absent.

Reproductive organization. Plants bisexual, with bisexual spikelets; with hermaphrodite florets.

Inflorescence. *Inflorescence of spicate main branches (when closely examined, of numerous lateral racemes, these short, imbricate and appressed)*; contracted (spiciform); *non-digitate*. Primary inflorescence branches 25–40 (or more?). Inflorescence espatheate; not comprising 'partial inflorescences' and foliar organs. Spikelet-bearing axes persistent. *Spikelets* solitary; secund (more or less); *subsessile*.

Female-fertile spikelets. *Spikelets* 3–3.5 mm long; adaxial; compressed laterally; *disarticulating between the glumes (the G₁ persistent, the G₂ falling with the florets)*; not disarticulating between the florets; with conventional internode spacings. Rachilla prolonged beyond the uppermost female-fertile floret; rachilla hairless (glabrous). The rachilla extension with incomplete florets. Hairy callus present. *Callus* short; blunt.

Glumes two; more or less equal; about equalling the spikelets; long relative to the adjacent lemmas (much longer); lateral to the rachis; hairless (scaberulous); shortly awned (with aristules to 1.5 mm long); non-carinate (rounded on the back); similar (narrowly oblong, membranous, the G₁ acute or obtuse, the G₂ truncate to emarginate). Lower glume 1 nerved. Upper glume 1 nerved. *Spikelets* with incomplete florets. The incomplete florets distal to the female-fertile florets. The distal incomplete florets 1, or 2 (the second floret male or sterile, the third reduced to an awn); incomplete florets awned.

Female-fertile florets 1. *Lemmas* less firm than the glumes to similar in texture to the glumes (thinly membranous); not becoming indurated; entire; pointed; *awned*. Awns 1; median; apical; non-geniculate; flexuous (slender); hairless (scaberulous); much longer than the body of the lemma. Lemmas hairy (loosely pilose above); non-carinate (rounded on the back); without a germination flap; 3 nerved. *Palea* present; relatively long; minutely apically notched; awnless, without apical setae (ciliolate); not indurated (hyaline); 2-nerved; 2-keeled (folded). *Lodicules* present; 2; free; fleshy; glabrous. *Stamens* 3. Anthers 1–1.5 mm long; not penicillate; without an apically prolonged connective. *Ovary* glabrous. Styles free to their bases. Stigmas 2.

Fruit, embryo and seedling. *Fruit* free from both lemma and palea (but embraced); small (1.3–1.6 mm long); fusiform; compressed dorsiventrally to not noticeably compressed (slightly flattened on the hilar side). Hilum short. Pericarp fused (?). Embryo large.

Abaxial leaf blade epidermis. *Costal/intercostal zonation* conspicuous. *Papillae* absent. *Long-cells* similar in shape costally and intercostally; of similar wall thickness costally and intercostally (walls of medium thickness). Mid-intercostal long-cells rectangular; having markedly sinuous walls. *Microhairs* present; elongated; clearly two-celled; chloridoid-type. Microhair apical cell wall of similar thickness/rigidity to that of the basal cell. Microhairs 18–21–22.5 microns long. Microhair basal cells 12 microns long. Microhairs (6.6–)7.5–9(–9.6) microns wide at the septum. Microhair total length/width at septum 2–2.8. Microhair apical cells 6.6–7.5–9 microns long. Microhair apical cell/total length ratio 0.33–0.5. *Stomata* common; 18–19.5–21 microns long. Subsidiaries mostly dome-shaped (often high). Guard-cells overlapping to flush with the interstomatals. *Intercostal short-cells* common; not paired (solitary); not silicified. Intercostal silica bodies absent. *Costal short-cells* conspicuously in long rows. Costal silica bodies present in alternate cell files of the costal zones; 'panicoid-type'; mostly long to short dumb-bell shaped; not sharp-pointed.

Transverse section of leaf blade, physiology. C₄; XyMS+ (mestome sheath often double, thick-walled). PCR sheaths of the primary vascular bundles interrupted; interrupted abaxially only. PCR sheath extensions absent. Mesophyll traversed by columns of colourless mesophyll cells (probably, though in most bundles the colourless cells fall short of the abaxial epidermis). *Leaf blade* adaxially flat. *Midrib* not readily distinguishable; with one bundle only. Bulliforms present in discrete, regular adaxial groups (very large groups, between all bundles); associated with colourless mesophyll cells to form deeply-penetrating fans (these probably sometimes linked by colourless cells with the abaxial epidermis). All the vascular bundles accompanied by sclerenchyma. Combined sclerenchyma girders present (with all the bundles); forming 'figures' (I's, in all bundles). Sclerenchyma all associated with vascular bundles.

Taxonomy. Chloridoideae; main chloridoid assemblage.

Ecology, geography, regional floristic distribution. 1 species. Southern tropical Africa. Helophytic, or mesophytic (streamsides); glycophytic.

Paleotropical. African. Sudano-Angolan. South Tropical African.

References, etc. Morphological/taxonomic: Hubbard 1940b. Leaf anatomical: this project.

Pohlidium Davidse, Soderstrom & Ellis

Habit, vegetative morphology. Perennial; caespitose. The flowering culms leafy. *Culms* 7–19 cm high; herbaceous; branched above. Culm nodes glabrous. Young shoots intravaginal. *Leaves* not basally aggregated; auriculate (the sheath being extended upwards into a single, median, prominent auricle, with which the ligule is marginally adnate); without auricular setae (the auricle glabrous or sparsely hairy). *Leaf blades* ovate to elliptic; broad; 8–16 mm wide; flat; conspicuously *pseudopetiolate*; cross veined; persistent (seemingly); rolled in bud. *Ligule* present to absent.

Reproductive organization. *Plants monoecious with all the fertile spikelets unisexual*; without hermaphrodite florets. *The spikelets* of sexually distinct forms on the same plant (male and female); *female-only and male-only (the males fewer, peripheral in the inflorescence)*. The male and female-fertile spikelets on different branches of the same inflorescence to segregated, in different parts of the same inflorescence branch (the males terminating the main axis and the major branches, or the lower branches sometimes exclusively female). The spikelets overtly heteromorphic.

Inflorescence. Inflorescence paniculate; open; espatheate; not comprising 'partial inflorescences' and foliar organs. Spikelet-bearing axes persistent. Spikelets not secund; pedicellate (the pedicels 3–14 mm long).

Female-sterile spikelets. Male spikelets 3–6.4 mm long, sometimes with a linear upper glume, 1–4 flowered, lemmas glabrous, stamens 2 or 3, or 2 plus a staminode, anthers 1.5–2 mm long, not penicillate, unappendaged; lodicules 2, fleshy, inconspicuously vascularised. The male spikelets with glumes (one), or without glumes; 1–4 floreted. Male florets 2 staminate, or 3 staminate (or two plus a staminode).

Female-fertile spikelets. *Spikelets* unconventional (being without glumes); 2.3–3 mm long; falling entire. Rachilla terminated by a female-fertile floret (?). Hairy callus absent. *Callus* absent.

Glumes absent. *Spikelets* with female-fertile florets only.

Female-fertile florets 1. Lemmas herbaceous; not becoming indurated; entire, or incised; when incised, minutely 3 lobed; not deeply cleft; awnless; hairy (between the nerves); non-carinate (seemingly— slightly sulcate between the nerves); without a germination flap; 5 nerved, or 7 nerved; with the nerves non-confluent. *Palea* present; relatively long (slightly shorter than the lemma); apically notched; awnless; without apical setae; textured like the lemma; not indurated; 2-nerved; 2-keeled (hairy between the keels). *Lodicules* present; 2; joined to free; fleshy; glabrous; not or scarcely vascularized (obscurely vascularised). *Stamens* 0 (sometimes with two minute staminodes). *Ovary* glabrous. Styles free to their bases. Stigmas 2.

Fruit, embryo and seedling. Fruit small (about 1.8 mm long); compressed laterally. *Hilum short (apparently)*. Embryo large (about 1/3 of the caryopsis length).

Abaxial leaf blade epidermis. *Papillae* present. Intercostal papillae over-arching the stomata; several per cell (2 to 3 per intercostal long-cell, sometimes branched, large, covering over much of the epidermis). Intercostal zones without typical long-cells (these being generally only slightly longer than wide). Mid-intercostal long-cells having markedly sinuous walls. *Microhairs* present; panicoid-type. *Stomata* common. Subsidiaries dome-shaped and triangular. Costal silica bodies 'panicoid-type', or oryzoid and 'panicoid-type' (?); cross shaped and dumb-bell shaped (? —described as irregular dumb-bells 'with somewhat angular and indented ends', and 'very small, narrow crosses'); not sharp-pointed.

Transverse section of leaf blade, physiology. C_3; XyMS+. *Mesophyll* with non-radiate chlorenchyma; with adaxial palisade; without arm cells; without fusoids. *Leaf blade* adaxially flat (apart from a conspicuous *adaxial* mid rib). *Midrib* conspicuous; with one bundle only. Bulliforms not in discrete, regular adaxial groups (the entire epidermis being bulliform). All the vascular bundles accompanied by sclerenchyma. Combined sclerenchyma girders present (in all the primary bundles save the midrib, which has an adaxial strand). Sclerenchyma all associated with vascular bundles.

Cytology. Chromosome base number, $x = 12$. $2n = 24$. 2 ploid.

Taxonomy. Bambusoideae; Oryzodae; Centotheceae.

Ecology, geography, regional floristic distribution. 1 species. Panama. Helophytic; shade species.

Neotropical. Caribbean.

References, etc. Morphological/taxonomic: Davidse, Soderstrom and Ellis 1986.

Poidium Nees

~ *Poa*

Habit, vegetative morphology. Perennial; caespitose. *Culms* 50–150 cm high (?); herbaceous; unbranched above. Young shoots intravaginal. Leaves non-auriculate. The fibrous remains of the sheaths persisting. Leaf blades narrow; flat, or folded; without cross venation; persistent. *Ligule an unfringed membrane*.

Reproductive organization. Plants bisexual, with bisexual spikelets; with hermaphrodite florets. Exposed-cleistogamous, or chasmogamous.

Inflorescence. *Inflorescence paniculate*; open; espatheate; not comprising 'partial inflorescences' and foliar organs. Spikelet-bearing axes persistent. Spikelets not secund; pedicellate.

Female-fertile spikelets. Spikelets compressed laterally; disarticulating above the glumes; disarticulating between the florets; with a distinctly elongated rachilla internode above the glumes. Rachilla prolonged beyond the uppermost female-fertile floret. The rachilla extension with incomplete florets. *Hairy callus absent*.

Glumes two; *more or less equal*; about equalling the spikelets; long relative to the adjacent lemmas; hairless; glabrous; awnless; carinate; similar (linear lanceolate or narrowly lanceolate). Lower glume 1 nerved, or 3 nerved. Upper glume 3 nerved. *Spikelets* with incomplete florets. The incomplete florets distal to the female-fertile florets. The distal incomplete florets 1; *incomplete florets* merely underdeveloped (rudimentary).

Female-fertile florets *2–4*. *Lemmas lanceolate*; not becoming indurated; entire; awnless; hairless; scabrous; carinate; without a germination flap; 5 nerved. *Palea* present; relatively long (narrow); awnless, without apical setae; not indurated; 2-nerved; 2-keeled (sulcate between the keels). *Lodicules* present; 2; free; membranous; glabrous; not toothed (acute). **Stamens** *1*. *Ovary* glabrous. Stigmas 2.

Fruit, embryo and seedling. *Fruit* longitudinally grooved (ventrally sulcate); *not noticeably compressed to trigonous (subtrigonous)*; hairy on the body. Hilum short.

Abaxial leaf blade epidermis. *Costal/intercostal zonation* conspicuous. Papillae absent. *Long-cells* markedly different in shape costally and intercostally (the costals much narrower); of similar wall thickness costally and intercostally. Mid-intercostal long-cells rectangular; having markedly sinuous walls (these conspicuously pitted). *Microhairs* absent. *Stomata* absent or very rare (in places), or common (in places, but then confined to single files adjacent to the costae). Subsidiaries parallel-sided. Guard-cells overlapped by the interstomatals (but only very slightly so). *Intercostal short-cells* common; in cork/silica-cell pairs; silicified. Intercostal silica bodies crescentic. Large, round prickle bases fairly common. *Crown cells* absent. *Costal short-cells* neither distinctly grouped into long rows nor predominantly paired (mostly in pairs and short rows). Costal silica bodies horizontally-elongated crenate/sinuous (short versions of the crenate type predominating), or 'panicoid-type' (if a few deeply crenate forms are so interpreted); if classifiable as panicoid type, nodular; not sharp-pointed.

Transverse section of leaf blade, physiology. C_3; XyMS+. *Mesophyll* with non-radiate chlorenchyma; without adaxial palisade. *Leaf blade* with distinct, prominent adaxial ribs; with the ribs more or less constant in size (flat- to round-topped). *Midrib* not readily distinguishable; with one bundle only. *The lamina* symmetrical on either side of the midrib. Bulliforms present in discrete, regular adaxial groups (in all the furrows); in simple fans. All the vascular bundles accompanied by sclerenchyma. Combined sclerenchyma girders present; forming 'figures' (most bundles with a large T — the adaxial girders generally being the more massive). Sclerenchyma all associated with vascular bundles.

Special diagnostic feature. Lemmas not as in *Briza* (q.v.). Female-fertile lemma not as in *Lombardochloa* (q.v.).

Taxonomy. Pooideae; Poodae; Poeae.

Ecology, geography, regional floristic distribution. 2 species. Tropical Brazil. Neotropical. Amazon and Central Brazilian.

References, etc. Morphological/taxonomic: Nicora and Rúgolo de Agrasar 1981. Leaf anatomical: this project.

Polevansia de Winter

Habit, vegetative morphology. Perennial; rhizomatous and stoloniferous (mat-forming, with long decumbent stems). *Culms* 4–45 cm high; herbaceous; unbranched above. Culm nodes glabrous. Culm internodes hollow. Plants unarmed. Young shoots intravaginal. *Leaves* mostly basal (the culm leaves reduced); non-auriculate. Leaf blades linear to linear-lanceolate; narrow; 1.5–2.5 mm wide (and 1–18 cm long); flat; without abaxial multicellular glands; without cross venation; persistent. *Ligule a fringed membrane (minutely fimbriate)*; truncate; to 0.4 mm long. *Contra-ligule* absent.

Reproductive organization. Plants bisexual, with bisexual spikelets; with hermaphrodite florets.

Inflorescence. *Inflorescence of spicate main branches (of appressed racemes, 2–3 cm long)*; contracted; espatheate; not comprising 'partial inflorescences' and foliar organs. **Spikelet-bearing axes *'racemes'***. The racemes without spikelets towards the base. Spikelet-bearing axes persistent. *Spikelets* solitary; *secund (on the racemes, at maturity deflexed upwards from the G_1, which remains erect against the rachis)*; pedicellate; imbricate.

Female-fertile spikelets. *Spikelets* 3.5–4.5 mm long; adaxial; *compressed dorsiventrally*; planoconvex; disarticulating above the glumes; with conventional internode spacings. Rachilla terminated by a female-fertile floret. Hairy callus present (small, the hairs short). *Callus* short; blunt.

Glumes two; very unequal; (the upper) long relative to the adjacent lemmas; dorsiventral to the rachis; hairless; glabrous; pointed (G_2), or not pointed (G_1); awnless; non-carinate (G_2 may be slightly keeled towards apex); very dissimilar (G_1 hyaline-membranous, nerveless, obtuse, G_2 lanceolate, firmly membranous, 1 nerved). *Lower glume* 0.5 times the length of the upper glume; shorter than the lowest lemma; 0 nerved. *Upper glume 1 nerved*. Spikelets with female-fertile florets only.

Female-fertile florets 1. *Lemmas* similar in texture to the glumes (like the G_2); not becoming indurated; entire; pointed, or blunt; shortly *mucronate*; hairy (on the lateral nerves, below), or hairless; non-carinate; having the margins tucked in onto the palea; without a germination flap; 3 nerved; with the nerves confluent towards the tip. *Palea* present; relatively long (narrowly elliptic, nearly equalling the lemma); tightly clasped by the lemma; entire (truncate), or apically notched; awnless, without apical setae; thinner than the lemma; not indurated (subhyaline); 2-nerved; 2-keeled. Palea keels wingless; glabrous. *Lodicules* present; 2; free; fleshy; glabrous. *Stamens* 3. Anthers 2 mm long; not penicillate; without an apically prolonged connective. *Ovary* glabrous. Styles free to their bases. Stigmas 2; white.

Abaxial leaf blade epidermis. *Costal/intercostal zonation* conspicuous. *Papillae* absent. *Long-cells* markedly different in shape costally and intercostally (the costals much narrower); of similar wall thickness costally and intercostally. Mid-intercostal long-cells rectangular; having markedly sinuous walls. *Microhairs* present; more or less spherical to elongated; clearly two-celled; chloridoid-type. Microhair apical cell wall of similar thickness/rigidity to that of the basal cell. Microhairs 27.6–33 microns long. Microhair basal cells 18–21 microns long. Microhairs 11.4–15 microns wide at the septum. Microhair total length/width at septum 2.6. Microhair apical cells 10.5–12.9 microns long. Microhair apical cell/total length ratio 0.4. *Stomata* common. Subsidiaries dome-shaped and triangular (mostly dome-shaped). Guard-cells overlapping to flush with the interstomatals. *Intercostal short-cells* common; in cork/silica-cell pairs; silicified. Intercostal silica bodies present and perfectly developed; tall-and-narrow, crescentic, and saddle shaped. *Costal short-cells* conspicuously in long rows (in places), or neither distinctly grouped into long rows nor predominantly paired. Costal silica bodies present in alternate cell files of the costal zones; predominantly saddle shaped; not sharp-pointed.

Transverse section of leaf blade, physiology. C_4; XyMS+. PCR sheath outlines even. PCR sheaths of the primary vascular bundles interrupted; interrupted abaxially only. PCR sheath extensions absent. PCR cell chloroplasts centripetal. *Mesophyll* with radiate chlorenchyma; without adaxial palisade; traversed by columns of colourless mesophyll cells (in places), or not traversed by colourless columns. *Leaf blade* adaxially flat. *Midrib* conspicuous (by virtue of its wide, rounded abaxial keel, contrasting with the flattened 'keels' of the other main bundles); with one bundle only. *The lamina* symmetrical on either side of the midrib. Bulliforms present in discrete, regular adaxial groups; associated with colourless mesophyll cells to form deeply-penetrating fans (these sometimes linked with traversing colourless columns). All the vascular bundles accompanied by sclerenchyma. Combined sclerenchyma girders present; forming 'figures' (most bundles with Is or anchors). Sclerenchyma all associated with vascular bundles. The lamina margins with fibres.

Taxonomy. Chloridoideae; main chloridoid assemblage.

Ecology, geography, regional floristic distribution. 1 species. South Africa. Mesophytic; species of open habitats; glycophytic. Mountain grassland.

Paleotropical. African. Sudano-Angolan. South Tropical African.

References, etc. Morphological/taxonomic: de Winter 1966b. Leaf anatomical: this project.

Special comments. Fruit data wanting.

Polliniopsis Hayata

Habit, vegetative morphology. Decumbent (culm bases creeping, with rooting nodes). *Culms* 30–40 cm high; herbaceous. Leaf blades lanceolate; narrow; to 5 mm wide; without cross venation. Ligule truncate.

Reproductive organization. *Plants bisexual, with bisexual spikelets*; with hermaphrodite florets. *The spikelets all alike in sexuality*; homomorphic.

Inflorescence. *Inflorescence* of spicate main branches, or paniculate (?— 'of paired racemes'); *non-digitate*; espatheate; not comprising 'partial inflorescences' and foliar organs. **Spikelet-bearing axes** 'racemes'; paired (slender, recurved, about 8 cm long); *persistent (?—rachis tough)*. 'Articles' linear. Spikelets paired; pedicellate; consistently in 'long-and-short' combinations; unequally pedicellate in each combination. Pedicels of the 'pedicellate' spikelets free of the rachis. The 'shorter' spikelets hermaphrodite. The 'longer' spikelets hermaphrodite.

Female-fertile spikelets. *Spikelets* 5 mm long; *compressed dorsiventrally*; falling with the glumes. Rachilla terminated by a female-fertile floret. Hairy callus present.

Glumes two; long relative to the adjacent lemmas; *awned; very dissimilar (the G_1 2-aristate with 1 mm aristules, the G_2 acuminate into an 8 mm awn)*. Lower glume flattened on the back; not pitted; relatively smooth. *Spikelets* with incomplete florets. The incomplete florets proximal to the female-fertile florets. *Spikelets with proximal incomplete florets*. *The proximal incomplete florets* 1; epaleate; sterile. *The proximal lemmas hyaline, linear, 2-toothed*; awned (the awn long, geniculate, from the sinus); more or less equalling the female-fertile lemmas; similar in texture to the female-fertile lemmas.

Female-fertile florets 1. Lemmas linear, similar to the L_1 in texture and form; less firm than the glumes (hyaline); incised; 2 lobed; not deeply cleft (bidentate); awned. Awns 1; median; from a sinus; geniculate; hairy; much longer than the body of the lemma (about 10 mm long). Lemmas hairless; non-carinate; without a germination flap. **Palea** present, or absent (?); if present, *very reduced*. *Lodicules* present; 2; free; fleshy; glabrous. *Ovary* glabrous (?). Stigmas 2.

Fruit, embryo and seedling. Hilum short.

Taxonomy. Panicoideae; Andropogonodae; Andropogoneae; Andropogoninae.

Ecology, geography, regional floristic distribution. 1 species. Formosa.

Paleotropical. Indomalesian. Indo-Chinese.
References, etc. Morphological/taxonomic: Hayata 1918.
Special comments. Fruit data wanting. Anatomical data wanting.

Polypogon Desf.

Chaetotropis Kunth, *Nowodworskya* Presl., *Raspailia* Presl., *Santia* Savi

Habit, vegetative morphology. Annual, or perennial; stoloniferous, or caespitose. *Culms* 2–120 cm high; herbaceous. Culm nodes glabrous. Culm internodes hollow. *Leaves* not basally aggregated; non-auriculate. Leaf blades linear to linear-lanceolate; narrow; 1–10 mm wide; usually flat; without cross venation; persistent. Ligule an unfringed membrane; truncate, or not truncate; 2–15 mm long.

Reproductive organization. Plants bisexual, with bisexual spikelets; with hermaphrodite florets; inbreeding.

Inflorescence. *Inflorescence paniculate*; bristly, contracted; more or less ovoid, or spicate, or more or less irregular; espatheate; not comprising 'partial inflorescences' and foliar organs. Spikelet-bearing axes persistent. Spikelets not secund; pedicellate.

Female-fertile spikelets. *Spikelets* 1.5–3 mm long; somewhat compressed laterally; *falling with the glumes (and the pedicel, or part of it). Rachilla terminated by a female-fertile floret*. Hairy callus absent. *Callus* very short.

Glumes present; two; relatively large; more or less equal; exceeding the spikelets; long relative to the adjacent lemmas; hairless; scabrous; pointed (entire to bilobed); usually awned (apically), or awnless; non-carinate; similar (chartaceous, scabrid). Lower glume 1 nerved. *Upper glume 1 nerved.* **Spikelets** with female-fertile florets only; *without proximal incomplete florets*.

Female-fertile florets 1. Lemmas less firm than the glumes (hyaline); not becoming indurated; entire to incised (truncate, finely toothed via excurrent nerves); awnless, or awned (usually). Awns 1; median; from a sinus, or dorsal; if dorsal, from near the top; non-geniculate, or geniculate; hairless; much shorter than the body of the lemma to about as long as the body of the lemma; entered by one vein; deciduous. *Lemmas* hairless; non-carinate; *5 nerved*. Palea present; relatively long, or conspicuous but relatively short (*Chaetotropis*); entire (truncate), or apically notched; awnless, without apical setae; not indurated (hyaline); 2-nerved; 2-keeled. *Lodicules* present; 2; free; membranous; glabrous; toothed, or not toothed. **Stamens 3**. Anthers 0.2–0.6 mm long; not penicillate. *Ovary* glabrous. Styles free to their bases. Stigmas 2; white.

Fruit, embryo and seedling. *Fruit* adhering to lemma and/or palea (usually); small; fusiform (*Chaetotropis*), or ellipsoid; longitudinally grooved; not noticeably compressed. *Hilum short*. Embryo large (rarely), or small; not waisted. Endosperm liquid in the mature fruit, or hard; with lipid; containing compound starch grains. Embryo with an epiblast; without a scutellar tail; with a negligible mesocotyl internode. Embryonic leaf margins meeting.

Seedling with a long mesocotyl. First seedling leaf with a well-developed lamina. The lamina narrow; erect; 3–5 veined.

Abaxial leaf blade epidermis. *Costal/intercostal zonation* conspicuous. *Papillae* absent. *Long-cells* similar in shape costally and intercostally; of similar wall thickness costally and intercostally. Mid-intercostal long-cells rectangular, or fusiform; having straight or only gently undulating walls. *Microhairs* absent. *Stomata* common; (31.5–)33–36(–37.5) microns long. Subsidiaries parallel-sided. Guard-cells overlapped by the interstomatals. *Intercostal short-cells* absent or very rare; when present, not paired (mostly solitary); not silicified. *Costal short-cells* neither distinctly grouped into long rows nor predominantly paired. Costal silica bodies horizontally-elongated crenate/sinuous (mostly), or horizontally-elongated smooth (a few); not sharp-pointed.

Transverse section of leaf blade, physiology. C_3; XyMS+. *Mesophyll* with non-radiate chlorenchyma; without adaxial palisade. *Leaf blade* with distinct, prominent adaxial ribs, or 'nodular' in section (rarely); with the ribs more or less constant in size. *Midrib* not readily distinguishable; with one bundle only. Bulliforms present in discrete, regular adaxial groups (in the furrows); in simple fans. All the vascular bundles accompanied by sclerenchyma. Combined sclerenchyma girders present, or absent (rarely); nowhere forming 'figures'. Sclerenchyma all associated with vascular bundles.

Culm anatomy. *Culm internode bundles* in one or two rings.

Phytochemistry. Leaves without flavonoid sulphates (1 species).

Cytology. Chromosome base number, $x = 7$. $2n = 14, 28, 42, 50$, and 60. 2, 4, 6, 7, and 8 ploid.

Taxonomy. Pooideae; Poodae; Aveneae.

Ecology, geography, regional floristic distribution. 18 species. Mediterranean, southwest Asia. Commonly adventive. Helophytic to mesophytic; species of open habitats; halophytic, or glycophytic. In moist ground.

Holarctic, Paleotropical, Neotropical, Cape, Australian, and Antarctic. Boreal and Tethyan. African and Indomalesian. Euro-Siberian, Eastern Asian, and Atlantic North American. Macaronesian, Mediterranean, and Irano-Turanian. Saharo-Sindian, Sudano-Angolan, Namib-Karoo, and Ascension and St. Helena. Indian, Indo-Chinese, and Papuan. Caribbean, Central Brazilian, Pampas, Andean, and Fernandezian. South-West Australian. Antarctic and Subantarctic. European. Canadian-Appalachian and Central Grasslands. Somalo-Ethiopian. South Temperate Oceanic Islands.

Hybrids. Intergeneric hybrids with *Agrostis* (×*Agropogon* P. Fourn.).

Rusts and smuts. Rusts — *Puccinia*. Taxonomically wide-ranging species: *Puccinia graminis* and *Puccinia coronata*. Smuts from Tilletiaceae. Tilletiaceae — *Entyloma*.

Economic importance. Significant weed species: *P. monspeliensis*, *P. viridis*.

References, etc. Leaf anatomical: Metcalfe 1960; this project.

Polytoca R.Br.

Cyathorhachis Steud.

Habit, vegetative morphology. Tall, robust perennial, or annual; caespitose. **Culms 120–250 cm high**. Culm nodes hairy, or glabrous. Culm internodes solid (sometimes compressed). *Leaves* not basally aggregated; auriculate, or non-auriculate. Leaf blades broad; 8–60 mm wide; flat; pseudopetiolate, or not pseudopetiolate; without cross venation; persistent. *Ligule* present; an unfringed membrane.

Reproductive organization. *Plants monoecious with all the fertile spikelets unisexual*; without hermaphrodite florets. The spikelets of sexually distinct forms on the same plant; female-only, male-only, and sterile. The male and female-fertile spikelets in different inflorescences and segregated, in different parts of the same inflorescence branch, or segregated, in different parts of the same inflorescence branch (male spikelets in the slender, upper part of the raceme, the females below, sometimes with some racemes all male, the latter often grouped in a terminal panicle). The spikelets overtly heteromorphic, or overtly heteromorphic and homomorphic (male, female and rudimentary, the pedicellate spikelets in both male and female parts of the inflorescence sometimes with the G_1 awned); in both homogamous and heterogamous combinations and all in heterogamous combinations, or all in heterogamous combinations.

Inflorescence. Inflorescence of terminal and axillary spicate racemes; digitate and non-digitate (the terminal, male racemes digitate or in a racemose panicle), or non-digitate (lateral, a single raceme). Rachides hollowed. Inflorescence spatheate; not comprising 'partial inflorescences' and foliar organs. *Spikelet-bearing axes* 'racemes' (peduncled); solitary; with substantial rachides; disarticulating (at least in the female part); disarticulating at the joints (in the lower, female part). *'Articles'* (of female segments) non-linear (widened and hollowed above); with a basal callus-knob. Spikelets paired; secund; sessile and pedicellate; consistently in 'long-and-short' combinations; in pedicellate/sessile combinations. Pedicels of the 'pedicellate' spikelets discernible, but fused with the rachis (wholly or in part). The 'shorter' spikelets female-only (in the lower part of the raceme), or male-only (above and in male racemes). The 'longer' spikelets sterile, or female-only and sterile (in the female part of the raceme), or male-only (?), or sterile (in the male part of the raceme, and male racemes).

Female-sterile spikelets. Pedicelled partners of the (sessile) female spikelets well developed but barren, or reduced. Male spikelets with membranous glumes and two male florets. The male spikelets with glumes; 2 floreted (both fertile).

Female-fertile spikelets. Spikelets compressed dorsiventrally; falling with the glumes. Rachilla terminated by a female-fertile floret. Hairy callus absent.

Glumes two; relatively large; more or less equal; long relative to the adjacent lemmas; awnless; *very dissimilar (G₁ obtuse or emarginate, indurated below, G₂ acuminate).* Lower glume not two-keeled (often winged above); convex on the back; 17–19 nerved. Upper glume 8–19 nerved. *Spikelets* with incomplete florets. The incomplete florets proximal to the female-fertile florets. *Spikelets with proximal incomplete florets. The proximal incomplete florets* 1; epaleate; sterile. The proximal lemmas awnless; 3–7 nerved; not becoming indurated (thin, acuminate).

Female-fertile florets 1. Lemmas less firm than the glumes (thinly membranous); not becoming indurated; awnless; hairless; non-carinate; 3 nerved. *Palea* present; textured like the lemma; not indurated; 2-nerved. *Lodicules* absent. *Stamens* 0. *Ovary* glabrous. Styles basally fused. Stigmas 2; red pigmented.

Fruit, embryo and seedling. *Fruit longitudinally grooved (and hollowed at the base). Hilum short (in the basal hollow).* Endosperm containing only simple starch grains. Embryo without an epiblast; with a scutellar tail; with an elongated mesocotyl internode. Embryonic leaf margins overlapping.

Abaxial leaf blade epidermis. *Costal/intercostal zonation* conspicuous. *Papillae* absent. *Long-cells* similar in shape costally and intercostally; of similar wall thickness costally and intercostally. Mid-intercostal long-cells rectangular; having markedly sinuous walls. *Microhairs* present; panicoid-type; (57–)66–75(–76.5) microns long; (9–)9.6–10.5(–11.4) microns wide at the septum. Microhair total length/width at septum 5.9–7.8. Microhair apical cells (39–)42–52.5(–54) microns long. Microhair apical cell/total length ratio 0.64–0.75. *Stomata* common; (45–)48–54(–60) microns long. Subsidiaries dome-shaped (mostly), or triangular. Guard-cells overlapping to flush with the interstomatals. *Intercostal short-cells* common; in cork/silica-cell pairs; silicified. *Costal short-cells* predominantly paired, or neither distinctly grouped into long rows nor predominantly paired. Costal silica bodies 'panicoid-type'; cross shaped and dumb-bell shaped; not sharp-pointed.

Transverse section of leaf blade, physiology. C₄; XyMS–. *Mesophyll* with non-radiate chlorenchyma; with arm cells (apparently some, but not completely convincing), or without arm cells (?). *Leaf blade* adaxially flat. *Midrib* conspicuous; having a conventional arc of bundles; with colourless mesophyll adaxially. Bulliforms present in discrete, regular adaxial groups; in simple fans. Many of the smallest vascular bundles unaccompanied by sclerenchyma. Combined sclerenchyma girders present; forming 'figures'. Sclerenchyma all associated with vascular bundles.

Cytology. 2n = 20 and 40.

Taxonomy. Panicoideae; Andropogonodae; Maydeae.

Ecology, geography, regional floristic distribution. 2 species. Indomalayan region. Mesophytic. Forest margins.

Paleotropical. Indomalesian. Indian, Indo-Chinese, and Papuan.

References, etc. Leaf anatomical: this project.

Polytrias Hackel

Aethonopogon Kuntze
~ *Eulalia* (*E. amaura*)

Habit, vegetative morphology. Creeping, mat-forming perennial; stoloniferous. *Culms* 10–30 cm high; herbaceous. Culm nodes glabrous. Culm internodes hollow. *Leaves* mostly basal; non-auriculate. Leaf blades narrow; without cross venation; persistent. Ligule a fringed membrane (short).

Reproductive organization. Plants bisexual, with bisexual spikelets; with hermaphrodite florets. *The spikelets usually all alike in sexuality*; more or less homomorphic (the pedicelled smaller, sometimes male-only).

Inflorescence. Inflorescence a solitary 'raceme'; espatheate; not comprising 'partial inflorescences' and foliar organs. *Spikelet-bearing axes* 'racemes' (many-jointed); solitary; with very slender rachides; disarticulating; disarticulating at the joints. *'Articles'* linear; not appendaged; disarticulating obliquely; densely long-hairy (the hairs brown). *Spikelets* in triplets; not secund; sessile and pedicellate; **consistently in 'long-and-short' combinations; in pedicellate/sessile combinations (two sessile, one pedicelled)**. Pedicels of the 'pedicellate' spikelets free of the rachis. The 'shorter' spikelets hermaphrodite. The 'longer' spikelets hermaphrodite (usually), or male-only.

Female-fertile spikelets. Spikelets 3.5–4.5 mm long; compressed dorsiventrally; falling with the glumes (the pedicelled falling from its pedicel, the sessile with the internode and pedicel). Rachilla terminated by a female-fertile floret. Hairy callus absent.

Glumes two; more or less equal; long relative to the adjacent lemmas; hairy; without conspicuous tufts or rows of hairs; awnless; very dissimilar (truncate, the G_1 cartilaginous and 2-keeled, the G_2 longer and naviculate). *Lower glume two-keeled*; flattened on the back to concave on the back; not pitted; relatively smooth; 4 nerved (with no median). Upper glume 3 nerved. **Spikelets** with female-fertile florets only; **without proximal incomplete florets**.

Female-fertile florets 1. Lemmas less firm than the glumes (thinly membranous); not becoming indurated; incised; 2 lobed; deeply cleft (the lobes narrow, hairy); awned. Awns 1; median; from a sinus; geniculate; much longer than the body of the lemma. Lemmas hairy (on the lobes); non-carinate; 1 nerved. *Palea* present; 2-nerved. **Lodicules absent**. *Stamens* 3. Anthers not penicillate. *Ovary* glabrous. Styles free to their bases. Stigmas 2; red pigmented.

Fruit, embryo and seedling. Hilum short. Embryo large.

Abaxial leaf blade epidermis. *Costal/intercostal zonation* conspicuous. *Papillae* present. Intercostal papillae over-arching the stomata; several per cell (finger-like). *Long-cells* similar in shape costally and intercostally; of similar wall thickness costally and intercostally (thin walled). Mid-intercostal long-cells rectangular; having markedly sinuous walls. *Microhairs* present; panicoid-type; (30–)31.5–36(–42) microns long; 4.2–5.4–6 microns wide at the septum. Microhair total length/width at septum 5.8–7.1. Microhair apical cells 16.5–19.5(–24) microns long. Microhair apical cell/total length ratio 0.52–0.57. *Stomata* common; 22.5–24 microns long. Subsidiaries triangular. Guard-cells overlapping to flush with the interstomatals. *Intercostal short-cells* common; not paired (solitary); silicified. *Costal short-cells* conspicuously in long rows. Costal silica bodies 'panicoid-type'; not sharp-pointed.

Transverse section of leaf blade, physiology. C_4; XyMS–. *Mesophyll* with radiate chlorenchyma. *Leaf blade* 'nodular' in section; with the ribs more or less constant in size. *Midrib* not readily distinguishable; with one bundle only. Bulliforms present in discrete, regular adaxial groups; in simple fans. All the vascular bundles accompanied by sclerenchyma. Combined sclerenchyma girders present; forming 'figures'. Sclerenchyma all associated with vascular bundles.

Taxonomy. Panicoideae; Andropogonodae; Andropogoneae; Andropogoninae.

Ecology, geography, regional floristic distribution. 1 species. Java. Species of open habitats. Waste land, roadsides.

Paleotropical and Australian. Indomalesian and Polynesian. Indo-Chinese, Malesian, and Papuan. Fijian. North and East Australian. Tropical North and East Australian.

Rusts and smuts. Rusts — *Puccinia*. Taxonomically wide-ranging species: '*Uromyces*' *schoenanthi*. Smuts from Ustilaginaceae. Ustilaginaceae — *Ustilago*.

Economic importance. Lawns and/or playing fields: in the humid tropics.

References, etc. Leaf anatomical: this project.

Pommereulla L.f.

Habit, vegetative morphology. Perennial; stoloniferous. *Culms* 5–15 cm high; herbaceous. Leaf blades linear; narrow; 2–3 mm wide; flat, or folded; exhibiting multicellular glands abaxially. The abaxial glands on the blade margins. *Leaf blades* without cross venation; *disarticulating from the sheaths (apparently)*. Ligule a fringed membrane to a fringe of hairs (a pubescent ridge).

Reproductive organization. Plants bisexual, with bisexual spikelets; with hermaphrodite florets.

Inflorescence. Inflorescence a single raceme (the raceme occasionally forked). Rachides flattened. Inflorescence espatheate (but the raceme usually partially enclosed in the uppermost, spathe-like sheath); not comprising 'partial inflorescences' and foliar organs. Spikelet-bearing axes persistent. Spikelets solitary (close or distant); secund (?); pedicellate.

Female-fertile spikelets. Spikelets more or less *morphologically 'conventional' (but the lemmas spirally arranged, forming an inverted cone, and changing form acropetally)*; 8 mm long; compressed dorsiventrally; disarticulating above the glumes; not disarticulating between the florets. Rachilla prolonged beyond the uppermost female-fertile floret. The rachilla extension with incomplete florets. *Callus* long (at the base of the lowermost floret); pointed.

Glumes two; (the upper) about equalling the spikelets (the lower shorter); (the upper) long relative to the adjacent lemmas; awnless; non-carinate; similar (membranous, persistent, amplexicaul at the base, the G_2 larger). Lower glume 1 nerved. Upper glume 3 nerved. *Spikelets* with incomplete florets. The incomplete florets both distal and proximal to the female-fertile florets. *The distal incomplete florets* merely underdeveloped. **The proximal incomplete florets 2** *(fan-shaped, embracing the fertile lemmas)*; epaleate; sterile. *The proximal lemmas 4-lobed, the lobes acute to short-awned, inner lobes narrower, the lemma with a slender awn from middle of back; awned (dorsally, and mucronate or aristulate from the four lobes)*; 7–9 nerved; more or less equalling the female-fertile lemmas to decidedly exceeding the female-fertile lemmas.

Female-fertile florets 2–3. Lemmas *conspicuously non-distichous*; similar to the proximal lemmas, becoming smaller acropetally; incised; 4 lobed (the outer lobes much longer); deeply cleft; awned. Awns 1; median; dorsal; from well down the back (from about middle); non-geniculate; hairless; much shorter than the body of the lemma. Lemmas hairy; non-carinate; 7–9 nerved. *Palea* present; relatively long (flat); entire to apically notched; awnless, without apical setae; 2-nerved; 2-keeled. *Lodicules* present; 2; free; glabrous; not or scarcely vascularized. *Stamens* 3. Anthers short; not penicillate; without an apically prolonged connective. *Ovary* glabrous. Styles free to their bases. Stigmas 2.

Fruit, embryo and seedling. Fruit ellipsoid. Embryo with an epiblast; with a scutellar tail; with an elongated mesocotyl internode. Embryonic leaf margins meeting.

Abaxial leaf blade epidermis. *Costal/intercostal zonation* conspicuous. *Papillae* present; intercostal. Intercostal papillae not over-arching the stomata; commonly consisting of one oblique swelling per cell (sometimes cells with two). Mid-intercostal long-cells rectangular; having markedly sinuous walls (and outer walls pitted). *Microhairs* present; more or less spherical; clearly two-celled (the basal cell sunken); chloridoid-type. Microhair apical cell wall of similar thickness/rigidity to that of the basal cell. Microhair basal cells 6–9 microns long (but sunken). Microhair total length/width at septum 1.3. Microhair apical cell/total length ratio 0.6. *Stomata* common. Subsidiaries markedly triangular. *Intercostal short-cells* fairly common; in cork/silica-cell pairs and not paired (some solitary); silicified. Intercostal silica bodies absent to imperfectly developed; rounded (narrowly oval), or crescentic. *Costal short-cells* conspicuously in long rows. Costal silica bodies present in alternate cell files of the costal zones; saddle shaped (to rectangular); not sharp-pointed.

Transverse section of leaf blade, physiology. C_4; XyMS+. PCR sheaths of the primary vascular bundles interrupted; interrupted both abaxially and adaxially. PCR sheath extensions absent. *Mesophyll* with radiate chlorenchyma; traversed by columns of colourless mesophyll cells. *Leaf blade* with distinct, prominent adaxial ribs (towards midrib), or adaxially flat (outside); with the ribs more or less constant in size (low). *Midrib* conspicuous; having a conventional arc of bundles (large median, several small strands either side). Bulliforms present in discrete, regular adaxial groups (including a large, median group in a groove over midrib); associated with colourless mesophyll cells to form deeply-penetrating fans (these linked to the abaxial epidermis by girders of colourless cells). All the vascular bundles accompanied by sclerenchyma. Combined sclerenchyma girders present (nearly all bundles). Sclerenchyma forming a hypodermal plate adaxially in midrib.

Taxonomy. Chloridoideae; main chloridoid assemblage (*'Pommereulleae'*).

Ecology, geography, regional floristic distribution. 1 species. Eastern Asia. Species of open habitats.

Paleotropical. Indomalesian. Indian.

References, etc. Leaf anatomical: Metcalfe 1960; Van den Borre 1994.

Special comments. Fruit data wanting.

Porteresia Tateoka

Sclerophyllum Griff., *Indoryza* Henry & Roy
~ *Oryza* (*O. coarctata*)

Habit, vegetative morphology. Perennial. Culms herbaceous. *Leaf blades* narrowly linear; *leathery (with tuberculate prickly margins)*; without cross venation.

Reproductive organization. Plants bisexual, with bisexual spikelets; with hermaphrodite florets.

Inflorescence. Inflorescence narrowly paniculate; open; espatheate; not comprising 'partial inflorescences' and foliar organs. Spikelet-bearing axes persistent. Spikelets not secund; pedicellate.

Female-fertile spikelets. *Spikelets* unconventional (owing to extreme reduction of the glumes, cf. *Oryza*); strongly compressed laterally; disarticulating above the glumes (if the pedicel cup be interpreted as glumes); with conventional internode spacings (i.e. the floret not stipitate). Rachilla terminated by a female-fertile floret. Hairy callus absent.

Glumes *present to absent (discernable only as obscure lobing of the pedicel tip)*; if considered present, two; *minute*; more or less equal; shorter than the adjacent lemmas; joined; awnless. Lower glume 0 nerved. Upper glume 0 nerved. *Spikelets* with incomplete florets. The incomplete florets proximal to the female-fertile florets. **The proximal incomplete florets** 2; epaleate; *sterile (much reduced, cf.* **Oryza***)*. The proximal lemmas awnless; exceeded by the female-fertile lemmas (less than half the length of the spikelet).

Female-fertile florets 1. Lemmas leathery; not becoming indurated; shortly aristulate; strongly carinate; 5–10 nerved. *Palea* present; relatively long (resembling the lemma); awnless, without apical setae; textured like the lemma; several nerved; one-keeled. *Lodicules* present; 2; membranous. *Stamens* 6. *Ovary* glabrous. Styles free to their bases. Stigmas 2.

Fruit, embryo and seedling. Hilum long-linear. Embryo large. Endosperm hard. Embryo with an epiblast (large); with a scutellar tail; with a negligible mesocotyl internode; with one scutellum bundle. Embryonic leaf margins overlapping.

Abaxial leaf blade epidermis. *Papillae* present. *Microhairs* absent. Costal silica bodies not sharp-pointed.

Transverse section of leaf blade, physiology. *Leaf blades* seemingly *consisting of midrib (with two superposed bundles in each rib, the adaxial one being inverted, and large air spaces in the mesophyll)*. C_3; XyMS+. Mesophyll with arm cells; without fusoids. Midrib having complex vascularization (regardless of whether or not the blade is interpreted as midrib, since every rib has superposed bundles). Bulliforms present in discrete, regular adaxial groups.

Cytology. Chromosome base number, $x = 12$. $2n = 48$. 4 ploid.

Taxonomy. Bambusoideae; Oryzodae; Oryzeae.

Ecology, geography, regional floristic distribution. 1 species. India, Burma. Helophytic; halophytic (in the brackish waters of deltas and tidal swamps).

Paleotropical. Indomalesian. Indian and Indo-Chinese.

References, etc. Morphological/taxonomic: Tateoka 1965a.

Potamophila R.Br.

Habit, vegetative morphology. Perennial; rhizomatous and caespitose. The flowering culms leafy. *Culms* 30–150 cm high; herbaceous. Culm nodes glabrous. Culm internodes hollow. Young shoots extravaginal. *Leaves* not basally aggregated; auriculate, or non-auriculate; without auricular setae. Leaf blades linear; narrow; 4–6 mm wide; not pseudopetiolate; without cross venation; disarticulating from the sheaths, or persistent; rolled in bud. Ligule an unfringed membrane; not truncate (acute or lacerate); 4–15 mm long. *Contra-ligule* absent.

Reproductive organization. Plants bisexual, with bisexual spikelets; with hermaphrodite florets. *The spikelets of sexually distinct forms on the same plant (hermaphrodite and unisexual)*; hermaphrodite and female-only, or hermaphrodite and male-only; homomorphic.

Inflorescence. Inflorescence paniculate; open (to loosely contracted); with capillary branchlets; espatheate; not comprising 'partial inflorescences' and foliar organs. Spikelets not secund; pedicellate.

Female-fertile spikelets. Spikelets *unconventional (by virtue of reduced glumes, cf.* **Oryza***)*; 3–5.5 mm long; strongly compressed laterally; disarticulating above the glumes (i.e., above the tiny cupule representing them); with conventional internode spacings (i.e. the floret not stipitate). Rachilla terminated by a female-fertile floret. Hairy callus absent.

Glumes present to absent (vestigial); if considered present, two; *minute (reduced to a cupular or bilobed rim)*; more or less equal; shorter than the adjacent lemmas; joined; awnless. Lower glume 0 nerved. Upper glume 0 nerved. *Spikelets* with incomplete florets. The incomplete florets proximal to the female-fertile florets. *The proximal incomplete florets* 2; epaleate; sterile. *The proximal lemmas* ovate; awnless; 0 nerved; *exceeded by the female-fertile lemmas (very small, less than 1/8 to 1/4 the spikelet length)*; less firm than the female-fertile lemmas to similar in texture to the female-fertile lemmas (membranous); not becoming indurated.

Female-fertile florets 1. Lemmas membranous or thinly chartaceous; not becoming indurated; entire; pointed to blunt; awnless; hairless (often scaberulous on the sides); non-carinate ('sub-keeled'); 5 nerved. *Palea* present; relatively long; not clasped by the lemma; entire; awnless, without apical setae; textured like the lemma; not indurated; several nerved (three nerved); one-keeled. *Lodicules* present; 2; free; membranous; glabrous. **Stamens 6**. *Ovary* glabrous. Styles free to their bases. Stigmas 2.

Fruit, embryo and seedling. Fruit compressed dorsiventrally to not noticeably compressed. Hilum long-linear. Embryo small; with an epiblast; without a scutellar tail; with a negligible mesocotyl internode. Embryonic leaf margins meeting.

Abaxial leaf blade epidermis. *Costal/intercostal zonation* conspicuous. *Papillae* absent. *Long-cells* similar in shape costally and intercostally; of similar wall thickness costally and intercostally. Mid-intercostal long-cells rectangular; having markedly sinuous walls. *Microhairs* present; panicoid-type; (34.5–)39–45(–51) microns long; 7.5–8.4–9 microns wide at the septum. Microhair total length/width at septum 4.3–6.1. Microhair apical cells (12–)16.5–21(–22.5) microns long. Microhair apical cell/total length ratio 0.35–0.47. *Stomata* common; 19.5–21 microns long. Subsidiaries dome-shaped (mostly), or dome-shaped. Guard-cells overlapping to flush with the interstomatals. *Intercostal short-cells* common; in cork/silica-cell pairs, or not paired; silicified, or not silicified. *Costal short-cells* predominantly paired, or neither distinctly grouped into long rows nor predominantly paired (rarely). Costal silica bodies oryzoid to 'panicoid-type' (mostly), or tall-and-narrow (a few); not sharp-pointed.

Transverse section of leaf blade, physiology. C_3; XyMS+. *Mesophyll* with non-radiate chlorenchyma; without adaxial palisade; with arm cells; with fusoids. The fusoids external to the PBS (short). *Leaf blade* with distinct, prominent adaxial ribs; with the ribs more or less constant in size, or with the ribs very irregular in sizes. *Midrib* conspicuous; having complex vascularization. Bulliforms present in discrete, regular adaxial groups; in simple fans. All the vascular bundles accompanied by sclerenchyma. Combined sclerenchyma girders present; forming 'figures'. Sclerenchyma all associated with vascular bundles.

Taxonomy. Bambusoideae; Oryzodae; Oryzeae.

Ecology, geography, regional floristic distribution. 1 species. Australia. Helophytic; glycophytic.

Australian. North and East Australian. Temperate and South-Eastern Australian.

References, etc. Leaf anatomical: this project.

Pringleochloa Scribner

Habit, vegetative morphology. Small perennial; stoloniferous (and mat forming). Culms herbaceous; unbranched above. Culm nodes glabrous. Plants unarmed. Young shoots intravaginal. *Leaves* mostly basal; non-auriculate. Leaf blades narrow; about 2 mm wide; not setaceous (but enrolling); without abaxial multicellular glands; without cross venation; persistent. *Ligule a fringe of hairs*.

Reproductive organization. *Plants monoecious with all the fertile spikelets unisexual*; without hermaphrodite florets. The spikelets of sexually distinct forms on the same plant; female-only and male-only. *The male and female-fertile spikelets in different inflorescences (the staminate spikelets in 4 or 5 short spikes on a long projecting peduncle, the pistillate spikelets clustered in the radical rosette)*.

Inflorescence. *Inflorescence of spicate main branches (male), or a false spike, with spikelets on contracted axes (contracted, headlike, with 2–5 glomerules, female)*. Inflorescence with axes ending in spikelets. Inflorescence espatheate; not comprising 'partial inflorescences' and foliar organs. Spikelet-bearing axes disarticulating; falling entire (the glomerules shed). Spikelets 3–5 spikelets in each glomerule; secund (male inflorescence), or not secund (female inflorescence).

Female-sterile spikelets. Male spikelets 2–4 mm long, 1-flowered, rachilla not prolonged, stamens 3. Rachilla of male spikelets terminated by a male floret. The male spikelets with glumes; without proximal incomplete florets; 1 floreted. Male florets 1; 3 staminate.

Female-fertile spikelets. *Spikelets* 4 mm long (excluding the awns); *compressed dorsiventrally*; falling with the glumes (in the glomerules). *Rachilla prolonged beyond the uppermost female-fertile floret*; rachilla hairless. The rachilla extension with incomplete florets.

Glumes two; more or less equal; long relative to the adjacent lemmas; hairy; pointed; awned (the G_2 acuminate into its short awn, the G_1 reduced almost to its awn); non-carinate; very dissimilar (the G_1 very narrow, the G_2 lanceolate and tapering into its awn). Lower glume 1 nerved. Upper glume 1 nerved. *Spikelets* with incomplete florets. The incomplete florets distal to the female-fertile florets (the empty lemmas each hairy, more dissected than L_1, the lowermost having about 15 awns, the uppermost 3). *The distal incomplete florets* 2–3; **incomplete florets** *clearly specialised and modified in form (constituting a multi-awned tip to the spikelet)*.

Female-fertile florets 1. Lemmas decidedly firmer than the glumes (membranous or cartilaginous); not becoming indurated; incised; 3 lobed; deeply cleft (to 1/3 or more of its length, at each side); awned. Awns 2, or 3; median and lateral, or lateral only (two short lateral awns and a median mucro); the median (when present) similar in form to the laterals (but smaller); non-geniculate; hairless (scabrid). The lateral awns exceeding the median (the median 'awn' reduced to a mucro, the laterals longer than the median, but shorter than the body of the lemma). Lemmas sparsely hairy; non-carinate; without a germination flap; 3 nerved. *Palea* present (sparsely hairy apically between the keels); relatively long; apically notched; awnless, without apical setae; not indurated (membranous); 2-nerved; 2-keeled. *Lodicules* present; 2; free; fleshy; glabrous; not or scarcely vascularized. *Stamens* 0 (3 staminodes only). *Ovary* glabrous. Styles free to their bases. Stigmas 2.

Fruit, embryo and seedling. *Fruit* free from both lemma and palea; small (about 2 mm long); ellipsoid; compressed dorsiventrally. Hilum short. Pericarp free. Embryo large.

Abaxial leaf blade epidermis. *Costal/intercostal zonation* conspicuous. *Papillae* absent. *Long-cells* similar in shape costally and intercostally; of similar wall thickness costally and intercostally. Mid-intercostal long-cells rectangular; having markedly sinuous walls (and pitted). *Microhairs* present; elongated; clearly two-celled; panicoid-type. Microhair apical cell wall thinner than that of the basal cell and often collapsed. Microhairs 39–42–51 microns long. Microhair basal cells 24 microns long. Microhairs 7.5–8.4–9.6 microns wide at the septum. Microhair total length/width at septum 5–5.6. Microhair apical cells (15–)16.5–19.5(–21) microns long. Microhair apical cell/total length ratio 0.38–0.46. *Stomata* common; 25.5–30 microns long. Subsidiaries dome-shaped (mostly), or triangular. Guard-cells overlapping to flush with the interstomatals. *Intercostal short-cells* common; in cork/silica-cell pairs; silicified. Intercostal silica bodies present and perfectly developed; tall-and-narrow, acutely-angled, and oryzoid-type. *Costal short-cells* conspicuously in long rows. Costal silica bodies present in alternate cell files of the costal zones; saddle shaped to crescentic; not sharp-pointed.

Transverse section of leaf blade, physiology. C_4; XyMS+. PCR sheaths of the primary vascular bundles interrupted; interrupted abaxially only. PCR sheath extensions absent. Mesophyll traversed by columns of colourless mesophyll cells. *Leaf blade* adaxially flat. *Midrib* not readily distinguishable; with one bundle only. Bulliforms present in discrete, regular adaxial groups; associated with colourless mesophyll cells to form deeply-penetrating fans (these linked with traversing colourless columns). All the vascular bundles accompanied by sclerenchyma. Combined sclerenchyma girders absent (mostly with adaxial and abaxial strands). Sclerenchyma all associated with vascular bundles. The lamina margins with fibres.

Taxonomy. Chloridoideae; main chloridoid assemblage.

Ecology, geography, regional floristic distribution. 1 species. Mexico. Species of open habitats. Dry calcareous plains.

Neotropical. Caribbean.

References, etc. Leaf anatomical: this project.

Prionanthium Desv.

Chondrolaena Nees, *Prionachne* Nees

Habit, vegetative morphology. Slender *annual*; caespitose. The flowering culms leafless. *Culms* 4–43 cm high; herbaceous; branched above, or unbranched above. Culm nodes glabrous. Culm internodes hollow. Plants unarmed; with multicellular glands (stalked or sessile, on the glumes of all three species, not on the leaves). Young shoots intravaginal. *Leaves* not basally aggregated; non-auriculate; without auricular setae (but with a tuft of hairs at the mouth of the sheath). Leaf blades linear (or filiform); narrow; 0.5–2(–3) mm wide (less than 8 cm long); flat, or rolled; without cross venation; persistent. Ligule a fringed membrane to a fringe of hairs; 0.3–1.5 mm long. *Contra-ligule* absent.

Reproductive organization. Plants bisexual, with bisexual spikelets; with hermaphrodite florets.

Inflorescence. *Inflorescence a single spike, or a single raceme (spike-like)*; contracted (3–8 cm long, the axis curved beside each spikelet); espatheate; not comprising 'partial inflorescences' and foliar organs. Spikelet-bearing axes persistent. *Spikelets* solitary, or paired; *secund*; biseriate; sessile to subsessile; consistently in 'long-and-short' combinations, or not in distinct 'long-and-short' combinations.

Female-fertile spikelets. Spikelets 3–7 mm long; adaxial (the upper glume gaping widely at anthesis); compressed laterally; disarticulating above the glumes; disarticulating between the florets (but the disarticulated florets remaining for some time within the persistent glumes). Rachilla prolonged beyond the uppermost female-fertile floret. The rachilla extension naked.

Glumes two; more or less equal (the lower slightly longer); about equalling the spikelets; long relative to the adjacent lemmas (exceeding them); dorsiventral to the rachis (in *P. pholiuroides*), or lateral to the rachis; hairy, or hairless (keels and sometimes the nerves tuberculate or pectinate); *not pointed (rounded)*; awnless; carinate (strongly, often asymmetrically); *similar (navicular, rigid, leathery with membranous margins enfolding the floret, usually with multicellular glands)*. Lower glume about equalling the lowest lemma, or much exceeding the lowest lemma; 5–8 nerved. Upper glume 5–8 nerved. *Spikelets* with female-fertile florets only, or with incomplete florets. The incomplete florets distal to the female-fertile florets. *The distal incomplete florets* merely underdeveloped; awnless.

Female-fertile florets 2. Lemmas acute, lanceolate or lanceolate-oblong; less firm than the glumes; not becoming indurated (thinly membranous); entire; pointed; awnless, or mucronate; hairy, or hairless. The hairs not in tufts (when present, evenly distributed). Lemmas non-carinate (abaxially rounded); without a germination flap; 3–5 nerved. *Palea* present (sub-linear); relatively long; apically notched; awnless, without apical setae; textured like the lemma; not indurated (hyaline, with thickened keels); 2-nerved; 2-keeled. *Lodicules* present (minute); 2; free; fleshy; glabrous. *Stamens* 3. Anthers 1.8–3.9 mm long; not penicillate; without an apically prolonged connective. *Ovary* glabrous. Styles free to their bases. Stigmas 2.

Fruit, embryo and seedling. Fruit narrowly lanceolate in outline; longitudinally grooved. Hilum long-linear. Embryo small. Endosperm containing only simple starch grains.

Abaxial leaf blade epidermis. *Costal/intercostal zonation* conspicuous. *Papillae* absent. Mid-intercostal long-cells rectangular and fusiform; having markedly sinuous walls. *Microhairs* present; panicoid-type; (63–)69–75(–81) microns long; 7.5–8.4–10.5 microns wide at the septum. Microhair total length/width at septum 6.9–10. Microhair apical cells 28.5–31.5 microns long. Microhair apical cell/total length ratio 0.38–0.46. *Stomata* common; 31.5–36 microns long. Subsidiaries parallel-sided to dome-shaped (mostly domes-shaped). Guard-cells overlapping to flush with the interstomatals. *Intercostal short-cells* common; in cork/silica-cell pairs (and solitary); silicified. *Costal short-cells* conspicuously in long rows. Costal silica bodies 'panicoid-type'; dumb-bell shaped and nodular; not sharp-pointed.

Transverse section of leaf blade, physiology. C_3; XyMS+. *Mesophyll* with non-radiate chlorenchyma; without adaxial palisade. *Leaf blade* with distinct, prominent adaxial ribs; with the ribs more or less constant in size. *Midrib* not readily distinguishable; with one bundle only. Bulliforms present in discrete, regular adaxial groups; in simple fans. All the vascular bundles accompanied by sclerenchyma. Combined sclerenchyma girders absent (small abaxial girders only, but with a wide adaxial strand at the top of each rib). Sclerenchyma all associated with vascular bundles.

Cytology. Chromosome base number, $x = 7$. $2n = 14$. 2 ploid (all three species).

Taxonomy. Arundinoideae; Danthonieae.

Ecology, geography, regional floristic distribution. 3 species. South Africa. Helophytic (in seasonally wet places); species of open habitats; glycophytic.

Cape.

References, etc. Morphological/taxonomic: Davidse, G. (1988). *Bothalia* **18**, 143–153. Leaf anatomical: this project.

Prosphytochloa Schweickerdt

~ *Potamophila* (*P. prehensilis*)

Habit, vegetative morphology. Perennial; rhizomatous (the rhizomes horizontal, with cataphylls). The flowering culms leafy. **Culms** 1000 cm high (or more); herbaceous; *scandent (by retrorse hairs on the leaf blade margins)*; branched above. Culm nodes hairy, or glabrous. Culm internodes hollow. Plants unarmed (but leaf margins and main veins spiny, with minute retrorse barbs). Young shoots intravaginal. *Leaves* not basally aggregated; auriculate (sheaths slightly auricled); without auricular setae. Leaf blades linear-lanceolate to lanceolate; narrow to broad; 4–15 mm wide; flat; without cross venation; persistent; rolled in bud. Ligule an unfringed membrane; not truncate (rounded, the margin lacerate-fimbriate); 1–1.5 mm long. *Contra-ligule* absent.

Reproductive organization. Plants bisexual, with bisexual spikelets; with hermaphrodite florets.

Inflorescence. Inflorescence paniculate (terminating main culm and laterals, the branchlets fine and stiff); open; espatheate; not comprising 'partial inflorescences' and foliar organs. Spikelet-bearing axes persistent. Spikelets solitary; not secund; pedicellate.

Female-fertile spikelets. *Spikelets* unconventional (by virtue of the reduced glumes); 6–9 mm long; slightly compressed laterally; disarticulating above the glumes (i.e. above the rudimentary glumes); not disarticulating between the florets. Rachilla terminated by a female-fertile floret. Hairy callus absent. *Callus* absent.

Glumes present, or absent; *two (reduced to a bilobed to entire hyaline cup)*; minute; more or less equal; shorter than the spikelets; shorter than the adjacent lemmas; joined; hairless; glabrous; not pointed (blunt); awnless. Lower glume 0 nerved. Upper glume 0 nerved. *Spikelets* with incomplete florets. The incomplete florets proximal to the female-fertile florets. **The proximal incomplete florets** 2; epaleate; *sterile (subulate, edged with minute hyaline spines, variable in size)*. The proximal lemmas awnless; 1 nerved; exceeded by the female-fertile lemmas (very short).

Female-fertile florets 1. Lemmas acuminate; chartaceous or leathery; entire; pointed; awnless; hairless; scabrous; carinate (the nerves with prickles); without a germination flap; 5 nerved; with the nerves confluent towards the tip. *Palea* present (similar to the lemma, which clasps it); relatively long; tightly clasped by the lemma; entire (acuminate); awnless, without apical setae; textured like the lemma; leathery; several nerved (3); one-keeled. *Lodicules* present; 2; free; membranous (above, but fleshy below); glabrous; not toothed; heavily vascularized. *Stamens* 6. Anthers 5 mm long; not penicillate; without an apically prolonged connective. *Ovary* glabrous. Styles free to their bases. Stigmas 2 (plumose); white.

Fruit, embryo and seedling. *Fruit* free from both lemma and palea; medium sized (5 to 6 mm long, brown); fusiform; longitudinally grooved; not noticeably compressed; longitudinally ribbed. Hilum long-linear. Embryo small. Endosperm containing compound starch grains. Embryo with an epiblast; without a scutellar tail.

Abaxial leaf blade epidermis. *Costal/intercostal zonation* conspicuous. *Papillae* present. Intercostal papillae several per cell (one or two rows of small, circular papillae on each long-cell, and the guard-cells overtopped by four small papillae, two from each subsidiary). *Long-cells* markedly different in shape costally and intercostally (the costals relatively long and narrow); of similar wall thickness costally and intercostally (walls of medium thickness). Mid-intercostal long-cells rectangular; having markedly sinuous walls. *Microhairs* present; panicoid-type; 45–48 microns long; 5.4–6 microns wide at the septum. Microhair total length/width at septum 7.5–8.9. Microhair apical cells (19.5–)21–22(–22.5) microns long. Microhair apical cell/total length ratio 0.41–0.49. *Stomata* common; 21–22.5 microns long. Subsidiaries papillate; dome-shaped to triangular. Guard-cells overlapped by the interstomatals (slightly). *Intercostal short-cells* absent or very rare. *Costal short-cells* conspicuously in long rows. Costal silica bodies oryzoid; not sharp-pointed.

Transverse section of leaf blade, physiology. C_3; XyMS+. *Mesophyll* with non-radiate chlorenchyma; without adaxial palisade; with arm cells; with fusoids (i.e. with lateral sheath extensions), or without fusoids (if these not so interpreted). The fusoids an integral part of the PBS. *Leaf blade* adaxially flat. *Midrib* conspicuous; having complex vascularization (there being a small bundle adaxial to the main one); with colourless mesophyll adaxially. *The lamina* symmetrical on either side of the midrib. Bulliforms present in discrete, regular adaxial groups; in simple fans (the fans broad). All the vascular bundles accompanied by sclerenchyma. Combined sclerenchyma girders present; forming 'figures'. Sclerenchyma all associated with vascular bundles.

Special diagnostic feature. *Scandent via leaf blades with retrorsely scabrid margins*.

Cytology. Chromosome base number, $x = 12$. $2n = 24$. 2 ploid.

Taxonomy. Bambusoideae; Oryzodae; Oryzeae.

Ecology, geography, regional floristic distribution. 1 species. South Africa. Helophytic; shade species; glycophytic.

Paleotropical. African. Sudano-Angolan. South Tropical African.

References, etc. Morphological/taxonomic: de Winter 1951. Leaf anatomical: this project.

Psammagrostis C. Gardner & C.E. Hubb.

Habit, vegetative morphology. Annual; decumbent (to geniculate ascending). Plants to 50 cm in diameter. *Culms* 10–20 cm long; herbaceous; branched above (and below, with all the branch systems producing inflorescences). Culm nodes glabrous. *Leaves* not basally aggregated; non-auriculate. The sheaths inflated, constricted at the throat, striate.

Leaf blades not all greatly reduced (but becoming reduced acropetally on the culms); linear-lanceolate to lanceolate (but tightly rolling when dry); narrow; 1–4 mm wide; setaceous (when dry), or not setaceous; flat, or rolled; exhibiting multicellular glands abaxially (at the base of macrohairs). The abaxial glands intercostal. Leaf blades without cross venation; disarticulating from the sheaths; rolled in bud. Ligule a fringe of hairs. *Contra-ligule* absent.

Reproductive organization. Plants bisexual, with bisexual spikelets; with hermaphrodite florets.

Inflorescence. *Inflorescence* falsely *paniculate (towards the culm tips, by proliferation of the axillary fascicles which occur in simpler form with the lower leaves)*; spatheate (each of the ultimate fascicles associated with a spatheole, within the axillant sheath); *a complex of 'partial inflorescences' and intervening foliar organs*. Spikelet-bearing axes *very much reduced (each to a deciduous, slightly clavate, 3–10 mm peduncle, with a terminal spikelet and often with one or two sessile, subsessile or more or less pedicellate lateral spikelets)*; clustered (2–4 per axil); *disarticulating*; falling entire (the striated peduncles disarticulating at the base). Spikelets not secund; sessile, subsessile, and pedicellate.

Female-fertile spikelets. *Spikelets* 4–8 mm long; somewhat *compressed laterally*; falling with the glumes (and with their companions on the peduncle); not disarticulating between the florets; with distinctly elongated rachilla internodes between the florets (up to 2 mm long, clavate). Rachilla prolonged beyond the uppermost female-fertile floret; rachilla hairless. The rachilla extension with incomplete florets.

Glumes two; very unequal; shorter than the spikelets; shorter than the adjacent lemmas; lateral to the rachis (in so far as the orientation is ascertainable); hairless; slightly scabrous; not pointed (blunt or erose); awnless; carinate; very dissimilar to similar (the lower being flimsier, sometimes hyaline throughout). Lower glume 0 nerved, or 1 nerved. Upper glume 3 nerved. *Spikelets* with incomplete florets. The incomplete florets distal to the female-fertile florets. *The distal incomplete florets* merely underdeveloped; awnless.

Female-fertile florets 3–7. Lemmas elliptical to broadly ovate; decidedly firmer than the glumes (leathery or cartilaginous, with hyaline margins); not becoming indurated; entire to incised (obtuse, truncate or slightly emarginate); not deeply cleft; mucronate; hairless; scabrous; carinate; without a germination flap; 3 nerved; with the nerves non-confluent. *Palea* present; relatively long; tightly clasped by the lemma; entire (the hyaline apex rounded); awnless, without apical setae; textured like the lemma (leathery to cartilaginous between the keels); 2-nerved; 2-keeled. Palea keels wingless; scabrous. *Lodicules* present; 2; free; fleshy; glabrous. *Stamens* 3. Anthers not penicillate; without an apically prolonged connective. *Ovary* glabrous. Styles free to their bases.

Fruit, embryo and seedling. Fruit small; trigonous; sculptured. Hilum short. Pericarp free. Embryo large.

Abaxial leaf blade epidermis. *Costal/intercostal zonation* conspicuous. *Papillae* absent. *Long-cells* markedly different in shape costally and intercostally (the costals relatively longer and narrower); of similar wall thickness costally and intercostally. Mid-intercostal long-cells rectangular; having markedly sinuous walls. *Microhairs* present; very elongated; clearly two-celled; chloridoid-type to *Enneapogon*-type (the slender basal cell long, relatively thick-walled, the short, rounded apical cell quite thin-walled). Microhair apical cell wall of similar thickness/rigidity to that of the basal cell. Microhairs (117–)126.6(–141) microns long. Microhair basal cells 111 microns long. Microhairs (12–)12.6(–14.4) microns wide at the septum. Microhair total length/width at septum (8.1–)10.1(–11.75). Microhair apical cells 16.5–20.7–24 microns long. Microhair apical cell/total length ratio 0.14–0.16–0.19. *Stomata* common; (27–)28.5(–30) microns long. Subsidiaries dome-shaped and triangular. Guard-cells overlapping to flush with the interstomatals. *Intercostal short-cells* absent or very rare (only one or two seen); not paired (the few seen solitary). Intercostal silica bodies absent. *Costal short-cells* conspicuously in long rows. Costal silica bodies present in alternate cell files of the costal zones; 'panicoid-type'; rather consistently 'smoothed out' dumb-bell shaped; not sharp-pointed.

Transverse section of leaf blade, physiology. C_4; XyMS+. PCR sheath outlines even. PCR sheaths of the primary vascular bundles interrupted; interrupted abaxially only.

PCR sheath extensions absent. PCR cell chloroplasts centripetal (seemingly, judging from the poor material seen). *Mesophyll* with radiate chlorenchyma. *Leaf blade* adaxially flat. *Midrib* not readily distinguishable; with one bundle only. *The lamina* symmetrical on either side of the midrib. Bulliforms present in discrete, regular adaxial groups; in simple fans (the median fan cells deeply penetrating). All the vascular bundles accompanied by sclerenchyma. Combined sclerenchyma girders present (all the major bundles with small girders top and bottom). Sclerenchyma all associated with vascular bundles. The lamina margins with fibres.

Taxonomy. Chloridoideae; main chloridoid assemblage.

Ecology, geography, regional floristic distribution. 1 species. Australia. Xerophytic; species of open habitats. Sandhills.

Australian. Central Australian.

References, etc. Morphological/taxonomic: Gardner and Hubbard 1938. Leaf anatomical: this project.

Psammochloa A. Hitchc.

Habit, vegetative morphology. Robust perennial; rhizomatous to stoloniferous. *Culms* 20–120 cm high (? — 'resembling *Ammophila* in appearance'); herbaceous. Culm nodes hidden by leaf sheaths. Leaves non-auriculate. *Leaf blades broad (loosely inrolled); not pseudopetiolate.* Ligule an unfringed membrane; up to 5–10 mm long.

Reproductive organization. Plants bisexual, with bisexual spikelets; with hermaphrodite florets.

Inflorescence. *Inflorescence paniculate*; open (diffuse); espatheate; not comprising 'partial inflorescences' and foliar organs. Spikelet-bearing axes persistent. Spikelets not secund; pedicellate.

Female-fertile spikelets. *Spikelets not noticeably compressed*; disarticulating above the glumes. Rachilla terminated by a female-fertile floret. Hairy callus absent.

Glumes two; more or less equal; somewhat shorter than the spikelets to exceeding the spikelets; long relative to the adjacent lemmas (sometimes slightly shorter than them); awnless; non-carinate; similar. Lower glume 5–7 nerved. Upper glume 5–7 nerved. **Spikelets** with female-fertile florets only; *without proximal incomplete florets*.

Female-fertile florets 1. Lemmas not convolute; decidedly firmer than the glumes; not becoming indurated (chartaceous); entire to incised (square to slightly notched); not deeply cleft; awned. Awns 1; median; from a sinus to dorsal; non-geniculate; straight; deciduous. Lemmas densely hairy; non-carinate; 7–9 nerved. **Palea** present; relatively long (closely resembling the lemma); entire (acute); textured like the lemma; several nerved (5–9 nerved, hairy); *keel-less*. *Lodicules* present; 3; free; membranous; ciliate (near apex); not toothed; not or scarcely vascularized. *Stamens* 3. Anthers penicillate, or not penicillate. *Ovary* glabrous. Styles free to their bases. Stigmas 2; white.

Fruit, embryo and seedling. *Fruit* free from both lemma and palea; medium sized, or large (8–11 mm long); not noticeably compressed. Hilum long-linear. *Pericarp free.* Embryo small; not waisted. Endosperm hard; containing compound starch grains. Embryo with an epiblast; without a scutellar tail; with a negligible mesocotyl internode. Embryonic leaf margins meeting.

Abaxial leaf blade epidermis. *Costal/intercostal zonation* conspicuous. *Papillae* absent. Mid-intercostal long-cells having markedly sinuous walls. *Microhairs* absent. *Stomata* common; (31.5–)34.5–36(–39) microns long. Subsidiaries low to high dome-shaped, or triangular. *Intercostal short-cells* common; not paired (arranged like the costals); silicified. Intercostal silica bodies cubical. *Costal short-cells* neither distinctly grouped into long rows nor predominantly paired. Costal silica bodies horizontally-elongated smooth, rounded, and tall-and-narrow (or cubical); not sharp-pointed.

Transverse section of leaf blade, physiology. C_3; XyMS+. *Mesophyll* with non-radiate chlorenchyma. *Leaf blade* with distinct, prominent adaxial ribs; with the ribs very irregular in sizes. *Midrib* not readily distinguishable; with one bundle only. Bulliforms present in discrete, regular adaxial groups (at the bases of the furrows); in simple fans. All the vascular bundles accompanied by sclerenchyma. Combined sclerenchyma girders present (with all the bundles); forming 'figures'. Sclerenchyma not all bundle-associated.

The 'extra' sclerenchyma in abaxial groups (in the form of small abaxial strands midway between the bundles).

Taxonomy. Arundinoideae; Stipeae.

Ecology, geography, regional floristic distribution. 1 species. Gobi Desert. Xerophytic; species of open habitats. A robust, *Ammophila*-like sandbinder.

Holarctic. Tethyan. Irano-Turanian.

Rusts and smuts. Rusts — *Puccinia*.

References, etc. Morphological/taxonomic: Bor 1951; Macfarlane and Watson 1980. Leaf anatomical: Metcalfe 1960; this project.

Psathyrostachys Nevski

Habit, vegetative morphology. *Perennial (with short, creeping rhizomes)*. Culms 15–100 cm high; herbaceous; unbranched above. Culm internodes hollow. Young shoots intravaginal. *Leaves* mostly basal, or not basally aggregated; auriculate (usually, at least some of them), or non-auriculate. Sheath margins joined (ramal leaves), or free (cauline leaves). Leaf blades narrowly linear; narrow; 1–4.5 mm wide; folded, or rolled; without cross venation; persistent. Ligule an unfringed membrane; truncate; 0.1–1 mm long. *Contra-ligule* absent.

Reproductive organization. Plants bisexual, with bisexual spikelets; with hermaphrodite florets. *The spikelets all alike in sexuality (?)*. Plants outbreeding.

Inflorescence. *Inflorescence a false spike, with spikelets on contracted axes (linear, erect)*; espatheate; not comprising 'partial inflorescences' and foliar organs. *Spikelet-bearing axes disarticulating; disarticulating at the joints*. Spikelets paired, or in triplets, or paired and in triplets; not secund; distichous (the groups distichously arranged on the rachis); sessile.

Female-fertile spikelets. Spikelets 6–16 mm long; compressed laterally to compressed dorsiventrally (?); falling with the glumes; not disarticulating between the florets (the rachilla without joints). Rachilla prolonged beyond the uppermost female-fertile floret; rachilla hairy (bristly), or hairless. The rachilla extension with incomplete florets.

Glumes two; more or less equal; free; displaced (borne side by side); densely hairy or scabrid; subulate (subulate-setiform); awned (awn-like); non-carinate; similar (subulate). Lower glume 1 nerved. Upper glume 1 nerved. *Spikelets* with incomplete florets (?). The incomplete florets distal to the female-fertile florets (above the perfect floret(s)). The distal incomplete florets 1; *incomplete florets merely underdeveloped*.

Female-fertile florets 1, or 2. Lemmas entire; acuminate pointed; mucronate to awned. Awns 1; median; apical; non-geniculate; much shorter than the body of the lemma to much longer than the body of the lemma (up to 8 mm long). Lemmas non-carinate (dorsally broadly rounded); without a germination flap; 5–7 nerved; with the nerves confluent towards the tip. *Palea* present; relatively long; 2-nerved; 2-keeled. *Lodicules* present; 2; free; membranous; ciliate; not toothed (entire); not or scarcely vascularized. *Stamens* 3. Anthers 1–6 mm long. **Ovary hairy**. Styles free to their bases. Stigmas 2.

Fruit, embryo and seedling. *Fruit* adhering to lemma and/or palea; small to medium sized (3–7.5 mm long); longitudinally grooved; compressed dorsiventrally; with hairs confined to a terminal tuft. Hilum long-linear. Embryo small.

Seedling with a loose coleoptile.

Abaxial leaf blade epidermis. *Costal/intercostal zonation* conspicuous. *Papillae* absent. Long-cells differing markedly in wall thickness costally and intercostally (costal walls thicker). Mid-intercostal long-cells rectangular; having straight or only gently undulating walls. *Microhairs* absent. *Stomata* common. Subsidiaries parallel-sided. Guard-cells overlapped by the interstomatals (sunken). *Intercostal short-cells* absent or very rare (their presence excluded by the abundance of prickles). Prickles abundant. *Costal short-cells* neither distinctly grouped into long rows nor predominantly paired. Costal silica bodies horizontally-elongated crenate/sinuous (large), or horizontally-elongated smooth; not sharp-pointed.

Transverse section of leaf blade, physiology. C_3; XyMS+. *Leaf blade* with distinct, prominent adaxial ribs; with the ribs more or less constant in size. *Midrib* not readily distinguishable; with one bundle only. Bulliforms present in discrete, regular adaxial groups;

in simple fans (these large, in the furrows). Many of the smallest vascular bundles unaccompanied by sclerenchyma. Combined sclerenchyma girders absent (abaxial only). Sclerenchyma all associated with vascular bundles.

Phytochemistry. Tissues of the culm bases with little or no starch. Fructosans predominantly short-chain.

Cytology. Chromosome base number, $x = 7$. $2n = 14$. 2 ploid. Haplomic genome content N.

Taxonomy. Pooideae; Triticodae; Triticeae.

Ecology, geography, regional floristic distribution. 8 species. Asia and Southeast Europe. Xerophytic; species of open habitats. Steppe and semi-desert regions, stony slopes.

Holarctic. Boreal and Tethyan. Euro-Siberian and Eastern Asian. Irano-Turanian. Siberian.

Hybrids. Intergeneric hybrids with *Leymus* (×*Leymostachys* Tsvelev). See also ×*Elymostachys* Tsvelev.

References, etc. Morphological/taxonomic: Löve 1984. Leaf anatomical: Metcalfe 1960 and this project.

Pseudanthistiria (Hackel) Hook.f.

Habit, vegetative morphology. Annual, or perennial (short-lived); culms prostrate or ascending, often very slender. Culms herbaceous. Leaf blades without cross venation. Ligule an unfringed membrane; short.

Reproductive organization. Plants bisexual, with bisexual spikelets; with hermaphrodite florets. *The spikelets of sexually distinct forms on the same plant*; hermaphrodite and male-only; overtly heteromorphic; all in heterogamous combinations.

Inflorescence. Inflorescence falsely paniculate (with short 'racemes' in numerous axillary, short-peduncled fascicles); spatheate; a complex of 'partial inflorescences' and intervening foliar organs. **Spikelet-bearing axes *very much reduced (the 'racemes' usually comprising a basal sessile-bisexual/pedicellate-male pair, and a terminal triad of one sessile-bisexual and two pedicellate-male spikelets)*;** solitary (in spatheolate fascicles); with very slender rachides; disarticulating; disarticulating at the joints (the short 'raceme' disarticulating beneath each sessile spikelet). '*Articles*' linear; not appendaged; disarticulating transversely. *Spikelets* unaccompanied by bractiform involucres, not associated with setiform vestigial branches (i.e., no involucral spikelets), in triplets (terminally), or paired (basally); sessile and pedicellate; consistently in 'long-and-short' combinations; in pedicellate/sessile combinations. Pedicels of the 'pedicellate' spikelets free of the rachis. The 'shorter' spikelets hermaphrodite. *The 'longer' spikelets male-only*.

Female-sterile spikelets. The pedicelled male spikelets longer, the callus narrowly linear to oblong, lacking the palea and sometimes also the lemma.

Female-fertile spikelets. Spikelets compressed dorsiventrally; falling with the glumes. Rachilla terminated by a female-fertile floret. Hairy callus present (minute, shortly bearded). *Callus* short; blunt.

Glumes two; more or less equal; long relative to the adjacent lemmas; hairless (glabrous, smooth); awnless; very dissimilar (membranous, the G_1 truncate, dorsally subconvex or concave, margins apically inflexed, broadly involute below, the G_2 acute, lanceolate, keeled). Lower glume not two-keeled; convex on the back (slightly), or flattened on the back, or concave on the back; not pitted; relatively smooth; 4–7 nerved. Upper glume 3 nerved. **Spikelets** with female-fertile florets only; *without proximal incomplete florets*.

Female-fertile florets 1. Lemmas stipitate; not becoming indurated; incised; 2 lobed (these minute, pointed); not deeply cleft; awned. Awns 1; median; from a sinus; geniculate; much longer than the body of the lemma. Lemmas non-carinate; without a germination flap; 1 nerved. **Palea *absent (in material seen)*.** *Lodicules* present; 2; free; fleshy; glabrous. *Ovary* glabrous. Stigmas 2.

Abaxial leaf blade epidermis. *Costal/intercostal zonation* conspicuous. *Papillae* present. Intercostal papillae not over-arching the stomata; several per cell (a single row per long-cell). Mid-intercostal long-cells rectangular; having markedly sinuous walls (thin; remotely sinuous). *Microhairs* present; panicoid-type; 36–48(–58) microns long; (3.9–)4.5–5.4 microns wide at the septum. Microhair total length/width at septum 6.7–10. Microhair apical cells (10.5–)12–20(–24) microns long. Microhair apical cell/total length ratio 0.29–0.41. *Stomata* common; 21–24 microns long. Subsidiaries low dome-shaped and triangular (not conspicuously pointed). Guard-cells overlapping to flush with the interstomatals. *Intercostal short-cells* absent or very rare (infrequent); silicified (a few). Intercostal silica bodies cross-shaped. *Costal short-cells* conspicuously in long rows (a few solitary). Costal silica bodies 'panicoid-type'; nodular (mostly), or dumb-bell shaped (a few); not sharp-pointed.

Transverse section of leaf blade, physiology. C_4 (vascular bundles 'crowded'); XyMS–. Mesophyll without fusoids (but 'often...a (sheath) cell on either side of each vascular bundle projecting into the adjacent mesophyll'). *Leaf blade* adaxially flat. *Midrib* not readily distinguishable; with one bundle only. Bulliforms present in discrete, regular adaxial groups to not in discrete, regular adaxial groups (irregularly grouped, tending to occupy the entire epidermis in parts); in places in simple fans. Many of the smallest vascular bundles unaccompanied by sclerenchyma. Combined sclerenchyma girders present (in the ribs). Sclerenchyma all associated with vascular bundles.

Culm anatomy. *Culm internode bundles* in one or two rings (1–2 more or less distinct circles).

Cytology. $2n = 20$.

Taxonomy. Panicoideae; Andropogonodae; Andropogoneae; Andropogoninae.

Ecology, geography, regional floristic distribution. 4 species. India, Ceylon to Thailand. Species of open habitats. Hillsides and disturbed ground.

Paleotropical. Indomalesian. Indian and Indo-Chinese.

Rusts and smuts. Smuts from Ustilaginaceae. Ustilaginaceae — *Sorosporium*.

References, etc. Leaf anatomical: Metcalfe 1960; this project.

Special comments. Fruit data wanting.

Pseudarrhenatherum Rouy

Thorea Rouy, *Thoreochloa* Holub
~ *Arrhenatherum*

Habit, vegetative morphology. Perennial; caespitose. **Culms** 30–120(–150) cm high; *herbaceous*; unbranched above; not tuberous. Culm nodes hairy, or glabrous. Leaves non-auriculate. *Leaf blades* narrow; *0.6–4 mm wide (with prominent adaxial ribs, by contrast with Arrhenatherum)*; setaceous, or not setaceous; flat, or rolled (convolute); without cross venation; persistent. **Ligule** present; *an unfringed membrane (to finely ciliate).*

Reproductive organization. Plants bisexual, with bisexual spikelets; with hermaphrodite florets.

Inflorescence. Inflorescence paniculate; open (to rather dense); espatheate; not comprising 'partial inflorescences' and foliar organs. Spikelet-bearing axes persistent. Spikelets not secund; pedicellate.

Female-fertile spikelets. Spikelets 5–10 mm long; compressed laterally; disarticulating above the glumes; not disarticulating between the florets. Rachilla prolonged beyond the uppermost female-fertile floret.

Glumes two; very unequal; long relative to the adjacent lemmas; pointed (acuminate); *awnless*; similar (thin). Lower glume 1 nerved, or 3 nerved. Upper glume 3 nerved. *Spikelets* with incomplete florets. The incomplete florets proximal to the female-fertile florets, or both distal and proximal to the female-fertile florets. *The distal incomplete florets* (when present) merely underdeveloped (a single rudiment). *Spikelets with proximal incomplete florets. The proximal incomplete florets* 1; paleate; male. *The proximal lemmas awned (the awn geniculate, dorsal, from the middle or above)*; 5–9 nerved (?); similar in texture to the female-fertile lemmas (but apically bifid).

Female-fertile florets 1. Lemmas acute; decidedly firmer than the glumes; not becoming indurated; entire; pointed; awnless, or awned (?). Awns 1; median; dorsal; from near the top; non-geniculate. Lemmas non-carinate; without a germination flap; 5–9 nerved (?). *Palea* present; relatively long; awnless, without apical setae; not indurated; 2-nerved; 2-keeled. Palea keels wingless. *Lodicules* present; 2; free; membranous. *Stamens* 3. Anthers 2.5–5 mm long. **Ovary** *hairy*. Styles free to their bases. Stigmas 2.

Fruit, embryo and seedling. *Fruit* small (2.5–3 mm long); *longitudinally grooved*; compressed dorsiventrally. Hilum long-linear (but at 1/3 of the grain length, shorter than in *Arrhenatherum*). Embryo small.

Abaxial leaf blade epidermis. *Costal/intercostal zonation* conspicuous. *Microhairs* presumably absent (see Zarco 1985). Costal silica bodies not sharp-pointed.

Transverse section of leaf blade, physiology. C_3; XyMS+. *Mesophyll* with non-radiate chlorenchyma; without adaxial palisade. *Leaf blade* with distinct, prominent adaxial ribs (the ribs more prominent than in *Arrhenatherum*); with the ribs more or less constant in size (*P. pallens*), or with the ribs very irregular in sizes (*P. longifolia*). *Midrib* conspicuous to not readily distinguishable; with one bundle only. *The lamina* symmetrical on either side of the midrib. Bulliforms present in discrete, regular adaxial groups; in simple fans. All the vascular bundles accompanied by sclerenchyma. Combined sclerenchyma girders present; forming 'figures'. Sclerenchyma all associated with vascular bundles (*P. longifolium*), or not all bundle-associated (*P. pallens*). The 'extra' sclerenchyma of *P. pallens* in a continuous abaxial layer.

Cytology. Chromosome base number, $x = 7$. $2n = 14$ (+/- 1B). 2 ploid.

Taxonomy. Pooideae; Poodae; Aveneae.

Ecology, geography, regional floristic distribution. 2 species. Western Europe. Species of open habitats. In dry grassland, calcicole.

Holarctic. Boreal and Tethyan. Euro-Siberian. Mediterranean. European.

References, etc. Morphological/taxonomic: Zarco 1985. Leaf anatomical: Zarco 1985.

Pseudechinolaena Stapf

Loxostachys Peter
Excluding *Perulifera*

Habit, vegetative morphology. Annual; decumbent. *Culms* 10–60 cm high; herbaceous; branched above. Culm nodes glabrous. Culm internodes solid. *Leaves* not basally aggregated; non-auriculate. *Leaf blades* lanceolate (acuminate); broad; 6–15 mm wide; not cordate, not sagittate (but asymmetrical at the base); flat; pseudopetiolate; *cross veined*; persistent. Ligule a fringed membrane; 1.8 mm long.

Reproductive organization. Plants bisexual, with bisexual spikelets; with hermaphrodite florets. The spikelets of sexually distinct forms on the same plant, or all alike in sexuality; hermaphrodite, or hermaphrodite and male-only, or hermaphrodite and sterile (some variously incomplete).

Inflorescence. Inflorescence of spicate main branches (spiciform racemes); espatheate; not comprising 'partial inflorescences' and foliar organs. Spikelet-bearing axes persistent. Spikelets paired, or solitary (via suppression of one of the pair); secund; pedicellate; consistently in 'long-and-short' combinations, or not in distinct 'long-and-short' combinations.

Female-fertile spikelets. Spikelets 4.6 mm long; adaxial; compressed laterally; falling with the glumes. Rachilla terminated by a female-fertile floret. Hairy callus absent.

Glumes two; more or less equal (or G_1 shorter); long relative to the adjacent lemmas; dorsiventral to the rachis; pointed; awnless; *very dissimilar (the lower smooth, the upper gibbous with translucent intercostal glands and often with hooked spines)*. Lower glume 3 nerved. *Upper glume* 7 nerved; *prickly*. *Spikelets* with incomplete florets. The incomplete florets proximal to the female-fertile florets. *Spikelets with proximal incomplete florets. The proximal incomplete florets* 1; paleate. Palea of the proximal incomplete florets fully developed. The proximal incomplete florets male, or sterile. The proximal lemmas awnless; 3 nerved; decidedly exceeding the female-fertile lemmas (equalling the spikelet); similar in texture to the female-fertile lemmas; not becoming indurated (chartaceous, with thin margins and a delicate hyaline area at the base, laterally compressed).

Female-fertile florets 1. **Lemmas** lanceolate; similar in texture to the glumes to decidedly firmer than the glumes (papery); smooth; not becoming indurated; entire; pointed; awnless; *hairless (sometimes with hooks)*; non-carinate (convex on the back); having the margins tucked in onto the palea; with a clear germination flap; 3–5 nerved. *Palea* present; entire; awnless, without apical setae; textured like the lemma; not indurated; 2-nerved. *Lodicules* present; 2; free; fleshy; glabrous. *Stamens* 3. Anthers not penicillate. *Ovary* glabrous. Styles fused, or free to their bases. Stigmas 2; red pigmented.

Fruit, embryo and seedling. Fruit small (1.5 mm long); ellipsoid; compressed dorsiventrally (ventrally flattened). Hilum short. Embryo large; waisted. Endosperm hard; without lipid.

Abaxial leaf blade epidermis. *Costal/intercostal zonation* conspicuous. *Papillae* absent. Long-cells of similar wall thickness costally and intercostally (thin walled). Intercostal zones with typical long-cells (though these rather short and broad). Mid-intercostal long-cells rectangular; having markedly sinuous walls. *Microhairs* present; panicoid-type; (36–)39–42(–48) microns long; (3.3–)3.9–4.5(–5.4) microns wide at the septum. Microhair total length/width at septum 6.7–11.8. Microhair apical cells 13.5–15–19.5 microns long. Microhair apical cell/total length ratio 0.32–0.42. *Stomata* common; about 27 microns long. Subsidiaries triangular. Guard-cells overlapping to flush with the interstomatals. *Intercostal short-cells* absent or very rare; not paired (those seen usually solitary); silicified, or not silicified. *Costal short-cells* conspicuously in long rows. Costal silica bodies 'panicoid-type'; not sharp-pointed.

Transverse section of leaf blade, physiology. C_3; XyMS+. *Mesophyll* with radiate chlorenchyma; *Isachne*-type. *Leaf blade* 'nodular' in section; with the ribs more or less constant in size. *Midrib* conspicuous; with one bundle only; with colourless mesophyll adaxially (associated with the median bundle). Bulliforms present in discrete, regular adaxial groups; in simple fans (these wide, occupying most of the shallow intercostal depressions). All the vascular bundles accompanied by sclerenchyma. Combined sclerenchyma girders present; forming 'figures'. Sclerenchyma all associated with vascular bundles.

Cytology. $2n = 36$.

Taxonomy. Panicoideae; Panicodae; Paniceae.

Ecology, geography, regional floristic distribution. 6 species. 5 in Madagascar, 1 pantropical. Mesophytic; shade species; glycophytic. In forest.

Paleotropical, Neotropical, and Cape. African, Madagascan, and Indomalesian. Sudano-Angolan and West African Rainforest. Indian, Malesian, and Papuan. Caribbean, Venezuela and Surinam, Amazon, Central Brazilian, Pampas, and Andean. Sahelo-Sudanian, Somalo-Ethiopian, South Tropical African, and Kalaharian.

References, etc. Leaf anatomical: this project.

Pseudobromus K. Schum.

~ *Festuca*

Habit, vegetative morphology. Perennial; caespitose. **Culms** 40–200 cm high; *herbaceous*; unbranched above. Culm nodes glabrous. Culm internodes hollow. *Leaves* not basally aggregated; auriculate (from the base of the blade). *Leaf blades* linear to linear-lanceolate; broad; 6–15 mm wide ('to 15 mm wide'); flat, or rolled; *cross veined*; persistent; rolled in bud. Ligule an unfringed membrane (with lacerate margins); not truncate; 3–8.5 mm long. *Contra-ligule* absent.

Reproductive organization. Plants bisexual, with bisexual spikelets; with hermaphrodite florets.

Inflorescence. Inflorescence paniculate; open; espatheate; not comprising 'partial inflorescences' and foliar organs. Spikelet-bearing axes persistent. Spikelets not secund; pedicellate.

Female-fertile spikelets. *Spikelets* 8–10 mm long; *not noticeably compressed*; disarticulating above the glumes; disarticulating between the florets. Rachilla prolonged beyond the uppermost female-fertile floret (even when there is only one floret); rachilla hairless (minutely scabrid). Hairy callus present, or absent. *Callus* short; blunt.

Glumes two; very unequal; shorter than the spikelets; shorter than the adjacent lemmas; hairless; glabrous; pointed (acute or acuminate); awnless; more or less *carinate*; similar. Lower glume shorter than the lowest lemma; 1–3 nerved. Upper glume 3 nerved. *Spikelets* with female-fertile florets only, or with incomplete florets. The incomplete florets distal to the female-fertile florets. *The distal incomplete florets* merely underdeveloped; when present, awned.

Female-fertile florets 1–2. *Lemmas* similar in texture to the glumes; not becoming indurated; *incised*; not deeply cleft (only notched); awned. Awns 1; median; dorsal, or apical; when apical, from near the top; non-geniculate, or geniculate; hairless (scabrid); much longer than the body of the lemma; entered by one vein. Lemmas hairy (with a tuft above the callus), or hairless; non-carinate; without a germination flap; 3–5 nerved. *Palea* present; relatively long; entire; awnless; without apical setae (glabrous); textured like the lemma; not indurated (membranous); 2-nerved; 2-keeled. *Lodicules* present; 2; free; membranous; glabrous; toothed, or not toothed; not or scarcely vascularized. **Stamens** 3. Anthers 3 mm long; not penicillate; without an apically prolonged connective. *Ovary* hairy. Styles free to their bases. Stigmas 2; white.

Fruit, embryo and seedling. Fruit medium sized (6–7 mm long); longitudinally grooved; compressed dorsiventrally; with hairs confined to a terminal tuft. *Hilum long-linear.* Embryo small. Endosperm hard; without lipid; containing compound starch grains. Embryo with an epiblast; without a scutellar tail; with a negligible mesocotyl internode. Embryonic leaf margins meeting.

Abaxial leaf blade epidermis. *Costal/intercostal zonation* conspicuous. *Papillae* absent. *Long-cells* markedly different in shape costally and intercostally (costals longer and narrower); of similar wall thickness costally and intercostally. Mid-intercostal long-cells fusiform; having straight or only gently undulating walls. *Microhairs* absent. *Stomata* common; 40.5–42–45 microns long. Subsidiaries parallel-sided. Guard-cells overlapped by the interstomatals (sunken). *Intercostal short-cells* absent or very rare (prickles only). *Costal short-cells* neither distinctly grouped into long rows nor predominantly paired. Costal silica bodies horizontally-elongated crenate/sinuous, or horizontally-elongated smooth; not sharp-pointed.

Transverse section of leaf blade, physiology. C_3; XyMS+. *Mesophyll* with non-radiate chlorenchyma. *Leaf blade* 'nodular' in section; with the ribs more or less constant in size. *Midrib* conspicuous; with one bundle only. Bulliforms present in discrete, regular adaxial groups (between the bundles); in simple fans. All the vascular bundles accompanied by sclerenchyma. Combined sclerenchyma girders present. Sclerenchyma all associated with vascular bundles.

Cytology. Chromosome base number, $x = 7$. $2n = 28$.

Taxonomy. Pooideae; Poodae; Poeae.

Ecology, geography, regional floristic distribution. 3 species. East tropical and South Africa, Madagascar. Mesophytic; shade species; glycophytic. In forests.

Paleotropical and Cape. African. Sudano-Angolan. Sahelo-Sudanian, Somalo-Ethiopian, and South Tropical African.

Economic importance. Important native pasture species: *P. engleri, P. sylvaticus.*

References, etc. Leaf anatomical: Metcalfe 1960 and this project.

Pseudochaetochloa A. Hitchc.

~ Pennisetum

Habit, vegetative morphology. Perennial; caespitose. **Culms** *50–150 cm high*; herbaceous; branched above. Culm nodes hairy (usually), or glabrous. Culm internodes solid. Leaves non-auriculate. Leaf blades narrow; 2–5 mm wide; without cross venation; persistent. Ligule a fringe of hairs.

Reproductive organization. *Plants dioecious (seemingly — only male material as yet formally described)*; without hermaphrodite florets. The spikelets all alike in sexuality (on the male plants, at least); female-only, or male-only. Plants outbreeding.

Inflorescence. Inflorescence paniculate; contracted (the spikelets in clusters of 1–5, with reduced-branch bristles); spicate. Inflorescence axes not ending in spikelets. Inflorescence espatheate; not comprising 'partial inflorescences' and foliar organs. Spikelet-bearing axes persistent. **Spikelets *with 'involucres' of 'bristles'***. The 'bristles' relatively slender, not spiny; deciduous with the spikelets. Spikelets not secund; pedicellate. Pedicel apices discoid.

Female-sterile spikelets. Male spikelets about 6 mm long, not compressed, falling with the glumes. Rachilla not prolonged. Glumes 2, unequal, membranous, much shorter than the adjacent lemmas, 3-nerved. Two male florets per spikelet, the lemmas 5-nerved. Palea conventionally eu-panicoid, 2-nerved. Lodicules 2, fleshy. Rachilla of male spikelets terminated by a male floret. The male spikelets with glumes; without proximal incomplete florets; 2 floreted. The lemmas awnless. Male florets 2.

Female-fertile spikelets. *Spikelets* as yet not formally described — only male plants known.

Abaxial leaf blade epidermis. *Costal/intercostal zonation* conspicuous. *Papillae* absent. *Long-cells* similar in shape costally and intercostally; of similar wall thickness costally and intercostally (quite thick walled). Mid-intercostal long-cells rectangular; having markedly sinuous walls. *Microhairs* present; panicoid-type; (72–)75–78 microns long; 6–7.5 microns wide at the septum. Microhair total length/width at septum 10.12. Microhair apical cells (37.5–)39–40.5(–42) microns long. Microhair apical cell/total length ratio 0.52–0.56. *Stomata* common; 36–42 microns long. Subsidiaries triangular. Guard-cells overlapping to flush with the interstomatals. *Intercostal short-cells* absent or very rare. Prickles abundant. *Costal short-cells* conspicuously in long rows. Costal silica bodies horizontally-elongated crenate/sinuous, or 'panicoid-type'; not sharp-pointed.

Transverse section of leaf blade, physiology. C_4; XyMS–. PCR sheath outlines uneven. PCR sheath extensions present. Maximum number of extension cells 1. *Mesophyll* with radiate chlorenchyma. *Leaf blade* with distinct, prominent adaxial ribs, or 'nodular' in section; with the ribs very irregular in sizes. *Midrib* conspicuous; having a conventional arc of bundles; with colourless mesophyll adaxially. Bulliforms present in discrete, regular adaxial groups; in simple fans. All the vascular bundles accompanied by sclerenchyma. Combined sclerenchyma girders present, or absent; nowhere forming 'figures'. Sclerenchyma all associated with vascular bundles.

Taxonomy. Panicoideae; Panicodae; Paniceae.

Ecology, geography, regional floristic distribution. 1 species. Australia. Species of open habitats. Stony slopes.

Australian. North and East Australian. Tropical North and East Australian.

References, etc. Leaf anatomical: this project.

Special comments. *Only male plants have yet been formally described inder this genus. It is now established that* Pennisetum arnhemicum *F. Muell. is the male component (Lazarides and Macfarlane, pers. comm.), but full details and the revised generic circumscription are not yet formally published*. Fruit data wanting.

Pseudocoix A. Camus

~ *Hickelia*

Habit, vegetative morphology. Perennial. The flowering culms leafy. *Culms* reaching 40–60 m long; woody and persistent (but weak); scandent (on trees and shrubs); branched above. Primary branches/mid-culm node several, in verticils. Rhizomes pachymorph. *Leaves* not basally aggregated. Leaf blades linear-lanceolate to lanceolate; narrow; 7–9 mm wide (to 15 cm long); not cordate, not sagittate (abruptly attenuate); (presumably) pseudopetiolate; disarticulating from the sheaths (presumably). *Ligule* present ('oblong').

Reproductive organization. Plants bisexual, with bisexual spikelets; with hermaphrodite florets.

Inflorescence. *Inflorescence a fascicle of single, bracteate spikelets, 'pendulous, subspiciform, recalling...(that)...of Coix by its ventricose lemmas'*; spatheate; a complex of 'partial inflorescences' and intervening foliar organs.

Female-fertile spikelets. Spikelets 15–16 mm long; not noticeably compressed (? ventricose); falling with the glumes. Rachilla prolonged beyond the uppermost female-fertile floret.

Glumes two to several (2–3, membranous); pointed (acuminate or cuspidate); awnless; similar. Lower glume many nerved. Upper glume many nerved. *Spikelets* with incomplete florets. The incomplete florets proximal to the female-fertile florets. *The proximal incomplete florets* 1, or 2 (their lemmas cuspidate, glabrous, shining, becoming ventricose). The proximal lemmas many-nerved; exceeded by the female-fertile lemmas; becoming indurated.

Female-fertile florets 1. **Lemmas** glossy; becoming *saccate (ventricose)*; decidedly firmer than the glumes; becoming indurated; entire; pointed; awned (long-cuspidate). Awns 1; median; apical; non-geniculate. Lemmas hairless; non-carinate; many-nerved. *Palea* present; relatively long; awnless, without apical setae; indurated (and inflated); 'parinervia'; 2-keeled (?— 'dorsally 'sulcate'). *Lodicules* present (large); 3; free. *Stamens* 6. **Ovary** hairy; *with a conspicuous apical appendage*. The appendage broadly conical, fleshy. Stigmas 3.

Fruit, embryo and seedling. Pericarp fleshy (apically rostrate); free.

Abaxial leaf blade epidermis. *Costal/intercostal zonation* conspicuous (mainly because of the macrohairs and prickles). *Papillae* present; costal and intercostal. Intercostal papillae over-arching the stomata (mostly with four papillae over-arching each stomatal apparatus); several per cell (mostly 5—8 in a median longitudinal row, large, circular, thickened). *Long-cells* similar in shape costally and intercostally; of similar wall thickness costally and intercostally. Mid-intercostal long-cells rectangular; having markedly sinuous walls. *Microhairs* present; elongated; clearly two-celled; panicoid-type; 60–72 microns long; 4.2–6 microns wide at the septum. Microhair total length/width at septum 10–16.4. Microhair apical cells 28.5–39 microns long. Microhair apical cell/total length ratio 0.48–0.54. *Stomata* common; 18–21 microns long. Subsidiaries non-papillate. The costal zones with large, bulbous-based macrohairs and occasional prickles. *Costal short-cells* predominantly paired (but in places the intervening 'long-cells' are quite short). Costal silica bodies saddle shaped (predominating), or oryzoid (a few, intergrading with saddles); not sharp-pointed.

Transverse section of leaf blade, physiology. C_3; XyMS+. *Mesophyll* with non-radiate chlorenchyma; with adaxial palisade; with arm cells; with fusoids. The fusoids external to the PBS. *Leaf blade* adaxially flat (except for the midrib). *Midrib* conspicuous; with one bundle only. *The lamina* symmetrical on either side of the midrib. Bulliforms present in discrete, regular adaxial groups (between the laterally adjacent pairs of fusoids); in simple fans. All the vascular bundles accompanied by sclerenchyma. Combined sclerenchyma girders present (with all the bundles); forming 'figures' (most bundles with anchors). Sclerenchyma all associated with vascular bundles.

Taxonomy. Bambusoideae; Bambusodae; Bambuseae.

Ecology, geography, regional floristic distribution. 1 species. Madagascar. Mesophytic.

Paleotropical. Madagascan.

References, etc. Leaf anatomical: this project.

Pseudodanthonia Bor & C.E. Hubb.

Excluding *Sinochasea*

Habit, vegetative morphology. Perennial; caespitose. *Culms herbaceous*; unbranched above. Young shoots intravaginal. Leaves non-auriculate. Leaf blades narrow; setaceous; without cross venation. *Ligule an unfringed membrane*; not truncate; 7 mm long.

Reproductive organization. Plants bisexual, with bisexual spikelets; with hermaphrodite florets.

Inflorescence. Inflorescence few spikeleted, or many spikeleted; a single raceme, or paniculate; open; espatheate; not comprising 'partial inflorescences' and foliar organs. Spikelet-bearing axes persistent. Spikelets not secund; pedicellate.

Female-fertile spikelets. Spikelets to 25 mm long; compressed laterally; disarticulating above the glumes; disarticulating between the florets. Rachilla prolonged beyond the uppermost female-fertile floret.

Glumes present; two; relatively large; very unequal to more or less equal; about equalling the spikelets; (the longer) long relative to the adjacent lemmas; pointed (acuminate); awnless; non-carinate. Lower glume 5–7 nerved. Upper glume 7–9 nerved.

Female-fertile florets *4–6*. **Lemmas** similar in texture to the glumes; not becoming indurated; *incised*; 2 lobed; not deeply cleft (notched); awned. Awns 1; median; from a sinus (flanked by the shortish, narrow acuminate lobes); geniculate; hairless (scabrid); much longer than the body of the lemma. Lemmas hairy; non-carinate; without a germination flap; 7–9 nerved. *Palea* present; conspicuous but relatively short; apically notched; awnless, without apical setae; 2-nerved; 2-keeled. **Lodicules** present; *2*; free; fleshy; glabrous; not or scarcely vascularized. *Stamens* 3. Anthers 4.5–5 mm long. **Ovary** *hairy (all over)*. Styles free to their bases. *Stigmas 3*; white.

Abaxial leaf blade epidermis. *Costal/intercostal zonation* lacking. *Papillae* absent. Mid-intercostal long-cells having markedly sinuous walls. *Microhairs* absent. *Stomata* absent or very rare. Intercostal short-cells in cork/silica-cell pairs. *Costal short-cells* predominantly paired. Costal silica bodies rounded and tall-and-narrow; not sharp-pointed.

Transverse section of leaf blade, physiology. C_3; XyMS+. *Leaf blade* with distinct, prominent adaxial ribs; with the ribs very irregular in sizes. *Midrib* not readily distinguishable; with one bundle only. All the vascular bundles accompanied by sclerenchyma. Combined sclerenchyma girders present; forming 'figures'. Sclerenchyma not all bundle-associated. The 'extra' sclerenchyma in a continuous abaxial layer.

Taxonomy. Pooideae, or Arundinoideae; if pooid, Poodae; Aveneae; if arundinoid, Danthonieae (?).

Ecology, geography, regional floristic distribution. 1 species. Himalayas. Species of open habitats. Mountain slopes.

Paleotropical. Indomalesian. Indian.

References, etc. Morphological/taxonomic: Macfarlane and Watson 1980. Leaf anatomical: this project.

Special comments. Fruit data wanting.

Pseudodichanthium Bor

Habit, vegetative morphology. Slender annual. *Culms* 60 cm high. Leaves non-auriculate. Leaf blades linear (attenuate); narrow; 4–6 mm wide; without cross venation. *Ligule an unfringed membrane (?)*; not truncate (ovate); 1 mm long.

Reproductive organization. Plants bisexual, with bisexual spikelets; with hermaphrodite florets. The spikelets of sexually distinct forms on the same plant; hermaphrodite and male-only, or hermaphrodite and sterile. The male and female-fertile spikelets mixed in the inflorescence. *The spikelets* more or less *homomorphic*; usually in both homogamous and heterogamous combinations (with the basal 2–3 pairs homogamous-sterile).

Inflorescence. *Inflorescence* terminal and axillary a single raceme; *non-digitate*; *espatheate (?)*; not comprising 'partial inflorescences' and foliar organs (?). *Spikelet-bearing axes* 'racemes' (curved, 1–1.75 cm long); solitary; with very slender rachides; disarticulating; disarticulating at the joints. *'Articles'* linear. Spikelets sessile and pedicellate; consistently in 'long-and-short' combinations (but the pedicels shorter at the base of the raceme); in pedicellate/sessile combinations. Pedicels of the 'pedicellate' spikelets free of the rachis. The 'shorter' spikelets hermaphrodite, or sterile (at the base of the raceme). The 'longer' spikelets male-only, or sterile.

Female-sterile spikelets. The pedicelled member larger, membranous, winged.

Female-fertile spikelets. Spikelets compressed dorsiventrally. Rachilla terminated by a female-fertile floret. Callus blunt.

Glumes two; long relative to the adjacent lemmas; free; hairless; glabrous; without conspicuous tufts or rows of hairs; awnless; very dissimilar (G_1 cartilaginaous, 2-keeled, winged, G_2 lanceolate, acute, papery). *Lower glume two-keeled (and broadly winged above)*; broadly *convex on the back*; not pitted; relatively smooth; 9–11 nerved (7–9 nerved between the keels). Upper glume 3 nerved. *Spikelets* with incomplete florets. The incomplete florets proximal to the female-fertile florets. *The proximal incomplete florets* 1; sterile. The proximal lemmas awnless; 0 nerved, or 1 nerved; not becoming indurated (hyaline).

Female-fertile florets 1. Lemmas stipitate, passing into the awn; not becoming indurated; entire; not deeply cleft; awned. Awns 1; median; apical; geniculate; hairy. Lemmas without a germination flap. **Palea** *absent*. **Stamens** 3. **Ovary** glabrous. Stigmas 2.

Fruit, embryo and seedling. Fruit small (3 mm long); compressed dorsiventrally.

Taxonomy. Panicoideae; Andropogonodae; Andropogoneae; Andropogoninae.

Ecology, geography, regional floristic distribution. 1 species. Bombay. Shade species.

Paleotropical. Indomalesian. Indian.

Special comments. Fruit data wanting. Anatomical data wanting.

Pseudopentameris Conert

~ *Danthonia* sensu lato

Habit, vegetative morphology. Perennial; caespitose. **Culms** 30–120 cm high; *herbaceous*; branched above, or unbranched above. Culm nodes glabrous. Culm internodes hollow. Plants unarmed. Young shoots intravaginal. **Leaves** not basally aggregated; *auriculate (from the base of the blade)*. Leaf blades linear; narrow (to 8 mm wide at the base); 3–8 mm wide; setaceous (at the tips), or not setaceous; flat, or rolled (usually convolute); without cross venation; persistent (old basal leaves with curled blades in *P. brachyphylla*). **Ligule** *a fringe of hairs*; about 0.5 mm long. *Contra-ligule* absent.

Reproductive organization. Plants bisexual, with bisexual spikelets; with hermaphrodite florets.

Inflorescence. Inflorescence few spikeleted; paniculate (4–25 cm long); contracted (but the central axis visible); without capillary branchlets; espatheate; not comprising 'partial inflorescences' and foliar organs. Spikelet-bearing axes persistent. Spikelets solitary; not secund; pedicellate.

Female-fertile spikelets. *Spikelets 35–55 mm long*; compressed laterally; disarticulating above the glumes; disarticulating between the florets; with conventional internode spacings. Rachilla prolonged beyond the uppermost female-fertile floret. The rachilla extension with incomplete florets. Hairy callus present. *Callus* long; pointed.

Glumes present; two; relatively large (35–55 mm long); more or less equal; about equalling the spikelets to exceeding the spikelets; *long relative to the adjacent lemmas (exceeding them)*; hairless; glabrous; pointed (acuminate); awnless; carinate; similar (lanceolate, membranous). Lower glume much exceeding the lowest lemma; 3 nerved, or 5(–7) nerved. Upper glume 3 nerved, or 5(–7) nerved. *Spikelets* with incomplete florets. The incomplete florets distal to the female-fertile florets. *The distal incomplete florets* merely underdeveloped.

Female-fertile florets 2, or 3. *Lemmas* decidedly firmer than the glumes (leathery); incised; basically *2 lobed (the triangular lobes themselves bifid)*; deeply cleft; awned. Awns 3; median and lateral; the median different in form from the laterals; from a sinus; geniculate; hairless (scabrid); much longer than the body of the lemma. The lateral awns shorter than the median (non-geniculate, loosely twisted, 10–16 mm long, from the inner margins of the lobes). Lemmas hairy (villous); non-carinate; having the margins tucked in onto the palea; without a germination flap; 9 nerved; with the nerves non-confluent. **Palea** *present (glabrous, by contrast with* **Pentameris***)*; relatively long (exceeding the lemma lobes); apically notched; awnless, without apical setae; textured like the lemma (leathery); 2-nerved; 2-keeled. Palea back glabrous. Palea keels glabrous. *Lodicules* present; free; fleshy; ciliate (or at least ciliolate), or glabrous; heavily vascularized. *Stamens* 3. Anthers about 8 mm long; not penicillate; without an apically prolonged connective. *Ovary* glabrous. Styles fused (the stigmatic hairs joining over the ovary apex). Stigmas 2; brown.

Fruit, embryo and seedling. Fruit medium sized (about 6 mm long); narrowly ellipsoid; longitudinally grooved; not noticeably compressed. Hilum long-linear (more than half the grain length).

Abaxial leaf blade epidermis. *Costal/intercostal zonation* conspicuous. *Papillae* absent. Intercostal zones with typical long-cells (though sometimes these are relativey short). Mid-intercostal long-cells rectangular, or fusiform, or rectangular and fusiform; having markedly sinuous walls, or having straight or only gently undulating walls. *Microhairs* absent (but present adaxially); panicoid type. *Stomata* absent or very rare, or common (seen only in one specimen of *P. macrantha*). Subsidiaries low dome-shaped. *Intercostal short-cells* common; in cork/silica-cell pairs; silicified. *Costal short-cells* conspicuously in long rows, or neither distinctly grouped into long rows nor predominantly paired (with numerous relatively long 'short-cells', or with numerous intervening long-cells). Costal silica bodies 'panicoid-type'; cross shaped to dumb-bell shaped; not sharp-pointed.

Transverse section of leaf blade, physiology. C_3; XyMS+. *Mesophyll* with non-radiate chlorenchyma; without adaxial palisade. *Leaf blade* with distinct, prominent adaxial ribs (with rounded sides and flat tops); with the ribs more or less constant in size to with the ribs very irregular in sizes (this being more noticeable in *P. brachyphylla*). *Midrib* not readily distinguishable (except by its position); with one bundle only. Bulliforms present in discrete, regular adaxial groups (in all the furrows); in simple fans. All the vascular bundles accompanied by sclerenchyma. Combined sclerenchyma girders present; forming 'figures' (all the main bundles with I's or T's).

Cytology. Chromosome base number, $x = 6$. $2n = 12$. 2 ploid.

Taxonomy. Arundinoideae; Danthonieae.

Ecology, geography, regional floristic distribution. 2 species. South Africa. Mesophytic; species of open habitats; glycophytic. Mountain Fynbos.

Paleotropical. African. Namib-Karoo.

References, etc. Morphological/taxonomic: Conert 1971. Leaf anatomical: Ellis 1985.

Pseudophleum M. Dogan

~ *Phleum* (*P. gibbum* Boiss.)

Habit, vegetative morphology. *Annual. Culms* 5–20 cm high; herbaceous; unbranched above. Culm nodes glabrous. The shoots not aromatic. *Leaves* not basally aggregated; non-auriculate; without auricular setae. Sheath margins free. Leaf blades narrow; 0.8–1.5 mm wide (0.8–5 cm long); not setaceous (usually convolute); usually rolled (convolute); not pseudopetiolate; without cross venation; persistent; rolled in bud. *Ligule* present; an unfringed membrane; not truncate (acute); 2.5–4 mm long.

Reproductive organization. Plants bisexual, with bisexual spikelets; with hermaphrodite florets.

Inflorescence. *Inflorescence* determinate; without pseudospikelets; paniculate; contracted; more or less ovoid to spicate; without capillary branchlets; non-digitate; espatheate; not comprising 'partial inflorescences' and foliar organs. Spikelet-bearing axes persistent. *Spikelets not secund*; shortly pedicellate, or subsessile.

Female-fertile spikelets. Spikelets 2.5–3 mm long (pale greenish); compressed laterally; disarticulating above the glumes. Rachilla terminated by a female-fertile floret (?). Hairy callus absent.

Glumes two; relatively large; *very unequal (or at least, supposedly more so than in* **Phleum)**; shorter than the spikelets; shorter than the adjacent lemmas; *joined (shortly, at base)*; hairless (rather leathery); pointed (acute); awnless; carinate; without a median keel-wing; *similar (gibbous, narrow at base)*. Lower glume 3 nerved. Upper glume 3 nerved. *Spikelets* with female-fertile florets only.

Female-fertile florets 1. **Lemmas** similar in texture to the glumes (leathery); not becoming indurated; *mucronate (terminally, the mucro to 0.6 mm long)*; hairy (dorsally covered with setules); carinate (in upper half), or non-carinate; without a germination flap; 5 nerved. *Palea* present; relatively long; apically notched; awnless, without apical setae; not indurated; 2-nerved; 2-keeled. Palea keels hairy (ciliate). *Lodicules* present; 2; free; membranous; glabrous; not toothed. *Stamens* 3. Anthers not penicillate; without an apically prolonged connective. *Ovary* glabrous; without a conspicuous apical append-age. Styles free to their bases. Stigmas 2.

Fruit, embryo and seedling. Fruit small (1.3 mm long). *Hilum short (elliptical)*. Embryo small.

Taxonomy. Pooideae; Poodae; Aveneae.

Ecology, geography, regional floristic distribution. 1 species. Turkey. Species of open habitats.

Holarctic. Tethyan. Mediterranean.

Special comments. Anatomical data wanting.

Pseudopogonatherum A. Camus

~ Eulalia

Habit, vegetative morphology. Annual, or perennial; caespitose. Culm nodes gla-brous. Culm internodes hollow. *Leaves* mostly basal; non-auriculate. Leaf blades very narrow; without cross venation; persistent. *Ligule a fringed membrane (short)*.

Reproductive organization. Plants bisexual, with bisexual spikelets; with hermaph-rodite florets. The spikelets homomorphic.

Inflorescence. *Inflorescence* of spicate main branches (having several spike-like 'racemes' on a short axis); open; *digitate*; espatheate; not comprising 'partial inflores-cences' and foliar organs. *Spikelet-bearing axes* 'racemes' (shortly pedunculate); persist-ent (tough, this being the main distinction from *Eulalia*), or disarticulating (in some forms); when disarticulating, disarticulating at the joints. 'Articles' glabrous on the back, with villous-ciliate edges. Spikelets paired; not secund; sessile and pedicellate, or subses-sile and pedicellate; consistently in 'long-and-short' combinations; in pedicellate/sessile combinations, or unequally pedicellate in each combination. The 'shorter' spikelets her-maphrodite. The 'longer' spikelets hermaphrodite.

Female-fertile spikelets. Spikelets compressed dorsiventrally; falling with the glumes (falling from the pedicels, or persistent on pedicels and internodes). Rachilla ter-minated by a female-fertile floret. Hairy callus present. Callus blunt.

Glumes present; two; *more or less equal*; long relative to the adjacent lemmas; with-out conspicuous tufts or rows of hairs; *awned*. Lower glume two-keeled, or not two-keeled; convex on the back, or concave on the back (slightly); not pitted; relatively smooth; 2 nerved. Upper glume 1 nerved. **Spikelets** *with female-fertile florets only (the L_1 often absent), or with incomplete florets*. The incomplete florets (when present) proxi-mal to the female-fertile florets. *The proximal incomplete florets* (when present) 1; sterile.

Female-fertile florets 1. Lemmas less firm than the glumes; incised; 2 lobed; deeply cleft (bifid), or not deeply cleft (emarginate); awned. Awns 1; median; from a sinus; gen-iculate; hairy (villous-ciliate at the edges of the column). *Lemmas* hairless; non-carinate; *0 nerved*. Palea present; nerveless. *Lodicules* present; 2; free; fleshy; glabrous; toothed, or not toothed. *Stamens* 3. Anthers not penicillate. *Ovary* glabrous. Styles free to their bases. Stigmas 2.

Fruit, embryo and seedling. Fruit compressed dorsiventrally to not noticeably com-pressed. Hilum short. Embryo large.

Abaxial leaf blade epidermis. *Costal/intercostal zonation* conspicuous. *Papillae* absent. *Long-cells* similar in shape costally and intercostally; of similar wall thickness costally and intercostally. Mid-intercostal long-cells rectangular; having markedly sinuous walls. *Microhairs* present; panicoid-type; (39–)40.5–45 microns long; 6–6.6–7 microns wide at the septum. Microhair total length/width at septum 5.6–6.8. Microhair apical cells 24–26–27 microns long. Microhair apical cell/total length ratio 0.57–0.65. *Stomata* common; (21–)22.5–25.5(–27) microns long. Subsidiaries triangular. Guard-cells overlapping to flush with the interstomatals. *Intercostal short-cells* common; not paired (solitary); not silicified. *Costal short-cells* conspicuously in long rows. Costal silica bodies 'panicoid-type'; not sharp-pointed.

Transverse section of leaf blade, physiology. C_4; XyMS–. *Mesophyll* with radiate chlorenchyma. *Leaf blade* 'nodular' in section, or adaxially flat. *Midrib* conspicuous; having a conventional arc of bundles; with colourless mesophyll adaxially (notably between the median bundle and its adaxial strand). Bulliforms present in discrete, regular adaxial groups; in simple fans (these large, occupying most of the epidermis). All the vascular bundles accompanied by sclerenchyma. Combined sclerenchyma girders present; forming 'figures'. Sclerenchyma all associated with vascular bundles.

Cytology. $2n = 40$.

Taxonomy. Panicoideae; Andropogonodae; Andropogoneae; Andropogoninae.

Ecology, geography, regional floristic distribution. 2 species. Tropical Asia. Paleotropical and Australian. Indomalesian. Indo-Chinese, Malesian, and Papuan. North and East Australian. Tropical North and East Australian.

References, etc. Leaf anatomical: this project.

Pseudoraphis Griff.

Habit, vegetative morphology. Perennial; *decumbent*. Culms herbaceous; branched above. Culm nodes glabrous. Culm internodes hollow. *Leaves* not basally aggregated; non-auriculate. Sheath margins free. The sheaths flattened. Leaf blades narrow; not setaceous; flat (glabrous); without cross venation; disarticulating from the sheaths, or persistent. Ligule an unfringed membrane.

Reproductive organization. Plants bisexual, with bisexual spikelets; with hermaphrodite florets (rarely), or without hermaphrodite florets.

Inflorescence. Inflorescence of spicate main branches (the slender flexuous branches sometimes reduced to one or two spikelets); non-digitate. *Inflorescence axes not ending in spikelets (the 'racemes' each terminated by a stiff bristle at least as long as the uppermost spikelet)*. Inflorescence espatheate; not comprising 'partial inflorescences' and foliar organs. *Spikelet-bearing axes* very much reduced (with only one spikelet), or 'racemes' (with a compressed peduncle, the few spikelets distant); persistent. Spikelets solitary; not secund; shortly pedicellate. Pedicel apices truncate, or discoid.

Female-fertile spikelets. Spikelets 6–10 mm long; lanceolate; abaxial; compressed dorsiventrally; falling with the glumes; with distinctly elongated rachilla internodes between the florets. The upper floret conspicuously stipitate. The stipe beneath the upper floret not filiform. Rachilla terminated by a female-fertile floret.

Glumes two; very unequal; (the longer) long relative to the adjacent lemmas; awned, or awnless (G_2 acuminate to awnlike above); very dissimilar (lower tiny, semiorbicular, rectangular or broadly obovate, the upper lanceolate, acuminate or tapering into an awn). Lower glume 0 nerved. Upper glume 5–9 nerved. **Spikelets *with incomplete florets (the lower floret male, the upper female or rarely hermaphrodite)***. The incomplete florets proximal to the female-fertile florets. *The proximal incomplete florets* 1; paleate. Palea of the proximal incomplete florets fully developed. The proximal incomplete florets male. The proximal lemmas awnless, or awned (sometimes awnlike above); 7–13 nerved; decidedly exceeding the female-fertile lemmas (similar to the G_2 in shape); less firm than the female-fertile lemmas to similar in texture to the female-fertile lemmas; not becoming indurated.

Female-fertile florets 1. **Lemmas** similar in texture to the glumes to decidedly firmer than the glumes; smooth; becoming indurated (slightly), or not becoming indurated; white in fruit; entire; awnless; hairless; non-carinate; having the margins lying flat and exposed on the palea; *0 nerved*. *Palea* present (oblong); relatively long; 'obscurely nerved'. *Lodicules* present; 2; joined, or free; fleshy; glabrous; not toothed. *Stamens* 3, or 0 (usually rudimentary or missing). Anthers not penicillate. *Ovary* glabrous (shortly stipitate). Styles fused, or free to their bases (rarely).

Fruit, embryo and seedling. Fruit small; compressed dorsiventrally. Hilum short. Embryo large.

Abaxial leaf blade epidermis. *Costal/intercostal zonation* conspicuous. *Papillae* present; costal and intercostal. Intercostal papillae over-arching the stomata; consisting of one symmetrical projection per cell. *Long-cells* similar in shape costally and intercostally; of similar wall thickness costally and intercostally. Intercostal zones exhibiting many atypical long-cells to without typical long-cells (the long-cells mostly short-rectangular). Mid-intercostal long-cells rectangular; having markedly sinuous walls. *Microhairs* present; elongated; clearly two-celled; panicoid-type; (31.5–)33–39 microns long (*P. paradoxa*), or 22–24 microns long (*P. spinescens*); (6–)6.3–7.2(–8.4) microns wide at the septum (*P. paradoxa*), or 6.6–7.5 microns wide at the septum (*P. spinescens*). Microhair total length/width at septum 4.6–6.5 (*P. paradoxa*), or 3–3.6 (*P. spinescens*). Microhair apical cells (14.4–)15.6–18.6(–19.5) microns long (*P. paradoxa*), or 10.2–10.5 microns long (*P. spinescens*). Microhair apical cell/total length ratio 0.46–0.59 (*P. paradoxa*), or 0.4–0.47 (*P. spinescens*). *Stomata* common; 19–29 microns long. Subsidiaries low to high dome-shaped. Guard-cells overlapped by the interstomatals. *Intercostal short-cells* absent or very rare. *Costal short-cells* neither distinctly grouped into long rows nor predominantly paired. Costal silica bodies horizontally-elongated crenate/sinuous (a few, in *P. spinescens*), or 'panicoid-type' (predominating, mostly rather angular, with truncated 'points'); not sharp-pointed.

Transverse section of leaf blade, physiology. C_4; XyMS–. *Mesophyll* with radiate chlorenchyma. *Leaf blade* adaxially flat. *Midrib* conspicuous to not readily distinguishable; having a conventional arc of bundles; with colourless mesophyll adaxially to without colourless mesophyll adaxially. Bulliforms present in discrete, regular adaxial groups, or not in discrete, regular adaxial groups (the epidermis irregularly bulliform, e.g. *P. paradoxa*); in *P. spinescens* in simple fans (these large, *Zea*-type, over the minor bundles). Many of the smallest vascular bundles unaccompanied by sclerenchyma. Combined sclerenchyma girders present; nowhere forming 'figures'. Sclerenchyma all associated with vascular bundles.

Cytology. Chromosome base number, $x = 8$. $2n = 16$.

Taxonomy. Panicoideae; Panicodae; Paniceae.

Ecology, geography, regional floristic distribution. 6 species. India & Japan to Australia. Hydrophytic to helophytic; species of open habitats. Marshes.

Holarctic, Paleotropical, and Australian. Boreal. Indomalesian. Eastern Asian. Indian, Indo-Chinese, Malesian, and Papuan. North and East Australian. Tropical North and East Australian and Temperate and South-Eastern Australian.

Rusts and smuts. Smuts from Ustilaginaceae. Ustilaginaceae — *Sorosporium* and *Sphacelotheca*.

References, etc. Morphological/taxonomic: Vickery 1952; Webster 1987. Leaf anatomical: this project.

Pseudoroegneria (Nevski) A. Löve

~ *Elymus, Elytrigia*

Habit, vegetative morphology. Perennial; *caespitose*. Culms 30–90 cm high; *herbaceous*; unbranched above. Young shoots intravaginal. *Leaves* not basally aggregated; auriculate. Leaf blades linear; narrow; flat, or rolled (involute or convolute); without cross venation; persistent. Ligule an unfringed membrane; truncate. *Contra-ligule* absent.

Reproductive organization. Plants bisexual, with bisexual spikelets; with hermaphrodite florets; outbreeding.

Inflorescence. *Inflorescence a single spike (the spikelets appressed)*. Rachides hollowed (against the spikelet). Inflorescence espatheate; not comprising 'partial inflorescences' and foliar organs. Spikelet-bearing axes usually persistent; fragile or not in *P. stipifolia*, where the spikelet may fall with the internode above. *Spikelets* solitary; not secund; distichous; sessile to subsessile; *distant (at least, the spikes lax by contrast with* **Elymus***)*.

Female-fertile spikelets. *Spikelets* 18–22 mm long; compressed laterally; disarticulating above the glumes; *not disarticulating between the florets*. Rachilla prolonged beyond the uppermost female-fertile floret; rachilla hairless (glabrous). The rachilla extension with incomplete florets. *Hairy callus present (inconspicuous)*. *Callus* short; blunt.

Glumes two; very unequal to more or less equal; shorter than the spikelets; shorter than the adjacent lemmas to long relative to the adjacent lemmas; lateral to the rachis; hairless (sometimes with a few spinules apically); pointed to not pointed (lanceolate or narrowly lanceolate); not subulate; awnless (but sometimes acuminate); non-carinate; similar. Lower glume 5–7 nerved. Upper glume 5–7 nerved. *Spikelets* with incomplete florets. The incomplete florets distal to the female-fertile florets. *The distal incomplete florets* merely underdeveloped.

Female-fertile florets 3–5. *Lemmas* linear-lanceolate; *similar in texture to the glumes*; entire; pointed; awnless, or awned. Awns when present, 1; apical; non-geniculate; recurving (widely divergent at maturity); much shorter than the body of the lemma to much longer than the body of the lemma (up to 30 mm long); entered by several veins. Lemmas hairless; glabrous, or scabrous; non-carinate; without a germination flap; 5 nerved; with the nerves confluent towards the tip. *Palea* present; relatively long (about equalling the lemma); awnless, without apical setae; thinner than the lemma; not indurated; 2-nerved; 2-keeled. Palea keels wingless. *Lodicules* present; 2; free; membranous; ciliate; not toothed. *Stamens* 3. Anthers 4–7 mm long. *Ovary* hairy. Styles free to their bases. Stigmas 2; white.

Fruit, embryo and seedling. *Fruit* adhering to lemma and/or palea; medium sized; longitudinally grooved; compressed dorsiventrally; with hairs confined to a terminal tuft. Hilum long-linear. Embryo small.

Abaxial leaf blade epidermis. *Costal/intercostal zonation* conspicuous. *Papillae* absent. *Long-cells* similar in shape costally and intercostally; of similar wall thickness costally and intercostally (walls of medium thickness). Mid-intercostal long-cells rectangular; having markedly sinuous walls (and pitted). *Microhairs* absent. *Stomata* common; (36–)37–39(–42) microns long. Subsidiaries predominantly parallel-sided. Guard-cells overlapped by the interstomatals (mostly, slightly). *Intercostal short-cells* common (abundant); in cork/silica-cell pairs and not paired (commonly solitary). *Costal short-cells* predominantly paired. Costal silica bodies rounded (mostly), or saddle shaped to crescentic (few, intergrading with the rounded forms); not sharp-pointed.

Transverse section of leaf blade, physiology. C_3; XyMS+. *Mesophyll* with non-radiate chlorenchyma; without adaxial palisade. *Leaf blade* with distinct, prominent adaxial ribs; with the ribs very irregular in sizes (small, round topped alternating with large, flat topped). *Midrib* not readily distinguishable; with one bundle only. Bulliforms present in discrete, regular adaxial groups (in all the furrows); in simple fans. All the vascular bundles accompanied by sclerenchyma. Combined sclerenchyma girders present (with most of the major bundles); forming 'figures' (I's with all the major bundles). Sclerenchyma all associated with vascular bundles.

Cytology. Chromosome base number, $x = 7$. $2n = 14$ and 28. 2 and 4 ploid. Haplomic genome content S.

Taxonomy. Pooideae; Triticodae; Triticeae.

Ecology, geography, regional floristic distribution. About 16 species. Middle East, central Asia, northern China, western North America. Xerophytic; species of open habitats. Drought tolerant.

Holarctic. Boreal and Tethyan. Euro-Siberian, Eastern Asian, Atlantic North American, and Rocky Mountains. Irano-Turanian. European and Siberian. Canadian-Appalachian.

Rusts and smuts. Rusts — *Puccinia*.

References, etc. Morphological/taxonomic: Löve 1984. Leaf anatomical: this project.

Pseudosasa Makino

Yadakeya Mak.
~ *Arundinaria*

Habit, vegetative morphology. Shrubby perennial; rhizomatous. The flowering culms leafy. *Culms* 200–500 cm high; woody and persistent; to 1.5 cm in diameter; branched above. *Primary branches/mid-culm node 1 (or 2–3 at upper nodes).* Culm sheaths persistent. Culm internodes hollow. Rhizomes pachymorph and leptomorph (metamorph type I). Plants unarmed. **Leaves** not basally aggregated; *with auricular setae.* Leaf blades broad; 10–50 mm wide ('to 5 cm'); pseudopetiolate; cross veined; disarticulating from the sheaths; rolled in bud. *Contra-ligule* present.

Reproductive organization. Plants bisexual, with bisexual spikelets; with hermaphrodite florets.

Inflorescence. Inflorescence paniculate (terminal). Inflorescence with axes ending in spikelets. Inflorescence spatheate; a complex of 'partial inflorescences' and intervening foliar organs. *Spikelet-bearing axes* 'racemes' (seemingly); persistent. Spikelets not secund; pedicellate.

Female-fertile spikelets. Spikelets 15–45 mm long; compressed laterally to not noticeably compressed; disarticulating above the glumes (?); disarticulating between the florets. Rachilla prolonged beyond the uppermost female-fertile floret. Hairy callus absent.

Glumes two; very unequal; shorter than the adjacent lemmas; pointed; awnless; similar. Lower glume 5 nerved (in material seen). Upper glume 9 nerved (in material seen). *Spikelets* with female-fertile florets only, or with incomplete florets. The incomplete florets distal to the female-fertile florets (assuming the distal florets may be imperfect).

Female-fertile florets 2–7. *Lemmas* entire; pointed; awnless, or mucronate (?); non-carinate; *17 nerved (in material seen)*. *Palea* present; relatively long; apically notched; awnless, without apical setae; several nerved (8–13, 13 in material seen); 2-keeled. *Lodicules* present; 3; free; membranous; ciliate; not toothed; heavily vascularized. **Stamens** *3(–4)*. Anthers not penicillate; without an apically prolonged connective. **Ovary** glabrous; *without a conspicuous apical appendage (the apex not swollen)*. Styles fused. *Stigmas 3*.

Fruit, embryo and seedling. Pericarp thin (seemingly). Embryo small; with an epiblast; with a scutellar tail; with a negligible mesocotyl internode. Embryonic leaf margins overlapping.

Abaxial leaf blade epidermis. *Costal/intercostal zonation* conspicuous. *Papillae* present; costal and intercostal. Intercostal papillae over-arching the stomata (and obscuring them); several per cell (abundant, narrow-elongated). Mid-intercostal long-cells rectangular; having markedly sinuous walls. *Microhairs* present; panicoid-type; without 'partitioning membranes' (in *Pseudosasa japonica*); (42–)49–51(–60) microns long; (4.2–)4.5–5.4(–6) microns wide at the septum. Microhair total length/width at septum 7.8–14.3. Microhair apical cells (21–)22.5–25.5(–27) microns long. Microhair apical cell/total length ratio 0.43–0.51. *Stomata* common; 27–28.5–30 microns long. *Intercostal short-cells* common; in cork/silica-cell pairs; silicified. Intercostal silica bodies crescentic, or saddle shaped, or tall-and-narrow (smallish). *Costal short-cells* conspicuously in long rows. Costal silica bodies saddle shaped (predominantly, very large), or oryzoid (a few approaching this); not sharp-pointed.

Transverse section of leaf blade, physiology. C_3; XyMS+. *Mesophyll* with non-radiate chlorenchyma; with adaxial palisade (in places); with arm cells; with fusoids. The fusoids external to the PBS. *Leaf blade* adaxially flat. *Midrib* conspicuous (and with the usual bamboo asymmetry); having complex vascularization. Bulliforms present in discrete, regular adaxial groups (these large); in simple fans. All the vascular bundles accompanied by sclerenchyma. Combined sclerenchyma girders present (with most bundles — a few with strands only); forming 'figures' (sometimes). Sclerenchyma all associated with vascular bundles.

Cytology. Chromosome base number, $x = 12$. $2n = 48$. 4 ploid.

Taxonomy. Bambusoideae; Bambusodae; Bambuseae.

Ecology, geography, regional floristic distribution. 8 species. Eastern Asia. Holarctic and Paleotropical. Boreal. Indomalesian. Eastern Asian. Indo-Chinese.

Rusts and smuts. Rusts — *Puccinia*. Taxonomically wide-ranging species: *Puccinia longicornis* and *Puccinia kusanoi*.

References, etc. Leaf anatomical: this project.

Pseudosorghum A. Camus

Habit, vegetative morphology. Erect or decumbent *annual*. *Culms* 40–100 cm high; herbaceous; branched above, or unbranched above. Culm internodes solid. *Leaves* not basally aggregated; non-auriculate. Leaf blades linear; narrow; 5–10 mm wide; without cross venation.

Reproductive organization. Plants bisexual, with bisexual spikelets; with hermaphrodite florets. The spikelets of sexually distinct forms on the same plant (heterogamous); hermaphrodite and sterile, or hermaphrodite and sterile; overtly heteromorphic; in both homogamous and heterogamous combinations (the lowest pair of some or all racemes sterile).

Inflorescence. *Inflorescence paniculate (all-sidedly branched, of few-to-many simple or branched spikelike racemes)*; contracted (relatively dense); espatheate; not comprising 'partial inflorescences' and foliar organs. *Spikelet-bearing axes* 'racemes' (spiciform, peduncled); the spikelet-bearing axes with 4–5 spikelet-bearing 'articles' to with more than 10 spikelet-bearing 'articles' (few-to-many-jointed — the racemes fairly long, by contrast with *Sorghum*); with very slender rachides; disarticulating; disarticulating at the joints. *'Articles'* linear; not appendaged; disarticulating transversely; densely long-hairy to somewhat hairy. *Spikelets* in triplets (a terminal triad), or paired (the rest); secund (all the pedicelled spikelets one one side); sessile and pedicellate; *consistently in 'long-and-short' combinations*; in pedicellate/sessile combinations. Pedicels of the 'pedicellate' spikelets free of the rachis. The 'shorter' spikelets hermaphrodite (except in lowest pair, where they may be male or sterile). The 'longer' spikelets male-only, or sterile.

Female-sterile spikelets. The basal barren spikelets lacking a palea.

Female-fertile spikelets. Spikelets 4–5.25 mm long; compressed dorsiventrally; falling with the glumes (with adjacent joint and pedicel). Rachilla terminated by a female-fertile floret. Hairy callus present (minute). *Callus* short; blunt.

Glumes two; more or less equal; long relative to the adjacent lemmas; glumes smooth, hairy or not; awnless; carinate (G_2), or non-carinate (G_1); very dissimilar (papery, the G_1 truncate and 2-keeled, the G_2 cymbiform and acute). *Lower glume two-keeled*; not pitted; relatively smooth; 9–13 nerved. *Upper glume 7 nerved. Spikelets* with incomplete florets. The incomplete florets proximal to the female-fertile florets. *The proximal incomplete florets* 1; epaleate; sterile. The proximal lemmas awnless; 2 nerved; similar in texture to the female-fertile lemmas (thinly membranous); not becoming indurated.

Female-fertile florets 1. **Lemmas** less firm than the glumes; incised; 2 lobed; deeply cleft (to about halfway); *awned*. Awns 1; median; from a sinus; geniculate; hairless (ciliolate along margins); much longer than the body of the lemma. Lemmas hairy (long-ciliate); non-carinate; without a germination flap. **Palea** present (but absent in the barren basal spikelet); *relatively long*; awnless, without apical setae; not indurated. *Lodicules* present; 2; free; fleshy; glabrous. *Stamens* 3. *Ovary* glabrous. Styles free to their bases. Stigmas 2.

Cytology. $2n = 20$.

Taxonomy. Panicoideae; Andropogonodae; Andropogoneae; Andropogoninae.

Ecology, geography, regional floristic distribution. 2 species. Indo-Malaysian. Mesophytic; shade species and species of open habitats.

Paleotropical. Indomalesian. Indian, Indo-Chinese, and Malesian.

References, etc. Morphological/taxonomic: Camus 1920.

Special comments. Fruit data wanting. Anatomical data wanting.

Pseudostachyum Munro

~ *Schizostachyum*

Habit, vegetative morphology. Perennial; rhizomatous. The flowering culms leafy. *Culms* 500–1600 cm high; woody and persistent; to 2.5 cm in diameter; scandent (above), or not scandent; branched above (near their tops). Culm sheaths truncate-triangular. Culm internodes hollow (the walls thin). Rhizomes culms arising singly from long creeping rhizome. Plants unarmed. *Leaves* not basally aggregated; with auricular setae (but these deciduous). Leaf blades lanceolate to elliptic (ending in a long, twisted, scabrous point); broad; 20–45 mm wide (8–30 cm long); not cordate, not sagittate (but asymmetric at the base); pseudopetiolate (the 'petiole' rather long, thick); cross veined; disarticulating from the sheaths; rolled in bud. *Ligule* present; short.

Reproductive organization. Plants bisexual, with bisexual spikelets; with hermaphrodite florets.

Inflorescence. *Inflorescence* indeterminate; *with pseudospikelets*; paniculate (large, with drooping branches); non-digitate; spatheate (the panicle leafy, the units bracteate); a complex of 'partial inflorescences' and intervening foliar organs. *Spikelet-bearing axes* 'racemes', or spikelike, or paniculate; persistent. Spikelets not secund; seemingly sessile and pedicellate.

Female-fertile spikelets. Spikelets 5 mm long; disarticulating above the glumes. Rachilla prolonged beyond the uppermost female-fertile floret. The rachilla extension with incomplete florets. Hairy callus absent.

Glumes one per spikelet; shorter than the adjacent lemmas; hairless; mucronate; awnless. Upper glume 7 nerved. *Spikelets* with incomplete florets. The incomplete florets distal to the female-fertile florets. *The distal incomplete florets* merely underdeveloped (the terminal rachilla bearing glumes or an imperfect floret).

Female-fertile florets 1. Lemmas similar in texture to the glumes (and similar in form); not becoming indurated; mucronate; usually finely ciliate above. **Palea *present***; relatively long; *convolute*; not indurated (thin); 2-keeled (the keels ciliate). *Lodicules* present (large, persistent); 3–5; free; membranous; ciliate; not toothed (acute, rounded or truncate); heavily vascularized. *Stamens* 6. Anthers with the connective apically prolonged (apiculate). *Ovary* glabrous; with a conspicuous apical appendage. The appendage long, stiff and tapering. Styles fused (into one long, rigid style). Stigmas 2 (short).

Fruit, embryo and seedling. *Fruit* fruit supported by persistent glume, lemma, palea and lodicules. Pericarp thick and hard (crustaceous); free.

Abaxial leaf blade epidermis. *Costal/intercostal zonation* conspicuous. *Papillae* present; intercostal. Intercostal papillae over-arching the stomata (and small — not seen elsewhere). Mid-intercostal long-cells having markedly sinuous walls (thin). *Microhairs* present; panicoid-type. *Stomata* common (obscured by papillae). Subsidiaries probably low dome-shaped. *Intercostal short-cells* common; in cork/silica-cell pairs; silicified. Intercostal silica bodies saddle shaped (narrower than the costals), or tall-and-narrow. *Costal short-cells* predominantly paired (plus a few short rows). Costal silica bodies saddle shaped; not sharp-pointed.

Transverse section of leaf blade, physiology. C_3; XyMS+. Mesophyll with arm cells; with fusoids (large). *Leaf blade* adaxially flat (except near midrib). *Midrib* conspicuous; having complex vascularization. Bulliforms present in discrete, regular adaxial groups; in simple fans. All the vascular bundles accompanied by sclerenchyma. Combined sclerenchyma girders present (with all the bundles); forming 'figures' (in many bundles).

Taxonomy. Bambusoideae; Bambusodae; Bambuseae.

Ecology, geography, regional floristic distribution. 1 species. Eastern Himalaya, Assam and Upper Burma. Shade species. Under large trees.

Paleotropical. Indomalesian. Indo-Chinese.

References, etc. Leaf anatomical: Metcalfe 1960.

Special comments. Fruit data wanting.

Pseudovossia A. Camus

~ Phacelurus, = Phacelurus (Vossia) cambogiensis

Habit, vegetative morphology. Culms herbaceous; unbranched above. Leaf blades without cross venation.

Reproductive organization. *Plants bisexual, with bisexual spikelets*; with hermaphrodite florets. *The spikelets of sexually distinct forms on the same plant*; hermaphrodite and male-only; overtly heteromorphic (the pedicelled spikelet laterally flattened); all in heterogamous combinations.

Inflorescence. *Inflorescence digitate*; espatheate; not comprising 'partial inflorescences' and foliar organs. **Spikelet-bearing axes** 'racemes'; *solitary*; *with substantial rachides (the joints relatively long by contrast with* **Vossia***)*; disarticulating; disarticulating at the joints. *'Articles' not appendaged*; disarticulating transversely. Spikelets paired (the pairs relatively remote, by contrast with *Vossia*); sessile and pedicellate; consistently in 'long-and-short' combinations; in pedicellate/sessile combinations (the pedicels long and articulated in the middle). Pedicels of the 'pedicellate' spikelets free of the rachis. The 'shorter' spikelets hermaphrodite (and not embedded in the rachis). The 'longer' spikelets male-only.

Female-sterile spikelets. *The pedicellate spikelets with a long callus, by contrast with* **Phacelurus.** *Male, laterally compressed, with a dorsally keeled G₁.*

Female-fertile spikelets. Spikelets compressed dorsiventrally; falling with the glumes. Rachilla terminated by a female-fertile floret. Hairy callus absent.

Glumes two; long relative to the adjacent lemmas; hairless; pointed; awnless; non-carinate; very dissimilar (the G_1 leathery, 2-keeled, the G_2 thinner, naviculate with a recurved apex). Lower glume two-keeled; flattened on the back; not pitted; relatively smooth. *Spikelets* with incomplete florets (hyaline). The incomplete florets proximal to the female-fertile florets. *Spikelets with proximal incomplete florets.* **The proximal incomplete florets** 1; paleate; *male*. The proximal lemmas awnless ('sub-aristate'); similar in texture to the female-fertile lemmas (hyaline).

Female-fertile florets 1. **Lemmas** less firm than the glumes (hyaline); entire; pointed, or blunt; *awnless*; non-carinate; without a germination flap; faintly 3 nerved. *Palea* present; awnless, without apical setae; 2-nerved. *Stamens* 3. *Ovary* glabrous. Stigmas 2.

Taxonomy. Panicoideae; Andropogonodae; Andropogoneae; Rottboelliinae.

Ecology, geography, regional floristic distribution. 1 species. Indochina. Paleotropical. Indomalesian. Indo-Chinese.

Special comments. Fruit data wanting. Anatomical data wanting.

Pseudoxytenanthera Soderstrom and Ellis

~ *Dendrocalamus*, *Oxytenanthera* (*Dendrocalamus monadelphus* Thwaites, *Oxytenanthera thwaitesii* Munro, *Oxytenanthera monadelpha* (Thwaites) Alston)

Habit, vegetative morphology. Perennial. The flowering culms leafy. **Culms** 400–800 cm high; *woody and persistent (stout and soft below, bending over and becoming thin and whiplike towards their tips)*; *to 1.5 cm in diameter*; 'vinelike'; branched above (the large central bud remains dormant while the laterals produce numerous basal branches, then subsequently develops into a long, whiplike branch producing clusters of branches at each node). Culm sheaths deciduous in their entirety. Pluricaespitose. Rhizomes pachymorph. Plants unarmed. Young shoots extravaginal. *Leaves* not basally aggregated; with auricular setae. Leaf blades linear-lanceolate (acuminate); broad; 20–30 mm wide (and 12–20 cm long); flat; pseudopetiolate; without cross venation; disarticulating from the sheaths; rolled in bud. Ligule in the form of a hard ridge with a curved, denticulate margin; truncate; 1.3–1.7 mm long. *Contra-ligule* present (in the form of a ciliolate ridge).

Reproductive organization. Plants bisexual, with bisexual spikelets; with hermaphrodite florets.

Inflorescence. Inflorescence *indeterminate (the spikelets each with two gemmiferous bracts beneath the glumes)*; *with pseudospikelets*; a spicate, leafless branch with sessile clusters of pseudospikelets; spatheate. *Spikelet-bearing axes* capitate (the clusters spaced along the main axis). *Spikelets* (i.e. the pseudospikelets) not secund; *sessile*.

Female-fertile spikelets. Spikelets 13–16 mm long; compressed laterally to not noticeably compressed (?); falling with the glumes; not disarticulating between the florets (the spikelet falling entire). Rachilla prolonged beyond the uppermost female-fertile floret.

Glumes two; very unequal (the upper longer); shorter than the spikelets; shorter than the adjacent lemmas; hairless; glabrous (but ciliate along the middle and upper margins); pointed; awnless; similar (obovate-triangular, becoming thickened towards the pointed tips). Lower glume 10–11 nerved. Upper glume 10–11 nerved. *Spikelets* with incomplete florets. The incomplete florets distal to the female-fertile florets. *The distal incomplete florets* merely underdeveloped; awnless.

Female-fertile florets 1–2 (i.e. 1–3 florets, with acropetal reduction). Lemmas obovate-triangular, thickened above; similar in texture to the glumes; entire; pointed; awnless to mucronate; hairless; glabrous; without a germination flap; 13–16 nerved (with transverse veinlets). *Palea* present; relatively long (a little shorter than the lemma); entire; awnless, without apical setae; thinner than the lemma; not indurated; several nerved (7–3 between the keels and one on each side); 2-keeled. Palea keels wingless; hairy (ciliate). *Lodicules* absent. **Stamens** 6; *monadelphous*. Anthers about 3 mm long (orange); penicillate; with the connective apically prolonged. *Ovary* glabrous (but the style pilose); with a conspicuous apical appendage (and a hollow style). The appendage long, stiff and tapering. Styles fused. Stigmas 2, or 3.

Fruit, embryo and seedling. Fruit medium sized (about 5 mm long); fusiform; not noticeably compressed (plano-convex). Hilum long-linear (dark, almost the length of the fruit). Pericarp thick and hard (above, but thinner and easily separable from the seed below); loosely adherent (below), or fused (above). Embryo small.

Abaxial leaf blade epidermis. *Costal/intercostal zonation* conspicuous. *Papillae* present; costal and intercostal. Intercostal papillae over-arching the stomata (and obscuring them); several per cell (with one to two rows of 8–12 per long-cell). Mid-intercostal long-cells rectangular; having markedly sinuous walls. *Microhairs* present; elongated; clearly two-celled; panicoid-type. *Stomata* common. *Intercostal short-cells* common; in cork/silica-cell pairs; silicified. Intercostal silica bodies tall-and-narrow. *Costal short-cells* conspicuously in long rows. Costal silica bodies predominantly saddle shaped (a large version of this form); not sharp-pointed.

Transverse section of leaf blade, physiology. C_3; XyMS+. Mesophyll with adaxial palisade (of arm-cells); with arm cells; with fusoids. The fusoids external to the PBS. *Leaf blade* adaxially flat (except near the midrib). *Midrib* conspicuous; having complex vascularization. *The lamina* distinctly asymmetrical on either side of the midrib. Bulliforms present in discrete, regular adaxial groups; in simple fans. All the vascular bundles accompanied by sclerenchyma. Combined sclerenchyma girders present; forming 'figures'. Sclerenchyma all associated with vascular bundles.

Taxonomy. Bambusoideae; Bambusodae; Bambuseae.

Ecology, geography, regional floristic distribution. 1 species. Southern India and Sri Lanka.

Paleotropical. Indomalesian. Indian.

References, etc. Morphological/taxonomic: Soderstrom and Ellis 1988. Leaf anatomical: Soderstrom and Ellis 1988.

Pseudozoysia Chiov.

Habit, vegetative morphology. Perennial. Culms herbaceous. Leaf blades narrow; setaceous; rolled (involute); without cross venation.

Reproductive organization. Plants bisexual, with bisexual spikelets; with hermaphrodite florets. *The spikelets all alike in sexuality*.

Inflorescence. Inflorescence a false spike, with spikelets on contracted axes (cylindrical, of short 'racemelets', basally enclosed in the upper leaf sheatrh); espatheate; not comprising 'partial inflorescences' and foliar organs. **Spikelet-bearing axes** very much reduced (to two contiguous spikelets); *disarticulating (?)*; falling entire (the racemelets falling, cf. *Tragus* etc.?). Spikelets paired; not secund; consistently in 'long-and-short' combinations. The 'shorter' spikelets hermaphrodite. The 'longer' spikelets hermaphrodite.

Female-fertile spikelets. *Spikelets not noticeably compressed (?)*; falling with the glumes (with the racemelet?). Rachilla terminated by a female-fertile floret. Hairy callus absent.

Glumes two; relatively large; exceeding the spikelets; long relative to the adjacent lemmas; hairless; not pointed; awnless; non-carinate; *very dissimilar (indurated, tuberculate, acute, the lower ovate, the upper subglobose)*. Lower glume and the upper tuberculate. **Upper glume** *distinctly saccate (enclosing the floret)*. *Spikelets* with female-fertile florets only.

Female-fertile florets 1. Lemmas less firm than the glumes (scarious); not becoming indurated; entire; blunt; awnless; hairless; non-carinate; without a germination flap. *Palea* present; conspicuous but relatively short (about half as long as the lemma); not indurated.

Abaxial leaf blade epidermis. *Costal/intercostal zonation* conspicuous. *Microhairs* present; chloridoid-type. *Stomata* common. Subsidiaries dome-shaped and triangular. *Intercostal short-cells* common (fairly); not paired (mainly solitary). *Costal short-cells* conspicuously in long rows. Costal silica bodies 'panicoid-type'; dumb-bell shaped; not sharp-pointed.

Transverse section of leaf blade, physiology. C_4; XyMS+. Bulliforms present in discrete, regular adaxial groups; probably associated with colourless mesophyll cells to form deeply-penetrating fans (?). All the vascular bundles accompanied by sclerenchyma. Combined sclerenchyma girders present (with all the bundles); forming 'figures'. Sclerenchyma all associated with vascular bundles.

Taxonomy. Chloridoideae; main chloridoid assemblage.

Ecology, geography, regional floristic distribution. 1 species. Somalia. Xerophytic; species of open habitats; halophytic. In coastal sand dunes.

Paleotropical. African. Sudano-Angolan. Somalo-Ethiopian.

Special comments. Fruit data wanting. Anatomical data wanting.

Psilathera Link

~ *Sesleria*

Habit, vegetative morphology. Perennial. *Culms* 3–10 cm high; herbaceous. Leaves non-auriculate. Leaf blades narrow; 1–1.5 mm wide; setaceous, or not setaceous; without cross venation. Ligule an unfringed membrane; not truncate; 1–2 mm long.

Reproductive organization. *Plants bisexual, with bisexual spikelets*; with hermaphrodite florets. *The spikelets of sexually distinct forms on the same plant*; hermaphrodite and sterile; *overtly heteromorphic (the sterile spikelets represented by ovate to lanceolate bracts at the base of the inflorescence).*

Inflorescence. *Inflorescence paniculate*; contracted; capitate, or more or less ovoid (6–10 x 4–6 mm); espatheate (but with bracts at base of inflorescence); not comprising 'partial inflorescences' and foliar organs. Spikelets not secund.

Female-sterile spikelets. The sterile spikelets vestigial, represented by bracts at the base of the inflorescence.

Female-fertile spikelets. *Spikelets 2 mm long*; *not noticeably compressed*; disarticulating above the glumes; disarticulating between the florets. Rachilla prolonged beyond the uppermost female-fertile floret. Hairy callus present.

Glumes two; more or less equal; long relative to the adjacent lemmas; awned (the awns to 2 mm long, straight or hooked), or awnless (but mucronate); carinate; similar (ovate, abruptly contracted above). Lower glume 1 nerved. Upper glume 1 nerved.

Female-fertile florets 1–2. Lemmas decidedly firmer than the glumes; not becoming indurated; incised; 5 lobed (toothed); not deeply cleft; awned. Awns 5; median and lateral (via the excurrent veins); the median similar in form to the laterals; non-geniculate; about as long as the body of the lemma to much longer than the body of the lemma. The lateral awns shorter than the median (the median to 1.5 mm long, the laterals about 1 mm). Lemmas non-carinate (dark blue, broadly ovate); without a germination flap; 5 nerved. *Palea* present; relatively long (exceeding the lemma); awned (with two 1–1.5 mm awns); 2-nerved; 2-keeled. *Lodicules* present; 2; free; membranous; glabrous; not toothed; not or scarcely vascularized. *Stamens* 3. **Ovary hairy.** Stigmas 2.

Fruit, embryo and seedling. Fruit small (2 mm long); with hairs confined to a terminal tuft. Hilum short. Embryo small.

Abaxial leaf blade epidermis. *Costal/intercostal zonation* fairly conspicuous. *Papillae* absent. *Long-cells* fairly markedly different in shape costally and intercostally (the costals narrower and more consistently rectangular); of similar wall thickness costally and intercostally (fairly thick walled). Mid-intercostal long-cells fusiform, or rectangular and fusiform (in places); having straight or only gently undulating walls (the walls heavily pitted). *Microhairs* absent. *Stomata* absent or very rare (none seen). *Inter-*

costal short-cells absent or very rare. With a few costal prickles. *Costal short-cells* neither distinctly grouped into long rows nor predominantly paired (solitary and a few short rows). Costal silica bodies present and well developed (but scarce); horizontally-elongated crenate/sinuous and horizontally-elongated smooth.

Transverse section of leaf blade, physiology. C_3; XyMS+. *Leaf blade* with distinct, prominent adaxial ribs; with the ribs more or less constant in size (round topped, one ber bundle). *Midrib* conspicuous (by its position on the fold of the blade, and the pointed keel); with one bundle only. *The lamina* symmetrical on either side of the midrib. Bulliforms present in discrete, regular adaxial groups, or not in discrete, regular adaxial groups (? — not clear in the material seen); if present, in simple fans. Many of the smallest vascular bundles unaccompanied by sclerenchyma (the sclerenchyma confined to a small abaxial strand in the keel, and marginal fibre groups). Combined sclerenchyma girders absent. Sclerenchyma all associated with vascular bundles.

Taxonomy. Pooideae; Poodae; Seslerieae.

Ecology, geography, regional floristic distribution. 1 species. Southern & central Europe.

Holarctic. Boreal and Tethyan. Euro-Siberian. Mediterranean. European.

References, etc. Leaf anatomical: this project.

Psilolemma Phillips

Habit, vegetative morphology. Perennial; stoloniferous (a wiry mat-grass). **Culms** 7–35 cm high; herbaceous; *tuberous*. *Leaves* not basally aggregated; conspicuously distichous; non-auriculate. Leaf blades linear (stiff, pungent); narrow; to 2 mm wide; rolled (convolute); without abaxial multicellular glands; without cross venation. Ligule a fringed membrane to a fringe of hairs.

Reproductive organization. *Plants bisexual, with bisexual spikelets*; with hermaphrodite florets.

Inflorescence. Inflorescence of spicate main branches (with few-spiculate racemes appressed to the central axis, the upper racemes often reduced to single spikelets), or paniculate (the lower racemes sometimes branched); contracted (spiciform); spicate; nondigitate; espatheate; not comprising 'partial inflorescences' and foliar organs. Spikelet-bearing axes persistent. Spikelets pedicellate (the pedicels to 3 mm long).

Female-fertile spikelets. Spikelets 8–15 mm long; compressed laterally (slightly), or not noticeably compressed; disarticulating above the glumes; disarticulating between the florets. Rachilla prolonged beyond the uppermost female-fertile floret.

Glumes two; very unequal; shorter than the spikelets; shorter than the adjacent lemmas; awnless; *non-carinate*; similar. Lower glume 1 nerved. Upper glume 1 nerved.

Female-fertile florets 4–14. Lemmas similar in texture to the glumes to decidedly firmer than the glumes (membranous); not becoming indurated; entire; blunt; awnless; hairless; glabrous; non-carinate (abaxially rounded); 3 nerved. *Palea* present; relatively long (glabrous); awnless, without apical setae; not indurated; 2-nerved; 2-keeled. Palea keels glabrous (smooth). Stigmas 2.

Fruit, embryo and seedling. Fruit small (1.1–1.2 mm long); ellipsoid; not noticeably compressed. Hilum short. *Pericarp loosely adherent (removable when soaked)*. Embryo large.

Abaxial leaf blade epidermis. *Costal/intercostal zonation* conspicuous. *Papillae* absent. *Long-cells* similar in shape costally and intercostally (or the costals rather narrower); differing markedly in wall thickness costally and intercostally (the costals thinner walled, and less conspicuously pitted). Mid-intercostal long-cells rectangular; having markedly sinuous walls. *Microhairs* present; more or less spherical to elongated; clearly two-celled; chloridoid-type. Microhair apical cell wall of similar thickness/rigidity to that of the basal cell. Microhairs 36.6–40.5 microns long. Microhair basal cells 27 microns long. Microhairs 18–21 microns wide at the septum. Microhair total length/width at septum 1.8–2. Microhair apical cells 12–16.5 microns long. Microhair apical cell/total length ratio 0.31–0.41. *Stomata* common; 22.5–33 microns long. Guard-cells overlapped by the interstomatals (conspicuously so, the entire stomatal apparatus sunken and the seemingly dome-shaped subsidiaries obscured). *Intercostal short-cells* common;

in cork/silica-cell pairs and not paired (some solitary); silicified. Intercostal silica bodies imperfectly developed; rounded and oryzoid-type. *Costal short-cells* predominantly paired. Costal silica bodies present throughout the costal zones; rounded to saddle shaped (intergrading); not sharp-pointed.

Transverse section of leaf blade, physiology. C_4; XyMS+. PCR sheath outlines even. PCR sheaths of the primary vascular bundles interrupted; interrupted abaxially only. PCR sheath extensions absent. *Mesophyll* with radiate chlorenchyma. *Leaf blade* with distinct, prominent adaxial ribs; with the ribs more or less constant in size (round topped). Midrib with one bundle only. *The lamina* symmetrical on either side of the midrib. Bulliforms present in discrete, regular adaxial groups; in simple fans (these large, in at least some of the furrows). All the vascular bundles accompanied by sclerenchyma. Combined sclerenchyma girders absent (all the bundles with large fibre groups, with girders or strands abaxially but seemingly only strands adaxially). Sclerenchyma all associated with vascular bundles. The lamina margins with fibres.

Taxonomy. Chloridoideae; main chloridoid assemblage.

Ecology, geography, regional floristic distribution. 1 species. East Africa. Species of open habitats; halophytic (coloniser of soda mud-flats).

Paleotropical. African. Sudano-Angolan. Somalo-Ethiopian and South Tropical African.

References, etc. Morphological/taxonomic: Phillips 1973. Leaf anatomical: this project.

Psilurus Trin.

Asprella Host

Habit, vegetative morphology. Slender annual; caespitose (or single culms). *Culms* 5–45 cm high; herbaceous. Culm nodes glabrous. Culm internodes solid, or hollow. Leaves non-auriculate. Sheaths not keeled. Leaf blades linear (filiform); narrow; about 0.2–1 mm wide; setaceous; rolled (convolute); without cross venation; persistent. **Ligule** present; *an unfringed membrane*; truncate; 0.2–0.3 mm long.

Reproductive organization. Plants bisexual, with bisexual spikelets; with hermaphrodite florets. The spikelets of sexually distinct forms on the same plant, or all alike in sexuality; hermaphrodite, or hermaphrodite and sterile (the basal spikelets sometimes reduced).

Inflorescence. *Inflorescence a single spike (thin, wiry, ultimately disarticulating, the spikelets initially sunken, ultimately spreading).* Rachides hollowed. Inflorescence espatheate; not comprising 'partial inflorescences' and foliar organs. Spikelet-bearing axes tardily disarticulating; disarticulating at the joints. Spikelets solitary; not secund; distichous; sessile; distant.

Female-fertile spikelets. *Spikelets* 3.5–6.8 mm long; *adaxial (the single ('upper') glume on the outside)*; compressed laterally; disarticulating above the glumes (the terminal spikelet), or falling with the glumes (the lower spikelets); with distinctly elongated rachilla internodes between the florets. Rachilla prolonged beyond the uppermost female-fertile floret. The rachilla extension with incomplete florets, or naked. Hairy callus absent. *Callus* short; blunt (glabrous).

Glumes one per spikelet (except for the terminal spikelet, with two); (the upper, only glume) *shorter than the adjacent lemmas (very small)*; dorsiventral to the rachis; pointed; awnless; non-carinate. Upper glume 1 nerved. *Spikelets* with female-fertile florets only, or with incomplete florets. The incomplete florets when present, distal to the female-fertile florets. *The distal incomplete florets* merely underdeveloped (rudimentary).

Female-fertile florets 1. Lemmas linear-lanceolate; membranous to leathery; not becoming indurated; entire, or incised; not deeply cleft; awned. Awns 1; median; from a sinus, or apical; non-geniculate; straight; hairless (scabrous); much shorter than the body of the lemma to much longer than the body of the lemma (2.3–5.5 mm long); entered by one vein. Awn bases flattened. Lemmas hairless; carinate; 3 nerved. *Palea* present; relatively long (as long as the lemma); tightly clasped by the lemma; entire; awnless, without apical setae; thinner than the lemma (membranous); not indurated; 2-nerved; 2-keeled.

Palea keels wingless. *Lodicules* present; 2; membranous; toothed. **Stamens** *1*. Anthers 0.4–2.2 mm long; not penicillate. *Ovary* glabrous. Styles free to their bases. Stigmas 2; white.

Fruit, embryo and seedling. *Fruit* slightly adhering to lemma and/or palea; small (about 4 mm long); linear, or ellipsoid; longitudinally grooved to not grooved; compressed laterally; scabrous. Hilum long-linear. Embryo small; not waisted. Endosperm hard; with lipid; containing compound starch grains. Embryo with an epiblast.

Seedling with a short mesocotyl; with a tight coleoptile. First seedling leaf with a well-developed lamina. The lamina narrow; erect; 3 veined.

Abaxial leaf blade epidermis. *Costal/intercostal zonation* conspicuous. *Papillae* absent. *Long-cells* similar in shape costally and intercostally; of similar wall thickness costally and intercostally. Mid-intercostal long-cells fusiform; having markedly sinuous walls. *Microhairs* absent. *Stomata* absent or very rare. *Intercostal short-cells* common; in cork/silica-cell pairs, or not paired (solitaries); silicified (when paired), or not silicified. *Costal short-cells* predominantly paired, or neither distinctly grouped into long rows nor predominantly paired. Costal silica bodies horizontally-elongated crenate/sinuous, horizontally-elongated smooth, rounded, and tall-and-narrow (and some rather square); not sharp-pointed.

Transverse section of leaf blade, physiology. C_3; XyMS+. *Mesophyll* with non-radiate chlorenchyma; without adaxial palisade. *Leaf blade* 'nodular' in section, or adaxially flat; with the ribs more or less constant in size. *Midrib* conspicuous; with one bundle only. Bulliforms poorly develo̧ ed, not in discrete, regular adaxial groups. Many of the smallest vascular bundles unaccompanied by sclerenchyma. Combined sclerenchyma girders absent. Sclerenchyma all associated with vascular bundles.

Cytology. Chromosome base number, $x = 7$. $2n = 14$ and 28. 2 and 4 ploid.

Taxonomy. Pooideae; Poodae; Poeae.

Ecology, geography, regional floristic distribution. 1 species. Mediterranean to Afghanistan. Commonly adventive. Species of open habitats. Dry places.

Holarctic. Boreal and Tethyan. Euro-Siberian. Mediterranean and Irano-Turanian. European.

Rusts and smuts. Rusts — *Puccinia*. Taxonomically wide-ranging species: *Puccinia graminis* and *Puccinia hordei*.

References, etc. Leaf anatomical: this project.

Pterochloris A. Camus

~ *Chloris*

Habit, vegetative morphology. Perennial; stoloniferous and decumbent. **Culms** about 10–40 cm high; *herbaceous*; unbranched above; 3 noded. Culm nodes exposed; glabrous. Culm leaves present. Upper culm leaf blades fully developed. Culm internodes solid. *Leaves* not basally aggregated; non-auriculate. The sheaths keeled. *Leaf blades* linear; narrow; 4–6 mm wide; folded; without abaxial multicellular glands; *pseudopetiolate*; without cross venation; disarticulating from the sheaths, or persistent (?). *Ligule a fringe of hairs*. *Contra-ligule* absent.

Reproductive organization. Plants bisexual, with bisexual spikelets; with hermaphrodite florets. The spikelets all alike in sexuality; hermaphrodite.

Inflorescence. Inflorescence a single spike, or of spicate main branches; digitate to non-digitate. Primary inflorescence branches 1–2(–4). Inflorescence espatheate; not comprising 'partial inflorescences' and foliar organs. Spikelet-bearing axes persistent. Spikelets solitary; secund (on one side of the rachis); biseriate; imbricate.

Female-fertile spikelets. *Spikelets* about *3 mm long*; detectably adaxial (but twisted to lie flatwise, cf. *Chloris*); compressed laterally; disarticulating above the glumes; not disarticulating between the florets. Rachilla prolonged beyond the uppermost female-fertile floret; rachilla hairless. The rachilla extension with incomplete florets. Hairy callus present (of short, white hairs). Callus pointed.

Glumes two; more or less equal; shorter than the spikelets; shorter than the adjacent lemmas; hairless; scabrous (along the midvein); pointed; awnless (but the lower conspicuously mucronate); carinate; very dissimilar (the lower much larger). Lower glume 1 nerved. Upper glume 1 nerved. *Spikelets* with incomplete florets. The incomplete florets distal to the female-fertile florets. The distal incomplete florets 1; *incomplete florets* red, male or sterile; incomplete florets awned.

Female-fertile florets 1. **Lemmas becoming broadly winged above, the wings membranous, bilobed, spreading out at maturity**; similar in texture to the glumes; incised; 2 lobed; not deeply cleft; awned. Awns 1; median; from a sinus to dorsal (from between the lobes, or just behind them); from near the top; non-geniculate (slender); flexuous; hairless; much longer than the body of the lemma; entered by one vein; persistent. Awn bases not twisted; not flattened. Lemmas hairless; glabrous, or scabrous (along the margins); carinate; without a germination flap; 3 nerved (below), or 5–7 nerved (above, the laterals branching into the wings); with the nerves non-confluent. *Palea* present; relatively long; entire; awnless, without apical setae; textured like the lemma; not indurated; 2-nerved; 2-keeled. Palea back glabrous. Palea keels winged; hairy. **Lodicules absent.** *Stamens* 3. Anthers about 1 mm long; not penicillate; without an apically prolonged connective. *Ovary* glabrous. Styles free to their bases; free. Style bases widely separated. Stigmas 2; red pigmented.

Fruit, embryo and seedling. Fruit ellipsoid; compressed dorsiventrally, or trigonous; glabrous; smooth. Hilum short. **Pericarp** thin; **loosely adherent**. Embryo large; not waisted. Endosperm hard.

Abaxial leaf blade epidermis. *Costal/intercostal zonation* conspicuous. *Papillae* absent. *Long-cells* similar in shape costally and intercostally; of similar wall thickness costally and intercostally. Intercostal zones with typical long-cells. Mid-intercostal long-cells rectangular; having markedly sinuous walls (coarsely so). *Microhairs* present; more or less spherical; clearly two-celled; chloridoid-type. Microhair apical cell wall of similar thickness/rigidity to that of the basal cell. Microhair basal cells 9 microns long. Microhair total length/width at septum about 1.4. Microhair apical cell/total length ratio about 0.6. *Stomata* common. Subsidiaries non-papillate; triangular. Guard-cells overlapping to flush with the interstomatals. *Intercostal short-cells* common; in cork/silica-cell pairs; silicified and not silicified. Intercostal silica bodies imperfectly developed; rounded, crescentic, tall-and-narrow, cross-shaped, and oryzoid-type. *Costal short-cells* conspicuously in long rows (but the 'short-cells' mostly quite long), or neither distinctly grouped into long rows nor predominantly paired (in places). Costal silica bodies present and well developed; present in alternate cell files of the costal zones; saddle shaped.

Transverse section of leaf blade, physiology. C_4; XyMS+. PCR sheath outlines even. PCR sheaths of the primary vascular bundles interrupted; interrupted both abaxially and adaxially. PCR sheath extensions absent. Mesophyll not *Isachne*-type. *Leaf blade* 'nodular' in section; with the ribs more or less constant in size. *Midrib* conspicuous; with one bundle only; with colourless mesophyll adaxially (some colourless tissue adjacent to a large median bulliform group). *The lamina* symmetrical on either side of the midrib. Bulliforms present in discrete, regular adaxial groups (in places), or not in discrete, regular adaxial groups (in places constituting most of the epidermis); commonly in simple fans. All the vascular bundles accompanied by sclerenchyma (but the smallest usually with only one or two fibres abaxially). Combined sclerenchyma girders present; forming 'figures' (the lateral primaries with I's). Sclerenchyma all associated with vascular bundles. The lamina margins with fibres.

Taxonomy. Chloridoideae; main chloridoid assemblage.

Ecology, geography, regional floristic distribution. 1 species. Madagascar. Paleotropical. Madagascan.

References, etc. Morphological/taxonomic: Camus 1957. Leaf anatomical: this project.

Puccinellia Parl.

Atropis (Trin.) Griseb.

Habit, vegetative morphology. Annual (a few), or perennial; rhizomatous, or stoloniferous, or caespitose, or decumbent. **Culms** 4–100 cm high; **herbaceous**. Culm nodes glabrous. Culm internodes hollow. Young shoots extravaginal, or intravaginal. *Leaves* not basally aggregated; non-auriculate. **Sheath margins** generally *free (rarely closed basally to almost one third their length)*. Leaf blades linear; narrow; 0.5–5 mm wide; setaceous, or not setaceous; flat, or folded, or rolled; without cross venation; persistent; rolled in bud. *Ligule an unfringed membrane*; truncate, or not truncate; 0.5–4.5 mm long.

Reproductive organization. Plants bisexual, with bisexual spikelets; with hermaphrodite florets; outbreeding and inbreeding. Exposed-cleistogamous, or chasmogamous.

Inflorescence. *Inflorescence paniculate (rarely reduced and racemelike)*; open (usually), or contracted; espatheate; not comprising 'partial inflorescences' and foliar organs. Spikelet-bearing axes persistent. Spikelets not secund; pedicellate.

Female-fertile spikelets. *Spikelets 2–13 mm long*; compressed laterally; disarticulating above the glumes; disarticulating between the florets. Rachilla prolonged beyond the uppermost female-fertile floret; rachilla hairless (glabrous). The rachilla extension with incomplete florets. *Hairy callus absent (but the lemma base often hairy).*

Glumes two; very unequal; shorter than the spikelets; shorter than the adjacent lemmas; pointed, or not pointed (oblong); awnless; *non-carinate*; similar. *Lower glume 1 nerved. Upper glume 3 nerved. Spikelets* with incomplete florets. The incomplete florets distal to the female-fertile florets.

Female-fertile florets *2–10. Lemmas similar in texture to the glumes (membranous, thinner apically)*; not becoming indurated; entire (or erose, often ciliolate); pointed, or blunt; awnless; hairy (often only at the base), or hairless; non-carinate (or rarely carinate); 5 nerved. *Palea* present; relatively long; apically notched (emarginate); awnless, without apical setae; 2-nerved; 2-keeled. Palea keels scabrous (or scaberulous), or hairy. *Lodicules* present; 2; free; membranous; glabrous; toothed. *Stamens* 3. Anthers 0.3–2.5 mm long; not penicillate. *Ovary* glabrous. Styles free to their bases. Stigmas 2.

Fruit, embryo and seedling. *Fruit free from both lemma and palea*; small; longitudinally grooved, or not grooved; compressed dorsiventrally. *Hilum short (round to oval).* Embryo small; not waisted. Endosperm hard; without lipid; containing compound starch grains. Embryo with an epiblast; without a scutellar tail; with a negligible mesocotyl internode. Embryonic leaf margins meeting.

Seedling with a loose coleoptile, or with a tight coleoptile. First seedling leaf with a well-developed lamina. The lamina narrow; 1–3 veined.

Abaxial leaf blade epidermis. *Costal/intercostal zonation* conspicuous. *Papillae* present. Intercostal papillae over-arching the stomata; consisting of one oblique swelling per cell. *Long-cells* similar in shape costally and intercostally (but the *costals* tending to be fusiform); of similar wall thickness costally and intercostally (thick walled). Mid-intercostal long-cells rectangular; having markedly sinuous walls. *Microhairs* absent. *Stomata* common; in *P. stricta* 24–27 microns long. Subsidiaries parallel-sided, or dome-shaped (slightly). Guard-cells overlapped by the interstomatals (in some species the stomata are sunken, with access to the exterior only via a triangular pore). *Intercostal short-cells* common; in cork/silica-cell pairs, or not paired (some solitary, some triplets); silicified, or not silicified. Intercostal silica bodies rounded (elliptical), or crescentic. *Costal short-cells* predominantly paired, or neither distinctly grouped into long rows nor predominantly paired. Costal silica bodies rounded (predominating), or crescentic; not sharp-pointed.

Transverse section of leaf blade, physiology. C_3; XyMS+. *Mesophyll* with non-radiate chlorenchyma; without adaxial palisade. Midrib with one bundle only. Bulliforms present in discrete, regular adaxial groups, or not in discrete, regular adaxial groups (sometimes of small cells, irregular in size, at the bases of furrows — cf. *Ammophila*); when regularly grouped, in simple fans. Many of the smallest vascular bundles unaccompanied by sclerenchyma. Combined sclerenchyma girders present. Sclerenchyma all associated with vascular bundles.

Culm anatomy. *Culm internode bundles* in one or two rings.

Cytology. Chromosome base number, $x = 7$. $2n = 14, 28, 35, 42, 49, 56, 70$, and 77. 2–11 ploid.

Taxonomy. Pooideae; Poodae; Poeae.

Ecology, geography, regional floristic distribution. About 80 species. North temperate. Commonly adventive. Helophytic, or mesophytic; usually halophytic.

Holarctic, Paleotropical, Neotropical, Cape, Australian, and Antarctic. Boreal, Tethyan, and Madrean. African. Arctic and Subarctic, Euro-Siberian, Eastern Asian, Atlantic North American, and Rocky Mountains. Macaronesian, Mediterranean, and Irano-Turanian. Sudano-Angolan and Namib-Karoo. Pampas and Andean. North and East Australian. New Zealand and Patagonian. European. Canadian-Appalachian. South Tropical African. Temperate and South-Eastern Australian.

Hybrids. Intergeneric hybrids with *Phippsia* — ×*Pucciphippsia* Tsvelev.

Rusts and smuts. Rusts — *Puccinia*. Taxonomically wide-ranging species: *Puccinia graminis, Puccinia coronata, Puccinia striiformis*, and '*Uromyces*' *dactylidis*. Smuts from Tilletiaceae and from Ustilaginaceae. Tilletiaceae — *Entyloma* and *Urocystis*. Ustilaginaceae — *Ustilago*.

Economic importance. Cultivated fodder: *P. airoides*. Important native pasture species: *P. airoides*.

References, etc. Leaf anatomical: Metcalfe 1960; this project.

Puelia Franch.

Atractocarpa Franch.

Habit, vegetative morphology. *Perennial (sometimes with root tubers)*; rhizomatous. The flowering culms leafless, or leafy (i.e. sometimes with separate fertile and vegetative culms). Culms herbaceous; unbranched above. Culm leaves present. Upper culm leaf blades fully developed. Culm internodes hollow. Rhizomes pachymorph. Plants unarmed. *Leaves* not basally aggregated; non-auriculate; without auricular setae. Leaf blades ovate-lanceolate; broad; 20–70 mm wide; flat; pseudopetiolate; cross veined; persistent. *Ligule a fringed membrane*. *Contra-ligule* present.

Reproductive organization. *Plants* bisexual, with bisexual spikelets; *without hermaphrodite florets (the spikelets with proximal male florets and a terminal female)*.

Inflorescence. Inflorescence narrowly paniculate; contracted; espatheate, or spatheate (in that the branches are sometimes subtended by small bracts); not comprising 'partial inflorescences' and foliar organs. *Spikelet-bearing axes* short 'racemes'; persistent. Spikelets not secund; pedicellate; imbricate.

Female-fertile spikelets. *Spikelets* 12–15 mm long; lanceolate to ovate; compressed laterally; disarticulating above the glumes; not disarticulating between the florets (the males falling together with the terminal female); *with distinctly elongated rachilla internodes between the florets (having a 1 mm internode beneath the female floret, bearing a fleshy outgrowth embracing the base of the floret)*. Rachilla terminated by a female-fertile floret (seemingly). Hairy callus absent.

Glumes two; very unequal, or more or less equal; shorter than the adjacent lemmas; ciliate on the margins; pointed (the upper), or not pointed; awnless; carinate; similar. Lower glume much shorter than half length of lowest lemma; 4 nerved, or 5 nerved. Upper glume 7 nerved. *Spikelets* with incomplete florets. The incomplete florets proximal to the female-fertile florets. *Spikelets with proximal incomplete florets*. *The proximal incomplete florets* 3–5; paleate. Palea of the proximal incomplete florets fully developed. *The proximal incomplete florets male (the stamens 6, monadelphous, at least sometimes with penicillate anthers)*. The proximal lemmas lanceolate; awnless; 11 nerved; exceeded by the female-fertile lemmas.

Female-fertile florets 1. *Lemmas convolute*; decidedly firmer than the glumes; smooth; not becoming indurated (pallid, softly leathery below, becoming cartilaginous towards the apex); entire; blunt; awnless; hairy (except towards the tip); non-carinate (rounded on the back); without a germination flap; 9–11 nerved. *Palea* present; relatively long; convolute; entire; awnless, without apical setae (apically fringed); thinner than the lemma; not indurated (but cartilaginous towards the apex); several nerved (5 nerved in the material seen); keel-less (abaxially rounded). Palea back hairy. *Lodicules* present; 3; free; membranous; ciliate; not toothed. *Stamens* 0. *Ovary* glabrous; with a conspicuous apical appendage. The appendage broadly conical, fleshy (and very long). Styles fused (into one). Stigmas 2–3; brown.

Fruit, embryo and seedling. Hilum long-linear. Pericarp free. Embryo small.

Abaxial leaf blade epidermis. *Costal/intercostal zonation* conspicuous. *Papillae* absent. *Long-cells* markedly different in shape costally and intercostally (the costals much narrower); of similar wall thickness costally and intercostally (thin walled). Intercostal zones with typical long-cells. Mid-intercostal long-cells rectangular; having markedly sinuous walls (the sinuosity deep, rather irregular). *Microhairs* seemingly absent (none seen in either of the species examined). *Stomata* common. Subsidiaries non-papillate; dome-shaped to triangular (the triangles often apically truncated). Guard-cells overlapping to flush with the interstomatals. *Intercostal short-cells* common; in cork/silica-cell pairs; silicified. Intercostal silica bodies when properly developed crescentic and oryzoid-type. *P. ciliata* with abundant small microhairs and prickles intercostally. Costal zones with short-cells. *Costal short-cells* conspicuously in long rows. Costal silica bodies present and well developed; consistently saddle shaped.

Transverse section of leaf blade, physiology. C_3; XyMS+. Mesophyll without adaxial palisade; with arm cells; with fusoids. The fusoids external to the PBS. *Leaf blade* adaxially flat. *Midrib* conspicuous; having complex vascularization (with an abaxial arc of three (a large median with small laterals) embedded in sclerenchyma, and a small adaxial sclerenchyma mass containing 1–3 small bundles); with colourless mesophyll adaxially. *The lamina* symmetrical on either side of the midrib. Bulliforms present in discrete, regular adaxial groups; in simple fans (these large and wide). All the vascular bundles accompanied by sclerenchyma. Combined sclerenchyma girders present (with all the bundles); forming 'figures' (I's and 'anchors'). Sclerenchyma all associated with vascular bundles.

Cytology. Chromosome base number, $x = 12$. $2n = 24$. 2 ploid.

Taxonomy. Bambusoideae; Bambusodae; Puelieae.

Ecology, geography, regional floristic distribution. 6 species. Tropical Africa. Shade species.

Paleotropical. African. Sudano-Angolan and West African Rainforest. South Tropical African.

References, etc. Leaf anatomical: Metcalfe 1960; this project.

Pyrrhanthera Zotov

Habit, vegetative morphology. Low, mat-forming perennial; rhizomatous and caespitose. **Culms *1–3 cm high*; *herbaceous***. Culm nodes glabrous. Culm internodes solid. *Leaves* mostly basal; non-auriculate. Leaf blades narrow; setaceous to not setaceous; folded; without cross venation; persistent.

Reproductive organization. Plants bisexual, with bisexual spikelets; with hermaphrodite florets.

Inflorescence. Inflorescence reduced to a single spikelet, or few spikeleted (1–3); a single raceme, or paniculate (much reduced); espatheate; not comprising 'partial inflorescences' and foliar organs. Spikelet-bearing axes persistent.

Female-fertile spikelets. *Spikelets* compressed laterally; disarticulating above the glumes; *disarticulating between the florets*. Rachilla prolonged beyond the uppermost female-fertile floret; rachilla hairy, or hairless. Hairy callus absent. *Callus* very short.

Glumes present; two; more or less equal; exceeding the spikelets (enclosing it); long relative to the adjacent lemmas; hairless; glabrous; awnless; similar (ovate, leathery). Lower glume 7–9 nerved. Upper glume 7–9 nerved. *Spikelets* with female-fertile florets only, or with incomplete florets. The incomplete florets distal to the female-fertile florets. *The distal incomplete florets* merely underdeveloped.

Female-fertile florets *2–3*. Lemmas similar in texture to the glumes (leathery); not becoming indurated; incised; minutely 3 lobed (3-toothed); not deeply cleft; mucronate (on the teeth); hairy (minutely pilose). The hairs not in tufts; not in transverse rows. Lemmas non-carinate; 9 nerved. *Palea* present; 2-nerved. *Lodicules* present; 2; free; fleshy; ciliate. *Stamens* 3. Anthers not penicillate. *Ovary* glabrous. Styles free to their bases. Stigmas 2.

Fruit, embryo and seedling. Fruit small; compressed dorsiventrally. Hilum short. *Pericarp thick and hard*; *free (the fruit a tiny nut)*. Embryo large; waisted; without an epiblast; with a scutellar tail.

Abaxial leaf blade epidermis. *Costal/intercostal zonation* conspicuous. *Papillae* absent. *Long-cells* similar in shape costally and intercostally; of similar wall thickness costally and intercostally. Mid-intercostal long-cells rectangular; having markedly sinuous walls. *Microhairs* present; panicoid-type to chloridoid-type; (45–)51–63(–66) microns long; (22.5–)24–25.5(–26.4) microns wide at the septum. Microhair total length/ width at septum 1.8–2.6. Microhair apical cells (13.5–)18–24(–25.5) microns long. Microhair apical cell/total length ratio 0.3–0.45. *Stomata* common; (39–)40–47(–48) microns long. Subsidiaries dome-shaped, or triangular. Guard-cells overlapped by the interstomatals, or overlapping to flush with the interstomatals. *Intercostal short-cells* common; in cork/silica-cell pairs; silicified. *Costal short-cells* conspicuously in long rows, or neither distinctly grouped into long rows nor predominantly paired. Costal silica bodies rounded, crescentic, and 'panicoid-type'; often cross shaped; not sharp-pointed.

Transverse section of leaf blade, physiology. C_3; XyMS+. *Mesophyll* with non-radiate chlorenchyma; without adaxial palisade. *Leaf blade* adaxially flat. *Midrib* conspicuous; with one bundle only. Bulliforms not in discrete, regular adaxial groups (confined to a simple fan over the midrib). Many of the smallest vascular bundles unaccompanied by sclerenchyma. Combined sclerenchyma girders present; forming 'figures'. Sclerenchyma all associated with vascular bundles.

Taxonomy. Arundinoideae; Danthonieae.

Ecology, geography, regional floristic distribution. 1 species. New Zealand. Species of open habitats; glycophytic. Upland grassy plains.

Antarctic. New Zealand.

References, etc. Morphological/taxonomic: Zotov 1963. Leaf anatomical: this project.

Racemobambos Holttum

Microcalamus Gamble, *Neomicrocalamus* Keng f.

Habit, vegetative morphology. Perennial. The flowering culms leafy. Culms woody and persistent; to 1 cm in diameter; scandent (or scrambling, slender); branched above. Primary branches/mid-culm node 3–20 (?—'many'). Culm sheaths deciduous in their entirety. Rhizomes pachymorph. Plants unarmed. *Leaves* not basally aggregated; auriculate (the auricles raised); with auricular setae (these slender). *Leaf blades narrow*; 4–10 mm wide ('to 1 cm'); pseudopetiolate; disarticulating from the sheaths; rolled in bud.

Reproductive organization. Plants bisexual, with bisexual spikelets; with hermaphrodite florets.

Inflorescence. *Inflorescence* determinate, or indeterminate (?); seemingly without pseudospikelets; a single raceme, or paniculate (of small terminal racemes or panicles, each spikelet subtended by a small bract); spatheate (not foliate, the bracts with or without a rudimentary blade); a complex of 'partial inflorescences' and intervening foliar organs. *Spikelet-bearing axes* 'racemes'; solitary; persistent. *Spikelets* solitary; *secund ('all spikelets twisted to side of blade opposite subtending leaf')*; more or less sessile; not in distinct 'long-and-short' combinations.

Female-fertile spikelets. Spikelets 20 mm long; compressed laterally; disarticulating above the glumes; disarticulating between the florets. Rachilla prolonged beyond the uppermost female-fertile floret. The rachilla extension with incomplete florets. Hairy callus absent.

Glumes two, or several (2–3); very unequal (G_1 shorter); shorter than the adjacent lemmas; free; pointed; awnless; non-carinate (G_2 slightly keeled near tip); *very dissimilar* (*G_1 much narrower*). Lower glume 1–3 nerved. Upper glume 9–11 nerved. *Spikelets* with incomplete florets. The incomplete florets distal to the female-fertile florets. The distal incomplete florets 1; *incomplete florets* merely underdeveloped.

Female-fertile florets 2–3. Lemmas similar in texture to the glumes; entire; pointed; awned (the apex setiform). Awns 1; median; apical; non-geniculate; much shorter than the body of the lemma (the setum about 3 mm long). Lemmas with fringed margins; non-carinate; 9–11 nerved (in material seen). *Palea* present; relatively long (but shorter than lemmas); apically notched; several nerved (9 in material seen); 2-keeled. Palea keels scabrous (fringed near the apex). *Lodicules* present; 3; free; membranous; ciliate, or glabrous. *Stamens* 6. Anthers 5 mm long; not penicillate; without an apically prolonged connective. **Ovary** hairy; *with a conspicuous apical appendage*. The appendage broadly conical, fleshy. Styles fused. Stigmas 3 (the ovary apex swollen, no hollow style).

Abaxial leaf blade epidermis. *Costal/intercostal zonation* conspicuous. *Papillae* present (costally and around the stomata only). Intercostal papillae over-arching the stomata; several per cell (often quite thick-walled, rather irregular). Long-cells of similar wall thickness costally and intercostally. Mid-intercostal long-cells rectangular; having markedly sinuous walls. *Microhairs* present; panicoid-type; (30–)39–45(–47) microns long; (6–)6.6–8.4(–9) microns wide at the septum. Microhair total length/width at septum 3.6–7.8. Microhair apical cells (18–)21–25.5(–28.5) microns long. Microhair apical cell/total length ratio 0.5–0.63. *Stomata* common (bordering the veins only, completely covered by papillae); 27–30–33 microns long. *Intercostal short-cells* common; in cork/silica-cell pairs (and solitary); silicified. Intercostal silica bodies crescentic and saddle shaped. Bulbous prickles with tiny points common. *Crown cells* absent. *Costal short-cells* conspicuously in long rows. Costal silica bodies saddle shaped (predominating), or oryzoid (quite common); not sharp-pointed.

Transverse section of leaf blade, physiology. C_3; XyMS+. Mesophyll without adaxial palisade; with arm cells; with fusoids. The fusoids external to the PBS. *Leaf blade* the lamina flat to low-ribbed. *Midrib* conspicuous (via the large bundle and the prominent adaxial rib); with one bundle only. Bulliforms present in discrete, regular adaxial groups; in simple fans. All the vascular bundles accompanied by sclerenchyma. Combined sclerenchyma girders present (with all the bundles); forming 'figures' (all bundles). Sclerenchyma all associated with vascular bundles.

Taxonomy. Bambusoideae; Bambusodae; Bambuseae.

Ecology, geography, regional floristic distribution. 18 species. Southern Asia, Malay Peninsula, Borneo.

Paleotropical. Indomalesian. Malesian and Papuan.

References, etc. Morphological/taxonomic: Chao and Renvoize 1989. Leaf anatomical: this project.

Special comments. Fruit data wanting.

Raddia Bertol.

Hellera Doell., *Strephium* Nees

Habit, vegetative morphology. Perennial; caespitose, or caespitose and rhizomatous. The flowering culms leafy. *Culms* 10–50 cm high; herbaceous; unbranched above (or 'nearly simple'). Plants unarmed. *Leaves* few, not basally aggregated; without auricular setae. Leaf blades lanceolate, or elliptic (apiculate); broad, or narrow; 4–12 mm wide (and 1–7 cm long); rounded at base; pseudopetiolate; cross veined; rolled in bud. *Ligule* obsolete or fimbriate.

Reproductive organization. *Plants monoecious with all the fertile spikelets unisexual*; without hermaphrodite florets. The spikelets of sexually distinct forms on the same plant; female-only and male-only. *The male and female-fertile spikelets in different inflorescences (comprising lateral synflorescences of female racemes, and terminal male panicles)*. The spikelets overtly heteromorphic.

Inflorescence. **Inflorescence** *indeterminate (a synflorescence)*; a complex of 'partial inflorescences' and intervening foliar organs. *Spikelet-bearing axes* paniculate; persistent. Spikelets not secund; pedicellate (the pedicels short, those of female spikelets longer and apically expanded).

Female-sterile spikelets. Male spikelets shorter-pedicelled, florets with 2–3 free stamens. The male spikelets without glumes; 1 floreted. Male florets 2 staminate, or 3 staminate.

Female-fertile spikelets. Spikelets 7–8 mm long; elliptic, or ovate; compressed dorsiventrally; disarticulating above the glumes. Rachilla terminated by a female-fertile floret. Hairy callus absent. *Callus* absent.

Glumes two; more or less equal; long relative to the adjacent lemmas; free; pointed (acuminate); similar (firmly membranous, with cartilaginous, undulating margins, cf. *Sucrea*). *Lower glume 5 nerved*. Upper glume 3 nerved. *Spikelets* with female-fertile florets only.

Female-fertile florets 1. Lemmas gibbous; decidedly firmer than the glumes (leathery); becoming indurated; awnless, or mucronate (?); non-carinate; 5 nerved. *Palea* present; awnless, without apical setae; indurated; obscurely 2-nerved. *Stamens* 0.

Fruit, embryo and seedling. *Fruit* free from both lemma and palea (but enclosed); small, or medium sized (2–7 mm long).

Seedling with a short mesocotyl. First seedling leaf without a lamina (or the blade much reduced).

Abaxial leaf blade epidermis. *Costal/intercostal zonation* conspicuous. *Papillae* present. Intercostal papillae not over-arching the stomata; several per cell (small, thick-walled). *Long-cells* markedly different in shape costally and intercostally (costals much narrower); of similar wall thickness costally and intercostally (medium thin walled). Mid-intercostal long-cells rectangular; having markedly sinuous walls. *Microhairs* present; panicoid-type; (45–)49–54(–57) microns long; (3.4–)4.5–5.4(–6) microns wide at the septum. Microhair total length/width at septum 9.2–11.5. Microhair apical cells 18–22.5 microns long. Microhair apical cell/total length ratio 0.38–0.42. *Stomata* common; 18–19.5–21 microns long. Subsidiaries predominantly triangular. Guard-cells overlapping to flush with the interstomatals. *Intercostal short-cells* common; in cork/silica-cell pairs (and solitary); silicified. *Costal short-cells* conspicuously in long rows. Costal silica bodies oryzoid (predominating), or 'panicoid-type' (a few of the oryzoid form more or less isodiametric; when isodiametric, cross shaped; not sharp-pointed.

Transverse section of leaf blade, physiology. C_3; XyMS+. Mesophyll without adaxial palisade; with arm cells (very conspicuous); with fusoids (very long and thin). The fusoids external to the PBS. *Leaf blade* adaxially flat. *Midrib* conspicuous; having a conventional arc of bundles (a large median with a smaller lateral on each side). Bulliforms present in discrete, regular adaxial groups (these large, wide); in simple fans. All the vascular bundles accompanied by sclerenchyma. Combined sclerenchyma girders present (with all the bundles); forming 'figures' (with most bundles). Sclerenchyma all associated with vascular bundles.

Cytology. Chromosome base number, $x = 11$. $2n = 22$. 2 ploid.

Taxonomy. Bambusoideae; Oryzodae; Olyreae.

Ecology, geography, regional floristic distribution. 5 species. Central and South America, West Indies. Shade species.

Neotropical. Caribbean, Venezuela and Surinam, Amazon, Central Brazilian, and Andean.

References, etc. Morphological/taxonomic: Soderstrom and Zuloaga 1988. Leaf anatomical: this project.

Special comments. Fruit data wanting.

Raddiella Swallen

Habit, vegetative morphology. Delicate perennial (*R. esenbeckii* and perhaps *R. minima*), or annual (mostly); caespitose, or decumbent (sometimes mat-forming). The flowering culms leafy. *Culms* 5–30(–40) cm high; herbaceous. Rhizomes pachymorph. Plants unarmed. *Leaves* not basally aggregated; auriculate, or non-auriculate; without auricular setae. *Leaf blades* ovate, or elliptic (to ovate-triangular, often stringly asymmetrical and apiculate); flimsy (membranous), or neither leathery nor flimsy; *narrow*; 2.4–8(–11) mm wide (and only 4–15(–22 mm long); slightly to much asymmetric at their bases; *pseudopetiolate*; cross veined. Ligule an unfringed membrane, or a fringed membrane (?—'membranous or membranous-ciliate'); truncate; 0.2–0.5 mm long.

Reproductive organization. *Plants monoecious with all the fertile spikelets unisexual*; without hermaphrodite florets. The spikelets of sexually distinct forms on the same plant; female-only and male-only. The male and female-fertile spikelets in different inflorescences, or mixed in the inflorescence (in partial (syn)inflorescences which are axillary and all-female, terminal and all-male, or mixed). The spikelets overtly heteromorphic.

Inflorescence. *Inflorescence* determinate, or indeterminate (some species with synflorescences); few spikeleted; a scanty, racemelike panicle, sometimes reduced to two spikelets. *Spikelet-bearing axes persistent*. Spikelets not secund; pedicellate.

Female-sterile spikelets. Male spikelets shorter pedicelled than females, hyaline and early deciduous. Rachilla of male spikelets terminated by a male floret. The male spikelets without glumes; without proximal incomplete florets; 1 floreted. Male florets 1; 3 staminate.

Female-fertile spikelets. *Spikelets 1.3–2.3(–2.7) mm long*; elliptic; falling with the glumes (mostly), or disarticulating above the glumes (the glumes persistent in two species); with a characteristic, thickened and indurate interglumal internode. Rachilla terminated by a female-fertile floret. *Callus* absent.

Glumes two; more or less equal; about equalling the spikelets; long relative to the adjacent lemmas; hairless; pointed (acute); similar (membranous-herbaceous, with basal pulvini). Lower glume 3 nerved. Upper glume 3 nerved, or 5 nerved. *Spikelets* with female-fertile florets only.

Female-fertile florets 1. Lemmas decidedly firmer than the glumes (thinly leathery); smooth to papillose; becoming indurated (bony); entire; awnless; non-carinate; having the margins tucked in onto the palea; with a clear germination flap. *Palea* present; tightly clasped by the lemma; awnless, without apical setae. *Stamens* 0. Styles fused. Stigmas 2.

Fruit, embryo and seedling. Fruit small; ellipsoid. *Hilum short (punctiform, elliptic or short-linear)*. Embryo small.

Abaxial leaf blade epidermis. *Costal/intercostal zonation* conspicuous. *Papillae* present; costal and intercostal (abundant). Intercostal papillae several per cell (most longcells with a single median row). *Long-cells* markedly different in shape costally and intercostally (the costals much narrower); of similar wall thickness costally and intercostally (thin walled). Mid-intercostal long-cells rectangular; having markedly sinuous walls (the sinuosity coarse). *Microhairs* present; elongated; clearly two-celled; panicoid-type; 34.5–42 microns long; 5.4–6.3 microns wide at the septum. Microhair total length/width at septum 6–7.8. Microhair apical cells 16.5–18 microns long. Microhair apical cell/total length ratio 0.4–0.5. *Stomata* common; 19.5–21 microns long. Subsidiaries papillate (two on each, but thin walled and relatively inconspicuous); predominantly triangular. Guardcells overlapped by the interstomatals (slightly), or overlapping to flush with the interstomatals. *Intercostal short-cells* common; in cork/silica-cell pairs; silicified. Intercostal silica bodies vertically elongated-nodular. Short, thick-walled macrohairs with simple bases common. *Crown cells* absent. *Costal short-cells* conspicuously in long rows. Costal silica bodies consistently oryzoid; not sharp-pointed.

Transverse section of leaf blade, physiology. C_3; XyMS+. Mesophyll with adaxial palisade; insufficiently well preserved to score for arm-cells, in the material seen; with fusoids. The fusoids external to the PBS. *Leaf blade* adaxially flat (and thin). Bulliforms present in discrete, regular adaxial groups (a large, wide group in every intercostal zone); in simple fans. All the vascular bundles accompanied by sclerenchyma. Combined sclerenchyma girders present; forming 'figures' (every bundle having a small I or 'anchor'). Sclerenchyma all associated with vascular bundles.

Cytology. Chromosome base number, $x = 10$.

Taxonomy. Bambusoideae; Oryzodae; Olyreae.

Ecology, geography, regional floristic distribution. 7 species. West Indies and tropical America. Helophytic; glycophytic. Damp depressions and wet rocks near waterfalls.

Neotropical. Caribbean, Venezuela and Surinam, Amazon, and Central Brazilian.

References, etc. Morphological/taxonomic: Maguire *et al.* 1965; Zuloaga and Judziewicz 1991. Leaf anatomical: this project.

Special comments. Fruit data wanting.

Ratzeburgia Kunth

Aikinia Wall.

Habit, vegetative morphology. Perennial; stoloniferous. *Culms* 30–50 cm high; herbaceous. Young shoots extravaginal. Leaves non-auriculate. Leaf blades narrow; 1–4 mm wide; without cross venation.

Reproductive organization. Plants bisexual, with bisexual spikelets; with hermaphrodite florets. The spikelets of sexually distinct forms on the same plant; hermaphrodite and male-only, or hermaphrodite and sterile; overtly heteromorphic.

Inflorescence. *Inflorescence of one linear, flattened, rather fragile spicate 'raceme'*. Rachides flattened. Inflorescence espatheate; not comprising 'partial inflorescences' and foliar organs. **Spikelet-bearing axes** *spikelike*; the spikelet-bearing axes with 6–10 spikelet-bearing 'articles' ('8–10 noded'); solitary; with substantial rachides; disarticulating; disarticulating at the joints. *'Articles'* non-linear; with a basal callus-knob; disarticulating transversely; somewhat hairy. *Spikelets in triplets*; secund (the rachis dorsiventral, the spikelets on one side); sessile and pedicellate; *consistently in 'long-and-short' combinations*; *in pedicellate/sessile combinations (two opposite sessile spikelets appressed back to back, one pedicellate)*. Pedicels of the 'pedicellate' spikelets free of the rachis (but closely opposed). The 'shorter' spikelets hermaphrodite. The 'longer' spikelets sterile.

Female-sterile spikelets. The pedicellate spikelet sterile, vestigial.

Female-fertile spikelets. Spikelets 5.6 mm long; abaxial; compressed dorsiventrally; falling with the glumes (with the adjacent joint and pedicel). Rachilla terminated by a female-fertile floret. Hairy callus absent.

Glumes two; more or less equal; long relative to the adjacent lemmas; free; dorsiventral to the rachis; awnless; *very dissimilar (the G_1 leathery or crustaceous, reticulate-ribbed, pitted, tuberculate and truncately winged at the tip, the G_2 thinner and glabrous)*. *Lower glume two-keeled (the keels apically broadly, truncately winged)*; lacunose with deep depressions (with 5–6 vertical rows of pits). *Spikelets* with incomplete florets. The incomplete florets proximal to the female-fertile florets. *The proximal incomplete florets* 1; epaleate; sterile. The proximal lemmas awnless; more or less equalling the female-fertile lemmas; similar in texture to the female-fertile lemmas; not becoming indurated.

Female-fertile florets 1. Lemmas less firm than the glumes (scarious); not becoming indurated; entire; pointed, or blunt; awnless; hairless; non-carinate; without a germination flap. **Palea** present; *conspicuous but relatively short, or very reduced*; awnless, without apical setae; not indurated (scarious). *Lodicules* present; free; glabrous. *Stamens* 3. Anthers not penicillate; without an apically prolonged connective. *Ovary* glabrous. Styles free to their bases. Stigmas 2.

Taxonomy. Panicoideae; Andropogonodae; Andropogoneae; Rottboelliinae.

Ecology, geography, regional floristic distribution. 1 species. Burma. Species of open habitats. Short grassland.

Paleotropical. Indomalesian. Indo-Chinese.

Special comments. Fruit data wanting. Anatomical data wanting.

Redfieldia Vasey

Habit, vegetative morphology. Arenicolous perennial; rhizomatous. *Culms* 50–120 cm high; herbaceous. *Leaves* not basally aggregated; non-auriculate. Leaf blades linear (to filiform); narrow; tapering to a fine point; rolled; without abaxial multicellular glands; without cross venation. **Ligule** present; *a fringe of hairs*; 1 mm long.

Reproductive organization. Plants bisexual, with bisexual spikelets; with hermaphrodite florets.

Inflorescence. *Inflorescence paniculate (up to half the length of the culm)*; open; *with capillary branchlets (these flexuous)*; espatheate; not comprising 'partial inflorescences' and foliar organs. Spikelet-bearing axes persistent. Spikelets not secund; pedicellate.

Female-fertile spikelets. Spikelets 5–8 mm long; cuneate; compressed laterally; disarticulating above the glumes; disarticulating between the florets. Rachilla prolonged beyond the uppermost female-fertile floret. *Hairy callus present (silky-pilose)*.

Glumes two; more or less equal (to somewhat unequal); shorter than the spikelets; shorter than the adjacent lemmas; hairless; glabrous; pointed; awnless; similar (narrow lanceolate, acuminate). Lower glume not two-keeled; 1 nerved. Upper glume 1 nerved.

Female-fertile florets 2–5. *Lemmas* not becoming indurated (papery); entire, or incised (sometimes tridentate); when incised not deeply cleft; awnless, or mucronate (the nerves sometimes excurrent, forming 3 minute teeth); *hairless (but with a basal, hairy callus)*; *carinate (compressed)*; without a germination flap; 3 nerved. *Palea* present; relatively long (as long as the lemma); 2-nerved. *Lodicules* present; 2; fleshy. *Stamens* 3 (?). *Ovary* glabrous. Stigmas 2.

Fruit, embryo and seedling. *Fruit* free from both lemma and palea; ellipsoid. Hilum short. Pericarp fused. Embryo large; with an epiblast; with a scutellar tail; with an elongated mesocotyl internode. Embryonic leaf margins overlapping.

Abaxial leaf blade epidermis. *Costal/intercostal zonation* conspicuous (the intercostal zones hidden in deep grooves). *Papillae* absent. Long-cells differing markedly in wall thickness costally and intercostally (the intercostals thinner walled). Mid-intercostal long-cells having straight or only gently undulating walls. *Microhairs* present; elongated; clearly two-celled; chloridoid-type. Microhair apical cell wall of similar thickness/rigidity to that of the basal cell. Microhairs 42–45 microns long. Microhair basal cells 33 microns long. Microhairs 9.6–12 microns wide at the septum. Microhair total length/width at septum 3.6–4.4. Microhair apical cells 6.9–10.5 microns long. Microhair apical cell/total length ratio 0.2–0.24. *Stomata* common; 18–21 microns long. Subsidiaries triangular. Guard-cells overlapping to flush with the interstomatals. *Intercostal short-cells* seemingly absent or very rare. Intercostal silica bodies absent. *Costal short-cells* conspicuously in long rows. Costal silica bodies present in alternate cell files of the costal zones; abundant, predominantly saddle shaped; not sharp-pointed.

Transverse section of leaf blade, physiology. C_4. The anatomical organization somewhat unusual, in that the PCR sheaths are often bifurcated in the outer parts of the adaxial ribs, by intrusion of the adaxial sclerenchyma girder. XyMS+. PCR sheath outlines even. PCR sheaths of the primary vascular bundles interrupted; interrupted laterally. PCR sheath extensions present. Maximum number of extension cells 12 (or more, not uniseriate but in the form of an adaxial block of PCR cells which is either partially or completely interrupted by the sclerenchyma girder). PCR cell chloroplasts centripetal. Mesophyll traversed by columns of colourless mesophyll cells (in each intercostal zone). *Leaf blade* 'nodular' in section; with the ribs more or less constant in size (tall, somewhat hollowed at the sides). *Midrib* not readily distinguishable; with one bundle only. *The lamina* symmetrical on either side of the midrib. Bulliforms present in discrete, regular adaxial groups; associated with colourless mesophyll cells to form deeply-penetrating fans (these linking with the traversing columns of colourless cells). All the vascular bundles accompanied by sclerenchyma. Combined sclerenchyma girders present; forming 'figures' (all the bundles with a large I or 'anchor', though in many the adaxial sclerenchyma mass is 'interrupted' by PCR tissue). Sclerenchyma all associated with vascular bundles.

Taxonomy. Chloridoideae; main chloridoid assemblage.

Ecology, geography, regional floristic distribution. 1 species. Central U.S.A. Xerophytic; species of open habitats. Inland sandhills.

Holarctic and Paleotropical. Boreal and Madrean. Madagascan. Atlantic North American and Rocky Mountains. Central Grasslands.

Rusts and smuts. Rusts — *Puccinia*. Smuts from Tilletiaceae. Tilletiaceae — *Tilletia*.

References, etc. Leaf anatomical: this project.

Reederochloa Soderstrom & H.F. Decker

Habit, vegetative morphology. Perennial; conspicuously stoloniferous and caespitose. *Culms* 3–11 cm high; herbaceous; unbranched above. Culm nodes sparsely or densely hairy. *Leaves* mostly basal. Leaf blades narrow; 0.2–0.5 mm wide (and 1.5–4 cm long); rolled (involute); without abaxial multicellular glands; not pseudopetiolate; without cross venation. Ligule present; *an unfringed membrane (apically erose)*.

Reproductive organization. *Plants dioecious*; without hermaphrodite florets. The spikelets all alike in sexuality (on the one plant); female-only, or male-only. Plants outbreeding.

Inflorescence. *Inflorescence* few spikeleted; *a cluster of 2–4 appressed spikelets, the males long exserted, the females sessile among the upper leaf sheaths*; digitate, or non-digitate (?); espatheate; not comprising 'partial inflorescences' and foliar organs. Spikelet-bearing axes persistent.

Female-sterile spikelets. Male spikelets 2–4 appressed, in exserted inflorescences 0.5–1.5 cm long. Spikelets 5–13 mm long, laterally compressed, glabrous, with 3–8 florets; palea glabrous, equalling the lemma; stamens 3, unequal. Rachilla of male spikelets prolonged beyond the uppermost male floret. The male spikelets with glumes; 3–8 floreted. Male florets 3 staminate.

Female-fertile spikelets. Spikelets 5–11 mm long; compressed laterally; tardily disarticulating above the glumes; disarticulating between the florets. Rachilla prolonged beyond the uppermost female-fertile floret. The rachilla extension with incomplete florets.

Glumes two; very unequal; shorter than the spikelets; shorter than the adjacent lemmas; hairy (at the base only — glabrous above); pointed; awnless; similar (obovate-lanceolate). Lower glume indistinctly 2–8 nerved. *Upper glume* indistinctly *8 nerved*. *Spikelets* with incomplete florets. The incomplete florets distal to the female-fertile florets. *The distal incomplete florets* merely underdeveloped.

Female-fertile florets 3–8. Lemmas broad at the base, narrowing above; not becoming indurated; entire; pointed; awnless; hairy (densely pilose at the base, glabrous above); 10–13 nerved. *Palea* present; relatively long (broadened and hairy at the base); awnless; without apical setae; not indurated. *Lodicules* present; fleshy (small); glabrous (?). *Stamens* 0. Ovary glabrous. Stigmas 2.

Fruit, embryo and seedling. Fruit ellipsoid; hairy on the body ('hairs about 2/5 as long as the fruit'); smooth (?). Hilum short. Pericarp fused.

Abaxial leaf blade epidermis. *Papillae* present; costal and intercostal. Intercostal papillae consisting of one symmetrical projection per cell. Mid-intercostal long-cells rectangular. *Microhairs* present; more or less spherical; ostensibly one-celled; chloridoid-type (often somewhat sunken). *Stomata* common. Intercostal short-cells not paired. Intercostal silica bodies absent. *Costal short-cells* neither distinctly grouped into long rows nor predominantly paired. Costal silica bodies present throughout the costal zones to confined to the central file(s) of the costal zones; rounded to saddle shaped; not sharp-pointed.

Transverse section of leaf blade, physiology. C_4; XyMS+. PCR sheaths of the primary vascular bundles complete. PCR sheath extensions absent. *Mesophyll* with radiate chlorenchyma; traversed by columns of colourless mesophyll cells. *Leaf blade* with distinct, prominent adaxial ribs. *Midrib* not readily distinguishable; with one bundle only. Bulliforms not in discrete, regular adaxial groups. All the vascular bundles accompanied by sclerenchyma. Combined sclerenchyma girders absent (there being no adaxial sclerenchyma). Sclerenchyma not all bundle-associated. The 'extra' sclerenchyma in abaxial groups (there being some sclerification of colourless cells in the abaxial furrows). The lamina margins with fibres.

Special diagnostic feature. Panicle loose, or if dense then interrupted, neither cylindrical nor ovoid: awns usually present, usually twisted, usually distinctly dorsal, conspicuous if inflorescence compact.

Cytology. $2n = 38$.

Taxonomy. Chloridoideae; main chloridoid assemblage.

Ecology, geography, regional floristic distribution. 1 species. Mexico. Xerophytic; species of open habitats; halophytic (alkali flats).

Holarctic. Madrean.

References, etc. Morphological/taxonomic: Soderstrom and Decker 1964. Leaf anatomical: Van den Borre 1994.

Rehia Fijten

Bulbulus Swallen

Habit, vegetative morphology. Perennial; caespitose (the slender culms in small dense tufts). The flowering culms leafy. **Culms** 12–19 cm high; herbaceous; unbranched above; *tuberous*. Culm nodes hairy. Plants unarmed. *Leaves* not basally aggregated; without auricular setae. Leaf blades broad; 7–16 mm wide (and 2–4.5 cm long, acute); somewhat cordate, or not cordate, not sagittate (but rounded at the base); pseudopetiolate; cross veined.

Reproductive organization. *Plants monoecious with all the fertile spikelets unisexual*; without hermaphrodite florets. The spikelets of sexually distinct forms on the same plant; female-only and male-only. The male and female-fertile spikelets mixed in the inflorescence. The spikelets overtly heteromorphic (the males smaller, glabrous).

Inflorescence. Inflorescence indeterminate (with synflorescences); few spikeleted, or many spikeleted; *paniculate (panicles terminal, small, borne singly or several together, each of 2–5 spikelet pairs or triplets, scarcely exserted or partially enclosed in a sheath)*; spatheate; a complex of 'partial inflorescences' and intervening foliar organs. Spikelet-bearing axes disarticulating. Spikelets in triplets (sometimes 1 female with 2 male), or paired; pedicellate; consistently in 'long-and-short' combinations; unequally pedicellate in each combination. The 'shorter' spikelets female-only. The 'longer' spikelets male-only.

Female-sterile spikelets. Male spikelets 5 mm long, the floret with 3 stamens. Rachilla of male spikelets terminated by a male floret. The male spikelets without glumes; without proximal incomplete florets; 1 floreted. Male florets 1; 3 staminate.

Female-fertile spikelets. Spikelets 7–8 mm long; lanceolate; with conventional internode spacings. Rachilla terminated by a female-fertile floret. *Callus* absent.

Glumes two; more or less equal; long relative to the adjacent lemmas; pointed (acute or acuminate, somewhat twisted); awnless; similar (papery). Lower glume 7 nerved. Upper glume 7 nerved. *Spikelets* with female-fertile florets only.

Female-fertile florets 1. Lemmas decidedly firmer than the glumes (?); becoming indurated; entire; pointed; awnless, or mucronate (?); hairy (pilose); having the margins tucked in onto the palea. *Palea* present. *Stamens* 0 (but sometimes with staminodes).

Fruit, embryo and seedling. Fruit medium sized (5.5 mm long). Endosperm containing compound starch grains.

Abaxial leaf blade epidermis. *Costal/intercostal zonation* conspicuous (and the intercostal zones each with a median astomatal zone). *Papillae* present; intercostal. Intercostal papillae over-arching the stomata (from the adjoining long-cells and interstomatals); several per cell (two or three irregularly shaped papillae, of various sizes, per cell in the stomatal zone). *Long-cells* markedly different in shape costally and intercostally (the costals narrower); of similar wall thickness costally and intercostally (fairly thin walled). Mid-intercostal long-cells rectangular; having markedly sinuous walls (the sinuosity fairly coarse). *Microhairs* present; elongated; clearly two-celled; panicoid-type; 63–69 microns long; 5.4–6.9 microns wide at the septum. Microhair total length/width at septum 9.6–12.1. Microhair apical cells 28.5–33 microns long. Microhair apical cell/total length ratio 0.4–0.5. *Stomata* common (in broad intercostal bands adjacent to the costal zones); 19.5–21 microns long. Subsidiaries papillate (two on each, large and conspicuous); predominantly high dome-shaped, or triangular. *Intercostal short-cells* common (in the astomatal zones); in cork/silica-cell pairs; silicified. Intercostal silica bodies vertically elongated-nodular. Prickles and macrohairs absent in the material seen. *Crown cells* absent. *Costal short-cells* conspicuously in long rows. Costal silica bodies oryzoid (mainly), or 'panicoid-type' (a few); sometimes approaching Maltese cross shaped; not sharp-pointed.

Transverse section of leaf blade, physiology. C_3; XyMS+. Mesophyll without arm cells (? — seemingly, in the distorted material seen); with fusoids. The fusoids external to the PBS. *Leaf blade* adaxially flat. *Midrib* conspicuous (with a large, round-topped adaxial projection); having a conventional arc of bundles (3 or 5); with colourless mesophyll adaxially. *The lamina* symmetrical on either side of the midrib. Bulliforms present in discrete, regular adaxial groups (a large, wide group in each intercostal zone); in simple fans. All the vascular bundles accompanied by sclerenchyma. Combined sclerenchyma girders present; forming 'figures' (all the bundles with an I or an 'anchor'). Sclerenchyma all associated with vascular bundles.

Cytology. Chromosome base number, $x = 10$. $2n = 20$ (*'Bulbulus nervatus'*).

Taxonomy. Bambusoideae; Oryzodae; Olyreae.

Ecology, geography, regional floristic distribution. 1 species. Brazil. Mesophytic; shade species. In forest.

Neotropical. Amazon.

References, etc. Leaf anatomical: this project.

Special comments. Fruit data wanting.

Reimarochloa A. Hitchc.

Habit, vegetative morphology. Perennial; caespitose. Culms herbaceous. Culm nodes glabrous. Culm internodes solid. *Leaves* mostly basal, or not basally aggregated; non-auriculate. Leaf blades narrow; not cordate, not sagittate; flat; without cross venation; persistent.

Reproductive organization. Plants bisexual, with bisexual spikelets; with hermaphrodite florets.

Inflorescence. *Inflorescence* of spicate main branches (slender racemes); open; *digitate*. Primary inflorescence branches 3–6. Rachides flattened. Inflorescence espatheate; not comprising 'partial inflorescences' and foliar organs. Spikelet-bearing axes persistent. Spikelets solitary; secund (on one side of the flattened rachis); biseriate; subsessile; rather distant.

Female-fertile spikelets. Spikelets *unconventional (either there is no L_1 and only one glume, or both glumes are missing: the occasional presence of an apparent glume with terminal spikelets suggests that most spikelets 'lack' both glumes)*; 2–5 mm long; lanceolate; abaxial; compressed dorsiventrally; planoconvex; falling with the glumes. The upper floret not stipitate. Rachilla terminated by a female-fertile floret. Hairy callus absent.

Glumes *absent*. *Spikelets* with incomplete florets (though the L_1 could be mistaken for a glume). The incomplete florets proximal to the female-fertile florets. *The proximal incomplete florets* 1; epaleate; sterile. *The proximal lemmas* pointed to acuminate; awnless; 3 nerved; *decidedly exceeding the female-fertile lemmas*; less firm than the female-fertile lemmas to similar in texture to the female-fertile lemmas; not becoming indurated.

Female-fertile florets 1. Lemmas acuminate; not saccate; membranous to thinly leathery; striate (or scarcely so); entire; pointed; awnless; hairless; non-carinate; having the margins lying flat and exposed on the palea (except at the base); 3 nerved. *Palea* present; relatively long; gaping (free for nearly half its length); awnless, without apical setae; textured like the lemma; 2-nerved; 2-keeled (flat). *Lodicules* present; 2; free; fleshy; glabrous; not or scarcely vascularized. *Stamens* 3. Anthers not penicillate; without an apically prolonged connective. *Ovary* glabrous. Styles free to their bases. Stigmas 2; brown (dried).

Fruit, embryo and seedling. Fruit small. Hilum short. Embryo large.

Abaxial leaf blade epidermis. *Costal/intercostal zonation* conspicuous. *Papillae* absent. Mid-intercostal long-cells having markedly sinuous walls. *Microhairs* present; panicoid-type. *Stomata* common. Subsidiaries dome-shaped (mostly), or triangular (rather indistinct). Guard-cells overlapping to flush with the interstomatals. *Intercostal short-cells* common; in cork/silica-cell pairs. *Costal short-cells* conspicuously in long rows. Costal silica bodies 'panicoid-type'; mostly nodular; not sharp-pointed.

Transverse section of leaf blade, physiology. C_4; XyMS–. *Leaf blade* 'nodular' in section to adaxially flat; with the ribs more or less constant in size. *Midrib* not readily distinguishable; with one bundle only. Bulliforms not in discrete, regular adaxial groups (indistinguishable from macrohair cushions, etc.). Many of the smallest vascular bundles unaccompanied by sclerenchyma. Combined sclerenchyma girders absent. Sclerenchyma all associated with vascular bundles.

Taxonomy. Panicoideae; Panicodae; Paniceae.

Ecology, geography, regional floristic distribution. 4 species. Southern U.S.A. to Argentina. Helophytic (mainly); glycophytic.

Holarctic and Neotropical. Madrean. Caribbean, Amazon, and Central Brazilian.

Rusts and smuts. Rusts — *Puccinia*. Taxonomically wide-ranging species: *Puccinia levis*.

References, etc. Morphological/taxonomic: Hitchcock 1909. Leaf anatomical: this project.

Reitzia Swallen

Habit, vegetative morphology. Perennial; densely caespitose. The flowering culms leafy. *Culms* about 10–25 cm high (?); herbaceous; unbranched above. Rhizomes pachymorph. Plants unarmed. *Leaves* not basally aggregated; without auricular setae. *Leaf blades broad*; 12–20 mm wide (4–7.5 cm long, acute); not cordate, not sagittate ('subtruncate' and slightly asymmetrical at base); pseudopetiolate.

Reproductive organization. *Plants monoecious with all the fertile spikelets unisexual*; without hermaphrodite florets. The spikelets of sexually distinct forms on the same plant; female-only and male-only. The male and female-fertile spikelets mixed in the inflorescence (each panicle with a few fmale spikelets, each accompanied by 1–2 males). The spikelets overtly heteromorphic; all in heterogamous combinations.

Inflorescence. Inflorescence *indeterminate (a synflorescence)*; of reduced axillary and terminal racemelike panicles; a complex of 'partial inflorescences' and intervening foliar organs. Spikelet-bearing axes persistent. Spikelets paired, or in triplets; not secund (?); pedicellate; consistently in 'long-and-short' combinations; in pedicellate/sessile combinations (each female with 1–2 males). The 'shorter' spikelets male-only. The 'longer' spikelets female-only.

Female-sterile spikelets. Male spikelets 3 mm long, without glumes, lemma thin and 3-nerved, stamens 3. Rachilla of male spikelets terminated by a male floret. The male spikelets without glumes; without proximal incomplete florets; 1 floreted. Male florets 1; 3 staminate.

Female-fertile spikelets. Spikelets about 6 mm long; lanceolate; compressed dorsiventrally; disarticulating above the glumes; with conventional internode spacings. Rachilla terminated by a female-fertile floret. *Callus* absent.

Glumes two; pointed (acuminate to attenuate); non-carinate; similar (leathery). *Lower glume 3 nerved*. Upper glume 3 nerved. *Spikelets* with female-fertile florets only.

Female-fertile florets 1. Lemmas becoming mottled with purple; becoming indurated (presumably); awnless; hairless; non-carinate; 3 nerved (the nerves white — according to Swallen), or 5 nerved (Calderón & Soderstrom). *Palea* present; indurated (presumably). *Stamens* 0. Stigmas 2 (with long processes, plumose, not curling).

Abaxial leaf blade epidermis. *Costal/intercostal zonation* conspicuous. *Papillae* present; costal and intercostal (but largely absent from the mid-intercostal regions). Intercostal papillae over-arching the stomata; several per cell (small, circular, numerous, in rows on the costal long-cells, scattered over the intercostals bordering the costae, and concentrated in rings around the stomatal crypts). *Long-cells* markedly different in shape costally and intercostally (the costals much narrower); of similar wall thickness costally and intercostally to differing markedly in wall thickness costally and intercostally (the costals thicker walled in places). Mid-intercostal long-cells rectangular; having markedly sinuous walls (the sinuosity coarse, fairly irregular). *Microhairs* present; elongated; clearly two-celled; panicoid-type. *Stomata* common. Subsidiaries papillate (two per cell, overlying the guard-cells); dome-shaped and triangular. Guard-cells seemingly overlapped by the interstomatals (the stomata in depressions between the interstomatals, overhung by the small papillae). *Intercostal short-cells* common; in cork/silica-cell pairs; silicified. Intercostal silica bodies crescentic, tall-and-narrow, cross-shaped, and vertically elongated-nodular (mainly a slim version of the latter). With numerous intercostal prickles. *Costal short-cells* conspicuously in long rows (over the main veins), or predominantly paired (over minor veins). Costal silica bodies present and well developed; 'panicoid-type'; cross shaped to dumb-bell shaped (large); sharp-pointed.

Transverse section of leaf blade, physiology. C₃; XyMS+. Mesophyll with arm cells; with fusoids. The fusoids external to the PBS. *Leaf blade* adaxially flat. *Midrib* conspicuous; having a conventional arc of bundles (a large median, with a small bundle on each side); with colourless mesophyll adaxially. *The lamina* symmetrical on either side of the midrib. Bulliforms present in discrete, regular adaxial groups; in simple fans (these wide, of large cells). All the vascular bundles accompanied by sclerenchyma. Combined sclerenchyma girders present (with all the bundles save the median); forming 'figures' (I's or 'anchors', in places). Sclerenchyma all associated with vascular bundles.

Taxonomy. Bambusoideae; Oryzodae; Olyreae.

Ecology, geography, regional floristic distribution. 1 species. Brazil. Shade species.

Neotropical. Central Brazilian and Pampas.

References, etc. Morphological/taxonomic: Swallen 1956. Leaf anatomical: this project.

Special comments. Fruit data wanting.

Relchela Steud.

Habit, vegetative morphology. Perennial; rhizomatous and stoloniferous. *Culms* 20–60 cm high; herbaceous; unbranched above. *Leaves* not basally aggregated; non-auriculate. *Leaf blades* linear; narrow; about 5 mm wide; flat; *not pseudopetiolate*; without cross venation; persistent. Ligule an unfringed membrane; toothed. *Contra-ligule* absent.

Reproductive organization. Plants bisexual, with bisexual spikelets; with hermaphrodite florets. Exposed-cleistogamous and chasmogamous.

Inflorescence. *Inflorescence paniculate*; contracted; more or less irregular; espatheate; not comprising 'partial inflorescences' and foliar organs. Spikelets not secund; pedicellate; imbricate.

Female-fertile spikelets. *Spikelets 4–5 mm long*; compressed laterally; disarticulating above the glumes; disarticulating between the florets; with conventional internode spacings. Rachilla prolonged beyond the uppermost female-fertile floret; rachilla hairy (between the florets, and above the upper one). The rachilla extension naked. *Hairy callus present.* Callus short; blunt.

Glumes two; very unequal to more or less equal; *about equalling the spikelets*; long relative to the adjacent lemmas; hairless; with scabrid keels; pointed; awnless; carinate; similar (lanceolate, membranous). Lower glume about equalling the lowest lemma; 1 nerved. Upper glume 3 nerved. **Spikelets** with female-fertile florets only; *without proximal incomplete florets*.

Female-fertile florets 1–2(–3). *Lemmas* oblong; *decidedly firmer than the glumes (leathery or crustaceous)*; entire to incised; blunt; when incised, slightly 2 lobed; not deeply cleft; awnless; hairless; scabrous (above, shining below); non-carinate (dorsally rounded); without a germination flap; 5 nerved, or 7 nerved. *Palea* present; relatively long; tightly clasped by the lemma; entire to apically notched; awnless, without apical setae; thinner than the lemma (membranous); not indurated; 2-nerved; 2-keeled. Palea keels wingless; hairy (ciliate). *Lodicules* present; 2; free; membranous; glabrous (?); toothed. *Stamens* 3. **Ovary** apically *hairy*. Stigmas 2.

Fruit, embryo and seedling. Fruit small (about 1.75 mm long); ellipsoid; longitudinally grooved; somewhat compressed dorsiventrally, or not noticeably compressed; with hairs confined to a terminal tuft. Hilum short (but elongated, almost a third as long as the fruit). Embryo small. Endosperm hard.

Abaxial leaf blade epidermis. *Costal/intercostal zonation* conspicuous. *Papillae* absent. Mid-intercostal long-cells fusiform; having straight or only gently undulating walls. *Microhairs* absent. *Stomata* common; 42–48 microns long. Subsidiaries parallel-sided. Guard-cells overlapped by the interstomatals. *Costal short-cells* conspicuously in long rows (in places), or neither distinctly grouped into long rows nor predominantly paired (in other places). Costal silica bodies horizontally-elongated crenate/sinuous (crenate, some short and approaching dunb-bells); not sharp-pointed.

Transverse section of leaf blade, physiology. C_3; XyMS+. Mesophyll without adaxial palisade. *Leaf blade* with distinct, prominent adaxial ribs. Midrib with one bundle only. *The lamina* symmetrical on either side of the midrib. Many of the smallest vascular bundles unaccompanied by sclerenchyma. Combined sclerenchyma girders present. Sclerenchyma all associated with vascular bundles.

Taxonomy. Pooideae; Poodae; Aveneae (?).

Ecology, geography, regional floristic distribution. 1 species. Chile and Argentina. Mesophytic; shade species. At margins of woods.

Neotropical. Pampas and Andean.

References, etc. Morphological/taxonomic: data from Nicora and Rúgolo de Agrasar 1987. Leaf anatomical: this project.

Rendlia Chiov.

~ *Microchloa*

Habit, vegetative morphology. Perennial; densely caespitose (from a cushion of old, fibrous leaf sheaths). *Culms* 5–35 cm high; herbaceous; unbranched above. Culm nodes glabrous. Culm internodes solid. Plants unarmed. Young shoots intravaginal. *Leaves* mostly basal; non-auriculate. The basal sheaths persistent, densely woolly, fibrous when old. *Leaf blades* linear; narrow; to *0.7 mm wide*; setaceous; folded (at the base, the adaxial surfaces adnate); without abaxial multicellular glands; without cross venation; persistent (but usually burned off annually). *Ligule* present; a fringed membrane (the 'membrane' unusually firm); 0.2–0.3 mm long. *Contra-ligule* absent.

Reproductive organization. Plants bisexual, with bisexual spikelets; with hermaphrodite florets. The spikelets of sexually distinct forms on the same plant, or all alike in sexuality; hermaphrodite, or hermaphrodite and sterile (the uppermost 2–3 spikelets reduced).

Inflorescence. Inflorescence a single spike (to 5 cm long, rarely a pair of spikes). *Inflorescence axes not ending in spikelets (their tips obtuse).* Rachides flattened and winged (convex on back, with membranous, hairy margins). Inflorescence espatheate; not comprising 'partial inflorescences' and foliar organs. *Spikelet-bearing axes* spikes; persistent. Spikelets solitary; secund; alternately biseriate; sessile; imbricate.

Female-fertile spikelets. *Spikelets* 4–5.5 mm long; adaxial; with the G_1 compressed obliquely, the G_2 compressed dorsiventrally; *disarticulating between the glumes (disarticulating above the persistent G_1, the G_2 falling with and enveloping the florets)*; not disarticulating between the florets; with conventional internode spacings. Rachilla prolonged beyond the uppermost female-fertile floret; rachilla hairless. The rachilla extension with incomplete florets. Hairy callus present (small). *Callus* short; pointed.

Glumes two; more or less equal; exceeding the spikelets; long relative to the adjacent lemmas (both longer than the florets); lateral to the rachis; hairy (lanose, G_2), or hairless (glabrous, G_1); without conspicuous tufts or rows of hairs; obtuse, lanceolate-oblong; awnless; carinate (G_1), or non-carinate (G_2); very dissimilar (firmly membranous, the G_1 obliquely laterally compressed, the G_2 dorsally compressed). Lower glume 1 nerved. Upper glume 1 nerved. *Spikelets* with incomplete florets. The incomplete florets distal to the female-fertile florets. The distal incomplete florets 1; *incomplete florets* male or sterile, banana-shaped, hairless, the lemma shorter and thinner than the L_1, the palea reduced or absent; incomplete florets awnless.

Female-fertile florets 1. Lemmas similar in texture to the glumes; not becoming indurated; incised; shortly 2 lobed; not deeply cleft; awnless; hairy (with dense, silky hairs on the keels); non-carinate (but rather, 3 keeled: laterally compressed, the margins beyond the lateral keels narrow, hyaline, infolded); without a germination flap; 3 nerved. *Palea* present; relatively long (slightly exceeding the lemma); apically notched; awnless, without apical setae (glabrous); thinner than the lemma; not indurated (hyaline-membranous); 2-nerved; 2-keeled. Palea keels wingless; minutely scabrous. *Lodicules* present (briefly joined to the base of the palea); 2; free; fleshy; glabrous. *Stamens* 3. Anthers 2.5 mm long; not penicillate; without an apically prolonged connective. *Ovary* glabrous (suppressed in the upper floret). Styles free to their bases. Stigmas 2; brown.

Fruit, embryo and seedling. *Fruit* free from both lemma and palea; small (2 mm long); ellipsoid; compressed dorsiventrally, or trigonous. Hilum short. Embryo small (seemingly, judging from immature material).

Abaxial leaf blade epidermis. *Costal/intercostal zonation* conspicuous. *Papillae* absent. Long-cells of similar wall thickness costally and intercostally (medium thick walled). Mid-intercostal long-cells rectangular; having markedly sinuous walls (and pitted). *Microhairs* present; more or less spherical; ostensibly one-celled (15 to 18 microns in diameter); chloridoid-type. *Stomata* common; 27–28.5–30 microns long. Subsidiaries high dome-shaped, or triangular. Guard-cells overlapping to flush with the interstomatals. *Intercostal short-cells* common; not paired (solitary); not silicified. Intercostal silica bodies absent. *Costal short-cells* conspicuously in long rows. Costal silica bodies confined to the central file(s) of the costal zones; almost exclusively saddle shaped (large); not sharp-pointed.

Transverse section of leaf blade, physiology. Leaf blades consisting of midrib (there being only a flange of lamina and two bundles on each side, these increasingly reduced acropetally). C_4; XyMS+. PCR sheaths of the primary vascular bundles interrupted; interrupted both abaxially and adaxially. PCR sheath extensions absent. PCR cell chloroplasts centripetal. Mesophyll without arm cells (although with some hints of wall intrusions in the material seen). Midrib having a conventional arc of bundles (a large arc of about 16 bundles, with the median slightly larger than the rest); with colourless mesophyll adaxially (the entire adaxial area of the leaf being 'colourless'). All the vascular bundles accompanied by sclerenchyma. Combined sclerenchyma girders present (with the large lamina-flange bundles only); forming 'figures' (the lamina-flange bundles only). Sclerenchyma all associated with vascular bundles. The lamina margins with fibres.

Taxonomy. Chloridoideae; main chloridoid assemblage.

Ecology, geography, regional floristic distribution. 1 species. Eastern tropical and southern Africa. Mesophytic; species of open habitats; glycophytic. Shallow soils in grasslands.

Paleotropical. African. Sudano-Angolan. South Tropical African.

References, etc. Leaf anatomical: this project.

Reynaudia Kunth

Habit, vegetative morphology. Perennial; caespitose. **Culms *15–40 cm high***; herbaceous; unbranched above. Culm nodes hairy. *Leaves* mostly basal; non-auriculate. Leaf blades narrow (slender); rolled (involute); without cross venation; persistent.

Reproductive organization. Plants bisexual, with bisexual spikelets; with hermaphrodite florets.

Inflorescence. *Inflorescence paniculate*; contracted (rather narrow, 3–6 cm long); espatheate; not comprising 'partial inflorescences' and foliar organs. Spikelet-bearing axes persistent. Spikelets pedicellate.

Female-fertile spikelets. *Spikelets* 3–4 mm long; *compressed laterally*; falling with the glumes. Rachilla terminated by a female-fertile floret. Hairy callus absent (?).

Glumes present; two; very unequal; (the upper) shorter than the spikelets to about equalling the spikelets (the upper about half as long); the upper shorter than the adjacent lemmas to long relative to the adjacent lemmas (sometimes little shorter than the spikelet); hairless; *awned (both long-awned from between the lobes of the bifid apex)*; carinate; firmly membranous. Lower glume strongly 3 nerved, or 5 nerved. Upper glume strongly 5 nerved. *Spikelets* with incomplete florets. The incomplete florets proximal to the female-fertile florets. *The proximal incomplete florets* 1; sterile. The proximal lemmas awned (shortly, from the apical notch), or awnless (but then mucronate); 3 nerved; decidedly exceeding the female-fertile lemmas; decidedly firmer than the female-fertile lemmas (similar to the glumes).

Female-fertile florets 1. Lemmas less firm than the glumes; thin; not deeply cleft; awnless, or mucronate (the nerve excurrent from the keel); carinate; without a germination flap; 5 nerved (with 2 pairs of obscure laterals). **Palea** *absent*. *Lodicules* absent (?). **Stamens** *2*.

Fruit, embryo and seedling. Fruit ellipsoid; compressed dorsiventrally.

Abaxial leaf blade epidermis. *Costal/intercostal zonation* conspicuous. *Papillae* absent. *Long-cells* similar in shape costally and intercostally (but the costals smaller and narrower); of similar wall thickness costally and intercostally (walls of medium thickness). Mid-intercostal long-cells rectangular; having markedly sinuous walls (pitted). *Microhairs* present; panicoid-type; 48–54–60 microns long; (9–)9.3–9.9(–10.2) microns wide at the septum. Microhair total length/width at septum 4.8–6.7. Microhair apical cells (27–)33–36(–39) microns long. Microhair apical cell/total length ratio 0.56–0.67. *Stomata* common; 27(–30) microns long. Subsidiaries triangular; including both triangular and parallel-sided forms on the same leaf. Guard-cells overlapping to flush with the interstomatals. *Intercostal short-cells* common; in cork/silica-cell pairs; silicified. Intercostal silica bodies crescentic. *Costal short-cells* predominantly paired. Costal silica bodies rounded (a few), or tall-and-narrow (very few), or crescentic (predominating); not sharp-pointed.

Transverse section of leaf blade, physiology. C_4; XyMS–. Mesophyll without adaxial palisade. *Leaf blade* with distinct, prominent adaxial ribs; with the ribs more or less constant in size (round-topped). *Midrib* not readily distinguishable; with one bundle only; with colourless mesophyll adaxially (each rib with an adaxial group of large colourless cells). Bulliforms present in discrete, regular adaxial groups; in simple fans (these small, in each furrow). All the vascular bundles accompanied by sclerenchyma. Combined sclerenchyma girders absent (each bundle with a wide, narrow adaxial strand and an abaxial girder). Sclerenchyma all associated with vascular bundles.

Taxonomy. Panicoideae; Panicodae; Paniceae (Arthropogoneae).

Ecology, geography, regional floristic distribution. 1 species. West Indies. Mesophytic; species of open habitats. Savanna, palm barrens, bushy slopes. Neotropical. Caribbean.

References, etc. Leaf anatomical: this project.

Special comments. Fruit data wanting.

Rhipidocladum McClure

Habit, vegetative morphology. Perennial. The flowering culms leafy. *Culms woody and persistent (slender, arborescent or scrambling)*; scandent to not scandent; branched above. *Primary branches/mid-culm node 1*. Culm internodes hollow. Unicaespitose. Rhizomes pachymorph. Plants unarmed. *Leaves* not basally aggregated; with auricular setae. Leaf blades broad; pseudopetiolate; cross veined, or without cross venation; disarticulating from the sheaths; rolled in bud.

Reproductive organization. Plants bisexual, with bisexual spikelets; with hermaphrodite florets.

Inflorescence. *Inflorescence* spatheate; *a complex of 'partial inflorescences' and intervening foliar organs*. *Spikelet-bearing axes* lax, spikes, or 'racemes'; persistent. *Spikelets* solitary (usually), or paired; secund (the racemes unilateral), or not secund (the racemes bilateral, zigzag); *sessile, or subsessile*; distant.

Female-fertile spikelets. Spikelets 20–60 mm long; compressed laterally (?); disarticulating above the glumes; disarticulating between the florets. Rachilla prolonged beyond the uppermost female-fertile floret.

Glumes two, or several (?); very unequal; shorter than the adjacent lemmas; awnless. *Spikelets* with incomplete florets. The incomplete florets distal to the female-fertile florets, or both distal and proximal to the female-fertile florets. *The distal incomplete florets* merely underdeveloped. *The proximal incomplete florets* 1. The proximal lemmas awnless; exceeded by the female-fertile lemmas; similar in texture to the female-fertile lemmas; not becoming indurated.

Female-fertile florets 2–8 (? 'few to several'). Lemmas similar in texture to the glumes; not becoming indurated; entire; pointed, or blunt; awnless. *Palea* present; entire; awnless, without apical setae; not indurated; 2-keeled (and sulcate). *Lodicules* present; 3 (rarely 0–2); free; membranous; ciliate. **Stamens** *3*. Anthers not penicillate; without an apically prolonged connective. **Ovary** glabrous; ***with a conspicuous apical appendage***. The appendage broadly conical, fleshy. Styles basally fused. Stigmas 2.

Fruit, embryo and seedling. Fruit longitudinally grooved, or not grooved. Hilum long-linear. Embryo small. Endosperm containing compound starch grains.

Abaxial leaf blade epidermis. *Costal/intercostal zonation* conspicuous. *Papillae* present; costal and intercostal (conspicuous, especially around the stomata and around the costal short-cells). Intercostal papillae over-arching the stomata; several per cell (most cells with one or two rows of circular papillae). *Long-cells* similar in shape costally and intercostally; of similar wall thickness costally and intercostally. Mid-intercostal long-cells rectangular; having markedly sinuous walls. *Microhairs* present; elongated; clearly two-celled; panicoid-type; 39–51 microns long; 6–8.4 microns wide at the septum. Micro-hair total length/width at septum 5.9–6.3. Microhair apical cells 21–27 microns long. Microhair apical cell/total length ratio 0.47–0.54. *Stomata* common; 24–28.5 microns long. Subsidiaries perhaps papillate, or it being hard to distinguish whether they come from the adjoining cells. Guard-cells overlapped by the interstomatals (the stomata and subsidiaries sunken, obscured by papillae). *Intercostal short-cells* common; not paired (in *R. racemiflorum*, and seemingly solitary but obscured by special quartets and rosettes of papillae in *R. pittieri*). *Costal short-cells* neither distinctly grouped into long rows nor predominantly paired (mostly seemingly solitary, obscured in *R. pittieri* by special papillae). Costal silica bodies absent, or poorly developed (in the material of the two species seen); (or at least, the silica cells) tall-and-narrow.

Transverse section of leaf blade, physiology. C_3; XyMS+. *Mesophyll* with non-radiate chlorenchyma; with adaxial palisade to without adaxial palisade; without arm cells; with fusoids (conspicuous in both *R. pittieri* and *R. racemiflorum*, though less so in the latter and more easily detected towards the edges of the lamina). The fusoids external to the PBS. *Leaf blade* adaxially flat (except for a single adaxial rib near one margin). *Midrib* conspicuous; with one bundle only. *The lamina* symmetrical on either side of the midrib (unless the marginal rib be interpreted as a highly asymmetric 'midrib'). Bulli-forms present in discrete, regular adaxial groups; in simple fans. All the vascular bundles accompanied by sclerenchyma. Combined sclerenchyma girders present. Sclerenchyma not all bundle-associated. The 'extra' sclerenchyma in abaxial groups and in adaxial groups; abaxial-hypodermal, the groups isolated and adaxial-hypodermal, contiguous with the bulliforms (there being groups of abaxial hypodermal fibres opposite the bulli-forms, and adaxial groups or isolated fibres lining or adjoining them).

Taxonomy. Bambusoideae; Bambusodae; Bambuseae.

Ecology, geography, regional floristic distribution. 11 species. Central & South America.

Neotropical. Caribbean.

References, etc. Leaf anatomical: this project.

Rhizocephalus Boiss.

Habit, vegetative morphology. Dwarf annual. **Culms *1–3 cm high (i.e., almost without culms)***; herbaceous. *Leaves* mostly basal; non-auriculate. **Sheath margins free**. Leaf blades linear-lanceolate; narrow; 1–4 mm wide; flat, or folded; without cross vena-tion. Ligule an unfringed membrane; not truncate; 2 mm long.

Reproductive organization. Plants bisexual, with bisexual spikelets; with hermaph-rodite florets. The spikelets all alike in sexuality.

Inflorescence. *Inflorescence paniculate*; contracted; capitate; espatheate (but the head-like panicles partly concealed among the leaves); not comprising 'partial inflores-cences' and foliar organs. Spikelet-bearing axes persistent. Spikelets not secund (?); shortly pedicellate.

Female-fertile spikelets. *Spikelets* 3–7 mm long; ***not noticeably compressed***; falling with the glumes. Rachilla terminated by a female-fertile floret. Hairy callus absent.

Glumes two; more or less equal; shorter than the adjacent lemmas; shortly **hairy (with clavate hairs)**; pointed (acute); awnless; non-carinate; similar (thinly learthery). Lower glume 3 nerved. Upper glume 3 nerved. **Spikelets** with female-fertile florets only; ***without proximal incomplete florets***.

Female-fertile florets 1. Lemmas acuminate; decidedly firmer than the glumes; not becoming indurated; entire; pointed; awnless, or mucronate; hairy; carinate; without a germination flap; 5 nerved (the lateral nerves much thickened). *Palea* present; relatively long; entire (lanceolate); awnless, without apical setae; thinner than the lemma; not indurated (hyaline); 2-nerved; 2-keeled, or keel-less. *Lodicules* absent. **Stamens 2**. Anthers 0.5–0.8 mm long; not penicillate; without an apically prolonged connective. *Ovary* glabrous. Styles fused. Stigmas 2.

Fruit, embryo and seedling. Fruit small (about 2 mm long); ellipsoid (rostrate); compressed dorsiventrally. Hilum short. Embryo small.

Abaxial leaf blade epidermis. *Costal/intercostal zonation* conspicuous. *Papillae* absent. *Long-cells* fairly similar in shape costally and intercostally (the costals smaller); of similar wall thickness costally and intercostally to differing markedly in wall thickness costally and intercostally (the costals somewhat thicker walled). Mid-intercostal long-cells rectangular and fusiform; having markedly sinuous walls to having straight or only gently undulating walls (heavily pitted, the sinuosity fine, more apparent at lower levels of focus). *Microhairs* absent. *Stomata* common (in quite deep depressions between the interstomatals, the guard-cells quite sausage-shaped); (33–)36–42(–45) microns long. Subsidiaries non-papillate; parallel-sided. Guard-cells overlapped by the interstomatals. *Intercostal short-cells* absent or very rare. With a few blunt costal prickles. *Costal short-cells* neither distinctly grouped into long rows nor predominantly paired (mostly solitary). Costal silica bodies present and well developed; horizontally-elongated crenate/sinuous and horizontally-elongated smooth (numerous, often quite irregular).

Transverse section of leaf blade, physiology. C_3; XyMS+. *Leaf blade* with distinct, prominent adaxial ribs; with the ribs more or less constant in size. *Midrib* conspicuous to not readily distinguishable (fairly obvious from the infolding of the blade, but scarcely so from bundle size or sclerenchyma layout); with one bundle only. *The lamina* symmetrical on either side of the midrib. Bulliforms not in discrete, regular adaxial groups (no bulliforms apparent). All the vascular bundles accompanied by sclerenchyma. Combined sclerenchyma girders absent (each bundle limited to small abaxial and adaxial strands). Sclerenchyma all associated with vascular bundles.

Taxonomy. Pooideae; Poodae; Aveneae.

Ecology, geography, regional floristic distribution. 2 species. Southwest and central Asia. Xerophytic; species of open habitats. Arid places.

Holarctic. Tethyan. Irano-Turanian.

References, etc. Leaf anatomical: this project.

Rhomboelytrum Link

Rhombolytrum Link
Excluding *Gymnachne*

Habit, vegetative morphology. Perennial; rhizomatous, or caespitose. *Culms* about 10–50 cm high; herbaceous; unbranched above. *Leaves* not basally aggregated; non-auriculate. Leaf blades linear; narrow; flat, or rolled; without cross venation; persistent. *Ligule an unfringed membrane.*

Reproductive organization. Plants bisexual, with bisexual spikelets; with hermaphrodite florets. Exposed-cleistogamous, or chasmogamous.

Inflorescence. *Inflorescence paniculate*; contracted; spicate to more or less irregular; espatheate; not comprising 'partial inflorescences' and foliar organs. Spikelet-bearing axes persistent. Spikelets not secund; pedicellate; imbricate.

Female-fertile spikelets. Spikelets 4.5–7 mm long; rhomboidal or oblong; compressed laterally (but the florets dorsiventrally flattened); disarticulating above the glumes; disarticulating between the florets; with conventional internode spacings. Rachilla prolonged beyond the uppermost female-fertile floret; rachilla hairless. The rachilla extension with incomplete florets. Hairy callus present. *Callus* short.

Glumes two; more or less equal; shorter than the spikelets; long relative to the adjacent lemmas; hairless; scabrous; awnless; carinate (the keel scabrid); similar (lanceolate or naviculate). Lower glume about equalling the lowest lemma; 3 nerved, or 5 nerved.

Upper glume 3 nerved, or 5 nerved. *Spikelets* with incomplete florets. The incomplete florets distal to the female-fertile florets. *The distal incomplete florets* merely underdeveloped; awnless.

Female-fertile florets 4–10. *Lemmas* broadly lanceolate or rhomboid, with thickened margins in the lower third; **becoming indurated (on the lower third of the margins, thin towards the tip)**; entire, or incised; pointed; when incised, 2 lobed; when incised, not deeply cleft (bidentate); awnless to mucronate; hairy (especially on the margins below); non-carinate; without a germination flap; 5 nerved, or 7 nerved, or 9 nerved. *Palea* present; relatively long (lanceolate); tightly clasped by the lemma; entire to apically notched; awnless, without apical setae; thinner than the lemma (papery); not indurated; 2-nerved; 2-keeled (flat between the hairy or fringed keels). Palea keels slightly winged; hairy. *Lodicules* present; 2; free; membranous; glabrous; not toothed. **Stamens 1.** Anthers very small in cleistogamous florets. *Ovary* glabrous. Styles free to their bases (subsessile). Stigmas 2.

Fruit, embryo and seedling. *Fruit* adhering to lemma and/or palea (to the palea); small; ellipsoid; compressed dorsiventrally. Hilum short. Embryo small. Endosperm hard.

Abaxial leaf blade epidermis. *Costal/intercostal zonation* conspicuous. *Papillae* absent. *Long-cells* markedly different in shape costally and intercostally (the costals narrow, rectangular); of similar wall thickness costally and intercostally (fairly thick walled). Mid-intercostal long-cells mostly fusiform; having markedly sinuous walls. *Microhairs* absent. *Stomata* common. Subsidiaries parallel-sided to triangular (many approaching the 'triangular with truncated apices' configuration). Guard-cells overlapped by the interstomatals. *Intercostal short-cells* common; in cork/silica-cell pairs; silicified. Intercostal silica bodies tall-and-narrow. *Costal short-cells* neither distinctly grouped into long rows nor predominantly paired (mostly in short rows to paired). Costal silica bodies horizontally-elongated crenate/sinuous (a few, but short and irregular so as sometimes to approach panicoid type crosses etc.), or rounded (a few, variously distorted), or crescentic (common, but variously distorted), or 'panicoid-type' (a few); a few assignable as cross shaped; not sharp-pointed.

Transverse section of leaf blade, physiology. C_3; XyMS+. *Mesophyll* with non-radiate chlorenchyma; without adaxial palisade. *Leaf blade* 'nodular' in section; with the ribs more or less constant in size (mostly round topped). *Midrib* not readily distinguishable; with one bundle only. *The lamina* symmetrical on either side of the midrib. Bulliforms present in discrete, regular adaxial groups (a large group in each furrow); in simple fans. All the vascular bundles accompanied by sclerenchyma. Combined sclerenchyma girders present (with all the bundles); forming 'figures' (slender to heavy I's). Sclerenchyma all associated with vascular bundles.

Special diagnostic feature. Lemmas not as in *Briza* (q.v.).

Taxonomy. Pooideae; Poodae; Poeae.

Ecology, geography, regional floristic distribution. 2 species. South America. Species of open habitats. Stony slopes.

Neotropical and Antarctic. Andean. Patagonian.

References, etc. Morphological/taxonomic: Nicora and Rúgolo de Agrasar 1981. Leaf anatomical: this project.

Rhynchelytrum Nees

Monachyron Parl.
~ *Melinis*; cf. *Tricholaena* — Fosberg and Sachet 1981

Habit, vegetative morphology. Annual (rarely), or perennial; rhizomatous, or stoloniferous, or caespitose, or decumbent. **Culms** 20–120 cm high; **herbaceous**; branched above, or unbranched above. Culm nodes hairy. Culm internodes hollow. *Leaves* not basally aggregated; auriculate (rarely), or non-auriculate. Leaf blades linear; broad (rarely), or narrow; flat, or rolled; without cross venation; persistent; rolled in bud. Ligule a fringed membrane to a fringe of hairs. *Contra-ligule* absent.

Reproductive organization. *Plants* bisexual, with bisexual spikelets; **with hermaphrodite florets**.

Inflorescence. Inflorescence paniculate, or of spicate main branches (often decompound, rarely composed of secund racemes); open, or contracted; with capillary branchlets, or without capillary branchlets; espatheate; not comprising 'partial inflorescences' and foliar organs. Spikelet-bearing axes persistent. Spikelets not secund; pedicellate. Pedicel apices discoid (often hairy).

Female-fertile spikelets. *Spikelets* 2–11 mm long; elliptic, or ovate; *compressed laterally (often asymmetric)*; disarticulating above the glumes (the fruiting floret falling first), or falling with the glumes (falling from the pedicel); with conventional internode spacings, or with a distinctly elongated rachilla internode between the glumes. Rachilla terminated by a female-fertile floret. Hairy callus present, or absent.

Glumes two; very unequal; (the upper) long relative to the adjacent lemmas; hairy (G_2 usually villous or pubescent below, rarely glabrous); awned (G_2 only, sometimes awned or mucronate or beaked upwards), or awnless; *very dissimilar (G_1 a scale up to 1/3 spikelet length, or reduced to a vestige or rim, G_2 longer, apically emarginate or bifid, and awned or mucronate from the sinus, firmly membranous or papery, curved on the back)*. Lower glume 0–1 nerved. Upper glume 5–7 nerved (often gibbous below). *Spikelets* with incomplete florets. The incomplete florets proximal to the female-fertile florets. *Spikelets with proximal incomplete florets*. *The proximal incomplete florets* 1; paleate, or epaleate (rarely). Palea of the proximal incomplete florets when present, fully developed to reduced. The proximal incomplete florets male, or sterile. *The proximal lemmas awned*; 5 nerved; decidedly exceeding the female-fertile lemmas (similar to the G_2, or narrower and less gibbous); *similar in texture to the female-fertile lemmas*; not becoming indurated.

Female-fertile florets 1. *Lemmas* less firm than the glumes to similar in texture to the glumes (membranous to papery); smooth (shiny); not becoming indurated; white in fruit; entire (truncate), or incised (emarginate or minutely two-lobed); when incised, 2 lobed; not deeply cleft; *awnless*; hairless (usually glabrous, rarely ciliate); non-carinate; having the margins lying flat and exposed on the palea; without a germination flap; 5 nerved. *Palea* present; relatively long; apically notched; 2-nerved. *Lodicules* present; 2; free; fleshy; glabrous. *Stamens* 3. Anthers not penicillate. *Ovary* glabrous. Styles fused, or free to their bases. Stigmas 2; brown.

Fruit, embryo and seedling. *Fruit* free from both lemma and palea; small; ellipsoid; compressed laterally, or not noticeably compressed. Hilum short. Embryo large; waisted, or not waisted; without an epiblast; with a scutellar tail; with an elongated mesocotyl internode. Embryonic leaf margins overlapping.

Seedling with a long mesocotyl. First seedling leaf with a well-developed lamina. The lamina broad; curved; 18 veined.

Abaxial leaf blade epidermis. *Costal/intercostal zonation* conspicuous. *Papillae* absent. *Long-cells* similar in shape costally and intercostally; of similar wall thickness costally and intercostally. Mid-intercostal long-cells rectangular; having markedly sinuous walls. *Microhairs* present; panicoid-type; 60–75–87 microns long; (4–)4.5–5.4(–6) microns wide at the septum. Microhair total length/width at septum 12–19.3. Microhair apical cells (35–)40–47(–48) microns long. Microhair apical cell/total length ratio 0.55–0.67. *Stomata* common; (27–)28.5–30(–36) microns long. Subsidiaries triangular. Guard-cells overlapping to flush with the interstomatals. *Intercostal short-cells* common; in cork/silica-cell pairs and not paired (solitaries); silicified (when paired), or not silicified. Intercostal silica bodies cross-shaped. *Costal short-cells* conspicuously in long rows. Costal silica bodies 'panicoid-type'; cross shaped to dumb-bell shaped, or nodular; not sharp-pointed.

Transverse section of leaf blade, physiology. C_4; biochemical type PCK (*R. repens*); XyMS+. PCR sheath outlines uneven. PCR sheath extensions absent. *PCR cell chloroplasts* ovoid; with well developed grana; centrifugal/peripheral. *Mesophyll* with radiate chlorenchyma. *Leaf blade* adaxially flat. *Midrib* conspicuous, or not readily distinguishable; with one bundle only, or having a conventional arc of bundles (rarely); with colourless mesophyll adaxially, or without colourless mesophyll adaxially. Bulliforms present in discrete, regular adaxial groups; in simple fans and associated with colourless mesophyll cells to form deeply-penetrating fans. Many of the smallest vascular bundles unaccompanied by sclerenchyma. Combined sclerenchyma girders present; forming 'figures'. Sclerenchyma all associated with vascular bundles.

Culm anatomy. *Culm internode bundles* in three or more rings.

Phytochemistry. Leaves without flavonoid sulphates (2 species). Leaf blade chlorophyll *a:b* ratio 3.63.

Cytology. Chromosome base number, $x = 9$. $2n = 36$. 4 ploid.

Taxonomy. Panicoideae; Panicodae; Paniceae (Melinideae).

Ecology, geography, regional floristic distribution. 14 species. Tropical Africa, Madagascar, Arabia to Indochina. Commonly adventive. Species of open habitats; glycophytic. Savanna and grassland, often in disturbed ground.

Holarctic, Paleotropical, and Neotropical. Boreal and Tethyan. African, Madagascan, and Indomalesian. Atlantic North American. Macaronesian and Irano-Turanian. Saharo-Sindian, Sudano-Angolan, West African Rainforest, and Namib-Karoo. Indo-Chinese, Malesian, and Papuan. Caribbean. Central Grasslands. Sahelo-Sudanian, Somalo-Ethiopian, South Tropical African, and Kalaharian.

Rusts and smuts. Rusts — *Puccinia*. Taxonomically wide-ranging species: *Puccinia levis*. Smuts from Tilletiaceae. Tilletiaceae — *Entyloma*.

Economic importance. Significant weed species: *R. repens*.

References, etc. Leaf anatomical: Metcalfe 1960 and this project.

Rhynchoryza Baillon

Habit, vegetative morphology. Perennial. Culms herbaceous. Culm nodes glabrous. *Leaves* not basally aggregated; non-auriculate; without auricular setae. Leaf blades linear; narrow; 5–10 mm wide; flat; seemingly pseudopetiolate (see illustration by Nicora and Rúgolo de Agrasar 1987); without cross venation; persistent. Ligule an unfringed membrane; not truncate (acute); large.

Reproductive organization. Plants bisexual, with bisexual spikelets; with hermaphrodite florets.

Inflorescence. Inflorescence paniculate; open; espatheate; not comprising 'partial inflorescences' and foliar organs. Spikelet-bearing axes persistent. Spikelets not secund; pedicellate.

Female-fertile spikelets. *Spikelets* unconventional (owing to extreme reduction of the glumes, cf. *Oryza*); about 15–25 mm long; compressed laterally; disarticulating above the glumes (in so far as these can be said to be present). Rachilla terminated by a female-fertile floret. Hairy callus absent. *Callus* absent.

Glumes *present to absent (vestigial, cf.* Oryza*)*; if considered present, two; minute; more or less equal; shorter than the spikelets; shorter than the adjacent lemmas; joined; awnless. Lower glume 0 nerved. Upper glume 0 nerved. *Spikelets* with incomplete florets. The incomplete florets proximal to the female-fertile florets. **The proximal incomplete florets** 2 (the lower larger); epaleate; *sterile (much reduced, cf. Oryza)*. *The proximal lemmas broadly ovate, cuspidate*; awnless; 3 nerved; *exceeded by the female-fertile lemmas (only about 1/10 the length of the spikelet)*; not becoming indurated.

Female-fertile florets 1. **Lemmas leathery below, with the tip beneath the awn in the form of a beak extending beyond the palea, containing transversely septate aerenchyma and specialised as a flotation device**; becoming indurated; entire; pointed; awned. Awns 1; apical; non-geniculate; hairless. Lemmas hairless; carinate; without a germination flap; 5 nerved. *Palea* present; relatively long; textured like the lemma; several nerved; one-keeled. *Lodicules* present; 2; membranous. *Stamens* 6. *Ovary* glabrous. Styles free to their bases. Stigmas 2.

Fruit, embryo and seedling. Fruit not noticeably compressed ('cylindrical to somewhat hexagonal in section'). Hilum long-linear. Embryo small. Endosperm hard.

Abaxial leaf blade epidermis. *Costal/intercostal zonation* conspicuous. *Papillae* present; intercostal (in the broad stomatal zones — adjoining the costae – only). Intercostal papillae over-arching the stomata (often from the interstomatals, as well as from the subsidiaries); several per cell (round, rather irregular in size and distribution on each cell). *Long-cells* similar in shape costally and intercostally; of similar wall thickness costally and intercostally. Mid-intercostal long-cells rectangular; having markedly sinuous walls.

Microhairs present; elongated; ostensibly one-celled; panicoid-type. *Stomata* common. Subsidiaries papillate; predominantly triangular. *Intercostal short-cells* common (especially in the astomatal regions); in cork/silica-cell pairs; silicified. Intercostal silica bodies narrowly oryzoid-type. *Costal short-cells* predominantly paired (and a few short rows). Costal silica bodies predominantly oryzoid (these abundant), or tall-and-narrow (a few, representing poorly developed oryzoids); not sharp-pointed.

Transverse section of leaf blade, physiology. C_3; XyMS+. *Mesophyll* with non-radiate chlorenchyma; without adaxial palisade; with arm cells (conspicuous); with fusoids. The fusoids external to the PBS (these unusual, in that they are often in pairs contiguous on their long axes, and sometimes with a third contiguous member inside, projecting further into the intercostal region). *Leaf blade* adaxially flat. *Midrib* very conspicuous (being much enlarged and aerenchymatous, with conspicuous groups of stellate chlorenchyma); having complex vascularization (with a complete, peripheral ring of bundles, the adaxial series being smaller but not 'inverted', and with a few small 'disorientated' bundles in the internal trabeculae between the air cavities); with colourless mesophyll adaxially (and large lacunae). Bulliforms present in discrete, regular adaxial groups; in simple fans. All the vascular bundles accompanied by sclerenchyma. Combined sclerenchyma girders present; forming 'figures' (most bundles with I's). Sclerenchyma all associated with vascular bundles.

Special diagnostic feature. *Female-fertile lemma with its tip extended beyond the palea as a conical, herbaceous beak (flotation device) composed of aerenchyma with transverse septa, tapering into an awn.*

Taxonomy. Bambusoideae; Oryzodae; Oryzeae.

Ecology, geography, regional floristic distribution. 1 species. Paraguay to Argentina. Helophytic; glycophytic.

Neotropical. Pampas.

References, etc. Leaf anatomical: this project.

Rhytachne Desv.

Lepturopsis Steud.

Habit, vegetative morphology. Annual (rarely), or perennial; caespitose. *Culms* 25–120 cm high; herbaceous; unbranched above; few noded. Culm nodes glabrous. Culm internodes solid. Young shoots intravaginal. *Leaves* not basally aggregated; auriculate (from the apex of the sheath), or non-auriculate (but with long hairs in the auricular position). Leaf blades linear; narrow (to filiform); 1–9 mm wide; setaceous to not setaceous (often rolled); flat, or folded, or rolled, or acicular; without cross venation; persistent. *Ligule an unfringed membrane*; truncate; 0.75–2 mm long. *Contra-ligule* absent.

Reproductive organization. Plants bisexual, with bisexual spikelets; with hermaphrodite florets. The spikelets nearly always of sexually distinct forms on the same plant; hermaphrodite and male-only, or hermaphrodite and sterile; overtly heteromorphic (the pedicellate spikelets nearly always much reduced); all in heterogamous combinations.

Inflorescence. *Inflorescence a single raceme (of single, usually terminal 'racemes' these cylindrical and themselves culm-like until the embedded spikelets open).* Rachides hollowed (the articles hollowed to take the contiguous sessile spikelets). Inflorescence spatheate, or espatheate; not comprising 'partial inflorescences' and foliar organs. **Spikelet-bearing axes** *spikelike; the spikelet-bearing axes with more than 10 spikelet-bearing 'articles' (12–30);* solitary; with substantial rachides; *disarticulating;* disarticulating at the joints. 'Articles' non-linear (clavate, as long as or longer than the sessile spikelet); with a basal callus-knob; appendaged, or not appendaged (the summit of the joint concave to cupular); disarticulating transversely; somewhat hairy (at the apex and onto the pedicels), or glabrous. Spikelets paired (but the pedicellate member sometimes reduced to a sclae-tipped pedicel); secund (the 'racemes' somewhat dorsiventral, the sessile spikelets in two alternating rows on one side); sessile and subsessile, or sessile and pedicellate; consistently in 'long-and-short' combinations; in pedicellate/sessile combinations. *Pedicels of the 'pedicellate' spikelets free of the rachis.* The 'shorter' spikelets hermaphrodite. The 'longer' spikelets male-only, or sterile (variously reduced, sometimes suppressed), or hermaphrodite (*R. perfecta*).

Female-sterile spikelets. *The pedicelled member usually suppressed, vestigial or represented by an awn, occasionally well developed (even bisexual, in* **R. perfecta***). The pedicel usually foliaceous.*

Female-fertile spikelets. Spikelets 2–8 mm long; abaxial; compressed dorsiventrally; planoconvex, or biconvex; falling with the glumes; with conventional internode spacings. Rachilla terminated by a female-fertile floret. Hairy callus present, or absent. *Callus* absent, or short; blunt.

Glumes two; more or less equal; about equalling the spikelets; long relative to the adjacent lemmas (exceeding them); dorsiventral to the rachis; hairless; glabrous; pointed; awned (G_1 and/or G_2, sometimes), or awnless; non-carinate; very dissimilar (G_1 leathery, convex, often transversely rugulose, G_2 membranous or hyaline, with or without a terminal subule). Lower glume two-keeled (and sometimes obscurely winged), or not two-keeled (then the sides rounded); convex on the back; not pitted; relatively smooth, or rugose (or longitudinally ribbed); obscurely 5 nerved. Upper glume 3–5 nerved. *Spikelets* with incomplete florets. The incomplete florets proximal to the female-fertile florets. *The proximal incomplete florets* 1; paleate, or epaleate. Palea of the proximal incomplete florets reduced. The proximal incomplete florets male, or sterile (rarely). The proximal lemmas awnless; more or less equalling the female-fertile lemmas; similar in texture to the female-fertile lemmas (hyaline); not becoming indurated.

Female-fertile florets 1. *Lemmas* linear-lanceolate to oblong; less firm than the glumes (hyaline); not becoming indurated; entire; awnless; hairless; non-carinate; without a germination flap; *1–2 nerved (rarely 3)*. *Palea* present; relatively long, or conspicuous but relatively short, or very reduced (shorter than the lemma, or very short); entire; awnless, without apical setae; textured like the lemma; not indurated (hyaline); 2-nerved, or nerveless; keel-less. *Lodicules* present; 2; free; fleshy. *Stamens* 3. Anthers 1.5–2.5 mm long; not penicillate; without an apically prolonged connective. *Ovary* glabrous. Stigmas 2; brown.

Fruit, embryo and seedling. Fruit compressed dorsiventrally.

Abaxial leaf blade epidermis. *Costal/intercostal zonation* conspicuous, or lacking. *Papillae* absent. *Long-cells* similar in shape costally and intercostally, or markedly different in shape costally and intercostally (with the costals much smaller); of similar wall thickness costally and intercostally. Mid-intercostal long-cells rectangular; having markedly sinuous walls. *Microhairs* present, or absent; when present, panicoid-type. *Stomata* common, or absent or very rare. Subsidiaries triangular. *Intercostal short-cells* common; in cork/silica-cell pairs (abundant); silicified. Intercostal silica bodies rounded (to elliptical). *Costal short-cells* predominantly paired. Costal silica bodies rounded; not sharp-pointed.

Transverse section of leaf blade, physiology. Leaf blades consisting of midrib, or 'laminar'. C_4; XyMS– (seemingly). PCR cell chloroplasts centrifugal/peripheral. *Leaf blade* with distinct, prominent adaxial ribs, or 'nodular' in section, or adaxially flat. *Midrib* conspicuous (or the blade acicular and reduced to the midrib); having a conventional arc of bundles; with colourless mesophyll adaxially. Bulliforms present in discrete, regular adaxial groups, or not in discrete, regular adaxial groups (typical bulliforms sometimes absent, the epidermis of large cells); sometimes in simple fans. Many of the smallest vascular bundles unaccompanied by sclerenchyma. Sclerenchyma all associated with vascular bundles.

Cytology. $2n = 20$.

Taxonomy. Panicoideae; Andropogonodae; Andropogoneae; Rottboelliinae.

Ecology, geography, regional floristic distribution. 12 species. Tropical and southern Africa, Madagascar, tropical South America. Helophytic (pans and riversides), or mesophytic (grasslands); species of open habitats; glycophytic.

Paleotropical. African and Madagascan. Sudano-Angolan and West African Rainforest. Somalo-Ethiopian and South Tropical African.

Rusts and smuts. Smuts from Ustilaginaceae. Ustilaginaceae — *Sorosporium*, *Sphacelotheca*, and *Tolyposporella*.

References, etc. Morphological/taxonomic: Clayton 1978. Leaf anatomical: Metcalfe 1960.

Special comments. Fruit data wanting.

Richardsiella Elffers & Kennedy O'Byrne

Habit, vegetative morphology. Annual (with filiform culms); caespitose (or with solitary culms). *Culms* 7–18 cm high. Plants unarmed. *Leaves* not basally aggregated. Leaf blades linear; narrow; to 1.5 mm wide (1–6.5 cm long); flat, or rolled (involute); exhibiting multicellular glands abaxially (at the base of hairs). The abaxial glands intercostal. Leaf blades without cross venation; persistent. Ligule a fringe of hairs; 1 mm long (with longer lateral tufts). *Contra-ligule* absent.

Reproductive organization. Plants bisexual, with bisexual spikelets; with hermaphrodite florets.

Inflorescence. *Inflorescence of spicate main branches*; contracted (to 7 cm long, with few-to-many 5–12 mm long incurved racemes); *non-digitate*. Primary inflorescence branches 3–11 (rarely 1–2). *Inflorescence axes not ending in spikelets (the main axis and the lateral racemes ending in a naked scaberulous bristle)*. Inflorescence espatheate; not comprising 'partial inflorescences' and foliar organs. *Spikelet-bearing axes* 'racemes'; persistent. Spikelets solitary (alternate, contiguous); secund; biseriate; pedicellate (the pedicels short, bristly).

Female-fertile spikelets. Spikelets 1.6–2 mm long; compressed laterally; disarticulating above the glumes; disarticulating between the florets. Rachilla prolonged beyond the uppermost female-fertile floret. The rachilla extension with incomplete florets. Hairy callus not recorded: the 'callus' referred to in the original description is the pedicel, beneath the persistent glumes.

Glumes two; very unequal; (the upper) long relative to the adjacent lemmas (equalling or exceeding the spikelets); hairy (with spreading tubercle-based hairs on the keel, fewer of them on G_1); pointed (acutely acuminate); awned (aristate, aristate-acuminate or caudate-acuminate), or awnless (acutely acuminate); carinate (somewhat); similar (deciduous, lanceolate to narrowly ovate and acutely acuminate, membranous). Lower glume shorter than the lowest lemma to about equalling the lowest lemma; 1 nerved. Upper glume 1 nerved. *Spikelets* with incomplete florets. The incomplete florets distal to the female-fertile florets. *The distal incomplete florets merely underdeveloped*.

Female-fertile florets 6–12. Lemmas similar in texture to the glumes to decidedly firmer than the glumes (membranous to hyaline); not becoming indurated; entire, or incised; not deeply cleft (entire or shallowly emarginate); awnless to mucronate (apiculate or shortly mucronate); hairy (ciliate, with stiff, white hairs on the margins); carinate to non-carinate; 1 nerved. *Palea* present; relatively long; entire (truncate, oblong); awnless, without apical setae; 2-nerved; 2-keeled (but convex and gibbous below between the rigid keels). Palea keels hairy (pectinate-ciliate). *Lodicules* present ('minute'), or absent; when present, 2; glabrous. *Stamens* 2 (anthers 0.25–0.4 mm long, broadly oblong). Anthers 0.25–0.3 mm long; not penicillate; without an apically prolonged connective. *Ovary* glabrous. Styles free to their bases (slender). Stigmas 2.

Fruit, embryo and seedling. *Fruit* free from both lemma and palea; small (0.3–0.4 mm long); ellipsoid; not noticeably compressed (terete). Hilum short. Pericarp free. Embryo large; waisted.

Abaxial leaf blade epidermis. *Costal/intercostal zonation* conspicuous. *Papillae* absent. *Long-cells* similar in shape costally and intercostally; of similar wall thickness costally and intercostally (the walls rather thick). Mid-intercostal long-cells rectangular; having markedly sinuous walls. *Microhairs* present; elongated; clearly two-celled; chloridoid-type. Microhair apical cell wall thinner than that of the basal cell but not tending to collapse. Microhairs (27–)31–32(–33) microns long. Microhair basal cells 12 microns long. Microhairs (6–)6.6–7.8(–9) microns wide at the septum. Microhair total length/width at septum 3.7–5.3. Microhair apical cells 18–19.5–21 microns long. Microhair apical cell/total length ratio 0.59–0.64. *Stomata* common; 18–27 microns long. Subsidiaries dome-shaped and triangular. Guard-cells overlapping to flush with the interstomatals. *Intercostal short-cells* common; not paired (mostly solitary); silicified. Intercostal silica bodies absent. *Costal short-cells* predominantly paired, or neither distinctly grouped into long rows nor predominantly paired (paired, solitary and in short rows). Costal silica bodies present in alternate cell files of the costal zones; saddle shaped (predominating, a rather irregular small version), or tall-and-narrow to crescentic (few); not sharp-pointed.

Transverse section of leaf blade, physiology. C_4; XyMS+. PCR sheaths of the primary vascular bundles interrupted; interrupted abaxially only. PCR sheath extensions absent. PCR cell chloroplasts centrifugal/peripheral. *Leaf blade* with distinct, prominent adaxial ribs; with the ribs very irregular in sizes (there being a big, flat-topped rib towards each margin). *Midrib* conspicuous (via its large, round-topped adaxial rib and large bundle); with one bundle only. Bulliforms present in discrete, regular adaxial groups (and sometimes in irregular groups); sometimes in simple fans, or associated with colourless mesophyll cells to form deeply-penetrating fans (?). Many of the smallest vascular bundles unaccompanied by sclerenchyma. Combined sclerenchyma girders present; forming 'figures' (primary bundles). Sclerenchyma all associated with vascular bundles.

Taxonomy. Chloridoideae; main chloridoid assemblage.

Ecology, geography, regional floristic distribution. 1 species. Southeast tropical Africa. Mesophytic, or xerophytic; species of open habitats; glycophytic. Dry sandy places.

Paleotropical. African. Sudano-Angolan. South Tropical African.

References, etc. Morphological/taxonomic: Elffers and Kennedy-O'Byrne 1949. Leaf anatomical: Metcalfe 1960; this project.

Robynsiochloa Jacques-Félix

~ *Rottboellia*

Habit, vegetative morphology. Stoloniferous. *Culms* 100–150 cm high; herbaceous (erect); branched above. Culm nodes exposed; hairy. Culm leaves present. Upper culm leaf blades fully developed. Culm internodes hollow. *Leaves* not basally aggregated; non-auriculate. Sheaths with tuberculate hairs. Leaf blades linear; narrow; 4–6 mm wide; flat (rarely), or folded; without cross venation; persistent. *Ligule an unfringed membrane*; somewhat truncate; 1.5–2 mm long. *Contra-ligule* absent.

Reproductive organization. Plants bisexual, with bisexual spikelets; with hermaphrodite florets. The spikelets of sexually distinct forms on the same plant; hermaphrodite and male-only. The male and female-fertile spikelets mixed in the inflorescence. *The spikelets overtly heteromorphic (and the sessile female spikelets crustaceous, the pedicellate males herbaceous)*; all in heterogamous combinations.

Inflorescence. Inflorescence a solitary 'raceme' terminating the culm; *espatheate*; not comprising 'partial inflorescences' and foliar organs (the 'raceme' more or less disengaging from the uppermost, non-spathiform sheath). *Spikelet-bearing axes* 'racemes' (spiciform), or spikelike; the spikelet-bearing axes with more than 10 spikelet-bearing 'articles' (15 to 20); solitary; with substantial rachides; disarticulating; disarticulating at the joints. *'Articles'* non-linear; with a basal callus-knob; not appendaged; disarticulating obliquely; glabrous. *Spikelets* paired; secund (the 'raceme' strongly dorsiventral, the sessile spikelets in two alternating rows on one side of the rachis); sessile and pedicellate; *consistently in 'long-and-short' combinations*; in pedicellate/sessile combinations. Pedicels of the 'pedicellate' spikelets discernible, but fused with the rachis (to a variable extent). The 'shorter' spikelets hermaphrodite. The 'longer' spikelets male-only.

Female-sterile spikelets. Pedicellate spikelets herbaceous, asymmetrically lanceolate; florets 1–2, the lower male, the upper male or sterile. The male spikelets with glumes; with proximal incomplete florets, or without proximal incomplete florets; 1 floreted, or 2 floreted. The lemmas awnless. Male florets 3 staminate.

Female-fertile spikelets. *Spikelets* 6 mm long; orange; abaxial; *compressed dorsiventrally (thick, obconical)*; planoconvex; falling with the glumes (and the pedicel and rachis segment); not disarticulating between the florets; with conventional internode spacings. Rachilla terminated by a female-fertile floret. Hairy callus absent. *Callus* short; blunt (dumpy).

Glumes two; more or less equal; long relative to the adjacent lemmas; free; dorsiventral to the rachis; hairless; glabrous; not pointed (G_1 bidentate, G_2 blunt); awnless; very dissimilar (G_1 crustaceous and rounded on the back, G_2 leathery on the back with hyaline margins, naviculate and keel-winged above). *Lower glume* two-keeled (the keels winged above); convex on the back; not pitted; *muricate (shining between the tiny projections)*; about 17 nerved. *Upper glume* about *15 nerved*. *Spikelets* with incomplete florets. The incomplete florets proximal to the female-fertile florets. *The proximal incomplete florets* 1; paleate. Palea of the proximal incomplete florets fully developed (indurated below, with hyaline wings). The proximal incomplete florets male. The proximal lemmas ovate-lanceolate; awnless; 3 nerved; more or less equalling the female-fertile lemmas; less firm than the female-fertile lemmas to similar in texture to the female-fertile lemmas (hyaline); not becoming indurated.

Female-fertile florets 1. *Lemmas* less firm than the glumes (membranous); not becoming indurated; entire; pointed; awnless; hairless; glabrous; *carinate (conspicuously naviculate)*; without a germination flap. *Palea* present; relatively long (linear); entire; awnless, without apical setae; textured like the lemma; not indurated; 2-nerved (the margins narrow, inflexed); 2-keeled. Palea back glabrous. Palea keels wingless; glabrous. *Lodicules* present; 2; free; fleshy; glabrous. *Stamens* 3. *Ovary* glabrous. Styles free to their bases; free. Style bases adjacent. Stigmas 2.

Fruit, embryo and seedling. Fruit small; golden yellow; subglobose; not noticeably compressed; glabrous. Hilum short. Embryo large. Endosperm hard; containing only simple starch grains.

Abaxial leaf blade epidermis. *Costal/intercostal zonation* conspicuous. *Papillae* absent. *Long-cells* markedly different in shape costally and intercostally (the costals much narrower, more regularly rectangular); of similar wall thickness costally and intercostally (the walls of medium thickness). Intercostal zones with typical long-cells. Mid-intercostal long-cells rectangular to fusiform; having markedly sinuous walls to having straight or only gently undulating walls. *Microhairs* present; elongated; clearly two-celled; panicoid-type. *Stomata* common. Subsidiaries non-papillate; rather consistently triangular. Guard-cells overlapping to flush with the interstomatals (consistently very slightly overlapping them). *Intercostal short-cells* common; in cork/silica-cell pairs and not paired (often solitary); silicified. Intercostal silica bodies tall-and-narrow, cross-shaped, and oryzoid-type. Neither macrohairs no prickles seen. Costal zones with short-cells. *Costal short-cells* conspicuously in long rows. Costal silica bodies present and well developed; 'panicoid-type'; cross shaped to dumb-bell shaped.

Transverse section of leaf blade, physiology. C_4; XyMS–. PCR sheath extensions absent. *Leaf blade* adaxially flat. *Midrib* conspicuous; having a conventional arc of bundles (with one large median); with colourless mesophyll adaxially. *The lamina* symmetrical on either side of the midrib. Bulliforms present in discrete, regular adaxial groups (in places), or not in discrete, regular adaxial groups (for the most part, the epidermis more or less irregularly bulliform); occasionally in simple fans (of the *Zea* type). Combined sclerenchyma girders present (with all the primaries and some of the other main bundles, some others with abaxial girders only). Sclerenchyma all associated with vascular bundles.

Taxonomy. Panicoideae; Andropogonodae; Andropogoneae; Rottboelliinae.

Ecology, geography, regional floristic distribution. 1 species. Tropical Africa. Paleotropical. African and Indomalesian. Sudano-Angolan and West African Rainforest. Indian. South Tropical African.

References, etc. Leaf anatomical: this project.

Rottboellia L.f.

Stegosia Lour.
Excluding *Robynsiochloa*

Habit, vegetative morphology. Annual; caespitose. *Culms* 30–300 cm high; herbaceous; branched above. Culm nodes glabrous. Culm internodes solid. *Leaves* not basally aggregated; non-auriculate. Leaf blades broad; flat; without cross venation; persistent. *Ligule an unfringed membrane (?), or a fringed membrane.*

Reproductive organization. Plants bisexual, with bisexual spikelets; with hermaphrodite florets. *The spikelets* of sexually distinct forms on the same plant; hermaphrodite and male-only, or hermaphrodite and sterile; *overtly heteromorphic (but all similar in texture)*; in both homogamous and heterogamous combinations (with terminal, homogamous imperfect pairs and heterogamous pairs below). Plants inbreeding. Exposed-cleistogamous, or chasmogamous.

Inflorescence. Inflorescence a single raceme, or paniculate (with terete, spike-like 'racemes', terminating the culms and branches, or axillary, solitary or in fascicles). *Inflorescence axes not ending in spikelets (the upper homogamous spikelet pairs increasingly reduced, forming a tapered, tail-like terminal appendage to the rachis).* Rachides hollowed. *Inflorescence* spatheate; *a complex of 'partial inflorescences' and intervening foliar organs. Spikelet-bearing axes* spikelike (cylindrical, with embedded spikelets); solitary and clustered (fascicled); with substantial rachides; disarticulating; disarticulating at the joints. *'Articles'* non-linear (flattened below, cupular above); with a basal callus-knob; disarticulating transversely, or disarticulating obliquely (and the pedicelled spikelets disarticulating obliquely, leaving shallow crescentic scars); glabrous. *Spikelets* paired; secund (the sessile spikelets in two alternating rows, on one side of the rachis); all sessile, but recognisably incorporating fused pedicels; *consistently in 'long-and-short' combinations*; in pedicellate/sessile combinations. Pedicels of the 'pedicellate' spikelets discernible, but fused with the rachis. The 'shorter' spikelets hermaphrodite. The 'longer' spikelets male-only, or sterile.

Female-sterile spikelets. The pedicelled spikelets male or sterile, striate, compressed, herbaceous.

Female-fertile spikelets. Spikelets compressed dorsiventrally (trigonous); falling with the glumes (and with the joint, the pedicelled spikelets falling separately). Rachilla terminated by a female-fertile floret. Hairy callus absent.

Glumes two; more or less equal; long relative to the adjacent lemmas; awnless; very dissimilar (the lower flat-backed and 2-keeled above, the upper naviculate, winged). Lower glume two-keeled (narrowly winged at the apex); convex on the back to flattened on the back; not pitted; relatively smooth (scabridulous); 11–13 nerved. Upper glume 11–13 nerved. *Spikelets* with incomplete florets. The incomplete florets proximal to the female-fertile florets. *The proximal incomplete florets* 1; paleate. Palea of the proximal incomplete florets fully developed. The proximal incomplete florets male. The proximal lemmas awnless; 3 nerved; more or less equalling the female-fertile lemmas; similar in texture to the female-fertile lemmas (hyaline); not becoming indurated.

Female-fertile florets 1. Lemmas less firm than the glumes; not becoming indurated; entire; entire; awnless; hairless; non-carinate; 3 nerved. *Palea* present; relatively long; entire; awnless, without apical setae; not indurated (hyaline); 2-nerved. *Lodicules* present; 2; free; fleshy; glabrous. *Stamens* 3. Anthers not penicillate. *Ovary* glabrous. Styles free to their bases. Stigmas 2.

Fruit, embryo and seedling. *Fruit* free from both lemma and palea; small; compressed dorsiventrally. Hilum short. Embryo large; waisted. Endosperm hard; without lipid.

Abaxial leaf blade epidermis. *Costal/intercostal zonation* conspicuous. *Papillae* absent. *Long-cells* similar in shape costally and intercostally; of similar wall thickness costally and intercostally (thin walled). Mid-intercostal long-cells rectangular; having markedly sinuous walls. *Microhairs* present; panicoid-type; (31–)36–42(–48) microns long; (4.2–)4.5–5.1(–5.4) microns wide at the septum. Microhair total length/width at septum 5.8–9.3. Microhair apical cells (15–)18–24(–29) microns long. Microhair apical cell/total length ratio 0.48–0.57. *Stomata* common; (22.5–)24–28.5(–30) microns long. Subsidiaries low dome-shaped and triangular. Guard-cells overlapping to flush with the interstomatals. *Intercostal short-cells* common; not paired (usually); silicified, or not silicified. Intercostal silica bodies cross-shaped, or vertically elongated-nodular. *Costal short-cells* conspicuously in long rows. Costal silica bodies 'panicoid-type'; cross shaped to dumb-bell shaped; not sharp-pointed.

Transverse section of leaf blade, physiology. C_4; XyMS–. PCR cells with a suberised lamella. PCR cell chloroplasts with reduced grana; centrifugal/peripheral. *Mesophyll* with radiate chlorenchyma. *Leaf blade* 'nodular' in section, or adaxially flat. *Midrib* conspicuous; having a conventional arc of bundles; with colourless mesophyll adaxially. Bulliforms present in discrete, regular adaxial groups (also irregularly grouped); in simple fans. Many of the smallest vascular bundles unaccompanied by sclerenchyma. Combined sclerenchyma girders present; forming 'figures'. Sclerenchyma all associated with vascular bundles.

Phytochemistry. Leaves without flavonoid sulphates (1 species).

Special diagnostic feature. Lower glume of female-fertile spikelet flattish, not pitted; 'pedicellate' spikelets reduced, herbaceous.

Cytology. Chromosome base number, $x = 9$ and 10. $2n = 20$, 36, 40, and 54. 2, 4, and 6 ploid.

Taxonomy. Panicoideae; Andropogonodae; Andropogoneae; Rottboelliinae.

Ecology, geography, regional floristic distribution. 4 species. Tropical and subtropical Africa, Asia. Commonly adventive. Helophytic to mesophytic; shade species, or species of open habitats; glycophytic. Woodland, swamps, often in disturbed ground or a weed of cultivated ground.

Holarctic, Paleotropical, Neotropical, and Australian. Tethyan. African, Madagascan, Indomalesian, and Neocaledonian. Macaronesian. Saharo-Sindian, Sudano-Angolan, West African Rainforest, and Namib-Karoo. Indian, Indo-Chinese, Malesian, and Papuan. Caribbean, Pampas, and Andean. North and East Australian. Sahelo-Sudanian, Somalo-Ethiopian, South Tropical African, and Kalaharian. Tropical North and East Australian.

Rusts and smuts. Rusts — *Puccinia*. Taxonomically wide-ranging species: *Puccinia microspora* and *Puccinia levis*. Smuts from Ustilaginaceae. Ustilaginaceae — *Sorosporium*, *Sphacelotheca*, and *Ustilago*.

Economic importance. Significant weed species: *R. cochinchinensis*, *R. exaltata*.

References, etc. Morphological/taxonomic: Veldkamp, de Konig and Sosef 1986. Leaf anatomical: Metcalfe 1960 and this project.

Rytidosperma Steud.

Notodanthonia Zotov

~ *Danthonia* sensu lato

Excluding *Chionochloa*, *Erythranthera*, *Karroochloa*, *Merxmuellera*, *Monachather*, *Monostachya*, *Plinthanthesis*

Habit, vegetative morphology. Perennial; caespitose. **Culms** 10–100 cm high; *herbaceous*; unbranched above. Culm nodes glabrous. Culm internodes solid, or hollow. *Leaves* mostly basal, or not basally aggregated (more often); non-auriculate. Leaf blades linear; narrow; setaceous, or not setaceous; flat, or folded, or rolled; exhibiting multicellular glands abaxially. The abaxial glands intercostal. Leaf blades without cross venation; persistent; once-folded in bud. *Ligule a fringed membrane (very short), or a fringe of hairs*; 0.5–5 mm long. *Contra-ligule* present, or absent.

Reproductive organization. *Plants bisexual, with bisexual spikelets*; with hermaphrodite florets. The spikelets all alike in sexuality. Plants inbreeding. Exposed-cleistogamous, or chasmogamous. Plants with hidden cleistogenes, or without hidden cleistogenes. The hidden cleistogenes in the leaf sheaths (mainly associated with the upper sheaths, and not highly modified, by contrast with American *Danthonia*).

Inflorescence. Inflorescence rather few spikeleted (usually), or many spikeleted; paniculate (sometimes almost a raceme); open to contracted (frequently open at anthesis, then closing when fruiting); espatheate; not comprising 'partial inflorescences' and foliar organs. Spikelet-bearing axes persistent. Spikelets not secund; pedicellate.

Female-fertile spikelets. Spikelets 4–25 mm long; compressed laterally; disarticulating above the glumes; disarticulating between the florets (the disarticulations oblique); with conventional internode spacings. Rachilla prolonged beyond the uppermost female-fertile floret. The rachilla extension with incomplete florets. Hairy callus present. *Callus* short to long; pointed, or blunt.

Glumes two; more or less equal; about equalling the spikelets, or exceeding the spikelets; long relative to the adjacent lemmas; pointed (acute); awnless; keeled above; similar (papery, persistent). *Lower glume 5–7 nerved*. Upper glume 5–7 nerved. *Spikelets* with female-fertile florets only, or with incomplete florets. The incomplete florets when present, as is usual, distal to the female-fertile florets. *The distal incomplete florets* merely underdeveloped. *Spikelets without proximal incomplete florets*.

Female-fertile florets *3–20*. *Lemmas* similar in texture to the glumes to decidedly firmer than the glumes (soft to leathery); not becoming indurated; incised; 2 lobed (the lobes acute, acuminate or setaceous); deeply cleft (often), or not deeply cleft; *awned (or occasionally only mucronate)*. *Awns* 1, or 3; median, or median and lateral; the median different in form from the laterals (when laterals present); from a sinus; *geniculate*; much shorter than the body of the lemma to much longer than the body of the lemma; entered by one vein. The lateral awns (when present) shorter than the median to exceeding the median (straight, terminating the lobes). Lemmas nearly always hairy (though there are a few exceptions). *The hairs* when present (as is usual) in tufts (or at least, the hairs of markedly different lengths); *in transverse rows (usually with transverse rows of tufts)*. Lemmas non-carinate (rounded on the back); without a germination flap; 7–15 nerved. *Palea* present; relatively long to conspicuous but relatively short; entire (truncate), or apically notched (minutely bilobed); awnless, without apical setae; not indurated (membranous, or firm towards the base); 2-nerved; 2-keeled. Palea keels wingless. **Lodicules** present; 2; free; fleshy, or membranous; *ciliate*. *Stamens* 3. Anthers 0.5–3 mm long; not penicillate. *Ovary* glabrous. Styles free to their bases. Stigmas 2; white, or red pigmented, or brown.

Fruit, embryo and seedling. **Fruit** free from both lemma and palea; small; ellipsoid; *longitudinally grooved (or somewhat hollowed)*; compressed dorsiventrally. *Hilum short (but at least sometimes linear)*. Embryo large; waisted. Endosperm hard; without lipid; containing compound starch grains. Embryo without an epiblast; with a scutellar tail; with an elongated mesocotyl internode. Embryonic leaf margins meeting.

Seedling with a short mesocotyl, or with a long mesocotyl. First seedling leaf with a well-developed lamina. The lamina narrow; curved; 3–5 veined.

Abaxial leaf blade epidermis. *Costal/intercostal zonation* conspicuous, or lacking. *Papillae* absent. *Long-cells* similar in shape costally and intercostally; of similar wall thickness costally and intercostally (usually). Mid-intercostal long-cells rectangular; having markedly sinuous walls. *Microhairs* present, or absent (but usually present somewhere on the leaf); elongated; clearly two-celled; panicoid-type. Microhair apical cell wall thinner than that of the basal cell and often collapsed. Microhairs without 'partitioning membranes' (in *R. linkii*). Microhair basal cells 24–36 microns long. *Stomata* common. Subsidiaries parallel-sided, or dome-shaped, or triangular. Guard-cells overlapping to flush with the interstomatals. *Intercostal short-cells* common, or absent or very rare; in cork/silica-cell pairs, or not paired. Intercostal silica bodies present and perfectly developed. *Costal short-cells* conspicuously in long rows, or predominantly paired. Costal silica bodies present in alternate cell files of the costal zones; rounded, or tall-and-narrow, or crescentic (to cuboid), or 'panicoid-type'; often cross shaped, or dumb-bell shaped; not sharp-pointed.

Transverse section of leaf blade, physiology. C_3; XyMS+. *Mesophyll* with radiate chlorenchyma, or with non-radiate chlorenchyma; without adaxial palisade. *Leaf blade* with distinct, prominent adaxial ribs; with the ribs more or less constant in size, or with the ribs very irregular in sizes. *Midrib* not readily distinguishable; with one bundle only. Bulliforms present in discrete, regular adaxial groups (in the furrows), or not in discrete, regular adaxial groups (sometimes of small, irregularly sized cells confined to the furrows); sometimes in simple fans. All the vascular bundles accompanied by sclerenchyma. Combined sclerenchyma girders present; forming 'figures'. Sclerenchyma all associated with vascular bundles, or not all bundle-associated. The 'extra' sclerenchyma (when present) in abaxial groups, or in a continuous abaxial layer. The lamina margins with fibres.

Phytochemistry. Tissues of the culm bases with abundant starch. Leaves containing flavonoid sulphates ('*Notodanthonia racemosa* (R.Br.) Zotov').

Cytology. Chromosome base number, $x = 6$. $2n = 24, 48, 72, 96$, and 120. 4–20 ploid.

Taxonomy. Arundinoideae; Danthonieae.

Ecology, geography, regional floristic distribution. About 60 species (in Austalasia, plus a few elsewhere in the Southern hemisphere?). Australasian region and South America?. Mesophytic, or xerophytic; mostly species of open habitats; glycophytic. Open grasslands, including uplands, and in sparse woodlands.

Paleotropical, Neotropical, Australian, and Antarctic. Indomalesian, Polynesian, and Neocaledonian. Papuan. Fijian. Pampas and Andean. North and East Australian, South-West Australian, and Central Australian. New Zealand and Patagonian.

Rusts and smuts. Rusts — *Puccinia*.

Economic importance. Lawns and/or playing fields: several species with potential.

References, etc. Morphological/taxonomic: Vickery 1956; Zotov 1963; Connor and Edgar 1979; Jacobs 1982. Leaf anatomical: this project.

Special comments. The Australasian component of *Danthonia*. Note that taxonomically and nomenclaturally, '*Danthonia*' is a mess. The characters which are supposed to distinguish genera here (e.g., lemma hairs, hilum shape) need surveying at world level.

Saccharum L.

Rhipidium Trin., *Saccharifera* Stokes
Excluding *Erianthus, Miscanthus, Lasiorhachis, Narenga*

Habit, vegetative morphology. Robust perennial (cane grasses); **rhizomatous**. *Culms* 150–1200 cm high; woody and persistent, or herbaceous. Culm nodes glabrous. Culm internodes solid. *Leaves* not basally aggregated; auriculate, or non-auriculate. Leaf blades usually linear; broad, or narrow; 10–40(–60) mm wide (to 60 cm in *S. officinarum*); flat; rolled in bud. **Ligule a fringed membrane**.

Reproductive organization. Plants bisexual, with bisexual spikelets; with hermaphrodite florets. **The spikelets all alike in sexuality**; homomorphic. Plants outbreeding. Apomictic, or reproducing sexually.

Inflorescence. *Inflorescence* paniculate; **open (plumose, silvery)**; espatheate; not comprising 'partial inflorescences' and foliar organs. **Spikelet-bearing axes** spiciform 'racemes'; with very slender rachides; **disarticulating**; disarticulating at the joints. '*Articles*' linear; without a basal callus-knob; not appendaged; disarticulating transversely; densely long-hairy to somewhat hairy. Spikelets paired; not secund; sessile and pedicellate; consistently in 'long-and-short' combinations; in pedicellate/sessile combinations. Pedicels of the 'pedicellate' spikelets free of the rachis. The 'shorter' spikelets hermaphrodite. The 'longer' spikelets hermaphrodite.

Female-fertile spikelets. Spikelets 2–6 mm long (the pedicellate and sessile members similar); compressed laterally, or not noticeably compressed, or compressed dorsiventrally; falling with the glumes (the pedicelled falling from the pedicel, the sessile falling with the joint and pedicel). Rachilla terminated by a female-fertile floret. Hairy callus present.

Glumes two; more or less equal; long relative to the adjacent lemmas; without conspicuous tufts or rows of hairs; awnless; very dissimilar (the sessile spikelet with the lower bicarinate, the upper imparinerved), or similar (the pedicellate spikelet with both imparinerved). Lower glume more or less two-keeled (in sessile spikelets), or not two-keeled (in pedicellate spikelets); convex on the back, or flattened on the back; not pitted; relatively smooth. *Spikelets* with incomplete florets. The incomplete florets proximal to the female-fertile florets. *The proximal incomplete florets* 1; paleate, or epaleate. Palea of the proximal incomplete florets fully developed, or reduced. The proximal incomplete florets sterile. The proximal lemmas awnless; not becoming indurated.

Female-fertile florets 1. Lemmas membranous, or reduced to a linear stipe; less firm than the glumes; not becoming indurated; entire; pointed, or blunt; mucronate, or awned. Awns 1 (bristle-like); median; apical; non-geniculate. **Lemmas hairy**; 0–1 nerved. **Palea** present, or absent; when present, relatively long, or conspicuous but relatively short, or very reduced; awnless, without apical setae; not indurated (hyaline); **nerveless**. *Lodicules* present; free; fleshy; ciliate; toothed (with a blunt tooth, or one on each margin). *Stamens* 3. *Ovary* glabrous. Styles fused. Stigmas 2; red pigmented.

Fruit, embryo and seedling. *Fruit* free from both lemma and palea; not noticeably compressed (terete). Hilum short. Embryo large, or small.

Seedling with a long mesocotyl.

Abaxial leaf blade epidermis. *Papillae* present, or absent. Intercostal papillae overarching the stomata. Mid-intercostal long-cells having markedly sinuous walls. *Microhairs* present; panicoid-type; (48–)51–72(–78) microns long; (4.5–)6–9(–9.6) microns wide at the septum. Microhair total length/width at septum 8.5–12. Microhair apical cells (16–)20–53(–63) microns long. Microhair apical cell/total length ratio 0.4–0.7. *Stomata* common; (33–)35–42(–45) microns long. Subsidiaries low dome-shaped to triangular. Guard-cells overlapping to flush with the interstomatals. *Intercostal short-cells* common, or absent or very rare; when present, silicified. Intercostal silica bodies when present, cross-shaped. *Costal short-cells* predominantly paired, or neither distinctly grouped into long rows nor predominantly paired. Costal silica bodies 'panicoid-type'; cross shaped to nodular; not sharp-pointed.

Transverse section of leaf blade, physiology. C_4; biochemical type NADP–ME (*S. officinarum*); XyMS–. PCR cell chloroplasts with reduced grana; centrifugal/peripheral. *Mesophyll* with radiate chlorenchyma; traversed by columns of colourless mesophyll cells, or not traversed by colourless columns. *Leaf blade* with distinct, prominent adaxial ribs, or 'nodular' in section, or adaxially flat; with the ribs very irregular in sizes. *Midrib* conspicuous; having a conventional arc of bundles; with colourless mesophyll adaxially. Bulliforms present in discrete, regular adaxial groups; in simple fans, or associated with colourless mesophyll cells to form deeply-penetrating fans (sometimes linked with traversing colourless tissue); associating with colourless mesophyll cells to form arches over small vascular bundles, or nowhere involved in bulliform-plus-colourless mesophyll arches. Many of the smallest vascular bundles unaccompanied by sclerenchyma. Combined sclerenchyma girders present; forming 'figures'. Sclerenchyma all associated with vascular bundles.

Culm anatomy. *Culm internode bundles* scattered.

Phytochemistry. Leaves containing flavonoid sulphates (5 species), or without flavonoid sulphates (*S. officinarum*).

Cytology. Chromosome base number, $x = 10$ and 12. $2n = 40$, or 60, or 68, or 76–78, or 80, or 90, or 46–128, or 110, or 112, or 116–117, or 144. 4, 6, 8, 9, and 12 ploid. Haploid nuclear DNA content 1.1–1.3 pg (3 species, one 6x and two 8x).

Taxonomy. Panicoideae; Andropogonodae; Andropogoneae; Andropogoninae.

Ecology, geography, regional floristic distribution. 5 species. Tropical, subtropical. Helophytic, or mesophytic; shade species, or species of open habitats. Mostly riversides and valleys, some on open hillsides.

Holarctic, Paleotropical, and Neotropical. Boreal, Tethyan, and Madrean. African, Madagascan, Indomalesian, and Polynesian. Eastern Asian. Mediterranean and Irano-Turanian. Saharo-Sindian, Sudano-Angolan, and West African Rainforest. Indian, Indo-Chinese, Malesian, and Papuan. Hawaiian and Fijian. Caribbean, Venezuela and Surinam, Central Brazilian, Pampas, and Andean. Sahelo-Sudanian, Somalo-Ethiopian, and South Tropical African.

Hybrids. Intergeneric hybrids with *Erianthus, Imperata, Miscanthidium, Miscanthus, Narenga, Sclerostachya, Sorghum*; a claim involving *Zea* is probably erroneous, and one involving *Bambusa* has been proven so.

Rusts and smuts. Rusts — *Puccinia*. Taxonomically wide-ranging species: *Puccinia miscanthae*. Smuts from Ustilaginaceae. Ustilaginaceae — *Sorosporium, Sphacelotheca*, and *Ustilago*.

Economic importance. Cultivated fodder: *S. officinarum*. Stems of *S. spontaneum* used for light construction work.

References, etc. Leaf anatomical: Metcalfe 1960; this project.

Sacciolepis Nash

Rhampholepis Stapf

Habit, vegetative morphology. Annual, or perennial; rhizomatous, or stoloniferous, or caespitose, or decumbent. *Culms* 10–200 cm high; herbaceous; branched above. Culm nodes glabrous. Culm internodes hollow. *Leaves* not basally aggregated; auriculate, or non-auriculate. Leaf blades linear to linear-lanceolate; narrow; flat, or rolled (convolute); cross veined, or without cross venation; persistent; rolled in bud. *Ligule an unfringed membrane to a fringed membrane.*

Reproductive organization. Plants bisexual, with bisexual spikelets; with hermaphrodite florets. The spikelets of sexually distinct forms on the same plant (e.g. *S. cingularis*), or all alike in sexuality; when of different kinds, hermaphrodite and sterile (the sterile spikelets, when present, at the base of the inflorescence).

Inflorescence. *Inflorescence paniculate (with filiform branches)*; narrowly contracted (usually), or open (rarely); usually spicate; espatheate; not comprising 'partial inflorescences' and foliar organs. Spikelet-bearing axes persistent. Spikelets not secund; shortly pedicellate. Pedicel apices discoid.

Female-fertile spikelets. *Spikelets* 0.8–5.2 mm long; elliptic, or lanceolate, or ovate; *compressed laterally to not noticeably compressed (gibbous, often oblique)*; falling with the glumes. Rachilla terminated by a female-fertile floret. Hairy callus absent.

Glumes two; very unequal; (the upper about equalling the spikelets (the lower 1/5 to 3/4 as long); (the longer) long relative to the adjacent lemmas; awnless; very dissimilar (both membranous or hyaline, prominently ribbed, the upper generally more or less inflated and gibbous). Lower glume 3–7 nerved. **Upper glume** *distinctly saccate*; 5–13 nerved. *Spikelets* with incomplete florets. The incomplete florets proximal to the female-fertile florets. *Spikelets with proximal incomplete florets. The proximal incomplete florets* 1; paleate, or epaleate. Palea of the proximal incomplete florets when present, fully developed to reduced. The proximal incomplete florets male (rarely), or sterile. The proximal lemmas less gibbous than the upper glume, otherwise more or less resembling it, or occasionally with a transverse row of hairs, and *S. fenestrata* with two basal 'windows'; awnless; 7 nerved; decidedly exceeding the female-fertile lemmas (about equalling G_2); less firm than the female-fertile lemmas; not becoming indurated.

Female-fertile florets 1 (laterally compressed). Lemmas acute, strongly convex; decidedly firmer than the glumes (papery to subcrustaceous); smooth; becoming indurated to not becoming indurated; white in fruit; entire; pointed; awnless; hairless (glossy); non-carinate; having the margins lying flat and exposed on the palea (but the margins not hyaline), or having the margins tucked in onto the palea; with a clear germination flap; 3 nerved, or 5 nerved (obscurely so). *Palea* present; relatively long; entire; awnless, without apical setae; textured like the lemma; indurated, or not indurated; 2-nerved. *Lodicules* present; 2; free; fleshy; glabrous. *Stamens* 3. Anthers not penicillate. *Ovary* glabrous. Styles free to their bases. Stigmas 2.

Fruit, embryo and seedling. Fruit small; compressed dorsiventrally. *Hilum short.* Embryo large; waisted. Endosperm hard; without lipid. Embryo without an epiblast; with a scutellar tail; with an elongated mesocotyl internode. Embryonic leaf margins overlapping.

Seedling with a long mesocotyl. First seedling leaf with a well-developed lamina. The lamina broad; supine.

Abaxial leaf blade epidermis. *Costal/intercostal zonation* conspicuous. *Papillae* absent. *Long-cells* markedly different in shape costally and intercostally; of similar wall thickness costally and intercostally. Intercostal zones with typical long-cells to exhibiting many atypical long-cells. Mid-intercostal long-cells rectangular; having markedly sinuous walls, or having straight or only gently undulating walls. *Microhairs* present; panicoid-type (but the apical cell sometimes quite broad). *Stomata* common. Subsidiaries low dome-shaped, or triangular. Guard-cells overlapping to flush with the interstomatals. *Intercostal short-cells* common, or absent or very rare; not paired, or in cork/silica-cell pairs and not paired; silicified, or not silicified. *Costal short-cells* conspicuously in long rows, or conspicuously in long rows and predominantly paired. Costal silica bodies tall-and-narrow, or 'panicoid-type' (mostly), or acutely-angled (sometimes); mostly nodular, or dumb-bell shaped; sharp-pointed, or not sharp-pointed (rarely).

Transverse section of leaf blade, physiology. C_3; XyMS+. *Mesophyll* with radiate chlorenchyma (e.g. *S. myosuroides*), or with non-radiate chlorenchyma (e.g *S. indica*); *Isachne*-type (e.g. *S. indica, S. africana*), or not *Isachne*-type (e.g. *S. chevalieri*); without fusoids (but lacunae sometimes present between the veins near the midrib). *Leaf blade* with distinct, prominent adaxial ribs. *Midrib* conspicuous; with one bundle only. Bulliforms often present in discrete, regular adaxial groups, or not in discrete, regular adaxial groups (confined to the adaxial groove of the midrib in *S. chevalieri*); when grouped, in simple fans. Many of the smallest vascular bundles unaccompanied by sclerenchyma. Combined sclerenchyma girders absent.

Phytochemistry. Leaves without flavonoid sulphates (2 species).

Cytology. Chromosome base number, $x = 9$. $2n = 16, 18, 36$, and 45. 2, 4, and 5 ploid.

Taxonomy. Panicoideae; Panicodae; Paniceae.

Ecology, geography, regional floristic distribution. 30 species. Tropical and subtropical. Hydrophytic to helophytic; species of open habitats; glycophytic. In or near water or in wet places.

Holarctic, Paleotropical, Neotropical, and Australian. Boreal. African, Madagascan, Indomalesian, Polynesian, and Neocaledonian. Eastern Asian. Saharo-Sindian, Sudano-Angolan, West African Rainforest, and Namib-Karoo. Indian, Indo-Chinese, Malesian, and Papuan. Polynesian. Caribbean, Venezuela and Surinam, Central Brazilian, Pampas, and Andean. North and East Australian. Sahelo-Sudanian, Somalo-Ethiopian, South Tropical African, and Kalaharian. Tropical North and East Australian.

Rusts and smuts. Rusts — *Puccinia*. Taxonomically wide-ranging species: *Puccinia emaculata*. Smuts from Ustilaginaceae. Ustilaginaceae — *Sphacelotheca* and *Ustilago*.

Economic importance. Important native pasture species: *S. africana, S. myosuroides*.

References, etc. Leaf anatomical: Metcalfe 1960; this project.

Sartidia de Winter

Habit, vegetative morphology. Perennial; densely caespitose. *Culms* 80–200 cm high; herbaceous; unbranched above. Culm nodes glabrous. Culm internodes solid. Plants unarmed. Young shoots intravaginal. *Leaves* not basally aggregated; non-auriculate. *Leaf blades* linear; narrow; 2–4 mm wide; rolled; *not pseudopetiolate*; without cross venation; persistent; rolled in bud. Ligule a fringed membrane, or a fringe of hairs; truncate; 0.3–1 mm long. *Contra-ligule* present (as a line of hairs, in *S. jucunda*), or absent.

Reproductive organization. Plants bisexual, with bisexual spikelets; with hermaphrodite florets.

Inflorescence. Inflorescence paniculate (erect, narrow, often interrupted); open; espatheate; not comprising 'partial inflorescences' and foliar organs. Spikelet-bearing axes persistent. Spikelets solitary; not secund; pedicellate.

Female-fertile spikelets. *Spikelets* 12–30 mm long; *not noticeably compressed*; disarticulating above the glumes. Rachilla terminated by a female-fertile floret. Hairy callus present (densely but shortly bearded). Callus pointed to blunt (shallowly bilobed).

Glumes two; more or less equal; long relative to the adjacent lemmas; free; hairless; glabrous; pointed (acute to acuminate); awned, or awnless; non-carinate (rounded on the back); similar (narrow, the nerves evanescent). Lower glume about equalling the lowest lemma; usually 3 nerved. Upper glume 3 nerved, or 5 nerved. **Spikelets** with female-fertile florets only; *without proximal incomplete florets*.

Female-fertile florets 1. *Lemmas subcylindrical, with involute margins, scarcely narrowed above*; not convolute (involute); decidedly firmer than the glumes; becoming indurated to not becoming indurated (leathery to cartilaginous); not deeply cleft; awned (cf. *Aristida*). *Awns triple or trifid, commonly with a basal column (or at least with the three spreading awns twisted together basally)*; apical; non-geniculate; hairless (glabrous or scabrid); about as long as the body of the lemma to much longer than the body of the lemma; persistent. Lemmas hairless (glabrous or scabrid); non-carinate; without a germination flap; 3 nerved; with the nerves non-confluent. **Palea** present; *conspicuous but relatively short (small, scale-like)*; tightly clasped by the lemma; entire (obtuse); awnless, without apical setae (glabrous); thinner than the lemma; not indurated (leathery below); 2-nerved; 2-keeled (grooved). Palea keels wingless; glabrous. *Lodicules* present (about equalling or exceeding the palea); 2; free; membranous; glabrous; toothed; heavily vascularized. *Stamens* 3. Anthers 5–6 mm long; not penicillate; without an apically prolonged connective. *Ovary* glabrous. Styles free to their bases. Stigmas 2 (plumose); white.

Fruit, embryo and seedling. *Fruit* free from both lemma and palea (but tightly enclosed by the lemma); medium sized (8–10 mm long); fusiform; longitudinally grooved; compressed dorsiventrally. *Hilum long-linear*. Embryo small (no more than 1/4 grain length). Endosperm containing compound starch grains. Embryo without an epiblast; without a scutellar tail; with an elongated mesocotyl internode. Embryonic leaf margins meeting.

First seedling leaf with a well-developed lamina. The lamina narrow; erect; 5 veined.

Abaxial leaf blade epidermis. *Costal/intercostal zonation* conspicuous. *Papillae* absent. Mid-intercostal long-cells rectangular; having markedly sinuous walls. *Microhairs* present; panicoid-type. *Stomata* common. Subsidiaries dome-shaped, or dome-shaped and triangular. *Costal short-cells* conspicuously in long rows. Costal silica bodies 'panicoid-type'; not sharp-pointed.

Transverse section of leaf blade, physiology. C_3; XyMS+. *Mesophyll* with non-radiate chlorenchyma. *Leaf blade* with distinct, prominent adaxial ribs; with the ribs very irregular in sizes. *Midrib* not readily distinguishable; with one bundle only. Bulliforms present in discrete, regular adaxial groups. All the vascular bundles accompanied by sclerenchyma. Combined sclerenchyma girders present (with most bundles); forming 'figures' (with most bundles). Sclerenchyma not all bundle-associated. The 'extra' sclerenchyma in abaxial groups; abaxial-hypodermal, the groups isolated (opposite the bulliforms).

Cytology. Chromosome base number, $x = 11$. $2n = 22$.

Taxonomy. Arundinoideae; Aristideae.

Ecology, geography, regional floristic distribution. 4 species. South Africa. Mesophytic; glycophytic.

Holarctic and Paleotropical. Boreal. African and Madagascan. Eastern Asian. Sudano-Angolan and Namib-Karoo. South Tropical African.

Sasa Makino & Shibata

Neosasamorpha Tatewaki, *Nipponobambusa* Muroi, *Sasaella* Mak., *Sasamorpha* Nakai

Habit, vegetative morphology. Small to medium, shrubby perennial. The flowering culms leafy. *Culms* 100–400 cm high; woody and persistent; branched above. *Primary branches/mid-culm node 1 (rarely 3, with one dominant)*. Culm sheaths persistent. Culm internodes hollow. *Rhizomes leptomorph*. Plants unarmed. **Leaves** not basally aggregated; *with auricular setae. Leaf blades* broadly *lanceolate, or elliptic (large)*; broad (acuminate); pseudopetiolate; cross veined; disarticulating from the sheaths; rolled in bud. Ligule an unfringed membrane.

Reproductive organization. Plants bisexual, with bisexual spikelets; with hermaphrodite florets.

Inflorescence. Inflorescence paniculate; open; spatheate (the long peduncle covered by sheaths); a complex of 'partial inflorescences' and intervening foliar organs. Spikelet-bearing axes persistent. Spikelets not secund; pedicellate.

Female-fertile spikelets. Spikelets compressed laterally; disarticulating above the glumes; disarticulating between the florets. Rachilla prolonged beyond the uppermost female-fertile floret. The rachilla extension with incomplete florets.

Glumes present; two; shorter than the adjacent lemmas; similar (scarious). *Spikelets* with incomplete florets. The incomplete florets distal to the female-fertile florets. *The distal incomplete florets* merely underdeveloped.

Female-fertile florets 3–13. Lemmas decidedly firmer than the glumes (with tessellate venation); not becoming indurated; non-carinate; more than 5-nerved. *Palea*

present; 2-keeled. *Lodicules* present; 3; free. **Stamens 6. Ovary** glabrous; *without a conspicuous apical appendage.* Styles fused. Stigmas 3 (plumose).

Fruit, embryo and seedling. Fruit longitudinally grooved. Endosperm containing compound starch grains.

Abaxial leaf blade epidermis. *Costal/intercostal zonation* conspicuous. *Papillae* present (abundant); costal and intercostal (present over minor bundles, lacking over the large ones). Intercostal papillae over-arching the stomata (and largely obscuring them); several per cell (large, circular, thick walled and refractory, a single row per cell). Mid-intercostal long-cells rectangular; having markedly sinuous walls (the sinuosity fairly fine, even). *Microhairs* present; elongated; clearly two-celled (the basal cells very long); panicoid-type. *Stomata* common. Subsidiaries papillate; dome-shaped. *Intercostal short-cells* not apparent in this highly papillate epidermis. With costal and intercostal, scattered prickles. *Costal short-cells* conspicuously in long rows. Costal silica bodies present and well developed; saddle shaped (large).

Transverse section of leaf blade, physiology. C_3; XyMS+. Mesophyll with adaxial palisade; with arm cells; with fusoids. The fusoids external to the PBS. *Leaf blade* adaxially flat. *Midrib* conspicuous; having complex vascularization. *The lamina* symmetrical on either side of the midrib. Bulliforms present in discrete, regular adaxial groups; in simple fans (these large). All the vascular bundles accompanied by sclerenchyma. Combined sclerenchyma girders present (with all the bundles); forming 'figures' (I's and T's). Sclerenchyma all associated with vascular bundles.

Cytology. Chromosome base number, $x = 12$. $2n = 48$ (sample including *Sasamorpha* and *Sasaella*). 4 ploid.

Taxonomy. Bambusoideae; Bambusodae; Bambuseae.

Ecology, geography, regional floristic distribution. About 50 species. Eastern Asia. Holarctic. Boreal. Euro-Siberian and Eastern Asian. Siberian.

Hybrids. May hybridize with *Semiarundinaria* (×*Hibanobambusa* Maruyama and Okamura).

Rusts and smuts. Rusts — *Stereostratum* and *Puccinia*. Taxonomically wide-ranging species: *Stereostratum corticoides*, *Puccinia longicornis*, and *Puccinia kusanoi*. Smuts from Ustilaginaceae. Ustilaginaceae — *Ustilago*.

References, etc. Leaf anatomical: this project.

Special comments. Fruit data wanting.

Saugetia A. Hitchc. & Chase

~ *Enteropogon*

Habit, vegetative morphology. Wiry perennial; caespitose. *Culms* 40–80 cm high; herbaceous; conspicuously branched above. Culm nodes glabrous. Culm internodes solid. Plants unarmed. Young shoots intravaginal. *Leaves* not basally aggregated; non-auriculate; without auricular setae (but hairy in the auricle positions). Leaf blades narrow; 0.2–1.5 mm wide; setaceous (or filiform); without abaxial multicellular glands; without cross venation. Ligule a fringed membrane; very short. *Contra-ligule* absent.

Reproductive organization. Plants bisexual, with bisexual spikelets; with hermaphrodite florets.

Inflorescence. *Inflorescence of spicate main branches (the branches 12 cm or more long), or a single raceme (the spikelets appressed to the hollows in the rachis, more or less contiguous)*; when branched, digitate. Primary inflorescence branches 1–2(–3). Inflorescence with axes ending in spikelets (but these imperfect in material seen). Rachides hollowed. Inflorescence espathaceate; not comprising 'partial inflorescences' and foliar organs. Spikelet-bearing axes slender; persistent. Spikelets solitary; secund (on one side of rachis); biseriate; shortly pedicellate, or subsessile.

Female-fertile spikelets. *Spikelets* 3 mm long; adaxial; compressed dorsiventrally; *disarticulating above the glumes*; *with a distinctly elongated rachilla internode above the glumes and with distinctly elongated rachilla internodes between the florets (the first floret stipitate, the second on an elongate, slender rachilla joint).* Rachilla prolonged beyond the uppermost female-fertile floret; rachilla hairless (glabrous). The rachilla extension with incomplete florets. Hairy callus present. *Callus* rather long (hairy at the base).

Glumes present; two; very unequal; shorter than the adjacent lemmas; dorsiventral to the rachis; hairless; glabrous; pointed; awnless; non-carinate (rounded on back); similar in shape and texture, thinly membranous to linear acuminate/setiform. Lower glume much shorter than half length of lowest lemma (about one-sixth); 0 nerved, or 1 nerved. Upper glume 1 nerved. *Spikelets* with incomplete florets. The incomplete florets distal to the female-fertile florets (terminating the rachilla). *The distal incomplete florets* 1; **incomplete florets** merely underdeveloped (reduced to a minute lemma); *incomplete florets awnless.*

Female-fertile florets 1. Lemmas similar in texture to the glumes to decidedly firmer than the glumes (membranous); smooth; not becoming indurated; incised; 2 lobed; not deeply cleft (bidentate); delicately awned. Awns 1; median; from a sinus; non-geniculate; flexuous (the tip thin and tortuous); hairless (scabrid); much longer than the body of the lemma (more than twice as long); entered by one vein; persistent. Lemmas hairless; glabrous; non-carinate (rounded on the back); without a germination flap; 3 nerved. *Palea* present; relatively long; apically notched (slightly 2-dentate); awnless, without apical setae (glabrous); thinner than the lemma (hyaline); not indurated; 2-nerved; 2-keeled (accommodating the rachilla). *Lodicules* present; 2; free; fleshy (clearly so, and long); glabrous; not toothed; not or scarcely vascularized. Anthers not penicillate; without an apically prolonged connective. *Ovary* glabrous. Styles free to their bases. Stigmas 2; red pigmented.

Fruit, embryo and seedling. Hilum short. Pericarp free (?).

Abaxial leaf blade epidermis. *Costal/intercostal zonation* conspicuous. *Papillae* absent. *Long-cells* similar in shape costally and intercostally; of similar wall thickness costally and intercostally. Mid-intercostal long-cells rectangular; having markedly sinuous walls (thick, pitted). *Microhairs* present; more or less spherical to elongated; clearly two-celled; chloridoid-type (basal cell short). Microhair apical cell wall of similar thickness/rigidity to that of the basal cell. Microhair basal cells 6 microns long. Microhair total length/width at septum 2. Microhair apical cell/total length ratio 0.5. *Stomata* common. Subsidiaries dome-shaped and triangular. Guard-cells overlapping to flush with the interstomatals. *Intercostal short-cells* common; in cork/silica-cell pairs; silicified. Intercostal silica bodies absent, or imperfectly developed; narrowly saddle shaped, or crescentic. *Costal short-cells* predominantly paired. Costal silica bodies present throughout the costal zones; crescentic (common), or tall-and-narrow, or saddle shaped (numerous, but in reduced form — almost tall-and-narrow, intergrading with crescents); not sharp-pointed.

Transverse section of leaf blade, physiology. *Leaf blades* largely *consisting of midrib*. C_4; XyMS+. PCR sheaths of the primary vascular bundles interrupted; interrupted both abaxially and adaxially. PCR sheath extensions absent. *Leaf blade* with distinct, prominent adaxial ribs (the section crescentic, with a large broad flat-topped midrib and only 1–2 small ribs on each side). *Midrib* conspicuous (the blade reduced to relatively small flanges on either side); having a conventional arc of bundles (a large median, and three small ones on either side); with colourless mesophyll adaxially. Bulliforms in conspicuous midrib 'hinges'; in deeply penetrating 'midrib hinge' fans. All the vascular bundles accompanied by sclerenchyma. Combined sclerenchyma girders present (with the lamina bundles); forming 'figures' (in the lamina bundles). Sclerenchyma all associated with vascular bundles. The 'extra' sclerenchyma in an adaxial layer. The lamina margins with fibres.

Taxonomy. Chloridoideae; main chloridoid assemblage.

Ecology, geography, regional floristic distribution. 2 species. West Indies. Glycophytic.

Neotropical. Caribbean.

References, etc. Leaf anatomical: this project.

Special comments. Fruit data wanting.

Schaffnerella Nash

Schaffnera Benth.

Habit, vegetative morphology. Slender annual; caespitose. *Culms* 6–12 cm high; herbaceous. Culm nodes glabrous. **Leaves** *mostly basal*; non-auriculate; without auricular setae. Sheath margins free (hyaline at the edges). Leaf blades narrow; about 0.2–1 mm wide; rolled (convolute); without abaxial multicellular glands; without cross venation (but these numerous and conspicuous under the transmission microscope). *Ligule an unfringed membrane*; not truncate (rounded-jagged); 0.5–1 mm long. *Contra-ligule* absent.

Reproductive organization. Plants bisexual, with bisexual spikelets; with hermaphrodite florets. The spikelets of sexually distinct forms on the same plant (the ultimate branch of the cymose fascicle with a sterile spikelet), or all alike in sexuality (?); hermaphrodite, or hermaphrodite and sterile (?).

Inflorescence. *Inflorescence complex paniculate (with small cymose fascicles of spatheate peduncles, each peduncle bearing two or three spikelets, each fascicle subtended and partly enclosed by a broad-sheathed spathe with a well developed lamina)*; spatheate (and spatheolate); *a complex of 'partial inflorescences' and intervening foliar organs*. Spikelet-bearing axes persistent. Spikelets not secund.

Female-fertile spikelets. Spikelets 5 mm long; compressed laterally; falling with the glumes; with conventional internode spacings. Rachilla terminated by a female-fertile floret. Hairy callus present (small). *Callus* short; blunt.

Glumes present; two; (the upper) relatively large; very unequal (the lower minute); shorter than the adjacent lemmas; hairy (G$_2$ hairy on the nerves); (the upper) *awned (with three or five awns)*; *very dissimilar (the lower minute and vestigial, the upper 3- or 5-aristate with long, straight, spreading, scabridulous awns, often laterally hyaline-appendiculate)*. Upper glume 3 nerved, or 5 nerved. *Spikelets* with female-fertile florets only.

Female-fertile florets 1. Lemmas similar in texture to the glumes (membranous); smooth; not becoming indurated; entire to incised; when incised, 2 lobed; not deeply cleft (bilobed); awned. Awns 1; median; dorsal; from near the top; non-geniculate; somewhat recurving; hairless (scabrid); much shorter than the body of the lemma (about half as long); entered by one vein. Lemmas hairy; carinate; without a germination flap; 3 nerved. *Palea* present (lanceolate); relatively long; entire (blunt); awnless, without apical setae (glabrous); not indurated (hyaline); 2-nerved; keel-less (the margins embracing the flower). **Lodicules** *absent*. Stamens 3; with free filaments (these short). Anthers relatively very long; not penicillate; without an apically prolonged connective. *Ovary* glabrous. Styles fused (the ovary attenuate into one long style). Stigmas 2.

Fruit, embryo and seedling. Fruit small; ellipsoid; compressed laterally. Pericarp fused.

Abaxial leaf blade epidermis. *Costal/intercostal zonation* conspicuous. *Papillae* present; intercostal (mainly on the interstomatals). Intercostal papillae not over-arching the stomata (but almost); consisting of one oblique swelling per cell (large). Mid-intercostal long-cells rectangular; having markedly sinuous walls (coarsely so). *Microhairs* present (very scarce); chloridoid-type (basal cell somewhat elongated). *Stomata* common; 15–18 microns long. Subsidiaries high dome-shaped, or triangular. Guard-cells overlapping to flush with the interstomatals. *Intercostal short-cells* absent or very rare. Intercostal silica bodies absent. *Costal short-cells* conspicuously in long rows. Costal silica bodies present in alternate cell files of the costal zones; rounded (a few, round to almost rectangular), or 'panicoid-type' (mostly); mostly rather rectangular cross shaped and butterfly shaped; not sharp-pointed.

Transverse section of leaf blade, physiology. C$_4$; XyMS+. PCR sheaths of the primary vascular bundles complete. PCR sheath extensions absent. *Leaf blade* 'nodular' in section. *Midrib* conspicuous (with a fairly heavy sclerenchymatous keel); with one bundle only. Bulliforms not in discrete, regular adaxial groups (the adaxial epidermis apparently being mainly of papillate and bulliform cells). Many of the smallest vascular bundles unaccompanied by sclerenchyma (most bundles unaccompanied by sclerenchyma: a few of the larger laterals with small abaxial strands, the midrib with a large abaxial strand (or girder?)). Combined sclerenchyma girders absent. Sclerenchyma all associated with vascular bundles.

Taxonomy. Chloridoideae; main chloridoid assemblage (superficially of andropogonoid appearance).

Ecology, geography, regional floristic distribution. 1 species. Central America. Holarctic. Madrean.

References, etc. Morphological/taxonomic: Bentham 1882. Leaf anatomical: this project.

Schedonnardus Steud.

Spirochloe Lunell

Habit, vegetative morphology. Perennial; caespitose. *Culms* 20–45 cm high; herbaceous; unbranched above. Culm internodes solid, or hollow. Leaves non-auriculate. Leaf blades narrow; 0.7–2 mm wide (to 5 cm long); flat (or involute); without abaxial multicellular glands; without cross venation. *Ligule an unfringed membrane.*

Reproductive organization. Plants bisexual, with bisexual spikelets; with hermaphrodite florets.

Inflorescence. *Inflorescence* of spicate main branches (stiff, slender divergent spikes, remote along a common axis); *deciduous in its entirety (the central axis first elongating and coiling into a loose spiral)*; open. Inflorescence with axes ending in spikelets. Inflorescence espatheate; not comprising 'partial inflorescences' and foliar organs. *Spikelets* solitary; secund; biseriate (appressed to the concave sides of the triquetrous rachis); *sessile*; somewhat imbricate.

Female-fertile spikelets. Spikelets 3–4 mm long; compressed laterally; disarticulating above the glumes. Rachilla terminated by a female-fertile floret. Hairy callus absent.

Glumes two; very unequal; (the upper) long relative to the adjacent lemmas (only slightly shorter than them); dorsiventral to the rachis (?); hairless; pointed; awnless; similar (lanceolate or acuminate, stiff, slightly divergent). Lower glume 1 nerved. Upper glume 1 nerved. *Spikelets* with female-fertile florets only.

Female-fertile florets 1. Lemmas acuminate; similar in texture to the glumes to decidedly firmer than the glumes (firmly membranous, rigid); entire; pointed; awnless to mucronate; hairless; glabrous to scabrid); carinate to non-carinate; 3 nerved. *Palea* present; relatively long (glabrous); entire; awnless, without apical setae; textured like the lemma; rigid; 2-nerved. *Lodicules* present; 2; free; fleshy; glabrous. *Stamens* 3. *Ovary* glabrous. Stigmas 2.

Fruit, embryo and seedling. Fruit small (2.5–3.5 mm long); fusiform. Hilum short (?). Pericarp fused (?). Embryo large; with an epiblast; with a scutellar tail; with an elongated mesocotyl internode. Embryonic leaf margins meeting.

Abaxial leaf blade epidermis. *Costal/intercostal zonation* conspicuous. *Papillae* present; intercostal. Intercostal papillae over-arching the stomata (at one end); consisting of one oblique swelling per cell (at one end of each interstomatal, and on a few of the long-cells as well). Long-cells of similar wall thickness costally and intercostally (walls of medium thickness). Mid-intercostal long-cells rectangular; having markedly sinuous walls. *Microhairs* present; elongated; clearly two-celled; chloridoid-type (the basal cell somewhat the longer). Microhair apical cell wall of similar thickness/rigidity to that of the basal cell. Microhairs 51–52.5–57 microns long. Microhair basal cells 36 microns long. Microhairs 18 microns wide at the septum. Microhair total length/width at septum 2.9. Microhair apical cells (15–)18–19.5(–21) microns long. Microhair apical cell/total length ratio 0.29–0.41. *Stomata* common; (21–)22.5–24(–27) microns long. Subsidiaries predominantly triangular. *Intercostal short-cells* common; not paired (mostly solitary); not silicified. Intercostal silica bodies absent. *Costal short-cells* conspicuously in long rows (but the files often interrupted by longish short-cells). Costal silica bodies present in alternate cell files of the costal zones; almost exclusively saddle shaped (a rather rectangular version); not sharp-pointed.

Transverse section of leaf blade, physiology. C_4; XyMS+. PCR sheaths of the primary vascular bundles interrupted; interrupted abaxially only. PCR sheath extensions absent. PCR cells without a suberised lamella. PCR cell chloroplasts with well developed grana; centripetal. *Mesophyll* with radiate chlorenchyma; traversed by columns of colourless mesophyll cells. *Leaf blade* with distinct, prominent adaxial ribs to 'nodular' in section; with the ribs more or less constant in size (round topped). *Midrib* conspicuous (via its marked abaxial keel, and the bulliform adaxial epidermis); with one bundle only. Bulliforms present in discrete, regular adaxial groups; associated with colourless mesophyll cells to form deeply-penetrating fans (these linked with traversing columns of colourless cells). All the vascular bundles accompanied by sclerenchyma. Combined sclerenchyma girders present (with the primaries only); forming 'figures' (some forming I's or 'anchors'). Sclerenchyma all associated with vascular bundles. The lamina margins with fibres.

Cytology. Chromosome base number, $x = 10$. $2n = 30$.

Taxonomy. Chloridoideae; main chloridoid assemblage.

Ecology, geography, regional floristic distribution. 1 species. Canada, U.S.A. to Argentina. Species of open habitats. Prairie.

Holarctic and Neotropical. Boreal and Madrean. Rocky Mountains. Pampas.

Rusts and smuts. Rusts — *Puccinia*. Taxonomically wide-ranging species: *Puccinia schedonnardi*.

Economic importance. Significant weed species: *S. paniculatus*.

References, etc. Leaf anatomical: Metcalfe 1960; this project.

Schenckochloa J.J. Ortíz

Diplachne barbata Hack.; *Gouinia barbata* (Hack.) Swallen

Habit, vegetative morphology. Perennial; caespitose. *Culms* 70–90 cm high; herbaceous; unbranched above; 2–4 noded. Culm nodes hidden by leaf sheaths. Young shoots intravaginal. **Leaves *mostly basal*;** non-auriculate. Leaf blades linear; narrow; 1–2 mm wide (10–30 cm long); flat, or rolled (convolute or involute); without cross venation. Ligule a fringed membrane. *Contra-ligule* absent.

Reproductive organization. Plants bisexual, with bisexual spikelets; with hermaphrodite florets. The spikelets hermaphrodite.

Inflorescence. *Inflorescence of spicate main branches (2–30 cm long, the lower branches 10–12 cm long, the rachis not grooved).* Inflorescence with axes ending in spikelets. Inflorescence espatheate; not comprising 'partial inflorescences' and foliar organs. The racemes spikelet bearing to the base. Spikelet-bearing axes persistent. Spikelets solitary; shortly pedicellate (the pedicels 0.5–3 mm long); imbricate to distant.

Female-fertile spikelets. *Spikelets 9–11 mm long*; cuneate, or lanceolate; compressed laterally; disarticulating above the glumes; disarticulating between the florets. *Rachilla* prolonged beyond the uppermost female-fertile floret; *rachilla hairless (sulcate-triangular in section towards the tip)*. Hairy callus present. Callus pointed.

Glumes two; more or less equal; shorter than the spikelets; shorter than the adjacent lemmas; hairless; glabrous; pointed; awnless; similar (lanceolate, memranous to papery). Lower glume 1 nerved. Upper glume 1 nerved. *Spikelets* with incomplete florets. The incomplete florets distal to the female-fertile florets. The distal incomplete florets 1–2 (?); *incomplete florets* merely underdeveloped; incomplete florets awned.

Female-fertile florets 4–6. *Lemmas* linear-lanceolate, laterally flattened; similar in texture to the glumes (membranous to papery); not becoming indurated; incised; shortly 2 lobed; not deeply cleft; *awned*. Awns 1; median; from a sinus; non-geniculate; straight, or flexuous; hairless (antrorsely scabrid); about as long as the body of the lemma to much longer than the body of the lemma (2–7 mm long); entered by one vein. *Awn bases* not twisted; *not flattened. Lemmas hairy (scabrid on the median nerve, the lateral nerves pectinate-ciliate, pilose at the base and between the nerves)*; carinate; without a germination flap; 3 nerved; with the nerves non-confluent. *Palea* present; relatively long, or conspicuous but relatively short (4–4.5 mm long); entire; awnless, without apical setae; textured like the lemma; not indurated; 2-nerved; 2-keeled. Palea back glabrous. *Palea keels* *scabrous*. *Stamens* 3. Anthers about 1 mm long (yellow). *Ovary* glabrous (?). Stigmas 2 (?).

Fruit, embryo and seedling. Fruit small (about 3 mm long); narrowly ellipsoid; compressed laterally and trigonous. Hilum short.

Abaxial leaf blade epidermis. *Papillae* absent. Intercostal zones with typical long-cells. Mid-intercostal long-cells rectangular; having markedly sinuous walls. *Microhairs* absent (chloridoid type present adaxially). *Stomata* absent or very rare. *Intercostal short-cells* common; in cork/silica-cell pairs and not paired. *Costal short-cells* predominantly paired, or neither distinctly grouped into long rows nor predominantly paired. Costal silica bodies present and well developed; 'panicoid-type'; dumb-bell shaped.

Transverse section of leaf blade, physiology. C_4. The anatomical organization conventional. XyMS+. PCR sheath outlines seemingly even. PCR sheath extensions absent. *Mesophyll* with radiate chlorenchyma; not *Isachne*-type; without fusoids. *Leaf blade* with distinct, prominent adaxial ribs; with the ribs more or less constant in size. *Midrib* not readily distinguishable; with one bundle only; without colourless mesophyll adaxially.

Taxonomy. Chloridoideae; main chloridoid assemblage.

Ecology, geography, regional floristic distribution. 1 species. North eastern Brazil. Species of open habitats. Coastal dunes and sandy soils.

Neotropical. Central Brazilian.

References, etc. Morphological/taxonomic: Ortíz 1991. Leaf anatomical: Ortíz 1991.

Special comments. Fruit data wanting. Anatomical data somewhat wanting.

Schismus P. Beauv.

Electra Panz., *Hemisacris* Steud.

Habit, vegetative morphology. Annual, or perennial (weakly, infrequently); caespitose (rarely), or decumbent (low). **Culms** 3–40 cm high; *herbaceous*; unbranched above. Culm nodes glabrous. Culm internodes solid, or hollow. Leaves non-auriculate. *Leaf blades* linear to linear-lanceolate; narrow; 0.5–2.5 mm wide; setaceous, or not setaceous; flat, or rolled (convolute); without abaxial multicellular glands; without cross venation; *persistent*. *Ligule a fringe of hairs*.

Reproductive organization. Plants bisexual, with bisexual spikelets; with hermaphrodite florets.

Inflorescence. Inflorescence paniculate; contracted; loosely spicate; espatheate; not comprising 'partial inflorescences' and foliar organs. Spikelet-bearing axes persistent. Spikelets solitary; not secund; pedicellate.

Female-fertile spikelets. Spikelets 4–8 mm long; slightly compressed laterally; falling with the glumes (falling entire), or disarticulating above the glumes; when disarticulating above the glumes, disarticulating between the florets (the upper florets florets falling singly, then the lower florets, glumes and pedicel falling together); with conventional internode spacings. Rachilla prolonged beyond the uppermost female-fertile floret. The rachilla extension with incomplete florets. Hairy callus present, or absent (or the hairs few and short).

Glumes two; more or less equal; shorter than the spikelets to about equalling the spikelets; *long relative to the adjacent lemmas*; free (widely separated); awnless; similar (herbaceous-membranous, the margins hyaline). Lower glume 5–7 nerved. Upper glume 3–7 nerved. *Spikelets* with incomplete florets. The incomplete florets distal to the female-fertile florets. *The distal incomplete florets* merely underdeveloped.

Female-fertile florets *5–10*. Lemmas similar in texture to the glumes (herbaceous, the lobes and margins hyaline); not becoming indurated; incised (to merely emarginate); 2 lobed; deeply cleft (bifid), or not deeply cleft (emarginate); awnless, or mucronate (from the sinus), or awned. *Awns* when present, 1; from a sinus; *non-geniculate*; much shorter than the body of the lemma to about as long as the body of the lemma; entered by one vein. Lemmas hairy (pilose dorsally or on the margins); non-carinate; 7–9 nerved. *Palea* present; relatively long; not indurated (membranous); 2-nerved; 2-keeled. *Lodicules* present; 2; joined, or free; fleshy; ciliate, or glabrous. *Stamens* 3. Anthers 0.2–0.5 mm long; not penicillate. *Ovary* glabrous. Styles free to their bases. Stigmas 2.

Fruit, embryo and seedling. *Fruit* free from both lemma and palea; compressed dorsiventrally. *Hilum short*. *Embryo large*. Endosperm hard; without lipid; containing compound starch grains.

Abaxial leaf blade epidermis. *Costal/intercostal zonation* conspicuous. *Papillae* absent. *Long-cells* similar in shape costally and intercostally; of similar wall thickness costally and intercostally. Mid-intercostal long-cells rectangular; having markedly sinuous walls. *Microhairs* present; elongated; clearly two-celled; panicoid-type. Microhair apical cell wall thinner than that of the basal cell and often collapsed. Microhairs 36–39–45 microns long. Microhair basal cells 18–21 microns long. Microhairs (6.9–)7.5–8.4(–9) microns wide at the septum. Microhair total length/width at septum 4–5.6. Microhair apical cells (19.5–)21–22.5(–24) microns long. Microhair apical cell/total length ratio 0.53–0.62. *Stomata* common; (19–)21–22(–24) microns long. Subsidiaries dome-shaped. Guard-cells overlapping to flush with the interstomatals. *Intercostal short-cells* common; not paired (solitary); not silicified. Intercostal silica bodies imperfectly developed. *Costal short-cells* conspicuously in long rows. Costal silica bodies present in alternate cell files of the costal zones; 'panicoid-type', or saddle shaped (or almost so), or horizontally-elongated crenate/sinuous (rarely); often (?) dumb-bell shaped; not sharp-pointed.

Transverse section of leaf blade, physiology. C_3; XyMS+. *Mesophyll* with radiate chlorenchyma, or with non-radiate chlorenchyma. *Leaf blade* with distinct, prominent adaxial ribs, or 'nodular' in section; with the ribs more or less constant in size. *Midrib* not readily distinguishable; with one bundle only. Bulliforms not in discrete, regular adaxial groups (bulliforms ill defined and small, save for the 'midrib hinges'). All the vascular bundles accompanied by sclerenchyma. Combined sclerenchyma girders present, or absent; forming 'figures'. Sclerenchyma all associated with vascular bundles. The lamina margins with fibres.

Special diagnostic feature. Female-fertile lemmas awnless, mucronate or with a short straight awn.

Cytology. Chromosome base number, $x = 6$. $2n = 12$. 2 ploid.

Taxonomy. Arundinoideae; Danthonieae.

Ecology, geography, regional floristic distribution. 5 species. Africa, Mediterranean to northwest India. Commonly adventive. Xerophytic; species of open habitats.

Holarctic, Paleotropical, and Cape. Boreal, Tethyan, and Madrean. African and Indomalesian. Euro-Siberian. Macaronesian, Mediterranean, and Irano-Turanian. Saharo-Sindian, Sudano-Angolan, and Namib-Karoo. Indian. European and Siberian. Somalo-Ethiopian.

Rusts and smuts. Rusts — *Puccinia*. Taxonomically wide-ranging species: *Puccinia hordei*. Smuts from Ustilaginaceae. Ustilaginaceae — *Ustilago*.

Economic importance. Important native pasture species: *S. arabicus*, *S. barbatus*.

References, etc. Leaf anatomical: Metcalfe 1960 and this project.

Schizachne Hackel

Habit, vegetative morphology. Perennial; caespitose (with short rhizomes). *Culms* 30–100 cm high; herbaceous; unbranched above. Culm nodes glabrous. Culm internodes hollow. *Leaves* not basally aggregated; non-auriculate. *Sheath margins joined*. Leaf blades linear; narrow; 1–4 mm wide; flat, or rolled (convolute); without cross venation; rolled in bud. *Ligule* present; an unfringed membrane; truncate (at back); 0.5–3 mm long.

Reproductive organization. Plants bisexual, with bisexual spikelets; with hermaphrodite florets.

Inflorescence. Inflorescence few spikeleted; paniculate; open (up to 10 cm). Inflorescence with axes ending in spikelets. Inflorescence espatheate; not comprising 'partial inflorescences' and foliar organs. Spikelet-bearing axes persistent. *Spikelets secund (in the lax panicle)*; long pedicellate.

Female-fertile spikelets. Spikelets 9–25 mm long; compressed laterally; disarticulating above the glumes; disarticulating between the florets. Rachilla prolonged beyond the uppermost female-fertile floret; rachilla hairless (glabrous). The rachilla extension with incomplete florets. **Hairy callus present (long-pilose)**. *Callus* short; blunt.

Glumes two; very unequal; shorter than the spikelets; shorter than the adjacent lemmas; pointed (acute); awnless; carinate, or non-carinate; similar (membranous, ovate-lanceolate). Lower glume (1–)3 nerved. Upper glume 5 nerved. *Spikelets* with incomplete florets. The incomplete florets distal to the female-fertile florets. The distal incomplete florets 1–2; *incomplete florets* merely underdeveloped.

Female-fertile florets 3–7. Lemmas decidedly firmer than the glumes (becoming leathery); not becoming indurated; incised; 2 lobed; not deeply cleft (incised to about 1/4); awned. Awns 1; median; dorsal; from near the top (behind the sinus); hairless; about as long as the body of the lemma to much longer than the body of the lemma; entered by one vein. Lemmas hairless; carinate to non-carinate; without a germination flap; 7–13 nerved. *Palea* present; relatively long, or conspicuous but relatively short, or very reduced; not indurated (membranous); 2-nerved; 2-keeled. Palea keels hairy (ciiate above). **Lodicules** present; 2; joined; *fleshy*; glabrous; not or scarcely vascularized. *Stamens* 3. Anthers 1–2 mm long. *Ovary* glabrous. Styles free to their bases. Stigmas 2; white.

Fruit, embryo and seedling. *Fruit* free from both lemma and palea; small, or medium sized (about 4 mm long). Hilum long-linear. Embryo small. Endosperm hard; without lipid; containing compound starch grains. Embryo with an epiblast; without a scutellar tail; with a negligible mesocotyl internode. Embryonic leaf margins meeting.

Abaxial leaf blade epidermis. *Costal/intercostal zonation* conspicuous. *Papillae* absent. *Long-cells* similar in shape costally and intercostally; of similar wall thickness costally and intercostally. Mid-intercostal long-cells rectangular; having markedly sinuous walls. *Microhairs* absent. *Stomata* absent or very rare. *Intercostal short-cells* common; not paired; not silicified. *Costal short-cells* neither distinctly grouped into long rows nor predominantly paired. Costal silica bodies horizontally-elongated crenate/sinuous (a few), or 'panicoid-type'; mostly elongated, iregularly nodular; not sharp-pointed.

Transverse section of leaf blade, physiology. C_3; XyMS+. *Mesophyll* with non-radiate chlorenchyma. *Leaf blade* 'nodular' in section; with the ribs more or less constant in size. Bulliforms present in discrete, regular adaxial groups; in simple fans. All the vascular bundles accompanied by sclerenchyma. Combined sclerenchyma girders present; forming 'figures'. Sclerenchyma all associated with vascular bundles.

Cytology. Chromosome base number, $x = 10$. $2n = 20$. 2 ploid.

Taxonomy. Pooideae; Poodae; Meliceae.

Ecology, geography, regional floristic distribution. 1 species. North Eurasia, Japan, North America. Mesophytic; shade species. In woods.

Holarctic. Boreal, Tethyan, and Madrean. Euro-Siberian, Eastern Asian, Atlantic North American, and Rocky Mountains. Mediterranean. European and Siberian. Canadian-Appalachian.

Rusts and smuts. Rusts — *Puccinia*. Taxonomically wide-ranging species: *Puccinia coronata*.

References, etc. Leaf anatomical: this project.

Schizachyrium Nees

Pithecurus Kunth, *Schizopogon* Spreng.
Excluding *Ystia*

Habit, vegetative morphology. Annual, or perennial; rhizomatous, or stoloniferous, or caespitose, or decumbent. *Culms* (5–)30–320 cm high; woody and persistent (rarely, and never tall), or herbaceous; branched above. Culm internodes solid, or hollow. *Leaves* not basally aggregated; non-auriculate. Leaf blades linear; narrow; without cross venation; persistent; once-folded in bud. **Ligule** present; *a fringed membrane (short)*.

Reproductive organization. Plants bisexual, with bisexual spikelets; with hermaphrodite florets. *The spikelets* of sexually distinct forms on the same plant; hermaphrodite and male-only, or hermaphrodite and sterile; overtly heteromorphic; *all in heterogamous combinations*. Exposed-cleistogamous, or chasmogamous.

Inflorescence. Inflorescence a single raceme, or paniculate (of single racemes, these sometimes solitary but usually in a spatheate false panicle). Rachides flattened. *Inflorescence* spatheate; *a complex of 'partial inflorescences' and intervening foliar organs*. **Spikelet-bearing axes** peduncled 'racemes'; *solitary (in their spathes, but often fascicled)*; *with substantial rachides*; disarticulating; disarticulating at the joints. 'Articles' non-linear (compressed, thickened above with a cupular, usually fringed apex); with toothed apices. Spikelets paired; secund; sessile and pedicellate; consistently in 'long-and-short' combinations; in pedicellate/sessile combinations. Pedicels of the 'pedicellate' spikelets free of the rachis. The 'shorter' spikelets hermaphrodite. The 'longer' spikelets male-only, or sterile.

Female-sterile spikelets. The pedicellate (male or sterile) members broader and flatter or reduced, the lower glume sometimes awned.

Female-fertile spikelets. Spikelets compressed dorsiventrally (or subterete below and compressed between the internode and the pedicel, lanceolate to linear); falling with the glumes (and the joint). Rachilla terminated by a female-fertile floret. Hairy callus present. *Callus* short; blunt.

Glumes present; two; more or less equal; long relative to the adjacent lemmas; awned (the upper, rarely), or awnless; very dissimilar (the lower bicarinate, the upper thinner and naviculate). *Lower glume two-keeled (the keels not winged, but excurrent as teeth or mucros)*; convex on the back to flattened on the back; not pitted; relatively smooth; 3–13 nerved (with several between the keels). Upper glume 1–3 nerved. *Spikelets* with incomplete florets. The incomplete florets proximal to the female-fertile florets. *Spikelets with proximal incomplete florets*. The proximal incomplete florets 1; epaleate; sterile. The proximal lemmas awnless; 2 nerved; similar in texture to the female-fertile lemmas (hyaline, often purple); not becoming indurated.

Female-fertile florets 1. **Lemmas** often stipitiform; less firm than the glumes (hyaline); not becoming indurated; usually *incised (rarely merely prolonged into the awn, without teeth)*; when incised, 2 lobed; deeply cleft (often deeply bifid), or not deeply cleft; awned. Awns 1; median; from a sinus, or apical; geniculate; hairless (glabrous); about as long as the body of the lemma, or much longer than the body of the lemma (usually short). Lemmas hairless; 1–3 nerved. *Palea* present, or absent (usually); when present, very reduced (to a minute, hyaline scale); not indurated (hyaline); nerveless. *Lodicules* present (tiny); 2; fleshy; ciliate, or glabrous. *Stamens* 2–3. Anthers not penicillate. *Ovary* glabrous. Styles free to their bases. Stigmas 2.

Fruit, embryo and seedling. *Fruit* free from both lemma and palea; compressed dorsiventrally (or subterete). Hilum short. Embryo large. Endosperm hard; without lipid. Embryo without an epiblast; with a scutellar tail; with an elongated mesocotyl internode. Embryonic leaf margins overlapping.

Abaxial leaf blade epidermis. *Costal/intercostal zonation* conspicuous. *Papillae* absent. Long-cells of similar wall thickness costally and intercostally (quite thin walled). Mid-intercostal long-cells rectangular; having markedly sinuous walls. *Microhairs* present; panicoid-type; (36–)42–50(–60) microns long; 6–8.4 microns wide at the septum. Microhair total length/width at septum 5.2–7. Microhair apical cells (18–)19–26(–30) microns long. Microhair apical cell/total length ratio 0.46–0.52. *Stomata* common; (25–)27–29(–30) microns long. Subsidiaries low to high dome-shaped to triangular. Guard-cells overlapping to flush with the interstomatals. *Intercostal short-cells* absent or very rare. *Costal short-cells* conspicuously in long rows. Costal silica bodies 'panicoid-type'; dumb-bell shaped (mostly, long to short), or nodular (some); not sharp-pointed.

Transverse section of leaf blade, physiology. C_4; biochemical type NADP–ME (1 species); XyMS–. PCR sheath outlines uneven. PCR cell chloroplasts centrifugal/peripheral. *Mesophyll* with radiate chlorenchyma, or with non-radiate chlorenchyma. *Leaf blade* adaxially flat. *Midrib* conspicuous; having a conventional arc of bundles; with colourless mesophyll adaxially, or without colourless mesophyll adaxially. Bulliforms not in discrete, regular adaxial groups (the adaxial epidermis of some species with scattered bulliform cells, *S. fragile* having the adaxial epidermis largely bulliform and a large fan of bulliforms over the midrib). Many of the smallest vascular bundles unaccompanied by sclerenchyma. Combined sclerenchyma girders present; forming 'figures' (some of the primary laterals with 'anchors' or I's). Sclerenchyma all associated with vascular bundles.

Culm anatomy. *Culm internode bundles* in three or more rings.

Phytochemistry. Leaves without flavonoid sulphates (3 species).

Cytology. Chromosome base number, $x = 5$ and 10. $2n = 20$, 30, 40, and 50.

Taxonomy. Panicoideae; Andropogonodae; Andropogoneae; Andropogoninae.

Ecology, geography, regional floristic distribution. About 60 species. Tropical. Helophytic, or mesophytic, or xerophytic; species of open habitats; halophytic (rarely), or glycophytic. Savanna, rarely sandy beaches or dunes.

Holarctic, Paleotropical, Neotropical, Australian, and Antarctic. Boreal, Tethyan, and Madrean. African, Madagascan, Indomalesian, and Neocaledonian. Eastern Asian and Atlantic North American. Irano-Turanian. Saharo-Sindian, Sudano-Angolan, West African Rainforest, and Namib-Karoo. Indian, Indo-Chinese, Malesian, and Papuan. Caribbean, Amazon, Central Brazilian, Pampas, and Andean. North and East Australian. Patagonian. Southern Atlantic North American and Central Grasslands. Sahelo-Sudanian, Somalo-Ethiopian, South Tropical African, and Kalaharian. Tropical North and East Australian.

Rusts and smuts. Rusts — *Puccinia*. Taxonomically wide-ranging species: '*Uromyces*' *clignyi*. Smuts from Ustilaginaceae. Ustilaginaceae — *Sorosporium* and *Sphacelotheca*.

Economic importance. Significant weed species: *S. condensatum*, *S. scoparium*. Important native pasture species: *S. brevifolium*, *S. sanguineum*.

References, etc. Leaf anatomical: Metcalfe 1960; this project.

Schizostachyum Nees

Leptocanna Chia & Fung
Excluding *Cephalostachyum*, *Dendrochloa*, *Pseudostachyum*, *Teinostachyum*

Habit, vegetative morphology. Perennial (the young culm internodes uniformly covered with adpressed white hairs). The flowering culms leafy. *Culms* 400–1200 cm high; woody and persistent; to 5 cm in diameter; scandent, or not scandent. Primary branches/mid-culm node several to many. Culm internodes solid, or hollow. Rhizomes pachymorph. Plants unarmed. Young shoots extravaginal. *Leaves* not basally aggregated; auriculate (the auricles sometimes small, often bristly); with auricular setae (curved). Sheath margins free. Leaf blades broad; 15–90 mm wide (15–50 cm long); pseudopetiolate; cross veined (and often with pellucid dots or dashes), or without cross venation; disarticulating from the sheaths. Ligule an unfringed membrane, or a fringed membrane.

Reproductive organization. Plants bisexual, with bisexual spikelets; with hermaphrodite florets.

Inflorescence. Inflorescence indeterminate; *with pseudospikelets*; a false spike, with spikelets on contracted axes, or paniculate (with alternate clusters of pseudospikelets in terminal spike-like or paniculate inflorescences, which are leafless or leafy below); spatheate; a complex of 'partial inflorescences' and intervening foliar organs. **Spikelet-bearing axes** *spikelike, or paniculate*; persistent. Spikelets not secund.

Female-fertile spikelets. *Spikelets 12–50 mm long*; compressed laterally to not noticeably compressed (subterete); disarticulating above the glumes; disarticulating between the florets. Rachilla usually prolonged beyond the uppermost female-fertile floret (with a rudimentary floret). The rachilla extension with incomplete florets.

Glumes present, or absent (depending on interpretation); if considered present, one per spikelet to several (3 or 4 'bracts' with buds, in lieu of glumes). *Spikelets* with incomplete florets. The incomplete florets distal to the female-fertile florets. *The distal incomplete florets* merely underdeveloped. *Spikelets without proximal incomplete florets.*

Female-fertile florets 1, or 3, or 4. Lemmas similar in texture to the glumes; awned (shortly spine-tipped), or awnless to mucronate (apiculate, with overlapping and convoluted margins). Awns when present, 1; apical; much shorter than the body of the lemma (to 3 mm long). Lemmas non-carinate; 11–17 nerved (in material seen). **Palea** *present*; relatively long (usually exceeding the lemma); apically notched; shortly 2-

pointed; several nerved (11–15 in material seen); 2-keeled (grooved between them, to accommodate the rachilla). *Lodicules* present, or absent (sometimes exhibiting transitions between stamens and lodicules); when present, 1–10; free; membranous; ciliate. *Stamens* usually 6; usually with free filaments. Anthers 5–14 mm long; penicillate, or not penicillate; with the connective apically prolonged, or without an apically prolonged connective. **Ovary** glabrous; *with a conspicuous apical appendage. The appendage long, stiff and tapering*. Styles fused (the single style hollow). Stigmas 3 (simple); white, or red pigmented.

Fruit, embryo and seedling. Fruit compressed laterally. Hilum long-linear. Pericarp thick and hard (leathery to crustaceous); free. Embryo small.

Abaxial leaf blade epidermis. *Costal/intercostal zonation* conspicuous. *Papillae* present (costally and intercostally). Intercostal papillae over-arching the stomata (and completely obscuring them); several per cell (large, one row per long-cell). *Long-cells* similar in shape costally and intercostally; of similar wall thickness costally and intercostally (walls of medium thickness). Mid-intercostal long-cells rectangular; having markedly sinuous walls. *Microhairs* present; panicoid-type; (54–)57–63(–69) microns long; (8.4–)9 microns wide at the septum. Microhair total length/width at septum 6–7.7. Microhair apical cells (28–)30–36(–39) microns long. Microhair apical cell/total length ratio 0.53–0.57. *Stomata* common (outlines obscured by papillae); 19.5–21–24 microns long. *Intercostal short-cells* common; in cork/silica-cell pairs; silicified. Intercostal silica bodies tall-and-narrow. *Crown cells* present (at least, having large umbonate or tiny-pointed 'prickles'). *Costal short-cells* predominantly paired. Costal silica bodies rounded, or saddle shaped, or tall-and-narrow, or crescentic, or oryzoid (predominantly tall-and-narrow, overlapping into ovals, crescents and near-oryzoids); not sharp-pointed.

Transverse section of leaf blade, physiology. C_3; XyMS+. Mesophyll with adaxial palisade; with arm cells; with fusoids. The fusoids external to the PBS. *Leaf blade* adaxially flat (save alongside the midrib). *Midrib* conspicuous; having complex vascularization. Bulliforms present in discrete, regular adaxial groups; in simple fans. All the vascular bundles accompanied by sclerenchyma. Combined sclerenchyma girders present; forming 'figures' (the larger bundles). Sclerenchyma all associated with vascular bundles.

Taxonomy. Bambusoideae; Bambusodae; Bambuseae.

Ecology, geography, regional floristic distribution. 35 species. Eastern Asia. Halophytic (*S. pulchellum*), or glycophytic. Mostly savanna, rarely coastal sand.

Paleotropical. Indomalesian. Indo-Chinese and Malesian.

Economic importance. Pliant, tough splints of *S. dumetorum* constitute thonging, and the whole culms are pounded and twisted into ropes. *S. jaculans* is the blowpipe bamboo of lowland Malaya.

References, etc. Leaf anatomical: Metcalfe 1960; this project.

Schmidtia Steud.

Antoschmidtia Boiss.

Habit, vegetative morphology. Annual, or perennial (usually viscid); caespitose to decumbent. *Culms* 15–100 cm high; herbaceous; branched above. Culm nodes hairy, or glabrous. Culm internodes hollow. Plants unarmed. Young shoots intravaginal. **Leaves** not basally aggregated; non-auriculate; *with auricular setae*. Leaf blades linear to linear-lanceolate; broad, or narrow; 4–13 mm wide; flat, or rolled; exhibiting multicellular glands abaxially (at the base of macrohairs). The abaxial glands intercostal. *Leaf blades not pseudopetiolate*; without cross venation; persistent. Ligule a fringe of hairs; truncate. *Contra-ligule* absent.

Reproductive organization. Plants bisexual, with bisexual spikelets; with hermaphrodite florets.

Inflorescence. Inflorescence paniculate; open, or contracted; espatheate; not comprising 'partial inflorescences' and foliar organs. Spikelet-bearing axes persistent. Spikelets solitary; not secund; subsessile to pedicellate.

Female-fertile spikelets. Spikelets 7–10 mm long; slightly compressed laterally; disarticulating above the glumes; not disarticulating between the florets; with conventional internode spacings. Rachilla prolonged beyond the uppermost female-fertile floret; rachilla hairy. The rachilla extension with incomplete florets. Hairy callus present. *Callus* short.

Glumes two; very unequal to more or less equal; about equalling the spikelets; long relative to the adjacent lemmas; hairless; pointed; awnless; slightly carinate, or non-carinate; similar (lanceolate, membranous, usually green or grey). **Lower glume** about 0.6–0.7 times the length of the upper glume; much exceeding the lowest lemma; *7–11 nerved*. Upper glume 7–11 nerved. *Spikelets* with incomplete florets. The incomplete florets distal to the female-fertile florets. The distal incomplete florets 1, or 2; *incomplete florets* merely underdeveloped; incomplete florets awned.

Female-fertile florets 3–9. *Lemmas* decidedly firmer than the glumes (subleathery); not becoming indurated; *incised*; 6 lobed; deeply cleft; awned. *Awns 5 (one awn from each sinus, and the lobes sometimes mucronate as well)*; median and lateral; the median similar in form to the laterals; non-geniculate; hairless (scabrid); about as long as the body of the lemma; entered by one vein. The lateral awns about equalling the median. Lemmas hairy (villous below); non-carinate (dorsally rounded); without a germination flap; 9 nerved; with the nerves non-confluent. *Palea* present; relatively long (longer than the body of the lemma); tightly clasped by the lemma; entire (pointed); awnless, without apical setae; thinner than the lemma; not indurated; 2-nerved; 2-keeled. Palea keels wingless; hairy (stiffly ciliate). *Lodicules* present; 2; fleshy; ciliate (sometimes glandular), or glabrous. *Stamens* 3. Anthers 3–4 mm long; not penicillate; without an apically prolonged connective. *Ovary* glabrous. Styles free to their bases. Stigmas 2; white.

Fruit, embryo and seedling. *Fruit* free from both lemma and palea; small (about 2.5 mm long); ellipsoid; compressed dorsiventrally. Hilum short. Pericarp fused. Embryo large; waisted; with an epiblast; with a scutellar tail; with an elongated mesocotyl internode. Embryonic leaf margins overlapping.

Abaxial leaf blade epidermis. *Costal/intercostal zonation* conspicuous. *Papillae* absent. *Long-cells* similar in shape costally and intercostally; of similar wall thickness costally and intercostally. Mid-intercostal long-cells rectangular; having markedly sinuous walls. *Microhairs* present; elongated; clearly two-celled; *Enneapogon*-type (very variable in size, at least in *S. pappophoroides*). Microhair apical cell wall thinner than that of the basal cell but not tending to collapse. Microhairs (54–)81–360(–426) microns long. Microhair basal cells 45–150–405 microns long. Microhairs (5.4–)6–9.6(–10.5) microns wide at the septum. Microhair total length/width at septum 10–34.3. Microhair apical cells (19.5–)24–51(–63) microns long. Microhair apical cell/total length ratio 0.15–0.36. *Stomata* common; (21–)24–30(–33) microns long. Guard-cells overlapping to flush with the interstomatals. *Intercostal short-cells* common; not paired (mainly solitary); not silicified. Intercostal silica bodies absent. *Costal short-cells* conspicuously in long rows. Costal silica bodies present in alternate cell files of the costal zones; 'panicoid-type'; cross shaped, dumb-bell shaped, and nodular; not sharp-pointed.

Transverse section of leaf blade, physiology. C_4; XyMS+. PCR sheath outlines uneven. PCR sheaths of the primary vascular bundles interrupted; interrupted abaxially only. PCR sheath extensions absent. PCR cell chloroplasts centrifugal/peripheral. *Mesophyll* with radiate chlorenchyma. *Leaf blade* adaxially flat. *Midrib* not readily distinguishable; with one bundle only. Bulliforms present in discrete, regular adaxial groups (between each bundle pair); in simple fans (the median cells deeply penetrating), or associated with colourless mesophyll cells to form deeply-penetrating fans. All the vascular bundles accompanied by sclerenchyma. Combined sclerenchyma girders absent (most bundles with strands only), or present (a few). Sclerenchyma all associated with vascular bundles.

Special diagnostic feature. Female-fertile lemmas 6-lobed and 5-awned, with an awn arising between each pair of lobes.

Cytology. Chromosome base number, $x = 9$. $2n = 36$.

Taxonomy. Chloridoideae; Pappophoreae.

Ecology, geography, regional floristic distribution. 2 species. Tropical and southern Africa, Cape Verde Is., Pakistan. Xerophytic; species of open habitats; glycophytic. Woods and bushland, on dry sandy soils.

Holarctic and Paleotropical. Tethyan. African. Macaronesian. Saharo-Sindian, Sudano-Angolan, and Namib-Karoo. Sahelo-Sudanian, Somalo-Ethiopian, South Tropical African, and Kalaharian.

References, etc. Leaf anatomical: this project; photos of *S. kalihariensis* and *S. pappophoroides* provided by R.P. Ellis.

Schoenefeldia Kunth

Habit, vegetative morphology. Annual, or perennial; caespitose. *Culms* 70–120 cm high; herbaceous. *Leaves* not basally aggregated; non-auriculate. Leaf blades linear; narrow; without abaxial multicellular glands; without cross venation. Ligule a fringed membrane (short).

Reproductive organization. Plants bisexual, with bisexual spikelets; with hermaphrodite florets.

Inflorescence. *Inflorescence* of spicate main branches (usually 2–6 sessile, flexuous spikes); *digitate*. Primary inflorescence branches (1–)2–6. Inflorescence espatheate; not comprising 'partial inflorescences' and foliar organs. Spikelet-bearing axes persistent. *Spikelets* solitary; secund (on one side of the rachis); biseriate; *sessile*; imbricate.

Female-fertile spikelets. *Spikelets* strongly *compressed laterally*; disarticulating above the glumes. Rachilla prolonged beyond the uppermost female-fertile floret. Hairy callus present.

Glumes two; very unequal; exceeding the spikelets; long relative to the adjacent lemmas (exceeding it); free; hairless; pointed; awned (G₁, sometimes), or awnless; carinate; *similar (persistent, narrow or setaceous, subhyaline, divergent)*. Lower glume 1 nerved. Upper glume 1 nerved. *Spikelets* with female-fertile florets only, or with incomplete florets. The incomplete florets when present, distal to the female-fertile florets. The distal incomplete florets when present, 1; incomplete florets awned.

Female-fertile florets *1*. Lemmas completely covering the palea; decidedly firmer than the glumes (often blackened at maturity); incised; 2 lobed; not deeply cleft (apically bifid); awned. *Awns* 1; median; from a sinus; non-geniculate, or geniculate; *much longer than the body of the lemma (very long, flexuous, tangling one another)*. Lemmas hairy (with appressed hairs); non-carinate (rounded on the back); 3 nerved. *Palea* present; 2-nerved; 2-keeled. *Lodicules* present; 2; free; fleshy; glabrous. *Stamens* 2–3. *Ovary* glabrous. Styles free to their bases. Stigmas 2.

Fruit, embryo and seedling. *Fruit* free from both lemma and palea; ellipsoid; compressed laterally. Hilum short. *Pericarp free*. Embryo large.

Abaxial leaf blade epidermis. *Costal/intercostal zonation* conspicuous. *Papillae* present; intercostal. Intercostal papillae not over-arching the stomata; consisting of one oblique swelling per cell. Mid-intercostal long-cells rectangular; having markedly sinuous walls. *Microhairs* present; more or less spherical to elongated; clearly two-celled; chloridoid-type. Microhair apical cell wall of similar thickness/rigidity to that of the basal cell. Microhairs 21–24 microns long. Microhair basal cells 15–18 microns long. Microhairs 9–10–10.2 microns wide at the septum. Microhair total length/width at septum 2.2–2.7. Microhair apical cells 7.5–9.6–10.5 microns long. Microhair apical cell/total length ratio 0.36–0.46. *Stomata* common; 19.5–21 microns long. Subsidiaries triangular. Guard-cells overlapping to flush with the interstomatals. *Intercostal short-cells* absent or very rare; not paired. Intercostal silica bodies absent. Large prickles present costally. *Costal short-cells* conspicuously in long rows. Costal silica bodies present in alternate cell files of the costal zones; exclusively saddle shaped; not sharp-pointed.

Transverse section of leaf blade, physiology. C₄; XyMS+. PCR sheaths of the primary vascular bundles interrupted; interrupted both abaxially and adaxially. PCR sheath extensions absent. *Mesophyll* with radiate chlorenchyma. *Leaf blade* adaxially flat. *Midrib* not readily distinguishable; with one bundle only. Bulliforms present in discrete, regular adaxial groups; in simple fans and associated with colourless mesophyll cells to form deeply-penetrating fans. All the vascular bundles accompanied by sclerenchyma. Combined sclerenchyma girders present (with all the bundles); forming 'figures' (in all the bundles). Sclerenchyma all associated with vascular bundles. The lamina margins with fibres.

Taxonomy. Chloridoideae; main chloridoid assemblage.

Ecology, geography, regional floristic distribution. 2 species. Tropical Africa, Asia. Species of open habitats. Savanna, hardpans and seasonally flooded flats.

Paleotropical. African, Madagascan, and Indomalesian. Saharo-Sindian, Sudano-Angolan, and West African Rainforest. Indian. Sahelo-Sudanian, Somalo-Ethiopian, and South Tropical African.

References, etc. Leaf anatomical: this project.

Sclerachne R.Br.

Habit, vegetative morphology. Erect to prostrate annual. *Culms* 10–90 cm high; herbaceous. Culm internodes solid. Leaves non-auriculate. Leaf blades linear-lanceolate (with broad base and acute apex); broad; without cross venation.

Reproductive organization. *Plants monoecious with all the fertile spikelets unisexual*; without hermaphrodite florets. The spikelets of sexually distinct forms on the same plant; female-only, male-only, and sterile. The male and female-fertile spikelets segregated, in different parts of the same inflorescence branch (in terms of the reduced 'racemes'), or mixed in the inflorescence (in terms of the panicle as a whole).

Inflorescence. Inflorescence paniculate (a complex pseudo-panicle: see Backer 1968); spatheate; a complex of 'partial inflorescences' and intervening foliar organs. Spikelet-bearing axes *very much reduced (to peduncled, spatheate, spike-like reduced 'racemes' of 2(–3) spikelets: the first female, the terminal male, the middle if present male or female)*; paired to clustered (fascicled); with substantial rachides; disarticulating; falling entire (the peduncle apex widened, articulated with the hollowed base of the first spikelet). 'Articles' non-linear; disarticulating obliquely. Spikelets solitary, or paired; consistently in 'long-and-short' combinations. Pedicels of the 'pedicellate' spikelets discernible, but fused with the rachis. The 'shorter' spikelets female-only (basal), or male-only (terminal). The 'longer' spikelets sterile.

Female-sterile spikelets. The male spikelet terminal, with two equal herbaceous, acute, many-nerved glumes; two lemmas with male flowers, each having three stamens and two large lodicules. Rachilla of male spikelets terminated by a male floret. The male spikelets with glumes; without proximal incomplete florets; 2 floreted. Male florets 2; 3 staminate.

Female-fertile spikelets. Spikelets falling with the glumes (and associated structures). Rachilla terminated by a female-fertile floret. Hairy callus absent.

Glumes two; long relative to the adjacent lemmas; hairless; awnless; *very dissimilar (the G_1 sheath-like, embracing base of raceme, indurated, 2-keeled, with one or two oblique transverse constrictions, emarginate and winged, the G_2 ovate acuminate, transparent, indurated, grooved on each side of mid-nerve, its apex appressed to the keels of G_1). Lower glume two-keeled*. Spikelets with incomplete florets. The incomplete florets proximal to the female-fertile florets. *Spikelets with proximal incomplete florets*. *The proximal incomplete florets* 1; sterile. The proximal lemmas awnless; 3 nerved; similar in texture to the female-fertile lemmas (hyaline, but ovate, long-acuminate); not becoming indurated.

Female-fertile florets 1. Lemmas acute; less firm than the glumes (hyaline); not becoming indurated; entire; pointed; awnless; hairless; non-carinate; without a germination flap; 1 nerved (narrow). *Palea* present; entire (acute); awnless, without apical setae; 2-nerved. *Lodicules* absent. *Stamens* 0.

Fruit, embryo and seedling. *Disseminule consisting of the disarticulated spikelet-bearing inflorescence unit (a false fruit, held together by the G_1 of the basal, female spikelet). Fruit longitudinally grooved (on the back)*. Hilum short.

Abaxial leaf blade epidermis. *Costal/intercostal zonation* conspicuous. *Papillae* absent. *Long-cells* markedly different in shape costally and intercostally (the costals narrower, more regularly rectangular); of similar wall thickness costally and intercostally to differing markedly in wall thickness costally and intercostally (the costals somewhat thicker walled). Mid-intercostal long-cells rectangular and fusiform; having markedly sinuous walls and having straight or only gently undulating walls (pitted, the sinuosity where present fairly fine). *Microhairs* present; elongated; clearly two-celled; panicoid-type. *Stomata* common. Subsidiaries non-papillate; low dome-shaped to triangular (their outer walls unusually thick). Guard-cells overlapping to flush with the interstomatals. *Intercostal short-cells* common; in cork/silica-cell pairs (but the raised cells calling for careful focusing); silicified. Intercostal silica bodies cross-shaped. No macrohairs or prickles seen. *Costal short-cells* conspicuously in long rows (over some veins), or predominantly paired (over others). Costal silica bodies present and well developed; 'panicoid-type'; cross shaped to butterfly shaped, or dumb-bell shaped (short).

Transverse section of leaf blade, physiology. C_4. The anatomical organization conventional. XyMS–. *Midrib* conspicuous; having a conventional arc of bundles; with colourless mesophyll adaxially. *The lamina* symmetrical on either side of the midrib. Bulliforms not reliably recordable in the material seen, but probably in the form of simple fans. Many of the smallest vascular bundles unaccompanied by sclerenchyma. Combined sclerenchyma girders present (only with the primaries, the sclerenchyma scanty save at the blade margins and abaxially in the midrib); forming 'figures' (small I's or 'anchors' with the primaries). Sclerenchyma all associated with vascular bundles.

Cytology. $2n = 20$.

Taxonomy. Panicoideae; Andropogonodae; Maydeae.

Ecology, geography, regional floristic distribution. 1 species. Indo-Southeast Asia. Forest margins.

Paleotropical. Indomalesian. Malesian.

References, etc. Leaf anatomical: this project.

Sclerochloa P. Beauv.

Amblychloa Link, *Crassipes* Swallen

Habit, vegetative morphology. *Annual*; caespitose. *Culms* 4–30 cm high; herbaceous. Culm nodes glabrous. Culm internodes solid. Leaves non-auriculate. Sheath margins joined to free. Leaf blades linear; narrow; flat, or rolled; without cross venation; persistent; once-folded in bud. *Ligule an unfringed membrane*; not truncate; 0.5–1.5 mm long.

Reproductive organization. Plants bisexual, with bisexual spikelets; with hermaphrodite florets.

Inflorescence. Inflorescence one-sided, a single raceme, or paniculate (with short, stout branches); contracted (narrow, rigid); espatheate; not comprising 'partial inflorescences' and foliar organs. Spikelet-bearing axes persistent. *Spikelets secund*; pedicellate (the pedicels stout, to 1 mm long).

Female-fertile spikelets. *Spikelets* 6–15 mm long; compressed laterally; *falling with the glumes (and with the pedicel)*; tardily disarticulating between the florets, or not disarticulating between the florets. Rachilla prolonged beyond the uppermost female-fertile floret; rachilla hairless. The rachilla extension with incomplete florets. Hairy callus absent.

Glumes two; very unequal; shorter than the spikelets; shorter than the adjacent lemmas; dorsiventral to the rachis; hairless; glabrous; not pointed (obtuse to emarginate); awnless; non-carinate; similar (herbaceous with membranous margins, oblong-ovate, somewhat asymmetric). Lower glume 3–5 nerved. Upper glume 7–9 nerved. *Spikelets* with incomplete florets. The incomplete florets distal to the female-fertile florets.

Female-fertile florets *3–8.* Lemmas decidedly firmer than the glumes (leathery, the margins membranous); not becoming indurated; entire, or incised; blunt; not deeply cleft (rounded to emarginate); awnless; hairless; glabrous; non-carinate; 5–7 nerved. *Palea* present; relatively long; not indurated (hyaline); 2-nerved; 2-keeled. *Lodicules* present; 2; free; membranous; glabrous; toothed. Anthers 0.6–1.5 mm long; not penicillate. *Ovary* glabrous. Styles free to their bases. Stigmas 2.

Fruit, embryo and seedling. *Fruit* free from both lemma and palea; small (2.5–3.5 mm long); trigonous. Hilum short (round). Embryo small; not waisted. Endosperm hard; with lipid, or without lipid. Embryo with an epiblast; without a scutellar tail; with a negligible mesocotyl internode. Embryonic leaf margins meeting.

Abaxial leaf blade epidermis. *Costal/intercostal zonation* conspicuous. *Papillae* absent. *Long-cells* similar in shape costally and intercostally; of similar wall thickness costally and intercostally. Mid-intercostal long-cells rectangular to fusiform; having straight or only gently undulating walls. *Microhairs* absent. *Stomata* common; 21–27 microns long (*S. dura*), or 33–39 microns long (*S. rigida*). Subsidiaries parallel-sided. Guard-cells overlapped by the interstomatals. *Intercostal short-cells* common; not paired (solitary); not silicified. *Costal short-cells* neither distinctly grouped into long rows nor predominantly paired. Costal silica bodies horizontally-elongated crenate/sinuous, or horizontally-elongated smooth (predomonating), or rounded (some, oval); not sharp-pointed.

Transverse section of leaf blade, physiology. C_3; XyMS+. *Mesophyll* with non-radiate chlorenchyma; without adaxial palisade. *Leaf blade* 'nodular' in section; with the ribs more or less constant in size. *Midrib* conspicuous; with one bundle only. Bulliforms not in discrete, regular adaxial groups (apart from the 'midrib hinges'). All the vascular bundles accompanied by sclerenchyma. Combined sclerenchyma girders present; forming 'figures'. Sclerenchyma all associated with vascular bundles.

Cytology. Chromosome base number, $x = 7$. $2n = 14$. 2 ploid.

Taxonomy. Pooideae; Poodae; Poeae.

Ecology, geography, regional floristic distribution. 2 species. Southern Europe to western Asia. Commonly adventive. Mesophytic to xerophytic; species of open habitats; halophytic to glycophytic (in dry weedy places and saline soils).

Holarctic. Boreal and Tethyan. Euro-Siberian and Atlantic North American. Mediterranean and Irano-Turanian. European. Central Grasslands.

Rusts and smuts. Rusts — *Puccinia*. Taxonomically wide-ranging species: '*Uromyces*' *dactylidis*.

References, etc. Leaf anatomical: this project.

Sclerodactylon Stapf

Arthrochlaena Laena

Habit, vegetative morphology. Perennial; stoloniferous and caespitose. *Culms* 30–80 cm high (rigid); unbranched above. Culm nodes glabrous. Culm internodes solid. Plants conspicuously armed (the rigid leaves sharp-pointed). *Leaves* mostly basal; non-auriculate. *Leaf blades acicular (junciform, circular in section, longitudinally striate, with a central 'pith')*; *hard, woody, needle-like*; without abaxial multicellular glands. Ligule a fringe of hairs (short, dense). *Contra-ligule* absent.

Reproductive organization. Plants bisexual, with bisexual spikelets; with hermaphrodite florets. The spikelets all alike in sexuality (except for reduced members at the tips of the spikes).

Inflorescence. *Inflorescence* of spicate main branches; *digitate*. Primary inflorescence branches 2–4. Inflorescence with axes ending in spikelets (but these more or less abortive), or axes not ending in spikelets (ending in a point). Inflorescence espatheate; not comprising 'partial inflorescences' and foliar organs. *Spikelet-bearing axes* spikes (3–15 cm long); persistent. Spikelets solitary; secund; biseriate; sessile; imbricate.

Female-fertile spikelets. Spikelets 10–21 mm long (by 4–5 mm wide); strongly compressed laterally; disarticulating above the glumes; disarticulating between the florets; with conventional internode spacings. Rachilla prolonged beyond the uppermost female-fertile floret; rachilla hairless (glabrous). The rachilla extension with incomplete florets. Hairy callus absent.

Glumes two; very unequal (G_1 about half the length of G_2); shorter than the spikelets; shorter than the adjacent lemmas (G_2 reaching about the middle of L_1); lateral to the rachis (i.e. the spikelet edgewise to the rachis); hairless; glabrous; pointed (acute); awnless; carinate; similar (ovate-acute, leathery, the upper sometimes pungent tipped). Lower glume 1 nerved. Upper glume 1 nerved. *Spikelets* with incomplete florets. The incomplete florets distal to the female-fertile florets. *The distal incomplete florets* merely underdeveloped. *Spikelets without proximal incomplete florets*.

Female-fertile florets 6–20. Lemmas similar in texture to the glumes (leathery); smooth; not becoming indurated; entire to incised; not deeply cleft (no more than minutely incised); awnless (but mucronulate), or mucronate (minutely, from the median nerve); hairless; glabrous; strongly carinate; without a germination flap; 3 nerved. *Palea* present; relatively long; minutely apically notched; awnless, without apical setae (glabrous); not indurated (but almost cartilaginous); 2-nerved; 2-keeled. Palea keels winged; minutely ciliolate. *Lodicules* present; 2; free; membranous; glabrous; toothed; very heavily vascularized (at the base). *Stamens* 3; with free filaments (these long). *Ovary* glabrous. Styles free to their bases (the bases swollen). Stigmas 2.

Fruit, embryo and seedling. Fruit small; ellipsoid; not noticeably compressed. Hilum short. Pericarp free (very flimsy). Embryo large; waisted.

Abaxial leaf blade epidermis. *Costal/intercostal zonation* conspicuous. *Papillae* present; intercostal. Intercostal papillae over-arching the stomata; consisting of one symmetrical projection per cell (the intercostal grooves filled with short, blunt, thick-walled, unicellular papilla-hairs). Intercostal zones details of intercostal zones invisible. *Microhairs* present (*Chloris* type, hard to find among the papilla-hairs in the leaf grooves); more or less spherical; ostensibly one-celled; chloridoid-type; (33–)39–42(–45) microns long. Microhair total length/width at septum 1.6–2. *Stomata* common (in the grooves); (16.5–)18–21(–27) microns long. Subsidiaries triangular. *Costal short-cells* neither distinctly grouped into long rows nor predominantly paired. Costal silica bodies present throughout the costal zones; rounded and saddle shaped (intergrading); not sharp-pointed.

Transverse section of leaf blade, physiology. C_4 (vascular bundles in a single ring at the periphery of the 'pith', linked with the epidermis by tall columns of sclerenchyma; rows of PCR cells adjacent to the sclerenchyma blocks, not sheathing the bundles (sometimes the PCR rows link internally between the bundles; cf. *Triodia*); blocks of PCA tissue sandwiched and enclosed between the PCR cell rows). The anatomical organization unconventional. Organization of PCR tissue *Triodia* type. XyMS+. Mesophyll with arm cells (abundant, very clear). All the vascular bundles accompanied by sclerenchyma (all bundles with elongated, 'outer' girders). Sclerenchyma not all bundle-associated. The 'extra' sclerenchyma in adaxial groups (morphologically so, being 'internal' to the ring of bundles).

Taxonomy. Chloridoideae; main chloridoid assemblage.

Ecology, geography, regional floristic distribution. 1 species. Madagascar, Aldabra, Assumption. Xerophytic (extreme); species of open habitats; halophytic (coral rocks, salt pans, mangrove swamps).

Paleotropical. Madagascan.

References, etc. Leaf anatomical: this project.

Special comments. Rows of 'crypts' containing crystalline material (NaCl?) occur within the grooves if the leaf blade.

Scleropogon Phil.

Lesourdia Fourn.

Habit, vegetative morphology. Perennial; stoloniferous, or stoloniferous and caespitose. **Culms *10–20 cm high***; herbaceous. *Leaves* mostly basal; non-auriculate. Leaf blades linear; narrow; 1.5–2 mm wide; flat to folded (folded and arcuate above); without cross venation. Ligule a fringe of hairs (very short).

Reproductive organization. *Plants monoecious with all the fertile spikelets unisexual, or dioecious*; without hermaphrodite florets. The spikelets of sexually distinct forms on the same plant (monoecious, the male and female spikelets mixed or in separate panicles), or all alike in sexuality (dioecious); female-only, or male-only, or female-only and male-only. The male and female-fertile spikelets in different inflorescences, or mixed in the inflorescence. Plants outbreeding (at least when dioecious).

Inflorescence. Inflorescence few spikeleted; a single raceme (spike-like, with few spikelets), or paniculate (contracted, with few spikelets); contracted; *spatheate (the glumes of female spikelets subtended by glume-like bracts)*; not comprising 'partial inflorescences' and foliar organs. Spikelet-bearing axes persistent. Spikelets not secund; shortly pedicellate, or subsessile; imbricate.

Female-sterile spikelets. Male spikelets with equal, lanceolate glumes separated by a short internode; 6–15 flowered; lemmas attenuate into short awns; 2 fleshy lodicules. Rachilla of male spikelets prolonged beyond the uppermost male floret. The male spikelets with glumes; 6–15 floreted. The lemmas shortly awned. Male florets 5–14 (?).

Female-fertile spikelets. *Spikelets* 25–30 mm long; compressed laterally to not noticeably compressed (narrowly cylindrical); *disarticulating above the glumes; not disarticulating between the florets (the florets falling together as a cylindric, many-awned unit)*. Rachilla prolonged beyond the uppermost female-fertile floret. Hairy callus present (sharp-bearded, at the base of the lowermost floret and of the unit as shed).

Glumes two; very unequal; shorter than the spikelets; long relative to the adjacent lemmas (?); hairless; pointed; awnless; similar (hyaline, lanceolate-acuminate). Lower glume 1 nerved, or 3 nerved. Upper glume 3 nerved. *Spikelets* with incomplete florets (basal florets male, those above female or rarely hermaphrodite). The incomplete florets proximal to the female-fertile florets. *Spikelets with proximal incomplete florets. The proximal incomplete florets* 1–2; paleate. Palea of the proximal incomplete florets fully developed. The proximal incomplete florets male. The proximal lemmas shortly awned; 3 nerved; exceeded by the female-fertile lemmas.

Female-fertile florets 1–3. Lemmas narrow; decidedly firmer than the glumes (cartilaginous); not becoming indurated; incised; 3 lobed, or 4 lobed (with a shallow median sinus, and a somewhat deeper cleft on either side); not deeply cleft; awned. *Awns 3 (70–140 mm long extensions from the nerves, arising between small lobes)*; median and lateral; the median similar in form to the laterals; non-geniculate; hairless (scabrous); much longer than the body of the lemma. The lateral awns long. Lemmas hairless; non-carinate (rounded on the back); 3 nerved. *Palea* present (narrow, hairy at the base); tightly clasped by the lemma; entire (obtuse); awned (with 2 short awns); 2-nerved (the nerves near the margins); 2-keeled. *Lodicules* absent. *Stamens* 0. *Ovary* glabrous (?). Stigmas 2.

Fruit, embryo and seedling. Disseminule consisting of the abscised spikelet. Fruit medium sized (about 4.8 mm long); linear; not noticeably compressed. Hilum short. Pericarp fused. Embryo large; with an epiblast; with a scutellar tail; with an elongated mesocotyl internode. Embryonic leaf margins overlapping.

Abaxial leaf blade epidermis. *Costal/intercostal zonation* conspicuous. *Papillae* present; intercostal (on the long-cells and interstomatals). Intercostal papillae overarching the stomata (both at the ends and laterally); mostly consisting of one oblique swelling per cell. *Long-cells* markedly different in shape costally and intercostally (the costals narrower and more regularly rectangular); of similar wall thickness costally and intercostally. Mid-intercostal long-cells rectangular; having markedly sinuous walls. *Microhairs* present; elongated; clearly two-celled; chloridoid-type (the basal cells relatively long); 57–60–66 microns long; 15–18–19.5 microns wide at the septum. Microhair total length/width at septum 2.9–4.4. Microhair apical cells 18–21 microns long. Microhair apical cell/total length ratio 0.3–0.35. *Stomata* common; 21–24–27 microns long. Subsidiaries high dome-shaped (mostly), or triangular. Guard-cells overlapped by the interstomatals (very slightly), or overlapping to flush with the interstomatals. *Intercostal short-cells* common; in cork/silica-cell pairs and not paired (mostly paired, but often superposed and sometimes appearing solitary); silicified. *Costal short-cells* conspicuously in long rows. Costal silica bodies 'panicoid-type'; cross shaped, butterfly shaped, and dumb-bell shaped; not sharp-pointed.

Transverse section of leaf blade, physiology. C_4; XyMS+. PCR sheath outlines even. PCR sheath extensions absent. *Mesophyll* with radiate chlorenchyma. *Leaf blade* 'nodular' in section to adaxially flat (with slight, rounded abaxial ribs). *Midrib* conspicuous (via its prominent, rounded abaxial keel); with one bundle only. Bulliforms present in discrete, regular adaxial groups; in simple fans and associated with colourless mesophyll cells to form deeply-penetrating fans. All the vascular bundles accompanied by sclerenchyma. Combined sclerenchyma girders present (with all the bundles); forming 'figures' (all the bundles with 'anchors'). Sclerenchyma all associated with vascular bundles (apart from small marginal fibre groups).

Cytology. Chromosome base number, $x = 10$. $2n = 28$ and 40.

Taxonomy. Chloridoideae; main chloridoid assemblage.

Ecology, geography, regional floristic distribution. 1 species. Southwest U.S.A. and northern Mexico, Chile and Argentina. Xerophytic; species of open habitats. Dry grassy plains.

Holarctic and Neotropical. Boreal and Madrean. Atlantic North American. Pampas and Andean. Central Grasslands.

Rusts and smuts. Rusts — *Puccinia*.

Economic importance. Important native pasture species: *S. brevifolius*.

References, etc. Leaf anatomical: Metcalfe 1960; this project.

Sclerostachya A. Camus

~ *Saccharum* (subgenus *Sclerostachyum*), *Miscanthus*

Habit, vegetative morphology. Robust perennial; often forming large clumps. *Culms* 200–300 cm high; herbaceous. *Leaf blades* linear-lanceolate (rigid, often very long); broad, or narrow; flat; *pseudopetiolate*; without cross venation. Ligule an unfringed membrane.

Reproductive organization. Plants bisexual, with bisexual spikelets; with hermaphrodite florets. The spikelets homomorphic.

Inflorescence. Inflorescence paniculate; open (large, narrow, decompound, with whorled or fascicled branches); without capillary branchlets (the branches not slender); espatheate; not comprising 'partial inflorescences' and foliar organs. *Spikelet-bearing axes persistent (but the branches jointed)*. *Spikelets* paired; not secund; all pedicellate; *consistently in 'long-and-short' combinations*; unequally pedicellate in each combination. Pedicels of the 'pedicellate' spikelets free of the rachis. The 'shorter' spikelets hermaphrodite. The 'longer' spikelets hermaphrodite.

Female-fertile spikelets. Spikelets compressed dorsiventrally; falling with the glumes (falling from the pedicels). Rachilla terminated by a female-fertile floret. Hairy callus present (short, the silky hairs up to half the length of the spikelet).

Glumes two; more or less equal; long relative to the adjacent lemmas; hairless (firm, smooth); awnless; carinate (G_2), or non-carinate (G_1); very dissimilar (the G_2 flat-backed, with two ciliate submarginal keels, notched, the G_2 naviculate). *Lower glume two-keeled*; 3 nerved. Upper glume 3 nerved. *Spikelets* with incomplete florets. The incomplete florets proximal to the female-fertile florets. *The proximal incomplete florets* 1; sterile. The proximal lemmas awnless; similar in texture to the female-fertile lemmas (with hyaline, narrow, ciliate margins); not becoming indurated.

Female-fertile florets 1. Lemmas not incised; less firm than the glumes (hyaline); not becoming indurated; entire; awnless; hairy; non-carinate; without a germination flap. *Palea* present; conspicuous but relatively short; awnless, without apical setae; not indurated (hyaline); keel-less. *Lodicules* present; 2; free; glabrous. *Stamens* 3. *Ovary* glabrous. Styles free to their bases. Stigmas 2.

Fruit, embryo and seedling. Fruit brown; smooth.

Phytochemistry. Leaves without flavonoid sulphates (1 species).

Cytology. $2n = 30$, 48, and 96.

Taxonomy. Panicoideae; Andropogonodae; Andropogoneae; Andropogoninae.

Ecology, geography, regional floristic distribution. 2–3 species. India to Malaysia. Paleotropical. Indomalesian. Indian, Indo-Chinese, and Malesian.

Hybrids. Intergeneric hybrids with *Narenga*, *Saccharum*.

Rusts and smuts. Rusts — *Puccinia*. Smuts from Ustilaginaceae. Ustilaginaceae — *Ustilago*.

Special comments. Anatomical data wanting.

Scolochloa Link

Fluminia Fries

Habit, vegetative morphology. *Perennial*; rhizomatous (with succulent rhizomes). *Culms* 70–200 cm high. Culm internodes hollow. *Leaves* not basally aggregated; non-auriculate. *Leaf blades* linear; broad, or narrow; 4–12 mm wide; flat; *not pseudopetiolate*; without cross venation. *Ligule an unfringed membrane*; truncate; 3–10 mm long.

Reproductive organization. Plants bisexual, with bisexual spikelets; with hermaphrodite florets.

Inflorescence. *Inflorescence paniculate*; open (to 30 cm long); espatheate; not comprising 'partial inflorescences' and foliar organs. Spikelet-bearing axes persistent. Spikelets not secund; pedicellate.

Female-fertile spikelets. Spikelets 7–11 mm long; compressed laterally; disarticulating above the glumes; disarticulating between the florets. Rachilla prolonged beyond the uppermost female-fertile floret. The rachilla extension naked. Hairy callus present. *Callus* short; pointed.

Glumes two; very unequal; (the longer) *long relative to the adjacent lemmas*; hairless; pointed (acute to acuminate); awnless; non-carinate; similar. Lower glume 1–5 nerved. Upper glume 3–7 nerved. *Spikelets* with female-fertile florets only.

Female-fertile florets 3–4. *Lemmas* similar in texture to the glumes to decidedly firmer than the glumes (somewhat leathery); not becoming indurated; indistinctly incised; 3 lobed (or lacerate); not deeply cleft; *awnless, or mucronate*; hairless (save for the villous callus); non-carinate; without a germination flap; 5–9 nerved. *Palea* present; relatively long; 2-nerved; 2-keeled. *Lodicules* present; free; membranous; glabrous; toothed, or not toothed; not or scarcely vascularized. *Stamens* 3. Anthers 2–4 mm long. **Ovary** apically *hairy*. Styles free to their bases. Stigmas 2.

Fruit, embryo and seedling. Fruit small (about 2 mm long); compressed dorsiventrally. *Hilum long-linear*. Embryo small (but nearly 1/3 of the grain length). Endosperm hard; without lipid; containing compound starch grains. Embryo with an epiblast; without a scutellar tail; with a negligible mesocotyl internode. Embryonic leaf margins meeting.

Abaxial leaf blade epidermis. *Costal/intercostal zonation* conspicuous. *Papillae* absent. *Long-cells* similar in shape costally and intercostally; of similar wall thickness costally and intercostally. Mid-intercostal long-cells rectangular; having markedly sinuous walls. *Microhairs* absent. *Stomata* common; 33–39 microns long. Subsidiaries parallel-sided and dome-shaped. Guard-cells overlapped by the interstomatals, or overlapping to flush with the interstomatals. Intercostal short-cells in cork/silica-cell pairs; silicified. *Costal short-cells* predominantly paired. Costal silica bodies horizontally-elongated crenate/sinuous (short), or rounded and crescentic (or more or less rectangular); not sharp-pointed.

Transverse section of leaf blade, physiology. C_3; XyMS+. *Mesophyll* with non-radiate chlorenchyma. Bulliforms present in discrete, regular adaxial groups (at the bases of the furrows); in simple fans. Sclerenchyma all associated with vascular bundles.

Cytology. Chromosome base number, $x = 7$. $2n = 28$. 4 ploid.

Taxonomy. Pooideae; Poodae; Poeae.

Ecology, geography, regional floristic distribution. 2 species. North temperate. Helophytic, or mesophytic. Lakes, rivers, wet meadows.

Holarctic. Boreal, Tethyan, and Madrean. Arctic and Subarctic, Euro-Siberian, and Rocky Mountains. Mediterranean. European and Siberian.

Rusts and smuts. Rusts — *Puccinia*. Taxonomically wide-ranging species: *Puccinia coronata* and *Puccinia recondita*. Smuts from Ustilaginaceae. Ustilaginaceae — *Ustilago*.

Economic importance. Important native pasture species: *S. festucacea*.

References, etc. Leaf anatomical: this project; Metcalfe 1960.

Scribneria Hackel

Habit, vegetative morphology. *Annual*; caespitose. *Culms* 7–30 cm high; herbaceous; unbranched above. Culm internodes hollow. *Leaves* not basally aggregated; non-auriculate. Leaf blades linear; narrow; about 0.8–1.6 mm wide; almost filiform; rolled (involute); without cross venation; persistent. Ligule an unfringed membrane; not truncate (acuminate, becoming lacerate); 1.5–4 mm long. *Contra-ligule* absent.

Reproductive organization. Plants bisexual, with bisexual spikelets; with hermaphrodite florets.

Inflorescence. Inflorescence a single spike (slender), or a single raceme (with short pedicels), or paniculate (rarely, having two spikelets per node, on different pedicels or on one branched peduncle); contracted; spicate. Rachides hollowed. Inflorescence espatheate; not comprising 'partial inflorescences' and foliar organs. Spikelet-bearing axes persistent. *Spikelets* solitary (mostly), or paired (occasionally); not secund; distichous; *subsessile to pedicellate (the pedicels embedded with the spikelets the rachis hollows, becoming shorter towards the inflorescence tip)*; imbricate.

Female-fertile spikelets. *Spikelets* 4–7 mm long; *compressed laterally (flatwise against the axis)*; disarticulating above the glumes. Rachilla prolonged beyond the uppermost female-fertile floret; rachilla hairy. The rachilla extension naked. Hairy callus present. *Callus* short (oblique).

Glumes two; more or less equal; exceeding the spikelets; long relative to the adjacent lemmas (exceeding them); free; lateral to the rachis; hairless; glabrous (minutely scabrous on the nerves); pointed; awnless; non-carinate (but keeled on the outer nerves); similar (usually purplish-tinged, both strongly keeled on the outer nerve with the side adjoining the rachis nerveless, one — assumed to be the G_1 — markedly narrower). Lower glume two-keeled (G_2 also); 2 nerved. *Upper glume 4 nerved. Spikelets* with female-fertile florets only.

Female-fertile florets 1. Lemmas less firm than the glumes (membranous); not becoming indurated; incised; 2 lobed; not deeply cleft (shortly bifid); awned. Awns 1; median; from a sinus; non-geniculate; hairless; much shorter than the body of the lemma to much longer than the body of the lemma; entered by one vein. Lemmas hairless; glabrous; carinate; without a germination flap; 3–5 nerved. *Palea* present; relatively long; tightly clasped by the lemma; briefly apically notched; awnless, without apical setae; thinner than the lemma (hyaline, glabrous); not indurated; 2-nerved; keel-less (the nerves closely apposed). *Lodicules* present; 2; free; membranous; glabrous; not toothed; not or scarcely vascularized. **Stamens 1**. *Ovary* glabrous. Styles free to their bases. Stigmas 2; white.

Fruit, embryo and seedling. *Fruit* free from both lemma and palea; small (2.5 mm long); fusiform; not noticeably compressed. *Hilum short (punctiform)*. Embryo small (but fairly large for a pooid grass — about 1/4 of the grain length). Endosperm hard; with lipid; containing compound starch grains.

Abaxial leaf blade epidermis. *Costal/intercostal zonation* conspicuous. *Papillae* absent. Mid-intercostal long-cells rectangular; having straight or only gently undulating walls. *Microhairs* absent. *Stomata* common; (31–)34–38(–39) microns long. Subsidiaries parallel-sided. Guard-cells overlapped by the interstomatals (slightly). *Intercostal short-cells* absent or very rare. *Costal short-cells* predominantly paired. Costal silica bodies absent (in material seen), or present and well developed; when present, horizontally-elongated smooth, or rounded.

Transverse section of leaf blade, physiology. C_3; XyMS+. *Mesophyll* with non-radiate chlorenchyma; without adaxial palisade. *Leaf blade* with slight, rounded adaxial ribs; with the ribs more or less constant in size. *Midrib* conspicuous (by its association with a small abaxial keel, and by the bundle possessing an abaxial strand rather than a girder); with one bundle only. *The lamina* symmetrical on either side of the midrib. Bulliforms present in discrete, regular adaxial groups; in simple fans. All the vascular bundles accompanied by sclerenchyma. Combined sclerenchyma girders present; forming 'figures' (all the main bundles with I's). Sclerenchyma all associated with vascular bundles.

Cytology. Chromosome base number, $x = 13$.

Taxonomy. Pooideae; Poodae; Aveneae.

Ecology, geography, regional floristic distribution. 1 species. Western U.S.A. Holarctic. Madrean.

Rusts and smuts. Rusts — *Puccinia*.

References, etc. Leaf anatomical: this project.

Scrotochloa Judziewicz

~ Leptaspis

Habit, vegetative morphology. Perennial; decumbent. *Culms* 30–100 cm high; herbaceous; unbranched above. Culm nodes glabrous. Culm internodes solid to hollow. *Leaves* not basally aggregated; non-auriculate; without auricular setae. *Leaf blades* broad; 10–70 mm wide; *pseudopetiolate (becoming inverted by twisting of the 'petiole')*; pinnately veined (the laterals diverging obliquely from the midrib); cross veined; persistent; rolled in bud. **Ligule** present; *an unfringed membrane (minute)*. *Contra-ligule* absent.

Reproductive organization. Plants monoecious with all the fertile spikelets unisexual; without hermaphrodite florets. The spikelets of sexually distinct forms on the same plant; female-only and male-only. The male and female-fertile spikelets mixed in the inflorescence. The spikelets overtly heteromorphic.

Inflorescence. Inflorescence paniculate (1–noded); open; non-digitate. Primary inflorescence branches 4–8. Inflorescence espatheate (the spikelet branchlets without subtending bracts); not comprising 'partial inflorescences' and foliar organs. Spikelet-bearing axes disarticulating; falling entire (the inflorescence separating below the single node). Spikelets paired; pedicellate (the female pedicels clavate); not in distinct 'long-and-short' combinations (the female spikelets tending to be accompanied by pedicelled males, the pedicels of similar length).

Female-sterile spikelets. The male spikelets much smaller, stamens 6, anthers non-penicillate, lodicules absent, the floret caducous. The male spikelets with glumes; 1 floreted. Male florets 6 staminate.

Female-fertile spikelets. Spikelets 4–9 mm long; compressed laterally; falling with the glumes. Rachilla terminated by a female-fertile floret; rachilla hairless. Hairy callus absent.

Glumes two; relatively large; more or less equal; long relative to the adjacent lemmas; pointed (acute); awnless; very dissimilar (both ovate, caducous, purplish brown). Lower glume 5–7 nerved. Upper glume 5–7 nerved. *Spikelets with female-fertile florets only.*

Female-fertile florets 1. **Lemmas urceolate-scrotoform, with connate margins and a terminal pore through which the style emerges**; decidedly firmer than the glumes; becoming indurated; awnless; densely hairy (uncinate-pubescent); non-carinate; 9 nerved. *Palea* present (linear); relatively long; awnless, without apical setae; indurated; 2-nerved; 2-keeled. *Lodicules* absent. *Stamens* 0 (with 3 minute staminodes). Styles fused (into one). Stigmas 3.

Fruit, embryo and seedling. *Fruit* free from both lemma and palea.

Abaxial leaf blade epidermis. *Costal/intercostal zonation* conspicuous (the zones conspicuously sinuous, in the material seen). *Papillae* absent. *Long-cells* markedly different in shape costally and intercostally (the costals much narrower — reminiscent of epidermal fibres). Intercostal zones exhibiting many atypical long-cells (in the sense that they are very variable in width, and of divers shapes). Mid-intercostal long-cells often rectangular (but frequently with one or both ends oblique or irregular); having markedly sinuous walls. *Microhairs* absent. *Stomata* absent or very rare (in places), or common. Subsidiaries mostly low to high dome-shaped (frequently accompanied by small interstomatal or long-cells almost constituting additional subsidiaries), or parallel-sided. Guard-cells overlapping to flush with the interstomatals. *Intercostal short-cells* absent or very rare. *Costal short-cells* neither distinctly grouped into long rows nor predominantly paired. Costal silica bodies rounded to upright-ovoid, with a raised prickly belt around the equator (small, abundant, the form as yet unknown elsewhere); sharp-pointed (i.e. around the middle).

Transverse section of leaf blade, physiology. C_3; XyMS+. Mesophyll without arm cells; with fusoids. *Midrib* conspicuous; with colourless mesophyll adaxially. Bulliforms present in discrete, regular adaxial groups; in simple fans (these wide).

Special diagnostic feature. *Having female spikelets, with shell- or urn-shaped lemmas which are closed save for an apical pore.*

Taxonomy. Bambusoideae; Oryzodae; Phareae.

Ecology, geography, regional floristic distribution. 2 species. Ceylon and Southeast Asia to New Guinea and the Solomon Islands. Shade species. In forest.

Paleotropical and Australian. Indomalesian. Indian, Indo-Chinese, Malesian, and Papuan. North and East Australian. Tropical North and East Australian.

References, etc. Morphological/taxonomic: Judziewicz 1984. Leaf anatomical: this project.

Special comments. Fruit data wanting.

Scutachne A. Hitchc. & Chase

Habit, vegetative morphology. Perennial; caespitose. *Culms* 40–70 cm high; herbaceous. Culm nodes glabrous. **Leaves *mostly basal*;** non-auriculate. Leaf blades narrow; 4–6(–10) mm wide; without cross venation; persistent.

Reproductive organization. Plants bisexual, with bisexual spikelets; with hermaphrodite florets. The spikelets of sexually distinct forms on the same plant, or all alike in sexuality (the short-pedicelled members poorly-developed or missing); hermaphrodite and sterile.

Inflorescence. *Inflorescence* a single spike (contracted about the primary branches), or paniculate; *open*; espatheate; not comprising 'partial inflorescences' and foliar organs. Spikelet-bearing axes persistent. Spikelets not secund; pedicellate; consistently in 'long-and-short' combinations, or not in distinct 'long-and-short' combinations; sometimes unequally pedicellate in each combination. The 'shorter' spikelets sterile. The 'longer' spikelets hermaphrodite.

Female-fertile spikelets. *Spikelets* 5 mm long; *compressed dorsiventrally*; falling with the glumes; with a distinctly elongated rachilla internode between the glumes. Rachilla terminated by a female-fertile floret. Hairy callus absent.

Glumes two; relatively large; very unequal; (the upper) about equalling the spikelets (the lower half as long); (the upper) long relative to the adjacent lemmas; (the upper) hairy; pointed; awnless; non-carinate; very dissimilar (the lower small, hyaline and appressed to the rachilla, the upper large and firm). Lower glume 7–9 nerved. *Upper glume 7 nerved*. *Spikelets* with incomplete florets. The incomplete florets proximal to the female-fertile florets. *Spikelets with proximal incomplete florets*. *The proximal incomplete florets* 1; paleate. Palea of the proximal incomplete florets not becoming conspicuously hardened and enlarged laterally. The proximal incomplete florets male. *The proximal lemmas* awnless; 5 nerved; *more or less equalling the female-fertile lemmas*; less firm than the female-fertile lemmas (leathery, like the upper glume).

Female-fertile florets 1. *Lemmas acuminate to a puberulous point; similar in texture to the glumes to decidedly firmer than the glumes*; pointed; awnless; hairless; non-carinate; *having the margins lying flat and exposed on the palea*; with a clear germination flap; 5 nerved. *Palea* present; relatively long; 2-nerved; 2-keeled. *Lodicules* present; 2; free; fleshy; glabrous; not or scarcely vascularized. *Stamens* 3. Anthers not penicillate; without an apically prolonged connective. *Ovary* glabrous. Styles free to their bases. Stigmas 2; red pigmented.

Abaxial leaf blade epidermis. *Costal/intercostal zonation* conspicuous. *Papillae* absent. Mid-intercostal long-cells having markedly sinuous walls. *Microhairs* present; panicoid-type; 60–66 microns long; 6–8.4 microns wide at the septum. Microhair total length/width at septum 7.1–11. Microhair apical cells 28.5–36 microns long. Microhair apical cell/total length ratio 0.45–0.55. *Stomata* common; 27–33 microns long. Subsidiaries dome-shaped (mostly), or triangular. Guard-cells overlapping to flush with the interstomatals. *Intercostal short-cells* common; in cork/silica-cell pairs. *Costal short-cells* conspicuously in long rows. Costal silica bodies horizontally-elongated smooth and 'panicoid-type'; mostly cross shaped and nodular; not sharp-pointed.

Transverse section of leaf blade, physiology. C_4; XyMS+. PCR sheath outlines uneven. *Mesophyll* with radiate chlorenchyma. *Leaf blade* 'nodular' in section; with the ribs more or less constant in size. *Midrib* not readily distinguishable; with one bundle only. Bulliforms present in discrete, regular adaxial groups; in simple fans. All the vascular bundles accompanied by sclerenchyma. Combined sclerenchyma girders present (with the bigger bundles); forming 'figures'. Sclerenchyma all associated with vascular bundles.

Taxonomy. Panicoideae; Panicodae; Paniceae.

Ecology, geography, regional floristic distribution. 2 species. Cuba. Species of open habitats. Rocky slopes.

Neotropical. Caribbean.

References, etc. Leaf anatomical: this project.

Special comments. Fruit data wanting.

Secale L.

Habit, vegetative morphology. *Annual (rarely perennial)*; caespitose (or the culms solitary). *Culms* 20–150 cm high; herbaceous; unbranched above. Culm nodes glabrous. Culm internodes hollow. *Leaves* not basally aggregated; auriculate. Leaf blades linear; broad, or narrow; 2.5–20 mm wide; flat, or rolled (convolute); without cross venation; persistent; rolled in bud. Ligule an unfringed membrane; truncate.

Reproductive organization. Plants bisexual, with bisexual spikelets; with hermaphrodite florets; outbreeding and inbreeding. Exposed-cleistogamous, or chasmogamous.

Inflorescence. *Inflorescence a single spike (laterally compressed, distichous)*. Rachides hollowed. Inflorescence espatheate; not comprising 'partial inflorescences' and foliar organs. Spikelet-bearing axes disarticulating, or persistent (in cultivated forms); disarticulating at the joints. Spikelets solitary; not secund; distichous; sessile.

Female-fertile spikelets. *Spikelets* 10–18 mm long; compressed laterally; *falling with the glumes (and the joint), or not disarticulating (in cultivated forms)*. Rachilla prolonged beyond the uppermost female-fertile floret; rachilla hairless. Hairy callus absent. *Callus very short*.

Glumes two; very unequal to more or less equal; shorter than the adjacent lemmas; free; lateral to the rachis; subulate; acuminate to awned; carinate (sharply keeled to the base); similar (membranous). Lower glume 1 nerved. *Upper glume 1 nerved*. *Spikelets* with female-fertile florets only, or with incomplete florets. The incomplete florets distal to the female-fertile florets. The distal incomplete florets 1; *incomplete florets* merely underdeveloped.

Female-fertile florets 2–3. Lemmas lanceolate, tapered to the awn; less firm than the glumes to similar in texture to the glumes; entire; pointed; awned. Awns 1; median; apical; non-geniculate; hairless (scabrid); much longer than the body of the lemma; entered by several veins. Lemmas hairless; carinate (the keel with rigid, pectinate cilia); 5 nerved; with the nerves non-confluent. *Palea* present; relatively long; apically notched; not indurated (hyaline); 2-nerved; 2-keeled. *Lodicules* present; 2; free; membranous; ciliate. *Stamens* 3. Anthers 2.3–12 mm long; not penicillate. *Ovary* hairy. Styles free to their bases. Stigmas 2; white.

Fruit, embryo and seedling. *Fruit* free from both lemma and palea; medium sized to large; ellipsoid; longitudinally grooved; slightly compressed dorsiventrally to not noticeably compressed; with hairs confined to a terminal tuft. Hilum long-linear. Embryo large to small (to 1/3 the length of the caryopsis). Endosperm hard; without lipid; containing only simple starch grains. Embryo without an epiblast; without a scutellar tail; with a negligible mesocotyl internode. Embryonic leaf margins meeting.

Seedling with a short mesocotyl; with a tight coleoptile. First seedling leaf with a well-developed lamina. The lamina narrow; erect; 9–15 veined (?).

Abaxial leaf blade epidermis. *Costal/intercostal zonation* conspicuous. *Papillae* absent. *Long-cells* similar in shape costally and intercostally; of similar wall thickness costally and intercostally. Mid-intercostal long-cells rectangular; having markedly sinuous walls. *Microhairs* absent. *Stomata* common; 48–49–51 microns long. Subsidiaries low dome-shaped, or parallel-sided. Guard-cells overlapped by the interstomatals. *Intercostal short-cells* common; in cork/silica-cell pairs; silicified. Intercostal silica bodies rounded (elliptical), or crescentic. *Costal short-cells* predominantly paired. Costal silica bodies rounded and crescentic; not sharp-pointed.

Transverse section of leaf blade, physiology. C_3; XyMS+. *Mesophyll* with non-radiate chlorenchyma; without adaxial palisade. *Leaf blade* with distinct, prominent adaxial ribs, or 'nodular' in section; with the ribs very irregular in sizes. *Midrib* not readily distinguishable; with one bundle only. Bulliforms present in discrete, regular adaxial groups; in simple fans. All the vascular bundles accompanied by sclerenchyma. Combined sclerenchyma girders present; nowhere forming 'figures'. Sclerenchyma all associated with vascular bundles.

Culm anatomy. *Culm internode bundles* in one or two rings.

Phytochemistry. Leaves without flavonoid sulphates (1 species).

Cytology. Chromosome base number, $x = 7$. $2n = 14$. 2 ploid. Haplomic genome content R. Haploid nuclear DNA content 7.2–9.5 pg (5 species, mean 8.3). Mean diploid 2c DNA value 16.8 pg (6 species, 14.8–19.0).

Taxonomy. Pooideae; Triticodae; Triticeae.

Ecology, geography, regional floristic distribution. 5 species. Mediterranean, eastern Europe to central Asia, and South Africa. Commonly adventive. Mesophytic, or xerophytic; species of open habitats. Sandy soils and dry hillsides.

Holarctic. Boreal and Tethyan. Euro-Siberian. Mediterranean and Irano-Turanian. European.

Hybrids. Intergeneric hybrids with *Triticum* (×*Triticosecale* Wittmack), *Agropyron*, *Aegilops* (×*Aegilosecale* Ciferri & Giacom.), *Hordeum* (×*Hordale* Ciferri & Giacom.), *Elytrigia*. ×*Agrotrisecale* Ciferri & Giacom. = *Agropyron* × *Secale* × *Triticum*.

Rusts and smuts. Rusts — *Puccinia*. Taxonomically wide-ranging species: *Puccinia graminis, Puccinia striiformis, Puccinia recondita, 'Uromyces' turcomanicum*, and *'Uromyces' fragilipes*. Smuts from Tilletiaceae and from Ustilaginaceae. Tilletiaceae — *Tilletia* and *Urocystis*. Ustilaginaceae — *Ustilago*.

Economic importance. Grain crop species: *S. cereale* (Rye).

References, etc. Morphological/taxonomic: Löve 1984. Leaf anatomical: Metcalfe 1960; this project.

Sehima Forssk.

Hologamium Nees

Habit, vegetative morphology. Annual, or perennial; caespitose. *Culms* 20–100 cm high; herbaceous; branched above, or unbranched above. Culm nodes hairy, or glabrous. Culm internodes solid. *Leaves* not basally aggregated; non-auriculate. Leaf blades linear; narrow; setaceous (rarely), or not setaceous; without cross venation; persistent. *Ligule a fringed membrane, or a fringe of hairs*.

Reproductive organization. Plants bisexual, with bisexual spikelets; with hermaphrodite florets. The spikelets of sexually distinct forms on the same plant; hermaphrodite and male-only, or hermaphrodite and sterile; overtly heteromorphic; all in heterogamous combinations.

Inflorescence. *Inflorescence a single raceme (a single, curved, culm-like 'raceme' with embedded spikelets)*. Rachides hollowed and flattened. *Inflorescence espatheate*; not comprising 'partial inflorescences' and foliar organs. **Spikelet-bearing axes** spikelike (laterally compressed, curved); solitary; with substantial rachides; *disarticulating*; disarticulating at the joints. *'Articles'* stoutly linear to subclavate; not appendaged; densely long-hairy (with white hairs ventrally). Spikelets paired; secund; sessile and pedicellate; consistently in 'long-and-short' combinations; in pedicellate/sessile combinations. Pedicels of the 'pedicellate' spikelets free of the rachis. The 'shorter' spikelets hermaphrodite. The 'longer' spikelets male-only, or sterile.

Female-sterile spikelets. *Pedicelled spikelets male or neuter, strongly dorsally compressed, flat, often with the G_1 large and strongly nerved, the lemmas awnless*. The lemmas awnless.

Female-fertile spikelets. Spikelets slightly compressed dorsiventrally, or compressed laterally and compressed dorsiventrally (commonly more or less square in section); falling with the glumes. Rachilla terminated by a female-fertile floret. Hairy callus present.

Glumes two; more or less equal; long relative to the adjacent lemmas; without conspicuous tufts or rows of hairs; *awned (the upper with an apical bristle-like awn, the lower bidentate or 2-mucronate)*; very dissimilar (the lower 2-keeled and 2-winged, the upper naviculate-subulate and awned). *Lower glume* two-keeled (scarcely winged); concave on the back, or sulcate on the back; not pitted; relatively smooth; *3–6 nerved*. Upper glume 3–6 nerved. *Spikelets* with incomplete florets. The incomplete florets proximal to the female-fertile florets. *The proximal incomplete florets* 1; paleate. Palea of the proximal incomplete florets fully developed. The proximal incomplete florets male. The proximal lemmas awnless; similar in texture to the female-fertile lemmas (hyaline); not becoming indurated.

Female-fertile florets 1. Lemmas less firm than the glumes (hyaline); not becoming indurated; incised; 2 lobed; awned. Awns 1; median; from a sinus; geniculate; hairless to hairy (puberulous to ciliate along its coils); much longer than the body of the lemma. Lemmas hairy (above), or hairless (glabrous); non-carinate; 2–3 nerved. *Palea* present; relatively long; not indurated (hyaline); 2-nerved. *Lodicules* present; 2; free; fleshy; glabrous. *Stamens* 3. Anthers not penicillate. *Ovary* glabrous. Styles free to their bases. Stigmas 2.

Fruit, embryo and seedling. *Fruit* free from both lemma and palea; concave on one side; compressed dorsiventrally. Hilum short. Embryo large.

Abaxial leaf blade epidermis. *Costal/intercostal zonation* conspicuous. *Papillae* absent. Mid-intercostal long-cells rectangular; having markedly sinuous walls (conspicuously pitted). *Microhairs* present; panicoid-type (unusually slender); (63–)72–75(–78) microns long; 3.6–3.9–4.5 microns wide at the septum. Microhair total length/width at septum 14–20.8. Microhair apical cells (24–)30–33(–35) microns long. Microhair apical cell/total length ratio 0.38–0.44. *Stomata* common; 24–27–30 microns long. Subsidiaries predominantly triangular. Guard-cells overlapping to flush with the interstomatals. *Intercostal short-cells* absent or very rare. *S. nervosa* with abundant prickles. *Crown cells* absent. *Costal short-cells* conspicuously in long rows. Costal silica bodies 'panicoid-type'; short to long dumb-bell shaped (in *S. nervosa*); not sharp-pointed.

Transverse section of leaf blade, physiology. C_4; XyMS–. PCR sheath extensions absent. PCR cell chloroplasts centrifugal/peripheral. *Mesophyll* with radiate chlorenchyma. *Leaf blade* 'nodular' in section; with the ribs very irregular in sizes (major bundles with large adaxial and abaxial ribs, the rest unribbed or with small ribs, in *S. nervosa*). *Midrib* not readily distinguishable; with one bundle only. Bulliforms present in discrete, regular adaxial groups (these large and wide); in simple fans. All the vascular bundles accompanied by sclerenchyma. Combined sclerenchyma girders present; forming 'figures' (all the primaries with heavy T's). Sclerenchyma all associated with vascular bundles.

Phytochemistry. Leaves without flavonoid sulphates (1 species).

Cytology. Chromosome base number, $x = 10$, 17, and 20. $2n = 34$ and 40.

Taxonomy. Panicoideae; Andropogonodae; Andropogoneae; Andropogoninae.

Ecology, geography, regional floristic distribution. 5 species. Warm Africa, India, Australia. Helophytic to mesophytic; species of open habitats; glycophytic. Savanna, sometimes on heavy clay.

Holarctic, Paleotropical, and Australian. Tethyan. African and Indomalesian. Macaronesian. Saharo-Sindian, Sudano-Angolan, and Namib-Karoo. Indian, Indo-Chinese, Malesian, and Papuan. North and East Australian. Sahelo-Sudanian, Somalo-Ethiopian, South Tropical African, and Kalaharian. Tropical North and East Australian.

Rusts and smuts. Smuts from Tilletiaceae and from Ustilaginaceae. Tilletiaceae — *Entyloma*. Ustilaginaceae — *Sphacelotheca*.

References, etc. Leaf anatomical: this project.

Semiarundinaria Makino

Brachystachyum Keng

Habit, vegetative morphology. Shrubby perennial (or small tree); rhizomatous. The flowering culms leafy. *Culms* 300–1000 cm high; woody and persistent; to 4 cm in diameter; cylindrical; branched above. Primary branches/mid-culm node 3–8 (?). Culm sheaths deciduous in their entirety (leathery). Culm internodes hollow. Pluricaespitose. Rhizomes leptomorph (metamorph type I). Plants unarmed. *Leaves* not basally aggregated; auriculate; without auricular setae. Leaf blades broad; pseudopetiolate; cross vened; disarticulating from the sheaths; rolled in bud.

Reproductive organization. Plants bisexual, with bisexual spikelets; with hermaphrodite florets.

Inflorescence. **Inflorescence** indeterminate; *with pseudospikelets (apparently)*; compound paniculate (in spatheolate racemes od 1–3 spikelets in fascicles at the nodes of the branches); spatheate; *a complex of 'partial inflorescences' and intervening foliar organs (and each spikelet or raceme with an outer, basal bract with a reduced lamina)*. *Spikelet-bearing axes* very much reduced, or 'racemes' (of 1–3 spikelets); persistent. Spikelets not secund.

Female-fertile spikelets. *Spikelets 5–10 mm long*; compressed laterally; disarticulating above the glumes (?); disarticulating between the florets (?). Rachilla prolonged beyond the uppermost female-fertile floret. The rachilla extension with incomplete florets. Hairy callus absent.

Glumes two; relatively large; shorter than the adjacent lemmas; pointed; awnless; similar. *Spikelets* with incomplete florets (presumably). The incomplete florets distal to the female-fertile florets (assuming terminal florets may be imperfect).

Female-fertile florets 3–4. Lemmas ovate-lanceolate, acuminate; entire; pointed; awnless, or mucronate, or awned. Awns 1; median; apical; non-geniculate; much shorter than the body of the lemma. Lemmas non-carinate; 7–11 nerved. *Palea* present; relatively long (about equalling the lemma); apically notched; several nerved; 2-keeled. *Lodicules* present; 3; free; membranous; ciliate; not toothed; heavily vascularized. **Stamens 3**. Anthers not penicillate; without an apically prolonged connective. *Ovary* glabrous; with a conspicuous apical appendage. The appendage broadly conical, fleshy. Styles fused (into one long style). Stigmas 3 (feathery).

Abaxial leaf blade epidermis. *Costal/intercostal zonation* conspicuous. *Papillae* present (especially in the stomatal bands). Intercostal papillae over-arching the stomata. Mid-intercostal long-cells having markedly sinuous walls (thin). *Microhairs* present; panicoid-type (but variable in shape). *Stomata* common. Subsidiaries low to high dome-shaped. *Intercostal short-cells* common; not paired (but paired with hooks). *Costal short-cells* predominantly paired (and solitary). Costal silica bodies saddle shaped; not sharp-pointed.

Transverse section of leaf blade, physiology. C_3; XyMS+. Mesophyll with arm cells; with fusoids. *Leaf blade* with distinct, prominent adaxial ribs (slight ribs and furrows), or adaxially flat; with the ribs more or less constant in size. *Midrib* conspicuous; having complex vascularization. Bulliforms present in discrete, regular adaxial groups; in simple fans, or in simple fans and associated with colourless mesophyll cells to form deeply-penetrating fans. All the vascular bundles accompanied by sclerenchyma. Combined sclerenchyma girders present (with all the bundles); forming 'figures' (with most bundles).

Cytology. Chromosome base number, $x = 12$. $2n = 48$. 4 ploid.

Taxonomy. Bambusoideae; Bambusodae; Bambuseae.

Ecology, geography, regional floristic distribution. About 5 species. China, Japan, Vietnam.

Holarctic and Paleotropical. Boreal. Indomalesian. Eastern Asian. Indo-Chinese.

Hybrids. May hybridize with *Sasa* (×*Hibanobambusa* Maruyama and Okamura).

Rusts and smuts. Rusts — *Stereostratum* and *Puccinia*. Taxonomically wide-ranging species: *Stereostratum corticoides* and *Puccinia kusanoi*.

References, etc. Leaf anatomical: Metcalfe 1960.

Special comments. Fruit data wanting.

Sesleria Scop.

Diptychum Dulac

Excluding *Oreochloa*, *Psilathera*, *Sesleriella*

Habit, vegetative morphology. Perennial; usually caespitose. **Culms** 10–70 cm high; herbaceous. Leaves non-auriculate. **Sheath margins joined**. Leaf blades linear; narrow; 0.4–6 mm wide; flat, or folded, or rolled (convolute); without cross venation; persistent. Ligule an unfringed membrane; truncate, or not truncate; 0.1–1 mm long.

Reproductive organization. Plants bisexual, with bisexual spikelets; with hermaphrodite florets. *The spikelets of sexually distinct forms on the same plant; hermaphrodite and sterile (the sterile spikelets reduced to a pair (usually) of bractlike scales (representing their glumes?) at the base of the inflorescence)*; overtly heteromorphic.

Inflorescence. *Inflorescence paniculate*; contracted (often bluish, greyish or whitish); more or less ovoid, or spicate, or more or less irregular; espatheate; not comprising 'partial inflorescences' and foliar organs. Spikelet-bearing axes persistent. Spikelets not secund.

Female-sterile spikelets. The sterile spikelets vestigial, represented by bracts at the base of the inflorescence.

Female-fertile spikelets. Spikelets 3.5–9 mm long; compressed laterally; disarticulating above the glumes; disarticulating between the florets. Rachilla prolonged beyond the uppermost female-fertile floret; rachilla hairless. Hairy callus absent.

Glumes two; more or less equal; shorter than the adjacent lemmas, or long relative to the adjacent lemmas (rarely); pointed (acuminate); shortly awned, or awnless; *carinate*; similar (membranous). Lower glume 1 nerved. Upper glume 1 nerved. *Spikelets* with female-fertile florets only, or with incomplete florets. The incomplete florets distal to the female-fertile florets.

Female-fertile florets 2–6. *Lemmas similar in texture to the glumes (membranous)*; not becoming indurated; incised; 2–5 lobed; not deeply cleft (with 2–5 teeth); mucronate to awned (the teeth usually aristulate). Awns when present, 1, or 3, or 5; median, or median and lateral; the median similar in form to the laterals (when laterals present); apical; non-geniculate; much shorter than the body of the lemma to about as long as the body of the lemma. The lateral awns when present, shorter than the median. Lemmas hairy, or hairless; non-carinate; 3–5 nerved. *Palea* present; relatively long; awnless, without apical setae, or with apical setae, or awned; 2-nerved; 2-keeled. *Lodicules* present; 2; free; membranous; glabrous; toothed; not or scarcely vascularized. *Stamens* 3. Anthers 2.5–4 mm long; not penicillate. *Ovary* hairy. Styles fused. Stigmas 2; white.

Fruit, embryo and seedling. *Fruit* free from both lemma and palea; small; compressed dorsiventrally; hairy on the body. Hilum short. Embryo small; not waisted. Endosperm hard; without lipid. Embryo with an epiblast; without a scutellar tail; with a negligible mesocotyl internode. Embryonic leaf margins meeting.

Seedling with a loose coleoptile. First seedling leaf with a well-developed lamina. The lamina narrow; 3 veined.

Abaxial leaf blade epidermis. *Costal/intercostal zonation* conspicuous. *Papillae* absent. *Long-cells* similar in shape costally and intercostally; differing markedly in wall thickness costally and intercostally. Mid-intercostal long-cells rectangular; having markedly sinuous walls. *Microhairs* absent. *Stomata* absent or very rare. *Intercostal short-cells* common, or absent or very rare; not paired. *Costal short-cells* predominantly paired. Costal silica bodies horizontally-elongated crenate/sinuous, or horizontally-elongated smooth, or rounded; not sharp-pointed.

Transverse section of leaf blade, physiology. C_3; XyMS+. *Mesophyll* with non-radiate chlorenchyma; seemingly without adaxial palisade (but see Metcalfe 1960). *Midrib* conspicuous, or not readily distinguishable; with one bundle only. Bulliforms present in discrete, regular adaxial groups, or not in discrete, regular adaxial groups (sometimes *Ammophila*-type, or confined to 'midrib hinges'); sometimes in simple fans. All the vascular bundles accompanied by sclerenchyma. Combined sclerenchyma girders present; forming 'figures'. Sclerenchyma all associated with vascular bundles, or not all bundle-associated (rarely).

Phytochemistry. Leaves without flavonoid sulphates (1 species).

Cytology. Chromosome base number, $x = 7$. $2n = 14$ (rarely), or 28, or 42, or 56. 2 ploid (rarely), or 4 ploid, or 6 ploid, or 8 ploid (often $2n = 56$).

Taxonomy. Pooideae; Poodae; Seslerieae.

Ecology, geography, regional floristic distribution. 25 species. Europe, western Asia. Mesophytic, or xerophytic; species of open habitats. Uplands and mountain rocks.

Holarctic. Boreal and Tethyan. Euro-Siberian. Mediterranean and Irano-Turanian. European.

Rusts and smuts. Rusts — *Puccinia*. Taxonomically wide-ranging species: *Puccinia graminis* and *Puccinia coronata*. Smuts from Tilletiaceae and from Ustilaginaceae. Tilletiaceae — *Entyloma*, *Tilletia*, and *Urocystis*. Ustilaginaceae — *Ustilago*.

References, etc. Morphological/taxonomic: Deyl 1946. Leaf anatomical: Metcalfe 1960; this project.

Sesleriella Deyl

~ *Sesleria*

Habit, vegetative morphology. Perennial; caespitose. **Culms 5–15 cm high**; herbaceous. Leaves non-auriculate. **Sheath margins free**. Leaf blades narrow; 1–2 mm wide; sagittate (e.g. *S. appendiculata*), or not cordate, not sagittate; not setaceous (convolute or flat); flat, or rolled (convolute); without cross venation. **Ligule an unfringed membrane**; not truncate; 0.5 mm long.

Reproductive organization. Plants bisexual, with bisexual spikelets; with hermaphrodite florets. *The spikelets of sexually distinct forms on the same plant*; **hermaphrodite and sterile (the sterile spikelets reduced to bractlike scales at the base of the inflorescence)**; overtly heteromorphic.

Inflorescence. Inflorescence paniculate; contracted; capitate, or more or less ovoid; espatheate; not comprising 'partial inflorescences' and foliar organs. Spikelet-bearing axes persistent. Spikelets not secund.

Female-sterile spikelets. The sterile spikelets vestigial, represented by bracts at the base of the inflorescence.

Female-fertile spikelets. Spikelets compressed laterally; disarticulating above the glumes. Rachilla prolonged beyond the uppermost female-fertile floret. The rachilla extension with incomplete florets.

Glumes two; more or less equal; shorter than the adjacent lemmas, or long relative to the adjacent lemmas; pointed; shortly awned, or awnless (then mucronate); carinate; similar (ovate, membranous). Lower glume 1 nerved. Upper glume 1 nerved. *Spikelets with incomplete florets*. The incomplete florets distal to the female-fertile florets.

Female-fertile florets 3–4. *Lemmas* decidedly firmer than the glumes (silvery or grey); not becoming indurated; *incised*; not deeply cleft (toothed); awnless, or mucronate (the teeth shortly aristulate); hairy (proximally); *carinate*; without a germination flap; 5 nerved. *Palea* present; relatively long; 2-nerved; 2-keeled. *Lodicules* present; 2; free; membranous; glabrous; toothed; not or scarcely vascularized. *Stamens* 3. *Ovary* glabrous. Styles fused. Stigmas 2.

Fruit, embryo and seedling. Fruit small (about 1.5 mm long). Embryo small.

Abaxial leaf blade epidermis. *Costal/intercostal zonation* conspicuous. *Papillae* absent. *Long-cells* costal long-cells smaller; of similar wall thickness costally and intercostally. Mid-intercostal long-cells rectangular; having markedly sinuous walls (and pitted). *Microhairs* absent. *Stomata* absent or very rare. *Intercostal short-cells* common; in cork/silica-cell pairs (and solitary); silicified (a few only), or not silicified. *Costal short-cells* neither distinctly grouped into long rows nor predominantly paired (many paired). Costal silica bodies horizontally-elongated crenate/sinuous (often short), or rounded (often irregular, integrading with the sinuous/crenate forms); not sharp-pointed.

Transverse section of leaf blade, physiology. C_3; XyMS+. *Mesophyll* with non-radiate chlorenchyma. *Leaf blade* with distinct, prominent adaxial ribs, or 'nodular' in section; with the ribs more or less constant in size. *Midrib* conspicuous (by the larger bundle); with one bundle only. Bulliforms not in discrete, regular adaxial groups (bulliforms not apparent). All the vascular bundles accompanied by sclerenchyma. Combined sclerenchyma girders present; forming 'figures' (I's). Sclerenchyma all associated with vascular bundles.

Cytology. Chromosome base number, $x = 7$.

Taxonomy. Pooideae; Poodae; Seslerieae.

Ecology, geography, regional floristic distribution. 2 species. Central and southern Europe.

Holarctic. Boreal and Tethyan. Euro-Siberian. Mediterranean. European.

References, etc. Leaf anatomical: this project.

Setaria P. Beauv.

Acrochaete Peter, *Chaetochloa* Scribn., *Miliastrum* Fabric., *Tansaniochloa* Rauschert
Excluding *Camusiella, Cymbosetaria*

Habit, vegetative morphology. Annual, or perennial; rhizomatous, or stoloniferous, or caespitose, or decumbent. *Culms* 10–320 cm high; herbaceous; branched above, or unbranched above. Culm internodes solid, or hollow. Young shoots extravaginal, or intravaginal, or extravaginal and intravaginal. *Leaves* not basally aggregated; non-auriculate; without auricular setae. *Leaf blades* broad, or narrow; usually *not cordate, not sagittate (occasionally sagittate or hastate, then perhaps referable to* **Cymbosetaria***)*; flat, or folded; without abaxial multicellular glands; pseudopetiolate (occasionally), or not pseudopetiolate; pinnately veined to palmately veined (rarely), or parallel veined; without cross venation; persistent; rolled in bud, or once-folded in bud, or folded like a fan in bud (in Sect. Ptychophyllum). *Ligule* present; a fringed membrane (narrow), or a fringe of hairs. *Contra-ligule* absent.

Reproductive organization. *Plants* bisexual, with bisexual spikelets; *with hermaphrodite florets*. The spikelets of sexually distinct forms on the same plant, or all alike in sexuality; hermaphrodite, or hermaphrodite and male-only, or hermaphrodite and sterile (when clustered, often not all fully developed). Plants outbreeding and inbreeding. Exposed-cleistogamous, or chasmogamous. Apomictic, or reproducing sexually. Viviparous (sometimes), or not viviparous.

Inflorescence. Inflorescence of spicate main branches (not uncommonly so — e.g., in Sect. Ptychophyllum — though this is ignored in many published keys), or a false spike, with spikelets on contracted axes, or paniculate; open, or contracted; when contracted more or less ovoid, or spicate, or more or less irregular. Inflorescence axes not ending in spikelets (produced into 'bristles' beyond the spikelets). Inflorescence espatheate; not comprising 'partial inflorescences' and foliar organs. *Spikelet-bearing axes persistent.* **Spikelets** *with 'involucres' of 'bristles', or subtended by solitary 'bristles' (e.g., in Sect. Ptychophyllum)*. The 'bristles' relatively slender, not spiny; persisting on the axis. Spikelets secund (when the spikelets are on clear primary branches, as in Sect. Ptychophyllum), or not secund; subsessile, or pedicellate. Pedicel apices discoid. Spikelets not in distinct 'long-and-short' combinations.

Female-fertile spikelets. *Spikelets* 2–4 mm long; elliptic, or ovate, or obovate; *compressed dorsiventrally*; falling with the glumes, or not disarticulating (in cultivated forms). Rachilla terminated by a female-fertile floret. Hairy callus absent.

Glumes two; relatively large; very unequal; shorter than the adjacent lemmas, or long relative to the adjacent lemmas; awnless; membranous. *Lower glume 3–6 nerved*. Upper glume 3–9 nerved. *Spikelets* with incomplete florets. The incomplete florets proximal to the female-fertile florets. *Spikelets with proximal incomplete florets*. The proximal incomplete florets 1; paleate, or epaleate. *Palea of the proximal incomplete florets* fully developed to reduced; *not becoming conspicuously hardened and enlarged laterally*. The proximal incomplete florets male, or sterile. The proximal lemmas awnless; 5 nerved, or 7 nerved (rarely); more or less equalling the female-fertile lemmas to decidedly exceeding the female-fertile lemmas; less firm than the female-fertile lemmas (membranous); not becoming indurated.

Female-fertile florets 1. Lemmas decidedly firmer than the glumes; rugose; becoming indurated (crustaceous); yellow in fruit; entire; pointed; awnless (usually apiculate); hairless; usually non-carinate (but cymbiform in species perhaps referable to *Cymbosetaria*); having the margins tucked in onto the palea; with a clear germination flap; 1–5 nerved. *Palea* present; relatively long; entire; awnless, without apical setae; textured like the lemma; indurated; 2-nerved. *Lodicules* present; 2; free; fleshy; glabrous. *Stamens* 3. Anthers not penicillate. *Ovary* glabrous. Styles free to their bases. Stigmas 2; white, or red pigmented.

Fruit, embryo and seedling. Fruit small; ellipsoid to subglobose; compressed dorsiventrally; sculptured, or smooth. Hilum short. Embryo large; waisted, or not waisted. Endosperm hard; without lipid; containing only simple starch grains. Embryo without an epiblast; with a scutellar tail; with an elongated mesocotyl internode. Embryonic leaf margins overlapping.

Seedling with a long mesocotyl; with a loose coleoptile. First seedling leaf with a well-developed lamina. The lamina broad; curved; veins 'many'.

Abaxial leaf blade epidermis. *Costal/intercostal zonation* conspicuous. *Papillae* absent. *Long-cells* similar in shape costally and intercostally (costals narrower); of similar wall thickness costally and intercostally. Mid-intercostal long-cells rectangular; having markedly sinuous walls. *Microhairs* present; elongated; clearly two-celled; panicoid-type. Microhair apical cell wall thinner than that of the basal cell and often collapsed. Microhairs (34–)38–96(–102) microns long. Microhair basal cells 21 microns long. Microhairs 7.5–8.4–9 microns wide at the septum. Microhair total length/width at septum 9.3–11.6. Microhair apical cells (16–)18–59(–62) microns long. Microhair apical cell/total length ratio (0.52–)0.57–0.64(–0.69). *Stomata* common; 39–42 microns long. Subsidiaries low to high dome-shaped, or triangular, or dome-shaped and triangular. Guard-cells overlapping to flush with the interstomatals. *Intercostal short-cells* common, or absent or very rare; not paired (solitary); when present, silicified, or not silicified. Intercostal silica bodies present and perfectly developed (few); when present, cross-shaped, or vertically elongated-nodular. *Costal short-cells* conspicuously in long rows, or neither distinctly grouped into long rows nor predominantly paired (rarely). Costal silica bodies present in alternate cell files of the costal zones; 'panicoid-type'; cross shaped, or butterfly shaped, or dumb-bell shaped, or nodular; not sharp-pointed.

Transverse section of leaf blade, physiology. C_4; biochemical type NADP–ME (5 species); XyMS–. PCR sheath outlines uneven. PCR sheaths of the primary vascular bundles complete. PCR sheath extensions absent. PCR cells with a suberised lamella. PCR cell chloroplasts with reduced grana; centrifugal/peripheral. *Mesophyll* with radiate chlorenchyma. *Leaf blade* 'nodular' in section, or adaxially flat; with the ribs more or less constant in size. *Midrib* conspicuous; with one bundle only, or having a conventional arc of bundles; with colourless mesophyll adaxially, or without colourless mesophyll adaxially. Bulliforms present in discrete, regular adaxial groups, or not in discrete, regular adaxial groups (somtimes the epidermis largely, irregularly bulliform); commonly in simple fans. Many of the smallest vascular bundles unaccompanied by sclerenchyma. Combined sclerenchyma girders present, or absent; nowhere forming 'figures'. Sclerenchyma all associated with vascular bundles. The lamina margins with fibres (few cells).

Culm anatomy. *Culm internode bundles* scattered.

Phytochemistry. Leaves containing flavonoid sulphates (*S. chevalieri*), or without flavonoid sulphates (10 species).

Cytology. Chromosome base number, $x = 9$ and 10. $2n = 18, 36, 54, 63$, and 72, or 36–54. 2, 4, 6, and 8 ploid. Nucleoli persistent.

Taxonomy. Panicoideae; Panicodae; Paniceae.

Ecology, geography, regional floristic distribution. About 110 species. Tropical and warm temperate. Commonly adventive. Generally mesophytic; shade species (e.g. *S. palmifolia*), or species of open habitats. In woodland, grassland, weedy places.

Holarctic, Paleotropical, Neotropical, Australian, and Antarctic. Boreal, Tethyan, and Madrean. African, Madagascan, Indomalesian, Polynesian, and Neocaledonian. Euro-Siberian, Eastern Asian, and Atlantic North American. Macaronesian, Mediterranean, and Irano-Turanian. Saharo-Sindian, Sudano-Angolan, West African Rainforest, and Namib-Karoo. Indian, Indo-Chinese, Malesian, and Papuan. Fijian. Caribbean, Venezuela and Surinam, Amazon, Central Brazilian, Pampas, and Andean. North and East Australian and Central Australian. Patagonian. European and Siberian. Southern Atlantic North American and Central Grasslands. Sahelo-Sudanian, Somalo-Ethiopian, South Tropical African, and Kalaharian. Tropical North and East Australian.

Rusts and smuts. Rusts — *Puccinia*. Taxonomically wide-ranging species: *Puccinia chaetochloae, Puccinia dolosa, Puccinia graminis, Puccinia substriata, 'Uromyces' setariae-italicae*, and *Puccinia esclavensis*. Smuts from Tilletiaceae and from Ustilaginaceae. Tilletiaceae — *Tilletia*. Ustilaginaceae — *Sorosporium, Sphacelotheca, Tolyposporium*, and *Ustilago*.

Economic importance. Significant weed species: *S. adhaerens, S. barbata, S. faberi, S. glauca, S. gracilis, S. italica, S. pallide-fusca, S. palmifolia, S. paniculifera, S. poiretiana, S. sphacelata, S. verticillata, S. verticilliformis, S. viridis*. Cultivated fodder: *S. sphacelata* (Nandi). Important native pasture species: e.g. *S. incrassata, S. kagarensis, S. longiseta, S. macrostachya, S. sphacelata*. Grain crop species: *S. italica* (Foxtail Millet) — as a major (China) or minor cereal, and as birdseed. *S. palmifolia* sometimes cultivated as a vegetable.

References, etc. Morphological/taxonomic: Clayton 1979. Leaf anatomical: Metcalfe 1960; this project.

Setariopsis Scribner ex Millsp.

Habit, vegetative morphology. Annual, or perennial; rhizomatous, or decumbent. *Culms* 15–50 cm high; herbaceous; branched above. Culm nodes glabrous. Culm internodes solid. *Leaves* not basally aggregated. Leaf blades broad to narrow; flat; without cross venation; persistent. **Ligule a fringed membrane**. *Contra-ligule* absent.

Reproductive organization. Plants bisexual, with bisexual spikelets; with hermaphrodite florets. The spikelets of sexually distinct forms on the same plant; hermaphrodite and sterile (the clusters with undeveloped spikelets at the base).

Inflorescence. Inflorescence a false spike, with spikelets on contracted axes, or of spicate main branches (the basal 'clusters' being quite elongated in *S. auriculata*); contracted. Inflorescence axes not ending in spikelets (produced into long bristles beyond the spikelets). Inflorescence espatheate; not comprising 'partial inflorescences' and foliar organs. Spikelet-bearing axes persistent. **Spikelets *subtended by solitary 'bristles' (these longer than the spikelets)***. The 'bristles' persisting on the axis. Spikelets not secund; not two-ranked; pedicellate. Pedicel apices discoid.

Female-fertile spikelets. Spikelets 2–3 mm long (gibbous); compressed dorsiventrally; falling with the glumes; with conventional internode spacings. The upper floret not stipitate. Rachilla terminated by a female-fertile floret. Hairy callus absent. *Callus* absent.

Glumes two; very unequal; (the upper) about equalling the spikelets; (the upper) long relative to the adjacent lemmas; hairless; glabrous; pointed (G_2), or not pointed (G_1 blunt); awnless; non-carinate; *very dissimilar (the G_1 small, broad and cordate, the G_2 large, very broad, saccate, auriculate and becoming indurated)*. **Lower glume 7–9 nerved**. *Upper glume* distinctly saccate (and auriculate); 11–19 nerved. *Spikelets* with incomplete florets. The incomplete florets proximal to the female-fertile florets. *The proximal incomplete florets* 1; paleate; sterile (seemingly). The proximal lemmas broad, lyrate with two swellings at the base; awnless; 7–11 nerved; decidedly exceeding the female-fertile lemmas; less firm than the female-fertile lemmas (cartilaginous, harder along the margins); becoming indurated.

Female-fertile florets 1. Lemmas not saccate; decidedly firmer than the glumes; rugose; becoming indurated; entire; pointed (and apiculate); awnless (though apiculate); hairless; glabrous; non-carinate; having the margins tucked in onto the palea; with a clear germination flap; 5 nerved. *Palea* present; relatively long; entire; awnless, without apical setae; textured like the lemma; indurated; 2-nerved; 2-keeled. Palea keels wingless. *Lodicules* present; 2; free; fleshy; glabrous; not or scarcely vascularized. Anthers about 0.5 mm long; not penicillate; without an apically prolonged connective. *Ovary* glabrous. Styles free to their bases. Stigmas 2.

Fruit, embryo and seedling. *Fruit* free from both lemma and palea; small; compressed dorsiventrally. Hilum short.

Abaxial leaf blade epidermis. *Costal/intercostal zonation* conspicuous. *Papillae* absent. *Long-cells* markedly different in shape costally and intercostally (the costals much narrower); of similar wall thickness costally and intercostally (thin walled). Mid-intercostal long-cells rectangular; having markedly sinuous walls (irregularly so). *Microhairs* present; panicoid-type; (45–)48–51(–54) microns long; 5.1–5.4–6 microns wide at the septum. Microhair total length/width at septum 8.3–10. Microhair apical cells (30–)33–34.5(–39) microns long. Microhair apical cell/total length ratio 0.67–0.73. *Stomata* common; 21–24 microns long. Subsidiaries medium to high dome-shaped, or

870

triangular (occasionally). Guard-cells overlapping to flush with the interstomatals. *Intercostal short-cells* common (in places); not paired (solitary); not silicified. *Costal short-cells* conspicuously in long rows. Costal silica bodies 'panicoid-type'; dumb-bell shaped and nodular; not sharp-pointed.

Transverse section of leaf blade, physiology. C_4; XyMS–. PCR sheath outlines even. *Leaf blade* adaxially flat. *Midrib* conspicuous; having a conventional arc of bundles (the large median bundle flanked on either side by two smaller ones); with colourless mesophyll adaxially. Bulliforms not in discrete, regular adaxial groups (the epidermis largely bulliform, and complicated by cushion hair bases). Many of the smallest vascular bundles unaccompanied by sclerenchyma. Combined sclerenchyma girders present (with the primaries only); forming 'figures'. Sclerenchyma all associated with vascular bundles.

Cytology. Chromosome base number, $x = 9$. $2n = 19$. 2 ploid.

Taxonomy. Panicoideae; Panicodae; Paniceae.

Ecology, geography, regional floristic distribution. 2 species. Mexico. Species of open habitats. Weedy places.

Holarctic and Neotropical. Madrean. Caribbean.

References, etc. Leaf anatomical: this project.

Special comments. Description based mainly on *S. auriculata*.

Shibataea Makino

Habit, vegetative morphology. Small, shrubby perennial; rhizomatous. The flowering culms leafy. *Culms* 20–100 cm high; woody and persistent; to 0.5 cm in diameter; flattened on one side; not scandent (zigzag); branched above. Culm nodes 2 ridged. Primary branches/mid-culm node 2–6. Culm sheaths deciduous in their entirety (papery). Pluricaespitose. Rhizomes leptomorph (rhizomes metamorph type I). Plants unarmed. *Leaves* not basally aggregated; non-auriculate. Leaf blades broad; 12–25 mm wide; pseudopetiolate; cross veined; disarticulating from the sheaths; rolled in bud. Ligule an unfringed membrane; truncate. *Contra-ligule* present.

Reproductive organization. Plants bisexual, with bisexual spikelets; with hermaphrodite florets.

Inflorescence. Inflorescence indeterminate (?); *with pseudospikelets (apparently — but the description seen unclear)*; with few-spikeleted racemes in lateral, spatheate clusters; spatheate; *a complex of 'partial inflorescences' and intervening foliar organs (but spikelets without basal outer bracts)*. Spikelet-bearing axes very much reduced (of few spikelets); persistent. Spikelets not secund.

Female-fertile spikelets. Spikelets 15–18 mm long; compressed laterally; disarticulating above the glumes (?). Rachilla prolonged beyond the uppermost female-fertile floret.

Glumes two, or several (?); very unequal; shorter than the adjacent lemmas; pointed; awnless; similar (ovate-lanceolate). *Spikelets* with female-fertile florets only, or with incomplete florets (?).

Female-fertile florets 2. Lemmas ovate-lanceolate; entire; pointed; awnless; hairless; non-carinate; 9–10 nerved. *Palea* present; relatively long (about equalling the lemma); entire (pointed); several nerved; 2-keeled. Palea back glabrous. *Lodicules* present; 3; free; membranous; ciliate; not toothed; heavily vascularized. **Stamens** *3*. Anthers not penicillate; without an apically prolonged connective. *Ovary* glabrous; with a conspicuous apical appendage. The appendage broadly conical, fleshy. Styles fused (into one, trifid above). Stigmas 3 (feathery).

Fruit, embryo and seedling. Fruit medium sized (7 mm long); not noticeably compressed (cylindric).

Abaxial leaf blade epidermis. *Costal/intercostal zonation* conspicuous. *Papillae* present. Intercostal papillae over-arching the stomata, or not over-arching the stomata; several per cell (one or more than one row per cell). Long-cells differing markedly in wall thickness costally and intercostally (costals thicker). Mid-intercostal long-cells having markedly sinuous walls (thin). *Microhairs* present; panicoid-type (but variable in shape). *Stomata* common. Subsidiaries low to high dome-shaped. *Intercostal short-cells* common; in cork/silica-cell pairs, or not paired. *Costal short-cells* conspicuously in long rows (but frequent long short-cells). Costal silica bodies saddle shaped, or 'panicoid-type' (or cuboid); often cross shaped, or butterfly shaped, or dumb-bell shaped; not sharp-pointed.

Transverse section of leaf blade, physiology. C_3; XyMS+. Mesophyll with arm cells; with fusoids (but small and inconspicuous). *Leaf blade* with distinct, prominent adaxial ribs (low), or adaxially flat; with the ribs more or less constant in size. *Midrib* conspicuous; having complex vascularization. Bulliforms present in discrete, regular adaxial groups; in simple fans. All the vascular bundles accompanied by sclerenchyma. Combined sclerenchyma girders present (with all the bundles); forming 'figures' (the large bundles).

Cytology. Chromosome base number, $x = 12$. $2n = 48$. 4 ploid.

Taxonomy. Bambusoideae; Bambusodae; Bambuseae.

Ecology, geography, regional floristic distribution. 3 species. Eastern Asia. Holarctic. Boreal. Eastern Asian.

References, etc. Leaf anatomical: Metcalfe 1960.

Special comments. Description very poor. Fruit data wanting.

Sieglingia Bernh.

~ *Danthonia* sensu lato

Habit, vegetative morphology. Perennial; caespitose. **Culms** 10–60 cm high; ***herbaceous***; tuberous, or not tuberous. Culm nodes glabrous. Culm internodes hollow. *Leaves* mostly basal; non-auriculate. Leaf blades narrow; flat, or rolled; without cross venation; persistent; rolled in bud, or once-folded in bud. ***Ligule a fringe of hairs***. *Contra-ligule* present (of hairs).

Reproductive organization. *Plants bisexual, with bisexual spikelets*; with hermaphrodite florets; inbreeding. Exposed-cleistogamous, or chasmogamous. Plants with hidden cleistogenes, or without hidden cleistogenes. The hidden cleistogenes (when present) in the leaf sheaths (sometimes basal and very modified).

Inflorescence. *Inflorescence few spikeleted*; a single raceme, or paniculate; open, or contracted; espatheate; not comprising 'partial inflorescences' and foliar organs. Spikelet-bearing axes persistent. Spikelets secund, or not secund; pedicellate.

Female-fertile spikelets. *Spikelets* 6–12(–14) mm long; compressed laterally; disarticulating above the glumes; disarticulating between the florets; ***with conventional internode spacings***. Rachilla prolonged beyond the uppermost female-fertile floret; rachilla hairless. The rachilla extension with incomplete florets. Hairy callus present.

Glumes two; more or less equal; about equalling the spikelets; long relative to the adjacent lemmas; pointed, or not pointed; awnless; similar (rounded below, keeled above, lanceolate to ovate). Lower glume 3–5 nerved. Upper glume 3–5 nerved. *Spikelets* with incomplete florets. The incomplete florets distal to the female-fertile florets.

Female-fertile florets 4–6. *Lemmas* similar in texture to the glumes to decidedly firmer than the glumes; not becoming indurated; incised; not deeply cleft; ***mucronate (via the central tooth)***; hairy (short-haired on the lower margins), or hairless. ***The hairs not in tufts***; *not in transverse rows*. Lemmas non-carinate; 7–9 nerved. *Palea* present; relatively long; entire; awnless, without apical setae; 2-nerved; 2-keeled. Palea keels wingless (but each enlarging and progressively thickening towards the base — by contrast with *Danthonia* s. str.?). *Lodicules* present; fleshy; glabrous. *Stamens* 3. Anthers 0.2–0.4 mm long; not penicillate. *Ovary* glabrous. Styles free to their bases. Stigmas 2.

Fruit, embryo and seedling. *Fruit* free from both lemma and palea; small; longitudinally grooved. ***Hilum long-linear***. *Embryo large (compared with* **Danthonia** *sensu stricto)*; waisted. Endosperm hard; without lipid. Embryo without an epiblast; with a scutellar tail; with an elongated mesocotyl internode. Embryonic leaf margins overlapping.

Seedling with a long mesocotyl; with a loose coleoptile. First seedling leaf with a well-developed lamina. The lamina narrow; 5 veined.

Abaxial leaf blade epidermis. *Costal/intercostal zonation* conspicuous. *Papillae* absent. *Long-cells* markedly different in shape costally and intercostally (the costals narrower); of similar wall thickness costally and intercostally (rather thick walled, and heavily pitted anticlinally and on the surface). Mid-intercostal long-cells rectangular;

having markedly sinuous walls. *Microhairs* present; panicoid-type (large); $(78-)81-99(-102)$ microns long; $(12-)12.6-15(-18)$ microns wide at the septum. Microhair total length/width at septum $4.3-7.9$. Microhair apical cells $(36-)40.5-46.5(-52.5)$ microns long. Microhair apical cell/total length ratio $0.44-0.52$. *Stomata* absent or very rare (scarce in material seen), or common (see Metcalfe); $31-34$ microns long. Subsidiaries dome-shaped and triangular (of the truncated type, in the material seen). Guard-cells overlapped by the interstomatals (very slightly), or overlapping to flush with the interstomatals. *Intercostal short-cells* common; in cork/silica-cell pairs (and solitary); silicified. Intercostal silica bodies mostly tall-and-narrow, or crescentic (a few cubical or saddle-shaped). *Costal short-cells* conspicuously in long rows. Costal silica bodies 'panicoid-type'; short dumb-bell shaped, or cross shaped; not sharp-pointed.

Transverse section of leaf blade, physiology. C_3; XyMS+. *Mesophyll* with non-radiate chlorenchyma; without adaxial palisade; without 'circular cells' (but often with a region of translucent, spongy tissue in the mid-interveinal regions of the mesophyll). *Leaf blade* with distinct, prominent adaxial ribs to adaxially flat (the adaxial ribs very slight); with the ribs more or less constant in size. *Midrib* conspicuous (by virtue of the large associated hinge groups); with one bundle only. Bulliforms present in discrete, regular adaxial groups (these conspicuous, in all the slight furrows); in simple fans. All the vascular bundles accompanied by sclerenchyma. Combined sclerenchyma girders present; forming 'figures' (the midrib with an anchor, the rest of the bundles with I's). Sclerenchyma all associated with vascular bundles.

Cytology. Chromosome base number, $x = 6$, or 9 (?). $2n = 24$, 36, and 124. 3, 4, and 14 ploid.

Taxonomy. Arundinoideae; Danthonieae.

Ecology, geography, regional floristic distribution. 1 species. Madeira, Algeria, Europe, Asia Minor. Commonly adventive.

Holarctic. Boreal and Tethyan. Euro-Siberian. Mediterranean and Irano-Turanian.

Hybrids. Intergeneric hybrids with *Danthonia* (×*Danthosieglingia* Domin).

Rusts and smuts. Rusts — *Puccinia*. Taxonomically wide-ranging species: *Puccinia brachypodii*.

References, etc. Leaf anatomical: Metcalfe 1960; this project.

Silentvalleya Nair, Sreekumar, Vajravelu & Bhargavan

Habit, vegetative morphology. Perennial; caespitose. **Culms** $40-100$ cm high; *herbaceous*; unbranched above. Culm nodes glabrous. Leaves non-auriculate. Leaf blades linear (with filiform tips); narrow; $0.4-0.6$ mm wide; without abaxial multicellular glands; without cross venation. Ligule an unfringed membrane (according to the original description); truncate.

Reproductive organization. Plants bisexual, with bisexual spikelets; with hermaphrodite florets.

Inflorescence. *Inflorescence of spicate main branches*; non-digitate (the slender 'racemes' scattered along a main axis). Primary inflorescence branches $6-10$. Inflorescence espatheate; not comprising 'partial inflorescences' and foliar organs. The racemes spikelet bearing to the base. Spikelet-bearing axes persistent. Spikelets solitary (at each node); secund (obscurely so?); biseriate; subsessile; imbricate.

Female-fertile spikelets. Spikelets about 8 mm long; compressed laterally; disarticulating above the glumes; disarticulating between the florets. Rachilla prolonged beyond the uppermost female-fertile floret; rachilla hairless (except for the callus at the base of each floret). The rachilla extension with incomplete florets. Hairy callus present. *Callus* short.

Glumes two; very unequal (the G_1 considerably the shorter); shorter than the spikelets; (the upper) long relative to the adjacent lemmas; hairless; glabrous; without conspicuous tufts or rows of hairs; pointed (G_1 acute, G_2 acuminate); awnless. **Lower glume** 0.5 times the length of the upper glume; shorter than the lowest lemma; *0 nerved*. Upper glume 3 nerved (the laterals faint). *Spikelets* with incomplete florets. The incomplete florets distal to the female-fertile florets. *The distal incomplete florets* merely underdeveloped.

Female-fertile florets *6–10 (sometimes purplish)*. *Lemmas* lanceolate, acuminate; not becoming indurated; entire; pointed; *awned*. Awns 1; median; apical (from the acuminate lemma tip); non-geniculate; hairless (scabrid); about as long as the body of the lemma to much longer than the body of the lemma. Lemmas hairless; glabrous; without a germination flap; 3 nerved (the laterals faint). *Palea* present; relatively long; entire; awnless, without apical setae; 2-nerved; 2-keeled. Palea keels winged; hairy (ciliate). *Lodicules* present; 2; free; fleshy (seemingly); glabrous. *Stamens* 3. Anthers 1.25 mm long (the filaments relatively short); not penicillate; without an apically prolonged connective. *Ovary* glabrous. Styles free to their bases. Stigmas 2.

Fruit, embryo and seedling. Fruit small (about 2 mm long, stipitate); not noticeably compressed. Hilum short (?). Pericarp fused (seemingly). Embryo large (?).

Abaxial leaf blade epidermis. *Costal/intercostal zonation* conspicuous. *Papillae* absent. Mid-intercostal long-cells rectangular. *Microhairs* present (but very scarce); elongated; clearly two-celled; chloridoid-type. Microhair apical cell wall of similar thickness/ rigidity to that of the basal cell. Microhair basal cells 18 microns long. Microhair total length/width at septum 2.5. Microhair apical cell/total length ratio 0.4. *Stomata* fairly common. Intercostal short-cells in cork/silica-cell pairs; silicified. Intercostal silica bodies present and perfectly developed; tall-and-narrow. *Costal short-cells* conspicuously in long rows. Costal silica bodies present in alternate cell files of the costal zones; rounded.

Transverse section of leaf blade, physiology. C_4; XyMS+. PCR sheath outlines uneven. PCR sheaths of the primary vascular bundles interrupted; interrupted both abaxially and adaxially. PCR sheath extensions present. Maximum number of extension cells 2. Mesophyll traversed by columns of colourless mesophyll cells. *Leaf blade* with distinct, prominent adaxial ribs. *Midrib* not readily distinguishable; with one bundle only. Bulliforms present in discrete, regular adaxial groups. All the vascular bundles accompanied by sclerenchyma. Combined sclerenchyma girders present. Sclerenchyma all associated with vascular bundles.

Taxonomy. Chloridoideae; main chloridoid assemblage (cf. *Diplachne*).

Ecology, geography, regional floristic distribution. 1 species. India. Paleotropical. Indomalesian. Indian.

References, etc. Morphological/taxonomic: Nair, Sreekumar, Vajravelu and Bhargavan 1982b. Leaf anatomical: Van den Borre 1994.

Simplicia Kirk

Habit, vegetative morphology. *Perennial*; caespitose. Culms herbaceous. Culm nodes glabrous. *Leaves* not basally aggregated; non-auriculate. Leaf blades narrow; without cross venation; persistent. Ligule an unfringed membrane; not truncate; 2–3 mm long.

Reproductive organization. Plants bisexual, with bisexual spikelets; with hermaphrodite florets.

Inflorescence. Inflorescence paniculate; open, or contracted; espatheate; not comprising 'partial inflorescences' and foliar organs. Spikelet-bearing axes persistent. Spikelets not secund; pedicellate.

Female-fertile spikelets. *Spikelets 2.5 mm long*; compressed laterally; disarticulating above the glumes. *Rachilla prolonged beyond the uppermost female-fertile floret*; rachilla hairy. The rachilla extension naked.

Glumes two; *minute*; very unequal to more or less equal; shorter than the spikelets; shorter than the adjacent lemmas (the upper less than 1/4 the length of the floret); pointed; awnless; non-carinate; similar. Lower glume 1 nerved. Upper glume 1–3 nerved. *Spikelets* with female-fertile florets only.

Female-fertile florets 1–2. Lemmas decidedly firmer than the glumes (membranous); not becoming indurated; pointed (acute); awnless, or awned (with or without an awnlet). Awns if present, 1; median; dorsal to apical; from near the top (subapical); non-geniculate. Lemmas hairless (but tuberculate-scabrid); carinate; 3 nerved. *Palea* present; relatively long; 1-nerved, or 2-nerved; one-keeled. *Lodicules* present; 2; free; membranous; glabrous; toothed. *Stamens* 2–3. Anthers not penicillate. *Ovary* glabrous. Styles free to their bases. Stigmas 2.

Fruit, embryo and seedling. Fruit small; compressed laterally. Hilum short. Embryo not waisted. Endosperm containing compound starch grains. Embryo with an epiblast.

Abaxial leaf blade epidermis. *Costal/intercostal zonation* conspicuous. *Papillae* absent. Long-cells of similar wall thickness costally and intercostally. Mid-intercostal long-cells fusiform; having straight or only gently undulating walls. *Microhairs* absent. *Stomata* absent or very rare. Subsidiaries parallel-sided. *Intercostal short-cells* absent or very rare. Costal silica bodies predominantly horizontally-elongated smooth, or horizontally-elongated crenate/sinuous; not sharp-pointed.

Transverse section of leaf blade, physiology. C_3; XyMS+. *Mesophyll* with non-radiate chlorenchyma. *Leaf blade* adaxially flat. *Midrib* conspicuous; with one bundle only. All the vascular bundles accompanied by sclerenchyma. Combined sclerenchyma girders present; forming 'figures'. Sclerenchyma all associated with vascular bundles.

Cytology. Chromosome base number, $x = 7$. $2n = 28$.

Taxonomy. Pooideae; Poodae; Poeae.

Ecology, geography, regional floristic distribution. 2 species. New Zealand. Species of open habitats. Grassland.

Antarctic. New Zealand.

References, etc. Leaf anatomical: this project.

Sinarundinaria Nakai

Ampelocalamus Chen, Wen and Sheng, *Burmabambusa* Keng f., *Chimonocalamus* Hsueh and Yi, *Drepanostachyum* Keng f.; interpreted by Clayton and Renvoize 1986 to include *Yushania* and *Otatea* as well

Included in *Fargesia* by Soderstrom and Ellis 1987

Excluding *Otatea, Yushania*

Habit, vegetative morphology. Perennial; rhizomatous. The flowering culms leafy. *Culms woody and persistent*; branched above. Culm nodes 1 ridged. *Primary branches/ mid-culm node 3–7*. Culm sheaths persistent. Culm internodes solid, or hollow. Rhizomes pachymorph. Plants conspicuously armed (*Chimonocalamus*), or unarmed. *Leaves* not basally aggregated. *Leaf blades linear-lanceolate*; broad to narrow; to 12 mm wide (in *S. nitida*); *pseudopetiolate*; cross veined; disarticulating from the sheaths. Ligule a fringed membrane; very short.

Reproductive organization. Plants bisexual, with bisexual spikelets; with hermaphrodite florets.

Inflorescence. *Inflorescence* determinate; without pseudospikelets; a single raceme, or paniculate; spatheate; *not comprising 'partial inflorescences' and foliar organs*. Spikelets pedicellate.

Female-fertile spikelets. Spikelets compressed laterally; disarticulating above the glumes; disarticulating between the florets. Rachilla prolonged beyond the uppermost female-fertile floret. The rachilla extension often with incomplete florets.

Glumes two; very unequal; shorter than the spikelets; shorter than the adjacent lemmas; similar. Upper glume 7 nerved. *Spikelets* often with incomplete florets. The incomplete florets distal to the female-fertile florets. *The distal incomplete florets* merely underdeveloped. *Spikelets without proximal incomplete florets*.

Female-fertile florets 1–5. Lemmas not becoming indurated; awnless to awned. Awns 1; median; apical; non-geniculate. *Lemmas* hairless; *7 nerved*. Palea present; relatively long; awnless, without apical setae; several nerved (between and outside the keels); 2-keeled. *Lodicules* present; 3; free; membranous; ciliate. Stamens 3. *Ovary without a conspicuous apical appendage. Stigmas 2*.

Abaxial leaf blade epidermis. *Costal/intercostal zonation* conspicuous. *Papillae* present. Intercostal papillae over-arching the stomata; several per cell (small, usually on row per long-cell). Mid-intercostal long-cells having markedly sinuous walls. *Microhairs* present; clearly two-celled; panicoid-type; 55–88 microns long. Microhair apical cells 18–29 microns long. *Stomata* common. Subsidiaries low to high dome-shaped, or triangular. *Intercostal short-cells* common (paired with hook bases), or absent or very rare. *Costal short-cells* conspicuously in long rows (e.g. *S. murieli*), or neither distinctly grouped into long rows nor predominantly paired (mostly in rows of 3–5 in *S. nitida*). Costal silica bodies saddle shaped, or 'panicoid-type'; when panicoid type, cross shaped, or cross shaped to dumb-bell shaped; not sharp-pointed.

Transverse section of leaf blade, physiology. XyMS+. Mesophyll with arm cells; with fusoids. The fusoids external to the PBS. *Leaf blade* with distinct, prominent adaxial ribs; with the ribs more or less constant in size (broad, low). Midrib having complex vascularization. Bulliforms present in discrete, regular adaxial groups; in simple fans and associated with colourless mesophyll cells to form deeply-penetrating fans. All the vascular bundles accompanied by sclerenchyma. Combined sclerenchyma girders present.

Taxonomy. Bambusoideae; Bambusodae; Bambuseae.

Ecology, geography, regional floristic distribution. About 50 species. Asia and Madagascar. Shade species and species of open habitats. Woodland and open places, low and high altitudes.

Holarctic and Paleotropical. Boreal and Tethyan. Indomalesian and Polynesian. Euro-Siberian and Eastern Asian. Indian, Indo-Chinese, and Malesian.

References, etc. Leaf anatomical: Metcalfe 1960, for '*Arundinaria*' *murieli* and '*A.*' *nitida*.

Special comments. Clayton and Renvoize (1986) and Soderstrom and Ellis (1987) present very different generic interpretations of the species associated with *Sinarundinaria, Otatea, Yushania, Fargesia* and *Thamnocalamus* (q.v.), without providing generic descriptions adequate for the present purpose, and the data compiled here are inevitably very unsatisfactory. Fruit data wanting.

Sinobambusa Makino

Neobambus Keng f.

Habit, vegetative morphology. Perennial (shrub or small tree); rhizomatous. The flowering culms leafy. *Culms* 200–800 cm high; woody and persistent (with long internodes); to 3.5 cm in diameter; with grooved internodes; branched above. Primary branches/mid-culm node 3 (but these branching again). Culm sheaths deciduous in their entirety (leathery). Culm internodes hollow. Pluricaespitose. Rhizomes leptomorph (metamorph type I). *Leaves* not basally aggregated; with auricular setae, or without auricular setae. Leaf blades broad; 10–35 mm wide; pseudopetiolate; cross veined, or without cross venation; disarticulating from the sheaths; rolled in bud.

Reproductive organization. Plants bisexual, with bisexual spikelets; with hermaphrodite florets.

Inflorescence. Inflorescence indeterminate; **with pseudospikelets (seemingly: the spikelets in tufts, each with a bract)**; lateral, the spikelets in clusters at the nodes; spatheate; a complex of 'partial inflorescences' and intervening foliar organs. *Spikelet-bearing axes* paniculate (clustered); persistent. Spikelets not secund; more or less sessile.

Female-fertile spikelets. Spikelets 30–100 mm long; compressed laterally to not noticeably compressed; disarticulating above the glumes; disarticulating between the florets (?). Rachilla prolonged beyond the uppermost female-fertile floret. Hairy callus absent.

Glumes two, or several (?); very unequal; shorter than the adjacent lemmas; awnless; similar.

Female-fertile florets 4–25. Lemmas ovate; entire; pointed; awnless; non-carinate; 11–15 nerved. *Palea* present; relatively long (almost as long as the lemma); entire (pointed); awnless, without apical setae; several nerved (5-nerved); 2-keeled. *Lodicules* present; 3; free; membranous; glabrous; not toothed; heavily vascularized. **Stamens 3**. **Ovary** glabrous; **without a conspicuous apical appendage**. Styles fused (into one, short). Stigmas 3.

Fruit, embryo and seedling. Fruit longitudinally grooved.

Abaxial leaf blade epidermis. *Costal/intercostal zonation* conspicuous. *Papillae* present; costal and intercostal (very abundant). Intercostal papillae over-arching the stomata; several per cell (smallish, in one or two rows per cell, sometimes clustered or in branched-pairs). *Long-cells* similar in shape costally and intercostally; of similar wall thickness costally and intercostally. Mid-intercostal long-cells rectangular; having markedly sinuous walls (the sinuosity fine to coarse). *Microhairs* present; elongated; clearly two-celled; panicoid-type. *Stomata* common. Subsidiaries non-papillate; mostly low, parallel-sided, or dome-shaped, or triangular; including both triangular and parallel-sided forms on the same leaf. *Intercostal short-cells* common; in cork/silica-cell pairs and not paired (some solitary). Numerous prickles with very reduced points costally, and a few macrohairs both costally and intercostally. *Costal short-cells* conspicuously in long rows (in places), or predominantly paired (in places), or neither distinctly grouped into long rows nor predominantly paired (in places, irregularly grouped). Costal silica bodies present and well developed; saddle shaped (large).

Transverse section of leaf blade, physiology. C₃. Mesophyll probably with adaxial palisade (but obscure in the poor material seen); with arm cells; with fusoids. The fusoids external to the PBS. *Leaf blade* with distinct, prominent adaxial ribs to 'nodular' in section (slightly ribbed over the primaries). *Midrib* conspicuous (by virtue of the large midrib bundle and abaxial keel, and the adaxial and abaxial sclerenchyma masses); with one bundle only; without colourless mesophyll adaxially. *The lamina* symmetrical on either side of the midrib. Bulliforms present in discrete, regular adaxial groups; in simple fans (these large and broad, in the shallow, wide furrows). All the vascular bundles accompanied by sclerenchyma. Combined sclerenchyma girders present (with the larger bundles); forming 'figures' (anchors, with the primaries). Sclerenchyma all associated with vascular bundles.

Cytology. Chromosome base number, x = 12. 2n = 48. 4 ploid.

Taxonomy. Bambusoideae; Bambusodae; Bambuseae.

Ecology, geography, regional floristic distribution. 17 species. Eastern Asia. Holarctic and Paleotropical. Boreal. Indomalesian. Eastern Asian. Indian and Indo-Chinese.

Rusts and smuts. Rusts — *Puccinia*. Taxonomically wide-ranging species: *Puccinia kusanoi*.

References, etc. Morphological/taxonomic: Chao and Renvoize 1989. Leaf anatomical: this project.

Sinochasea Keng

~ *Pseudodanthonia*

Habit, vegetative morphology. Perennial; caespitose. Culms herbaceous. Leaves non-auriculate. Leaf blades narrow; setaceous; without cross venation. *Ligule an unfringed membrane*; truncate; 0.5–1 mm long.

Reproductive organization. Plants bisexual, with bisexual spikelets; with hermaphrodite florets.

Inflorescence. Inflorescence paniculate; contracted; espatheate; not comprising 'partial inflorescences' and foliar organs. Spikelet-bearing axes persistent. Spikelets not secund; pedicellate.

Female-fertile spikelets. Spikelets 10–12 mm long; compressed laterally to not noticeably compressed; disarticulating above the glumes. *Rachilla prolonged beyond the uppermost female-fertile floret*. The rachilla extension naked. Hairy callus present.

Glumes two; more or less equal; long relative to the adjacent lemmas (exceeding them); pointed (acute); awnless; carinate; similar. Lower glume 5(–6) nerved. Upper glume (4–)5 nerved. *Spikelets* with female-fertile florets only.

Female-fertile florets 1. *Lemmas* similar in texture to the glumes (firm); not becoming indurated; *incised (deeply notched)*; awned. Awns 1; median; from a sinus; geniculate; entered by one vein. Lemmas hairy; non-carinate; without a germination flap; 5 nerved. *Palea* present; relatively long; 2-nerved. *Lodicules* present; 2; free; membranous; glabrous; not toothed; not or scarcely vascularized. **Stamens 3**. *Ovary* glabrous. Styles free to their bases. *Stigmas 3*.

Taxonomy. Pooideae; Poodae; Aveneae.

Ecology, geography, regional floristic distribution. 1 species. China. Holarctic. Boreal. Eastern Asian.

Special comments. Fruit data wanting. Anatomical data wanting.

Sitanion Raf.

~ Elymus

Habit, vegetative morphology. Perennial; caespitose. *Culms* 11–100 cm high; herbaceous. Culm internodes hollow. Young shoots intravaginal. *Leaves* not basally aggregated; auriculate. *Leaf blades* narrow; to *6 mm wide*; flat, or folded, or rolled (convolute); without cross venation. Ligule an unfringed membrane; truncate; 0.3–1 mm long.

Reproductive organization. Plants bisexual, with bisexual spikelets; with hermaphrodite florets. The spikelets of sexually distinct forms on the same plant; hermaphrodite and sterile (sterile spikelets accompanying the female-fertile ones, either irregularly in the clusters, or the central spikelet fertile and the laterals sterile).

Inflorescence. *Inflorescence a single spike, or a false spike, with spikelets on contracted axes (with 1–4, but usually 2(-3), spikelets per node, very bristly via acicular glumes, awned lemmas and reduced spikelets)*; espatheate; not comprising 'partial inflorescences' and foliar organs. *Spikelet-bearing axes disarticulating*; disarticulating at the joints. **Spikelets** associated with bractiform involucres (constituted by glumes), or unaccompanied by bractiform involucres, not associated with setiform vestigial branches; usually, fairly consistently *paired*; not secund; sessile.

Female-sterile spikelets. Sometimes with 1(-2)of the spikelets in each group sterile and reduced to groups of awns.

Female-fertile spikelets. *Spikelets* 5–15 mm long; *not noticeably compressed*; falling with the glumes (the pairs/clusters falling with the rachis segment); disarticulating between the florets (the rachilla often also fragile), or not disarticulating between the florets. *Rachilla prolonged beyond the uppermost female-fertile floret.* Hairy callus absent.

Glumes two; relatively large; more or less equal; shorter than the adjacent lemmas, or long relative to the adjacent lemmas; displaced (borne side by side); awned (the awns scabrid, straight or flexuous); non-carinate; very dissimilar, or similar (cartilaginous basally, long, entire, bifid or deeply cleft into one to several awns). Lower glume 1–4 nerved. Upper glume 1–4 nerved. *Spikelets* usually with incomplete florets. The incomplete florets distal to the female-fertile florets (*S. longifolium*), or both distal and proximal to the female-fertile florets (usually, in at least one of the spikelets of a group). The distal incomplete florets 1; *incomplete florets merely underdeveloped. The proximal incomplete florets* when present, 1; epaleate; sterile. The proximal lemmas awned (reduced to an awn).

Female-fertile florets 1–6. Lemmas lanceolate; less firm than the glumes to similar in texture to the glumes; not becoming indurated (firm); incised; slightly 2 lobed; not deeply cleft (slightly bidentate); awned. Awns 1, or 3, or 5; median, or median and lateral (the lateral nerves sometimes excurrent as bristles); the median similar in form to the laterals (when laterals present); from a sinus, or apical; non-geniculate; much longer than the body of the lemma; entered by several veins. Lemmas hairy, or hairless; non-carinate; without a germination flap; 3–5 nerved. *Palea* present; relatively long; awnless, without apical setae, or awned; 2-nerved (these sometimes extending into awns to 5 mm long); 2-keeled. *Lodicules* present; 2; free; membranous; ciliate, or glabrous; toothed, or not toothed; not or scarcely vascularized. *Stamens* 3. Anthers 1–2 mm long. *Ovary* hairy. Styles free to their bases. Stigmas 2.

Fruit, embryo and seedling. *Fruit* adhering to lemma and/or palea; medium sized (5–6 mm long); longitudinally grooved; compressed dorsiventrally; with hairs confined to a terminal tuft. Hilum long-linear. Embryo small. Endosperm hard; without lipid. Embryo with an epiblast; without a scutellar tail; with a negligible mesocotyl internode. Embryonic leaf margins meeting.

Abaxial leaf blade epidermis. *Costal/intercostal zonation* conspicuous. *Papillae* absent. *Long-cells* similar in shape costally and intercostally (the costals somewhat narrower); of similar wall thickness costally and intercostally (rather thick walled). Mid-intercostal long-cells rectangular; having markedly sinuous walls (conspicuously pitted). *Microhairs* absent. *Stomata* common; (33–)34–36 microns long. Subsidiaries low dome-shaped (exclusively). Guard-cells overlapped by the interstomatals. *Intercostal short-cells* common; in cork/silica-cell pairs (and solitary); silicified. Short prickles common. *Crown* cells absent (but the prickles with crown-cell like bases). *Costal short-cells* neither distinctly grouped into long rows nor predominantly paired (solitary, paired and in short rows). Costal silica bodies tall-and-narrow, or crescentic; not sharp-pointed.

Transverse section of leaf blade, physiology. C₃; XyMS+. Mesophyll without adaxial palisade. *Leaf blade* with distinct, prominent adaxial ribs; with the ribs very irregular in sizes. *Midrib* not readily distinguishable; with one bundle only. Bulliforms present in discrete, regular adaxial groups (in the furrows); in simple fans. All the vascular bundles accompanied by sclerenchyma. Combined sclerenchyma girders present (with all the largest bundles); forming 'figures' (conspicuous I's). Sclerenchyma all associated with vascular bundles.

Cytology. Chromosome base number, $x = 7$. $2n = 28$. 4 ploid. Haplomic genome content H and S.

Taxonomy. Pooideae; Triticodae; Triticeae.

Ecology, geography, regional floristic distribution. 4 species. Western temperate North America. Xerophytic.

Holarctic. Madrean.

Hybrids. Integeneric hybrids with *Agropyron* (×*Agrositanion* Bowden), *Elymus*, *Hordeum* (×*Sitordeum* Bowden), *Lophopyrum*. See also ×*Elysitanion* Bowden.

Rusts and smuts. Rusts — *Puccinia*. Taxonomically wide-ranging species: *Puccinia graminis*, *Puccinia striiformis*, *Puccinia montanensis*, and *Puccinia recondita*. Smuts from Ustilaginaceae. Ustilaginaceae — *Ustilago*.

References, etc. Morphological/taxonomic: Löve 1984. Leaf anatomical: this project.

Snowdenia C.E. Hubb.

Beckera Fresen.

Habit, vegetative morphology. Annual, or perennial; stoloniferous, or decumbent. *Culms* 15–180 cm high; herbaceous. Culm nodes glabrous. *Leaves* not basally aggregated; non-auriculate. Leaf blades narrow; without cross venation; persistent. Ligule membranous; 1 mm long.

Reproductive organization. Plants bisexual, with bisexual spikelets; with hermaphrodite florets. The spikelets all alike in sexuality.

Inflorescence. Inflorescence of spicate main branches (slender spiciform racemes), or a single raceme; digitate, or non-digitate (when of a single raceme); espatheate; not comprising 'partial inflorescences' and foliar organs. Spikelet-bearing axes persistent. Spikelets solitary; secund; pedicellate.

Female-fertile spikelets. *Spikelets* 2–4 mm long; adaxial; *compressed dorsiventrally*; falling with the glumes. Rachilla terminated by a female-fertile floret. Hairy callus absent.

Glumes present; two; *minute*; more or less equal; shorter than the adjacent lemmas; dorsiventral to the rachis; not pointed; awnless; *similar (reduced to small, rotund scales)*. Lower glume 0 nerved. Upper glume 0 nerved. *Spikelets* with incomplete florets. The incomplete florets proximal to the female-fertile florets. *The proximal incomplete florets* 1; sterile. The proximal lemmas usually awned; decidedly exceeding the female-fertile lemmas (equalling the spikelet).

Female-fertile florets 1. Lemmas not becoming indurated (membranous); entire; pointed; awnless, or mucronate, or awned. Awns when present, 1; apical; non-geniculate; much shorter than the body of the lemma. Lemmas hairless; non-carinate; having the margins lying flat and exposed on the palea; 3–5 nerved. **Palea** present; *conspicuous but relatively short (less than half the floret length)*. *Lodicules* present; 2; free; fleshy; glabrous. **Stamens 3.** Anthers not penicillate. *Ovary* glabrous. Stigmas 2.

Fruit, embryo and seedling. *Fruit* free from both lemma and palea; small; compressed dorsiventrally. Hilum short. Embryo large.

Abaxial leaf blade epidermis. *Costal/intercostal zonation* conspicuous. *Papillae* present. Mid-intercostal long-cells having markedly sinuous walls. *Microhairs* present; panicoid-type; 28–45 microns long. Microhair apical cells 16–27 microns long. Microhair apical cell/total length ratio 0.55. *Stomata* common. Subsidiaries high dome-shaped and triangular. *Intercostal short-cells* absent or very rare. *Costal short-cells* neither distinctly grouped into long rows nor predominantly paired. Costal silica bodies infrequent, 'panicoid-type'; nodular; not sharp-pointed.

Transverse section of leaf blade, physiology. C_4; XyMS–. *Mesophyll* with radiate chlorenchyma. *Leaf blade* adaxially flat. *Midrib* conspicuous; having a conventional arc of bundles; with colourless mesophyll adaxially. Bulliforms present in discrete, regular adaxial groups (the epidermis partly, irregularly bulliform). Many of the smallest vascular bundles unaccompanied by sclerenchyma. Combined sclerenchyma girders present. Sclerenchyma all associated with vascular bundles.

Culm anatomy. *Culm internode bundles* scattered.

Taxonomy. Panicoideae; Panicodae; Paniceae (Arthropogoneae).

Ecology, geography, regional floristic distribution. 4 species. Tropical East Africa. Forest margins.

Paleotropical. African. Sudano-Angolan. Sahelo-Sudanian, Somalo-Ethiopian, and South Tropical African.

Rusts and smuts. Rusts — *Puccinia*.

Economic importance. Important native pasture species: *S. polystachya*.

References, etc. Leaf anatomical: Metcalfe 1960.

Soderstromia Morton

Fourniera Scribn.

Habit, vegetative morphology. Slender annual, or perennial; stoloniferous. *Culms* 20 cm high; herbaceous; branched above. Culm nodes glabrous. *Leaves* not basally aggregated; non-auriculate. Leaf blades narrow; 1–2 mm wide (2 to 6 cm long); without abaxial multicellular glands; without cross venation. Ligule a fringed membrane (short, densely fringed). *Contra-ligule* absent.

Reproductive organization. Plants monoecious with all the fertile spikelets unisexual; without hermaphrodite florets. The spikelets all alike in sexuality (i.e., all male or all female, when dioecious), or of sexually distinct forms on the same plant (when monoecious); female-only, or male-only, or female-only and male-only. The male and female-fertile spikelets in different inflorescences (in separate tufts), or on different branches of the same inflorescence. The spikelets overtly heteromorphic.

Inflorescence. Inflorescence a false spike, with spikelets on contracted axes (when each spikelet is interpreted as a reduced cluster, cf. *Bouteloua*: see below). Inflorescence with the main rachis terminating in a short abortive apex. Inflorescence espatheate; not comprising 'partial inflorescences' and foliar organs. **Spikelet-bearing axes** *very much reduced (each cluster comprising 3 basal, shortly-pedicellate, flattened, several-nerved, dentate, opaque 'bracts' (reduced spikelets), and a central, longer-pedicellate, glumeless fertile spikelet)*; disarticulating; falling entire (i.e., the clusters falling from the persistent main axis). **Spikelets** *associated with bractiform involucres (comprising reduced spikelets)*. The involucres shed with the fertile spikelets. Spikelets shortly pedicellate (including the 'bracts').

Female-sterile spikelets. Male spikelets with two fertile florets and glumelike basal bracts, the 'cluster' suggesting a single, bizarre spikelet; male flowers 3-staminate. The male spikelets 2 floreted. Male florets 2; 3 staminate.

Female-fertile spikelets. Spikelets unconventional (because of the complicating 'bracts'); 4–5 mm long; *compressed dorsiventrally*; the clusters disarticulating. Rachilla prolonged beyond the uppermost female-fertile floret; rachilla hairless. The rachilla extension with incomplete florets.

Glumes *absent*. *Spikelets* with incomplete florets. The incomplete florets distal to the female-fertile florets. *The distal incomplete florets* 1; **incomplete florets** *clearly specialised and modified in form (the sterile floret consisting of 3 long, straight awns)*.

Female-fertile florets 1. Lemmas not becoming indurated (membranous); incised; 3 lobed; deeply cleft (trifid, the lobes subacuminate, the central one longer); awnless; hairless; non-carinate; without a germination flap; 3 nerved. *Palea* present; relatively long; entire; awnless, without apical setae (glabrous); not indurated (membranous); 2-nerved;

2-keeled. *Lodicules* present; 2; free; membranous; glabrous; not or scarcely vascularized. *Stamens* 0. Anthers relatively long.

Fruit, embryo and seedling. Fruit small; ellipsoid; compressed dorsiventrally. Hilum short. Pericarp fused. Embryo large.

Abaxial leaf blade epidermis. *Costal/intercostal zonation* conspicuous. *Papillae* absent. *Long-cells* markedly different in shape costally and intercostally (the costals narrower, more rectangular); of similar wall thickness costally and intercostally (thick, pitted). Mid-intercostal long-cells rectangular to fusiform; having markedly sinuous walls. *Microhairs* present; elongated; clearly two-celled; panicoid-type to chloridoid-type (apical cells narrowing to the thick-walled tip, which is more or less apiculate). Microhair apical cell wall thinner than that of the basal cell but not tending to collapse. Microhairs (30–)31–39 microns long; (7.2–)7.5–8.4(–9) microns wide at the septum. Microhair total length/width at septum 3.6–4.3. Microhair apical cells 16.5–18–19.5 microns long. Microhair apical cell/total length ratio 0.42–0.65. *Stomata* common; 24–25–27 microns long. Subsidiaries dome-shaped (mostly), or triangular (in some files). Guard-cells overlapping to flush with the interstomatals. *Intercostal short-cells* common; in cork/silica-cell pairs; silicified. Intercostal silica bodies present and perfectly developed; tall-and-narrow. *Costal short-cells* conspicuously in long rows (mostly), or predominantly paired to neither distinctly grouped into long rows nor predominantly paired (paired or in short rows over some veins). Costal silica bodies present in alternate cell files of the costal zones; saddle shaped (predominantly), or crescentic; not sharp-pointed.

Transverse section of leaf blade, physiology. C_4; XyMS+. PCR sheaths of the primary vascular bundles interrupted; interrupted both abaxially and adaxially. PCR sheath extensions absent. *Leaf blade* 'nodular' in section to adaxially flat (with rounded, low abaxial ribs). *Midrib* not readily distinguishable; with one bundle only. Bulliforms present in discrete, regular adaxial groups; associated with colourless mesophyll cells to form deeply-penetrating fans. All the vascular bundles accompanied by sclerenchyma (but small bundles with only tiny strands). Combined sclerenchyma girders present (with the primaries); forming 'figures' (the primaries). Sclerenchyma all associated with vascular bundles.

Taxonomy. Chloridoideae; main chloridoid assemblage.

Ecology, geography, regional floristic distribution. 1 species. Mexico, Central America. Species of open habitats. Short grassland.

Neotropical. Caribbean.

References, etc. Leaf anatomical: this project.

Sohnsia Airy Shaw

Calamochloa Fourn.

Habit, vegetative morphology. Perennial; caespitose (forming tough clumps). *Culms* 30–100 cm high; herbaceous; unbranched above. Culm nodes hairy (or at least, hairy below the nodes). Leaves auriculate (from the sheath), or non-auriculate. Leaf blades linear (glaucous); narrow; flat, or rolled (becoming involute when dry); without abaxial multicellular glands; without cross venation; disarticulating from the sheaths (the lower sheaths persistent). Ligule a fringed membrane to a fringe of hairs (a ciliate rim).

Reproductive organization. *Plants dioecious*; without hermaphrodite florets (the stamens of pistillate plants seemingly sterile). The spikelets all alike in sexuality; female-only, or male-only. Plants outbreeding.

Inflorescence. *Inflorescence of spicate main branches (the males and females similar)*; contracted. Primary inflorescence branches 6–12. Inflorescence with axes ending in spikelets (?). Inflorescence espatheate; not comprising 'partial inflorescences' and foliar organs. The racemes spikelet bearing to the base. Spikelet-bearing axes persistent. Spikelets pedicellate.

Female-sterile spikelets. The male spikelets 3–5 flowered, the rachilla not disarticulating; glumes equal, lemmas 3-awned; stamens 3, well developed, the pistil rudimentary. Rachilla of male spikelets prolonged beyond the uppermost male floret. The male spikelets with glumes; 3–5 floreted. The lemmas 3 awned. Male florets 3–4 (?); 3 staminate.

Female-fertile spikelets. Spikelets 10–12 mm long; abaxial; compressed laterally, or not noticeably compressed (?); disarticulating above the glumes; not disarticulating between the florets (the florets usually falling together); with conventional internode spacings (the rachilla joints short). Rachilla prolonged beyond the uppermost female-fertile floret; rachilla hairless. The rachilla extension with incomplete florets. *Hairy callus present. Callus* short.

Glumes two; more or less equal ('subequal'); shorter than the spikelets; shorter than the adjacent lemmas; dorsiventral to the rachis; hairless (scaberulous on the keel and towards the tip, otherwise glabrous); pointed; awnless; carinate; similar. Lower glume 1 nerved. Upper glume 1 nerved. *Spikelets* with incomplete florets. The incomplete florets distal to the female-fertile florets. The distal incomplete florets 1 (the third, terminal floret being incomplete); *incomplete florets* merely underdeveloped.

Female-fertile florets 2. Lemmas not becoming indurated; incised; 3 lobed, or 5 lobed (?); deeply cleft; awned. Awns 3; median and lateral; the median similar in form to the laterals; dorsal (from between two hyaline lobes); from near the top (from the back of the median lobe); non-geniculate; recurving; hairless (antrorsely scabrous); about as long as the body of the lemma; entered by one vein. The lateral awns shorter than the median to about equalling the median. Lemmas hairy (pilose marginally, and on each side of the median nerve above the base of the diverging median awn); 3 nerved, or 5 nerved (or more?). *Palea* present; relatively long (almost equalling the lemma); obscurely apically notched; awnless, without apical setae; 2-nerved; 2-keeled. Palea keels narrowly winged. *Lodicules* present; 2; free; fleshy (seemingly cuneate); glabrous. *Stamens* 3 (but apparently sterile). *Ovary* glabrous. Styles free to their bases. Stigmas 2.

Fruit, embryo and seedling. *Fruit* free from both lemma and palea. Pericarp fused.

Abaxial leaf blade epidermis. *Costal/intercostal zonation* conspicuous. *Papillae* absent. *Long-cells* markedly different in shape costally and intercostally (the costals much narrower); of similar wall thickness costally and intercostally (of medium thickness). Mid-intercostal long-cells rectangular; having markedly sinuous walls (the sinuosity coarse, irregular). *Microhairs* present; elongated; clearly two-celled; chloridoid-type. Microhair apical cell wall of similar thickness/rigidity to that of the basal cell. Microhair basal cells 30 microns long. Microhair total length/width at septum 3.5. Microhair apical cell/total length ratio 0.3. *Stomata* common. Subsidiaries non-papillate; dome-shaped to triangular (often rather flat topped). Guard-cells overlapping to flush with the interstomatals. *Intercostal short-cells* absent or very rare; not paired. Intercostal silica bodies absent. Bulbous prickles with reduced points abundant over the veins. *Costal short-cells* conspicuously in long rows. Costal silica bodies present and well developed; present in alternate cell files of the costal zones; saddle shaped.

Transverse section of leaf blade, physiology. C_4. The anatomical organization conventional. XyMS+. PCR sheath outlines even. PCR sheaths of the primary vascular bundles interrupted; interrupted both abaxially and adaxially. PCR sheath extensions absent. PCR cell chloroplasts centripetal (very clearly so, even in the dried material seen). *Mesophyll* with radiate chlorenchyma; traversed by columns of colourless mesophyll cells (the broad columns of large cells, 2–3 cells wide, between all adjacent veins). *Leaf blade* with distinct, prominent adaxial ribs to 'nodular' in section; with the ribs more or less constant in size (large, broad, flat to broadly round topped, one over each bundle). *Midrib* not readily distinguishable; with one bundle only. Bulliforms not in discrete, regular adaxial groups (in that they are not distinguishable from the colourless columns). All the vascular bundles accompanied by sclerenchyma. Combined sclerenchyma girders present (with every bundle); forming 'figures' (large I's, which interrupt the PCR sheaths top and bottom). Sclerenchyma all associated with vascular bundles. The lamina margins with fibres.

Taxonomy. Chloridoideae; main chloridoid assemblage.

Ecology, geography, regional floristic distribution. 1 species. Mexico. Xerophytic; species of open habitats. Dry hillsides.

Holarctic. Madrean.

References, etc. Morphological/taxonomic: Sohns 1956. Leaf anatomical: this project.

Special comments. Fruit data wanting.

Sorghastrum Nash

Dipogon Steud., *Poranthera* Raf.

Habit, vegetative morphology. Annual, or perennial; caespitose. *Culms* 70–150 cm high; herbaceous; branched above, or unbranched above. *Leaves* not basally aggregated; auriculate, or non-auriculate. Leaf blades without cross venation; rolled in bud. Ligule an unfringed membrane to a fringed membrane.

Reproductive organization. Plants bisexual, with bisexual spikelets; with hermaphrodite florets. The spikelets of sexually distinct forms on the same plant; hermaphrodite and sterile; overtly heteromorphic (in that the sterile spikelets are reduced to pedicels), or homomorphic (rarely the pedicellate spikelets are well developed and simiar to the sessile ones). Plants outbreeding.

Inflorescence. Inflorescence paniculate (narrowly elongated, more or less unilateral panicles of much-reduced, capillary 'racemes'); open (usually narrow); with capillary branchlets; subdigitate to non-digitate. *Inflorescence axes not ending in spikelets (in that the axis of each inflorescence unit ends in a bristle resembling a sterile pedicel)*. Inflorescence espatheate; not comprising 'partial inflorescences' and foliar organs. *Spikelet-bearing axes* very much reduced (the ultimate units with very few spikelets, often only one accompanied by the sterile pedicel); disarticulating (but the disarticulating units much reduced); falling entire (when reduced to one joint), or disarticulating at the joints. *Spikelets* nearly always *paired (but ostensibly solitary, by virtue of the 'pedicelled' member being reduced to its pedicel — by contrast with* **Sorghum***)*; secund (the inflorescence one-sided), or not secund; sessile and pedicellate (but usually ostensibly solitary by suppression); *consistently in 'long-and-short' combinations (but the sterile member of each combination is nearly always reduced to its pedicel)*. Pedicels of the 'pedicellate' spikelets free of the rachis. The 'shorter' spikelets hermaphrodite. The 'longer' spikelets sterile (reduced to pedicels).

Female-fertile spikelets. Spikelets 5–8 mm long; compressed dorsiventrally (plump); falling with the glumes (and the joint). Rachilla terminated by a female-fertile floret. Hairy callus present.

Glumes two; more or less equal; long relative to the adjacent lemmas; free; hairy (G_1); without conspicuous tufts or rows of hairs; pointed; awnless; very dissimilar (the lower flattened and often hairy on the back, the upper glabrous and slightly keeled above). Lower glume not two-keeled; convex on the back to flattened on the back; not pitted; relatively smooth; 9 nerved. Upper glume 5 nerved. *Spikelets* with incomplete florets. The incomplete florets proximal to the female-fertile florets. *The proximal incomplete florets* 1; epaleate; sterile. The proximal lemmas 2-lobed; awnless; 2 nerved; more or less equalling the female-fertile lemmas to decidedly exceeding the female-fertile lemmas; not becoming indurated (hyaline).

Female-fertile florets 1. Lemmas less firm than the glumes; linear, almost reduced to the awn, the margins narrow and thin; incised; awned. Awns 1; median; from a sinus; geniculate; hairless (glabrous); much longer than the body of the lemma. Lemmas hairy (sometimes long-ciliate), or hairless; non-carinate; *1 nerved (?)*. *Palea* present, or absent; when present, conspicuous but relatively short, or very reduced; not indurated. *Lodicules* present; 2; free; fleshy; glabrous. *Stamens* 3; with free filaments. Anthers not penicillate; without an apically prolonged connective. *Ovary* glabrous. Styles fused, or free to their bases. Stigmas 2.

Fruit, embryo and seedling. *Fruit* free from both lemma and palea; small. Hilum short. Embryo large. Endosperm hard. Embryo without an epiblast; with a scutellar tail; with an elongated mesocotyl internode. Embryonic leaf margins overlapping.

Abaxial leaf blade epidermis. *Costal/intercostal zonation* conspicuous. *Papillae* present; costal, or intercostal. Mid-intercostal long-cells rectangular; having markedly sinuous walls. *Microhairs* present; panicoid-type. *Stomata* common. Subsidiaries triangular. *Intercostal short-cells* common. *Costal short-cells* conspicuously in long rows, or predominantly paired. Costal silica bodies horizontally-elongated crenate/sinuous, or tall-and-narrow, or 'panicoid-type'; not sharp-pointed.

Transverse section of leaf blade, physiology. C_4; biochemical type NADP–ME (*S. nutans*); XyMS–. PCR sheath outlines uneven. PCR sheath extensions present, or absent. Maximum number of extension cells 1. PCR cell chloroplasts with reduced grana; centrifugal/peripheral. *Mesophyll* with radiate chlorenchyma. *Leaf blade* with distinct, prominent adaxial ribs, or 'nodular' in section, or adaxially flat; when ribbed with the ribs more or less constant in size, or with the ribs very irregular in sizes. *Midrib* conspicuous; having a conventional arc of bundles; with colourless mesophyll adaxially. Bulliforms present in discrete, regular adaxial groups; associated with colourless mesophyll cells to form deeply-penetrating fans and combining with colourless mesophyll cells to form narrow groups penetrating into the mesophyll; associating with colourless mesophyll cells to form arches over small vascular bundles (all the bulliforms involved in these configurations). Combined sclerenchyma girders present; forming 'figures'. Sclerenchyma all associated with vascular bundles.

Phytochemistry. Tissues of the culm bases with abundant starch.

Special diagnostic feature. Spikelets ostensibly solitary, each accompanied by a barren pedicel.

Cytology. Chromosome base number, $x = 10$. $2n = 20, 40$, and 60. 2, 4, and 6 ploid.

Taxonomy. Panicoideae; Andropogonodae; Andropogoneae; Andropogoninae.

Ecology, geography, regional floristic distribution. About 20 species. Mainly tropical and subtropical Africa and America. Commonly adventive. Helophytic to mesophytic; shade species, or species of open habitats; glycophytic. Savanna and woodland margins, often in wet places.

Holarctic, Paleotropical, and Neotropical. Boreal and Madrean. African and Madagascan. Atlantic North American. Sudano-Angolan, West African Rainforest, and Namib-Karoo. Caribbean, Central Brazilian, Pampas, and Andean. Canadian-Appalachian, Southern Atlantic North American, and Central Grasslands. Sahelo-Sudanian, South Tropical African, and Kalaharian.

Rusts and smuts. Rusts — *Puccinia*. Taxonomically wide-ranging species: '*Uromyces*' *clignyi*. Smuts from Ustilaginaceae. Ustilaginaceae — *Sphacelotheca* and *Tolyposporella*.

Economic importance. Important native pasture species: *S. arundinaceum* (sometimes cyanogenic when young); *S. nutans*.

References, etc. Leaf anatomical: photos of *S. friesii* and *S. stipoides* provided by R.P. Ellis.

Sorghum Moench

Blumenbachia Koel., *Sarga* Ewart & White

Habit, vegetative morphology. Annual, or perennial; rhizomatous, or stoloniferous, or caespitose, or decumbent. *Culms* 60–300 cm high; herbaceous; branched above (rarely), or unbranched above. Culm nodes hairy (e.g. in subgenera *Parasorghum* and *Stiposorghum*), or glabrous. Culm internodes solid (or sometimes hollow below). *Leaves* not basally aggregated; non-auriculate. Leaf blades linear to lanceolate; broad, or narrow; not setaceous; usually flat; without cross venation; persistent; rolled in bud. *Ligule* present; an unfringed membrane to a fringed membrane, or a fringe of hairs (rarely). *Contra-ligule* absent.

Reproductive organization. Plants bisexual, with bisexual spikelets; with hermaphrodite florets. **The spikelets of sexually distinct forms on the same plant**; hermaphrodite and male-only, or hermaphrodite and sterile; overtly heteromorphic; all in heterogamous combinations. Plants outbreeding and inbreeding. Exposed-cleistogamous, or chasmogamous. Apomictic, or reproducing sexually.

Inflorescence. Inflorescence paniculate, or of spicate main branches (usually large, the branched or simple primary branches usually whorled); open, or contracted; with capillary branchlets, or without capillary branchlets; *espatheate*; not comprising 'partial inflorescences' and foliar organs. **Spikelet-bearing axes** very much reduced, or 'racemes' (short racemes with 1–6(-8) articles only); *the spikelet-bearing axes* with only one spikelet-bearing 'article', or with 2–3 spikelet-bearing 'articles', or with 4–5 spikelet-bearing 'articles', or with 6–10 spikelet-bearing 'articles' (rarely); *with very slender rachides*; disarticulating, or persistent (in cultivated forms); falling entire (when reduced to one joint), or disarticulating at the joints. *'Articles'* linear (flattened); without a basal callus-knob; disarticulating transversely. *Spikelets* paired, or in triplets (terminal); not secund; sessile and pedicellate; *consistently in 'long-and-short' combinations*; in pedicellate/sessile combinations. Pedicels of the 'pedicellate' spikelets free of the rachis. The 'shorter' spikelets hermaphrodite. *The 'longer' spikelets male-only, or sterile (but not reduced to the pedicel, by contrast with* **Sorghastrum***).*

Female-sterile spikelets. The pedicellate spikelet male or sterile, much narrower and awnless or reduced to a glume, very rarely suppressed (*S. angustum.*

Female-fertile spikelets. *Spikelets compressed dorsiventrally*; falling with the glumes, or not disarticulating (in cultivated forms). Rachilla terminated by a female-fertile floret. Hairy callus present, or absent. Callus pointed, or blunt.

Glumes two; more or less equal; long relative to the adjacent lemmas; without conspicuous tufts or rows of hairs; awnless, or awned (the upper, sometimes); very dissimilar (the lower flat or rounded on the back save at the summit, upper naviculate). Lower glume not two-keeled (below, but becoming two-keeled and winged above); convex on the back to flattened on the back; not pitted; relatively smooth; 5–11 nerved. Upper glume 3–11 nerved. *Spikelets* with incomplete florets. The incomplete florets proximal to the female-fertile florets. *Spikelets with proximal incomplete florets. The proximal incomplete florets* 1; epaleate; sterile. The proximal lemmas awnless; 0 nerved, or 2 nerved; similar in texture to the female-fertile lemmas (hyaline, ciliate); not becoming indurated.

Female-fertile florets 1. **Lemmas** less firm than the glumes (hyaline, ciliate); not becoming indurated; *incised*; awnless to mucronate (rarely), or awned. Awns when present, 1; from a sinus; geniculate; hairless (glabrous); much shorter than the body of the lemma to much longer than the body of the lemma. Lemmas hairless; non-carinate; 1–3 nerved. *Palea* present, or absent; when present, relatively long, or conspicuous but relatively short, or very reduced; not indurated (hyaline); nerveless, or 2-nerved. **Lodicules** present; 2; free; more or less fleshy; usually *ciliate*. **Stamens** 3. Anthers penicillate, or not penicillate. **Ovary** glabrous, or hairy (occasionally with a terminal hair tuft, e.g. *S. intrans*); *without a conspicuous apical appendage*. Styles free to their bases. Stigmas 2; red pigmented.

Fruit, embryo and seedling. *Fruit* free from both lemma and palea; small, or medium sized, or large; compressed dorsiventrally, or not noticeably compressed. Hilum short. Embryo large. Endosperm hard; without lipid; containing only simple starch grains. Embryo without an epiblast; with a scutellar tail; with an elongated mesocotyl internode. Embryonic leaf margins overlapping.

Seedling with a long mesocotyl. First seedling leaf with a well-developed lamina. The lamina broad; curved; 21–30 veined.

Abaxial leaf blade epidermis. *Costal/intercostal zonation* conspicuous. *Papillae* present, or absent. Intercostal papillae over-arching the stomata; several per cell (the costal papillae consisting of fingerlike projections, the intercostal ones of larger, oblique swellings). *Long-cells* similar in shape costally and intercostally; of similar wall thickness costally and intercostally. Mid-intercostal long-cells rectangular; having markedly sinuous walls. *Microhairs* present; panicoid-type; 34–78 microns long; 6.6–9.6 microns wide at the septum. Microhair total length/width at septum 6.3–11.8. Microhair apical cells (15–)21–30(–36) microns long. Microhair apical cell/total length ratio 0.39–0.62. *Stomata* common; 27–36 microns long. Subsidiaries usually at least somewhat triangular. Guard-cells overlapping to flush with the interstomatals. *Intercostal short-cells* absent or very rare (rare); when present, in cork/silica-cell pairs, or not paired (solitary); silicified. Intercostal silica bodies when present, tall-and-narrow, or crescentic. *Costal short-cells* conspicuously in long rows. Costal silica bodies 'panicoid-type'; usually cross shaped to dumb-bell shaped, or nodular; not sharp-pointed.

Transverse section of leaf blade, physiology. C_4; biochemical type NADP–ME (3 species); XyMS–. PCR sheath outlines uneven. PCR cell chloroplasts with reduced grana;

centrifugal/peripheral. *Mesophyll* with radiate chlorenchyma. *Leaf blade* with distinct, prominent adaxial ribs, or adaxially flat. *Midrib* conspicuous; having a conventional arc of bundles; with colourless mesophyll adaxially. Bulliforms present in discrete, regular adaxial groups, or not in discrete, regular adaxial groups (often irregularly grouped and/ or occupying most of the epidermis); sometimes in simple fans, or associated with colourless mesophyll cells to form deeply-penetrating fans. Many of the smallest vascular bundles unaccompanied by sclerenchyma. Combined sclerenchyma girders present; nowhere forming 'figures'. Sclerenchyma all associated with vascular bundles.

Culm anatomy. *Culm internode bundles* scattered.

Phytochemistry. Tissues of the culm bases with abundant starch. Leaves without flavonoid sulphates (4 species).

Special diagnostic feature. Spikelets paired, all the pedicels spikelet-bearing.

Cytology. Chromosome base number, $x = 5$. $2n = 10$, 20, and 40 (and aneuploids — 26, 33, 38–39 etc.). Haploid nuclear DNA content 1.2–2.7 pg (8 $4x$ species, mean 2.3). Nucleoli persistent.

Taxonomy. Panicoideae; Andropogonodae; Andropogoneae; Andropogoninae.

Ecology, geography, regional floristic distribution. About 30 species. Tropical and subtropical. Commonly adventive. Mesophytic; shade species, or species of open habitats; glycophytic. Savanna and forest margins, alluvial plains and disturbed ground.

Holarctic, Paleotropical, Neotropical, Australian, and Antarctic. Boreal, Tethyan, and Madrean. African, Madagascan, Indomalesian, and Neocaledonian. Euro-Siberian, Eastern Asian, and Atlantic North American. Mediterranean and Irano-Turanian. Saharo-Sindian, Sudano-Angolan, West African Rainforest, and Namib-Karoo. Indian, Indo-Chinese, Malesian, and Papuan. Caribbean, Central Brazilian, Pampas, and Andean. North and East Australian and Central Australian. Patagonian. European. Canadian-Appalachian. Sahelo-Sudanian, Somalo-Ethiopian, South Tropical African, and Kalaharian. Tropical North and East Australian and Temperate and South-Eastern Australian.

Hybrids. Intergeneric hybrids with *Saccharum*.

Rusts and smuts. Rusts — *Puccinia*. Taxonomically wide-ranging species: *Puccinia nakanishikii* and *Puccinia levis*. Smuts from Ustilaginaceae. Ustilaginaceae — *Sorosporium*, *Sphacelotheca*, *Tolyposporium*, and *Ustilago*.

Economic importance. Significant weed species: *S. almum*, *S. bicolor*, *S. halepense*. Cultivated fodder: *S. halepense* (Johnson), and hybrids involving *S. arundinaceum*, *S. bicolor* and *S. halepense* (e.g. Sudan). Grain crop species: *S. bicolor* (Sorghum), with many cultivars.

References, etc. Leaf anatomical: Metcalfe 1960; this project.

Spartina Schreber

Chauvinia Steud., *Limnetis* Rich., *Ponceletia* Thours, *Psammophila* Schult., *Solenachne* Steud., *Trachynotia* Michaux, *Tristania* Poir.

Habit, vegetative morphology. Perennial; rhizomatous to stoloniferous, or caespitose. *Culms* 20–300 cm high; herbaceous. Culm nodes glabrous. Culm internodes solid, or hollow. *Leaves* not basally aggregated; non-auriculate. *Leaf blades* linear; tough; broad, or narrow; flat, or rolled; without abaxial multicellular glands; without cross venation; *disarticulating from the sheaths*; rolled in bud. *Ligule* present; a fringe of hairs.

Reproductive organization. Plants bisexual, with bisexual spikelets; with hermaphrodite florets. Exposed-cleistogamous, or chasmogamous.

Inflorescence. *Inflorescence of spicate main branches (with 2 to many long or short spikes, borne racemosely on the main axis). Inflorescence axes not ending in spikelets (their slender, naked tips often prolonged).* Inflorescence espatheate; not comprising 'partial inflorescences' and foliar organs. Spikelet-bearing axes persistent. Spikelets solitary; secund (the rachides dorsiventral, triquetrous); biseriate (appressed or pectinate); sessile.

Female-fertile spikelets. Spikelets 6–18 mm long; abaxial; strongly compressed laterally; falling with the glumes. Rachilla terminated by a female-fertile floret. Hairy callus absent.

Glumes two; very unequal (the upper longer); (the upper) long relative to the adjacent lemmas (often exceeding it); free; dorsiventral to the rachis; hairy, or hairless; pointed; shortly awned, or awnless; similar (leathery or membranous). Lower glume 1 nerved. Upper glume 1–3 nerved. *Spikelets* with female-fertile florets only.

Female-fertile florets 1. Lemmas not becoming indurated; entire, or incised; when entire pointed, or blunt; not deeply cleft (minutely bidentate); awnless; hairy (shortly), or hairless (glabrous); carinate; 1–3 nerved. *Palea* present; relatively long; awnless, without apical setae; 2-nerved; 2-keeled. **Lodicules *absent*.** *Stamens* 3. Anthers 3–13 mm long (relatively long); not penicillate. *Ovary* glabrous. Styles fused. Stigmas 2.

Fruit, embryo and seedling. *Fruit* free from both lemma and palea; medium sized; fusiform. Hilum short. Pericarp loosely adherent, or fused. Embryo large. Endosperm hard; without lipid; containing compound starch grains. Embryo with an epiblast, or without an epiblast; with a scutellar tail; with an elongated mesocotyl internode. Embryonic leaf margins meeting.

Seedling with a loose coleoptile. First seedling leaf with a well-developed lamina. The lamina narrow; erect; 'many'.

Abaxial leaf blade epidermis. *Costal/intercostal zonation* conspicuous, or lacking. *Papillae* absent. *Long-cells* similar in shape costally and intercostally; of similar wall thickness costally and intercostally (walls very thick and pitted), or differing markedly in wall thickness costally and intercostally. Mid-intercostal long-cells rectangular; having markedly sinuous walls. *Microhairs* present, or absent; chloridoid-type (sunken in 'crypts' in the epidermis); with 'partitioning membranes' (in *S. anglica*, *S. foliosa*). The 'partitioning membranes' in the basal cell. *Stomata* absent or very rare, or common. Subsidiaries dome-shaped, or triangular. Guard-cells overlapping to flush with the interstomatals. *Intercostal short-cells* common; in cork/silica-cell pairs (a few), or not paired (mainly solitary); silicified (usually), or not silicified. Intercostal silica bodies absent, or imperfectly developed; when present, tall-and-narrow, or cubical. *Costal short-cells* neither distinctly grouped into long rows nor predominantly paired. Costal silica bodies present throughout the costal zones; horizontally-elongated crenate/sinuous, or horizontally-elongated smooth, or rounded, or tall-and-narrow (or more or less rectangular); not sharp-pointed.

Transverse section of leaf blade, physiology. C_4; biochemical type PCK (*S. anglica*); XyMS+. PCR sheath outlines uneven. PCR sheaths of the primary vascular bundles complete. PCR sheath extensions present. Maximum number of extension cells 7–10. PCR cells without a suberised lamella. PCR cell chloroplasts with well developed grana; centrifugal/peripheral. *Mesophyll* with radiate chlorenchyma. *Leaf blade* with distinct, prominent adaxial ribs; with the ribs very irregular in sizes. *Midrib* not readily distinguishable; with one bundle only. Bulliforms not in discrete, regular adaxial groups (arranged inconspicuously at the bases of the furrows, cf. *Ammophila*). All the vascular bundles accompanied by sclerenchyma. Combined sclerenchyma girders present; nowhere forming 'figures'. Sclerenchyma all associated with vascular bundles. The lamina margins with fibres.

Culm anatomy. *Culm internode bundles* in one or two rings.

Phytochemistry. Tissues of the culm bases with abundant starch. Leaves without flavonoid sulphates (4 species).

Cytology. Chromosome base number, $x = 7$ and 10. $2n = 28, 40, 42, 60, 62, 84, 120, 122,$ and 124. 3, 4, 6, 8, and 12 ploid.

Taxonomy. Chloridoideae; main chloridoid assemblage.

Ecology, geography, regional floristic distribution. 16 species. Temperate America, coastal Europe, Africa, Tristan da Cunha. Commonly adventive. Hydrophytic to helophytic; species of open habitats; halophytic.

Holarctic; Paleotropical, Neotropical, Cape, and Antarctic. Boreal, Tethyan, and Madrean. African. Euro-Siberian, Atlantic North American, and Rocky Mountains. Macaronesian and Mediterranean. Namib-Karoo. Caribbean, Amazon, and Pampas. Patagonian and Antarctic and Subantarctic. European. Canadian-Appalachian and Central Grasslands. South Temperate Oceanic Islands.

Rusts and smuts. Rusts — *Puccinia*.

Economic importance. Significant weed species: *S. alterniflora*, *S. cynosuroides*. Important native pasture species: *S. pectinata*.

References, etc. Leaf anatomical: Metcalfe 1960; this project.

Spartochloa C.E. Hubb.

Habit, vegetative morphology. *Juncus*-like *perennial (a switch plant, with reduced leaf blades and green culms)*; caespitose. *Culms* 30–60 cm high; herbaceous; unbranched above; culm bases rather swollen within the cataphylls. Culm internodes solid. Plants unarmed. *Leaves* mostly basal; non-auriculate. **Leaf blades** *greatly reduced*; narrow (tiny, reduced); without cross venation; disarticulating from the sheaths. Ligule a fringe of hairs. *Contra-ligule* present (of sparse hairs).

Reproductive organization. Plants bisexual, with bisexual spikelets; with hermaphrodite florets.

Inflorescence. Inflorescence paniculate; contracted; spicate; espatheate; not comprising 'partial inflorescences' and foliar organs. Spikelet-bearing axes persistent. Spikelets not secund; pedicellate; consistently in 'long-and-short' combinations; unequally pedicellate in each combination. Pedicels of the 'pedicellate' spikelets free of the rachis. The 'shorter' spikelets hermaphrodite. The 'longer' spikelets hermaphrodite.

Female-fertile spikelets. Spikelets 4–5 mm long; compressed laterally; disarticulating above the glumes; disarticulating between the florets. Rachilla prolonged beyond the uppermost female-fertile floret; rachilla hairless. The rachilla extension with incomplete florets. Hairy callus absent.

Glumes two; more or less equal; shorter than the adjacent lemmas; lateral to the rachis (the spikelets tending to be flatwise against the rachis); hairless; pointed; awnless (but mucronate); carinate; similar (membranous). Lower glume 1–3 nerved. Upper glume 3 nerved. *Spikelets* with incomplete florets. The incomplete florets distal to the female-fertile florets. *The distal incomplete florets* merely underdeveloped.

Female-fertile florets 3–6. Lemmas similar in texture to the glumes (firmly membranous); not becoming indurated; entire, or incised; not deeply cleft; awnless, or mucronate (from a minute sinus); hairless; glabrous; somewhat carinate; without a germination flap; 5–9 nerved. *Palea* present; conspicuous but relatively short (about a third of the lemma length); apically notched (hairy at the tip); awnless, without apical setae; not indurated; 2-nerved; 2-keeled. *Lodicules* present; 2; free; fleshy; glabrous; not or scarcely vascularized. *Stamens* 3. Anthers not penicillate; without an apically prolonged connective. *Ovary* glabrous. Styles free to their bases. Stigmas 2; red pigmented.

Fruit, embryo and seedling. Fruit medium sized (about 5 mm long); black; slightly longitudinally grooved; sculptured (pitted). Hilum short. Pericarp thick and hard (black, with minute pits). Embryo small; not waisted. Endosperm hard; containing compound starch grains.

Abaxial leaf blade epidermis. *Costal/intercostal zonation* conspicuous. *Papillae* absent. Mid-intercostal long-cells rectangular; having markedly sinuous walls. *Microhairs* present; panicoid-type; about 45 microns long; 7.5 microns wide at the septum. Microhair total length/width at septum about 6. Microhair apical cells about 22.5 microns long. Microhair apical cell/total length ratio 0.5. *Stomata* common; 27–35 microns long. Subsidiaries medium dome-shaped. Guard-cells overlapping to flush with the interstomatals. *Intercostal short-cells* common; in cork/silica-cell pairs; silicified. *Costal short-cells* conspicuously in long rows. Costal silica bodies 'panicoid-type'; not sharp-pointed.

Transverse section of leaf blade, physiology. C_3; XyMS+. *Mesophyll* with non-radiate chlorenchyma. *Leaf blade* with distinct, prominent adaxial ribs. *Midrib* not readily distinguishable; with one bundle only. Bulliforms present in discrete, regular adaxial groups. All the vascular bundles accompanied by sclerenchyma. Combined sclerenchyma girders present. Sclerenchyma all associated with vascular bundles.

Special diagnostic feature. *Rush-like, with reduced leaf blades*.

Taxonomy. Arundinoideae; Spartochloeae.

Ecology, geography, regional floristic distribution. 1 species. Australia. Xerophytic; species of open habitats. Arid places.

Australian. South-West Australian and Central Australian.

References, etc. Morphological/taxonomic: Hubbard 1952; Macfarlane and Watson 1980. Leaf anatomical: this project.

Spathia Ewart

Habit, vegetative morphology. Annual; caespitose. The flowering culms leafy. *Culms* 20–75 cm high; herbaceous; branched above. Culm nodes hairy. Culm internodes solid. *Leaves* not basally aggregated; non-auriculate. The sheaths usually inflated. Leaf blades narrow; 2–4 mm wide (reducing in length on the upper culm, where the sheaths become enlarged to constitute spathes); without cross venation; disarticulating from the sheaths. Ligule a fringed membrane.

Reproductive organization. Plants bisexual, with bisexual spikelets; with hermaphrodite florets. The spikelets of sexually distinct forms on the same plant; hermaphrodite and male-only. The male and female-fertile spikelets mixed in the inflorescence. The spikelets overtly heteromorphic (the pedicellate spikelets less hairy, awnless); in both homogamous and heterogamous combinations (each raceme with one basal, homogamous pair). Exposed-cleistogamous.

Inflorescence. *Inflorescence* of spicate main branches (of sessile or subsessile 'racemes'); *digitate*; *spatheate (with one broad, membranous spathe enclosing the inflorescence)*; a complex of 'partial inflorescences' and intervening foliar organs (up to 3 partial inflorescences), or not comprising 'partial inflorescences' and foliar organs. *Spikelet-bearing axes* 'racemes'; clustered (3–5); with very slender rachides; disarticulating; disarticulating at the joints. *'Articles'* linear; not appendaged; disarticulating obliquely; densely long-hairy (with brown hairs). Spikelets paired; not secund; sessile and pedicellate, or subsessile and pedicellate; consistently in 'long-and-short' combinations. Pedicels of the 'pedicellate' spikelets free of the rachis. The 'shorter' spikelets hermaphrodite. The 'longer' spikelets male-only.

Female-sterile spikelets. The less hairy, awnless pedicellate spikelet with two glumes and a floret containing 2 stamens, the callus short and oblong. Rachilla of male spikelets terminated by a male floret. The male spikelets with glumes (two); 1 floreted. The lemmas awnless. Male florets 2 staminate.

Female-fertile spikelets. Spikelets 5.5–6.2 mm long; compressed dorsiventrally; falling with the glumes. Rachilla terminated by a female-fertile floret. Hairy callus present (with long dense brown hairs). Callus blunt.

Glumes two; more or less equal; free; hairy (the hairs brown); awnless; very dissimilar (the lower bicarinate and blunt, the upper cymbiform and pointed). Lower glume two-keeled; convex on the back; 7–9 nerved. Upper glume 3 nerved. *Spikelets* with incomplete florets. The incomplete florets proximal to the female-fertile florets. *The proximal incomplete florets* 1; sterile. *The proximal lemmas* awnless; 0 nerved; *less firm than the female-fertile lemmas*; not becoming indurated.

Female-fertile florets 1. Lemmas stipe-like beneath the awn; less firm than the glumes; not becoming indurated; entire; awned. Awns 1; median; apical; geniculate; hairless (glabrous); much longer than the body of the lemma. Lemmas non-carinate; 0–1 nerved. **Palea absent**. Lodicules present; 2; fleshy; glabrous; not or scarcely vascularized. *Stamens* 3. Anthers about 0.6 mm long; not penicillate; without an apically prolonged connective. *Ovary* glabrous. Styles free to their bases. Stigmas 2; red pigmented.

Fruit, embryo and seedling. Fruit small; compressed dorsiventrally. Hilum short. Embryo large; waisted. Endosperm hard; containing only simple starch grains. Embryo without an epiblast; with a scutellar tail; with an elongated mesocotyl internode. Embryonic leaf margins overlapping.

Abaxial leaf blade epidermis. *Costal/intercostal zonation* conspicuous. *Papillae* present. Intercostal papillae not over-arching the stomata; consisting of one oblique swelling per cell. *Long-cells* similar in shape costally and intercostally; of similar wall thickness costally and intercostally. Mid-intercostal long-cells rectangular; having markedly sinuous walls. *Microhairs* present; panicoid-type; 48–60 microns long; (3.9–)5.4–5.7(–6) microns wide at the septum. Microhair total length/width at septum 9.4–13. Microhair apical cells (16.5–)19.5–22.5(–24) microns long. Microhair apical cell/total length ratio 0.32–0.47. *Stomata* common; (22.5–)24–25.5(–27) microns long.

Subsidiaries triangular. Guard-cells overlapped by the interstomatals. *Intercostal short-cells* absent or very rare. *Costal short-cells* conspicuously in long rows. Costal silica bodies 'panicoid-type' (large); not sharp-pointed.

Transverse section of leaf blade, physiology. C_4; XyMS–. *Mesophyll* with radiate chlorenchyma. *Leaf blade* adaxially flat. *Midrib* conspicuous; with one bundle only, or having a conventional arc of bundles (rarely); with colourless mesophyll adaxially. Bulliforms present in discrete, regular adaxial groups to not in discrete, regular adaxial groups; sometimes more or les in simple fans. Many of the smallest vascular bundles unaccompanied by sclerenchyma. Combined sclerenchyma girders present; forming 'figures'. Sclerenchyma all associated with vascular bundles.

Taxonomy. Panicoideae; Andropogonodae; Andropogoneae; Andropogoninae.

Ecology, geography, regional floristic distribution. 1 species. Northern Australia. Species of open habitats. Grassy plains.

Australian. North and East Australian. Tropical North and East Australian.

References, etc. Leaf anatomical: this project.

Sphaerobambos S. Dransfield

Habit, vegetative morphology. Perennial. The flowering culms leafless, or leafy. **Culms** 400–600 cm high (or up to 10 m long); *woody and persistent*; to 0.4 cm in diameter, or 5 cm in diameter; cylindrical (?); scandent, or not scandent; branched above. Primary branches/mid-culm node 3–25 ('few to many'). Culm sheaths deciduous in their entirety. Culm internodes hollow. Rhizomes pachymorph. Plants unarmed. *Leaves* not basally aggregated; auriculate; with auricular setae. Leaf blades lanceolate to ovate-lanceolate; broad; 10–60 mm wide (and 100–230 mm long); pseudopetiolate; disarticulating from the sheaths; rolled in bud.

Reproductive organization. Plants bisexual, with bisexual spikelets; with hermaphrodite florets.

Inflorescence. *Inflorescence* indeterminate; with pseudospikelets; a compound panicle; spatheate; a complex of 'partial inflorescences' and intervening foliar organs. *Spikelet-bearing axes* paniculate. Spikelets not secund; sessile to subsessile.

Female-fertile spikelets. Spikelets 6–8 mm long (*S. subtilis*), or 15 mm long; compressed laterally; disarticulating above the glumes; disarticulating between the florets. Rachilla prolonged beyond the uppermost female-fertile floret. The rachilla extension with incomplete florets.

Glumes two to several (usually 3); very unequal; shorter than the spikelets; shorter than the adjacent lemmas; pointed to not pointed; awnless (but mucronate); similar. Lower glume of *S. hirsuta* 7 nerved. Upper glume of *S. hirsuta* 9 nerved. *Spikelets* with incomplete florets. The incomplete florets distal to the female-fertile florets. *The distal incomplete florets* merely underdeveloped. *Spikelets* seemingly **with proximal incomplete florets**.

Female-fertile florets 3–5. Lemmas similar in texture to the glumes; not becoming indurated; entire; pointed; awnless, or mucronate; hairless; glabrous (or glabrescent); without a germination flap; 5–9 nerved. *Palea* present; relatively long (exceeding the lemma); tightly clasped by the lemma; entire; awnless, without apical setae; several nerved (in *S. subtilis* — two between the keels, two outside on each side); 2-keeled. Palea keels winged; hairy (to ciliolate). *Lodicules* not mentioned in the available descriptions. *Stamens* 6. Anthers 2.3–2.5 mm long; not penicillate; with the connective apically prolonged. **Ovary hairy**; with a conspicuous apical appendage. The appendage broadly conical, fleshy. Styles fused (into one). Stigmas 3.

Fruit, embryo and seedling. Fruit large (to about 6 mm diameter, where known); subglobose; not noticeably compressed. Hilum short. Pericarp fleshy. Embryo large. *Seed 'non-endospermic'*.

Taxonomy. Bambusoideae; Bambusodae; Bambuseae.

Ecology, geography, regional floristic distribution. 3 species. Malesia. Mesophytic.

Paleotropical. Indomalesian. Malesian.

References, etc. Morphological/taxonomic: Dransfield, S. (1989). *Kew Bull.* **44**, 425–434.

Special comments. Anatomical data wanting.

Sphaerocaryum Nees ex Hook.f.

Graya Steud., *Steudelella* Honda

Habit, vegetative morphology. Annual; stoloniferous and decumbent. *Culms* 5–15 cm high; herbaceous. Culm nodes hairy. Culm internodes hollow. *Leaves* not basally aggregated; non-auriculate. *Leaf blades* ovate; broad, or narrow; *3–10 mm wide (small but relatively broad,* Commelina-*like); cordate (amplexicaul)*; obscurely cross veined; persistent. *Ligule a fringe of hairs*.

Reproductive organization. Plants bisexual, with bisexual spikelets; with hermaphrodite florets.

Inflorescence. *Inflorescence* paniculate; open; *with capillary branchlets*; espatheate; not comprising 'partial inflorescences' and foliar organs. Spikelet-bearing axes persistent. Spikelets not secund; pedicellate.

Female-fertile spikelets. **Spikelets** seemingly morphologically 'conventional'; 0.8–1.4 mm long; *not noticeably compressed*; falling with the glumes (but the glumes often deciduous). Rachilla terminated by a female-fertile floret. Hairy callus absent.

Glumes present; assumed to be two (since the inflorescence is not contracted); relatively large; more or less equal; shorter than the adjacent lemmas; not pointed (obtuse); awnless; similar (cymbiform, hyaline). Lower glume 0 nerved. Upper glume 1 nerved. *Spikelets* with female-fertile florets only (i.e., a necessary corollary, assuming there are 2 glumes).

Female-fertile florets 1. Lemmas similar in texture to the glumes (hyaline); not becoming indurated; entire; pointed to blunt; hairy; non-carinate (cymbiform); having the margins lying flat and exposed on the palea; without a germination flap; 1 nerved, or 2 nerved. *Palea* present; entire (oblong,obtuse); awnless, without apical setae; textured like the lemma; not indurated; 2-nerved. *Lodicules* present (minute); 2. *Stamens* 3. Anthers not penicillate. *Ovary* glabrous. Styles fused. Stigmas 2.

Fruit, embryo and seedling. *Fruit* free from both lemma and palea; small (about 0.5 mm long). Hilum short.

Abaxial leaf blade epidermis. *Costal/intercostal zonation* conspicuous. *Papillae* present (many intercostal cells papillate). Intercostal papillae over-arching the stomata; consisting of one oblique swelling per cell to consisting of one symmetrical projection per cell. Intercostal zones without typical long-cells (cf. *Isachne*). Mid-intercostal long-cells having straight or only gently undulating walls. *Microhairs* present; chloridoid-type (the apical cell somewhat pointed, but thick-walled and broader than long); 16.5 microns long; 7.5 microns wide at the septum. Microhair total length/width at septum 2.2. Microhair apical cells 9 microns long. Microhair apical cell/total length ratio 0.55. *Stomata* common. Subsidiaries parallel-sided to triangular (rather irregular forms, owing to bulbously-intruding surrounding cells); including both triangular and parallel-sided forms on the same leaf. Guard-cells overlapping to flush with the interstomatals. *Intercostal short-cells* absent or very rare (simply not detectable in this kind of epidermis). *Costal short-cells* conspicuously in long rows. Costal silica bodies 'panicoid-type' to acutely-angled; acutely angled cross shaped; sharp-pointed (acutely angled crosses).

Transverse section of leaf blade, physiology. C_3; XyMS+. *Mesophyll* with radiate chlorenchyma; *Isachne*-type. *Leaf blade* with distinct, prominent adaxial ribs; with the ribs more or less constant in size. *Midrib* not readily distinguishable; with one bundle only. Bulliforms present in discrete, regular adaxial groups (in the shallow furrows); in simple fans. All the vascular bundles accompanied by sclerenchyma. Combined sclerenchyma girders present; forming 'figures'. Sclerenchyma all associated with vascular bundles.

Cytology. Chromosome base number, $x = 10$. $2n = 20$.

Taxonomy. Panicoideae; Panicodae; Isachneae.

Ecology, geography, regional floristic distribution. 1 species. India to southern China, Formosa, Malay Peninsula, Banka. Helophytic to mesophytic. Paleotropical. Indomalesian. Indian, Indo-Chinese, and Malesian.

References, etc. Leaf anatomical: Metcalfe 1960; this project.

Spheneria Kuhlm

Habit, vegetative morphology. Small, delicate perennial; caespitose. Culms unbranched above. Culm nodes glabrous. Culm internodes solid. *Leaves* mostly basal; non-auriculate. Leaf blades narrow; setaceous (pilose); without cross venation; persistent. **Ligule** present; *an unfringed membrane (jagged, small, with a conspicuous fringe adjoining it on the lamina)*; 0.25 mm long.

Reproductive organization. Plants bisexual, with bisexual spikelets; with hermaphrodite florets.

Inflorescence. *Inflorescence of spicate main branches*; with thin, delicate rachides. Primary inflorescence branches few. Inflorescence espatheate; not comprising 'partial inflorescences' and foliar organs. Spikelet-bearing axes persistent. *Spikelets* solitary; secund; biseriate; *pedicellate (the pedicels appearing bipartite — inside glabrous, outside pilose)*.

Female-fertile spikelets. **Spikelets** unconventional (lacking an organ, assumed here to be the lower glume); *1 mm long*; turbinate; abaxial; compressed dorsiventrally; falling with the glumes (the pedicel splitting lengthways). Rachilla terminated by a female-fertile floret. *Hairy callus present (constituted by the length of 'pedicel' which comes away)*.

Glumes present; *one per spikelet*; the upper, only glume long relative to the adjacent lemmas; dorsiventral to the rachis; hairy (pilose with white hairs, and a subterminal transverse fringe); awnless; non-carinate. Upper glume faintly 5 nerved. *Spikelets* with incomplete florets. The incomplete florets proximal to the female-fertile florets. *The proximal incomplete florets* 1; sterile. *The proximal lemmas membranous, pilose with white hairs and a subterminal fringe, like the glume, but apparently with is a basal, pilose appendage fused to the outside of the pedicel — cf.* **Centrochloa**); awnless; faintly 3 nerved; more or less equalling the female-fertile lemmas; less firm than the female-fertile lemmas; not becoming indurated.

Female-fertile florets 1. **Lemmas** decidedly firmer than the glumes; becoming indurated (thinly); brown in fruit (shiny); entire; pointed, or blunt; *awnless (but abruptly beaked, the beak hooked)*; hairless; non-carinate; having the margins lying flat and exposed on the palea; with a clear germination flap; faintly 3 nerved. *Palea* present; relatively long; 2-nerved (thinly indurated); 2-keeled. *Lodicules* present; 2; free; fleshy; glabrous; not or scarcely vascularized. *Stamens* 3. *Ovary* glabrous. Styles free to their bases. Stigmas 2.

Fruit, embryo and seedling. *Fruit* free from both lemma and palea; small (about 1 mm long); compressed dorsiventrally. Hilum short. Embryo large. Endosperm hard.

Abaxial leaf blade epidermis. *Costal/intercostal zonation* lacking. *Papillae* absent. Mid-intercostal long-cells having markedly sinuous walls. *Microhairs* present; panicoid-type; (42–)45–51(–54) microns long; 5.4–5.7–6.9 microns wide at the septum. Microhair total length/width at septum 6.5–9.5. Microhair apical cells (22–)33–35(–38) microns long. Microhair apical cell/total length ratio 0.53–0.7. *Stomata* absent or very rare. *Intercostal short-cells* common; in cork/silica-cell pairs. Large macrohairs present, with complex cushion bases. *Costal short-cells* predominantly paired (all short-cells in cc/sc pairs). Costal silica bodies rounded (a few), or tall-and-narrow, or 'panicoid-type' (basically this type, commonly 'reduced'); often one-sided cross shaped (often with sharp points); sharp-pointed (often has sharp points on the crosses).

Transverse section of leaf blade, physiology. C_4; XyMS–. *Leaf blade* with distinct, prominent adaxial ribs; with the ribs very irregular in sizes. *Midrib* conspicuous (a larger rib); having a conventional arc of bundles (1 large and 2 small bundles). Bulliforms not in discrete, regular adaxial groups. All the vascular bundles accompanied by sclerenchyma. Combined sclerenchyma girders absent. Sclerenchyma not all bundle-associated. The 'extra' sclerenchyma in a continuous abaxial layer.

Taxonomy. Panicoideae; Panicodae; Paniceae.

Ecology, geography, regional floristic distribution. 1 species. Tropical South America. Savanna.
Neotropical. Amazon.
References, etc. Leaf anatomical: this project.

Sphenopholis Scribner

Colobanthium (Reichenb.) Taylor, *Colobanthus* (Trin.) Spach, *Reboulea* Kunth

Habit, vegetative morphology. Perennial (usually), or annual; caespitose. *Culms* 30–120 cm high; herbaceous. Culm internodes hollow. *Leaves* not basally aggregated; non-auriculate. *Sheath margins free. Leaf blades* narrow; mostly flat; *not pseudopetiolate*; without cross venation; rolled in bud. *Ligule an unfringed membrane*; truncate; 0.5–4 mm long.

Reproductive organization. Plants bisexual, with bisexual spikelets; with hermaphrodite florets.

Inflorescence. *Inflorescence paniculate*; open, or contracted; with capillary branchlets, or without capillary branchlets; espatheate; not comprising 'partial inflorescences' and foliar organs. Spikelet-bearing axes persistent. Spikelets not secund; pedicellate.

Female-fertile spikelets. *Spikelets* 1.5–5 mm long; compressed laterally; *falling with the glumes*; not disarticulating between the florets. *Rachilla prolonged beyond the uppermost female-fertile floret (as a bristle)*. The rachilla extension naked. Hairy callus absent.

Glumes two; very unequal to more or less equal; long relative to the adjacent lemmas; awnless; carinate; *very dissimilar (rather leathery with thin margins, the G_1 narrow-lanceolate and acute, the G_2 broader and oblanceolate or obovate)*. Lower glume 1–3 nerved. Upper glume 3–5 nerved. *Spikelets* with female-fertile florets only, or with incomplete florets. The incomplete florets distal to the female-fertile florets.

Female-fertile florets (1–)2(–3). Lemmas similar in texture to the glumes; not becoming indurated; entire; pointed; awnless, or mucronate, or awned. Awns when present, 1; dorsal; from near the top; geniculate; hairless (scabrid); much shorter than the body of the lemma to about as long as the body of the lemma. Lemmas hairless; non-carinate (dorsally rounded); without a germination flap; 3–5 nerved. *Palea* present; relatively long; gaping; thinner than the lemma; not indurated (hyaline); 2-nerved; 2-keeled. *Lodicules* present; free; membranous; glabrous; toothed, or not toothed; not or scarcely vascularized. *Stamens* 3. Anthers 0.3–0.7 mm long. *Ovary* glabrous. Styles free to their bases. Stigmas 2.

Fruit, embryo and seedling. Embryo small. Endosperm liquid in the mature fruit. Embryo with an epiblast; without a scutellar tail; with a negligible mesocotyl internode. Embryonic leaf margins meeting.

Abaxial leaf blade epidermis. *Costal/intercostal zonation* conspicuous. *Papillae* absent. *Long-cells* markedly different in shape costally and intercostally (costals shorter); of similar wall thickness costally and intercostally. Mid-intercostal long-cells rectangular and fusiform; having straight or only gently undulating walls. *Microhairs* absent. *Stomata* common (in files adjoining the veins); 27–33 microns long. Subsidiaries low dome-shaped, or parallel-sided. Guard-cells overlapped by the interstomatals. *Intercostal short-cells* absent or very rare. Rows of prickles present costally. *Costal short-cells* neither distinctly grouped into long rows nor predominantly paired (the rows interrupted by prickles). Costal silica bodies abundant, horizontally-elongated crenate/sinuous; not sharp-pointed.

Transverse section of leaf blade, physiology. C_3; XyMS+. *Mesophyll* with non-radiate chlorenchyma. *Leaf blade* with distinct, prominent adaxial ribs; with the ribs more or less constant in size. Bulliforms present in discrete, regular adaxial groups; in simple fans. All the vascular bundles accompanied by sclerenchyma. Combined sclerenchyma girders present. Sclerenchyma all associated with vascular bundles.

Cytology. Chromosome base number, $x = 7$. $2n = 14$. 2 ploid.

Taxonomy. Pooideae; Poodae; Aveneae.

Ecology, geography, regional floristic distribution. 5 species. North America, West Indies. Helophytic, or mesophytic, or xerophytic; shade species, or species of open habitats; glycophytic. Woodland, marshes, prairies.

Holarctic. Boreal and Madrean. Atlantic North American and Rocky Mountains. Canadian-Appalachian, Southern Atlantic North American, and Central Grasslands.

Hybrids. Intergeneric hybrids with *Trisetum*.

Rusts and smuts. Rusts — *Puccinia*. Taxonomically wide-ranging species: *Puccinia graminis* and *Puccinia striiformis*.

References, etc. Leaf anatomical: this project.

Sphenopus Trin.

Habit, vegetative morphology. Annual; caespitose. *Culms* (4–)7–30 cm high; herbaceous; 1–3 noded. Culm nodes exposed, or hidden by leaf sheaths; glabrous. Culm internodes hollow. Young shoots intravaginal. *Leaves* not basally aggregated; non-auriculate. Sheaths not keeled, terete. Leaf blades linear; narrow; 0.3–0.6 mm wide (in *S. divaricatus*); flat to folded (to almost filiform); without cross venation; persistent. Ligule an unfringed membrane; not truncate; 0.6–3.4 mm long.

Reproductive organization. Plants bisexual, with bisexual spikelets; with hermaphrodite florets.

Inflorescence. *Inflorescence paniculate (with numerous, very small spikelets)*; open; *with conspicuously divaricate branchlets*; with capillary branchlets to without capillary branchlets; espatheate; not comprising 'partial inflorescences' and foliar organs. Spikelet-bearing axes persistent. Spikelets not secund; long pedicellate (the pedicels claviform).

Female-fertile spikelets. Spikelets 1.5–2.8 mm long; elliptic; compressed laterally; disarticulating above the glumes; disarticulating between the florets. Rachilla prolonged beyond the uppermost female-fertile floret; rachilla hairless (glabrous or minutely aculeolate). The rachilla extension naked. Hairy callus absent. *Callus* short; blunt (glabrous).

Glumes two; minute to relatively large (G_1 0.2–0.4 mm long, G_2 0.5–0.9 mm in *S. divaricatus*); very unequal; shorter than the spikelets; shorter than the adjacent lemmas; not pointed (emarginate, truncate or rounded); awnless; carinate; very dissimilar to similar (hyaline to membranous, rounded to emarginate, the lower smaller). *Lower glume 0 nerved*. Upper glume 1 nerved. *Spikelets* with female-fertile florets only.

Female-fertile florets 2–7. Lemmas similar in texture to the glumes (membranous, with hyaline margins wider above); not becoming indurated; entire; pointed, or blunt; awnless; hairless; glabrous; non-carinate (but keeled on all three veins); 3 nerved; with the nerves non-confluent. *Palea* present; relatively long; tightly clasped by the lemma; entire, or apically notched; thinner than the lemma to textured like the lemma (hyaline); 2-nerved; 2-keeled. Palea keels wingless. *Lodicules* present; 2; free; membranous; glabrous; not toothed. *Stamens* 3. Anthers 0.2–0.6 mm long; not penicillate. *Ovary* glabrous. Styles free to their bases. Stigmas 2; white.

Fruit, embryo and seedling. *Fruit* adhering to lemma and/or palea, or free from both lemma and palea; small (1–1.2 mm long in *S. divaricatus*); oblong or ellipsoid; shallowly ventrally longitudinally grooved; compressed laterally, or not noticeably compressed. Hilum short. Endosperm liquid in the mature fruit, or hard; with lipid. Embryo with an epiblast; without a scutellar tail; with a negligible mesocotyl internode. Embryonic leaf margins meeting.

Abaxial leaf blade epidermis. *Costal/intercostal zonation* conspicuous, or lacking. *Papillae* absent. *Long-cells* similar in shape costally and intercostally; of similar wall thickness costally and intercostally. Mid-intercostal long-cells rectangular; having markedly sinuous walls, or having straight or only gently undulating walls. *Microhairs* absent. *Stomata* absent or very rare. *Intercostal short-cells* common; in cork/silica-cell pairs (usually), or not paired (solitary); silicified (when paired), or not silicified. *Costal short-cells* predominantly paired. Costal silica bodies horizontally-elongated smooth, or tall-and-narrow, or crescentic; not sharp-pointed.

Transverse section of leaf blade, physiology. C_3; XyMS+. *Mesophyll* with non-radiate chlorenchyma; without adaxial palisade. *Leaf blade* with distinct, prominent adaxial ribs; with the ribs more or less constant in size. *Midrib* conspicuous (rarely), or not readily distinguishable; with one bundle only. Bulliforms not in discrete, regular adaxial groups (not apparent in the large-celled epidermis, save for the 'midrib hinges'). Many of the smallest vascular bundles unaccompanied by sclerenchyma. Combined sclerenchyma girders absent. Sclerenchyma all associated with vascular bundles.

Cytology. Chromosome base number, $x = 6$ and 7. $2n = 12$ and 24. 2 and 4 ploid.

Taxonomy. Pooideae; Poodae; Poeae.

Ecology, geography, regional floristic distribution. 2 species. Mediterranean to western Asia. Commonly adventive. Species of open habitats; halophytic. Saline soils and maritime sand.

Holarctic and Paleotropical. Boreal and Tethyan. African. Euro-Siberian. Macaronesian, Mediterranean, and Irano-Turanian. Saharo-Sindian. European.

Rusts and smuts. Smuts from Tilletiaceae. Tilletiaceae — *Tilletia*.

References, etc. Leaf anatomical: this project.

Special comments. Fruit data wanting.

Spinifex L.

Ixalum Forst.

Habit, vegetative morphology. Perennial; rhizomatous and caespitose. *Culms* 50–100 cm high; herbaceous. Culm nodes hairy, or glabrous. Culm internodes solid, or hollow. *Leaves* not basally aggregated; non-auriculate. Leaf blades linear; narrow; hard, woody, needle-like, or not needle-like; without cross venation; persistent; rolled in bud. Ligule a fringe of hairs.

Reproductive organization. Plants dioecious; with hermaphrodite florets, or without hermaphrodite florets. The spikelets all alike in sexuality (on the same plant); female-only, or hermaphrodite (? — the female florets often containing well formed stamens), or male-only. Plants outbreeding.

Inflorescence. *Inflorescence very peculiar — the female-fertile (hermaphrodite) spikelets solitary at the bases of long, bare rachides, which are bristle-like and clustered in dense spatheate umbels: the latter fall entire and are blown about; male spikelets in rigid spikes clustered in spatheate umbels*; the umbel deciduous in its entirety. Inflorescence axes not ending in spikelets (the female-fertile spikelets solitary at the bases of the long, pointed rachides). Inflorescence spatheate. Spikelet-bearing axes persistent. *Spikelets* associated with bractiform involucres (consisting of spathes). The involucres shed with the fertile spikelets. Spikelets subsessile. Pedicel apices cupuliform.

Female-sterile spikelets. *The male spikelets on separate plants, in spatheate umbels of rigid spikes, the spikelets falling from their pedicels; rachilla not prolonged, no hairy callus. Glumes unequal, shorter than the adjacent lemma, convex on the back, awnless, 3–7 nerved. Lemmas 2, similar, awnless, each with a staminate flower. Paleas 2-nerved, about equalling the lemmas. Lodicules 2, often united, glabrous. Stamens 3, ovary rudimentary*. Rachilla of male spikelets terminated by a male floret. The male spikelets with glumes; without proximal incomplete florets; 2 floreted (both fertile). The lemmas awnless. Male florets 2.

Female-fertile spikelets. Spikelets elliptic, or lanceolate, or ovate; compressed dorsiventrally; falling with the glumes (female-fertile spikelets falling with the umbel, tardily disarticulating from their pedicels). Rachilla terminated by a female-fertile floret. Hairy callus absent.

Glumes two; more or less equal; long relative to the adjacent lemmas; hairy, or hairless; awnless; non-carinate; similar (acute, papery, entire). Lower glume 7–11 nerved. Upper glume 7–11 nerved. *Spikelets* with incomplete florets. The incomplete florets proximal to the female-fertile florets. *The proximal incomplete florets* 1; epaleate; sterile. The proximal lemmas awnless; 5 nerved; exceeded by the female-fertile lemmas to decidedly exceeding the female-fertile lemmas; similar in texture to the female-fertile lemmas; not becoming indurated.

Female-fertile florets 1. Lemmas decidedly firmer than the glumes (finally); smooth; becoming indurated to not becoming indurated; yellow in fruit; awnless; hairless; non-carinate; with a clear germination flap; 3–11 nerved. *Palea* present; relatively long; textured like the lemma; indurated; 2-nerved. *Lodicules* present; 2; free; membranous; glabrous; not toothed. *Stamens* 3, or 0 (then 3 staminodes). *Ovary* glabrous. Styles fused. Stigmas 2; white.

Fruit, embryo and seedling. Fruit medium sized; ellipsoid; compressed dorsiventrally. Hilum short. Embryo large; waisted.

Seedling with a long mesocotyl. First seedling leaf with a well-developed lamina. The lamina curved.

Abaxial leaf blade epidermis. *Costal/intercostal zonation* conspicuous (but commonly obscured by the macrohairs). *Papillae* absent. *Long-cells* similar in shape costally and intercostally; of similar wall thickness costally and intercostally. Mid-intercostal long-cells rectangular; having markedly sinuous walls to having straight or only gently undulating walls. *Microhairs* present; panicoid-type; 108–120 microns long. Microhair apical cells 72–80 microns long. Microhair apical cell/total length ratio 0.67. *Stomata* common; 24–30 microns long. Subsidiaries parallel-sided, or dome-shaped, or triangular (basically, triangles with the apices truncated to varying extents). Guard-cells overlapped by the interstomatals (the stomata in deep pits). *Intercostal short-cells* common; in cork/silica-cell pairs, or not paired (then solitary); silicified (rarely), or not silicified. Intercostal silica bodies tall-and-narrow, or cubical. Numerous large, thick-walled macrohairs present. *Costal short-cells* conspicuously in long rows, or predominantly paired, or neither distinctly grouped into long rows nor predominantly paired. Costal silica bodies present and well developed, or poorly developed (or few), or absent; horizontally-elongated smooth (sometimes, a few), or tall-and-narrow (or more or less cuboid, with points), or 'panicoid-type'; sometimes cross shaped, or dumb-bell shaped; sharp-pointed (i.e., the cuboid forms), or not sharp-pointed.

Transverse section of leaf blade, physiology. C_4; biochemical type NADP–ME (1 species); XyMS– (or 'XyMS variable'). PCR sheath outlines uneven. PCR sheath extensions present. Maximum number of extension cells 3–10. PCR cells with a suberised lamella. PCR cell chloroplasts with reduced grana; centrifugal/peripheral. *Mesophyll* with radiate chlorenchyma. *Leaf blade* with distinct, prominent adaxial ribs; with the ribs very irregular in sizes. *Midrib* conspicuous, or not readily distinguishable; with one bundle only, or having a conventional arc of bundles (rarely). Bulliforms present in discrete, regular adaxial groups (the groups sometimes poorly defined); in simple fans. Many of the smallest vascular bundles unaccompanied by sclerenchyma. Combined sclerenchyma girders present; forming 'figures'. Sclerenchyma all associated with vascular bundles.

Culm anatomy. *Culm internode bundles* scattered.

Special diagnostic feature. *Female inflorescence a large, deciduous globular head of sessile, bristle-tipped racemes*.

Cytology. Chromosome base number, $x = 9$. $2n = 18$. 2 ploid.

Taxonomy. Panicoideae; Panicodae; Paniceae.

Ecology, geography, regional floristic distribution. 4 species. Eastern Asia, Indomalayan region, Pacific, Australia. Xerophytic; species of open habitats; halophytic. Binding coastal sand dunes.

Holarctic, Paleotropical, Australian, and Antarctic. Boreal. Indomalesian and Neocaledonian. Eastern Asian. Indian, Indo-Chinese, Malesian, and Papuan. North and East Australian and South-West Australian. New Zealand. Tropical North and East Australian and Temperate and South-Eastern Australian.

Rusts and smuts. Smuts from Tilletiaceae and from Ustilaginaceae. Tilletiaceae — *Tilletia*. Ustilaginaceae — *Ustilago*.

References, etc. Leaf anatomical: Metcalfe 1960; this project.

Spodiopogon Trin.

Excluding *Eccoilopus*

Habit, vegetative morphology. Annual, or perennial. *Culms* 150 cm high; herbaceous. Culm internodes solid. Leaf blades linear to lanceolate; cordate to sagittate, or not cordate, not sagittate; flat; pseudopetiolate to not pseudopetiolate; without cross venation. Ligule present; *an unfringed membrane (bearded at its base)*.

Reproductive organization. Plants bisexual, with bisexual spikelets; with hermaphrodite florets. *The spikelets all alike in sexuality (homogamous)*; overtly heteromorphic (the pedicelled members with imperfect lemma awns, but sometimes with awned glumes).

Inflorescence. *Inflorescence paniculate*; open (much branched); *espatheate*; not comprising 'partial inflorescences' and foliar organs. *Spikelet-bearing axes* 'racemes' (usually short); with very slender rachides; disarticulating; disarticulating at the joints. *'Articles'* non-linear (clavate, with cupular pubescent apices); not appendaged; disarticulating transversely. *Spikelets* in triplets, or paired (all paired, or with a terminal triad); sessile and pedicellate; *consistently in 'long-and-short' combinations*; in pedicellate/sessile combinations. Pedicels of the 'pedicellate' spikelets free of the rachis. The 'shorter' spikelets hermaphrodite. The 'longer' spikelets hermaphrodite.

Female-fertile spikelets. *Spikelets compressed dorsiventrally*; falling with the glumes (the sessile members falling with adjacent joint and pedicel, the pedicellate members falling from their pedicels). Rachilla terminated by a female-fertile floret. Hairy callus present (bearded). *Callus* short.

Glumes two; G_2 larger; long relative to the adjacent lemmas; hairy, or hairless; awned (sometimes shortly so in pedicellate spikelets), or awnless; *very dissimilar (the lower glume chartaceous, pallid, with raised nerves, the upper cymbiform)*. Lower glume not two-keeled (the nerves equally ribbed); convex on the back; not pitted; relatively smooth; 5–9 nerved (and ridged). Upper glume 3–9 nerved. *Spikelets* with incomplete florets. The incomplete florets proximal to the female-fertile florets. *The proximal incomplete florets* 1; male, or sterile. The proximal lemmas awnless; 3 nerved; as long as spikelet, similar to G_2; similar in texture to the female-fertile lemmas (hyaline); not becoming indurated.

Female-fertile florets 1. **Lemmas** less firm than the glumes (hyaline); not becoming indurated; *incised*; 2 lobed; deeply cleft (1/3–3/4); awned (but those of the pedicellate spikelets imperfectly so). Awns 1; median; from a sinus; geniculate. Lemmas non-carinate; without a germination flap; 3 nerved. *Palea* present; relatively long; deeply apically notched; awnless, without apical setae (margins ciliate); not indurated (hyaline). **Lodicules** present; 2; free; *ciliate (on the larger lobe)*; toothed (unequally 2-lobed). *Stamens* 3. *Ovary* glabrous. Stigmas 2.

Fruit, embryo and seedling. Fruit fusiform; compressed laterally. Hilum short. Embryo large.

Seedling with a long mesocotyl. First seedling leaf with a well-developed lamina. The lamina broad; curved; about 41 veined.

Abaxial leaf blade epidermis. *Costal/intercostal zonation* conspicuous. *Papillae* present; costal and intercostal. Intercostal papillae not over-arching the stomata (for the most part); several per cell (large, irregular, somewhat oblique, mostly in a single row per cell). *Long-cells* markedly different in shape costally and intercostally (the costals narrower). Mid-intercostal long-cells rectangular; having markedly sinuous walls. *Microhairs* present; elongated; clearly two-celled; panicoid-type; 27–36 microns long; (4.5–)4.8–5.4(–6) microns wide at the septum. Microhair total length/width at septum 5–6.7. Microhair apical cells 15–21 microns long. Microhair apical cell/total length ratio 0.5–0.58. *Stomata* common; 22.5–30 microns long. Subsidiaries non-papillate; mostly high to low dome-shaped. Guard-cells overlapping to flush with the interstomatals. *Intercostal short-cells* absent or very rare. *Costal short-cells* conspicuously in long rows. Costal silica bodies 'panicoid-type'; consistently nodular (elongated); not sharp-pointed.

Transverse section of leaf blade, physiology. C_4; XyMS–. PCR sheath outlines even. PCR sheath extensions absent. PCR cell chloroplasts seemingly centripetal. *Midrib* very conspicuous; having a conventional arc of bundles (three large bundles, with small ones between and peripheral to them); with colourless mesophyll adaxially. *The lamina* symmetrical on either side of the midrib. Bulliforms present in discrete, regular adaxial groups to not in discrete, regular adaxial groups (epidermis mostly irregularly bulliform); in places in simple fans. Many of the smallest vascular bundles unaccompanied by sclerenchyma. Combined sclerenchyma girders present; forming 'figures' (the primaries and some smaller bundles with conspicuous I's). Sclerenchyma all associated with vascular bundles.

Cytology. $2n = 40$ and 42.

Taxonomy. Panicoideae; Andropogonodae; Andropogoneae; Andropogoninae.

Ecology, geography, regional floristic distribution. 10 species. Mainly temperate Asia, Middle East. Grassy hillsides.

Holarctic and Paleotropical. Boreal and Tethyan. Indomalesian. Euro-Siberian and Eastern Asian. Irano-Turanian. Indian, Indo-Chinese, and Malesian. Siberian.

Rusts and smuts. Rusts — *Puccinia*. Taxonomically wide-ranging species: *Puccinia miyoshiana*. Smuts from Ustilaginaceae. Ustilaginaceae — *Sorosporium, Sphacelotheca,* and *Ustilago*.

References, etc. Leaf anatomical: Metcalfe 1960; this project.

Sporobolus R.Br.

Agrosticula Raddi, *Bauchea* Fourn., *Cryptostachys* Steud., *Diachyrium* Griseb., *Spermachiton* Llanos, *Triachyrum* A. Br.

Habit, vegetative morphology. Annual (rarely), or perennial; rhizomatous, or stoloniferous, or caespitose, or decumbent. **Culms** 5–160(–300) cm high; *herbaceous*. Culm nodes glabrous. Culm internodes solid (usually), or hollow (rarely). Plants with multicellular glands (rarely, e.g. on the pedicels in *S. heterolepis*), or without multicellular glands. Leaves non-auriculate. *Leaf blades* linear; narrow; setaceous, or not setaceous; flat, or folded, or rolled, or acicular (rarely solid-cylindrical); without abaxial multicellular glands; without cross venation; *persistent*; rolled in bud. Ligule a fringed membrane (narrow), or a fringe of hairs. *Contra-ligule* present (of hairs), or absent.

Reproductive organization. *Plants bisexual, with bisexual spikelets*; with hermaphrodite florets; inbreeding. Exposed-cleistogamous, or chasmogamous.

Inflorescence. *Inflorescence a false spike, with spikelets on contracted axes (rarely), or a single raceme (rarely), or paniculate*; open to contracted (rarely with the lowest branches sterile — *S. panicoides*); spicate, or more or less irregular; with capillary branchlets, or without capillary branchlets. Rachides hollowed, or flattened, or winged, or neither flattened nor hollowed, not winged. Inflorescence espatheate; not comprising 'partial inflorescences' and foliar organs. Spikelet-bearing axes persistent. Spikelets secund, or not secund; subsessile, or pedicellate.

Female-fertile spikelets. *Spikelets 0.8–3.5(–6) mm long (i.e. usually small, but rarely to 6 mm)*; often fusiform; *compressed laterally to not noticeably compressed*; disarticulating above the glumes. Rachilla prolonged beyond the uppermost female-fertile floret (rarely, Section *Chaetorachis*), or terminated by a female-fertile floret. The rachilla extension when present, with incomplete florets, or naked. *Hairy callus absent. Callus* absent, or short; (when present) blunt.

Glumes two; very unequal (G_1 often very short), or more or less equal; shorter than the adjacent lemmas, or long relative to the adjacent lemmas; hairless; glabrous; awnless (occasionally mucronate); carinate (slightly), or non-carinate (convex); very dissimilar to similar (persistent or subpersistent, thinly membranous or hyaline, the upper usually resembling the lemma). Lower glume 1 nerved. *Upper glume 1 nerved. Spikelets* with female-fertile florets only, or with incomplete florets (rarely). The incomplete florets when present, distal to the female-fertile florets. *Spikelets without proximal incomplete florets*.

Female-fertile florets *1 (species with two or more florets being rather arbitrarily excluded* — cf. **Eragrostis, Thellungia**). *Lemmas* usually similar to G_2; usually thinly membranous, rarely papery; not becoming indurated; entire; pointed, or blunt; *awnless* (**S. molleri** *subulate-tipped*); hairless; *glabrous (usually shiny, often olive or grey)*; carinate to non-carinate; *1(–3) nerved*. **Palea** present; *relatively long*; entire (seemingly in all species, when young), or apically notched to deeply bifid (often splitting as the grain develops); textured like the lemma (delicate); not indurated; 2-nerved; 2-keeled (but the back often induplicate, bringing them into contiguity). *Lodicules* present, or absent; when present, 2; free; fleshy; glabrous. *Stamens* (1–)2–3. Anthers 0.2–2.5 mm long; not penicillate. *Ovary* glabrous. Styles free to their bases. Stigmas 2; white, or brown.

Fruit, embryo and seedling. *Fruit* free from both lemma and palea; small (0.3–2 mm long); compressed laterally, or compressed dorsiventrally, or not noticeably compressed (variously globular or compressed). Hilum short. **Pericarp free (commonly swelling and mucilaginous when wet, forcibly ejecting the seed).** Embryo large; not waisted. Endosperm hard; without lipid; containing compound starch grains. Embryo with an epiblast; with a scutellar tail; with an elongated mesocotyl internode. Embryonic leaf margins meeting.

Seedling with a long mesocotyl. First seedling leaf with a well-developed lamina. The lamina broad (rarely), or narrow; erect; 3–5 veined.

Abaxial leaf blade epidermis. *Costal/intercostal zonation* conspicuous. *Papillae* absent. *Long-cells* similar in shape costally and intercostally; of similar wall thickness costally and intercostally. Mid-intercostal long-cells rectangular; having markedly sinuous walls. *Microhairs* present; usually more or less spherical; ostensibly one-celled (usually), or clearly two-celled (e.g. *S. wrightii*); chloridoid-type. Microhair apical cell wall thinner than that of the basal cell but not tending to collapse. Microhairs with 'partitioning membranes'. The 'partitioning membranes' in the basal cell. Microhairs 9–20 microns long, or 30–37 microns long (*S. wrightii*). Microhair basal cells 24 microns long. Microhair total length/width at septum 2. Microhair apical cell/total length ratio 0.4–0.5. *Stomata* common; 24–30 microns long. Subsidiaries low dome-shaped, or triangular. Guard-cells overlapping to flush with the interstomatals. *Intercostal short-cells* common; in cork/silica-cell pairs and not paired; silicified (when paired), or not silicified. Intercostal silica bodies present and perfectly developed; crescentic, or rounded, or saddle shaped, or tall-and-narrow. *Costal short-cells* conspicuously in long rows, or predominantly paired. Costal silica bodies present in alternate cell files of the costal zones; rounded, or saddle shaped, or tall-and-narrow, or crescentic; not sharp-pointed.

Transverse section of leaf blade, physiology. C_4; biochemical type PCK (6 species), or NAD–ME (4 species); XyMS+. PCR sheath outlines uneven, or even. PCR sheaths of the primary vascular bundles interrupted; interrupted abaxially only. PCR sheath extensions present, or absent. Maximum number of extension cells when present, 2–5. PCR cells with a suberised lamella. *PCR cell chloroplasts* ovoid, or elongated; with well developed grana; centrifugal/peripheral, or centripetal. *Mesophyll* with radiate chlorenchyma. *Leaf blade* with distinct, prominent adaxial ribs; with the ribs more or less constant in size. *Midrib* conspicuous, or not readily distinguishable; with one bundle only, or having a conventional arc of bundles; with colourless mesophyll adaxially (usually), or without colourless mesophyll adaxially. Bulliforms present in discrete, regular adaxial groups; associated with colourless mesophyll cells to form deeply-penetrating fans (usually), or in simple fans and associated with colourless mesophyll cells to form deeply-penetrating fans (sometimes), or combining with colourless mesophyll cells to form narrow groups penetrating into the mesophyll (e.g. *S. wrightii*). All the vascular bundles accompanied by sclerenchyma. Combined sclerenchyma girders present; forming 'figures'. Sclerenchyma all associated with vascular bundles. The lamina margins with fibres.

Culm anatomy. *Culm internode bundles* in one or two rings, or in three or more rings, or scattered.

Phytochemistry. Tissues of the culm bases with abundant starch. Leaves without flavonoid sulphates (4 species). Leaf blade chlorophyll *a:b* ratio 3.28–3.69 (PCK), or 3.75–3.89 (NAD-ME).

Cytology. Chromosome base number, $x = 9$ and 10. $2n = 18$, 24, 36, 38, 54, 72, 80, 88, 90, 108, and 126. 2, 4, 6, 8, 9, 10, 12, and 13 ploid. Nucleoli persistent.

Taxonomy. Chloridoideae; main chloridoid assemblage.

Ecology, geography, regional floristic distribution. About 160 species. Tropical and warm temperate. Commonly adventive. Mesophytic, or xerophytic; halophytic, or glycophytic. In diverse habitats, including coastal sand dunes.

Holarctic, Paleotropical, Neotropical, Cape, Australian, and Antarctic. Boreal, Tethyan, and Madrean. African, Madagascan, Indomalesian, Polynesian, and Neocaledonian. Euro-Siberian, Eastern Asian, Atlantic North American, and Rocky Mountains. Macaronesian, Mediterranean, and Irano-Turanian. Saharo-Sindian, Sudano-Angolan, West African Rainforest, Namib-Karoo, and Ascension and St. Helena. Indian, Indo-Chinese, Malesian, and Papuan. Fijian. Caribbean, Venezuela and Surinam, Amazon, Central Brazilian, Pampas, and Andean. North and East Australian, South-West Australian, and Central Australian. New Zealand. European. Canadian-Appalachian, Southern Atlantic North American, and Central Grasslands. Sahelo-Sudanian, Somalo-Ethiopian, South Tropical African, and Kalaharian. Tropical North and East Australian and Temperate and South-Eastern Australian.

Rusts and smuts. Rusts — *Puccinia*. Taxonomically wide-ranging species: *Puccinia schedonnardi*. Smuts from Tilletiaceae and from Ustilaginaceae. Tilletiaceae — *Entyloma*, *Melanotaenium*, and *Tilletia*. Ustilaginaceae — *Sorosporium*, *Sphacelotheca*, *Tolyposporella*, and *Ustilago*.

Economic importance. Significant weed species: *S. airoides*, *S. africanus*, *S. cryptandrus*, *S. diander*, *S. elongatus*, *S. fertilis*, *S. indicus*, *S. neglectus*, *S. poiretii*, *S. pyramidalis*, *S. tremulus*, *S. vaginiflorus*, *S. virginicus*, etc. Important native pasture species: *S. airoides*, *S. cryptandrus*, *S. elongatus*, *S. helvolus*, *S. interruptus*, *S. ioclados*, *S. pyramidalis* (in arid places), *S. wrightii*.

References, etc. Leaf anatomical: Metcalfe 1960; this project.

Special comments. Fairly arbitrarily but mostly readily separable from *Eragrostis*; see also *Thellungia*.

Steinchisma Raf.

~ *Panicum* (Subgenus *Steinchisma*, including (e.g.) *Panicum hians* = *P. milioides*

Habit, vegetative morphology. Perennial. *Culms herbaceous.* Leaf blades without cross venation.

Reproductive organization. *Plants bisexual, with bisexual spikelets*; with hermaphrodite florets.

Inflorescence. *Inflorescence paniculate*; more or less contracted; espatheate; not comprising 'partial inflorescences' and foliar organs. Spikelet-bearing axes persistent. Spikelets not secund; pedicellate. Pedicel apices cupuliform.

Female-fertile spikelets. *Spikelets compressed dorsiventrally*; falling with the glumes; with conventional internode spacings. The upper floret not stipitate. Rachilla terminated by a female-fertile floret. Hairy callus absent.

Glumes two; *very unequal (the lower short)*; (the upper) long relative to the adjacent lemmas (almost as long as the spikelet); without conspicuous tufts or rows of hairs; awnless; non-carinate. Lower glume shorter than the lowest lemma. *Spikelets* with incomplete florets. The incomplete florets proximal to the female-fertile florets. *Spikelets with proximal incomplete florets. The proximal incomplete florets* 1; paleate. *Palea of the proximal incomplete florets* fully developed; *becoming conspicuously hardened and enlarged laterally. The proximal lemmas* awnless; *3 nerved; less firm than the female-fertile lemmas.*

Female-fertile florets 1. *Lemmas* not saccate; decidedly firmer than the glumes; smooth; *becoming indurated*; entire; awnless; non-carinate; having the margins tucked in onto the palea; with a clear germination flap. *Palea* present; awnless, without apical setae. *Ovary* glabrous. Stigmas 2.

Fruit, embryo and seedling. Hilum short. Embryo large.

Abaxial leaf blade epidermis. *Costal/intercostal zonation* conspicuous. *Papillae* absent. Long-cells differing markedly in wall thickness costally and intercostally (the costals thinner walled). Mid-intercostal long-cells rectangular; having markedly sinuous walls. *Microhairs* present; elongated; clearly two-celled; panicoid-type; of *S. hians* 54–57 microns long; 5.4–6 microns wide at the septum. Microhair total length/width at septum 9–10. Microhair apical cells 27–34.5 microns long. Microhair apical cell/total length ratio 0.5–0.65. *Stomata* common; in *S. hians* 21–24 microns long. Subsidiaries mostly high dome-shaped, or triangular. Guard-cells overlapping to flush with the interstomatals. *Intercostal short-cells* common; in cork/silica-cell pairs and not paired (mostly ostensibly solitary, but a few obvious pairs); not silicified (in the material of *S. hians* seen). Small intercostal prickles abundant. *Crown cells* absent. *Costal short-cells* conspicuously in long rows. Costal silica bodies abundant, 'panicoid-type'; cross shaped, butterfly shaped, and dumb-bell shaped (short); not sharp-pointed.

Transverse section of leaf blade, physiology. C_3 to C_4 (intermediate in all three species tested, with C_3-like mesophyll layout). The anatomical organization unconventional (in that there are some organelles in the outer sheath). XyMS+. PCR cell chloroplasts centripetal. *Mesophyll* with non-radiate chlorenchyma; without adaxial palisade. *Leaf blade* with distinct, prominent adaxial ribs to 'nodular' in section; with the ribs more or less constant in size (round topped). *Midrib* not readily distinguishable; with one bundle only. *The lamina* symmetrical on either side of the midrib. Bulliforms present in discrete, regular adaxial groups (in all the furrows); in simple fans. All the vascular bundles accompanied by sclerenchyma. Combined sclerenchyma girders present; forming 'figures' (all the bundles with I's or 'anchors'). Sclerenchyma all associated with vascular bundles.

Phytochemistry. Leaves containing flavonoid sulphates, or without flavonoid sulphates.

Special diagnostic feature. *Plants not as in* **Dichanthelium** *(q.v.)*.

Taxonomy. Panicoideae; Panicodae; Paniceae.

Ecology, geography, regional floristic distribution. 4 species. Southern U.S.A. to Argentina. Commonly adventive. Mesophytic. Damp grassland.

Holarctic and Neotropical. Boreal and Madrean. Caribbean, Venezuela and Surinam, Amazon, Central Brazilian, and Pampas.

Economic importance. Significant weed species: *S. hians*.

References, etc. Leaf anatomical: this project.

Steirachne Ekman

Habit, vegetative morphology. Perennial; caespitose. *Culms* 120 cm high; herbaceous; unbranched above. Culm nodes glabrous. Young shoots intravaginal. Leaves non-auriculate. Leaf blades linear (acuminate, setaceous-tipped); narrow; 0.5–2 mm wide; without abaxial multicellular glands; without cross venation; persistent. Ligule a fringed membrane, or a fringe of hairs. *Contra-ligule* absent.

Reproductive organization. Plants bisexual, with bisexual spikelets; with hermaphrodite florets. The spikelets all alike in sexuality. *Plants with hidden cleistogenes*. The hidden cleistogenes in the leaf sheaths.

Inflorescence. Inflorescence paniculate; open (compound, the primary branches spiralled); with capillary branchlets; espatheate; not comprising 'partial inflorescences' and foliar organs. Spikelet-bearing axes persistent. Spikelets not secund; pedicellate.

Female-fertile spikelets. *Spikelets* 7–11 mm long; compressed laterally; disarticulating above the glumes; disarticulating between the florets; *with distinctly elongated rachilla internodes between the florets*. Rachilla prolonged beyond the uppermost female-fertile floret; rachilla hairy. The rachilla extension with incomplete florets. Hairy callus present. *Callus* short.

Glumes present; two; very unequal; shorter than the spikelets; shorter than the adjacent lemmas; hairless; pointed (acute); awnless; carinate; similar (chaffy). Lower glume 1 nerved. Upper glume 1 nerved. *Spikelets* with incomplete florets. The incomplete florets distal to the female-fertile florets. *The distal incomplete florets* several; **incomplete florets** *merely underdeveloped (some larger than the perfect florets, but not conspicuously different in shape, by contrast with* Ectrosia*)*.

Female-fertile florets 7–8. Lemmas lanceolate-ovate; similar in texture to the glumes (membranous); not becoming indurated; entire; pointed (very acute); mucronate to awned (acuminate-aristate). Awns 1; median; apical; non-geniculate; hairless; much shorter than the body of the lemma. *Lemmas* hairless; glabrous; *carinate (save at the broadly rounded base)*; without a germination flap; 3 nerved, or 5 nerved (the laterals not excurrent). *Palea* present; relatively long; apically notched; awnless, without apical setae, or with apical setae (via slightly excurrent nerves); not indurated (membranous); 2-nerved; 2-keeled. Palea keels winged (sulcate and 2-winged); serrate-scabrid. *Lodicules* present; 2; free (minute); fleshy; glabrous. *Stamens* 2; with free filaments (short). Anthers 0.8 mm long; not penicillate (greyish); without an apically prolonged connective. *Ovary* glabrous. Styles free to their bases. Stigmas 2.

Fruit, embryo and seedling. Fruit ellipsoid. Hilum short. Pericarp fused. Embryo large; with an epiblast; with a scutellar tail; with an elongated mesocotyl internode. Embryonic leaf margins meeting.

Abaxial leaf blade epidermis. *Costal/intercostal zonation* conspicuous. *Papillae* absent. *Long-cells* similar in shape costally and intercostally; of similar wall thickness costally and intercostally (rather thick walled). Mid-intercostal long-cells rectangular; having markedly sinuous walls (and pitted). *Microhairs* present; elongated; clearly two-celled; panicoid-type (but distal cells quite broad). Microhair apical cell wall thinner than that of the basal cell and often collapsed. Microhairs 36–39 microns long. Microhair basal cells 15 microns long. Microhairs 8.4–9.3 microns wide at the septum. Microhair total length/width at septum 3.9–4.6. Microhair apical cells 19.5–24 microns long. Microhair apical cell/total length ratio 0.54–0.67. *Stomata* common; 24–30 microns long. Subsidiaries low dome-shaped. Guard-cells overlapping to flush with the interstomatals. *Intercostal short-cells* common; not paired (solitary); not silicified. Intercostal silica bodies absent. *Costal short-cells* predominantly paired. Costal silica bodies present throughout the costal zones; saddle shaped, tall-and-narrow, crescentic, and oryzoid (but predominantly broad to narrow saddles); not sharp-pointed.

Transverse section of leaf blade, physiology. C_4; XyMS+. PCR sheath outlines uneven. PCR sheaths of the primary vascular bundles interrupted; interrupted abaxially only. PCR sheath extensions present. Maximum number of extension cells 3–4. *Leaf blade* with distinct, prominent adaxial ribs (round-topped). *Midrib* not readily distinguishable (a somewhat larger bundle and rib); with one bundle only. Bulliforms present in discrete, regular adaxial groups (in every adaxial groove); in simple fans. All the vascular bundles accompanied by sclerenchyma. Combined sclerenchyma girders present (with all the bundles); forming 'figures' (nearly all the bundles). Sclerenchyma all associated with vascular bundles. The lamina margins with fibres.

Taxonomy. Chloridoideae; main chloridoid assemblage.

Ecology, geography, regional floristic distribution. 2 species. Brazil. Species of open habitats.

Neotropical. Amazon and Central Brazilian.

References, etc. Morphological/taxonomic: Ekman 1911. Leaf anatomical: this project.

Stenotaphrum Trin.

Diastemenanthe Steud., *Ophiurinella* Desv.

Habit, vegetative morphology. Annual, or perennial; rhizomatous, or stoloniferous, or caespitose. *Culms* 10–60 cm high; herbaceous; branched above. Culm nodes glabrous. *Leaves* not basally aggregated; non-auriculate. The sheaths compressed. *Leaf blades lanceolate to elliptic*; broad, or narrow; flat, or folded (when young); pseudopetiolate, or not pseudopetiolate; without cross venation; disarticulating from the sheaths, or persistent (rarely); once-folded in bud. Ligule present; *a fringed membrane*.

Reproductive organization. *Plants bisexual, with bisexual spikelets*; with hermaphrodite florets. The spikelets all alike in sexuality.

Inflorescence. *Inflorescence of spicate main branches, or a false spike, with spikelets on contracted axes (the spikelets 1 to several, in very short spike-like racemes embedded in hollows of the corky or foliaceous common axis, or in longer racemes closely appressed to it). Inflorescence axes not ending in spikelets (minutely baretipped, when not coalesced with the main axis).* Rachides hollowed. Inflorescence spatheate (the small racemes subtended/enclosed by spathes which are laterally adnate to the rachis), or espatheate; a complex of 'partial inflorescences' and intervening foliar organs, or not comprising 'partial inflorescences' and foliar organs. *Spikelet-bearing axes very much reduced (sometimes reduced to one spikelet with no rachis extension, or coalesced with the main axis); disarticulating; falling entire (the free racemes falling with the joint* of the main axis), or disarticulating at the joints (when the 'spikelet bearing unit' consists of a coalesced main axis and branches). **Spikelets *unaccompanied by bractiform involucres, not associated with setiform vestigial branches*; *secund*.**

Female-fertile spikelets. Spikelets elliptic, or lanceolate, or ovate; abaxial; compressed dorsiventrally; falling with the glumes. Rachilla terminated by a female-fertile floret. Hairy callus absent.

Glumes two; very unequal; (the longer) long relative to the adjacent lemmas, or shorter than the adjacent lemmas; dorsiventral to the rachis; awnless; very dissimilar (lower minute, scale-like, upper large, substantial), or similar (both small, scale-like). *Lower glume 0 nerved*. Upper glume 5–9 nerved. *Spikelets* with incomplete florets. The incomplete florets proximal to the female-fertile florets. *The proximal incomplete florets* 1; paleate, or epaleate. Palea of the proximal incomplete florets when present, fully developed. The proximal incomplete florets male, or sterile (rarely). The proximal lemmas awnless; 7–9 nerved; similar in texture to the female-fertile lemmas, or decidedly firmer than the female-fertile lemmas (leathery or papery); not becoming indurated.

Female-fertile florets 1. Lemmas decidedly firmer than the glumes (papery to subleathery); smooth to striate; not becoming indurated; yellow in fruit; entire; pointed; awnless; hairless; non-carinate; having the margins lying flat and exposed on the palea; with a clear germination flap; 3–5 nerved. *Palea* present; relatively long; entire (pointed); awnless, without apical setae; textured like the lemma; not indurated; 2-nerved. *Lodicules* present; 2; fleshy. *Stamens* 3. Anthers not penicillate. *Ovary* glabrous. Styles fused. Stigmas 2; white, or red pigmented.

Fruit, embryo and seedling. Disseminule comprising the rachis segment and associated structures, or consisting of the disarticulated spikelet-bearing inflorescence unit, or constituted by the complete, deciduous inflorescence. Fruit small; ellipsoid; compressed dorsiventrally. Hilum short. Embryo large; not waisted.

Abaxial leaf blade epidermis. *Costal/intercostal zonation* conspicuous. *Papillae* absent. *Long-cells* similar in shape costally and intercostally; of similar wall thickness costally and intercostally. Mid-intercostal long-cells rectangular; having markedly sinuous walls. *Microhairs* present; panicoid-type; (45–)54–78(–84) microns long; 9–12(–13.5) microns wide at the septum. Microhair total length/width at septum 6.2–7.7. Microhair apical cells 32–62 microns long. Microhair apical cell/total length ratio 0.61–0.76. *Stomata* common; (27–)28.5–30(–33) microns long. Subsidiaries low dome-shaped, or triangular, or dome-shaped and triangular. Guard-cells overlapping to flush with the interstomatals. *Intercostal short-cells* common; in cork/silica-cell pairs and not paired (solitary); silicified, or not silicified. Intercostal silica bodies when present, tall-and-narrow. *Costal short-cells* conspicuously in long rows. Costal silica bodies 'panicoid-type'; cross shaped to dumb-bell shaped, or nodular; not sharp-pointed.

Transverse section of leaf blade, physiology. C_4; XyMS–. PCR sheath outlines uneven. PCR sheath extensions absent. PCR cell chloroplasts with reduced grana; centrifugal/peripheral. *Mesophyll* with radiate chlorenchyma. *Leaf blade* adaxially flat; with the ribs more or less constant in size. *Midrib* conspicuous; having a conventional arc of bundles; without colourless mesophyll adaxially (but with colourless cells *abaxially* instead, here and sometimes with other main bundles). Bulliforms not in discrete, regular adaxial groups (absent or irregularly disposed, apart from a group over the midrib). Many of the smallest vascular bundles unaccompanied by sclerenchyma. Combined sclerenchyma girders absent. Sclerenchyma all associated with vascular bundles.

Cytology. Chromosome base number, $x = 9$. $2n = 18$, 20, and 36. 2 and 4 ploid.

Taxonomy. Panicoideae; Panicodae; Paniceae.

Ecology, geography, regional floristic distribution. 7 species. Tropical and subtropical. Commonly adventive. Mesophytic; species of open habitats; halophytic, or glycophytic. Usually in sandy soils near the coast, sometimes inland.

Holarctic, Paleotropical, Neotropical, Cape, and Australian. Boreal. African, Madagascan, Indomalesian, Polynesian, and Neocaledonian. Atlantic North American. Sudano-Angolan. Indian, Indo-Chinese, Malesian, and Papuan. Fijian. Caribbean, Central Brazilian, Pampas, and Andean. North and East Australian. Southern Atlantic North American and Central Grasslands. Sahelo-Sudanian, Somalo-Ethiopian, and South Tropical African. Tropical North and East Australian.

Rusts and smuts. Rusts — *Puccinia*. Taxonomically wide-ranging species: *Puccinia stenotaphri* and '*Uromyces*' *setariae-italicae*. Smuts from Ustilaginaceae. Ustilaginaceae — *Sorosporium, Sphacelotheca*, and *Ustilago*.

Economic importance. Significant weed species: *S. dimidiatum, S. secundatum*. Important native pasture species: *S. dimidiatum, S. secundatum*. Lawns and/or playing fields: *S. secundatum* (in warm coastal regions).

References, etc. Morphological/taxonomic: Saur 1972. Leaf anatomical: Metcalfe 1960; this project.

Stephanachne Keng

Pappagrostis Roshev.

Habit, vegetative morphology. Perennial; caespitose. Culms herbaceous. Leaves non-auriculate. Leaf blades without cross venation. *Ligule an unfringed membrane*; not truncate; 1–2 mm long.

Reproductive organization. Plants bisexual, with bisexual spikelets; with hermaphrodite florets. The spikelets all alike in sexuality.

Inflorescence. Inflorescence paniculate; contracted (to spiciform); espatheate; not comprising 'partial inflorescences' and foliar organs. Spikelet-bearing axes persistent. Spikelets not secund; pedicellate.

Female-fertile spikelets. *Spikelets 4.5–7 mm long*; *not noticeably compressed*; disarticulating above the glumes. *Rachilla prolonged beyond the uppermost female-fertile floret*. The rachilla extension naked. Hairy callus present.

Glumes two; more or less equal; about equalling the spikelets; long relative to the adjacent lemmas; pointed (acuminate); awnless; carinate, or non-carinate; similar (membranous or herbaceous). Lower glume 1–3 nerved. Upper glume 3 nerved. *Spikelets with female-fertile florets only*.

Female-fertile florets 1. Lemmas decidedly firmer than the glumes (thinly leathery); not becoming indurated; incised; 2 lobed; deeply cleft to not deeply cleft (notched, or cleft to a third of its length); awned. Awns 1, or 3 (by extension from the acuminate lobes); median, or median and median and lateral; the median different in form from the laterals; from a sinus; geniculate. Lemmas hairy. *The hairs in tufts; in transverse rows (with a long transverse tuft at the base of each glabrous lobe)*. Lemmas non-carinate; without a germination flap; 5 nerved. **Palea** present; relatively long; 2-nerved; *keel-less*. Palea back hairy. *Lodicules* present; 2 (*Pappagrostis*), or 3; free; membranous; glabrous; not toothed; not or scarcely vascularized. **Stamens 3**. *Ovary* glabrous. Styles free to their bases. Stigmas 2.

Fruit, embryo and seedling. Fruit small (about 3 mm long). *Hilum long-linear*. Embryo small.

Abaxial leaf blade epidermis. *Costal/intercostal zonation* conspicuous. *Papillae* absent. Long-cells differing markedly in wall thickness costally and intercostally (costals thicker-walled). Mid-intercostal long-cells rectangular; having markedly sinuous walls. *Microhairs* absent. *Stomata* common; 27–30 microns long. Subsidiaries parallel-sided. Guard-cells overlapped by the interstomatals. *Intercostal short-cells* fairly common; not paired (mostly solitary); not silicified. *Costal short-cells* neither distinctly grouped into long rows nor predominantly paired. Costal silica bodies absent, or poorly developed (in the material seen).

Transverse section of leaf blade, physiology. C_3; XyMS+. Mesophyll without adaxial palisade. *Leaf blade* with distinct, prominent adaxial ribs to 'nodular' in section; with the ribs very irregular in sizes. *Midrib* not readily distinguishable; with one bundle only. Bulliforms present in discrete, regular adaxial groups; in simple fans. All the vascular bundles accompanied by sclerenchyma. Combined sclerenchyma girders present (with the primaries); forming 'figures'. Sclerenchyma all associated with vascular bundles.

Cytology. Chromosome base number, $x = 12$ (or 6). $2n = 24$.

Taxonomy. Pooideae; Poodae; Aveneae.

Ecology, geography, regional floristic distribution. 2 species. China, eastern Russia.

Holarctic. Boreal. Eastern Asian.
References, etc. Leaf anatomical: this project.

Stereochlaena Hackel

Chloridion Stapf
Excluding *Baptorhachis*
Habit, vegetative morphology. Annual, or perennial; stoloniferous, or caespitose. *Culms* 60–150 cm high; herbaceous; branched above, or unbranched above. Culm nodes glabrous. Culm internodes hollow. Young shoots intravaginal. *Leaves* not basally aggregated; non-auriculate. Leaf blades linear to linear-lanceolate; narrow; 2–8(–11) mm wide; flat; without cross venation; persistent. *Ligule* present; a fringed membrane; 0.2–0.7 mm long.

Reproductive organization. Plants bisexual, with bisexual spikelets; with hermaphrodite florets.

Inflorescence. *Inflorescence* of spicate main branches (slender spike-like racemes); paired or *digitate*. Primary inflorescence branches 2–8. Rachides winged (with narrow, fringed wings). Inflorescence espatheate; not comprising 'partial inflorescences' and foliar organs. *Spikelet-bearing axes* 'racemes'. The racemes spikelet bearing to the base, or without spikelets towards the base. *Spikelet-bearing axes persistent*. Spikelets paired; secund; biseriate; shortly pedicellate, or subsessile; consistently in 'long-and-short' combinations (but homogamous). The 'shorter' spikelets hermaphrodite. The 'longer' spikelets hermaphrodite.

Female-fertile spikelets. *Spikelets* 2–4.5 mm long; abaxial; *compressed dorsiventrally*; flattened dorsally and ventrally; falling with the glumes; with conventional internode spacings. Rachilla terminated by a female-fertile floret. Hairy callus absent. *Callus absent*.

Glumes one per spikelet, or two; minute, or relatively large (the G_1 minute or absent, the G_2 minute to almost as long as the spikelet); when both present, very unequal; (the upper) shorter than the spikelets to about equalling the spikelets; shorter than the adjacent lemmas to long relative to the adjacent lemmas; dorsiventral to the rachis; hairless; awned (G_2 only, sometimes), or awnless; non-carinate; when both present, very dissimilar. Lower glume when present, 0 nerved. Upper glume 1–3 nerved (?). *Spikelets* with incomplete florets. The incomplete florets proximal to the female-fertile florets. *The proximal incomplete florets* 1; paleate, or epaleate (?). Palea of the proximal incomplete florets when present, reduced. The proximal incomplete florets sterile. The proximal lemmas awned (the terminal awn from 3–30 mm long); 5 nerved, or 7 nerved (in *S. tridentata*, the lateral keels form a tooth on either side of the awn); more or less equalling the female-fertile lemmas (excluding the awn); less firm than the female-fertile lemmas (membranous, scabrid, 2-keeled or not); not becoming indurated.

Female-fertile florets 1. Lemmas decidedly firmer than the glumes (papery); not becoming indurated; brown in fruit; entire; pointed; awnless (sometimes apiculate); hairless; having the margins lying flat and exposed on the palea; with a clear germination flap; faintly 3 nerved. *Palea* present; relatively long; 2-nerved. **Lodicules** *absent*. **Stamens** 3. Anthers about 2 mm long; not penicillate; without an apically prolonged connective. *Ovary* glabrous. Styles fused. Stigmas 2; red pigmented.

Fruit, embryo and seedling. Fruit small (about 1.7 mm long); elongate ellipsoid; compressed dorsiventrally. Hilum short (punctiform). Embryo large (about 1/3 the fruit length).

Abaxial leaf blade epidermis. *Costal/intercostal zonation* conspicuous. *Papillae* absent. Intercostal zones with typical long-cells and exhibiting many atypical long-cells (varying from place to place). Mid-intercostal long-cells rectangular (elongated to almost square); having markedly sinuous walls. *Microhairs* present; panicoid-type; 78–84 microns long; 4.5–6 microns wide at the septum. Microhair total length/width at septum 13–18.7. Microhair apical cells 51–52.5 microns long. Microhair apical cell/total length ratio 0.61–0.67. *Stomata* common; 30–32(–33) microns long. Subsidiaries parallel-sided (in some parts of the blade), or dome-shaped, or triangular; including both triangular and parallel-sided forms on the same leaf. Guard-cells overlapping to flush with the interstomatals. *Intercostal short-cells* absent or very rare (the apparent examples being prickle bases). Sometimes exhibiting macrohairs with complex cushion bases. *Costal short-cells* conspicuously in long rows. Costal silica bodies 'panicoid-type'; nearly all cross shaped; not sharp-pointed.

Transverse section of leaf blade, physiology. C_4; XyMS–. PCR sheath outlines uneven. PCR cell chloroplasts centrifugal/peripheral. *Mesophyll* with radiate chlorenchyma; traversed by columns of colourless mesophyll cells (apparently, in places). *Leaf blade* adaxially flat. *Midrib* conspicuous; having a conventional arc of bundles (one large and up to 8 small bundles); with colourless mesophyll adaxially. Bulliforms not in discrete, regular adaxial groups (bulliform cells constituting most of the epidermis). Many of the smallest vascular bundles unaccompanied by sclerenchyma. Combined sclerenchyma girders absent. Sclerenchyma all associated with vascular bundles.

Taxonomy. Panicoideae; Panicodae; Paniceae.

Ecology, geography, regional floristic distribution. 5 species. Tropical east Africa. Mesophytic; species of open habitats; glycophytic. Savanna grasslands.

Paleotropical and Cape. African. Sudano-Angolan and West African Rainforest. Sahelo-Sudanian, Somalo-Ethiopian, South Tropical African, and Kalaharian.

Rusts and smuts. Rusts — *Puccinia*. Taxonomically wide-ranging species: *Puccinia stenotaphri*. Smuts from Ustilaginaceae. Ustilaginaceae — *Ustilago*.

References, etc. Morphological/taxonomic: Stapf 1900b; Clayton 1978b. Leaf anatomical: this project.

Steyermarkochloa Davidse and Ellis

Habit, vegetative morphology. Perennial; caespitose (in dense clumps). *The flowering culms leafless (the culms dimorphic. Vegetative, with internodes not elongated, erect, with bladeless sheaths at the base, the uppermost of these to 20 cm long and clasping the solitary, terminal blade-bearing leaf. Reproductive culms with the internodes short below and much elongated above, all with bladeless sheaths).* Culms 50–350 cm high; herbaceous; unbranched above. Culm sheaths persistent. Culm internodes of the fertile culms hollow. Plants unarmed. *Leaves not clearly differentiated into sheath and blade (in that the single developed leaf has no ligule, and the long, terete petiolar region gradually opens into the flattened laminar part)*; non-auriculate. Sheath margins joined, or free (depending on interpretation of the highly peculiar 'sheath' of the solitary leaf, which is solid, cylindrical and stemlike, with two concentric rings of vascular bundles and numerous lacunae). Leaf blades linear; narrow; 2.8–6.5 mm wide; flattened, with plano-convex laminae flanking a narrow midrib, the distal 1–2 cm of the blades fused into a blunt, navicular tip; pseudopetiolate (or petiolate!); without cross venation; persistent. *Ligule* absent (i.e., the single complete leaf eligulate).

Reproductive organization. Plants bisexual, with bisexual spikelets; with hermaphrodite florets. The spikelets of sexually distinct forms on the same plant; hermaphrodite, female-only, and male-only (male or bisexual at the base of the inflorescence, bisexual in the middle, female towards the tip). The male and female-fertile spikelets segregated, in different parts of the same inflorescence branch. The spikelets overtly heteromorphic (see below).

Inflorescence. Inflorescence a single raceme (7–50 cm long, exserted-pedunculate, spicate, the spikelets densely arranged in irregular whorls, irregularly spiralled near the base); espatheate; not comprising 'partial inflorescences' and foliar organs. Spikelet-bearing axes solitary; persistent. Spikelets not secund; shortly pedicellate; imbricate.

Female-sterile spikelets. Basal male spikelets dorsally compressed to terete, 4.5–7.7 mm long, with 2 or 3 florets, 2 functionally male, the uppermost rudimentary. Rachilla of male spikelets prolonged beyond the uppermost male floret. The male spikelets without proximal incomplete florets; 2–3 floreted. Male florets 2.

Female-fertile spikelets. Spikelets 4.5–7.5 mm long (bisexual spikelets), or 9–17 mm long (female spikelets); not noticeably compressed to compressed dorsiventrally; falling with the glumes. Rachilla prolonged beyond the uppermost female-fertile floret. Hairy callus absent.

Glumes two; more or less equal; shorter than the adjacent lemmas; hairless; not pointed (obtuse or truncate); awnless; non-carinate; very dissimilar (G_1 2-keeled, hardened and thickened, G_2 rounded on the back or slightly flattened). Lower glume two-keeled; 3–7 nerved. Upper glume 3–6 nerved. *Spikelets* with incomplete florets. The incomplete florets distal to the female-fertile florets, or both distal and proximal to the female-fertile florets (always with a distal rudimentary floret, the lower floret usually male in bisexual spikelets, sterile in female spikelets). *The distal incomplete florets* merely underdeveloped. *The proximal incomplete florets* 1; male (in bisexual spikelets), or sterile (sterile in female spikelets). The proximal lemmas awnless; 3–7 nerved (in bisexual spikelets), or 5–9 nerved (in female spikelets); similar in texture to the female-fertile lemmas (herbaceous); not becoming indurated.

Female-fertile florets 1 (or occasionally L_1 and L_2 both bisexual in the bisexual spikelets). Lemmas not becoming indurated; entire; pointed (broadly acute in bisexual spikelets), or blunt (in female spikelets); awnless; hairless; non-carinate; 3–7 nerved (in bisexual spikelets), or 5–11 nerved (in female spikelets). *Palea* present; relatively long (about equalling the lemma in bisexual spikelets. In female spikelets much exceeding it, curved, somewhat twisted and convolute, the apex forming a distinct oriface for the stigmas); awnless, without apical setae; membranous in bisexual spikelets, conspicuously spongy-thickened except for the herbaceous tip and membranous apex in female spikelets; 2-nerved (or more, in bisexual spikelets), or several nerved (5–11 in female spikelets); 2-keeled, or keel-less. Palea keels wingless. *Lodicules* absent. *Stamens* 2 (and posterior, in bisexual florets), or 0 (female florets with two posterior staminodes); with free filaments to monadelphous (the filaments free or fused along their length). Anthers 2.2–3.8 mm long; not penicillate; without an apically prolonged connective. *Ovary* glabrous. Styles fused (the ovary attenuate into a short joint style in bisexual florets, into a long one in female florets). Stigmas 2.

Fruit, embryo and seedling. Fruit medium sized (about 5 mm long); fusiform; not noticeably compressed. Hilum long-linear. Embryo small (about 1/5 as long as the grain).

Abaxial leaf blade epidermis. *Costal/intercostal zonation* conspicuous. *Papillae* absent. Mid-intercostal long-cells rectangular; having markedly sinuous walls (the anticlinal walls thick and pitted). *Microhairs* absent. *Stomata* common. Subsidiaries dome-shaped. *Intercostal short-cells* common; in cork/silica-cell pairs; silicified. Intercostal silica bodies irregularly rounded. *Costal short-cells* predominantly paired. Costal silica bodies present and well developed (in *S. cameroni*), or poorly developed (*S. uniflora*); 'panicoid-type' (*S. cameroni*), or crescentic and oryzoid (*S. uniflora*, but unclearly defined); not sharp-pointed.

Transverse section of leaf blade, physiology. C_3; XyMS+. *Mesophyll* with non-radiate chlorenchyma; without adaxial palisade; without arm cells; without fusoids (but interrupted by a series of large, symmetrically-disposed lacunae, which separate the two rows of vascular bundles into vertically aligned pairs of bundles). *Leaf blade* adaxially flat. *Midrib* conspicuous (comprising a narrow median zone, separated from the thickened laminar flanges by large bulliform 'hinges'); with one bundle only. Bulliforms absent, except for the midrib 'hinges'. All the vascular bundles accompanied by sclerenchyma.

Special diagnostic feature. *Culms dimorphic, the fertile culms leafless, the vegetative culms each with a single developed leaf, this being eligulate and with a terete, culm-like 'sheath'*.

Taxonomy. Bambusoideae, or Arundinoideae; Steyermarkochloeae.

Ecology, geography, regional floristic distribution. 2 species. Venezuela and Colombia. Helophytic. Stream margins in savanna.

Neotropical. Venezuela and Surinam.

References, etc. Morphological/taxonomic: Davidse and Ellis 1984. Leaf anatomical: this project.

Stiburus Stapf

~ *Eragrostis*

Habit, vegetative morphology. *Annual*; caespitose. *Culms* 10–63 cm high; herbaceous; unbranched above. Plants unarmed. Young shoots intravaginal. *Leaves* mostly basal; non-auriculate. Leaf blades narrow; setaceous (in the upper part), or not setaceous (but narrow); exhibiting multicellular glands abaxially (at the base of macrohairs). The abaxial glands intercostal. Leaf blades without cross venation; persistent. *Ligule a fringed membrane (very narrow), or a fringe of hairs*; ligule very short.

Reproductive organization. Plants bisexual, with bisexual spikelets; with hermaphrodite florets.

Inflorescence. *Inflorescence paniculate*; contracted; spicate (purplish); non-digitate; espatheate; not comprising 'partial inflorescences' and foliar organs. Spikelet-bearing axes persistent. Spikelets not secund; pedicellate.

Female-fertile spikelets. *Spikelets 4 mm long*; compressed laterally; disarticulating above the glumes; disarticulating between the florets. Rachilla prolonged beyond the uppermost female-fertile floret; rachilla hairless (save at the nodes). The rachilla extension with incomplete florets. *Hairy callus present (but minute). Callus* short.

Glumes two; very unequal to more or less equal; shorter than the adjacent lemmas, or long relative to the adjacent lemmas; hairy (with abundant tubercle based hairs); pointed (acute or acuminate); awnless; carinate; similar. Lower glume longer than half length of lowest lemma; 1 nerved. *Upper glume 1 nerved. Spikelets* with incomplete florets. The incomplete florets distal to the female-fertile florets. *The distal incomplete florets* merely underdeveloped.

Female-fertile florets 1–5. *Lemmas* similar in texture to the glumes (thin); not becoming indurated; entire; pointed; *mucronate (excurrent into the mucro)*; hairy; carinate, or non-carinate (*S. conrathii*); without a germination flap; 3 nerved. **Palea** present; relatively long, or conspicuous but relatively short; *entire*; awnless, without apical setae; not indurated (thin); 2-nerved; 2-keeled. **Lodicules** present; 2; free; *fleshy (tiny)*; glabrous; not or scarcely vascularized. *Stamens* 3. Anthers minute; not penicillate; without an apically prolonged connective. *Ovary* glabrous. Styles free to their bases. Stigmas 2.

Fruit, embryo and seedling. *Fruit* free from both lemma and palea; small (about 2 mm long); not noticeably compressed. Hilum short. Pericarp fused (probably).

Abaxial leaf blade epidermis. *Costal/intercostal zonation* conspicuous. *Papillae* absent. *Long-cells* similar in shape costally and intercostally; of similar wall thickness costally and intercostally. Mid-intercostal long-cells rectangular; having markedly sinuous walls (and pitted). *Microhairs* absent. *Stomata* common; 16–21 microns long. Subsidiaries dome-shaped. Guard-cells overlapped by the interstomatals. *Intercostal short-cells* common; in cork/silica-cell pairs and not paired (some solitary); silicified (sometimes). Intercostal silica bodies absent. With abundant large cushion based macrohairs. *Costal short-cells* conspicuously in long rows. Costal silica bodies present in alternate cell files of the costal zones; 'panicoid-type'; cross shaped, or dumb-bell shaped (mostly two-lobed); not sharp-pointed.

Transverse section of leaf blade, physiology. C_4; XyMS+ (the ms cells very large, larger than the PCR cells, with very thick walls). PCR sheaths of the primary vascular bundles interrupted; interrupted both abaxially and adaxially. PCR sheath extensions absent. PCR cell chloroplasts centrifugal/peripheral. Mesophyll traversed by columns of colourless mesophyll cells (very wide columns, cf. *Aristida*). *Leaf blade* with distinct, prominent adaxial ribs to 'nodular' in section; with the ribs more or less constant in size (primary bundles in bigger ribs). *Midrib* not readily distinguishable; with one bundle only. Bulliforms present in discrete, regular adaxial groups; associated with colourless mesophyll cells to form deeply-penetrating fans (these linked with the wide traversing columns of colourless cells). All the vascular bundles accompanied by sclerenchyma. Combined sclerenchyma girders present (with all bundles — the fibre groups interrupting the PCR sheath); forming 'figures'. Sclerenchyma all associated with vascular bundles. The lamina margins with fibres.

Taxonomy. Chloridoideae; main chloridoid assemblage.

Ecology, geography, regional floristic distribution. 2 species. Southern Africa.

Paleotropical. African. Sudano-Angolan. South Tropical African.
References, etc. Leaf anatomical: this project.

Stilpnophleum Nevski

~ Deyeuxia, Calamagrostis

Habit, vegetative morphology. Perennial; caespitose. Culms herbaceous. Leaves non-auriculate. Leaf blades narrow; without cross venation. **Ligule** present; *an unfringed membrane*; not truncate; 3–8 mm long.

Reproductive organization. Plants bisexual, with bisexual spikelets; with hermaphrodite florets.

Inflorescence. *Inflorescence paniculate*; *contracted*; capitate to spicate; espatheate; not comprising 'partial inflorescences' and foliar organs. Spikelet-bearing axes persistent. Spikelets not secund; pedicellate.

Female-fertile spikelets. *Spikelets* 5–7 mm long; *not noticeably compressed*; disarticulating above the glumes. Rachilla prolonged beyond the uppermost female-fertile floret; rachilla hairy. The rachilla extension naked. Hairy callus present.

Glumes two; more or less equal; long relative to the adjacent lemmas; pointed (acute to acuminate); awnless; carinate; similar (lanceolate or lanceolate-ovate). Lower glume 1 nerved. *Upper glume 1 nerved*. Spikelets with female-fertile florets only.

Female-fertile florets 1. *Lemmas similar in texture to the glumes*; not becoming indurated; *incised*; not deeply cleft (toothed?); awned. Awns 1; median; dorsal; usually from well down the back; geniculate. Lemmas hairless; non-carinate; without a germination flap; 5 nerved. *Palea* present; relatively long; 2-nerved; 2-keeled, or keel-less. *Lodicules* present; 2; free; membranous; glabrous; usually toothed; not or scarcely vascularized. **Stamens 3**. *Ovary* glabrous. Styles free to their bases. Stigmas 2.

Fruit, embryo and seedling. Fruit longitudinally grooved. Hilum short. Embryo small.

Abaxial leaf blade epidermis. *Costal/intercostal zonation* conspicuous. *Papillae* absent. Mid-intercostal long-cells fusiform (mostly); having markedly sinuous walls (some, slightly), or having straight or only gently undulating walls. *Microhairs* absent. *Stomata* common; (33–)36–42(–45) microns long. Subsidiaries parallel-sided. Guard-cells overlapped by the interstomatals (clearly sunken). *Intercostal short-cells* absent or very rare. Prickles present over the veins. *Costal short-cells* neither distinctly grouped into long rows nor predominantly paired. Costal silica bodies horizontally-elongated crenate/sinuous and horizontally-elongated smooth; not sharp-pointed.

Transverse section of leaf blade, physiology. C_3; XyMS+. Mesophyll without adaxial palisade. *Leaf blade* with distinct, prominent adaxial ribs to 'nodular' in section; with the ribs very irregular in sizes (large alternating with small). *Midrib* not readily distinguishable; with one bundle only. Bulliforms present in discrete, regular adaxial groups; in simple fans. All the vascular bundles accompanied by sclerenchyma. Combined sclerenchyma girders present (with the primaries); forming 'figures'. Sclerenchyma all associated with vascular bundles.

Taxonomy. Pooideae; Poodae; Aveneae.

Ecology, geography, regional floristic distribution. 1 species. Russia. Holarctic. Tethyan. Irano-Turanian.

References, etc. Leaf anatomical: this project.

Stipa L.

Achnatherum P. Beauv. p.p., *Aristella* Bertol., *Jarava* Ruiz & Pavon, *Lasiagrostis*, *Macrochloa* Kunth, *Orthoraphium* Nees, *Patis* Ohwi, *Ptilagrostis* Griseb., *Sparteum* P. Beauv, *Timouria* Roshev.

Excluding *Anemanthele, Lorenzochloa, Nassella, Orthachne, Trikeraia*

Habit, vegetative morphology. Perennial (rarely annual, e.g. *S. capensis, S. parvula*); caespitose. *Culms* 10–250 cm high; woody and persistent (rarely, with persistent canes), or herbaceous; branched above, or unbranched above. Culm nodes hairy, or glabrous. Culm internodes solid (rarely), or hollow. *Leaves* mostly basal, or not basally aggregated; auriculate, or non-auriculate. **Leaf blades** greatly reduced (occasionally), or not all greatly reduced (usually); linear; nearly always narrow; setaceous, or not setaceous; flat (e.g. *S. dregeana*, *S. sibirica*), or folded, or rolled; *not pseudopetiolate*; without cross venation; persistent; rolled in bud (e.g., *Achnatherum*), or oncefolded in bud. Ligule an unfringed membrane, or a fringed membrane. *Contra-ligule* present, or absent.

Reproductive organization. *Plants bisexual, with bisexual spikelets*; with hermaphrodite florets; inbreeding. Exposed-cleistogamous, or chasmogamous. Plants with hidden cleistogenes, or without hidden cleistogenes. The hidden cleistogenes (when present) in the leaf sheaths (sometimes basal and very modified).

Inflorescence. Inflorescence few spikeleted to many spikeleted; paniculate; not deciduous; open, or contracted; with capillary branchlets, or without capillary branchlets; espatheate; not comprising 'partial inflorescences' and foliar organs. Spikelet-bearing axes persistent. Spikelets not secund; pedicellate.

Female-fertile spikelets. Spikelets 3–12 mm long (narrow); compressed laterally to not noticeably compressed; disarticulating above the glumes. Rachilla terminated by a female-fertile floret. Hairy callus present (long and sharp-pointed, except in *Ptilagrostis*). *Callus* short (*Ptilagrostis*), or long; pointed, or blunt.

Glumes two; very unequal to more or less equal; about equalling the spikelets to exceeding the spikelets; nearly always long relative to the adjacent lemmas; pointed (acute or acuminate); awnless, or awned (sometimes aristate); similar. Lower glume not two-keeled (linear or lanceolate, membranous or papery); 1–4 nerved. Upper glume 3–6 nerved. **Spikelets** with female-fertile florets only; *without proximal incomplete florets*.

Female-fertile florets 1. *Lemmas* nearly always convolute, or not convolute (occasionally); not saccate; *decidedly firmer than the glumes (narrow, hiding the palea)*; *becoming indurated (usually, horny)*; entire, or incised (shortly 2-toothed in *Ptilagrostis*); when incised, usually not deeply cleft (deeply cleft in *S. gigantea*); awned. *Awns* 1; median; from a sinus (*Ptilagrostis*), or apical; *geniculate (or sometimes bigeniculate, subulate and non-geniculate in S. saltensis)*; hairless, or hairy, or long-plumose; much shorter than the body of the lemma to much longer than the body of the lemma (sometimes very long, up to 50 cm in *S. pulcherrima*); entered by several veins (3); deciduous (infrequently), or persistent (usually, though often with a basal 'articulation'). Lemmas hairy, or hairless (rarely); non-carinate (terete); without a germination flap; 3–7 nerved. **Palea** usually present (enclosed by the lemma); relatively long (usually), or conspicuous but relatively short to very reduced (rarely); awnless, without apical setae; *thinner than the lemma (translucent)*; indurated (more or less, at least the exposed part); 2-nerved, or nerveless (rarely); *keel-less*. Lodicules present; 2 (rarely), or 3; free; fleshy, or membranous (stipoid); glabrous; not toothed. *Stamens* 3. Anthers 1.2–9 mm long; penicillate, or not penicillate. *Ovary* glabrous. Styles free to their bases. Stigmas 2, or 3, or 4 (3–4 in Section Barbata); white.

Fruit, embryo and seedling. *Fruit* free from both lemma and palea; small, or medium sized, or large; fusiform; compressed laterally, or not noticeably compressed. **Hilum long-linear.** Embryo small; not waisted. Endosperm hard; without lipid; containing only simple starch grains, or containing compound starch grains. Embryo with an epiblast; without a scutellar tail; with a negligible mesocotyl internode. Embryonic leaf margins meeting.

Seedling with a short mesocotyl, or with a long mesocotyl. First seedling leaf with a well-developed lamina. The lamina narrow; erect; 3–5 veined.

Abaxial leaf blade epidermis. *Costal/intercostal zonation* conspicuous, or lacking. *Papillae* present (rarely), or absent; when present, intercostal. Intercostal papillae not over-arching the stomata; several per cell. Mid-intercostal long-cells rectangular (occasionally with a few fusiform); having markedly sinuous walls. *Microhairs* present (probably, sometimes — found to date only adaxially), or absent; elongated; ostensibly one-celled; hard to find and study, seemingly peculiar —perhaps one-celled. *Stomata* absent or very rare, or common; (22–)24–45(–48) microns long. Subsidiaries non-papillate; low to high dome-shaped, or triangular. Guard-cells overlapped by the interstomatals, or overlapping to flush with the interstomatals (rarely). *Intercostal short-cells* common; in cork/silica-cell pairs; silicified. Intercostal silica bodies tall-and-narrow, or crescentic, or rounded (to elliptical). *Costal short-cells* conspicuously in long rows, or predominantly paired, or neither distinctly grouped into long rows nor predominantly paired. Costal silica bodies horizontally-elongated crenate/sinuous (sometimes, but short with only one or two crenations), or horizontally-elongated smooth, or rounded, or saddle shaped, or tall-and-narrow (or cuboid), or crescentic, or 'panicoid-type' (i.e., exhibiting most of the silica body forms, in various combinations from species to species); often cross shaped, or dumb-bell shaped; not sharp-pointed.

Transverse section of leaf blade, physiology. C_3; XyMS+. *Mesophyll* with non-radiate chlorenchyma; without adaxial palisade. *Leaf blade* with distinct, prominent adaxial ribs, or 'nodular' in section; with the ribs more or less constant in size, or with the ribs very irregular in sizes. *Midrib* conspicuous, or not readily distinguishable; with one bundle only, or having a conventional arc of bundles. Bulliforms at the bases of the furrows, present in discrete, regular adaxial groups, or not in discrete, regular adaxial groups (the bulliform cells small and/or inconspicuous, cf. *Ammophila*, in many species); sometimes in simple fans. All the vascular bundles accompanied by sclerenchyma. Combined sclerenchyma girders present, or absent; forming 'figures', or nowhere forming 'figures'. Sclerenchyma all associated with vascular bundles, or not all bundle-associated. The 'extra' sclerenchyma when present, in abaxial groups, or in a continuous abaxial layer.

Phytochemistry. Tissues of the culm bases with abundant starch. Leaves without flavonoid sulphates (3 species).

Special diagnostic feature. *Spikelet not as in* **Nassella** *(q.v.)*.

Cytology. Chromosome base number, $x = 9$, 10, 11, 12, and 22. $2n = 22$, 28, 40, 44, 48, 68, and 96. 2, 4, and 8 ploid (and aneuploids). Nucleoli disappearing before metaphase.

Taxonomy. Arundinoideae; Stipeae.

Ecology, geography, regional floristic distribution. 300 species. Tropical and temperate. Commonly adventive. Mesophytic to xerophytic.

Holarctic, Paleotropical, Neotropical, Cape, Australian, and Antarctic. Boreal, Tethyan, and Madrean. African. Euro-Siberian, Atlantic North American, and Rocky Mountains. Macaronesian, Mediterranean, and Irano-Turanian. Saharo-Sindian and Sudano-Angolan. Caribbean, Pampas, and Andean. North and East Australian, South-West Australian, and Central Australian. New Zealand and Patagonian. European and Siberian. Canadian-Appalachian, Southern Atlantic North American, and Central Grasslands. Sahelo-Sudanian, Somalo-Ethiopian, and South Tropical African. Tropical North and East Australian and Temperate and South-Eastern Australian.

Hybrids. Intergeneric hybrids with *Oryzopsis* — ×*Stiporyzopsis* B.L. Johnson & Rogler.

Rusts and smuts. Rusts — *Puccinia*. Taxonomically wide-ranging species: *Puccinia graminella*, *Puccinia striiformis*, and *Puccinia monoica*. Smuts from Tilletiaceae and from Ustilaginaceae. Tilletiaceae — *Tilletia* and *Urocystis*. Ustilaginaceae — *Sorosporium*, *Sphacelotheca*, *Tolyposporium*, and *Ustilago*.

Economic importance. Significant weed species: the pointed callus of many species (e.g. *S. comata*, *S. spartea*) sometimes injuring livestock. Important native pasture species: many important in dry climates, e.g. *S. barbata*, *S. capensis*, *S. lessingiana*. *S. tenacissima* (Halfa, Esparto grass) used for papermaking mats and cordage.

References, etc. Morphological/taxonomic: de Winter 1965; Barkworth 1983; Freitag 1985; Vickery Jacobs and Everett 1986; Barkworth and Everett 1987. Leaf anatomical: mainly Metcalfe 1960; this project.

Stipagrostis Nees

Schistachne Fig. & De Not

Habit, vegetative morphology. Annual (rarely), or perennial; caespitose. *Culms* 10–200 cm high; herbaceous; branched above, or unbranched above. Culm nodes hairy, or glabrous. Culm internodes solid, or hollow. Plants unarmed. *Leaves* mostly basal, or not basally aggregated. Leaf blades narrowly linear; narrow; setaceous, or not setaceous; flat (rarely), or folded, or rolled (or subterete); without cross venation; disarticulating from the sheaths, or persistent; once-folded in bud. *Ligule a fringe of hairs*.

Reproductive organization. Plants bisexual, with bisexual spikelets; with hermaphrodite florets.

Inflorescence. *Inflorescence paniculate*; open, or contracted; espatheate; not comprising 'partial inflorescences' and foliar organs. Spikelet-bearing axes persistent. Spikelets solitary; not secund; pedicellate.

Female-fertile spikelets. Spikelets 7–20 mm long (?); compressed laterally to not noticeably compressed; disarticulating above the glumes. Rachilla terminated by a female-fertile floret. Hairy callus present, or absent. **Callus** *long (well developed, bearded or not)*; pointed (rarely rounded or minutely bifid).

Glumes two; very unequal to more or less equal; (the longer) long relative to the adjacent lemmas (usually exceeding it); pointed to not pointed (acuminate to obtuse); awnless; non-carinate; similar (scarious). Lower glume usually 3 nerved. Upper glume 3(–11) nerved. **Spikelets** with female-fertile florets only; *without proximal incomplete florets*.

Female-fertile florets *1*. Lemmas cylindrical; convolute, or not convolute; decidedly firmer than the glumes (leathery, the glumes membranous); becoming indurated; not deeply cleft; awned. *Awns* usually *triple or trifid, commonly with a basal column, or not of the triple/trifid, basal column type (S.* **anomala**); apical; non-geniculate (at least, not geniculate in the usual sense); usually *long-plumose (at least on the median branch), or hairless (S.* **anomala**); much longer than the body of the lemma; entered by several veins (3 veins in the column); deciduous (at the base of the column, or near the middle of the lemma — rarely persistent). Lemmas hairless (usually glabrous or scabrid); non-carinate; with a clear germination flap; 3 nerved. **Palea** present; *conspicuous but relatively short (usually less than half lemma length)*; entire to apically notched; awnless; without apical setae; thinner than the lemma; *indurated*; 2-nerved; keel-less. *Lodicules* present, or absent; when present, 2; free; membranous; glabrous; not toothed. **Stamens** *3*. Anthers 2–5.5 mm long; not penicillate; without an apically prolonged connective. *Ovary* glabrous. Styles free to their bases. Stigmas 2.

Fruit, embryo and seedling. *Fruit* free from both lemma and palea (but tightly enclosed in the lemma); fusiform; longitudinally grooved (often, shallowly); compressed dorsiventrally to not noticeably compressed (?). *Hilum long-linear*. Embryo large; waisted, or not waisted. Endosperm containing compound starch grains. Embryo without an epiblast; with a scutellar tail; with an elongated mesocotyl internode; with one scutellum bundle. Embryonic leaf margins meeting.

First seedling leaf with a well-developed lamina. The lamina narrow; erect; 5 veined.

Abaxial leaf blade epidermis. *Costal/intercostal zonation* conspicuous. *Papillae* absent. Long-cells of similar wall thickness costally and intercostally. Mid-intercostal long-cells rectangular; having markedly sinuous walls. *Microhairs* present; elongated; clearly two-celled; panicoid-type; 45–51–57 microns long; 6–7.5 microns wide at the septum. Microhair total length/width at septum 6–8.6. Microhair apical cells 24–28–35 microns long. Microhair apical cell/total length ratio 0.53–0.6. *Stomata* common; (30–)34.5–45(–48) microns long. Subsidiaries dome-shaped and triangular. Guard-cells overlapping to flush with the interstomatals. *Intercostal short-cells* common; in cork/silica-cell pairs; silicified. *Crown cells* present (in the form of numerous bulbous-based and pitted, short pointed or umbonate 'prickles'). *Costal short-cells* neither distinctly grouped into long rows nor predominantly paired. Costal silica bodies rounded (predominantly, in some species?), or saddle shaped (of the broad, almost round type, predominant in (e.g.) *S. ciliata*); not sharp-pointed.

Transverse section of leaf blade, physiology. C_4; XyMS+ (and PCR sheath single, by contrast with *Aristida*). Mesophyll traversed by columns of colourless mesophyll cells. *Leaf blade* with distinct, prominent adaxial ribs to 'nodular' in section; with the ribs very irregular in sizes. *Midrib* not readily distinguishable; with one bundle only. Bulliforms present in discrete, regular adaxial groups; associated with colourless mesophyll cells to form deeply-penetrating fans (these linking with traversing colourless columns). All the vascular bundles accompanied by sclerenchyma. Combined sclerenchyma girders present (with the primaries); forming 'figures' (anchors and I's). Sclerenchyma all associated with vascular bundles.

Phytochemistry. Leaves containing flavonoid sulphates (*S. garubensis*).

Cytology. Chromosome base number, $x = 11$. $2n = 22$ and 44. 2 and 4 ploid.

Taxonomy. Arundinoideae; Aristideae.

Ecology, geography, regional floristic distribution. 50 species. Africa, southwest Asia, northwest India. Xerophytic; species of open habitats. Desert and semidesert, sometimes dunes — e.g. *S. ciliata*, a sandbinder.

Holarctic and Paleotropical. Boreal and Tethyan. African. Euro-Siberian. Macaronesian. Saharo-Sindian, Sudano-Angolan, and Namib-Karoo. European. Sahelo-Sudanian, Somalo-Ethiopian, South Tropical African, and Kalaharian.

Economic importance. Cultivated fodder: *S. ciliata*, *S. uniplumis*. Important native pasture species: e.g. *S. ciliata*, *S. obtusa*, *S. plumosa*.

References, etc. Morphological/taxonomic: de Winter 1965. Leaf anatomical: this project.

Streblochaete Hochst.

Koordersiochloa Merr., *Pseudostreptogyne* A. Camus

Habit, vegetative morphology. Perennial; caespitose. *Culms* 30–100 cm high; herbaceous; unbranched above. Leaves non-auriculate. ***Sheath margins joined***. Leaf blades linear-lanceolate; rolled; without cross venation. Ligule an unfringed membrane; not truncate; 3–12 mm long.

Reproductive organization. Plants bisexual, with bisexual spikelets; with hermaphrodite florets.

Inflorescence. Inflorescence paniculate; open (narrow); espatheate; not comprising 'partial inflorescences' and foliar organs. Spikelet-bearing axes persistent. Spikelets secund to not secund (the panicle sometimes unilateral), or not secund; pedicellate.

Female-fertile spikelets. *Spikelets* 16–28 mm long; ***not noticeably compressed***; disarticulating above the glumes; disarticulating between the florets. Rachilla prolonged beyond the uppermost female-fertile floret. The rachilla extension with incomplete florets. Hairy callus present. *Callus* long; pointed.

Glumes present; two; very unequal; shorter than the spikelets; shorter than the adjacent lemmas; hairless; pointed; awnless; non-carinate (rounded dorsally); similar (narrow, membranous-herbaceous with hyaline margins). Lower glume 3 nerved. ***Upper glume 5 nerved (with transverse linking)***. *Spikelets* with incomplete florets. The incomplete florets distal to the female-fertile florets. The distal incomplete florets several, in a lanceolate cluster; *incomplete florets merely underdeveloped* (male or sterile).

Female-fertile florets 2–6. Lemmas similar in texture to the glumes to decidedly firmer than the glumes (herbaceous); not becoming indurated (or hardening slightly); shortly incised (to nearly entire); slightly 2 lobed; not deeply cleft; awned. ***Awns 1***; median; dorsal; from near the top; ***geniculate (the very long, filiform awns coiling and intertwining with one another, so that the spikelet is dispersed as a unit)***; hairless (scabrid); much longer than the body of the lemma (distally filiform); entered by one vein. Lemmas hairless; non-carinate (dorsally rounded); without a germination flap; 7 nerved. **Palea** present; ***conspicuous but relatively short (about half the lemma length)***; apically notched; not indurated (herbaceous); 2-nerved; 2-keeled. *Lodicules* present; 2; free; membranous; ciliate, or glabrous; not toothed; not or scarcely vascularized. *Stamens* 3. *Ovary* glabrous. Styles free to their bases. Stigmas 2.

Fruit, embryo and seedling. Fruit longitudinally grooved; compressed dorsiventrally. Hilum short. Embryo small. Endosperm hard.

Abaxial leaf blade epidermis. *Costal/intercostal zonation* conspicuous. *Papillae* absent. *Long-cells* the costal long-cells narrower; differing markedly in wall thickness costally and intercostally (the costals thicker). Mid-intercostal long-cells rectangular; having markedly sinuous walls. *Microhairs* absent. *Stomata* absent or very rare. *Intercostal short-cells* common; not paired; not silicified. Prickles present over the veins. *Costal* short-cells neither distinctly grouped into long rows nor predominantly paired. Costal silica bodies horizontally-elongated crenate/sinuous; not sharp-pointed.

Transverse section of leaf blade, physiology. C_3; XyMS+. Sclerenchyma all associated with vascular bundles.

Cytology. Chromosome base number, $x = 10$.

Taxonomy. Pooideae; Poodae; Meliceae.

Ecology, geography, regional floristic distribution. 1 species. Tropical Africa, Java, Lombok, Philippines; montane. Mesophytic; shade species; glycophytic. Montane forest glades.

Paleotropical. African, Madagascan, and Indomalesian. Sudano-Angolan and West African Rainforest. Malesian. Sahelo-Sudanian, Somalo-Ethiopian, and South Tropical African.

References, etc. Morphological/taxonomic: Tateoka 1965b. Leaf anatomical: this project.

Streptochaeta Schrad.

Lepideilema Trin.

Habit, vegetative morphology. Perennial; rhizomatous (the internodes crowded). The flowering culms leafy. Culms 35–105 cm high; *woody and persistent*; to 0.3 cm in diameter; branched above, or unbranched above. Culm nodes glabrous. Culm internodes hollow. Pluricaespitose. Rhizomes pachymorph. Plants unarmed. *Leaves* not basally aggregated; spirally disposed; auriculate (from the sheath); with auricular setae. The juncture of blade and sheath a smooth, dark band of tissue covered by cilia abaxially. *Leaf blades* lanceolate to ovate; broad; (5–)10–95 mm wide; pseudopetiolate; *palmately veined to pinnately veined (in that the veins converge into the midrib from a distance of several cm above the base of the lamina)*; cross veined; disarticulating from the sheaths. Ligule a fringe of hairs. *Contra-ligule* absent.

Reproductive organization. Plants bisexual, with bisexual spikelets; with hermaphrodite florets. Not viviparous.

Inflorescence. *Inflorescence* determinate; with pseudospikelets, or without pseudospikelets (depending on interpretation: buds sometimes present in the 'bract' axils, but these seemingly not proliferating); a single raceme (or a raceme of spikes, if the 'spikelet' be so interpreted). Inflorescence with axes ending in spikelets, or axes not ending in spikelets (sometimes ending in a tuft of hairs). Inflorescence espatheate; not comprising 'partial inflorescences' and foliar organs. *Spikelet-bearing axes* spikes; persistent (tough). Spikelets solitary; not secund; not two-ranked (the raceme many-sided); pedicellate. Pedicel apices cupuliform.

Female-fertile spikelets. Spikelets *unconventional (with 5 spirally arranged, dentate 'bracts' (?glumes, sometimes with axillary buds), a sixth produced into a very long, coiled awn, 7 and 8 side by side opposing 6, and 9–11 whorled to form a central cone, lemma, palea and lodicules absent)*; 10–20 mm long; not noticeably compressed; falling with the glumes (all the pseudospikelets of the inflorescence often shed as a unit, entangled via the long awns of their G_6's which are deflected by the central cones into the cleft between bracts 7 and eight). Rachilla terminated by a female-fertile floret. Hairy callus absent.

Glumes present; several (in the form of 11 or twelve 'bracts', as described elsewhere, the lowermost to 4 mm long, the uppermost to 15 mm); very unequal; hairless; awned (the G_6 with a long, coiled, terminal awn), or awnless (the rest); non-carinate. *Spikelets* with female-fertile florets only.

Female-fertile florets 1. **Lemmas absent, or at least unrecognisable.** *Palea* absent. *Lodicules* absent. *Stamens* 6; monadelphous. Anthers not penicillate; without an apically prolonged connective. *Ovary* glabrous. Styles fused. Stigmas 3.

Fruit, embryo and seedling. *Fruit* free from both lemma and palea; linear; not grooved; not noticeably compressed (terete). Hilum long-linear. Pericarp loosely adherent to fused. Embryo small. Endosperm hard; containing compound starch grains. Embryo without an epiblast; with a scutellar tail; with a negligible mesocotyl internode; with more than one scutellum bundle. Embryonic leaf margins overlapping.

Seedling with a short mesocotyl; with a loose coleoptile. First seedling leaf without a lamina.

Abaxial leaf blade epidermis. *Costal/intercostal zonation* conspicuous. *Papillae* absent. Long-cells of similar wall thickness costally and intercostally (walls of medium thickness). Mid-intercostal long-cells rectangular and fusiform (but the latter forms asymmetric, reflecting oblique end-walls); having markedly sinuous walls. *Microhairs* present; panicoid-type; (84–)90–96(–99) microns long; (6–)7.5–9 microns wide at the septum. Microhair total length/width at septum 9.3–13.2. Microhair apical cells (43.5–)58.5–66(–69) microns long. Microhair apical cell/total length ratio 0.52–0.72. *Stomata* common; 24–33 microns long. Subsidiaries high dome-shaped, or parallel-sided. Guard-cells overlapping to flush with the interstomatals. *Intercostal short-cells* absent or very rare. *Costal short-cells* neither distinctly grouped into long rows nor predominantly paired. Costal silica bodies exclusively saddle shaped; not sharp-pointed.

Transverse section of leaf blade, physiology. C_3; XyMS+. *Mesophyll* with non-radiate chlorenchyma; without arm cells (or at least without conspicuous arm cells, in the material seen); with fusoids. The fusoids external to the PBS. *Leaf blade* adaxially flat. *Midrib* conspicuous (via a T-shaped adaxial rib and a small, rounded, abaxial keel); having complex vascularization (there being a small bundle adaxial to the main one); with colourless mesophyll adaxially. Bulliforms present in discrete, regular adaxial groups (these large and wide); in simple fans. All the vascular bundles accompanied by sclerenchyma. Combined sclerenchyma girders present; forming 'figures'. Sclerenchyma all associated with vascular bundles.

Cytology. Chromosome base number, $x = 11$. $2n = 22$. 2 ploid.

Taxonomy. Bambusoideae; Bambusodae; Streptochaeteae.

Ecology, geography, regional floristic distribution. 3 species. Mexico to Argentina. Shade species.

Neotropical. Caribbean, Amazon, Central Brazilian, and Andean.

References, etc. Morphological/taxonomic: Judziewicz and Soderstrom 1989. Leaf anatomical: Metcalfe 1960; this project.

Streptogyna P. Beauv.

Streptia Doell

Habit, vegetative morphology. Perennial; stoloniferous to caespitose. The flowering culms leafy. *Culms* 60–100 cm high; herbaceous; to 0.5 cm in diameter; unbranched above. Rhizomes pachymorph. Plants unarmed. *Leaves* mostly basal (*S. americana*), or not basally aggregated (*S. crinita*); auriculate (at the top of the sheath — inconspicuous or absent in *S. americana*); with auricular setae (better developed in *S. crinita*). Leaf blades linear to lanceolate; broad; 13–36 mm wide (15–25 cm long); pseudopetiolate, or not pseudopetiolate; cross veined (few laterals, above the base); disarticulating from the sheaths. Ligule a fringed membrane; truncate. *Contra-ligule* present (in the form of a hard rim).

Reproductive organization. *Plants bisexual, with bisexual spikelets*; with hermaphrodite florets.

Inflorescence. Inflorescence a single raceme (inilateral, spiciform); espatheate; not comprising 'partial inflorescences' and foliar organs. *Spikelet-bearing axes* 'racemes'; persistent. *Spikelets secund*; pedicellate.

Female-fertile spikelets. *Spikelets* 25–35 mm long; compressed laterally to not noticeably compressed; disarticulating above the glumes; *disarticulating between the florets (the rachilla joint at the base of each floret forming a hook, aiding in dispersal)*; with distinctly elongated rachilla internodes between the florets. Rachilla prolonged beyond the uppermost female-fertile floret; rachilla hairless. The rachilla extension with incomplete florets. Hairy callus present, or absent.

Glumes two; very unequal (G_2 much longer); shorter than the adjacent lemmas; hairless; awnless; papery. Lower glume 1–3(–5) nerved. Upper glume (5–)7–17 nerved.

Spikelets with incomplete florets. The incomplete florets distal to the female-fertile florets (with glabrous lemmas). *The distal incomplete florets* merely underdeveloped.

Female-fertile florets 2–4. **Lemmas convolute**; similar in texture to the glumes to decidedly firmer than the glumes (rigidly leathery); not becoming indurated; incised; 2 lobed; not deeply cleft (bidentate); awned. Awns 1; median; from a sinus; non-geniculate; straight; hairless (antrorsely scabrous); much shorter than the body of the lemma to about as long as the body of the lemma. Lemmas hairy (silky villous below the middle); non-carinate; 7–13 nerved. *Palea* present (narrowly linear); relatively long; apically notched; awnless, without apical setae; not indurated; 2-nerved (these median, contiguous). *Lodicules* present; 3; free; membranous; ciliate (only inconspicuously so in *S. americana*); not toothed; heavily vascularized. *Stamens* 2. Anthers not penicillate; without an apically prolonged connective. *Ovary* glabrous, or hairy. Styles fused (only at the base in *S. americana*). Stigmas 2 (*S. crinita*), or 3 (*S. americana* — in both species, they elongate after anthesis to entangle with one another and with the lemma awns).

Fruit, embryo and seedling. *Fruit* free from both lemma and palea; large; linear; compressed laterally; when present, with hairs confined to a terminal tuft. Hilum long-linear. Embryo small; not waisted. Endosperm hard; containing only simple starch grains, or containing compound starch grains (some, in *S. crinita*). Embryo with an epiblast; with a scutellar tail; with a negligible mesocotyl internode. Embryonic leaf margins overlapping.

Seedling with a short mesocotyl; with a loose coleoptile (with a short mucro). First seedling leaf with a well-developed lamina. The lamina narrow; erect.

Abaxial leaf blade epidermis. *Costal/intercostal zonation* conspicuous. *Papillae* absent. *Long-cells* similar in shape costally and intercostally; differing markedly in wall thickness costally and intercostally (the costals rather thicker walled). Mid-intercostal long-cells rectangular (mostly); having markedly sinuous walls. *Microhairs* absent. *Stomata* common; 18–24 microns long. Subsidiaries dome-shaped (in *S. americana*), or dome-shaped and triangular (in *S. crinita*). Guard-cells overlapping to flush with the interstomatals. *Intercostal short-cells* common; in cork/silica-cell pairs (mostly); silicified. Intercostal silica bodies narrowly saddle shaped. *Costal short-cells* conspicuously in long rows (*S. crinita*), or predominantly paired (*S. americana*). Costal silica bodies saddle shaped (predominating), or tall-and-narrow to crescentic; not sharp-pointed.

Transverse section of leaf blade, physiology. C_3; XyMS+. *Mesophyll* with non-radiate chlorenchyma; without adaxial palisade; with arm cells (*S. crinita*), or without arm cells (no trace of these in *S. americana*); with fusoids. The fusoids external to the PBS. *Leaf blade* adaxially flat. *Midrib* conspicuous; having complex vascularization (the large keel with an abaxial arc, plus one small adaxial bundle); with colourless mesophyll adaxially. Bulliforms present in discrete, regular adaxial groups (these large); in simple fans. All the vascular bundles accompanied by sclerenchyma (or a few without, in *S. crinita*). Combined sclerenchyma girders present (with most bundles); forming 'figures' (a few with I's and T's). Sclerenchyma all associated with vascular bundles.

Cytology. Chromosome base number, $x = 12$. $2n = 24$.

Taxonomy. Bambusoideae; Bambusodae; Streptogyneae.

Ecology, geography, regional floristic distribution. 2 species. Africa, India, Central and South America. Shade species; glycophytic. In dry forests.

Paleotropical and Neotropical. African and Indomalesian. Sudano-Angolan and West African Rainforest. Indian. Caribbean, Amazon, and Central Brazilian. Sahelo-Sudanian and South Tropical African.

References, etc. Morphological/taxonomic: Soderstrom and Judziewicz 1987. Leaf anatomical: Metcalfe 1960; this project.

Streptolophus Hughes

Habit, vegetative morphology. Rambling annual; decumbent (?). *Culms* 100–200 cm high; herbaceous; branched above. Culm nodes hairy. *Leaves* not basally aggregated. *Leaf blades* lanceolate; broad; 15–20 mm wide; **cordate to sagittate**; pseudopetiolate (with long petioles); without cross venation. Ligule a fringed membrane.

Reproductive organization. Plants bisexual, with bisexual spikelets; with hermaphrodite florets. The spikelets of sexually distinct forms on the same plant, or all alike in sexuality; hermaphrodite and sterile (sterile at the tip of main inflorescence axis).

Inflorescence. *Inflorescence a false spike, with spikelets on contracted axes (the lateral branches comprising spikelets and bristles organized in cuneate, peduncled burrs, the burrs racemosely and loosely borne on a slender, wavy rachis)*; a raceme of burrs, by contrast with the sessile burrs of *Cenchrus*. Inflorescence axes not ending in spikelets (the main rachis and the burr-like branches terminating in spiny bristles). Inflorescence espatheate; not comprising 'partial inflorescences' and foliar organs. *Spikelet-bearing axes* very much reduced (to pedunculate burrs of spikelets-plus-bristles); disarticulating; falling entire (the burrs falling entire with their peduncles). **Spikelets** *with 'involucres' of 'bristles' (the spinescent bristles dichotomising from two basal stumps, forming a basketlike support for the spikelets)*. The 'bristles' spiny, markedly coalescent basally; deciduous with the spikelets. Spikelets solitary; not secund.

Female-fertile spikelets. *Spikelets* 3.5 mm long; *compressed dorsiventrally*; falling with the glumes (in the burr). Rachilla terminated by a female-fertile floret. Hairy callus absent.

Glumes two; very unequal to more or less equal; shorter than the adjacent lemmas; hairless; awnless. Lower glume 0 nerved. Upper glume 3 nerved. *Spikelets* with incomplete florets. The incomplete florets proximal to the female-fertile florets. *The proximal incomplete florets* 1; epaleate; sterile. The proximal lemmas awnless.

Female-fertile florets 1. Lemmas decidedly firmer than the glumes; striate (minutely verrucose); becoming leathery; awnless, or mucronate (?); hairless; non-carinate; 5 nerved. *Palea* present; 2-nerved. *Lodicules* present; 2; free. *Stamens* 3. Anthers not penicillate. *Ovary* glabrous. Styles fused. Stigmas 2.

Fruit, embryo and seedling. Fruit small (2 mm long); compressed dorsiventrally. Hilum short. Embryo large.

Abaxial leaf blade epidermis. *Costal/intercostal zonation* conspicuous. *Papillae* absent. Mid-intercostal long-cells rectangular; having markedly sinuous walls. *Microhairs* present; panicoid-type; (54–)60–78 microns long; (5.4–)5.7–6(–7.2) microns wide at the septum. Microhair total length/width at septum 8.3–14.4. Microhair apical cells (34–)37.5–49.5(–54) microns long. Microhair apical cell/total length ratio 0.63–0.69. *Stomata* common; 22.5–27 microns long. Subsidiaries dome-shaped (mostly), or triangular. Guard-cells overlapping to flush with the interstomatals. *Intercostal short-cells* absent or very rare. *Costal short-cells* conspicuously in long rows (the rows interrupted by large prickles). Costal silica bodies 'panicoid-type'; mostly cross shaped and dumb-bell shaped (short); not sharp-pointed.

Transverse section of leaf blade, physiology. C_4; XyMS–. *Midrib* conspicuous (via its keel); with one bundle only. Bulliforms not in discrete, regular adaxial groups (epidermis seemingly extensively bulliform). Many of the smallest vascular bundles unaccompanied by sclerenchyma. Combined sclerenchyma girders present (primaries only — most larger bundles with abaxial girders only, or with abaxial girders and adaxial strands); nowhere forming 'figures'. Sclerenchyma all associated with vascular bundles.

Taxonomy. Panicoideae; Panicodae; Paniceae.

Ecology, geography, regional floristic distribution. 1 species. Angola. Mesophytic. Paleotropical. African. Sudano-Angolan. South Tropical African.

References, etc. Morphological/taxonomic: Hughes 1923. Leaf anatomical: this project.

Streptostachys Desv.

Habit, vegetative morphology. Perennial; decumbent. **Culms** 25–150(–200) cm high; *herbaceous*; branched above, or unbranched above. Culm nodes hairy, or glabrous. Culm internodes solid, or hollow. *Leaves* not basally aggregated; non-auriculate. **Leaf blades** broadly lanceolate, or ovate-lanceolate; *broad*; 7–30(–43) mm wide; cordate, or not cordate, not sagittate; not pseudopetiolate; without cross venation; persistent. **Ligule** present, or absent; when present, *a fringe of hairs (sometimes with scattered hairs in the ligule position, and marginal tufts)*.

Reproductive organization. Plants bisexual, with bisexual spikelets; with hermaphrodite florets. The spikelets all alike in sexuality.

Inflorescence. *Inflorescence* large, leafy, *paniculate (its limits hard to define)*; open. Rachides sinuous. Inflorescence spatheate, or espatheate (?); a complex of 'partial inflorescences' and intervening foliar organs, or not comprising 'partial inflorescences' and foliar organs (? — with narrow secondary panicles and racemes). **Spikelet-bearing axes** 'racemes', or spikelike, or paniculate; *persistent*. Spikelets solitary, or paired; not secund; subsessile to pedicellate.

Female-fertile spikelets. *Spikelets* 3.6–9.3 mm long; narrowly oblong; *compressed dorsiventrally*; falling with the glumes; with a distinctly elongated rachilla internode between the glumes, with a distinctly elongated rachilla internode above the glumes, and with distinctly elongated rachilla internodes between the florets (the internodes thickened). The upper floret conspicuously stipitate. The stipe beneath the upper floret filiform (straight); homogeneous. Rachilla terminated by a female-fertile floret. Hairy callus absent.

Glumes two; very unequal, or more or less equal; (at least the upper) long relative to the adjacent lemmas (more or less equalling the lower lemma); dorsiventral to the rachis; sparsely hairy (on the back); not pointed (blunt); awnless; non-carinate; very dissimilar to similar (herbaceous, the lower having its thickened base forming a downwardly projecting rim). Lower glume with short, sparse hairs; (1–)3–5(–7) nerved. *Upper glume* somewhat hooded apically; 5 nerved, or 7 nerved. *Spikelets* with incomplete florets. The incomplete florets proximal to the female-fertile florets. **Spikelets with proximal incomplete florets**. *The proximal incomplete florets* 1; paleate, or epaleate. Palea of the proximal incomplete florets (when present) fully developed to reduced. The proximal incomplete florets male, or sterile. *The proximal lemmas* awnless; (3–)5 nerved, or 7 nerved; *more or less equalling the female-fertile lemmas*; less firm than the female-fertile lemmas (membranous-herbaceous, sparsely hairy above); not becoming indurated.

Female-fertile florets 1. *Lemmas* decidedly firmer than the glumes (shiny); becoming indurated (thinly); entire; pointed; awnless; hairless; non-carinate; *having the margins lying flat and exposed on the palea*; with a clear germination flap; 5 nerved. *Palea* present; relatively long; entire; thinly indurated; indurated; 2-nerved; 2-keeled. Palea keels wingless. *Lodicules* present; 2; free; fleshy; glabrous; not or scarcely vascularized. *Stamens* 3. Anthers not penicillate; without an apically prolonged connective (the anthers divaricate). *Ovary* glabrous. Styles fused (*S. asperifolia*), or free to their bases. Stigmas dark.

Fruit, embryo and seedling. *Hilum long-linear*. Embryo 'less than half the length of the caryopsis'.

Abaxial leaf blade epidermis. *Costal/intercostal zonation* conspicuous. *Papillae* absent. Mid-intercostal long-cells rectangular; having markedly sinuous walls. *Microhairs* present; panicoid-type; (36–)39–48 microns long; (4.8–)5.1–6.6(–6.9) microns wide at the septum. Microhair total length/width at septum 6.1–9.4. Microhair apical cells (18–)19–27(–33) microns long. Microhair apical cell/total length ratio 0.5–0.73. *Stomata* common; (28.5–)30–42 microns long. Subsidiaries triangular. Guard-cells overlapping to flush with the interstomatals. *Intercostal short-cells* common; in cork/silica-cell pairs. *Costal short-cells* conspicuously in long rows. Costal silica bodies 'panicoid-type'; mostly cross shaped; not sharp-pointed.

Transverse section of leaf blade, physiology. C_4 (*S. macrantha, S. ramosa*), or C_3 (*S. asperifolia*); XyMS+. *Mesophyll* with radiate chlorenchyma, or with non-radiate chlorenchyma; with adaxial palisade (seemingly, in *S. asperifolia*), or without adaxial palisade; *Isachne*-type (perhaps, somewhat, in *S. asperifolia*), or not *Isachne*-type; with arm cells (a few); with fusoids (in *S. asperifolia*), or without fusoids. The fusoids external to the PBS. *Leaf blade* adaxially flat. *Midrib* not readily distinguishable; with one bundle only. Bulliforms present in discrete, regular adaxial groups; in simple fans. Many of the smallest vascular bundles unaccompanied by sclerenchyma (in the C_4 species), or all the vascular bundles accompanied by sclerenchyma (*S. asperifolia*). Combined sclerenchyma girders present; forming 'figures' (I's in *S. macrantha*, T's in *S. ramosa*), or nowhere forming 'figures' (*S. asperifolia*). Sclerenchyma all associated with vascular bundles.

Taxonomy. Panicoideae; Panicodae; Paniceae.

Ecology, geography, regional floristic distribution. 3 species. North tropical South America, Trinidad. Savanna.

Neotropical. Caribbean, Venezuela and Surinam, Amazon, and Central Brazilian.

References, etc. Morphological/taxonomic: Morrone and Zuloaga 1991. Leaf anatomical: this project; Morrone and Zuloaga 1991.

Special comments. *Streptostachys* as carefully re-circumscribed by Morrone and Zuloaga (i.e. with '*S. aucuminata*' referred to *Urochloa*, and '*S. robusta*' excluded) continues to present a wonderful blend of panicoid and bambusoid features, and is very unusual in including both C_3 and C_4 species. Fruit data wanting.

Styppeiochloa de Winter

Crinipes Hochst.

Habit, vegetative morphology. Perennial; densely caespitose (the hard, fibrous basal sheaths forming tough, fire-resistant mats). **Culms** 10–70 cm high; *herbaceous*; unbranched above (wiry, the nodes hidden at the base). Culm sheaths persistent (tomentose, splitting into fibres). Culm internodes solid. Plants unarmed. Young shoots intravaginal. *Leaves* mostly basal, or not basally aggregated; non-auriculate. **Leaf blades** linear; narrow; to *1 mm wide*; setaceous (resembling the culms); rolled (convolute); without cross venation; persistent. **Ligule a fringe of hairs**; about 0.4 mm long. *Contra-ligule* absent.

Reproductive organization. Plants bisexual, with bisexual spikelets; with hermaphrodite florets.

Inflorescence. Inflorescence paniculate; contracted (scanty, the spikelets appressed to the panicle branches); espatheate; not comprising 'partial inflorescences' and foliar organs. Spikelet-bearing axes persistent. Spikelets pedicellate.

Female-fertile spikelets. Spikelets compressed laterally; disarticulating above the glumes; disarticulating between the florets; with conventional internode spacings. Rachilla prolonged beyond the uppermost female-fertile floret. The rachilla extension with incomplete florets. Hairy callus present. *Callus* short; blunt (truncate).

Glumes two; very unequal, or more or less equal; shorter than the spikelets; shorter than the adjacent lemmas; hairless; glabrous; pointed; short awned (or aristate from the excurrent mid-nerve), or awnless; carinate; similar (lanceolate, their apices 3-lobed or acute/entire). Lower glume shorter than the lowest lemma; 1 nerved. Upper glume 1–3 nerved. *Spikelets* with incomplete florets. The incomplete florets distal to the female-fertile florets. *The distal incomplete florets* merely underdeveloped; awned, or awnless.

Female-fertile florets 2–5. Lemmas similar in texture to the glumes (membranous); not becoming indurated; incised; apically 3 lobed; not deeply cleft; mucronate to awned (the three lobes with awns or mucros). Awns when present, 3; median, or median and lateral (via shortly excurrent nerves); the median similar in form to the laterals (when laterals present); apical; non-geniculate; hairless (scaberulous); much shorter than the body of the lemma; entered by one vein. The lateral awns shorter than the median. Lemmas hairy (at the margins, near the base); carinate; without a germination flap; 3–5(–7) nerved; with the nerves non-confluent. **Palea** present (narrowly lanceolate); relatively long (equalling the lemma); apically notched; *with apical setae (via the excurrent nerves)*; textured like the lemma (membranous); not indurated; 2-nerved; 2-keeled (concave between the keels). *Lodicules* present; heavily vascularized (2–5 nerved?). *Stamens* 3. Anthers about 2.5 mm long; not penicillate; without an apically prolonged connective. *Ovary* glabrous; without a conspicuous apical appendage (but the style bases knob-like). Styles free to their bases. Stigmas 2.

Fruit, embryo and seedling. *Fruit* free from both lemma and palea; small (about 2 mm long); fusiform; not noticeably compressed (nearly terete). *Hilum long-linear*. Embryo small; waisted.

Abaxial leaf blade epidermis. *Costal/intercostal zonation* the costal and intercostal zonation fairly indistinct. *Papillae* absent. *Long-cells* similar in shape costally and intercostally; of similar wall thickness costally and intercostally. Mid-intercostal long-cells rectangular; having markedly sinuous walls. *Microhairs* present; panicoid-type. *Stomata* absent or very rare. *Intercostal short-cells* common. *Costal short-cells* conspicuously in long rows. Costal silica bodies 'panicoid-type'; dumb-bell shaped; not sharp-pointed.

Transverse section of leaf blade, physiology. C_3; XyMS+. *Mesophyll* with non-radiate chlorenchyma; without adaxial palisade; not *Isachne*-type (the cells isodiametric). *Leaf blade* with distinct, prominent adaxial ribs; with the ribs very irregular in sizes. *Midrib* not readily distinguishable; with one bundle only. Bulliforms present in discrete, regular adaxial groups (in all the furrows); in simple fans. All the vascular bundles accompanied by sclerenchyma. Combined sclerenchyma girders present (with all the large bundles); forming 'figures' (the large bundles). Sclerenchyma not all bundle-associated. The 'extra' sclerenchyma in a continuous abaxial layer.

Taxonomy. Arundinoideae; Danthonieae.

Ecology, geography, regional floristic distribution. 2 species. South and southeastern tropical African mountains. Helophytic to mesophytic; species of open habitats; glycophytic. Where there is impeded drainage, in mountains.

Paleotropical. African and Madagascan. Sudano-Angolan. South Tropical African.

References, etc. Morphological/taxonomic: de Winter 1966a. Leaf anatomical: this project; photos provided by R.P. Ellis.

Sucrea Soderstrom

Habit, vegetative morphology. Perennial; rhizomatous and caespitose. **Culms** 30–100 cm high; herbaceous; *unbranched above*. Culm nodes hairy, or glabrous. Culm sheaths persistent. Plants unarmed. *Leaves* mostly basal, or not basally aggregated; without auricular setae. *Leaf blades* broadly *ovate*; broad; 25–100 mm wide; pseudopetiolate; cross veined; rolled in bud. *Ligule* present, or absent.

Reproductive organization. *Plants monoecious with all the fertile spikelets unisexual*; without hermaphrodite florets. The spikelets of sexually distinct forms on the same plant; female-only and male-only. *The male and female-fertile spikelets segregated, in different parts of the same inflorescence branch (the female spikelets terminal, the males below).* The spikelets overtly heteromorphic.

Inflorescence. Inflorescence paniculate; open (terminal); spatheate (the terminal sheath constituting a spathe); not comprising 'partial inflorescences' and foliar organs. *Spikelet-bearing axes* paniculate; persistent. Spikelets not secund; pedicellate (the pedicels of female spikelets thickened apically); not in distinct 'long-and-short' combinations.

Female-sterile spikelets. Male spikelets without glumes; lemma membranous, 3-nerved; palea membranous, 2-nerved; lodicules 3, fleshy; stamens 3, free, non-penicillate. Rachilla of male spikelets terminated by a male floret. The male spikelets without glumes; without proximal incomplete florets; 1 floreted. Male florets 1; 3 staminate. The staminal filaments free.

Female-fertile spikelets. Spikelets 6.5–10 mm long; disarticulating above the glumes. Rachilla terminated by a female-fertile floret. Hairy callus absent. *Callus* absent.

Glumes two; more or less equal; long relative to the adjacent lemmas (exceeding them); hairy (finely pubescent); pointed (ovate-acuminate to shortly subulate); shortly *awned (with a terminal scabrid subule)*; non-carinate; *similar (with cartilaginous margins, twisting with age, tessellate)*. Lower glume 5 nerved. Upper glume 3 nerved. *Spikelets* with female-fertile florets only.

Female-fertile florets 1. **Lemmas fusiform, completely embracing the palea**; decidedly firmer than the glumes; pitted; becoming indurated (leathery when young); awnless; hairless; glabrous; non-carinate; having the margins tucked in onto the palea; without a germination flap (apparently); 5 nerved (tessellate). *Palea* present; relatively long (slightly shorter than the lemma); entire; awnless, without apical setae; textured like the lemma; indurated; 2-nerved (? '2 strong nerves and transverse veinlets'). *Lodicules* present; 3; free; fleshy; glabrous; heavily vascularized. *Stamens* 0 (3 staminodes). *Ovary* glabrous. Styles fused (the ovary attenuate into one long style). Stigmas 2.

Fruit, embryo and seedling. Hilum long-linear. Embryo small; waisted.

Abaxial leaf blade epidermis. *Costal/intercostal zonation* conspicuous. *Papillae* present. Intercostal papillae not over-arching the stomata; several per cell (round, thick walled, in 2–3 irregular rows on long-cells and interstomatals, the costals with mostly only one row). *Long-cells* markedly different in shape costally and intercostally (the intercostals much broader); of similar wall thickness costally and intercostally (quite thin walled). Mid-intercostal long-cells rectangular; having markedly sinuous walls (coarsely so). *Microhairs* present; panicoid-type; (45–)48–51(–54) microns long; 5.4–6.3 microns wide at the septum. Microhair total length/width at septum 7.1–10. Microhair apical cells 19.5–25.5 microns long. Microhair apical cell/total length ratio 0.43–0.5. *Stomata* common; 21–24 microns long. Subsidiaries mostly high dome-shaped. Guard-cells overlapping to flush with the interstomatals. *Intercostal short-cells* common; in cork/silica-cell pairs (mostly). *Costal short-cells* conspicuously in long rows. Costal silica bodies saddle shaped, oryzoid, and 'panicoid-type' (forming an almost continuous series); the panicoid form cross shaped; not sharp-pointed.

Transverse section of leaf blade, physiology. C_3; XyMS+. Mesophyll with adaxial palisade; with arm cells; with fusoids. The fusoids external to the PBS. *Leaf blade* adaxially flat (except for the midrib). *Midrib* conspicuous; having a conventional arc of bundles (a large median, and a small one on either side); with colourless mesophyll adaxially. Bulliforms present in discrete, regular adaxial groups (these wide); in simple fans. All the vascular bundles accompanied by sclerenchyma. Combined sclerenchyma girders present (with most of the main bundles); forming 'figures' (most of the main bundles with I's or 'anchors').

Cytology. Chromosome base number, $x = 11$. $2n = 22$. 2 ploid.

Taxonomy. Bambusoideae; Oryzodae; Olyreae.

Ecology, geography, regional floristic distribution. 3 species. Brazil. Shade species. In forest.

Neotropical. Central Brazilian.

References, etc. Morphological/taxonomic: Soderstrom 1981c. Leaf anatomical: this project.

Suddia Renvoize

Habit, vegetative morphology. Perennial; rhizomatous. *Culms* 140–300 cm high; herbaceous; to 2.5 cm in diameter; unbranched above. Culm internodes solid (spongy). *Leaves* not basally aggregated (3–5 laminate leaves per culm, with several bladeless sheaths at the base); without auricular setae. *Leaf blades* lanceolate; broad; 35–105 mm wide (22–112 cm long); *sagittate*; flat; *pseudopetiolate (the 'petiole' 16–100 cm long)*; *pinnately veined*; cross veined; with a conspicuous articulation at the base. *Ligule* present; an unfringed membrane; not truncate; 35–80 mm long.

Reproductive organization. Plants probably bisexual, with bisexual spikelets (but known only from material with smutted, imperfect spikelets — i.e., as yet inadequately described); probably with hermaphrodite florets.

Inflorescence. Inflorescence paniculate; open (much branched, up to 35 cm long, the lower branches often in verticils); espatheate; not comprising 'partial inflorescences' and foliar organs. Spikelet-bearing axes persistent (apparently). Spikelets not secund; pedicellate.

Female-fertile spikelets. Spikelets 3.5–4.3 mm long. Rachilla terminated by a female-fertile floret (seemingly).

Glumes two; long relative to the adjacent lemmas; pointed; awnless; similar. Lower glume 5–7 nerved. Upper glume 7–9 nerved.

Female-fertile florets 1 (seemingly). Lemmas entire; pointed; awnless; 3 nerved.

Abaxial leaf blade epidermis. *Costal/intercostal zonation* conspicuous. *Papillae* absent. Mid-intercostal long-cells rectangular; having markedly sinuous walls. *Microhairs* present; panicoid-type. *Stomata* common. Subsidiaries seemingly dome-shaped and triangular. *Intercostal short-cells* common; not paired (seemingly solitary). *Costal short-cells* conspicuously in long rows. Costal silica bodies 'panicoid-type'; not sharp-pointed.

Transverse section of leaf blade, physiology. C_3; XyMS+. *Mesophyll* with non-radiate chlorenchyma; without arm cells; with fusoids. The fusoids external to the PBS. *Leaf blade* adaxially flat. *Midrib* conspicuous (very much so, with large air spaces); having complex vascularization. Bulliforms present in discrete, regular adaxial groups.

All the vascular bundles accompanied by sclerenchyma. Combined sclerenchyma girders present. Sclerenchyma all associated with vascular bundles.

Taxonomy. Bambusoideae; Oryzodae; Phareae.

Ecology, geography, regional floristic distribution. 1 species. Sudan. Hydrophytic to helophytic.

Paleotropical. African. Saharo-Sindian.

References, etc. Morphological/taxonomic: Renvioze 1984.

Special comments. Morphological description very incomplete. Fruit data wanting.

Swallenia Soderstrom & Decker

Ectosperma Swallen

Habit, vegetative morphology. Perennial; rhizomatous. *Culms* about 10–60 cm high (?); herbaceous; branched above (with sterile and fertile branches). Plants unarmed. *Leaves* not basally aggregated; non-auriculate. Leaf blades narrow; 3–6 mm wide (4–14 cm long); flat; without abaxial multicellular glands; without cross venation. *Ligule a fringe of hairs. Contra-ligule* absent.

Reproductive organization. *Plants bisexual, with bisexual spikelets*; with hermaphrodite florets.

Inflorescence. Inflorescence paniculate; contracted (sparse, 4–10 cm long); spicate (the branches short, appressed, with 1–3 spikelets); espatheate; not comprising 'partial inflorescences' and foliar organs. Spikelet-bearing axes persistent. Spikelets not secund; pedicellate.

Female-fertile spikelets. *Spikelets 10–15 mm long (and almost as wide)*; *compressed laterally*; *not disarticulating (in material seen, even in the presence of fruit)*; not disarticulating between the florets; with conventional internode spacings. Rachilla prolonged beyond the uppermost female-fertile floret. The rachilla extension with incomplete florets. Hairy callus present.

Glumes two; more or less equal; long relative to the adjacent lemmas (almost as long as the spikelet, spreading); hairless; glabrous; pointed (acuminate); awnless; similar (thinly membranous). Lower glume 5–7 nerved. Upper glume 7–11 nerved. *Spikelets* with incomplete florets. The incomplete florets distal to the female-fertile florets. *The distal incomplete florets* merely underdeveloped.

Female-fertile florets 3–7. Lemmas very broad; similar in texture to the glumes to decidedly firmer than the glumes (membranous or papery); not becoming indurated; entire; pointed (acute), or blunt (but mucronate); awnless, or mucronate; hairy (densely villous below, on the margins and sometimes intercostally); non-carinate (dorsally rounded); without a germination flap; 5–7 nerved. *Palea* present; relatively long (often exceeding the lemma); entire to apically notched (erose-truncate); with apical setae to awned (the 2 nerves shortly excurrent); not indurated (thin, villous below and marginally); 2-nerved; 2-keeled. *Lodicules* present (large); 2; fleshy; glabrous; each with a thin, auriculate lobe; heavily vascularized. *Stamens* 3. Anthers 3.5 mm long. *Ovary* glabrous. Styles free to their bases. Stigmas 2; probably white.

Fruit, embryo and seedling. *Fruit* free from both lemma and palea (and falling free readily); small to medium sized (about 4 mm long); ellipsoid; compressed dorsiventrally (rectangular in section); smooth (brown). Hilum short. Pericarp fused. Embryo large; with an epiblast; with a scutellar tail; with an elongated mesocotyl internode. Embryonic leaf margins meeting.

Abaxial leaf blade epidermis. *Costal/intercostal zonation* conspicuous. *Papillae* present; intercostal. Intercostal papillae papilliform cells on the sides of the intercostal grooves extensively obscuring the stomata; many intercostal cells swollen-papilliform. *Long-cells* markedly different in shape costally and intercostally (costal long-cells 'normal'); differing markedly in wall thickness costally and intercostally (costals thicker-walled). Intercostal zones exhibiting many atypical long-cells (these being rather short). Mid-intercostal long-cells having straight or only gently undulating walls. *Microhairs* present; elongated; clearly two-celled; chloridoid-type (large, with distending basal cell — but inconspicuous, because both cells thin-walled and tending to collapse); 27–33 microns long; 19.5–21 microns wide at the septum. Microhair total length/width at septum 1.38–1.57. Microhair apical cells 7.2–7.5 microns long. Microhair apical cell/ total length ratio 0.22–0.28. *Stomata* common (but mostly obscured by prickles and distended intercostal cells); 21–27 microns long. Subsidiaries where visible, high dome-shaped to triangular. *Intercostal short-cells* absent or very rare (infrequent pairs); in cork/ silica-cell pairs (but rare). Intercostal silica bodies absent. Prickles common alongside the intercostal zones. *Costal short-cells* conspicuously in long rows. Costal silica bodies present in alternate cell files of the costal zones; horizontally-elongated smooth and rounded (these predominating, including potato shapes and large, supine crescents), or 'panicoid-type' (a few); the panicoid form somewhat nodular; not sharp-pointed.

Transverse section of leaf blade, physiology. C_4; XyMS+. PCR sheaths of the primary vascular bundles complete. PCR sheath extensions absent. *Mesophyll* with radiate chlorenchyma. *Leaf blade* 'nodular' in section (opposing, rounded adaxial and abaxial ribs); with the ribs more or less constant in size. *Midrib* not readily distinguishable; with one bundle only. Bulliforms present in discrete, regular adaxial groups (and in irregular groups); sometimes in simple fans. All the vascular bundles accompanied by sclerenchyma. Combined sclerenchyma girders absent (most bundles with heavy adaxial and abaxial strands only). Sclerenchyma all associated with vascular bundles. The lamina margins with fibres.

Taxonomy. Chloridoideae; main chloridoid assemblage.

Ecology, geography, regional floristic distribution. 1 species. California. Xerophytic; species of open habitats. Sand dunes.

Holarctic. Madrean.

References, etc. Morphological/taxonomic: Swallen 1956. Leaf anatomical: Metcalfe 1960; this project.

Swallenochloa McClure

~ *Chusquea*

Habit, vegetative morphology. Perennial. The flowering culms leafy. *Culms* woody and persistent; *scandent*; branched above. Primary branches/mid-culm node (1–)3–7 (i.e. few). Culm sheaths persistent (the blades caducous). Culm internodes hollow ('or pith breaking down'). Unicaespitose. Rhizomes pachymorph. Plants unarmed. Young shoots intravaginal. *Leaves* not basally aggregated; auriculate (the auricles small), or non-auriculate; without auricular setae. Leaf blades broad; pseudopetiolate; cross veined; disarticulating from the sheaths; rolled in bud. *Contra-ligule* present, or absent.

Reproductive organization. Plants bisexual, with bisexual spikelets; with hermaphrodite florets.

Inflorescence. Inflorescence paniculate (terminating the culm in addition to leafy or leafless lateral shoots); contracted; spicate (the lateral branches appressed); *spatheate*; a complex of 'partial inflorescences' and intervening foliar organs. *Spikelet-bearing axes* paniculate; persistent. Spikelets not secund; pedicellate.

Female-fertile spikelets. Spikelets 5–9 mm long; compressed laterally (?); disarticulating above the glumes; disarticulating between the florets (beneath the fertile lemma). Rachilla terminated by a female-fertile floret.

Glumes two; minute, or relatively large; very unequal; shorter than the adjacent lemmas; not pointed (blunt); awnless; non-carinate; similar. Lower glume much shorter than half length of lowest lemma. *Spikelets* with incomplete florets. The incomplete florets proximal to the female-fertile florets. *Spikelets with proximal incomplete florets. The proximal incomplete florets* 2; sterile. The proximal lemmas awnless (with a very short subule); exceeded by the female-fertile lemmas, or exceeded by the female-fertile lemmas to more or less equalling the female-fertile lemmas; similar in texture to the female-fertile lemmas; not becoming indurated.

Female-fertile florets 1. Lemmas not becoming indurated; entire; pointed; mucronate. *Palea* present; relatively long; apically notched; awnless, without apical setae, or with apical setae; not indurated; several nerved; 2-keeled to keel-less (being keeled only near tip). *Lodicules* present; 3; free; membranous; ciliate; not toothed; heavily vascularized.

Stamens 3. Anthers not penicillate; without an apically prolonged connective. **Ovary** glabrous; *without a conspicuous apical appendage.* Styles fused (the ovary attenuate). Stigmas 2.

Abaxial leaf blade epidermis. *Costal/intercostal zonation* conspicuous (and the intercostal zones with a median astomatal zone). *Papillae* present; costal and intercostal (but conspicuously absent from the median, astomatal intercostal regions). Intercostal papillae over-arching the stomata; several per cell (12–20 small, thick walled, lobed or branching papillae per long-cell, in two or three iregular rows). *Long-cells* similar in shape costally and intercostally to markedly different in shape costally and intercostally (those the costal zones much smaller); of similar wall thickness costally and intercostally. Mid-intercostal long-cells rectangular; having markedly sinuous walls. *Microhairs* present; elongated; clearly two-celled; panicoid-type; 39–45 microns long; 6.6–10.5 microns wide at the septum. Microhair total length/width at septum 3.7–6.8. Microhair apical cells 18–24 microns long. Microhair apical cell/total length ratio 0.45–0.55. *Stomata* common; 27–30 microns long. Subsidiaries papillate (often with two on each). *Intercostal short-cells* common (in the astomatal regions); in cork/silica-cell pairs; silicified. Intercostal silica bodies crescentic. Macrohairs and prickles lacking in the material seen. *Crown cells* absent. *Costal short-cells* conspicuously in long rows. Costal silica bodies predominantly saddle shaped and oryzoid; not sharp-pointed.

Transverse section of leaf blade, physiology. C_3; XyMS+. Mesophyll with adaxial palisade; with arm cells; with fusoids. The fusoids external to the PBS. *Leaf blade* with distinct, prominent adaxial ribs; with the ribs more or less constant in size (broad, flat topped). *Midrib* conspicuous; having complex vascularization. *The lamina* symmetrical on either side of the midrib. Bulliforms present in discrete, regular adaxial groups (a conspicuous group in each furrow); in simple fans. All the vascular bundles accompanied by sclerenchyma. Combined sclerenchyma girders present; forming 'figures' (the minor bundles with slender I's, the primary bundles with 'anchors'). Sclerenchyma all associated with vascular bundles.

Cytology. $2n = 40$.

Taxonomy. Bambusoideae; Bambusodae; Bambuseae.

Ecology, geography, regional floristic distribution. 5 species. Bolivia and Brazil to Costa Rica.

Neotropical. Caribbean.

References, etc. Morphological/taxonomic: Soderstrom 1978; Soderstrom and Calderón 1978b. Leaf anatomical: this project.

Special comments. Fruit data wanting.

Symplectrodia Lazarides

Habit, vegetative morphology. Perennial (robust or slender); caespitose (and rhizomatous). *Culms* 30–135 cm high. Culm nodes hairy (*S. gracilis*), or glabrous (*S. lanosa*). Culm internodes solid. The shoots not aromatic. *Leaves* not basally aggregated; non-auriculate; with hair tufts in the ligule region. Sheath margins free. The basal sheaths sometimes reddish brown or with woolly hairs. *Leaf blades* narrow; acicular; **hard, woody, needle-like (cf. Triodia)**; without abaxial multicellular glands; without cross venation. Ligule a fringe of hairs. *Contra-ligule* absent.

Reproductive organization. Plants bisexual, with bisexual spikelets; with hermaphrodite florets.

Inflorescence. Inflorescence paniculate; open to contracted. Inflorescence with axes ending in spikelets. Inflorescence espatheate; not comprising 'partial inflorescences' and foliar organs. Spikelet-bearing axes persistent. Spikelets solitary; not secund; long pedicellate.

Female-fertile spikelets. *Spikelets not noticeably compressed;* disarticulating above the glumes; disarticulating between the florets; *with distinctly elongated rachilla internodes between the florets (the internode above the fertile floret up to 5 mm long and adnate to the palea, and the upper internodes also elongated at maturity, with the sterile florets ultimately disarticulating individually).* Rachilla prolonged beyond the uppermost female-fertile floret. The rachilla extension with incomplete florets. Hairy callus present. Callus pointed (curved or oblique).

Glumes two; very unequal; shorter than the spikelets to about equalling the spikelets; long relative to the adjacent lemmas; hairless; usually glabrous; pointed; awned to awnless (often aristulate); non-carinate (rounded or flattened dorsally); similar (cartilaginous, lanceolate-elliptic, acuminate). Lower glume 1–5 nerved. Upper glume 3–7 nerved. *Spikelets* with incomplete florets. The incomplete florets distal to the female-fertile florets (without paleas, often reduced). The distal incomplete florets 2–5; *incomplete florets* clearly specialised and modified in form (empty, cartilaginous, 3-lobed, 3-nerved, unequally 3-awned, finally becoming distant and prominently exserted by elongation of the internodes); incomplete florets awned.

Female-fertile florets 1. Lemmas similar in texture to the glumes (cartilaginous); entire; pointed; awned. Awns 1; median; apical; non-geniculate; hairless; much shorter than the body of the lemma to about as long as the body of the lemma. Lemmas hairy (marginally and on the midnerve), or hairless; without a germination flap; 3 nerved. *Palea* present; relatively long; entire (acute); awnless, without apical setae; hardened and adnate to the rachilla below; 2-nerved; 2-keeled. *Lodicules* present; 2; fleshy; glabrous; not or scarcely vascularized. *Stamens* 3. Anthers not penicillate; without an apically prolonged connective. *Ovary* glabrous. Styles free to their bases. Stigmas 2.

Fruit, embryo and seedling. *Fruit* small to medium sized (4–4.5 mm long); ellipsoid; *longitudinally grooved (on the hilar face)*; compressed dorsiventrally. Hilum short. Pericarp fused. Embryo large.

Abaxial leaf blade epidermis. *Costal/intercostal zonation* conspicuous (the intercostal zones sunken in narrow grooves, inaccessible for observation in surface view). *Papillae* present; intercostal, or costal and intercostal. Intercostal papillae not overarching the stomata; consisting of one oblique swelling per cell, or consisting of one symmetrical projection per cell, or several per cell. *Microhairs* present; elongated; clearly two-celled; chloridoid-type. Microhair apical cell wall of similar thickness/rigidity to that of the basal cell. Microhairs about 39 microns long (where recordable); 15 microns wide at the septum. Microhair total length/width at septum 2.6. Microhair apical cells 7.5 microns long. Microhair apical cell/total length ratio 0.19. *Stomata* not observable; 18–21 microns long. Subsidiaries papillate, or non-papillate. *Costal short-cells* conspicuously in long rows. Costal silica bodies present in alternate cell files of the costal zones; 'panicoid-type'; dumb-bell shaped (large and sometimes very asymmetric); not sharp-pointed.

Transverse section of leaf blade, physiology. C$_4$. The anatomical organization unconventional. Organization of PCR tissue *Triodia* type (PCR 'sheaths' lateral only or 'draping'). XyMS+. PCR sheath outlines even, or uneven to even. Mesophyll traversed by columns of colourless mesophyll cells (linking the adaxial and abaxial grooves); with arm cells (cf. *Sclerodactylon, Triodia*). *Leaf blade* 'nodular' in section; with the ribs very irregular in sizes. *Midrib* not readily distinguishable; with one bundle only. Bulliforms present in discrete, regular adaxial groups; associated with colourless mesophyll cells to form deeply-penetrating fans (incorporated in or linked with the colourless girders). All the vascular bundles accompanied by sclerenchyma. Combined sclerenchyma girders present; forming 'figures' (large anchors). Sclerenchyma not all bundle-associated. The 'extra' sclerenchyma in a continuous abaxial layer.

Taxonomy. Chloridoideae; Triodieae.

Ecology, geography, regional floristic distribution. 2 species. Australia. Xerophytic; species of open habitats.

Australian. North and East Australian. Tropical North and East Australian.

References, etc. Morphological/taxonomic: Lazarides 1985. Leaf anatomical: this project.

Taeniatherum Nevski

~ Elymus

Habit, vegetative morphology. *Annual. Culms* 5–60 cm high; herbaceous. Culm internodes hollow. *Leaves* not basally aggregated; auriculate (the auricles small). Leaf blades linear; narrow; 0.3–4 mm wide; flat, or rolled (convolute); without cross venation. **Ligule** present (short); *an unfringed membrane*; truncate; 0.2–0.5 mm long.

Reproductive organization. Plants bisexual, with bisexual spikelets; with hermaphrodite florets. The spikelets of sexually distinct forms on the same plant; hermaphrodite and sterile (sterile at the tip of the rachis). Viviparous (sometimes), or not viviparous.

Inflorescence. Inflorescence a false spike, with spikelets on contracted axes (the clusters reduced to spikelet pairs); espatheate; not comprising 'partial inflorescences' and foliar organs. Spikelet-bearing axes persistent. Spikelets paired; not secund; distichous; sessile.

Female-fertile spikelets. *Spikelets* 8–15 mm long; *compressed dorsiventrally*; disarticulating above the glumes. Rachilla prolonged beyond the uppermost female-fertile floret; rachilla hairless (slightly scabrid, flattened between the florets). The rachilla extension with incomplete florets. Hairy callus absent. *Callus* long; pointed (with only a few marginal spicules).

Glumes two; very unequal to more or less equal; *joined*; hairless; scabrous; subulate; awned (awn-like); similar (indurated at the base). Lower glume 1 nerved, or 3 nerved. Upper glume 1 nerved, or 3 nerved. *Spikelets* with incomplete florets. The incomplete florets distal to the female-fertile florets. The distal incomplete florets usually 1; *incomplete florets* merely underdeveloped.

Female-fertile florets 1. Lemmas attenuate to the awn; less firm than the glumes to similar in texture to the glumes; entire; pointed; awned. Awns 1; median; apical; non-geniculate; recurving; hairless (scabrid); much longer than the body of the lemma (4–12 cm long); entered by several veins. Lemmas hairless; scabrous; non-carinate; without a germination flap; 5 nerved. *Palea* present; relatively long (about equalling the lemma); entire (obtuse), or apically notched (slightly); 2-nerved; 2-keeled (scabrid on the keels). *Lodicules* present; 2; free; membranous; ciliate; toothed; not or scarcely vascularized. *Stamens* 3. Anthers 0.6–1.4 mm long (relatively short). **Ovary hairy.** Styles free to their bases. Stigmas 2.

Fruit, embryo and seedling. *Fruit* somewhat adhering to lemma and/or palea; medium sized (6–9 mm long); longitudinally grooved; compressed dorsiventrally; with hairs confined to a terminal tuft. Hilum long-linear. Embryo small. Endosperm hard; without lipid; containing only simple starch grains. Embryo with an epiblast, or without an epiblast; without a scutellar tail; with a negligible mesocotyl internode. Embryonic leaf margins meeting.

Abaxial leaf blade epidermis. *Costal/intercostal zonation* conspicuous. *Papillae* absent. Long-cells of similar wall thickness costally and intercostally (quite thick walled). Mid-intercostal long-cells rectangular and fusiform; having markedly sinuous walls, or having straight or only gently undulating walls. *Microhairs* absent. *Stomata* common; 28–33 microns long. Subsidiaries dome-shaped (mostly, low), or parallel-sided. Guard-cells overlapped by the interstomatals (but at the most very slightly), or overlapping to flush with the interstomatals. *Intercostal short-cells* common; in cork/silica-cell pairs (mostly); silicified. Intercostal silica bodies rounded to crescentic. *Crown cells* present. Costal short-cells predominantly paired, or neither distinctly grouped into long rows nor predominantly paired. Costal silica bodies horizontally-elongated crenate/sinuous, or horizontally-elongated smooth, or rounded, or crescentic; not sharp-pointed.

Transverse section of leaf blade, physiology. C_3; XyMS+. *Mesophyll* with non-radiate chlorenchyma; without adaxial palisade. *Leaf blade* with distinct, prominent adaxial ribs; with the ribs more or less constant in size (wide, rounded). *Midrib* not readily distinguishable; with one bundle only. Bulliforms present in discrete, regular adaxial groups; in simple fans. Many of the smallest vascular bundles unaccompanied by sclerenchyma, or all the vascular bundles accompanied by sclerenchyma. Combined sclerenchyma girders present; forming 'figures' (in *T. caput-medusae*). Sclerenchyma all associated with vascular bundles.

Cytology. Chromosome base number, $x = 7$. $2n = 14$. 2 ploid. Haplomic genome content T. Haploid nuclear DNA content 4.4 pg (1 species). Mean diploid 2c DNA value 8.8 pg.

Taxonomy. Pooideae; Triticodae; Triticeae.

Ecology, geography, regional floristic distribution. 2 species. Mediterranean to northwest India. Commonly adventive. Xerophytic.
Holarctic. Boreal and Tethyan. Euro-Siberian. Irano-Turanian. European.
Hybrids. Sterile hybrids made with *Hordeum* and *Aegilops*.
Rusts and smuts. Rusts — *Puccinia*. Taxonomically wide-ranging species: *Puccinia striiformis* and *Puccinia hordei*. Smuts from Tilletiaceae and from Ustilaginaceae. Tilletiaceae — *Tilletia*. Ustilaginaceae — *Ustilago*.
Economic importance. Important native pasture species: *T. asperum, T. crinitum*.
References, etc. Morphological/taxonomic: Löve 1984. Leaf anatomical: Metcalfe 1960; this project.

Taeniorhachis T.A. Cope

Habit, vegetative morphology. Perennial; rhizomatous and stoloniferous. *Culms* up to 8 cm high; herbaceous; 2–5 noded (?). Culm nodes hidden by leaf sheaths. *Leaves* not basally aggregated; conspicuously distichous; non-auriculate; without auricular setae. Leaf blades lanceolate to ovate; 'rigid, with white cartilaginous margins, densely pubescent especially below'; narrow; about 2–4 mm wide (?); flat, or folded; without cross venation (?); persistent (?). Ligule an unfringed membrane, or a fringed membrane (? — state indistinguishable from the illustration, not mentioned in the description); truncate (very short).

Reproductive organization. Plants bisexual, with bisexual spikelets; with hermaphrodite florets. *The spikelets all alike in sexuality*.

Inflorescence. *Inflorescence* of spicate main branches; *digitate*. Primary inflorescence branches 2 (the pair of conjugate 'racemes' 1.5–2.5 cm long, their rachides 2.5–3 mm wide). *Rachides hollowed, flattened, and winged (triquetrous, broadly winged laterally, narrowly winged ventrally down the middle)*. Inflorescence espatheate; not comprising 'partial inflorescences' and foliar organs. *Spikelet-bearing axes* (the conjugate 'racemes') paired; *disarticulating*; falling entire. *Spikelets* unaccompanied by bractiform involucres, not associated with setiform vestigial branches; paired; secund; biseriate (the pairs alternating on either side of the median ridge on the ventral side of the rachis); pedicellate; imbricate; consistently in 'long-and-short' combinations; unequally pedicellate in each combination. Pedicels of the 'pedicellate' spikelets free of the rachis. The 'shorter' spikelets hermaphrodite. The 'longer' spikelets hermaphrodite.

Female-fertile spikelets. Spikelets 6–6.5 mm long; elliptic to lanceolate (lanceolate-elliptic); abaxial (?); compressed dorsiventrally; planoconvex (flattened ventrally, convex dorsally); falling with the glumes (?). The upper floret not stipitate. Rachilla terminated by a female-fertile floret. Hairy callus absent.

Glumes two; (the upper) relatively large; very unequal (the lower a minute scale 0.5–0.6 mm long, the upper much larger and more substantial); shorter than the spikelets (the larger, upper glume being about 2/3 the length of the spikelet); *shorter than the adjacent lemmas*; hairy (the upper being pilose between the nerves); awnless; non-carinate; very dissimilar (the lower a minute, ovate scale, the upper lanceolate and membranous). *Lower glume 0 nerved*. Upper glume 3 nerved. *Spikelets* with incomplete florets. The incomplete florets proximal to the female-fertile florets. *The proximal incomplete florets* 1; epaleate; sterile. The proximal lemmas prominently nerved, villous in the outer interspaces; awnless; 9–11 nerved (the nerves close together); more or less equalling the female-fertile lemmas.

Female-fertile florets 1. *Lemmas* acuminate, chartaceous with hyaline margins, finely longitudinally striate; decidedly firmer than the glumes (chartaceous); finely striate; not becoming indurated; entire; pointed (acuminate); awnless; hairless (?); non-carinate; *having the margins lying flat and exposed on the palea (the hyaline margins enfolding and concealing most of the palea)*. *Palea* present; relatively long (?); tightly clasped by the lemma; awnless, without apical setae.

Fruit, embryo and seedling. Fruit 'pallid'; ellipsoid.

Taxonomy. Panicoideae; Panicodae; presumably Paniceae (seemingly a *Digitaria* relative, but not reliably assignable from the available description).

Ecology, geography, regional floristic distribution. 1 species (*T. repens* Cope). Collected only once, in Somalia northeast of Mogadishu. Xerophytic; species of open habitats; halophytic. On a coastal dune of white sand.

Paleotropical. African. Sudano-Angolan. Somalo-Ethiopian.

References, etc. Morphological/taxonomic: Cope 1993.

Special comments. These very inadequate data have been encoded, with some tentative guesswork, from the original very poor description and illustration. The former (including Latin diagnosis) occupies only one page, but includes the statement 'Caryopsis not seen', separated by 8 lines from 'fruit ellipsoid, pallid'. Ligule, spikelet orientation, female-fertile palea, lodicules, androecium (etc.) are neither dealt with in the text nor represented in the illustration. Fruit data wanting. Anatomical data wanting.

Tarigidia Stent

Habit, vegetative morphology. Glaucous perennial; caespitose. *Culms* 80–150 cm high; herbaceous; branched above, or unbranched above. Culm nodes glabrous. Culm internodes solid. Plants unarmed. Young shoots intravaginal. **Leaves *mostly basal (but cauline leaves conspicuous)*;** slightly auriculate, or non-auriculate. The sheath bases sometimes profusely hairy. Leaf blades linear; narrow; 3–6 mm wide; flat (with thickened margins); without cross venation; persistent. *Ligule an unfringed membrane;* truncate; 1–5 mm long. *Contra-ligule* absent.

Reproductive organization. Plants bisexual, with bisexual spikelets; with hermaphrodite florets. The spikelets of sexually distinct forms on the same plant, or all alike in sexuality; hermaphrodite, or hermaphrodite and sterile (in that some at the raceme bases may be sterile).

Inflorescence. *Inflorescence a false spike, with spikelets on contracted axes, or paniculate;* contracted; spicate (or the lower branches ascending). Inflorescence with axes ending in spikelets, or axes not ending in spikelets (rarely, the lower branches are shortly produced). Inflorescence espatheate; not comprising 'partial inflorescences' and foliar organs. **Spikelet-bearing axes** 'racemes' (at the base of the inflorescence), or capitate (at the apex); *disarticulating (at least the lower branches do so);* falling entire. Spikelets clustered or in pairs on lower panicle branches; secund; shortly pedicellate; imbricate.

Female-fertile spikelets. Spikelets 4–4.5 mm long; abaxial; compressed dorsiventrally; falling with the glumes; not disarticulating between the florets; with conventional internode spacings. Rachilla terminated by a female-fertile floret. Hairy callus absent. *Callus* absent.

Glumes two; very unequal (the lower sometimes much reduced), or more or less equal (usually); shorter than the spikelets; *shorter than the adjacent lemmas;* free; hairy (sometimes densely so); pointed (acuminate); awnless; non-carinate. Lower glume shorter than the lowest lemma; 1 nerved. Upper glume 3 nerved. *Spikelets* with incomplete florets. The incomplete florets proximal to the female-fertile florets. *Spikelets with proximal incomplete florets. The proximal incomplete florets* 1; epaleate; sterile. The proximal lemmas awnless; 5–7 nerved; more or less equalling the female-fertile lemmas; similar in texture to the female-fertile lemmas (papery, with long spreading hairs between nerves and at the margins); not becoming indurated.

Female-fertile florets 1. Lemmas elliptical-obtuse; similar in texture to the glumes; not becoming indurated; entire; pointed; awnless; hairless; non-carinate; having the margins lying flat and exposed on the palea; without a germination flap. *Palea* present; relatively long; tightly clasped by the lemma; entire; awnless, without apical setae; textured like the lemma; not indurated; 2-nerved. *Lodicules* present; 2; free; fleshy; glabrous; not or scarcely vascularized. *Stamens* 3. Anthers 2.5 mm long; not penicillate; without an apically prolonged connective. *Ovary* glabrous. Styles free to their bases. Stigmas 2; red pigmented.

Abaxial leaf blade epidermis. *Costal/intercostal zonation* conspicuous. *Papillae* absent. Mid-intercostal long-cells having markedly sinuous walls and having straight or only gently undulating walls (i.e., mixed). *Microhairs* present; panicoid-type; (69–)78–93(–99) microns long; 4.5–5.1 microns wide at the septum. Microhair total length/width at septum 13.5–20.7. Microhair apical cells (37.5–)42–50(–51) microns long. Microhair apical cell/total length ratio 0.5–0.56. *Stomata* common; 30–33 microns long. Subsidiaries high dome-shaped (mostly), or triangular. Guard-cells overlapping to flush with the interstomatals. *Intercostal short-cells* absent or very rare. *Costal short-cells* conspicuously in long rows. Costal silica bodies 'panicoid-type'; cross shaped to butterfly shaped; not sharp-pointed.

Transverse section of leaf blade, physiology. C_4; XyMS–. PCR cell chloroplasts centrifugal/peripheral. *Mesophyll* with radiate chlorenchyma. *Leaf blade* adaxially flat. *Midrib* conspicuous; having a conventional arc of bundles (one large bundle, about 10 small); with colourless mesophyll adaxially. Bulliforms present in discrete, regular adaxial groups; in simple fans. Many of the smallest vascular bundles unaccompanied by sclerenchyma. Combined sclerenchyma girders absent. Sclerenchyma all associated with vascular bundles.

Taxonomy. Panicoideae; Panicodae; Paniceae.

Ecology, geography, regional floristic distribution. 1 species. Southern Africa. Mesophytic to xerophytic; species of open habitats; glycophytic. Dry grassland.

Paleotropical and Cape. African. Sudano-Angolan and Namib-Karoo. South Tropical African.

Hybrids. A taxon sharing features of *Digitaria* and *Anthephora*: originally suggested as an intergeneric hybrid — which accords with absence of fruit in material seen, and the scattered distribution.

References, etc. Morphological/taxonomic: Stent 1932; Loxton 1974. Leaf anatomical: this project.

Special comments. Fruit data wanting.

Tatianyx Zuloaga and Soderstrom

Habit, vegetative morphology. Perennial; caespitose (with densely villous cataphylls). *Culms* 40–100 cm high; herbaceous; unbranched above. Culm nodes hairy. Young shoots intravaginal. *Leaves* not basally aggregated; non-auriculate. Leaf blades linear-lanceolate (acuminate, those of the upper leaves reduced); narrow; 2–4 mm wide; flat, or rolled; without cross venation; persistent. *Ligule a fringe of hairs*; 0.3–0.4 mm long.

Reproductive organization. Plants bisexual, with bisexual spikelets; with hermaphrodite florets.

Inflorescence. *Inflorescence paniculate*; open (pyramidal, many-flowered); *with capillary branchlets*; espatheate; not comprising 'partial inflorescences' and foliar organs. Spikelet-bearing axes persistent. Spikelets not secund; pedicellate (the pedicels flexuous). Pedicel apices oblique.

Female-fertile spikelets. *Spikelets* 4–4.5 mm long (silvery-villous); *compressed dorsiventrally*; falling with the glumes; with a distinctly elongated rachilla internode between the glumes and with distinctly elongated rachilla internodes between the florets. The upper floret not stipitate (at least, not like *Ichnanthus*). Rachilla terminated by a female-fertile floret. Hairy callus present (via the falcate pedicel tip). *Callus* short.

Glumes present; two; very unequal; (the upper) long relative to the adjacent lemmas; hairy (silvery-villous); without conspicuous tufts or rows of hairs; pointed (acute); awnless; non-carinate; similar. Lower glume longer than half length of lowest lemma; 3–5 nerved. Upper glume 5–7 nerved. *Spikelets* with incomplete florets. The incomplete florets proximal to the female-fertile florets. *The proximal incomplete florets* 1; paleate. Palea of the proximal incomplete florets fully developed (two winged, hyaline); not becoming conspicuously hardened and enlarged laterally. The proximal incomplete florets male, or sterile. The proximal lemmas awnless; 5–7 nerved; more or less equalling the female-fertile lemmas to decidedly exceeding the female-fertile lemmas; less firm than the female-fertile lemmas (herbaceous, hairy, similar to the glumes); not becoming indurated.

Female-fertile florets 1. Lemmas ellipsoid; decidedly firmer than the glumes; smooth (shining, pale); becoming indurated; entire; pointed, or blunt; awnless; hairless; glabrous; non-carinate; having the margins tucked in onto the palea; with a clear germination flap (?). *Palea* present; relatively long; entire; awnless, without apical setae; textured like the lemma; indurated; 2-nerved; keel-less. *Lodicules* present; 2; free; fleshy; glabrous. *Stamens* 3. Anthers 1–1.4 mm long; not penicillate; without an apically prolonged connective. *Ovary* glabrous. Stigmas 2.

Fruit, embryo and seedling. Fruit small (2.2 mm long); longitudinally grooved; compressed dorsiventrally. *Hilum long-linear*. Embryo large.

Abaxial leaf blade epidermis. *Costal/intercostal zonation* conspicuous. *Papillae* absent. *Long-cells* markedly different in shape costally and intercostally (the costals more regularly rectangular); of similar wall thickness costally and intercostally. Mid-intercostal long-cells rectangular (though mostly somewhat narrowed at the ends); having markedly sinuous walls. *Microhairs* present (and abundant); of peculiar form, not reliably interpretable in surface view: the apical cell relatively thick walled and darkly pigmented, usually more or less bent over and often pyriform, the basal cell apparently more or less sunken. *Stomata* common; 48–51 microns long. Subsidiaries low dome-shaped (mostly), or triangular. Guard-cells overlapping to flush with the interstomatals. *Intercostal short-cells* common; in cork/silica-cell pairs; silicified. Intercostal silica bodies crescentic and oryzoid-type. *Costal short-cells* predominantly paired. Costal silica bodies rounded, or tall-and-narrow (common), or crescentic (commonly broadly D-shaped); not sharp-pointed.

Transverse section of leaf blade, physiology. C_4. The anatomical organization somewhat unconventional (in exhibiting numerous small, peculiarly positioned vascular bundles). XyMS+. PCR sheath outlines even. PCR sheath extensions absent. Mesophyll without adaxial palisade. *Leaf blade* 'nodular' in section (peculiarly so: each half-blade with flat bottomed abaxial ribs corresponding with each main bundle, the adaxial surface with a wide, triangular marginal rib and internal to it one broad rib incorporating two costal zones and the intercostal region between them). Midrib and blade having complex vascularization (there being numerous small bundles, including some aligned vertically adjoining the abaxial girders of the primary bundles, and some occurring adaxially to the two primaries which are incorporated in the blade margin). Many of the smallest vascular bundles unaccompanied by sclerenchyma. Combined sclerenchyma girders present; forming 'figures' (all the primaries having a short adaxial girder and an elongated abaxial one, the combination constituting a massive 'I'). Sclerenchyma all associated with vascular bundles.

Taxonomy. Panicoideae; Panicodae; Paniceae (probably).

Ecology, geography, regional floristic distribution. 1 species. Brazil. Species of open habitats. Savanna.

Neotropical. Central Brazilian.

References, etc. Leaf anatomical: this project: described here from poorly preserved material, with a disintegrating midrib which could not be observed. The peculiar vascular bundle pattern and the unusual microhairs need further, detailed study.

Teinostachyum Munro

~ *Schizostachyum*

Habit, vegetative morphology. Perennial; caespitose. The flowering culms leafy. *Culms* 400–1000 cm high; woody and persistent; to 3 cm in diameter; not scandent (shrubby or arborescent); branched above. Primary branches/mid-culm node many. Young shoots extravaginal. *Leaves* not basally aggregated. Leaf blades broad; 10–50 mm wide (15–30 cm long); cordate, or not cordate, not sagittate; pseudopetiolate (the blade recurved); without cross venation (but with bridging by pellucid glands); disarticulating from the sheaths; rolled in bud. Ligule an unfringed membrane.

Reproductive organization. Plants bisexual, with bisexual spikelets; with hermaphrodite florets.

Inflorescence. Inflorescence indeterminate; *with pseudospikelets (?)*; paniculate (spikelets in spike-like panicles); spatheate; a complex of 'partial inflorescences' and intervening foliar organs. *Spikelet-bearing axes* paniculate (the spike-like panicles on leafy branches); persistent. Spikelets not secund.

Female-fertile spikelets. Spikelets 15–75 mm long; disarticulating above the glumes; disarticulating between the florets. Rachilla prolonged beyond the uppermost female-fertile floret. The rachilla extension with incomplete florets.

Glumes one per spikelet, or two; shorter than the adjacent lemmas; awnless. *Spikelets* with incomplete florets. The incomplete florets both distal and proximal to the female-fertile florets. *The distal incomplete florets* merely underdeveloped. **Spikelets with proximal incomplete florets**. The proximal incomplete florets 1. The proximal lemmas shortly awned, or awnless (but mucronate); similar in texture to the female-fertile lemmas.

Female-fertile florets 2–10 (? 'few to many'). Lemmas similar in texture to the glumes; not becoming indurated; entire; pointed; mucronate to awned. Awns when present, 1; apical; much shorter than the body of the lemma. Lemmas 9–11 nerved. *Palea* present; several nerved (5–9); 2-keeled. Palea keels hairy (ciliate). *Lodicules* present; 3; free; membranous; ciliate, or glabrous; not toothed; heavily vascularized (with 3–9 nerves). *Stamens* 6. Anthers not penicillate; with the connective apically prolonged, or without an apically prolonged connective. **Ovary glabrous**; with a conspicuous apical appendage. The appendage long, stiff and tapering. Styles fused. Stigmas 2, or 3.

Fruit, embryo and seedling. *Fruit* free from both lemma and palea. Pericarp thick and hard (crustaceous); free.

Abaxial leaf blade epidermis. *Costal/intercostal zonation* conspicuous. *Papillae* present (numerous, costally and intercostally). Intercostal papillae over-arching the stomata; several per cell (tall, variable in size and irregular in shape, thick-walled). Long-cells of similar wall thickness costally and intercostally (thin walled). Mid-intercostal long-cells rectangular; having markedly sinuous walls. *Microhairs* present; clearly two-celled, or uniseriate (occasionally three-celled); panicoid-type (some with 2 basal cells); (57–)63–66(–69) microns long; (5.1–)6–6.6(–7.8) microns wide at the septum. Microhair total length/width at septum 8.8–12.9. Microhair apical cells (27–)28.5–30(–33) microns long. Microhair apical cell/total length ratio 0.43–0.53. *Stomata* common (in bands alongside the veins); 24–27 microns long. *Intercostal short-cells* common; in cork/silica-cell pairs; silicified. Intercostal silica bodies tall-and-narrow, or crescentic, or saddle shaped (the saddles mostly imperfect). Numerous, bulbous-based, tiny-pointed 'prickles' present. *Costal short-cells* predominantly paired (plus a few short rows). Costal silica bodies saddle shaped (abundant), or tall-and-narrow (a few), or oryzoid (a few); not sharp-pointed.

Transverse section of leaf blade, physiology. C_3; XyMS+. Mesophyll with adaxial palisade; with arm cells; with fusoids. The fusoids external to the PBS. *Leaf blade* adaxially flat (except beside midrib). *Midrib* conspicuous (large); having complex vascular-ization. Bulliforms present in discrete, regular adaxial groups; in simple fans. All the vascular bundles accompanied by sclerenchyma. Combined sclerenchyma girders present (with all the bundles); forming 'figures' (all bundles). Sclerenchyma all associated with vascular bundles.

Taxonomy. Bambusoideae; Bambusodae; Bambuseae.

Ecology, geography, regional floristic distribution. 3 species. India. Paleotropical. Indomalesian. Indian.

References, etc. Leaf anatomical: this project.

Special comments. Fruit data wanting.

Tetrachaete Chiov.

Habit, vegetative morphology. Wiry annual; loosely caespitose. *Culms* 4–25 cm high; herbaceous. Culm nodes glabrous. *Leaves* not basally aggregated; non-auriculate. Leaf blades narrow; setaceous to not setaceous; flat, or rolled (convolute); without abaxial multicellular glands; without cross venation; persistent. Ligule a fringed membrane, or a fringe of hairs.

Reproductive organization. Plants bisexual, with bisexual spikelets; with hermaph-rodite florets.

Inflorescence. *Inflorescence a false spike, with spikelets on contracted axes (from an inflated sheath, a short spike of twin-spikelet glomerules, each glomerule comprising a pair of spikelets subtended by an involucre of four glumes)*. Inflorescence with axes ending in spikelets (or at least, in terminal glomerules). Inflorescence spatheate (via an inflated upper sheath with reduced lamina); not comprising 'partial inflorescences' and foliar organs. *Spikelet-bearing axes* very much reduced (to the glomerules); disarticulating; falling entire (i.e., the glomerules disarticulate from the persistent main axis). Spikelets paired; not secund (the glomerules spiralled on the slender rachis); sessile.

Female-fertile spikelets. Spikelets 3–4 mm long; compressed dorsiventrally; falling with the glumes (in the glomerules). Rachilla minutely prolonged beyond the uppermost female-fertile floret. The rachilla extension naked. Hairy callus present (at the base of the glomerule).

Glumes two; relatively large (awn-like, to 1.5 cm long, bending out-wards); more or less equal; exceeding the spikelets; long relative to the adjacent lemmas (far exceeding it); slightly *joined*; *lateral to the rachis (relative to the spikelet, both being behind the lemma)*; hairy (densely pilose around the spikelets, scabrid above); awned (awn-like); similar (identical). Lower glume 1 nerved. Upper glume 1 nerved. *Spikelets* with female-fertile florets only.

Female-fertile florets 1. **Lemmas saccate (gibbous)**; less firm than the glumes (firmly membranous); not becoming indurated; entire; pointed; awned (attenuate into a 4 mm awn). Awns 1; median; apical; non-geniculate; much longer than the body of the lemma. Lemmas hairy (pilose, especially on the veins); strongly carinate; without a germination flap; 3 nerved. *Palea* present; relatively long; 2-nerved (acuminate, minutely pubescent); 2-keeled. *Lodicules* absent (seemingly). *Stamens* 3. Anthers not penicillate; without an apically prolonged connective. *Ovary* glabrous.

Fruit, embryo and seedling. *Fruit* free from both lemma and palea; small (2 mm long); sub-triquetrous; sculptured. Hilum short. Pericarp fused. Embryo large.

Abaxial leaf blade epidermis. *Costal/intercostal zonation* conspicuous. *Papillae* absent. *Long-cells* similar in shape costally and intercostally; of similar wall thickness costally and intercostally. Mid-intercostal long-cells rectangular; having markedly sinuous walls. *Microhairs* present; more or less spherical to elongated; clearly two-celled; chloridoid-type. Microhair apical cell wall of similar thickness/rigidity to that of the basal cell. Microhairs 21–27 microns long. Microhair basal cells 15 microns long. Microhairs 6–8.4 microns wide at the septum. Microhair total length/width at septum 2.7–3.6. Microhair apical cells 6.6–9 microns long. Microhair apical cell/total length ratio 0.31–0.4. *Stomata* common; 18–21 microns long. Subsidiaries triangular (mostly with truncated apices). Guard-cells overlapping to flush with the interstomatals. *Intercostal short-cells* common; in cork/silica-cell pairs (small); silicified. Intercostal silica bodies present and perfectly developed; panicoid type. *Costal short-cells* conspicuously in long rows. Costal silica bodies present in alternate cell files of the costal zones; 'panicoid-type'; cross shaped and dumb-bell shaped; not sharp-pointed.

Transverse section of leaf blade, physiology. C_4; XyMS+. PCR sheaths of the primary vascular bundles interrupted; interrupted abaxially only. PCR sheath extensions absent. *Leaf blade* with distinct, prominent adaxial ribs to 'nodular' in section; with the ribs more or less constant in size. *Midrib* not readily distinguishable; with one bundle only. Bulliforms present in discrete, regular adaxial groups; in simple fans. All the vascular bundles accompanied by sclerenchyma. Combined sclerenchyma girders present; forming 'figures'. Sclerenchyma all associated with vascular bundles. The lamina margins with fibres.

Taxonomy. Chloridoideae; main chloridoid assemblage.

Ecology, geography, regional floristic distribution. 1 species. Eritrea, Arabia. Xerophytic; species of open habitats. Dry stony slopes.

Paleotropical. African and Indomalesian. Saharo-Sindian and Sudano-Angolan. Indian. Somalo-Ethiopian and South Tropical African.

References, etc. Leaf anatomical: this project.

Tetrachne Nees

Habit, vegetative morphology. Perennial; forming large tufts, the shoots crowded on a short, oblique rhizome. **Culms** 30–100 cm high; *herbaceous*; to 0.3 cm in diameter; branched above to unbranched above. Culm nodes glabrous. Culm internodes solid. Plants unarmed. Young shoots intravaginal. *Leaves* mostly basal; non-auriculate. Leaf blades linear; narrow; 1–4 mm wide; setaceous; usually rolled; without abaxial multicellular glands; without cross venation; persistent. *Ligule a fringe of hairs (dense)*; about 1 mm long. *Contra-ligule* absent.

Reproductive organization. Plants bisexual, with bisexual spikelets; with hermaphrodite florets. Viviparous, or not viviparous.

Inflorescence. *Inflorescence of spicate main branches (the few to many branches appressed, short, dense)*. Inflorescence with axes ending in spikelets. Inflorescence espatheate; not comprising 'partial inflorescences' and foliar organs. The racemes spikelet bearing to the base. Spikelet-bearing axes persistent. Spikelets solitary; secund; biseriate (on one side of rachis, crowded); very shortly pedicellate; densely imbricate.

Female-fertile spikelets. *Spikelets* 4–6 mm long; adaxial; compressed laterally; *falling with the glumes*; not disarticulating between the florets; with conventional internode spacings. Rachilla prolonged beyond the uppermost female-fertile floret; rachilla hairless. The rachilla extension with incomplete florets. Hairy callus absent. *Callus* absent.

Glumes two; *more or less equal*; shorter than the adjacent lemmas; lateral to the rachis; hairless (scabrid on the keels); pointed; awnless; carinate; the keels somewhat winged; similar (thin, acute, the lower smaller). Lower glume shorter than the lowest lemma; 1 nerved. Upper glume 1 nerved. *Spikelets* with incomplete florets. The incomplete florets both distal and proximal to the female-fertile florets. *The distal incomplete florets* merely underdeveloped; awnless. *The proximal incomplete florets* 2; epaleate; sterile. The proximal lemmas awnless; 1–3 nerved; exceeded by the female-fertile lemmas; similar in texture to the female-fertile lemmas (membranous); not becoming indurated (similar to the glumes, but rather larger).

Female-fertile florets 3–5. Lemmas similar in texture to the glumes; not becoming indurated; entire; pointed; awnless; hairless; glabrous (scabrous on the keels); carinate. The keel slightly winged (cf. the glumes). Lemmas having the margins lying flat and exposed on the palea; without a germination flap; 5 nerved. *Palea* present; relatively long (equalling the lemma); entire; awnless, without apical setae; textured like the lemma; not indurated (thin); 2-nerved; 2-keeled. Palea keels winged; scabrous. *Lodicules* present; 2; free; fleshy; glabrous; heavily vascularized. *Stamens* 3. Anthers 2 mm long; not penicillate; without an apically prolonged connective. *Ovary* glabrous. Styles free to their bases. Stigmas 2; white.

Fruit, embryo and seedling. *Fruit* free from both lemma and palea; small (2.5 mm long); fusiform; slightly compressed laterally. Hilum short. Pericarp fused, or loosely adherent (removable with difficulty after soaking). Embryo large (about 2/3 the length of the fruit); not waisted.

Abaxial leaf blade epidermis. *Costal/intercostal zonation* conspicuous. *Papillae* absent. *Long-cells* similar in shape costally and intercostally; of similar wall thickness costally and intercostally. Mid-intercostal long-cells rectangular; having markedly sinuous walls. *Microhairs* present; elongated; clearly two-celled; panicoid-type and chloridoid-type (material seen has a mixture of panicoid type and more or less chloridoid type — but the apical cells are thin walled and often collapsed or missing). Microhair apical cell wall thinner than that of the basal cell but not tending to collapse. Microhairs (24–)27–30 microns long. Microhair basal cells 15 microns long. Microhairs 6–6.9 microns wide at the septum. Microhair total length/width at septum 3.9–5. Microhair apical cells (9–)12–13.5 microns long. Microhair apical cell/total length ratio 0.38–0.45. *Stomata* common; 21–24 microns long. Subsidiaries dome-shaped and triangular (mostly domes). Guard-cells overlapping to flush with the interstomatals. *Intercostal short-cells* common; in cork/silica-cell pairs; silicified. Intercostal silica bodies present and perfectly developed; saddle shaped. *Costal short-cells* predominantly paired. Costal silica bodies present throughout the costal zones; saddle shaped (small), or crescentic (a few, and intermediates with saddles); not sharp-pointed.

Transverse section of leaf blade, physiology. C_4; XyMS+. PCR sheath outlines fairly even. PCR sheaths of the primary vascular bundles interrupted; interrupted both abaxially and adaxially. PCR sheath extensions absent. PCR cell chloroplasts centripetal. *Mesophyll* with radiate chlorenchyma; traversed by columns of colourless mesophyll cells (the columns often wide, cf. *Aristida*). *Leaf blade* with distinct, prominent adaxial ribs to 'nodular' in section; with the ribs more or less constant in size (low, round-topped). *Midrib* conspicuous to not readily distinguishable; with one bundle only, or having a conventional arc of bundles; with colourless mesophyll adaxially, or without colourless mesophyll adaxially. Bulliforms present in discrete, regular adaxial groups; associated with colourless mesophyll cells to form deeply-penetrating fans (these linking with colourless columns). All the vascular bundles accompanied by sclerenchyma. Combined sclerenchyma girders present (with the primaries— the others with strands only); forming 'figures'. Sclerenchyma not all bundle-associated. The 'extra' sclerenchyma in abaxial groups. The lamina margins with fibres.

Cytology. Chromosome base number, $x = 10$. $2n = 20$. 2 ploid.

Taxonomy. Chloridoideae; main chloridoid assemblage.

Ecology, geography, regional floristic distribution. 1 species. South Africa and Pakistan. Mesophytic (often in alluvial soil); species of open habitats; glycophytic. In high altitude grassland.

Paleotropical and Cape. African. Sudano-Angolan. South Tropical African.

References, etc. Leaf anatomical: this project.

Special comments. Some cultivated specimens hint at hybridization with *Fingerhuthia*.

Tetrapogon Desf.

Codonachne Steud., *Cryptochloris* Benth., *Lepidopironia* A. Rich.

Habit, vegetative morphology. Annual, or perennial; stoloniferous, or caespitose. *Culms* 13–85 cm high; herbaceous. Culm nodes glabrous (pale or dark). *Leaves* not basally aggregated; non-auriculate. Sheath margins free. The sheaths keeled. Leaf blades linear (tapered); narrow; usually folded; without abaxial multicellular glands; without cross venation. Ligule a fringed membrane; very narrow.

Reproductive organization. Plants bisexual, with bisexual spikelets; with hermaphrodite florets. Exposed-cleistogamous.

Inflorescence. *Inflorescence a single spike, or of spicate main branches (comprising 1–3 upright racemes or spikes, of which 2 may be partly or completely fused along their backs)*; digitate, or subdigitate, or non-digitate. Primary inflorescence branches 1–3. Inflorescence espatheate; not comprising 'partial inflorescences' and foliar organs. Spikelet-bearing axes persistent. Spikelets solitary, or paired; secund (the rachis one-sided); biseriate; subsessile.

Female-fertile spikelets. Spikelets 2.5–12 mm long; cuneate; compressed laterally; disarticulating above the glumes (the glumes persistent); not disarticulating between the florets, or disarticulating between the florets (but under the fertile florets only). Rachilla prolonged beyond the uppermost female-fertile floret. The rachilla extension with incomplete florets. Hairy callus present.

Glumes two; relatively large; more or less equal; *long relative to the adjacent lemmas*; hairless; glabrous; pointed (acute or acuminate); awned (awn-tipped), or awnless; carinate; similar (lanceolate, long-pointed, subhyaline). Lower glume 1 nerved. Upper glume 1 nerved. *Spikelets* with incomplete florets. The incomplete florets distal to the female-fertile florets. The distal incomplete florets 1–4; incomplete florets awned.

Female-fertile florets 2–7. *Lemmas* decidedly firmer than the glumes (herbaceous, leathery, the margins hyaline); not becoming indurated; entire (truncate), or incised; when entire blunt; not deeply cleft; *awned*. *Awns* 1; median; *dorsal*; from near the top; non-geniculate; much longer than the body of the lemma. Lemmas hairy (on the back, the margins glabrous); carinate; 3–5 nerved. *Palea* present; relatively long; apically notched; awnless; without apical setae; not indurated (membranous, sometimes hairy); 2-nerved; 2-keeled. Palea keels hairy (ciliate). *Lodicules* present; 2; free; fleshy; glabrous. *Stamens* 3. Anthers 0.5–0.7 mm long; not penicillate. *Ovary* glabrous. Stigmas 2.

Fruit, embryo and seedling. *Fruit* free from both lemma and palea; small (1.5–3 mm long); ellipsoid; compressed laterally (in one species), or compressed dorsiventrally. Hilum short. *Pericarp free*. Embryo large.

Abaxial leaf blade epidermis. *Costal/intercostal zonation* conspicuous. *Papillae* present; intercostal. Intercostal papillae small, consisting of one oblique swelling per cell. Mid-intercostal long-cells rectangular; having markedly sinuous walls (coarsely so). *Microhairs* present; elongated; clearly two-celled; chloridoid-type. Microhair apical cell wall of similar thickness/rigidity to that of the basal cell. Microhairs 18–21 microns long. Microhair basal cells 15 microns long. Microhairs (6–)7.2–8.4 microns wide at the septum. Microhair total length/width at septum 2.1–3. Microhair apical cells (7.5–)9–10.5 microns long. Microhair apical cell/total length ratio 0.42–0.58. *Stomata* common; 13.5–15 microns long. Subsidiaries triangular. Guard-cells overlapping to flush with the interstomatals. *Intercostal short-cells* common; not paired (solitary); not silicified. Intercostal silica bodies absent. *Costal short-cells* conspicuously in long rows. Costal silica bodies present in alternate cell files of the costal zones; exclusively saddle shaped; not sharp-pointed.

Transverse section of leaf blade, physiology. C_4; XyMS+. PCR sheaths of the primary vascular bundles interrupted; interrupted both abaxially and adaxially. PCR sheath extensions absent. PCR cell chloroplasts centripetal. *Mesophyll* with radiate chlorenchyma. *Leaf blade* 'nodular' in section to adaxially flat; with the ribs more or less constant in size. *Midrib* conspicuous; with one bundle only; with colourless mesophyll adaxially. Bulliforms present in discrete, regular adaxial groups; in simple fans (mostly, with deeply penetrating median cells), or associated with colourless mesophyll cells to form deeply-penetrating fans. All the vascular bundles accompanied by sclerenchyma. Combined sclerenchyma girders present (with the primaries); forming 'figures' (in the primaries). Sclerenchyma all associated with vascular bundles. The lamina margins with fibres.

Cytology. Chromosome base number, $x = 10$. $2n = 20$. 2 ploid.

Taxonomy. Chloridoideae; main chloridoid assemblage.

Ecology, geography, regional floristic distribution. 5–6 species. Mediterranean to India, tropical and South Africa. Helophytic to mesophytic; shade species, or species of open habitats; glycophytic. Savanna.

Holarctic and Paleotropical. Tethyan. African and Indomalesian. Macaronesian and Irano-Turanian. Saharo-Sindian, Sudano-Angolan, and Namib-Karoo. Indian. Sahelo-Sudanian, Somalo-Ethiopian, and South Tropical African.

Economic importance. Important native pasture species: *T. tenellus* (and other species potentially useful), in dry places.

References, etc. Morphological/taxonomic: Launert 1970. Leaf anatomical: this project.

Tetrarrhena R.Br.

~ Ehrharta

Habit, vegetative morphology. Perennial; stoloniferous and decumbent. *Culms* woody and persistent to herbaceous; scandent (often), or not scandent (wiry, often long and scrambling, 'sometimes capable of entangling a horse'); ***branched above***. Culm nodes glabrous. Culm internodes hollow. Young shoots extravaginal. *Leaves* not basally aggregated; non-auriculate. Leaf blades narrow; flat (or concave), or rolled; not pseudopetiolate; cross veined (rarely), or without cross venation; persistent. Ligule an unfringed membrane to a fringed membrane; truncate; short. *Contra-ligule* absent.

Reproductive organization. Plants bisexual, with bisexual spikelets; with hermaphrodite florets.

Inflorescence. *Inflorescence* few spikeleted; *a single raceme (spike-like, the axis flexuous)*; espatheate; not comprising 'partial inflorescences' and foliar organs. Spikelet-bearing axes persistent. *Spikelets* solitary; not secund; ***subsessile***.

Female-fertile spikelets. Spikelets 4.8–7 mm long; compressed laterally; disarticulating above the glumes. Rachilla terminated by a female-fertile floret. Hairy callus absent.

Glumes two; very unequal; shorter than the adjacent lemmas; not pointed (truncate); awnless; similar (leathery to scarious). Lower glume 1 nerved. Upper glume 5 nerved.

Spikelets with incomplete florets. The incomplete florets proximal to the female-fertile florets. *The proximal incomplete florets* 2; epaleate; sterile. The proximal lemmas awnless; faintly 7 nerved; more or less equalling the female-fertile lemmas; similar in texture to the female-fertile lemmas (tough); not becoming indurated.

Female-fertile florets 1. Lemmas similar in texture to the glumes to decidedly firmer than the glumes (leathery); not becoming indurated; entire; blunt; awnless; hairless; carinate to non-carinate; 7 nerved. **Palea** present; relatively long, or conspicuous but relatively short; entire (acute); awnless, without apical setae; thinner than the lemma (membranous); not indurated; *1-nerved*; *one-keeled (laterally compressed)*. Lodicules present (large); 2; membranous; ciliate; heavily vascularized. **Stamens 4 (usually), or 2** **(T. oreophila**). Anthers 2–3 mm long; not penicillate. *Ovary* glabrous. Styles free to their bases.

Fruit, embryo and seedling. Fruit compressed laterally. Hilum short. Embryo small. Endosperm containing compound starch grains. Embryo with an epiblast; with a scutellar tail; with a negligible mesocotyl internode.

Seedling with a short mesocotyl. First seedling leaf with a well-developed lamina. The lamina narrow; erect; 5 veined.

Abaxial leaf blade epidermis. *Costal/intercostal zonation* conspicuous. *Papillae* absent. *Long-cells* similar in shape costally and intercostally; of similar wall thickness costally and intercostally. Mid-intercostal long-cells rectangular; having markedly sinuous walls. *Microhairs* present; panicoid-type; (30–)34–60(–63) microns long; 4.5–5.1 microns wide at the septum (*T. oreophila*), or 9.6–18 microns wide at the septum. Microhair total length/width at septum 2.1–7.1, or 10 (in *T. oreophila*). Microhair apical cells (10.5–)12–35(–36) microns long. Microhair apical cell/total length ratio 0.31–0.71. *Stomata* absent or very rare; 22.5–45 microns long. Subsidiaries dome-shaped, or triangular. Guard-cells overlapping to flush with the interstomatals. *Intercostal short-cells* common; in cork/silica-cell pairs; silicified. Intercostal silica bodies tall-and-narrow, or rounded (or oval), or crescentic. *Costal short-cells* conspicuously in long rows. Costal silica bodies rounded, or 'panicoid-type'; not sharp-pointed.

Transverse section of leaf blade, physiology. C_3; XyMS+. *PBS cells* without a suberised lamella. *Mesophyll* with radiate chlorenchyma, or with non-radiate chlorenchyma; without adaxial palisade; without arm cells; without fusoids. *Leaf blade* with distinct, prominent adaxial ribs, or 'nodular' in section, or adaxially flat; with the ribs more or less constant in size. *Midrib* not readily distinguishable; with one bundle only. Bulliforms present in discrete, regular adaxial groups; in simple fans. All the vascular bundles accompanied by sclerenchyma. Combined sclerenchyma girders present; forming 'figures', or nowhere forming 'figures'. Sclerenchyma all associated with vascular bundles.

Taxonomy. Bambusoideae; Oryzodae; Ehrharteae.

Ecology, geography, regional floristic distribution. 5 species. Australia. Shade species.

Australian. North and East Australian and South-West Australian. Tropical North and East Australian and Temperate and South-Eastern Australian.

Rusts and smuts. Rusts — *Puccinia*.

References, etc. Morphological/taxonomic: Vickery 1975; Willemse 1982. Leaf anatomical: Metcalfe 1960 and this project.

Thamnocalamus Munro

Fargesia Franch., *Himalayacalamus* Keng f.

Habit, vegetative morphology. Perennial; caespitose. The flowering culms leafy. **Culms** 100–500 cm high; *woody and persistent*; to 2 cm in diameter; branched above. Culm nodes glabrous. Primary branches/mid-culm node 5–8. Unicaespitose. Rhizomes pachymorph. Plants unarmed. Young shoots intravaginal. *Leaves* not basally aggregated; non-auriculate; without auricular setae. Leaf blades broad; pseudopetiolate; cross veined; disarticulating from the sheaths; rolled in bud. Ligule a fringed membrane; truncate; 1.5 mm long. *Contra-ligule* present.

Reproductive organization. Plants bisexual, with bisexual spikelets; with hermaphrodite florets.

Inflorescence. *Inflorescence* determinate; without pseudospikelets; a single raceme, or paniculate; contracted; spatheate; a complex of 'partial inflorescences' and intervening foliar organs. Spikelets solitary; not secund; pedicellate.

Female-fertile spikelets. *Spikelets* 15–18 mm long; *not noticeably compressed*. Rachilla prolonged beyond the uppermost female-fertile floret; rachilla hairless. The rachilla extension with incomplete florets. Hairy callus absent.

Glumes two; more or less equal; *long relative to the adjacent lemmas*; hairless; pointed; awnless (the upper pointed); non-carinate; similar. Lower glume 8–9 nerved. Upper glume 9–13 nerved. *Spikelets* with incomplete florets. The incomplete florets distal to the female-fertile florets. *The distal incomplete florets* merely underdeveloped.

Female-fertile florets 2–8. Lemmas similar in texture to the glumes; smooth; not becoming indurated; entire; pointed; awnless, or mucronate (?); hairless; non-carinate; without a germination flap; 10–11 nerved. *Palea* present; relatively long; apically notched; awnless, without apical setae; not indurated; several nerved; 2-keeled. *Lodicules* present; 3; free; membranous; ciliate; not toothed; heavily vascularized. *Stamens* 3. Anthers not penicillate; without an apically prolonged connective. **Ovary** glabrous; *without a conspicuous apical appendage*. Styles free to their bases. Stigmas 3.

Abaxial leaf blade epidermis. *Costal/intercostal zonation* conspicuous. *Papillae* present; costal and intercostal. Intercostal papillae over-arching the stomata (from the interstomatals); several per cell (small, thickened, one row per cell on the long-cells of astomatal zones, irregular on the interstomatals). *Long-cells* markedly different in shape costally and intercostally. Mid-intercostal long-cells rectangular; having markedly sinuous walls. *Microhairs* present; panicoid-type. *Stomata* common. Subsidiaries parallel-sided, or dome-shaped, or parallel-sided and dome-shaped. Guard-cells overlapped by the interstomatals. *Intercostal short-cells* common; in cork/silica-cell pairs and not paired (solitary and paired). *Costal short-cells* neither distinctly grouped into long rows nor predominantly paired. Costal silica bodies saddle shaped and 'panicoid-type'; not sharp-pointed.

Transverse section of leaf blade, physiology. C_3; XyMS+. *Mesophyll* with non-radiate chlorenchyma; with adaxial palisade; with arm cells; with fusoids. The fusoids external to the PBS. *Leaf blade* adaxially flat (the ribbing at most slight). *Midrib* conspicuous; usually with one bundle only (complex in *T. aristatus*). *The lamina* symmetrical on either side of the midrib. Bulliforms present in discrete, regular adaxial groups; in simple fans, or associated with colourless mesophyll cells to form deeply-penetrating fans. All the vascular bundles accompanied by sclerenchyma. Combined sclerenchyma girders present; forming 'figures'. Sclerenchyma all associated with vascular bundles.

Cytology. Chromosome base number, $x = 12$.

Taxonomy. Bambusoideae; Bambusodae; Bambuseae.

Ecology, geography, regional floristic distribution. 6 species. Eastern Asia, South Africa. Helophytic, or mesophytic; glycophytic.

Holarctic and Paleotropical. Boreal. African. Eastern Asian. Sudano-Angolan. South Tropical African.

References, etc. Leaf anatomical: Metcalfe 1960 (*'Arundinaria tessellata'*; Soderstrom and Ellis 1982.

Special comments. See Clayton and Renvoize (1986) and Soderstrom and Ellis (1987) for completely different generic interpretations of the species in this circle of affinity: there are no available generic descriptions adequate for the present purpose. Fruit data wanting.

Thaumastochloa C.E. Hubb.

Habit, vegetative morphology. Annual, or perennial. Culms herbaceous. Culm nodes hairy, or glabrous. Culm internodes solid. Leaves non-auriculate. Leaf blades narrow; flat, or rolled (convolute); without cross venation; persistent. *Ligule a fringed membrane to a fringe of hairs*.

Reproductive organization. Plants bisexual, with bisexual spikelets; with hermaphrodite florets. *The spikelets of sexually distinct forms on the same plant*; hermaphrodite and sterile; overtly heteromorphic (the pedicelled spikelets greatly reduced).

Inflorescence. Inflorescence usually compound, of single, pedunculate, cylindrical, dorsiventral racemes. Rachides hollowed. Inflorescence spatheate, or espatheate (depending on interpretation); a complex of 'partial inflorescences' and intervening foliar organs, or not comprising 'partial inflorescences' and foliar organs. **Spikelet-bearing axes** very much reduced, or spikelike (peduncled, slender, with one or few spikelets); solitary; with substantial rachides (hollowed); disarticulating; *disarticulating at the joints (the rachis fragile, but the lowest spikelet falling with the peduncle, which becomes stiff, curved and pointed to function in dispersal), or falling entire (the only spikelet falling with the specialised peduncle)*. 'Articles' non-linear (concavo-convex, compressed or subterete); with a basal callus-knob; disarticulating transversely to disarticulating obliquely. Spikelets unaccompanied by bractiform involucres, not associated with setiform vestigial branches; paired (if the very reduced 'pedicelled spikelet' is interpreted as such); *secund (the sessile spikelets in two alternating rows, on one side of the rachis)*; sessile and pedicellate (the pedicelled member vestigial); *consistently in 'long-and-short' combinations*; in pedicellate/sessile combinations. *Pedicels of the 'pedicellate' spikelets discernible, but fused with the rachis*. The 'shorter' spikelets hermaphrodite. The 'longer' spikelets sterile (and vestigial).

Female-sterile spikelets. The pedicelled spikelet reduced to a glume, or represented only by its fused pedicel.

Female-fertile spikelets. Spikelets compressed dorsiventrally; falling with the glumes (and with the joint). Rachilla terminated by a female-fertile floret. Hairy callus absent.

Glumes two; relatively large; more or less equal; long relative to the adjacent lemmas; awnless; very dissimilar (the lower indurated and sometimes rugulose, the upper hyaline and cymbiform). *Lower glume two-keeled (wingless)*; convex on the back to concave on the back; not pitted; transversely rugose, or relatively smooth; 5–9 nerved. Upper glume 3–5 nerved. *Spikelets* with incomplete florets. The incomplete florets proximal to the female-fertile florets. *The proximal incomplete florets* 1; epaleate; sterile. *The proximal lemmas* awnless; *2 nerved (towards the margins)*; more or less equalling the female-fertile lemmas to decidedly exceeding the female-fertile lemmas; similar in texture to the female-fertile lemmas to decidedly firmer than the female-fertile lemmas (hyaline).

Female-fertile florets 1. Lemmas less firm than the glumes (hyaline); not becoming indurated; entire; blunt; awnless; hairless; non-carinate; 0–3 nerved. *Palea* present; 1-nerved, or 2-nerved, or nerveless. *Lodicules* present; 2; fleshy. *Stamens* 3. Anthers not penicillate. *Ovary* glabrous. Styles free to their bases.

Fruit, embryo and seedling. Fruit compressed dorsiventrally. Hilum short. Embryo large. Endosperm containing only simple starch grains. Embryo without an epiblast; with a scutellar tail; with an elongated mesocotyl internode. Embryonic leaf margins overlapping.

Abaxial leaf blade epidermis. *Costal/intercostal zonation* conspicuous. *Papillae* absent. *Long-cells* similar in shape costally and intercostally (but the costals much smaller); of similar wall thickness costally and intercostally. Intercostal zones with typical long-cells (with a few short ones). Mid-intercostal long-cells rectangular; having markedly sinuous walls. *Microhairs* present; panicoid-type; 45–51 microns long; 4.5–7.5 microns wide at the septum. Microhair total length/width at septum 6–11.3. Microhair apical cells 25–33 microns long. Microhair apical cell/total length ratio 0.57–0.65. *Stomata* common; 37–42 microns long. Subsidiaries triangular. Guard-cells overlapping to flush with the interstomatals. *Intercostal short-cells* absent or very rare. Cushion-based macro-hairs present, prickle bases numerous. *Costal short-cells* conspicuously in long rows. Costal silica bodies 'panicoid-type'; not sharp-pointed.

Transverse section of leaf blade, physiology. C_4; XyMS–. *Mesophyll* with radiate chlorenchyma. *Leaf blade* adaxially flat. *Midrib* not readily distinguishable; with one bundle only. Bulliforms present in discrete, regular adaxial groups (in places, but the epidermis mostly irregularly bulliform); occasionally in simple fans (these large-celled, *Zea*-type). Many of the smallest vascular bundles unaccompanied by sclerenchyma.

Combined sclerenchyma girders present; nowhere forming 'figures'. Sclerenchyma all associated with vascular bundles.

Special diagnostic feature. *Spikelets not arranged as in* **Manisuris** *(q.v.).*

Taxonomy. Panicoideae; Andropogonodae; Andropogoneae; Rottboelliinae.

Ecology, geography, regional floristic distribution. 7 species. Northeast India, Southeast Asia to Formosa, Philippines, Marianne & Caroline Is., Moluccas, Australia. Woodland.

Paleotropical and Australian. Indomalesian. Indian, Indo-Chinese, Malesian, and Papuan. North and East Australian. Tropical North and East Australian.

References, etc. Morphological/taxonomic: Hubbard 1936. Leaf anatomical: this project.

Thelepogon Roth.

Rhiniachne Steud.

Habit, vegetative morphology. Rather stout, erect or decumbent annual (often with prop roots). *Culms* 10–150 cm high; herbaceous; branched above, or unbranched above. Culm nodes glabrous. Culm internodes solid. Young shoots intravaginal. *Leaves* not basally aggregated; non-auriculate. *Leaf blades* lanceolate (from the amplexicaul base); broad; to 30 mm wide; *cordate*; flat. Ligule an unfringed membrane, or a fringed membrane; 0.5–0.8 mm long. *Contra-ligule* absent.

Reproductive organization. Plants bisexual, with bisexual spikelets; with hermaphrodite florets. *The spikelets of sexually distinct forms on the same plant*; hermaphrodite and sterile (fertile/sterile, but the latter reduced to their pedicels).

Inflorescence. *Inflorescence* of spicate main branches (of long, brittle golden 'racemes'); *subdigitate (the racemes on a short common axis)*. Primary inflorescence branches 2–20. Inflorescence espatheate; not comprising 'partial inflorescences' and foliar organs. *Spikelet-bearing axes* 'racemes' (with numerous spikelets). The racemes without spikelets towards the base. *Spikelet-bearing axes* clustered; *with substantial rachides*; disarticulating; disarticulating at the joints. 'Articles' non-linear (curved, concave, compressed below and clavate upwards); without a basal callus-knob; not appendaged; disarticulating transversely; somewhat hairy (i.e., the margins ciliate). Spikelets paired (but the pedicellate one reduced to the pedicel); sessile and pedicellate (the pedicelled component vestigial); consistently in 'long-and-short' combinations; in pedicellate/sessile combinations. *Pedicels of the 'pedicellate' spikelets* basally *free of the rachis (the pedicel and joint separated below, contiguous above)*. The 'shorter' spikelets hermaphrodite. The 'longer' spikelets sterile (and vestigial).

Female-sterile spikelets. *The pedicellate member absent, represented only by the flattened, linear pedicel.*

Female-fertile spikelets. Spikelets 5–13 mm long; abaxial; compressed dorsiventrally; falling with the glumes (deciduous with the joint and the sterile pedicel); with conventional internode spacings. Rachilla terminated by a female-fertile floret. Hairy callus present (annular). *Callus* short; blunt.

Glumes two; more or less equal; exceeding the spikelets; long relative to the adjacent lemmas; dorsiventral to the rachis; hairless; glabrous; pointed; awnless; non-carinate (rounded on the back); very dissimilar (both wingless, more or less rugose, the lower firmer). Lower glume much exceeding the lowest lemma; not two-keeled; convex on the back; not pitted; strongly rugose, or muricate, or tuberculate; 7–9 nerved. Upper glume 1 nerved (to 3?). *Spikelets* with incomplete florets. The incomplete florets proximal to the female-fertile florets. *The proximal incomplete florets* 1; paleate. Palea of the proximal incomplete florets fully developed. The proximal incomplete florets male. The proximal lemmas awnless; 2 nerved; more or less equalling the female-fertile lemmas; similar in texture to the female-fertile lemmas (hyaline); not becoming indurated.

Female-fertile florets 1. Lemmas less firm than the glumes (hyaline); not becoming indurated; incised; deeply cleft; awned. Awns 1; median; from a sinus; geniculate; hairless (glabrous); much longer than the body of the lemma. Lemmas hairless; non-carinate; without a germination flap; 3 nerved, or 5 nerved. *Palea* present; relatively long; entire;

awnless, without apical setae; not indurated (hyaline); 2-nerved; keel-less. *Lodicules* present; 2; free; fleshy; glabrous. *Stamens* 3. *Ovary* glabrous. Styles free to their bases. Stigmas 2.

Fruit, embryo and seedling. *Fruit* free from both lemma and palea; small (about 3 mm long); ellipsoid; compressed dorsiventrally. Hilum short. Embryo large (about half the length of the fruit).

Abaxial leaf blade epidermis. *Costal/intercostal zonation* conspicuous. *Papillae* absent. *Long-cells* markedly different in shape costally and intercostally (the costals narrower); of similar wall thickness costally and intercostally. Mid-intercostal long-cells rectangular; having markedly sinuous walls (but fairly thin). *Microhairs* present; panicoid-type; 57–61.5 microns long; 9–10.5 microns wide at the septum. Microhair total length/width at septum 5.4–6.7. Microhair apical cells 33–37.5 microns long. Microhair apical cell/total length ratio 0.55–0.63. *Stomata* common; 45–51 microns long. Subsidiaries triangular. Guard-cells overlapping to flush with the interstomatals. *Intercostal short-cells* common; in cork/silica-cell pairs (usually); silicified. Intercostal silica bodies crescentic and oryzoid-type. *Costal short-cells* neither distinctly grouped into long rows nor predominantly paired (pairs, triplets and solitaries). Costal silica bodies 'panicoid-type'; mainly cross shaped and dumb-bell shaped; not sharp-pointed.

Transverse section of leaf blade, physiology. C_4; XyMS–. *Leaf blade* adaxially flat. *Midrib* conspicuous; with one bundle only. Bulliforms not in discrete, regular adaxial groups (constituting most of the adaxial epidermis). Many of the smallest vascular bundles unaccompanied by sclerenchyma. Combined sclerenchyma girders present; nowhere forming 'figures' (but almost so in the midrib). Sclerenchyma all associated with vascular bundles.

Taxonomy. Panicoideae; Andropogonodae; Andropogoneae; Andropogoninae.

Ecology, geography, regional floristic distribution. 1 species. Tropical Africa, Asia. Helophytic to mesophytic; species of open habitats; glycophytic. Seasonally wet, heavy soils and disturbed ground.

Paleotropical. African and Indomalesian. Sudano-Angolan and West African Rain-forest. Indian, Indo-Chinese, and Malesian. Sahelo-Sudanian, Somalo-Ethiopian, and South Tropical African.

References, etc. Leaf anatomical: this project.

Thellungia Stapf

~ *Eragrostis*

Habit, vegetative morphology. Robust perennial; caespitose. **Culms 80–150 cm high**; *herbaceous*. Culm nodes glabrous. *Leaves* mostly basal; non-auriculate. Leaf blades narrow; 2–5 mm wide; flat; without abaxial multicellular glands; without cross venation; persistent. Ligule a fringe of hairs.

Reproductive organization. Plants bisexual, with bisexual spikelets; with hermaphrodite florets. Exposed-cleistogamous.

Inflorescence. *Inflorescence paniculate*; contracted (the primary branches short, erect-appressed); spicate; espatheate; not comprising 'partial inflorescences' and foliar organs. Spikelet-bearing axes persistent. *Spikelets* not secund; *subsessile*.

Female-fertile spikelets. Spikelets about 4 mm long; compressed laterally; disarticulating above the glumes; disarticulating between the florets. Rachilla prolonged beyond the uppermost female-fertile floret, or terminated by a female-fertile floret (all the florets hermaphrodite?); rachilla hairless. The rachilla extension (when present) with incomplete florets, or naked. Hairy callus absent.

Glumes two; very unequal; shorter than the adjacent lemmas; pointed; awnless; similar (membranous, narrow). Lower glume 1 nerved. Upper glume 1 nerved. **Spikelets** with female-fertile florets only, or with incomplete florets; *without proximal incomplete florets*.

Female-fertile florets (1–)3–4. Lemmas similar in texture to the glumes (membranous); not becoming indurated; entire; pointed; awnless; hairless; glabrous; carinate. The keel wingless. **Lemmas 1 nerved.** **Palea** present; *conspicuous but relatively* *short (about one half the lemma length)*; entire (pointed); awnless, without apical setae; not indurated; 2-nerved. *Stamens* 3. Anthers not penicillate. *Ovary* glabrous. Stigmas 2.

Fruit, embryo and seedling. Fruit small; compressed laterally. Hilum short. Pericarp free. Embryo large.

Abaxial leaf blade epidermis. *Costal/intercostal zonation* conspicuous. *Papillae* absent. *Long-cells* markedly different in shape costally and intercostally (the costals much narrower); of similar wall thickness costally and intercostally (walls of medium thickness). Mid-intercostal long-cells rectangular; having markedly sinuous walls. *Microhairs* present; more or less spherical to elongated; clearly two-celled; chloridoid-type (the basal and apical cells both short). Microhair apical cell wall thinner than that of the basal cell and often collapsed. Microhairs (21–)22.5–24(–27) microns long. Microhair basal cells 18 microns long. Microhairs 11.4–14.4 microns wide at the septum. Microhair total length/width at septum 1.7–2.1. Microhair apical cells 10.5–13.5 microns long. Microhair apical cell/total length ratio 0.44–0.56. *Stomata* common; 27–36 microns long. Subsidiaries dome-shaped and triangular. Guard-cells overlapping to flush with the interstomatals. *Intercostal short-cells* common; in cork/silica-cell pairs and not paired (some solitary, a few threes). Intercostal silica bodies absent. *Costal short-cells* conspicuously in long rows (but in places the 'short-cells' are quite long). Costal silica bodies present in alternate cell files of the costal zones; saddle shaped and 'panicoid-type', or tall-and-narrow (a few, bordering the veins); the panicoid type cross shaped (but saddles and intermediates predominating); not sharp-pointed.

Transverse section of leaf blade, physiology. C_4; XyMS+. PCR sheath outlines even. PCR sheaths of the primary vascular bundles interrupted; interrupted both abaxially and adaxially. PCR sheath extensions absent. PCR cell chloroplasts centripetal. *Mesophyll* with radiate chlorenchyma. *Leaf blade* 'nodular' in section; with the ribs more or less constant in size. *Midrib* not readily distinguishable; with one bundle only. Bulliforms present in discrete, regular adaxial groups (in each groove); in simple fans (these predominating, with large, deeply penetrating median cells), or associated with colourless mesophyll cells to form deeply-penetrating fans (a few). All the vascular bundles accompanied by sclerenchyma. Combined sclerenchyma girders present (with the primary bundles only — the rest mostly with strands adaxially); forming 'figures' (all the primary bundles with massive 'anchors'). Sclerenchyma all associated with vascular bundles. The lamina margins with fibres.

Taxonomy. Chloridoideae; main chloridoid assemblage.

Ecology, geography, regional floristic distribution. 1 species. Australia. Australian. North and East Australian. Tropical North and East Australian.

References, etc. Leaf anatomical: this project.

Special comments. Bridging *Sporobolus* and *Eragrostis*, and incompletely separable from the latter in terms of the current descriptions.

Themeda Forssk.

Androscepia Brong., *Anthistiria* L. f., *Aristaria* Jungh., *Heterelytron* Jungh., *Perobachne* Presl

Habit, vegetative morphology. Annual, or perennial; caespitose (coarse, very rarely stoloniferous). *Culms* 30–310 cm high; herbaceous; branched above, or unbranched above. Culm nodes hairy, or glabrous. Culm internodes solid, or hollow. *Leaves* not basally aggregated; non-auriculate. *Leaf blades* linear; *narrow*; flat, or folded; without cross venation; persistent; rolled in bud, or once-folded in bud. Ligule an unfringed membrane to a fringed membrane. *Contra-ligule* absent.

Reproductive organization. *Plants bisexual, with bisexual spikelets*; with hermaphrodite florets. The spikelets of sexually distinct forms on the same plant; hermaphrodite and male-only, or hermaphrodite and sterile; overtly heteromorphic; in both homogamous and heterogamous combinations (the lower pairs homogamous, homomorphic, forming an involucre under the terminal triad). Apomictic, or reproducing sexually.

Inflorescence. Inflorescence paniculate (leafy, comprising short racemes in spatheate, hard-to-interpret clusters); open; spatheate; *a complex of 'partial inflores-*

cences' and intervening foliar organs (composed of short racemes in spatheate clusters: each cluster terminated by 1–3 pairs of spikelets, one of each pair sessile and bisexual, the other pedicelled and male-or-sterile (or a triplet of 1 terminal sessile spikelet with 2 pedicellate ones), the whole surrounded by a whorl of 4 male or sterile, sessile spikelets constituting an involucre). **Spikelet-bearing axes** very much reduced; **clustered (the racemes solitary in their spatheoles, these units in groups of three or more in short capituliform glomerules)**; disarticulating; disarticulating at the joints (each raceme disarticulating at the level of the female-fertile spikelets). **Spikelets** *associated with bractiform involucres (constituted by the four imperfect spikelets)*. The involucres persistent on the rachis. Spikelets paired and in triplets; the clusters secund, or not secund; sessile and pedicellate; consistently in 'long-and-short' combinations; in pedicellate/sessile combinations. Pedicels of the 'pedicellate' spikelets free of the rachis. The 'shorter' spikelets hermaphrodite. The 'longer' spikelets male-only, or sterile.

Female-fertile spikelets. Spikelets not noticeably compressed to compressed dorsiventrally; falling with the glumes (the clusters disarticulating immediately above the involucres). Rachilla terminated by a female-fertile floret. *Hairy callus present (usually acute to pungent).* Callus pointed.

Glumes two; more or less equal; long relative to the adjacent lemmas; awnless; with the keel conspicuously winged, or without a median keel-wing; leathery. Lower glume not two-keeled; convex on the back; not pitted; relatively smooth; 7–11 nerved. Upper glume 1–5 nerved. *Spikelets* with incomplete florets. The incomplete florets proximal to the female-fertile florets. *The proximal incomplete florets* 1; epaleate; sterile. The proximal lemmas awnless; 0 nerved; similar in texture to the female-fertile lemmas (hyaline); not becoming indurated.

Female-fertile florets 1. Lemmas hyaline, stipitate beneath the awn; less firm than the glumes; not becoming indurated; usually entire; not deeply cleft; awned. Awns 1; median; usually apical; geniculate; hairy; much shorter than the body of the lemma to much longer than the body of the lemma. *Lemmas* hairless; non-carinate; *1 nerved. Palea* present, or absent; when present, conspicuous but relatively short, or very reduced; not indurated (hyaline); nerveless. *Lodicules* present; 2; free; fleshy; glabrous. *Stamens* 3. Anthers not penicillate. *Ovary* glabrous. Styles free to their bases. Stigmas 2; red pigmented.

Fruit, embryo and seedling. *Fruit* free from both lemma and palea; small; longitudinally grooved (channelled down one side); compressed dorsiventrally. Hilum short. Embryo large. Endosperm hard; without lipid; containing compound starch grains. Embryo without an epiblast; with a scutellar tail; with an elongated mesocotyl internode. Embryonic leaf margins overlapping.

Seedling with a long mesocotyl. First seedling leaf with a well-developed lamina. The lamina broad; curved; 21–30 veined.

Abaxial leaf blade epidermis. *Costal/intercostal zonation* conspicuous. *Papillae* present. Intercostal papillae over-arching the stomata; several per cell (finger-like projections). *Long-cells* similar in shape costally and intercostally; differing markedly in wall thickness costally and intercostally (the costals thicker-walled). Mid-intercostal long-cells rectangular; having markedly sinuous walls. *Microhairs* present; panicoid-type; (26–)38–64(–72) microns long; 4.2–6.6 microns wide at the septum. Microhair total length/width at septum 7.8–14.3. Microhair apical cells (10–)13–19.5(–22.5) microns long. Microhair apical cell/total length ratio 0.35–0.44. *Stomata* common; 24–30 microns long. Subsidiaries papillate; low dome-shaped and triangular, or triangular. Guard-cells overlapping to flush with the interstomatals. *Intercostal short-cells* common, or absent or very rare; in cork/silica-cell pairs and not paired; silicified (rarely), or not silicified. Intercostal silica bodies tall-and-narrow, or vertically elongated-nodular, or cross-shaped. *Costal short-cells* conspicuously in long rows. Costal silica bodies 'panicoid-type'; cross shaped to dumb-bell shaped, or nodular; not sharp-pointed.

Transverse section of leaf blade, physiology. C_4; XyMS–. PCR sheath outlines uneven. PCR cell chloroplasts with reduced grana; centrifugal/peripheral. *Mesophyll* with radiate chlorenchyma. *Leaf blade* 'nodular' in section, or adaxially flat; with the ribs more or less constant in size. *Midrib* conspicuous; with one bundle only, or having a conventional arc of bundles. Bulliforms not in discrete, regular adaxial groups (constituting most of the epidermis, irregular save for a large fan over the keel). Many of the smallest vascular bundles unaccompanied by sclerenchyma. Combined sclerenchyma girders present; forming 'figures', or nowhere forming 'figures'. Sclerenchyma all associated with vascular bundles.

Culm anatomy. *Culm internode bundles* in one or two rings.

Phytochemistry. Tissues of the culm bases with abundant starch. Leaves without flavonoid sulphates (1 species).

Cytology. Chromosome base number, $x = 5$ and 10. $2n = 20, 40, 60,$ and 80 (and aneuploids).

Taxonomy. Panicoideae; Andropogonodae; Andropogoneae; Andropogoninae.

Ecology, geography, regional floristic distribution. 18 species. Warm Africa, Asia, Australia. Commonly adventive. Mesophytic; species of open habitats; glycophytic. Savanna.

Holarctic, Paleotropical, Cape, and Australian. Boreal and Tethyan. African, Madagascan, Indomalesian, and Neocaledonian. Eastern Asian. Macaronesian and Irano-Turanian. Saharo-Sindian, Sudano-Angolan, West African Rainforest, and Namib-Karoo. Indian, Indo-Chinese, Malesian, and Papuan. North and East Australian, South-West Australian, and Central Australian. Sahelo-Sudanian, Somalo-Ethiopian, South Tropical African, and Kalaharian. Tropical North and East Australian and Temperate and South-Eastern Australian.

Rusts and smuts. Rusts — *Phakopsora* and *Puccinia*. Taxonomically wide-ranging species: *Phakopsora incompleta, Puccinia versicolor,* and *'Uromyces' clignyi.* Smuts from Ustilaginaceae. Ustilaginaceae — *Sorosporium, Sphacelotheca, Tolyposporium,* and *Ustilago.*

Economic importance. Significant weed species: *T. arguens, T. villosa.* Important native pasture species: *T. anathera, T. australis, T. cymbaria, T. triandra.*

References, etc. Leaf anatomical: Metcalfe 1960; this project.

Thinopyrum A. Löve

~ *Elytrigia, Elymus*

Habit, vegetative morphology. Rigid, erect, glaucous perennial; rhizomatous. *Culms* 25–70 cm high; herbaceous; unbranched above. Young shoots extravaginal. *The shoots aromatic.* Leaves not basally aggregated; non-auriculate. Leaf blades linear; narrow; 6–10 mm wide; rolled (involute); without cross venation; persistent. Ligule an unfringed membrane; truncate. *Contra-ligule* absent.

Reproductive organization. Plants bisexual, with bisexual spikelets; with hermaphrodite florets; outbreeding, or inbreeding.

Inflorescence. Inflorescence a single spike (the spikelets usually appressed). Rachides flattened (on the side facing the spikelets, and smooth on the main angles). Inflorescence espatheate; not comprising 'partial inflorescences' and foliar organs. Spikelet-bearing axes disarticulating (fragile); disarticulating at the joints (the spikelets falling with the internode below). Spikelets solitary; not secund; distichous; sessile to subsessile; usually imbricate.

Female-fertile spikelets. Spikelets compressed laterally; falling with the glumes. Rachilla prolonged beyond the uppermost female-fertile floret. The rachilla extension with incomplete florets. Hairy callus absent. *Callus* short (glabrous); pointed.

Glumes two; more or less equal (subequal); shorter than the spikelets; shorter than the adjacent lemmas; lateral to the rachis; pointed to not pointed (obtuse, acute or truncate); not subulate; awnless; non-carinate; similar. Lower glume 4–12 nerved. Upper glume 4–12 nerved. *Spikelets* with incomplete florets. The incomplete florets distal to the female-fertile florets. *The distal incomplete florets* merely underdeveloped.

Female-fertile florets 2–10. Lemmas similar in texture to the glumes (leathery); entire; blunt; awnless; hairless; glabrous; non-carinate (except towards the tip); without a germination flap; 5 nerved; with the nerves confluent towards the tip. *Palea* present; relatively long; awnless, without apical setae; 2-nerved; 2-keeled. *Lodicules* present; 2;

free; membranous; ciliate; 'usually one-lobed'. *Stamens* 3. Anthers 4–12 mm long. *Ovary* hairy. Styles free to their bases. Stigmas 2; white.

Fruit, embryo and seedling. *Fruit* free from both lemma and palea; medium sized to large; longitudinally grooved; compressed dorsiventrally; with hairs confined to a terminal tuft. *Hilum long-linear*. Embryo small. Endosperm hard; without lipid.

Abaxial leaf blade epidermis. *Costal/intercostal zonation* lacking. *Papillae* absent. *Long-cells* similar in shape costally and intercostally (indistinguishable); of similar wall thickness costally and intercostally (fairly thick walled). Mid-intercostal long-cells rectangular; having markedly sinuous walls (and pitted). *Microhairs* absent. *Stomata* absent or very rare; 29–39 microns long. *Intercostal short-cells* common; not paired (solitary). *Costal short-cells* neither distinctly grouped into long rows nor predominantly paired (solitary). Costal silica bodies absent, or poorly developed; tall-and-narrow (sometimes present).

Transverse section of leaf blade, physiology. C_3; XyMS+. *Mesophyll* with non-radiate chlorenchyma; without adaxial palisade. *Leaf blade* with distinct, prominent adaxial ribs; with the ribs very irregular in sizes (large, flat topped ribs alternating with small round topped ones). *Midrib* not readily distinguishable; with one bundle only. Bulliforms present in discrete, regular adaxial groups (in the furrows). All the vascular bundles accompanied by sclerenchyma. Combined sclerenchyma girders present (with the primaries, in the large ribs); forming 'figures' (massive anchors with the main bundles). Sclerenchyma not all bundle-associated (there being a continuous abaxial layer, linking the 'feet' of the 'anchors'). The 'extra' sclerenchyma in a continuous abaxial layer.

Cytology. Chromosome base number, $x = 7$. $2n = 14$, 28, and 42. 2, 4, and 6 ploid. Haplomic genome content J. Haploid nuclear DNA content 5.5 pg (1 species).

Taxonomy. Pooideae; Triticodae; Triticeae.

Ecology, geography, regional floristic distribution. 5 species. Coasts of Europe. Commonly adventive. Xerophytic; species of open habitats; halophytic. Coastal sands. Holarctic and Cape. Boreal and Tethyan. Euro-Siberian. Irano-Turanian. European.

Hybrids. Intergeneric hybrids with *Leymus* and *Elytrigia*.

References, etc. Morphological/taxonomic: Löve 1984. Leaf anatomical: Metcalfe 1960; this project.

Thrasya Kunth

Habit, vegetative morphology. Perennial; caespitose. *Culms herbaceous*; unbranched above. Culm nodes hairy, or glabrous. **Leaves** *mostly basal*. Leaf blades narrow; without cross venation; persistent. Ligule an unfringed membrane; not truncate.

Reproductive organization. Plants bisexual, with bisexual spikelets; with hermaphrodite florets. The spikelets all alike in sexuality.

Inflorescence. *Inflorescence a single raceme (spicate, with a winged rachis which partially embraces the spikelets)*. *Inflorescence axes not ending in spikelets (the spikes apparently bare-tipped)*. Rachides winged (partially embracing the spikelets). Inflorescence espatheate; not comprising 'partial inflorescences' and foliar organs. Spikelet-bearing axes persistent. *Spikelets* solitary, or paired; *secund (often in back to back pairs, in one row along the midrib of the rachis)*; not two-ranked (in one row); pedicellate; consistently in 'long-and-short' combinations (in *T. mosquitiensis*), or not in distinct 'long-and-short' combinations; in *T. mosquitiensis* unequally pedicellate in each combination (the upper pedicel here being longer, though partially adnate to the rachis). Pedicels of the 'pedicellate' spikelets (when in unequally-pedicellate pairs) discernible, but fused with the rachis. The 'shorter' spikelets hermaphrodite. The 'longer' spikelets hermaphrodite.

Female-fertile spikelets. *Spikelets* 2–4 mm long; alternately *abaxial and adaxial*; compressed laterally, or not noticeably compressed, or compressed dorsiventrally; falling with the glumes. Rachilla terminated by a female-fertile floret. Hairy callus absent (but with a glabrous swelling involving the glume bases, at the base of the spikelet).

Glumes present; two; very unequal; shorter than the adjacent lemmas, or long relative to the adjacent lemmas; dorsiventral to the rachis; hairy (the upper), or hairless; not pointed; awnless (but the upper sometimes mucronate with produced nerves); non-carinate;

very dissimilar (the lower minute and scarious, the upper conspicuous but thin). **Lower glume 0 nerved**. Upper glume 3–7 nerved. *Spikelets* with incomplete florets. The incomplete florets proximal to the female-fertile florets. *The proximal incomplete florets* 1; paleate. Palea of the proximal incomplete florets fully developed. The proximal incomplete florets male. **The proximal lemmas furrowed or bifid**; awnless (sometimes mucronate); 5–7 nerved; more or less equalling the female-fertile lemmas; similar in texture to the female-fertile lemmas (firm); not becoming indurated (subindurate).

Female-fertile florets 1. Lemmas decidedly firmer than the glumes; striate; becoming indurated (thinly so); entire; pointed; awnless; hairy (on the apical margin only), or hairless; non-carinate; having the margins tucked in onto the palea; with a clear germination flap; faintly 3–5 nerved. *Palea* present; relatively long; textured like the lemma (firm); 2-nerved; 2-keeled. *Lodicules* present; 2; free; fleshy; glabrous; not or scarcely vascularized. *Stamens* 3. Anthers not penicillate; without an apically prolonged connective (the anthers divaricate). *Ovary* glabrous. Styles free to their bases. Stigmas 2; red pigmented.

Fruit, embryo and seedling. *Fruit* free from both lemma and palea; compressed dorsiventrally. Hilum short to long-linear (elongated, up to half the grain length). Embryo large.

Abaxial leaf blade epidermis. *Costal/intercostal zonation* conspicuous. *Papillae* present, or absent. Intercostal papillae over-arching the stomata; consisting of one symmetrical projection per cell. *Long-cells* similar in shape costally and intercostally. Mid-intercostal long-cells rectangular; having markedly sinuous walls, or having straight or only gently undulating walls. *Microhairs* present; panicoid-type; (33–)39–66(–69) microns long; (6–)7.5–9.6(–11.4) microns wide at the septum. Microhair total length/width at septum 3.5–4.2 (in *T. reticulata*), or 7.5–11 (in *T. petrosa* and *T. thrasyoides*). Microhair apical cells (19.5–)21–45(–46.5) microns long. Microhair apical cell/total length ratio 0.54–0.73. *Stomata* common; (24–)26–39(–42) microns long. Subsidiaries triangular, or dome-shaped and triangular (*T. reticulata*), or parallel-sided, dome-shaped, and triangular (*T. petrosa*); including both triangular and parallel-sided forms on the same leaf (e.g. *T. petrosa*), or not including both parallel-sided and triangular forms on the same leaf (e.g. *T. reticulata*, *T. thrasyoides*). Guard-cells overlapping to flush with the interstomatals. *Intercostal short-cells* common; in cork/silica-cell pairs (e.g. *T. thrasyoides*), or not paired (*T. petrosa*). *Costal short-cells* predominantly paired to neither distinctly grouped into long rows nor predominantly paired. Costal silica bodies present and well developed to absent (a few in *T. reticulata* and *T. thrasyoides*, none in *T. petrosa*); 'panicoid-type' (e.g. in *T. reticulata*, and reduced forms in *T. thrasyoides*); not sharp-pointed.

Transverse section of leaf blade, physiology. C_4; XyMS–. *Mesophyll* with radiate chlorenchyma. *Leaf blade* 'nodular' in section to adaxially flat; with the ribs more or less constant in size. *Midrib* conspicuous; having a conventional arc of bundles (1 large and about 20 small bundles); with colourless mesophyll adaxially. Bulliforms present in discrete, regular adaxial groups; in simple fans. Many of the smallest vascular bundles unaccompanied by sclerenchyma. Combined sclerenchyma girders present; forming 'figures', or nowhere forming 'figures'. Sclerenchyma all associated with vascular bundles.

Taxonomy. Panicoideae; Panicodae; Paniceae.

Ecology, geography, regional floristic distribution. About 20 species. Tropical South America, Trinidad. Helophytic; species of open habitats. Savanna, on wet sands.

Neotropical. Caribbean, Venezuela and Surinam, Central Brazilian, and Andean.

Rusts and smuts. Rusts — *Puccinia*. Taxonomically wide-ranging species: *Puccinia levis*.

References, etc. Morphological/taxonomic: Davidse and Burman 1987. Leaf anatomical: this project.

Thrasyopsis L. Parodi

Habit, vegetative morphology. Perennial; caespitose. **Culms herbaceous**; unbranched above. Culm nodes hairy. *Leaves* mostly basal; non-auriculate. Leaf blades narrow; without cross venation; persistent. **Ligule an unfringed membrane**; not truncate; 1 mm long.

Reproductive organization. Plants bisexual, with bisexual spikelets; with hermaphrodite florets.

Inflorescence. *Inflorescence a single spike, or of spicate main branches (with short plump spikes)*; digitate (the material of *T. repanda* seen having 1–2 spikes), or non-digitate (when consisting of only one spike). Primary inflorescence branches 1–2. Rachides flattened (broad and flat). Inflorescence espatheate; not comprising 'partial inflorescences' and foliar organs. Spikelet-bearing axes persistent. Spikelets solitary; secund; biseriate; subsessile.

Female-fertile spikelets. *Spikelets* 4–5 mm long; *compressed dorsiventrally*; biconvex; falling with the glumes. Rachilla terminated by a female-fertile floret. Hairy callus absent.

Glumes present; two; more or less equal; long relative to the adjacent lemmas; lateral to the rachis; hairless; awnless; non-carinate; *similar (herbaceous)*. Lower glume 6–7 nerved. *Upper glume 9 nerved*. *Spikelets* with incomplete florets. The incomplete florets proximal to the female-fertile florets. *Spikelets with proximal incomplete florets. The proximal incomplete florets* 1; paleate; *male*. The proximal lemmas awnless; 7–9 nerved; more or less equalling the female-fertile lemmas; less firm than the female-fertile lemmas (membranous, glabrous); not becoming indurated.

Female-fertile florets 1. Lemmas decidedly firmer than the glumes (rigid, shining); becoming indurated (thinly); entire; pointed; awnless; hairless (shiny); non-carinate; having the margins lying flat and exposed on the palea; inconspicuously 3 nerved. *Palea* present; relatively long (firm); 2-nerved; 2-keeled. *Lodicules* present; 2; free; fleshy; glabrous; not or scarcely vascularized. *Stamens* 3. Anthers not penicillate; without an apically prolonged connective (but the anthers divaricate). *Ovary* glabrous. Styles fused. Stigmas 2; dark.

Abaxial leaf blade epidermis. *Costal/intercostal zonation* conspicuous. *Papillae* present. Intercostal papillae not over-arching the stomata; consisting of one oblique swelling per cell to consisting of one symmetrical projection per cell. Mid-intercostal long-cells having markedly sinuous walls. *Microhairs* present; panicoid-type; (33–)39–42 microns long; 5.4–6.6 microns wide at the septum. Microhair total length/width at septum 6.1–7. Microhair apical cells (15–)22.5–24(–26) microns long. Microhair apical cell/total length ratio 0.45–0.61. *Stomata* common; 24–28.5 microns long. Subsidiaries triangular. Guard-cells overlapping to flush with the interstomatals. *Intercostal short-cells* absent or very rare. Numerous cushion-based macrohairs present. *Costal short-cells* conspicuously in long rows. Costal silica bodies 'panicoid-type'; cross shaped, butterfly shaped, dumb-bell shaped, and nodular (all forms often with points); sharppointed (with points on representatives of the various forms).

Transverse section of leaf blade, physiology. C_4; XyMS–. *Mesophyll* with radiate chlorenchyma. *Leaf blade* adaxially flat. *Midrib* conspicuous; having a conventional arc of bundles (3 large, 8 small bundles); with colourless mesophyll adaxially. Bulliforms in broad 'fans', extending all the way from one bundle to the next. Many of the smallest vascular bundles unaccompanied by sclerenchyma. Combined sclerenchyma girders present; forming 'figures'. Sclerenchyma all associated with vascular bundles.

Taxonomy. Panicoideae; Panicodae; Paniceae.

Ecology, geography, regional floristic distribution. 2 species. Brazil. Savanna. Neotropical. Central Brazilian and Pampas.

References, etc. Leaf anatomical: this project.

Special comments. Fruit data wanting.

Thuarea Pers.

Ornithocephalochloa Kurz, *Microthuareia* Thouars, *Thouarsia* Kuntze

Habit, vegetative morphology. Perennial; decumbent (creeping, mat-forming). *Culms* 5–30 cm high; herbaceous; unbranched above. Culm nodes glabrous. Culm internodes solid. Plants unarmed. Young shoots intravaginal. *Leaves* not basally aggregated; non-auriculate. Leaf blades linear-lanceolate (broad-based); broad, or narrow; without cross venation; persistent; rolled in bud. Ligule a fringe of hairs.

Reproductive organization. Plants bisexual, with bisexual spikelets; with hermaphrodite florets, or without hermaphrodite florets. The spikelets of sexually distinct forms on the same plant; hermaphrodite and male-only, or hermaphrodite, female-only, and male-only, or female-only and male-only (the lower 1–2 spikelets hermaphrodite or female, the upper 2–6 male-only). The male and female-fertile spikelets segregated, in different parts of the same inflorescence branch (the male spikelets distal to the female-fertile ones on the rachis). The spikelets homomorphic.

Inflorescence. *Inflorescence a peculiar spiciform 'raceme', terminal, shortly peduncled, broad and winglike in the female-fertile part, narrow and beak-like in the upper male part*. Rachides flattened (below). Inflorescence spatheate (the young inflorescence enclosed in the spathaceous blade of the uppermost culm leaf); not comprising 'partial inflorescences' and foliar organs. *Spikelet-bearing axes* spikelike; solitary; persistent. *Spikelets* discernably paired; secund; *not two-ranked (in one row, on the channelled side of the rachis)*; discernably sessile and pedicellate (though the pedicels fused); consistently in 'long-and-short' combinations. Pedicels of the 'pedicellate' spikelets discernible, but fused with the rachis.

Female-sterile spikelets. Male spikelets in the upper part of the raceme, articulated with their bulbous pedicels, deciduous. Both or only the upper of the two florets male, with 3 stamens. Rachilla of male spikelets terminated by a male floret. The male spikelets with proximal incomplete florets, or without proximal incomplete florets; 2 floreted (both or only one fertile). Male florets 1, or 2; 3 staminate.

Female-fertile spikelets. *Spikelets* elliptic, or lanceolate, or ovate; adaxial; compressed dorsiventrally; *not disarticulating (after anthesis the upper (male) spikelets fall, then the axis bends to enclose the developing fruit, and the flowering branch bends down to thrust the ripened seed into the sand)*. Rachilla terminated by a female-fertile floret. Hairy callus absent.

Glumes one per spikelet to two (the G_1 when present minute, hyaline); shorter than the adjacent lemmas; dorsiventral to the rachis; the upper softly hairy; not pointed (obtuse); awnless; non-carinate; very dissimilar (when both present, the lower vestigial). Lower glume when present, 0 nerved. Upper glume 5 nerved. *Spikelets* with incomplete florets. The incomplete florets proximal to the female-fertile florets. *The proximal incomplete florets* 1; paleate. Palea of the proximal incomplete florets fully developed. The proximal incomplete florets male. The proximal lemmas awnless; 5–7 nerved; more or less equalling the female-fertile lemmas; less firm than the female-fertile lemmas; not becoming indurated.

Female-fertile florets 1. Lemmas decidedly firmer than the glumes; smooth; papery; yellow in fruit; entire; blunt; awnless; hairy (apically, otherwise glabrous); non-carinate; having the margins lying flat and exposed on the palea; with a clear germination flap; 5–9 nerved. *Palea* present; relatively long; entire; awnless, without apical setae; textured like the lemma; 2-nerved; 2-keeled. *Lodicules* present; 2; free; fleshy; glabrous. *Stamens* 3. Anthers not penicillate; without an apically prolonged connective. *Ovary* glabrous. Styles fused, or free to their bases. Stigmas 2; white.

Fruit, embryo and seedling. Fruit small to medium sized (3–4 mm long); compressed dorsiventrally. Hilum short. Embryo large; without an epiblast; with a scutellar tail; with an elongated mesocotyl internode; with one scutellum bundle. Embryonic leaf margins overlapping.

Abaxial leaf blade epidermis. *Costal/intercostal zonation* conspicuous. *Papillae* absent. Intercostal zones without typical long-cells (except for occasional typical long-cells mid-way between the more widely spaced veins). Mid-intercostal long-cells having straight or only gently undulating walls. *Microhairs* present (but sometimes scarce, and hard to find among the macrohairs); elongated; clearly two-celled; panicoid-type; 42–55–67.5 microns long; 5.4–6.2–6.9 microns wide at the septum. Microhair total length/width at septum 7.7–10. Microhair apical cells 30–37–45 microns long. Microhair apical cell/total length ratio 0.59–0.73. *Stomata* common; 21–27 microns long. Subsidiaries dome-shaped. Guard-cells overlapping to flush with the interstomatals. *Intercostal short-cells* absent or very rare. Macrohairs abundant. *Costal short-cells* conspicuously in long rows, or neither distinctly grouped into long rows nor predominantly paired. Costal silica bodies poorly developed, or absent (in the material seen); when present, probably 'panicoid-type' (judging from the silica cell shapes); presumably dumb-bell shaped; not sharp-pointed.

Transverse section of leaf blade, physiology. C_4; XyMS+. PCR sheath outlines uneven. PCR sheath extensions absent. PCR cell chloroplasts centrifugal/peripheral. *Mesophyll* with radiate chlorenchyma; traversed by columns of colourless mesophyll cells (rarely), or not traversed by colourless columns. *Leaf blade* adaxially flat; with the ribs more or less constant in size. *Midrib* not readily distinguishable; with one bundle only. Bulliforms present in discrete, regular adaxial groups; in simple fans. Many of the smallest vascular bundles unaccompanied by sclerenchyma. Combined sclerenchyma girders absent. Sclerenchyma all associated with vascular bundles.

Phytochemistry. Leaves without flavonoid sulphates (*T. involuta*).

Special diagnostic feature. Spikelets not borne on a broad, leaflike rachis (the flattened rachis not leaflike). *Flowering culms ultimately bending over, so as to enclose the ripening fruit.*

Taxonomy. Panicoideae; Panicodae; Paniceae.

Ecology, geography, regional floristic distribution. 2 species. 1 in Madagascar, 1 Indomalaya, North Australia, New Guinea. Commonly adventive. Species of open habitats; halophytic. Seashore sand.

Paleotropical and Australian. Madagascan, Indomalesian, Polynesian, and Neocaledonian. Indian, Indo-Chinese, Malesian, and Papuan. Fijian. North and East Australian. Tropical North and East Australian.

Economic importance. Significant weed species: *T. involuta*. Important native pasture species: *T. involuta*.

References, etc. Leaf anatomical: this project.

Thyridachne C.E. Hubb.

Tisserantiella Mimeur

Habit, vegetative morphology. Annual. Culms herbaceous. Leaf blades linear; narrow; without cross venation. Ligule membranous.

Reproductive organization. Plants bisexual, with bisexual spikelets; with hermaphrodite florets.

Inflorescence. Inflorescence a single raceme, or paniculate (narrow, slender, spiciform); contracted; espatheate; not comprising 'partial inflorescences' and foliar organs. Spikelet-bearing axes persistent. Spikelets mostly paired; sessile and pedicellate. Pedicel apices minutely cupuliform. *Spikelets consistently in 'long-and-short' combinations*; in pedicellate/sessile combinations. The 'shorter' spikelets hermaphrodite. The 'longer' spikelets hermaphrodite.

Female-fertile spikelets. *Spikelets* abaxial; *compressed dorsiventrally (falcate in lateral view)*; falling with the glumes; with conventional internode spacings. Rachilla terminated by a female-fertile floret. Hairy callus absent.

Glumes two; very unequal; (the upper) long relative to the adjacent lemmas; hairless; awnless; non-carinate; *very dissimilar (the G_1 small, broad, sub-orbicular and thinly membranous, the G_2 gibbous and enveloping the spikelet by its inflexed margins, oblong-elliptical, obtusely bi- to tri-lobed at the summit, basally membranous, leathery below but hyaline at the base of the median line).* Lower glume much shorter than half length of lowest lemma (about a quarter as long as G_2); 0 nerved. **Upper glume** *distinctly saccate (gibbous)*; 5 nerved (obscurely). *Spikelets* with incomplete florets. The incomplete florets proximal to the female-fertile florets. *The proximal incomplete florets* 1; paleate; male, or sterile. *The proximal lemmas apically obscurely trilobed, channelled down the back, the channel membranous and enlarging below to form a 'window', the margins and apex also membranous, leathery elsewhere*; awnless; obscurely 3 nerved; more or less equalling the female-fertile lemmas to decidedly exceeding the female-fertile lemmas (equalling G_2); thinly leathery.

Female-fertile florets 1. Lemmas broadly oval; less firm than the glumes to similar in texture to the glumes (somewhat leathery or cartilaginous, with membranous apices); not becoming indurated; entire; blunt; awnless; hairless; non-carinate; 3 nerved. *Palea* present; relatively long; entire; awnless, without apical setae; textured like the lemma; not indurated; 2-nerved; keel-less (or flat). **Lodicules** *absent*. *Stamens* 3; with free filaments. Anthers not penicillate; without an apically prolonged connective. *Ovary* glabrous. Styles free to their bases. Stigmas 2.

Fruit, embryo and seedling. Fruit compressed dorsiventrally (plano-concave to plano-convex). Hilum short. Embryo large.

Abaxial leaf blade epidermis. *Costal/intercostal zonation* conspicuous. *Papillae* absent. *Long-cells* markedly different in shape costally and intercostally (the costals much more regularly rectangular); of similar wall thickness costally and intercostally (fairly thick walled). Intercostal zones with typical long-cells. Mid-intercostal long-cells more or less rectangular; having markedly sinuous walls (the sinuosity very coarse, irregular). *Microhairs* present, or absent (very scarce — commoner adaxially); panicoid-type; 21–27 microns long; 5.7–6 microns wide at the septum. Microhair total length/width at septum 3.5–4.7. Microhair apical cells 12.6–21 microns long. Microhair apical cell/total length ratio 0.6–0.78. *Stomata* common; 33–39 microns long. Subsidiaries non-papillate; rather low triangular. Guard-cells overlapping to flush with the interstomatals. *Intercostal short-cells* common; in cork/silica-cell pairs (the pairs tiny, superposed); silicified. Intercostal silica bodies rounded. No macrohairs or prickles seen. *Costal short-cells* predominantly paired. Costal silica bodies rounded (predominating), or crescentic, or tall-and-narrow (a few).

Transverse section of leaf blade, physiology. C_3 (with large mesophyll lacunae); XyMS+. *Leaf blade* with distinct, prominent adaxial ribs (the ribs topped by clusters of large, thin walled macrohairs); with the ribs more or less constant in size. *Midrib* not readily distinguishable; with one bundle only. Bulliforms not in discrete, regular adaxial groups. All the vascular bundles accompanied by sclerenchyma. Combined sclerenchyma girders present (if the large, thin walled 'colourless' tissue between the bundles and the ad- and abaxial fibre groups is interpreted as sclerenchyma). Sclerenchyma all associated with vascular bundles (apart from large fibre groups in the distended blade margins).

Taxonomy. Panicoideae; Panicodae; Paniceae.

Ecology, geography, regional floristic distribution. 1 species. Tropical Africa. Helophytic (in shallow pools); glycophytic.

Paleotropical. African. Sudano-Angolan. Sahelo-Sudanian.

References, etc. Morphological/taxonomic: Hubbard 1949. Leaf anatomical: this project.

Thyridolepis S. T. Blake

Habit, vegetative morphology. Perennial (with basal scaly, woolly cataphylls); caespitose, or decumbent. *Culms* 15–50 cm high; woody and persistent, or herbaceous; tuberous, or not tuberous. Culm nodes hairy, or glabrous. Culm internodes solid. Young shoots extravaginal and intravaginal. *Leaves* not basally aggregated; non-auriculate. Leaf blades linear to ovate; narrow; 1.5–4.5 mm wide; not setaceous; somewhat pseudopetiolate; without cross venation; disarticulating from the sheaths; rolled in bud. **Ligule** present; *a fringe of hairs*.

Reproductive organization. Plants bisexual, with bisexual spikelets; with hermaphrodite florets. The spikelets of sexually distinct forms on the same plant; hermaphrodite and sterile (the lowermost spikelets being reduced). Exposed-cleistogamous, or chasmogamous.

Inflorescence. *Inflorescence a single raceme (spikelike, bristly)*; espatheate; not comprising 'partial inflorescences' and foliar organs. Spikelet-bearing axes persistent. Spikelets solitary; not secund; shortly pedicellate. Pedicel apices oblique, or discoid.

Female-fertile spikelets. Spikelets 4–7.5 mm long; oblong, or elliptic, or lanceolate; abaxial; compressed dorsiventrally; falling with the glumes. Rachilla terminated by a female-fertile floret. Hairy callus present.

Glumes two; more or less equal; long relative to the adjacent lemmas; dorsiventral to the rachis; hairy; awnless; non-carinate; *very dissimilar (both leathery, the lower blunt, with a transverse row of tubercle-based bristles along the top of a rectangular, semi-transparent or pigmented 'window', the upper broader, rostrate, with tufts of tubercle-based bristles along the margins).* Lower glume 7–11 nerved. Upper glume 7–11 nerved. *Spikelets* with incomplete florets. The incomplete florets proximal to the female-fertile florets. *Spikelets with proximal incomplete florets. The proximal incomplete florets* 1; paleate, or epaleate. Palea of the proximal incomplete florets when present, reduced. The proximal incomplete florets male, or sterile. The proximal lemmas awnless; 5 nerved; more or less equalling the female-fertile lemmas; becoming indurated (not gibbous).

Female-fertile florets 1. Lemmas similar in texture to the glumes; not becoming indurated (thinly rigid as in *Paraneurachne*, by contrast with *Neurachne*); brown in fruit; entire; pointed; awnless; hairless; non-carinate; having the margins lying flat and exposed on the palea; with a clear germination flap; 3–5 nerved. *Palea* present; relatively long; entire; awnless, without apical setae; textured like the lemma; not indurated; 2-nerved. *Lodicules* present, or absent; when present, 2; free; fleshy; glabrous. *Stamens* 3. Anthers very short; not penicillate. *Ovary* glabrous. Styles free to their bases. Stigmas 2 (and only 2 styles, no appendage on the grain).

Fruit, embryo and seedling. Fruit small; compressed dorsiventrally. Hilum short. Embryo large; waisted. Endosperm containing only simple starch grains. Embryo without an epiblast; with a scutellar tail; with a negligible mesocotyl internode.

Seedling with a short mesocotyl. First seedling leaf with a well-developed lamina. The lamina broad; curved.

Abaxial leaf blade epidermis. *Costal/intercostal zonation* conspicuous. *Papillae* absent. *Long-cells* similar in shape costally and intercostally. Mid-intercostal long-cells rectangular; having markedly sinuous walls. *Microhairs* present; panicoid-type; (36–)48–72(–75) microns long; 5.4–9.6 microns wide at the septum. Microhair total length/width at septum 4.3–10.6. Microhair apical cells (21–)25–38(–42) microns long. Microhair apical cell/total length ratio 0.47–0.64. *Stomata* common; 21–39 microns long. Subsidiaries triangular, or dome-shaped and triangular (*T. xerophila*), or parallel-sided, dome-shaped, and triangular (*T. multiculmis*); including both triangular and parallel-sided forms on the same leaf, or not including both parallel-sided and triangular forms on the same leaf. Guard-cells overlapping to flush with the interstomatals. *Intercostal short-cells* common, or absent or very rare; in cork/silica-cell pairs, or not paired; silicified, or not silicified. Two of the species with cushion-based macrohairs and/or prickles. *Crown cells* present, or absent. *Costal short-cells* conspicuously in long rows (but sometimes also short rows; solitaries; pairs). Costal silica bodies 'panicoid-type'; cross shaped, butterfly shaped, dumb-bell shaped, and nodular; not sharp-pointed.

Transverse section of leaf blade, physiology. C_3; XyMS+. *PBS cells* without a suberised lamella. *Mesophyll* with radiate chlorenchyma; *Isachne*-type. *Leaf blade* with distinct, prominent adaxial ribs, or 'nodular' in section, or adaxially flat; with the ribs more or less constant in size. *Midrib* not readily distinguishable; with one bundle only. Bulliforms present in discrete, regular adaxial groups (sometimes associated with/comprising hair cushions); often in simple fans. All the vascular bundles accompanied by sclerenchyma. Combined sclerenchyma girders absent (combining strong abaxial girders with adaxial strands, the latter linked with the bundles by vertically-elongated colourless cells). Sclerenchyma all associated with vascular bundles (apart from strong marginal groups).

Special diagnostic feature. *Lower glume with a rectangular window, surmounted by bristles.*

Cytology. Chromosome base number, $x = 9$. $2n = 18$ and 36. 2 and 4 ploid.

Taxonomy. Panicoideae; Panicodae; Neurachneae.

Ecology, geography, regional floristic distribution. 3 species. Australia. Xerophytic; species of open habitats. Dry grassland and scrub.

Australian. North and East Australian and Central Australian. Tropical North and East Australian.

Economic importance. Important native pasture species: *T. mitchelliana* (drought tolerant).

References, etc. Morphological/taxonomic: Blake 1972b. Leaf anatomical: Hattersley *et al.* 1982; this project.

Thyrsia Stapf

~ *Phacelurus*

Habit, vegetative morphology. Stout, reed-like, tall *annual*. Culms herbaceous. Leaf blades narrow; not setaceous; flat (long, hard); without cross venation. *Ligule an unfringed membrane*.

Reproductive organization. Plants bisexual, with bisexual spikelets; with hermaphrodite florets. *The spikelets* of sexually distinct forms on the same plant (heterogamous); hermaphrodite and male-only, or hermaphrodite and sterile (or the pedicellate spikelet rarely hermaphrodite); overtly heteromorphic, or homomorphic; *in both homogamous and heterogamous combinations*.

Inflorescence. *Inflorescence of spicate main branches (with 'racemes' borne in terminal racemes or panicles)*. Rachides hollowed. Inflorescence espatheate; not comprising 'partial inflorescences' and foliar organs. **Spikelet-bearing axes** *spikelike (rather thick, cylindrical)*; with substantial rachides; disarticulating; disarticulating at the joints. *'Articles'* non-linear (constricted and dilated); not appendaged; disarticulating transversely; glabrous. *Spikelets* paired; secund (the 'racemes' more or less dorsiventral); sessile and pedicellate; *consistently in 'long-and-short' combinations*; in pedicellate/sessile combinations. Pedicels of the 'pedicellate' spikelets free of the rachis (but contiguous). The 'shorter' spikelets hermaphrodite. *The 'longer' spikelets* usually *male-only, or sterile (rarely hermaphrodite)*.

Female-sterile spikelets. The pedicellate spikelets usually sterile or male-only, sometimes reduced to their pedicels.

Female-fertile spikelets. Spikelets abaxial; compressed dorsiventrally; falling with the glumes (and the joint). Rachilla terminated by a female-fertile floret. Hairy callus absent.

Glumes two; more or less equal; long relative to the adjacent lemmas; dorsiventral to the rachis; hairless (glabrous, smooth); pointed; awnless; carinate (G_2), or non-carinate (G_1); very dissimilar (leathery or papery, the upper naviculate). Lower glume two-keeled; flattened on the back; not pitted; relatively smooth (shining). *Spikelets* with incomplete florets. The incomplete florets proximal to the female-fertile florets. *The proximal incomplete florets* 1; epaleate; sterile. The proximal lemmas awnless; 2 nerved; more or less equalling the female-fertile lemmas; similar in texture to the female-fertile lemmas (hyaline); not becoming indurated.

Female-fertile florets 1. Lemmas less firm than the glumes (hyaline); not becoming indurated; entire; pointed; awnless; hairless; non-carinate; without a germination flap; 3 nerved. *Palea* present; relatively long; entire; awnless, without apical setae; not indurated; 2-nerved. *Lodicules* present; 2; free; fleshy; glabrous. *Stamens* 3. *Ovary* glabrous. Stigmas red pigmented.

Fruit, embryo and seedling. Fruit compressed dorsiventrally.

Abaxial leaf blade epidermis. *Costal/intercostal zonation* conspicuous. *Papillae* absent. *Long-cells* markedly different in shape costally and intercostally (the costals narrower and more regularly rectangular, the intercostals unusually large); of similar wall thickness costally and intercostally (of medium thickness). Mid-intercostal long-cells rectangular to fusiform; having markedly sinuous walls (the sinuosity quite fine, conspicuously pitted). *Microhairs* present; elongated; clearly two-celled; panicoid-type (rather large). *Stomata* common. Subsidiaries non-papillate; conspicuously triangular. Guard-cells overlapping to flush with the interstomatals. *Intercostal short-cells* common; in cork/silica-cell pairs; silicified. Intercostal silica bodies crescentic to tall-and-narrow. Bulbous prickles with tiny points common over the main veins. *Costal short-cells* predominantly paired. Costal silica bodies present and well developed to poorly developed; tall-and-narrow.

Transverse section of leaf blade, physiology. C_4. The anatomical organization conventional. XyMS–. PCR sheath extensions absent. *Mesophyll* with radiate chlorenchyma. *Leaf blade* more or less adaxially flat (with very slight ribs over the primary bundles). *Midrib* conspicuous; having a conventional arc of bundles (a large, median primary with several smaller bundles on either side); with colourless mesophyll adaxially. *The lamina* symmetrical on either side of the midrib. Bulliforms not in discrete, regular adaxial groups (the epidermis consisting of rather regular, large cells). Many of the smallest vascular bundles unaccompanied by sclerenchyma. Combined sclerenchyma girders present (with the primaries only); forming 'figures' (slight I's, in places, but the sclerenchyma scanty everywhere except in the large abaxial keel of the midrib). Sclerenchyma all associated with vascular bundles.

Taxonomy. Panicoideae; Andropogonodae; Andropogoneae; Rottboelliinae.

Ecology, geography, regional floristic distribution. 3–4 species. Tropical Africa, Asia.

Paleotropical. African and Indomalesian. Sudano-Angolan and West African Rainforest. Indian, Indo-Chinese, and Malesian. Somalo-Ethiopian and South Tropical African.

References, etc. Morphological/taxonomic: Stapf 1922. Leaf anatomical: this project.

Thyrsostachys Gamble

Habit, vegetative morphology. Arborescent perennial; caespitose. The flowering culms leafy. *Culms* 800–1000 cm high; woody and persistent; to 6 cm in diameter; branched above. Primary branches/mid-culm node several. Culm sheaths persistent. Rhizomes pachymorph. Plants unarmed. *Leaves* not basally aggregated; auricles very small; with auricular setae (these small), or without auricular setae. Leaf blades broad; about 10–15 mm wide (by 7–18 cm long); pseudopetiolate; disarticulating from the sheaths; rolled in bud. Ligule an unfringed membrane.

Reproductive organization. Plants bisexual, with bisexual spikelets; with hermaphrodite florets.

Inflorescence. Inflorescence indeterminate; *with pseudospikelets*; paniculate (large, compound, thyrsoid, the branch nodes bearing sessile, spatheate clusters each of few pseudospikelets); spatheate (the spikelet groups in the axils of short sheaths); a complex of 'partial inflorescences' and intervening foliar organs. *Spikelet-bearing axes* paniculate; persistent. Spikelets not secund.

Female-fertile spikelets. Spikelets 10–25 mm long; disarticulating above the glumes; disarticulating between the florets. Rachilla prolonged beyond the uppermost female-fertile floret. The rachilla extension with incomplete florets.

Glumes one per spikelet, or two; very unequal; shorter than the spikelets; shorter than the adjacent lemmas; hairy; pointed; awnless; similar. Upper glume about 9 nerved ('about 4 on either side'). *Spikelets* with incomplete florets. The incomplete florets both distal and proximal to the female-fertile florets. **The distal incomplete florets** *merely underdeveloped (differing from* **Bambusa** *in the more reduced terminal floret). Spikelets with proximal incomplete florets.* **The proximal incomplete florets** 1, or 2; *paleate (the palea of the lowest floret deeply bifid).* Palea of the proximal incomplete florets fully developed. The proximal incomplete florets male. The proximal lemmas awnless; exceeded by the female-fertile lemmas; similar in texture to the female-fertile lemmas.

Female-fertile florets *1.* Lemmas papery; not becoming indurated; entire; pointed; awnless, or mucronate (?); hairy; many nerved. **Palea** *present*; relatively long; apically notched (less cleft thn in the proximal floret); several nerved; 2-keeled (less clearly so than in the proximal florets). *Lodicules* present, or absent; when present, 1–3; free; membranous (narrow); ciliate; not toothed. *Stamens* 6. Anthers not penicillate; shortly with the connective apically prolonged. *Ovary* glabrous; with a conspicuous apical appendage. The appendage broadly conical, fleshy. Styles fused (into one, the base not hollow - thick, forming beak in fruit). Stigmas 3.

Fruit, embryo and seedling. Fruit medium sized (5 to 10 mm long); not noticeably compressed (cylindrical); smooth (glabrous). Hilum long-linear. *Pericarp thick and hard (or at least, crustaceous)*; free, or fused (?). Embryo small (prominent). Endosperm containing compound starch grains.

Abaxial leaf blade epidermis. *Costal/intercostal zonation* conspicuous. *Papillae* present (papillae small, variously shaped). Intercostal papillae over-arching the stomata. Mid-intercostal long-cells having markedly sinuous walls (thin). *Microhairs* present; panicoid-type. *Stomata* common (obscured by papillae). *Intercostal short-cells* common; in cork/silica-cell pairs; silicified. Intercostal silica bodies saddle shaped. *Costal short-cells* conspicuously in long rows. Costal silica bodies saddle shaped; not sharp-pointed.

Transverse section of leaf blade, physiology. C_3; XyMS+. Mesophyll with arm cells; with fusoids. *Leaf blade* 'nodular' in section. *Midrib* conspicuous; having complex vascularization. Bulliforms present in discrete, regular adaxial groups; in simple fans and associated with colourless mesophyll cells to form deeply-penetrating fans. All the vascular bundles accompanied by sclerenchyma. Combined sclerenchyma girders present; forming 'figures' (in the large bundles).

Cytology. Chromosome base number, $x = 12$. $2n = 72$. 6 ploid.

Taxonomy. Bambusoideae; Bambusodae; Bambuseae.

Ecology, geography, regional floristic distribution. 2 species. Burma, Siam. Rain forest.

Paleotropical. Indomalesian. Indian and Indo-Chinese.

References, etc. Leaf anatomical: Metcalfe 1960.

Thysanolaena Nees

Myriachaeta Moritzi

Habit, vegetative morphology. Tufted perennial; reedlike. *Culms* 150–400 cm high; woody and persistent; branched above (shrubby). Culm internodes solid. *Leaves* not basally aggregated; auriculate, or non-auriculate. *Leaf blades* lanceolate (-acuminate); somewhat *leathery*; broad; (30–)40–70(–100) mm wide (up to 60 cm long); somewhat *cordate (amplexicaul)*; flat; pseudopetiolate; cross veined; disarticulating from the sheaths; rolled in bud. Ligule an unfringed membrane, or a fringed membrane (minutely ciliolate); truncate (cartilaginous). *Contra-ligule* present.

Reproductive organization. Plants bisexual, with bisexual spikelets; with hermaphrodite florets.

Inflorescence. *Inflorescence paniculate (large, with numerous tiny spikelets)*; open (contracted on the primary branches); espatheate; not comprising 'partial inflorescences' and foliar organs. Spikelet-bearing axes disarticulating (pedicels and ultimate branchlets disarticulating). Spikelets secund; pedicellate.

Female-fertile spikelets. Spikelets 1.2–1.8 mm long; compressed laterally (and somewhat asymmetric); disarticulating above the glumes and falling with the glumes (falling with the pedicels, but also disarticulating above the glumes); tardily disarticulating between the florets. Rachilla prolonged beyond the uppermost female-fertile floret (ending in a flattened process 0.5 mm long, with a flattened tip). Hairy callus absent.

Glumes two; relatively large; very unequal to more or less equal; shorter than the spikelets; shorter than the adjacent lemmas; not pointed (obtuse); awnless; non-carinate; similar (broadly oval, hyaline). Lower glume 0–1 nerved. Upper glume 0–1 nerved. *Spikelets* with incomplete florets. The incomplete florets proximal to the female-fertile florets, or both distal and proximal to the female-fertile florets (proximal incomplete floret always present, distal rudiment present or absent). *The distal incomplete florets* when present, merely underdeveloped. *The proximal incomplete florets* 1; epaleate; sterile. The proximal lemmas acuminate, glabrous; awnless (acuminate); 1–3 nerved; decidedly exceeding the female-fertile lemmas (equalling the spikelet); less firm than the female-fertile lemmas (membranous); not becoming indurated.

Female-fertile florets 1. Lemmas decidedly firmer than the glumes; becoming indurated (firmer than the L_1); entire; pointed; awnless, or mucronate; hairy; non-carinate; without a germination flap; 3 nerved. *Palea* present; conspicuous but relatively short; apically notched; awnless, without apical setae; not indurated (thin); 2-nerved; 2-keeled. *Lodicules* present; 2; free; fleshy; glabrous; not or scarcely vascularized. *Stamens* 2–3. Anthers 0.8 mm long; not penicillate. *Ovary* glabrous. Styles free to their bases. Stigmas 2; red pigmented.

Fruit, embryo and seedling. *Fruit* free from both lemma and palea; small; ellipsoid, or subglobose; not noticeably compressed. Hilum short. Embryo large; not waisted.

Abaxial leaf blade epidermis. *Costal/intercostal zonation* conspicuous. *Papillae* absent. *Long-cells* similar in shape costally and intercostally; of similar wall thickness costally and intercostally. Mid-intercostal long-cells rectangular; having markedly sinuous walls. *Microhairs* present (but very scarce in material seen: variable, cf. *Phragmites*?); panicoid-type. *Stomata* common. Subsidiaries low dome-shaped, or triangular. Guard-cells overlapping to flush with the interstomatals. *Intercostal short-cells* common; in cork/silica-cell pairs; silicified. Intercostal silica bodies tall-and-narrow, or crescentic, or vertically elongated-nodular, or oryzoid-type. *Costal short-cells* conspicuously in long rows. Costal silica bodies 'panicoid-type'; cross shaped, butterfly shaped, and dumb-bell shaped; not sharp-pointed.

Transverse section of leaf blade, physiology. C_3; XyMS+. *Mesophyll* with non-radiate chlorenchyma; without adaxial palisade; with arm cells. *Leaf blade* 'nodular' in section; with the ribs more or less constant in size. *Midrib* conspicuous; having a conventional arc of bundles. Bulliforms present in discrete, regular adaxial groups (between each bundle pair); consistently in simple fans (in the material seen, the fans large with a deeply penetrating median cell). All the vascular bundles accompanied by sclerenchyma. Combined sclerenchyma girders present; forming 'figures' (most of the bundles with a smallish I or T). Sclerenchyma all associated with vascular bundles.

Cytology. Chromosome base number, $x = 11$, or 12 (?).

Taxonomy. Arundinoideae; Arundineae.

Ecology, geography, regional floristic distribution. 1 species. Tropical Asia. Species of open habitats; glycophytic. On mountains.

Paleotropical. Indomalesian. Indian, Indo-Chinese, Malesian, and Papuan.

Economic importance. Important native pasture species: *T. maxima*.

References, etc. Leaf anatomical: Metcalfe 1960; this project.

Torreyochloa Church

Habit, vegetative morphology. Perennial; stoloniferous, or caespitose. Culms 20–50 cm high; *herbaceous*. Culm internodes hollow. Leaves non-auriculate. Leaf blades linear; 1–2.5 mm wide; flat; rolled in bud. *Ligule an unfringed membrane*; not truncate; 0.5–5 mm long.

Reproductive organization. Plants bisexual, with bisexual spikelets; with hermaphrodite florets.

Inflorescence. *Inflorescence paniculate*; open; espatheate; not comprising 'partial inflorescences' and foliar organs. Spikelet-bearing axes persistent. Spikelets not secund; pedicellate.

Female-fertile spikelets. Spikelets 2–6 mm long; compressed laterally to not noticeably compressed; disarticulating above the glumes; disarticulating between the florets. Rachilla prolonged beyond the uppermost female-fertile floret. Hairy callus absent.

Glumes two; very unequal; shorter than the adjacent lemmas; *not pointed (apically rounded)*; awnless; *non-carinate*; similar. Lower glume 1 nerved, or 3 nerved. Upper glume 1–3 nerved. *Spikelets* with female-fertile florets only, or with incomplete florets. The incomplete florets distal to the female-fertile florets.

Female-fertile florets 3–7. *Lemmas decidedly firmer than the glumes*; not becoming indurated; entire, or incised; when entire, blunt; when incised, not deeply cleft (denticulate); awnless; hairless; non-carinate; without a germination flap; *5–7 nerved (these prominent, scaberulous)*. Palea present; relatively long; 2-nerved; 2-keeled. *Lodicules* present; 2; free; membranous; glabrous; toothed, or not toothed; not or scarcely vascularized. Stamens 3. *Anthers 0.3–0.6 mm long*. Ovary glabrous (rarely), or hairy. Styles free to their bases. Stigmas 2; white.

Fruit, embryo and seedling. Fruit free from both lemma and palea; *compressed laterally*; with hairs confined to a terminal tuft. *Hilum short*. Embryo small. Endosperm hard; containing compound starch grains. Embryo with an epiblast; without a scutellar tail; with a negligible mesocotyl internode. Embryonic leaf margins meeting.

Abaxial leaf blade epidermis. *Costal/intercostal zonation* conspicuous. *Papillae* absent. *Long-cells* markedly different in shape costally and intercostally; of similar wall thickness costally and intercostally. Mid-intercostal long-cells rectangular, or fusiform; having straight or only gently undulating walls. *Microhairs* absent. *Stomata* common; 27–38 microns long. Subsidiaries parallel-sided. Guard-cells overlapped by the interstomatals. *Intercostal short-cells* absent or very rare. *Costal short-cells* neither distinctly grouped into long rows nor predominantly paired. Costal silica bodies horizontally-elongated crenate/sinuous (mostly), or horizontally-elongated smooth (a few), or 'panicoid-type' (a few); sometimes nodular; not sharp-pointed.

Transverse section of leaf blade, physiology. C_3; XyMS+. *Mesophyll* with non-radiate chlorenchyma; without adaxial palisade. *Midrib* not readily distinguishable; with one bundle only. Bulliforms present in discrete, regular adaxial groups (in the furrows); in simple fans (the fans wide). All the vascular bundles accompanied by sclerenchyma. Combined sclerenchyma girders present. Sclerenchyma all associated with vascular bundles.

Cytology. Chromosome base number, $x = 7$. $2n = 14$. 2 ploid.

Taxonomy. Pooideae; Poodae; Poeae.

Ecology, geography, regional floristic distribution. 4 species. Northern Asia, North America. Helophytic; species of open habitats; glycophytic. Wet meadows and in shallow water.

Holarctic. Boreal. Euro-Siberian, Eastern Asian, Atlantic North American, and Rocky Mountains. Siberian. Canadian-Appalachian.

References, etc. Leaf anatomical: this project.

Tovarochloa T.D. Macfarlane and P. P.-H. But

Habit, vegetative morphology. Very diminutive alpine *annual*. Culms *0.3–1 cm high*; herbaceous. Culm nodes hidden by leaf sheaths (the internodes condensed). Leaves non-auriculate. Sheaths green or hyaline, with broad membranous margins. Leaf blades linear to ovate; narrow (but *relatively* broad, and short); flat, or folded; without cross venation. *Ligule* present (lower leaves), or absent (upper leaves); an unfringed membrane; not truncate; 0.3 mm long.

Reproductive organization. Plants bisexual, with bisexual spikelets; with hermaphrodite florets. *The spikelets all alike in sexuality*.

Inflorescence. *Inflorescence paniculate (usually shorter than the leaves)*; contracted; capitate (nearly concealed); espatheate; not comprising 'partial inflorescences' and foliar organs. Spikelet-bearing axes persistent. Spikelets not secund; pedicellate.

Female-fertile spikelets. Spikelets 3–3.3 mm long; compressed laterally; disarticulating above the glumes. *Rachilla terminated by a female-fertile floret*. Hairy callus present, or absent.

Glumes two; more or less equal; shorter than the adjacent lemmas, or long relative to the adjacent lemmas; hairless; glabrous; pointed (apiculate, the apiculum sometimes recurved); awnless; carinate; very dissimilar to similar (membranous except along the vein, the lower sometimes apically lobed or shouldered). Lower glume 1 nerved. *Upper glume 1 nerved. Spikelets* with female-fertile florets only.

Female-fertile florets 1. Lemmas similar in texture to the glumes (slightly firmer); not becoming indurated (membranous); entire; pointed; awnless (but apiculate), or mucronate (the awnlet less than 1mm long); uniformly hairy; non-carinate; without a germination flap; 1 nerved, or 3 nerved, or 5 nerved (sometimes with one or two pairs of short laterals); with the nerves non-confluent. *Palea* present; relatively long; awnless, without apical setae, or with apical setae (with one or two points, according to whether one or two veined); textured like the lemma; not indurated (membranous); 1-nerved, or 2-nerved; one-keeled (when 1 nerved), or keel-less. Palea back hairy. *Lodicules* present; 2; free; membranous; glabrous; toothed; not or scarcely vascularized. *Stamens* 2, or 3 (?). Anthers 0.4–0.6 mm long. *Ovary* glabrous. Styles free to their bases. Stigmas 2; white.

Fruit, embryo and seedling. *Fruit* free from both lemma and palea; small (1.5–1.7 mm long); pale brown; compressed dorsiventrally. Hilum short. Embryo small. Endosperm hard; with lipid; containing compound starch grains.

Abaxial leaf blade epidermis. *Costal/intercostal zonation* lacking. *Papillae* absent. Mid-intercostal long-cells fusiform; having straight or only gently undulating walls. *Microhairs* absent. *Stomata* common. Subsidiaries parallel-sided and dome-shaped. *Intercostal short-cells* common to absent or very rare; not paired (solitary); not silicified.

Costal zones with short-cells (but few silica cells). *Costal short-cells* predominantly paired. Costal silica bodies not sharp-pointed.

Transverse section of leaf blade, physiology. C_3; XyMS+. *Leaf blade* adaxially flat. *Midrib* not readily distinguishable; with one bundle only.

Taxonomy. Pooideae; Poodae; Aveneae.

Ecology, geography, regional floristic distribution. 1 species. Peru. Species of open habitats. High Andes.

Neotropical. Andean.

References, etc. Morphological/taxonomic: Macfarlane and But 1982.

Trachypogon Nees

Homopogon Stapf

Habit, vegetative morphology. Slender perennial (very rarely annual); caespitose. *Culms* 30–200 cm high; herbaceous; unbranched above. Culm nodes hairy (with white hairs). Leaves non-auriculate. Leaf blades linear; narrow; not setaceous; rolled (usually, convolute), or flat (sometimes). *Ligule an unfringed membrane*; not truncate.

Reproductive organization. Plants bisexual, with bisexual spikelets; with hermaphrodite florets. The spikelets of sexually distinct forms on the same plant; hermaphrodite and male-only, or hermaphrodite and sterile; more or less overtly heteromorphic (the male or neuter spikelets without a callus, often awnless); all in heterogamous combinations.

Inflorescence. Inflorescence of spicate main branches, or a single raceme; digitate, or non-digitate (when unbranched); espatheate; not comprising 'partial inflorescences' and foliar organs. **Spikelet-bearing axes** 'racemes' (long, terminating the culms); solitary, or paired, or clustered (up to 5 'racemes'); *persistent (but the joints articulated and usually shortly bearded)*. 'Articles' not appendaged; oblique. *Spikelets* paired; pedicellate; *consistently in 'long-and-short' combinations (in which the usual pattern of sexuality is inverted)*; unequally pedicellate in each combination. Pedicels of the 'pedicellate' spikelets free of the rachis. The 'shorter' spikelets male-only, or sterile. The 'longer' spikelets hermaphrodite.

Female-sterile spikelets. The short-pedicelled male or neuter spikelets persistent, sometimes dorsally flattened. Without a callus, often awnless. L_1 sterile. Rachilla of male spikelets terminated by a male floret. The male spikelets with proximal incomplete florets. The lemmas awnless. Male florets 1.

Female-fertile spikelets. *Spikelets not noticeably compressed (cylindrical)*; falling with the glumes (falling from the pedicels). Rachilla terminated by a female-fertile floret. Hairy callus present (in the pedicelled members). Callus of the pedicelled members pointed (attached obliquely to the pedicel).

Glumes two; relatively large; more or less equal; long relative to the adjacent lemmas; hairy; without conspicuous tufts or rows of hairs; awnless; carinate (G_2), or non-carinate (G_1); very dissimilar (the G_1 firmer, convolute and 2-keeled, the G_2 thinner, channelled on each side of the rounded keel). *Lower glume two-keeled*; not pitted; relatively smooth; 7–11 nerved (the nerves inconspicuous between the keels, anastomosing above). Upper glume 3 nerved. *Spikelets* with incomplete florets. The incomplete florets proximal to the female-fertile florets. *The proximal incomplete florets* 1; epaleate; sterile. The proximal lemmas awnless; 2 nerved; more or less equalling the female-fertile lemmas; hyaline; not becoming indurated (ciliate above).

Female-fertile florets 1. Lemmas hyaline basally, but becoming stipitate-cartilaginous above; less firm than the glumes; not becoming indurated; entire; not deeply cleft; awned. Awns 1; median; apical; geniculate; hairy to long-plumose; much longer than the body of the lemma. Lemmas hairy; non-carinate; without a germination flap; 3 nerved. *Palea* present, or absent; when present, very reduced; not indurated (hyaline). *Lodicules* present; 2; free; fleshy; glabrous. *Stamens* 3. *Ovary* glabrous. Stigmas 2.

Fruit, embryo and seedling. Embryo large.

Abaxial leaf blade epidermis. *Costal/intercostal zonation* conspicuous. *Papillae* absent. Mid-intercostal long-cells rectangular; having markedly sinuous walls. *Microhairs* present; panicoid-type; (39–)42–45(–48) microns long; 5.4–6 microns wide at the septum. Microhair total length/width at septum 7–8.9. Microhair apical cells (18–)24–26(–27) microns long. Microhair apical cell/total length ratio 0.46–0.57. *Stomata* common; 22.5–27 microns long. Subsidiaries mostly triangular. Guard-cells overlapping to flush with the interstomatals. *Intercostal short-cells* absent or very rare. Bulbous-based costal prickles abundant. Costal zones with short-cells. *Costal short-cells* conspicuously in long rows. Costal silica bodies 'panicoid-type'; cross shaped (some), or dumb-bell shaped (mostly); not sharp-pointed.

Transverse section of leaf blade, physiology. C_4; XyMS– (but many individual bundles have intervening mestome on one side, or are approaching the XyMS+ condition, and the main midrib bundle is XyMS+). PCR cell chloroplasts centrifugal/peripheral. *Mesophyll* with radiate chlorenchyma; traversed by columns of colourless mesophyll cells (these being the 'arms' of bulliform-plus-colourless cell arches). *Leaf blade* with distinct, prominent adaxial ribs (low, round-topped, over primary bundles). *Midrib* conspicuous; having a conventional arc of bundles (a large median, and three small laterals on either side); with colourless mesophyll adaxially. Bulliforms present in discrete, regular adaxial groups; associated with colourless mesophyll cells to form deeply-penetrating fans (linking via colourless cells with the abaxial epidermis); associating with colourless mesophyll cells to form arches over small vascular bundles. All the vascular bundles accompanied by sclerenchyma (even the smallest having minute abaxial strands). Combined sclerenchyma girders present (with the primary laterals); forming 'figures' (the primary laterals). Sclerenchyma all associated with vascular bundles.

Phytochemistry. Leaves containing flavonoid sulphates (1 species, doubtfully), or without flavonoid sulphates (3 species).

Cytology. Chromosome base number, $x = 5$, or 10. $2n = 20$ and 40.

Taxonomy. Panicoideae; Andropogonodae; Andropogoneae; Andropogoninae.

Ecology, geography, regional floristic distribution. About 13 species. Tropical America and Africa, Madagascar. Mesophytic; species of open habitats; glycophytic. Savanna.

Holarctic, Paleotropical, Neotropical, and Cape. Boreal and Madrean. African. Atlantic North American. Sudano-Angolan, West African Rainforest, and Namib-Karoo. Caribbean, Venezuela and Surinam, Central Brazilian, Pampas, and Andean. Southern Atlantic North American and Central Grasslands. Sahelo-Sudanian, Somalo-Ethiopian, South Tropical African, and Kalaharian.

Rusts and smuts. Rusts — *Puccinia.* Taxonomically wide-ranging species: *Puccinia eritraeensis* and *Puccinia versicolor.* Smuts from Ustilaginaceae. Ustilaginaceae — *Sphacelotheca* and *Ustilago.*

References, etc. Leaf anatomical: this project.

Trachys Pers.

Trachyozus Reichenb., *Trachystachys* Dietr.

Habit, vegetative morphology. Diffuse annual; decumbent. *Culms* 15–30 cm high; herbaceous. Culm nodes hairy. *Leaves* not basally aggregated; non-auriculate. Leaf blades narrow, or broad; 6–12 mm wide (by 2.5–5 cm long); without cross venation; persistent. *Ligule a fringe of hairs*; 3 mm long.

Reproductive organization. Plants bisexual, with bisexual spikelets; with hermaphrodite florets. The spikelets of sexually distinct forms on the same plant; hermaphrodite and sterile, or hermaphrodite, male-only, and sterile (the clusters of complete spikelets surrounded by incomplete ones); overtly heteromorphic (each cluster with proximal barren spikelets, the outermost reduced to thickened, rigid bracts or recurved spines).

Inflorescence. *Inflorescence of spicate main branches (of one-sided, winged, spikelike racemes, the underside of each bearing alternating, short-peduncled glomerules, the latter containing 2–3 bisexual spikelets plus 'bracts' and spines)*; usually *digitate*. Primary inflorescence branches (1–)2–3. Rachides winged. *Spikelet-bearing axes* very much reduced (to glomerules); disarticulating; falling entire (the glomerules constituting burrs, which fall with the rachis joint, i.e. the main rachides also disarticulate). **Spikelets** *associated with bractiform involucres (spiny, representing the* *variously reduced sessile/subsessile spikelets surrounding the clusters).* The involucres shed with the fertile spikelets. *Spikelets secund (the clusters from the midrib on one side of the broad, flat, jointed rachis)*; subsessile.

Female-fertile spikelets. *Spikelets* unconventional (or hard to interpret, because of the associated involucral spikelets); 5–6 mm long; compressed dorsiventrally; falling with the glumes (within the cluster). Rachilla terminated by a female-fertile floret. Hairy callus absent.

Glumes two; very unequal; shorter than the spikelets; shorter than the adjacent lemmas; the upper hairy; pointed (acuminate); awnless; non-carinate; very dissimilar (the lower small, subulate-lanceolate and glabrous, the upper membranous and lanceolate, hairy). Lower glume 0 nerved. Upper glume 3 nerved. *Spikelets* with incomplete florets. The incomplete florets proximal to the female-fertile florets. *The proximal incomplete florets* 1; paleate. Palea of the proximal incomplete florets reduced (tiny). The proximal incomplete florets sterile. The proximal lemmas expanded to form a broadly ovate, involucral scale; awnless (acute); 11–17 nerved; decidedly exceeding the female-fertile lemmas; decidedly firmer than the female-fertile lemmas (leathery); becoming indurated.

Female-fertile florets 1. Lemmas ovate-lanceolate to linear-oblong; similar in texture to the glumes (membranous or papery); not becoming indurated; entire; pointed; awnless; hairless; non-carinate; having the margins lying flat and exposed on the palea; with a clear germination flap; 3 nerved. *Palea* present; relatively long; 2-nerved (membranous); 2-keeled. *Lodicules* present (tiny); 2; free; fleshy; glabrous; not or scarcely vascularized. *Stamens* 3. Anthers penicillate; without an apically prolonged connective. *Ovary* glabrous. Styles fused. Stigmas 2; red pigmented.

Fruit, embryo and seedling. *Fruit* free from both lemma and palea; small (3 mm long); compressed dorsiventrally. Hilum short. Embryo large; waisted.

Abaxial leaf blade epidermis. *Costal/intercostal zonation* conspicuous. *Papillae* absent. Intercostal zones exhibiting many atypical long-cells. Mid-intercostal long-cells having markedly sinuous walls and having straight or only gently undulating walls (mixed). *Microhairs* present; panicoid-type; 51–57 microns long; 5.1–5.4 microns wide at the septum. Microhair total length/width at septum 9.4–11.2. Microhair apical cells 25.5–27 microns long. Microhair apical cell/total length ratio 0.45–0.5. *Stomata* common; 23.5–28–33 microns long. Subsidiaries mostly low dome-shaped (sometimes tending to parallel). Guard-cells overlapping to flush with the interstomatals. *Intercostal short-cells* absent or very rare. With cushion-based macrohairs. Costal zones with short-cells. *Costal short-cells* conspicuously in long rows (but rather ambiguous, because the costal short-cells rather long!). Costal silica bodies 'panicoid-type'; mostly dumb-bell shaped; not sharp-pointed.

Transverse section of leaf blade, physiology. C_4; XyMS–. *Mesophyll* with radiate chlorenchyma. *Leaf blade* adaxially flat (?). *Midrib* conspicuous; having a conventional arc of bundles (1 large plus 8 small bundles); with colourless mesophyll adaxially. Bulliforms not in discrete, regular adaxial groups (constituting most of the epidermis). Many of the smallest vascular bundles unaccompanied by sclerenchyma. Combined sclerenchyma girders absent. Sclerenchyma all associated with vascular bundles.

Taxonomy. Panicoideae; Panicodae; Paniceae.

Ecology, geography, regional floristic distribution. 1 species. Southern India & Burma, especially coastal. Species of open habitats; halophytic, or glycophytic. Mainly in coastal sand.

Paleotropical. Indomalesian. Indian and Indo-Chinese.

References, etc. Leaf anatomical: this project.

Tragus Haller

Habit, vegetative morphology. Annual, or perennial; stoloniferous, or decumbent (usually creeping). *Culms* 5–65 cm high; herbaceous. Culm nodes glabrous. Culm internodes solid. *Leaves* not basally aggregated; non-auriculate. Leaf blades narrow; slightly cordate, or not cordate, not sagittate; not setaceous (somewhat rigid, the margins pectinate); flat; without abaxial multicellular glands; without cross venation; persistent; rolled in bud. *Ligule* present; a fringed membrane (very narrow), or a fringe of hairs.

Reproductive organization. *Plants bisexual, with bisexual spikelets*; with hermaphrodite florets. The spikelets of sexually distinct forms on the same plant, or all alike in sexuality; hermaphrodite, or hermaphrodite and sterile (with one or more members of the cluster reduced).

Inflorescence. *Inflorescence a false spike, with spikelets on contracted axes (a spicate raceme of crowded glomerules, the latter very shortly- or rarely long- peduncled, each with 2–5 spikelets)*; espatheate; not comprising 'partial inflorescences' and foliar organs. **Spikelet-bearing axes** very much reduced (to the glomerules); *disarticulating*; falling entire (the clusters falling whole). Spikelets not secund; sessile to subsessile.

Female-fertile spikelets. *Spikelets* 2–5 mm long; adaxial; *compressed dorsiventrally*; falling with the glumes (in the cluster). Rachilla terminated by a female-fertile floret. Hairy callus absent.

Glumes one per spikelet, or two; very unequal (the G_1 much reduced or absent); (the G_2) long relative to the adjacent lemmas (equalling the spikelet); free; dorsiventral to the rachis; pointed (acute or acuminate); awnless; non-carinate; *very dissimilar (the lower tiny, scarious or absent, the upper large, hard, with 5 rows of hooked spines on the back)*. Lower glume 0 nerved. *Upper glume* 5–7 nerved; *prickly. Spikelets* with female-fertile florets only.

Female-fertile florets 1. Lemmas lanceolate, acute or acuminate; less firm than the glumes (membranous); not becoming indurated; entire; pointed; awnless; hairy (with minute spinous bristles, all over or only centrally); non-carinate; 3 nerved. *Palea* present; entire (pointed); not indurated (hyaline, glabrous); 2-nerved. *Lodicules* present; 2; free; fleshy; glabrous; toothed. *Stamens* 3. Anthers 0.4–0.7 mm long; not penicillate. *Ovary* glabrous. Styles free to their bases. Stigmas 2; white.

Fruit, embryo and seedling. Fruit small; compressed dorsiventrally, or not noticeably compressed. Hilum short. Pericarp fused. Embryo large; not waisted. Endosperm hard; without lipid; containing compound starch grains. Embryo with an epiblast; with a scutellar tail; with an elongated mesocotyl internode; with one scutellum bundle. Embryonic leaf margins meeting.

Seedling with a short mesocotyl. First seedling leaf with a well-developed lamina. The lamina broad; supine; 6–12 veined.

Abaxial leaf blade epidermis. *Costal/intercostal zonation* conspicuous. *Papillae* present, or absent; intercostal. Intercostal papillae not over-arching the stomata; consisting of one oblique swelling per cell. Mid-intercostal long-cells having markedly sinuous walls. *Microhairs* present, or absent; when present, elongated; clearly two-celled; when present, chloridoid-type. Microhair apical cell wall of similar thickness/rigidity to that of the basal cell. Microhair basal cells 30 microns long. Microhair total length/width at septum 2.5. Microhair apical cell/total length ratio 0.3. *Stomata* common; 25–30 microns long. Subsidiaries low dome-shaped, or dome-shaped and triangular. Guard-cells overlapping to flush with the interstomatals. *Intercostal short-cells* common (rarely), or absent or very rare; silicified, or not silicified. Intercostal silica bodies absent, or imperfectly developed; when present, crescentic. Costal zones with short-cells. *Costal short-cells* conspicuously in long rows. Costal silica bodies present in alternate cell files of the costal zones; saddle shaped (mostly), or crescentic (a few); not sharp-pointed.

Transverse section of leaf blade, physiology. C_4; XyMS+. PCR sheath outlines even. PCR sheaths of the primary vascular bundles interrupted; interrupted abaxially only. PCR sheath extensions present. Maximum number of extension cells 1. PCR cell chloroplasts with well developed grana; centripetal. *Mesophyll* with radiate chlorenchyma. *Leaf blade* 'nodular' in section, or adaxially flat; with the ribs more or less constant in size. *Midrib* not readily distinguishable; with one bundle only. Bulliforms present in discrete, regular adaxial groups; in simple fans and associated with colourless mesophyll cells to form deeply-penetrating fans. All the vascular bundles accompanied by sclerenchyma. Combined sclerenchyma girders present; forming 'figures'. Sclerenchyma all associated with vascular bundles. The lamina margins with fibres.

Culm anatomy. *Culm internode bundles* in one or two rings, or in three or more rings.

Cytology. Chromosome base number, $x = 10$. $2n = 20$ and 40. 2 and 4 ploid.

Taxonomy. Chloridoideae; main chloridoid assemblage.

Ecology, geography, regional floristic distribution. 7 species. 6 in warm Africa, 1 pantropical. Commonly adventive. Mesophytic to xerophytic; species of open habitats; glycophytic. Often in disturbed ground.

Holarctic, Paleotropical, Neotropical, Cape, and Australian. Boreal, Tethyan, and Madrean. African, Madagascan, and Neocaledonian. Euro-Siberian and Atlantic North American. Macaronesian, Mediterranean, and Irano-Turanian. Saharo-Sindian, Sudano-Angolan, and Namib-Karoo. Caribbean, Central Brazilian, Pampas, and Andean. North and East Australian and Central Australian. European. Southern Atlantic North American. Sahelo-Sudanian, Somalo-Ethiopian, South Tropical African, and Kalaharian. Tropical North and East Australian and Temperate and South-Eastern Australian.

Rusts and smuts. Rusts — *Puccinia*. Smuts from Ustilaginaceae. Ustilaginaceae — *Sphacelotheca* and *Ustilago*.

Economic importance. Significant weed species: *T. berteronianus*, *T. racemosus*, *T. roxburghii*.

References, etc. Leaf anatomical: Metcalfe 1960; this project.

Tribolium Desv.

Brizopyrum Stapf, *Lasiochloa* Kunth, *Plagiochloa* Adamson and Sprague

Habit, vegetative morphology. Annual, or perennial; rhizomatous, or stoloniferous, or caespitose. **Culms** 2–60 cm high; *herbaceous*; branched above, or unbranched above. Culm nodes glabrous. Culm internodes hollow. Plants unarmed. Young shoots intravaginal. *Leaves* mostly basal, or not basally aggregated; non-auriculate. Leaf blades narrow; 0.3–4 mm wide; setaceous, or not setaceous; flat, or rolled; without abaxial multicellular glands; without cross venation; persistent. *Ligule a fringed membrane to a fringe of hairs*; 0.3–2 mm long (the fringe sometimes double, the row of short hairs interspersed with longer ones). *Contra-ligule* absent.

Reproductive organization. Plants bisexual, with bisexual spikelets; with hermaphrodite florets.

Inflorescence. Inflorescence few spikeleted to many spikeleted; a single spike, or a single raceme, or paniculate; contracted; capitate, or more or less ovoid, or spicate (sometimes interrupted); espatheate; not comprising 'partial inflorescences' and foliar organs. Spikelet-bearing axes persistent. Spikelets solitary; secund, or not secund; biseriate, or not two-ranked; very shortly pedicellate, or subsessile, or sessile; imbricate.

Female-fertile spikelets. Spikelets 2–10 mm long; broadly cuneate, or suborbicular; compressed laterally, or not noticeably compressed; disarticulating above the glumes; tardily disarticulating between the florets; with conventional internode spacings. Rachilla prolonged beyond the uppermost female-fertile floret; rachilla hairless. The rachilla extension with incomplete florets. Hairy callus absent. *Callus* short.

Glumes two; relatively large; very unequal to more or less equal; shorter than the spikelets to exceeding the spikelets; shorter than the adjacent lemmas, or long relative to the adjacent lemmas; dorsiventral to the rachis (when orientation ascertainable); hairy (with glandular, often tubercle-based hairs), or hairless; glabrous, or scabrous; pointed (acute, acuminate or subulate-caudate); shortly awned, or awnless; carinate, or non-carinate; similar (naviculate, membranous to chartaceous). Lower glume shorter than the lowest lemma to much exceeding the lowest lemma; (3–)5(–7) nerved. *Upper glume 5(–7) nerved. Spikelets* with incomplete florets. The incomplete florets distal to the female-fertile florets. *The distal incomplete florets merely underdeveloped; awnless.*

Female-fertile florets *2–9(–14)*. *Lemmas* less firm than the glumes to similar in texture to the glumes (membranous to chartaceous); not becoming indurated; entire; pointed to blunt; not deeply cleft; awnless to mucronate; *hairy (usually with clavate hairs), or hairless*; carinate (usually), or non-carinate (sometimes); without a germination flap; 5–9 nerved. *Palea* present; relatively long; awnless, without apical setae; thinner than the lemma to textured like the lemma; not indurated (membranous); 2-nerved; 2-keeled. Palea keels winged, or wingless; scabrous, or hairy. *Lodicules* present; 2; free; fleshy; ciliate, or glabrous; heavily vascularized, or not or scarcely vascularized. *Stamens* 3. Anthers

1–2.5 mm long; not penicillate; without an apically prolonged connective. *Ovary* glabrous. Styles free to their bases. Stigmas 2; white, or brown.

Fruit, embryo and seedling. *Fruit* free from both lemma and palea; small (1–1.2 mm long); compressed dorsiventrally. *Hilum short*. *Pericarp* fairly *loosely adherent*. Embryo small; waisted; without an epiblast; with a scutellar tail; with an elongated mesocotyl internode. Embryonic leaf margins meeting.

Abaxial leaf blade epidermis. *Costal/intercostal zonation* conspicuous. *Papillae* absent. *Long-cells* similar in shape costally and intercostally, or markedly different in shape costally and intercostally (when the costals are much narrower); of similar wall thickness costally and intercostally. Mid-intercostal long-cells rectangular to fusiform; having markedly sinuous walls. *Microhairs* present; elongated; clearly two-celled; panicoid-type. Microhair apical cell wall thinner than that of the basal cell and often collapsed. Microhairs (58.5–)64–84(–99) microns long. Microhair basal cells 30–36 microns long. Microhairs 9.6–16.5 microns wide at the septum. Microhair total length/width at septum 3.8–8.3. Microhair apical cells (30–)34–50(–60) microns long. Microhair apical cell/ total length ratio 0.48–0.66. *Stomata* absent or very rare, or common; 21–33 microns long. Subsidiaries high dome-shaped, or dome-shaped and triangular. Guard-cells overlapped by the interstomatals (slightly), or overlapping to flush with the interstomatals. *Intercostal short-cells* common; in cork/silica-cell pairs (but often apparently solitary, through overlapping); silicified. Intercostal silica bodies present and perfectly developed. Costal zones with short-cells. *Costal short-cells* conspicuously in long rows. Costal silica bodies present throughout the costal zones; rounded (few), or 'panicoid-type'; predominantly cross shaped and dumb-bell shaped (short); not sharp-pointed.

Transverse section of leaf blade, physiology. C_3; XyMS+. *Mesophyll* with non-radiate chlorenchyma; without adaxial palisade. *Leaf blade* with distinct, prominent adaxial ribs to adaxially flat (the ribs low); with the ribs more or less constant in size. *Midrib* not readily distinguishable; with one bundle only. Bulliforms present in discrete, regular adaxial groups (in each of the slight furrows); in simple fans. Many of the smallest vascular bundles unaccompanied by sclerenchyma, or all the vascular bundles accompanied by sclerenchyma. Combined sclerenchyma girders present, or absent (but then with strong adaxial and abaxial strands); forming 'figures' (most major bundles with narrow I's). Sclerenchyma all associated with vascular bundles, or not all bundle-associated. The 'extra' sclerenchyma in abaxial groups. The lamina margins with fibres.

Phytochemistry. Tissues of the culm bases with abundant starch, or with little or no starch.

Cytology. Chromosome base number, $x = 6$. $2n = 12$. 2 ploid.

Taxonomy. Arundinoideae; Danthonieae.

Ecology, geography, regional floristic distribution. 11 species. South Africa. Commonly adventive. Mesophytic to xerophytic (winter rainfall); species of open habitats; glycophytic. Fynbos and Karoo.

Paleotropical and Cape. African. Namib-Karoo and Ascension and St. Helena.

Rusts and smuts. Smuts from Ustilaginaceae. Ustilaginaceae — *Ustilago*.

References, etc. Morphological/taxonomic: Renvoize 1985c. Leaf anatomical: this project.

Tricholaena Schrad.

Excluding *Rhynchelytrum*

Habit, vegetative morphology. Annual (rarely), or perennial; caespitose, or decumbent. **Culms** 10–120 cm high; herbaceous; *unbranched above*. Culm internodes hollow. Leaf blades often glaucous-inrolled, rigid; without cross venation. Ligule a fringed membrane (very narrow), or a fringe of hairs.

Reproductive organization. Plants bisexual, with bisexual spikelets; with hermaphrodite florets. Apomictic, or reproducing sexually.

Inflorescence. *Inflorescence paniculate*; open, or contracted; with capillary branchlets (these flexuous); espatheate; not comprising 'partial inflorescences' and foliar organs. Spikelet-bearing axes persistent. Spikelets not secund; pedicellate.

Female-fertile spikelets. *Spikelets 2–3.5 mm long*; *compressed laterally*; with a distinctly elongated rachilla internode between the glumes (having G_1 slightly separated from G_2). Rachilla terminated by a female-fertile floret. Hairy callus absent.

Glumes present; one per spikelet, or two; very unequal; (the longer) long relative to the adjacent lemmas; free; hairy (the upper and/or both sometimes variously hairy), or hairless; awnless (the upper sometimes mucronate); very dissimilar (the lower often reduced to a tiny scale, hairy or glabrous), or similar (rarely). **Lower glume** about 0.1–0.2 times the length of the upper glume; *0–1 nerved*. Upper glume indistinctly 3 nerved, or 5 nerved (thinly membranous, emarginate to acute). *Spikelets* with incomplete florets. The incomplete florets proximal to the female-fertile florets. *Spikelets with proximal incomplete florets*. *The proximal incomplete florets* 1; paleate. Palea of the proximal incomplete florets fully developed. *The proximal incomplete florets male*. The proximal lemmas awnless; 3–7 nerved; decidedly exceeding the female-fertile lemmas; less firm than the female-fertile lemmas (often hairy, resembling the upper glume); not becoming indurated.

Female-fertile florets 1 (dorsally compressed). *Lemmas decidedly firmer than the glumes (cartilaginous to sub-crustaceous)*; smooth; becoming indurated to not becoming indurated; entire to incised; not deeply cleft (obtuse to emarginate); awnless; hairless (shiny); non-carinate; having the margins lying flat and exposed on the palea; 3–5 nerved. *Palea* present; relatively long; tightly clasped by the lemma; entire; textured like the lemma; 2-nerved. *Lodicules* present; 2; free; glabrous; not or scarcely vascularized. *Stamens* 3. *Ovary* glabrous. Styles fused, or free to their bases.

Fruit, embryo and seedling. *Fruit* free from both lemma and palea (but enclosed); compressed dorsiventrally. *Hilum short*. Embryo large.

Abaxial leaf blade epidermis. *Costal/intercostal zonation* conspicuous. *Papillae* absent. Mid-intercostal long-cells having markedly sinuous walls. *Microhairs* present; panicoid-type; 51–66 microns long; 5.4–6 microns wide at the septum. Microhair total length/width at septum 8.5–12.2. Microhair apical cells 24–39 microns long. Microhair apical cell/total length ratio 0.47–0.59. *Stomata* common; 30–36 microns long. Subsidiaries triangular. Guard-cells overlapping to flush with the interstomatals. *Intercostal short-cells* common; in cork/silica-cell pairs. Costal zones with short-cells. *Costal short-cells* mixture on the one epidermis. Costal silica bodies 'panicoid-type'; cross shaped (some), or nodular (mostly); not sharp-pointed.

Transverse section of leaf blade, physiology. C_4; XyMS+. PCR sheath outlines uneven. PCR cell chloroplasts centrifugal/peripheral. *Mesophyll* with radiate chlorenchyma. *Leaf blade* 'nodular' in section to adaxially flat; with the ribs more or less constant in size. *Midrib* not readily distinguishable; with one bundle only to having a conventional arc of bundles (depending on delimitation of mid-rib). Bulliforms present in discrete, regular adaxial groups; in simple fans. Many of the smallest vascular bundles unaccompanied by sclerenchyma. Combined sclerenchyma girders present; forming 'figures'. Sclerenchyma all associated with vascular bundles.

Cytology. Chromosome base number, $x = 9$. $2n = 36$. 4 ploid.

Taxonomy. Panicoideae; Panicodae; Paniceae (Melinideae).

Ecology, geography, regional floristic distribution. 12 species. Africa, Madagascar, Canaries, Mediterranean. Xerophytic; species of open habitats; glycophytic. Sandy and stony soil, sometimes ruderal.

Holarctic, Paleotropical, Neotropical, and Cape. Tethyan. African and Madagascan. Macaronesian, Mediterranean, and Irano-Turanian. Saharo-Sindian, Sudano-Angolan, and Namib-Karoo. Andean. Sahelo-Sudanian, Somalo-Ethiopian, South Tropical African, and Kalaharian.

Rusts and smuts. Rusts — *Physopella*. Smuts from Ustilaginaceae. Ustilaginaceae — *Sphacelotheca* and *Ustilago*.

Economic importance. Important native pasture species: *T. teneriffae*.

References, etc. Leaf anatomical: this project.

Trichoneura Anderss.

Crossotropis Stapf

Habit, vegetative morphology. Annual, or perennial (xeromorphic). **Culms** 12–100 cm high; *herbaceous*; branched above (often), or unbranched above. Culm nodes glabrous. Culm internodes solid. *Leaves* not basally aggregated; non-auriculate. Leaf blades linear (pointed); narrow; setaceous, or not setaceous; usually flat; without abaxial multicellular glands; without cross venation. *Ligule an unfringed membrane*; truncate.

Reproductive organization. Plants bisexual, with bisexual spikelets; with hermaphrodite florets.

Inflorescence. *Inflorescence of spicate main branches (the racemes scattered along a central axis)*; open. Primary inflorescence branches 8–40. Rachides hollowed. Inflorescence espatheate; not comprising 'partial inflorescences' and foliar organs. Spikelet-bearing axes persistent. Spikelets solitary; secund; biseriate (on two faces of the trigonous rachis); subsessile.

Female-fertile spikelets. *Spikelets* 5.3–14 mm long; adaxial; *compressed laterally*; disarticulating above the glumes; *disarticulating between the florets*. Rachilla prolonged beyond the uppermost female-fertile floret. The rachilla extension with incomplete florets. Hairy callus present.

Glumes two; more or less equal; about equalling the spikelets to exceeding the spikelets; *long relative to the adjacent lemmas*; dorsiventral to the rachis; pointed; awned, or awnless (tapered into a mucro or short awn); carinate; similar (narrowly lanceolate, membranous, persistent). Lower glume much exceeding the lowest lemma; 1 nerved. Upper glume 1 nerved. *Spikelets* with incomplete florets. The incomplete florets distal to the female-fertile florets.

Female-fertile florets *2–8*. Lemmas similar in texture to the glumes (membranous); not becoming indurated; incised; 2 lobed; not deeply cleft (bluntly 2-toothed); mucronate, or awned. Awns when present, 1; from a sinus; non-geniculate; much shorter than the body of the lemma. Lemmas hairy (conspicuously pectinate-ciliate on the lateral nerves); non-carinate (rounded on the back); 3 nerved. *Palea* present; relatively long; entire to apically notched; awnless, without apical setae; 2-nerved; 2-keeled. Palea keels scabrous, or hairy. *Lodicules* present; 2; free; fleshy; glabrous. *Stamens* 3. Anthers not penicillate. *Ovary* glabrous. Styles free to their bases. Stigmas 2.

Fruit, embryo and seedling. **Fruit** free from both lemma and palea; small; *compressed dorsiventrally*. *Hilum short*. Pericarp fused. Embryo large; waisted.

Abaxial leaf blade epidermis. *Costal/intercostal zonation* conspicuous. *Papillae* absent. *Long-cells* markedly different in shape costally and intercostally (costals narrower); of similar wall thickness costally and intercostally. Mid-intercostal long-cells rectangular; having markedly sinuous walls (deeply). *Microhairs* present; more or less spherical; clearly two-celled; chloridoid-type (basal cell globular, apical cell small). Microhair apical cell wall thinner than that of the basal cell but not tending to collapse. Microhairs 25–27(–30) microns long. Microhair basal cells 24 microns long. Microhairs (11.4–)12–13.5(–15) microns wide at the septum. Microhair total length/width at septum 1.9–2.4. Microhair apical cells (7.5–)8.4–9(–10.5) microns long. Microhair apical cell/total length ratio 0.28–0.38. *Stomata* common; 22–26 microns long. Subsidiaries dome-shaped (mostly), or triangular. Guard-cells overlapping to flush with the interstomatals. *Intercostal short-cells* absent or very rare. Intercostal silica bodies absent. Numerous small, papilla-like prickles present. Costal zones with short-cells. *Costal short-cells* conspicuously in long rows (but short cells often relatively long). Costal silica bodies present in alternate cell files of the costal zones; 'panicoid-type'; dumb-bell shaped (mostly), or cross shaped (few), or nodular (few); not sharp-pointed.

Transverse section of leaf blade, physiology. C_4; XyMS+. PCR sheaths of the primary vascular bundles interrupted; interrupted abaxially only. PCR sheath extensions absent. PCR cell chloroplasts centripetal. *Leaf blade* 'nodular' in section to adaxially flat. *Midrib* not readily distinguishable; with one bundle only. Bulliforms present in discrete, regular adaxial groups; in simple fans and associated with colourless mesophyll cells to form deeply-penetrating fans. All the vascular bundles accompanied by sclerenchyma. Combined sclerenchyma girders present (with all the bundles); forming 'figures' (I's or anchors in all bundles). Sclerenchyma all associated with vascular bundles. The lamina margins with fibres.

Cytology. Chromosome base number, $x = 10$. $2n = 20$. 2 ploid.

Taxonomy. Chloridoideae; main chloridoid assemblage.

Ecology, geography, regional floristic distribution. 7 species. America, tropical Africa. Xerophytic; species of open habitats. In sandy or stony soil.

Holarctic, Paleotropical, and Neotropical. Boreal. African. Atlantic North American. Saharo-Sindian, Sudano-Angolan, West African Rainforest, and Namib-Karoo. Andean. Southern Atlantic North American and Central Grasslands. Sahelo-Sudanian, Somalo-Ethiopian, South Tropical African, and Kalaharian.

Rusts and smuts. Rusts — *Puccinia*.

References, etc. Morphological/taxonomic: Phillips 1973. Leaf anatomical: this project.

Trichopteryx Nees

Habit, vegetative morphology. Annual, or perennial (with slender culms); caespitose, or decumbent. *Culms* 2–90 cm high; herbaceous; branched above, or unbranched above. Culm nodes hairy. Culm internodes hollow. Plants unarmed. Young shoots intravaginal. *Leaves* not basally aggregated; non-auriculate. Leaf blades linear-lanceolate to lanceolate; narrow; 2–6 mm wide (the base rounded or contracted); without cross venation. *Ligule a fringe of hairs*; 0.3 mm long.

Reproductive organization. Plants bisexual, with bisexual spikelets; with hermaphrodite florets.

Inflorescence. *Inflorescence* paniculate; open, or contracted; *with capillary branchlets*; espatheate; not comprising 'partial inflorescences' and foliar organs. Spikelet-bearing axes persistent. *Spikelets solitary, or paired*; not secund; pedicellate; consistently in 'long-and-short' combinations, or not in distinct 'long-and-short' combinations; when long-and-short unequally pedicellate in each combination.

Female-fertile spikelets. Spikelets 2.5–6 mm long; brown; compressed laterally to not noticeably compressed; disarticulating above the glumes; disarticulating between the florets (disarticulating readily between L_1 and L_2, less readily between G_2 and L_1); with conventional internode spacings. Rachilla terminated by a female-fertile floret; rachilla hairless. Hairy callus present. *Callus* short; blunt.

Glumes two; relatively large; very unequal (G_1 one third to one half spikelet length); (the upper) about equalling the spikelets; (the upper) long relative to the adjacent lemmas; hairy, or hairless; pointed (G_1 rarely obtuse); awnless (though the G_1 can be aristulate and the G_2 acuminate); non-carinate; similar (membranous or papery, G_1 narrower). Lower glume shorter than the lowest lemma; 3 nerved. *Upper glume 3 nerved*. Spikelets with incomplete florets. The incomplete florets proximal to the female-fertile florets. *Spikelets with proximal incomplete florets*. The proximal incomplete florets 1; paleate. Palea of the proximal incomplete florets fully developed (two keeled, thin). The proximal incomplete florets male, or sterile. The proximal lemmas awnless; (1–)3 nerved; decidedly exceeding the female-fertile lemmas; less firm than the female-fertile lemmas to similar in texture to the female-fertile lemmas; not becoming indurated (resembling the G_2 in form and texture).

Female-fertile florets 1. Lemmas similar in texture to the glumes to decidedly firmer than the glumes (membranous, hardening to leathery); incised; 2 lobed; deeply cleft; awned. Awns 1, or 3; median, or median and lateral (with the lobes terminating in awns additional to the median); the median different in form from the laterals (when laterals present); from a sinus; geniculate; hairless; much longer than the body of the lemma; entered by one vein; persistent. The lateral awns when present, shorter than the median (straight). Lemmas hairy. *The hairs in tufts (with a sub-marginal tuft of erect hairs, in the middle on each side)*. Lemmas non-carinate; having the margins tucked in onto the palea (the palea embraced and almost enclosed); without a germination flap; 5–7 nerved; with the nerves non-confluent. *Palea* present; relatively long; minutely apically notched; awnless, without apical setae; not indurated (hyaline); 2-nerved; 2-keeled. Palea keels wingless. *Lodicules* present; 2; free; fleshy. **Stamens 2**. Anthers about 1 mm long; not

964

penicillate; without an apically prolonged connective. *Ovary* glabrous. Styles free to their bases. Stigmas 2; white.

Fruit, embryo and seedling. Fruit longitudinally grooved. Hilum long-linear. Embryo large.

Abaxial leaf blade epidermis. *Costal/intercostal zonation* conspicuous. *Papillae* absent. *Long-cells* markedly different in shape costally and intercostally (the costals narrower and more regularly rectangular in *T. stolziana* and *T. dregeana*); differing markedly in wall thickness costally and intercostally (the costals thinner walled). Mid-intercostal long-cells rectangular and fusiform; having markedly sinuous walls. *Microhairs* present; panicoid-type; (36–)39–42(–45) microns long; (5.1–)5.7–6(–6.6) microns wide at the septum. Microhair total length/width at septum 5.9–8.2. Microhair apical cells 21–25.5(–28.5) microns long. Microhair apical cell/total length ratio 0.54–0.68. *Stomata* common; 31–36 microns long. Subsidiaries triangular, or dome-shaped and triangular. Guard-cells overlapping to flush with the interstomatals. *Intercostal short-cells* common (sometimes paired with macrohair bases); not silicified. Costal zones with short-cells. *Costal short-cells* conspicuously in long rows. Costal silica bodies 'panicoid-type'; mostly dumb-bell shaped; not sharp-pointed.

Transverse section of leaf blade, physiology. C_4. The anatomical organization conventional, or unconventional. Organization of PCR tissue when unconventional *Arundinella* type. XyMS–. PCR sheath outlines uneven. PCR cell chloroplasts centrifugal/peripheral. Mesophyll exhibiting 'circular cells', or without 'circular cells'. *Leaf blade* adaxially flat. *Midrib* not readily distinguishable (at least in the two species seen); with one bundle only. Bulliforms present in discrete, regular adaxial groups, or not in discrete, regular adaxial groups (*T. stolziana* having a largely, irregularly bulliform epidermis). Many of the smallest vascular bundles unaccompanied by sclerenchyma. Combined sclerenchyma girders present, or absent (*T. stolziana* with little sclerenchyma, as strands only); when present, forming 'figures'. Sclerenchyma all associated with vascular bundles.

Cytology. Chromosome base number, $x = 12$ (?). $2n = 24$. 2 ploid (?).

Taxonomy. Panicoideae; Panicodae; Arundinelleae.

Ecology, geography, regional floristic distribution. 5 species. Southern and tropical Africa, Madagascar. Helophytic, or mesophytic; shade species, or species of open habitats; glycophytic. Streambanks, grasslands and forest margins.

Paleotropical and Neotropical. African and Madagascan. Sudano-Angolan and West African Rainforest. Andean. Sahelo-Sudanian, Somalo-Ethiopian, and South Tropical African.

References, etc. Leaf anatomical: Metcalfe 1960 and this project; photos of *T. dregeana* provided by R.P. Ellis.

Tridens Roem. & Schult.

Antonella Caro, *Gossweilerochloa* Renvoize, *Tricuspis* P. Beauv., *Windsoria* Nutt Excluding *Erioneuron*

Habit, vegetative morphology. *Perennial*; rhizomatous to stoloniferous (rarely), or caespitose. **Culms** 5–160 cm high; *herbaceous*; branched above, or unbranched above. Culm nodes glabrous. Culm internodes hollow. Young shoots intravaginal. *Leaves* mostly basal, or not basally aggregated; non-auriculate. Leaf blades linear, or linear-lanceolate; narrow; setaceous, or not setaceous; flat, or folded, or rolled; without abaxial multicellular glands; without cross venation; rolled in bud. **Ligule** present; *a fringed membrane, or a fringe of hairs*.

Reproductive organization. *Plants bisexual, with bisexual spikelets*; with hermaphrodite florets. Exposed-cleistogamous. *Plants without hidden cleistogenes*.

Inflorescence. *Inflorescence paniculate*; open to contracted (very variable); capitate to spicate; with capillary branchlets, or without capillary branchlets; espatheate; not comprising 'partial inflorescences' and foliar organs. Spikelet-bearing axes persistent. Spikelets not secund; pedicellate.

Female-fertile spikelets. Spikelets 4–15 mm long; compressed laterally; disarticulating above the glumes; disarticulating between the florets; with conventional internode spacings. Rachilla prolonged beyond the uppermost female-fertile floret. The rachilla extension with incomplete florets. Hairy callus present, or absent. *Callus* short; blunt.

Glumes two; *more or less equal*; shorter than the spikelets; shorter than the adjacent lemmas, or long relative to the adjacent lemmas; pointed (acute to acuminate); awnless; carinate; similar (membranous, often thin). Lower glume 1 nerved. Upper glume 1–3(–7) nerved. *Spikelets* with incomplete florets. The incomplete florets distal to the female-fertile florets. *The distal incomplete florets* merely underdeveloped.

Female-fertile florets 3–12. *Lemmas* not becoming indurated; *incised*; 2 lobed; deeply cleft, or not deeply cleft (from minutely emarginate or toothed, to deeply and obtusely 2-lobed); mucronate to awned (the midnerve usually excurrent as a mucro or short awn, the laterals also often excurrent). Awns when present, 1, or 3; median, or median and lateral; the median similar in form to the laterals (when laterals present); from a sinus; non-geniculate; much shorter than the body of the lemma; entered by one vein. The lateral awns shorter than the median. Lemmas hairy (on the nerves, at least below); carinate to non-carinate; without a germination flap; *3 nerved. Palea* present; relatively long; apically notched; awnless, without apical setae; not indurated; 2-nerved; 2-keeled. Palea keels wingless; glabrous to hairy. *Lodicules* present; 2; free; fleshy; glabrous. *Stamens* 3. Anthers not penicillate; without an apically prolonged connective. *Ovary* glabrous. Styles free to their bases. Stigmas 2; red pigmented.

Fruit, embryo and seedling. Fruit small to medium sized; dull brown; ellipsoid; compressed dorsiventrally. *Hilum short (less than half the fruit length).* Pericarp fused. *Embryo small (about 2/5 of the grain length).* Endosperm hard; without lipid. Embryo with an epiblast; with a scutellar tail; with an elongated mesocotyl internode. Embryonic leaf margins meeting.

Abaxial leaf blade epidermis. *Costal/intercostal zonation* conspicuous. *Papillae* absent. *Long-cells* similar in shape costally and intercostally (the costals smaller, narrower); of similar wall thickness costally and intercostally. Mid-intercostal long-cells rectangular; having markedly sinuous walls. *Microhairs* present; elongated; clearly two-celled; chloridoid-type (but peculiar, the basal cell seemingly expanded below into a 'foot'). Microhair apical cell wall of similar thickness/rigidity to that of the basal cell. Microhairs 18–21 microns long. Microhair basal cells 15 microns long. Microhairs 9–10.5 microns wide at the septum. Microhair total length/width at septum 1.7–2.3. Microhair apical cells 5.4–7.5 microns long. Microhair apical cell/total length ratio 0.26–0.42. *Stomata* common; 24–29 microns long. Subsidiaries dome-shaped (mostly), or triangular (a few, almost). Guard-cells overlapping to flush with the interstomatals. *Intercostal short-cells* common (fairly); not paired (solitary); not silicified. Intercostal silica bodies absent. A few intercostal prickles present. *Crown cells* absent. Costal zones with short-cells. *Costal short-cells* conspicuously in long rows. Costal silica bodies present in alternate cell files of the costal zones; saddle shaped (abundant), or 'panicoid-type'; when panicoid type, cross shaped, or dumb-bell shaped, or nodular (variously incomplete); not sharp-pointed.

Transverse section of leaf blade, physiology. C_4; biochemical type PCK; XyMS+. PCR sheath outlines uneven. PCR sheaths of the primary vascular bundles interrupted; interrupted abaxially only. PCR sheath extensions absent. PCR cells with a suberised lamella. *PCR cell chloroplasts* ovoid; with well developed grana; centrifugal/peripheral. *Mesophyll* with radiate chlorenchyma; not traversed by colourless columns (though some of the bulliform-plus-colourless groups come close to traversing). *Leaf blade* adaxially flat. *Midrib* conspicuous (by virtue of a larger bundle with heavy abaxial sclerenchyma, and adaxial colourless tissue); with one bundle only; with colourless mesophyll adaxially. Bulliforms present in discrete, regular adaxial groups; associated with colourless mesophyll cells to form deeply-penetrating fans (these deeply penetrating). All the vascular bundles accompanied by sclerenchyma. Combined sclerenchyma girders present (with all the main bundles); forming 'figures' (midrib with an anchor, the rest with anchors or I's). Sclerenchyma all associated with vascular bundles. The lamina margins with fibres.

Phytochemistry. Leaf blade chlorophyll *a:b* ratio 3.83–3.87.

Cytology. Chromosome base number, $x = 8$, or 10 (?). $2n = 16, 32, 40,$ and 72. 2, 4, 5, and 9 ploid.

Taxonomy. Chloridoideae; main chloridoid assemblage.

Ecology, geography, regional floristic distribution. 18 species. Eastern and southern U.S.A. and Mexico. Species of open habitats. Meadows, plains, open woodland. Holarctic and Neotropical. Boreal and Madrean. Atlantic North American. Central Brazilian and Pampas. Southern Atlantic North American and Central Grasslands. **Rusts and smuts.** Rusts — *Puccinia*. Taxonomically wide-ranging species: *Puccinia aristidae*. Smuts from Ustilaginaceae. Ustilaginaceae — *Ustilago*. **References, etc.** Leaf anatomical: Metcalfe 1960; this project.

Trikeraia Bor

Stipa sect. Lasiagrostis (Link) Hackel, *Achnatherum* sect. Trikeraia (Bor) Tsvelev (see Freitag 1985)

~ *Stipa*

Habit, vegetative morphology. Perennial; rhizomatous. *Culms* 7.5–20 cm high (*T. oreophila*), or 60–70 cm high (*T. hookeri*); herbaceous. Leaves non-auriculate. Leaf blades narrow; 1–1.5 mm wide; rolled; without cross venation; persistent. *Ligule an unfringed membrane*; 0.4 mm long (*T. oreophila*), or 2 mm long (*T. hookeri*).

Reproductive organization. Plants bisexual, with bisexual spikelets; with hermaphrodite florets.

Inflorescence. Inflorescence paniculate; open, or contracted; espatheate; not comprising 'partial inflorescences' and foliar organs. Spikelet-bearing axes persistent. Spikelets not secund; pedicellate.

Female-fertile spikelets. *Spikelets* slightly *compressed dorsiventrally, or not noticeably compressed*; disarticulating above the glumes. *Rachilla terminated by a female-fertile floret*. Hairy callus present. Callus blunt.

Glumes two; more or less equal; about equalling the spikelets, or exceeding the spikelets; long relative to the adjacent lemmas; pointed; awnless; non-carinate; similar. Lower glume 3 nerved (*T. hookeri*), or 5 nerved (*T. oreophila*). Upper glume 3 nerved. **Spikelets with female-fertile florets only; *without proximal incomplete florets*.**

Female-fertile florets 1. **Lemmas** not convolute (covering only the sides of the palea); decidedly firmer than the glumes (membranous); not becoming indurated; *incised*; 1 lobed (dissected into two shortly awned lobes); awned. *Awns 3*; median and lateral (with two short straight lateral awns in addition to the median one); the median somewhat different in form from the laterals; from a sinus; non-geniculate; *tending to be straight, but slightly twisted basally*; entered by several veins (3). The lateral awns shorter than the median. Awn bases somewhat twisted. Lemmas hairy (on the back, the lobes glabrous); non-carinate; 3–7 nerved. *Palea* present; relatively long; entire (acute); 2-nerved (hairy down the middle like the lemma); keel-less. *Lodicules* present; 3. *Third lodicule present*. Lodicules free; membranous (two 'stipoid' and fleshy at the base); glabrous; toothed; not or scarcely vascularized. **Stamens 3.** Anthers 1–1.5 mm long (*T. oreophila*), or 4–4.5 mm long (*T. hookeri*). *Ovary* glabrous. Styles free to their bases. Stigmas 2.

Abaxial leaf blade epidermis. *Costal/intercostal zonation* conspicuous. *Papillae* present; costal and intercostal. Intercostal papillae over-arching the stomata (and obscuring them); several per cell. Mid-intercostal long-cells having markedly sinuous walls. *Microhairs* absent (also absent adaxially). *Stomata* common; 33–39 microns long. Subsidiaries non-papillate. *Intercostal short-cells* common; in cork/silica-cell pairs. Costal zones with short-cells. *Costal short-cells* neither distinctly grouped into long rows nor predominantly paired. Costal silica bodies 'panicoid-type'; dumb-bell shaped to nodular; not sharp-pointed.

Transverse section of leaf blade, physiology. C_3; XyMS+. *Mesophyll* with non-radiate chlorenchyma. *Leaf blade* with distinct, prominent adaxial ribs. *Midrib* not readily distinguishable; with one bundle only. Bulliforms present in discrete, regular adaxial groups (restricted to the bases of the furrows, small and irregular, cf. *Ammophila*). All the vascular bundles accompanied by sclerenchyma. Combined sclerenchyma girders present; forming 'figures'. Sclerenchyma not all bundle-associated (sometimes has a continuous abaxial layer). The 'extra' sclerenchyma in a continuous abaxial layer.

Taxonomy. Arundinoideae; Stipeae.

Ecology, geography, regional floristic distribution. 2 species. Pakistan to Tibet. Species of open habitats; glycophytic. Montane, by water. Holarctic. Tethyan. Irano-Turanian.

References, etc. Morphological/taxonomic: Bor 1955; Macfarlane and Watson 1980; Freitag 1985; Cope 1988. Leaf anatomical: Metcalfe 1960; this project.

Special comments. Fruit data wanting.

Trilobachne Schenk ex Henrard

Habit, vegetative morphology. Annual. *Culms* 20–100 cm high (or more?); herbaceous. Culm nodes hairy. Plants unarmed. *Leaves* not basally aggregated. Leaf blades linear-lanceolate, or lanceolate; narrow; 50–75 mm wide (to 70 cm long).

Reproductive organization. *Plants monoecious with all the fertile spikelets unisexual;* without hermaphrodite florets. The spikelets of sexually distinct forms on the same plant; female-only and male-only. The male and female-fertile spikelets in different inflorescences (at least, in different partial inflorescences). The spikelets overtly heteromorphic.

Inflorescence. *Inflorescence of distinct male and female partial inflorescences mixed in fascicles in the upper leaf axils;* spatheate; a complex of 'partial inflorescences' and intervening foliar organs (with a hierarchy of spathes and spatheoles). *Spikelet-bearing axes* 'racemes'; clustered (panicled); with substantial rachides; disarticulating; disarticulating at the joints (swollen at base of fertile spikelets). *'Articles'* non-linear; disarticulating transversely. Spikelets sessile and pedicellate, or subsessile and pedicellate; consistently in 'long-and-short' combinations (in both male and female inflorescences); in pedicellate/sessile combinations, or unequally pedicellate in each combination. Pedicels of the 'pedicellate' spikelets free of the rachis (to somewhat fused at their bases). The 'shorter' spikelets female-only (female inflorescence), or male-only (male inflorescence). The 'longer' spikelets sterile.

Female-sterile spikelets. Male spikelets each with 2 male florets, each with 2 fleshy, glabrous lodicules and 3 free stamens. Rachilla of male spikelets terminated by a male floret. The male spikelets without proximal incomplete florets; 2 floreted (both fertile). Male florets 2; 3 staminate.

Female-fertile spikelets. Spikelets 7–8 mm long; compressed dorsiventrally; falling with the glumes. Rachilla terminated by a female-fertile floret. Hairy callus absent.

Glumes two; more or less equal; long relative to the adjacent lemmas; G_1 hairy; awnless; non-carinate; *very dissimilar (G_1 crustaceous, cupular at the base, trilobed with the central lobe largest, the G_2 ovate-acute, thinner).* Lower glume not two-keeled. *Spikelets* with incomplete florets. The incomplete florets proximal to the female-fertile florets. *Spikelets with proximal incomplete florets. The proximal incomplete florets* 1; sterile. The proximal lemmas awnless; 13–15 nerved; decidedly exceeding the female-fertile lemmas; not becoming indurated.

Female-fertile florets 1. Lemmas less firm than the glumes (hyaline); not becoming indurated; entire; pointed, or blunt; awnless; hairless; non-carinate; without a germination flap. *Palea* present (but small); conspicuous but relatively short (about 1/3 of the lemma length); entire to apically notched; awnless, without apical setae; not indurated (hyaline); nerveless; keel-less. *Lodicules* absent. *Stamens* 0. *Ovary* glabrous. Stigmas 2; red pigmented.

Fruit, embryo and seedling. Fruit slightly longitudinally grooved; compressed dorsiventrally. *Hilum long-linear (elongated, in the sulcus).*

Special diagnostic feature. *Spikelets not borne on a broad, leaflike rachis.*

Cytology. $2n = 20$.

Taxonomy. Panicoideae; Andropogonodae; Maydeae.

Ecology, geography, regional floristic distribution. 1 species. India. Forest margins. Paleotropical. Indomalesian. Indian.

References, etc. Morphological/taxonomic: Stapf 1894.

Special comments. Anatomical data wanting.

Triniochloa A. Hitchc.

Habit, vegetative morphology. Perennial; caespitose. Culms herbaceous. Leaves non-auriculate. *Sheath margins joined.* Leaf blades without cross venation. Ligule an unfringed membrane; not truncate; 8–9 mm long.

Reproductive organization. Plants bisexual, with bisexual spikelets; with hermaphrodite florets.

Inflorescence. Inflorescence paniculate; open; espatheate; not comprising 'partial inflorescences' and foliar organs. Spikelet-bearing axes persistent. Spikelets not secund; pedicellate.

Female-fertile spikelets. Spikelets 10–12 mm long; not noticeably compressed, or compressed dorsiventrally; disarticulating above the glumes. *Rachilla terminated by a female-fertile floret.* Hairy callus present. Callus blunt.

Glumes present; two; relatively large; very unequal; shorter than the spikelets, or about equalling the spikelets; shorter than the adjacent lemmas, or long relative to the adjacent lemmas; pointed (acute or acuminate); awnless; non-carinate; similar (thin, membranous, papery). Lower glume 1 nerved. Upper glume 1 nerved. *Spikelets* with female-fertile florets only.

Female-fertile florets 1. Lemmas decidedly firmer than the glumes (subleathery); not becoming indurated; incised; setaceously 2 lobed; not deeply cleft; awned. Awns 1; median; dorsal; from near the top, or from well down the back; geniculate; entered by several veins. Lemmas hairless; non-carinate; without a germination flap; 5(–7) nerved. *Palea* present; relatively long; 2-nerved; closely 2-keeled (sulcate between). *Lodicules* present; 2; joined; fleshy; glabrous; not or scarcely vascularized. *Stamens* 3. *Ovary* glabrous. Styles free to their bases. Stigmas 2.

Fruit, embryo and seedling. Fruit medium sized (about 5 mm long); compressed dorsiventrally. *Hilum long-linear.* Embryo small. Endosperm hard. Embryo with an epiblast.

Abaxial leaf blade epidermis. *Costal/intercostal zonation* conspicuous. *Papillae* absent. *Long-cells* similar in shape costally and intercostally; of similar wall thickness costally and intercostally. Mid-intercostal long-cells rectangular and fusiform; having markedly sinuous walls and having straight or only gently undulating walls. *Microhairs* absent. *Stomata* absent or very rare. *Intercostal short-cells* common; not paired (nearly all solitary); not silicified. A few prickles present over the veins. Costal zones with short-cells. *Costal short-cells* neither distinctly grouped into long rows nor predominantly paired. Costal silica bodies horizontally-elongated crenate/sinuous and 'panicoid-type' (a continuous series of elongated-crenate, elongated-nodular and crosses, even a few dumb-bells); when panicoid type, cross shaped, or dumb-bell shaped, or nodular; not sharp-pointed.

Transverse section of leaf blade, physiology. C_3; XyMS+. *Mesophyll* with non-radiate chlorenchyma; without adaxial palisade. Leaf blade with the ribs very irregular in sizes. *Midrib* conspicuous (via its large bundle and a prominent abaxial keel); with one bundle only. *The lamina* symmetrical on either side of the midrib. All the vascular bundles accompanied by sclerenchyma. Combined sclerenchyma girders present (with all the bundles); forming 'figures' (the large bundles with T's, the smaller bundles with I's). Sclerenchyma all associated with vascular bundles.

Cytology. Chromosome base number, $x = 8$. $2n = 32$. 4 ploid.

Taxonomy. Pooideae; Poodae; Meliceae.

Ecology, geography, regional floristic distribution. 4–5 species. Mexico to Ecuador and Peru. Xerophytic; species of open habitats. Stony hillsides.

Neotropical. Caribbean and Andean.

References, etc. Leaf anatomical: this project.

Triodia R.Br.

Triodon Baumg.

Habit, vegetative morphology. Prickly *perennial ('porcupine grass')*; caespitose. *Culms* 20–120(–240) cm high. Culm nodes glabrous. The shoots aromatic (often resinous), or not aromatic. Leaves auriculate, or non-auriculate. Sheath margins free. The sheaths often viscid. ***Leaf blades*** linear; narrow; not setaceous; acicular; ***hard, woody, needle-like***; without abaxial multicellular glands; without cross venation; persistent; once-folded in bud. *Ligule* present; a fringe of hairs.

Reproductive organization. Plants bisexual, with bisexual spikelets; with hermaphrodite florets.

Inflorescence. Inflorescence a single spike (*T. spicata*), or paniculate; open, or contracted; when contracted, spicate, or more or less irregular; with capillary branchlets (notably, in *T. pascoeana*), or without capillary branchlets; espatheate; not comprising 'partial inflorescences' and foliar organs. Spikelet-bearing axes persistent. Spikelets solitary; not secund; subsessile to pedicellate.

Female-fertile spikelets. Spikelets compressed laterally; disarticulating above the glumes; disarticulating between the florets; with conventional internode spacings (by contrast with *Symplectrodia*). Rachilla prolonged beyond the uppermost female-fertile floret; rachilla hairless. Hairy callus present. *Callus* short; blunt.

Glumes two; more or less equal; ***shorter than the spikelets (to subequal)***; shorter than the adjacent lemmas, or long relative to the adjacent lemmas; hairless; glabrous; awnless; carinate, or non-carinate. Lower glume 1–7(–20) nerved. Upper glume 1–7(–20) nerved. *Spikelets* with female-fertile florets only, or with incomplete florets. The incomplete florets distal to the female-fertile florets. ***Spikelets without proximal incomplete florets.***

Female-fertile florets 4–15. ***Lemmas becoming indurated (usually, in the body)***; usually incised; 3 lobed (or 3-toothed, rarely entire); ***not deeply cleft (at least by contrast with* Plectrachne, *the lobes usually being shorter than the body)***; awnless, or mucronate; hairy (usually, on the body — the lobes often glabrous); carinate to non-carinate; 3–20 nerved. *Palea* present; relatively long; 2-nerved; 2-keeled. Palea keels winged, or wingless. *Lodicules* present; 2; free; fleshy, or membranous; glabrous; heavily vascularized. *Stamens* 3. Anthers not penicillate. *Ovary* glabrous. Styles free to their bases. Stigmas 2.

Fruit, embryo and seedling. Fruit small; ellipsoid. Hilum short. Pericarp fused. Embryo large; not waisted. Endosperm containing compound starch grains. Embryo with an epiblast; with a scutellar tail; with an elongated mesocotyl internode. Embryonic leaf margins meeting.

Abaxial leaf blade epidermis. *Costal/intercostal zonation* conspicuous, or conspicuous and lacking (the abaxial grooves lacking across the leaf centre). *Papillae* absent (but abundant adaxially). *Microhairs* present (probably, but well hidden in the deep, narrow and hairy abaxial grooves — common enough adaxially); chloridoid-type; with 'partitioning membranes' (in *T. scariosa*). The 'partitioning membranes' in the basal cell. *Stomata* absent or very rare (or if present abaxially, confined to the grooves and unseen). Costal zones with short-cells. *Costal short-cells* conspicuously in long rows. Costal silica bodies present in alternate cell files of the costal zones; 'panicoid-type'; dumb-bell shaped; not sharp-pointed.

Transverse section of leaf blade, physiology. C_4. The anatomical organization unconventional. Organization of PCR tissue *Triodia* type (i.e. the 'sheaths' lateral-only to the main bundles, and at least in the lateral parts of the blade, draping over adjacent bundles, so that many PCR cells are very distant from the vascular tissue). Biochemical type NAD–ME (*T. scariosa*); XyMS+. PCR sheath outlines even. PCR cells without a suberised lamella. *PCR cell chloroplasts* ovoid; with well developed grana; centrifugal/peripheral. *Mesophyll* with radiate chlorenchyma; traversed by columns of colourless mesophyll cells (linking the adaxial and abaxial grooves, at least in the middle part of the blade where the PCR cells do not 'drape'); with arm cells. *Leaf blade* 'nodular' in section (with narrow abaxial grooves), or with distinct, prominent adaxial ribs and 'nodular' in section (sometimes lacking the abaxial grooves towards the margins); with the ribs very irregular in sizes. *Midrib* conspicuous, or not readily distinguishable; with one bundle only, or having a conventional arc of bundles; without colourless mesophyll adaxially (but the blade often with *abaxial* blocks of colourless tissue towards the margins). Bulliforms present in discrete, regular adaxial groups (but incorporated in the colourless tissue); associated with colourless mesophyll cells to form deeply-penetrating fans (incorporated in or linked with the colourless girders). All the vascular bundles accompanied by sclerenchyma. Combined sclerenchyma girders present, or absent; forming 'figures'. Sclerenchyma all associated with vascular bundles, or not all bundle-associated. The 'extra' sclerenchyma when present, in abaxial groups, or in a continuous abaxial layer.

Phytochemistry. Leaf blade chlorophyll a:b ratio 4.52–4.91.

Taxonomy. Chloridoideae; Triodieae.

Ecology, geography, regional floristic distribution. 35 species. Australia. Extreme xerophytic; species of open habitats. On arid, sandy or stony soils.

Australian. North and East Australian and Central Australian. Tropical North and East Australian and Temperate and South-Eastern Australian.

Rusts and smuts. Smuts from Ustilaginaceae. Ustilaginaceae — *Ustilago*.

References, etc. Morphological/taxonomic: Burbidge 1953, 1960; Jacobs 1971. Leaf anatomical: Metcalfe 1960 and this project.

Triplachne Link

Habit, vegetative morphology. *Annual*. *Culms* 5–25 cm high; herbaceous. Culm nodes glabrous. Leaves non-auriculate. Leaf blades linear; narrow; 2–3 mm wide; flat; without cross venation; persistent. Ligule an unfringed membrane; truncate; 2–3 mm long.

Reproductive organization. Plants bisexual, with bisexual spikelets; with hermaphrodite florets.

Inflorescence. Inflorescence paniculate (1–5 cm long); contracted; spicate, or more or less ovoid; espatheate; not comprising 'partial inflorescences' and foliar organs. Spikelet-bearing axes persistent. Spikelets not secund; pedicellate.

Female-fertile spikelets. Spikelets 3–4.5 mm long; compressed laterally; disarticulating above the glumes. Rachilla prolonged beyond the uppermost female-fertile floret. The rachilla extension naked. Hairy callus present.

Glumes two; more or less equal; exceeding the spikelets; long relative to the adjacent lemmas; hairless (shiny, scabrid on the keel); pointed; awnless; carinate; similar (membranous, lanceolate, acute). Lower glume 1 nerved. Upper glume 1 nerved. *Spikelets* with female-fertile florets only.

Female-fertile florets 1. **Lemmas** truncate; somewhat *saccate (gibbously ovate in profile)*; less firm than the glumes (scarious); not becoming indurated; entire; *awned*. Awns 3; median and lateral (the two outer veins excurrent, plus the median); the median different in form from the laterals; dorsal; from well down the back; geniculate; much longer than the body of the lemma; entered by one vein. The lateral awns shorter than the median (straight, via the outer, excurrent veins). Lemmas hairy (with long appressed hairs); non-carinate; obscurely *5 nerved*. *Palea* present; relatively long; tightly clasped by the lemma; 2-nerved. *Lodicules* absent. *Stamens* 3. Anthers 0.5 mm long; not penicillate. *Ovary* glabrous. Styles free to their bases. Stigmas 2.

Fruit, embryo and seedling. Fruit small. Hilum short. Embryo small.

Abaxial leaf blade epidermis. *Costal/intercostal zonation* conspicuous. *Papillae* absent. *Long-cells* markedly different in shape costally and intercostally (the costals long and narrow, the intercostals somewhat diamond-shaped and inflated); of similar wall thickness costally and intercostally. Mid-intercostal long-cells fusiform (strongly so); having straight or only gently undulating walls. *Microhairs* absent. *Stomata* common; 37–45 microns long. Subsidiaries dome-shaped, or triangular. Guard-cells overlapped by the interstomatals. *Intercostal short-cells* common; in cork/silica-cell pairs and not paired (solitary); silicified (when paired), or not silicified. Costal zones with short-cells. *Costal short-cells* conspicuously in long rows, or neither distinctly grouped into long rows nor predominantly paired. Costal silica bodies horizontally-elongated crenate/sinuous, or 'panicoid-type'; not sharp-pointed.

Transverse section of leaf blade, physiology. C_3; XyMS+. *Mesophyll* with non-radiate chlorenchyma. *Leaf blade* with distinct, prominent adaxial ribs, or 'nodular' in section; with the ribs more or less constant in size. *Midrib* conspicuous; with one bundle only. Bulliforms present in discrete, regular adaxial groups; in simple fans. All the vascular bundles accompanied by sclerenchyma. Combined sclerenchyma girders present; forming 'figures'. Sclerenchyma all associated with vascular bundles.

Taxonomy. Pooideae; Poodae; Aveneae.

Ecology, geography, regional floristic distribution. 1 species. Mediterranean, Canaries. Xerophytic; species of open habitats. Near the sea.

Holarctic and Paleotropical. Tethyan. African. Macaronesian and Mediterranean. Saharo-Sindian.

References, etc. Leaf anatomical: this project.

Triplasis P. Beauv.

Diplocea Raf., *Merisachne* Steud., *Uralepis* Nutt

Habit, vegetative morphology. Annual, or perennial; caespitose. *Culms* 30–100 cm high; herbaceous; unbranched above. Culm nodes hairy. *Leaves* not basally aggregated; non-auriculate (but with hair tufts at the auricle regions). Leaf blades narrow; flat, or rolled (involute); without abaxial multicellular glands; without cross venation; rolled in bud. Ligule a fringe of hairs.

Reproductive organization. Plants bisexual, with bisexual spikelets; with hermaphrodite florets. Exposed-cleistogamous, or chasmogamous. *Plants with hidden cleistogenes*. The hidden cleistogenes in the leaf sheaths (the culms breaking at the nodes, leaving the cleistogenes within the sheaths; small panicles in upper sheaths, reduced to modified spikelets in the lower ones).

Inflorescence. Inflorescence few spikeleted; paniculate (short, purple); open; espatheate; not comprising 'partial inflorescences' and foliar organs. Spikelet-bearing axes persistent. Spikelets not secund; pedicellate.

Female-fertile spikelets. *Spikelets* 6–10 mm long (linear); compressed laterally; disarticulating above the glumes; disarticulating between the florets; *with a distinctly elongated rachilla internode above the glumes and with distinctly elongated rachilla internodes between the florets*. Rachilla prolonged beyond the uppermost female-fertile floret. Hairy callus present.

Glumes two; very unequal to more or less equal; shorter than the spikelets; shorter than the adjacent lemmas; hairless; pointed, or not pointed; awnless; carinate; similar. Lower glume 1 nerved. Upper glume 1 nerved. *Spikelets* with female-fertile florets only, or with incomplete florets. The incomplete florets distal to the female-fertile florets.

Female-fertile florets 2–4. Lemmas incised; 2 lobed; mucronate, or awned. Awns when present, 1; from a sinus (or mucronate from the notch); non-geniculate; hairy (silky tomentose); much shorter than the body of the lemma to about as long as the body of the lemma (to slightly longer). Lemmas hairy (the nerves silky-villous); *carinate*; without a germination flap; 3 nerved. *Palea* present; 2-nerved; 2-keeled. Palea keels hairy (densely long-villous, glabrous between them). **Lodicules** *present*; 2; fleshy. *Ovary* glabrous. Styles fused. Stigmas 2; red pigmented.

Fruit, embryo and seedling. Fruit small; dull brown; ellipsoid. Pericarp thin; fused (?). Embryo with an epiblast; with a scutellar tail; with an elongated mesocotyl internode. Embryonic leaf margins meeting.

Abaxial leaf blade epidermis. *Costal/intercostal zonation* conspicuous (the intercostal zones in deep grooves). *Papillae* absent. *Long-cells* markedly different in shape costally and intercostally (the costals much narrower); of similar wall thickness costally and intercostally. Intercostal zones with typical long-cells, or exhibiting many atypical long-cells (with short members, in places). Mid-intercostal long-cells rectangular; having markedly sinuous walls. *Microhairs* present; more or less spherical to elongated; clearly two-celled; chloridoid-type (but the apical cells unusually thin walled, often collapsed). Microhair apical cell wall thinner than that of the basal cell and often collapsed to thinner than that of the basal cell but not tending to collapse. Microhairs of *T. pupurea* 24–27 microns long. Microhair basal cells 30 microns long. Microhairs 11.4–13.5 microns wide at the septum. Microhair total length/width at septum 1.9–2.3. Microhair apical cells 5.4–8.4 microns long. Microhair apical cell/total length ratio 0.2–0.3. *Stomata* common (in deep grooves, overarched by the costal prickles); 16.5–27 microns long. Subsidiaries low dome-shaped (predominating), or triangular. Guard-cells overlapped by the interstomatals to overlapping to flush with the interstomatals. *Intercostal short-cells* common; not paired (sometimes paired with prickles, sometimes ostensibly solitary). Intercostal silica bodies absent. Large costal and small intercostal prickles abundant. Costal zones with short-cells. *Costal short-cells* conspicuously in long rows, or conspicuously in long rows, predominantly paired, and neither distinctly grouped into long rows nor predominantly paired (i.e. varying from place to place in *T. americana*). Costal silica bodies present in alternate cell files of the costal zones; 'panicoid-type'; large, dumb-bell shaped (short to long); not sharp-pointed.

Transverse section of leaf blade, physiology. C_4; XyMS+. PCR sheath outlines even. PCR sheaths of the primary vascular bundles interrupted; interrupted both abaxially and adaxially. PCR sheath extensions absent. PCR cell chloroplasts centripetal (in *T. americana*). Mesophyll traversed by columns of colourless mesophyll cells. *Leaf blade* 'nodular' in section; with the ribs more or less constant in size (round to flat topped). *Midrib* not readily distinguishable; with one bundle only. Bulliforms present in discrete, regular adaxial groups (in every furrow); associated with colourless mesophyll cells to form deeply-penetrating fans (these continued into the colourless columns). All the vascular bundles accompanied by sclerenchyma. Combined sclerenchyma girders present (with all the bundles); forming 'figures' (every bundle with a conspicuous I or 'anchor'). Sclerenchyma all associated with vascular bundles. The lamina margins without fibres.

Cytology. Chromosome base number, $x = 10$.

Taxonomy. Chloridoideae; main chloridoid assemblage.

Ecology, geography, regional floristic distribution. 2 species. Southeast U.S.A. Xerophytic; species of open habitats. Dry sandy soils.

Holarctic and Neotropical. Boreal. Atlantic North American. Caribbean. Canadian-Appalachian, Southern Atlantic North American, and Central Grasslands.

Rusts and smuts. Rusts — *Puccinia*. Taxonomically wide-ranging species: *Puccinia schedonnardi*. Smuts from Ustilaginaceae. Ustilaginaceae — *Ustilago*.

References, etc. Leaf anatomical: this project.

Special comments. Fruit data wanting.

Triplopogon Bor

Habit, vegetative morphology. Large, stilt-rooted annual. *Culms* 50–180 cm high (or more?); herbaceous. Culm nodes glabrous. *Leaves* not basally aggregated; non-auriculate. *Leaf blades* lanceolate (acuminate); broad; 10–35 mm wide; *pseudopetiolate (attenuate to the base)*. Ligule an unfringed membrane; 1.5 mm long.

Reproductive organization. *Plants bisexual, with bisexual spikelets*; *without hermaphrodite florets (the L_2 having a female-only floret)*. The spikelets of sexually distinct forms on the same plant; hermaphrodite and sterile; overtly heteromorphic (glumes of the pedicelled spikelets not hair-tufted); in both homogamous and heterogamous combinations (the lowermost pair both sterile).

Inflorescence. Inflorescence falsely paniculate (the numerous culm branches terminating in pedunculate 'racemes'); spatheate; a complex of 'partial inflorescences' and intervening foliar organs. *Spikelet-bearing axes* 'racemes' (6 or more noded); solitary; disarticulating; disarticulating at the joints. *'Articles'* linear (to slightly widened upwards, flattened); not appendaged; disarticulating transversely; densely long-hairy (densely silky on margins). Spikelets paired; sessile and pedicellate; consistently in 'long-and-short' combinations; in pedicellate/sessile combinations. Pedicels of the 'pedicellate' spikelets free of the rachis. *The 'shorter' spikelets hermaphrodite, or sterile (the lowermost sessile spikelet being rudimentary, and the second pair often reduced)*. The 'longer' spikelets sterile (?).

Female-sterile spikelets. Pedicelled spikelets sterile(?), to 5 mm long.

Female-fertile spikelets. Spikelets 8–10 mm long; compressed laterally; falling with the glumes. Rachilla terminated by a female-fertile floret. Hairy callus present.

Glumes one per spikelet; relatively large; long relative to the adjacent lemmas; hairy; *with distinct hair tufts (the G_1 with two tufts, the G_2 with one)*; awnless; very dissimilar (the G_1 lanceolate, with a median groove and incurved margins and a hair tuft on either side above the middle, the G_2 complanate, not grooved, with one median tuft). Lower glume not two-keeled (except towards the tip, and wingless); sulcate on the back; not pitted; relatively smooth; 'many-nerved'. Upper glume 'about 9-nerved'. *Spikelets* with incomplete florets. The incomplete florets proximal to the female-fertile florets. *The proximal incomplete florets* 1; paleate; male (the floret with 3 stamens). The proximal lemmas awnless; 3 nerved; more or less equalling the female-fertile lemmas to decidedly exceeding the female-fertile lemmas; similar in texture to the female-fertile lemmas (thin); not becoming indurated.

Female-fertile florets 1. Lemmas less firm than the glumes (thin); not becoming indurated; incised; not deeply cleft; awned. Awns 1; median; from a sinus; geniculate; hairless (scabrid); much longer than the body of the lemma. Lemmas hairless; glabrous; non-carinate; without a germination flap; 3 nerved. *Palea* present; conspicuous but relatively short; entire; awnless, without apical setae; not indurated (thin); 2-nerved; keel-less. *Lodicules* present; 2; free; fleshy; glabrous. *Stamens* 0. *Ovary* glabrous. Styles free to their bases. Stigmas 2.

Fruit, embryo and seedling. Fruit small (3.5 mm long); longitudinally grooved (adaxially); strongly compressed laterally (and 3-angled). Hilum short. Embryo large.

Abaxial leaf blade epidermis. *Costal/intercostal zonation* conspicuous. *Papillae* present. Intercostal papillae over-arching the stomata; several per cell (a row per long-cell). Intercostal zones with typical long-cells to exhibiting many atypical long-cells (some of them short). Mid-intercostal long-cells having markedly sinuous walls (thin). *Microhairs* present; panicoid-type; (24–)31–44 microns long. Microhair apical cells (13–)16–29 microns long. Microhair apical cell/total length ratio 0.6. *Stomata* common. Subsidiaries tall dome-shaped and triangular. *Intercostal short-cells* absent or very rare (infrequent). Costal zones with short-cells. *Costal short-cells* conspicuously in long rows, or neither distinctly grouped into long rows nor predominantly paired (confused by long 'short-cells'). Costal silica bodies 'panicoid-type'; cross shaped to dumb-bell shaped, or nodular (and intermediates); not sharp-pointed.

Transverse section of leaf blade, physiology. C_4 (vb's 'crowded'); XyMS−. *Mesophyll* with radiate chlorenchyma. *Leaf blade* adaxially flat. *Midrib* conspicuous (keel rounded); having a conventional arc of bundles (3 large median bundles and several smaller laterals); with colourless mesophyll adaxially. Bulliforms not in discrete, regular adaxial groups (in irregular groups, cf. *Ammophila*). Many of the smallest vascular bundles unaccompanied by sclerenchyma. Combined sclerenchyma girders present. Sclerenchyma as a hypodermal adaxial plate in the keel.

Taxonomy. Panicoideae; Andropogonodae; Andropogoneae; Andropogoninae.

Ecology, geography, regional floristic distribution. 1 species. Bombay. Forest margins.

Paleotropical. Indomalesian. Indian.

Rusts and smuts. Smuts from Ustilaginaceae. Ustilaginaceae — *Sphacelotheca*.

References, etc. Morphological/taxonomic: Bor 1954. Leaf anatomical: Metcalfe 1960.

Tripogon Roem. & Schult.

Archangelina Kuntze, *Kralikia* Coss. & Dur., *Kralikiella* Batt. & Trab., *Plagiolytrum* Nees

Excluding *Oropetium*

Habit, vegetative morphology. Annual, or perennial; caespitose. Culms 4–65 cm high; *herbaceous*; unbranched above. Culm nodes glabrous. Culm internodes hollow. *Leaves* mostly basal; non-auriculate. Leaf blades linear (often filiform); narrow; setaceous, or not setaceous; without abaxial multicellular glands; without cross venation; persistent. Ligule an unfringed membrane to a fringe of hairs.

Reproductive organization. Plants bisexual, with bisexual spikelets; with hermaphrodite florets.

Inflorescence. *Inflorescence a single spike (slender)*. Rachides hollowed. Inflorescence espatheate; not comprising 'partial inflorescences' and foliar organs. Spikelet-bearing axes persistent. *Spikelets* solitary; not secund; alternately distichous; *sessile*; distant.

Female-fertile spikelets. Spikelets 3–25 mm long; adaxial; compressed laterally; disarticulating above the glumes; disarticulating between the florets. Rachilla prolonged beyond the uppermost female-fertile floret; rachilla hairless (glabrous). The rachilla extension with incomplete florets. Hairy callus present (minute). *Callus* short.

Glumes two; very unequal to more or less equal; shorter than the spikelets; shorter than the adjacent lemmas to long relative to the adjacent lemmas; dorsiventral to the rachis; hairless; glabrous; awnless; carinate; very dissimilar, or similar (membranous, narrow, the G_1 often asymmetric). Lower glume 1 nerved. Upper glume 1 nerved, or 3 nerved, or 5 nerved. *Spikelets* with incomplete florets. The incomplete florets distal to the female-fertile florets (or rarely, the L_1 also neuter). *The distal incomplete florets* merely underdeveloped.

Female-fertile florets *3–20*. Lemmas not becoming indurated (scarious with hyaline margins); 2 lobed, or 4 lobed; not deeply cleft (the teeth small); mucronate, or awned (usually awned or mucronate from a median sinus or behind the apex, the lobes sometimes awned or mucronate). Awns when present, 1, or 3, or 5; median, or median and lateral (via mucronate to awned lobes); the median similar in form to the laterals (when laterals present); from a sinus, or apical; non-geniculate; much shorter than the body of the lemma to much longer than the body of the lemma. *Lemmas* hairless; glabrous; *carinate*; *1–3 nerved*. *Palea* present; entire (truncate), or apically notched; awnless, without apical setae; not indurated (hyaline); 1-nerved, or 2-nerved; 2-keeled. Palea keels usually winged (below). *Lodicules* present; 2; free; fleshy; glabrous. *Stamens* 2, or 3. Anthers 0.8–1.3 mm long; not penicillate. **Ovary** glabrous; *without a conspicuous apical appendage*. Styles free to their bases. Stigmas 2.

Fruit, embryo and seedling. *Fruit* free from both lemma and palea; small (0.8–2.2 mm long); not noticeably compressed (terete), or trigonous. Hilum short. Pericarp fused. Embryo large, or small (1/3 the length of the fruit or somewhat less). Endosperm hard; without lipid; containing compound starch grains. Embryo with an epiblast; with a scutellar tail; with an elongated mesocotyl internode. Embryonic leaf margins meeting.

First seedling leaf with a well-developed lamina.

Abaxial leaf blade epidermis. *Costal/intercostal zonation* conspicuous. *Papillae* absent. *Long-cells* similar in shape costally and intercostally; of similar wall thickness costally and intercostally. Mid-intercostal long-cells rectangular; having markedly sinuous walls. *Microhairs* present; more or less spherical to elongated; clearly two-celled; chloridoid-type. Microhair apical cell wall of similar thickness/rigidity to that of the basal cell. Microhairs 16.5–21 microns long. Microhair basal cells 2.4–3.6 microns long. Microhairs 9–12 microns wide at the septum. Microhair total length/width at septum 1.6–2. Microhair apical cells 10.5–12(–13.5) microns long. Microhair apical cell/total length ratio 0.58–0.67. *Stomata* common; 18–21 microns long. Subsidiaries dome-shaped, or triangular. Guard-cells overlapping to flush with the interstomatals. *Intercostal short-cells* absent or very rare; in cork/silica-cell pairs; silicified. Intercostal silica bodies imperfectly developed. Costal zones with short-cells. *Costal short-cells* conspicuously in long rows. Costal silica bodies present in alternate cell files of the costal zones; saddle shaped and tall-and-narrow (or rectangular); sharp-pointed.

Transverse section of leaf blade, physiology. C_4; XyMS+. PCR sheath outlines even. PCR sheaths of the primary vascular bundles interrupted; interrupted abaxially only. PCR sheath extensions absent. PCR cell chloroplasts centripetal. *Mesophyll* with radiate chlorenchyma; traversed by columns of colourless mesophyll cells, or not traversed by colourless columns. *Leaf blade* 'nodular' in section, or adaxially flat; with the ribs more or less constant in size. *Midrib* conspicuous, or not readily distinguishable; with one bundle only. Bulliforms associated with colourless mesophyll cells to form deeply-penetrating fans (these sometimes linking with traversing columns of colourless cells). All the vascular bundles accompanied by sclerenchyma. Combined sclerenchyma girders present; forming 'figures'. Sclerenchyma all associated with vascular bundles. The lamina margins with fibres.

Phytochemistry. Leaves without flavonoid sulphates (1 species).

Cytology. Chromosome base number, $x = 10$. $2n = 20$.

Taxonomy. Chloridoideae; main chloridoid assemblage.

Ecology, geography, regional floristic distribution. About 30 species. Tropical Africa, Asia, Australia. Helophytic to xerophytic; species of open habitats; glycophytic. Holarctic, Paleotropical, Neotropical, and Australian. Boreal and Tethyan. African and Indomalesian. Euro-Siberian, Eastern Asian, and Atlantic North American. Macaronesian and Irano-Turanian. Saharo-Sindian, Sudano-Angolan, West African Rainforest, and Namib-Karoo. Indian, Indo-Chinese, and Papuan. Central Brazilian, Pampas, and Andean. North and East Australian and Central Australian. Siberian. Southern Atlantic North American and Central Grasslands. Sahelo-Sudanian, Somalo-Ethiopian, South Tropical African, and Kalaharian. Tropical North and East Australian and Temperate and South-Eastern Australian.

Rusts and smuts. Rusts — *Puccinia*.

References, etc. Morphological/taxonomic: Phillips and Launert 1971. Leaf anatomical: Metcalfe 1960; this project.

Tripsacum L.

Dactylodes Kuntze, *Digitaria* Adans.

Habit, vegetative morphology. Perennial; forming clumps, from thick knotty rhizomes. *Culms* 70–400 cm high. *Leaves* not basally aggregated; non-auriculate. Leaf blades broad, or narrow; flat. Ligule an unfringed membrane to a fringed membrane; short.

Reproductive organization. *Plants monoecious with all the fertile spikelets unisexual*; without hermaphrodite florets. The spikelets of sexually distinct forms on the same plant; female-only and male-only. The male and female-fertile spikelets segregated, in different parts of the same inflorescence branch. The spikelets overtly heteromorphic. Apomictic, or reproducing sexually.

Inflorescence. Inflorescence of spicate main branches (terminal and axillary inflorescences of 1–5 racemes, the male spikelets distal to the females on the same rachis); open; digitate, or non-digitate (when unbranched), or subdigitate. Primary inflorescence branches 1–5. Rachides hollowed (in association with the female spikelets). Inflorescence espatheate; not comprising 'partial inflorescences' and foliar organs. **Spikelet-bearing axes** spikelike; *with substantial rachides (the female part becoming bony)*; disarticulating; disarticulating at the joints (in the lower, female part — the male section falling whole). 'Articles' non-linear (the female spikelets sunken in the hollows). Spikelets solitary (female spikelets, sometimes accompanied by a rudiment), or paired (males); secund (above: the male spikelets on one side of rachis), or not secund (below: female spikelets opposite); sessile (female), or sessile and pedicellate (male); consistently in 'long-and-short' combinations (male), or not in distinct 'long-and-short' combinations (female). The 'shorter' spikelets female-only, or male-only. The 'longer' spikelets male-only.

Female-sterile spikelets. Male spikelets 5–12 mm long, with 2 membranous glumes and 2 florets. The male spikelets with glumes (two); 2 floreted.

Female-fertile spikelets. *Spikelets* morphologically 'conventional' (but the inner parts hyaline); 5–10 mm long; abaxial (?); compressed laterally, or not noticeably compressed, or compressed dorsiventrally (?); falling with the glumes (and the joint). Rachilla terminated by a female-fertile floret. Hairy callus absent.

Glumes two; *very unequal (indurate, fused with the rachis)*; long relative to the adjacent lemmas (enclosing the female spikelets); dorsiventral to the rachis; hairless; awnless; non-carinate. *Spikelets* with incomplete florets. The incomplete florets proximal to the female-fertile florets. *Spikelets with proximal incomplete florets*. *The proximal incomplete florets* 1; paleate, or epaleate; sterile. The proximal lemmas awnless; 0 nerved; more or less equalling the female-fertile lemmas to decidedly exceeding the female-fertile lemmas; similar in texture to the female-fertile lemmas (hyaline); not becoming indurated.

Female-fertile florets 1. **Lemmas** less firm than the glumes (hyaline); not becoming indurated; awnless; hairless; non-carinate; without a germination flap; *0 nerved. Palea* present (hyaline); awnless, without apical setae; not indurated (hyaline). *Lodicules* absent. *Stamens* 0. *Ovary* glabrous. Styles fused. Stigmas 2.

Fruit, embryo and seedling. *Hilum short.* Embryo large. Endosperm hard; without lipid. Embryo without an epiblast; with a scutellar tail; with an elongated mesocotyl internode. Embryonic leaf margins overlapping.

Abaxial leaf blade epidermis. *Costal/intercostal zonation* conspicuous. *Papillae* absent. *Long-cells* similar in shape costally and intercostally; of similar wall thickness costally and intercostally (walls of medium thickness). Mid-intercostal long-cells rectangular; having markedly sinuous walls. *Microhairs* present; panicoid-type (but often balanoform and blunt tipped); 49–78 microns long (*T. dactyloides*), or 81–90 microns long (*T. laxum*); 12–13.5 microns wide at the septum (in *T. laxum*). Microhair total length/width at septum 6.2–7.5 (*laxum*). Microhair apical cells 25–45 microns long (*T. dactyloides*), or 55–66 microns long (*T. laxum*). Microhair apical cell/total length ratio 0.54 (*T. dactyloides*), or 0.69–0.73 (*laxum*). *Stomata* common; 57–60 microns long. Subsidiaries dome-shaped to triangular. Guard-cells overlapping to flush with the interstomatals. *Intercostal short-cells* common; in cork/silica-cell pairs (and solitary); silicified. Intercostal silica bodies cross-shaped, or crescentic, or vertically elongated-nodular. Costal zones with short-cells. *Costal short-cells* conspicuously in long rows (in places), or neither distinctly grouped into long rows nor predominantly paired. Costal silica bodies tall-and-narrow (few), or 'panicoid-type'; cross shaped to dumb-bell shaped; not sharp-pointed.

Transverse section of leaf blade, physiology. C_4; XyMS–. PCR cell chloroplasts with reduced grana. *Mesophyll* with non-radiate chlorenchyma. *Leaf blade* adaxially flat. *Midrib* conspicuous (keeled); having a conventional arc of bundles (a large median, and a few to many small laterals); with colourless mesophyll adaxially. Bulliforms present in discrete, regular adaxial groups; in simple fans (sometimes the groups large celled — *Zea*-type). Many of the smallest vascular bundles unaccompanied by sclerenchyma. Combined sclerenchyma girders present (with all but the smaller bundles).

Phytochemistry. Leaves containing flavonoid sulphates (1 species), or without flavonoid sulphates (1 species).

Cytology. Chromosome base number, $x = 9$. $2n = 36, 72, 90$, and 108. $4, 8, 10$, and 12 ploid. Nucleoli persistent.

Taxonomy. Panicoideae; Andropogonodae; Maydeae.

Ecology, geography, regional floristic distribution. About 12 species. Warm America. Commonly adventive. Helophytic, or mesophytic; shade species and species of open habitats. Open woodland and damp places.

Holarctic, Paleotropical, and Neotropical. Boreal and Madrean. Indomalesian. Atlantic North American. Malesian. Caribbean, Venezuela and Surinam, Central Brazilian, and Andean. Southern Atlantic North American and Central Grasslands.

Hybrids. Intergeneric hybrids with *Zea*.

Rusts and smuts. Rusts — *Physopella* and *Puccinia*. Smuts from Ustilaginaceae. Ustilaginaceae — *Ustilago*.

Economic importance. Significant weed species: *T. laxum*. Cultivated fodder: several, notably *T. andersonii* (Guatemala), *T. dactyloides, T. laxum*.

References, etc. Leaf anatomical: Metcalfe 1960; this project.

Triraphis R.Br.

Habit, vegetative morphology. Annual, or perennial; caespitose (mostly small xeromorphs). **Culms** (1–)4–140 cm high; *herbaceous*. Culm nodes glabrous. Culm internodes solid. *Leaves* mostly basal; non-auriculate. Leaf blades narrow; setaceous, or not setaceous; flat, or rolled (or junciform); without abaxial multicellular glands; without cross venation; persistent. *Ligule a fringed membrane, or a fringe of hairs*.

Reproductive organization. *Plants bisexual, with bisexual spikelets*; with hermaphrodite florets.

Inflorescence. *Inflorescence paniculate*; open, or contracted (rarely spiciform); when contracted, spicate to more or less irregular; with capillary branchlets, or without capillary branchlets; espatheate; not comprising 'partial inflorescences' and foliar organs. Spikelet-bearing axes persistent. Spikelets not secund; pedicellate.

Female-fertile spikelets. Spikelets compressed laterally; disarticulating above the glumes; disarticulating between the florets. Rachilla prolonged beyond the uppermost female-fertile floret; rachilla hairless. The rachilla extension with incomplete florets. Hairy callus present.

Glumes two; relatively large; very unequal (rarely), or more or less equal; shorter than the spikelets; *shorter than the adjacent lemmas*; pointed, or not pointed (often bidentate); awned (or mucronate, from the sinus), or awnless; carinate; similar (narrow, persistent). Lower glume 1 nerved. *Upper glume 1 nerved. Spikelets* with incomplete florets. The incomplete florets distal to the female-fertile florets. *The distal incomplete florets* merely underdeveloped.

Female-fertile florets 5–10. *Lemmas* not becoming indurated (membranous); *incised; 3 lobed, or 4 lobed (the central lobe bidentate)*; deeply cleft (on either side of the central lobe); awned. Awns 3; median and lateral (the lateral lobes setiform-awned or mucronate); the median similar in form to the laterals; from a sinus (of the central lobe); non-geniculate (setiform). The lateral awns shorter than the median. Lemmas hairy (villous on the lateral nerves); non-carinate (3-keeled), or carinate (when the lateral keels are near the margins); 3 nerved. *Palea* present; shorter than the lemma; apically notched, or deeply bifid; not indurated (hyaline); 2-nerved; 2-keeled. *Lodicules* present; 2; free; fleshy, or membranous; glabrous. *Stamens* 3. Anthers not penicillate. *Ovary* glabrous. Styles free to their bases. Stigmas 2; white.

Fruit, embryo and seedling. Fruit free from both lemma and palea; small; *linear*; trigonous. Hilum short. Pericarp fused. Embryo large; waisted. Endosperm containing compound starch grains. Embryo with an epiblast; with a scutellar tail; with an elongated mesocotyl internode; with one scutellum bundle. Embryonic leaf margins meeting.

First seedling leaf with a well-developed lamina. The lamina broad (fairly); curved.

Abaxial leaf blade epidermis. *Costal/intercostal zonation* conspicuous. *Papillae* absent. *Long-cells* similar in shape costally and intercostally; of similar wall thickness costally and intercostally. Mid-intercostal long-cells rectangular; having markedly sinuous walls. *Microhairs* present; elongated; clearly two-celled; panicoid-type. Microhair apical cell wall thinner than that of the basal cell and often collapsed. Microhairs without 'partitioning membranes' (in *T. mollis*); (40–)50–72(–75) microns long; (6–)6.6–7.5(–8.4) microns wide at the septum. Microhair total length/width at septum 6.8–11. Microhair apical cells (27–)30–42(–46) microns long. Microhair apical cell/total length ratio 0.44–0.63. *Stomata* common; (24–)25–31.5(–36) microns long. Subsidiaries parallel-sided and dome-shaped, or dome-shaped and triangular, or parallel-sided, dome-shaped, and triangular; including both triangular and parallel-sided forms on the same leaf. Guard-cells overlapping to flush with the interstomatals. *Intercostal short-cells* absent or very rare; not paired (solitary); not silicified. Intercostal silica bodies absent. Costal zones with short-cells. *Costal short-cells* conspicuously in long rows, or neither distinctly grouped into long rows nor predominantly paired. Costal silica bodies present in alternate cell files of the costal zones; 'panicoid-type'; cross shaped to butterfly shaped, or dumb-bell shaped; not sharp-pointed.

Transverse section of leaf blade, physiology. C_4; biochemical type NAD–ME (*T. mollis*); XyMS+. PCR sheath outlines uneven. PCR sheaths of the primary vascular bundles complete. PCR sheath extensions present. Maximum number of extension cells 3. PCR cells with a suberised lamella. *PCR cell chloroplasts* ovoid; with well developed grana; centrifugal/peripheral. *Mesophyll* with radiate chlorenchyma. *Leaf blade* 'nodular' in section; with the ribs more or less constant in size, or with the ribs very irregular in sizes. *Midrib* conspicuous, or not readily distinguishable; with one bundle only. Bulliforms present in discrete, regular adaxial groups; in simple fans, or associated with colourless mesophyll cells to form deeply-penetrating fans (in *T. pumilio*, according to Metcalfe 1960). All the vascular bundles accompanied by sclerenchyma. Combined sclerenchyma girders present; forming 'figures'. Sclerenchyma all associated with vascular bundles.

Phytochemistry. Leaf blade chlorophyll *a:b* ratio 4–4.01.

Cytology. Chromosome base number, $x = 10$. $2n = 20$. 2 ploid.

Taxonomy. Chloridoideae; main chloridoid assemblage.

Ecology, geography, regional floristic distribution. 7 species. Tropical and southern Africa, Australia. Mesophytic to xerophytic; species of open habitats; glycophytic. Savanna, in sandy or rocky soil.

Paleotropical and Australian. African. Saharo-Sindian, Sudano-Angolan, and Namib-Karoo. North and East Australian and Central Australian. South Tropical African and Kalaharian. Tropical North and East Australian and Temperate and South-Eastern Australian.

References, etc. Leaf anatomical: Metcalfe 1960; this project.

Triscenia Griseb.

Habit, vegetative morphology. Perennial; caespitose. *Culms* 20–50 cm high; herbaceous; unbranched above. Culm internodes hollow. *Leaves* mostly basal. Leaf blades narrow; setaceous (or filiform); acicular (reduced to the midrib); without cross venation.

Reproductive organization. Plants bisexual, with bisexual spikelets; with hermaphrodite florets.

Inflorescence. *Inflorescence of spicate main branches*; contracted (the few-spikeleted, ascending branches distant). Primary inflorescence branches 3–5. Inflorescence espatheate; not comprising 'partial inflorescences' and foliar organs. *Spikelet-bearing axes* 'racemes' (the spikelets appressed); solitary; persistent. Spikelets secund (?); shortly pedicellate (the pedicels slender).

Female-fertile spikelets. Spikelets 3 mm long; compressed dorsiventrally; falling with the glumes. Rachilla terminated by a female-fertile floret. Hairy callus absent.

Glumes two; relatively large; very unequal (G_1 about half as long as the G_2); (the upper) about equalling the spikelets; (the upper) long relative to the adjacent lemmas; hairless; pointed (G_1 acute, G_2 acuminate); awnless; non-carinate. *Lower glume* about 0.5 times the length of the upper glume; 1 nerved. Upper glume 3 nerved. *Spikelets* with incomplete florets. The incomplete florets proximal to the female-fertile florets. *The proximal incomplete florets* 1; epaleate; sterile. The proximal lemmas awnless (similar to G_2); 3 nerved; decidedly exceeding the female-fertile lemmas; less firm than the female-fertile lemmas to similar in texture to the female-fertile lemmas; not becoming indurated.

Female-fertile florets 1. **Lemmas narrow, rather acuminate, chartaceous with firm, flat margins**; less firm than the glumes to similar in texture to the glumes (thin, but papery); not becoming indurated; entire; pointed; awnless; hairless; glabrous; non-carinate; having the margins lying flat and exposed on the palea ('palea nearly enclosed in flat edges of lemma'); without a germination flap; 3 nerved. *Palea* present; relatively long; tightly clasped by the lemma; entire; awnless, without apical setae; textured like the lemma; not indurated; 2-nerved; keel-less. *Lodicules* present; 2; free; fleshy; glabrous; not or scarcely vascularized. *Stamens* 3. Anthers not penicillate; without an apically prolonged connective. *Ovary* glabrous. Styles free to their bases. Stigmas 2; white.

Fruit, embryo and seedling. Embryo small (see Pilger).

Abaxial leaf blade epidermis. *Costal/intercostal zonation* conspicuous. *Papillae* absent. *Long-cells* markedly different in shape costally and intercostally (the costals much smaller); of similar wall thickness costally and intercostally (rather thin walled). Mid-intercostal long-cells rectangular; having markedly sinuous walls. *Microhairs* present; panicoid-type (rather broad); 54–66 microns long; (7.5–)8.4–11.4(–11.7) microns wide at the septum. Microhair total length/width at septum 4.7–7.8. Microhair apical cells (28.5–)31.5–39(–42) microns long. Microhair apical cell/total length ratio 0.5–0.64. *Stomata* common; 36–40.5 microns long. Subsidiaries triangular (low, but consistently with points). Guard-cells overlapping to flush with the interstomatals (overlapping). *Intercostal short-cells* common; in cork/silica-cell pairs; silicified. Intercostal silica bodies oryzoid-type (apparently, though mostly incompletely developed in material seen). Costal zones with short-cells. *Costal short-cells* predominantly paired. Costal silica bodies mostly poorly developed; 'panicoid-type'; cross shaped, dumb-bell shaped, and nodular (mostly poorly developed, in the material seen); not sharp-pointed.

Transverse section of leaf blade, physiology. *Leaf blades consisting of midrib*. C_3; XyMS+. Mesophyll without adaxial palisade. Midrib having a conventional arc of bundles (the leaf having a deep, u-shaped arc comprising three primaries and four small bundles); with colourless mesophyll adaxially (the centre of the leaf blade occupied by colourless 'pith', in which the bundles are embedded, these being in contact with the peripheral chlorenchyma only via their abaxial sclerenchyma girders). All the vascular bundles accompanied by sclerenchyma. Combined sclerenchyma girders absent (no adaxial sclerenchyma). Sclerenchyma all associated with vascular bundles.

Taxonomy. Panicoideae; Panicodae; Paniceae.

Ecology, geography, regional floristic distribution. 1 species. Cuba. Streamsides. Neotropical. Caribbean.

References, etc. Leaf anatomical: this project.

Special comments. Fruit data wanting.

Trisetum Pers.

Acrospelion Schult., *Parvotrisetum* Chrtek, *Rupestrina* Prov., *Sennenia* Sennen, *Trisetaria* Forssk., *Trisetarium* Poir.

Excluding *Avellinia, Peyritschia*

Habit, vegetative morphology. Annual (*Trisetaria*), or perennial; rhizomatous, or caespitose. **Culms** 4–150 cm high; *herbaceous*. Culm internodes hollow. Young shoots extravaginal. Leaves non-auriculate. Leaf blades linear; narrow; 0.2–12 mm wide; flat, or rolled (convolute); without cross venation; persistent; rolled in bud. Ligule an unfringed membrane (sometimes puberulent or ciliolate); truncate, or not truncate; 1–1.5 mm long.

Reproductive organization. *Plants bisexual, with bisexual spikelets*; with hermaphrodite florets. The spikelets all alike in sexuality. Plants outbreeding and inbreeding. Exposed-cleistogamous, or chasmogamous.

Inflorescence. *Inflorescence paniculate*; open, or contracted; when contracted, *spicate (but interrupted), or more or less irregular*; espatheate; not comprising 'partial inflorescences' and foliar organs. Spikelet-bearing axes persistent. Spikelets not secund; pedicellate.

Female-fertile spikelets. Spikelets 2.4–9 mm long; compressed laterally; disarticulating above the glumes, or falling with the glumes, or not disarticulating. Rachilla prolonged beyond the uppermost female-fertile floret; rachilla hairy (usually), or hairless (rarely). *Hairy callus* usually *present. Callus* short.

Glumes two; very unequal to more or less equal; shorter than the spikelets, or about equalling the spikelets; shorter than the adjacent lemmas, or long relative to the adjacent lemmas; pointed (acute or acuminate); awnless; carinate; similar. Lower glume 1–3 nerved. Upper glume 1–5 nerved. *Spikelets* with female-fertile florets only, or with incomplete florets. The incomplete florets distal to the female-fertile florets.

Female-fertile florets (1–)2–5(–12). *Lemmas similar in texture to the glumes*; not becoming indurated; incised; 2 lobed (bifid or with two setae); usually conspicuously *awned (by contrast with* **Koeleria**, *but rarely awnless). Awns* when present, 1, or 3; median, or median and lateral (via setae from the lobes); the median different in form from the laterals (when laterals present); *dorsal (or 'subterminal')*; from near the top, or from well down the back; non-geniculate, or geniculate; much shorter than the body of the lemma to much longer than the body of the lemma; entered by one vein. The lateral awns when present, shorter than the median. *Lemmas* hairless; *carinate*; 3–7 nerved. **Palea** present; relatively long; *gaping; with apical setae (the two nerves ending as bristle tips)*; thinner than the lemma; 2-nerved; 2-keeled. *Lodicules* present; 2; free; membranous; ciliate, or glabrous; toothed, or not toothed; not or scarcely vascularized. *Stamens* 3. Anthers 0.3–4.5 mm long; not penicillate; without an apically prolonged connective. *Ovary* usually glabrous. Styles free to their bases. Stigmas 2; white.

Fruit, embryo and seedling. Fruit small, or medium sized; dull; slightly compressed laterally. Hilum short. Embryo small; not waisted. Endosperm liquid in the mature fruit; with lipid; containing compound starch grains. Embryo with an epiblast; without a scutellar tail; with a negligible mesocotyl internode. Embryonic leaf margins meeting.

Seedling with a long mesocotyl; with a tight coleoptile. First seedling leaf with a well-developed lamina. The lamina narrow; 3 veined.

Abaxial leaf blade epidermis. *Costal/intercostal zonation* conspicuous. *Papillae* absent. *Long-cells* similar in shape costally and intercostally, or markedly different in shape costally and intercostally; differing markedly in wall thickness costally and intercostally, or differing markedly in wall thickness costally and intercostally. Mid-intercostal long-cells rectangular, or fusiform; having markedly sinuous walls, or having straight or only gently undulating walls. *Microhairs* absent. *Stomata* common; (30–)33–41(–42) microns long. Subsidiaries parallel-sided. Guard-cells overlapped by the interstomatals. *Intercostal short-cells* common, or absent or very rare; not paired; not silicified. Costal zones with short-cells. *Costal short-cells* conspicuously in long rows, or predominantly paired, or neither distinctly grouped into long rows nor predominantly paired. Costal silica bodies horizontally-elongated crenate/sinuous, or horizontally-elongated smooth; not sharp-pointed.

Transverse section of leaf blade, physiology. C_3; XyMS+. *Mesophyll* with non-radiate chlorenchyma. Leaf blade with the ribs more or less constant in size, or with the ribs very irregular in sizes. *Midrib* not readily distinguishable; with one bundle only. Bulliforms present in discrete, regular adaxial groups; in simple fans. All the vascular bundles accompanied by sclerenchyma. Combined sclerenchyma girders present. Sclerenchyma all associated with vascular bundles.

Phytochemistry. Leaves without flavonoid sulphates (1 species).

Special diagnostic feature. *Panicle loose, or if dense then interrupted, neither cylindrical nor ovoid: awns usually present, usually twisted, usually distinctly dorsal, conspicuous if inflorescence compact.*

Cytology. Chromosome base number, $x = 6$ and 7. $2n = 12$, 14, 24, 28, 42, and 56. 2, 4, 6, and 8 ploid.

Taxonomy. Pooideae; Poodae; Aveneae.

Ecology, geography, regional floristic distribution. About 85 species. North &. South temperate. Commonly adventive. Mesophytic, or xerophytic; mostly species of open habitats; glycophytic. Meadows, mountain slopes, upland grasslands, weedy places.

Holarctic, Paleotropical, Neotropical, Australian, and Antarctic. Boreal, Tethyan, and Madrean. Indomalesian and Polynesian. Arctic and Subarctic, Euro-Siberian, Eastern Asian, Atlantic North American, and Rocky Mountains. Mediterranean and Irano-Turanian. Indo-Chinese and Papuan. Hawaiian. Caribbean, Andean, and Fernandezian. North and East Australian. New Zealand and Patagonian. European and Siberian. Canadian-Appalachian. Temperate and South-Eastern Australian.

Hybrids. Integeneric hybrids with *Koeleria* (×*Trisetokoeleria* Tsvelev), *Sphenopholis*.

Rusts and smuts. Rusts — *Puccinia*. Taxonomically wide-ranging species: *Puccinia graminis*, *Puccinia coronata*, *Puccinia striiformis*, *Puccinia brachypodii*, *Puccinia poarum*, *Puccinia hordei*, *Puccinia recondita*, '*Uromyces*' *dactylidis*, and *Puccinia monoica*. Smuts from Tilletiaceae and from Ustilaginaceae. Tilletiaceae — *Entyloma* and *Urocystis*. Ustilaginaceae — *Ustilago*.

Economic importance. Cultivated fodder: *T. flavescens*. Important native pasture species: *T. flavescens*, *T. spicatum*, *T. wolfii* etc.

References, etc. Morphological/taxonomic: Koch 1979. Leaf anatomical: Metcalfe 1960; this project.

Special comments. See description of *Koeleria* for comments on the unsatisfactory taxonomic situation around *Koeleria* and *Trisetum* (involving *Trisetaria*, *Graphephorum*, *Rostraria*, *Lophochloa* and *Peyritschia*).

Tristachya Nees

Apochaete (C. E. Hubbard) Phipps, *Dolichochaete* Phipps, *Loudetia* A. Br., *Monopogon* Presl, *Muantijamvella* Phipps, *Veseyochloa* Phipps
Excluding *Isalus*

Habit, vegetative morphology. Annual (rarely), or perennial; caespitose. *Culms* 15–270 cm high; herbaceous; tuberous, or not tuberous. Leaf blades broad, or narrow; flat, or rolled (then involute or convolute, often rigid). Ligule a fringe of hairs.

Reproductive organization. *Plants bisexual, with bisexual spikelets*; with hermaphrodite florets.

Inflorescence. Inflorescence paniculate, or a single raceme (of triads); open, or contracted; espatheate; not comprising 'partial inflorescences' and foliar organs. Spikelet-bearing axes persistent. *Spikelets* usually in triplets (the triads terminating the panicle branches, sometimes in unequally pedicelled groups of 2–3); not secund; *pedicellate (the pedicels within a triad usually connate)*.

Female-fertile spikelets. *Spikelets* 10–45 mm long (often large); brown; compressed laterally to not noticeably compressed (?); *disarticulating above the glumes*; disarticulating between the florets (disarticulating easily between L_1 and L_2, less easily between G_2 and L_1). *Rachilla terminated by a female-fertile floret (rarely prolonged?)*; rachilla hairless. Hairy callus present. **Callus** long; narrowly conical, *pointed (usually pungent, sometimes narrowly obtuse in* T. huillensis*)*.

Glumes two; *more or less equal*; long relative to the adjacent lemmas; free; hairy (sometimes with cushion-based hairs), or hairless (glabrous); pointed, or not pointed; awnless (obtuse, or lanceolate to acuminate, or rostrate); non-carinate (flattened or convex); similar. Lower glume 3 nerved, or 5 nerved (rarely). Upper glume 3 nerved, or 5 nerved (rarely). *Spikelets* with incomplete florets. The incomplete florets proximal to the female-fertile florets. *Spikelets with proximal incomplete florets. The proximal incomplete florets* 1; paleate. Palea of the proximal incomplete florets fully developed (narrow, two keeled). The proximal incomplete florets male. The proximal lemmas awnless (similar to G_2); (3–)5–9 nerved; exceeded by the female-fertile lemmas; less firm than the female-fertile lemmas to similar in texture to the female-fertile lemmas (papery to leathery); not becoming indurated.

Female-fertile florets 1. **Lemmas** similar in texture to the glumes to decidedly firmer than the glumes (leathery to cartilaginous); not becoming indurated; *incised*; 2 lobed (the lobes obtuse, pointed or aristulate); *not deeply cleft*; awned. Awns 1; median; from a sinus (from between the lobes); geniculate; hairless (scabrid), or hairy; much longer than the body of the lemma; deciduous (usually), or persistent (*T. bicrinita*). Lemmas hairy (usually), or hairless. The hairs in tufts (rarely, with tufts at the bases of the lobes), or not in tufts. Lemmas non-carinate; having the margins tucked in onto the palea (the latter enclosed, save at its summit); with a clear germination flap; 5–7 nerved. *Palea* present; awnless, without apical setae; 2-nerved; 2-keeled. *Palea keels wingless (thickened, but not wing-shaped)*. *Lodicules* present; 2; free; fleshy (narrowly cuneate). *Stamens* 3 (usually?). **Ovary hairy.** Styles free to their bases. Stigmas 2.

Fruit, embryo and seedling. *Fruit* free from both lemma and palea (but embraced by lemma and palea); longitudinally grooved. Hilum long-linear. Embryo large.

Abaxial leaf blade epidermis. *Costal/intercostal zonation* conspicuous. *Papillae* absent. Mid-intercostal long-cells rectangular; having markedly sinuous walls. *Microhairs* present; panicoid-type; (48–)54–87(–90) microns long; 6–7.2 microns wide at the septum. Microhair total length/width at septum 10.9–15. Microhair apical cells 23–30 microns long (*T. hispida*), or 40–47 microns long (*T. bequartii*). Microhair apical cell/total length ratio 0.45–0.54. *Stomata* absent or very rare (e.g. in *T. bequartii*), or common. Subsidiaries dome-shaped (mostly, low), or triangular. *Intercostal short-cells* common (fairly); in cork/silica-cell pairs (and solitary). Costal zones with short-cells. *Costal short-cells* conspicuously in long rows. Costal silica bodies 'panicoid-type'; mostly dumb-bell shaped; not sharp-pointed.

Transverse section of leaf blade, physiology. C_4; XyMS–. PCR cell chloroplasts centrifugal/peripheral. Mesophyll traversed by columns of colourless mesophyll cells, or not traversed by colourless columns (in *T. bequartii*, there are broad zones of colourless tissue both adaxially and abaxially, but the vascular bundles are embedded in a continuous central zone of photosynthetic tissue). *Leaf blade* with distinct, prominent adaxial ribs to adaxially flat. *Midrib* conspicuous, or not readily distinguishable; with one bundle only, or having a conventional arc of bundles; with colourless mesophyll adaxially. Bulliforms present in discrete, regular adaxial groups; in simple fans, or associated with colourless

982

mesophyll cells to form deeply-penetrating fans (sometimes traversing to the abaxial epidermis); associating with colourless mesophyll cells to form arches over small vascular bundles, or nowhere involved in bulliform-plus-colourless mesophyll arches. Many of the smallest vascular bundles unaccompanied by sclerenchyma. Combined sclerenchyma girders present, or absent (e.g. *T. bequartii*); forming 'figures' (when present). Sclerenchyma all associated with vascular bundles.

Special diagnostic feature. The lower glume exceeding the female-fertile lemma.

Cytology. Chromosome base number, $x = 10$ and 12. $2n = 24$ and 40.

Taxonomy. Panicoideae; Panicodae; Arundinelleae.

Ecology, geography, regional floristic distribution. About 20 species. Tropical and southern Africa, Madagascar, tropical America. Helophytic to xerophytic; shade species and species of open habitats; glycophytic. Grassland and savanna, woodland and floodplains, wet to dry soils.

Paleotropical and Neotropical. African. Sudano-Angolan, West African Rainforest, and Namib-Karoo. Caribbean, Central Brazilian, and Andean. Sahelo-Sudanian, Somalo-Ethiopian, South Tropical African, and Kalaharian.

Rusts and smuts. Rusts — *Puccinia*. Smuts from Ustilaginaceae. Ustilaginaceae — *Sphacelotheca* and *Tolyposporium*.

References, etc. Leaf anatomical: Metcalfe 1960; this project.

Triticum L.

Crithodium Link, *Deina* Alefeld, *Frumentum* Krause, *Gigachilon* Seidl, *Nivieria* Ser., *Spelta* Wolf

Habit, vegetative morphology. *Annual*; caespitose. *Culms* 40–170 cm high; herbaceous; unbranched above. Culm nodes hairy, or glabrous. Culm internodes solid, or hollow. *Leaves* not basally aggregated; auriculate. Leaf blades narrowly to broadly linear; broad to narrow; 2–20 mm wide; flat; without cross venation; persistent; rolled in bud. Ligule an unfringed membrane; truncate; 0.6–2 mm long.

Reproductive organization. Plants bisexual, with bisexual spikelets; with hermaphrodite florets. The spikelets of sexually distinct forms on the same plant, or all alike in sexuality; hermaphrodite, or hermaphrodite and sterile (often incomplete at the base of the spike). Plants inbreeding.

Inflorescence. *Inflorescence a single spike (elongated)*. Rachides hollowed. Inflorescence espatheate; not comprising 'partial inflorescences' and foliar organs. Spikelet-bearing axes with substantial rachides (these flattened or hollowed); disarticulating (*Crithodium*), or persistent; when fragile, disarticulating at the joints (above or below the spikelet). Spikelets solitary; not secund; distichous; sessile.

Female-fertile spikelets. Spikelets 9–16 mm long; compressed laterally; disarticulating above the glumes, or falling with the glumes, or not disarticulating (in cultivated forms); not disarticulating between the florets, or disarticulating between the florets. Rachilla prolonged beyond the uppermost female-fertile floret. The rachilla extension with incomplete florets. Hairy callus absent. *Callus* very short; blunt.

Glumes two; more or less equal; shorter than the adjacent lemmas, or long relative to the adjacent lemmas; lateral to the rachis; without conspicuous tufts or rows of hairs; not pointed (obtuse, truncate or bidentate via the lateral nerves); not subulate; awned (or mucronate), or awnless; *carinate (at least above, before the grain expands)*; with the keel conspicuously winged (and crested, in *Gigachilon*), or without a median keel-wing; similar (ovate or oblong, chartaceous or rarely membranous). *Lower glume 5–11 nerved*. Upper glume 5–11 nerved. *Spikelets* usually with incomplete florets. The incomplete florets distal to the female-fertile florets. The distal incomplete florets usually 1, or 2; *incomplete florets* merely underdeveloped.

Female-fertile florets 1 (*Crithodium*), or 2–6. Lemmas similar in texture to the glumes to decidedly firmer than the glumes; becoming indurated; entire to incised (1 to several dentate, or acuminate); awnless, or awned. Awns when present, 1; median; from a sinus, or apical; non-geniculate; much shorter than the body of the lemma to much longer than the body of the lemma; entered by several veins. Lemmas hairy, or hairless (but scabrid); non-carinate, or carinate (at least above); 5–11 nerved; with the nerves non-confluent. *Palea* present; relatively long; entire (not split or divided at maturity), or apically notched (*Crithodium*); 2-nerved; 2-keeled. Palea keels somewhat winged. *Lodicules* present; 2; free; membranous; ciliate; not toothed. *Stamens* 3. Anthers 2–4.5 mm long; not penicillate. **Ovary** *hairy*. Styles free to their bases. Stigmas 2; white.

Fruit, embryo and seedling. *Fruit* free from both lemma and palea (free threshing); medium sized to large (to 11 mm long); ellipsoid; longitudinally grooved; compressed dorsiventrally, or not noticeably compressed; with hairs confined to a terminal tuft. *Hilum long-linear.* Embryo large to small (to 1/3 the caryopsis length); not waisted. Endosperm hard; without lipid; containing only simple starch grains. Embryo with an epiblast; without a scutellar tail; with a negligible mesocotyl internode. Embryonic leaf margins meeting.

Seedling with a short mesocotyl; with a tight coleoptile. First seedling leaf with a well-developed lamina. The lamina narrow; erect; 'many'.

Abaxial leaf blade epidermis. *Costal/intercostal zonation* conspicuous. *Papillae* absent. Long-cells of similar wall thickness costally and intercostally. Mid-intercostal long-cells rectangular, or rectangular and fusiform; having markedly sinuous walls, or having straight or only gently undulating walls, or having markedly sinuous walls and having straight or only gently undulating walls. *Microhairs* absent. *Stomata* common; 63–69 microns long. Subsidiaries parallel-sided, or dome-shaped (low). Guard-cells overlapped by the interstomatals. *Intercostal short-cells* common (e.g. *T. polonicum*), or absent or very rare (usually); when present, mostly in cork/silica-cell pairs; silicified. Intercostal silica bodies rounded (or oval), or crescentic. *Crown cells* present. Costal zones with short-cells. *Costal short-cells* predominantly paired (*T. polonicum*), or neither distinctly grouped into long rows nor predominantly paired (or varying from vein to vein). Costal silica bodies horizontally-elongated crenate/sinuous, or horizontally-elongated smooth, or rounded, or crescentic; not sharp-pointed.

Transverse section of leaf blade, physiology. C_3; XyMS+. *PBS cells* without a suberised lamella. *Mesophyll* with non-radiate chlorenchyma. *Leaf blade* with distinct, prominent adaxial ribs; with the ribs more or less constant in size. *Midrib* conspicuous; with one bundle only, or having a conventional arc of bundles. *The lamina* symmetrical on either side of the midrib. Bulliforms present in discrete, regular adaxial groups (in the furrows); in simple fans. Many of the smallest vascular bundles unaccompanied by sclerenchyma. Combined sclerenchyma girders present (rarely), or absent; if present, nowhere forming 'figures'. Sclerenchyma all associated with vascular bundles.

Culm anatomy. *Culm internode bundles* in one or two rings.

Phytochemistry. Tissues of the culm bases with little or no starch. Leaves without flavonoid sulphates (1 species).

Cytology. Chromosome base number, $x = 7$. $2n = 42$, or 14 and 28 (*Crithodium*). 6 ploid, or 2 and 4 ploid (*Crithodium*). Haplomic genome content A (*Crithodium*), or A and B (*Gigichilon*), or A, B, and D (*Triticum* sensu stricto). Haploid nuclear DNA content 4.9–6.2 pg (8 species, mean 5.9). Mean diploid 2c DNA value 12.2 pg (3 species, 9.8–13.8).

Taxonomy. Pooideae; Triticodae; Triticeae.

Ecology, geography, regional floristic distribution. 8 species. Europe, Mediterranean, Western Asia. Commonly adventive. Mesophytic, or xerophytic; species of open habitats. Stony hillsides, dry grassland, weedy places.

Holarctic. Tethyan. Mediterranean.

Hybrids. Intergeneric hybrids with *Aegilops* (×*Aegilotriticum* Wagner ex Tschermak), *Agropyron* (×*Agrotriticum* Ciferri & Giacom.), *Elymus*, *Elytrigia* (×*Trititrigia* Tsvelev), *Hordeum* (×*Tritordeum* Aschers. & Graebn.), *Lophopyrum*, *Secale* (×*Triticosecale* Wittmack). See also ×*Elymotriticum* P. Fourn., ×*Agrotrisecale* Ciferri & Giacom. (*Agropyron* × *Secale* × *Triticum*), ×*Oryticum* Wang & Tang (supposedly *Oryza* × *Triticum*).

Rusts and smuts. Rusts — *Puccinia*. Taxonomically wide-ranging species: *Puccinia graminis*, *Puccinia striiformis*, and *Puccinia recondita*. Smuts from Tilletiaceae and from Ustilaginaceae. Tilletiaceae — *Tilletia* and *Urocystis*. Ustilaginaceae — *Ustilago*.

Economic importance. Grain crop species: several, but modern agriculture largely confined to *T. aestivum* (Bread Wheat) and *T. durum* (Macaroni Wheat).

References, etc. Morphological/taxonomic: Löve 1984. Leaf anatomical: Metcalfe 1960; this project.

Tsvelevia E. Alekseev

Poa kerguelensis, ~ *Poa*, *Festuca*

Habit, vegetative morphology. Perennial; densely caespitose. Culms 4–10 cm high; *herbaceous*; unbranched above. Young shoots intravaginal. Leaves non-auriculate. Sheaths scarious. Leaf blades linear; narrow; 0.5–0.7 mm wide; folded to rolled (folded to involute); without cross venation; persistent; once-folded in bud. *Ligule an unfringed membrane*; not truncate; 0.7–1.5 mm long. *Contra-ligule* absent.

Reproductive organization. Plants bisexual, with bisexual spikelets; with hermaphrodite florets.

Inflorescence. Inflorescence paniculate; contracted; more or less ovoid; espatheate; not comprising 'partial inflorescences' and foliar organs. Spikelet-bearing axes persistent. Spikelets not secund; pedicellate.

Female-fertile spikelets. *Spikelets 3.5–4 mm long*; compressed laterally; disarticulating above the glumes; disarticulating between the florets. Rachilla prolonged beyond the uppermost female-fertile floret; rachilla hairless. The rachilla extension with incomplete florets. Hairy callus absent.

Glumes two; very unequal (to almost equal); shorter than the spikelets to about equalling the spikelets; shorter than the adjacent lemmas; hairless; glabrous; pointed; not subulate; awnless; carinate; similar. *Lower glume* about 0.6–0.75 times the length of the upper glume; shorter than the lowest lemma to about equalling the lowest lemma; 1 nerved. Upper glume 3 nerved. *Spikelets* with incomplete florets. The incomplete florets distal to the female-fertile florets. *The distal incomplete florets* merely underdeveloped; awnless.

Female-fertile florets 2–3. Lemmas similar in texture to the glumes (thin); not becoming indurated; entire to incised; pointed; if incised, 2–3 lobed; not deeply cleft; awnless; hairy (especially on the nerves); *carinate*; without a germination flap; 3 nerved; with the nerves non-confluent. *Palea* present; relatively long; tightly clasped by the lemma; apically notched; awnless, without apical setae; textured like the lemma; not indurated; 2-nerved; 2-keeled. Palea keels wingless; hairy (below). *Lodicules* present; 2; free; membranous; glabrous; toothed. *Stamens* 3. Anthers 0.5–0.7 mm long; not penicillate; without an apically prolonged connective. *Ovary* glabrous. Styles free to their bases. Stigmas 2; white.

Fruit, embryo and seedling. Fruit free from both lemma and palea; small (about 2 mm long); *longitudinally grooved*; slightly compressed dorsiventrally. *Hilum long-linear*. Embryo small.

Abaxial leaf blade epidermis. *Costal/intercostal zonation* conspicuous. *Papillae* absent. Long-cells of similar wall thickness costally and intercostally (thick walled). Mid-intercostal long-cells rectangular; having markedly sinuous walls (heavily pitted). *Microhairs* absent. *Stomata* absent or very rare. *Intercostal short-cells* common; in cork/silica-cell pairs. Costal zones with short-cells. *Costal short-cells* predominantly paired (with only a few short rows). Costal silica bodies rounded and crescentic; not sharp-pointed.

Transverse section of leaf blade, physiology. C_3; XyMS+. *Mesophyll* with non-radiate chlorenchyma; without adaxial palisade. *Leaf blade* adaxially flat. *Midrib* conspicuous (by the rounded abaxial keel, and flanking 'hinges'); with one bundle only. Bulliforms not in discrete, regular adaxial groups (other than the midrib 'hinges'). All the vascular bundles accompanied by sclerenchyma. Combined sclerenchyma girders absent (each bundle with a small abaxial strand only). Sclerenchyma all associated with vascular bundles.

Taxonomy. Pooideae; Poodae; Poeae.

Ecology, geography, regional floristic distribution. 1 species. Chatham Islands. Xerophytic; halophytic. Maritime sand.

Antarctic. New Zealand.

References, etc. Morphological/taxonomic: Alekseev 1985. Leaf anatomical: this project.

Tuctoria J. Reeder

Habit, vegetative morphology. Annual; culms ascending or decumbent, not producing juvenile floating leaves. Culms herbaceous. Culm internodes solid. The shoots aromatic. **Leaves** not basally aggregated; *not clearly differentiated into sheath and blade*; non-auriculate. Sheath margins with little or no differentiation into sheath and blade. Leaf blades narrow; 5–12 mm wide; flat, or rolled; exhibiting multicellular glands abaxially. The abaxial glands intercostal. Leaf blades without cross venation; persistent. *Ligule* absent.

Reproductive organization. Plants bisexual, with bisexual spikelets; with hermaphrodite florets.

Inflorescence. Inflorescence dense, many sided, a single spike, or a single raceme; espatheate; not comprising 'partial inflorescences' and foliar organs. Spikelet-bearing axes persistent. Spikelets not secund (spiralled); sessile to subsessile.

Female-fertile spikelets. *Spikelets compressed laterally*; disarticulating above the glumes; disarticulating between the florets. Rachilla prolonged beyond the uppermost female-fertile floret. The rachilla extension with incomplete florets.

Glumes two; more or less equal; shorter than the adjacent lemmas; not pointed; awnless; similar (broad, 2–5 toothed). Lower glume 9–15 nerved. Upper glume 9–15 nerved. *Spikelets* with incomplete florets. The incomplete florets distal to the female-fertile florets.

Female-fertile florets 5–30(–40). *Lemmas* entire, or incised; *not deeply cleft (entire, erose or denticulate)*; awnless, or mucronate (usually, centrally); 11–17 nerved. *Palea* present; relatively long; 2-nerved; 2-keeled. Palea keels wingless; glabrous. *Lodicules* present (minute, often fused to the palea); 2; fleshy. *Stamens* 3. *Ovary* glabrous. Stigmas 2.

Fruit, embryo and seedling. *Fruit* not viscid; compressed laterally. Hilum short. Pericarp fused. Embryo large (clearly visible through the pericarp); with an epiblast; with a scutellar tail; with an elongated mesocotyl internode. Embryonic leaf margins meeting.

Abaxial leaf blade epidermis. *Costal/intercostal zonation* conspicuous. *Papillae* absent. Mid-intercostal long-cells rectangular; having markedly sinuous walls. *Microhairs* present; more or less spherical to elongated; ostensibly one-celled to clearly two-celled; chloridoid-type (of the sunken, 'button mushroom' type); (10.5–)12–15(–18) microns long; 9–12–13.5 microns wide at the septum. Microhair total length/width at septum 1–1.3. Microhair apical cells (6–)6.6–7.5(–10.5) microns long. Microhair apical cell/total length ratio 0.47–0.63. *Stomata* common; (21–)24–27(–29) microns long. Subsidiaries parallel-sided (broadly), or dome-shaped. Guard-cells overlapped by the interstomatals. *Intercostal short-cells* absent or very rare. Intercostal silica bodies absent. With costal prickles and long, thin intercostal macrohairs. Costal zones with short-cells. *Costal short-cells* conspicuously in long rows. Costal silica bodies present in alternate cell files of the costal zones; 'panicoid-type'; nodular (mostly, elongated), or dumb-bell shaped; not sharp-pointed.

Transverse section of leaf blade, physiology. C_4; XyMS+. PCR sheath extensions absent. *Midrib* seemingly not readily distinguishable; with one bundle only. Bulliforms present in discrete, regular adaxial groups (in the furrows, the details not observable in material seen). All the vascular bundles accompanied by sclerenchyma. Combined sclerenchyma girders present (only some of the major bundles). Sclerenchyma all associated with vascular bundles.

Cytology. Chromosome base number, $x = 10$. $2n = 24$ and 40. 2 and 4 ploid (and aneuploids).

Taxonomy. Chloridoideae; Orcuttieae.

Ecology, geography, regional floristic distribution. 3 species. California and Mexico. Helophytic.

Holarctic. Madrean.

References, etc. Morphological/taxonomic: Reeder 1965. Leaf anatomical: this project.

Uniola L.

Nevroctola Raf., *Trisiola* Raf. (1817), *Triunila* Raf.
Excluding *Leptochloöpsis*

Habit, vegetative morphology. Perennial; *rhizomatous, or stoloniferous.* Culms 100–220 cm high; herbaceous. Leaf blades harsh; narrow; without abaxial multicellular glands. Ligule a fringe of hairs.

Reproductive organization. Plants bisexual, with bisexual spikelets; with hermaphrodite florets. Exposed-cleistogamous, or chasmogamous.

Inflorescence. *Inflorescence of spicate main branches, or paniculate*; contracted (the crowded branches overlapping), or open; espatheate; not comprising 'partial inflorescences' and foliar organs. Spikelet-bearing axes persistent. Spikelets not secund; long to short pedicellate.

Female-fertile spikelets. *Spikelets* 10–25 mm long; compressed laterally; *falling with the glumes*; not disarticulating between the florets; with conventional internode spacings. Rachilla prolonged beyond the uppermost female-fertile floret.

Glumes two; shorter than the spikelets; pointed; awnless; carinate; similar (narrow, rigid). *Spikelets* with incomplete florets. The incomplete florets proximal to the female-fertile florets, or both distal and proximal to the female-fertile florets. *The distal incomplete florets* (when present) merely underdeveloped. *Spikelets with proximal incomplete florets.* **The proximal incomplete florets 3–5.** The proximal lemmas similar in texture to the female-fertile lemmas.

Female-fertile florets 7–15. Lemmas acute or acuminate; papery or leathery; not becoming indurated; entire; pointed; awnless; carinate (and compressed); 7–10 nerved. *Palea* present; relatively long; awnless, without apical setae; 2-nerved; 2-keeled. Palea keels winged; hairy (ciliate). *Lodicules* present; 2; free; fleshy; glabrous; heavily vascularized, or not or scarcely vascularized. *Stamens* 3.

Fruit, embryo and seedling. Fruit ellipsoid. Pericarp free. Embryo large; waisted; without an epiblast; with a scutellar tail; with an elongated mesocotyl internode; with more than one scutellum bundle. Embryonic leaf margins meeting.

First seedling leaf with a well-developed lamina. The lamina narrow; erect.

Abaxial leaf blade epidermis. *Costal/intercostal zonation* conspicuous. *Papillae* absent. *Long-cells* similar in shape costally and intercostally to markedly different in shape costally and intercostally (the costals narrower, more regularly rectangular); of similar wall thickness costally and intercostally (fairly thick walled). Mid-intercostal long-cells rectangular, or rectangular and fusiform (in *U. paniculata*); having markedly sinuous walls. *Microhairs* absent (but present adaxially); chloridoid type. *Stomata* common; 21–33 microns long. Subsidiaries low to high dome-shaped. Guard-cells overlapped by the interstomatals (and somewhat sunken, in *U. paniculata*), or overlapping to flush with the interstomatals (in *U. pittieri*). *Intercostal short-cells* common; in cork/silica-cell pairs and not paired; silicified (but no perfect silica bodies seen), or not silicified. Intercostal silica bodies present and perfectly developed. Macrohairs and prickles absent. *Crown cells* absent. Costal zones with short-cells. *Costal short-cells* neither distinctly grouped into long rows nor predominantly paired. Costal silica bodies poorly developed (and scarce, in the material of both species examined); present throughout the costal zones; where detectable rounded to saddle shaped (well developed saddles adaxially); sometimes sharp-pointed, or not sharp-pointed.

Transverse section of leaf blade, physiology. C_4; XyMS+. PCR sheath outlines even. PCR sheaths of the primary vascular bundles interrupted; interrupted both abaxially and adaxially. PCR sheath extensions absent. PCR cell chloroplasts with well developed grana; probably centrifugal/peripheral (? — judging from inadequate material). *Mesophyll* with radiate chlorenchyma; traversed by columns of colourless mesophyll cells (the columns wide). *Leaf blade* with distinct, prominent adaxial ribs; with the ribs more or less constant in size (round to flat topped, in *U. pittieri*), or with the ribs very irregular in sizes (tall and flat topped, with hollowed sides in *U. paniculata*). *Midrib* not readily distinguishable; with one bundle only. *The lamina* symmetrical on either side of the midrib. *Bulliforms* present in discrete, regular adaxial groups; associated with colourless mesophyll cells to form deeply-penetrating fans (these in every furrow, linked with the colourless columns). All the vascular bundles accompanied by sclerenchyma. Combined sclerenchyma girders present (*U. pittieri*), or absent (*U. paniculata*, where the upper region of each of the tall ribs is occupied by lightly lignified 'colourless tissue', containing a few scattered thick-walled fibres, linking the bundle with an adaxial strand); in *U. pitteri* forming 'figures' (every bundle being associated with a substantial I). Sclerenchyma not all bundle-associated. The 'extra' sclerenchyma in abaxial groups; abaxial-hypodermal, the groups continuous with colourless columns. The lamina margins with fibres.

Cytology. Chromosome base number, $x = 10$. $2n = 40$.

Taxonomy. Chloridoideae; main chloridoid assemblage.

Ecology, geography, regional floristic distribution. 2 species. North America, West Indies. Species of open habitats; halophytic. Sand dunes and salt flats.

Neotropical. Caribbean and Andean.

Economic importance. *U. paniculata* a significant sandbinder.

References, etc. Morphological/taxonomic: Yates 1966a. Leaf anatomical: this project.

Special comments. Fruit data wanting.

Uranthoecium Stapf

Habit, vegetative morphology. Annual (or short-lived perennial); caespitose. *Culms* 15–50 cm high; herbaceous; flattened; branched above. Culm nodes hairy. Culm internodes hollow. *Leaves* not basally aggregated; non-auriculate. Leaf blades linear; narrow; about 4 mm wide; flat (soft); without cross venation; persistent. Ligule a fringed membrane to a fringe of hairs; 0.75–1.5 mm long.

Reproductive organization. *Plants bisexual, with bisexual spikelets*; with hermaphrodite florets.

Inflorescence. *Inflorescence of spicate main branches (of short spikes, each of 2–4 spikelets, the spikes on a common, flattened cartilaginous axis)*; contracted. Inflorescence with axes ending in spikelets (but these rudimentary), or axes not ending in spikelets (often ending in a flattened bristle). Rachides flattened (i.e., the common axis). Inflorescence espatheate; not comprising 'partial inflorescences' and foliar organs. *Spikelet-bearing axes disarticulating*; falling entire (the short spikes falling with the adjacent internode of the fragile common axis). Spikelets solitary; not secund; sessile. Pedicel apices discoid.

Female-fertile spikelets. Spikelets 8–10 mm long; lanceolate; compressed dorsiventrally; falling with the glumes (and with the spike). Rachilla terminated by a female-fertile floret. Hairy callus absent.

Glumes two; relatively large; very unequal; shorter than the adjacent lemmas (the upper 1/2, the lower 1/4 its length); hairless; glabrous; not pointed; awnless; similar (firmly chartaceous, truncate). Lower glume 6–7 nerved. **Upper glume** *distinctly saccate (gibbous)*; 7 nerved. *Spikelets* with incomplete florets. The incomplete florets proximal to the female-fertile florets. *Spikelets with proximal incomplete florets*. *The proximal incomplete florets* 1; paleate. Palea of the proximal incomplete florets fully developed (2-keeled). The proximal incomplete florets sterile. The proximal lemmas membranous along the midline, leathery on the sides, caudate-acuminate; awnless (but caudate-acuminate); 5–7 nerved; not becoming indurated (but cartilaginous between the outer nerves).

Female-fertile florets 1. **Lemmas ovate-elliptic, subulate-caudate above**; decidedly firmer than the glumes; smooth; becoming indurated (thinly); yellow in fruit; entire; pointed; awned. Awns 1; median; apical; non-geniculate. Lemmas hairless; non-carinate; having the margins lying flat and exposed on the palea; 5 nerved. **Palea** present; *entire (apically prolonged, the tip caudate)*; with apical setae; textured like the lemma; indurated; 2-nerved. *Lodicules* present; 2; fleshy. *Stamens* 3. Anthers about 0.5 mm long. *Ovary* glabrous. Styles free to their bases. Stigmas 2.

Fruit, embryo and seedling. Fruit small. Hilum short. Embryo large.

Abaxial leaf blade epidermis. *Costal/intercostal zonation* conspicuous. *Papillae* absent. *Long-cells* similar in shape costally and intercostally; of similar wall thickness costally and intercostally. Mid-intercostal long-cells rectangular; having markedly sinuous walls. *Microhairs* present; panicoid-type; (66–)69–75(–81) microns long; 6–7.5 microns wide at the septum. Microhair total length/width at septum 9.2–11.9. Microhair apical cells (30–)33–42(–43.5) microns long. Microhair apical cell/total length ratio 0.43–0.57. *Stomata* common; 34.5–36 microns long. Subsidiaries triangular. Guard-cells overlapping to flush with the interstomatals. *Intercostal short-cells* common; in cork/silica-cell pairs; silicified. Intercostal silica bodies mainly cross-shaped. Prickles abundant. Costal zones with short-cells. *Costal short-cells* conspicuously in long rows. Costal silica bodies 'panicoid-type'; not sharp-pointed.

Transverse section of leaf blade, physiology. C_4; XyMS–. *Mesophyll* with radiate chlorenchyma. *Leaf blade* adaxially with numerous large, thin walled, blunt macrohairs; with the ribs more or less constant in size. *Midrib* not readily distinguishable; with one bundle only. Bulliforms not in discrete, regular adaxial groups (the epidermis extensively, irregularly bulliform). Many of the smallest vascular bundles unaccompanied by sclerenchyma. Combined sclerenchyma girders absent (the large bundles with heavy strands simulate combined girders in collapsed material, but there always appear to be thin walled, colourless cells intevening between the PCR sheath and the fibre groups). Sclerenchyma all associated with vascular bundles.

Taxonomy. Panicoideae; Panicodae; Paniceae.

Ecology, geography, regional floristic distribution. 1 species. Australia. Xerophytic; species of open habitats. Grassland on clay.

Australian. North and East Australian and Central Australian. Tropical North and East Australian.

References, etc. Leaf anatomical: this project.

Urelytrum Hackel

Habit, vegetative morphology. Annual (rarely), or perennial; caespitose. *Culms* 60–250 cm high; herbaceous (erect); unbranched above. Culm nodes glabrous. Culm internodes solid. Young shoots intravaginal. The shoots aromatic (and with a strong bitter taste), or not aromatic. **Leaves** mostly basal, or not basally aggregated; *auriculate (the auricles from the sheaths, glabrous or hairy)*. Leaf blades linear; broad, or narrow; flat, or rolled (convolute); without cross venation; persistent. Ligule an unfringed membrane; not truncate (apically rounded); 2–5 mm long.

Reproductive organization. Plants bisexual, with bisexual spikelets; with hermaphrodite florets. The spikelets of sexually distinct forms on the same plant; hermaphrodite and male-only, or hermaphrodite and sterile; overtly heteromorphic (the pedicellate spikelet usually with a long-awned G_1); all in heterogamous combinations.

Inflorescence. Inflorescence of spicate main branches, or a single raceme (of one to many long, rigid, spiciform, cylindrical or slightly flattened 'racemes'); digitate, or non-digitate (when the 'raceme' solitary). Rachides hollowed and flattened. Inflorescence espatheate; not comprising 'partial inflorescences' and foliar organs. *Spikelet-bearing axes* 'racemes' (these long, many-noded); the spikelet-bearing axes with more than 10 spikelet-bearing 'articles' (up to 35). The racemes spikelet bearing to the base. Spikelet-bearing axes solitary to clustered; with substantial rachides; disarticulating; disarticulating at the joints. **'Articles'** non-linear (clavate); with a basal callus-knob; *appendaged*; disarticulating obliquely; somewhat hairy, or glabrous. *Spikelets* paired; sessile and pedicellate (the pedicels resembling the internodes); *consistently in 'long-and-short' combinations*; in pedicellate/sessile combinations. Pedicels of the 'pedicellate' spikelets free of the rachis. The 'shorter' spikelets hermaphrodite. The 'longer' spikelets male-only (very rarely with one floret hermaphrodite), or sterile.

Female-sterile spikelets. *The pedicelled spikelets with a conspicuously long-awned G_1. Having two florets, usually both male, very occasionally one hermaphrodite.*

Sometimes sterile and reduced to the glumes. The male spikelets with glumes (the G_1 conspicuously awned); 2 floreted (usually both male).

Female-fertile spikelets. *Spikelets* 5–10 mm long; abaxial; *compressed dorsiventrally*; planoconvex; falling with the glumes (and with the adjacent joint and pedicel); not disarticulating between the florets; with conventional internode spacings. Rachilla terminated by a female-fertile floret. Hairy callus present. *Callus* long (but concealed by the joint until the raceme disarticulates); blunt.

Glumes two; more or less equal; exceeding the spikelets; long relative to the adjacent lemmas (larger than them); dorsiventral to the rachis; hairy (G_1), or hairless (G_2); without conspicuous tufts or rows of hairs; pointed; awned (G_1 occasionally bi-aristulate), or awnless; carinate (G_2), or non-carinate (G_1); very dissimilar (the G_1 leathery, dorsally flattened, 2-keeled, the G_2 thinner, naviculate-keeled). *Lower glume* much exceeding the lowest lemma; *two-keeled*; flattened on the back; not pitted; relatively smooth, or relatively smooth and muricate; obscurely 5–7 nerved. Upper glume 3 nerved. *Spikelets* with incomplete florets. The incomplete florets proximal to the female-fertile florets. *Spikelets with proximal incomplete florets. The proximal incomplete florets* 1; paleate. Palea of the proximal incomplete florets fully developed. *The proximal incomplete florets male.* The proximal lemmas awnless; 2 nerved; more or less equalling the female-fertile lemmas; similar in texture to the female-fertile lemmas (hyaline); not becoming indurated.

Female-fertile florets 1. Lemmas less firm than the glumes (hyaline); not becoming indurated; entire; pointed, or blunt; awnless; hairless; non-carinate; without a germination flap; 2–5 nerved. *Palea* present; relatively long; entire; awnless, without apical setae; textured like the lemma; not indurated (hyaline); 2-nerved; flat backed. *Lodicules* present; 2; free; fleshy; glabrous; not toothed. *Stamens* 3. Anthers 1.5–4 mm long; not penicillate; without an apically prolonged connective. *Ovary* glabrous. Styles free to their bases. Stigmas 2; red pigmented.

Fruit, embryo and seedling. *Fruit* free from both lemma and palea; small (about 3–4 mm long); ellipsoid. Hilum short. Embryo large.

Abaxial leaf blade epidermis. *Costal/intercostal zonation* conspicuous. *Papillae* present. Intercostal papillae over-arching the stomata; consisting of one oblique swelling per cell to consisting of one symmetrical projection per cell. Mid-intercostal long-cells rectangular; having markedly sinuous walls and having straight or only gently undulating walls. *Microhairs* present; panicoid-type; 57–66 microns long; (6.6–)7.5–8.4(–9) microns wide at the septum. Microhair total length/width at septum 6.7–9.1. Microhair apical cells 28.5–36 microns long. Microhair apical cell/total length ratio 0.5–0.55. *Stomata* common (most grooves with a single row); (42–)43–45(–51) microns long. Subsidiaries dome-shaped (mostly), or triangular. Guard-cells overlapping to flush with the interstomatals. *Intercostal short-cells* absent or very rare. Costal zones with short-cells. *Costal short-cells* conspicuously in long rows, or predominantly paired (especially over the midrib). Costal silica bodies 'panicoid-type'; dumb-bell shaped and nodular; not sharp-pointed.

Transverse section of leaf blade, physiology. C_4; XyMS–. PCR sheath outlines uneven. PCR cell chloroplasts centrifugal/peripheral. *Mesophyll* with radiate chlorenchyma; traversed by columns of colourless mesophyll cells. *Leaf blade* with distinct, prominent adaxial ribs (and small narrow abaxial ones); with the ribs more or less constant in size (becoming smaller away from the midrib). *Midrib* conspicuous; having a conventional arc of bundles (a large median, and one or two smaller laterals on either side, according to interpretation of the midrib limits); with colourless mesophyll adaxially. Bulliforms present in discrete, regular adaxial groups to not in discrete, regular adaxial groups (the groups very large); irregularly grouped,sometimes associated with colourless mesophyll cells to form deeply-penetrating fans; associating with colourless mesophyll cells to form arches over small vascular bundles (the arms of the arches sometimes traversing the mesophyll). All the vascular bundles accompanied by sclerenchyma. Combined sclerenchyma girders present (with the primaries); forming 'figures' (I's and T's). Sclerenchyma all associated with vascular bundles.

Special diagnostic feature. *The lower glume of the pedicellate spikelet with a 5–10 mm (or longer) awn.*

Cytology. Chromosome base number, $x = 10$. $2n = 20$.

Taxonomy. Panicoideae; Andropogonodae; Andropogoneae; Rottboelliinae.

Ecology, geography, regional floristic distribution. 7 species. South and tropical Africa, Madagascar. Mesophytic; species of open habitats; glycophytic. Savanna grassland.

Paleotropical. African and Madagascan. Sudano-Angolan, West African Rainforest, and Namib-Karoo. Sahelo-Sudanian, Somalo-Ethiopian, South Tropical African, and Kalaharian.

Rusts and smuts. Smuts from Ustilaginaceae. Ustilaginaceae — *Sorosporium*.

References, etc. Leaf anatomical: Metcalfe 1960; this project; photos of *U. agropyroides* provided by R.P. Ellis.

Urochlaena Nees

Habit, vegetative morphology. Annual; caespitose. *Culms* 7–20 cm high; herbaceous (glabrous); much branched from the base. Culm nodes glabrous. Culm internodes hollow. Plants unarmed. Young shoots intravaginal. *Leaves* not basally aggregated; non-auriculate; without auricular setae (but hair-tufted at the mouth of the sheath). *The uppermost sheath blade-bearing, broadly winged from the margins in the upper half and clasping the inflorescence*. Leaf blades linear; narrow; 0.5–2 mm wide (to 3 cm long); flat, or rolled; without cross venation; persistent. *Ligule a fringed membrane*; not truncate (acute); 0.5 mm long. *Contra-ligule* absent.

Reproductive organization. Plants bisexual, with bisexual spikelets; with hermaphrodite florets. The spikelets of sexually distinct forms on the same plant; hermaphrodite and sterile; overtly heteromorphic (those at the bases of the lower branches 1-flowered, or consisting of 2–4 empty glumes).

Inflorescence. *Inflorescence* paniculate; *deciduous in its entirety (the culm disarticulating at the uppermost node, complete with the inflorescence and the uppermost leaf)*; contracted (to 2.5 cm long); *more or less ovoid (small, embraced at the base by the sheath of the uppermost leaf and deciduous with it)*; not comprising 'partial inflorescences' and foliar organs. Spikelets solitary; not secund.

Female-fertile spikelets. Spikelets 4 mm long; slightly compressed laterally; falling with the glumes (and with the whole inflorescence, the adjacent node and its leaf); with conventional internode spacings. Rachilla prolonged beyond the uppermost female-fertile floret; rachilla hairy. The rachilla extension with incomplete florets. *Callus* absent.

Glumes two; relatively large; more or less equal; shorter than the spikelets; hairy (at least, the upper margins with tubercle based hairs); pointed; *awned (acuminate into scabrid 8–13 mm awns)*; non-carinate (dorsally rounded); similar (ovate-oblong, acuminate, membranous). Lower glume shorter than the lowest lemma; 4–6 nerved. Upper glume 5–7 nerved. *Spikelets* with incomplete florets. The incomplete florets distal to the female-fertile florets. *The distal incomplete florets* merely underdeveloped; awned.

Female-fertile florets *3–7*. Lemmas similar in texture to the glumes; not becoming indurated; entire; pointed; awned (tapering into the awn). Awns 1; median; apical; non-geniculate; recurving; much shorter than the body of the lemma to much longer than the body of the lemma (but shorter than the glume awns); entered by several veins. Lemmas hairy (with fine tubercle-based marginal hairs above, and club-shaped hairs on the mid-nerve); non-carinate (somewhat rounded on the back); without a germination flap; 7–9 nerved; with the nerves confluent towards the tip. *Palea* present (linear-oblong); relatively long (equalling the lemma); apically notched; awnless, without apical setae; thinner than the lemma (membranous); not indurated; 2-nerved; 2-keeled. Palea keels wingless; hairy. *Lodicules* present (minute); 2; fleshy; glabrous. *Stamens* 3. Anthers about 3 mm long; not penicillate; without an apically prolonged connective. *Ovary* glabrous. Styles free to their bases (short). Stigmas 2; white.

Fruit, embryo and seedling. *Fruit* free from both lemma and palea (enclosed by little-altered lemma and palea); small (1–2 mm long); compressed dorsiventrally (and obscurely concave on the front). Hilum short (but relatively large). Pericarp free. Embryo large; without an epiblast; with a scutellar tail; with an elongated mesocotyl internode. Embryonic leaf margins meeting.

Abaxial leaf blade epidermis. *Costal/intercostal zonation* conspicuous. *Papillae* absent. *Long-cells* similar in shape costally and intercostally to markedly different in shape costally and intercostally (the costals narrower); of similar wall thickness costally and intercostally. Mid-intercostal long-cells rectangular; having markedly sinuous walls. *Microhairs* present; panicoid-type. *Stomata* common. Subsidiaries dome-shaped (mainly), or triangular. *Intercostal short-cells* common. Costal zones with short-cells. *Costal short-cells* conspicuously in long rows. Costal silica bodies 'panicoid-type'; small, dumb-bell shaped and nodular; not sharp-pointed.

Transverse section of leaf blade, physiology. C_3; XyMS+. *Mesophyll* with radiate chlorenchyma (somewhat so); without adaxial palisade; tending to *Isachne*-type (loose). *Leaf blade* with distinct, prominent adaxial ribs; with the ribs more or less constant in size (low, round-topped). *Midrib* not readily distinguishable; with one bundle only. Bulliforms not in discrete, regular adaxial groups (the bulliforms in conspicuously irregular groups). Combined sclerenchyma girders absent. Sclerenchyma all associated with vascular bundles.

Cytology. Chromosome base number, $x = 7$.

Taxonomy. Arundinoideae; Danthonieae (?).

Ecology, geography, regional floristic distribution. 1 species. South Africa. Xerophytic; species of open habitats; glycophytic. In Karoo.

Cape.

References, etc. Leaf anatomical: photos provided by R.P. Ellis.

Urochloa P. Beauv.

Habit, vegetative morphology. Annual, or perennial; rhizomatous, or stoloniferous, or caespitose, or decumbent. *Culms* 20–170 cm high; herbaceous; branched above, or unbranched above; tuberous, or not tuberous. Culm nodes hairy. Culm internodes solid, or hollow. *Leaves* not basally aggregated; non-auriculate. Leaf blades linear to lanceolate; narrow; flat, or rolled; without cross venation; persistent; rolled in bud. Ligule a fringed membrane to a fringe of hairs.

Reproductive organization. Plants bisexual, with bisexual spikelets; with hermaphrodite florets. The spikelets of sexually distinct forms on the same plant (some spikelets reduced to disc-tipped pedicels), or all alike in sexuality; hermaphrodite, or hermaphrodite and sterile. Apomictic, or reproducing sexually.

Inflorescence. *Inflorescence of spicate main branches (these sessile or subsessile)*; digitate, or subdigitate, or non-digitate. Primary inflorescence branches 2–20(–25). Inflorescence espatheate; not comprising 'partial inflorescences' and foliar organs. Spikelet-bearing axes persistent. Spikelets solitary, or paired (or in fascicles of 3 to 4); secund; shortly pedicellate, or subsessile. Pedicel apices oblique, or discoid.

Female-fertile spikelets. *Spikelets* abaxial (when orientation ascertainable); *compressed dorsiventrally*; planoconvex; falling with the glumes. Rachilla terminated by a female-fertile floret.

Glumes two; very unequal, or more or less equal (rarely); (the upper) long relative to the adjacent lemmas; awnless; very dissimilar, or similar (membranous, the lower sometimes tiny). *Lower glume 3 nerved.* Upper glume 5–11 nerved. *Spikelets with incomplete florets.* The incomplete florets proximal to the female-fertile florets. *Spikelets with proximal incomplete florets.* The proximal incomplete florets 1; paleate, or epaleate. Palea of the proximal incomplete florets (when present) fully developed to reduced. The proximal incomplete florets male, or sterile. The proximal lemmas awnless; 5–7 nerved (rarely more); decidedly exceeding the female-fertile lemmas; less firm than the female-fertile lemmas.

Female-fertile florets 1. *Lemmas decidedly firmer than the glumes*; rugose; becoming indurated (crustaceous); yellow in fruit; entire; blunt; usually *awned (or at least strongly mucronate)*. Awns 1; median; apical; non-geniculate; hairless (scabrid or barbellate); much shorter than the body of the lemma. Lemmas hairless; non-carinate; having the margins tucked in onto the palea; with a clear germination flap; 5–7 nerved. *Palea* present; relatively long; tightly clasped by the lemma; entire; awnless, without apical setae; 2-nerved. *Lodicules* present; 2; free; fleshy; glabrous. *Stamens* 3. Anthers not penicillate. *Ovary* glabrous. Styles free to their bases. Stigmas 2.

Fruit, embryo and seedling. Fruit small; ellipsoid to subglobose; compressed dorsiventrally. Hilum short. Embryo large; waisted. Endosperm containing only simple starch grains.

Abaxial leaf blade epidermis. *Costal/intercostal zonation* conspicuous. *Papillae* absent. *Long-cells* similar in shape costally and intercostally; of similar wall thickness costally and intercostally. Mid-intercostal long-cells rectangular; having markedly sinuous walls. *Microhairs* present; panicoid-type; 52–75 microns long; 4.5–6 microns wide at the septum. Microhair total length/width at septum 12.5–16.7. Microhair apical cells 30–46.5 microns long. Microhair apical cell/total length ratio 0.6–0.63. *Stomata* common; 31.5–33 microns long. Subsidiaries triangular. Guard-cells overlapping to flush with the interstomatals. *Intercostal short-cells* common; in cork/silica-cell pairs and not paired (solitary); silicified (when paired), or not silicified. Intercostal silica bodies cross-shaped, or vertically elongated-nodular. Costal zones with short-cells. *Costal short-cells* conspicuously in long rows. Costal silica bodies 'panicoid-type'; cross shaped, or butterfly shaped, or dumb-bell shaped; not sharp-pointed.

Transverse section of leaf blade, physiology. C_4; biochemical type PCK (4 species); XyMS+. PCR sheath outlines uneven. PCR sheath extensions absent. PCR cell chloroplasts centrifugal/peripheral. *Mesophyll* with radiate chlorenchyma. *Leaf blade* 'nodular' in section; with the ribs more or less constant in size. *Midrib* conspicuous; having a conventional arc of bundles; with colourless mesophyll adaxially, or without colourless mesophyll adaxially. Bulliforms present in discrete, regular adaxial groups; in simple fans. All the vascular bundles accompanied by sclerenchyma. Combined sclerenchyma girders present, or absent; nowhere forming 'figures'. Sclerenchyma all associated with vascular bundles.

Special diagnostic feature. *No* **Eriochloa-*type* '*callus*'.

Cytology. Chromosome base number, $x = 7$, 9, and 15. $2n = 14$ (*U. reptans*), or 28, or 30, or 36, or 42, or 44, or 48. Nucleoli persistent.

Taxonomy. Panicoideae; Panicodae; Paniceae.

Ecology, geography, regional floristic distribution. 11 species. Tropical Africa, Asia. Commonly adventive. Mesophytic; shade species, or species of open habitats (usually); glycophytic. Savanna, often weedy.

Holarctic, Paleotropical, and Neotropical. Tethyan. African, Madagascan, and Indomalesian. Irano-Turanian. Saharo-Sindian, Sudano-Angolan, West African Rainforest, and Namib-Karoo. Indian and Indo-Chinese. Caribbean. Sahelo-Sudanian, Somalo-Ethiopian, South Tropical African, and Kalaharian.

Rusts and smuts. Rusts — *Puccinia*. Taxonomically wide-ranging species: '*Uromyces*' *setariae-italicae*. Smuts from Ustilaginaceae. Ustilaginaceae — *Sorosporium*, *Sphacelotheca*, and *Ustilago*.

Economic importance. Significant weed species: *U. panicoides*. Cultivated fodder: *U. mosambicensis*. Important native pasture species: e.g. *U. brachyura*, *U. mosambicensis*, *U. oligotricha*, *U. trichopus*.

References, etc. Leaf anatomical: Metcalfe 1960; this project.

Special comments. Unsatisfactorily delimited from other close allies of *Panicum*, in particular *Brachiaria*.

Urochondra C.E. Hubb.

Habit, vegetative morphology. Perennial; caespitose. **Culms** 40 cm high; **herbaceous**; unbranched above. Plants unarmed. *Leaves* not basally aggregated; non-auriculate. Leaf blades linear (acuminate, rigid, subulate-tipped); narrow; 2–7 mm wide; rolled (convolute), or acicular (towards the tip); without abaxial multicellular glands; without cross venation; persistent. *Ligule a fringe of hairs (dense)*. *Contra-ligule* absent.

Reproductive organization. Plants bisexual, with bisexual spikelets; with hermaphrodite florets.

Inflorescence. Inflorescence paniculate; contracted; spicate (5–10 cm long in material seen); espatheate; not comprising 'partial inflorescences' and foliar organs. Spikelet-bearing axes persistent. Spikelets not secund; very shortly pedicellate.
Female-fertile spikelets. Spikelets 2.5 mm long; compressed laterally; disarticulating above the glumes; with conventional internode spacings. Rachilla terminated by a female-fertile floret. Hairy callus absent.

Glumes two; more or less equal; shorter than the adjacent lemmas (or at least, somewhat shorter); hairy (slightly, on the upper third of the mid-nerve); pointed; awnless (not mucronate); carinate; similar (lanceolate, hyaline). Lower glume 1 nerved. Upper glume 1 nerved. *Spikelets* with female-fertile florets only.

Female-fertile florets 1. Lemmas acuminate; similar in texture to the glumes (hyaline); not becoming indurated; entire; pointed; mucronate (to aristulate); hairy, or hairless; carinate; without a germination flap; 1 nerved. *Palea* present; relatively long; apically notched; awnless, without apical setae (glabrous); not indurated (hyaline, delicate); 2-nerved (but the veins very scanty); 2-keeled. *Lodicules* absent. *Stamens* 3. Anthers relatively long; not penicillate; without an apically prolonged connective. **Ovary** hairy; *with a conspicuous apical appendage (beaked).* Styles free to their bases. Stigmas 2.

Fruit, embryo and seedling. Fruit small; ellipsoid. *Pericarp free (becoming swollen when wet, extruding the seed).*

Abaxial leaf blade epidermis. *Costal/intercostal zonation* conspicuous (the intercostal zones in deep grooves, their detailed histology inaccessible). *Papillae* present; intercostal. Intercostal papillae consisting of one symmetrical projection per cell (thick-walled). Long-cells differing markedly in wall thickness costally and intercostally (costals thicker-walled). Mid-intercostal long-cells rectangular; having markedly sinuous walls (and pitted, with abundant pits on outer walls). *Microhairs* present (in deep, narrow abaxial grooves, obscured by thick walled macrohairs and blunt prickles); more or less spherical to elongated; clearly two-celled; chloridoid-type. Microhair apical cell wall of similar thickness/rigidity to that of the basal cell. Microhairs 30–36 microns long. Microhair basal cells 2 microns long. Microhairs 13.5–19.5 microns wide at the septum. Microhair total length/width at septum 1.8–2.2. Microhair apical cells 15–18 microns long. Microhair apical cell/total length ratio 0.5–0.55. *Stomata* common; 15–18 microns long. Subsidiaries high dome-shaped and triangular. Guard-cells overlapping to flush with the interstomatals. Costal zones with short-cells. *Costal short-cells* predominantly paired. Costal silica bodies present throughout the costal zones; rounded, saddle shaped, tall-and-narrow, and crescentic; not sharp-pointed.

Transverse section of leaf blade, physiology. C_4; XyMS+. PCR sheaths of the primary vascular bundles interrupted; interrupted both abaxially and adaxially. PCR sheath extensions absent. Mesophyll traversed by columns of colourless mesophyll cells. *Leaf blade* 'nodular' in section (deep, narrow abaxial grooves opposite the broader adaxial furrows); with the ribs very irregular in sizes (large flat-topped ribs alternating with small pointed ones). *Midrib* not readily distinguishable; with one bundle only; with colourless mesophyll adaxially (exhibiting extensive blocks of colourless tissue adaxially and abaxially with all the bundles, each block bounded adaxially and abaxially by sclerenchyma and laterally by the rows of PCR cells). Bulliforms present in discrete, regular adaxial groups; associated with colourless mesophyll cells to form deeply-penetrating fans (these connected with the traversing columns of colourless cells). All the vascular bundles accompanied by sclerenchyma. Combined sclerenchyma girders present (with the primaries, the smaller bundles with only abaxial girders); forming 'figures' (heavily I-shaped in the main bundles). Sclerenchyma all associated with vascular bundles. The lamina margins with fibres.

Taxonomy. Chloridoideae; main chloridoid assemblage.

Ecology, geography, regional floristic distribution. 1 species. Northeast tropical Africa. Species of open habitats; halophytic. In coastal sand.

Holarctic and Paleotropical. Tethyan. African. Irano-Turanian. Saharo-Sindian and Sudano-Angolan. Somalo-Ethiopian.

References, etc. Leaf anatomical: this project.

Special comments. Fruit data wanting.

Vahlodea Fries

~ *Deschampsia*

Habit, vegetative morphology. Perennial; rhizomatous, or caespitose. **Culms** 15–80 cm high; *herbaceous*. Culm nodes glabrous. Culm internodes hollow. *Leaves* not basally aggregated; non-auriculate. Sheath margins free. Leaf blades linear; narrow; 1–5 mm wide; flat; without cross venation; persistent; rolled in bud. *Ligule an unfringed membrane*; truncate, or not truncate (dentate or lacerate); 2 mm long.

Reproductive organization. Plants bisexual, with bisexual spikelets; with hermaphrodite florets.

Inflorescence. *Inflorescence paniculate*; open; with capillary branchlets; espatheate; not comprising 'partial inflorescences' and foliar organs. Spikelet-bearing axes persistent. Spikelets not secund; pedicellate.

Female-fertile spikelets. Spikelets 2–5.5 mm long; compressed laterally to not noticeably compressed; disarticulating above the glumes; disarticulating between the florets. Rachilla prolonged beyond the uppermost female-fertile floret. The rachilla extension naked. Hairy callus present (the hairs 1/2 to 2/3 as long as the lemma). *Callus hairs present, more than 0.5 mm long*.

Glumes present; two; more or less equal; *exceeding the spikelets*; *long relative to the adjacent lemmas (about twice as long)*; pointed (acute); awnless; carinate; similar. Lower glume 1 nerved. Upper glume 3 nerved. *Spikelets* with female-fertile florets only.

Female-fertile florets *2–3*. *Lemmas* similar in texture to the glumes; not becoming indurated; *entire*; blunt; *awned*. *Awns* 1; median; *dorsal*; from near the top, or from well down the back; geniculate; proximally hairy. Lemmas hairy (above), or hairless (glabrous above the callus); non-carinate; without a germination flap; 5 nerved. *Palea* present; relatively long; 2-nerved; 2-keeled. *Lodicules* present; 2; free; membranous; glabrous; toothed, or not toothed; not or scarcely vascularized. *Stamens* 3. Anthers 0.4–0.8 mm long. *Ovary* glabrous. Styles free to their bases. Stigmas 2; white.

Fruit, embryo and seedling. *Fruit* adhering to lemma and/or palea, or free from both lemma and palea. Embryo small. Endosperm hard.

Abaxial leaf blade epidermis. *Costal/intercostal zonation* conspicuous. *Papillae* absent. *Long-cells* markedly different in shape costally and intercostally (costals more rectangular); differing markedly in wall thickness costally and intercostally (costals thicker-walled). Mid-intercostal long-cells fusiform; having straight or only gently undulating walls. *Microhairs* absent. *Stomata* common; (21–)24–33(–36) microns long. Subsidiaries parallel-sided. Guard-cells overlapped by the interstomatals. *Intercostal short-cells* absent or very rare. Prickle bases abundant. Costal zones with short-cells. *Costal short-cells* neither distinctly grouped into long rows nor predominantly paired. Costal silica bodies horizontally-elongated crenate/sinuous, or horizontally-elongated smooth (few); not sharp-pointed.

Transverse section of leaf blade, physiology. C_3; XyMS+. *Mesophyll* with non-radiate chlorenchyma; without adaxial palisade. *Leaf blade* 'nodular' in section to adaxially flat. *Midrib* not readily distinguishable; with one bundle only. Bulliforms present in discrete, regular adaxial groups; in simple fans (these wide, shallow). All the vascular bundles accompanied by sclerenchyma. Combined sclerenchyma girders present (with the primaries only); nowhere forming 'figures'. Sclerenchyma all associated with vascular bundles.

Cytology. Chromosome base number, $x = 7$. $2n = 14$. 2 ploid. Haploid nuclear DNA content 3.2 pg (1 species).

Taxonomy. Pooideae; Poodae; Aveneae.

Ecology, geography, regional floristic distribution. 3–4 species. Northeast Asia. Helophytic (calcifuge).

Holarctic and Antarctic. Boreal. Arctic and Subarctic and Euro-Siberian. Patagonian. European and Siberian.

References, etc. Leaf anatomical: this project.

Vaseyochloa A. Hitchc.

Habit, vegetative morphology. Perennial; rhizomatous and caespitose. **Culms** 40–100 cm high; *herbaceous*; unbranched above. Culm nodes glabrous. Plants unarmed. Leaves non-auriculate. Leaf blades narrow; 1–4 mm wide; flat, or rolled (loosely involute); without abaxial multicellular glands; without cross venation. *Ligule a fringe of hairs*. *Contra-ligule* present (partial, of hairs).

Reproductive organization. Plants bisexual, with bisexual spikelets; with hermaphrodite florets.

Inflorescence. *Inflorescence paniculate*; *open (5–20 cm long, with few branches)*; espatheate; not comprising 'partial inflorescences' and foliar organs. Spikelet-bearing axes persistent. Spikelets not secund; pedicellate.

Female-fertile spikelets. Spikelets 10–18 mm long; narrowly ovate; compressed laterally to not noticeably compressed; disarticulating above the glumes; disarticulating between the florets; with conventional internode spacings. Rachilla prolonged beyond the uppermost female-fertile floret; rachilla hairless (glabrous). The rachilla extension with incomplete florets. Hairy callus present (as the stipe-like base of the lemma).

Glumes two; very unequal; shorter than the spikelets; shorter than the adjacent lemmas; hairless; glabrous; pointed (G_1), or not pointed (G_2); awnless; carinate (G_1), or non-carinate (G_2); similar (rather firm, G_2 broader). Lower glume 3–5 nerved. Upper glume 7–9 nerved. *Spikelets* with incomplete florets. The incomplete florets distal to the female-fertile florets. *The distal incomplete florets* merely underdeveloped. *Spikelets without proximal incomplete florets*.

Female-fertile florets 6–12. *Lemmas* similar in texture to the glumes (membranous or thinly papery); not becoming indurated; *entire*; pointed to blunt; awnless; hairy (below); non-carinate (dorsally rounded); without a germination flap; distinctly *7–9 nerved*. *Palea* present; relatively long; probably apically notched (splitting at maturity); awnless, without apical setae; not indurated (membranous); 2-nerved; 2-keeled. Palea keels narrowly winged (and hollowed between the wings). *Lodicules* present; 2; free; fleshy; glabrous. *Stamens* 3; with free filaments (these very long). Anthers short; not penicillate; without an apically prolonged connective. *Ovary* glabrous. Styles free to their bases. Stigmas 2 (sometimes with a vestigial third style).

Fruit, embryo and seedling. Fruit small (2.5–3 mm long); black; ovate-rounded, ventrally deeply concave; compressed dorsiventrally (concavo-convex); sculptured to smooth (obscurely striate). *Hilum short*. Pericarp fused. Embryo large; with an epiblast; with a scutellar tail; with an elongated mesocotyl internode. Embryonic leaf margins meeting.

Abaxial leaf blade epidermis. *Costal/intercostal zonation* conspicuous. *Papillae* present; intercostal. Intercostal papillae over-arching the stomata (at one end); consisting of one oblique swelling per cell (at one end of interstomatals and some intercostal long-cells). *Long-cells* markedly different in shape costally and intercostally (the costals longer); of similar wall thickness costally and intercostally (medium-thin walled). Mid-intercostal long-cells rectangular; having markedly sinuous walls (and pitted). *Microhairs* present; elongated; clearly two-celled; chloridoid-type (unusually large). Microhair apical cell wall of similar thickness/rigidity to that of the basal cell. Microhairs (33–)36–39(–42) microns long. Microhair basal cells 24 microns long. Microhairs 10.5–12–14.4 microns wide at the septum. Microhair total length/width at septum 2.7–3.5. Microhair apical cells (7.5–)8.4–9(–12) microns long. Microhair apical cell/total length ratio 0.21–0.27. *Stomata* common; 21–22.5(–24) microns long. Subsidiaries dome-shaped. Guard-cells overlapping to flush with the interstomatals. *Intercostal short-cells* common; not paired (solitary); not silicified. Intercostal silica bodies absent. Costal zones with short-cells. *Costal short-cells* conspicuously in long rows (to solitary — the 'short-cells' mainly very long). Costal silica bodies present in alternate cell files of the costal zones; 'panicoid-type'; mostly dumb-bell shaped (some quite elongated); not sharp-pointed.

Transverse section of leaf blade, physiology. C_4; XyMS+. PCR sheaths of the primary vascular bundles interrupted; interrupted abaxially only. PCR sheath extensions absent. PCR cell chloroplasts centripetal. *Leaf blade* with distinct, prominent adaxial ribs to 'nodular' in section (the ribs round-topped); with the ribs more or less constant in size. *Midrib* not readily distinguishable; with one bundle only. Bulliforms present in discrete, regular adaxial groups (in all the adaxial furrows); seemingly exclusively in simple fans. All the vascular bundles accompanied by sclerenchyma. Combined sclerenchyma girders present (with all the bundles); forming 'figures' (all bundles). Sclerenchyma all associated with vascular bundles. The lamina margins with fibres.

Cytology. $2n = 60$.

Taxonomy. Chloridoideae; main chloridoid assemblage.

Ecology, geography, regional floristic distribution. 1 species. Texas. Species of open habitats. Sandy, open woods.

Holarctic. Boreal. Atlantic North American. Southern Atlantic North American.

References, etc. Leaf anatomical: this project.

Ventenata Koeler

Heteranthus Borkh., *Heterochaeta* Schult.

Excluding *Gaudiniopsis, Pilgerochloa*

Habit, vegetative morphology. *Annual. Culms* 10–70 cm high; herbaceous. Culm nodes glabrous. Culm internodes hollow. *Leaves* not basally aggregated; non-auriculate. Leaf blades linear; narrow; 1–2.5 mm wide; folded, or rolled (convolute); without cross venation; rolled in bud. *Ligule an unfringed membrane*; not truncate (acute, often lacerate); 2–4 mm long.

Reproductive organization. Plants bisexual, with bisexual spikelets; with hermaphrodite florets.

Inflorescence. Inflorescence paniculate, or a single raceme; open, or contracted; sometimes spicate; espatheate; not comprising 'partial inflorescences' and foliar organs. Spikelet-bearing axes persistent. Spikelets not secund; pedicellate.

Female-fertile spikelets. *Spikelets 8–15 mm long*; compressed laterally; disarticulating above the glumes; *disarticulating between the florets (the lowermost floret persisting, falling later with the glumes and pedicel)*. Rachilla prolonged beyond the uppermost female-fertile floret. *Hairy callus present. Callus* short; pointed (in the upper florets).

Glumes two; very unequal; *shorter than the adjacent lemmas*; pointed (acute to acuminate); awnless; *non-carinate*; similar (herbaceous, with wide, scarious margins). *Lower glume 3–7 nerved.* Upper glume 3–9 nerved. *Spikelets* with female-fertile florets only, or with incomplete florets. The incomplete florets distal to the female-fertile florets.

Female-fertile florets 2–7. *Lemmas* similar in texture to the glumes; not becoming indurated (papery, margins scarious); entire (the lowest), or incised (apically bidentate or bifid); *awned (the L_1 shorter and awnless but sometimes having an apical seta, the upper awned lemmas often bifid, with apical teeth or setae additional to the awn)*. Awns 1; median; dorsal; from well down the back; geniculate. Lemmas hairless; non-carinate; without a germination flap; 5 nerved. **Palea** present; *relatively long*; tightly clasped by the lemma; 2-nerved; 2-keeled. *Lodicules* present; 2; free; membranous; glabrous; toothed, or not toothed; not or scarcely vascularized. *Stamens* 3. Anthers 1.2–2.5 mm long. *Ovary* glabrous. Styles free to their bases. Stigmas 2; white.

Fruit, embryo and seedling. Fruit small (2.5 mm long); compressed laterally. Hilum short. Embryo small. Endosperm liquid in the mature fruit; with lipid.

Abaxial leaf blade epidermis. *Costal/intercostal zonation* conspicuous. *Papillae* absent. *Long-cells* similar in shape costally and intercostally; of similar wall thickness costally and intercostally. Mid-intercostal long-cells rectangular; having markedly sinuous walls. *Microhairs* absent. *Stomata* common (fairly); 31–39 microns long. Subsidiaries tending to parallel-sided, or dome-shaped. Guard-cells overlapped by the interstomatals (mostly, slightly). *Intercostal short-cells* common; in cork/silica-cell pairs; silicified. Costal zones with short-cells. *Costal short-cells* predominantly paired. Costal silica bodies horizontally-elongated crenate/sinuous (fairly short), or rounded (a few), or crescentic (a few); not sharp-pointed.

Transverse section of leaf blade, physiology. C_3; XyMS+. *Mesophyll* with non-radiate chlorenchyma; without adaxial palisade. *Leaf blade* with distinct, prominent adaxial ribs; with the ribs more or less constant in size (round topped). Midrib with one bundle only. *The lamina* symmetrical on either side of the midrib. Bulliforms seemingly not in discrete, regular adaxial groups (bulliforms not apparent in material seen). All the vascular bundles accompanied by sclerenchyma. Combined sclerenchyma girders absent (the main bundles with an abaxial girder and an adaxial strand). Sclerenchyma all associated with vascular bundles.

Cytology. Chromosome base number, $x = 7$. $2n = 14$. 2 ploid.

Taxonomy. Pooideae; Poodae; Aveneae.

Ecology, geography, regional floristic distribution. 5 species. Southern Europe. Xerophytic; species of open habitats.

Holarctic. Boreal and Tethyan. Euro-Siberian. Mediterranean. European.

Rusts and smuts. Rusts — *Puccinia.* Taxonomically wide-ranging species: *Puccinia graminis.*

References, etc. Leaf anatomical: this project.

Vetiveria Bory

Mandelorna Steud., *Lenormandia* Steud.

Habit, vegetative morphology. Perennial (often with aromatic roots); forming large clumps from stout rhizomes. Culms 50–300 cm high; herbaceous; *unbranched above.* Culm nodes glabrous. Culm internodes solid. *Leaves* mostly basal; non-auriculate. Sheath margins free. The lower sheaths compressed. Leaf blades linear; broad, or narrow; without cross venation; persistent. Ligule a fringed membrane to a fringe of hairs.

Reproductive organization. Plants bisexual, with bisexual spikelets; with hermaphrodite florets. *The spikelets of sexually distinct forms on the same plant*; hermaphrodite and male-only, or hermaphrodite and sterile; *homomorphic*; all in heterogamous combinations.

Inflorescence. Inflorescence of spicate main branches, or paniculate (a panicle with slender, whorled, simple or rarely compound racemes); open; espatheate; not comprising 'partial inflorescences' and foliar organs. *Spikelet-bearing axes* 'racemes'; the spikelet-bearing axes with 2–3 spikelet-bearing 'articles' to with 6–10 spikelet-bearing 'articles', or with more than 10 spikelet-bearing 'articles' (typically with many spikelet pairs); with very slender rachides; disarticulating; disarticulating at the joints. *'Articles'* linear; not appendaged; disarticulating transversely. Spikelets paired; secund (rarely), or not secund; sessile and pedicellate; consistently in 'long-and-short' combinations; in pedicellate/sessile combinations. Pedicels of the 'pedicellate' spikelets free of the rachis. The 'shorter' spikelets hermaphrodite. The 'longer' spikelets male-only, or sterile.

Female-sterile spikelets. The pedicelled, male spikelets similar to the sessile ones, or slightly smaller.

Female-fertile spikelets. *Spikelets* 4.5–10 mm long; *compressed laterally*; falling with the glumes (and with the joint and pedicel). Rachilla terminated by a female-fertile floret. Hairy callus present, or absent. Callus pointed to blunt.

Glumes two; more or less equal; long relative to the adjacent lemmas; awned (G_2, sometimes), or awnless; very dissimilar (the lower rounded on the back, the upper naviculate). Lower glume convex on the back; not pitted; spinulose; 5 nerved. Upper glume 3 nerved. *Spikelets* with incomplete florets. The incomplete florets proximal to the female-fertile florets. *The proximal incomplete florets* 1; epaleate; sterile. The proximal lemmas awnless; 2 nerved; similar in texture to the female-fertile lemmas (hyaline); not becoming indurated.

Female-fertile florets 1. Lemmas less firm than the glumes (hyaline); not becoming indurated; incised; not deeply cleft (bidentate); awnless, or mucronate, or awned. *Awns* when present, 1; *from a sinus*; geniculate; hairless (glabrous); much shorter than the body of the lemma to much longer than the body of the lemma. Lemmas hairless; non-carinate; 1–3 nerved. **Palea** present, or absent; when present, very reduced; apically notched; awnless, without apical setae; not indurated (hyaline); *nerveless. Lodicules* present; 2; free;

fleshy; glabrous. *Stamens* 3. Anthers not penicillate. *Ovary* glabrous. Styles free to their bases. Stigmas 2.

Fruit, embryo and seedling. Fruit small; not noticeably compressed. Hilum short. Embryo large. Endosperm containing only simple starch grains. Embryo without an epiblast; with a scutellar tail; with an elongated mesocotyl internode. Embryonic leaf margins overlapping.

First seedling leaf with a well-developed lamina. The lamina broad; curved; 21–30 veined.

Abaxial leaf blade epidermis. *Costal/intercostal zonation* conspicuous. *Papillae* absent. *Long-cells* similar in shape costally and intercostally; of similar wall thickness costally and intercostally. Mid-intercostal long-cells rectangular; having markedly sinuous walls (the sinuosities very tight in *V. elongata*). *Microhairs* present; panicoid-type (but often balanoform — the thin walled apical cells quite broad and blunt); (39–)48–51(–54) microns long; 9–12.6 microns wide at the septum. Microhair total length/width at septum 3.8–5.5. Microhair apical cells (25–)27–30(–33) microns long. Microhair apical cell/total length ratio 0.55–0.63. *Stomata* common; 27–33 microns long. Subsidiaries low dome-shaped, or triangular. Guard-cells overlapped by the interstomatals (the interstomatal end walls very thickened in *V. elongata*). *Intercostal short-cells* common; in cork/silica-cell pairs; silicified. Intercostal silica bodies tall-and-narrow, or cross-shaped. Costal zones with short-cells. *Costal short-cells* predominantly paired. Costal silica bodies tall-and-narrow (exclusively, in *V. elongata*), or 'panicoid-type'; cross shaped (in *V. zizanioides*); not sharp-pointed.

Transverse section of leaf blade, physiology. C_4; XyMS–. PCR sheath outlines even. PCR cell chloroplasts with reduced grana; centrifugal/peripheral. *Mesophyll* with radiate chlorenchyma. *Leaf blade* adaxially flat. *Midrib* conspicuous; having a conventional arc of bundles; with colourless mesophyll adaxially (the adaxial mesophyll of the rest of the blade also extensively colourless in *V. elongata*, and with large intercellular lacunae in *V. zizanioides*). Bulliforms not in discrete, regular adaxial groups (except in association with the midrib). Many of the smallest vascular bundles unaccompanied by sclerenchyma. Combined sclerenchyma girders present. Sclerenchyma all associated with vascular bundles.

Cytology. Chromosome base number, x = 5 and 10. $2n$ = 20 and 40. Nucleoli persistent.

Taxonomy. Panicoideae; Andropogonodae; Andropogoneae; Andropogoninae.

Ecology, geography, regional floristic distribution. 10 species. Tropical Africa, Asia, Australia. Helophytic; glycophytic. Floodplains and streambanks.

Paleotropical and Australian. African, Madagascan, and Indomalesian. Saharo-Sindian, Sudano-Angolan, West African Rainforest, and Namib-Karoo. Indian, Indo-Chinese, Malesian, and Papuan. North and East Australian. Sahelo-Sudanian, Somalo-Ethiopian, South Tropical African, and Kalaharian. Tropical North and East Australian.

Rusts and smuts. Smuts from Tilletiaceae and from Ustilaginaceae. Tilletiaceae — *Tilletia*. Ustilaginaceae — *Ustilago*.

Economic importance. Commercial essential oils: *V. zizanioides* (from the roots). *V. zizanioides* is valuable for hedging, and as a guard against soil erosion (it also 'repels pests such as rats and snakes' — O. Sattaur 1989, *New Scientist* **1664**, 16–17).

References, etc. Leaf anatomical: Metcalfe 1960; this project.

Vietnamochloa Veldkamp & Nowack

Habit, vegetative morphology. *Annual*; caespitose. *Culms* 25 cm high; herbaceous; unbranched above. *Leaves* mostly basal. Leaf blades linear; 1.5–2 mm wide; flat, or rolled; not needle-like; without abaxial multicellular glands; not pseudopetiolate. Ligule a fringe of hairs.

Reproductive organization. Plants bisexual, with bisexual spikelets; with hermaphrodite florets. The spikelets all alike in sexuality.

Inflorescence. *Inflorescence paniculate*; non-digitate; espatheate. Spikelet-bearing axes persistent. Spikelets solitary; not secund; pedicellate; distant.

Female-fertile spikelets. *Spikelets* 4–4.25 mm long; fusiform; *compressed dorsiventrally*; disarticulating above the glumes. *Rachilla prolonged beyond the uppermost female-fertile floret*; rachilla hairless. The rachilla extension naked. Hairy callus present.

Glumes two; more or less equal; *exceeding the spikelets*; long relative to the adjacent lemmas; hairless; glabrous; pointed; awnless; carinate; similar. Lower glume 1 nerved. Upper glume 1 nerved. *Spikelets* with female-fertile florets only.

Female-fertile florets 1. Lemmas similar in texture to the glumes; entire; not deeply cleft; awned. Awns not of the triple/trifid, basal column type; 1; median; non-geniculate; much shorter than the body of the lemma. Lemmas hairy; non-carinate; 3 nerved. *Palea* present; relatively long; apically notched; awnless, without apical setae; 2-nerved. Palea back hairy. *Stamens* 3. Anthers 1.5 mm long. Styles free to their bases. Stigmas 2.

Fruit, embryo and seedling. *Fruit* free from both lemma and palea; small; compressed dorsiventrally; smooth. *Hilum short*. Pericarp thin; fused. Embryo large.

Abaxial leaf blade epidermis. *Costal/intercostal zonation* conspicuous. *Papillae* absent. *Long-cells* similar in shape costally and intercostally (broader intercostally); of similar wall thickness costally and intercostally. Mid-intercostal long-cells rectangular; having markedly sinuous walls (irregular). *Microhairs* present; elongated; clearly two-celled; chloridoid-type. Microhair apical cell wall thinner than that of the basal cell but not tending to collapse. Microhairs 24–30 microns long. Microhair basal cells 18–21 microns long. Microhairs 6 microns wide at the septum. Microhair total length/width at septum 6. Microhair apical cells 9–12 microns long. Microhair apical cell/total length ratio 0.3. *Stomata* common; 15 microns long. Subsidiaries non-papillate; dome-shaped. Guard-cells overlapping to flush with the interstomatals. *Intercostal short-cells* common (few); mostly in cork/silica-cell pairs (when present); silicified (few). Intercostal silica bodies present and perfectly developed (but few); tall-and-narrow. Prickles and macrohairs present only on the leaf margins. *Costal short-cells* conspicuously in long rows. Costal silica bodies present and well developed; present in alternate cell files of the costal zones; saddle shaped.

Transverse section of leaf blade, physiology. C_4. The anatomical organization conventional. XyMS+. PCR sheath outlines even. PCR sheaths of the primary vascular bundles complete. PCR sheath extensions absent. Mesophyll without adaxial palisade. *Leaf blade* adaxially flat (but abaxially ribbed). *Midrib* not readily distinguishable; with one bundle only. *The lamina* symmetrical on either side of the midrib. Bulliforms present in discrete, regular adaxial groups; in simple fans (middle cell very large and deeply penetrating the chlorenchyma). All the vascular bundles accompanied by sclerenchyma. Sclerenchyma all associated with vascular bundles. The lamina margins with fibres (large).

Taxonomy. Chloridoideae; main chloridoid assemblage.

Ecology, geography, regional floristic distribution. 1 species. Vietnam. Paleotropical. Indomalesian. Indo-Chinese.

References, etc. Morphological/taxonomic: Nowack and Veldkamp 1994. Leaf anatomical: Van den Borre 1994.

Vietnamosasa Nguyen

~ *Arundinaria, Pseudosasa* — including *A. ciliata* A. Camus and *A. pusilla* Chevalier and Camus

Habit, vegetative morphology. Perennial; rhizomatous. The flowering culms leafy. *Culms* 50–150 cm high (small bamboos); woody and persistent; to 1 cm in diameter; cylindrical; branched above. Primary branches/mid-culm node not stated — 'branches fasciculate'. Culm internodes solid, or hollow. Rhizomes leptomorph. Plants unarmed. *Leaves* not basally aggregated; auriculate; *with auricular setae*. *Leaf blades* linear-lanceolate; narrow; 2–8(–10) mm wide; broadly *cordate ('subauriculate')*; pseudopetiolate; cross veined; disarticulating from the sheaths. *Ligule* present; membranous.

Reproductive organization. Plants bisexual, with bisexual spikelets; with hermaphrodite florets.

Inflorescence. *Inflorescence* determinate, or indeterminate (? —in *V. pusilla*, where the presence and number of glumes is variable, and the lower is sometimes gemmiparous); with pseudospikelets (?), or without pseudospikelets; spatheate, or espatheate. **Spikelet-bearing axes 'racemes', or spikelike (flowering branches 'simple')**; clustered (in fascicles); persistent. Spikelets sessile to pedicellate.

Female-fertile spikelets. Spikelets 30–70 mm long; compressed laterally; disarticulating above the glumes; disarticulating between the florets. Rachilla prolonged beyond the uppermost female-fertile floret.

Glumes present, or absent (sometimes, in *V. pusilla*); when present, one per spikelet to two (*V. pusilla*), or two. Lower glume shorter than the lowest lemma. *Spikelets* with incomplete florets. The incomplete florets distal to the female-fertile florets. *The distal incomplete florets* merely underdeveloped.

Female-fertile florets 6–12. Lemmas lanceolate; awnless; hairless; multinerved. *Palea* present; relatively long; entire, or apically notched; awnless, without apical setae; several nerved; 2-keeled. *Lodicules* present; 3; free; membranous; ciliate; not toothed; heavily vascularized (below). *Stamens* 6. Anthers 3.5–7 mm long (the filaments short). Ovary without a conspicuous apical appendage. Styles basally fused. Stigmas 3; red pigmented.

Fruit, embryo and seedling. Fruit longitudinally grooved.

Taxonomy. Bambusoideae; Bambusodae; Bambuseae.

Ecology, geography, regional floristic distribution. 3 species. Southern Vietnam. Paleotropical. Indomalesian. Indo-Chinese.

References, etc. Morphological/taxonomic: Nguyen, T.Q. (1990). New taxa of bamboos (Poaceae, Bambusoideae) from Vietnam. *Botanicheskii Zhurnal* **75**, 221–225. See Chevalier and Camus 1921, A. Camus 1919 and E.G. & A. Camus 1923 for descriptions of included *Arundinaria* (*Pseudosasa*) species.

Special comments. Description inadequate. Fruit data wanting. Anatomical data wanting.

Viguierella A. Camus

Habit, vegetative morphology. Slender annual; caespitose. *Culms* 10–40 cm high; herbaceous; branched above. Culm nodes glabrous. *Leaves* not basally aggregated; non-auriculate. Leaf blades linear; narrow; 2–3 mm wide; flat, or rolled; without abaxial multicellular glands; without cross venation. *Ligule* present; a fringe of hairs. *Contra-ligule* absent.

Reproductive organization. Plants bisexual, with bisexual spikelets; with hermaphrodite florets. *The spikelets* of sexually distinct forms on the same plant (theoretically), or all alike in sexuality; hermaphrodite, or hermaphrodite and sterile (by recognition of well disguised vestigial 'spikelets'); *overtly heteromorphic, or homomorphic (each manifest spikelet being subtended at its base by a tiny hyaline bract, (?)representing a reduced spikelet or branch system)*; all in heterogamous combinations.

Inflorescence. Inflorescence a single raceme (a bottlebrush); espatheate (but bracteate); not comprising 'partial inflorescences' and foliar organs. *Spikelet-bearing axes* spikelike; persistent. **Spikelets associated with bractiform involucres (or at least, each with one small 'bract')**. The involucres persistent on the rachis. Spikelets ostensibly solitary; shortly pedicellate (erect, the pedicel becoming the callus).

Female-fertile spikelets. Spikelets 4–7 mm long; compressed laterally; falling with the glumes (and with the pungent pedicel). Rachilla prolonged beyond the uppermost female-fertile floret; rachilla hairless. The rachilla extension with incomplete florets. Hairy callus present (1–1.2 mm long). *Callus* of the spikelet short; pointed.

Glumes two; more or less equal; shorter than the adjacent lemmas; hairy; awned (with long, straight, scabrid terminal awns); somewhat carinate; similar (hairy, asymmetric, truncate or bilobed). Lower glume 2 nerved. Upper glume 3 nerved. *Spikelets* with incomplete florets. The incomplete florets distal to the female-fertile florets. The distal incomplete florets 1–3; incomplete florets awned (reduced to a cluster of awns).

Female-fertile florets 1. Lemmas similar in texture to the glumes (membranous); not becoming indurated; entire, or incised; not deeply cleft; awned. Awns 1; median; from a sinus (this slight), or apical; non-geniculate; hairless (scabrid); much longer than the body of the lemma. Lemmas hairless; glabrous; non-carinate (dorsally rounded); without a germination flap; 3 nerved, or 5 nerved. *Palea* present; relatively long; entire (pointed); with apical setae (glabrous); not indurated (membranous); 2-nerved; 2-keeled. *Lodicules* present; free; fleshy; ciliate, or glabrous. Stamens with free filaments (these long). Anthers short; not penicillate; without an apically prolonged connective. *Ovary* glabrous. Styles free to their bases. Stigmas 2.

Fruit, embryo and seedling. Fruit banana-shaped. Embryo with an epiblast; with a scutellar tail; with an elongated mesocotyl internode. Embryonic leaf margins meeting.

Abaxial leaf blade epidermis. *Costal/intercostal zonation* conspicuous. *Papillae* absent. *Long-cells* similar in shape costally and intercostally; of similar wall thickness costally and intercostally (thick walled). Mid-intercostal long-cells rectangular; having markedly sinuous walls (and pitted). *Microhairs* present; elongated; clearly two-celled; panicoid-type; 36–39–45 microns long. Microhair basal cells 18 microns long. Microhairs (5.1–)5.4–5.7(–6) microns wide at the septum. Microhair total length/width at septum 6.3–7.6. Microhair apical cells 19.5–21 microns long. Microhair apical cell/total length ratio 0.43–0.54. *Stomata* common; 19–22.5 microns long. Subsidiaries mostly dome-shaped. Guard-cells overlapping to flush with the interstomatals. *Intercostal short-cells* common; not paired (solitary); not silicified. Intercostal silica bodies absent. Costal zones with short-cells. *Costal short-cells* predominantly paired. Costal silica bodies present throughout the costal zones; rounded to crescentic (predominantly more or less imperfect saddles, intergrading with the other forms); not sharp-pointed.

Transverse section of leaf blade, physiology. C_4; XyMS+. PCR sheath outlines uneven. PCR sheaths of the primary vascular bundles interrupted; interrupted abaxially only. PCR sheath extensions present. Maximum number of extension cells 1–2. Mesophyll with fusoids (in the sense that a PCR cell on either side of each bundle protrudes laterally, constituting a wing which approaches that of the adjacent bundle). The fusoids an integral part of the PBS. *Leaf blade* adaxially flat. *Midrib* conspicuous to not readily distinguishable (a slightly larger bundle and keel); with one bundle only. Bulliforms present in discrete, regular adaxial groups (between each bundle pair); in simple fans (the median cells deeply penetrating). All the vascular bundles accompanied by sclerenchyma. Combined sclerenchyma girders present (with all the bundles); forming 'figures' (all the bundles). Sclerenchyma all associated with vascular bundles. The lamina margins with fibres.

Special diagnostic feature. *The inflorescence a spicate 'raceme', with each spikelet subtended at its base by a tiny hyaline bract: Madagascar*.

Taxonomy. Chloridoideae; main chloridoid assemblage.

Ecology, geography, regional floristic distribution. 1 species. Madagascar. Xerophytic; species of open habitats.

Paleotropical. Madagascan.

References, etc. Leaf anatomical: this project.

Special comments. Fruit data wanting.

Vossia Wall. & Griff.

Habit, vegetative morphology. Aquatic perennial; rhizomatous. The flowering culms leafy. *Culms* 100–200 cm high (above the water, from floating culms up to 7 m long); herbaceous (propagating from stem fragments); to 1 cm in diameter. Culm leaves present. Upper culm leaf blades fully developed. Culm internodes solid. *Leaves* not basally aggregated; non-auriculate. *Leaf blades* linear to linear-lanceolate; *broad*; 6–25 mm wide (up to 1 m long); flat; without cross venation. Ligule present; *a fringed membrane*.

Reproductive organization. Plants bisexual, with bisexual spikelets; with hermaphrodite florets. *The spikelets* of sexually distinct forms on the same plant, or all alike in sexuality; hermaphrodite, or hermaphrodite and male-only; *homomorphic*; all in heterogamous combinations.

Inflorescence. *Inflorescence* of spicate main branches (rarely a single 'raceme'); usually *digitate*. Primary inflorescence branches several to many, rarely only one. Rach-

1002

ides hollowed. Inflorescence espatheate; not comprising 'partial inflorescences' and foliar organs. *Spikelet-bearing axes* 'racemes' (spiciform, subcylindrical or flattened, with 12 or more internodes); the spikelet-bearing axes with more than 10 spikelet-bearing 'articles' (12 or more); clustered; with substantial rachides; disarticulating (but rachis not very fragile); disarticulating at the joints. *'Articles'* non-linear (flattened dorsally, inflated above); not appendaged; disarticulating transversely; glabrous. Spikelets paired; not secund; distichous (in alternating pairs on the zigzag rachis); sessile and pedicellate; consistently in 'long-and-short' combinations; in pedicellate/sessile combinations. Pedicels of the 'pedicellate' spikelets free of the rachis. The 'shorter' spikelets hermaphrodite. The 'longer' spikelets hermaphrodite, or male-only.

Female-fertile spikelets. Spikelets 6–8 mm long; compressed dorsiventrally; planoconvex; falling with the glumes (and with the adjacent joint and pedicel). Rachilla terminated by a female-fertile floret. Hairy callus absent.

Glumes two; *more or less equal (when the apical convergence of the veins is considered, but the very long awn of the lower confuses the issue)*; (the longer) long relative to the adjacent lemmas; hairless (scabrid on margins and keels); pointed; *awned (or at least, the G_1 long-caudate, with a flat tail)*; carinate (G_2), or non-carinate (G_1); very dissimilar (the G_1 leathery, flat-backed, caudate-acuminate and 2-keeled, the G_2 thinner and naviculate). Lower glume two-keeled; flattened on the back; not pitted; relatively smooth; indistinctly many veined between the keels. Upper glume indistinctly several to many veined. *Spikelets* with incomplete florets. The incomplete florets proximal to the female-fertile florets. *The proximal incomplete florets* 1; paleate. Palea of the proximal incomplete florets fully developed. The proximal incomplete florets male. The proximal lemmas awnless; 2 nerved; more or less equalling the female-fertile lemmas; similar in texture to the female-fertile lemmas (hyaline); not becoming indurated.

Female-fertile florets 1. Lemmas less firm than the glumes (hyaline); not becoming indurated; entire; pointed; awnless; hairless; glabrous; non-carinate; without a germination flap; 3 nerved. *Palea* present; relatively long; tightly clasped by the lemma; entire; awnless, without apical setae; textured like the lemma; not indurated; 2-nerved; 2-keeled. *Lodicules* present; 2; free; fleshy; glabrous. *Stamens* 3. Anthers not penicillate; without an apically prolonged connective. *Ovary* glabrous. Stigmas 2.

Abaxial leaf blade epidermis. *Costal/intercostal zonation* conspicuous (in places). *Papillae* present (in places), or absent; where detected, intercostal. Intercostal papillae consisting of one oblique swelling per cell (where seen, at one end of the long-cells). *Long-cells* similar in shape costally and intercostally; of similar wall thickness costally and intercostally (of medium thickness). Mid-intercostal long-cells rectangular; having markedly sinuous walls. *Microhairs* present; elongated; clearly two-celled; panicoid-type. *Stomata* common. Subsidiaries non-papillate; dome-shaped to triangular. Guard-cells overlapping to flush with the interstomatals. *Intercostal short-cells* common; obscurely in cork/silica-cell pairs; silicified. Intercostal silica bodies tall-and-narrow (abundant, but small and seemingly poorly developed). With rows of large prickles on some of the costae. *Costal short-cells* predominantly paired. Costal silica bodies present and well developed to poorly developed; mostly tall-and-narrow (abundant, but mostly small and seemingly poorly developed).

Transverse section of leaf blade, physiology. C_4; XyMS–. PCR sheath outlines uneven. PCR cell chloroplasts centrifugal/peripheral. *Mesophyll* with radiate chlorenchyma. *Leaf blade* with distinct, prominent adaxial ribs to 'nodular' in section; with the ribs very irregular in sizes (emphatically so). *Midrib* very conspicuous; having a conventional arc of bundles (with a large median primary, a large primary bordering the midrib on each side, and small bundles in between); with colourless mesophyll adaxially (and an adaxial lignified hypodermal layer). *The lamina* distinctly asymmetrical on either side of the midrib. Bulliforms present in discrete, regular adaxial groups; in simple fans (these large, *Zea*-type, often situated over a minor bundle). All the vascular bundles accompanied by sclerenchyma. Combined sclerenchyma girders present (with the main bundles); sometimes forming 'figures' (slight I's only). Sclerenchyma all associated with vascular bundles.

Cytology. $2n = 20$.

Taxonomy. Panicoideae; Andropogonodae; Andropogoneae; Rottboelliinae.

Ecology, geography, regional floristic distribution. 1 species. Tropical Africa and Asia. Hydrophytic, or helophytic; species of open habitats; glycophytic. Swamps and river margins.

Paleotropical. African and Indomalesian. Sudano-Angolan, West African Rainforest, and Namib-Karoo. Indian and Indo-Chinese. Sahelo-Sudanian, Somalo-Ethiopian, South Tropical African, and Kalaharian.

References, etc. Leaf anatomical: this project, and photos provided by R.P. Ellis.

Special comments. Fruit data wanting.

Vulpia C. Gmelin

Chloammia Rafin., *Distomomischus* Dulac, *Festucaria* Link, *Loretia* Duval-Jouve, *Mygalurus* Link, *Narduretia* Villar, *Nardurus* (Bluff, Nees & Schauer) Reichenb., *Prosphysis* Dulac, *Zerna* Panzer
Excluding *Ctenopsis*

Habit, vegetative morphology. Annual, or perennial (rarely); caespitose. *Culms* 5–90 cm high; herbaceous; unbranched above. Culm nodes glabrous. Culm internodes solid, or hollow. Young shoots intravaginal. *Leaves* not basally aggregated; non-auriculate. Leaf blades linear; narrow; 0.5–3 mm wide; setaceous to not setaceous; flat, or rolled (convolute when dry); without cross venation; persistent; rolled in bud. *Ligule an unfringed membrane*; truncate, or not truncate; 0.2–6 mm long.

Reproductive organization. Plants bisexual, with bisexual spikelets; with hermaphrodite florets; inbreeding. Exposed-cleistogamous, or chasmogamous.

Inflorescence. *Inflorescence a single raceme (rarely), or paniculate*; open, or contracted; when contracted spicate, or more or less irregular; espatheate; not comprising 'partial inflorescences' and foliar organs. Spikelet-bearing axes persistent. *Spikelets* usually more or less *secund*; pedicellate.

Female-fertile spikelets. Spikelets 5–16 mm long; compressed laterally; disarticulating above the glumes (also, sometimes, at the base of the pedicel); disarticulating between the florets. Rachilla prolonged beyond the uppermost female-fertile floret; rachilla hairy, or hairless. The rachilla extension with incomplete florets. Hairy callus absent. *Callus* short.

Glumes two; very unequal; shorter than the spikelets; shorter than the adjacent lemmas, or long relative to the adjacent lemmas; free; dorsiventral to the rachis; pointed; awned (G_2, sometimes), or awnless; non-carinate; *very dissimilar (usually, the G_1 often minute, the G_2 acute to acuminate)*. Lower glume 0 nerved, or 1 nerved. Upper glume 1 nerved, or 3 nerved. *Spikelets* with incomplete florets. The incomplete florets distal to the female-fertile florets.

Female-fertile florets 2–15 (rarely only 1). Lemmas tapered; decidedly firmer than the glumes (chartaceous, with thin margins); not becoming indurated; entire; pointed; awned. Awns 1; median; apical; non-geniculate; much shorter than the body of the lemma to much longer than the body of the lemma; entered by one vein. Lemmas hairy, or hairless; carinate to non-carinate; 3–5 nerved. *Palea* present; relatively long; apically notched; awnless, without apical setae; not indurated; 2-nerved; 2-keeled. *Lodicules* present; 2; free; membranous; glabrous; toothed, or not toothed. **Stamens 1–2 (rarely 3)**. Anthers 0.3–5 mm long; not penicillate. *Ovary* glabrous, or hairy. Styles free to their bases. Stigmas 2.

Fruit, embryo and seedling. *Fruit* adhering to lemma and/or palea; small, or medium sized, or large; longitudinally grooved; compressed dorsiventrally; with hairs confined to a terminal tuft. Hilum long-linear. Embryo small; not waisted. Endosperm hard; with lipid; containing compound starch grains. Embryo with an epiblast.

Seedling with a tight coleoptile. First seedling leaf with a well-developed lamina. The lamina narrow; erect; 1–3 veined.

Abaxial leaf blade epidermis. *Costal/intercostal zonation* conspicuous. *Papillae* absent. *Long-cells* similar in shape costally and intercostally. Mid-intercostal long-cells rectangular, or fusiform; having markedly sinuous walls. *Microhairs* absent. *Stomata* absent or very rare, or common; when present, (36–)39–51(–54) microns long. Subsidiaries parallel-sided. Guard-cells overlapped by the interstomatals, or overlapping to flush with the interstomatals. *Intercostal short-cells* common; in cork/silica-cell pairs (mainly), or not paired (solitary); silicified (when paired). Costal zones with short-cells. *Costal short-cells* predominantly paired. Costal silica bodies horizontally-elongated crenate/sinuous, or horizontally-elongated smooth (a few), or rounded, or crescentic; not sharp-pointed.

Transverse section of leaf blade, physiology. C_3; XyMS+. *Mesophyll* with non-radiate chlorenchyma; without adaxial palisade. *Leaf blade* with distinct, prominent adaxial ribs; with the ribs more or less constant in size. *Midrib* conspicuous; with one bundle only. Bulliforms present in discrete, regular adaxial groups (at the bases of the furrows); in simple fans. All the vascular bundles accompanied by sclerenchyma. Combined sclerenchyma girders absent. Sclerenchyma all associated with vascular bundles.

Phytochemistry. Tissues of the culm bases with little or no starch.

Cytology. Chromosome base number, $x = 7$. $2n = 14$, 28, and 42. 2, 4, and 6 ploid.

Taxonomy. Pooideae; Poodae; Poeae.

Ecology, geography, regional floristic distribution. 23 species. Temperate. Commonly adventive. Mesophytic, or xerophytic; species of open habitats; halophytic (sometimes), or glycophytic. Dry places, including coastal sand.

Holarctic, Paleotropical, Neotropical, and Antarctic. Boreal, Tethyan, and Madrean. African. Euro-Siberian, Atlantic North American, and Rocky Mountains. Macaronesian, Mediterranean, and Irano-Turanian. Saharo-Sindian, Sudano-Angolan, and West African Rainforest. Caribbean and Pampas. Patagonian. European. Southern Atlantic North American and Central Grasslands. Sahelo-Sudanian, Somalo-Ethiopian, and South Tropical African.

Hybrids. Intergeneric hybrids with *Festuca* — ×*Festulpia* Melderis ex Stace & R. Cotton (several species involved).

Rusts and smuts. Rusts — *Puccinia*. Taxonomically wide-ranging species: *Puccinia graminis, Puccinia coronata, Puccinia brachypodii, Puccinia hordei, Puccinia recondita, 'Uromyces' fragilipes*, and '*Uromyces' dactylidis*. Smuts from Tilletiaceae and from Ustilaginaceae. Tilletiaceae — *Entyloma* and *Tilletia*. Ustilaginaceae — *Ustilago*.

Economic importance. Significant weed species: *V. bromoides, V. myuros*.

References, etc. Morphological/taxonomic: Cotton and Stace 1977, Stace 1981. Leaf anatomical: this project.

Vulpiella (Trabut) Burollet

Habit, vegetative morphology. Annual. *Culms* 10–40 cm high; herbaceous. Leaves non-auriculate. Leaf blades linear; narrow (?); without cross venation. *Ligule an unfringed membrane*; truncate; 2–4 mm long.

Reproductive organization. Plants bisexual, with bisexual spikelets; with hermaphrodite florets.

Inflorescence. *Inflorescence paniculate (2–10 cm long, with up to 4 branches at each node)*; open (the branches and pedicels pulvinate); *without conspicuously divaricate branchlets (by contrast with* **Cutandia***)*; espatheate; not comprising 'partial inflorescences' and foliar organs. *Spikelet-bearing axes disarticulating*; falling entire (branches disarticulating at the pulvini). Spikelets not secund; pedicellate.

Female-fertile spikelets. *Spikelets* 8–25(–40) mm long; compressed laterally; disarticulating above the glumes and falling with the glumes; *disarticulating between the florets (each lemma falling with the internode below, rather than the one above)*. Rachilla prolonged beyond the uppermost female-fertile floret. The rachilla extension with incomplete florets. Hairy callus absent.

Glumes two; very unequal; shorter than the adjacent lemmas; dorsiventral to the rachis; hairless (scabrid on the midribs); pointed (acuminate); shortly awned; non-carinate; similar. Lower glume 1 nerved. Upper glume 1–3 nerved. *Spikelets* with incomplete florets. The incomplete florets distal to the female-fertile florets.

Female-fertile florets *5–12(–18)*. Lemmas aculeolate, acuminate; similar in texture to the glumes; not becoming indurated (firm); entire; pointed; awned (the awn to 8 mm long). Awns 1; median; from a sinus, or apical; non-geniculate; much shorter than the body of the lemma to about as long as the body of the lemma; entered by one vein. Lemmas hairless; non-carinate; without a germination flap; 3 nerved. *Palea* present; relatively long; 2-nerved; 2-keeled. *Lodicules* present; 2; free; membranous; glabrous; toothed, or not toothed; not or scarcely vascularized. *Stamens* 3. *Ovary* glabrous. Styles free to their bases. Stigmas 2.

Fruit, embryo and seedling. *Fruit* adhering to lemma and/or palea (to the palea); small (3 mm long); longitudinally grooved; compressed dorsiventrally. Hilum short (oblong-linear). Embryo small. Endosperm hard; containing compound starch grains.

Abaxial leaf blade epidermis. *Costal/intercostal zonation* conspicuous. *Papillae* absent. *Long-cells* markedly different in shape costally and intercostally; of similar wall thickness costally and intercostally. Mid-intercostal long-cells fusiform; having straight or only gently undulating walls. *Microhairs* absent. *Stomata* common. Subsidiaries parallel-sided. Guard-cells overlapped by the interstomatals. *Intercostal short-cells* absent or very rare. Costal zones with short-cells. *Costal short-cells* conspicuously in long rows (mainly, sometimes paired). Costal silica bodies horizontally-elongated crenate/sinuous, or horizontally-elongated smooth (few); not sharp-pointed.

Transverse section of leaf blade, physiology. C_3. *Mesophyll* with non-radiate chlorenchyma; without adaxial palisade. *Leaf blade* with distinct, prominent adaxial ribs; with the ribs more or less constant in size. *Midrib* not readily distinguishable; with one bundle only. Bulliforms present in discrete, regular adaxial groups; in simple fans. Sclerenchyma all associated with vascular bundles.

Cytology. Chromosome base number, $x = 7$.

Taxonomy. Pooideae; Poodae; Poeae.

Ecology, geography, regional floristic distribution. 1 species. Mediterranean. Xerophytic; species of open habitats. Dry sandy places.

Holarctic. Boreal and Tethyan. Euro-Siberian. Mediterranean. European.

References, etc. Leaf anatomical: this project.

Wangenheimia Moench

Habit, vegetative morphology. Rather rigid *annual*. *Culms* 10–30 cm high; herbaceous. Leaves non-auriculate. Leaf blades linear; narrow; not setaceous; rolled (convolute); without cross venation. Ligule an unfringed membrane; not truncate; 1.5–2.5 mm long.

Reproductive organization. Plants bisexual, with bisexual spikelets; with hermaphrodite florets. The spikelets all alike in sexuality.

Inflorescence. *Inflorescence a single raceme (spike-like, the pedicels short)*; espatheate; not comprising 'partial inflorescences' and foliar organs. Spikelet-bearing axes persistent. Spikelets solitary; secund; subsessile to pedicellate (the pedicels 0.2 to 0.6 mm long); imbricate.

Female-fertile spikelets. Spikelets 3.5–8 mm long; compressed laterally; disarticulating above the glumes; tardily disarticulating between the florets. Rachilla prolonged beyond the uppermost female-fertile floret. The rachilla extension with incomplete florets. Hairy callus absent.

Glumes two (leathery, the midrib thickened); more or less equal; long relative to the adjacent lemmas; *displaced (the G_1 orientated parallel to the rachilla on the flattened side of the spikelet)*; pointed (acute); awnless; carinate; *very dissimilar (the lower linear, subulate, the upper lanceolate)*. Lower glume 1 nerved. Upper glume 2 nerved, or 3 nerved. *Spikelets* with incomplete florets. The incomplete florets distal to the female-fertile florets. *The distal incomplete florets* merely underdeveloped.

Female-fertile florets 3–11. Lemmas similar in texture to the glumes (leathery, the margins membranous); not becoming indurated; entire; pointed, or blunt (acuminate to obtuse-apiculate); awnless; hairy; more or less *carinate*; without a germination flap; 5 nerved. *Palea* present; relatively long; apically notched (shortly bifid); awnless, without apical setae; 2-nerved; 2-keeled. *Lodicules* present; 2; free; membranous; glabrous; toothed; not or scarcely vascularized. *Stamens* 3. Anthers 2–2.5 mm long. *Ovary* glabrous. Styles free to their bases. Stigmas 2.

Fruit, embryo and seedling. *Fruit* free from both lemma and palea; small (1.8–2.4 mm long); compressed dorsiventrally. *Hilum short*. Embryo small; with an epiblast; without a scutellar tail; with a negligible mesocotyl internode. Embryonic leaf margins meeting.

Abaxial leaf blade epidermis. *Costal/intercostal zonation* conspicuous (costal silica-bodies larger). *Papillae* absent. *Long-cells* similar in shape costally and intercostally; of similar wall thickness costally and intercostally. Mid-intercostal long-cells rectangular; having markedly sinuous walls (coarsely). *Microhairs* absent. *Stomata* absent or very rare. *Intercostal short-cells* common; nearly all in cork/silica-cell pairs; silicified. *Crown cells* present (over the veins). Costal zones with short-cells. *Costal short-cells* predominantly paired. Costal silica bodies horizontally-elongated crenate/sinuous (a few, short), or rounded (mostly, more or less irregular); not sharp-pointed.

Transverse section of leaf blade, physiology. C_3; XyMS+. *Mesophyll* with non-radiate chlorenchyma; without adaxial palisade. *Leaf blade* with distinct, prominent adaxial ribs; with the ribs more or less constant in size. *Midrib* not readily distinguishable; with one bundle only. Bulliforms present in discrete, regular adaxial groups; in simple fans. All the vascular bundles accompanied by sclerenchyma (but scanty sclerenchyma). Combined sclerenchyma girders absent (strands only). Sclerenchyma all associated with vascular bundles (save at the blade margins).

Cytology. Chromosome base number, $x = 7$. $2n = 14$. 2 ploid.

Taxonomy. Pooideae; Poodae; Poeae.

Ecology, geography, regional floristic distribution. 2 species. Spain, North Africa. Species of open habitats.

Holarctic. Boreal and Tethyan. Euro-Siberian. Mediterranean. European.

References, etc. Leaf anatomical: this project.

Whiteochloa C.E. Hubb.

Habit, vegetative morphology. *Annual (or short-lived perennials)*; caespitose to decumbent. *Culms* 30–105 cm high; herbaceous. Culm nodes hairy, or glabrous. Culm internodes solid. *Leaves* not basally aggregated; non-auriculate. Leaf blades narrow; 2–5.5 mm wide; loosely folded, or flat; without cross venation; persistent. Ligule a fringed membrane.

Reproductive organization. Plants bisexual, with bisexual spikelets; with hermaphrodite florets.

Inflorescence. *Inflorescence paniculate*; open (the branches solitary, usually distant); with capillary branchlets. Inflorescence with axes ending in spikelets (usually), or axes not ending in spikelets (the branches sometimes bristle-tipped in *W. cymbiformis*). Inflorescence espatheate; not comprising 'partial inflorescences' and foliar organs. Spikelet-bearing axes persistent. Spikelets solitary, or paired; secund to not secund; pedicellate. Pedicel apices discoid (rarely), or cupuliform. Spikelets not in distinct 'long-and-short' combinations (but often paired).

Female-fertile spikelets. *Spikelets* 2.25–4 mm long; oblong, or elliptic, or obovate; slightly *compressed laterally, or not noticeably compressed*; falling with the glumes; with distinctly elongated rachilla internodes between the florets. The upper floret conspicuously stipitate (shortly so). Rachilla terminated by a female-fertile floret. Hairy callus absent.

Glumes two; very unequal; (the upper) long relative to the adjacent lemmas; hairy (G_2 hispid or ciliate on the nerves, with tubercle-based hairs), or hairless; awnless (sometimes mucronate); *very dissimilar (membranous, the G_1 broadly ovate and enclosing the base of the spikelet, the G_2 larger, cymbiform, acuminate with a hardened apex)*. Lower glume 3–5 nerved. Upper glume 5–7 nerved (the nerves prominent). *Spikelets* with incomplete florets. The incomplete florets proximal to the female-fertile florets. *Spikelets with proximal incomplete florets*. The proximal incomplete florets 1; paleate. Palea of the proximal incomplete florets fully developed. The proximal incomplete florets male. The proximal lemmas awnless; 5 nerved; decidedly exceeding the female-fertile lemmas (with dorsal groove or depression); less firm than the female-fertile lemmas to similar in texture to the female-fertile lemmas (membranous to leathery); becoming indurated, or not becoming indurated.

Female-fertile florets 1. **Lemmas** decidedly firmer than the glumes; faintly but distictly rugose; becoming indurated (crustaceous); yellow in fruit; entire; pointed (acute or acuminate); *mucronate (with a slightly curved apiculum)*; hairy (apically puberulous), or hairless (glabrous); non-carinate; having the margins lying flat and exposed on the palea; 5 nerved. **Palea** present; relatively long; entire; awnless, without apical setae; *indurated (with indurated flaps)*; 2-nerved; 2-keeled. *Lodicules* present; free; fleshy; glabrous. *Stamens* 3. Anthers not penicillate. *Ovary* glabrous. Styles free to their bases. Stigmas 2; red pigmented.

Fruit, embryo and seedling. Fruit small (1.5–2.25 mm long); compressed dorsiventrally. Hilum short. Embryo large.

Abaxial leaf blade epidermis. *Costal/intercostal zonation* conspicuous. *Papillae* absent. *Long-cells* similar in shape costally and intercostally (but the intercostals much larger, and inflated); of similar wall thickness costally and intercostally. Mid-intercostal long-cells fusiform; having markedly sinuous walls, or having straight or only gently undulating walls. *Microhairs* present; panicoid-type; 54–60 microns long; 6.3–9 microns wide at the septum. Microhair total length/width at septum 6.7–7.6. Microhair apical cells 36–39 microns long. Microhair apical cell/total length ratio 0.63–0.69. *Stomata* common; 33–36 microns long. Subsidiaries triangular. Guard-cells overlapping to flush with the interstomatals. *Intercostal short-cells* absent or very rare; not paired (solitary); not silicified. Costal zones with short-cells. *Costal short-cells* conspicuously in long rows. Costal silica bodies 'panicoid-type'; not sharp-pointed.

Transverse section of leaf blade, physiology. C_4; XyMS–. PCR sheath outlines uneven. PCR sheath extensions absent. *Mesophyll* with radiate chlorenchyma; traversed by columns of colourless mesophyll cells. *Leaf blade* adaxially flat. *Midrib* conspicuous; having a conventional arc of bundles; with colourless mesophyll adaxially. Bulliforms present in discrete, regular adaxial groups, or not in discrete, regular adaxial groups (then the epidermis irregularly bulliform); when clearly grouped, in simple fans (these large-celled, *Zea*-type). Many of the smallest vascular bundles unaccompanied by sclerenchyma. Combined sclerenchyma girders present; nowhere forming 'figures'. Sclerenchyma all associated with vascular bundles.

Taxonomy. Panicoideae; Panicodae; Paniceae.

Ecology, geography, regional floristic distribution. 5 species. Australia and Moluccas. Species of open habitats. Sandy alluvial soils in savanna.

Australian. North and East Australian. Tropical North and East Australian.

References, etc. Leaf anatomical: this project.

Willkommia Hackel

Willbleibia Herter

Habit, vegetative morphology. Annual, or perennial; stoloniferous, or caespitose. *Culms* 20–40 cm high; herbaceous; unbranched above. Culm nodes glabrous. Culm internodes hollow. Plants unarmed. *Leaves* not basally aggregated; non-auriculate. Leaf blades linear; narrow; 2–4 mm wide; flat; exhibiting multicellular glands abaxially. The abaxial glands on the blade margins and intercostal. Leaf blades without cross venation; persistent. *Ligule a fringe of hairs.* Contra-ligule absent.

Reproductive organization. Plants bisexual, with bisexual spikelets; with hermaphrodite florets. Exposed-cleistogamous, or chasmogamous (?).

Inflorescence. *Inflorescence of spicate main branches*; non-digitate (the spikes scattered along a central axis). Primary inflorescence branches 3–8. Rachides flattened and winged. Inflorescence espatheate; not comprising 'partial inflorescences' and foliar organs. Spikelet-bearing axes persistent. Spikelets secund (on one side of the rachis); biseriate; very shortly pedicellate.

Female-fertile spikelets. *Spikelets* about 4 mm long; adaxial; *compressed dorsiventrally*; disarticulating above the glumes; with a distinctly elongated rachilla internode between the glumes and with a distinctly elongated rachilla internode above the glumes. Rachilla terminated by a female-fertile floret. Hairy callus absent. Callus pointed.

Glumes two; very unequal (G_1 about two-thirds the length of G_2); (the upper) long relative to the adjacent lemmas (slightly exceeding the spikelet); dorsiventral to the rachis; hairless (glabrous to scabrid); pointed (somewhat, in G_2), or not pointed (blunt to slightly notched, G_1); awnless; non-carinate (rounded); *similar (thin, the G_1 flimsier)*. *Lower glume 0 nerved.* Upper glume 1 nerved. **Spikelets** with female-fertile florets only; *without proximal incomplete florets*.

Female-fertile florets 1. Lemmas less firm than the glumes to similar in texture to the glumes; not becoming indurated; entire; pointed; awnless (but acuminate), or mucronate to awned. Awns when present, 1; apical; non-geniculate; hairless; much shorter than the body of the lemma; entered by one vein. Lemmas hairy; non-carinate (but the median nerve slightly prominent); without a germination flap; 3 nerved. *Palea* present; relatively long (glabrous or silky-hairy); entire to apically notched (minutely trilobed); awnless, without apical setae; not indurated (thinly membranous, flimsy); 2-nerved; 2-keeled. *Lodicules* present; 2; free; fleshy; glabrous; not or scarcely vascularized. **Stamens 3**. Anthers relatively long; not penicillate; without an apically prolonged connective. *Ovary* glabrous. Styles free to their bases. Stigmas 2.

Fruit, embryo and seedling. Fruit ellipsoid. Hilum short. Embryo large; with an epiblast; with a scutellar tail; with an elongated mesocotyl internode. Embryonic leaf margins meeting.

Abaxial leaf blade epidermis. *Costal/intercostal zonation* conspicuous. *Papillae* present; intercostal. Intercostal papillae not over-arching the stomata (or scarcely so); consisting of one oblique swelling per cell, or consisting of one symmetrical projection per cell. *Long-cells* markedly different in shape costally and intercostally (costals much longer, much narrower). Intercostal zones exhibiting many atypical long-cells to without typical long-cells (the long-cells short to very short). Mid-intercostal long-cells rectangular; having markedly sinuous walls to having straight or only gently undulating walls. *Microhairs* present; more or less spherical; clearly two-celled; chloridoid-type (but the apical cell thin-walled). Microhair apical cell wall thinner than that of the basal cell and often collapsed. Microhairs 27–30 microns long. Microhair basal cells 15 microns long. Microhairs 17.4–19.5 microns wide at the septum. Microhair total length/width at septum 1.5–1.7. Microhair apical cells 9–12 microns long. Microhair apical cell/total length ratio 0.3–0.42. *Stomata* common; 18–21 microns long. Subsidiaries dome-shaped and triangular. Guard-cells overlapping to flush with the interstomatals. *Intercostal short-cells* absent or very rare. Intercostal silica bodies absent. Costal zones with short-cells. *Costal short-cells* conspicuously in long rows. Costal silica bodies present in alternate cell files of the costal zones; predominantly saddle shaped; not sharp-pointed.

Transverse section of leaf blade, physiology. C_4; XyMS+. PCR sheaths of the primary vascular bundles interrupted; interrupted abaxially only. PCR sheath extensions absent. *Leaf blade* 'nodular' in section to adaxially flat (with slight abaxial ribs). *Midrib* conspicuous (via a larger bundle and sclerenchyma mass); with one bundle only. Bulliforms present in discrete, regular adaxial groups; in simple fans (mostly), or associated with colourless mesophyll cells to form deeply-penetrating fans (few of these). All the vascular bundles accompanied by sclerenchyma. Combined sclerenchyma girders present (with the large bundles); forming 'figures' (the large bundles). Sclerenchyma all associated with vascular bundles. The lamina margins with fibres.

Taxonomy. Chloridoideae; main chloridoid assemblage.

Ecology, geography, regional floristic distribution. 2 species. 1 in southern U.S.A., 3 in southern Africa. Xerophytic. Open habitats in savanna, usually halophytic.

Holarctic and Neotropical. Boreal. Atlantic North American. Pampas. Southern Atlantic North American.

References, etc. Leaf anatomical: this project.

Special comments. Fruit data wanting.

Xerochloa R.Br.

Kerinozoma Steud.

Habit, vegetative morphology. Annual (rarely), or perennial; caespitose (often rush-like), or decumbent. Culms woody and persistent, or herbaceous. Culm nodes glabrous. Culm internodes solid. *Leaves* not basally aggregated; non-auriculate. Leaf blades narrow; setaceous, or not setaceous; flat, or rolled (involute); without cross venation; persistent. Ligule a fringed membrane (very reduced).

Reproductive organization. Plants bisexual, with bisexual spikelets; with hermaphrodite florets, or without hermaphrodite florets (the upper floret of 'perfect' spikelets sometimes (?) being female, with 2–3 staminodes). The spikelets of sexually distinct forms on the same plant (heteromorphous, with some of the spikelets 'reduced'); hermaphrodite and sterile, or hermaphrodite, female-only, and sterile.

Inflorescence. *Inflorescence of globose, spathaceous units, fascicled and interspersed with prophylls, the fascicled units solitary and terminating the culm, or arranged in a leafy panicle.* Inflorescence axes not ending in spikelets (the rachis produced beyond the uppermost spikelet). Inflorescence spatheate; a complex of 'partial inflorescences' and intervening foliar organs. **Spikelet-bearing axes** very much reduced (to very short spikes, with tough rachides); *clustered (with groups of 3–5 in a spathe)*; disarticulating; falling entire, or disarticulating at the joints (the ultimate units breaking up, or falling entire). Spikelets solitary; not secund; sessile; not in distinct 'long-and-short' combinations.

Female-sterile spikelets. The literature suggests that some spikelets are variously 'reduced'.

Female-fertile spikelets. Spikelets 4.8–12 mm long; oblong, or elliptic, or lanceolate, or ovate, or obovate; abaxial; compressed laterally to compressed dorsiventrally; falling with the glumes (with the spikelet bearing axes falling, and a second zone of disarticulation beneath the glumes); with conventional internode spacings. Rachilla terminated by a female-fertile floret. Hairy callus absent.

Glumes one per spikelet, or two; very unequal; shorter than the adjacent lemmas; hairy, or hairless; awnless; non-carinate. Lower glume 0–1 nerved. Upper glume 2–5 nerved. *Spikelets* with incomplete florets. The incomplete florets proximal to the female-fertile florets. *Spikelets with proximal incomplete florets. The proximal incomplete florets* 1; paleate. Palea of the proximal incomplete florets fully developed. The proximal incomplete florets male, or sterile. The proximal lemmas awnless; 3–9 nerved; more or less equalling the female-fertile lemmas to decidedly exceeding the female-fertile lemmas; similar in texture to the female-fertile lemmas (membranous to cartilaginous, with a hyaline area at the base).

Female-fertile florets 1. Lemmas acuminate; decidedly firmer than the glumes (membranous to cartilaginous); smooth to rugose (or muricate); becoming indurated to not becoming indurated; white in fruit, or yellow in fruit, or brown in fruit; entire; awnless; hairless; glabrous; non-carinate; without a germination flap; 2 nerved. *Palea* present; textured like the lemma; 2-nerved. **Lodicules** *absent (or vestigial)*. Stamens 3, or 0 (or staminodal?). Anthers not penicillate. *Ovary* glabrous. Styles fused. Stigmas 2; red pigmented.

Fruit, embryo and seedling. Fruit compressed dorsiventrally. Hilum short. Embryo large. Endosperm containing only simple starch grains. Embryo without an epiblast; with a scutellar tail; with an elongated mesocotyl internode.

Abaxial leaf blade epidermis. *Costal/intercostal zonation* conspicuous. *Papillae* absent. Long-cells of similar wall thickness costally and intercostally. Mid-intercostal long-cells rectangular, or fusiform; having markedly sinuous walls. *Microhairs* present (often arising from specialized epidermal cells); panicoid-type; (48–)51–72(–78) microns long; (10.5–)11.4–14.4(–15) microns wide at the septum. Microhair total length/width at septum 3.6–6.5. Microhair apical cells (27–)30–51(–57) microns long. Microhair apical cell/total length ratio 0.53–0.75. *Stomata* common; (30–)33–51 microns long. Subsidiaries dome-shaped. Guard-cells overlapping to flush with the interstomatals. *Intercostal short-cells* common, or absent or very rare; in cork/silica-cell pairs (usually); silicified. Costal zones with short-cells. *Costal short-cells* neither distinctly grouped into long rows nor predominantly paired. Costal silica bodies rounded (predominating), or 'panicoid-type'; not sharp-pointed.

Transverse section of leaf blade, physiology. *Leaf blades* almost entirely *consisting of midrib (obviously so in* **X. barbata** *and* **X. imberbis,** *less obviously so in* **X. laniflora** *where the quite broad 'midrib' is recognisable as such by the reduced lateral flanges and with reference the other species).* C_4; XyMS–. PCR sheath outlines uneven. PCR sheath extensions absent. *Mesophyll* with radiate chlorenchyma; traversed by columns of colourless mesophyll cells, or not traversed by colourless columns. Midrib (i.e. the bulk of the blade) having a conventional arc of bundles; with colourless mesophyll adaxially. Bulliforms not in discrete, regular adaxial groups (the adaxial epidermis being largely bulliform); associating with colourless mesophyll cells to form arches over small vascular bundles. Many of the smallest vascular bundles unaccompanied by sclerenchyma, or all the vascular bundles accompanied by sclerenchyma. Combined sclerenchyma girders absent. Sclerenchyma all associated with vascular bundles.

Special diagnostic feature. Rush-like, with reduced leaf blades, or not rush-like.

Taxonomy. Panicoideae; Panicodae; Paniceae.

Ecology, geography, regional floristic distribution. 4 species. Siam, Java, Australia. Xerophytic (extreme); species of open habitats. Dry stream beds and clay flats. Paleotropical and Australian. Indomalesian. Malesian. North and East Australian. Tropical North and East Australian.

References, etc. Morphological/taxonomic: Webster 1987. Leaf anatomical: this project.

Yakirra Lazarides & R. Webster

Ichnanthus australiensis and relatives

Habit, vegetative morphology. Annual, or perennial; caespitose. **Culms** 8–75 cm high; *herbaceous*; branched above, or unbranched above. Culm nodes hairy, or glabrous. Culm internodes solid, or hollow. *Leaves* not basally aggregated; non-auriculate. Leaf blades linear to lanceolate; narrow; 1.5–7 mm wide (3–30 cm long); flat, or rolled; without cross venation. Ligule present; *a fringe of hairs.*

Reproductive organization. *Plants bisexual, with bisexual spikelets*; with hermaphrodite florets.

Inflorescence. *Inflorescence paniculate (rarely, reduced to racemes — e.g. in* **Y. australiensis***)*; open, or contracted; espatheate; not comprising 'partial inflorescences' and foliar organs. Spikelet-bearing axes persistent. Spikelets solitary; not secund; pedicellate. *Pedicel apices cupuliform.*

Female-fertile spikelets. *Spikelets* 3.2–6.8 mm long; oblong, or elliptic, or obovate, or oblanceolate; adaxial (when appressed); *compressed dorsiventrally*; falling with the glumes; *with a distinctly elongated rachilla internode between the glumes and with distinctly elongated rachilla internodes between the florets (the latter being a straight, swollen elaiosome).* The upper floret conspicuously stipitate. The stipe beneath the upper floret not filiform; straight and swollen; homogeneous. Rachilla terminated by a female-fertile floret. Hairy callus absent.

Glumes two; very unequal; (the upper) long relative to the adjacent lemmas; hairless (glabrous, save for adaxial hairs at the apex of G_2); pointed (acuminate to acute, ovate, elliptic to oblanceolate); awnless; non-carinate; membranous. Lower glume 3–9 nerved. Upper glume 5–9 nerved. *Spikelets* with incomplete florets. The incomplete florets proximal to the female-fertile florets. *Spikelets with proximal incomplete florets. The proximal incomplete florets* 1; paleate. Palea of the proximal incomplete florets fully developed to reduced. The proximal incomplete florets sterile. *The proximal lemmas* awnless; *5–9 nerved (elliptic to oblanceolate, acute to acuminate)*; longer than the fertile lemmas; *less firm than the female-fertile lemmas (membranous)*; not becoming indurated.

Female-fertile florets 1. *Lemmas* acute; decidedly firmer than the glumes; smooth; becoming indurated; yellow in fruit, or brown in fruit; entire; pointed; awnless; hairless; glabrous; non-carinate (dorsally rounded); *having the margins tucked in onto the palea*; with a clear germination flap. *Palea* present; relatively long; entire; awnless, without apical setae; indurated; 2-nerved. *Lodicules* present; 2; fleshy; glabrous. *Stamens* 3. *Ovary* glabrous. Styles free to their bases. Stigmas 2; red pigmented.

Fruit, embryo and seedling. Fruit small; compressed dorsiventrally. *Hilum short*. Embryo large.

Abaxial leaf blade epidermis. *Costal/intercostal zonation* conspicuous. *Papillae* absent. Mid-intercostal long-cells having markedly sinuous walls. *Microhairs* present; panicoid-type; (39–)42–51 microns long; (4.5–)5.1–7.5 microns wide at the septum. Microhair total length/width at septum 5.9–10.7. Microhair apical cells (18–)19.5–28.5(–30) microns long. Microhair apical cell/total length ratio 0.46–0.62. *Stomata* common; (33–)36–48(–51) microns long. Subsidiaries dome-shaped and triangular. Guard-cells overlapping to flush with the interstomatals. *Intercostal short-cells* common. Costal zones with short-cells. *Costal short-cells* conspicuously in long rows. Costal silica bodies 'panicoid-type'; not sharp-pointed.

Transverse section of leaf blade, physiology. *C₄*; XyMS+. PCR sheath outlines even. PCR sheath extensions present, or absent. Maximum number of extension cells when present, 1. PCR cell chloroplasts centripetal. *Mesophyll* with radiate chlorenchyma. *Leaf blade* 'nodular' in section. *Midrib* not readily distinguishable; with one bundle only. Bulliforms present in discrete, regular adaxial groups; in simple fans. All the vascular bundles accompanied by sclerenchyma. Combined sclerenchyma girders present; nowhere forming 'figures'. Sclerenchyma all associated with vascular bundles.

Taxonomy. Panicoideae; Panicodae; Paniceae.

Ecology, geography, regional floristic distribution. 7 species. Australia, Burma. Mesophytic to xerophytic.

Paleotropical and Australian. Indomalesian. Indo-Chinese. North and East Australian and Central Australian. Tropical North and East Australian.

References, etc. Morphological/taxonomic: Lazarides and Webster 1984. Leaf anatomical: this project.

Ystia P. Compère

~ *Schizachyrium (S. kwiluense)*

Habit, vegetative morphology. Perennial. *Culms* about 50–150 cm high; herbaceous; branched above. *Leaves* not basally aggregated. Leaf blades narrowly linear; narrow; without cross venation. *Ligule an unfringed membrane*; short.

Reproductive organization. Plants bisexual, with bisexual spikelets; with hermaphrodite florets. *The spikelets all alike in sexuality*; overtly heteromorphic (the pedicellate spikelets with awns reduced or lacking).

Inflorescence. Inflorescence paniculate (a false panicle of small 'racemes'); spatheate; *a complex of 'partial inflorescences' and intervening foliar organs*. Spikelet-bearing axes 'racemes'; the spikelet-bearing axes with 6–10 spikelet-bearing 'articles' (with 6–7 spikelet pairs); solitary (in the spatheoles); with very slender rachides (flexuose); disarticulating; disarticulating at the joints. 'Articles' scabrid. *Spikelets* paired and in triplets (with terminal triad); sessile and pedicellate; *consistently in 'long-and-short' combinations*; in pedicellate/sessile combinations. Pedicels of the 'pedicellate' spikelets free of the rachis. The 'shorter' spikelets hermaphrodite. The 'longer' spikelets hermaphrodite.

Female-fertile spikelets. Spikelets 4–5 mm long; falling with the glumes. Rachilla terminated by a female-fertile floret. Hairy callus present. *Callus* short; blunt.

Glumes present; two; long relative to the adjacent lemmas; hairless (glabrous below, scabrid above); awnless; very dissimilar (the G₁ papery with involute margins and apically bidentate, the G₂ thinner, naviculate, shortly subulate or acuminate). Lower glume not two-keeled (except at the tip); not pitted; relatively smooth; (5–)7(–9) nerved. Upper glume 3 nerved. *Spikelets* with incomplete florets. The incomplete florets proximal to the female-fertile florets. *The proximal incomplete florets* 1; epaleate (reduced to a hyaline lemma); sterile. The proximal lemmas awnless.

Female-fertile florets 1. Lemmas less firm than the glumes (hyaline); not becoming indurated; incised; deeply cleft (to about a third); awned (in the sessile spikelet), or awnless to mucronate (in the pedicellate spikelet). Awns 1; median; from a sinus; geniculate; shortly hairy, or hairless (scabrid); about as long as the body of the lemma to much longer than the body of the lemma (4–5 mm long, in the sessile spikelet). Lemmas non-carinate; without a germination flap. **Palea** *absent*. *Lodicules* present; 2; fleshy. *Stamens* 3. Stigmas 2.

Taxonomy. Panicoideae; Andropogonodae; Andropogoneae; Andropogoninae.
Ecology, geography, regional floristic distribution. 1 species. Tropical Africa. Paleotropical. African. Sudano-Angolan and West African Rainforest. Somalo-Ethiopian and South Tropical African.
References, etc. Morphological/taxonomic: Compère 1963.
Special comments. Fruit data wanting. Anatomical data wanting.

Yushania Keng

~ Sinarundinaria

Habit, vegetative morphology. Perennial; rhizomatous. The flowering culms leafy. *Culms* 100–400 cm high; woody and persistent; to 2 cm in diameter; branched above. *Primary branches/mid-culm node 3*. Culm sheaths persistent. Culm internodes hollow. Pluricaespitose. Rhizomes metamorph type II. Plants unarmed. **Leaves** not basally aggregated; auricles inconspicuous; *with auricular setae (short brown bristles)*. Leaf blades lanceolate; broad, or narrow; 5–13 mm wide (4–18 cm long); pseudopetiolate; cross veined; disarticulating from the sheaths; rolled in bud. *Ligule* present. *Contra-ligule* present.

Reproductive organization. Plants bisexual, with bisexual spikelets; with hermaphrodite florets. Not viviparous.

Inflorescence. Inflorescence paniculate (with few spikelets per cluster); *spatheate*; a complex of 'partial inflorescences' and intervening foliar organs. *Spikelet-bearing axes* spikelike, or paniculate; persistent. Spikelets not secund; pedicellate.

Female-fertile spikelets. Spikelets compressed laterally; disarticulating above the glumes; disarticulating between the florets. Rachilla prolonged beyond the uppermost female-fertile floret; rachilla hairy (at the swollen tips of the joints). The rachilla extension with incomplete florets. Hairy callus absent.

Glumes two; shorter than the adjacent lemmas; pointed; awnless (acute to acuminate-tipped); carinate; similar. *Spikelets* with incomplete florets. The incomplete florets distal to the female-fertile florets. *The distal incomplete florets* merely underdeveloped.

Female-fertile florets 2–7. Lemmas similar in texture to the glumes; not becoming indurated; entire; pointed; mucronate to awned. Awns 1; median; apical; non-geniculate; much shorter than the body of the lemma. Lemmas carinate; 7 nerved, or 9 nerved. *Palea* present; relatively long; apically notched (acutely bifid); not indurated; several nerved (2 between the keels, 2 between each keel and the margin); 2-keeled. **Lodicules** present; 3; *joined*; membranous; ciliate; not toothed; heavily vascularized. **Stamens** *3*. Anthers not penicillate; without an apically prolonged connective. *Ovary* glabrous; without a conspicuous apical appendage. Styles fused. Stigmas 2.

Fruit, embryo and seedling. Fruit longitudinally grooved; compressed dorsiventrally. Hilum long-linear. Embryo small.

Taxonomy. Bambusoideae; Bambusodae; Bambuseae.
Ecology, geography, regional floristic distribution. 1 species. China including Taiwan. Paleotropical. Indomalesian. Indo-Chinese and Malesian.
Special comments. See Clayton and Renvoize (1986) and Soderstrom and Ellis (1987) for very different generic interpretations of species in this circle of affinity: there are no available generic descriptions adequate for the present purpose. Anatomical data wanting.

Yvesia A. Camus

Habit, vegetative morphology. Slender *annual*. The flowering culms leafless. *Culms* 15–22 cm high; herbaceous; unbranched above. Culm nodes exposed; glabrous. *Leaves* mostly basal (almost rosette-forming); non-auriculate. Sheaths glabrous, striate. Leaf blades lanceolate to ovate-lanceolate; narrow (but relatively broad); 2–4 mm wide; 'basi rotundata'; flat (rigid, sub-erect); without cross venation; persistent. Ligule a fringed membrane; short. *Contra-ligule* absent.

Reproductive organization. Plants bisexual, with bisexual spikelets; with hermaphrodite florets.

Inflorescence. Inflorescence paniculate (delicate); open; with capillary branchlets; espatheate; not comprising 'partial inflorescences' and foliar organs. Spikelet-bearing axes persistent. Spikelets solitary; secund to not secund (the branches indistinctly unilateral); shortly to long pedicellate. Pedicel apices cupuliform.

Female-fertile spikelets. Spikelets about 2.5 mm long; lanceolate; adaxial (insofar as orientation detectable); slightly compressed dorsiventrally; biconvex; falling with the glumes. Rachilla terminated by a female-fertile floret. *Hairy callus present (with long, dense white hairs, to about 0.5 mm long).* Callus blunt.

Glumes one per spikelet; (the upper, only glume) about equalling the spikelets; long relative to the adjacent lemmas; conspicuously hairy (with long, white pubescence); pointed (acute); awnless; non-carinate. Upper glume 3 nerved. *Spikelets with incomplete florets. The incomplete florets proximal to the female-fertile florets. Spikelets with proximal incomplete florets. The proximal incomplete florets* 1; epaleate; sterile. The proximal lemmas conspicuously pubescent, similar to the glume; awnless; 5 nerved (the laterals inconspicuous); decidedly exceeding the female-fertile lemmas; less firm than the female-fertile lemmas; not becoming indurated.

Female-fertile florets 1. *Lemmas* obovate-oblong; decidedly firmer than the glumes (becoming cartilaginous, distinctly hardened); striate; entire; *mucronate (or at least, mucronulate)*; hairless; glabrous; non-carinate; having the margins tucked in onto the palea; with a clear germination flap; 5 nerved; with the nerves confluent towards the tip. *Palea* present; relatively long; tightly clasped by the lemma; entire; awnless, without apical setae; textured like the lemma (and glabrous); not indurated; 2-nerved; keel-less. Palea back glabrous. *Lodicules* present; 2; free; fleshy; glabrous; not or scarcely vascularized. *Stamens* 3. *Ovary* glabrous. Styles free to their bases. Stigmas 2; red pigmented.

Fruit, embryo and seedling. *Fruit* free from both lemma and palea (but enclosed); small (1.2 mm long); yellow; not noticeably compressed; glabrous. Hilum short. Embryo large; waisted. Endosperm hard.

Abaxial leaf blade epidermis. *Costal/intercostal zonation* conspicuous. *Papillae* absent. *Long-cells* markedly different in shape costally and intercostally (the costals much narrower); of similar wall thickness costally and intercostally (fairly thin walled). Intercostal zones with typical long-cells, or exhibiting many atypical long-cells (often short to almost isodiametric). Mid-intercostal long-cells rectangular and fusiform; having markedly sinuous walls and having straight or only gently undulating walls. *Microhairs* present; panicoid-type (only the bases seen). *Stomata* common. Subsidiaries non-papillate; low dome-shaped and triangular. Guard-cells overlapping to flush with the interstomatals. *Intercostal short-cells* fairly common; in cork/silica-cell pairs; silicified. Intercostal silica bodies cross-shaped and oryzoid-type. No microhairs or prickles seen. Costal zones with short-cells. *Costal short-cells* conspicuously in long rows. Costal silica bodies present and well developed; 'panicoid-type'; cross shaped, butterfly shaped, dumb-bell shaped, and nodular.

Transverse section of leaf blade, physiology. C_4; XyMS+. PCR sheath outlines even. PCR sheath extensions absent. PCR cell chloroplasts seemingly centripetal (fairly clearly so, in the poorish material seen). *Leaf blade* adaxially flat. *Midrib* not readily distinguishable; with one bundle only. *The lamina* symmetrical on either side of the midrib. Bulliforms present in discrete, regular adaxial groups; in simple fans (their median cells quite large and deeply penetrating, separating all the bundles). Many of the smallest vascular bundles unaccompanied by sclerenchyma (or with very little, some seemingly with none). Combined sclerenchyma girders present; forming 'figures' (the sclerenchyma quite scanty, forming slight I's with larger bundles). Sclerenchyma all associated with vascular bundles.

Taxonomy. Panicoideae; Panicodae; Paniceae.

Ecology, geography, regional floristic distribution. 1 species. Madagascar. Streamside rocks.
Paleotropical. Madagascan.

References, etc. Morphological/taxonomic: Camus 1927. *Bull. Soc. Bot. Fr.* **73**, 687–690. Leaf anatomical: this project.

Zea L.

Mays Miller, *Mayzea* Raf., *Reana* Brignoli, *Thalysia* Kuntze. = *Zea mays* subsp. *mays* Excluding *Euchlaena*, Teosinte (see comments)

Habit, vegetative morphology. Robust annual. *Culms* 200–450 cm high; herbaceous. Culm nodes glabrous. Culm internodes solid. *Leaves* not basally aggregated; non-auriculate. Leaf blades broad; flat; without cross venation; persistent; rolled in bud. Ligule a fringed membrane.

Reproductive organization. Plants monoecious with all the fertile spikelets unisexual; without hermaphrodite florets. The spikelets of sexually distinct forms on the same plant (male and female); female-only and male-only. The male and female-fertile spikelets in different inflorescences. The spikelets overtly heteromorphic. Plants outbreeding.

Inflorescence. *Inflorescence peculiar: the female axillary, comprising a stout, spicate spadix with spikelets in few to several longitudinal rows, terminating in a tuft of long pendulous styles ('silks'); the male spikelets in terminal panicles of spiciform 'tassels'*; spatheate; not comprising 'partial inflorescences' and foliar organs. Spikelet-bearing axes persistent. Spikelets paired; not secund; consistently in 'long-and-short' combinations (male), or not in distinct 'long-and-short' combinations (female).

Female-sterile spikelets. Male spikelets in pairs, two-flowered, with many-nerved, membranous glumes. Lemmas and paleas hyaline, the florets with three stamens and two cuneate lodicules. Rachilla of male spikelets terminated by a male floret. The male spikelets with glumes (two); without proximal incomplete florets; 2 floreted. The lemmas awnless. Male florets 2; 3 staminate.

Female-fertile spikelets. Spikelets not noticeably compressed to compressed dorsiventrally; not disarticulating. Rachilla terminated by a female-fertile floret. Hairy callus absent.

Glumes two; more or less equal; long relative to the adjacent lemmas; not pointed; awnless; similar (basally fleshy, hyaline above). Lower glume not pitted; relatively smooth; 0 nerved. Upper glume 0 nerved. *Spikelets* with incomplete florets. The incomplete florets proximal to the female-fertile florets. *The proximal incomplete florets* 1; paleate, or epaleate. Palea of the proximal incomplete florets when present, fully developed to reduced. The proximal incomplete florets sterile. The proximal lemmas awnless; 0 nerved; more or less equalling the female-fertile lemmas to decidedly exceeding the female-fertile lemmas; similar in texture to the female-fertile lemmas (hyaline); not becoming indurated.

Female-fertile florets 1. Lemmas less firm than the glumes; not becoming indurated; awnless; hairless; non-carinate; 3 nerved. *Palea* present; relatively long; apically notched; with apical setae; not indurated (papery at base, hyaline above). *Lodicules* absent. *Stamens* 0. *Ovary* glabrous. Styles fused (fused nearly to tip). Stigmas 2 (at tip of style); white to red pigmented (or green).

Fruit, embryo and seedling. Fruit medium sized; compressed dorsiventrally, or not noticeably compressed. Hilum short. Embryo large. Endosperm hard; without lipid; containing only simple starch grains. Embryo without an epiblast; with a scutellar tail; with an elongated mesocotyl internode. Embryonic leaf margins overlapping.

Seedling with a long mesocotyl. First seedling leaf with a well-developed lamina. The lamina broad; curved.

Abaxial leaf blade epidermis. *Papillae* absent. Mid-intercostal long-cells having markedly sinuous walls. *Microhairs* present; panicoid-type; without 'partitioning membranes' *(Zea mays)*; (39–)42–51(–55) microns long; (8.4–)9.6–11.4(–12) microns wide at the septum. Microhair total length/width at septum 3.4–5.4. Microhair apical cells (21–)32–38 microns long. Microhair apical cell/total length ratio 0.5–0.73. *Stomata* common; (39–)42–48(–60) microns long. Subsidiaries triangular. Guard-cells overlapping to flush with the interstomatals. *Intercostal short-cells* common; silicified. Intercostal silica bodies vertically elongated-nodular, or cross-shaped. Costal zones with short-cells. *Costal short-cells* conspicuously in long rows, or neither distinctly grouped into long rows nor predominantly paired. Costal silica bodies 'panicoid-type'; cross shaped; not sharp-pointed.

Transverse section of leaf blade, physiology. C_4; biochemical type NADP–ME; XyMS–. PCR sheath outlines uneven. PCR cells with a suberised lamella. PCR cell chloroplasts with reduced grana; centrifugal/peripheral. *Mesophyll* with radiate chlorenchyma. *Leaf blade* adaxially flat. *Midrib* conspicuous; having a conventional arc of bundles; with colourless mesophyll adaxially. Bulliforms present in discrete, regular adaxial groups; in simple fans (the groups large-celled — '*Zea*-type'). Many of the smallest vascular bundles unaccompanied by sclerenchyma. Combined sclerenchyma girders present; nowhere forming 'figures'. Sclerenchyma all associated with vascular bundles.

Culm anatomy. *Culm internode bundles* scattered.

Phytochemistry. Leaves containing flavonoid sulphates (*Z. mays*). Leaf blade chlorophyll *a:b* ratio 4.37.

Special diagnostic feature. *Fruiting inflorescence a massive, axillary, spatheate cob with a spongy axis, the fruits in many rows.*

Cytology. Chromosome base number, $x = 10$. $2n = 20$. 2 ploid. Haploid nuclear DNA content 2.4–5.5 pg (6 values for *Z. mays*, mean 4.4). Mean diploid 2c DNA value 5.2 pg (values from 4.4 to 11.0 given for *Zea mays* by different authors). Nucleoli persistent.

Taxonomy. Panicoideae; Andropogonodae; Maydeae.

Ecology, geography, regional floristic distribution. 1 species. Unknown in the wild: originated in tropical America, cultivated in all warm to tropical regions of both hemispheres. Mesophytic.

Neotropical. Caribbean, Venezuela and Surinam, Amazon, Pampas, and Andean.

Hybrids. Intergeneric hybrids with *Euchlaena* (×*Euchlaezea* Janaki ex Bor), *Saccharum*, *Tripsacum*.

Rusts and smuts. Rusts — *Physopella* and *Puccinia*. Smuts from Ustilaginaceae. Ustilaginaceae — *Sphacelotheca* and *Ustilago*.

Economic importance. Cultivated fodder: *Zea mays*. Grain crop species: *Z. mays* (Maize, Corn, Mealies).

References, etc. Leaf anatomical: Metcalfe 1960; this project.

Special comments. The descriptions of maize relatives require updating, to conform with modern taxonomic views (e.g. Doebley and Iltis 1980, Iltis 1987). The older and convenient but artificial treatment, separating the readily distinguishable *Zea mays* subsp. *mays* from the rest, is retained here pending acqisition of detailed comparative morphological data.

Zenkeria Trin.

Habit, vegetative morphology. Perennial; caespitose. **Culms** 60–130 cm high; *herbaceous*. *Leaf blades* broad, or narrow; 5–25 mm wide; flat, or rolled (convolute or involute); *pseudopetiolate (the pseudopetiole stiff)*; without cross venation. *Ligule a fringe of hairs*.

Reproductive organization. Plants bisexual, with bisexual spikelets; with hermaphrodite florets.

Inflorescence. Inflorescence paniculate; open, or contracted; espatheate; not comprising 'partial inflorescences' and foliar organs. Spikelet-bearing axes persistent. Spikelets not secund; pedicellate.

Female-fertile spikelets. Spikelets 2.5–4 mm long; compressed laterally; disarticulating above the glumes. Rachilla very shortly prolonged beyond the uppermost female-fertile floret, or terminated by a female-fertile floret; rachilla hairy. The rachilla extension (when present) naked.

Glumes two; very unequal (G_2 longer), or more or less equal; shorter than the spikelets; shorter than the adjacent lemmas; awnless; carinate; similar (spreading, ovate). Lower glume much shorter than half length of lowest lemma, or longer than half length of lowest lemma; 1 nerved. Upper glume 1 nerved. *Spikelets* with female-fertile florets only.

Female-fertile florets 2 *(similar)*. *Lemmas* similar in texture to the glumes to decidedly firmer than the glumes; not becoming indurated (leathery); awnless (acuminate in *Z. elegans*); *hairy (below the middle)*; 5–7 nerved. *Palea* present; awnless, without apical setae, or with apical setae; 2-keeled. Palea keels hairy (long-ciliate). *Lodicules* present; 2; free; membranous. *Stamens* 3.

Fruit, embryo and seedling. Endosperm containing compound starch grains.

Abaxial leaf blade epidermis. *Costal/intercostal zonation* conspicuous. *Papillae* absent. *Long-cells* markedly different in shape costally and intercostally; of similar wall thickness costally and intercostally. Mid-intercostal long-cells rectangular; having markedly sinuous walls (and pitted; but the sinuations very fine; by contrast with the coarse sinuations of the costals). *Microhairs* present; panicoid-type (large, the apical cell as long as the basal cell). *Stomata* common. Subsidiaries dome-shaped. Guard-cells overlapping to flush with the interstomatals. *Intercostal short-cells* common; not paired (solitary (square-ish)); not silicified. Costal zones with short-cells. *Costal short-cells* conspicuously in long rows (though the short-cells sometimes rather long). Costal silica bodies oryzoid (very regular, cf. *Pheidochloa*); not sharp-pointed.

Transverse section of leaf blade, physiology. C_3; XyMS+. *Mesophyll* with nonradiate chlorenchyma; without adaxial palisade; without arm cells (but the cell walls frequently very sinuous). *Leaf blade* with distinct, prominent adaxial ribs; with the ribs very irregular in sizes (large, flat-topped and a few small round-topped). *Midrib* not readily distinguishable; with one bundle only. Bulliforms present in discrete, regular adaxial groups (large groups, regularly disposed between the ribs); in simple fans and associated with colourless mesophyll cells to form deeply-penetrating fans. All the vascular bundles accompanied by sclerenchyma. Combined sclerenchyma girders present (with all the bundles); forming 'figures' (all bundles). Sclerenchyma all associated with vascular bundles.

Taxonomy. Arundinoideae; Danthonieae.

Ecology, geography, regional floristic distribution. 4 species. India, Ceylon, Burma. Species of open habitats. Upland grassland.

Paleotropical. Indomalesian. Indian.

References, etc. Leaf anatomical: this project.

Special comments. Fruit data wanting.

Zeugites P. Browne

Despretzia, Galeottia Mart. & Gal., *Krombholzia* Fourn., *Senites* Adans.

Habit, vegetative morphology. Perennial; stoloniferous, or decumbent. The flowering culms leafy. *Culms* 30–100 cm high; herbaceous; clambering or trailing; branched above. Primary branches/mid-culm node 1. Plants unarmed. *Leaves* not basally aggregated; non-auriculate. Leaf blades lanceolate to ovate; broad, or narrow; pseudopetiolate (the pseudopetioles long and slender); cross veined; rolled in bud.

Reproductive organization. *Plants bisexual, with bisexual spikelets; without hermaphrodite florets (the lowermost florets in each spikelet female, the upper male)*.

Inflorescence. Inflorescence paniculate; open; with capillary branchlets; espatheate; not comprising 'partial inflorescences' and foliar organs.

Female-fertile spikelets. Spikelets falling with the glumes; not disarticulating between the florets, or disarticulating between the florets (the upper, male part sometimes shed separately). Rachilla prolonged beyond the uppermost female-fertile floret; rachilla hairless. The rachilla extension with incomplete florets.

Glumes one per spikelet; shorter than the adjacent lemmas; *not pointed (truncate to dentate, with cross nerves)*; awnless. Lower glume several-nerved. Upper glume 5 nerved. *Spikelets* with incomplete florets. The incomplete florets distal to the female-fertile florets (male). The distal incomplete florets 1–14. *Spikelets without proximal incomplete florets.*

Female-fertile florets 1. Lemmas slightly gibbous; entire to incised; not deeply cleft (acute to truncate or dentate); awnless, or mucronate (?); hairless; 7–13 nerved. *Palea* present; awnless, without apical setae. *Lodicules* present; 2; free; fleshy. *Stamens* 0 (3 in the male florets). Stigmas 2.

Fruit, embryo and seedling. Hilum short. Embryo small; with an epiblast; with a scutellar tail; with a negligible mesocotyl internode. Embryonic leaf margins overlapping.

Abaxial leaf blade epidermis. *Costal/intercostal zonation* conspicuous. *Papillae* absent. *Long-cells* markedly different in shape costally and intercostally (the costals narrower). Mid-intercostal long-cells rectangular; having markedly sinuous walls (deeply). *Microhairs* present; panicoid-type; 30–31–57 microns long; 3.6–5.7 microns wide at the septum. Microhair total length/width at septum 6.9–10. Microhair apical cells (12–)13–31.5(–33) microns long. Microhair apical cell/total length ratio 0.38–0.58. *Stomata* common; 21–33 microns long. Subsidiaries dome-shaped and triangular. Guard-cells overlapping to flush with the interstomatals. *Intercostal short-cells* absent or very rare (but abundant microhairs and (small) microhair bases in *Z. pittieri*). Costal zones with short-cells. *Costal short-cells* conspicuously in long rows, or neither distinctly grouped into long rows nor predominantly paired (in *Z. pittieri*, the rows are interrupted by prickles). Costal silica bodies 'panicoid-type'; cross shaped and nodular; not sharp-pointed.

Transverse section of leaf blade, physiology. C_3; XyMS+. *Mesophyll* with non-radiate chlorenchyma; with adaxial palisade; without arm cells; without fusoids. *Leaf blade* adaxially flat. *Midrib* conspicuous (a larger bundle); with one bundle only. Bulliforms present in discrete, regular adaxial groups; in simple fans (the fans large, occupying about half the width of the mesophyll). All the vascular bundles accompanied by sclerenchyma. Combined sclerenchyma girders present. Sclerenchyma all associated with vascular bundles.

Cytology. $2n = 46$.

Taxonomy. Bambusoideae; Oryzodae; Centotheceae.

Ecology, geography, regional floristic distribution. About 12 species. Mexico to Venezuela, West Indies. Shade species; glycophytic. Ravines and bushy hillsides.

Neotropical. Caribbean and Venezuela and Surinam.

References, etc. Leaf anatomical: this project.

Zingeria P. Smirnov

Zingeriopsis Probat.

Habit, vegetative morphology. *Annual*. Culms 20–50 cm high; herbaceous. Leaves non-auriculate. Leaf blades narrowly linear; narrow; 0.5–2 mm wide; flat (usually), or folded, or rolled; without cross venation. *Ligule an unfringed membrane*; not truncate; 1–4 mm long.

Reproductive organization. Plants bisexual, with bisexual spikelets; with hermaphrodite florets.

Inflorescence. *Inflorescence* delicate, *paniculate*; open. Inflorescence with axes ending in spikelets. Inflorescence espatheate; not comprising 'partial inflorescences' and foliar organs. Spikelet-bearing axes persistent. Spikelets not secund; pedicellate.

Female-fertile spikelets. *Spikelets* 1–3 mm long; slightly *compressed dorsiventrally*; *disarticulating above the glumes*. Rachilla terminated by a female-fertile floret. Hairy callus present. *Callus* very short.

Glumes two; relatively large; very unequal (*Z. verticillata*), or more or less equal; (the longer) shorter than the spikelets to about equalling the spikelets; long relative to the adjacent lemmas; hairless; glabrous; pointed (acute); awnless; non-carinate; similar (ovate, membranous). Lower glume 1 nerved. Upper glume 3 nerved. *Spikelets* with female-fertile florets only.

Female-fertile florets 1. **Lemmas** decidedly firmer than the glumes; not becoming indurated (papery in fruit); entire; blunt; awnless; **hairy *(with clavate hairs)***; non-carinate; without a germination flap; 3 nerved. *Palea* present (similar to the lemma); relatively long; textured like the lemma; not indurated; 2-nerved; 2-keeled. *Lodicules* present; 2; free; membranous; glabrous; toothed, or not toothed; not or scarcely vascularized. *Stamens* 3. Anthers 0.5–1.3 mm long. *Ovary* glabrous. Styles free to their bases. Stigmas 2.

Fruit, embryo and seedling. *Fruit* adhering to lemma and/or palea (often), or free from both lemma and palea; small (0.3–1.7 mm long); compressed dorsiventrally. Hilum short. **Pericarp thick and hard**; free. Embryo small.

Abaxial leaf blade epidermis. *Costal/intercostal zonation* conspicuous. *Papillae* present; costal and intercostal. Intercostal papillae over-arching the stomata (slightly, in *Z. trichopoda*), or not over-arching the stomata; consisting of one oblique swelling per cell (costally), or consisting of one symmetrical projection per cell (intercostally). Long-cells differing markedly in wall thickness costally and intercostally (costals thicker). Mid-intercostal long-cells rectangular, or rectangular and fusiform (*Z. kochii*); having straight or only gently undulating walls. *Microhairs* absent. *Stomata* common; (25–)27–30(–32) microns long. Subsidiaries low dome-shaped, or parallel-sided. Guard-cells overlapped by the interstomatals. *Intercostal short-cells* absent or very rare. Costal zones with short-cells. *Costal short-cells* neither distinctly grouped into long rows nor predominantly paired. Costal silica bodies horizontally-elongated crenate/sinuous, or horizontally-elongated smooth (a few), or rounded (very few); not sharp-pointed.

Transverse section of leaf blade, physiology. C_3; XyMS+. Mesophyll without adaxial palisade. *Midrib* not readily distinguishable; with one bundle only. *The lamina* symmetrical on either side of the midrib. All the vascular bundles accompanied by sclerenchyma. Combined sclerenchyma girders present (with the largest bundles). Sclerenchyma all associated with vascular bundles.

Cytology. Chromosome base number, $x = 2$. $2n = 4$, 8, and 12. 2 ploid ($2n = 4$), or 4 ploid, or 6 ploid.

Taxonomy. Pooideae; Poodae; Aveneae.

Ecology, geography, regional floristic distribution. 4 species. Southeast Russia, Western Asia. Mesophytic; species of open habitats; glycophytic. Meadows and streamsides.

Holarctic. Boreal and Tethyan. Euro-Siberian. Irano-Turanian. European.

References, etc. Leaf anatomical: this project.

Zizania L.

Ceratochaete Lunell, *Elymus* Mitchell, *Fartis* Adans., *Hydropyrum* Link, *Melinum* Link

Habit, vegetative morphology. Tall, aquatic annual, or perennial. *Culms* 100–300 cm high; herbaceous. Culm nodes hairy, or glabrous. Culm internodes hollow. Leaves non-auriculate. Leaf blades linear-lanceolate; broad (to 3 cm), or narrow; 5–30 mm wide; flat; pseudopetiolate, or not pseudopetiolate; without cross venation (?). *Ligule an unfringed membrane*; 3–11 mm long.

Reproductive organization. *Plants monoecious with all the fertile spikelets unisexual*; without hermaphrodite florets. The spikelets of sexually distinct forms on the same plant; female-only and male-only. The male and female-fertile spikelets on different branches of the same inflorescence (the lower, pendent branches male, the upper, ascending ones female). Plants inbreeding.

Inflorescence. Inflorescence paniculate (large, terminal); open; espatheate; not comprising 'partial inflorescences' and foliar organs. Spikelet-bearing axes persistent. Spikelets not secund; pedicellate.

Female-sterile spikelets. Male spikelets pendent, lemmas membranous, acute or short-awned, 5-nerved; palea 3-nerved; 6 free stamens. The male spikelets without glumes; 1 floreted. The lemmas awnless to awned. Male florets 6 staminate. The staminal filaments free.

Female-fertile spikelets. Spikelets unconventional (the glumes 'obsolete'); *10–25 mm long (ascending)*; compressed laterally to not noticeably compressed; falling entire. Rachilla terminated by a female-fertile floret. Hairy callus absent.

Glumes *absent (but perhaps represented by a small collar-like ridge on the pedicel)*. Spikelets with female-fertile florets only.

Female-fertile florets 1. Lemmas acuminate into the long, acuminate lemma; not becoming indurated (papery); entire; pointed; awned. Awns 1; median; apical (from the into a long, slender awn); non-geniculate; about as long as the body of the lemma to much longer than the body of the lemma. Lemmas non-carinate; 3–5 nerved. *Palea* present; relatively long; 2-nerved; one-keeled. *Stamens* 0. *Ovary* glabrous. Styles free to their bases. Stigmas 2.

Fruit, embryo and seedling. Fruit medium sized, or large (10–20 mm long); not noticeably compressed (cylindrical). Hilum long-linear. Endosperm hard; without lipid. Embryo with an epiblast; with a scutellar tail; with a negligible mesocotyl internode. Embryonic leaf margins overlapping.

Seedling with a long mesocotyl; with a loose coleoptile (margins free, lamina well developed).

Abaxial leaf blade epidermis. *Costal/intercostal zonation* conspicuous. *Papillae* present. Intercostal papillae over-arching the stomata; several per cell (each long-cell with one large and many small papillae). Mid-intercostal long-cells having markedly sinuous walls (the walls thin). *Microhairs* present; panicoid-type (tapering to base and apex); 26–36 microns long. Microhair apical cells 14–18 microns long. Microhair apical cell/total length ratio 0.48. *Stomata* common. Subsidiaries triangular. *Intercostal short-cells* common; in cork/silica-cell pairs; silicified. Intercostal silica bodies narrowly oryzoid-type, or vertically elongated-nodular. Costal zones with short-cells. *Costal short-cells* predominantly paired, or neither distinctly grouped into long rows nor predominantly paired. Costal silica bodies oryzoid; not sharp-pointed.

Transverse section of leaf blade, physiology. C_3; XyMS+. *Mesophyll* with non-radiate chlorenchyma; with arm cells; with fusoids. The fusoids external to the PBS. *Midrib* conspicuous (with a characteristic system of intercellular spaces); having complex vascularization. Bulliforms present in discrete, regular adaxial groups; in simple fans, or associated with colourless mesophyll cells to form deeply-penetrating fans. All the vascular bundles accompanied by sclerenchyma. Combined sclerenchyma girders present.

Cytology. Chromosome base number, $x = 15$. $2n = 30$ and 34. 2 ploid. Haploid nuclear DNA content 2.2 pg (1 species). Mean diploid 2c DNA value 4.4 pg (*Z. aquatica*).

Taxonomy. Bambusoideae; Oryzodae; Oryzeae.

Ecology, geography, regional floristic distribution. 3 species. North America and Eurasia. Hydrophytic to helophytic.

Holarctic and Paleotropical. Boreal. Indomalesian. Euro-Siberian, Eastern Asian, and Atlantic North American. Indian and Indo-Chinese. Siberian. Canadian-Appalachian and Central Grasslands.

Rusts and smuts. Smuts from Tilletiaceae and from Ustilaginaceae. Tilletiaceae — *Entyloma*. Ustilaginaceae — *Ustilago*.

Economic importance. Grain crop species: breeding of *Zizania aquatica* (Wildrice, long gathered from wild stands) is yielding non-shattering forms of potential economic value. *Z. latifolia* cultivated for its edible young shoots.

References, etc. Leaf anatomical: Metcalfe 1960 and this project.

Zizaniopsis Doell & Aschers.

Habit, vegetative morphology. Robust perennial; rhizomatous and caespitose. The flowering culms leafy. *Culms* 100–350 cm high; herbaceous; unbranched above. Pluricaespitose. Plants unarmed. *Leaves* not basally aggregated. Leaf blades linear; broad; 10–30 mm wide (and to 1.2 m long); flat; without cross venation. Ligule an unfringed membrane; not truncate (ovate-lanceolate); 15–80 mm long.

Reproductive organization. *Plants monoecious with all the fertile spikelets unisexual*; without hermaphrodite florets. The spikelets of sexually distinct forms on the same plant; female-only and male-only. The male and female-fertile spikelets on different branches of the same inflorescence, or segregated, in different parts of the same inflorescence branch (the males below the females on the same panicle branches). The spikelets overtly heteromorphic.

Inflorescence. Inflorescence paniculate (large, to 100 cm or more long); open; espatheate; not comprising 'partial inflorescences' and foliar organs. Spikelet-bearing axes persistent. Spikelets not secund; pedicellate.

Female-sterile spikelets. Male spikelets sometimes larger, with 7-nerved lemma, 2-nerved palea and 6 free, non-penicillate stamens. The male spikelets without glumes; 1 floreted. Male florets 6 staminate. The staminal filaments free.

Female-fertile spikelets. Spikelets unconventional (through lacking glumes); *6–8 mm long*; subcylindrical; compressed laterally to not noticeably compressed; disarticulating from the pedicels. Rachilla terminated by a female-fertile floret. Hairy callus absent.

Glumes *absent*. *Spikelets* with female-fertile florets only.

Female-fertile florets 1. Lemmas lanceolate, tapered into the awn; entire; pointed; awned. Awns 1; median; apical; non-geniculate; hairless; much shorter than the body of the lemma to much longer than the body of the lemma. Lemmas 5–7 nerved. *Palea* present; relatively long; entire (narrow, pointed); several nerved (3); one-keeled. *Stamens* 0. *Ovary* glabrous. Styles fused (into one long style). Stigmas 2.

Fruit, embryo and seedling. *Fruit* free from both lemma and palea; medium sized (6–7 mm long); not noticeably compressed (cylindrical). Hilum long-linear. Endosperm hard; without lipid.

Abaxial leaf blade epidermis. *Costal/intercostal zonation* conspicuous. *Papillae* present; intercostal. Intercostal papillae over-arching the stomata (in places), or not over-arching the stomata (elsewhere); several per cell (most cells in *Z. bonariense* with a median row of thick walled, circular or bifurcated papillae, and often with a single very large papilla in addition; long-cells in *Z. miliacea* with 2 or 3 irregular rows of small, thick walled, circular papillae, which tend to associate with those of the interstomatals to encircle the stomata). *Long-cells* similar in shape costally and intercostally to markedly different in shape costally and intercostally (the costals more regularly rectangular); of similar wall thickness costally and intercostally (fairly thick walled). Mid-intercostal long-cells rectangular; having markedly sinuous walls (the sinuosity coarse to fairly fine). *Microhairs* present; elongated; ostensibly one-celled; of peculiar form — small to medium sized, thin walled, pyriform with pointed tips, and bent over. *Stomata* common; 24–27 microns long (*Z. miliacea*), or 36–39 microns long (*Z. bonariensis*). Subsidiaries non-papillate; dome-shaped and triangular (mostly low domes, in *Z. bonariensis*). Guard-cells overlapped by the interstomatals (slightly). *Intercostal short-cells* common; in cork/silica-cell pairs; silicified, or not silicified. Intercostal silica bodies oryzoid-type, or vertically elongated-nodular (in *Z. miliacea*, ill-defined in *Z. bonariensis*). Costal zones with short-cells. *Costal short-cells* conspicuously in long rows (*Z. miliacea*), or predominantly paired (*Z. bonariensis*). Costal silica bodies oryzoid; not sharp-pointed.

Transverse section of leaf blade, physiology. C_3; XyMS+ (the mestome sheath heavily lignified, often double). Mesophyll without adaxial palisade; without arm cells (seemingly, in the poor material seen), or without arm cells (?); with fusoids. The fusoids external to the PBS. *Leaf blade* with distinct, prominent adaxial ribs to 'nodular' in section. *Midrib* conspicuous; having complex vascularization; with colourless mesophyll adaxially (and with large areas of stellate aerenchyma). *The lamina* symmetrical on either side of the midrib. Bulliforms present in discrete, regular adaxial groups; in simple fans (the fans large). All the vascular bundles accompanied by sclerenchyma. Combined sclerenchyma girders present; forming 'figures' (every bundle with an I or an 'anchor'). Sclerenchyma all associated with vascular bundles.

Cytology. Chromosome base number, $x = 12$. $2n = 24$. 2 ploid. Nucleoli persistent.

Taxonomy. Bambusoideae; Oryzodae; Oryzeae.

Ecology, geography, regional floristic distribution. 5 species. Southern U.S.A., South America. Helophytic.

Holarctic and Neotropical. Boreal. Atlantic North American. Caribbean and Pampas. Southern Atlantic North American and Central Grasslands.
References, etc. Leaf anatomical: this project.

Zonotriche Phipps

Mitwabochloa Phipps, *Piptostachya* (C. E. Hubbard) Phipps

Habit, vegetative morphology. Perennial; caespitose. *Culms* 60–150 cm high; herbaceous. *Leaves* not basally aggregated. Leaf blades linear; broad to narrow; flat; without cross venation. *Ligule a fringe of hairs*.

Reproductive organization. Plants bisexual, with bisexual spikelets; with hermaphrodite florets.

Inflorescence. Inflorescence paniculate; open to contracted; espatheate; not comprising 'partial inflorescences' and foliar organs. *Spikelet-bearing axes disarticulating; falling entire (the villous peduncle developing a fracture zone, so that the spikelet triads fall entire). Spikelets* in triplets (compact); not secund; *pedicellate (the pedicels connate in the triplet); not in distinct 'long-and-short' combinations.*

Female-fertile spikelets. Spikelets 10–28 mm long; brown; disarticulating above the glumes and falling with the glumes; disarticulating between the florets. *Rachilla terminated by a female-fertile floret.* Hairy callus present (beneath the L_2). *Callus* short; blunt.

Glumes two; very unequal to more or less equal; (the longer) long relative to the adjacent lemmas; hairy (with tubercle-based hairs); without conspicuous tufts or rows of hairs; pointed; awnless; similar (narrowly lanceolate, papery to leathery). Lower glume 3 nerved. Upper glume 3 nerved. *Spikelets* with incomplete florets. The incomplete florets proximal to the female-fertile florets. *The proximal incomplete florets* 1; paleate; male. The proximal lemmas 5 nerved; exceeded by the female-fertile lemmas (2/3 to 3/4 as long as the spikelet).

Female-fertile florets 1. Lemmas thinly leathery; not becoming indurated; incised; 2 lobed; deeply cleft; awned. Awns 1, or 3; median (the lateral lobes muticous), or median and lateral (the lateral lobes aristate); the median different in form from the laterals (when laterals present); from a sinus; geniculate; hairless; much longer than the body of the lemma; persistent. Lemmas hairy (variously so). The hairs in tufts, or not in tufts; in transverse rows, or not in transverse rows. Lemmas non-carinate; 5–9 nerved. *Palea* present; 2-nerved; 2-keeled. Palea keels winged, or wingless. *Lodicules* present; fleshy. *Stamens* 3.

Fruit, embryo and seedling. Fruit with hairs confined to a terminal tuft (or hairless). *Hilum long-linear*. Embryo large.

Abaxial leaf blade epidermis. *Costal/intercostal zonation* conspicuous. *Papillae* absent. *Long-cells* markedly different in shape costally and intercostally (the costals much narrower); of similar wall thickness costally and intercostally (of medium wall thickness). Mid-intercostal long-cells rectangular; having markedly sinuous walls (these with conspicuous pits). *Microhairs* present; elongated; clearly two-celled; panicoid-type; (the only complete example seen) 57 microns long; 6 microns wide at the septum. Microhair total length/width at septum 9.5. Microhair apical cells 27 microns long. Microhair apical cell/total length ratio 0.47. *Stomata* common; 27–33 microns long. Subsidiaries low to high dome-shaped and triangular. Guard-cells overlapping to flush with the interstomatals. *Intercostal short-cells* common; in cork/silica-cell pairs and not paired (sometimes seemingly solitary); silicified. A few small intercostal prickles present. *Crown cells* absent. Costal zones with short-cells. *Costal short-cells* conspicuously in long rows. Costal silica bodies 'panicoid-type'; dumb-bell shaped (elongated); not sharp-pointed.

Transverse section of leaf blade, physiology. C_4. The anatomical organization somewhat unconventional. Seemingly XyMS+ (the primaries certainly double-sheathed, and the inner sheath seemingly empty, in the fairly well preserved material seen). PCR sheath outlines even. PCR sheath extensions absent. *Leaf blade* with only low, round-topped ribs over the primary bundles. *Midrib* not readily distinguishable; with one bundle only. Bulliforms present in discrete, regular adaxial groups (large groups between the primary bundles); in simple fans. All the vascular bundles accompanied by sclerenchyma.

Combined sclerenchyma girders present (with the primaries); forming 'figures' (I's and T's). Sclerenchyma all associated with vascular bundles.

Taxonomy. Panicoideae; Panicodae; Arundinelleae.

Ecology, geography, regional floristic distribution. 3 species. Tropical Africa. Savanna woodland.

Paleotropical. African. Sudano-Angolan. Somalo-Ethiopian and South Tropical African.

References, etc. Leaf anatomical: this project.

Zoysia Willd.

Brousemichea Bal., *Matrella* Pers., *Osterdamia* Necker ex Kuntze

Habit, vegetative morphology. Mat forming perennial; rhizomatous. *Culms* 5–50 cm high; herbaceous. Culm nodes glabrous. Culm internodes solid. *Leaves* not basally aggregated; conspicuously distichous; non-auriculate. *Leaf blades* very narrow; flat, or rolled; without abaxial multicellular glands; *not pseudopetiolate*; without cross venation; persistent; rolled in bud. *Ligule a fringed membrane to a fringe of hairs*.

Reproductive organization. *Plants bisexual, with bisexual spikelets*; with hermaphrodite florets.

Inflorescence. *Inflorescence* reduced to a single spikelet (in *Z. minima*), or few spikeleted to many spikeleted; *a false spike, with spikelets on contracted axes, or a single raceme*; espatheate; not comprising 'partial inflorescences' and foliar organs. Spikelet-bearing axes persistent. Spikelets solitary; not secund; shortly pedicellate.

Female-fertile spikelets. Spikelets 2.3–3.4 mm long; adaxial; compressed laterally, or not noticeably compressed, or compressed dorsiventrally; falling with the glumes (falling singly from the persistent pedicels). Rachilla terminated by a female-fertile floret. Hairy callus absent.

Glumes present; one per spikelet, or two; very unequal; (the upper) long relative to the adjacent lemmas; awned (or mucronate, via the excurrent midnerve), or awnless; very dissimilar (the lower when present minute and scarious). Lower glume 0 nerved. Upper glume 1 nerved. *Spikelets* with female-fertile florets only.

Female-fertile florets 1. **Lemmas** less firm than the glumes (hyaline); not becoming indurated; entire; blunt; *mucronate*; hairless; non-carinate; 1 nerved. **Palea** present, or absent; when present, conspicuous but relatively short, or very reduced; not indurated; *1-nerved, or nerveless*. **Lodicules absent**. Stamens 2, or 3. Anthers 1.3–3 mm long; not penicillate. *Ovary* glabrous. Styles fused, or free to their bases. Stigmas 2; white.

Fruit, embryo and seedling. *Fruit* free from both lemma and palea; small; compressed laterally. Hilum short. Pericarp fused. Embryo large; not waisted. Endosperm hard; without lipid; containing compound starch grains. Embryo with an epiblast; with a scutellar tail; with an elongated mesocotyl internode. Embryonic leaf margins meeting.

Seedling with a long mesocotyl. First seedling leaf with a well-developed lamina. The lamina broad; erect.

Abaxial leaf blade epidermis. *Costal/intercostal zonation* conspicuous. *Papillae* absent (but abundant adaxially). *Long-cells* similar in shape costally and intercostally; of similar wall thickness costally and intercostally. Mid-intercostal long-cells rectangular; having markedly sinuous walls. *Microhairs* present; more or less spherical; clearly two-celled; chloridoid-type. Microhair apical cell wall of similar thickness/rigidity to that of the basal cell. Microhairs with 'partitioning membranes' (in *Z. macrantha*). The 'partitioning membranes' in the basal cell. Microhairs (27–)28.5–30(–31.5) microns long. Microhair basal cells 15 microns long. Microhairs (13.5–)15.5–16.5 microns wide at the septum. Microhair total length/width at septum 1.5–2.2. Microhair apical cells 10.5–12 microns long. Microhair apical cell/total length ratio 0.37–0.47. *Stomata* common; (18–)19.5–21(–22.5) microns long. Subsidiaries dome-shaped, or dome-shaped and triangular. Guard-cells overlapping to flush with the interstomatals. *Intercostal short-cells* common; in cork/silica-cell pairs and not paired (some solitary); silicified. Intercostal silica bodies present and perfectly developed; crescentic. Costal zones with short-cells. *Costal short-cells* predominantly paired, or neither distinctly grouped into long rows nor predominantly paired. Costal silica bodies present throughout the costal zones; rounded, saddle shaped, and crescentic; not sharp-pointed.

Transverse section of leaf blade, physiology. C_4; biochemical type PCK (*Z. japonica*); XyMS+. PCR sheath outlines uneven. PCR sheaths of the primary vascular bundles interrupted; interrupted abaxially only to interrupted both abaxially and adaxially. PCR sheath extensions present. Maximum number of extension cells 4–6. PCR cell chloroplasts with well developed grana; centrifugal/peripheral. *Mesophyll* with radiate chlorenchyma; traversed by columns of colourless mesophyll cells. *Leaf blade* adaxially flat. *Midrib* not readily distinguishable; with one bundle only. Bulliforms present in discrete, regular adaxial groups; associated with colourless mesophyll cells to form deeply-penetrating fans (these often linked with the abaxial epidermis by columns of colourless cells). All the vascular bundles accompanied by sclerenchyma. Combined sclerenchyma girders present; forming 'figures'. Sclerenchyma all associated with vascular bundles. The lamina margins with fibres.

Cytology. Chromosome base number, $x = 10$. $2n = 40$. 4 ploid.

Taxonomy. Chloridoideae; main chloridoid assemblage.

Ecology, geography, regional floristic distribution. 10 species. Mascarene Is. to New Zealand. Commonly adventive. Xerophytic; species of open habitats; halophytic. Coastal sands and hinterland.

Holarctic, Paleotropical, Neotropical, Australian, and Antarctic. Boreal. Madagascan, Indomalesian, and Neocaledonian. Eastern Asian and Atlantic North American. Indian, Indo-Chinese, Malesian, and Papuan. Caribbean. North and East Australian. New Zealand. Canadian-Appalachian. Tropical North and East Australian.

Rusts and smuts. Rusts — *Puccinia*.

Economic importance. Significant weed species: *Z. matrella*. Cultivated fodder: *Z. macrantha* (in coastal sand and brackish places). Lawns and/or playing fields: *Z. japonica*, *Z. matrella*, *Z. tenuifolia*.

References, etc. Leaf anatomical: Metcalfe 1960; this project.

Zygochloa S. T. Blake

Habit, vegetative morphology. Shrubby perennial ('cane-grass'); rhizomatous (and tussock-forming). *Culms* 80–200 cm high; woody and persistent; to 0.9 cm in diameter (near the base); grooved on one side; branched above. Culm nodes glabrous. Culm internodes solid. Young shoots intravaginal. *Leaves* not basally aggregated; non-auriculate. Leaf blades broad to narrow; 4–10 mm wide (to 30 cm long); flat (short, stiff); without cross venation; disarticulating from the sheaths. Ligule a fringe of hairs.

Reproductive organization. Plants dioecious; without hermaphrodite florets. The spikelets all alike in sexuality (on the same plant); female-only, or male-only. Plants outbreeding.

Inflorescence. *Inflorescence a spatheate panicle of 'bracteate' heads*; open. Inflorescence axes not ending in spikelets (interpreting one of the 'bracts' as a modified axis tip). Inflorescence spatheate (and 'bracteate'); a complex of 'partial inflorescences' and intervening foliar organs. **Spikelet-bearing axes *very much reduced (to a single spikelet accompanied by two basal bracts and the small, naked, bractlike axis tip, the units grouped into capitate heads)*; disarticulating; falling entire (the heads falling). Spikelets *associated with bractiform involucres (each spikelet associated with the three chaffy, rigid-tipped 'bracts')*. The involucres shed with the fertile spikelets. Spikelets solitary; not secund; subsessile. Pedicel apices discoid. Spikelets not in distinct 'long-and-short' combinations.

Female-sterile spikelets. Male spikelets on separate individuals, in small bracteate heads 1–2 cm in diameter. Glumes similar, 5–7 nerved; lemmas 2, similar, awnless, 5 nerved, each with a male floret. Paleas conspicuous, with winged keels. 3 stamens, anthers 4 mm long, no gynoecium. Rachilla of male spikelets terminated by a male floret. The male spikelets with glumes; without proximal incomplete florets; 2 floreted (both fertile). The lemmas awnless. Male florets 2; 3 staminate.

Female-fertile spikelets. Spikelets 6–10 mm long; lanceolate, or ovate; compressed dorsiventrally; falling with the glumes. Rachilla terminated by a female-fertile floret. Hairy callus absent.

Glumes two; more or less equal; shorter than the adjacent lemmas; pointed, or not pointed; awnless; similar (papery). Lower glume 5–9 nerved. Upper glume 5–9 nerved. *Spikelets* with incomplete florets. The incomplete florets proximal to the female-fertile florets. *The proximal incomplete florets* 1; sterile. The proximal lemmas awnless; 5 nerved; becoming indurated.

Female-fertile florets 1. Lemmas abruptly acuminate; decidedly firmer than the glumes; smooth, or striate; becoming indurated (crustaceous); yellow in fruit; entire; pointed; awnless; hairless; non-carinate; having the margins tucked in onto the palea; 5 nerved. *Palea* present; entire (ovate, abruptly acuminate); awnless; without apical setae; textured like the lemma; 2-nerved. *Lodicules* present; 2; fleshy. *Stamens* 0 (3 staminodes). *Ovary* glabrous. Styles fused. Stigmas 2.

Fruit, embryo and seedling. Fruit small (about 3 mm long). Hilum short.

Abaxial leaf blade epidermis. *Costal/intercostal zonation* conspicuous. *Papillae* absent. *Long-cells* similar in shape costally and intercostally; of similar wall thickness costally and intercostally. Mid-intercostal long-cells rectangular; having markedly sinuous walls. *Microhairs* present, or absent (not seen on male material); chloridoid-type; about 66 microns long; 9 microns wide at the septum. Microhair total length/width at septum 7.3. Microhair apical cells about 42 microns long. Microhair apical cell/total length ratio 0.64. *Stomata* common; 42–57 microns long. Subsidiaries parallel-sided, dome-shaped, and triangular; including both triangular and parallel-sided forms on the same leaf. Guard-cells overlapping to flush with the interstomatals. *Intercostal short-cells* common; in cork/silica-cell pairs and not paired (solitary); silicified (when paired), or not silicified. Costal zones with short-cells. *Costal short-cells* conspicuously in long rows. Costal silica bodies 'panicoid-type'; not sharp-pointed.

Transverse section of leaf blade, physiology. C_4; XyMS–. PCR sheath outlines uneven. *Mesophyll* with radiate chlorenchyma. *Leaf blade* 'nodular' in section; with the ribs more or less constant in size. *Midrib* not readily distinguishable; with one bundle only. Bulliforms present in discrete, regular adaxial groups (the epidermis extensively, irregularly bulliform). Many of the smallest vascular bundles unaccompanied by sclerenchyma. Combined sclerenchyma girders present; forming 'figures'. Sclerenchyma all associated with vascular bundles.

Special diagnostic feature. *Stems cane-like, spikelets in bracteate, globular 1–3.5 cm heads*.

Taxonomy. Panicoideae; Panicodae; Paniceae.

Ecology, geography, regional floristic distribution. 1 species. Australia. Xerophytic; species of open habitats. Arid sandy and rocky places.

Australian. Central Australian.

References, etc. Morphological/taxonomic: Vickery 1975; Webster 1987. Leaf anatomical: this project.

Special comments. Fruit data wanting.

REFERENCES AND SOURCES OF DATA

The references listed here have mostly been used in preparing the generic descriptions, either as sources of descriptive, floristic, nomenclatural, economic or ethnic data, in connection with generic circumscriptions and/or nomenclature, or as sources of background information in relation to preparing the character list. A few have not yet been used directly, but are informative on aspects which merit inclusion in the future. Note that 'indirect' descriptive data of the kind involving extrapolation from tribal or subfamily descriptions or deductions from printed keys have not in general been accepted. Some references relevant to software are also included.

The references are annotated with 'subject areas', mainly to aid searching via computer editors, but assignment of many of the articles is somewhat arbitrary and some appear under more than one heading.

M Sources of comparative morphological and/or floristic data for the family, or for large suites of genera

T General taxonomic and descriptive references pertaining to particular genera, or to small series of genera (type descriptions of genera are listed here only when they were found at some stage to yield data directly useful in the present context)

A Sources of comparative anatomical data for the family, or for suites of genera, or of background information on anatomical characters and character states

C Classification of the family, phylogeny, classificatory methodology

P Data sources for comparative biochemistry, physiology, pathogens

G Chromosomes, breeding systems, hybridization

S Software and relevant background

I Sources of illustrations

Inclusion in generic descriptions of particular taxonomic references does not necessarily mean that these have provided much of the descriptive data. Nomenclatural inconsistencies between genera described here and the references pertaining to them result from taxonomic realignments. Many of the works listed are largely or in part compilations, which themselves list primary data sources. The latter are given here only when they have been consulted directly.

Adams, C.D. (1972). 'Flowering Plants of Jamaica.' University of the West Indies, Mona. *M*

Aiken, S.G. (1983–4). Unpublished morphological data on Canadian grass genera. *M*

Aiken, S.G. and Darbyshire, S.J. (1983). 'Grass Genera of the Western Canadian Cattle Rangelands.' Biosystematics Research Institute Monograph No. 29. (Agriculture Canada: Ottawa.) *M*

Aiken, S.G., Darbyshire, S.J. and Lefkovitch, L.P. (1985). Restricted taxonomic value of leaf sections in Canadian narrow-leaved *Festuca* (Poaceae). *Can. J. Bot.* **63**(6), 995–1007. *T*

Al-Aish, M. and Brown, W.V. (1958). Grass germination responses to I.P.C. and classification. *Amer. J. Bot.* **45**, 16–23. *P*

Allred, K.W. (1982). Describing the grass inflorescence. *J. Range Management* **35**, 672–675. *M*

Alekseev, E.B. (1976). *Austrofestuca* E. Alekseev, Comb. Nov. — A new genus of the family Poaceae from Australia. *Byull. Mosk. Obsh. Ispyt. Prirody* **81**(5), 55–60. *T*

Alekseev, E.B. (1985). New genera of grasses. *Byull. Mosk. Obsh. Ispyt. Prirody* **90**, 102–109. (*Tsvelevia, Festucella, Hookerochloa, Parafestuca*). *T*

Alekseev, E.B. (1984). Genus *Festuca* L. (Poaceae) in Mexico et America Centrali. *Novitates Systematicae Plantarum Vascularum* **21**, 25–58. *T*

Amarasinghe, V. and Watson, L. (1988). Comparative ultrastructure of microhairs in grasses. *Bot. J. Linn. Soc.* **98**, 303–319. [A]

Amarasinghe, V. and Watson, L. (1989). Variation in salt secretory activity of microhairs in grasses. *Aust. J. Plant Physiol.* **16**, 219–229. [P]

Anderson, D.E. (1961). Taxonomy and distribution of the genus *Phalaris*. *Iowa State J. Sci.* **36**, 1–96. [T]

Anderson, D.E. (1974). Taxonomy of the genus *Chloris* (Gramineae). *Brigham Young Univ. Sci. Bull. Biol.* Ser. **19**, 1–133. [T]

Anton, A.M. and Cocucci, A.E. (1984). The grass megagametophyte and its possible phylogenetic implications. *Pl. Syst. Evol.* **146**, 117–121. [AG]

Appels, R., Scholes, G. and Chapman, G.D. (1987). The nature of change in nuclear DNA in the evolution of grasses. In 'Grass Systematics and Evolution', pp. 73–87. (Eds T.R. Soderstrom, K.W. Hilu, C.S. Campbell and M.E. Barkworth.) (Smithsonian Inst.: Washington.) [P]

Artucio, P.I. and Laguardia, A. (1987). Un Nuevo Enfoque Hacia la Definicion del Fruto de las Graminas. *Bol. Invest. Uraguay Fac. Agron.* **3**. [AT]

Atkins, R.J., Barkworth, M.E. and Dewey, R.D. (1984). A taxonomic study of *Leymus ambiguus* and *L. salinus* (Poaceae: Triticeae). *Syst. Bot.* **9**(3), 279–294. [T]

Auquier, P. (1963). Critères anciens at modernes dans la systematique de Graminées. *Nature Mosana* **16**, 1–63. [C]

Auquier, P. (1980). *Hubbardochloa*, a new genus of Grasses from Rwanda and Burundi. *Bull. Jard. Bot. Nat. Belg.* et *Bull. Nat. Plantentium Belg.* **50**, 241–247. [T]

Auquier, P. and Somers, Y. (1967). Récherches histotaxonomiques sur le chaume de Poaceae. *Bull. Soc. Roy. Bot. Belgique* **100**, 95–140. [A]

Avdulov, N.P. (1931). Karyo-systematische Untersuchung der Familie Gramineen. *Bull. Appl. Bot. Pl. Breed.* Supp. **44**. [G]

Baaijens, G.J. and Veldkamp, J.F. (1991). *Sporobolus* (Gramineae) in Malesia. *Blumea* **35**, 393–458. [T]

Backer, C.A. and Bakjuizen van den Brink, R.C. (1968). 'Flora of Java.' Vol. III. (Wolters-Noordhoff N.V.: Grongingen.) [M]

Barker, N.P. (1986). The shape and ultrastructure of the caryopsis of *Pentameris* and *Pseudopentameris*. *Bothalia* **16**, 65–69. [T]

Barkworth, M.E. (1983). *Ptilagrostis* in North America and its relationship to other Stipeae (Gramineae). *Syst. Bot.* **8**(4), 395–419. [T]

Barkworth, M.E. and Atkins, R.J. (1984). *Leymus* Hochst. Gramineae: Triticeae in North America: taxonomy and distribution. *Amer. J. Bot.* **71**(5), 609–625. [T]

Barkworth, M.E. and Dewey, D.R. (1985). Genomically based genera in the perennial Triticeae of North America: identification and membership. *Amer. J. Bot.* **72**(5), 767–776. [CGT]

Barkworth, M.E., Dewey, D.R. and Atkins, R.J. (1983). New generic concepts in the Triticeae of the Intermountain Region: key and comments. *Great Basin Naturalist* **43**(4), 561–572. [T]

Barkworth, M.E. and Everett, J. (1987). Evolution in the Stipeae: identification and relationships of its monophyletic taxa. In 'Grass Systematics and Evolution', pp. 251–264. (Eds T.R. Soderstrom, K.W. Hilu, C.S. Campbell and M.E. Barkworth.) (Smithsonian Inst.: Washington.) [T]

Baum, B.R. (1968). *Danthonia*. Plant Res. Inst. Contr. No. 676. Canadian Dept Agric., Ottawa. *Taxon* **17**, 444–446. [T]

Baum, B.R. (1973). The genus *Danthoniastrum*, about its circumscription, past and present status, and some taxonomic principles. *Österr. Bot. Z.* **122**, 51–77. [T]

Baum, B.R. (1977). Oats, wild and cultivated — a monograph of the genus *Avena* L. (Poaceae). (Dept Agric.: Ottawa.) [T]

Baum, B.R. (1978b). Taxonomy of the tribe Triticeae (Poaceae) using various numerical techniques. II. Classification. *Can. J. Bot.* **56**, 27–56. [C]

Baum, B.R. (1987). Numerical taxonomic analyses of the Poaceae. In 'Grass Systematics and Evolution', pp. 334–342. (Eds T.R. Soderstrom, K.W. Hilu, C.S. Campbell and M.E. Barkworth.) (Smithsonian Inst.: Washington.) [C]

Baum, B.R., Estes, J.R. and Gupta, P.K. (1987). Assessment of the genomic system of classification in the Triticeae. *Amer. J. Bot.* **74**, 1388–1395. *CG*

Beetle, A.A. (1955). The four subfamilies of the Gramineae. *Bull. Torrey Bot. Club* **82**, 196–197. *C*

Bennett, M.D. and Smith, J.B. (1976). Nuclear DNA amounts in angiosperms. *Philosophical Transactions, Royal Society of London*, series B, **274**, 227–274. *P*

Bennett, M.D., Smith, J.B. and Heslop-Harrison, J.S. (1982). Nuclear DNA amounts in Angiosperms. *Philosophical Transactions, Royal Society of London*, series B, **216**, 179–199. *P*

Bentham, G. (1882). *Schaffnera gracilis* Benth. *Hook. Ic. Pl.* **14**, t.1378. (*Schaffnerella.*) *T*

Bentham, G. and Hooker, J.D. (1862). 'Genera Plantarum.' Vol. 1. (Reeve: London.) *C*

Bentham, G. and Hooker, J.D. (1873). 'Genera Plantarum.' Vol. 2. (Reeve: London.) *C*

Bentham, G. and Hooker, J.D. (1883). Gramineae. In 'Genera Plantarum', Vol. 3. (Reeve: London.) *CM*

Bentham, G. and von Mueller, F. (1878). 'Flora Australiensis.' Vol. VII. (Reeve: London.) *CM*

Bjorkman, S.O. (1960). Studies in *Agrostis* and related genera. *Symb. Bot. Uppsal.* **17**, 1–112. *T*

Black, G.A. (1950). Novas espécies de Paniceae (Gramineae) do Brazil. *Boll. Tech. Inst. Agron. Norte* **20**, 29. (*Froesiochloa*, Olyreae.) *T*

Black, J.M. (1978). Lycopodiaceae-Orchidaceae. In 'Flora of South Australia', Part I, 3rd edition. (Govt Printer: South Australia.) *M*

Blake, S.T. (1941a). Studies of Queensland grasses. II. *Ectrosia, Sporobolus, Tragus, Arundinella, Eriochloa, Thaumastochloa, Sorghum*. Papers, Department of Biology, University of Queensland No. 1(18), pp. 1–22. *T*

Blake, S.T. (1941b). New genera of Australian grasses. *Dimorphochloa, Ancistrachne, Zygochloa*. Papers, Department of Biology, University of Queensland No. 1(19), pp. 1–12. *T*

Blake, S.T. (1944a). On *Streptachne* R.Br. and *Pheidochloa* gen. nov., two genera of grasses from Queensland. *Proc. Roy. Soc. Qld* **56**, 11–22. *T*

Blake, S.T. (1944b). Monographic studies in the Australian Andropogoneae, Part 1. *Bothriochloa, Capillipedium, Chrysopogon, Iseilema*. Papers, Department of Biology, University of Queensland No. 3, pp. 1–62. *T*

Blake, S.T. (1946). Two new grasses from New Guinea. *Blumea Suppl.* **3**(56), 56–62. (*Ancistragrostis, Buergersiochloa*.) *T*

Blake, S.T. (1972a). *Plinthanthesis* and *Danthonia* and a review of the Australian species of *Leptochloa* (Gramineae). *Contr. Qld. Herb.* **14**. (Mentions *Chionochloa, Sieglingia, Notodanthonia, Monachather* and *Erythranthera*.) *T*

Blake, S.T. (1972b). *Neurachne* and its allies (Gramineae). *Contr. Qld Herb.* **13**, 1–53. *T*

Blake, S.T. (1973). Taxonomic and nomenclatural studies in the Gramineae, No. 3. *Proc. Roy. Soc. Qld* **84**(3), 61–70. (*Digitaria, Ectrosia, Paspalidium*.) *T*

Bolkhovskikh, Z. *et al.* (1969). 'Chromosome Numbers of Flowering Plants.' (U.S.S.R. Academy of Sciences: Leningrad.) *G*

Bor, N.L. (1940). Gramineae. In 'Flora of Assam', Vol. V. (Ed. Kanjilal *et al.*) (Assam Government: Shillong.) *M*

Bor, N.L. (1948). *Bhidea* Stapf.: A new genus of Indian grasses. *Kew Bull.* **3**, 445–447. *T*

Bor, N.L. (1949). *Pogonachne* Bor.: A new genus of Indian grasses. *Kew Bull.* **1**, 176–178. *T*

Bor, N.L. (1951). *Timouria* Roshev. and *Psammochloa* Hitchc. *Kew Bull.* 1951: 186–192. *T*

Bor, N.L. (1952). Notes on Asiatic grasses: I. *Danthonia* in India. *Kew Bull.* 1951: 75–81. *T*

Bor, N.L. (1954). Notes on Asiatic Grasses: XV. *Triplopogon* Bor, a new genus of Indian grasses. *Kew Bull.* **1**, 51–56. *T*

Bor, N.L. (1954 publ. 1955). Notes on Asiatic grasses. XXII. *Trikeraia* Bor, a new genus of Stipeae. *Kew Bull.* 1954, 555–557. *T*

Bor, N.L. (1957a). Notes on Asiatic Grasses: XXVIII. *Kew Bull.* 1 412–413. (*Eulaliopsis.*) *T*

Bor, N.L. (1957b). Notes on Asiatic Grasses: XXXII. *Kew Bull.* 1. (*Indopoa.*) *T*

Bor, N.L. (1960). 'Grasses of Burma, Ceylon, India and Pakistan.' (Pergamon: Oxford.) (Excluding Bambuseae.) *CM*

Bor, N.L. (1968a). Gramineae. In 'Flora of Iraq', Vol. 9. (Ministry of Agric.: Baghdad.) *M*

Bor, N.L. (1968b). Studies in the Flora of Thailand (43). Gramineae: some new taxa. *Dansk Bot. Arkiv* 23(4), 467–471. (*Chumsriella, Ischaemum, Tripogon.*) *T*

Bor, N.L. (1970). Gramineae. In 'Flora Iranica', No. 70. (Ed. K.H. Rechinger.) (Akademische Druck- und Verlagsanstalt: Graz) *M*

Bosser, J. (1965). Notes sur les Graminées de Madagascar. II. Sur l'identité des genres *Boivinella* A. Camus et *Cyphochlaena* Hack. *Adansonia* n.s., 411–413. *T*

Bosser, J. (1969). 'Graminées des paturages et des cultures a Madagascar.' (Orstom: Paris.) *M*

Brandenburg, D.M., Blackwell, W.H. and Thieret, J.W. (1991). Revision of the genus *Cinna* (Poaceae). *Sida* 14, 581–596. *T*

Brown, W.V. (1958). Leaf anatomy in grass systematics. *Bot. Gaz.* 119, 170–178. *AC*

Brown, W.V. (1960). A cytological difference between the Eu-panicoideae and the Chlorideae (Gramineae). *Southw. Nat.* 5, 7–11. *A*

Brown, W.V. (1975). Variations in anatomy, associations, and origins of Kranz tissue. *Am. J. Bot.* 62, 395–402.

Brown, W.V. (1977). The Kranz syndrome and its subtypes in grass systematics. *Mem. Torrey Bot. Club* 23, 1–97, 126–130. *ACP*

Brown, W.V., Harris, W.F. and Graham, J.D. (1959). Grass morphology and systematics. I. The internode. *Southw. Nat.* 4. *M*

Brown, W.V. and Emery, W.H.P. (1957). Persistent nucleoli in grass systematics. *Am. J. Bot.* 44, 585–590.

Burbidge, N.T. (1941). A revision of the Australian species of *Enneapogon* Desf. *Proc. Linn. Soc. Lond.* Session 153, 52–91. *T*

Burbidge, N.T. (1953). The genus *Triodia* R.Br. (Gramineae). *Aust. J. Bot.* 1, 121–184. *T*

Burbidge, N.T. (1960). Further notes on *Triodia* R.Br. (Gramineae), with descriptions of five new species and one variety. *Aust. J. Bot.* 8, 381–395. *T*

Burbidge, N.T. and Gray, M. (1970). 'Flora of the Australian Capital Territory.' (Australian National University Press: Canberra.) *MT*

But, P.P.–H., Chia, L., Fung, H. and Hu, S.-Y. (1985). 'Hong Kong Bamboos.' (Urban Council: Hong Kong.) *M*

Butzin, F. (1966). *Mezochloa*, eine neue Paniceen-Gattung von Madagaskar. *Willdenowia* 4(2), 209–214. *T*

Butzin, F. (1970). Die systematische Gliederung der Paniceae. *Willdenowia* 6(1), 179–192. *CT*

Butzin, F. (1971). Der Umfang der Melinideae (Gramineae, Panicoideae) und die neue Gatung *Mildbraediochloa*. *Willdenowia* 6(2), 285–289. *T*

Butzin, F. (1973). Subfamilial nomenclature of Poaceae. *Willdenowia* 7, 113–168. *CM*

Butzin, F. (1979a). Synopsis der neuen Gramineengattungen der letzten 25 Jahre. *Willdenowia* 8, 471–480. *C*

Butzin, F. (1979b). Apikale Reduktionen im Infloreszenzberich der *Gramineae*. *Willdenowia* 9, 161–167. *M*

Cabrera, A.L. *et al.* (1970). Gramineae. In 'Flora de la Provincia de Buenos Aires', Part 2. (Coleccion Cientifica del I.N.T.A.: Buenos Aires.) *M*

Calderón, C.E. and Soderstrom, T.R. (1967). Las Gramineas Tropicales afines a *Olyra* L. *Atas do Simp. sôbre a Biota Amazônica* 4, 67–76. *T*

Calderón, C.E. and Soderstrom, T.R. (1973). 'Morphological and Anatomical Considerations of the Grass Subfamily Bambusoideae, Based on the New Genus *Maclurolyra*.' Smithsonian Contr. Bot. No. 11. (Smithsonian Inst.: Washington.) *T*

Calderón, C.E. and Soderstrom, T.R. (1980). 'The Genera of Bambusoideae (Poaceae) of the American Continent: Keys and Comments.' Smithsonian Contr. Bot. No. 44. (Smithsonian Inst.: Washington.) *CMT*

Campbell, S.C. (1985). The subfamilies and tribes of Gramineae (Poaceae) in the Southeastern United States. *J. Arnold Arboretum* **66**, 123–199. *C*

Campbell, C.S., Garwood, P.E. and Specht, L.P. (1986). Bambusoid affinities of the north temperate genus *Brachyelytrum* (Gramineae). *Bull. Torrey Bot. Club* **113**, 135–141. *T*

Campbell, C.S. and Kellogg, E.A. (1987). Sister group relationships of the Poaceae. In 'Grass Systematics and Evolution', pp. 217–224. (Eds T.R. Soderstrom, K.W. Hilu, C.S. Campbell and M.E. Barkworth.) (Smithsonian Inst.: Washington.) *C*

Campbell, S.C., Quinn, J.A., Cheplick, G.P. and Bell, T.J. (1983). Cleistogamy in Grasses. *Ann. Rev. Ecol. Syst.* **14**, 411–441. *GM*

Camus, E.G. (1913). 'Les Bambusées.' (Lechevalier: Paris.) *M*

Camus, A. (1919). *Arundinaria ciliata*. *Bull. Mus. Paris* 1919, 672. (*Vietnamosasa*.) *T*

Camus, A. (1920). Note sur le genre *Pseudosorghum* A. Camus. et Note sur le genre *Neohusnotia* A. Camus. *Mus. d'Hist. Nat. de Paris Bull.* **26**, 662–664. *T*

Camus, A. (1923). Le genre *Leptosaccharum* (Hackel) A. Camus. *Bull. Soc. Bot. France* **70**, 736–738. *T*

Camus, A. (1924). Genres nouveaux de Bambusées malgaches. *Compt. Rend. Acad. Sci.* **179**, 478–79. (*Pseudocoix, Hickelia*.) *T*

Camus, A. (1925a). *Hitchcockella*, genre nouveau de Bambusées malgaches. *Compt. Rend. Acad. Sci.* **181**, 253–255. *T*

Camus, A. (1925b). *Boivinella*, genre nouveau de Graminées: *Lecomtella* genre nouveau de Graminées malgaches. *Compt. Rend. Acad. Sci.*. **181**, 174–177, 567–568. *T*

Camus, A. (1927a). *Yvesia*, genre nouveau et espèces nouvelles de Graminées malgaches. *Bull. Soc. Bot. France* **73**, 687–691. (*Oriza, Brachiaria*.) *T*

Camus, A. (1927b). *Perulifera*, genre nouveau de la tribu des *Boivinelleae*. *Bull. Soc. Bot. France* **74**, 889–893. *T*

Camus, A. (1930). *Pseudostreptogyne*, genre nouveau de Graminées. *Bull. Soc. Bot. France* **77**, 476–479. *T*

Camus, A. (1957). *Pterochloris* (Graminées), genre nouveau de Madagascar. *Bull. du Mus. Nat. d'Hist. Naturelle*. 2e série, t.XXIX, no.4, 349–350. *T*

Camus, E.G. and Camus, A. (1923). Graminées. In 'Flore Général de L'Indo-Chine', Vol. 7. (Eds Lecompte, M.H. and Humbert, H.) (Masson: Paris.) *MT*

Canfield, R.H. (1934). Stem structure of grasses on the Jornada experimental range. *Bot. Gaz.* **95**, 636–648. *A*

Carolin, R.C., Jacobs, S.W.L. and Vesk, M. (1973). The structure of the cells of the mesophyll and parenchymatous bundle sheath of the Gramineae. *Bot. J. Linn. Soc.* **66**, 259–275. *A*

Chao, Chi-son and Chu, Cheng-de (1983). A study on the bamboo genus *Indosasa* of China. *Acta. Phytotax. Sinica* **21**, 60–75. *T*

Chao, Chi-son and Renvoize, S.A. (1989). A revision of the species described under *Arundinaria* (Gramineae) in Southeast Asia. *Kew Bull.* **44**, 349–367. *T*

Chaianan, Chumsri (1972). A revision of *Germainia* Balansa & Poitrasson (Gramineae). *Thai Forest Bull. (Botany)* **6**, 29–59. *T*

Chase, A. (1908). Notes on genera of Paniceae. III. *Proc. Biol. Soc. Washington* **21**, 185–186. (*Mniochloa*.) *T*

Chase, M.W. *et al.* (1993). Phylogenetics of seed plants: an analysis of nucleotide sequences from the plastid gene *rbc*L. *Ann. Missouri Bot. Gard.* **80**, 528–580. *C*

Chevalier, M.A. and Camus, A. (1921). Deux bambous nouveaux de l'Annam. *Bull. Mus. Paris* 1921, 450–452. (*Arundinaria pusilla = Vietnamosasa*.) *T*

Chia, L., Fung, H. and Yang, Y.L. (1988). *Monocladus, genus novum bambusoidearum* (Poaceae). *Acta Phytotaxonomica Sinica* **26**, 211–216. *T*

Chippindall, L.K.A. (1955). 'The Grasses and Pastures of South Africa.' (CNA: Johannesburg.) *M*

Chu, Cheng-dee and Chao, Chi-sen (1979). *Acidosasa*. *J. Nanjing Coll. For. Prod.* **142**. *T*

Clayton, W.D. (1965) Introduction to Gramineae. In 'The Gramineae, a Study of Cereal, Bamboo and Grass'. (Ed. A. Arber.) (Reprint edition.) (Cramer: Weinheim.) *C*

Clayton, W.D. *et al.* (1970–1982). Gramineae, Part 1. In 'Flora of Tropical East Africa'. (Ed. E. Milne-Redhead and R.M. Polhill.) (Crown Agents: London.) *M*

Clayton, W.D. (1966). Studies in Gramineae: X. Andropogoneae, the genus *Anadelphia* Hack. *Kew Bull.* **20**, 275–285. *T*

Clayton, W.D. (1966). Studies in Gramineae: XI. Adropogoneae, the genus *Elymandra* Hack. *Kew Bull.* **20**, 287–293. *T*

Clayton, W.D. (1966). Studies in the Gramineae: XII. *Parahyparrhenia, Hyperthelia, Exotheca. Kew Bull.* **20**, 433–449. *T*

Clayton, W.D. (1967–8.) Studies in Gramineae: XIII. Chlorideae: XIV. Arundineae. (*Phragmites* Adans.): XV. Arundinelleae: XVI. A remarkable new genus from Tanzania. (*Farrago*). *Kew Bull.* **21**, 99–117, 119–23, 125–127. *T*

Clayton, W.D. (1969). 'A Revision of the Genus *Hyparrhenia*.' Kew Bull. Add. Series II. (H.M. Stationery Office: London.) *T*

Clayton, W.D. (1970). Studies in the Gramineae: XXII. A curious new genus from Tanzania. *Kew Bull.* **24**, 461–463. (*Chlorocalymma*.) *T*

Clayton, W.D. (1971). Studies in the *Gramineae:* XXVI. Numerical taxonomy of the Arundinelleae. *Kew Bull.* **26**(44), 111–123. *T*

Clayton, W.D. (1972a). Gramineae. In 'Flora of West Tropical Africa', Vol. 3(2). (Ed. F.N. Hepper.) (Crown Agents: London.) *M*

Clayton, W.D. (1972b). Studies in the Gramineae: XXVI. (*Leptothrium, Tragus, Eragrostis.*) *Kew Bull.* **27**(1), 151–153. *T*

Clayton, W.D. (1972c). XXXI. The awned genera of Andropogoneae: XXXIX. *Kew Bull.* **27**(1), 447–449. *M*

Clayton, W.D. (1972d). XXXIX Andropogoneae, Centotheceae. *Kew Bull.* **27**(1), 457–471. *T*

Clayton, W.D. (1973). Studies in the Gramineae: XXXII. The tribe Zoysieae Miq. XXXIII. The awnless genera of the Andropogoneae. *Kew Bull.* **28**, 37–58. *MT*

Clayton, W.D. (1976). The chorology of African mountain grasses. *Kew Bull.* **31**(2), 274–288. *C*

Clayton, W.D. (1977). New species and combinations in African and Indian grasses. Studies in the Gramineae: XLIII. *Kew Bull.* **32**(4), 579–581. (*Chaetopoa, Dichanthium, Echinochloa, Loxodera, Pennisetum, Schizachyrium.*) *T*

Clayton, W.D. (1978a). The genus *Rhytachne* (Gramineae): The genus *Holcolemma* (Gramineae). *Kew Bull.* **32**(4), 767–74. *T*

Clayton, W.D. (1978b). The genus *Phacelurus* (Gramineae): The genus *Stereochlaena* (Gramineae). *Kew Bull.* **33**(2), 175–179, 295–297. (*Stereochlaena foliosa = Baptorhachis.*) *T*

Clayton, W.D. (1979). Notes on *Setaria* (Gramineae). *Kew Bull.* **33**(3), 501–609. *T*

Clayton, W.D. (1981a). Notes on the tribe *Andropogoneae* (Gramineae). *Kew Bull.* **35**(4), 823–828. *T*

Clayton, W.D. (1981b). Evolution and distribution of grasses. *Ann. Missouri Bot. Gard.* **68**, 5–14. *C*

Clayton, W.D. (1982). Notes on subfamily Chloridoideae (Gramineae). *Kew Bull.* **37**(3), 417–420. *T*

Clayton, W.D. (1985). Miscellaneous notes on pooid grasses. *Kew Bull.* **40**(4), 727–729. (*Danthoniastrum, Metcalfia, Neuropoa.*) *T*

Clayton, W.D. (1987). Miscellaneous notes on panicoid grasses. *Kew Bull.* **42**, 401-403. *T*

Clayton, W.D. and Hepper, F.N. (1974). Computer-aided chorology of W. African grasses. *Kew Bull.* **29**(1), 213–234. *C*

Clayton, W.D. and Panigrahi, G. (1974). Computer-aided chorology of Indian grasses. *Kew Bull.* **29**(4), 669–686. *C*

Clayton, W.D., Phillips, S.M. and Renvoize, S.A. (1974). Gramineae, Part 2. In 'Flora of Tropical East Africa', pp. 1–450. (Eds E. Milne-Redhead and R.M. Polhill.) (Crown Agents: London.) *M*

Clayton, W.D. and Renvoize, S.A. (1982). Gramineae, Part 3. In 'Flora of East Tropical Africa', pp. 1–898. (Ed. R.M. Polhill.) (A.A. Balkema: Rotterdam.) *M*

Clayton, W.D. and Renvoize, S.A. (1986). 'Genera Graminum. Grasses of the World.' Kew Bull. Add. Series XII. (H.M. Stationery Office: London.) *CM*

Clayton, W.D. and Richardson, F.R. (1973). The tribe Zoysieae Miq.: Studies in the Gramineae. XXXII. *Kew Bull.* **28**, 37–48. *AM*

Clifford, H.T. (1964). The systematic position of the grass genus *Micraira* F. Muell. *Univ. Queensland Papers, Dept of Botany* **4**, 87–93. *T*

Clifford, H.T. (1965). The classification of Poaceae: a statistical study. *Papers, Dept of Biology, Univ. of Qld.* **4**(15), 243–253. *C*

Clifford, H.T. (1967). A contribution to the leaf-anatomy of *Hubbardia heptaneuron* Bor (Gramineae). *Kew Bull.* **21**(1), 169–174. *T*

Clifford, H.T. (1968). Attribute correlations in the Poaceae (grasses). *Bot. J. Linn. Soc.* **62**, 59–67. *C*

Clifford, H.T. (1972–1988). Assorted unpublished observations on grass morphology, especially on seedling structure. *M*

Clifford, H.T. (1987). Spikelet and floral morphology. In 'Grass Systematics and Evolution', pp. 21–30. (Eds T.R. Soderstrom, K.W. Hilu, C.S. Campbell and M.E. Barkworth.) (Smithsonian Inst.: Washington.) *GM*

Clifford, H.T. and Goodall, D.W. (1967). A numerical contribution to the classification of Poaceae. *Aust. J. Bot.* **15**, 499–519. *ACM*

Clifford, H.T. and Watson, L. (1977). 'Identifying Grasses: Data, Methods and Illustrations.' (Queensland University Press: Brisbane.) *ACM*

Clifford, H.T., Williams, W.T. and Lance, G.N. (1969). A further numerical contribution to the classification of Poaceae. *Aust. J. Bot.* **17**, 119–131. *C*

Compère, P. (1963). *Ystia*, Nouveau Genre de Gramineae au Congo. *Bull. Jard. Bot.* (Etat, Bruxelles) **33**, 399–401. *T*

Conert, H.J. (1957). Beitrage zur Monographie der Arundinelleae. *Bot. Jahrb.* **77**(2/3), 226–354. *AM*

Conert, J.H. (1959). Über die Stellung der Gattung *Phaenosperma* im System der Gramineae. *Bot. Jahrb. Syst.* **78**, 195–207. *T*

Conert, H.J. (1960). *Metcalfia*, eine neue Gattung der Gramineen. *Willdenowia* **2**, 417–419. *T*

Conert, H.J. (1962). Über die Gramineen-Gattung *Asthenatherum* Nevski. *Senck. Biol.* **43**(4), 239–266. *T*

Conert, H.J. (1965). Über den Verwandshaftskreis der *Danthonia curva*. *Senck. Biol.* **46**(2), 175–182. *T*

Conert, H.J. (1969). *Karroochloa*, eine neue Gattung der Gramineen. *Senck. Biol.* **50**(3/4), 289–318. *T*

Conert, H.J. (1970). *Merxmuellera*, eine neue Gattung der Gramineen. (Poaceae: Arundinoideae.) *Senck. Biol.* **51**(1/2), 129–133. *T*

Conert, H.J. (1971). The genus *Danthonia* in Africa. *Mitt. Bot. Staats München* **10**, 299–308. (*Pseudopentameris, Merxmuellera.*) *T*

Connor, H.E. (1979). Breeding systems in grasses: a survey. *N.Z. J. Bot.* **17**, 547–574. *G*

Connor, H.E. (1981). Evolution of reproductive systems in the Gramineae. *Ann. Missouri Bot. Gard.* **68**, 48–74. *G*

Connor, H.E. (1984). Breeding systems in New Zealand grasses IX. Sex ratios in dioecious *Spinifex sericeus*. *N.Z. J. Bot.* **22**, 569–574. *G*

Connor, H.E. (1987). Reproductive biology in the grasses. In 'Grass Systematics and Evolution', pp. 117–132. (Eds T.R. Soderstrom, K.W. Hilu, C.S. Campbell and M.E. Barkworth.) (Smithsonian Inst.: Washington.) *G*

Connor, H.E. (1991). *Chionochloa* Zotov (Gramineae) in New Zealand. *New Zealand J. Bot.* **29**, 219–282. *T*

Connor, H.E. and Edgar, E. (1979). *Rytidosperma* Steudel (*Notodanthonia* Zotov) in New Zealand. *N. Z. J. Bot.* **17**, 311–337. *T*

Connor, H.E. and Edgar, E. (1986). Australasian alpine grasses: diversification and specialization. In 'The Flora and Fauna of Alpine Australasia', pp. 413–434. (Ed. B.A. Barlow.) (CSIRO: Melbourne) *G*

Cooke, and Stapf, O. (1908). *Pseudodichanthelium. Kew Bull.* 1908, 450–451 (as *Andropogon (Dichanthium?) serrafalcoides*). *T*

Cope, T.A. (1982). Poaceae. In 'Flora of West Pakistan'. (Eds E. Nasir and S.I. Ali.) (The Editors, sponsored by U.S. Department of Agriculture: Karachi.) *M*

Cope, T.A. (1988). A new species of *Trikeraia* (Gramineae) from Tibet and Bhutan. *Kew Bull.* **42**, 350. *T*

Cope, T.A. (1993). *Taeniorhachis*: a new genus of *Gramineae* from Somalia. *Kew Bull.* **48**, 403–405. *T*

Cotton, R. and Stace, C.A. (1977). Morphological and anatomical variation in *Vulpia* (Gramineae). *Bot. Not.* **130**, 173–188. *T*

Cronquist, A. (1968). 'The evolution and classification of flowering plants.' (Nelson: London.) *C*

Cronquist, A. (1981). 'An integrated system of classification of flowering plants.' (Columbia University Press: New York.) *C*

Cronquist, A., Holmgren, A.H., Holmgren, N.H., Reveal, J.L. and Holmgren, P.K. (1977). 'Intermountain Flora.' (Columbia University Press: New York.) *M*

Cummins, G.B. (1971). 'The Rust Fungi of Cereals, Grasses and Bamboos.' (Springer-Verlag: New York.) *P*

Dallwitz, M.J. (1974). A flexible computer program for generating identification keys. *Syst. Zool.* **23**, 50–7. *S*

Dallwitz, M.J. (1980). A general system for coding taxonomic descriptions. *Taxon* **29**, 41–6. *S*

Dallwitz, M.J. (1984). Automatic typesetting of computer-generated keys and descriptions. In 'Databases in Systematics', Systematics Association Special Volume No. 26, pp. 279–90. (Eds R. Allkin and F.A. Bisby.) (Academic Press: London.) *S*

Dallwitz, M.J. (1988a). Representation of extreme and normal values in DELTA. *DELTA Newsletter* **1**, 3. *S*

Dallwitz, M.J. (1988b). Notes on INTKEY. *DELTA Newsletter* **2**, 3–4. *S*

Dallwitz, M.J. (1989a). Diagnostic descriptions from INTKEY and CONFOR. *DELTA Newsletter* **3**, 8–13. *S*

Dallwitz, M.J. (1989b). Diagnostic descriptions for groups of taxa. *DELTA Newsletter* **4**, 8–13. *S*

Dallwitz, M.J. (1992). DELTA and INTKEY. In 'Proceedings of Workshop on Artificial Intelligence and Modern Computer Methods for Systematic Studies in Biology'. (Ed. R. Fortuner.) (University of California: Davis.) *S*

Dallwitz, M.J., Paine, T.A., and Zurcher, E.J. (1993). 'User's Guide to the DELTA System: a General System for Processing Taxonomic Descriptions.' 4th edition. 136pp. (CSIRO Division of Entomology: Canberra.) *S*

Dallwitz, M.J. and Zurcher, E.J. (1988). 'User's Guide to TYPSET: a Computer Type-setting Program.' 2nd edition. CSIRO Aust. Div. Entomol. Rep. No. 18, 1–28. *S*

Damanakis, M.E. and Economoy, G. (1986). 'The Grasses of Greece. Records of Occurrence and Distribution.' (Benaki Phytopathological Institute: Athens.) *M*

Darbyshire, S.J., Cayouette, J. and Warwick, S.I. (1992). The intergeneric hybrid origin of *Poa labradorica* Steudel (Poaceae). *Pl. Syst. Evol.* (in press). *T*

Darbyshire, S.J. and Cayouette, J. (1992). An examination of the holotype of ×*Dupontia dezhvevii* Prob. (Poaceae). *Taxon* **41**, 737–743. *T*

Davidse, G. (1978). A systematic study of the genus *Lasiacis* (Gramineae, Paniceae). *Ann. Missouri Bot. Gard.* **65**, 1133–1161. *T*

Davidse, G. (1987). Four new species of *Axonopus* (Poaceae: Paniceae) from tropical America. *Ann. Miss. Bot. Gard.* **74**, 416–423. *T*

Davidse, G. (1988). A revision of the genus *Prionanthium* (Poaceae–Arundineae). *Bothalia* **18**, 143–153. *T*

Davidse, G. and Burman, A.G. (1987). A new species of *Thrasya* (Poaceae: Panicoideae) from the Mosquitia of Nicaragua and Honduras. *Ann. Missouri Bot. Gard.* **74**, 434–436. *T*

Davidse, G. and Ellis, R.P. (1984). *Steyermarkochloa unifolia*, a new genus from Venezuela and Colombia (Poaceae: Arundinoideae: Steyermarkochloeae). *Ann. Missouri Bot. Gard.* **71**, 994–1012. *T*

Davidse, G., Soderstrom, T.R. and Ellis, R.P. (1986). *Pohlidium petiolatum* (Poaceae: Centotheceae), a new genus and species from Panama. *Syst. Bot.* **11**(1), 131–144. *T*

Davidse, G. and Ellis, R.P. (1987). *Arundoclaytonia*, a new genus of the Steyermark-ochloeae (Poaceae: Arundinoideae) from Brazil. *Ann. Missouri Bot. Gard.* **74**, 479–490. *T*

Davis, P.H. (1985). 'Flora of Turkey and the East Aegean Islands.' Vol. 9. (Edinburgh University Press) *M*

Dayanandan, P., Hebard, F.V., Baldwin, D. Van and Kaufman, P.B. (1977). Structure of gravity-sensitive sheath and internodal pulvini in grass shoots. *Amer. J. Bot.* **64**, 1189–1199. *M*

Decker, H.F. (1964a). An anatomic-systematic study of the classical tribe Festuceae (Gramineae). *Amer. J. Bot.* **51**, 453–463. *AC*

Decker, H.F. (1964b). Affinities of the grass genus *Ampelodesmos*. *Brittonia* **16**, 76–79. *T*

de Koning, R. and Sosef, M.S.M. (1985). The Malesian Species of *Paspalum* L. (Gramineae). *Blumea* **30**, 279–318. *T*

Dengler, N.G., Dengler, R.E. and Hattersley, P.W. (1985). Differing ontogenetic origins in leaf blades of C_4 grasses (Poaceae). *Amer. J. Bot.* **72**(2), 284–302. *A*

Deshpande, U.R., Prakash, Ved and Singh, N.P. (1989). *Bhidea borii*, a new species of Poaceae from India. *Current Science* **58**, 1094–1095. *T*

de Wet, J.M.J. (1954). The genus *Danthonia* in grass phylogeny. *Amer. J. Bot.* **41**, 204–201. *T*

de Wet, J.M.J. (1987). Hybridization and polyploidy in the Poaceae. In 'Grass Systematics and Evolution', pp. 188–194. (Eds T.R. Soderstrom, K.W. Hilu, C.S. Campbell and M.E. Barkworth.) (Smithsonian Inst.: Washington.) *G*

Dewey, D.R. (1984). The genomic system of classification as a guide to intergeneric hybridization with the perennial Triticeae. In 'Gene Manipulation in Plant Improvement', pp. 209–279. (Ed. J.P. Gustafson.) (Plenum Press: New York.) *CGT*

de Winter, B. (1951a). A morphological, anatomical and cytological study of *Potamophila prehensilis* (Nees) Benth. *Bothalia* **6**, 117–134. (*Prosphytochloa*.) *T*

de Winter, B. (1951b). *Pseudobromus*. *Bothalia* **6**, 139–149. *T*

de Winter, B. (1960). A new genus of Gramineae. *Bothalia* **7**, 387–390. (*Diandrochloa*.) *T*

de Winter, B. (1961). *Kaokochloa*. *Bothalia* **7**, 479–480. *T*

de Winter, B. (1965). The South African Stipeae and Aristideae (Gramineae). *Bothalia* **8**(3), 201–404. *T*

de Winter, B. (1966a). *Styppeiochloa* de Winter, gen. nov. *Bothalia* **9**(1), 134–137. *T*

de Winter, B. (1966b). Gramineae: *Polevansia* de Winter, gen. nov. *Bothalia* **9**(1), 130–134. *T*

Deyl, M. (1946). Study of the genus *Sesleria*. *Opera Botanica Czvechica* vol. **3**. *T*

Doebley, J.F. and Iltis, H.H. (1980). Taxonomy of *Zea* (Gramineae). I. A subgeneric classification, with key to taxa. *Amer. J. Bot.* **67**, 982–993. *C*

Dore, W.G. (1956). Some grass genera with liquid endosperm. *Bull. Torrey Bot. Club* **83**(5), 335–337. *A*

Dore, W.G. and McNeill, J. (1980). 'Grasses of Ontario.' Monograph 26. (Research Branch, Agriculture Canada: Ottawa.) *M*

Dransfield, S. (1989). *Sphaerobambos*, a new genus of bamboo (*Gramineae–Bambusoideae*) from Malesia. *Kew Bull.* **44**, 425–434. *T*

Du Plessis, H. and Spies, J.J. (1992). Chromosome numbers in the genus *Pentaschistis* (Poaceae, Danthonieae). *Taxon* **41**, 709–720. *G*

Dyer, R.A. (1976). 'The Genera of Southern African Flowering Plants.' Vol. 2. (Dept Agric. & Technical Services: Pretoria.) *M*

Ekman, (1911). Neue Brasillianische Gräser. *Arkiv För. Bot.* **10**(17), 37–39. (*Steirachne*.) *T*

Elffers, J. and Kennedy-O'Byrne, J. (1957). Notes on African Grasses: XXIV. *Richardsiella*, a new genus of grasses from Tropical Africa. *Kew Bull.* **11**, 455–462. *T*

Ellis, R.P. (1976). A procedure for standardizing comparative leaf anatomy in the Poaceae. I. The blade as viewed in transverse section. *Bothalia* **12**(1), 65–109. *A*

Ellis, R.P. (1977a). Distribution of the Kranz syndrome in the Southern African Eragrostoideae and Panicoideae, according to bundle sheath anatomy and cytology. *Agroplanta* **9**, 73–110. *AP*

Ellis. R.P. (1977b). Leaf anatomy of the South African Danthonieae. I. The genus *Dregeochloa*. *Bothalia* **12**, 209–213. *T*

Ellis, R.P. (1979). A procedure for standardizing comparative leaf anatomy in the Poaceae. II. The epidermis as seen in surface view. *Bothalia* **12**(4), 641–672. *A*

Ellis, R.P. (1980–1983). Leaf anatomy of the South African Danthonieae. II–VII (*Merxmuellera*). *Bothalia* **13–14**. *T*

Ellis, R.P. (1984a). Leaf anatomy of the South African Danthonieae (Poaceae). IX. *Asthenatherum glaucum*. *Bothalia* **15**(1/2), 153–159. *T*

Ellis, R.P. (1984b). *Eragrostis walteri* — a first record of non-Kranz leaf anatomy in the sub-family Chloridoideae (Poaceae). *Sth. Afr. J. Bot.* **3**(6), 380–386. *PT*

Ellis, R.P. (1985). Leaf anatomy of the South African Danthonieae (Poaceae). X. *Pseudopentameris*: XI. *Pentameris longiglumis* and *Pentameris* sp. nov.: XII. *Pentameris thuarii*: XIII. *Pentameris macrocalycina* and *P. obtusifolia*. *Bothalia* **15**(3/4), 561–571, 573–585. *T*

Ellis, R.P. (1986a). Leaf anatomy of the South African Danthonieae (Poaceae). XIV. *Pentameris dregeana*. *Bothalia* **16**, 235–241. (*Pentaschistis*.) *T*

Ellis, R.P. (1986b). Leaf anatomy of the South African Danthonieae. XV. The genus *Elytrophorus*. *Bothalia* **16**, 243–249. *T*

Ellis, R.P. (1987a). Unpublished information on leaf blade anatomy of southern African species, provided in the form of photos of original preparations. *A*

Ellis, R.P. (1987b). Leaf anatomy of the genus *Ehrharta* (Poaceae) in southern Africa: the Setacea group. *Bothalia* **17**(1), 75–89. *T*

Ellis, R.P. (1987c). Leaf anatomy of the genus *Ehrharta* (Poaceae) in southern Africa: the Villosa group. *Bothalia* **17**, 195–204. *T*

Ellis, R.P. (1988a). Leaf anatomy of the South African Danthonieae. XVI. The genus *Urochlaena*. *Bothalia* **18**, 101–104. *T*

Ellis, R.P. (1988b). Leaf anatomy in the South African Danthonieae. XVII. The genus *Chaetobromus*. *Bothalia* **18**, 195–209. *T*

Ellis, R.P. (1988c). Leaf anatomy and systematics of *Panicum* (Poaceae: Panicoideae) in southern Africa. *Monogr. Syst. Bot. Missouri Bot. Gard.* **25**, 129–156. *T*

Ellis, R.P. (1989). Leaf anatomy of the Southa African Danthonieae. XVIII. *Centropodia mossamedensis*. *Bothalia* **19**, 41–43. *T*

Esen, A. and Hilu, K.W. (1989). Immunological affinities among subfamilies of the Poaceae. *Amer. J. Bot.* **76**, 196–203. *C*

Esen, A. and Hilu, K.W. (1991). Electrophoretic and immunological studies of prolamins in Poaceae: II. Phylogenetic affinities of the Aristideae. *Taxon* **40**, 5–17. *C*

Fairbrothers, D.E. and Johnson, M.A. (1961). The precipitation reaction as indication of relationships in some grasses. *Recent Advances in Botany* **1**, 116–120. *P*

Farris, J.S. (1988). 'Hennig86.' Version 1.5. (Port Jefferson Station: New York) *S*

Faruqi, S.A., Quraish, H.B. and Halai, N. (1979). Chromosome number and morphological characteristics of some *Andropogoneae* of Pakistan. *Cytologia* **44**, 585–605. *G*

Felsenstein, J. (1993). 'PHYLIP (Phylogeny Inference Package).' Version 3.5c. Distributed by the author. (Department of Genetics, University of Washington: Seattle.) *S*

Fijten, F. (1975). A taxonomic revision of *Buergersiochloa* Pilg. (Gramineae). *Blumea* **22**, 415–418. *T*

Filgueiras, T.S. (1982). Taxonomia e distribuicão de *Arthropogon* Nees (Gramineae). *Bradea (Bol. do Herbarium Bradeanum)* **3**, 303–322. *T*

Filgueiras, T.S. (1989). Revisão de *Mesosetum* Steudel (Gramineae: Paniceae). *Acta Amazonica* **19**, 47–114. *T*

Filgueiras, T.S., Davidse, G. and Zuloaga, F.O (1993). *Ophiochloa*, a new endemic serpentine grass genus (Poaceae: Paniceae) from the Brazilian cerrado vegetatio. *Novon* **3**, 360–366. *T*

Fosberg, F.R. and Sachet, M.-H. 1982). Micronesian Poaceae: critical and distributional notes. *Micronesica* **18**, 69–80. (*Leptutopetium.*) *T*

Fournier, E. (1836). *Calamochloa, Gynerium, Gouinia.* In *Mexicanas Plantas* **2**, 102–103. *T*

Frean, M.L. and Marks, E. (1988). Chromosome numbers of C_3 and C_4 variants within *Alloteropsis semialata* (R. Br.) Hitchc. (Poaceae). *Bot. J. Linn. Soc.* **97**, 255–259. *T*

Frederiksen, S. and von Bothmer, R. (1989). Intergeneric hybridization between *Taeniatherum* and different genera of Triticeae, Poaceae. *Nord. J. Bot.* **9**, 229–240, HOL 122. *G*

Freitag, H. (1975). The genus *Piptatherum* (Gramineae) in Southwest Asia. *Notes, Royal Bot. Gard. Edinburgh* **33**(3), 341–408. *T*

Freitag, H. (1985). The genus Stipa (Gramineae) in Southwest and South Asia. *Notes, Royal Bot. Gard. Edinburgh* **42**(3), 355–489. *T*

Garcia, A.T.R., Blanca, G. and Torres, C.M. (1987). *Linkagrostis*, un género nuevo de la familia Poaceae. *Candollea* **42**, 379–388. *T*

Gardner, C.A. (1952). Gramineae. In 'Flora of Western Australia', Vol. 1, part 1. (Govt Printers: Perth.) *MI*

Gardner, C.A. and Hubbard, C.E. (1938). *Psammagrostis wiseana* Gardner & Hubbard. *Hook Ic. Pl.* **34**, t. 3361, 1–3. *T*

Gibbs A., Ding Shouwei, Howe J., Keese P., Mackenzie A., Skotnicki M., Srifah P. and Torronen M. (1990). Old v. new characters for systematics: cautionary tales from virology. *Aust. Syst. Bot.* **3**, 159–163. *CP*

Gibbs Russell, G.E. (1987). Taxonomy of the genus *Ehrharta* (Poaceae) in southern Africa: the Setacea group. *Bothalia* **17**(1), 67–73. *T*

Gibbs Russell, G.E. (1987–1988). Unpublished original morphological observations on southern African grasses (especially endemic genera). *MT*

Gibbs Russell, G.E. and Ellis, R.P. (1987). Species groups in the genus *Ehrharta* (Poaceae) in southern Africa. *Bothalia* **17**(1), 51–65. *T*

Gibbs Russell, G.E. and Ellis, R.P. (1988). Taxonomy and leaf anatomy of the genus *Ehrharta* (Poaceae) in southern Africa: the Dura group. *Bothalia* **18**, 165–171. *T*

Gibbs Russell, G.E., Reid, C., Vanrooy, J. and Smook, L. (1985). 'List of Species of Southern African Plants.' 2nd edition. Part 1. Memoirs of the Botanical Survey of South Africa No. 51. (Botanical Research Institute: Pretoria.) *MT*

Gibbs Russell, G.E., Watson, L., Koekemoer, M., Smook, L., Barker, N.P., Anderson, H.M. and Dallwitz, M.J. (1990). 'Grasses of Southern Africa. An Identification Manual with Keys, Descriptions, Classification and Automated Identification and Information Retrieval from Computerized Data.' Memoirs of the Botanical Survey of South Africa No. 58. 437 pp. (Botanical Research Institute: Pretoria.) *SI*

Gilliland, H.B. *et al.* (1971). Grasses of Malaya. In 'A Revised Flora of Malaya', Vol. 3. (Govt Printer, Botanic Gard.: Singapore.) *M*

Goldblatt, P. (ed.) (1979–1985). 'Index to Plant Chromosome Numbers.' (Missouri Botanical Garden: St. Louis.) *G*

Gould, F.W. (1973). A systematic treatment of *Garnotia. Kew Bull.* **27**(3), 515–561. *T*

Gould, F.W. (1974). Nomenclatural changes in the Poaceae. *Brittonia* **26**, 59–60. *T*

Gould, F.W. (1978). Poaceae. In 'Flora of Lesser Antilles', Vol. 3. (Ed. R.A. Howard.) (Arnold Arboretum: Harvard.) *M*

Gould, F.W. (1979a). The genus *Bouteloua* (Poaceae). *Ann. Missouri Bot. Gard.* **66**, 348–416. *T*

Gould, F.W. (1979b). 'A Key to the Genera of Mexican Grasses.' MP 1422. (Texas Agricultural Experiment Station) *M*

Gould, F.W. (1983). 'Grass Systematics.' 2nd edition. (McGraw-Hill: New York.) *CM*

Gould, F.W. and Clark, C.A. (1978). *Dichanthelium* (Poaceae) in the United States and Canada. *Ann. Missouri Bot. Gard.* **65**, 1088–1132. *T*

Gould, F.W. and Soderstrom, T.R. (1967). Chromosome numbers of some tropical American grasses. *Amer. J. Bot.* **54**, 676–683. *GT*

Griffiths, D. (1912). The Gramma grasses. *Contr. U.S. Nat. Herb.* **14**(3), 358–363. (*Bouteloua, Cathestecum.*) *T*

Grime, J.P., Shacklock, J.M. and Band, S.R. (1985). Nuclear DNA contents, shoot phenology and species co-existence in a limestone grassland community. *New Phytol.* **100**, 435–445. *P*

Groves, E.W. (1981). Vascular plant collections from the Tristan da Cunha group of islands. *Bull. Br. Mus. nat. Hist. (Bot.)* **8**, 333– 420. (*Parodiochloa.*) *T*

Guerra, L.C. (1981). A new genus of the *Poaceae. Folia Geobot. et Phytotaxon.* **16**, 439–440. (*Aristopsis.*) *T*

Gutierrez, M., Edwards, G.E. and Brown, W.V. (1976). PEP carboxykinase containing species in the *Brachiaria* group of the subfamily Panicoideae. *Biochem. Syst. Ecol.* **4**, 47–49. *P*

Gutierrez, M., Gracen, V.E. and Edwards, G.E. (1974a). Biochemical and cytological relationships in C₄ plants. *Planta Berl.* **119**, 279–300. *P*

Gutierrez, M., Gracen, V.E. and Edwards, G.E. (1974b). Biochemical and cytological bases for classifying C₄ plants into groups. *Plant Physiol. Suppl.* **27**. *P*

Hackel, E. (1887). Gramineae. In 'Die Naturlichen Pflanzenfamilien', Teil II, Abteilung 2. (Eds A. Engler and K. Prantl.) (Engelmann: Leipzig.) *CM*

Häfliger, E. and Scholz, H. (1980, 1981). 'Grass Weeds.' Vols 1 and 2. (CIBA-GEIGY Ltd: Basle.) *M*

Hamby, K.R. and Zimmer, E.A. (1988). Ribosomal RNA sequences for inferring phylogeny within the grass family (Poaceae). *Plant Syst. Evol.* **160**, 29–37. *CP*

Harberd, D.J. (1972). A note on the mesocotyl in the systematics of the Gramineae. *Ann. Bot.* **36**, 599–603. *M*

Harborne, J.B. and Williams, C.A. (1976). Flavonoid patterns in leaves of the Gramineae. *Biochem. Syst. Ecol.* **4**, 267–280. *P*

Harris, S.A. and Ingram, R.I. (1991). Chloroplast DNA and biosystematics: the effects of intraspecific diversity on plastid transmission. *Taxon* **40**, 393–412. *CP*

Hatch, M.D. and Kagawa, T. (1974). NAD-malic enzyme in leaves with C₄-pathway photosynthesis and its role in C₄ acid decarboxylation. *Arch. Biochem. Biophys.* **160**, 346–349. *P*

Hatch, M.D., Kagawa, T. and Craig, S. (1975). Subdivision of C₄ pathway species based on differing C₄ acid decarboxylating systems and ultrastructural features. *Aust. J. Plant Physiol.* **2**, 111–128. *P*

Hattersley, P.W. (1987). Variations in photosynthetic pathway. In 'Grass Systematics and Evolution', pp. 49–64. (Eds T.R. Soderstrom, K.W. Hilu, C.S. Campbell and M.E. Barkworth.) (Smithsonian Inst.: Washington.) *P*

Hattersley, P.W. (1992). C₄ photosynthetic pathway variation in grasses: (Poaceae): its significance for arid and semi-arid lands. In 'Desertified Grasslands: their Biology and Management', pp. 182–212. (Ed. G.P. Chapman.) (Linnean Society of London.) *AP*

Hattersley, P.W. and Browning, A.J. (1981). Occurrence of the suberized lamella in leaves of grasses of different photosynthetic types. *Protoplasma* **109**, 371–401. *AP*

Hattersley, P.W. and Long, C. (1990). Systematics, biogeography, and photosynthetic pathway variation in Indo-Malayan/African *Alloteropsis* J.S. Presl ex C. Presl (Poaceae). *ASBS Symposium 1990: Indo-Pacific Biogeography*, 14. *AP*

Hattersley, P.W. and Roksandic, Z. (1983). δ¹³C Values of C₃ and C₄ species of Australian *Neurachne* and its allies (Poaceae). *Aust. J. Bot.* **31**, 317–321. *T*

Hattersley, P.W. and Watson, L. (1975). Anatomical parameters for predicting photosynthetic pathways in grass leaves. *Phytomorphology* **25**, 325–333. *AP*

Hattersley, P.W. and Watson, L. (1976). C₄ grasses: an anatomical criterion for distinguishing between NADP-malic enzyme species and PCK or NAD-malic enzyme species. *Aust. J. Bot.* **24**, 297–308. *AP*

Hattersley. P.W. and Watson, L. (1993). Diversificaton of photosynthesis in grasses. In 'Grass Evolution and Domestication', Ed. G.P. Chapman. (Cambridge University Press.) *AP*

Hattersley, P.W., Watson, L. and Johnston, C.R. (1982). Remarkable leaf anatomical variations in *Neurachne* and its allies (Poaceae) in relation to C₃ and C₄ photosynthesis. *Bot. J. Linn. Soc.* **84**, 265–272. *AT*

Hattersley, P.W., Watson, L. and Osmond, C.B. (1977). *In situ* immunofluorescent labelling of ribulose-1, 5–bisphosphate carboxylase in C_3 and C_4 plants. *Aust. J. Pl. Physiol.* **4**, 523–539. **P**

Hattersley, P.W., Wong, Suan-Chin, Perry, S. and Roksandic, Z. (1986). Comparative ultrastructure and gas exchange characteristics of the C_3-C_4 intermediate *Neurachne minor* S.T. Blake (Poaceae). *Pl. Cell Environ.* **9**(3), 217–233. **T**

Hayata, (1918). *Ic. Pl. Formosa* **7** (76: f45), 76–77. (*Pollinia, Polliniopsis, Saccharum.*) **T**

Henty, E.E. (1969). 'A Manual of the Grasses of New Guinea.' (Dept of Forests: Lae.) **M**

Hilu, K.W. (1981). Taxonomic status of the disputable *Eleusine compressa* (Gramineae). *Kew Bull.* **36**(3), 559–563. (*Ochthochloa.*) **T**

Hilu, K.W. and Johnson. J.L. (1991). Chloroplast DNA reassociation and grass phylogeny. *Plant Syst. Evol.* **176**, 21–31. **C**

Hilu, K.W. and Wright, K. (1982). Systematics of Gramineae: a cluster analysis study. *Taxon* **31**, 9–36. **C**

Hitch, P.A. and Sharman, B.C. (1971). The vascular pattern of festucoid grass axes with particular reference to nodal plexi. *Bot. Gaz.* **132**, 38–56. **A**

Hitchcock, A.S. (1909). Catalogue of the grasses of Cuba. *Contr. U.S. Nat. Herb.* (Washington.) **12**, 183–189. (*Reimarochloa.*) **T**

Hitchcock, A.S. (1922). Grasses of British Guiana. *Contr. U.S. Nat. Herb.* (Washington) **22**(6), 439–514. **MT**

Hitchcock, A.S. (1927). 'The Grasses of Ecuador, Peru and Bolivia.' *Contr. U.S. Nat. Herb.* (Washington) **24**(8), 291–556. **M**

Hitchcock, A.S. (1930). The grasses of Central America. *Contr. U.S. Nat. Herb.* (Washington) **24**(9), 557–762. **MT**

Hitchcock, A.S. (1933). *Orinus* Hitchc., gen. nov. *J. Washington Acad. Sci.* **23**(3), 136. **T**

Hitchcock, A.S. (1936). Manual of the grasses of the West Indies. U.S. Dept of Agric. Misc. Publ. No. 243, (Govt Printing Office: Washington.) **M**

Hitchcock, A.S. and Chase, A. (1910). The North American species of *Panicum. Contr. U.S. Nat. Herb.* (Washington) **15**, 1–396. **T**

Hitchcock, A.S. and Chase, A. (1915). Tropical North American species of *Panicum. Contr. U.S. Nat. Herb.* (Washington) **17**(6), 459–539. **T**

Hitchcock, A.S. and Chase, A. (1920a). Revisions of North American grasses. *Contr. U.S. Nat. Herb.* (Washington) **22**(1). (*Ichnanthus, Laciacis, Brachiaria, Cenchrus.*) **T**

Hitchcock, A.S. and Chase, A. (1920b). Revisions of North American grasses. *Contr. U.S. Nat. Herb.* (Washington) **22**(3). (*Isachne, Oplismenus, Echinochloa, Chaetochloa.*) **T**

Hitchcock, A.S. and Chase, A. (1950). 'Manual of the Grasses of the United States.' U.S. Dept of Agric., Misc. Publ. No. 200. (Govt Printing Office: Washington.) **CM**

Holttum, R.E. (1956). The classification of bamboos. *Phytomorphology* 74–90. **C**

Holttum, R.E. (1967). The bamboos of New Guinea. *Kew Bull.* **21**, 263–292. **M**

Hooker, J.D. (1896). *Ischnochloa. Hook. Ic. Pl.* **25**, t. 2466 **T**

Hooker, J.D. (1897). Gramineae. In 'Flora of British India', Vol. VIII, CLXXIII. (Reeve: London) **M**

Hoshikawa, K. (1969). Underground organs of the seedlings and the systematics of Gramineae. *Bot. Gaz.* **130**, 192–203. **M**

Hoshino, T. and Davidse, G. (1988). Chromosome numbers of grasses (Poaceae) from southern Africa. *Ann. Missouri Bot Gard.* **75**, 866–873. **G**

Hsu, Chien-Chang (1978). Poaceae. In 'Flora of Taiwan', Vol. 5. (Epoch: Taipei.) **M**

Hubbard, C.E. (1929). Notes on African Grasses: XI. A new genus of grasses from Bechuanaland. *Kew Bull.* 319–323. (*Megaloprotachne.*) **T**

Hubbard, C.E. (1933a). *Thellungia advena* Stapf. *Hook. Ic. Pl.* **32**, t.3184, 103. **T**

Hubbard, C.E. (1933b). *Cleistochloa subjuncea* C.E. Hubbard. *Hook. Ic. Pl.* **33**, t.3209, 1–6. **T**

Hubbard, C.E. (1933c). *Calyptochloa gracillima* C.E. Hubbard. *Hook. Ic. Pl.* **33**, t.3210, 1–3. **T**

Hubbard, C.E. (1934). *Homopholis belsonii* C.E. Hubbard. *Hook. Ic. Pl.* **33**, t.3231, 1–2. *T*

Hubbard, C.E. (1935). *Echinopogon intermedius* C.E. Hubbard (with key to *Echinopogon* spp.). *Hook. Ic. Pl.* **33**, t.3261, 1–10. *T*

Hubbard, C.E. (1935). *Heterachne abortiva* (R.Br.) Druce (and key to the species of *Heterachne*). *Hook. Ic. Pl.* **33**, t.3283, 1–4. *T*

Hubbard, C.E. (1936a). *Thaumastochloa rariflora* (Bailey) C.E. Hubbard and *Thaumastochloa brassii* C.E. Hubbard (and key to the genera of Rottboelliastrae in Australia). *Hook. Ic. Pl.* **33**, t.3313, 3314. *T*

Hubbard, C.E. (1936b). *Apochiton* (and key to the genera of the Eragrosteae). *Hook. Ic. Pl.* **34**, t.3319. *T*

Hubbard, C.E. (1936c). A new genus from North East Africa. *Kew Bull.* 1936: 329. (*Phaenanthoecium*.) *T*

Hubbard, C.E. (1937a). *Danthonidium. Hook. Ic. Pl.* **34**, t. 3331. *T*

Hubbard, C.E. (1937b). *Chaetostichium minimum* Hochst. *Hook. Ic. Pl.* **34**, t. 3341. *T*

Hubbard, C.E. (1939a). *Humbertochloa greenwayi. Hook Ic. Pl.* **34**, t.3387. *T*

Hubbard, C.E. (1939b). Notes on African Grasses. XXII. *Kew Bull.* **10**, 643–655 (*Leptagrostis*.) *T*

Hubbard, C.E. (1940a). *Alloeochaete andongensis* Rendle. *Hook. Ic. Pl.* **35**, t.3418. *T*

Hubbard, C.E. (1940b). *Pogonochloa greenwayi* C.E. Hubbard. *Hook. Ic. Pl.* **35**, t.3421. *T*

Hubbard, C.E. (1943). *Limnopoa meeboldi* (C.E.C. Fischer). *Hook. Ic. Pl.* **35**, t. 3432. *T*

Hubbard, C.E. (1947). *Hydrothauma* C.E. Hubbard. *Hook. Ic. Pl.* **35**, t. 3458. *T*

Hubbard, C.E. (1949). *Thryidachne. Kew Bull.* **4**, 363. *T*

Hubbard, C.E. (1951). *Kerriochloa. Hook. Ic. Pl.* **35**, t.3494 *T*

Hubbard, C.E. (1952). Gramineae Australiensis: IV. *Spartochloa*, a new genus of Australian grasses. *Kew Bull.* **3**, 307–308. *T*

Hubbard, C.E. (1956). *Heteropholis. Hook. Ic. Pl.* **36**, t.3548 *T*

Hubbard, C.E. (1957). Notes on African Grasses: XXV. *Nematopoa*, a new genus from Southern Rhodesia: XXVI. *Piptophyllum*, a new genus from Angola: XXVII. *Crinipes* Hochst., a genus of grasses from North East Tropical Africa. *Kew Bull.* **12**, 51–58. *T*

Hubbard, C.E. (1962). *Maltebrunia gabonensis. Hook. Ic. Pl.* **36**, t. 3595. *T*

Hubbard, C.E. (1966). Gramineae. In 'A Dictionary of Flowering Plants and Ferns'. (Ed. J.C. Willis.) 8th edition, revised by H.K. Airy Shaw. (Cambridge University Press.) *C*

Hubbard, C.E. (1967a). *Habrochloa bullockii* (Gramineae). *Hook. Ic. Pl.* **37**, t.3645. *T*

Hubbard, C.E. (1967b). *Chaetopoa taylori. Hook. Ic. Pl.* **37** t.3646, t.3646. *T*

Hubbard, C.E. (1968). 'Grasses.' (Penguin: Harmondsworth.) *M*

Hubbard, C.E. and Schweickerdt, H.G. (1936). XXXII. Notes on African Grasses: XIX. *Oryzidium*, a new genus from South West Africa. *Kew Bull.* **5**, 326–329. *T*

Hughes, D.K. (1923). *Streptolophus, Orthachne* and *Streptachne. Kew Bull.* **5**, 177–181, 301–303. *T*

Hunziker, J.H., Wulff, A.F. and Soderstrom, T.R. (1982). Chromosome studies on the Bambusoideae (Gramineae). *Brittonia* **34**(1), 30–35. *G*

Hunziker, J.H., Wulff, A.F. and Soderstrom, T.R. (1989). Chromosome studies on *Anomochloa* and other Bambusoideae (Gramineae). *Darwiniana* **29**, 41–45. *G*

Hunziker, J.H. and Stebbins, G.L. (1987). Chromosomal evolution in the Gramineae. In 'Grass Systematics and Evolution', pp. 179–187. (Eds T.R. Soderstrom, K.W. Hilu, C.S. Campbell and M.E. Barkworth.) (Smithsonian Inst.: Washington.) *G*

Hutchinson, J. (1973). 'The Families of Flowering Plants Arranged According to a New System Based on their Probable Phylogeny.' 3rd edition. (Clarendon: Oxford.) *C*

Ibrahim, K.M. and Kabuye, C.H.S. (1987). 'An Illustrated Manual of the Grasses of Kenya.' (F.A.O., United Nations: Rome.) *M*

Iltis, H.H. and Doebley, J.F. (1980). Taxonomy of *Zea*. II. Subspecific categories in the *Zea mays* complex and a generic synopsis. *Amer. J. Bot.* **67**, 994–1004. *C*

Innes, R. Rose (1977). 'A Manual of Ghana Grasses.' (Land Resources Division, Ministry of Overseas Development: Surbiton, England.) *M*

Jacobs, S.W.L. (1971). Systematic position of the genera *Triodia* R.Br. and *Plectrachne* Henr. (Gramineae). *Proc. Linn. Soc. N.S.W.* **96**(3), 175–185. *T*

Jacobs, S.W.L. (1982). Comment on Proposal 520: Conservation of *Notodanthonia* Zotov (Gramineae). *Taxon* **31**, 737–743. *T*

Jacobs, S.W.L. (1984). A new grass genus from Australia. *Kew Bull.* **40**(3), 659–661. (*Monodia.*) *T*

Jacobs, S.W.L. (1990). Notes on Australian grasses (Poaceae). *Austrofestuca, Festucella, Hookerochloa, Australopyrum. Telopea* **3**, 601–603. *T*

Jacobs, S.W.L. and Highet, J. (1988). Re-evaluation of the characters used to distinguish *Enteropogon* from *Chloris* (Poaceae). *Telopea* **3**, 217–222. *T*

Jacobs, S.W.L. and Lapinpuro, L. (1986). The Australian species of *Amphibromus* (Poaceae). *Telopea* **2**, 715–729. *T*

Jacques-Félix, H. (1954). Notes sur les Graminées d'Afrique tropicale. VI. (Série Bambusoïde). *J.A.T.B.A.* **1**(1–4), 40–60. *T*

Jacques-Félix, H. (1962). Les Graminées (Poaceae) d'Afrique tropicale. I. Généralités, classification, description des genres. *Bull. Sci. Inst. des Recherches Agronomiques tropicales* **8**. (I.R.A.T.: Paris.) *ACM*

Jain, S.K. (1970). The genus *Manisuris* L. (Poaceae) in India. *Bull. Bot. Surv. India* **12**(1–4), 6–17. (Most of the species described now referred to *Glyphochloa.*) *T*

Jansen, P. (1952). *Ectrosiopsis* (Ohwi) Jansen. *Act. Bot. Meerl.* **1**, 474–475. *T*

Jauhar, P.P. and Crane, C.F. (1989). An evaluation of Baum *et al.*'s assessment of the genomic system of classification in the Triticeae. *Amer. J. Bot.* **76**, 571–576. *C*

Jirásek, V. (1966). Über die systematische Einordnung der Gattung *Molinia* Schrank (Poaceae). *Preslia* **38**, 23–65. *T*

Jirasek, V. and Jozifova, M. (1968). Morphology of lodicules, their variability and importance in the taxonomy of the Poaceae. *Bol. Soc. Arg. Bot.* **12**, 324–349. *M*

Johnston, C.R. and Watson, L. (1977). Microhairs: A universal characteristic of non-festucoid grass genera. *Phytomorphology* **26**(3), 297–301. *A*

Johnston, C.R. and Watson, L. (1981). Germination flaps in grass lemmas. *Phytomorphology* **31**(1&2), 78–85. *M*

Joshi, Y.C. *et al* (1983). Salt excretion by glands in *Diplachne fusca. Indian J. Plant Physiol.* **26**(2), 203. *P*

Judziewicz, E.J. (1984). *Scrotochloa*, a new genus of paleotropical pharoid grasses. *Phytologia* **56**(4), 299–304. *T*

Judziewicz, E.J. and Soderstrom, T.R. (1989). 'Morphological, Anatomical and Taxonomic Studies in *Anomochloa* and *Streptochaeta* (Poaceae: Bambusoideae).' Smithsonian Contr. Bot. No. 68. (Smithsonian Inst.: Washington.) *T*

Kabuye, C.H.S. and Renvoize, S.A. (1975). The genus *Alloeochaete*, tribe Danthonieae (Gramineae). *Kew Bull.* **30**(3), 569–577. *T*

Kakudidi, E.K.Z., Lazarides, M. and Carnahan, J.A. (1989). A revision of *Enneapogon* (Poaceae, Pappophoreae) in Australia. *Aust. J. Syst. Bot.* **1**, 325–353. *T*

Kammacher, P. et. al. (1973). Nombres Chromosomiques des Graminés de Côte-d'Ivoire. *Candollea* **28**, 194–217. *G*

Kartesz, J. and Kartesz, R. (1980). 'A Synonymized Checklist of the Vascular Flora of the United States, Canada and Greenland.' (University of North Carolina Press: Chapel Hill.) *M*

Kawanabe, S. (1979). Relationship between germination temperature and the classification of grasses. *Jap. J. Trop. Agric.* **23**(4), 173–180. *P*

Kellogg, E.A. (1985–6). Unpublished, morphological observations, notably on germination flaps of lemmas. *M*

Kellogg, E.A and Campbell, C.S. (1987). Phylogenetic analyses of the Gramineae. In 'Grass Systematics and Evolution', pp. 310–324. (Eds T.R. Soderstrom, K.W. Hilu, C.S. Campbell and M.E. Barkworth.) (Smithsonian Inst.: Washington.) *C*

Kellogg, E.A. (1989). Comments on genomic genera in the Triticeae (Poaceae). *Amer. J. Bot.* **76**, 796–805. *C*

King, G.J. and Ingrouille, M.J. (1987). Genome heterogeneity and classification of the Poaceae. *New Phytol.* **107**, 633–644. *CP*

Knobloch, I.W. (1968). 'A Check List of Crosses in the Gramineae.' (Dept of Botany and Plant Pathology, Michigan State University: E. Lansing.) (Comprehensive references to 1967 — not a source of data. The original references not yet entered here.) *G*

Koch, S.D. (1978). Notes on the genus *Eragrostis* (Gramineae) in the southeastern United States. *Rhodora* **80**, 390–403. (*Diandrochloa, Neeragrostis.*) *T*

Koch, S.D. (1979). The relationships of three Mexican Aveneae and some new characters for distinguishing *Deschampsia* and *Trisetum* (Gramineae). *Taxon* **28**, 225–235. (Also deals with *Peyritschia.*) *T*

Koechlin, J. (1962). Famille des graminées. In 'Flore du Gabon'. (Eds Aubréville, A., *et al.*) (Governement de la Republique du Gabon: Aubreville.) *M*

Koidzum, (1925). Contributiones ad cognitionem. *Bot. Mag. (Tokyo)* Vol. XXXIX, 23–24. (*Chikusichloa.*) *T*

Korthoff, H.M. and Veldkamp, J.F. (1984). A revision of *Aniselytron* with some new combinations in *Deyeuxia* in SE. Asia (Gramineae). *Gardens' Bull.* **37**(2), 213–223. *T*

Kuwabara, Y. (1960). The first seedling leaf in grass systematics. *J. Jap. Bot.* **35**, 139–145. *M*

Kuwabara, Y. (1961). On the shape and the direction of leaves of grass seedlings. *J. Jap. Bot.* **36**, 368–373. *M*

Laegaard, S. (1987). The genus *Aciachne. Nord. J. Bot.* **7**, 662–672. *T*

Launert, E. (1961). *Loxodera* and *Lepargochloa*, two new genera (Gramineae) from South Tropical Africa. *Bol. da Soc. Brot.* **35**, 79–89. *T*

Launert, E. (1963). *Bol. Soc. Brot.* **37**, 80. (*Loxodera.*) *T*

Launert, E. (1965a). Pro domo: *Loxodera. Senck. Biol.* **46**(2), 121–122. (Gramineae, Andropogoneae, Rottboelliineae, Rottboelliininae.) *T*

Launert, E. (1965b). *Eccoptocarpha*, a new genus from Zambia. *Senck. Biol.* **46**(2), 123–128. *T*

Launert, E. (1970). Miscellaneous taxa of Gramineae from South West Africa. *Mitt. Bot. München* Band **VIII**, 147–163. (*Pseudobrachiaria, Psilochloa* and *Acroceras, Tetrapogon, Leucophrys.*) *T*

Lazarides, M. (1970). 'The Grasses of Central Australia.' (Australian National University Press: Canberra). *M*

Lazarides, M. (1972). A revision of Australian Chlorideae. *Aust. J. Bot. Suppl. Series*, No. 5. *T*

Lazarides, M. (1978). The genus *Whiteochloa* C.E.Hubbard (Poaceae, Paniceae). *Brunonia* **1**(1), 69–93. *T*

Lazarides, M. (1979). *Micraira* F. Muell. (Poaceae, Micrairoideae): *Hygrochloa*, a new genus of aquatic grasses from the Northern Territory. *Brunonia* **2**, 67–84, 85–91. *T*

Lazarides, M. (1980a). The genus *Leptochloa* Beauv. (Poaceae, Eragrostideae) in Australia and Papua New Guinea. *Brunonia* **3**, 247–269. *T*

Lazarides, M. (1980b). *Aristida* L. (Poaceae) in Australia. *Brunonia* **3**, 271–333. *T*

Lazarides, M. (1980c). 'Tropical grasses of Southeast Asia.' Phanerogamarum, Monographiae XII. (Cramer: Vaduz.) *M*

Lazarides, M. (1981). Gramineae (Poaceae). In 'Flora of Central Australia'. (Ed. J.P. Jessop.) (A.H. & A.W. Reed: Sydney.) *M*

Lazarides, M. (1985). New taxa of tropical Australian grasses. *Nuytsia* **5**, 273–303. (*Plectrachne, Symplectrodia, Austrochloris, Arthragrostis, Oxychloris, Micraira.*) *T*

Lazarides, M., Hacker, J.B. and Andrew, M.H. (1991). Taxonomy, cytology and ecology of indigenous Australian sorghums (*Sorghum* Moench: Andropogoneae: Poaceae). *Aust. Syst. Bot.* **4**, 591–635. *T*

Lazarides, M., Lenz, J. and Watson, L. (1991). *Clausospicula*, a new Australian genus of grasses (Poaceae, Andropogoneae). *Aust. Syst. Bot.* **4**, 391–405. *T*

Lazarides, M. and Watson, L. (1986). *Cyperochloa*, a new genus in the Arundinoideae Dumortier (Poaceae). *Brunonia* **9**, 215–221. *T*

Lazarides, M. and Webster, R.D. (1984). *Yakirra* (Paniceae, Poaceae), a new genus for Australia. *Brunonia* **7**, 289–296. *T*

Lemée, A. (1929–1951). 'Dictionnaire descriptif et synonymique des genres de plantes phanérogames.' (Imprimerie Commerciale et Administrative: Brest.) *M*

Lin Wan-Tao. (1988). New taxa and combinations of Bambusoideae from China. *Acta Phytotaxonomica Sinica* **26**, 144–149. (*Metasasa*.) *T*

Linder, P.H. and Ellis, R.P. (1990). A revision of *Pentaschistis* (Arundineae: Poaceae). *Contr. Bolus Herbarium* **12**, 1–124. *T*

Londoño, X., and Peterson, P.M. (1992). *Guadua chacoensis* (Poaceae: Bambuseae), and its taxonomic identity, morphology and affinities. *Novon* **2**, 41–47. *T*

Löve, A. (1984). Conspectus of the Triticeae. *Feddes Repert.* **95**, 425–521. *CGM*

Löve, A. and Connor, H.E. (1982). Relationships and taxonomy of New Zealand wheat-grasses. *N.Z. J. Bot.* **20**, 169–186. (Triticeae, *Elymus, Agropyron, Cockaynea*.) *G*

Löve, A. and Löve, D. (1978). IOPB Chromosome number reports LXI. *Taxon* **27**, 375. *G*

Loxton (1974). *Tarigidia. Bothalia* **11**, 285. *T*

Macfarlane, T.D. (1979). 'A Taxonomic Study of the Pooid Grasses.' (Ph.D. Thesis, Australian National University: Canberra.) *ACM*

Macfarlane, T.D. (1987). Poaceae subfamily Pooideae. In 'Grass Systematics and Evolution', pp. 265–276. (Eds T.R. Soderstrom, K.W. Hilu, C.S. Campbell and M.E. Barkworth.) (Smithsonian Inst.: Washington.) *C*

Macfarlane, T.D. and But, P.P-H. (1982). *Tovarochloa* (Poaceae: Pooideae). A new genus from the High Andes of Peru. *Brittonia* **34**(4), 478–481. *T*

Macfarlane, T.D. and Watson, L. (1980). The circumscription of Poaceae subfamily Pooideae, with notes on some controversial genera. *Taxon* **29**, 645–666. *ACM*

Macfarlane, T.D. and Watson, L. (1982). The classification of Poaceae subfamily Pooideae. *Taxon* **31**(2), 178–203. (Poanae, Triticanae.) *C*

Maddison, W.P., and Maddison, D.R. (1992). 'MacClade: Analysis of Phylogeny and Character Evolution.' Version 3. 398pp. (Sinauer Associates: Sunderland, Massachusetts.) *S*

Maguire, B. *et al.* (1965). The Botany of the Guayana Highland. VI. *Mem. NY Bot. Gard.* **12**(3), 1–7. (*Raddiella*.) *T*

Maire, R. (1952, 1953, 1955). Gramineae. In 'Flore de L'Afrique du Nord', Vols 1–3. (Lechevalier: Paris.) *M*

Martin, A.C. (1947). Comparative internal morphology of seeds. *Amer. Midl. Nat.* **36**, 513–660. *A*

Matthei, O.R. (1975). Der *Briza*-Komplex in Südamerika: *Briza, Calotheca, Chascolytrum, Poidium* (Gramineae). *Willdenowia* Beiheft **8**, 168 pp. *T*

McClure, F.A. (1961). Toward a fuller description of Bambusoideae (Gramineae). *Kew Bull.* **15**, 321–324. *M*

McClure, F.A. (1973). 'Genera of Bamboos Native to the New World (Gramineae: Bambusoideae).' (Smithsonian Inst.: Washington.) *M*

Metcalfe, C.R. (1960). 'Anatomy of the Monocotyledons. I. Gramineae.' (Clarendon Press: Oxford.) *A*

Michael, P.W. (1980). A new perennial species of *Echinochloa* from New Guinea. *Telopea* **2**(1), 31–33. *T*

Moore, D.M. (1982). 'Flora Europaea Check-List and Chromosome Index.' (Cambridge University Press.) *G*

Moore, R.J. (ed.) (1973). 'Index to Plant Chromosome Numbers 1967–71.' Regnum Vegetabile Vol. 90. (I.A.P.T.: Utrecht.) *G*

Moore, R.J. (ed.) (1974). 'Index to Plant Chromosome Numbers 1972.' Regnum Vegetabile Vol. 91. (Oosthoek, Scheltema and Holkema: Utrecht.) *G*

Moore, R.J. (ed.) (1977). 'Index to Plant Chromosome Numbers for 1973–4.' Regnum Vegetabile Vol. 96. (Bohn, Scheltema and Holkema: Utrecht.) *G*

Morat, P. (1981). Note sur les Graminées de la Nouvelle-Calédonie. VI : *Lepturopetium*, genre nouveau endémique. *Adansonia*, sér. 2, **20**(4), 377–381. *T*

Morrone, O, Filgueiras, T.S., Zuloaga, F.O. and Dubcovsky, J (1993). A revision of *Anthaenantiopsis* (Poaceae: Paniceae). *Systematic Botany* **18**, 434–453. *T*

Morrone, O. and Zuloaga, F. (1991). Revision del genero *Streptostachys* (Poaceae-Panicoideae), su posicion sistematica dentro de la tribu Paniceae. *Ann. Missouri Bot. Gard.* **78**, 359–376. *T*

Muller, F.M. (1978). 'Seedlings of the North-Western European Lowland.' (Junk: The Hague.) *M*

Nair, V.J., Ramachandran, V.S. and Sreekumar, P.V. (1982a). *Chandrasekharania*. A new genus of Poaceae from Kerala, India. *Proc. Indian Acad. Sci. (Plant Sci.)* **91**, 79–82. *T*

Nair, V.J., Sreekumar, P.V., Vajravelu, E. and Bhargavan, P. (1982b). *Silentvalleya*. A new genus of Poaceae from Kerala, India. *J. Bombay Nat. Hist. Soc.* **79**, 654–656. *T*

Napper, D.M. (1963). Notes on East African grasses. *Kirkia* **3**, 117, 131–4. (*Pogononera, Coelachyrum.*) *T*

Napper, D.M. (1965). Grasses of Tanganyika. Ministry of Agriculture, Forests and Wildlife, Tanzania. Bulletin. No. 18. *M*

Nguyen, T.Q. (1990). New taxa of bamboos (Poaceae, Bambusoideae) from Vietnam. *Botanicheskii Zhurnal* **75**, 221–225. *T*

Nicora, E.G. (1962). Revalidacion de género d Gramineas 'Neeragrostis' de la flora Norteamericana. *Revista Argentina de Agronomia* **29**, 1–11. *T*

Nicora, E.G. and Rúgolo de Agrasar (1981). Los géneros sudamericanos afines a *Briza* L. *Darwiniana* **23**, 279–309. *T*

Nicora, E.G. and Rúgolo de Agrasar, Z.E. (1987). 'Los generos de Gramineas de America Austral.' (Editorial Hemisferio sur S.A.: Buenos Aires.) *M*

Ohwi, J. (1965). Gramineae. In 'Flora of Japan', English edition. (Eds F.G. Meyer and E.H. Walker.) (Smithsonian Inst.: Washington.) *M*

Ortíz, J.J. (1991). *Schenckochloa* (Poaceae, Chloridoideae, Eragrostidae), un género nuevo de Noreste Brasil. *Candollea* **46**, 241–249. *T*

Ortíz, J.J. (1993). Estudio sistematico del genero *Gouinia* (Gramineae, Chloridoideae). *Acta Bot. Mexico* **23**, 1–33. *T*

Pankhurst, R.J. (1986). A package of computer programs for handling taxonomic databases. *Comput. Applic. Biosc.* **2**, 33–9. *S*

Parkinson, C.E. (1933). A new Burmese bamboo. *Indian Forester* **59**, 707–709. (*Dendrochloa.*) *T*

Parodi, L.R. (1937). *Oplismenopsis. Not. Mus. la Plata Bot.* **2**. *T*

Parodi, L.R. (1938). Gramineas Austroamericanas nuevas o criticas. *Not. Mus. la Plata Bot.* Vol. 8. **40**, 75–100. (*Erianthecium, Anthaenantiopsis, Lasiacis, Setaria.*) *MT*

Parodi, L.R. (1961). La taxonomia de las Gramineas Argentinas a la luz de las investigaciones mas recentes. *Recent Adv. Bot.* **1**, 125–129. *CT*

Parodi, L.R. (1969). Gramineas: la familia botánica de los Pastos. In 'Flora ilustrada de Entre Ríos (Argentina)', Vol. 6, part 2, pp. 270–272. (Ed. Arturo Burkart.) (INTA: Buenos Aires.) (*Leptocoryphium.*) *CM*

Parodi, L.R. and Freir, F. (1945). Observaciones taxonómicas sobre las gramíneas estipeas. *Cienciae i investigoción* **1**(3), 144–146. (*Stipeae.*) *T*

Partridge, T.R., Dallwitz, M.J., and Watson, L. (1993). 'A Primer for the DELTA System.' 3rd edition. 15pp. (CSIRO Division of Entomology: Canberra.) *S*

Payne, R.W. (1975). GENKEY: a program for constructing diagnostic keys. In 'Biological Identification with Computers', pp. 65–72. (Ed. R.J. Pankhurst.) (Academic Press, London.) *S*

Peterson, P.M. (1989). A re-evaluation of *Bealia mexicana* (Poaceae–Eragrostideae). *Madroño* **36**, 260–265. *T*

Peterson, P.M. and Annable, C.R. (1990). A revision of *Blepharoneuron* (Poaceae: Eragrostideae). *Systematic Botany* **15**, 515–525. *T*

Peterson, P.M., Annable, C.R. and Franceschi, V.R. (1989). Comparative leaf anatomy of the annual *Muhlenbergia* (Poaceae). *Nord. J. Bot.* **8**, 575–583. *T*

Philipson, M.N. and Connor, H.E. (1984). Haustorial synergids in danthonoid grasses. *Bot. Gaz.* **145**(1), 78–82. *A*

Phillips, S.M. (1974). Studies in the *Gramineae*: XXXV. *Kew Bull.* **29**(2), 267–270. (*Psilolemma, Drake-Brockmania, Dactyloctenium, Heterocarpha, Trichoneura.*) *T*

Phillips, S.M. (1974). A review of the genus *Oropetium* (Gramineae). *Kew Bull.* **30**(3), 467–470. *T*

Phillips, S.M. (1982). A numerical analysis of Eragrostideae (Gramineae). *Kew Bull.* **37**, 133–162. *TC*

Phillips, S.M. and Launert, E. (1971). A revision of the African species of *Tripogon* Roem. & Schult. *Kew Bull.* **25**(2), 301–322. *T*

Phipps, J.B. (1967). Studies in the Arundinelleae (Gramineae). XVI. *Kirkia*, 418–427. (*Pleioneura, Danthoniopsis, Gilgiochloa*.) *T*

Pierce (1978). *Griffithsochloa. Bull. Torrey Bot. Club* **105**, 134. *T*

Pilger, R. (1954). Das System der Gramineae unter Ausschlus der Bambusoideae. *Bot. Jahrb. Syst., Pflanzengeschichte und Pflanzengeographie* **76**(3), 281–384. *C*

Potzal, E. (1964). Graminales. In 'A. Engler's Syllabus der Pflanzenfamilien', 12th edition. (Ed. H. Melchior.) (Gebrüder Borntrager: Berlin.) *C*

Prat, H. (1960). Revue d'Agrostologie; vers une classification naturelle des Graminées. *Bull. Soc. Bot. Fr.* **107**, 32–79. *C*

Prendergast, H.D.V. (1987). 'Structural, Biochemical and Geographical Relationships in Australian C_4 Grasses.' (Ph.D. Thesis, Australian National University: Canberra.) *AP*

Prendergast, H.D.V. and Hattersley, P.W. (1985). Distribution and cytology of Australian *Neurachne* and its allies (Poaceae), a group containing C_3, C_4 and C_3-C_4 intermediate species. *Aust. J. Bot.* **33**, 317–336. *GT*

Prendergast, H.D.V. and Hattersley, P.W. (1987). Australian C_4 grasses: leaf blade anatomical features in relation to C_4 acid decarboxylation types. *Aust. J. Bot.* **35**, 355–382. *AP*

Prendergast, H.D.V., Hattersley, P.W. and Stone, N.E. (1987). New structural/biochemical associations in leaf blades of C_4 grasses (Poaceae). *Aust. J. Pl. Physiol.* **14**, 403–420. *AP*

Probatov, N.S. and Yurtsev. B.A. (1984). New taxa of Poaceae from the North-east of the U.S.S.R. *Botanicheskii Zhurnal* **69**, 688–692. (Description of *Dupontopoa*.) *G*

Quraish, H.B. and Faruqi, S.A. (1979). Cytogenetics of *Cymbopogon* species from Pakistan. *Caryologia* **32**(3), 311–327. *G*

Reeder, J.R. (1957). The embryo in grass systematics. *Amer. J. Bot.* **44**, 756–768. *AC*

Reeder, J.R. (1962). The bambusoid embryo: a reappraisal. *Amer. J. Bot.* **49**, 639–641. *A*

Reeder, J.R. (1965). The tribe Orcuttieae and the subtribes of the Pappophoreae (Gramineae). *Madroño* **18**, 19–28. *T*

Reeder, J.R. (1982). Systematics of the tribe Orcuttieae (Gramineae) and the description of a new segregate genus, *Tuctoria. Amer. J. Bot.* **69**, 1082–1095. *T*

Reeder, J.R. and Reeder, C.G. (1963). Notes on Mexican grasses II. *Cyclostachya*, a new diocecious genus. *Bull. Torrey Bot. Club* **90**(2), 193–201. *T*

Reeder, J.R. and Reeder, C.G. (1968). *Parodiella*, A new genus of grasses from the High Andes. *Bol. de la Soc. Argent. de Bot.* **12**, 268–283. *T*

Reeder, J.R., Reeder, C.G. and Rzedowski, J. (1965). Notes on Mexican grasses. III. *Buchlomimus*, another diocecious genus. *Brittonia* **17**, 26–33. *T*

Reeder, J.R. and Toolin, L.J. (1989). Notes on *Pappophorum. Systematic Botany* **14**, 349–358. *T*

Reichert, E.T. (1913). 'The Differentiation and Specificity of Starches in Relation to Genera, Species etc.' Part 1. (Carnegie Inst.: Washington.) *AP*

Rendle, A.B. (1899). *Pogonarthria falcata. Hook. Ic. Pl.* (Plate 2610.) *T*

Rendle, A.B. (1922). *Leucophrys mesocoma. Hook. Ic. Pl.* **VI**(4). t.3095 *Kew Bull.* **31**(4), 844. *T*

Renvoize, S.A. (1979). A new genus in the tribe Arundineae (Gramineae). *Kew Bull.* **33**(3), 525–527. (*Gossweilerochloa*.) *T*

Renvoize, S.A. (1981). The sub-family Arundinoideae and its position in relation to a general classification of the Gramineae. *Kew Bull.* **36**(1), 85–102. *C*

Renvoize, S.A. (1982a). A survey of leaf-blade anatomy in grasses. I. Andropogoneae. *Kew Bull.* **37**(2), 315–321. *A*

Renvoize, S.A. (1982b). A new genus and several new species of grasses from Bahia (Brazil). *Kew Bull.* **37**(2), 323–332. *T*

Renvoize, S.A. (1982c). A survey of leaf-blade anatomy in grasses. II. Arundinelleae: III. *Garnotieae. Kew Bull.* **37**(3), 489–95, 497–500. [A]

Renvoize, S.A. (1983). A survey of leaf-blade anatomy in grasses. IV. *Eragrostideae. Kew Bull.* **37**(4), 469–478. [A]

Renvoize, S.A. (1985a). A survey of leaf-blade anatomy in grasses V. The bamboo allies. *Kew Bull.* **40**(3), 509–535. [A]

Renvoize, S.A. (1985b). A survey of leaf-blade anatomy in grasses: VI. Stipeae. A survey of leaf-blade anatomy in grasses: VII. Pommereulleae, Orcuttieae and Pappophoreae. *Kew Bull.* **40**(4), 731–744. [A]

Renvoize, S.A. (1985c). A review of *Tribolium. Kew Bull.* **40**, 795–799. [T]

Renvoize, S.A. (1987a). A survey of leaf-blade anatomy in grasses: X. Bambuseae. *Kew Bull.* **42**(1), 201. [A]

Renvoize, S.A. (1987b). A survey of leaf-blade anatomy in grasses: XI. Paniceae *Kew Bull.* **42**(3), 739–768. [A]

Renvoize, S.A. and Clayton, W.D. (1983). A new species of *Lophacme* (Gramineae) from Zambia. *Kew Bull.* **38**(1), 61–62. [T]

Renvoize, S.A. and Zuloaga, F. (1984). The genus *Panicum*, group *Lorea* (Gramineae). *Kew Bull.* **39**(1), 185–202. [T]

Richel, T., Basaula, N., Lejoly, J. and Vande Plassche, D. (1985). Analyse phytogeographique d'une liste provisoire des Poacées Zaïroises. *Bull. Soc. Roy. Bot. Belgique* **118**, 212–232. [AM]

Romero Zarco, C. (1984). Revisión del Género *Helictotrichon* Bess. ex Schultes & Schultes Fil. (Gramineae) en la Península Ibérica. I. Estudio Taxonómico. *Anales Jar. Bot. de Madrid* **41** 97–124. [T]

Romero Zarco, C. (1985). Revisión. II. Estudios Experimentales. *Anales Jar. Bot. de Madrid* **42**(1), 133–154. (*Helictotrichon.*) [T]

Romero Zarco, C. and Cabezudo, B. (1983). *Micropyropsis*, Genero Nuevo de Gramineae. *Lagascalia* **11**(1), 94–99. [T]

Rosengurtt, B. and Arrillaga de Maffei, B.R. (1979). *Lombardochloa*, Nuevo Genero de Gramineae. *Anales de la Facultad de Quirnica*, (Montevideo Uraguay) **9**, 255–268. [T]

Rosengurtt, B. and Laguardia, A. (1987). La Tendencia Envolvente del Escudete en el Embrion de Panicoideae (Gramineae). *Bol. Fac. Agron., Univ. Repub., Uruguay* **1**. [A]

Rosengurtt, B., Laguardia, A. and Arrillaga de Maffei, B.R. (1972). El carácter lipids del endosperma central en especies de gramineas. *Bol. Fac. Agron., Univ. Repub., Uruguay* **124**, 3–43. [AP]

Roshevits, R.Yu. (1937). 'Grasses. An Introduction to the Study of Fodder and Cereal Grasses.' Transl. 1980. (INSDOC: New Delhi-110012.) [CM]

Rost, T.L. and Lersten, N.R. (1973). A Synopsis and Selected Bibliography of Grass Caryopsis Anatomy and Fine Structure. *Iowa State J. Res.* **48**(1), 47–87. [A]

Rudall, P. and Dransfield, S. (1989). Fruit structure and development in *Dinochloa* and *Ochlandra* (Gramineae-Bambusoideae). *Ann. Bot.* **63**, 29–38. [T]

Rúgolo de Agrasar, Z.E. (1982). Revalidacion de genero *Bromidium* Nees et Meyen emend Pilger (Gramineae). *Darwinia* **24**, 187–216. [T]

Runemark, H. and Heneen, W.K. (1968). *Elymus* and *Agropyron*, a problem of generic delimitation. *Bot. Not.* **121**, 51–79. [T]

Sanchez, E. (1983). *Dasyochloa* Willdenow ex Rydberg (Poaceae). *Lilloa* **36**(1), 131–138. [T]

Saur, J.D. (1972). Revision of *Stenotaphrum* (Gramineae: Paniceae) with attention to its historical geography. *Brittonia* **24**, 202–222. [T]

Schmid, M. (1958). Flore agrostologique de l'Indochine. *Agron. Trop.* **13**, 7–51. [M]

Scholz, H. (1982a). Eine neue *Danthoniastrum*-Art (Gramineae) aus Albanien. *Brachiaria deflexa* und *B. multispiculata* sp. nova (Gramineae). *Willdenowia* **12**, 47–49, 287–289. [T]

Schouten, Y. and Veldkamp, J.F. (1985). A revision of *Anthoxanthum* including *Hierochloë* (Gramineae) in Malesia and Thailand. *Blumea* **30**, 319–351. [T]

Schwab, C. (1972). Floral structure and embryology of *Diarrhena* (Gramineae). *Diss. Abstr. Int.* B. **32**, 3812–3813. *T*

Schweickerdt, (1936). *Cymbosetaria sagittifolia* (A. Rich.) *Hook. Ic. Pl.* t.3320. *T*

Scoggan, H.J. (1978). Poaceae. In 'Flora of Canada', Vol. 2. (Nat. Mus. Canada: Ottawa.) *M*

Seberg, O., Frederiksen, S., Baden, C. and Linde-Lausen, IB. (1991). *Peridictyon*, a new genus from the Balkan peninsula, and its relationship with *Festucopsis* (Poaceae). *Willdenowia* **21**, 87–104. *T*

Sekine, Y. (1959). Ligule patterns of the Poaceae. *J. Jap. Bot.* **34**, 129–138. *M*

Senaratna, S.D.J.E. (1956). 'The Grasses of Ceylon.' Peradeniya Manual No. 8. (Govt Press: Ceylon.) *M*

Sendulsky, T. and Soderstrom, T.R. (1984). 'Revision of the South American Genus *Otachyrium* (Poaceae: Panicoideae).' Smithsonian Contr. Bot. No. 57. (Smithsonian Inst.: Washington.) *T*

Sharma, H.C. and Gill, B.S. (1982). Variability in spiklet disarticulation in *Agropyron* species. *Can. J. Bot.* **60**, 1771–1775. *T*

Simon, B.K. (1971). Rhodesian and Zambian grass lists. *Kirkia* **8**, 3–83. *M*

Simon, B.K. (1972). A revision of the genus *Sacciolepis* in the 'Flora Zambesiaca' area. *Kew Bull.* **27**, 387–406. *T*

Simon, B.K. (1978). A preliminary checklist of Australian grasses. Queensl. Dep. Primary Ind. Bot. Branch Tech. Bull. No. 3. *M*

Simon, B.K. (1984). New taxa of and nomenclatural changes in *Aristida* L. (Poaceae) in Australia. *Austrobailya* **2**(1), 86–102. *T*

Simon, B.K. (1986). *Planichloa* (Poaceae, Chloridoideae, Eragrostideae), a new grass genus from northern Queensland. *Austrobaileya* **2**, 211–216. *T*

Simon, B.K. (1987). Unpublished computer file of data on the grass genera of the world, comprehensively recorded for floristic distributions in terms of the Regions and Sub-regions of Takhtajan (1969). *M*

Simon, B.K. (1990). 'A Key to Australian Grasses.' Queensland Department of Primary Industries Information Series QI89019. (Queensland Department of Primary Industries: Brisbane.) *M*

Simon, B.K. (1992). Studies in Australian grasses 6. *Alexfloydia*, *Cliffordiochloa* and *Dallwatsonia*, three new panicoid grass genera from Eastern Australia. *Austrobaileya* **3**, 669–691. *T*

Smith, B.N. and Brown, W.V. (1974). The Kranz syndrome in the Gramineae as indicated by carbon isotopic ratios. *Amer. J. Bot.* **60**, 505–513. *P*

Smith, D. (1968). Classification of several native North American grasses as starch or fructosan accumulators in relation to taxonomy. *J. Br. Grassland Soc.* **23**, 306–309. *P*

Smith, D. (1973). The non-structural carbohydrates. In 'The Chemistry and Biochemistry of Herbage', pp. 105–155. (Eds G.W. Butler and R.W. Bailey.) (Academic Press: London.) *P*

Smith, P. (1970). Taxonomy and nomenclature of the Brome grasses (*Bromus* s.l.). *Notes from the Royal Botanic Gardens, Edinburgh* **30**, 361–375. *T*

Smouter, H. and Simpson, R.J. (1989). Occurrence of fructans in the Gramineae (Poaceae). *New Phytol.* **111**, 359–368. *P*

Sneath, P.H.A. (1988). The phenetic and cladistic approaches. In 'Prospects in Systematics', pp. 252–277. (Ed. D.L. Hawksworth.) (Clarendon Press: Oxford.) *C*

Soderstrom, T.R. (1967). Taxonomic study of Subgenus *Podosemum* and Section *Epicampes* of Muhlenbergia (Gramineae). *Contr. U.S. Nat. Herb.* **34**(4), 75–189. *T*

Soderstrom, T.R. (1969). Botany of the Guayana Highland. VIII. *Mem. NY Bot. Gard.* **18**(2), 11–22. (*Neurolepis*.) *T*

Soderstrom, T.R. (1978). *Chusquea* and *Swallenochloa* (Poaceae: Bambusoideae): Generic relationships and new species. *Brittonia* **30**(3), 297–312. *T*

Soderstrom, T.R. (1980). A new species of *Lithachne* (Poaceae: Bambusoideae) and remarks on its sleep movements. *Brittonia* **32**(4), 495–501. *T*

Soderstrom, T.R. (1981a). Some evolutionary trends in the Bambusoideae (Poaceae). *Ann. Missouri Bot. Gard.* **68**, 15–47. *MT*

1046

Soderstrom, T.R. (1981b). The grass subfamily Centostecoideae. *Taxon* **30**, 614–616. *T*

Soderstrom, T.R. (1981c). *Sucrea* (Poaceae: Bambusoideae), a new genus from Brazil. *Brittonia* **33**(2), 198–210. *T*

Soderstrom, T.R. (1981d). Observations on a fire-adapted Bamboo of the Brazilian Cerrado, *Actinocladum verticillatum* (Poacea: Bambusoideae). *Amer. J. Bot..* **68**(9), 1200–1211. *T*

Soderstrom, T.R. (1981e). *Olmeca*, a new genus of Mexican bamboos with fleshy fruits. *Amer. J. Bot.* **68**(10), 1361–1373. *T*

Soderstrom, T.R. (1982a). *Cryptochloa dressleri* (Poaceae), a new bambusoid grass from Panama. *Brittonia* **34**(1), 25–29. *T*

Soderstrom, T.R. (1982b). New species of *Cryptochloa* and *Piresia* (Poaceae: Bambusoidea). *Brittonia* **34**(2), 199–209. *T*

Soderstrom, T.R. and Calderón, C.E. (1978a). The species of *Chusquea* (Poaceae: Bambusoideae) with verticillate buds. *Brittonia* **30**(2), 154–164. *T*

Soderstrom, T.R. and Calderón, C.E. (1978b). *Chusquea* and *Swallenochloa* (Poaceae: Bambusoideae): Generic relationships and new species. *Brittonia* **30**(3), 297–312. *T*

Soderstrom, T.R. and Calderón, C.E. (1979b). *Arberella* (Poaceae: Bambusoideae): a new genus from tropical America. *Brittonia* **31**(4), 433–445. *T*

Soderstrom, T.R. and Decker, H.F. (1964). *Reederochloa*, a new genus of dioecious grasses from Mexico. *Brittonia* **16**, 334–339. *T*

Soderstrom, T.R. and Decker, H.F. (1965). *Allolepis*: a new segregate of *Distichlis* (Gramineae). *Madroño* **18**(2), 33–39. *T*

Soderstrom, T.R. and Decker, H.F. (1973). *Calderonella*, a new genus of grasses, and its relationships to the centostecoid genera. *Ann. Missouri Bot. Gard.* **60**(2), 427–441. *T*

Soderstrom, T.R. and Ellis, R.P. (1982). Taxonomic status of the endemic South African bamboo, *Thamnocalamus tessellatus*. *Bothalia* **14**, 53–67. *T*

Soderstrom, T.R. and Ellis, R.P. (1987). The position of bamboo genera and allies in a system of grass classification. In 'Grass Systematics and Evolution', 225–238. (Eds T.R. Soderstrom, K.W. Hilu, C.S. Campbell and M.E. Barkworth.) (Smithsonian Inst.: Washington.) *C*

Soderstrom, T.R. and Ellis, R.P (1988). 'The Woody Bamboos (Poaceae: Bambuseae) of Sri Lanka: a Morphological-Anatomical Study.' Smithsonian Contr. Bot. No. 72. 75 pp. (Smithsonian Inst.: Washington.) *T*

Soderstrom, T.R., Ellis, R.P. and Judziewicz, E.J. (1987). 'The Phareae and Streptogyneae of Sri Lanka: a Morphological-Anatomical Study.' Smithsonian Contr. Bot. No. 65. 27 pp. (Smithsonian Inst.: Washington.) *T*

Soderstrom, T.R. and Londoño, X. (1988). A morphological study of *Alvimia* (Poaceae: Bambuseae), a new Brazilian bamboo with fleshy fruits. *Amer. J. Bot.* **75**, 819–839. *T*

Soderstrom, T.R. and Zuloaga, F.O. (1988). New Species of Grasses in *Arberella, Cryptochloa*, and *Raddia* (Poaceae: Bambusoideae: Olyreae). *Brittonia* **37**(1), 22–35 *T*

Soderstrom, T.R. and Zuloaga, O. (1989). 'A Revision of the Genus *Olyra* and the New Segregate Genus *Parodiolyra* (Poaceae: Bambusoideae: Olyreae).' Smithsonian Contr. Bot. No. 69. (Smithsonian Inst.: Washington.) *T*

Soenarko, S. (1977). The genus *Cymbopogon* Sprengel (Gramineae). *Reinwardtia* **9**, 225–375. *T*

Sohns, (1956). *Calamochloa*: A Mexican Grass. *J. Washington Acad. Sci.* **46**(4), 109–112. (=*Sohnsia*.) *T*

Soreng, R.J., Davis, J.I. and Doyle, J.J. (1990). A phylogenetic analysis of chloroplast DNA restriction site variation in Poaceae subfam. Pooideae. *Plant Syst. Evol.* **172**, 83–97. *C*

Stace, C.A. (1978a). Changing concepts of the genus *Nardurus* Reichenb. (Gramineae). *Bot. J. Linn. Soc.* **76**, 344–350. *T*

Stace, C.A. (1978b). Notes on *Cutandia* and related genera. *Bot. J. Linn. Soc.* **76**, 350–353. *T*

Stace, C.A. (1989). IOPB chromosome data 1. *International Organization of Plant Biosystematists Newsletter* **13**, 15–21. *G*

Stanfield, D.P. (1970). 'The Flora of Nigeria.' (University Press: Ibadan.) *M*

Stapf, O. (1894). *Trilobachne. Hook. Ic. Pl.* **4** (part II). *T*

Stapf, O. (1895). *Cyathopus sikkimensis. Hook. Ic. Pl.* Plate 2395. *T*

Stapf, O. (1896). *Halopyrum mucronatum. Hook. Ic. Pl.* Plate 2448. *T*

Stapf, O. (1900a). Gramineae. In 'Flora Capensis', Vol. 7. (Ed. W.T. Thiselton-Dyer.) (Reeve: Ashford.) *M*

Stapf, O. (1900b). *Chloridion cameroni. Hook. Ic. Pl.* 4th series, **7**(4), Plate 2640. *T*

Stapf, O. (1911). *Dignathia. Hook. Ic. Pl.* t.2950. *T*

Stapf, O. (1915). *Homozeugos fragile. Hook. Ic. Pl.* t.3033. *T*

Stapf, O. (1916). *Odontelytrum. Hook. Ic. Pl.* **31**, t.3074. *T*

Stapf, O. (1917). Gramineae. In 'Flora of Tropical Africa', Vol. 9, CLVII. (Reeve: London.) *M*

Stapf, O. (1922). *Diheterpogon, Odyssea mucronata, Thyrsia. Hook. Ic. Pl.* t.3078, t.3093, t.3100. *T*

Stapf, O. (1927). *Lasiorrhachis hildebrandtii* Stapf. *Hook. Ic. Pl.* t.3124. *T*

Stapf, O. and Hubbard, C.E. (1929). XLIII. A new genus of grasses. *Kew Bull.* **8**, 244–247. (*Holcolemma.*) *T*

Stapf, O. and Hubbard, C.E. (1935). *Sclerandrium* (and a key to the genera of Polliniastrae). *Hook. Ic. Pl.* **33**, t.3262. *T*

Stebbins, G.L. and Crampton, B. (1961). A suggested revision of the grass genera of temperate North America. *Advances in Botany (Lectures and Symposia, IX Int. Bot. Congr.)* **1**, 133–145. *C*

Stent, S.M. (1923). South African Gramineae. A new genus and seven new species. *Bothalia* **1** 170. (*Mosdenia.*) *T*

Stent, S.M. (1924). *Bothalia* **1**, 292–293. (*Dactyloctenium, Lophacme, Enneapogon, Schmidtia.*) *T*

Stent, S.M. (1932). XVIII. Notes on African Grasses: XII. a new genus from the Orange Free State. *Kew Bull.* 1932, 151–153. (*Tarigidia.*) *T*

Stewart, D.R.M. (1964). Stalked glandular hairs in the Pappophoreae (Gramineae). *Ann. Bot.* **28**(112), 565–567. *T*

Stieber, M.T. (1987). Revision of *Ichnanthus* sect. Foveolatus (Gramineae: Panicoideae). *Syst. Bot.* **12**(2), 187–216. *T*

Swallen, J.R. (1935). The grass genus *Gouinia. Amer. J. Bot.* **22**, 31–41. *T*

Swallen, J.R. (1950). *Ectosperma*, a new genus of grasses from California. *Washington Acad. Sci.* **40**, 19–21. (*Swallenia.*) *T*

Swallen, J.R. (1956). New Grasses from Santa Catarina. *Sellowia — Anias Bot. do Hbr.* **7**, 7–12. (*Reitzia.*) *T*

Swallen, J.R. (1964). Two new genera of Olreae from South America. *Phytologia* **11**(3), 153–154. (*Bulbulus, Piresia.*) *T*

Swallen, J.R. (1968). *Acostia*, a new genus of grasses from Ecuador. *Bol. Soc. Arg. de Bot.* **12**, 109–110. *T*

Swofford, D.L. (1984). 'Phylogenetic analysis using parsimony.' Version 2.2. (Illinois Natural History Survey: Champaign.) *S*

Swofford, D.L. (1991). 'Phylogenetic analysis using parsimony (PAUP).' Version 3.0s. (Illinois Natural History Survey: Champaign.) *S*

Taira, H. (1968). Amino acid patterns of grass seed and systematics. *Proc. Jap. Soc. Pl. Taxon.* **2**, 14–17. *P*

Takhtajan, A. (1969). 'Flowering Plants: Origin and Dispersal.' Transl. C.A. Jeffrey. (Oliver & Boyd: Edinburgh.) (Source of the phytogeographical system employed.) *M*

Takhtajan, A.L. (1980). Outline of the classification of flowering plants (Magnoliophyta). *Bot. Rev.* **46**, 225–359. *C*

Taleisnik, E.L. and Anton, A.M. (1988). Salt glands in *Pappophorum* (Poaceae). *Ann. Bot.* **62**, 383–388. *P*

Tateoka, T. (1954). On the systematic significance of starch grains of seeds in Poaceae. *J. Jap. Bot.* **29**, 341–346. *AP*

Tateoka, T. (1955). Further studies of starch grains of seeds in Poaceae from the viewpoint of systematics. *J. Jap. Bot.* **30**, 199–207. *AP*

Tateoka, T. (1956). Notes on some grasses. I. Systematic position of the genus *Phaenosperma*. *Bot. Mag. (Tokyo)* **69**, 311–313. [T]

Tateoka, T. (1957a). Proposition of a new phylogenetic system of Poaceae. *J. Jap. Bot.* **32**, 275–287. [CT]

Tateoka, T. (1957b). Notes on some grasses. III. 5. Affinities of the genus *Brylkinia*. 6. Systematic position of the genus *Diarrhena*. *Bot. Mag. (Tokyo)* **70**, 8–12. [T]

Tateoka, T. (1959). Notes on some Grasses. IX. Systematic significance of Bicellular Microhairs of Leaf Epidermis. *Bot. Gaz.* **121**, 80–91. [A]

Tateoka, T. (1961). Notes on some grasses. XI. Leaf structure of *Eriachne*. *Bull. Torrey Bot. Club* **88**(1), 11–20. [T]

Tateoka, T. (1962). Starch grains of endosperm in grass systematics. *Bot. Mag. (Tokyo)* **75**, 377–383. [AP]

Tateoka, T. (1964a). Notes on some grasses. XVI. Embryo structure of the genus *Oryza* in relation to the systematics. *Amer. J. Bot.* **51**(5), 539–543. [T]

Tateoka, T. (1964b). Notes on some grasses. XVII. *Metcalfia*, a primative genus of the tribe Aveneae. *Bot. Mag. (Tokyo)* **77**, 66–72. [T]

Tateoka, T. (1965a). *Portersia*, a new genus of Gramineae. *Bull. Nat. Sci. Mus. Tokyo*, **8**(3),405–406. [T]

Tateoka, T. (1965b). Notes on some grasses. XVIII. Affinities of the genus *Streblochaete*. *Bot. Mag. (Tokyo)* **78**, 289–293. [T]

Tateoka, T. (1967). Lodicules of the tribes Ehrharteae and Aristideae. *Bull. Nat. Sci. Mus., Tokyo* **10**, 444–453. [T]

Tateoka, T. and Takagi, T. (1967). Notes on some grasses. XIX. Systematic significance of microhairs of lodicule epidermis. *Bot. Mag. (Tokyo)* **80**(952), 394–403. [A]

Terrell, E.E. (1971). Survey of the occurrence of liquid or soft endosperm in grass genera. *Bull. Torr. Bot. Cl.* **98**, 264–271. [AMP]

Terrell, E.E. and Peterson, P.M. (1993). Caryopsis morphology and classification in the Triticeae (Pooideae: Poaceae). *Smithsonian Contributions to Botany* **83**, 25 pp. (Smithsonian Inst.: Washington.) [T]

Thomasson, J.R. (1987). Fossil grasses: 1820–1986 and beyond. In 'Grass Systematics and Evolution', pp. 159–167. (Eds T.R. Soderstrom, K.W. Hilu, C.S. Campbell and M.E. Barkworth.) (Smithsonian Inst.: Washington.) [C]

Torres, I. Hernández and Koch, S.D. (1987). The status of the genus *Peyritschia* (Gramineae: Pooideae). *Phytologia* **61**, 453–455. [T]

Tran Van Nam (1971). La ligule dorsale des Graminées. *Bull.Soc. bot. Fr.* **118**, 639–658. [M]

Tsvelev, N.N. (1976). 'Grasses of the Soviet Union.' Vols I and II. (Nanka: Leningrad). English translation 1983. (Oxonion Press: New Delhi.) [CM]

Tsvelev, N.N. (1989). The system of grasses (Poaceae) and their evolution. *Bot. Rev.* **55**, 141–204. [C]

Türpe, A.M. (1970). Sobre la anatomía foliar de *Jansenella griffithiana* (C. Mueller) Bor. *Senck. Biol.* **51**(3/4), 277–285. [T]

Tutin, T.G., Heywood, V.H. *et al.* (eds) (1980). 'Flora Europaea.' Vol. 5. Poaceae. (Cambridge University Press.) [CM]

Valdès, J., Morden, C.W. and Hatch, S.L. (1986). *Gouldochloa*, a new genus of centhothecoid grasses from Tamaulipas, Mexico. *Syst. Bot.* **11**(1), 112–119 [T]

Van den Borre, A. (1993–4). Extensive anatomical data on Chloridoideae, as yet unpublished. [A]

Van den Borre, A., and Watson, L. (1994). The infrogeneric classification of *Eragrostis* (Poaceae). *Taxon* (in press). [AMT]

Van Eck-Borsboom, M.H.J. (1980). A Revision of *Eriachne* R. Br. (Gramineae) in Asia and Malesia. *Blumea* **26**(1), 127–138. [T]

Van Welzen, P.C. (1981). A taxonomic revision of the genus *Arthraxon* Beauv. (Gramineae). *Blumea* **27**, 255–300. [T]

Veldkamp, J.F. (1974). A taxonomic revision of *Dichelachne* Endl. (Gramineae), with some combinations in *Stipa* L. and *Oryzopsis* Michx. *Blumea* **22**, 5–12. [T]

Veldkamp, J.F. (1980). Conservation of *Notodanthonia* Zotov (Gramineae). *Taxon* **29**(2/3), 293–298. [T]

Veldkamp, J.F. (1985). *Anemanthele* Veldk. (Gramineae: Stipeae), a new genus from New Zealand. *Acta Bot. Neerl.* **34**(1), 105–109. *T*

Veldkamp, J.F., De Konig, R. and Sosef, M.S.M. (1986). Generic delimitation of *Rottboellia* and related genera (Gramineae). *Blumea* **31**, 281–307. *T*

Veldkamp, J.E.F. and Nowack, R. (1994). *Vietnamochloa. Adansonia* (in press). *T*

Veldkamp, J.F. and Schleindelin, H.J. van. (1989). *Australopyrum, Brachypodium* and *Elymus* (Gramineae) in Malesia. *Blumea* **34**, 61–76. *T*

Vickery, J.W. (1939). Revision of the indigenous species of *Festuca* Linn. in Australia. *Contr. N.S.W Nat. Herb.* **1**, 5–15. *T*

Vickery, J.W. (1940). A revision of the Australian species of *Deyeuxia* Clar. ex Beauv., with notes on the status of the genera *Calamagrostis* and *Deyeuxia. Contr. N.S.W. Nat. Herb.* **1**, 43–82. *T*

Vickery, J.W. (1941). A revision of the Australian species of *Agrostis* L. *Contr. N.S.W. Nat. Herb.* **1**, 101–119. *T*

Vickery, J.W. (1950). The species of *Amphipogon* R.Br. (Gramineae). *Contr. N.S.W. Nat. Herb.* **1**, 281–295. *T*

Vickery, J.W. (1952). *Pseudoraphis spinescens* (R.Br.) n. comb. *Proc. Roy. Soc. Qld.* **62**, 69–72. *T*

Vickery, J.W. (1956). A revision of the Australian species of *Danthonia* DC. *Contr. N.S.W. Nat. Herb.* **2**(3). *T*

Vickery, J.W. (1961). Gramineae, Part 1. Contr. N.S.W. Herbarium. Flora series, No. 19. (N.S.W. Dept of Agric.: Sydney.) *M*

Vickery, J.W. (1963). *Dryopoa*, a new grass genus allied to *Poa. Contr. N.S.W. Nat. Herb.* **3**, 195–197. *T*

Vickery, J.W. (1970). A taxonomic study of the genus *Poa* L. in Australia. *Contr. N.S.W. Nat. Herb.* **4**(4). (*Poa fax* = *Neuropoa* Clayton.) (See *Kew Bull.* **40**, (1985) 727–729.) *T*

Vickery, J.W. (1975a). Gramineae, Part 2. Flora of N.S.W. No. 19. (N.S.W. Dept of Agric.: Sydney.) *M*

Vickery, J.W. (1975b). Contributions to the taxonomy of Australian grasses. III. *Paspalidium, Setaria, Microlaena. Telopea* **1**, 40–43. *T*

Vickery, J.W. and Jacobs, S.W.L. (1980). *Nassella* and *Oryzopsis* (Poaceae) in New South Wales. *Telopea* **2**(1), 17–23. *T*

Vickery, J.W., Jacobs, S.W.L. and Everett, J. (1986). Taxonomic studies in *Stipa* (Poaceae) in Australia. *Telopea* **3**(1), 1–132. *T*

Wagnon, H.K. (1952). A revision of *Bromus* sect. *Bromopsis* of North America. *Brittonia* **7**, 415–480. *T*

Walsh, N.G. (1989). A new species of *Tetrarrhena* R.Br. (Poaceae) from Victoria. *Muelleria* **7**, 85–98. *T*

Walters, S.M. (1961). The shaping of Angiosperm taxonomy. *New Phytol.* **60**, 74–84. *C*

Walters, S.M. (1986). The name of the rose: a review of ideas on the European bias in Angiosperm classification. *New Phytol.* **104**, 537–546. *C*

Watson, L. (1971). Basic taxonomic data: the need for organization over presentation and accumulation. *Taxon* **20**, 131-136. *S*

Watson, L. (1972). Smuts on grasses: some general implications of the incidence of Ustilaginales on the genera of Gramineae. *Quart. Rev. Bot.* **47**(1), 46–62. *P*

Watson, L. (1977). Data for identifying Egyptian grass genera, and a summarised classification. *Publ. Cairo Univ. Herbarium* **7** and **8**, 231–254. *MI*

Watson, L. (1983). Taxonomic patterns in grass pollen antigens and allergens and their potential significance. In Baldo, B.A. and Howden, M.E.H. (eds.), '*Allergen standardization in Australia and current allergen research*'. *Proc. Sydney Allergen Group* **3**, 37–47. *P*

Watson, L. (1990). The Grass Family, Poaceae. Introductory Chapter, In: *Reproductive Versatility in the Grasses*. (Ed. G.P. Chapman). Cambridge University Press. *C*

Watson, L., Aiken, S.G., Dallwitz, M.J., Lefkovitch, L.P. and Dubé, M. (1986). Canadian Grass Genera: keys and descriptions in English and French from an automated data bank. *Can. J. Bot.* **64**, 55–70 (with 2 microfiches). *MS*

Watson, L. and Clifford, H.T. (1976). The major groups of Australasian grasses: a guide to sampling. *Aust. J. Bot.* **24**, 489–507. *C*

Watson, L., Clifford, H.T. and Dallwitz, M.J. (1985). The classification of Poaceae: subfamilies and supertribes. *Aust. J. Bot.* **33**, 433–484. *C*

Watson, L. and Dallwitz, M.J. (1980). 'Australian Grass Genera: Anatomy, Morphology and Keys.' 209 pp. (Research School of Biological Sciences, Australian National University: Canberra.) *ACMS*

Watson, L. and Dallwitz, M.J. (1981). An automated data bank for grass genera. *Taxon* **30**, 424–429. With 2 microfiches comprising 'Grass Genera: Descriptions', 2nd (1980) edition. *ACMS*

Watson, L. and Dallwitz, M.J. (1985). 'Australian Grass Genera: Anatomy, Morphology, Keys and Classification.' 2nd edition. 165 pp. (Research School of Biological Sciences, Australian National University: Canberra.) *ACMS*

Watson, L. and Dallwitz, M.J. (1988). 'Grass Genera of the World: Illustrations of Characters, Descriptions, Classification, Interactive Identification, Information Retrieval.' (Research School of Biological Sciences, Australian National University: Canberra.) *ACMS*

Watson, L. and Dallwitz, M.J. (1989). Grass genera of the world: interactive identification and information retrieval. *Flora Online* **22**. *ACMS*

Watson, L. and Dallwitz, M.J. (1991). Grass genera of the world: an INTKEY package for automated identification and information retrieval of data including synonyms, morphology, anatomy, physiology, cytology, classification, pathogens, world and local distribution, and references. 2nd edition. *Flora Online* **22**. *ACMS*

Watson, L. and Dallwitz, M.J. (1992). The Grass Genera of the World. 1038 pp., CAB International, Cambridge. (Generated automatically from the database.) *ACMS*

Watson, L. and Dallwitz, M.J. (1994). 'The Grass Genera of the World: Interactive Identification and Information Retrieval.' Floppy disks for MS-DOS. (CSIRO Publications: Melbourne.) *ACMS*

Watson, L., Dallwitz, M.J., Gibbs, A.J. and Pankhurst, R.J. (1988). Automated taxonomic descriptions. In 'Prospects in Systematics', pp. 292–304. (Ed. D.L. Hawksworth.) (Clarendon Press: Oxford.) *S*

Watson, L., Damanakis, M. and Dallwitz, M.J. (1988). 'The Grass Genera of Greece'. 231 pp. In Greek. (University of Crete: Heraklion.) *MS*

Watson, L. and Gibbs, A.J. (1974). Taxonomic patterns in the host ranges of viruses among grasses, and suggestions on generic sampling for host-range studies. *Ann. appl. Biol.* **77**, 23–32. *P*

Watson, L., Gibbs Russell, G.E. and Dallwitz, M.J. (1989). Grass genera of southern Africa: interactive identification and information retrieval from an automated data bank. *S. Afr. J. Bot.* **55**, 452–63. *S*

Watson, L. and Johnston, C.R. (1978). Taxonomic variation and stomatal insertion among grass leaves. *Aust. J. Bot.* **26**, 235–238. *A*

Watson, L. and Knox, R.B. (1976). Pollen wall antigens and allergens: taxonomically-ordered variation among grasses. *Ann. Bot.* **40**, 399–408. *P*

Watson, L. and Milne, P.W. (1972). A flexible system for automatic generation of special purpose keys and its application to Australian grass genera. *Aust. J. Bot.* **20**, 331-352. *S*

Weatherwax, P. (1939). The morphology and phylogenetic position of the genus *Jouvea* (Gramineae). *Bull. Torrey Bot. Club* **66**, 315–325. *T*

Webster, R.D. (1983). A revision of the genus *Digitaria* Haller (Paniceae: Poaceae) in Australia. *Brunonia* **6**, 131–216. *T*

Webster, R.D. (1987). 'The Australian Paniceae (Poaceae).' (J. Cramer: Berlin & Stuttgart.) *M*

Wen, T.H. and He, X.L. (1989). The morphology of fruits and starches in bamboos, and its relation to systematic position. *Acta Phytotaxonomica Sinica* **27**, 365–377. *T*

Wheeler, D.E.B., Jacobs, S.W.L. and Norton, B.E.L. (1982). 'Grass Genera of New South Wales.' Monograph No. 3. (University of New England Publishing Unit: Armidale.) *M*

Willemse, L.P.M. (1982). A discussion of the Ehrharteae (Gramineae) with special reference to the Malesian taxa formerly included in *Microlaena*. *Blumea* **28**, 181–194. *T*

Willis, J.H. (1970). Ferns, conifers and monocotyledons. In 'A Handbook to Plants in Victoria', Vol. 1. (Melbourne University Press.) *M*

Yates, H.O. (1966a). Morphology and cytology of *Uniola* (Gramineae). *Southw. Nat.* **11**(2), 145–189. *T*

Yates, H.O. (1966b). Revision of grasses traditionally referred to *Uniola*. II. *Chasmanthium*. *Southw. Nat.* **11**(4), 415–455 *T*

Yen, C. and Yang, J-L. (1990). *Kengyilia gobicola*, a new taxon from west China. *Can. J. Bot.* **68**, 1894–1897. *T*

Yeoh, Hock-Hin and Watson, L. (1981). Systematic variation in amino acid compositions of grass caryopses. *Phytochem.* **20**, 1041–1051. *P*

Yeoh, Hock-Hin and Watson, L. (1982a). Variations in free protein amino acid compositions of grass leaves. *Biochem. Sys. Ecol.* **10**, 55–63. *P*

Yeoh, Hock-Hin and Watson, L. (1982b). Taxonomic variation in total leaf protein amino acid compositions of grasses. *Phytochem.* **21**, 615–626. *P*

Yeoh, Hock-Hin, Badger, M.R. and Watson, L. (1980). Variations in K_mCO_2 of ribulose-1,5–biphosphate carboxylase among grasses. *Pl. Physiol.* **66**, 1110–1112. *P*

Yeoh, Hock-Hin, Badger, M.R. and Watson, L. (1981). Variations in kinetic properties of ribulose-1,5–bisphosphate carboxylases among plants. *Pl. Physiol.* **67**, 1151–1155. *P*

Yeoh, Hock-Hin, Stone, N.E. and Watson, L. (1981). Taxonomic variation in amino acid compositions of ribulose-1,5–bisphosphate carboxylases from grasses. *Biochem. Sys. Ecol.* **9**(4), 307–312. *P*

Yeoh, Hock-Hin, Stone, N.E. and Watson, L. (1982). Taxonomic variation in the subunit amino acid compositions of RuBP carboxylases from grasses. *Phytochem.* **21**(1), 71–80. *P*

Young, D.J. and Watson, L. (1970). The classification of Dicotyledons: a study of the upper levels of the hierarchy. *Aust. J. Bot.* **18**, 387–433. *C*

Zarco, C. Romero (1985). Estudio taxonomico del genero *Pseudarrhenatherum* Rouy (Gramineae) en la peninsula Iberica. *Lagascalia* **13**, 255–273. *T*

Zuloaga, F.O. (1981). Las especies argentinas del género *Ichnanthus* (Gramineae). El género *Panicum* (Gramineae) en la República Argentina. II. *Darwiniana* **23**(1), 189–221, 233–256. *T*

Zuloaga, F.O. (1987). Estudio exomorfologico e histofoliar de las especies americanas del genero *Acroceras* (Poaceae: Paniceae). *Darwiniana* **28**, 191–217. *T*

Zuloaga, F.O. (1987). Systematics of New World species of *Panicum* (Poaceae: Paniceae). In 'Grass Systematics and Evolution', pp. 287–306. (Eds T.R. Soderstrom, K.W. Hilu, C.S. Campbell and M.E. Barkworth.) (Smithsonian Inst.: Washington.) *T*

Zuloaga, F.O. and Morrone, O. (1993). *Gerritea*, a new genus of Paniceae (Poaceae: Panicoideae) from South America. *Novon* **3**, 213–219. *T*

Zuloaga, F.O. and Judziewicz, E.J. (1991). A revision of *Raddiella* (Poacaea: Bambusoideae: Olyreae). *Ann. Miss. Bot. Gard.* **78**, 928–941. *T*

Zuloaga, F.O. and Judziewicz, E.J. (1993). *Agnesia*, a new genus of Amazonian herbaceous bamboos (Poaceae: Bambusoidieae: Olyreae). *Novon* **3**, 306–309. *T*

Zuloaga, F.O., Morrone, O. and Judziewicz, E.J. (1993). Endemic herbaceous bamboo genera of Cuba (Poaceae: Bambusoideae: Olyreae). *Ann. Miss. Bot. Gard.* **80**, 846–861. *MT*

Zuloaga, F.O. and Soderstrom, T.R. (1985). Classification of the outlying species of New World *Panicum* (Poaceae: Paniceae). *Smithsonian Contrib. Bot.* **59**. (Smithsonian Inst.: Washington.) *T*

APPENDIX

Species Sampled for Anatomy

Leaf Blades

The following species have been examined for leaf blade anatomy in the course of constructing the data bank, and are represented on the Taxonomy Lab's slide collection. Details of individual specimens are obtainable from the Collectors' Notebooks in the Taxonomy Lab. Leaf anatomical information in the descriptions has been taken from divers sources (see References), but largely represents a combination of observations on this material with those of Metcalfe (1960). This is a list of permanent slides: it does not include species examined via temporary preparations from living material, thus somewhat underestimating the extent of sampling.

In future undertakings of this kind (and perhaps in future versions of this package), the kind of information conveyed here will be incorporated directly into the DELTA data files. Meanwhile, please note that while every effort has been made to ensure that the anatomical data *entered in the DELTA generic descriptions* conform with the generic alignments employed there, this is not true of the present list. Names of species sampled were entered down the years under generic alignments in use at the time, and some obvious conflicts reflect the low priority awarded to the very laborious and relatively unimportant task of updating them nomenclaturally.

Acamptoclados sessilispica Nash, *Achlaena piptostachya* Griseb., *Aciachne pulvinata* Benth., *Acrachne verticillata* Wight & Arn., *Acrachne racemosa* Ohwi, *Acritochaete volkensii* Pilger, *Acroceras macrum* Stapf, *Acroceras munroanum* (Balansa) Henrard, *Actinocladum verticillatum* (Nees) McClure, *Aegilops cylindrica* Host, *Aegopogon cenchroides* Willd., *Aeluropus littoralis* (Gouan) Parl., *Afrotrichloris martinii* Chiov., *Agenium villosum* Nees, *Agropyron cristatum* (L.) J. Gaertner, *Agropyron scabrum* (Labill.) Beauv. (= *Elymus*), *Agropyropsis lolium* (Batt. & Trab.) A. Camus, *Agrostis aemula* R. Br., *Agrostis australiensis* Mez, *Agrostis avenacea* Gmelin, *Agrostis billardieri* R.Br., *Agrostis tenuis* Sibth. (= *A. capillaris* L.), *Aira cupiana* Guss., *Aira elegans* Willd., *Airopsis tenella* (Cav.) Ascherson & Graebner, *Alexfloydia repens* B.K. Simon, *Alloeochaete geniculata* (Rendle) C.E. Hubb., *Allolepis texana* (Vasey) Soderstrom & Decker, *Alloteropsis cimicina* (L.) Stapf, *Alloteropsis semialata* (R.Br.) Hitchc., *Alopecurus pratensis* L., *Alvimia auriculata* Soderstrom & Londoño, *Amblyopyrum muticum* (Jaub. & Spach) Eig, *Ammochloa involucrata* Murbeck, *Ammochloa palestina* Boiss., *Ammophila arenaria* (L.) Link, *Ampelodesmos mauritanicus* (Poir.) Dur. & Schinz., *Amphibromus* 'sp.', *Amphibromus neesii* Steud., *Amphicarpum muhlenbergianum* (Schult.) Hitchc., *Amphipogon amphipogonoides* (Steud) Vickery, *Amphipogon avenaceus* R. Br., *Amphipogon caricinus* F. Muell., *Amphipogon debilis* R. Br., *Amphipogon laguroides* R. Br., *Amphipogon strictus* R. Br., *Amphipogon turbinatus* R. Br., *Anadelphia bigeniculata* W.D. Clayton, *Anadelphia leptocoma* Pilger, *Anadelphia pumila* Jac.-Fél., *Anadelphia trepidaria* (Stapf) Stapf, *'Anadelphia scyphofera'* W.D. Clayton, *Ancistrachne uncinulata* (R. Br.) Blake, *Ancistragrostis uncinioides* S.T. Blake, *Andropogon virginicus* L., *Andropterum variegatum* Stapf (= *A. stolzii*), *Aniselytron treutleri* (= *Aulacolepis*), *Anisopogon avenaceus* R. Br., *Anthaenantiopsis trachystachya* (Nees) Pilger, *Anthenantia rufa* (Ell.) Schult., *Anthephora acuminata* (Rendle) Stapf & Hubbard, *Anthephora pubescens* Nees, *Anthochloa lepidula* Nees & Meyen, *Anthoxanthum odoratum* L., *Antinoria agrostidea* (DC.) Parl., *Apera spica-venti* (L.) Beauv., *Aphanelytrum procumbens*, *Apluda mutica* L., *Apochiton burttii* Hubbard, *Apoclada cannavierira* (Alvarado de Silviera) McClure, *Apocopis courtallumensis* (Steud.) Henr., *Apocopis mangalorensis*

(Hochst.) Henr., *Arberella dressleri* Soderstrom & Calderon, *Arctagrostis latifolia* (R. Br.) Griseb., *Arctophila fulva* (Trin.) N.J. Anderss., *Aristida adscensionis* L., *Aristida aemulans* Melderis, *Aristida brainii* Melderis, *Aristida congesta* Roem. & Schult., *Aristida caput-medusae* Domin, *Aristida hordeacea* Kunth, *Aristida kenyensis* Henr., *Aristida meridionalis* Henr., *Aristida ramosa* R. Br., *Aristida rhiniochloa* Hochst., *Aristida scabrivalvis* Hack., *Aristida stipitata* Hack., *Aristida vestita* Thunb., *Arrhenatherum elatius* (L.) Beauv., *Arthragrostis aristispicula* B.K. Simon, *Arthragrostis deschampsioides* (Domin) Lazarides, *Arthraxon hispidus* (Thunb.) Makino, *Arthropogon villosus* Nees, *Arundinaria alpina* K. Schuman, *Arundinaria* sp., *Arundinella nepalensis* Trin, *Arundo donax* L., *Asthenatherum forskahlei* (= *Centropodia*), *Asthenochloa tenera* Buse, *Astrebla pectinata* (Lindl.) F. Muell. ex Benth., *Athroostachys capitata* (Hook.) Benth., *Atractantha falcata* McClure, *Aulacolepis treutleri* (O. Ktze.) Hack. (= *Aniselytron*), *Aulonemia fulgor* McClure, *Australopyrum retrofractum* (J. W. Vickery) A. Löve (ssp. *velutinum*), *Australopyrum pectinatum* (Labill.) A. Löve, *Austrochloris dichanthoides* (Everist) Lazarides, *Austrofestuca littoralis* (Labill.) A. Alexeev, *Avellinia michelii* (Savi) Parl., *Avena fatua* L., *Axonopus affinis* Chase

Bambusa arnhemica F. Muell., *Bealia mexicana* Scribn. ex Beal, *Beckmannia syzigachne* (Steud.) Fernald, *Bellardiochloa variegata* (Lam.) Kerguélen, *Bewsia biflora* (Hack.) Goossens, *Bhidea burnsiana* Bor, *Blepharidachne bigelovii* (S. Watts) Hack., *Blepharidachne kingii* (S. Watts) Hack., *Blepharoneuron tricholepis* (Torr.) Nash, *Boissiera squarrosa* (Banks & Soland.) Nevski, *Bothriochloa macra* (Steudel) Blake, *Bouteloua curtipendula* (Michx.) Torr., *Brachiaria decumbens* Stapf, *Brachiaria eruciformis* (J.E. Sm.) Griseb., *Brachiaria foliosa* R.Br., *Brachyachne convergens* (F. Muell.) Stapf, *Brachychloa schiemanniana* Phillips, *Brachyelytrum erectum* Beauv., *Brachypodium distachyon* (L.) Beauv., *Brachypodium pinnatum* (L.) Beauv., *Briza maxima* L., *Briza minor* L., *Briza triloba* (= *B. subaristata*), *Bromus diandrus* Roth, *Bromus erectus* Huds. (= *Zerna*), *Bromus mollis* L. (= *B. hordeaceus* L.), *Bromus unioloides* Kunth, *Brylkinia caudata* (Munro) Fr. Schmidt, *Buchloë dactyloides* (Nutt.) Engelm., *Buchlomimus nervatus* (Swallen) Reeder & Rzedowski, *Buergersiochloa bambusoides* Pilger

Calamagrostis epigejos (L.) Roth, *Calamovilfa longifolia* (Hook.) Hack., *Calderonella sylvatica* Soderstrom & Decker, *Calosteca brizoides* (Lam.) Mattei, *Calyptochloa gracillima* Hubbard, *Camusiella vatkeana* (Schum.) Bosser, *Capillipedium spicigerum* Blake, *Castellia tuberculosa* (Moris) Bor, *Catabrosa aquatica* (L.) Beauv., *Catabrosella humilis* (Bieb.) Tsvel., *Catalepis gracilis* Stapf & Stent, *Catapodium loliaceum* (Hudson) Link (= *Desmazeria marina*), *Catapodium rigidum* (L.) Hubbard, *Cathestechum erectum* Vasey & Hack., *Cenchrus australis* R. Br., *Cenchrus biflorus* Roxb., *Cenchrus brownii* Roem. & Schultes, *Cenchrus ciliaris* L., *Cenchrus echinatus* L., *Cenchrus elymoides* F. Muell., *Cenchrus incertus* M.A. Curtis, *Cenchrus longispinus* (Hack.) Fern., *Cenchrus setigerus* Vahl., *Centotheca* sp., *Centotheca lappacea* (L.) Desvaux, *Centrochloa singularis* Swallen, *Centropodia glauca* (Nees) T.A. Cope (*Asthenatherum*), *Cephalostachyum pergracile* Munro, *Chaetium cubanum* Hitchc., *Chaetobromus dregeanus* Nees, *Chaetopoa taylori* C.E. Hubbard, *Chaetopogon fasciculatus* Link., *Chaetostichium minimum* (Hochst.) C.E. Hubb., *Chamaeraphis hordeacea* R. Br., *Chasechloa madagascariensis* (Baker) Camus, *Chasmanthium latifolium* (Michx.) Yates, *Chasmanthium laxum* (L.) Yates, *Chevalierella congoensis* A. Camus (= *C. dewildemanii*), *Chevalierella dewildemanii* (Vand.) Van der Veken, *Chikusichloa aquatica* Koidz., *Chimonobambusa densifolia* Nakai, *Chionachne cyathopoda* (Muell.) Muell. ex Benth., *Chionochloa conspicua* (Forst.) Zotov, *Chionochloa frigida* (Vickery) Conert, *Chionochloa pallida* (R. Br.) Conert, *Chloachne oplismenoides* (Hack.) Stapf ex Robyns, *Chloris gayana* Kunth, *Chloris truncata* R. Br., *Chlorocalymma cryptacanthum* W. Clayton, *Chrysochloa hubbardiana* Germain & Risopoulos, *Chrysochloa orientalis* (Hubb.) Swallen, *Chrysopogon fallax* Blake, *Chusquea longiligulata* (Soderstrom & Calderon) Clark (= *Swallenochloa*), *Cinna latifolia* (Trev.) Griseb., *Cladoraphis spinosa* (L. f.) S.M. Phillips, *Clausospicula extensa* M. Lazarides, *Cleistochloa subjuncea* Hubbard, *Cliffordiochloa parvispicula*

B.K. Simon, *Cockaynea gracilis* (Hook. f.) Zotov, *Coelachne pulchella* R. Br., *Coelachyropsis lagopoides* (Burm. f.) Bor, *Coelachyrum poaeflorum* Chiov., *Coelorachis rottboellioides* (R. Br.) Camus, *Coix lacryma-jobi* L., *Colanthelia cingulata* (McClure & Smith) McClure, *Coleanthus subtilis* (Tratt.) Seidel., *Colpodium altaicum* Trin. (= *Paracolpodium*), *Colpodium chionogeiton* Pilg. (= *Keinochloa*), *Commelinidium gabunense* (Hack.) Stapf, *Cornucopiae cucullatum* L., *Cortaderia archboldii* (Hitchc.) Connor & Edgar, *Cortaderia bifida* Pilger, *Cortaderia selloana* (Schultes & Schultes) Aschers & Graebner, *Corynephorus canescens* (L.) Beauv., *Cottea pappophoroides* Kunth, *Craspedorhachis africana* Benth., *Crinipes abyssinicus* (A. Rich.) Hochst., *Crithopsis delineana* (Schultes) Roshev., *Crypsis aculeata* (L.) Ait., *Cryptochloa capillata* (Trin.) Soderstrom, *Ctenium polystachyum* Balansa, *Ctenopsis patens* (Boiss.) Meld., *Ctenopsis pectinella* (Del.) De Not, *Cutandia dichotoma* (Forsk.) Trab., *Cutandia memphitica* (Spreng.) Richt., *Cutandia philistaea* Benth., *Cyclostachya stolonifera* (Scribn.) Reeder & Reeder, *Cymbopogon refractus* (R. Br.) Camus, *Cymbosetaria sagittifolia* (A. Rich.) Schweick., *Cynodon dactylon* (L.) Pers., *Cynosurus cristatus* L., *Cynosurus echinatus* L., *Cyperochloa hirsuta* Lazarides & Watson, *Cyphochlaena madagascariensis* Hack., *Cypholepis yemenica* (Schweinf.) Chiov., *Cyrtococcum* sp.

Dactylis glomerata L., *Dactyloctenium aegyptium* (L.) Beauv., *Daknopholis boivinii* (A. Camus) Clayton, *Dallwatsonia felliana* B.K. Simon, *Danthonia* aff. *alpicola* Vickery (= *Rytidosperma*), *Danthonia caespitosa* Gaudich. (= *Rytidosperma*), *Danthonia carphoides* F. Muell. ex Benth. (= *Rytidosperma*), *Danthonia dimidiata* Vickery (= *Rytidosperma*), *Danthonia linkii* Kunth (= *Rytidosperma*), *Danthonia nivicola* Vickery (= *Rytidosperma*), *Danthonia nudiflora* Morris (= *Rytidosperma*), *Danthonia procera* Vickery (= *Rytidosperma*), *Danthonia 'robusta'* (= *Chionochloa pallida*), *Danthonia setacea* R. Br. (= *Rytidosperma*), *Danthoniastrum compactum* (Holub) Holub, *Danthonidium gammiei* C.E. Hubb., *Danthoniopsis barbata* (Nees) C.E. Hubb., *Danthoniopsis occidentalis* Jacques-Félix, *Dasyochloa pulchella* (H.B.K.) Willd. ex Rydb., *Dasypoa scaberula* Pilger, *Dasypyrum villosum* (Coss. & Dur.) Dur., *Decaryochloa diadelpha* A. Camus, *Dendrocalamus gigantus* Munro, *Deschampsia caespitosa* (L.) Beauv., *Desmazeria marina* (L.) Druce, *Desmostachya bipinnata* (L.) Stapf, *Deyeuxia affinis* M. Gray, *Deyeuxia crassiuscula* Vickery, *Deyeuxia quadriseta* (Labill.) Benth., *Diandrochloa confertiflora* (J.M. Black) de Winter, *Diandrostachya chrysothrix* (= *Loudetiopsis*), *Diarrhena americana* P. Beauv., *Diarrhena japonica* Franch. & Savat., *Diarrhena mandshurica* Maxim. (= *Neomolinia*), *Dichaetaria wightii* Nees, *Dichanthelium angustifolium* (Ell.) Gould, *Dichanthelium clandestinum* (L.) Gould, *Dichanthelium commutatum* (Schult.) Gould, *Dichanthelium dichotomum* (L.) Gould, *Dichanthelium lanuginosum* (Ell.) Gould, *Dichanthelium lindheimeri* (Nash) Gould, *Dichanthelium linearifolium* (Scribn.) Gould, *Dichanthelium oligosanthes* (Schult.) Gould, *Dichanthelium scoparium* (Lam.) Gould, *Dichanthelium sphaerocarpon* (Ell.) Gould, *Dichanthium annulatum* (Forssk.) Stapf, *Dichelachne micrantha* (Cav.) Domin, *Diectomis fasigiata* (Sw.). P.Beauv., *Dielsiochloa floribunda* (Pilg.) Pilg., *Digastrium fragile* (Hack.) A. Camus, *Digitaria aequiglumis* (Hackel ex Arechav.) Parodi, *Digitaria ammophila* (F. Muell.) Hughes, *Digitaria bicornis* (Lam.) Roem. & Schult., *Digitaria breviglumis* (Domin) Henr., *Digitaria brownii* (Roemer & Schultes) Hughes, *Digitaria coenicola* (F. Muell.) Hughes, *Digitaria divaricatissima* (R. Br.) Hughes, *Digitaria milanjiana* (Rendle) Stapf, *Digitaria parviflora* (R. Br.) Hughes, *Digitaria sanguinalis* (L.) Scop., *Digitariopsis redheadi* Hubbard, *Dignathia villosa* Hubbard, *Dilophotriche pobeguinii* Jacques-Félix, *Dimeria* sp., *Dimorphochloa rigida* Blake, *Dinebra retroflexa* (Vahl) Panzer, *Dinochloa pubiramea* Gamble, *Diplachne parviflora* (R. Br.) Benth., *Diplopogon setaceus* R.Br., *Dissanthelium minimum* Pilger, *Dissochondrus biflorus* Kuntze, *Distichlis distichophylla* (Labill.) Fassett, *Drake-Brockmania somalensis* Stapf, *Dregeochloa calviniensis* Conert, *Dryopoa dives* Vickery, *Dupontia pilosantha* Rupr., *Duthiea bromoides* Hack. *Dybowskia seretii* (De Wild) Stapf

Eccoptocarpha obconiciventris Launert, *Echinaria capitata* (L.) Desf., *Echinochloa colona* (L.) Link, *Echinochloa crus-galli* (L.) Beauv., *Echinolaena gracilis* Swallen,

Echinolaena inflexa Chase, *Echinopogon cheelii* Hubbard, *Echinopogon ovatus* (Forst.)
Beauv., *Ectrosia agrostoides* Benth., *Ectrosia eragrostoides* Domin (= *E. lasioclada*
(Merr.) S.T. Blake), *Ectrosia leporina* R. Br., *Ectrosiopsis lasioclada* (Ohwi) Jansen,
Ehrharta barbinodis Nees ex Trin., *Ehrharta brevifolia* Schrad. var *brevifolia, Ehrharta
capensis* Thunb., *Ehrharta calycina* J.C. Sm., *Ehrharta dura* Nees ex Trin., *Ehrharta
delicatula* (Nees) Stapf, *Ehrharta erecta* Lam. subsp. *natalensis, Ehrharta longiflora*
Smith, *Ehrharta melicoides* Thunb., *Ehrharta pusilla* Nees ex Trin., *Ehrharta ramosa*
Thunb., *Ehrharta rehmannii* Stapf, *Ehrharta rehmanii* Stapf var. *filiformis, Ehrharta
rupestris* Nees ex Trin., *Ehrharta setacea* Nees ex Trin., *Ehrharta subspicata* Stapf,
Ehrharta uniflora Burch. ex Stapf, *Ehrharta triandra* Nees ex Trin., *Ehrharta villosa*
Schult., *Ekmanochloa aristata* Ekman, *Eleusine indica* (L.) Gaertn., *Elionurus citreus*
(R. Br.) Munro ex Benth., *Elymandra androphila* (Stapf) Stapf, *Elymandra
archaelymandra* (Jac.-Fél.) W.D. Clayton, *Elymandra grallata* (Stapf) Clayton,
Elymandra subulata Jac.-Fél., *Elymus scaber* (Labill.) Löve (= *Agropyrum scabrum*),
Elytrigia elongata (Host) Nevski (= *Lophopyrum*), *Elytrigia junceiformis* A. & D. Löve
(= *Thinopyrum*), *Elytrigia pycnantha* (Godron) A. Löve, *Elytrigia repens* (L.) Nevski,
Elytrophorus spicatus (Willd.) A. Camus, *Elytrostachys clavigera* McClure,
Enneapogon avenaceus (Lindl.) C.E. Hubb., *Enneapogon brachystachyus* (Jaub. &
Spach.) Stapf, *Enneapogon caerulescens* (Gaudich.) N.T. Burbidge, *Enneapogon
cenchroides* (Roem. & Schult.) C.E. Hubb., *Enneapogon clelandii* N.T. Burbidge,
Enneapogon cylindricus N.T. Burbidge, *Enneapogon desvauxii* Beauv., *Enneapogon
elegans* (Nees) Stapf, *Enneapogon glaber* N.T. Burbidge, *Enneapogon gracilis* (R. Br.)
Beauv., *Enneapogon nigricans* (R. Br.) Beauv., *Enneapogon oblongus* N.T. Burbidge,
Enneapogon planifolius N.T. Burbidge, *Enneapogon polyphyllus* (Domin) N.T.
Burbidge, *Enneapogon purpurascens* (R. Br.) N.T. Burbidge, *Enneapogon scoparius*
Stapf, *Enteropogon acicularis* (Lindl.) Lazarides, *Entolasia imbricata* Stapf, *Entolasia
marginata* (R. Br.) Hughes, *Entolasia olivacea* Stapf, *Entolasia stricta* (R. Br.) Hughes,
Entolasia whiteana C.E. Hubb., *Entoplocamia aristulata* Stapf, *Eragrostiella bifaria*
(Vahl) Bor, *Eragrostis chloromelas* Steudel, *Eragrostis cilianensis* (All.) Lutati,
Eragrostis confertiflora J.M. Black (= *Diandrochloa*), *Eragrostis benthamii* Mattei,
Eragrostis elongata (Willd.) Jacq., *Eragrostis eriopoda* Benth., *Eragrostis
neomexicana* Vasey, *Eragrostis nindensis* Fical. & Hiern, *Eragrostis pallens* Hack.,
Eragrostis parviflora (R. Br.) Trin., *Eragrostis walteri* Pilg., *Eremochloa bimaculata*
Hackel, *Eremopoa altaica* (Trin.) Roshev., *Eremopogon foveolatus* (Del.) Stapf,
Eremopyrum bonaepartis (Spreng.) Nevski *Eremopyrum distans* (C. Koch) Nevski,
Eremopyrum orientale (L.) Jaub. & Spach, *Eremopyrum triticeum* (Gaertn.) Nevski,
Eriachne anomala Hartley, *Eriachne aristidea* F. Muell., *Eriachne glabrata* (Maiden)
Hartley, *Eriachne mucronata* R. Br., *Eriachne nervosa* Ewart & Cookson, *Eriachne
obtusa* R. Br., *Eriachne ovata* Nees, *Eriachne pulchella* Domin, *Erianthecium
bulbosum* Parodi, *Erianthus fastigiatus* (Nees ex Steud.) Stapf, *Eriochloa australiensis*
Stapf ex Thell., *Eriochloa crebra* S.T. Blake, *Eriochloa decumbens* Bailey, *Eriochloa
procera* (Retz.) C.E. Hubb., *Eriochloa pseudoacrotricha* (Stapf ex Thell.) Black,
Eriochrysis cayennensis Beauv., *Erioneuron pilosum* (Buckley) Nash, *Erythranthera
australis* (Petrie) Zotov, *Eulalia fulva* (R. Br.) Kuntze, *Eulaliopsis binata* (Retz.) C.E.
Hubb., *Eustachys distichophylla* (Lag.) Nees *Exotheca abyssinica* (A. Rich.) Anderss.,

Festuca arundinacea Schreb., *Festuca asperula* Vickery, *Festuca hookeriana* F. Muell.
ex Hook. f. (= *Hookerochloa*), *Festuca leptopogon* Stapf, *Festuca littoralis* Labill. (=
Austrofestuca), *Festuca rubra* L., *Festucella eriopoda* (Vickery) E. Alexeev,
Fingerhuthia africana Lehm.

Garnotia stricta Brongn., *Gastridium ventricosum* (Gouan) Schinz & Thell., *Gaudinia
fragilis* (L.) Beauv., *Gaudinia maroccana* Trab. ex Pitard, *Gaudiniopsis macra* (Bieb.)
Eig, *Germainia truncatiglumis* (Muell. ex Benth.) Chaianan, *Gilgiochloa indurata*
Pilger, *Glyceria australis* C.E. Hubb., *Glyceria septentionalis* Hitchc., *Gouinia
guatemalensis* (Hack.) Swallen (= *G. latifolia* (Griseb.) Vasey var.*guatemalensis*
(Hack.) Ortiz), *Gouinia longiramea* Swallen (= *G. virgata* (Presl.) Scribn.), *Guaduella
foliosa* Pilg., *Guaduella marantifolia* Franch., *Guaduella zenkeri* Pilg., *Gymnachne
koelerioides* (Trin.) Parodi, *Gymnopogon foliosus* (Willd.) Nees *Gynerium sagittatum*

(Aubl.) Beauv.,

Habrochloa bullocki C.E. Hubb., *Hackelochloa granularis* (L.) Kuntze, *Hainardia cylindrica* (Willd.) Greuter, *Hakonechloa macra* (Monro) Makino, *Halopyrum mucronatum* (L.) Stapf, *Harpachne schimperi* A. Rich., *Harpochloa fallax* (L.f.) Kuntze, *Helictotrichon pubescens* (Huds.) Pilger, *Helleria fragilis* Luces, *Helleria livida* Fourn. ex Hemsl., *Hemarthria uncinata* R. Br., *Henrardia persica* (Boiss.) C.E. Hubb., *Heterachne abortiva* (R. Br.) Druce, *Heteranthelium piliferum* (Banks and Soland.) Hochst., *Heteropholis nigrescens* (Yhw.) C.E. Hubb., *Heteropogon triticeus* (R. Br.) Stapf, *Hierochloë rariflora* Hook. f., *Hierochloë redolens* (Vahl) Roemer & Schultes, *Hilaria jamesii* Torr., *Holcolemma canaliculatum* (Steud.) Stapf & Hubbard, *Holcus lanatus* L., *Homolepis aturensis* Chase, *Homolepis glutinosa* (Swartz) Zuloaga & Soderstrom, *Homolepis isocalycia* (Meyer) Chase, *Homolepis longispicula* (Doell) Chase, *Homopholis proluta* (F. Muell.) R. Webster, *Homozeugos eylesii* C.E. Hubb., *Hookerochloa hookeriana* (F. Muell. ex Hook. f.) E. Alexeev, *Hordelymus europaeus* (L.) Hartz, *Hordeum glaucum* Steud., *Hordeum leporinum* Link, *Hordeum vulgare* L., *Hubbardochloa gracilis* Auquier, *Humbertochloa greenwayi* C.E. Hubbard, *Hydrochloa caroliniensis* P. Beauv., *Hydrothauma manicatum* C.E. Hubbard, *Hygrochloa aquatica* Lazarides, *Hygrochloa craveni* Lazarides, *Hygroryza aristida* (Retz.) Nees, *Hylebates cordatus* Chippindall, *Hymenachne amplexicaulis* (Rudge) Nees, *Hyparrhenia filipendula* (Hochst.) Stapf, *Hyperthelia dissoluta* (Steud.) Clayton, *Hypseochloa cameroonensis* C.E. Hubb., *Hystrix patula* Moench

Ichnanthus vicinus (F.M. Bailey) Merr., *Imperata cylindrica* (L.) Beauv., *Indocalamus debilis* (Thwait.) Alston, *Indopoa pauperculus* (Stapf) Bor, *Isachne confusa* Ohwi, *Isachne globosa* (Thunb.) Kuntze, *Isachne pulchella* Roth. ex R. & S., *Isalus isalensis* (Lam) Phipps, *Isalus humbertii* (A. Camus) Phipps, *Ischaemum fragile* R. Br., *Ischnurus pulchellus* Balf. f., *Iseilema vaginiflorum* Domin, *Ixophorus unisetus* (Presl) Schlecht.

Jardinea gabonensis Steud., *Jouvea pilosa* (Presl) Scribn., *Jouvea straminea* Fourn.

Kampochloa brachyphylla W. Clayton, *Kengia serotina* (L.) Packer (subsp. *serotina*), *Keniochloa cheinogeiton* (= *Colpodium*), *Koeleria australiensis* Domin

Lagurus ovatus L., *Lamarckia aurea* (L.) Moench., *Lamprothyrsus peruvianus* Hitchcock, *Lasiacis ligulata* Hitchc. & Chase, *Lasiacis nigra* Davidse, *Lasiacis procerrima* Hitchc. ex Chase, *Lasiacis sorghoides* Hitchc. & Chase, *Lasiochloa echinata* (Thunb.) Henr. (= *Tribolium*), *Lasiorhachis hildebrandtii* (Hack.) Stapf, *Lasiurus hirsutus* (Forrsk.) Boiss., *Latipes senegalensis* (= *Leptothrium*), *Lecomtella madagascariensis* A. Camus, *Leersia hexandra* Swartz, *Leptaspis banksii* (R. Br.), *Leptocarydion vulpiastrum* (De Not) Stapf, *Leptochloa decipiens* (R. Br.) Stapf ex Maiden, *Leptochloa digitata* (R. Br.) Domin, *Leptochloa filiformis* (Lam.) Beauv., *Leptochloa neesii* (Thw.) Benth., *Leptochloa panicea* (Retz.) Ohwi, *Leptochloa uniflora* Hochst., *Leptochloa virgata* (L.) P. Beauv., *Leptochloöpsis virgata* (Poir.) Yates, *Leptocoryphium lanatum* Nees, *Leptoloma cognatum* (Schult.) Chase, *Leptosaccharum filiforme* Camus, *Leptothrium rigidum* Kunth, *Leptothrium senegalensis* Kunth, *Lepturidium insulare* Hitchc. & Ekman, *Lepturopetium marshallense* Fosberg & Sachet, *Lepturella aristata* Stapf, *Lepturus repens* (Forst.) R. Br., *Leucopoa 'spadicea'*, *Leymus arenarius* (L.) Hochst., *Libyella cyrenaica* (Dur. & Barr.) Pamp., *Limnas stelleri* Trin., *Limnodea arkansana* (Nutt.) Dewey, *Limnopoa meeboldii* (Fischer) Hubbard, *Lindbergella sintenisii* (Lindb.) Bor, *Lintonia nutans* Stapf, *Lithachne pauciflora* Beauv., *Littledalea racemosa* Keng, *Loliolum subulatum* (Banks and Soland.) Eig, *Lolium perenne* L., *Lolium rigidum* Gaud., *Lophacme digitata* Stapf, *Lophatherum gracile* Brongn., *Lophochloa cristata* (L.) Hylander (= *Koeleria*, *Rostraria*), *Lophochloa phleoides* (Vill.) Reichb., *Lophochloa pumila* (Desf.) Bor (= *Trisetum*), *Lopholepis ornithocephala* Steud., *Lophopyrum elongatum* (Host) A. Löve (= *Elytrigia*), *Lorenzochloa erectifolia* (Swallen) J. & C. Reeder, *Loudetia simplex* (Nees) C.E. Hubb., *Loudetiopsis chrysothrix* (Nees) Conert, *Louisiella fluitans* Hubbard & Leonard, *Loxodera caespitosa* (C.E. Hubb.) Clayton, *Loxodera ledermannii* (Pilger)

Clayton, *Luziola pittieri* Luces, *Lycurus phleoides* H. B. & K., *Lygeum spartum* L.
Maillea crypsoides (D'Urv.) Boiss., *Malacurus lanatus* (Korsh.) Nevski, *Maltebrunia leersioides* Kunth, *Manisuris myuros* L., *Megalachne berteroniana* Steud., *Megaloprotachne albescens* Hubbard, *Melanocenchris abyssinica* (R. Br.) Hochst., *Melica argentata* E. Desv., *Melica argyrea* Hack., *Melica californica* Scribn., *Melica ciliata* L., *Melica uniflora* Retz., *Melinis minutiflora* Domin, *Melocanna bambusoides* Trin., *Merostachys 'anceps'*, *Mesosetum altum* Swallen, *Mesosetum loliiforme* (Hochst. ex Steud.) Hitchc., *Mesosetum pittieri* Hitchc., *Metcalfia mexicana* Conert, *Mibora minima* (L.) Desv., *Mibora verna* Beauv. (= *M. minima*, *Micraira* sp. nov. Lazarides, *Micraira subulifolia* F. Muell., *Microbriza poaemorpha* (Presl.) Parodi, *Microcalamus barbinoides* Franch., *Microchloa indica* (L.f.) Beauv., *Microlaena stipoides* (Labill.) R. Br., *Micropyrum tenellum* L., *Microstegium spectabile* (Trin.) A. Camus, *Mildbraediochloa reynaudioides* Butzin, *Milium effusum* L., *Miscanthidium sorghum* (Nees) Stapf, *Miscanthus sinensis* Anderss., *Mnesithea laevis* (Retz.) Kunth, *Mnesithea mollicoma* (Hance) A. Camus, *Molineriella minuta* (L.) Rouy (= *Periballia minuta* (L.) Aesch. & Graebn.), *Molinia caerulea* (L.) Moench, *Moliniopsis japonica* Hayata, *Monachather paradoxa* Steud., *Monanthochloë littoralis* Engelm., *Monelytrum luderitzianum* Hack., *Monerma cylindrica* (= *Hainardia*), *Monium funereum* Jac-Fél., *Monium macrochaetum* Stapf, *Monium trichaetum* Reznik, *Monocymbium ceresiiforme* (Nees) Stapf, *Monodia stipoides* S.W.L. Jacobs, *Monostachya oreoboloides* (F. Muell.) Hitchc., *Muhlenbergia arisanensis* Hayata, *Muhlenbergia asperifolia* Nees & Meyen ex Trin., *Muhlenbergia minuscula* H. Scholz, *Munroa squarrosa* (Nutt.) Torr., *Myriocladus paludicolus* Swallen, *Myriostachya wightiana* (Nees) Hook. f.

Narduroides salzmannii (Boiss.) Rouy, *Nardurus maritimus* (L.) Murbeck (= *Vulpia unilateralis*), *Nardus stricta* L., *Nassella pubiflora* (Trin. & Rupr.) Desv., *Nassella trichotoma* (Nees) Hackel ex Arechav., *Nastus hooglandii* Holttum, *Nastus obtusus* Holttum, *Neesiochloa barbata* (Trin.) Pilger, *Nematopoa longipes* (Stapf & C.E. Hubb.) C.E. Hubb., *Neobouteloua lophostachya* (Griseb.) Gould, *Neomolinia mandshurica* (Maxim.) Honda (= *Diarrhena*), *Neostapfia colusana* Davy, *Neostapfiella humbertiana* A. Camus, *Neostapfiella perrieri* A. Camus, *Nephelochloa orientalis* Boiss., *Neurachne alopecuroidea* R. Br., *Neurachne lanigera* S.T. Blake, *Neurachne minor* S.T. Blake, *Neurachne munroi* (F. Muell.) F. Muell., *Neurachne queenslandica* S.T. Blake, *Neurachne tenuifolia* Blake, *Neyraudia arundinacea* (L.) Henr., *Neyraudia neyraudiana* (Kunth) Keng, *Notochloë microdon* (Benth.) Domin

Ochlandra stridula Thw., *Ochthochloa compressa* Edgew., *Odyssea mucronata* Stapf, *Olmeca recta* Soderstrom, *Olyra latifolia* L., *Ophiuros megaphyllas* Stapf ex Haines, *Opizia stolonifera* Presl., *Oplismenopsis najada* (Hack. & Arech.) Parodi, *Oplismenus aemulus* (R. Br.) Roem. & Schult., *Oplismenus burmanii* (Retz.) P. Beauv., *Oplismenus compositus* P. Beauv., *Oplismenus hirtellus* (L.) P. Beauv., *Oplismenus undulatifolius* (Ard.) Roem. & Schult., *Orcuttia californica* Vasey, *Orcuttia inaequalis* Hoover, *Oreochloa disticha* (Wulf.) Link, *Orinus kokonorica* (Hao) Keng, *Orinus thoroldii* A. Hitchc., *Oropetium thomaeum* (L. f.) Trin., *Orthachne breviseta* A. Hitchc., *Orthoclada laxa* Beauv., *Oryza australiensis* Domin, *Oryza sativa* L., *Oryzidium barnardii* Hubbard & Schweickerdt, *Oryzopsis miliacea* (L.) Benth. & Hook. f. (= *Piptatherum*), *Otachyrium succisum* (Swallen) Sendulsky & Soderstrom, *Otatea acuminata* (Munro) Calderon & Soderstrom, *Ottochloa gracillima* Hubbard, *Oxychloris scariosa* (F. Muell.) Lazarides *Oxyrhachis gracillima* (Bak.) C.E. Hubb., *Oxytenanthera monadelphia* (Thwait.) Alston

Panicum antidotale Retz, *Panicum bulbosum* H.B.K., *Panicum buncei* F. Muell. ex Benth., *Panicum cambogiense* Balansa, *Panicum coloratum* L., *Panicum effusum* R. Br., *Panicum incomtum* Trin., *Panicum laevifolium* Hack., *Panicum maximum* Jacq., *Panicum miliaceum* L., *Panicum mitchellii* Benth., *Panicum paludosum* Roxb., *Panicum pygmaeum* R. Br., *Panicum simile* Domin, *Panicum subxerophyllum* Domin, *Panicum trachyrhachis* Benth., *Pappophorum sub-bulbosum* Arech, *Parafestuca albida* (Lowe) E. Alexeev, *Parahyparrhenia annua* (Hack.) W.D. Clayton, *Paraneurachne muelleri* (Hackel) Blake, *Parapholis incurva* (L.) Hubbard, *Paratheria prostrata* Griseb.,

Parectenium novae-hollandiae Beauv., *Parodiolyra lateralis* (Presl) Soderstrom & Zuloaga, *Parodiolyra luetzelburgii* (Pilger) Soderstrom & Zuloaga, *Parodiolyra ramosissima* (Trin.) Soderstrom & Zuloaga, *Pascopyrum smithii* (Rydb.) A. Löve, *Paspalidium flavidum* Stapf, *Paspalidium gracile* (R. Br.) Hughes, *Paspalidium jubiflorum* (Trin.) Hughes, *Paspalidium radiatum* Vickery, *Paspalidium rarum* (R. Br.) Hughes, *Paspalum conjugatum* Berg., *Paspalum dilatatum* Poiret, *Paspalum distichum* L., *Paspalum vaginatum* Swartz., *Pennisetum alopecuroides* (L.) Sprengel, *Pennisetum villosum* R. Br., *Pentapogon quadrifidus* (Labill.) Baillon, *Pentarrhaphis scabra* H.B.K., *Pentaschistis airoides* (Nees) Stapf, *Pereilema brasilianum* Trin., *Pereilema crinitum* J. & C. Presl, *Periballia minuta* (L.) Aesch. & Graebn., *Peridictyon sanctum* (Janka) Seberg, Frederiksen and Baden, *Perotis rara* R. Br., *Perulifera madagascariensis* A. Camus, *Peyritschia koelerioides* (Peyr.) Fourn. (= *Trisetum altijugum* (Fourn.) Scribn.), *Peyritschia pringlei* Scribn. (= *Trisetum kochianum* I. Holz. T.), *Phacelurus huillensis* (Rendle) Stapf, *Phaenanthoecium koestlinii* (Hochst. ex A. Rich.) C.E. Hubb., *Phaenosperma globosa* Munro, *Phalaris aquatica* L., *Phalaris arundinacea* L., *Pharus latifolius* L., *Pheidochloa gracilis* Blake, *Pheidochloa vulpioides* Blake, *Phippsia algida* (Solander) R. Br., *Phleum pratense* L., *Pholiurus pannonicus* (Host) Trin., *Phragmites australis* (Cav.) Trin. ex Steud., *Pilgerochloa blanchei* (Boiss.) Eig, *Piptochaetium brevicalyx* Ricker ex Hitchc., *Piptochaetium fimbriatum* (H.B.K.) Hitchc., *Piresia goeldii* Swallen, *Piresiella strephioides* (Griseb.) Zuloaga, Morrone & Judziewicz, *Plagiantha tenella* Renvoize, *Plagiochloa acutiflora* (Nees) Adamson & Sprague (= *Tribolium*), *Plagiosetum refractum* (F. Muell.) Benth., *Planichloa nervilemma* B.K. Simon, *Plectrachne melvillei* C.E. Hubb., *Plectrachne schinzii* Henrard, *Pleiadelphia gossweileri* Stapf, *Pleuropogon californicus* (Nees) Benth., *Pleuropogon sabinii* R. Br., *Plinthanthesis paradoxa* (R. Br.) S. T. Blake, *Plinthanthesis rodwayi* (C. E. Hubbard) S. T. Blake, *Poa annua* L., *Poa clivicola* Vickery, *Poa ensiformis* Vickery, *Poa exilis* Vickery (= *P. meionectes* Vickery), *Poa helmsii* Vickery, *Poa labillaridieri* Steud., *Poa phillipsiana* Vickery, *Poa pratensis* L., *Poa queenslandica* C.E. Hubb., *Poa saxicola* R.Br., *Poa sieberana* Nees, *Poagrostis pusilla* Stapf, *Pobeguinea afzeliana* (Rendle) Jac.-Fél., *Pobeguinea hamata* (Stapf) Jac.-Fél., *Podophorus bromoides* Phil., *Poecilostachys festucaceus* (Mez) Camus, *Poecilostachys viguieri* Camus, *Pogonarthria fleckii* Hack., *Pogonarthria squarrosa* Pilger, *Pogonatherum paniceum* (Lam.) Hackel, *Pogoneura biflora* Napper, *Pogonochloa greenwayi* Hubbard, *Poidium itatiaiae* (Ekman) Nicora & Rugolo, *Polevansia stricta* de Winter, *Polypogon monspeliensis* (L.) Desf., *Polytoca macrophylla* Benth., *Polytrias amura* (Buse) Kuntze, *Pommereulla cornucopia* L., *Potamophila parviflora* R. Br., *Pringleochloa stolonifera* (Fourn.) Scribn., *Prionanthium pholiuroides* Stapf, *Prosphytochloa prehensilis* (Nees) Schweick., *Psammagrostis wiseana* C. Gardner & C.E. Hubbard, *Psammochloa villosa* (Trin.) Bor, *Psathyrostachys junceus* (Fischer) Nevski, *Pseudanthistiria umbellata* (Hack.) Hook. f., *Pseudechinolaena polystachya* (H.B.K.) Stapf, *Pseudobromus sylvaticus* K. Schum., *Pseudochaetochloa australiensis*, *Pseudocoix perrieri* A. Camus, *Pseudodanthonia himalaica* Bor & C.E. Hubb., *Pseudopogonatherum irritans* Hitchc., *Pseudoraphis abortiva* (R.Br.) Pilger, *Pseudoraphis paradoxa* (R.Br.) Pilger, *Pseudoraphis spinescens* Vickery, *Pseudoroegneria spicata* (Pursh) A. Löve, *Pseudosasa japonica* (Sieb. & Zucc. ex Steud.) Makino ex Nakai, *Psilathera tenella* (Host) Link, *Psilolemma jaegeri* (Pilger) S.M. Phillips, *Psilurus incurvus* (Gouan) Schinz & Thell., *Pterochloris humbertiana* A. Camus, *Puccinellia stricta* (Hook.f.) Blom, *Puelia ciliata* Franch., *P. dewevrei* Wild. & Dur., *Pyrrhanthera exigua* (Kirk) Zotov

Racemobambos gibbsiae (Stapf) Holttum, *Raddia brasiliensis* A. Bertoloni, *Raddiella esenbeckiana* (Steud.) Calderon & Soderstrom, *Redfieldia flexuosa* (Thurb.) Vasey, *Reederochloa eludens* Soderstrom & Decker, *Rehia nervata* (Swallen) Fitjen, *Reimaria brasiliensis* Schlecht. (= *Paspalum*), *Reimarochloa brasiliensis* Hitchc., *Reitzia smithii* Swallen, *Relchela panicoides* Steud., *Rendlia altera* (Rendle) Chiov., *Reynaudia filiformis* Kunth., *Rhipidocladum pittieri* (Hack.) McClure, *Rhipidocladum racemiflorum* (Steud.) McClure, *Rhizocephalus orientalis* Boiss., *Rhomboelytrum monandrum* (Hack.) Nicora & Rugolo, *Rhynchelytrum repens* (Willd.) Hubbard,

Rhynchoryza suloulata (Nees) Baillon, *Richardsiella eruciformis* Elffers & Kennedy O'Byrne, *Robynsiochloa purpurascens* (Robyns) Jac.-Fél., *Roegneria canina* (L.) Nevski (= *Elymus caninus*), *Rottboellia exaltata* L. f., *Rytidosperma* — see *Danthonia*

Saccharum robustum Brandes & Jeswiet, *Saccharum spontaneum* L., *Sacciolepis indica* (L.) Chase, *Sasa kurilensis* (Rupr.) Makino & Shibata, *Saugetia pleiostachya* Hitchc. & Eckman, *Schaffnerella gracilis* (Benth.) Nash, *Schedonnardus paniculatus* (Nutt.) Trel., *Schismus barbatus* (L.) Thell., *Schizachne purpurascens* (Torr.) Swallen, *Schizachyrium fragile* (R. Br.) A. Camus, *Schizostachyum lima* Merril, *Schmidtia pappophoroides* Steud. ex Schmidt, *Schoenefeldia transiens* Chiov., *Sclerachne punctata* R. Br., *Sclerandrium truncatiglumis* (F. Muell. ex Benth.) Stapf & C.E. Hubb. (= *Germainia*), *Sclerochloa dura* (L.) Beauv., *Sclerodactylon macrostachyum* (Benth.) A. Camus, *Scleropoa rigida* (L.) Griseb. (= *Catapodium*), *Scleropogon brevifolius* Phil., *Scolochloa festucacea* (Willd.) Link, *Scribneria bolanderi* (Thurb.) Hack., *Scrotochloa urceolata* (Roxb.) Judziewicz, *Scutachne dura* (Griseb.) Hitchc. & Chase, *Secale cereale* L., *Sehima nervosa* (Rottb.) Stapf, *Sesleria caerulea* (L.) Ard., *Sesleriella (Sesleria) sphaerocephala*, *Setaria geniculata* (Lam.) Beauv., *Setaria palmifolia* (Koenig) Stapf, *Setaria verticillata* (L.) Beauv., *Setariopsis auriculata* Scribn., *Sieglingia decumbens* (L.) Bernh., *Silentvalleya naivii* Nair, Sreekumar, Vajravdu & Bhargavan, *Simplicia buchananii* Zotov, *Sinobambusa tootsik* Makino, *Sitanion hystrix* (Nutt.) J.G. Smith, *Snowdenia polystachya* (Fresen) C.E. Hubbard, *Soderstromia mexicana* (Scribn.) Morton, *Sohnsia filifolia* (Fourn.) Airy Shaw, *Sorghum halepense* (L.) Pers., *Sorghum leiocladum* (Hackel) Hubbard, *Spartina maritima* (Curtis) Fernald, *Spartochloa scirpioides* Hubbard, *Spathia neurosa* Ewart & Archer, *Sphaerocaryum malacense* (Trin.) Pilger, *Spheneria kegelii* (C. Muell.) Pilger, *Sphenopholis obtusata* (Michx.) Scribner, *Sphenopus divaricatus* (Gouan) Reichenb., *Sphenopus* sp., *Spinifex hirsutus* Labill., *Spinifex longifolius* R. Br., *Spinifex sericeus* R. Br., *Spodiopogon lacei* Hole, *Sporobolus crebar* de Nardi, *Sporobolus elongatus* R. Br., *Sporobolus pulchellus* R. Br., *Steinchisma hians* (*Panicum hians* Ell.), *Steirachne diandra* Ekman, *Stenotaphrum micranthum* (Desv.) C.E. Hubb., *Stenotaphrum secundatum* (Walter) Kuntze, *Stephanachne* (*Pappagrostis*) *pappophorea*, *Stereochlaena cameronii* (Stapf) Pilger, *Steyermarkochloa uniflora* Davidse & Ellis, *Stiburus alopecuroides* Stapf, *Stilpnophleum anthoxanthoides*, *Stipa aristiglumis* F. Muell., *Stipa bigeniculata* Hughes, *Stipa blackii* Hubbard, *Stipa columbiana* Maccun, *Stipa comata* Trin. & Rupr., *Stipa densiflora* Hughes, *Stipa elegantissima* Labill., *Stipa falcata* Hughes, *Stipa falcata/scabra*, *Stipa filiculmis* Delile, *Stipa filifolia* Nees, *Stipa gigantea* Link, *Stipa ichu* Kunth, *Stipa juncoides* Speg., *Stipa leucotricha* Trin. & Rupr., *Stipa neomexicana* (Thurb.) Scribn., *Stipa nervosa* Vickery, *Stipa pubescens* R. Br., *Stipa ramosissima* (Trin.) Trin., *Stipa teretifolia* Steudel., *Stipa verticillata* Nees & Sprengel, *Stipagrostis ciliata* (Desf.) de Winter, *Stipagrostis uniplumis* (Licht.) de Winter, *Streblochaete longiarista* Pilger, *Streptochaeta spicata* Schrad. ex Nees, *Streptogyna americana* C.E. Hubb., *Streptolophus sagittifolius* Hughes, *Streptostachys asperifolia* Desf., *Sucrea maculata* Soderstrom, *Swallenia alexandrae* (Swallen) Soderstrom & Decker, *Swallenochloa longiligulata* Soderstrom & Calderon, *Symplectrodia lanosa* M. Lazarides, *Symplectrodia gracilis* M. Lazarides

Taeniatherum caput-medusae (L.) Nevski, *Tarigidia aequiglumis* (Goossens) Stent, *Tatianyx arnactites* (Trin.) Zuloaga & Soderstrom, *Teinostachyum attenuatum* Munro, *Tetrachaete elionuroides* Chiov., *Tetrachne dregei* Nees, *Tetrapogon tenellus* Chiov., *Tetrarrhena distichophylla* R. Br., *Tetrarrhena juncea* R. Br., *Tetrarrhena laevis* R. Br., *Tetrarrhena oreophila* D.I. Morris, *Thaumastochloa* sp., *Thelepogon elegans* Roth, *Thellungia advena* Stapf, *Themeda australis* (R. Br.) Stapf, *Thinopyrum* spp. — see *Elytrigia*, *Thrasya petrosa* (Trin.) Chase, *Thrasya reticulata* Swallen, *Thrasya thrasyoides* (Trin.) Chase, *Thrasyopsis repanda* (Nees) Parodi, *Thuarea involuta* (Forster f.) R. Br. ex Roem. & Schultes, *Thyridachne tisserantii* C.E. Hubb., *Thyridolepis mitchelliana* (Nees) Blake, *Thyridolepis multiculmis* (Pilger) Blake, *Thyridolepis xerophila* (Domin) Blake, *Thyrsia inflata* Stapf, *Thyrsia undulatifolia* (Chiov.) Robyns, *Thysanolaena maxima* (Roxb.) O.K., *Torreyochloa pallida* Church, *Trachynia distachya* (L.) Link (= *Brachypodium*), *Trachypogon plumosus* Nees,

Trachys muricata Steud., *Tragus australianus* Blake, *Tribolium* (see *Lasiochloa, Plagiochloa*), *Tricholaena monachne* (Trin.) Stapf & Hubbard, *Tricholaena teneriffae* Parl., *Trichoneura arenaria* (Hochst. & Steud.) Ekman, *Trichopteryx stolziana* Henrard, *Tridens braziliensis* Nees, *Trikeraia hookeri* (Stapf) Bor, *Triniochloa stipoides* (H.B.K.) Hitchcock, *Triodia basedowii* Pritzel, *Triodia clelandii* N.T. Burbidge, *Triodia concinna* N.T. Burbidge, *Triodia hubbardii* N.T. Burbidge, *Triodia irritans* R. Br., *Triodia longiceps* J.M. Black, *Triodia marginata* N.T. Burbidge, *Triodia pungens* R. Br., *Triodia racemigera* Gardner, *Triodia secunda* N.T. Burbidge, *Triodia spicata* N.T. Burbidge, *Triplachne nitens* (Guss.) Link, *Triplasis americana* Beauv., *Triplasis purpurea* (Walt.) Chapm., *Tripogon loliiformis* (Muell.) Hubbard, *Tripsacum laxum* Nash, *Triraphis mollis* R. Br., *Triscenia ovina* Griseb., *Trisetum pumilum* (Desf.) Kunth (= *Lophochloa pumila* (Desf.) Bor), *Trisetum spicatum* (L.) Richter, *Tristachya bequartii* de Willd., *Triticum aestivum* L., *Tsvelevia kerguelensis* (Hook. f.) E. Alexeev, *Tuctoria greenei* (Vasey) J. Reeder

Uniola paniculata L., *Uniola pittieri* Hack., *Uranthoecium truncatum* (Maiden & Betche) Stapf, *Urelytrum squarrosum* Hack., *Urochloa maxima* (Jacq.) R. Webster (= *Panicum maximum*), *Urochloa mosambicensis* (Hackel) Dandy, *Urochondra setulosa* (Trin.) Hubbard

Vahlodea atropurpurea (Wahl.) Fries, *Vaseyochloa multinervosa* (Vasey) Hitchc., *Ventenata dubia* (Leers) Coss., *Vetiveria elongata* (R. Br.) Stapf ex C.E. Hubbard, *Viguierella madagascariensis* A. Camus, *Vossia cuspidata* Wall. & Griff., *Vulpia bromoides* (L.) Gray, *Vulpia incrassata* (Salz. ex Loesel.) Parl., *Vulpia inops* Hack., *Vulpia megalura* (Nutt.) Rudb., *Vulpia myuros* (L.) Gmelin, *Vulpia unilateralis* (L.) Stace (= *Vulpia* Sect. Nardurus)

Wangenheimia lima (L.) Trin., *Whiteochloa cymbiformis* (Hughes) B.K. Simon, *Whiteochloa semitonsa* (Muell. ex Benth.) Hubbard, *Willkommia sarmentosa* Hack.

Xerochloa barbata R. Br., *Xerochloa imberbis* R. Br., *Xerochloa laniflora* Benth.

Yakirra australiensis (Domin) Lazarides & Webster, *Yakirra majuscula* (F. Muell. ex Benth.) Lazarides & Webster, *Yakirra muelleri* (Hughes) Lazarides & Webster, *Yakirra nulla* Lazarides & Webster *Yakirra pauciflora* (R.Br.) Lazarides & Webster, *Yvesia madagascariensis* A. Camus

Zea mays L., *Zenkeria elegans* Trin., *Zeugites mexicana* (Kunth) Trin. ex Steud., *Zeugites pittieri* Hack., *Zingeria kochii* (Mez) Tsvelev, *Zingeria trichopoda* (Boiss.) P. Smirn., *Zizaniopsis miliacea* (Michx.) Döll & Aschers, *Zizaniopsis bonariensis* (Bal. & Poitr.) Spegazz., *Zonotriche inamoena* (K. Schum.) Phipps, *Zoysia macrantha* Desvaux, *Zygochloa paradoxa* (R. Br.) Blake

Embryos

Embryos of the following few species have been sectioned in the Taxonomy Lab, where the preparations are housed.

Dichaetaria wightii Nees, *Eriachne ovata* Nees, *Homolepis aturensis* Chase, *Triodia basedowii* Pritzel, *Triraphis mollis* R. Br.

INDEX OF NAMES

Amblychloa Link = *Sclerochloa* P.
Beauv.
Amblyopyrum Eig
Amblytes Dulac. = *Molinia* Schrank
×*Ammocalamagrostis* P. Fourn., see
Ammophila Host, *Calamagrostis*
Adans.
Ammochloa Boiss.
Ammophila Host
Ampelocalamus Chen, Wen and Sheng =
Sinarundinaria Nakai
Ampelodesmos Link
Amphibromus Nees
Amphicarpum Kunth
Amphidonax Nees = *Arundo* L.
Amphigenes Janka = *Festuca* L.
Amphilophis Nash = *Bothriochloa*
Kuntze
Amphipogon R.Br.
Amphochaeta Anderss. = *Pennisetum*
Rich.
Anachortus Jirásek and Chrtek =
Corynephorus P. Beauv.
Anachyris Nees = *Paspalum* L.
Anadelphia Hackel
Anadelphia scyphofera W.D. Clayton
Anastrophus Schlecht. = *Axonopus* P.
Beauv.
Anatherum Nabelek = *Festuca* L.
Anatherum P. Beauv. = *Andropogon* L.
Ancistrachne S. T. Blake
Ancistragrostis S. T. Blake
Ancistrochloa Honda = *Calamagrostis*
Adans.
Andropogon L.
Andropterum Stapf
Androscepia Brong. = *Themeda* Forssk.
Anelytrum Hack. = *Avena* L.
Anemagrostis Trin. = *Apera* Adans.
Anemanthele Veldk.
Aneurolepidium Nevski = *Leymus*
Hochst.
Anisachne Keng = *Calamagrostis*
Adans.
Anisantha Koch = *Bromus* L.
Aniselytron Merr.
Anisopogon R.Br.
Anomalotis Steud. = *Agrostis* L.
Anomochloa Brongn.
Anoplia Steud. = *Leptochloa* P. Beauv.
Anthaenantia P. Beauv. = *Anthenantia* P.
Beauv.
Anthaenantiopsis Pilger
Anthenantia P. Beauv.
Anthephora Schreber
Anthipsimus Raf. = *Muhlenbergia*
Schreber
Anthistiria L. f. = *Themeda* Forssk.

Anthochloa Nees & Meyen ex Nees
Anthopogon Nutt. = *Gymnopogon* P.
Beauv.
Anthosachne Steud. = *Elymus* L.
Anthoxanthum L.
Antichloa Steud. = *Bouteloua* Lag.
Antinoria Parl.
Antitragus Gaertn. = *Crypsis* Aiton
Antonella Caro = *Tridens* Roem. &
Schult.
Antoschmidtia Boiss. = *Schmidtia* Steud.
Apera Adans.
Aphanelytrum Hackel
Aplexia Faf. = *Leersia* Soland.
Aplocera Raf. = *Ctenium* Panzer
Apluda L.
Apochaete (C. E. Hubbard) Phipps =
Tristachya Nees
Apochiton C.E. Hubb.
Apoclada McClure
Apocopis Nees
Apogon Steud. = *Chloris* O. Swartz
Apogonia Nutt = *Coelorachis* Brongn.
Arberella Soderstrom & Calderón
Archangelina Kuntze = *Tripogon* Roem.
& Schult.
Arctagrostis Griseb.
×*Arctodupontia* Tsvelev, see *Arctophila*
Rupr. ex Andersson, *Dupontia* R.Br.
Arctophila Rupr. ex Andersson
Arctopoa (Griseb.) Probat. = *Poa* L.
Argillochloa Weber = *Festuca* L.
Argopogon Mimeur = *Ischaemum* L.
Aristaria Jungh. = *Themeda* Forssk.
Aristavena Albers & Butzin =
Deschampsia P. Beauv.
Aristella Bertol. = *Stipa* L.
Aristida L.
Aristidium (Endl.) Lindley = *Bouteloua*
Lag.
Aristopsis Catasus = *Aristida* L.
Arrhenatherum P. Beauv.
Arrozia Kunth = *Luziola* A.L. Juss.
Arthragrostis Lazarides
Arthratherum P. Beauv. = *Aristida* L.
Arthraxon P. Beauv.
Arthrochlaena Laena = *Sclerodactylon*
Stapf
Arthrochloa Lorch = *Acrachne* Wright &
Arn. ex Chiov.
Arthrochloa R. Br. = *Holcus* L.
Arthrochortus Lowe = *Lolium* L.
Arthrolophis (Trin.) Chiov. =
Andropogon L.
Arthropogon Nees
Arthrostachya Link = *Gaudinia* P.
Beauv.
Arthrostachys Desv. = *Andropogon* L.

Arthrostylidium Rupr.
Arundarbor Kuntze = *Bambusa* Schreber
Arundinaria Mich.
Arundinella Raddi
Arundo L.
Arundoclaytonia Davidse & Ellis
Asperella Humb. = *Elymus* L.
Asprella Host = *Psilurus* Trin.
Asprella Schreb. = *Leersia* Soland.
Aspris Adans. = *Aira* L.
Asthenatherum Nevski = *Centropodia* Reichenb.
Asthenochloa Buese
Astrebla F. Muell.
Ataxia R. Br. = *Hierochloë* (Gmel.) R.Br.
Athernotus Dulac = *Calamagrostis* Adans.
Atherophora Steud. = *Aegopogon* Humb. & Bonpl. ex Willd.
Atheropogon Willd. = *Bouteloua* Lag.
Athroostachys Bentham
Atractantha McClure
Atractocarpa Franch. = *Puelia* Franch.
Atropis (Trin.) Griseb. = *Puccinellia* Parl.
Aulacolepis Hack. = *Aniselytron* Merr.
Aulaxanthus Elliott = *Anthenantia* P. Beauv.
Aulaxia Nutt = *Anthenantia* P. Beauv.
Aulonemia Goudot
Australopyrum (Tsvelev) A. Löve
Austrochloris Lazarides
Austrofestuca (Tsvel.) E.B. Alekseev
Avellinia Parl.
Avena L.
Avena Scop. = *Lagurus* L.
Avenaria Fabrich. = *Bromus* L.
Avenastrum Opiz = *Helictotrichon* Besser ex Roem. & Schult.
Avenella Parl. = *Deschampsia* P. Beauv.
Avenochloa Holub = *Helictotrichon* Besser ex Roem. & Schult.
Avenula (Dumort.) Dumort. = *Helictotrichon* Besser ex Roem. & Schult.
Axonopus P. Beauv.
Balansochloa Kuntze = *Germainia* Bal. & Poitr.
Baldingera Gaertn., Meyer & Scherb. = *Phalaris* L.
Baldomiria Herter = *Leptochloa* P. Beauv.
Bambusa Schreber
Baptorhachis Clayton & Renvoize
Bashania Keng f. & Yi = *Arundinaria* Mich.

Batratherum Nees = *Arthraxon* P. Beauv.
Bauchea Fourn. = *Sporobolus* R.Br.
Bealia Scribner
Beckera Fresen. = *Snowdenia* C.E. Hubb.
Beckeria Bernh. = *Melica* L.
Beckeropsis Figari & de Not.
Beckmannia Host
Beesha Kunth = *Melocanna* Trin.
Beesha Munro = *Ochlandra* Thwaites
Bellardiochloa Chiov.
Berchtoldia Presl = *Chaetium* Nees
Berghausia Endl. = *Garnotia* Brongn.
Bewsia Goossens
Bhidea Stapf ex Bor
Biatherium Desv. = *Gymnopogon* P. Beauv.
Bifaria (Hack.) Kuntze = *Mesosetum* Steud.
Blakeochloa Veldkamp = *Plinthanthesis* Steud.
Blepharidachne Hackel
Blepharochloa Endl. = *Leersia* Soland.
Blepharoneuron Nash
Bluffia Nees = *Alloteropsis* Presl
Blumenbachia Koel. = *Sorghum* Moench
Blyttia Fries = *Cinna* L.
Boissiera Hochst. & Steud.
Boivinella A. Camus
Bonia Balansa = *Bambusa* Schreber
Boriskerella Terekhov = *Eragrostis* N. M. Wolf
Bothriochloa Kuntze
Bouteloua Lag.
Brachatera Desv. = *Danthonia* DC.
Brachiaria Griseb.
Brachyachne (Benth.) Stapf
Brachyathera Kuntze = *Danthonia* DC.
Brachychloa Phillips
Brachyelytrum P. Beauv.
Brachypodium P. Beauv.
Brachystachyum Keng = *Semiarundinaria* Makino
Brachystylus Dulac = *Koeleria* Pers.
Bracteola Swallen = *Chrysochloa* Swallen
Brandtia Kunth = *Arundinella* Raddi
Brasilocalamus Nakai = *Merostachys* Spreng.
Brevipodium A. & D. Löve = *Brachypodium* P. Beauv.
Briza L.
Brizochloa Jirásek & Chrtek = *Briza* L.
Brizopyrum Link = *Desmazeria* Dumort.
Brizopyrum Stapf = *Tribolium* Desv.
Bromelica (Thurber) Farw. = *Melica* L.
Bromidium Nees = *Agrostis* L.

×*Bromofestuca* Prodan., see *Festuca* L., *Bromus* L.

Bromopsis (Dumort.) Fourr. = *Bromus* L.

Bromuniola Stapf & C.E. Hubb.

Bromus L.

Brousemichea Bal. = *Zoysia* Willd.

Brylkinia Schmidt

Bucetum Parnell = *Festuca* L.

Buchloë Engelm.

Buchlomimus Reeder, Reeder & Rzedowski

Buchmannia Nutt = *Beckmannia* Host

Buergersiochloa Pilger

Bulbilis Raf. = *Buchloë* Engelm.

Bulbulus Swallen = *Rehia* Fijten

Burmabambusa Keng f. = *Sinarundinaria* Nakai

Butania Keng f. = *Arundinaria* Mich.

Cabrera Lag. = *Axonopus* P. Beauv.

Calamagrostis Adans.

Calamina P. Beauv. = *Apluda* L.

Calamochloa Fourn. = *Sohnsia* Airy Shaw

Calamochloe Reichenb. = *Arundinella* Raddi

×*Calamophila* O. Schwartz, see *Calamagrostis* Adans.

Calamovilfa Hackel

Calanthera Hook. = *Buchloë* Engelm.

Calderonella Soderstrom & Decker

Callichloea Steud. = *Elionurus* Humb. & Bonpl.

Calosteca Desv.

Calotheca P. Beauv. = *Calosteca* Desv.

Calotheria Wight and Arn. = *Enneapogon* Desv. ex P. Beauv.

Calycodon Nutt = *Muhlenbergia* Schreber

Calyptochloa C.E. Hubb.

Campeiostachys Drob. = *Elymus* L.

Campella Link = *Deschampsia* P. Beauv.

Campuloa Desv. = *Ctenium* Panzer

Campulosus Desv. = *Ctenium* Panzer

Camusia Lorch = *Acrachne* Wright & Arn. ex Chiov.

Camusiella Bosser

Candollea Steud. = *Agrostis* L.

Capillipedium Stapf

Capriola Adans. = *Cynodon* Rich.

Caryochloa Spreng. = *Piptochaetium* Presl

Caryochloa Trin. = *Luziola* A.L. Juss.

Caryophyllea Opiz = *Aira* L.

Casiostega Galeotti = *Opizia* J. & C. Presl

Castellia Tineo.

Catabrosa P. Beauv.

Catabrosella (Tzvelev) Tzvelev

Catalepis Stapf & Stent

Catapodium Link

Catatherophora Steud. = *Pennisetum* Rich.

Cathestechum J. Presl

Cenchrus L.

Centosteca Desv. = *Centotheca* Desv.

Centotheca Desv.

Centrochloa Swallen

Centrophorum Trin. = *Chrysopogon* Trin.

Centropodia Reichenb.

Cephalochloa Coss. & Dur. = *Ammochloa* Boiss.

Cephalostachyum Munro

Ceratochaete Lunell = *Zizania* L.

Ceratochloa P. Beauv. = *Bromus* L.

Cerea Schlecht. = *Paspalum* L.

Ceresia Pers. = *Paspalum* L.

Ceytosis Munro = *Crypsis* Aiton

Chaboissaea Fourn.

Chaetaria P. Beauv. = *Aristida* L.

Chaetium Nees

Chaetobromus Nees

Chaetochloa Scribn. = *Setaria* P. Beauv.

Chaetopoa C.E. Hubb.

Chaetopogon Janchen

Chaetostichium (Hochst.) C.E. Hubb.

Chaetotropis Kunth = *Agrostis* L.

Chaetotropis Kunth = *Polypogon* Desf.

Chaeturus Link = *Chaetopogon* Janchen

Chalcoelytrum Lunell = *Chrysopogon* Trin.

Chamaecalamus Meyen = *Calamagrostis* Adans.

Chamaedactylis T. Nees = *Aeluropus* Trin.

Chamaeraphis R.Br.

Chamagrostis Borkh. = *Mibora* Adans.

Chandrasekharania V.J. Nair, V.S. Ramachandran & P.V. Sreekumar

Chascolytrum Desv. = *Briza* L.

Chasea Nieuw. = *Panicum* L.

Chasechloa A. Camus

Chasmanthium Link

Chasmopodium Stapf

Chauvinia Steud. = *Spartina* Schreber

Chennapyrum Löve = *Aegilops* L.

Chevalierella A. Camus

Chikusichloa Koidz.

Chilochloa P. Beauv. = *Phleum* L.

Chimonobambusa Makino

Chimonocalamus Hsueh and Yi = *Sinarundinaria* Nakai

Chionachne R.Br.

Chionochloa Zotov

Chloachne Stapf
Chloammia Raf. = *Vulpia* C. Gmelin
Chloothamnus Büse = *Nastus* Juss.
Chloridion Stapf = *Stereochlaena*
 Hackel
Chloridopsis Hack. = *Chloris* O. Swartz
Chloris O. Swartz
Chlorocalymma W. Clayton
Chloroides Regel = *Eustachys* Desf.
Chloropsis Kuntze = *Chloris* O. Swartz
Chlorostis Raf. = *Chloris* O. Swartz
Chondrachyrum Nees = *Briza* L.
Chondrolaena Nees = *Prionanthium*
 Desv.
Chondrosum Desv. = *Bouteloua* Lag.
Chrysochloa Swallen
Chrysopogon Trin.
Chrysurus Pers. = *Lamarckia* Moench
 mut. Koeler
Chumsriella Bor
Chusquea Kunth
Cinna L.
Cinnagrostis Griseb. = *Calamagrostis*
 Adans.
Cladoraphis Franch.
Claudia Opiz = *Melica* L.
Clausospicula M. Lazarides
Clavinodum Wen = *Arundinaria* Mich.
Cleachne Roland. ex Rottb. = *Paspalum*
 L.
Cleistachne Benth.
Cleistochloa C.E. Hubb.
Cleistogenes Keng = *Kengia* Packer
Cliffordiochloa B.K Simon
Clinelymus (Griseb.) Nevski = *Elymus*
 L.
Clomena P. Beauv. = *Muhlenbergia*
 Schreber
Cockaynea Zotov
Codonachne Steud. = *Tetrapogon* Desf.
Coelachne R.Br.
Coelachyropsis Bor
Coelachyrum Hochst. & Nees
Coelarthron Hook.f. = *Microstegium*
 Nees
Coeleochloa Steud. = *Coelachyrum*
 Hochst. & Nees
Coelorachis Brongn.
Coelorhachis Brongn. = *Coelorachis*
 Brongn.
Coix L.
Colanthelia McClure & Smith
Coleanthus Seidl.
Coleataenia Griseb. = *Panicum* L.
Collardoa Cav. = *Ischaemum* L.
Colobachne P. Beauv. = *Alopecurus* L.
Colobanthium (Reichenb.) Taylor =
 Sphenopholis Scribner

Colobanthus (Trin.) Spach =
 Sphenopholis Scribner
Colpodium Trin.
Commelinidium Stapf
Comopyrum Löve = *Aegilops* L.
Coridochloa Nees = *Alloteropsis* Presl
Cornucopiae L.
Cortaderia Stapf
Corynephorus P. Beauv.
Costia Willkom = *Agropyron* Gaertn.
Cottea Kunth
Craepalia Schrank = *Lolium* L.
Craspedorhachis Benth.
Crassipes Swallen = *Sclerochloa* P.
 Beauv.
Criciuma Soderstrom & Londoño =
 Bambusa Schreber
Crinipes Hochst.
Critesion Raf. = *Hordeum* L.
Critho Meyer = *Hordeum* L.
Crithodium Link = *Triticum* L.
Crithopsis Jaub & the Spach
Crossotropis Stapf = *Trichoneura*
 Anderss.
Crypsinna Fourn. = *Muhlenbergia*
 Schreber
Crypsis Aiton
Cryptochloa Swallen
Cryptochloris Benth. = *Tetrapogon*
 Desf.
Cryptostachys Steud. = *Sporobolus*
 R.Br.
Crypturus Link = *Lolium* L.
Ctenium Panzer
Ctenopsis De Not
Curtopogon P. Beauv. = *Aristida* L.
Cutandia Willk.
Cuviera Koel. = *Hordelymus* (Jessen)
 C.O. Harz
Cyathopus Stapf
Cyathorhachis Steud. = *Polytoca* R.Br.
Cyclichnium Dulac = *Gaudinia* P.
 Beauv.
Cyclostachya J. & C. Reeder
Cycloteria Stapf = *Coelorachis* Brongn.
Cylindropyrum (Jaub. & Spach) Löve =
 Aegilops L.
Cymbanthelia Anderss. = *Cymbopogon*
 Spreng.
Cymbopogon Spreng.
Cymbosetaria Schweick.
Cymotochloa Schlecht. = *Paspalum* L.
×*Cynochloris* Clifford & Everist, see
 Chloris O. Swartz, *Cynodon* Rich.
Cynodon Rich.
Cynosurus L.
Cyperochloa Lazarides & L. Watson
Cyphochlaena Hackel

Cypholepis Chiov.
Cyrtococcum Stapf
Czerniaevia Ledeb. = *Deschampsia* P.
　Beauv.
Czernya Presl = *Phragmites* Adans.
Dactilon Vill. = *Cynodon* Rich.
Dactylis L.
Dactyloctenium Willd.
Dactylodes Kuntze = *Tripsacum* L.
Dactylogramma Link = *Muhlenbergia*
　Schreber
Daknopholis W. Clayton
Dallwatsonia B.K. Simon
Dalucum Adans. = *Melica* L.
Danthonia DC.
Danthoniastrum (J. Holub) J. Holub
Danthonidium C.E. Hubb.
Danthoniopsis Stapf
Danthorhiza Ten. = *Helictotrichon*
　Besser ex Roem. & Schult.
×*Danthosieglingia* Domin, see
　Danthonia DC., *Sieglingia* Bernh.
Dasyochloa Willd. ex Rydberg
Dasypoa Pilger
Dasypyrum (Cosson & Durieu) Durand
Davidsea Soderstrom and Ellis
Davyella Hack. = *Neostapfia* Davy
Decandolea Batard = *Agrostis* L.
Decaryella A. Camus
Decaryochloa A. Camus
Deina Alefeld = *Triticum* L.
Dendragrostis Jackson = *Chusquea*
　Kunth
Dendrocalamopsis (Chia & Fung) Keng
　f. = *Bambusa* Schreber
Dendrocalamus Nees
Dendrochloa C.E. Parkinson
Deschampsia P. Beauv.
Desmazeria Dumort.
Desmostachya (Hook. f.) Stapf
Despretzia = *Zeugites* P. Browne
Devauxia Kunth = *Glyceria* R.Br.
Deyeuxia Clarion ex P. Beauv.
Diachroa Nutt = *Leptochloa* P. Beauv.
Diachyrium Griseb. = *Sporobolus* R.Br.
Diacisperma Kuntze = *Leptochloa* P.
　Beauv.
Diandrochloa de Winter
Diandrolyra Stapf
Diandrostachya Jacq.-Fél.
Diarina Raf. = *Diarrhena* P. Beauv.
Diarrhena P. Beauv.
Diastemenanthe Steud. = *Stenotaphrum*
　Trin.
Dichaetaria Nees
Dichanthelium (A. Hitchc. & Chase)
　Gould
Dichanthium Willem.

Dichelachne Endl.
Dichromus Schlecht. = *Paspalum* L.
Dictyochloa (Murbeck) E.G. Camus =
　Ammochloa Boiss.
Didactylon Zoll. & Mor. = *Dimeria*
　R.Br.
Didymochaeta Steud. = *Agrostis* L.
Diectomis Kunth
Dielsiochloa Pilger
Digastrium (Hackel) A. Camus
Digitaria Adans. = *Tripsacum* L.
Digitaria Fabric. = *Paspalum* L.
Digitaria Haller
Digitariella De Winter = *Digitaria*
　Haller
Digitariopsis C.E. Hubb.
Dignathia Stapf
Digraphis Trin. = *Phalaris* L.
Diheteropogon (Hack.) Stapf
Dilepyrum Michaux = *Muhlenbergia*
　Schreber
Dilepyrum Raf. = *Oryzopsis* Michx.
Dileucaden (Raf.) Steud. = *Panicum* L.
Dilophotriche Jacq.-Fél.
Dimeiostemon Raf. = *Andropogon* L.
Dimeria R.Br.
Dimeria Raf. = *Hierochloë* (Gmel.)
　R.Br.
Dimorphochloa S. T. Blake
Dimorphostachys Fourn. = *Paspalum* L.
Dinebra Jacq.
Dinochloa Buese
Diplachne P. Beauv.
Diplasanthum Desv. = *Dichanthium*
　Willem.
Diplax Bennett = *Ehrharta* Thunb.
Diplocea Raf. = *Triplasis* P. Beauv.
Diplopogon R.Br.
Dipogon Steud. = *Sorghastrum* Nash
Dipogonia P. Beauv. = *Diplopogon*
　R.Br.
Dipterum Desv. = *Mnesithea* Kunth
Diptychum Dulac = *Sesleria* Scop.
Disakisperma Steud. = *Leptochloa* P.
　Beauv.
Disarrenum Labill. = *Hierochloë*
　(Gmel.) R.Br.
Dissanthelium Trin.
Dissochondrus (Hillebr.) Kuntze
Distichlis Raf.
Distomomischus Dulac = *Vulpia* C.
　Gmelin
Doellochloa Kuntze = *Gymnopogon* P.
　Beauv.
Dolichochaete Phipps = *Tristachya* Nees
Donacium Fries = *Arundo* L.
Donax P. Beauv. = *Arundo* L.
Drake-Brockmania Stapf

Dregeochloa Conert
Drepanostachyum Keng f. =
　Sinarundinaria Nakai
Drymochloa Holub = *Festuca* L.
Drymonaetes Fourr. = *Festuca* L.
Dryopoa Vick.
Dupontia R.Br.
×*Dupontopoa* N.S. Probatova, see
　Dupontia R.Br., *Poa* L.
Duthiea Hackel
Dybowskia Stapf
Eatonia Raf. = *Panicum* L.
Eccoilopus Steud.
Eccoptocarpha Launert
Echinalysium Trin. = *Elytrophorus* P.
　Beauv.
Echinaria Desf.
Echinaria Fabric. = *Cenchrus* L.
Echinochloa P. Beauv.
Echinolaena Desv.
Echinopogon P. Beauv.
Ectosperma Swallen = *Swallenia*
　Soderstrom & Decker
Ectrosia R.Br.
Ectrosiopsis (Ohwi) Jansen
Ehrharta Thunb.
Ehrhartia Weber = *Leersia* Soland.
Ekmanochloa Hitchcock
Electra Panz. = *Schismus* P. Beauv.
Eleusine Gaertn.
Elionurus Humb. & Bonpl.
×*Elyleymus* Baum, see *Leymus* Hochst.
×*Elyhordeum* Zizan & Petrowa, see
　Hordeum L.
Elymandra Stapf
×*Elymostachys* Tsvelev, see
　Psathyrostachys Nevski
×*Elymotriticum* P. Fourn., see *Triticum*
　L.
Elymus L.
Elymus Mitchell = *Zizania* L.
×*Elysitanion* Bowden, see *Sitanion* Raf.
Elytrigia Desv.
Elytroblepharum (Steud.) Schlecht. =
　Digitaria Haller
Elytroblepharum Steud. = *Digitaria*
　Haller
×*Elytrohordeum* Hylander, see *Elytrigia*
　Desv., *Hordeum* L.
Elytrophorus P. Beauv.
Elytrostachys McClure
Endallex Raf. = *Phalaris* L.
Endodia Raf. = *Leersia* Soland.
Enneapogon Desv. ex P. Beauv.
Enodium Gaud. = *Molinia* Schrank
Enteropogon Nees
Entolasia Stapf
Entoplocamia Stapf

Ephebopogon Steud. = *Microstegium*
　Nees
Epicampes Presl = *Muhlenbergia*
　Schreber
Eragrostiella Bor
Eragrostis N. M. Wolf
Eremitis Doell = *Pariana* Aub.
Eremocaulon Soderstrom & Londoño =
　Bambusa Schreber
Eremochloa Buese
Eremochloe S. Wats. = *Blepharidachne*
　Hackel
Eremopoa Roshev.
Eremopogon Stapf
Eremopyrum (Ledeb.) Jaub. & Spach
Eriachne Phil. = *Digitaria* Haller
Eriachne R.Br.
Erianthecium L. Parodi
Erianthus Michx.
Erioblastus Honda = *Deschampsia* P.
　Beauv.
Eriochaeta Fig. & De Not = *Pennisetum*
　Rich.
Eriochloa Kunth
Eriochrysis P. Beauv.
Eriocoma Nutt. = *Oryzopsis* Michx.
Eriolytrum Kunth = *Panicum* L.
Erioneuron = *Dasyochloa* Willd. ex
　Rydberg
Erioneuron Nash
Eriopodium Hochst. = *Andropogon* L.
Erochloe Raf. = *Eragrostis* N. M. Wolf
Erosion Lunell = *Eragrostis* N. M. Wolf
Erucaria Cerv. = *Bouteloua* Lag.
Erythranthera Zotov
Euchlaena Schrad.
×*Euchlaezea* Janaki ex Bor, see
　Euchlaena Schrad., *Zea* L.
Euclasta Franch.
Eudonax Fries = *Arundo* L.
Eulalia Kunth
Eulaliopsis Honda
Euraphis (Trin.) Lindley = *Boissiera*
　Hochst. & Steud.
Eustachys Desf.
Euthryptochloa Cope
Eutriana Trin. = *Bouteloua* Lag.
Exagrostis Steud. = *Eragrostis* N. M.
　Wolf
Exotheca Anderss.
Exydra Endl. = *Glyceria* R.Br.
Falmiria Reichenb. = *Gaudinia* P.
　Beauv.
Falonia Adans. = *Cynosurus* L.
Fargesia Franch.
Farrago W. Clayton
Fartis Adans. = *Zizania* L.
Fendleria Steud. = *Oryzopsis* Michx.

Ferrocalamus Hsueh & Keng f. =
　Indocalamus Nakai
Festuca L.
Festucaria Fabric. = *Festuca* L.
Festucaria Link = *Vulpia* C. Gmelin
Festucella E. Alekseev
Festucopsis (C.E. Hubb.) Melderis
×*Festulolium* Aschers. & Graebn., see
　Festuca L., *Lolium* L.
×*Festulpia* Melderis ex Stace & R.
　Cotton, see *Festuca* L., *Vulpia* C.
　Gmelin
Fibichia Koel. = *Cynodon* Rich.
Filipedium Raiz. & Jain = *Capillipedium*
　Stapf
Fingerhuthia Nees
Fiorinia Parl. = *Aira* L.
Flavia Fabric. = *Anthoxanthum* L.
Fluminia Fries = *Scolochloa* Link
Foenodorum Krause = *Anthoxanthum* L.
Forasaccus Bub. = *Bromus* L.
Fourniera Scribn. = *Soderstromia*
　Morton
Froesiochloa G.A. Black
Frumentum Krause = *Triticum* L.
Fussia Schur = *Aira* L.
Galeottia Mart. & Gal. = *Zeugites* P.
　Browne
Gamelythrum Nees = *Amphipogon* R.Br.
Garnotia Brongn.
Garnotiella Stapf = *Asthenochloa* Buese
Gastridium P. Beauv.
Gastropyrum (Jaub. & Spach) Löve =
　Aegilops L.
Gaudinia P. Beauv.
Gaudiniopsis (Boiss.) Eig
Gazachloa Phipps = *Danthoniopsis*
　Stapf
Gelidocalamus Wen = *Indocalamus*
　Nakai
Genea (Dumort.) Dumort. = *Bromus* L.
Geopogon Steud. = *Chloris* O. Swartz
Germainia Bal. & Poitr.
Gerritea Zuloaga & Morrone
Gigachilon Seidl = *Triticum* L.
Gigantochloa Kurtz ex Munro
Gilgiochloa Pilger
Ginannia Bub. = *Holcus* L.
Glandiloba (Raf.) Steud. = *Eriochloa*
　Kunth
Glaziophyton Franch.
Glyceria R.Br.
Glyphochloa W. D. Clayton
Gnomonia Lunell = *Festuca* L.
Goldbachia Trin. = *Arundinella* Raddi
Gossweilerochloa Renvoize = *Tridens*
　Roem. & Schult.
Gouinia Fourn.

Goulardia Husn. = *Elymus* L.
Gouldochloa Valdés, Morden & Hatch
Gracilea Hook. f. = *Melanocenchris*
　Nees
Gramen Krause = *Festuca* L.
Gramerium Desv. = *Digitaria* Haller
Graminiastrum Krause = *Dissanthelium*
　Trin.
Graphephorum Desv.
Graya Steud. = *Sphaerocaryum* Nees ex
　Hook.f.
Greenia Nutt = *Limnodea* Dewey ex
　Coult.
Greslania Bal.
Griffithsochloa G.J. Pierce
Guadua Kunth = *Bambusa* Schreber
Guaduella Franch.
Gymnachne L. Parodi
Gymnandropogon (Nees) Duthie =
　Bothriochloa Kuntze
Gymnanthelia Schweinf. = *Cymbopogon*
　Spreng.
Gymnopogon P. Beauv.
Gymnostichum Schreb. = *Elymus* L.
Gymnotrix P. Beauv. = *Pennisetum* Rich.
Gynerium Humb. & Bonpl. = *Gynerium*
　P. Beauv.
Gynerium P. Beauv.
Habrochloa C.E. Hubb.
Habrurus Hochst. = *Elionurus* Humb. &
　Bonpl.
Hackelochloa Kuntze
Hainardia Greuter
Hakonechloa Makino
Halochloa Griseb. = *Monanthochloë*
　Engelm.
Halopyrum Stapf
Haplachne Presl = *Dimeria* R.Br.
Harpachne Hochst.
Harpochloa Kunth
Haynaldia Schur = *Dasypyrum* (Cosson
　& Durieu) Durand
Hekaterosachne Steud. = *Oplismenus* P.
　Beauv.
Heleochloa Fries = *Glyceria* R.Br.
Heleochloa Hort ex Roem. = *Crypsis*
　Aiton
Heleochloa P. Beauv. = *Phleum* L.
Helictotrichon Besser ex Roem. &
　Schult.
Hellera Doell. = *Raddia* Bertol.
Helleria Fourn.
Hellerochloa Rauschert = *Festuca* L.
Helopus Trin. = *Eriochloa* Kunth
Hemarthria R.Br.
Hemibromus Steud. = *Glyceria* R.Br.
Hemigymnia Stapf = *Ottochloa* Dandy
Hemimunroa Parodi = *Munroa* J. Torr.

Hemisacris Steud. = *Schismus* P. Beauv.
Hemisorghum C.E. Hubb.
Henrardia C.E. Hubb.
Hesperochloa Rydb. = *Leucopoa* Griseb.
Heterachne Benth.
Heteranthelium Hochst.
Heteranthoecia Stapf
Heteranthus Borkh. = *Ventenata* Koeler
Heterelytron Jungh. = *Themeda* Forssk.
Heterocarpha Stapf & C.E. Hubb.
Heterochaeta Schult. = *Ventenata*
    Koeler
Heterochloa Desv. = *Andropogon* L.
Heterolepis Boiss. = *Chloris* O. Swartz
Heteropholis C.E. Hubb.
Heteropogon Pers.
Heterosteca Desv. = *Bouteloua* Lag.
Heuffelia Schur = *Helictotrichon* Besser
    ex Roem. & Schult.
Hexarrhena Presl = *Hilaria* Kunth
×*Hibanobambusa* Maruyama and
    Okamura, see *Sasa* Makino &
    Shibata, *Semiarundinaria* Makino
Hickelia A. Camus
Hierochloë (Gmel.) R.Br.
Hilaria Kunth
Himalayacalamus Keng f. =
    Thamnocalamus Munro
Hippagrostis Kuntze = *Oplismenus* P.
    Beauv.
Hitchcockella A. Camus
Holboellia Hook. = *Lopholepis* Decne.
Holcolemma Stapf & C.E. Hubb.
Holcus L.
Hologamium Nees = *Sehima* Forssk.
Holosetum Steud. = *Alloteropsis* Presl
Holttumochloa K.M. Wong = *Bambusa*
    Schreber
Homalachna Kuntze = *Holcus* L.
Homalocenchrus Mieg = *Leersia* Soland.
Homoeatherum Nees = *Andropogon* L.
Homoiachne Pilger = *Deschampsia* P.
    Beauv.
Homolepis Chase
Homopholis C.E. Hubb.
Homoplitis Trin. = *Pogonatherum* P.
    Beauv.
Homopogon Stapf = *Trachypogon* Nees
Homozeugos Stapf
Hookerochloa E. Alekseev
×*Hordale* Ciferri & Giacom., see
    Hordeum L., *Secale* L.
Hordelymus (Jessen) C.O. Harz
Hordeum L.
Houzeaubambus Mattei = *Oxytenanthera*
    Munro
Hubbardia Bor
Hubbardochloa Auquier

Humbertochloa A. Camus & Stapf
Hyalopoa (Tzvelev) Tzvelev
Hydrochloa Hartm. = *Glyceria* R.Br.
Hydrochloa P. Beauv.
Hydropoa (Dumort.) Dumort. = *Glyceria*
    R.Br.
Hydropyrum Link = *Zizania* L.
Hydrothauma C.E. Hubb.
Hygrochloa Lazarides
Hygroryza Nees
Hylebates Chippindall
Hymenachne P. Beauv.
Hymenothecium Lag. = *Aegopogon*
    Humb. & Bonpl. ex Willd.
Hyparrhenia Anderss.
Hyperthelia W. Clayton
Hypogynium Nees
Hypseochloa C.E. Hubb.
Hypudaerus A. Br. = *Anthephora*
    Schreber
Hystericina Steud. = *Echinopogon* P.
    Beauv.
Hystrix Moench
Ichnanthus P. Beauv.
Imperata Cyr.
Indocalamus Nakai
Indochloa Bor = *Euclasta* Franch.
Indopoa Bor
Indoryza Henry & Roy = *Porteresia*
    Tateoka
Indosasa McLure
Ipnum Phil. = *Leptochloa* P. Beauv.
Irulia Bedd. = *Ochlandra* Thwaites
Isachne R.Br.
Isalus J. Phipps
Ischaemopogon Griseb. = *Ischaemum* L.
Ischaemum L.
Ischnanthus Roem. & Schult. =
    Ichnanthus P. Beauv.
Ischnochloa J.D. Hook.
Ischnurus Balf.
Ischurochloa Büse = *Bambusa* Schreber
Iseilema Anderss.
Ixalum Forst. = *Spinifex* L.
Ixophorus Schlechtd.
Jacquesfelixia Phipps = *Danthoniopsis*
    Stapf
Jansenella Bor
Jarava Ruiz & Pavon = *Stipa* L.
Jardinea Steud.
Joachima Ten. = *Beckmannia* Host
Joannegria Chiov. = *Lintonia* Stapf
Jouvea Fourn.
Kampochloa W. Clayton
Kaokochloa de Winter
Karroochloa Conert & Túrpe
Kengia Packer
Kengyilia Yen & J.L Yang

Keniochloa Melderis = *Colpodium* Trin.
Kerinozoma Steud. = *Xerochloa* R.Br.
Kerriochloa C.E. Hubb.
Kielboul Adans. = *Aristida* L.
Kiharapyrum Löve = *Aegilops* L.
Kinabaluchloa K.M. Wong = *Bambusa*
Schreber
Klemachloa Parker = *Dendrocalamus*
Nees
Knappia Sm. = *Mibora* Adans.
Koeleria Pers.
Koordersiochloa Merr. = *Streblochaete*
Hochst.
Korycarpus Lag. = *Diarrhena* P. Beauv.
Kralikella Coss. and Dur. = *Oropetium*
Trin.
Kralikia Coss. & Dur. = *Tripogon*
Roem. & Schult.
Kralikiella Batt. & Trab. = *Tripogon*
Roem. & Schult.
Kratzmannia Opiz = *Agropyron* Gaertn.
Krombholzia Fourn. = *Zeugites* P.
Browne
Ktenosachne Steud. = *Koeleria* Pers.
Lachnagrostis Trin. = *Agrostis* L.
Lacryma Medik = *Coix* L.
Lacryma-jobi Ort. = *Coix* L.
Lacrymaria Fabric. = *Coix* L.
Laertia Gromov = *Leersia* Soland.
Lagurus L.
Lamarckia Moench mut. Koeler
Lamprothyrsus Pilger
Langsdorffia Regel = *Eustachys* Desf.
Lappogopsis Steud. = *Axonopus* P.
Beauv.
Lasiacis A. Hitchc.
Lasiagrostis = *Stipa* L.
Lasiochloa Kunth = *Tribolium* Desv.
Lasiolytrum Steud. = *Arthraxon* P.
Beauv.
Lasiorhachis (Hack.) Stapf
Lasiorrachis Stapf = *Lasiorhachis*
(Hack.) Stapf
Lasiostega Benth. = *Buchloë* Engelm.
Lasiotrichos Lehm. = *Fingerhuthia* Nees
Lasiurus Boiss.
Latipes Kunth = *Leptothrium* Kunth
Lecomtella A. Camus
Leersia Soland.
Leiopoa Ohwi = *Festuca* L.
Leleba Nakai = *Bambusa* Schreber
Lenormandia Steud. = *Vetiveria* Bory
Lepargochloa Launert
Lepeocercis Trin. = *Dichanthium*
Willem.
Lepideilema Trin. = *Streptochaeta*
Schrad.

Lepidopironia A. Rich. = *Tetrapogon*
Desf.
Lepidurus Janchen = *Parapholis* C.E.
Hubb.
Lepitoma Steud. = *Pleuropogon* R.Br.
Lepiurus Dum. = *Lepturus* R.Br.
Leptagrostis C.E. Hubb.
Leptaspis R.Br.
Leptatherum Nees = *Microstegium* Nees
Leptocanna Chia & Fung =
Schizostachyum Nees
Leptocarydion Stapf
Leptocercus Raf. = *Lepturus* R.Br.
Leptochloa P. Beauv.
Leptochloöpsis Yates
Leptochloris Kuntze = *Chloris* O. Swartz
Leptocoryphium Nees
Leptoloma Chase
Leptophyllochloa Cald. = *Koeleria* Pers.
Leptopogon Roberty = *Andropogon* L.
Leptosaccharum (Hackel) A. Camus
Leptostachys Meyer = *Leptochloa* P.
Beauv.
Leptothrium Kunth
Lepturella Stapf
Lepturidium Hitchc. and Ekman
Lepturopetium Morat
Lepturopsis Steud. = *Rhytachne* Desv.
Lepturus R.Br.
Lepyroxis Fourn. = *Muhlenbergia*
Schreber
Lerchenfeldia Schur = *Deschampsia* P.
Beauv.
Lesourdia Fourn. = *Scleropogon* Phil.
Leucophrys Rendle
Leucopoa Griseb.
×*Leymopyron* Tsvelev, see *Agropyron*
Gaertn., *Leymus* Hochst.
×*Leymostachys* Tsvelev, see *Leymus*
Hochst., *Psathyrostachys* Nevski
×*Leymotrigia* Tsvelev, see *Elytrigia*
Desv., *Leymus* Hochst.
Leymus Hochst.
Libertia Lejeune = *Bromus* L.
Libyella Pamp.
Limnas Trin.
Limnetis Rich. = *Spartina* Schreber
Limnodea Dewey ex Coult.
Limnopoa C.E. Hubb.
Lindbergella Bor
Lingnania McClure = *Bambusa* Schreber
Linkagrostis Garcia, Blanca & Torres
Linospartum Adans. = *Lygeum* Loefl ex
L.
Lintonia Stapf
Lithachne P. Beauv.
Littledalea Hemsley

Lodicularia P. Beauv. = *Hemarthria* R.Br.
Lojaconoa Gand. = *Festuca* L.
Loliolum Krecz. & Bobr.
Lolium L.
Lombardochloa Rosengurtt & Arillaga
Lophacme Stapf
Lophatherum Brongn.
Lophochlaena Nees = *Pleuropogon* R.Br.
Lophochloa Reichenb. = *Koeleria* Pers.
Lopholepis Decne.
Lophopogon Hackel
Lophopyrum A. Löve
Lorenzochloa J. & C. Reeder
Loretia Duval-Jouve = *Vulpia* C. Gmelin
Loudetia A. Br. = *Tristachya* Nees
Loudetia Hochst.
Loudetiopsis Conert
Louisiella C.E. Hubb. & Léonard
Loxodera Launert
Loxostachys Peter = *Pseudechinolaena* Stapf
Loydia Delile = *Pennisetum* Rich.
Lucaea Kunth = *Arthraxon* P. Beauv.
Ludolphia Willd. = *Arundinaria* Mich.
Luziola A.L. Juss.
Lycochloa Samuelsson
Lycurus Kunth
Lygeum Loefl ex L.
Maclurochloa K.M. Wong = *Bambusa* Schreber
Maclurolyra Calderón and Soderstrom
Macroblepharus Philippi = *Eragrostis* N. M. Wolf
Macrobriza (Tsvel.) Tsvel. = *Briza* L.
Macrochaeta Steud. = *Pennisetum* Rich.
Macrochloa Kunth = *Stipa* L.
Macronax Raf = *Arundinaria* Mich.
Macrostachya A. Rich. = *Enteropogon* Nees
Maillea Parl.
Maizilla Schlecht. = *Paspalum* L.
Malacurus Nevski
Maltebrunia Kunth
Mandelorna Steud. = *Vetiveria* Bory
Manisuris L.
Mapira Adans. = *Olyra* L.
Massia Bal. = *Eriachne* R.Br.
Matrella Pers. = *Zoysia* Willd.
Matudacalamus Mackawa = *Aulonemia* Goudot
Mays Miller = *Zea* L.
Mayzea Raf. = *Zea* L.
Megalachne Steud.
Megaloprotachne C.E. Hubb.
Megastachya P. Beauv.
Melanocenchris Nees

Melica L.
Melinis P. Beauv.
Melinum Link = *Zizania* L.
Melocalamus Benth.
Melocanna Trin.
Meoschium P. Beauv. = *Ischaemum* L.
Merathrepta Raf. = *Danthonia* DC.
Meringurus Murbeck = *Gaudinia* P. Beauv.
Merisachne Steud. = *Triplasis* P. Beauv.
Merostachys Spreng.
Merxmuellera Conert
Mesosetum Steud.
Metasasa W.T. Lin
Metcalfia Conert
Mezochloa Butzin = *Alloteropsis* Presl
Mibora Adans.
Micagrostis Juss. = *Mibora* Adans.
Michelaria Dumort. = *Bromus* L.
Micraira F. Muell.
Microbambus K. Schum. = *Guaduella* Franch.
Microbriza Parodi ex Nicora et Rúg.
Microcalamus Franch.
Microcalamus Gamble = *Racemobambos* Holttum
Microchloa R.Br.
Microlaena R.Br.
Micropogon Pfeiffer = *Microchloa* R.Br.
Micropyropsis Zarco & Cabezudo
Micropyrum Link
Microstegium Nees
Microthuareia Thouars = *Thuarea* Pers.
Miegia Pers. = *Arundinaria* Mich.
Mildbraediochloa Butzin
Milearium Moench = *Milium* L.
Miliastrum Fabric. = *Setaria* P. Beauv.
Milium Adans. = *Panicum* L.
Milium L.
Miphragtes Nieuwland = *Phragmites* Adans.
Miquelia Arn. & Nees = *Garnotia* Brongn.
Miscanthidium Stapf
Miscanthus Anderss.
Mitwabochloa Phipps = *Zonotriche* Phipps
Mnesithea Kunth
Mniochloa Chase
Moenchia Steud. = *Paspalum* L.
Molineria Parl. = *Periballia* Trin.
Molineriella Rouy = *Periballia* Trin.
Molinia Schrank
Moliniopsis Gand. = *Kengia* Packer
Moliniopsis Hayata = *Molinia* Schrank
Monachather Steud.
Monachne P. Beauv. = *Panicum* L.

Monachyron Parl. = *Rhynchelytrum* Nees
Monanthochloë Engelm.
Monathera Raf. = *Ctenium* Panzer
Monelytrum Hackel
Monerma (Willd.) Coss. & Dur. = *Hainardia* Greuter
Monerma P. Beauv. = *Lepturus* R.Br.
Monium Stapf
Monocera Elliott = *Ctenium* Panzer
Monochaete Doell. = *Gymnopogon* P. Beauv.
Monocladus Chia, Fung & Yang
Monocymbium Stapf
Monodia S.W.L. Jacobs
Monopogon Presl = *Tristachya* Nees
Monostachya Merr.
Monostemon Henr. = *Microbriza* Parodi ex Nicora et Rúg.
Moorea Lemaire = *Cortaderia* Stapf
Mosdenia Stent
Moulinsia Raf. = *Aristida* L.
Muantijamvella Phipps = *Tristachya* Nees
Muhlenbergia Schreber
Munroa J. Torr.
Mygalurus Link = *Vulpia* C. Gmelin
Myriachaeta Moritzi = *Thysanolaena* Nees
Myriocladus Swallen
Myriostachya J.D. Hook.
Nabelekia Roshev. = *Festuca* L.
Narduretia Villar = *Vulpia* C. Gmelin
Narduroides Rouy
Nardurus (Bluff, Nees & Schauer) Reichenb. = *Vulpia* C. Gmelin
Nardus L.
Narenga Bor
Nassella Desv.
Nastus Juss.
Nastus Lunell = *Cenchrus* L.
Natschia Bub. = *Nardus* L.
Navicularia Radd. = *Ichnanthus* P. Beauv.
Neeragrostis Bush
Neesiochloa Pilger
Negria Chiov. = *Lintonia* Stapf
Nemastachys Steud. = *Microstegium* Nees
Nematopoa C.E. Hubb.
Neoaulacolepis Rauschert = *Aniselytron* Merr.
Neobambus Keng f. = *Sinobambusa* Makino
Neobouteloua Gould
Neohouzeaua A. Camus
Neohusnotia A. Camus = *Acroceras* Stapf

Neomicrocalamus Keng f. = *Racemobambos* Holttum
Neomolinia Honda = *Diarrhena* P. Beauv.
Neosasamorpha Tatewaki = *Sasa* Makino & Shibata
Neoschischkinia Tsvelev = *Agrostis* L.
Neosinocalamus Keng f. = *Dendrocalamus* Nees
Neostapfia Davy
Neostapfiella A. Camus
Nephelochloa Boiss.
Nestlera Steud. = *Bouteloua* Lag.
Neurachne R.Br.
Neurolepis Meissner
Neuropoa Clayton = *Poa* L.
Nevroctola Raf. = *Uniola* L.
Nevroloma Raf. = *Glyceria* R.Br.
Nevskiella Krecz & Vved. = *Bromus* L.
Neyraudia Hook.f.
Nipponobambusa Muroi = *Sasa* Makino & Shibata
Nipponocalamus Nakai = *Arundinaria* Mich.
Nivieria Ser. = *Triticum* L.
Normanboria Butzin = *Acrachne* Wright & Arn. ex Chiov.
Nothoholcus Nash = *Holcus* L.
Notholcus Hitchc. = *Holcus* L.
Notochloë Domin.
Notodanthonia Zotov = *Rytidosperma* Steud.
Notonema Raf. = *Agrostis* L.
Nowodworskya Presl = *Polypogon* Desf.
Ochlandra Thwaites
Ochthochloa Edgwe.
Odontelytrum Hackel
Odyssea Stapf
Oedipachne Link = *Eriochloa* Kunth
Oligostachyum Wang & Ye = *Arundinaria* Mich.
Olmeca Soderstrom
Olyra L.
Omeiocalamus Keng f. = *Arundinaria* Mich.
Onoea Franch. & Sav. = *Diarrhena* P. Beauv.
Ophiochloa Filgueiras, Davidse & Zuloaga
Ophiurinella Desv. = *Stenotaphrum* Trin.
Ophiuros Gaertn. f.
Opizia J. & C. Presl
Oplismenopsis L. Parodi
Oplismenus P. Beauv.
Orcuttia Vasey
Oreiostachys Gamble = *Nastus* Juss.
Oreobambos K. Schum.

Oreocalamus Keng = *Chimonobambusa* Makino
Oreochloa Link
Oreopoa Grand. = *Poa* L.
Orinus A. Hitchc.
Ornithocephalochloa Kurz = *Thuarea* Pers.
Ornithospermum Dumoulin = *Echinochloa* P. Beauv.
Oropetium Trin.
Orrhopygium Löve = *Aegilops* L.
Orthachne Nees ex Steud.
Orthoclada P. Beauv.
Orthopogon R. Br. = *Oplismenus* P. Beauv.
Orthoraphium Nees = *Stipa* L.
×*Oryticum* Wang & Tang, see *Oryza* L., *Triticum* L.
Oryza L.
Oryzidium C.E. Hubb. & Schweick.
Oryzopsis Michx.
Osterdamia Necker ex Kuntze = *Zoysia* Willd.
Otachyrium Nees
Otatea McClure & Smith
Ottochloa Dandy
Oxyanthe Steud. = *Phragmites* Adans.
Oxychloris Lazarides
Oxydenia Nutt = *Leptochloa* P. Beauv.
Oxyrhachis Pilger
Oxytenanthera Munro
Padia Moritzi = *Oryza* L.
Pallasia Scop. = *Crypsis* Aiton
Paneion Lunell = *Poa* L.
Panicastrella Moench = *Echinaria* Desf.
Panicum L.
Pantathera Phil. = *Megalachne* Steud.
Pappagrostis Roshev. = *Stephanachne* Keng
Pappophorum Schreber
Paracolpodium Tsvelev. = *Colpodium* Trin.
Paractaenium P. Beauv. = *Parectenium* P. Beauv. corr. Stapf
Parafestuca E. Alekseev
Parahyparrhenia A. Camus
Paraneurachne S. T. Blake
Parapholis C.E. Hubb.
Paratheria Griseb.
Parectenium P. Beauv. corr. Stapf
Pariana Aub.
Parodiella J. & C. Reeder = *Lorenzochloa* J. & C. Reeder
Parodiochloa C.E. Hubb. = *Poa* L.
Parodiolyra Soderstrom and Zuloaga
Parvotrisetum Chrtek = *Trisetum* Pers.
Pascopyrum A. Löve
Paspalanthium Desv. = *Paspalum* L.

Paspalidium Stapf
Paspalum L.
Patis Ohwi = *Stipa* L.
Patropyrum Löve = *Aegilops* L.
Pechea Lapeyr. = *Crypsis* Aiton
Pectinaria (Benth.) Hack. = *Eremochloa* Buese
Peltophorus Desv. = *Manisuris* L.
Penicillaria Willd. = *Pennisetum* Rich.
Peniculus Swallen = *Mesosetum* Steud.
Pennisetum Rich.
Pentacraspedon Steud. = *Amphipogon* R.Br.
Pentameris P. Beauv.
Pentapogon R.Br.
Pentarrhaphis Kunth
Pentaschistis (Nees) Spach
Pentastachya Steud. = *Pennisetum* Rich.
Pentatherum Nabelek = *Agrostis* L.
Pereilema J. & C. Presl
Periballia Trin.
Peridictyon O. Seberg, S. Frederiksen & C. Baden
Perlaria Fabric. = *Aegilops* L.
Perobachne Presl = *Themeda* Forssk.
Perotis Aiton
Perrierbambus A. Camus
Perulifera A. Camus
Petriella Zotov
Petrina Phipps = *Danthoniopsis* Stapf
Peyritschia Fourn.
Phacellaria Steud. = *Chloris* O. Swartz
Phacelurus Griseb.
Phaenanthoecium C.E. Hubb.
Phaenosperma Munro ex Benth.
Phalarella Boiss. = *Phleum* L.
Phalaridantha St-Lager = *Phalaris* L.
Phalaridium Nees & Meyen = *Dissanthelium* Trin.
Phalaris L.
Phalaroides Wolf = *Phalaris* L.
Phanopyrum (Raf.) Nash = *Panicum* L.
Pharus P.Browne
Pheidochloa S. T. Blake
Phippsia R.Br.
Phleum L.
Pholiurus Trin.
Phragmites Adans.
Phyllorhachis Trimen
Phyllostachys Sieb & Zucc.
Pilgerochloa Eig
Piptatherum P. Beauv.
Piptochaetium Presl
Piptophyllum C.E. Hubb.
Piptostachya (C. E. Hubbard) Phipps = *Zonotriche* Phipps
Piresia Swallen

Piresiella Judziewicz, Zuloaga &
   Morone
Pithecurus Kunth = *Schizachyrium* Nees
Plagiantha Renvoize
Plagiarthron Duv. = *Loxodera* Launert
Plagiochloa Adamson and Sprague =
   Tribolium Desv.
Plagiolytrum Nees = *Tripogon* Roem. &
   Schult.
Plagiosetum Benth.
Planichloa B. Simon
Planotia Munro = *Neurolepis* Meissner
Plantinia Bub. = *Phleum* L.
Platonia Kunth = *Neurolepis* Meissner
Plazerium Kunth = *Eriochrysis* P.
   Beauv.
Plectrachne Henrard
Pleiadelphia Stapf
Pleioblastus Nakai = *Arundinaria* Mich.
Pleiodon Reichenb. = *Bouteloua* Lag.
Pleioneura (C. E. Hubb.) Phipps =
   Danthoniopsis Stapf
Pleopogon Nutt = *Lycurus* Kunth
Pleuraphis Torrey = *Hilaria* Kunth
Pleuroplitis Trin. = *Arthraxon* P. Beauv.
Pleuropogon R.Br.
Plinthanthesis Steud.
Plotia Steud. = *Glyceria* R.Br.
Poa L.
Poagrostis Raf. = *Poa* L.
Poagrostis Stapf
Poarion Reichenb. = *Koeleria* Pers.
Pobeguinea Jacques-Félix
Podagrostis (Griseb.) Scribn. = *Agrostis*
   L.
Podinapus Dulac = *Deschampsia* P.
   Beauv.
Podophorus Phil.
Podopogon Raf. = *Piptochaetium* Presl
Podosemum Desv. = *Muhlenbergia*
   Schreber
Poecilostachys Hackel
Pogochloa S. Moore = *Gouinia* Fourn.
Pogonachne Bor
Pogonarthria Stapf
Pogonatherum P. Beauv.
Pogoneura Napper
Pogonochloa C.E. Hubb.
Pogonopsis Presl = *Pogonatherum* P.
   Beauv.
Pohlidium Davidse, Soderstrom & Ellis
Poidium Nees
Polevansia de Winter
Pollinia Spreng. = *Chrysopogon* Trin.
Pollinidium Haines = *Eulaliopsis* Honda
Polliniopsis Hayata
Polyneura Peter = *Panicum* L.
Polyodon Kunth = *Bouteloua* Lag.

Polypogon Desf.
Polyraphis (Trin.) Lindley =
   Pappophorum Schreber
Polyschistis Presl = *Pentarrhaphis*
   Kunth
Polytoca R.Br.
Polytrias Hackel
Pommereulla L.f.
Ponceletia Thours = *Spartina* Schreber
Poranthera Raf. = *Sorghastrum* Nash
Porroteranthe Steud. = *Glyceria* R.Br.
Porteresia Tateoka
Potamochloa Griff. = *Hygroryza* Nees
Potamophila R.Br.
Preissia Opiz = *Avena* L.
Pringleochloa Scribner
Prionachne Nees = *Prionanthium* Desv.
Prionanthium Desv.
Prosphysis Dulac = *Vulpia* C. Gmelin
Prosphytochloa Schweickerdt
Psamma P. Beauv. = *Ammophila* Host
Psammagrostis C. Gardner & C.E.
   Hubb.
Psammochloa A. Hitchc.
Psammophila Schult. = *Spartina*
   Schreber
Psathyrostachys Nevski
Pseudanthistiria (Hackel) Hook.f.
Pseudarrhenatherum Rouy
Pseudechinolaena Stapf
Pseudobrachiaria Launert = *Brachiaria*
   Griseb.
Pseudobromus K. Schum.
Pseudochaetochloa A. Hitchc.
Pseudocoix A. Camus
Pseudodanthonia Bor & C.E. Hubb.
Pseudodichanthium Bor
Pseudolasiacis A. Camus = *Lasiacis* A.
   Hitchc.
Pseudopentameris Conert
Pseudophacelurus (Steud.) A. Camus =
   Phacelurus Griseb.
Pseudophleum M. Dogan
Pseudopogonatherum A. Camus
Pseudoraphis Griff.
Pseudoroegneria (Nevski) A. Löve
Pseudoryza Griff. = *Leersia* Soland.
Pseudosasa Makino
Pseudosecale (Godron) Degen =
   Dasypyrum (Cosson & Durieu)
   Durand
Pseudosorghum A. Camus
Pseudostachyum Munro
Pseudostreptogyne A. Camus =
   Streblochaete Hochst.
Pseudovossia A. Camus
Pseudoxytenanthera Soderstrom and
   Ellis

Pseudozoysia Chiov.
Psilantha (K. Koch) Tzvelev = *Eragrostis* N. M. Wolf
Psilathera Link
Psilochloa Launert = *Panicum* L.
Psilolemma Phillips
Psilopogon Hochst. = *Microstegium* Nees
Psilostachys Steud. = *Dimeria* R.Br.
Psilurus Trin.
Pterium Desv. = *Lamarckia* Moench mut. Koeler
Pterochlaena Chiov. = *Alloteropsis* Presl
Pterochloris A. Camus
Pteropodium Steud. = *Calamagrostis* Adans.
Pterostachyum Steud. = *Dimeria* R.Br.
Ptilagrostis Griseb. = *Stipa* L.
Ptiloneilema Steud. = *Melanocenchris* Nees
Puccinellia Parl.
×*Pucciphippsia* Tsvelev, see *Phippsia* R.Br., *Puccinellia* Parl.
Puelia Franch.
Puliculum Haines = *Eulalia* Kunth
Pyrrhanthera Zotov
Quiongzhuea Hsueh & Yi = *Chimonobambusa* Makino
Rabdochloa P. Beauv. = *Leptochloa* P. Beauv.
Racemobambos Holttum
Raddia Bertol.
Raddia Mazziari = *Crypsis* Aiton
Raddiella Swallen
Ramosia Merr. = *Centotheca* Desv.
Raphis Lour. = *Chrysopogon* Trin.
Raram Adans. = *Cenchrus* L.
Raspailia Presl = *Polypogon* Desf.
Rattraya Phipps = *Danthoniopsis* Stapf
Ratzeburgia Kunth
Reana Brignoli = *Zea* L.
Reboulea Kunth = *Sphenopholis* Scribner
Redfieldia Vasey
Reederochloa Soderstrom & H.F. Decker
Rehia Fijten
Reimaria Fluegge = *Paspalum* L.
Reimarochloa A. Hitchc.
Reitzia Swallen
Relchela Steud.
Rendlia Chiov.
Rettbergia Raddi = *Chusquea* Kunth
Reynaudia Kunth
Rhachidospermum Vasey = *Jouvea* Fourn.
Rhampholepis Stapf = *Sacciolepis* Nash
Rhiniachne Steud. = *Thelepogon* Roth.

Rhipidium Trin. = *Saccharum* L.
Rhipidocladum McClure
Rhizocephalus Boiss.
Rhomboelytrum Link
Rhombolytrum Link = *Rhomboelytrum* Link
Rhynchelytrum Nees
Rhynchoryza Baillon
Rhytachne Desv.
Richardsiella Elffers & Kennedy O'Byrne
Riedelia Kunth = *Arundinella* Raddi
Ripidium Trin. = *Erianthus* Michx.
Robynsiochloa Jacques-Félix
Roegneria C. Koch = *Elymus* L.
Roemeria Roem. & Schult. = *Diarrhena* P. Beauv.
Roshevitzia Tsvelev = *Eragrostis* N. M. Wolf
Rostraria Trin. = *Koeleria* Pers.
Rothia Borkh. = *Mibora* Adans.
Rottboellia L.f.
Roylea Steud. = *Melanocenchris* Nees
Rupestrina Prov. = *Trisetum* Pers.
Rytidosperma Steud.
Rytilix Raf. = *Hackelochloa* Kuntze
Sabsab Adans. = *Paspalum* L.
Saccharifera Stokes = *Saccharum* L.
Saccharum L.
Sacciolepis Nash
Salmasia Bub. = *Aira* L.
Sanguinaria Bub. = *Digitaria* Haller
Sanguinella Gleichen = *Digitaria* Haller
Santia Savi = *Polypogon* Desf.
Sarga Ewart & White = *Sorghum* Moench
Sartidia de Winter
Sasa Makino & Shibata
Sasaella Mak. = *Sasa* Makino & Shibata
Sasamorpha Nakai = *Sasa* Makino & Shibata
Saugetia A. Hitchc. & Chase
Savastana Schrank = *Hierochloë* (Gmel.) R.Br.
Schaffnera Benth. = *Schaffnerella* Nash
Schaffnerella Nash
Schedonnardus Steud.
Schedonorus Beauv. = *Austrofestuca* (Tsvel.) E.B. Alekseev
Schellingia Steud. = *Aegopogon* Humb. & Bonpl. ex Willd.
Schenckochloa J.J. Ortíz
Schismus P. Beauv.
Schistachne Fig. & De Not = *Stipagrostis* Nees
Schizachne Hackel
Schizachyrium Nees

Schizopogon Spreng. = *Schizachyrium*
  Nees
Schizostachyum Nees
Schmidtia Steud.
Schmidtia Tratt. = *Coleanthus* Seidl.
Schnizleina Steud. = *Boissiera* Hochst.
  & Steud.
Schoenanthus Adans. = *Ischaemum* L.
Schoenefeldia Kunth
Schultesia Spreng. = *Eustachys* Desf.
Sciadonardus Steud. = *Gymnopogon* P.
  Beauv.
Scirpobambus Kuntze = *Oxytenanthera*
  Munro
Sclerachne R.Br.
Sclerachne Trin. = *Limnodea* Dewey ex
  Coult.
Sclerandrium Stapf & Hubbard =
  Germainia Bal. & Poitr.
Sclerochlaena A. Camus =
  Cyphochlaena Hackel
Sclerochloa P. Beauv.
Sclerodactylon Stapf
Sclerodeuxia Pilger = *Calamagrostis*
  Adans.
Sclerodeyeuxia (Stapf) Pilger =
  Deyeuxia Clarion ex P. Beauv.
Scleropelta Buckley = *Hilaria* Kunth
Sclerophyllum Griff. = *Porteresia*
  Tateoka
Scleropoa Griseb. = *Catapodium* Link
Scleropogon Phil.
Sclerostachya A. Camus
Scolochloa Link
Scolochloa Mert. & Koch = *Arundo* L.
Scribneria Hackel
Scrotochloa Judziewicz
Scutachne A. Hitchc. & Chase
Secale L.
Secalidium Schur = *Dasypyrum* (Cosson
  & Durieu) Durand
Sehima Forssk.
Semeiostachys Drob. = *Elymus* L.
Semiarundinaria Makino
Senisetum Koidz. = *Agrostis* L.
Senites Adans. = *Zeugites* P. Browne
Sennenia Sennen = *Trisetum* Pers.
Sericrostis Raf. = *Muhlenbergia*
  Schreber
Sericura Hassk. = *Pennisetum* Rich.
Serrafalcus Parl. = *Bromus* L.
Sesleria Scop.
Sesleriella Deyl
Setaria P. Beauv.
Setariopsis Scribner ex Millsp.
Setiacis S.L. Chen & Y.X. Jin =
  Panicum L.
Setosa Ewart = *Chamaeraphis* R.Br.

Shibataea Makino
Sieglingia Bernh.
Silentvalleya Nair, Sreekumar,
  Vajravelu & Bhargavan
Simplicia Kirk
Sinarundinaria Nakai
Sinoarundinaria Ohwi = *Phyllostachys*
  Sieb & Zucc.
Sinobambusa Makino
Sinocalamus McClure = *Dendrocalamus*
  Nees
Sinochasea Keng
Sitanion Raf.
Sitopsis (Jaub. & Spach) Löve =
  Aegilops L.
×*Sitordeum* Bowden, see *Hordeum* L.,
  Sitanion Raf.
Snowdenia C.E. Hubb.
Soderstromia Morton
Soejatmia K.M. Wong = *Bambusa*
  Schreber
Sohnsia Airy Shaw
Solenachne Steud. = *Spartina* Schreber
Solenophyllum Baillon = *Monanthochloë*
  Engelm.
Sorghastrum Nash
Sorghum Adans. = *Holcus* L.
Sorghum Moench
Sparteum P. Beauv. = *Stipa* L.
Spartina Schreber
Spartochloa C.E. Hubb.
Spartum P. Beauv. = *Lygeum* Loefl ex L.
Spathia Ewart
Spelta Wolf = *Triticum* L.
Spermachiton Llanos = *Sporobolus*
  R.Br.
Sphaerella Bub. = *Airopsis* Desv.
Sphaerium Kuntze = *Coix* L.
Sphaerobambos S. Dransfield
Sphaerocaryum Nees ex Hook.f.
Spheneria Kuhlm
Sphenopholis Scribner
Sphenopus Trin.
Spinifex L.
Spirochloe Lunell = *Schedonnardus*
  Steud.
Spirotheros Raf. = *Heteropogon* Pers.
Spodiopogon Trin.
Sporobolus R.Br.
Stapfia Davy = *Neostapfia* Davy
Stapfiola Kuntze = *Desmostachya*
  (Hook. f.) Stapf
Stegosia Lour. = *Rottboellia* L.f.
Steinchisma Raf.
Steirachne Ekman
Stelephuros Adans. = *Phleum* L.
Stemmatospermum P. Beauv. = *Nastus*
  Juss.

Stenochloa Nutt = *Dissanthelium* Trin.
Stenofestuca (Honda) Nakai = *Bromus* L.
Stenotaphrum Trin.
Stephanachne Keng
Stereochlaena Hackel
Steudelella Honda = *Sphaerocaryum* Nees ex Hook.f.
Steyermarkochloa Davidse and Ellis
Stiburus Stapf
Stilpnophleum Nevski
Stipa L.
Stipagrostis Nees
Stipavena Vierh. = *Helictotrichon* Besser ex Roem. & Schult.
×*Stiporyzopsis* B.L. Johnson & Rogler, see *Oryzopsis* Michx., *Stipa* L.
Streblochaete Hochst.
Strephium Nees = *Raddia* Bertol.
Streptachne R. Br. = *Aristida* L.
Streptia Doell = *Streptogyna* P. Beauv.
Streptochaeta Schrad.
Streptogyna P. Beauv.
Streptolophus Hughes
Streptostachys Desv.
Strombodurus Steud. = *Pentarrhaphis* Kunth
Sturmia Hoppe = *Mibora* Adans.
Styppeiochloa de Winter
Suaria Schrank = *Melinis* P. Beauv.
Sucrea Soderstrom
Suddia Renvoize
Swallenia Soderstrom & Decker
Swallenochloa McClure
Syllepis Fourn. = *Imperata* Cyr.
Symbasiandra Steud. = *Hilaria* Kunth
Symplectrodia Lazarides
Synaphe Dulac = *Catapodium* Link
Syntherisma Walt. = *Digitaria* Haller
Taeniatherum Nevski
Taeniorhachis T.A. Cope
Tansaniochloa Rauschert = *Setaria* P. Beauv.
Tarigidia Stent
Tatianyx Zuloaga and Soderstrom
Teinostachyum Munro
Tema Adans. = *Echinochloa* P. Beauv.
Terellia Lunell = *Elymus* L.
Tetrachaete Chiov.
Tetrachne Nees
Tetragonocalamus Nakai = *Bambusa* Schreber
Tetrapogon Desf.
Tetrarrhena R.Br.
Thalysia Kuntze. = *Zea* L.
Thamnocalamus Munro
Thaumastochloa C.E. Hubb.
Thelepogon Roth.

Thellungia Stapf
Themeda Forssk.
Thinopyrum A. Löve
Thorea Rouy = *Pseudarrhenatherum* Rouy
Thoreochloa Holub = *Pseudarrhenatherum* Rouy
Thouarsia Kuntze = *Thuarea* Pers.
Thrasya Kunth
Thrasyopsis L. Parodi
Thrixgyne Keng = *Duthiea* Hackel
Thuarea Pers.
Thurberia Benth. = *Limnodea* Dewey ex Coult.
Thyridachne C.E. Hubb.
Thyridolepis S. T. Blake
Thyridostachyum Nees = *Mnesithea* Kunth
Thyrsia Stapf
Thyrsostachys Gamble
Thysanachne Presl = *Arundinella* Raddi
Thysanolaena Nees
Timouria Roshev. = *Stipa* L.
Tinaea Garzia = *Lamarckia* Moench mut. Koeler
Tisserantiella Mimeur = *Thyridachne* C.E. Hubb.
Torgesia Bornm. = *Crypsis* Aiton
Torresia Ruiz & Pavon = *Hierochloë* (Gmel.) R.Br.
Torreyochloa Church
Tosagris P. Beauv. = *Muhlenbergia* Schreber
Tovarochloa T.D. Macfarlane and P. P.-H. But
Tozzettia Savi = *Alopecurus* L.
Trachynia Link = *Brachypodium* P. Beauv.
Trachynotia Michaux = *Spartina* Schreber
Trachyozus Reichenb. = *Trachys* Pers.
Trachypoa Bub. = *Dactylis* L.
Trachypogon Nees
Trachys Pers.
Trachystachys Dietr. = *Trachys* Pers.
Tragus Haller
Tragus Panzer = *Brachypodium* P. Beauv.
Tremularia Fabric. = *Briza* L.
Triachyrum A. Br. = *Sporobolus* R.Br.
Triaena Kunth = *Bouteloua* Lag.
Trianthium Desv. = *Chrysopogon* Trin.
Triarrhena (Maxim.) Nakai = *Miscanthus* Anderss.
Triatherus Raf. = *Ctenium* Panzer
Triavenopsis Candargy = *Duthiea* Hackel
Tribolium Desv.

Trichachne Nees = *Digitaria* Haller
Trichloris Benth. = *Chloris* O. Swartz
Trichochloa DC. = *Muhlenbergia*
Schreber
Trichodium Michaux = *Agrostis* L.
Tricholaena Schrad.
Trichoneura Anderss.
Trichoon Roth = *Phragmites* Adans.
Trichopteryx Nees
Trichopyrum (Nevski) Löve = *Elytrigia*
Desv.
Tricuspis P. Beauv. = *Tridens* Roem. &
Schult.
Tridens Roem. & Schult.
Triglossum Roem. & Schult. =
Arundinaria Mich.
Trikeraia Bor
Trilobachne Schenk ex Henrard
Triniochloa A. Hitchc.
Triniusa Steud. = *Bromus* L.
Triodia R.Br.
Triodon Baumg. = *Triodia* R.Br.
Triphlebia Stapf = *Eragrostis* N. M.
Wolf
Triplachne Link
Triplasis P. Beauv.
Triplathera (Endl.) Lindley = *Bouteloua*
Lag.
Triplopogon Bor
Tripogon Roem. & Schult.
Tripsacum L.
Triraphis R.Br.
Triscenia Griseb.
Trisetaria Forssk. = *Trisetum* Pers.
Trisetarium Poir. = *Trisetum* Pers.
Trisetobromus Nevski = *Bromus* L.
×*Trisetokoeleria* Tsvelev, see *Koeleria*
Pers., *Trisetum* Pers.
Trisetum Pers.
Trisiola Raf. (1817) = *Uniola* L.
Trisiola Raf. (1825) = *Distichlis* Raf.
Tristachya Nees
Tristania Poir. = *Spartina* Schreber
Tristegis Nees = *Melinis* P. Beauv.
×*Triticosecale* Wittmack, see *Secale* L.,
Triticum L.
Triticum L.
×*Trititrigia* Tsvelev, see *Elytrigia* Desv.,
Triticum L.
×*Tritordeum* Aschers. & Graebn., see
Hordeum L., *Triticum* L.
Triunila Raf. = *Uniola* L.
Trixostis Raf. = *Aristida* L.
Trochera L. Rich. = *Ehrharta* Thunb.
Tschompskia Aschers. & Graebn. =
Arundinaria Mich.
Tsvelevia E. Alekseev
Tuctoria J. Reeder

Turraya Wall. = *Leersia* Soland.
Typhoides Moench = *Phalaris* L.
Uniola L.
Urachne Trin. = *Oryzopsis* Michx.
Uralepis Nutt = *Triplasis* P. Beauv.
Uranthoecium Stapf
Urelytrum Hackel
Urochlaena Nees
Urochloa P. Beauv.
Urochondra C.E. Hubb.
Vahlodea Fries
Valota Adans. = *Digitaria* Haller
Vaseya Thurber = *Muhlenbergia*
Schreber
Vaseyochloa A. Hitchc.
Ventenata Koeler
Verinea Merino = *Melica* L.
Veseyochloa Phipps = *Tristachya* Nees
Vetiveria Bory
Vietnamochloa Veldkamp & Nowack
Vietnamosasa Nguyen
Viguierella A. Camus
Vilfa Adans. = *Agrostis* L.
Vilfa Beauv. = *Phippsia* R.Br.
Vilfagrostis Doell = *Eragrostis* N. M.
Wolf
Vossia Wall. & Griff.
Vulpia C. Gmelin
Vulpiella (Trabut) Burollet
Wangenheimia Moench
Wasatchia M. E. Jones = *Festuca* L.
Whiteochloa C.E. Hubb.
Wiesta Boiss. = *Boissiera* Hochst. &
Steud.
Wilhelmsia Koch = *Koeleria* Pers.
Wilibald-Schmidtia Conrad = *Danthonia*
DC.
Wilibalda Roth = *Coleanthus* Seidl.
Willbleibia Herter = *Willkommia* Hackel
Willkommia Hackel
Windsoria Nutt = *Tridens* Roem. &
Schult.
Wirtgenia Doell = *Paspalum* L.
Woodrowia Stapf = *Dimeria* R.Br.
Xanthonanthus St-Lager =
Anthoxanthum L.
Xenochloa Roem. & Schult. =
Phragmites Adans.
Xerochloa R.Br.
Xerodanthia Phipps = *Danthoniopsis*
Stapf
Xiphagrostis Cov. = *Miscanthus*
Anderss.
Xystidium Trin. = *Perotis* Aiton
Yadakeya Mak. = *Pseudosasa* Makino
Yakirra Lazarides & R. Webster
Ystia P. Compère
Yushania Keng

Yvesia A. Camus
Zea L.
Zeia Lunell = *Elymus* L.
Zenkeria Trin.
Zeocrithon P. Beauv. = *Hordeum* L.
Zeocriton Wolf = *Hordeum* L.
Zerna Panzer = *Vulpia* C. Gmelin
Zeugites P. Browne
Zingeria P. Smirnov
Zingeriopsis Probat. = *Zingeria* P.
    Smirnov
Zizania L.
Zizaniopsis Doell & Aschers.
Zonotriche Phipps
Zoysia Willd.
Zygochloa S. T. Blake